ADDITIVES IN POLYMERS

Industrial Analysis and Applications

ADDITIVES IN POLYMERS
Industrial Analysis and Applications

Jan C.J. Bart

DSM Research, The Netherlands

John Wiley & Sons, Ltd

Copyright © 2005   John Wiley & Sons Ltd, The Atrium, Southern Gate, Chichester,
West Sussex PO19 8SQ, England

Telephone (+44) 1243 779777

Email (for orders and customer service enquiries): cs-books@wiley.co.uk
Visit our Home Page on www.wiley.com

Other Wiley Editorial Offices

John Wiley & Sons Inc., 111 River Street, Hoboken, NJ 07030, USA

Jossey-Bass, 989 Market Street, San Francisco, CA 94103-1741, USA

Wiley-VCH Verlag GmbH, Boschstr. 12, D-69469 Weinheim, Germany

John Wiley & Sons Australia Ltd, 33 Park Road, Milton, Queensland 4064, Australia

John Wiley & Sons (Asia) Pte Ltd, 2 Clementi Loop #02-01, Jin Xing Distripark, Singapore 129809

John Wiley & Sons Canada Ltd, 22 Worcester Road, Etobicoke, Ontario, Canada M9W 1L1

Wiley also publishes its books in a variety of electronic formats. Some content that appears
in print may not be available in electronic books.

Library of Congress Cataloging-in-Publication Data

Bart, Jan C. J.
   Additives in polymers : industrial analysis and applications / Jan C.J. Bart.
      p. cm.
   Includes bibliographical references and index.
   ISBN 0-470-85062-0 (acid-free paper)
   1. Polymers – Additives. 2. Polymers – Analysis. I . Title.
   TP1142.B37 2005
   668.9 – dc22

                          2004015411

British Library Cataloguing in Publication Data

A catalogue record for this book is available from the British Library

ISBN 0-470-85062-0

Typeset in 9/11pt Times by Laserwords Private Limited, Chennai, India
Printed and bound in Great Britain by Antony Rowe Ltd, Chippenham, Wiltshire
This book is printed on acid-free paper responsibly manufactured from sustainable forestry
in which at least two trees are planted for each one used for paper production.

The author and publishers wish to thank C. Gerhardt, GmbH & Co. KG for providing the cover image:
'Soxtherm' an original painted by Douglas Swan (1997).

Contents

Foreword

Loss of knowledge is an acute threat to companies. The crucial question is how existing knowledge and new technologies can be harnessed as a corporate resource. A major problem facing industry is retaining knowledge within the company, in particular in times of acceleration of innovation. Moreover, in industrial research there is an unmistakable shift from generating knowledge and solving problems by experimental work towards detecting, selecting and absorbing knowledge from the external knowledge infrastructure and adapting it to specific situations. This book contributes a great deal to preserving and critically evaluating knowledge in the field of the analytics of polymer additives.

Additives play a leading role in the success of commercial plastics, elastomers, rubbers, coatings and adhesives. Without additives, many polymers would simply be of limited use. Although polymer additive analysis claims a history of use spanning at least half a century it is, nevertheless, still a continuously evolving research area with new and modified procedures related to increasingly sophisticated products. In many ways, this has led to a plethora of traditional and new chemical, physico-chemical and physical techniques and applications that are confusing to the specialist and beginner alike. An overview of developments across all areas of polymer additive analysis is lacking and a unified approach should therefore be of considerable assistance. This work shows that industrially relevant polymer additive analysis has developed into a very broad and complex field, in retrospect at the limit for one single author and problem holder. Also, despite the many advances direct polymer additive analysis has not yet displaced conventional wet chemical routes.

In this respect, current state-of-the-art ends up in a draw. This book makes a substantial contribution to the current literature on the analytics of polymer additives, follows up an earlier industrial tradition and lays a foundation for the future. It will be of great value to a broad readership comprising industrial and academic (analytical) chemists, polymer scientists and physicists, technologists and engineers, and other professionals involved in R&D, production, use and re-use of polymers and additives in all areas of application, including manufacturers, formulators, compounders, end users, government legislators and their staff, forensic scientists, etc.

With a rapidly developing field as this one, this book can only be considered as a work in progress. Hopefully, this monograph will help users to avoid reinventing the classical analytical wheel, and abandon obsolete, old practices, to redirect their efforts eventually towards more appropriate, though sometimes complicated equipment, to become sufficiently proficient to solve real-life analytical problems efficiently and with confidence, or even to devise innovative and challenging new directions. Certainly, this book will save significant time and effort for those analysts faced with cracking complex polymer additive cocktails. As nothing holds true for ever, it will be most appropriate to review the field again within the next decade.

Jos Put
Vice President R&D Materials
DSM Research
Geleen
The Netherlands

Preface

Whenever textbooks on polymer chemistry deal with polymer analytical aspects, macromolecular characterisation is usually overemphasised giving the unsuspecting reader the incorrect impression that polymers and formulated polymeric materials are one and the same thing. This treatise, which attempts to remedy such an oversight, is concerned with the characterisation of additives embedded in a broad variety of polymeric matrices. The topic is particularly relevant in view of the impressive growth in the use of synthetic polymeric materials and significant analytical advances in terms of sample preparation, chromatography, detection systems, hyphenation and computation in the last two decades. In every field of science and engineering, it is convenient to have at one's disposal an up-to-date handbook to provide specialists with a broad collection of technical details about the individual elements of the field. This has now come true for polymer/additive analysis.

The purpose of this monograph, the first to be dedicated exclusively to the analytics of additives in polymers, is to evaluate critically the extensive problem-solving experience in the polymer industry. Although this book is not intended to be a treatise on modern analytical tools in general or on polymer analysis *en large*, an outline of the principles and characteristics of relevant instrumental techniques (without hands-on details) was deemed necessary to clarify the current state-of-the-art of the analysis of additives in polymers and to accustom the reader to the unavoidable professional nomenclature. The book, which provides an in-depth overview of additive analysis by focusing on a wide array of applications in R&D, production, quality control and technical service, reflects the recent explosive development of the field. Rather than being a compendium, cookery book or laboratory manual for qualitative and/or quantitative analysis of specific additives in a variety of commercial polymers, with no limits to impractical academic exoticism (analysis for its own sake), the book focuses on the fundamental characteristics of the arsenal of techniques utilised industrially in direct relation to application in real-life polymer/additive analysis. The analyst requires *expert knowledge*, i.e. understanding of the strengths, weaknesses and limits of application of each technique and how they relate to practical problems. Therefore, the chapters are replete with selected and more common applications illustrating why particular additives are analysed by a specific method. By understanding the underlying principles, the mystery of the problem disappears. Expertise, of course, requires more than understanding of the principles alone. Consequently this book does not serve to become overnight expert in the area of polymer/additive analysis. Rather, it helps the emerging generation of polymer analysts to obtain a rapid grasp of the material in minimal time but is no substitute for personal experience.

Additives in Polymers: Industrial Analysis and Applications fulfils a need and provides information not currently available from another single literature source. This book is different from other books on polymer analysis in a number of ways. Instrumental methods are categorised according to general deformulation principles; there is more emphasis on effective problem solving and promoting understanding than on factual information or instrumental capabilities without focus on any specific analyte or polymer class. The tools of the trade are introduced when appropriate in the deformulation strategy, not on the basis of their general properties only. In particular, the author has tried to emphasise the importance of employing rational methods to laboratory, *in situ* and on-line polymer/additive analysis. The present text is an appraisal of the literature and methodology currently available (tool description), from which the inexperienced 'deformulator' can select those means necessary to tackle his own problem and finally write his own recipe and clear procedures in compliance with local instrumental possibilities. The critical evaluation of methods also indicates what still needs to be done. From an industry perspective, it is clear that above all there is a need to improve the quantitative aspects of the methods.

Although wide-ranging, the author does not claim to present a collection of 10 comprehensive reviews. Instead, *illustrative* examples, drawn from closely related fields (polymers, rubbers, coatings, adhesives), are given to outline the ranges of applicability. The value of the book stays in the applications. No book is perfect and no doubt equally deserving papers have been omitted and some undeserving ones have been included. However, with the number of techniques much greater than originally planned the text should be kept within reasonable bounds. The reader may keep in mind the lines

For what there was none cared a jot.
But all were wroth with what was not.

Theory and practice of polymer/additive analysis are not a regular part of analytical education, and usually require on-the-job training. The intention in writing this text was to appeal to as wide an audience as possible. Using an instructional approach, this reference book helps orienting chemists and technicians with little or no background in polymer/additive analysis who would like to gain rapidly a solid understanding of its fundamentals and industrial practice. Seasoned analysts of polymer formulations may use the text to quickly understand terms and techniques which fall outside of their immediate experience. The author has attempted to bring together many recent developments in the field in order to provide the reader with valuable insight into current trends and thinking. Finally, this book can also serve as a modern textbook for advanced undergraduate and graduate courses in many disciplines including analytical chemistry, polymer chemistry and industrial chemistry.

In planning this book the author has chosen a monograph in decathlon fashion. This allows critical comparisons between methods and has the advantage of a unified structure. The disadvantage is that no individual can have specialist knowledge in all fields equal to that of the sum of the experts. To overcome this drawback extensive peer review has been built in. For each individual technique more excellent textbooks are available, properly referenced, albeit with less focus on the analysis of additives in polymers. However, the steep growth curve during the past two decades has made reporting on this subject an almost elusive target.

Each chapter of this monograph is essentially self-contained. The reader can consult any subchapter individually. Together they should give a good grounding of the basic tools for dealing with the subject matter.

The reader is well advised to read the two introductory chapters first, which define the analytical problem area and general deformulation schemes. The next chapters tackle polymer/additive deformulation strategically in an ever-increasing order of sophistication in analytical ingenuity. Conventional, indirect, polymer/additive analysis methods, mainly involving wet chemistry routes, are described in Chapters 3 to 9. The book is concluded with prospects in Chapter 10. Extensive appendices describe additive classes; a glossary of symbols, and databases. To facilitate rapid consultation the text has been provided with *eye-catchers*. Each chapter concludes with up-to-date references to the primary literature (no patent literature). Contributions from many of the top industrial research laboratories throughout the world are included in this book, which represents the most extensive compilation of polymer/additive analysis ever. Once more it comes true that most research is being carried out beyond one's own R&D establishment.

The author has not tried to include a complete *ab-initio* literature search in any particular area. The majority of references in the text are from recent publications (1980–2003). This is not because excellent older references are no longer relevant. Rather, these are frequently no longer used because: (i) more recent work is a fine-tuned extension of prior work; (ii) the 'classic' texts list extensive work up to 1980; and (iii) older methods are frequently based on inferior or obsolete technology and thus direct transfer of methods may be difficult or impossible. Readers familiar with the 'classics' in the field will find that almost everything has changed considerably.

As most (industrial) practitioners have access to rapid library search facilities, it is recommended that a literature search on the analysis of a *specific* additive in a given polymer be carried out at the time, in order to generate the most recent references. Consequently, the author does not apologise for omitting references to specific analyses. However, every effort has been made to keep the book up-to-date with the latest methodological developments. Each chapter comprises a critical list of recommended general reading (books, reviews) for those who want to explore the subjects in greater depth.

This book should convince even the most hardened of the 'doubting Thomases' that polymer/additive analysis has gone a long way. With a developing field such as this one, any report represents only work in progress and is not the last word.

Geleen
December 2003

About the Author

Jan C.J. Bart (PhD Structural Chemistry, University of Amsterdam) is a senior scientist with broad interest in materials characterisation, heterogeneous catalysis and product development who spent an industrial carrier in R&D with Monsanto, Montedison and DSM Research in various countries. The author has held several teaching assignments and researched extensively in both academic and industrial areas; he authored over 250 scientific papers, including chapters in books. Dr Bart has acted as a Ramsay Memorial Fellow at the Universities of Leeds (Colour Chemistry) and Oxford (Material Science), a visiting scientist at Institut de Recherches sur la Catalyse (CNRS, Villeurbanne), and a Meyerhoff Visiting Professor at WIS (Rehovoth), and held an Invited Professorship at USTC (Hefei). He is currently a Full Professor of Industrial Chemistry at the University of Messina.

He is also a member of the Royal Society of Chemistry, Royal Dutch Chemical Society, Society of Plastic Engineers and The Institute of Materials.

Acknowledgements

This book summarises the enormous work done and published by many scientists who believe in polymer analysis. It is humbling to notice how much collective expertise is behind the current state-of-the-art in polymer/additive analysis and how little is at the command of any individual. The high degree of creativity and ingenuity within the international scientific community is inspiring. The size of the book shows the high overall productivity. Even so, only a fraction of the pertinent literature was cited.

Any project has its supporters and opponents, ranging from those faithful who repeatedly encourage to others who actively discourage. The author wishes to thank DSM from CTO to operational managers at DSM Research BV for providing foresight and generous resources for monitoring developments in this field of interest, for stimulating the work and granting permission for publication. This monograph was finalised during a sabbatical year granted only half-heartedly by the Faculty of Science of the University of Messina. The end-product may convince academic sceptics that a book marks a more permanent contribution to transfer of know-how from industry to academia than a standard one-semester course for ever-dwindling flocks of students.

The author thanks colleagues (at DSM Research) and former colleagues (now at SABIC Euro Petrochemicals) for taking on the difficult job of critically reading various chapters of the book. Reviewing means lots of work and not much appreciation from the general public. Information Services at DSM Research have been crucial in providing much needed help in literature search. Each chapter saw many versions, which needed seemingly endless word-processing. Without the expert help and patience of Mrs Coba Hendriks, who cared repeatedly about every dot and dash, it would not have been possible to complete this work successfully. A special word of thanks goes to Mihaela and David for their hospitality and endurance during the many years of preparation of this text.

The author expresses his gratitude to peer reviewers of this project for recommendation to the publisher and thanks editor and members of staff at John Wiley & Sons, Ltd for their professional assistance and guidance from manuscript to printed volume. The kind permission granted by journal publishers, book editors and equipment producers to use illustrations and tables from other sources is gratefully acknowledged. The exact references are given in figures and table captions. Every effort has been made to contact copyright holders of any material reproduced within the text and the author apologises if any have been overlooked. The author and publisher wish to thank C. Gerhardt GmbH & Co KG for providing the cover image, from an original painted by Douglas Swan.

Jan C.J. Bart
Geleen
December 2003

Disclaimer

CHAPTER 1

Search before Research

Introduction

The successful use of plastic materials in many applications, such as in the automotive industry, the electronics sector, the packaging and manufacturing of consumer goods, is substantially attributable to the incorporation of additives into virgin (and recycled) resins. Polymer industry is impossible without additives. Additives in plastics provide the means whereby processing problems, property performance limitations and restricted environmental stability are overcome. In the continuous quest for easier processing, enhanced physical properties, better long-term performance and the need to respond to new environmental health regulations, additive packages continue to evolve and diversify.

Additives can mean ingredients for plastics but they play a crucial role also in other materials, such as coatings, lacquers and paints, printing inks, photographic films and papers, and their processing. In this respect there is a considerable overlap between the plastics industry and the textiles, rubber, adhesives and food technology industries. For example, pigments can be used outside the plastics industry in synthetic fibres, inks, coatings, and rubbers, while plasticisers are used in energetic materials formulations (polymeric composite explosives and propellants). Additives for plastics are

therefore to be seen in the larger context of *specialty chemicals*. 'Specialties' are considered to be chemicals with specific properties tailored to niche markets, special segments or even individual companies. Customers purchase these chemicals to achieve a desired performance. Polymer and coatings additives are ideal specialty chemicals: very specific in their application and very effective in their performance, usually with a good deal of price inelasticity. The corresponding business is associated with considerable innovation and technical application knowledge. Research and development are essential and global operation is vital in this area.

Plastics additives now constitute a highly successful and essential sector of the chemical industry. Polymer additives are a growing sector of the specialty chemical industry. Some materials that have been sold for over 20 years are regarded today as commodity chemicals, particularly when patents covering their use have expired. Others, however, have a shorter life or have even disappeared almost without trace, e.g. when the production process cannot be made suitably economic, when unforeseen toxicity problems occur or when a new generation of additive renders them technically obsolete.

Additives In Polymers: Industrial Analysis And Applications   J. C. J. Bart
© 2005 John Wiley & Sons, Ltd   ISBN: 0-470-85062-0

1.1 ADDITIVES

It is useful at this point to consider the definition of an additive as given by the EC: an additive is a substance which is incorporated into plastics to achieve a technical effect in the finished product, and is intended to be an essential part of the finished article. Some examples of additives are antioxidants, antistatic agents, antifogging agents, emulsifiers, fillers, impact modifiers, lubricants, plasticisers, release agents, solvents, stabilisers, thickeners and UV absorbers. Additives may be either organic (e.g. alkyl phenols, hydroxybenzophenones), inorganic (e.g. oxides, salts, fillers) or organometallic (e.g. metallocarboxylates, Ni complexes, Zn accelerators). Classes of commercial plastic, rubber and coatings additives and their functionalities are given in Appendices II and III.

Since the very early stages of the development of the polymer industry it was realised that useful materials could only be obtained if certain additives were incorporated into the polymer matrix, in a process normally known as 'compounding'. Additives confer on plastics significant extensions of properties in one or more directions, such as general durability, stiffness and strength, impact resistance, thermal resistance, resistance to flexure and wear, acoustic isolation, etc. The steady increase in demand for plastic products by industry and consumers shows that plastic materials are becoming more performing and are capturing the classical fields of other materials. This evolution is also reflected in higher service temperature, dynamic and mechanical strength, stronger resistance against chemicals or radiation, and odourless formulations. Consequently, a modern plastic part often represents a high technology product of material science with the material's properties being not in the least part attributable to additives. Additives (and fillers), in the broadest sense, are essential ingredients of a manufactured polymeric material. An additive can be a primary ingredient that forms an integral part of the end product's basic characteristics, or a secondary ingredient which functions to improve performance and/or durability. Polypropylene is an outstanding example showing how polymer additives can change a vulnerable and unstable macromolecular material into a high-volume market product. The expansion of polyolefin applications into various areas of industrial and every-day use was in most cases achieved due to the employment of such speciality chemicals.

Additives may be monomeric, oligomeric or high polymeric (typically: impact modifiers and processing aids). They may be liquid-like or high-melting and therefore show very different viscosity compared to the polymer melt in which they are to be dispersed.

Selection of additives is critical and often a proprietary knowledge. Computer-aided design is used for organic compounds as active additives for polymeric compositions [1]. An advantage of virtual additives is that they do not require any additive analysis!

Additives are normally present in plastics formulations intentionally for a variety of purposes. There may also be unintentional additives, such as water, contaminants, caprolactam monomer in recycled nylon, stearic acid in calcium stearate, compounding process aids, etc. Strictly speaking, substances which just provide a suitable medium in which polymerisation occurs or directly influence polymer synthesis are not additives and are called polymerisation aids. Some examples are accelerators, catalysts, catalyst supports, catalyst modifiers, chain stoppers, cross-linking agents, initiators and promoters, polymerisation inhibitors, etc. From an analytical point of view it is not relevant for which purpose substances were added to a polymer (intentionally or not). Therefore, for the scope of this book an *extended definition* of 'additive' will be used, namely anything in a polymeric material that is not the polymer itself. This therefore includes catalyst residues, contaminants, solvents, low molecular components (monomers, oligomers), degradation and interaction products, etc. At most, it is of interest to estimate on beforehand whether the original substance added is intended to be transformed (as most polymerisation aids).

Additives are needed not only to make resins processable and to improve the properties of the moulded product during use. As the scope of plastics has increased, so has the *range of additives*: for better mechanical properties, resistance to heat, light and weathering, flame retardancy, electrical conductivity, etc. The demands of packaging have produced additive systems to aid the efficient production of film, and have developed the general need for additives which are safe for use in packaging and other applications where there is direct contact with food or drink.

The number of additives in use today runs to many thousands, their chemistry is often extremely complex and the choice of materials can be bewildering. Most commercial additives are single compounds, but some are oligomeric or technical mixtures. Examples of polymer additives containing various components are Irgafos P-EPQ, Anchor DNPD [2], technical grade glycerylmonostearate [3] and various HAS oligomers [4]. Polymeric hindered amine light stabilisers are very important constituents of many industrial formulations. In these formulations, it is often not just one component that is of interest. Rather, the overall identity, as determined by the presence and distribution of the individual components, is critical. The processing stabiliser Irgafos P-EPQ consists of a mixture of seven compounds and the antistatic agent N,N-bis-(2-hydroxyethyl) alkylamine contains five components [5]. Similarly, the antistat Atmos 150 is composed of glycerol mono- and distearate. Ethoxylated alcohols consist of polydisperse mixtures. 'Nonyl phenol' is a mixture of monoalkyl phenols with branched side-chains and an average molecular weight of 215 [6]. Commercial calcium stearate is composed of 70% stearate and 30% palmitate. Also dialkylphthalates are technical materials as well as the high-molecular weight (MW) release agent pentaerythritoltetrastearate (PETS). Flame retardants are often also mixtures, such as polybromodiphenyl ethers (PBDEs) or brominated epoxy oligomers (BEOs). Surfactants rarely occur as pure compounds.

It is also to be realised that many additives are commercialised under a variety of *product names*. Appendix III shows some examples for a selection of stabilisers, namely a phenolic antioxidant (2,2'-methylene-bis-(6-*tert*-butyl-4-methylphenol)), an aromatic amine (N-1,3-dimethyl-butyl-N'-phenyl-paraphenylene-diamine), a phosphite (trisnonylphenylphosphite), a thiosynergist (dilaurylthiodipropionate), a UV-absorber (2-hydroxy-4-n-octoxybenzophenone), a nickel-quencher ((2,2'-thio-bis-(4-*tert*-octylphenolato)-n-butylamine)-nickel), a low-MW hindered amine light stabiliser or HALS (di-(2,2,6,6-tetramethyl-4-piperidinyl)-sebacate) and a polymeric HALS compound (Tinuvin 622). Various commercial additive products are binary or ternary blends. Examples are Irganox B225 (Irganox 1010/Irgafos 168, 1:1), Ultranox 2840 (Ultranox 276/Weston 619, 3:2), and Tinuvin B75 (Irganox 1135/Tinuvin 765/Tinuvin 571, 1:2:2).

It may be seen from Appendix II that the tertiary *literature* about polymer additives is vast. Books on the subject fall into one of two categories. Some provide commercial information, in the form of data about the multitude of additive grades, or about changes in the market. Others are more concerned with accounts of the scientific and technical principles underlying current practice. This book gives higher priority to promoting understanding of the principles of polymer/additive deformulation than to just conveying factual information.

1.1.1 Additive Functionality

Additives used in plastics materials are normally classified according to their intended *performance*, rather than on a chemical basis (cf. Appendix II). For ease of survey it is convenient to classify them into groups with similar functions. The main functions of polymer additives are given in Table 1.1.

Generally, polymer modification by additives provides a cost-effective and flexible means to alter polymer properties. Traditionally, however, the use of an additive is very property-specific in nature, with usually one or two material enhancements being sought. An additive capable of enhancing one property often does so at the cost of a separate trait. Today many additives are *multifunctional* and combine different additive functionalities such as melt and light stabilisation (e.g. in Nylostab® S-EED) or metal deactivation and antioxidation (e.g. in Lowinox® MD24) (cf. Table 10.14). Dimethyl methyl phosphonate (DMMP) is a multifunctional molecular additive acting as an antiplasticiser, processing aid and flame retardant in cross-linked epoxies. In a variety on the theme, some multifunctional antioxidants, such as the high-MW Chimassorb 944, combine multiple functions in one molecule. Adhikari *et al.* [7] have presented a critical analysis of seven categories of multifunctional rubber additives having various combinations of antidegradant, activator, processing aid, accelerator, antioxidant, retarder, curing agent, dispersant, and mould release agent functions.

In analogy to plastics additives, paper coating additives are distinguished in as many as twenty-one functional property categories (for dispersion, foam and air entrainment control, viscosity modification, levelling and evening, water retention, lubricity, spoilage control, optical brightness improvement, dry pick improvement, dry nub improvement and abrasion resistance, wet pick improvement, wet rub improvement, gloss-ink hold-out, grease and oil resistance, water resistance, plasticity, fold endurance, electroconductivity, gloss improvement, organic solvent coating additives, colouring), even excluding those materials whose primary function is as a binder, pigment or vehicle [8].

Typical technology questions raised by plastic producers and manufacturers and directed at the additive supplier are given in Table 1.2, as exemplified in the application of injection moulding of polyamides. These problems may be tackled with appropriate addition of chain extenders and cross-linking agents, nucleating agents and lubricants, release agents, reinforcements, etc.

There are now far more categories of additives than a few decades ago. The corresponding changes in additive technology are driven partly by the desire to produce plastics which are ever more closely specified for particular purposes. The *benefits* of plastics additives are not marginal. As outlined before, they are not simply optional extras but essential ingredients, which make all

Table 1.1 Main functions of polymer additives

Polymerisation/chemical modification aids

Accelerators	Cross-linking agents
Chain growth regulators	Promoters
Compatibilisers	

Improvement in processability and productivity (transformation aids)

Defoaming and blowing agents	Release agents
Flow promoters	Surfactants
Plasticisers	Thixotropic agents, thickening agents
Processing aids	Wetting agents
Slip agents and lubricants (internal and external)	

Increased resistance to degradation during processing or application

Acid scavengers	Metal deactivators
Biostabilisers	Processing/thermal stabilisers
Light/UV stabilisers	

Improvement/modification of mechanical properties

Compatibilisers	Impact modifiers (elastomers)
Cross-linking agents	Nucleating agents
Fibrous reinforcements (glass, carbon)	Plasticisers or flexibilisers
Fillers and particle reinforcements	

Improvement of product performance

Antistatic agents	Friction agents
Blowing agents	Odour modifiers
EMI shielding agents	Plasticisers
Flame retardants	Smoke suppressants

Improvement of surface properties

Adhesion promoters	Lubricants
Antifogging agents	Slip and antiblocking agents
Antistatic agents	Surfactants
Antiwear additives	Wetting agents
Coupling agents	

Improvement of optical properties

Nucleating agents	Pigments and colorants
Optical brighteners	

Reduction of formulation cost

Diluents and extenders	Particulate fillers

Table 1.2 Technology questions related to injection moulding of polyamides

- Short cycle times
- Better mould release
- Plate-out and deposits on moulds and plastics surfaces
- Feeding problems
- Increased dimensional stability, less shrinkage
- Processing protection against depolymerisation and yellowing
- Better melt flow
- Improved surface of glass-reinforced parts
- Better strength of flow lines in moulded parts
- Higher molecular weight
- Rise of impact strength and elongation at break

the difference between success and failure in plastics technology. Typically, PVC is a material whose utility is greatly determined by plasticisers and other additives. The bottom line on the use of any additive is a desired level of performance. The additive package formulation needs to achieve cost effectively the performance required for a given application. In this respect we recall that early plastics were often unsatisfactory, partly because of inadequate additive packages. In the past, complaints about plastics articles were common. Use of additives brings along also some potential *disadvantages*. Many people have been influenced by a widespread public suspicion of chemicals in general (and additives in particular, whether in foods or plastics). Technological actions must take place within an increasingly (and understandably) strict environment which regulates the potential hazards of chemicals in the workplace, the use of plastics materials in contact with foodstuffs, the possible side-effects of additives as well as the long-term influence of the additives on the environment when the product is recycled or otherwise comes to final disposal.

Concerns are expressed by legislation and regulations, such as:

- General Health & Safety   Fitness for purpose (food/water contact materials, toys, medical)
- Montreal Protocol   Blowing agents for foams
- EU Directives   Food contact
- Landfill Directives   Disposal, recycling
- Life Cycle Analysis   Realistic evaluation of product use (flame retardants, volatiles, etc.).

All additives are subject to some form of regulatory control through general health and safety at work legislation. From an environmental and legislative point of view three additive types in particular experience pressures, namely halogen-containing flame retardants (actions pending), heavy metals (as used in pigments and PVC stabiliser systems), and plasticisers. The trend towards the incineration of plastics, which recovers considerable energy for further use, leads to concern and thought about the effects of any additives on the emissions produced. Environmental issues often have beneficial consequences. The toxicity of certain pigments, both in plastics and in paints, has been a driving force for the development of new, safer pigments with applications in wider areas than those originally envisaged. Where food contacts are the issue, the additives used must be rigorously tested to avoid any tainting of the contents of the packaging. On the whole, the benefits of additives far outweigh the disadvantages.

1.2 PLASTICS FORMULATIONS

Plastic additives are a diverse group of specialty chemicals that are either incorporated into the plastic product prior to or during processing, or applied to the surface of the product when processing has been completed. To a great extent, the selection of the appropriate additive is the responsibility of the plastic processor or the compounder carrying out the modification. Scheme 1.1 illustrates the use of typical additives in the process from polymerisation to product manufacturing.

Figure 1.1 describes the interrelationships between the players in plastic materials manufacturing, which is considerably more complex than for the coating industry. The product performance specifications are defined by the end-users. Specialty additives demand is nowadays migrating to compounders, converters and distributors.

The *rubber industry* was the original user of additives. Rubber is a thermosetting polymer, which classically requires curing (peroxides), in a reaction which must be controlled by initiators (e.g. sulfur compounds), accelerators (e.g. aniline), retarders, etc. The whole compounding and moulding process is to be controlled by antioxidants and antiscorch agents to prevent decomposition. Plasticisers are added to improve processability, and adhesion promoters may be added to improve the bonding with reinforcement. To protect cured rubber products during lifetime, other additives are introduced into the compound to confer resistance to ozone, ultraviolet and internal heat build-up (hysteresis) as the compound is stressed. Other vital components of a final rubber compound are fillers as reinforcing agents, pigments, and extenders (essentially low-cost fillers).

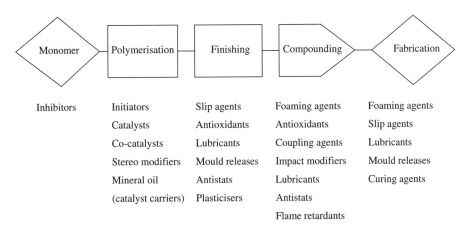

Scheme 1.1 Exemplified application of additives in various stages of the production process of a polymeric material

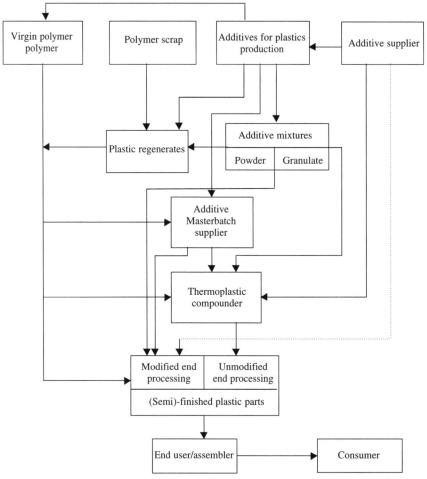

Figure 1.1 Methods of manufacturing plastic materials. After Titzschkau [9]. Reproduced by permission of Intertech Corporation, Portland, MN

The compounded rubber is therefore a highly complex chemical system, difficult to analyse (cf. Section 2.2).

Table 1.3 shows the build-up of a typical recipe for PP grades. It is important to take into account possible incompatibilities, such as co-additive interactions leading to undesired effects.

Typical additive packages for *engineering thermoplastics* have been described by Titzschkau [9], such as processing aids for PA, PP, or PET/PBT, three-component additive packages for polyamides and polyesters (nucleating agent, lubricant and process heat stabiliser) and coated copper stabilisers for polyamides. Additive packages or combinations of up to five or more additives are quite common. A typical white window PVC profile formulation comprises an acrylic impact modifier, TiO_2, $CaCO_3$, calcium stearate, a

Table 1.3 Basic additive formulation for polypropylene

- Long-term stabiliser (always for Z/N PP, usually phenolic AOs)
- Melt stabiliser (phosphite or phosphonite)
- Acid scavenger (always for Z/N PP)
- Slip and antiblocking agents (for film)
- Nucleators, clarifiers, antistatics (for specific injection moulding applications)
- Specific antioxidants (for fibres; nonvolatiles, gas-fading)
- UV absorbers (for automotive)

processing aid, polyethylene wax, oxidised polyethylene wax, an external/internal lubricant and lead stabilisers. Not surprisingly, the additives largely determine the cost price of PVC. Typical fibre formulations comprise primary and secondary process stabilisers, a neutraliser, UV additive, pigment, optical brighteners and a flame retardant.

1.2.1 Supply Forms

Various physical supply forms for product formulation exist: powders, irregular flakes, beads/prills, granulate (highly extruded or compacted), lenses, pastilles, spheres, emulsions and liquids. The majority of the additives are *solids*. Product shape is strongly influenced by the production method of the additive, typically extrusion, (strand) pelletising, grinding, spraying, flaking, or pastillating. The main concern of the additive producer is always to have a defined throughput (kg/h) of pellets with a specific average diameter (mm) from a given material. A current trend is the re-working of traditional workhorse grades of some additive classes into environmentally more acceptable product forms, which offer greater safety and are easier to handle and to mix. Traditional additives in powder form emit dust and tend to flow erratically in feeder equipment causing worker hygiene and handling problems [10,11]. Priority challenges in the field of product form performance of additives are dust reduction, dosing optimisation and dispersion improvement. Conventional approaches to meet these goals are based on mechanical compaction or mechanical treatments, using large compression forces and significant amounts (approximately 20–60 %) of processing aids causing secondary deterioration effects. Additives in the *ideal physical form* have a spherical product shape ($d_{50} = 500$–$1500\,\mu$m), exhibit the same performance as the original powder, ensure high homogeneity and dispersibility rate, are mechanically resistant, show no segregation in the polymer and are more suitable for feeding, dosing and blending. Some relevant milestones in additive development in the past 25 years have been the introduction of dust-free formulations of light stabilisers (1979), free-flowing antioxidants, light stabilisers and compounds (1983), free-flowing beads of oligomer light stabilisers (1989), free-flowing/dust-free oligomer light stabilisers and antioxidants (1991), durable dust-free antioxidants and compounds (1995) and customised additives (one-packs, in powder form, as dust-free compacted granules or as masterbatches). Free-flowing silica fillers have been created by dispersion of siloxane gums [12]. Additive concentrates are also available in granulate form (e.g. Morstille 18, a pastille form of DSTDP from Morton Performance Chemicals). Compared to masterbatches, these formulations have the advantage that they can be prepared at very low temperatures and the additives are thus likely to be virtually intact. Some innovative spherical particle systems with narrow size distribution have recently been introduced, such as drop process pelletising with industrial applications for waxes, saponified fatty acids, metal stearates, metal soaps, stabilisers and colour concentrates [13], and continuous fluidised bed (FB) spray granulation, as demonstrated for carbon-black (CB), TiO_2, flame retardants (FR), colour pigments, organic based stabilisers (OBS) and light stabilisers [14]. Drop process pelletising of low-viscosity plastics and additives is applicable to materials available as liquids or melts with viscosities below 500 cP.

The last 15 years have witnessed a constantly increasing impact of *additive masterbatches* (concentrates containing a higher level of additives dispersed in the parent polymer), e.g. for antistatics [15], foaming agents, flame retardants, impact modifiers, antimicrobials, modifier masterbatches for surface improvement and shear reduction, colour masterbatches [16], etc. The use of concentrated additive masterbatches and sophisticated material delivery systems gives high confidence in polymer compounding. Other important reasons for choosing additive masterbatches instead of pure additives are the physical form, dosability, ease of handling, homogeneous mixing, safety, additive protection and improvement of performance, influence of carrier system, supplier experience and cost. Porous polyolefin carriers offer masterbatch suppliers an inexpensive and simple way of producing high concentrates without having to use an extruder. Pure additives usually require specific handling. In fact, some additives have to be dispersed like pigments to avoid agglomeration; some others need to be intensively kneaded. It is difficult to choose processing conditions that offer simultaneously an optimum on mixing/dispersing/kneading/dissolving efficiency as required for processing of additives with very different properties. Masterbatches go some way to overcome these problems. An additive producer or a masterbatch supplier may carry out additive selection and production of the mixture.

Blending and/or custom blending is another current trend. *One-pack systems* may offer antioxidant activity, processing aid and lubrication or anticorrosive activity in one package, usually in a low- or nondusting product form. A proprietary database [17] mentions already some 140 commercial binary and ternary phenol-phosphite blends, HALS-containing blends and miscellaneous blends. As most polymer processing requires both primary and secondary antioxidant addition, 'one-pack' blends containing these components are another obvious development. Antioxidant blends are combinations of primary hindered phenolic and secondary organophosphite antioxidants, which synergistically act together to provide excellent performance in the prevention of thermo-oxidative degradation of the polymer. Examples are Ciba Specialty Chemicals Irganox B series additive blends (e.g. Irganox B561 is Irganox 1010 and Irgafos 168 in 1:4 parts) and Great Lakes Chemical No Dust Blends (NDB) [18–20]. The move to multicomponent packages takes away the risk of operator error, leads to productivity benefits, aids ISO protocols and good housekeeping. Other *advantages* are ease of dosage, reduction in concentration variability during polymer production (quality control, less off-spec product), in logistics and in analytical costs (analysis only of the easiest detectable component). By controlling the composition of the NDB with analytical instruments the precision has been found to be of the same order of magnitude as the tolerance of the used analytical methods. However, because the weighing operation is carried out by an electronic balance, the achieved precision level is always higher than the one detectable with common analyses. Dosing/homogeneity accuracy affecting stabiliser additivation has been addressed by Sasselli [18].

For the purpose of cost reduction, it is sometimes dangerous practice to limit analysis of the components of dry-blends or other mixtures to the determination of one 'critical' component only. As shown by Pahl and Grosse-Aschhoff [21] various degrees of dispersion may easily invalidate such conventional assumptions. Techniques do exist (e.g. near-infrared spectroscopy) which simultaneously determine all components and can therefore cope with problems of heterogeneity. Main disadvantages of one-pack systems are loss of flexibility and price. A trend towards uniformity and streamlining of the product range nowadays applies especially to producers of polyolefins and PVC; however, the situation is different for engineering thermoplastics where it is virtually impossible to avoid producing tailor-made product modifications.

Only a few additives are *liquids* (e.g. Vitamin E), which require different handling. A recent development is incorporation of the (viscous) liquid and low-melting additives in high concentration in high-porosity carriers, such as LDPE (Stamypor®, $d = 925 \, \text{kg/m}^3$) [22]. Such nonhygroscopic holey beads can successfully be used for the production of polyolefin concentrates with liquid and low-melting polar and nonpolar additives, such as antistats, anticondensation agents, slip agents, mould-release agents, lubricants, antioxidants, UV stabilisers, pigments, polyisobutylene, pastes and fragrances; temperature-sensitive reactants such as silanes, peroxides and chain extenders offer safety and efficiency improvements. Due to its spherical shape, Stamypor® remains a free-flowing product even after

high liquid loading (exceeding 50 %), allowing good dispersion. The loading process just requires a low-speed mixer. Similarly, AKZO's microporous carriers (Accurel), based on PP, HDPE, LDPE, EVA, PC, ABS, SAN and nylon, have a load capacity of up to 70 %.

The *concentration* of additives in a polymer depends on the intended function. Each additive has specific concentration ranges in which it does not affect the properties of the matrix. Additive levels amount to at least some 100 ppm (e.g. Vitamin E as a melt processing stabiliser), although catalyst residues and unintentional contaminants may show lower levels. Typical antioxidant levels to inhibit thermal oxidation in polyolefins are of the order of 0.1 wt%. However, in applications of LDPE, LLDPE, HDPE and PP calling for very good toughness or high deformability of the material the filler content easily amounts to 30–40 wt% [23]. There are many nominally organic plastics articles which actually consist of considerably less than 50 % organic polymer, the remainder being largely inorganic additives. For example, in general spumific flame retardant additives are less efficient in polyamides than either halogenated or intumescent additives and much higher loading levels, typically 50–60 wt%, are then required in order to prevent dripping and to obtain the same levels of flame retardancy that can be achieved with typically 10–25 wt% of a halogenated or an intumescent additive. Some other highly filled compounds are cross-linked PMMA/72 wt% SiO_2 (Silgranit), PMMA/62 wt% Si (Silacron), PMMA/62 wt% $Al(OH)_3$ (Corian). The Japanese manufacturer Kanebo Gosen has developed a heavily metal-filled PA resin for production of electronic and automotive components with a specific gravity of 13 g/cm^3 (compared with 11.3 of lead). Composite polymeric materials containing high percentages of nonpolymeric materials are used extensively in the fabrication and engineering industries. Twin-screw systems are configured to continuously mix very high levels of metal fillers (90 wt% +) with various polymer binders. Obviously, at such high loading levels a distortion of the balance of mechanical properties of the base polymer results, but this may be acceptable (e.g. 70 wt% of fused silica properly balances the coefficient of thermal expansion in an epoxy moulding resin) [24]. Similarly, a large proportion of automotive dashboard skins has a high plasticiser content (ca. 70 phr). PVC is almost unique in its ability to accept addition of very high plasticiser levels (up to 100 phr and above) while still retaining useful mechanical properties. Also for processing of cellulose acetate it is necessary to incorporate relatively high levels of a polar plasticiser (typically

50 phr) to lower the softening temperature below the decomposition temperature.

1.2.2 Additive Delivery

Additives can be incorporated into the polymer at several stages: (i) during the polymerisation stage (directly during production of the plastic in the reactor); (ii) addition to the finished granulate in a subsequent processing (compounding/mixing) stage, or in the processing machine itself. Finally, additives may be applied to the finished part surface. Much depends on the type of additive and polymer. Automated powder sampling systems have been described [25] and handling of solid additives has been reviewed [26]. A crucial aspect is to obtain a completely homogeneous mixture of polymer and additive – a difficult technological target, as shown by microscopy and chemiluminescence studies. Additives such as stabilisers can be introduced at the raw material manufacturing stage, whereas performance-critical additives (such as flame retardants) are introduced at the compounding stage. Additives to confer special technical properties are usually introduced in a secondary compounding stage.

To facilitate in-plant compounding, most suppliers have developed systems which efficiently and reproducibly deliver a controlled additive 'package' to a compound, using either a specialised concentrate or a masterbatch formulation. Some of the polymer manufacturers have also made available *advanced additive delivery* systems, which they have often developed originally for their own use (e.g. Eastman, Montell).

In the most sophisticated operations, there are facilities for *reactive compounding*, in which reactive additives are chemically bound as an integral part of the polymeric structure. Thus it is possible to produce hundreds of very differentiated modified plastics from very few basic plastic types and the range of recipes and possible varieties is virtually inexhaustible.

1.3 ECONOMIC IMPACT OF POLYMER ADDITIVES

Plastic and rubber additives are both commodity chemicals and specialties. The *Handbook of Plastic and Rubber Additives* [27] mentions over 13 000 products; antioxidants and antiozonants amount to more than 1500 trade name products and chemicals [28], flame retardants to some 1000 chemicals [29] and antimicrobials to over 1200 products [30].

In the past decades, polymer materials have been continuously replacing more traditional materials such as paper, metal, glass, stone, wood, natural fibres and natural rubber in the fields of clothing industry, E&E components, automotive materials, aeronautics, leisure, food packaging, sports goods, etc. Without the existence of suitable polymer materials progress in many of these areas would have been limited. Polymer materials are appreciated for their chemical, physical and economical qualities including low production cost, safety aspects and low environmental impact (cf. life-cycle analysis).

Plastic additives account for 15–20 wt% of the total volume of plastic products marketed. Estimates of the size of the *world additives market* vary considerably according to classification. Table 1.4 shows

Table 1.4 The global additives business (various estimates and forecasts)

	Global				USA	Western Europe		
	1996[a] (%)	1996[a] (kt)	1996[a] (US$m)	1996[b] (kt)	1995[b] (US$m)	1996[b] (US$m)	2001[b] (US$m)	2000[c] (kt)
Fillers	n/a					1020	1060	
Plasticisers						1930	1960	
Colorants						1200	1370	
Flame retardants	31	843			718	580	670	375
Heat/light stabilisers	17	462		295		530	620	
Impact modifiers	17	462						
Lubricants	16	435						
Antioxidants	7	190		88		370	480	
Others	12	326				1000	1190	
Europe	26		4100					
North America	23		3700					
Asia/Pacific	39		6200					
Rest of world	12		1900					
Total		2720	15 900			6630	7350	

Source: [a] Phillip Townsend Associates; [b] Business Communications Co.; [c] Schmidt [31].

some marked differences in the estimates of the global additives business volume (depending upon definition of 'additive'). According to Phillip Townsend Associates Inc. [32] the world market for performance additives (modifiers, property extenders and processing agents), thus excluding commodity materials such as fillers and pigments but including plasticisers, was worth nearly US$15.9 billion or 7.9 mt/yr in 1996. Recent figures for 2001 are 14.6 billion (by region: US 28 %, EU 26 %, AP 38 %, ROW 10 %) or 8.0 mt/yr [33] denoting a reduction in the growth rate, shrinking value (Asian crisis, WTC effect), and margin compression for material suppliers and compounders. Plastics additives are a highly competitive business.

Table 1.5 is a breakdown of the consumption by additive class. Total EU additive consumption is reported as 6989 kt (1997) growing up to 9031 kt (2002), with fillers 4346 kt, plasticisers 940 kt and colourants 728 kt (in 1997) being the main classes. Additive consumption by polymer classification for Europe is given in Table 1.6.

Worldwide consumption of performance additives (excluding plasticisers) grew from just over 2.7 mt in 1996 to 3.6 mt in 2001. Flame retardants make up 31 % of the volume and stabilisers, impact modifiers and lubricants each account for around 16–17 %. Flame retardant markets (construction, E&E devices, automotive) are headed for unprecedented development and change, being threatened by environmental, health and safety issues. The global demand for mineral-based FR compounds will increase dramatically.

The total bulk volume of additives derives from modifiers, while the value comes from relatively small volumes of increasing high-performance chemicals, for stabilising, curing/cross-linking, colouring and flame-retarding various types of plastics, both thermoplastics and thermosets. More precisely, a breakdown of the 1999 world market (totalling 7.6 mt and US$15.0 billion) shows modifiers (coupling agents, impact modifiers, nucleating agents, organic peroxides, chemical blowing agents, plasticisers) at 69 % of total volume and 51 % of total value, property extenders (antioxidants, preservatives, light and heat stabilisers, antistatic agents, flame retardants) at 23 % of total volume and 41 % of total value, processing aids (mould release agents, lubricants, antiblocking agents, slip agents) at 8 % of total volume and 8 % of total value [35].

The plastic additive market is characterised by a highly fragmented global market. Nevertheless, global customers, a maturing technology and expiring patents are fuelling the field. Each of the additive classes is favoured by a different customer group. Modifiers are largely purchased by fabricators, who account for 69 % of the modifier consumption (volume). Resin manufacturers or captive compounders capture 16 % and merchant compounders purchased the remaining 15 % of the modifiers. In 1994, resin manufacturers consumed 1.9 billion pounds of property extenders, merchant compounders 28 %, and fabricators the remaining 7 %. The processing aids are the most evenly consumed class of polymer additives. Fabricators lead with 44 % of total volume, followed by resin manufacturers with 33 % and merchant compounders consuming the remaining 23 %. The average cost of the polymer additive classes varies widely.

Demand for the different classes of polymer additives varies by resin. Modifiers and processing aids rely heavily on PVC while the property extenders are primarily used in non-PVC resins. PVC is by far the largest *consuming resin* for polymer additives (excluding fillers), accounting for some 80 % of the world-wide volume or 60 % in total value. Polyolefins are a distant second accounting for 8 % and 17 %, respectively [36].

The European consumption of plasticisers (as the main modifier) is gradually increasing, as shown in Table 1.7, with an expected growth of 2.7 % for 2001–2006. The total European market for flame-retardant chemicals (percent of revenues by product type – forecast for the year 2003) is as follows:

Table 1.5 Consumption of plastic additives by type (1998)

Plasticisers	32 %	Organic peroxides	6 %
Flame retardants	14	Lubricants/mould release agents	6
Heat stabilisers	12	Light stabilisers	3
Impact modifiers/ processing aids	10	Others	8
Antioxidants	9	Total	US$14.9 billion

Source: Townsend's Polymer Services & Information (T-PSI). Reproduced by permission.

Table 1.6 Additive consumption by polymer classification in Europe

Class	1997	2002	1997	2002
Commodity thermoplastics	4500 kt	5500 kt	62.2 %	59.5 %
Engineering plastics	1050	1900	15.3	20.7
Thermosets	1500	1600	22.5	19.8

After Dufton [34]. Reproduced by permission of Rapra Technology Ltd.

Table 1.7 European consumption of plasticisers (kt)

Plasticiser type	1993	1997	2000
DOP	484	522	562
DINP/DIDP	371	454	490
Other phthalates	78	91	100
Other plasticisers	147	171	184
Total	1080	1239	1336

Source: CDC.

Table 1.8 Plastic additives – expected global growth rates 1999–2004[a]

Highest growth (6–7 %)	*Medium growth (4–5 %)*
Coupling agents	Antiblocking agents[b]
Light stabilisers[b]	Antioxidants[b]
Nucleating/clarifying agents	Antistatic agents
	Chemical blowing agents
Lowest growth (3 % or less)	Flame retardants[b]
	Heat stabilisers[b]
Biocides	Impact modifiers/processing aids[b]
Plasticisers[b]	Lubricants/mould release agents[b]
	Organic peroxides
	Slip agents

[a] Not corrected for WTC effect.
[b] Additives most affected by environmental.
After Galvanek *et al.* [35]. Reproduced by permission of Rapra Technology Ltd.

phosphorous-based chemicals (38.2 %), inorganic compounds and melamine (36.2 %), halogen-based chemicals (25.6 %). The volume of halogenated flame retardants in Europe has not declined. The European market for additives (plasticisers, light and heat stabilisers, flame retardants and antioxidants) is expected to grow from US$2.2 billion (1998) to 2.6 billion (2005) [37]. *Global growth rates* for plastics additives are given in Table 1.8, with some performance additives showing 'above-average' potential.

Light stabilisers are the fastest-growing sector of the US additives market. Large amounts of stabilisers are also used for the protection of various petroleum products, foods, sanitary goods, cosmetics, and pharmaceuticals. The most extensive development, however, is addressed to the field of polymer stabilisation. The global consumption of light stabilisers in plastics in 1996 amounted to 24.8 kt world-wide, namely PP 45 %, PE 29 %, styrenics 5 %, EP 7 %, PVC 9 % and other polymers 5 %. Similar figures for antioxidants are 206.5 kt world-wide with PP 40 %, PE 25 %, styrenics 15 %, EP 10 %, PVC 5 % and other polymers 5 % (source: Phillip Townsend Associates Inc.). More than 200 users worldwide consume over US$400 million of light stabilisers. Much lower growth is predicted for

Table 1.9 Factors affecting plastic additives growth

- Resin demand/mix changes
- End use demand and requirements
- New technologies
- Interpolymer competition
- Regional growth patterns
- Substitution of traditional materials
- Lifetime shortening
- Miniaturisation
- Drive for shareholder value
- Focus on the customer ('one stop shopping')
- Environmental regulatory issues

plasticisers [38]. For the seven main types of plasticiser – phthalates, aliphatics, epoxidised vegetable oils (EPOs), phosphates, trimellitates, citrates and polymerics – the predicted growth rate is 2.8 % for the period 1999–2004 for a global demand of 4.6 mt in 1999. Factors affecting plastic additive growth are given in Table 1.9.

On the whole, the amount of additive per pound weight of resin is decreasing as more efficient materials are developed, cost reduction is attempted and, in some cases the concentration of potentially toxic substances is cut. Excluding the filler market (largest in size: 50 vol%; 15 % of total value), there are over 400 suppliers of performance polymer additives (antiblocking/slip agents, antioxidants, antistatics, coupling agents, chemical blowing agents, flame retardants, heat stabilisers, impact modifiers, light stabilisers, lubricants, mould-release agents, nucleating agents, organic peroxides, plasticisers and preservatives) worldwide, including already over 200 producers of colour masterbatches only in Europe [32].

Figure 1.1 shows that the methods of manufacturing (semi-)finished plastic parts involve various players: equipment manufacturers, polymer producers, additive suppliers, compounders and final processors. It can be safely assumed that the compounder will continue to be the main customer for additives and additive concentrates also in the future. Finally, the recently established Plastics Additive Museum (Lingen, Bavaria), by a pioneer in PVC additives (Bärlocher GmbH), shows that the business is coming to age.

1.4 ANALYSIS OF PLASTICS

In contrast to low-MW substances, which are composed of identical molecules (eventually apart from isomers), macromolecules constitute a statistical assembly of molecules of different molecular weight, composition, chain architecture, branching, stereoregularities (tacticity), geometric isomerism, etc. Examination of polymer systems requires determination of several types of polydispersity, such as molecular weights $\langle M_n \rangle$, $\langle M_w \rangle$, molecular weight distribution (MWD), compositional homogeneity, functionality distribution, etc. Various chromatographies, such as size-exclusion chromatography (SEC), high-performance liquid chromatography (HPLC) and thin-layer chromatography (TLC), are helpful in these analyses. Yet, even with this much of effort a polymer is not fully characterised as other 'details' are of great practical importance, such as rest monomer (e.g. styrene in PS), oligomers, or volatiles (such as water in nylons). Catalyst residues are another inherent, and important, impurity in a polymer, especially in relation to stability. Consequently, full characterisation of an unknown polymer is a challenging task. However, this is more child's play in comparison to the requirements of extensive chemical analysis of a *polymeric material*, constituted of a formulation of the aforementioned statistical assembly of macromolecules with organic and/or inorganic additives, fillers, etc. *Textbooks* on various aspects of the determination of the complex structure of polymers (in particular macromolecular characterisation in terms of molar mass, chemical composition, functionality and architecture) [39–57] outnumber those covering *analysis of additives* in polymers [41,50,54,55,58–63] or textbooks dealing with the in-service aspects of the materials [58,60]. Actually, in industrial practice these problems are usually treated separately as different interests are addressed. This does not mean to say that no polymer/additive sample will ever be examined both to characterise the polymer and the additive composition. However, frequently the chemical nature of the polymeric matrix of a formulated polymeric material is already known (but usually not for rubbers). Eventually, for additive analysis only the nature of the polymer needs to be assessed (mainly for solvent choice), but not its polydispersity or other structural details. Consequently, and in view of the considerable spread of the analytical topics, it is not surprising that few authors dare deal in depth with molecular characterisation of polymers and polymer/additive analysis in one monograph [41,50,52,54,55]; the latter are also fairly dated. The required *level of analysis* is often not merely that of the identification of the additives of Appendix II (relatively simple), but a full analysis of all active ingredients present in a polymeric matrix, both qualitatively (not straightforward) and quantitatively (difficult), and sometimes even in a spatially resolved fashion (very difficult). Representative sampling is

Table 1.10 Basic needs in polymer/additive analysis

- Qualitative identification of a particular additive in a sample
- Quantitative determination of the additive concentration
- Reliability, accuracy
- Sensitivity (down to 0.01 wt% or less)
- Short analysis time (e.g. simultaneous analyses, automation)

obviously of immediate concern. Basic needs in polymer/additive analysis are given in Table 1.10.

Industrial analytical laboratories search for methodologies that allow high quality analysis with enhanced sensitivity, short overall analysis times through significant reductions in sample preparation, reduced cost per analysis through fewer man-hours per sample, reduced solvent usage and disposal costs, and minimisation of errors due to analyte loss and contamination during evaporation. The experience and criticism of analysts influence the economical aspects of analysis methods very substantially.

The ability to reproducibly determine the additive package present in polymers is of major concern to resin manufacturers, converters (compounders), end-users, regulators and others. Qualitative and/or quantitative knowledge of compounding ingredients, to be obtained by additive analysis, may be needed in various stages of a *product lifecycle* (Table 1.11).

Analytical support is required throughout for base polymers, compounds, additives, polymer-based products and manufacturing sector products and components. Product development (e.g. surface active additives, such as antistatics, slip and antiblocking additives) leading to better performing products requires an in-depth understanding of the mechanism of action. Polymer/additive analysis contributes to this understanding. Apart from polymer microstructural analysis, polymer/additive analysis is the only way to investigate the effects of processing conditions on a polymer at the molecular level. The determination of factors affecting additive consumption can lead to an improved understanding of how to process polymers both cost effectively and with maintenance of final product properties as a goal. However, in order to determine additive consumption and draw valid conclusions, the technologist requires reliable and reproducible methods for additive level determination.

It is equally important for the manufacturer and regulator to know the level of additives in a polymer material to ensure that the product is fit for its intended purpose. Additive analysis marks sources of supply, provides a (total) process signature and may actually be used as a *fingerprint* of a polymeric material, in particular as molecular characterisation of the polymer

Table 1.11 Analytical product lifecycle

Development
- Materials selection
- Product development (structure/property relationships)
- Improved product design specification
- Improvement of product quality
- Compound formulation
- Testing (stability, flame retardancy, etc.)
- Process development
- Vendor and competitor additive package analysis (reverse engineering)
- Countertyping
- Research applications
- Application development
- Food contact (toxicology)
- Migration studies (compliance with regulations, blooming, staining; medical)
- Service life prediction

Production
- Assessment of raw materials (purity; 'hidden' ingredients; supplier monitoring)
- Quality control of intermediate and end-products (plant support; process deviations)
- Manufacturing problems (compounding or processing errors)
- Spy on production process (via low-MW by-products)
- Deposit compositions (dosing problems, caking, etc.)
- Contamination
- Process improvements
- Technical specification compliance
- Standardisation of semimanufactured and intermediate products (SPC, GLP)
- Additive depletion during polymer processing
- Industrial troubleshooting (defect and failure analysis at polymer, additive and masterbatch producers, or at polymer processors)
- Occupational safety and hygiene

In-service
- Failure diagnosis (degradation of product performance)
- Customer support
- Product recall (complaints, substandard batches)
- Compliance testing
- Taste, odour, discoloration problems (yellowing, mystery contaminants)
- Emission of VOCs and decomposition products
- Additive depletion or degradation during materials lifetime
- Post-mortem analysis
- Materials changes (ageing, dynamic loading, etc.)
- Grade detection (ownership; materials recognition, markers)
- Government regulations
- Environmental
- Claims, litigation
- Expert witness service
- Forensic investigation
- Recycling (characterisation, restabilisation)

is often less selective in fair discrimination. Important problems, both for the additive producer and user,

are the determination and control of impurities in additives (determination via chromatographic methods). It is equally important to be able to determine the concentration of adventitious volatiles in polymers (arising from the manufacturing method), which usually range from a few tens of ppm to several hundred ppm. Volatile residues may affect the processability and mechanical properties of polymers, or may cause tainting in case of foodstuff- or beverage-packaging grades of polymers. Residual volatiles are often indicative of the production process (and therefore play a role in product protection). Typically, oligomer extracts often provide a fingerprint permitting to establish the origin of a polymeric material (e.g. of polycarbonate). In both cases a broad knowledge of competitor products is required. Obviously, for a detailed insight in the differences of the fingerprints further identification (e.g. by means of LC-MS) is necessary. 'Tracers', based on uncommon elements (e.g. strontium stearates), are increasingly being used by polymer manufacturers as a rapid means of screening 'complaint' polymer samples, in order to ascertain the ownership of the material.

Apart from *routine quality control* actions, additive analysis is often called upon in relation to testing additive effectiveness as well as in connection with food packaging and medical plastics, where the identities and levels of potentially toxic substances must be accurately known and controlled. Food contact plastics are regulated by maximum concentrations allowable in the plastic, which applies to residual monomers and processing aids as well as additives [64–66]. Analytical measurements provide not only a method of quality control but also a means of establishing the loss of stabilisers as a function of material processing and product ageing.

Additive analysis is also beneficial in the identification of reaction or transformation products, as well as of odorants or irritants that evolve from a polymeric material during processing or use, in dealing with problems involving migration and diffusion phenomena, in the *deformulation* of unknown additive cocktails, and in solving the origin of complaints (e.g. regarding discoloured or early aged materials), etc. Moreover, inadvertent contaminants may often pose considerable practical problems as sources of yellowing or discoloration. Consequently, some applications require a quick semiquantitative analysis to support production while in other circumstances the analytical chemist must act as a detective to determine which additives are incorporated into a sample and then quantify them. Unless the analyst is sufficiently familiar with the type of sample to be tested, it makes good sense to apply a specific test for

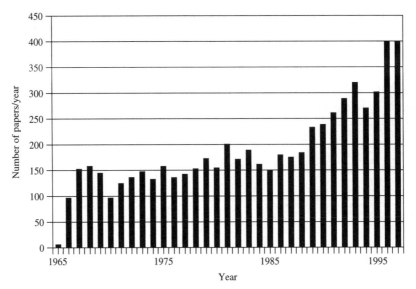

Figure 1.2   Number of scientific publications per year on polymer/additive analysis. Source: *Chem. Abstr.*

the presence of specific additives before embarking upon the involved determination of something that might not be there. Gabriel and Mulley [67] have detailed a range of such tests for anionic surfactants. However, in some cases it is imperative that a complete characterisation of a system of additives in a polymer is made, e.g. if a product is not meeting performance expectations.

Nothing is more difficult than screening for the general unknown (nontargeted analysis). Identification and/or verification and quantification of a complete additive package, consisting in the best case of fairly low-MW organics in à polymeric matrix, is a considerable analytical challenge. The detection of an additive in a polymer is determined by the following parameters: (i) chemical nature of additive and polymer matrix; (ii) concentration; and (iii) thermal stability (fragmentation). The analytical problem becomes even considerably more complex for *polymeric* additives, in case of interaction between additive and polymer backbone (grafted functionalities), or in the presence of degradation phenomena. Analysis of polymeric additives such as Chimassorb 944 is notoriously difficult (cf. Sections 4.4.2.2 and 4.4.2.3). The same analytical misfortune may be bestowed on a polymer material supplier who stealthily tries to build up some knowledge of competitor products (either in a 'me too' approach or with more refined objectives). The task is the more demanding as it is not acceptable from an analytical point of view that fragmentation occurs during analysis. This may easily be the case in handling thermally labile compounds. For instance, both mass spectrometry and

thermal analysis indicate initial fragmentation resulting in the loss of 2-hydroxybenzophenone from a 2-hydroxybenzophenone based phosphite, or of 4-amino tetramethylpiperidine from a novel phosphite stabiliser based on a bis-hindered phenolic moiety coupled to a 4-aminotetramethylpiperidine chromophore [68].

The need for complete compositional analysis of additive packages in industrial plastics for both research and quality control applications has led to the development of numerous analyte-specific test procedures in recent years.

Table 1.12 and Figure 1.2 give a fair idea of the vast amount of polymer/additive analysis *literature* published in the last three decades and the effort spent on each technique. It is clearly not the purpose of this book to deal with analytical methods for any class of additives in particular, even less so to describe all reported analytical procedures for a given additive. Rather, those general analytical procedures are highlighted which have been used in the past and are still being applied nowadays, and especially those which may be expected to have an increasing impact in the near future. Over 90 % of all reported polymer/additive investigations have been concerned with polyolefins and polyesters.

Table 1.13, which lists the main techniques used for polymer/additive analysis, allows some interesting observations. Classical extraction methods still score very high amongst sample preparation techniques; on the other hand, not unexpectedly, inorganic analysis methods are not in frequent use; for separation purposes

Table 1.12 Scientific publications on analysis and determination of additives and additive classes in polymers and plastics

Class	No. entries	Class	No. entries
Additives	1647	Lubricants	437
Antioxidants	829	Metal deactivators	4
Antiozonants	34	Nucleating agents	47
Antistatics	86	Optical brighteners	10
Blowing agents	127	Photoinitiators	113
Cross-linking agents	442	Plasticisers	713
Emulsifiers	186	Quenchers	43
Fire/flame retardants	132	Stabilisers	530
Free radical scavengers	22	Surfactants	917
Hydroperoxide decomposers	16	UV-absorbers	100
Impact modifiers	32	Total	5792

Source: *Chem. Abstr.* 1968–1997.

Table 1.13 Scientific publications on analytical methods for the qualitative and quantitative determination of additives and additive classes of Table 1.12

Sample preparation techniques: extraction (776), Soxhlet (33), ultrasonics (24), microwave (21), SFE (22), Soxtec® (−), ASE® (2)
Chromatographic techniques: chromatography (1024), GC (424), HS-GC (6), GC-MS (102), (HP)LC (321), (HP)LC-FTIR (31), (HP)LC-MS (30), SEC (39), SFC (34), SFE-SFC (14), CE/CZE (32)
Spectroscopic techniques: spectroscopy (567), UV (478), (FT)IR (727), NIRS (15), NMR (247), SFE-SFC-FTIR (2)
Mass spectrometric techniques: mass spectrometry (387), CI (20), EI (7), ESP-MS (12), FAB (15), FD (11), FTICR-MS (44), MALDI-ToFMS (12), others (APCI, DCI, DI, DP, DT, FI, LSIMS, PB, PD, PSP, TSP) (4)
Thermal techniques: thermal analysis (357), pyrolysis (201), PyGC (97), PyMS (65), PyGC-MS (42)
Inorganics: AAS (16), AED (3), ICP (8), XRF (23)
Various: direct analysis (29), LD (24), ToF-SIMS (32), ISE (49)

Source: *Chem. Abstr.* 1968–1997.

chromatography (especially GC and HPLC) and thermo-analytical techniques are being relied upon; for detection much trust is laid upon (FT)IR, UV, NMR and MS, with the latter detection method being divided up into a bewildering number of subdisciplines. Hyphenated techniques are holding the future. Direct (in-polymer) analytical methods are still rather few but constitute a growth area.

Despite continuing improvements in instrumental methods for chemical analysis, the reliable analysis of organic (and inorganic) additives in polymers remains a formidable task because of the complexity of commercial polymer formulations [69,70]. Frequently more than one method may perform an analysis. An example of such a *multiple choice* is the quantitative analysis of a nonpolymeric component in a polymer matrix, such as dioctylphthalate (DOP) in PVC. If DOP is the only carbonyl containing material present, IR is feasible with suitable calibration. An alternative is quantitative analysis by GC, LC or SEC. Usually a multitechnique approach is necessary. A good *multidisciplinary analysis* is the study of performance of antistatic and slip additives in polyolefins, as studied by means of XPS, PA-FTIR, IR, TLC, NMR, potentiometry and chemiluminescence [71]. It is the added value of the analyst to apply those techniques that are most likely to provide a rapid answer.

In compliance with EURACHEM/CITAC Guide 2 [72] polymer/additive analysis can be considered as a collection of discrete subtasks (Figure 1.3), each consisting of a number of unit processes, themselves composed of modules containing routine unit operations. The unit processes are characterised as being separated by natural dividing lines at which work can be interrupted and the test portion can be stored without detriment before the next step.

Critical expert forums for all aspects related to the analytics of additives in polymers are the ACS Analytical Division, SPE Polymer Analysis and Polymer Modifiers & Additives Analysis Divisions or the German Arbeitskreis Polymeranalytik (cf. homepage DKI).

1.4.1 Regulations and Standardisation

The additive content of polymers needs to be monitored for quality and regulatory reasons. Examples of regulations with limits are food-contact rubber articles intended for repeated use (21 CFR 177.2600) and food-contact packaging for irradiated food (21 CFR 179.45). Unfortunately, the additive composition is not usually disclosed to the analysts of the food packaging industries, nor to those of food surveillance programmes. In the framework of control of materials and articles a systematic approach to such control has been elaborated in The Netherlands to meet *Dutch Regulations* [73] in existence before Directive 90/128/EEC. Practical application of this approach for over a decade has shown that analysis of the type of polymer used in a given food-contact situation, and of the additives and other constituents that might be present requires great experience. The Dutch test method is subject of discussion by the CEN working

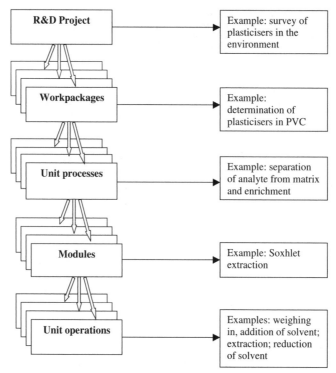

Figure 1.3 Breakdown of a polymer/additive analysis project into unit operations

group WG 2 of Technical Committee TC 194; a CEN standard is in preparation [74]. The Dutch method is applied in its original or slightly modified form by a number of government laboratories in Denmark, Greece, Norway, Sweden and Switzerland and in industrial test laboratories (especially by converters).

It should be mentioned that the Food Additives Analytical Manual (FAAM) [75] provides analysts with FDA evaluated methodology (partly subjected to collaborative study) needed to determine compliance with food additive regulations, including procedures for indirect food additives, such as butylated hydroxy-anisole (BHA), butylated hydroxytoluene (BHT), *t*-butylhydroxyquinone (TBHQ), dilaurylthiopropionate (DLTDP), fatty acid methyl esters (FAME), sodium benzoate, sorbitol, and others.

Analysts in industry prefer in many cases to maintain consistent methods for their analyses. Recommended ASTM analytical procedures are quite well developed in the rubber and polymer industry. As an example, we mention the standard test method for determination of phenolic antioxidants and erucamide slip additives in LDPE using liquid chromatography [76]. However, the current industry standard test methods (ASTM, AOAC, IUPAC, etc.) use a large number of solvents in vast

quantities whereas spent solvent waste stream disposal has become an important issue.

National and *supranational (EEC) regulations* are being enforced to exercise control on the use of a list of additives used for the production of food contact plastics (Synoptic Document N.7 of the Commission of the European Communities [77]), but often without adequate analytical support. Moreover, industry has generated its own *company-specific standards* and (validated) analytical procedures for polymer/additive analysis. For reasons of compliance the polymer producers (chemical industry) are well advised to take good notice of company-specific norms of their end-users, e.g. document D 40 5271 (PSA) regulates extraction of plasticisers and additives, instruction D 40 1753 (PSA) [78] concerns the quantitative evaluation of the principal components of vulcanised rubbers by means of thermogravimetric analysis, document PV 3935 (Volkswagen/Seat/Škoda/Audi) [79] describes an analytical method for the determination of polymer type and additives by means of pyrolysis - gas chromatography/mass spectrometry (PyGC-MS), whereas instruction PB VWT 709 (Daimler Benz) [80] regulates the determination of gaseous and condensable emissions in car interiors by means of thermodesorption.

For standard or proprietary polymer additive blends there is the need for analytical certification of the components. Blend technology has been developed for two- to six-component polymer additive blend systems, with certified analytical results [81]. Finally, there exist physical collections of reference additive samples, both public [82] and proprietary. The Dutch Food Inspection Service reference collection comprises 100 of the most important additives used in food contact plastics [83–85]. Reference compounds of a broad range of additives used in commercial plastics and rubber formulations are generally also available from the major additive manufacturers. These additive samples can be used as reference or calibration standards for chromatographic or spectroscopic analysis. DSM Plastics Reference Collection of Additives comprises over 1400 samples.

1.4.2 Prior Art

As shown in Appendix II there is a multitude of books concerning various aspects of additives in polymers (either commercial information or technical/scientific principles). However, in the past, only a fairly restricted number of contributions in *textbooks* has devoted attention to the analysis of additives in plastics. The field of polymer/additive analysis has grown steadily since its inception in the late 1950s.

Leadership in many branches of chemistry resides outside the traditional academic boundaries. In many areas, including polymer/additive analytics, industrial laboratories have assumed the leadership. This is not surprising because of the geography of the problem. Haslam *et al.*'s classic textbook [54] on the identification and analysis of plastics, reflecting the considerable analytical experience of the ICI Plastics Division, has been *the* established working reference source for industrial chemists concerned with plastics analysis in the 1965–1983 period (though limited to literature up to 1970 only). However, the authors deal mostly with the molecular characterisation of (co)polymers, such as vinyl resins, polyesters, nylons, polyolefins, fluorocarbon polymers, rubbers, thermosetting rubbers, and natural rubbers, with limited attention to the analysis of plasticisers, fillers and solvents. Whereas in the first edition (1965) attention was restricted mainly to IR methods, the state-of-the-art techniques in 1970 had broadened considerably to include UV/VIS, IR, NMR, GC, PyGC, GC-IR, AAS, AES, XRF and automated titrations. Although now dated and obviously light on modern instrumental methodology this book contains a wealth of information on polymer/additive analysis. In the German language area early books on polymer analysis by Schröder *et al.* [55] and on rubber analysis by Ostromow [61] should be mentioned. Krause *et al.* [86] have also briefly described chemical analysis of additives (limited to fillers, stabilisers, dyes and pigments), and have compiled an extensive and useful index to ASTM and DIN standards for analysis and characterisation of polymers, resins, rubbers, plastics and fibres. In the past, chromatographic, spectral, mass spectrometric, electrochemical and radioisotopic methods were most widely employed for the determination of additives [87].

Crompton [88,89] has regularly provided detailed accounts of the scientific principles underlying current practice, targeted mainly at the experienced industrial technologist. An extensive review on polymer/additive analyses (period 1960–1980) is contained in Crompton [52]. More recently, the same author [41] has described polymer analysis (polymer microstructure, copolymer composition, molecular weight distribution, functional groups, fractionation) together with polymer/additive analysis (separation of polymer and additives, identification of additives, volatiles and catalyst residues); the monograph provides a single source of information on polymer/additive analysis techniques up to 1980. Crompton described *practical analytical methods* for the determination of classes of additives (by functionality: antioxidants, stabilisers, antiozonants, plasticisers, pigments, flame retardants, accelerators, etc.). Mitchell [53] has covered many aspects of polymer analysis and characterisation, including analysis of additives, residual monomers and oligomers, moisture and adventitious impurities. Gooch [59] has recently addressed the analysis and deformulation of a variety of polymeric materials (paints, plastics, adhesives and inks).

In a later manual of plastics analysis Crompton [50] deals with *all* aspects of polymer analysis, including the polymer structure (compositional analysis), as well as deliberately added nonpolymeric processing chemicals used during manufacture, chemicals added to improve the polymer properties during service life and impurities, such as water and processing solvents, unreacted monomers, etc. If any criticism is allowed, only a modest and very selective share (20 %) of references *post* 1980 is included in this textbook, which therefore cannot claim to be an up-to-date or complete review of the world literature on the subject of polymer/additive analysis. Some 108 detailed experimental procedures, again largely dated (1950s 3 %, 1960s 27 %, 1970s 30 %, 1980s 7 %, previously unpublished 33 %), were included. In this respect the *current text*, organised by clusters of analytical techniques dressed up with applications, is a more rational rather than a descriptive approach, apart from being dedicated *exclusively* to polymer/additive analysis and more up-to-date (up to end 2002). Scheirs [58] has described a range of techniques and strategies for the compositional and failure analysis of polymeric materials and products, with applications of analytical methods for troubleshooting industrial problems. This practical description (with literature coverage up to 1997) is complementary to the current more analytical approach. Forrest [63] has recently touched on techniques and methods used to characterise and carry out QC work on plastics, deformulation of plastic compounds and investigation of failure of plastic products (both molecular characterisation and additive analysis). The same author [90] has described analytical aspects of the characterisation of rubber polymers, with emphasis on the determination of the principal components in rubber compounds and product deformulation.

The recent literature also contains a number of *reviews* more limited in scope and frequently outside the English language area (Table 1.14). Several papers are worth mentioning. Squirrell [139] has reviewed ICI's approach towards the analysis of additives and process residues in plastics materials (state-of-the-art 1981); Scrivens and Jackson [111] have described current MS practice. Lattimer and Harris [105] have put emphasis on the extraction of additives from polymers, MS analysis of the extracts, GC-MS and LC-MS as well as on direct mass spectral analysis of polymer additives, including thermal desorption and pyrolysis. Developments in techniques, instrumentation and problem solving in applied polymer analysis and characterisation up to 1987 have been described by Mitchell [134]. Later Foster [140] has addressed analytical methods in relation to testing of oxidation inhibition, and Rotschová and Pospíšil [130] have published an excellent review covering both indirect and direct analysis methods of stabilisers (state-of-the art 1989). A fairly comprehensive overview of the literature on additive analysis (restricted to some protective agents, such as antioxidants and UV stabilisers) has been published by Freitag [133] in 1993 with emphasis on sample preparation (Soxhlet and reflux extraction, dissolution/precipitation), TLC, HPLC, GC and various spectroscopic techniques. Other reviews are more specific. Munteanu [98] has reported an excellent account of the analysis of antioxidants and light stabilisers in polyolefins by HPLC. Newton [94] has reviewed wet chemical analysis of additives in various polymers (PP, PVC, PTFE and polyamides). Thomas [131] has presented a general overview of the analytical techniques available to qualitate and quantify the primary and secondary additive stabilisers (antioxidants, processing and

Table 1.14 Selected reviews concerning polymer/additive analysis

Topic(s)	Reference(s)
Sample preparation (SFE, chromatographic analysis)	[91,92]
Extractions	[93]
Separation and analysis of additives in polymers	[94]
Qualitative/quantitative analysis (GC)	[95]
Qualitative/quantitative analysis (surface micro analysis)	[96]
Vulcanised rubbers (HS-GC)	[97]
Polyolefins, antioxidants and light stabilisers (LC)	[98]
TLC applications	[99]
SFC applications	[100]
Pyrolysis-GC applications	[101]
PP pellets, additives (NIRS)	[102]
PVC, plasticisers, stabilisers, fillers (IR, GC)	[103]
Volatile additives (direct mass spectrometry)	[104]
Organic additives (qualitative MS analysis, TD, PyMS, GC-MS, LC-MS)	[105]
LC-MS applications	[106]
FAB-MS applications	[107,108]
LDI applications	[109]
LD FTICR-MS applications	[110]
Direct and indirect analysis (FD-MS, MALDI, LSIMS, TD-GC-MS)	[111]
Inorganic additives	[112]
Thermal analysis, applications	[24,113,114]
Thermal evolution techniques, applications	[115]
Blends of monomeric and polymeric additives, separations	[116]
Polymer/additive analysis; general; softeners	[96]
Additives in plastics and rubbers (instrumental analysis)	[117–124]
HALS (chromatographic and spectroscopic methods)	[125]
Low-MW organic additives (extraction)	[126]
Radioactive tracers in stabilisation technology	[127]
Packaging materials, residual monomers	[95]
Cable insulating compounds, additives	[128]
Surface modifying additives, siloxane surfactants	[129]
Identification and determination of stabilisers of oxidation processes	[130]
Stabilisers (chromatography, analytical artefacts)	[6]
Polyolefins, antioxidants, processing and light stabilisers	[131–133]
Additives, impurities, degradation products	[134]
Rubbers, sulfur-containing additives	[135]
Antimicrobials	[136]
Antioxidants	[137,138]

light stabilisers) in polyolefin substrates. Berger [100] has examined the use of SFC in the analysis of polymer additives. Bataillard *et al.* [132] have recently given broad coverage of polymer/additive analytics. Gouya [141] has recently reviewed dispersion and analytical methods of a variety of polymer additives (plasticisers, lubricants, stabilisers, coupling agents, photocatalysts, surfactants, vulcanisation and cross-linking agents). Finally, Hummel [62] has described the application of vibrational (FTIR, UV, Raman) and mass spectrometries for identification and structure elucidation of plastics additives (mainly antioxidants, stabilisers, plasticisers, pigments, fillers, rubber chemicals). Other reviews are mentioned under the specific headings of the following chapters. Many analytical tools are more capable of compound *class* identification than of specific compound analysis.

Progress in the field of polymer/additive analysis in the last three decades can best be illustrated by an old recipe for the direct determination of organotin stabilisers in PVC [142]:

PVC (200 mg) was dissolved in 20 mL THF and precipitated with 50 mL EtOH. Several drops of 0.1 % pyrocatechol violet solution were added to the heated filtrate until a blue colour appeared. This solution was titrated with 0.001 M EDTA until the change via green to yellow. In the presence of Mg, Ca, and Zn, Eriochrome Black was added before titration.

Physico-chemical instrumental analysis nowadays has greatly suppressed such chemical handwork. An internet website disseminates methods of analysis and supporting spectroscopic information on monomers and additives used for food contact materials (principally packaging).

1.4.3 Databases

Access to databases provides one of the most critical tools for polymer/additive analysis, as it greatly determines the efficiency (and cost) of the overall operations (cf. Table 1.15). The literature provides a wealth of *spectral information*, available in bound volumes of illustrations or in a computer format (e.g. on CD ROMs), such as the Bio-Rad Sadtler, Aldrich, Bruker-Merck, Nicolet and Rapra collections of FTIR and/or NMR spectra of organic compounds, additives (including plasticisers, flame retardants, surfactants), rubber chemicals, polymers and inorganics (including minerals). Typically, on-line IR (and Raman) data files comprise approximately 200 000 spectra of pure compounds (Sadtler with 1740 polymer additives, 1480 plasticisers, 590 flame retardants, 3070 dyes and pigments, 700 curing agents, 850 basic surfactants as specific products),

7000 ('Hummel'), 18 500 (Aldrich) [143], 3000 (Merck) [144], 3000 (coatings), 1000 (inorganics) and 576 spectra (Rapra Collection of Infrared Spectra of Rubbers, Plastics and Thermoplastic Elastomers) [145]. Over 50 000 FTIR, NIR and Raman spectra of polymers and related compounds are available. Important support has been provided by Scholl [146] in 1981 with a spectroscopic atlas of fillers and processing additives and later more extensively (analytical methods and spectroscopy) by Hummel. The 'Hummel/Scholl infrared databases' (now Chemical Concepts) include fillers, processing aids and surfactants [146–150], with typically 1520 auxiliaries and additives, and 1570 monomers and low-MW substances. The atlas of industrial surfactants comprises 1082 FTIR spectra [149]. Infrared spectra of plasticisers have also been collected [151]. Library searching of SFC/FTIR spectra can be carried out with reference to the Sprouse Library of Polymer Additives [152], as distributed by Nicolet. This library contains 325 reference spectra recorded either neat, as chloroform casts, or as nujol mulls. A recent atlas of plastics additives contains 772 FTIR spectra [62]. However, e-databases are more efficient than those in printed format. As to the distribution of reference spectroscopic databases, cf. Davies [153], Hummel [154] has illustrated the possibilities and limitations of computer-based searching with special FTIR libraries (additives, surfactants, monomers, etc.).

As public libraries are far from being complete as to (newly introduced) commercial polymer additives various *proprietary databases* have been created. In this respect, ICI keeps a company database on polyolefin stabilisers with over 4000 records [155]. The Hoechst IR and Raman spectroscopic database (80 000 entries) contains some 500 internal reference data for polymer additives [156]. DSM Research has developed a more general customised polymer additive reference database for analytical purposes running on Windows 98 and NT and comprising over 1000 of the most common industrially used additives with molecular and structural formula, molecular weight, systematic chemical name, and commercial brand names, CAS Registry number, commercial names, GC, HPLC, NMR, UV, IR, Raman, MS and ToF-SIMS data, some chemical and physical properties (such as melting point, density, etc.) and safety data [157]. DSM also disposes of a proprietary trade mark/chemical structure database for over 250 antioxidants, light and heat stabilisers (with some 1100 entries [17], cf. Appendix III). It is noticed that this database has grown considerably in the last 5 years. Ciba-Geigy has issued positive lists of additives for plastics, elastomers and synthetic fibres [158] and of additives used in the PVC industry [159], both for food contact applications.

Additive analysis also greatly benefits from the new NIST 02 release of the NIST/EPA/NIH Mass Spectral Library, containing 143 000 verified spectra and almost complete coverage with chemical structures [160]; the Wiley library contains 230 000 mass spectra. However, all mass spectra are not created equal. Consequently, secure comparison of mass spectra for positive compound identification requires attention to the ionisation mode. In relation to PyGC-MS analysis the Shimadzu/VW additive mass spectra library is of interest [161]. The 1993 publication *Spectra for the Identification of Monomers in Food Packaging* [162,163] presents FTIR and MS information on monomeric substances and other starting substances listed in Directive 90/128/EEC [65], which restricts the range of compounds that can be used for the production of plastics materials and articles intended for food contact applications. Reference [164] contains other spectra for the identification of additives in food packaging. The handbook *Spectra for the Identification of Additives in Food Packaging* [84] compiled with EC funding under the SM&T programme, contains a collection of spectra for the identification of 100 of the most important additives used in plastics packaging and coatings. Infrared and mass spectra are presented, together with ^{1}H NMR spectra and GC data. This file is accessible via an internet website (*http://cpf.jrc.it/smt/*) [82]. In another recent compilation nearly 400 fully assigned NMR spectra of some 300 polymers and polymer additives are collected [165].

Obviously, use of such databases often fails in case of interaction between additives. As an example we mention additive/antistat interaction in PP, as observed by Dieckmann *et al.* [166]. In this case analysis and performance data demonstrate chemical interaction between glycerol esters and acid neutralisers. This phenomenon is pronounced when the additive is a strong base, like synthetic hydrotalcite, or a metal carboxylate. Similar problems may arise after ageing of a polymer. A common request in a technical support analytical laboratory is to analyse the additives in a sample that has prematurely failed in an exposure test, when at best an unexposed control sample is available. Under some circumstances, heat or light exposure may have transformed the additive into other products. *Reaction product identification* then usually requires a general library of their spectroscopic or mass spectrometric profiles. For example, Bell *et al.* [167] have focused attention on the degradation of light stabilisers and antioxidants in chemical and photo-oxidising environments. HPLC-UV/VIS, FTIR, and GC-MS were found to be suitable techniques to follow the degradation chemistry of the additives. On the other hand, the study of the chemistry of benzotriazoles turned out to be more difficult because of the insolubility of the resinous degradation products.

The importance of adequate support by database reference material is well illustrated with the following case. After chromatographic separation (TLC, CC), the combination of ^{1}H/^{13}C NMR, DI-MS (EI), FTIR and HPLC (UV/VIS, DAD and MS) a flame retardant in a Japanese polypropylene TV cabinet on the European market was identified as tetrabromobisphenol-*S*-bis-(2,3-dibromopropyl ether) (TBBP-S) [168]. The result was verified by synthesis of reference material; the product was finally identified as Non Nen #52 from Marubishi Oil Chemical Co., Ltd (Osaka), not registered in any spectral database.

1.4.4 Scope

As the structure of additives for polymers becomes evermore complex, there is an increasing need for good reliable analysis of additives to meet more exacting performance demands. High-quality analytical methods are needed for high-quality products. As the rate of accumulation of scientific knowledge accelerates, it becomes increasingly difficult for scientists to find relevant information quickly and effectively and to put it in the proper context. This certainly holds for such a broad field as the analytics of polymer additives. This book provides comprehensive coverage of the current status of the (qualitative and quantitative) analysis techniques for additive determination in commercial polymers at the lowest level of analytical sophistication (bulk analysis). No technique will suit every need. Emphasis is laid on understanding and applicability. As additive analysis is typically an *industrial problem solving* area particular attention is paid to cost effective, real-life, analytical approaches. In particular, the prospects of analysis conducted after separation of the additives from the polymer are compared. Recent years have seen an almost quantum increase in the range of analytical techniques, particularly involving hyphenated chromatography, spectroscopy and mass spectrometry. This book draws them all together, in a comprehensively up-dated version with specific applications. Limitations of current additive analysis methodology are indicated.

The main goal of this book is to set the scene for 'Analytical Excellence' in the field of polymer/additive analytics, not unlike Manufacturing or Operational Excellence programs in industry. Table 1.15 shows the necessary ingredients. For product analysis a mix of analytical technique specialists and product specialists is ideal. Some industrial analytical departments are structured in this fashion [169]. The key to the successful analysis of additives in a polymer for a specific application not only requires a comprehensive understanding of commercial additives but also knowledge of the polymer matrix and its targeted application, as well as the required tests to be passed for that specific application [170,171]. For in-polymer additive analysis already Crompton [52] had stressed the importance of familiarity with the chemical and physical properties of additives. For this purpose the reader is referred to Appendix II.

An adequate measurement technique is not sufficient for good analytical results. Analyses are operated under ISO 9001 and carried out according to validated analytical protocols. Deviations from such protocols are to be given as amendments (a priori) or as deviations (a

posteriori). Moreover, under no conditions the analyst should change the analyte.

It is clearly not the *purpose* of this monograph to deal with analytical methods for any class of additives in particular. Also, this book is not intended to be a manual for reported analytical procedures for a given additive. Yet, in order to comply with the primary value of reporting particular method developments and applications in the literature, namely that the reader knows that at least someone was successful with a particular analyte in a particular matrix, extensive referencing is included. Literature was covered as comprised in *CA Selects^{SM} Plastics Additives* up to the end of 2002. In this text those analytical procedures in particular are highlighted which have been used in the (recent) past and are still being applied today and especially those which may reasonably be expected to have an increasing impact in the near future. Relevant techniques are detailed up to the point that the applications may be rationalised and understood. Although the devil is often in the detail, a deeper level of abstraction of the numerous analytical tools utilised (Figure 1.4) would have led to an unmanageable amount of experimental data for which the original literature is the most suitable source of information. By describing many real applications, the author tries to alert the reader to the opportunities of the various techniques. For the selection of the many citations in this monograph, next to their information content, their topicality and availability also played a role. No claim is made to comprehensive coverage.

This highly specific book is not a collection of independent reviews, but promotes understanding and emphasises development of problem-solving ability. It is just the tip of the iceberg of the field. Expecting to use cookbook methods does not necessarily work. In fact, such methods are unavailable for most analyte–matrix pairs anyway, and actually are pretty dangerous for the unaware. This book just provides the routing (Figure 1.5). Those applying the calibration equation coefficients of a published method and expecting to get immediately good quantitative results may easily get disillusioned. Such an approach only works well when the application is very well defined and adhered to.

It should be understood that the reported practices of polymer/additive analysis, being the focus of this book, equally well apply to additive analysis of rubbers, textile fibres, surface coatings, paints, resins, adhesives, paper and food, but specific product knowledge gives the edge. Both fresh and aged materials may be analysed, as well as those of both industrial and forensic origin.

Table 1.15 Requirements for Analytical Excellence

Feature	Action(s)
Analytical strategy	Clear-cut corporate strategy
Product analysis	Technique and product specialists
Operator competence	Analyte specific physico-chemical background
Standardisation	Norms, validated analytical procedures, certification
State-of-the-art	Instrumentation, method development, current awareness
Multidisciplinary approach	Standard Operating Procedures
Quality control	Calibration, (certified) reference materials
Efficiency	Databases, benchmarking, 'Best Practice'
Profitability	Excellent knowledge infrastructure, automation, low cost
Analytical information management	Central repository for analytical data
Analytical sample management	LIMS
Quality assurance	ISO 9001
Reproducibility	SPC
Reliance	No dependence on one analysis technique
Bureaucracy	Fewer analytical chemists than analytical problems
Proficiency testing	Round-robins
Health, safety, and environment	Responsible Care

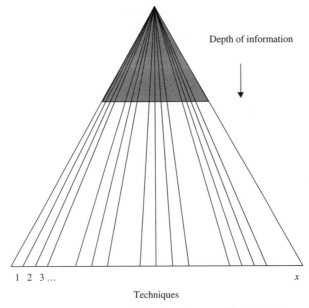

Depth of information

1 2 3...   x

Techniques

Figure 1.4 Level of abstraction of analytical tools on additives in polymers with delimitations adopted in the text (shaded)

Figure 1.5 Road map of polymer/additive analysis. Reprinted with permission from *LC.GC Europe*, Vol. 14, Number 8, August 2001. LC.GC Europe is a copyrighted publication of Advanstar Communications Inc. All rights reserved

This book, being the first one *entirely* dedicated to analysis of additives in polymers, is closely targeted to R&D units, manufacturers, compounders, leading end-users, universities and colleges, government/independent testing and certification bodies. It is expected to contribute to the development of recognised analytical procedures for the determination of constituents of plastics, as aspired to by various organisations ranging from forensic institutions to Greenpeace [172] and others.

Many polymer companies have not maintained a cadre of experts on the analysis of additives in polymers. Consequently, there is a need to train a new generation of people about additives and methods of deformulating them. Outsourcing of polymer/additive analysis is usually not an option for the reasons mentioned (unfamiliarity with the underlying chemistry). Today, there are also more sophisticated compounders who formulate their own products. Along with the increased popularity of blended, pelletised and compacted mixtures of additives, this suggests that there should be broad interest in the topics of this book among formulators of concentrates and additive producers.

1.4.5 Chapter Overview

Methods of analysis are either chemical or physical in nature. Chemical methods of analysis are based on the selective interaction of materials (chromatographic media, etc.) with analytes. Physical methods of analysis are based on the direct interaction of materials (analyte) with electromagnetic waves; corpuscular beams, such as electrons and neutrons; or electric, magnetic, and gravimetric fields. It is important to appreciate that various methods of instrumental analysis, including all kinds of spectroscopies, have completely changed the nature of chemical research. Physical and chemical methods are not clearly separated. The goal of Chapter 2 is to familiarise the reader with general reverse engineering schemes proposed for polymers and rubbers. Industrial practice and developments are illustrated for the 1980–2002 period.

Conventional polymer/additive analysis by wet chemical means is described in seven chapters. Chapter 3 describes the separation of the polymer from the additive components. Consequently, emphasis is both on classical liquid–solid extractions and modern pressurised extraction methods. Digestion techniques and depolymerisation approaches to polymer/additive analysis are also treated. Chapter 4 provides a critical review of modern chromatographic separation techniques for additive analysis, based on the solvent extracts according to Chapter 3, or evaporative losses. The main techniques and applications covered are GC, HS-GC, HPLC, SEC, GPEC®, TLC and CE. The power of specific detection modes is evaluated. Different spectroscopic methods (UV/VIS, mid-IR, luminescence and NMR), applicable to additives in extracts before and after chromatographic separation, are described in Chapter 5 and are critically evaluated. Chapter 6 summarises the power of the most important organic mass spectrometric techniques for direct and indirect additive analysis. In Chapter 7 use and power of a great variety of (multi)hyphenated and multidimensional techniques (sample preparation–chromatography–detection) for polymer/additive analysis of extracts are assessed. The applications of multidimensional spectroscopy are outlined. The reader is given detailed insight in the possibilities and limitations of these sophisticated analytical techniques for the specific applications of interest. This chapter reflects the explosion of procedures in the last few years. Both conventional and more modern element analytical protocols for extract and in-polymer additive analysis are illustrated in Chapter 8, including inorganic MS and radioanalytical methods. The limited usefulness of electrochemical techniques for this purpose is pointed out. Chapter 9 addresses the prospects for direct chromatographic, spectroscopic and mass spectrometric methods of deformulation of polymer/additive dissolutions without prior separation of the components.

The final chapter summarises the book with special emphasis on the future of polymer/additive analysis. The methods, results and their evaluation presented in this chapter encompass all material developed in the book's previous chapters. Three appendices contain lists of symbols, describe the functionality of common additives (as a reminder) and show an excerpt of an industrial polymer additive database.

1.5 BIBLIOGRAPHY

1.5.1 Plastics Additives

J.S. Murphy (ed.), *Additives for Plastics*, Elsevier Advanced Technology, Oxford (2001).

H. Zweifel (ed.), *Plastics Additives Handbook*, Hanser Publishers, Munich (2000).

M. Ash and I. Ash, *Handbook of Industrial Chemical Additives*, Synapse Information Resources, Endicott, NY (2000).

G. Pritchard, *Plastics Additives. An A–Z Reference*, Chapman & Hall, London (1998).

M. Ash and I. Ash, *Handbook of Plastic and Rubber Additives*, Gower Publishing Ltd, Aldershot (1995).

J. Edenbaum (ed.), *Plastics Additives and Modifiers Handbook*, Van Nostrand Reinhold, New York, NY (1992).

R. Gächter and H. Müller (eds), *Plastics Additives Handbook*, Hanser Publishers, Munich (1990).

J.T. Lutz (ed.), *Thermoplastic Polymer Additives: Theory and Practice*, M. Dekker, New York, NY (1989).

I.P. Maslov (ed.), *Chemical Additives in Polymers, Handbook* (in Russian), Khimiya, Moscow (1981).

1.5.2 Processing Technologies

W. Hoyle (ed.), *Developments in Handling and Processing Technologies*, The Royal Society of Chemistry, Cambridge (2001).

N. Harnby, M.F. Edwards and A.W. Nienow, *Mixing in the Process Industries*, Butterworth-Heinemann Ltd, Oxford (1992).

1.5.3 Instrumental Analysis

H. Günzler and A.W. Williams (eds), *Handbook of Analytical Techniques*, Wiley-VCH, Weinheim (2001).

K.A. Rubinson and J.F. Rubinson, *Contemporary Instrumental Analysis*, Prentice Hall, Upper Saddle River, NJ (2000).

F. Rouessac and A. Rouessac, *Chemical Analysis: Modern Instrumentation Methods and Techniques*, John Wiley & Sons, New York, NY (2000).

R. Kellner, J.-M. Mermet, M. Otto and H.M. Widmer (eds), *Analytical Chemistry*, Wiley-VCH, Weinheim (1998).

A. Settle (ed.), *Handbook of Instrumental Techniques in Analytical Chemistry*, Prentice Hall, Upper Saddle River, NJ (1997).

J.R.J. Paré and J.M.R. Bélanger, *Instrumental Methods in Food Analysis*, Elsevier, Amsterdam (1997).

1.5.4 Polymer Analysis

H. Lobo and J. Bonilla, *Handbook of Plastics Analysis*, M. Dekker, New York, NY (2003).

B. Stuart, *Polymer Analysis*, John Wiley & Sons, Ltd, Chichester (2002).

J. Scheirs, *Compositional and Failure Analysis of Polymers. A Practical Approach*, John Wiley & Sons, Ltd, Chichester (2000).

D. Campbell, R.A. Pethrick and J.R. White, *Polymer Characterization*, International Specialized Book Services, Portland, OR (2000).

R.A. Pethrick and J.V. Dawkins (eds), *Modern Techniques for Polymer Characterisation*, John Wiley & Sons, Ltd, Chichester (1999).

S.J. Spells (ed.), *Characterization of Solid Polymers – New Techniques and Developments*, Kluwer, Dordrecht (1994).

B.J. Hunt and M.I. James (eds), *Polymer Characterisation*, Blackie A&P, London (1993).

G. Kämpf, *Characterization of Plastics by Physical Methods*, Hanser Gardner, Cincinnati, OH (1986).

The Japan Society for Analytical Chemistry, *Handbook of Polymer Analysis (Kobunshi Bunseki Handobukku)*, Asakura Publ. Co., Tokyo (1985).

1.5.5 Polymer/additive Analysis

D.O. Hummel, *Atlas of Plastics Additives: Analysis by Spectrometric Methods*, Springer-Verlag, Berlin (2002).

M.F. Forrest, *Analysis of Plastics, Rapra Review Report No. 149*, Rapra Technology Ltd, Shawbury (2002).

T.R. Crompton, *Manual of Plastics Analysis*, Plenum Press, New York, NY (1998).

T.R. Crompton, *Practical Polymer Analysis*, Plenum Press, New York, NY (1993).

T.R. Crompton, *Analysis of Polymers. An Introduction*, Pergamon Press, Oxford (1989).

J. Mitchell (ed.), *Applied Polymer Analysis and Characterization: Recent Developments in Techniques, Instrumentation, Problem Solving*, Hanser Publishers, Munich (1987).

A. Krause, A. Lange and M. Ezrin, *Plastics Analysis Guide – Chemical and Instrumental Methods*, Hanser Verlag, Munich (1983).

1.6 REFERENCES

1. I.V. Germashev, V.E. Derbisher, M.N. Tsapleva and E.V. Derbisher, *Khim. Promyshl.* (6), 43–8 (2002).
2. P. Molander, R. Trones, K. Haugland and T. Greibrokk, *Analyst* **124**, 1137–41 (1999).
3. R. Trones, T. Andersen, I. Hunnes and T. Greibrokk, *J. Chromatogr.* **A814**, 55–61 (1998).
4. R. Trones, T. Andersen, T. Greibrokk and D.R. Hegna, *J. Chromatogr.* **A874**, 65–71 (2000).
5. T. Bücherl, A. Gruner and N. Palibroda, *Packag. Technol. Sci.* **7**, 139–54 (1994).
6. B.S. Middleditch, *Analytical Artifacts GC, MS, HPLC, TLC and PC*, *J. Chromatogr. Library Series No. 44*, Elsevier, Amsterdam (1989).
7. B. Adhikari, T.K. Khamra and S. Maiti, *J. Sci. Ind. Res.* **51** (6), 429–43 (1992).
8. P. Pothier (ed.), *Additives for Paper Coating*, TAPPI Press, Atlanta, GA (1998).
9. K. Titzschkau, *Proceedings Polymer Additives 95* (Chicago, IL), Intertech Corporation, Portland, MN (1995).
10. R. Weinekötter and H. Gericke, *Mischen von Feststoffen*, Springer-Verlag, Berlin (1995).
11. H. Gericke (ed.), *Dosieren von Feststoffen (Schüttgütern)*, Gericke GmbH, Rielasingen (1989).
12. N. Singh, J. Anostario, A. Tate and R. Dean, *Polym. Prepr. (Am. Chem. Soc., Div. Polym. Chem.)* **39** (1), 518–9 (1998).
13. H. Müller and R. Herbert, *Proceedings Addcon World 2002*, Budapest (2002), Paper 22.
14. M. Bauer, *Proceedings Addcon World 2002*, Budapest (2002), Paper 13.
15. U. Dietrich, *Kunststoffe* **88** (6), 858–60 (1998).
16. R. Steiner, *Proceedings Masterbatch 98*, Düsseldorf (1998).
17. P. Gijsman, *Stabiliser Trade Name-Chemical Structure Index*, DSM Research, Geleen (2004).
18. G. Sasselli, *Proceedings Polypropylene '98 (7th Annual World Congress)*, Maack Business Services, Zürich (1998).
19. C. Callierotti, R.E. Lee, O.I. Kuvshinnikova, K. Keck-Antoine, B. Johnson and J.W. Kim, *Proceedings Polyolefins 2001*, Houston, TX (2001), pp. 447–53.
20. B.W. Johnson, K. Keck-Antoine, R.E. Lee and C. Callierotti, *Plast. Addit. Compd.* **3** (6), 14–21 (2001).
21. M.H. Pahl and M. Grosse-Aschhoff, *Kunststoffe* **87** (7), 872–4 (1997).
22. C. Bruens, R. Nieland and D. Stanssens, *Polym. Polym. Comp.* **7** (8), 581–7 (1999).
23. H.P. Schlumpf, in *Plastics Additives Handbook* (R. Gächter and H. Müller, eds), Hanser Publishers, Munich (1990), pp. 525–91.
24. H.E. Bair, in *Thermal Characterization of Polymeric Materials* (E.A. Turi, ed.), Academic Press, San Diego, CA (1997), Ch. 10.
25. L.G. Meisinger, *Chem. Plants Process.* (3), 46–7 (2000).
26. M. Hubis, in *Plastics Additives Handbook* (H. Zweifel, ed.), Hanser Publishers, Munich (2000), pp. 1112–22.
27. M. Ash and I. Ash, *Handbook of Plastic and Rubber Additives*, Gower Publishing Ltd, Aldershot (1995).
28. M. Ash and I. Ash, *The Index of Antioxidants and Antiozonants*, Gower Publishing Ltd, Aldershot (1997).

29. M. Ash and I. Ash, *The Index of Flame Retardants*, Gower Publishing Ltd, Aldershot (1997).

30. M. Ash and I. Ash, *The Index of Antimicrobials*, Gower Publishing Ltd, Aldershot (1996).

31. R. Schmidt, *Proceedings Addcon World '98*, London (1998), Paper 17.

32. Phillip Townsend Associates Inc., *Chemical Additives for Plastics: Apparent Supplier Strategies*, Houston, TX (1997).

33. L.N. Kattas and F. Gastrock, *Proceedings Addcon World 2002*, Budapest (2002), Paper 1.

34. P. Dufton, *Functional Additives for the Plastics Industry*, Rapra Technology Ltd, Shawbury (1998).

35. T. Galvanek, T. Gastrock and L.N. Kattas, *Proceedings Addcon World 2000*, Basel (2000), Paper 6.

36. L.N. Kattas, *Proceedings Addcon World '98*, London (1998), Paper 1.

37. Frost & Sullivan, *The European Market for Plastic Additives*, Frankfurt (1999).

38. Chemical Market Resources Inc., *Plasticizers, Markets, Technologies & Trends 1999–2004*, Houston, TX (1999).

39. R.A. Pethrick and J.V. Dawkins (eds), *Modern Techniques for Polymer Characterisation*, John Wiley & Sons, Ltd, Chichester (1999).

40. H. Pasch and B. Trathnigg, *HPLC of Polymers*, Springer-Verlag, Berlin (1997).

41. T.R. Crompton, *Practical Polymer Analysis*, Plenum Press, New York, NY (1993).

42. M. Potschka and P.L. Dubin (eds), *Strategies in Size Exclusion Chromatography*, ACS Symp. Series 635, American Chemical Society, Washington, DC (1996).

43. S.R. Holding and E. Meehan, *Molecular Weight Characterisation of Synthetic Polymers, Rapra Review Report No. 83*, Rapra Technology Ltd, Shawbury (1995).

44. Th.P. Wampler (ed.), *Applied Pyrolysis Handbook*, M. Dekker, New York, NY (1995).

45. B. Stuart, *Polymer Analysis*, John Wiley & Sons, Ltd, Chichester (2002).

46. T. Provder (ed.), *Chromatography of Polymers*, ACS Symp. Series 521, American Chemical Society, Washington, DC (1993).

47. R. White, *Chromatography/Fourier Transform Infrared Spectroscopy and Its Applications*, M. Dekker, New York, NY (1990).

48. M.P. Stevens, *Polymer Chemistry. An Introduction*, Oxford University Press, New York, NY (1990).

49. J.I. Kroschwitz (ed.), *Polymers: Polymer Characterization and Analysis*, John Wiley & Sons, Inc., New York, NY (1990).

50. T.R. Crompton, *Manual of Plastics Analysis*, Plenum Press, New York, NY (1998).

51. A.R. Cooper (ed.), *Determination of Molecular Weight*, John Wiley & Sons, Inc., New York, NY (1989).

52. T.R. Crompton, *Analysis of Polymers. An Introduction*, Pergamon Press, Oxford (1989).

53. J. Mitchell (ed.), *Applied Polymer Analysis and Characterization, Recent Developments in Techniques, Instrumentation, Problem Solving*, Hanser Publishers, Munich (1987).

54. J. Haslam, H.A. Willis and D.C.M. Squirrell, *Identification and Analysis of Plastics*, John Wiley & Sons, Ltd, Chichester (1983, and earlier editions).

55. E. Schröder, J. Franz and E. Hagen, *Ausgewählte Methoden zur Plastanalytik*, Akademie Verlag, Berlin (1976), p. 31.

56. D. Campbell, R.A. Pethrick and J.R. White, *Polymer Characterization*, International Specialized Book Services, Portland, OR (2000).

57. G. Kämpf, *Characterization of Plastics by Physical Methods*, Hanser Gardner, Cincinnati, OH (1986).

58. J. Scheirs, *Compositional and Failure Analysis of Polymers. A Practical Approach*, John Wiley & Sons, Ltd, Chichester (2000).

59. J.W. Gooch, *Analysis and Deformulation of Polymeric Materials, Paints, Plastics, Adhesives and Inks*, Plenum Press, New York, NY (1997).

60. G. Lawson, *Chemical Analysis of Polymers, Rapra Review Report No. 47*, Rapra Technology Ltd, Shawbury (1991).

61. H. Ostromow, *Analyse von Kautschuken und Elastomeren*, Springer-Verlag, Berlin (1981).

62. D.O. Hummel, *Atlas of Plastics Additives: Analysis by Spectrometric Methods*, Springer-Verlag, Berlin (2002).

63. M.F. Forrest, *Analysis of Plastics, Rapra Review Report No. 149*, Rapra Technology Ltd, Shawbury (2002).

64. EEC 89/109, *Off. J. Eur. Comm.* **L40**, 38 (1989).

65. EEC 90/128, *Off. J. Eur. Comm.* **L349**, 26–47 (1990).

66. EEC 92/39, *Off. J. Eur. Comm.* **L168**, 21 (1992).

67. D.M. Gabriel and V.J. Mulley, in *Anionic Surfactants – Chemical Analysis* (J. Cross, ed.), M. Dekker, New York, NY (1977).

68. N.S. Allen, R.A. Ortiz, G.J. Anderson and M. Sasaki, *Polym. Degrad. Stab.* **46** (1), 75–84 (1994).

69. R. Gächter and H. Müller (eds), *Plastics Additives Handbook*, Hanser Publishers, Munich (1993).

70. E.W. Flick, *Plastic Additives: An Industrial Guide*, Noyes Publications, Park Ridge, NJ (1986).

71. A.H. Sharma, F. Mozayeni and J. Alberts, *Proceedings Polyolefins XI*, Houston, TX (1999), pp. 679–703.

72. EURACHEM Guidance Document No. CITAC Guide 2: *QA Best Practice for Research and Development and Non-routine Analysis*, Working Draft 40, April 1997. Available from the EURACHEM Secretariat, Dept of Chemistry and Biochemistry, University of Lisbon, Portugal.

73. Nederlandse Warenwet (Dutch Food and Commodity Act), *Packaging and Food Utensils Regulations* (VGB), SDU Uitgeverij, The Hague (1997).

74. D. van Battum and J.B.H. van Lierop, *Materials and Articles in Contact with Foodstuffs, Guide for Examination of Plastic Food Contact Materials*. CEN TC 194/SC1/WG2 Document N118 (1997).

75. C. Warner, J. Modderman, T. Fazio, M. Beroza, G. Schwartzman and K. Fominaya, *Food Additives Analytical Manual*, AOAC International, Arlington, VA (1993).

76. ASTM D 1996–97, *Standard Test Method for Determination of Phenolic Antioxidants and Erucamide Slip Additives in Low Density Polyethylene Using Liquid Chromatography (LC), Annual Book of ASTM Standards*, ASTM, West Conshohocken, PA (1997), Vol. 08.01.

77. Commission of the European Communities, *Synoptic Document No. 7. Draft of Provisional List of Monomers and Additives Used in the Manufacture of Plastics and Coatings Intended to Come into Contact with Foodstuffs*, Commission Document CS/PM/2356, Brussels (15 May, 1994).

78. Groupes PEUGEOT SA et RENAULT, *Quantitative Evaluation of Constituents in Vulcanised Rubbers by Means of TG, Méthode d'Essai D40/1753* (March 1991).

79. VOLKSWAGEN AG, SEAT SA, ŠKODA Automobilova AS, AUDI AG, *Py-GC/MS für Kunststoffe und Elastomeren, Zentralnorm PV 3935*, Wolfsburg (December 1997).

80. DAIMLER BENZ, *Arbeitsvorschrift zur Bestimmung von gasförmigen und kondensierbaren Emissionen aus Fahrzeuginnenausstattungsmaterialien mit Thermodesorption, Prüfanweisung PB VWT 709* (1997).

81. G. Harm, J. Wittenauer and P. Roscoe, *Proceedings Polymer Additives '95*, Chicago, IL (1995).

82. J. Gilbert, C. Simoneau, D. Cote and A. Boenke, *Food Addit. Contam.* **17** (10), 889–93 (2000).

83. *Plastics Reference Collection of Additives*, Food Inspection Service, Utrecht, The Netherlands.

84. B. van Lierop, L. Castle, A. Feigenbaum and A. Boenke, *Spectra for the Identification of Additives in Food Packaging*, Kluwer Academic Publishers, Dordrecht (1998).

85. B. van Lierop, L. Castle, A. Feigenbaum, K. Ehlert and A. Boenke, *Food Addit. Contam.* **15** (7), 855–60 (1998).

86. A. Krause, A. Lange and M. Ezrin, *Plastics Analysis Guide – Chemical and Industrial Methods*, Hanser Verlag, Munich (1983), p. 267.

87. J. Rotschová and J. Pospíšil, *Identification and Determination of Stabilizers in Polymers*, Edition Macro, Ser. R-Revue, Vol. R-3, ČSAV, Prague (1980).

88. T.R. Crompton, *Chemical Analysis of Additives in Plastics*, Pergamon Press, Elmsford, NY (1971).

89. T.R. Crompton, *Chemical Analysis of Additives in Plastics*, Pergamon Press, Oxford (1977).

90. M.J. Forrest, *Rubber Analysis in Polymers, Compounds and Products, Rapra Review Report No. 139*, Rapra Technology Ltd, Shawbury (2001).

91. T.P. Lynch, *J. Chromatogr. Libr.* **56**, 269–303 (1995).

92. F. David, *Chem. Mag. (Ghent)* **17** (4), 12–15 (1991).

93. H.J. Vandenburg and A.A. Clifford, in *Extraction Methods in Organic Analysis* (A.J. Handley, ed.), Sheffield Academic Press, Sheffield (1999), pp. 221–42.

94. I.D. Newton, in *Polymer Characterization* (B.J. Hunt and M.I. James, eds), Blackie A&P, Glasgow (1993), pp. 8–36.

95. R. Yoda, *Bunseki* (10), 731–7 (1978).

96. H. Takeguchi and S. Takayama, *Bunseki* (4), 309–14 (1997).

97. N. Oguri, *Nippon Gomu Kyokaishi* **67** (11), 768–80 (1994).

98. D. Munteanu, in *Mechanisms of Polymer Degradation and Stability* (G. Scott, ed.), Elsevier, London (1990), pp. 211–314.

99. R. Amos, *Talanta* **20** (12), 1231–60 (1973).

100. C. Berger, in *Supercritical Fluid Chromatography with Packed Columns* (K. Anton and C. Berger, eds), M. Dekker, New York, NY (1997), pp. 301–48.

101. F.C.-Y. Wang, *J. Chromatogr.* **A843** (1/2), 413–23 (1999).

102. K. Tatsubayashi and F.A. DeThomas, *Purasuchikkusu Eji* **40** (6), 157–62 (1994).

103. J. Pavlov, *Polimeri (Zagreb)* **5** (10), 241–3 (1984).

104. S. Foti and G. Montaudo, *Mass Spectrometry of Large Molecules, Comm. Eur. Communities Rept. EUR 9421* (1985), pp. 283–302.

105. R.P. Lattimer and R.E. Harris, *Mass Spectrom. Rev.* **4**, 369 (1985).

106. T. Kashiwagi, O. Hiroaki, M. Katoh and H. Takigawa, *Sumitomo Kagaku (Osaka)* (2), 71–85 (1993).

107. J. Čermák, *Chem. Listy* **83** (5), 488–505 (1989).

108. R.L. Cochran, *Appl. Spectrosc. Rev.* **22** (2–3), 137–87 (1986).

109. K.R. Lykke, P. Wurz, D.H. Parker and M. Pellin, *Appl. Opt.* **32** (6), 857–66 (1993).

110. B. Asamoto, in *FT-ICR/MS: Analytical Applications of Fourier Transform Ion Cyclotron Resonance Mass Spectrometry* (B. Asamoto, ed.), VCH, New York, NY (1991), pp. 157–85.

111. J.H. Scrivens and A.T. Jackson, *NATO ASI Ser.* **C521**, 201–34 (1999).

112. H. Narasaki, *Bunseki* (12), 913–17 (1982).

113. K.G.H. Raemaekers and J.C.J. Bart, *Thermochim. Acta* **295**, 1–58 (1997).

114. H.E. Bair, in *Thermal Characterization of Polymeric Materials* (E.A. Turi, ed.), Academic Press, New York, NY (1981), pp. 845–909.

115. J. Chiu and E.F. Palermo, *Anal. Chim. Acta* **81** (1), 1–19 (1976).

116. S. Affolter and S. Hofstetter, *Kautsch. Gummi Kunststoffe* **48** (3), 173–9 (1995).

117. N. Asada, *Bunseki* (9), 697–703 (1991).

118. M. Kojima, *Bunseki* (9), 665–70 (1988).

119. S. Takayama, *Bunseki* (4), 239–44 (1987).

120. R. Yoda, *Bunseki* (1), 75–82 (1984).

121. G.R. Coe, *Plast. Compd.* **1** (3), 41–50 (1978).

122. S. Hirayanagi, *Bunseki* (12), 1008–13 (1989).

123. R.W. Crecely and C.E. Day, in *Plastics Additives. An A–Z Reference* (G. Pritchard, ed.), Chapman & Hall, London (1998), pp. 26–31.

124. P. Meng and L. Li, *Zhongguo Suliao* **13** (8), 16–20 (1999).

125. B. Marcato and M. Vianello, *Proceedings HPLC '99*, Granada (1999).

126. T.A. Fadeeva, V.M. Ryabikova, G.S. Popova and L.V. Toporkova, *Plast. Massy* (10), 72–4 (1990).

127. T.J. Henman and S.R. Oldland, *Dev. Polym. Stab.* **7**, 275–302 (1984).

128. L. Veres, *Kunststoffe* **66** (8), 493–6 (1976).

129. I. Yilgor, E. Yilgor and B. Gruening, *Tenside, Surfactants, Deterg.* **30** (3), 158–64 (1993).

130. J. Rotschová and J. Pospíšil, in *Oxidation Inhibition in Organic Materials* (J. Pospíšil and P.P. Klemchuk, eds), CRC Press, Boca Raton, FL (1989), Vol. II, pp. 347–68.

131. M.P. Thomas, *J. Vinyl Addit. Technol.* **2** (4), 330–8 (1996).

132. P. Bataillard, L. Evangelista and M. Thomas, in *Plastics Additives Handbook* (H. Zweifel, ed.), Hanser Publishers, Munich (2000), pp. 1047–83.

133. W. Freitag, in *Plastics Additives Handbook* (R. Gächter and H. Müller, eds), Hanser Publishers, Munich (1993), pp. 909–49.

134. J. Mitchell, in *Applied Polymer Analysis and Characterization: Recent Developments in Techniques, Instrumentation, Problem Solving* (J. Mitchell, ed.), Hanser Publishers, Munich (1987), pp. 3–81.

135. H.V. Drushel, in *The Analytical Chemistry of Sulfur and Its Compounds* (J. Karchmer, ed.), Wiley-Interscience, New York, NY (1970), Pt II, Ch. 7.

136. E. Lück and M. Jager, *Antimicrobial Food Additives. Characteristics, Uses, Effects*, Springer-Verlag, Berlin (1997).

137. D.A. Wheeler, *Talanta* **15**, 1315–34 (1968).

138. T. Tikuisis and V. Dang, in *Plastics Additives. An A–Z Reference* (G. Pritchard, ed.), Chapman & Hall, London (1998), pp. 80–94.

139. D.C.M. Squirrell, *Analyst* **106**, 1042–56 (1981).

140. G.N. Foster, in *Oxidation Inhibition in Organic Materials* (J. Pospíšil and P.P. Klemchuk, eds), CRC Press, Boca Raton, FL (1989), Vol. II, pp. 299–346.

141. H. Gouya, *Kobunshi no Hyomen Kaishitsu to Oyo*, 251–72 (2001).

142. K.G. Bergnes, U. Rudt and D. Mack, *Dtsch. Lebensm.-Rdsch.* **63** (6), 180 (1967).

143. J. Behnke, *The Aldrich Library of FT-IR Spectra*, Aldrich Chemical Co., Milwaukee, WI (1997).

144. K.G.R. Pachler, F. Matlok and H.-U. Gremlich (eds), *Merck FT-IR Atlas*, Wiley-VCH, Weinheim (1988).

145. J.A. Sidwell, *The Rapra Collection of Infrared Spectra of Rubbers, Plastics and Thermoplastic Elastomers*, Rapra Technology Ltd, Shawbury (1997).

146. F. Scholl, *Atlas der Polymer- und Kunststoffanalyse*, Vol. 3. *Zusatzstoffe und Verarbeitungshilfsmittel*, Verlag Chemie, Weinheim (1981).

147. D.O. Hummel (ed.), *Atlas of Polymer and Plastics Analysis*, Vol. 1, Wiley-VCH, Weinheim (1991).

148. D.O. Hummel, *Atlas of Polymer and Plastics Analysis: Analytical Methods, Spectroscopy, Characteristic Absorptions, Description of Compound Classes*, Vol. 2, John Wiley & Sons, Inc., New York, NY (1988).

149. D.O. Hummel, *Analysis of Surfactants, Atlas of FTIR-Spectra with Interpretations*, Hanser/Gardner, Cincinnati, OH (1996).

150. D.O. Hummel, *Atlas der Polymer- und Kunststoffanalyse*, Bd. 2: *Kunststoffe, Fasern, Kautschuk, Harze; Ausgangs- und Hilfsstoffe, Abbauprodukte*, C. Hanser Verlag, Munich (1984).

151. C.D. Craver (ed.), *Infrared Spectra of Plasticizers and Other Additives*, The Coblentz Society Inc., Kirkwood, MO (1982).

152. *Sprouse Library of Polymer Additives*, Sprouse Scientific Systems, Paoli, PA.

153. A.N. Davies, *Spectrosc. Europe* **12** (4), 22–6 (2000).

154. D.O. Hummel, *Macromol. Symp.* **119**, 65–77 (1997).

155. T.J. Henman (ed.), *World Index of Polyolefine Stabilizers*, Kogan Page, London (1982).

156. E. Zeisberger and T. Fröhlich, *Spectrosc. Europe* **5** (3), 16–9 (1993).

157. R. Jennissen, *Techn. Report DSM Research*, Geleen (1998).

158. U. Schönhausen, *Positive List I. Additives for Plastics, Elastomers and Synthetic Fibres for Food Contact Applications*, Ciba-Geigy Ltd, Basel (1993).

159. U. Schönhausen, *Positive List II. Additives Used in the PVC-Industry for Food Contact Applications*, Ciba-Geigy Ltd, Basel (1990).

160. NIST 02, National Institute of Standards and Technology, Gaithersburg, MD (2002).

161. M. Geissler, *Kunststoffe* **87** (2), 194–6 (1997).

162. J. Bush, J. Gilbert and X. Goenaga, *Spectra for the Identification of Monomers in Food Packaging*, Kluwer Academic Publishers, Dordrecht (1993).

163. J. Gilbert, J. Bush, A. Lopez de Sa, J.B.H. van Lierop and X. Goenaga, *Food Addit. Contam.* **11**, 71–4 (1994).

164. Bureau Communautaire de Référence (BCR), *Spectra for the Identification of Additives in Food Packaging*, (DG 12, MAT 1 CT 930037), Geel (1993).

165. A.J. Brandolini and D.D. Hills, *NMR Spectra of Polymers and Polymer Additives*, M. Dekker, New York, NY (2000).

166. D. Dieckmann, W. Nyberg, D. Lopez and P. Barnes, *J. Vinyl Addit. Technol.* **2** (1), 57–62 (1996).

167. B. Bell, D.E. Beyer, N.L. Maecker, R.R. Papenfus and D.B. Priddy, *J. Appl. Polym. Sci.* **54** (11), 1605–12 (1994).

168. F.T. Dettmer, H. Wickmann, J. de Boer and M. Bahadir, *Chemisphere* **39** (9), 1523–32 (1999).

169. G. Stegeman and E. Venema (GE Plastics), *Personal communication* (1999).

170. F.C.-Y. Wang, *J. Chromatogr.* **A886**, 225–35 (2000).

171. F.C.-Y. Wang, *J. Chromatogr.* **A883**, 199–210 (2000).

172. R. Stringer, I. Labunska, D. Santillo, P. Johnston, J. Siddorn and A. Stephenson, *Environ. Sci., Pollut. Res.* **7** (1), 27–36 (2000).

CHAPTER 2

If you don't know where you're going, you'll never know when you get there

Deformulation Principles

The analysis of an unknown number of unknown additives in unknown concentration in an unknown polymeric matrix is a demanding task for the analytical chemist for a variety of circumstances (Table 2.1). Primary analytical needs include the identification of the additives, the quantification of the additive levels, and the examination of additive stability. Obviously, the experimental analytical conditions must be such that no measurable polymer degradation or additive loss occurs during analysis.

Table 2.1 Factors affecting polymer/additive analysis

- Nature of the polymer matrix (physical and chemical properties of the polymer determine how the additives can be separated, if at all)
- Wide coverage of chemical materials (both organic and inorganic, varying greatly in molecular weight, volatility and polarity)
- Complex mixtures of compounds (frequently of completely unknown type and concentration in an unknown matrix)
- Wide additive concentration range (restricting the analytical choice of method)
- Purity (many additives are technical grade substances, (isomeric) mixtures, e.g. fatty amides or epoxidised vegetable oils)
- High reactivity and low thermal stability of many additives (especially antioxidants)
- Mutual interferences of the components (presence of transformation and degradation products)
- Trend towards polymeric and grafted additive functions
- Residual monomers and oligomers

Successful analytical methodologies not only must distinguish the number of mixture components but should also provide characteristic structural information about each additive. For identification purposes various public or proprietary databases for low-MW additives can be used.

Most methods for the determination of additives in plastics come essentially under two headings, namely with or without sample preparation. The following eight analytical categories are thus distinguished:

- (i) Solvent extraction (cf. Chapter 3).
- (ii) Dissolution methods (cf. Chapters 3.7 and 9).
- (iii) Digestion techniques (cf. Chapter 8.2).
- (iv) Depolymerisation methods, e.g. hydrolysis (cf. Chapter 3.8).
- (v) Heat extraction: examination of volatiles released (destructive testing by thermal methods, pyrolysis, laser desorption, photolysis).
- (vi) Nondestructive or *in situ* analysis of the polymeric material (spectroscopy, microscopy).
- (vii) In-process analysis, i.e. polymer melt sampling.
- (viii) Miscellaneous (chemical reactions) (cf. Chapter 2.5).

Developing a method for analysis of polymer additives is dependent upon several factors, such as the nature of the matrix and the volatility, molecular weight, solubility,

Additives In Polymers: Industrial Analysis And Applications   J. C. J. Bart
© 2005 John Wiley & Sons, Ltd   ISBN: 0-470-85062-0

stability and concentration of the analyte(s). Other factors influencing the choice are instrumental, such as detector solvent compatibility. It is particularly necessary to distinguish between a generic identification analytical problem and quantitative analysis. The latter requires considerably more time and effort. If the question relates to the determination of the presence of a particular additive, a minimum detection limit must be established. Usually the primary goal is to identify the additives and thus to quantify them with the least difficulty, i.e. according to the simplest deformulation scheme. The challenge is then to develop methods which are not prone to interferences from other components in the polymer matrix. In the R&D environment, at variance to quality control, analyte specific protocols are usually not the issue. Any method selected must be validated.

Accurate analysis of a multicomponent formulated material is quite complicated. Plastic formulations can be mixtures of materials of widely varying concentrations and 5–15 ingredients are not uncommon. Some practical examples are given in Tables 2.3 and 2.8.

There are thousands of commercially available additives of diverse chemical classes and with masses ranging from a few hundred to several thousand Daltons (cf. also Appendix II). Deformulation means *reverse engineering*, with subsequent analysis of each separated component. Product deformulation may hint towards the process of origin. Deformulation will combine several

techniques to build up a profile of the polymer base and additives and is frequently used in reformulation. The strategy for polymer/additive analysis has varied widely, as will be illustrated in this chapter. Modern polymer/additive deformulation is essentially carried out according to four different approaches, in increasing order of sophistication, namely analysis of the analytes separated from the polymer (typically an extract), of analytes and polymer in solution, directly in the solid state or in the melt. Understanding of the physico-chemical aspects of polymers is an asset.

2.1 POLYMER IDENTIFICATION

Principles and Characteristics A first step in additive analysis is the identification of the matrix. In this respect the objective for most polymer analyses for R&D purposes is merely the definition of the most appropriate extraction conditions (solvent choice), whereas in rubber or coatings analysis usually the simultaneous characterisation of the polymeric components and the additives is at stake. In fact, one of the most basic tests to carry out on a rubber sample is to determine the base polymer. Figure 2.1 shows the broad variety of additive containing polymeric matrices.

Identification of the chemical nature of a plastics sample is a problem facing processors and users of

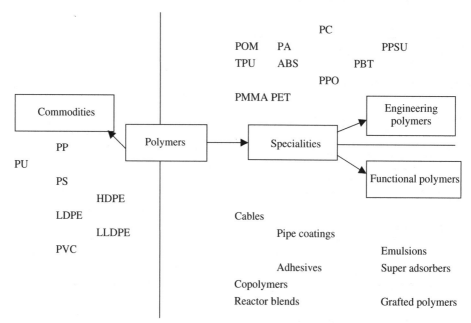

Figure 2.1 Polymeric matrices

plastics. Identification of plastics is formally regulated by ISO 1043 (plastics-symbols). More practically, a variety of identification methods of plastics has been given in the literature [1–10].

Polymer analysis and characterisation ranges from a quick identification/fingerprinting approach based on some simple physical and chemical tests to a more detailed microstructural elucidation requiring the full range of chromatographic and spectroscopic studies. It is the case to distinguish between the need for identification of samples coming from R&D projects and technical service, etc., and rapid plastics identification for recycling purposes.

In identifying an unknown polymer it is useful to compare some of its physical characteristics with properties based on information obtained from known polymers. The simplest physical and chemical tests such as elemental analysis, saponification number or solvent solubility, ignition or burning/heating tests are of limited value only, in particular in case of copolymers. Moreover, nonflame grades will not burn. Braun [2] has described simple methods for identification of plastics, such as use, appearance, examination of microtome sections, burning/heating tests (odours produced on ignition), colour reactions and other specific identification reactions, testing for heteroatoms and other preliminary tests (solubility, density). Colour reactions for homopolymers are rather laborious and best regarded as confirmatory tests for polymer identifications achieved by other means. Classification by solubility is equally of limited value, especially when applied to copolymers. Many materials or copolymers cannot be identified with such simple means. More successful methods of identifying an unknown polymer consist in the comparison of its physical characteristics with those of spectroscopic (FTIR, NMR), mass spectrometric (PyMS, PyGC-MS), or thermal (T_g, T_m) libraries.

Infrared spectroscopy is a major tool for polymer and rubber identification [11,12]. *Infrared analysis* usually suffices for identification of the plastic material provided absence of complications by interferences from heavy loadings of additives, such as pigments or fillers. As additives can impede the unambiguous assignment of a plastic, it is frequently necessary to separate the plastic from the additives. For example, heavily plasticised PVC may contain up to 60 % of a plasticiser, which needs to be removed prior to attempted identification of the polymer. Also an ester plasticiser contained in a nitrile rubber may obscure identification of the polymer. Because typical rubber compounds only contain some 50 % polymer direct FTIR analysis rarely provides a definitive answer. It is usually necessary first to effectively remove the interference of various other major constituents, e.g. by means of solvent extraction. Reference may be made to the Hummel/Sadtler Polymer Database, which contains 1920 spectra. Identification of components in a mixture can be performed by using: (i) curve fitting techniques [13]; and (ii) knowledge-based systems [14]. Methods based on curve-fitting require that the composite spectrum can be approximated by a linear combination of component spectra, which is not always true for solid mixtures such as polymer blends or copolymers. In knowledge-based systems the human interpretation process is more closely approached.

Composition and structure of macromolecules can also be elucidated through analysis of volatile pyrolysis products (PyGC, PyMS) that are large enough to possess the substantial structural elements of the original polymer. Fingerprint identification of polymers can take advantage of PyGC-MS, which relies on a comparison of mass spectra of evolved gases of unknown and reference samples (library data); identification of peaks is not necessary. For polymer identification pyrolysis temperatures kept between 500 °C and 800 °C for 10 s are usually recommended. Pyrolysis GC enables the polymer type within a rubber sample to be identified. The method can be applied either to the extracted portion of the sample or to the sample as received. For identification of polymers by PyGC an ISO standard is in preparation (ISO/DIS 7270). A comparative evaluation of optical processes for rapid polymer identification and those based on detection of pyrolysis products has appeared [15].

As to thermal techniques, DSC can also answer many questions and is useful as a rapid method of checking the nature of polymers (T_g, T_m and enthalpy of fusion). Identification of HDPE/PA6 plastic films (cooking bags), LDPE/PA6 composite films (packaging material for sausages and cheese) and LDPE/EVAL/PA6 (used in the food sector) by DSC have been reported [16]. However, the results may again be inconclusive because of the wide range of polymers manufactured especially taking into account copolymers, and in view of additive interferences. For quantification of the polymer component within a rubber, the easiest technique to use is TGA.

Rapid identification of plastics by spectroscopic and x-ray methods were reviewed [17]. Several books are available [2,11].

Applications Rapid industrial polymer identification systems have been developed to sort plastic components in cars, plastics used in the building and construction industry and plastic films. In recycling of plastics material *automated identification* is crucial, regardless of colour, degree of contamination, and additive and moisture content. The quality of recycled material depends on the purity of the starting material and, therefore, on the performance of the identification/sorting techniques.

Different sorting and identification methods are known, using e.g. properties such as density, electrical, magnetical, or chemical separation. However, there are similar polymers, like copolymers or polymer blends, as well as materials with different additives that cannot be separated by these methods. Therefore, optical measurements, especially spectroscopic techniques have been developed. Various types of identification techniques have been evaluated [18]. Currently available methods for the identification of waste plastics to facilitate separation include colour recognition systems, optical methods (readout of codes), bar code reading systems, XRF, NIRS, FTIR, Raman spectroscopy, MS of gaseous decomposition products, laser desorption-ion mobility spectrometry (LD-IMS), fluorescent dye markers, thermo-optical methods (based on differences in dielectric constants of plastics after dielectric heating), and ultrasonic and photoacoustic methods [19]. Among these techniques, NIR, MIR and Raman are those with the best reliability for identification of plastic materials. Chemometric approaches determining spectral distances between a sample spectrum and reference spectra are in use for identification of plastic materials by optical measurements. For example, identification of plastic materials such as flame retarded ABS, HIPS, SAN, PP, PE, PA, POM, PMMA, PC and PPO by spectroscopic (FTIR) measurements has taken advantage of the determination of spectral distances with reference spectra (using simple PLS, PCA, PDA or more complex methods like neural networks and Mahalanobis distances) and can be achieved with high reliability (correct identification rate of 95 to 98 %) within 2 s [20].

Fast non-contacting NIRS (absorption and reflection) with powerful software techniques is indicated as the method of choice whatever plastics originating from household waste are to be identified with a subsecond ID cycle time [21–24]. High-throughput NIRS with InGaAs detection and PCA analysis has been proposed for on-line identification of post-consumer packages of different polymers (PE, PET, PP, PS, PVC) on an industrial conveyer belt [25]. Near-infrared using an acousto-optical tuneable filter (AOTF) can distinguish in real-time between the chemical signatures of over 30 types of plastic with a reported 100 % success rate [26]. Advantages of the NIR spectral range compared to mid-IR are: NIR photodetectors like Ge, InAs or InGaAs

photodiodes with fast response and higher detectivity, and quartz fibre optics with low attenuation.

A hand-held laser probe using Raman inelastic light scattering has been described for rapid automotive plastic identification (for recycling purposes; ID cycle time ca. 1 s), also suitable for darkly pigmented, textured and reinforced polymer grades [27,28]; the system is sensitive to the presence of CB. Also LD-IMS has been proposed for sorting and recycling of polymer samples, including PVC, PA6.6, HDPE, LDPE and ABS [29].

Automatic identification and sorting of waste plastics may also rely on the incorporation of fluorescent tracers into the materials, giving each a unique fingerprint [30]. Alternatively, a small amount of an additive such as Li_2CO_3 or $Li_2B_4O_7$ may be used in plastics for easy identification of the type of resin (e.g. by flame or spectral analysis) during recycling. For engineering plastics, various methods have to be combined because of the wide variety of materials and additives to be identified.

Identification and sorting of plastics in waste materials were reviewed [19,31]. Garbassi [32] has stressed the important role played by polymer analysis and characterisation in plastics recycling.

2.2 ADDITIVE ANALYSIS OF RUBBERS: 'BEST PRACTICE'

In the rubber field it is not only the polymer that determines the properties of an elastomer, but many accompanying substances, like fillers, pigments, plasticisers, curing agents, antioxidants, stabilisers and processing aids (cf. Table 2.2). With rubbers the possible compositional permutations are numerous. In fact, already within the additive group of CBs there are more than 30 different possible products.

Table 2.3 shows a typical rubber composition. Some rubber components are amongst the most difficult compounds to analyse. For example, vulcanisation chemicals are changed by incorporation into the polymer

Table 2.2 Components of practical rubber formulations

- Elastomers
- Curatives (sulfur, accelerators, peroxides, sulfur-donor systems, etc.)
- Cure-system activators (ZnO, stearic acid, lauric acid, etc.)
- Fillers and reinforcing agents (CB, mineral fillers, and short fibres)
- Process oil, plasticisers
- Additives (antioxidants, antiozonants, softeners, tackifiers, peptisers, scorch inhibitors, colorants, flame retardants, blowing agents, process aids, etc.)

Table 2.3 Typical rubber composition

Component(s)	Analytical data
Polymers	Nitrile rubber 55: PVC 45 blend, 39 % total polymer content
Plasticiser	DEHP
CB	2 %
Organic flame retardants	Penta BDE (13 %), EHDPP
Inorganics	ZnO 2 %; Al(OH)$_3$ 20 %, Sb$_2$O$_3$ 2 %
Antidegradant	Diphenylamine/acetone condensation product
Accelerators	CBS, TMTD, MBT/MBTS
Cure system (species)	Sulfur-based; phthalimide (PVI)

Table 2.4 General analytical procedures

ASTM D 297	Chemical analysis of rubber (specific gravity; extract, filler, CB, polymer and sulfur analysis; antidegradant and plasticiser analysis)
ASTM D 1278	Analysis of natural rubber
ASTM D 1416	Analysis of synthetic rubber
ASTM D 2007	Oils–characterisation–CC
ASTM D 2702	Rubber chemicals–Infrared
ASTM D 2703	Rubber chemicals–Ultraviolet
ASTM D 3156	Antidegradant–TLC
ASTM D 3452	Rubber–Pyrolysis GC
ASTM D 3677	Rubber identification by IR spectrophotometry
ASTM E 442	Cl$_2$, Br$_2$, I$_2$ by oxygen flask

After American Society for Testing and Materials [39].

and through chemical reactions. Therefore, a complete analysis is very laborious and time-consuming [10,33]. A comprehensive list of materials used in the industrial elastomer formulations and their sources is available [34].

The earliest comprehensive publication on analysis of rubbers by Tuttle [35] was published in 1922. British Standard Methods for the analysis of rubbers were first published in 1950, ASTM Standard Methods in 1958. The British Standards were later supplemented by Wake [36]. Burger [37] of the United States Rubber Company has reported a comprehensive bibliography from Tuttle through 1958. Rubber analysis is complex and possible interferences are numerous. Some rubbers contain more than 15 ingredients, some at a low (<1 %) level below the solubility limit, and others are very volatile, thermally labile or will have undergone chemical reactions during vulcanisation making accurate original quantifications impossible to deduce. Even with the greatest effort it is unlikely that more than 90–95 % of a complex rubber formulation can be completely elucidated by analytical means. Fundamental techniques for the quantitative analysis of vulcanised rubber compounds are normally conducted according to ASTM D 297 and supplementary testing methods

(Table 2.4) or proprietary test methods. Forrest [38] has listed 94 international *rubber analysis standards* (ISO) and 20 ISO standards in preparation referring to latices, carbon-black-filled compositions, raw and compounded rubbers.

Conventional rubber compound analysis requires several instrumental techniques, in addition to considerable pretreatment of the sample to isolate classes of components, before these selected tests can be definitive. Table 2.5 lists some general analytical tools. Spectroscopic methods such as FTIR and NMR often encounter difficulties in the analysis of vulcanised rubbers since they are insoluble and usually contain many kinds of additives such as a curing agent, plasticisers, stabilisers and fillers. Pyrolysis is advantageous for the practical analysis of insoluble polymeric materials.

Identification of rubber compounding ingredients using TLC was repeatedly reviewed (pre-1971) [41,42]. The most useful technique for the analysis of cured samples is GC-MS, either by solution injection or dynamic headspace. Headspace techniques provide a means of identifying the largest range of cure system breakdown products. Diagnostic fragment data are

Table 2.5 Analytical tools for the characterisation of rubbers

Material	Main components	Minor components	Structural aspects
Raw material	TGA, S(qual.)[a], XRF	S(quant.)[a], EGA[a], colour reactions[a]	XPS[c], ToF-SIMS[c]
Solvent extract	FTIR	TGA, TLC[a], GC[a], HPLC[a]	NMR[d]
Residue of solvent extract	TGA, FTIR	S(quant.)[a], PyGC[a,b], NMR[b]	NMR[d,e], RS[b,e], XANES[e]
Ash	FTIR	XRF	XRD[d]

[a] Characterisation of the vulcanisation system.
[b] Analysis of blends and copolymers.
[c] Surface properties.
[d] Molecular structure.
[e] Type of cross-linking bridges.
After Affolter [40]. From S. Affolter, *Macromolecular Symposia*, **165**, 133–142 (2001), © Wiley-VCH, 2001. Reproduced by permission of Wiley-VCH.

Table 2.6 Characterisation of rubber components (synopsis)

Component(s)	Standard(s)	Technique(s)
Base polymer	ISO 4650, ISO 5945, ISO 7270, ISO 9924	FTIR (after extraction), PyGC, PyFTIR, PyGC-MS, ^{13}C NMR, ^{1}H NMR, TGA
Residual monomers	ISO/DIS 17052	TD-GC-MS, HS-GC-MS
Plasticiser/oil, organic flame retardants	ISO 1407	Solvent extraction, FTIR, HPLC, TGA
CB	ISO 1408, ISO 12245, ISO/DAM/1304	TGA, Table 2.7
Inorganic fillers	ISO 247, ISO 9924.1	TGA, XRF, furnace ashing, FTIR, XRF, ICP
Antidegradants	ISO 11089, ISO 4645.2, ASTM D 3156	HS-GC-MS, solvent extraction, GC, HPLC, TLC, UV, DP-MS
Cure system additives	ISO 10398, ISO 11235 ISO 11236	GC-MS, DHS-GC-MS, HPLC, TLC
Blowing agents	–	GC-MS
Inorganic flame retardants	–	TGA, XRF
Peroxide co-agents	–	GC-MS, HS-GC-MS
Sulfur	ISO 6528	Combustion flask, furnace combustion

included in the books by Willoughby [43] and Forrest [38]. The use of HPLC for the analysis of accelerators of uncured samples has also been described [44] and a critical comparison of GC and HPLC in the analysis of a CB filled NMR vulcanisate was reported [45]. GC does not reveal various volatile or thermally labile ingredients. HPLC conceals substances with low UV absorption; with dithiocarbamates interaction with heavy metals from metallic parts of the equipment were noted.

Table 2.6 summarises the main techniques and standards used for the determination of specific rubber components. ISO 9924-1 (2000) describes the use of TGA for the determination of the composition of butadiene, ethylene-propylene, butyl, isoprene and SBR rubbers.

Table 2.7 lists techniques used to characterise *carbon-blacks*. Analysis of CB in rubber vulcanisates requires recovery of CB by digestion of the matrix followed by filtration, or by nonoxidative pyrolysis. Dispersion of CB within rubber products is usually assessed by the Cabot dispersion test, or by means of TEM. Kruse [46] has reviewed rubber microscopy, including the determination of the microstructure of CB in rubber compounds and vulcanisates and their qualitative and quantitative determination. Analysis of free CB features measurements of: (i) particulate and aggregate size (SEM, TEM, XRD, AFM, STM); (ii) total surface area according to the BET method (ISO 4652), iodine adsorption (ISO 1304) or cetyltrimethylammonium bromide (CTAB) adsorption (ASTM D 3765); and (iii) external surface area, according to the dibutylphthalate (DBP) test (ASTM D 2414). TGA is an excellent technique for the quantification of CB in rubbers. However, it is very limited in being able to distinguish the different types of

Table 2.7 Techniques used to characterise CBs

Technique(s)	Application(s)	Reference(s)
SEM, TEM, XRD	Particulate and aggregate size	[48]
BET, CTAB	Total surface area	[49]
Neutron scattering	Particle structure	[50]
RS and x-ray scattering, Cabot dispersion test, TEM	Particle microstructure, dispersion	[46,51,52]
GC-MS, SIMS, XPS, IGC	Surface chemistry	[53]
AFM, STM	Surface structure	[54]
IGC	Surface energies and thermodynamic parameters	[55,56]

CB. Bertrand and Weng [47] have reported CB surface characterisation by ToF-SIMS and XPS. The sensitivity of the former (ToF-SIMS) is much higher than that of the latter (XPS) but quantification is much better with XPS.

TGA is also convenient for the quantification of *inorganic fillers* such as barites, silica and silicates. The limitation of the technique is that the total inorganic content of the rubber is obtained, with no indication of the relative proportion if a blend of fillers is present. Since most rubber compounds contain zinc oxide as part of the cure system (sulfur systems) or as an acid acceptor (peroxide cure systems), it is common practice to complement the TGA data with at least a semi-quantitative elemental technique, such as XRF, in order to obtain qualitative as well as quantitative data. Furnace ashing at 550 °C (ISO 247 (1990)) is another technique commonly used for the isolation and quantification of inorganics. Dry ashing may be unsuitable for halogen-containing rubbers and acid ashing is not recommended for raw rubbery polymers. Qualitative information about the composition of the ash can be obtained by FTIR, XRF or ICP.

Plasticiser/oil in rubber is usually determined by solvent extraction (ISO 1407) and FTIR identification [57]; TGA can usually provide good quantifications of plasticiser contents. *Antidegradants* in rubber compounds may be determined by HS-GC-MS for volatile species (e.g. BHT, IPPD), but usually solvent extraction is required, followed by GC-MS, HPLC, UV or DP-MS analysis. Since cross-linked rubbers are insoluble, more complex extraction procedures must be carried out. The determination of antioxidants in rubbers by means of HPLC and TLC has been reviewed [58]. The TLC technique for antidegradants in rubbers is described in ASTM D 3156 and ISO 4645.2 (1984). Direct probe EIMS was also used to analyse antioxidants (hindered phenols and aromatic amines) in rubber extracts [59]. ISO 11 089 (1997) deals with the determination of N-phenyl-β-naphthylamine and poly-2,2,4-trimethyl-1,2-dihydroquinoline (TMDQ) as well as other generic types of antiozonants such as N-alkyl-N'-phenyl-p-phenylenediamines (e.g. IPPD and 6PPD) and N-aryl-N'-aryl-p-phenylenediamines (e.g. DPPD), by means of HPLC.

There are two principal types of *cure systems* used in rubber products: sulfur and peroxide. Affolter [40] has presented analytical methods leading to the characterisation of the vulcanisation system of unknown elastomers (sulfur or peroxide cross-linked). It is to be considered, however, that sulfur in rubbers may originate not only from the vulcanisation process (bound or combined sulfur) but also from additives such as accelerator reaction products, organic plasticisers (e.g. alkylpolyether-thioether), organic stabilisers (e.g. DSTDP), organic modifiers, organic fillers (e.g. ebonite powder), inorganics (e.g. $BaSO_4$, free elemental sulfur, metallic sulfide) or inorganic pigments (e.g. C.I. Pigment Blue 29, C.I. Pigment Orange 20). Depending on how much information is needed, sulfur analysis can be complex. In order to make reliable statements about the vulcanisation system of unknown elastomeric material, the results of several analytical methods have to be evaluated, in particular thermal degradation, pyrolysis and chromatography (frequently using FPD and AED detection). Total sulfur content in rubbers may be determined by the oxygen combustion flask method (ISO 6528-1 (1992) and BS 7164-23.1 (1993)), sodium peroxide fusion (ISO 6528-2 (1992)) or furnace combustion (ISO 6528-3 (1988)). The oxygen flask combustion technique can also be used for the determination of total halogens (titration method)

or quantification of specific halogens (ion chromatography method) in a rubber. Cure state studies may be carried out by means of DSC [60], TMA and DTMA. Andersson *et al.* [61] used PyGC to study sulfur bridges in filled NR vulcanisates. Some sulfur cure system accelerators (e.g. TMTD) produce nitrosamines. Analysis of rubber compounds for nitrosamines usually takes the form of extraction followed by GC analysis using a specific detector (e.g. ELCD). Nitrosamines in rubber were reviewed [62]. Peroxide co-agents (e.g. TAC, TAIC) can be detected in solvent extracts by (headspace) GC-MS as they are reasonably volatile and thermally stable. A comprehensive review of the use of chemical analysis techniques to investigate the cure of rubbers and other polymer systems is available [63]. Analysis for accelerators can be difficult depending on the complexity of the accelerator system.

Standardisation of EPDM characterisation tests (molecular composition, stabiliser and oil content) for QC and specification purposes was reported [64,65]. Infrared spectroscopy (rather than HPLC or photometry) is recommended for the determination of the stabiliser content (hindered phenol type) of EP(D)M [65]. Determination of the oil content of oil-extended EPDM is best carried out by Soxhlet extraction using MEK as a solvent [66]. A round robin test was reported that evaluated the various techniques currently used in the investigation of unknown rubber compounds (passenger tyre tread stock formulations) [67].

Leyden and Rabb [68] have illustrated the *rubber formula reconstruction* process. Table 2.8 shows some of the difficulties encountered in the reconstruction of a complex wire and cable jacket compound. Forrest [38] has illustrated the reverse engineering of the reasonably complex highly flame retardant elastomer compound of

Table 2.8 Actual and reconstructed formulations compared

Reconstructed formula		Actual formula	
Material	phr	Material	phr
SBR 1503	100.0	SBR 1503	100.0
N-774 black	55.9	N-330 black	15.0
Hard kaolin clay	16.7	N-774 black	40.0
HI-aromatic oil/resin	19.1	Cumar MH 2½	20.0
Zinc oxide	4.8	Suprex clay	20.0
Stearic acid	2.0	Zinc oxide	5.0
Petroleum wax	2–3	Stearic acid	2.0
Wingstay 300	2	Petroleum wax	3.0
Sulfur	1.5	Santoflex 13	1.0
Accelerator system	2	Sulfur	1.85
		Beutene	1.00
		Monex	0.75

After Leyden and Rabb [68]. Reproduced by permission of Communication Channels Inc. (Argus Press Holdings Inc.).

Table 2.3 as a completely worked out example using quantitative solvent extraction, ash content determination, TGA, FTIR, XRF, GC-MS, HS-GC-MS, PyFTIR, ICP, and s-NMR. Information on the cure and antidegradant systems was obtained (assigned species/possible origin), as follows: cyclohexane thiol/CBS accelerator; benzothiazole/MBT, MBTS or CBS accelerators; *N*, *N*-dimethylformamide/TMTD accelerator; phthalimide/Santoguard PVI; and *N*-phenylbenzene amine/possibly a diphenyl/acetone amine antioxidant.

It is of interest to examine the development of the *analytical toolbox* for rubber deformulation over the last two decades and the role of emerging technologies (Table 2.9). *Bayer technology* (1981) for the qualitative and quantitative analysis of rubbers and elastomers consisted of a multitechnique approach comprising extraction (Soxhlet, DIN 53 553), wet chemistry (colour reactions, photometry), electrochemistry (polarography, conductometry), various forms of chromatography (PC, GC, off-line PyGC, TLC), spectroscopy (UV, IR, off-line PyIR), and microscopy (OM, SEM, TEM, fluorescence) [10]. Reported applications concerned the identification of plasticisers, fatty acids, stabilisers, antioxidants, vulcanisation accelerators, free/total/bound sulfur, minerals and CB. *Monsanto* (1983) used direct-probe MS for *in situ* quantitative analysis of additives and rubber and made use of ^{31}P NMR [69].

The conventional approach to rubber analysis employed by *BFGoodrich* (and others) is shown in Scheme 2.1. Pausch [82] has reviewed the analytical approach utilised for determining the AO composition. Pausch *et al.* [70] have initially experimented with direct compound analysis using ATR-FTIR, NMR, PyMS, and TGA and have subsequently focused on a wide

variety of pyrolysis and mass spectrometric methods (EIMS, CIMS, FDMS, FIMS, AC-MS, MS/MS, Py-FIMS, TPPy-MS, PyGC-MS) [70,76,77,83–87]. Py-FIMS was used as a screening technique to obtain an overview of the chemical composition.

Lussier [71] has given an overview of *Uniroyal Chemical's* approach to the analysis of compounded elastomers (Scheme 2.2). Uncured compounds are first extracted with ethanol to remove oils for subsequent analysis, whereas cured compounds are best extracted with ETA (ethanol/toluene azeotrope). Uncured compounds are then dissolved in a low-boiling solvent (chloroform, toluene), and filler and CB are removed by filtration. When the compound is cured, extended treatment in *o*-dichlorobenzene (ODCB) (b.p. 180 °C) will usually suffice to dissolve enough polymer to allow its separation from filler and CB via hot filtration. Polymer identification was based on IR spectroscopy (key role), CB analysis followed ASTM D 297, filler analysis (after direct ashing at 550–600 °C in air) by means of IR, AAS and XRD. Antioxidant analysis proceeded by IR examination of the nonpolymer ethanol or ETA organic extracts. For unknown AO systems (preparative) TLC was used with IR, NMR or MS identification. Alternatively GC-MS was applied directly to the preparative TLC eluent.

Deformulation of vulcanised rubbers and rubber compounds at *Dunlop* (1988) is given in Scheme 2.3. Schnecko and Angerer [72] have reviewed the effectiveness of NMR, MS, TG and DSC for the analysis of rubber and rubber compounds containing curing agents, fillers, accelerators and other additives. PyGC has been widely used for the analysis of elastomers, e.g. in the determination of the vulcanisation mode (peroxide or sulfur) of natural rubbers.

Table 2.9 Tools for rubber deformulation

Company	Analytical toolbox	Reference(s)	Year
Bayer	Soxhlet, wet chemistry, electrochemistry, PC, GC, PyGC, TLC, UV, IR, PyIR, OM, SEM, TEM	[10]	1981
Monsanto	DP-MS, ^{31}P NMR	[69]	1983
BFGoodrich	ATR-FTIR, NMR, PyMS, TGA	[70]	1983
Uniroyal Chemical	Soxhlet, Underwriters extraction, GC-MS, TLC, HPLC, UV/VIS, IR, NMR, AAS, XRD, TGA	[71]	1983
Dunlop	US, GC, PyGC, HPLC, IR, ATR-FTIR, PyIR, s-NMR, PyMS, TG, DTA, XRF, XRD	[72]	1988
Polysar	UV/VIS, IR, NMR, GC, HS-GC, HPLC, SEC, IC, DSC, TG, TG-MS, PyGC-MS, AAS, OM	[73]	1988
Goodyear	LDMS	[74,75]	1991–1992
BFGoodrich	GC-MS, LC, TLC, IR, NMR, XRF, MS and Py methods	[76,77]	1993–1995
Ameripol Sympol	SFE, PyGC-MS	[78,79]	1996
Akron Rubber	Reflux extraction, GC, TLC, HPLC, GC-MS, FTIR, Py, SEM, TEM, AAS	[80]	1999
Bayer	Soxhlet, MAE, US, SFE, PFE, GC, HPLC, TLC, UV, IR, R, NMR, MS, TG, TG-FTIR, GC-MS, PyGC-MS, Py-FTIR, TD-GC-MS, AAS, ICP, OM, TEM, ATR-FTIR	[81]	2000

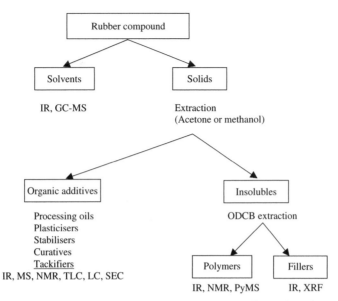

Scheme 2.1 BFGoodrich rubber separation and analysis scheme. After Lattimer [76]. Reprinted from *Journal of Analytical and Applied Pyrolysis*, **26**, R.P. Lattimer, 65–92, Copyright (1993), with permission from Elsevier

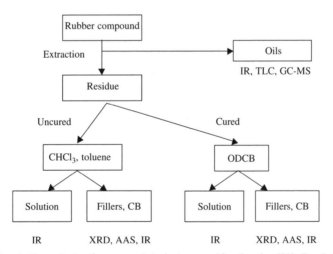

Scheme 2.2 Uniroyal Chemical's analysis of compounded elastomers. After Lussier [71]. Reprinted with permission from *Proceedings of the 122nd Meeting of the ACS Rubber Division*, Paper No. 27. Copyright © (1983), Rubber Division, American Chemical Society, Inc.

The logical approach to problem solving for rubber analysis at *Polysar Ltd* was described by Chu [73] (cf. Schemes 2.4 and 2.5). Systematic analysis involves sampling, elimination of interference and measurement. Methods employed include chromatography (GC, HS-GC, HPLC, SEC, IC), spectroscopy (AAS, UV/VIS, IR, NMR), MS, microscopy and thermal analysis. The specific role of each of these techniques for the analysis of rubber compounds with or without

compounding ingredients, cured rubber products, latexes and monomers is outlined in Table 2.10.

Typical applications at Polysar included the quantification of residual solvents and monomers in finished rubber products (e.g. styrene in SBR), quality control of feedstocks such as benzene or ethyl benzene as impurities in styrene monomer, and the analysis of samples collected from environmental monitoring programs.

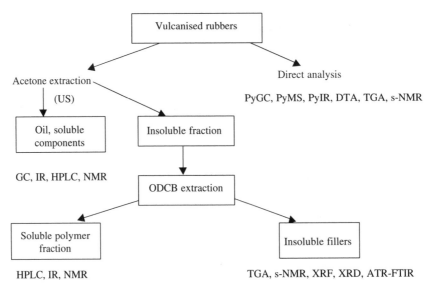

Scheme 2.3 Dunlop's deformulation of vulcanised rubbers. After Schnecko and Angerer [72]. Reproduced by permission of Hüthig GmbH

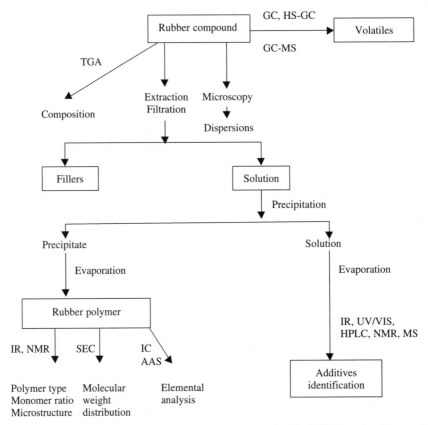

Scheme 2.4 Rubber compounds with or without compounding ingredients. After Chu [73]. Reproduced by permission of Hüthig GmbH

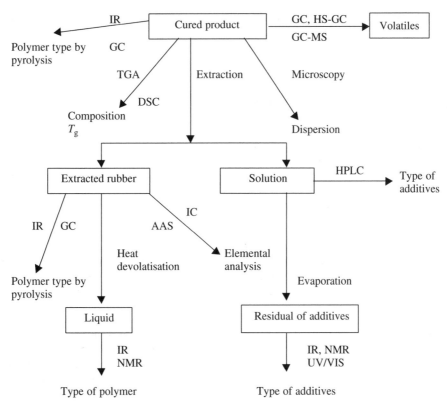

Scheme 2.5   Cured rubber products. After Chu [73]. Reproduced by permission of Hüthig GmbH

Table 2.10   Multitechnique analysis of rubber products

Technique(s)	Applications
IR	Identification of rubber, polymer type (quantitative analysis), additives
UV/VIS	Additives (quantitatively)
NMR	Polymer type, monomer sequence and ratio, isomeric nature
PyGC-MS/TG-MS	Residual monomer/solvent identification, feedstock analysis
GC	Residual monomers/solvents (quantification), QC, impurity tracing
HS-GC	Trace volatile organics, QC
HPLC	Nonvolatile organics
SEC	MWD
IC	Organic acids analysis; Cl⁻, Br⁻, SO₄²⁻ (quantification)
OM	Identification of compounding ingredients, particle size, dispersion, surface blooming, image analysis
TGA	Composition of rubber compounds, filler and rubber type
DSC	T_g
AAS	Catalyst residues, trace metals

After Chu [73]. Reproduced by permission of Hüthig GmbH.

It is quite clear from Schemes 2.1–2.5 that in rubbers polymer identification and additive analysis are highly interlinked. This is at variance to procedures used in polymer/additive analysis. The methods for qualitative and quantitative analysis of the composition of rubber products are detailed in ASTM D 297 'Rubber Products–Chemical Analysis' [39].

At *Goodyear* laser-desorption MS has been used for direct analysis of rubber additives (e.g. antioxidants, antiozonants, vulcanising agents, processing oils, silica fillers, etc.), *in situ* at the surface of an elastomeric vulcanisate [74,75].

The main concern when selecting an analytical test method for characterising polymers is the potential interference from rubber additives such as oils, organic acids, antioxidants, cross-linking agents, resins, binders, accelerators, CB, and other fillers. *Ameripol Synpol* uses SFE for the extraction of extender oil, organic acids and antioxidants from rubbers [78]. Total extractables are often determined by weighing the sample or the extraction thimble plus sample before and after extraction. For characterisation of rubbers PyGC-MS was used [79].

Formula reconstruction of rubbers at *Akron Rubber* [80] can be divided into the following steps (Scheme 2.6):

(i) Extraction of rubber compound (2 g, 16 h reflux).
(ii) Polymer identification (ASTM D 3677 'Rubber Identification by Infrared Spectrophotometry') and quantification (cast film).
(iii) CB type (pyrolysis at 800 °C to 900 °C, followed by TEM analysis according to ASTM D 1765 'Carbon Blacks Used in Rubber Products') and concentration (electron microscope surface area measurement).
(iv) Identification and quantification of inorganic components (after ashing at 900 °C).
(v) Analytical evaluation of the extract to identify and quantify such components as oils and plasticisers (ASTM D 2007), antidegradants, accelerators, antiozonants, accelerator fragments and curatives (ASTM D 3156), etc.; quantification by external standard method.
(vi) Determination of sulfur level (Leco sulfur determinator; pyrolysis at 1350 °C, SO_2 measurement).

Following this scheme, Coz and Baranwal [80] have reported the reverse engineering of four unknown cured rubber compounds, representing a radial passenger tyre tread, a radiator hose, an oil pan seal and an engine gasket.

Finally, Brück [81] has illustrated the current *Bayer* approach to the analysis of rubbers and vulcanisates (Scheme 2.7). It is interesting to notice the great analogy to previous (older) deformulation schemes (Schemes 2.1–2.6), but for the specific analytical tools. *In situ* analysis uses mainly ^{13}C s-NMR, PyGC-MS, TG and FTIR. Various FTIR techniques are used to identify the polymer also in CB-filled rubbers. Solid-state ^{13}C NMR is used mainly for the quantitative determination of the polymer composition and cross-linking structure. Near-infrared Raman spectroscopy (Nd:YAG) can be used in case of CB-free vulcanisates for establishing the polymer composition. TD-GC-MS is applied for the identification of accelerators and antioxidants. Direct analysis of rubber compounds thus yields simultaneously information about polymer and additives. TGA is highly suited for quantitative analysis of rubbers: polymer, additives, CB and fillers. Comparison of PyGC-MS results of rubbers at low and high temperature (e.g. 300 and 750 °C) yields information about low-MW additives. High-performance TLC (HPTLC) is considered to be well suited to the analysis of rubber extracts and allows identification of oils, antioxidants and plasticisers in less than 10 min. NMR, UV and IR spectroscopic analysis of extracts is usually possible only after separation of the components. Coupling of GC, HPLC, SEC with MS or FTIR is advantageous for identification of the soluble components of a vulcanisate of unknown composition. Infrared spectroscopy holds a

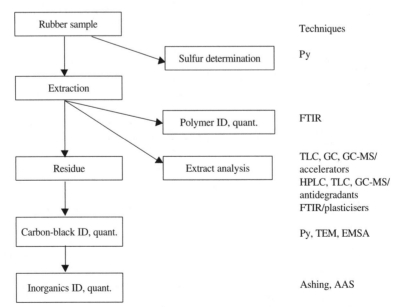

Scheme 2.6 Rubber deformulation. After Coz and Baranwal [80]. Reproduced by permission of *Rubber World Magazine* (Lippincott)

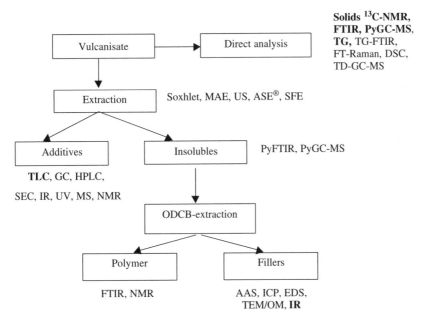

Scheme 2.7 Analysis of rubbers and vulcanisates (bold: main techniques). After Brück [81]. Reproduced by permission of Deutsches Institut für Kautschute Technologie, Hannover

key position for the determination of the polymer components in vulcanisates, either as liquid pyrolysis products [88], ODCB soluble components or thin sections of the extracted vulcanisate. Carbon-black-filled vulcanisates can be investigated by means of ATR-FTIR or PA-FTIR. When filler absorptions dominate in the IR spectra polymer analysis is often impossible by this technique.

Brück [81] has illustrated the analysis of antioxidants in a CB-free vulcanisate of unknown composition according to Scheme 2.7. Some components detected by off-line TD-GC-MS (cyclohexylamine, aniline and benzothiazole) were clearly indicative of the CBS accelerator; other TD components were identified as the antioxidants BHT, 6PPD, Vulcanox BKF and the antiozonant Vulkazon AFS. In the methanol extract also the stabiliser ODPA was identified. The presence of an aromatic oil was clearly derived from the GC-MS spectra of the thermal and methanol extracts. The procedure is very similar to that of Scheme 2.3.

Okumoto [89] has reported an analytical scheme (Scheme 2.8) for automotive rubber products (ENB-EPDM vulcanisates). For high-resolution PyGC analysis, organic additives are first removed from the rubber/(CB, inorganics) formulation. Carbon-black and inorganic material hardly interfere with pyrolysis. For the analysis of the additives the extracted soluble substances are characterised by FTIR, TLC, HPLC and GC.

Deformulation of complex materials, such as rubbers, is a typical example of the need for a multitechnique approach, which requires a well-equipped analytical toolbox (cf. Schemes 2.1–2.8). The preferred methods of choice give simultaneously information about polymer and additives. Overall, the general approach to compound analysis has not really changed over the years, although some new instrumental methods have been adopted in the past few years. At present there is a great deal of in situ analysis: s-NMR, FTIR, TG, with a growing tendency to solve analytical problems by coupling a separation technique with an identification technique, e.g. PyGC-MS, TD-GC-MS, TG-FTIR or TG-MS. This trend is particularly clear by comparing Bayer technology 1981 and 2000. It is equally interesting that some 'old-fashioned' techniques, such as Soxhlet and TLC are still high on the user's priority list. As rubbers are very heterogeneous materials, which fail to be extracted with ease, solid-state analysis requires some special caution.

Analytical techniques used in troubleshooting and formulation experimentation available to the rubber compounder were reviewed [90]. Various textbooks deal with the analysis of rubber and rubber-like polymers [10,38,91]. Forrest [38] has illustrated the use of wet chemistry, spectroscopic, chromatographic, thermal, elemental and microscopy techniques in rubber analysis.

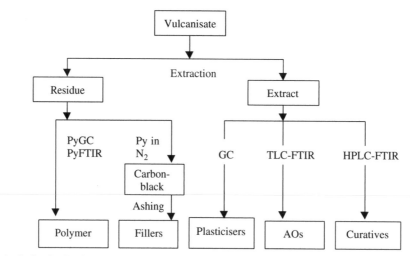

Scheme 2.8   Analysis of vulcanised rubbers. After Okumoto [89]. Reproduced by permission of the author

2.3   POLYMER EXTRACT ANALYSIS

In the past many schemes for the analysis of additives have been developed, mainly in industrial research laboratories. Many of these procedures are matrix specific (i.e. developed for polyolefins, PVC, PTFE, ABS, PS, acrylics, polyamides, etc. [5,9,92,93]) and only some are analyte-class specific (e.g. the identification of sulfur-containing dialkyl stabilisers, or dialkyltin carboxylated and hemi-ester stabilisers in PVC [94]), but most require a considerable amount of chemical handling (extraction, centrifugation, dissolution, filtration, precipitation, decantation, ashing, titration, distillation, gravimetry, etc.). Abundant use of chemicals for solvent extraction, dissolution, chromatography, and identification (e.g. with spray reagents) contrast with direct (*in situ*) examination of the materials. The current trend is to minimise chemical handling and to generalise the procedures.

Gooch [4] has reported general deformulation schemes for solid paints and coatings, liquid paints, solid plastics, liquid plastic specimens, solid and liquid adhesives, in which preliminary examination (by OM, SEM, EDXRA) is followed by separation of the individual components and their subsequent identification (Scheme 2.9). Gooch employs different deformulation schemes for solid and liquid specimens, but essentially identical approaches for plastics, paints, adhesives and inks (Scheme 2.10).

In this context, liquid specimens are polymers dissolved in solvent, in dispersion, or of very low molecular weight. The specimen may contain pigments and additives. If the specimen contains obvious colour and turbidity, then it must be prepared for complete deformulation by separating components, as shown in Scheme 2.11.

In the author's laboratory a wide range of polymers (PE, PP, EPDM, ABS, SAN, SMA, PC, PET, PBT, PA6/6.6/4.6) and copolymers is examined routinely on additive content (qualitatively and quantitatively). It proves being good practice to start analysis of a polymeric material for unknown additives with rapid stereomicroscopic examination in order to gain information about heterogeneity and additive shape (e.g. chopped fibres). It is equally a good habit to carry out qualitative elemental screening (mainly Ca, Br, Ni, P, Sb, Zn, etc.) using 2 mm thick pressed plaques (no grinding in order to avoid contamination). Elemental analysis should be of sufficient sensitivity that important elements are not missed. Optical emission spectroscopy and x-ray fluorescence spectrometry are suitable methods for the concentration ranges of interest (>10 ppm). XRF is applicable to all elements with $Z > 12$. Nitrogen is determinable at the desired levels by micro Kjeldahl digestion techniques. Elemental analysis reduces the possibility of overlooking hetero-element containing additives. Any element identified must be accounted for in the subsequent deformulation, be it that the elements identified are originating from the polymer, are indicative of the additive system or represent adventitious impurities or catalyst residues. Detection of elements such as N, S, P, Si, Sb and halogens at relatively high concentrations may point to additives in the polymer. Ashing a sample on the end of a spatula is a simple way to confirm the presence of an inorganic residue.

For the purpose of IR analysis, in order to establish the nature of the polymer matrix, the sample is pressed

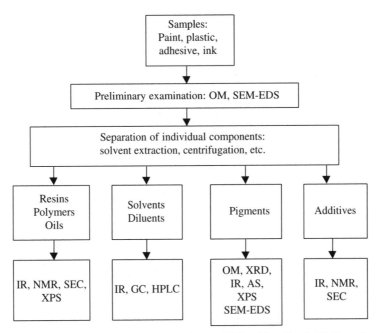

Scheme 2.9 General deformulation for paints, plastics, adhesives and inks. After Gooch [4]. Reproduced from J.W. Gooch, *Analysis and Deformulation of Polymeric Materials, Paints, Plastics, Adhesives and Intes*, Plenum Press, New York, NY (1997), by permission of Kluwer Academic/Plenum Publishers

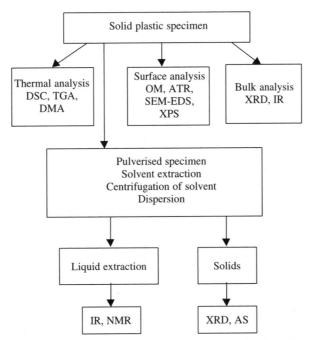

Scheme 2.10 Scheme for deformulation of solid plastic specimens. After Gooch [4]. Reproduced from J.W. Gooch, *Analysis and Deformulation of Polymeric Materials, Paints, Plastics, Adhesives and Intes*, Plenum Press, New York, NY (1997), by permission of Kluwer Academic/Plenum Publishers

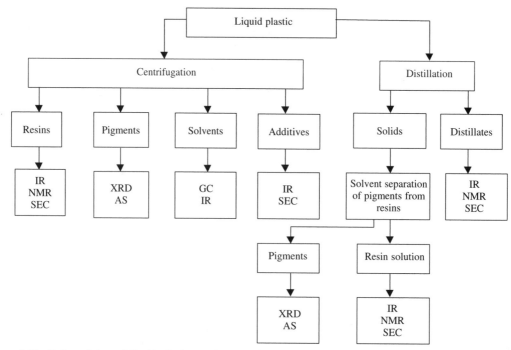

Scheme 2.11 Deformulation of liquid plastic specimens. After Gooch [4]. Reproduced from J.W. Gooch, *Analysis and Deformulation of Polymeric Materials, Paints, Plastics, Adhesives and Intes*, Plenum Press, New York, NY (1997), by permission of Kluwer Academic/Plenum Publishers

into a 100–150 μm thick film. This identification step not only helps in defining the extraction liquid but also the optimal temperature for forming an XRF disk by hot pressing. Preliminary in-polymer examination is terminated with UV spectrophotometry. In case of a polyolefin matrix *in situ* UV analysis may give a first impression of the presence of phenolic antioxidants (absorption at about 280 nm) or UV absorbers (absorption at 330–340 nm) down to 25 ppm levels. UV spectroscopy cannot be used for the detection of additives in polymers which absorb above about 250 nm (e.g. ABS).

Polymer/additive analysis then usually proceeds by separation of polymer and additives (cf. Scheme 2.12) using one out of many solvent extraction techniques (cf. Chapter 3). After extraction the residue is pressed into a thin film to verify that all extractables have been removed. UV spectroscopy is used for verification of the presence of components with a chromophoric moiety (phenolic antioxidants and/or UV absorbers) and IR spectroscopy to verify the absence of IR bands extraneous to the polymer. The XRF results before and after extraction are compared, especially when the elemental analysis does not comply with the preliminary indications of the nature of the additive package. This may occur for example in PA6/PA6.6 blends where

H_3PO_4 is being used for transamidation; in such a case XRF determines higher phosphorous concentrations than justified on the basis of the phosphite additives. Ion chromatography may then be required.

Analytes in the extract are frequently identified/ verified by means of standard (HT)GC-MS methods for volatile components, by HPLC-PDA for less volatile organics and by TLC for nonvolatiles. Where identification, particularly of minor organic components is required, some separation from the plastic compound is usually a must. This is often the case of surfactants and fatty acids, which are frequently multicomponent formulations. Some commercial additives such as several hindered amine light stabilisers (HALS) (e.g. Tinuvin 622 and Chimassorb 944) are oligomeric. In the past various rather exotic procedures have been described for their determination, including direct UV spectroscopic analysis of a polyolefin extract [95], PyGC of an extract, SEC, degradation by saponification followed by LC, etc. They cannot be analysed satisfactorily by LC; Trones *et al.* [96] have recently successfully implemented HTLC for this purpose. For unknown components HPLC-FTIR, preparative TLC, GC-MS or NMR may be used. A complete complex mixture analysis (especially

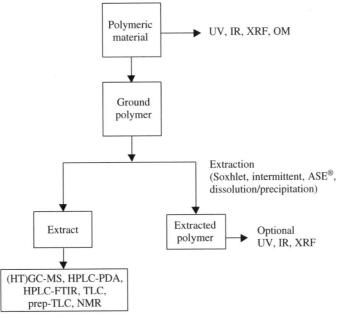

Scheme 2.12   Polymer extract analysis

including quantitative determinations) is rarely solvable by one particular technique, and MS investigations coupled with IR and NMR techniques tend to be the prime tools for many industrial deformulation investigations of the organic components of an 'unknown' formulated product. For sufficiently volatile components, separation, identification and quantification can often be carried out in one analytical process, such as (HT)GC-MS, HPLC-PDA and HPLC-FTIR. As extra identification power LC-MS (either LC-(PB)-MS or LC-APCI-MS) may be used. Completeness of extraction is verified by means of UV or XRF analysis of the residual polymer.

To assure consistency and speed in multidisciplinary structure analysis of low-MW compounds involving various techniques (IR, NMR, MS, etc.) most industrial laboratories use a Standard Operating Procedure (SOP). In such schemes IR analysis is frequently used as a cheap 'filter' for a quick starting control and as a means for *verification*. As IR detects only structural units identification of an unknown compound on the basis of IR is difficult. Mass spectrometry is used as the prime *identification* tool and is especially important in the determination of the exact mass and gross formulae. While structural prognostication on the basis of MS is difficult for the non-expert, a posteriori interpretation is quite feasible. ^{1}H NMR is both easy and cheap, however requires greater sample quantities than either

IR or MS and does not easily distinguish between a monomer and a dimer. NMR is found to be most precious for the identification of P-containing additives. It should be realised that high investments are needed for a universal artificial intelligence structural identification software package based on NMR, MS and IR (Chemical Concepts/Spec Info).

Several qualifying features for polymer extract analysis are summarised in Table 2.11. Quantitative separation of polymer and (thermolabile and/or volatile) additives without decomposition of the analyte(s) is difficult for thermoplasts, but even more difficult for

Table 2.11   Characteristics of polymer extract analysis

Advantages

- Analysis after isolation of the additive
- Analytes easily accessible to chromatographic and spectroscopic techniques

Disadvantages

- Selection of the extraction solvent (difficult for unknowns)
- Yield is function of time and temperature
- Treatment of thermolabile, oxidation or hydrolysis sensitive additives
- Chemical handling
- Use of solvents
- Analyst time and cost
- Additional sample preparation steps often required

thermosetting polymers and vulcanised rubbers due to the compounding ingredients being locked into the matrix by CB and cross-linking of the rubber polymer. An additional problem inherent in most rubber mixtures comes from many of the additives being already fragmented due to the curing and cross-linking process, which combine to new products. Existing extraction techniques produce fragments. If an additive is of a reactive type and copolymerised into the polymer backbone or grafted onto the polymer separation and detection by GC or LC may be difficult or impossible.

After extraction and before analysis additional sample preparation steps may be needed, but these are definitely undesirable:

(i) Concentration step (often necessary after Soxhlet extraction).
(ii) Oligomer clean-up (problem for HTGC), provided the extraction solvent has been chosen properly.
(iii) Derivatisation (e.g. for polar or thermolabile components, to induce UV absorptivity, etc.).

Recent attention has focused on MS for the direct analysis of polymer extracts, using soft ionisation sources to provide enhanced molecular ion signals and less fragment ions, thereby facilitating spectral interpretation. The direct MS analysis of polymer extracts has been accomplished using fast atom bombardment (FAB) [97,98], laser desorption (LD) [97,99], field desorption (FD) [100] and chemical ionisation (CI) [100].

2.4 *IN SITU* POLYMER/ADDITIVE ANALYSIS

Unfortunately, extraction procedures are often elaborate and labour intensive since many of the polymer matrices are poorly soluble or insoluble. For this reason, substantial efforts have been directed towards additive analysis without prior separation from the polymer. Chapter 9 deals with *direct* methods in which such separation of polymer and additive can be omitted. Yet, this direct protocol still requires sample pretreatment (dissolution) of the polymer/additive system as before.

Even more interest exists in direct *in-polymer analysis* of intact bulk samples as delivered (powder, granulate, sheet, film, etc.). This is by no means a simple matter. In fact, as plastic materials usually contain several components, analysis of the levels whilst still in the plastic is usually difficult. Some additives can be analysed without extraction

by using various spectroscopic techniques, such as NMR [101], UV spectroscopy [102] and UV desorption/mass spectrometry [103], but often the complexity of most plastics precludes this. Several other direct methods of analysis, such as IR, fluorescence, phosphorescence and XRF have been reported [104–106]. However, as these methods generally lack specificity, determination of additives often needs a preliminary solid–liquid extraction.

Alternative approaches consist in heat extraction by means of thermal analysis, thermal volatilisation and (laser) desorption techniques, or pyrolysis. In most cases mass spectrometric detection modes are used. Early MS work has focused on thermal desorption of the additives from the bulk polymer, followed by electron impact ionisation (EI) [98,100], CI [100,107] and field ionisation (FI) [100]. These methods are limited in that the polymer additives must be both stable and volatile at the higher temperatures, which is not always the case since many additives are thermally labile. More recently, soft ionisation methods have been applied to the analysis of additives from bulk polymeric material. These ionisation methods include FAB [100] and LD [97,108], which may provide qualitative information with minimal sample pretreatment. A comparison with FAB [97] has shown that LD Fourier transform ion cyclotron resonance (LD-FTICR) is superior for polymer additive identification by giving less molecular ion fragmentation. While PyGC-MS is a much-used tool for the analysis of rubber compounds (both for the characterisation of the polymer and additives), as shown in Section 2.2, its usefulness for the *in situ* in-polymer additive analysis is equally acknowledged.

The deformulation schemes presented in this chapter show that polymer/additive analytics is a rapidly developing field. The analogy of related fields (paper, textiles, resins, coatings, rubbers, paints, food, adhesives and inks) is evident. The same or quite similar principles are being used, with some variations due to the specific nature of the additives and the matrix (cf. deformulation of rubbers in Section 2.2 and of plastics in Sections 2.3 and 2.4). An interesting comparison can be made with *paper chemistry*. Paper is composed chiefly of cellulose, but contains sizing agents, wet- and dry-strength resins, surface treatment agents, pigments, and fixing agents. Sizing agents, such as higher fatty acid salts, rosin, and alkyl ketene dimers, are usually analysed by IR spectroscopy, absorption spectrophotometry, and GC after conventional solvent extraction from paper matrices. Also in this case there is a drive towards direct determination methods [109]. Le [110] has recently described a scheme for the analysis of paper additives by means of spot tests, FTIR and TLC methods.

The increase in scientific production on polymer deformulation is considerably higher than the inflation in scientific papers (3.4 % as from 1989). Bourgeois [111] has recently described reverse engineering of polyolefins and polyolefin blends.

2.5 CLASS-SPECIFIC POLYMER/ADDITIVE ANALYSIS

While most polymer/additive analysis procedures are based on solvent or heat extraction, dissolution/precipitation, digestions or nondestructive techniques generally suitable for various additive classes and polymer matrices, a few class-selective procedures have been described which are based on specific chemical reactions. These wet chemical techniques are to be considered as isolated cases with great specificity.

In the past, the reduction of metallic ions and certain diazotised organic compounds by *hindered phenols* has served as the basis for colorimetric methods of analysis [112–114].

Kellum [115] has described a class-selective oxidation chemistry procedure for the quantitative determination of *secondary antioxidants* in extracts of PE and PP with great precision (better than 1 %). Diorgano sulfides and tertiary phosphites can be quantitatively oxidised with *m*-chloroperoxybenzoic acid to the corresponding sulfones and phosphates with no interference from other stabilisers or additives. Hindered phenols, benzophenones, triazoles, fatty acid amides, and stearate salts show no sensitivity to the oxidation reaction. The basis for the procedure was reported by Paquette [116], namely:

$$R_2S + 2ClC_6H_4COO_2H \longrightarrow R_2SO_2 + 2ClC_6H_4COOH$$

The oxidation reaction proceeds to completion within 30 to 40 min; the oxidant consumed is precisely determined on an excess of reagent by means of an iodometric procedure. Polyethylene samples containing varying amounts of 4,4′-thiobis(6-*t*-butyl-*m*-cresol) and other synergists were analysed with results close to the expected values. The method has been applied to numerous commercially available polymers.

The use of model chain reactions of oxidation of liquid hydrocarbons for quantitative characterisation of stabilisers is not new [118]. It has been demonstrated that determination of the state of *phenolic antioxidants* in PP films can be executed by using a model chain reaction of initiated oxidation of cumene [119,120]. The method utilises the inherent property of antioxidants to deactivate certain oxidatively active molecules of the initiated oxidative model reaction. The procedure has been used for the analysis of sterically hindered phenolic stabilisers in engineering HDPE materials (pipe and geogrid material) [121] and PP non-woven [117,121]. Schroeder *et al.* [117] have used the measurement of the steady-state rate of post-induction oxygen consumption during the model reaction of initiated cumene oxidation for the simultaneous quantitative determination of a hindered phenolic antioxidant (Irganox 1330) and 2,2,6,6-tetramethylpiperidine derived HAS-antioxidants, Chimassorb 119FL and Chimassorb 944, even in the presence of organophosphites in ground PP samples. The

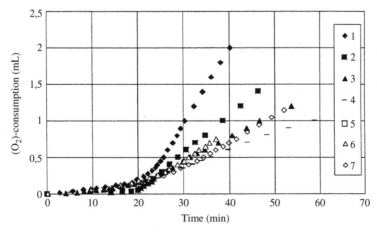

Figure 2.2 Kinetic curves of oxygen consumption for cumene initiated oxidation in the presence of Chimassorb 119FL (1, 0; 2, 1.5×10^{-4}; 3, 2.6×10^{-4}; 4, 4.4×10^{-4} mol/L) or Chimassorb 944 (1, 0; 5, 2.4×10^{-4}; 6, 4.8×10^{-4}; 7, 7.2×10^{-4} mol/L) and Irganox 1330 (1.3×10^{-5} mol/L) at 60 °C. After Schroeder *et al.* [117]. Reproduced by permission of Rapra Technology Ltd

rate of the post-induction oxidation reaction depends linearly on the reciprocal square root of the concentration of the HAS-antioxidant type for a sufficiently wide concentration range (Figure 2.2). The analytical method for the determination of sterically hindered phenol stabilisers by this model reaction of initiated oxidation of cumene outperforms or amends other methods in several respects. The execution is extremely simple and inexpensive. The analyses can be carried out directly with an appropriate ground fraction (particle size $< 100\,\mu$m), without previous extraction or separation. The model reaction is carried out under mild conditions ($60\,^{\circ}$C) in a well-known system. The basic functional principle of the model reaction is a rather specific property of some compounds to deactivate $RO_2^{\bullet}$ peroxy radicals. As a functional chemical reactivity criterion reigns the analysis additives not deactivating the initiating system or $RO_2^{\bullet}$ peroxy radicals should not disturb. A disadvantage of the procedure is the limited uncertainty of the inhibition coefficient f, representing the number of $RO_2^{\bullet}$ peroxy radicals deactivated per antioxidising functional group of one molecule of inhibitor or how many oxidation chains are terminated by one antioxidising group of one molecule of an inhibitor (one stabiliser equivalent). Mika *et al.* [122] have described indirect methods for determination of HALS in polyolefins based on quantitative microanalysis to nitrogen by hydrogenolytic transformation of nitrogen into ammonia. Erroneous results are generated in case of the presence of other nitrogen-containing additives.

2.6 BIBLIOGRAPHY

2.6.1 Polymer Identification

G.A.L. Verleye, N.P.G. Roeges and M.O. de Moor, *Easy Identification of Plastics and Rubbers*, Rapra Technology Ltd, Shawbury (2001).

R.D. Pascoe, *Sorting of Waste Plastics for Recycling, Rapra Review Report No. 124*, Rapra Technology Ltd, Shawbury (2000).

D. Brown, *Simple Methods for Identification of Plastics*, Hanser-Gardner Publishers, Cincinnati, OH (1999).

2.6.2 Deformulation of Rubbers

M.J. Forrest, *Rubber Analysis–Polymers, Compounds and Products, Rapra Review Report No. 139*, Rapra Technology Ltd, Shawbury (2001).

M.J.R. Loadman, *Analysis of Rubber and Rubber-like Polymers*, Kluwer Academic Publishers, Dordrecht (1999).

B.G. Willoughby, *Cure Assessment by Physical and Chemical Techniques, Rapra Review Report. No. 68*, Rapra Technology Ltd, Shawbury (1993).

H. Ostromow, *Analyse von Kautschuken und Elastomeren*, Springer Verlag, Berlin (1981).

2.6.3 Deformulation of Polymers

M.F. Forrest, *Analysis of Plastics, Rapra Review Report No. 149*, Rapra Technology Ltd, Shawbury (2002).

T.R. Crompton, *Manual of Plastics Analysis*, Plenum Press, New York, NY (1998).

J.W. Gooch, *Analysis and Deformulation of Polymeric Materials, Paints, Plastics, Additives and Inks*, Plenum Press, New York, NY (1997).

T.R. Crompton, *Practical Polymer Analysis*, Plenum Press, New York, NY (1993).

2.7 REFERENCES

1. J. Scheirs, *Compositional and Failure Analysis of Polymers, A Practical Approach*, John Wiley & Sons, Ltd, Chichester (2000).
2. D. Braun, *Simple Methods for Identification of Plastics*, Hanser-Gardner Publishers, Cincinnati, OH (1999).
3. T.R. Crompton, *Manual of Plastics Analysis*, Plenum Press, New York, NY (1998).
4. J.W. Gooch, *Analysis and Deformulation of Polymeric Materials, Paints, Plastics, Adhesives and Inks*, Plenum Press, New York, NY (1997).
5. T.R. Crompton, *Practical Polymer Analysis*, Plenum Press, New York, NY (1993).
6. T.R. Crompton, *Analysis of Polymers. An Introduction*, Pergamon Press, Oxford (1989).
7. J. Mitchell (ed.), *Applied Polymer Analysis and Characterization. Recent Developments in Techniques, Instrumentation, Problem Solving*, Hanser Publishers, Munich (1987).
8. J. Haslam, H.A. Willis and D.C.M. Squirrell, *Identification and Analysis of Plastics*, John Wiley & Sons, Ltd, Chichester (1983).
9. A. Krause, A. Lange, M. Ezrin, *Plastics Analysis Guide. Chemical and Instrumental Methods*, Hanser Publishers, Munich (1983).
10. H. Ostromow, *Analyse von Kautschuken und Elastomeren*, Springer-Verlag, Berlin (1981).
11. G.A.L. Verleye, N.P.G. Roeges and M.O. de Moor, *Easy Identification of Plastics and Rubbers*, Rapra Technology Ltd, Shawbury (2001).
12. H.L. Dinsmore and D.C. Smith, *Anal. Chem.* **20**, 11–24 (1948).
13. P. Saarinen and J. Kauppinen, *Appl. Spectrosc.* **45**, 953 (1991).
14. J.A. de Koeijer, H.J. Luinge, J.H. van der Maas, J.M. Chalmers and P.J. Tayler, *Vib. Spectrosc.* **4**, 285–99 (1993).

15. K. Vornberger and B. Willenberg, *Kunstst. Plast. Europe* **84** (5), 17–9 (1994).

16. Mettler-Toledo, *Collected Applications Thermal Analysis: Food*, Schwerzenbach (1998), pp. 47–8.

17. A.K. Bledzki and D. Kardasz, *Intl. Polym. Sci. Technol.* **25** (4), T/90–6 (1998).

18. D.F. Gentle, in *Plastics in Automotive Engineering* (H.G. Haldenwanger and L. Vollrath, eds), C. Hanser, Munich (1994), pp. 261–71.

19. R.D. Pascoe, *Sorting of Waste Plastics for Recycling, Rapra Review Report No. 124*, Rapra Technology Ltd, Shawbury (2000).

20. E. Zoidis (to Sony Intl.), *Eur. Pat. Appl.* 1,072,882 (July 29, 1999).

21. N. Eisenreich and Th. Rohe, *Kunststoffe* **86** (2), 222–4 (1996).

22. H. Ritzmann, *Proceedings Recycle '93*, Davos (1993), Paper 11.

23. N. Eisenreich, J. Herz, H. Kull, W. Mayer and T. Rhoe, *Proceedings SPE ANTEC '96*, Indianapolis, IN (1996), pp. 3131–5.

24. T. Amano, *Proceedings Recycle '95*, Davos (1995), Paper 68.

25. R. Feldhoff, D. Wienke, K. Cammann and H. Fuchs, *Appl. Spectrosc.* **51** (3), 362–8 (1997).

26. S. Moore, *Modern Plastics Int.* **29** (3), 37 (1999).

27. E.R. Grant, E.L. Pomerlean and A.J. Brooks, *Proceedings ARC '97*, Chicago, IL (1997), pp. 61–70.

28. M. Knights, *Plastics Technol.* **43** (12), 24–5 (1997).

29. M. Simpson, D.R. Anderson, C.W. McLeod and M. Cooke, *Analyst* **118** (4), 449–51 (1993).

30. B. Simmons, R. Ahmad and B. Overton, *Br. Plast. Rubb.* (6), 4–10 (1998).

31. A. Bledzki and W. Nowaczek, *Polim. Tworz. Wielk.* **38** (11), 511–8 (1993).

32. F. Garbassi, *Polymer News* **17** (12), 375–7 (1992).

33. W.C. Wake, B.K. Tidd and M.J.R. Loadman, *Analysis of Rubbers and Rubber-like Polymers*, Applied Science Publishers, London (1983).

34. J. Lippincott, *Rubber World Blue Book: Materials, Compounding Ingredients, and Machinery for Rubber* (Annual), Bell Communications, New York, NY (1993).

35. J.B. Tuttle, *The Analysis of Rubber*, The Chemical Catalog Company, New York, NY (1922).

36. W.C. Wake, *The Analysis of Rubber and Rubber-like Polymers*, MacLean & Sons Ltd, London (1958); Wiley-Interscience, Chichester (1969).

37. V.L. Burger, *Rubber Chem. Technol.* **32**, 1452 (1959).

38. M.J. Forrest, *Rubber Analysis–Polymers, Compounds and Products, Rapra Review Reports No. 139*, Rapra Technology Ltd, Shawbury (2001).

39. American Society for Testing and Materials, *Annual Book of ASTM Standards*, Vol. 09.01 (Rubber) and Vol. 09.02 (Rubber Products), ASTM, Philadelphia, PA (1990).

40. S. Affolter, *Macromol. Symp.* **165**, 133–142 (2001).

41. J.G. Kreiner and W.C. Warner, *J. Chromatogr.* **44**, 315–30 (1969).

42. J.G. Kreiner, *Rubber Chem. Technol.* **44**, 381–96 (1971).

43. B.G. Willoughby, *Rubber Fume: Ingredient/Emission Relationships*, Rapra Technology Ltd, Shawbury (1994).

44. D. Gross and K. Strauss, *Kautsch. Gummi Kunstst.* **32** (1), 18–22 (1979).

45. G. Stein, E. Wunstel and C. Held, *Kautsch. Gummi Kunstst.* **52** (2), 124–33 (1999).

46. J. Kruse, *Rubber Chem. Technol.* **46**, 653–785 (1973).

47. P. Bertrand and L.T. Weng, *Rubber Chem. Technol.* **72** (2), 384–97 (1999).

48. W.M. Hess and C.R. Herd, in *Carbon Black Science and Technology* (J.-B. Donnet, R.C. Bansal and M.-J. Wang, eds), M. Dekker, New York, NY (1993), pp. 89–173.

49. T.G. Lamond and C.R. Gillingham, *Rubber J.* **152**, 65 (1970).

50. R.P. Hjelm, W.A. Wampler, P.A. Seeger and M. Gerspacher, *J. Mater. Res.* **9**, 3210 (1994).

51. T.C. Gruber, T.C. Zerda and M. Gerspacher, *Carbon* **31**, 1209 (1993).

52. T.C. Gruber, T.C. Zerda and M. Gerspacher, *Carbon* **32**, 1377 (1994).

53. J.A. Ayala, W.M. Hess, A.O. Dotsan and G.A. Joyce, *Rubber Chem. Technol.* **63**, 747 (1990).

54. H. Raab, J. Froehlich and D. Goeritz, *Proceedings IRC '97*, Kuala Lumpur (1997), pp. 171–3.

55. M.-J. Wang, S. Wolff and J.-B. Donnet, *Rubber Chem. Technol.* **64**, 714 (1991).

56. M.-J. Wang and S. Wolff, *Rubber Chem. Technol.* **65**, 890–907 (1992).

57. F. Scholl, *Atlas der Polymer- und Kunststoffanalyse*, Vol. 3, *Zusatzstoffe und Verarbeitungshilfsmittel*, Verlag Chemie, Weinheim (1981).

58. A.B. Sullivan, G.H. Kuhls and R.H. Campbell, *Rubber Age* **108**, 41–9 (1976).

59. D.W. Carlson, M.W. Hayes, H.C. Ransaw, R.S. McFadden and A.G. Altenau, *Anal. Chem.* **43**, 1874–6 (1971).

60. D.W. Brazier, *Rubber Chem. Technol.* **53**, 457 (1980).

61. E.M. Andersson, I. Ericsson and L. Trojer, *J. Appl. Polym. Sci.* **27**, 2527–37 (1982).

62. B.G. Willoughby and K.W. Scott, *Nitrosamines in Rubber*, Rapra Technology Ltd, Shawbury (1997).

63. B.G. Willoughby, *Cure Assessment by Physical and Chemical Techniques, Rapra Review Report No. 68*, Rapra Technology Ltd, Shawbury (1993).

64. J.W.M. Noordermeer, *Rubber World* **216** (2), 18–22 (1997).

65. J.W.M. Noordermeer, *Rubber World* **216** (4), 16–20 (1997).

66. J. Fourreau, *Matér. Techn.* **4–5**, 80 (1992).

67. G.J. Frisone, D.L. Schwarz and R.A. Ludwigsen, *Proceedings ITEC '96*, Akron, OH (1996), pp. 21–32.

68. J.J. Leyden and J.M. Rabb, *Elastomerics* **112** (4), 34–7 (1980).

69. J. Pierre and J. van Bree, *Kunststoffe* **73**, 319 (1983).

70. J.B. Pausch, R.P. Lattimer and H.L.C. Meuzelaar, *Rubber Chem. Technol.* **56**, 1031–44 (1983).

71. F.E. Lussier, *Proceedings 122nd Meeting ACS Rubber Division*, Toronto (1983), Paper No. 27.

72. H. Schnecko and G. Angerer, *Kautsch. Gummi Kunstst.* **41** (2), 149–53 (1988).
73. C.Y. Chu, *Kautsch. Gummi Kunstst.* **41** (1), 33–9 (1988).
74. W.H. Waddell, K.A. Benzing, L.R. Evans, S.K. Mowdood, D.A. Weil, J.M. McMahon, R.H. Cody and J.A. Kinsinger, *Rubber Chem. Technol.* **64**, 622–34 (1991).
75. W.H. Waddell, K.A. Benzing, L.R. Evans and J.M. McMahon, *Rubber Chem. Technol.* **65**, 411–26 (1992).
76. R.P. Lattimer, *J. Anal. Appl. Pyrol.* **26**, 65–92 (1993).
77. R.P. Lattimer, *J. Anal. Appl. Pyrol.* **31**, 203 (1995).
78. J.K. Sekinger, G.N. Ghebremeskel and L.H. Concienne, *Rubber Chem. Technol.* **69**, 851–7 (1996).
79. G.N. Ghebremeskel, J.K. Sekinger, J.L. Hoffpauir and C. Hendrix, *Rubber Chem. Technol.* **63**, 874–84 (1996).
80. D. Coz and K. Baranwal, *Rubber World*, 30–5 (January 1999).
81. D. Brück, *Fortbildungsseminar Moderne Prüfmethoden – von Rohstoffen über Mischungen bis zum Produkt*, Deutsches Institut für Kautschuk Technologie, Hannover (May 2000).
82. J.B. Pausch, *Anal. Chem.* **54**, 89A–96A (1982).
83. R.P. Lattimer and R.E. Harris, *Mass Spectrom. Rev.* **4**, 369 (1985).
84. R.P. Lattimer, *Rubber Chem. Technol.* **61**, 658 (1988).
85. R.P. Lattimer, H. Muenster and H. Budzikiewicz, *Rubber Chem. Technol.* **63**, 298 (1990).
86. H.-R. Schulten, B. Plage and R.P. Lattimer, *Rubber Chem. Technol.* **62**, 698 (1989).
87. R.P. Lattimer and R.E. Harris, *Rubber Chem. Technol.* **62**, 548 (1989).
88. ISO/DIS 4650.2, *Rubber Identification IR Spectrometric Method*.
89. T. Okumoto, *Proceedings Int. Symposium on Analytical and Applied Pyrolysis*, Nagoya (1994), Paper L-19.
90. P.R. Dlunzneski, *Preprints 158th ACS Rubber Division Meeting*, Cincinnati, OH (2000), Paper 101.
91. M.J.R. Loadman, *Analysis of Rubber and Rubberlike Polymers*, Kluwer Academic Publishers, Dordrecht (1999).
92. D.C.M. Squirrell, *Analyst* **106**, 1042–56 (1981).
93. I.D. Newton, in *Polymer Characterization* (B.J. Hunt and M.I. James, eds), Blackie A&P, Glasgow (1993), pp. 8–36.
94. J. Udris, *Analyst (London)* **96**, 130 (1971).
95. W. Freitag, *Fresenius' Z. Anal. Chem.* **316**, 495 (1983).
96. R. Trones, T. Andersen, T. Greibrokk and D.R. Hegna, *J. Chromatogr.* **A874**(1), 65–71 (2000).
97. C.L. Johlman, C.L. Wilkins, J.D. Hogan, T.L. Donovan, D.A. Laude Jr and M.J. Youssefi, *Anal. Chem.* **62**, 1167–72 (1990).
98. T. Yoshikawa, K. Ushimi, K. Kimura and M. Tamura, *J. Appl. Polym. Sci.* **15**, 2065 (1971).
99. B. Asamoto, J.R. Young and R.J. Citerin, *Anal. Chem.* **62**, 61–70 (1990).
100. R.P. Lattimer, R.E. Harris, C.K. Rhee and H.-R. Schulten, *Anal. Chem.* **58**, 3188–95 (1986).
101. F.C. Schilling and V.J. Kuck, *Polym. Degrad. Stab.* **31**, 141–52 (1991).
102. B. Brauer, T. Funke and H. Schulenberg-Schell, *Dtsch. Lebensm. Rdsch.* **91**, 381–5 (1995).
103. S.J. Wright, M.J. Dale, P.R.R. Langridge-Smith, Q. Zhan and R. Zenobi, *Anal. Chem.* **68**, 3585–94 (1996).
104. H.L. Spell and R.D. Eddy, *Anal. Chem.* **32**, 1811 (1960).
105. H.V. Drushel and A.L. Sommers, *Anal. Chem.* **36**, 836 (1964).
106. A. Wheeler, *Talanta* **15**, 1315 (1968).
107. P. Rudewicz and B. Munson, *Anal. Chem.* **58**, 358 (1986).
108. A.T. Hsu and A.G. Marshall, *Anal. Chem.* **60**, 932 (1988).
109. T. Yano, H. Ohtani, S. Tsuge and T. Obokata, *Tappi J.* **74** (2), 197–201 (1991).
110. P.-C. Le, *PTS-MS 2022 Report (Aktuelle Analyseverfahren für die Untersuchung von Papier, Kreislauf und Abwasser)*, PTS Analytik-Tage (2000), Paper 3.1–16.
111. Y. Bourgeois, in *Polymer Characterisation 2002*, Rapra Technology Ltd, Shawbury (2000).
112. K. Metcalfe and R.F. Tomlinson, *Plastics (London)* **25**, 336 (1960).
113. C.L. Hilton, *Anal. Chem.* **32**, 383 (1960).
114. C. Szalkowski and J. Garber, *J. Agr. Food Chem.* **10**, 490 (1962).
115. G.E. Kellum, *Anal. Chem.* **43** (13), 1843–7 (1971).
116. L.A. Paquette, *J. Am. Chem. Soc.* **86**, 4085–92 (1964).
117. H.F. Schroeder, E.B. Zeynalov, H. Bahr and Th. Rybak, *Polym. Polym. Compos.* **10** (1), 73–82 (2002).
118. G. Scott, *Atmospheric Oxidation and Antioxidants*, Elsevier, Amsterdam (1965).
119. V.F. Tsepalov, A.A. Kharitonova, E.B. Zeinalov and G.P. Gladyshev, *Azerb. Khim. Zhurn.* **4**, 113 (1981).
120. A.A. Kharitonova, V.F. Tsepalov, G.P. Gladyshev, C. de Jonge and V.G. Mies, *Kinet. Kataliz* **21**, 1600 (1980).
121. E.B. Zeinalov, H.F. Schroeder and H. Bahr, *Proceedings Addcon World 2000*, Basel (2000), Paper 3.
122. V. Mika, L. Preisler and J. Sodomka, *Polym. Degrad. Stab.* **28**, 215–25 (1990).

CHAPTER 3

Sorting the wheat from the chaff

Sample Preparation Perspectives

Additives In Polymers: Industrial Analysis And Applications J. C. J. Bart
© 2005 John Wiley & Sons, Ltd ISBN: 0-470-85062-0

Analytical techniques for the quantitative determination of additives in polymers generally fall into two classes: *indirect* (or *destructive*) and *direct* (or *nondestructive*). Destructive methods require an irreversible alteration to the sample so that the additive can be removed from the plastic material for subsequent detention. This chapter separates the additive 'wheat' from the polymer 'chaff', and deals with sample preparation techniques for indirect analysis.

Total sample preparation is recognised as an integral and critical part of the analysis process because it precedes the use of any particular instrumental method. Both *sampling* and sample preparation are important steps in any analytical procedure. During sampling, special attention has to be paid to the question of a representative average of the sample. Proper sampling of potentially nonhomogeneous materials, such as polymeric blends and compounds, is a challenge in analytical chemistry [1]. The ever-increasing precision of analytical techniques puts more stringent requirements on methods to provide representative samples. A large fraction of the analysis error is caused by poor sampling techniques.

According to a recent survey analytical time-keeping in an industrial laboratory is as follows: sample collection 6 %, sample preparation 61 %, analysis 6 %, and data management 27 % [2]. Traditionally, sample preparation has been a 'manual' technology, which was viewed as 'low-tech'. Market drivers for interest in new sample preparation technologies meet the analytical challenges facing industrial laboratories. (Table 10.34) However, new sample preparation technologies suffer from slow acceptance for various reasons such as general inertia of users (learning curve), time required to validate new methods (equivalency testing), a confusing abundance of techniques, slow acceptance by regulatory agencies, payback of capital investment, and others. Automation of sample preparation has lagged behind automation of other analytical manipulations. However, this situation has recently changed as sample preparation

had become the logistic bottleneck. Emphasis is now on the integration of sampling and sample preparation.

The *ideal sample preparation* would be no sample preparation at all, as in the case of direct solid sampling (cf. Section 8.3.4); NIR spectroscopy is typically such a technology. Whether one must solubilise (extraction-GC or HPLC), vaporise (GC, MS), or surface analyse (MS, FTIR) usually some means of sampling must be developed. The sampling step is dominated by instrumentation designed to aid removal of components from the solid polymer materials by a selective 'separation' method. There are many preparation techniques that can be used individually or sequentially according to the complexity of the sample, the nature of the matrix, the analytes, and the instrumental techniques available (Table 3.1). *Classical sample preparation methods* (typically batch adsorption, centrifugation, dialysis, distillation, filtration, liquid–liquid extraction, liquid–solid extraction, lyophilisation, open-column chromatography, precipitation, etc.) are perceived as rate determining in the modern analytical cycle, being most tedious. Traditional methods are also solvent consuming, may result in the loss of volatile components and are the source of much of the imprecision and inaccuracy of the overall analysis. Therefore, the importance of sample preparation in analytical chemistry cannot be overemphasised.

In general, new sample preparation technologies are faster, more efficient and cost effective than traditional sample preparation techniques. They are also safer, more easily automated, use smaller amounts of sample and less organic solvent, provide better target analyte recovery with enhanced precision and accuracy. Attention to the sample preparation steps has also become an important consideration in reducing contamination. A useful general guide to sample preparation has been published [3]. A recent review on sample preparation methods for polymer/additive analysis is also available [4].

The preponderance of analytical sample preparation methods employs some type of *extraction*. Traditionally, these methods were liquid–liquid, liquid–solid, and hot

Table 3.1 Sample preparation processes in active use for polymer/additive analysis

Washing/soaking[a]	Grinding (homogenisation)
Shake-flask extraction[a]	Film pressing/casting
In-vial LSE[a]	Microtoming
Reflux extraction	Weighing
Soxhlet/Soxtec®/Soxtherm® extraction	Mixing
Solid–fluid–vortex extraction[a]	pH change
Intermittent extraction[a]	Liquid handling
Dissolution/precipitation	Solvent exchange
Sonication	Centrifugation
SFE	Dilution
MAE	Evaporation/concentration
PFE (ASE®/ESE®)	Heating/cooling
SFE/ESE®[a]	Column switching
SPE, SPME, HS-SPME	Derivatisation
Hydrolysis	Filtration
Decomposition/digestion	Drying
Thermal extraction (HS, TD, LD)	Glass-ware cleaning (trace analysis)
Pyrolysis	

[a] Limited use.

solvent extraction (e.g. Soxhlet). The area of analytical scale extractions has recently received more attention because it is one of the least developed stages in the assemblage of steps taken to determine compounds in complex matrices, such as polymers, fuel oils, food, blood, soil, etc. [5]. The recent great emphasis on sample preparation techniques has led to a plethora of new approaches such as SFE, MAE, PFE, SPE, SPME, SBSE, HSSE, SLM, MMLLE, MESI, TD, LD, etc.

Most of the polymer/additive samples that have to be analysed are complex, which calls for some form of *chromatography*. The principal objectives of sample preparation for chromatographic analysis are dissolution of the analytes in a suitable solvent and removal from the solution of as many interfering compounds as possible. For a polymer/additive matrix this can be achieved by solvent extraction of the additive and dissolution or hydrolysis of the polymer. The major constraints on sample preparation are low organic analyte concentrations in polymer/additive analysis and the presence of large amounts of unwanted interfering substances in the matrix (polymer, oligomers, fillers, etc.) from which the analyte must be separated before introduction into the analytical column. In most cases the low-MW soluble polymer fraction interferes with the subsequent detection and quantitation of the extracted additives. Filtration of the solvent extract may eventually remove a wax fraction. The actual extraction step should not degrade, modify or adulterate the additives in any way. If the extracts must be concentrated, evaporation should be performed carefully because of the volatility of low-MW

additives (e.g. BHT). In principle, quantitative analysis of an additive is best carried out by *dissolution* (100 % recovery) rather than by extraction, especially when the procedure does not require additional handling (evaporation, preconcentration, redissolution, etc.). In fact, in theory partitioning of the analyte between polymer and solvent prevents complete extraction. However, as the quantity of extracting solvent is much larger than that of the polymeric material, and the partition coefficients usually favour the solvent, in practice at equilibrium very low levels in the polymer will result. A problem may arise though when the additive partitions with the polymer in the dissolution step [6].

The ideal extraction method should be rapid, simple, and inexpensive to perform, yield a quantitative recovery of target analytes without loss or transformation, yield a sample that is immediately ready for analysis without additional concentration or class fractionation steps, and generate no additional laboratory wastes. Unfortunately, the different extraction techniques frequently fail to meet all of these goals. Analytically spoken extraction is far from an ideal sample preparation procedure as the extraction yield varies as a function of the experimental conditions and from sample to sample. Furthermore, many additives are high-MW compounds, thermally labile, or grafted to the polymeric backbone. Analytical extraction and recovery (100 % of the theoretical loading) is therefore often difficult to achieve. However, for industrial analytical purposes the ability to reproducibly extract these materials out of polymer matrices at a high level (>90 %) is usually sufficient to ensure that the correct amount of material is in the formulation to afford the necessary protection during processing and end use. Problems do occur though. Some additives have the tendency to bind to fillers, such as CB [6]. This has caused great difficulty in extractions for systems containing high levels of CB where the extraction recovery may be as low as 50 %. Special problems have also been reported for the higher-MW HALS.

When using any solvent extraction system, one of the most important decisions is the selection of the solvent to be used. The properties which should be considered when choosing the appropriate solvent are: selectivity; distribution coefficients; insolubility; recoverability; density; interfacial tension; chemical reactivity; viscosity; vapour pressure; freezing point; safety; and cost. A balance must be obtained between the efficiency of extraction (the yield), the stability of the additive under the extraction conditions, the (instrumental and analyst) time required and cost of the equipment. Once extracted the functionality is lost and only the chemistry of the additive is important for its qualitative/quantitative detection.

Polymer extraction procedures using organic solvents do not extract all types of organic additives from polymers, and many inorganic compounds and metal inorganic compounds (e.g. calcium stearate) are insoluble. Most types of organic polymer additives, however, can be readily extracted from polymers with organic solvents of various types. For additives of high-MW or of low solubility in the extracting medium, polymer dissolution/reprecipitation or PyGC-MS may be methods of choice.

As scientists strive for ever lower detection limits, sample preparation techniques must inevitably continue to improve. Some future directions in sample preparation for chromatography can be delineated as follows:

- High speed analysis requirements which influence sample preparation demands (being the rate determining step of the analytical process).
- Miniaturisation (lower sample mass, less solvent).
- Organic solvent-less techniques (e.g. subcritical water extraction, headspace SPME).
- On-line coupled sample preparation/separation/identification systems (e.g. SFE-GC, PFE/automated evaporation/HPLC).

Given that approximately two-thirds of a typical analytical chromatographic procedure is spent on sample preparation, it can contribute markedly to the overall measurement uncertainty (reportedly 30% of the total error bar). Techniques to decrease the measurement uncertainty due to sample preparation are: (i) adequate working techniques (e.g. weighing instead of volumetric operations, robots instead of manual handling); (ii) minimising working steps; (iii) choosing dilution steps carefully (preferably large volumes); (iv) making sample and reference measurements in close time proximity on the same instrument; (v) parallel working principles; (vi) internal standards and artificial matrices or use of CRMs; and (vii) increasing the number of determinations [7]. Advances in manual and automated sample preparation of solid and liquid samples have recently critically been reviewed [8]. Several monographs deal with sampling and sample preparation.

3.1 SOLVENTS

Solvents play an important role in relation to polymer science and technology, starting from media for polymerisation reactions, etc. When polymers are used as constituents of coatings, paints, and lacquers they require solvents as dispersing agents [9]. Also in industrial processes, such as the spinning and finishing processes in fibre manufacturing and the casting and finishing processes in membrane manufacturing, the penetration of small molecules into a polymer solid or liquid plays an important role.

For polymer/additive analysis solvents and solvent properties are often crucial in relation to dissolution, solvent extraction, chromatography and spectroscopy. In all cases (different) restrictions on use can be formulated. In extraction the quality of solvents is important. All solvents should be specified as HPLC grade when ordered. It is critical that the solvents be dry (free of water) and that they contain no stabilisers, peroxides or other oxidising materials. If either condition exists, immediate partial or complete hydrolysis or oxidation, e.g. of phosphites, may take place. Marcus [10] has recently presented reliable data on the chemical and physical properties of solvents and has addressed such issues as solvent extraction, solubility and partitioning, and UV and IR spectroscopy windows. For further general consultation on solvents the reader is referred to references [9,11–17]. Appropriate precautions should be taken when handling (toxic) solvents. For safety procedures inherent to a particular solvent, the Material Safety Data (MSD) sheet should be consulted [18]. Various *solvent classification schemes* have been presented, which are based on polarity [15,19,20], acidity and basicity [21], or a mix of hydrogen bonding, electron pair donating abilities, polarity, and the extent of self-association [22]. A common solvent classification scheme [23] is:

(i) nonpolar solvents (such as hexane and tetrachloromethane);
(ii) solvents of low polarity (such as toluene and chloroform);
(iii) aprotic dipolar solvents (such as acetone and N, N-dimethylformamide);
(iv) protic and protogenic solvents (such as ethanol and nitromethane);
(v) basic solvents (such as pyridine and 1,2-diaminoethane);
(vi) acidic solvents (such as 3-methylphenol and butanoic acid).

The term polarity includes parameters such as dipole moment, hydrogen bonding, polarisability, entropy and enthalpy. Since the term polarity cannot be defined unambiguously in physical terms, a classification on this basis is not meaningful.

For the purpose of polymer/additive analysis it is useful to rank solvents according to their boiling point ranges into low boilers (b.p. <100 °C), medium boilers (b.p. 100–150 °C) and high boilers (b.p. >150 °C). Based on their evaporation numbers, solvents can be subdivided into four groups: high volatility (<10), moderate volatility (10–35), low volatility (35–50) and very low volatility (>50).

For selection of alternative solvents (non-ozone depleting) for separation processes (extraction and HPLC mobile phase optimisation) references [24,25] are very useful.

3.1.1 Polymer Solubility Criteria

Polymers may be distinguished on the basis of their solubility:

(i) Polymers completely soluble in organic solvents (e.g. PS).
(ii) Polymers which do not completely dissolve in organic solvents but swell sufficiently to allow extraction (e.g. LDPE, PP, ABS, cross-linked styrene-butadiene copolymer).
(iii) Polymers almost completely insoluble in organic solvents (e.g. PTFE).
(iv) Water-soluble polymers.

In order to study the conditions for polymer dissolution molecular interactions present in polymers have to be considered. The configuration of a polymer in solution is dependent on the balance between the interaction of the segments with each other and the interaction of the segments with the solvent. The most common configuration of synthetic polymers is a random coil. One of the reasons for conformational differences of polymers in solution is the 'thermodynamic quality' of the solvents. In 'good' solvents, which exert stronger secondary forces with the polymer segments, the polymer exhibits the extended chain conformation. The solubility of polymers not only depends on the solvent type, but also on the duration and temperature of exposure, concentration, the molar mass, microstructure (crystallinity, morphology) and macrostructure of the polymer. Some *rules of thumb* of polymer solubility were developed before the advance of theoretical studies:

(i) The 'like dissolves like' principle applies to polymers. A polymer frequently dissolves best in its own monomer or in a structurally similar solvent.
(ii) Solubility normally increases with rising temperature.

(iii) The solubility of chemically related compounds decreases with increasing molecular mass since the intermolecular forces of interaction increase. This is the principle of solution fractionation.
(iv) Branching increases the solubility compared to a linear polymer of the same molecular weight.
(v) Solubility decreases with increasing melting points. Crystalline polymers dissolve to a lesser extent and frequently become soluble only close to their melting point. For example, PE dissolves in various solvents only when heated.

Polymer–solvent interactions and solubility testing were summarised by Staal [26].

3.1.2 Solubility Parameters

Solvency is the interacting force (strength) of a solvent (or additive) for a designated polymer. The free energy of mixing for a polymer–solvent system can be expressed as:

$$\Delta G_{mix} = \Delta H_{mix} - T \Delta S_{mix} \qquad (3.1)$$

where ΔG_{mix} is the Gibbs free energy, ΔH_{mix} the enthalpy and ΔS_{mix} the entropy of mixing. Dissolution of a polymer in a solvent is a matter of a balance between two forces. The first is due to the entropy of mixing, which is low for polymers in solution. Entropy is always acting in favour of mixing. The second force is due to the enthalpy of mixing, which is a measure of the interaction energy between a polymer segment as compared to the interaction energy between segments and solvent molecules alone. This latter force is normally positive and hence opposing the mixing of the two components. This opposing force normally decreases with temperature since it is counteracted by the thermal energy. For mixing/solubility ΔG_{mix} must be negative. If ΔG_{mix} is minimised, the solvent/polymer interaction is most favourable and the polymer will be dissolved or swelled by the solvent. Dissolution of a material is possible only if the free enthalpy of the solution decreases with respect to the sum of the pure components. If the polymer–solvent interaction is strong, and ΔH negative, dissolution is always possible. If ΔG_{mix} is large, polymer–solvent interaction is unfavourable and the solvent will not affect the polymer.

The value of the *solubility parameter* δ (in $cal^{1/2}$ $cm^{-3/2}$ units) can be calculated from the evaporation enthalpy of a liquid at a given temperature:

$$\delta = \left(\frac{\Delta H_p - RT}{V} \right)^{1/2} \qquad (3.2)$$

where ΔH_p is the latent heat of vaporisation and V is the molar volume [27]. The solubility parameter, which is the square root of the cohesive energy density (CED), is a specific temperature-dependent parameter for each solvent and represents a direct thermodynamic measure of the forces of attraction in solvents. Cohesive energy density is the energy needed to remove a molecule from its nearest neighbours. Both direct and indirect methods of evaluation of evaporation enthalpy are available [28]. It is not possible to determine solubility parameters of polymers by direct measurement of the evaporation enthalpy. For this reason, all methods are indirect. The underlying principles of these methods are based on the theory of regular solutions that assume that the best mutual dissolution of substances is observed at equal values of solubility parameters. Methods of experimental evaluation and calculation of solubility parameters of polymers have been reviewed [28].

According to Hildebrand and Scott [27] the mixing heat, ΔH_{mix}, is given by $\Delta H_{mix} = V_{mix} (\delta_1 - \delta_2)^2 \varphi_1 \varphi_2$, where V_{mix} is the total volume, φ_i are the volume fractions, δ_i^2 are the CEDs of solvent and polymer, and δ is commonly referred to as the *Hildebrand solubility parameter*. The predictive value of solubility, which requires a means of determining CEDs, is limited. The CED values are derived from energies of vaporisation at 25 °C for liquids; for solids, the CED forces are estimated according to Small's correlations [29]. Hansen [29a] proposed that the total solubility parameter of a solvent is made up of contributions from dispersive or interactive (London) forces, permanent dipole–dipole interactions, and hydrogen bonding forces, such that $\delta_o^2 = \delta_d^2 + \delta_p^2 + \delta_h^2$, where δ_o is the total solubility parameter and δ_d, δ_p and δ_h are the dispersive, polar and hydrogen solubility parameters, respectively. Solubility parameters vary with temperature in a given solvent. The CED or solubility parameter of polymers cannot be calculated via the heat of vaporisation. An indirect method is to find the best solvents for the polymer. The solubility parameter of the polymer would then equal that of the solvent. The solubility parameters only indicate whether mixing is favourable or unfavourable but gives no measure of the degree of mixing or the equilibrium solubility. Hildebrand parameters of solvents (Table 3.2) and polymers are widely available from a variety of published sources [30–32], including practical examples relating to coatings and fibres [33]. Wright [34] has given a list of experimentally determined solubility parameters with an indication of the degree of hydrogen bonding. Krauskopf [35,36] has described the quantitative prediction of plasticiser solvency in PVC formulations using

Table 3.2   Solubility parameters of common solvents

Solvent	Solubility parameter ($MPa^{1/2}$)	Solvent	Solubility parameter ($MPa^{1/2}$)
Hexane	14.9	Acetone	20.3
Cyclohexane	16.8	2-Propanol	23.5
Ethyl acetate	18.6	Acetonitrile	24.3
Chloroform	19.0	Ethanol	26.0
Dichloromethane	19.8	Methanol	29.7

After Vandenburg and Clifford [4]. Reproduced by permission of Sheffield Academic Press/CRC Press.

Hansen solubility parameters. A convenient scheme for the evaluation of solvency is the use of the Hansen plot, a 3D diagram positioning δ_d, δ_p and δ_h of polymer and solvent.

A polymer will be more soluble in a solvent if the solubility parameters are similar. Also, at a fixed temperature, a solvent with a closer solubility parameter swells the polymer more, and thus gives faster extractions. Consequently, for extractions from PP ($\delta = 16.6\,MPa^{1/2}$) at 120 °C, the extraction rate increases with decreasing difference in solubility parameter [37]. Acetone ($\Delta\delta = 3.7\,MPa^{1/2}$) gives the fastest extractions, followed by 2-propanol ($\Delta\delta = 6.9\,MPa^{1/2}$) with acetonitrile being the slowest ($\Delta\delta = 7.7\,MPa^{1/2}$). If the solubility parameter is too close, as for chloroform, the polymer softens and dissolves.

Experimental methods for virtually all types of solubility measurements have critically been evaluated [38].

3.1.3 Polymer Solutions

Dissolving a polymer is unlike dissolving low-MW compounds because of the vastly different dimensions of solvent and polymer molecules. For high-MW compounds (e.g. UHMWPE, $>10^6$ a.m.u.) dissolution occurs in two stages. Initially the solvent molecules, which are more mobile than the polymer chains, penetrate into and diffuse through the polymer matrix to form a *swollen, solvated mass* called a *gel*. The polymer solution then becomes highly viscous and sticky. The next step in the dissolution process is that the polymer chains disentangle from the gel and *diffuse* into the solvent. This is a slow process since the polymer chain dynamics, which is dependent on the polymer MW, is rate determining. Lower MW polymer chains disentangle spontaneously. Dissolution of a polymer can sometimes be a problem.

How does one choose a solvent? A review on the strategy for solvent selection in order to achieve extraction or liquid chromatography has appeared [25].

The solubility of polymers in solvents is guided by the use of the solubility parameters. Roughly, the closer the solubility parameters, the more polymer will dissolve in a solvent. The solubility parameter is thus based on the assumption that 'like dissolves like'. The suitability of solvents for polymer swelling and dissolution has been widely documented [39]. There are many sources for the empirically determined solubility parameters of polymers and solvents, and listings of solvent/nonsolvent systems [10,31,40,41] (cf. Section 3.7). The solubility parameters are generally determined at or near room temperature but not above the boiling points of the solvents. As the temperature rises, the miscibility range of the polymer will increase. Horie [42] has published solubility charts of common synthetic polymers, including the swelling properties of some network resins. In most of the original sources no standard definition for solvent/nonsolvent systems has been used. A standard practice for preparing polymer solutions is available with information on solvents, concentration, time, pressure, mixing time and heating for 75 typical polymers [43].

True (active) solvents dissolve a given substance at room temperature; the solubility parameters of the solvents and the solute are similar. Polystyrene, for example, has a solubility parameter of 9.1 $cal^{1/2}$ $cm^{-3/2}$ and suitable solvents are cyclohexane ($\delta = 8.2$), benzene ($\delta = 9.2$) and methylethyl ketone ($\delta = 9.3$), while *n*-hexane ($\delta = 7.3$) and ethanol ($\delta = 12.7$) are nonsolvents, i.e. do not dissolve the polymer. Conventional low impact polystyrene is soluble in cold and hot toluene, whereas high-density polyethylene or propylene have little or no solubility in this solvent. However, if the polystyrene contains some copolymerised butadiene, as occurs in the case of high impact polystyrenes, then due to the presence of cross-linked gel, the polymer would not completely dissolve in hot toluene. So even in the case of simple polymers solubility tests are of limited value and for them to provide any useful information requires detailed knowledge. Studies of polymer solubilities using thermodynamic principles have led to semi-empirical relationships for predicting solubility. Van Duin [44] has constructed a solubility plot on the basis of experimental solubility parameters of 40 selected solvents [40], solubility parameters of 50 selected polymers calculated according to Coleman *et al.* [45], and data on (non)solvents for polymer/solvent combinations [40] in order to define solubility regions for copolymers on the basis of composition (e.g. PP/SMA or PP/PS/rubber blends).

Nonsolvents are unable to dissolve the substance in question. Their solubility parameters and hydrogen bond parameters lie outside the solubility regions of the substances to be dissolved. A polymer is not soluble in certain liquids due to a large difference in the interaction energy between segments of the polymer and solvent molecules as compared to the interaction energy between segment–segment and solvent–solvent molecules. Thus, in order to achieve some solubility the segment–solvent interaction energy should be as close as possible to the interaction energy between the segment–segment and solvent–solvent molecules. Certain combinations of two or more solvents may become nonsolvents. Conversely mixtures of two or more nonsolvents may sometimes become solvents. The classification of a certain compound as a nonsolvent does not necessarily imply ability to act as a precipitant since this is influenced also by the nature of the particular solvent of a solvent–nonsolvent pair. However, most nonsolvents combine both properties.

Whereas polymers of sufficiently high molecular weight may be soluble in the common solvents with some difficulty, network polymers do not dissolve, even at elevated temperature. They usually swell depending on the nature and cross-link density. Marcus [10] described the swelling of polystyrene cross-linked by divinylbenzene.

Polymer solutions may undergo phase separation. According to the Flory–Huggins theory for polymer solutions, a polymer solution with a high molecular weight is less stable compared with a solution of the same polymer of lower molecular weight. Hence, when a polymer solution separates, the high molecular weight species will separate in one phase. This phenomenon is used in fractionation of polymer samples with respect to molecular weight. Phase separation can also be achieved by adding a nonsolvent to the polymer–solvent system.

For polymer/additive analysis complete dissolution is not a prerequisite. Rather, the solvent should at least swell the polymer by diffusion, which allows the physically blended additives to dissolve. True dissolution occurs predominantly when polymer chain lengths are small, on the order of 5000–10 000 Da. Solvent choice for dissolution or extraction should take into account *restrictions* imposed by further analysis steps (compatibility with chromatographic and/or spectroscopic requirements). When microwave extraction of additives from a polymer is followed by HPLC analysis, the solvent must be compatible with the HPLC mobile phase so that solvent exchange is not required before analysis.

3.2 EXTRACTION STRATEGY

Quantitative or qualitative analysis of an analyte usually requires analyte isolation from matrix components, concentration, class fractionation or a combination of these actions. The extent of isolation, purification and concentration of the analyte is determined by the matrix (complexity, composition), the concentration of the analyte in the matrix, the selectivity and sensitivity required in the analysis and the analytical objectives (e.g. screening, quantitative or qualitative analysis). Analyte isolation may consist of many different techniques, which can be classified according to their function, as follows:

(i) release of analytes from a matrix;
(ii) removal of contaminants (LLE or SPE);
(iii) liquid handling (dilution, evaporation, dissolution, etc.);
(iv) pre- and post-column derivatisation (selectivity, sensitivity).

It is the difficult task of the analytical chemist to select the sample preparation technique best-suited for the problem at hand. The more tools there are in the toolkit, the larger the chances of finding a sample preparation technique that offers the desired characteristics. The goal of any extraction technique is to obtain extraction efficiency for the analyte which meets the analytical requirements in the shortest possible time. In some analytical procedures little sample handling is needed [46–49].

The role of extraction in sample preparation consists in analyte concentration, clean-up and change of physical form. Extraction procedures, though time-consuming, have these advantages:

(i) Isolation of organic additives from polymer and fillers facilitates both chromatographic and spectroscopic analysis.
(ii) Higher molecular weight (or more labile) additives can be detected more readily in the isolated extract, since desorption ionisation techniques (e.g. FD and FAB) can be used with the extract but not with the compounded polymer.
(iii) Quantitative procedures are facilitated.

Contaminated solvents and glassware are a very well known problem in analysis involving extraction. The major problem in the use of solvents is contamination with plasticisers, especially DEHP. After sample extraction usually enrichment of the analytes is required prior to the analysis.

Extraction is not a simple, almost trivial process. The selection of an extraction procedure depends on the nature of the additive, the polymer matrix, and the requirements of the extraction. Various factors need to

be considered: (i) initial particle size of the sample; (ii) nature of additive (MW, polarity, shape); (iii) stability of the analyte; (iv) chemistry of the additive (interactions between additive and polymeric substrate); (v) solubility of the additive in the extraction solvent; (vi) diffusion rate of the solvent into the polymer; and (vii) conditions (T, p, t, pH) of the extraction process. The actual extraction step should not degrade, modify or adulterate the additives in any way.

To obtain representative samples from nonhomogeneous sample materials, such as polymer compounds, *particle-size reduction* techniques need often to be applied (not for film) [50]. Also, for destructive in-polymer additive analysis it is advantageous to change the physical state of solid samples to provide a larger surface area per unit mass. Complete extraction is sometimes achieved only after grinding the sample. Typically, Perlstein [51] has reported recoveries of only 59 % for extraction of Tinuvin 320 from unground PVC after 16 h of Soxhlet extraction with diethyl ether while recoveries rise to 97 % for ground polymer.

A wide variety of methods for reducing sample particle size are available, such as blending, chopping, crushing, cutting, grinding, homogenising, macerating, milling, mincing, pressing, pulverising and sieving [52]. The choice of device is determined by the following parameters:

- Type of material based on its hardness (Mohs' scale number for hard materials or degree of softness for samples, such as plastics, rubber, paper and leather).
- Initial particle size (chunks or powder).
- Final desired particle size (millimetres or micrometres).
- Required sample quantity or throughput.
- Possible contamination that can interfere with the subsequent analysis.

All particle size reduction involves some sort of abrasion, so contamination by the grinding tools is a constant threat. Common laboratory grinders can granulate most plastic materials into a powder-like form with a particle size ranging from less than 100 up to 1000 μm. During this sample preparation step alteration or loss of the desired (volatile) analyte may occur (e.g. BHT, m.p. 69 °C). Cryogrinding effectively mitigates volatility loss. Malleable or elastic samples, such as rubber or plastic must be cooled before grinding to make them more brittle. Not all samples that are difficult to grind at room temperature can be ground successfully in a freezer mill. Certain polymers, although easily ground in pellet form, remain flexible at liquid nitrogen temperatures when they are in a fibrous or thin-film form [53].

It is necessary to take the trouble to select the most appropriate *extraction solvent* for a specific polymer/additive problem. For this purpose various important points have to be considered: selectivity (the solvent should extract only the components of interest, in order to avoid additional separation processes), and suitability (the extracting agent should be removable under soft conditions without losing the components of interest). In addition to extracting the additives of interest, a portion of the polymer matrix may also be solvated during the extraction step. In most cases the soluble polymer fraction interferes with the subsequent detection and quantitation of the extracted additives. The choice of the ideal solvent for a given polymer additive pair cannot be defined without taking into consideration the limitations imposed by the proposed extraction technique. For example, microwave-assisted extraction requires a solvent (or at least a solvent component) with a high dielectric loss coefficient. Harmonisation of extraction/leaching tests for the purpose of polymer/additive analysis therefore appears to be a futile action.

An increase in the *temperature* can cause the polymer to undergo a transition from the glassy to the rubbery form at the glass transition temperature T_g, with diffusion being much faster in the rubbery than in the crystalline glassy form. A sharp rise in the extraction rate would therefore be expected at T_g, which is defined as the onset of long-range, coordinated molecular motion of the repeating unit of a polymer. A higher temperature should therefore be chosen to maximise diffusion in order to achieve a more complete extraction. The best strategy for fast extractions is to select a solvent that can be used at as high a temperature as possible, namely just below that at which the polymer dissolves in that solvent. To achieve maximum extraction rates with a given solvent, the extraction should be carried out in conditions causing the maximum *swelling* without dissolving the polymer. Principal factors in swelling interaction are the polymer chemical composition, crystallinity, molecular weight and temperature. Swelling the polymer with solvent has a significant effect on the diffusion rate. Diffusion of ethylbenzene through polystyrene swollen with CO_2 has been reported to be 10^6 times faster than through the unswollen polymer [54]. Risch [55] studied the effect of swelling on antioxidant extraction.

Solubility parameters can be a useful guide to solvent selection, but precise quantitative relationships between solvent properties and extraction rates are not yet possible [37]. As an illustrative example we mention extraction of Irganox 1010 from PP [37]. Freeze-ground PP was extracted at 120 °C with 2-propanol,

acetone, and acetonitrile and at 62 °C with chloroform. Chloroform could not be used at 120 °C as the polymer dissolved. It was observed that the greater the solubility parameter differences between solvent and PP, the slower the extraction. The different activation energies obtained using different solvents indicate that the process being measured is not simply diffusion through PP but diffusion through the polymer swollen by the solvent. The use of solvent mixtures (one for swelling the polymer, the other to enhance the solubility of the additive) has also been reported. For example, for extraction of antioxidants from polymers Nielson [56] has used cyclohexane (as swelling agent)/2-propanol (as extractant). The solvent (mixture) should also be selected such that it dissolves the polymer below the melting point of the polymer or the temperature at which the analyte breaks down.

Finally, it is very difficult to keep standard solutions for monitoring or recalibration purposes. Control batches should be purged and kept with a dry, inert gas. As long as results are reproduced, the system is considered to be in calibration. The relatively long extraction times usually prohibit the use of these methods for quality control analysis applications in a plastics manufacturing plant.

Liquid (solvent) extraction is not the only way of sample preparation, but stands along with various forms of heat extraction (headspace, thermal desorption, pyrolysis, etc.) and with laser desorption techniques.

3.3 CONVENTIONAL EXTRACTION TECHNOLOGIES

Liquid extraction technologies span a century. Extraction can be directly conducted as static, dynamic, static–dynamic, or dynamic–cryotrap. Cryotrapping enables analytes to be removed from a matrix and deposited onto a technically more inert surface. For analytes with complex matrices, this method is highly advantageous. For pure analytes or for analytes with simple matrices, however, cryotrapping can be overkill, as it effectively means two sample extractions for a given run.

Extraction techniques for polymeric matrices can be divided into 'traditional' and 'new'. The traditional techniques include Soxhlet extraction, boiling under reflux, shaking extraction and sonication. All these methods are at atmospheric pressure. When the sample is added to a solvent, which is boiled under reflux (i.e. at the highest possible temperature without applying an external pressure) extractions tend to be much faster than Soxhlet extractions. Examples are the Soxtec®,

Soxtherm®, FEXTRA® and intermittent extractions. The newer methods (SFE, ASE®, MAE) invariably operate at pressures above atmospheric. Micromethods of solvent extraction have also been described [57,58].

ISO 6427 (1992) describes the determination of matter extractable from plastics by organic solvents (conventional methods). Solvent extraction has been reviewed [59]. Few books are dedicated to the multifarious extraction methods in organic analysis [60].

3.3.1 Liquid–Liquid Extraction

Solvent extraction is defined as the process of separating one constituent from a mixture by dissolving it into a solvent in which it is soluble but in which the other constituents of the mixture are not, or are at least less soluble. The three main reasons for using solvent extraction are: (i) to isolate a component or analyte of interest; (ii) to remove potential interferents from a matrix; and (iii) to preconcentrate an analyte prior to measurement.

When the term solvent extraction is used, it is often taken as meaning extraction involving two or more liquids, i.e. liquid–liquid extraction (LLE). Although not directly relevant for in-polymer additive analysis, analysis of aqueous solutions is important for environmental reasons. To separate and preconcentrate the analyte component by solvent extraction, the sample can be contacted with an organic phase. Conventional solvent extractions may be applied in various ways. Usually, this method of separation is carried out in the laboratory by shaking a solution of a solute in one solvent with another solvent which is immiscible with the first. The two solvents are then allowed to separate. The simplest scheme extracts only the component to be determined. In a reversed scheme, interfering and matrix elements are removed by extraction; the analyte component in the aqueous phase is then not preconcentrated. Another method may be used which includes both extraction and back-extraction of the component to be determined. Yet, LLE, which is the traditional method of treating samples prior to GC injection, is generally no longer acceptable for a variety of reasons (tedious procedures, difficult automation, need for relatively large sample volume and expensive, highly purified solvents, large volumes, safety). A flow-chart for LLE is available [37]. Recently, a micro volume LLE system has been developed that requires small volumes of aqueous sample and organic extracting solvent [61]. Also, intensive periodic extraction by means of a minute amount of extractant, spread on a suitable support material, is

feasible [62]. Other methods which reduce or eliminate the use of (conventional) solvents include static and dynamic headspace (for volatile compounds) and SFE and SPME (for semivolatiles). In many cases, some form of (conventional) solvent-containing analyte reduction of volume is required. This nowadays often takes the form of solid-phase extraction (SPE). These approaches, as well as reducing the solvent volume and hence preconcentrating the analyte, may also include inherent clean-up processes. Although LLEs are less time-consuming than LSE, the additives are subjected to more losses due to volatilisation and/or degradation because of the (usually) higher temperatures employed. Despite the development of alternative methods of sample preparation, such as SPE, liquid–liquid (solvent) extraction remains a very popular technique (e.g. in the bioanalytical field).

Isolation of the products from complex matrixes (e.g. polymer and water, air, or soil) is often a demanding task. In the process of stability testing (10 days at 40 °C, 1 h at reflux temperature) of selected plastic additives (DEHA, DEHP and Irganox 1076) in EU aqueous simulants, the additive samples after exposure were simply extracted from the aqueous simulants with hexane [63]. A sonication step was necessary to ensure maximum extraction of control samples. Albertsson *et al.* developed several sample preparation techniques using headspace-GC-MS [64], LLE [65] and SPE [66–68]. A practical guide to LLE is available [3].

3.3.2 Liquid–Solid Extraction

In liquid–solid extraction (LSE) the analyte is extracted from the solid by a liquid, which is separated by filtration. *Numerous extraction processes*, representing various types and levels of energy, have been described: steam distillation, simultaneous steam distillation–solvent extraction (SDE), passive hot solvent extraction, forced-flow leaching, (automated) Soxhlet extraction, shake-flask method, mechanically agitated reflux extraction, ultrasound-assisted extraction, γ-ray-assisted extraction, microwave-assisted extraction (MAE), microwave-enhanced extraction (Soxwave®), microwave-assisted process (MAP™), gas-phase MAE, enhanced fluidity extraction, hot (subcritical) water extraction, supercritical fluid extraction (SFE), supercritical assisted liquid extraction, pressurised hot water extraction, enhanced solvent extraction (ESE®), solution/precipitation, etc. The most successful systems are described in Sections 3.3.3–3.4.6. Other, less frequently used extractors are the Underwriters extraction apparatus [69], which is more effective than Soxhlet, because of the faster rate of solvent recycling, tightly capped bottles in which the sample and solvent are heated under pressure [70], Wiley extractors [71–73], and flasks in which the sample is merely steeped in solvent [74]. The Wiley extractor was used in a method for the determination of 2,6-di-t-butyl-p-cresol in polyolefins [73]. Sevini and Marcato [75] mentioned use of a Kumagawa-type extractor and Fiorenza *et al.* [76] an ASTM extraction device. These more exotic extractors have had little follow-up.

Classical solvent extraction is a phase transfer of solute from the solid phase to solution. All analytical extractions from solid matrices undergo *three processes*: (i) transport of the analyte through the sample particle, which may involve overcoming the energy of interaction between analyte and sample matrix and/or analyte diffusion through the sample particle; (ii) transport of the analyte from the surface of the sample particle into the bulk extraction fluid, which may involve overcoming the adsorption energy at the particle surface; and (iii) transport of the extracting fluid/analyte solution away from the sample particle. These transport processes are dependent on a variety of physical properties, such as diffusion, viscosity, partitioning, solubility, and surface tension. These properties generally become more favourable at higher temperatures.

Essentially, extraction of an analyte from one phase into a second phase is dependent upon two main factors: solubility and equilibrium. The principle by which solvent extraction is successful is that 'like dissolves like'. To identify which solvent performs best in which system, a number of chemical properties must be considered to determine the efficiency and success of an extraction [77]. Separation of a solute from solid, liquid or gaseous sample by using a suitable solvent is reliant upon the relationship described by Nernst's distribution or partition law. The traditional distribution or partition coefficient is defined as $K_D = C_s/C_l$, where C_s is the concentration of the solute in the solid and C_l is the species concentration in the liquid. A small K_D value stands for a more powerful solvent which is more likely to accumulate the target analyte. The shape of the partition isotherm can be used to deduce the behaviour of the solute in the extracting solvent. In theory, partitioning of the analyte between polymer and solvent prevents complete extraction. However, as the quantity of extracting solvent is much larger than that of the polymeric material, and the partition coefficients usually favour the solvent, in practice at equilibrium very low levels in the polymer will result.

The *efficacy and rate* of the extraction process of additives from polymers are influenced by numerous factors including:

(i) the molecular weight of the additive;
(ii) the migration rate of the additive in the polymer (high molecular weight, bulky additives diffuse more slowly);
(iii) the diffusion rate of the solvent in the polymer (a function of density, degree of crystallinity and other factors; HDPE is more difficult to extract than LDPE);
(iv) the solid particle size (polymer surface/solvent ratio);
(v) the analyte solubility in the extracting fluid (nature of the solvent);
(vi) analyte–matrix interactions;
(vii) extraction conditions (T, t, p, pH).

The *extraction time* has been observed to vary linearly with polymer density and decreases with smaller particle size [78,79]. The extraction time varies considerably for different solvents and additives. Small particle sizes are often essential to complete the extraction in reasonable times, and the solvents must be carefully selected to swell the polymer to dissolve the additives quantitatively. By powdering PP to 50 mesh size, 98 % extraction of BHT can be achieved by shaking at room temperature for 30 min with carbon disulfide. With iso-octane the same recovery requires 125 min; Santonox is extractable quantitatively with iso-octane only after 2000 min. The choice of solvent significantly influences the duration of the extraction. For example talc filled PP can be extracted in 72 h with chloroform, but needs only 24 h with THF [80]. pH plays a role in extracting weakly acidic and basic organic solutes, but is rarely addressed explicitly as a parameter.

It can be shown that multistep extraction is advisable, i.e. it is always better to use several small portions of solvent (e.g. $5 \times 20\,cm^3$) to extract a sample than to extract with one large portion (e.g. $1 \times 100\,cm^3$) [77]. As mentioned already, for general purposes a *recovery* of greater than 90 % is usually considered acceptable in polymer/additive analysis; no analytical recovery is required. A flow-chart for LSE is available [3].

3.3.3 Classical Solvent Extractions of Additives from Polymers

The first step in destructive polymer/additive deformulation is usually transfer of the additives into the liquid phase in order to render them accessible to a wide range of chromatographic, spectroscopic and spectrometric methods. Methods adapted for separation of polymer and additives are typically based on LSE, dissolution/reprecipitation or hydrolysis [81–83]. Of practical concern in extraction is the volatility of additives and the transformation of thermally labile components.

Conventional methods of polymer extraction use large quantities of solvents as in shake-flask extraction or a Soxhlet extraction apparatus. For all classical extraction methods, solvent selectivity, in general, is low, i.e. solvents with high capacity tend to have low selectivity. In reflux extractions, which are still quite popular in polymer applications, the polymer is refluxed with a hot solvent, which disperses it to provide a solvent phase containing additives. In these conditions solvents are at their atmospheric boiling point. These methods are lengthy and labour intensive. *Fractional extraction* is based on solvents with increasing solvent power (cf. also [81]).

The classical methods of solvent extraction of polymers can be conveniently divided into those for which *heat* is required (Soxhlet/Soxtec®), and those methods for which no heat is added, but which utilise some form of *agitation*, i.e. shaking or sonication (Table 3.3). Other LSE procedures consist in soaking the polymer in boiling solvents [84,85] and cold LSE [80,86]. These methods are also time-consuming, use large amounts of solvents which are scheduled to be restricted in the future, and exhibit other limitations when analytes are present in small quantities, where they may actually be lost in concentration steps following extraction. Many norms are still based on such standard procedures [87,88].

More recently, some novel *pressurised procedures* for extraction of additives from polymers have been developed (Table 3.3). The principal objectives of all these techniques, such as SFE [89], MAE [90] and pressurised fluid extraction (PFE), is to replace the conventional extraction methods by shortening the extraction time, reduction in solvent use and automation.

Table 3.4 summarises the main characteristics of a variety of sample preparation modes for in-polymer additive analysis. Table 3.5 is a short literature evaluation of various extraction techniques. Majors [91] has recently reviewed the changing role of extraction in preparation of solid samples. Vandenburg and Clifford [4] and others [6,91–95] have reviewed several sample preparation techniques, including polymer dissolution, LSE and SFE, microwave dissolution, ultrasonication and accelerated solvent extraction.

Table 3.3 Essential conditions of liquid–solid extraction methods

Extraction principle	Conditions			
	Hot	Cold	Agitation	Pressure
Fractional	+	–	–	–
Washing/soaking	+	+	–	–
Shake-flask	+	+	+	–
In-vial (hot block)	+	–	–/+[a]	–
Parr bomb	+	–	–	+
Soxhlet (reflux)	+	–	–	–
Soxtec®/Soxtherm®	+	–	–	–
Solid–fluid–vortex	+	–	+	+[h]
Intermittent	+	–	+	+[h]
Distillation–extraction (SDE)	+	–	–	–
Forced-flow leaching	+	–	–	+[i]
Cold extraction	–	+	+	–
Sonication (US)	+[d]	+	+	–
Supercritical fluid (SFE)	+	+	–	+
Pressurised hot water (PHWE)	+	–	–	+
Microwave-assisted (MAE)	+	–	–	+[b], –[c]
Soxwave® (FMW)	+	–	+	–
Microwave-assisted (MAP™)	+[e]	–[f]	–	+[b], –[c]
Pressurised fluid (PFE)[g]	+	–	–	+

[a] Upon cooling;
[b] closed-vessel;
[c] open-vessel;
[d] usually ambient –40 °C;
[e] analyte in matrix;
[f] solvent;
[g] accelerated solvent extraction (ASE®) and enhanced solvent extraction (ESE®);
[h] filtration pressure and/or vacuum;
[i] high-pressure flow.

3.3.3.1　Off-line vs. On-line Extraction Methods

In *off-line extraction* the extracted analytes are collected and isolated independently from any subsequent analytical technique, which is to be employed next. For example, the extracted analyte can be collected in a solvent or on a solid sorbent. The choice of the collection method affects the possibilities for further analysis. The extracts may be used for final direct measurements (i.e. without further separation), e.g. UV and IR analysis. More usually, however, extraction is a pre-separation technique for chromatography, either off-line (the most common mode of SFE) or on-line (e.g. SFE-GC, SFE-LC-FTIR, etc.). The solvents used in extraction may affect subsequent chromatography.

In *on-line extraction* the process is coupled directly ('hyphenated') to the analytical technique used for further analysis of the extract (either spectroscopy or, more frequently, chromatography, because of the limited selectivity of extraction). Common examples include SFE-GC, SFE-SFC, SFE-HPLC, SFE-FTIR,

Table 3.4 Summary of polymer/additive dissolution and extraction techniques[a]

	Dissolution	Soxhlet	Soxtec®	Intermittent	Ultrasonication	SFE	MAE	PFE
Solvent	Any	Any	Any	Àny	Any	(Modified) CO_2	Solvent with microwave absorbing component	Any
Typical sample size	1–10 g	1–10 g	1–5 g	1–5 g	10–50 g	50–100 mg[b], 1–10 g[c]	1–5 g	1–10 g
Analysis time	40 min –1 h	6–48 h	2 h	2 h	40 min –1 h	20 min –2 h	10 min –1 h	15 min
Solvent volume	1–5 w/w %	50–500 mL	15 mL	75 mL	50–500 mL	10–50 mL (modifier)	30 mL	15–50 mL
Quantitative	Yes	No	No	No	No	No	No	No
Experimental parameters	S, T, t	S, T, t	S, T, t	S, T, t	S, T, t	S, M, T, p, t, f	S, T, p, t	S, T, p, t
Compatibility with on-line chromatography	Yes	No	No	No	No	Yes	No	No
Advantages	Low cost; No grinding; Broad applicability; High b.p. solvent contamination of analyte	Low investment; Simple equipment; Simultaneous extractions in series	Low investment; Simple equipment; Rapid; Economic solvent use; Good reproducibility	Low investment; Simple equipment	Economical; Simple equipment; Not traumatic	Almost solvent free; Concentrated analyte; Rapid; Low temperatures	Rapid; Automated; Simultaneous extraction; Low solvent use	Rapid; User friendly; Automated; Sequential extractions; Not analyst labour intensive

(*continued overleaf*)

Table 3.4 (*continued*)

	Dissolution	Soxhlet	Soxtec®	Intermittent	Ultrasonication	SFE	MAE	PFE
Disadvantages	Analyst labour intensive	Solvent intensive	Automated	Not commercial	(Cryo)grinding	Variable extract collection modes Automation, hyphenation Tuneable solvent power		Limited solvent use
	Difficult to automate	Slow		(Cryo)grinding	Not always effective	High investment	Modest investment	Modest investment
	Insolubility of cross-linked polymers	(Cryo)grinding	(Cryo)grinding			Modifier (for polar extractants)	Dielectric solvents	Off-line
	Volatility of additives	Concentration step	Concentration step	Concentration step	Labour intensive	(Cryo)grinding	Safety precautions	Pressure device
	Transformation of additives	Loss of volatiles			No automation	Requires optimisation	Micronisation	(Cryo)grinding
					Clean-up needed	Requires high-purity extractants	No uniform absorption of microwave energy	Requires optimisation
	Not universal	Thermally labile component unfriendly				Trouble-prone	Requires optimisation	Limited experience for polymer/additives
References	[4,6]	[4,92]	[4]	–	[4,56,92]	[4,6,89,92]	[4,6,92,96,97]	[4,92]

[a] M = modifiers, S = solvent, T = temperature, f = flow-rate, p = pressure, t = time;
[b] on-line;
[c] off-line.

Table 3.5 Scientific publications on extraction techniques for the qualitative and quantitative determination of additives and additive classes of Appendix II

Technique	No. entries
Extraction	776
Accelerated solvent extraction (ASE®)	2
Microwave (MAE)	21
Supercritical fluid extraction (SFE)	25
Soxhlet extraction	33
Soxtec®	–
Ultrasonics	24

Source: *Chem. Abstr.* 1968–1997.

SPE-HPLC and SPME-GC (cf. Sections 7.1 and 7.2). For meaningful results it is crucial to dispose of the correct interface between the various combined techniques.

Hyphenation offers several *advantages* including the total transfer of the extract, which results in maximum sensitivity. Sample handling and exposure to the atmosphere can be eliminated between the extraction step and the analysis, which reduces analyte contamination and degradation. For example, with a special guard column and multi-solvent gradient liquid chromatography (HPLC), extraction and analysis of monomers, additives, oligomers, and polymer can be performed in one step on-line [98]. The principle of this method is to dissolve the sample in a good solvent and precipitate or suspend the polymer sample in the HPLC with a nonsolvent. By adding a solvent to the adsorbed suspension with a gradient system, the components will redissolve and elute from the column (cf. Section 4.4.2.4). Combined chromatographic–spectroscopic techniques allow complex multicomponent data to be obtained from a single experiment.

3.3.3.2 Fractional Extraction

Principles and Characteristics Fractional solution procedures usually consist of consecutive extractions with solvents of increasing solvent power. These labour intensive methods benefit from a larger surface area to mass ratio. Other methods for fractionation by solubility rely on fractional precipitation through addition of a nonsolvent, lowering the temperature or solvent volatilisation (Section 3.7).

Applications References [99–102] have reported fractional extraction of cresolic and phenolic antioxidants from PE. Conventional analysis of PVC compounds requires the subsequent use of various solvents

for extraction of the different organic components used as additives, e.g. plasticisers, lubricants and stabilisers [103]. Braun and Bezdadea [104] have reported a procedure for the analysis of plate-out formed during PVC extrusion based on dissolution in THF. However, THF is a nonsolvent for some organic components of PVC compounds, e.g. outer lubricants of very low polarity. On the other hand, 1,2,4-trichlorobenzene (TCB) acts as a solvent for nearly all the organic components used in PVC compounds including Pd, Cd, Ba and Zn salts of fatty acids, ester waxes, DEHP, epoxidised soya oil, stearic acid, di-*n*-butyltin mercaptide and glycerol mono-oleate. Fractional extraction of PVC with diethylether, methanol and THF, as proposed by Crompton [81], is no longer to be considered an economically viable analysis procedure. Using successive Soxhlet extractions with *n*-heptane, *n*-octane and xylene polypropylenes may be fractionated [105]. By carefully selecting extraction conditions class-selective SFE is possible (Section 3.4.2.3).

3.3.3.3 Washing/soaking

Principles and Characteristics In this most simple liquid–solid interaction a resin is washed with a solvent to (selectively) remove external components. This additive isolation procedure can be used to show that a component is applied on the surface of a polymer pellet, as opposed to blending throughout the polymer. Other LSE procedures consist in soaking the polymer in boiling solvent and cold liquids.

Applications Kumar [106] has reported ethanol washing of styrenic polymers to remove external stearic acid for FTIR identification. The conventional polymer dissolution/reprecipitation method, in which stearic acid is extracted from the bulk *and* from the surface of styrenic polymers, fails to establish whether stearic acid is used as an external or internal lubricant. Similar washing procedures are being applied in diffusion and blooming studies. Extraction of additives from rubbers or plastics was also carried out by placing the testing material in a porous PTFE bag (pore diameter 10–100 μm); the additives were extracted by feeding a solvent to the bag [107]. Also soaking the polymer in boiling solvents [84,85] and cold LSE have been reported [80,86].

3.3.3.4 Shake-flask Extraction

Principles and Characteristics Conventional LSE in the form of shake-flask extraction is carried out by placing a sample into a suitable glass container, adding a suitable organic solvent, and then agitating or shaking (rocking or circular action) in simple laboratory devices for a determined time period to allow the analytes to dissolve into the surrounding liquid until they are completely removed. This method works well when the analyte is very soluble in the extracting solvent and the sample is quite porous. To get more effective solid–liquid contact, samples must first be in a finely divided state. Heating or refluxing the sample in hot solvent may speed up the extraction process. Equilibrium of the analyte between the solid and solvent should strongly favour the solvent. After extraction, the solvent-containing analyte needs to be separated from the insoluble matrix by means of centrifugation, decantation, and/or filtration. The shake-flask (or shake-filter) method can be performed in batches, which helps the overall sample throughput. Multiple extractions can easily be carried out. It may be advisable to repeat the process several times with fresh solvent and then combine all extracts. Shake-flask methods are probably the least efficient of the solid extraction methods and generally produce the lowest yields for difficult to extract matrices.

Mixing, wrist shaking, or tumbling combinations offer a simple, effective, but time-consuming and not highly reproducible method of extraction. Sample-to-solvent ratios are similar to Soxhlet ratios. Although sample manipulation is reduced, this technique requires nearly as much time as the Soxhlet method. Shake-flask extraction is low cost.

Applications   Shake-flask extraction nowadays finds only limited application in polymer/additive analysis. Carlson *et al.* [108] used this technique to extract antioxidants from rubber vulcanisates for identification purposes (NMR, IR, MS). Wrist-action shaking at room temperature was also used as the sample preparation step for the UV and IR determination of Ionol CP, Santonox R and oleamide extracted from pelletised polyethylene using different solvents [78]. BHT could be extracted in 98 % yield from powdered PP by shaking at room temperature for 30 min with carbon disulfide.

In the framework of the phthalate controversy Wilkinson and Lamb [109] used various *in vitro* methods in which known amounts of soft PVC materials were shaken, stirred, impacted, or otherwise mechanically agitated in some type of simulated saliva under controlled conditions (T, t), and the saliva extracted into hexane for analysis. Shaking-flask (liquid–liquid) extraction was also used for the solvent extraction of nonionic surfactants of the general type $RO(CH_2CH_2O)_nH$ (where R

is an alkyl or alkylphenyl group and n is the number-average degree of polymerisation) [110].

3.3.3.5   In-vial LSE

Principles and Characteristics   In-vial or hot block extraction uses a solid block of aluminium with bored chambers for the sample vials. These chambers are only slightly larger than the vial diameters and deep enough that only about one half of the cap height protrudes above the top of the block. The temperature of the block is controlled by the use of a heater module, which holds the block(s) snugly on all sides plus the bottom. The system has thermocouple leads placed into ports also bored into the aluminium block. The hot block extraction method uses the same type of sample vials, septa, solvents and internal standards as in the microwave extraction method. In this procedure about 1 g of sample is placed in a 20 mL headspace vial; after adding some 10 mL of solvent the vial is capped and placed in the hot block for 15 min (sometimes with additional repeated manual shaking). After removing the vial from the hot block it is mechanically shaken until cooling. The required aliquot is then removed for analysis.

Advantages of the block heater are simplicity, ease of use, exact temperature control, automation, total extraction of additives (when – as usual – complete dissolution of the polymer takes place), reasonably short extraction times, the provision to extract many samples (up to 18 or more) simultaneously and the fact that at least 3 g of sample can be used, thus reducing errors due to heterogeneity of the sample. A *shortcoming* of this system is that the block becomes extremely hot. This extraction method can therefore be quite traumatic to the polymer if either long extraction times or extreme temperatures are used [23]. Another drawback is poor chromatographic resolution. Care must be taken to avoid contact with the block when adding or removing vials. There is also the potential, with this open system, for the explosion of a vial. The use of proper safety equipment is required when handling the hot vials.

Applications   The in-vial LSE of additives is used in several laboratories (e.g. GE, DSM Plant Laboratories) but other extraction methods are more popular. Use may be envisaged in particular for specific routine applications in plant services at low cost. Desrosiers [23] has reported that up to 3 g of either ground or thin disk strips may be put completely into solution at 140 °C and an extraction period of 15 min. Ashton [111] used hot block extraction for the determination of bis-(2,4-di-*tert*-butylphenyl) pentaerythritol diphosphite (and its associated breakdown products) in PP.

3.3.3.6 Parr Bomb Extraction

Principles and Characteristics In the Parr bomb extraction technique (see also Section 8.2.1.2), a Teflon-tetrafluoroethylene fluorocarbon resin lined vessel is used for sample containment. A sample (typically 1 g) is heated in a small volume of solvent at T, t, p of choice.

Applications The determination and quantification of oligomers from PET has been carried out using various techniques: Soxhlet extraction followed by gravimetric analysis with identification by HPLC-DAD, selective precipitation of the polymer (from a trifluoroacetic acid solution), and chloroform extraction under pressure in a sealed Parr bomb [112]. Heating of a 1 g sample in 20 mL chloroform at 100 °C for 2 h allows a precision of ±5 %.

Costley *et al.* [113] have evaluated the use of a range of organic solvents (dichloromethane, water, acetone, hexane, xylene) in the microwave extraction of oligomers from PET and have compared MAE to alternative extraction approaches (Soxhlet, Parr bomb).

3.3.3.7 Reflux Extraction

Principles and Characteristics In boiling under reflux procedures a small amount of ground polymer (typically 3 g) is placed in a headspace jar (typically 100 mL) and solvent (typically 30 mL) is added. After sealing, the jar is placed in an oven at a temperature where the solvent slowly refluxes. The solvent is, therefore, at the highest temperature possible without applying an external pressure. Consequently, reflux extractions tend to be much faster than Soxhlet extractions. Examples are Soxtec®, Soxtherm®, FEXTRA® and intermittent extraction. Whilst, in theory, partitioning of the analyte between the polymer and solvent prevents complete extraction, this hardly ever constitutes a problem in practice. As the quantity of solvent is much larger than that of the polymer, and the partition coefficients usually favour the solvent, very low additive levels in the polymer result at equilibrium. Any solvent or solvent mixture can be used.

The *advantages* of solvent reflux are that the low cost equipment is very simple and the extractions

are faster than Soxhlet extraction with high recovery for all additives; automation is possible. A *disadvantage* is that, in theory, complete extraction can never be attained (despite large solvent use) and solvents, which give rapid extractions, often also dissolve lower oligomers. The method is also less suited for thermolabile analytes.

Applications Caceres *et al.* [114] compared various methods for extraction of Tinuvin 770 and Chimassorb 944 from HDPE pellets, namely room temperature diffusion in $CHCl_3$ (20 % extraction), ultrasonication (20 % extraction), Soxtec® extraction with DCM (nonsolvent) (50 % extraction), dissolution (dichlorobenzene)/precipitation (2-propanol) (65–70 % recovery) and boiling under reflux with toluene (solvent) at 160 °C (95 % extraction). By changing conditions (nature of solvent, T, t) similar comparisons do not have much added value. Table 3.6 compares the results of reflux extraction and MAE for additives in LDPE [115].

Reflux extraction of additives and wax from polyolefins was reported [116]. Subsequently, the additives were adsorbed onto an adsorbent (Florisil) and the wax was removed from the extract before chromatography. Boiling extractions of SBR are described in ASTM D 1416-89; 1 g of rubber is extracted by boiling in two 100 mL portions of 75/25 vol% isopropyl alcohol/toluene. Reflux heating with strong solvents, such as THF, dichloromethane or chloroform has been reported [117]. Reflux extraction has also been used for the 3 h extraction of caprolactam and oligomers from PA6 in boiling methanol.

The advantages of automation of extraction under reflux *versus* manual sample preparation are well illustrated for HDPE/(Irganox 1010, Irgafos 168) (Table 3.7) [118]. No spikes were observed in the former case, as opposed to the latter. Boiling under reflux is considered by some [4] as the best 'all round' conventional extraction method.

Table 3.6 Relative efficiencies of microwave-assisted and reflux extraction techniques for additives in LDPE[a]

Additive	Slip agent		Antioxidant	
Method	Microwave	Reflux	Microwave	Reflux
Conc. (ppm)	819	843	329	343
RSD (%)	1.38	3.52	1.87	2.78

[a] Conditions: 5 g of ground polymers in 50 mL isopropyl alcohol, at 125 °C for 10 min, followed by GC-FID analysis.
After Scheirs [115]. From J. Scheirs, *Compositional and Failure Analysis of Polymers. A Practical Approach.* Copyright © 2000 John Wiley & Sons, Limited. Reproduced with permission.

Table 3.7 Extraction under reflux of HDPE/(Irganox 1010, Irgafos 168)

	Automated sample preparation[a]		Manual sample preparation[b]	
	Irganox 1010	Irgafos 168	Irganox 1010	Irgafos 168
Max. variability (%)	11	13	51	39
RSD (%)	2.3	2.3	5.3	3.2

[a] Forty-six observations.
[b] Ninety observations.

3.3.3.8 Soxhlet Extraction

Principles and Characteristics For solid samples, the *most widely used extraction method* is Soxhlet extraction. In fact, for more than a century Soxhlet extractors have been in frequent use in chemical and pharmaceutical laboratories. Soxhlet extraction is mentioned by some 33 % of the respondents of a recent survey as a sample preparation in regular use [119]. Named after Baron Franz von Soxhlet (born 1848), the Soxhlet principle is based on continuous extraction of the solid by repeated boiling–condensation cycles of a solvent in such a way that the extraction fluid is continuously refreshed [120]. During operation target analytes are exposed to pure, clean hot solvent on each pass. In this method, the solid sample is placed in a Soxhlet thimble, which is a disposable, porous container made of stiffened filter paper. The thimble is placed in the Soxhlet apparatus, in which refluxing extraction solvent condenses into the thimble and the soluble components leach out. (The process may also be termed a form of reflux extraction.) The Soxhlet apparatus is designed to siphon the solvent with the extracted components after the inner chamber holding the thimble is filled to a specific volume with solution. The siphoned solution containing the dissolved analytes is then returned to the boiling flask, and the process is repeated until the analyte is successfully removed from the solid sample. Two variations of the device are possible. The difference between the two designs is whether the solvent vapour is allowed to cool slightly by passing to the outside of the (standard Soxhlet) apparatus or whether it remains within the body of the apparatus in contact with the sample-containing thimble (so-called 'cold' and 'warm' Soxhlets). As the extracted analyte will normally have a higher boiling point than the solvent it is preferentially retained in the flask and fresh solvent recirculates. A disadvantage of this approach is that the organic solvent is below its boiling point when it passes through the sample contained in the thimble. This contributes to long extraction times.

Soxhlet extractions typically employ solid–liquid ratios ranging from 1:10 to 1:50. This allows even minimally soluble analytes to be dissolved. However, even under the most favourable solution conditions, a target analyte might not easily be desorbed due to physical problems, inaccessibility and the inability of even the best solvent to compete with tightly bound solute [121], which limits the solvent extraction efficiency. Soxhlet extraction requires some method development. The efficacy of the procedure depends mainly on the choice of solvent. *Soxhlet solvents* for polymers are normally good swelling agents for the polymer to be extracted [122]. The ideal solvent would not dissolve any of the polymeric matrices, but would selectively remove all the desired analytes. This is hardly ever achieved and the resulting extract usually consists of a mixture of additives and other ingredients of the composition (monomers and oligomers). During Soxhlet extraction the solvent is the vapour condensate, which will only have the same composition as a mixture of solvents if an azeotropic mixture is used. For example, for extraction of ABS, the azeotropic mixture cyclohexane/benzene/ethanol (58.8:30.4:10.8 vol% ratio) is used, in which benzene acts as the swelling agent and ethanol as the nonsolvent. Soxhlet solvents are usually low boiling point liquids that evaporate easily during analyte recovery steps. An extra concentration step is usually required to make chromatographic quantification of low levels of additives possible. Because the Soxhlet system operates at atmospheric pressure and at a temperature of the extracting solution below the solvent boiling point such open-vessel extractions may easily take up to 16–20 h to reach an acceptable level of solute recovery. The extraction dynamics is limited by an azeotrope of solvent mixtures; boiling point of the azeotrope and the relative ratio of the solvents may restrict the analyte solubility. Fortunately, many methods have been published, so a quick literature search for similar sample types may save time.

Advantages and disadvantages of Soxhlet extraction are collected in Table 3.8. Soxhlet extraction is cheap and best for freely flowing powders. An important advantage of this process is that the polymer is constantly extracted with condensed solvent containing no extracted analyte and, therefore, total extraction is theoretically possible without equilibrium being set up between polymer and solvent. Conventional extraction methods – Soxhlet and sonication – also have some severe limitations because they lack selectivity and yield extracts containing species that can degrade column performance. Moreover, Soxhlet extraction suffers from the solubility dependence of the additives. Because the dissolved analyte is allowed to accumulate in

Table 3.8 Main characteristics of conventional Soxhlet extraction

Advantages	Disadvantages
• Low cost	• Time-consuming
• Unattended operation	• Relative nonselectivity
• Automation (cf. Section 3.3.3.9)	• (Cryo) grinding required
• High recovery (>90%)	• High (toxic) solvent use
• Excellent reproducibility	• Potential for solvent loss
• Large sample sizes	• Loss of volatile compounds
• Broad experience	• Less suitable for thermolabile analytes
• Not overly traumatic	• Clean-up strategy required
• Any solvent	
• No reverse diffusion of additives back into polymer	

the boiling flask, the analyte must be stable at the boiling point of the extraction solvent. For a long period of time, Soxhlet extraction has been the method of choice for forced migration of additives out of a polymer matrix [123]. For many applications using Soxhlet extraction, subsequent clean-up strategies (such as SPE), solvent concentration–replacement schemes or dilutions may be necessary to obtain an extract that is suitable for analysis. Moreover, high-resolution, chromatographic and spectroscopic methods are often necessary to separate the target analytes from co-extracted analytes or classes of analytes that are not of analytical interest.

It has been reported that substances extracted from paper Soxhlet thimbles include 2-cyclohexen-1-ol, 2-cyclohexen-1-one, 2,4-di-*tert*-butylthiophene, 2,5-di-*tert*-butylthiophene, 3,4-dimethoxyacetophenone, 1,2-epoxycyclohexane, eicosane, dodecane, tridecane, pentadecane, hexadecane, heptadecane, octadecane, nonadecane and several plasticisers (DEHP, DBP, DEP, DIBP, DOP) [124]. However, a major problem with this method (as in principle with most extraction methods) is that quantitative recovery is not assured, not even with a long extraction time. The extraction rate is slow, ranging typically from 6 h [125] to 48 h [84,126], and extraction efficiencies are not always particularly favourable. This is because this procedure does not result in complete dissolution of the polymer, as with a heating block, or makes use of the addition of a second energy source, as does microwave digestion or ultrasonic digestion. Long extraction times are needed in particular for the higher molecular weight antioxidants, such as Irganox 1010. For a research laboratory the *Soxhlet extraction time* (overnight) is less crucial than

for plant laboratory services. The time-consuming nature of Soxhlet extraction [84,125] is generally imputed to slow diffusion of the analytes from the sample matrix into the extraction fluid and/or to slow desorption of the components from the sample matrix. Small-volume Soxhlet extractors and thimbles can accommodate mg-scale sample amounts. ISO 1407 (1992) describes the use of Soxhlet extraction for the determination of extractables from rubbers and elastomers (similar norms are not available for most other extraction methods). The US Environmental Protection Agency (EPA) Method 3540 (Soxhlet extraction) is an approved method for the extraction of semivolatile organics from solids. A guide to choosing Soxhlet extraction conditions has recently been published [3].

The classical Soxhlet extraction technique has seen some improvements, mainly in the submersion of the whole extraction thimble into the boiling extraction solvent, degree of automation, and in reduction of solvent volume. In a recently introduced universal extraction system (Büchi) four SLE methods are contained in one device: Soxhlet Standard, Soxhlet warm, hot extraction and continuous flow. It is possible to use solvents with boiling points of up to 150 °C; inert gas can be supplied during the extraction process.

Automated Soxhlet extraction [127] can be faster (2–4 h) and uses less solvent (50–100 mL) than conventional Soxhlet extractions because the sample is immersed in a boiling solvent during the first portion of the extraction set-up. Soxhlet extraction is now gradually being replaced by more efficient techniques. For example, the conventional Soxhlet method, where only the hydrostatic pressure of the extractive fluid operates (ca. 0.005–0.01 bar), is less effective than intermittent or solid–fluid–vortex extraction. Various alternative sample preparation techniques, such as MAE [128,129], sonication extraction [130,131], SFE [132,133] and enhanced solvent extraction [134] have been developed in the last few years as competitors to the Soxhlet procedure. Compared to Soxhlet extraction, each of the *alternative techniques* reduces the amount of solvent required and/or shortens the sample preparation time. By the introduction of microwave or sonication, by using supercritical fluids, or by extracting the components at elevated temperatures, the rates of diffusion and desorption can be increased significantly. Increased temperature during extractions gives favourable gains relative to time and solvent usage. In microwave extraction systems times are shortened compared to those in Soxhlet extraction because the solvent can be heated above its boiling point with the use of pressurisable vessels.

Applications Despite various negative aspects and emerging competitive techniques Soxhlet extraction is still widely used. The greatest strength of the procedure is the fact that it allows for large sample weights. Thus, homogeneity problems become less of a concern. The Soxhlet system for the extraction of compounds of interest from solid matrices prior to further analysis has been so well established over the years that the number of individual applications of the technique runs into the thousands. Soxhlet extraction has been reported in matrix/additive systems, such as: PVC/(Tinuvin 320, Cyasorb UV9, Uvinul N-539) [75], PE/Cyasorb UV531 [135], LDPE/(DSTDP, Irganox 1035, Santonox R, Vulcup) [136], PP/(Irganox 1010/1330/3114, Irgafos 168, Atmer 163, Tinuvin 326) [79], PP/(Tinuvin 144/770, Hostavin TMN 20) [75], PP/(DLTDP, DSTDP, TNPP, BHT, Goodrite 3114, Weston 618, Topanol CA, Irganox 1010/1076, Cyasorb UV531, oleamide, erucamide, Ethyl 330, stearamide) [84], PP/(Irganox 1010/3114, Irgafos 168, Tinuvin 440/770, Tinuvin P, erucamide) [137] and PVC/commercial tin maleate stabiliser [138], etc. Dilettato *et al.* [139] used Soxhlet extraction of ground PE pellets followed by TLC, cSFC and GC-MS, and of PP pellets followed by cSFC. Dettmer *et al.* [140] reported Soxhlet extraction (toluene) of the flame retardant tetrabromobisphenol-*S*-bis-(2,3-dibromopropyl ether) followed by precipitation of co-extracted PP by means of ethyl acetate (nonsolvent). Soxhlet extraction of dashboard foils was followed by preparative NPLC and RPLC; the fractions were characterised by IR methods [141].

The additives in common use today exhibit a varying range of solubility or miscibility with 'safe' solvents. The analyst then must design with some care a solvent system that will extract both lower and higher molecular weight additives, less polar and more polar organics while at the same time not solubilising the polymer. For example, Kozlowski and Gallagher [142] have standardised on a two-component solvent, a 1:1 blend of methanol and hexane for the Soxhlet extraction of octyl and phthalate ester plasticisers from flexible PVC compounds. Robertson and Rowley [143] have published a detailed description of methods for the solvent extraction of plasticisers from PVC and other polymers prior to further analysis by GC. In the analysis quantitative separation of plasticisers from the other ingredients is the first and most important step. The efficacies of ether, CCl_4 and CH_3OH, in the Soxhlet extraction of a number of PVC plasticiser compositions were compared and binary azeotropes of CH_3OH with CCl_4, $CHCl_3$, 1,2-dichlorethane, and acetone, and of

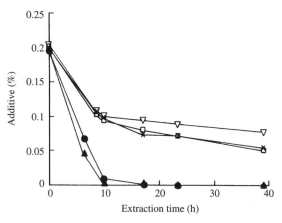

Figure 3.1 Loss of stabilisers from 0.5 mm HDPE plaques through Soxhlet extraction with *n*-heptane as a function of extraction time. ●, Anox 20; ▲, Anox PP18; ▽, Silanox ($n = 10$); □, Silanox ($n = 8.5$); x, Silanox ($n = 7$). After Gray and Lee [144]. Reproduced by permission of Blackwell Science

diethyl ether with 1,2-epoxypropane, were evaluated. The results show that considerable care should be exercised in claiming *quantitative extraction* for a given solvent and polymer/additive matrix. Similar reports are quite frequent and suggest the need for careful definition of the Soxhlet extraction conditions where quantitative recovery is at stake. Gray and Lee [144] compared the Soxhlet extractability of several low- and high-MW antioxidants in HDPE plaques (Figure 3.1). Cramers *et al.* [122,145] have compared Soxhlet, SFE and ASE® extractions of monomers and oligomers of PA6 and PBT. Soxhlet extracted brominated FRs from sediments gave extraction yields of less than 36 % compared with PHWE [146]. Such comparisons are frequently encountered in the literature.

Oligomers in HDPE are readily determined gravimetrically by using the so-called 'cyclohexane extractables' test. In this method, a small quantity of PE resin is Soxhlet extracted with cyclohexane and determined gravimetrically after drying. The test does not give the MW(D) of the extracted species.

As to artefacts, various plasticisers (DEP, DBP, DIBP, DHP) have been found as a constituent of paper Soxhlet thimbles [124].

An important *QC analysis* in the fibre and textile industry is the surface finish determination by Soxhlet extraction (AATCC Test Method 94–1992). Solvent extraction is used on textile materials to determine naturally occurring oily and waxy materials that have not been completely removed from the fibres (ASTM Method D 2257-96). Meanwhile, environmental, safety and cost concerns have resulted in an increased interest in alternative analytical approaches, such as NIRA, s-NMR, SFE and ASE®.

3.3.3.9 Fast Soxhlet Extractions

Principles and Characteristics A number of devices have been developed in the last three decades which modify the traditional Soxhlet vapour condensate extraction design and adapt the extraction temperature. These systems are known as automated Soxhlet, Soxtec® or Soxtherm®. Emphasis is nowadays on miniaturisation and extraction apparatus emitting only small amounts of solvent to the environment [147]. Modern fast Soxhlet extractors, such as the manual *Soxtec*® apparatus, use a three-step process that combines rapid LSE in boiling solvent followed by the traditional reflux approach. In the initial boiling stage, a thimble containing the sample is immersed in the boiling solvent for approximately 1 h. In the subsequent vapour condensate extraction stage the thimble is then raised manually above the boiling solvent, which now contains most of the extractable material. This step resembles the traditional Soxhlet extraction process but is much shorter (about 1 h). In the last step (10–15 min), the distilled solvent is collected for possible reuse or disposal (Figure 3.2). At the same time, this final action concentrates the analyte in the boiling flask during the solvent removal process. As compared with the traditional Soxhlet extraction, in which condensed solvent far below boiling temperature is allowed to drip into the extraction thimble during the entire extraction process, analytes in the Soxtec® thimble are exposed to boiling solvent. This provides quicker initial extraction. The Soxtec® technique requires only 20 % of the volumes of solvents used in the Soxhlet or sonication methods, is fast (approximately 2 h per sample), and less labour-intensive than the sonication method. The extracts obtained by Soxtec® extraction do not require filtration and are concentrated directly in the Soxtec® apparatus. Both sample size and solvent quantity are smaller, with matrix-to-solvent ratios similar to ordinary Soxhlet ratios. By weighing the aluminium thimble before and after extraction the total amount extracted can be determined. The procedure was evaluated by Lopez-Avila *et al.* [127] for 64 basic, neutral and acidic organic compounds (amongst which the phthalates DMP, DBP, DEHP and DOA) from soils.

Using the original Soxtec® system, the operational steps are manual, but the One-Touch Soxtec® and

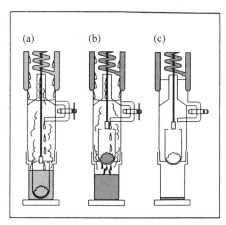

Figure 3.2 The three-step Soxhlet extraction process of the Soxtec® extraction system: (a) solubilisation of extractable matter from sample immersed in boiling solvent; (b) rinsing of extracted solid; and (c) concentration of extracted sample and collection of distilled solvent for reuse or disposal. Reproduced by permission of FOSS Analytical A/S, Hillerød

Soxtherm® *extractors* automate them. The patented Soxtherm® process consists of five (automatic) stages:

(i) The sample is immersed in a *boiling solvent*; an equilibrium is established between the analyte in solution and on the sample surface.
(ii) The level of the solvent is lowered to below the extraction thimble and excess solvent is collected in a solvent recovery tank.
(iii) The material is extracted by the refluxing, condensed solvent and collected in the solvent, below the extraction thimble.
(iv) The bulk of the solvent is distilled over into a storage tank for recovery.
(v) Some of the residual solvent is removed via convection heating.

Soxtherm® is up to five times faster than the traditional Soxhlet method thus providing considerable time saving while producing highly precise results. Its fully automatic analysis and unattended running guarantees ultimate precision and reduces solvent consumption by 80 %. The reproducibility is comparable to Soxhlet (1 % or better). Soxtherm® can be operated for all common solvents (temperatures up to 200 °C or 300 °C) and samples and allows up to six simultaneous determinations. In order to avoid contamination the apparatus is equipped with glass beakers and Viton sealing rings or other suitable nonextractable devices (silicone O-rings for acetone, Paraflour O-rings for dichloromethane or solvent mixtures). Soxtec® and

Soxtherm® can be considered as Soxhlet with *hot extraction* (at the boiling point of the solvent in the first stage).

Applications   Although Soxtec® combines the best qualities of reflux and Soxhlet extractions up to now fairly little evidence has been reported concerning the efficacy of this system for polymer and rubber analysis. Nevertheless, it appears that oligomers and other reaction residues, softeners, antioxidants (e.g. BHT) and several other additives used to modify polymers are easily extracted from PVC, PP, PE, PS, rubber and many other polymeric materials. Also, some leading international plastic, rubber and packaging companies have made Soxtec® an integral part of their quality control routines. Some application examples where Soxtec® has proved successful are [148]:

- Extraction of plasticisers in PVC-tubings for medical equipment. Quality control.
- Extraction of antioxidants in polypropylene production. Process control.
- Extraction of plastics and resins in electronic devices.
- Extraction of plasticisers and additives in plastic parts used in car production.

For the extraction of rubber and rubber compounds a wide variety of solvents (ethyl acetate, acetone, toluene, chloroform, carbon tetrachloride, hexane) have been used [149]. Soxtec® extraction has also been used for HDPE/(Tinuvin 770, Chimassorb 944) [114] and has been compared to ultrasonic extraction, room temperature diffusion, dissolution/precipitation and reflux extraction. The relatively poor performance of the Soxtec® extraction (50 % after 4 h in DCM) as compared with the reflux extraction (95 % after 2–4 h in toluene at 60 °C) was described to the large difference in temperature between the boiling solvents. Soxtec® was also used to extract oil finish from synthetic polymer yarn (calibration set range of 0.18–0.33 %, standard error 0.015 %) as reference data for NIRS method development [150].

Soxtherm® can be used for a variety of substrates, such as plasticisers, and additives in plastics and rubber, fibre and textile coatings, colorants on textiles, fat in food, oils and lubricants, fertiliser coatings, etc. Specific reports are scarce. EPA method 3541 has adopted automated Soxhlet extraction [151].

3.3.3.10   Solid–Fluid–Vortex Extraction

Principles and Characteristics   The principle of solid–fluid–vortex extraction, a recent development [152], is based on the creation of a relatively high filtration pressure as a result of cooling off a vapour chamber in a boiler vessel in such a way that there is (ideally) complete condensation and the extractive fluid is forced through a filter and/or extraction material at nearly one atmosphere in the case of open extractor systems and at more than one atmosphere in the case of closed extractor systems (cf. hydrostatic pressures up to 0.01 bar in Soxhlet).

The intermittent solid–fluid–vortex extraction method [153] consists of four stages:

(i)   A weighed extractable sample (1–20 g) and part of the solvent are fed into an extraction tube, whereas the remaining solvent fills the basic boiling vessel (total solvent ca. 100 mL).
(ii) Upon boiling the solvent vapours pass a PTFE membrane filter (10–20 μm), penetrate the sample and condense on the cooler. The continuous stream of solvent vapour vigorously fluidises the sample/solvent mixture heated at boiling temperature. This fluidised bed technique renders extraction particularly effective.
(iii) After a preset heating time has elapsed the boiler vessel and its contents are cooled rapidly.
(iv) Cooling and condensation creates underpressure in the basic vessel; consequently, the extractive solution flows back through the membrane filter into the basic vessel.

This cycle of vaporisation of the solvent, condensation, extraction, and vacuum-filtration may be repeated any number of times in a solid–fluid serial extractor. The occurrence of an *extractive material fluid bed* as a result of the flow of boiling hot vapour provides for effective extraction, while pressure filtration provides for short cycle times. This functional principle makes it possible to achieve filtration pressures which are 50–100 times more effective than when using the Soxhlet method, where only the low hydrostatic pressure of the extractive fluid operates. Solid–fluid–vortex extraction according to the proprietary FEXTRA® (Feststoff Extraktion) principle is low cost.

Applications   Redeker [154] has reported applications for the following systems: PE/pigments (0.1–0.4 mm particles or film), BR/fatty acids, PVC/plasticiser and rubber/filler. Using 2–20 g and ethanol/toluene (for butadiene rubber), acetone (for PVC), xylene (for filled rubber) and a mixture (for PE) as solvents, total extraction times of 140–300 min (10–20 cycles) were considerably shorter (though not 50–100 times) than in conventional Soxhlet extraction. Quantitatively the results are comparable to Soxhlet. The method is also suited for the determination of residual monomer levels.

Other reported applications are in the field of agrochemistry, e.g. the determination of oil content in oilseeds [155], and in such diverse areas as environmental chemistry, soil samples, foodstuffs, natural products, animal tissues, pharmaceutical drugs, metabolism research, etc. *Warning:* The author has noticed that the seals of the commercial apparatus (fexIKA) contain unwanted extractable material.

3.3.3.11 *Intermittent Extraction*

Principles and Characteristics Intermittent extraction, a proprietary DSM development, is based on similar principles as solid–fluid–vortex extraction. In this technique an all-glass apparatus (Figure 3.3) contains a sintered filter separating the extraction and boiler vessels. A sample (5 g) is loaded in the extraction vessel and ca. 75 mL of solvent in the boiler vessel. The solvent is heated intermittently by means of a glass enclosed heating coil of adjustable current. Cooling with nitrogen alternates a heating cycle. During the heating and/or boiling stage a turbulent flow is created around the extractable sample; cooling creates a pressure drop in the boiler vessel which induces back-flow of the extractive solution. The cycle time is regulated in such a way that the sample is completely immersed at the end of the cycle (ca. 30 mL).

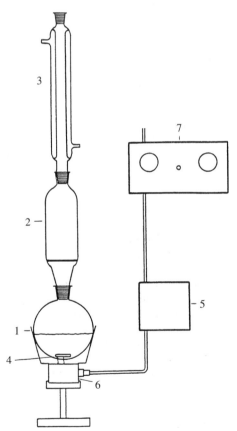

Figure 3.3 Apparatus for intermittent extraction of solids: 1, boiler vessel; 2, extraction vessel; 3, cooler; 4, heating coil; 5, transformer; 6, contact-block; 7, timer. Reproduced by permission of DSM Research, Geleen

Advantages of intermittent extraction are similar to those of solid–fluid–vortex extraction. A disadvantage is caking of the polymer in the case of a high polymer concentration in the extract or by loss of extraction fluid. Interlaboratory comparison is restricted.

Applications The method is in use for the determination of water extractable organics in PA6 and PA4.6, and for alkane extraction of waxes from HDPE (in nitrogen atmosphere to prevent oxidation) [156]. Ethylene-bis-stearamide (EBA) can be extracted from ABS in 30 min using intermittent extraction; in this case quantitative Soxhlet extraction was not possible. Nelissen [157] has used intermittent extraction with MTBE for the analysis of the flame retarder system of Tribit 1500 GN30.

3.3.3.12 Distillation–Extraction Techniques

Principles and Characteristics Steam distillation can sometimes be applied as an alternative approach for purification and sample preparation of heat-sensitive samples. This technique is rather gentle in that the distilled materials are never heated to temperatures greater than that of steam. Brock and Louth [158] and Likens and Nickerson [159] have described distillation–extraction apparatus. Although the apparatus is typically charged with a small volume of extracting solvent, it often requires additional concentration prior to analysis and some loss of highly volatile compounds is common. A modified Likens–Nickerson apparatus for steam distillation–extraction was described [160] and has also been adapted for microscale preparations [161]. Seidel and Lindner [162] recently proposed a redesigned simultaneous steam distillation–solvent extraction (SDE) apparatus as a universal sample enrichment technique. The technique is based on extraction using nonpolar organic solvents, which are lighter than water, e.g. petroleum ether (40–60 °C). One of the main advantages of SDE is the high purity of the obtained organic extract, which eliminates further clean-up prior to GC-ECD analysis. In Seidel and Lindner [162] the redesigned SDE apparatus was compared with existing competitive distillation–extraction apparatus, Soxhlet extraction and SFE. SFE has replaced SDE and LLE operations.

Applications Simultaneous steam distillation–solvent extraction has been in use for many years [163]. Steam distillation combined with continuous liquid extraction is an efficient technique for the removal and isolation of volatile compounds in various matrices (environmental,

food, natural products) and has been used extensively in flavour investigations. Brock and Louth [158] have described an unusual procedure for the identification of accelerators and antioxidants in rubbers that takes advantage of the tendency of accelerators to decompose during mixing or vulcanisation, or upon extraction from compounded stocks. By refluxing vulcanised and unvulcanised rubbers in EtOH/1 N HCl decomposition of unstable accelerators was promoted; guanidine and antioxidants were recovered unchanged. The extracted components were then separated using steam distillation and LLE techniques. Data on 24 accelerators and 12 antioxidants were obtained in a simple, rapid and unambiguous way. The accelerators used could be determined from the decomposition behaviour of known compounds. The antioxidants were identified by UV and colour tests. Tetramethylsuccinonitrile (TMSN), a decomposition product of 2,2′-azobisisobutyronitrile (AIBN), was extracted from dichloromethane soluble plastics such as PSF, PMMA and MABS by steam distillation [164]. Haslam *et al.* [82] have reported distillation *in vacuo* to recover volatile antioxidants from PE (e.g. 4-methyl 2,6-di-*t*-butylphenol and phenyl-α-naphthylamines). Squirrell [165] has used steam/solvent (water–diethyl ether) distillation of packaging films for the determination of odour- and taint-forming additives. Examination of volatile compounds extraction from PP sheets by hot water (steam distillation–extraction), according to Likens and Nickerson [159], has allowed assessing the organoleptic consequences of the temperature of sheets injection [166]. To obtain more precise information on the chemical origin of undesirable odour of PP cups, aromatic extracts were made from sheets, similarly prepared, using an extraction technique involving a steam distillation associated with a co-distillation of pentane. Steam distillation was chosen because it simulates a number of drastic interactions between food and plastic material, as met when making coffee in an electric coffee machine, rehydrating lyophilised food in a plastic bowl, or heating food in its package in a microwave oven. Steam distillation is frequently used for the analysis of food additives [167] and has also been employed for the separation and concentration of accelerants in fire debris samples [168]. The recently introduced SDE technique has not yet found application for polymer additive analysis.

3.3.3.13 Forced-flow Leaching

Principles and Characteristics Less well-characterised extraction methods include solvent leaching at high temperature with forced-solvent flow. In the latter approach based on liquid–solid chromatography a sample is packed in a (stainless steel) flow-through tube, which is heated to a temperature near the solvent's boiling point with solvent flowing through it [169]. The method is suitable for particulate samples; the solvent can be pumped or pushed through with high-pressure nitrogen gas. Although a smaller volume of solvent is used than Soxhlet, the yield is similar and the preparation time is reduced considerably.

Applications Mangani *et al.* [169] introduced forced-flow leaching for the direct extraction with hot toluene (at 100 °C) of low-MW PAHs from ashes. The method does not appear to have been used for polymer/additive analysis.

3.3.3.14 Cold Extraction

Principles and Characteristics In still another device, the equilibrium extraction unit (Equilibrium Extractor) is designed for the cold extraction (working temperature: +2 °C to +30 °C) of materials which are heat and/or light sensitive. The system comprises a series of specially designed glass extraction vessels mounted in a rack, which is inserted in a programmable tumbling device. Each extraction vessel is comprised of two glass parts. The lower part is essentially a large test tube into which the solvent of choice is placed. The upper part is specially designed so that as the complete vessel rotates the solvent becomes trapped in the upper section and can only return by passing through the sample, which is contained in a cellulose thimble attached to the upper part. As the solvent returns to the lower vessel elution of the extractable matter is achieved. By correctly selecting the number of cycles (from 1 to 99; time per cycle: 1 min) equilibrium between the sample (maximum weight: 22 g) and solvent (maximum volume: 200 cm^3) is achieved. The apparatus allows efficient and reproducible extractions, features reduced time demand on the analyst, no metallic or plastic contamination of extract (all glass construction of extractor, sample in contact only with solvent, glass and the cellulose thimble) and free choice of solvent. Reported extraction times are up to 3 h. Granules can be extracted without any further size reduction. If grinding is possible, the extraction time may be reduced further.

Applications The extractor is designed for use in any application where heat and/or light sensitive materials are to be extracted. In addition, by purging the vessel with an inert gas, such as oxygen-free nitrogen, materials sensitive to aerial oxidation can be successfully extracted. The characteristics of this latter LSE technique make it more suitable for food applications than for polymers. Nevertheless, additives may be extracted with ether from cryogenically milled HDPE at 20 °C for 16 h. In this process low-MW fractions, which often determine the organoleptic properties of the material, are not co-extracted.

3.3.4 Sonication

Sound propagation is considered to cover the frequency range from 0.01 to 2000 MHz. The highest frequency perceptible to a human ear is approximately 18 kHz. *Diagnostic ultrasound* (US) technologies use sounds at low power and very high frequencies (typically 5–10 MHz). These sound waves have no permanent effect on the physical or chemical character of the material being examined. However, if sound waves with a lower frequency of between 20 kHz and 100 kHz and a higher power are applied to a material, *physical changes* are induced; from about 300 kHz to 2 MHz *chemical effects* prevail.

The power levels used in *high-intensity* applications are so large (typically in the range 10–1000 W cm^{-2}) that they can provoke a number of changes within the propagating medium – mainly through the phenomenon of *cavitation*. This process ('making holes in liquids') can be likened to the process of boiling (bubble formation). As a sound wave passes through a liquid, the molecules in the wave path suffer alternate periods of compression (positive acoustic pressure) and depression (negative acoustic pressure). If the intensity of the wave is large enough, the negative pressure produced during the rarefaction half of the cycle can be sufficient to cause the intermolecular separation to exceed a critical value. In this case the liquid is literally pulled apart, and cavities or voids are formed [170]. Upon subsequent implosion (10^{-7} s) the gas in the voids is compressed (up to 1000 atm) and consequently the temperature is raised (up to 4000 °C). This phenomenon, which occurs with the frequency of the acoustic waves, i.e. at least 20 000 times per second, is generally thought to be the mechanism contributing to degradation of polymers in solution. The intense pressure, shear and temperature gradients generated by ultrasonic waves within a material can physically disrupt its structure ('*sonoprocessing*'), or promote certain chemical reactions ('*sonochemistry*'). At sufficiently high levels ultrasonics have deleterious (biological, medical and chemical) effects.

Typical mechanical effects are mass and heat transfer, outgassing, diffusion, mixing, crystallisation, etc. Not surprisingly, therefore, high-intensity ultrasonics finds application in such diverse areas as food and chemical processing, polymer science and technology, organic synthesis, paint production, etc., for the purpose of promoting crystallisation processes, pigment dispersion, chemical reactions, extraction, cleaning, welding of plastics, plating, preparations of catalysts and nano-sized powders, facilitation of (ultra) filtration, dialysis and reverse-osmosis processes by continuously 'cleaning' the interface, regeneration of waste ion-exchange resins, etc. Ultrasound comes in particularly useful because it gives efficient emulsification, particle deagglomeration and dispersion. Ultrasonic techniques are finding increasing use in (physical or chemical) modification of the properties of foods, for example to disrupt cells, promote chemical reactions, inhibit enzymes, and tenderise meat; other applications are oxidation–reduction in aqueous solutions, alcoholic beverage ageing, oil hydrogenation, extraction of enzymes and proteins [170,171]. Cavitation is used to advantage to generate emulsions of immiscible liquids, such as oil and water. Ultrasonically produced emulsions are used widely in the food, paint and polish, cosmetic and pharmaceutical industries, up to coal slurries.

In ultrasonic cleaning systems (e.g. for surgical instruments) an ultrasonic generator converts 220 V 60 Hz power into 1500 V 40 000 Hz and thus produces a high-frequency alternating electrical signal, which is applied to a transducer. The transducer expands when exposed to a positive electrical signal and contracts at a negative signal, creating sound waves. The effect of the implosion of bubbles created during sonication is commonly used either to clean or to erode solid surfaces. Micro explosions (cavitation) produce the high levels of cleanliness associated with ultrasonics. In a heterogeneous solid–liquid system collapsing cavitation bubbles have significant mechanical effects. Being based on sound, US cleaning is omnidirectional and effective cleaning occurs anywhere the ultrasonics penetrate. The ultrasonic field in an ultrasonic cleaner is not homogeneous. Hilgert [172] has specified the characteristics of ultrasonic cleaning systems.

Sonochemistry started in 1927 when Richards and Loomis [173] first described chemical reactions brought about by ultrasonic waves, but rapid development of ultrasound in chemistry really only began in the 1980s. Over the past decades there has been a remarkable expansion in the use of ultrasound as an energy source to produce bond scission and to promote or modify chemical reactivity. Although acoustic cavitation plays a major role in most sonication-sensitive reactions there is no general consensus on the mechanism of sonochemistry [174–176]. It is extremely important that the acoustic systems giving rise to sonochemistry be well characterised in order to allow reproducibility and duplication. The results of a sonochemical reaction largely depend on the placement of the reaction vessel. The processes promoted by ultrasound, such as efficient mixing, cavitation and microstreaming all aid organic reactions. Ultrasound is established as an important technique in organic synthesis, e.g. ultrasound assisted oxidations and sonochemical hydrogenation [174,177–180]. The use of ultrasound can be viewed as an alternative to phase transfer catalysts (PTC) [181]. High-intensity ultrasonics may initiate (co)polymerisation but can also easily cause depolymerisation (e.g. of acrylic polymers) and degradation. Ultrasound techniques can be used to monitor the course of a polymerisation reaction and characterise the thermomechanical properties of polymer solids and melts. Pethrick [182] has reported ultrasonic studies of polymeric solids and solutions. Curing of polymers and resins using ultrasonic conditions is very promising. The problem of real-time determination of the state of cure of composite materials, with particularly large structures being cured in autoclaves, is of considerable industrial importance. Mitra and Booth [183] reported first all-optical remote (noncontact) measurements of ultrasonic propagation times in an epoxy sample throughout the entire curing process. Similarly, resonant ultrasound spectroscopy has been applied to the problem of monitoring the curing of carbon fibre reinforced epoxy composite material in a noncontact manner [184]. Cure monitoring of carbon fibre reinforced epoxy composites is necessary to reduce the amount of energy invested in the cured part.

Use of ultrasounds in catalyst preparation leads to higher penetration of the active metal inside the pores of the support and greatly increases the metal dispersion on the support [185]. Major advances in ultrasonic technology have increased the acoustic power and sensitivity of transducers.

For general aspects on sonochemistry the reader is referred to references [174,180], and for cavitation to references [175,186]. Cordemans [187] has briefly reviewed the use of (ultra)sound in the chemical industry. Typical applications include thermally induced polymer cross-linking, dispersion of TiO₂ pigments in paints, and stabilisation of emulsions. High power ultrasonic waves allow rapid *in situ* copolymerisation and compatibilisation of immiscible polymer melt blends. Roberts [170] has reviewed high-intensity ultrasonics, cavitation and relevant parameters (frequency, intensity,

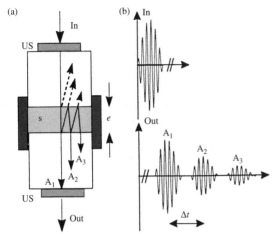

Figure 3.4 (a) Schematic of ultrasonic technique. S, Polymer sample of thickness e; US, piezoelectric transducer. (b) Ultrasonic echo patterns for evaluation of sound velocity and attenuation. After Sahnoune *et al.* [189]. Reproduced by permission of the Society of Plastics Engineers (SPE)

liquid properties and temperature); the most recent book on the subject is due to Abramov [188].

Commercially available *low-power ultrasonic* devices allow measurements for a wide variety of *diagnostic purposes*: (i) thickness of materials; (ii) interface detection; (iii) flow-rate, velocity; (iv) level measurements; (v) detection of extraneous matter and defect structures (damage); (vi) composition, phase transitions, particle size; (vii) (medical) imaging; (viii) echo ranging; (ix) US spectroscopy; (x) nebulisation; and (xi) sonochemical analysis. Most ultrasonic instruments utilise either pulsed or continuous-wave ultrasound [190]. The simplest and most widely used technique for carrying out ultrasonic measurements is called the pulse-echo technique (Figure 3.4). Modern noncontact ultrasonic testing equipment allows nondestructive inspection, providing high resolution and high penetration. Unlike optical methods, acoustic testing allows the study of opaque liquids and solids. Laser-acoustics can be used for testing thin films [191]. The method is ultrasonic and based on surface acoustic waves induced by short laser pulses. The technique is useful for nondestructively testing coated materials. Transmission of an acoustic wave through a material is characterised by velocity and attenuation (the equivalent of absorption in optical spectroscopy).

Low-intensity ultrasound uses power levels (typically $<1\,\mathrm{W\,cm^{-2}}$) that are considered to be so small that the ultrasonic wave causes no physical or chemical alterations in the properties of the material through which the wave passes, i.e. it is nondestructive. However,

it has been reported that although ultrasound action with a frequency of 10–100 MHz, an intensity of $0.05–15\,\mathrm{W/cm^2}$, and a duration of 0.1–20 min is sufficient for the purpose of chemical analysis, even under these mild conditions, ultrasonic vibrations do initiate and intensify oxidation and reduction, hydrolysis, polymerisation, molecular rearrangement, metal electrodeposition, and solvation of minerals and soils; they also change the states of ions in solution, etc. Low-intensity ultrasound is fairly inexpensive, is capable of rapid and precise measurements, can be used on-line, and provides information about physico-chemical properties [171]. The use of low-intensity ultrasonic waves in medicine is well known [192], e.g. in diagnostic US imaging for obstetrical/gynaecological applications [193], and for therapeutic reasons (cancer therapy). Here, a pulse of US is used to 'probe' the sample under investigation; comparison of the pulse shape before and after transmission and a measurement of the transit time in the sample provide information on many physical parameters.

The main *advantages* of US are that it is relatively inexpensive, rapid, precise, nondestructive and noninvasive and can be applied off-line or on-line to systems that are concentrated and optically opaque. One of the major *disadvantages* of US techniques is the attenuation of US by small gas bubbles.

Alig *et al.* [194] have reviewed new ultrasonic methods for characterisation of polymeric materials.

3.3.4.1 *Ultrasonically Assisted Extraction*

Principles and Characteristics Analytical extraction can be speeded up by efficient agitation, such as sonication. Ultrasound seems to be especially useful with small extraction cells where conventional stirring methods cannot be used. Gentle heating generated during sonication further aids the extraction process. Experiments concerning ultrasonically assisted extraction may be carried out in various ways (Figure 3.5):

(i) indirect sonication using an ultrasonic bath;
(ii) direct sonication using an ultrasonic bath; and
(iii) direct sonication using an ultrasonic reactor provided with a horn.

On a laboratory scale, generally an ultrasonic probe (horn) and an ultrasonic cleaner are used. The ultrasonic field in an ultrasonic cleaner is not homogeneous. Sonication extraction uses ultrasonic frequencies to disrupt or detach the target analyte from the matrix. Horn type sonic probes operate at pulsed powers of 400–600 W in the sample solvent container. Ultrasonic extraction works by agitating the solution and producing cavitation in the

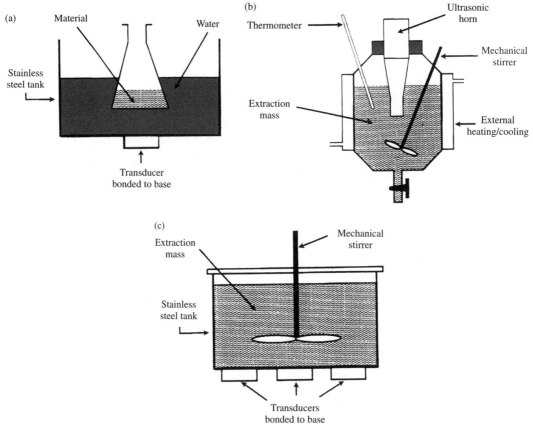

Figure 3.5 (a) Indirect sonication using an ultrasonic bath. (b) Direct sonication using an ultrasonic horn. (c) Direct sonication using an ultrasonic bath. After Vinatoru *et al.* [195]. Reproduced from *Ultrasonics Sonochemistry*, **4**, M. Vinatoru *et al.*, 135–139, Copyright (1997), with permission from Elsevier

liquid. The most probable mechanism for the ultrasonic enhancement of extraction is intensification of mass transfer and easier access of the solvent to the analyte. Similarly, the rate of transfer across the polymer/liquid boundary layer may be enhanced, but not the diffusion of additives within the polymer. Analogous considerations hold for solid–fluid–vortex and intermittent extractions. Sonication is rapid and effective for certain situations because cavitation raises the temperature at the particulate surface creating a localised superheating even though bulk heating is minimal [175]. Extractions occur in minutes versus hours. Sample throughput is low because only one sample at a time is processed. Sample sizes may average 30 g and total solvent volumes typically range from 150 to 300 mL. Depending on the level of the additives polymer samples as small as 2 g in 10–30 mL of solvent can be used, especially for screening.

Desrosiers [23] considers ultrasonic bath extraction to be one of the best methods of in-polymer extraction.

Since relatively low temperatures are employed, it is the least traumatic to the polymer matrix compared to Soxhlet, hot block or microwave extractions. Oligomers are still being generated but at a vastly reduced level and the generation of oxidation/degradation products is completely eliminated. Total or near total extractions of all additives are possible without putting the sample totally into solution. Extraction times are longer than those utilised in hot block and microwave digestion procedures. However, US extraction from polymers is more efficient than simple flask-shaking [78]. Table 3.9 summarises the *main characteristics* of US extraction. Table 3.4 compares ultrasonication to other sample preparation methods.

Sonication of polychlorinated biphenyls (PCBs) from soils has been the subject of an interlaboratory study [196] with 129 participants. Two standard methods of extraction were evaluated: EPA method 3540 for Soxhlet extraction and EPA method 3550 for sonication

Table 3.9 Main characteristics of ultrasonic extraction

Advantages
- Any solvent
- Rapid, reproducible
- Low cost
- No risk for hydrolysis breakdown (solvent dependent)
- Mild extraction (least traumatic)

Disadvantages
- Low recovery for some additives (high-MW)
- Needs grinding
- No automation

extraction [197,198]. The results indicated that Soxhlet extraction is more reliable and gives more accurate results than sonication, which requires more care and expertise to obtain correct results. Sonication extraction requires much less time than Soxhlet extraction but still requires large amounts of solvents [127]; the technique is quite labour intensive. In both cases, the sample extracts usually require clean-up and concentration.

Not all of the classical extraction processes are suitable for ultrasonic enhancement. For example, among the existing techniques used to obtain bio-active extracts from plant material (direct distillation, water steam distillation, organic solvent extraction, maceration, cold/hot fat extraction, etc.) [195] the water steam distillation is not amenable to ultrasonic enhancement.

Applications Extraction is typically accomplished by refluxing the polymer in an appropriate solvent for 1–48 h [84,199]. In many cases, ultrasonic exposure reduces the extraction time [90,200]. According to Table 3.5 there are several reports of US extraction from polymers. Ultrasonic extraction has been used for HDPE/(BHT, Irganox antioxidants, Isonox, Cyasorb, Am 340, MD 1024, Irgafos 168) [56], LDPE/Chimassorb 81 [201], SBR/tri(nonylphenyl) phosphite [200], HDPE/(Tinuvin 770, Chimassorb 944) [114], etc. Nielson [90] compared the recoveries obtained for a variety of analytes from PP, LDPE and HDPE with Soxhlet, ultrasonic bath and microwave oven. For all samples, the ultrasonic extraction could be achieved within 1 h. For LDPE and PP most compounds (except Irganox 1010) were extracted within 10 min. Further experiments by Nielson [56] on extraction from HDPE confirmed these results. Where phosphite antioxidants (such as Irgafos 168) are present the use of the solvent mixture DCM–cyclohexane was preferred as it prevented hydrolysis of the phosphite by extraction solvents such as alcohols [56]. Similarly, phosphite esters also undergo hydrolysis

during RPLC if exposed to water for too long a period. US was also used for chloroform extraction of 150 μm thick LDPE/Chimassorb 940 film and of 25 μm MDPE/(Irganox 1010, Irgafos 168, calcium and zinc stearate) film. Fast and total recovery was achieved in 15 min (Irgafos 168), 45 min (Irganox 1010), and 60 min (Chimassorb 944) at 60 °C [202]. Extractions of LDPE/Chimassorb 81 and EVA/Chimassorb 81 both with DCM were also reported [201]. Migration of antioxidants (Irganox 1010, Irgafos 168) from MDPE film in natural environments was studied by an ultrasonic extraction technique using $CHCl_3$ as extraction solvent to recover the nonmigrated antioxidants from the polymeric matrix, followed by quantitative HPLC-UV analysis [203].

Recent studies show that usually 30–60 min of sonication is required to quantitatively extract common *antioxidants* from PE but this method is not as efficient as 60 min reflux heating according to the ASTM extraction method [199,204]. Monteiro and Matos [205] have reported ultrasonic extraction of antioxidants, UVAs and slip agents from PP and HDPE in 30–45 min. Caceres *et al.* [114] have applied ultrasonication to extract HDPE/(Tinuvin 770, Chimassorb 944). The additives could only be extracted at less than 20 % recoveries from pellets using ultrasonic extractions of up to 5 h. Métois *et al.* [206] compared Soxhlet and US (iso-octane and dichloromethane) extractions of antioxidants (Irganox 1010, Irgafos 168) and bis(ethanolamine) antistatics from HDPE and PP film. In this case the most efficient extraction method is iso-octane (b.p. 100 °C) Soxhlet reflux, probably on account of the much higher temperature than in an ultrasonic bath. UV/VIS was used for quantification of PP, HDPE/(Irganox 1010, Irgafos 168, Armostat, Lutostat) extracts. US and MAE were used and compared in the study of degradation products from degradable HDPE/PP blends with biodegradable additives aged in soil [207]. Meng *et al.* [208] determined light stabilisers in PVC qualitatively and quantitatively by means of HPLC after ultrasonic extraction. Ultrasonic extraction of PVC/(DEHP, DINP) sheets and toys was compared with shaking and sucking experiments [209].

Table 3.10 shows the recovery from PP of Irgafos 168 and its oxidised and hydrolysed by-products by various extraction procedures. As may be observed, One-Step Microwave-Assisted Extraction (OSM) and US lead both to negligible hydrolytic additive degradation. The measured additive decay (by oxidation) is essentially due to the antioxidant activity during the processing (extrusion) step of the polymer and not to the US or microwave heating treatment.

Table 3.10 Recovery from PP of Irgafos 168 and its oxidised and hydrolysed by-products[a] by different extraction procedures

Extraction mode	Product analysis (ppm)[b]		
	Irgafos 168	Irgafos 168 phosphate	2,4-DTBP
One-step MAE (OSM)	810 ± 20	195 ± 4	5
US (1 h, anhydrous n-hexane, r.t.)	800 ± 20	190 ± 4	6
Reflux (6 h, HCCl$_3$)	780 ± 19	210 ± 4	25
Dissolution (xylene)/ precipitation (methanol)	720 ± 18	280 ± 6	32

[a] Original stabiliser concentration: 1000 ppm.
[b] HPLC analysis.
After Marcato and Vianello [210]. Reprinted from *Journal of Chromatography*, **A869**, B. Marcato and M. Vianello, 285–300, Copyright (2000), with permission from Elsevier.

As to Irgafos 168 the reader is advised to notice the results of a round-robin involving PP/(Irganox 1076, Irgafos 168) [209a]. Ultrasonication at room temperature with anhydrous n-hexane or acetone is a suitable soft extraction mode for the determination of aromatic phosphites and phosphonites, such as Ultranox 626 and Sandostab P-EPQ, which easily degrade in heating extraction procedures [210].

Brandt [200] has extracted tri(nonylphenyl) phosphite (TNPP) from a styrene-butadiene polymer using iso-octane. Brown [211] has reported US extraction of acrylic acid monomer from polyacrylates. Ultrasonication was also shown to be a fast and efficient extraction method for organophosphate ester *flame retardants* and *plasticisers* [212]. Greenpeace [213] has recently reported the concentration of phthalate esters in 72 toys (mostly made in China) using shaking and sonication extraction methods. Extraction and analytical procedures were carefully quality controlled. QC procedures and acceptance criteria were based on USEPA method 606 for the analysis of phthalates in water samples [214]. Extraction efficiency was tested by spiking blank matrix and by standard addition to phthalate-containing samples. For removal of fatty acids from the surface of EVA pellets a 1 min ultrasonic bath treatment in isopropanol is sufficient [215]. It has been noticed that the experimental ultrasonic extraction conditions are often ill defined and do not allow independent verification.

Manufactured aramid fibres are usually treated by a binder such as oils and surfactants to protect the fibre surface and improve the ability to handle. In order to remove the binder the knitted fabric is immersed in

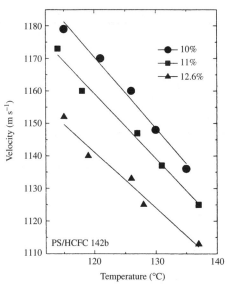

Figure 3.6 Sound velocity in polystyrene/HCFC 142b before phase separation as a function of temperature. The blowing agent concentrations (wt%) are indicated. After Sahnoune *et al.* [189]. Reproduced by permission of the Society of Plastic Engineers

cyclohexane under ultrasonic waves [216]. Ultrasonic velocity and attenuation measurements also provide important information about the state of a blowing agent within the polymer matrix and may be used for the determination of the influence of *blowing agents* on the rheology of polymers [189,217]. As shown by Figure 3.6 the sound velocity is very sensitive to the blowing agent (HCFC 142b) concentration in polystyrene. Ultrasonics is therefore a powerful and practical tool to investigate polymer foams.

Ultrasonically assisted extraction is also widely used for the isolation of effective medical components and bioactive principles from plant material [195]. The most common application of low-intensity ultrasound is as an analytical technique for providing information about the physico-chemical properties of foods, such as in the analysis of edible fats and oils (oil composition, oil content, droplet size of emulsions, and solid fat content) [171,218]. Ultrasonic techniques are also used for fluids characterisation [219].

A wealth of other data can be obtained from the use of US as an analytical method. *Sonoelectrochemical analysis* of trace metals [220] and organic compounds [221] has been reported. Ultrasonic atomisation [222] is used in many fields where a dispersion of liquid particles is required. Ultrasonic nebulisation (USN) is used for analysis of organic solutions in conjunction with ICP-AES/MS [223,224] and MIP-AES [225].

High-intensity low frequency US waves (~400 kHz to 2.5 MHz) can effect polymerisation [174,226,227] but may also induce *degradation* [226]. The extent of degradation increases with decreasing frequency. As mentioned already, ultrasonic methods have been successfully applied to monitoring of polymer processing, chemical reactions (polymerisation or curing of thermosets), film formation, glue processes, and crystallisation in polymers. Although the relationships between material properties and acoustical parameters have been studied for a long time [228], ultrasonic devices are not really frequently used for polymeric material characterisation [194]. When propagated in polymeric materials, acoustic waves are influenced both by the structure and by molecular relaxations. From the velocity and attenuation of longitudinal and/or shear waves it is possible to estimate the visco-elastic and other polymeric material properties [189,229]. For decades, ultrasonic techniques have been excellent tools for *nondestructive testing* (defect detection with high sensitivity) and for characterisation of microstructures or imaging of materials. Molinero [230] has described the physical basis, performance and limits. Stanullo *et al.* [231] have reported US signal analysis to monitor damage development in short fibre-reinforced polymers. The orientation of short glass fibres in PP has been determined by means of nondestructive ultrasonic backscattering [232]. Alig and Lellinger [229] have recently summarised ultrasonic methods for characterising polymeric materials.

Noncontact industrial ultrasonic sensors using echo-pulse technology have been developed to measure the time-of-flight of acoustical transmission reflected from a liquid, slurry, granular, or free-flowing powder surface, including plastics. Various aspects of the injection moulding process can be monitored on-line with US pulse-echo methods [233]. Ultrasonic sensors measure the velocity and attenuation of an ultrasonic wave through the polymeric mixture during extrusion [234,235]. Ultrasonic in-line monitoring of copolymer co-extrusion of HDPE and EPDM [235] and of filler content and viscosity in PE extrusion [229,236] have been reported. Also Brown *et al.* [237] have described the use of ultrasound for monitoring of polyolefin melt process variables, including polymer composition additives (e.g. Irganox 1010) and fillers, for extrusion and injection moulding processes. Noninvasive ultrasound measurements are currently not yet widely used, due to limitations in the high-temperature transducer technology and real-time signal processing.

Singh [238] has recently reviewed emerging ultrasonic techniques for process sensing and control in manufacturing (for structure–property relationships), such as analytical ultrasonics and ultrasonic spectroscopy, etc. Ultrasonic spectroscopy is being used for the characterisation of materials [239]. For general applications of the use of ultrasonics and cavitation, Crawford [240] may be consulted.

3.4 HIGH-PRESSURE SOLVENT EXTRACTION METHODS

As shown in Table 3.3 in most extractions temperature is used as a fundamental parameter. Only in a few conventional extraction methods pressure also plays a (minor) role but then mostly at normal boiling point temperature. A number of new extraction methods have been developed which share a single commonality, namely pressure. These are supercritical fluid extraction (SFE), closed-vessel microwave-assisted extraction (MAE), accelerated solvent extraction (ASE®) or enhanced solvent extraction (ESE®) (subsequently both referred to as pressurised fluid extraction, PFE) and pressurised hot (subcritical) water extraction (PHWE or SWE). If sufficient pressure is exerted on the solvent during extraction, temperatures above the normal boiling point of the extracting solvent can be used. Solubility, mass transfer effects and disruption of surface equilibria are the main reasons for enhanced extraction performance of liquid solvents at elevated temperatures and pressures. Higher temperature increases the capacity of solvents to solubilise analytes, faster diffusion rates occur and kinetics improves. Also, increased temperatures can disrupt strong solute–matrix interactions and decrease the viscosity of liquid solvents, thus allowing better penetration of matrix particles. Pressure also facilitates extractions from samples in which the analytes have been trapped in matrix pores. All these factors concur in improving the extraction yields. For samples in which the analytes are already readily accessible, pressure gives an advantage in analogy to liquid chromatography. Most high-pressure solvent extractions are performed in the 75–150 °C range (SFE, MAE, PFE), but some may reach 250–300 °C (PHWE, SWE).

3.4.1 Supercritical Fluid Technology

3.4.1.1 Supercritical fluids

When a substance is brought above a particular critical temperature, T_c, and pressure, p_c, and is unable to be condensed to a liquid by pressure alone, it exists in a condition called the supercritical fluid (SCF) state

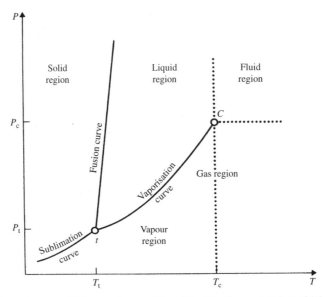

Figure 3.7 *PT* diagram for a substance which expands on melting. After Van Ness and Abbott [241]. Reproduced by permission of McGraw-Hill, New York

(Figure 3.7) [241]. Some consider the SCF state to be more extended and comprising the area of the phase diagram above T_c independent of p_c [242]. Critical temperature and pressure are usually defined as the maximum temperature at which a gas can be converted to a liquid by an increase in pressure, and the maximum pressure at which a liquid can be converted to a gas by an increase in temperature, respectively. In a *PT* diagram the vaporisation curve ends at the critical point. At a temperature above the critical point, the vapour and liquid have the same density. The *critical parameters* for some common fluids in analytical studies are listed in Table 3.11, but others may be found elsewhere [243], in particular, $T_c = 31.3\,°C$ and $p_c = 7.38\,MPa$ for the most common SCF (CO_2). Supercritical CO_2 ($scCO_2$) is widely used because of its convenient critical parameters, low cost, and safety aspects (low toxicity, nonexplosive).

Baron Cagniard de la Tour [244] made the first reported observation of the occurrence of a supercritical phase in 1822. Tables 3.12 and 3.13 summarise some of the main useful features of SCFs. Several properties of SCFs make them ideal candidates as solvents for industrial extraction processes [245,246].

The physical properties of a SCF are intermediate between those of a typical gas or liquid. For example, the diffusivity of a SCF is intermediate between a liquid and a gas and the viscosity is similar to a gas. The density of a SCF can be changed by varying the applied pressure on the fluid and can range between that exhibited by a gas to liquid-like values when the fluid is compressed

Table 3.11 Critical parameters for some fluids used in analytical supercritical fluid techniques (data taken from SF Solver™ package, Isco, USA)

Fluid	Critical temperature T_c (°C)	Critical pressure p_c (atm)
Xenon	16.7	57.6
Carbon dioxide	31.3	72.9
Nitrous oxide	36.6	71.5
Sulfur hexafluoride	45.7	37.1
Propane	96.8	41.9
Ammonia	132.4	112.0
Pentane	196.6	33.3
Water	374.1	218.3

After Lynch [89]. Reprinted from *Journal of Chromatography Library*, **56**, T.P. Lynch, 269–303, Copyright (1995), with permission from Elsevier.

Table 3.12 Density, viscosity, and diffusion coefficient of gas, liquid and supercritical fluids

Aggregation state	Density (kg m^{-3})	Viscosity (cP, 10^{-3} Ns m^{-2})	Diffusion coefficient (cm^{2}s^{-1})
Gas	1	0.01	10^{-1}
Liquid	1000	0.5–1.0	10^{-5}
SCF	200–700	0.05–0.1	10^{-4}–10^{-3}

After Harvala *et al.* [247]. Reprinted from *Extraction '87*, I. Chem. E Symposium Series, No. 103, T. Harvala *et al.*, 233–243, Copyright (1987), with permission from Elsevier.

at high pressures. As the density of a SCF is typically 100–1000 times higher than that of a gas and more

Table 3.13 Useful features of supercritical fluids

- High solvating power, tuneable via mechanical compression
- Solute diffusivities considerably larger than liquid solvents
- Fast extraction fluxes
- Low temperature extraction/analysis conditions
- Utilisation of nontoxic extraction media
- Facilitation of recycle of extraction fluid
- Solubility enhancement by addition of organic co-solvents
- Ease of removal of dissolved analytes

comparable to that of a liquid, molecular interactions can be strong owing to short intermolecular distances. Consequently, the solvating properties are similar to those of liquids, but with significantly lower viscosities and higher diffusion coefficients. The 10–100 times lower viscosities and the 10–100 times higher diffusion coefficients in SCFs compared with liquids result in a significantly enhanced mass transfer of solutes in extractions with SCFs than in conventional extractions with liquids [248]. As the *solvent strength* of an SCF is directly related to its density, the solvating ability of a SCF towards a particular species can easily be modified by changing the extraction pressure (and, to a lesser extent, the temperature) [249]. Higher pressure at constant temperature increases density and solvating power, whereas higher temperature at constant pressure decreases density and hence solvating power. In short, a SCF is essentially as a solvent with continuously adjustable solvent power. As a solvent, $scCO_2$ emulates hexane at low pressures (densities) and methylene chloride–acetone–chloroform at higher pressures. The 'tuneable' solvent strengths of SCFs make selective extraction possible. The ability to vary the density of the fluid, and therefore the solubility of the solute, by adjusting temperature and pressure, is not easily done in conventional solvent extraction. The ability of a SCF to solubilise solids was first reported by Hannay and Hogarth [250] in 1879. It was later established that the solubilities of low volatile organic solutes in SCFs are orders of magnitude higher than expected. The low viscosity and surface tension of SCFs allow the fluids to enter a porous matrix, dissolve a given component and exit without any of the problems normally associated with liquids.

The use of solvents above their normal conditions of temperature and pressure, up to and including the supercritical state, expands the range of analytical methods exploiting an overall spectrum of solubility, polarity and volatility properties of solvents and mobile phases. The fundamentals and applications of SCFs have been reviewed [243] and described in numerous books [248,251–256].

3.4.1.2 Emergence Pattern for SCF Technologies

The growing interest in SCFs as attractive industrial solvents for chemical engineering process applications is highlighted by many monographs and reviews which have appeared in the literature since 1985. Supercritical fluid technology has become a truly interdisciplinary field studied not only by chemists and chemical engineers but also by scientists and engineers concerned with foods, pharmaceuticals, soils, polymers, pollutants, propellants, etc. This is testified by international symposia on SFE/SFC, which are held at regular intervals (Park City, UT, 1988; Snowbird, UT, 1989; Park City, UT, 1991; Cincinnati, OH, 1992; Baltimore, MD, 1994; Uppsala, 1995; Indianapolis, IN, 1996; St Louis, MO, 1998; Munich, 2000; Myrthe Beach, SC, 2001). There is currently great interest in SFE for both process engineering [257,258] and analytical applications [132].

The oil industry has been a major force in the early and further development and application of SCFs. Petroleum-based applications cover exploration and production of the crude oil, processing in the refinery and use in automotive engines. Lynch [89] has reviewed the use of SFE in the petroleum industry. Petroleum-based patents referring to SCFs date from as early as 1936 [257]. A process which takes advantage of the dramatic changes in properties, which can occur around the critical point, is the technique of propane deasphalting for lube oil refining and purified light oil – wax – asphalt – naphthenes fractions. Later, there has been strong emphasis on SFE both as an alternative process extraction technology and an energy savings technology. Such a technology is the Residual Oil Supercritical Extraction (ROSE) process developed by Kerr-McGee Refining (USA) for the removal of lighter products from the residue of the commercial distillation of crude oil [259]. Levy [260] has reviewed fossil fuel applications of SFE and SFC with examples including determination of polymer additives.

Numerous other process applications were developed for SFE, such as in the area of materials processing, food and raw material modification [248]. For example, the Solexol process concerns the purification of vegetable and fish oils with SCF-propane and concentration of polyunsaturated triglycerides. The food and pharmaceutical industries have become another focus for SFE in the 1970s. Many patents resulted from early studies, covering the SFE of coffee, tea, tobacco and spices. In 1979, HAG AG (D) built the first large-scale production plant using SFE to remove caffeine from green coffee beans [261]. One of the most prolific applications for SFE today is the extraction of fats and oils in food samples [262]. Also the extraction (with $scCO_2$) of essential oils, flavours, and fragrances from plant materials, e.g. hop, tobacco and wood pulping, is now being developed on an industrial scale [247,263]. Principles and applications of SFE in food analysis were reviewed recently [242].

Maturation of SCF technology has only recently begun to be recognised. Its applications in chemical processing, cleaning, purification, extraction, separation, fractionation, impregnation, swelling, sorption, drying, foaming, micronisation and organic synthesis are now well established. Supercritical fluids have been used in *extraction* of several materials and analytes, such as: (i) fossil fuel; (ii) food and food products (fats, essential oils, vitamins); (iii) botanicals and natural products; (iv) biotic (animal) fluids and tissues; (v) polymers (additives); (vi) aqueous samples; and (vii) environmental samples, soil (pesticides). SFE is being used for regeneration of activated charcoal, extraction of pollutants from contaminated soils, and extraction of pollutants from polluted waste waters. As to the environmental analysis of soils, sediments, sludges and fly ashes more polar components, e.g. chlorophenols and triamines are extracted using methanol as a modifier in CO_2. Other applications include, e.g., the selective extraction of PCBs from eggs.

Supercritical fluid solvents have recently also been the focus of active R&D programmes in the area of polymer chemistry and processing technology. Examples include use of $scCO_2$ as a continuous phase for polymer synthesis (e.g. fluoropolymers, butyl acrylate), synthesis within CO_2 swollen polymers, SCF impregnation, SCF-assisted dyeing, processing of polymer solutions into porous particles and porous films, extractions from polymers, and as a blowing agent for foam production or as a polymer processing aid. To effectively utilise these processes, it is imperative to understand the solubility of CO_2 in polymers and its effect upon their properties [264].

The solubility of $scCO_2$ in various polymers has been reported [265]. The quartz crystal microbalance technique reported by Aubert [264] for measurement of the solubility of CO_2 in polymers is of value to investigations into the use of CO_2 as an agent to infuse chemicals into polymers. In such cases, $scCO_2$ acts as a plasticiser for glassy polymers and allows 'solvent-free' incorporation of additives, fragrances, and pharmaceutical principles (e.g. ascorbic acid, ascorbil palmitate, BHA, propyl and lauryl gallate) [266,267]. Infusion with $scCO_2$ achieves much higher loadings of the polymer

Table 3.14 Analytical techniques utilising supercritical fluids

Supercritical fluid chromatography	Thermo-optical absorption
Thin-layer chromatography	Immunoassay
Atomic absorption spectroscopy	Flow injection analysis
Nuclear magnetic resonance spectroscopy	Inductively coupled plasma
Mass spectrometry	Electrochemical
Fourier transform infrared spectrometry	Electrophoresis
Ion mobility spectrometry	Extraction
Field flow fractionation	Sample preparation (trace analysis)

compared to those achieved by static diffusion. Impregnation of polymers with SCFs is of considerable importance in relation to sustained delivery devices [268]. Bruna has reviewed the use of SFE in polymer technology [269]. Books have appeared on SCF technology [270] and cleaning [271]. References [256,272,273] have reported industrial uses of SFE. Although the use of SFE has previously generally been confined to relatively large-scale chemical processing applications [248,257,259,274], there is also scope for broad *analytical applicability* of SCFs (Table 3.14). The success of SFE depends vastly on the matrix.

3.4.1.3 Polymer–SCF Interaction

Potential interactions of an SCF, such as CO_2, with a polymeric material may include the following: (i) sorption; (ii) swelling; (iii) dissolution; (iv) plasticisation and depression of T_g; (v) crystallisation and increase of T_m and melting enthalpy; (vi) increased diffusion rate; (vii) changes in solubility of additives; (viii) weight loss; (ix) change of mechanical properties; (x) change of surface properties; and (xi) nucleation of voids within the polymer structure [275]. These phenomena are quite important in certain applications, such as the formation of polymer foams, impregnation of polymers with chemical additives, extraction of low-MW species from polymers, and separation of gas mixtures, using polymer membranes. The plasticising effect of CO_2 is not only a purely physical phenomenon but it also reflects its ability to interact with the basic sites (electron donating functional groups) in polymers [276], which reduces chain–chain interactions and increases the mobility of polymer segments. Plasticisation of polymer substrates with CO_2 results in orders of magnitude higher diffusion rates [277,278]. Only a limited number of polymers is soluble in $scCO_2$ (fluoropolymers and polydimethylsiloxane) at accessible temperatures and pressures (less than $100\,^{\circ}\mathrm{C}$ and 50 MPa). The effects of scCO$_2$ on the mechanical, thermal and interaction properties of 20 polymers were determined.

Extraction of low-MW solutes by SCFs is often rate-limited by the interaction of the solute with a polymer. SFE from polymers is usually a diffusion-driven process. The success of such a process hinges upon the ability of the SCF to swell the polymer, possibly plasticising it, and thereby increasing the diffusivity of the solute in the polymer phase during extraction. Swelling of a polymer by the SCF, and its effects on solute diffusivity, are of paramount importance in understanding and modelling practical SFE of a typical low-MW solute from a polymer. In cases where the SCF swells the polymer strongly (e.g. CO$_2$/ethylbenzene/PS), as is often the case at high pressure, the extractions are neither close to equilibrium nor completely diffusion limited [54]. This results in unusual variations in the amounts of ethylbenzene extracted versus temperature and pressure. Linear expansion measurements in an LVDT apparatus can be used to evaluate the dissolution of CO$_2$ in the polymer solid state [279]. The method has been applied to study swelling behaviour of Viton™ and butyl rubber O-rings [279], and of PVDF and PTFE [280].

Kazarian *et al.* [281–283] have used various spectroscopic techniques (including FTIR, time-resolved ATR-FTIR, Raman, UV/VIS and fluorescence spectroscopy) to characterise polymers processed with scCO$_2$. FTIR and ATR-FTIR spectroscopy have played an important role in developing the understanding and *in situ* monitoring of many SCF processes, such as drying, extraction and impregnation of polymeric materials.

3.4.2 Analytical SFE

Principles and Characteristics Supercritical fluid extraction uses the principles of traditional LSE. Recently SFE has become a much studied means of analytical sample preparation, particularly for the removal of analytes of interest from solid matrices prior to chromatography. SFE has also been evaluated for its potential for extraction of in-polymer additives. In SFE three interrelated factors, solubility, diffusion and matrix, influence recovery. For successful extraction, the solute must be sufficiently soluble in the SCF. The timescale for diffusion/transport depends on the shape and dimensions of the matrix particles. Mass transfer from the polymer surface to the SCF extractant is very fast because of the high diffusivity in SCFs and the layer of stagnant SCF around the solid particles is very thin. Therefore, the rate-limiting step in SFE is either

diffusion in the polymer particles or solubility in the SCF [278]. Which step actually controls the extraction kinetics can be identified by monitoring the extraction yields at different temperatures or different solvent flow-rates. Other matrix effects (such as strong adsorption of analyte molecules on surface sites, trapping of molecules in the polymer chains) are held responsible for the very slow final stage in some applications [284]. This matrix effect is the least well understood. Because of these many phenomena, extractions from polymers show unusual rate-limited behaviour that often cannot be predicted by simpler desorption models.

Various models of SFE have been published, which aim at understanding the *kinetics* of the processes. For many dynamic extractions of compounds from solid matrices, e.g. for additives in polymers, the analytes are present in small amounts in the matrix and during extraction their concentration in the SCF is well below the solubility limit. The *rate of extraction* is then not determined principally by solubility, but by the rate of mass transfer out of the matrix. Supercritical gas extraction usually falls very clearly into the class of purely diffusional operations. Gere *et al.* [285] have reported the physico-chemical principles that are the foundation of theory and practice of SCF analytical techniques. The authors stress in particular the use of intrinsic solubility parameters (such as the Hildebrand solubility parameter δ), in relation to the solubility of analytes in SCFs and optimisation of SFE conditions.

Bartle *et al.* [286] described a simple model for *diffusion-limited extractions* from spherical particles (the so-called 'hot-ball' model). The model was extended to cover polymer films and a nonuniform distribution of the extractant [287]. Also the effect of solubility on extraction was incorporated [288] and the effects of pressure and flow-rate on extraction have been rationalised [289]. In this idealised scheme the matrix is supposed to contain small quantities of extractable materials, such that the extraction is not solubility limited. The model is that of diffusion out of a homogeneous spherical particle into a medium in which the extracted species is infinitely dilute. The ratio of mass remaining (m) in the particle of radius r at time t to the initial amount (m_0) is given by:

$$m/m_0 = (6/\pi^2) \sum_{n=1}^{\infty} (1/n^2) \exp(-n^2 \pi^2 Dt/r^2)$$

$$(3.3)$$

where n is an integer and D is the diffusion coefficient. The corresponding *extraction profile* with time falls off steeply initially, becoming eventually linear at the scaled time $t_r \approx 0.5$ (Figure 3.8). The physical explanation of

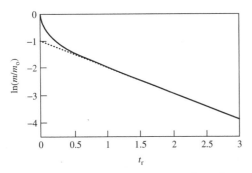

Figure 3.8 $Ln(m/m_0)$ vs. scaled time $t_r(= \pi^2 Dt/r^2)$ for the hot-ball model, including the effect of particle shape. After Bartle *et al.* [286]. Reproduced from *Journal of Supercritical Fluids*, **3**, K.D. Bartle *et al.*, 143–149, Copyright (1990), with permission from Elsevier

Table 3.15 Extrapolation to total mass of BHT in PP pellets[a,b]

Time (min)	Wt extracted (μg)	Time (min)	Wt extracted (μg)
0–20	7.1		
20–60	25.0		
60–120	45.7		
120–180	36.8		
180–240	26.8	0–240	141.4 (m_1)
240–300	16.4		
300–360	17.1	240–360	33.5 (m_2)
360–480	27.8	360–480	27.8 (m_3)
Total	202.7		
Given total	356.8		
Total from equation (3.4)			338.3

[a] 178.4 mg PP/0.2 wt% BHT.
[b] Extraction conditions: $scCO_2$, 50 °C at 400 atm; flow-rate 0.45 mL min^{-1}.
After Bartle *et al.* [286]. Reproduced from *Journal of Supercritical Fluids*, **3**, K.D. Bartle *et al.*, 143–149, Copyright (1990), with permission from Elsevier.

the shape of the curve is that the additive near the surface is rapidly extracted until a smooth falling concentration gradient is established across the particles. The extraction rate is then completely controlled by the rate at which the additive diffuses to the surface. Correspondingly, extraction is always facilitated by reducing one dimension of the extractable matrix (particles, films or foams). For this reason it is usually necessary to grind the polymer before extraction to provide a high surface area and a short distance for the analyte to diffuse. The model has shown its validity for a number of real systems and has been applied to the extraction of polymer additives from beads and polymer films (using the geometry of the infinite slab) [286,290,291].

A characteristic of SCF extractions is that the majority of the analytes is removed during a short period at the beginning of the extraction. An initial fall to a straight-line portion, predicted by the hot-ball model [286], was observed in all cases. In this period (up to $t_r = 0.5$), some 63 % of the material is lost from the 'hot-ball' sphere. The time required to extract 99 % of the material, however, corresponds to $t_r = 4.1$, i.e. almost 10 times that needed to remove the first 50 %. The long extraction tail, although involving the minority of solute, is important for quantitative results. However, the use of the 'hot-ball' model may be able to provide a method of extrapolation, which considerably reduces the time needed to obtain quantitative results [286]. If extraction is carried out at least as long as the initial nonlinear period to obtain an extracted mass m_1, followed by extraction over two subsequent equal time periods to obtain masses m_2 and m_3, then it can be readily shown, using Equation (3.3) that m_0, the total mass in the sample, is given by:

$$m_0 = m_1 + m_2^2/(m_2 - m_3) \qquad (3.4)$$

In the case of extraction from polymers, there is an advantage in working with sample pellets (no loss during grinding). However, a fairly exhaustive extraction of polymer pellets of a few mm in diameter is likely to take 80 h. The extrapolation procedure (Equation 3.4) can then be applied. Table 3.15 gives data for extraction of BHT from standard PP cylinders of ca. 3 mm (both length and diameter). Although extraction was carried out for 8 h with only 57 % of the additive extracted, an estimate of the final amount was made using the above equation which is only 5.2 % below the given value.

In *solubility limited extractions* the rate is reduced at the beginning of the extraction. Models for SFE predict that these extractions should be carried out at as high a temperature and pressure as possible. However, the polymer will eventually soften and melt, and the particles will coalesce.

Numerous reviews cover the state of analytical SFE [89,132,242,259,273,286,292–305]. Monographs on (analytical) SFE are references [253,257,306–311]. Modelling of SFE has been treated by reference [312]. A practical guide to SFE is available [3].

3.4.2.1 SFE Equipment

The use of SCF as extracting media and mobile phases for chromatography is now commonplace, and SFE and SFC occupy established niches in analytical chemistry. Commercially available instrumentation for SFE and SFC have been available since the mid 1980s. Basic requirements for analytical scale SFE equipment that can perform selective removal or solubilisation of target analytes or analyte classes consist of a

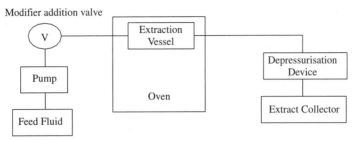

Figure 3.9   Schematic representation of a supercritical fluid extraction system

supply of fluid at the required pressure, a pressure-safe sample vessel (for pressures up to 680 atm), a means of controlling the extraction temperature, a depressurisation device and a means of collecting the extract on depressurisation (Figure 3.9). For design features, see references [313,314]. SFE is a deceptively simple technique. Samples are placed in a high-pressure high-temperature cartridge, and the SCF is passed through the sample and is then depressurised. After expansion of the SCF the extracted components are no longer soluble in gaseous CO_2 and are collected off-line (in a collector) or on-line (GC, SFC or LC). The collector may consist of an analyte-soluble solvent trap, a cooled trap or an absorbent trap. *Post-extraction analyte collection* is an essential part of SFE equipment, for which various options are available. Two methods are based on precipitating the components from the extract by reduction of the fluid density, i.e. via (i) temperature increase (isobaric method) or (ii) pressure reduction (isothermal method). The third technique is based on the adsorption of the solute on an appropriate stationary phase [248]. The possibility of loss of analytes at the trapping stage must be verified. Hirata and Okamoto [315] have noted that in order to trap polymer additives after decompression of an extract-laden SCF the restrictor had to be connected to two trapping tubes in series. Most of the analytes were trapped efficiently in the first (empty) trap, but about one-third of the analytes were trapped on the silica in the second trap. Internal standards may be used to check the efficiency of the trapping system [89].

In proper experimentation important considerations are the chemical nature of the analyte and the proposed analytical technique for further analysis of the extract. When a modifier is used it is best to use the modifier solvent as the trapping solvent. A disadvantage of solid-based traps is that most subsequent separation techniques (e.g. GC or HPLC) require a solution; consequently, it is then necessary to carry out a (small scale) solvent extraction to remove the analytes from

the trap into solution. It is possible to collect *selective fractions* of the extractable components by operating variation of the experimental conditions. One of the main claimed advantages of the SFE technique is that it works best at low temperatures (up to 100 °C) and is therefore considered useful for thermally labile substances. However, this does not always apply, as temperatures in the region of 200 °C are often needed to obtain maximum analyte recovery.

An SFE instrument can be designed as a single standalone instrument performing a range of manual processes: extraction, fractionation, concentration, solvent exchange, reconstitution and derivatisation.

Lynch [316] has recently critically reviewed the future development in analytical instrumental SCF technology. Modern instruments include manual and automated extractors that can handle large sample volumes (100 mL for manual SFE and 10 mL for automated SFE). From an experimental point of view, in using SFE attention should be paid to contamination from the seals of the SCF extractor.

3.4.2.2   SFE Techniques

SFE can be carried out in three different ways. In a *static extraction* (no flow-rate), the extraction vessel is pressurised to the desired pressure with the extracting fluid and then simply left for a certain length of time. The main benefit of this method is that the fluid has time to penetrate the matrix. It is most applicable when the analyte has a high affinity for the solvent and a low affinity for the matrix and also when the solubility limit of the analyte in the fluid is much higher than the actual level reached during the extraction [89]. This method was popular in early SFE experiments but has declined in favour of *dynamic SFE*. Here, fresh SCF is continuously passed over the sample, extracting soluble compounds and depositing them in a suitable solvent or on a solid trap. The dynamic mode is particularly useful when the concentration of the solute in the sample is high relative to its solubility in the fluid or when the solute has a high affinity for the sample matrix. Many modern instruments can operate in both dynamic and static modes and often the best approach is to combine them in a pulsed extraction mode. *Recirculating extraction* combines some of the features of both static and dynamic extraction.

SFE can be operated in off-line and on-line mode. *Off-line extraction* refers to methods where the extracted analytes are collected and isolated in some way not directly coupled to the subsequent analytical techniques, which are to be employed. This is the simplest way to perform SFE. While a significant part of the early work was aimed at utilising the compatibility of SFE to chromatographic separation, the inflexibility of such systems has probably limited their continued usage. The situation is exactly the opposite for off-line SFE. The flexibility of off-line operation allows the analyst to focus on the sample preparation only. Optimisation of SFE to maximise analyte recovery may be carried out without fear of overloading a chromatographic column and also leaves a certain amount of freedom of choice pertaining to the analysis, e.g. GC, HPLC, IR, etc. [272]. Depending on the nature of the target analytes to be extracted it is possible to select three basic modes of off-line collection: (i) empty vial collection (for nonvolatile or high-MW analytes); (ii) glass vial filled with a suitable liquid solvent (for analytes with intermediate volatility); and (iii) cryogenically cooled solid-phase adsorbent trap (for volatile analytes). The most well-known and frequently used method of analyte trapping during off-line SFE is trapping into a liquid solvent. Adsorbents for SFE solid-phase trapping are C_{18}-modified silica (30–45 μm), C_{18}-glass beads, C_{18}-unibeads, cyano-modified silica unibeads (30 μm), fused-silica beads (80–100 mesh), silanised glass beads (80–100 mesh), stainless steel balls (1200 μm) or unibeads (80–100 mesh) [317]. Szumski and Buszewski [318] have used on-line SFE-SPE coupling as a selective trap for extract collection and have described a new trapping device for off-line dynamic SFE based on a specially prepared SPE column with different types of adsorbents. For on-line SFE techniques, see Section 7.1.1.

3.4.2.3   Use of SFE for Polymer/Additives

SFE of polymers is a viable extraction method and generally much faster than the aforementioned traditional methods (Section 3.3.3). For example, Hirata and Okamoto [315] reported that additives could be extracted at 90 % recoveries from PE and PP films using $scCO_2$ at 25.4 MPa and 35 °C within 2 h as compared

with 24 h for Soxhlet extraction. Cotton *et al.* [319] showed that Irganox 1010, Irgafos 168 and erucamide can be extracted from ground PE and Tinuvin 770 from PP at 92–95 % recoveries in 15 min at 42.9 MPa and 60 °C using SFE. The optimum extraction conditions generally need to be defined.

Tables 3.4 and 3.16 show the *main characteristics* of SFE. The technique has important advantages especially in cases where loss or degradation of target analytes is likely to occur. The generation of oxidation products will not present any problems. Some authors have claimed that selectivity can be introduced in modern SFE instrumentation [320]. Extraction selectivity and efficiency can be controlled by the nature of the supercritical medium (Selectivity 1). Another opportunity for selectivity concerns the density of the

Table 3.16   Characteristics of supercritical fluid extraction

Advantages

- Extraction selectivity (tuneable solvent strength: p, T, modifier)
- Extractability of nonpolar to moderately polar compounds
- Speed and efficiency (favourable diffusion coefficients and surface tension)
- Closed system instrumentation
- Reduction in chemical exposure (oxygen-free environment, little sample degradation)
- Not traumatic (extraction of thermally labile and volatile analytes at low temperature)
- 'Green chemistry' processing (no environmental restrictions, virtually no organic solvents)
- Cost-effective extraction solvents
- Wide applicability (polymers, food, environmental)
- Ease of separating analytes from SCFs
- Solid or liquid trapping for added selectivity
- Simplicity of control of extraction conditions
- Potential of fractionation during collection
- Quantitation (internal standard needed)
- Automated (potential for multiple concurrent analysis)
- Post-extraction manipulation (minimisation of sample clean-up procedures)
- On-line potential (SFE-SFC, SFE-FTIR, etc.).

Disadvantages

- Complicated technique (SCF)
- Technological drawbacks (high p, T)
- Strong matrix dependency (physical conditions of sample influence SCF penetration)
- Limited polarity (modifiers)
- Low recovery for oligomeric additives
- Careful control of variables to obtain reproducibility
- Poor robustness (prone to operational problems)
- Extensive optimisation required for each system (large number of variables)
- Expensive: high investment and operating costs (need for high-purity $scCO_2$)
- Not matured (developing knowledge base)

supercritical medium and the temperature (Selectivity 2). After leaching of the sample, the extract is collected on a solid trap with a nonpolar or polar adsorbent, which can be selected according to the application (Selectivity 3). The trap is then rinsed with a solvent, the polarity of which can be chosen to desorb the solutes of interest in a selective way (Selectivity 4) and an adsorbent can be added in the extraction thimble (Selectivity 5). By carefully selecting extraction conditions class-selective SFE has been indicated as a new direction in off-line analytical SFE [321]. Class-selective extractions using SFE would then have the potential to yield extracts free of matrix components that may interfere with the analysis of the target analytes. The choice of SFE conditions to achieve class-selective extractions appears to be highly matrix dependent. However, the statement that SFE provides the selectivity required for even the most complex samples deserves some comment. In fact, the supposed enhanced selectivity of SFE is a myth: the claim of selectivity of SFE is unjustified. In extraction there is always a trade off between a reproducible complete extraction and selectivity. Only techniques that apply mild extraction conditions or use some type of very specific chemistry can be selective. Mild extraction techniques, however, are restricted in their range of application and are likely to show matrix-dependent extraction yields. Even the Soxhlet method can be made more selective by using a series of extraction solvents of increasing strength. Applying mild extraction conditions can perform selective SFE, but under these conditions the technique is likely to suffer from a strong matrix-dependency of the extraction yields, just as any other mild extraction method would [77].

While reduced solvent use is often promoted as the driving force behind SFE, in actuality, the performance attributes (greater selectivity, reduced time, quantitative yields, lower cost per extraction, new capabilities) are what has driven the technology. An additional advantage of SFE is the reduced use or elimination of organic solvents that are often required to remove the analyte from its matrix. The most common solvents used in SFE are gases at room temperature and below and as such, on depressurisation from the extraction system, are easily removed from even a cryogenically cooled analyte collection vessel without the need for distillation or the application of heat. This feature has made on-line SFE techniques particularly attractive to industries whose current sample preparation techniques require solvents that are undergoing increasing EPA regulation (i.e. chlorinated solvents). The CO_2 used in SFE is from fermentation processes and is therefore not a net contributor to the greenhouse effect. Further, most extraction methods using $scCO_2$ are designed so that the gas may be recycled.

Major *disadvantages* that slow the growth and acceptance of the field are the need for extensive optimisation (11 parameters in the extraction stage and 7 parameters in the analyte collection stage, cf. ref. [322]) and the fact that the knowledge base necessary for evaluating new applications is still developing. Determination of solubilities in SCFs underpins much SFE work. As each sample is extracted in turn rather than simultaneously, extraction time is longer than that of some other methods. Up to 5 g of sample can be successfully extracted in 45–60 min with temperatures below 100 °C. However, it has also been reported that some components still show very low, unacceptable extraction coefficients [23]. Poor extraction efficiency may sometimes be misinterpreted as being the result of poor extraction solubilities; analyte loss during the collection stage may occur. Another significant problem involves plugging.

A problem that has been encountered in the use of single component SCFs is that those with easily attainable critical parameters are relatively nonpolar and the compounds which can be solubilised are therefore also limited by their polarity. This problem can be partially overcome by using 'solvent modifiers'. The mechanism by which these modifiers work is complex and still subject of research.

SFE and SFC require a high-purity feedstock of liquid CO_2 (electron capture impurities below 100 ppt, and mass responsiveness impurities below 10 ppb). Impurities can be detrimental to the use of SFE in trace analysis. Hinz and Wenclawiak [323] have investigated SFE/SFC grade CO_2 by means of GC with FID, ECD and MS detection. Quantification of the impurities, using FID or ECD, was achieved introducing an internal standard into the CO_2 flow.

Generally SFE has proved a greater success than SFC. However, the need for successful automation is a significant restriction in many routine applications. SFE has been promoted as the ideal technique for sample preparation for chromatography. Meanwhile it is clear that this is far too optimistic [77,292]. As shown in Section 3.4.2.7, SFE does not guarantee quantitative analysis. Before any technique can be fully accepted, it should be capable of generating reproducible results. This is clearly not the case in SFE. Also, sample sizes of (on-line) SFE tend to be much smaller than in other methods, such as MAE or ASE® (Table 3.4), which raises the risk of nonrepresentative sampling. There is a need for SFE to be carried out on reference materials of known composition determined by an alternative technique.

SFE instrument development has greatly been stimulated by the desire of the Environmental Protection Agency (EPA) to replace many of their traditional liquid–solvent extraction methods by SFE with carbon dioxide. In the regulatory environment, EPA and FDA approved SFE and SFC applications are now becoming available. Yet, further development requires interlaboratory validation of methods. Several reviews describe analytical SFE applied to polymer additives [89,92,324].

3.4.2.4 Factors Affecting SFE from Polymer/Additive Matrices

Several authors [92,292,317] have discussed a number of factors affecting SFE from polymers. All classic and new extraction techniques require *pre-extraction procedures* to ensure that appropriate solvent contact is maximised for solid and semisolid matrices. The pre-extraction strategies for SFE are given in Table 3.17.

SFE usually requires pre-extraction manipulation in the form of cryogenic grinding, except in cases where analytes are sorbed only on the surface or outer particle periphery. The optimum particle diameter is about $10-50\,\mu m$. Diatomaceous earth is used extensively in SFE sample preparation procedures. This solid support helps to disperse the sample evenly, allowing the SCF to solvate the analytes of interest efficiently and without interference from moisture.

In extraction from a polymer/additive solid matrix the rate-determining step in the extraction process is governed by the interaction of the solvent of sufficient dissolution power with the matrix and the removal of the analyte (cf. Section 3.4.1.3). There appears to exist a direct relationship between degree of swelling and efficiency of extraction. The amount of CO_2 absorbed depends on temperature, pressure and the polymer concerned. Crystalline polymers are–not surprisingly–plasticised less

Table 3.18 Pressure and equivalent solvent data

Pressure at 35 °C (bar)	Pressure at 55 °C (bar)	Equivalent solvent
78	100	Hydrogen
133	235	Methane
155	265	n-Pentane
180	300	n-Hexane
220	350	n-Octane
360	520	Cyclohexane
480	>660	CCl₄

than the amorphous polymers [325,326]. CO_2 is not soluble in the crystalline regions of polymers [327].

With conventional solvent extraction, the only variable which can be readily altered to change solubility significantly is temperature. However, with an SCF, the solvent power can be adjusted over a wide range by varying *pressure* and therefore density. At higher pressure the density of the SCF increases and hence the solubility of the extractant. Higher pressure may increase the extraction rate in amorphous polymers, even when solubility does not limit the extraction as the plasticisation of the polymer increases with absorbed CO_2. This means that the solvent power and hence the selectivity of the solvent can be fine tuned to target the analytes of interest (Table 3.18). Typical solubility curves as a function of pressure for extractions not limited by the solubility of the additives in the SCF show a threshold pressure below which the analytes are not significantly soluble and an almost linear increase in solubility until a plateau is reached at higher pressure [278].

Increased temperatures often exert opposite effects on the solute diffusivity in the polymer (typically increasing exponentially) and equilibrium solubility in the SCF (typically passing through a maximum). Higher *extraction temperatures* may result in: (i) a decrease in the density of CO_2, therefore decreasing its solvation power; (ii) an increase in diffusion coefficients for additives in the polymer, so increasing mass transfer into the extracting solvent; and (iii) polymer melting or softening [278]. The decrease in density at higher temperature is almost linear. This is of greatest importance where the solubility of the extract in the SCF completely limits the extraction. Extraction at temperatures above T_g is generally much faster than below T_g. However, plasticisation alters the T_g and softening points, and therefore the temperature and pressure selected for extraction can be lower than would be expected from physical data available on the polymer. The temperature is usually chosen just

Table 3.17 Pre-extraction strategies for SFE

Strategy	Actions
Moisture handling	Oven drying using drying agents such as Hydromatrix, sodium sulfate and magnesium sulfate
Physical manipulation	Grinding, cutting or milling the sample
Dispersant addition	Adding Hydromatrix, celite, sand or glass beads
Modifier addition	Static soaking, adding derivatising agents or adding secondary modifiers

above T_g, where the polymer undergoes a transition from a glassy to a rubbery form. As diffusion in the rubbery state is much faster than in the crystalline state a sharp rise in extraction rate results at T_g. In many cases the optimum extraction temperature is approximately 120 °C.

Diffusion from the polymer is generally the rate-limiting process in particular during the later stages of extraction. The rate of diffusion in polymers follows an exponential Arrhenius curve with temperature, rate $\propto A \exp(-E/RT)$, where E is the activation energy, R is the gas constant and T is the absolute temperature. The process is slow with diffusion coefficients of the order of 10^{-10} cm^2 s^{-1} at 40 °C. Consequently, at higher temperature the diffusion rate increases steeply. High pressure is also needed for rapid extractions, both to increase solubility of the analyte and to plasticise the polymer. However, as high pressures may lower the softening point of the polymer, the optimum conditions need to be determined experimentally. The extraction temperature is also chosen in relation to the thermal stability of the additives.

The *flow-rate* affects only solubility-limited extractions. Increasing the flow-rate has a similar effect to increasing pressure. Baner *et al.* [328] observed a marked reduction in time to determine total extractables from the biodegradable Biopol polymer with increased flow rate.

A drawback of the use of most single component SCFs with easily attainable critical parameters is that these are relatively nonpolar; therefore the type of components which can be solubilised are limited by polarity. For example, CO_2 is essentially a nonpolar solvent with a solubility parameter similar to hexane at low pressures, and therefore it readily dissolves many common nonpolar low-MW materials. The low polarity of CO_2 makes it a poor solvent for most polymers. Consequently, provided scCO$_2$ allows for sufficient swelling of a polymer, nonpolar additives may easily be extracted from this matrix. Similarly, the polarity of a dye has great influence on the solubility in scCO$_2$ in supercritical dyeing. Yet, SFE has also potential for quite polar compounds. For extraction of *polar compounds* by means of scCO$_2$ a first impression about the solubility of the additives can be obtained on the basis of the Flory–Huggins interaction parameter χ and the Hildebrand solubility parameters δ_1 (component) and δ_2 (CO_2) [329]:

$$\chi = V_2(\delta_1 - \delta_2)^2/RT + \chi_s \qquad (3.5)$$

in which V_2 is the molar volume of CO_2 at extraction conditions, and χ_s is the entropic contribution to the interaction parameter. For extraction of a component which is soluble in the extraction solvent the experimental conditions need to be chosen in such a way that:

$$\chi < \chi_c = [1 + (V_1/V_2)^{0.5}]^2/2(V_1/V_2) \qquad (3.6)$$

where V_1 is the molar volume of the analyte.

The solvating power of SCFs can be adjusted by adding *solvent modifiers*. In the case of scCO$_2$, methanol is often added to increase polarity, aliphatic hydrocarbons to decrease it, toluene to impart aromaticity and tributyl phosphate to enhance solvation of metal complexes. For those additives which display a low solubility in dense CO_2, the use of polar co-solvents, or modifiers with different dipole moments (up to $\mu = 4.5$ D) is highly recommended. Solvent modifiers, which may be nonpolar, aromatic or polar, can act by swelling the polymer, thereby increasing diffusion and extraction rate even when solubility is not a limiting factor. Solvent modifiers can act to displace analyte molecules, which are adsorbed on active sites of the sample matrix. Methanol has been found useful when extracting from polar nylons, and aromatic modifiers from nonpolar poly(olefins). Methanol-modified extracts may span a wide polarity range and may contain highly polar components. Since the influence of modifier on extraction efficiencies can be dependent on the sample matrix and target analyte, choosing a modifier for a particular application can require trying modifiers with different polarities. As indicated by a systematic study in which over 30 modifiers were tested for extraction of a given analyte, a higher modifier dipole moment is a necessary but not a sufficient requirement for effective extraction [330]. Clifford [255] listed some compounds which are commonly used as modifiers. Critical parameters for CO_2-based modifiers have been reported [317,331]. Levy *et al.* [332] have discussed the use of various modifiers, including methanol, hexane, acetone, chloroform, dichloromethane, toluene and tributylphosphate, in on-line and off-line SFE with cryogenic adsorbent trapping. Page *et al.* [331] have given a comprehensive review of modifiers.

The concept of supercriticality is more complex if a two-component fluid is used. For most mixtures used, SFE must be carried out above a certain pressure to ensure that the fluid is in one phase. For MeOH–CO_2 mixtures at 50 °C the fluid is in one phase and can be described as supercritical above 95 bar, whatever the composition [284]. Compounds may also be added to the supercritical phase as a reactant rather than as a simple modifier.

The *nature of the extractant* affects the extraction in both the solubility and diffusion limiting cases.

Diffusion in a polymer is slower for larger molecules and hence also the extraction. High-MW organic compounds are less soluble in $scCO_2$ and consequently solubility will also limit the extraction more for larger molecules. For example, extraction of DIOP from PVC at 45 MPa and 90 °C was almost complete after 20 min, whereas only 50 % of the polar Topanol CA was extracted [333]. Irganox 1010 has proved difficult to extract in a number of cases when smaller compounds were easily extracted [291,334]. Extraction of high-MW additives, such as oligomeric HALS, has proved to be poor. Extraction of flame retardants at 30.4 MPa for 10 min at 60 °C was complete for low-MW molecules (up to 472 Da), but the largest molecule of 571 Da was only extracted for 77 % [335]. In the case of interaction of an additive to a matrix sorption site smaller molecules may be extracted more slowly than larger molecules [336].

As diffusion to the surface of a polymer is one of the limiting steps in extraction, the *particle size* or film thickness of a sample is also important [278,333,337–340]. With the typical diffusion coefficients of additives in polymers a particle diameter of about 0.3 mm is required for an extraction time of about 1000 s at 40 °C. An exception to this is the extraction of thin films and foams, for which the shortest dimension is small. It is not surprising that no more than 50 % of antioxidants could be extracted from PP pellets as opposed to 90 % recoveries from the same polymer extruded into film [341]. Grinding of the polymer is usually an essential step before extraction. Care should be taken to avoid loss of volatile additives owing to the heat generated in such processes. Therefore, cryogrinding is preferred.

3.4.2.5 Optimised Operational Parameters in Analytical SFE

Published evidence highlights the efficacy of SFE. However, the method is highly matrix and analyte dependent and must be optimised for each combination of material and analyte. Interaction between analyte and matrix is often difficult to predict and optimisation of the extraction procedure is not simple. Understanding of the processes that occur during SFE has lagged behind instrumental developments. The results obtained from SFE are highly dependent on the operational parameters used during the extraction (Table 3.19).

Optimisation of these conditions is a most difficult step and of great concern in SFE, since it has a direct impact on the added value of the technique. Optimisation of the experimental conditions for extraction

Table 3.19 Parameters affecting extraction efficiencies in SFE

- Extraction cell volume
- Matrix properties (nature, particle size, pore structure, water content, adsorptive strength)
- Sample size
- Pre-extraction strategy (drying agents, dispersants)
- Sample introduction mode (extraction thimble loading)
- Fluid composition (SCF, co-solvents and concentration)
- Extraction temperature
- System pressure (density)
- Flow-rate
- Temperature of oven
- Extraction time (static and dynamic)
- Static or dynamic mode
- Analyte collection/trapping scheme (packed bed or solvent trapping)
- Analyte collection temperatures (restrictor, trap, elution)
- Solute parameters (molecular weight, polarity, volatility)

has been the subject of a large number of papers in the literature [77,292,322,342–344] and an *optimisation strategy* for polymeric samples has recently been proposed [145]. According to Salafranca *et al.* [322] the main variables in SFE are extraction temperature, system pressure, dynamic time, percentage modifier and various temperature settings in the collection section. All these variables influence the extraction performance, with the most significant ones being pressure and temperature; the latter showed a nonlinear influence. Gere and Derrico [343] consider the nature of the analyte and matrix, SCF choice, pressure and temperature as being most critical. The selectivity of the analysis depends greatly on the choice of the fluid composition [345]. A rule of thumb for optimum extraction conditions is for dynamic extraction, with the temperature just below the softening point of the polymer under experimental conditions, the pressure as high as possible, and with about 10 % addition of a swelling solvent as modifier. For the application of analytical SCF methodology, a commercial software package, SF-Solver [346], allows calculating and searching the various parameters (pressure vs. density isotherm, Hildebrand solubility index, libraries for solubility data, etc.) useful for working with SCFs.

The task of embarking on *method development* for a new polymer/additive sample can be intimidating due to the number of parameters that can be varied during an extraction. Because 'cook-book' methods are unavailable for most analyte–matrix pairs, analysts may feel they are condemned to use a trial-and-error approach to optimise extraction–collection conditions for SFE. However, a flow sheet considering the most important parameters in the SFE process is available [3]. There are several approaches to method development in SFE. They range from an exhaustive study of the effects and interactions of as many as 18 parameters to a simple heuristic, empirical approach. One approach to an SFE method development strategy has been outlined in reference [347]. A more rigorous experimental design has been outlined by Otero-Keil [348]. Cramers *et al.* [145,349] have also given some guidelines that describe a strategy for method development for SFE of polymeric samples. The consideration is that SFE extraction of polymeric materials basically involves three subsequent steps:

(i) diffusion of the solutes from the core of the particles to the surface;
(ii) transfer of the components from the particle surface into the extraction fluid (with the distribution coefficient of the solute between the matrix and the SCF being the control parameter); and
(iii) elution of the components out of the extraction cell by the flow of supercritical extractant.

The actual SFE extraction rate is determined by the slowest of these three steps. Identification of the rate-determining step is an important aspect in method development for SFE. The extraction kinetics in SFE may be understood by changing the extraction flow-rate. Such experiments provide valuable information about the nature of the limiting step in extraction, namely *thermodynamics* (i.e. the distribution of the analytes between the SCF and the sample matrix at equilibrium), or *kinetics* (i.e. the time required to approach that equilibrium). A general strategy for optimising experimental parameters in SFE of polymeric materials is shown in Figure 3.10.

As for polymers the rate-limiting step in SFE is generally diffusion, conditions should be selected that enhance *diffusion* inside the particles. This means conditions of maximum swelling of the material, selection of modifiers that strongly partition into the polymer and high temperatures (below the softening or melting point of the material). Modifiers are more effective at low temperatures. Maximum enhancement is obtained when the polarity of the modifier matches that of the polymer.

Also Richter *et al.* [350] have described a method development paradigm for logically optimising the extraction parameters for a given matrix. Gere and Derrico [351] have presented a heuristic approach to preliminary experiments and method development in SFE and have produced a generic guide for SFE method development. This process is considered the minimum effort for developing a robust, repeatable, quantitative sample preparation, assuming that little or no method is previously available. The robustness of SFE has not been widely evaluated for polymer/additive matrices. Also a full-factorial design approach for the optimisation of SFEs has been described [322]. The multitude of parameters that can affect the extraction process in SFE in combination with the lack of fundamental knowledge about how these parameters affect the extraction, are to blame for the fact that method development in SFE is still largely empirical even though significant advances have been reported in recent years as to the various mechanisms underlying SFE.

In order to increase the overall extraction efficiency during SFE sonication has been applied [352]. Ultra-sound creates intense sinusoidal variations in density and pressure, which improve solute mass transfer. Development of an SFE method is a time-consuming process. For new methods, analysts should refer the results to a traditional sample preparation method such as Soxhlet or LLE.

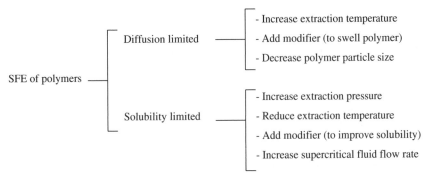

Figure 3.10 Optimisation strategy for supercritical fluid extraction of polymeric samples. After Lou *et al.* [145]. Reproduced from the *Journal of Chromatographic Science* by permission of Preston Publications, A Division of Preston Industries, Inc.

Table 3.20 Usefulness of chemical derivatisation in SFE

- Decreases the polarity of polar analytes (beneficial for conventional extraction)
- Increases the solubility of analytes in $scCO_2$
- Enhances polar analyte volatility (for GC analysis)
- Increases the analyte's thermal stability
- Enables coupling of SFE and various chromatographic methods of separation and detection
- Improves detectability (beneficial for FID, ECD, NPD and MS in GC, and for PDA in HPLC)
- Promotes analyte extractability from aqueous and solid samples (by reducing polarity)
- Integrates the sample preparation step (to reduce analysis time and costs)
- Deactivates active matrix sites (to facilitate the release of analytes)

3.4.2.6 Derivatisation Reactions

An approach that has been developed in response to the desire to extend the utility of SCFs to polar compound analysis is coupling of chemical derivatisation reactions to SFE. Table 3.20 lists situations when derivatisation may be useful [353].

Combining derivatisation with SFE first emerged in 1990 as an alternative approach for the determination of polar compounds in complex solid matrices. A wide array of polar compounds, including fatty acids, phenols, organometals and surfactants have been determined using methods where derivatisation occurs either prior to extraction, under *in situ* conditions, or after extraction under injection port conditions. In particular, four principal types of derivatisation reactions common to analytical chemistry have been coupled with SFE, namely esterification, alkylation (of organotin compounds), silylation and acylation (of phenols) reactions. Field [354] has reviewed coupling chemical derivatisation reactions with SFE for the determination of trace levels of organic compounds in solid matrices. Quite obviously, the need for derivatisation and modifiers reduces the usefulness of SFE in routine in-polymer additive analysis.

3.4.2.7 Quantitative Analytical SFE

Principles and Characteristics A good extraction method, in addition to quantitative recoveries, ensures a high degree of selectivity. The demands for quantitative and selective extraction are difficult to meet simultaneously. Quantitative extraction with recoveries being independent of the matrix requires a strong extraction fluid. On the other hand, selective extraction is usually associated with fairly mild extraction conditions where only the components of interest are dissolved

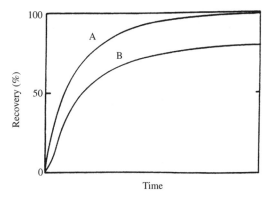

Figure 3.11 Schematic curves of percentage extracted vs. time for SFE without (A) and with (B) substantial matrix effects. After Clifford [284]. Reprinted from *Encyclopedia of Analytical Science* (A. Townshend, ed.), A.A. Clifford, pp. 4724–4729, Copyright (1995), with permission from Elsevier

from the matrix. Extraction by a supercritical (or any) fluid is never complete in finite time. It is relatively rapid initially, but is followed by a long tail in the curve of percentage analyte extracted vs. time (curve A of Figure 3.11). In a typical situation 50 % is extracted in a few minutes, but it may take hours before ca. 99 % is extracted. To achieve *quantitative extraction* for analytical purposes, a full kinetic investigation of extraction is required for any particular application. In simple terms, the fact that in two subsequent extractions much less solute is obtained in the second extraction does not mean that extraction is essentially complete. For SFE to be of use to the analytical chemist it must be made quantitative. For that purpose it is useful to add a reference standard to the extraction thimble. Internal standards added to the sample in the cell may be used both as a means of quantification of the analyte and for the purpose of checking the efficiency of the trapping system [89]. In the determination of oligomers an internal standard may be composed of three *n*-hydrocarbons. For example, where oligomers are all composed of even carbon numbers all the odd carbon number compounds are available for use as internal standards. Thus, a solution of known concentration of three normal hydrocarbons of odd carbon number, covering the analytical range of interest, e.g. nC_9, nC_{13}, and nC_{19}, is then added to the sample in the extraction cell. The resulting extract is analysed and the standards are used to calculate the concentration of the other components in the extract.

In order to overcome one of the main disadvantages of SFE, Salafranca *et al.* [322] have proposed the use of full-factorial design with the objective of attaining optimum extraction conditions. It was considered that

SFE involves numerous different variables and consequently this extraction technique requires a careful optimisation process to attain quantitative results. The factorial design procedure allows a minimum number of experiments (but still 24 in the specific case), and permits evaluation of the interactions between different variables. Quite obviously, the established optimal conditions are valid only within the selected ranges. It was concluded that each polymeric sample has to be considered individually and that both the physical appearance and sample size significantly affect the extraction efficiency. SFE optimisation by full-factorial design for the determination of additives in polyolefins thus shows that SFE can hardly be considered a routine tool for reliable quantitative determinations. Thilén and Shishoo [355] carried out another optimisation study for quantification of PP/(Irganox 1010, Irgafos 168) using SFE combined with HPLC-UV (cf. Figures 7.7 and 7.8). In the optimum conditions (pressure 384 bar, temperature 120 °C, methanol as modifier) quantitative extractions were found to be significantly faster than those reported earlier in the literature. A fast and quantitative extraction by means of SFE has also been reported for phthalate plasticisers in PVC when the surface/volume ratio of the sample is increased [356].

Supercritical extractions may be quantified gravimetrically or by using HPLC-ELSD and UV/VIS detection [357]. *Gravimetric weight loss methods* involve evaporation of the extraction or recovery solvent followed by drying, equilibration and weighing. These steps may add as much as 2 to 4 h to the total analysis time. Certain finish analysis applications do not even lend themselves to simple gravimetric analysis, as prescribed by some established procedures (AATCC Test Method 94–1992; ASTM Method D 2257-89), especially in the case of the small sample sizes used typically in SFE. For example, a 0.2 % level of finish on fibre, which is relatively high for polyester staple fibre, corresponds to a weight loss of only 4 mg from a 2 g fibre sample. A complicating problem of low finish levels is the quantity of low-MW oligomers, which are almost always co-extracted with the finish. These problems may be overcome by on-line SFE-FTIR [358] or by off-line SFE with HPLC-ELSD and UV/VIS detection for quantitative analysis [357].

A real breakthrough of analytical SFE for in-polymer analysis is still uncertain. The expectations and needs of industrial researchers and routine laboratories have not been fulfilled. SFE presents some severe drawbacks (optimisation, quantification, coupling, and constraints as to polarity of the extractable analytes), which cannot easily be overcome by instrumental breakthroughs but

are inherent to the matrix−analyte pair. This hampers routine use of the technique. It is therefore not surprising to notice a gradual shift away from SFE in a number of industrial research laboratories. It is also noticeable that in a recent round-robin involving extraction techniques SFE was not amongst the techniques of choice of some 24 participating European laboratories [209a].

Applications The majority of SFE applications involves the extraction of dry solid matrices. Supercritical fluid extraction has demonstrated great utility for the extraction of organic analytes from a wide variety of solid matrices. The combination of fast extractions and easy solvent evaporation has resulted in numerous applications for SFE. Important areas of analytical SFE are: environmental analysis (41 %), food analysis (38 %) and polymer characterisation (11 %) [292]. Determination of additives in polymers is considered attractive by SFE because: (i) the SCF can more quickly permeate throughout the polymer matrix compared to conventional solvents, resulting in a rapid extraction; (ii) the polymer matrix is (generally) not soluble in SCFs, so that polymer dissolution and subsequent precipitation are not necessary; and (iii) organic solvents are not required, or are used only in very small quantities, reducing preparation time and disposal costs [359].

Various authors have been concerned with the proper choice of *SFE parameters* in order to enhance the extraction yield. Lou *et al.* [291] have thoroughly investigated the effects of a number of parameters for extracting additives in PE and developed a two-film extraction theory considering mass transfer across a phase boundary. Cotton *et al.* [278] have described the rate and extent of dynamic SFE with scCO$_2$ of Irgafos 168, Irganox 1010 and Tinuvin 770 from commercial PP, with particular attention to diffusion, solubility and matrix effects. Higher pressure leads to increased extraction rates at 50 °C up to a given value (30.4 MPa), after which solubility no longer limits the extraction. The rate of extraction increased sharply between 20 °C and 140 °C, but decreased at higher temperatures with the onset of melting when the surface area decreases. The authors proposed a diffusion-limited extraction model. The effect of pressure and solubility on the kinetics of extraction was also studied for the system PP/Irgafos 168 at 70 °C with pressures of 75 to 400 bar [301]. Nerín *et al.* [360] used full factorial experimental design in SFE of several antioxidants, plasticisers, lubricants and related additives from both virgin and recycled PET samples with different physical matrix characteristics (pellets, flakes, pressed pellets and frozen ground

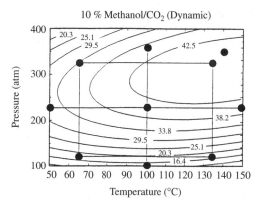

10 % Methanol/CO$_2$ (Dynamic)

Figure 3.12 Optimisation of supercritical fluid extraction of plasticisers (wt%) in PVC. After Bicking and Levy [359]. Reproduced by permission of Supercritical Conferences

plastic). The SFE was optimised and compared to total dissolution. The Simplex method has been used to optimise p, T, t parameters in SFE of liquid crystalline 4,4'-dibutylazobenzene from a PS matrix [361]. Bicking and Levy [359] have optimised the determination of plasticisers in PVC samples by SFE. As shown in Figure 3.12 there is no single optimum set of conditions for complete extraction. Rather, there are minimum temperature and pressure conditions that are necessary to achieve maximum recovery. Any combination of conditions above these minima, up to the operating limits of the equipment, provide good recovery.

Polymer applications of SFE generally fall into one of two categories: extraction of the oligomeric material or removal of additives from the polymer matrix. SFE can uniquely address these areas because of the high solubility of many additives in scCO$_2$ and because of the high diffusion (at low temperatures) available with supercritical extracting fluids. Consequently, the technique has been applied to extraction of a very wide range of low-MW compounds from an equally wide range of polymers, including polyolefins, nylons, PET, PBT, PVC, PS, etc. SFE has thus found applications in extraction of monomers in PET [362], aromatic amines in rubber [363], stabilisers in PP [364], organotin stabilisers [338], light stabilisers and AOs in polymers used in packaging [365]. The majority of analytical SFE studies reported to date involved extraction of relatively nonpolar compounds from samples smaller than 10 g. Specifically, SFE has been used for the following selection of polymer matrix/additive systems: PVC/(DIOP, CPE, Topanol) [333], PVC/stabilisers [366], PVC/Tinuvin P [367], PVC/DEHP [269,333,336,368], PVC/DOP [269], PVC/DIOP [333], PVC/(DBP, DOP) [336], PVC/organotin stabiliser [338], PVC/(DIOP, DEHP,

BHT, Tinuvin P, tributyltin chloride, vinylchloride) [369], PAG/additives [370], PE/erucamide acid amide [371], HDPE/Ethanox 330 [372], PE/(Chimassorb, Tinuvin 144, Irganox 1010/1076/1098, Naugard 524, DSTDP, Ionox 330) [373], PE/additives [374], PE/(Irganox 1010/1076, Irgafos 168) [375,376], PE and PP/(Irganox 1010, BHT, erucamide, Tinuvin 770, Irgafos 168, Isonox 129, DLTDP) [377], PE and PS/(BHT, BHEB, Isonox 129, Irganox 1010/1076, Irgafos 168, Cyasorb UV3346, Cyanox 1790, stearyl stearamide, HBCD, erucamide) [378], PP/(BHT, Tinuvin 326, Seenox DM, Irganox 1010) [334], PP/(Irganox 1010/1076) [328], PP/(Irganox 1076, Irgafos 168, Hostanox SE-2) [344], PP/(Irganox PS 800, Hostanox O3) [344], PP/(Hostanox SE2, Hostanox O3) [344], PP/(Irgafos 168, Hostanox O3) [344], PP/(Irgafos 168, Irganox 1010) [344], PE and PP/(Cyasorb UV531, Topanol OC, Irganox 1010/1330/1076) [315], PU/(Amgard TCEP, Amgard TMCP, Amgard V6, Thermolin 101) [367], PU/(TPP, BHT, Irganox 1010/1076) [369], PU/(BHT, plasticisers) [379], POM/Tinuvin 234 [380], etc. SFE was used by Tikuisis and Cossar [381] in the determination of the concentration of organophosphites in polymeric materials. Due to the slow mass transfer kinetics polymer pellets must be sliced or ground to smaller particles in order to give acceptable yields with SFE [378,382]. SFE of phenolic compounds from active matrices, such as CB, is difficult [383].

Some typical applications in SFE of polymer/additive analysis are illustrated below. Hunt *et al.* [333] found that supercritical extraction of DIOP and Topanol CA from ground *PVC* increased with temperature up to 90 °C at 45 MPa, then levelled off, presumably as solubility became the limiting factor. The extraction of DOP and DBP plasticisers from PVC by scCO$_2$ at 52 MPa increased from 50 to 80 °C, when extraction was almost complete in 25 min [336]. At 70 °C the amount extracted increased from 79 to 95 % for pressures from 22 to 60 MPa. SFE has the potential to shorten extraction times for traces (<20 ppm) of additives (DBP and DOP) in flexible PVC formulations with similar or even better extraction efficiencies compared with traditional LSE techniques [384]. Marín *et al.* [336] have used off-line SFE-GC to determine the detection limits for DBP and DOP in flexible PVC. The method developed was compared with Soxhlet liquid extraction. At such low additive concentrations a maximum efficiency in the extractive process and an adequate separative system are needed to avoid interferences with other components that are present at high concentrations in the PVC formulations, such as DINP. Results obtained

Table 3.21 Extraction of plasticiser (DOP, DINP) from flexible PVC[a,b]

Plasticiser conc. (phr)	SFE (%)	Soxhlet (%)	Nominal (%)
70	40.0 ± 0.8	41.1 ± 2.0	41.30
50	32.0 ± 3.0	33.0 ± 2.0	33.31
30	19.3 ± 0.2	20.2 ± 0.4	20.02
10	6.5 ± 0.5	6.3 ± 0.3	6.68

[a] Extraction conditions: 48.263 MPa, 95 °C, 25 min.
[b] Results based on three replicate extractions.
After Marín *et al.* [336]. Reprinted from *Journal of Chromatography*, **A750**, M.L. Marín *et al.*, 183–190, Copyright (1996), with permission from Elsevier.

with SFE are comparable to those obtained with the Soxhlet extraction method, as shown in Table 3.21. Off-line SFE-GC-FID was also used for the analysis of citrates and benzoates in PVC [385]; results were compared with conventional Soxhlet extraction. Marín *et al.* [356] have optimised the variables for quantitative extraction of phthalate plasticisers (DBP, DIHP, DOP, DINP, DIDP) from PVC used in toy manufacturing. Important factors were T, t, p and sample morphology (particle size and structure). SFE of phthalates used as plasticisers in PVC derivatives present some advantages in comparison to traditional extraction methods such as Soxhlet. Plasticisers in SFE extracts of PVC were determined gravimetrically [317]. Recovery of dialkyl organotin stabilisers from freeze-ground, unplasticised PVC at 17.7 MPa has been reported to increase with temperature up to 90 °C, and then levels off with higher temperature, as solubility becomes limiting [338]. Hunt *et al.* [333] have observed that only 50 % of Topanol CA is extractable from PVC at 45 MPa and 90 °C within 30 min; after addition of a methanol modifier the extraction yield increased.

Extraction efficiency of BHT and Irganox 1010 from freeze-ground *PP* has been found to increase from 30 to 90 °C at 30.4 MPa [334]. At 90 °C extraction was completed in 60 min except for Irganox 1010. Extraction of erucamide from PE at 45 °C increased by raising the pressure from 15.2 MPa to 20.3 MPa, when also an Irganox-type antioxidant was extractable. The extraction efficiency for six antioxidants in PP did not exceed 75 % with a particularly low level (30 %) for Irganox 1010 and Hostanox O3 [344]. *HDPE* was reported to be more efficiently extracted by Soxhlet extraction, while several compounded polymers gave high yields with SFE [382]. Extractions were performed at 90–120 °C at 350 bar, with scCO₂ or 5 % of 2-propanol in CO₂. Higher pressures had no effect on yield. By increasing the extraction time compared to the sample size, HDPE can also be extracted with good yields [324]. *LDPE*

could be extracted in good yields of some of the common additives [340]. Extraction of TNPP, Irganox 1076 and Weston 618 from LDPE film was measured at 45.6 MPa at different temperatures [378]. At 60 °C, only 40–60 % was extractable. At 150 °C, i.e. 10 °C above the melting point, surprisingly >95 % was recovered. Usually, heating above the melting point reduces the extraction rate as melting reduces the surface area.

Calcium montanate (a mixture of aliphatic hydrocarbons, fatty acid esters of C_{28}–C_{32} acids, the calcium salts of these fatty acids, and the acids themselves) can be selectively extracted with SCFs (Section 3.6.2) [386]. Following the procedure for the analysis of calcium montanate [386], Swagten-Linssen [387] has carried out extraction of LDPE/(Irganox B220, zinc stearate) with pure scCO₂ and modified by 10 % MeOH or 10 % MeOH/citric acid (complexing agent). Although zinc stearate can also be determined by means of XRF, this technique fails to give information about oxidation of the fatty acids. Because of the difficulties associated with the chromatography of extracted fatty acids, it is generally necessary to derivatise *fatty acids* to methyl esters. SFE with liquid collection offers the ability not only to remove the fatty acids from the matrix, but also to form the easily chromatographable alkyl esters in a single step. Daniel and Taylor [388] have reported formation of the alkyl esters of a series of fatty acids during SFE using both traditional reagents such as BF₃ in methanol and/or methanol-modified CO₂ in the absence of derivatising reagents.

The polymer matrix affects the extraction efficiency, as can be illustrated for the case of LDPE and HDPE charged with Chimassorb 944, distearyl 3,3′-thiodipropionate (DSTDP), Irganox 1010, Irganox 1076, Tinuvin 144, Irganox 1098, Ionox 330 and oxidised Naugard 524. Extraction at 60.8 MPa and 40 °C requires 30 min for LDPE and 5 h for HDPE. All additives were extracted from LDPE, whereas some additives would not extract from HDPE even with 3 % methanol modifier [373]. Torkos *et al.* [389] used SFE for the specification of polymer structures and the dosage of additives to PE at 50 °C and at 2000–5000 psi using again scCO₂ as the extraction medium. The effectiveness of SFE was compared with Soxhlet extraction. GC and GC-MS methods were used to determine the extracted compounds.

The polymer manufacturing industry monitors extractables in their products for a variety of purposes. These include not only additives but also mineral oils and *waxes* as well as incompletely polymerised dimers and trimers that have a pronounced effect in the finished product. Myer *et al.* [390] have compared the efficiency of SFE and Soxhlet extraction and analysis of Mekon wax and Polywax 1000, mineral oil, dimers, and trimers in PS pellets. SFE proved to give comparable recoveries to those of Soxhlet extraction but with dramatically shortened extraction times. This is the general experience. Under the appropriate conditions, SFE provides faster more selective or cleaner extracts for a variety of analytes in numerous matrices. SFE has also been employed to determine caprolactam, *oligomers* and cyclic trimers in nylons [145,319,339,391], dimer/trimer and volatiles in PBT [145,337], cyclic trimer in PET [287,362,392], CCl_4 in poly(isoprene) [393], alkylbenzenes and styrene oligomers in PS [54,121,391], CFCs in PU [394], and paraffins and olefins in LDPE [369]. Sekinger *et al.* [395] determined total extractables in styrene-butadiene rubber with SFE. The technique has also been used for the extraction of extender oil, organic acids and antioxidants from rubbers [395] and of aromatic amines from tyre rubber [363]. The total amount of solvent extracts from seven vulcanised rubbers of different polarity were determined by SFE and Soxhlet (DIN ISO 1407), with off-line HPLC analysis [396]. The results are comparable, but SFE was considered to offer more advantages. One-step SFE with denitrosation to secondary amines and derivatisation with GC-NPD detection was used for the determination of *N*-nitrosamines in latex [397]. For a wide number of analytes the recoveries are often comparable or better than conventional organic solvent extraction procedures. Consequently, SFE has been proposed as the sample preparation method of choice for many applications.

EC-harmonisation has caused many legislative changes concerning food contact materials and articles. New limits for global and specific migration have been set. New analytical methods are needed to examine whether *packaging materials* meet these requirements. Solvent-based migration and extraction methods can be used to gravimetrically determine the amount of potential migrants. For the measurement of specific migration limits it is necessary to analyse the composition of the extract. Bücherl and Piringer [398] collect the extract in a cryo trap and use SFC separation. A Global-SFE method has been developed for the determination of migratable substances in food contact materials; polymer samples are extracted at 37.8 MPa and 80 °C with scCO₂ and the extract is measured by flame ionisation detection (FID) [398]. As FID can only identify substances by their retention times, which can be shifted considerably by changes in the CO₂ flowrate, an MS-interface is often coupled to SFC; this tool is indispensable for the attribution of unknown peaks.

SFE has widely been applied to the study of *fibre and textile finishes* [357,358,399–403]. Many of the components used in fibre and textile finishing are low polarity, high-MW compounds (i.e. waxes, surfactants, oils) and are therefore well suited to SCF-based analysis. Using 100 % scCO₂ as the extraction fluid, a variety of wax and oil based textile finishes can be extracted from nylon, as beads, fibres or threads, in a relatively short time [399]. By altering the composition of the supercritical solvent and extraction conditions, the nature of the extraction can be changed from extracting only the surface components of the polymer, to extracting the components of the polymer itself. Caprolactam and its oligomers are extracted using CO₂ modified with 15 mol% MeOH, at 600 atm at 100 °C. Figure 3.13 shows the recovery of total extractables of nylon 6 by SFE as a function of the extraction time. Off-line SFE has also been applied to the analysis of fibre finishes on other fibre/textile matrices, such as PU/3 wt% poly(dimethylsiloxane) oil, PA/0.5–1 wt% (glycerol triesters, alcohol ethoxylates, alcohol PO-EO blocked surfactants, phosphites, PEG derivatives, fatty acid soaps) and Kevlar/0.5–1 wt% (substituted phenol, $C_8–C_{18}$ triglycerides, sorbitol derivatives) [358]. Calibration curves were established for these three finishes. Although off-line dynamic extraction showed the finishes to be over 89 % extractable with pure CO₂, some finish components, notably the sorbitol-based components, have low recoveries due to their low solubility in CO₂.

Drews *et al.* [403] have focused on the use of SFE for quantitative analysis of fibre finishes as a replacement of current fibre analysis methods (Soxhlet) requiring chlorinated solvents to remove the finish from the fibre. It was observed that while the extraction

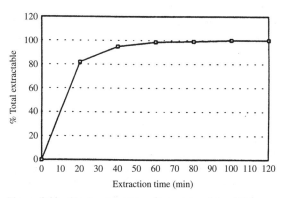

Figure 3.13 Total extractables from nylon-6 by SFE as a function of extraction time. After Ezzell and Thompson [399]. Reproduced by permission of Supercritical Conferences

of the finish components appears to be solubility limited, other analytes such as oligomers and polymer additives are both solubility and diffusion limited. Drews *et al.* [401] have also compared SFE extraction with freon, 1,1,1-trichloroethane and petroleum ether Soxhlet extractions of fibre and fabric finishes from commercial PET filament yarns, nylon yarns and PP staple fibres. The SFE conditions employed for each sample were chosen on the basis of calculated Hildebrand solubility parameters, the boiling points of the solvents used and the time required to reach specific extraction cell volume changes. In all cases, the correlation between the SFE and Soxhlet extraction results has been very good. The results show that SFE can be successfully used to replace Soxhlet extractions for a wide range of textile and fibre *QC applications*. In addition to reducing or eliminating organic solvent consumption the use of SFE significantly reduces the time required for a gravimetric finish analysis procedure.

Swagten-Linssen [387] has reported quantitative extraction of crystalline PA6/(Irganox 245, Ultranox 626, 2,4-DTBP) with scCO$_2$ as an alternative to dissolution (HFIP)/precipitation (MTBE).

As may be seen from Table 3.22, MTBE does not extract additives from PA6, as opposed to dissolution in the expensive solvent HFIP. It is also evident that in these conditions intact Ultranox 626 is not observed; the hydrolysis product 2,4-di-*t*-butylphenol (2,4-DTBP) is observed instead. ^{31}P NMR confirms hydrolysis of Ultranox 626. The results do not discriminate between hydrolysis during mixing or analysis. As also SFE does not detect Ultranox 626 hydrolysis is likely to occur in the mixing step. Dissolution with HFIP and SFE (after optimisation) give identical results. In this case the added value of SFE extraction consists in a considerable cost reduction.

SFE has also been used to analyse phenanthrene in PET and PMMA [404]. Nazem and Taylor [405] used SFE to determine antioxidant (Irganox 1010/1076, Irgafos 168) recovery from variously TEGDMA *cross-linked polymethacrylates* (PMMA, PEMA, PBMA).

Low percent recoveries were observed especially for the Irganoxes (<50 %) as opposed to nearly quantitative Irgafos 168 recovery. SFE extraction of the multicomponent additive Hoechst Wachs S (for mould deposit prevention) from Stapron S can be performed with higher yields than by suspension/precipitation [387]. Extraction of *flame retardants* from PU was not possible at 60 °C and 10.1 MPa; the recovery was low at 20.3 MPa and quantitative within 10 min at 30.4 MPa [335]. In the framework of polymer recycling and/or disposal SFE has separated the halogenated flame retardants TBBA, TBPA and HBCD from ABS composites [406]; IR spectroscopy was used to identify the organic flame retardants and EDX analysis to determine the total FR content (Br). SFE has also found application as an alternative approach to conventional liquid ion-pair extraction for recovering sulfonated linear alkylbenzenesulfonate *surfactants*, aliphatic secondary alkane sulfonate surfactants, and stilbene-based fluorescent whitening agents from sewage sludges and sediments by adding tetrabutylammonium hydrogen sulfate directly to the SFE extraction cells [407]. Extracts obtained by ion-pair/SFE contained the tetrabutylammonium ion-pairs of both classes of sulfonate surfactants and the whitening agent. In other more unique uses of SFE in polymer applications, Roston *et al.* [408] assayed a polymeric controlled-release drug formulation with SFE and Nerín *et al.* [409] quantitatively determined pesticides in post-consumer recycled plastics.

SFE has been used extensively in the analysis of *solid* polymers. Supercritical fluid extraction of *liquid* samples is undertaken less widely because dissolution or entrainment of the matrix can occur. As illustrated elsewhere SFE has also been applied for the analysis of liquid poly(alkylene glycol) (PAG) lubricants and sorbitan ester formulations [370]. The analysis of PAG additives (antioxidants, biocides and anticorrosion, antiwear and antifoaming agents) is hindered by the presence of the low molecular weight PAG matrix (liquid) and therefore a method for the selective separation of additives from PAG is required. The PAG

Table 3.22 Comparison of extraction methods for polyamide samples

Sample	Irganox 245				Ultranox 626				2,4-DTBP	
	Extraction[a]	Dissolution[b]	SFE[c]	Nominal[c]	Extraction[a]	Dissolution[b]	SFE[c]	Nominal[c]	Dissolution[b]	SFE[c]
1	20	0.33	0.32	0.50	–	–	–	–	–	–
2	10	0.18	0.18	0.25	<10	<10	–	0.25	0.19	0.17

[a] Intermittent extraction with MTBE; in ppm.
[b] Dissolution (HFIP)/precipitation (MTBE); in %.
[c] In %.
After Swagten-Linssen [387].

matrix was immobilised on silica and the extracted analytes were passed through a silica column positioned in-line in the SCF stream.

Supercritical fluid extraction cannot only be used for the separation of additives from the polymeric matrix but also for *fractionation of polymers* [257,410]. The polymer (considered to be an ideal mixture of oligomers) is extracted for a number of successive periods of time. In each period the pressure is constant and higher than in the previous period. Extraction is continued in each period until very little polymer is being extracted at that pressure. Using this method, PE has been separated into narrow fractions using $scCO_2$ [411]. A requirement for application of the method is that the polymer fractions are soluble in the SFE fluid. As polyisobutene is not very soluble in CO_2 only low oligomers can be extracted. For poly(dimethylsiloxane), which is much more soluble in CO_2, the fractions obtained had higher polydispersities [410].

3.4.3 Subcritical Water Extraction

Principles and Characteristics Water is an interesting alternative for an extraction fluid because of its unique properties and nontoxic characteristics. Two states of water have so far been used in the continuous extraction mode, namely subcritical (at $100\,°C < T < 374\,°C$ and sufficient pressure to maintain water in the liquid state) and supercritical ($T > 374\,°C$, $p > 218$ bar). Unfortunately, supercritical water is highly corrosive, and the high temperatures required may lead to thermal degradation of less stable organic compounds. However, water is also an excellent medium for extraction below its critical temperature [412]. Subcritical water exhibits lower corrosive effects.

The outstanding feature of both sub- and supercritical water as a leaching agent is its capacity for altering its dielectric constant as a function of temperature. As water is heated under moderate pressure, the dielectric constant (and hence, the effective polarity) drops to the point where it behaves as an organic-like solvent. The dielectric constant (ε) of water is primarily an inverse function of temperature and only mildly dependent on pressure. Extractions are effective using water without the need to go to the supercritical state. At room temperature, ε of water is near 80 making it much too polar to solvate hydrophobic organic compounds. However, at $300\,°C$ and 500 psi, ε of liquid water drops to approximately 20, quite similar to ethanol ($\varepsilon = 24$) and methanol ($\varepsilon = 33$) at $25\,°C$. Many organic compounds are sufficiently soluble to

be extracted by water under subcritical conditions. Instrumentation for Subcritical (or Superheated) Water Extraction (SWE), also termed *Pressurised Hot Water Extraction* (PHWE), has been developed [413,414]. PHWE is defined as both water and steam extraction (cf. expresso coffee machine). In SWE analytes are extracted into subcritical water, which flows through an extraction cell (dynamic mode), normally at $200\,°C$ and 750 psi of pressure. In many applications, temperatures above $200\,°C$ must be used to achieve efficient extraction and this requires special extraction vessels, valves and sealing materials. After extraction, the water is cooled and the extracted analytes are collected either in an organic solvent or a solid-phase trap. Unfortunately, commercial PWHE equipment is not available. Hot subcritical water extraction has been combined with SPME [415].

The conditions in PHWE are typically harsh and, therefore, the method is not suitable for thermolabile compounds. Analytes may also react with each other or with the water molecules during the extraction. From an analytical point of view the most salient negative factors of SWE in the continuous mode are co-extraction of undesirable components of the matrix (usually polar components) and dilution of the analyte in the extract. This calls for a clean-up and concentration step prior to individual separation and detection of the target compounds.

Miller and Hawthorne [416] have developed a chromatographic method that allows subcritical (hot/liquid) water to be used as a mobile phase for packed-column RPLC with solute detection by FID, UV or F; also PHWE-LC-GC-FID couplings are used. Before LC elution the extract is dried in a solid-phase trap to remove the water. In analogy to SFE-SFC, on-line coupled superheated water extraction-superheated water chromatography (SWE-SWC) has been proposed [417]. On-line sample extraction, clean-up and fractionation increases sensitivity, avoids contamination and minimises sources of error.

With the emergence of SWE as an alternative to SFE opportunities exist for combining derivatisation reactions with aqueous extractions. Although extractions using superheated and supercritical water yield pleasing results, many instrumental problems will have to be overcome before this technique is ready to leave the (academic) research laboratories [77]. This approach might play a significant role in future analytical extractions.

Applications Subcritical water extractions with suitable adjustments to the temperature (up to $250\,°C$)

allow the selective extraction of polar, low-polarity and nonpolar organic compounds [418]. Miller and Hawthorne [416] have recently demonstrated that water is a strong enough mobile phase to elute alcohols, hydroxy-substituted benzenes, and amino acids from a reversed-phase polystyrene stationary phase when heated above 40 °C. SWE has been used to extract organic pollutants such as polynuclear aromatic hydro-carbons (PAHs) and polychlorinated biphenyls (PCBs) from sediments and industrial soil, reducing the use of organic solvents in chemical analysis [413,419]. Other applications involve extractions from plant matrices. No polymer additive analysis by means of SWE or PHWE has been reported so far. A model matrix with antioxi-dants was SWE extracted and the components were cold trapped and subsequently released by raising the temper-ature in stages and eluted with superheated water using a thermal gradient [417].

3.4.4 Microwave Technology

3.4.4.1 Microwave field interactions with matter

The *microwave region* of the electromagnetic spec-trum lies between IR radiation and radio frequencies and extends from wavelengths of 1 cm to 1 m (fre-quencies of 30 GHz to 300 MHz, respectively). Wave-lengths between 1 cm and 25 cm are extensively used for radar transmissions and the remaining wavelength range for telecommunications. In order to avoid inter-ferences the principal frequencies allocated for indus-trial use (in domestic and industrial microwave heaters) are either 12.2 cm (2.45 GHz) or 33.3 cm (900 MHz). Domestic microwave ovens generally generate up to 8 kW of continuous power at 2.45 GHz, eventually up to 100 kW at a frequency of 900 MHz. A microwave generator or magnetron converts DC electrical energy into microwaves. The device consists essentially of a cylindrical diode, in an axial magnetic field, with a ring of cavities, which become resonant or excited in a way that makes it a source for the oscillations of microwave energy. Microwaves can be used both to study molecular characteristics and properties and to enhance chemical reactions, including sample digestion and separation.

Microwave spectroscopy is generally defined as the high-resolution absorption spectroscopy of molecular rotational transitions in the gas phase. Microwave spec-troscopy observes the transitions between the quantised rotational sublevels of a given vibrational state in the electronic ground state of free molecules. Molecular

rotational resonance (MRR) spectra are measured in the 3 to 1000 GHz range (from $\lambda = 10$ cm to $\lambda = 0.3$ mm). The range between 8 and 40 GHz is particularly impor-tant because the rotational transitions with low quan-tum numbers are usually found there for molecules with small to medium molecular masses. With its sharp bands, microwave spectroscopy provides an excellent fingerprinting technique for molecules in the gas phase. MRR spectroscopy is eminently suitable for the high-precision determination of molecular structures.

In liquids and solids, where the molecules are generally not free to rotate independently, the spectra are too broad to be observed. It is for these phases that *microwave dielectric loss heating* effects are relevant and need to be distinguished from the spectroscopic effects. The microwave dielectric heating effect uses the ability of some liquids and solids to transform electromagnetic energy into heat and thereby drive chemical reactions. The energy involved in microwave cleaning exceeds that of microwave spectroscopy.

Microwave-enhanced chemistry is an emerging field [420]. Microwave radiation can be used in organic, organometallic and inorganic synthesis, as well as in catalysis. A vast number of organic reactions has been studied under microwave conditions. Reduction of reac-tion time, improved yield, and decreased side reactions (purity) are beneficial effects. Reaction selectivity is also affected. Wan *et al.* [421] have contributed significantly to the development of microwave catalysis. In general, evidence exists that catalysed reactions, such as the conversion of cyclohexene to benzene, catalytic crack-ing, decomposition of organic halides, decomposition of methane to ethane and hydrogen, and steam reform-ing, are accelerated by microwave heating when com-pared with conventional heating. Bond and Moyes [422] have advanced some tentative explanations. The future of microwaves in catalysis lies in the area of specialty chemicals where the energy costs are insignificant com-pared with the value of the product. Many uses of microwave heating are to be noted also in biochemi-cal and biomedical applications, (in)organic chemistry and polymers [423,424]. Microwave heating is widely used in rubber vulcanisation. Curing of epoxy resins, SMC and coating are other commercial applications. The application of microwave energy to organic syn-thesis is also quite recent and is characterised by fast growth [425,426]. Simple thermal effects can explain all microwave-accelerated organic reactions. Applica-tions of microwave dielectric heating effects to synthetic problems in chemistry have been reviewed [427].

The use of microwaves for heating chemical reactors has increased dramatically over the past decade [428].

Table 3.23 Vessel types for microwave applications

Vessel type	Description
HP closed	High-pressure closed vessel (>80 atm)
MP closed	Medium-pressure closed vessel (10–80 atm)
LP closed	Low-pressure closed vessel (<10 atm)
Open	Atmospheric pressure vessel (1 atm)
Flow through	Flowing stream of reaction mixture passing through a microwave field
Stopped flow	Flowing stream of reaction mixture that is stopped when it is contained within a microwave field

After Walker *et al.* [430]. Reprinted with permission from P.J. Walker *et al.*, in *Microwave-Enhanced Chemistry* (H.M. Kingston and S.J. Haswell, eds), American Chemical Society, Washington, DC (1997), pp. 55–121. Copyright (1997) American Chemical Society.

The microwave technique is widely applied to process polymer materials, e.g. in microwave cure [429]. *Microwave processing* is a developing technology.

Table 3.23 gives an overview of the vessel types in use for microwave applications. It is especially important to distinguish between open vessel (as used in Soxwave®) and closed vessel (pressurised) *microwave heating systems* (as in MAE). Both open-vessel and closed-vessel microwave systems use direct absorption of microwave radiation through essentially microwave transparent vessel materials (Teflon, PC).

Applications of microwave radiation relevant to the analytical chemist cover:

(i) microwave sample preparation;
(ii) acceleration to pre-concentration of analytical samples under a microwave field;
(iii) microwave-induced evolved gas analysis;
(iv) thermal desorption of solid traps by microwave energy; and
(v) microwave thermal analysis.

Thermal desorption of solid traps by microwave energy is unsuitable for thermally labile compounds. In microwave thermal analysis [431] the (solid) sample is heated directly via interactions of the microwaves with the sample, providing more even heating and reduction of temperature gradients in comparison to heating with electrical furnaces. By passing air over a microwave-heated volatile sample evolved gases may be collected [432].

In *microwave sample preparation* the following processes may be distinguished:

• microwave drying;
• microwave digestion/(acid) dissolution (from atmospheric to high pressure, Table 3.23) for elemental analysis (cf. Section 8.2.2.2);

• microwave ashing (cf. Section 8.2.2.2);
• microwave solvent extraction (e.g. MAE, Soxwave® and MAP™).

Table 3.24 summarises the main microwave-assisted solid sampling modes. LeBlanc [433] has described microwave-accelerated dissolution of polymers. Microwave energy has also been used to assist the one-step *in situ* extraction from samples and headspace sampling on SPME fibres to develop a simple, fast, and solventless process [434]. Technology for microwave-digestion flow systems is available for sample digestion, dissolution and detection as well as organic synthesis. Burguera *et al.* [435] described coupling of microwave sample dissolution to FAAS by using flow-injection analysis (FIA) methodology. The impetus behind such on-line developments is to further reduce the labour-intensive process of batch microwave sample dissolution by using a flow system, thus increasing sample throughput and facilitating automation. On-line microwave sample dissolution has also been coupled with ICP-AES/MS [436]. Microwave cavities are also used for microwave-induced plasma spectrometry (MIP-AES and MIP-MS) (cf. Section 8.3.2.6). In this chapter we deal with the relatively new technique of microwave-assisted extraction.

3.4.4.2 Microwave Dielectric Heating

Principles and Characteristics Microwave and radio-frequency (RF) heating are to be opposed to 'classical' means, namely thermal conductance and convection and may excel where these traditional methods of heating are inefficient or ineffective. Microwave energy is more expensive than conventional energy and/or heat. Interaction of matter with microwaves leads to macroscopic or microscopic thermal effects. Unlike microwave spectroscopy the microwave heating effect does not result from well-spaced discrete quantised energy states. Electric energy is converted into heat as a result of the physical principle of dielectric loss. Typically, microwave energy is lost to a sample by two primary mechanisms:

Table 3.24 Microwave-assisted solid sampling

Characteristic	Mode
Type of microwaves	Dispersed, focused
Type of treatment	Digestion, leaching
Type of solvent/extractant	Organic, aqueous
Pressure	High, low

ionic conduction and dipole rotation. Ionic conduction is the conductive migration of dissolved ions in the applied electromagnetic field. Dipole rotation refers to the alignment, in an electric field, of molecules in the sample that have permanent or induced dipole moments. As the electric field of the microwave energy increases, polarised molecules are aligned; as the field decreases, thermally induced disorder is restored and thermal energy is released. In most cases ionic conduction and dipole rotation take place simultaneously. In the microwave and RF region the rate of heating is dependent only on the material *dielectric properties* at the specific frequency and intensity of the electromagnetic field at a given point in space. The extent to which a material is heated when subjected to microwave radiation depends on the dielectric constant or electric permittivity (ε') and the dielectric loss factor (ε''). The dielectric constant describes the ease with which a material is polarised by an electric field, while the dielectric loss factor is the physical parameter that measures the efficiency of conversion of the electromagnetic radiation to heat. Both ε' and ε'' vary with frequency, with the latter reaching a maximum, as the dielectric constant falls, in the microwave band of the electromagnetic spectrum. The ratio of the two gives the dielectric loss tangent or dissipation factor:

$$\tan \delta = \varepsilon''/\varepsilon' \qquad (3.7)$$

which defines the ability of a material to convert electromagnetic energy into thermal energy at a given temperature and frequency. A substance that absorbs microwave energy strongly is called a sensitiser. A major advantage of microwave and RF heating is that a medium is heated throughout, i.e. homogeneously. This holds true for many media, even poor heat conductors. Most organic media can be heated throughout and heated extremely quickly. Not all materials with high dielectric permittivity or dielectric constant ε' exhibit a high loss factor and high absorption of microwave energy. One other important parameter of electromagnetic heating is the *penetration depth*, which varies with frequency and dielectric properties, but is in general limited, restricting the technique to small-volume vessels. Each compound has its own characteristic frequency at which it will absorb energy most efficiently. For example, water will absorb energy most efficiently at around 20 GHz. Other forms of electromagnetic radiation have a much restricted penetration depth, and thermal conductivity is the limiting factor in heating by IR or shorter wavelengths of electromagnetic energy. Consequently, in those cases uneven heating may occur.

Microwave radiation does not have sufficient energy to be classified as ionising radiation. The energy content of microwaves (0.00001 eV) is much lower than that of x-rays (124 000 eV) and several orders of magnitude below the energy necessary to disrupt covalent bonds or hydrogen bonds of common organic molecules. The origin of the heating effect produced by the high frequency electromagnetic waves arises from the ability of an electric field to exert a force on charged particles. Microwave heating involves a special form of dielectric heating. Here frictional heat is produced in a nonconducting or only slightly conducting body by a high frequency electromagnetic field. This frictional heating is a result of the fact that molecules with an intrinsic dipole structure continually seek to align themselves to the alternating field.

Irradiation of matter with an electromagnetic field of appropriate frequencies (in the microwave and RF) facilitates penetration of energy into the bulk of the material. Microwave energy is deposited directly in the heated material, so the interior of the object can be heated without the mediation of conductive heating. Microwave energy is believed to provide uniform and fast heating due to the heat generated directly from the interaction between the material's molecules and microwave electromagnetic field. Entire bulk regions of a dielectric material can then be heated simultaneously, without any major temperature gradient. When microwave processed materials are not homogeneous this leads to complications in predicting the distribution of microwave energy inside the material.

Microwave (dielectric) heating in solution may distinguish three situations, which determine heating characteristics:

- a single solvent or mixture of solvents that have high dielectric loss coefficients;
- solvent mixtures of high and low dielectric loss;
- a susceptible sample that has a high dielectric loss in a low dielectric loss solvent.

In organic microwave sample preparation, the polarisability of the solvent molecules depends on the nature of the solvent and its relative permittivity. Therefore, the greater the relative permittivity, the more thermal energy is released and the more rapid the heating is for a given frequency. Polar molecules and ionic solutions will absorb microwave energy strongly in comparison to nonpolar molecules. In microwave field processing of materials that are poor absorbers of microwave energy a single-mode resonant cavity has to be used. A single-mode microwave oven allows a sample to be placed in a much higher electric cavity power (three orders of magnitude) than obtained in a multimode oven. This allows microwave heating of relatively low-loss materials like glasses or polymers using low power.

The *advantages* of microwave heating of solvents are manifold; the ability to rapidly reach elevated temperatures and hold them for consistent times is most important. Magnetrons are simple, reliable, and relatively inexpensive. However, the application of microwave energy to flammable organic compounds, such as solvents, can pose serious *hazards*. As the chemical and physical principles underlying the technology are deceptively simple, personnel dealing with microwave experiments must exercise an extraordinary level of safety and attention. Only approved laboratory equipment and scientifically sound procedures should be used. For laboratory microwave safety, see reference [437].

Microwaves, microwave-enhanced chemistry and microwave sample preparation have been described in several reviews [92] and books [420,438–440]. Zlotorzynski [441] recently described the fundamental principles of electromagnetic field interaction with matter. Many workers have developed the theory of microwave heating; Mingos and Baghurst [427] reviewed the origins of microwave dielectric heating. Duquesne's SamplePrep Web (*http://www.sampleprep.duq.edu/sampleprep*) is a source of information for standard microwave methods, acid decomposition chemistry and safety [442].

Applications Applications of microwave heating are both *technological* and *analytical*. Renoe [443] mentions various areas, such as the (petro)chemical industry for polymers and additives to polymers in process monitoring and quality control schemes; the environmental laboratory for priority pollutants, total petroleum hydrocarbons, and pesticides; industrial applications for extractions of lubricants and additives to process materials in fibres and contaminant analyses; the agricultural industry for pesticides and additives to seeds, feeds, plants, and fertilisers; health-care products, e.g. the ability to solvent-extract colorants, emulsifiers, active and inactive ingredients from cleaning products, lotions, and food supplements; the food industry, e.g. analyses of natural materials in food such as fats and oils and analyses of additives and preservatives; and the pharmaceutical industry, which involves a range of applications for process and quality control of active and inert ingredients. The effect of microwave vs. conventional heating on the migration of dioctyl adipate (DOA) and acetyltributyl citrate (ATBC) plasticisers from food grade PVC and PVDC/PVC films into fatty foodstuffs has been studied

using indirect GC after saponification of the ester-type plasticisers [444,445].

Most polymers are effectively transparent to microwaves, and therefore polymer/microwave interactions are unlikely. Polymers with low dielectric constant do not absorb microwave energy at all. The introduction of polar groups increases the dielectric constant. On the other hand, microwave heating may be used as a rapid technique for dissolving polyolefins prior to GPC analysis. Molecular weight of the polyolefins is not affected by the microwave heating process in TCB at 2.45 GHz [446]. Whether microwave heating of a polymer/additive matrix affects dissolution of the polymer or extraction of the additive is greatly determined by the nature of the solvent. For extraction purposes the choice of solvent is somehow restricted as it must contain a component with a high relative permittivity to be heated by microwaves. Microwave transparent media do not possess a significant dielectric constant, e.g. hexane (1.9), CCl_4 (2.2), and liquid CO_2 (1.6 at 0 °C and 50 atm), as opposed to large dielectric constant-possessing substances such as water (80.4). Pure hydrocarbons and other commonly used extraction solvents like toluene or cyclohexane, which have no dipole moment, do not absorb microwave energy. Therefore, choosing a solvent for its microwave absorbing character is important in order for these extractions to succeed.

Industrial applications of microwaves have been described [447]. Zlotorzynski [441] has described the latest advances in the application of microwave energy to analytical chemistry.

In the nonchemical field cooking, drying, sterilisation and pasteurisation are areas where microwave heating is applied on a large scale.

3.4.5 Microwave-assisted Extractions

Microwave-assisted extraction (MAE) of analytes from various matrices using organic solvents has been operative since 1986 [128]. In this process microwave energy is used to heat solvents in contact with a solid sample uniformly and to partition compounds of analytical interest from the sample matrix into the solvent. The way in which microwaves enhance extraction is not fully understood. The main factors to consider include improved transport properties of molecules, molecular agitation, the heating of solvents above their boiling points and, in some cases, product selectivity.

Microwave-assisted solvent extraction is usually carried out in one of the following modes:

(i) In a closed microwave-transparent vessel with an extraction solvent of high dielectric constant (cf. Section 3.4.5.1).

(ii) In an open microwave-transparent vessel with a microwave absorbing solvent (cf. Section 3.4.5.2).

(iii) In a microwave-transparent open or closed vessel with a nonmicrowave absorbing solvent (i.e. with a low dielectric constant) (cf. Section 3.4.5.3).

A few other variations on the microwave extraction theme have been described. In *intermittent MAE*, in order to avoid the effect of high temperatures during the microwave extraction procedure, samples are irradiated for only 30 s at a time and are not allowed to boil [128]. Craveiro *et al.* [432] have passed air over a microwave-heated volatile sample (fresh plant material) and collected the evolved water and oils in a condenser outside the oven to extract the compound without using an extraction solvent. The result of this *microwave-induced evolved gas* sample preparation is qualitatively not different from steam distillation, but much faster. In *dynamic microwave-assisted extraction* (DMAE) a dipolar solvent flows continuously through a temperature controlled extraction cell, which is positioned in a regulated microwave field. The course of extraction can be monitored, by leading the extracted material through an HPLC-type detector [448]. DMAE can be coupled on-line to a C_{18}-solid phase pre-concentration step in order to extract, concentrate and pre-clean samples simultaneously. The latter techniques have not been applied to in-polymer additive analysis.

3.4.5.1 Pressurised MAEs

Principles and Characteristics The approach is a direct descendant of closed-vessel microwave acid digestions and solvent extractions of organic analytes from solid samples. In microwave solvent extraction, the solvent is in constant contact with the solute and the matrix surface. The process is a partitioning of the analytes from the sample matrix into the solvent, with the kinetics driven by the elevated temperatures and the choice of solvent or solvent mixtures [443]. The solvent usually contains a component with a high dielectric loss coefficient, which can be heated by microwaves. The microwave-compatible solvent (characterised by large dielectric-loss tangents and loss coefficients) couples with the electromagnetic (EM) field, and through dielectric relaxation mechanisms transfers heat to the solvent medium. In MAE, the polymeric matrix does not

generally absorb microwave energy. Both temperature and pressure influence the extraction. High temperature extractions accomplished in closed vessels may result in pressures up to 200 psi ($\sim$14 bar) within the containers. In a homogeneous polar solvent, dielectric heating may result in partitioning at a relatively low temperature (50–200 °C) depending on the solvent susceptibility to microwave energy. Partitioning of compounds in extraction (solute into the solvent) consists of desorption at the matrix–solvent interface followed by diffusion of the analyte into the solvent [121].

In approaching the MAE applications it is necessary to understand some important instrument design constraints and the parameters to control for consistent solvent extractions of samples with microwave heating. In the closed pressure-resistant MAE vessels, equipped with temperature and pressure control and safety valves, the solvent is allowed to be heated under pressure above its normal boiling point and remain liquid. Microwave instrumentation achieves rapid sample dissolution through the elevation of reaction rate due to increases in temperature and pressure. Elevations in pressure are the combined result of the vapour pressure of the digestion mixture and of gases evolved by the dissolution reactions occurring. The balance between these two gas pressures can determine the temperature achievable in a closed container or vessel of fixed volume during a particular sample digestion. Pressure and temperature feedback are important control parameters in MAE operations [97].

In *closed-vessel MAE* the sample is in direct contact with the solvent mixture, not with a distilled azeotrope (as in case of a Soxhlet). Analyte solubility can then be maximised by fixing the solvent's mixture ratio and controlling the temperature. In a closed vessel, the solvent can be heated by microwave energy to any temperature, limited only by the vessel's pressure specifications. Table 3.25 shows the temperatures that can be reached with common laboratory solvents using open and closed vessels. In the latter case temperatures are often some 100 °C higher than boiling point. Closed vessel technology takes advantage of a chemically resistant polyfluoro alkoxy alkane (PFA)-Teflon lining inside a vessel shell constructed of high-mechanical strength polymer material that is also microwave transparent. Because of the instrumental configuration, MAE is well-suited for extracting large numbers of similar samples.

MAE equipment allows a high level of convenience, speed and automation.

The key to successful MAE is proper selection of the organic solvent on the basis of solvent polarity

Table 3.25 Solvents commonly used in microwave-assisted extraction

Solvent	Dielectric constant	Boiling point (°C)	Closed-vessel temperature (°C)
Hexane	1.89	68.7	–
Methanol	32.63	64.7	151
Acetone	20.7	56.2	164
Acetonitrile	37.5	81.6	194

After Hasty and Revesz [449]. Reproduced by permission of International Scientific Communications, Inc.

Table 3.26 Characteristics of closed-vessel microwave-assisted extraction

Advantages

- Fast (10–30 min)
- Selective heating
- High throughput (multivessel, up to 14 samples simultaneously)
- Low solvent volumes (10–50 mL)
- Low solvent exposure
- High efficiency (total extraction)
- No loss of volatile analytes
- No matrix dependency
- Matrix dissolution not necessary
- Not labour intensive (polymer grinding not strictly necessary)

Disadvantages

- New technique
- Medium to high cost
- Special nonmicrowave absorbing extraction vessels required
- Need for definition of extraction parameters (optimisation required for every polymer/additive combination)
- Restricted solvent choice (dipole moment)
- Risk of analyte decomposition (thermolabile products)
- Filtration of sample required after cooling
- Safety precautions (pressurised conditions: <200 psi; flammable solvents)

and dielectric compatibility. *Solvent choice* is dictated by the solubility of the analytes of interest in the solvent, the interaction of the solvent and the matrix (swelling) and the microwave absorbing properties of the solvent. The magnitude of the solvent dipole moment is the main factor that correlates with the microwave-heating characteristics of the organic solvent. The larger the dipole moment the more vigorously the solvent molecules will oscillate in the microwave field. For microwave extraction to be effective, the solvent must be able to absorb microwave radiation and pass it on in the form of heat to other molecules in the system. Equation (3.7) measures how well a certain solvent will pass on energy to others. Nonpolar solvents with low relative permittivities (such as hexane or toluene) are hardly affected by microwave energy and therefore require polar co-solvents (e.g. water or acetone), if they are to be used as solvents in MAE. If extraction between nonpolar molecules is required, then the choice of solvent is the main factor to consider.

The *main characteristics* of closed-vessel MAE are given in Table 3.26. Microwave extraction is a clean process technology. The volume of solvent used for MAE is typically 30 mL, in comparison to as much as 500 mL for conventional methods. Another advantage compared with Soxhlet extraction is that in principle any composition of solvent mixtures can be used. MAE is more selective than Soxhlet extraction, which is advantageous for a specific additive but disadvantageous in the case of screening. In many ways, the advantages of MAE are similar to those found in ASE®, i.e. that the benefits of high temperatures can be used with low boiling solvents through the use of pressure. High temperatures may favour swelling of the polymer owing to enhanced solvent–polymer interaction, provided microwave absorbing solvent is also present. However, as a consequence of the high temperatures attained, one can expect MAE to be traumatic to the polymer, where trauma to the polymer is defined as the likelihood of wax or oligomer discharge from the polymer matrix and that of oxidation/degradation reactions

taking place under the conditions of use. Thus despite the fact that MAE does not dissolve the polymer and that therefore no precipitating step is needed, the presence of low-MW oligomers in the solution cannot entirely be avoided. The difference in heating mechanism between MAE and ASE® has some other effect. With a solvent of high dielectric constant, heating will be faster than that of an oven. However, for a solvent of low dielectric constant heating will be slow or nonexistent. On the other hand, if the sample itself efficiently absorbs microwave energy, the use of non-absorbing solvent allows extractions in cool solvents to be carried out, thus reducing the risks of artefacts and analyte decomposition. Total extraction of the additives of interest may be expected when MAE produces complete or nearly complete dissolution of the polymer. Temperature-controlled MAEs lead to low RSD values [90]. According to Thomas [6] a limitation of MAE is that an extraction method must be developed for every polymer/additive combination. However, as shown by Marcato and Vianello [210] two procedures, OSM and TSM, are sufficient (at least for polyolefins).

Table 3.4 contains a *comparison* of microwave extraction with other sample preparation procedures. MAE compares favourably with Soxhlet/Soxtec® and

Table 3.27 In-polymer additive analysis by means of closed-vessel MAE

Matrices	Additives	Solvents[a]	References
HDPE, LDPE, PP	Irgafos 168, Chimassorb 81, Irganox 1010	Acetone/heptane (1:1); trichloroethane	[96]
HDPE	BHT, Irganox 1010, Irganox 1076	IPA/cyclohexane (1:1); DCM/IPA (98:2)	[90]
LDPE	BHT, BHEB, Isonox 129, Irganox 1010, Irganox 1076	IPA/cyclohexane (1:1); DCM/IPA (98:2)	[90]
LDPE	Erucamide	IPA/cyclohexane (1:1); DCM/IPA (98:2)	[90]
PP	Cyasorb UV531, Irganox 3114; Irgafos 168, Irganox 1010; Ultranox 626, Irganox 3114; BHT, Irganox 1010	IPA/cyclohexane (1:1); DCM/IPA (98:2)	[90]
PP	Irganox 1010	IPA	[457]
PET	Cyclic trimer	DCM, water, acetone, hexane/acetone (1:1), acetone/DCM (1:1)	[113]
MDPE	Irganox 1010, Irgafos 168	n.d.	[202,457]
LDPE	Chimassorb 944	Various binary solvent mixtures	[458]

[a] IPA, isopropanol; DCM, dichloromethane.

sonication extraction techniques [450]. The yields of the compounds obtained by MAE at high frequency (2.45 GHz) are comparable with those obtained by the traditional Soxhlet or shaking extraction, but MAE is much faster than conventional methods. Due to considerable reduction in exposure time the method is therefore more suitable for extraction of thermally labile compounds. Extracts obtained by MAE have none of the artefacts that can be found in the corresponding Soxhlet extracts.

The polymer/additive system in combination with the proposed extraction technique determines the preferred solvent. In ASE® the solvent must swell but not dissolve the polymer, whereas MAE requires a high dielectric solvent or solvent component. This makes solvent selection for MAE more problematical than for ASE®. Therefore, MAE may be the preferred method for a plant laboratory analysing large numbers of similar samples (e.g. nonpolar or polar additives in polyolefins [210]). At variance to ASE®, in MAE dissolution of the polymer will not block any transfer lines. Complete dissolution of the sample leads to rapid extractions, the polymer precipitating when the solvent cools. However, partial dissolution and softening of the polymer will result in agglomeration of particles and a reduction in extraction rate.

The popularity of MAE methods for in-polymer additive analysis is reflected in a limited list of reported applications. This is both on account of the former lack of dedicated microwave equipment designed specially for small analytical samples and the relatively recent commercial introduction of the technique. Microwave extraction for analytical purposes is a relatively new growth area [441].

Reviews on (the applications of) microwave-assisted sample preparation have recently been published

[443,451–454], but all with little emphasis on in-polymer additive analysis. Kingston *et al.* [455] have indicated future areas of development in microwave sample preparation and instrument development.

Applications The broad industrial analytical applicability of microwave heating was mentioned before (see Section 3.4.4.2). The chemical industry requires extractions of additives (antioxidants, colorants, and slip agents) from plastic resins or vulcanised products. So far there have been relatively few publications on microwave-assisted solvent extraction from polymers (Table 3.5). As may be seen from Tables 3.27 and 3.28, most MAE work has concerned *polyolefins*.

In order to perform extraction of additives or dissolution of polymers, solvents that absorb microwave energy are necessary. This is more important than direct absorption of microwave energy by the polymer or additives. When microwave extraction of additives

Table 3.28 Comparison of microwave and reflux extraction solvent efficiencies for additives in LLDPE

Additive	Microwave (mg L^{-1})[a]		Reflux (mg L^{-1})
	Isopropanol	Isobutanol	Isobutanol
BHT	382 (2.28)	372 (1.59)	394
Erucamide	501 (4.92)	435 (4.56)	423
Isonox 129	198 (1.25)	196 (0.92)	188
Cyasorb UV531	421 (2.73)	420 (0.56)	426
Irganox 1010	405 (2.27)	407 (1.21)	397
Irganox 1076	443 (1.73)	422 (0.46)	414
Irgafos 168	458 (2.58)	449 (0.50)	454
n	6	6	2

[a] Parenthetical values are % RSD for n extractions.
After Jassie *et al.* [454]. Reprinted with permission from L. Jassie *et al.*, in *Microwave-Enhanced Chemistry* (H.M. Kingston and S.J. Haswell, eds), American Chemical Society, Washington, DC (1997), pp. 569–609. Copyright (1997) American Chemical Society.

from a polymer is followed by HPLC analysis the solvent should preferably be compatible with the HPLC mobile phase so that solvent exchange is not required before analysis. Common extraction solvents such as alcohols, especially 2-propanol, are good reflux extraction solvents especially for additives in LDPE and also heat rapidly in the microwave. In contrast, solvents typically used to extract additives from HDPE, such as hexane, heptane, cyclohexane, toluene, and iso-octane, are less polar and heat poorly in the microwave. Thus, for HDPE, a binary solvent mixture is required consisting of a nonpolar solvent to swell the resin and a polar solvent that heats rapidly in the microwave [454].

Suitability of microwave solvent extraction for polymer/additive extraction was first shown by Freitag and John [96] for the quantitative determination of antioxidants (Irganox 1010 and Irgafos 168) and a light stabiliser (Chimassorb 81) in HDPE, LDPE and PP. Fairly quantitative extraction of the stabilisers (>90 %) from powdered polymer (1 g) was achieved within 3 to 6 min using 30 mL of 1,1,1-trichloroethane (Figure 3.14) or a 1:1 mixture of acetone and n-heptane (Figure 3.15). An important aim was to minimise thermal degradation of the antioxidants during extraction. Nielson [56,90] has published detailed reports of the extraction and subsequent separation and quantitation of various additives in HDPE, LDPE and PP using different extraction techniques (MAE, US, Soxhlet). The microwave

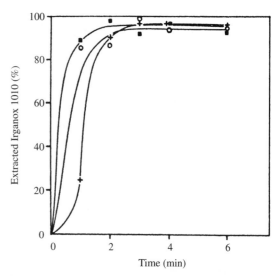

Figure 3.14 Microwave extraction of 0.1 % Irganox 1010 from ground (20 mesh) polyolefins (■, PP; ○, LDPE; +, HDPE) with 1,1,1-trichloroethane. After Freitag and John [96]. From W. Freitag and O. John, *Angewandte Macromoleculare Chemie*, **175**, 181–185 (1990). © Wiley-VCH, 1990. Reproduced by permission of Wiley-VCH

oven provides a very fast means of extracting the additives from ground resin. Depending upon the components present in the formulation this technology can be performed on unground resins as well. MAE and US extractions of various HDPE, LDPE and PP

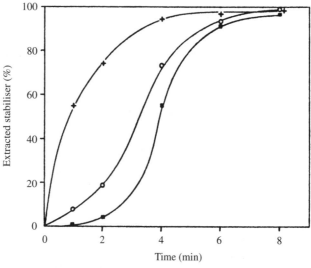

Figure 3.15 Microwave extraction of 81 Irganox 1010 (■), Irgafos 168 (○) and Chimassorb 0.1% (+) from powdered (20 mesh) HDPE with acetone/n-heptane (1:1 v/v). After Freitag and John [96]. From W. Freitag and O. John, *Angewandte Macromoleculare Chemie*, **175**, 181–185 (1990). © Wiley-VCH, 1990. Reproduced by permission of Wiley-VCH

pellets were compared [90]. MAE has also been reported for HDPE/(BHT, Irgafos 168, Irganox 1010, Irganox 1076) [456].

In many cases the use of a solvent mixture, consisting of a solvent causing swelling and another favouring migration of the additives out of the polymer matrix, is advantageous. Table 3.27 gives several examples of this approach.

Jassie *et al.* [454] have reviewed MAE applications for polymeric materials. Common antioxidants like BHT (small and volatile), Irganox 1076 (medium-size) and Irganox 1010 (large and difficult-to-extract) are efficiently microwave extracted from HDPE in 20 min. Irganox 1010 (molecular mass 1178 g mol^{-1}) is the bulkiest and migrates slowly. If the extracts contain small polar antioxidants and UV stabilisers and are analysed by RPLC, then solvent exchange is required because heptane is not miscible with the acetonitrile:water gradient typically used to separate antioxidants. Thus, common additives such as BHT, BHEB, Irganox MD 1024, and Tinuvin P normally cannot be analysed without a solvent-exchange step.

In comparing microwave with reflux extractions of LLDPE/(BHT/BHEB, erucamide, Isonox 129, Irganox 1010, Irganox 1035, Irganox 1076, TNPP) (I) in different mesh size, and of LLDPE/(BHT, erucamide, Isonox 129, Cyasorb UV531, Irganox 1010, Irganox 1076, Irgafos 168) (II), Jassie *et al.* [454] encountered some

typical problems (Table 3.28). For example, 20 mesh polyethylene swells slightly in 50 mL isobutanol when heated for 60 min at reflux conditions. By using similar solvents in microwave heating, such as isopropanol and isobutanol for LLDPE (and also cyclohexane for HDPE), the polymer melts at the elevated pressure and temperature conditions and resolidifies on cooling. The problem was overcome by modifying the solvent-to-polymer ratio. Equivalent extraction efficiencies in I were obtained in 10 min for both 20- and 30-mesh grinds. Ten-minute extractions are sufficient to ensure that all additives have been extracted, but it is brief enough that smaller, more volatile additives are not lost. Recovery and precision data for several other additives in II were also reported. Extraction results with microwave heating in isobutanol or isopropanol compared with a 60 min reflux extraction also show good reproducibility (<3 % RSD for all additives analysed).

By using *experimental design techniques*, the optimal conditions that give high extraction efficiencies of all additives may be determined in a given sample. Figure 3.16 indicates quantitative extraction from HDPE of five additives within 9 min. This result corresponds to having optimum conditions in the vessel at the maximum pressure and temperature for about 4 min. No additives were lost or degraded during the extraction procedure. The procedure is rugged enough to extract 11 additives (Tinuvin P, BHT, BHEB, Cyasorb UV531,

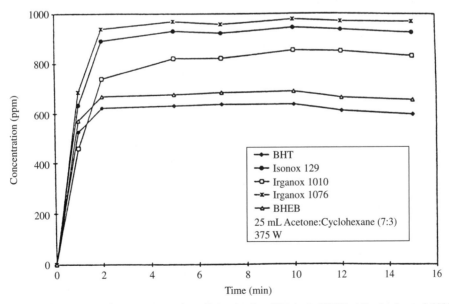

Figure 3.16 Influence of time on microwave-extraction efficiencies for additives in HDPE. After Jassie *et al.* [454]. Reprinted with permission from L. Jassie *et al.*, in *Microwave-Enhanced Chemistry* (H.M. Kingston and S.J. Haswell, eds), American Chemical Society, Washington, DC (1997), pp. 569–609. Copyright (1997) American Chemical Society

Topanol CA, Irganox 1035, Isonox 129, Irganox 3114, Ultranox 626, Irganox 1010, Irganox 1076) with acetone–cyclohexane (7:3) from HDPE resins of different crystallinity and melt indices. Recovery and precision data for these HDPE additives show that the microwave method has an RSD of <4% [454]. As expected, the solvents used for extracting additives from PE behave differently during microwave heating. At extraction temperatures near 140 °C, the acetone–cyclohexane mixture reaches pressures in excess of 100 psig, as the result of the high vapour pressures of the solvent. At the same temperature, isopropanol and isobutanol, in contrast, reach only 78 psig and ~35 psig, respectively [454].

Haider and Karlsson [202] compared MAE and US of MDPE/(Irganox 1010, Irgafos 168, calcium/zinc stearate). Irganox 1010 and Irgafos 168 were determined on the basis of peak areas obtained by HPLC and UV absorbance measurements at 245 and 280 nm. Using three different binary solvent mixtures, total additive recoveries from LDPE/Chimassorb 944 (film) and MDPE/(Irganox 1010, Irgafos 168) film were achieved after 30–60 min at temperatures from 70 to 120 °C [458]. Vandenburg *et al.* [457] compared extractions of Irganox 1010 from PP with ASE®, MAE and conventional techniques. Extraction using MAE gave the same recovery as exhaustive refluxing with chloroform with 5 min extraction at 150 °C and 3 min warm-up time. MAE and ASE® gave equal recoveries. Dean [272] has compared MAE with Soxhlet, sonication and SFE for a large number of compounds, including phthalate esters. Application note [459] describes MAE of the *total extractables* of polypropylene using acetone–hexane (4:1) (extraction time 30 min); application note [460] refers to microwave extraction of DOP from PVC.

To avoid the time-consuming grinding of plastics samples, Freitag and John [96] have tried dissolving thick samples, i.e. pellets of 3 mm diameter, instead of extracting them. The *dissolution*, performed with a 1:1 mixture of toluene and 1,2-dichlorobenzene at 100% microwave output during 5 min, yielded satisfactory results (95% of expected concentration) only for PP, whereas the recovery from the polyethylenes was below 85%.

Many of the promising features noticed by Freitag and John [96], Nielson [90] and Jassie *et al.* [454] have further been developed by Marcato and Vianello [210,461], who have described a unique *total concept* for in-polyolefin additive analysis based on MAE-HPLC-ELSD/UV. In this procedure Montell has aimed at setting up a total solution scheme for different kinds of unground (granular or Spheripol form) polyolefinic materials (iPP, HDPE, LLDPE, HECO, PB) using solvent mixtures with the lowest toxicity, heated through microwave energy. Two distinct microwave-assisted processes (MAPs) are sufficient to obtain additive extraction with minimal solubilisation of the oligomeric fraction, yielding extracts ready for direct HPLC injection. A one-step microwave-assisted extraction (OSM) is useful for a wide variety of low-medium polarity additives, such as stabilisers, flame retardants, antistatics, antislip and processing agents. A two-step microwave-assisted solvent extraction (TSM) addresses medium-high polarity additives, such as organic salts, anti-gasfading, antiacid or nucleating agents, or high molecular mass additives, such as polymeric HALS. All the OSM and TSM extraction solutions were monitored by RPLC with either UV or ELSD. The procedures were tested on various commercial products. Analytical results are satisfactory.

The combination of MAE and HPLC-UV/ELSD is a convenient way for systematic, complete and accurate analysis of organic additives in polyolefins. *Advantages* are direct extraction from pelletised matrices (no grinding), direct HPLC injection of extraction solutions (no coagulation, washing, post-concentration), extraction of a very wide dipolarity range of organic analytes, virtually negligible degradation of stabilisers, reduced time of sample treatment, reduced solvent volume, avoidance of chlorinated or aromatic solvents, fast and total recovery of either 'difficult' or highly polar compounds and appreciable accuracy.

Successful extraction of additives from polymeric matrices requires a proper selection of organic solvents. Solvent choice was based on the following solvent properties:

(i) the solubilisation capacity for the additives of interest;
(ii) the microwave absorption (ability to convert electromagnetic energy to thermal energy) and the following objectives;
(iii) the lowest toxicity;
(iv) a proper degree of polymer swelling at high temperature;
(v) dissolution of the smallest quantity of polymer at room temperature.

In the resulting binary solvent mixtures of either OSM or TSM micronising step procedures, one apolar component (*n*-hexane or *n*-heptane) serves to give high swelling-melting power to the polymer but does not heat under microwave irradiation. The second polar component (ethyl acetate, acetone, and isopropyl alcohol) has a sufficient dipole moment to facilitate heating in the microwave field and produces a shrinkage effect on the polymer macrostructure, preventing its solvation.

For *one-step microwave-assisted extraction* (OSM) the following parameters were optimised:

(i) ratio between polymer amount and solvent mixture volume;
(ii) extraction temperature; and
(iii) heating time.

Efficient extractions for various kinds of polyolefins were obtained with 2.5 g of polymer and 25 mL of solvent extraction mixture. These quantities allow representative sampling of the polymer, do not produce an excessive dilution of the extracted additives to compromise the HPLC sensitivity limits, and allow direct HPLC injection of the extraction liquids without further concentration. Table 3.29 summarises the OSM conditions for a variety of polymer matrices loaded with the same additive package (Irganox 1010, Irganox PS 802, Irganox 1076, Irgafos 168, GMS 40, CaSt), and for PB/(Irganox 1010, Irganox 1076, Irgafos 168).

OSM extracts were analysed by HPLC-UV/ELSD; each compound present in the sample solution was identified by comparing its retention time with that of

Table 3.29 OSM conditions

Sample amount	2.5 g (pelletised polymer)
Solvent mixture volume	25 mL ethyl acetate/n-hexane (3:1)
Microwave power	1 kW
Extraction time	15 min
Extraction temperature	80–125 °C[a]

[a] Depending on polymer structure and composition: iPP, 125 °C; HDPE, 125 °C; LLDPE, 100 °C; heterophasic C_2/C_3 copolymer (HECO), 125 °C or 110 °C; amorphous copolymer C_2/C_3 (Supersoft®), 110 °C; PB, 80 °C.

the corresponding peak in a standard solution. The OSM procedure was evaluated using some common additives, each with peculiar analytical problems. For example, Irganox 1010 shows poor solubility in hydrocarbon solvents and therefore needs a high swelling grade of the polymeric matrix to be completely extracted. Irgafos 168 suffers a fast degradation at $T > 140$ °C and hydrolysis to 2,4-DTBP. Irganox PS 802, with a chemical structure similar to polyolefinic matrices, shows difficult recovery. Quantification of the UV absorbers Irganox 1010 and Irgafos 168 was carried out by comparison with an external standard solution of the pure additives. Quantification of Irganox PS 802 (non-UV absorbent) was carried out using a suitable calibration curve obtained by analysis of several solutions with different concentrations or pure additive. Recoveries of over 95 % were achieved. Table 3.10 compares the recovery of Irgafos 168 and its oxidised and hydrolysed by-products from PP by means of OSM, US, reflux and dissolution/precipitation. Both US and OSM lead to negligible hydrolytic additive degradation with the best recovery for OSM. The measured additive decay is essentially due to the AO activity during the processing (extrusion) step of the polymer and not to the microwave heating treatment.

Table 3.30 shows the excellent performance of OSM for the extraction of the medium polar additive Irganox 1010. The reader may compare the results with a round-robin comprising the analysis of PP/(Irganox 1010, Irgafos 168) [209a].

The OSM MAP can be effectively applied to most of the organic additives for polyolefins. Its validity has been tested by comparing the OSM with traditional reflux extraction procedures for primary AOs (phenols), secondary AOs (aliphatic and aromatic phosphites;

Table 3.30 Comparison of extraction procedures for PP/Irganox 1010[a]

Extraction procedure	Sample		Solvents for additive extraction and PP oligomer precipitation		Time (h)
	Morphology	Amount (g)	Nature	Volume (mL)	
Solvent reflux under stirring	Pellets	5–10	Chlorinated solvents, acetone, aliphatic alcohols, ethyl acetate	80–100	4–6
Soxhlet	Pellets	5–10	Chlorinated solvents, acetone, aliphatic alcohols, ethyl acetate	100–250 or 80–100	≥6
Soxtec®	Pellets	5–10	Chlorinated solvents, acetone, aliphatic alcohols, ethyl acetate	80–100 or 10–20	≥6
Ultrasonication	Ground (≤1 mm)	2.5–5	Chlorinated hydrocarbons, acetone	50–100	≥6
OSM	Pellets	2.5	Mixture of n-hexane, ethyl acetate	18.75 6.25	0.25

[a] Recoveries ≥95 %.
After Marcato and Vianello [210]. Reprinted from *Journal of Chromatography*, **A869**, B. Marcato and M. Vianello, 285–300, Copyright (2000), with permission from Elsevier.

aliphatic thioesters), UVAs (benzotriazoles and ben-zophenones), light stabilisers (monomeric HALS), anti-statics (glycerol monoacyl esters and aliphatic amines), slip agents (fatty amides) and flame retardants (halogen derivatives) [210].

Organic polar compounds, e.g. sodium benzoate or Irganox 1425, are very difficult to extract quantita-tively from polyolefinic matrices, because of the reduced 'swelling' and the consequent poor diffusion effects of the solvents in which they are soluble. To overcome this problem Montell prepares polymer samples with a very high specific superficial area so as to increase the diffu-sion of polar solvents into the polymeric macrostructure. This can generally be achieved via cryogenic grinding, solubilising and coagulating, microwave dissolution at high temperature in closed vessel and recrystallisation. According to Marcato and Vianello [210] the latter proce-dure (called micronisation) gives the best results in terms of polymer surface area produced, recovery of additives, time and volume of solvent consumed. In the *two-step microwave-assisted solvnet extraction* (TSM) the first (micronising) step thus determines complete dissolution by microwave heating, and by cooling recrystallisation of the polymer as very small particles or lamellar stacks (<1 μm); in the second (leaching or reactive) step the additives are dissolved via addition of a suitable amount of an organic polar solvent to the extraction vessel. Typ-ical TSM conditions for micronisation are 2.0 g of pel-letised polymer, 25 mL solvent mixture volume, 1 kW microwave power, and time/temperature depending on the polymer (Table 3.31).

With this new morphology, the polymer releases polar additives more easily into the solvent mixture. Polar solvents, such as methanol (25 mL), added to the vessel containing the micronised mixture, easily solubilise polar organic additives that are suspended in the previous solvent mixture. Thus a good recovery can easily be obtained by short manual shaking (leaching) for 10–15 min. Solubilisation or the recovery of polar organic additives can also be achieved after a chemical

Table 3.31 TSM conditions for micronisation

Polymer	Solvent mixture (vol/vol)	Temperature (°C)	Time (min)
iPP	Acetone/*n*-hexane (1:3)	130	30
HDPE	Acetone/*n*-heptane (1:4)	135	40
LLDPE	Isopropanol/*n*-heptane (1:3)	125	30
HECO	Acetone/*n*-hexane (1:1)	125	30
Supersoft®	Acetone/*n*-hexane (1:1)	125	30
PB	Ethyl acetate/*n*-hexane (3:1)	140	30

reaction. In fact, good extraction of some organic additives may require oxidation (e.g. by 2,6-di-*tert*-butylhydroperoxide), saponification (e.g. with TBAH), hydrolysis (e.g. with phosphoric acid) or derivatisation (e.g. with 1,4-dibromoacetophenone). TSM procedures give excellent recovery results for the following medium to high polar additives, extracted using (i) the *leaching mode*: nucleating agents (aromatic phosphonate salts, sorbitol derivatives), light stabilisers (polymeric HALS); (ii) the *hydrolytic mode*: antiacid agents (fatty acid salts); (iii) the *saponifying mode*: light stabilisers (polymeric HALS); or (iv) the *oxidising mode*: stabilisers (aromatic phosphite and phosphite–phosphonite mixtures) [210].

A good example of analysis by TSM extraction, coupled with chemical reaction, is given by the analysis of fatty acid salts (e.g. calcium stearate). Fatty acid salts may be transformed completely into the corresponding acids. The free fatty acids (mainly stearic, palmitic and myristic) are completely solubilised in the top layer of the organic solvent mixture inside the microwave vessel and can be determined quantitatively by HPLC-ELSD [462]. TSM extraction tests were carried out on PP/(Irganox 1010/1076/1425, Irgafos 168, sodium benzoate), and TSM procedures were tested using the nucleating agent sodium benzoate and the anti-gas fading agent Irganox 1425 as very polar organic additives. Recoveries for sodium benzoate and Irganox 1425 were better than 95 %. Table 3.32 shows that recoveries of the highly polar anti-gas fading additive Irganox 1425 (MW 695) from 2 g of PP sample as high as those of TSM extractions can be obtained only after polymer dissolution with xylene, its coagulation with methanol, an intermediate washing with acetone (solvent with intermediate dipolarity between xylene and methanol) and multi-washings with methanol of the coagulated polymer. Clearly, TSM has advantages over competing extraction procedures.

The fact that low-MW oligomers can be extracted from the polymer during cooking and storage of food-stuffs is of particular concern to the polymer manufac-turer and the food retailer. This concern over plastic contact materials and issues relating to the recycling of waste plastics have made it necessary to investigate the extractability of oligomers from PET. For the determina-tion and quantification of *oligomers* from PET a number of methods have been proposed to differentiate between the species. MAE is useful for exhaustive extractions using organic solvents (with a dipole moment). Cost-ley *et al.* [113] have evaluated various organic solvents (dichloromethane, water, acetone, hexane, xylene) in microwave extraction of oligomers from PET and have compared MAE to alternative extraction approaches

Table 3.32 Comparison of extraction procedures for PP/Irganox 1425

Extraction procedure	Sample morphology	Solvent		Irganox 1425 recovery range (%)[a]	Time (h)
		Nature	Volume (mL)		
Two-step MAE (TSM)	Pellets	Acetone	6.25	99–103	1.5–2
		n-Hexane	18.75		
		Methanol	25.00		
Solvent reflux 6 h; post-concentration	Ground (<1 mm)	Methanol	80–100	60–65	7–8
Dissolution (xylene)/crystallisation, washings, filtering, concentration	Pellets	Xylene	80	75–86	4.5
		Methanol	160		
Dissolution (xylene)/precipitation (methanol), washings, filtering, concentration	Pellets	Xylene	80	95–102	5–6
		Acetone	80		
		Methanol	160		

[a] HPLC analysis.
After Marcato and Vianello [210]. Reprinted from *Journal of Chromatography*, **A869**, B. Marcato and M. Vianello, 285–300, Copyright (2000), with permission from Elsevier.

(Soxhlet, Parr bomb). The main parameters that influence extractability in MAE are temperature, pressure, time of extraction and solvent selection. Microwave extraction of PET with dichloromethane (DCM) at 120 °C for 2 h gave comparable results to Soxhlet extraction for 24 h. At 125 °C the microwave extraction time is reduced to 0.5 h, but the polymer fuses above 120 °C and PET degradation phenomena occur. MAE using hexane–acetone (1:1), water, acetone and acetone–DCM (1:1) all gave much lower recoveries than DCM at 120 °C. This would be expected from the solubility parameters. At the same temperature, DCM would swell the polymer most. The major extractable component of the low-MW oligomers is the cyclic trimer; fractions up to the heptamer, nonamer and decamer have been reported. The effects of chemicals migrating by microwaving foods in various plastics have been addressed elsewhere [463].

MAE has also been used for the extraction of adipate plasticisers from PVC [464]. The efficiency of MAE depends on the kind of solvent, the temperature achieved and the heating time. The final temperature reached depends on the microwave power, number of vessels and irradiation time. Higher recovery values than SFE were reported for both phthalate and adipate. Other reports on microwave-assisted solvent extraction have appeared [465–467].

MAE of additives from polymeric matrices has clearly established good records. Some additional studies may be needed in order to validate this approach for analytical sample preparation. Microwave heating has also been applied to dissolve polymers for molecular weight determination [446].

Work is in progress to validate the MAE method, proposed for EPA, in a *multi-laboratory evaluation* study. Nothing similar has been reported for additives in polymeric matrices. Dean *et al.* [452] have reviewed microwave-assisted solvent extraction in environmental organic analysis. Chee *et al.* [468] have reported MAE of phthalate esters (DMP, DEP, DAP, DBP, BBP, DEHP) from marine sediments. The focus to date has centred on extractions from solid samples. However, recent experience suggests that MAE may also be important for extractions from liquids.

3.4.5.2 Focused MAEs at Atmospheric Pressure

Principles and Characteristics Despite its widespread application the Soxhlet technique suffers from several disadvantages, amongst which long extraction times and relatively large solvent volumes. For this reason various modifications to the basic technique have been proposed to accelerate the process (cf. Section 3.3.3.9). Recently an instrument (Soxwave®, Prolabo) for carrying-out *focused microwave-enhanced extraction* (FMW) at atmospheric pressure has been introduced which results in considerable time savings (Figure 3.17). Focused microwave-assisted Soxhlet extraction (FMASE) is a hybrid between MAE and conventional Soxhlet extraction. The extraction process is a conventional reflux extraction at the boiling point of the solvent at atmospheric pressure in an open vessel. In contrast to the closed systems where the microwave energy is dispersed throughout the oven, the commercially available atmospheric pressure systems are based

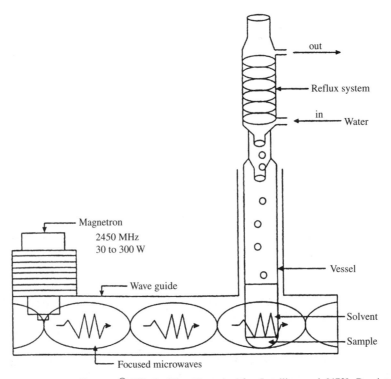

Figure 3.17 Microwave apparatus (Soxwave® 100, Prolabo, France). After Letellier *et al.* [472]. Reprinted with permission from *LC.GC International*, Vol. 12, Number 4, April 1999, pp. 222–225. LC.GC International is a copyrighted publication of Advanstar Communications Inc. All rights reserved

on the use of microwaves that are focused into the vessel. Sample and solvent are introduced into a glass container, which is fitted with a condenser to prevent loss of volatiles and solvent, and placed within a protective glass sheath. Microwave radiation (2.45 GHz, 30 to 300 W) is allowed to interact with the solvent in the bottom of the vessel and is focused precisely to increase reproducibility [469–471]. The organic solvent is refluxed through the sample. Magnetic stirring of the samples allows the efficient extraction of even large amounts of samples. Typical operating conditions are: 30 mL solvent for extraction of 1–30 g of sample. In the open-style system, individual sample vessels are heated sequentially. Focused microwave-assisted extracts need to be reduced to a small volume (a few mL) using a rotary evaporator.

The principle of FMW involves the heating of both the solvent and the matrix by wave/matter interactions. The microwave energy is converted into heat by two mechanisms: dipole rotation and ionic conductance. The heating is, therefore, selective with only polar or moderately polar compounds susceptible. Due to the use of low microwave energy the structure of target molecules remains intact.

Advantages of the FMW method are [471,472]:

(i) homogeneous and reproducible sample treatment (owing to the focusing principle);
(ii) rapid heating to the boiling temperature (1 min at 30 W);
(iii) secure (atmospheric pressure, no risk of explosion);
(iv) no compound degradation (boiling point extraction);
(v) not significantly matrix-dependent;
(vi) no cooling required after extraction;
(vii) low solvent requirements;
(viii) significant time savings (typical extraction time 10 min).

Recovery yields are equivalent to those of conventional systems but the solvent volume requirements are much lower (by about 10 times) meaning a considerable cost saving. Focused microwave-atmospheric pressure technology allows the possibility of adding reagents at any time during the microwave-assisted process. Even though the principle of focused microwaves is efficient in terms of energy transfer, it only allows for the use of one flask at the time. Careful optimisation of the extraction medium, power applied, and exposure time is required to preserve the stability of the organic compounds to be separated.

The open-vessel MAE approach can be contrasted to closed-vessel MAE (Section 3.4.5.1). Focused microwave-atmospheric pressure technology was reviewed by Mermet [473].

Applications With focused microwaves additives from polymers can be extracted with 100 % recovery in 30 min (as compared to 8 h by Soxhlet) with 30 mL solvent consumption. The extraction of organic compounds in polymeric matrices by using focused microwave technology at atmospheric pressure is at an early stage of exploration. The method has been applied to the separation of organotin compounds (in sediments) [473] and has been validated for the extraction of PAHs using a certified matrix, the 'Standard Reference Material, SRM 1941a'. Rocca *et al.* [474] have reported the analysis of essential oils in chewing gums. Specific literature reports regarding application to polymer/additive analysis are still to appear, quite at difference to the more established use of closed-vessel MAE.

3.4.5.3 *Microwave-assisted Process*™

Principles and Characteristics Paré *et al.* [475] have patented another approach to extraction, the Microwave-Assisted Process (MAP™). In MAP™ the microwaves (2.45 GHz, 500 W) *directly* heat the material to be extracted, which is immersed in a microwave transparent solvent (such as hexane, benzene or iso-octane). MAP™ offers a radical change from conventional sample preparation work in the analytical laboratory. The technology was first introduced for liquid-phase extraction but has been extended to gas-phase extraction (headspace analysis). MAP™ constitutes a relatively new series of technologies that relate to novel methods of enhancing chemistry using microwave energy [476].

MAP™ makes use of physical phenomena that are fundamentally different compared to those applied in current sample preparation techniques. Previously, application of microwave energy as a heat source, as opposed to a resistive source of heating, was based upon the ability to heat selectively an extractant over a matrix. The *fundamental principle* behind MAP™ is just the opposite. It is based upon the fact that different chemical substances absorb microwave energy

to different extents. In this case the microwaves travel freely through the microwave-transparent solvent and reach the matrix where they interact selectively with (and heat) molecules of high dielectric constant. The localised heating disrupts the matrix structure and allows very rapid extractions. Normally the extractant medium is selected to be transparent to the microwave and therefore remains cool, which is of importance for desired labile or volatile components, reduces the risk of artefacts and analyte decomposition. However, if some heating of the medium is permissible, it may be partially transparent. Obviously, the solvent must be selected for its ability to solubilise the desired product. As a result of the selective application of energy right into the matrix, MAP™ is the only extraction process that effects a *direct migration* of the desired components out of the matrix. Other extraction processes make use of a common action mechanism, namely the random transfer of energy to a container, the extractant and matrix. The extractant, then, diffuses into the matrix and solubilises the various constituents (e.g. Soxhlet) or entrains them (e.g. steam distillation) out of the matrix. Therefore, most extraction processes are based on the fundamental importance of diffusivity. The very basic nature of such processes prevents to a large extent speciation, or selective extraction.

The MAP™ process thus features a partitioning mechanism in which the sample (often a biological material) is a good dielectric in the presence of a low dielectric, poorly heating solvent. Intrinsic moisture is often a desirable component in MAP™ because water superheats and eventually leads to structural rupture. The dispersed moisture content for this process is preferably 40–90 % [475], but materials containing as little as 20 % moisture are amenable to the process. For matrices that are relatively poor microwave absorbers, the desired thermal energy transfer may be achieved by simply impregnating them, or by dispersing within them some strong microwave-absorbing components. MAP™ methods are rugged, as evidenced by the small variations in procedures required to obtain excellent recoveries and precision from a wide product range.

The *advantages* associated with the MAP™ technology as compared to conventional and automatic Soxhlet methods are considerable (Table 3.33). In MAP™ high sensitivity and selectivity by fractionation are achieved using different extraction media with similar, or better, linearity and reproducibility parameters. One of the principle features of the process is the lower temperatures observed in the microwave-extracted materials in contrast to volumetric heating usually experienced in traditional solvent procedures. These lower temperatures

Table 3.33   Characteristics of MAP™ technology

Advantages

- Large tolerance to sample size (from subgram to multiple-gram)
- Low solvent consumption and reduced waste management costs
- Cool solvents (reduced analyte decomposition)
- Selective extraction
- Reduced processing time (within minutes)
- Limited pre-processing (e.g. homogenisation)
- No diffusion limitation (direct extraction)
- Liquid-phase and gas-phase extractions

Disadvantages

- Safety precautions
- Need for microwave-absorbing matrix

and the migration of the analyte into the cooler surrounding solvent, where heat is dissipated, are supposed to account for the fact that little degradation of analytes is observed [475]. A high level of safety precautions is required, whether or not operation is done in an open vessel at low pressure or in a closed vessel at high pressure. However, given the insensitivity of MAP™ to the diffusivity of solvents, or to diffusion processes associated with analytes for that matter, it is generally not necessary to operate under high-pressure conditions.

Paré *et al.* [477] have indicated two main areas of extractions based on MAP™ technology, namely liquid-phase and gas-phase extraction. In *liquid-phase extraction*, provided that there is an adequate amount of high dielectric constant material in the matrix, it is possible to design a MAP™ extraction procedure with almost any of the organic solvents (of smaller dielectric constant) routinely used in chemical analysis, bearing in mind that the solvent must be selected for its ability to solubilise the desired product. The greater the difference in dielectric properties, the better the efficiency of MAP™. Although the MAP™ concept was originally introduced as a liquid-phase extraction, it applies equally well to extraction in any type of fluid, liquid, liquefied gas, gas *per se*, or SCF. Diffusivity being a parameter of minor importance in MAP™, it is possible to imagine liquefied gas extraction using liquid carbon dioxide, instead of scCO$_2$. This offers the advantage of eliminating the high pressure and temperature requirements for conventional SFE operation. Liquid carbon dioxide offers good solubility parameters but suffers from a poor diffusivity coefficient. The latter characteristic makes the use of liquid CO$_2$ as a solvent in current extraction processes virtually impossible within reasonable extraction time boundaries. However, diffusivity is not a problem in MAP™. The technique has also

been used under sub- and supercritical fluid conditions (*MAP™-SFE*) [478]. Results obtained for operation of MAP™-SFE hyphenation lead to higher yields than conventional SFE and shorter extraction times. Among the main advantages of MAP™-SFE is its capacity to be able to operate at much lower temperatures and pressures than currently used.

The principles behind MAP™ liquid-phase and *gas-phase extractions* are fundamentally similar and rely on the use of microwaves to selectively apply energy to a matrix rather than to the environment surrounding it. MAP™ gas-phase extractions (MAP-HS) give better sensitivity than the conventional static headspace extraction method. MAP-HS may also be applied in dynamic applications. This allows the application of a prolonged, low-power irradiation, or of a multi-pulse irradiation of the sample, thus providing a means to extract all of the volatile analytes from the matrix [477].

Applications   MAP™ provides a technique whereby intact organic and organometallic compounds can be extracted more selectively and more rapidly with similar or better recovery when compared to conventional extraction processes. The applications of MAP™ extend far beyond extraction and range from analytical-scale to industrial processing for commercial purposes. The fields of application of MAP™ in the liquid phase are very diversified and include the environmental (water), agricultural (natural products), food, biomedical and pharmaceutical areas, consumer products, cosmetics, as well as process monitoring and control [477]. The patented technology applies to the extraction of a variety of chemical substances from a wide range of matrices such as water, soils, animal and plant tissues, and a variety of man-made products [477]. Examples are the 'extraction of oils', aromas and flavours from plant material, analytes from sediments, etc. The *residual moisture content* of matrices has been the single most important factor in the development of MAP™ analytical methods, irrespective of their nature. The microwaves interact selectively with the free water molecules contained in the matrix and cause localised heating that gives rise to a sudden non-uniform elevation in temperature with more pronounced effects on systems that can not accommodate the high internal pressures that are created (e.g. in plant and animal tissues, foams). These rupture spontaneously, allowing the organic contents to flow freely toward the relatively cool surrounding solvent that solubilises them rapidly. Extraction is often achieved in less than 2 min.

The applicability of MAP™ to dehydrated or dry materials is not precluded. In fact, most matrices can either be rehydrated, impregnated or swollen with a substance that possesses a relatively high dielectric constant in order to achieve the same microwave effects as would be obtained from moist samples [475,479]. For the purpose of extraction of additives from polymers a higher rate of success might be expected in the application to engineering plastics (e.g. polyamides) than to polyolefins with a low dielectric constant. No such specific applications have been given yet. However, MAP™ has been applied in cleaning polyurethane foams (PUFs), which are used to monitor and assess air quality. Use of MAP™ for the extraction of PAHs from PUFs is considerably more economic than Soxhlet extraction, at similar or even higher yields [480].

Microwave-assisted gas-phase extraction (MAP-HS) has been used to liberate volatile organic compounds (VOCs) from water samples [481,482]. In MAP™ applied to headspace analysis the microwave energy is imparted selectively to the water because the latter absorbs microwaves preferentially to the surrounding medium. The energy is then released to the neighbouring low-absorption species (VOCs) which are, in turn, vaporised selectively and rapidly in relation to their vapour pressures and their heat capacity. The VOCs are subsequently introduced into a GC with FID. Within 1 min of sample preparation time the MAP™ approach gives higher detector responses, with better precision and more rapidly than the conventional 30 min static headspace sampling approach.

3.4.6 Pressurised Fluid Extraction

Principles and Characteristics The terminology pressurised fluid extraction (PFE), pressurised liquid extraction (PLE), pressurised solvent extraction (PSE) or hot solvent extraction stands for the development of extraction devices coined as Accelerated Solvent Extraction (ASE®, Dionex) or Enhanced Solvent Extraction (ESE®, Isco-Suprex) for commercialisation purposes that significantly streamline sample preparation with liquid solvents relative to the traditional Soxhlet. Pressurised fluid extraction (PFE, i.e. ASE® or ESE®) is a relatively new LSE process performed at elevated temperatures, usually between 50 °C and 200 °C and pressures of 6.9–13.8 MPa (but below supercritical conditions), for short periods of time (5–10 min) using low amounts of solvents (eventually up to 100 mL) [134,483]. Figure 3.18 shows a schematic diagram of PFE (ASE®). The extraction procedure consists of a combination of dynamic and static flow of the solvent through a heated extraction cell containing

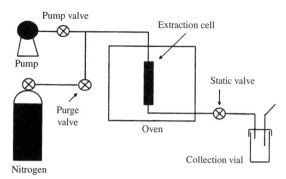

Figure 3.18 Schematic representation of a pressurised fluid extraction (PFE) system. After Ezzell [486]. Reproduced by permission of Sheffield Academic Press/CRC Press

the sample. In principle, any type of solvent or solvent mixture may be used, and solvent volumes of less than 15 mL are required for a 10-g sample. The solid sample eventually mixed with an inert support (e.g. silica of small particle size or Hydromatrix™ [484]) to avoid sample agglomeration at high *p* and *T* is placed in a cartridge and the hot pressurised liquid solvent is passed through the sample. A period of static extraction follows, and the solvent is then purged into a collecting vial. A similar volume of solvent is flushed through the cell to transfer all the extracted material to the collecting vial. This has the advantage of being able to use any solvent at high temperatures to achieve the benefits of high dissolving power and swelling of the polymer. The sequential technique is applicable to a variety of matrices, including polymers, plant and animal tissue, foods, and environmental samples [485].

PFE is based on the adjustment of known extraction conditions of traditional solvent extraction to higher temperatures and pressures. The main reasons for enhanced extraction performance at elevated temperature and pressure are: (i) solubility and mass transfer effects; and (ii) disruption of surface equilibria [487]. In PFE, a certain *minimum pressure* is required to maintain the extraction solvent in the liquid state at a temperature above the atmospheric boiling point. High pressure elevates the boiling point of the solvent and also enhances penetration of the solvent into the sample matrix. This accelerates the desorption of analytes from the sample surface and their dissolution into the solvent. The final result is improved extraction efficiency along with short extraction time and low solvent requirements. While pressures well above the values required to keep the extraction solvent from boiling should be used, no influence on the ASE® extraction efficiency is noticeable by variations from 100 to 300 bar [122].

The efficiency of extraction is mainly dependent on *temperature* as it influences physical properties of the sample and its interaction with the liquid phase. The extraction is influenced by the surface tension of the solvent and its penetration into the sample (i.e. its viscosity) and by the diffusion rate and solubility of the analytes: all parameters that are normally improved by a temperature increase. High temperature increases the rate of extraction. Lou *et al.* [122] studied the *kinetics of mass transfer* in PFE of polymeric samples considering that the extraction process in PFE consists of three steps:

(i) diffusion from the core of the polymeric particles to the surface;
(ii) transfer from the surface into the extraction fluid; and
(iii) elution of the solutes out of the extraction cell (solubility).

Steps (i) and (ii) are controlled by molecular diffusion. Higher operating temperatures can improve the kinetics of mass transfer in all three steps. Vandenburg *et al.* [37] described the kinetics of PFE extraction using the 'hot-ball' model [286] derived for SFE extractions.

In PFE the effects of *solvent flow-rate* on the extraction efficiency can be used to determine whether extraction is limited by diffusion or by solubility. Compared to SFE, in PFE a slightly more complicated situation prevails. In PFE, mass transfer from the surface of the polymer particles through the layer of stagnant solvent around the particles into the extraction solvent is much slower because of the lower solute diffusivity in the extraction solvent and the relatively thick layer of stagnant solvent around the particles, Hence, in PFE extraction of polymeric samples, the rate-limiting step can be diffusion inside the polymer particles, transfer from the surface into the extractant, or solubility in the extractant. If extraction rates are controlled by diffusion inside the polymer particles and/or transfer through the layer of stagnant extractant around the polymer particles, the solvent flow-rate will show little influence on the extraction recovery [122]. But, if extraction rates are controlled by solubility, larger solvent flow-rates will yield higher extraction recoveries. The solubility of additives is expected to be high in the solvents at elevated temperatures, and therefore solubility-limiting behaviour should be rare in ASE®. This was indeed found to be the case for extraction of Irganox 1010 from PP [37]. Extraction of additives from polymers tends to be *diffusion-limited*. Not surprisingly then, a small sample size is preferred provided that the requirements for sample homogeneity and sensitivity are satisfied. With the PFE extraction rate being limited by diffusion of the analytes in the polymer particles, the extraction rate can be greatly increased by decreasing the particle size.

The static ASE® mode may lead to incomplete extraction because of the limited volume of solvent used. Dynamic ASE® yields faster extractions by continuously providing fresh extraction solvent to the sample, but obviously requires a larger solvent volume than static ASE® and is therefore less suited for trace analysis. In practice, a combination of static and dynamic extraction is often considered to be the best choice.

Method development in PFE is considered by some workers [134] as being relatively straightforward, but this opinion is questionable [122]. Ezzell [486] described PFE (ASE®) method development. Pre-extraction steps comprise grinding, dispersing and drying. In principle, extraction parameters to be considered are sample size, particle size, extractant temperature, pressure, number of cycles, time, static vs. dynamic extraction and flow-rate. Fortunately, the number of critical experimental parameters that really need to be optimised is relatively small, certainly in comparison to SFE. *Optimisation* is, perhaps, the area of the greatest concern in method development in PFE, which has not yet reached maturity. Lou *et al.* [122] have faced systematically the effect of most of the aforementioned parameters on the extraction efficiency of ASE® for polymeric samples. Apart from particle size, extraction temperature and type of solvent used were found to be the most critical. Generally, if a particular solvent works well in a conventional procedure, it will also work well in PFE. Compatibility with the post-extraction analytical technique, the need for extract concentration (solvent volatility) and the cost of the solvent should all be considered. If a traditional extraction approach requires an extract clean-up step prior to quantification, the PFE extract will probably require the same manipulation. For preliminary experimentation, a starting temperature of 100 °C is usually recommended. An increase in temperature generally has a positive impact on extraction but only up to the point where analyte or matrix degradation begins to occur. These problems, however, are not generally observed in the normal operating ranges of PFE. Most PFE extractions are performed at a standard pressure of 10.3 MPa. The effect of static time should always be explored in conjunction with static cycles, in order to produce a complete extraction in the most efficient way possible. Analysts should use a *method validation* technique to evaluate the success of method modifications. One way to validate extraction efficiency is to re-extract the same sample and check for remaining analyte(s). The commercial extractor allows the analyst to automate re-extraction for easy method validation [488].

Selection of a suitable extraction solvent is probably the most difficult step in optimising PFE for polymers [122]. *Solvent selection* used in PFE of any particular analyte can be done in several ways:

(i) if the matrix/analyte has a standard extraction procedure, the same solvent system may be used;
(ii) analytes known to be soluble in a particular solvent or mixture of solvents may be extracted using that solvent system;
(iii) using the Hildebrand solubility parameters;
(iv) on the basis of polarity: analytes with known polarity will be extracted using a solvent with a compatible polarity index;
(v) on the basis of dielectric constant;
(vi) on the basis of dipole moment.

The problem in selecting a solvent for an extraction is how to determine which one will swell but not dissolve the polymer at high temperatures and solubilise the analyte. Unfortunately, the positive effects of solvents, which swell the polymer in Soxhlet and reflux methods, cannot fully be exploited because, at the elevated temperatures used, these solvents will (partially) dissolve or soften the polymer [122]. Softening causes agglomeration problems similar to SFE. Moreover, dissolved polymer reprecipitates in the transfer lines of the instrument, causing plugging of the system tubing, valves or restrictor. Therefore, the advantages of faster diffusion gained at high temperatures cannot be achieved with swelling solvents. However, solvents which do not interact at all with the polymer will not swell the polymer even at temperatures at which the polymer will melt or the additive may decompose. Therefore, the solvent must be selected to swell, but not dissolve, the polymer significantly at high temperatures. Consequently, solvents used in conventional extractions are not necessarily good solvents for PFE [122]. Moreover, as almost no data are available on the solubility of polymers or solutes in high temperature solvents under PFE conditions solvent selection on this basis is still largely empirical. This situation is quite different for environmental samples where selection of the solvent is relatively straightforward (same solvent as employed in conventional extraction methods) [134].

The promising approach taken by Vandenburg *et al.* [37,489] is to use initially a solvent with a Hildebrand solubility parameter several $MPa^{1/2}$ different from the polymer (i.e. a 'poor', nonswelling solvent for the polymer) to determine experimentally the maximum temperature at which the polymer can be extracted. The optimal extraction temperature depends on the polymer (melting point under PFE conditions; degradation), the extraction solvent, and the target analytes (and their thermostability). Preferably the solvent is also compatible with the follow-up analytical separation technique, e.g. HPLC. A 'stronger' solvent with a solubility parameter similar to the polymer is then added in increments. This will cause the polymer to swell and hence increase diffusion and extraction rates. Eventually the solvent mixture will dissolve the polymer, causing reduced surface area and slower extractions, as well as possibly blocking transfer lines. In this way an optimum 'strength' of solvent can be determined at the optimum temperature for each polymer and additive.

Vandenburg *et al.* [37,489] have described the use of Hildebrand solubility parameters in a simple and fast solvent selection procedure for PFE of a variety of polymers. Hildebrand parameters for several common solvents and polymers are presented in Tables 3.2 and 3.34, respectively. When the proper solvent mixture for the polymer was determined, PFE resulted in essentially the same recoveries as the traditional extraction methods, but used much less time and solvent. PFE can be used to give very fast extractions and appears to offer the greatest flexibility of solvents and solvent mixtures. The method is ideal for a laboratory which analyses a large number of different polymers.

Table 3.35 lists the main *advantages and disadvantages* of PFE techniques. PFE is generally considered to be an exhaustive extraction: under the appropriate conditions all of the extractable content of a matrix will be solubilised. Indeed, poor selectivity is usually encountered in PFE since the solvents that are chosen have good solvating power for a wide range of analytes. A clean-up step following extraction is therefore normally required. For this reason (i.e. high solvating power) PFE method development appears to be less matrix dependent than SFE for a specific sample type. At least for the

Table 3.34 Solubility parameters (δ, $MPa^{1/2}$) of some common polymers

Polymer	Solubility parameter
PP	16.6
PVC	19.5
PET	20.5
PA6.6	28.0
PMMA	19.0

After Vandenburg *et al.* [489]. From H.J. Vandenburg *et al.*, *Analyst*, **124**, pp. 1707–1710 (1999). Reproduced by permission of The Royal Society of Chemistry.

Table 3.35 Main characteristics of pressurised fluid extraction

Advantages

- Rapid extractions (15 min) above boiling points
- Minimal solvent usage
- Principles well understood (similar to LSE)
- Limited method development (Hildebrand solubility parameters for solvent choice)
- Flexibility of solvents and solvent mixtures
- Some selectivity
- Sequential operation (fully automated; up to 24 samples)
- Ease of use
- Suitable for R&D samples, up to 50 g
- Automatable

Disadvantages

- New technique (few applications)
- High capital cost (but entry-level ASE® system available)
- Optimisation required for each system (but limited method development)
- Analyte stability problems (at high temperature)
- Exhaustive (rather than selective) extraction
- Low recovery for high-MW HALS
- Grinding recommended
- Uses conventional oven heating; thermal lag

Table 3.36 ASE® saves solvent and time

Extraction technique	Average solvent used per sample (mL)	Average extraction time per sample
Soxhlet	200–500	4–48 h
Automated Soxhlet	50–100	1–4 h
Sonication	100–300	30 min–1 h
Closed-vessel MAE	20–50	30 min–1 h
SFE	10–50	30 min–2 h
ASE®	15–40	12–18 min

extraction of environmental samples, no matrix dependency of recoveries was observed [134]. Nevertheless, by applying knowledge of the sample matrix and its constituents, some degree of selectivity may be achieved. PFE extraction parameters, which most affect selectivity in extraction, are the choice of solvent and extraction temperature [486]. If, in a given set of conditions, the target compounds are recovered with a high degree of co-extractable materials, a shift in solvent polarity or a reduction in extraction temperature may result in a reduction of the contaminant material. As the sample is diluted further concentration is required. Large volume injection can eventually avoid the concentration step.

While the decrease in extraction time is favourable for laboratories in general, it can be critical when laboratory analyses are used in feedback control of production cycles and quality control of manufacturing processes. The volume of solvents used in PFE can be some 10 times less than traditional extraction methods (cf. Table 3.36). PFE cuts solvent consumption by up to 95 %. Because so little solvent is used, final clean-up and concentration are fast; direct injection in analytical devices is often possible. Automated PFE systems can extract up to 24 sample cells.

The use of the Hildebrand solubility parameter approach to aid solvent selection with a few simple experiments, starting from the liquid solvents used in traditional extraction methods, limits the efforts needed in method development. As for other extraction techniques that do not dissolve the polymer, the extraction efficiency is proportional to the surface area of the sample matrix; therefore, grinding is often recommended prior to PFE. Stability problems with analytes and matrices are limited in the 75–150 °C range, but more common at 200 °C. Ultranox 626 is not hydrolysed in PFE conditions. So far, evidence produced indicates reliable quantitation by means of PFE, but further confirmation is needed. PFE is to be considered as a promising technique for polymer/additive sample preparation despite the fact that the procedure may need modifications to optimise extraction speed and efficiency, as each polymer formulation is unique.

Table 3.4 compares the use of PFE with other sample preparation techniques. ASE® technology [134] has quickly become an attractive alternative to other traditional and new extraction techniques. PFE is suitable for exhaustive extractions using organic solvents and is the most similar to common standard methods. The technique was developed to overcome the most important disadvantages of SFE, such as matrix-dependency of extraction yields and restrictor plugging. In part, PFE is a competitor to SFE; in part the techniques are complementary. From the experimental results, it appears that PFE offers significant advantages over SFE with scCO$_2$ alone for extraction of compounds with a low solubility in this medium. However, addition of modifiers to scCO$_2$ can result in a similar performance for both techniques. PFE puts to good use the same advantages as SFE (increased extraction rate at higher p, T), but is not limited in solvent selection, as is the case in practice with SFE (merely confined to scCO$_2$, restricting extraction to certain additive classes). It is perhaps for this reason that PFE most recently is enjoying popularity.

Dean [272] has compared various extraction methods. The use of ASE® has made a large initial impact due to its acceptance as an EPA method. Inter-extraction competition is expected to favour the technique even further. Vandenburg *et al.* [457] have compared PFE and MAE with atmospheric pressure methods for extraction of Irganox 1010 from freeze-ground PP. Accelerated solvent extraction has been reviewed by Ezzell [486].

Applications Pressurised fluid extraction (ASE®, ESE®) is still in its early stage of development, both for polymeric and other samples. At present, most applications are found in the environmental, food, pharmaceutical and nutraceutical areas. Few reports describe the application of PFE to the extraction of monomers, oligomers and additives from polymers and most work is very recent. An application note [488] has provided some *guiding principles* for ASE® applied to additives in polymers, namely as follows:

- Disperse the ground polymer sample with an inert matrix (to prevent agglomeration of the particles).
- Define the main 'weak' solvent (that dissolves the additives but not the polymer).
- Add a small amount of a 'strong' solvent (that causes swelling of the polymer).
- Use a cellulose thimble in the extraction cell (to prevent plugging).
- Use multiple static cycles (to maximise concentration gradients).
- Use sample re-extraction (to confirm completeness of extraction).

A more explicit approach to solvent selection for ASE® of polymers, based on Hildebrand solubility

parameters, has been given [37,489]. Test extractions have shown that this method of solvent selection can be used for a wide variety of polymers. The method has been used to determine optimum conditions for ASE® of additives from ground PP/Irganox 1010 (cf. Figure 3.19), PVC/DOP pellets and PA6.6/dimer; optimum extraction conditions were also set for PET and PMMA. Table 3.37 shows the Hildebrand solubility parameters for these systems.

Irganox 1010 could be extracted from PP in 18 min, using propan-2-ol–cyclohexane (97.5:2.5). For plasticiser extraction from PVC, optimum conditions were hexane–ethyl acetate (60:40) at 170 °C, giving effectively complete extractions after 13 min. For nylon, the optimum conditions were hexane–ethanol (60:40) at 170 °C, and extractions from ground material were 95 % complete after 16 min, including warm-up time. Irganox 1330 was used as an internal standard in PP, n-C_{20} in

Table 3.37 Hildebrand solubility parameters (δ, MPa$^{1/2}$)

Matrix	Polymer	Strong solvent		Weak solvent	
PP/0.15 % Irganox 1010	16.6	Cyclohexane	16.8	Propan-2-ol	23.8
PVC/38 % DOP	19.5	Ethyl acetate	18.6	Hexane	14.9
PA6.6/dimer	28.0	Ethanol	26.0	Hexane	14.9
PET	20.5	Ethyl acetate	18.6	Hexane	14.9
PMMA	19.0	Ethyl acetate	18.6	Hexane	14.9

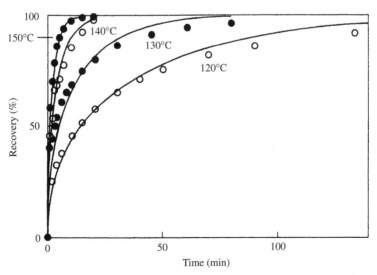

Figure 3.19 Accelerated solvent extraction of Irganox 1010 from PP with 2-propanol at various temperatures. Points, experimental data; solid lines, curve fitted using the 'hot-ball' model. After Vandenburg *et al.* [37]. Reprinted with permission from H.J. Vandenburg *et al.*, *Analytical Chemistry*, **70**, 1943–1948 (1998). Copyright (1998) American Chemical Society

PVC and nylon 6.6; calibration curves for Irganox 1076 and DOP were applied.

Solvents that are good swelling agents, and hence give fastest extractions during Soxhlet extraction, tend to dissolve the polymer at the high temperatures used during PFE. Macko *et al.* [490,491] who used high pressures to dissolve PE in heptane at 160 °C exploited this. The polymer is not soluble in heptane at room temperature and therefore precipitates on cooling. Hexane can be readily removed by evaporation and the analytes redissolved in RPLC solvents.

Accelerated solvent extraction has been exploited for *in-polyolefin additive analysis*. Hinman *et al.* [492] have used ASE® for extraction of antioxidants (BHT, BHEB, Isonox 129, Irganox 1010, Irganox 1076) from LDPE, and have compared ASE® and SFE in the extraction of antioxidants from LDPE. Swagten-Linssen [387] has extracted HDPE/(Irganox 1010, Irgafos 168) by means of ASE® (solvent acetonitrile), followed by quantification by HPLC-PDA. Irganox 1010 was obtained in 94 % yield. UV control of the extracted solid confirmed that the extraction was not complete. Post-extraction verification by means of UV is impossible if the material to be extracted is dispersed with Hydromatrix™. Similarly, also the analysis of HDPE/Irganox B220 by means of ASE® was optimised. Carlson *et al.* [493] used acetonitrile–ethyl acetate (1:1) or isopropyl alcohol (IPA)–cyclohexane (95:5) to extract BHT, Irganox 1010, Irganox 1076 and Irgafos 168 from ground HDPE at 125 °C (Table 3.38).

Waldebäck *et al.* [494] have developed a simple, accurate and fast ASE® method for the determination of Irganox 1076 in LLDPE before and after γ-irradiation. The specific aim was to find proper extraction conditions for the balance between high diffusion rate of the analyte vs. solvation of the plastic material. The extraction yield for the different combinations of solvents and temperatures is shown in Table 3.39, from which ethyl acetate was chosen. Extraction yields detected were always less than 100 % of the added antioxidants (200–300 ppm) probably due to Irganox 1076 losses during the manufacturing, processing and/or storage of

Table 3.38 Extraction of polymer additives from ground HDPE[a]

Compound	BHT	Irganox 1010	Irganox 1076	Irgafos 168
% Recoverage	110.9	89.7	87.1	93.3
RSD (%) ($n = 5$)	15.9	3.1	0.6	2.3

[a] Extraction at 140 °C with acetonitrile:ethyl acetate (1:1).
After Carlson *et al.* [493]. From R. Carlson *et al.*, in *Proceedings of the 10th International Symposium on Polymer Analysis and Characterization*, Toronto (1997). Reproduced by permission.

Table 3.39 Extraction yield (in μg g^{-1}) of Irganox 1076 from LLDPE

Temperature (°C)	Methanol	Acetonitrile	Ethyl acetate	Propan-2-ol	THF	MIBK
50	13	11	34	–	–	–
75	43	53	111	48	123	37
100	128	125	173	54	–	57

After Waldebäck *et al.* [494]. From M. Waldebäck *et al.*, *Analyst*, **123**, 1205–1207 (1998). Reproduced by permission of The Royal Society of Chemistry.

the plastic. At temperatures above 80 °C LLDPE ground granules start melting, but whole granules do withstand fast extractions at 100 °C. In this way the grinding step could be omitted, simplifying and accelerating the method. Moreover, any risk of antioxidant degradation during grinding was eliminated. The extracts did also not require further concentration before LC analysis, again reducing losses of Irganox 1076. Mixing ethyl acetate with hexane as a swelling agent raised the extraction yield of the antioxidant. Reproducible results were obtained with RSD values of 9.7–12.5 %. After γ-irradiation (2 × 25 kGy) 96 % of Irganox 1076 could no longer be extracted. Similarly, Yagoubi *et al.* [495] observed a 95 % loss in Irganox 1076 content, after polyethylene vinyl acetate was radio-treated by 25 kGy. The difference in Irganox 1076 yield is probably on account of scission of this molecule and an increase of cross-linking within the polymer due to irradiation.

Use of a 'strong' solvent for PP (cyclohexane) during PFE has resulted in melting or dissolving at moderate temperatures and low extraction rates. The best solvent for extraction from PP using PFE was found to be propan-2-ol at 150 °C (within 5 min) although acetone at 140 °C gave only slightly slower extractions [37]. Addition of 2.5 % cyclohexane to propan-2-ol gives slightly faster extractions with ASE® than using pure propan-2-ol. ASE® of PP/(Irganox 1010, Irganox 1076, Irgafos 168) was described in reference [488] with quantification relative to an internal standard (Irganox 1330) added to the finished extract.

PFE has also been used for additive analysis of *engineering plastics and rubbers*. Lou *et al.* [122], who discussed the mechanisms of ASE® extraction of polymeric samples, have given guidelines for the optimisation of ASE® extractions of polymeric matrices, as illustrated for low-MW organics, such as dimers and trimers from PBT and caprolactam from PA6. Hexane was used for these extractions, even though this solvent gives poor Soxhlet recoveries. Total extractables from styrene-butadiene rubbers, including naphthenic oils, aromatic oils and organic acids, with IPA at 185 °C gave results comparable to ASTM D 1416 involving reflux in ethanol–toluene (70:30) [493]. The application of ASE® for the determination of the BFRs 1,2-bis(tribromophenoxy) ethane (TBPE), polybrominated diphenylether (PBDE) and polybrominated biphenyl (PBB) in ABS and PS is time- and solvent-saving in comparison to Soxhlet extraction, but the retrieval rates are lower [496]. It is also possible to extract PEG 400 DME quantitatively from PET by means of ASE® within 90 min, thus avoiding the use of HFIP in a dissolution/precipitation procedure. ASE® results obtained for PVC/(DOA, TOP, DOP, TOTM) were comparable with the conventional Soxhlet method (ASTM D 2124) but are obtained with reduced solvent consumption and extraction times [497]. Drews *et al.* [357] have used ASE® extracts with HPLC-ELSD and UV/VIS detection in *QC analysis* of butyl stearate surface finish on nylon, PP and polyester yarn and staple fibres using MeCl$_2$ or MeOH as a solvent.

Although PFE lacks a proven total concept for in-polymer analysis, as in the case of closed-vessel MAE (though limited to polyolefins), a framework for method development and optimisation is now available which is expected to be an excellent guide for a wide variety of applications, including non-polyolefinic matrices. Already, reported results refer to HDPE, LDPE, LLDPE, PP, PA6, PA6.6, PET, PBT, PMMA, PS, PVC, ABS, styrene-butadiene rubbers, while others may be added, such as the determination of oil in EPDM, the quantification of the water-insoluble fraction in nylon, as well as the determination of the isotacticity of polypropylene and of heptane insolubles. Thus PFE seems to cover a much broader polymer matrix range than MAE and appears to be quite suitable for R&D samples.

ASE® is recognised as an official extraction method in US EPA SW-846 Method 3545 and has shown to be equivalent to standard EPA extraction technology in terms of recovery and precision.

3.4.6.1 Hybrid Supercritical Fluid Extraction/ Enhanced Solvent Extraction

Principles and Characteristics In an attempt to develop a *unified sample preparation system* for extraction of various matrix/analyte combinations Ashraf-Khorassani *et al.* [498] have described a hybrid supercritical fluid extraction/enhanced solvent extraction (SFE/ESE®) system to remove both polar and nonpolar analytes from various matrices. The idea is that a single instrument that can perform extractions via pure CO$_2$ solvent, and all gradients thereof affords

economical advantages over systems that can handle only one extraction medium. The hybrid SFE/ESE® process proceeds in the same way as ESE®. A solvent is allowed to enter the extraction chamber containing an extraction cartridge. The vessel is pressurised and after attaining equilibrium (p,T) a static extraction time is initiated. The extracted material is then washed out of the extraction vessel (rinse step). This is then followed by a pressurised CO$_2$ flush.

A single SFE/ESE® instrument may perform (i) pressurised CO$_2$ (SFE), (ii) pressurised CO$_2$/modifier and (iii) pressurised modifier (i.e. ASE®/ESE®, organic solvent) extractions. The division between SFE and ASE®/ESE® blurs when high percentages of modifier are used. Each method has its own unique advantages and applications. ESE® is a viable method to conduct matrix/analyte extraction provided a solvent with good solvating power for the analyte is selected. Sample clean-up is necessary for certain matrix/analyte combinations. In some circumstances studied [498], SFE may offer a better choice since recoveries are comparable but the clean-up step is not necessary.

Applications SFE/ESE® is a very recent development, which has been used to extract additives from LDPE/(BHT, BHEB, Isonox 129, Irganox 1010, Irganox 1076) with EtOAc/CH$_3$CN (1:1) (ESE®) in combination with scCO$_2$ (SFE) [498].

3.5 SORBENT EXTRACTION

Many of the classical techniques used in the preparation of samples for chromatography are labour-intensive, cumbersome, and prone to sample loss caused by multistep manual manipulations. During the past few years, miniaturisation has become a dominant trend in analytical chemistry. At the same time, work in GC and HPLC has focused on improved injection techniques and on increasing speed, sensitivity and efficiency. Separation times for both techniques are now measured in minutes. Miniaturised sample preparation techniques in combination with state-of-the-art analytical instrumentation result in faster analysis, higher sample throughput, lower solvent consumption, less manpower in sample preparation, while maintaining or even improving limits.

Analysis can be improved if *clean-up* and/or *preconcentration* of sample components is achieved. Traditionally, analytes are preconcentrated from liquid samples using LLE. This technique is characterised by a low number of theoretical plates, is laborious, uses vast amounts of expensive and toxic organic solvents of high purity (to avoid chromatographic inter-ferences after working up), is difficult to automate and requires additional preconcentration steps prior to analysis. Modern alternatives to LLE as laboratory ana-lyte preconcentration or sampling methods are solid-phase extraction (SPE) and solid-phase microextraction (SPME), which overcome some of the disadvantages encountered with LLE. SPE methods can handle small samples, reduce solvent consumption, have fewer steps, save time and labour, provide better efficiency, prevent emulsions, enable easy sample collection and are more amenable to automation.

In sorptive extraction methods compounds are partitioned from the sample into the bulk of a polymeric retaining phase. The devices used for this purpose exhibit bulk retention instead of surface adsorption. Sorption is a well known chromatographic process with linear sorption isotherms and superior performance for polar and/or reactive compounds and is performed in the absence of catalytic degradation reactions (high inertness). Solvent-free sorptive extraction techniques have evolved from Open Tubular Trapping (OTT) to SPME (cf. Section 3.5.2) and packed PDMS tubes to the recently introduced Stir Bar Sorptive Extraction (SBSE) (cf. Scheme 3.1). The sensitivity of these tech-niques is primarily determined by the amount of sor-bent employed. The most commonly applied sorbent is poly(dimethylsiloxane) (PDMS), which offers some desirable properties: (i) best GC stationary phase (apo-lar); (ii) decomposition products very specific and not related to solutes of interest; (iii) retention indices avail-able for many compounds; and (iv) similar H_2O/PDMS and H_2O/octanol distributions. PDMS traps based on 100 % PDMS particles for analyte trapping are similar to SPE or air adsorption tubes. The large amount of sor-bent (250 mg) allows high enrichment and sensitivity,

high maximum flow-rates and rapid sampling, but loss of volatiles is eventually a drawback.

Classical LLEs have also been replaced by mem-brane extractions such as SLM (supported liquid mem-brane extraction), MMLLE (microporous membrane liq-uid–liquid extraction) and MESI (membrane extraction with a sorbent interface). All of these techniques use a nonporous membrane, involving partitioning of the ana-lytes [499]. SLM is a sample handling technique which can be used for selective extraction of a particular class of compounds from complex (aqueous) matrices [500]. Membrane extraction with a sorbent interface (MESI) is suitable for VOC analysis (e.g. in a MESI-µGC-TCD configuration) [501,502].

3.5.1 Solid-phase Extraction

Principles and Characteristics   Solid-phase extrac-tion (SPE) is a very popular sample preparation and clean-up technique. In SPE solutes are extracted from a liquid (or gaseous) phase into a solid phase. Substances that have been extracted by the solid particles can be removed by washing with an appropriate liquid eluent. Usually, the volume of solvent needed for complete elu-tion of the analytes is much smaller (typically <1 mL) than the original sample volume. A concentration of the analytes is thus achieved.

Although SPE can be done in a batch equilibration similar to that used in LLE, it is much more common to use a small tube (minicolumn) or cartridge packed with the solid particles. SPE is often referred to as LSE, bonded phase or sorbent extraction; SPE is a refinement of open-column chromatography. The mechanisms of retention include reversed phase, normal phase, and ion exchange.

Packed-bed SPE was introduced as a *sample prepa-ration technique* in the early 1970s but did not start

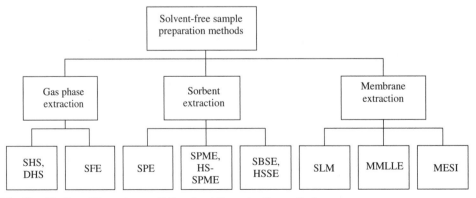

Scheme 3.1 Classification of the main essentially solvent-free extraction methods

to be commonly used until 1985. The last decade has witnessed a great increase in the use of SPE as a powerful technique for rapid, selective preparation in the overall analytical process. The SPE procedure consists of: (i) conditioning; (ii) loading and/or adsorption; (iii) washing; and (iv) elution. Before adsorption of analytes by the stationary phase can begin, the sorbent bed must be made compatible with the liquid solution. The necessary pretreatment involves the use of a mediating solvent that will promote better surface contact between the phases (wetting). With modern solid extractants, in the loading stage equilibrium is rapidly attained and the analytes tend to be extracted in a zone near the top of the SPE column. A carefully chosen wash liquid affords the opportunity to remove coadsorbed matrix materials from the SPE column. In the elution step the adsorbed analytes are selectively removed from the solid extractant by a strong solvent and are returned to a liquid phase that is appropriate for instrumental analysis. The final result is that the analytes are concentrated in a relatively small volume of clean solvent; the highly purified extract is ready for more effective chromatography without any additional sample work-up. For use in GC, the eluting solvent should be free from impurities and fairly low boiling so that the large solvent peak will not interfere with the sample peaks.

The sample volume initially introduced onto the sorbent, the choice of sorbent and solvent system and careful control of the amount of solvent used are of paramount importance for effective pre-concentration and/or clean-up of the analyte in the sample. The number of theoretical plates in an SPE column is low ($N = 10$–25). SPE is a multistage separation method and as such requires only a reasonable *difference* in extractability to separate two solutes. In SPE concentration factors of 1000 or more are possible, as compared to up to 100 for LLE with vortex mixing.

High selectivity (i.e. the ability to separate analytes from matrix interferences) is one of the most powerful aspects of SPE. This highly selective nature of SPE is based on the extraction sorbent chemistry, on the great variety of possible sorbent/solvent combinations to effect highly selective extractions (more limited in LLE where immiscible liquids are needed) and on the choice of SPE operating modes. Consequently, SPE solves many of the most demanding sample preparation problems.

Both SPE and LLE involve a partitioning of compounds between two phases: solid and liquid for SPE and two immiscible liquids for LLE. As modern SPE is a technique in which the basic principles of liquid chromatography are used to isolate the compound(s) of interest from the sample matrix it is placed between the classical LSE and column liquid chromatography, in full accord with the IUPAC definition of chromatography [503]. SPE offers several important advantages over LLE: (i) faster; (ii) easier manipulation; (iii) use of much smaller amounts of liquid organic solvents; (iv) less stringent requirements for separation; (v) higher concentration factors. In liquid chromatography the intent is to separate various sample components from one another on the basis of different rates of migration through the column. In SPE the sample solution itself is the mobile phase. SPE in a column provides multiple equilibrations and therefore requires only a reasonable *difference* in extractability to separate two solutes. Ideally, the capacity (or retention) factors of the sample components in SPE are very high (100, 1000, or more) as compared to LC (1 to 10). Under these conditions the sample components will be retained as a single, tight band on the column. Fritz [504] has compared LLE, SPE and LC. There is little difference between short LC columns and SPE cartridges. HPLC and SPE have many similarities in terms of retention mechanisms and solvent types; also essentially the same packings are used (mainly C_{18} phases) with different particle packing sizes, although Majors [505] noticed some phase differences. However, these techniques are performed under quite different experimental conditions. Major differences are the nature of the elution characteristics, the plates/column and cost (Table 3.40).

The *sorbents* used for SPE are available in three basic formats, namely disks, prepacked cartridges, and syringe barrels, each with certain advantages. Solid media employed in SPE may be classified as nonpolar, polar and ion-exchange phases with mixed retention mechanisms (Table 3.41). The large sorbent selection ranges also

Table 3.40 A comparison of HPLC and SPE under normal operating conditions

Factor	HPLC	SPE
Format	Stainless steel column	Plastic cartridge
Particle size (μm)	5 (or 3)	40
Particle shape	Spherical	Irregular
Plates/column	20–25 000	<100
Separation	Continuous elution	Digital elution
Cost of column	US\$ 250–300	US\$ 1–2
Operation	Reusable	Used once
Separation modes	Many	Many
Cost of operation	Moderate–high	Low
Cost of equipment	High	Low

Table 3.41　Solid-phase extraction mechanisms and phases

Mechanism	Sorbent	Analyte type	Matrix type	Analyte eluent
Nonpolar extraction (reversed phase)	C_{18}, C_8, C_2, phenyl, cyclohexyl, cyanopropyl	Nonpolar functional groups, such as alkyl and aromatic	Polar solutions (aqueous buffers)	Polar solvents such as methanol, acetonitrile and water
Polar extraction (normal bonded phase or adsorption)	Silica, diol, cyano, amino, diamino	Polar functional groups, such as amine and hydroxyl	Nonpolar solvents, oils	Nonpolar solvents, such as hexane and methylene chloride
Cation exchange	Strong (sulfonic acid) or weak (carboxylic acid)	Positively charged functional groups, such as amines	Aqueous, low ionic strength	Buffers such as acetate, citrate and phosphate
Anion exchange	Strong (tetraalkylammonium), weak (DEAE, amino)	Negatively charged functional groups, such as organic acids	Aqueous, low ionic strength	Buffers such as phosphate and acetate

After Majors *et al.* [3]. Reprinted with permission from *Supplement to LC.GC Europe*, Vol. 13, Number 11, November 2000. LC.GC Europe is a copyrighted publication of Advantar Communications Inc. All rights reserved.

comprise mixed mode types and polymer-based versions. Activated charcoal, alumina, silica gel, magnesium silicate (Florisil), chemically bonded and modified silica phases (which allow extraction from both aqueous and nonaqueous solvents) and polymers, e.g. styrene divinylbenzene copolymers, are being used. Baltussen *et al.* [506] have evaluated PDMS as a solid phase. The PDMS-based extraction technique shows resemblances to other techniques such as SPME and OTT.

The selectivity of a sorbent (or wash and eluting solvent) is determined by the difference in affinity for the analyte compared to the matrix compounds. The availability of different chemically modified surfaces ensures maximum *extraction selectivity*. By choosing the proper surface chemistry, users can optimise their extraction procedure to maximise cleanliness by adjusting chemical selectivity. SPE only requires the identification of a retention mechanism that would extract the analyte selectively and allows other matrix components to flow through the bed. The average silica particle size is typically 30 to 60 μm and allows easy elution under low pressure. The small mean pore diameter (typically 6 nm) excludes molecules of MW higher than 15 000–20 000 a.m.u., and the large surface area (typically 500–600 m² g⁻¹) confers a high sample loading capacity.

Table 3.42 lists the main factors influencing optimisation of SPE. When considering a specific extraction problem, many different aspects influence column selection, including: nature of the analytes and of the sample matrix; degree of purity required; nature of major contaminants in the sample; and final analytical procedure. Reversed-phase sorbents have nonpolar functional groups and preferentially retain nonpolar compounds. Thus, for a nonpolar analyte, to remove polar interferences using a polar sorbent phase, the sample

Table 3.42　Factors influencing optimisation of solid-phase extraction

Choice of phase type	Cartridge conditioning solvent
Phase weight	Wash solvent polarity, pH
Particle size	Elution solvent polarity, pH
pH of sample	flow-rate
Dilution of sample	

After Stevenson *et al.* [507]. Reproduced by permission of Sheffield Academic Press/CRC Press.

environment should be made as nonpolar as possible. Normal-phase sorbents with polar functional groups are more likely to retain polar compounds, such as analytes with hydroxyl or amine groups (e.g. chloroform extract containing polyamines). Ion-exchange sorbents have either cationic or anionic functional groups.

The choice of solvent directly influences the retention of the analyte on the sorbent and its subsequent elution, whereas the solvent polarity determines the solvent strength (or ability to elute the analyte from the sorbent in a smaller volume than a weaker solvent). Dean [272] gave solvent strengths for normal- and reversed-phase sorbents. The elution solvent should be one in which the analytes are soluble and should ideally be compatible with the final analysis technique. For example, for HPLC analysis, a solvent similar to the mobile phase is a good choice of elution solvent. For the elution step it is also important to consider the volume of the solvent. A minimum volume of elution solvent (typically 250 μL per 100 mg of sorbent) allows maximum concentration of the analytes.

Differentiation of SPE methods can be made according to the following criteria: (i) the type of sorbent; (ii) the mode of performing SPE (static or dynamic); (iii) (in case of dynamic SPE) off-line (cartridges) or on-line;

Table 3.43 Characteristics of SPE sample preparation techniques

Characteristic	Off-line	On-line
System	Open	Closed
Analysis	Batch-wise	Serial
Reuse of SPE column	No	Yes
Loss of sample	Yes	No

and (iv) the type of interactions (apolar, polar, ion-exchange, covalent, mixed). On-line SPE processes, also known as pre-column concentration techniques, overcome some of the drawbacks associated with off-line SPE (Table 3.43). In the absence of sample manipulation between the pre-concentration and the analysis steps, analyte loss and risk of contamination are reduced and the detection limits and reproducibility are improved. As the whole sample extract enters the analytical column the sample volume can be smaller, the consumption of organic solvents is lower, and the potential for automation is improved.

Numerous SPE methods have been developed, which differ mainly in the way the adsorbent is used to isolate the analytes from the matrix and materials used. Recently, emphasis is on the use of short (1 cm long) pre-concentration columns, which provide rapid desorption, a need for a small sample volume and easy coupling with LC analysis. Although these methods generally show adequate performance the extraction selectivity is usually poor. This can be a problem in low-level target compound analysis but not for screening procedures. Basically, only by changing the solid-phase material or increasing the amount of the retaining phase can lack of retention be overcome.

There are several ways in which SPE can prove useful:

(i) Removal of interfering compounds. In sample isolation, the compound of interest can be selectively sorbed onto the solid phase while matrix interferences are allowed to pass through the SPE column. In the reverse manner, matrix isolation can be used to bind interferences to the solid phase, allowing the sample of interest to pass through and be collected.

(ii) Pre-concentration of the analyte to a level suitable for the sensitivity of the analytical technique of choice. For trace enrichment a large volume of dilute sample is passed over the stationary phase. The enriched sample is stripped quantitatively by displacement with a small volume of an appropriate strong eluent, eliminating the need for an additional evaporation step.

(iii) Fractionation of the sample into different (groups of) compounds, as in classical column chromatography, eluting each fraction with a different liquid phase.

(iv) Storage of analytes that are unstable in a liquid medium or with relatively high volatility.

(v) Derivatisation reactions between reactive groups of the analyte(s) and on the adsorbent surface.

SPE offers many advantages over LLE, as shown in Table 3.44. The main disadvantage is the need for method development. There will never be a universal SPE method because the sample pretreatment depends strongly on the analytical demand.

SPE is an alternative rapid sample preparation mode for GC and HPLC (cf. Sections 7.1.1.2 and 7.1.1.6). Combination of SPE and SFE provides a solvent-less preparation technique. In this approach, the sample can

Table 3.44 Characteristics of solid-phase extraction

Advantages

- Very simple and safe to use
- Speed (10–20 min; few operational steps)
- Versatile
- Ease of automation
- Low solvent, reagent, and apparatus costs (relatively inexpensive)
- Variable sample volumes (from 100 µL to 1 L)
- Small volumes of elution solvent (<1 mL)
- Effective concentration of analytes (over wide range of volatility and polarity)
- High analyte recoveries as a result of few sample transfers
- Minimal evaporation
- Minimal exposure to environment of labile samples
- No formation of emulsions
- Excellent selectivity (wide choice of stationary phases)
- Ability to simultaneously extract analytes of wide polarity range
- Highly purified extracts
- Use of disposable extraction columns (in a multitude of column sizes)
- No cross-contamination
- Prolonged column life-time (particulate matter and strongly retained compounds can be removed)
- Compatibility with instrumental analysis (precolumn hyphenation)
- Availability of many product formats
- Alternative to LLE

Disadvantages

- Method development required (choice of sorbent and optimisation of sorbent selectivity)
- Significant background interferences (high blank values)
- Labour intensive (unless automated)
- Poor reproducibility (batch-to-batch sorbent performance)
- Variable recovery
- Plugging of cartridges, blocking of pores

be collected or isolated on an SPE disk or cartridge, and the device can be placed in the extraction thimble of an SFE instrument. Using scCO$_2$ as an extractant, analysts can remove compounds from the disk selectively. On the other hand, during SFE, the analytes extracted from the matrix are transported through the restrictor to the trapping device (off-line SFE) or focused in the chromatographic system (on-line SFE). Methods of analyte trapping during off-line SFE are trapping into a liquid solvent or collection on solid surface. Szumski and Buszewski [318] have described a trapping device for off-line dynamic SFE based on a specially prepared SPE column with different types of adsorbents. Also microwave-assisted SPE is being practised [465].

Miniaturisation of SPE has also been described [504]. Thurman and Mills [508] discussed the history and future of SPE. The technique will continue to replace LLE. More on-line use of both LC and GC are prospected. As instruments such as GC-MS and HPLC-MS become more sensitive, smaller sample sizes may be used. New phases will continue to be introduced to take full advantage of specific interactions. It is expected that at last sample handling and SPE will reach the level of sophistication that its relatives in LC have reached, and perhaps go beyond.

Principles and practice of SPE have been described in several recent monographs [504,508–512] and reviews [505,513–516]. A practical guide to SPE is available [3].

Applications   The application of SPE technology to the isolation of a specific analyte or class of analytes from various matrices has grown tremendously in the last decade. Much of this growth has been due to the relative ease of operation with samples being poured directly onto the column with little or no prior preparation, and the wide range of polymer phases bound to solid supports currently available. The versatility of SPE allows it to be used for a number of purposes, such as clean-up, concentration of nonvolatile and semi-volatile analytes from liquid samples, solvent exchange (analytes being transferred from one particular matrix environment into another, e.g. aqueous to organic) before chromatographic analysis, desalting, derivatisation (analytes are retained on a sorbent, derivatised, then eluted), class fractionation (separation of a sample into different compound groups sharing common properties), minimisation of matrix effects, and optimisation of extraction efficiency.

Liquid samples, in particular aqueous in nature, are the most frequently encountered (waste water, food and beverages, biological fluids, soil and agricultural samples). SPE is a well-established technique, which is much used in clinical chemistry for drugs-of-abuse and general biological/biotech samples, in biochemical and pharmaceutical analysis, and in aqueous environmental samples for the determination of organic pollutants. Most SPE methods are developed for a single compound, possibly with an internal standard in the case of quantitative determinations.

Junk *et al.* [517] have observed C$_8$–C$_{28}$ alkanes and alkenes (residues of the polymerisation process), alkylphthalate (plasticisers), BHT and quinone oxidation products (AOs) as interference compounds by extracting SPE cartridges and by analysing these extracts by GC-MS. Consequently, in the past processing samples using SPE may have resulted in contamination of the extract with artefacts from the SPE device (composed of medical-grade polyolefins, glass and virgin PTFE). Medical-grade PP contains a low level of extractables that may easily appear when performing sensitive GC with ECD, FID or MS detectors. Modern SPE devices show marked improvement in extractable reduction. As the levels of extractables are decreased, more gas chromatographers will adopt SPE as an extraction technique.

SPE has been applied to phthalate esters (plasticisers in PVC), polar pesticides (agricultural usage) and for other continuous pollution monitoring problems and environmental analyses [272]. For these applications SPE has largely displaced LLE as the preferred technique for the preparation of liquid samples, e.g. EPA method 506 is concerned with the determination of phthalates and adipate esters in drinking water.

SPE has also found applications for dyes [518–521]. In another application SPE was also used to achieve concentration [157]. Extraction of rubber (containing 50 % oil) does not easily allow additive determination in view of the low additive content. However, the additives may be collected on the SPE column by treating the rubber with *n*-heptane removing the oil. Additives are then eluted by means of CHCl$_3$ or ether and can easily be determined. Sarbach *et al.* [522] used SPE for sampling in a study of migration of impurities from a multilayer plastics container into a parenteral infusion fluid. 4-Alkylphenols in water samples, which are degradation products of polyethoxylates widely used as nonionic surfactants, were determined quantitatively using SPE on poly(styrene-divinylbenzene) and HPLC-UV (at 278 nm) [523]; the method was validated. SPE has also been used for the extraction of antioxidants in ointments [524].

In a typical application of SPE as a *clean-up* device in polymer/additive analysis Nelissen [157] has examined a PA6 compound with calcium stearate adhered on the surface by means of paraffin oil.

Extraction with CHCl$_3$ separates paraffin oil, nylon oligomers and stearic acid from calcium stearate; after concentration to dryness the residue was taken up in *n*-heptane and passed over an SPE cartridge, which retained the polar components, namely oligomers and fatty acid, removing the apolar paraffin oil. In planning SPE experiments solvent choice is greatly aided by preliminary TLC sampling to verify elution and/or retention of analytes. SPE can be used for clean-up of the extract output by SFE. SPE is also used for clean-up of analytical samples prior to the determination of various trace metal ions, e.g. by means of ICP-MS.

SPE is a useful device for working up of polymer additive dissolutions: the apolar polymer is retained on the C$_{18}$ sorbent, while analytes may be eluted. In the fractionation of dissolutions it is advantageous to make use of the differences in polarity and affinity of the components with the sorbent. SPE of applied samples may be done with cartridges or disks, either off- or on-line. A flow-chart for the use of SPE has been published [3]. Applications of SPE have been described in several monographs [511,512].

3.5.2 Solid-phase Microextraction

Principles and Characteristics Solid-phase microextraction (SPME) is a patented microscale adsorption/desorption technique developed by Pawliszyn *et al.* [525–531], which represents a recent development in sample preparation and sample concentration. In SPME analytes partition from a sample into a polymeric stationary phase that is thin-coated on a fused-silica rod (typically 1 cm × 100 µm). Several configurations of SPME have been proposed including fibre, tubing, stirrer/fan, etc. SPME was introduced as a solvent-free sample preparation technique for GC.

The most common *SPME device* (commercially introduced by Supelco Inc. in 1992) consists of a fused silica fibre (uncoated or coated with PDMS, polyimide, liquid crystal polyacrylate, Carbowax or graphite), protected in a micro-syringe assembly (Figure 3.20). Solid-phase microextraction is a *two-stage process* whereby an organic analyte is first extracted from a liquid (usually aqueous) sample, from headspace above a liquid or solid or from the atmosphere (liquid and headspace sampling; concentration step) and then adsorbed onto the surface of a phase-coated silica fibre. The basic approach is to use a small amount of the extracting phase, usually less than 1 µL; the sample volume can be very large. Adsorption of the analytes on the SPME fibre is allowed for a controlled time

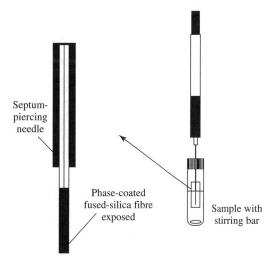

Septum-piercing needle

Phase-coated fused-silica fibre exposed

Sample with stirring bar

Figure 3.20 Solid-phase microextraction apparatus. Reproduced by permission of Supelco Inc. (Sigma-Aldrich)

period preselected according to their affinity toward the fibre coating. The extracting phase can be either high-MW polymeric liquid, similar in nature to stationary phases in chromatography, or a solid sorbent, typically of high porosity to increase the surface area available for adsorption. The amount of analyte extracted by the coating at equilibrium is determined by the magnitude of the partition coefficient (distribution ratio) of the analyte between sample matrix and coating material. Adsorption equilibria are usually attained in 2–30 min, after which time the fibre is retracted within the protective needle. In the desorption step, the microfibre is inserted in the hot injector of a gas chromatograph or other suitable instrument for separation and quantitation.

The most important step of the two-stage process is adsorption of the analyte onto a suitable stationary phase. The choice of sorbent is essential, in that it must have a strong affinity for the target analyte so that preconcentration can occur from either dilute liquid samples or the gas phase. Mani [532] has described the properties of commercial SPME coatings. The sensitivity of the method and the linear range are dependent upon the volume of the stationary phase and the partition coefficient K between fibre material and analyte in question. Several fibre materials with different polarities are commercially available. The most reported phase-coating for SPME is PDMS, which is similar to the medium apolar reversed-phase sorbents in chromatography and stable at high temperatures. The introduction of SPME fibres with varying polarity provides selectivity for different groups of analytes. Most of the fibres are compatible with

HPLC solvents (PDMS-coated fibres are incompatible with hexane). PDMS fibres are more selective towards nonpolar compounds and polyacrylate fibres towards polar compounds such as acids, alcohols, phenols and aldehydes. Another feature of SPME fibre selectivity is discrimination towards high-MW volatiles. SPME has successfully been applied to the analysis of both polar and nonpolar analytes from solid, liquid or gas phases. Li and Weber [533] have addressed the issue of selectivity in SPME.

The main principle of operation of SPME is partitioning of analytes between a liquid or gaseous sample and the polymeric stationary phase according to their partition coefficients K [531]. The theory for the dynamics of the absorption process has been developed [534]. SPME is essentially an *equilibrium process* in which equilibria are established among the concentrations of an analyte in the liquid or solid sample, in the headspace above the sample, and in the phase on the fused silica fibre. SPME is thus not a total extraction technique. The amount of analyte extracted at a given time is dependent upon the mass transfer of an analyte through the liquid phase [535]. The extraction efficiency of the fibre is a combination of extraction time, thickness of the stationary phase, and magnitude of the partition coefficient for the stationary phase. The distribution constant generally increases with increasing molecular weight and boiling point of the analyte. Volatile compounds require a thick phase-coat; a thin coat is most effective for adsorbing/desorbing semi-volatile analytes. Also the sensitivity and the linear range of the method are dependent upon these parameters.

It obviously takes a certain sampling time for an analyte to reach partition equilibrium between the sample matrix and the SPME fibre. The equilibrium time ranges from a few minutes to several hours depending on the sampling conditions and nature of the analyte. In practice, analytes are therefore usually not completely extracted from the matrix. Efficient sample agitation is essential to attain a reasonably fast equilibrium. Without any stirring, sample analytes reach the fibre surface only by diffusion, which is relatively slow. Sonication may help to accelerate partitioning of analytes over the phases according to their partition coefficients. For high accuracy and precision in SPME full equilibration is not even necessary, but consistent sampling time and other sampling parameters are quite essential. The extraction time in the pre-equilibrium period is a critical parameter in the SPME sampling process. All movements of the SPME fibre from adsorption to desorption are precisely timed for optimum precision. Extraction time is determined by the time required to

obtain precise extractions for the analyte with the highest distribution constant.

Two basic methods are used for SPME: direct immersion of the fibre into the sample and headspace sampling. Experimental parameters comprise the polarity of the sample matrix and coating material, solvent and salting-out. Other parameters for optimisation of SPME conditions include desorption time, injector port temperature and initial oven temperature.

Without calibration SPME is only a representative sampling method whereas SFE is truly quantitative. Originally, the SPME technique was considered as a screening method, but reliable *quantitation* of analytes down to the ppt level, particularly in water analysis, is possible [536]. SPME can be very reproducible and quantitative with linear response over several orders of magnitude provided good procedures are followed. Quantification of analytes can be carried out by using external or internal standard calibration (and isotopic dilution), standard addition or application of a known distribution constant. Homogeneous type matrices require only external calibration standards to achieve reproducible quantitation. Internal standards or standard addition work best with more heterogeneous samples.

The ideal way of carrying out a quantitative analysis with a sampling technique is to completely transfer an analyte from the sample matrix to the analytical apparatus (e.g. in dissolution/precipitation methods, cf. Section 3.7). Quantitative transfer is highly improbable for SPME. In exhaustive extraction, selectivity is sacrificed to obtain a quantitative transfer of target analytes into the extracting phase. One advantage of this approach is that, in principle, it does not require calibration, since all the analytes of interest are transferred to the extracting phase. On the other hand, an equilibrium approach in which only a small fraction of each analyte is actually extracted usually requires careful calibration for quantitative work. Besides, equilibrium methods are more selective because they take full advantage of the differences in extracting-phase/matrix distribution constants to separate target analytes from interferents. When the partition equilibrium is reached, the extracted amount of an analyte is proportional to its initial concentration in the sample matrix phase. Ai [537] has shown that quantitative SPME analysis is feasible no matter whether the partition equilibrium is retained or not. Quantitative aspects of SPME were reviewed by Pawliszyn [538]. Calibration parameters in SPME (distribution coefficient K, diffusion coefficient D_s, reaction rate constant k or volume V_s) may be determined theoretically, empirically or experimentally. Provided proper calibration strategies are followed, SPME can thus yield quantitative data, and excellent precision, reproducibility and linearity (over a wide range) have been demonstrated. Various factors, such as sample volume, headspace volume, sampling time, temperature, fibre coating, sample matrix components and others, affect precision in SPME. Quantification of higher MW volatiles by SPME is hindered by the slow transport of analytes into the gaseous phase, which results in long equilibration times and headspace depletion of analyte during sampling. Obviously, for quantitation it is essential that the extraction time be monitored carefully. As with any other extraction/concentration technique, it is best to use multiple internal standards in SPME methods, and to treat the standards and the analytes in an identical manner. SPME can attain detection limits of $15 \, ng \, L^{-1}$ for both volatile and nonvolatile compounds. The technique is rugged, and fibres can usually be used for 100 or more samplings.

The most common approach for SPME is its (online) coupling to GC or GC-MS (cf. Section 7.1.1.3), although its use for HPLC (cf. Section 7.1.1.7) and SFC has also been reported. For analysis the fibre is forced into a chromatograph capillary injector where the analytes are either thermally desorbed for a particular time (usually in $1-2$ min) and separated on the GC column or solvent desorbed (HPLC). The small dimensions of the fibre ($100-300 \, \mu m$) are convenient for on-column injections. Although the amount of analytes recovered by SPME is relatively small compared to several other methods, analyte losses due to sample handling are nil and the entire extraction is desorbed into the injection port of the gas chromatograph. As SPME is a microextraction technique coupling may be envisaged to microseparation techniques (μHPLC, CE).

Desorption of an analyte from the SPME fibre depends on the boiling point of the analyte, the thickness of the coating on the fibre, and the temperature of the injection port. The fibre can immediately be used for a successive analysis. Some modifications of the GC injector or addition of a desorption module are required. It is possible to automate SPME for routine analysis of many compounds by either GC-MS or HPLC. A significant advantage of SPME over LLE is the absence of the solvent peak in SPME chromatograms. SPME eliminates the separate concentration step from the SPE and LLE methods because the analytes diffuse directly into the coating of the SPME device and are concentrated there.

Derivatisation/SPME of polar analytes in the sample matrix is the simplest way to improve an analyte's partition coefficient and enhance SPME performance [539]. With the continued introduction of new SPME fibres, the development of derivatisation methods, and the possible application of μHPLC and capillary LC columns to in-tube SPME, SPME and related techniques hold promise for increasing accuracy and throughput in sample preparation for a wide variety of analytes and matrices [540].

SPME meets various important analytical requirements: analysis at the spot, elimination of solvent use, integration with a sampling step and hyphenation (to GC, HPLC or IR). For spot sampling, the fibre is exposed to a sample matrix until the partitioning equilibrium is reached between sample matrix and coating material. In time-averaging mode, the fibre remains in the needle during exposure of the SPME device to the sample. The method is experimentally simpler than thermal desorption from solid-phase packings. This convenient, solvent-free process facilitates sharp injection bands and rapid separations [541]. These features of SPME result in integration of the first step in the analytical process: sampling, sample preparation, and introduction of the extracted mixture to the analytical instrument.

Table 3.45 lists the *main characteristics* of SPME. The technique is sensitive, reduces analyte loss and can successfully be applied to the analysis of both polar and nonpolar volatile and nonvolatile analytes from solid or liquid and in the gas phase [535]. Room temperature operation of SPME favours thermolabile compounds (only heating during injection into GC). Method

Table 3.45 Characteristics of solid-phase microextraction

Advantages

- Relatively easy to handle
- Speed ($10-60$ min)
- Simple device; automated systems available
- Solventless; no chromatographic solvent interferences
- Not traumatic to analytes
- Economical (more than 50 extractions per fibre on average)
- Sample volume ($1 \, mL$ to $1 \, L$)
- Particularly useful for small aqueous samples
- Theoretical principles available (adsorption on an external phase; total desorption in an injector)
- Screening method
- Very low detection limits (ppt)
- Versatile (adapts to any GC or HPLC system)
- Suitable for on-site analysis and process monitoring
- Rugged, robust

Disadvantages

- Method development required
- New technology (gaining in popularity)
- Fragile fibres
- Quantitatively exacting
- Equilibrium process
- Selected suppliers only
- Not suitable for organic liquids

development is the main disadvantage of SPME. Optimisation parameters include coating material, extraction mode (direct and headspace), extraction time and temperature, ionic strength, pH, agitation conditions and sample volume. Unfortunately, SPME is also not (yet) suitable for organic solutions although Nafion-based systems are under development.

Table 3.46 compares SPME and SPE. Although SPME has in common with SPE that the analytes are concentrated by adsorption into a solid phase, SPE involves absorbing the analyte from the sample onto a modified solid support. In practice, the two techniques are quite different. SPME differs from conventional SPE in that SPE isolates the majority of the analyte from a sample (>90%) but injects only about 1 to 2% of the sample onto the GC. SPME isolates a much smaller quantity of analyte (2–20%), but that entire sample is injected into the GC. SPME is easy-to-perform and often significantly more rapid and simpler than SPE, but its quantitative aspect is exacting. Both conventional SPE and SPME minimise the use of solvents for sample preparation and free analysts from tedious sample clean-up. Where SPE can replace LLE

in many instances, SPME can be used instead of LLE and static headspace. Solid-phase microextraction onto chemically modified fused silica fibres with thermal desorption eliminates the problems associated with SPE while retaining the advantages; solvents are completely eliminated, blanks are greatly reduced, and extraction time can be reduced to a few minutes. The time efficiency, portability, precision, and detection limit as well as low cost of the SPME technique have been significantly improved in comparison to traditional SPE. SPME and membrane SPE performed on a semimicro- or miniaturised scale (SM-SPE and M-SPE) are really complementary techniques. Valor *et al.* [542] compared SPME with headspace analysis.

Solid-phase microextraction eliminates many of the drawbacks of other sample preparation techniques, such as headspace, purge and trap, LLE, SPE, or simultaneous distillation/extraction techniques, including excessive preparation time or extravagant use of high-purity organic solvents. SPME ranks amongst other solvent-free sample preparation methods, notably SBSE (Section 3.5.3) and PT (Section 4.2.2) which essentially operate at room temperature, and DHS (Section 4.2.2),

Table 3.46 Comparison of sorbent technologies

	SPE	SPME
Description of system	Analyte retained on solid adsorbent; extraneous sample material washed from sorbent.	Analyte retained on a sorbent-containing fibre attached to a silica support.
	Desorption of analyte using organic solvent	Thermal desorption
Function/objective	Sample clean-up	Sampling method
	Pre-concentration	Concentration
	Class fractionation	Fast extraction
	Solvent exchange	
Acceptability	Widely accepted	New technology
Sample matrices	Liquids (aqueous or organic), gases	Liquids, solids, gases
Sample size	1 mL–1 L	1 mL–1 L
Sample pretreatment	Occasional matrix clean-up	Occasional matrix clean-up; not for headspace sampling
Analytes	Semivolatiles and slightly volatile compounds	Volatiles and semivolatiles
Recovery of analytes	Total extraction	Equilibrium method
Quantitation	Poor reproducibility	Comparison to spiked blank matrix or standard addition
Hyphenation	On-line GC, HPLC, SPE-SFE, SFE-SPE	GC (most common), HPLC.
		Direct thermal desorption.
		Transfer of all analytes
Extraction time	10–20 min	10–60 min (optimisation required)
Solvent usage	Solvent for wetting and elution of analyte (10–20 mL)	None
Cost	Relatively low	Relatively low
	Disposable cartridges	
Ease of operation	Automated and robotic systems	Automated systems
		Fragility of fibres
Regulatory approval	Several EPA methods	None at present
Main disadvantage	Method development	Method development
		Not suitable for organic solvents
Suppliers	Commonly available	Selected suppliers (Supelco, Varian)

TD and LD which are thermally more traumatic to the analyte. SPME has an advantage over PT techniques by not requiring expensive thermal desorption equipment. In the absence of any solvent, the adsorbed components desorb rapidly; consequently detection limits of SPME are very low (ppb–ppt range) and solvent interferences are excluded. SPME is relatively new and awaits approval by government regulatory agencies.

For recent advances of SPME as an analyte extraction and analytical instrument introduction technique the reader is referred to several reviews [531,543,544,544a] and books [272,545].

Applications SPME may be classified with other *concentration methods* such as CIS, CT, SFE in the gas phase and solvent evaporation, LVI in the liquid phase. Because it is simple, fast, inexpensive, and requires no solvents, SPME is potentially a very useful sample preparation method for analysis of components in solid or liquid samples. SPME is a universal tool for isolation and pre-concentration of pollutants for different matrices. The compact nature of the sampling device and the elimination of solvents allow SPME to be easily adapted for automation and provide an excellent opportunity to perform mobile sampling, such as indoor air monitoring (IAM), VOC concentration detection in chemical storage rooms and field analysis (SPME-portable GC). Other important application areas are in food chemistry (flavour and fragrance analysis) [546], in forensic science, toxicology (biological fluids) and environmental science (pesticides). The choice of the proper SPE device for a given application depends on sample volume, degree of contamination, complexity of the sample matrix, quantities of compounds of interest, type and solvent strength of the sample matrix.

SPME-IR has been applied to VOCs in soil samples [547]. Industrial applications to in-process streams can well be envisaged. SPME has not yet extensively been explored for polymers, but the determination of residual volatiles, semi-volatiles and degradation products in polymers has been reported [548]. It is equally well possible to use SPME for plasticiser analysis in various matrices (water, milk, blood, processed food, etc.).

In-fibre derivatisation/SPME has been reported for the analysis of polar analytes. Derivatisation allows target analytes to be converted to less polar and more volatile species prior to GC analysis. In-fibre derivatisation with diazomethane was applied to long-chain (C_{16}, C_{18}) fatty acids in aqueous solutions. Initially, the polyacrylate fibre was placed in an aqueous sample containing the fatty acids. After sufficient extraction time,

the fibre was withdrawn and inserted into the headspace of another vial containing diazomethane/diethyl ether to carry out the derivatisation. Limits of detection of ng/L were achieved [539].

For further applications the reader is referred to Pawliszyn [549]. Application of SPME in water analysis has been reviewed [536].

3.5.3 Stir Bar Sorptive Extraction

Principles and Characteristics Stir bar sorptive extraction (SBSE) is a novel extraction technique for the analysis of volatile and semi-volatile micropollutants from aqueous samples [550,551]. In this technique, 50–300 µL PDMS coatings are applied to 1- to 5-cm-long bars consisting of magnetic core material, sealed in glass. The amount of PDMS on the stir bar amounts to 10 to 350 mg, thus greatly increasing the sensitivity in comparison to SPME. The amount of PDMS used in SPME is typically in the order of 0.5 µL or less. The extraction mechanism of SBSE is similar to that of SPME but the enrichment factor is about 100–1000 times higher [550]. Thermal desorption of the stir bar allows total transfer to the GC; refocusing is necessary. As an alternative the stir bars can be desorbed by liquid extraction followed by conventional or large volume injection. The same principle can be applied for LC analysis.

Another recently developed technique is *headspace sorptive extraction* (HSSE) with PDMS stir bars [552]. HSSE-GC was compared with SHS and HS-SPME. SBSE and HSSE extract organic analytes from aqueous or vapour samples. In SBSE, the stir bar is inserted into the aqueous sample and extraction takes place during stirring whereas in HSSE the glass rod is suspended within the headspace volume and sampling takes place during headspace equilibration. New trends are the development of selective sorbents.

Applications PDMS coated stir bars (Twister™, Gerstel) are profitably used in SBSE-TDS-CGC-MS (or AED, PFPD, ICP-MS) configurations for product control analysis or otherwise as a powerful tool for the extraction and analysis of organic compounds in aqueous matrices (drinking or waste water, body fluids, beverages, dairy products, and processed foods). For example, using SBSE-TDS-CGC-PFPD a complete range of technical nonylphenols in 10–50 ppb range was observed in red wine bottled with a plastic stopper. SBSE has also been used for the determination of ppq-level traces of organotin compounds in environmental samples with TD-CGC-ICP-MS [553].

3.6 METHODOLOGICAL COMPARISON OF EXTRACTION METHODS

The main characteristics of the ideal extraction method are given in Table 3.47, which at the same time are also criteria for comparison of sample preparation techniques. It is unlikely that a unique best method can be defined, which is analyte and matrix independent. Extraction is affected by polymer functionality, molecular weight and cross-linking. Selective extraction of some additives is basically not possible. Hence, the goal of an ideal extraction would be the complete extraction of all additives from the polymer for subsequent chromatographic separation.

Experimental comparisons may suffer from a lack of optimal conditions for all methods considered or may be based on biased evaluation. It is frequently noticed that results quoted by the preferred extraction technique compare extremely favourably with existing extraction technology. Also, lack of prospects of using CRMs is not helpful for comparisons. However, it appears that for a given infrastructure (R&D vs. plant laboratory) and need (routine vs. occasional operations), and depending on the mix of polymeric matrices to be handled, some preferences may clearly be expressed.

If the focus is on a wide variety of additives in polyolefins only the OSM/TSM procedure of Marcato and Vianello [210] appears to be far reaching, despite the fact that this *total design approach* must have required considerable effort in terms of method development. On the other hand, in comparison considerably more method development and optimisation appears to be necessary for SFE of a single polymer/additive matrix, but even then still without the demonstrated benefit of a total approach [322]. Where analyte integrity is of greater concern recourse should be taken to less

Table 3.47 Requirements for the ideal extraction method

- Rapid in routine operation (low overall and analyst time)
- Simple (no need for frequent method development/optimisation)
- Selective
- Low cost (investment, consumables)
- Broad analyte coverage and integrity (no analyte degradation, transformation)
- Low detection limits
- Total recoverage (quantitation; no target analyte loss)
- Practical linear dynamic range
- Yields a sample ready for direct further analysis (no concentration or clean-up steps)
- Generates no additional laboratory wastes
- Portable

traumatic methods and/or conditions (e.g. cold extractions, US, OSM, SFE). Ultrasonication is considered as a *soft approach*. The prospects of selective extraction are very limited (cf. Section 3.6.2); at present SFE is the technique which lays claim to most selective features. Table 3.4 compares the main characteristics of extraction and dissolution methods, as applicable to polymer/additive systems. Traditional extraction methods at atmospheric pressure, such as Soxhlet extraction, boiling under reflux, shake-flask, sonication and dissolution/precipitation are well documented and use simple inexpensive equipment, which requires little attention, but give long extraction times. Moreover, they can usually not be automated and, in the case of dissolution methods, often require subsequent clean-up steps. Therefore, these methods are probably most suitable for laboratories with only occasional requirements for polymer analysis, with boiling under reflux being considered as the best 'all round' performer [4]. The newer methods of extraction, such as SFE, MAE and PFE, all require an outlay of significant funds, all employ elevated temperature and almost invariably also high pressures (cf. Table 3.3). There appears to be no single method which is preferable in all cases.

Soxhlet extraction is well established, and generally exhaustively extracts all additives. The selection of extraction solvent can make large differences to the extraction time. The generally long extraction times followed by concentration steps may determine losses of volatile or thermally labile components. Because this form of extraction is one of the oldest and still widely used in industry, it is the standard to which many of the newer extraction technologies (which are likely to determine future applications) are referred. However, it should be realised that extraction mechanisms may be different, and thus comparisons are sometimes irrelevant.

It is not uncommon that extraction techniques are unfairly compared. Appropriate interlaboratory studies are few. Soxhlet and *sonication extraction* (EPA methods 3540 and 3550, respectively) were compared in an interlaboratory study (129 participants) for PCBs in soil. Results from laboratories using Soxhlet extraction were significantly more accurate than those obtained using sonication, especially at higher concentrations, but with equal precision [196]. This is rationalised by the observation that the Soxhlet procedure presents the sample with fresh solvent so that the extraction solvent is never saturated, unlike the sonication procedure. Sonication is very sensitive to the solvent polarity, nonpolar solvents producing considerably less accurate results than polar solvents. It is not as sensitive to clean-up procedures as it does not extract as much interfering material as the Soxhlet procedure. However, although Soxhlet extraction is far less sensitive to solvent polarity, it is still affected by the polarity of the extraction solvent.

The use of high pressure offers the fastest, automated extraction methods. SFE, MAE and PFE all significantly improve on extraction rates achieved with conventional methods but require the purchase of special equipment. Again, choosing between these techniques is not a simple matter. The greatest amount of published evidence concerns the efficacy of *supercritical fluid extraction*. In the early 1990s SFE was advertised as the extraction panacea. In fact, the use of SFE in many analytical applications has become of such great interest that it has become a powerful alternative to conventional extraction methods. The advantages of SFE as a sample preparation technique result from the physical properties of SCFs [554]. SFE is much faster than Soxhlet extraction (an order of magnitude, according to Dean [251]), achieving mostly rapid and quantitative extraction provided ground samples, thin films or foams are used, Nevertheless, SFE of polyolefins has little added value as compared to chloroform Soxhlet extraction. The response times of SFE are smaller but this is often not a crucial issue in a research environment. The extraction efficiencies are comparable with chloroform Soxhlet extraction.

SFE might appear as a more interesting technique for the extraction of additives from engineering plastics, due to the complicated and hazardous techniques used for extracting/dissolving of engineering plastics. However, one should consider that in SFE the polymer is not actually dissolved and the extraction is a function of the ability of the SCF to permeate the sample. Then, most difficulty is likely to be experienced with extraction of high-MW and/or polar compounds. In SFE this poses the need for a modifier to enhance the extraction rate. For extractions that still take an excessively long time, an extrapolation procedure can be used to determine the analyte concentration in the plastic without total extraction. Equipment exists which can extract several samples simultaneously and which can be programmed to extract up to 28 samples consecutively.

SFE has now been available long enough to allow an evaluation of its prospects for polymer/additive extraction. SFE is still around, but EPA and FDA approved SFE methods are still wanting. The main problem is strong matrix effects. SFE is not a cookbook method for one's matrix. Not unlike microwave extraction, SFE requires that a specific method be developed to optimise the recovery for each polymer/additive system. Therefore, the success of SFE depends on the polymer

matrix and its additives. It has turned out that optimisation of the extractions is not simple. Many factors have to be considered, in particular choice of temperature, pressure and modifier; method optimisation can be quite time-consuming. Sample sizes tend to be much smaller than in conventional methods, which raises the possibility of nonrepresentative samples being used. An additional drawback of using SCFs in extraction is the requirement for very high-purity extractants. On the other hand, SFE is an efficient stand-alone sample preparation/fractionation method generating extracts that are totally free of residual solvent. It can be used to concentrate analytes from a bulk matrix. This is often important when using capillary SFC because of the limited dynamic range and volume capacity of the technique. Lynch [89] and David [376] have recently reviewed the use of supercritical extraction for sample preparation for chromatographic analysis. As SFE can easily be interfaced to SFC many SFE-SFC applications have been described in the field of additive analysis. However, optimisation of on-line SFE-SFC is very troublesome.

Accelerated solvent extraction, which is quite effective in extraction from environmental samples, appears to have great potential for extraction from polymers. The main experimental problem is the selection of the extraction solvent, which should not (partially) dissolve the polymer at high temperature. To this end a *rational approach* is available based on Hildebrand solubility parameters (cf. Section 3.4.6). Besides, temperature is the most important parameter to be optimised. In ASE®, extraction temperatures well above the normal boiling point of the solvent are used, as opposed to Soxhlet extraction. ASE® thus offers great flexibility of solvents and solvent mixtures. Experimentally, up to 24 consecutive analyses can be programmed independently and some initial experiments are required to optimise the method to give rapid extractions without dissolving the polymer. Due to the flexibility of ASE® in solvent selection, some additional selectivity can be introduced by using different solvents. ASE® gives recoveries comparable to those obtained with Soxhlet extraction at a much reduced solvent consumption. ASE® would appear to be ideal for a laboratory which analyses a large number of different polymers [4]. More development is required in this field; reported applications are still few. Although SFE and ASE® differ in many respects, the extraction mechanisms are quite similar. Important differences between the two techniques are the high p, T for ASE® as compared to high p and low T for SFE. Higher p, T lead to greater solubility of the analyte in the extraction medium, lower viscosity of the extraction medium and higher diffusion of the extraction medium in the polymer matrix. ASE® allows a wide choice in solvents, where in practice SFE is greatly confined to the use of scCO$_2$. In the extraction of polymeric samples, ASE® is more effective than SFE with pure CO$_2$, particularly in the extraction of components having a poor solubility in scCO$_2$ and/or when high extraction temperatures are necessary to increase diffusion of the analytes in the polymer particles. It appears that ASE® is becoming much more popular than SFE.

In MAEs the heat source is microwave energy instead of conventional heat. The two extreme approaches adopted consist in heating of the bulk solvent (MAE) and selective heating of the sample matrix (MAP™). In *microwave-assisted extraction* (MAE), which is very similar to PFE, the solvents used must contain a component which absorbs microwaves strongly. This makes the selection of solvents more problematical than for ASE®, as for example pure hydrocarbons cannot be used. Extraction times are similar to ASE® but, as 14 samples can be extracted simultaneously, the overall extraction time per sample can be much less for replicate extractions. Therefore, MAE may be the preferred method for a laboratory analysing large numbers of similar samples. According to Raynie and Innis [555] SFE, ASE® and MAE are best viewed as complementary, rather than competitive techniques. Each method relies on the controlled manipulation of temperature and pressure to achieve the desired results. ASE® and SFE are similar with the exception that, in SFE, temperatures near or greater than the critical temperature are used so that the unique properties of SCFs can be exploited for extraction. ASE® and MAE are similar with the exception that ASE® uses an applied pressure and is instrumentally configured as a static/dynamic process, while pressures in MAE are the result of heating in a closed, static system. Each technique, therefore, offers some advantages and disadvantages, the final choice is dependent on the requirements of the user and the particular application. With careful thought, as one gains experience, each of the three methods can be used in similar applications.

Desrosiers [23] has dared ranking extracting methods as follows (in order of preference): SFE, US, hot block or MAE, Soxhlet (to be phased out as quickly as possible). Munteanu [556] has evaluated extraction techniques for additives from polymers prior to chromatographic analysis (up to 1990). The analytical extraction of additives from polymers has recently critically been reviewed with emphasis on SFE, MAE and ASE® [92]. Dean [272] compared modern extraction techniques, with focus on environmental analysis.

Validation of extraction procedures is frequently lacking. A good assessment of quality assurance implies that the extraction recoveries are verified, e.g. by spiking of standard addition. A major drawback is that the spike is not always bound the same way as the compounds of interest. For the development of good extraction methods, materials with an incurred analyte (i.e. bound to the matrix in the same way as the unknown), which is preferably labelled (radioactive labelling would allow verification of the recovery), would be necessary. Such materials not being available, the extraction method used should be validated by other independent methods, e.g. by verification against 'known' samples and by use of a recovery SPC chart. A mere comparison of extraction methods is no validation.

3.6.1 Experimental Comparisons

Table 3.48 shows some formulations for which extractions have been compared. Soxhlet extraction is frequently taken as the reference to measure progress.

Here we report only a small selection of experimental results. There are few systematic studies of experimental results of new and conventional techniques for extraction from polymers. Vandenburg *et al.* [457] have recently compared PFE and MAE with atmospheric pressure methods for extraction of PP/(Irganox 1010, Irganox 1330).

Amongst the traditional atmospheric pressure extraction methods the chloroform reflux method is the fastest, with 'complete' extraction after about 40 min (Table 3.49). The times taken to reach 90 % extraction for PFE using propan-2-ol at 150 °C and acetone at 140 °C were 5 and 6 min, respectively. For effectively complete extraction, from loaded extraction vessel to extract ready for analysis, PFE required 15 min, MAE 28 min and reflux with chloroform 45 min. Microwave extraction offers faster sample analysis for large numbers of identical analyses, largely because multiple samples can be extracted simultaneously and the warm up time is shorter. There is no risk of blockages in the equipment. For PFE (i.e. ASE®/ESE®), analysis for a single sample is faster than for MAE, as the sample can be analysed without waiting for it to cool down. There is no restriction on solvents which can be used (as in case of MAE), but they must be carefully selected so as not to dissolve the polymer and thus cause blockages. Propan-2-ol was found to be the best single solvent for PFEs for PP.

Vandenburg *et al.* [92] have also recently published accounts of laboratory methods for the extraction of

Table 3.48 Comparison of extraction performance of polymer/additive systems

Matrix	Extraction methods[a]									Reference
	S	S®	R	SF	US	SFE	MAE	PFE	D/P	
PP/(Irganox 1010, Irganox 1330)	+		+	+	+		+	+		[457]
PP/(Irgafos 168)			+		+		+		+	[210]
HDPE/(BHT, Irganox 1010, Irganox 1076)	+				+		+			[90]
PP/(Irganox 3114, Cyasorb UV531, AM 340)					+		+			[90]
PP/(Irganox 1010, Irgafos 168)					+		+			[90]
PP/(BHT, Irganox 1010)					+		+			[90]
LDPE/(BHT, BHEB, Isonox 129, Irganox 1010, Irganox 1076)					+		+			[90]
LDPE/erucamide					+		+			[90]
HDPE/(Tinuvin 770, Chimassorb 944)		+	+		+				+	[114]
LDPE/Slip agent			+				+			[115]
LDPE/Antioxidant			+				+			[115]
PVC/DOP						+				[384]
MDPE/(Irganox 1010, Irgafos 168)					+		+			[202]
PA6/ε-caprolactam	+					+				[145]
PBT/dimer (trimer)						+				[145]
LDPE/antioxidants	+					+		+		[492]
PS/(AOs, UVAs, FRs, antistats, lubricants)	+					+			+	[557]

[a] S, Soxhlet; S®, Soxtec; R, reflux; SF, shake-flask; US, ultrasonics; SFE, supercritical fluid extraction; MAE, microwave-assisted extraction; PFE, pressurised fluid extraction (ASE®, ESE®); D/P, dissolution/precipitation.

Table 3.49 Approximate time for 90 % extraction from freeze-group PP/(Irganox 1010, Irganox 1330) (atmospheric pressure extraction)

Extraction technique	Time to 90 % extraction (min)
Reflux (chloroform)	24
Reflux (cyclohexane–propan-2-ol, 1:1)	38
Ultrasonic (chloroform)	78
Shake-flask (chloroform)	86
Soxhlet (chloroform)	84

After Vandenburg et al. [457]. From H.J. Vandenburg et al., *Analyst*, **124**, 397–400 (1999). Reproduced by permission of The Royal Society of Chemistry.

additive and low-MW organics from over 75 polymer systems. SFE and Soxhlet extraction have been compared in PA6/ε-caprolactam extraction [145], in PBT/oligomer extraction [145], in PET/cyclic trimer extraction [287], in PVC/phthalates extraction [336] and in PE/additives extractions [291,373]. On the other hand, SFE and LSE have been used for the same polymer/low-MW organic systems, namely PET/cyclic trimer [392], PVC/(DIOP, plasticiser) [333] and PP/(BHT, Tinuvin 326, Seenox DM, Irganox 1010) [334]. Marín et al. [384] have observed an increase in extraction efficiency for phthalate plasticisers in paints, coatings and PVC products used in toys when SFE is used instead of the traditional Soxhlet method. Affolter and Hofstetter [558] have compared Soxhlet and SFE extraction of natural rubber products. Figure 3.21 shows the correspondence between the total

extractables (extender oil, organic acids and antioxidants) from SBR, as determined by means of SFE and the ASTM D 1416 boiling extraction method [395]. Werthmann et al. [396] have compared SFE of seven vulcanised rubbers with different polarity with conventional Soxhlet extraction according to DIN ISO 1407 [559]. The results are similar but SFE offers more advantages.

Wieboldt et al. [560] have described SFE-SFC-FTIR analysis of hindered phenol primary antioxidants and phosphite secondary antioxidants in PE. SFE is more selective for the lower-range low-MW polymer than Soxhlet-type extraction. This yields a chromatogram with less interference from low-MW polymer peaks in the region where the additive components elute. As a result, SFE appears to be a better choice than Soxhlet-type extraction for the selective removal of additives from flaked polymer. SFE and dissolution/precipitation methods were compared for a PVC/stabiliser system [366].

Kirschner et al. [358] have observed a lower percent finish-on-yarn (FOY) for SFE as compared to solvent extraction of various fibre/textile matrices. This is rationalised as organic solvents tend to extract components from a matrix more vigorously than scCO2 and thus remove more of the oligomer and organic components present in the fibre. SFE is a potentially 'softer' extraction technique since it removes less of the polymer from the fibre matrix than liquid solvent extraction.

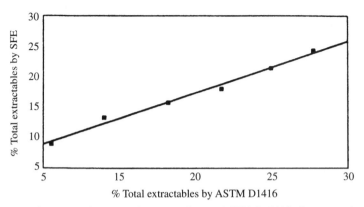

Figure 3.21 Comparison of SFE and reflux extraction (according to ASTM D 1416) for styrene-butadiene rubbers. After Sekinger *et al.* [395]. Reprinted with permission from *Rubber Chemistry and Technology*. Copyright © (1996), Rubber Division, American Chemical Society, Inc.

Caceres *et al.* [114] have compared several methods for extraction of Tinuvin 770 and Chimassorb 944 from unground HDPE pellets. Room temperature diffusion into chloroform and ultrasonication gave less than 20 % extraction. Soxtec® extraction with DCM for 4 h resulted in 50 % extraction. Dissolution of the polymer in DCB at 160 °C for 1 h followed by reprecipitation of the polymer with propan-2-ol gave 65–70 % recovery. The most successful method was boiling under reflux with toluene at 160 °C for 2–4 h, which extracted 95 % of both additives. The relatively poor performance of the Soxtec® extraction compared with the reflux extraction is probably due to the large difference in temperature between the boiling solvents. Nielson [90] performed both microwave extraction with ultrasonication for HDPE, PP and LDPE. BHT, Irganox 1010 and Irganox 1076 could be extracted at >90 % recoveries from ground HDPE in 20 min.

Freitag and John [96] studied rapid separation of stabilisers from plastics. Fairly quantitative extraction (>90 % of the expected content) of stabilisers from a powdered polymer was achieved by MAE within 3 to 6 min, as compared to 16 h of Soxhlet extraction for the same recovery. MAE and Soxhlet extraction have also been compared in the analysis of cyclic trimer in PET [113]. On the other hand, Ganzler *et al.* [128] compared the extraction yields for various types of compounds from nonpolymeric matrices for microwave irradiation with those obtained by the traditional Soxhlet or shake-flask extraction methods. Microwave extraction was more effective than the conventional methods, in particular in the case of polar compounds. As expected, the efficiency of the former is high especially when the extraction solvents contain water. With the high dipole moment of water, microwave heating is more

effective for desorption and extraction processes of polar compounds (up to about 100 times).

Hinman *et al.* [492] have compared SFE and ASE® in the extraction of antioxidants from LDPE. Comparable extraction yields were obtained with both techniques. However, sample clean-up was necessary after ASE®, while with SFE the extract could be analysed directly without any post-extraction clean-up. Supercritical fluid extraction of 15 polymer additives (AOs, UVAs, process lubricants, flame retardants and antistatic agents) from eight PS formulations was compared to dissolution/precipitation extractions [557]. Additive recoveries were comparable. Numerous additional comparisons can be found under the specific headings of the extraction techniques (Sections 3.3 and 3.4).

3.6.2 Extraction Selectivity

Principles and Characteristics A sample can contain a great number of compounds, but analysts are usually interested only in the qualitative presence (and the quantitative amount) of a small number of the total compounds. Selectivity is an important parameter in analytical separations. The total analytical process clearly benefits from selectivity enhancement arising from appropriate sample preparation strategies. Selective separation of groups or compound classes can simplify a mixture of analytes before analysis, which in turn enhances analytical precision and sensitivity. Selective fractionation, in some cases, allows easier resolution of the compounds of interest, so analysts can avoid the extreme conditions of high-resolution columns.

Selectivity in extraction can be defined as the ability to extract only the analyte(s) of interest, while leaving other compounds behind. Co-extractives are additional components of a sample, which are extracted along with those of interest, and which may interfere with the analysis. In polymer/additive analysis this is typically the case of rest monomers, solvents and oligomers contained in the polymer, which are co-extracted together with the additives. The presence of these co-extractable materials may necessitate a post-extraction clean-up step to remove compounds, which interfere with chromatographic analysis. Complete selectivity, extraction of solely target compounds, is normally difficult to accomplish due to the closely related solubility characteristics of compounds in a sample matrix. In some cases, the polarity and solubility of the target compounds may be so closely related to those of the co-extractives as to make selective isolation at the extraction step improbable. In many cases, selective fractionation simply removes groups of compounds that most analysts would consider impurities or contaminants in an analytical column or separation.

Several techniques allow a certain level of *target analyte selectivity* as a function of temperature, such as thermal desorption, headspace sampling and SPME. In all of these techniques, the extraction mechanism is governed primarily by analyte volatility. In SPE class fractionation is based on the nature of the solid phase material. Microwave extractions, accelerated solvent extraction and SFE have recently been used or modified to achieve some degree of selective class or target analyte extractions. SFE is the only sample preparation technique that incorporates thermal- and solubility-driven mechanisms. Compared with other LSE techniques, such as Soxhlet and ASE®, that are used for universal sample treatment, SFE qualifies as being most appropriate for selective extraction of analytes and analyte classes. SFE using 100 % CO_2 has been reported to be more selective for the lower-range low-MW polyethylene than Soxhlet-type extraction [560]. The unique control of volatility–solubility relationships in SFE provides the basis for enhancing the extraction selectivity. By judiciously adjusting the tuneable solubilising power of the SCF in SFE, analysts can selectively reduce the number of compounds present [343]. Selectivity enhancement in SFE involves the adjustment of pressure, temperature, fluid composition, fluid flow and time to isolate specific target compounds from complex sample matrices. The physical SFE parameters provide a higher level of selectivity control than for other LSE techniques [317]. Nevertheless, as argued in Section 3.4.2.3 the claim of selectivity of SFE is largely overdone. While obtaining a pure analyte from complex matrices using SFE or SWE is unlikely, sequential extractions of different compound

classes have been demonstrated [132,418]. Membrane techniques for chromatography (SML, MMLLE, MESI) provide selective enrichment. As was pointed out by Holden [77], mild extraction techniques are likely to be the most selective. Soft extractions are 'cold' extractions at room temperature, sonication or standard Soxhlet extraction. However, these techniques are restricted in their range of application and are likely to show matrix-dependent extraction yields.

Selective extractions are not only of interest to solvent extraction, but also to thermal extractions. For example, selective *in situ* detection of polymer additives is possible using laser mass spectrometry, notably UV laser desorption/MS [561]. The proper matching of extraction technique to a sample determines the success of the operation and enhances the precision and accuracy of the analysis.

Applications Work by Salafranca *et al.* [562], who used ultrasonication for extraction of recycled HDPE, shows the power of the technique (170 compounds detected) but at the same time illustrates the poor selectivity of this extraction method. Few selective extractions have been reported for in-polymer additive analysis. Gere and Derrico [351] have described selective fractionation by means of SFE illustrating the adjustable nature of the SFE process through various combinations of temperature, pressure and time periods. Venema and Jelink [386] have described selective $scCO_2$ extraction from PA6.6 of the various fractions of the external lubricant calcium montanate, which is a technical product composed of C_{28}–C_{32} fatty acids, fatty acid esters and calcium salts, with contaminations of aliphatic hydrocarbons. The aliphatic hydrocarbons are extractable by means of $scCO_2$, the free fatty acids by $scCO_2$/MeOH and the calcium salts by $scCO_2$/MeOH/citric acid, whereby the complexing agent citric acid binds the cation and sets the acids free.

Antioxidants, biocides, anticorrosion, antiwear and antifoaming agents in low molecular mass liquid poly(alkylene glycols) (PAGs) could selectively be extracted by means of SFE by immobilising PAG on a silica adsorbent, despite the fact that both the additives and free PAG lubricants tend to be soluble in $scCO_2$ [370]. Clifford *et al.* [410] have reported fractionation of polyisobutene and PDMS using SFE; the oligomers detected can be used to calculate the number-averaged molar mass and polydispersity. Fractionation of polymers using SFE has recently been reviewed [257].

Analysis by ^{13}C NMR has shown that the 7 and 10 MPa $scCO_2$ extracts of sorbitan ester (a complex mixture of sugars) contain mainly cyclic monoesters, whereas the 15 and 20 MPa scCO$_2$ extracts contain cyclic di- and triesters with some fatty acid, and the 10 % MeOH–scCO$_2$ extracts are composed of linear esters [370]. This particular example shows the value of SFE as a sample preparation technique for the spectroscopic characterisation of technical mixtures. Very sharp class fractionation with SFE is applied industrially in the flavour and fragrance industries.

With suitable adjustments to the temperature, also subcritical water extraction (SWE) or pressurised hot water extraction (PHWE) allows selective extraction of polar (chlorinated phenols), low-polarity (PCBs and PAHs) and nonpolar (alkanes) organic compounds from industrial soils [418].

3.6.3 'Nonextractable' Additive Analysis

Principles and Characteristics Modern technology is making ever-increasing demands on the durability of plastics, rubbers and coatings. Highly demanding operating conditions together with a critical attitude of the consumer have contributed to an increasing demand for higher *persistence* of polymer additives. Retention of additive functionality in polymers is therefore gaining importance. Oligomers and polymers prepared from reactive functionalised low-MW compounds and having properties of various additive classes (such as AO, MD, LS, FR and BIO-S) have been developed. The high cost/performance ratio allows the exploitation of such systems only in very demanding niche applications for specialty polymers. Low physical persistence therefore limits the exploitation of the inherent chemical efficiency of additives in polymers used under demanding conditions in domestic, engineering or medical applications. Synthetic polymeric materials are increasingly being used in environments which are hostile to their long-term durability. High temperatures lead to the rapid loss of many antioxidants and stabilisers from polymers, but equally important is the leaching action of aggressive media such as lubricating oils in engine components, and dry cleaning solvents or detergents in fibres. Other examples of conditions which increase the *physical losses* of additives include contact of rubber articles with hot oils, contact with gasolines containing polar additives (e.g. methyl *tert*-butyl ether) and hydraulic fluids, contact of polyolefins with streaming hot water, engineering applications of polymers in hot air atmospheres or under space conditions. For example, in some demanding applications (fibre production, extractive environments,

etc.) low-MW HALS stabilisers fail to meet expectations. More 'persistent' high-MW and polymeric stabilisers were introduced to the market in the beginning of the 1980s as the solution for demanding applications. Polymer stabilisers having $\langle M_n \rangle$ 800 Da, which have sufficient resistance against extraction by components of (processed) foods and do not enter into the metabolic cycle, may usually overcome food extractability problems. Nonextractability is equally important for medical devices. Cattell [563] has described polymeric materials in chemically resistant applications. *Extraction resistance* or nonextractability is important for a wide variety of outdoor applications, such as stadium seats, car bumpers, outdoor furniture and containers [564], gas and hot water pipes [565], tanks, telecommunication wire and cables, agricultural films, geotextiles and geomembranes.

Various additives show considerable extraction resistance, such as impact modifiers (polyacrylates and polyblends: PVC/EVA, PVC/ABS, etc.), high-polymeric processing aids (PMMA-based), elastomers as high-MW plasticisers, reactive flame retardants (e.g. tetrabromobisphenol-A, tetrabromophthalic anhydride, tetrabromophthalate diol, dibromostyrene). Direct measurement of additives by UV and IR spectroscopy of moulded films is particularly useful in analysing for additives that are difficult to extract, although in such cases the calibration of standards may present a problem and interferences from other additives are possible.

Polarity of organic solvents and their ability to swell the polymer matrix increase losses of additives due to leaching. In this respect, water is generally considered to be less serious than organic solvents. Nevertheless, serious damage may be caused during long-term contact with water. For example, extraction of aromatic amines from tyres [566] or of phenolic antioxidants from PE [567,568] has been reported. The ease with which water leaches various phenylenediamine derivatives from NR and SBR changes inversely with the MW of the phenylenediamine and with the leachant acidity [566]. Intensive extraction of stabilisers can take place during laundering of textiles with aqueous solutions of detergents. Following up the extended application of hindered amine light stabilisers (HALS) in stabilisation of foils and fibres, a great deal of attention has also been paid to HALS extractability by water and aqueous solutions of detergents or solvents used in dry cleaning [569].

Determination of extractability of stabilisers is a substantial part of the rating of stabiliser efficiency where leaching is a common deterioration effect or where legislation requires quantitative data dealing with extractability of stabilisers from packaging materials by food simulants. In this respect, it can be stated that if an ingredient of a plastic material cannot be extracted by foodstuff with which it is in contact, it does not constitute a toxic hazard.

High compatibility of the additive with the polymer matrix usually extends performance but at the expense of extractability. Lack of extractability is of analytical concern, e.g. in case of functionalised oligomer and polymeric additives, grafted and reactive additive functionalities, and chemisorbed and absorbed additives. More generally, additive recovery levels significantly lower than the target level may arise from various situations:

(i) Cross-linked polymer, inhibiting extraction of additives from the polymer (remedy: more aggressive extraction mode).
(ii) Binding of additive to polymer, e.g. reaction of the thioether group of Irganox 1520 with rubbers or styrenics, or of HALS in weathered samples (occasional remedy: acid/base titration).
(iii) Grafted additive functionalities (remedy: *in situ* additive analysis).
(iv) Adsorption of the additive by CB in a polymer, e.g. in masterbatch samples (remedy: displacement of the additive by adding an amine).
(v) High compatibility with the polymer; comparable solubility of the additive in the polymer and the extraction solvent (remedy: different solvent).
(vi) Occlusion of the additive by precipitating polymer (remedy: addition of an internal standard co-additive).
(vii) Transformation of the additive during analysis (remedy: milder extraction conditions: lower temperature, inert atmosphere, less aggressive solvents).
(viii) Transformation of the additive during processing (remedy: LC-MS analysis of all transformation products).
(ix) Oligomers (remedy: dissolution followed by SEC).
(x) Insolubility (remedy: *in situ* additive analysis).

Thermosetting polymers such as vulcanised rubber are more difficult to analyse than thermoplastic polymers due to the compounding ingredients being locked into the matrix by CB and cross-linking of the rubber polymer. For example, the antidegradant (or its reaction products) in NR/IPPD becomes nonextractable after several months of storage at room temperature [570]. A further problem inherent in most thermosetting rubber mixtures arises from many of the additives being already

fragmented due to the curing and cross-linking process. Stearic acid in rubbers can only be determined directly and not in extracts, as it is not totally extractable [571]. A gradual diminution in the extractable levels of Irganox 1076/1010 in food contact polymers was observed as γ-irradiation progresses [572].

Dealing with incomplete extraction is particularly challenging when analysing polymers containing unknown additive levels. A common strategy is to perform multiple extractions. When incomplete extraction is suspected, it is also useful to apply an alternative analysis technique, such as spectroscopy or elemental analysis. It is good practice to compare the IR spectrum of the polymer before and after extraction to verify the presence of absorbance bands related to the additive.

It is of course critical to discount that the additive undergoes a thermal or chemical conversion during the extraction step. To verify interaction between an additive and other compounds in the polymer, a sample of the polymer during the reflux step may be spiked with a known amount of the same additive.

When the analyte is present in the polymer at very low concentrations some special precautions are needed to enhance the sensitivity of the extraction process, i.e. to lower the detection limit. The sample may be concentrated prior to analysis by SCF or solvent evaporation (at as low a temperature as possible to avoid degradation or partial loss of volatile analytes). Alternatively, a larger amount of polymer sample may be extracted (followed by LVI). Samples may also be concentrated or matrix effects minimised by using SPE [573,574].

3.6.3.1 Case Studies of Extraction Resistance

Nonbound and polymer-bound hindered phenolic antioxidants

The driving force for replacement of BHT (MW 220) for hindered phenolics such as Irganox 1010 and/or Anox 20 (MW 1178) as well as of the low-MW hindered amine Lowilite 77 (MW 481) for polymeric HALS such as Uvasil 299 (MW 1800) has been the greater permanence in the polymer matrix.

In more recent developments siloxane additives (Figure 3.22) exhibit good resistance to extraction, as shown in Figure 3.1 for Soxhlet extraction of HDPE plaques. While nonpolymeric antioxidants (Anox 20 and Anox PP18) were completely extracted after 10 h, at the same time more than 40 % of the initial concentration

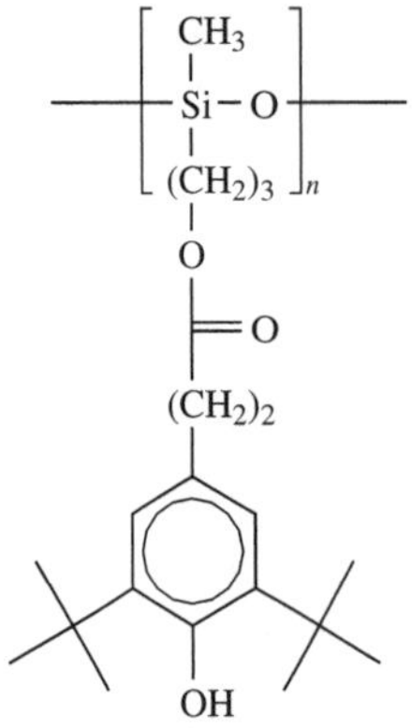

Figure 3.22 Polysiloxane-based antioxidant (Silanox). After Gray and Lee [144]. Reproduced by permission of Blackwell Science

MW siloxane oligomers, while the highest remain almost unextracted even at 40 h of extraction. The low extractability of siloxane-based additives has further been enhanced by the inclusion of graftable pendant groups on the polysiloxane backbone [575]. Examples of graftable moieties are shown in Figure 3.23. Figure 3.24 compares extraction results of graftable and nongraftable siloxane-based phenolic antioxidants with monomeric traditional antioxidants. The siloxane-based products outperform both Anox 20 and Anox PP18. A marked improvement was achieved when the siloxane graft ($n/m = 5$) was evaluated after grafting the material to LDPE.

It has been observed that complete immobilisation of the stabiliser through a graft leads to deactivation. However, proper selection of the ratio of phenolic to graftable groups leads to a polymer-bound product which retains sufficient mobility to provide a high level of antioxidant activity. An n/m ratio of 5–10 provides an optimal balance of graftability and antioxidant activity [144].

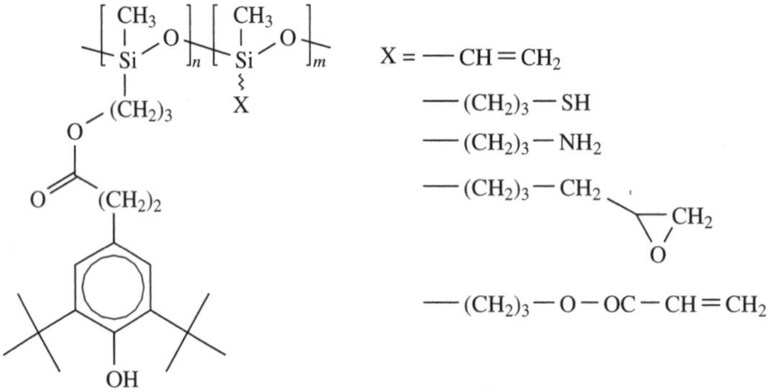

of Silanox ($n = 7–10$) is still present in the plaques. Figure 3.1 demonstrates early extraction of the lower

Figure 3.23 Graftable polysiloxane structures. After Gray and Lee [144]. Reproduced by permission of Blackwell Science

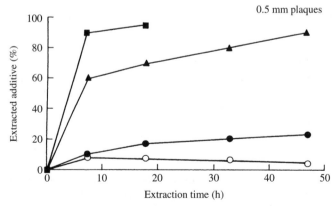

Figure 3.24 Extraction resistance for graftable Silanox. ▲, Anox 20; ■, Anox PP18; ●, Silanox (as shown in Figure 3.22); ○, Silanox graft $n/m = 5$. After Gray and Lee [144]. Reproduced by permission of Blackwell Science

The high flexibility of the polysiloxane chemistry allows synthesis of copolymers containing different functional groups as compatibilisers, phenolics, HALS, phosphites, etc., thus extending the low extractability to the full stabilising package. Although polyolefin stabilisation by grafting represents an important development the polymer industry still prefers to achieve stabilisation by the traditional means of melt blending with additives as specialty chemicals.

Polymer-bound additive functionalities

As already shown, it is technically possible to incorporate additive functional groups within the structure of a polymer itself, thus dispensing with easily extractable small-molecular additives. However, the various attempts of incorporation of additive functionalities into the polymer chain, by copolymerisation or free radical initiated grafting, have not yet led to widespread practical use, mainly for economical reasons. Many macromolecular stabiliser-functionalised systems and reactive stabiliser-functionalised monomers have been described (cf. ref. [576]). Examples are bound-in chromophores, e.g. the benzotriazole moiety incorporated into polymers [577,578], but also copolymerisation with special monomers containing an inhibitor structural unit, leading to the incorporation of the antioxidant into the polymer chain. Copolymers of styrene and benzophenone-type UV stabilisers have been described [579]. Chemical combination of an antioxidant with the polymer leads to a high degree of resistance to (oil) extraction.

A somewhat different approach consists in *chemical grafting* onto conventional polymers e.g. in the case of DBBA [580]. Grafting is an important approach to the synthesis of polymeric additives, in particular stabilisers, and implies reactions of conventional polymers with functionalised molecules. Grafted polyolefins are of interest due to a relatively good compatibility with the host polymer. Clariant Huningue SA used patented photo-reactive chemistry in the development of Sanduvor PR-31. The technology grafts HALS molecules to the matrix polymer, forming an outer layer which absorbs UV and improves the long-term weathering resistance [581]. Volatility, extraction and migration are all reduced. The stabiliser is especially suitable for pigmented and unpigmented polyolefins and also for selected engineering resins. Polymer supported 2-hydroxybenzophenones and 2-(2-hydroxyphenyl)-2H-benzotriazoles have useful properties in articles used under high exposure of UV light;

examples are transparent durable acrylic films, outdoor paints, etc.

Kim *et al.* [582] have described maleimide-based antioxidants melt grafted onto low-MW PE. IR spectroscopic methods and titration were used for the quantitative determination of the extent of grafting of the monomeric antioxidant. Smedberg *et al.* [583] have characterised polymer-bound stabilisers by FTIR and NMR. The binding of antioxidants and photostabilisers to polyurethanes was verified by UV/VIS spectroscopy [584].

Further developments in the range of halogen-containing flame retardants are reactive systems in which the polymer chain is modified by grafting *flame retardant molecules* within the framework of the polymerisation process. Incorporation of the organophosphorous FR functionality *within* the polymeric structure, in line with the current drive for the development of less volatile molecules that reduce the risk of the evolution of toxic phosphorous-containing species within the smoke from burning materials, also leads to high extraction resistance [585]. Flame retardants based on reactive molecules/monomers, such as tetrabromobisphenol A (TBBPA) derivatives, tetrabromophthalate diols, dibromostyrene (DBS) and tribromophenol (TBP), chemically bound to the polymer (mainly in thermosets) possess excellent resistance to migration. Brominated styrene grafted onto PET is capable of acting as a nonvolatile FR for PET fabrics [586]. Other functionalised polymers will probably be commercialised in the future due to an increased pressure of legislative requirements against physically nonpersistent additives used for stabilisation of polymeric materials in contact with food or regulations for handling toxic rubber chemicals.

In cases where additive molecules have been grafted to the polymer all of the methods that require extraction are less than quantitative. Graft copolymers acting as interfacial agents [587] are quite inaccessible to extraction.

Al-Malaika [588] has reviewed antioxidant grafting on polymers.

Additive Adsorption

The performance of stabilisers in respect of their physical persistency can also be improved by physical adsorption on surfaces of reinforcing fillers, e.g. of CB or amorphous microground silica [589]. Mineral fillers are well known to adsorb polymer additives, especially stabilisers necessary for processing and long-term stability. Not surprisingly therefore, mineral fillers affect the chemical stability of polymers in processing and, even more so, in application. The main reason for the effect on the long-term properties is adsorption of part of the stabilisation system by the filler particles, reducing the mobility and availability of these substances in the polymer matrix. It should be stressed, however, that synthesis of organic macromolecular stabilisers is a more efficient approach to develop persistent stabilisers having a relatively high content of active moieties.

From an analytical point of view the strong adsorption phenomena may be disturbing. Already Squirrell [165] has noticed that strong adsorption or chemisorption of additives onto the polymer–filler matrix may invalidate quantitative extraction. Haacke *et al.* [590] have reported strong adsorption of UV light stabilisers on pigment particles in polymeric coatings, leading to incomplete extraction. SFE of phenolic compounds from a highly active matrix, such as CB, is nearly impossible [383]. Wolfschwenger *et al.* [591] have examined the additive adsorption for a variety of other systems, such as Irganox 1010 and Irgafos 168 on talc, calcium carbonate, China clay and wollastonite, and of Irganox 3114, Irganox 1010, Irganox 1330, Hostanox SE 10 and DSTDP on talc. Analytically, the adsorption can be measured by mixing a solution of the additive with the filler and monitoring relative loss of additive in relation to the fresh solution. Extraction of additives strongly adsorbed or chemisorbed onto the *polymer–filler matrix* must be carefully watched by the analyst, as a change in the method of manufacture of, for example, the filler or in the method of compounding the plastic formulation can also markedly alter the degree of adsorption bonding produced and hence invalidate an established quantitative extraction procedure. The use of 'stronger' extraction reagents can cause complications at the measurement stage and hence each system must be carefully screened and frequently checked.

High-MW Additives

Synthesis of compounds having an increased molecular weight represents the most natural way of producing physically persistent additives in demanding applications, which resist extraction or leaching. Pospíšil [576] has described such functionalised oligomers and polymers as stabiliser systems for conventional polymers. Antioxidants of higher MW have been developed for enhanced extraction resistance. Pan *et al.* [592] have recently reported several new monomeric antioxidants containing hindered phenol, which can be copolymerised with vinyl monomer by free radical initiation for the preparation of polymeric antioxidants. UV stabiliser monomers of the benzophenone and benzotriazole type were copolymerised with styrene and methyl methacrylate in order to overcome the diffusion problem [593]. Other recent developments in primary antioxidants include hindered phenolics with low volatility and extractability, such as Ralox AC, and Lankromark LE 314, which are superior to conventional phenolics in PP fibres. On the other hand, *oligomeric antioxidants*, such as the commercial product Flectol H, which is partially polymeric, go some way toward a solution to the problem of antioxidant volatility but are not substantive under conditions of extraction by hot oils or dry cleaning solvents. Ciba's Tinuvin 791 and 783 *blends* exploit the advantages of combining low- and high-MW to give superior light stabilisation in thick PP sections and talc-filled grades. Low-MW Tinuvin gives low mobility and superior stabilisation; high-MW Tinuvin and Chimassorb give good compatibility, extraction resistance and long-term thermal stability. The same holds for FS-812 (Ciba), which is a mixture of high-MW and low-MW HALS plus a benzotriazole UV absorber and hydroxylamine processing stabiliser. Tinuvin 783 is an FDA-approved high performance stabiliser for low-colour applications and good volatility for rotational moulding; Tinuvin 111 has good extraction resistance, low gas-fade and low pigment interaction, especially for film and fibre.

According to Malik and Ligner [594] 'optimal molecular weight' and 'optimal mobility' of light stabilisers determine best efficiency results for additives having $\langle M_n \rangle \approx 2700$ Da. To improve extraction stability polymeric antioxidants with MW above 3000 Da have been developed [595]. Taking into account the character of the host polymer one notices that for purposes of chemical efficiency and physical persistence *polymeric additives* (usually antioxidants) having MWs in broad limits of $\langle M_n \rangle$ 3000 to 20 000 Da are deemed acceptable according to the character of the host polymer. Average molecular weight ($\langle M_n \rangle$) of commercial stabilisers are Modanox 2000 $\sim$ 1100 Da, Tinuvin 622 $\sim$ 3600 Da, Chimassorb 944 > 2500 Da and Cyasorb UV2126 $\sim$ 50 000 Da. Polymeric HALS (such as Tinuvin 622, Chimassorb 944, Hostavin N 30 and Cyasorb

UV3346) offer superior compatibility, low volatility, excellent resistance to extraction and contribute to heat stability. The nonextractable nature of these additives makes their quantification challenging. The quantification of Chimassorb 944 in polyolefins is possible using an UV absorption method after dissolution of the polymer [596].

High-MW hindered amine thermal stabiliser (HATS) formulations are designed for advanced extraction resistant long-term stabilisation, i.e. for use in extractive environments such as polyolefin pipes, fulfilling the stringent requirement of guaranteeing product lifetime of more than 50 years [565]. These systems offer much better gas and hot water resistance than the low-MW phenolic antioxidant systems. Ethanox 330

(Albemarle) has shown unique resistance to extraction from polyolefins. No radioactivity was detected in water run through a MDPE pipe stabilised with ^{14}C-labelled Ethanox 330 for 10 months at ambient conditions, or in an accelerated 3-month test at 80°C (limit of detection 25 ppb).

The commercial evolution of hindered amine stabilisers is best illustrated by the sequence of the monomeric Tinuvin 770, oligomeric Tinuvin 622, Chimassorb 944 and Chimassorb 119, and most recently Chimassorb 2020 (Figure 3.25) with increased control on volatility, migration, extractability and co-additive interactions [597]. The hindered amine thermal stabilisers Chimassorb 944 and Chimassorb 119, outperform Cyasorb UV3346 as to extraction resistance [565].

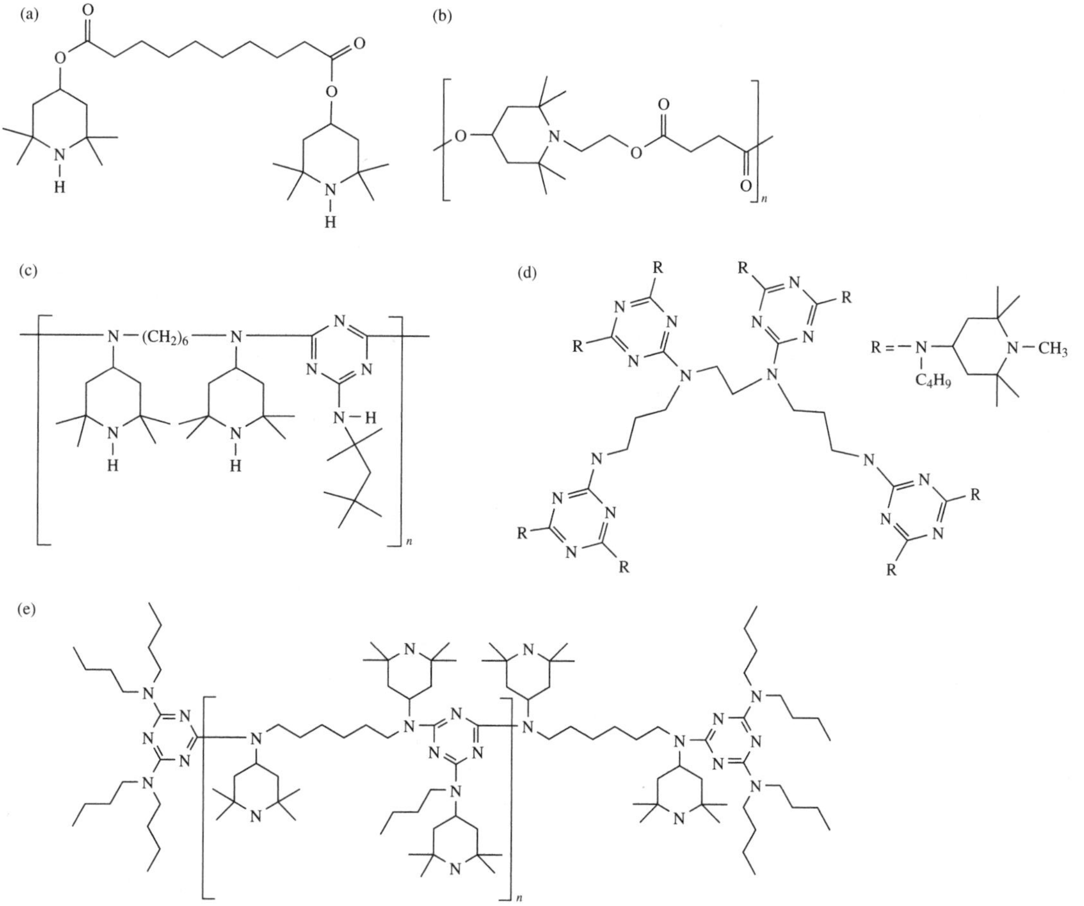

Figure 3.25 Chemical formulae of (a) Tinuvin 770, (b) Tinuvin 622, (c) Chimassorb 944, (d) Chimassorb 119 and (e) Chimassorb 2020

Other polymeric additive types are also quite inaccessible to analytical extraction. This holds for butadiene-based block copolymers as toughening additives and Goodyear's acrylate terpolymer Sunigum P 7395, etc. Similarly, oligomeric and *polymeric flame retardants* such as Saytex HP 7010 (Albemarle) (brominated polystyrene for GFR PBT), Fire Guard 7500 (Sankyo Organic Chemicals Co), a functionalised oligocarbonate for ABS prepared from 4,4′-isopropylidenebis(dibromophenol) and FR 1025, poly(pentabromobenzyl acrylate) (Technion), TBBPA carbonate oligomers, TBBPA/BPA carbonate oligomers, TBBPA carbonate dimers, brominated poly(phenylene oxide), poly-DBS (high–medium and low-MW), DBS-polypropylene graft, DBS-latex copolymer and DBS-styrene copolymer [598] possess excellent resistance to migration. As the practical application of oligomeric (polymeric) additive functionalities (antioxidants, flame retardants) is expected to increase this will have analytical consequences. Drastic solvent extraction methods, such as total dissolution or hydrolysis techniques, or heat extractions are required for the quantitative determination of high-MW additives.

Insoluble Additives

In most rubber compounds, extractability of lead-based additives (acting as HCl scavengers) can be reduced to very low levels by replacement of litharge or red lead with lower solubility stabilisers, such as dibasic lead phthalate or phosphite and tribasic lead sulfate. The most common reaction product of dibasic lead phthalate in scavenging chloride is the dichloro derivative, which is highly insoluble (i.e. reduced extractability by aqueous media). Although generally nonmigratory, such additives when extracted under vigorous conditions, as in the EPA Toxic Characteristic Leaching Procedure, may still yield sufficient soluble lead that product scrap would be classed as hazardous waste. Grossman [599,600] has recently presented modified lead stabilisers such that the complex reaction products arising from stabilisation are extremely insoluble (classified in EPA Toxicity Category IV, i.e. nontoxic). Because of their low solubility, migration or extractability of these additives, once mixed into a polymer, is very low. In the specific case, elemental analysis is an excellent tool for the determination of total lead compounds.

Ultramarines are completely insoluble in aqueous media and organic solvents. They are also nonmigratory when used in pigmented resin systems.

3.7 POLYMER/ADDITIVE DISSOLUTION METHODS

Principles and Characteristics For a variety of reasons, LSE does not always yield a sufficient additive recovery level. In general, this extraction mode is not adequate for isolation of high-MW additives, nor for additives which are not highly soluble in the extraction fluid. To overcome the disadvantage of the slow rate of LSE, some procedures for the separation of additives from polymers make use of the quicker and more efficient LLEs which involve the complete dissolution of the sample as an effective way of liberating the additives from the host polymer. However, this is not always possible. Having to dissolve polymers is a severe bottleneck for many analytical methods. Some polymers, such as nylons, are insoluble in most solvents. Although there are solvents that can dissolve them, high temperatures are often required, the viscosity of the final solution may be so high that extraction is practically impossible, or the solvent is very corrosive and presents safety hazards. Although LLEs are less time-consuming than LSE, the additives are subjected to more losses due to volatilisation and/or degradation because of the higher temperatures employed.

Polymer solubility criteria and polymer solutions have already been described (Sections 3.1.1–3.1.3). Follow-up steps after dissolution may comprise SPE, LLE evaporation, precipitation or cooling with the latter two being most common. The *dissolution/antisolvent precipitation method* (D/P) involves dissolution of the organic phase of the polymer composition in a large volume of a suitable (usually boiling) solvent, followed by precipitation of the polymeric constituents, often in a finely divided form in suspension with the inorganic fillers, by means of addition of a nonsolvent for the polymer, leaving the additives and low-MW oligomers dissolved in the solvent–nonsolvent mixture [596,601,602]. A clean-up step may be needed. If necessary, the obtained suspension is filtered and the additive is concentrated in the filtrate by evaporation, which is then analysed. Reversed-phase chromatography is the most common method for additive analysis, and some organic solvents must be removed before analysis, which is difficult in the case of solvents with a high boiling point, such as decalin. Some solvents, which are frequently used as a precipitant, are chemically reactive. For this reason methanol is often replaced by the less reactive acetonitrile.

A limited list of *solvent/nonsolvent combinations* for dissolution/precipitation is given in Table 3.50. More extensive listings may be found elsewhere

Table 3.50 Possible solvent/nonsolvent combinations for common polymers

Polymer	Solvents[a]	Antisolvents
Polyethylene (HDPE)	Xylene[b], decaline[b], TCB[b]	Acetone, MeOH, ether
LDPE	Toluene	MeOH, acetonitrile
Polytetrafluoroethylene (PTFE)	Insoluble	–
Polypropylene (PP)	Toluene	MeOH, acetonitrile
Atactic	General hydrocarbons	Ethyl acetate, iPrOH
Isotactic	Xylene[b], decaline[b], TCB[b]	Acetone, MeOH, ether
Polybutadiene	Hydrocarbons, benzene	Gasoline, alcohols, esters, ketones
Polyisobutylene	Gasoline, ether	Alcohols, esters
Polyisoprene	Benzene	Gasoline, alcohols, esters, ketones
Polyoxymethylene (POM)	HFIP/toluene (1:3)	MeOH, EtOH
	DMF (hot); benzyl alcohol (hot)	MeOH, ether
Polyamides (PA6, PA6.6, PA12, PA4.6)	Trifluoroethane/toluene (1:1)	MeOH, EtOH
	HFIP	MeOH, acetonitrile
	Formic acid, DMF, m-cresol	MeOH, ether
Polyurethanes (uncured)	Formic acid, DMF	MeOH, ether
Polyesters (except PET)	Toluene	MeOH, EtOH, iPrOH
	Chloroform, benzene	MeOH, ether
PET	HFIP/DCM/toluene (9.9:0.1:1)	Dioxane
	THF	Acetonitrile, MTBE
	m-Cresol, o-chlorophenol	MeOH, acetone
PBT	HFIP/toluene (1:3)	MeOH, EtOH
Polycarbonate (PC)	THF	MeOH, EtOH
	DCM	MeOH, acetonitrile
Poly(methyl methacrylate) (PMMA)	Toluene	MeOH, EtOH
	Chloroform, acetone, THF	MeOH, petroleum ether
	Acetone	MeOH, acetonitrile
Poly(vinyl chloride) (PVC)	Toluene, THF, DMF	MeOH, EtOH, hexane, acetone
Poly(vinylidene chloride) (PVDC)	THF, dioxane, ketones, butylacetate	Hydrocarbons, alcohols, phenols
Poly(vinyl alcohol) (PVOH)	H₂O, formamide	Gasoline, aromatic hydrocarbons, alcohols
Poly(vinyl acetal)	THF, ketones, esters, ethylene glycol	Methanol, aliphatic hydrocarbons
Poly(vinyl acetate) (PVAc)	Aromatic and chlorinated hydrocarbons, MeOH, ketones	Gasoline, ethylene glycol, cyclohexanol
Poly(vinyl butyral)	Ketones, esters	MeOH, aliphatic hydrocarbons
Polystyrene (PS)	Toluene, chloroform, cyclohexanone, DCM	MeOH, EtOH, acetonitrile
Styrenics (ABS, AES, MBS, SAN)	Toluene	MeOH, EtOH
	Acetone	MeOH, acetonitrile
Polyacrylonitrile (PAN)	DMF, DMSO	MeOH, ether, H₂O
Polyketones	Ketones, esters	Gasoline, alcohols
Polydimethylsiloxane (PDMS)	Chloroform, ether, heptane	MeOH, EtOH
Polysulfone	THF	THF–H₂O gradient
Rubbers (NBR, BR, EPDM, SIS, ESBR, SSBR, EPR)	Toluene	MeOH, acetonitrile
	Chlorinated hydrocarbons	MeOH, ketones, esters
Cellulose esters	Acetone, esters	Aliphatic hydrocarbons
Alkyd resins	Aromatics, chlorinated hydrocarbons, lower alcohols, ketones, esters	Hydrocarbons
Natural resins	Alcohol, benzene, chlorinated hydrocarbons, esters, ether	Aliphatic hydrocarbons
Epoxy resins (uncured)	Alcohol, esters, ketones, dioxane	H₂O, hydrocarbons
Polyacrylamide	H₂O	Alcohols, esters, hydrocarbons
Polyglycols	H₂O, chlorinated hydrocarbons, alcohols	Gasoline

[a] DCM, dichloromethane; DMF, dimethylformamide; DMSO, dimethylsulfoxide; HFIP, hexafluoroisopropanol; MTBE, methyl-t-butylether; TCB, trichlorobenzene; THF, tetrahydrofuran.
[b] $T > 130\,°C$.
Adapted from references [16,81,83,93,115,603,604].

[40,93,603–605]. In reality, finding a suitable solvent is not as easy as simply matching the polymer's solubility parameter (δ value). It is also important to take into account the effects of polymer crystallinity (as in the case of aPP and iPP, LDPE and HDPE). Because of their various chemical structures, it may be necessary to experiment with solvent, temperature, and time conditions to optimise the extraction strategy.

Table 3.51 lists the *main characteristics* of the two-step extraction for in-polymer additive analysis. The dissolution method has the advantage of being in principle analytically a better approach for characterisation of an unknown additive mixture than solvent extraction. The supernatant obtained is extremely useful for either direct analysis of known additives or for subsequent separation of their components by one of a number of (multidimensional) chromatographic techniques prior to examination of the individual separated compounds by spectroscopy for identification and for quantitation. Reprecipitation avoids some other disadvantages of extraction methods. It is suitable for additives of high MW, is time saving, and does not require grinding. The main asset of the technique is quantitation. The good performance of the dissolution/precipitation procedure was confirmed by the results of a round-robin for the determination of plasticisers and antioxidants in polymeric matrices [209a]. If chemisorbed constituents are not completely released from the polymer – filler matrix they are often left in a form very vulnerable to attack by the analytical reagent(s) used in the analysis. For example, titration of amine/amide slip additives with perchloric acid in nonaqueous media can be carried out

in the presence of suspended solids. Dissolving the polymer is a useful technique to isolate *inorganic fillers*. Filtering and centrifuging can then be used to collect the insoluble solids.

As with all techniques, the dissolution/precipitation method also has its disadvantages. Although the method is very effective and applicable to a broad class of polymers, it is not universal: e.g. Teflon and cross-linked polymers do not dissolve. While there is no possibility for some target analyte remaining bound in the polymer network, inclusion in the reprecipitated polymer can occur. However, such events may be controlled by addition (before precipitation) of an internal standard. The dissolution/reprecipitation procedure is time-consuming. The procedure depends upon the extraction temperature, solvent used, matrix and additive being extracted. Exhaustive extraction may require several dissolution/precipitation steps, extended extraction times, or changes in solvent and temperature to obtain quantitative results. The numerous handling steps contribute to error. Other disadvantages are the co-separation of oligomers and low-MW waxes [125], which may need to be removed before further analysis, and the possibility of chemical transformations or interactions. The large excess of solvent used for dissolution might interfere with the interpretation of the chromatogram, obscuring some of the peaks of interest. The procedure also suffers from the solubility dependence of the additives and risk of volatilisation, which may yield erroneous results.

Several dissolution modes (boiling solvent, hot block, Parr bomb, high-pressure dissolution) are possible. The configuration of the main procedures for fractionation by solubility and their chromatographic follow-up is given in Table 3.52.

Table 3.51 Main characteristics of dissolution/precipitation methods

Advantages

- Wide applicability
- No grinding of polymer required
- Exhaustive extraction
- Quantitative additive recovery (unless coprecipitation)
- Relatively low cost
- Allowance for fractional reprecipitation

Disadvantages

- Analyst intensive
- Solvent intensive
- Not universal
- Poor sensitivity
- Additive volatility and/or transformation
- Difficult to automate
- Unsuited for production quality control (time-consuming)
- Unsuited for thermoset cross-linked polymers
- Possible coprecipitation of analyte with polymer
- Clean-up steps frequently required (separation of nonelutable oligomers, heavy waxes)

Table 3.52 Dissolution-based polymer/additive analysis procedures

One-/two-step extraction mode		Chromatography/ spectroscopy
Dissolution	Reprecipitation (antisolvent)	RPLC
Dissolution	Cooling	RPLC
High-pressure dissolution	Cooling	RPLC
Dissolution	Precipitation, solvent evaporation, redissolution	RPLC, FTIR, NMR
Dissolution	Liquid–liquid extraction	RPLC
Dissolution	–	GPEC®, SEC, SEC-GC, SEC-HPLC, TLC

The polymer may be dissolved first and then extracted with another solvent (LLE) [79]. Addition of an antisolvent is not always necessary. The high-MW polymer may precipitate out as the solution is cooled [490,491,596,606], as in the case of PE and PP (reflux in decaline/cooling), or dissolution/precipitation may be followed by evaporation to dryness and redissolution in order to allow for a more suitable chromatographic or spectroscopic solvent [106]. Crompton [81] has reported many practical examples of such actions. According to Schabron *et al.* [606–608] only dissolution/cooling leads to good recovery of additives. Recently, various *alternative dissolution-based polymer-additive procedures* have been advanced. In the multiple-solvent gradient method the sample is dissolved in a good solvent and is precipitated on top of a HPLC column with a nonsolvent [98]. By adding a solvent to the adsorbed suspension in gradient mode monomers, additives and polymers will redissolve and elute from the column.

Usually high boiling solvents are required for dissolving the polymer. The solvent is then difficult to remove after precipitation of the polymer. This problem was addressed by Macko *et al.* [490,491], who described pressurised dissolution/precipitation of MDPE/(Irganox 1010, Irgafos 168), MDPE/Irganox 1330 and UHMWPE/α-tocopherol. The polymeric matrix was dissolved in a UV-transparent low-boiling solvent, i.e. pure *n*-heptane or *n*-heptane/isopropanol in an autoclave at 160–170 °C (well above the normal boiling point) under elevated pressure and in inert atmosphere. As the polymer is not soluble in heptane at room temperature it precipitates on cooling. In the simplest procedure the supernatant solution obtained is then analysed directly by isocratic HPLC with UV detection. No additional manipulation with the solution (evaporation, preconcentration, dissolution, etc.) is usually necessary. Otherwise, heptane can readily be removed by evaporation and replaced by an RPLC solvent. Advantages of the procedure are the short analysis time (ca. 2 h), good quantitative recovery and reproducibility of the results ($\sigma = 2$–5%) and a low detection limit. The approach can be applied for quantitative analysis of all additives soluble in an aliphatic solvent which is able to dissolve the crystalline matrix polymers at high temperature and pressure and can be used as an eluent in HPLC or evaporated in vacuum at low temperature.

Table 3.4 compares the performance of dissolution and various extraction procedures. Chapter 9 addresses in more detail direct methods of deformulation of polymer/additive dissolutions, i.e. without separation of polymer and additives. Dissolution/precipitation is quite the opposite of solvent-free methods for polymer/additive analysis, such as thermal desorption (TD).

Applications Dissolution/reprecipitation is claimed to be the most widespread approach to polymer/additive analysis [603], but recent round-robins cast some doubt on this statement. Dissolution appears to be practised much less than LSEs. However, in cases where exhaustive extraction is difficult, e.g. for polyolefins containing high-MW (polymeric) additives, a dissolution/precipitation method is preferred.

A common technique used for polyolefin samples is to dissolve the sample using solvents such as xylene, decalin, toluene and di- or trichlorobenzene heated to temperatures as high as 130–150 °C. After the plastic sample has been solvated, the polymeric component is precipitated by cooling and/or by adding a cold nonsolvent such as acetone, methanol or isopropanol. Polypropylene does not completely dissolve in toluene under reflux for 0.5 to 1 h with magnetic stirring (typically, 2 g of polymer in 40 mL of toluene), yet the additives may be extracted [603]. In addition to additives, most solvents also extract some low-MW polymer with subsequent contamination of the extract. To overcome this a procedure for obtaining polymer-free additive extracts from PE, PP and PS has been described based on low-temperature extraction with *n*-hexane at 0 °C [100].

Dissolution (hot decalin)/cooling was used for the determination of hindered phenol antioxidants (BHT, Irganox 1010/1076) in polyolefins [609]. Conditions for other common dissolution procedures are: PP/Tinuvin 770 (dissolution in tetrachloroethane or HCCl$_3$, cooling), PA4.6/Irganox 1098 (dissolution: HCOOH; precipitation: acetonitrile/water) and PE/Irganox B220 (dissolution in xylene, followed by centrifugation and liquid phase XRF of phosphorous). Gelbin and Jackson [610] have described a simple dissolution (toluene)/precipitation (methanol) method that allows to analyse for TNPP in LLDPE with regard to its integrity (hydrolysis, oxidation and nonylphenol formation) and load level in the polymer resin. After evaporation to dryness the residue was taken up in cyclohexane and used as in HPLC-ELSD/UV and/or ^{31}P NMR analysis. Sharma *et al.* [611] have used dissolution (decalin)/precipitation (methanol) for the analysis of simple and complex mixtures of ethoxylated amines, ethoxylated amides, primary and secondary amides and glycerol esters in HDPE-based masterbatches by nonaqueous potentiometric titration.

The pressurised dissolution/cooling procedure of Macko *et al.* [490], which uses a UV-transparent low-boiling point solvent, is fast and simple as no additional evaporation of the solvent, preconcentration or redissolution of the additive is necessary. Macko *et al.* [491] have given an extensive listing of HPLC analyses of aromatic antioxidants and UVAs which can be separated with *n*-heptane and *n*-hexane as the main component of the mobile phase. The method was also used for HPLC quantification of thioether antioxidants (Santonox R, Chimox 14 and Irganox PS 802) in MDPE [612].

To avoid time-consuming grinding of polyolefin samples, Freitag and John [96] have dissolved thick samples, i.e. pellets of 3 mm diameter in a 1:1 mixture of toluene and 1,2-dichlorobenzene using microwaves during 5 min. Satisfactory results (95 % of expected concentration) were obtained only for PP, whereas recovery from polyethylenes was below 85 %. The *polymeric light stabiliser* Chimassorb 944 (MW 2500–4000 Da) can be quantitatively determined in PP after dissolution (hot decalin)/cooling, followed by extraction of the suspension by diluted sulfuric acid and UV determination [596]. The extraction step is essential, as interfering additives remain in the organic layer. High-MW HALS stabilisers such as Chimassorb 119/944 and Cyasorb UV3346 in polyolefins may be analysed by dissolution (toluene)/precipitation (methanol, triethylamine) followed by isocratic HPLC-UV (244 nm) with THF–diethanolamine as the mobile phase (SD 0.4 %, LOD 100 ppm for Chimassorb 944) [613,614]. Dissolution (xylene)/precipitation (2-propanol) followed by TLC and size-exclusion nonaqueous RPLC was used as another mode of quantitative analysis of Chimassorb 944 in polyolefins [615]. Interferences from other polymer additives may be removed by the use of a silica cartridge pretreatment.

Tinuvin 622 (MW > 3100–4000 Da) in polyolefins can be determined by dissolution (toluene)/derivatisation (TBAH)/precipitation (alcohol) which saponifies the polymeric light stabiliser to the diol 4-hydroxy-2,2,6,6-tetramethyl-1-piperidine ethanol which is then quantified using either NPLC or RPLC with UV detection [616]. The advantage of the gradient system is separation of any potential matrix interference. Pigments do not interfere. However, in this procedure any information on the oligomer distribution of Tinuvin 622 is lost, as opposed to the μHTLC-ELSD procedure adopted by Trones [617]. Roberson and Patonay [618] reported recoveries of 94.2 %, 97.4 % and 95.6 % of Tinuvin 622, Chimassorb 944 and Sandostab P-EPQ from PP following the dissolution (toluene)/precipitation (methylene chloride) mode.

PP/Uvasil 299 fibres were analysed by ^{1}H NMR after dissolution (toluene)/precipitation (methanol) and PP/Uvasil 299 pellets after reflux extraction with precipitation (methanol) of the oligomers [619]. The dried extracts were redissolved in $CDCl_3$, containing the DMP internal standard. The method allows quantification of the high-MW HALS at very low levels (0.044 ± 0.001 %).

Wang and Buzanowski [620] have used dissolution (toluene/chloroform)/precipitation (methanol) for isolation of a lubricant (technical-grade stearic acid; Spartech 14 575) from PE. Fink *et al.* [621] have developed a method for analysis of metal additives in recycled thermoplasts from *electronic waste*, based on dissolving the samples in an organic solvent and subsequent analysis of the corresponding solutions or suspensions by TXRF. The procedure is considerably less time-consuming than the conventional digestion of the polymer matrix.

Dissolution ($CHCl_3$)/precipitation (*n*-hexane) can be used for polycarbonate. For quantitative measurements Irganox 1063 is added as an internal standard. Dissolution/precipitation has also been applied to PET/(butyric acid, Malathion, Diazinon) [601]. Some polymers are soluble in few solvents only, e.g. PET, which has been dissolved in HFIP/DCM, and the polymer is precipitated by addition of acetone or methanol [601,622]. Various extraction approaches have been utilised for the analysis of oligomers from PET including selective precipitation of the polymer from a trifluoroacetic solution and chloroform extraction under pressure in a Parr bomb [112]. In the selective precipitation approach the successive addition of chloroform, acetone, distilled water and concentrated NH_4OH allows isolation of oligomers. This approach has the advantage of oligomer isolation at or below room temperature. The major extractable component of the low-MW oligomers is cyclic trimer, but also fractions up to the heptamer, nonamer and decamer have been reported [113]. In another similar study of PET the cyclic trimer and other oligomers up to the heptamer have been determined by means of dissolution/precipitation [622]. Begley and Hollifield [623] determined residual reactants and reaction by-products in PET by dissolution methods.

Crompton [81] has reported in detail fractional precipitation procedures for various polymers, such as polyolefins, acrylic polymers, PS, PVC, PUR, rubbers, polyarylamide, and various copolymers. Metcalf and Tomlinson [624] have reported fractional precipitation of antioxidants from PE. Separation of low-MW species from the polymer by repeated fractional precipitation [625] is unsatisfactory for routine use because of the time and manipulative skills required.

For engineering plastics various operational protocols are in use, such as dissolution (HFIP/TFE)/precipitation (cyclohexane) [626] or dissolution (HFIP)/precipitation (MTBE) [157]. Crompton [83] has reported dissolution methodology for the determination of residual isobutane, n- and iso-pentane, neo- and iso-hexane (10 ppm) in expandable and expanded polystyrene by GC (accuracy ± 5 % after internal standard addition). Also the determination of 50 ppm styrene monomer and other volatiles (benzene, toluene, ethylbenzene, m/p-xylene, cumene, o-xylene, n-propylbenzene, m/p-ethyltoluenes) in PS and styrene-butadiene copolymers (expandable, self extinguishing, crystal, high impact grade and foodstuff packaging grades) was carried out by dissolution in propylene oxide in the presence of n-undecane as an internal standard in GC-FID analysis [627]. Similarly, Steichen [628] has reported the determination of vinyl chloride, butadiene, acrylonitrile, styrene and 2-ethylhexyl acrylate monomers in polymers by dissolution procedures followed by headspace analysis. For the analysis of organic additives in polyamides (except for lubricants, oligomeric HALS and nucleating agents) the polymer is dissolved (HFIP) and reprecipitated (MTBE). Extraction with $HCCl_3$ gives quantitative access to external lubricants, such as EBA after elimination of polyamide oligomers [157].

Dissolution (toluene)/precipitation (methanol) of EPS beads (atactic PS foams with ca. 2 wt.% HBCD and $BaSO_4$) with TLC separation and IR identification allowed detection of hexabromocyclododecane; for quantification XRF analysis (Br) was used [629]. Dissolution/precipitation of PS/lubricants with evaporation to dryness and redissolution was followed by FTIR analysis [106].

N-alkyl-N'-aryl-p-phenylenediamine and N,N'-diaryl-p-phenylene-diamine in oil-extended black and non-black SBR masterbatches and vulcanisates were determined by dissolution (benzene)/precipitation (methanol) [630]. In case of a polymer containing percentage levels of inorganic metal oxides, dissolution of the polymer in an organic solvent followed by extraction with an aqueous acid may suffice to separate the polymer from the metallic constituent. Separation of gross polymer from additives and fillers is important for obtaining fractions for further examination. Cook and Lehrle [631] have separated oil/additive mixtures by dissolution (in n-hexane), followed by precipitation of the polymer using methanol.

In a *comparative qualitative study* El Mansouri *et al.* [632] examined dissolution/precipitation and Soxhlet extraction of phosphorous antioxidants in PP. The dissolution (refluxing toluene)/precipitation

Table 3.53 Quantification of PP additives

Additives	Dissolution/precipitation		Soxhlet	
	Amount (ppm)	RSD (%)	Amount (ppm)	RSD (%)
Irganox 1010	750	2	880	1.1
Irgafos 168	880[a]	1.1	41	2.3
Ultranox 626	41	2.3	340	1.3

[a] As oxidised Irgafos 168.
After El Mansouri *et al.* [632]. Reproduced by permission of the authors and Friedr. Vieweg & Sohn.

(methanol) extract contained DBS, Ultranox 626, Irganox 1010, Irgafos 168, oxidised Irgafos 168 (not in Soxhlet!) and DTBP (a hydrolysis product of Ultranox 626). The Soxhlet extract (CH_2Cl_2, 50 °C) comprised DBS, Ultranox 626, Irganox 1010, Irgafos 168, DTBP and a second degradation product of Ultranox 626 (not in dissolution!), but no Irgafos 168. Table 3.53 summarises the average additive concentrations as determined by means of dissolution/precipitation and Soxhlet extraction. The level of Irganox 1010 was approximately equal in both extraction procedures, whereas the quantity of phosphite esters in the Soxhlet extract is larger than in the dissolution/precipitation extract. Ultranox 626 is more sensitive to heat than Irgafos 168. The different results are ascribed to the high temperature (100 °C) of the dissolution/precipitation extraction on the one hand, and hydrolysis on the other hand.

In the framework of migration testing to food simulants (iso-octane) propylene oligomers were isolated from the polymer matrix by dissolution/precipitation and Soxhlet methods and characterised by UV, FTIR, HPLC and SEC [633]. The average molecular weights differed according to the extraction method. The fraction obtained by dissolution/precipitation was homogeneous in MW, whereas the one obtained by Soxhlet extraction had a higher polydispersity index (M_w/M_n).

Cortes *et al.* [634] have recently used μSEC-GC/LC in a *comparative quantitative study* of dissolution and dissolution/precipitation of PC/(2,4-di-t-butylphenol, nonylphenol isomers, Tinuvin 329, Irgafos 168) and ABS/(nonylphenol isomers, Tinuvin P, BBP, Vanox 2246, Tinuvin 328/770, Topanol CA, Acrawax). For the ABS sample the dissolution approach determined a four-fold higher concentration for Vanox 2246 than by dissolution/precipitation of the sample, indicating that precipitation can yield low (incorrect) results for additives which exhibit solubility dependence. Using both sample preparations equivalent concentrations were observed for the additives of the PC sample, except for Tinuvin 329.

Dissolution followed by evaporation to dryness and redissolution has been applied for PS/lubricants [106] and dissolution followed by LLE in polyamides/lubricants [79]. In some instances the polymer does not need to be precipitated out of the solution. For example, polyurethane samples can be readily dissolved using hot DMF and the solution can then be analysed directly for antioxidant content by high-temperature gel chromatography, employing an in-line guard column to filter out the polymer. Staal *et al.* [98] have reported dissolution of polycarbonate and polysulfone in THF and precipitation onto a C_{18} guard column. A gradient from water–THF (50:50 vol%) as a nonsolvent to 100 % THF successively enabled elution of the additives, oligomers and finally the polymer itself.

A British Standard method [635] describes dissolution of polymers in refluxing toluene, with reprecipitation of the polymer by addition of ethanol.

In an industrial application dissolution/reprecipitation technology is used to separate and recover nylon from carpet waste [636]. Carpets are generally composed of three primary polymer components, namely polypropylene (backing), SBR latex (binding) and nylon (face fibres), and calcium carbonate filler. The process involves selective dissolution of nylon (typically constituting more than 50 wt% of carpet polymer mass) with an 88 wt% liquid formic acid solution and recovery of nylon powder with $scCO_2$ antisolvent precipitation at high pressure. Papaspyrides and Kartalis [637] used dimethylsulfoxide as a solvent for PA6 and formic acid for PA6.6, and methylethylketone as the nonsolvent for both polymers.

3.8 HYDROLYSIS

Principles and Characteristics In principle, both classical methods of chemical and physico-chemical analysis (burning and staining tests, density, solubility, softening point, refractive index, acid number and saponification value, iodine, hydroxyl and carbonyl values, qualitative determination of hetero atoms, ash determination) and modern instrumental analytical methods (IR, NMR) play a role in the practical deformulation of plastics [54]. For example, dye investigations of textiles can be divided into two groups: distinguishing between classes of dyes and identification of individual dyestuffs. Drawing firm conclusions about the identity of individual dyes or finishes on the basis of wet chemical analytical tests (spot tests) requires considerable skill, experience and reference material. Determining the class

of dye is generally a much simpler process than identifying an individual dye, which often necessitates the use of instrumental analytical methods. Identification of the dye class is easier if the fibres are identified first, because this limits the classes to be looked for. Hurwitz *et al.* [638] described many preliminary and confirmatory tests grouped according to fibres.

Sophisticated instrumental techniques are continually being developed and gradually replace the classical wet chemistry analytical methods. *Wet chemical analysis* is destructive: the sample is dissolved or altered. Nowadays the analyst is highly focused on instrumental methods and chemometrics. Yet, chemical work-up methods (e.g. hydrolysis with alcoholic alkali, alkali fusion, aminolysis, and transesterification, etc.) and other 'wet laboratory' skills should not be forgotten.

For the purpose of the identification and quantification of additives (broadly defined) in polymeric materials extraction and dissolution methods are favoured (Sections 3.3–3.7). However, additives are also made accessible analytically by digestion of the sample matrix (cf. Section 8.2). Such wet chemical techniques, that remove the sample matrix first, are often limited to mg amounts because of pressure build-up in destruction vessels. Another reactive extraction approach to facilitate additive analysis is *depolymerisation* by acid hydrolysis or saponification, sometimes under pressure. This is then frequently followed by chemical methods such as titrimetry or photometry for final identification and quantification.

Although polymers in-service are required to be resistant toward hydrolysis and solar degradation, for polymer deformulation purposes hydrolysis is an asset. Highly crystalline materials such as compounded polyamides are difficult to extract. For such materials hydrolysis or other forms of chemolysis render additives accessible for analysis. Polymers, which may profitably be depolymerised into their monomers by hydrolysis include PET, PBT, PC, PU, PES, POM, PA and others. Hydrolysis occurs when moisture causes chain scissions to occur within the molecule. In polyesters, chain scissions take place at the ester linkages (R-CO-O-R'), which causes a reduction in molecular weight as well as in mechanical properties. Polyesters show their susceptibility to hydrolysis with dramatic shifts in molecular weight distribution. Apart from access to the additives fraction, hydrolysis also facilitates molecular characterisation of the polymer. In this context, it is noticed that condensation polymers (polyesters, -amides, -ethers, -carbonates, -urethanes) have also been studied much

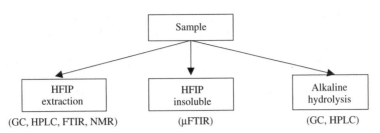

Figure 3.26 Deformulation of flame retardants in polyesters

less extensively by PyGC than polyolefins or vinylpoly-mers (e.g. PVC), which is partly due to the fact that these polymers can be chemically degraded and are, therefore, more readily studied by conventional analytical techniques.

Applications In the two-step microwave-assisted solvent extraction (TSM) procedure the first micronising step, which determines complete dissolution by microwave heating and subsequent recrystallisation of the polymer as very small particles by cooling, is followed by a second (leaching or reactive) step for dissolution of the additives. TSM procedures give excellent recovery results for the following medium to highly polar additives using: (i) a leaching mode for nucleating agents and light stabilisers; (ii) a hydrolytic mode (e.g. with phosphoric acid) for antiacid agents; (iii) a saponifying mode (e.g. with TBAH) for light stabilisers; or (iv) an oxidising mode (e.g. by 2,6-di-*tert*-butylhydroxyperoxide) for aromatic phosphites and phosphite–phosphonite mixtures [210].

Hydrolysis can be used to examine the structure of cross-linked polyester resins and to degrade the polymeric matrix for quantitative analytical access to impact modifiers or inorganic additives. (Pressurised) hydrolysis of *polyesters* leads to aromatic diacids (which can be determined directly by means of HPLC) and diols (determination after derivatisation to the corresponding acetates); polyesters may also be subjected to direct methanolysis. Carbodiimide-based hydrolysis stabilisers such as Stabaxol I, P and P100, which have interacted with the carboxyl endgroups of polyesters under formation of a urethane, can be GC analysed as aromatic amines after hydrolysis with methanolic KOH followed by acetylation [157]. Flame retardants in polyesters are often analysed after basic hydrolysis of the polymer. In the presence of ester groupings in the FR partial or total hydrolysis may occur, which impairs identification. For example, for the separation of PBT from Tribit 1500GN30 (Samyang Korea) alkaline hydrolysis was needed (1 N ethanolic KOH at 140 °C, 4 h); the FR,

poly(pentabromobenzyl)acrylate, was not hydrolysis stable [157].

Nelissen [157] has adopted a standard approach for the determination of FRs in polyester compounds (Figure 3.26). The analysis is complicated by spectral interferences of PET/PBT and by the complexity of FR structures, notably DBDPO, Saytex BT 93 W (ethylene-bis-tetrabromophthalimide), PDBS 80, Pyrochek 68PB, Saytex HP7010, Saytex 8010, FR1808, FR 1025, F 2400, BC 52 and BC 58 (brominated polycarbonate oligomers).

Direct spectroscopic additive analysis of PET/PBT compounds is often limited. An IR window from 2800 to 1730 cm^{-1} can be used to detect the presence of FRs. PET/PBT IR absorption bands in the 1730–1650 cm^{-1} range may obscure FR absorptions (e.g. in the case of BC 52, BC 58 and Saytex BT 93 W). Similarly, ^{1}H NMR spectroscopy is of limited use for the analysis of FRs with few protons. A useful exception is FR 1025 with its 5.5 ppm CH$_2$ resonance of the brominated benzylester. Unless a database is available PyGC-MS is also of limited use in the identification of FRs in compounds. However, brominated polystyrenes such as Pyrochek 68B, PDBS 80 and Saytex HP 7010 can fairly easily be distinguished by this technique. Wang [639] has developed PyGC-MS into a routine tool for analysis of FRs. A usual extraction procedure for additives from crystalline polyesters is dissolution (HFIP)/precipitation (MTBE) followed as a matter of course by HPLC-DAD, in view of the UV chromophores of the FRs. However, for this analysis method the solubility of the FRs in the eluent acetonitrile is quite important with FR 1808 being soluble, DBDPO, BC 52 and BC 58 poorly soluble and Saytex 8010, Saytex BT 93 W and F 2400 insoluble. GC analysis is less indicated as most FRs used in polyesters are high-MW and have low volatility. After concentration of the extract the FRs may be identified by FTIR after correction for the presence of polyester oligomers. Not surprisingly, identification of additives in the extract is favoured with respect to direct analysis in the compound. Given the low solubility of many FRs in HFIP, FTIR microspectroscopy can be used for their detection in the HFIP insoluble fraction. Alkaline hydrolysis (KOH/MeOH) of such polyester compounds in an autoclave at 140 °C can be used for compositional analysis, i.e. the determination of the contents of diacid (HPLC) and diol (GC, after acetylation). At the same time, hydrolysable FRs (e.g. BC 52, BC 58, F 2400) are depolymerised into their monomers and can be identified and quantified. Brominated polystyrenes such as Pyrochek 68PB, PDBS 80, Saytex HP7010 and Saytex 8010 are hydrolysis stable, but can thereupon easily be analysed in view of their solubility in chloroform.

Hydrolysis of *polyamide-based formulations* with 6 N HCl followed by TLC allows differentiation between α-aminocaproic acid (ACA) and hexamethylenediamine (HMD) (hydrolysis products of PA6 and PA6.6, respectively), even at low levels. The monomer composition (PA6/PA6.6 ratio) can be derived after chromatographic determination of the adipic acid (AA) content. Extraction of the hydrolysate with ether and derivatisation allow the quantitative determination of fatty acids (from lubricants) by means of GC (Figure 3.27). Further HCl/HF treatment of the hydrolysis residue, which is composed of mineral fillers, CB and nonhydrolysable polymers (e.g. impact modifiers) permits determination of total IM and CB contents; CB is measured quantitatively by means of TGA [157]. Acid hydrolysis of flame retarded polyamides allows to determine the adipic acid content (indicative of PA6.6) by means of HPLC, HCN content (indicative of melamine cyanurate) and fatty acid (indicative of a stearate) by means of GC [640]. Determination of ethylene oxide-based antistatic agents

in nylon-6 and similar polymers may be carried out by hydrolysis with HCl, precipitation of the antistatic agent by addition of a known amount of $K_4Fe(CN)_6$ and titration of the excess $Fe(CN)_6^{4-}$ [641]. The method gives accurate results for 1–3 % antistat in the polymer. Li and Hu [642] have studied the discoloration mechanism of PA6 using acid hydrolysis and HPLC to separate a low-MW yellow substance from other hydrolysis products. Characterisation of the chromophore was carried out by IR, FAB, MS, UV, ^{1}H NMR and ^{13}C NMR.

While additive analysis of polyamides is usually carried out by dissolution in HFIP and hydrolysis in 6 N HCl, *polyphthalamides* (PPAs) are quite insoluble in many solvents and very resistant to hydrolysis. The highly thermally stable PPAs can be adequately hydrolysed by means of high pressure microwave acid digestion (at 140–180 °C) in 10 mL Teflon vessels. This procedure allows simultaneous analysis of polymer composition and additives [643]. Also the polymer, oligomer and additive composition of *polycarbonates* can be examined after hydrolysis. However, it is necessary to optimise the reaction conditions in order to avoid degradation of bisphenol A. In the procedures for the analysis of dialkyltin stabilisers in *PVC*, described by Udris [644], in some instances the methods can be put on a quantitative basis, e.g. the GC determination of alcohols produced by hydrolysis of ester groups.

Various classes of additives may undergo chemical reaction in hydrolytic conditions. For example, hydrolysis of aromatic phosphites forms phenolic compounds [645]. Additives may also undergo reaction with the polymer chain, which causes extraction difficulties.

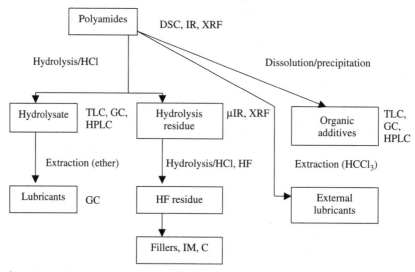

Figure 3.27 Deformulation of polyamide blend formulations

An analytical solution for molecules with alkaline functionality is *acid/base titration*. In this technique, the polymer is dissolved, but not precipitated prior to analysis. In this way, the additive, even if polymer-bound, is still in solution and titratable. This principle has also been applied for the determination of 0.01 % stearic acid and sodium stearate in SBR solutions. The polymer was diluted with toluene/absolute ethanol mixed solvent and stearic acid was determined by titration with 0.1 M ethanolic NaOH solution to the *m*-cresol purple endpoint; similarly, sodium stearate was titrated with 0.05 M ethanolic HCl solution [83]. Also long-chain acid lubricants (e.g. stearic acid) in acrylic polyesters were quantitatively determined by titration of the extract.

Instrumental methods for the determination of water in polymeric materials often rely on heat release of water from the polymer matrix. However, in some cases (e.g. PET) the polymer is hydrolysed and a simple Karl Fischer method is then preferred. Small quantities of water ($10\,\mu g$–15 mg) of water in polymers (e.g. PBT, PA6, PA4.6, PC) can be determined rapidly and accurately by means of a coulometric titration after heating at 50 to 240 °C with a detection limit in the order of 20 ppm.

The accurate determination of CB or thermally sensitive pigments and *fillers* in polymers, such as PET and PA, can be achieved by digestion of the polymer matrix using acid- or base-catalysed hydrolysis. Nylon additives such as CB, MoS_2 and glass fibre, which are not decomposed by acid hydrolysis, may be determined by direct gravimetry [82]. Boric acid, formed by hydrolysis of GFR EPs, is determined after acetylation. Wet chemical analysis is traditionally also used for quantification of carbohydrates in wood and pulps, namely acid hydrolysis followed by reduction, acetylation and GC analysis. More recently, PyGC-MS is being used in the latter application [646].

In a typical example of *reactive extraction* a fatty acid ester was easily formed by adding a small amount of an alcohol to a fatty acid metal salt and extracting at 315 °C [647]. This method was applied to PP/calcium stearate and PP/zinc stearate containing pellets; the presence of the fatty acid metal salts was confirmed by GC-MS.

Oligomeric *hindered amine light stabilisers*, such as Tinuvin 622 and Chimassorb 944, resist satisfactory analysis by conventional HPLC and have required direct UV spectroscopic analysis of a polyolefin extract [596], PyGC of an extract [618,648], or SEC of an extract [649]. Freitag *et al.* [616] determined Tinuvin 622 in LDPE, HDPE and PP by saponification of the polymer dissolution in hot toluene via addition of an alcoholic solution of TBAH to form 4-hydroxy-2,2,6,6-tetramethyl-1-piperidine ethanol with simultaneous precipitation of the polyolefin matrix and quantification by means of HPLC. The method is specific, accurate and reproducible and yields good recoveries (100–105 %). Kurihara *et al.* [650] have reported a qualitative GC-MS analysis after trimethylsilylation of a HALS by alcohol added thermal extraction in KOH/MeOH solution at 315 °C.

Industrial processes for recycling have been developed which take advantage of reactively separating hydrolysable polymers from nonhydrolysable waste plastics [651].

3.9 BIBLIOGRAPHY

3.9.1 Sampling and sample preparation

J. Pawliszyn (ed.), *Sampling and Sample Preparation for Field and Laboratory*, Elsevier, Amsterdam (2003).

S. Mitra and J.D. Winefordner, *Sample Preparation Techniques in Analytical Chemistry*, Wiley-Interscience, New York, NY (2003).

P. Gy, *Sampling for Analytical Purposes*, John Wiley & Sons, Ltd, Chichester (1998).

M. Stoeppler (ed.), *Sampling and Sample Preparation. Practical Guide for Analytical Chemists*, Springer Verlag, Berlin (1997).

G.E. Baiulescu, P. Dumitrescu and P.Gh. Zugrăvescu, *Sampling*, Ellis Horwood, Chichester (1991).

J.W. Weaver (ed.), *Analytical Methods for a Textile Laboratory*, American Association of Textile Chemists and Colorists, Research Triangle Park, NC (1984).

3.9.2 Solvents/solubility

R.P.T. Tomkins, G. Hefter and C. Young, *The Experimental Determination of Solubilities*, John Wiley & Sons, Ltd, Chichester (2003).

I. Teraoka, *Polymer Solutions*, John Wiley & Sons, Ltd, Chichester (2002).

M. McParland and N. Bates (eds), *Toxicology of Solvents*, Rapra Technology Ltd, Shawbury (2002).

G. Wypych (ed.), *Handbook of Solvents*, ChemTec Publishing, Toronto (2001).

Y. Marcus, *The Properties of Solvents*, John Wiley & Sons, Inc., New York, NY (1998).

D. Stoye and W. Freitag (eds), *Paints, Coatings and Solvents*, Wiley-VCH, Weinheim (1998).

A.F.M. Barton, *CRC Handbook of Solubility Parameters and Other Cohesion Parameters*, CRC Press, Boca Raton, FL (1991).

3.9.3 Extraction methods

S.A. Scheppers Wercinski (ed.), *Solid Phase Microextraction: A Practical Guide*, M. Dekker, New York, NY (1999).

M. Caude and D. Thiébaut (eds), *Practical Supercritical Fluid Chromatography and Extraction*, Harwood, Amsterdam (1999).

J.S. Fritz, *Analytical Solid-phase Extraction*, Wiley-VCH, New York, NY (1999).

J. Pawliszyn (ed.), *Applications of Solid Phase Microextraction*, The Royal Society of Chemistry, Cambridge (1999).

A.J. Handley (ed.), *Extraction Methods in Organic Analysis*, Sheffield Academic Press, Sheffield (1998).

J.R. Dean, *Extraction Methods for Environmental Analysis*, John Wiley & Sons, Ltd, Chichester (1998).

T. Clifford, *Fundamentals of Supercritical Fluids*, Oxford University Press, Oxford (1998).

E.M. Thurman and M.S. Mills, *Solid Phase Extraction: Principles in Practice*, John Wiley & Sons, Inc., New York, NY (1998).

W. Simpson, *Solid Phase Extraction: Principles, Strategies and Applications*. M. Dekker, New York, NY (1997).

M.A. Abraham and A.K. Sunol (eds), *Supercritical Fluids, Extraction and Pollution Prevention*, ACS Symposium Series 670, American Chemical Society, Washington, DC (1997).

R.M. Smith and S.B. Hawthorne (eds), *Supercritical Fluids in Chromatography and Extraction*, Elsevier, Amsterdam (1997).

H.M. Kingston and S.J. Haswell (eds) *Microwave-enhanced Chemistry. Fundamentals, Sample Preparation and Applications*, American Chemical Society, Washington, DC (1997).

J. Pawliszyn, *Solid Phase Microextraction. Theory and Practice*, John Wiley & Sons, Inc., New York, NY (1997).

L.T. Taylor, *Supercritical Fluid Extraction*, John Wiley & Sons, Ltd, Chichester (1996).

P.D. McDonald and E.S.P. Bouvier, *Solid Phase Extraction Applications Guide and Bibliography, A Resource for Sample Preparation Methods Development*, Waters, Milford, MA (1995).

M.D. Luque de Castro, M. Valcárcel and M.T. Tena, *Analytical Supercritical Fluid Extraction*, Springer Verlag, Heidelberg (1994).

N. Simpson and K.C. van Horne (eds), *Varian Sorbent Extraction Technology Handbook*, Varian Sample Preparation Products, Harbor City, CA (1993).

H.M. Kingston and L.B. Jassie (eds), *Introduction to Microwave Sample Preparation*, ACS Professional Reference Book, American Chemical Society, Washington, DC (1988).

M. Zief and R. Kiser, *Sorbent Extraction for Sample Preparation*, J.T. Baker, Phillipsburg, NJ (1988).

K.C. van Horne, *Sorbent Extraction Technology*, Analytichem International, Harbor City, CA (1985).

3.10 REFERENCES

1. P.M. Gy, *Sampling for Analytical Purposes*, John Wiley & Sons, Ltd, Chichester (1998).

2. N. Pipkin, *Am. Lab.*, 40D (November 1990).

3. R.J. Majors, K.P. Kelly, J. Ezzell and M.T. Paulus, *Supplement to LC.GC Europe* **13** (11) (2000).

4. H.J. Vandenburg and A.A. Clifford, in *Extraction Methods in Organic Analysis* (A.J. Handley, ed.), Sheffield Academic Press/CRC Press, Sheffield (1999), pp. 221–42.

5. R.E. Majors, *LC.GC* **13**, 742 (1995).

6. M.P. Thomas, *J. Vinyl Addit. Technol.* **2** (4), 330–8 (1996).

7. V.R. Meyer, *LC.GC Europe* **15** (7), 398–404 (2002).

8. R.E. Majors, *Proceedings of Solutions for Scientist Symposium*, Advanstar Communications, London (1999).

9. D. Stoye and W. Freitag (eds), *Paints, Coatings and Solvents*, Wiley-VCH, Weinheim (1998).

10. Y. Marcus, *The Properties of Solvents*, John Wiley & Sons, Inc., New York, NY (1998).

11. E.W. Flick, *Industrial Solvents Handbook*, Noyes Data Corp., Park Ridge, NJ (1991).

12. J. Roire, *Les Solvants*, Erec, Puteaux (1989).

13. M. Ash and I. Ash, *Dispersants, Solvents and Solubilizers*, E. Arnold, London (1988).

14. G. Wypych (ed.), *Handbook of Solvents*, ChemTec Publishing, Toronto (2001).

15. C. Reichardt, *Solvents and Solvent Effects in Organic Chemistry*, Wiley-VCH, Weinheim (2003).

16. O. Fuchs, in *Polymer Handbook* (J. Brandrup and E.H. Immergut, eds), Wiley-Interscience, New York, NY (1989), pp. VII 379–407.

17. E.A. Grulke, in *Polymer Handbook* (J. Brandrup, E.H. Immergut and E.A. Grulke, eds), Wiley-Interscience, New York, NY (1999), p. VII–675.

18. M. McParland and N. Bates, *Toxicology of Solvents*, Rapra Technology Ltd, Shawbury (2002).

19. L.R. Snyder, *J. Chromatogr. Sci.* **16**, 223 (1978).

20. A. Tímár-Balázsy and D. Eastop, *Chemical Principles of Textile Conservation*, Butterworths-Heinemann, Oxford (1998).

21. A.J. Parker, *Chem. Rev.* **69**, 1 (1969).

22. M. Chastrette, *Tetrahedron* **35**, 1441 (1979).

23. D. Desrosiers, *Proceedings SPE PMAD RETEC*, Atlantic City, NJ (1996), pp. 145–76.

24. L. Snyder, *Tech. Chem.* **12**, 25 (1978).

25. V.J. Barwick, *Trends Anal. Chem.* **16** (6), 293–309 (1997).

26. W.J. Staal, *PhD Thesis*, Technical University of Eindhoven (1996).

27. J.H. Hildebrand and R.L. Scott, *The Solubility of Nonelectrolytes*, Reinhold Publ., New York, NY (1964).

28. V.Yu. Senichev and V.V. Tereshatov, in *Handbook of Solvents* (G. Wypych, ed.), ChemTec Publishing, Toronto (2001), pp. 243–52.

29. P.A. Small, *J. Appl. Chem.* **3**, 71 (1953).

29a. C.M. Hansen, *I & EC Prod. Res. Dev.* **8**, 2 (1969).

30. A.F.M. Barton, *Handbook of Solubility Parameters and Other Cohesion Parameters*, CRC Press, Boca Raton, FL (1983).

31. A.F.M. Barton, *Handbook of Polymer–Liquid Interaction Parameters and Solubility Parameters*, CRC Press, Boca Raton, FL (1990).

32. V.Yu. Senichev and V.V. Tereshatov, in *Handbook of Solvents* (G. Wypych, ed.), ChemTec Publishing, Toronto (2001), pp. 101–24.

33. C. Hansen, *Hansen Solubility Parameters. A User's Handbook*, CRC Press, Boca Raton, FL (2000).

34. D. Wright, *Environmental Stress Cracking of Plastics*, Rapra Technology Ltd, Shawbury (1996).

35. L.G. Krauskopf, *Proceedings SPE ANTEC '99*, New York, NY (1999), pp. 3512–25.

36. L.G. Krauskopf, *J. Vinyl Addit. Technol.* **5** (2), 101–6 (1999).

37. H.J. Vandenburg, A.A. Clifford, K.D. Bartle, S. Zhu, J. Carroll, I.D. Newton and L.M. Garden, *Anal. Chem.* **70**, 1943–8 (1998).

38. R.P.T. Tomkins, G. Hefter and C. Young, *The Experimental Determination of Solubilities*, John Wiley & Sons, Ltd, Chichester (2002).

39. W. Hofmann, *Rubber Technology Handbook*, Hanser Publishers, Munich (1996).

40. J. Brandrup, E.H. Immergut and E.A. Grulke (eds), *Polymer Handbook*, John Wiley & Sons, Inc., New York, NY (1999).

41. G. Bodor, *Structural Investigation of Polymers*, Ellis Horwood, New York, NY (1991).

42. C.V. Horie, *Materials for Conservation*, Butterworths, London (1987), pp. 193–223.

43. ASTM D 5226-98. *Standard Practice for Dissolving Polymer Materials*.

44. M. van Duin, *Techn. Rept DSM Research*, Geleen (1998).

45. M.M. Coleman, J.F. Graf and P.C. Painter, *Specific Interactions and the Miscibility of Polymer Blends*, Technomic Publishing Inc., Lancaster (1991).

46. R. Sehan and W. Kimmer, *Jena Rev.* **35**, 31 (1990).

47. H. Winterberg, *Plaste Kautsch.* **37**, 185 (1990).

48. G. Meszlényi, M. Sipos, E. Juhasz and M. Lelkes, *J. Appl. Polym. Sci.: Appl. Polym. Symp.* **48**, 411 (1991).

49. X.Z. Xiang, K. Dahlgren, W.P. Enlow and A.G. Marshall, *Anal. Chem.* **64**, 2862 (1992).

50. R.F. Cross and R.E. Majors, *LC.GC Europe* **13** (10), 727–33 (2000).

51. P. Perlstein, *Anal. Chim. Acta* **149**, 21–7 (1983).

52. R.E. Majors, *LC.GC Intl.* **11** (7), 434–9 (1998).

53. Spex CertiPrep Inc., *Handbook of Sample Preparation and Handling*, Metuchen, NJ (1997), p. 116.

54. K.M. Dooley, D. Launey, J.M. Becnel and T. Caines, *ACS Symposium Series* **608**, 269–80 (1995).

55. B.G. Risch, *Proceedings SPE ANTEC '99*, New York, NY (1999), pp. 3361–5.

56. R.C. Nielson, *J. Liq. Chromatogr.* **16**, 1625–38 (1993).

57. W. Dunges, *Pre-chromatographic Micromethods*, Huethig, Heidelberg (1987).

58. T.S. Ma and V. Horak, *Microscale Manipulations in Chemistry*, John Wiley & Sons, Inc., New York, NY (1976).

59. T.C. Lo, in *Encyclopedia of Physical Science and Technology* (R.A. Meyers, ed.), Academic Press, San Diego, CA (2001), Vol. 15.

60. A.J. Handley (ed.), *Extraction Methods in Organic Analysis*, Sheffield Academic Press/CRC Press, Sheffield (1999).

61. K. Carlsson and B. Karlberg, *Anal. Chim. Acta* **415**, 1 (2000).

62. A. Dolev, E. Kehat and R. Lavie, *Ind. Eng. Chem. Res.* **38** (4), 1618–24 (1999).

63. C. Simoneau and P. Hannaert, *Food Addit. Contam.* **16** (5), 197–206 (1999).

64. A.-C. Albertsson, C. Barenstedt and S. Karlsson, *Polym. Degrad. Stab.* **37**, 163 (1992).

65. A.-C. Albertsson, C. Barenstedt and S. Karlsson, *Acta Polym.* **45**, 97 (1994).

66. A.-C. Albertsson, C. Barenstedt and S. Karlsson, *J. Chromatogr.* **A690**, 207 (1995).

67. M. Hakkarainen, A.-C. Albertsson and S. Karlsson, *Polym. Degrad. Stab.* **52**, 283 (1996).

68. M. Hakkarainen, A.-C. Albertsson and S. Karlsson, *J. Chromatogr.* **A741**, 251 (1996).

69. F.E. Lussier, *ACS Rubber Div. 122nd Mtg* (1983), Paper No. 27.

70. H.V. Drushel and A.L. Sommers, *Anal. Chem.* **36**, 836 (1964).

71. C.L. Hilton, *Anal. Chem.* **32**, 1554 (1960).

72. C. Stafford, *Anal. Chem.* **32**, 1811 (1960).

73. C. Stafford, *Anal. Chem.* **34**, 794–6 (1962).

74. R.F. van der Heide and O. Wouters, *Z. Lebensm. Unters. Forsch.* **117**, 129 (1962).

75. F. Sevini and B. Marcato, *J. Chromatogr.* **260**, 507–12 (1983).

76. A. Fiorenza, G. Bonomi and A. Saredi, *Mater. Plast. Elast.* **10**, 1045–50 (1965).

77. A.J. Holden, in *Extraction Methods in Organic Analysis* (A.J. Handley, ed.), Sheffield Academic Press/CRC Press, Sheffield (1999), pp. 5–53.

78. H.L. Spell and R.D. Eddy, *Anal. Chem.* **32**, 1811–4 (1960).

79. I.D. Newton, in *Polymer Characterisation* (B.J. Hunt and M.I. James, eds), Blackie, Glasgow (1993), pp. 8–36.

80. A.M. Wims and S.J. Swarin, *J. Appl. Polym. Sci.* **19**, 1243–56 (1975).

81. T.R. Crompton, *Practical Polymer Analysis*, Plenum Press, New York, NY (1993).

82. J. Haslam, H.A. Willis and D.C.M. Squirrell, *Identification and Analysis of Plastics*, John Wiley & Sons, Ltd, Chichester (1983).

83. T.R. Crompton, *Manual of Plastics Analysis*, Plenum Press, New York, NY (1998).

84. M.A. Haney and W.A. Dark, *J. Chromatogr. Sci.* **18**, 655–9 (1980).

85. U. Gasslander and H. Jaegfeldt, *Anal. Chim. Acta* **166**, 243–51 (1984).

86. J.D. Vargo and K.L. Olson, *Anal. Chem.* **57**, 672–5 (1985).

87. ISO 6427. *Kunststoffe Bestimmung der extrahierbaren Bestandteile durch organische Lösungsmittel (Standardverfahren)* (1998).
88. ISO 1407. *Kautschuk und Elastomere. Bestimmung der mit Lösemitteln extrahierbaren Bestandteile* (1992).
89. T.P. Lynch, *J. Chromatogr. Libr.* **56**, 269–303 (1995).
90. R.C. Nielson, *J. Liq. Chromatogr.* **14**, 503–19 (1991).
91. R.E. Majors, *LC.GC* **14** (2), 88–96 (1996).
92. H.J. Vandenburg, A.A. Clifford, K.D. Bartle, J. Carroll, I. Newton, L.M. Garden, J.R. Dean and C.T. Costley, *Analyst* **122**, 101–15R (1997).
93. W. Freitag, in *Kunststoff-Additive* (R. Gächter and H. Müller, eds), Hanser, Munich (1990), pp. 909–46.
94. T.A. Fadeeva, V.M. Ryabikova, G.S. Popova and L.V. Toporkova, *Plast. Massy* (10), 72–4 (1990).
95. R.E. Majors, *LC.GC* **13**, 82 (1995).
96. W. Freitag and O. John, *Angew. Makromol. Chem.* **175**, 181–5 (1990).
97. E.E. King and D. Barclay, *Intl. Lab.* **28** (1), 7–11 (1998).
98. W.J. Staal, P. Cools, A.M. van Herk and A.L. German, *Chromatographia* **37** (3–4), 218–20 (1993).
99. R.H. Campbell and R.W. Wize, *J. Chromatogr.* **12**, 178 (1963).
100. D.F. Slonaker and D.C. Sievers, *Anal. Chem.* **36**, 1130 (1964).
101. R.F. van der Heide and O. Wouters, *Z. Lebensm. Unters. Forsch.* **117**, 129 (1962).
102. E. Schröder and G. Rudolph, *Plaste Kautsch.* **10**, 22 (1963).
103. E. Schröder, J. Franz and E. Hagen, *Ausgewälte Methoden zur Plastanalytik*, Akademie-Verlag, Berlin (1976), p. 31.
104. D. Braun and E. Bezdadea, *Angew. Makromol. Chem.* **113**, 77–90 (1983).
105. R. Paukkeri and A. Lehtinen, *Polymer* **35** (8), 1673–9 (1994).
106. T. Kumar, *Analyst* **115**, 1319–22 (1990).
107. I. Tsujikaido and Y. Matsuda, *Jpn. Kokai Tokkyo Koho* JP 86-36209 (to Sumitomo Electric Industries Ltd), February 20, 1986.
108. D.W. Carlson, M.W. Hayes, H.C. Ransaw, R.S. McFadden and A.G. Altenau, *Anal. Chem.* **43** (13), 1874–6 (1971).
109. C.F. Wilkinson and J.C. Lamb, *Regul. Toxic. Pharmacol.* **30** (2), 140–55 (1999).
110. K. Tôei, S. Motomizu and T. Umano, *Talanta* **29**, 103–6 (1982).
111. H.C. Ashton, *Angew. Makromol. Chem.* **261/262**, 9–23 (1998).
112. W.R. Hudgins, K. Theurer and T. Mariani, *J. Appl. Polym. Sci.: Appl. Polym. Symp.* **34**, 145–55 (1978).
113. C.T. Costley, J.R. Dean, I.D. Newton and J. Carroll, *Anal. Commun.* **34**, 89–91 (1997).
114. A. Caceres, F. Ysambert, J. Lopez and N. Marquez, *Sep. Sci. Technol.* **31**, 2287–98 (1996).
115. J. Scheirs, *Compositional and Failure Analysis of Polymers. A Practical Approach*, John Wiley & Sons, Ltd, Chichester (2000).

116. M. Nagata and Y. Kishioka, *Bunseki Kagaku* **38** (6), T75–80 (1989).
117. G. Matz, *J. Chromatogr.* **587**, 205 (1991).
118. L. Evangelista and V. Jeanguyot, *Symposium on Additives for Polymers*, Certech, Namur (2001).
119. *LG.GC Magazine Survey* (1997).
120. F. Soxhlet and J. Szombathy, *Dinglers Polytech. J.* **232**, 461 (1879).
121. S.B. Hawthorne, A.B. Galy, V.O. Schnitt and D.J. Miller, *Anal. Chem.* **67**, 2723–32 (1995).
122. X.W. Lou, H.-G. Janssen and C.A. Cramers, *Anal. Chem.* **69**, 1598–1603 (1997).
123. R.G. Lichtenthaler and F.J. Ranfelt, *J. Chromatogr.* **149**, 553 (1978).
124. H.G. Nowicki, C.A. Kieda, R.F. Devine, V. Current and T.H. Schaefers, *Anal. Lett.* **A12**, 769–76 (1979).
125. T.R. Crompton, *Eur. Polym. J.* **4**, 473 (1968).
126. R.E. Majors, *J. Chromatogr. Sci.* **8**, 339–45 (1970).
127. V. Lopez-Avila, K. Bauer, J. Milanes and W.F. Beckert, *J. Assoc. Off. Anal. Chem.* **76**, 864–80 (1993).
128. K. Ganzler, A. Salgó and K. Valkó, *J. Chromatogr.* **371**, 299–306 (1986).
129. V. Lopez-Avila, R. Young, R. Kim and W.F. Beckert, *J. Chromatogr. Sci.* **33**, 481–4 (1995).
130. A.S.Y. Chau and L.J. Babjak, *J. Assoc. Off. Anal. Chem.* **62**, 107–13 (1979).
131. F.M. Dunnivant and A.W. Elzerman, *J. Assoc. Off. Anal. Chem.* **71**, 551–6 (1988).
132. S.B. Hawthorne, *Anal. Chem.* **62**, 633–42A (1990).
133. T.L. Chester, J.D. Pinkston and D.E. Raynie, *Anal. Chem.* **64**, 153–70R (1992).
134. B.E. Richter, B.A. Jones, J.L. Ezzell, N.L. Porter, N. Avdalovic and C. Pohl, *Anal. Chem.* **68**, 1033–9 (1996).
135. J. Lehotay, J. Daněček, O. Liška, O.J. Leško and E. Brandšteterová, *J. Appl. Polym. Sci.* **25**, 1943 (1980).
136. R.E. Majors and E.L. Johnson, *J. Chromatogr. Sci.* **167**, 17 (1978).
137. M.W. Raynor, K.D. Bartle, I.L. Davies, A. Williams, A.A. Clifford, J.M. Chalmers and B.W. Cook, *Anal. Chem.* **60**, 427 (1988).
138. J.-L. Gardette and J. Lemaire, *Polym. Degrad. Stab.* **34**, 135–67 (1991).
139. D. Dilettato, P.J. Arpino, K. Nguyen and A. Bruchet, *J. High Resolut. Chromatogr.* **14** (5), 335–42 (1991).
140. F.T. Dettmer, H. Wickmann, J. de Boer and M. Bahadir, *Chemosphere* **39** (9), 1523–32 (1999).
141. U.D. Standt, *Angew. Makromol. Chem.* **150**, 13–9 (1987).
142. R.R. Kozlowski and T.K. Gallagher, *J. Vinyl Add. Technol.* **3**, 249–55 (1997).
143. M.M. Robertson and R.H. Rowley, *Br. Plastics*, 26–9 (January 26, 1960).
144. R.L. Gray and R.E. Lee, in *Chemistry and Technology of Polymer Additives* (S. Al-Malaika, A. Golovoy and C.A. Wilkie, eds), Blackwell Science, Oxford (1999), pp. 21–35.
145. X.W. Lou, H.-G. Janssen and C.A. Cramers, *J. Chromatogr. Sci.* **34**, 282–90 (1996).

146. K. Kuosmanen, T. Hyötyläinen, K. Hartonen and M.-L. Riekkola, *J. Chromatogr.* **A943**, 113–22 (2002).

147. G. Persson and F. Alstin (to Tecator AB), *WO97/00109* (January 3, 1997).

148. *Tecator Application Short Note* ASTN 52/88 (1988).

149. *Tecator Application Short Note* ASTN 53/88 (1988).

150. K. Michalski and W.B. Mroczyk, *J. Near Infrared Spectrosc.* **6**, 247–51 (1998).

151. U.S. Environmental Protection Agency (EPA), *Method 3541, Automated Soxhlet Extraction* (September 1994).

152. J. Redeker, *German Patent* 19612037 (October 2, 1997).

153. H. Siegel, *GIT Fachz. Lab.* **41**, 486–7 (1997).

154. J. Redeker, *LaborPraxis* **21** (4), 60–4 (1997).

155. B. Matthäus, *Bioforum* **21** (1/2), 26–9 (1998).

156. P.C.E. Tackx, H. Nelissen and H. Kolnaar, *Techn. Rept DSM Research*, Geleen (1998).

157. H. Nelissen, *Techn. Rept DSM Research*, Geleen (2000).

158. M.J. Brock and G.D. Louth, *Anal. Chem.* **27**, 1575–80 (1955).

159. S.L. Likens and G.B. Nickerson, *Proc. Am. Soc. Brew. Chem.*, 5–13 (1964).

160. T.H. Schultz, R.A. Flath, T.R. Mon, S.B. Eggling and R. Teranishi, *J. Agric. Food Chem.* **25**, 446–9 (1977).

161. M. Godefroot, M. Stechele, P. Sandra and M. Verzele, *J. High Resolut. Chromatogr., Chromatogr. Commun.* **5**, 75 (1982).

162. V. Seidel and W. Lindner, *Anal. Chem.* **65**, 3677 (1993).

163. R.E. Majors, *LC.GC Intl*, 19–24 (January 1999).

164. H. Ishiwata, T. Inoue, M. Yamamoto and K. Yoshihara, *J. Agric. Food Chem.* **36**, 1310–3 (1988).

165. D.C.M. Squirrell, *Analyst* **106**, 1042–56 (1981).

166. P. Rebeyrolle-Bernard and P. Etiévant, *J. App. Polym. Sci.* **49**, 1159–64 (1993).

167. E. Lück and M. Jager, *Antimicrobial Food Additives. Characteristics, Uses, Effects*, Springer, Berlin (1997).

168. ASTM E 1385-90, *Standard Practice for Separation and Concentration of Flammable or Combustible Liquid Residues from Fire Debris Samples by Steam Distillation*, American Society for Testing and Materials, Philadelphia, PA (1991).

169. F. Mangani, A. Cappiello, G. Crescentini, F. Bruner and L. Bonfanti, *Anal. Chem.* **59**, 2066–9 (1987).

170. R.T. Roberts, *Spec. Publ.-R. Soc. Chem.* **106**, 287–97 (1992).

171. D.J. McClements, *Trends Food Sci. Technol.* **6** (9), 293–9 (1995).

172. J. Hilgert, *Met. Finish.* **95** (4), 54–6 (1997).

173. W.T. Richards and A.L. Loomis, *J. Am. Chem. Soc.* **49**, 3086 (1927).

174. T.J. Mason and J.P. Lorimer, *Sonochemistry: Theory, Applications and Uses of Ultrasound in Chemistry*, Ellis Horwood, Chichester (1988).

175. L.A. Crum, *Ultrasonics Sonochem.* **2**, S147–52 (1995).

176. J.-L. Luche, *Adv. Sonochem.* **3**, 85–124 (1993).

177. T.J. Mason, *Practical Sonochemistry*, Ellis Horwood, Chichester (1991).

178. T.J. Mason, *The Uses of Ultrasound in Chemistry*, The Royal Society of Chemistry, Cambridge (1990).

179. S.V. Ley and C.M.R. Low, *Ultrasound in Synthesis*, Springer Verlag, Berlin (1989).

180. J.-L. Luche (ed.), *Synthetic Organic Sonochemistry*, Plenum Press, New York, NY (1998).

181. T.J. Mason, *Educ. Chem.* **24**, 104 (1987).

182. R.A. Pethrick, *Adv. Sonochem.* **2**, 65–133 (1991).

183. B. Mitra and D.J. Booth, *Ultrasonics* **35**, 569–72 (1998).

184. D.R. Bauer, *J. Coat. Technol.* **66**, 57 (1994).

185. C.L. Bianchi, E. Gotti, L. Toscano and V. Ragaini, *Ultrasonics Sonochem.* **4**, 317–20 (1997).

186. F.R. Young, *Cavitation*, McGraw-Hill, London (1989).

187. E. Cordemans, *Chem. Mag. (Leuven)* (2), 12–4 (1999).

188. O.V. Abramov, *High-Intensity Ultrasonic: Theory and Industrial Applications*, Gordon & Breach, New York, NY (1998).

189. A. Sahnoune, L. Piché, A. Hamel, R. Gendron, L.E. Daigneault and L.-M. Caron, *Proceedings SPE ANTEC '97*, Toronto (1997), pp. 2259–63.

190. L.C. Lynworth, *Ultrasonic Measurements for Process Control: Theory, Techniques, Applications*, Academic Press, New York, NY (1989).

191. D. Schneider, *Materials World*, 11–12 (April 2001).

192. F.A. Duck, A.C. Baker and H.C. Starritt (eds), *Ultrasound in Medicine*, Institute of Physics Publ., Philadelphia, PA (1998).

193. W.D. O'Brien, *Jpn J. Appl. Phys. Part. 1*, **37** (5B), 2781–8 (1998).

194. I. Alig, D. Lellinger and S. Tadjbakhsch, *Proceedings ACS Div. Polym. Mater. Sci. Engng* **79**, 31–2 (1998).

195. M. Vinatoru, M. Toma, O. Radu, P.I. Filip, D. Lazurca and T.J. Mason, *Ultrasonics Sonochem.* **4**, 135–9 (1997).

196. D.E. Kimbrough, R. Chin and J. Wakakuwa, *Analyst* **119**, 1283–92 (1994).

197. U.S. Environmental Protection Agency (EPA), *Method 3540C, Soxhlet Extraction* (January 1995).

198. U.S. Environmental Protection Agency (EPA), *Method 3550, Ultrasonic Extraction* (January 1995).

199. ASTM STP D 1996, *Test Method for Determination of Phenolic Antioxidants and Erucamide Slip Additives in LDPE Using Liquid Chromatography, Annual Book of ASTM Standards* (D.L. Dolmage, ed.), American Society for Testing and Materials, Philadelphia, PA (1993), Vol. 08.01.

200. H.J. Brandt, *Anal. Chem.* **33**, 1390–1 (1961).

201. C. Nerín, J. Salafranca and J. Cacho, *Food Addit. Contam.* **13**, 243–50 (1996).

202. N. Haider and S. Karlsson, *Analyst* **124**, 797–800 (1999).

203. N. Haider and S. Karlsson, *Biomacromol.* **1**, 481–7 (2001).

204. G.J. DeMenna and W.J. Edison, *Novel Sample Preparation Techniques for Chemical Analysis – Microwave and Pressure, Dissolution, Chemical Analysis of Metals*, ASTM STP 994 (F.T. Coyle, ed.), American Society for Testing and Materials, Philadelphia, PA (1987), p. 45.

205. M.G. Monteiro and V.F. Matos, *Proc. Intl. GPC Symp. '87*, Millipore Corp., Milford, MA (1987), pp. 437–46.

206. P. Métois, D. Scholle, J. Bouquant and A. Feigenbaum, *Food Addit. Contam.* **15**(1), 100–11 (1998).

207. L. Contat-Rodrigo, N. Haider, A. Ribes-Greus and S. Karlsson, *J. Appl. Polym. Sci.* **79** (6), 1101–12 (2001).

208. P. Meng, L. Li and G. Song, *Suliao Gongye* **25** (5), 104–5 (1997).

209. F. Fiala, I. Steiner and K. Kubesch, *Dtsch. Lebensm.- Rdsch.* **96** (2), 51–7 (2000).

209a. J.C.J. Bart, M. Schmid and S. Affolter, *Anal. Sci.*, **17**, i729–32 (2001).

210. B. Marcato and M. Vianello, *J. Chromatogr.* **A869**, 285–300 (2000).

211. L. Brown, *Analyst (London)* **104**, 1165 (1979).

212. H. Carlsson, U. Nilsson, G. Becker and C. Oestman, *Environm. Sci. Technol.* **31**(10), 2931–6 (1997).

213. R. Stringer, I. Labunska, D. Santillo, P. Johnston, J. Siddorn and A. Stephenson, *Envir. Sci. Pollut. Res.* **7** (1), 27–36 (2000).

214. USEPA, *Method 606 – Phthalate Ester.* Federal Register **409** (209), 73–80 (1984).

215. M.G. Botros, *J. Plast. Film Shtg.* **11**, 326–37 (1995).

216. K. Kitagawa, S. Hayasaki and Y. Ozaki, *Vibr. Spectrosc.* **15**, 43–51 (1997).

217. A. Sahnoune, A. Hamel and L. Piché, *Proceedings SPE ANTEC '98*, Atlanta, GA (1998), pp. 1927–31.

218. D.J. McClements and M.J.W. Povey, *Ultrasonics* **30** (6), 383–8 (1992).

219. M.J.W. Povey, *Ultrasonic Techniques for Fluids Characterization*, Academic Press, San Diego, CA (1997).

220. A.M. Oliveira Brett, C.M.A. Brett, F.-M. Matysik and S. Matysik, *Ultrasonics Sonochem.* **4**, 123–4 (1997).

221. A.M. Oliveira Brett and F.-M. Matysik, *Ultrasonics Sonochem.* **4**, 125–6 (1997).

222. A. Morgan, *Adv. Sonochem.* **3**, 145–64 (1993).

223. R. Ma, C.W. McLeod and K.C. Thompson, in *Plasma Source Mass Spectrometry, Developments and Applications* (G. Holland and S.D. Tanner, eds), The Royal Society of Chemistry, Cambridge (1997), pp. 192–201.

224. Y.C. Wang and W.B. Yang, *Chin. Anal. Chem.* **19**, 102 (1991).

225. Q.H. Jin, H.Q. Zhang and S.R. Yu, *Spectrosc. Spectrum Anal.* **9**, 31 (1989).

226. B. Hartmann, *J. Appl. Polym. Sci.* **19**, 3241 (1975).

227. T.J. Mason, *Chemistry with Ultrasound*, Elsevier Applied Science, Oxford (1990).

228. W.P. Mason (ed.), in *Physical Acoustics*, Academic Press, New York, NY (1964), Vol. I, Pt A.

229. I. Alig and D. Lellinger, *Chem. Innov.* **30** (2), 12–8 (2000).

230. I. Molinero, *Vide, Couches Minces* **251** (Suppl.), 21–8 (1990).

231. J. Stanullo, S. Bojinski, N. Gold, S. Shapiro and G. Busse, *Ultrasonics* **36** (1–5), 455–60 (1998).

232. G. Bechtold, K. Friedrich, K.M. Gaffney and J. Botsis, in *Verbundwerkstoffe Werkstoffverbunde* (K. Friedrich, ed.), DGM Informationsgesellschaft, Oberursel (1997), pp. 845–50.

233. S.-S.L. Wen, C.-K. Jen, K.T. Nguyen, A. Derdouri and Y. Simard, *Proceedings SPE ANTEC '99*, New York, NY (1999), pp. 636–40.

234. Y. Thomas, K.C. Cole, L.E. Daigneault and L.-M. Caron, *Proceedings SPE ANTEC '96*, Indianapolis, IN (1996), pp. 1855–9.

235. D.R. Franca, C.-K. Jen, K.T. Nguyen and R. Gendron, *Polym. Eng. Sci.* **40**(1), 82–94 (2000).

236. I. Alig, D. Lellinger and S. Tadjbakhsch, *Abstracts Advanced Methods of Polymer Characterization: New Developments and Applications in Industry*, Mainz (1999), Paper O3.

237. E.C. Brown, T.L.D. Collins, A.J. Dawson, P. Olley and P.D. Coates, *Proceedings SPE ANTEC '98*, Atlanta, GA (1998), pp. 335–9.

238. S. Singh, *Proceedings 14th World Conference on Nondestructive Testing*, New Dehli (1996), Vol. 2, pp. 581–7.

239. J. Wu, *Proc. SPIE-Int. Soc. Opt. Eng.* **3341**, 143–53 (1998).

240. A.E. Crawford, *Ultrasonic Engineering*, Butterworths, London (1955).

241. H.C. Van Ness and M.M. Abbott, *Classical Thermodynamics of Nonelectrolytic Solutions*, McGraw-Hill Co., New York, NY (1982).

242. D.R. Gere, L.G. Randall and D. Callahan, in *Instrumental Methods in Food Analysis* (J.R.J. Paré and J.M.R. Bélanger, eds), Elsevier, Amsterdam (1997), pp. 421–84.

243. A.K. Sunol and S.G. Sunol, in *Handbook of Solvents* (G. Wypych, ed.), ChemTec Publishing, Toronto (2001), pp. 1419–59.

244. C. de la Tour, *Ann. Chim. Phys.* **21**, 127 (1822).

245. A. Clifford and K.D. Bartle, *Chem. Brit.* (6), 499 (1993).

246. M. Poliakoff and S. Howdle, *Chem. Brit.* (2), 118 (1995).

247. T. Harvala, M. Alkio and V. Komppa, in *Extraction '87, I. Chem. E Symposium Series* No. 103, Pergamon Press (1987), pp. 233–43.

248. E. Stahl, K.W. Quirin and D. Gerard, *Verdichtete Gase zur Extraktion und Raffination*, Springer, Berlin (1987).

249. S.B. Hawthorne and D.J. Miller, *J. Chromatogr. Sci.* **24**, 258 (1986).

250. J.B. Hannay and J. Hogarth, *Proc. R. Soc. London* **29**, 324 (1879).

251. J.R. Dean (ed.), *Applications of Supercritical Fluids in Industrial Analysis*, Blackie A&P, Boca Raton, FL (1993).

252. O. Kaiser (ed.), *Supercritical Fluids in Separation Sciences*, Incom, Düsseldorf (1997).

253. R.M. Smith and S.B. Hawthorne (eds), *Supercritical Fluids in Chromatography and Extraction*, Elsevier, Amsterdam (1997).

254. B. Wenclawiak (ed.), *Analysis with Supercritical Fluids: Extraction and Chromatography*, Springer-Verlag, Berlin (1992).

255. T. Clifford, *Fundamentals of Supercritical Fluids*, Oxford University Press, Oxford (1998).

256. M.A. Abraham and A.K. Sunol (eds), *Supercritical Fluids Extraction and Pollution Prevention*, ACS Symposium

Ser. 670, American Chemical Society, Washington, DC (1997).

257. M.A. McHugh and V.J. Krukonis, *Supercritical Fluid Extraction – Principles and Practice*, Butterworth-Heinemann, Boston, MA (1996).

258. D.A. Moyler, in *Distilled Beverage Flavors – Recent Developments* (J.R. Piggott and A. Patterson, eds), Ellis Horwood, Chichester (1989), pp. 319–28.

259. G. Nicolaon, *Rev. Energ.* **375**, 283–92 (1985).

260. J.M. Levy, *J. High Resolut. Chromatogr.* **17** (4), 212–6 (1994).

261. S.S. Rizvi, A.L. Benado, J.A. Zollweg and J.A. Daniëls, *Food Technol.* **40**, 55–65 (1986).

262. J.M. Levy, V. Danielson, R. Ravey and L. Dolata, *LC.GC* **12** (12), 920–3 (1994).

263. P. Hubert and O. Vitzthum, *Angew. Chem. Intl. Ed.* **17**, 710–15 (1978).

264. J.H. Aubert, *J. Supercrit. Fluids* **11**, 163–72 (1998).

265. J.F. Parcher and P.S. Wells, *Proceedings 8th Intl. Symposium on Supercritical Fluid Chromatography and Extraction*, St Louis, MO (1998), Paper L-29.

266. A. Cortesi, I. Kikic, P. Alessi, G. Turtoi and S. Garnier, *J. Supercrit. Fluids* **14** (2), 139–44 (1999).

267. S.J. Avison, D.A. Gray, G.M. Davidson and A. Taylor, *J. Agric. Food Chem.* **49**(1), 270–5 (2001).

268. P.G. Debenedetti, J.W. Tom, S.-D. Yeo and G.-B. Lim, *J. Controlled Release* **24**, 27–44 (1993).

269. J.M. Bruna, *Rev. Plast. Mod.* **69**, 448–51 (1995).

270. T.J. Bruno and J.F. Ely (eds), *Supercritical Fluid Technology: Reviews in Modern Theory and Applications*, CRC Press, Boca Raton, FL (1991).

271. J. McHardy and S.P. Sawan (eds), *Supercritical Fluid Cleaning. Fundamentals, Technology and Applications*, Noyes Publications, Westwood, NJ (1998).

272. R.D. Dean, *Extraction Methods for Environmental Analysis*, John Wiley & Sons, Ltd, Chichester (1998).

273. C.L. Phelps, N.G. Smart and C.M. Wai, *J. Chem. Educ.* **73**, 1163–8 (1996).

274. G.G. Hoyer, *ChemTech* **15**, 440 (1985).

275. S.P. Sawan, Y.-T. Shieh, J.-H. Su, G. Manivannan and W.D. Spall, in *Supercritical Fluid Cleaning* (J. McHardy and S.P. Sawan, eds), Noyes Publications, Westwood, NJ (1998), pp. 121–61.

276. S.G. Kazarian, M.F. Vincent, F.V. Bright, C.L. Liotta and C.E. Eckert, *J. Am. Chem. Soc.* **118**, 1729–36 (1996).

277. A.R. Berens, G.S. Huvard, R.W. Korsmeyer and F.W. Kunig, *J. Appl. Polym. Sci.* **46**, 231 (1992).

278. N.J. Cotton, K.D. Bartle, A.A. Clifford and C.J. Dowle, *J. Appl. Polym. Sci.* **48**, 1607–19 (1993).

279. D.H. Ender, *ChemTech* **16**, 52 (1986).

280. K.J. Wynne, S. Shenoy, T. Fujiwara, D. Woerdeman and S. Irie, *Preprints IUPAC Polymer Conference PC 2002*, Kyoto (2002), p. 428.

281. S.G. Kazarian, B.J. Briscoe, D. Coombs and G. Bulter, *Spectrosc. Europe* **11** (3), 10–6 (1999).

282. M. Poliakoff, S.M. Howdle and S.G. Kazarian, *Angew. Chem. Int. Ed. Engl.* **34**, 1275 (1995) and references therein.

283. S.G. Kazarian, *Proceedings SPE ANTEC '99*, New York, NY (1999), pp. 2433–5.

284. A.A. Clifford, in *Encyclopedia of Analytical Science* (A. Townshend, ed.), Academic Press, London (1995), pp. 4724–9.

285. D.R. Gere, C.R. Knipe, D.C. Messer and L.T. Taylor, *Proceedings 5th Intl. Symposium on Supercritical Fluid Chromatography and Extraction*, Baltimore, MD (1994), paper A-16.

286. K.D. Bartle, A.A. Clifford, S.B. Hawthorne, J.J. Langenfeld, D.J. Miller and R. Robinson, *J. Supercrit. Fluids* **3**, 143–9 (1990).

287. K.D. Bartle, T. Boddington, A.A. Clifford, N.J. Cotton and C.F. Dowle, *Anal. Chem.* **63**, 2371–7 (1991).

288. K.D. Bartle, T. Boddington, A.A. Clifford and S.B. Hawthorne, *J. Supercrit. Fluids* **5**, 207–12 (1992).

289. A.A. Clifford, K.D. Bartle and S.A. Zhu, *Anal. Proc.* **32**, 227–30 (1995).

290. Z.Y. Liu, P.B. Farnsworth and M.L. Lee, *J. Microcol. Sep.* **4**, 199–208 (1992).

291. X.W. Lou, H.-G. Janssen and C.A. Cramers, *J. Microcol. Sep.* **7**, 303–17 (1995).

292. H.-G. Janssen and X.W. Lou, in *Extraction Methods in Organic Analysis* (A.J. Handley, ed.), Sheffield Academic Press/CRC Press, Sheffield (1999), pp. 100–45.

293. R.M. Smith, *J. Chromatogr.* **A856**, 83–115 (1999).

294. R.W. Vannoort, J.-P. Chervet, H. Lingeman, G.J. de Jong and U.A.Th. Brinkman, *J. Chromatogr.* **505**, 45–77 (1990).

295. M. Natangelo, D. Giavini and E.B. Fanelli, *Lab2000*, **10**, 46–52 (1996).

296. A. Kot, P. Sandra and A.M. Kolodziejczyk, *Pol. J. Environ. Stud.* **5**, 5–15 (1996).

297. J.W. King, *Trends Anal. Chem.* **14**, 474–81 (1995).

298. X. Chaudot, A. Tambute and M. Caude, *Analusis* **25**, 81–96 (1997).

299. B.A. Charpentier and M.R. Sevenants (eds), *Supercritical Fluid Extraction and Chromatography: Techniques and Applications*, ACS Symposium Series 366, American Chemical Society, Washington, DC (1988).

300. T. Berger and T. Greibrokk, *Chromatography: Principles and Practice (Practical Supercritical Fluid Chromatography and Extraction)* (2), 107–48 (1999).

301. A.A. Clifford, in *Supercritical Fluid Extraction and Its Use in Chromatographic Sample Preparation* (S.A. Westwood, ed.), Blackie A&P, London (1993), pp. 1–38.

302. S.B. Hawthorne, in *Supercritical Fluid Extraction and Its Use in Chromatographic Sample Preparation* (S.A. Westwood, ed.), Blackie A&P, London (1993), pp. 39–64.

303. S.B. Hawthorne, D.J. Miller and J.J. Langenfeld, in *Hyphenated Techniques in Supercritical Fluid Chromatography and Extraction* (K. Jinno, ed.), Elsevier, Amsterdam (1992), Ch. 12.

304. M.E. Paulaitis, V.J. Krukonis, R.T. Kurnik and R.C. Reid, *Rev. Chem. Eng.* **1**(2), 179–250 (1983).

305. V. Berry, *LC.GC* **8**, 734 (1990).

306. M.D. Luque de Castro, M. Valcárcel and M.T. Tena, *Analytical Supercritical Fluid Extraction*, Springer Verlag, Heidelberg (1994).

307. L. Taylor, *Supercritical Fluid Extraction*, John Wiley & Sons Ltd, Chichester (1996).

308. K. Jinno (ed.), *Hyphenated Techniques in Supercritical Fluid Chromatography and Extraction*, Elsevier, Amsterdam (1992).

309. L.G. Randall, W.S. Miles, F. Rowland and C.R. Knipe, *Designing a Sample Preparation Method Which Employs Supercritical Fluid Extraction (SFE)*, Hewlett-Packard Publ. 435091-2102E, Wilmington, DE (1994).

310. M.L. Lee and K.E. Markides (eds), *Analytical Supercritical Fluid Chromatography and Extraction*, Chromatography Conferences Inc., Provo, UT (1990).

311. M. Caude and D. Thiebaut (eds), *Practical Supercritical Fluid Chromatography and Extraction*, Harwood Academic Publishers, Amsterdam (1999).

312. C.B.C. Ohanuzue and J.H. Nobbs, *Indian J. Eng. Mater. Sci.* **6** (1), 43–7 (1999).

313. R.B. Schlake, *Proceedings 5th Intl Symposium on Supercritical Fluid Chromatography and Extraction*, Baltimore, MD (1994), paper D-25.

314. F. Mellor, U. Just and Th. Strumpf, *J. Chromatogr.* **A679** (1), 147–52 (1994).

315. Y. Hirata and Y. Okamoto, *J. Microcol. Sep.* **1**, 46 (1989).

316. T.P. Lynch, in *Applications of Supercritical Fluids in Industrial Analysis* (J.R. Dean, ed.), Blackie, Glasgow (1993), pp. 188–218.

317. J.M. Levy, *LC.GC Europe* **13** (3), 174–81 (2000).

318. M. Szumski and B. Buszewski, *Intl. Lab.*, 20–24 (November 1999).

319. N.J. Cotton, K.D. Bartle, A.A. Clifford, S. Ashraf, R. Moulder and C.J. Dowle, *HRC-J. High Resolut. Chromatogr.* **14**, 164–8 (1991).

320. A. Medvedovici, A. Kot, F. David and P. Sandra, in *Supercritical Fluid Chromatography with Packed Columns* (K. Anton and C. Berger, eds), M. Dekker, New York, NY (1998).

321. J.E. France and J.W. King, *J. Agric. Food Chem.* **39**, 1874 (1991).

322. J. Salafranca, J. Cacho and C. Nerín, *J. High Resol. Chromatogr.* **22** (10), 553–8 (1999).

323. D.C. Hinz and B.W. Wenclawiak, *Proceedings 8th Intl. Symposium on Supercritical Fluid Chromatography and Extraction*, St Louis, MO (1998), paper A-02.

324. T. Greibrokk, *J. Chromatogr.* **A703**, 523–35 (1995).

325. Y.T. Shieh, J.H. Su, G. Manivannan, P.H.C. Lee, S.P. Sawan and W.D. Spall, *J. Appl. Polym. Sci.* **59**, 695 (1996).

326. Y.T. Shieh, J.H. Su, G. Manivannan, P.H.C. Lee, S.P. Sawan and W.D. Spall, *J. Appl. Polym. Sci.* **59**, 707–17 (1996).

327. A.S. Michaels and H.J. Bixler, *J. Polym. Sci.* **50**, 393–412 (1961).

328. L. Baner, T. Bücherl, J. Ewender and R. Franz, *J. Supercrit. Fluids* **5**, 213 (1992).

329. J.W. King, *J. Chromatogr. Sci.* **27**, 355–64 (1989).

330. J.B. Morris, M.A. Schroeder, R.A. Pesce-Rodriguez, K.L. McNesby and R.A. Fifer, *Proceedings 8th Intl. Symposium on Supercritical Fluid Chromatography and Extraction*, St. Louis, MO (1998), paper D-20.

331. S.H. Page, S.R. Sumpter and M.L. Lee, *J. Microcol. Sep.* **4** (2), 91–122 (1992).

332. J.M. Levy, L. Dolata, R.M. Ravey, E. Storozynsky and K.A. Holowczak, *J. High Resolut. Chromatogr.* **16**, 368–71 (1993).

333. T.P. Hunt, C.J. Dowle and G. Greenway, *Analyst* **116**, 1299–304 (1991).

334. H. Daimon and Y. Hirata, *Chromatographia* **32**, 549–54 (1991).

335. G.A. Mackay and R.M. Smith, *Analyst* **118**, 741–5 (1993).

336. M.L. Marín, A. Jiménez, J. Lopez and J. Vilaplana, *J. Chromatogr.* **A750**, 183–90 (1996).

337. S. Schmidt, L. Blomberg and T. Wannman, *Chromatographia* **28**, 400–4 (1989).

338. J.W. Oudsema and C.F. Poole, *HRC-J. High Resolut. Chromatogr.* **16**, 198–202 (1993).

339. A. Venema, H.J.F.M. Vandeven, F. David and P. Sandra, *HRC-J. High Resolut. Chromatogr.* **16**, 522–4 (1993).

340. M. Ashraf-Khorassani and J.M. Levy, *HRC-J. High Resolut. Chromatogr.* **13**, 742–7 (1990).

341. J.A. Garde, J. Galotto, R. Catalá and R. Gavara, *Proceedings ILSI Symposium on Food Packaging: Ensuring the Quality and Safety of Foods*, Budapest (1996).

342. B.W. Wenclawiak, F. Eisenbeiss, T. Hees, M. Krappe, G. Maio, T. Paschke, C. Rathmann and J. Schipke, *Proc. 5th Intl. Symposium on Supercricital Fluid Chromatography and Extraction*, Baltimore, MD (1994), pp. 24–5.

343. D.R. Gere and E.M. Derrico, *LC.GC Intl.* **7** (6), 325–31 (1994).

344. J.A. Garde, R. Catalá and R. Gavara, *Food Addit. Contam.* **15** (6), 701–8 (1998).

345. K.D. Bartle and A.A. Clifford, *Proceedings 7th Intl. Symposium on Supercritical Fluid Chromatography and Extraction*, Indianapolis, IN (1996), paper L-01.

346. SF-Solver™, *Software for Supercritical Fluid Analysis*, Isco Inc., Lincoln, NB (1991).

347. C.R. Knipe, W.S. Miles, F. Rowland and L.G. Randall, *Designing a Sample Preparation Method That Employs Supercritical Fluid Extraction*, Hewlett-Packard Company, Part No. 07680-90400 (1993).

348. Z. Otero-Keil, *Pittcon Conference*, New Orleans, LA (1992), paper 677.

349. H.-G. Janssen, X.W. Lou and C.A. Cramers, *Proceedings 8th Intl. Symposium on Supercritical Fluid Chromatography and Extraction*, St. Louis, MO (1998), paper L-50.

350. B.E. Richter, J.L. Ezzell, N.L. Porter and D.E. Knowles, *Proceedings 5th Symposium on Supercritical Fluid Chromatography and Extraction*, Baltimore, MD (1994), pp. 25–6.

351. D.R. Gere and E.M. Derrico, *LC.GC Intl.* **7** (7), 370–5 (1994).

352. B.W. Wright, J.L. Fulton, A.J. Kopriva and R.D. Smith, *ACS Symp. Ser.* **366**, 44 (1988).

353. D.R. Knapp, *Handbook of Analytical Derivatisation Reactions*, John Wiley & Sons Inc., New York, NY (1979).

354. J.A. Field, *J. Chromatogr.* **A785**, 239–49 (1997).

355. M. Thilén and R. Shishoo, *J. Appl. Polym. Sci.* **76** (6), 938–46 (2000).

356. M.L. Marín, A. Jiménez, V. Berenguer and J. López, *J. Supercrit. Fluids* **12**, 271–7 (1998).

357. M.J. Drews, K. Ivey, J. Helvey and F. van Lenten, *Proceedings 8th Intl. Symposium on Supercritical Fluid Chromatography and Extraction*, St. Louis, MO (1998), paper E-20.

358. C.H. Kirschner, S.L. Jordan, L.T. Taylor and P. Seemuth, *Anal. Chem.* **66**, 882–7 (1994).

359. M.K.L. Bicking and J.M. Levy, *Proceedings 5th Intl. Symposium on Supercritical Fluid Chromatography and Extraction*, Baltimore, MD (1994), paper F-13.

360. C. Nerín, E. Asensio, C. Fernandez and R. Battle, *Quim. Anal. (Barcelona)* **19** (4), 205–12 (2000).

361. M.-D. Bermudez, F.-J. Carrion-Vilches, G. Martinez-Nicolas and M. Pagan, *J. Supercrit. Fluids* **23** (1), 59–63 (2002).

362. S.T. Küppers, *Chromatographia* **33**, 434 (1992).

363. V. Janda, J. Kriz, J. Vejrosta and K.D. Bartle, *J. Chromatogr.* **A669**, 241 (1994).

364. R. Moulder, J.P. Kithinji, M.W. Raynor, K.D. Bartle and A.A. Clifford, *J. High Resolut. Chromatogr.* **12**, 688 (1989).

365. T. Bücherl, A. Gruner and N. Palibroda, *Packag. Technol. Sci.* **7**, 139 (1994).

366. B. Brauer, T. Funke and H. Schulenbergschell, *Dtsch. Lebensm. - Rdsch.* **91**, 381 (1995).

367. G.A. Mackay and R.M. Smith, *HRC-J. High Resolut. Chromatogr.* **18**, 607–9 (1995).

368. K. Ventura, R. Kovar, M. Mikesova, P. Karasek and J. Vejrosta, *Collect. Czech. Chem. Commun.* **60**, 1109 (1995).

369. J.H. Braybrook and G.A. Mackay, *Polym. Intl.* **27**, 157–64 (1992).

370. T.P. Hunt, C.J. Dowle and G. Greenway, *Analyst* **118**, 17–22 (1993).

371. H. Engelhardt, J. Zapp and P. Kolla, *Chromatographia* **32**, 527 (1991).

372. A.M. Pinto and L.T. Taylor, *J. Chromatogr.* **A811** (1 + 2), 163–70 (1998).

373. C.G. Juo, S.W. Chen and G.R. Her, *Anal. Chim. Acta* **311**, 153–64 (1995).

374. Y. Hirata, F. Nakata and M. Horihata, *J. High Resolut. Chromatogr.* **11**, 81–4 (1988).

375. Y. Ito, T. Takeuchi, D. Ishii and M. Goto, *J. Chromatogr.* **346**, 161 (1985).

376. F. David, *Chem. Mag. (Ghent)* **17** (4), 12–5 (1991).

377. T.W. Ryan, S.G. Jocklovich, J.C. Watkins and E.J. Levy, *J. Chromatogr.* **505**, 273–82 (1990).

378. M. Ashraf-Khorassani, D.S. Boyer and J.M. Levy, *J. Chromatogr. Sci.* **29**, 517–21 (1991).

379. G.A. Mackay and R.M. Smith, *J. Chromatogr. Sci.* **32**, 455–60 (1994).

380. Suprex Corporation, *Application Note SFE-72* (1989).

381. T. Tikuisis and M. Cossar, *Proceedings SPE ANTEC '93*, New Orleans, LA (1993), p. 270.

382. S. Ourén, T. Greibrokk. D.R. Hegna and K. Kleveland, in *The 2nd European Symposium on Analytical Supercritical Chromatography and Extraction, Riva del Garda* (P. Sandra and K. Markides, eds), Huethig, Heidelberg (1993), p. 308.

383. M. Ashraf-Khorassani, S. Gidanian and. Y. Yamini, *J. Chromatogr. Sci.* **33**, 658–62 (1995).

384. M.L. Marín, J. López, A. Sánchez, J. Vilaplana and J. Jiménez, *Bull. Environ. Contam. Toxicol.* **60** (1), 68–73 (1998).

385. R.M. Guerra, M.L. Marín, A. Sánchez and A. Jiménez, *J. Chromatogr.* **A950** (1–2), 31–9 (2002).

386. A. Venema and T.J. Jelink, *J. High Resolut. Chromatogr.* **16**, 166–8 (1993).

387. J. Swagten-Linssen, *Techn. Rept DSM Research*, Geleen (2000).

388. L.H. McDaniel and L.T. Taylor, *Proceedings 8th Intl. Symposium on Supercritical Fluid Chromatography and Extraction*, St Louis, MO (1998), paper C-03.

389. K. Torkos, P. Horvatovich and J. Borossay, *Olaj, Szappan, Kozmet.* **45** (spec. issue), 62–5 (1996).

390. L.J.D. Myer, J.H. Damian, P.B. Liescheski and J. Tehrani, *ACS Symp. Ser.* **488**, 221–36 (1992).

391. S.L. Jordan, L.T. Taylor, P.D. Seemuth and R.J. Miller, *Text. Chem. Color.* **29**, 25 (1997).

392. N.J. Cotton, K.D. Bartle, A.A. Clifford and C.J. Dowle, *J. Chromatogr. Sci.* **31**, 157–61 (1993).

393. A.N. Burgess and K. Jackson, *J. Appl. Polym. Sci.* **46**, 1395–9 (1992).

394. G. Filardo, A. Galia, S. Gambino, G. Silvestri and M. Poidomani, *J. Supercrit. Fluids* **9**, 234 (1996).

395. J.K. Sekinger, G.N. Ghebremeskel and L.H. Concienne, *Rubber Chem. Technol.* **69**, 851–7 (1996).

396. B. Werthmann, V. Neyen, F. Milczewski and R. Borowski, *Kautsch. Gummi Kunstst.* **51** (2), 118–21 (1998).

397. F. Reche, M.C. Garrigós, M.L. Marín and A. Jiménez, *Abstracts 7th Intl. Symposium on Hyphenated Techniques in Chromatography and Hyphenated Chromatographic Analyzers (HTC-7)*, Bruges (2002), paper P132.

398. T. Bücherl and O.G. Piringer, *Proceedings 5th Intl. Symposium on Supercritical Fluid Chromatography and Extraction*, Baltimore, MD (1994), pp. 31–2.

399. J. Ezzell and B. Thompson, *Proceedings 5th Intl. Symposium on Supercritical Fluid Chromatography and Extraction*, Baltimore, MD (1994), paper E-4.

400. S.L. Jordan, L.T. Taylor and P.D. Seemuth, *Proceedings 5th Intl Symposium on Supercritical Fluid Chromatography and Extraction*, Baltimore, MD (1994), paper E-9.

401. M.J. Drews, J. Ivey, C. Lam and S. Feng, *Proceedings 5th Intl Symposium on Supercritical Fluid Chromatography and Extraction*, Baltimore, MD (1994), paper F-7.

402. S.G. Yocklovich, S.F. Sarner and J.M. Levy, *Am. Lab.* **21**, 26 (1989).

403. M.J. Drews, K. Ivey and J. Helvey, *Proceedings 7th Intl Symposium on Supercritical Fluid Chromatography and Extraction*, Indianapolis, IN (1996), paper L-33.

404. M.D. Johnson, E.J. LeBœuf, T.M. Young and W.J. Weber, *Proceedings 7th Intl. Symposium on Supercritical Fluid Chromatography and Extraction*, Indianapolis, IN (1996), paper D-17.

405. N. Nazem and L.F. Taylor, *J. Chromatogr. Sci.* **40**, 181–6 (2002).

406. E. Marioth, G. Bunte and Th. Härdle, *Polym. Recycl.* **2** (4), 303–8 (1996).

407. J.A. Field and W. Giger, *Proceedings 5th Intl. Symposium on Supercritical Fluid Chromatography and Extraction*, Baltimore, MD (1994), pp. 19–20.

408. D.A. Roston, J.J. Sun, P.W. Collins, W.E. Perkins and S.J. Tremont, *J. Pharm. Biomed. Anal.* **13**, 1513–20 (1995).

409. C. Nerín, R. Batlle and J. Cacho, *Anal. Chem.* **69**, 3304–13 (1997).

410. A.A. Clifford, K.D. Bartle, I. Gelebart and S. Zhu, *J. Chromatogr.* **A785**, 395–401 (1997).

411. J.C. Via, C.L. Braue and L.T. Taylor, *Anal. Chem.* **66**, 603 (1994).

412. Y. Yang, S. Bøwadt, S.B. Hawthorne and D.J. Miller, *Anal. Chem.* **67**, 4571–6 (1995).

413. K. Hartonen, K. Inkala, M. Kangas and M.-L. Riekkola, *J. Chromatogr.* **A785**, 219–26 (1997).

414. T.E. Young, S.C. Ecker, R.E. Synovec, N.T. Hawley, J.P. Lomber and C.M. Wai, *Talanta* **45**, 1189–99 (1998).

415. H. Daimon and J. Pawliszyn, *Anal. Commun.* **33**, 421–4 (1996).

416. J.D. Miller and S.B. Hawthorne, *Anal. Chem.* **69**, 623–7 (1997).

417. R. Tajuddin and R.M. Smith, *Analyst* **127** (7), 883–5 (2002).

418. S.B. Hawthorne, Y. Yang and D.J. Miller, *Anal. Chem.* **66**, 2912–20 (1994).

419. T.L. Chester, J.D. Pinkston and D.E. Raynie, *Anal. Chem.* **68**, 487R (1996).

420. H.M. Kingston and S.J. Haswell (eds), *Microwave-enhanced Chemistry, Fundamentals, Sample Preparation and Applications*, American Chemical Society, Washington, DC (1997).

421. J. Wan, M. Tse, H. Husby and M. Depew, *J. Microwave Power* **25**, 32 (1990).

422. G. Bond and R.B. Moyes, in *Microwave-enhanced Chemistry* (H.M. Kingston and S.J. Haswell, eds), American Chemical Society, Washington, DC (1997), pp. 551–68.

423. G. Majetich and K. Wheless, in *Microwave-enhanced Chemistry* (H.M. Kingston and S.J. Haswell, eds), American Chemical Society, Washington, DC (1997), pp. 455–505.

424. D.R. Baghurst and D.M.P. Mingos, in *Microwave-enhanced Chemistry* (H.M. Kingston and S.J. Haswell, eds), American Chemical Society, Washington (1997), pp. 523–50.

425. D.M. Mingos and D.R. Baghurst, *Chem. Soc. Rev.* **20**, 1 (1991).

426. A. Loupy (ed.), *Microwaves in Organic Synthesis*, Wiley-VCH, Weinheim (2002).

427. D.M.P. Mingos and D.R. Baghurst, in *Microwave-enhanced Chemistry* (H.M. Kingston and S.J. Haswell, eds), American Chemical Society, Washington, DC (1997), pp. 3–53.

428. G. Majetich, R. Hicks and E. Neas, in *Microwave-enhanced Chemistry* (H.M. Kingston and S.J. Haswell, eds), American Chemical Society, Washington, DC (1997), pp. 507–22.

429. X. Fang and D.A. Scola, *Proceedings ACS Div. Polym. Mater.: Sci. Engng.*, Anaheim, CA (1999), Vol. 80, pp. 322–3.

430. P.J. Walker, S. Chalk and H.M. Kingston, in *Microwave-enhanced Chemistry* (H.M. Kingston and S.J. Haswell, eds), American Chemical Society, Washington, DC (1997), pp. 55–121.

431. G.M.B. Parkes, P.A. Barnes, E.L. Charsley and G. Bond, *J. Therm. Anal. Calorim.* **56**, 723–31 (1999).

432. A.A. Craveiro, F.J.A. Matos and J.W. Alencar, *Flavour Fragance J.* **4**, 43 (1989).

433. G.N. LeBlanc, *Am. Lab.* **32** (18), 32–7 (2000).

434. Y.I. Chen, Y.-S Su and J.-F. Jen, *Abstracts 7th Intl. Symposium on Hyphenated Techniques in Chromatography and Hyphenated Chromatographic Analyzers (HTC-7)*, Bruges (2002), paper P25.

435. M. Burguera, J.L. Burguera and O.M. Alarcón, *Anal. Chim. Acta* **179**, 351 (1986).

436. R.E. Sturgeon, S.N. Willie, B.A. Methven, J.W.H. Lam and H. Matusiewicz, *J. Anal. At. Spectrom.* **10**, 981 (1995).

437. H.M. Kingston, P.J. Walter, W.G. Engelhart and P.J. Parsons, in *Microwave-Enhanced Chemistry* (H.M. Kingston and S.J. Haswell, eds), American Chemical Society, Washington, DC (1997), pp. 697–742.

438. E.H. Grant (ed.), *Microwaves: Industrial, Scientific and Medical Applications*, Artech House, Norwood, MA (1992).

439. A.J. Baden-Fuller, *Microwaves: An Introduction to Microwave Theory and Techniques*, Pergamon Press, Oxford (1990).

440. H.M. Kingston and L.B. Jassie (eds), *Introduction to Microwave Sample Preparation*, American Chemical Society, Washington, DC (1988).

441. A. Złotorzynski, *Crit. Rev. Anal. Chem.* **25**, 43–76 (1995).

442. S.J. Chalk, H.M. Kingston, P.J. Walter, K. McQuillin and J. Brown, in *Microwave-Enhanced Chemistry* (H.M.

Kingston and S.J. Haswell, eds), American Chemical Society, Washington, DC (1997), pp. 667–95.

443. B.W. Renoe, *Am. Lab.* 34–40 (August 1994).

444. A.B. Badeka and M.G. Kontominas, *Z. Lebensm.-Unters. Forsch.* **A208** (1), 69–73 (1999).

445. A.B. Badeka, K. Pappa and M.G. Kontominas, *Z. Lebensm.-Unters. Forsch.* **A208** (5–6), 429–33 (1999).

446. L.B. Gilman and W.A. Dark, *Proceedings Intl GPC Symposium '89*, Newton, MA (1989).

447. J. Thuéry, *Microwaves: Industrial, Scientific and Medical Applications* (E.H. Grant, ed.), Artech House, Norwood, MA (1992).

448. M. Ericsson and A. Colmsjo, *Abstracts 6th Intl. Symposium on Hyphenated Techniques in Chromatography and Hyphenated Chromatographic Analyzers (HTC-6)*, Bruges (2000), paper 16.2.

449. E. Hasty and R. Revesz, *Am. Lab.* **27**, 66–73 (1995).

450. V. Lopez-Avila, *Sonication and Soxhlet Extraction in Environmental Analysis: Methods Comparison. EPA Report 600/X-93/010*, US Environmental Protection Agency: Environmental Monitoring Systems Laboratory, Las Vegas, NV (February 1993).

451. F.E. Smith and E.A. Arsenault, *Talanta* **43**, 1207–68 (1996).

452. J.R. Dean, L. Fitzpatrick and C. Heslop, in *Extraction Methods in Organic Synthesis* (A.J. Handley, ed.), Sheffield Academic Press/CRC Press, Sheffield (1999), pp. 166–93.

453. T.-B. Hsu, *Huaxue* **56** (4), 285–94 (1998).

454. L. Jassie, R. Revesz, T. Kierstead, E. Hasty and S. Matz, in *Microwave-enhanced Chemistry* (H.M. Kingston and S.J. Haswell, eds), American Chemical Society, Washington, DC (1997), pp. 569–609.

455. H.M. Kingston, P.J. Walter, S. Chalk, E. Lorentzen and D. Link, in *Microwave-enhanced Chemistry* (H.M. Kingston and S.J. Haswell, eds), American Chemical Society, Washington, DC (1997), pp. 223–349.

456. *CEM Application Note E012*, CEM Corporation, Matthews, NC (n.d.).

457. H.J. Vandenburg, A.A. Clifford, K.D. Bartle, J. Carroll and I.D. Newton, *Analyst* **124**, 397–400 (1999).

458. W. Camacho and S. Karlsson, *Polym. Degrad. Stab.* **71** (1), 123–34 (2001).

459. *CEM Application Note E007*, CEM Corporation, Matthews, NC (1988).

460. *CEM Application Note E010*, CEM Corporation, Matthews, NC (n.d.).

461. B. Marcato and M. Vianello, *Proceedings HPLC '99*, Granada (1999).

462. G. Cecchin and B. Marcato, *J. Chromatogr.* **730**, 83 (1996).

463. Anon., *Plastics in the Microwave: A Common Sense Approach*, American Plastics Council, Washington, DC (1999).

464. J. Cano, M.L. Marín, A. Sanchez and V. Hernadis, *J. Chromatogr.* **A963** (1–2), 401–9 (2002).

465. K.K. Chee, M.K. Wong and K.K. Lee, *Anal. Chim. Acta* **330**, 217–27 (1996).

466. W.-H. He and S.-J. Hsieh, *Anal. Chim. Acta* **428**, 111 (2001).

467. V. Lopez-Avila, R. Young, J. Benedicto, P. Ho, R. Kim and W.F. Beckert, *Anal. Chem.* **67**, 2096–2102 (1995).

468. K.K. Chee, M.K. Wong and H.K. Lee, *Chromatographia* **42**, 378–84 (1996).

469. H. Budzinski, A. Papineau, P. Baumard and P. Garrigues, *C.R. Acad. Sci. (Paris), Ser. IIb*, **321** (2), 69–79 (1995).

470. H. Budzinski, P. Baumard, A. Papineau, S. Wise and P. Garrigues, *Polycyclic Aromatic Compounds* **9** (1–4), 225–32 (1996).

471. M. Letellier, H. Budzinski, P. Garrigues and S. Wise, *Spectroscopy* **13**, 71–80 (1997).

472. M. Letellier, H. Budzinski and P. Garrigues, *LC.GC Intl.* **12** (4), 222–5 (1999).

473. J.M. Mermet, in *Microwave-enhanced Chemistry* (H.M. Kingston and S.J. Haswell, eds), American Chemical Society, Washington, DC (1997), pp. 371–400.

474. B. Rocca, C. Arzouyan and J. Estienne, *J. Ann. Fals. Exp. Chim.* **911**, 347 (1992).

475. J.R.J. Paré, M. Segouin and J. Lapointe (to Dept. of Environment of Canada), *US Pat.* 5,002,784 (March 26, 1991).

476. J.R.J. Paré and J.M.R. Bélanger, in *Instrumental Methods in Food Analysis* (J.R.J. Paré and J.M.R. Bélanger, eds), Elsevier, Amsterdam (1997), pp. 395–420.

477. J.R.J. Paré, J.M.R. Bélanger and S.S. Stafford, *Trends Anal. Chem.* **13**, 176–84 (1994).

478. J.M.R. Bélanger, K. Li, Y.Y. Shu, R.D. Turpin, R. Singhvi, M.M. Punt and J.R.J. Paré, *Proceedings 7th Intl. Symposium on Supercritical Fluid Chromatography and Extraction*, Indianapolis, IN (1996), paper L-20.

479. J. Lapointe, J. Paré, M. Sigouin, J.J.R. Paré and J.R.J. Paré (to Canadian Ministry of Environment), *US Pat.* 5,458,897 (October 17, 1995).

480. Y.Y. Shu, R.C. Lao, J.M.R. Bélanger, M.F. Fingas and J.R.J. Paré, *Proceedings 12th Technical Seminar on Chemical Spills*, Edmonton (1995), pp. 165–71.

481. J.R.J. Paré, J.M.R. Bélanger, K. Li and S.S. Stafford, *J. Microcol. Sep.* **7**, 37–40 (1995).

482. J.R.J. Paré, J.M.R. Bélanger, D.E. Thornton, K. Li, M.L. Llompart, M. Fingas and S.A. Blenkinsopp, *Spectroscopy* **13**, 23–32 (1996/97).

483. B.E. Richter, C.A. Pohl, N.L. Porter, B.A. Jones, J.L. Ezzell and N. Avdalovic (to Dionex Corp.), *US Pat.* 5,843,311 (December 1, 1998).

484. M.J. Hirsch, *Varian Chromatography News* (1), 4–5 (2002).

485. B. Richter, J. Ezzell, D. Felix, K. Roberts and D. Later, *Am. Lab.*, 24–8 (February 1995).

486. J.L. Ezzell, in *Extraction Methods in Organic Analysis* (A.J. Handley, ed.), Sheffield Academic Press/CRC Press, Sheffield (1999), pp. 146–65.

487. U.A.Th. Brinkman and R.J.J. Vreuls, *LC.GC Intl.* **8**, 694 (1995).

488. *Dionex Application Note 331*, Dionex Corporation, Sunnyvale, CA (1998).

489. H.J. Vandenburg, A.A. Clifford, K.D. Bartle, R.E. Carlson, J. Carroll and I.D. Newton, *Analyst* **124**, 1707–10 (1999).

490. T. Macko, R. Siegl and K. Lederer, *Angew. Makromol. Chem.* **227**, 179–91 (1995).

491. T. Macko, B. Furtner and K. Lederer, *J. Appl. Polym. Sci.* **62**, 2201–7 (1996).

492. S.A. Hinman, L.T. Taylor, M. Ashraf-Khorassani and D.C. Messer, *Proceedings 8th Intl. Symposium on Supercritical Fluid Chromatography and Extraction*, St Louis, MO (1998), paper D-08.

493. R. Carlson, J. Clark, J. Ezzell and R. Joyce, *Proceedings 10th Intl. Symposium on Polymer Analysis and Characterization*, Toronto (1997).

494. M. Waldebäck, C. Jansson, F.J. Señoráns and K.E. Markides, *Analyst* **123**, 1205–7 (1998).

495. N. Yagoubi, A. Baillet, F. Pellerin and D. Ferrier, *Nucl. Instrum. Methods Phys. Res., Sect. B*, **105**, 340 (1995).

496. M. Riess, T. Ernst, G. Biermann and R. van Eldik, *GIT Labor-Fachz.* **42**(10), 1008–9 (1998).

497. F. Hofler, *Chem. Labor Biotech* **51** (2), 56–8 (2000).

498. M. Ashraf-Khorassani, S. Hinman and L.T. Taylor, *J. High Resolut. Chromatogr.* **22** (5), 271–5 (1999).

499. J.Å. Jönsson and L. Mathiasson, *Trends Anal. Chem.* **18**, 318–25 (1999).

500. N. Megersa, T. Solomon and J.Å. Jönsson, *J. Chromatogr.* **A830**, 203–10 (1999).

501. Y.Z. Luo and J. Pawliszyn, *Anal. Chem.* **72**, 1058–63 (2000).

502. A. Segal, T. Górecki, P. Mussche, J. Lips and J. Pawliszyn, *J. Chromatogr.* **A873**, 13–27 (2000).

503. L.S. Ettre, *Chromatographia* **38**, 521 (1994).

504. J.S. Fritz, *Analytical Solid-phase Extraction*, Wiley-VCH, New York, NY (1999).

505. R.E. Majors, *LC.GC Intl.* **11** (8), 8–15 (1998).

506. E. Baltussen, H. Snijders, H.-G. Janssen, P. Sandra and C.A. Cramers, *J. Chromatogr.* **A802**, 285–95 (1998).

507. D. Stevenson, S. Miller and I.D. Wilson, in *Extraction Methods in Organic Analysis* (A.J. Handley, ed.), Sheffield Academic Press/CRC Press, Sheffield (1999), pp. 194–220.

508. E.M. Thurman and M.S. Mills, *Solid-phase Extraction. Principles and Practice*, John Wiley & Sons Inc., New York, NY (1998).

509. K.C. van Horne, *Sorbent Extraction Technology*, Analytichem International, Harbor City, CA (1985).

510. M. Zief and R. Kiser, *Solid Phase Extraction for Sample Preparation*, J.T. Baker, Phillipsburg (1988).

511. P.D. McDonald and E.S.P. Bouvier, *Solid Phase Extraction Applications Guide and Bibliography, A Resource for Sample Preparation Methods Development*, Waters, Milford, MA (1995).

512. N. Simpson, *Solid Phase Extraction: Principles, Strategies and Applications*, M. Dekker, New York, NY (1997).

513. G. Font, J. Mañes, J.C. Moltó and Y. Picó, *J. Chromatogr.* **642**, 135 (1993).

514. L.A. Berrueta, B. Gallo and F. Vicente, *Chromatographia* **40**, 474–83 (1995).

515. Z.E. Penton, *Adv. Chromatography* **38**, 205–36 (1997).

516. M.-C. Hennion, *J. Chromatogr.* **A856**, 3–54 (1999).

517. G.A. Junk, M.J. Avery and J.J. Richard, *Anal. Chem.* **60** (13), 1347–50 (1988).

518. S.M. Dugar, J.N. Leibowitz and R.H. Dyer, *J. AOAC Int.* **77**, 1335 (1994).

519. E. Milanova and B.B. Sithole, *Tappi J.* **80**, 121–8 (1997).

520. K. Ohto, Y. Tanaka and K. Inoue, *Chem. Lett.* **7**, 647 (1997).

521. G. Malofeeva, O. Petrukhin, L. Rozhkova, B. Spivakov, G. Genkina and T. Mastryukova, *Russ. J. Anal. Chem.* **51**, 1061 (1996).

522. Ch. Sarbach, N. Yagoubi, J. Sauzières, Ch. Renaux, D. Ferrier and E. Postaire, *Int. J. Pharm.* **140**, 169–74 (1996).

523. A.M. Kvistad, E. Lundanes and T. Greibrokk, *Chromatographia* **48**, 707–13 (1998).

524. T.T. Nguyen, R. Kringstad and K.E. Rasmussen, *J. Chromatogr.* **366**, 445–50 (1986).

525. R.G. Belardi and J. Pawliszyn, *Water Pollut. Res. J. Can.* **25**, 179 (1989).

526. C.L. Arthur and J. Pawliszyn, *Anal. Chem.* **62**, 2145–8 (1990).

527. C.L. Arthur, L.M. Killam, K.D. Buchholz, J. Pawliszyn and J.R. Berg, *Anal. Chem.* **64**, 1960 (1992).

528. Z. Zhang and J. Pawliszyn, *Anal. Chem.* **65**, 1843–52 (1993).

529. C.L. Arthur, D.W. Potter, K.D. Buchholz, S. Motlagh and J. Pawliszyn, *LC.GC Intl.* **5** (10), 8 (1992).

530. J. Pawliszyn, *Trends Anal. Chem.* **14** (3), 113 (1995).

531. Z. Zhang, M.J. Yang and J. Pawliszyn, *Anal. Chem.* **66**, 844A (1994).

532. V. Mani, in *Applications of Solid Phase Microextraction* (J. Pawliszyn, ed.), The Royal Society of Chemistry, Cambridge (1999), pp. 57–72.

533. S. Li and S.G. Weber, in *Applications of Solid Phase Microextraction* (J. Pawliszyn, ed.), The Royal Society of Chemistry, Cambridge (1999), pp. 49–56.

534. D. Louch, S. Motlagh and J. Pawliszyn, *Anal. Chem.* **64**, 1187–99 (1992).

535. S. Magdic and J.B. Pawliszyn, *J. Chromatogr.* **A723**, 111–22 (1996).

536. C. Grote and K. Levsen, in *Applications of Solid Phase Microextraction* (J. Pawliszyn, ed.), The Royal Society of Chemistry, Cambridge (1999), pp. 169–87.

537. J. Ai, in *Applications of Solid Phase Microextraction* (J. Pawliszyn, ed.), The Royal Society of Chemistry, Cambridge (1999), pp. 22–37.

538. J. Pawliszyn, in *Applications of Solid Phase Microextraction* (J. Pawliszyn, ed.), The Royal Society of Chemistry, Cambridge (1999), pp. 3–21.

539. L. Pan and J. Pawliszyn, *Anal. Chem.* **69**, 196–205 (1997).

540. J. Chen and J. Pawliszyn, *Anal. Chem.* **67**, 2530–3 (1995).

541. T. Gorecki and J. Pawliszyn, *Anal. Chem.* **67**, 3265–74 (1995).

542. I. Valor, C. Cortada and J.C. Molto, *J. High Resolut. Chromatogr.* **19**, 472 (1996).

543. H.L. Lord and J. Pawliszyn, *LC.GC Intl.* **11** (12), 776–85 (1998).

544. Y. Luo and J. Pawliszyn, in *Extraction Methods in Organic Synthesis* (A.J. Handley, ed.), Sheffield Academic Press/CRC Press, Sheffield (1999), pp. 75–99.

544a. J.V. Hinshaw, *LC.GC Europe* **16** (2), 803–7 (2003).

545. J. Pawliszyn, *Solid Phase Microextraction, Theory and Practice*, John Wiley & Sons Inc., New York, NY (1997).

546. T.J. Braggins, C.C. Grimm and F.R. Visser, in *Applications of Solid Phase Microextraction* (J. Pawliszyn, ed.), The Royal Society of Chemistry, Cambridge (1999), pp. 407–22.

547. D.C. Stahl and D.C. Tilotta, in *Applications of Solid Phase Microextraction* (J. Pawliszyn, ed.), The Royal Society of Chemistry, Cambridge (1999), pp. 625–37.

548. Z. Penton, *SPME Application Note 7*, Varian Chromatography Systems, Walnut Creek, CA (1995).

549. J. Pawliszyn, (ed.), *Applications of Solid Phase Microextraction*, The Royal Society of Chemistry, Cambridge (1999).

550. E. Baltussen, P. Sandra, F. David and C. Cramers, *J. Microcol. Sep.* **11** (10), 737–47 (1999).

551. A. Hoffmann, R. Bremer, P. Sandra and F. David, *LaborPraxis* **24** (2), 60–2 (2000).

552. C. Bicchi, C. Cordero, C. Iori, P. Rubiolo and P. Sandra, *J. High Resolut. Chromatogr.* **23** (9), 539–46 (2000).

553. J. Vercauteren, C. Peres, C. Devos, P. Sandra, F. Vanhaecke and L. Moens, *Anal. Chem.* **73** (7), 1509–14 (2001).

554. J.S. Rowlinson, *Fluid Phase Equilibria* **10**, 135 (1983).

555. D.E. Raynie and D.P. Innis, *Proceedings 7th Intl. Symposium on Supercritical Fluid Chromatography and Extraction*, Indianapolis, IN (1996), paper L-21.

556. D. Munteanu, in *Mechanisms of Polymer Degradation and Stabilization* (G. Scott, ed.), Elsevier Applied Science, London (1990), pp. 211–314.

557. S.H. Smith and L.T. Taylor, *Chromatographia* **56** (3/4), 165–9 (2002).

558. S. Affolter and S. Hofstetter, *Kautsch. Gummi Kunstst.* **48** (3), 173–9 (1995).

559. DIN ISO 1407, *Kautschuk und Elastomere, Bestimmung der mit Lösemitteln extrahierbaren Bestandteile* (1996).

560. R.C. Wieboldt, K.D. Kempfert and D.L. Dalrymple, *Appl. Spectrosc.* **44**, 1028–34 (1990).

561. S.J. Wright, M.J. Dale, P.R.R. Langridge-Smith, Q. Zhan and R. Zenobi, *Anal. Chem.* **68**, 3585–94 (1996).

562. J. Salafranca, J. Cacho and C. Nerín, *Polymer Recycling* **4** (1), 13–25 (1999).

563. D. Cattell, *Polymers in Chemically Resistant Applications, Rapra Review Reports* 39, Rapra Technology, Shawbury (1991).

564. T. Provder (ed.), *Chromatography of Polymers, ACS Symposium Series* 521, American Chemical Society, Washington, DC (1993).

565. Th. Schmutz, E. Kramer, H. Zweifel and G. Dörner, *J. Elastomers Plast.* **30** (1), 55–67 (1998).

566. E.J. Latos and A.K. Sparks, *Rubber J.* **151** (6), 18 (1969).

567. G. Pfahler and K. Loetzsch, *Kunststoffe-German Plast.* **78**, 142 (1988).

568. T.P. Gendek, T.A. Hatton and R.C. Reid, *I & EC Res.* **28**, 1036 (1989).

569. R.J. Tucker and P.V. Susi, *Polym. Prepr.* **25** (1), 34 (1984).

570. A.J. Aarts and K.M. Baker, *Kautsch. Gummi Kunstst.* **37**, 497–500 (1984).

571. R.P. Lattimer, R.E. Harris, C.K. Rhee and H.-R. Schulten, *Anal. Chem.* **58**, 3188–95 (1986).

572. D.W. Allen, D.A. Leathard, C. Smith and J.D. McGuinness, *Chem. Ind. (London)*, 198–9 (1987).

573. M. Moors, D.L. Massart and R.D. McDowall, *Pure Appl. Chem.* **66**, 277–304 (1994).

574. L.A. Berrueta, B. Gallo and F. Vicente, *Chromatographia* **40**, 474–83 (1995).

575. R.L. Gray and C. Neri, *Proceedings 19th Ann. Intl. Conference on Advances in the Stabilization and Degradation of Polymers*, Luzern (1997), pp. 63–79.

576. J. Pospíšil, *Adv. Polym. Sci.* **101**, 65–167 (1991).

577. P.M. Gomez, S.K. Fu, A. Gupta and O. Vogl, *Polym. Prepr.* **26** (1), 100 (1985).

578. R. Liu, S. Wu, S. Li, F. Xi and O. Vogl, *Polym. Bull.* **20**, 59 (1988).

579. R.J. Parmar, S. Saxena and J.S. Parmar, *Angew. Makromol. Chem.* **259**, 1–5 (1998).

580. J. Pospíšil, in *Oxidation Inhibition in Organic Materials* (J. Pospíšil and P.P. Klemchuk, eds), CRC Press, Boca Raton, FL (1989), Vol. I, pp. 193–224.

581. J. Malik, G. Ligner and L. Avár, *Polym. Degrad. Stab.* **60** (1), 205–13 (1998).

582. T.H. Kim, H.-K. Kim, D.R. Oh, M.S. Lee, K.H. Chae and S.-Y. Kaang, *J. Appl. Polym. Sci.* **77** (13), 2968–73 (2000).

583. A. Smedberg, T. Hjertberg and B. Gustafsson, *Proceedings 19th Intl. Conference on Advances in the Stabilization and Degradation of Polymers*, Luzern (1997), p. 293.

584. C. Bolcu, *Ann. West Univ. Timişoara, Ser. Chem.* **5**, 11–16 (1996).

585. J.W. Wheeler, Y. Zhang and J.C. Tebby, *Proceedings 6th European Meeting on Fire Retardancy of Polymeric Materials*, Lille (1997), pp. 90–1.

586. M. Day, T. Suprunchuk, J.D. Cooney and D.M. Wiles, *J. Appl. Polym. Sci.* **33**, 2041 (1987).

587. R. Greco, in *Plastics Additives. An A-Z Reference* (G. Pritchard, ed.), Chapman & Hall, London (1998), pp. 375–85.

588. S. Al-Malaika, in *Chemistry and Technology of Polymer Additives* (S. Al-Malaika, A. Golovoy and C.A. Wilkie, eds), Blackwell Science, Oxford (1999), pp. 1–20.

589. A. Vidal, M. Feder and E. Papirer, *Proceedings Intl. Conference Rubber '89*, Prague (1989), Vol. II, p. 54.

590. G. Haacke, E. Longordo, F.F. Andrawes and B.H. Campbell, *Progr. Org. Coat.* **34** (1–4), 75–83 (1998).

591. J. Wolfschwenger, A. Hauer, M. Gahleitner and W. Neissl, *Proceedings Eurofillers '97*, British Plastics Federation, London (1997), pp. 375–7.

592. J.-Q. Pan, N.C. Liu and W.W.Y. Lau, *Polym. Degrad. Stab.* **62**, 315–22 (1998).

593. H. Pasch, K.F. Shuhaibar and S. Attari, *J. Appl. Polym. Sci.* **42**, 263–71 (1991).

594. J. Malik and G. Ligner, *Proceedings Intl. Conf. Adv. Plast. Technol. APT '97*, Gliwice (1997), pp. 11/1–11/8.

595. F. Gugumus, in *Oxidation Inhibition in Organic Materials* (J. Pospíšil and P.P. Klemchuk, eds), CRC Press, Boca Raton, FL (1989), Vol. I, pp. 61–172.

596. W. Freitag, *Fresenius Z. Anal. Chem.* **316**, 495–6 (1983).

597. P. Holbein, M. Bonora and D. Horsey, *Proceedings Addcon World '98*, London (1998), P8.

598. J.S. Murphy, *The Additives for Plastics Handbook*, Elsevier Advanced Technology, Oxford (2001).

599. R.F. Grossman, *Rubber World* **220** (5), 20–1 (1999).

600. R.F. Grossman, *Gummi Fasern Kunstst.* **52** (6), 454–6 (1999).

601. V. Komolprasert, A.R. Lawson and W.A. Hargreaves, *J. Agric. Food Chem.* **43**, 1963–5 (1995).

602. T.J. Nielson, I.M. Jagerstad, R.E. Hoste and B.T.G. Sivik, *J. Agric. Food Chem.* **39**, 1234 (1991).

603. P. Bataillard, L. Evangelista and M. Thomas, in *Plastics Additives Handbook* (H. Zweifel, ed.), Hanser Publishers, Munich (2000), pp. 1047–83.

604. D. Braun, *Simple Methods for Identification of Plastics*, Hanser Publishers, Munich (1996).

605. A. Krause, A. Lange and M. Ezrin, *Plastics Analysis Guide. Chemical and Instrumental Methods*, Hanser Publishers, Munich (1983).

606. J.F. Schabron and L.E. Fenska, *Anal. Chem.* **52**, 1411–5 (1980).

607. J.F. Schabron and C.Z. Bradfield, *J. Appl. Polym. Sci.* **26**, 2479 (1981).

608. J.F. Schabron, V.J. Smith and J.L. Ware, *J. Liq. Chromatogr.* **5**, 613 (1982).

609. J.F. Schabron (to Phillips Petroleum Co.), *US Pat.* 4,576,917A (March 18, 1986).

610. M.E. Gelbin and K. Jackson, *Proceedings SPE ANTEC 2002*, San Francisco, CA (2002), pp. 2025–9.

611. A.H. Sharma, F. Mozayeni and J. Alberts, *Proceedings SPE Polyolefins XI*, Houston, TX (1999), pp. 679–703.

612. T. Macko, B. Furtner and K. Lederer, *Int. J. Polym. Anal. Charact.* **3**, 369–79 (1997).

613. R. Verstappen, *Techn. Rept DSM Research*, Geleen (1994).

614. P. Matuska, *J. Chromatogr.* **606**, 136–40 (1992).

615. B. Marcato, C. Fantazzini and F. Sevini, *J. Chromatogr.* **553**, 415–22 (1991).

616. W. Freitag, R. Wurster and N. Mady, *J. Chromatogr.* **450**, 426–9 (1988).

617. R. Trones, *PhD Thesis*, University of Oslo (1999).

618. M.A. Roberson and G. Patonay, *J. Chromatogr.* **505**, 375–84 (1990).

619. S. Narayan, R.E. Lee, D. Hallberg and V. Malatesta, *Proceedings SPE ANTEC 2000*, Orlando, FL (2000), pp. 2178–9.

620. F.C.-Y. Wang and W.C. Buzanowski, *J. Chromatogr.* **A891**, 313–24 (2000).

621. H. Fink, U. Panne, M. Theisen, R. Niessner, T. Probit and X. Lin, *Fresenius' Z. Anal. Chem.* **368** (2–3), 235–9 (2000).

622. K.A. Barns, A.P. Damant, J.R. Startin and L. Castle, *J. Chromatogr.* **A712**, 191–9 (1995).

623. T.H. Begley and H.C. Hollifield, *J. Assoc. Off. Anal. Chem.* **72** (3), 468–70 (1989).

624. K. Metcalf and R. Tomlinson, *Plastics (London)* **25**, 319 (1960).

625. L. Robinson, *Ann. Ist. Super. Sanità* **8**, 542 (1972).

626. G. Stegeman and E. Venema (GE Plastics), *personal communication* (1999).

627. T.R. Crompton and L.W. Myers, *Eur. Polym. J.* **4**, 355 (1968).

628. R.J. Steichen, *Anal. Chem.* **48**, 1398 (1976).

629. G.v.d. Velden, *Techn. Rept DSM Research*, Geleen (1997).

630. R.H. Campbell and E.J. Young, *Rubber Age (NY)*, **100** (3), 71–5 (1968).

631. S. Cook and R. Lehrle, *Eur. Polym. J.* **29** (1), 1–8 (1993).

632. H. El Mansouri, N. Yagoubi and D. Ferrier, *Chromatographia* **48**, 491–6 (1998).

633. H. El Mansouri, N. Yagoubi, D. Scholler, A. Feigenbaum and D. Ferrier, *J. Appl. Polym. Sci.* **71** (3), 371–5 (1999).

634. H.J. Cortes, G.E. Bormett and J.D. Graham, *J. Microcol. Sep.* **4** (1), 51–7 (1992).

635. *British Standard 2782, Part 4, Method 405D* (1965).

636. A.T. Griffith, Y. Park and C.B. Roberts, *Polym. Plast. Technol. Engng.* **38** (3), 411–31 (1999).

637. C.D. Papaspyrides and C.N. Kartalis, *Polym. Engng. Sci.* **40** (4), 979–84 (2000).

638. M.D. Hurwitz, V.S. Salvin and R.L. McConnel, in *Analytical Methods for a Textile Laboratory* (J.W. Weaver, ed.), American Association of Textile Chemists and Colorists, Research Triangle Park, NC (1984), pp. 129–54.

639. F.C.-Y. Wang, *J. Chromatogr.* **A886**, 225–35 (2000).

640. A. Schaafsma, *Techn. Rept DSM Research*, Geleen (1998).

641. A. Milosz, *Wlokna Chem.* **9** (1), 9–18 (1983).

642. R. Li and X. Hu, *Polym. Degrad. Stab.* **62**, 523–8 (1998).

643. J. Cosemans-Craeghs, *Techn. Rept DSM Research*, Geleen (1999).

644. J. Udris, *Analyst (London)* **96**, 130 (1971).

645. D. Stein and D. Stevenson, *J. Vinyl Addit. Technol.* **6** (3), 129–37 (2000).

646. M. Kleen. G. Lindblad and S. Backa, *J. Anal. Appl. Pyrol.* **25**, 209–27 (1993).

647. K. Kurihara and F. Tanoue, *Bunseki Kagaku* **49** (4), 265–8 (2000).

648. P. Perlstein and P. Orme, *J. Chromatogr.* **325**, 87–93 (1985).

649. S.G. Gharfeh, *J. Chromatogr.* **389**, 211 (1987).

650. K. Kurihara, F. Tsuchiya and T. Shoji, *Bunseki Kagaku* **51** (8), 647–52 (2002).

651. R.Y. Saleh and W.E. Wellman (to Exxon Research & Engineering Co.), *US Pat.* 5,325,791 (July 5, 1994).

CHAPTER 4

Lost information cannot be retrieved (D. Taupin, 1988)

Separation Techniques

Separation methods occupy a central position in organic analyses. Chromatographic, optical, magnetic, electrostatic, or dynamic fields are of interest from the viewpoints of separation science. Single molecule separation and detection are a final goal of the analytical separation. Chromatography in general comprises all separation techniques in which analytes partition between different phases that move relative to each other or where the analytes have different migration velocities. A chromatographic system consists of a fixed phase (liquid or solid stationary phase) which retains and a moving phase (gaseous or liquid mobile phase) which transports. The mixture to be analysed is introduced into the system via the mobile phase; the affinity of a solute for one phase over the other governs its separation from the other components. Retention is based on various

attraction forces which give rise to different modes of separation. Those substances distributed preferentially in the moving phase pass through the chromatographic system faster than those that are distributed preferentially in the stationary phase. Thus the substances are eluted in the inverse order of the magnitude of their distribution coefficients with respect to the stationary phase. The development of both separation efficiency and new detection systems has led to continuous improvements in selectivity and sensitivity, two integrated parameters in chromatography.

In the previous chapter on sample preparation for chromatographic analysis the principal objective has been to secure dissolution of analytes in a suitable solvent and removal from the solution of as many interfering compounds as possible. General *sample handling*

Additives In Polymers: Industrial Analysis And Applications J. C. J. Bart
© 2005 John Wiley & Sons, Ltd ISBN: 0-470-85062-0

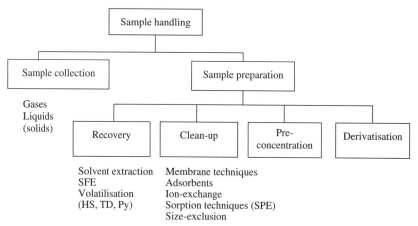

Scheme 4.1   Sample handling in chromatography

procedures for chromatography are given in Scheme 4.1. Most of the samples that have to be analysed by chromatography are too complex, too dilute, or, in their original state, incompatible with the chromatographic system. In these cases, direct injection cannot be used and sample pretreatment prior to injection of the sample into the chromatography system is required. The actual techniques used depend on the aggregation state of the sample to be analysed and the nature of the analytes (e.g. volatility and thermal stability): volume measurement for gases; volume measurement, dilution and solvent changeover for liquids; weighing, dissolution, leaching for solids. The main reasons why sample pretreatment is required are: (i) to protect the chromatographic system from contamination by nonelutable heavy material; (ii) to reduce sample complexity by selectively removing components that are not relevant but could elute with the target analyte; and (iii) to preconcentrate the components of interest to bring them into the working range of the chromatographic system. The relative importance of these three factors is significantly different for GC (relevant contamination) and LC (complexity reduction). Drawbacks of sample preparation are: (i) complexity; (ii) source of random and systematic errors; (iii) time-consuming; (iv) human participation; (v) variability; (vi) difficult control; and (vii) hazardous [1]. Sample preparation is often the rate-limiting step of chromatographic analysis.

Boundaries in chromatography and extraction are blurring, as evident from the relation between GC, SFC and HPLC, the use of superheated/subcritical water for extraction and chromatography, and the role of enhanced fluidity solvents and pressurised fluid extractions [2]. Extraction is an extreme form of chromatography. *Separation science* recognises that there is unity in the

separation method and that a continuum exists from gases to liquids. The same basic principles apply all over the pressure/temperature spectrum. The operation of each individual separation method is governed by the physical properties (such as density, viscosity and diffusion rates) of the mobile phase or solvent used. These properties define shape and size of the column and stationary phase and the types of analytes that can best be separated.

Chromatography is one of the most important analytical techniques: approximately 50 % of the total industrial analytical effort passes through chromatography. It allows the separation and subsequently the qualitative and quantitative analysis of complex mixtures, as long as the samples are volatile or soluble in a suitable solvent. Solutes must have appreciable solubility in a fluid for that fluid to be an effective chromatographic mobile phase. An inherent assumption in the following discussion is a general familiarity with basic analytical separation techniques. In chromatographic techniques a great number of variables play a role (Table 4.1). Most important are:

- physical state of the mobile phase (with T, p adjustable physical properties);
- geometrical arrangements of the stationary phase;
- type of stationary phase;
- basis of separation.

The physical properties of the mobile phase, mainly viscosity, diffusivity and solubility, affect the flow characteristics, column efficiency (kinetics), and retention (thermodynamics) in the chromatographic process. These physical properties are affected by temperature. Chromatographic techniques, although basically simple in

Table 4.1 Major variants in chromatography

Variable of chromatographic system	Variations in chromatography
Mobile phase	Gas, SCF, liquid, ionic solution Pressure, density Single vs. multiple component Polar vs. nonpolar
Stationary phase	Liquid, solid Ionic gel, cross-linked liquid Bonded phase Polar vs. nonpolar
Separation mechanism	Adsorption, partition, ion-exchange Size-exclusion, affinity
Development mode	Zonal, displacement, frontal
Isotherm linearity	Linear vs. nonlinear
Geometry	Column, (packed, capillary, multicapillary, cast) Layer (paper, thin layer) Planar vs. nonplanar
Flow	Capillary forces, gravitation, forced flow Electroosmosis
Elution	Isocratic Gradient (programmed temperature, chromathermography, programmed flow)
Chromatographic dimensionality	1D, 2D, multidimensional Continuous vs. discrete
Physical scale	Microbore – macrobore Nanoscale – preparative scale
Analytical use	Qualitative vs. quantitative

Table 4.2 Operating variables of some separation techniques

Operating variable	GC	LC	SFC	SFE
Temperature	x	x	x	x
Pressure/density			x	x
Mobile phase composition		x	x	x
Stationary phase	x	x	x	

After Davies *et al.* [3]. Reprinted from I.L. Davies *et al.*, in *Multidimensional Chromatography* (H.J. Cortes, ed.), Marcel Dekker Inc., New York, NY (1990), pp. 301–330, by courtesy of Marcel Dekker Inc.

Time-dominated processes inherently govern chromatography. The horizontal axis of a chromatogram is time (and not energy as in spectroscopy). To describe the quality of a chromatographic system the concepts of the height equivalent to a theoretical plate, HETP or H, and the number of theoretical plates N are used (Equation 4.1):

$$H = L/N \qquad (4.1)$$

where L is the length of the chromatographic bed. Important aspects in chromatography are the distribution coefficient (expressing the degree of preference of the analyte for one of the phases) and the length of the chromatographic bed over which equilibrium is reached (theoretical plate). If the components of a mixture have different distribution coefficients in a given chromatographic system, they can be separated if the number of plates is high enough.

There are various ways of *classification* of separation techniques. Interactions between sample components and mobile and stationary phases can be classified as either adsorption or absorption. In adsorption chromatography (also referred to as liquid–solid chromatography, LSC) the sample is attracted to the surface of the phase. In absorption or partition chromatography the sample diffuses into the interior of the stationary phase. Most chromatographic separations are a combination of adsorption and absorption phenomena. Other separation modes are ion-exchange chromatography (IEC) and size-exclusion chromatography (SEC). Solvents used in adsorption chromatography are arranged in eluotropic series; the eluting action of solvents increases with increasing polarity: pentane, petroleum ether, hexane, heptane, cyclohexane, tetrachloromethane, trichloroethylene, benzene, dichloromethane, trichloromethane, diethyl ether, ethyl acetate, pyridine, acetone, propanol, ethanol, methanol, and water.

In practice, chromatography takes place on a layer or in a tube. Meyer [4] has compared analytical column-type chromatographic methods. A column can be an open capillary or a packed tube. In the first case the mobile phase is coated as a thin film on the inner wall of the capillary. In most cases GC is used as open-tubular

principle, involve complex physical chemical concepts. For the effective use of the techniques, the chromatographic process must be well understood so that the separation can be optimised. Chromatography is based on the interaction of molecules with molecules (not with a field). Physical or chemical interaction of the mobile and stationary phases with the sample molecules is based on transport, partitioning, adsorption, solvation and polarity phenomena (in a very broad sense). A mixture can be separated if its compounds are differentially retained. The main parameters for retention control are temperature in GC, solvent strength in LC, and solvent density/strength and temperature in SFC. SEC and CE are examples of chromatographies in which there may be negligible interaction between the analyte and the stationary phase. With the broad range of mobile phases and extremely large number of stationary phases, it is almost a certainty that a compound can be separated from a bulk mixture. Solubility parameters are crucial in the selection of chromatographic mode and column. Various *operating parameters* may be varied to achieve the optimum chromatographic selectivity and retention (Table 4.2).

GLC, whereas LC is performed in packed columns. SFC is intermediate between GC and LC and it is no surprise that it can be used equally well with open capillaries and packed columns. A layer can be a sheet of paper (paper chromatography) or an adsorbent that is fixed onto a sheet of aluminium or glass (TLC). If the chromatographic separation is performed on a layer (PC or TLC), the separated bands are not eluted but remain in the layer. The supplementary column and layer liquid chromatographic techniques have always been characteristically developed in constant mutual interaction (cf. Table 4.3). The advent of high-performance liquid chromatography (HPLC) has entailed fundamental renewal of the most popular planar layer liquid chromatographic technique, TLC. High-performance TLC (HPTLC) is based on the use of chromatoplates coated with fine particles of narrow particle size distribution sorbent. The greatly increased developing time on a HPTLC chromatoplate has made it necessary to employ forced flow, which is also used in rotation planar chromatography (RPC) and in high-speed TLC (HSTLC) using electroosmosis to force the eluent. Overpressured layer chromatography (OPLC), using a pump system for admission of the eluent, is a real planar version of HPLC [5].

Chromatography can be used for both qualitative and quantitative analysis. The retention time, t_R, and in planar chromatography the R_f value, are physicochemical characteristics for a pure compound in a given chromatographic system. Nowadays also separation on a microscale is perfectly possible (e.g. by means of µGC, µLC or electrophoresis). Analytically, the problem is detection on the µL or nL level. Some typical

Table 4.3 Some milestones in separation science

Year	Development
1903–1940	Adsorption/desorption [6]
1938/1958	Thin-layer chromatography [7,8]
1941	Partition chromatography [9]
1944	Paper chromatography [10]
1952	Gas chromatography [11]
1958	Capillary columns for GC [12]
1958	Headspace sampling [13]
1962	Supercritical fluid chromatography [14]
1964	Size-exclusion chromatography [15]
1967/1981	Capillary electrophoresis [16]
1970	High-performance liquid chromatography [16a]
1975	Ion chromatography [17]
1979	Fused-silica capillary columns [18]
>1985	Hyphenations
1989	Unified chromatography [2,19]
1995	Injectors with temperature programming for GC
1996	Multicapillary columns for GC [20]
2000	Monolithic sorbents for LC [21,22]

Table 4.4 Main features of microchromatography

Advantages

- Speed (GC, CEC)
- Low eluent consumption (LC)
- Good mass detection limits (all techniques)
- Good heat dissipation (CEC)
- Small sample need (LC, CEC)
- Applicable for expensive stationary phases (LC, CEC)

Disadvantages

- Less robust (all techniques)
- Reduced sample loadability (GC)
- Poor concentration detection limits (all techniques)
- High pressure drop (GC)
- Critical injection band width (GC, LC)
- Critical system volumes (LC, CEC)

injection volumes of chromatographic techniques are: cSFC, < 0.1 µL; cGC, 1 µL; HPLC, 10 µL. Table 4.4 shows the main features of microchromatography.

A chromatographic separation step provides various advantages to the analytical procedure: (i) each component is isolated from the others (which facilitates identification); (ii) minor components in mixtures may be detected more readily than by direct analysis techniques; (iii) the chromatographic retention parameter provides additional confirmation that a particular component is present or absent; and (iv) quantitative analysis. However, chromatography alone does not provide information on the identity of a totally unknown sample.

Selecting the appropriate chromatographic technique for a given separation problem amongst those indicated in Scheme 4.2 is a challenging task. Table 4.5 sets the general criteria. Whether a given problem should be tackled by GC, SFC, HPLC or some electrophoretic method depends on many parameters. Generally, whenever capillary GC can be used, it should be preferred over any of the other chromatographic techniques as it offers a resolving power, separation speed and a userfriendliness unsurpassed by competitive methods. In fact, the *first rule of chromatography* states: 'If you *can* perform a separation by GC, you *should* use it' [23]. A major problem of GC is that it is limited to the separation

Table 4.5 Criteria for choice of separation techniques[a]

- Ability to solve real-life problems
- Availability of commercial instrumentation
- Understanding of the fundamentals of the technique
- Figures of merit (capacity, efficiency, speed, resolution, selectivity, LOD, LOQ, etc.)
- Ease of method optimisation and validation and transfer
- Ability to reduce labour, time and cost per analysis

[a] Including pre- and post-chromatographic requirements.

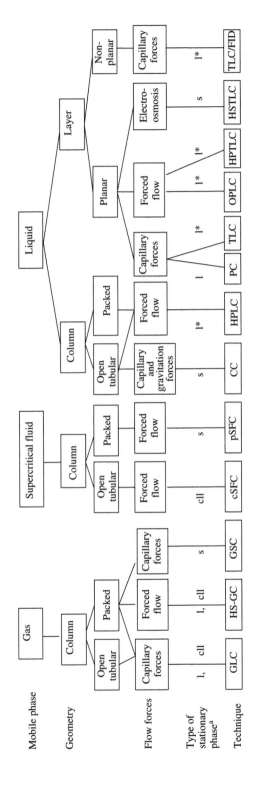

Scheme 4.2 Classification of chromatographic methods

[a]l, liquid; s, solid; cll, cross-linked liquid; l*, liquid (on gel support)

of only a small fraction of all known compounds (i.e. modest molecular weight, thermally stable, volatile and semivolatile compounds): of the more than 9×10^6 compounds assigned CAS numbers, only 1.4×10^5 are amenable to GC analysis [24]. GC outperforms most other separation techniques in terms of speed, resolution and sensitivity but its application area is relatively small. High-temperature capillary GC (HTGC) nowadays allows analysis of components that previously could only be analysed using HPLC, pSFC or cSFC. Separations that are performed at oven temperatures below 325 °C are usually considered as 'normal' GC and are applicable to relatively volatile materials (up to about 600 Da). HTGC ($T_{\text{oven}}^{\max} = 450$ °C) extends the MW range of action to about 1500 Da. Surprisingly, the thermal stability of organic compounds in a fully inert system, i.e. properly deactivated fused-silica column, inert carrier gas and a highly purified stationary phase, is much greater than generally accepted [25]. With the development of HTGC the practical applicability of GC for the analysis of high-MW materials such as polymer additives, surfactants, hydrocarbon oils, etc., has increased significantly. However, sample introduction in HTGC is extremely difficult. For thermally labile analytes that do not volatilise at temperatures up to 250 °C and are strongly polar GC is not the technique of choice.

Dry column chromatography and HPLC allow rapid analysis, but at a higher cost than GC because high-quality pump and column packing are necessary. Classical gravity-flow and preparative-scale work is time-consuming. Proper choice of conditions, however, allows virtually any organic compound (other than, e.g., salts) to be analysed by LC. In many cases LC can obtain the separation of organic compounds with high molecular weights and polar functional groups, but there is a lack of sensitive universal LC detectors. In these cases, SFC is an alternative which replaces NPLC, expands the GC application range towards higher MW, and has advantages in oligomer analysis over SEC. There is a continual shift in many applications back and forth between GC, SFC and LC following up subtle changes in technology.

It is difficult to compare separation techniques in any general way. Comparison may be based on the traditional *figures of merit*, such as resolution R_s (including column efficiency N, selectivity, retention, and peak capacity), chromatographic speed, sample capacity, sensitivity, detection and column impedance, as well as breadth of application. Usually a trade-off between these attributes is found. Berger [26] has compared GC, pSFC, cSFC, LC and CE on the basis

of these specifications. Middleditch *et al.* [27,28] have listed an impressive range of stabilisers and additives which can be found in commercial solvents and which cause artifacts in chromatography. There is a great deal of difference between separating a set of model compounds and separating even the same compounds in a real-life sample.

Chromatographic methods are essential to separate and resolve the multiple components of polymer/additive extracts or dissolutions. Current chromatographic methodology for *additive analysis* comprises GC (including HTGC, HSGC, HS-GC, TD-GC), SFC (both cSFC and pSFC), various forms of column LC (NPLC, RPLC, µHPLC, µHTLC, HSLC, SEC, GPEC®, IC) and planar LC (TLC, HPTLC), all of which can be coupled ('hyphenated') to either prechromatographic sample preparation techniques or postchromatographic identification methods (cf. Chapter 7). Capillary electrophoretic techniques (HPCE, CZE) only play a minor role. All these procedures have their advantages and limitations, as will be shown in subsequent paragraphs. One of these limitations (now partly overcome) is that for chromatographic techniques generally quite a large amount of sample is still necessary for a good separation. This differs from direct additive analysis procedures involving volatilisation of solid material (TA, Py, TD and LD techniques). A recent breakdown for techniques used for additive analysis in polymers gives the following estimate: RPLC 60 %, NPLC 5 %, GC-FID 14 %, GC-MS 2 %, TLC 6 %, UV 2 %, EA 6 %, and others (DSC, TGA, NMR, MS) 5 % [29].

For the validation of chromatographic methods a good starting point can be found in the EURACHEM/WELAC document on Accreditation of Chemical Laboratories [30]. ASTM Committee E19 on Chromatography covers all forms of chromatography including gas, liquid, ion and supercritical chromatography.

Chromatography is the widest used technique (especially in the pharmaceutical industry) that enables separation of molecular components of interest. Generally, however, chromatographic data alone are insufficient evidence to substantiate the presence of a specific component in a mixture. For the chromatographer, separation is not the final definitive procedure, since monitoring and identification of eluted chromatographic peaks is vital to the effectiveness of separation. For analytical purposes (identification and selective quantification) specific detectors play a crucial role. Given their general applicability for all forms of chromatography we start this chapter by introducing these detectors.

4.1 ANALYTICAL DETECTORS

The overall performance of a separation method is intrinsically linked to that of the detector used as part of the system. A detector is a device that monitors, in the dimensions of space or time, the presence of the components of a mixture that has been subjected to a chromatographic process. The detection methods provide evidence concerning the quality of the separation and serve especially to increase sensitivity and selectivity. The quantitative aspects of chromatographic analyses are dependent upon the detector capabilities.

Detectors are composed of a sensor and associated electronics. Design and performance of any detector depends heavily on the column and chromatographic system with which it is associated. Because of the complexity of many mixtures analysed and the limitation in regard to resolution, despite the use of high-resolution capillary columns and multicolumn systems, specific detectors are frequently necessary to gain selectivity and simplify the separation system. Many detectors have been developed with sensitivities toward specific elements or certain functional groups in molecules. Those detectors that exhibit the highest sensitivity are often very specific in response, e.g. the electron capture detector in GC or the fluorescence detector in LC. Because

of their specific response, the noise level is always relatively low. Scott [31] distinguishes some 12 detector criteria, amongst which linear dynamic range, response, noise level, sensitivity (or minimum detectable concentration), time constant, flow, pressure and temperature sensitivity, ruggedness, etc. Some detectors generate 2D data, i.e. data representing signal strength as a function of time. Others, including fluorescence and diode-array UV/VIS detectors, generate 3D data including not only signal strength but also spectral data for each point in time. *Design principles* of the main chromatographic detectors are given in Table 4.6.

Detectors have been classified in a number of different ways, although no classification is ideal:

(i) Specific and nonspecific detectors.
(ii) Element specific detectors.
(iii) Bulk property and solute property detectors.
(iv) Mass sensitive and concentration sensitive detectors.
(v) Spectroscopic detectors.
(vi) Non-spectroscopic detectors.

The first classification is based on the nature of the detector response. Table 4.7 ranks several chromatographic detectors as specific and nonspecific. A *nonspecific or universal detector* responds to all solutes present in the mobile phase and this performance makes it a

Table 4.6 Design principles of chromatographic detectors

Acronym	Full name	Design principle
FID	Flame ionisation detector	Burning of compound in flame
O-FID	Oxygen-specific FID	Cracking, methanisation
NPD (FTD)	Nitrogen–phosphorous detector	Thermoionic emission
FPD	Flame photometric detector	Chemiluminescence
ECD	Electron capture detector	Molecular capturing of low-energy electrons
ELCD, HECD	Electrolytic conductivity (Hall) detector	Thermal generation of inorganic ions
SCD	Sulfur chemiluminescence detector (flame and flameless)	Ozone-induced CL
RCD	Redox chemiluminescence detector	Ozone-induced CL
CLND	Chemiluminescence nitrogen detector	Pyro-chemiluminescence
TCD	Thermal conductivity detector (katharometer)	Differences in thermal conductance of analyte vs. carrier gas
FLD	Fluorescence detector	Excitation/fluorescence
UVD (DAD)	UV/VIS detector[a] (diode-array detector)	UV absorbance
AED	Atomic emission detector[b]	Atomic emission spectroscopy
ICP-MS	Inductively coupled plasma[b]	Argon plasma discharge
MIP-MS	Microwave-induced (helium) plasma[b]	Microwave-induced electrical discharge
ELSD	Evaporative light scattering detector	Raleigh scattering of atomised eluent
RID	Refractive index detector	Differential refractive index
MS	Mass spectrometer	Fragmentation and ionisation
FTIR	Fourier transform infrared	Infrared absorption
PID	Photoionisation detector	Photoinduced (UV) ionisation
LIF	Laser-induced fluorescence	Fluorescence

[a] Fixed wavelength.
[b] Tandem.

Table 4.7 Some specific and nonspecific chromatographic detectors

Detector type	GC	HPLC	SFC	TLC
Specific	FPD, NPD, RCD, CLND	FLD, RCD, CLND	FPD, RCD, CLND, SCD	FLD
Nonspecific	FID, TCD	RID, ELSD	FID	

very useful and popular type of detector [31]. However, because general chromatographic detectors such as FID respond to most or all of the organic species present in a sample, producing complicated traces, it is often advantageous to use more specific detectors in the analyses of complex mixtures. *Specific detectors* are not genuinely substance-specific but respond only to a particular type of compound or a particular chemical group (e.g. RCD for alcohols, ketones, acids; PID for olefins and aromatics). A selective detector can alleviate both problems, i.e. separation and identification by responding only to the analyte of interest regardless of other constituents of the matrix. With a few exceptions (ELCD, FSCD, PID, MS, RCD) all selective detectors are element and not organic compound selective. The ideal, selective, multielement detection system (for GC) has the following properties [32]:

(i) measures simultaneously all elements emerging from the column;
(ii) LOD of $< 100\,\mathrm{pg\,s^{-1}}$ for each element;
(iii) wide linear dynamic range ($10\,\mathrm{pg\,s^{-1}}$ to $10\,\mu\mathrm{g\,s^{-1}}$);
(iv) fast response;
(v) insensitive towards all other elements; and
(vi) response independent of molecular structure.

Spectrometric detection systems based on measurement of atomic weight and atomic emission can potentially fulfil these requirements.

Table 4.8 shows some frequently used *element selective chromatographic detectors* (ESDs) and Table 4.9

Table 4.8 Main element specific chromatographic detectors (ESDs)

ESD	Element specificity[a]	ESD	Element specificity
NPD/FTD	N, P, X	ICP-MS	Multiple
FPD	S, P, N, Sn	MIP-MS/AED	Multiple
ECD	X, carbonyl, nitro	SCD	S, N
ELCD	X, N, S	O-FID	O
AED	Multiple	CLND	N, NO

[a] X = halogen.

Table 4.9 Comparison of several commercially available sulfur-selective detectors

Detector	Minimum detectability (pg S s^{-1})	Selectivity (S/C)[a]	Linear response
FPD	2–50	10^4–10^5	No
ELCD	2–50	10^4–10^5	Yes
AED	1–2	1.5×10^5	Yes
Flame SCD	0.5–5	$>10^7$	Yes
Flameless SCD	0.05–0.5	$>10^7$	Yes

[a] Selectivity relative to an equal mass of carbon.
After Shearer [33]. Reprinted from R.L. Shearer, in *Selective Detectors* (R.E. Sievers, ed.), John Wiley & Sons, Inc., New York, NY, pp. 35–69, Copyright © (1995, John Wiley & Sons, Inc.). Copyright © 1995 John Wiley & Sons, Inc. This material is used by permission of John Wiley & Sons, Inc.

compares various sulfur-selective detectors. A truly selective detector such as the sulfur-selective FPD detector essentially requires no separation of a single sulfur-containing analyte from other compounds; nevertheless, it is often convenient to couple a chromatographic step with selective detection. One can rely on the discrimination of the detector to distinguish and measure analytes not fully separated from other nonsulfur-containing compounds. Analyte identification is considerably enhanced by the use of a selective detection scheme, as the chromatographic trace is simplified to include only specific (i.e. sulfur-containing) compounds. The widespread adoption of these detectors shows the value of element-selective detection, but most are limited in elemental scope in comparison with a full spectral technique such as atomic emission. Also, plasma source mass spectrometry offers selective detection with excellent sensitivity and is used for the analysis and speciation of trace elements. In comparison to AED, detection limits are generally two to three orders of magnitude lower for plasma MS determinations. Additionally, isotope abundance information may be obtained and multielement studies may be carried out. Speciation of trace elements is a popular research area.

Bulk property detectors function by measuring some bulk physical property of the mobile phase, e.g., thermal conductivity or refractive index. As a bulk property is being measured, the detector responses are very susceptible to changes in the mobile phase composition or temperature; these devices cannot be used for gradient elution in LC. They are also very sensitive to the operating conditions of the chromatograph (pressure, flowrate) [31]. Detectors such as TCD, while approaching universality in detection, suffer from limited sensitivity and inability to characterise eluate species.

In contrast, *solute property detectors* directly measure some physico-chemical property of the eluate species that the mobile phase does not possess, or to a minor extent only. An example of a solute property detector is the NPD in GC, which responds only to nitrogen or phosphorous, or the UV detector in LC, which detects only UV absorbing substances. Depending upon the property sensed, such detectors can show high sensitivity and also provide eluate identification and characterisation. Some solute property detectors give 'selective' information or, under optimal circumstances, 'specific' information on the eluates. This type of classification is confusing, as a given detector can exhibit both characteristics depending on the operating conditions. Also, this classification is of little use for SFC and TLC detectors (Table 4.10).

Some detectors respond to changes in solute concentration while others respond to the change in mass passing through the sensor per time unit (Table 4.11). *Concentration sensitive detectors* provide an output that is directly related to the concentration of the solute in the mobile phase passing through it. The *mass sensitive detector* responds to the mass of solute passing through it per unit time and is thus independent of the volume flow of mobile phase. Table 4.12 shows the sensitivity of some typical chromatographic detectors.

Spectroscopic detectors, which measure different spectral properties (absorption, fluorescence, scattering, etc.), may be element selective, structure or functionality selective, or property selective. The most common

Table 4.10 Bulk property and solute property chromatographic detectors

Detector type	GC	HPLC	SFC	TLC
Bulk property	TCD	RID	–	–
Solute property	NPD	UV	UV	UV

Table 4.11 Concentration and mass-sensitive detectors

Detector type	GC	HPLC	SFC	TLC
Concentration sensitive	TCD, ECD	UV	UV	UV
Mass sensitive	FID	ELSD	FID, ELSD	–

Table 4.12 Sensitivity of chromatographic detectors

Detector	Sensitivity[a]	Detector	Sensitivity[a]
TCD	10^{-3}–10^{-8}	PID	10^{-5}–10^{-12}
FID	10^{-3}–10^{-11}	ELCD	10^{-6}–10^{-11}
NPD	10^{-5}–10^{-12}	SCD	10^{-6}–10^{-14}
ECD	10^{-8}–10^{-13}	CLND	10^{-5}–10^{-13}
FPD	10^{-6}–10^{-10}	ELSD	10^{-7}–10^{-9}

[a] g of sample.

spectroscopic techniques employed to confirm substance identity and used for structure elucidation are UV, IR, Raman, MS, NMR, AAS and AES. There is a difference between a spectroscopic detector and a chromatograph-spectrometer coupling, the latter being a so-called *hyphenated* or *tandem system* (cf. Chapter 7). A spectroscopic detector is designed and constructed specifically as a chromatography detector, e.g. the diode array detector. The spectrometer used in tandem instruments, although packaged with a chromatograph, is usually a standard instrument that is interfaced to the chromatograph. Effective analysis must involve optimisation of both the separation and detection processes, as well as the interface between them.

Conventional single channel variable wavelength detectors deliver only quantitative information, whereas further qualitative information on component identity and peak purity are entirely missing. Peak purity testing in routine analysis (based on chemometric principles) adds important quality information to the analytical results. Multi-wavelength detectors allow simple purity checks by ratio calculation between the signals acquired at two different wavelengths. If the spectral difference between the main component and the chemically similar impurity is only visible at a spectral range different from the two selected wavelengths, the impurity becomes invisible for this detection technique. Only a diode array detector providing high-resolution spectra (typically better than 1 nm) adds significant information on component identity and peak purity [34].

Nonspectroscopic detection schemes are generally based on ionisation (e.g. FID, PID, ECD, MS) or thermal, chemical and (electro)chemical effects (e.g. CL, FPD, ECD, coulometry, colorimetry). Thermal detectors generally exhibit a poor selectivity. Electrochemical detectors are based on the principles of capacitance (dielectric constant detector), resistance (conductivity detector), voltage (potentiometric detector) and current (coulometric, polarographic and amperometric detectors) [35].

Advantages and disadvantages of chromatographic detectors and tandem systems are given in Table 4.13, whereas Table 4.14 gives a breakdown of the various detectors over the main chromatographic techniques. Simultaneous detection is possible, e.g. SCD/CLND.

Drushel [58] and others [31,59] have described the needs of the chromatographer in the area of detectors. Specific texts concern detection in quantitative GC [54], diode-array detection in HPLC [48], selective detectors [39] and element-specific chromatographic detection by AES [60], electrochemical detectors [61] and laser detectors [62].

Table 4.13 Advantages and disadvantages of chromatographic detectors

Detector	Advantages	Disadvantages	References
FID	General (all organic substances) Widest linear dynamic range High sensitivity Simple, easy to operate Inexpensive	Poor selectivity Destructive	[31,36,37]
O-FID	Oxygen specific High reproducibility	Destructive Low ppm range sensitivity	[38,39]
NPD (FTD)	Relatively high sensitivity Specific response	Smaller dynamic range than FID Susceptible to operating conditions	[31]
FPD (PFPD)	Relatively inexpensive High sensitivity Relatively robust	Not suited for quantitation in complex mixtures Destructive	[31,40,41]
ECD	Halogenated compounds Extremely sensitive to high-electron affinity compounds Ease of use Cheap	Limited to ion detection Poor selectivity Small linear range	[31]
ELCD	Good lower limit of detection Good selectivity Inexpensive	Several critical parameters Difficult to use	[42]
CLND	Total nitrogen specific Easy quantitation (single standard)	High linearity (10^5)	[43,44]
O-SCD	Higher selectivity than AED Excellent sensitivity Linear response Odorant analysis	Skilled operators (flame SCD)	[33,40,45]
TCD	Universal Gas analysis detector Rugged	Relatively insensitive Susceptible to operating conditions Questionable quantitative response	[31]
RCD	Selective for readily oxidisable compounds	Variable response factors	[46]
FLD	Very high sensitivity Simple operation Reliable	Few intrinsically fluorogenic compounds	[31,47]
UV (DAD)	High resolution Component identity Peak purity testing Robust	Limited to UV absorbing analytes Wide variability in compound absorptivities	[31,48]
AED	Simultaneous multielement Superior to FPD for quantitative analysis (S compounds) High sensitivity Versatile	Expensive Requires skilled analysts	[33]
ICP-MS	Multielement detection Trace elements Speciation	Expensive	[36]
MIP-MS	Multielement detection (especially nonmetals) Trace elements	Expensive Selectivity poor for many elements	[36,49,50]
ELSD	Quasi-universal No dependence on eluent conditions	Droplet size control Moderate sensitivity (low ng) Compound-dependent and non-linear detector response	[31,51–53]
RID	Quasi-universal MW, MWD information	Insensitive Very sensitive to temperature variations Limited dynamic range	[31]
MS	High sensitivity (low pg) High selectivity (MS^n mode)	Spectral interpretation	[54]
FTIR	Solute confirmation	Less informative than MS Limited use in structure elucidation	[31]
PID	Universal High sensitivity Compound selective Nondestructive	Partial selectivity	[55,56]
LIF	Very high sensitivity	Wavelength restrictions	[57]

Table 4.14 Characteristics of chromatographic detectors and tandem systems

Detector	GC	HPLC	SFC	TLC	SEC	Properties[a]	Quantification
FID	+		+		+	M, U	+
O-FID	+					E	+
NPD	+		+			E, S, SP	+
FPD	+		+			E, M, S	±
ECD	+	+	+			E, C, S	+
ELCD	+	+	+			E, C, M, S	+
CLND	+	+	+		+	S, M	+
O-SCD	+	+	+			E	+
TCD	+		+			BP, C, U	+
RCD	+	+	+			S	+
FLD		+	+	+		S, SD	+
UV (DAD)		+	+	+		C, S, SD, SP	+
AED	+	+	+			E, M, SD, U	+
ICP-MS	+	+	+			E, SD, U	+
MIP-MS	+	(+)	+			E, SD, U	+
LS[b]	+	+	+		+	M, U	+
RID	+	+	+		+	BP, C, U	−
DV					+	M	+
MSD	+	+	+	+	+	M, SD, U	−
FTIR	+	+	+	+	+	C, S, SP, U, SD	+
PID	+		+			C, S	−
LIF		+		+		S, SD	+

[a] S, specific; U, universal; BP, bulk property; SP, solute property; M, mass sensitive; C, concentration sensitive; E, element specific; SD, spectroscopic detector.
[b] ELSD, LALLS, RALLS.

4.2 GAS CHROMATOGRAPHY

Principles and Characteristics Gas chromatography (GC) is a physico-chemical technique of separation of substances from their mixtures based on different degrees of distribution of each substance between two heterogeneous phases, stationary and mobile, the mobile phase being gaseous. Gas chromatography offers separation of substances from their mixtures that have been converted to the gas phase and that, under given conditions, are sufficiently stable. The *basic principle* is simple. The sample is injected into a carrier gas and carried through a column of suitable length. Components in the mixture are separated by relative affinity of the compounds with the column packing material. A primary driving force for separation of organic components by GC is simple boiling point differences: higher boiling compounds are eluted more slowly. The times required by the components to reach the detector and the intensities of the responses from the detector are processed to provide qualitative and/or quantitative analysis. Liquid samples can be directly injected into the GC instrument; solids must be dissolved, pyrolysed, or subjected to thermal desorption. Upon injection, liquid samples are subjected to a very hot injection port, which immediately converts them to the gas phase. A chemically inert gas (usually He, Ar, N_2 or CO_2, depending on the detection system) transports the gases into a heated column, which is internally coated with a stationary phase. The gaseous mobile phase has no direct influence on the separation.

As mentioned before, there are two common types of GC: gas–liquid chromatography (GLC) and gas–solid chromatography (GSC), depending on the physical state of the stationary phase. GSC is seldom used. In GLC the analyte is partitioned between the mobile phase (gas) and a liquid phase, which is retained on an inert solid support. The liquid phase should ideally possess a low volatility (so that it does not volatilise with the analyte), be thermally stable and chemically inert, and have favourable solvent characteristics.

Gas chromatography, introduced in 1952 by James and Martin [11], now represents a mature analytical technique which has been enjoying major innovations [63]: introduction of FID [64] and high-resolution capillary columns with metal tubings [12] in 1958, development of a temperature-programmable GC in 1959, introduction of open-tubular capillary (OTC) columns (mid-1970s), almost generalised replacement of packed columns with fused-silica capillaries in the 1979–1985 period offering the inertness of glass with the flexibility of steel capillaries, development of detectors with improved selectivity and sensitivity, high-temperature GC [65], extensive hyphenation (GC-MS,

GC-FTIR, GC-AED, GC-ICP-MS, cf. Chapter 7), fast GC separations (1996) and most recently the development of sophisticated injectors with temperature-programming capability and high-resolution systems (GC-ToFMS). As a result, modern GC systems are quite advanced (Scheme 4.3) and GC is one of the most widely applied instrumental techniques.

The main *instrumental elements* of GC are a source of carrier gas, injector (for gas, liquid or solid samples of sufficient volatility), column (typically 25 m), oven programming (which influences the partial vapour pressures of the analytes), selection and control of chromatographic column and detector, and where appropriate, an autosampler and interface to a hyphenated technique. Detailed descriptions of each of these parts can be found in dedicated literature (e.g. ref. [66]) and may vary for different GC models. The dimensions of the injection system are important. GC sample volumes are normally between 0.5 and 5 μL. The GC oven provides a controlled temperature for the chromatographic column. In most GC systems, the set temperature can be kept within 0.1 °C, and a range between -100 °C to 400 °C can be achieved using either a cryogenic agent (liquid N_2 or CO_2) or electric heating. Also, GC ovens are commonly able to provide temperature gradients such that a sequence of isotherm and gradient portions is available.

Gas chromatography often requires some preliminary *sample preparation steps* (collection, storage, extraction, clean-up, preconcentration, derivatisation). Some samples may be ready for injection with no preparation, but most require some form of modification, even if this is only addition of a suitable solvent to a highly concentrated sample. Sample preparation is often the rate-limiting step of chromatographic analysis. Gas extraction techniques for sample preparation in GC are solvent extraction (cf. Chapter 3), heat extraction or photon extraction. These apply also to polymer/additive samples. In addition to classical techniques such as extraction and filtration, a number of instrumental sample preparation procedures exist (cf. Chapter 7). These techniques include purge-and-trap sampling, headspace sampling, thermal desorption, automated solid-phase microextraction, accelerated solvent microextraction and many others (cf. Scheme 4.3). SPME allows isolation of analytes of interest from very dilute solutions. In general, sample preparation alters analyte concentrations, may include analyte derivatisation, remove some impurities such as nonvolatile residues and transfer analytes into another solvent or remove the solvent entirely. Typical sample preparation steps are clean-up (e.g. SPE, retention gaps) and

preconcentration or enrichment (e.g. TD-CID, MESI). Nowadays, multipurpose samplers are available for standard injection, large volume injection, headspace injection, solid-phase microextraction (SPME) and automated sample preparation (dilution, extraction, derivatisation, internal standard addition).

Several sample preparation techniques are performed inside the inlet system. Large-volume injection can be carried out by a number of methods including programmed temperature vaporisation (PTV). Automated SPE may be interfaced to GC using a PTV injector for large volume injection. SPE-PTV-GC with on-column injection is suited to analysis of thermolabile compounds.

For direct gas sampling a diffusive sampler may be used, which is a device capable of taking samples of gas or vapour pollutants from the atmosphere at a rate controlled by a physical process such as diffusion through a static air layer or permeation through a membrane, but which does not involve the active movement of air through the sampler [67]. Adsorption by active charcoal is commonly used for collecting trace components prior to GC analysis. Rektorik [68] has described thermal desorption of solid-traps using microwave energy. The desorbed analyte is swept into the analytical column by a carrier gas [69].

There are basically three methods of liquid sampling in GC: direct sampling, solid-phase extraction and liquid extraction. The traditional method of treating liquid samples prior to GC injection is liquid–liquid extraction (LLE), but several alternative methods, which reduce or eliminate the use of solvents, are preferred nowadays, such as static and dynamic headspace (DHS) for volatile compounds and supercritical fluid extraction (SFE) and solid-phase extraction (SPE) for semivolatiles. The method chosen depends on concentration and nature of the substances of interest that are present in the liquid. Direct sampling is used when the substances to be assayed are major components of the liquid. The other two extraction procedures are used when the pertinent solutes are present in very low concentration. Modern automated on-line SPE-GC-MS is configured either for at-column conditions or rapid large-volume injection (RLVI).

Table 4.15 lists the many possibilities for solid sampling for GC analysis. In general, sample preparation should be considered in close conjunction with injection. Robotic sample processors have been introduced for automatic preparation, solvent extraction and injection of samples for GC and GC-MS analyses. Usually, facilities are included for solvent, reagent, and standard additions and for derivatisation of samples.

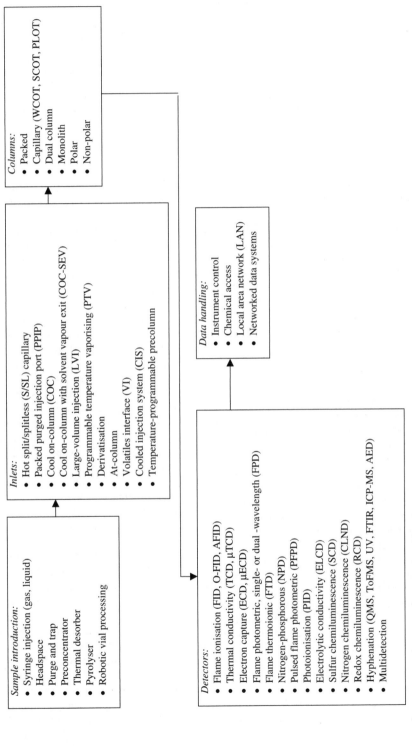

Scheme 4.3   Advanced gas chromatographic systems

Table 4.15 Sample preparation methods used in GC analysis

Gas sampling	Liquid sampling	Solid sampling
Diffusive sampler	Liquid–liquid extraction (LLE)	Conventional solvent extraction methods[a]
Membrane extraction (MESI)	Solid-phase extraction (SPE)	Pressurised solvent extraction methods[b]
	SPE-PTV-GC	Headspace GC (SHS, DHS)
	Solid-phase microextraction (SPME)	Thermal desorption (TD, DTD)
	Headspace GC (SHS, DHS)	Pyrolysis (Py)
	Large-volume injection (LVI)	Photolysis
	Coupled HPLC-GC	Photon extraction (LD)
	Membrane extraction (MESI)	Difficult matrix introduction (DMI)
	Difficult matrix introduction (DMI)	

[a] (Automated) Soxhlet extraction, reflux extraction, sonication, etc. (Section 3.3).
[b] SFE, MAE, ASE®/ESE® (Section 3.4).

Separation by GC is based on the vapour pressures of volatilised compounds and their affinity for the liquid stationary phase, which coats a solid support, as they pass down the column in a carrier gas. The practice of GC can be divided into two broad categories, packed and capillary or open-tubular column (OTC) based, the latter ascribed to Golay [70]. Hill and Simpson [71] have discussed the characteristics of the various types of chromatographic columns and GC support materials. Historically the first column types for GC were packed metal and glass columns, typically 3–6 mm in width and 1–5 m long. *Packed columns* contain solid particles of an inert support, such as diatomaceous earth coated with the stationary phase. Major drawbacks of traditional packed columns are peak broadening due to the presence of particles inside the column (eddy diffusion), low maximum achievable plate number and low speed. Advantages are high loadability and wide working range. As packed columns contain much more stationary phase peaks are moved further apart than in capillary columns which carry much less stationary phase; however, the dispersion is relatively large compared with that from a long capillary column. Use of regular packed columns is now limited to specific applications, while capillary columns are widely employed for 80 % of all GC analyses [72]. *Capillary columns* contain the stationary phase as a film on the inner wall. The stationary phase is commonly a liquid on a stationary surface, a cross-linked material, or a bonded phase. Special columns such as capillary columns containing fine solid particles coated with the stationary phase are also known. There are several advantages to capillary GC (compared to packed-column GC):

- Very accurate analyses are quickly attained on very small samples.
- Column 'bleed' is normally not a serious problem since capillary columns contain little stationary phase.

Capillary GC (CGC) is restricted to analytical procedures: usually only 1 μL of the analyte solution can be injected without overloading the column. Preparative-scale operations are normally carried out on packed columns. Capillary columns are commonly made from silica and have an outer coating (polyimide, aluminium, etc.) that improves their mechanical resistance. The primary advantages of fused-silica columns over glass and other materials used for OTCs are flexibility of the column and inherent inertness of the material [18]. Fused-silica columns are used in GC, microcolumn LC and SFC. Typical fused-silica columns such as CP-Sil 5CB (Chrompack; 25 m, 0.32 mm i.d., 0.12 μm film thickness) are in common use for GC up to 325 °C. Capillary columns can be made much longer (5 to 100 m) and thus produce many more theoretical plates (10^5+) than packed columns and achieve very high separation efficiency. The column length may be chosen in relation to the complexity of the sample; typically, 25 m for effective separation of most samples and 50 m for very complex samples containing many closely eluting components. The column internal diameter (i.d.) is in a range between 50 μm and 100 μm to 0.6 mm or even wider. Columns are classified as microbore for i.d. < 0.1 mm, minibore for i.d. = 0.18 mm, narrow-bore for i.d. = 0.25 mm, regular for i.d. = 0.32 mm, and megabore for i.d. = 0.53 mm. The internal diameter of the column is adapted to the concentration of the sample to be analysed. Small diameter columns, e.g. 0.25 mm, offer high efficiency and are useful for quantification of small amounts of sample, but are easily overloaded. Columns of i.d. 0.32 mm are recommended for on-column injections, while 0.53 mm columns are designed for high-volume injections. Major drawbacks of narrow-bore capillary columns are low sample capacity and limited working range. An advantage is high speed. Monoliths are chromatographic columns that are cast as continuous homogeneous phases rather than packed as individual particles. Multicapillary columns (over 900 capillaries with $40\,\mu m$ i.d.) were introduced by Cooke [20] in 1996.

Film thickness on the inner wall of the column is another very important parameter and varies between $0.1\,\mu m$ and $5\,\mu m$, typically $0.25\,\mu m$. Increasing resolution by increasing either the stationary phase loading on a packed column or length of the capillary column will always result in a proportional rise in retention time. The choice of the chemical nature of the *stationary phase* in a chromatographic column determines at least in part the range of compounds that can be analysed. Changing polarity of the stationary phase enables the chromatographer to change retention times of components relative to each other (selectivity). For capillary columns, numerous stationary phases are known but only about half a dozen are utilised very frequently because of their ability to separate a variety of compounds. Such phases are polyethylene glycol, dimethylpolysiloxane and dimethylpolysiloxane copolymer. There are three categories of stationary phase coatings: simple wall coatings (WCOT), coatings dispersed on porous solid-support (SCOT) to increase adsorbent surface area, and a dispersion of porous solid particles on the inside wall (PLOT). There are also two major classifications of columns: nonpolar and polar. Nonpolar columns separate components almost completely by boiling point differences. Polar columns separate components at least to some degrees on the basis of differences in dipole–dipole interactions between each component and the stationary phase. The stationary phase is thus a material that separates components by selective attraction (selective adsorption). Scott [73] has described methods of stationary phase assessment.

Trying to determine which column is ideal for a specific analysis can be difficult with over 1000 different columns on the market [74]. A proper choice implies a definition of parameters such as column material, stationary phase (polarity), i.d., film thickness and column length. Guides to column selection are available [74,75]. The most important consideration is the stationary phase. When selecting an i.d., sample concentration and instrumentation must be considered. If the concentration of the sample exceeds the column's capacity, then loss of resolution, poor reproducibility and peak distortion will result. Film thickness has a direct effect on retention and the elution temperature for each sample compound. Longer columns provide more resolving probe, increase analysis times and cost.

The coiled columns, through which the carrier gas passes, are placed in a temperature-controlled GC oven. Typical column temperature changes range from 50 to $300\,°C$ and should be just above the boiling point of the sample so that elution of the analytes occurs in a reasonable time. In GC, temperature gradients may be programmed to improve chromatographic resolution. The time taken for a component to travel from the injector end of the column to the detector end is called the *retention time*, and for a separation of two components to occur must be different. Increasing the temperature of a GC column reduces the retention time of a component and extends the range of analysis to substances of low volatility such as oils and waxes. In view of this temperature dependence, GC separations are always performed at controlled temperature conditions. One distinguishes isotherm separations, gradient separations, or sequences of isotherm and gradient portions. The temperature program is commonly chosen to achieve an acceptable separation of the components in a reasonable time span. While separation is usually better at lower temperatures, isothermal chromatography is not practical for complex samples. Gas chromatographic separation in temperature gradient is affected by various factors: variation in gas flow, distribution constants, and peak broadening. Relative peak areas are used to measure the amount of each compound.

Retention times are the primary means of *peak identification*. Far from being absolutely stable they vary from run-to-run, day-to-day and column-to-column. Some variation is normal. Hinshaw [76] has discussed problems with retention times. Retention time variability falls into three broad classifications: random fluctuations, drifting and sudden changes. A relatively high insensitivity toward experimental parameters may be achieved when retention is measured under isothermal conditions as 'relative retention times', that is, the retention of a compound is compared with that of one or more standard compounds. Such data are reported as retention indices. The highest precision in retention indices is obtained with nonpolar columns. A standard retention index library for capillary GC is available [77]. Gradual stationary-phase loss from the column causes a trend toward shorter retention times (drifting). Large retention time fluctuations may derive from carrier-gas leaks, oven temperature problems, operator error, poor injection technique or poor oven temperature repeatability. Strictly speaking, positive peak identification with nonselective detectors, such as FID, requires duplicate analyses with two different columns. It frequently occurs that more than one substance elutes simultaneously on a single column, but it is less likely that they will be eluted at the same time on two columns with different stationary phase chemistries. With the use of three different types of phases, the probability of peak coincidence in all three columns would be relatively low, at least when the solutes belong to a defined group of compounds. Selective detectors, such as FPD and ECD, help eliminating peaks that fall outside of a particular class, but they do not provide primary identification. Mass spectrometry detectors may be used to provide necessary confirmation of peak identity.

GC can relatively easily be accelerated. This requires improvement of the standard GC column injection (several hundreds of ms; band broadening) and fast data acquisition systems. The possibility of fast or *high-speed gas chromatography* (HSGC) with chromatograms in the time frame of 1 min or even less and with peak widths (FWHHs) ranging from 20 to 200 ms has been demonstrated [78]. General operating requirements for HSGC are deduced from Equation (4.2) which gives the separation time as the retention time t_R of the last component of interest

$$t_R = (L/u)(k_n + 1) \qquad (4.2)$$

in which L is the column length, u is the average carrier gas velocity in the column, and k_n is the retention factor of the last component of interest. The basic principle of decreasing analysis times in GC is miniaturisation: essentially a decrease of the particle size in packed columns or the column diameter in capillary GC. In practice, fast GC can be achieved by:

- shorter analytical columns (from 100 to 10 m);
- reduced column i.d. (from 0.32 to < 0.1 mm);
- multicapillary GC;
- optimising carrier gas type and flow-rate (from 1 to 3 mL min^{-1} He);
- temperature programming;
- low pressure (LPGC; 0.53 mm columns);
- cryofocusing inlets (from several hundreds of ms to 5 ms; narrow chromatographic peaks).

Fast analyses trade off speed, sensitivity, dynamic range and chromatographic resolution. Standard GC equipment can be upgraded with cryofocusing injection in 5 ms. The most common approach is to use shorter capillary columns with reduced internal diameters and reduced film thickness. Very fast chromatography with short (12–16 ft) columns was actually described a long time ago by Golay [70]. Short columns, higher-than-usual carrier gas velocities, and relatively small retention

factors can reduce separation times by one to two orders of magnitude. The use of short columns facilitates analysis of more thermolabile compounds on the basis that the compounds spend less time in the GC system. Short packed capillary columns containing small support particles (10–15 μm in diameter) are equally suitable for fast GC but require high inlet pressures [79–81]; the same practical problem exists for multicapillary columns [79]. Van Lieshout *et al.* [79] have compared capillary columns packed with 15 μm spherical octadecylsilica particles, a multicapillary column (900 capillaries, i.d. 40 μm) and 50 μm narrow-bore and conventional-bore columns for fast GC. Use of narrow-bore GC for additive analysis is still largely unexplored and not without problems. Fast run times also call for more efficient sample introduction and autosamplers. A combined on-column and PTV injector allowing injection volumes of up to 1 mL can achieve this. Sample focusing inlets are used. Liners with a large internal diameter give excessive line broadening in fast GC [82]. In order to perform fast GC, several requirements were defined [83]:

- Instrumentation allowing high inlet pressure (4–10 bar) and high split ratios (1:1000).
- Fast line and oven heating rates.
- Capillary columns with reduced i.d.
- Fast electronics for detection and data collection.

Temperature programming rates and time constants of detectors and sample introduction systems must be adapted to the required speed of analysis. For fast GC it is essential to tailor stationary phase selectivity for target separations. Because resolution is sacrificed for speed, achieving high column efficiency is particularly important in HSGC. Capillary columns with i.d. ≤100 μm are considered very suitable for fast GC. However, as the working range of such narrow columns is restricted to only a few ng due to limited sample capacity high sensitivity is required. It should be considered that proper high-speed analysis comprises the total analysis time from the beginning of one analysis to resetting for the next run. The instrumental measurement time is only part of the total process. Fast GC techniques thus require that sample preparation time is short enough to justify the technique itself. In fact, a laborious clean-up procedure would greatly impair the whole process time unless multiple sample preparation is possible.

Today, there are three experimental approaches to temperature programming for faster GC separation: isothermal GC, fast temperature programming with conventional ovens and by means of resistive heating. In the latter case the column is placed in a

5 m long resistively heated metal tube allowing for a temperature-programming rate of $30\,°C\,s^{-1}$ [84,85]. Such modular upgrade packages make fast temperature programming available for existing GC systems based on capillary microbore and megabore technology. Various approaches for obtaining fast GC separations, namely a narrow-bore column used under conventional GC operating conditions and under typical fast GC conditions (with high oven temperature programming) and resistive heating technology, have been compared [86]. Use of narrow-bore columns appears as the most straightforward way to reduce analysis times in GC without compromising resolution. For samples that are not too complex, resistive heating technology is promising, as shown in Figure 4.1 for fast GC separation (1.5 min) of a standard test extract containing some 18 polymer additives (the observed peak broadening illustrates the absence of electronic pressure control EPC for the GC used), accelerating chromatographic analysis by a factor of 20. Detectors for HSGC are FID and μTCD. For fast GC-MS scanning must be rapidly enough to ensure that the peak profile includes sufficient data points for determination. The benefits of high-speed GC are reduced manpower costs, smaller laboratory space requirements, faster sample turnaround times, higher sample throughput and more reliable data. However, the reliability of HSGC needs further improvement, in particular as to column reproducibility. As pointed out by Ettre [87], truly high-speed chromatography can only be introduced into routine analytical laboratories after complete redesign of present-day GC systems rather than by the use of upgrade kits only. For novel concepts for fast GC, see van Deursen [88].

The usefulness of *ultrafast GC* with separation of simple samples in a few seconds is limited. Extremely fast analysis in the ms range is only possible if the required number of theoretical plates is relatively low. GC-ToFMS allows ultrafast separations (within seconds).

Gas chromatographic analysis starts with introduction of the sample on the column, with or without sample preparation steps. The choice of inlet system will be dictated primarily by the characteristics of the sample after any preparation steps outside the inlet. Clearly, sample preparation has a profound influence on the choice of injection technique. For example, analysts may skip the solvent evaporation step after extraction by eliminating solvent in the inlet with splitless transfer into the column. *Sample introduction techniques* are essentially of two types: conventional and programmed temperature sample introduction. Vogt *et al.* [89] first described the latter in 1979. Injection of samples, which

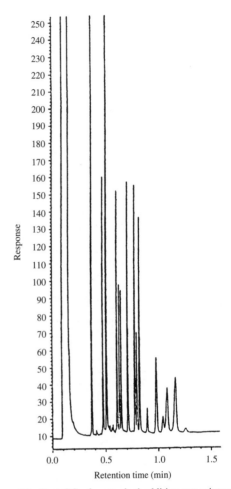

Figure 4.1 Fast GC of a standard additive test mixture (18 components)

is relatively straightforward with packed columns, made directly into the packing, becomes more critically important for capillary columns, which have less sample capacity and lower carrier gas flow-rates. Several methods of injection are available. There is no such thing as a universal injection system and there probably never will be. Introduction of liquid samples in GC is not trivial and has always been more difficult than in the case of HS-GC or TD-GC. Especially in CGC extraordinary demands are placed on the sample introduction device. An extremely small sample amount has to be introduced accurately and rapidly in a reproducible manner into the smallest possible gas volume. This gas plug must be transferred into the column without any losses by degradation, adsorption or discrimination. It is evident that this requires very sophisticated inlet devices.

Table 4.16 Common injection techniques

Technique	Description	Sample type	Band width	Typical injection size (μL)	Sample transfer (%)
Direct	Sample passes directly from syringe to hot inlet where it vaporises. Packed column inlets	Dilute samples Thermally stable	Fair, some focusing required	0.1–2	100
Hot split	Sample (usually liquid) passes from syringe into hot inlet; rapid vaporisation; initial oven temperature not critical; small fraction allowed to column (split ratios 1:20 to 1:400)	Concentrated samples	Narrow	0.1–2	0.01–10
Hot splitless	Liquid sample passes from syringe into hot inlet; initial oven temperature critical (ca. 10 °C lower than b.p. of solvent); bulk of sample enters column (splitless time of 10–40 s)	Dilute samples, containing heavy by-products	Very broad, focusing required	0.1–2	80–95
Cool on-column	Liquid sample is directly and totally passed from syringe into column or its extension. Cold injection followed by temperature program	Dilute, thermally labile samples; high-boiling components	Fair, some focusing required	0.1–1	100
Programmed-temperature split	Liquid sample from syringe into cold inlet; heat programming vaporises sample. Small fraction to column	Concentrated samples; thermally labile	Broad, some focusing required	0.1–2	0.01–10
Programmed-temperature splitless	Liquid sample from syringe into cold inlet; heat programming vaporises sample. Bulk of sample vapour enters column during splitless time venting (0.5–1 min)	Dilute samples; thermally labile	Very broad, focusing required	0.1–2	80–95
Solvent venting, without splitting	Liquid sample from syringe into cold inlet. Solvent is vented at low temperature, condensing nonvolatiles. Heat programming subsequently vaporises the residues, which enter column as in splitless injection	Dilute samples; thermally labile	Broad, some focusing required	1–1000	80–95

For routine analysis of liquid samples, four injection techniques are available: split, splitless, on-column and programmed temperature vaporising injection [90].

Injection systems of a capillary gas chromatography should fulfil two essential requirements: (i) the injected amount should not overload the column; (ii) the width of the injected sample plug should be small compared with band broadening due to the chromatographic separation. Good injection techniques are those which achieve optimum separation efficiency of the column, allow accurate and reproducible injection of representative amounts of sample, cause no change in sample composition (no discrimination) and have wide applicability (from undiluted samples to trace analysis). A simple injection classification scheme describes the injection process as a two-step process, with emphasis on temperature of the inlet system and on transfer of (part of the bulk of) the sample into the column [91]. The most popular conventional sample introduction techniques are split/splitless and cool on-column injection (cf. Tables 4.16 and 4.17).

Table 4.17 Characteristics of various syringe injection methods

Injection mode	Column type	Detection limit	Advantages	Disadvantages
Direct	Packed, SCOT, large bore WCOT	200 ng per component	Direct quantitation	Flash vaporisation
			Ease of operation	Decreased resolution (column requirements)
			High sample capacity	
Hot split	Column independent	50 ppm (FID)	Ease of operation	Flash vaporisation
			Independent of sample	Thermal degradation
			Rugged	Discrimination of higher boiling compounds possible
			Qualitative analysis	
				Not suitable for trace analysis
				Poor, indirect quantification
Hot splitless	WCOT	0.5 ppm (FID) without preconcentration	Lower injection temperature than split	Flash vaporisation
				Optimisation required (splitless time, oven temperature, solvent)
			Trace analysis	
			Handles 'dirty' samples	
			Automation	Limited number of solvents ('solvent effect')
				Thermal degradation possible
				Discrimination possible
				Poor direct quantification
				Unsuitable for very polar substances
Cool on-column	>250 μm column (i.d.)	1 ppm (FID)	Reduced thermal degradation and discrimination	Control of operational conditions (initial oven temperature)
			Wide range of analyte concentrations	Optimisation required
				Not applicable for polar solvents
			High sample capacity (LVI)	Column contamination by 'dirty matrices'
			Autosamplers	
			Direct quantification	Poor long term stability
			Excellent precision	
Cold split	Column independent	100-fold increase in sensitivity with respect to conventional GC injection	Reproducible and accurate sampling	Low volume of CIS liners (2–3 μL)
			Controlled evaporation	
			No discrimination on basis of b.p.	
			Rapid transfer of sample to column	
Cold splitless	WCOT; thick film	Idem	Reproducible and accurate sampling	Small sample size introduction (<2 μL)
			Controlled evaporation	Thermal degradation possible (long residence times of components in liner)
			No discrimination on basis of b.p.	
Solvent venting[a]	WCOT	1 ppm (FID)	Large volume injection	Limitations in solvent detector compatibility
			Low detection limits	
			Minimisation of solvent entering column	
			Allows polar solvents	
			Automation	

[a] Intermediate between cold split and splitless.

Splitting reduces the size of a sample to an amount compatible with the sample capacity of the capillary column. Split injection can be used to prevent column overload, but may give inaccurate results due to variations in sample volume, injector and detector temperatures and the way the syringe is handled. Owing to the difficulty in precisely injecting small quantities of solvent solutions, it is recommended that an internal standard method be employed for quantitative work. To use *splitless injection*, it is necessary to reduce the initial temperature of the oven to a value below the boiling point of the solvent, modify the GC temperature program, increase the helium flow-rate through the column, and use a column that is compatible with the solvent. In splitless injection reconcentration of the sample is necessary. Two reconcentration mechanisms are commonly considered: (i) reconcentration of high-boiling components by a cold trapping mechanism; and (ii) reconcentration of low-boiling components (b.p. about equal to solvent) by the 'solvent effect'. *Cool on-column* (COC) injection is theoretically the most ideal method of introducing a sample onto a column and is the method of choice for all samples containing high-boiling components that would not be quantitatively transferred to the column if an injection technique employing flash vaporisation were used. There are many opinions on the selection of the optimum injection temperature for on-column injection. On-column injection allows accurately known volumes to be very reproducibly introduced directly onto the column. In many respects on-column injection is superior to split/splitless injection, but the technique is less rugged as column contamination is common; use of a retention gap is advisable to increase column life. Which specific inlets and injection techniques are to be used is largely determined by the sample to be analysed. The need for different inlets can be circumvented by use of a programmed temperature vaporising inlet (PTV).

The advent of high-resolution capillary gas chromatography (HR-CGC) with on-column injection has resulted in improved GC analysis of polymer additives [92–94]. The solution of the additive mixture is injected directly into the cold end of the capillary column by means of a cold injector. Thus, sample discrimination, the instantaneous evaporation of the sample solvent, is avoided. The nonvaporising, on-column injection combined with very high resolution of the capillary columns allows accurate separation, identification and quantification of additives of complex mixtures. With the *solvent venting technique*, the sample is introduced into the column without splitting and sample concentrations

lower than 1 ppm can be analysed with reproducibility better than 5 % [95].

By introducing the sample at a low initial liner temperature, many of the disadvantages of the classical hot injection techniques can be circumvented. Because of these advantages, *programmed-temperature vaporisation* (PTV) multiple injection is generally considered to be the most universal inlet technology available [96]. Synonyms such as 'cooled injection system' (CIS) or 'cold/programmable injector' better reflect the main principle of the injection system. With PTV technology (ATAS OPTIC 2 or Gerstel injectors) volumes of up to $100 \mu L$ (or even $1000 \mu L$) may be introduced into a large-capacity, sorbent-packed liner. By temperature programming of the liner (up to final temperature of 400 to 450 °C), solvent is evaporated and analytes are transferred to the GC column. The capillary column is protected against contamination by trapping any involatile matrix compounds on the liner. Using this technique detection limits can be routinely improved by a factor of 100. A variety of temperature ramps and pressure programming options provide maximum injection flexibility to PTV/CIS. Such modern versatile, automatic and manual injection systems allow conventional on-column, split/splitless injection modes, large-volume and off-line sampling, combined multistep thermal desorption/pyrolysis operations of solid samples in connection with HTGC-FID [97] and expert injection modes (facilitating SPE-GC, SFE-GC, LC-GC, on-line cryo-trapping) with multiramp programmable temperature and pressure profiles. The cooling feature of PTV/CIS makes it a perfect interface for all techniques that require cold trapping, notably large-volume injection, static headspace sampling and thermal desorption. In the cooled injection modes thermal degradation is less likely to occur. PTV injectors extend discrimination-free injection to C_{100} compounds and beyond (100 % recovery of Irganox 1010). High-temperature operation (up to 600 °C) has other advantages, such as reducing carry-over between runs, and enabling derivatisations, reaction studies and sample pyrolysis. Nevertheless, some drawbacks still remain. Not all solvent/detector combinations are compatible, such as chlorine-containing solvents with NPD and halogen-containing solvents with ECD. In solid-sampling PTV injection quantification is limited (5–18 %) and needs further improvement [98].

The traditional sample volumes in CGC ($0.1-3 \mu L$) limit sample preparation possibilities. Specialised *large-volume injection (LVI) techniques* are designed to load more sample in the GC system (typically $150 \mu L$) by placing a length of uncoated fused-silica tubing in front of the analytical column. This procedure also provides enormous improvements in sensitivity by elimination of the highly volatile components from the sample (usually the solvent if the sample is injected as a dilute solution) maintaining a reduced sample load. LVI methods are employed effectively for CGC determination of trace analytes. Detection limits of trace compounds are primarily determined by the solubility of the sample in the solvent and the injection volume. The basic requirement for LVI is that injection and evaporation rates are equal. The ability to calculate the evaporation rate (venting) is an important aid in setting up a LVI method. Retention gaps with different polarities have stimulated LVI. Typical large-volume methods are on-column LVI, splitless PTV and at-column concentrating LVI. In *large-volume on-column injection* (LVI-COC), a long precolumn is employed to separate sample solvent and target compounds. The large-volume sample is injected onto the precolumn and the solvent is evaporated. However, this process is often accompanied by some loss of the target compounds. The precolumn is rapidly contaminated. This injection mode allows in-vial mini-volume extraction followed by direct injection, both for apolar and polar components. LVI-COC is a complex approach that requires careful optimisation. In using conventional PTV injection for rapid large-volume liquid sampling (RLVI) a liner is loaded with packing material. The large sample volume (150 μL) is injected in split mode at an injector temperature below the sample solvent boiling point. The sample solvent flows through the split vent and the analytes are retained on the packing material in the liner. After complete evaporation of the sample solvent the injector is switched to splitless mode and the temperature of the liner is increased; the target compounds in the liner are then transferred into the GC column. The main features of PTV for sample handling are shown in Table 4.18.

At-column concentration is a LVI (50 μL) technique which takes advantage of a temperature gradient in the liner whereby target compounds are concentrated at the inlet of the analytical GC capillary column [99]. Most of the sample solvent is evaporated in the empty liner and the target compounds are retained at a small range of the analytical column and transferred under cool conditions (below b.p. of the solvent). As no long precolumn and packing material in the liner are employed the previously mentioned drawbacks are avoided. At-column LVI combines the advantages of LVI-COC and LVI-PTV. This injection mode allows high sensitivity without degradation of the analyte. At-column LVI is suitable for analysis of 'difficult' samples, such as thermally labile or strongly polar additives at low concentration or reactive components without

Table 4.18 Characteristics of PTV for thermal sample pretreatment

Advantages

- High sample intakes (reduced homogeneity problems)
- Position of column relative to thermal zone (very short distance)
- No early loss of volatiles (no transfer lines)
- Discrimination-free quantitative transfer of products (up to C_{120})
- Flexibility of temperature selection (−50 to 630 °C)
- Possibility of multitemperature treatment
- Wide range of split flows
- Multipurpose injector
- Inexpensive
- Allowance for pyrolysis and chemolysis

Disadvantages

- Catalytic decomposition (due to packing material)
- Strong retention on packing in liner

discrimination or reaction product formation. At-column LVI is considered to be a more practical alternative to LVI-COC for labile and/or high-MW compounds.

Major advantages of LVI methods are higher sensitivity (compare the 100–1000 μL volume in LVI to the maximum injection volume of about 1 μL in conventional splitless or on-column injection), elimination of sample preparation steps (such as solvent evaporation) and use in hyphenated techniques (e.g. SPE-GC, LC-GC, GC-MS), which gives opportunities for greater automation, faster sample throughput, better data quality, improved quantitation, lower cost per analysis and fewer samples re-analysed. At-column is a very good reference technique for rapid LVI. Characteristics of LVI methods are summarised in Tables 4.19 and 4.20. Hankemeier [100] has discussed automated sample preparation and LVI for GC with spectrometric detection.

Using a programmable injector selective introduction of components onto a column can be achieved in split/splitless mode. *Selective exclusion/transfer* is a way of controlling which components enter the column. Transfer of selected groups of peaks to the column reduces analysis time of complex mixtures. Selective removal of major sample components (including solvent elimination) increases detection limits and resolution of trace analyte peaks. Removal of heavy, involatile compounds protects and prolongs the life of column and detector (especially MSD) and exclusion/transfer of labile compounds reduces degradation. Parameters to achieve positive discrimination are isothermal temperature and time. Transfer depends on volatility (however, for practical considerations boiling points are used). In order to check the viability of this approach the lowest injector temperature for total transfer of volatiles and

Table 4.19   Characteristics of LVI methods

Method	Advantages	Disadvantages
Rapid LVI	Good for volatiles Robust Easy to use Handles dirty samples	Limited by packing material Needs optimisation Thermal degradation possible Difficult to set-up and use
On-column LVI	Good for labile compounds	Loss of target compounds Not very robust High maintenance
At-column concentration LVI	Cool sample transfer to column No discrimination Little optimisation needed Simple to use, robust Handles 'difficult' samples Good repeatability High sensitivity Rugged No evidence of degradation Use in hyphenation Wide applicability	Not suitable for dirty samples

the highest temperature for total exclusion of involatiles need to be established. Advantages of selective exclusion are summarised in Table 4.21. Applications of selective exclusion are the exclusion of major components in highly concentrated samples and of late running components, and transfer of thermally labile compounds at the lowest injector temperature. Involatiles can be kept in the liner without decomposition.

Difficult matrix introduction (DMI) is another recently introduced way of automating trace analysis in complex and dirty matrices [101]. The technique may be used for both liquid and dirty solid samples. In DMI a sample extract or sample matrix (solid) is introduced directly into a microvial in the injector. Volatiles are desorbed directly

Table 4.20   Summary of LVI characteristics

	RLVI	At-column
Volatility	C_8-C_{40}	C_{12} upwards
Optimisation	Straightforward	Little needed
Robustness	Good	OK
General applicability	Similar to splitless injection	Similar to on-column injection

Table 4.21   Advantages of selective exclusion

- Cost effective
- Protects column and detector
- Decreases run times
- Simple to use (only two main parameters to optimise)
- Can enhance sensitivity for trace impurities
- Applicable to compounds with known or unknown volatilities

from the sample onto the head of the GC capillary column, whereas involatiles from the matrix are kept in the microvial which is disposed of after use. DMI allows for selective exclusion: volatiles (e.g. solvents) may be removed by venting under controlled conditions; compounds of interest may be transferred onto the column using the lowest possible final temperature (limits charring of matrix); involatiles are kept in the DMI microvial in the liner, in a controllable way. On the basis of the volatility (or b.p.) of the lightest and heaviest target analytes the initial and final temperature of the injector may be regulated in such a way as to maximise recoveries of target analytes and minimise transfer of matrix. The method is applicable to polymer additives. DMI permits analysis of trace contaminants in a wide range of different matrices with little if any sample preparation. The process can be modelled and automated and is characterised by low consumable costs.

Hinshaw [91] has addressed the problem of selection of the appropriate injection technique. Hundreds of liners are available per GC brand and application range. The most important injection techniques are listed in Table 4.16.

As GC is not only used as a separation medium but also as an analytical technique *detection* has an important function. Even if the column tolerates high-solute levels, detector requirements may determine the best injection technique or they may dictate adding a sample dilution step before injection to bring injected quantities within the optimal operating range. GC instruments accommodate an extremely wide range of solute concentrations. Minimum and maximum solute amounts in a sample must lie within the available dynamic ranges for inlet, column and detector. The correct choice of sample preparation and injection techniques, matched with an appropriate column and detector, helps ensure the reliability and longevity of the instrument components and quality of the analytical results.

The detector senses the presence of a component different from the carrier gas and generates an electrical signal preferably proportional with the (often very small) amount of analyte. Many GC detectors are utilised (Scheme 4.3), which are generally more sensitive than those used in LC. Thermal conductivity (katharometer) detectors were favoured in the early days of GC. Four detectors (FID, ECD, NPD, FPD) now account for over 95 % of all GC applications; of these, FID is a very versatile, sensitive and linear detector, which is used most. Because the FID response is proportional to weight of components per unit time, it is ideal for use with open tubular columns and small diameter packed columns. Even with the extraordinary ability of modern capillary columns to separate the components of a sample with general detectors (high-resolving power; ca. 10^5 theoretical plates), analysis of a single compound or group of compounds in complex samples can still be elusive and therefore selective detectors are needed (e.g. NPD, ECD, RCD, FPD, SCD, CLND, FTD, O-FID, ELCD). Some of these detectors are element-specific (cf. Section 4.1). Lower limits of detection of some representative GC detectors are given in Table 4.22. The electron capture detector is probably one of the most sensitive GC detectors available (minimum detectable concentration ca. 10^{-13} g mL^{-1}). Also various identification techniques have been hyphenated to GC and act as detectors (MS, FTIR, AED, ICP-MS, MIP-MS), as discussed in Chapter 7. Mass spectrometric detectors have sensitivity at least as good as FIDs, and facilitate characterisation of the components associated with the chromatographic peaks from their mass spectral fragmentation patterns. In addition, mass spectrometers can detect and measure permanent gases and other small molecules to which FID detectors are totally insensitive (e.g. HCl, CO_2, H_2O, NH_3). Identification of components is usually achieved by comparison of retention time and relative responses of standards to ECD/FID, and by directly using GC-MS (cf. Section 7.3.1.2). Gas chromatography is empowered by multidetection methods, e.g. PID/ELCD, PID/FID, PID/NPD, FTIR/FID, ECD/FID and SCD/CLND [102,103]. Multiple detection GC has been reviewed [103]. For more detailed descriptions on GC detectors, see Section 4.1 and the numerous literature references [46,73,104,105].

Table 4.22 Lower limits of detection (LLD) of some GC detectors

Detector	LLD	Detector	LLD
MIP-AED	1–100 pg s^{-1}	ELCD	ca. 10 pg
FPD	1 ng s^{-1}	ECD	fg–ng
SCD	<1 pg s^{-1}	TCD	400 pg mL^{-1}
NPD	ca. 0.1 pg	FID	2 pg s^{-1}
O-FID	0.1–1 ng	FTIR	0.2–40 ng
		MS	0.25–100 pg

Organic compounds for GC analysis may be classified in terms of polarity: very nonpolar (VNP), nonpolar (NP), polar (P) and very polar (VP). The principal requirements for analysis of such analytes by GC are volatility and stability to heat. Specifically, the compound must be stable enough to survive the conditions necessary to convert it to the gas phase. Polar organic compounds that contain active hydrogens (e.g. hydroxyl, amino and thiol groups) are typically of low volatility due to their tendency to self-associate or through hydrogen bond formation. *Derivatisation* (chemolysis) is common practice in GC and is conveniently carried out in a PTV injector attached to a HTGC-FID unit. Its purpose is increasing the volatility of the analytes by decreasing their polarity and capability of hydrogen bonding or to increase the analyte's thermal stability. Common derivatisations are permethylation and trimethylsilylation. However, depending on the nature of the compounds to be analysed (alcohols, thiols, phenols, amines, aldehydes, ketones, acids), numerous other derivatisation types may be utilised [106–108]. Robotic autosamplers are available for automatic derivatisation, extraction, dilution and injection (RLVI, TD, SPME, HS).

Gas chromatography is commonly used to analyse mixtures for *quantification*. A wide variety of special detectors with adequate linear response ranges are available for quantification of various classes of compounds (cf. Table 4.14). Quantification by direct injection may be used to determine additives, residual monomers and solvents in product formulations, coated films, and solid materials [109]. On the other hand, reliable quantification by means of solid-injection PTV-GC, HS-GC and PyGC techniques is not always trivial.

Preparative chromatography is a powerful tool to isolate working quantities of compounds from complex samples. *Preparative GC*, which is normally carried out on packed columns, is useful for components with a b.p. < 450 °C and levels > 100 ppm (impossible for < 10 ppm; questionable for 10–100 ppm). Total amount needed is 2 µL.

Key drivers for innovation in process GC are now leading to the development of micro gas chromatography

(μGC). The expected configuration is μGC-μTCD with various columns in series (if a component is not separated by the first column, then by the next). As FID measures weight of components, this detector is unsuitable for μ-streams; instead, the sensitivity of TCD is not size dependent (μTCD: filament length 1000 μm, thickness 0.2 μm, gold). μGC is expected to have the biggest impact for situations with homogeneous flows (process conditions) but its acceptance will be narrowed when heterogeneity comes into play.

The complete GC system must be optimised in relation to the complexity of the sample in order to obtain accurate results. Schoenmakers [110] has addressed the issue of *method development* for chromatography. Key parameters to be optimised are: (i) sample preparation (optimisation of volatility, thermal stability and polarity of target analytes; derivatisation); (ii) proper choice of column, mobile and stationary phase (column capacity, polarity, retention times, resolution); (iii) selection of most appropriate injection mode (sensitivity, thermal degradation, automation); and (iv) selection of detector (specificity, linearity range). For selection of carrier gases and conditions, cf. Hinshaw [111]. High-purity GC standards are available as reference substances [112]. The preparation of calibration standards for volatile organic compounds (VOCs) for GC analysis has been described [113]. Hinshaw [75] has recently addressed strategies for GC optimisation, In particular, the interrelationship between solute amounts and inlet, column and detector considerations were illustrated in a GC dynamic range nomogram. By following minimum and maximum solute amounts through the inlet system, column and detector, GC users can quickly evaluate the suitability of their instrument system for a particular analysis. Inlet parameter operating limits were suggested. Analysing the same sample on two columns of different polarity can increase both the qualitative and quantitative reliability. Instead of repeating the analysis a simple solution to improving analytical reliability without reducing sample throughput is to use a simultaneous dual-column technique. Method development for fatty acid analysis was discussed in depth by Craske and Bannon [114]. *Validated GC determinations* of additives in extracts from various matrices can be found in various compendia, e.g. ASTM [115] and FAAM [116].

Table 4.23 shows the *main characteristics* of advanced GC. The use of FID coupled to a high-efficiency capillary column is sufficient to perform routine additive analysis in the 20 ppm concentration range in solution. With adjustment of injection volume in an on-column injector the detection levels surpass

Table 4.23 Characteristics of advanced GC as an analytical method

Advantages

- Prime technique for thermally stable and semivolatile analytes (MW < 1200)
- Relatively simple in use; reliable
- Very small sample amounts (mg to tens of mg; μL); nondestructive
- No solvent eluent
- Inlet versatility
- Extremely high resolving power (capillary columns)
- Compatibility with a wide variety of sensitive and specific detectors; multidetection
- Sensitive (LOD <5 ppm)
- Adaptability to quantitative analysis (typical RSDs of 1–5 %)
- Fairly short analysis times (min to tens of min); ultrafast GC option
- Efficient, automation
- Cost effective operation and maintenance
- Standard commercial chromatographic equipment
- Various injection modes
- High-temperature option
- Hyphenation (MS, FTIR, AED, ICP-MS)
- Wide variety of ancillary techniques (HS-GC, TD-GC, PyGC, etc.)
- Access to multidisciplinary techniques
- Wide problem solving ability
- Miniaturisation (3 × 2 × 1 in.)

Disadvantages

- Stability of stationary phase at high temperature
- Not suitable for thermally labile samples
- Limited to volatile components (adequate partial pressure)
- Polarity restrictions
- Need for high-purity carrier gas (cost)
- Optimisation strategy desirable
- Destructive
- Sample preparation (can be difficult)
- Poor peak identity confirmation (requires spectroscopy)
- Difficult trapping of analytes from vapour state
- Poor prep-scale separations
- High response time for large molecules
- On-column fouling for high-MW additives

those of HPLC. Volatility is a serious limitation in the general applicability of GC as there are far more involatile materials than there are volatile. Two major factors limit the broad utilisation of GC for stabiliser analysis. There is a trend away from the application of volatile stabilisers. Moreover, there are also considerably more peaks chromatographed in GC than in LC, emerging from the polymeric material soluble in the extract. In fact, unless the injection samples are very clean and free from oligomeric material or soluble polymeric material, GC can prove to be quite expensive and frustrating to operate on a daily basis.

Different column coatings of different polarities can be used to achieve the separation of a wide variety of stabilisers. Irganox 1010 (MW 1178) can be gas chromatographed, but cannot be routinely quantified because of the on-column fouling that takes place.

In comparison to HPLC, GC is a more simplistic approach to separation, which requires no solvent eluents or gradient pumps. Gas chromatography is more robust than HPLC. In addition, GC columns have more theoretical plates (10^5 to 10^6) than HPLC columns, and therefore, may provide better analyte resolution in very complex mixtures. The sensitivity of GC detectors is generally at least two or three orders of magnitude greater than their HPLC counterparts. It follows that much less sample preparation and concentration is needed in GC analyses or, alternatively, the analysis can have a much wider dynamic range. By using derivatising techniques many highly polar and relatively high molecular weight compounds can still be separated by a GC system. Many functional groups (e.g. OH, NH, NH_2, SH, $CONH_2$, NOH) form trimethylsilyl and acylated derivatives. Poole and Poole [35] have described derivatisation techniques for GC.

Current trends in GC relate to miniaturisation, fast-GC, improved selectivity (mainly for short columns), stability of column stationary phases (reduction of bleeding) and increasing use of MS detection [117]. Finally, GC can be readily hyphenated with spectroscopic techniques without using involved interfaces and thus can easily provide unambiguous solute identification.

General texts on GC are numerous [118,119]; narrow-bore GC was addressed by van Es [120]. Sample introduction techniques and GC inlet systems have been reviewed [25,90]; and split/splitless [121] and on-column injection [122] were considered specifically. Stationary phases [123], multiple detection [103], derivatisation [124,125], and quantitative analysis in GC [109] have been described. High-speed GC has recently been reviewed [126]. For a compendium of GC terms and techniques, see Hinshaw [127].

Various ancillary GC techniques are: headspace GC (Section 4.2.2), thermal desorption GC, pyrolysis GC, hyphenated methods (Chapter 7), multidimensional techniques (Section 7.4.1) and process GC.

Applications Conventional GC is a workhorse in the qualitative and quantitative analysis of polymer additives in complex mixtures and has found numerous applications. Both GC and auxiliary techniques are particularly useful for characterisation of (semi)volatile constituents and additives ranging from gases to hydrocarbon waxes ($\leq C_{60}$), alcohols, fatty acids and their esters, plasticisers, antioxidants and stabilisers (lower MW amines and hindered phenols), surfactants, and vulcanisation accelerators contained in a wide variety of matrices (polyolefins, rubbers, polyesters, polycondensates, etc.). Analytes must have sufficient volatility in the operating range from ambient to about 325 °C. Detection is usually by means of FID; identification is based on retention times. The limit of detection for FID analysis of polymer solutions or extracts is often only 10–20 ppm. Since determinations of sub-ppm levels are frequently called for, solution or extraction methods usually may not be sensitive enough to obtain the desired analytical results. Thermally labile compounds such as some azo dyes cannot be analysed by GC. Trialkyl leads are also unstable at the temperatures used in GC; moreover, these compounds interact with the GC column [128]. Sensitivity ranges from ppb to % (sample dependent). Because of the wide applicability only a select account is given here to illustrate the potential of the technique rather than attempting comprehensive coverage.

Major types of *volatile constituents* in polymers include unreacted monomers, nonpolymerisable components of the original charge stock, residual polymerisation solvents, and water. Frequently, complex nonpolymerisable mixtures are present. The concentration of these substances may need to be determined for various reasons, such as the effects on materials properties and the risk of tainting in foodstuff- and beverage-packaging grades. For this purpose various GC methods are in regular use:

(i) Dissolution and/or dissolution/precipitation followed by GC.
(ii) Thermal desorption-gas chromatography (TD-GC).
(iii) Headspace analysis (HS-GC).

Extraction or dissolution almost invariably will cause low-MW material in a polymer to be present to some extent in the solution to be chromatographed. Solvent peaks interfere especially in trace analysis; solvent impurities also may interfere. For identification or determination of residual solvents in polymers it is mandatory to use solventless methods of analysis so as not to confuse solvents in which the sample is dissolved for analysis with residual solvents in the sample. Gas chromatographic methods for the analysis of some low-boiling substances in the manufacture of polyester polymers have been reviewed [129]. The contents of residual solvents (CH_2Cl_2, C_6H_5Cl) and monomers (bisphenol A, dichlorodiphenyl sulfone) in commercial polycarbonates and polysulfones were determined. Also residual monomers in PVAc latices were analysed by GC methods [130]. GC was also used to determine the migration of dimethylterephthalate (DMT) and dimethylisophthalate (DMIP) from plastics to food simulants [131].

Crompton [132] has given an extensive listing (some 50 references) of methods for the determination of unreacted monomers, nonpolymerisable feedstock components and volatiles in polystyrene, poly α-methylstyrene, styrene-butadiene, poly(meth)acrylates, styrene-ethyl acrylate, vinyl acrylate copolymer, rubber adhesives, polyester resin moldings, ABS, SAN, polycarbonates, PE, PP, etc. The same author [133] has described the determination of water and volatiles in LDPE, polyacrylics and polyvinyls. Styrene and a wide range of aromatic volatiles (down to 10 ppm) in PS were determined with 5 % accuracy by GC-FID after dissolving the sample in propylene oxide containing a known concentration of *n*-undecane as an internal standard [133]. When the PS solution was injected into the liner, the glass fibre retained the polymer. DMI techniques would be equally appropriate here. Values obtained by headspace analysis differed by up to 20 % from those obtained by the solution procedure, which are known to be correct. Lou *et al.* [134] have examined ASE® extracts of monomers and oligomers of PA6 and PBT by means of GC-FID. Janssen [25] has described analysis of low volatility samples using high-temperature PTV injection.

Various techniques have been used for the determination of *oligomers*, including GC [135], HPLC [136–138], TLC for polystyrene and poly α-methylstyrene [139] and SEC for polyesters [140,141]. GC and PyGC-MS can also profitably be used for the analysis of the compositions of volatile products formed using different flame retardants (FRs). Takeda [142] reported that volumes and compositions of the volatile products and morphology of the char were affected by FRs, polymers (PC, PPE, PBT) and their reactions from 300

to 600 °C. Both GC and MS techniques are profitably used to determine decomposition products in liquid and gas phase after polymerisation is finished. Such decomposition studies are important to determine the efficiency of the initiators used.

Gas chromatography is widely applied in the *paint and varnish* industry for analysis of solvents, oils, resins, plasticisers and polymers (by pyrolysis techniques). GC-TCD and GC-FID on capillary and packed columns are used for direct determination of solvents in paints [143,144]. DIN ISO 11890-2 describes an officially approved GC test procedure for solvents in paints and varnishes (VOCs 0.1–15 %).

GC (and GC-MS) are the most useful methods for *plasticiser analysis* (cf. Figure 4.2), although polyesters and chloroparaffins are not amenable to separation by GC. ASTM D 3465 describes a test method for purity of monomeric plasticisers by GC. Dimethylphthalate and its by-products have been determined by GC [145]. Due to their volatility, esters used as plasticisers in polymers such as flexible PVC can be determined by analysis of a solvent extract. Gas chromatography complemented by IR, HPLC and physical observations has been used to identify monomeric plasticisers in flexible PVC compounds [146]. For early successful GC analysis of plasticisers in polymers packed columns were used [147–149]. Diisoheptylphthalate, diisooctyladipate and diisooctylmaleate are plasticisers which exhibit peaks upon GC [150]. Wandel *et al.* [148] described GC analysis of plasticisers extracted from rubbers and elastomers. The determination of primary plasticisers in PVC calendered formulations has been reported [151]. Gas chromatography has also been used for quality control of the phthalate plasticiser content in PVC for wire and cable manufacturing [152]. High accuracy (±2 %) was achieved using THF solutions. The method is suitable for evaluating the insulation and shells of

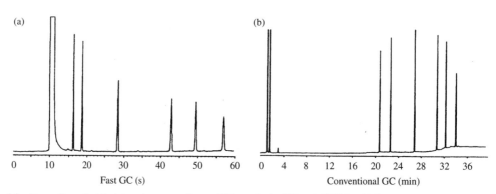

Figure 4.2   Separation of phthalate esters according to EPA methods 606 and 8060 by (a) fast temperature programming and (b) conventional CGC. Reproduced by permission of Thermedics Detection Inc.

wires and cables during service and for identification of plasticised PVC cable compositions. Pavlov [153] has reported analysis of plasticisers and other additives in PVC compounds by GC and other methods. Phthalates in PVC have been analysed by GC-FID [154] but for samples with very low levels GC-ECD is preferred [155]. GC is equally useful in the determination of the transfer of plasticisers from polymers to food products [156] as well as for the analysis of phthalate esters (up to 40 %) in plastic toys after Soxhlet extraction in dichloromethane [157]. Detection limits of DMP, DEP, DBP, DEHP, DOP, DNP, BBP and BOP were ca. 5 ppm, for DINP and DIDP ca. 200 ppm. EPA method 606 has compared conventional and high-speed GC-FID analysis for several phthalates (DMP, DEP, DBP, BBP, DEHP, DOP); flash GC separation was achieved in less than 60 s as compared to 34 min for conventional GC.

GC is extensively used to determine phenolic and amine *antioxidants*, UV light absorbers, stabilisers and organic peroxide residues, in particular in polyolefins, polystyrene and rubbers (cf. Table 61 of Crompton [158]). Ostromow [159] has described the quantitative determination of stabilisers and AOs in acetone or methanol extracts of rubbers and elastomers by means of GC. The method is restricted to analytes which volatilise between 160 °C and 300 °C without decomposition. A selection of 47 reports on GC analysis of AOs in elastomers (period 1959–1982) has been published [160]. Much of this work does not represent current practice particularly with regard to the stationary phases. The analysis of AOs and vulcanisation accelerators in rubber materials by GC, HPLC and TLC was reviewed [161]. Ali *et al.* [162] have reported development of a megabore capillary GC-FID method for the determination of AOs and UVAs in LDPE and PP. With the higher capacity of megabore columns, higher sample quantities can be injected, permitting detection of lower additive concentrations. High recovery of additives from the polymer was found (>99.0 %). The precision of the method is in the 3–5 % RSD range. Recently, GC-FID analysis of the AOs 2,5-di-*t*-butylhydroquinone and 2,5-di-*t*-amylhydroquinone in commercial rubber gloves was described [163]. Diorganosulfide secondary AOs in polyolefins were determined in extracts, also by GC-FID [133].

The content of 2,4-di-*t*-butylphenol in Irganox P-EPQ (MW 1035) may be determined by GC-FID using eicosane as an internal standard. Also for Tinuvin P (MW 225), Tinuvin 765 (MW 509), Tinuvin 770 (MW 481) and Irganox PS 800 (MW 514) GC-FID is the preferred technique.

By-products of Tinuvin 765 [bis(1,2,2,6,6-pentamethyl-4-piperidinyl)sebacate] are hydroxy-pentamethyl-piperidine (HPMP), dimethylsebacate (DMS), and C_{10}–C_{12} mono- and diesters. Also Tinuvin 770 [bis(2,2,6,6-tetramethyl-4-piperidyl) sebacate] contains C_{11}–C_{12} homologues and monoesters. Irganox PS 800 (dilauryl thiodipropionate) contains dodecanol as a by-product. The content of Irgafos 168 in Chimassorb B948 (Chimassorb 944/Irgafos 168 blend) can be determined by GC using octadecane as an internal standard; Chimassorb 944 does not elute. Unreacted Tinuvin 770 in PP is determined after dissolution of the resin in chloroform or tetrachloroethane. Gas chromatography is limited by the high MW and polar nature of many AOs and light stabilisers, which are designed to be reactive and decompose when exposed to heat [160]. Direct conventional GC analysis of the phenolic AO Irganox 1010 [pentaerythritol 3-(3,5-di-*t*-butyl-4-hydroxyphenyl)propionate], used in food packaging films, is difficult because of the high MW (1176 Da). However, after conversion to methyl 3-(3,5-di-*t*-butyl-4-hydroxyphenyl)propionate conventional GC analysis is possible [164]. Analysis of HALS based on 2,2,6,6-tetramethylpiperidine in polymers has been addressed by Rotschová and Pospíšil [165], using both direct determination and destructive methods, spectroscopic and chromatographic techniques. Most AOs degrade while in the polymer matrix (or in food). Consequently, analytical methods should preferably measure both the residual AOs and their degradation products. Scheirs *et al.* [166] have reported that GC analysis of PE/(Irganox 1076, Irgafos 168; 1:1 B-blend) leads apparently to a higher amount of Irganox 1076 than originally added and to a correspondingly lower Irgafos 168 amount and suggested that during compounding the phosphite decomposes to form a phosphate product that co-elutes with Irganox 1076. A CGC method was validated for the AOs Topanol A and Topanol O at concentrations of approximately 250 ppm in 11-bromoundecyl methacrylate and 50 ppm in methylmethacrylate [167]. Figge *et al.* [168] have reported GC analysis of an organotin stabiliser for PVC. Also nonionic surfactants may be determined by GC after extraction [169].

Chromatographic methods were developed for the systematic determination of five classes of additives in PE for food packaging [170]. In Soxhlet extractions phenolic AOs and acid amides were determined by HPLC and CGC, respectively. Thiodipropionic acid esters were determined by HRGC as higher alcohols obtained after saponification of the extracts with KOH. Glycerol fatty acid esters and stearates were determined by HRGC as their fatty acid methyl esters after methylation of the extracts. Retention times of Antage W-300 and W-400 were close. No additives could be detected in commercial PE packages of milk. Tetramethylsuccinonitrile (TMSN), a decomposition product of 2,2'-azobisisobutyronitrile (AIBN), was determined by GC-NPD in a large variety of food contact plastics [171]. In a rapid and precise method for the qualitative and quantitative analysis of additives in *polyolefins* (100–5000 ppm) by means of CGC-FID adsorbent pretreatment was mentioned explicitly [172]. The adsorbent (Florisil) was suitable for the separation of additives from wax in reflux extracts. Recoveries of the additives from the adsorbent were 92–103 %. The method was extended to other polymers containing additives [173]. A review has covered analysis of AOs, weatherproof agents, lubricants, and antistatic agents in PE and PP by means of GC, GC-MS and HPLC [174].

Mersch and Zimmer [175] have reported GC analysis of rubber extracts: aliphatic/naphthenic and aromatic oils, AOs and paraffin waxes. It is generally not possible to apply GC directly for *accelerator analysis* as most of these compounds are thermally unstable and decomposition may occur during analysis. Determination of accelerators in rubbers is an active research area, which is based on the identification of their degradation products; some accelerators such as TMTD and ZDMC yield the same degradation products, at variance to MBT. An amine-specific electrical conductivity detector enables direct GC determination of volatile (cyclo)aliphatic amines without the need for prior derivatisation [176]. EPA method 607 describes the GC separation of nitrosamines (direct and split injection). Nitrosamine analysis in (stored) rubbers is particularly challenging with legal limits currently set at <1 ppm for dimethylnitrosamine. No quite satisfactory standard methods are available and extraction, GC-TCD, HS-GS and pulsed NMR techniques are used. Extreme caution must be taken to avoid the risk of false-positive and false-negative results [177]. False positives can be obtained when nitrosamines are formed during sampling (e.g. by interaction of nitrites and amines).

Nonvolatile compounds cannot be analysed unless pyrolysis or derivatisation converts them to a condition amenable to GC. *Derivatisation GC* (or LC) has been used for several components such as erucamide (imidisation for volatility), fatty amines (aromatic amidation for UV detectability), and polyethylene oxides (esterification for both volatility and detectability) [178]. The surface concentration of erucamide on extruded LLDPE films was determined quantitatively by surface washings with ether, followed by evaporation, dissolution

in THF and GC analysis [179]. A correlation with the coefficient of friction (COF) was established. GLC is commonly used for the analysis of fatty acids as their volatile methyl esters [180]. Hoffman [181] has described a derivatisation GC method for the determination of water in polyesters, which consists of reaction of water with a mixture of hexamethyldisilazane and trimethylchlorosilane to form hexamethyldisiloxane. In most instances methods based on GC require organometallic compounds to be derivatised so as to confer thermal stability, whereas this is unnecessary with LC separations. Epoxidised soybean oil (ESBO) additive in PVC and PVDC was determined by GC using dioxolane derivatisation with tri(11,12-epoxy)eicosanion as an internal standard [182]. Also a fast and cost-effective procedure to quantify ESBO by an external standard method has been reported [183]. GC-FID analysis was carried out after conversion of ESBO into fatty acid methyl esters with TMAH. The procedure is applicable to commercial epoxidised oils, polymer additive packages and PVC-containing epoxidised oils. In PVC samples, ESBO was extracted with toluene and derivatised prior to GC analysis. The methyl esters of mono-, di- and triepoxyoctadecanoic acid were separated with a short capillary column. Using derivatisation reactions (such as silanisation) PTV injection can be used for less labour intensive analyses of several additives in polymeric matrices, e.g. acids. Stearic acid in PET can be determined by means of thermal desorption using PTV-GC. For the determination of (calcium) stearate the sample is hydrolysed, derivatised with acetic anhydride and GC analysed.

David *et al.* [184] have shown that *cool on-column* injection and the use of deactivated thermally stable columns in CGC-FID and CGC-FID-MS for quantitative determination of additives (antistatics, antifogging agents, UV and light stabilisers, antioxidants, etc.) in mixtures prevents thermal degradation of high-MW compounds. Perkins *et al.* [101] have reported development of an analysis method for 100 ppm polymer additives in a 500 μL SEC fraction in DCM by means of *at-column GC* (total elution time 27 min; repeatability 3–7 %). Requirements for the method were: (i) on-line; (ii) use of whole fraction (LVI); and (iii) determination of high-MW compounds (1200 Da) at low concentrations. *Difficult matrix introduction* (DMI) and selective extraction can be used for GC analysis of silicone oil contamination in paints and other complex analytical problems.

Many standard *compendium methods* (ASTM, EPA, FAAM) are based on GC analysis. Examples are the GC-FID determination of fatty acid methyl esters (FAME;

C_{12} to C_{20}, primarily C_{16} to C_{18}), used as surface lubricants in the manufacture of food-contact articles. The method, which uses ethyl palmitate (Eastman Chemicals No. 1575 Red Label) as an internal standard, has been validated at 200 ppm total FAME [185]. Other FAME standards (methyl palmitate, methyl stearate, methyl oleate, methyl linoleate and methyl linolenate) are available (Applied Science Laboratories) [116]. Worked out examples of additive determinations are given in the Food Additives Analytical Manual [116], which also describes a great many of indirect food additives, such as BHA, BHT, TBHQ, 1-chloro-2-propanol, DLTDP, fatty acid methyl esters, *n*-heptyl-*p*-hydroxybenzoate, propylgallate, sodium benzoate, sodium stearoyl-2-lactylate, sorbitol and phenolic antioxidants. EPA methods 606 and 8060 describe the CGC separation of phthalate esters (direct injection) (cf. Figure 4.2).

Other typical polymer/additive determinations which may be carried out by GC are the determination of slip agents (calcium, zinc, aluminium, lithium and sodium stearate, calcium montanate) in polyamide, of stearylstearate and nonylstearate in polyamide, of ERL 4221 in PBT, of 1,4-butane diol, terephthalic, isophthalic and 1,2,4-tribenzenecarboxylic acid in polyesters and polyetherester block copolymers, of Loxiol 862 in PC, etc. Gas chromatography is also used in health, safety and environmental monitoring related to the production and use of additive containing polymeric materials. Other problems of industrial relevance are the determination of isoprene, styrene and butadiene in SBR (using dissolution techniques or HS-GC), of ENB and DCPD in EPDM, and stabilisers and oil in rubbers. Not surprisingly, the two traditional methods of determination of oil content in rubbers (dissolution/precipitation and extraction) yield different results. Namely, in one case coprecipitation may occur and in the other co-extraction of rubber oligomers.

Van der Wal [186] has reported narrow-bore PTV-GC-FID of a standard test mixture (cf. Figure 4.1) of polymer additives (Irganox 245/1035/1076/1098/1330, Irgafos 168, Tinuvin 326/327/770/P, Ultranox 626, DMTDP, DSTDP, DBPC, Ionox 330, Kemamide E and tricaprine in MTBE); analysis cycle time was reduced from 1 h (conventional CGC) to 15 min with a minor reduction in peak capacity. A dynamic range of at least 0.3–100 ppm was demonstrated with CV < 10%. PTV injection also allows analysis of both volatile and high-boiling components without the need for time-consuming sample pretreatment such as solvent extraction. PTV injectors are employed to directly analyse *solid samples* by thermal desorption-pyrolysis within the injection liner. Lancaster *et al.* [187] have reported the MS detection (EI and CI) of off-odour oxygenated compounds in printed PE film for wrapping foodstuffs, the determination (EIMS) and quantification (FID, 'internal' standard *n*-eneicosane, *n*-C_{21}) of low molecular mass hydrocarbons (C_8 to C_{24}) at ppm level in PE reactor powder and pellet, and the identification of DUP and DIUP plasticisers in PVC. PTV analysis results in complete recovery of analyte molecules from polymeric matrices, as verified by making repeat analyses on the same sample, and can easily be made quantitative by the use of a suitable internal standard. The application of PTV in CGC analysis was reviewed [188].

Juvet *et al.* [189] have described *photolytic degradation* (irradiation of polymer films using a high-UV intensity mercury source at 360 nm) to elucidate structure and quantity of polymer additives from the nature of the photolysis products using GC-FID. The UV absorbance of the polymer influences the response to photolytic degradation. Without a sensitiser (as applicable with liquid samples) photolysis times of 30 min are needed for decomposition of polymers with low molar absorptivity in the UV region; irradiation times of some 5 min were typical in quantitative determination of polymer additives since these materials generally have greater cross-section for UV radiation and decompose more readily than the polymer itself. Photolysis products are most easily released by raising the temperature above T_g. Due to the control on input energy and a predictable manner of photolytical decomposition, photolytic degradation yields a rather simple decomposition pattern.

Table 4.24 summarises the main features of solid-state photolysis-GC for polymer/additive analysis. Photolytic degradation products obtained from PE/1% DLTDP and PE/1% DSTDP, PMMA/9% DNP and PVC/19% DOP were examined [189]. The method has found little follow-up.

General applications of CGC have been described [190]; recent reviews covering the application of GC to polymer analysis are relatively few [129,161,174,191,192] and all noncomprehensive.

Table 4.24 Main characteristics of solid-state photolysis-GC for polymer/additive analysis

Advantages
- Simple and reproducible decomposition patterns
- No need for prior separation
- Small sample amount (<1 mg)
- Rapid (5–30 min)

Disadvantages
- Restrictions as for GC
- Limited experience

4.2.1 High-temperature Gas Chromatography

Principles and Characteristics Major limitations of GC are the stability of the stationary phase and of the solute(s) at high temperature. At higher polarity and molecular weight of the solutes, a higher temperature is necessary to provide adequate solute partial pressure to allow GC separation to be realised. This must be matched by sufficient thermal stability. The use of temperature is of great importance as an elution parameter in GC. The term 'high-temperature gas chromatography' (HTGC) is ill defined, but it is generally accepted that HTGC starts at the working limit of conventional GC (about 330 °C), with an upper working temperature presently set at about 450 °C. Despite the fact that HTGC is simply an extension of GC to higher working temperatures, it is best treated separately as the technique needs special instrumentation.

Blum and Aichholz [193] first made *instrumental considerations*. A simple conventional GC system cannot be used for HTGC. Particular attention should be paid to the sample introduction system, carrier gases/carrier gas regulation, column and detector(s). Sample introduction in HTGC is extremely difficult. Conventional hot split and hot splitless injection cannot be used as these techniques suffer from severe discrimination of high-boiling components. On the other hand, on-line column injection and PTV are highly suitable for HTGC. For HTGC the PTV injector should operate up to about 500 to 600 °C in order to permit quantitative transfer of high-MW materials onto the column in a short time. Results of on-column injection and HT-PTV injection are at least qualitatively comparable [25]. For absolute peak areas, however, on-column injection yields better reproducibility. The main advantages of HT-PTV injection are discrimination-free sample introduction of high-MW materials and the fact that this technique is easier to automate than on-column injection. Another distinct advantage of HT-PTV injection over on-column injection is that PTV injection is relatively insusceptible to the presence of high-MW residues in the sample. Polymer extracts often contain a considerable amount of co-extracted oligomers that cannot be eluted from a GC column, not even at the maximum allowable operating temperature. With PTV injection the high-MW impurities from the sample are retained in the liner.

Hydrogen is the only acceptable carrier gas in HTGC. Three support materials have been used for HTGC, fused-silica, metal and borosilicate glass. Although polyimide-clad fused silica dominates in 'modern' GC, its suitability in HTGC is limited to about 350 °C [65,194]. Aluminum-clad fused-silica capillary columns for HTGC were introduced in 1986 [65] but did not meet expectation. The second support material, metal, is only available as coated capillary columns. Rohwer and Pretorius [194] have proposed the use of borosilicate capillary glass as an alternative to fused silica for high-temperature applications. Although a systematic comparison between various high-temperature stable supports has not been reported, borosilicate glass is favoured in most applications. For HTGC (and also GC-MS) low-bleed stationary phases are required. Elution of the stationary phase (column bleed) can cause misidentification of analytes, loss of sensitivity, and inaccurate quantitation. Column bleed is more easily detectable at higher temperatures. With the production of bonded-phase columns, column bleed is less of a problem. Most stationary phases in HTGC are polysiloxanes. The primary advantage of OH-terminated polysiloxane phases is the increase of inertness and thermostability (up to over 400 °C). Therefore, the use of this class of polymers is a prerequisite for any HTGC, whether glass- or metal-clad fused-silica columns are used. At present only a limited variety of more or less apolar coated capillary columns are commercially available and most HTGC applications are therefore carried out on laboratory-made glass columns or are custom-made, e.g. WCOT CP-SimDist Ultimetal (Varian) columns (10 m, 0.25 mm i.d., 0.15 μm film thickness). The very thin film is required for minimising column bleed at high temperatures. Some stationary phases can be used at remarkably high temperatures (e.g. Dexsil 300 up to 450 °C). It is then not surprising that HTGC is carried out only in a limited number of laboratories. HTGC detectors are MS, and modified FID and AFID.

Obviously many analysts are reluctant to use CGC for the separation of medium-mass molecules (ca. 1000 Da) even if the compounds are thermostable. It is actually amazing that on a highly inert, deactivated capillary column many compounds are unexpectedly thermally stable. Polar functional groups, which might suggest that the compound could be thermally labile, are in many cases sterically shielded and sufficiently protected against nucleophilic attacks. These facts may well be at the origin that early scepticism about the usefulness of HTGC [94] is now waived.

Gas chromatography at the maximum working temperature of the method is a borderline application. Many potential HPLC and SFC applications fall in the working range of HTGC. Nevertheless, with respect to selectivity and separation efficiency, an HTGC method has to be optimised in the same manner as in conventional GC. This can be difficult. In contrast to conventional GC the solution to most HTGC problems needs to be developed. If HTGC can be used it is a powerful separation technique for the determination of thermally stable, medium-mass molecules, without comparable alternatives [193].

The use of HTGC instead of alternative techniques has several *advantages*, not unlike those of conventional GC (excellent separation efficiency, selectivity and signal-to-noise ratio, ease of coupling to highly sensitive universal or specific detectors). Conventional GC allows analysis of additives with molecular weight up to about 600 Da. Expressed in masses of the analysed substrates, the difference between conventional GC and HTGC is about 600–1000 atomic mass units, depending on the nature and volatility of the compounds. Another advantage is that the presence of oxidation/degradation peaks is much less of an issue in HTGC than in HPLC. A single column can be chosen which successfully resolves most additives of interest, as well as their primary by-products. In many cases analysis times are some 25 % shorter than for HPLC. In general, the chromatograms themselves do not require manual correction of the baselines and peak resolutions, as often occurs with HPLC [195]. The method has good sensitivity and additives can be quantified with confidence at very low levels. If HTGC can be employed, it is the method of choice for quantitative as well as qualitative trace analysis of complex mixtures. A *weakness* of the method is the fact that some oligomers are always present, appear in the chromatogram and may have to be properly accounted for in the case of interference with peaks representing very low levels of additives or their by-products.

According to Blum and Damasceno [94] the real importance of HTGC (as well as cSFC) lies in the combination with MS. Coupling of HTGC to MS is discussed in Blum and Aichholz [193].

Applications On a comparative basis, HTGC is a relatively new tool and extremely valuable for the analyses of extracted polymer additives, as shown by industrial problem solving. For satisfactory analysis of in-polymer additives by HTGC two specific conditions are to be met. The instrument should be equipped with a cool on-column injection port to better preserve some of the additives and/or their by-products that may be thermally labile. The instrument must also have electronic pressure control so that some of the very high-boiling components, such as Irganox 1010, are not permanently adsorbed onto the column coating, i.e. they are never eluted under the conditions of the analysis.

In the literature several HTGC methods for additive analysis have been reported [94,193,195–198], differing in column type and temperature (usually not exceeding 380–420 °C), and molar mass range of the additives eluted. Attention should be paid to column bleeding phenomena and premature ageing. Verdurmen *et al.* [197] have developed a screening HTGC-FID method at the lower limit of the HTGC range (325 °C) in order to achieve good separation between frequently encountered additives in polymer materials. The method aimed at being a verification of the type of additive in a stand-alone technique without the aid of identifying detectors and the separation step in GC-MS analysis. In the experimental conditions a maximum molar mass of 700 Da elutes; a test mixture of 19 additives is well separated. Quantification of the components may be performed utilising calibration factors relative to an internal standard. Blum and Damasceno [94] have used high-temperature glass capillary gas chromatography with OH-terminated polysiloxane stationary phases. In particular, 21 AOs and UVAs, including Tinuvin 234/320/326/327, Irganox 245/259/1010/1019/1076/1098/1330/3114/3125, Chimassorb 966 and other stabilisers with MW ranging from 315 to 1196 Da were separated in 4 min by means of HTGC-FID in the 190–420 °C range.

The importance of using sufficiently high temperatures when using PTV injection in HTGC may be illustrated by the observation that in analysis of a polymer additive mixture Irganox 1010 was completely absent at a final PTV injector temperature of only 400 °C, as opposed to 600 °C. Apparently the vapour pressure of this compound at 400 °C is too low to allow transfer to the column. For some additives, e.g. Irgafos 168, thermal degradation can occur during PTV sample introduction.

The compositions of the antistatic agents Ampacet 100134, Atmer 1013 and Atmer 7105 were studied by means of HTGC-FID and HTGC-MS after acetylation [199]. The former two products are mainly composed of GMS/GMP mixtures in various ratios while Atmer 7105 contains two alkyl (C_{14}, C_{16}) diethanolamines. Cold injection HTGC-FID and HTGC-MS were reported for routine determination of additives (antistatics and antifogging agents, UVAs, AOs) in polymer formulations [198]. Differences in *cis/trans* ratio of oleamide (Kemamide U) and erucamide (Unislip 1752/3) batches were detected using HTGC-MS [199a].

4.2.2 Headspace Gas Chromatography

Principles and Characteristics Headspace sampling is one of the major sample handling procedures in GC. Headspace analysis is essentially an extraction procedure in which a gas is used instead of a liquid as the solvent; a gas is an ideal 'solvent' for highly volatile compounds. In this case the analytes will be distributed between the condensed phase and gas phase and conditions are adjusted to favour the latter. *Gas extraction techniques* can be carried out in several variants: as a single step (static headspace), by stepwise repeating of the extraction (multiple headspace extraction) and also by stripping the volatiles (dynamic headspace) by a continuous flow of an inert purge gas (total trapping).

Headspace gas chromatography (HS-GC) is a modification of regular GC. In *static HS-GC* (SHS-GC), often called equilibrium headspace sampling, a sample is usually a solution sealed in a closed vial (typically 20 mL volume) held at a constant temperature for a given time until thermodynamic equilibrium in the binary heterogeneous system has been reached between vapour and solution phase. SHS-GC initiated in the 1958–1964 period with significant contributions by Machata [200]. Headspace sampling relies totally on volatilisation to separate analytes from a sample matrix. The higher vapour pressure of the analyte as compared to that of the rest of the sample enriches the headspace with the analyte even though the molar concentration of the analyte is small. According to Raoult's law (Equation 4.3), the partial pressure (P_a), or molar concentration, of analyte in the headspace is directly proportional to its molar concentration in the liquid sample (X_a) by a factor (VP_a), the vapour pressure of the analyte at the temperature of the sample.

$$P_a = (X_a).(VP_a) \qquad (4.3)$$

The headspace chamber is heated to take advantage of the increase of vapour pressure with temperature. An aliquot of the vapour phase (headspace) is then injected into the column to proceed as regular GC. In this way, HS-GC is used to analyse low concentrations of components with a high vapour pressure in a low vapour pressure matrix.

Important factors influencing the analyte volatilisation process are related to diffusion, porosity, and surface area (for solids). To obtain reproducible results it is necessary to control storage temperature and time strictly. The temperature of the sample is very important because of the specific boiling points of the various analytes. The partition coefficient, K, at equilibrium, is influenced by temperature and matrix. The first commercial headspace sampling system was introduced over 35 years ago. In advanced, fully automated, GC instrumentation syringe (gas, liquid) injection is now combined with a variety of other sample introduction systems such as static headspace, thermal desorption (TD) and pyrolysis (Py); split/splitless or on-column injection modes are available. Multipurpose samplers nowadays allow headspace sampling for on-line enrichment of (semi)volatile analytes from liquids and large volume sampling (LVS), to 1000 µL. Detectors used in HS-GC are various, including ELCD, ECD, FID, FPD, FTIR, MS, PID, TCD and NPD. Headspace sampling can also be used in hyphenated systems (e.g. HS-GC-FTIR and HS-GC-MS).

HS-GC can be quantitated in several ways [201]. *Quantitative analysis* depends on calibration. Headspace methods often use external standardisation [202–204]. For accurate quantitative determination the solution–vapour (or solid–vapour) partition coefficient for each analyte must be known. Determination of such partition coefficients is often difficult in practice. Use of an external standard also implies that chromatographic response factors have not been determined on the actual system to be analysed. For quantitative analysis it is important that the sample matrix is matched to standards, otherwise the proportion of the analyte distributed between the phases may be different and results inaccurate. Assuming the matrix is matched and equilibrium reached, the concentration of the analyte originally in the sample is proportional to the concentration in the headspace. Matrix effects can be overcome by employing the standard addition method. Although this works well with liquid samples in HS-GC, solid samples present problems being difficult to obtain a homogeneous mixture of standard and sample [205]. Furthermore, the analytes are highly volatile, which makes handling difficult. These problems are overcome by the multiple headspace extraction (MHE) method [206]. With MHE (or stepwise dynamic gas extraction) the quantitation limit depends on the number of extractions. Analysts must achieve a response of three times the noise level for the last extraction, when most of the analyte has already been extracted. Quantitative results require accurate control of the sample temperature during the headspace procedure. For overviews of methods for quantitative analysis (internal normalisation, internal and external standard methods, standard addition method, multiple headspace extraction, analysis of solid samples) and method development in HS-GC, see cf. Kolb and Ettre [207].

Table 4.25 Main features of headspace sampling

Advantages

- Little or no sample preparation
- Ease of use; reliable
- Excellent quantitative results
- Matrix-independent quantitation using MHE
- Automation (high sample throughput)
- No solvent consumption
- Mature technology (8th generation)

Disadvantages

- Limited to volatile analytes
- Limited sensitivity (in particular for solution headspace)

Parameters influencing the performance of headspace methods include sample preparation, sample temperature, equilibration time, carrier gas pressure, pressurisation time, sampling time and transfer line temperature. For validation of headspace instrumentation, see Kolb and Ettre [207].

Table 4.25 lists the *main characteristics* of headspace sampling. In HS-GC sample preparation is very often limited to placing a sample in a vial. Sample extraction, clean-up and preconcentration are not necessary. Elimination of solvents in preparing samples for GC has several benefits:

(i)   Increased selectivity (thermal desorption of the sample compounds to a GC permits analysis of 100 % of the sample, instead of an aliquot).

(ii) Elimination of the solvent peak.

(iii) No environmental concerns (no evaporating solvents, no waste disposal).

Volatile compounds in almost any sample matrix can be extracted simply and highly reproducibly by use of headspace techniques. Headspace sampling uses much more material than most thermoanalytical techniques (e.g. TGA), which is an advantage in sample representativity. A disadvantage of thermal volatilisation and desorption of additives from polymeric matrices compared to solvent extraction routes is that solvent swelling favours the desorption process. The most significant limitation of SHS sampling is lack of sensitivity. The sensitivity depends on the total mass of the analyte transferred to the column. Cryofocusing in HS-GC-MS greatly increases the sensitivity (detection limits: ppt).

With regard to the solution approach, it is imperative that the solvent used be of the highest possible purity. *Solution headspace* is applicable to a much wider range of samples than the solid approach. When working with sample solutions, headspace equilibrium is more readily attained and the calibration procedure simplified. The sensitivity of the solution method depends on the vapour pressure of the analyte and its solubility in the solvent phase. The greatest detection sensitivities using solution headspace analysis are obtained for monomers having relatively low boiling points. Greater sensitivities and shorter analysis times were reported with the headspace analysis method than for direct injection of polymer solutions into a GC [133]. However, the headspace sensitivity of the solution approach is reduced compared to direct analysis of the solid sample. In solvents the partition coefficient is usually large in comparison to a solid-sample matrix. Although solid headspace sampling provides about 10-fold more sensitivity than the solution headspace method this procedure may be applied only to sample systems where equilibration with headspace is rapid and complete. Whenever a solid sample is soluble in an organic solvent or in water, the determination can be simplified by using the solution approach. The equilibration time will then depend on the viscosity of the resulting solution and will not necessarily be shorter than it would be if the solid sample were analysed directly. Furthermore, even if equilibration between solid and headspace is obtained, the partition coefficient must also be determined for the component of interest in each type of sample matrix.

Advantages and disadvantages of HS-GC over regular GC are summarised in Table. 4.26. HS-GC fingerprinting chromatograms obviously include only the *volatile* components present and do not provide a complete picture of sample composition; on the other hand, when solvent extraction is used, all the soluble sample constituents are removed, including also those having no appreciable vapour pressure at the equilibration temperature. Headspace analysis enhances the peaks of volatile trace components.

Table 4.26 Advantages and disadvantages of HS-GC over conventional GC

Advantages

- Selective on more volatile components
- More sensitive
- No chromatographic disturbance of nonvolatile sample components
- Simplicity: easier sample preparation

Disadvantages

- More complicated quantitation (matrix effects except for MHE)
- Sometimes long equilibration times

Static headspace may also be carried out by substituting the heating step by a microwave treatment. In this procedure the material is immersed in a solvent that is transparent to microwaves relative to the sample in order to impart most, if not all, of the microwave energy to the sample [208]. Another configuration of MAP gas-phase extraction relates to dynamic headspace sampling.

Wahlroos [209] has first suggested continuous gas extraction. *Dynamic headspace* (DHS) (purge-and-trap, PT) involves bubbling a (high-purity) inert gas, such as N_2 or He at a specified flow-rate and time, through an aqueous sample (typically 5 mL; solids must be suspended in water) at ambient temperature. This liberates volatile organic compounds, which are efficiently transferred from the aqueous to the vapour phase. During this purge step, the inert gas flow sweeps the vapour through a trap containing adsorbent materials, which retain the VOCs (Figure 4.3). Next, the trap is rapidly heated to desorb the trapped VOCs in a narrow band. The procedure is particularly useful for concentrating VOCs that are insoluble or poorly soluble in water and have boiling points below 200 °C. The tubes used in PT analyses are generally packed with multiple beds of various sorbent materials, so that a broad range of high- and low-MW compounds, polar and nonpolar, can be trapped in a single tube. Selective trapping uses an adsorbent having high affinity for the analyte of interest. Use of PT in combination with thermal stripping is advantageous

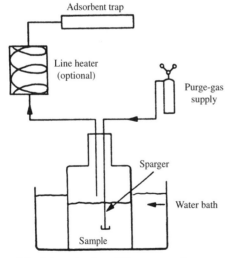

Figure 4.3 Schematic of an off-line purge-and-trap apparatus. After Cole and Woolfenden [208a]. Reprinted with permission from *LC.GC*, Vol. 10, Number 2, February 1992, pp. 76–82. LC.GC is a copyrighted publication of Advantar Communications Inc. All rights reserved

for several reasons. Purge-and-trap sample preparation is extremely simple and fast, thermally stripped samples deliver sharp, symmetrical peaks, and, because the sample delivered to the GC column is undiluted, sensitivity is enhanced considerably compared to solvent extraction and solvent desorbed samples. A theoretical model for quantitative determination of volatile compounds in solid polymers by DHS sampling has been presented which correlates desorption parameters (temperature and flow-rate) with component-polymer properties (diffusion and distribution coefficients) [210]. DHS-GC has more parameters to be tailored and optimised for specific sample properties than SHS-GC, such as selecting the various adsorbents for filling a trap. Purge-and-trap is the method of choice for extracting and concentrating VOCs from almost any matrix (food, polymers, pharmaceuticals, sediments, water samples, etc.).

Headspace analysis was reviewed [132], including the development of the PT technique [211]. Applied HS-GC has been dealt with in several monographs [207,212].

Applications Sampling from headspace has been explored extensively for many years [212,213]. The technique coupled with GC has been used for quantitative analysis of *volatile organic compounds* in gaseous, liquid or solid samples in areas as diverse as food, beverage, polymers, clinical biochemistry, and environmental monitoring. Determination of residual volatile chemicals in polymers by SHS-GC is common practice [214]. For product safety purposes it is necessary to monitor exposure to possible toxic substances, undesirable odours and other types of irritants that may result from volatiles or decomposition products of polymers. HS-GC is used both for R&D and QC purposes. Equilibrium HS-GC is used to analyse a wide variety of liquids (aqueous samples, simulants, oils, emulsions, gels, ointments, etc.). Solution headspace sampling for GC for additives in polymers is less important than direct sampling for solid polymers. However, most of the officially recommended headspace procedures utilise the solution approach.

Typical areas of application are identification of trace (ppm or ppb level) volatile organics in complex mixtures (e.g. olfactory principles) and monitoring of residual monomers in polymeric materials. Apart from HS-GC, analysis of volatiles can also be carried out by a variety of other methods, including hydrodistillation, SFE, US, adsorption trapping and SPME.

SHS-GC is amongst the best methods of choice for volatile *monomer analysis* and effectively gives both isolation and preconcentration of the analyte in one step. Headspace methods are used extensively for determination of residual monomers and other residues in polymer compositions after dissolution or dispersion in a suitable solvent and equilibration in a sealed vial at constant temperature. SHS-GC is routinely being applied for the determination of C_2 in LDPE, C_6, C_8 and toluene in LLDPE, C_2, C_7 and C_8 in PP powder, decaline and TEA in UHMWPE, 1,3-butadiene in PC/ABS blends, and for total emission analysis in PP. A method involving HS-GC has been described for determination of monomers and volatiles in PVC [215]. Quantitation limits (defined as the monomer concentration necessary to produce a peak at least three times the baseline noise or 3 % of full scale) for residual monomers as low as 1 ppm have been reported for direct solution injection GC, as compared to 0.05 ppm for solution headspace [158]. For the analysis of 0.05–0.5 ppm of vinyl chloride, butadiene, acrylonitrile, styrene and 2-ethylhexyl acrylate monomers in (co)polymers a N, N'-dimethyl acetamide solution of the polymer was subject to HS-GC analysis [216]. FDA has accepted an official SHS-GC method for the analysis of vinyl chloride in PVC food packaging [217]. *Vinyl chloride monomer* was also measured directly in an aqueous suspension of PVC using headspace membrane enrichment followed by on-line process GC-FID determination [218].

HS-GC methods have equally been used for chromatographic analysis of residual volatile substances in PS [219]. In particular, various methods have been described for the determination of *styrene monomer* in PS by solution headspace analysis [204,220]. Residual styrene monomer in PS granules can be determined in about 100 min in DMF solution using n-butylbenzene as an internal standard; for this monomer solid headspace sampling is considerably less suitable as over 20 h are required to reach equilibrium [204]. Shanks [221] has determined residual styrene and butadiene in polymers with an analytical sensitivity of 0.05 to 5 ppm by SHS analysis of polymer solutions. The method development for determination of residual styrene monomer in PS samples and of residual solvent (toluene) in a printed laminated plastic film by HS-GC was illustrated [207]. Less volatile monomers such as styrene (b.p. 145 °C) and 2-ethylhexyl acrylate (b.p. 214 °C) may not be determined using headspace techniques with the same sensitivities realised for more volatile monomers. Steichen [216] has reported a 600-fold increase in headspace sensitivity for the analysis of residual 2-ethylhexyl acrylate by adding water to the solution in dimethylacetamide.

Solution SHS-GC-FID was used for the determination of dichloromethane (DCM) in PC [222] and solid SHS-GD for the analysis of residual solvents in transdermal drug-delivery systems [223] and rest solvents in packaging films [224]. Kolb *et al.* [225] have applied MHE to the determination of residual monomers in polymers and residual solvents in other solid samples. The multiple headspace extraction method was illustrated for residual ethylene oxide (EO) in surgical PVC [207]. Various ASTM methods (F 151-91, D 4526-85, D 3749-87, D 4322-91, D 4443-89, D 4740-93) deal with the determination of residual solvents in flexible packaging materials (cf. ref. [226]).

For *odour analysis* the most valuable detectors are FID, AED and MSD. NPD and ECD detectors are very sensitive, but limited in use to nitrogen or halogen containing compounds; TCD and FTIR detectors are not sensitive enough for odour analysis by direct headspace autosampling.

Concentrations and recovery of hexane, tridecane and BHT from PP were predicted on the basis of a quantitative theoretical model for DHS sampling [210]. The insolubility of thermoset polyesters in organic solvents means that determination of volatile components can only be reliably carried out using a PT device [227–229]. Solvent extraction SHS-GC-MS was used to analyse the concentration of benzene (originating from the t-butyl perbenzoate initiator) in powdered samples of thermoset polyester compounded for the manufacture of plastic cookware [230]. This approach yields minimum levels of volatiles present as there is no guarantee of total extraction into the solvent. In this case the use of a TD technique for monitoring of total volatiles had previously failed to indicate the presence of benzene, possibly because of its high volatility in relation to the other compounds being measured.

In addition to the direct analysis of a sample for its quantitative and/or qualitative composition, HS-GC can be used for *physico-chemical measurements*, such as the determination of vapour pressures.

4.3 SUPERCRITICAL FLUID CHROMATOGRAPHY

Principles and Characteristics Klesper *et al.* [14] have introduced supercritical fluid chromatography (SFC). The oil industry has been a major force in the development of many aspects of the application of supercritical fluids. Much of the pioneering development of SFC was carried out by Sie and Rijnders [231,232], who also coined the term 'supercritical fluid chromatography' [233].

In SFC the same type of supercritical fluid (i.e. a fluid above its critical temperature and critical pressure) is used as the mobile phase to chromatograph additives as described in the SFE section (Section 3.4.2). As pointed out by Berger [26] SFC is a misnomer as there is no physical or chemical characteristic important to the separations that is unique to supercritical fluids. Nonetheless, the scope of SFC is essentially limited by the inherent physical chemistry of the fluid. The *theoretical basis* of SFC is well established [234]. SFC separations involve normal-phase partition or adsorption. SFC combines the advantages of both GC and HPLC into a hybrid technique. However, it is important to realise that conditions do not exist where a supercritical fluid can have *both* the solvation capacity of the liquid-like phase and the high diffusion rate of the gas-like phase at the same time, i.e. SFC does not have all the advantages of GC and LC. The normal-phase characteristics of SFC offer separations not available on reversed-phase HPLC [235].

The *basic SFC system* comprises a mobile phase delivery system, an injector (as in HPLC), oven, restrictor, detector and a control/data system. In SFC the mobile phase is supplied to the LC pump where the pressure of the fluid is raised above the critical pressure. Pressure control is the primary variable in SFC. In SFC temperature is also important, but more as a supplementary parameter to pressure programming. Samples are introduced into the fluid stream via an LC injection valve and separated on a column placed in a GC oven thermostatted above the critical temperature of the mobile phase. A postcolumn restrictor ensures that the fluid is maintained above its critical pressure throughout the separation process. Detectors positioned either before or after the postcolumn restrictor monitor analytes eluting from the column. The key feature differentiating SFC from conventional techniques is the use of the significantly elevated pressure at the column outlet. This allows not only to use mobile phases that are either impossible or impractical under conventional LC and GC conditions but also to use more ordinary

fluids at temperatures off limits at ambient pressure. These capabilities provide truly significant advances in speed and selectivity. Roberts [236] described *SFC instrumentation*; injection methods have been reviewed [237] and restrictors for SFC have been discussed [238]. Commercial SFC instruments were first introduced in the 1980s for packed and then for capillary chromatography. Adjusting the pressure in SFC (up to 400 bar) is used to achieve gradient elution; mobile phase gradients and temperature gradients determine an optimum separation. A desirable feature of SFC is the ability to elute more polar solutes by raising the pressure. Obviously, changing the pressure is far easier than changing the composition of a fluid, such as in gradient elution LC (cf. Section 4.4.2.4). In SFC the main pressure drop takes place abruptly in the back-pressure regulator, whereas in the HPLC system the pressure drops gradually as the mobile phase solvent passes through the column to atmospheric pressure.

Implementation of SFC has initially been hampered by instrumental problems, such as back-pressure regulation, need for syringe pumps, consistent flow-rates, pressure and density gradient control, modifier gradient elution, small volume injection (nL), poor reproducibility of injection, and miniaturised detection. These difficulties, which limited sensitivity, precision or reproducibility in industrial applications, were eventually overcome. Because instrumentation for SFC is quite complex and expensive, the technique is still not widely accepted. At the present time few SFC instrument manufacturers are active. Berger and Wilson [239] have described packed SFC instrumentation equipped with FID, UV/VIS and NPD, which can also be employed for open-tubular SFC in a pressure-control mode. Column technology has been largely borrowed from GC (for the open-tubular format) or from HPLC (for the packed format). Open-tubular coated capillaries (50–100 μm i.d.), packed capillaries (100–500 μm i.d.), and packed columns (1–4.6 mm i.d.) have been used for SFC (Table 4.27).

Table 4.27 Columns used for supercritical fluid chromatography

Parameter	Packed column			Open-tubular column
	Conventional	Narrow-bore	Micro	
Internal diameter (mm)	2–4.6	1.0	<0.5	0.05–0.1
Length (cm)	10–30	10–30	50–150	300–2000
Column material	Stainless steel	Stainless steel	Fused silica	Fused silica
Particle/film (μm)	$3–10^a$	$3–10^a$	$3–10^a$	$0.05–0.5^b$
Stationary phase	Chemically bonded silica, alumina, polymeric resins			Polysiloxanes
Pressure drop (MPa)	7	–	–	0.01

[a] Average diameter of particles packed into column.
[b] Thickness of film coated onto the inner wall surface of capillary column.
After Robards *et al.* [240]. Reprinted from K. Robards *et al.*, *Principles and Practice of Modern Chromatographic Methods*, Academic Press, Copyright (1994), with permission from Elsevier.

The selection of the column type is mainly determined by the composition of the sample. In general open-tubular (capillary) columns are preferred for low-density (gas-like) SFC, whereas packed columns are most useful for high-density (liquid-like) SFC. Open-tubular columns can provide a much larger number of theoretical plates than packed columns for the same pressure drop. Volumetric flow-rates are much higher in packed column SFC (pSFC) than in open-tubular column SFC (cSFC), which makes injection and flow control less problematic.

For a proper evaluation of the prospects of SFC for polymer/additive analysis it is imperative to understand the essential differences between cSFC and pSFC. *Capillary (open-tubular) columns* in SFC were first reported in 1981 [241]. Commercial cSFC instrumentation was introduced in 1985 and dominated the marketplace until 1992. Open-tubular columns for SFC differ from those in GC in that column diameters are generally much smaller (50–100 μm i.d., 3–20 m length), to maintain high efficiency. This leads to low flow-rates. In cSFC precolumn splitting ratios from 1:5 to 1:50 are commonly used to prevent solvent overloading. Methylpolysiloxanes are the most popular phases in both cSFC and GC, but dedicated phases have been designed for cSFC. The inertness of capillary columns provides better chromatography for polar compounds without the need for polar modifiers. Open-tubular columns are commonly used with neat $scCO_2$, isothermal pressure or density programming, FID or MS detection, and are usually prescribed when GC and HPLC do not work. Injection is commonly done in a split mode with injected quantities in the nL range. This limits the sensitivity of cSFC to 10–20 ppm. The sensitivity limitation can be overcome by large-volume injection ($>100 \mu L$) coupled with a solvent elimination step and analyte refocusing [242]. With such an injection system the major limitations associated with cSFC injection are overcome. Advantages of LVI-cSFC over conventional cSFC technology are therefore: (i) improved sensitivity and specificity; (ii) direct injection of liquids (including water); (iii) improved injection precision; and (iv) automation. Samples may be introduced into cSFC also by on-line SFE coupling (cf. Section 7.1.1.4) or by using microcolumn LC (cf. Section 7.4.3). The low flow-rate used in cSFC provides great detection flexibility and interfacing to hyphenated techniques. The most important attribute of cSFC is the use of FID as the universal, constant response factor detector missing in LC. Interlaboratory reproducibility of cSFC-FID has been reported [243]. Capillary SFC columns can be used at higher

Table 4.28 Main features of open-tubular SFC

Advantages

- High efficiency (theoretical plates m^{-1} > 10^5)
- Low mobile phase flow-rates ($< 0.5\,mL\,min^{-1}$)
- High resolution
- Inert
- Universal detection
- Excellent quantitation (in absence of organic modifiers)
- Low operating costs
- Unique applications

Disadvantages

- Low speed (analysis time >60 min)
- Low sensitivity (analyte solution concentration > 10 ppm)
- Selectivity only conferred by (limited range of commercial) stationary phases
- Small sample injection volumes (50–100 nL)
- Low injection precision
- Nonvariable restriction
- Difficult reproducibility of absolute retention times
- Time-consuming hardware changes
- Limited application range

temperature than packed columns. Table 4.28 lists the main characteristics of cSFC.

The main interest in cSFC comes from the high efficiency that can be obtained for involatile samples. The capillary column has the advantage of being able to chromatograph many analytes without additional solvent modifiers. cSFC generates two to three orders of magnitude more theoretical plates for a given separation than a typical packed column of 5 μm particles.

Although cSFC shows relatively poor figures of merit (speed, sensitivity, detection dynamic range and sample capacity) as well as a limited application area, its applications tend to be *unique*. These include solutes that can be solvated with pure $scCO_2$ and quantified with FID. Linear density programs typical in cSFC are ideal for homologous series found in surfactants, many prepolymers, etc. Selectivity in cSFC, which can be achieved by mobile phase density and temperature programming, relies on selective interactions with the stationary phase. Quantitative analysis in cSFC may be rendered difficult by small injected volumes; the use of internal standards is recommended.

As compared to HPLC, cSFC shows higher efficiency, universal and selective detection, minimal derivatisation for separation and the ability to separate thermally labile organic compounds. Often, cSFC analyses are also considerably faster. This arises because higher mobile phase diffusion coefficients translate directly into higher optimum velocities. However, sensitivity, detection dynamic range and sample capacity are lower. cSFC may also be characterised as an extension of GC to larger, low volatility, but mostly thermally stable molecules. In the traditional figures of merit, speed, resolution, and sensitivity, cSFC does not compete well with GC. In principle, the most important niche of cSFD is elution of larger, moderately polar solutes using pure CO_2 and detection with FID. Capillaries tend to work best as an extension of GC into higher molecular weights. A capillary column is the right choice when *all* of the following conditions apply:

 (i) major/minor component analysis;
 (ii) if FID (or other GC detectors) are required;
 (iii) if pressure programming of pure fluids provides an adequate range of solvent strength;
 (iv) if the available instrumentation only allows pressure control.

Although SFC has been promoted as a replacement for HTGC [244], cSFC has essentially lost out against HTGC and pSFC and now seems somewhat ignored.

In *packed column SFC*, with equipment common to HPLC, the SCF has a higher diffusion rate and lower viscosity than most liquid solvents. The separations are therefore more efficient and can be operated at a higher linear flow-rate than LC. Packed columns are commonly used with CO_2 and modifier combinations, gradient elution and UV or ELSD, and are usually prescribed for time and cost savings vs. conventional HPLC. Due to the low polarity of CO_2, pSFC is limited to a normal-phase role. Packed columns (1.0–4.6 mm i.d., 6–25 cm length) are generally similar to HPLC columns with 3–10 μm porous particles, capable of withstanding high pressures. In pSFC very many stationary phases are available from HPLC. The most common stationary phases are silica-based chemically bonded phases; other materials are polar adsorbents such as silica, alumina and polymeric materials. Nearly all recent work on stationary SFC phases has been for packed columns. Derivatisation may be required to reduce the polarity.

A model based on multicomponent solubility parameters may explain retention in pSFC. Selectivity in pSFC can be achieved by adjusting relative retention by balancing the stationary phase polarity against solubility in the mobile phase. Interactions between the stationary phase and the sample solutes are of more significance for packed than capillary columns. This provides a major advantage for pSFC over cSFC in terms of the selectivity towards specific solutes. A mobile and stationary phase selection guide for pSFC is available [26]. Parameters influencing the selectivity in pSFC are pressure and gradients. Packed columns pose much

less of a sample injection problem than open-tubular columns owing to their greater loadability, and direct injection methods have, in fact, proven highly effective and reproducible for packed column use. Commercial instruments with flow-independent automated pressure control, designed for use in pSFC, have re-emerged in the early 1990s, with emphasis on more polar solutes [26]. This is in line with the increasing trend to micro-column chromatography.

The most important area for packed column use involves modified mobile phases (MPs). Consequently, pSFC needs detection systems in which the MP modifier and possible additive(s) do not interfere, and in which detection of low or non-UV-absorbing molecules is possible in combination with pressure/modifier gradients. The disadvantage of adding even small amounts of modifier is that FID can no longer be used as a detector. In the presence of polar modifiers in pSFC the detection systems are restricted basically to spectroscopic detection, namely UVD, LSD, MSD (using PB and TSP interfaces as in LC). ELSD can substitute FID and covers the quasi-universal detection mode, while NPD and ECD cover the specific detection mode in pSFC on a routine basis. As ELSD detects non-UV absorbing molecules dual detection with UV is an attractive option.

The main features of packed column SFC are given in Table 4.29. A few major attributes of pSFC are: independent dynamic pressure and flow controls, common use of binary and tertiary MPs, composition programming preferred over pressure programming, elution of much more polar solutes, trace vs. major–minor component analysis, and UV, ECD, NPD, and (occasional) FID detection. In practice, pump seals in pSFC had to be selected to be resistant to the extraction of additives. Low viscosity leads to low-pressure drop per unit length, which leads to long columns and high-plate counts, when desired. The low viscosity of the mobile phase combined with increased diffusivity often translates to improved resolution and shorter analysis times in pSFC. Other reasons for using pSFC instead of LC include faster column equilibration and method development, lower pressure drop across the column, long lifetime of the column, ease of solvent removal and possibilities for semipreparative fraction collection.

As compared to GC, pSFC possesses inferior figures of merit (efficiency, speed, sensitivity, detection, resolution), but allows a greater sample capacity and is more widely applicable. pSFC can be used as an orthogonal separation technique to verify the accuracy of GC methods. pSFC should be thought of as an extension of LC because of similarities in equipment and approach. The

Table 4.29 Main features of packed column SFC

Advantages

- High sample capacity (20 µL direct injection)
- Reproducible direct injections
- Low mobile phase flow-rate
- Variable restrictor (programming of density and pressure, independent of flow-rate)
- Mobile phase composition programming
- Ease of solvent removal
- Tuneable and unique selectivities
- Full range of detectors – GC and HPLC (volumetric flow compatible with HPLC detectors)
- High detection selectivity
- Rapid analysis (up to 3–5× faster than HPLC)
- Up to 70 % cost-savings vs. HPLC
- Robust equipment
- Potential replacement of NPLC
- Hardware and maintenance as HPLC
- Linked or longer columns for greater resolution and selectivity

Disadvantages

- Relatively low column efficiencies (28 000 theoretical plates m^{-1}, but multicolumns)
- Adsorptive packings (highly active)
- Few stationary phases optimised for pSFC (use of conventional HPLC stationary phases)
- Small range of eluents
- Modifiers required with most available columns
- No FID with most modifiers (ELSD as a poor substitute)
- Difficult routine working level
- Significant pressure drops
- Applicability to small number of substance groups

greatest difference is simply the need to hold the outlet pressure above ambient to prevent expansion of the fluid. In practice it is important to solve the problem of automating the pressure gradient without changing the flow-rate. Optimum flow-rate is about five times higher than in HPLC.

Irrespective of the nominal nature of the stationary phase, SFC is essentially a normal-phase separation mode in which a primary retention mechanism is the interaction of the analyte with the stationary phase. Thus any assay where the reversed-phase mode of separation is preferable, which includes the great majority of analytes, will probably not transfer easily to SFC. As a normal-phase method pSFC cannot compete with the widespread dominance of RPLC, pSFC thus rivals NPLC and is actually superior to LC in terms of speed, efficiency, selectivity, and detection options. CO_2 separations are generally superior to many existing normal-phase methods. Extra benefits inherent to pSFC as compared to HPLC are: less organic solvent, lower running costs, easier coupling with MS and FTIR spectrometers and with light-scattering detectors. Quantification values

(linearity, reproducibility, detection limits, etc.) of pSFC are comparable to those in HPLC. Quantitative reproducibility of pSFC compares favourably to that of GC or LC. Hartkopf [178] has reported the time and area reproducibility (as RSD) of Isonox 129 determined by LC (time 0.5 %, area 5 %) and pSFC (time 0.2 %, area 6 %). pSFC has a great potential in areas where quantitative analysis is needed such as migration studies, process development, in-process and quality control.

Packed columns are the right choice for (moderately) polar solutes, for solutes containing multiple polar functional groups, when modifiers are required or trace analysis is desired, and when high precision or high speed are needed. There is a niche role for pSFC, namely as an easier and more robust method to carry out normal-phase separations than with nonpolar organic eluents. SFC is well suited for oligomers and 'difficult' additives such as phosphites (hydrolysis) and fatty amines. The advantage of SFC is that it encompasses many different additives (e.g. hindered phenolics, nitrogen heterocycles containing amide or ester linkages, thiodipropionates, aromatic phosphites, fatty amide lubricants) on one column with no need for derivatisation.

Both pSFC and cSFC compete with LC, and to a lesser extent, GC. Packed capillary column SFC provides remarkable economy in SFC operation, requiring a few orders of magnitude lower mobile and stationary phase consumptions as compared with conventional packed column SFC. The choice between packed and capillary columns in SFC has been addressed [26]. The question of the 'best' column type for SFC does not arise. pSFC and cSFC tend to best solve different problems. pSFC has little in common and does not compete with cSFC. Many of the approaches used in cSFC are among the worst things to try with pSFCs (and vice versa) [26]. While many workers have extolled the virtues of both open-tubular and packed columns for SFC, few experimental comparisons are available wherein typical operating conditions and identical parameters (where feasible) are employed. However, Schwartz *et al.* [245] have compared the two types of columns based on theoretical considerations. Because of its higher capacity (sample loadability), faster elution times, better reproducibility, easier column handling and ruggedness, pSFC is generally preferred over cSFC. In SFE-SFC-MS applications cSFC is the dominant technique because it easily couples to MS. However, pSFC on microbore columns and pcSFC are good alternatives to improve the limit of detection and analyse additives present in very low amounts (as in the case of migration studies).

SFC requires polar *stationary phases*. Bare silica is widely used. The most appropriate bonded phases include cyanopropyl, aminopropyl, diol, and ion-exchange columns. The availability of a wide range of bonded phases and column types has extended the applications of SFC to many different chemical systems. The most commonly used *mobile phase* for SFC is $scCO_2$, although ethane, pentane, dichlorodifluoromethane, diethylether and THF have also been employed. In SFC the mobile phase is kept under supercritical conditions by means of a restrictor. CO_2 does have some limitations, such as reaction with primary and secondary amines to form ureas and carbamates. The quality of CO_2 that is needed depends on the type of detector used and sensitivity required (99.9995 % for FID, 99.95 % for UVD). The critical properties and chromatographic behaviour of scN_2O are very similar to $scCO_2$. Use of scN_2O offers no particular advantages in detection; reaction with primary amines is avoided. Because of the complications of using single-component polar solvents, organic *modifiers* are used to enhance the solvent strength of low-polarity fluids, such as CO_2. Successfully used modifiers are methanol, ethanol, THF and acetonitrile; hexane and MTBE are less frequently used [246]. The older literature has also mentioned dichloromethane, propylene carbonate, formic acid, water, 1-propanol and 2-propanol. Blackwell *et al.* [247] determined the relative eluotropic strengths of various modifiers in SFC. The interesting point about supercritical fluids is the possibility to change their density, and consequently their solvation power, simply by modifying the pressure. A pressure gradient in SFC is equivalent to an elution gradient in HPLC or a temperature gradient in GC. Programming methods for cSFC and pSFC are extremely versatile [248] and involve temperature, composition, pressure, density and flow. Of these possibilities pressure or density gradients and simultaneous density–temperature or pressure–temperature programming are the more practical. In this way it is possible to fine tune the retention of analytes, thus modifying the selectivity. In pSFC gradient elution is most common (usually isothermal and isobaric, but may be combined with pressure programming: HPLC like). It is important to remember, however, that supercritical fluids can have the properties of *either* gases or liquids according to temperature and pressure, but never of both simultaneously. The combination of low polarity mobile phase with a relatively polar stationary phase means that most SFC separations involve normal-phase partition or adsorption. In the absence of specific interactions, elution time is then a function of molecular mass and polarity.

Retention in SFC is not as easily rationalised as in GC or LC. It is dependent upon temperature, pressure, mobile phase density and composition of the stationary and mobile phases. Many of these experimental variables are interactive and do not change in a simple or easily predictable manner. For a given column, *solute retention* is affected by four major factors: nature of supercritical fluid (solubilising power), pressure, temperature, and gradient profile. The most confusing single aspect of retention in SFC involves comparing retentions on capillary and packed columns. Typical packed columns are 10 to 100 times more retentive than capillaries. Using the same fluids for both types of columns will therefore produce very different retention characteristics. Berger [26] has addressed the effects of the instrumental variables on retention, selectivity and efficiency in SFC. These parameters (in order of decreasing importance) are: percentage modifier, p, T, flow (for retention); T, p, percentage modifier, flow (for selectivity); and flow, p, T, percentage modifier (for efficiency).

A powerful advantage of SFC is that more detectors can be interfaced with SFC than with any other chromatographic technique (Table 4.30). There are only a few *detectors* which operate under supercritical conditions. Consequently, as the sample is transferred from the chromatograph to the detector, it must undergo a phase change from a supercritical fluid to a liquid or gas before detection. Most detectors can be made compatible with both cSFC and pSFC if flow and pressure limits are taken into account appropriately. GC-based detectors such as FID and LC-based detectors such as UVD are the most commonly used, but the detection limits of both still need to be improved to reach sensitivity for SFC compatible with that in LC and GC. Commercial cSFC-FID became available in 1985 and pSFC-PDA in 1992. When CO_2 is used as the mobile phase, FID provides excellent quantification

Table 4.30 Detectors in SFC[a]

	Available	Developmental
GC type	FID (10), FPD (1000), ECD (1), TID/NPD (50), CLND (60), SCD (−), RCD (−)	SCD (75), PID (100), IMS (50), ICP-MS (?), NMR (?)
LC type	UVD (100), RI (10 000)	FLD (10), ELSD (5000)
Coupled	DAD (100), evaporative IR (2000), on-column IR (10 000), MS (20), AED (10)	

[a] Approximate detection limit (pg) in parentheses.

due to the direct relationship between detector response and mass of the analyte. UVD is particularly popular for those conditions incompatible with FID, i.e. with wide-bore columns and for mobile phases containing organic solvent modifiers. The most difficult problem associated with SFC detection is mobile phase compatibility. scCO$_2$ is compatible with both FID and UV absorption detection. Interferences occur with other detection systems. For example, with IR detection CO$_2$ obscures broad areas of the usable spectra. As FID responds to virtually all organic solvents, addition of a modifier into the MP leads to baseline variations and poor sensitivity.

Universal SFC detectors are FID, PID and ELSD. SFC-ELSD is a promising technique for analysing polar compounds that have no response in the UV region [249,250]. ELSDs work very well with SFC instruments because of the ease of evaporating the mobile phase and can be used with a high mole fraction of modifier. Among the spectroscopic detectors (UVD, FLD, MSD, FTIR, NMR) MS and FTIR are also frequently used (Section 7.3.2). Spectral detectors for SFC fall into categories of ion detectors (MS, IMS) and optical detectors and are operated at high (e.g. FTIR), ambient (e.g. FTIR) or low (e.g. MS) pressures. Optical detectors must have very small volumes to be able to withstand high pressures. Plasma emission spectrometry is also used for detection in SFC, such as ICP [251,252], RF discharge [253] and various forms of microwave induced plasma [254–260], but merely represents a developmental stage. Plasma-based detectors show good detection limits, precision and/or accuracy, wide dynamic ranges, and capabilities for simultaneous multielement analysis. In addition, complete chromatographic resolution is not required because of their elemental selectivity. Most SFC-ICP-MS papers deal with the separation of organometallic compounds.

Selective SFC detectors are either (prevalently) single-element, such as O-FID, TID/NPD, FPD, SCD, ECD, IMS, or multielement/plasma detectors such as ICP-AES/MS and MIP-MS. Few selective detectors have been widely studied. As an SFC detection method, ion mobility spectrometry (IMS) has several unique advantages, including: (i) nonselective (FID-like) responses for most compounds; (ii) selective (ECD-like) responses for electronegative compounds; (iii) selective detection of compounds not containing heteroatoms; (iv) detection of compounds not containing chromophores; and (v) detection of compounds contained in mobile phases in the presence of organic modifiers [261,262]. Moreover, ion mobility data can be used to suggest unknown identities. When matched with

chromatographic and ion mobility data from standards, confirmation of identifications can be achieved. Detection limits, mobile phase compatibility, and response versatility of FT-IMS exceed most inexpensive detection methods available for SFC today. No fully optimised SFC-FTIMS instrument is available. Selective detectors, besides UVD, are appropriate when attempting to characterise multicomponent mixtures. Dual detection such as SFC-UV-ICP has also been described [263]. Electrochemical or additional separation methods may also be used following SFC to obtain qualitative information about the separated components.

With recent emphasis shifting strongly to packed columns it is not surprising to notice a shift in detector development from GC- to LC-like detectors. Several authors [36,46,236,238,264] have discussed detection in SFC.

Method development in SFC needs to consider the choice of the mobile phase (MP), column and detector type, stationary phase (SP), selectivity tuning, and involves pressure/density, temperature and MP (including modifier) as the main variables. In general, to minimise time and costs, conditions need to be selected that: (i) achieve high selectivity; (ii) minimise mobile phase viscosity; (iii) maximise the diffusion coefficient; and (iv) use the lowest modifier concentrations and highest temperatures possible, being mindful of selectivity, thermal stability and peak shapes. Matching of the two most appropriate phases (MP and SP) for a separation is not, necessarily, trivial. Many practical guidelines allow prediction of near optimum phases [26]. For *performance testing* Grob's mix will not work in SFC because it is not retained well enough. An alternative is anthracene, *n*-eicosanol, *n*-docosanoic acid, *n*-triacontane and tridodecylamine at 1 mg mL^{-1} (dissolved in 10 % MEK in toluene) for 0.1 µL direct injections.

General characteristics of SFC are given in Table 4.31. An important *advantage* of SFC is the rapid column re-equilibration with both normal- and bonded-phase systems following a programme run. Since separations are carried out at low temperatures, SFC is considered as a separation technique that allows analysing thermally labile and nonvolatile compounds not amenable to GC and with greater efficiency and detector compatibility than LC. The technique is also perceived as a method for nonpolar analytes but with poor reproducibility and numerous operational problems even though the latter have now largely been removed. As a rule of thumb substances with some solubility in methanol or a less polar organic solvent are likely to be separated by CO$_2$-based mobile phases. As a corollary, substances requiring aqueous conditions to

Table 4.31 General features of supercritical fluid chromatography

Advantages

- Analysis of thermally labile and nonvolatile compounds
- Easily adjustable selectivity
- High sample loadings (pSFC)
- Preparative analysis tool
- Variable solvation power
- Detector friendly (wide choice)
- High resolution (cSFC)
- More versatile than HPLC
- Straightforward method development (less time-comsuming than for HPLC or GC)
- High throughput
- Interfacing to hyphenated techniques

Disadvantages

- Normal-phase technique
- Limited choice of mobile and stationary phases
- Difficult application to highly polar systems
- Instrumentally more complicated than HPLC
- Plugging at restrictor
- Poor injection precision (cSFC)
- Long analysis times (cSFC)
- Retention prediction
- Expensive equipment
- Limited choice of surviving suppliers

dissolve are poor candidates for separation by CO_2-based SFC. The low viscosity of the mobile phase permits connection of several columns in a series. This explains why several studies using SFC are directed at analyses of oligomers. Because it is more rapid than HPLC, SFC can be used for process analysis. With small injection volumes and split injection mode used in SFC the analytical precision valves are about 3–5 % relative. The precision obtained in SFC is a significant *limitation* of the technique. Adoption of SFE and SFC as official methods by regulatory authorities has been slow.

Supercritical fluid techniques (SFE and SFC) have the potential to fill the analytical grey area between GC and LC. Supercritical fluids should be regarded as a unifying point in chromatography, bridging liquid and gas chromatographic MPs. The solubility properties, physical properties (viscosity and diffusion coefficients) and detector compatibility of the SCF may combine to make SFC more suitable for a given application than either GC or HPLC; the techniques are complementary. In LC and SFC, volatility is not a prerequisite and separation is influenced by differences in analyte affinity for the SPs and MPs. The differences between SFC and GC or LC (the limiting cases of unified chromatography) are more practical than theoretical. SFC possesses inferior figures of merit compared to GC but SFC (in particular pSFC) is more widely applicable. On the

other hand, from the physical properties cSFC can never compete with GC in speed. SFC provides the high resolution of GC along with the ability to analyse less volatile compounds as done by HPLC. In comparison to LC, SFC possesses superior figures of merit but is less applicable than LC. The differences between the techniques have often been exaggerated. The primary differences are higher diffusivity and lower viscosity than liquids, allowing faster analysis and/or higher resolution than LC. A major advantage of SFC over HPLC is that SFC is truly 'detector friendly'. Another advantage of SFC over HPLC is the possibility of using modifier and density gradients or a combination of both. Several SFC-LC comparisons have appeared [265–269]. Contrary to common belief, SFC is not displacing HPLC and has not reached the once expected level of broad acceptance; SFC is not a popular analytical instrument. At most the normal-phase SFC method may be expected to reach the level of application of NPLC. Frey [270] has discussed the application of packed-column SFC as a 'universal' SFC/HPLC system.

Although SFC fills a niche in what can be considered as a continuum of separation eluents from gases to liquids, it cannot claim a unique status: subcritical water extraction (SWE, cf. Section 3.4.3) and pressurised fluid extraction (PFE, cf. Section 3.4.6) are other examples of eluents where altering the conditions cause a useful change in the solvation properties.

The future of SFC will focus most on packed columns, modified CO_2, and a variety of either universal, element specific, or spectrometric detectors. Consequently, SFC will become more viewed like HPLC and may, in fact, replace HPLC in a selected number of applications because SCFs have both better mass transport properties than liquids and are less harmful to the environment [245]. There remains a significant group of applications in both chromatography and extraction for which supercritical fluids are the most suitable solvents. Complex mixtures, which cannot be suitably analysed by either GC (because of involatility or instability) or HPLC (because of complexity and long retention) may be more effectively analysed by SFC. SFC seems doomed to a *niche status* between GC and HPLC with applicability to a restricted number of substance groups in which the technique is superior to GC or HPLC [271]. In general, SFC is supposed to perform especially well in case of additives that have poor solvent solubility, no chromophore, high-MW (if soluble in $scCO_2$) or are nonvolatile or thermally labile. A decline in SFC activities is noticeable since the early 1990s. The major future of the supercritical fluid methodology is most likely as an extraction technique for sample preparation (cf. Section 3.4.2) rather than in chromatography.

SFC has been widely reviewed [272–283], with particular emphasis on pSFC [284–286] and cSFC [287], SFC principles and techniques [246], instrumentation [272,288,289], columns and phases [283,290], detection [236,291], past and future [292]. Various recent monographs deal with supercritical fluids, SFE and/or SFC [293–303], some specifically with pSFC [26,304].

Applications SFC now accounts for less than 5 % of the HPLC market. Areas of greatest impact of SFC are petrochemical separations, agrochemical and soil analysis, food analysis, and analysis of natural products and synthetic polymers. Most drug compounds are too polar for routine SFC. For chiral chromatography and metal complex analysis and applications, which favour the normal-phase mode, SFC offers particular advantages over existing methods. Most other applications concern samples that are poorly handled by current GC or HPLC technology. In some cases, SFC is favoured by absence of UV absorbing chromophores in the analyte, which complicates detection in HPLC. In many instances, however, analytes are so complex that it may require considerable effort to develop a suitable HPLC method.

Use of SFC in the polymer industry has been described as one of the growth areas of application of cSFC [305]. The most important applications are molecular mass distribution of low-MW polymers and characterisation of monomeric residuals, polymer additives and stabilisers. Generally, HTGC provides higher resolution in a shorter analysis time, but SFC expands the analysis range towards higher-MW, non-volatile components, enabling elution of thermally labile or moderately polar components. Such SFC analyses are preferably performed at the highest temperature

that samples can withstand without risk of degradation. Discrimination of high-MW vs. low-MW components in SFC is controversial and still needs to be addressed.

Some typical areas for SFC comprise waxes, surfactants and dyes. Marked advantages of SFC over GC in the analysis of surfactants (e.g. ethoxylates) have been reported [246]. It is arguably the best chromatographic method for the separation of *nonionic surfactants*, as reviewed by Cserhati and Forgacs [306]. pSFC-UV-ELSD has been used for separation of oligomers of the nonionic surfactant Triton X-100 [249].

Oligomers are perhaps best analysed by SFC as it is applicable for polymers that lack the volatility necessary for GC analysis and which prove difficult to characterise by HPLC. SFC is best suited for the analysis of *oligomers* with molecular masses less than 1000 Da; higher-MW (up to 10 kDa) oligomers are generally amenable to analysis by LDMS [307]. A direct comparison was reported on low-MW PDMS oligomers in the mass range from 1000 to 10 000 Da by means of SFC and ToF-SIMS [308]. Results fit well up to mass 3000 Da although remarkable differences were noticed for higher masses due to mass discrimination effects which are more pronounced for SFC than for ToF-SIMS. In contrast to NMR spectroscopy both methods SFC and ToF-SIMS give information on number as well as weight average molecular weights of PDMS oligomers, and therefore the polydispersity of a molecular weight distribution can be assessed.

In the analysis of synthetic lower-MW oligomers SFC offers considerable advantages over SEC [309]. Examples are polysiloxanes (MW < 25 kDa) [310] (cf. also Figure 4.4), polyethers (nonionic surfactants), polyesters (up to 10 kDa) [311], PEGs [312] and poly(propyleneglycol)s [246]. Selective IMS detection of nonhetero-containing compounds has been accomplished for oligomers in various polymeric compounds, including Triton X-100/114/305 [261,313],

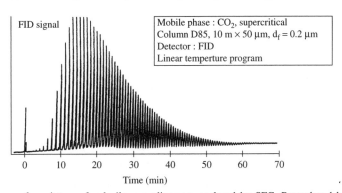

Figure 4.4 Chromatogram of a mixture of polysiloxane oligomers analysed by SFC. Reproduced by permission of Thermo Electron Corporation

PS-800 [313] and PDMS [314]. pSFC elutes polydecenes up to C_{140} and other poly(alpha-olefins) (PAOs), in use as mineral oil substitutes in lubricants, up to about 3000 Da [178]. Conventional GC can elute only the lower-MW portion of most PAOs. Lynch [315] and others [316,317] have reported separation of polystyrene oligomers by pSFC employing UVD and LSD. pSFC can determine all relevant oligomers in commercial polystyrene (up to tetramers and pentamers), whereas GC is usually limited to dimers and trimers. Resolution is better than typically achieved by LC [178]. SFC-FID and SFC-UV (276 nm) have been used for analysis of oligomers of PMMA and other acrylic resins, potential sources of odour problems and migration into foodstuffs from acrylic kitchenware [318].

Many SFC applications in the literature concern the analysis of UV-active *additives in polymers* and coatings after extraction [319–335]. On-line SFE-SFC has also been used [319,321–325,333–335] (cf. Section 7.1.1.4). cSFC is often a matter of choice because every kind of polymer additive can be detected with FID [326,328,330,331,335–337]. Some publications show interesting examples of pSFC with UV detection [329,332,337–339]. Figure 4.5 shows the pSFC separation developed for routine formulation control of

a polymer additive mixture [338]. Although the control is usually done by HPLC, special chromatographic conditions are needed for the calcium salt Irganox 1425. A pSFC method allows analysis of all components in the mixture in one run within 6 min.

Berger [340] has examined the use of pSFC in polymer/additive analysis. As many polymer additives are moderately polar and nonvolatile SFC is an appropriate separation technique at temperatures well below those at which additives decompose [300,341,342]. SFC is also a method of choice for additives which hydrolyse easily. Consequently, Raynor *et al.* [343] and others [284,344] consider that SFC (especially in combination with SFE) is the method of choice for analysing polymer additives as a relatively fast and efficient sample preparation method. Characterisation of product mixtures of nonpolar to moderately polar components encompassing a wide range of molecular masses can be accomplished by cSFC-FID. Unknown polymer additives may be identified quite adequately by means of cSFC-FID by comparison with retention times of standards [343]. However, identification by this method tends to be time-consuming and requires that all the candidate compounds are on hand. SFC-FID of some low-to-medium polarity additives on reversed-phase packed columns

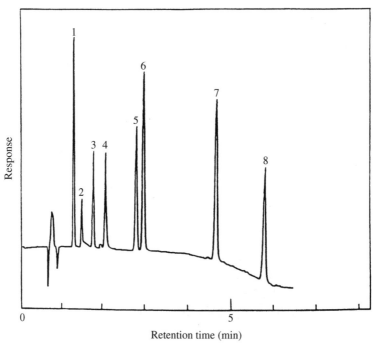

Figure 4.5 pSFC-UV chromatogram (278 nm) of a synthetic mixture of polymer additives. 1, Metilox; 2, Irgafos 168; 3, Irganox 1076; 4, Irganox 1425; 5, Irganox 259; 6, Irganox 1330; 7, Irganox 1010; 8, Irganox 1098. After Giorgetti *et al.* [338]. Reproduced from the *Journal of Chromatographic Science* by permission of Preston Publications, A Division of Preston Industries, Inc.

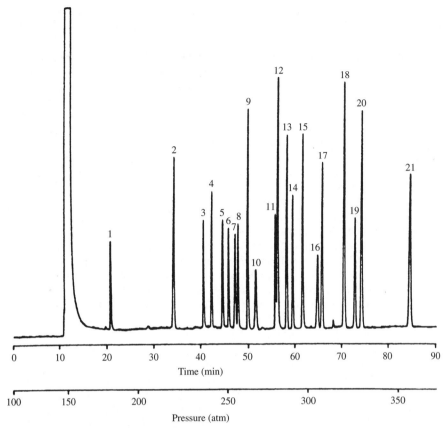

Figure 4.6 cSFC-FID chromatogram of a synthetic mixture of polymer additives. 1–21, Topanol OC, Tinuvin P/292/320/326 /328, Chimassorb 81, erucamide, Tinuvin 770/440, Irgafos 168, Tinuvin 144, Irganox PS 800/1076/MD 1025/245/1035/3114/PS 802/1330/1010, in this order. For conditions see Raynor *et al.* [343]. Reprinted with permission from Raynor *et al.*, *Analytical Chemistry*, **60**, 427–433 (1988). Copyright (1988) American Chemical Society

and open-tubular capillary columns using both scCO$_2$ and scN$_2$O have been compared [337]. Greibrokk *et al.* [345] have described polymer/additives assays by cSFC.

Figure 4.6 shows an example of cSFC-FID characterisation of a complex mixture of polymer additives, analysed without apparent degradation of any of the additives. In view of the limited number of additives in a given polymer the greater speed of analysis afforded by lower resolution pSFC is often preferred to cSFC with high theoretical plate numbers and chromatographic resolution. More polar compounds are usually analysed by pSFC-UV. pSFC of Isonox 129, erucamide, Irgafos 168, BHT and Irganox 1010/1076 clearly illustrates the problems that may be encountered with basic and/or highly polar compounds on packed columns, namely strong absorption of polar compounds such as aliphatic amines and amides on the residual silanols of bonded phases for packed column SFC; while peak

shape is poor, polar compounds can be eluted [178]. Bartle *et al.* [332,346] have reported the separation of as many as 21 additives (mainly AOs and light stabilisers with a wide range of molecular weights and boiling points) on various columns coated with methyl-, octyl- and biphenyl-substituted polysiloxane stationary phases using both pSFC-FID and pSFC-UV operated with (modified) scCO$_2$ at 140 °C. Certain additives, e.g. erucamide and Irganox PS 800, require methanol modification for elution. No one packed column permitted analysis of all of the additives present in the samples of commercial PP with CO$_2$ as MP. The information obtained with SFC-FID, SFC-UV or HPLC-UV is usually not specific enough for accurate qualitative identification in the case of complex additive mixtures. In certain circumstances retention time data can result in erroneous identification (cf. Raynor *et al.* [343]). For example, on a methylpolysiloxane-coated column Tinuvin 144 can be confused with a degradation product of Irgafos 168. For positive identification SFC-FTIR is needed; SFC can use dual FID and FTIR or MS detection.

As most of the additives in *polyolefins* are of medium polarity and soluble in $scCO_2$, SFC is suitable for their analysis. Apart from Raynor *et al.* [343], the high potential of cSFC for determination of plastic additives was also demonstrated by Arpino *et al.* [347] for the synthetic mixture of BHT, Irganox 1010/1076/1330/PS 800/PS 802/2246, Tinuvin 329/326/P and Irgafos 168. The FID detector permits a uniform response for all the model additives, including non-UV/VIS absorbing substances. In real extraction conditions co-extracted oligomers disturb FID detection [348]. It was reported that before SFC it is necessary to precipitate and filtrate co-extruded propylene oligomers from packaging films [331]. The large number of peaks observed in pSFC-FID analysis of a LLDPE/(BHT, Irganox 1010/1076) extract was attributed to oligomer and/or degraded additives [178]. Quantitation of Irgafos 168 by SFC has been complicated by its degradation and hydrolysis products [343].

A comparison was also made between the separation of *AOs and UV stabilisers* used in the manufacturing of food packaging materials by NPLC under isocratic or gradient elution and by SFC with $scCO_2$ [347]. SFC proved to be more efficient for separation and determination of additives with a wide range of boiling points and for substances lacking a UV chromophoric moiety. Dilettato *et al.* [330] described the use of Soxhlet extraction of ground PE pellets followed by TLC, cSFC and GC-MS, and of PP pellets followed by cSFC. cSFC-FID is perfectly suited for characterisation of PE/(Irganox 1010, BHT, stearic acid) and PP/(Atmos 150, Irganox PS 802/1010) extracts [330]. On the other hand, HPLC-RID did not provide satisfactory results for these materials owing to poor resolving power of the technique and low sensitivity of the RI detector (used for lack of any UV chromophoric moiety in the glycerides and Irganox PS 802). Identification of co-extruded low-mass oligomers and degradation products of Irganox 1010 was carried out by GC-MS. cSFC of PP/(Irgafos 168, Irganox 1010, Tinuvin 770) was also reported [349]. LVI-cSFC was applied to a 10 μL injection of the hindered phenols Irganox 1010/1076/1330 at a 1–2 ppm concentration level with RSD = 0.16 % [242]. By using isopropanol as the solvent, SFC-ELSD has quantified Irgafos 168 down to 5 ng levels [350]. pSFC has been used to separate a synthetic mixture of polymer additives (Metilox, Irgafos 168, Irganox 259/1010/1076/1098/1330/1425) in 6 min [338]. The calcium salt Irganox 1425 cannot usually be

eluted with other Irganox family members by HPLC. The separation of the hydroxylamine additive Irgastab FS042 is an instance where SFC presents a distinct advantage over other chromatographic methods. This hydroxylamine is a process stabiliser that undergoes transformation during processing. The compounds to which it transforms depend on the polymer type, residual catalyst concentration, and processing temperature. SFC has been shown to be the only chromatographic technique capable of analysing not only the original additive, but also the transformation products [351].

Berg *et al.* [328] have used SFC-FID and GC-MS for the characterisation and identification of migrants (Irganox 1010, Irgafos 168, N,N'-bis(2-hydroxyethyl)-C_{12}-C_{14} amine, GMS, GDS and calcium stearate) from PP to an acid-based food simulant; SFC-FID was utilised for determination of the mass balance and Irgafos 168, and GC-MS for the derivatised amine. As shown by SFC, Irgafos P-EPQ, used as a processing stabiliser, is a mixture of seven compounds, namely Irgafos 168 (26.2 %), 4,3'-mono-P-EPQ (1.3 %), 4,4'-mono-P-EPQ (8.8 %), oxidised 4,4'-mono-P-EPQ (0.8 %), 3.3'-P-EPQ (1.3 %), 4,3'-P-EPQ (15.4 %), 4,4'-P-EPQ (27.4 %), and BHT (18.8 %), with the latter volatile compound being identified by GC [352]. The quantitative composition of the mixture was calculated by the peak areas of the SFC-FID chromatogram. Chimassorb 944 is not readily soluble in $scCO_2$, even in the presence of a modifier, and cannot easily be chromatographed by means of SFC. pSFC-FID of diorganotin compounds, used as heat and light stabilisers in rigid plastics such as PVC, was also reported [353].

Hunt *et al.* [354] used cSFC for the separation of extracts of poly(alkylene glycol) *lubricants* and sorbitan ester formulations. Doehl *et al.* [337] have compared the performance of cSFC-FID and pSFC-FID with both $scCO_2$ and scN_2O in the analysis of the *antiblocking agents* oleamide and erucamide, the antistatic Armostat 400 and antioxidant Hostanox SE-10, none of which can be detected by UV absorption. By using open-tubular capillary columns, PAs as well as (un)substituted heavy carboxylic acids ($\geq C_{18}$) can be eluted.

SFC has played an important role in the extraction and isolation of *fatty acids* [355,356]. Underivatised fatty acids and methyl esters of fatty acids are surprisingly easy to elute using a bonded phase or a silica based packed column and pure CO_2, probably due to the long hydrocarbon tails on the molecules [357]. On the other hand, most aromatic and polysubstituted acids will not elute. Triglycerides with saturated fatty acids can be analysed faster with pSCF-ELSD than with GC-FID and do not require sample preparation [358]. Using cSFC, Moulder *et al.* [331] avoided the derivatisation step by extracting metal salts with acidified ethanol, which yields ethyl esters of the acids directly. Long chain fatty amines cannot be eluted using pure CO_2 and cSFC [359].

SFC-FID is widely used for the analysis of (non-volatile) *textile finish components*. An application of SFC in fuel product analysis is the determination of *lubricating oil additives*, which consist of complex mixtures of compounds such as zinc dialkylthiophosphates, organic sulfur compounds (e.g. nonylphenyl sulfides), hindered phenols (e.g. 2,6-di-*t*-butyl-4-methylphenol), hindered amines (e.g. dioctyldiphenylamines) and surfactants (sulfonic acid salts). Classical TLC, SEC and LC analysis are not satisfactory here because of the complexity of such mixtures of compounds, while their lability precludes GC determination. Both cSFC and pSFC enable analysis of most of these chemical classes [305]. Rather few examples have been reported of thermally unstable compounds analysed by SFC; an example of *thermally labile polymer additives* are fire retardants [360]. pSFC has been used for the separation of a mixture of methylvinylsilicones and peroxides (thermally labile analytes) [361].

Some selected applications (chemical class, mobile phase, stationary phase, column, detector) are as follows: fatty acids, CO_2, PEG or CN-PS, OTC, FID; organotin, CO_2, PhMe-PS, OTC, FPD [305]. The pSFC-ICP system is an effective tool for *speciation* of metal-containing compounds [362], even in the presence of modifiers to control the retention of components.

Supercritical fluid chromatography has some inherent restrictions. As the polarity of CO_2 is low, many analytes of interest are simply insoluble and cannot be analysed. Also compounds with acid and basic functional groups in their structure, and compounds which are water-soluble only cannot be satisfactorily analysed using pSFC techniques [363].

General SFC applications were reviewed [305]. Berger [364] has reviewed the separation of 'polar' solutes (mainly of pharmaceutical interest) by pSFC; here 'polar' stands for solutes that do not elute or elute with poor peak shapes from packed columns, using pure CO_2 as the mobile phase. A recent review has focused on the use of pSFC in the analysis of additives in plastics, rubbers, textiles and other polymeric materials, with emphasis on process development (Irganox 1010, Chimassorb 944, Wingstay L), troubleshooting of production processes, and development of new polymer additives (Tinuvin 360/400/1577) [284]. No examples were found where it would have made sense to replace an exciting GC method by a pSFC method. When GC did not work, as, for example, in the separation of Tinuvin P, cSFC is probably the best choice. When cSFC has some limitation, as in the separation of Tinuvin 1577, pSFC can bring a real improvement. In most cases pSFC was used as an orthogonal separation technique to validate or improve existing methods (HPLC, GC), to 'de-bottleneck' the capacity of the laboratory in taking advantage of much shorter equilibration, to reduce method development and/or analysis time, and to reduce the consumption of combustible organic solvent.

Monographs dealing with SFC applications have appeared [304,342]. In spite of all the various advertised advantages, SFC techniques take a long time to take off in industrial additive analysis (cf. Table 1.13). Few robust SFC applications were developed.

4.4 LIQUID CHROMATOGRAPHY TECHNIQUES

Principles and Characteristics Liquid chromatography is the generic name used to describe any chromatographic procedure in which the mobile phase is a liquid. It may be classified according to the mechanism of retention in adsorption, partition, size-exclusion, affinity and ion-exchange (Scheme 4.4). These mechanisms form the basis for the chromatographic modes of

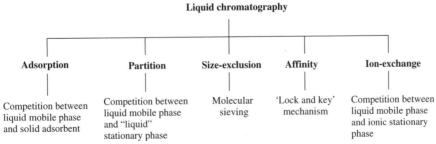

Scheme 4.4 Classification of liquid chromatography according to the retention mechanism. After Weston and Brown [365]. Reprinted from A. Weston and P.R. Brown, *HPLC and CE. Principles and Practice*, Academic Press, Copyright (1997), with permission from Elsevier

normal-phase (NP) and reversed-phase (RP) and make possible the separation of a wide variety of solutes, including ionic, ionisable, polar, nonpolar and polymeric compounds. Partition chromatography, which has become the most widely used of all liquid chromatographic procedures, is conveniently divided into liquid–liquid chromatography (LLC) and bonded-phase chromatography (BPC). The difference between the two lies in the method by which the stationary phase is held on the packing. In liquid–liquid partition packings, retention is by physical adsorption, while in bonded-phase packings, covalent bonds are involved. Liquid chromatography comprises all of these techniques, including recently introduced developments, such as solid-phase extraction (SPE).

Liquid chromatography occurs basically in two forms, as layer chromatography, where the stationary phase is arranged in the form of a planar bed, or as column liquid chromatography, where the stationary phase is confined in a tubular column. In planar chromatography the mobile phase normally is driven by capillary action, while in column liquid chromatography the mobile phase is pumped through the bed of a stationary phase. Table 4.32 compares planar and column chromatographies.

The major attribute that distinguishes planar techniques from column chromatography is that in the former separation and detection are discontinuous ('off-line'). In column chromatography analytes are carried through the entire column and monitored at the end, usually by flow-through detectors measuring changes in some physical characteristics of the effluent (optical refraction, UV absorption, etc.). A comparison between planar and column chromatography has been given [366].

Applications Figure 4.7 shows the general areas of application of liquid chromatography techniques.

4.4.1 Planar Chromatographies

Principles and Characteristics Planar chromatographic methods include *paper chromatography* (PC), thin- and thick-layer chromatography, and electrochromatography, all with rather simple instrumental requirements. Each method makes use of a flat, relatively thin layer of material that is either self-supporting or is coated on a glass, plastic, or metal surface. The mobile phase moves through the stationary phase by capillary action (controlled by capillary pressure Δp), sometimes assisted by gravity or an electrical potential. Planar chromatography has several advantages over column separations (Table 4.33).

Although the interest in, and application of layer chromatography has historically resulted from the development of PC, it was soon replaced by *thin-layer chromatography* (TLC). In PC, only one stationary phase matrix is available (cellulose), at variance to TLC (silica, polyamide, ion-exchange resins, cellulose). Using a silica-gel plate, separation of a sample can be accomplished in approximately 1 h as compared with many hours on paper. The plate size is much smaller than the necessary paper size. Also, more samples can be spotted

Table 4.32 Complementary features of planar and column chromatographies

Feature	Planar chromatography	Column chromatography
Conditions	Atmospheric pressure (capillary forces)	External force (pressure)
Theoretical plate number	Low	High
Separation performance	Limited	Relatively large
Stationary phase	Accessible (during/after development)	Inaccessible
Separations	Simultaneous (including standards, calibration)	Sequential
Separation mechanisms	Few	Many
Prevailing separation mode	Normal-phase	Reversed-phase
Sample throughput	High	Low
Use of stationary phase	Single	Multiple
Unit operating costs	Low	High
Matrix contamination	No problem	Problematic
Sample preparation	Low	High
Separation mode	Development mode	Elution mode
Detection	In presence of stationary phase	In presence of mobile phase
Detection strategy	Highly flexible (sequential detection)	On-line
Sample integrity	Guaranteed	Implied
Automation	Low level	High level
Detection possibilities	Low	High

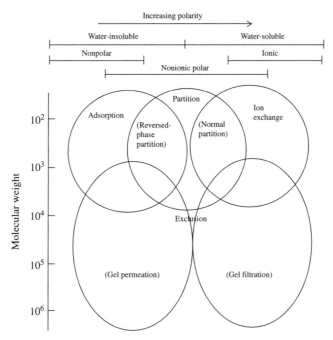

Figure 4.7 Applications of liquid chromatography. After Saunders [367]. Reproduced by permission of International Thomson Publishing

Table 4.33 Advantages of planar chromatography

- Simple to use
- Small sample sizes
- Reproducible R_f values
- Simultaneous separation of several samples
- 2D development
- Detection by specific/sensitive colour reagents
- Specific contact detection (x-ray film, digital autoradiography)
- Visual detection of UV-absorbing compounds

on a single TLC plate than on the corresponding paper. Another advantage of TLC is the availability of plates containing fluorescent reagents that allow direct evaluation of developed plates under UV light. In PC the chromatogram has to be sprayed with a reagent to make colourless analytes visible [366]. Paper chromatography has lost favour compared to TLC because the latter technique is faster, more efficient, and allows more versatility in the choice of stationary and mobile phases.

4.4.1.1 Paper Chromatography

Principles and Characteristics Paper chromatography, which was developed in 1944 [10], is the simplest and cheapest form of chromatography and has been used quite extensively in the past. For most applications the method has now been replaced by TLC and various forms of electrophoresis. It is covered here mainly for historical reasons.

Paper chromatography is essentially a liquid–liquid partitioning technique between two (usually immiscible) phases, namely a stationary polar solvent and a mobile, less polar organic solvent. Although the chromatographic bed is composed of a sheet of paper (i.e. highly purified cellulose), the stationary phase essentially consists of water adsorbed to the cellulose as well as of the polymer itself. Other liquids can be made to displace water, however, thus providing a different type of stationary phase. For example, paper treated with silicone or paraffin oil permits RP-PC, in which the mobile phase is a polar solvent. Also available commercially are special papers that contain an adsorbent or an ion-exchange resin, thus permitting adsorption and ion-exchange paper chromatography. However, in normal operation, most solvents used in PC contain some water. Consequently, components that are highly water-soluble or exhibit the greatest hydrogen-bonding capacity move more slowly. PC is performed on sheets or strips of paper (typically Whatman No. 1 or No. 3 filter paper). Low-porosity paper will produce a slow rate of movement of the developing solvent while thick papers have increased sample capacity. The sample, dissolved in a volatile solvent, is applied as a spot near the end of the paper (optimum spot diameter 2 mm, maximum sample quantity 500 µg). Separation takes place in a closed chamber using either ascending flow (capillary action) or descending mobile flow (capillary and gravity action) with the bottom and/or top of the paper immersed into the mobile phase. When the mobile phase has almost reached the opposite end of the sheet the paper is removed from the developing tank, the solvent front is marked and the paper is dried (2–4 h). Coloured substances are observed visually; fluorescing compounds can be detected under a UV lamp and labelled (radioactive) components by an appropriate counter or scanner. Other invisible substances may be visualised by chemical means. It is possible to obtain reproducible and accurate quantitative results in PC using standardised conditions. In cases of difficult separation 2D PC can be used, which consists of developing a square sheet of paper subsequently in two perpendicular directions. The technique takes advantage of the different separation abilities of various mobile and stationary phase systems.

The *main features* of PC are: low cost, need for small sample amount, high level of resolution, ease of detection and quantitation, simplicity of apparatus and use, difficult reproducibility (because of variation in fibres) and susceptibility to chemical attack. Identification of the separated components is facilitated by the reproducible R_f values. Detection methods in PC have been reviewed [368]. Fluorescence has been used for many years as a means of locating the components of a mixture separated by PC or TLC. However, also ATR-IR and SERS are useful. Preparative PC is unsuitable for trace analysis because filter paper inevitably contains contaminants (e.g. phthalate esters, plasticisers) [369]. For that purpose an acceptable substitute is glass-fibre paper [28].

Applications Despite its advantages, PC has not enjoyed lasting popularity. The procedure is too slow for most applications. This form of chromatography has commonly been used for highly polar compounds such as sugars, amino acids and dyes and pigments.

For the purpose of polymer/additive analysis most applications refer to *vulcanisate analysis*. Weber [370] has determined various vulcanisation accelerators (Vulkazit Thiuram/Pextra N/Merkapto/AZ/DM) in rubbers using PC. Similarly, Zijp [371] has described application of PC for identification of various vulcanisation accelerator classes (guanidines, dithiocarbaminates, thiuramsulfides, mercapto-substituted heterocyclic compounds, thioureas, etc.). The same author has also identified accelerators and AOs in rubber of unknown composition using PC. Gaczyński and Steplien [372] have published a PC method for identification of the six most commonly used accelerators (Poland) and N-phenyl-2-naphthylamine. Auler [373,374] examined PC for *antioxidants*, antiozonants and accelerators. Many early stabiliser analyses were performed by PC [160,375,376], but the technique is nowadays practically no longer used for this purpose. Paper chromatography for the separation of additives, including metal-free stabilisers, typically requires 16 h [377–379]. Comprehensive schemes for the systematic identification of AOs and accelerators by PC, using R_f values and specific colour reactions, have been reported [371,373,380]. Table 7 of Vimalasiri *et al.* [160] lists 18 paper chromatographic separations of accelerators and AOs in elastomers in the 1952–1965 period. Paper chromatography has quite recently still been used for identification of some AOs of the gallate type [381] and for phenolic AOs in PE [382].

Both PC and TLC have been used as analytical methods for QC of oleamide production from urea and oleic acid [383]. Polyamide-impregnated PC is effective in separating fatty amides of different carbon contents [384].

Cyclic *oligomers* of PA6 can be separated by PC [385,386]; also PET and linear PET oligomers were separated by this technique [387]. Similarly, PC has been used for the determination of PEGs, but was limited by its insensitivity and low repeatability [388]. PC was also used in the determination of Cd, Pb and Zn salts of fatty acids [389]. ATR-IR has been used to identify the plasticisers DEHP and TEHTM separated by PC [390]. Although this combined method is inferior in sensitivity and resolution to modern hyphenated separation systems it is simple, cheap and suitable for routine analysis of components like polymer additives. However, the applicability of ATR-IR for *in situ* identification of components separated by PC is severely restricted by background interference.

Tran [391,392] has reported ng detection of *dyes* on filter paper by SERS. Silver colloidal hydrosols stabilised by filter supports enhance the Raman scattering of adsorbed dyes. Typical detection limits are: 500 pg (crystal violet), 7 ng (1,1′,9-trimethyl-2,2′-cyanine perchlorate), 15 ng (3,3′-diethylthiacarbocyanine chloride) and 240 ng (methyl red) using a 3 mW He-Ne laser.

The analysis of AOs by a variety of methods including PC has been reviewed [376]. Nowadays, PC hardly finds application for polymer/additive analysis.

4.4.1.2 Thin-layer Chromatography

Principles and Characteristics High-performance thin-layer chromatography (HPTLC), also known as planar chromatography, is an analytical technique with separation power and reproducibility superior to conventional TLC, which was first used in 1938 [7] and modified in 1958 [8]. HPTLC is based on the use of precoated TLC plates with small particle sizes (3–5 μm) and precise instruments for each step of the chromatographic process.

In TLC the stationary phase is pre-wet by volatile components in the mobile phase present in the vapour phase of the chromatographic chamber. The mobile phase is at the bottom of the developing chamber and advances on the stationary phase; its movement depends on capillary forces. The stationary phase is equilibrated by the mobile phase front during its movement. Separations obtained under capillary flow controlled conditions are limited to a maximum of about 5000 theoretical plates. Forced-flow development requires an external force to move the mobile phase through the layer.

The fundamental separation principle underlying TLC depends on interaction of the sample molecules with the mobile and stationary phases while still being influenced by the vapour phase in the developing chamber. The great number of combinations of adsorbents (a large choice of precoated plates) and solvents (from aqueous to lipophilic, acidic to basic) gives the technique its high flexibility. In contrast to other chromatographic techniques, the solvents can be selected independently of the requirements of the detection system, since the solvent is evaporated from the TLC plate prior to evaluation. Thus solvents can be used which cause interference in the case of HPLC UV detection. The technique allows the simultaneous running of standards and samples and has reduced needs for sample clean-up due to the one-time use of the stationary phase. Carry-over of material from one sample to another, due to sample components being retained in the stationary phase of a column, cannot occur in TLC. The technique is undemanding as far as cost of consumables is concerned. Only 10 mL of solvent are required in a modern horizontal developing chamber for the development of up to 72 samples.

Automated devices have been introduced for the three main steps of the chromatographic process, namely sample application, chromatogram development and evaluation. Appropriate *sample application*, i.e. its deposition on the plate as a small start zone, without damage to the solid-phase layer, is critical to the success of TLC. Sample application modes include spotting with the help of disposable glass capillaries, micropipettes, syringes and capillary dispersers. Computer controlled automatic sample application devices use highly reproducible μL syringes and offer adjustable rates for deposition of the sample as well as a choice between spotwise or bandwise sample application. On-line sample application is possible.

The TLC process is an off-line process. A number of samples are chromatographed simultaneously, side-by-side. HPTLC is fast (5 min), allows simultaneous separation and can be carried out with the same carrier materials as HPLC. Silica gel and chemically bonded silica gel sorbents are used predominantly in HPTLC; other stationary phases are cellulose-based [393]. Separation mechanisms are either NPC (normal-phase chromatography), RPC (reversed-phase chromatography) or IEC (ion-exchange chromatography). RPC on hydrophobic layers is not as widely used in TLC as it is in column chromatography. The resolution capabilities of TLC using silica gel absorbent as compared to C_{18} reversed-phase absorbent have been compared for 18 commercially available plasticisers, and 52 amine and 36 phenolic AOs [394].

In TLC the choice of mobile phase depends primarily on the additive in question. Gedeon *et al.* [394] have listed mobile phases for the separation of AOs and plasticisers. Bataillard *et al.* [351] have reported R_f values for various solvents and visualisation modes for a great variety of primary and secondary AOs, UVAs, HALS and metal deactivators.

Automatic instruments allow reproducible control of all aspects of *chromatogram development*, including pre-conditioning time, composition of the mobile phase and drying parameters. The user can also preset the required migration distance. In linear development, samples are applied along one edge of the plate and by allowing the mobile phase to migrate by capillary action towards the opposite edge separation is performed. This is the standard method of development. Advances in instrumental TLC techniques have aimed at increasing the separation efficiency, such as gradient elution in normal-phase, multidimensional separations by coupling gradient-elution RPLC with automated multiple development (AMD) normal-phase gradient elution [395]. For mixtures of components spanning a wide retention range, some form of gradient development is required to obtain even spacing of the components in the chromatogram.

Multiple developments have been widely used for planar chromatography. The common feature of multiple development techniques is the use of successive, repeated developments of the layer, with removal of the mobile phase each time. These techniques offer several advantages for separation of complex mixtures [393]. Multiple development improves separation efficiency, but its use is limited by generation of artefact spots and also by irreversible adsorption. Instruments for AMD use a combination of multiple stepwise development with gradient separation. In the AMD procedure the chromatogram is developed repeatedly in the same direction; each partial run goes over a longer solvent migration distance than the one before. Each next partial run uses a solvent of lower elution strength. In this way, a stepwise elution gradient is formed. With AMD mixtures of up to 40 substances can be resolved in a chromatogram covering as little as 80 mm. A separation capacity of several hundred spots is possible by 2D TLC. The acceptance of 2D planar chromatography for quantitative analysis is very much dependent on providing a convenient method for *in situ* quantitative detection and data analysis, such as optical imaging systems.

Modern TLC is thus an effective tool for separation of complex solutions into their components. Analysis can be qualitative or quantitative. When the composition or the analysed solution is known approximately, identification of the substance contained in a particular spot by its positional coordinates is frequently adequate. HPTLC provides the means not only for flexible screening procedures but also for demanding quantitative determinations. HPTLC features highly sensitive scanning densitometry and video technology for rapid *chromatogram evaluation* and documentation.

Three types of detection methods for TLC may be distinguished, namely physical, microchemical and biological–physiological. The more common detection methods for polymer additives in TLC are given in Figure 4.8. Detection is an off-line process, thus several detection techniques may be used one after the other.

Detection should preferably be carried out immediately after chromatogram development, in order to reduce loss of volatile constituents. *Physical methods* include photometric absorption and fluorescence and phosphorescence inhibition, and the detection of radioactively labelled substances by means of autoradiographic techniques, scintillation procedures or other radiometric methods (including NAA). These methods are nondestructive. Radiometry is a sensitive method of detection. The analyte of interest may be made radioactive by attaching an isotope before application to the plate or after development. A quick way of applying a radioactive tracer to developed spots is to expose the plate to radioactive iodine. Radioimaging detectors are used in 2D TLC for monitoring of analytes containing radioactive elements [396,397]. X-ray methods may be used to monitor radiolabelled compounds.

In situ scanning of TLC plates employing optical instrumentation (spectrodensitometer) is relatively recent but is now considered essential for accurate spot location and precise quantitation. The surface of the plate can be examined using either reflected (most commonly), transmitted or fluorescent light. Spectral resolutions range from 25 to 200 μm. Spectra recording is fast (up to $100\,\mathrm{nm\,s^{-1}}$). High-performance densitometers for *in situ* photometric evaluation cover the wavelength range from 190 to 800 nm. Common light sources are deuterium (190 to 400 nm), tungsten filament (350 to 800 nm) and low-pressure mercury vapour lamps (254 nm, 578 nm). For the theory of spectrodensitometry, see Wall and Wilson [398].

Post-chromatographic detection in TLC usually proceeds according to one of three procedures: (i) fluorescence detection (many organic substances exhibit natural fluorescence or can be derivatised to form fluorescent compounds) [399]; (ii) UV absorption; and (iii) visually (many chemical procedures – reagent spray, derivatisation – render the spots on the TLC plate visible). There is no difficulty in detecting coloured substances

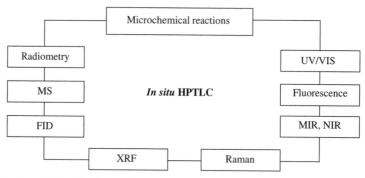

Figure 4.8 *In situ* detection and identification methods

or compounds with intrinsic fluorescence on TLC chromatograms. Many TLC separations are only rendered visible with UV illumination and, therefore, must be photographed under these conditions to obtain an image. For visible radiation and UV TLC photography the reader is referred to Heinz and Vitek [400] and Vitek [401], respectively.

The standard device of detection is a UV densitometer. Two types of UV light are required for inspecting thin-layer chromatograms: long- and short-wave UV light (at 366 and 254 nm, respectively). For short-wave UV light, eye protection against direct radiation is required. Under 254 nm UV substances absorbing at that wavelength become visible, provided the TLC layer contains a fluorescent indicator. These substances appear as dark spots on a bright background. Under 366 nm UV substances with either inherent or reagent induced fluorescence or phosphorescence appear as bright spots, often differently coloured, on a dark background. The sensitivity of this detection method increases with the intensity of the long-wave UV light and also as more visible light is eliminated [368]. UV/VIS analysis of HPTLC allows both qualitative and quantitative analysis of a wide range of AOs, light stabilisers and soluble rubber components [351,402]. In conjunction with R_f values, automatic comparison of measured UV/VIS spectra with a library of standard spectra can help in identification of the resolved components. It should be considered, however, that the thin layer adsorbents can cause either bathochromic or hypsochromic shifts in UV/VIS spectra. The paucity of bands in the UV spectrum means that structurally similar substances cannot generally be distinguished. A chromophore is, furthermore, needed for detection based on UV absorption. Thin-layer chromatographic grades of silica gel usually contain traces of organic impurities. These may interfere in the interpretation of the eluted plate particularly when they absorb below 250 μm in the UV region [158]. The adsorbent impurities usually do not have an appreciable absorption in the IR region. Quantitative assessment of resolved components can be achieved by plate readers, i.e. densitometry in either absorbance or fluorescence mode. Comparison of in situ absorption spectra is only possible if there are nearly equal amounts of substance in the spots [403].

Substances that cannot be directly measured in UV/VIS or F mode can generally be quantitated after chemical derivatisation [404]. Transfer of derivatising reagents onto the plate can be accomplished by spraying or by immersing the plate using an automatic immersion device. Such a device controls the immersion rate, dipping time and depth of immersion with high precision. Dipping and evaporation procedures give better precision than manual spraying of chromatograms [405]. Derivatisation to coloured, fluorescent, or UV absorbing compounds can be carried out both pre- or post-chromatographically. In situ pre-chromatographic derivatisation often involves oxidation and reduction reactions, hydrolysis, halogenation, nitration, esterification, etc. In situ post-chromatographic derivatisation techniques improve the selectivity of the separation, increase the sensitivity of detection and enhance the precision of the subsequent quantitative analysis. Microchemical reactions may be universal, i.e. nonspecific, directed towards particular functional groups, allowing group-specific identifications to be made. Substance-specific derivatisations are virtually impossible. Kovar and Morlock [406] have listed pre-chromatographic and post-chromatographic in situ derivatisations with functional group specificity. Other lists of chromogenic (spray) reagents for the recognition of functional groups (group-specific reagents) are given in the literature [133,407–409]. Detection reagents are also reported in Zweig and Sherma [410]. Hundreds of specific and sensitive colour reagents are available. Crompton [158] has reported general spray reagents for TLC as well as specific spray reagents for various additive classes (phenolic and amine AOs, dialkylthiodipropionates, phthalate ester plasticisers, acids, bases and carboxylic acids, alkanolamines, alkyl phenols, carbonyl compounds and organic peroxides). The aim of microchemical investigations on TLC/HPTLC plates is to provide information concerning identity and/or purity. Unequivocal identification of a substance requires combined information (R_f values, colour reactions, UV/VIS, IR, Raman, mass spectra, etc.). The specificity of detection can be increased by combination of several detection methods (e.g. physical and microchemical).

Modern TLC systems are equipped with video scan/store capabilities and offer 254 nm short-wave UV (direct light), 366 nm long-wave UV (direct light) and white light (direct light and transmitted light). Chromatogram evaluation by video densitometry is gaining importance. Advantages are ease, speed of analysis, electronic archiving (e.g. for retroactive quantification), image analysis and low operating costs. Contemporary video imaging technology uses the visible part of the spectrum only. In comparison with classical densitometry limitations are that the UV region, which is exceptionally productive for planar chromatography, is only indirectly accessible. Fluorescing substances are excited by UV, resulting in emitted visible light, which is measured, whereas UV absorbing substances can only be detected via the fluorescence quenching of an indicator incorporated in the layer.

On TLC, any material(s) left at the origin may be detected, if not resolved. Thus a good 'mass balance' may be achieved. On the other hand, in GC or LC there is always a possibility that some of the unknown mixture may remain on the column. QTLC may be based on absorption, fluorescence (quenching), FTIR or radioactivity determinations, eventually using scanners. When *quantitative evaluation* is performed, the reflection mode is generally used. The Kubelka–Munk theory is applied. All optical methods for the quantitative evaluation of thin-layer chromatograms are based upon measuring the difference in optical response between blank portions of the medium and regions where a separated substance is present. Multiwavelength evaluation allows each individual fraction to be quantitied at its optimum wavelength. The vast majority of quantitative evaluations is based on a calibration curve using a series of standards spanning the concentration range of the analyte to be determined. Conventional methods of quantitation of fractions resolved by TLC using *in situ* spectrophotometry or photodensitometry are of limited utility to substances that contain weak or no chromophoric groups. Substances separated on a TLC plate may be consecutively vaporised by laser pyrolysis and transported to an FID or ECD for quantitation [411]. Also the technique of TLC using chromatotubes (tubular TLC) can be used in conjunction with FID (Iatroscan) without the need for colour reagents. TLC-FID is useful for quantitative determination of analytes that do not absorb UV or visible light, do not fluoresce or are not volatile enough for GC. Quantitative TLC suffers from inaccuracy and irreproducibility of sample application and spot quantification. Additional operations are involved when the spots are re-extracted from the sorbent with a suitable solvent and determined in the solution. With complete computer control of data acquisition and manipulation, the reproducibility of densitometric measurements is easily reduced below 1 % RSD. Ebel [412] extensively reviewed the methodology of TLC and HPTLC quantification.

TLC plates are of particular interest as substrates for spectroscopy: (i) as a storage device for off-line spectroscopic analysis; (ii) for efficient *in situ* detection and identification; and (iii) for exploitation of spectroscopic techniques that cannot be used in HPLC. Thin-layer chromatography combined with HR MAS (NMR) can be used for compound identification without the need for elution from the stationary phase [413]. Recently also TLC-XRF was found suitable for *in situ* TLC imaging of elements [414]. The combination

of TLC with XRF was accomplished without any interface. It enables direct, nondestructive visualisation of elements developed on a TLC plate. In addition to inorganic compounds, organic compounds including electronegative elements (e.g. flame retardants) can be detected simultaneously that cannot always be measured satisfactorily by competitive techniques. Hyphenated TLC techniques are described in Section 7.3.5.

Practical aspects of TLC *method development* comprise: (i) searching for a suitable developing solvent; (ii) optimising the visualisation and evaluation process; and (iii) method validation. Table 4.34 lists the main features of HPTLC.

Table 4.34 Characteristics of TLC/HPTLC

Advantages

- Minimal sample requirements and sample preparation
- Analysis of 'dirty' samples (e.g. dissolved polymeric materials)
- Simultaneous sample clean-up and separation
- Precoated layers; sorbents with wide range of sorption properties for selectivity optimisation (HPTLC)
- Disposable stationary phase (no regeneration needed)
- Nonconventional sample application and development techniques (HPTLC)
- Unsurpassed flexibility (in chromatographic conditions)
- Monitoring of progress of separation
- Multidimensional separation
- High sample throughput (70 samples and standards in parallel); HTS device (screening technique)
- Static detection (post-chromatographic, sequential, multiple evaluation)
- Temporally distinct development and detection processes
- Easy post-chromatographic visualisation/derivatisation techniques
- Observation of all sample components in the chromatogram (except for highly volatiles)
- Selectivity of detection
- Low-cost chromatography (TLC); time efficiency, automatisation (HPTLC)
- Ease of use (TLC)
- Low mobile phase consumption
- Highly reproducible
- Rapid method development and validation
- No post-chromatographic clean-up
- Storage device ('analytical diskette', archiving) (HPTLC)
- Photographic documentation; video technology (HPTLC)
- *In situ* recording and quantitation (automated scanning densitometers) (HPTLC)
- Low detection limits (low-pg range)
- Robust

Disadvantages

- Restricted separation efficiency
- Restricted plate length (low plate number)
- Influence of environmental conditions on reproducibility of R_f values
- Difficult control on mobile phase velocity
- Cumbersome quantitation procedures

There is almost no limitation to the composition of the mobile phase; a wide choice of stationary and mobile phase combinations is available. Chromatographic conditions can be changed within minutes. Detection in HPTLC takes place in the absence of the mobile phase. Multiple subsequent detection of the same chromatogram is possible (densitometry, FTIR, R, MS, etc.). The information stored in a thin-layer chromatogram can be used for different detection and identification methods, because the processes of chromatographic development and detection or identification are independent both in time and space. In TLC it is more appropriate to define separation in space rather than in time, as in HPLC. If post-chromatographic derivatisation is employed, substance classes can be selectively detected without interference of others. Multiple samples can be analysed on one plate. Sample application, plate development, equilibration time and densitometric evaluation are all matters of minutes [415]. The speed derives from the ability to simultaneously spot, develop, and automatically scan a number of separations on a single TLC plate. Typically, 15–16 samples and standards (1–5000 ng) are separated on a single chromatographic plate with rapid developments (3–20 min) over short migration distances (2–7 cm). TLC makes it possible to work with ng and μg quantities of substances and provides absolute limits of detection at levels of 10^{-8} to 10^{-7} μg. This sensitivity makes TLC comparable with physical methods such as AAS, MS and XRF, etc. TLC can be used quite successfully for separation of practically any mixture of cations and anions.

The main limitation of TLC is its restricted separation efficiency. The separating efficiency (in terms of plates per metre) decreases rapidly over long development distances. That is, highest efficiencies are only achievable within a development distance of approximately 4–7 cm. Therefore, the total number of theoretical plates achievable on an HPTLC plate is limited (about 5000) and inferior to long LC or GC columns. Consequently, complex separations of many compounds are usually not achievable by means of HPTLC. This method is most useful for quantitating only a few components in simple or complex sample matrices. The efficiencies can also be reduced if the plate is overloaded, in an attempt to detect very trace components in a sample.

The use of conventional photographic systems for data acquisition, printing and archiving results of planar chromatography [401] has now largely been superseded by video *documentation systems*. A sample method for documentation of TLC plates has been described by a combination of computer, scanner, and digital colour thermal printer resulting in a very rapid, inexpensive and accurate retrieval system of TLC separations in the form of monochrome or colour images [416]. The use of a thermoprinter enables an accurate colour image of a TLC plate, which allows later semiquantitative determinations. Such documentation is of great importance for validation of the analytical procedure.

As many samples and reference substances may be chromatographed alongside each other, TLC is an excellent general *screening method* for additive content. In fact, in some laboratories [417] almost every incoming sample without a complete history is first subjected to a TLC screen to identify which additives are present. In most cases TLC is universal. A select few sets of conditions have been developed for identification of additives by using schemes that are specific for various additive classes. Certain classes of additives (e.g. antioxidants) may be separated from each other by using silica gel plates with a solvent combination of methylene chloride and hexane. Different classes of additives may also be differentiated and analysed by spotting the samples onto several plates and immersing them into several developing solvent conditions. More than one R_f value per analyte is important for confirmation, especially where matrix or other extractants interfere [351]. Alternatively, the TLC, GC and HPLC R_f values may be used to identify an unknown component. With a little experience and practice, additives may be unequivocally identified by a simple TLC test alone. Screening of additives in polymer extracts by means of TLC is frequently more rewarding than by means of in-polymer PyGC-MS. TLC as a screening tool was demonstrated in the separation of 54 amine antioxidants and/or antiozonants in elastomers [418]. Polymeric HALS tends to exhibit elongated streaks because of their molecular weight distribution.

TLC is more rapid (ca. 10-fold) in comparison with PC and provides 10 to 100 times lower determination limits. The wide variety of detection methods for TLC is the great advantage of the method over column techniques (HPLC, GE, SFC, CZE). Compared with GC and HPLC, conventional TLC is a less powerful separation technique. The smaller particle size of the sorbents allows better resolution in HPTLC than in conventional TLC. The average efficiency for the HPTLC layer is better than that of the conventional TLC layer, but not by much. The virtue of the HPTLC layer is that it requires shorter migration distances to achieve a given efficiency, resulting in faster separations and more compact zones that are easier to detect by scanning densitometry. Overpressured layer chromatography (OPLC) enhances the separation power, and its combination with densitometry allows the use of TLC for selective and sensitive quantitative determination of minor components of samples.

The main characteristic features of HPTLC (use of fine particle layers for fast separations, sorbents with a wide range of sorption properties, high degree of automation for sample application, development and detection) are the exact opposite of conventional TLC. Expectations in terms of performance, ease of use and quantitative information from the two approaches to TLC are truly opposite [419]. Modern TLC faces an uncertain future while conventional TLC is likely to survive as a general laboratory tool.

HPLC should be used in those cases where the number of components to be separated exceeds the separation capacity of TLC. HPTLC and HPLC are now being considered as complementary rather than competitive. Modern HPTLC is no longer inferior to other liquid chromatographic techniques with respect to precision and sensitivity. In comparison to HPLC TLC offers higher sample throughput, ability to analyse crude samples, wider choice of solvents, selective reagent sprays, ability to see irreversibly absorbed fractions, low cost and low solvent use. Other distinct advantages of TLC over HPLC are greater detection possibilities, use of disposable plates, and easier sample preparation [420]. TLC is especially useful to analyse components in cases when HPLC has difficulties and has proved to be very effective for those additives too involatile to be determined by GC or HPLC. In particular, separations of substances with strongly differing polarities are performed better using TLC. The technique has been used with great advantage over HPLC and spectroscopy in the identification of rubber compounding ingredients [394]. TLC can also assist HPLC in the investigation of the best mobile phase for complete separation of an unknown additive mixture, since changing the mobile phase in HPLC is more complex and time-consuming than in TLC. A comparison of HPTLC and HPLC analyses has been published [421].

TLC of larger quantities of materials (10 to 1000 mg) on thick layers (1–5 mm), for the purpose of isolating separated substances for further analysis or use, is called *preparative layer chromatography* (PLC). Most preparative applications are carried out on 20 × 20 silica gel or alumina plates with a layer containing a fluorescent indicator to facilitate nondestructive detection.

The main *developments*, which have ensured survival of TLC as a technique, may be summarised as follows:

(i) HPTLC using stationary sorbent layers with fine particles.

(ii) Computer-controlled automatic sample application devices (improved reproducibility).

(iii) Forced-flow TLC (FFTLC) – also called overpressured layer chromatography (OPLC) – with mobile phase movement propagated by a pump [419].

(iv) High-speed TLC, with the mobile phase being propagated by electroosmosis.

(v) Thin-layer electrochromatography (TLE) [422].

(vi) Gradient techniques.

(vii) Smaller TLC plate size (increase in elution speed, decrease in solvent consumption).

(viii) Automated multiple development (AMD).

(ix) On-line detection methods.

(x) Densitometer for quantitative evaluation of plates.

(xi) Special TLC methods: high-pressure planar liquid chromatography (HPPLC) [423], rotational planar chromatography (RPC) [424], tubular TLC-FID, hot TLC [425], radio-TLC.

The latest innovation is the introduction of ultra-thin silica layers. These layers are only $10\,\mu m$ thick (compared to $200–250\,\mu m$ in conventional plates) and are not based on granular adsorbents but consist of monolithic silica. *Ultra-thin layer chromatography* (UTLC) plates offer a unique combination of short migration distances, fast development times and extremely low solvent consumption. The absence of silica particles allows UTLC silica gel layers to be manufactured without any sort of binders, that are normally needed to stabilise silica particles at the glass support surface. UTLC plates will significantly reduce analysis time, solvent consumption and increase sensitivity in both qualitative and quantitative applications (Table 4.35). Miniaturised planar chromatography will rival other microanalytical techniques.

Selectivity of the separation in TLC is achieved by various of the aforementioned techniques (e.g. multiple development, gradient elution, sequence TLC, AMD, HPPLC or OPLC). Multidimensional TLC methods are described in Section 7.4.4.

Numerous reviews on (HP)TLC have recently appeared [366,393,419,426–429], some being quite specific: basic TLC technique [430], quantitative and

Table 4.35 Typical performance characteristics of TLC, HPTLC and UTLC plates

Performance characteristic	TLC	HPTLC	UTLC
Sample application volume (nL) (spot-wise)	1000–5000	100–500	5–20
Migration distance (cm)	10–15	3–7	1–3
Analysis time (min)	15–200	5–30	1–6
Solvent consumption (mL)	100	20	1–4
Detection limit (pg)	1000	100	10

qualitative (HP)TLC and hyphenated techniques [426, 431–433], visualisation techniques [408], theory and mechanism [434] and TLC-FID [435,436]. Other more dated papers review TLC of rubber compounding ingredients [394,437] and detection methods in TLC [368]. There are also recent books dealing with TLC [438–442]. Reagents and detection methods in TLC were treated in Jork *et al.* [407] and physical and chemical detection methods in Jork *et al.* [399]. It should be mentioned that in 1996 Sherma and Fried [443] excluded a chapter on polymers and oligomers because of a lack of sufficient new information on this topic since 1989. To a great extent this conclusion still holds true today.

Applications Conventional TLC was the most successful separation technique in the 1960s and early 1970s for identification of components in plastics. Amos [409] has published a comprehensive review on the use of TLC for various additive types (antioxidants, stabilisers, plasticisers, curing agents, antistatic agents, peroxides) in polymers and rubber vulcanisates (1973 status). More recently, Freitag [429] has reviewed TLC applications in additive analysis. TLC has been extensively applied to the determination of additives in polymer extracts [444,445].

Major applications of modern TLC comprise various sample types: biomedical, pharmaceutical, forensic, clinical, biological, environmental and industrial (product uniformity, impurity determination, surfactants, synthetic dyes); the technique is also frequently used in food science (some 10 % of published papers) [446]. Although polymer/additive analysis takes up a small share, it is apparent from deformulation schemes presented in Chapter 2 that (HP)TLC plays an appreciable role in industrial problem solving even though this is not reflected in a flood of scientific papers. TLC is not only useful for polymer additive extracts but in particular for direct separations based on dissolutions.

Use of TLC in industrial organic analysis falls into three main categories, namely monitoring chemical processes, identification of unknown components (in particular heat-sensitive and nonvolatile components) and in quantitative analysis. HPTLC is especially suitable for comparison of samples based on fingerprints, as a screening method, and as the method of choice in detection of degradation and by-products. TLC should be considered for those applications where many samples requiring minimal sample preparation are to be analysed, so that its use permits a decrease in the number of sample preparation steps, and where post-chromatographic reactions are required to detect separated analytes. TLC is sometimes employed as

a scout technique for rapid screening and to quickly determine separating conditions for HPLC analysis.

Plastic foils used for wrapping TLC plates were once reported to contain amides of oleic, stearic, or palmitic acid as antiblocking agents which may migrate from the wrapping material [28]. Similarly, Amos [447] has encountered interferences from contaminants in solvents in the TLC analysis of phenolic AOs in turbine oils.

TLC has been used extensively in *rubber and elastomer analysis* in many laboratories [394]. An early review was published on identification by TLC of rubber compounding ingredients including 36 amine AOs and antiozonants, 31 phenolic AOs, 3 guanidines, 28 accelerators, and 10 amines obtained from accelerators [448]. Also Plekhotkina [449] has reviewed the use of TLC in the rubber vulcanisate industry (determination of organic compounds and free sulfur). TLC analysis is suitable to identify accelerators [450], processing oils and soluble sulfur in vulcanisates [451], thiobisphenolic AOs [452] and mercaptobenzothiazole derivatives [453]. A more general review on chromatographic analysis (CC, PC, TLC, GC, HPLC, SEC) of elastomer antidegradants (amines, phenols, sulfur compounds) and accelerators (thiazoles, sulfenamides, thiuram sulfides, thioureas, dithiocarbamates, aldehyde amines, guanidines, xanthates) has appeared [160] reporting 61 TLC papers (1955–1983 period). Ostromow [159] has described identification of some 25 plasticisers in rubbers and elastomers by means of TLC.

Migration of additives in (un)vulcanised kneaded rubber compositions (such as 1 mol% IPPD containing vulcanised natural rubber sheet) to the surface has been analysed qualitatively and/or quantitatively by pressing the samples on TLC plates, followed by measuring the additives adsorbed on the plates with densitometers to obtain UV/VIS spectra or after adding colour formers [454]. Most of the work done using TLC for identification of rubber processing ingredients has been focused on determination of antidegradants using silica gel absorbents. ASTM D 3156 gives a widely used procedure for identification of AOs [455]. ISO 4645 (1984) describes TLC methods for identification of antidegradants in rubber and rubber products. Analysis of rubber extracts for AOs and plasticisers by means of HPTLC can be achieved in 10 min [402]. TLC and FTIR analysis have shown that the main additives in rubber sealants included dioctyl decanedioate plasticiser, rosin soap emulsifier and a poly(ammonium acrylate) thickening agent [456]. Nishiguchi [457] has conducted qualitative and quantitative analyses of antiaging agents and vulcanisation accelerators in vulcanised rubber by TLC and GC. The additives were identified by their R_f values and coloration using 2,6-dichloroquinonechlorimide and p-diazobenzenesulfonic acid as colouring agents. Most recently, less published activity on TLC analysis of elastomers has been noticed.

In the area of process monitoring TLC has been used for the study of the thermal decomposition of zinc di-isopropyl dithiophosphate (antiwear additive in lubricating oils) [458]. TLC analysis has been reported as a quality control tool for analysis of dispersing agents (alkylsalicylates, thioalkylphenolates), AOs (dithiophosphates, dialkyldithiophosphates) and their intermediates in lubricating oil (UV detection, 10 ng sensitivity) [459]. TLC is most successful in *lubricating oil analysis* [409] for determination of AOs, detergents, dispersants, antiwear, corrosion inhibitors, viscosity index improvers, and pour point depressants. An analysis scheme, based on TLC, of a complex lubricant has been published [409]. TLC is applicable for following additive depletion rates in used oils during service or laboratory simulated engine tests. A review on the application of TLC to the analysis of additives in oil products has appeared [460].

TLC is especially suited for detection of stabilisers and AOs [455] and was used for separation of 30 *antioxidants* using nine different eluents [461]. Milligen [462] has described colour reactions of over 50 AOs, accelerators and peptisers. A wide variety of polymer/additive analyses with TLC have been reported, including EVA/DLTP [463], ABS/DLTDP [464] and styrene butadiene copolymer/antioxidants [463,465,466]. Urbanski *et al.* [467] and Crompton [158] have given examples of analytical applications of TLC with polymer additive systems, including phenolic and dilaurylthiodipropionate AOs, UV light stabilisers, organic peroxide residues and organotin heat stabilisers. TLC was also used to separate 4-methyl 2,6-di-t-butylphenol, tris(p-nonyl)phosphite, 2-hydroxy-4-n-octoxybenzophenone, DLTDP, stearic acid and oleamide in polyolefins with subsequent identification of the separated compounds by UV, IR, NMR and other methods [468]. Most reported TLC work on polymer/stabiliser analysis stems from the 1980s. Ostromow [159] has described TLC analysis of stabilisers and AOs according to DIN 53622.

Squirrell [469] has mentioned TLC of commercial polymer additives (Topanol OC, DLTDP, DSTDP, Ionox 330, Irganox 259/288/1010/1076, Nonox WSP, Polygard, Santonox R, Tinuvin 326/327/328, Topanol CA, UV531, Hoechst D 55, oleamide, erucamide, Ethomeen T12, stearic acid, and others). A TLC procedure has also been described for identification of

the monomeric HALS photostabilisers Tinuvin 770 and Hostavin TMN20 in polyolefins [470] but a different, more modern, approach (MAE-HPLC-ELSD) is now being used instead [471].

In the 1960s and early 1970s TLC formed an essential part of the analytical methods of Vitamin E (a collective term for tocopherols and tocotrienols, a series of potent AOs structurally derived from 6-chromanol). Differentiation of the most common antioxidants in a single TLC system based on silica gel is possible, provided the polarity range is not too wide. Thus, α-tocopherol, BHT, BHA, gallate esters, and ascorbyl palmitate have been concurrently chromatographed [472,473]. However, when still more polar AOs, e.g. free ascorbic acid, are included, simultaneous determination is no longer feasible. It is then more convenient to develop two systems, one for 'polar' substances (ascorbic acid, ascorbyl palmitate, gallate esters) and one for 'nonpolar' antioxidants (α-tocopherol, BHT, BHA). By using HPTLC plates, Alary *et al.* [474] were able to reduce the time to separate five polar and six nonpolar AOs, respectively, to 5–10 min. Few recent papers deal with TLC of Vitamin E; HPLC has apparently taken over. TLC in conjunction with older colorimetric or GC methods is retrograde in this area.

A TLC method has also been developed to identify 57 phenols and 6 organic phosphites for stabilisation of elastomers [475]. TLC analysis of 54 amines (including diphenylamines, naphthylamines, and p-phenylenediamines) used as AOs and antiozonants for elastomers has highlighted the numerous minor components of these very impure industrial compounds [418]. Airaudo *et al.* [476] have identified phenolic AOs in 102 medical and pharmaceutical rubber articles by means of TLC (conditions: Soxhlet acetone extracts; mobile phases: benzene, benzene–hexane or benzene–ethyl acetate–acetone; spray reagent: N-chloro-2,6-dichloro-(p-benzoquinone monoimine)). Despite the great number of phenols proposed as AOs for elastomers, only eight compounds were found in current use, namely BHT, Cyanox 2246, Cyanox 425, Permanax WSP, Naugawhite, Wingstay T, Naugard SP and Vulkanox CS. HPTLC was also used in a study of the migration of impurities (ε-caprolactam and a phthalate) from a multilayer plastics container into a parenteral infusion liquid [477]. Other reports on the analysis of AOs and lubricating agents in *pharmaceutical packagings* by planar chromatography have appeared [478,479]. Dreassi *et al.* [480] have evaluated the release of Irganox 1330 from HDPE to food simulants by means of HPTLC. Detection and quantification limits were suitable for quality control. Crompton [133] has described detailed procedures for the determination of Santonox R phenolic AOs (down to 0.002%) in HDPE, six phenolic AOs in polyalkenes, 32 phenolic and amine AOs, UVAs and organotin stabilisers in a variety of polymeric matrices, Cyasorb UV531 in PE, *p-tert*-butylperbenzoate and benzoyl peroxide residues in PS, accelerators and AOs in nonvulcanised rubber compounds, traces (1 ppm) of 2,4- and 2,6-diaminotoluenes in flexible polyurethane foams (fluorimetric method) and additive mixtures (Topanol CA, DLTDP, Ionox 330; Cyasorb UV531 and Ionol CP). Also the reproducibility of recovery from PP of Ionox 330 (λ_{max} 277 μm) and di-*n*-butylphthalate (λ_{max} 222 μm) from TLC plate was reported [158]. The content of the light stabiliser Sandovur EPU in PP was determined by TLC in a CHCl₃ extract with direct (*in situ*) evaluation of the reflectance of the spots [481]. Using fluorescent indicator-containing layers, the presence of small amounts of UV light-absorbing antioxidants in polymers can directly be determined.

A small number of papers has dealt with the detection of *plasticisers* by means of TLC [482,483]. The quantitative evaluation by means of HPTLC (densitometry at 254 nm) of DEHP and DINP migrated from PVC articles has been reported [484]. Apart from the separation of plasticisers and extenders (phthalate esters, aliphatic esters, phosphate esters, epoxy esters and chlorinated hydrocarbons) from PVC [469] TLC has equally been used for the determination of lubricants and stabilisers in PVC [485]. In the former case the separated bands were marked by UV, removed from the plate and extracted for IR and/or MS examination. TLC was also instrumental in separating the four components of a rosin diglycol plasticiser for PVC [486]. Affolter [483] has reviewed methods of characterisation of polymeric and monomeric polyester plasticisers. Zitko [487] has determined the fungicide 10,10'-oxybisphenoarsine (OBPA) in PVC liners and brominated flame retardants (BFRs) in styrofoam and plastics in use in facilities holding aquatic fauna. HPTLC has also been used for screening of 16 nonionic and anionic surfactants in latex products [488]. Some of these surfactants do not absorb in the UV range.

Lawrence and Ducharme [489] have described a fast, simplified method for the detection of fluorescent *whiteners* in polymers, in which the polymer dissolution was applied directly to the thin layer. Also the separation of optical brighteners (Leucopur EGM, Azur 4, Azur 5, Hostalux ABC, Uvitex OB, Eastobrite OB) from plastics and migration into water and olive oil was studied by HPTLC on RP-18 silica using various mobile phase mixtures and UV detection [490].

TLC is used nowadays for screening for PP antioxidants, analysis of ZBEC and ZDEC vulcanisation accelerators in SBR gel foams, screening for the presence of Wingstay L and Flextol antioxidants in latices, analysis of yellowing problems (e.g. BHT), analysis of allergenic dyes (eco-labelling schemes), etc. [491].

Identification of dyes on dyed textiles is traditionally carried out by destructive techniques [493]. TLC is an outstanding technique for identification of extracted *dyestuffs* and examination of inks. Figure 4.9 shows HPTLC/SERRS analysis of acridine orange [492]. Wright *et al.* [494] have described a simple and rapid TLC-videodensitometric method for *in situ* quantification of lower halogenated subsidiary colours (LHSC) in multiple dye samples. The results obtained by this method were compared with those obtained by an indirect TLC-spectrophotometric method and those from HPLC. The total time for the TLC-videodensitometric assay of five standards and four samples applied to each plate was less than 45 min. The method is applicable for use in routine batch-certification analysis. Loger *et al.* [495,496] have chromatographed 19 basic dyes for PAN fibres on alumina on thin-layer with ethanol−water (5:2) and another 11 dyes on silica gel G with pyridine−water

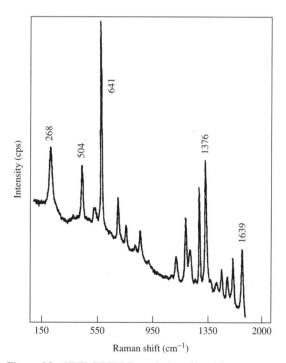

Figure 4.9 HPTLC/SERRS analysis of acridine orange. Laser excitation line, 488 nm; laser power, 15 mW. After Koglin [492]. Reproduced by permission of the Research Institute for Medicinal Plants, Budakalasz, Hungary

(1:2) as the solvent system. Also the separation of 26 basic dyes, suitable for dyeing acrylic fibres and silk with reversed-phase TLC, has been reported [497]. Laing *et al.* [498] has reported standardisation of TLC systems for comparison of fibre dyes. Dyes in textiles from archaeological excavations have also been characterised by TLC [499]. Burns and McGuigan [500] have described TLC identification and determination of cationic dyes in dye liquors for acrylic fibre and extracted from finished products. Qualitative HPTLC with post-chromatographic derivatisation and quantitative AMD HPTLC have been used for identification and determination of 20 carcinogenic amines derived from azo dyes for coloring textiles [501]. HPTLC is a reliable and low cost tool for analysis of textiles and leather in order to ensure absence of carcinogenic amines derived from certain azo dyes and is more attractive than HPLC, which requires expensive ion pairing agents in many cases. Detection of prohibited azo dyes is based on TLC analysis of the amine cleavage products [502,503]. Various mobile phases have been tested for TLC of 18 commercial dyes (e.g. for polyester fabrics) on polyamide sheets [504]. The use of TLC for the analysis of synthetic dyes has been reviewed [505]. Not surprisingly, *colour body problems* frequently make good use of HPTLC analysis. TLC, GC-MS and UV spectroscopy were used to ascribe yellowing of a radiation curable coating to degradation of BHT stabiliser to diphenochinon [506].

Various additives in PE (Santonox, Nonox DPPD, Neozone A, Ionol and Agerite White) were determined by conventional TLC [507]. Other additives in PE, studied by means of TLC, were Tinuvin P 120/326/327/770, Cyasorb UV531, Anti UV P (2-hydroxy-4-*n*-octyloxybenzophenone), Irganox 1076, Sanduvor EPU, AO-4 and Dastib 242/263 [508]. TLC has also been used in the analysis of additives in polyurethanes [509,510] as well as of slip additives (ethoxylated amines and amides) in HDPE extracts [511].

Multiple direction elution TLC has been used for the separation of PET *oligomers* or prepolymer [512]. The oligomer distribution is highly dependent on the method of isolation of the oligomers (TLC, Soxhlet or dissolution/precipitation). Jung and Lee [513] have determined a dimer of 2-phenylbenzoxazole in a polyester resin by means of TLC. A comparative study of the separation of styrene oligomers (up to dodecamers) by TLC and OPLC has appeared [514].

Reversed-phase TLC has been applied to identify antioxidants and plasticisers [394], zinc dialkyldithiophosphates and polymer additives in lubrication oil [515] and optical brighteners in plastics [490]. Organic peroxides were quantified by RPTLC (detection limit of 40 ng after post-chromatographic derivatisation) [516]. Fairly little work has been done on the separation of antidegradants using C_{18} reversed-phase absorbents. Reversed-phase HPTLC and UV densitometry have been used for the determination of sunscreen octyl salicylate in cosmetics [517].

Organotin stabilisers in PVC compositions were determined by means of TLC [518]. The determination of *inorganic species* (down to ppb range) by TLC has been reviewed on several occasions [519,520]. Instrumental methods successful in the quantitative estimation of inorganic TLC zones directly on the plate are densitometry, fluorimetry, radiometry, planimetry and visual methods. In the framework of recycling studies, in particular as to metal stabilisers in PVC, Braun and Richter [521] have carried out inorganic cation separation on cellulose TLC, using the classical experimental conditions established by paper chromatography. TRXRF would appear to be a valid alternative for this problem [522].

Apart from styrene oligomers [514], it appears that OPLC analysis of polymer additives has not been reported. However, the technique has been used for analysis of food antioxidants (BHA, BHT, NDGA and propyl, octyl, and dodecyl gallate) on silica with five different solvent mixtures and densitometric detection [479].

Thin-layer chromatography is employed in many areas of QC and routine monitoring of product quality [458]. Fluorescence scanning, densitometry or videodensitometry are used for quantification. Not all polymer additives are amenable to TLC analysis. Some fatty acid amides are virtually insoluble in organic solvents and cannot be isolated by thin-layer or column chromatography.

Several TLC related reviews have appeared, namely on polymers and oligomers [523], and on the toxicity of polymer additives [524]. Wilson and Morden [525] have reported practical applications of TLC. Applications of TLC have been described in a monograph [526].

4.4.2 Column Chromatographies

Principles and Characteristics Column liquid chromatography is the parent of all other types of chromatography. The technique used by Tswett is now called classical open-column liquid chromatography or simply LC. In column chromatography the stationary phase is contained in a column and the mobile phase flows through the (pre-wetted) stationary phase. Column liquid chromatography is an on-line process. Samples are chromatographed sequentially. Adsorbents are commonly silica gel and alumina. Silica gel is more commonly used and is especially favourable for sensitive organic compounds. Mobile phase flow results from either its weight (gravity), as in *classical liquid chromatography*, or is regulated by pumps (in HPLC). In an isocratic process the same solvent of the sample is used to elute the mixture; when an increase or decrease in solvent polarity is required, it is called solvent gradient elution. A variety of chromatography columns are available commercially:

(i) slurry packed columns, in which solvent is used to pack the column with adsorbent;
(ii) dry columns, prepared without the use of solvents;
(iii) microscale columns [527].

Dry column chromatography [528] provides several improvements over traditional column chromatography, such as better resolution and high speed. Another important characteristic is the near-quantitative applicability of TLC results in dry column analysis. Knowledge of the TLC characteristics of a sample is useful before column chromatography is employed. Careful control of the moisture content of the adsorbent is crucial to the dry column as well as other types of chromatography.

Column chromatography is directly applicable to preparative-scale separations. However, this approach usually requires many hours with the classical gravity flow columns. Flash chromatography is a type of column chromatography in which air pressure is applied to the top of the column to push the solvent through the column at a much faster rate. With this method, mixtures can be separated in a very short period of time. Flash chromatography has typically been a low pressure (10–20 psi) adsorption method that handles mg–g separations (preparative NPLC); pressure-driven apparatus (100 psi) enhances the speed of the separation process. The development of *high-performance liquid chromatography* (HPLC) has marked an important advance in analytical science. Since the sample size is commonly 1–10 μL of a solution containing perhaps 1–10 % of a mixture of many constituents, the detector system has to be capable of responding to less than 1 μg of material. The success of HPLC is therefore critically dependent on the efficiency of the detector. Detection takes place in the presence of the mobile phase. Therefore the mobile phase must be selected with a view to not interfering. Together with lifetime considerations this has resulted in the fact that the vast majority of column chromatographic separations are carried out

in the reversed-phase mode. A main characteristic of modern HPLC as compared with classic, gravity-flow LC is the shorter analysis time.

A particular column can be used for different types of LC by changing the eluent components. For example, a column packed with RP-18 bonded silica gel can be used for SEC with THF, NPLC with *n*-hexane, and RPLC with aqueous acetonitrile. When separation cannot be achieved by improving the theoretical plate number of a column, it may be achieved by selection of an appropriate stationary phase material and/or eluent.

4.4.2.1 Classical Liquid Chromatography

Principles and Characteristics In classical open-column chromatography a glass tube is packed with a filling consisting of the adsorbent. The sample, dissolved in a solvent, is added to the top of the adsorbent bed, and an elution solvent is passed through the chromatographic column under gravity. By carefully adjusting a few parameters, almost any mixture of analytes can be separated. Important parameters include the adsorbent chosen (particle size 150–200 μm), the polarity of solvent(s), and size of column (typically i.d. 1–2 cm, length 50–500 mm). The flow-rate of the solvent is significant in the effectiveness of the separation: the mixture to be separated has to remain on the column long enough to allow an equilibrium between stationary and mobile phases. The choice of adsorbent and solvent is similar to the rules governing TLC. A polar adsorbent will have better retention capabilities for a polar analyte whereas the polar analyte will elute more quickly when the adsorbent is nonpolar. To allow for maximum separation of the analytes it is necessary to optimise the different polarities of the adsorbent/solvent combination. Performing classical column chromatography (CC) is a simple process. Before use, the column should be conditioned by washing with a series of solvents. For elution of compounds of different polarities, more than one solvent might be required in the eluotropic order, less polar solvents first followed by more polar ones. Large extract volumes (sometimes 5 to 100 L) are required to obtain the sample loading needed for the separation.

Detection in CC may be visually for coloured compounds. Different methods can be used to monitor colourless compounds (collecting fractions; addition of an inorganic phosphor to the column adsorbent). The detector choice is quite limited, with UV, RI and molecular fluorescence (F) emission being the most popular. A fluorescent column adsorbent is extremely useful for monitoring the development of bands of colourless compounds. Observing the column under a hand-held UV lamp allows monitoring of band progress. A glass column cannot be used with a fluorescent column adsorbent since glass blocks UV light. UV transparent nylon tubings may be used instead.

Although CC is probably the most exacting chromatographic technique to perform, it has the *advantage* of being able to handle relatively large samples (50–500 mg). Only 500 mg of dry adsorbent is used in microscale columns. CC can be used qualitatively and quantitatively, is selective and nondestructive, and a good preparation tool with a wide application range. A *disadvantage* is the rather poor chromatographic efficiency. Band resolution problems such as streaking and unevenness can occur. Possible causes are impure solvents, uneven adsorbent packing, contaminated adsorbent, or decomposition of the sample on the column. If the column has been packed too tightly, the elution flow may not reach the desired flow of one drop per second. Traditional CC typically takes many hours, quantification is not satisfactory and is not ideal for trace analysis.

In *fast chromatography*, the column is pressurised with a flow control adaptor attached to the top of the column to pressure the column. Flash chromatography, which is a low-pressure LC technique that uses large-particle packings generally with an average particle diameter of 40 μm packed into glass or plastic large-bore preparative columns, is making a comeback. Flash chromatography allows separations of samples weighing 0.01–10.0 g in less than 15 min elution time [529].

The relation between CC and TLC is a matter of interest. The less time consuming TLC method can profitably be used for finding suitable adsorbents and development solvents for achieving satisfactory separation in CC. These conditions can usually be translated to a column without difficulty. Similarly, TLC is useful as a rapid method of monitoring the purity of fractions obtained in column separations. TLC usually provides enough of each of the individual polymer additives in a sufficiently high state of purity to enable them to be identified by spectroscopic techniques. Eventually, larger quantities in the 50–500 mg range can be obtained by scaling up the preparation on a column. Apart from preparative purposes, today the traditional column LC is entirely replaced by modern LC techniques. However, open-column chromatography still finds use in the modern laboratory but more for sample pretreatment than for analysis.

Applications Open-column chromatography was used for polymer/additive analysis mainly in the 1950–1970 period (cf. Vimalasiri *et al.* [160]). Examples are the application of CC to styrene-butadiene copolymer/(additives, low-MW compounds) [530] and rubbers/(accelerators, antioxidants) [531]. Column chromatography of nine plasticisers in PVC with various elution solvents has been reported [44], as well as the separation of CHCl$_3$ solvent extracts of PE/(BHT, Santonox R) on an alumina column [532]. Similarly, Santonox R and Ionol CP were easily separated using benzene and Topanol CA and dilaurylthiodipropionate using cyclohexane: ethyl acetate (9:1 v/v) [533]. CC on neutral alumina has been used for the separation of antioxidants, accelerators and plasticisers in rubber extracts [534]. Column chromatography of polymer additives has been reviewed [160,375,376].

4.4.2.2 High-performance Liquid Chromatography

Principles and Characteristics High-performance liquid chromatography (HPLC) is a dominant analytical tool which continues to expand. It is estimated that worldwide some 150 000 HPLC systems are in use (sales to reach US$260 million by 2005). Modern HPLC started with the preparation of high-efficiency columns from 10 μm particles and modification of the silica surface via silanisation in 1973. This has created sufficient separation power to tackle a broad range of real-life problems within acceptable analysis time. HPLC evolved from preparative column chromatography and its performance (efficiency, resolution) has been greatly enhanced by the use of elaborate stationary phases composed of spherical particles with small diameters (now typically between 2 μm and 5 μm). Because of the small particles head pressure, typically of the order of 100 atm, is needed to force the mobile phase through the packing in short, small diameter columns. The method is therefore also often called *high-pressure liquid chromatography*. Forced migration of a liquid phase through a stationary phase is encountered in many chromatographic techniques. One of the aspects particular to HPLC is that of the partition mechanisms between analyte, mobile phase and stationary phase. These are based on coefficients of adsorption or partition. Ettre [535] has described the evolution of LC.

Efficient separation of analytes in HPLC requires an appropriate solubility of the analytes in the system and neither too strong nor too weak binding/solubility of the analytes in the solvated stationary phase. The distribution of analytes in the HPLC phases can be described by the equilibrium constant (K_d) for the concentration of the analyte in the mobile phase (C_m) and in the solvated stationary phase (C_s):

$$K_d = C_s / C_m \qquad (4.4)$$

Modular *HPLC instruments* consist of a solvent (mobile phase) reservoir, degassing system, solvent delivery system (or pump), a sample introduction system (or injector), thermostated column, detector(s) and data system. The reproducibility of the chromatography is improved if the column is located in an oven at constant temperature. Similarly, recent progress in packing material technology allows for better control of the particle size down to sub-2 μm, improving again the resolution [535a]. Instrumental developments such as microbore columns and spectrophotometric PDA detection coupled with efficient data-processing systems now offer great potential to the determination of additives even in complex samples. Injection of liquid samples (usually 10–20 μL, minimum 1 μL) is less problematic for HPLC than for GC, and an injection valve fitted with an injection loop is often used. Autosamplers constitute a major uncertainty factor in HPLC analysis precision [536]. It is impossible to run all autosamplers of a routine laboratory at a relative standard deviation (RSD) better than 0.5–1 %.

Liquid chromatography has a number of different *configurations* with regard to technical (instrumental) as well as separation modes. The HPLC system can be operated in either isocratic mode, i.e. the same mobile phase composition throughout the chromatographic run, or by gradient elution (GE), i.e. the mobile phase composition varies with run time. The choice of operation depends largely on the number of analytes to be separated and the speed required. For routine work involving analysis of a few components, it is nearly always possible to develop an isocratic separation which allows both identification and quantification of individual components. Isocratic HPLC separation of polymer additives is generally preferred because of better precision and simplicity compared with the gradient elution mode of separation. Regardless of the mode of chromatography, gradient elution involves using a weak eluting solvent at the start of the chromatographic run and adding increasing proportions of a strong eluting solvent over the course of the separation. Gradient elution frequently improves separation efficiency, just as temperature programming helps in both GC and HPLC. GE-HPLC is preferred for the separation of complex mixtures of varying polarity, is also used for rapid screening of unknown additives and for pilot work to find a suitable mobile phase for an isocratic separation.

Normal-phase (NP) and reversed-phase (RP) liquid chromatography are simple divisions of the LC techniques based on the relative polarities of the mobile and stationary phases (Figure 4.10). Both NPLC and RPLC analysis make use of either the isocratic or gradient elution modes of separation (i.e. constant or variable composition of the mobile phase, respectively). Selection from these four available separation techniques depends on many variables but basically on the number and chemical structure of the compounds to be separated and on the scope of the analysis.

When the predominant functional group of the stationary phase is more polar than the commonly used mobile phases, the separation technique is termed *normal-phase HPLC* (NPLC), formerly also called adsorption liquid chromatography. In NPLC, many types

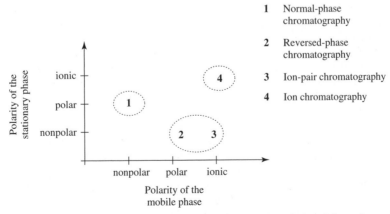

1 Normal-phase chromatography

2 Reversed-phase chromatography

3 Ion-pair chromatography

4 Ion chromatography

Figure 4.10 Classification of liquid chromatographic methods depending on the polarity of the stationary and mobile phases. After Schäffer *et al.* [537]. Reproduced by permission of Metrohm AG, Herisau, Switzerland

of interaction have been described but the basic molecular interaction is electrostatic forces. The specificity of NPLC is the result of the direct formation of a strong molecular interaction (hydrogen bonding) between sample molecules and the adsorbent. Although normal-phase chromatographs can be performed using either partition or adsorption mechanisms, the dominant retention mechanism is adsorption. The typical stationary phases employed in normal-phase chromatography are common porous adsorbents, such as silica or alumina with polar hydroxyl groups on the surface. However, as these columns lack selectivity nowadays bonded stationary phases are more popular. The use of varying bonded phase types, i.e. cyanopropyl (CN), amino (NH_2) and diol, can result in large differences in selectivity. The main mechanism of NPLC is based on displacement, i.e. the competition of solute and solvent molecules for active sites. The mobile phase consists of a nonpolar solvent such as a hydrocarbon (the weak solvent) which may be mixed with a small amount (usually <5 vol%) of a more polar solvent (the strong solvent). Weak solvents are those which interact very little with the stationary phase. In NPLC the weak solvents are nonpolar hydrocarbons such as n-hexane, n-heptane, iso-octane and cyclohexane. The strong solvents in the NPLC separation mode are more polar compounds such as CH_2Cl_2, $CHCl_3$, CH_2Cl-CH_2Cl or 2-propanol. In NPLC the more polar analytes are more strongly retained on the polar stationary phase and, consequently, the less polar analytes elute first at shorter retention times. Separation basically depends on packing type, polarity of the additives and mobile phase strength, and needs to be optimised. The isocratic mode of HPLC separation is not ideally suited to analysis of complex mixtures containing stabilisers with a wide range of polarity. During GE chromatography in normal-phase systems the concentration of one or more polar solvent(s) in a nonpolar solvent is increased. A disadvantage of this technique with respect to reversed-phase gradient elution is the possible preferential adsorption of the more polar solvent(s) on the surface of the polar adsorbent, which may upset the gradient profile. GE-HPLC separation is used when complex mixtures of additives with very different polarity have to be analysed. As the composition of the mobile phase is varied during the separation this provides an additional parameter compared to the isocratic mode. By proper choice of the mobile phase strengths at the beginning and end of the separation and gradient time and shape, the separation may be drastically improved in terms of peak resolution, detection and analysis time. The most practical method for finding a good eluent for NPLC is a trial experiment using TLC. Not

surprisingly, NPLC is most commonly applied to analysis of samples that are soluble in nonpolar solvents. Normal-phase chromatography is especially useful for analysis of water-sensitive samples, analytes that are very insoluble in water, and certain geometric isomers difficult to separate by reversed-phase chromatography. NPLC exhibits a unique ability to distinguish between solutes with different number of electronegative atoms, such as oxygen or nitrogen, or molecules with different functional groups. Adsorption chromatography is therefore widely used for class separations. In contrast to RPLC, NPLC is not very effective for the separation of members of a homologous series. Organic solvents used in normal-phase separations are more MS-friendly than some of the buffers typically used in HPLC. Isocratic elution (NPLC) is ineffective for the complete separation of different model additives (AOs, UVAs) in a single run [347]. Normal gradient-elution separation is excluded for the analysis of non-UV absorbing additives which require the use of RI detection and, therefore, an isocratic separation. For example, commonly used additives such as Irganox PS 800 and PS 802 do not have a chromophoric moiety and cannot be determined using a UV detector, while an RI detector only functions under isocratic conditions and is not very sensitive. Consequently, NPLC appears limited to the routine target compound determination of a finite number of additives, but investigation of complex mixtures from various plastic extracts is difficult and time-consuming.

Development of HPLC was slow when still a NP method but exploded with the arrival of RP methods. NPLC has fallen in disfavour with many chromatographers because of some of the complexities involved. Under certain circumstances, lengthy equilibration times or long-term retention time reproducibility problems are encountered, which are due largely to the sensitivity of the technique to the presence of small concentrations of polar contaminants in the mobile phase. Another major disadvantage of NPLC is the effect of water adsorption, which leads to a decrease in solute retention and adsorbent deactivation. If these problems are controlled, the technique typically gives chromatograms superior to reversed-phase methods, due to the low viscosity of the commonly used solvents. Probably the greatest drawback of NPLC is the lack of separation selectivity between different packing materials: virtually all compounds elute in the same order regardless of the column selected. Normal-phase chromatography relies on 'bare' silica or silica functionalised with diol, cyanopropyl or amino groups. In modern chromatography NPLC is by far the least commonly employed of the major separation modes. The main reason is the

Table 4.36 Main features of NPLC

Advantages

- Simplicity of operation
- Good resolution
- Routine target compounds determination
- Separation of analytes soluble in nonpolar solvents
- Recommended for water-sensitive samples
- Separation of geometric isomers
- Allowance for class separations
- MS friendly

Disadvantages

- Volatile solvents
- Preferential adsorption of polar solvents (water)
- Lack of selectivity (isocratic mode)
- Long-term irreproducibility of retention times
- Long equilibration periods
- Difficult for complex mixtures
- No effective separation of homologous series
- Limited detector choice

fact that RP chromatography readily allows for greater variation of separation selectivity for the same types of solutes used in normal-phase chromatography. However, for the compound classes indicated NPLC still has its place in the separations laboratory. Table 4.36 lists the main characteristics of NPLC.

Reversed-phase liquid chromatography (RPLC) is characterised by the use of nonpolar stationary phases and eluents which are more polar than the stationary phase. In RPLC the stationary phase is usually a hydrophobic bonded phase such as octadecylsilane (RP-18) or octylsilane (RP-8) whereas the mobile phase usually consists of polar solvents such as water and water-soluble organic solvents. Three organic solvents, acetonitrile, methanol and THF, dominate in RPLC. Increasing the concentrations of the organic modifier decreases the overall retention times, but changes in relative retention times depend on the properties of the analytes. Hanai [538] has discussed eluent selection. The mobile phase should be free of particles and, especially in gradient chromatography, free of UV-absorbing impurities. Retention is driven by hydrophobic interactions with the stationary phase and polar interactions with the mobile phase. Less polar sample components are more strongly retained on the nonpolar stationary phase than the more polar components which present little interaction with the stationary phase. For low-MW solutes, retention is mainly governed by a partitioning process, rather than by adsorption. However, a complete theory of the retention mechanism of reversed-phase chromatography must also include partitioning and adsorption [539]. In RPLC the obtainable separation

of analytes is mainly determined by the characteristics of the mobile phase, although appreciable differences in selectivity exist between different bonded phase columns. Important parameters to consider in RPLC are type and amount of organic modifier concentration of salts and impurities, pH and temperature. A polar component will largely transfer to the mobile phase and exit the column first, whereas a nonpolar component will partition more to the stationary phase and not move until the solvent strength of the mobile phase increases sufficiently. The nonpolar components of a sample interact more with the relatively nonpolar hydrocarbon column packing and thus elute later than polar components. The elution order of solutes in RPLC is thus that of decreasing polarity, i.e. increasing hydrophobicity, while in NPLC the least polar compound elutes first. NPLC and RPLC are complementary in elution order. Compared with isocratic analysis the gradient elution mode of separation of complex mixtures with many additives of very different polarity improves the peak resolution and detection and reduces the analysis times. The main parameter to control in these separations is also the variation of the strength of the mobile phase.

RPLC is today by far the most popular HPLC technique. About 80 % of all analytical separations of low-MW samples is carried out using an RP method. RPLC owes its popularity mainly to the fact that good results are obtained for many compounds with few technical complications despite the fact that its mechanism is complex. RPLC is particularly well suited for separation of mixtures of small hydrophobic molecules and molecules with medium polarity. GE-RPLC is the workhorse of the industrial laboratory. The reason is that reproducibility depends mainly on eluent properties (so far the columns are reproducible). Method development is easy. For reliable methods in HPLC correct quantitation is more important than constant retention time. However, method transfer is a big problem. While in isocratic work retention times are constant, if method transfer takes place in gradient mode there is often a considerable retention time difference between the results prescribed by the method and those found by experiment. There are many variations in RPLC in which various mobile phase additives are used to impart a different selectivity. For example, for RPLC of anions, the addition of a buffer and tetraalkylammonium salt would allow ion pairing to occur and effect separations that rival ion-exchange chromatography. The volume injected is usually between 10 and 100 µL in RPLC, although for nonpolar substances in water injection of up to 10 mL has been used for trace analysis. When large volumes are injected, nonpolar components will concentrate in a narrow band at the top of the column with a predominantly aqueous mobile phase, and only start moving down the column at higher solvent concentrations. Temperature control is essential for achieving reliable and reproducible HPLC separations. Increasing the effective operating temperature generally decreases retention, improves the separation efficiency and may alter the separation selectivity. RPLC separations generally use aqueous mobile phases that are more viscous than typical NPLC mobile phases. Therefore, for RPLC it is advantageous to operate at higher temperatures. Temperature control may be used as an adjustable parameter, i.e. an additional variable that can be used to improve the analysis. In reversed-phase, HPLC operating conditions must sometimes be modified dramatically to allow compatibility with MS detection. Table 4.37 lists the main features of RPLC.

Selection of columns and mobile phases is determined after consideration of the chemistry of the analytes. In HPLC, the mobile phase is a liquid, while the stationary phase can be a solid or a liquid immobilised on a solid. A stationary phase may have chemical functional groups or compounds physically or chemically bonded to its surface. Resolution and efficiency of HPLC are closely associated with the active surface area of the materials used as stationary phase. Generally, the efficiency of a column increases with decreasing particle size, but back-pressure and mobile phase viscosity increase simultaneously. Selection of the *stationary phase* material is generally not difficult when the retention mechanism of the intended separation is understood. The fundamental behaviour of stationary phase materials is related to their solubility–interaction

Table 4.37 Main features of RPLC

Advantages

- Ease of use, simplicity, versatility and scope
- Few technical complications
- High column efficiencies
- Reproducible retention times
- Fairly reproducible and stable bonded-phase columns
- Quantitation
- Easy method development
- Considerable variation in separation selectivity
- Fast equilibration
- Predictable elution order
- Suitable for separation of small hydrophobic and medium polarity molecules
- Effective separation of homologous series

Disadvantages

- Analysis time
- Method transfer
- Limited detector choice
- MS unfriendliness

properties. Particle types used in HPLC columns are: (i) superficially porous particles (pellicular materials, e.g. $1-2\,\mu m$ porous outer layer of a coated $30\,\mu m$ i.d. stationary phase); (ii) very small totally porous particles (microparticulate $3-30\,\mu m$); (iii) totally porous particles (e.g. $30\,\mu m$); and (iv) monolithic sorbents (bimodal pore structure). The majority of HPLC columns today are packed with spherical, microparticulate ($3-10\,\mu m$ diameter) materials, which results in stable, high-efficiency columns which can be used for relatively large sample loadings. In addition to the physical nature of the particle (i.e. microporous or pellicular) and its size and shape, the particle material also significantly affects the performance of the HPLC packing. Column packing alternatives include the use of rigid solids (most commonly silica), resins (usually PS/DVB, methacrylate and polyvinylalcohol) and soft gels. HPLC is moving into new column and stationary phase design and miniaturisation (HPLC-chip).

In NPLC, which refers to the use of adsorption, i.e. liquid–solid chromatography (LSC), the surface of microparticulate silica (or other adsorbent) constitutes the most commonly used polar stationary phase; normal bonded-phase chromatography (N-BPC) is typified by nitrile- or amino-bonded stationary phases. Silica columns with a broad range of properties are commercially available (with standard particle sizes of 3, 5 and $10\,\mu m$, and pore sizes of about $6-15\,nm$). A typical HPLC column is packed with a stationary phase of a pore size of $10\,nm$ and contains a surface area of between 100 and $150\,m^2\,mL^{-1}$ of mobile phase volume.

One of the most important advances with HPLC was the introduction of bonded phases on silica where the active silanol groupings are reacted with a silylating reagent with a nonpolar tail to give a nonpolar coated phase, such as octadecylsilane (ODS). Bonded phases with easily adjustable polarity constitute the basis of reversed-phase partition chromatography. The performance of a bonded phase silica column is determined by: (i) the base silica material and its pretreatment; (ii) the type of stationary phase bonded onto the silica; (iii) the amount of stationary phase material (or carbon load) bonded onto the silica; and (iv) any secondary bonding (or end-capping) reactions. RP columns are named by the nature of the bonded R group; C_{18} is the most common functional group on bonded RP columns. Other functional groups include C_8, phenyl, C_6, C_4, C_2, CN, diol, NH_2 and NO_2. Bonded silica phases, such as RP-8 (dimethyloctylsilane), RP-18 or ODS (dimethyloctadecylsilane), are usually of relatively low polarity. The nature of the functional group controls selectivity, while the chain length controls column efficiency. For most RPLC separations, a good general-purpose column is typically 15–30 cm in length, packed with C$_{18}$ functionalised 3–5 μm spherical silica and a carbon load of 7–10 %. The column should be end-capped and have an efficiency of 5000–10 000 theoretical plates. In RP chromatography it is possible to select a column out of over 600 available stationary phases with different selectivities and efficiencies. Only few systematic approaches have been published to characterise stationary phases. Engelhardt [540] has discussed commonly used column test procedures.

The characteristics of some common bonded RPLC columns in terms of particle size/shape, pore size, degree of end-capping and carbon load are described [240]. Silica-based matrices with bonded phases have some inherent problems in RPLC which can lead to deteriorating separation performance. Residual acidic silanol groups on silica-based packings interact strongly with basic compounds, resulting in peak tailing and poor resolution. Aqueous acidic eluent conditions also result in gradual loss of both the alkyl bonded phase, creating more free silanols, and in dissolution of the base silica itself. Column lifetime is shortened. Although most stationary phases in RPLC are silica-based bonded phases, polymeric phases, phases based on inorganic substrates other than silica, and graphitised carbon are applied as well. Polymeric packings are completely insoluble and will not contaminate isolated fractions with leachables.

The recent introduction of monolithic sorbents [21,22,541,542] in which phases are cast rather than packed, with silica and polymeric chemistries, are a breakthrough which eliminates problems associated with high column back-pressures. The material is comprised of a silica skeleton with a bimodal pore structure of 13 nm mesopores and 2 μm macropores. The latter serve as through-pores and enable analyte transport under low pressure. The activated surface for chromatographic separation made available by the mesopores is ca. 300 m^2 g^{-1}. The overall porosity of the monolithic silica matrix exceeds 80 % (as compared to 65 % for conventional particulate silica packings). Monolithic columns, which come with C$_{18}$ surface modification and end-capping, are comparable in selectivity to conventional RP columns. This allows for easy transfer of existing methods from particulate to monolithic columns. The separation time is decreased by the same factor as the increased flow-rate from 1 to 9 mL min^{-1} (flow-rates of 2 to 3 mL min^{-1} can rarely be exceeded for particulate columns). Challenging applications may be envisaged using a combination of flow and solvent gradients. By coupling of several monolith columns beds in excess of 10 000 plates are possible; up to 10 × 10 mm Chromolith^{TM} columns have been successfully linked.

Standardisation of RP columns is an issue of debate. RPLC stationary phases were reviewed and compared and the ideal properties were defined [543]. For the development of RP packing materials, see Dolan [544].

In HPLC the *mobile phase* is forced past the stationary phase under pressure and the solutes partition between the two phases according to their distribution coefficients and rates of mass transfer. The most important parameter for control of LC is the composition of the eluent. Liquid chromatography presents unlimited possibilities of eluent selection. The type of stationary phase influences the choice of mobile phase used and the retention characteristics of the solutes. In eluent selection the crucial factor is to control the solubility of the analytes in the eluent. HPLC eluents are classified in solvents, nonsolvents and eluents showing partial solubility. For miscibility of solvents, see Unger [545]. Other properties of an ideal mobile phase are nonflammability, nontoxicity, readily availability in a high and consistent state of purity, absence of detection interferences (such as UV/VIS absorbance), low viscosity and high diffusion rate, ability to alter the solvent strength, environmental friendliness and low cost. An additional experimental variable is the flow-rate of the mobile phase. The choice of the mobile phase solvent also varies with each type of HPLC (NPLC vs. RPLC, isocratic vs. gradient). Resolution in LC is controlled primarily by the mobile phase.

A good index of solvent strength for adsorption chromatography is the experimental adsorption *solvent strength* parameter $\varepsilon°$, as tabulated by Snyder and Kirkland [546] and others [240]. While solvent mixtures of equivalent eluting strengths, termed isoeluotropic, give similar total elution times, they may differ in solvent selectivity, i.e. changes in the relative elution of the peaks. Perhaps the greatest practical problem associated with operation of NP chromatography is the effect of water on the activity of polar adsorbents. Solvent strength in RPLC can be expressed as a function of polarity, and the eluting strength of an RPLC solvent is generally inversely related to its polarity. Once a mixture of the desired solvent strength is defined for a given separation, changing the nature of the organic modifier, while maintaining constant eluotropic strength, can alter the *solvent selectivity*. Transfer rules may be used for calculating the organic modifier content of isoeluotropic mobile phases [547]. Other factors which affect retention in RPLC are temperature, pH and mobile phase additives. Temperature affects viscosity and/or column back-pressure, sample solubility, retention time stability and bonded phase degradation. The ease with which gradients can be used with RPLC is one of the reasons for its widespread use. However, solvent gradient elution is not simple with columns of small inner diameters. Superheated-water between 100 °C and 240 °C has recently been advanced as an eluent for RPLC [548]. Superheated-water separations are comparable to reversed-phase methods so are recommended for the separation of medium- and high-polarity analytes.

Scheme 4.5 illustrates *HPLC phase selection*. Column manufacturers may have an applications database from which they can recommend a column and a method. Specific methods have been established for quite a large number of analytes, such as additives (e.g. antioxidants). Column selection and column technology have been reviewed [549]. Contrary to GC, and with the exception of SEC, selectivity in HPLC is determined not by the column alone but also by the mobile phase. There is therefore no one-for-one assignment between an analytical problem and the 'best' column for this problem.

The chromatographic column is the heart of the HPLC system. Generally, the longer the column, the higher its efficiency and resolution; the shorter the column, the faster the separation. Also, the larger the column diameter, the greater its loading capacity, while the narrower the column, the greater its mass

sensitivity. Most analytical columns range from 50 to 300 mm in length.

Most of today's analytical separations using HPLC are performed routinely on columns with i.d. 1.0–4.6 mm with flow-rates of $50 \, \mu L \, min^{-1}$ up to $2 \, mL \, min^{-1}$ (Table 4.38). Precise flow-rates (<1 % variation) can be delivered. An advantage of the application of conventional LC over the use of microbore or capillary LC systems is that commercially available flow cells with high-sample capacity can be used. *Column technology* has advanced significantly into various directions: (i) reproducibility of the column packing operation and the mechanical stability of the packed bed; (ii) chemistry of HPLC packings; and (iii) smaller particle sizes and column diameters [550–552]. Polymeric RP packings replace conventional silica materials providing similar resolution without the inherent disadvantages of silica bonded phases. There is nowadays a trend towards smaller columns in HPLC. The theoretical and practical difficulties in downscaling from standard-bore HPLC to small-bore HPLC were discussed by Weston and Brown [365]. Main drivers for use of narrow-bore, microbore and packed capillary column technology are hyphenation with MS detection and faster analysis. External constraints also force the user to employ micro-HPLC, such as increasing waste disposal costs and the limit of determination which can be obtained using conventional

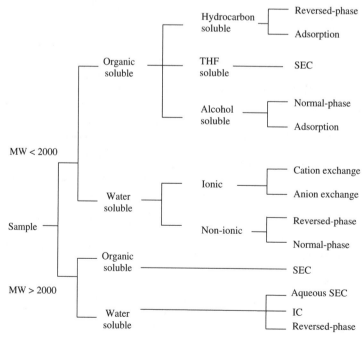

Scheme 4.5   Phase selection process

Table 4.38 Nomenclature for HPLC columns (cylindrical formats)

Description	Dimension	Typical flow-rate[a]
Preparative column HPLC	i.d. > 10 mm	> 20 mL min^{-1}
Semipreparative column HPLC	5 mm $<$ i.d. ≤ 10 mm	5.0–40 mL min^{-1}
Normal-bore column HPLC	4 mm $\leq$ i.d. ≤ 5 mm	1.0–10 mL min^{-1}
Narrow (small)-bore column HPLC	2.1 mm $<$ i.d. < 4 mm	0.3–3.0 mL min^{-1}
Microbore column HPLC	1 mm $\leq$ i.d. ≤ 2.1 mm	50–1000 μL min^{-1}
Capillary column HPLC	100 μm $<$ i.d. < 1 mm	0.4–200 μL min^{-1}
Nanobore column HPLC	25 μm $\leq$ i.d. ≤ 100 μm	25–4000 nL min^{-1}
Open-tubular liquid chromatography	< 25 μm i.d.	< 25 nL min^{-1}

[a] Velocity 1–10 mm s^{-1}.

methods. More impressive than the reduction in solvent use is the 95 % reduction in sample size.

Microbore LC requires special injection systems, detectors and capillaries, while narrow-bore systems can be operated with normal equipment. Detector technology has been the principle problem preventing widespread use of microcolumns in HPLC. Lower flow-rates in microcolumn HPLC (10–50 μL min^{-1}) allow introduction of the total effluent into the ion source. This increases the sensitivity of the method. Microbore columns are also suitable for performing LC-ICP-MS separations as lower organic solvent concentrations are transferred to the ICP-MS. A possible limitation is that only small samples (1–10 nL) can be analysed because of the limited loadability of these columns. In general, it is more difficult to achieve the same high efficiencies and high sensitivity on a small column than on a large column. Packed column SFC offers an attractive alternative. The main characteristics of microcolumn HPLC are summarised in Table 4.39.

Packed capillary columns in HPLC have received greater attention over the last few years as a result of the increased availability of instrumentation for miniaturised HPLC. By reducing the column i.d. from conventional size of 4.6 mm to 0.32 mm, the mass limit of detection (mLOD) can be improved theoretically by a factor of 200, if the same mass of analyte is injected and both the conventional and capillary size detector cell have similar response. An injection volume of 0.06 μL is typically used in packed capillary LC, when phase focusing is not performed; phase focusing allows for injection volumes of up to 200 μL. Packed capillary LC offers several advantages over conventional LC, such as reduced consumption of mobile and stationary phases, increased mass sensitivity, less demanding coupling requirements, suitability of using temperature programming for retention control, and availability of long columns to enhance the resolution of complex mixtures and overall efficiency

Table 4.39 Main features of microcolumn HPLC

Advantages

- Ideal for analysis of very small samples
- Reduced sample dilution on-column
- Minimum sample loss
- Reduced solvent and stationary phase consumption
- Possibility of using 'exotic' solvents
- Increased compatibility with spectroscopic/spectrometric techniques (e.g. MS, NMR)
- Dramatic increase in mass sensitivity of concentration-sensitive detectors
- High column efficiency and resolution
- More economical column manufacturing

Limitations

- More stringent instrumental design
- Adequate dimensions of connecting capillaries and detector volume
- Higher back-pressure for smaller particles
- Plate height determining the length of the column for a desired resolution
- Development of widely applicable instrumentation
- Complicated maintenance
- Low overall public acceptance

($>10^5$ plates m^{-1}) [553]. Unlike narrow-bore columns, the use of 1 mm and smaller i.d. columns requires either the use of a specially designed *micro-HPLC system*, or modification of a standard HPLC instrument, in order to handle the micro-volume sample and solvent fluidics. Conventional HPLC systems can now be converted into a micro-HPLC system within minutes. Recently a dedicated μHPLC has been developed that fulfills all the requirements to perform micro-, capillary- and nanoscale LC separations.

Microbore and packed capillary HPLC column technology has not yet met the requirements for breakthrough of new technologies [554]. On commercial instruments in general efficiency and detectability with microbore columns are lower than with normal-bore columns. Microbore and capillary HPLC suffer from complicated maintenance. For recent developments in LC column technology, see references [554–556].

Advances in packings, column technology and HPLC instrumentation have resulted in development of fast LC, which can be either in high-speed or high-resolution mode. Microcolumn and *high-speed LC* (HSLC) techniques use one of two possibilities to improve the column efficiency, i.e. decrease of packing particle size for HSLC and decrease of column internal diameter for microcolumn LC. Both these improvements result in an increase in theoretical plates per unit column (up to 15 000 plates per column), i.e. higher efficiency due to the decrease of peak equivalent theoretical plates. Fast LC performs rapid analysis using short (3–10 cm), 4–6 mm i.d. columns packed with small particles (3 μm), usually operating in the 1–3 mL min⁻¹ range. In contrast to conventional 10–15 cm columns, fast LC can enhance laboratory productivity considerably without sacrificing resolution, reliability, and convenience. Routine fast LC analysis times are 1–3 min for isocratic and 5–10 min for gradient elution separations. Recently 50 mm HPLC monolith columns for fast analysis have been introduced. High-speed μHPLC columns (33 × 7 mm) reduce analysis time by some 70–80 %. High-throughput screening (HTS) with capillary HPLC is fast and sensitive. Fast LC or rapid-resolution LC columns are finding increased use in QC/QA and HTS applications such as LC-MS/MS. Due to its advantages, the use of HSLC in QA is widespread in the plastics industry, including the analysis of polymer additives. Dong and DiCesare [557] performed rapid analysis of 5 AOs in only 70 s and for 9 plasticisers in 150 s. Concepts, performance, requirements and applications of fast LC are widely reported [558].

Once the chromatographic separation on the column has been conducted, the composition of the eluent must be determined using a detector. Absolute retention times are difficult to reproduce even within one laboratory on a single system. Most HPLC separations are carried out by direct comparison of an analyte with retention of a standard sample on the same column and instrument or by comparison with an internal standard in a system calibrated using an authentic sample (relative retention factors). Apart from being very time-consuming this requires a collection of suspected compounds. By providing a *retention value*, which is largely independent of the exact operating conditions and which can be reproduced in different laboratories, retention indices enable results to be transferred between laboratories and for identifications to be assigned from library values, even in the absence of authentic samples. Smith [559] has discussed application of retention indices that are largely independent of the composition of the mobile phase (but still dependent on the stationary phase). Identification of compounds from their retention indexes in gradient RPLC has been described [560] and reviewed for HPLC [559]. Problems associated with the difficulty of ensuring reproducibility of the retention properties and selectivities from one commercial reversed-phase to another and even from one batch to another of the same product have also been discussed [560]. If good reproducibility of retention times is required, the column temperature should be kept constant. Although retention times are more reproducible in RPLC than in NPLC identification of compounds in additive mixtures should not be based only on peak retention times. Another factor complicating identification is that many of the additives are not pure, but mixtures. Matching to spectral libraries is indicated to avoid misidentification.

Development of *HPLC detectors* has been slow and all generally show a lower sensitivity than their GC counterparts. Detection in HPLC is far more difficult than in GC as a solute in a liquid modifies the overall properties of the carrier only to a minor extent as compared to that of the same concentration of solution in a gas. The detector response will thus be close to the detector noise and the sensitivity likely to be very limited. In order to obtain maximum sensitivity it is necessary to choose the right mobile phase. In most HPLC detectors, the eluent flows through a measuring cell, where the change of a physical or chemical property with elution time is detected. The flow cell requires a small volume to eliminate band spreading. In general, the detector should be smaller than 100 μL for most conventional packed columns; the detector volume should range from 1 μL for microbore columns to 0.1 μL for packed microcapillaries and to 10 nL for open microtubular capillaries. Requirements to HPLC detecting systems which operate in a dynamic manner are given in Table 4.40. These requirements are difficult to meet.

Detector selectivity is much more important in LC than in GC since, in general, separations must be performed with a much smaller number of theoretical plates, and for complex mixtures both column separation and detector discrimination may be equally significant in obtaining an acceptable result. Sensitivity is important for trace analysis and for compatibility with the small sizes and miniaturised detector volumes associated with microcolumns in LC. The introduction of small bore packed columns in HPLC with reduced peak volume places an even greater strain on LC detector design. It is generally desirable to have a nondestructive detector; this allows coupling several detectors in series (dual

Table 4.40 Requirements to a HPLC detector

- On-line
- High sensitivity (1 part of solute or less in 10^6 parts of eluent)
- Low noise levels
- Large linear dynamic range (10^3–10^6 for straightforward quantitative analysis)
- Insensitivity to variations in operating parameters (p, T, flow-rate, etc.)
- Gradient compatibility
- Universal and/or selective
- Capable of providing information on the identity of the solute
- Fast response time
- Small detector volume
- Sample preservation
- No remixing of components upon passing the detector
- Long-term operational stability
- Ease of operation and reliability
- Low investment and operating costs

detection) to obtain complementary information in a single chromatographic run.

Specifications for modern detectors in HPLC are given by Hanai [538] and comprise spectroscopic detectors (UV, F, FTIR, Raman, RID, ICP, AAS, AES), electrochemical detectors (polarography, coulometry, (pulsed) amperometry, conductivity), mass spectrometric and other devices (FID, ECD, ELSD, ESR, NMR). None of these detectors meets all the requirement criteria of Table 4.40. The four most commonly used HPLC detectors are UV (80 %), electrochemical, fluorescence and refractive index detectors. As these detectors are several orders of magnitude less sensitive than their GC counterparts, sensor contamination is not so severe, and generally less maintenance is required. Most detectors employed in HPLC are concentration dependent.

HPLC detectors are often classified (in a misleading way) as bulk property or solute property detectors (Figure 4.11), or as nonselective and selective detectors [561]. Bulk property detectors are mostly adversely affected by small changes in the mobile phase composition and temperature, which usually precludes the use of such techniques as flow programming or gradient elution. *Solute property detectors* respond to a physical or chemical characteristic of the solute which, ideally, is independent of the mobile phase. Selective detectors in HPLC are UV, FTIR, F, ECD. Among the photometric detectors, the UV detector is the most frequently used. The popularity of UV detection in HPLC is primarily due to its sensitivity towards a large number of constituents in a range of 210–280 nm. Actually, chromatographers demand the option of monitoring at wavelengths as low as 190 nm so as to pick up compounds lacking chromophores. UV detectors can be used with gradient elution provided solvents do not absorb significantly over the wavelength range that is being used for detection. UV cut-off values (at which transmission falls to 10 %) of some typical HPLC solvents (nm) are given in Table 5.4. In RPLC the most commonly used organic solvents are methanol and acetonitrile, since these do not absorb appreciably above 200 nm; ethanol and propanol are also relatively transparent to 200 nm. Unstabilised THF may be useful above about 240 nm. In NP operation, solvent selection requires more care as many solvents appropriate for chromatography absorb UV light very strongly. Solvents that can be used in NPLC (e.g. with a polar stationary phase such as

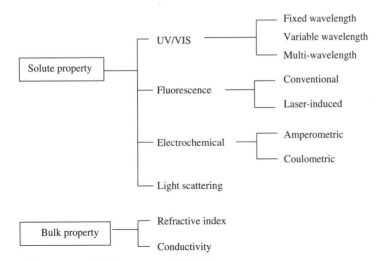

Figure 4.11 Schematic classification of different types of HPLC detectors

Table 4.41   Characteristics of common HPLC detectors

Detector	Type[a]	Maximum sensitivity[b]	Flow-rate sensitivity	Temperature sensitivity	Gradient compatibility
UV/VIS	SP	10^{-9}	No	Low	Yes[d]
Electrochemical	SP	10^{-12}	Yes	1.5 % per °C	Yes
Fluorescence/LIF	SP	$10^{-11}/10^{-14}$	No	Moderate	Yes
Light scattering (ELSD)	SP	10^{-8}	[c]	Limited	Yes
Refractive index	BP	10^{-7}	No	$\pm 10^{-4}$ per °C	No
Conductivity (ELCD)	BP	10^{-8}	Yes	2 % per °C	No

[a] SP, solute property; BP, bulk property.
[b] $g\ mL^{-1}$
[c] Not relevant.
[d] UV transparency required.

silica gel) are *n*-paraffins, methylene dichloride, THF and (small quantities of) aliphatic alcohols [31]. *Fixed wavelength UV detectors* operate with light of a single wavelength (typically 280, 254, 229 or 214 nm). The choice of wavelength makes the response of the UV detector close to that of a universal detector. Most HPLC separations of polymer additives have been monitored with UV detectors (at 280 or at 254 nm). Some additives, such as fatty acid amides, do not absorb in the usual UV detection range and require RI detection. RI detection is less sensitive than the UV detector and excludes the gradient elution mode of separation. With the *variable-wavelength UV detector* the optimum wavelength can be set for the type of compounds being examined. Consequently, this detector is most commonly used for trace analysis in HPLC with detection levels down to about 0.1 ng of substances containing strongly absorbing UV chromophores, such as aromatic rings. *Multiwavelength UV detectors*, either the dispersion or diode array detector, require a broad emission light source, such as a deuterium or xenon lamp. Common use of the multiwavelength detection consists in selecting a wavelength that is characteristically absorbed by a particular component or components of a mixture in order to enhance the sensitivity to those particular solutes. The resolution of a PDA detector ($\Delta\lambda$) depends on the number of diodes in the array and on the wavelength range covered (state-of-the-art 1.2 nm). PDA can be used to check the purity of a given solute. Diode arrays extending into the visible are valuable as analysis tools for colour problems. For example, quinones, which are often responsible for yellow discoloration in polymers, have unique and strong absorbances at visible wavelengths and can therefore be identified based upon their absorbance bands without isolation. The HPLC-PDA procedure for additive analysis consists of separation of the components, peak purity verification, identification of the type of component utilising relative retention times and matching with a UV

spectra library, followed by quantification utilising calibration factors [197,562]. However, for unambiguous identification of unknown components the selectivity resolution of a UV spectrum is usually unreliable. Moreover, when UV spectra of a class of compounds exhibit only small differences (as for phenolic AOs) the UV technique cannot even be used for reliable verification purposes. RPLC-PDA with spectral sort software has been described [563]. Diode-array detection in HPLC is described in George [564]. IR detectors are very useful, but limited to certain mobile phases which do not absorb at the detection wavelength (cf. Section 7.3.3.1).

In order to achieve detection limits below the ng mL^{-1} range only amperometric, chemiluminescence, radiometric, or conventional fluorescence (CF) can be applied (Table 4.41). *Fluorescence detectors* are generally about 100 times more sensitive and more selective than UV detectors. The selectivity of fluorescence detection is due to the fact that only aromatic and conjugated molecules can be analysed, and by applying specific excitation and emission wavelengths the selectivity can even be increased. Pre- or postcolumn derivatisation in HPLC is a technique that is most commonly performed prior to UV absorption or fluorescence detection [565]. In LIF detection systems, excitation power may be increased up to six orders of magnitude compared to CF detection. Most LC-LIF detection concerns underivatised polynuclear aromatic hydrocarbons (PAHs) and fluorescing dyes (e.g. polymethines). Because only a limited number of analytes possess native fluorescence, derivatisation of the analyte before detection is normally required in trace analysis of organic solutes by means of LIF detection. LIF detection in HPLC was reviewed [566]. Apart from LIF, other laser-based HPLC detectors are LS and Raman. Laser spectroscopic methods for detection in LC have been reviewed [567].

Electrochemical (amperometric) detection is used especially for μHPLC and offers high sensitivity and selectivity for analytes that are electroactive at a modest potential (e.g. quinones, nitro and nitroso compounds by reduction, phenols and aromatic amines by oxidation). *Electrochemical detection* in HPLC has been reviewed [568]. HPLC detection may also be achieved by *light scattering* modes for nonchromophoric molecules. ELSD can be employed with solvent gradients, as opposed to RI detection. ELSD is readily interfaced to HPLC, μHPLC, SFC and HT-SEC. HPLC-ELSD is useful as a first screening (verification, no identification) in polymer/additive analysis and for quantitative evaluation. In the MAE-HPLC-PDA-ELSD method for the analysis of additives in polyolefins [471] the proposed gradient composition (narrow UV window, $\lambda > 230$ nm; measurement at 273 nm) is not quite ideal as many additives exhibit high sensitivity between 200 nm and 230 nm. ELSD operation is also not optimal for analytes eluting at different gradient compositions (response linearity problems). Drawbacks of ELSD are relatively low sensitivity and poor detection limits but these problems can be overcome by analyte-specific optimisation. At this point, however, it is questionable whether HPLC-ELSD with analyte-specific optimisation is more advantageous than reliance on universal HTGC and HPLC methods for routine analysis of unknown samples. Replacement of MS detection by the inexpensive ELSD during the optimisation stage of a separation method is convenient since the elution condition requirements are the same for both detections. HPLC-ELSD detection has been reviewed [569] and applications were described [570].

Bulk property detectors generally have neither the sensitivity nor the linear dynamic range of solute property detectors and, as a consequence, are less frequently used in modern HPLC analyses. Furthermore, none can be used with gradient elution, flow or temperature programming and therefore they place considerable restrictions on the choice of chromatographic system. In analogy to the equally non highly sensitive ELSD, the RI detector is used for substances that do not absorb UV light, or those which are nonconducting and do not fluoresce [31]. Essentially, RID will detect all substances with a refractive index differing from that of the mobile phase. The universal RI detector only operates in isocratic elution, as opposed to ELSD which can be applied also in gradient elution conditions. A niche application for RID is in molecular characterisation. Basically, there are four types of *universal detectors* which can be applied to HPLC: refractive index, conductivity, density and evaporative detectors. Unlike FID for GC, a sensitive universal detector for HPLC does not exist. The most obvious need in chromatography is a good mass-type detector for HPLC. Despite the enormous

increase of HPLC applications, detection of solutes without a suitable chromophore is still a limiting factor. In particular, there is no sensitive, nonspecific detector available suitable also for gradient elution. ELCD is used mainly for ion detection in ion chromatography. Less frequently used HPLC detectors are SCD (quantitation of unknown compounds), ECD and PDECD (no halogenated solvents), NPD (no nitrogen-containing solvents) and PDPID. HPLC-MS couplings have been described in Section 7.3.3.2.

Derivatives are prepared in LC primarily to improve the response of an analyte to a particular detector and less frequently to improve the stability of the analyte in a particular separation system, or to improve the chromatographic separation of a mixture containing overlapping peaks [35,565].

HPLC (in both NP and RP modes) is quite suitable for *speciation* by coupling to FAAS, ETAAS, ICP-MS and MIP-MS [571,572]. Coupling of plasma source mass spectrometry with chromatographic techniques offers selective detection with excellent sensitivity. For HPLC-ICP-MS detection limits are in the sub-ng to pg range [36]. Metal ion determination and speciation by LC have been reviewed [573,574] with particular regard to ion chromatography [575].

The object of LC is rarely to determine the global composition of a sample. It is used more frequently to precisely measure the concentration of a species present. In *quantitative analysis*, use of a universal detector is not necessary. For accurate quantitative measurement selective detectors are needed or detectors able to remove the eluent from the solute before monitoring. This limits the choice mainly to UV/VIS, FTIR, amperometric detectors, fluorimeters and laser light-scattering detectors. In the related field of food analysis three types of in-line HPLC detector are used to determine additives such as vitamins: absorbance, fluorescence, and electrochemical. Quantitative HPLC evaluation of each separated additive is generally carried out by the external standard method (average analytical error 2 % or better, excluding the variability of the extraction step). If more precise results are required, internal standard techniques may be applied. The use of an internal standard such as 5′-(octadecyloxy)-*m*-terphenyl-2′-ol is advantageous because it can take into account additive losses during sample preparation. The limits of additive detection in polyolefins are usually in the 10–60 ppm range [576]. This is quite sufficient for the analysis at additive levels of about 500 ppm, which are typical for most commercial polyolefins. HPLC is well suited for the quantitative determination of complete groups of closely related compounds.

Table 4.42 Main characteristics of HPLC

Advantages

- Only small sample required (<1 mg)
- Nondestructive
- Detection of additives of high polarity or high-MW in complex matrices
- Large specificity
- Accomodates volatile, nonvolatile and thermally unstable substances
- High technical level, selectivity and versatility
- Wide range of stationary phases and detectors
- Possibility of using gradient elution
- Quantitative analysis with sufficient precision (0.3 % at best)
- Low detection limits (LOD <5 ppm)
- Rugged, robust
- Fairly straightforward preparative scale
- Wide applicability
- Automation

Disadvantages

- Time-consuming sample preparation (by extraction or dissolution/reprecipitation)
- Slow (ca. 1 h)
- Restricted resolving power
- Lack of a highly sensitive universal detector
- Unreliable interlaboratory method transfer
- Regular maintenance requirements (e.g. eluent supply)
- Column ageing
- Use of large amounts of solvents (preparative scale)
- Analytical skills required
- Relatively high cost

The current status and further prospects of HPLC detectors have widely been reviewed [35,577–580].

Table 4.42 summarises the *main characteristics* of HPLC. The main assets of modern HPLC instrumentation are gradient reproducibility, flow-rates from $50 \, \text{nL} \, \text{min}^{-1}$ to $10 \, \text{mL} \, \text{min}^{-1}$, highly efficient handling of minor amounts of sample, fast equilibration characteristics for fast gradients, high UV sensitivity, flexible use of PDA and FLD, theoretical plates $>100\,000 \, \text{m}^{-1}$. HPLC separation does not depend on the volatility of the analytes. Selectivity in HPLC is obtained by choosing optimal columns and setting chromatographic conditions, such as mobile phase composition, column temperature and detector wavelength. On-line sorptive preconcentration may be used in connection with HPLC. The main drawbacks of HPLC are the lack of a high-sensitivity universal detector and the insufficient chromatographic efficiency to separate many of the complex polymer extraction mixtures. The chemical similarity of many additives and the limited chromatographic efficiency of packed columns, even when employing special column packings and gradient schemes, is still a major problem with HPLC methods. HPLC is probably the most

general quantitative polymer/additive analytical method in use at present. There is no single column that will successfully separate all of the components of interest if one has a broad based analytical interest in additives. However, two analytical columns, such as a phenyl type and C_8 type column, cover a broad range of polymer additives. A small percentage of organic compounds may react with the stationary phase of some columns. Proper choice of conditions, in order to prevent undesirable side reactions, allows virtually any organic compound to be analysed by HPLC. Problems encountered when high-MW polymer is injected onto a HPLC column include: irreversible absorption onto the stationary phase (or slow resorption), precipitation, or co-elution with desired components. Usually sample preparation steps are needed to separate the polymer from the lower-MW components prior to actual chromatography. The use of guard columns (very short precolumns that act as a filter for any minute particles in the sample) is optional but usually recommended.

Advanced *method development* in HPLC is well understood [581]. Optimisation of the main experimental parameters relate to: (i) stationary phase chemistry; (ii) reversed vs. normal-phase option; (iii) mobile phase (composition, gradient, flow-rate); (iv) temperature programming; and (v) detector choice. Understanding of the fundamental descriptors for isocratic and gradient elution are crucial to a successful method development strategy [582]. There is general consensus about the critical parameters for the HPLC process. HPLC suffers from a lack of long-term reproducibility, although the repeatability of retention times can be excellent. The transfer of methods between laboratories is often unreliable and hence the ability to identify analytes other than by direct comparison is limited. HPLC method development is a challenging task by the multitude of stationary phases available and solvent choice. As to the latter, two factors need to be considered: (i) solvent chemical and physical properties; and (ii) effect of these properties on the chromatographic process. A HPLC solvent guide is available [583]. Development of an HPLC method starts with the determination of the solubility of the sample components. Scheme 4.5 summarises appropriate separation methods. Since the mobile phase governs solute–stationary phase interactions, its choice is critical. Strategies for optimising mobile phases for HPLC have been discussed [583a, 584]. The mobile phase should not degrade. The fastest way of developing a new method is to use a mobile phase gradient (from a weak solvent to higher solvent strength). Column kits that provide different NP and RP chemistries are ideal for method development. Rational selection of optimised experimental conditions for chromatographic

Table 4.43 Detector options

Detector	Solvent requirements
UV/VIS, F	UV-grade non UV-absorbing solvents
RI	No mobile phase gradients
ECD, ELCD	Conducting mobile phase
ELSD, MS	Volatile solvents and volatile buffers

Table 4.44 Parameters influencing selectivity

Parameter	HPLC	SFC	CZE
Stationary phase	++	+	−
pH	++	−	++
Ionic strength	+	−	−
Modifier	++	+	−
Ion pairing agent	+	+	+
Pressure	−	++	−
Temperature	+	+	−
Gradients	++	++	−

separation of analytes is realised nowadays with the help of specialised computer programs which incorporate chromatographic theory and real data to facilitate optimisation of resolution and run time [585–589]. Stability of test and standard solutions should be assessed at least over a period of time covering preparation and analysis. Peak symmetry in HPLC gives an indication of column performance. Detector choice is related to solvent requirements (Table 4.43). For method development in HPLC, see references [581,590].

The final step of method development is *validation* of the HPLC method. Optimisation of chromatographic selectivity [110], performance verification testing of HPLC equipment [591], validation of computerised LC systems [592] and validation of analysis results using HPLC-PDA [34] were reported. The feasibility of automated validation of HPLC methods has been demonstrated [593]. Interlaboratory transfer of HPLC methods has been described [594].

Since liquids can solvate and elute perhaps 90 % of all compounds, HPLC is a nearly universal separation technique. Analysis of extracts by HPLC-PDA has one inherent advantage over GC. The extract chromatogram is usually much simpler than that of a GC-FID analysis since most co-extractives do not contain UV chromophores. Most organic compounds do, however, respond in a FID. As a result, quantitation from a UV chromatogram may involve less error due to more reliable integration. Unfortunately, the figures of merit of HPLC (speed, efficiency, resolution, detection) are relatively poor. Nonetheless, HPLC is not a less powerful technique than GC because modifying the composition of the mobile phase can increase the resolution. On the other hand, the range of applicability of HPLC is the widest of any chromatographic technique. Applications include most of the compounds for which GC is a poor choice, including nonvolatile, thermally labile, high-MW (from about 50 Da to several thousands), and multifunctional polar molecules. HPLC is particularly useful when medium size compounds such as antioxidants, curatives, plasticisers are to be studied. As most HPLC separations are operated at ambient temperature decomposition of samples, which is always a concern with GC, rarely occurs. Highly accurate, almost

universal detectors, such as UV/VIS, make quantitation far easier than with TLC.

Packed column SFC and CE are both able to make inroads into the application area served by HPLC, but from opposite extremes of polarity and with little overlap. CE is likely to be more efficient and faster, but mostly applicable to very polar molecules and ions. SFC qualifies as a more reproducible, trace technique, with greater selectivity and multiple detection options. HPLC and CE have been compared [365]. Owing to their orthogonality, CZE and SFC are worth developing, not in competition or as an alternative to HPLC, but as an additional method in order to augment the information obtained from the analysis. With the broad scope of possible eluents and stationary phases, HPLC has fewer constraints than SFC and CZE. The parameters influencing selectivity may be used as a guide to optimisation (Table 4.44).

HPLC should be used when a small number of components have to be separated in a short time. When a high number of plates is required, SFC and CZE are preferred. In contrast to HPLC and SFC, CZE can only be used in a miniaturised form, resulting in constraints in terms of loading capacity and sensitivity. CZE has a limited application in trace analysis. The greatest advantage of HPLC over SFC is that a large number of stationary phases and mobile phases are available with a wider range of polarities. Organic modifier gradients can be run in HPLC and SFC, but not in CZE. In SFC the density (pressure gradients) can be used to change the selectivity rapidly. In CZE selectivity changes are brought about by changes in pH.

HPLC is an excellent tool for quality control, process development, troubleshooting and research. HPLC has recently been reviewed [240,595], in particular with regard to detection [561,596], RPLC [597] and analysis of metal complexes [598]. Recent technological developments in HPLC were highlighted [599], including ultra performance liquid chromatography [535a]. Monographs dealing with HPLC are numerous [365,600], some being more specific: microscale HPLC [553],

HPLC detection [578,601], gradient elution [602], RPLC [603], HPLC columns [549] and derivatisation reactions [604]. The *Journal of Chromatography A* **656** (1993) deals extensively with reversed-phase chromatography.

Applications Chromatography is a preferred technique for additive analysis as it allows both separation of additives in a mixture and subsequent quantitation. Despite the developments in GC, this technique cannot separate many polymer additives. Even with its lower efficiency in comparison to GC, HPLC is today one of the cornerstones in a polymer additive laboratory. Judging by the number of publications in recent years, HPLC is first among analytical methods for additives (confirmation/identification/quantification). Most additives may be analysed by HPLC if they can be dissolved in an HPLC solvent and absorb UV light. Typical polymer/additive analyses are carried out using LPE followed by HPLC with UV or RI detection [605–611]. Verification of the identity of an analyte is then based on a combination of retention time, UV and RI evidence. RPLC is used most frequently for polymer/additive analysis, but normal-phase and SEC are also used. Consequently, techniques for additive analysis by HPLC are legion.

HPLC methods of determining the amounts of different additives in polymeric materials are preceded by an extraction process or dissolution of the polymer matrix. Although extraction-HPLC is often observed to be superior to the traditional spectroscopic techniques (UV and IR) in analysing additives, it is frequently difficult to obtain reproducible results in view of the variability of the extraction yield. On the other hand, it is equally difficult to obtain quantitative data in the dissolution/reprecipitation-HPLC method because of entrapment of analytes in the polymer precipitate and the potential for high absorption of the additives on the polymer surface.

Normal-phase chromatography is still widely used for the determination of nonpolar additives in a variety of commercial products and pharmaceutical formulations, e.g. the separation of nonpolar components in the nonionic surfactant Triton X-100. Most of the NPLC analyses of polymer additives have been performed in isocratic mode [576]. However, isocratic HPLC methods are incapable of separating a substantial number of industrially used additives [605,608,612–616]. Normal-phase chromatography of Irgafos 168, Irganox 1010/1076/3114 was shown [240]. NPLC-UV has been used for quantitative analysis of additives in PP/(Irganox 1010/1076, Irgafos 168) after Soxhlet extraction (88 %

recovery in 7 h). To prevent hydrolysis of Irgafos 168 use of dried mobile phase is required. Mobile phase optimisation for the separation of four non or partially aqueous soluble AOs (BHT, BHA, Irganox 565/1010) and two UVAs (Tinuvin 327/P) was reported [608]. Direct analyses of anhydrides, based on spectroscopic techniques (e.g. IR and NMR) to monitor functional groups with poor quantitative precision at low (<0.1 %) concentrations, are often difficult due to their susceptibility to hydrolysis. Indirect techniques such as titration, colorimetry and GC involve hydrolysis and derivatisation. For analytes susceptible to hydrolysis, the use of nonaqueous solvent systems is essential and NPLC is the preferred choice of analysis. Patterson and Escott [617] have developed direct NPLC analysis using both isocratic and gradient systems with UV detection of a wide range of (di)anhydrides including phthalic anhydride (used in both composite resin formulations and the plasticiser industry) and maleic anhydride. Acid impurities at 5 ppm level could be detected. NPLC-ELSD has been used for the determination of low concentration (>8 ppm) paraffin waxes in food packaging materials free of polyolefin oligomer [618]; the paraffin type was identified by LC-GC-FID and FTIR. HPLC offers considerable advantages over GC for the analysis of Vitamin E structures, such as greater chromatographic and detection selectivity, less complicated sample preparation and no need for derivatisation. Ball [619] has used NPLC to determine Vitamin E in food extracts. NPLC is capable of separating isocratically all of the eight tocopherols and tocotrienols. One of the more important uses of NPLC is in fact the separation of isomers. Al-Malaika *et al.* [620] have described an extensive study of HPLC separation of stereoisomers and spectroscopic (FTIR, UV, NMR, MS) characterisation of other oxidation products of *dl*-α-tocopherol (trimers, dihydroxydimers and spirodimers) formed in PE and PP during melt processing. Munteanu [576] has given examples of various polymer additives and solvent extracts that have been separated by gradient-elution NPLC. Early NPLC work used for separation of additives from polyolefins [621–623] did not prove to be as selective as RPLC [624]. However, Nielson [625] used both NPLC and RPLC for the analysis of HDPE containing (BHT, BHEB, Irganox 1010/1076), (BHT, Irganox 1010/1076), (MD-1024, Isonox 129, Irgafos 168) and (BHEB, Isonox 129, Irgafos 168). The results obtained by means of NPLC for HDPE/(Irgafos 168, Isonox 129, Irganox 1010/1076) were nearly as good as for RPLC.

Method development starts from an understanding of the physical and chemical properties of the sample molecule and/or impurities present. If the sample is of low-MW (<2000 Da) and soluble in reversed-phase (water/organic) solvents, then RPLC is one of the methods of choice (Scheme 4.5). *Reversed-phase chromatography* is the most common method for additive analysis. Many RPLC analyses of polymer additives have been performed in isocratic mode [576]. In some cases very complex mixtures, with over 20 compounds of different polarity, were separated. The main challenge in these analyses was to define the optimum strength of the mobile phase for separation of the majority of mixture components. Also GE-RPLC is extensively being used for analysis of polymer additives, as is quite apparent from the review by Munteanu [576]. It is common to employ a solvent gradient when analysing an unknown mixture or when the various components require different solvent composition for elution. RPLC separation of polymer additives combined with PDA detection and spectral sort software has been described [563]. RPLC-ELSD has been used for detection of various additives, such as Irganox 109/1076, Irgafos 12/38/168, Chimassorb 119 and Tinuvin 622, FS 210 FF and Tinuvin 622 [626]. The hydrolytically and thermally stable diphosphite Doverphos S-9228 [bis-(2,4-dicumylphenyl)spiropentaerythritoldiphosphite] and Doverphos 4-HP (tris-nonylphenylphosphite) in PP and HDPE film (at 600 to 1500 ppm level) were extracted (US, 2 h) and analysed by RPLC (water/ACN) [627]. ^{31}P NMR and DP-FIMS were also utilised in verifying the structures of the oxidation products. Oxidation of phosphites to mono- and diphosphates in PP was followed as a function of temperature and number of extrusion passes. Analysis of BHA and BHT in chewing gum by means of RPLC-UV or EC detection has been described [628]. Also the autoxidation of the AO 6-ethoxy-1,2-dihydro-2,2,4-trimethylquinoline (EDTMQ) was studied by RPLC [629]. RPLC is equally a very useful technique for analysis of trace levels of monomers and additives in aqueous food simulants, provided the substance contains a UV chromophore.

Many industrial laboratories conducting significant amounts of additive analyses have developed a *'universal' HPLC method* which may be used to separate most of the additives of interest. Thomas [417] has reported a method that can separate over 20 common primary and secondary stabilisers. Verdurmen *et al.* [197] employ a gradient ranging from 60 % acetonitrile/40 % water to 100 % acetonitrile; subsequently, all components are eluted off the column in isocratic mode. Irganox 1063 is used as a suitable internal standard since this compound is not frequently encountered in commercial polymers, elutes without overlap to other additives and shows good UV absorbency. In order to develop a universal method from the standpoint of sensitivity, reproducibility, spectra matching, and resolution, certain criteria need to be met. RPLC is the most reproducible mode. Skelly *et al.* [563] have developed spectra sort software in connection with a universal RPLC-PDA method for polymer additives. For that purpose a database was set up consisting of response factors and UV peak areas $\mu g^{-1}\,mL^{-1}$ at 210 and 280 nm (partial listing for 51 PS and ABS additives in Skelly *et al.* [563]). For greatest sensitivity, a monitoring wavelength of 210 nm is recommended, which requires use of a UV transparent solvent such as acetonitrile. A highly efficient 3-μm reversed-phase column imposes itself for resolution of a vast number of polymer additives. For separation of additives with a wide range of polarities gradient elution (from 40 % acetonitrile–water to 100 % acetonitrile) rather than isocratic elution is recommended. Changes in retention time which occur due to differences in gradient formation (acetonitrile-based) or column degradation were accounted for. In a synthetic mixture containing 38 additives at nominally 10 $\mu g\,mL^{-1}$ in solution, examined by gradient RPLC, the spectra software correctly identified 35 of the components. Extracts of various polymer matrices (PS/(nonylphenol, Ionol, DBDPO); PE/(Irgafos 168, Irganox 1076, Ultranox 626) and PC/(Tinuvin 234/350D)) were also analysed correctly, besides showing the presence of monomers, oligomers and hydrolysis products. Smith and Taylor [630] have reported HPLC method development for a solvent mixture of 15 polymer additives (AOs, UVAs, process lubricants, FRs and antistatic agents). Only nine of the additives could be assayed by means of RPLC. Attempts to develop a 'universal' method for PS additive packages failed. Water-soluble compounds such as acrylamide, methacrylamide and methacrylonitrile have sufficient lipophilic character in order to be retained and separated in RPLC columns using water as the eluent [631,632]. Acrylic acid monomer in polyacrylates was determined by US extraction and HPLC [633]. GC and GE-HPLC were used both for the analysis of additives and free *monomers* (e.g. diisocyanates, diols, epoxy compounds, diketones, diacids) affecting the performance of coatings; in particular, GE-HPLC was used to determine toluene diisocyanate and isophthalic acid [634]. HPLC is also the method of choice for the analysis of terephthalic, isophthalic and 1,2,4-tribenzene carboxylic acid in PBT. GC and HPLC techniques were found to be in good agreement in determining the content of free phenol and alkylphenol isomers in technical and formulated arylphosphates (plasticisers and FRs) at concentration levels below 0.1 % [635]. HPLC has also been used to analyse traces of methylene bis(aniline) in polyurethanes [636].

HPLC is the most widely used technique for characterising oligomers (other techniques are SFC, SEC, LDMS, TLC). Some examples of oligomers separated by HPLC are PEG [637], PC [638], PS [639], PA6 [640] and PET [641–643]. The high resolution of modern HPLC columns and high sensitivity of UV detection allow individual *oligomers* (including positional isomers and secondary compounds) to be separated and identified. The peak distribution pattern often permits identification of commercial products. In fact, HPLC analysis of polycarbonate extracts gives besides information about the additives also a fingerprint of the oligomers of PC. As the oligomer patterns of various manufacturers differ, this may be used to attribute ownership; even grades of different MFIs can be distinguished [644]. For determination and quantification of oligomers from PET various methods have been proposed to differentiate between the species, i.e. extraction (Soxhlet, MAE, Parr) followed by gradient HPLC-UV [642,645]. GE-HPLC has been used for determination of oligomers in dissolved PET [141,646] and epoxy resins [646]. HPLC can also be applied to resins without UV-active groups by using mass detectors. Styrene oligomers have been analysed by RPLC [647]. NPLC with a selective eluent gives separations according to the number of oligomer units or to the stereoisomers of individual oligomers [138]. Isotactic, syndiotactic, and atactic isomers of polystyrene oligomers from an 800 MW monodisperse PS standard were resolved on a C_8 column (RI) using a 100 % propylene carbonate mobile phase [648].

A simple RPLC system was developed for the determination of PEGs with $M_r < 600$ Da [637]. On the other hand, NPLC using an amino column, derivatisation by 3,5-dinitrobenzoate and gradient elution, gives better resolution for oligomers of PEGs of average M_r 400 to 2000 Da. Other HPLC-UV studies of oligomers of PEG and MPEG have appeared [649,650]. Cyclic and linear PA6 oligomers have recently successfully been detected by RPLC-ELSD on a C_{18} modified silica-based column in a mobile phase of 65–95 % formic acid in water [651]. GE-HPLC (25–75 % MeOH/H_2O) with CLND detection was used for quantitative oligomer analysis of PA6 extracts. RPLC using a C_{18} ODS column, combined with SPE, has been used for the separation and characterisation of alkylphenols (APs) and alkylphenol polyethoxylates (APEOs) in real samples [652]; these oligomers have been difficult to separate using NPLC methods.

Oligomeric additives with broad MWD tend to be a problem in conventional HPLC conditions. In cases where no interest exists in the oligomer distribution it is common practice to solve the problem by creating a uniform structural unit useful for analysis. For example, isocratic (or gradient) LC-UV was used for the determination of the polymeric light stabiliser Tinuvin 622 in polyolefins using dissolution (toluene)/derivatisation (TBAH)-precipitation (alcohol); the diol formed was quantitatively determined by NPLC [653].

Because of the reactivity, polarity, thermolability, and volatility of certain additives the chromatographic separation of polyolefin extracts is usually performed by HPLC [557,654]. HPLC analysis of *antioxidants, light stabilisers* and other polymer additives are carried out only by LSC and BPC, rather than by LLC. The advent of stable, reproducible, chemically bonded phases has made bonded-phase chromatography the most widely used separation technique for the analysis of stabilisers. The functionality of the most widely used bonded phases in NP-HPLC are amino and cyanonitrile. NPLC but more frequently RPLC has been used for the separation of AOs. The isocratic mode of separation is preferred for simplicity in the analysis of polymer extracts containing a single AO or a relatively simple mixture of AOs. Separation of a larger number of analytes of very different polarity is generally achieved by GE-HPLC. A large number of AOs and UVAs can be analysed by HPLC using a mobile phase containing *n*-hexane or *n*-heptane as the main component [655]. These combinations of solvents are UV transparent in a broad range of wavelengths and allow detection of very small concentrations of UV absorbing additives. If *n*-heptane (or *n*-hexane) is evaporated from the obtained solution of additives, then the additives may be redissolved in another solvent (e.g. in acetonitrile), thus expanding the scope of the method. Since the majority of additives function as light stabilisers or contain some degree of unsaturation, most additives can be quantitated by the use of HPLC. Munteanu *et al.* [605] obtained reliable data by simultaneously monitoring the elution of 22 stabilisers with RI and UV at 254 and 280 nm. Both UV/RI and especially UV_{254}/UV_{280} response ratios depend on the chemical structure of stabilisers and may be used together with retention time values for identifying the stabilisers and to detect overlapping peaks. The UV_{254}/UV_{280} ratio provides very useful information about the chemical structure of the analytes; PE/(Irganox 1010/1076, BHT) down to 10 ppm has been analysed by rapid extraction followed by HPLC [622].

The concentration of the UV stabiliser 2 hydroxy-4-octoxybenzophenone in *LDPE film* was measured by performing HPLC-UV on a chloroform extract [656]. ASTM D 1996 is a standard test method for determination of phenolic AOs and erucamide slip additives in LDPE using HPLC [657]. Antioxidant concentration levels of *LLDPE formulations* were determined by SFE-HPLC analysis [657,658]. Haider and Karlsson [659] determined Irganox 1010 and Irgafos 168 extracted from *MDPE matrix* on the basis of peak areas obtained by HPLC and absorbance measurements registered by UV spectroscopy at 245 and 280 nm. A HPLC method has been reported for determination of Tinuvin 770 and Hostavin TMN 20 in PP. No interference was observed for BHT, Irganox 1010/1076, Irgafos 168, Ionox 330, Cyasorb UV531, Tinuvin 120/326/327, fatty acids, fatty acid salts and fatty acid amides [470]. RPLC with quaternary gradient elution has separated five chemical groups of additives from *PP extracts*, namely lower-MW di-t-butylphenol (DTBP), UVA (Tinuvin 326), hindered phenolic AOs (Irganox 1010) and phosphorous AOs (Irgafos 168 and Ultranox 626) with their degradation products [660]. Several polymer additives were separated on an aminopropyl column with ELSD detection [345]: Tinuvin 327/770 (isocratic methanol/DCM mobile phase), Atmer 129, Radiamuls 142, GMS (chloroform/methanol gradient). Some typical separations on a C_{18} column have concerned Irgafos P-EPQ and Irgafos 168 (acetonitrile/chloroform; $\lambda = 220$ nm) [345]; Irganox 1010/1076, Similizer BHT (acetonitrile/acetone); Santowhite, Irganox 1010/1330/3114 (acetonitrile; $\lambda = 280$ nm) [661]; Irganox 245/259/565/1010/1035, Anox-3114, Tinuvin P/234/320/326/327/328 (methanol/water) [606]; Hostanox SE-10 (acetone; $\lambda = 220$ nm or ELSD) [345]. Typical HPLC assays were developed for Irganox 245 (MW 586.8), Irganox 1010 (MW 1177.6), Irganox 1076 (MW 530.9), Irganox 1098 (MW 637.0), Irgafos 168 (MW 646.9), Tinuvin 329 (MW 323.4), Irganox 1520 (MW 424.8), Irganox B220 (Irgafos 168: Irganox 1010 3:1), Irganox B215 (Irgafos 168: Irganox 1010 2:1), Irganox B225 (Irgafos 168: Irganox 1010 1:1) and Irganox B900 (Irgafos 168: Irganox 1010 4:1) [662].

Monomeric HALS have been determined by HPLC [470,663]. Excellent separation was achieved for HALS-type samples (Tinuvin 770 and Chimassorb 944) with NPLC-PDA (230 nm) using an amino column with acetonitrile/water as the mobile phase; RPLC using C_{18} or cyano columns was not effective [664].

Munteanu [576] has given numerous examples of HPLC studies of *chemical transformations* of stabilisers in polymers. Such analyses are very useful for various purposes: (i) for determining material balances on production processes: (ii) to verify the purity

of production batches; (iii) to study decomposition pathways; and (iv) for FDA approval and, generally, for a better understanding of the behaviour and performance of these additives. Various chemical transformations of additives during polymer processing or the long-term life of the finished articles have been studied by HPLC analysis of solvent extracts [616,665–667]. Examples are degradation of Goodrite 3114 [666,668], thermal degradation of Irgafos 168, transformation of 2,2,6,6-tetramethyl-4-hydroxypiperidine and 2,2,6,6 tetramethyl-4-hydroxypiperidine-*N*-oxyl [665], and oxidation of Irganox 1010 [669]. Thermal degradation of Irganox 1076 at 250 °C results in mono- and didealkylated derivatives by means of a retro-Friedel–Crafts reaction and formation of isobutylene [670]. The products are spectrally distinct and therefore analysis by HPLC-UV is very convenient [671].

Werthmann *et al.* [672] have used HPLC with UV detection at selective wavelengths for the quantitative determination of AOs (Vulkanox BKF/4020/4010 NA/HS and Irganox 1010/1076) in vulcanised *rubbers* (CR, NBR, NR, NR-SBR, NBR-SBR and EPDM). Light stabilisers in US extracted *PVC* were determined qualitatively and quantitatively by gradient HPLC-UV [673]. PVC/(DBP, DDP, DeBP) and PE/(Irganox 1076, CAO-40, Santonox R) were analysed by Soxhlet extraction followed by HPLC-UV with an accuracy of ±5 % down to 0.01 % [674]. Materials separated from *polycarbonate* by extraction or dissolution/precipitation may be a complex mixture containing monomers, oligomers, UVAs and AOs. RPLC-PDA detected bisphenol A, Irganox 1010 and Cyasorb UV5411 in optical grade PC [675]. The same technique allowed determination of 1 ppm of the hydroxybenzophenone MARK LA-51, a proposed tracer in PC [676]. Dissolution/precipitation extractions and SFE of *polystyrene* resins (containing Irganox 1076, Wytox, Tinuvin P/770, Hostatat) were similar as compared by HPLC using dual detectors (UV, ELSD). Mady *et al.* [509] have reported GC, HPLC, SEC and TLC procedures for the determination of UVAs and AOs in polyurethanes (before and after processing and in service). Contaminants of PUR, diphenylmethane-4,4'-diisocyanate and toluene diisocyanate, were derivatised and determined on a RP-18 column in buffered acetonitrile/water [677].

RPLC-ELSD has been used for analysis of *polyolefin additives* [678]. Improvements in sensitivity by use of narrow-bore columns were reported. MAE-RPLC-ELSD/UV (273 nm) has been proposed as a universal one-step and/or two-step procedure for the analysis of additives in polyolefins (Section 3.4.5.1), with the reference mixture of standard additives for the analysis of OSM extracts being composed of 2,4-DTBP, BHT, palmitamide, monopalmitine, oleamide, monostearine, stearic acid, erucamide, Goodrite 1114, Weston 626, Irganox 1010/1076/1330, DLTDP, Irgafos 168 phosphite/phosphate and DSTDP (sulfoxide) [471]. PO/glycerides may be analysed by means of HPLC-ELSD instead of GC. Yagoubi *et al.* [661] have compared three detection systems (UVD, ECD, LSD) for four phenolic AOs (Santowhite, Goodrite 3114, Irganox 1010/1330) in packaging materials by means of isocratic RPLC. Quantitative analysis of AOs, UVAs and slip agents extracted from PP and HDPE was accomplished by isocratic HPLC [679]. RPLC of three AOs (a benzoate, BHT and Irganox 1076) and erucamide slip agent in an extract of a polyolefin formulation was reported [680].

HPLC can be used to identify soluble residues of *accelerators* and other organic substances in extracts of vulcanisates [681,682]. Vimalasiri *et al.* [160] have summarised some 25 papers on the application of HPLC for the determination of accelerators and AOs in elastomers (1965–1982 period). Günter and Jürgen [683] determined MBT in rubber baby bottle nipples by HPLC. HPLC-CLND/UV (in series and with postcolumn split) has been used for the analysis of amines and other nitrogen-containing compounds without UV chromophores. For *N*-nitroso compounds (NOC) or nitrosamines (and/or nitrosamides, which are potent carcinogens, with or without metabolic activation) various methods have been developed such as TLC, GC, GC-MS and HPLC. For analytical purposes, *N*-nitrosamines are roughly divided into volatile and nonvolatile compounds. A variety of detectors has been used for the HPLC determination of NOC, including UV, F, EC, TEA and postcolumn denitrosation detectors [684]. Since most NOC absorb strongly in the UV region (220–235 and 330–375 nm) UV detectors are very useful. Since most NOC can be reduced or oxidised, electrochemical detectors give a highly sensitive response to these compounds. Of all the detectors, the thermal energy detector (TEA) is most widely used for GC and HPLC determination of NOC. A sensitive HPLC method with fluorescence detection has been described for measuring H_2O_2 and organic peroxides such as dealkylperoxides, diacyl peroxides, peresters, etc. [685]. After HPLC separation a postcolumn photoreactor converts various organic peroxides to H_2O_2 and hydroperoxides, which then react with the derivatisation agents *p*-hydroxyphenylacetic acid (PHPAA) and horseradish peroxidase (HRP) to form the fluorescent PHPAA dimer. ISO 11089 (1997) describes the determination of *antidegradants* in rubbers by HPLC. HPLC-CED

(Coulometric electrochemical detection) has been used for the analysis of AOs in resins and the detection of peroxides [686].

Gross and Strauss [687] have described HPLC of *plasticisers* extracted from rubbers and elastomers. NPLC-RI has been used for the analysis of plasticisers (DIOP, DBP, DEP, DMP) in PVC [688]. Khan *et al.* [689] have successfully separated the high-priority esters DMP, DEP, DBP, DAP, BBP, again used as plasticisers, with isocratic HPLC. However, DEHP could be separated satisfactorily only with gradient elution. HPLC-ELSD has been used for a variety of problems related to the monomeric plasticisers chlorinated paraffin and DIOP, and a DIOA/DBP/DIDP mixture in PVC [690]. Reversed-phases with extended polar selectivity were used to reduce analysis time for wide-polarity-range samples including phthalate esters (DMP, DBP, DOP) [691]. RPLC-UV of six phthalates (DMP, DEP, BBP, DBP, DOP, DNOP) was reported; the same phthalates were separated by GC-FID in 8 min using a temperature program [692]. The contents of plasticisers such as triacetin, triethylcitrate, DMP, DEP, DBP in controlled release coating membranes was determined by RPLC at 217 and 260 nm [693].

Identification of a large variety of technical *flame retardants* in polymeric waste materials (from TV sets to personal computers) was achieved in a single RPLC-UV run [694]. Luyk [695] used semipreparative RPLC analysis to describe the degradation of decabromodiphenyl ethers (DBDPO). Detection in RPLC with UV absorbance or fluorescence is based upon aromaticity rather than the number of bromine substituents. Therefore, RPLC was used as a qualitative analysis instrument to determine relative increases in the yield of hepta- and octabromodibenzofurans after each extrusion cycle of HIPS/(DBDPO, Sb_2O_3) [695]. FR 1808 (octabromotrimethylphenylindane) in PBT has been determined after dissolution (HFIP)/precipitation (acetonitrile) by means of a universal HPLC method using Irganox 1063 as an internal standard [696].

HPLC-UV is a popular technique to analyse *textile dyes* extracted from polyester fibres [697], acidic dyes from wool fibres [698] and basic dyes from acrylic fibres [699]. HPLC provides better sensitivity and resolution than TLC [697–699]. GE-RPLC has been used for the determination of 18 disperse dyes (e.g. Navy D-2G-133, Orange CB, Yellow D-3R and Red D-2G) extracted from polyester [700]. Compared with the traditional TLC method, HPLC offers lower detection limits, better observation of contaminant peaks, and reproducible quantitative results. HPLC has also been used to determine azo dyes [701,702].

Aromatic amines formed from the reduction of azo colorants in toy products were analysed by means of HPLC-PDA [703]. Drews *et al.* [704] have applied HPLC/ELSD and UV/VIS detection for quantifying SFE and ASE extracts of butyl stearate finish on various commercial yarns. From the calibrated ELSD response the total extract (finish and polyester trimer) is obtained and from the UV/VIS response the trimer only. Representative SFE-ELSD/UV *finish analysis* data compare satisfactorily to their corresponding SFE gravimetric weight recovery results. GC, HPLC and SEC are also used for characterisation of low-MW compounds (e.g. curing agents, plasticisers, by-products of curing reactions) in epoxy resin *adhesives*.

Ethoxylated alcohols such as polyoxyethylene glycols, $HO(CH_2CH_2O)_nH$ offer a real challenge to analytical chemists because they are usually composed of complex mixtures of many components, reaching even a quantity of a dozen (polydisperse mixtures). Thirty-one separate peaks were generated during the analysis of nonionic octylphenol polyether alcohol *surfactants* by isocratic RPLC-PDA [705]. Ethoxylated alkylphenols were analysed by RPLC [583]. Nonylphenol ethoxylate surfactants up to 14 ethoxy units were analysed by isocratic RPLC-UV ($\lambda = 229$ nm) [706]. Detection limits of $1 \mu g L^{-1}$ for individual surfactants were reported. HPLC methods for the analysis of nonionic surfactants on normal and reversed-phases, molecular sieves and ion exchangers with UV (preferred), RI, FID, fluorimetric, conductometric, MS and ELSD detection were reviewed [707]. The application of conventional HPLC to surfactants is limited to a molecular weight of about 2000 Da. Higher-MW components require temperature programming HPLC, high-resolution SEC or LD-FTMS analysis. HPLC is also the most convenient technique for the determination of cationic surfactants which are widely used in industrial and commercial formulations, including textile softeners. The method is more suitable than most others, such as GC (derivatisation required), photometry or potentiometry (interferences).

For HPLC analysis of *preservatives* (antimicrobials such as polar organic acids, or antioxidants such as BHA, BHT, PG, TBHQ and tocopherols) a variety of stationary phases, mobile phases and detectors can be used [711]. Common antibacterials such as carbadox, thiamphenicol, furazolidone, oxolinic acid, sulfadimethoxine, sulfaquinoxaline, nalidixic and piromidic acid can be analysed by GE-RPLC-UV (at 254 nm). Collaborative studies have been reported for the HPLC determination of the antimicrobial sodium benzoate in aqueous solutions [712]. Plastics devices used for field collection of water samples may contain polymer additives (such as resorcinol monobenzoate, 2,4-dihydroxybenzophenone or bisphenol A) or cyanobacterial microcystins [713].

HPLC has also emerged as one of the most powerful methods for the instrumental analysis of *food additives*. The analytical problem of the HPLC identification and quantification of AOs in edible oils and fats resembles the determination of AOs in polymers. GE-HPLC-UV/ECD was used for the analysis of 12 food antioxidants; detection limits of 61 pg (UV; MHB) and 360 pg (ECD; BHA) were reported [714]. Common food additives (such as *p*-hydroxybenzoic acid, sorbic acid, *p*-toluic acid and their esters) can be analysed by gradient RPLC-UV (at 230 nm). Food AOs examined by HPLC were BHT, BHA, PG, THBP, TBHQ, NDGA, ACP, Ionox-100, DG and TPA [715]. The use of GE-RPLC to determine quantitatively five AOs (BHA, BHT, PG, OG and DG) in fat or oil extracts has been reported [716]. HPLC-UV was also used in a study on the stability of five commercial AOs in food simulants [717]. ESA [718] has developed a direct gradient HPLC method for the determination of synthetic phenolic AOs in food and oils using a coulometric electrochemical array detector which avoids any concentration step, in contrast to AOAC method 983.15. Food-grade phenol AOs (BHT, BHA, PG, TBHQ) were also determined by HPLC under gradient elution with amperometric detection at a nickel phthalocyanine polymer modified electrode (PME) [719]. This PME can also be used as an amperometric detector under flow-injection conditions [720]. Ritter [721] has addressed migration of the epoxy resin bisphenol-A-diglycidylether (BADGE) and its hydrolysis products in packaging. Analysis was carried out by HPLC-UV after Soxhlet extraction; more sensitive measurements were carried out with fluorescence detection. Flak and Pilsbacher [715] have reviewed HPLC determination of food additives and preservatives. A monograph on food analysis by HPLC has appeared [722].

HPLC has also been used for analysing fatty acid mixtures [708] and for the characterisation of *fatty acids* and their derivatives [709]. Fatty acids are commonly analysed on polymeric RPLC columns. Only multiple unsaturated fatty acids can be detected by UV in HPLC; the others require derivatisation into UV-absorbing or fluorescing derivatives. Simultaneous determination of saturated and unsaturated fatty acids ($C_{12}-C_{24}$) by means of RPLC has been reported [710]. Derivatisation is necessary.

Eleven zinc dialkyldithiophosphates (ZDDPs) in *lubricating oil additives* were separated by NPLC [723]; eight ZDDPs were separated on an ODS column using methanol–ethanol–water with linear gradient elution [724]. RPLC of mixtures of ZDDPs is less successful as the zinc complexes are not very soluble in aqueous-based eluents. Tetraalkyltin compounds were separated in normal-phase mode [725]. Astruc *et al.* [726] described on-line discontinuous detection in HPLC by GFAAS and its application to butyltin compounds at trace level. The chromatographic procedure has been applied to tetrabutyl, tributyl-, dibutyl- and monobutyltin speciation in water.

RPLC-PDA is frequently used for *quality control*, such as the determination of free Irganox 1098 in PA4.6 (at 278 nm; after dissolution/precipitation), of free Irganox 1010/1076 in PP (at 278 nm; after extraction with MTBE, thus avoiding dissolution of polymer waxes), of Luperco 802 in PP (at 218 nm, after extraction with HCCl₃), and of Tinuvin 122 in HDPE (at 225 nm; as diol). The advantages of the use of HSLC over conventional LC in QC of plastics and additives have been demonstrated, e.g. for AOs in PE, mixed phthalate esters and residual terephthalic acid in PET and partially cured epoxy resins [557].

HPLC is expected to become even more important as more additives are being produced with lower volatile properties. HPLC applications demanding columns with i.d. <1 mm are constantly increasing providing the user with access to higher sensitivity and resolution with the capillary and micro techniques.

An excellent and comprehensive review has covered HPLC analysis of AOs and light stabilisers up to 1990 [576]. Normal vs. reversed-phase and isocratic vs. gradient-elution HPLC separation of synthetic mixtures of additives and of solvent extracts from polymers were discussed.

Temperature-programmed Liquid Chromatography

Principles and Characteristics Most HPLC analyses of polymer additives have been performed at room temperature, although isocratic and gradient RPLC of polyolefin additives at 40–60 °C have been reported in the past [557]. The use of temperature as an elution parameter in HPLC has long been realised, both for conventional 4.6 mm i.d. columns [727] and for capillary column dimensions (0.5 mm) [728]. Liquid chromatography with a dynamic temperature gradient was described in 1973 [729]. Jinno *et al.* [730] used temperature programming to obtain shorter analysis time and higher resolution in high-speed analysis ($r_t < 2$ min). Now that HPLC columns are available which can withstand temperatures as high as 200 °C

temperature has become an important variable to be optimised. For the present status of HTLC columns, see Marin *et al.* [730a]. Retention, efficiency, selectivity, column back-pressure and stability of analyte and stationary phase are all influenced by temperature in HPLC. The most important influence of increased temperature is a reduction in the viscosity of the mobile phase and an increase in diffusion rates. Recently high-temperature (HT) and temperature-programmed (TP) separations of polymer additives have been carried out (up to about 150 °C).

To achieve fast response throughout a column from the applied temperature capillary columns and small thermal mass are required. Accordingly, temperature programming can successfully be performed utilising packed-capillary columns, as an alternative to traditional gradient elution [728,731,732]. These columns allow for efficient heat transfer and minimised radial temperature gradients that will reduce the efficiency and show improved compatibility with some detectors. Using higher temperatures allows for the use of shorter, narrower columns in a manner that dramatically improves performance over wide-bore methods. With columns specially made for temperature programming, the repeatability of retention times and the lifetime of columns during temperature programming now approaches that of standard HPLC columns.

Various research groups have examined the possibilities of using elevated temperatures [733] or a temperature gradient in capillary LC [734,735]. Microcolumn separations were conducted by either superimposing a flow and temperature gradient [734] or by combining temperature programming and gradient elution [735]. Replacing gradient elution with temperature programming is possible for many (but not all) reversed-phase applications.

Temperature-programmed packed-capillary LC (TPLC) is a robust and accessible technique which has the potential of making available the separation of complex samples containing relatively high-MW compounds of low water solubility. The variety of stationary and mobile phases available, in addition to the relatively easy way of adjusting the elution strength and solubility of the solutes in the mobile phase utilising temperature programming, makes this technique capable of filling the gap between SFC, GC, HTSEC and conventional HPLC with regard to a niche of applications. TP-μHPLC has been coupled to various detectors (UV, ELSD, ICP-MS). The use of mobile phases with relatively high UV cut-off limits use of an UV detector. Trones *et al.* [736,737] have developed μHTLC-ELSD.

A drawback of the method is the relatively poor sensitivity of ELS detection.

Advantages of temperature programming are: (i) improved separation efficiency at elevated temperatures; (ii) use of selectivity effects as a $f(T)$; (iii) alternative to instrumental problems with solvent gradients in miniaturised systems; and (iv) reduced viscosity of the mobile phase. The benefits of HTLC on packed capillaries are demonstrated by peak shape and analysis time (rapid method development). Temperature programming is not only an alternative to solvent gradient elution, but is required for success in niche applications. Exploiting temperature programming extends the application range of LC. However, elevated temperatures cause concern for decomposition of analytes [737a]. HTLC with packed-capillary columns and nonaqueous mobile phases is a technique suitable for analysing some polymer additives and technical waxes which are often not adequately separated by GC, SFC or conventional LC methods [738].

When large volumes of sample are to be focused in LC, the analytes are traditionally dissolved in a solvent of noneluting properties. In practice, also on-column focusing large-volume injections to packed-capillary columns is possible [739,740]. Molander *et al.* [740] have recently introduced the concept of subambient temperature-promoted large-volume solute enrichment in packed-capillary LC, where the analytes are injected at low column temperatures where elution is suppressed, prior to efficient temperature-programmed gradient elution of hydrophobic analytes. This procedure provides characteristics similar to that of solvent-assisted packed-capillary column switching in LC, allowing for sensitive trace determination of enlarged sample volumes of hydrophobic compounds (such as Irganox 1076) that are not easily dissolved in solvents that are compatible with solvent-assisted large-volume solute enrichment [741]. It was shown that simply by varying the operation temperature during sample preparation and analysis, small volumes (100 μL) of one single solvent can be used as extraction solvent, sample solvent and mobile phase, elegantly simplifying the total process.

Temperature programming in LC was recently reviewed [742].

Applications High-temperature liquid chromatography with packed-capillary columns, nonaqueous mobile phases, and ELSD and ICP-MS detection, has been developed specifically as a robust analytical tool for the analysis of high-MW polymer additives [731,738]. Dissolving such moderately polar, heavy compounds with low water solubility at ambient temperature usually requires a combination of nonaqueous relatively polar mobile phases and high temperatures (isothermal or temperature programming). It follows that the separation system should contain nonaqueous mobile phases and be temperature controlled. Temperature programming has successfully been used for partial characterisation and for purity testing of different polymer additives.

Fatty amides and esters were chromatographed with μHTLC-ELSD: technical grade glycerylmonostearate (70–100 °C range), oleamide, stearamide (stearic acid) and erucamide (80–150 °C range) [736] and GMS 40 (40–160 °C range) [737]; LOD of 3 ng. Trones [738] has reported μHTLC-ELSD separation in isothermal mode of Irganox 1076/3114, Irgafos 168 and chrysene; laser light scattering enables detection of ng amounts. Temperature-programmed μHTLC-ELSD separations were reported for the *antioxidants* Irganox 1076/3114 (50–150 °C range) [736] and μHTLC-UV programming for Tinuvin 327, Irganox 1076 and Irgafos 168 (75–150 °C range) [731]. The latter test mixture could be analysed with the temperature program in about 20 % of the time needed for isothermal analysis at 50 °C. Very low levels of Irganox 1076 (0.6 ppm) extracted (Soxhlet, MAE) from LDPE were determined by LVI-μHTLC-UV (at 280 nm) using a temperature program from 7 to 90 °C; LOD 3.3 ng [740].

The most spectacular results with temperature-programmed LC have been obtained for some notoriously difficult *polymeric additives*. Characterisation of the oligomeric HALS stabiliser poly [[6-[(1,1,3,3-tetramethylbutyl) amino]-1,3,5-triazine-2,4-diyl][(2,2,6,6-tetramethyl-4-piperidyl)imino]-1,6-hexanediyl [(2,2,6,6-tetramethyl-4-piperidyl)imino]] (I) (Figure 4.12) is difficult for several reasons; it has a broad MWD, may contain isomers, and has several amino groups that promote almost irreversible adsorption to silica based column packings in LC.

Nonaqueous packed-capillary temperature-programmed μHPLC-ELSD used for the analysis of commercial products of three different trademarks of (I) (Chimassorb 944, HALS 94 and Uvisol DL 449) showed that the products contained some 40 different homologues and other components [738]. HTLC-ELSD gives evidence that the HAS compound contains 1–20 monomeric units ($n = 1$–20); at the high end this is equivalent to a MW of approximately 9300 Da. This contrasts to the earlier suggestion of $n = 1$–6 based on SEC. HALS 94 and Uvisol DL 449 appear to be almost identical products and differ significantly from Chimassorb 944 (Figures 4.13 and 4.14). Temperature programming enables differentiation between various commercial products, regarding the total number

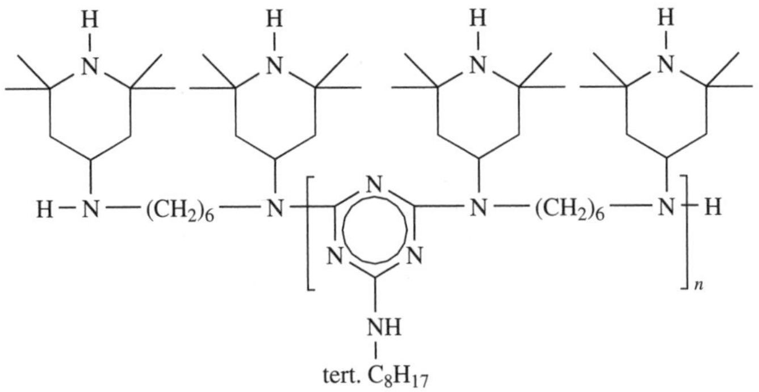

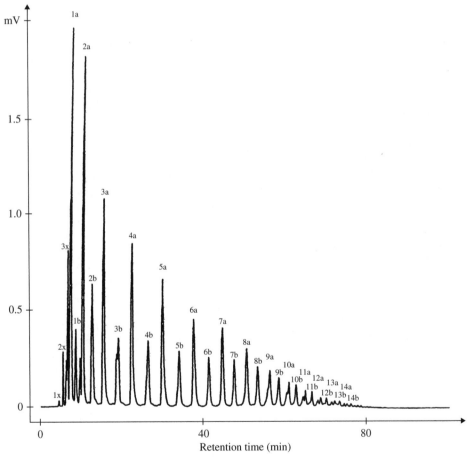

Figure 4.12 HAS compound with IUPAC name poly[[6-[(1,1,3,3-tetramethylbutyl)amino]-1,3,5-triazine-2,4-diyl][(2,2,6,6-tetramethyl-4-piperidyl)imino]-1,6-hexanediyl[(2,2,6,6-tetramethyl-4-piperidyl) imino]]

Figure 4.13   10 µg Chimassorb 944 (dissolved in mobile phase) analysed by use of µHTLC-ELSD. The xa numbers indicate the assumed main units in the oligomer. 1x, 2x, 3x and xb are assumed to be by-products or impurities. After Trones [738]. Reproduced by permission of R. Trones

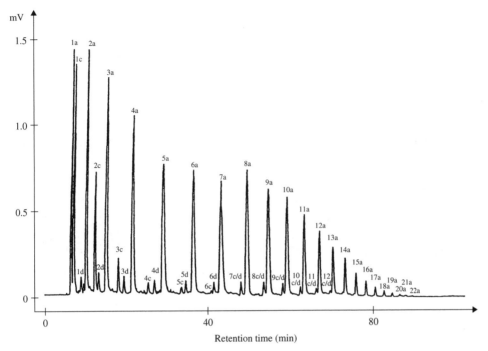

Figure 4.14 10 μg Uvisol DL 449 (dissolved in mobile phase) analysed by use of μHTLC-ELSD. The *x*a numbers indicate the assumed main units in the oligomer. *x*c and *x*d are assumed to be by-products or impurities. After Trones [738]. Reproduced by permission of R. Trones

of components and the quantity of each component [738,743].

Chromatographic separations of different HAS trade products of (I) (eluted as one peak), that were extracted from polymer matrixes, have been performed by PyGC [744], HPLC-FID/UV [745–748] and SEC-UV [749,750]. Freitag [747] has reported an RPLC-UV (at 240 nm) method for the determination of 0.05–0.5 % Chimassorb 944 in polyolefins (RSD 2.2 %). Matuska [746] has described an isocratic HPLC method using dissolution (toluene)/precipitation (methanol) and two LiChrogel SEC columns with THF–diethanol as an eluent and UV detection at 244 nm. Verstappen [751] reported an isocratic HPLC method in which Irganox 1010 (in concentrations not exceeding Chimassorb 944), Irgafos 168 (phosphate), Irganox 1076 and Tinuvin 326

do not interfere. Gharfeh [745] has used size-exclusion nonaqueous reversed-phase chromatography for the separation of Chimassorb 944 from other additives in a PE sample dissolved in decaline. The analysis (with FID detection) was carried out with a step gradient from toluene to 0.1 M piperidine in toluene and can be applied for concentrations ranging from 0.05 to 10 %. Chimassorb 944 is adsorbed to the nonpolar PS/DVB column but elutes in piperidine/toluene. These studies have not resulted in any information of the *oligomer distribution* or possible isomers in the trade products.

Poly-(N-β-hydroxyethyl-2,2,6,6-tetramethyl-4-hydroxypiperidyl succinate) (II) (Figure 4.15) is an oligomer where *n* is assumed to be 2–14, whereas the majority of the product is believed to be $n = 11$–14. The average molecular weight, determined by SEC,

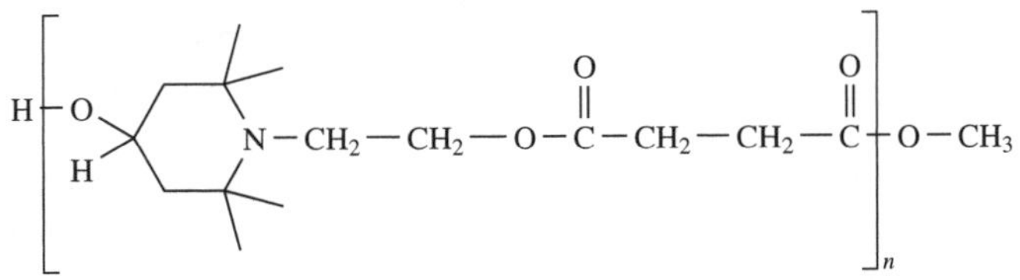

Figure 4.15 HAS compound with IUPAC name poly-(N-β-hydroxyethyl-2,2,6,6-tetramethyl-4-hydroxypiperidyl succinate)

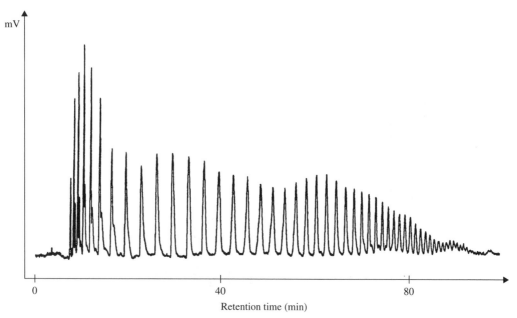

Figure 4.16 12.5 μg Lowilite 62 (dissolved in toluene) analysed by use of μHTLC-ELSD. After Trones [738]. Reproduced by permission of R. Trones

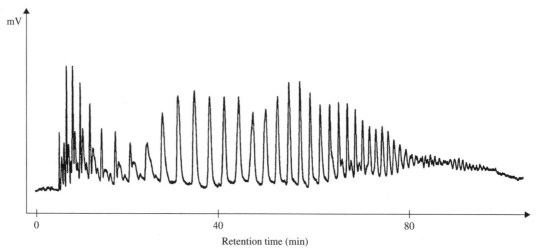

Figure 4.17 12.5 μg Tinuvin 622 LD (dissolved in toluene) analysed by use of μHTLC-ELSD. After Trones [738]. Reproduced by permission of R. Trones

is reported to be 3100–4000 Da [752]. Analysis of this product by GC and HPLC requires derivatisation (using TBAH) [753]. However, this procedure reduces the oligomer to small molecular units (MW ≈ 200), with no information about the oligomer distribution as a result.

Three different trademark products of (II), namely Tinuvin 622 LD, Lowilite 62 and Uvisol DL 226 were examined by μHTLC-ELSD with thermal programming [738,743]. The observed peak distribution pattern with more than 50 different homologues is characteristic of each commercial product (Figures 4.16–4.18). Also in this case the number of monomeric units is significantly higher than originally suggested by the manufacturers; the high end of the MW range appears to be in the vicinity of about 15 000 Da.

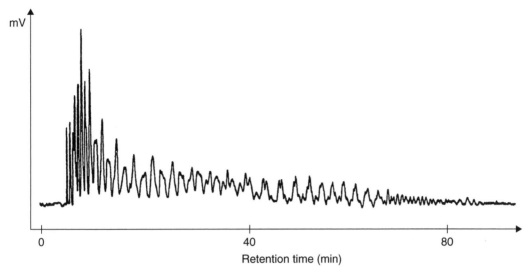

Figure 4.18 12.5 µg Uvisol DL 226 (dissolved in toluene) analysed by use of µHTLC-ELSD. After Trones [738]. Reproduced by permission of R. Trones

Temperature-programmed µHPLC of Irgafos P-EPQ clearly showed that this AO contains five different components: three diphosphonites, one monophosphonite and one phosphite compound (Irgafos 168). TP-µHPLC is clearly superior to SFC [345] for this analysis. The effect of using high isothermal temperatures and temperature programming above ambient (from 50 to 175 °C) in packed reversed-phase capillary LC columns using nonaqueous mobile phases has been used for partial characterisation and for *purity testing* of Irganox 1076, Irganox 1010, Irgafos 168, Irgafos P-EPQ, Irgafos 168-phosphate, Anchor DNPD, and Chimassorb 119 [754]. Also the analysis of polyoxyethylenic *surfactants* by HPLC using temperature programming has been reported [755]. Temperature-programmed packed-capillary HTLC has clearly shown its potential for the characterisation of relatively high-MW polymer additives used as AOs. These are compounds that are not adequately characterised by conventional separation techniques. Further work in this direction might benefit from µHTLC-ESI-MS.

Temperature programming packed-capillary and open capillary HTLC-ICP-MS (up to 200 °C) has been reported [738]. µHTLC-ICP-MS instrumental coupling has been applied to organotin (tetraethyltin, tributyltin chloride and triphenyltin chloride) and organolead (tetramethyllead and tetraethyllead) compounds [756]. HTLC-ICP-MS can be used for the determination of *organometallic compounds* at low concentrations. The observed limit of detection (LOD) was 5 pg for Pb in

tetraethyllead, and 9 pg for Sn in tetraethyltin using HTLC-ICP-MS [738].

4.4.2.3 Size-exclusion Chromatography

Principles and Characteristics The development of gel permeation chromatography (GPC) or gel filtration [15] in the mid 1960s by fractionation of macromolecules according to molecular size by LC has revolutionised the determination of molecular weight distributions (MWDs) in polymers. Since size exclusion (or steric exclusion) by a gel composed of small porous particles is the dominant fractionation mechanism the technique is commonly termed high-performance size-exclusion chromatography (HPSEC) or simply size-exclusion chromatography (SEC). Unlike other modes of LC, such as adsorption, partition or ion exchange, where enthalpic interactions (dipole–dipole, hydrogen bonds, London forces, ion–ion) with the stationary phase or packing surfaces are predominant, entropic effects govern the separation in SEC. SEC separates on the basis of *molecular hydrodynamic volume* or size (Stokes' radius), i.e. on the basis of physical sieving processes (Figure 4.19) and not on chemical phenomena. The size of any polymer molecule dissolved in a solvent is dependent on its molecular weight, composition, branching, microstructure and solvent quality. The separation mechanism is, however, not the only difference between these techniques. The main difference between SEC and other modes of LC is that the information about

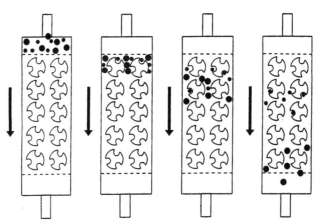

Figure 4.19 Pictorial view of size-exclusion chromatography. After Wu [757]. Reprinted from C.-S. Wu, ed., *Handbook of Size Exclusion Chromatography*, Marcel Dekker Inc., New York, NY (1995), by courtesy of Marcel Dekker Inc.

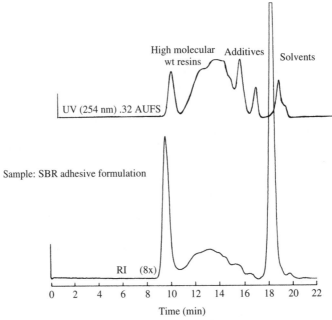

Figure 4.20 Size-exclusion chromatogram of SBR adhesive formulation in THF using UV and RI detectors. After Alfredson and Tschida [760]. Reproduced by permission of Varian, Inc.

the samples is coded into the *absolute* peak position (while in liquid adsorption chromatography the relative elution order is sufficient) and peak shape. In SEC, retention is a function of changes in conformational entropy of macromolecules diffusing into or out of pores with diameter similar to the size of the analyte molecules. Also at variance to other modes of LC, SEC separation is determined exclusively by particle size and pore size distribution of the stationary phase, while the mobile phase should have no effect (good solvent required). Any chemical or electrostatic, attractive or repulsive interactions between polymer and stationary phase have to be prevented. If the retention of a particular compound is influenced by molecular interactions between the solute and the mobile and stationary phases the results are likely to be in error. Whereas for polymer analysis the full exclusion mechanism is crucial, this is not necessary for separation of additives.

In SEC, the polymer is dissolved in an appropriate solvent and is injected into a column packed with chemically inert porous particles of fairly defined pore size. In SEC, solvent quality must be good. Selection of a *stationary phase* is critical. The most frequently used column packings for nonaqueous organic solvents are either based on porous silica or on semi-rigid (highly cross-linked) organic gels, in most cases PS/DVB-based, packed in a stainless steel chromatographic column (5–8 mm i.d., 1 ft long) and swollen by a mobile phase. High-performance column packing materials are still referred to as 'gels' although the highly cross-linked, macroporous polymer beads exhibit high rigidity. Some specialised (silica) columns are available for HFIP applications and plasma treated PS/DVB columns for light scattering detection. For chromatographic reasons, the packing material generally consists of spherical beads with an average diameter ranging between 3 μm and 20 μm and a narrow size distribution. For retention mechanism reasons, the sizes of the pores should have the same order of magnitude as that of the macromolecules to be separated in solution. For technical reasons, it is only possible to synthesise packings with a limited range of pore sizes, and the SEC column is generally an assembly of several columns in series, packed with several gels of different porosities. Column selection involves matching the pore size of the packing to the molecular size of the sample molecules. Since the columns are defined in terms of their molecular weight exclusion limits, it might be necessary to connect several columns (with various pore size, e.g. 60, 100, 250 and 1000 Å) in series to perform a separation of a sample consisting of species of widely different molecular sizes. Alternatively, mixed-gel columns (packed with a blend of pore size media) may be used. SEC packings are available in pore sizes that allow separation over a wide molecular weight range of about 10^2–10^7 Da. Thus, packings with small pores (e.g. less than 100 Å) allow separation of relatively small molecules, with molecular weights up to about 1000 Da. The pore size distribution should be such as to provide sufficient resolution between the lowest molecular weight sample components and the various peaks due to low molecular mass species, such as AOs, added internal markers, etc. In practice, modern SEC for polymer characterisation is generally performed with two to four 1-ft columns of 30 000 to 50 000 theoretical plates with a separation time of generally less than 45 min.

Improved *resolution* in SEC can be readily achieved by: (i) increasing the column length or number of columns used in series (at the expense of additional analysis time and cost); (ii) smaller particle size (down to 3 μm); or (iii) an increase in pore volume of the packing media. Indeed, increasing pore volume and decreasing particle size of porous PS/DVB materials lead to increased resolution compared to conventional media [758]. The current 3-μm columns exhibit extremely high efficiency ($>10^5$ plates m^{-1}). These have particular application in the very low molecular weight SEC separation range, competing with other types of HPLC for resolution of small molecules.

The SEC *mobile phase* is generally the same solvent as used to dissolve the polymer. Requirements for mobile phases in SEC are given in Pasch and Trathnigg [759]. Molecules that are too large to penetrate the pores of the packing elute in the interstitial or void volume of the column (exclusion limit). Smaller molecules penetrate into the pores, access greater pore volume and, as a result, elute at a later time (Figure 4.20). Solvent molecules are normally the smallest molecules in the eluent system and are thus eluted at a retention time known as the 'dead time', t_0. The largest elution volume in SEC is the total mobile phase volume in the column: the sum of pore volume and interstitial volume; this restricts the separation space. The smallest elution volume is the interstitial volume. The entire separation occurs within the volume of the pores, which typically equals only 30–40 % of the total column volume.

The SEC mechanism demands only an isocratic (constant composition) solvent system with normally a single solvent. The most frequently used organic solvents are THF, chloroform, toluene, esters, ketones, DMF, etc. The key *solvent parameters* of interest in SEC are: (i) solubility parameter; (ii) refractive index; (iii) UV/IR absorbance; (iv) viscosity; and (v) boiling point. Sample solutions are typically prepared at concentrations in the region of 0.5–5 mg mL^{-1}. In general an injection volume of 25–100 μL per 300 × 7.5 mm column should be employed. For SEC operation with polyolefins chlorinated solvents (for detector sensitivity and increased boiling point) and elevated temperatures (110 to 150 °C) are required to dissolve olefin polymer. HFIP is the preferred solvent for SEC analysis of polyesters and polyamides.

Instrumentation requirements for SEC are somewhat simpler than those of other modes of HPLC, since mobile phase gradients are not used; however, adequate computer support for data acquisition and processing is essential. *Method development* involves finding a suitable solvent for the sample and choosing a mixed bed column or, more often, a set of columns in series to match the pore size of the column(s) with the size distribution of the sample.

In SEC analysis a restricted number out of the numerous HPLC detectors is usually applied (Table 4.14). *SEC detectors* may be divided in three categories: concentration-based (DRI, ELSD, FID, UV/VIS, IR), molecular weight and structure-sensitive (DV, MS, LS) and chemical composition-sensitive detectors (UV, IR, FL, MS, NMR). When SEC is only equipped with a concentration detector, just molar masses relative to the polymer used for calibration can be obtained. On-line coupling of SEC-DV provides absolute molar mass values for the unknown (calibration needed). SEC-LS directly provides the molar mass of the eluting species without the need of calibrating the system. The differential laser refractometer, which measures the differential refractive index (DRI) between solvent and column effluent, is the preferred concentration detector for most SEC applications. This detector provides a signal proportional to the polymer concentration but does not give structural information. Fixed-wavelength or tuneable UV and IR detectors are also available for SEC of polymers containing suitable chromophores. Both are limited by the need for finding a solvent that is transparent to the incident light. With a solvent frequently used in SEC, tetrahydrofuran, problems with UV transmission are likely to arise from UV absorption due to added stabiliser. Either UV (DAD) or RI detectors are commonly used for SEC analysis of additives, oligomers, or polymers (i.e. when the mobile phase is non-UV absorbing, and no solubility limitations exist). SEC-DV is less appropriate for additive analysis. The column effluent can be collected whenever desired (RI and UV detectors are nondestructive) so as to obtain pure fractions for identification by IR, MS, NMR, etc. A successful approach to combining the separating power of SEC with the identification capability of IR is in using systems where there is a solvent removal step between SEC and IR. The use of an evaporative light scattering detector (ELSD) allows observation of both UV absorbing and non-UV absorbing polymer samples. ELSD does not require the eluent to be UV transparent and thus permits a much wider choice of eluent systems. ELSD is used when RI and UV detection are impossible. This detector is quasi universal being molar mass and composition dependent. Various static laser light scattering detectors (LALLS, DALLS, TALLS, MALLS, RALLS) are used for measurement of absolute molar masses (but cf. Clarke [761]) and in branching studies. An LS detector is more sensitive to the high-MW end than a concentration detector, and less sensitive to the low-MW end.

The trend in modern SEC is to couple one or several other detectors (in particular molecular weight sensitive) to the concentration detector providing complementary information on the polymer, for example the triple detector (DRI/DV/LS) combination [762]. However, in most cases the choice of a suitable detector is dependent on the specific polymer being characterised. Advantages of *multiple detection* are: (i) purity analysis; (ii) quantitative analysis of mixtures (when response factors are known), including additives in polymers; (iii) chemical composition distribution; (iv) absolute MWD (structural information, degree of branching); and (v) branching distribution. SEC-RI/DV for determination of a polymer MWD is becoming routine even though this analytical system is of limited use for the analysis of low-MW polymers (<10 000 Da) [763]. Multidetector SEC was recently summarised by Clarke [764].

SEC is a *secondary method* for determination of molecular masses, i.e. generally establishes *relative* rather than *absolute* molecular weights. The relation between log M and V (elution volume) has to be established by calibration with polymer standards of known molecular weight. An on-line light scattering (LS) detector allows measurement of absolute molar masses without the need for SEC column calibration. In the absence of an absolute molecular weight detector, it is necessary to calibrate the SEC system in order to generate MWD information. Calibration can be carried out with commercially available mono- and polydisperse standards (e.g. PS, PMMA or polyisoprene) [51], leading to 'polymer equivalent' molecular weights. However, whilst conventional SEC calibration plots of log M vs. retention time t_R vary according to polymer type (different sizes in solution), on a plot of log M $[\eta]$ vs. t_R, where M is the molecular weight and $[\eta]$ is the intrinsic viscosity, most polymer types can be fitted in a single *universal calibration* mode [765]. Consequently, it is now well established that for most polymers, by using SEC coupled to viscometry and concentration detectors, the 'true' molecular weight of polymer types other than the calibrants can be obtained. When SEC is combined with molecular weight sensitive detectors (light scattering and viscometry) allowances can be made for differences in chemical type (only if structure effect is present) and structure (branching). An accurate elution volume is critical for molecular weight determination by using universal SEC calibration.

In conjunction with molecular weight sensitive detection systems, SEC can be used for determining various *molecular weight parameters*, such as molecular size, conformation and branching, as a function of MW. Furthermore, by interfacing SEC with spectrometry, polymer compositional heterogeneity can be determined.

Even if SEC has invaded practically all polymer analytical laboratories, there are still many problems which cannot be solved by this technique despite the fact that the number and types of detectors at the end of the column have been multiplied. An alternative to polymer characterisation by means of SEC is the use of laser-based MS methods, in particular MALDI-ToFMS and LD-FTMS, provided low polydispersity (PD <1.05) [763]. The high resolving power achievable with the FTMS instrument allows some molecular information to be inferred from a single mass spectrum. The strengths of MS are in the molecular weight regions where SEC performance tails off or when detailed molecular structures are needed. There are prospects of coupling SEC directly to FTMS by means of the electrospray method (Section 7.3.4.2).

Table 4.45 shows the *main features* of SEC. This technique has become an indispensable tool for polymer characterisation. SEC has some advantages over other LC methods, such as the predictability of the end of a chromatographic run and of the retention times in a calibrated chromatographic system. SEC is an attractive technique for prefractionation or sample clean-up prior to a more sensitive RPLC technique. This intermediate step is especially interesting for experimental purposes whenever polymer matrix interference cannot be separated from the peak of interest [647]. Disadvantages are that the whole separation must be eluted within the excluded and dead volumes. A limitation of SEC occurs when the analyte shows adsorption effects on the column material or when the molecules form aggregates in solution.

In principle, any type of sample can be analysed by SEC provided that it can be solubilised and that there are no enthalpic interactions between sample and packing material. By definition then, this technique cannot be carried out on vulcanisates and even unvulcanised fully compounded rubber samples can present problems due to filler–rubber interactions. The primary use of SEC is to determine the whole MWD of polymers and the various averages (number, viscosity, weight, z-average) based on a calibration curve and to allow qualitative comparisons of different samples. Many commercial polymers have a broad MWD leading to strong peak overlap in the chromatography of complex multicomponent systems.

Although conventional SEC is usually reserved for industrial polymers and biopolymers, it can also be used for screening of unknown mixtures composed of polymeric, oligomeric and low-MW components which do not interfere with other resin constituents. SEC is used to separate polymer from additives. In particular, low-resolution SEC is well suited for *groupswise additive analysis*. The method simplifies additive characterisation since components of the mixtures are grouped in the chromatogram based on molecular weight or hydrodynamic volume, and not boiling point as with GC, or polarity as with HPLC. As a nonadsorptive, isocratic method, the selectivity of SEC relative to that of other high-performance chromatographies is poor. Since separation of homologues differing by one or two carbon atoms is usually easily accomplished by reversed-phase or normal-phase chromatography, SEC has not been seen to be advantageous. Consequently, liquid adsorption chromatography (HPLC, SFC) is generally considered to be a more practical form of chromatography for the purpose of additive analysis; this has limited use of SEC in this area to some special applications. However, with the current high-efficiency columns SEC is now much better equipped as a primary separation technique for low-MW additives than in the past.

Miniaturised SEC uses small fused-silica packed-capillary columns (0.32–1 mm i.d., 30–200 cm) instead of relatively large metal columns. Miniaturisation puts stringent requirements on the quality of SEC columns. Advantages of μSEC are: (i) much smaller amounts of (toxic, expensive) solvents; (ii) smaller samples; (iii) better and easier temperature control; (iv) increased detector compatibility (e.g. MS); and (v) greatly reduced

Table 4.45 Main characteristics of size-exclusion chromatography

Advantages

- Key tool for molecular characterisation (MWD, MMD, FTD, MAD)
- Routinely applicable to all soluble polymers (or extracts)
- High efficiency (automated sample injection carousel)
- Small and predictable electron volumes
- Normally no sample analyte loss or on-column reactions
- Many on-line detectors (IR, MS, LS, UV, NMR, DV)
- Multidimensional chromatography (GC, LC)
- Sample clean-up
- Commercial equipment
- Miniaturisation (under development)

Disadvantages

- Apparent simplicity
- Limited separation efficiency (not for 3-μm columns)
- Limited peak capacity
- Risk of polymer degradation (high-shear forces)
- Need for solubility in an organic solvent
- Need for high-quality solvents (particulate matter-free)
- Molecular separation according to size (instead of mass)
- Calibration essential (not for SEC-LS)
- Critical data acquisition and processing
- Not applicable to samples of similar hydrodynamic size
- Not applicable to analytes that interact with column packing material

costs (columns, and eventually hardware). Such developments in separation require similar advancements in detection techniques, e.g. miniaturised ELSD [737]. Fast μSEC-ELSD/UV using $250\,\mu\text{m} \times 15\,\text{cm}$ columns is currently being evaluated for at-line MW determinations. For practical high-throughput screening (HTS) of large combinatorial libraries fast SEC-ELS methods are available, requiring between 40 s and 2 min per sample [766]. Other developments are electrically (rather than pressure) driven SEC. However, ED-SEC is not likely to replace conventional pressure driven SEC (PD-SEC) in the near future. Suitable detectors for size-exclusion electrochromatography (SEEC) are UV, F, LS and DV.

Aqueous SEC is widely used for the determination of MWDs of a variety of synthetic and naturally occurring water soluble polymers, as well as for separations of small molecules. The column requirements for aqueous SEC are very demanding to eliminate ionic and hydrophobic effects.

SEC can also be combined with other techniques in order to obtain additional information on composition and structure of polymers. Hyphenated and multidimensional SEC techniques are described in Section 7.3.4 and 7.4.3.1, respectively.

Several reviews [767,768] and books [51,757,767, 769] deal with SEC in relation to the molecular weight characterisation of synthetic polymers (see also Bibliography). Trends in the development of column technology, detectors and data handling for SEC have recently been discussed [770,771]. The field produces some 1200 papers per year.

Applications The primary application of SEC is to define the MW and MWD of polymeric materials. The method can be used to investigate the use of regrind in a sample. The technique also finds utility in the analysis of additives in polymeric resins. For the separation of low-MW compounds conventional SEC suffers from poor resolution, so that it is not able to separate the individual components from complex polymer additive mixtures. However, this disadvantage turns into an advantage for groupswise additive analysis where identification and/or quantitation of individual components is not desired. With the recently greatly increased efficiency, SEC finds some interesting *niche applications*. It is intuitive that the strengths of SEC in the polymer/additive field concern oligomers (e.g. polymeric HALS) and some analyses that are not amenable to HPLC, such as the direct analysis of polymer/additive dissolutions (e.g. for screening purposes) or polymer-bound additives. SEC and GPEC® are the only LC techniques that allow direct analysis of additives in polymers, without the need for a preliminary extraction step; to that extent, SEC is more performing. Applicability of the SEC technique assumes existence of a solvent compatible with sample, separation media, and equipment.

There are essentially two methods of analysing additives in polymers by means of SEC: (i) direct injection of the polymer solution into a SEC system equipped with a high-efficiency column for small molecules; and (ii) extraction using an appropriate solvent, followed by separation of the extract. In the former case the peaks appearing after the polymer peak are allowed to flow into a second column (SEC or another LC mode) to separate and analyse the additives (Section 9.1.1). The size-exclusion mode can be used satisfactorily when the molecular sizes of the additives are different, but other separation modes should be used when molecular sizes are similar. Despite several disadvantages of SEC in comparison with HPLC, *direct determination of analytes* in polymers as well as in other materials such as lubricating and vegetable oils, is of considerable use provided that the matrix is of a sufficiently high-MW to be almost completely resolved from the additive components of the mixture.

In SEC analysis of additive extracts from polymers, the effect of the extraction solvent on the mobile phase is less critical than in HPLC analysis. The extraction solvents typically employed generally do not interfere with the SEC mobile phases. Moreover, the same solvents are often used both as extraction solvent and as mobile phase. Therefore, there is no need to evaporate the extract to dryness prior to analysis and then to redissolve it in a suitable solvent. Typical extraction procedures often produce extracts that generally contain a small amount of wax. Frequently, removal of such oligomers from an extract is necessary, e.g. by means of precipitation, centrifuging, precolumn filtration or protection (use of a reversed-phase guard column). In SEC separations the presence of polyolefin wax does not usually disturb provided that the MW of the wax is higher than that of the analysed compounds.

Howard [772] has been amongst the first to show the usefulness of conventional SEC for polymer/additive systems. Čoupek *et al.* [773] have also reported results with this technique in an early stage; their work was limited to synthetic mixtures of additives. The use of open-column SEC in the analysis of plastics additives has been reported [774]. Qualitative analysis of additives has been performed by stopped-flow SEC with IR detection [775]. Polypropylene oligomers were isolated from a PP/(Irganox 1010, Irgafos 168, DBS) matrix by dissolution (toluene)/precipitation (methanol) and Soxhlet (methylene chloride) methods and were characterised by FTIR, UV, HPLC and SEC [776]. Interestingly enough, the average molecular weights are different depending upon the quality of extraction with the fraction obtained by the dissolution/precipitation method being homogeneous in molecular weight and the one obtained by Soxhlet extraction having a higher polydispersity index ($\langle M_w/M_n \rangle$). The fraction obtained by Soxhlet extraction was relatively polydisperse because its MWD appeared bimodal.

In other experiments, PVC/plasticiser extracts (*n*-hexane) were separated by SEC using THF or chloroform as the mobile phase. Similarly, PE film was immersed in THF for several hours and the extract was concentrated by a factor of 20 prior to injection into a SEC system [51]. However, use of extraction techniques followed by injection into a SEC system for separation of low-MW additives is not the most obvious analytical approach in view of the relatively low resolution of conventional SEC in the low molecular mass range. For this purpose efficient column packing materials with small pore sizes are to be used.

Oligomeric species are present in most commercial polymers and play an important role in terms of surface properties (adhesion, printability, lubrication), rheology, processability and extractability (odour, taste). SEC is one of the preferred techniques for characterising oligomers. Other techniques are (temperature-programmed) HPLC, SFC, GC-MS, TLC and LDMS. In a typical SEC output oligomers are positioned between polymer components and the additive region on the basis of molecular weight. Oligomers separated by SEC have originated from PS [777], PVC [778], epoxy [779], PVAL [780] and other polymers [781]. Snyder and Breder [782] have described a rapid SEC method (15 to 20 min) for oligomer analysis of polyolefins. Campana *et al.* [783] have compared the analysis of PEG oligomers by means of SEC, LDI-ToFMS and LD-FTMS, with vastly improving resolution, in this order. The separation selectivity for oligomers according to the molar mass distribution is often better in interactive chromatography (NPLC, RPLC) than in SEC. Gradient elution is generally required to achieve adequate peak capacity in NPLC/RPLC of oligomers. The separation can be optimised by adjusting the initial mobile phase composition, gradient time and shape, and by temperature programming. Using high-resolution SEC columns with greatly increased pore volume ($>55\,000$ plates m^{-1}) oligomeric distributions of PS, PMMA, and epoxy resins (Epikote 828 and 1001) were analysed [758].

Diglycidyl ether of bisphenol A (DGEBA, MW 340 Da) and 4,4'-dihydroxy-diphenylmethane (DHDPM, MW 200 Da) were analysed by SEC-MALS [784]. DGEBA and DHDPM are the basic oligomers of epoxy resins and phenol–formaldehyde condensates, respectively, which are widely used in the electronic and automotive industries. Excellent reproducibility ($\pm 1\%$) and good accuracy (better than 10%) were observed. SEC has also been used for the determination of mineral oil in extended elastomers [785] and in PS [178]. With heptane containing 0.05% isopropanol as the mobile phase, mineral oil is completely unretained and elutes before the solvent via SEC; all other components in a PS extract are retained on silica and elute after the solvent peak.

Polymeric HALS (such as Chimassorb 944/2020, Tinuvin 622, Cyasorb UV3346) are difficult to analyse and a variety of methods have been employed, including UV spectroscopy, PyGC, NPLC, RPLC, TP-HPLC, TLC and SEC. In similar cases the capability of SEC for MW determination is an important advantage. Isolation of high-molecular HALS usually requires the dissolution method. In fact, it is not possible to isolate Chimassorb 944 from PE by means of SFE. Gharfeh [745] developed size-exclusion non-aqueous reversed-phase chromatography for separation of 0.05 to 10% Chimassorb 944 from other polyolefin additives (RSD of 3% in the concentration range of 0.1%). A step-gradient from toluene to 0.1 M piperidine in toluene was used for analysis. SEC-UV (239 nm) has also been used for the quantitative determination of Chimassorb 905/944, Cyasorb UV3346 and Tinuvin 770 in PP after dissolution (xylene)/precipitation (2-propanol). The MWD profile of HALS, as produced and after polymer extraction, was obtained [749]. Some other methods for determination of Chimassorb 944, such as Kjeldahl (nitrogen analysis) [786], UV ($\lambda_{max} = 244$ nm) [787] or FTIR (absorbance of the triazine group) [788] of the polymer extract lack accuracy or are subject to interferences from other additives. A semiquantitative PyGC method has been described based on polymer extracts. Chimassorb 944, Cyasorb UV3346 and Uvasorb HA 88 in polyolefins in the range of 100 ppm to 0.45 wt% were determined with a standard deviation of 0.4% by dissolution (toluene)/precipitation (methanol, 1% TEA) followed by quantification by means of an isocratic HPLC method using two size-exclusion columns, THF–diethanolamine as the mobile phase and UV detection at 244 nm [751,789]. Also direct analysis of Chimassorb 944 in the food simulant HB 307 without sample preparation was reported [790]. The limited resolution of SEC is taken to advantage in the quantification of Chimassorb 944 as the various oligomeric components are not separated. As quantitative SEC analysis is carried out with a UV detector, the sensitivity of the method does not exceed that of UV analysis. Also Tinuvin 622 may be analysed by SEC, but not necessarily so [417]. SEC does not provide single ethoxylate oligomer separation. HTLC has provided significantly more information about oligomeric additives than SEC [743]. It is the case to notice that also MALDI -ToFMS and LD-FTMS produce higher resolution data than that obtained by conventional SEC. Whereas SEC and ToF-MS data only give information about the molecular weight FTMS also provides the exact molecular mass and element composition. Although there is general qualitative agreement between the aforementioned mass spectrometric techniques, this is often at variance to SEC results [791]. It has also been shown that 500 μL SEC fractions of high-MW polymer additives (ca. 1200 Da) in DCM can be determined by at-column GC [101].

Polymeric plasticisers have been used as partial or total replacements for di(2-ethylhexyl)adipate (DEHA) in PVC cling film to reduce levels of plasticiser migration when used for food contact. Castle *et al.* [792] used SEC in combination with ^{1}H l-NMR and MS for the isolation and identification of seven individual oligomers in the most commonly employed polymeric plasticiser, poly(butylene adipate) (Reoplex R346). Both mass (RI) and specific ester moiety (UV) were being monitored (Figure 4.21). The oligomers were identified

as a series of diol-terminated units ranging from a trimer up to an 11-monomer unit, along with a cyclic tetramer, all in the molecular weight range of 300–1100 Da.

Using a SEC column packed with hydrophilic polyhydroxyethyl methacrylate gels and THF and methanol as mobile phases phthalate esters (DMP, DEP, DNBP, DNHP, DNNP and DNDP) and oligostyrenes were separated [793]. The size-exclusion effect was predominant when THF was used as the mobile phase and adsorption interactions between solutes and the gels superimposed the size-exclusion effect when methanol was the mobile phase. PVC/(tris(2-ethylhexyl)trimellitate and phthalate) was analysed by (HT)SEC-UV and/or SEC-RI in THF dissolution; other plasticisers (DEHP, DBP, DEP, DMP) did not interfere [794]. Also SEC-ELSD was used for the separation of a polymeric adipate–phthalate mixture [690]. This analysis is difficult by means of SEC-UV at 215 nm because of solvent (THF) absorption. Both SEC-UV (243 nm) and NPLC/RPLC-UV (224 nm) were used for the analysis of phthalate esters (DEHP, DNBP). Using 3-μm 100 Å high-efficiency PS/DVB columns dialkyl phthalates (DOP, DBP, DEP, DMP) were easily separated (Figure 4.22) [758]. Also a mixture of phthalate esters (methyl-, ethyl-, *n*-butyl-, *n*-octyl- and *n*-nonylphthalate esters) has been separated in SEC mode [795]. Using SEC columns with good resolution at low-MW for analysis of PVC/(plasticisers, epoxidised soybean oil), the separated components are pure enough to give distinctive spectra [796]. SEC has been successfully used for plasticisers which are present

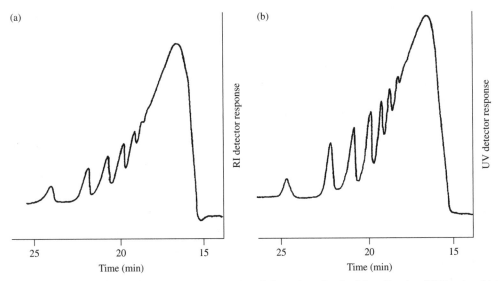

Figure 4.21 High-performance size-exclusion chromatograms of the polymeric plasticiser Reoplex R346 using (a) RI and (b) UV detection. After Castle *et al.* [792]. Reproduced from L. Castle *et al.*, *Food Addit. Contam.*, **8**, 565–576 (1991), by permission of Taylor & Francis Ltd (http://www.tandf.co.uk/journals)

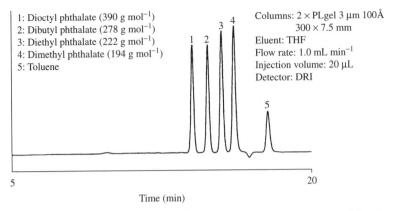

1: Dioctyl phthalate (390 g mol^{-1})
2: Dibutyl phthalate (278 g mol^{-1})
3: Diethyl phthalate (222 g mol^{-1})
4: Dimethyl phthalate (194 g mol^{-1})
5: Toluene

Columns: 2 × PLgel 3 µm 100Å
300 × 7.5 mm
Eluent: THF
Flow rate: 1.0 mL min^{-1}
Injection volume: 20 µL
Detector: DRI

Time (min)

Figure 4.22 Analysis of dialkyl phthalates using 3-µm 100 Å PS/DVB columns. After Meehan and Saunders [758]. Reproduced by permission of International Scientific Communications

in relatively large amounts and have UV-absorbing properties [797], but additives at low concentrations present more difficulties especially in the case of non-UV-absorbing species.

It is general experience that SEC can be used to elucidate the structures of complex low-MW mixtures in the range of 150–1000 Da [798] and to identify low-MW additives in polymers [623,773,799–802]. However, direct HPSEC for such compounds in plastics will not be suitable for all applications. Apart from the inherently limited resolving power of conventional SEC restrictions are solubility limitations (e.g. for polyolefins) and adequate sensitivity. One factor limiting the wider use of SEC in plastics analysis is the detection limit obtainable with the small sample loadings (<500 µg) which can be used on 8-mm analytical PS/DVB columns. Larger columns allow higher loadings. For good detection limits in SEC experiments it is important to determine the maximum loading of polymer which can be applied to the column whilst retaining complete separation of small molecules. Problems have been reported in the recovery of AOs [774] and organotin stabilisers [774,803]. The method is therefore not universal. Methods have been developed for 0.3 % TDI monomer and residual isocyanates in urethane adhesives [804,805], butyl-lithium in PS [806], methylacrylate in polyacrylates [807] and quantification of styrene in copolymers [808]. The use of SEC has also been extended to the analysis of surface coatings [809–811].

Shepherd and Gilbert [812] have used SEC analyses with 50-, 100- and 500-Å gel columns of TNPP and epoxidised oil (HCl scavenger) extracted from PVC sheeting that was subjected to a controlled heat treatment. They were able to identify rapid breakdown of the hydrolytically unstable TNPP at 170 °C

compounding conditions. Protivová and Pospíšil [813] have examined conventional SEC behaviour of some 10 amine antiozonants, *antioxidants* and 20 model compounds. SEC is capable of determining down to 0.02 % amine AOs and antidegradants in rubbers with an accuracy of ±5 %. Vimalasiri *et al.* [160] have discussed various other papers devoted to the application of SEC to elastomer antioxidant and accelerator analysis [623,773,801,813–815]. Although SEC is in principle a good method for separating antioxidants by molecular size, a complete separation takes over 3 h per sample, as compared to 30 min in HPLC separations. Nevertheless, both for HPLC and SEC (extraction mode) total analysis times are controlled mainly by the time to isolate the additives (>24 h when total Soxhlet extraction is required; the extraction times can be greatly reduced by SFE or ASE®). Gupta and Salovey [816] have reported characterisation of butadiene-acrylonitrile (BAN) elastomers containing chemically bound (inextractable) antioxidants (PgA) using THF dissolutions and SEC separation with RI and UV detection (254 nm for free AO, 280 nm for bound AO). From the UV absorption at 280 nm, the mass of antioxidant bound to the polymer was calculated (Figure 4.23). SEC-RI/UV was also suitable for characterisation of the chemical heterogeneity of UV stabiliser-containing polymers, prepared by copolymerisation of styrene and methyl methacrylate with polymerisable UV stabilisers or by grafting of UV stabiliser monomers (benzophenones, phenyl and naphthyl benzotriazoles) onto preactivated PVC and for the quantitative determination of the amounts of copolymers, oligomers and monomers, respectively [817]. Figure 4.20 shows SEC chromatograms for an adhesive formulation including SBR rubber, low-MW additives and solvents, for

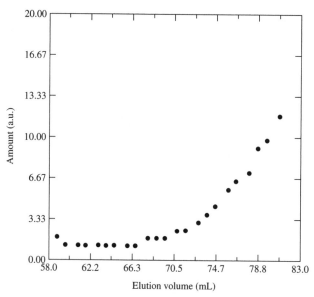

Figure 4.23 Antioxidant distribution in BAN copolymer as a function of elution volume. After Gupta and Salovey [816]. Reprinted with permission from *Rubber Chemistry and Technology*. Copyright © (1985), Rubber Division, American Chemical Society, Inc.

two different detectors [760]. The chromatogram contains information about molecular weight of additives which may aid in their identification.

For identification of additives with SEC a calibration curve using common stabilisers of known molecular weight as standards can be used [623]. The *quantitative determination* of additives in the solvent extract can be performed by means of linear calibration curves of peak height vs. stabiliser concentration in a standard solution [623,801]. Besides identification and quantification of polymer additives in solvent extracts, some earlier SEC analyses attempted to determine the purity of individual stabilisers or to show possible chemical transformation of stabilisers [801]. Conventional SEC offers the advantage that several species, varying little in molecular size but having commonality of structure, elute in a single peak. This allows for a simplified calibration procedure since only a single calibration curve needs to be determined. SEC has been used in this way to determine the plasticiser content in PS [818] and PVC compounds. Groupwise detection of dialkylphthalates (independent of alkyl group) takes advantage of the generally low separating power of SEC (one peak). Separation of single components requires HPLC or an appropriate choice of a SEC column. Dialkylphthalates (R = Me, Et, *n*-Bt, Oct) are easily distinguished by RI detection on a 5-μm 50 Å 7.5 × 300 mm column within 10 min using THF as an eluent. The technique has also been applied to the isolation of plasticisers from foods [819–822]. Similarly,

polychlorinated biphenyls (PCBs), used as plasticisers (Arachlor®), yield multicomponent peaks in GC (7–12 peaks in Arachlor® 1221–1226) making quantitation difficult; moreover, these PCBs have to be isolated from the polymer matrix with an appropriate solvent before injection into the GC system. A rapid analysis entailing minimal steps in sample preparation consists in dissolving the sample, followed by subsequent SEC separation of PCBs from the polymer matrix (PS) and finally quantitation [797]. Precision levels are as good as or better than classical GC methods.

Formulated lubricating oils contain a complex additives mixture including metallic detergents, antioxidants, ashless dispersants, viscosity index improvers, etc., in order to neutralise contaminants generated by engines, remove the heat and reduce the friction. In a stabilised lubricating oil the AOs Ethanox 702 (MW 436) and tricresyl phosphate (MW 368) were readily separated from each other and from the oil matrix [815]. Complex mixtures of low-MW *lubricant additives*, such as methylene bridged hindered *t*-butylphenolic AOs, sulfur bridged hindered *t*-butylphenolic AOs and alkylated diphenylamine AOs were analysed in groupwise fashion [798]. Chromatograms produced by HPLC for these complex mixtures contain a myriad of peaks rather than a simple chromatographic fingerprint, because HPLC separation is based on polarity and SEC separation on hydrodynamic volume. The simplicity of the SEC chromatograms makes them easier to interpret qualitatively than a corresponding HPLC chromatogram.

SEC-RI/UV has also been used to analyse some 26 thioorganotin compounds, organotin carboxylates and chlorides, essentially PVC stabilisers, and some of their main by-products and related compounds (thioesters and dithioesters, *n*-alkanes) [803]. Not all organotin chlorides were stable in the adopted analysis conditions. N, N'-ethylene-bis-stearamide and -oleamide in common plastics (ABS, SAN, PUR, LDPE, PA6.6) can be analysed by SEC after derivatisation with trifluoroacetic anhydride. SEC analysis of fatty alcohol ethoxylates (FAE), used as nonionic surfactants, has also been described [759].

SEC-UV/VIS may be used for locating chromophores in polymers. In an acrylate rubber (ASA) modified styrene-*co*-acrylonitrile (SAN) swimming pool accessory part, that had turned bright orange during use, SEC-UV/VIS analysis has shown that the visible absorbing portion of the polymer is located in small molecules (estimated MW 300–400) in the polymer. These were further identified by means of GC-MS and HPLC analysis. The discoloration was primarily due to interaction of stabilisers and swimming pool/spa chemicals [823]. By using very high loadings, sufficient quantities of contaminant fractions in SEC may be collected to permit identification by IR spectroscopy.

SEC has been claimed to be a very good routine monitoring technique providing qualitative and quantitative analyses for *quality control*. The technique is used as a QC tool in the medical plastics industry for fingerprinting polymer MWD and in analysing polymers, oligomers, monomers and most compound additives [824]. Similarly, SEC-RI/UV with a series of small pore SEC columns is used as a fingerprinting tool for confirming product composition of AO containing lubricants [798]. Ono [825] has reported SEC analysis of additives in fluoro rubbers for QC. Also for QC purposes batches of Chimassorb 944, Tinuvin 622, Irgafos 168, Irganox 1010 and Tinuvin 327/770 were analysed by 5-µm 50 Å PS/DVB columns and ELS detection using THF as the eluent with 0.1 % diethanolamine added to minimise interactions between samples and column packing [826]. SEC has allowed routine screening to verify supplied formulations and to determine the AO concentration in commercial PP and automotive moulded parts [623].

Despite general poor resolution and low sensitivity SEC has been used regularly for additive analysis, especially in the 1970–1985 period (for some 20 references, see Munteanu [576]). After the advent of HPLC its use has declined somewhat. However, SEC

solves various analytical problems not appropriate for HPLC. The technique finds also routine application in multidimensional mode in industrial laboratories (Section 7.4.3.1). On-line at-column GC analysis of high-MW compounds at low concentration (500 ppm) in 500 µL SEC fractions in DCM was described [101]; RSD values of 2–8 % were reported. The record of SEC in polymer/additive analysis is more impressive than might appear at first sight.

4.4.2.4 Gradient Elution Chromatography of Polymers

Principles and Characteristics Gradient elution is widely applied in high-performance liquid chromatography (GE-HPLC) to overcome certain general separation problems and to resolve more efficiently certain kinds of samples. By interactive LC, separation of polymers may be performed on the basis of differences in the chemical structure present along the chain molecules and independently from their MWDs. Depending on the molecular structures involved, different separation mechanisms can be exploited and various forms of LC can be applied in the characterisation of polymers, including size-exclusion chromatography (SEC) and liquid adsorption chromatography (LAC), eventually at critical conditions (LACCC). In a related group of LC techniques, *high-performance precipitation liquid chromatography* (HPPLC) in which the retention mechanism is dominated by solubility effects, solvent selection is based on using a nonsolvent such that the polymer is precipitated onto the column packing, and is subsequently eluted by applying a solvent gradient with nonsolvent/solvent. In HPPLC the sample solvent differs from the starting eluent. The great advantage of the HPPLC method for characterisation of polymers is the universal applicability, if a convenient precipitation/dissolution system exists. Eluent gradient polymer liquid adsorption chromatography is also known under the Waters' trade name of *Gradient Polymer Elution Chromatography* (GPEC®) [827]. GPEC® does not refer to a specific mechanism, but merely describes technique (gradient elution chromatography) and application (polymers).

Whereas SEC is the dominant technique for the characterisation of polymers, various nonexclusion liquid chromatographic (NELC) methods, such as GPEC® and LACCC offer equally valid possibilities for deformulation of complex polymer systems. In fact, molecular characterisation of polymers in the precipitation/adsorption mode (gradient HPLC) enables differences in chemical structure and composition to be detected – a task that is almost impracticable with SEC, although feasible in hyphenated SEC mode (Section 7.3.4).

The precipitation/adsorption mode of synthetic polymers is not as straightforward as size exclusion. In contrast to the isocratic elution of SEC, chromatography of polymers in the sorption mode usually requires a gradient of the mobile phase. This is because the differences in retention for polymers of varying sizes under isocratic conditions are often too large. Various gradient HPLC methods for the analysis of polymers have been described differing in the mode of precipitation/redissolution of the polymer and the prevailing separation mechanism.

The principle of GPEC® is to dissolve the sample in a good solvent and precipitate or suspend the polymer sample in a HPLC instrument with a nonsolvent. This precipitate or suspension will adsorb on a guard column with a special flow distributor to prevent column plugging. By adding a solvent to the adsorbed suspension with a gradient, the monomers, additives and polymers will redissolve and elute from the column according to molar mass *and* chemical composition. Retention is governed by solubility effects, size exclusion and sorption (adsorption and/or partitioning). The separation mechanism of GPEC® is based on a combination of a precipitation/redissolution mechanism (prevailing) and one controlled by column interactions (sorption and steric exclusion), but a thorough understanding of the technique is still lacking [828]. The contribution of each of the separation mechanisms depends on the combination of sample and separation system. The separation achieved by GPEC® depends on various factors, namely the applied solvent system (solvent/nonsolvent), type of column, column temperature, applied gradients (gradient curve, flow, initial and final conditions) and injection conditions (volume, concentration, and sample solvent). For several reasons, e.g. the occurrence of solubility effects, special adsorption features such as multisite attachment and conformational changes, as well as size-exclusion effects, GPEC® differs from gradient elution chromatography of low molar mass solutes.

In analogy to HPLC, different types of chromatography can be applied, namely normal-phase mode (NP-GPEC®) and reversed-phase mode (RP-GPEC®). RP-GPEC® is most common and applied with a nonpolar column in combination with an eluent decreasing in polarity. NP-GPEC® is performed with a combination of an eluent increasing in polarity and a polar column. The main characteristics of both modes are given in Table 4.46. RP-GPEC® can be used to determine the chemical composition distribution (CCD) of

Table 4.46 Characteristics of GPEC®

- RP-GPEC®

- Provides additional information beyond size-exclusion methods
- Information on chemical composition only for low molar masses
- Only qualitative information on microstructural differences
- Versatile fingerprinting technique

NP-GPEC®

- Quantitative information on end group distributions and CCDs (also for higher molar masses)
- Combination with SEC recommended
- Laborious method

multicomponent polymer systems; in this application the technique is not unique. RP-GPEC® must be considered mainly as a versatile, qualitative, fingerprinting tool rather than a method for quantitative evaluation of the chemical microstructure. In contrast, NP-GPEC® provides more and quantitative information on small microstructural differences, e.g. functional end-groups in polar molecules. Within certain limits GPEC® can also be used for the determination of the molar mass distribution (MMD).

Table 4.47 lists the main features of GPEC® as applied to additive analysis in polymers. Extraction is always complete (100 % recovery) in view of the strong solvent at the end of the gradient. Compared to SEC, gradient elution separation can be achieved at higher flow-rates in a much shorter time. Unfortunately, GPEC® is not universally applicable. As the GPEC® mechanism is highly dependent on the conditions used and on the applied polymer, development of a universal model or theory is difficult. The conditions have to be optimised for each new GPEC® separation. This makes GPEC® a time-consuming method

Table 4.47 Main virtues of GPEC® in polymer/additive analysis

Advantages

- Highly reproducible
- Complete extraction
- Suitable detectors (UV, ELSD, MS)
- Complete extraction
- Simultaneous identification and quantification of polymer and additives

Disadvantages

- Not universally applicable
- Mechanistically complex
- Difficult optimisation/method development
- Lack of applications

with art-like character. Moreover, the most important step in the application of gradient polymer elution chromatography is to find proper solvents and nonsolvents for polymers. A sample can only be injected if it is dissolved in a liquid. The possible combinations of polymer/solvent/nonsolvent/column are numerous and thus optimisation is not straightforward. Despite these drawbacks, in many cases GPEC® reveals differences between samples that cannot easily be obtained by other techniques.

The sensitivity of the RI detector to changes in mobile phase composition makes it unacceptable for use in gradient polymer analysis. In this respect, UV detection is much better suited. ELSD shows superior performance for gradient applications. This quasi-universal mass detector is essentially unaffected by nontransparent eluents and by changes in the mobile phase composition since the solvents are evaporated prior to detection. This combined with the excellent sensitivity for polymer samples, and the possibility to observe both UV absorbing and non-UV absorbing polymer samples make ELSD the best detector of choice for gradient analysis of polymers. By combining ELSD with a PDA, it is possible to detect and quantitate polymer additives and other molecules in traditional phase separations, determine peak purity, identify unknowns through library matching and carry out compositional analysis across the MWD of many copolymers [829]. For polymers a calibration line is needed for each different component; for low-molar mass additives calibration lines are less important. GPEC®-MS coupling (in particular with soft ionisation techniques) is necessary to provide the much needed peak identification, provided that peak resolution is high enough to accurately assign individual peaks. Both GPEC®-SEC and SEC-GPEC® multidimensional chromatography have been proposed [830].

Gradient HPLC of polymers has been described in a monograph [831], PhD theses [828,832] and a recent review [830].

Applications Van der Maeden *et al.* [646] first used GE-HPLC for the qualitative and quantitative analysis of *oligomeric mixtures*, such as low-MW resins (epoxy up to 16-mer, *o*-cresol novolak up to 16-mer, *p*-cresol novolak up to 13-mer), prepolymers (poly-(2,6-diphenyl-*p*-phenylene oxide) up to 20-mer), PET (up to 14-mer) and ethoxylated octaphenol surfactants (up to 19-mer). In many GE-HPLC separations of oligomeric mixtures, a compromise has to be found between sample loading, injection volume and compatibility of the sample solvent and the initial phase system. Therefore,

the gradient elution technique is somewhat limited and complicated. Gradient elution is usually a prerequisite for successful analyses of polymers and oligomers under sorptive conditions.

GPEC® can be used for the determination of the solubility of polymers (e.g. in the case of PB, PMMA and PS), and qualifies as a powerful technique for deformulation of the total composition of complex polymeric materials, including oligomers, additives and residual monomers. Separation of high-MW polymers is dominated by a precipitation–redissolution mechanism, and the separation efficiency is comparable to that in conventional RPLC or NPLC. Staal *et al.* [827] have described on-line extraction in one step of polymers, oligomers, additives and monomers by a multisolvent gradient on packed HPLC columns. However, no specific identification or quantification was reported. Additionally, polymer molecules can be separated according to molar mass, chemical composition and functional end groups. Despite not qualifying as a universal method, gradient HPLC-ELSD/PDA is well suited for deformulation applications. GPEC® has been applied to powder coatings (determination of oligomers and cyclic dimers) [833]. Oligomers in polyester samples were identified by hyphenation of GPEC® with ESI-MS and UV spectroscopy [834].

Alden and Woodman [829] have shown the use of gradient separations to the analysis of *polymer additives* using dual PDA and ELSD detection. Using a linear ternary H_2O/ACN/THF gradient various low-MW polymer additives were analysed by the RP mechanism, namely the UV stabilisers Tinuvin 328/440/900; the phthalate plasticisers DCHP, DOP (DEHP), DIDP, UDP; the slip agents oleamide and erucamide and antistat stearic acid (Figure 4.24); and the antioxidants Irgafos 168, Irganox 1076 (all within 38 min, using ELSD).

The antioxidants Tinuvin P, Irganox MD1024, BHT, BHEB, Cyasorb UV531, Isonox 129, Irganox 1010/1076/3114 and Irgafos 168 (in 100–400 ppm concentrations) were determined with an ACN/H_2O gradient within 9 min using UV detection ($\lambda = 230$ nm) with an average RSD of 0.4 % for 12 injections (Figure 4.25). The method is highly reproducible and sensitive.

On the whole, GPEC® remains a technique in search for polymer/additive applications with real added value [835]. Practical applications of GPEC® may be found in the analysis of polymer blends [836], laminates and packaging materials. For example, the technique can be used for determination of the impact modifier content in PS packaging material, which contains a soluble transparent rubber for transparent applications,

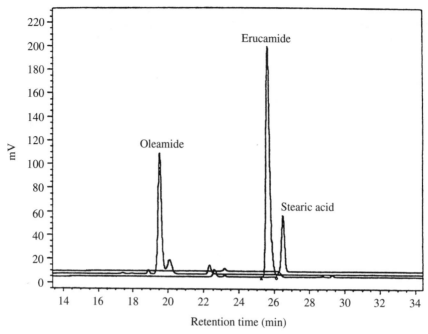

Figure 4.24 Gradient elution of antislips and antistats; ELSD detection. After Alden and Woodman [829]. Reproduced by permission of the authors

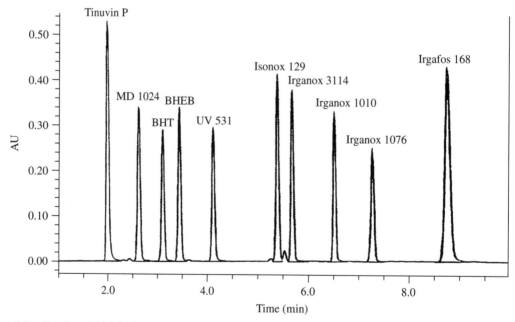

Figure 4.25 Overlay of 12 injections of antioxidant standards; gradient elution, UV detection at 230 nm. After Alden and Woodman [829]. Reproduced by permission of the authors

or an insoluble grafted styrene-butadiene copolymer for nontransparent application. At variance to IR, which determines total styrene, GPEC® differentiates between free and bonded styrene; RP-GPEC® is actually a fingerprint of the manufacturing process: batch-to-batch variations in the production of polymeric materials can be detected accurately. Minor differences can be identified by statistical methods like multivariate analysis (impurity profiling).

4.4.2.5 Ion Chromatography

Principles and Characteristics Ion chromatography (IC), also referred to as eluent suppressed ion chromatography [837] or conductometric chromatography, is an ion-exchange chromatographic technique related to HPLC. Small *et al.* [17] first described the conductometric approach to chromatographic analysis of ions in 1975. In a strict sense, IC is not a separate chromatographic technique, but rather a specialised application of established ion-exchange techniques. Nowadays the definition of IC has expanded greatly and can best be categorised in terms of the solutes separated (inorganic anions and cations, low-MW water-soluble organic acids and bases) rather than the manner in which the separation is achieved.

Separation of ionic species can occur by ion exchange, ion suppression or ion pairing. In IC an aqueous atomic or molecular ionic medium is injected into an eluent stream. Typical eluents are $Na_2CO_3/NaHCO_3$ for anions such as Cl^- and Br^-, dilute HCl for monovalent cations, and *m*-phenylene diamine/HCl for divalent cations. The *elution mechanism* is based on displacement of ionic equilibria The stationary phase is a solid used to conduct ion exchange. The ions are separated by their relative interactions with the column. The ion exchanger groups bound to the support material basically determine the separatory behaviour of an ion-exchange column. Stationary phases in IC are usually reticulated PS/DVB spheres (15 μm diameter) chemically modified as cation or anion exchangers, modified silica or film resins. The most important group of ion exchangers is that of the organic materials based on synthetic resins. Different columns are used for separation of anions and cations. In IC, ion-exchange materials of low capacity with particle sizes of 5–10 μm are employed. This makes it possible to separate and detect ions or polar compounds efficiently and rapidly with eluents of low concentration. The use of suitable combinations of support materials and ion-exchanger groups allows optimal conditions to be provided for special separation problems.

The combination of ion-exchange columns and conductivity detection represents the most important type of IC of which two techniques are used in practice. Since the eluents are ionic species, by necessity they must be eliminated or suppressed or they will overwhelm the detector. In *direct ion chromatography* (i.e. without chemical suppression), eluents with salts of organic acids in low concentration are employed on ion exchangers of very low capacity to achieve a relatively low-background conductivity, which can be suppressed directly by electronic means. This operating mode places high demands on the conductivity detector. Calibration curves obtained without chemical suppression remain linear for several decimal powers. Detection of ionic species in a sample is difficult when the analytes are present in low concentration in a mobile phase that contains high quantities of ions. This is why the detection mode based on conductivity needs a special arrangement. It requires a suppression device to be inserted between column and detector, which is used to eliminate the ions in the mobile phase via acid–base type reactions. The introduction of *chemical suppression* has been a major step forward in the applicability of IC. Chemical suppression is based on the use of salts of weakly dissociated acids (e.g. $NaHCO_3$) as eluents. By passing the eluent from an anion exchange column through a cation exchanger in the hydrogen form the salts that make up the sample peaks are converted into more highly conducting acids and the eluent is converted into weak acids or water. The suppressor reduces the conductivity when the mobile phase is highly ionic. The overall effect is a substantial increase in sensitivity. In IC with chemical suppression, the background conductivity is suppressed both chemically and electronically. If chemical suppression is used slightly curved calibration curves are usually obtained. Eluent conductivity suppression is no longer the exclusive detection method for anions by IC.

Many IC techniques are now available using single column or dual-column systems with various detection modes. *Detection methods* in IC are subdivided as follows [838]: (i) electrochemical (conductometry, amperometry or potentiometry); (ii) spectroscopic (UV/VIS, RI, AAS, AES, ICP); (iii) mass spectrometric; and (iv) postcolumn reaction detection (AFS, CL). The mainstay of routine IC is still the nonspecific conductometric detector. A significant disadvantage of suppressed conductivity detection is the fact that weak to very weak acid anions (e.g. silicate, cyanide) yield poor sensitivity. IC combined with potentiometric detection techniques using ISEs allows quantification of selected analytes even in complex matrices. The main drawback of ISEs in ion chromatography is the slow response of many electrodes. Amperometric and potentiometric detectors play a minor role for routine applications. Other detection techniques also provide some information about identity, structure or elemental composition of the analytes. Direct UV/VIS detection is a straightforward technique in IC for monitoring analytes of sufficient absorptivity in the presence of a high concentration of ions without UV absorption. Indirect UV/VIS detection has universal application in direct IC for use with eluents with high UV absorption (e.g. phthalate buffers). UV/VIS multiwavelength detection uses diode arrays or fast scanning detectors. Apart from straightforward applications for analytes with native fluorescence properties postcolumn reactions allow for fluorescence or chemiluminescence (CL) detection. 8-Hydroxyquinoline-5-sulfonic acid (HQS) is a suitable postcolumn reagent for fluorescence detection that responds selectively to metal ions like Mg^{2+}, Ca^{2+}, Zn^{2+}, Cd^{2+}, Al^{3+}, Ga^{3+} and In^{3+}. Indirect organic and inorganic cationic IC with fluorometric detection has also been reported [839]. A specific application of CL detection in IC is analysis of oxalates.

Various forms of off- and on-line AES/AAS can achieve element specific detection in IC. The majority of atomic emission techniques for detection in IC are based on ICP. In the field of speciation analysis both IC-ICP-AES and IC-ICP-MS play an important role. Besides the availability of the ICP ion source for elemental MS analysis, structural information can be provided by interfaces and ion sources like particle beam or electrospray.

In contrast to the determination of organic analytes where the use of high-performance separations is almost inevitable in complex matrices, inorganic analysis has powerful tools in highly selective and sensitive spectroscopic methods and thus a separation step is often unnecessary. This is especially true for determination of inorganic cations. In the determination of *inorganic cations* by IC various spectrophotometric and electrochemical methods of detection can be employed. Transition metal ions are particularly suited to spectrophotometric detection but also to such electrochemical methods as coulometry at controlled potential, conductivity measurements, amperometry and indirect potentiometry. Typical lower cation detection limits by IC are 1–2 ppt (NH_4^+, Ca^{2+}, Li^+, Mg^{2+}, K^+, Na^+, Co^{2+}, Cu^{2+}, Fe^{2+}, Mn^{2+}, Ni^+, Zn^{2+}). IC competes on a realistic basis with atomic spectrometric techniques. For many applications, spectroscopic techniques such as AAS are preferred for the determination of simple inorganic ions. For complex mixtures, for samples derived from complex matrices,

or in cases when differentiation of oxidation states is required, ion exchange is usually superior.

On the other hand, the number of methods for the determination of *inorganic anions* is limited. Spectrophotometric procedures and methods based on ion-selective electrode potentiometry are usually not sufficiently sensitive for trace analyses and are mostly capable of determining only a single analyte. Therefore, high-performance separations are often needed in the analysis of anions. The greatest utility of IC is thus for the analysis of anions for which there are no other rapid analytical methods. IC is the only technique available to monitor several anions during the same test run. IC can be used for simple profiling of anions and cations in aqueous-based media. Standard IC with injections of 20 µL of sample and separations of 5–15 min can be used to determine sub-0.1 ppm levels of anions and cations in aqueous solution. IC can be used both for analytical and preparative purposes. Large sample volumes, up to 1300 µL, can be injected to determine trace anions and cations and to attain detection limits of 10–400 ng L^{-1}. Using chemical suppression anion detection limits in IC range from 5 ppb (F^-, Cl^-, NO_3^-, PO_4^{3-}, NO_2^-, Br^-, SO_4^{2-}) to 5–10 ppm (acetate, formate, oxalate, hexafluorosilicate). IC-APCI-MS has been used for analysis of low-MW organic and inorganic anions [840].

Both IC and CE are separation techniques used to examine aqueous samples for anions, cations and trace levels of organic acids. CE produces separations by a different mechanism than IC, and produces sharper peaks. Therefore, CE can easily perform some separations that are difficult by IC. The determination of inorganic anions by IC and CE was compared [841].

Ion chromatography has been reviewed [61,842–844], including ion-exchange columns [845] and detection techniques [838,842]; various monographs covering this chromatographic method have appeared [846–848] (Bibliography). Gradient elution in IC has been dealt with specifically [849].

Applications   Applications of IC extend beyond the measurement of anions and cations that initially contributed to the success of the technique. Polar organic and inorganic species can also be measured. Ion chromatography can profitably be used for the analysis of *ionic degradation products*. For example, IC permits determination of the elemental composition of additives in polymers from the products of pyrolysis or oxidative thermal degradation. The lower detection limit for additives in polymers are 0.1 % by PyGC and 0.0001–0.01 % by IC. The latter technique determines the anion and cation compositions of the degradation products. Tikuisis and Dang [348] used both wet ashing and destruction of HDPE formulations in an oxygen combustion bomb to oxidise the hydrocarbon matrix of the polymer, leaving behind various ionic species in the acidic aqueous solution. Sulfur and phosphorous ions generated from thioester and/or phosphite type AOs could be quantified using specific ion electrodes or IC. Nitrogenous-based antioxidants (e.g. aryl amines) present in the plastic sample could be measured by traditional Kjeldahl analysis or pyrolysis/chemiluminescence using an automated nitrogen analyser. Sotnikov and Bubchikova [850] have analysed PC containing tetrabromobisphenol-A flame retardant and EPDM with N-, S- and Br- containing additives. Stutts [851] has used IC with UV detection (at 254 nm) for the quantitative analysis of dithionite and thiosulfate (detection limits of 10 and 2 ppm, respectively). The latter is a known degradation product of the bleaching agent sodium dithionite and is detrimental in the paper processing industry. IC has also been used to examine the hydrolysis resistance of PUR [852]. This process leads to the formation of various ionic products: carboxylic acids, protonated amines, hydrolysis products of additives, such as F^- from foam additives and PO_4^{3-} from flame retardants, which can be determined by IC. IC has also been used to determine endgroups in polyamides [853].

The *chloride ion* is one of the most frequently analysed by IC, e.g. following up combustion of polymers [854,855]; similar analyses were reported for the bromide ion [854,855] and nitrite [855]. Analysis of polyester resins for halogens or phosphorous components may be carried out via conversion to halides and phosphates, respectively.

Mono-, di- and tributyltin ion species (in water) have been determined by cation exchange IC-ICP-MS at 0.2 ng detection limits [856]. IC is also particularly useful for HSE purposes, such as the determination of acid gases in the workplace. Applications of IC have been reviewed [857].

4.5 CAPILLARY ELECTROPHORETIC TECHNIQUES

Principles and Characteristics In electrophoresis the separation of electrically charged particles or molecules in a conductive liquid medium, usually aqueous, is achieved under the influence of a high electric field. This differs from chromatographic separations which imply a distribution between a mobile and a stationary phase. Instead of migration or retention times, which are chromatographic descriptors, CE focuses on mobility. Capillary electrophoresis was first described by Hjertén [16] in 1967 and is based on the same principles as traditional gel electrophoresis. The technique is suitable for the separation of ionic and neutral compounds.

Mixtures of chemical compounds can be separated by electrophoresis based on differences in the electrophoretic mobility which are principally determined by charge-to-mass ratio and physical dimensions of the analyte, viscosity of the medium and interaction of the analyte with a buffer. In *high-performance capillary electrophoresis* (HPCE) all components are largely miniaturised with a small total volume of about 1 μL. In this technique a buffer-filled narrow-bore fused-silica capillary (i.d. 50–75 μm, 30–100 cm long) is placed between two buffer reservoirs, and an electric field (up to 100 μA, 30 kV) is applied across the capillary. This results in an acceleration of ions in the capillary. Both size and shape of the analytes affect the electrophoretic mobility. The most noticeable contribution of the capillary wall, electroosmotic flow (EOF), is a very important feature in HPCE. As the solutes migrate through the capillary, a detector positioned just before the outlet of the separating capillary detects them. On-column detectors monitor a finite length of the capillary. The same capillary can be used for separations of polar and nonpolar compounds, small and large molecules, at ambient pressure without the need for a liquid pump.

Capillary electrophoresis offers several useful methods for (i) fast, highly efficient separations of ionic species; (ii) fast separations of macromolecules (biopolymers); and (iii) development of small volume separations-based sensors. The very low-solvent flow (1–10 nL min⁻¹) CE technique, which is capable of providing exceptional separation efficiencies, places great demands on injection, detection and the other processes involved. The total volume of the capillaries typically used in CE is a few microlitres. CE instrumentation must deliver nL volumes reproducibly every time. The peak width of an analyte obtained from an electropherogram depends not only on the bandwidth of the analyte in the capillary but also on the migration rate of the analyte.

There are three main *electrophoretic methods of separation*: (i) zone electrophoresis, where the components are separated on a basis of relative mobilities; (ii) isotachophoresis, where the separation is again based on relative mobilities but where the solutes are sandwiched between leading and terminating electrolytes;

and (iii) isoelectric focusing, where the solutes are separated according to their isoelectric points. In all electrophoretic systems, movement of charged species is always accompanied to some extent by electroosmotic flow. Since there are several different modes of capillary electrophoresis available on a standard CE instrument, namely capillary zone electrophoresis (CZE), capillary gel electrophoresis (CGE), micellar electrokinetic capillary chromatography (MEKC or MECC), capillary electrochromatography (CEC), capillary isoelectric focusing (CIEF), and capillary isotachophoresis (CITP), the technique is applicable to many different types of samples [858]. HPCE has perhaps the greatest molecular weight dynamic range of all the known separation techniques. Each separation mode differs in the mechanism or combination of mechanisms.

Free solution (FS) capillary zone electrophoresis, which is by far the most popular of the CE modes, separates analytes based on differences between their charge-to-size ratios in an aqueous medium. Zone electrophoresis refers to migration of molecules as zones, which do not undergo zone spreading due to diffusion. CZE uses a homogeneous buffer system (constant pH) over the entire separation path and time, which leads to simplified method development compared to that of the more complex technique of CITP. CZE may be used to separate a series of charged solutes. CITP uses a discontinuous buffer system and CIEF is used exclusively for amphoteric substances. In MEKC a micellar solution is employed as the electrophoretic medium [859]. Its separation principle is based on chromatography, i.e. the difference in the distribution between solvent and micelles of ionic surfactants. MEKC distinguishes itself from CE in that it can separate neutrals as well as ionic analytes. Capillary electrochromatography relies on the phenomenon of electroosmotic flow rather than high pressure to force the mobile phase through a fused-silica capillary filled with reversed-phase packing material [860,861]. CEC combines some of the advantages of HPLC (selectivity) and CE (efficiency) but is not simply to be considered as a hybrid technique of HPLC and CE as it works only for neutral compounds and is quite insensitive [862]. CEC is not yet mature. A comparable separation technique has been developed for macromolecular compounds [863]. Techniques suitable for separation of small molecules are CZE, MEKC and ITP. By proper selection of the mode of electrophoresis, anionic, cationic and neutral small molecules can be separated, often during the course of one run. Separating neutral compounds by CE is impossible using simple buffers. However, such separations are routinely achieved by using ionic additives such as the surfactant sodium dodecyl sulfate (SDS).

Modes of *detection* in both HPLC and CE are genuinely similar. However, given the small capillary dimensions in CE and the minuscule zone volumes produced (5–50 nL) sensitive and specific detection directly on the separating column is a major challenge. Detector types for CE (in decreasing order of use) are: UV/VIS (direct or indirect; PDA), fluorescence (FLNS, LIF), electrochemical (conductivity, amperometry), MS, Raman, radiometric, refractive index detection. For a comparison of different detection modes in CE, see Guzman and Majors [864]. The detectors employed are mostly concentration dependent. Due to the short light pathlength with on-line UV detection in CE (the inner diameter of the capillary), concentration limits of detection are usually well above 10^{-6} mol L^{-1}. Other types of detection may give better sensitivity. Indirect UV detection is widely used in CE for analysing compounds with limited UV activity, such as metal ions, simple anions and small organic acids (e.g. maleic and succinic acid). A UV-absorbing ion is added to the electrolyte to generate a high UV background signal, e.g. a phthalate for the separation of a range of small organic acids [865]. The concentration sensitivity of optical detection techniques for HPLC tends to be better than that of CE. Raman microprobe spectroscopy is a useful tool for probing structures of molecules within an electrophoresis capillary [866]. Laser-induced fluorescence (LIF) detection offers the highest sensitivity in CE (10^{-11} M with minimum detectable quantities in the zeptomole region), but is expensive and lacks universality [867,868]. Moreover, degradation of the analytes in intense laser beams cannot be excluded. The application of voltammetric detectors for CE is affected by the limited range of elements which provide appropriate electrochemical signals [869–871]. CE systems have also been coupled to miniaturised ICP-MS detection systems [872,873], and to PIXE [874] for element detection. Developments towards miniaturisation of ELSD are under way. In some cases, two or more detectors are connected in series. Capillary electrophoresis has been interfaced to ESI-MS and FAB/FIB-MS. Commercial CE-MS interfaces generally involve electrospray ionisation. The principles of CE are little affected by such couplings, but due attention should be given to the high voltages involved with capillary electrophoresis. Hyphenated CE techniques are described in Section 7.3.6.

CE can be used for both qualitative and quantitative analysis. Good reproducibility in migration times (qualitative aspects) and peak integration (quantitative aspects) is necessary. Qualitative analysis is usually based on comparisons of migration times between standards and samples. Quantitative analysis in CE is performed in essentially the same way as in HPLC since peak heights/areas are directly proportional to concentration when a concentration-dependent detector (e.g. UV/VIS, F) is used [546]. Baker [875] has gone into considerable detail as to the several parameters affecting the reproducibility and accuracy of *quantitative analysis* by CE and quotes reproducibility in the range of 1–2 % RSD. The precision of quantitative results is superior to most analytical techniques. Improved precision has mainly been achieved by using internal standards [876].

Table 4.48 summarises the *main characteristics* of CE. One of the main advantages of CE is that it requires only simple instrumentation (basically a high-voltage power supply, two buffer reservoirs, a capillary and a detector). Commercial instruments for CE have been available since 1988. Solvent compatibility is the main bottleneck for electrokinetic methods as something has to be ionised, either the polymer itself, some component of the separation medium, or the stationary phase material (in SEEC). Solvents with a relatively high dielectric constant must be used, and application of electrokinetic separation techniques is limited to those polymer types that are soluble in solvents such as water, DMF, acetonitrile, and HFIP. Non-aqueous capillary electrophoresis (NACE) offers advantages such as increased solubility for compounds that are not readily soluble in water and reduced sorption of hydrophobic substances onto capillary walls. Resolution, selectivity, efficiency and analysis time can be readily adjusted by selecting the right solvent (mixture).

Another limiting factor for applying electrokinetic separations to synthetic polymer analysis is the availability of suitable detectors. Although a UV detector is used, many interesting polymers do not have an appreciable absorbance in a suitable wavelength range. Also the sensitivity of the current HPCE detectors is inadequate. The tiny injection and detection volumes (on the order of a few pL to about 5 nL) used in CE undoubtedly set the sensitivity limits to a level which is approximately 50 times poorer than that of HPLC. The LOD requires improvement. In order to overcome the poor concentration limits of detection, sample introduction techniques are used that provide on-column focusing and allow the use of large sample volumes. Sensitivity enhancement in CE has recently been discussed; many devices or methods developed for preconcentration and enhancement of concentration sensitivity detection in CE were listed [864]. In some cases concern has been expressed

Table 4.48 Principle features of capillary electrophoresis

Advantages

- Relatively simple, inexpensive instrumentation
- Exceptionally high efficiency
- Fast separations without the need to employ gradient elution
- High sample throughput
- Flexibility in separation mechanism
- High-resolution separations; elimination of band broadening by convection
- Applicability to a wide variety of sample types (ionic, polar ionic/nonionic, nonpolar nonionic, high-molecular) and complex mixtures
- Low-volume technique; minute sample volume requirements (typically 1–10 nL)
- Greatly reduced analyte losses (small surface area)
- High mass detectability (zeptomole range); ultramicroanalysis
- Low reagent consumption (environmental friendly) and equipment maintenance (long lasting capillary columns)
- Reduced sample preparation time
- Direct injection of aqueous media; use of extremely dilute samples
- Ambient temperature (minimisation of sample decomposition)
- Efficient heat dissipation
- High precision; reliable
- Moderate method development effort
- Ease of operation
- Suitable for routine analytical separations
- Automated sampling and fraction collection
- Recognised by regulatory authorities (FDA, etc.)
- Commercial instrumentation; miniaturisation

Disadvantages

- High voltages
- Temperature control
- Solvent compatibility
- Limited loading capacity of analyte solutions
- Retention time repeatability and reproducibility
- Limited detector choice (need for extremely sensitive detection systems)
- Poor detection sensitivity
- Limited dynamic range
- Quantitation questionable (CZE)
- Requirement for sample homogeneity; clogging by large particles
- Unfit for separation and collection of large amounts of sample (not a preparative technique)
- Limited literature sources

about the reproducibility of CE. The techniques that can be used for peak purity assessment in CE are similar to those used in HPLC (spectral suppression, absorbance ratio, etc.). CE is not yet mature as a quantification tool.

The small injection volume of CE is desirable only when a small amount of sample is available. With materials which are heterogeneous on a microscale the need for extraction/dissolution of representative amounts of which only minute quantities are used in CE experiments leads easily to sensitivity problems. For application to polymeric materials the representativity of the minute sample volumes is quite relevant. Important factors for routine application of CE to synthetic polymers are also cleaning or regeneration procedures for the capillary and size or mass calibration.

In most situations analysts can achieve a rapid reasonable separation of compounds using an appropriate standard CE method with generic operating conditions [877]. This eliminates or reduces dramatically the need for *method development*. Major instrumental error sources in CE are detection, integration and injection. General guidelines for validation of CE methods are available and similar to those of HPLC [878]. Validated CE methods often perform the same as, or better than, the corresponding HPLC methods.

Traditional electrophoresis and capillary electrophoresis are competitive techniques as both can be used for the analysis of similar types of samples. On the other hand, whereas HPLC and GC are complementary techniques since they are generally applicable to different sample types, HPLC and CE are more competitive with each other since they are applicable to many of the same types of samples. Yet, they exhibit different selectivities and thus are very suitable for cross-validation studies. CE is well suited for analysis of both polar and nonpolar compounds, i.e. water-soluble and water-insoluble compounds. CE may separate compounds that have been traditionally difficult to handle by HPLC (e.g. polar substances, large molecules, limited size samples).

Comparisons of CE and HPLC on similar analytical problems have been reported [879–883]. In contrast to CE and HPLC, GC is best suited for analysis of nonpolar, lower MW, volatile compounds. HPLC and GC have detection limits roughly 100–1000 times lower than CE, while traditional electrophoresis has detection limits comparable to CE. As already mentioned, the inferior detection sensitivity and precision of CE when compared with HPLC are caused by the technique's nanoscale.

Because HPLC and HPCE are based on different physico-chemical principles, HPCE may be expected to address areas in which HPLC has shortcomings [884]. One such area is time of separation. In terms of speed of analysis, selectivity, quantitation, methods to control separation mechanism, orthogonality, CE performs better than conventional electrophoresis and varies from HPLC (Table 4.49). CE has very high efficiency compared to HPLC (up to two orders of magnitude) or GC. For typical capillary dimensions 10^5–10^6 theoretical plates are common in CE compared to 20 000 for a conventional HPLC column and

Table 4.49 Capability and potential of CE relative to HPLC

Criteria	CE	HPLC
Principles	Proven	Proven
Separation efficiency	Very high	Moderate
Reproducibility	Relatively high	High
Sensitivity		
Mass	Good	Good
Concentration	Moderate	Excellent
Affordability	Economical	Economical
Automation	Easy	Easy
Speed	Very rapid	Moderately rapid
Quantitation	Relatively easy	Easy
Sample capacity	nL range	>μL
Methods to control separation mechanism	Developing	Developed
Selectivity		
Diversity	Excellent	Excellent
Implementation	Easy	Relatively easy
Developed methods	Few	Many
Literature sources	Fair	Extensive
Orthogonality to HPLC	Orthogonal	–
Safety precautions	High voltage	High pressure
Equipment maintenance	Simple	Moderately simple
Environmental friendliness (solvent usage/waste)	Little waste and solvent	Solvent/waste
Consumable costs	Low	Relatively low
Prospects for growth	Maturing	Mature

150 000 for a capillary GC column. Sample preparation requirements for CE are those for HPLC (e.g. SPE). HPLC and CE are both robust methods. In CE capillary temperature control is crucial. Comparison of robustness implies a variation of different parameters, such as the mobile phase composition, the buffer pH and molarity, temperature, flow-rate and sample solvent [883]. Comparison of validated CE and HPLC methods shows that HPLC is about a factor of two better than CE for all quantitative parameters. Standard deviations of CZE analytical results are typically one full order of magnitude greater than those obtained with HPLC. The relative properties of HPLC and CE were discussed [885,886], also in relation to food grade antioxidants [887]. Although initially it was thought that CE might replace HPLC [875], this has proven not yet to be the case. The view has been expressed that future CE will be superior to LC not only in terms of separation performance, but also in precision [888]. CE is cost effective compared with HPLC; some 1000–1500 CE systems are sold per year.

CE and IC are complementary techniques and their use in tandem is advantageous [886]. The determination of inorganic anions by CE and IC was critically compared [841]; CE is usually more cost effective. As compared to ion-exchange chromatography with conductivity detection, CE ion analysis is faster and offers cost savings. For a comparison of CE to other separation techniques, see Baker [875].

Capillary electrophoresis is still in a state of evolution and a technique of choice for certain applications (chiral analysis; small ion analysis in food and beverage industries; bioanalysis). The number of reviews [365,858,884,889–896], books and manuals [365,897–903] published on (HP)CE/CEC/CZE in the last decade is overwhelming, in particular in relation to the importance of the technique (see also Bibliography). CE-LIFS has been reviewed [904].

Applications Capillary electrophoresis is applied mainly in the pharmaceuticals (vitamins, amino acids, nucleotides) and protein–peptide fields, in clinical and forensic science (drug screening), for agrochemicals and fine chemicals as well as for environmental purposes. HPCE allows separation of almost any type of compound, regardless of the chemical nature, size, conformation, or charge. Major application types are assay, impurity determination and identity confirmation, determination of small (in)organic ions and low-MW organics, and trace analysis (within the limits of the insufficient CE sensitivity). CE excels in the rapid analysis of highly charged polar analytes and ions. CE makes routine separations possible at nano or atto levels.

Apart from paints, electrokinetic separations find limited application for synthetic polymers [905], mainly because of solvent compatibility (CE is mostly an aqueous technique) and competition of SEC (reproducibility). Reasons in favour of the use of CE-like methods for polymer analysis are speed, sample throughput and low solvent consumption. Nevertheless, CE provides some interesting possibilities for polymer separation. Electrokinetic methods have been developed based on differences in ionisation, degree of interaction with solvent constituents, and molecular size and conformation.

Since the early 1990s an increasing number of papers has been devoted to the application of CE for the analysis of both *inorganic cations* [906–915] and *low-molecular-mass anions* [915–922]. Standard CE methods have been developed and validated for determining inorganic anions (e.g. chloride, sulfate and nitrate), small carboxylic acids and metal ions that all have limited or no UV absorbance. In those situations, short UV wavelengths (190 nm) or indirect UV detection should be used. Such methods might be extended to metallic

stearates used as acid scavengers or lubricants. Applications of CE analysis to inorganic anions are numerous but very limited in the field of polymer/additive analysis. It is expected that applications of CE to the determination of inorganic anions will partially replace IC.

Braun and Richter [923] have described an application of CE in additive analysis, namely quantitative analysis of *heat stabilisers* in PVC, such as Irgastab 17A and 18 MOK-N, which are metal-based (in the past usually Cd, Ba and Pb, now nontoxic Ca, Zn and Sn). Quantitative metal analysis is of interest for PVC recycling purposes. Various alternative approaches are possible for such quantitative analysis, such as XRF [924], polarography [925] and AAS [923]. The performance of AAS, CE and complexometric titrations in the analysis of the heavy metal content in PVC was compared [923]. For all methods investigated the metals must be separated from the polymer and transferred into an aqueous phase.

CE is widely applied to the analysis of *organic acids* using indirect UV detection. The equivalence of CE and IC determinations of organic acids (up to ppm level) has been shown [926]. The need for removal of organic matrices by SPE, as required by IC, was eliminated. In this respect, CE is a superior technique for organic acid analysis. Saturated and unsaturated *fatty acids* (C_{12}–C_{24}) in use as lubricants, softeners and textile auxiliary substances, have been separated simultaneously by CE with indirect UV detection at 214 nm (using *p*-hydroxybenzoate) and by RPLC [710]. The advantages of CE for the analysis of fatty acids include fast separation, rapid method development, low sample and solvent consumption and no derivatisation. The disadvantages are the higher limit of detection and lower reproducibility than HPLC.

Only few applications have been reported to determine *antioxidants* in rubbers or polymers by using electrochemical methods [927,928]. Sawada *et al.* [929] reported successful separations by coupling the antioxidants with *p*-diazobenzene sulfonic acid before electrophoresis. Amine AOs were coupled in acetic acid and phenolic AOs in NaOH-ethanol were analysed by CE methods. MEKC separation of the four major food grade antioxidants (PG, BHA, BHT, TBHQ) was completed within 6 min with pmole amount detection using UV absorption [930]. RPLC was not as efficient and required larger sample amounts and longer separation times.

The different classes of compounds that have been successfully separated by MEKC include, among others, phenols [932,933], antioxidants [930,934], vitamins [935], phthalate esters [931,936] (Figure 4.26) and charged and neutral dyestuffs, including some dispersive dyes [937–939]. The separation of various priority

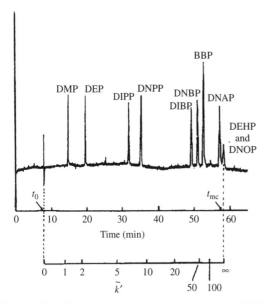

Figure 4.26 Chromatogram of phthalate esters in MEKC with methanol mixed solution. After Takeda *et al.* [931]. Reprinted with permission from S. Takeda *et al.*, *Anal. Chem.* **65**, 2489–2492 (1993). Copyright (1993) American Chemical Society

phthalate esters (DMP, DEP, DBP, DAP, BBP, DEHP), used as *plasticisers* in polymer formulations, by MEKC was reported [936]. The amount of sample injected was typically about 1.5 nL. MEKC with methanol and urea as buffer modifiers was also employed for separation of benzothiazole sulfenamides (TBBS and CBS), used as *accelerators* in manufacturing of rubber products [940]. These compounds tend to have poor stability in polar solvents. Kok *et al.* [905] have used MEKC for the oligomeric separation of PEG 600 after derivatisation with phenylisocyanate (PIC). CGE has been used for the separation of alkylbenzoates, alkylphenones, alkylbenzenes and oxidation inhibitors [941]. CZE in aqueous H_3PO_4 with HFIP using UV detection at 190 nm can be used for the selective separation of the linear oligomers of PA6 [651].

CE is also potentially a useful alternative analytical tool for monitoring of chemicals (dyes, flame retardants and lubricants) involved in various steps of the *textile fibre* manufacturing process. In this area, CE compares favourably with existing techniques. CZE-MS^n was used for the analysis of sulfonated azo dyes [942]. A variety of fluorescent analytes including thiazole orange dyes have been characterised by CE-FLNS [943].

SPE-NACE-UV has been employed for the determination of phenols (in environmental samples as by-products in the production of plastics and dyes and in the pulp industry) [944]. Various other methods, such as GC-MS, HPLC-PDA, HPLC-ECD, HPLC-PB-MS and SPE-HPLC-APCI-MS have been used for the same purpose [945].

CE has been used for the analysis of anionic *surfactants* [946,947] and can be considered as complementary to HPLC for the analysis of cationic surfactants with advantages of minimal solvent consumption, higher efficiency, easy cleaning and inexpensive replacement of columns and the ability of fast method development by changing the electrolyte composition. Also the separation of polystyrene sulfonates with polymeric additives by CE has been reported [948]. Moreover, CE has also been used for the analysis of *polymeric water treatment additives*, such as acrylic acid copolymer flocculants, phosphonates, low-MW acids and inorganic anions. The technique provides for analyst time-savings and has lower detection limits and improved quantification for determination of anionic polymers, compared to HPLC.

Recently a decreased level of CE activity has been noticed with a shift of attention towards other separation techniques such as electrochromatography. CE is apparently not more frequently used partly because of early instrumental problems associated with lower sensitivity, sample injection, and lack of precision and reliability compared with HPLC. CE has slumped in many application areas with relatively few accepted routine methods and few manufacturers in the market place. While the slow acceptance of electrokinetic separations in polymer analysis has been attributed to conservatism [905], it is more likely that as yet no unique information has been generated in this area or eventually only the same information has been gathered in a more efficient manner than by conventional means. The applications of CE have recently been reviewed [949,950]; metal ion determination by CE was specifically addressed by Pacaková *et al.* [951].

4.6 BIBLIOGRAPHY

4.6.1 General Texts

C.F. Poole, *Essence of Chromatography*, Elsevier, Amsterdam (2002).

J. Cazes (ed.), *Encyclopedia of Chromatography*, M. Dekker, New York, NY (2001).

H.J. Issaq (ed.), *A Century of Separation Science*, M. Dekker, New York, NY (2001).

C.W. Gehrke, R.L. Wixom and E. Bayer, *Chromatography: A Century of Discovery, 1900–2000*, Elsevier, Amsterdam (2001).

C.E. Meloan, *Chemical Separations: Principles, Techniques & Experiments*, Wiley-Interscience, New York, NY (2000).

T. Toyo'oka, *Modern Derivatization Methods for Separation Sciences*, John Wiley & Sons Ltd, Chichester (1999).

A. Felinger (ed.), *Data Analysis and Signal Processing in Chromatography*, Elsevier, Oxford (1998).

E. Forgács and T. Cserháti, *Molecular Basis of Chromatographic Separation*, CRC Press, Boca Raton, FL (1997).

R. Kaliszan, *Structure and Retention in Chromatography. A Chemometric Approach*, Harwood Academic Publishers, Amsterdam (1997).

B. Baars, J.v.d. Berg, H.-G. Janssen, P. Schoenmakers, R. Tijssen and N. Vonk, *Chromatografie in de praktijk*, Ten Hagen & Stam, Den Haag (1997) (in Dutch).

K. Robards, P.R. Haddad and P.E. Jackson, *Principles and Practice of Modern Chromatographic Methods*, Academic Press, San Diego, CA (1994).

B.S. Middleditch, *Analytical Artifacts GC, MS, HPLC, TL and PC*, Elsevier, Amsterdam (1992).

4.6.2 Detectors

R.P.W. Scott, *Chromatographic Detectors*, M. Dekker, New York, NY (1996).

R.E. Sievers (ed.), *Selective Detectors. Environmental, Industrial, and Biomedical Applications*, John Wiley & Sons Inc., New York, NY (1995).

D. Stiles, A. Calokerinos and A. Townshend (eds), *Flame Chemiluminescence Analysis by Molecular Emission Cavity Detection*, John Wiley & Sons, Ltd, Chichester (1994).

P.C. Uden (ed.), *Element-Specific Chromatographic Detection by Atomic Emission Spectroscopy*, ACS Symposium Series, Vol. 479, American Chemical Society, Washington, DC (1992).

4.6.3 Gas Chromatography

J.V. Hinshaw, *Getting the Best Results from Your Gas Chromatograph*, Wiley-Interscience, New York, NY (2003).

A.J. Handley and E.R. Adlard (eds), *Gas Chromatographic Techniques and Applications*, Sheffield Academic Press, Sheffield (2001).

R.P.W. Scott, *Introduction to Analytical Gas Chromatography*, M. Dekker, New York, NY (1998).

H.M. McNair and J.M. Miller, *Basic Gas Chromatography*, John Wiley & Sons, Inc., New York, NY (1998).

H.-G. Janssen, *Sample Introduction Techniques for Capillary Gas Chromatography* (Reference GI-12981), Gerstel Inc., Mülheim a.d. Ruhr (1998).

W. Jennings, E. Mittlefehldt and P. Stremple, *Analytical Gas Chromatography*, Academic Press, San Diego, CA (1997).

B. Kolb and L.S. Ettre, *Static Headspace-Gas Chromatography. Theory and Practice*, Wiley-VCH, New York, NY (1997).

D.W. Grant, *Capillary Gas Chromatography*, John Wiley & Sons, Inc., New York, NY (1996).

H. Hachenberg and K. Beringer, *Die Headspace-Gaschromatographie als Analysen- und Messmethode*, Springer-Verlag, Berlin (1996).

R.L. Grob, *Modern Practice of Gas Chromatography*, John Wiley & Sons, Inc., New York, NY (1995).

I.A. Fowlis, *Gas Chromatography*, John Wiley & Sons, Ltd, Chichester (1995).

L.S. Ettre and J.V. Hinshaw, *Basic Relationships of Gas Chromatography*, Advanstar Communications, Eugene, OR (1993).

4.6.4 Supercritical Fluid Chromatography

K. Anton and C. Berger, *Supercritical Fluid Chromatography with Packed Columns – Techniques and Applications*, M. Dekker, New York, NY (1998).

R.M. Smith and S.B. Hawthorne (eds), *Supercritical Fluids in Chromatography and Extraction*, Elsevier, Amsterdam (1997).

F.A. Settle (ed.), *Supercritical Fluid Chromatography and Extraction*, Prentice Hall, Upper Saddle River, NJ (1997).

M.S. Verrall (ed.), *Supercritical Fluid Extraction and Chromatography*, John Wiley & Sons, Ltd, Chichester (1996).

T.A. Berger, *Packed Column SFC*, The Royal Society of Chemistry, Cambridge (1995).

S.A. Westwood (ed.), *Supercritical Fluid Extraction and Its Use in Chromatographic Sample Preparation*, Blackie A&P, London (1993).

4.6.5 Thin-layer Chromatography

B. Fried and J. Sherma, *Thin-Layer Chromatography: Techniques and Applications*, M. Dekker, New York, NY (1999).

E. Soczewinski (ed.), *Planar Chromatography (TLC) – Handbook of Thin Layer Chromatography*, Harwood Academic Publishers, Amsterdam (1998).

J. Sherma and B. Fried (eds), *Handbook of Thin-layer Chromatography*, M. Dekker, New York, NY (1996).

L. Kraus, A. Koch and S. Hofstetter-Kuhn, *Dünnschichtchromatographie*, Springer-Verlag, Berlin (1996).

H. Jork, W. Funk, W. Fischer and H. Wimmer, *Thin-layer Chromatography*, Vol. 1b, *Reagents and Detection Methods*, VCH, Weinheim (1994).

H.P. Frey and K. Zieloff, *Qualitative and quantitative Dünnschicht-chromatographie, Grundlagen und Praxis*, VCH, Weinheim (1992).

J.C. Touchstone, *Practice in Thin Layer Chromatography*, John Wiley & Sons, Inc., New York, NY (1992).

H. Jork, W. Funk, W. Fischer and H. Wimmer, *Thin-layer Chromatography*, Vol. Ia, *Physical and Chemical Detection Methods: Fundamentals, Reagents*, VCH, Weinheim (1990).

F. Geiss, *Fundamentals of Thin-layer Chromatography*, Hüthig Verlag, Heidelberg (1987).

4.6.6   Liquid Chromatography

P.C. Sadek, *The HPLC Solvent Guide*, John Wiley & Sons, Inc., New York, NY (2002).

J. Swadesh (ed.), *HPLC, Practical and Industrial Applications*, CRC Press, Boca Raton, FL (2000).

S. Kromidas, *Practical Problem Solving in HPLC*, John Wiley & Sons, Ltd, Chichester (2000).

T. Hanai, *HPLC. A Practical Guide*, The Royal Society of Chemistry, Cambridge (1999).

V.R. Meyer (ed.), *Practical High-performance Liquid Chromatography*, John Wiley & Sons, Ltd, Chichester (1999).

W.R. LaCourse, *Pulsed Electrochemical Detection in High Performance Liquid Chromatography*, John Wiley & Sons, Inc., New York, NY (1997).

H. Pasch and B. Trathnigg, *HPLC of Polymers*, Springer-Verlag, Berlin (1997).

A. Weston and P.R. Brown, *HPLC and CE. Principles and Practice*, Academic Press, San Diego, CA (1997).

U.D. Neue, *HPLC Columns, Theory, Technology and Practice*, John Wiley & Sons, Ltd, Chichester (1997).

L.R. Snyder, J.J. Kirkland and J.L. Glajch, *Practical HPLC Method Development*, John Wiley & Sons, Inc., New York, NY (1997).

R.P.W. Scott, *Liquid Chromatography for the Analyst (Chromatographic Science Series, Vol. 67)*, M. Dekker, New York, NY (1994).

L. Huber and S. George, *Diode-array Detection in High-performance Liquid Chromatography*, M. Dekker, New York, NY (1993).

S. Ahuja, *Trace and Ultratrace Analysis by HPLC*, John Wiley & Sons, Inc., New York, NY (1991).

A.T. Rhys Williams, *Fluorescence Detection in Liquid Chromatography*, Perkin-Elmer, Beaconsfield (1986).

4.6.7   Size-exclusion Chromatography

C.-S. Wu, *Handbook of Size Exclusion Chromatography*, Academic Press, New York, NY (1999).

S. Mori and H.G. Barth, *Size Exclusion Chromatography*, Springer-Verlag, Berlin (1999).

4.6.8   Ion Chromatography

Anon, *The Whole World of Ion Chromatography*, Metrohm AG, Herisau (2004).

C. Eith, M. Kolb, A. Seubert and K.H. Viewweger, *Practical Ion Chromatography*, Metrohm AG, Herisau (2001).

J.S. Fritz and D.T. Gjerde, *Ion Chromatography*, Wiley-VCH, Weinheim (2000).

H. Schäffer, M. Laeubli and A. Doerig, *Theory of Ion Chromatography*, Metrohm AG, Herisau (2000).

P.R. Haddad and P.E. Jackson, *Ion Chromatography (J. Chromatogr. Libr., Vol. 46)*, Elsevier, Amsterdam (1990).

4.6.9   Capillary Electrophoretic Techniques

Z. Deyl and F. Švec, *Capillary Electrochromatography*, Elsevier Science, Amsterdam (2001).

K.D. Bartle and P. Myers (eds), *Capillary Electrochromatography*, The Royal Society of Chemistry, Cambridge (2001).

D. Heiger, *High Performance Capillary Electrophoresis, Agilent Technologies Primer, Publ. no. 5968-9963E*, Wilmington, DE (2000).

R. Weinberger, *Practical Capillary Electrophoresis*, Academic Press, Orlando, FL (2000).

H. Sørensen, S. Sørensen, C. Bjerjegaard and S. Michaelsen, *Chromatography and Capillary Electrophoresis in Food Analysis*, The Royal Society of Chemistry, Cambridge (1999).

M. Khaledi (ed.), *High Performance Capillary Electrophoresis: Theory, Techniques and Applications*, John Wiley & Sons, Inc., New York, NY (1998).

P. Camilleri (ed.), *Capillary Electrophoresis: Theory and Practice*, CRC Press Inc., Boca Raton, FL (1998).

K.D. Altria (ed.), *Capillary Electrophoresis Guidebook: Principles, Operation and Applications*, Humana Press, Totowa, NJ (1996).

H. Shintani and J. Polonsky (eds), *Handbook of Capillary Electrophoresis Applications*, Chapman & Hall, London (1996).

E. Jackim and L. Watts Jackim (eds), *Capillary Electrophoresis Procedures Manual: A Laboratory User's Aid for Quick Starts*, Elsevier Science Publishers, Amsterdam (1996).

J.P. Landers (ed.), *Handbook of Capillary Electrophoresis*, CRC Press Inc., Boca Raton, FL (1996).

H. Engelhardt, *Capillary Electrophoresis*, F. Vieweg & Sohn, Wiesbaden (1995).

D.R. Baker, *Capillary Electrophoresis*, John Wiley & Sons, Inc., New York, NY (1994).

R. Kuhn and S. Hofstetter-Kuhn, *Capillary Electrophoresis Principles and Practice*, Springer-Verlag, Berlin (1993).

N.A. Guzman (ed.), *Capillary Electrophoresis Technology*, M. Dekker, New York, NY (1993).

P. Jandik and G. Bonn, *Capillary Electrophoresis of Small Molecules and Ions*, VCH Publishers, New York, NY (1993).

R. Weinberger, *Practical Capillary Electrophoresis*, Academic Press, San Diego, CA (1993).

J. Vindevogel and P. Sandra, *Introduction to Micellar Electrokinetic Chromatography*, Hüthig Verlag, Heidelberg (1992).

P.D. Grossman and J.C. Colburn (eds), *Capillary Electrophoresis: Theory and Practice*, Academic Press, San Diego, CA (1992).

S.F.Y. Li, *Capillary Electrophoresis, Practice and Applications*, Elsevier Science Publishers, Amsterdam (1992).

4.7 REFERENCES

1. S. Cardenas, M. Gallego and M. Valcárcel, *LC.GC Europe* **14** (3), 174–9 (2001).
2. T.L. Chester, *Anal. Chem.* **69**, 165A–9A (1997).
3. I.L. Davies, K.E. Markides, M.L. Lee and K.D. Bartle, in *Multidimensional Chromatography* (H.J. Cortes, ed.), M. Dekker, New York, NY (1990), pp. 301–30.
4. V.R. Meyer, in *Encyclopedia of Analytical Science* (A. Townshend, ed.), Academic Press, London (1995), pp. 720–9.
5. E. Mincsovics, K. Ferenczi-Fodor and E. Tyihák, in *Handbook of Thin-layer Chromatography* (J. Sherma and B. Fried, eds), M. Dekker, New York, NY (1996), Ch. 7.
6. M.S. Tswett, *Tr. Varshav. Obshch. Estestvoispyt., Otd. Biol.* **14**, 20–39 (1903).
7. N.A. Izmailov and H.S. Shraiber, *Farmatsiya (Sofia)* **3**, 1 (1938).
8. E. Stahl, *Chem. Ztg.* **82**, 323–9 (1958).
9. A.J.P. Martin and R.L.M. Synge, *Biochem. J.* **35**, 1358 (1941).
10. R. Consden, A.H. Gordon and A.J.P. Martin, *Biochem. J.* **38**, 224 (1944).
11. A.T. James and A.J.P. Martin, *Analyst* **77**, 915–32 (1952).
12. M.J.E. Golay, in *Gas Chromatography* (V.J. Coates, ed.), Academic Press, New York, NY (1958), pp. 1–13.
13. L. Bovijn, J. Pirotte and A. Berger, in *Gas Chromatography 1958 (Amsterdam Symposium)*, D.H. Desty (ed.), Butterworths, London (1958), pp. 310–20.
14. E. Klesper, A.H. Corwin and D.A. Turner, *J. Org. Chem.* **27**, 700–1 (1962).
15. J.C. Moore, *J. Polym. Sci.* **A2**, 835 (1964).
16. S. Hjertén, *Chromatogr. Rev.* **9** (2), 122–219 (1967).
16a. C. Horváth, *Pittsburgh Conference* (1970).
17. H. Small, T.S. Stevens and W.C. Bauman, *Anal. Chem.* **47**, 1801–9 (1975).
18. R.D. Dandeneau and H.E. Zerenner, *J. High Resolut. Chromatogr., Chromatogr. Commun.* **2**, 351–6 (1979).
19. D. Ishii and T. Takeuchi, *J. Chromatogr. Sci.* **27**, 71–4 (1989).
20. W.S. Cooke, *Today's Chemist at Work* **1**, 16 (1996).
21. T. Cabrera, D. Lubda, H.-M. Eggenweiler, H. Minakuchi and K. Nakanishi, *J. High Resolut. Chromatogr.* **23** (1), 93–9 (2000).
22. C. Schäfer, K. Cabrera, D. Lubda, K. Sinz and D. Cunningham, *Intl. Lab.* **31** (2A), 14–15 (April 2001).
23. H.M. NcNair, *Chromatographia* **7**, 161–5 (1974).
24. M.J.I. Mattina, *J. Chromatogr. Libr.* **59**, 325–44 (1996).
25. H.-G. Janssen, *ATAS Chromatography Technical Notes No. 5*, Veldhoven (1993).
26. T.A. Berger, *Packed Column SFC*, The Royal Society of Chemistry, Cambridge (1995).
27. B.S. Middleditch and A. Zlatkis, *J. Chromatogr. Sci.* **25**, 547 (1987).
28. B.S. Middleditch, *Analytical Artifacts GC, MS, HPLC, TLC and PC, J. Chromatogr. Library Series No. 44*, Elsevier, Amsterdam (1989).
29. L. Evangelista and V. Jeanguyot, *Symposium on Additives for Polymers*, Certech, Namur (2001).
30. *Accreditation for Chemical Laboratories: Guidance on the Interpretation of the EN 45000 Series of Standards and ISO Guide 25, Section 15: Validation, Eurachem Guidance Document No. 1*, WELAC Guidance Document No. WGD2 (April 1993).
31. R.P.W. Scott, *Chromatographic Detectors*, M. Dekker, New York, NY (1996).
32. A. de Wit and J. Beens, *J. Chromatogr. Libr.* **56**, 159–200 (1995).
33. R.L. Shearer, in *Selective Detectors* (R.E. Sievers, ed.), John Wiley & Sons, Inc., New York, NY (1995), pp. 35–69.
34. L. Jaufmann, *Analusis Mag.* **26** (2), M22–4 (1998).
35. C.F. Poole and S.K. Poole, *Chromatography Today*, Elsevier, Amsterdam (1991).
36. F.A. Byrdy and J.A. Caruso, in *Selective Detectors* (R.E. Sievers, ed.), John Wiley & Sons, Inc., New York, NY (1995), pp. 171–207.
37. L.S. Ettre, *LC.GC Europe* **15** (6), 364–73 (2002).
38. U.K. Gökeler, in *Selective Detectors* (R.E. Sievers, ed.), John Wiley & Sons, Inc., New York, NY (1995), pp. 99–126.
39. R.E. Sievers (ed.), *Selective Detectors*, John Wiley & Sons, Inc., New York, NY (1995).
40. T.B. Ryerson and R.E. Sievers, in *Selective Detectors* (R.E. Sievers, ed.), John Wiley & Sons, Inc., New York, NY (1995), pp. 1–33.
41. S.O. Farwell and C.J. Barinaga, *J. Chromatogr. Sci.* **24**, 483 (1986).
42. R.C. Hall, *J. Chromatogr. Sci.* **12**, 152 (1974).
43. H. Shi, J. Thompson, B. Strode, L.T. Taylor and E.M. Fujinari, in *Instrumental Methods in Food and Beverage Analysis* (D.L.B. Wetzel and G. Charalambous, eds), Elsevier, Amsterdam (1998), Chap. 6.
44. E.M. Fujinari and L.O. Courthaudon, *J. Chromatogr.* **592**, 209 (1992).
45. R.L. Shearer, *Am. Lab.* **12**, 24 (1994).
46. A.J. Dunham and R.E. Sievers, in *Selective Detectors* (R.E. Sievers, ed.), John Wiley & Sons, Inc., New York, NY (1995), pp. 71–97.
47. A. Hulshoff and H. Lingeman, in *Molecular Luminescence Spectroscopy. Methods and Applications* (S.G. Schulman, ed.), John Wiley & Sons, Inc., New York, NY (1985), Pt I, pp. 621–716.
48. L. Huber and S. George (eds), *Diode-array Detection in High-performance Liquid Chromatography*, M. Dekker, New York, NY (1993).
49. T.H. Risby and Y. Talmi, *CRC Crit. Rev. Anal. Chem.* **14** (3), 231 (1983).
50. P.C. Uden, in *Selective Detectors* (R.E. Sievers, ed.), John Wiley & Sons, Inc., New York, NY (1995), pp. 143–69.

51. S. Mori and H.G. Barth, *Size Exclusion Chromatography*, Springer-Verlag, Berlin (1999).

52. C.S. Young and J.W. Dolan, *LC.GC Europe* **16** (3), 132–7 (2003).

53. M.J. Wilcox, *Abstracts 22nd Intl Symposium on High Performance Liquid Phase Separations and Related Techniques (HPLC '98)*, St Louis, MO (1998).

54. P.C. Uden, in *Quantitative Analysis Using Chromatographic Techniques* (E. Katz, ed.), John Wiley & Sons, Ltd, Chichester (1987), pp. 99–155.

55. J.N. Davenport and E.R. Adlard, *J. Chromatogr.* **290**, 13–32 (1984).

56. W. Haag and C. Wrenn, *Handbook of Theory and Applications of Direct-reading Photoionization Detectors (PIDs)*, RAE Systems Inc., Sunnyvale, CA (2002).

57. G.R. van Hecke and K.K. Karukstis, *A Guide to Lasers in Chemistry*, Jones and Bartlett Publ., Boston, MA (1998).

58. H.V. Drushel, *J. Chromatogr. Sci.* **21**, 375 (1983).

59. H.M. McNair and J.M. Miller, *Basic Gas Chromatography*, John Wiley & Sons, Inc., New York, NY (1998).

60. P.C. Uden, *Element-specific Chromatographic Detection by Atomic Emission Spectroscopy*, ACS Symposium Series, Vol. 479, American Chemical Society, Washington, DC (1992).

61. J.K. Swadesh, in *HPLC. Practical and Industrial Applications* (J.K. Swadesh, ed.), CRC Press, Boca Raton, FL (1996), pp. 171–243.

62. R.B. Green, *Anal. Chem.* **55**, 20A (1983).

63. J.V. Hinshaw, *LC.GC* **13**, 536, 944 (1995).

64. I.G. McWilliam and R.A. Dewar, *Nature* **181**, 760 (1958).

65. S.R. Lipsky and M.L. Duffy, *J. High Resolut. Chromatogr., Chromatogr. Commun.* **9**, 376, 725 (1986).

66. R.L. Grob, *Modern Practice of Gas Chromatography*, John Wiley & Sons, Inc., New York, NY (1995).

67. R.H. Brown, R.P. Harvey, C.J. Purnell and K.J. Saunders, *Am. Ind. Hyg. Assoc. J.* **45** (2), 67–75 (1984).

68. J. Rektorik, *Eur. Pat. Appl.* EP 73,176 (June 7, 1982).

69. J. Rektorik, in *Sample Introduction in Capillary Gas Chromatography* (P. Sandra, ed.), A. Hüthig Verlag, Heidelberg (1985), Vol. 1, p. 217.

70. M.J.E. Golay (to Perkin Elmer), *US Pat.* 2,920,476 (June 24, 1957).

71. H.H. Hill and G. Simpson, in *Encyclopedia of Analytical Science* (A. Townshend, ed.), Academic Press, London (1995), pp. 1766–71.

72. L.S. Ettre, *Intl Lab.*, 4–6 (January 1999).

73. R.P.W. Scott, *Introduction to Analytical Gas Chromatography*, M. Dekker, New York, NY (1998).

74. Restek Product Guide, Restek Corporation, Bellefonte, PA (1999).

75. J.V. Hinshaw, *LC.GC Europe* **14** (3), 155–60 (2001).

76. J.V. Hinshaw, *LC.GC Intl* **11**, 234–9 (April 1998).

77. J.F. Sprouce and A. Varano, *Intl. Lab.*, 54 (November/December 1984).

78. R. Sacks, H. Smith and M. Nowak, *Anal. Chem.* **70**, 29–37A (1998).

79. M.P.M. van Lieshout, M. van Deursen, R. Derks, H.-G. Janssen and C.A. Cramers, *J. Microcol. Sep.* **11**, 155–62 (1999).

80. R.J. Jonker, H. Poppe and J.F.K. Huber, *Anal. Chem.* **54**, 2447 (1982).

81. Y. Shen and M.L. Lee, *J. Microcol. Sep.* **9**, 21 (1997).

82. M.P.M. van Lieshout, M. van Deursen, R. Derks, H.-G. Janssen and C.A. Cramers, *J. High Resolut. Chromatogr.* **22**, 116–8 (1999).

83. L. Mondello, G. Zappia, G. Errante, P. Dugo and G. Dugo, *LC.GC Europe*, 495–502 (July 2000).

84. *Flash-2D-GC^{TM} Brochure*, Thermedics Detection Inc., Chelmsford, MA (1996).

85. S.J. MacDonald and D. Wheeler, *Intl Lab.*, 21C–23C (September 1998).

86. M.P.M. van Lieshout, R. Derks, H.-G. Janssen and C.A. Cramers, *J. High Resolut. Chromatogr.* **21**, 583–6 (1998).

87. L.S. Ettre, *LC.GC Intl* **10**, 782–4 (December 1997).

88. M.M. van Deursen, *PhD Thesis*, Technical University of Eindhoven (2002).

89. W. Vogt, K. Jacob and H.W. Obwexer, *J. Chromatogr.* **174**, 437 (1979).

90. H.-G. Janssen, *Sample Introduction Techniques for Capillary Gas Chromatography*, Gerstel Inc., Mülheim a.d. Ruhr (1998), Reference GI-12981.

91. J.V. Hinshaw, *LC.GC Intl* **12** (2), 83–6 (1999).

92. G. Di Pasquale and M. Galli, *J. High Resolut. Chromatogr., Chromatogr. Commun.* **7**, 484 (1984).

93. G. Di Pasquale, L. Giambelli, A. Soffientini and R. Paiello, *J. High Resolut. Chromatogr., Chromatogr. Commun.* **8**, 618 (1985).

94. W. Blum and L. Damasceno, *J. High Resolut. Chromatogr.* **10**, 472–6 (1987).

95. B.E. Berg, A.M. Flaaten, J. Paus and T. Greibrokk, *J. Microcol. Sep.* **4**, 227 (1992).

96. E.S. Poy, F. Visani and F. Terrosi, *J. Chromatogr.* **217**, 81 (1981).

97. M.P.M. van Lieshout, H.-G. Janssen, C.A. Cramers and G.A. van den Bosch, *J. Chromatogr.* **A764**, 73–84 (1997).

98. M.P.M. van Lieshout, H.-G. Janssen, C.A. Cramers, M.J.J. Hetem and H.J.P. Schaik, *J. High Resolut. Chromatogr.* **19**, 193 (1996).

99. R. Sasano, Y. Hiramatsu, M. Kurano and M. Furuno, *Proceedings 22nd Intl Symposium on Capillary Chromatography*, Gifu (1999).

100. T. Hankemeier, *PhD Thesis*, Vrije Universiteit Amsterdam (2000).

101. R. Perkins, D. Nicolas and R. Sasano, *www.ATAS-int.com* (2000).

102. J.-F. Borny and J. Werner, *Abstracts 6th Intl Symposium on Hyphenated Techniques in Chromatography and Hyphenated Chromatographic Analyzers (HTC-6)*, Bruges (2000), P13.

103. I.S. Krull, M. Swartz and J.N. Driscoll, *Adv. Chromatogr.* **24**, 247–316 (1984).

104. E.R. Adlard, *CRC Crit. Rev. Anal. Chem.* **5** (1), 1–11, 13–36 (1975).

105. S.A. Liebman and T.P. Wampler, in *Pyrolysis and GC in Polymer Analysis* (S.A. Liebman and E.J. Levy, eds), M. Dekker, New York, NY (1985), pp. 53–148.

106. S.C. Moldoveanu, *Analytical Pyrolysis of Natural Organic Polymers*, Elsevier, Amsterdam (1998).

107. S. Ahuja, in *Ultratrace Analysis of Pharmaceuticals and Other Compounds of Interest* (S. Ahuja, ed.), John Wiley & Sons, Inc., New York, NY (1986), pp. 19–90.

108. V.G. Berezkin, *Chemical Methods in Gas Chromatography*, Elsevier, Amsterdam (1981).

109. J. Novák, *Quantitative Analysis by Gas Chromatography*, M. Dekker, New York, NY (1987).

110. P.J. Schoenmakers, *Optimization of Chromatographic Selectivity: A Guide to Method Development*, Elsevier Science Publishers, Amsterdam (1986).

111. J.V. Hinshaw, *LC.GC Europe* **15** (2), 80–4 (2002).

112. Riedel-de Haën, *High Purity Standards*, Seelze, BRD (1998).

113. J. McKinley and R.E. Majors, *LC.GC Europe* **13** (12), 892–901 (2000).

114. J.D. Craske and C.D. Bannon, *J. Am. Oil Chem. Soc.* **64**, 1413 (1987).

115. *ASTM Annual Book of ASTM Standards*, ASTM, West Conshohocken, PA (2000).

116. C. Warner, J. Modderman, T. Fazio, M. Beroza, G. Schwartzman and K. Fominaya, *Food Additives Analytical Manual*, AOAC Intl, Arlington, VA (1993).

117. L.S. Ettre, *LC.GC Europe* **14** (2), 72–3 (2001).

118. V.G. Berezkin, *Gas Liquid Solid Chromatography*, M. Dekker, New York, NY (1991).

119. K.J. Hyver (ed.), *High-resolution Gas Chromatography*, Hewlett-Packard Co., Avondale, PA (1989).

120. A. van Es, *High Speed Narrow Bore Capillary Gas Chromatography*, Hüthig Publishers, Heidelberg (1992).

121. K. Grob, *Split and Splitless Injection in Capillary GC*, Hüthig Publishers, Heidelberg (1993).

122. K. Grob, *On-column Injection in Capillary Gas Chromatography*, Hüthig Publishers, Heidelberg (1987).

123. H. Rotzsche, *Stationary Phases in Gas Chromatography*, Elsevier, Amsterdam (1991).

124. K. Blau and J.M. Halket, *Handbook of Derivatives in Chromatography*, John Wiley & Sons, Inc., New York, NY (1993).

125. T. Toyo'oka, *Modern Derivatization Methods for Separation Sciences*, John Wiley & Sons, Ltd, Chichester (1999).

126. C.A. Cramers, H.-G. Janssen, M.M. van Deursen and P.A. Leclercq, *J. Chromatogr.* **A856**, 315–29 (1999).

127. J.V. Hinshaw, *LC.GC Europe* **16** (3), 140–6 (2003).

128. J.W. Robinson and E.D. Boothe, *Spectrosc. Lett.* **17**, 689 (1984).

129. E.E. Sotnikov, L.A. Bubchikova and T.G. Lanskova, *Plast. Massy* (3), 53–8 (1988).

130. A. Maltha, *Proceedings Organic Peroxides Symposium*, AKZO, Apeldoorn (1998).

131. T. Stareczek, G. Kaminska, J. Ermel and K. Kortylewska, *Polim. Tworz. Wielkoczast.* **44** (3), 218–21 (1999).

132. T.R. Crompton, in *Polymer Devolatilization* (R.J. Albalak, ed.), M. Dekker, New York, NY (1996), pp. 575–643.

133. T.R. Crompton, *Manual of Plastics Analysis*, Plenum Press, New York, NY (1998).

134. X. Lou, H. Janssen and C.A. Cramers, *Anal. Chem.* **69**, 1598–1603 (1997).

135. D.F. Utterback, D.S. Millington and A. Gold, *Anal. Chem.* **56**, 470 (1984).

136. J.F. Ludwig and A.G. Bailie, *Anal. Chem.* **56**, 2081 (1984).

137. T.H. Mourey, *Anal. Chem.* **56**, 1777–81 (1984).

138. T.H. Mourey, G.A. Smith and L.R. Snyder, *Anal. Chem.* **56**, 1773–7 (1984).

139. E.S. Gankina, M.D. Val'chikhina and B.G. Belen'kii, *Vysokomol. Soedin. Ser.* **A18** (5), 1170 (1976).

140. S.J. Shiono, *Polym. Sci. A-1* **17**, 4120 (1979).

141. L.M. Zaborsky, *Anal. Chem.* **49**, 1166 (1977).

142. K. Takeda, *Proceedings 10th Annual BCC Conference on Flame Retardancy*, Stamford, CT (1999).

143. ASTM D 3271–93, *Practice for Direct Injection of Solvent-reducible Paints into a Gas Chromatograph for Solvent Analysis, Annual Book of ASTM Standards*, ASTM, West Conshohocken, PA (1998), Vol. 06.01.

144. ASTM D 268–96, *Guide for Sampling and Testing Volatile Solvents and Chemical Intermediates for Use in Paint and Related Coatings and Materials, Annual Book of ASTM Standards*, ASTM, West Conshohocken, PA (1998), Vol. 06.04.

145. *Varian Chrompack Application Note 19-GC*, Varian Inc., Walnut Creek, CA (n.d.).

146. R.R. Kozlowski and T.K. Gallagher, *J. Vinyl Addit. Technol.* **3** (3), 249–55 (1997).

147. A. Krishen, *Anal. Chem.* **43**, 1130 (1971).

148. M. Wandel, H. Tengler and H. Ostromow, *Die Analyse von Weichmachern*, Springer Verlag, Berlin (1967).

149. L. Jacque and G. Guiochon, *Chim. Anal. (Paris)* **49** (1), 3–10 (1967).

150. J.D. Ramsey, T.D. Lee, M.D. Osselton and A.C. Moffat, *J. Chromatogr.* **184**, 185–206 (1980).

151. D.T. Burns, W.P. Hayes and P. Steele, *J. Chromatogr.* **103**, 339 (1975).

152. A.A. Krychkov and A.S. Rysev, *Plast. Massy* (1), 71–3 (1990).

153. J. Pavlov, *Polimeri (Zagreb)* **5** (10), 241–3 (1984).

154. D.T. Burns, W.P. Hayes and P. Steele, *J. Chromatogr.* **103**, 241 (1975).

155. C.S. Giam, H.S. Chan and G.S. Neff, *Anal. Chem.* **47**, 2225 (1975).

156. V.D. Feofanov, *Gig. Sanit.* **36** (6), 75–80 (1971).

157. S.C. Rastogi, *Chromatographia* **47** (11/12), 724–6 (1998).

158. T.R. Crompton, *Analysis of Polymers. An Introduction*, Pergamon Press, Oxford (1989).

159. H. Ostromow, *Analyse von Kautschuken und Elastomeren*, Springer-Verlag, Berlin (1981).

160. P.A.D.T. Vimalasiri, J.K. Haken and R.P. Burford, *J. Chromatogr.* **300**, 303–65 (1984).

161. K. Watanabe, *Gomu Genzairyo, Seihin no Bunseki to Kenkyu Kaihatsu, Genba eno Oyo, Gomu Gijutsu Shinpojumu, 41st*, Nippon Gomu Kyokai Kenkyukai, Tokyo (1995).

162. S. Ali, B.D. Bhatt and R.N. Nigam, in *Polymer Science* (I.S. Bhardwaj, ed.), Appl. Publishers, New Dehli (1994), Vol. 2, pp. 922–6.

163. I. Matsunaga, H. Nakashima and N. Miyano, *Osaka-furitsu Koshu Eisei Kenkyusho Kenkyu Hokoku* **36**, 149–57 (1998).

164. S. Tan and T. Okada, *Shokuhin Eiseigaku Zasshi* **24** (2), 207–12 (1983).

165. J. Rotschová and J. Pospíšil, *Plaste Kautsch.* **36** (9), 289–93 (1989).

166. J. Scheirs, J. Pospíšil, M.C. O'Connor and S.W. Bigger, *Polym. Prep. Am. Chem. Soc. Div. Polym. Chem.* **34**, 203 (1993).

167. I.D. Smith and D.G. Waters, *J. Chromatogr.* **596** (2), 290–3 (1992).

168. K. Figge, J. Koch and H. Lubba, *J. Chromatogr.* **131**, 317 (1977).

169. B. Stancher, F. Tunis and L. Favretto, *J. Chromatogr.* **131**, 309–16 (1977).

170. S. Tan, S. Tadenuma, M. Hojo, Y. Yoshida, Y. Matsuzawa, T. Tatsuno and T. Okada, *Shokuhin Eiseigaku Zasshi* **27** (3), 229–37 (1986).

171. H. Ishiwata, T. Inoue, M. Yamamoto and K. Yoshihara, *J. Agric. Food Chem.* **36**, 1310–3 (1988).

172. M. Nagata and Y. Kishioka, *Bunseki Kagaku* **38** (6), T75–80 (1989).

173. M. Nagata and Y. Kishioka, *J. High Resolut. Chromatogr.* **14** (9), 639–42 (1991).

174. M. Yuasa, *Idemitsu Giho* **34** (2), 190–8 (1991).

175. F. Mersch and R. Zimmer, *Kautsch. Gummi Kunstst.* **39**, 427–32 (1986).

176. J. Ewender and O. Piringer, *Dtsch. Lebensm. Rdsch.* **87** (1), 5–7 (1991).

177. I.S. Krull, T.Y. Fan and D.H. Fine, *Anal. Chem.* **50**, 698–701 (1978).

178. A. Hartkopf, *Proceedings 8th Intl Symposium on Supercritical Fluid Chromatography and Extraction*, St Louis, MO (1998).

179. M.X. Ramirez, D.E. Hirt, B. Roberts, M. Havens and N. Miranda, *Proceedings SPE ANTEC 2000*, Orlando, FL (2000), pp. 2873–6.

180. D.L. Carpenter and H.T. Silver, *J. Chromatogr. Sci.* **14**, 405 (1976).

181. E.R. Hoffman, *Anal. Chem.* **48**, 445 (1976).

182. L. Castle, M. Sharman and J. Gilbert, *J. Chromatogr.* **437** (1), 274–80 (1988).

183. L.M.T. Han and G. Szajer, *J. Am. Oil Chem. Soc.* **71** (6), 669–70 (1994).

184. F. David, L. Vanderroost, P. Sandra and S. Stafford, *Tec. Lab.* **14** (176), 678–83 (1992).

185. A.E. Stafford, G. Fuller, H.R. Bolin and B.E. Mackey, *J. Agric. Food Chem.* **22**, 478–9 (1974).

186. S.v.d. Wal, *Techn. Rept DSM Research*, Geleen (1999).

187. J.S. Lancaster, T.P. Lynch and P.G. McDowell, *J. High Resolut. Chromatogr.* **23**, 479–84 (2000).

188. W. Engewald, J. Teske and J. Efer, *J. Chromatogr.* **A856**, 259–78 (1999).

189. R.S. Juvet, L.S. Smith and K.P. Li, *Anal. Chem.* **44**, 49 (1972).

190. W. Jennings and J. Nikelly, *Capillary Chromatography: The Applications*, Hüthig Verlag, Heidelberg (1991).

191. Y. Ishii, *Nippon Setchaku Kyokaishi* **16** (2), 71–8 (1980).

192. K. Hoshimoto, *Gomu Genzairyo, Seihin no Bunseki to Kenkyu Kaihatsu, Genba eno Oyo, Gomu Gijutsu Shinpojumu* **41**, 24–37 (1995).

193. W. Blum and R. Aichholz, in *Encyclopedia of Analytical Science* (A. Townshend, ed.), Academic Press, London (1995), pp. 1833–44.

194. E.R. Rohwer and V. Pretorius, *J. High Resolut. Chromatogr., Chromatogr. Commun.* **10**, 145 (1987).

195. D. Desrosiers, *Proceedings SPE PMAD RETEC*, Atlantic City, NJ (1996), pp. 145–76.

196. F. David, L. Vanderroost, P. Sandra and S. Stafford, *Hewlett Packard Appl. Note 228–149; Intl Labmate XVII (III)*, 13 (1992).

197. E. Verdurmen, R. Verstappen, J. Swagten, H. Nelissen, G. Heemels and J.C.J. Bart, *Abstracts 4th Intl Symposium on Hyphenated Techniques in Chromatography and Hyphenated Chromatographic Analyzers (HTC-4)*, Bruges (1996).

198. F. David, L. Vanderroost, P. Sandra and S. Stafford, *Lab. 2000* **12** (6), 74–81 (1999).

199. J. Swagten, E. Lazeroms and C. Bostoen, *Techn. Rept DSM Research*, Geleen (1998).

199a. J. Swagten, *Tech. Rept DSM Research*, Geleen (1996).

200. G. Machata, *Mikrochim. Acta* (2/4), 262–71 (1964).

201. L.S. Ettre, B. Kolb and S.G. Hurt, *Am. Lab.* **10** (5), 76–83 (1983).

202. S.J. Romano, J.A. Renner and P.M. Leitner, *Anal. Chem.* **45**, 2327–30 (1973).

203. A.R. Berens, L.B. Crider, C.J. Tomanek and J.M. Whitney, *J. Appl. Polym. Sci.* **19**, 3169–72 (1975).

204. L. Rohrschneider, *Fresenius' Z. Anal. Chem.* **255**, 345–50 (1971).

205. N. Kumar and J.G. Gow, *J. Chromatogr.* **667**, 235–40 (1994).

206. B. Kolb and L.S. Ettre, *Chromatographia* **32**, 505–13 (1991).

207. B. Kolb and L.S. Ettre, *Static Headspace-Gas Chromatography. Theory and Practice*, Wiley-VCH, New York, NY (1997).

208. J.R.J. Paré, J.M.R. Bélanger and S.S. Stafford, *Trends Anal. Chem.* **13**, 176–84 (1994).

208a. A. Cole and E. Woolfenden, *LC.GC* **10** (2), 76–82 (1992).

209. Ö. Wahlroos, *Ann. Acad. Sci. Fenn. Ser. A II, Chemica* **122**, 1 (1963).

210. A. Hagman and S. Jacobsson, *Anal. Chem.* **61**, 102 (1989).

211. S.M. Abel, A.K. Vickers and D. Decker, *J. Chromatogr. Sci.* **32**, 328 (1994).

212. B. Kolb (ed.), *Applied Headspace Gas Chromatography*, Heyden, London (1980).

213. B.V. Ioffe and A.G. Vitenberg, *Headspace Analysis and Related Methods in Gas Chromatography*, John Wiley & Sons, Inc., New York, NY (1984).

214. R.P. Lattimer and J.B. Pausch, *Am. Lab.*, 80–8 (August 1980).

215. A.R. Berens, *Proceedings 168th ACS Mtg*, Atlantic City, NJ (1974).

216. R.J. Steichen, *Anal. Chem.* **48**, 1398–402 (1976).

217. H.L. Dennison, C.V. Breder, T. McNeal, R.C. Snyder, J.A. Roach and J.A. Sphon, *J. Assoc. Off. Anal. Chem.* **61**, 813 (1978).

218. H. Wouters, J. van Pol, J. Gummersbach, P. Berckmoes and R. Smiths, *Abstracts 6th Intl Symposium on Hyphenated Techniques in Chromatography and Hyphenated Chromatographic Analyzers (HTC-6)*, Bruges (2000), Paper 11.3.

219. T. Sugita, H. Ishiwata, Y. Kawamura, T. Baba, T. Umehara, S. Morita and T. Yamada, *J. Food Hyg. Soc. Jpn.* **36** (2), 263–8 (1996).

220. R.A. Shanks, *Pye Unicam Newsletter* (1975).

221. R.A. Shanks, *Scan* **6**, 20–1 (1975).

222. C. Di Pasquale, G. Di Iorio and T. Capaccioli, *J. Chromatogr.* **152**, 538–41 (1978).

223. P. Klaffenbach, C. Coors, D. Kronenfeld, C. Brüse and H.-G. Schulz, *LC.GC Intl* 166–74 (March 1998).

224. B. Kolb, *Angewandte Gas-Chromatographie*, Bodenseewerk Perkin-Elmer & Co., GmbH, Überlingen (1968).

225. B. Kolb, P. Pospíšil and M. Auer, *Chromatographia* **19**, 113–22 (1984).

226. Committee of the Industrial Association for Food Technology and Packaging, *Verpack. Rdsch.* **40** (7), 56 (1989).

227. P. Werkhoff and W. Bretschneider, *J. Chromatogr.* **405**, 87–98 (1987).

228. S. Schmidt, L. Bromberg and T. Wannman, *J. High Resolut. Chromatogr., Chromatogr. Commun.* **11**, 242–7 (1988).

229. J.W. Long, R.W. Snyder and J. Spalik, *J. Chromatogr.* **450**, 394–8 (1988).

230. J.M. Jickells, C. Crews, L. Castle and J. Gilbert, *Food Addit. Contam.* **7** (2), 197–205 (1990).

231. S.T. Sie, W. van Beersum and G.W.A. Rijnders, *Separation Sci.* **1**, 469 (1966).

232. S.T. Sie and G.W.A. Rijnders, *Separation Sci.* **2**, 699, 729, 755 (1967).

233. S.T. Sie and G.W.A. Rijnders, *Anal. Chim. Acta* **38**, 31 (1967).

234. M.G. Rawdon, *Anal. Chem.* **56**, 831 (1984).

235. J.R. Wheeler and M.E. McNally, *Fresenius' Z. Anal. Chem.* **330**, 237 (1988).

236. I. Roberts, in *Chromatography in the Petroleum Industry* (E.R. Adlard, ed.), Elsevier Science, Oxford (1995), Ch. 11.

237. I.J. Koski and M.L. Lee, *J. Microcol. Sep.* **3**, 481–90 (1991).

238. H.H. Hill and D.A. Atkinson, in *Hyphenated Techniques in Supercritical Fluid Chromatography and Extraction* (K. Jinno, ed.), Elsevier, Amsterdam (1992), pp. 1–8.

239. T.A. Berger and W.H. Wilson, *Proceedings 4th Intl Symposium on Supercritical Fluid Chromatography and Extraction*, Cincinnati, OH (1992), pp. 7–8.

240. K. Robards, P.R. Haddad and P.E. Jackson, *Principles and Practice of Modern Chromatographic Methods*, Academic Press, London (1994).

241. M. Novotny, S.R. Springston, P.A. Peaden, J.C. Fjeldsted and M.L. Lee, *Anal. Chem.* **53**, 407A (1981).

242. R.M. Campbell, H.J. Cortes and L.S. Green, *Anal. Chem.* **64** (22), 2852–7 (1992).

243. J.D. Pinkston, D.P. Innis, R.T. Hentschel and S. Dressman, *Abstracts 8th Intl Symposium on Supercritical Fluid Chromatography and Extraction*, St Louis, MO (1998), E-19.

244. H.E. Schwartz, R.G. Brownlee, M.M. Boduszynski and F. Su, *Anal. Chem.* **59**, 1393 (1987).

245. H.E. Schwartz, P.J. Barthel, S.E. Moring and H.H. Lauer, *LC.GC* **5**, 490 (1987).

246. K.D. Bartle, in *Encyclopedia of Analytical Science* (A. Townshend, ed.), Academic Press, London (1995), pp. 4849–56.

247. J.A. Blackwell, G.O. Cantrell, J.D. Weckwerth and P.W. Carr, *Abstracts 7th Intl Symposium on Supercritical Fluid Chromatography and Extraction*, Indianapolis, IN (1996), A-06.

248. E. Klesper and F.P. Schmitz, *J. Chromatogr.* **402**, 1 (1987).

249. M. Takeuchi and T. Saito, in *Hyphenated Techniques in Supercritical Fluid Chromatography and Extraction* (K. Jinno, ed.), Elsevier, Amsterdam (1992), Ch. 4.

250. M. Lafosse, in *Practical Supercritical Fluid Chromatography and Extraction* (M. Caude and D. Thiébaut, eds), Harwood Academic Publishers, Amsterdam (1999), pp. 201–18.

251. J. Olesik and S. Olesik, *Anal. Chem.* **59**, 796 (1987).

252. C. Fujimoto, H. Yoshida and K. Jinno, *J. Chromatogr.* **411**, 213 (1987).

253. R.J. Shelton, P.B. Farnsworth, K.E. Markides and M.L. Lee, *Anal. Chem.* **61**, 1815 (1989).

254. G.L. Long, C.B. Motley and L.D. Perkins, *ACS Symp. Ser.* **479**, 242–56 (1992).

255. W.L. Shen, N.P. Vela, B.S. Sheppard and J.A. Caruso, *Anal. Chem.* **63**, 1491–6 (1991).

256. L.K. Olson and J.A. Caruso, *J. Anal. At. Spectrom.* **7**, 993–8 (1992).

257. D. Luffer, L. Galante, L. David, M. Novotny and G. Hieftje, *Anal. Chem.* **60**, 1365 (1988).

258. C.B. Motley, M. Ashraf-Khorassani and G.L. Long, *Appl. Spectrosc.* **43**, 737 (1989).

259. J.W. Carnahan, G.K. Webster and L. Zhang, *Proceedings 16th Annual Mtg of the Federation of Analytical Chemistry and Spectroscopist Societies*, Chicago, IL (1989), Paper 517.

260. Q. Jin, G. Hieftje, F. Wang and D.M. Chambers, *Proceedings Winter Conference on Plasma Spectrochemistry*, St Petersburg, FL (1990), Paper S-8.

261. H.H. Hill and E.E. Tarver, *J. Chromatogr. Libr.* **53**, 9–24 (1992).

262. M.X. Huang, K.E. Markides and M.L. Lee, *Chromatographia* **31** (3/4), 163–7 (1991).

263. K. Jinno, H. Mae and C. Fujimoto, *J. High Resolut. Chromatogr.* **13**, 13 (1990).

264. H.H. Hill and M.M. Gallagher, *J. Microcol. Sep.* **2**, 114 (1990).

265. G.A. Galuba and M. Gogolewski, *Chem. Anal. (Warsaw)* **41**, 737–41 (1996).

266. S. Johnson and E.D. Morgan, *J. Chromatogr.* **761**, 53–63 (1997).

267. R.C. Williams, M.S. Alasandro, V.L. Fasone, R.J. Boucher and J.F. Edwards, *J. Pharm. Biomed. Anal.* **14**, 1539–46 (1996).

268. J.L. Bernal, M.J. del Nazal, H. Velasco and L. Toribio, *J. Liq. Chromatogr. Relat. Technol.* **19**, 1579–89 (1996).

269. K.L. Williams, L.C. Sander and S.A. Wise, *J. Chromatogr.* **746**, 91–101 (1996).

270. C.R. Frey, *Proceedings 8th Intl Symposium on Supercritical Fluid Chromatography and Extraction*, St Louis, MO (1998), p. L–43.

271. C.M. Harris, *Anal. Chem.* **74**, 87A–91 (2002).

272. T. Berger and T. Greibrokk, *Chromatogr.: Princ. Pract.* **2**, 107–48 (1999).

273. T.L. Chester, J.D. Pinkston and D.E. Raynie, *Anal. Chem.* **68**, 487–514 (1996); **70** (12), 301R–19R (1998).

274. L. Taylor, in *Supercritical Fluid Chromatography and Extraction* (F.A. Settle, ed.), Prentice Hall, Upper Saddle River, NJ (1997), pp. 183–97.

275. A.A. Clifford, in *Supercritical Fluid Extraction and Chromatography* (M.S. Verrall, ed.), John Wiley & Sons, Ltd, Chichester (1996), pp. 241–57.

276. P.J. Sandra, A. Kot, A. Medvedovici, F. David and L. Toribio, in *Supercritical Fluids in Separation Sciences* (O. Kaiser, ed.), Incom, Düsseldorf (1997), pp. 279–89.

277. L.T. Taylor, *Chromatogr. Sci.* **35**, 374–82 (1997).

278. H.M. McNair, *Am. Lab. (Shelton, CT)* **29**, 10–13 (1997).

279. S.H.Y. Wong, in *Supercritical Fluid Chromatography* (S.H.Y. Wong and I. Sunshine, eds), CRC Press, Boca Raton, FL (1997), pp. 51–69.

280. I. Roberts, *J. Chromatogr. Libr.* **56**, 305–45 (1995).

281. H.-G. Janssen and C.A. Cramers, *Anal. Proc.* **30**, 89 (1993).

282. P.J. Schoenmakers and L.G.M. Uunk, *J. Chromatogr. Libr.* **51A**, A 339–91 (1992).

283. S.H. Page, S.R. Sumpter and M.L. Lee, *J. Microcol. Sep.* **4** (2), 91–122 (1992).

284. C. Berger, *Chromatogr. Sci. Ser.* **75**, 301–48 (1998).

285. H.-G. Janssen and X. Lou, in *Practical Supercritical Fluid Chromatography and Extraction* (M. Caude and D. Thiébaut, eds), Harwood Academic Publishers, Amsterdam (1999), pp. 15–52.

286. T.A. Berger, in *Supercritical Fluids in Chromatography and Extraction* (R.M. Smith and S.B. Hawthorne, eds), Elsevier Science, Amsterdam (1997).

287. M.W. Raynor, V. Sewram and M. Venayagomoorthy, in *Practical Supercritical Fluid Chromatography and Extraction* (M. Caude and D. Thiébaut, eds), Harwood Academic Publishers, Amsterdam (1999), pp. 53–106.

288. K. Anton, in *Encyclopedia of Analytical Science* (A. Townshend, ed.), Academic Press, London (1995), pp. 4856–62.

289. L.T. Taylor, in *Supercritical Fluids Extraction and Pollution Prevention* (M.A. Abraham and A.K. Sunol, eds), American Chemical Society, Washington, DC (1997), pp. 134–53.

290. C.R. Yonker, in *Encyclopedia of Analytical Science* (A. Townshend, ed.), Academic Press, London (1995), pp. 4863–9.

291. D. Thiébaut, in *Practical Supercritical Fluid Chromatography and Extraction* (M. Caude and D. Thiébaut, eds), Harwood Academic Publishers, Amsterdam (1999), pp. 149–60.

292. R.M. Smith, *J. Chromatogr.* **A856**, 83–115 (1999).

293. T. Clifford, *Fundamentals of Supercritical Fluids*, Oxford University Press, Oxford (1998).

294. M. Caude and D. Thiébaut (eds), *Supercritical Fluid Chromatography and Extraction*, Harwood Academic Publishers, Amsterdam (1999).

295. K.W. Hutchinson, *Innovations in Supercritical Fluids – Science and Technology*, ACS Symposium Series, Vol. 608, American Chemical Society, Washington, DC (1996).

296. M. Saito, Y. Yamauchi and T. Okuyama, *Fractionation by Packed-column SFC and SFE – Principles and Applications*, VCH, New York, NY (1994).

297. B. Wenclawiak (ed.), *Analysis with Supercritical Fluids: Extraction and Chromatography*, Springer-Verlag, Berlin (1992).

298. M.L. Lee and K.E. Markides (eds), *Proceedings Analytical Supercritical Fluid Chromatography and Extraction*, Chromatography Conferences, Provo, UT (1990).

299. M. Yoshioka, S. Parvez, T. Miyazaki and H. Parvez, *Supercritical Fluid Chromatography and Micro HPLC*, VSP, Utrecht (1989).

300. C.M. White (ed.), *Modern Supercritical Fluid Chromatography*, Hüthig Verlag, Heidelberg (1988).

301. B.A. Charpentier and M.R. Sevenants (eds), *Supercritical Fluid Extraction and Chromatography. Techniques and Applications*, ACS Symposium Series, Vol. 366, American Chemical Society, Washington, DC (1988).

302. R.M. Smith (ed.), *Supercritical Fluid Chromatography*, The Royal Society of Chemistry, Cambridge (1988).

303. K. Jinno (ed.), *Hyphenated Techniques in Supercritical Fluid Chromatography and Extraction*, Elsevier, Amsterdam (1992).

304. K. Anton and C. Berger (eds), *Supercritical Fluid Chromatography with Packed Columns,* M. Dekker, New York, NY (1998).

305. J.M. Bayona, in *Encyclopedia of Analytical Science* (A. Townshend, ed.), Academic Press, London (1995), pp. 4869–79.

306. T. Cserhati and E. Forgacs, *J. Chromatogr.* **774**, 265–79 (1997).

307. R.B. Cody, A. Bjarnason and D.A. Weil, *Oxford Ser. Opt. Sci.* **1**, 316 (1990).

308. B. Hagenhoff, A. Benninghoven, H. Barthel and W. Zoller, *Anal. Chem.* **63**, 2466–9 (1991).

309. D.R. Gere, *HP Application Note AN 800-3*, Hewlett-Packard, Avondale, PA (1983).

310. U. Just, F. Mellor and F. Keidel, *J. Chromatogr.* **683**, 105 (1994).

311. D. Leyendecker and D. Leyendecker, *LaborPraxis* **13** (Special), 29–34 (1989).

312. R.E.A. Ecott and N. Mortimer, *J. Chromatogr.* **553**, 423 (1991).

313. R.L. Eatherton, M.A. Morrissey, W.F. Siems and H.H. Hill, *J. High Resolut. Chromatogr., Chromatogr. Commun.* **9**, 154 (1986).

314. M.A. Morissey, W.F. Siems and H.H. Hill, *J. Chromatogr.* **505**, 215 (1990).

315. T.P. Lynch, in *Applications of Supercritical Fluids in Industrial Analysis* (J.R. Dean, ed.), Blackie, Glasgow (1993), pp. 188–218.

316. E. Klesper and W. Hartmann, *Eur. Polym. J.* **14**, 77 (1978).

317. Y. Hirata and F. Nakata, *J. Chromatogr.* **295**, 315 (1984).

318. W.J. Simonsick and L.L. Litty, *ACS Symp. Ser. (Supercritical Fluid Technology)* **488**, 288–303 (1992).

319. A.L. Baner, T. Bücherl, J. Ewender and R. Franz, *J. Supercrit. Fluids* **5**, 213–19 (1992).

320. W.J. Simonsick, L.L. Litty and L. Lance, *ACS Symp. Ser. (Supercritical Fluid Technology)* **488**, 288–303 (1992).

321. H. Daimon and Y. Hirata, *Chromatographia* **32**, 549–54 (1991).

322. M. Ashraf-Khorassani, D.S. Boyer and J.M. Levy, *J. Chromatogr. Sci.* **29**, 517–21 (1991).

323. M. Ashraf-Khorassani and J.M. Levy, *J. High Resolut. Chromatogr.* **13**, 742–7 (1990).

324. T.W. Ryan, S.G. Yocklovich, J.C. Watkins and E.J. Levy, *J. Chromatogr.* **505**, 273–82 (1990).

325. K. Anton, R. Menes and H.M. Widmer, *Chromatographia* **26**, 221–3 (1988).

326. B.E. Berg, D.R. Hegna, N. Orlien and T. Greibrokk, *Chromatographia* **37**, 271–6 (1993).

327. G.A. Mackay and R.M. Smith, *Anal. Proc.* **29**, 463–4 (1992).

328. B.E. Berg, D.R. Hegna, N. Orlien and T. Greibrokk, *J. High Resolut. Chromatogr.* **15**, 837–9 (1992).

329. T.P. Hunt, C.J. Dowle and G. Greenway, *Analyst* **116**, 1299–1304 (1991).

330. D. Dilettato, P.J. Arpino, N. Khoa and A. Bruchet, *J. High Resolut. Chromatogr.* **14**, 335–42 (1991).

331. R. Moulder, J.P. Kithinji, M.W. Raynor, K.D. Bartle and A.A. Clifford, *J. High Resolut. Chromatogr.* **12**, 688–91 (1989).

332. J.P. Kithinji, K.D. Bartle, M.W. Raynor and A.A. Clifford, *Analyst* **115**, 125–8 (1990).

333. T. Bücherl, A.L. Baner and O.-G. Piringer, *Dtsch. Lebensm. Rdsch.* **89**, 69–71 (1993).

334. N.J. Cotton, K.D. Bartle, A.A. Clifford, S. Ashraf, R. Moulder and C.F. Dowle, *J. High Resolut. Chromatogr.* **14**, 164–8 (1991).

335. J.W. Oudsema and C.F. Poole, *J. High Resolut. Chromatogr.* **16**, 198–202 (1993).

336. D. Dilettato and P.J. Arpino, *Proceedings 5th European Conference on Food Chemistry* (INRA, ed.), Versailles (1989), Vol. 1, pp. 73–7.

337. J. Doehl, A. Farbrot, T. Greibrokk and B. Iversen, *J. Chromatogr.* **392**, 175–84 (1987).

338. A. Giorgetti, N. Pericles, H.M. Widmer, K. Anton and P. Dätwyler, *J. Chromatogr. Sci.* **27**, 318–24 (1989).

339. M.A. Morrissey, A. Giorgetti, M. Polasek, N. Pericles and H.M. Widmer, *J. Chromatogr. Sci.* **29**, 237–42 (1991).

340. C. Berger, in *Supercritical Fluid Chromatography with Packed Columns* (K. Anton and C. Berger, eds), M. Dekker, New York, NY (1997), pp. 301–48.

341. R.C. Wieboldt, K.D. Kempfert, D.L. Dalrymple, *Appl. Spectrosc.* **44**, 1028–34 (1990).

342. K.E. Markides and M.L. Lee, *SFC Applications*, Brigham Young University Press, Provo, UT (1989).

343. M.W. Raynor, K.D. Bartle, I.L. Davies, A. Williams, A.A. Clifford, J.M. Chalmers and B.W. Cook, *Anal. Chem.* **60**, 427–33 (1988).

344. D. Later, B. Richter and M. Anderson, *LC.GC* **4**, 992 (1986).

345. T. Greibrokk, B.E. Berg, S. Hoffmann, H.R. Norli and Q. Ying, *J. Chromatogr.* **505**, 283 (1990).

346. K.D. Bartle, A.A. Clifford, N.J. Cotton, J.P. Kithinji, R. Moulder and M.W. Raynor, *Anal. Proc. (London)* **27**, 240–1 (1990).

347. P.J. Arpino, D. Dilettato, K. Nguyen and A. Bruchet, *J. High Resolut. Chromatogr.* **13** (1), 5–12 (1990).

348. T. Tikuisis and V. Dang, in *Plastics Additives. An A–Z Reference* (G. Pritchard, ed.), Chapman & Hall, London (1998), pp. 80–94.

349. N.J. Cotton, K.D. Bartle, A.A. Clifford and C.J. Dowle, *J. Appl. Polym. Sci.* **48** (9), 1607–19 (1993).

350. S. Hoffman and T. Greibrokk, *J. Microcol. Sep.* **1**, 35 (1989).

351. P. Bataillard, L. Evangelista and M. Thomas, in *Plastics Additives Handbook* (H. Zweifel, ed.), Hanser Publishers, Munich (2000), pp. 1047–83.

352. T. Bücherl, A. Gruner and N. Palibroda, *Packag. Technol. Sci.* **7** (3), 139–53 (1994).

353. C.F. Poole, J.W. Oudsema and K.G. Miller, *Abstracts 5th Intl Symposium on Supercritical Fluid Chromatography and Extraction*, Baltimore, MD (1994), pp. 18–9.

354. T.P. Hunt, C.J. Dowle and G. Greenway, *Analyst* **118**, 17–22 (1993).

355. Y. Ikushima, N. Saito, K. Hatakeda and S. Ito, in *Supercritical Fluid Processing of Food Biomaterials*

(S.S.H. Rizvi, ed.), Blackie, New York, NY (1994), pp. 244–54.

356. Y. Kadota, I. Tanaka, Y. Ohtsu and M. Yamaguchi, *Nihon Yukagakkaishi* **46**, 397–403 (1997).

357. K. Sakaki, *J. Chromatogr.* **648**, 451–7 (1993).

358. K. Anton, M. Bach, C. Berger, F. Walch, G. Jaccard and Y. Carlier, *J. Chromatogr. Sci.* **32** (10), 430–8 (1994).

359. F. David and P. Sandra, *J. High Resolut. Chromatogr.* **11**, 897 (1987).

360. G.A. MacKay and R.M. Smith, *Analyst* **118**, 741 (1993).

361. D.R. Gere, *HP Application Note AN 800–4*, Hewlett-Packard, Avondale, PA (1983).

362. J.W. Oudsema and C.F. Poole, *J. High Resolut. Chromatogr.* **16**, 198–202 (1993).

363. K. Anton, *Abstracts 7th Intl Symposium on Supercritical Fluid Chromatography and Extraction*, Indianapolis, IN (1996), L–10.

364. T.A. Berger, *J. Chromatogr.* **785**, 3–33 (1997).

365. A. Weston and P.R. Brown, *HPLC and CE. Principles and Practice*, Academic Press, San Diego, CA (1997).

366. H. Kalász and M. Báthori, *LC.GC Intl* 440–5 (July 1997).

367. D.L. Saunders, in *Chromatography* (E. Heftmann, ed.), Van Nostrand Reinhold, New York, NY (1975), p. 81.

368. G.C. Barrett, *Adv. Chromatogr.* **11**, 145–79 (1974).

369. H.G. Nowicki, C.A. Kieda, R.F. Devine, V. Current and T.H. Schaefers, *Anal. Lett.* **A12**, 769–76 (1979).

370. K. Weber, *Plaste Kautsch.* **1** (2), 38–9 (1954).

371. J.W.H. Zijp, *Rec. Trav. Chim. Pays Bas* **75**, 1129, 1155 (1956).

372. R. Gaczyński and M. Steplien, *Przemysl Chem.* **38** (9), 571 (1959).

373. H. Auler, *Rubber Chem. Technol.* **37**, 950 (1964).

374. H. Auler, *Gummi Asbest Kunstst.* **14**, 1024, 1081 (1960).

375. T.R. Crompton, *Chemical Analysis of Additives in Plastics*, Pergamon Press, Oxford (1977).

376. D.A. Wheeler, *Talanta* **15**, 1315–34 (1968).

377. E. Schröder and E. Hagen, *Plaste Kautsch.* **15**, 625 (1968).

378. E. Hagen, *Plaste Kautsch.* **14**, 158 (1967).

379. O. Korn and H. Woggon, *Plaste Kautsch.* **11**, 278 (1964).

380. F.B. Williamson, *Rubber J. Intl Plast.* **148**, 24 (1966).

381. R. ter Heide, *Fette, Seife, Anstrichmittel* **60**, 360 (1958).

382. E. Schröder and G. Rudolph, *Plaste Kautsch.* **10**, 22 (1963).

383. F. Johan, M. Zaharia, I. Onaca and F. Biaza, *Rev. Chim. (Bucharest)* **17** (10), 643–7 (1966).

384. K.-T. Wang, *J. Chin. Chem. Soc. (Taiwan)* **7**, 64–8 (1960).

385. M. Rothe, *Makromol. Chem.* **35**, 183–99 (1960).

386. I. Rothe and M. Rothe, *Makromol. Chem.* **68**, 206–10 (1963).

387. H. Zahn and R. Krzikalla, *Makromol. Chem.* **23**, 31 (1957).

388. G.L. Sedlen and J.H. Benedict, *J. Assoc. Off. Chem. Soc.* **45**, 652 (1968).

389. E. Schröder and K. Thinius, *Dtsch. Farben-Z.* **14**, 144–8, 189–93 (1960).

390. K. Sreenivasan, *Chromatographia* **22** (1–6), 199–200 (1986).

391. C.D. Tran, *Anal. Chem.* **56** (4), 824–6 (1984).

392. C.D. Tran, *J. Chromatogr.* **292**, 432 (1984).

393. C.F. Poole, in *Encyclopedia of Analytical Science* (A. Townshend, ed.), Academic Press, London (1995), pp. 5203–11.

394. B.J. Gedeon, T. Chu and S. Copeland, *Rubber Chem. Technol.* **56**, 1080–95 (1983).

395. D.E. Jänchen, *J. Chin. Assoc. Instr. Anal. (Fenxi Ceshi Tongbao)* **8**, 6–14 (1989).

396. J. Szunyog, E. Mincsovics, I. Hazal and I. Klebovich, *J. Planar Chromatogr.* **11**, 25 (1998).

397. H. Filthuth, in *Clinical and Environmental Applications of Quantitative Thin Layer Chromatography* (J. Touchstone, ed.), John Wiley & Sons, Inc., New York, NY (1980).

398. P. Wall and I.D. Wilson, in *Encyclopedia of Analytical Science* (A. Townshend, ed.), Academic Press, London (1995), pp. 5195–5203.

399. H. Jork, W. Funk, W. Fischer and H. Wimmer, *Thin-layer Chromatography*, VCH, Weinheim (1990), Vol. 1.

400. D.E. Heinz and R.K. Vitek, *J. Chromatogr. Sci.* **13**, 570–6 (1975).

401. R.K. Vitek, in *Handbook of Thin-layer Chromatography* (J. Sherma and B. Fried, eds), M. Dekker, New York, NY (1991), pp. 211–48.

402. D. Brück, *Fortbildungsseminar Moderne Prüfmethoden – von Rohstoffen über Mischungen zum Produkt*, DIK, Hannover (May 2000).

403. S. Ebel and J.S. Kang, *J. Planar Chromatogr.* **3**, 42–6 (1990).

404. B. Krumholz and K. Wenz, *J. Planar Chromatogr.* **4**, 370–2 (1991).

405. W. Funk and M. Heiligenthal, *GIT Fachz. Lab. Suppl. 5 (Chromatographie)*, 49–51 (1984).

406. K.-A. Kovar and G.E. Morlock, in *Handbook of Thin-layer Chromatography* (J. Sherma and B. Fried, eds), M. Dekker, New York, NY (1996), Ch. 8.

407. H. Jork, W. Funk, W. Fischer and H. Wimmer, *Thin-layer Chromatography: Reagents and Detection Methods*, VCH, Weinheim (1994), Vol. 1b.

408. H. Jork, in *Encyclopedia of Analytical Science* (A. Townshend, ed.), Academic Press, London (1995), pp. 5182–94.

409. R. Amos, *Talanta* **20** (12), 1231–60 (1973).

410. G. Zweig and J. Sherma, *Handbook of Chromatography – General Data and Principles*, CRC Press, Boca Raton, FL (1972), Vol. II, pp. 103–89.

411. J. Zhu and Y.S. Yeung, *J. Chromatogr.* **461**, 139–45 (1989).

412. S. Ebel, *J. Planar Chromatogr.* **2**, 410–9 (1989).

413. I.D. Wilson, M. Spraul and E. Humpfer, *J. Planar Chromatogr.* **10**, 217–9 (1997).

414. M. Ohnishi-Kameyama and T. Nagata, *Anal. Chem.* **70** (9), 1916–20 (1998).

415. W. Fischer, H.-E. Hauck and G. Wieland, *Labo* (6), 56–62 (1997).

416. J.K. Rózylo, R. Siembida and A. Jamrozek-Manko, *J. Planar Chromatogr.* **10**, 225–8 (1997).

417. M.P. Thomas, *J. Vinyl Addit. Technol.* **2** (4), 330–8 (1996).

418. C.B. Airaudo, A. Gayte-Sorbier, P. Aujoulat and V. Mercier, *J. Chromatogr.* **437**, 59–82 (1988).

419. C.F. Poole, *J. Chromatogr.* **A856**, 399–427 (1999).

420. R.E. Majors, *LC.GC Intl* **3**, 8 (1990).

421. B. Renger, *J. Planar Chromatogr.* **12**, 58–62 (1999).

422. A.G. Howard, T. Shafik, F. Moffatt and I.D. Wilson, *J. Chromatogr.* **A844** (1+2), 333–40 (1999).

423. R.E. Kaiser, *Einführung in die HPLC*, Hüthig Verlag, Heidelberg (1987).

424. M. Mazurek and Z. Witkiewicz, *Chem. Anal. (Warsaw)* **43**, 529 (1998).

425. D.W. Armstrong and X.F. Yang, *J. Chromatogr.* **456**, 440–3 (1988).

426. J.-P. Salo and H. Salomies, *Farmaseuttinen aikakaus-kirja-DOSIS* **12**, 152–64 (1996).

427. J. Sherma, *Anal. Chem.* **70**, 7R–26R (1998).

428. T. Cserhati and E. Forgacs, *J. Chromatogr. Sci.* **35**, 383–91 (1997).

429. W. Freitag, in *Kunststoff-Additive* (R. Gächter and H. Müller, eds), Hanser Verlag, Munich (1990), pp. 909–46.

430. J. Sherma, in *Handbook of Thin-layer Chromatography* (J. Sherma and B. Fried, eds), M. Dekker, New York, NY (1996), Ch. 1.

431. G. Herbst, *Labo* **2**, 8–14 (1997).

432. H. Jork, *LaborPraxis* **2**, 110 (1992).

433. A. Mohammad, *J. Planar Chromatogr.* **10**, 48–54 (1997).

434. T. Kowalska, *Chromatogr. Sci.* **55**, 43–69 (1991).

435. K.D. Mukherjee, *Chromatogr. Sci.* **55**, 339–50 (1991).

436. M. Ranný, *Chem. Listy* **76** (11), 1121–46 (1982).

437. J.G. Kreiner, *Rubber Chem. Technol.* **44**, 381–96 (1971).

438. E. Hahn-Deinstrop, *Dünnschicht-Chromatographie. Praktische Durchführung und Fehlervermeidung*, Wiley-VCH, Weinheim (1997).

439. B. Fried and J. Sherma, *Practical Thin-Layer Chromatography – A Multidisciplinary Approach*, CRC Press, Boca Raton, FL (1996).

440. N. Grinberg (ed.), *Modern Thin-layer Chromatography*, M. Dekker, New York, NY (1990).

441. K. Bauer, L. Gros and W. Sauer, *Thin-layer Chromatography – An Introduction*, Hüthig Verlag, Heidelberg (1991).

442. F. Geiss, *Fundamentals of Thin Layer Chromatography*, Hüthig Verlag, Heidelberg (1987).

443. J. Sherma and B. Fried, *Handbook of Thin-layer Chromatography*, M. Dekker, New York, NY (1996).

444. J.G. Cobler and C.D. Chow, *Anal. Chem.* **51**, 287R (1979).

445. H. Schweppe, in *The Analytical Chemistry of Synthetic Dyes* (K. Venkataraman, ed.), John Wiley & Sons, Inc., New York, NY (1977), p. 49.

446. E. Lück and M. Jager, *Antimicrobial Food Additives. Characteristics, Uses, Effects*, Springer-Verlag, Berlin (1997).

447. R. Amos, *J. Chromatogr.* **48**, 343–52 (1970).

448. J.G. Kreiner and W.C. Warner, *J. Chromatogr.* **44**, 315–30 (1969).

449. M.M. Plekhotkina, *Proizvod. Shin. Rezinotekh. Asbes-totekh. Izdelii, Nauch.-Tekh. Sb.* **1**, 29–31 (1973); *Ref. Zh. Khim.* 12S694.

450. T. Parys, D. Jaroszynska and I. Kochel, *Polim. Tworz. Wielk.* **28**, 284 (1983); *C.A.* **101**, 39 622h (1984).

451. D. Brück, *Kautsch. Gummi Kunstst.* **39**, 1165 (1986).

452. J. Rotschová, T.-T. Son and J. Pospíšil, *J. Chromatogr.* **246**, 346 (1982).

453. C.B. Airaudo, A. Gayte-Sorbier, R. Momburg and P. Laurent, *J. Chromatogr.* **354**, 341 (1986).

454. K. Akimoto, N. Shinohara, S. Tachi, T. Sako and Y. Yamamoto (to Ouchi Shinko Chemical Industrial Co.), *Jpn. Kokai Tokkyo Koho* 90–123150 (May 15, 1990).

455. ASTM Standard D 3156-81, *Thin-layer Chromatographic Analysis of Antidegradants (Stabilizers, Antioxidants and Antiozonants)* in *Raw and Vulcanized Rubbers, Annual Book of ASTM Standards*, ASTM, Philadelphia, PA (1990).

456. S. Wang, Y. Pang and D. Chen, *Huaxue Shijie* **31** (8), 361–3 (1990).

457. K. Nishiguchi, *Chiba-ken Kogyo Shikenjo Kenkyu Hokoku* **9**, 3–7 (1995).

458. F.C.A. Killer and R. Amos, *J. Inst. Petrol.* **52**, 315 (1966).

459. Zh. Zai, *Chinese J. Oil Refinement (Shiyou Lianzhi)* **8**, 37–40 (1984).

460. Y. Cheng, *Chin J. Chromatogr. (Se Pu)* **2**, 268–72 (1985).

461. J.H.v.d. Neut and A.C. Maagdenberg, *Plastics (London)* **31**, 66–7 (1966).

462. M.B. Milligen, *Anal. Chem.* **46**, 746 (1974).

463. J. Sedlar, E. Feniokova and J. Pac, *Analyst* **99**, 50 (1974).

464. H. Nelissen, *Tech. Rept DSM Research*, Geleen (1989).

465. C.A. Parks, *Anal. Chem.* **33**, 140 (1961).

466. D. Simpson and B.R. Currell, *Analyst (London)* **96**, 515 (1971).

467. J. Urbanski, W. Czerwinski, K. Janicka, F. Majewska and H. Zowall, *Handbook of Analysis of Synthetic Polymers and Plastics*, John Wiley & Sons, Inc., New York, NY (1977).

468. J. Haslam, H.A. Willis and D.C.M. Squirrell, *Identification and Analysis of Plastics*, John Wiley & Sons, Ltd, Chichester (1983).

469. D.C.M. Squirrell, *Analyst* **106**, 1042–56 (1981).

470. F. Sevini and B. Marcato, *J. Chromatogr.* **260**, 507–12 (1983).

471. B. Marcato and M. Vianello, *J. Chromatogr.* **A869**, 285–300 (2000).

472. C.H. Van Peteghem and D.A. Dekeyser, *J. Assoc. Off. Anal. Chem.* **64**, 1331 (1981).

473. C. Gertz and K. Herrmann, *Z. Lebensm. Unters. Forsch.* **177**, 186 (1983).

474. J. Alary, C. Grosset and A. Coeur, *Ann. Phar. Franç.* **40**, 301 (1982).

475. C.B. Airaudo, A. Gayte-Sorbier, P. Laurent and R. Creuseavu, *J. Chromatogr.* **314**, 349 (1984).

476. C.B. Airaudo, A. Gayte-Sorbier and R. Creusevau, *J. Chromatogr.* **392**, 407–14 (1987).

477. C. Sarbach, N. Yagoubi, J. Sauzières, C. Renaux, D. Ferrier and E. Postaire, *Int. J. Pharm.* **140**, 169–74 (1996).

478. P. Delvodre, H. El Mansouri, N. Yagoubi, C. Sarbach and D. Ferrier-Baylocq, *Proceedings 9th Intl Symposium on Planar Chromatography*, Interlaken (1997), pp. 70–7.

479. R. Siembida, *Proceedings 9th Intl Symposium on Planar Chromatography*, Interlaken (1997), pp. 321–4.

480. E. Dreassi, M. Bonifacio and P. Corti, *Food Addit. Contam.* **15** (4), 466–72 (1998).

481. J. Fahrnich, M. Popl and E. Vyborna, *Sci. Pap. Prague Inst. Chem. Technol.* **18**, 105 (1983).

482. M. Wandel, H. Tengler and H. Ostromow, *Die Analyse von Weichmachern*, Springer-Verlag, Berlin (1967).

483. S. Affolter, *Kautsch. Gummi Kunstst.* **44** (1), 1026–8 (1991).

484. F. Fiala, I. Steiner and K. Kubesh, *Dtsch. Lebensm. Rdsch.* **90**, 51–7 (2000).

485. H. Huber and J. Wimmer, *Kunststoffe* **58** (11), 786–8 (1968).

486. Y. Shi, H. Zhou and D. Wang, *Linchan Huaxue Yu Gongye* **12** (4), 293–7 (1992).

487. V. Zitko, *Chemosphere* **28**, 1211–5 (1994); **38**, 629–32 (1999).

488. M.L. Langhorst, *J. Planar Chromatogr.* **2**, 346–54 (1989).

489. A.H. Lawrence and D. Ducharme, *J. Chromatogr.* **194**, 434–6 (1980).

490. B. Iliano, A.-M. Oudar and J. Gosselé, *Dtsch. Lebensm. Rdsch.* **91**, 205–8 (1995).

491. E. Albrecht and J. Vindevogel, *Techn. Rept Centexbel*, Ghent (2002).

492. E. Koglin, *J. Planar Chromatogr. – Modern TLC* **3**, 117–20 (1990).

493. P.A. Martoglio, S.P. Bouffard, A.J. Sommer, J.E. Katon and K.A. Jakes, *Anal. Chem.* **62**, 1123A–8A (1990).

494. P.R. Wright, N. Richfield-Fratz, A. Rasooly and A. Weisz, *J. Planar Chromatogr. – Modern TLC* **10** (3), 157–62 (1997).

495. S. Loger, J. Perkavec and M. Perpar, *Mikrochim. Acta* 712 (1964).

496. S. Loger, J. Perkavec and M. Perpar, *J. Chromatogr.* **30**, 14 (1967).

497. T.A. Perenich, *Proceedings National Technical AATCC Conference* (1982), pp. 177–84.

498. D.K. Laing, L. Boughey and A.W. Hartshorne, *J. Forensic Sci. Soc.* **30**, 299 (1990).

499. P. Walton and G.W. Taylor, *Chromatography and Analysis* 5–7 (1991).

500. D.T. Burns and A.A. McGuigan, *Fresenius' Z. Anal. Chem.* **340**, 377–9 (1991).

501. *CAMAG Application Note A-64.5*, Muttenz (n.d.).

502. B. Kuster and U. Wahl, *Textilveredlung* **32**, 121–4 (1997).

503. G. Alemany, M. Akaarir, C. Rossello and A. Gamundi, *Biomed. Chromatogr.* **10**, 225–7 (1996).

504. V. Arsov, V. Kostova and B. Mesrob, *J. Chromatogr.* **267** (2), 448–54 (1983).

505. V.K. Gupta, in *Handbook of Thin-layer Chromatography* (J. Sherma and B. Fried, eds), M. Dekker, New York, NY (1996), Ch. 28.

506. G. Meyers and P. Gijsman, *Techn. Rept DSM Research*, Geleen (1998).

507. R.F.v.d. Heide and O. Woulters, *Z. Lebensm. Unters. Forsch.* **117**, 129–31 (1962).

508. J. Lehotay, J. Daněček, O. Líška, J. Leško and E. Brandšteterová, *J. Appl. Polym. Sci.* **25**, 1943–50 (1980).

509. N.H. Mady, R. Liu and J. Viczkus, *Proceedings 30th SPI Annu. Tech. Mark. Conf. (Polyurethanes)*, Toronto (1986), pp. 332–5.

510. M.J. Forrest, in *Developments in Polymer Analysis and Characterisation*, Rapra Technology Ltd, Shawbury (1999), Paper 4.

511. A.H. Sharma, F. Mozayeni and J. Alberts, *Proceedings SPE Polyolefins XI*, Houston, TX (1999), pp. 679–703.

512. W.R. Hudgins, K. Theurer and T. Mariani, *J. Appl. Polym. Sci.: Appl. Polym. Symp.* **34**, 145–55 (1978).

513. J.H. Jung and K.H. Lee, *Anal. Sci. Technol.* **12** (1), 84–8 (1999).

514. G. Katay, L. Litinova, E. Mincsovics, E. Melenewskaya and E. Tyihak, *J. Planar Chromatogr.* **12**, 340–4 (2000).

515. J. Kuniya, *Bunseki Kagaku* **37**, T87–91 (1988).

516. K. Wang. R. Zang and Z. He, *Fenxi Huaxue* **20**, 19–22 (1992).

517. E. Westgate and J. Sherma, *Intl Lab.* **30** (5), 36–40 (2000).

518. J. Udris, *Analyst* **96**, 130–9 (1971).

519. A. Mohammad, M. Ajmai, S. Anwar and E. Iraqui, *J. Planar Chromatogr.* **9**, 318–60 (1996).

520. M.P. Volynets and B.F. Myasoedov, *Crit. Rev. Anal. Chem.* **25** (4), 247–312 (1996).

521. D. Braun and E. Richter, *Kautsch. Gummi Kunstst.* **50**, 112 (1997).

522. U. Simmross, R. Fischer, F. Düwel and U. Müller, *Fresenius' Z. Anal. Chem.* **358**, 541–5 (1997).

523. E.S. Gankina and B.G. Belenkii, *Chromatogr. Sci.* **55**, 807–62 (1991).

524. V.A. Tsendrovskaya, *Gig. Sanit.* (3), 80–2 (1973).

525. I.D. Wilson and W. Morden, *LC.GC Intl* **12** (2), 72–80 (1999).

526. B. Fried and J. Sherma, *Thin-layer Chromatography*, M. Dekker, New York, NY (1999).

527. D.W. Mayo, R.M. Pike and P.K. Trumper, *Microscale Organic Laboratory with Multistep and Multiscale Syntheses*, John Wiley & Sons, Inc., New York, NY (1994), pp. 99–100.

528. J.M. Bohen, M.M. Jouillie, F.A. Kaplan and B. Loev, *J. Chem. Educ.* **50**, 367 (1973).

529. W. Still, M. Kahn and A. Mitra, *J. Org. Chem.* **43** (14), 2923 (1978).

530. G.J. Fallick, P.C. Talarico and R.R. McCough, *Proceedings SPE ANTEC '76*, Atlantic City, NJ (1976), p. 574.

531. C.A. Darker and J.M. Berriman, *Trans. Inst. Rubber Ind.* **28**, 279 (1952).

532. R.H. Campbell and R.W. Wize, *J. Chromatogr.* **12**, 178 (1963).

533. T.R. Crompton, *Eur. Polym. J.* **4**, 473–96 (1968).

534. A. Fiorenza, G. Bonomi and A. Saredi, *Mater. Plast. Elast.* **10**, 1045–50 (1965).

535. L.S. Ettre, in *HPLC* (C. Horváth, ed.), Academic Press, New York, NY (1980), Vol. 1, pp. 1–74.

535a. M.E. Swartz and B.J. Murphy, *LabPlus Intl* **18**(3), 6–9 (2004).

536. S. Küppers, B. Renger and V.R. Meyer, *LC.GC Europe* **13** (2), 114–18 (2000).

537. H. Schäffer, M. Lauebli and A. Doerig, *Theory of Ion Chromatography*, Metrohm AG, Herisau (2000).

538. T. Hanai, *HPLC: A Practical Guide*, The Royal Society of Chemistry, Cambridge (1999).

539. M. Jaroniec, *J. Chromatogr.* **A656**, 37–50 (1993).

540. H. Engelhardt, *Abstracts 22nd Intl Symposium on High Performance Liquid Phase Separation and Related Techniques HPLC '98*, St Louis, MO (1998), L–0401.

541. H. Zou, X. Huang, M. Ye and Q. Luo, *J. Chromatogr.* **A954**, 5–32 (2002).

542. N. Tanaka, H. Kobayashi, K. Nakanishi, H. Minakuchi and N. Ishizuka, *Anal. Chem.* **73** (15), 420A–9A (2001).

543. J.J. Pesek and E.J. Williamsen, *J. Chromatogr. Libr.* **57**, 371–401 (1995).

544. J.W. Dolan, *LC.GC Intl* 292–9 (May 1998).

545. K. Unger (ed.), *Handbuch der HPLC*, GIT Verlag, Darmstadt (1989), Vol. 1.

546. L.R. Snyder and J.J. Kirkland, *Introduction to Modern Liquid Chromatography*, John Wiley & Sons, Inc., New York, NY (1979).

547. P.J. Schoenmakers, H.A.H. Billiet and L. de Galan, *J. Chromatogr.* **205**, 13 (1981).

548. R.M. Smith, R.J. Burgess, O. Chienthavorn and J.R. Stuttard, *LC.GC Intl* **12** (1), 30–5 (1999).

549. U.D. Neue, *HPLC Columns. Theory, Technology and Practice*, John Wiley & Sons, Ltd, Chichester (1997).

550. R.E. Majors, *LC.GC Intl* 212–21 (April 1999).

551. R.E. Majors, *LC.GC Europe* 284–301 (2001).

552. Special Issue on *Recent Developments in LC Column Technology, LC.GC Europe* **16** (6a) (2003).

553. D. Ishii (ed.), *Introduction to Microscale High Performance Liquid Chromatography*, VCH, New York, NY (1988).

554. R. Stevenson, K. Mistry and I. Krull, *Am. Lab.* **30** (16), 16A (1998).

555. P. Myers and K. Bartle, in *Recent Applications in LC-MS, LC.GC Europe Spec. Edn* pp. 44–7 (November 2002).

556. J.P.C. Vissers, *J. Chromatogr.* **A856**, 117–43 (1999).

557. M.W. Dong and J.L. DiCesare, *Plast. Engng* **39** (2), 25–8 (1983).

558. J.R. Gant and M.W. Dong, *Pharm. Technol.* **11** (10), 44 (1987).

559. R.M. Smith, *J. Chromatogr. Libr.* **57**, 145–69 (1995).

560. P. Kuronen, *J. Chromatogr. Libr.* **57**, 209–33 (1995).

561. P.C. White, *Analyst* **109**, 677–97; 973–84 (1984).

562. Y. Kawamura, M. Miura, T. Sugita, T. Yamada and M. Takeda, *Shokuhin Eiseigaku Zasshi* **37**, 272–80 (1996).

563. N.E. Skelly, J.D. Graham, Z. Iskandarani and D. Priddy, *Polym. Mater. Sci. Eng.* **59**, 23–7 (1988).

564. S.A. George, in *Diode-array Detection in High-performance Liquid Chromatography* (L. Huber and S. George, eds), M. Dekker, New York, NY (1993), pp. 41–9.

565. G. Lunn, L. Hellwig and A. Cecchini, *Handbook of Derivatization Reactions of HPLC*, John Wiley & Sons, Inc., New York, NY (1998).

566. C.M.B. van den Beld and H. Lingeman, *Pract. Spectrosc.* **12**, 237 (1991).

567. E.S. Yeung, *Adv. Chromatogr.* **23**, 1–63 (1983).

568. I.N. Acworth and M. Bowers, in *Coulometric Electrode Array Detectors for HPLC* (I.N. Acworth, M. Naoi, H. Parvez and S. Parvez, eds), VSP, Utrecht (1997).

569. V.L. Cebolla, L. Membrado, J. Vela and A.C. Ferrando, *Semin. Food Anal.* **2** (3), 171–89 (1997).

570. M. Wilcox, *Intl. Labmate* **23** (6), 35–6 (1998).

571. S.J. Hill, M.J. Bloxham and P.J. Worsfold, *J. Anal. At. Spectrom.* **8**, 499 (1993).

572. M.J. Tomlinson, J. Wang and J. Caruso, *J. Anal. At. Spectrom.* **9**, 957 (1994).

573. C. Sarzanini and E. Mentasti, *J. Chromatogr.* **A789**, 301–21 (1997).

574. K. Robards, P. Starr and E. Patsalides, *Analyst* **116**, 1247–73 (1991).

575. P.K. Dasgupta, *Anal. Chem.* **64**, 775 (1992).

576. D. Munteanu, in *Mechanisms of Polymer Degradation and Stabilisation* (G. Scott, ed.), Elsevier Applied Science, London (1990), pp. 211–314.

577. S.R. Abbott and J. Tusa, *J. Liq. Chromatogr.* **6**, 77 (1983).

578. R.P.W. Scott, *Liquid Chromatography Detectors*, Elsevier Science Publishers, Amsterdam (1986).

579. E.S. Yeung, *Detectors for Liquid Chromatography*, John Wiley & Sons, Inc., New York, NY (1986).

580. T.M. Vickery (eds), *Liquid Chromatography Detectors*, M. Dekker, New York, NY (1983), pp. 1–434.

581. L.R. Snyder, J.J. Kirkland and J.L. Glajch, *Practical HPLC Method Development*, John Wiley & Sons, Ltd, Chichester (1998).

582. L.R. Snyder, in *Chromatography* (E. Heftmann, ed.), Elsevier, Amsterdam (1992), Ch. 1.

583. P.C. Sadek, *The HPLC Solvent Guide*, John Wiley & Sons, Inc., New York, NY (2002).

583a. J.C. Berridge, *Techniques for the Automated Optimization of HPLC Separations*, John Wiley & Sons, Inc., New York, NY (1985).

584. V.J. Barwick, *Trends Anal. Chem.* **16** (6), 293–309 (1997).

585. T. Baczek, R. Kaliszan, H.A. Claessens and M.A. van Straten, *LC.GC Europe* 304–13 (2001).

586. M. Pfeffer, K. Bürkle and N. Ruszkowki, *Proceedings Advances in Process Analytics and Control Technology (APACT 03)*, York (2003), P13.

587. M. Pfeffer and H. Windt, *Fresenius' Z. Anal. Chem.* **369**, 36–41 (2001).

588. K. Bürkle and J. Paschlau, *LaborPraxis* **7/8**, 56–8 (2002).

589. J. Dolan, *Proceedings Solutions for Scientists Symposium*, Advanstar Communications, London (1999).

590. P.J. Schoenmakers and M. Mulholland, *Chromatographia* **25** (8), 737 (1988).

591. D. Parriott, *LC.GC* **12** (2), 134 (1994).

592. W.B. Furman, T.P. Layloff and R.F. Tetzlaff, *J. AOAC Intl* **77** (5), 1314 (1994).

593. M. Pfeffer and R. Golejewski, *Proceedings Advances in Process Analytics and Control Technology (APACT 03)*, York (2003), P14.

594. J.J. Kirschbaum, *J. Pharm. Biomed. Anal.* **7** (7), 813 (1989).

595. F. Rouessac and A. Rouessac, *Chemical Analysis. Modern Instrumental Methods and Techniques*, John Wiley & Sons, Ltd, Chichester (2000).

596. J. Wang, in *HPLC Detection: Newer Methods* (G. Patoney, ed.), VCH Publishers, New York, NY (1992), Ch. 5.

597. R.D. Shah and C.A. Maryanoff, in *HPLC, Practical and Industrial Applications* (J. Swadesh, ed.), CRC Press, Boca Raton, FL (1996), pp. 111–69.

598. P. Wang and H.K. Lee, *J. Chromatogr.* **A789**, 437–51 (1997).

599. W.R. LaCourse, *Anal. Chem.* **74**, 2813–32 (2002).

600. S. Lindsay, *High Performance Liquid Chromatography*, John Wiley & Sons, Ltd, Chichester (1992).

601. G. Patoney (ed.), *HPLC Detection: Newer Methods*, VCH Publishers, New York, NY (1992).

602. P. Jandera and J. Churacek, *Gradient Elution in Column Chromatography*, Elsevier Science Publishers, New York, NY (1985).

603. G. Szepesi, *How to Use Reverse-phase HPLC*, VCH Publishers, New York, NY (1992).

604. G. Lunn and L. Hellwig, *Handbook of Derivatization Reactions for HPLC*, John Wiley & Sons, Ltd, Chichester (1999).

605. D. Munteanu, A. Isfan, C. Isfan and I. Tincul, *Chromatographia* **23**, 7–14 (1987).

606. K. Jinno and Y. Yokayama, *J. Chromatogr.* **550**, 325 (1991).

607. R.C. Nielson, *J. Liq. Chromatogr.* **14**, 503–19 (1991).

608. E. Lesellier, P. Saint Martin and A. Tchapla, *LC.GC Intl* **5** (11), 38–43 (1992).

609. E. Lesellier and A. Tchapla, *Chromatographia* **36**, 135 (1993).

610. H.J. Cortes, G.E. Bormett and J.D. Graham, *J. Microcol. Sep.* **4**, 51 (1992).

611. C. Nerín, J. Salafranca, J. Cacho and C. Rubio, *J. Chromatogr.* **A690**, 230 (1995).

612. D.W. Allen, M.R. Clench, A. Crowson and D.A. Leathard, *Polym. Degrad. Stab.* **39**, 293–7 (1993).

613. M.T. Baker and J.F. Johnson, *Polym. Mater. Sci. Eng.* **56**, 194–7 (1987).

614. K. Sreenivasan, *Chromatographia* **32**, 285–6 (1991).

615. V.C. Francis, Y.N. Sharma and I.S. Bhardwaj, *Angew. Makromol. Chem.* **113**, 219–25 (1983).

616. P. Perlstein, *Anal. Chim. Acta* **149**, 21–7 (1983).

617. S.D. Patterson and R.E.A. Escott, *High Perform. Polym.* **2** (3), 197–207 (1990).

618. J. Simal-Gándara, M. Sarria-Vidal and R. Rijk, *J. AOAC Intl* **83** (2), 311–19 (2000).

619. G.F.M. Ball, in *Food Analysis by HPLC* (L.M.L. Nollet, ed.), M. Dekker, New York, NY (1992), pp. 275–340.

620. S. Al-Malaika, S. Issenhuth and D. Burdick, *Polym. Degrad. Stab.* **73**, 491–503 (2001).

621. J.F. Schabron, R.J. Hurtubise and H.F. Silver, *Anal. Chem.* **50**, 1911 (1978).

622. J.F. Schabron and L.E. Fenska, *Anal. Chem.* **52**, 1411–5 (1980).

623. A.M. Wims and S.J. Swarin, *J. Appl. Polym. Sci.* **19**, 1243–56 (1975).

624. *Liquid Chromatography Procedure for Polyolefin Additives, WAPP-100*, Waters, Milford, MA (1978).

625. R.C. Nielson, *J. Liq. Chromatogr.* **16** (7), 1625–38 (1993).

626. A.F. Coffey, E.A. Meadows and F.P. Warner, *Polymer Laboratories Paper PP-1380*, Church Stratton (n.d.).

627. M.R. Jakupca, D.R. Stevenson and D.L. Stein, *Proceedings Polyolefins X (SPE)*, Houston, TX (1997), pp. 627–52.

628. H.P. Nissen and H.W. Kreysel, *GIT Suppl.* **2**, 41 (1986).

629. S. Kato and K. Kanohta, *J. Chromatogr.* **324**, 462–6 (1985).

630. S.H. Smith and L.T. Taylor, *Chromatographia* **56** (3/4), 165–9 (2002).

631. N.E. Skelly and E.R. Husser, *Anal. Chem.* **50**, 1959 (1978).

632. E.R. Husser, R.H. Stehl and D.R. Price, *Anal. Chem.* **49**, 154 (1977).

633. L. Brown, *Analyst (London)* **104**, 1165 (1979).

634. J.H. van Dijk, P.C.G.M. Janssen and L.G.J. van der Ven, *Proceedings 16th FATIPEC*, Liège (1982), Vol. 1, pp. 253–65.

635. M. Petro and L. Bystrický, *Chem. Pap.* **40** (3), 357–62 (1986).

636. D.A. Ernes and D.T. Hanshumaker, *Anal. Chem.* **55**, 408 (1983).

637. C. Sun, M. Baird and J. Simpson, *J. Chromatogr.* **A800**, 231–8 (1998).

638. C. Bailly, *Polymer* **27**, 776 (1986).

639. P. Jandera, *J. Chromatogr.* **449**, 361 (1988).

640. J. Scheirs, *Compositional and Failure Analysis of Polymers. A Practical Approach*, John Wiley & Sons, Ltd, Chichester (2000).

641. V. Dulio, R. Po, R. Borrelli, A. Guarini and C. Santini, *Angew. Makromol. Chem.* **225**, 109 (1995).

642. W.R. Hudgins, K. Theurer and T. Mariani, *J. Appl. Polym. Sci.: Appl. Polym. Symp.* **34**, 145–55 (1978).

643. B. Lang and H. Makart, *Melliand Textilber. Int.* **56**, 647 (1975).

644. H. Nelissen, *Techn Rept DSM Research*, Geleen (1999).

645. C.T. Costley, J.R. Dean, I.D. Newton and J. Carroll, *Anal. Commun.* **34**, 89–91 (1997).

646. F.P.B. van der Maeden, M.E.F. Biemond and P.C.G.M. Janssen, *J. Chromatogr.* **149**, 539–52 (1978).

647. H. El Mansouri, N. Yagoubi and D. Ferrier, *J. Chromatogr.* **A771**, 111–8 (1997).

648. J.J. Lewis, L.B. Rogers and R.E. Pauls, *J. Chromatogr.* **264**, 339 (1983).

649. W.H. Leister, L.E. Weaner and D.G. Walker, *J. Chromatogr.* **A704**, 369–76 (1995).

650. R.E.A. Escott and N. Mortimer, *J. Chromatogr.* **553**, 423–32 (1991).

651. Y. Mengerink, S.v.d. Wal, H.A. Claessens and C.A. Cramers, *J. Chromatogr.* **A871** (1/2), 259–68 (2000).

652. T. Takasu, A. Iles and K. Hasebe, *Anal. Bioanal. Chem.* **372** (4), 554–61 (2002).

653. W. Freitag, R. Wurster and N. Mady, *J. Chromatogr.* **450**, 426–9 (1988).

654. D. Baylocq, C. Majcherczyk and F. Pellerin, *Ann. Pharm. Fr.* **43**, 329 (1985).

655. T. Macko, B. Furtner and K. Lederer, *J. Appl. Polym. Sci.* **62**, 2201–7 (1996).

656. P. Gijsman, J. Hennekens and K. Janssen, *Polym. Degrad. Stab.* **46**, 63–74 (1994).

657. ASTM D 1996–97, *Test Method for Determination of Phenolic Antioxidants and Erucamide Slip Additives in Low Density Polyethylene Using Liquid Chromatography (LC)*, Annual Book of ASTM Standards, ASTM, West Conshohocken, PA (1997), Vol. 08.01.

658. T. Tikuisis and M. Cossar, *Proceedings SPE ANTEC '93*, New Orleans, LA (1993), p. 270.

659. N. Haider and S. Karlsson, *Analyst* **124**, 797–800 (1999).

660. H. El Mansouri, N. Yagoubi and D. Ferrier, *Chromatographia* **48** (7/8), 491–6 (1998).

661. N. Yagoubi, A.E. Baillet, F. Pellerin and D. Baylocq, *J. Chromatogr.* **522**, 131–41 (1990).

662. Ciba-Geigy Ltd, *Analytical Methods*, Basel (n.d.).

663. J.F. Schabron and D.Z. Bradfield, *J. Appl. Polym. Sci.* **26**, 2479 (1981).

664. A. Caceres, F. Ysambert, J. Lopez and N. Marquez, *Sep. Sci. Technol.* **31**, 2287–98 (1996).

665. D.J. Carlsson, D.W. Grattan, T. Suprunchuk and D.M. Wiles, *J. Appl. Polym. Sci.* **22**, 2217 (1978).

666. M.A. Haney and W.A. Dark, *J. Chromatogr. Sci.* **18**, 655 (1980).

667. I. Vit, M. Popl and J. Fähnrich, *Chem. Prymysl.* **34**, 642 (1984).

668. F. Pellerin, C. Majcherczyk and D. Baylocq, *Talanta* **33**, 85 (1986).

669. R.O. Jonas, B.A.G. Parsons and R. Wilkinson, *Proceedings Intl Conference on Chemical and Physical Phenomena in the Ageing of Polymers*, Prague (1988), p. 84.

670. J. Pospíšil, *Polym. Degrad. Stab.* **39**, 103 (1993).

671. J. Scheirs, J. Pospíšil, M.J. O'Connor and S.W. Bigger, *Adv. Chem. Ser.* **249**, 359–74 (1996).

672. B. Werthmann, V. Neyen, F. Milczewski and R. Borowski, *Kautsch. Gummi Kunstst.* **51**, 118–21 (1998).

673. P. Meng, L. Li and G. Song, *Suliao Gongye* **25** (5), 104–5 (1997).

674. R.E. Majors, *J. Chromatogr. Sci.* **8**, 338 (1970).

675. S.-J. Kim, *Anal. Sci. Technol.* **13** (3), 282–90 (2000).

676. J. Cosemans, *Techn. Rept DSM Research*, Geleen (1993).

677. S.C. Rastogi, *Chromatographia* **28**, 15 (1989).

678. A.F. Coffey, E.A. Meadows and F.P. Warner, *Abstracts 22nd Intl Symposium on High Performance Liquid Phase Separations and Related Techniques HPLC '98*, St Louis, MO (1998), P–0507-W.

679. M.G. Monteiro and V.F. Matos, *Proceedings Intl GPC Symposium '87*, Millipore Corp., Milford, MA (1987), pp. 437–46.

680. T. Alfredson, *Reverse Phase Gradient Analysis of Polyolefin Polymer Additives, Publication No. LC-123*, Varian Instrument Group, Walnut Creek, CA (n.d.).

681. W. Müller, A. Zergiebel and R. Sourisseau, *Plast. Kautsch.* **35** (7), 249–51 (1988).

682. M.E. Quinn and W.W. McGee, *J. Chromatogr.* **350** (1), 187–203 (1985).

683. B. Günter and D.H. Jürgen, *Lebensmittelchem. Gerichtl. Chem.* **36**, 90 (1982).

684. N.P. Sen, in *Food Analysis by HPLC* (L.M.L. Nollet, ed.), M. Dekker, New York, NY (1992), pp. 673–96.

685. J. Hong, J. Maguhn, D. Freitag and A. Kettrup, *Fresenius' Z. Anal. Chem.* **361**, 124–8 (1998).

686. Y. Kageyama, A. Yoshida, K. Takahashi and T. Nishii, *Chemitopia* **14**, 10–9 (1994).

687. D. Gross and K. Strauss, *Kunststoffe* **67**, 426–32 (1977).

688. *Varian Chrompack Application Note 1158-HPLC*, Walnut Creek, CA (n.d.).

689. M.R. Khan, C.P. Ong, S.F.Y. Li and H.K. Lee, *J. Chromatogr.* **513**, 360 (1990).

690. B.E. Turner, *LC.GC Intl* **4** (7), 22–5 (1991).

691. I. Chappell, *LC.GC Europe* **15** (3), 156–64 (2002).

692. *Shimadzu Application Note HPLC-9*, Kyoto (n.d.).

693. L. Zhang, J.-H. Hu, J. Lian, Z. Li and Q. Zhu, *Zhongguo Yiyao Gongye Zazhi* **32** (10), 466–8 (2001).

694. M. Riess and R. van Eldik, *J. Chromatogr.* **A827** (1), 65–71 (1998).

695. R. Luyk, *PhD Thesis*, University of Amsterdam (1993).

696. H. Nelissen, *Techn. Rept DSM Research*, Geleen (1997).

697. B.B. Wheals, P.C. White and M.D. Paterson, *J. Chromatogr.* **350**, 295 (1985).

698. D.K. Laing, R. Gill and H.M. Bickley, *J. Chromatogr.* **442**, 187 (1988).

699. R.M.E. Griffin, S.J. Speers and T.G. Kee, *J. Chromatogr.* **A674**, 271 (1994).

700. J.C. West, *J. Chromatogr.* **208**, 47 (1981).

701. A. Shan, D. Harbin and C.W. Jameson, *J. Chromatogr. Sci.* **26**, 439 (1988).

702. R.J. Passarelli and E.S. Jacobs, *J. Chromatogr. Sci.* **13**, 153 (1975).

703. M.C. Garrigós, F. Reche, M.L. Marín and A. Jiménez, *Abstracts 7th Intl Symposium on Hyphenated Techniques in Chromatography and Hyphenated Chromatographic Analyzers (HTC-7)*, Bruges (2002), Paper P52.

704. M.J. Drews, K. Ivey, J. Helvey and F. van Lenten, *Proceedings 8th Intl Symposium on Supercritical Fluid Chromatography and Extraction*, St Louis, MO (1998), E–20.

705. M.Y. Ye, R.G. Walkup and K.D. Hill, *J. Liq. Chromatogr.* **18**, 2309 (1995).

706. M.J. Scarlett, J.A. Fisher, H. Zhang and M. Ronan, *Water Res.* **10**, 2109 (1994).

707. W. Miszkiewicz and J. Szymanowski, *Crit. Rev. Anal. Chem.* **25** (4), 203–46 (1996).

708. K. Lingeman, *PhD Thesis*, State University of Utrecht (1986).

709. M.J. Cooper and M.W.J. Anders, *J. Chromatogr. Sci.* **13**, 407 (1975).

710. K. Heinig, F. Hissner, S. Martin and C. Vogt, *Intl Lab.* 9–40 (September 1998).

711. *Restek HPLC Application Note # 59398*, Restek Corporation, Bellefonte, PA (2001).

712. B.B. Woodward, G.P. Heffelfinger and D.I. Ruggles, *J. AOAC* **62**, 1011–9 (1979).

713. M. Ikawa, N. Phillips, J.F. Haney and J.J. Sasner, *Toxicon* **37** (6), 923–9 (1999).

714. C. Grosset, D. Cantin, A. Villet and J. Alary, *Talanta* **37**, 301–6 (1990).

715. W. Flak and L. Pilsbacher, in *Food Analysis by HPLC* (L.M.L. Nollet, ed.), M. Dekker, New York, NY (1992), pp. 421–56.

716. K.J. Hammond, *J. Assoc. Off. Publ. Analysts* **16**, 17 (1978).

717. P.G. Demertzis and R. Franz, *Food Addit. Contam.* **15** (1), 93–9 (1998).

718. *ESA Application Note*, ESA Inc., Chelmsford, MA (1999).

719. M.A. Ruiz, E. García-Moreno, C. Barbas and J.M. Pingarrón, *Electroanalysis* **11** (7), 470–4 (1999).

720. M.A. Ruiz, M.G. Blázquez and J.M. Pingarrón, *Anal. Chim. Acta* **305**, 49 (1995).

721. A. Ritter, *Proceedings 3rd Würzburger Tage der instrumentellen Analytik in der Polymertechnik*, SKZ, Würzburg (1997).

722. L.M.L. Nollet (ed.), *Food Analysis by HPLC*, M. Dekker, New York, NY (1992).

723. N. Lambropoulos, T.J. Cardwell, D. Caridi and P.J. Marriott, *J. Chromatogr.* **A749**, 87–94 (1996).

724. H. Shigekuni and M. Shimizu, *Nisseki Rebyu* **33**, 126 (1991).

725. A. Praet, C. Dewaele, L. Verdonck and G.P. van der Kelen, *J. Chromatogr.* **507**, 427 (1990).

726. M. Astruc, A. Astruc and R. Pinel, *Mikrochim. Acta* **109**, 83 (1992).

727. R.E. Majors, *Analusis* **3**, 549 (1975).

728. T. Takeuchi, Y. Watanabe and D. Ishii, *J. High Resolut. Chromatogr.* **4**, 300 (1981).

729. M. Krejči and D. Kouřilová, *Abstracts 1st Intl Symposium on Column Liquid Chromatography*, Interlaken (1973).

730. K. Jinno, J.B. Phillips and D.P. Carney, *Anal. Chem.* **57**, 574 (1985).

730a. S.J. Marin, B.A. Jones, W.D. Felix and J. Clark, *Abstracts HPLC-2004*, Philadelphia, PA (2004).

731. R. Trones, A. Iveland and T. Greibrokk, *J. Microcol. Sep.* **7**, 505–12 (1995).

732. D. Ishii, in *Introduction to Microscale High Performance Liquid Chromatography* (D. Ishii, ed.), VCH, New York, NY (1988), p. 1.

733. G. Sheng, Y. Shen and M.L. Lee, *J. Microcol. Sep.* **9**, 63 (1997).

734. F. Houdiere, P.W.J. Fowler and N.M. Djordjevic, *Anal. Chem.* **69**, 2589 (1997).

735. M.H. Chen en C. Horváth, *J. Chromatogr.* **A788**, 51 (1997).

736. R. Trones, T. Andersen, I. Hunnes and T. Greibrokk, *J. Chromatogr.* **A814**, 55–61 (1998).

737. R. Trones, T. Andersen and T. Greibrokk, *J. High Resolut. Chromatogr.* **22** (5), 283–6 (1999).

737a. J. Clark, *LabPlus Intl* **18**(3), 10–3 (2004).

738. R. Trones, *PhD Thesis*, University of Oslo (1999).

739. P. Molander, S.J. Thommesen, I.A. Bruheim, R. Trones, T. Greibrokk, E. Lundanes and T.E. Gundersen, *J.High Resolut. Chromatogr.* **22** (9), 490–4 (1999).

740. P. Molander, K. Haugland, D.R. Hegna, E. Ommundsen, E. Lundanes and T. Greibrokk, *J. Chromatogr.* **A864**, (1), 103–9 (1999).

741. P. Molander, A. Holm, E. Lundanes, D.R. Hegna, E. Ommundsen and T. Greibrokk, *Analyst* **127**, 892–7 (2002).

742. T. Greibrokk and T. Andersen, *J. Sep. Sci.* **24** (12), 899–909 (2001).

743. R. Trones, T. Andersen, T. Greibrokk and D.R. Hegna, *J. Chromatogr.* **A874** (1), 65–71 (2000).

744. P. Perlstein and P. Orme, *Chromatographia* **17**, 576 (1985).

745. S.G. Gharfeh, *J. Chromatogr.* **389**, 211 (1987).

746. P. Matuska, *J. Chromatogr.* **606**, 136–40 (1992).

747. W. Freitag, *J. Chromatogr.* **450**, 430–2 (1988).

748. G. Ligner, *Analytical Procedure for Sanduvor 3944, Sandoz Polymer Additives, No. 5794–1*, Basel (n.d.).

749. B. Marcato, C. Fantazzini and F. Sevini, *J. Chromatogr.* **553**, 415–22 (1991).

750. L. Gaiani and D. Herzfeld, *Analytical Method for Chimassorb 944LD (KBC/3)*, Ciba-Geigy Additives, Basel (n.d.).

751. R. Verstappen, *Techn. Rept DSM Research*, Geleen (1994).

752. *Personal communication* Borealis, Stathelle (1999).

753. *Ciba Specialty Chemicals KC-158/1: Tinuvin 622LD Determination in Polyolefins*, Basel (n.d.).

754. P. Molander, R. Trones, K. Haugland and T. Greibrokk, *Analyst* **124**, 1137–41 (1999).

755. B. Desmazières and P.-L. Desbène, *Abstracts 23rd Intl Symposium on High Performance Liquid Phase Separations and Related Techniques HPLC '99*, Granada (1999).

756. R. Trones, A. Tangen, W. Lund and T. Greibrokk, *J. Chromatogr.* **A835**, 105–12 (1999).

757. C.-S. Wu (ed.), *Handbook of Size Exclusion Chromatography*, M. Dekker, New York, NY (1995).

758. E. Meehan and G. Saunders, *Intl Lab.* **31** (5), 27–9 (2001).

759. H. Pasch and B. Trathnigg, *HPLC of Polymers*, Springer-Verlag, Berlin (1997).

760. T. Alfredson and J. Tschida, *High Performance Size Exclusion Analysis of Coatings and Adhesive Polymer Formulations, Publ. No. LC-122*, Varian Instrument Group, Walnut Creek, CA (n.d.).

761. P.G. Clarke, *LC.GC Europe, Applications Book*, 44–6 (April 2003).

762. C. Jackson and H.G. Barth, *Trends Polym. Sci.* **2** (6), 203–7 (1994).

763. J.E. Campana, T. Havard and A. Jarrell, *Polym. Mater. Sci. Engng* **69**, 456–7 (1993).

764. P.G. Clarke, *Intl Lab. News* 8–9 (December 1999).

765. Z. Grubisic, P. Rempp and H. Benoit, *J. Polym. Sci.: Polym. Lett. Ed.* **B5**, 753 (1967).

766. R.B. Nielsen, A.L. Safir, M. Petro, T.S. Lee and P. Huefner, *Proceedings ACS Div. Polym. Mater.: Sci. Engng.* **80**, 92 (1999).

767. S.R. Holding and E. Meehan, *Molecular Weight Characterisation of Synthetic Polymers, Rapra Review Rept. No. 83*, Rapra Technology Ltd, Shawbury (1995).

768. R.G. Beri, L.S. Hacche and C.F. Martin, in *HPLC Practical and Industrial Applications* (J.K. Swadesh, ed.), CRC Press, Boca Raton, FL (1996), pp. 245–304.

769. B.J. Hunt and S.B. Holding (eds), *Size Exclusion Chromatography*, Chapman & Hall, New York, NY (1989).

770. E. Meehan, *LC.GC Europe* **14** (8), 432–3 (2001).

771. H.G. Barth, *LC.GC Europe* **16** (6a), 46–50 (2003).

772. I.M. Howard III, *J. Chromatogr.* **55**, 15 (1971).

773. J. Čoupek, S. Pokorný, J. Protivová, J. Holčik, M. Karvaš and J. Pospíšil, *J. Chromatogr.* **65**, 279 (1972).

774. M.J. Shepherd and J.J. Gilbert, *J. Chromatogr.* **178**, 435–41 (1979).

775. F.M. Mirabella, E.M. Barrall and J.F. Johnson, *Polymer* **17**, 17 (1976).

776. H. El Mansouri, N. Yagoubi, D. Scholler, A. Feigenbaum and D. Ferrier, *J. Appl. Polym. Sci.* **71** (3), 371–5 (1999).

777. Y. Kato, S. Kido, H. Watanabe, M. Yamamoto and T. Hashimoto, *J. Appl. Polym. Sci.* **19**, 629 (1975).

778. J.V. Dawkins, M.J. Forrest and M.J. Shepherd, *J. Liq. Chromatogr.* **13**, 3001 (1990).

779. G. Dallas and S.D. Abbott, in *Liquid Chromatographic Analysis of Food and Beverages* (G. Charalambous, ed.), Academic Press, New York, NY (1979), Vol. 2.

780. S. Mori, *J. Liq. Chromatogr.* **11**, 1205 (1988).

781. S. Nakamura, S. Ishiguro, T. Yamada and S. Moriizumi, *J. Chromatogr.* **84**, 279–88 (1973).

782. R.C. Snyder and C.V. Breder, *J. AOAC* **64** (4), 999 (1981).

783. J.E. Campana, L.-S. Sheng, S.L. Shew and B.E. Winger, *Trends Anal. Chem.* **13**, 239 (1994).

784. Wyatt Technology Corp., *LC.GC Europe Applications Book*, 25 (June 2001).

785. R.D. Mate and H.S. Lundstrom, *J. Polym. Sci.* **C21**, 317 (1968).

786. *Chimassorb 944 Determination in Polypropylene, High Density Polyethylene (HDPE) and Low Density Polyethylene (LDPE) by the Total Nitrogen Content Analytical Method, Code No. KC65/1*, Ciba-Geigy, Basel (1980).

787. *Quantitative Analysis of Chimassorb 944 in Polyolefins by Ultraviolet Spectroscopy. Analytical Method No. C-260*, Ciba-Geigy, Ardsley, NY (1982).

788. *Semiquantitative Determination of Chimassorb 944 in Low Density Polyethylene Polymer by Infrared Spectroscopy. Analytical Method No. C-259*, Ciba-Geigy, Ardsley, NY (1981).

789. R. Jennissen, P. Op den Camp and J. Swagten, *Techn. Rept DSM Research*, Geleen (1999).

790. P. Knape and R. Jennissen, *Techn. Rept DSM Research*, Geleen (1998).

791. L.-S. Sheng, S.L. Shew, B.E. Winger and J.E. Campana, *ACS Symp. Ser.* **581**, 55 (1994).

792. L. Castle, A.J. Mercer and J. Gilbert, *Food Addit. Contam.* **8**, 565–76 (1991).

793. Y. Mukoyama and S. Mori, *J. Liq. Chromatogr.* **12** (8), 1417–30 (1989).

794. K. Sreenivasan, *J. Liq. Chromatogr.* **9**, 2425 (1986).

795. S. Mori, *J. Chromatogr.* **129**, 53 (1976).

796. A. Krause, A. Lange and M. Ezrin, *Plastics Analysis Guide. Chemical and Instrumental Methods*, Hanser Publishers, Munich (1983).

797. J.M. Pacco and A.K. Mukherji, *J. Chromatogr.* **144**, 113–7 (1977).

798. S.V. Greene and V.J. Gatto, *Proceedings Intl GPC Symposium '98*, Phoenix, AZ (1998), pp. 612–32.

799. J.K. Swadesh, C.W. Stewart Jr and P.C. Uden, *Analyst* **118**, 1123 (1993).

800. F.N. Larsen, *Appl. Polym. Symp.* **8**, 111 (1969).

801. J. Protivová, J. Pospíšil and L. Zikmund, *J. Polym. Sci., Polym. Symp.* **40**, 233 (1973).

802. D.E. Hillman and C. Heathcote, in *Size Exclusion Chromatography* (B.J. Hunt and S.R. Holding, eds), Chapman & Hall, New York, NY (1989), Ch. 6.

803. J. Jirackova-Audouin, D. Ranceze and J. Verdu, *Analusis* **13** (2), 59–64 (1985).

804. F. Spagnolo and W.M. Malone, *J. Chromatogr. Sci.* **14**, 52 (1976).

805. U. Lotz, *Farbe Lack* **85**, 172 (1979).

806. J.H. Cox and R.G. Anthony, *J. Appl. Polym. Sci.* **19**, 821 (1975).

807. F. Eisenbeiss, E. Dumont and H. Henke, *Angew. Makromol. Chem.* **71**, 67 (1978).

808. J.M. Pacco, A.K. Mukherji and D.L. Evans, *Sep. Sci. Technol.* **13**, 277 (1978).

809. R.L. Bartosiewicz, *J. Polym. Sci., Part C*, 329 (1968).

810. H. Batzer and S.A. Zahir, *J. Appl. Polym. Sci.* **19**, 585 (1975).

811. A.F. Cunningham, G.C. Furneaux and D.E. Hilleman, *Anal. Chem.* **48**, 2192 (1976).

812. M.J. Shepherd and J. Gilbert, *J. Chromatogr.* **218**, 703 (1981).

813. J. Protivová and J. Pospíšil, *J. Chromatogr.* **88**, 99 (1974).

814. J. Protivová, J. Pospíšil and J. Holčik, *J. Chromatogr.* **92**, 361 (1974).

815. R.E. Majors and E.L. Johnson, *J. Chromatogr.* **167**, 17 (1978).

816. H.K. Gupta and R. Salovey, *Rubber Chem. Technol.* **58**, 295–303 (1985).

817. H. Pasch, A. Al-Mobasher, S. Attari. K.F. Shuhaibar and F.A. Rasoul, *J. Appl. Polym. Sci.: Appl. Polym. Symp.* **45**, 209–26 (1990).

818. D.F. Alliet and J.M. Pacco, *Sep. Sci.* **6**, 153 (1971).

819. T.R. Startin, M. Sharman, M.D. Rose, I. Parker, A.J. Mercer, L. Castle and J. Gilbert, *Food Addit. Contam.* **4**, 385–98 (1987).

820. A.J. Mercer, L. Castle, J.R. Startin and J. Gilbert, *Food Addit. Contam.* **4**, 399–406 (1987).

821. A.J. Mercer, L. Castle, J.R. Startin and J. Gilbert, *Food Addit. Contam.* **5**, 9–20 (1988).

822. L. Castle, J. Gilbert, S.M. Jickells and J.W. Gramshaw, *J. Chromatogr.* **437**, 281–6 (1988).

823. D.B. Priddy, *Trends Appl. Spectrosc.* **1**, 133 (1993).

824. J.P. Helfrich and E.C. Conrad, *Proceedings SPE ANTEC '85*, Washington, DC (1985), pp. 226–9.

825. S. Ono (to Shimadzu Corp.), *Jpn Pat.* 91–182254 (February 2, 1993).

826. *Polymer Laboratories Application Note*, Polymer Laboratories Ltd, Church Stretton (2002).

827. W.J. Staal, P. Cools, A.M. van Herk and A.L. German, *Sep. Technol.* 857–9 (1994); *Chromatographia* **37** (3/4), 218–20 (1993).

828. H.J.A. Philipsen, *PhD Thesis*, Technical University of Eindhoven (1998).

829. P.G. Alden and M. Woodman, *Proceedings Intl GPC Symposium '98*, Phoenix, AZ (1998), pp. 437–57.

830. B. Klumperman and H.J.A. Philipsen, *LC.GC Intl* 18–25 (January 1998).

831. G. Glöckner, *Gradient HPLC of Copolymers and Chromatographic Cross-fractionation*, Springer-Verlag, Heidelberg (1991).

832. W.J. Staal, *PhD Thesis*, Technical University of Eindhoven (1996).

833. P. Cools, *PhD Thesis*, Technical University of Eindhoven (1999).

834. F.A. Buytenhuys, H.J.F.M. van de Ven, M. Nielen, P.J.C.H. Cools, B. Klumperman and A.L. German, *Abstracts Intl GPC Symposium '98*, Phoenix, AZ (1998), p. 31.

835. W.J. Staal, *Proceedings Waters Intl GPC Symposium '96*, San Diego, CA (1996).

836. W.J. Staal, *Proceedings Intl Technical Symposium on GPC and LC Analysis of Polymers and Related Materials*, Waters Chromatography, Boston, MA (1989).

837. E.L. Johnson, in *Ion Chromatography* (J.G. Tarter, ed.), M. Dekker, New York, NY (1987), pp. 1–22.

838. P.R. Haddad and P. Jandik, in *Ion Chromatography* (J.G. Tarter, ed.), M. Dekker, New York, NY (1987), pp. 87–156.

839. J.H. Sherman and N.D. Danielson, *Anal. Chem.* **59**, 1483–5 (1987).

840. W. Ahrer and W. Buchberger, *Abstracts 15th Montreux Symposium on Liquid Chromatography/Mass Spectrometry*, Montreux (1998), p. 32.

841. V. Pacáková and K. Štulík, *J. Chromatogr.* **A789**, 169–80 (1997).

842. W.W. Buchberger and P.R. Haddad, *J. Chromatogr.* **A789**, 67–83 (1997).

843. H. Small, *J. Chromatogr.* **546**, 3–15 (1991).

844. P.K. Dasgupta, in *Ion Chromatography* (J.G. Tarter, ed.), M. Dekker, New York, NY (1987), pp. 191–367.

845. C. Pohl, *LC.GC Europe* **16** (6a), 51–4 (2003).

846. H. Small, *Ion Chromatography*, Plenum Press, New York, NY (1989).

847. J. Weiss, *Ionenchromatographie*, VCH, Weinheim (1991).

848. J.G. Tarter (ed.), *Ion Chromatography*, M. Dekker, New York, NY (1987).

849. R.D. Rocklin, C.A. Pohl and J.A. Schibler, *J. Chromatogr.* **411**, 107–19 (1987).

850. E.E. Sotnikov and L.A. Bubchikova, *Zh. Anal. Khim.* **42** (10), 1896–900 (1987).

851. K.J. Stutts, *Anal. Chem.* **59**, 543–4 (1987).

852. J. Von Unterrichter-Worthmann and F. Quella, *Kunststoffe* **74** (11), 682–4 (1984).

853. H. Müller, W. Nielinger and A. Horbach, *Angew. Makromol. Chem.* **108**, 1–8 (1982).

854. R.D. Holm and S.A. Barksdale, in *Ion Chromatographic Analysis of Environmental Pollutants* (E. Sawicki, J.D. Mulik and E. Wittgenstein, eds), Ann Arbor Science, Ann Arbor, MI (1978), pp. 99–110.

855. L.C. Speitel, J.C. Spurgeon and R.A. Filipczak, in *Ion Chromatographic Analysis of Environmental Pollutants* (E. Sawicki, J.D. Mulik and E. Wittgenstein, eds), Ann Arbor Science, Ann Arbor, MI (1978), Vol. 2, pp. 75–87.

856. J.I. Garcia-Alonso, A. Sanz-Medel and L. Ebdon, *Anal. Chim. Acta* **283**, 261 (1993).

857. J.G. Tarter, in *Ion Chromatography* (J.G. Tarter, ed.), M. Dekker, New York, NY (1987), pp. 369–415.

858. S. Swedberg, in *Instrumental Methods in Food Analysis* (J.R.J. Paré and J.M.R. Bélanger, eds), Elsevier, Amsterdam (1997), pp. 367–94.

859. S. Terabe, K. Otsuka and T. Ando, *Anal. Chem.* **56**, 111 (1984).

860. V. Pretorius, B.J. Hopkins and J.D. Schieke, *J. Chromatogr.* **99**, 23 (1974).

861. J.H. Knox and I.H. Grant, *Chromatographia* **24**, 135 (1987).

862. S. Kitagawa, H. Watanabe and T. Tsuda, *Electrophoresis* **20**, 9–17 (1999).

863. E. Venema, J.C. Kraak, R. Tyssen and H. Poppe, *J. Chromatogr.* **A837**, 3 (1999).

864. N.A. Guzman and R.E. Majors, *LC.GC Europe* **14** (6), 288–302 (2001).

865. K.D. Altria, K.H. Assi, S.M. Bryant and B.J. Clark, *Chromatographia* **44**, 367–71 (1997).

866. T.L. Rapp, W.K. Kowalcyk, K.L. Davis, E.A. Todd, K.L. Liu and M.D. Morris, *Anal. Chem.* **64**, 2434–7 (1992).

867. E. Gonzalez and J.J. Laserna, *Quim. Anal. (Barcelona)* **16** (1), 3–15 (1997).

868. F. Couderc, E. Caussé, N. Siméon, C. Bayle, M. Nertz and B. Feurer, *Intl Lab.* **31** (4A), 6–8 (2001).

869. R.A. Wallingford and A.G. Ewing, *Anal. Chem.* **59**, 1762 (1987).

870. A.G. Ewing, J.M. Mesaros and P.F. Gavin, *Anal. Chem.* **66**, 527 (1994).

871. M.J. Tomlinson, L. Lin and J.A. Caruso, *Analyst* **120**, 283 (1985).

872. J.W. Olesik, J.A. Kinzer and S.V. Olesik, *Ann. Chem.* **67**, 1 (1995).

873. Y. Liu, V. Lopez-Avila, J.J. Zhu, D.R. Wiederin and W.F. Becket, *Anal. Chem.* **67**, 2020 (1995).

874. C. Vogt, J. Vogt and H. Wittrisch, *J. Chromatogr.* **A727**, 301 (1996).

875. D.R. Baker, *Capillary Electrophoresis*, John Wiley & Sons, Ltd, Chichester (1995).

876. K.D. Altria, *LC.GC Europe* **15** (9), 588–94 (2002).

877. K.D. Altria, *LC.GC Europe* **13**, 320–9 (2000).

878. H. Fabre and K.D. Altria, *LC.GC Europe* **14**, 302–10 (2001).

879. A. Klockow, A. Paulus, V. Figueiredo, R. Amado and H.M. Widmer, *J. Chromatogr.* **A680**, 187 (1994).

880. Z.K. Shihabi and K.S. Oles, *Clin. Chem.* **40**, 1904 (1994).

881. M.C. Carneiro, L. Puignou and M.T. Galceran, *J. Chromatogr.* **A669**, 217 (1994).

882. S.J. Willimans, D.M. Goodall and K.P. Evans, *J. Chromatogr.* **629**, 379 (1993).

883. A. Kunkel, S. Günter, C. Dette and H. Wätzig, *J. Chromatogr.* **A781**, 445–55 (1997).

884. C.W. Demarest, E.A. Monnot-Chase, J. Jiu and R. Weinberger, in *Capillary Electrophoresis* (P.D. Grossman and J.C. Colburn, eds), Academic Press, San Diego, CA (1992), Ch. 11.

885. S.F.Y. Li, *Capillary Electrophoresis, Principles, Practice and Applications*, Elsevier, Amsterdam (1992).

886. P. Jandik and G. Bonn, *Capillary Electrophoresis of Small Molecules and Ions*, VCH Publishers, New York, NY (1993).

887. C.A. Hill III, A. Zhu and M.G. Zeece, *J. Agric. Food Chem.* **42**, 919 (1994).

888. A. Kunkel, M. Degenhardt, B. Schirm and H. Wätzig, *J. Chromatogr.* **A768**, 17–27 (1997).

889. A. Dermaux and P. Sandra, *Electrophoresis* **20**, 3027 (1999).

890. K.D. Altria, *J. Chromatogr.* **A856**, 443 (1999).

891. H. Sørensen, S. Sørensen, C. Bjerjegaard and S. Michaelsen, *Chromatography and Capillary Electrophoresis in Food Analysis*, The Royal Society of Chemistry, Cambridge (1999), Ch. 10.

892. H. Poppe, *Adv. Chromatogr.* **38**, 233 (1998).

893. R. Stevenson, K. Mistry and I.S. Krull, *Intl Lab.* 11C–22C (November 1998).

894. K. Benedek and A. Guttman, in *HPLC, Practical and Industrial Applications* (J. Swadesh, ed.), CRC Press, Boca Raton, FL (1996), pp. 305–46.

895. K.D. Altria and S.M. Bryant, *LC.GC Intl* 27–9 (January 1997).

896. W.G. Kuhr and C.A. Monnig, *Anal. Chem.* **64**, 389R (1992).

897. R.T. Kennedy (ed.), *Special Issue on Capillary Electrophoresis and Capillary Electrochromatography*, in *Analyst* **123** (7) (1998).

898. E. Jackim (ed.), *Capillary Electrophoresis Procedures Manual*, Elsevier Science, Amsterdam (1996).

899. E. Jackim and L. Watts-Jackim (eds), *Capillary Electrophoresis Procedures Manual: A Laboratory User's Aid for Quick Starts*, Elsevier Science Publishers, Amsterdam (1996).

900. I.S. Krull, R.L. Stevenson, K. Mistry and M.E. Swartz, *Capillary Electrochromatography and Pressurized Flow Capillary Electrochromatography*, HNB Publishing, New York, NY (2000).

901. K. Bartle and P. Myers (eds), *Capillary Electrochromatography*, The Royal Society of Chemistry, Cambridge (2000).

902. H. Engelhardt, *Capillary Electrophoresis*, F. Vieweg & Sons, Wiesbaden (1995).

903. D.N. Heiger, *High Performance Capillary Electrophoresis – An Introduction*, Hewlett-Packard Co., Wilmington, DE (1992).

904. G.R.v. Hecke and K.K. Karukstis, *A Guide to Lasers in Chemistry*, Jones and Bartlett Publ., Boston, MA (1998).

905. W.Th. Kok, R. Stol and R. Tyssen, *Anal. Chem.* **72**, 469A–76A (2000).

906. W. Beck and H. Engelhardt, *Chromatographia* **33**, 313 (1992).

907. P. Jandik, W.R. Jones and A.L. Heckenberg, *J. Chromatogr.* **593**, 289 (1992).

908. M. Koberda, M. Konkowski, P. Youngberg, W.R. Jones and A. Weston, *J. Chromatogr.* **602**, 235 (1992).

909. A. Weston, P.R. Brown, A.L. Heckenberg, P. Jandik and W.R. Jones, *J. Chromatogr.* **602**, 249 (1992).

910. A. Weston, P.R. Brown, P. Jandik, A.L. Heckenberg and W.R. Jones, *J. Chromatogr.* **608**, 395 (1992).

911. M. Chen and R.M. Cassidy, *J. Chromatogr.* **640**, 425 (1993).

912. T.-I. Lin, Y.-H. Lee and Y.C. Chen, *J. Chromatogr.* **A654**, 167 (1993).

913. C. Quang and M.C. Khaledi, *J. Chromatogr.* **A659**, 459 (1994).

914. E. Šimuničová, D. Kaniansky and K. Lokšíková, *J. Chromatogr.* **A665**, 203–9 (1994).

915. Z. Krivácsy, Á. Molnár, E. Tarjányi, A. Gelencsér, G. Kiss and J. Hlavay, *J. Chromatogr.* **A781**, 223–31 (1997).

916. J. Romano, P. Jandik, W.R. Jones and P.E. Jackson, *J. Chromatogr.* **546**, 411 (1991).

917. P. Jandik and W.R. Jones, *J. Chromatogr.* **546**, 431 (1991).

918. S.A. Oehrle, *J. Chromatogr.* **A671**, 383 (1994).

919. N.J. Benz and J.S. Fritz, *J. Chromatogr.* **A671**, 437 (1994).

920. R. Stahl, *J. Chromatogr.* **A686**, 143 (1994).

921. L. Song, Q. Qu, W. Yu and G. Xu, *J. Chromatogr.* **A696**, 307 (1995).

922. L. Gross and F.G. Yeung, *Anal. Chem.* **62**, 427–31 (1990).

923. D. Braun and E. Richter, *Kautsch. Gummi Kunstst.* **50**, 696 (1997).

924. W.J. Rühl, *Z. Anal. Chem.* **322**, 710 (1985).

925. E. Schröder and S. Malz, *Plast. Kautsch.* **5**, 416 (1958).

926. J. Chen, B.P. Preston and M.J. Zimmerman, *J. Chromatogr.* **A781**, 205–13 (1997).

927. M. Sawada, I. Yamaji and T. Yamashina, *Nippon Gomu Kyokaishi* **35**, 284–90 (1962).

928. I. Yamaji, M. Sawada and T. Yamashina, *Tokyo Kogyo Shikensho Hokoku* **60** (5), 176–82, 183–7 (1965).

929. M. Sawada, I. Yamaji and T. Yamashina, *Nippon Gomu Kyokaishi* **35**, 291–5 (1962).

930. C.A. Hall, A. Zhu and M.G. Zeece, *J. Agric. Food Chem.* **42**, 919–21 (1994).

931. S. Takeda, S. Wachida, M. Yamane, A. Kawahara and K. Higashi, *Anal. Chem.* **65**, 2489–92 (1993).

932. S. Terabe, K. Otsuka, K. Ichikawa, A. Tsuchiya and T. Ando, *Anal. Chem.* **56**, 113–6 (1984).

933. C.P. Ong, C.L. Ng, N.C. Chong, H.K. Lee and S.F.Y. Li, *J. Chromatogr.* **516**, 263–70 (1990).

934. W.G. Kuhr, in *Capillary Electrophoresis, Theory and Practice* (P. Camilleri, ed.), CRC Press, Boca Raton, FL (1993), pp. 66–116.

935. N. Nishi, N. Tsumagari, T. Kakimoto and S. Terabe, *J. Chromatogr.* **465**, 331–43 (1989).

936. C.P. Ong, H.K. Lee and S.F.Y. Li, *J. Chromatogr.* **542**, 473–81 (1991).

937. K.P. Evans and G.L. Beaumont, *J. Chromatogr.* **636**, 153 (1993).

938. S.M. Burkinshaw, D. Hinks and D.M. Lewis, *J. Chromatogr.* **640**, 413 (1993).

939. W.C. Brumley, C.M. Brownrigg and A.H. Crange, *J. Chromatogr.* **A680**, 635 (1994).

940. M.W.F. Nielen and M.J.A. Mensink, *J. High Resolut. Chromatogr.* **14**, 417 (1991).

941. M.R. Schure, R.E. Murphy, W.L. Klotz and W. Lau, *Anal. Chem.* **70** (23), 4985–95 (1998).

942. E.D. Lee, W. Mueck, J. Henion and T. Covey, *Biomed. Environ. Mass Spectrom.* **18**, 253, 844 (1989).

943. N. Milanovich, M. Shu, J.M. Hayes and G.J. Small, *Biospectroscopy* **2**, 125 (1996).

944. S. Morales and R. Cela, *J. Chromatogr.* **A896**, 95–104 (2000).

945. R. Wissiack, E. Rosenberg and M. Grasserbauer, *J. Chromatogr.* **A896**, 159–70 (2000).

946. K. Heinig, C. Vogt and G. Werner, *J. Chromatogr.* **A781**, 17–22 (1997).

947. J. Zweigenbaum, *Chromatography* **11**, 9 (1990).

948. J.B. Poli and M.R. Schure, *Anal. Chem.* **64** (8), 896–904 (1992).

949. J. Chapman and J. Hobbs, *LC.GC Intl* 266–79 (1999).

950. H. Shintani and J. Polonsky (eds), *Handbook of Capillary Electrophoresis Applications*, Chapman & Hall, London (1996).

951. V. Pacáková, P. Coufal and K. Štulík, *J. Chromatogr.* **A834**, 257 (1999).

CHAPTER 5

Adding colour to plastics

Polymer/Additive Analysis: The Spectroscopic Alternative

Spectroscopy is the study of the wavelength or frequency dependence of any optical process in which a substance gains or loses energy through interaction with radiation. Spectroscopy can be employed with pure substances for the purpose of obtaining more information on their molecular structure or can be utilised for the detection of particular chemical species in a sample (analytical application). The terminology 'spectroscopy' is most appropriate when spectra are studied for identifying peaks and chemical structures whereas 'spectrometry' refers to measuring and collecting spectra in numerical form for calculating physical or chemical properties, often without identifying peaks. The relative importance of the various spectroscopic techniques for the qualitative and quantitative determination of additives in polymers may be seen from Table 5.1.

In this chapter, spectroscopy is an umbrella term for a variety of complementary methods such as UV/VIS, IR, luminescence, and NMR, with the object of examining mainly polymer *additives in solution* after extraction but usually before a chromatographic separation. On-line spectroscopic detection hyphenated to chromatography is dealt with in Chapter 7.

The electromagnetic spectrum is shown in Figure 5.1 and most common spectroscopic methods are shown in Table 5.2. The spectroscopic regions are not exactly defined: slightly different boundaries are found in the literature.

The physical basis of spectroscopy is the interaction of light with matter. The main types of interaction of electromagnetic radiation with matter are absorption, reflection, excitation-emission (fluorescence, phosphorescence, luminescence), scattering, diffraction, and photochemical reaction (absorbance and bond breaking). Radiation damage may occur. Traditionally, spectroscopy is the measurement of light intensity

Table 5.1 Scientific publications on spectroscopic methods for the qualitative and quantitative determination of additives and additive classes of Table 1.12

Spectroscopic analysis	No. entries	Spectroscopic analysis	No. entries
Spectroscopy	567	NIRS	15
UV (PDA)	478	NMR	247
(FT)IR	727		

Source: *Chem. Abstr.*, 1968–1997.

Additives In Polymers: Industrial Analysis And Applications   J. C. J. Bart
© 2005 John Wiley & Sons, Ltd   ISBN: 0-470-85062-0

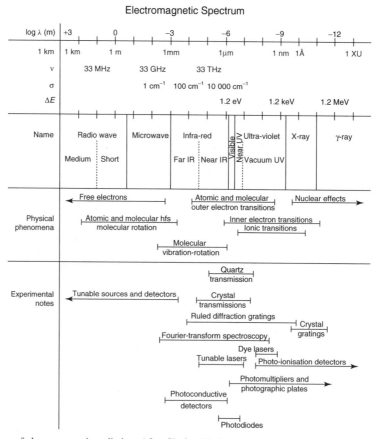

Figure 5.1 The range of electromagnetic radiation. After Siesler [1]. Reproduced by permission of the author

Table 5.2 Common spectroscopic methods based on electromagnetic radiation

Type of spectroscopy	Usual wavelength range[a]	Usual wavenumber range (cm^{-1})	Type of quantum transition
Gamma-ray emission	0.005–1.4 Å	–	Nuclear
X-ray absorption, emission, fluorescence, and diffraction	0.1–100 Å	–	Inner electron
Vacuum ultraviolet absorption	10–180 nm	$1 \times 10^6 – 5 \times 10^4$	Bonding electrons
Ultraviolet/visible absorption, emission, and fluorescence	180–780 nm	$5 \times 10^4 – 1.3 \times 10^4$	Bonding electrons
Infrared absorption and Raman scattering	0.78–300 μm	$1.3 \times 10^4 – 3.3 \times 10^1$	Rotation/vibration of molecules
Microwave absorption	0.75–3.75 cm	1.3–0.3	Rotation of molecules
Electron spin resonance	3 cm	0.33	Spin of electrons in a magnetic field
Nuclear magnetic resonance	0.6–10 m	$1.7 \times 10^{-2} – 1 \times 10^{-3}$	Spin of nuclei in a magnetic field

[a] 1 Å = 10^{-10} m = 10^{-8} cm.
1 nm = 10^{-9} m = 10^{-7} cm.
1 μm = 10^{-6} m = 10^{-4} cm.

(*1*) that is emitted, transmitted, scattered, or reflected from a sample, as a function of wavelength, at high spectral resolution – but without any spatial information. Spectroscopy is at the very heart of modern developments within polymer science, being used not only to characterise the microstructure of the macromolecular chains, but also to analyse the other essential components constituting a polymeric material.

A modern spectrophotometer (UV/VIS, NIR, mid-IR) consists of a number of essential components: source; optical bench (mirror, filter, grating, Fourier transform, diode array, IRED, AOTF); sample holder; detector (PDA, CCD); amplifier; computer; control. Important experimental parameters are the optical resolution (the minimum difference in wavelength that can be separated by the spectrometer) and the width of the light beam entering the spectrometer (the fixed entrance slit or fibre core). Modern echelle spectral analysers record simultaneously from UV to NIR.

Each spectroscopic technique (electronic, vibrational/rotational, resonance, etc.) has strengths and weaknesses, which determine its utility for studying polymer additives, either as pure materials or in polymers. The applicability depends on a variety of factors: the identity of the particular additive(s) (known/unknown); the amount of sample available; the analysis time desired; the identity of the polymer matrix; and the need for quantitation. The most relevant spectroscopic methods commonly used for studying polymers (excluding surfaces) are IR, Raman (vibrational), NMR, ESR (spin resonance), UV/VIS, fluorescence (electronic) and x-ray or electron scattering.

Absorption spectroscopy is based on the selectivity of the wavelengths of light absorbed by different chemical compounds. Absorption spectra in the *ultraviolet/visible* result from electronic transitions and so relate to electronic configurations, whereas absorption spectra in the IR region generally result from vibrational transitions in the sample (Figure 5.2). Energy differences involving rotational changes are even smaller with transitions in the far IR and microwave range. The advantage of vibrational techniques relative to many other analytical procedures is the absolute verification that is possible through the selectivity of molecular fingerprinting, if reference spectra are available. In comparison to *IR spectroscopy* qualitative additive analysis in the UV/VIS range plays a minor role. The latter technique is more indicated for quantitative analysis. However, as only some additives, e.g. unsaturated and aromatic stabilisers, exhibit specific absorption bands in the UV range this limits the general applicability of UV/VIS spectrophotometry in this field. Despite the similarities between the IR absorption and Raman (inelastic) scattering techniques, Raman spectroscopy is used only in the framework of in-polymer additive (micro)analysis. As will readily be apparent, other photophysics processes, such as *luminescence* (i.e. fluorescence/phosphorescence), are considerably less useful in polymer/additive deformulation. Despite the fact that *NMR spectroscopy* is a primary tool for structural analysis in solution, its impact on polymer/additive analysis has been limited to niche applications. Other techniques are usually preferred for routine analysis. Problems stand in relation to accessibility and the relatively low sensitivity of the technique in the past which has restricted applications for low concentration level additives, whereas for additives in high loadings (e.g. plasticisers or flame retardants) competitor techniques usually prevail. A notable exception is the use of ^{31}P liquid NMR for phosphorous-containing additives. However, NMR is a highly performing technique yielding simultaneous information on polymer and additives. A major advantage of NMR in comparison with other forms of spectroscopy is the possibility of manipulating and modifying the nuclear spin Hamiltonian almost without any restriction, and to adapt it to the special needs of the problem to be solved. At variance to UV and IR spectroscopy, in NMR it is very often possible to simplify complex spectra by modifying the Hamiltonian to an extent which permits a successful analysis. Because nuclear interactions are weak, it is possible to introduce competitive perturbations of sufficient strength to override certain interactions. In optical spectroscopy, the relevant interactions are of much higher energy, and similar manipulations are quite impossible.

Advanced computerisation and sensorisation and developments in the field of multielement optical detectors (CCD and PDA) and fibre optic remote spectroscopy have added modularity and flexibility. Silica–silica fibres used for spectroscopy applications are multimode with core diameters from 50 to 1000 μm. The application of new technologies to optical instrumentation (e.g. improved gratings in spectrographs, the use of

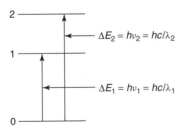

Figure 5.2 Absorption. Radiation can be absorbed by the analyte, producing a beam of diminished transmitted power, if the frequency of the incident beam ν_1 or ν_2 corresponds to the energy difference ΔE_1 or ΔE_2

highly reflecting dielectric coatings in interferometers, the development of multichannel analysers and image intensifiers) has greatly extended the sensitivity limits. New spectroscopic techniques such as Fourier transform spectroscopy, molecular beam spectroscopy and various kinds of double-resonance methods have contributed to further progress.

For polymer/additive analysis *solvent properties* are important in relation to solvent extraction, solubility, chromatography and spectroscopy (cf. Section 3.1). In general, spectroscopic solvents should be stable and transparent in the relevant range of wavelengths. They should be able to dissolve the analytes to be examined and not contain impurities affecting the stability of the substance or the validity of the method (selectivity, repeatability, limit of detection, analytical response). Theoretically, the solvent chosen should have minimal interaction with the solute. On the other hand, the so-called solvent effect can sometimes help in structure elucidation in UV, IR and NMR spectroscopies. Data regarding the cut-off points of solvents commonly in use in UV/VIS spectroscopy are readily available [2]. In the same way the range of transparency for IR solvents may be derived. Complete IR spectra of organic solvents can be found in the Sadtler *Infrared Spectra Handbook of Common Organic Solvents* [3]. Sources, detectors and window materials for UV/VIS, NIR and IR spectroscopy have equally been reported [4].

Decades of combined spectral and chemistry expertise have led to vast collections of searchable user *databases* containing over 300 000 UV, IR, Raman and NMR spectra, covering pure compounds, a broad range of commercial products and special libraries for applications in polymer chemistry (cf. Section 1.4.3). Spectral libraries are now on the hard disks of computers. Interpretation of spectra is frequently made only by computer-aided search for the nearest match in a digitised library. The spectroscopic literature has been used to establish computer-driven assignment programs (artificial intelligence).

The amount of information, which can be extracted from a spectrum, depends essentially on the attainable spectral or time resolution and on the detection sensitivity that can be achieved. Derivative spectra can be used to enhance differences among spectra, to resolve overlapping bands in qualitative analysis and, most importantly, to reduce the effects of interference from scattering, matrix, or other absorbing compounds in quantitative analysis. *Chemometric techniques* make powerful tools for processing the vast amounts of information produced by spectroscopic techniques, as a result of which the performance is significantly

enhanced. It is obviously particularly important to make the appropriate selection in data analysis in order to avoid generating bad results from good measuring data. Compounds in complex mixtures may be identified by spectral subtractions, multivariate analysis over the full or limited spectral range, etc. and instrumental measurements ('numbers') are translated into concentrations or masses to complete the analysis. The computations now routinely applied in modern spectroscopy are incomparably more complex than those traditionally used. In fact, the efficiency of a spectroscopic technique is currently greatly dictated by the chemometric procedure used to acquire the qualitative or quantitative information it provides.

5.1 ULTRAVIOLET/VISIBLE SPECTROPHOTOMETRY

Principles and Characteristics   UV/VIS spectrophotometry is concerned with the wavelength ranges of 180–380 nm (UV) and 380–780 nm (VIS) of the electromagnetic spectrum. Natural UV light is usually subdivided into three classes: near UV-A (315–380 nm), middle UV-B (280–315 nm) and far UV-C (180–280 nm). UV-A is the least harmful because it has the least energy and is used to induce fluorescent materials to emit visible light (dark glow). UV-B is typically the most destructive form of UV light, because it carries enough energy to damage biological tissues; short wavelength UV-C is almost completely absorbed in air. Extreme or vacuum UV is assigned to the 10–180 nm wavelength range. Ideal organic sunscreens (such as *p*-aminobenzoic acid and others [5]) should strongly absorb UV light in the 280–380 nm range. UV/VIS absorption spectroscopy is the simplest of the single-photon processes. In some molecules and atoms, photons of UV and visible light have enough energy to cause transitions between the different *electronic energy levels*. For atoms these transitions result in very narrow absorbance bands at wavelengths highly characteristic of the difference in energy levels of the absorbing species. However, for molecules, vibrational and rotational energy levels are superimposed on the electronic energy levels. Because many transitions with different energies can occur, the bands are broadened, even more so in solutions owing to solvent – solute interactions. Thus UV/VIS spectra obtained by measuring the variation in light intensity over the entire wavelength range generally show only a few *broad absorbance bands*. High-resolution UV spectroscopy is observed by seeding a few percent of molecules in a carrier gas of a supersonic beam which achieves molecular cooling without condensation [6].

Compared with techniques such as IR, which produces many narrow bands, UV/VIS spectroscopy normally provides a limited amount of qualitative information. UV/VIS spectrophotometry for organic analysis involves the promotion of electrons in σ, π and n orbitals from the ground state to higher energy states, i.e. $\sigma \to \sigma^*$, $n \to \sigma^*$, $n \to \pi^*$ and $\pi \to \pi^*$ transitions. Most absorption by organic compounds results from the presence of π (i.e. unsaturated) bonds. Because many compounds exhibit either very weak or no absorbance in the UV/VIS regions, a number of methods using chemical derivatisation have been developed. Such methods usually involve adding an organic reagent, which forms a complex with strong absorptivity. In indirect absorbance detection the mobile phase, rather than the analyte, contains a UV chromophore. Indirect detection is nonspecific.

A *spectrophotometer* for measuring the transmittance or absorbance of a sample as a function of wavelength consists of a source of broadband electromagnetic radiation (usually a deuterium arc and/or a tungsten–halogen lamp), a dispersion device, a sample area and a detector (Figure 5.3). As the monochromator is scanned through its wavelength range a spectrum is measured. Table 5.3 shows the main *light sources* for transmission, absorption and reflection spectroscopic studies. Tungsten–halogen

light sources provide a very stable output and are mostly used for the visible range and are suited for irradiance calibration. Deuterium light sources, with a high peak around 640 nm, are mostly used for UV absorption and reflection measurements. Pulsed xenon light sources are most suited for applications that require a long lifetime where a stable output is not priority. Xenon lamps are used only for applications in which high intensity is required, such as diffuse reflectance. Whenever a high power small wavelength range illumination is needed a LED can be efficiently coupled with fibre optics. A typical application is the use as an excitation light source for fluorescence applications.

Dispersion devices in most spectrophotometers are either holographic gratings or prisms with quartz prisms being replaced. Modern diffraction grating monochromators permit the selection of individual wavelengths in 1 nm increments between 200 nm and 800 nm. Diffraction gratings provide uniform spectral dispersion. Three types of UV detectors may be distinguished which differ in the way monochromatic light is obtained: (i) fixed-wavelength detectors (now replaced); (ii) variable-wavelength detectors; and (iii) diode array detectors. Spectrophotometers normally contain a photomultiplier tube (PMT) as a single detector or a photodiode array (PDA) detector. The latter, introduced in 1979, is an integrated circuit with an array of photosensitive sites,

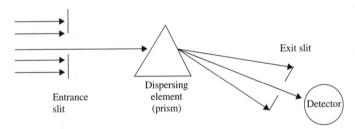

Figure 5.3 Schematics of a typical dispersive spectrometer. After Williams [7]. From R. Williams, *Spectroscopy and the Fourier Transform*, VCH Publishers, New York, NY. © Wiley-VCH, 1996. Reproduced by permission of Wiley-VCH

Table 5.3 Light sources

Application	Wavelength range (nm)	Type	Principle
Colour/VIS/NIR	360–1700	Tungsten-halogen	Continuous
UV	200–400	Deuterium	Continuous
UV/VIS/NIR refl./abs.	200–1700	Deuterium/halogen	Continuous
UV/VIS/NIR absorption	200–1700	Deuterium/halogen	Continuous
UV/VIS	200–1000	Xenon	Pulsed
Fluorescence blue	450–490	LED	Continuous
Wavelength calibration	253–922	Mercury-argon	Continuous
Irradiance calibration	360–1100	Tungsten-halogen	Continuous
Irradiance calibration	200–1100	Deuterium/halogen	Continuous

or pixels (typically 1024), linearly arranged on a single silicon chip. When a PDA is placed along the focal plane of a diffraction grating, each pixel on the PDA acts as a slit and detector for a given wavelength of dispersed radiation. The PDA detector (optical resolution ~1 nm) allows entire spectra to be collected very quickly because all wavelengths of dispersed radiation are measured simultaneously. More recent detectors include charge-coupled devices (CCDs). UV detectors are usually linear from a few milliabsorbance units (MAU) to 1 or 2 absorbance units (AU). PDA is linear from 0.0001 to 2.0 AU over the entire wavelength range and allows purity assessment and accurate quantitation of major and minor compounds.

Various *spectrophotometer configurations* are commercially available. Early UV/VIS equipment was single beam: blank and sample were measured consecutively, causing unwanted dependency on lamp drift. These spectrometers are well suited for measuring absorbance at a single point in the spectrum but less appropriate for measuring various compounds at different wavelengths or for obtaining spectra of samples. The conventional 'single measurement at a single wavelength' approach to obtaining results is insufficient for assuring optimum results. Double-beam spectrophotometers were originally developed to compensate for changes in lamp intensity, but suffer from reduced throughput and sensitivity. Recent advances in lamp and electronics design have improved the stability of single-beam spectrophotometers. Both conventional and diode array spectrophotometers are single-beam. Single-beam instruments are low cost, offer great ease of use, and the simple optical system allows high throughput and hence high sensitivity. Split-beam spectrophotometers resemble double-beam spectrophotometers but use a beam splitter instead of a chopper to send light along the blank and sample paths simultaneously to two separate but identical detectors. With a dual-wavelength spectrophotometer, two wavelengths can be measured simultaneously for special applications. These complex instruments are typically significantly more expensive than conventional spectrophotometers and have been largely replaced by diode array spectrophotometers, which are multiwavelength instruments. Diode array spectrophotometers offer considerable advantages over conventional scanning spectrophotometers, namely speed (parallel data acquisition and electronic scanning capabilities), especially useful for dynamic systems (e.g. FIA, HPLC); full spectra acquisition (for method development, multicomponent analysis, derivative spectroscopy and error detection); maximisation of dynamic range (through use of alternative wavelengths for high and low concentration ranges);

excellent wavelength resettability; high accuracy; higher throughput; lower noise levels; high reliability (mechanically simple, robust); and low cost of ownership.

PDA detection's real advantage – rapid collection of entire spectra – is exploited most successfully with HPLC and allows simultaneous monitoring of all wavelengths of interest as the HPLC separation progresses. UV spectrometry is the main detection technique for HPLC. Diode array detectors that extend into the visible wavelengths are valuable for colour body analysis. Disadvantages of PDA detectors are that resolution depends on the sampling interval of the diode array and that stray light is more problematic than with conventional spectrophotometers.

UV/VIS absorption spectroscopy, pioneered by Beckman (1941), is one of the oldest and most widely used instrumental techniques, despite being regarded by some analysts as obsolete. Recently there has been a renaissance in UV spectroscopy with many new techniques, instruments and data processing methods [8]. Modern highest specification UV/VIS absorption and fluorescence/phosphorescence spectrometer instruments extend their wavelength region from the far UV (175 nm) into the NIR region (1100 nm). Small footprint UV/VIS spectrometers (200–1100 nm) are now available. Paul [9] has traced the history of UV/VIS instrumental developments.

UV/VIS spectrophotometry can be used to determine many physico-chemical characteristics of compounds and thus can provide information as to the identity of a particular compound. Although UV/VIS spectra do not enable absolute *identification* of an unknown, they are frequently used to confirm the identity of a substance through comparison of the measured spectrum with a reference spectrum. However, UV spectrophotometry is not highly specific, and can obviously only be applied to polymer additives which are absorbers of UV radiation, i.e. contain chromophoric groups. Both UV and IR monitor functional entities rather than the entire molecular structure. A functional group's proximity to other electropositive or electronegative structures in a molecule affects the absorbance spectrum, allowing one to infer some details of molecular structure.

Most UV/VIS applications concern *single-component quantification*, which normally requires only relative measurements (e.g. the absorbance of an unknown concentration relative to the absorbance of a standard). Absorption and return to ground state are fast processes, and equilibrium is reached very quickly. Thus, absorption of UV/VIS light is quantitatively highly accurate. Method development involves selecting the wavelength(s) that yield the best results for the particular analysis. Quantitative analysis by solution spectrophotometry has been in continuous practice for more than a century [10]. A natural trend in spectrophotometry is an approach to *multicomponent analysis*. UV can offer the analytical scientist an alternative technique for separating and determining the concentrations of analytes in a mixture by mathematical rather than chromatographic separation [11]. Quantitation of compounds with highly overlapping spectra in a mixture is a difficult analytical problem, especially at greatly differing analyte concentration levels. Powerful signal processing analytical techniques, multiwavelength and derivative spectrophotometry have been developed. Multiple measurements at multiple wavelengths or (preferably) full spectra yield the best accuracy and precision of results.

Modern *multiwavelength analysis* utilises the reversed matrix representation of the Beer–Bouguer–Lambert law. It is applicable to the simultaneous determination of a larger number of components, even those with very close absorption maxima. The most important factors in analysis of spectrophotometric data are the selection of the wavelengths at which measurements are made and the extent of wavelength overdetermination that is necessary. The analysis of a solution containing p components using UV/VIS transmission spectroscopy with n calibration samples ('standards'), recorded at q wavelengths ('sensors'), and analysed by classical least squares methods has been exemplified in Massart *et al.* [12]. The standard software of modern UV/VIS-NIR spectrometers now includes a multicomponent quantitation mode for simultaneous quantitation of up to eight components as well as a multiwavelength mode for measurement up to six different wavelengths.

Difference spectroscopy involves the measurement of an absorbance difference between a liquid sample and a reference solution. Development of powerful signal-processing techniques like *derivative spectrophotometry* and multiwavelength analysis facilitates the resolution of complex mixtures with highly overlapped spectra [13]. Derivative spectroscopy involves plotting the nth order derivative of a spectrum with respect to wavelength rather than the spectrum itself. Derivative spectra are used to enhance differences among spectra, to resolve overlapping bands in qualitative analysis and, most importantly, to reduce the effects of interference from scattering, baseline drift, matrix, or broad absorbing compounds in quantitative analysis. For the resolution of components in a mixture, the use of second and fourth derivatives offers considerable advantages [14]. Resolution enhancement can greatly increase the fingerprinting use of UV spectra. Derivative spectrophotometry has become a standard in spectrophotometric laboratories.

It has proved to be very useful, providing both qualitative and quantitative information derived from mathematical processing of UV/VIS spectra. The principles of derivative spectrophotometry were discussed [15,16]. Obviously, derivatisation of spectra does not provide any additional information to that acquired during the measurement, but allows for easier interpretation. In particular, the possibility of resolving overlapping peaks makes derivative spectrophotometry a valuable tool for multicomponent analysis. Typically, derivative spectrophotometry is useful for the simultaneous determination of two additives in polymeric materials with very closely positioned absorption maxima. In quantitative analysis, derivative spectrophotometry leads to an increase in selectivity.

The accuracy of an analytical method should be confirmed by an independent method or by the analysis of certified reference materials. It is regrettable that detailed comparative studies of methods developed are still too uncommon in the analytical literature. Owen [17] has described the principles of UV/VIS *method development and validation*, regulatory requirements and standards. Method validation is required for determining whether performance characteristics of the method are suitable for the intended purpose. Some instrumental parameters may affect the accuracy and precision of measured absorbance values. Full performance verification of a spectrophotometer is time-consuming. Absolute wavelength accuracy and absolute photometric accuracy are very important in qualitative UV/VIS analysis, particularly for the identification and confirmation of unknowns. Nowadays standard instrument software comprises fully automated system validation for certain parameters (e.g. resolution or baseline stability) without the use of reference materials, while tests for photometric accuracy, wavelength reproducibility and stray light may be carried out with reference materials. NIST has given a range of solid and liquid standards for determining wavelength accuracy, photometric accuracy and stray light [18]. New CRMs for the determination of photometric accuracy in UV spectrophotometry were reported quite recently [19]. ASTM E 275-83 is the standard practice for describing and measuring performance of UV/VIS and NIR spectrophotometers.

Frequently recurring *analytical problems* are scattering, baseline shift, or unwanted broad absorbing components. Derivative spectroscopy is an excellent tool for reducing or eliminating such disturbances. In general, when measuring UV/VIS spectra we want only absorbance to occur. In the case of significant (Rayleigh or Tyndall) scattering quantitative analysis is seriously impaired. Occasionally, energy-rich low-wavelength UV causes unwanted photochemical reactions; some samples fluoresce.

UV/VIS spectroscopy, which is applied primarily to measure compounds in the vapour phase, liquids or solutions, is considered a routine technique and is used extensively in QA/QC laboratories. The transmission mode, which is restricted to the wavelength region in which the sample is transparent, is simpler and allows more accurate quantitative analysis than do reflectance measurements of solids. However, solvents exhibit a cut-off wavelength in the UV range below which they absorb too strongly for sample measurements to be performed. Table 5.4 allows an evaluation of common *solvents* for UV spectroscopic measurements. In planning UV spectrophotometry after extraction it should thus be considered that many solvents (aromatics, esters, ketones, DMF, $CHCl_3$, etc.) do not allow UV detection at low wavelengths (<250 nm); in ethers the UV cut-off may be dramatically influenced by contaminations, which are formed by oxidation, such as peroxides in THF, etc., or by stabilisers. Consequently, the choice of an extraction solvent is limited when UV/VIS spectrometry is to be employed next, without work-up. Most spectrophotometers can be used to measure colour. The colour of matter is related to its absorptivity or reflectivity.

Table 5.5 shows the *main characteristics* of UV spectrophotometry as applied to polymer/additive analysis. Growing interest in automatic sample processing looks upon spectrophotometry as a convenient detection technique due to the relatively low cost of the equipment and easy and cheap maintenance. The main advantage of UV/VIS spectroscopy is its extreme sensitivity, which permits typical absorption detection limits in solution of 10^{-5} M (conventional transmission) to 10^{-7} M (photoacoustic). The use of low concentrations of substrates gives relatively ideal solutions [20]. As UV/VIS spectra of analytes in solution show little fine structure, the technique is of relatively low diagnostic value; on the other hand, it is one of the most widely used for quantitative analysis. Absorption of UV/VIS light is quantitatively highly accurate. The simple linear relationship between

Table 5.4 Typical UV solvents and approximate UV cut-off wavelengths (nm)

Solvent	UV cut-off	Solvent	UV cut-off
Acetonitrile	190	Methanol	210
Distilled water	190	Cyclohexane	211
Iso-octane	195	Chloroform	246
n-Hexane	201	Dimethylsulfoxide	270
1,4-Dioxane	205	Xylene, benzene	280
Ethanol (95 vol%)	207	Acetone	331

Table 5.5 Main features of UV/VIS spectrophotometry

Advantages

- Routine technique
- High sensitivity
- Simple, low cost
- Fast analysis times (<2 min)
- Highly accurate (but nonspecific) quantitation
- Wide applicability

Disadvantages

- Lack of specificity (mixtures!)
- Limited qualitative information
- Interferences (by highly UV absorbing species)
- Time-consuming extractions
- Extraction required for filled/pigmented systems
- Low sensitivity for additives transparent in UV region

absorbance and concentration and the relative ease of measurement have made UV/VIS spectroscopy the basis for thousands of quantitative analytical methods, especially for rapid production and quality control.

Isolating additives from a polymer matrix simplifies UV/VIS analysis because interferences by the plastic decrease by several orders of magnitude. However, considering the time-consuming sample preparation, one may question if a chromatographic, and thus a specific quantification method should not be preferable. Nowadays, the drawback of poor selectivity can (partly) be overcome by powerful signal-processing techniques. These all make spectrophotometry a competitive technique, if not in ultratrace, at least in trace analysis due to its high sensitivity. Nevertheless, the use of UV/VIS spectrophotometry to analyse additives in extracts or in the solid state (in polymers or rubber compounds) is somewhat limited when compared with IR or NMR, because the analyte (or sample) has to possess a chromophore, such as conjugated dienes or unsaturated ketones in order to be observed. Also, most plastics are multicomponent additive systems which complicates analysis. The technique is considerably more sensitive and faster than NMR, but the problems of unambiguous peak assignment and quantitation remain.

With the development of more and more additives with closely similar absorbance bands UV analysis has lost its original role as a premier additive analysis technique. However, UV analysis can still be a very useful technique if one knows the exact additive package in a sample and there are no unknowns. Then the concentrations of the additives can be determined from the absorbance using Beer's law.

There is much effort in the international community in relation to the topic of UV measurement and the calibration of UV measuring devices. For standards and best practice in absorption spectrometry, see references [8,13] and Starna® Brand's CRMs for UV/VIS spectrophotometry (absorbance/transmission, wavelength, resolution, stray light). Under the right conditions, potassium dichromate solutions (prepared from NIST SRM 935a Potassium Dichromate, in accordance with NIST Special Publication SP260-54) sealed in cells are sufficiently stable, and have all the prerequisites to be calibrated as UV CRM [19].

A collection of UV spectra of plasticisers, fluorescent whitening agents (optical brighteners), UV absorbers, as well as of phenolic and aminic antioxidants was published by Hummel and Scholl [21]. UV absorbance data for isolated chromophores are listed elsewhere [22]. A general UV atlas of organic compounds is available [23].

Use of UV spectroscopy in hyphenation to separation techniques is described in Chapter 7. Łobiński and Marczenko [16] have described recent advances in UV/VIS spectrophotometry.

Applications Applications of UV/VIS spectrophotometry can be found in the areas of extraction monitoring and control, migration and blooming, polymer impregnation, in-polymer analysis, polymer melts, polymer-bound additives, purity determinations, colour body analysis and microscopy. Most samples measured with UV/VIS spectroscopy are in solution. However, in comparison to IR spectroscopy additive analysis in the UV/VIS range plays only a minor role as only a limited class of compounds exhibits specific absorption bands in the UV range with an intensity proportional to the additive concentration. Characteristic UV absorption bands of various common polymer additives are given in Scheirs [24].

Modern UV spectrophotometers are suitable to efficiently investigate the transmission and/or reflection of polymeric materials either in solution, as powder, plate, film or sheet. Typical UV detection wavelengths in the range of 180–350 nm can be utilised only in solvents or polymer films with a sufficiently low absorbance. The value of the data is considerably enhanced if the scanned wavelength range also covers the crucial spectral region between 170 nm and 190 nm. Extended analytical power can be achieved when UV spectrophotometry is carried out in the vapour phase (cf. GC-UV), thus avoiding the disadvantage of the liquid solvent carrier which frequently eliminates UV spectrophotometric details of a sample. UV spectroscopical behaviour (extinction coefficient, position of λ_{max}) is an important indication for the performance and in the analysis of any UV absorber.

In this paragraph we will mainly concentrate on UV/VIS applications in *solution*, i.e. after extraction of a polymer/additive sample. As indicated elsewhere [25], UV spectroscopy is less suitable for the determination of *monomers* in polymers due to lack of sensitivity (e.g. lower detection limit of about 200 ppm styrene in PS) and interference from other additives (in particular other aromatic hydrocarbons and antioxidants) in polymer formulations. Urbanski [26] has used UV spectroscopy to determine styrene monomer in chloroform extracts of polystyrene. UV spectroscopic methods for determining styrene cannot differentiate between the various volatile substances present in polystyrene [27]. Direct UV spectroscopic methods for the analysis of volatiles in PS are reliable only in case of samples not containing UV-absorbing antioxidants or any other type of strongly UV-absorbing additive. Hunt *et al.* [28] used UV absorbance to monitor CO_2 extracted solutes for detecting complete extraction of additives. Past research has mainly concentrated on additives that have UV absorbance above 260 nm and most efforts were directed towards the determination of a single additive. By improving the transmission efficiency of probes used in spectroscopic systems, simultaneous in-line quantitative monitoring of multiple additives can now be achieved. The UV approach requires standards or measurement of extinction coefficients in order to provide a quantitative determination of additives (or their degradation products). Yang [29] has given some examples of the limitation of Beer–Lambert's law in strong UV absorbing antioxidant CH_2Cl_2 solutions. The UV absorbance at two peak wavelengths plotted against AO concentration (Figure 5.4) with the concentration levels of 0.048 %, 0.100 %, 0.148 %, and 0.200 % at a pathlength of 2 mm shows that Beer's law is indeed obeyed by Irganox 3114/CH_2Cl_2 solutions. Instead Beer's law was not obeyed for BHT, although a linear relationship was held at the low UV absorbance range (less than 2.00 AUs).

As the majority of stabilisers has the structure of aromatics, which are UV-active and show a distinct UV spectrum, UV spectrophotometry is a very efficient analytical method for qualitative and quantitative analysis of stabilisers and similar substances in polymers. For *UV absorbers*, UV detection (before and after chromatographic separation) is an appropriate analytical tool. Haslam *et al.* [30] have used UV spectroscopy for the quantitative determination of UVAs (methyl salicylate, phenyl salicylate, DHB, stilbene and resorcinol monobenzoate) and plasticisers (DBP) in PMMA and methyl methacrylate-ethyl acrylate copolymers. From the intensity ratio

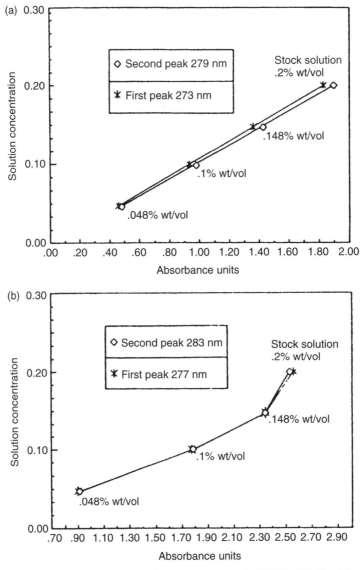

Figure 5.4 Peak UV absorbance vs. concentration of (a) Irganox 3114 and (b) BHT in CH$_2$Cl$_2$. After Yang [29]. Reproduced by permission of the author

at two wavelengths (e.g. UV$_{254}$/UV$_{280}$) it is often possible to derive the compound class [31]. Tojo and Matsumoto [32] have studied the identification of several hydroxybenzophenone UVAs and dialkylphthalate plasticisers in polymers by UV and other means. Other selected applications of spectrophotometry in the determination of organic compounds concern diethyldithiophosphates (at 293 nm in chloroform extract) [33], peroxides and surfactants [34]. Alkyldithiophosphates are used as additives to lubricants

and as anticorrosion agents. Also the trimethylquinoline content in PE was determined with UV spectroscopy [35]. Pern [36] has noticed that the loss rate of Cyasorb UV531 and the progress of discoloration of ethylene-vinylacetate (EVA) encapsulants from light yellow to brown follow a sigmoidal pattern.

UV/VIS spectrophotometry was the earliest technique to determine *antioxidants* [37–39]. UV/VIS has been used for the quantitative determination of BHT, stalite, etc. in rubbers [40]. Quantitative UV spectrophotometric analysis of polymer extracts has also been reported for polyolefins for a wide variety of stabilisers (Topanol OC, Binox M, Ionox 330, Ionol CP, BHT, Santonox R, phenyldiamines, Tinuvin 326, Irganox 1010, Irgafos 168), but also for antistatics (Armostat, Lutostat), and slip agents such as oleamide (for the older literature, see references [25,41–44]). Brandolini *et al.* [45] have attempted to study degradation during processing of several commercially available phosphorous-containing secondary antioxidants (Naugard 524, TNPP and Irgafos P-EPQ, see Figure 9.3). As the degradation products are rather weak UV absorbers (broad absorbance at 335 nm, not observable in the polymer samples), UV was not generally informative for these studies, contrary to ^{31}P NMR results. Rao *et al.* [46] have developed a method using UV/VIS quantification of BHT, Irganox 1076, Tinuvin 327 in PP and Irganox 1010/1076 in EP copolymer extracts. The total amount of additives was quantified using calibration plots in the appropriate ranges obtained for standards. The reliability of the UV method was shown by the recovery of the additives from the polymer standards. This method can be conveniently extended to the analysis of additives in commercial polyolefin samples.

Billingham *et al.* [47] determined 11 phenolic antioxidants in iPP by extraction of the additive into refluxing chloroform, followed by UV spectroscopy of the extract. The phenolic peak at 280 nm was used for the analysis and the appropriate extinction coefficients were determined separately. Care is required to avoid loss of small and volatile molecules during extraction experiments. UV spectroscopy has also been used to quantify two oxidation products of Topanol O (butylated hydroxytoluene), namely 3,3′,5,5′-tetra (*tert*-butyl)stilbenequinone and 3,3′,5,5′-tetra(*tert*-butyl)diphenoquinone [48]. The content of Cyasorb UV5411 in extracted optical grade PC was established as 0.12 wt% by quantitative analysis through the use of UV spectroscopy [49]. *Polymeric additives* are difficult to analyse by chromatographic techniques because of their molecular weight distribution. Here, spectroscopic methods offer a useful alternative. Chimassorb 944 (MW 2500–4000 Da) in PP can be determined quantitatively by dissolution (hot decalin)/cooling, followed by extraction with diluted sulfuric acid and UV determination [50]. UV spectroscopy was also used for quantitative analysis of chloroform extracts of LDPE/Chimassorb 944 [51].

Traditionally, Irganox B *blends* (Irganox 1010/1076 and Irgafos 168) are analysed by HPLC methods, which require a long sequence of preliminary actions: milling,

extraction, solvent evaporation and dissolution in acetonitrile with addition of an internal standard (Irganox 1063). Alternatively, the Irganox B blend content in the polymer may be monitored more rapidly (but rather inaccurately) on the basis of the XRF measured phosphorous content (assuming a constant blend component ratio, which only holds true macroscopically). Turning to spectroscopic methods, advantage can be taken of the fact that the UV spectra of Irganox 1010/1076 and Irgafos 168 are very much alike. As the phenyl groups are the main structural elements in these compounds, a deviation from the ideal composition of the blend has less consequence for UV analysis than one based on phosphorous measurement only. Consequently, for quantitative determination use can be made of small differences in the UV spectra of the various blend components and of multivariate calibration with multiwavelength detection. Heat stability testing of antioxidants and UV absorbers may be carried out by UV transmission spectroscopy in solution at a prescribed wavelength following thermal treatment. Ostromow [52] has used UV spectrophotometry to determine stabilisers with phenolic hydroxyl groups in extracts of *rubbers* and elastomers.

In order to control compliance with EC directives [53] of polyolefin *food packaging materials* containing aromatic antioxidants (Irganox 1010, Irgafos 168), bis(ethanolamine) antistatics (Armostat, Lutostat) and the internal lubricant erucamide, UV/VIS and ^{1}H NMR were evaluated as spectroscopic alternatives to the more time-consuming migration tests. Extraction was monitored using UV spectrophotometry (with chemical derivatisation). Optimisation relied on extraction kinetics, which included the demonstration that extraction is more severe than migration. Only a few hours are required to conclude whether a material complies with the regulations. UV/VIS thus also allows rate of migration studies for polymer additives.

For direct determination of *plasticisers* in PVC the sample was dissolved in THF, and phthalate or phosphate plasticisers were measured from absorbance at 240 and 275 nm or 257, 262 and 268 nm, respectively [54]. The organic tin compound OTC (heat stabiliser) was extracted from PVC by US extraction and analysed by UV spectrophotometry [55].

The UV method performs well in the absence of interferences of other additives or pigments but is liable to be in error owing to interference by highly absorbing *impurities* that may be present in the sample or in the solvent [56,57]. Commercial erucamides with variable light stability have been analysed by means of SEC-UV (254 nm)/RI and UV absorption in MeOH [58].

While pure erucamide does not exhibit absorption in the 250–280 nm range all commercial products showed UV absorption maxima at 230, 257, 267 and 278 nm (with variable intensities), indicative of the presence of oxidation products (up to 8 wt%) as impurities with chromophoric groups. Colourless impurities do not have an interference effect. UV/VIS was also used to evaluate discoloration of 50 wt/wt% mixtures of commercial erucamide and inorganic antiblock agents [59]. Apart from the interference by impurities from solvents chemical methods suffer from lengthy procedures.

Chemical derivatisation is sometimes exploited in the case of lack of UV absorbing moieties. Stafford [60] has developed an oxidation procedure for the spectrophotometric determination of Ionol (2,6-di-*t*-butyl-*p*-cresol) in polyolefins; oxidation of Ionol in alkaline isopropyl alcohol produces a solution with a strong absorption band at 365 nm. The procedure eliminates the interference of Santonox R (4,4′-thiobis-(6-*t*-butyl-*m*-cresol)). Similarly, in non-alkaline medium 2,6-di-*t*-butyl-4-methylphenol and 4-substituted 2,6-xylenol phenolic antioxidants have virtually identical UV spectra; in alkaline medium the spectra are sufficiently different to permit determination without mutual interference [61]. A multiwavelength approach might have been considered as an alternative to chemical derivatisation. Ruddle and Wilson [62] reported UV characterisation of PE extracts of three antioxidants (Topanol OC, Ionox 330 and Binox M), all with identical UV spectra and $\lambda_{max} = 277$ nm, after reaction with nickel peroxide in alkaline ethanolic solutions, to induce marked differentiation in different solvents and allow positive identification. Nonionic *surfactants* of the type $RO(CH_2CH_2O)_nH$ were determined by UV spectrophotometry after derivatisation with tetrabromophenolphthalein ethyl ester potassium salt [34]. Magill and Becker [63] have described a rapid and sensitive spectrophotometric method to quantitate the peroxides present in the surfactants sorbitan monooleate and monostearate. The method, which relies on the peroxide conversion of iodide to iodine, works also for Polysorbate 60 and other surfactants and is more accurate than a titrimetric assay.

A method suitable for quantification of the functional class of bis(ethanol)amine antistatics, which lack UV chromophores, consists of reaction with methyl orange [53]. Atmer 163 (alkyl-diethanol amine) has been determined as a yellow complex at 415 nm after interaction with a bromophenol/cresole mixture [64]. Hilton [65] coupled extracted phenolic antioxidants with diazotised *p*-nitroaniline in strongly acidic medium and carried out identification on the basis of the visible absorption spectrum in alkaline solution. The antioxidant Nonox CI in

Table 5.6 UV/VIS analyses of some selected antioxidants, light and heat stabilisers

Compound	Functionality	Solvent[a]	Absorption maxima (nm)[b]
Tinuvin 320	UV absorber	Toluene	346, 305
Tinuvin 326	UV absorber	Toluene	355, 312
Tinuvin 327	UV absorber	Toluene	353, 313
Irganox MD 1024	Metal deactivator	Methanol	276
Uvitex OB	Whitener	DMA	375, 356, 395
Chimassorb 944	HM-HALS	1 N H_2SO_4	247, 209
Goodrite 3034	LM-HALS	1 N H_2SO_4	202

[a] DMA, *N*,*N*-dimethylacetamide.
[b] First value: λ_{max}.

LDPE or HDPE extracts has been determined colorimetrically at 430 nm by oxidation with H_2O_2 in the presence of H_2SO_4 [66]. *p*-Phenylenediamine derivatives such as Flexzone 3C, used as antiozonants in rubber products, have been determined colorimetrically after oxidation to the corresponding Wurster salts [67]. A wide range of amine AOs in polyolefins has been determined by the *p*-nitroaniline spectrophotometric procedure [68]. Monoethanolamine (MEA) in a slip agent in PE film has been determined as a salicylaldehyde derivative by spectrophotometric quantification at 385 nm [69]. Table 5.6 contains additional examples of the use of UV/VIS spectrophotometry for the determination of additives in polymers.

As to *visible spectrophotometric methods*, most colorimetric procedures relate to PE/antioxidant systems (cf. ref. [44]), including derivatisations [65,66,68] and ashing-spectrophotometric methods for the determination of metal (catalyst residue) traces [42,70]. Treatment of fluxing benzene extracts of oil-extended black and nonblack SBR masterbatches and vulcanisates with benzoyl peroxide after methanol precipitation of the rubber develops a yellow hue proportional to the concentration of antidegradants such as *N*-alkyl-*N*′-aryl-phenylenediamines and *N*,*N*′-diaryl-phenylenediamine [71]. Koch [72] has reported the direct quantitative colorimetric determination of various di-*n*-octyltin compounds in the fat simulant HB 307 using complex formation with dithizone up to a detection limit of 0.75 ppm. Similarly, a direct colorimetric determination of sodium alkyl sulfates and sodium C_{12}-C_{16} alkyl sulfonates in HB 307 after reaction with methylene blue is possible up to a detection limit of 0.2–0.5 ppm. Diphenylpicrylhydrazyl (DPPH) has been used as a reagent for direct colorimetric determination of phenolic antioxidants and mercapto compounds in the concentration range around 1 ppm in the fat simulant HB 307 [73]. The applicability of the method is limited but may be used to determine the total specific migration, SML(T).

Phenolic antioxidants in rubber extracts were determined indirectly photometrically after reaction with Fe(III) salts which form a red Fe(II)-dipyridyl compound. The method was applicable to Vulkanox BKF and Vulkanox KB [52]. Similarly, aromatic amines (Vulkanox PBN, 4020, DDA, 4010 NA) were determined photometrically after coupling with Echtrotsalz GG (4-nitrobenzdiazonium fluoroborate). For qualitative analysis of *vulcanisation accelerators* in extracts of rubbers and elastomers colour reactions with dithiocarbamates (for Vulkacit P, ZP, L, LDA, LDB, WL), thiuram derivatives (for Vulkacit I), zinc 2-mercaptobenzthiazol (for Vulkacit ZM, DM, F, AZ, CZ, MOZ, DZ) and hexamethylene tetramine (for Vulkacit H30), were mentioned as well as PC and TLC analyses (according to DIN 53622) followed by IR identification [52]. 8-Hydroquinoline extraction of interference ions and alizarin-La^{3+} complexation were utilised for the spectrophotometric determination of fluorine in silica used as an antistatic agent in PE [74]. Also Polygard (trisnonylphenylphosphite) in styrene-butadienes has been determined by colorimetric methods [75,76]. Most procedures are fairly dated; for more detailed descriptions see references [25,42,44].

UV/VIS is a very simple method for identifying the principle natural and synthetic dyes [77]. Optical absorption spectroscopy has been used for the determination of acid *dyes* in textile dyeing formulations [78] and is particularly indicated for the identification of *extracted pigments*. For rapid screening and QC of solvent dyes for engineering thermoplastics, UV/VIS in solution using double-beam PDA instrumentation (for purity control), TLC (for troubleshooting and initial supplier qualification) and ICP (for organic and inorganic insoluble impurity profiling) are now used industrially, replacing the time- and resource-consuming traditional production of moulded chips for that purpose [79]. Differential spectroscopy was used to compare trial to standard samples. With correct sample preparation and selection of the optical path length, the strength of the colourant can be determined with ±3 % accuracy. Shakhnovich and Barren [79] have illustrated the detection of C.I. Solvent Green 3 as a low concentration

impurity in C.I. Solvent Violet 13. Haisch and Niessner [80] have described the use of photoacoustic spectroscopy (PA-UV) for the analysis of high concentrated textile dyestuff (5 g L^{-1}) where conventional transmission spectroscopy fails.

UV/VIS spectrophotometry can also be used for *extraction monitoring* on the basis of the chromophore of functional classes and to follow up polymer impregnation with additives in scCO$_2$ [81].

5.2 INFRARED SPECTROSCOPY

Principles and Characteristics Vibrational spectroscopic techniques such as IR and Raman are exquisitely sensitive to molecular structure. These techniques yield incisive results in studies of pure compounds or for rather simple mixtures but are less powerful in the analysis of complex systems. The IR spectrum of a material can be different depending on the state of the molecule (i.e. solid, liquid or gas). In relation to polymer/additive analysis it is convenient to separate discussions on the utility of FTIR for indirect analysis of extracts from direct *in situ* analysis.

Infrared radiation refers broadly to the part of the electromagnetic spectrum between the visible and microwave regions. It is convenient to divide the IR region into three parts (Table 5.7): far-IR, mid-IR and near-IR. The regions are not exactly defined: slightly different boundaries for the IR regions are found in the literature. Of greatest practical use for polymer/additive analysis are mid-IR and near-IR; the far-IR region is irrelevant for this purpose.

The vibrational and rotational motions of the chemically bound constituents of matter have frequencies in the IR region. Industrial IR spectroscopy is concerned primarily with *molecular vibrations*, as transitions between individual rotational states can be measured only in IR spectra of small molecules in the gas phase. Rotational – vibrational transitions are analysed by quantum mechanics. To a first approximation, the vibrational frequency of a bond in the mid-IR can be treated as a simple harmonic oscillator by the following equation:

$$v = \frac{1}{2\pi} \sqrt{\frac{K}{\mu}} \qquad (5.1)$$

Table 5.7 The IR regions of the spectrum

Region	λ (cm)	ν (cm^{-1})	ν (Hz)
Near-IR	$2.5 \times 10^{-4} - 7.8 \times 10^{-5}$	4000–12800	$1.2 \times 10^{14} - 3.8 \times 10^{14}$
Mid-IR	$5 \times 10^{-3} - 2.5 \times 10^{-4}$	200–4000	$6 \times 10^{12} - 1.2 \times 10^{14}$
Far-IR	$0.1 - 5 \times 10^{-3}$	10–200	$3 \times 10^{11} - 6 \times 10^{12}$

where v is the frequency of vibration, K is the force constant describing the bond between two atoms and μ is the reduced mass of the system. This simplified treatment of chemical bonds as harmonic oscillators explains the fundamental absorbance bands present in the mid-IR. However, real molecules do not behave totally harmonically. Interaction of IR radiation with a vibrating molecule is only possible if the electric vector of the radiation field oscillates with the same frequency as does the molecular dipole moment. A vibration is IR active only of the molecular dipole moment is modulated by the normal vibration (selection rule):

$$(\delta\mu/\delta q)_0 \neq 0 \qquad (5.2)$$

where μ is the molecular dipole moment and q stands for the normal coordinate describing the motion of the atoms during a normal vibration. The intensity of an IR absorption band is proportional to the square of the change in dipole moment, with respect to the change in the normal coordinate $(\partial\mu/\partial q)^2$ of the molecular vibration. Nature and quality of the observed spectrum will further depend on the chemical (e.g. solid, liquid or gas) and physical (e.g. path length, turbidity, surface condition, etc.) properties of the sample. Information concerning *all* aspects of the molecular structure and intermolecular interactions are encoded in the vibrational spectra.

A nonlinear molecule of N atoms with $3N$ degrees of freedom possesses $3N - 6$ normal vibrational modes, which not all are active. The prediction of the number of (absorption or emission) bands to be observed in the IR spectrum of a molecule on the basis of its molecular structure, and hence symmetry, is the domain of group theory [82]. Polymer molecules contain a very high number of atoms, yet their IR spectra are relatively simple. This can be explained by the fact that the polymer consists of identical monomeric units (except for the end-groups).

The interactions of photons with molecules are described by molecular cross-sections. For IR spectroscopy the cross-section is some two orders of magnitude smaller with respect to UV or fluorescence spectroscopy but about 10 orders of magnitude bigger than for Raman scattering. The peaks in IR spectra represent the excitation of vibrational modes of the molecules in the sample and thus are associated with the various chemical bonds and functional groups present in the molecules. The frequencies of the characteristic absorption bands lie within a relatively narrow range, almost independent of the composition of the rest of the molecule. The relative constancy of these group frequencies allows determination of the characteristic

functional groups in the molecule. The exact frequency values give further information about the neighbourhood of these functional groups. Thus structural details can be deduced. The IR spectrum of a compound is one of its most characteristic physical properties and can be regarded as its *'fingerprint'* in a similar way as the XRD pattern for crystalline materials. Infrared and Raman spectroscopy provide complementary images of molecular vibrations, because in these spectroscopic techniques the mechanisms of the interaction of light quanta with molecules are quite different. Mid-infrared and Raman show different sensitivities to functional groups (e.g. mid-IR is good for $-OH$, $C=O$, Raman is best for $-C=C-$), different sampling requirements (Raman is intrinsically easier), quantitation (neither technique is absolute, mid-IR generally shows higher sensitivity) and cost (mid-IR generally lower cost).

An *IR spectrometer* essentially consists of a source of continuous IR radiation, an absorption cell, a dispersing element, a detector and display. The most common *radiation sources* are thermal ones: quartz-halogen lamps (NIR), Nernst glower and glow bar/Globar (mid-IR) and mercury arc lamp (FIR); there are only few – and expensive – tuneable lasers available for the IR range. The IR region is particularly demanding for optical measurements because of the poor intensities of the available sources and the relatively poor detectors available. *Infrared detectors* are of various types: thermal (thermocouples), pyroelectric, quantum or photoacoustic. Thermal detectors are less sensitive and relatively slow; common pyroelectric detectors are TGS (triglycine sulfate) and DTGS (deuterated TGS). Quantum (or photon) detectors such as PbS, PbSe, InGaAs, InSb, PbSnTe and CdHgTe (or mercury cadmium telluride, MCT) are both sensitive and fast. At present the most commonly used mid-IR detectors are DTGS, for both dispersive and FT spectrometers, and MCT, for more sensitive and/or rapid FT applications. Chalmers and Dent [83] have reported the relative sensitivities of detectors used in IR and Raman spectroscopy.

Filters, prisms, gratings, monochromators, and interferometers are essential components of most IR instruments and enable measurement at selective wavelengths or over a narrow band. Discrete IR wavelength selection is only possible with a laser source. For many years, an IR spectrum was obtained by passing an IR beam through the sample and scanning the spectrum with a dispersion device (the familiar diffraction grating). *Dispersive IR spectrometers* are slow, as the absorption of each frequency is measured individually, and relatively insensitive. This contrasts with today's devices which might detect 0.01 % of a compound in some matrix.

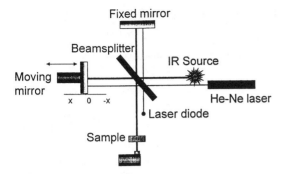

Figure 5.5 Michelson interferometer. Reproduced by permission of Thermo Electron, B.V.

Fourier-transform infrared spectrometers, which use similar sources, sampling devices and detectors, but work through the interference of two IR beams controlled by a mirror system, outperform dispersive instruments. While FTIR is steadily displacing dispersive IR in the workplace, there are nevertheless still many double beam dispersive instruments in use. Analysis of analytical signals such as IR, NMR, NQR and MS takes great benefits from Fourier-transform techniques, based on fast Fourier-transform (FFT) algorithms [84]. The underlying concepts that make Fourier spectroscopies attractive are primarily mathematical and electronic rather than spectroscopic [7]. The basic technique of a Fourier-transform spectrometer is light interference. The FTIR technique makes use of the Michelson interferometer [85] to examine the transmitted energy through a sample at all times. The light beam is split into two light beams by means of a beam splitter (Figure 5.5). Upon recombination, interference takes place depending on the difference in path length and wavelength. Measuring the intensity at varying path length differences produces an *interferogram*. By applying fast Fourier-transformation on the interferogram a spectrum of transmittance against wavenumber is generated which contains all the information required for multicomponent analysis. The FT spectrometer is particularly suited to the problems of the IR region and has revitalised IR spectroscopy.

In particular, using FTIR instrumentation instead of its dispersive equivalent the user benefits are:

- Multiplex advantage (Fellgett's): all wavelengths are measured simultaneously [86].
- Throughput advantage (Jacquinot's): higher energy throughput (larger apertures) [87].

- Frequency advantage (Conne's): internal calibration is derived from He-Ne laser (precision $0.01\,cm^{-1}$) [88].

Fourier-transform instruments provide very high resolution ($\leq 0.001\,cm^{-1}$) spectra quickly with large signal-to-noise ratios. As N frequencies are measured with one detector only the result is $\sqrt{N}$ less noise than its dispersively measured counterpart. More effective use is being made of the power of the radiation source. Modern instruments can now record a complete spectrum in less than $1/50\,s$, which is also the basis for several fast hyphenated techniques such as GC-FTIR and TG-FTIR. The multiplex advantage has also stimulated the use of several sampling accessories, such as diffuse reflection, photoacoustic, and attenuated reflection cells, which require up to 1000 scans to produce a high quality spectrum of even the most difficult samples. The throughput advantage arises because the interferometer has no gratings or slits to limit the optical throughput. Wavenumber precision (Conne's) and accuracy are not only important for validation purposes but equally necessary to generate precise information of chemical structures. A further advantage of FT spectrometers is the fact that the data are available in digital form. Spectroscopic databases can therefore be compiled either on the spectrometer computer or an external computer.

FTIR spectroscopy has become a routine technique. FTIR can often be conducted in a nondestructive manner, is a quick and easy method for quality control, trace analysis and polymer identification. Polymer samples can be analysed in all possible textures and excellent spectra can be obtained. FT spectrometers are a class of analytical instruments of significant relevance to *quality assurance* and quality control for the chemical, pharmaceutical, food and other industries. However, they are subject to many more possible types of systematic ordinate error than are grating instruments. Modern IQ/OQ/PQ validation packages support tests monitoring all of the critical performance factors of IR instruments, including noise level, resolution, linearity, and checks on both the wavelength and intensity scales to guarantee short- and long-term stability. In particular, where the spectrophotometer is used for regular transmittance or for absorbance measurements for quantitative purposes, the validity of the ordinate scale is of obvious relevance to the Quality Management System of the laboratory. Birch and Clarke [88a] have discussed the sources of ordinate error (i.e. transmittance) in FT spectroscopy and have identified as many as 50 different categories. Whereas uncertainties in transmittance and regular reflectance measurements on a grating instrument are only a few tenths of a percent, within FT spectrometers these are often over a percent, without even considering the additional errors in the reflectometer accessory. For these reasons the National Physical Laboratory (NPL) continues to use grating IR spectrophotometers for determining and supplying *IR standards* for the ordinate scales of various properties. NPL is the only national measurement standards laboratory that supplies an extensive range of IR standards: regular reflectance, hemispherical reflectance and wavenumber standards. NPL has also developed transmittance standards calibrated to a known absolute accuracy for checking mid-IR grating and FT spectrometer systems [89,90].

In most cases some form of *sample preparation* is required in order to obtain a good quality FTIR spectrum. Spectra of *gases* or low-boiling liquids may be obtained by expansion of the sample into an evacuated cell. Gas cells are available in lengths of a few centimetres up to 40 m. Standard IR spectrometer cells are not much longer than 10 cm; long paths are achieved by multiple reflection optics. In the vapour phase, the molecules are free to rotate; transitions between the various rotational levels are superimposed on the pure vibrational transition, leading to a spectrum with considerable structure.

Transmission spectroscopy remains the most commonly used and traditional IR measurement for samples that can be prepared in a transparent form. Various options exist for IR sampling of *liquids* (Table 5.8). Liquids in transmission may be measured neat or in solution in cells with fixed or demountable spacers, or as a film directly between IR-transmitting windows. A fixed path length in sealed cells is preferred for quantitative work. The choice of window material (KBr, ZnSe, CaF$_2$ or BaF$_2$) depends on solubility, chemical compatibility and the range needed [92]. Pressing a liquid sample between flat plates produces a film 0.01 mm or less in thickness, the plates being held together by capillary action. Generally, in order to obtain the IR spectrum of a pure liquid, a sample thickness of <0.015–0.025 mm is required. Samples of 1–10 mg are needed. Infrared spectra of a solid sample may be taken in transmission after dissolution in a suitable solvent. The solution method may pose a solubility problem. Soluble polymers are frequently studied in solution with compensation for solvent absorption. Unfortunately, a nonpolar liquid that is transparent across the entire mid-IR region and which dissolves all samples does not exist. As a consequence, most solid samples are run by other methods. Moreover, good quality spectra will not be obtained from samples containing particles whose dimensions are comparable

Table 5.8 Infrared sampling methods for liquids

Sample	Sampling method	Comments
Solvent extracts[a]	Cast directly on the KBr and evaporate solvent	–
	Cast on to silicon wafer	Large areas can be coated, no moisture absorption and no interfering peaks
Solutions	Fixed-path length liquid cell with KBr windows	Best for low-viscosity liquids
	Wick cell	For small volumes of liquid (1 µL)
On-line chromatographic fractions[b]	Evaporative FTIR	–
Liquid polymers and curatives	Capillary film of fluid pressed between KBr discs	For uncured matrix resins and oligomers
Liquids	Horizontal ATR	Wipe on/wipe off; single and multireflection technologies

[a] Common solvents used for IR analysis include chloroform, acetone, methanol and hexane.
[b] For review on sample preparation for IR hyphenated techniques, see McClure [91].

to or exceed the wavelength of the radiation passing through the sample, owing to light scattering effects.

Transmission methods impose fairly severe limitations on sample thickness (except for samples in the gas phase) because the amount of IR energy absorbed by the sample is proportional to its thickness. Beyond a certain thickness, a sample will not transmit any IR radiation in the regions of the spectrum where it is strongly absorbing. Repetitive analysis of liquid samples is made easy by the wipe on/wipe off sampling afforded by the horizontal ATR accessory. Liquid samples can be analysed easily by both single and multireflection technologies. Internal reflection spectroscopy can readily be used to identify solutes in volatile solvents since the solvent can easily be evaporated, leaving the solute as a thin layer on the surface of the internal reflection element (IRE). Yet, transmission measurements are easier to make than diffuse reflectance, they provide higher sensitivity, more precision and enable monofilament fibre optics to be used. There is no universal IR transmission cell that can be used easily to measure the broad range of various supercritical solvent systems. Diamond ATR-FTIR is an effective method for analysing supercritical fluids [93].

Infrared spectroscopy yields qualitative and quantitative information on chemical constitution, functional groups and impurities. Infrared data may be readily interpreted negatively so that one may definitely preclude the presence of certain functional groups, if the corresponding group vibrations are absent in the spectrum. *Qualitative analysis* may be based on a similarity match (e.g. for end product control), discriminant analysis (e.g. for the determination of a compound class) or search standards. The Sadtler IR library (Bio-Rad, Sadtler Division Philadelphia, PA) now contains some 185 000 IR spectra. The Basic Surfactants IR Database from Sadtler contains over 450 FTIR spectra of commercially available surfactants. Hummel [94] has described a computer-based search with special FTIR libraries for (industrial) polymers and resins, additives, monomers, pyrolysates and educts; the Hummel–Scholl compilation comprises an IR spectra library for over 1000 additives [21,95]. Bruker has developed a Plastics Identification System, a general-use polymer identification system based on mid-IR spectroscopy. The Rapra collection of IR spectra of some 1150 rubbers, plastics and thermoplastic elastomers [96] includes both spectra of films (transmission) and pyrolysates of cross-linked rubbers and thermoset materials (condensed phase), spectra of technically important rubber and plastic based materials. Nicolet Instrument Co. and Aldrich Chemical Co. have introduced an expanded condensed-phase library of high-resolution FTIR reference spectra (18 500 compounds) [97]. The Sprouse Library of Polymer Additives (Sprouse Scientific Systems, Paoli, PA), as distributed by Nicolet, contains 325 reference spectra recorded either neat, as chloroform casts, or as nujol mulls. Other comprehensive spectra catalogs of organic polymers, fibres, fillers, plasticisers, processing aids, surfactants and auxiliary materials are available [21,98–103]. FTIR databases of 1509 polymers, 290 plasticisers and additives, 90 PP additives and 637 surfactants are accessible through internet service running under *http://ftirsearch.com* [104]. In many instances in-house, laboratory-specific data collections have been installed which often better reflect the particular interests of a laboratory (Section 1.4.3). Vapour phase spectra are rather few (several thousands). Databanks facilitate archiving, and also allow quick comparisons of measured and library spectra. For computer-based search specialised FTIR, MS and ToF-SIMS industrial additive libraries have been developed or are now under development. In the absence of reference spectra valuable information about the molecular structure of the analysed compound may still be obtained by spectral interpretation.

Until fairly recently, IR spectroscopy was scarcely used in *quantitative analysis* owing to its many inherent shortcomings (e.g. extensive band overlap, failure to fulfil Beer's law over wide enough concentration ranges, irreproducible baselines, elevated instrumental noise, low sensitivity). The advent of FTIR spectroscopy, which overcomes some of these drawbacks, in addition to the development of powerful chemometric techniques for data processing, provides an effective means for tackling the analysis of complex mixtures without the need for any prior separation of their components.

Table 5.9 summarises the *main features* of FTIR spectroscopy as applied to extracts (separated or not). Since many additives have quite different absorbance profiles FTIR is an excellent tool for recognition. Qualitative identification is relatively straightforward for the different classes of additives. Library searching entails a sequential, point-by-point, statistical correlation analysis of the unknown spectrum with each of the spectra in the library. Fully automated analysis of

Table 5.9 Main characteristics of FTIR spectroscopy for polymer/additive extracts

Advantages

- Easy to operate, rapid (few seconds), reliable, versatile, low cost (atmospheric conditions)
- Relatively simple
- Small sample amounts
- Qualitative (chemical bonding) and quantitative information
- All elements (but not element specific)
- Specific and characteristic absorption bands
- Excellent reference databases (verification, identification)
- Simultaneous detection of different components of a mixture during one scan
- High absorption coefficients
- Simple, robust quantitative algorithms (peak height/area of isolated bands); standards usually needed
- Calibration transfer
- Mature technique and instrumentation (including desktop)
- Excellent wavenumber accuracy
- On-line hyphenated techniques available
- Wide applicability

Disadvantages

- Sample preparation needed (extraction, dissolution, separation)
- Transparent sample recommended
- Readily absorbed IR radiation
- Relatively poor analyte sensitivity
- Low specificity
- Interferences (co-additives with same functional groups)
- Difficult speciation of components in mixture analysis
- Energy-limited technique
- Low radiation intensity at detector
- Highly dependent on well-characterised calibration standards and sample preparation
- Few commercially available traceable standards

compounds that can be extracted from complex matrices requires a PCA algorithm for rapid discrimination [105]. Vibrational spectroscopy is basically employed to detect smallest amounts of a given analyte in a large excess of other compounds; contaminations in the ppm range can be detected. Recent improvements in FTIR detection technology have reduced detection limits to 100 pg or less for strong IR absorbers. FTIR is less sensitive than UV, but more specific. Advantages of FTIR over MS are that the interpretation is generally simpler and the measurement does not require a vacuum. FTIR spectroscopy is much faster (and cheaper) than NMR. As such, it holds promise for direct observation of additives and additive degradation products in polymers. However, the specificity of IR spectroscopy is not outstanding: e.g. various antioxidants, HALS stabilisers, plasticisers and lubricants absorb in the $1730\,cm^{-1}$ region. Thus the exact identification of members of a compound class may not be possible. Also, spectral overlaps, particularly for complex mixtures, may hinder identification of some components. Some drawbacks of IR stem from the fact that IR is an energy-limited technique. Situations frequently occur where there is not enough energy to measure accurately very weak or very strong bands necessary for an analysis. The bands could be intrinsically weak or arise from low concentrations or extremely small amounts of sample. Speciating individual components in complex mixtures is also limited. This disadvantage arises from the band overlap from the various components in a mixture. Although advances in multivariate data-handling methods (e.g. PLS, PCR or CLS) have improved this situation, it is still difficult to speciate more than a handful of components in a mixture. The obvious alternative is to employ a chromatographic technique first.

FTIR instrumentation is mature. A typical routine mid-IR spectrometer has KBr optics, best resolution of around $1\,cm^{-1}$, and a room temperature DTGS detector. Noise levels below $0.1\,\%\,T$ peak-to-peak can be achieved in a few seconds. The sample compartment will accommodate a variety of sampling accessories such as those for ATR (attenuated total reflection) and diffuse reflection. At present, IR spectra can be obtained with fast and very fast FTIR interferometers with microscopes, in reflection and microreflection, in diffusion, at very low or very high temperatures, in dilute solutions, etc. *Hyphenated IR techniques* such as PyFTIR, TG-FTIR, GC-FTIR, HPLC-FTIR and SEC-FTIR (Chapter 7) can simplify many problems and streamline the selection process by doing multiple analyses with one sampling. Solvent absorbance limits flow-through IR spectroscopy cells so as to make them impractical for polymer analysis. Advanced FTIR

spectroscopy allows IR imaging and time resolved spectroscopy (TRS). Dynamic FTIR characterisation is possible of time and temperature-dependent phenomena such as relaxation and exchange processes; light-fibre optics may be applied to process control and remote sensing. There are areas where improved performance is desirable, such as the sensitivity of chromatographic interfaces and the spatial resolution of IR microscopes.

General techniques of qualitative IR analysis and the identification of material by IR absorption spectroscopy are described in ASTM E 1252 and ASTM E 204, respectively. The theoretical aspects of FTIR spectroscopy and analysis have been described in many books [106,107] and reviews [108]. Coates [109] has reviewed sampling methods for IR spectroscopy.

Applications FTIR in combination with a cone calorimeter or in relation to various fire smoke toxicity tests has considerable potential for on-line analysis of *fire gases* from burning rubbers and plastics [110]. Fire gases from burning polyurethane foams are composed of HCN, NO_x, CO, CO_2, HCl and HBr (partly originating from FR additives) [111]. Infrared spectroscopy can also be used for sampling and quantitative analysis of *air and particulate matter* (e.g. around polymer processing operations) directly in a long-path cell. The advantage of IR here is its identification power, a disadvantage is occasional insufficient sensitivity. The detection limit for IR is about 0.1 to 1 ppm depending upon the compound [112]. For known analytes and many samples GC is more efficient for this application. GC-MS is generally more time-consuming than direct IR examination.

Many analytical techniques are in use for the qualitative and quantitative evaluation of *monomers and oligomers* extracted from PA6 (GC, differential refractometry, IR, PC, SEC, HPLC, RPLC, etc.). FTIR has been used for quantitative analysis of caprolactam oligomer content (extract %) in polyamide-6 [113]. The method, which involves a 3 h extraction in boiling methanol, is suitable for process control and plant environment. Kolnaar [114] has used FTIR characterisation of fractional extracts with pentane, hexane, and heptane of HDPE for blow moulding applications. Vinyl acetate in packaging film has similarly been determined by quantitative FTIR.

Composition and structure of newly developed additives are commonly examined by IR, NMR, MS and elemental analysis, e.g. recently developed higher MW antioxidants [115]. Infrared spectroscopy is also well suited to the direct verification of compound composition and quantitative determination of additives in polymers. Gray and Neri [116] have used Soxhlet extraction and IR measurements for the quantitative evaluation of the remaining antioxidant concentration in HDPE plaques. As shown in the polymer/additive deformulation scheme in Scheme 2.12, it is good practice to subject a sample to rapid IR screening as a thin film both before and after extraction in order to verify complete removal of extractables. Verification is based on comparison to reference spectra. However, additives of quite similar structure yield very similar IR spectra, which may limit unambiguous assignment. In view of the mostly low additive concentrations only the most intensive IR bands (e.g. carbonyl bands) can be used for quantitation. Calibration curves need to be available. The content of the BASF 5050 light stabiliser in PE/EVA copolymer was determined by IR spectroscopy (correlation coefficient 0.998) [117].

As well as being much faster than its dispersive ancestor, FTIR is more accurate because many spectra can be taken and averaged. As a result, FTIR has greatly expanded the use of IR analysis, including the analysis of extracts of polymer/additive matrices. These samples are particularly well suited for transmission measurements and can easily be kept within the boundaries of Beer's law to allow for quantitative measurements; where necessary attenuated total reflectance (ATR) measurements are made. In the *rubber industry* the FTIR transmission technique is generally accepted (ASTM D 3677-90). However, this procedure is preceded by a time-consuming and complicated extraction procedure. Blanco *et al.* [117a] have addressed the problem of the simultaneous determination of various organic compounds, namely stearic acid, *N*-phenyl-*N'*-isopropyl-*p*-phenylene diamine, aliphatic and aromatic processing oils, 1,2-dihydro-2,2,4-trimethylquinoline (Flextol H), DOS, dicumylperoxide, hindered bisphenol, nickel dimethyldithiocarbamate, trimethylolpropane trimethylacrylate and *N*-(cyclohexylthio)phthalimide (PVI), used as rubber additives by processing their FTIR spectroscopy data using partial least-squares regression for multivariate calibration. This procedure allows one to fully exploit the vast amount of information contained in the IR spectra for organic substances with a view to the simultaneous resolution of *complex mixtures.* The wavenumber range used should contain appreciable absorption of all the species analysed in order to improve the results obtained in the quantitation. Models including the wavenumber range from 1800 to 1650 cm^{-1} were the only ones that allowed the additives present in each sample to be quantified simultaneously. The proposed methodology was applied to experimental batches consisting of very different number of samples; quite satisfactory errors of prediction (2–5 %) were obtained.

A variety of *plasticisers* (aliphatic ester, alkyl phthalate, benzoate), dialkyl tin compounds and emulsifiers in PVC were determined by IR [44]. Examination of tin stabilisers containing sulfur (e.g. dibutyltin di-2-ethylhexyl thioglycollate) required a specific chemical test or XRF for the detection of sulfur, decomposition of the stabiliser and identification of 2-ethylhexanol by GC and IR methods, identification of the thioglycollic acid from the thioglycollate ester in the residue with IR or NMR, and of the oxide derivative in the acidified filtrate by IR or TLC, and gravimetric analysis of tin after wet digestion [44]. ASTM D 2124-99 describes a standard test method for analysis of components of PVC compounds using IR [118]. The PVC compound is solvent-extracted in order to separate the plasticiser from the compound. The resin is dissolved from the remaining compound and the inorganic fillers and stabilisers separated by centrifuging. By this technique, the compound is separated into plasticisers, resin, and inorganic stabilisers and fillers. Each may be analysed quantitatively by IR. PVC formulations are too varied to be covered adequately by a single test method. Complementary procedures, such as chromatographic and other separations, will be necessary to separate specific components and extend the applications of ASTM D 2124. Udris [119] has used a variety of techniques, such as GC, NMR and IR to identify different types of dialkyltin carboxylated and hemiester stabilisers and dialkyltin thiocompounds in PVC. Kozlowski and Gallagher [120] have recently described the identification of monomeric octylester plasticisers in flexible PVC compounds by means of IR and GC techniques. Infrared examination of the total plasticiser content in vinyl extracts containing co-additives, such as heat stabilisers, lubricants, emulsifying agents, monomers, etc. without previous chromatographic separation is less satisfactory and may well be misleading. There are currently very few methods which allow to determine migration of a whole group of substances at once. The total specific migration limit SML(T) of mixtures of plasticisers from PVC has been determined by means of FTIR by focusing attention on a common functional group, a so-called *group method* [121]. Table 5.10 shows that migration of mixtures of plasticisers can be determined if the results are expressed in molarity of ester group. Infrared was also used to determine the average carbon number in an alkyl monoester phthalate mixture [122].

As reported by Nelissen [58], unreacted carbodiimides (Stabaxol I, P and P100), used as *hydrolysis stabilisers* in e.g. polyesters, are readily detected by IR analysis on the basis of the $-N{=}C{=}N-$ absorption at 2140 cm^{-1}, but without further discrimination within

Table 5.10   IR spectra of aromatic plasticisers

Plasticiser[a]	ν_{max} (cm^{-1})[b]	ε[c]	ε/n[d]	Plasticiser[a]	ν_{max} (cm^{-1})[b]	ε[c]	ε/n[d]
TEHTM	1720	1356	452	DTMHP	1720	920	460
DEHP	1720	912	456	DIOP	1720	900	450
DBP	1720	920	460	DIUP	1720	894	447
MEHP	1720 (1705)	628	[e]	DCHP	1715	924	462

[a] See Appendix I.
[b] In chloroform.
[c] Molar absorptivity coefficient, L mol^{-1} cm^{-1}.
[d] Number of ester groups in the molecule.
[e] Superposition of two different absorptions.

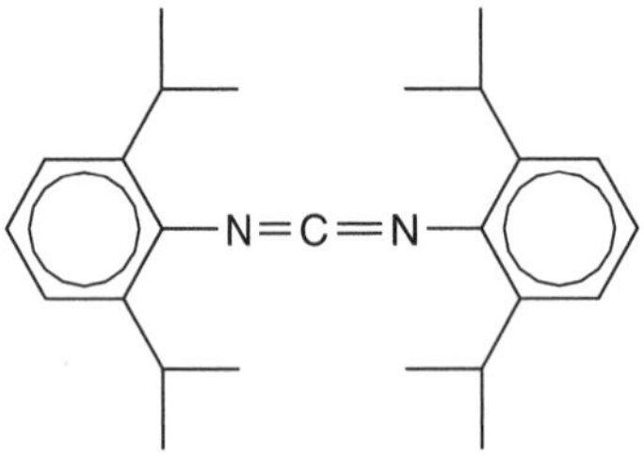

Scheme 5.1   Structural formula of Stabaxol I

the class (Scheme 5.1). Interference of heavy polymer absorption bands has excluded discrimination of brominated *flame retardants* such as PDBS 80, Pyrochek 68PB or Saytex HP 7010 in polyamide matrices by means of transmission FTIR spectroscopy. Prior extraction is necessary for that purpose [58]. As well known, fatty acids can function as external and internal *lubricants*. Obviously, conventional polymer dissolution/reprecipitation cannot distinguish between stearic acid from the bulk and the surface of styrenic resin pellets; an ethanol wash of the resin selectively removes external stearic acid. The accuracy of the FTIR determination is within 6 % with a detection limit of ca. 0.01 wt% [123]. FTIR spectroscopy has also been used to determine reliably trace amounts (ppm level) of entrained polydimethylsiloxane (PDMS) in plastic additives [124]. The additives were dissolved in DMF and the *silicones* were extracted with pentane using continuous liquid–liquid extraction. Interference resulting from remaining extracting solvents, additive residues or simultaneously extracted impurities were eliminated by spectral subtraction. Organic *binders* in paper coatings were analysed by FTIR [125].

IR absorption bands are sensitive to the local environment and therefore a spectral shift may occur when an additive is extracted from the polymer. This allows real-time monitoring of the extraction process *in situ*, as illustrated by Howdle *et al.* [126] using an organometallic complex. On the other hand,

FTIR has also been used to study vinyl trimethoxysilane grafting onto LDPE induced by dicumyl peroxide [127].

Infrared measurement of additive concentrations is a more complex analysis than initially expected, as some additives may undergo a variety of chemical reactions during processing, as shown by Reeder *et al.* [128] for the FTIR analysis of phosphites in polyolefins. Some further examples of IR work refer to PVC/metal stearates [129], and PE/Santonox R [68,130]. Klingbeil [131] has examined the decomposition of various organic peroxyesters (TBPB, TBPP, TBPA and TBPO) and a peroxidicarbonate (BOPD) as a function of pressure, temperature and solvent by means of quantitative FTIR using an optical high p, T reaction cell.

In conclusion, IR analysis of polymer/additive extracts before chromatographic separation takes advantage mainly of straightforward transmission measurements. Without separation it is often possible to make class assignments (e.g. in the reported examples on plasticisers and carbodiimide hydrolysis stabilisers); it may eventually be necessary to use multivariate techniques. Infrared detection of chromatographic effluents is dealt with in Chapter 7.

For more extensive references on the use of IR techniques the reader is referred to previous compilations [41,42,44]. The application of FTIR spectroscopy to the analysis of plastic additives has extensively been reviewed [95].

5.3   LUMINESCENCE SPECTROSCOPY

Principles and Characteristics   The term 'luminescence' describes the radiative evolution of energy other than blackbody radiation which may accompany the decay of a population of electronically excited chromophores as it relaxes to that of the thermally equilibrated ground state of the system. The frequency of the

Table 5.11 Classification of luminescence

Type of excitation	Name
Absorption of electromagnetic radiation	Photoluminescence (fluorescence, phosphorescence)
High-energy particles	Radioluminescence
Cathode rays/electron beams	Cathodoluminescence
Electric fields	Electroluminescence
Thermally activated ion recombination	Thermoluminescence
Chemical reaction (oxidation)	Chemiluminescence (CL), oxyluminescence (OL)
Cold plasma	Plasma-induced luminescence (PIL)
Electrochemical reactions	Electrochemical luminescence (ECL)
Biological processes (enzymatic)	Bioluminescence
Fiction/mechanical forces	Triboluminescence
Sound/ultrasound	Sonoluminescence

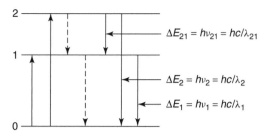

Figure 5.6 Fluorescence. Absorption of incident radiation from an external source causes excitation of the analyte to state 1 or 2. Excited species can dissipate the excess energy by emission of a photon or by radiationless processes (dashed lines). The frequencies emitted correspond to the energy differences between levels

emitted radiation depends on the difference in energy levels between the excited states and lower energy state. The probability of spontaneous luminescence is related to the probability of the electronic transition from the ground to the excited state. Luminescence covers all emissions of light from the near-IR to the near-UV regions. Many types of energy can excite luminescence in susceptible compounds (Table 5.11).

If a chemical reaction results in production of a molecular species in an excited electronic state that emits light, this phenomenon is termed chemiluminescence. *Chemiluminescence* usually occurs in the gas or liquid phase. In the case of *photoluminescence*, a molecule in the gas, liquid or solid phase absorbs light of wavelength λ_1 from an external light source, decays to a lower energy excited electronic state, and then emits light of wavelength λ_2 as it radiatively returns to its ground electronic state. Generally, the emission (em) wavelength is longer than the excitation (ex) wavelength but in resonance emission $\lambda_1 = \lambda_2$. Two spectrally and temporally distinct forms of luminescence exist: *fluorescence* and *phosphorescence*. Fluorescence, occurring between states of like multiplicity, is a quantum-mechanically 'allowed' transition (cf. Figure 5.6) and generally occurs at higher energies and upon shorter timescales than the 'lower probability' process of phosphorescence, which occurs between electronic states of differing multiplicity. Fluorescence is generally complete in about 10^{-5} s (or less) from the time of excitation whereas phosphorescence emission may extend for minutes or even hours after irradiation has ceased (up to

10^8 times longer than fluorescence). Chemiluminescence and bioluminescence (such as emission of VIS light by fireflies and glow-worms) are not considered to be fluorescence. In both phenomena, there is no transfer of light energy as in fluorescence, and the emission of light is only the result of a chemical reaction.

Although instrumentation for observing photoluminescence has many similarities to that used for measuring absorption in the UV region in the spectrum, the optical requirements of a fluorescence spectrometer are quite different from those of an absorption spectrometer. Therefore the use of fluorescence attachments for UV/VIS spectrometers is not usually successful. The characteristic right-angled optics of a fluorescence instrument (source-dispersion-sample/dispersion-detector) allow fluorescence to be detected in the absence of the transmitted beam and minimises Rayleigh scattering (Figure 5.7). Almost all fluorimeters are effectively single beam instruments.

Scanning *spectrofluorometers* (with horizontal or vertical beam geometry) contain both an excitation and an emission monochromator. In 'emission spectra', the ex wavelength is fixed (usually at a wavelength at which the sample has significant absorbance) and the em monochromator is scanned. In 'excitation spectra', on the other hand, the em wavelength is fixed and the ex monochromator is scanned. (Note that it is still fluorescence that is being detected!) The main luminescence-related measurements concern both emission and excitation spectra, luminescence kinetics (or lifetimes) and time-resolved emission spectra. Quite the most important feature of any luminescence is the *emission spectrum*. Luminescence is measured quantitatively on an axis perpendicular to the direction of the energy source. In addition to its wavelength and temporal characteristics, the luminescence observed from a given population of excited states is dependent

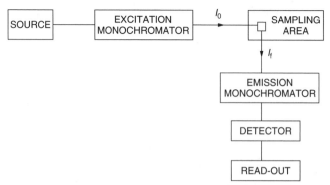

Figure 5.7 Layout of a typical fluorescence spectrometer. After Rendell [132]. Reprinted from D. Rendell, *Fluorescence and Phosphorescence Spectroscopy*. Copyright © 1987 John Wiley & Sons, Limited. Reproduced with permission

upon both the concentration of chromophores, dispersed in the medium under investigation, and their orientations with respect to the incident radiation and its polarisation characteristics. The intensity (I_L) of a luminescent material is measured using the relationship:

$$I_L = \beta I_o c \qquad (5.3)$$

where β is a proportionality constant, I_o is the intensity of the light source radiation, and c is the concentration of the luminescent atoms or molecules in a sample specimen. As the fluorescence intensity from a given sample is proportional to the incident light source intensity, fluorimetry thus requires bright but stable sources (usually Xe lamps). With appropriate calibration luminescence spectroscopy can be made quantitative.

Table 5.12 shows the *main features* of luminescence spectroscopy. The much higher sensitivity and specificity of luminescence techniques compared to absorption techniques is an obvious advantage for excitation spectra. In solution studies, pg mL^{-1} levels can often be determined, as compared to µg mL^{-1} levels in absorption spectroscopy. The greater sensitivity of luminescence techniques stems from the fact that the

Table 5.12 Main features of luminescence spectroscopy

Advantages

- Very high sensitivity (trace analysis)
- High specificity
- Nondestructive

Disadvantages

- Restricted quantitative use (with appropriate calibration; vulnerability to quenching)
- Low-light-level phenomenon
- Limited applicability
- Limited compound classes

luminescence intensity can be enhanced by increasing the intensity of the excitation source, which is not the case in absorption. The sensitivity of fluorimetry also derives from the fact that the technique detects a small emission signal above a zero background, rather than a small difference between two large numbers, as in absorptiometry. Luminescence is generally a low-light-level phenomenon, and in order to increase further the sensitivity of luminescence detection methods there is a continuous need for improved sensitive detection devices (such as CCD). Luminescence techniques are generally more selective than absorption methods, since two distinct wavelengths (those of absorption and emission) can be used to characterise a sample. High performance spectrophotometers provide a maximum spectral resolution of about 0.8 nm. Luminescence spectroscopy is generally regarded as nondestructive. Luminescence has recently been reviewed [133,134]; various books are available [132,135,136].

Fluorescence measurements for chromatographic applications are made with *fluorescence detectors* (FLD). These are amongst the most sensitive detectors in liquid chromatography but thorough method development is required [137]. Very low analyte detection limits (10^{-15}–10^{-4} mol L^{-1}) are reported. Until recently, however, FLDs have been built to provide single-wavelength information. In the past, fluorescence spectra had to be acquired under stop-flow conditions and data analysis was time-consuming. In modern fluorescence detector technology these drawbacks have been overcome and spectra are acquired on-line simultaneously with the chromatographic detector signal. In modern optical design a high power Xe flash lamp (with µs flash duration; 260–660 nm) is used for high intensity excitation in the UV range with holographic gratings as monochromators. The excitation (ex) and emission (em) monochromators can be operated in spectral and signal mode. The spectral mode is used to obtain multisignal or spectral information; the signal mode offers the lowest limits of detection available to LC since all data points are generated at a single ex and em wavelength but has low versatility. Fluorescence detectors often show limits of detection down to the ppt level. Chromatographic data can be displayed in three dimensions (em or ex wavelength, retention time and intensity). Spectra can be evaluated against spectra from customised spectral libraries for automated compound confirmation or to control peak purity. Various procedures may be followed to obtain complete information on a compound's fluorescence [137]. It is possible to use a FLD/DAD combination in order to acquire UV/VIS spectra (equivalent to excitation spectra) with DAD and emission spectra with FLD in a single run. During method development, one or two chromatographic runs can be sufficient to optimise wavelength settings for a series of analytes.

Fluorimetric methods of analysis make use of the natural fluorescence of the analyte, the formation of a fluorescent derivative or the quenching of the fluorescence of a suitable compound by the analyte. Fluorescence cannot occur unless there is light absorption, so that all *fluorescent molecules* absorb, but the reverse is not true; only a small fraction of all absorbing compounds exhibits fluorescence. The types of molecule most likely to show useful fluorescence are those with delocalised π-orbital systems. Often, the more rigid the molecule the stronger the fluorescence intensity. Naturally fluorescent compounds include Vitamin A, E (tocopherol).

Fluorescence results are more dependent upon the experimental conditions than absorption spectrometry [8]. Fluorescence depends on a variety of factors: molecular structure (aromatic, aliphatic, substituents, etc.); pH (of ionisable compounds); solvent (in case of interaction of solvent and analyte); temperature and concentration. A common problem with the use of fluorescence for quantitative analysis is that many compounds can affect the fluorescence (quenching). For example, dissolved oxygen is a well-known quencher of the fluorescence of aromatic compounds. Vulnerability to the presence of fluorescence quenchers restricts quantitative use to well-defined or carefully purified samples, conditions which often apply to the effluent of a chromatographic column. A critical issue in trace level fluorescence detection is to have an HPLC system free of fluorescence contamination, including suspended particles which increase scattered light. Solvents should not contribute a background signal. Most of the nonhalogen-containing solvents for HPLC can be used in fluorescence detection. The majority of HPLC solvents do not contain a fluorophore. In HPLC,

amines can be fluorescently labelled permitting very low detection limits to be achieved – in the order of attomoles (10^{-18} mol). Fluorescence based identification of related analytes in a mixture often requires a chromatographic separation of the components prior to detection. Quantification of additives, as opposed to pesticide residues, by fluorescence is not yet very popular despite its specificity and sensitivity.

Instrumentation for fluorescence spectroscopy has been reviewed [8]. For standards in fluorescence spectroscopy, see Miller [138]. Fluorescence detection in HPLC has recently been reviewed [137]. Phosphorescence detection of polymer/additive extracts is not being practised.

Applications Categorised as either fluorescent or phosphorescent, luminescent compounds are commonly added to coatings, paints, and plastics for special effects. Common uses include CRT screens, sports paraphernalia, toys, markings, and light tubes. Several kinds of fluorescent dyes such as Fluorescein, Rhodamine B, and Acridine Orange are used for morphology observation. Luminescence also attracts interest as a (macro)molecular probe of energy migration, chain dynamics and interactions [139].

Luminescence (fluorescence) emanating from PET has been associated with monomer units at short wavelengths and aggregated ground-state dimers at longer wavelengths [140,141]. Extensive studies have been undertaken on the luminescence characteristics of PET with a view of ascertaining the nature of the *oxidation products* and organic contaminants [142–144]. The fluorescence ex and em spectra of mono- and dihydroxyterephthalate chromophores have been characterised in both solution and solid poly(ethylene-*co*-1,4-cyclohexanedimethylene terephthalate) (PECT) [145]. Fluorescence analysis of manufactured crude terephthalic acid (CTA) showed that the major colour forming *impurity*, 4-carboxybenzaldehyde (4-CBA), dominates the spectrum as the primary component. According to Qi *et al.* [146] low-MW atactic PP plays an important role in the luminescence emission of commercial polypropylene. Burdon and Billingham [147] have used luminescence techniques for epoxy resin cure characterisation. Phillips and Carey [148] have recently dealt with polymer luminescence and photophysics and have focused attention arbitrarily on species that absorb in the spectral region from 250 nm to longer wavelengths, where luminescence may be an additional fate of photoexcited species. Luminescence techniques have been used both for indirect determination of additives in polymers (as a detector to chromatography, see also Section 7.3) and as a direct analysis tool.

Fluorescence is much more widely used for analysis than phosphorescence. Yet, the use of fluorescent detectors is limited to the restricted set of additives with fluorescent properties. Fluorescence detection is highly recommended for food analysis (e.g. vitamins), bioscience applications, and environmental analysis. As to polymer/additive analysis fluorescence and phosphorescence analysis of UV absorbers, optical brighteners, phenolic and aromatic amine antioxidants are most recurrent [25] with an extensive listing for 29 UVAs and AOs in an organic solvent medium at r.t. and 77 K by Kirkbright *et al.* [149].

Stabilisers are usually determined by a time-consuming extraction from the polymer, followed by an IR or UV spectrophotometric measurement on the extract. Most stabilisers are complex aromatic compounds which exhibit intense UV absorption and therefore should show luminescence in many cases. The fluorescence emission spectra of Irgafos 168 and its phosphate degradation product, recorded in hexane at an excitation wavelength of 270 nm, are not spectrally distinct. However, the fluorescence quantum yield of the phosphate greatly exceeds that of the phosphite and this difference may enable quantitation of the phosphate concentration [150]. The application of emission spectroscopy to additive analysis was illustrated for Nonox CI (N,N'-di-β-naphthyl-p-phenylene-diamine) [149] with fluorescence ex/em peaks at 392/490 nm and phosphorescence ex/em at 382/516 nm. Parker and Barnes [151] have reported the use of fluorescence for the determination of N-phenyl-1-naphthylamine and N-phenyl-2-naphthylamine in extracted vulcanised rubber. While pine tar and other additives in the rubber seriously interfered with the absorption spectrophotometric method this was not the case with the fluorometric method.

Fluorescence whitening agents (FWA) or *optical brighteners* are substances which strongly absorb the short-wave energy (340–400 nm) in the UV region of the spectrum and re-emit it almost quantitatively in the form of long-wave (400–440 nm), blue fluorescent light. FWAs are widely used in various industries: textile, laundry detergent and soap, pulp and paper, plastic and polymer. FWAs are important additives which improve the whiteness and brightness of various commercial products. Textile auxiliaries such as optical brighteners and *sizing agents* in aqueous solutions were analysed by means of 2D fluorescence spectroscopy [152]; on-line quantitative 2D fluorescence spectroscopy of substances added onto and adsorbed by polyamide

woven fabrics was reported. The concentration of the FWA 2,5-bis(5-*tert*-butyl-2-benzoxalyl)thiophene in olive oil food simulant was determined by fluorescence spectroscopy [153].

By their nature, many *UV absorbers* are amenable to analysis by fluorimetric analysis. In many instances visible fluorescence techniques are less subject to interference by other polymer additives in a polymer extract than are UV methods of analysis. In fluorescence analysis (ex at 367 nm, em at 400–440 nm) of a PS/Uvitex OB chloroform dissolution AOs such as Ionol CP, Ionox 330, Polygard and Wingstay T/W do not interfere; detection limit of 10 ppm [41].

Fluorescence spectroscopy is also particularly well-suited to clarify many aspects of *polymer/surfactant interactions* on a molecular scale. The technique provides information on the mean aggregation numbers of the complexes formed and measures of the polarity and internal fluidity of these structures. Such interactions may be monitored by fluorescence not only with extrinsic probes or labelled polymers, but also by using fluorescent surfactants. Schild and Tirrell [154] have reported the use of sodium 2-(N-dodecylamino) naphthalene-6-sulfonate (SDN6S) to study the interactions between ionic surfactants and poly(N-isopropylacrylamide).

Quite obviously, UV and *fluorescence detection in HPLC* is restricted to analytes with chromophoric groups in their structure, which many polymer additives do not possess. Verstappen [155] has recorded ex and em spectra of phenols (DBPC, Goodrite 3114, nonylphenol, Irganox 1010/1076, Topanol CA), phosphorous-containing compounds (Irgafos 168, Irgafos 168 phosphate, Mark C), amines (Permanax DPPD, Nonox CI), hydroxyalkoxybenzophenones (Cyasorb UV531), hydroxyphenylbenzotriazoles (Tinuvin P/326/328), phthalates (DBP, DOP), a fatty amide (Kemamide E), a whitening agent (Uvitex OB), and halogenated aromatics (Firemaster 680, DBBP, DBDPO) for the purpose of investigating the applicability of fluorescence detection for HPLC. It was noticed that the UVAs hydroxyalkoxybenzophenones and hydroxybenzotriazoles show no fluorescence. Although many additives (phenols, phosphorous compounds and phthalates) do fluoresce, they exhibit similar ex maxima (220 ± 5 nm) and em maxima (300 ± 10 nm); consequently, the selectivity is low and these additives cannot be distinguished by means of luminescence. Some fluorescing secondary amines, such as Permanax DPPD and Nonox CI, are no longer applied in polyolefins because of discolouring and toxicity problems. As expected, the aliphatic saturated fatty amide Kemamide-E does not fluoresce. The presence of halogens in flame retardants usually suppresses the intensity of a fluorescence signal. In many cases, fluorescence detection provides much higher sensitivity than UV detection. From a comparison of fluorescence and UV detection of various phenols (Irganox 1010/1076, DBPC and nonylphenol) and phosphorous-containing compounds (Irgafos 168 and Irgafos 168 phosphate) it appears that fluorescence detection based on the ex wavelength of 275 nm is usually more sensitive than UV detection at 280 nm (the second ex maximum for many additives) [155]. However, generally for these additives little advantage was found in the use of fluorescence instead of UV detection except for nonylphenol where fluorescence is 5 to 10 times more sensitive than UV detection. For many additives the selectivity of fluorescence detection is actually lower than UV detection. Fluorescence detection coupled to HPLC analysis of additives therefore usually has few advantages over UV detection except for some specific cases (e.g. organic whiteners). Fluorescence spectroscopic methods have found application for on-line and *in situ* measurements.

Fluorescence in UV radiation is a frequently used method for detection of *TLC spots*, e.g. of Tinuvin 326 [42]. The fluorescence emitted by optical brighteners under UV light on a thin-layer plate has been utilised as a means of analysing these compounds [42]. On the whole, the use of fluorescence detection in polymer/additive analysis of extracts is certainly not overwhelming. Applied fluorescence has been described in a monograph [156].

In principle, measurement of the *phosphorescence* characteristics of samples obtained after extraction of polymers with organic solvents may also yield useful information regarding the nature and concentration of the additives present. Parker and Hatchard [157] have examined the possibilities of phosphorescence measurements for *N*-phenyl-2-naphthylamine. Although it should be possible to determine various analytes simultaneously by correct choice of ex and em wavelengths and phosphorescence decay, no pertinent reports are available. Phosphorescence finds limited application for the direct determination of additives in polymers (without prior extraction).

5.4 HIGH-RESOLUTION NUCLEAR MAGNETIC RESONANCE SPECTROSCOPY

Principles and Characteristics No other physico-chemical method applied to liquids has come close to

rivalling nuclear magnetic resonance (NMR) with regard to versatility as a general, quantitative and detailed source of information at the molecular level. Among the features of NMR spectroscopy that have turned it into such a powerful collection of analytical methods is the freedom to manipulate independently both spin and spatial components of the nuclear spin interaction tensors. This imparts enormous flexibility and selectivity to the type of structural and dynamical information that can be obtained.

Radiofrequency spectroscopy (NMR) was introduced in 1946 [158,159]. The development of the NMR method over the last 30 years has been characterised by evolution in magnet design and cryotechnology, the introduction of computer-based operating systems and pulsed Fourier transform methods, which permit the performance of new types of experiment that control production, acquisition and processing of the experimental data. New pulse sequences, double-resonance techniques and gradient spectroscopy allow different experiments and have opened up the area of multidimensional NMR and NMRI.

The term NMR now covers a large diversity of structural and analytical techniques, which produce signals in the *frequency domain* (high-resolution continuous wave and Fourier-transform NMR spectroscopy), the *time domain* (low-resolution pulse NMR) and the *spatial domain* (magnetic resonance imaging) (Table 5.13). This has generated several stand-alone research fields, such as liquid-, solid-, low-resolution NMR, and NMR imaging and microscopy. Depending on the particular instrumental NMR technique, it is possible to gain structural information concerning a molecule, determine the concentrations of either major or minor constituents in solids and liquids, obtain information about the mobility and physical state of components, or analyse the spatial distribution and mobility of nuclei within a sample. Some of this information would be difficult, if not impossible, to obtain by any other method. NMR is now a mature spectroscopic technique and actually one of the most powerful spectroscopies available for the study of the structure and dynamics of condensed matter. The costs and constraints associated with the instrumentation required to obtain each type of information are very different and determine whether a particular technique may be applied to quality control measurements and process analysis, or to (polymer) research applications. The basic components of an *NMR spectrometer* consist of a superconducting magnet to generate a magnetic field B_0, a tuneable RF oscillator to generate a secondary magnetic field B_1 and/or B_2 in the transmitter coil(s), a sample cavity, a receiver coil and peripheral devices to detect,

Table 5.13 Typical uses of NMR in polymer analysis and characterisation

Technique	Properties studied	Matrix
Frequency domain		
FTNMR	Chemical composition	Liquids, solids (polymers)
	Molecular structure	Liquids, solids (polymers)
MAS NMR	Molecular structure	Solids (polymers)
Time domain		
FID	Rubber content	Polymer blends
	Fat content	
Spin echo	Oil and moisture	Rubbers
	Texture	
T_1 and T_2	Free and bound water	Swelling
	Oil and moisture	Elastomers
PFG-NMR	Diffusion (rate and distance)	Emulsions
Spatial domain		
NMR imaging	Diffusion (distribution, mobility)	Polymers
		Food packaging materials
	Local motion (dynamic processes)	Polymers (mass transport)
	Heterogeneity (voids, fillers)	Compounds, interfaces Foams
	Cross-link density	Elastomers
	Physical ageing	Polymers, elastomers

process and display the NMR signals. Field strengths of modern commercially available magnets reach ca. 20 T (for comparison: the earth's magnetic field is a mere 6×10^{-5} T).

The NMR phenomenon is associated with absorption and emission of energy (in this case RF radiation), similarly to other forms of molecular spectroscopy, such as UV or IR spectroscopy. The energy associated with NMR is very weak in comparison with the other common types of spectroscopy (of the order of $0.1 \, \mathrm{J \, mol^{-1}}$). The difference between NMR and other forms of spectroscopy is that under appropriate conditions the absorption of energy by nuclei takes place only in the presence of a strong *magnetic field* at frequencies governed by the characteristics of the sample. There are four basic steps in an *NMR experiment*: (i) generation of magnetisation; (ii) resonant energy absorption and coherence creation; (iii) detection; and (iv) signal processing. When a sample is placed between the poles of a magnet the nuclei

contained within the sample become polarised such that there exists a net magnetisation aligned parallel to the magnetic field ($\mathbf{B}_0$). The magnetic moments (μ) do not become totally parallel with the external field but are allowed to adopt only some quantised orientations (e.g. $\theta = 54°44'$ for ^{1}H nuclei). In this situation a supplementary movement takes place: the magnetic moment precesses around the direction of $\mathbf{B}_0$ with an angular speed ω_0 (Figure 5.8). The magnitude of this magnetisation can be measured by applying a second magnetic field ($\mathbf{B}_1$) perpendicular to the first in the form of a pulse of RF radiation which rotates the magnetisation into the plane perpendicular to the field (a '90° pulse'). The magnetisation rotates in this plane and induces a signal in the receiver coil of the spectrometer. The energy required ($E_0 = h\nu_0$) is in the RF range, the associated frequency having values between 10 MHz (or even less) and 900 MHz. The photon energy is absorbed by a nucleus in the lower energy spin state ($m = +1/2$), and the nucleus is flipped into its higher energy spin state ($m = -1/2$) (Figure 5.9). The absorption of energy stimulating transitions between the two levels of energy in a magnetic field is called magnetic resonance. Thus, the resonance condition obtained by equalling E_0 to ΔE can be expressed either as:

$$\nu_0 = \gamma \mathbf{B}_0 / 2\pi \quad \text{or} \quad \omega_0 = \gamma \mathbf{B}_0 \qquad (5.4)$$

where the proportionality constant γ is the gyromagnetic ratio and the angular frequency $\omega_0 = 2\pi\nu_0$. When the rotational frequency of $\mathbf{B}_1$ is exactly equal to ν_0 the *resonance phenomenon* is observed, and energy is absorbed. The rotation frequency ω_0 is the absorption frequency of a particular type of nucleus in a given magnetic field ($\mathbf{B}_0$). The gyromagnetic ratio γ describes how much the spin state energies of a given nucleus vary with changes in the external magnetic field. As the nuclei lose the absorbed energy from the radiowave, they line up again. By measuring the specific radiofrequencies

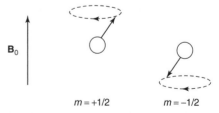

Figure 5.8 Precession of the magnetic moment in each of the two possible spin states of an $I = 1/2$ nucleus in an external magnetic field $\mathbf{B}_0$. After Macomber [160]. Reprinted from R.S. Macomber, *A Complete Introduction to Modern NMR Spectroscopy*, John Wiley & Sons, Inc., New York, NY, Copyright © (1998, John Wiley & Sons, Inc.). This material is used by permission of John Wiley & Sons, Inc.

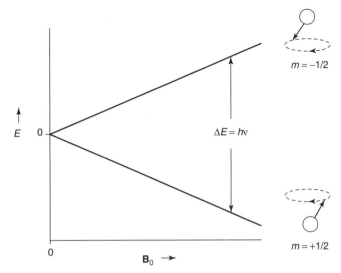

$m = -1/2$

E 0

$\Delta E = h\nu$

$m = +1/2$

0        $\mathbf{B_0}$ →

Figure 5.9   Relative energy of both spin states of an $I = 1/2$ nucleus as a function of the strength of the external magnetic field $\mathbf{B_0}$. After Macomber [160]. Reprinted from R.S. Macomber, *A Complete Introduction to Modern NMR Spectroscopy*, John Wiley & Sons, Inc., New York, NY, Copyright © (1998, John Wiley & Sons, Inc.). This material is used by permission of John Wiley & Sons, Inc.

that are emitted by the nuclei and the rate at which the realignment occurs detailed information is obtained about the molecular structure and motion of the sample.

In principle, one obtains an NMR spectrum from any chemical compound containing atoms whose nuclei have nonzero magnetic moments. A nucleus will have a nonzero nuclear spin quantum number I giving an NMR signal as long as the number of protons Z and neutrons A are not both even numbers. Almost every element has an *NMR active isotope* and can therefore be detected (although with different sensitivities). Isotopes such as ^{12}C, ^{16}O and ^{32}S have no magnetic moment and cannot therefore be studied by NMR. Most *common NMR nuclei* (^{1}H, ^{13}C, ^{19}F, ^{31}P) have the spin number (I) equal 1/2 and are named spin 1/2 nuclei. NMR spectra of practically all magnetically active isotopes have been recorded. Table 5.14 summarises those nuclei which are useful in studying vinyl stabilisers [161]. Other relevant nuclei for polymer/additive analysis are ^{23}Na and ^{29}Si.

Depending on how the secondary magnetic field is applied, there are two fundamentally different types of spectrometers, namely, continuous wave (CW) and pulse Fourier transform (PFT) spectrometers. The older *continuous wave NMR spectrometers* (the equivalent of dispersive spectrometry) were operated in one of two modes: (i) fixed magnetic field strength and frequency (ν_1) sweeping of $\mathbf{B_1}$ irradiation; or (ii) fixed irradiation frequency and variable field strength. In this way, when the resonance condition is reached for a particular type of nuclei ($\nu_1 = \nu_0$), the energy is absorbed and

Table 5.14   NMR nuclei in stabiliser analysis of vinyl polymers

Nucleus	Natural abundance (%)	Relative sensitivity	Ease of use
^{1}H	99.985	1.00	1.00
^{13}C	1.108	1.59×10^{-2}	1.76×10^{-4}
^{15}N	0.37	1.04×10^{-3}	3.85×10^{-6}
^{19}F	100.00	0.833	0.833
^{31}P	100.00	6.63×10^{-2}	6.63×10^{-2}
^{119}Sn	8.58	5.18×10^{-2}	4.45×10^{-3}
^{113}Cd	12.26	1.09×10^{-2}	1.34×10^{-3}
^{207}Pb	22.6	9.16×10^{-3}	2.07×10^{-3}

transitions take place. CW NMR spectrometers, which present many inconveniences and are insensitive, are no longer used.

In a *pulsed-mode NMR* experiment, which is performed at both constant magnetic field *and* constant RF frequency, RF radiation is supplied by a short intense computer-controlled pulse of RF current through the transmitter coil. This single-frequency pulse at an operating frequency ν_1 is characterised by a power (controlling the magnitude of $\mathbf{B_1}$) and a pulse width (t_p), with the duration of the pulse being measured in microseconds. The exact value of ν_1 is designed to be slightly offset from the range of nuclear precession frequencies to be examined (the spectral width). The signals collected represent the *difference* between the applied frequency ν_1 and the Larmor frequency ν_0 of each nucleus. The ultimate result of the pulse is a complex signal that is the superposition of the precession frequencies of *all* nuclei within the spectral width. The induced signal is monitored as a function of time. Such a pattern is called the *modulated free induction decay* (FID) signal (or *time-domain spectrum*). The decay is the result of spin–spin relaxation. After collecting data from one pulse, one must wait for the nuclei to relax to equilibrium. In a pulsed-mode NMR experiment the collection of FID data by a pulse data acquisition-delay sequence is repeated enough times to yield a time-averaged signal possessing the desired S/N ratio. From the FID data the frequency (Δv_i) and intensity of each component wave can be deduced by Fourier transformation. The Fourier transform of the FID is the NMR spectrum. The obtained *frequency-domain spectrum* generated in a pulsed-mode Fourier transform experiment contains all the same information as the spectrum obtained in a traditional continuous wave experiment, but in principle all the relevant spectroscopic information can be generated in just a few seconds from a single RF pulse (subject to S/N limitations) [158,162]. The resolution of an NMR spectrum obtained from an FTNMR spectrometer is limited by the amount of time spent recording the FID. The highest magnetic field commercially available is 900 MHz [163]. Increasing the magnetic field even further does not necessarily increase the resolution (due to chemical shift anisotropy). The type of NMR spectroscopy just described is termed *one-dimensional NMR* (1D NMR). Many ambiguities remain in 1D spectra such as overlapping resonances, the assignment of multiplets and the identification of connected transitions. FTNMR dominates in magnetic resonance spectrometry in the same way and for the same reasons as it does in the IR. However, the signal is generated in an entirely different fashion.

The main *NMR parameters* are: the gyromagnetic ratio γ; the nuclear shielding, σ, which describes the shielding of the nucleus from the applied magnetic field by the surrounding electrons and gives rise to chemical shifts; J, which relates to nuclear spin–spin coupling and depends upon relative nuclear orientations; and the times T_1 and T_2 which refer to the relaxation processes encountered by the nuclei excited in the NMR experiment. The *gyromagnetic ratio* γ is a physical property of the nuclei. Different chemical elements have different gyromagnetic ratios. These result in different energies and allow the study of different nuclei individually. On the basis of Equation (5.4) all nuclei with a given value of γ will produce a single absorption in the NMR spectrum. In such a case NMR spectroscopy would be of limited chemical interest only. In reality

the expression for the resonance condition needs to take into account that the magnetic field experienced by the resonating nuclei differs from $\mathbf{B}_0$ owing to shielding of the nucleus in a molecule by the surrounding electrons. More appropriately, the expression for the resonance condition is:

$$\omega_{\text{eff}} = \gamma \mathbf{B}_0 (1 - \sigma) \qquad (5.5)$$

where σ is the *nuclear shielding parameter*. Many NMR signals appear as multiplets, the structure of which arises from spin–spin coupling interactions with other nuclei in the molecule. The separation between adjacent members of a multiplet can give the value of J, the *spin coupling constant* (in Hz). In analogy to σ, the value of J depends upon the chemical environment of the nuclei concerned. Hence, values of J are of use in molecular structure determinations.

The difference in resonance NMR frequency of a chemically shielded nucleus measured relative to that of a suitable reference compound is termed *chemical shift* [164,165], and is a measure of the immediate electromagnetic environment of a nucleus. While the chemical shift depends on the $\mathbf{B}_0$ field, J does not. Chemical shifts, which cover a range of about 10 ppm for protons (i.e. 600 Hz in case of a 14.1 kG magnetic field) and 250 ppm for ^{13}C, are the substance of NMR.

Nuclear spins interact with a magnetic field (the chemical shift) and with each other (spin–spin coupling); NMR spectra are affected by chemical equilibria (chemical exchange) and molecular motion (spin-lattice and spin-spin relaxations). Several time-dependent phenomena affect the nuclei, such as *relaxation*, the relaxation related nuclear Overhauser effect and exchange processes. Resonance intensities are influenced by the 'relaxation times' T_1 and T_2, which are measures of time required by a particular nucleus to return to the equilibrium status after excitation by an RF field. Relaxation is a crucial process in the NMR technique. When immersed in a magnetic field, the spin states establish a new (Boltzmann) equilibrium distribution with a slight excess of nuclei in the lower energy state. The equilibration process is *not* infinitely fast. In fact, the rate at which the new equilibrium is established is governed by a quantity called the *spin-lattice (or longitudinal) relaxation time*, T_1. The exact relation involves exponential decay. T_1 governs the evolution of the longitudinal or z-magnetisation, M_z, toward the equilibrium value, M_0. The values of T_1 range broadly, depending on the particular type of nucleus, the location of the nucleus (atom) within a molecule, the size of the molecule, the physical state of the sample (solid or liquid), and temperature. For liquids or solutions, values of 10^{-2}–10^2 s are typical. For crystalline solids, T_1 values are much longer (minutes to days). A measurement of T_1 can provide an estimate of internuclear separation; T_1 is sensitive to motions of the chains.

There is a second relaxation process, called *spin-spin (or transverse) relaxation*, at a rate controlled by the spin–spin relaxation time T_2. It governs the evolution of the xy magnetisation toward its equilibrium value, which is zero. In the fluid state with fast motion and extreme narrowing T_1 and T_2 are equal; in the solid state with slow motion and full line broadening T_2 becomes much shorter than T_1. The so-called '180° pulse' which inverts the spin population present immediately prior to the pulse is important for the accurate determination of T_1 and the true T_2 value. The spin–spin relaxation time calculated from the experimental line widths is called T_2^*; the ideal NMR line shape is Lorentzian and its FWHH is controlled by T_2. Unlike chemical shifts and spin–spin coupling constants, relaxation times are not directly related to molecular structure, but depend on molecular mobility.

With the extremely small energy quanta ($\sim 10^{-6}$ eV) involved in NMR transitions the technique has inherently low *sensitivity*. Various measures can be taken for optimisation of the sensitivity, including: (i) use of Fourier techniques; (ii) high-field spectroscopy (sensitivity proportional to $\gamma^{5/2} B_0^{3/2}$; B_0 for 1-NMR 900 MHz; (iii) use of large sample volumes; (iv) cooling of the sample; and (v) enhancement of the magnetisation by the nuclear Overhauser effect, cross-polarisation in solids and liquids. Exciting all the nuclei of interest simultaneously, and collecting all of the signals simultaneously leads to higher sensitivity in pulsed FTNMR spectroscopy. FTNMR has made it feasible to obtain spectra for low sensitivity/low abundance nuclei such as ^{13}C. By far the simplest procedure for enhancing the polarisation of low-γ spins is the application of a saturating RF field to the high-γ spins. This process

of polarisation transfer and enhancement is called the *nuclear Overhauser effect* (NOE) [166,167]. NOE is inextricably associated with the dipole–dipole relaxation mechanism, occurs only during broadband decoupling and leads to enhancements in resonance intensities for the signals of the through-space neighbouring groups. The NOE in solution and dipolar splittings in solids yield inter- and intramolecular distances (0.1–1 nm).

Cross-polarisation (CP) in the rotating frame has been introduced as a means of transferring polarisation between different nuclear species in solids [168], and has become of central importance for obtaining spectra of rare spins with low gyromagnetic ratios such as ^{13}C, since a significant sensitivity enhancement may be achieved. Cross-polarisation can be used either for direct observation of low-sensitivity nuclei or for indirect detection of such nuclear species via high-sensitivity nuclei such as protons [169].

NMR spectra allow measurement of chemical shifts, signal intensities, coupling constants, line-widths and relaxation times. The corresponding chemical information is given in Table 5.15. Another important property of the magnetic active nuclei is their interaction through space. Through-space dipolar coupling is an effect of the local magnetic field generated by the neighbouring magnetic nuclei at the position of the observed nucleus. In solution, the splitting pattern of the dipolar coupling cannot be seen, but the NOE can be observed [171].

A wide variety of 1D and nD NMR techniques are available. In many applications of 1D NMR spectroscopy, the modification of the spin Hamiltonian plays an essential role. Standard techniques are double resonance for spin decoupling, multipulse techniques, pulsed-field gradients, selective pulsing, sample spinning, etc. Manipulation of the Hamiltonian requires an external perturbation of the system, which may either be time-independent or time-dependent. Time-independent

Table 5.15 Basic information in an NMR spectrum

Feature	NMR parameter	Symbol (units)	Chemical information
Peak position	Chemical shift	δ (ppm) (absolute Hz)	Functional group identification
Peak multiplicity	Coupling constant	J (Hz)	Number and type of nuclei within three bonds of the nucleus being detected
Peak intensity	Integrated area		Count of nuclei of given type
Line shape and line width	Spin-spin relaxation time	T_2 (s)	Lifetime of spin state determined by dynamic processes and local magnetic environment
Time dependence of intensity	Spin-lattice relaxation time and nuclear Overhauser effect (NOE)	T_1 (s), NOE	Molecular dynamics and magnetic interactions

perturbations are changes of temperature, pressure, solvents, or static magnetic field. Time-dependent perturbations are mechanical sample spinning and stationary or pulsed RF fields. Radiofrequency fields can be applied continuously, in the form of a periodic pulse train, or as aperiodic pulse sequences. Periodic multiple-pulse sequences are extremely versatile for the suppression of selected interactions, e.g. solvent (water) suppression. Depending on the information desired, experiments can be designed to measure, remove, or correlate one or more of the principal spin interactions, including the chemical shift, spin–spin scalar (J) coupling, electric quadrupole coupling, and magnetic dipole–dipole coupling, each of which depends on molecular orientation. Often one of the chief aims of such intervention is to enhance spectral resolution.

Decoupling is achieved by applying a second RF field at the resonance of the nucleus whose coupling is to be removed. The decoupling may be selective, involving a single group or even a single line within a group, or may be broad-banded and cover the full spectrum. Both homo- (same type of isotope being irradiated and observed) and heteronuclear decoupling are usual. Decoupling leads to a simplification of the spectrum and facilitates signal assignment.

Highly sophisticated pulse sequences have been developed for the extraction of the desired information from 1D and multidimensional NMR spectra [172]. The same techniques can be used for high-resolution l-NMR, s-NMR and NQR. *Pulse experiments* are commonly used for the measurement of relaxation times [173], for the study of diffusion processes [174] and for the investigation of chemical reactions [175]. Davies *et al.* [176] have described naming and proposed reporting of common NMR pulse sequences (IUPAC task group). An overview of pulse sequence experiments has been given [177].

The main applications of high-resolution NMR are based on studying individual signals in the frequency domain (the common NMR spectrum). Success of high-resolution NMR techniques was ensured originally by its extreme power in solving the structure of pure compounds. Relaxation measurements for individual signals (thus combining the study of the time and frequency domains) provide valuable information both in terms of structural assignments and mobility of various parts of the molecule. NMR relaxation is a major tool for the determination of *molecular dynamics* on the μs to ps time scale. Dynamic information that is not in any way accessible from other techniques is nowadays conveniently available from either NMR spin relaxation data

or from NMR-based multicomponent studies of molecular self-diffusion. The entire spectrum of molecular motions can be probed, including molecular rotation, translational self-diffusion, 'coherent' rotational motion, and the internal motion in nonrigid molecules. In order to quantify mobility, a parameter named *correlation time* is commonly used. Correlation time represents the time that a particular site (atom) in a molecule stays in the same position. Relaxation times are also used in the measurement and evaluation of the images obtained using *nuclear magnetic resonance imaging* (NMRI) methods.

Samples can be investigated by NMR spectroscopy in all states: gaseous, liquid, dispersions in liquid, gels, and solids. A compilation of fully assigned ^{13}C and ^{31}P *l-NMR spectra* of some 300 polymers and 54 polymer additives (mainly esters of (un)saturated acids, aromatic di- or triacid esters and phosphites and phosphates, used as plasticisers and AOs) is available [178]. For other spectral collections comprising monomers, catalysts, curing agents and other additives, see references [179–182]. In general, NMR spectroscopy does not rely on spectral matches; use of multipulse techniques and multidimensional NMR experiments is helpful in unambiguously elucidating the structure of an unknown analyte. Interpretation of NMR spectra relies on the concept of chemical shift equivalence, an understanding of which depends on stereochemical concepts. A key reference is available [183]

NMR provides one of the most powerful techniques for *identification* of unknown compounds based on high-resolution proton spectra (chemical shift: type; integration: relative numbers) or ^{13}C information (number of nonequivalent carbon atoms; types of carbon; number of protons at each C atom). *Structural information* may be obtained in subsequent steps from chemical shifts in single-pulse NMR experiments, homo- and heteronuclear spin–spin connectivities and corresponding coupling constants, from relaxation data such as NOEs, T_1s T_2s, or from even more sophisticated 2D techniques. In most cases the presence of a NOE enhancement is all that is required to establish the stereochemistry at a particular centre [167]. For a proper description of the microstructure of a macromolecule NMR spectroscopy has now overtaken IR spectroscopy as the analytical tool in general use.

One of the fastest growing areas in NMR over the past decade has been the use of *'pulsed field gradients'*, or PFG-NMR, for selective 1D and 2D experiments. The basic pulsed gradient spin-echo (PGSE) experiment [174] relies on the use of pulsed linear magnetic field gradients (of amplitude g, duration δ and separation Δ) that are applied during a *spin-echo experiment* [184], involving two or more RF pulses (in the simplest case separated in time by τ). Whereas NMR can be used to probe the molecular structure of a polymer, NMRI to obtain quantitative penetrant profiles in a polymer, PFG-NMR is used to measure the self-diffusivity of the penetrant within the polymer. This combination of techniques gives insights into how polymer structures restrict or promote *additive migration*.

Table 5.16 shows the *main features* of NMR. Although NMR spectroscopy is not as sensitive as other spectroscopic methods (for ^1H NMR concentrations $> 100\,\mu M$), the detail and lack of ambiguity of the information makes it the most effective tool for structural identification and elucidation in the liquid state. ^1H NMR is a technique that is capable of measuring a wide range of different organic compounds in a single sample without the need for any complex sample preparation. Furthermore, ^1H NMR provides a high resolution analysis, it is fast and quantitatively reliable. Depending on the strength of the magnetic field used it is possible to measure compounds down to a level of about 1 ppm. The strength of NMR spectroscopy lies in the application of multidimensional NMR experiments to an analytical problem. The superior spectral resolution of 1-NMR permits the study of large molecules with complicated structures. Liquid NMR allows the analysis of the microstructure of polymers (composition, tacticity, end-groups, chain branching, etc.). Because magnets

Table 5.16 Main characteristics of high-resolution pulse Fourier-transform 1-NMR spectroscopy

Advantages

- Multinuclear detectability
- High spectral resolution
- High specificity
- Ease of manipulation of nuclear spin Hamiltonians (spectral simplification)
- Supreme structural tool (conformation)
- Determination of molecular dynamics
- Reliable quantification without calibration (primary technique)
- Hyphenation
- Multidimensionality
- Mature technology

Disadvantages

- Sensitivity (magnetic field strength dependent)
- Fairly long data acquisition times (for some nuclei)
- Highly complex information
- Solvent dependency (limited to analytes soluble in NMR solvents)
- Deuterated, inert solvent system
- No separation involved
- Expensive equipment
- Laboratory-based technique
- Need for skilled operator

are sensitive to the environment, high-resolution NMR is a laboratory-based technique. A major drawback is that the information obtained from 1D NMR data is highly complex. In fact, no separation is involved; therefore, it is sometimes necessary to isolate an unknown analyte from the matrix or other co-additives prior to analysis. For several *hyphenated NMR spectroscopy techniques*, such as SFE-NMR, SFC-NMR and HPLC-NMR, see Chapter 7. Complex mixtures may also be unravelled by means of multidimensional NMR or PFG-NMR. On the other hand, also multivariate analysis applied to NMR spectra significantly lowers the information overload [185].

Chemical shifts (given in ppm) are always determined relative to the signal of a known standard, namely tetramethylsilane, $(CH_3)_4Si$ or TMS in ^1H NMR, which is set to 0 ppm. The chemical shift dispersion in *proton NMR* is small (ca. 10 ppm) and the resonance lines cannot always be unambiguously assigned. The sensitivity of 500 MHz ^1H NMR experiments is some 400-fold greater than ^{13}C NMR experiments. Stronger magnetic fields offer improved sensitivity, greater spread of chemical shifts making assignment easier for complicated molecules, and stronger interproton Overhauser enhancements. Quantitative evaluations are most readily obtained by ^1H NMR spectroscopy.

While ^1H NMR is an important tool, which requires some $10-100\,\mu g$, it yields organic structural information only indirectly (viewed through the hydrogen nuclei). In principle, the ^{13}C *nucleus* is the most informative probe for organic structure determination by means of FTNMR. The special advantages, which make ^{13}C NMR an attractive alternative to ^1H NMR for the solution of analytical problems, include:

- A broad chemical shift range of 250 ppm (i.e. about 20 times greater than that of ^1H NMR), which greatly facilitates resolution of individual resonances in complex mixtures.
- Simpler peak assignment than for ^1H NMR.
- ^{13}C labelling.
- Applications for liquid and solid samples.
- Signal enhancement in liquids (NOE) and solids (H–C cross-polarisation).
- Direct H–C dipole–dipole coupling dominating ^{13}C relaxation (for protonated carbons); information about molecular shapes and motion are contained in the experimental spin-lattice relaxation times.

A disadvantage in the utilisation of ^{13}C NMR is the intrinsic low relative sensitivity (Table 5.14). However, higher magnetic fields and better probe design have significantly increased the sensitivity for ^{13}C nuclei. Practical sample sizes are 5 mg/0.6 mL.

One-dimensional ^{1}H and ^{13}C NMR experiments usually provide sufficient information for the assignment and identification of additives. Multidimensional NMR techniques and other multipulse techniques (e.g. distortionless enhancement of polarisation transfer, DEPT) can be used, mainly to analyse complicated structures [186].

Another nucleus that is very amenable to NMR analysis of additives is ^{31}P. For this nucleus the major nuclear properties that enhance sensitivity in NMR spectroscopy are all highly favourable: nuclear spin ($I = 1/2$), natural abundance (essentially 100 %), and resonant frequency (higher than ^{13}C). Consequently, *^{31}P NMR* experiments are generally much easier to perform than ^{13}C NMR experiments, which are routine in modern NMR laboratories. Since the advent of multinuclear high-field PFT instruments in the late 1970s, ^{31}P NMR has been in widespread use. All that is usually necessary to perform ^{31}P NMR is to tune the appropriate NMR probe to the phosphorous frequency and carry out the necessary calibrations. Another advantage of ^{31}P NMR is its wide range of observed chemical shifts (~400 ppm), which are very dependent on oxidation state. This wide separation allows for the unambiguous identification of additives and additive degradation products. The use of homo- and heteronuclear coupling constants and multinuclear (e.g. ^{1}H, ^{13}C, ^{31}P) NMR data permit full determination of the structure of mono- and polyphosphorous compounds. ^{31}P NMR is also widely used in a quantitative fashion. For further reference, see Gorenstein [187].

Spectral width, dynamic range, resolution and sensitivity are expected to be pushed toward further limits. An emerging advancement in NMR spectroscopy is the DOSY technique (Section 5.4.1.1) which offers a separation capability as a function of the rates of steady state diffusion of molecules in solution.

The ASTM designation E 386-90 'Standard Practice for Data Presentation Relating to High-resolution Nuclear Magnetic Resonance (NMR) Spectroscopy' is the valid regulation procedure in NMR.

Basic principles of modern NMR spectroscopy are the subject of many textbooks [167,188–196], including pulse techniques [197]; for NMR of polymers, see Bodor [198]. A guide to multinuclear magnetic resonance is also available [199]. Several texts deal specifically with multidimensional NMR spectroscopy [169,197,200–202]. Ernst *et al.* [169] have reviewed the study of dynamic processes, such as chemical exchange

and cross-relaxation. A review on NMR in industry has also appeared [203].

Applications High-resolution NMR spectroscopy is relatively insensitive and fairly concentrated solutions (~10^{-1} M) must be used, but its great diagnostic character has made it the ideal spectroscopic technique for structure determinations, especially in organic systems as well as polymers. NMR is especially a valuable tool in conjunction with MS and IR as the mass spectrum indicates the groups of atoms present, IR describes functional groups and NMR reveals their interconnections. NMR is more important as a tool for fundamental research than for troubleshooting. Typically, NMR can be employed for the study of the mechanisms of oxidation and photodegradation of polyolefins [204]. Use of complementary methods (NMR, IR, MS) has proven to be most useful in the identification of additive transformation products and in that context are an integral part in the characterisation and registration of new additives. Computer-assisted identification of additive mixtures in polymers by combining MS, ^{1}H NMR, ^{31}P NMR and LC-MS data has been reported [205]. Industrial NMR analysis of polymers mainly aims at quantification; structural identification is required less frequently. In the past, the NMR technique has not been used in polymer/additive deformulation to the same extent as UV and FTIR spectroscopy (Table 5.1), not in the last place due to its lower sensitivity, occasionally long measurement times, high sample weights and overlapping of signals corresponding to characteristic groups in the combined spectrum. Many of these drawbacks are now overcome.

NMR finds major application in the determination of molecular structure (reaction) dynamics and on-line monitoring of manufacturing processes. Samples may be gases, liquids or solids, including polymers. Principle component analysis (PCA) provides a useful way of reducing the data size without considerably reducing the amount of information that can be derived from NMR data. Discriminant analysis can be applied to the compressed data to search through all of the principle components scores and classify the samples in groups. Once the group structure has been established, unknowns can be identified based solely on variations in their NMR spectrum. The information contained within the NMR spectrum is sufficient to identify the manufacturer.

Table 5.17 shows the use of NMR in polymer structural and dynamical studies. ^{13}C 1-NMR is successfully applied in detailed characterisation of homo- and

Table 5.17 NMR in structural and dynamical studies of polymers

Feature	Length scale	NMR method
Chemical groups	0.1–1 nm	1-NMR, MAS NMR
Inter-/intramolecular distances	0.1–1 nm	Nuclear Overhauser effect (l-NMR), dipolar splittings (s-NMR)
Orientations (fibres, films)	0.2–2 nm	Anisotropic interactions (s-NMR)
Torsion angles in chain	0.3–0.9 nm	Conformation dependent chemical shifts (s-NMR)
Configurational statistics	0.3–1.5 nm	l-NMR
Molecular mobility (dynamics)	0.5–500 nm	Relaxation times and line-shapes; multidimensional exchange experiments
Spatially resolved spectroscopy	20 μm–1 cm	Imaging (NMRI)

After Schmidt-Rohr [206]. Reprinted with permission from K. Schmidt-Rohr, *ACS Symposium Series*, **598**, 184–190 (1995). Copyright (1995) American Chemical Society.

copolymers: analysis of microstructure at a molecular level, comonomer composition, sequence distribution, functional and end-groups, tacticity, chain irregularities, polymer modifications (grafting, cross-linking), etc.

Suitable *solvents* are to be used to extract polymeric materials for direct 1-NMR measurements. The ideal NMR solvent should contain no protons and be inert, low boiling, and inexpensive. The properties of common deuterated solvents are listed elsewhere [207]. An improper solvent choice may lead to selective dissolution of the additives [208]. Generally accepted solvents for recording NMR spectra of polyolefins are 1,2,4-trichlorobenzene (TCB) and tetrachloroethane (TCE), often in combination with TCE-d_2, ODCB-d_4 or benzene-d_6 as a locking agent (for correction of instrumental drift). While TCE is thermolytically but not photolytically stable (HCl loss), benzene is volatile at higher temperatures. For rapid screening polyolefins may be contacted with CDCl$_3$ and the soluble fraction (low-MW additives) determined by 1-NMR. For more thorough work, a total extract in a non-NMR solvent is usually first evaporated to dryness and the ^{1}H NMR spectrum is then recorded in a NMR solvent.

^{1}H NMR spectra can be used in several ways, from a simple fingerprint of the potential migrants to an identification procedure. A great variety of 1-NMR analyses of additives extracted from polyolefins (and other polymers) has been reported. NMR suffers from the same limitations inherent to the extraction procedure as do GC and HPLC. However, for NMR studies extraction of the additives from the polymer is not a must (Section 9.2.1). In case of extruded polymer samples, such as granulate, it is best to sample several pellets in preparing the NMR sample so as to minimise sample to sample variations in additive concentration.

Successful additive analysis on *polymer extracts* by means of 1-NMR generally means:

(i) ^{1}H, ^{13}C or ^{31}P NMR, occasionally also ^{19}F or ^{119}Sn;

(ii) additives soluble in the NMR solvent (thus excluding metal soaps, such as calcium stearate, hydrotalcite, etc.);

(iii) no interaction of NMR solvent with additives;

(iv) separation of the signals of interest from those of the NMR solvent and the molecular oligomer fraction;

(v) room temperature measurements as opposed to elevated temperatures for polymer/additives in dissolution (cf. Section 9.2.1);

(vi) averaging over a large number of scans, especially for extracts with a complex spectrum; and

(vii) quantification referring to the polymer concentration.

Shift reagents can be used for separation of overlapping NMR signals [209]. For example, Rao *et al.* [46] used Eu(thd)$_3$ for the separation of signals in 1-NMR spectra of extracts of PP and EPM consisting of AO and UVA mixtures (BHT, Irganox 1010/1076, Tinuvin 327).

The following *polymer/additive related problems* have been subject of 1-NMR investigations:

(i) identification (fingerprinting) and quantification of extracted additives, with emphasis on antioxidants, antistatics, flame retardants, stabilisers, plasticisers, surfactants, vulcanisation accelerators, including oils from rubbers (often in a multidisciplinary fashion);

(ii) characterisation of conversion and degradation products;

(iii) discrimination of additives;

(iv) mixture analysis; and

(v) self-diffusion studies (using ^{1}H NMR pulsed gradient spin-echo, PGSE).

For detection of compounds in complex mixtures 2D methods are often needed. However, application of 2D correlation NMR experiments and advanced field gradient techniques is still fairly limited and awaits the possibility of quantitative evaluation.

NMR spectroscopy is most effective in qualitative analysis when the samples examined are substantially *pure compounds* and has been used to confirm the theoretically predicted low-energy conformations of the *N*-acylated hindered amine light stabiliser Tinuvin 440 [210]. Trace amounts of PDMS (quantification limit 0.1 ppm) in plastic additives, dyes and pigments were determined by ^{1}H NMR after Soxhlet extraction [211]. ^{1}H NMR was also used for the detection of octadecanol, an impurity in Irganox PS 802 (3,3′-dioctadecyl thiodipropionate). NMR has identified the nature of a supposedly UV stabiliser of empirical formula $C_{17}H_{18}N_3ClO$ [44] (Scheme 5.2).

Freitag and Lind [212] have used ^{1}H NMR for the quantitative determination of BHT, DLTDP, Irganox 1010/1076 in PP extracts, and of Tinuvin 622 in PE extracts (obtained by dissolution/precipitation). The method is useful for determination of complex *stabiliser mixtures* in a single analysis run (observed relative error 5–10 %) and also suitable for analysis of oligomeric or polymeric stabilisers (e.g. Tinuvin 622), which are difficult to determine otherwise. ^{1}H NMR was recently also used to characterise a series of new high-MW antioxidants synthesised by isocyanation of a functional hindered phenol [115]. Unstable intermediates in the oxidation of Irganox 1076 by means of $K_3Fe(CN)_6$ were monitored by using ^{1}H 1-NMR and ESR methods [213].

More problems are encountered with *plasticisers* because most extracts from polymer compositions are mixtures and, when separated by TLC, the amount of the individual fractions is often too small for convenient examination by *^{1}H 1-NMR spectroscopy*. Moreover, the original plasticisers themselves are often mixtures. For example, tricresyl (tritolyl)phosphate is based on mixed cresols, while most of the higher phthalate esters are based on complex mixtures of alcohols.

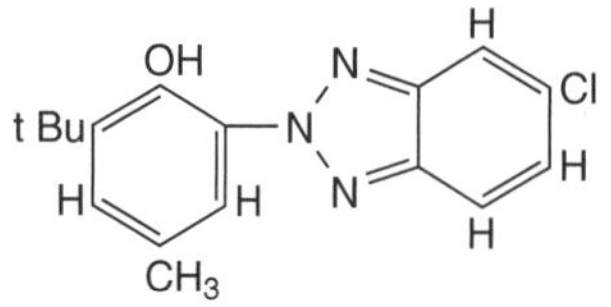

Nevertheless, NMR examination of plasticisers can often be rewarding, as pointed out by Haslam *et al.* [44]. Thus, in the case of phthalate plasticisers, the lower alkyl phthalates are substantially pure compounds which give excellent ^{1}H NMR spectra. Although it may not be possible to establish the precise chemical nature of the higher alkyl phthalates, the average chain length of the alkyl group can be deduced from the ratio of the areas of the resonances from aryl and alkyl protons, Also, the more complex phthalates (e.g. alkyl phthalyl, alkyl glycollates) give useful spectra [44,214]. NMR examination is particularly well suited for polyester plasticisers, which may often be recovered from a composition in a comparatively clean state by hot methanol extraction. Udris [119] has reported multidisciplinary (^{1}H NMR and IR) identification of sulfur-containing dialkyl stabilisers in PVC. PVC additives are accessible to analysis both by ^{1}H NMR spectroscopy [215] and ^{13}C NMR spectroscopy [216].

The compositions of copolymers of styrene, methyl methacrylate, acrylonitrile and acrylamide with diethyl vinyl phosphonate (S-DEVP, MMA-DEVP, AN-DEVP and AM-DEVP), with incorporated FR functionality, were analysed by means of ^{1}H 1-NMR in CDCl$_3$, DMSO-d$_6$ and D$_2$O [217].

^{1}H NMR is also a favoured analysis tool for Vitamin E, avoiding degradation during analysis. The nature of specific *transformation products* based on isolated isomeric forms of oxidative coupling products, aldehydes and a quinone, formed when vitamin E (*dl-α*-tocopherol) is incorporated by PE and PP during melt processing, was investigated by various techniques (HPLC, FTIR, UV, ^{1}H and ^{13}C NMR, and MS) [218]. ^{1}H NMR has also been applied to discriminate between various (unreacted) carbodiimides (low molecular Stabaxol I, and the oligomers Stabaxol P and P100), used as *hydrolysis stabilisers* in polyesters [58].

The NMR method offers various advantages for *quality tests*: (i) no special requirement for the samples: swollen gels, solid materials, emulsions, suspensions, granules and powders can be analysed; (ii) mineral fillers do not disturb; (iii) relatively fast analysis; (iv) measurement of volume average properties; and (v) no need for chemicals. Although the favoured analytical approach for identification of additives in food relies heavily on chromatographic techniques, in particular HPLC, ^{1}H NMR is equally a powerful tool for quality control of most food packaging plastics, especially in an industrial framework [219]. ^{1}H NMR fingerprinting was used for compliance testing and safety control of food packaging materials (polyolefins) by means

Scheme 5.2 Structural formula of 2-(3′-*t*-butyl-2′-hydroxy-5′–methylphenyl)-5-chlorobenzotriazole

of extract analysis. Feigenbaum *et al.* [220,221] have proposed a classification of EEC additives [222] on the basis of the NMR chemical shift, the reasoning being that the chemical shift of a given proton is an intrinsic property of the substance for a given solvent, concentration and temperature. Identical spectra can be assured months or years later with the same, or a similar, instrument. This is more reliable than retention times in chromatographic methods, which are strongly dependent on the age of the column and analytical conditions. The ^{1}H NMR fingerprint could thus allow batch-to-batch control, and spot such things as replacement of Irganox 1010 for Irganox 1076. This type of fingerprint is more sensitive than other approaches, such as IR spectroscopy. Taking into account that ^{1}H NMR can be very useful for identification of functional groups and structural units of extracted additives, it is not surprising that a classification of *food packaging additives* on the basis of similarities in functional groups and their corresponding chemical shift ranges has been proposed (Table 5.18) [220]. Such an approach may be applied for daily control of the quality of packaging materials in the food industries. It could equally well be applied to determine the total (or group) specific migration limits, which were recently introduced by the EEC. An example of the usefulness of this classification method is the case of phthalic or adipic esters, used as plasticisers. These complex products, synthesised from mixtures of oxo alcohols, have different chain lengths and branching. Such mixtures typically give very complex chromatograms, but relatively simple ^{1}H NMR spectra (class 2 of Table 5.18) and the structure of polymeric plasticisers was recently elucidated by this technique [223].

Métois *et al.* [53] have used ^{1}H NMR for compliance testing of polyolefin food packaging materials (HDPE and PP film) with EU regulations developing a quick analysis of two functional classes of plastic additives (aromatic antioxidants and antistatic agents). Identification of individual additives is facilitated if they belong to more than one functional group class; reference may be made to a compilation of ^{1}H NMR spectra of food packaging additives [224,225]. The spectra can be used to identify additives either pure, or as constituents in polymer extracts.

^{1}H NMR has frequently been used for the characterisation of complex polyetheresters both in relation to chemical composition (PBT: PTHF: EO/PO ratio) and additive concentration. Using calibration curves *stabilisers* such as Irganox 1330, Irganox 1098 (thermolabile) and Naugard 445 could be determined with ^{1}H 1-NMR in $C_2D_2Cl_4$ [226]. Antioxidants in a wide variety of

Table 5.18 Main functional classes from the EEC inventory of additives intended for use in Food Contact Plastics Materials (Commission of the European Communities, 1991) [222]

Functional additive classes	^{1}H NMR	Examples
1. Fatty saturated acids	1–2.5 ppm	Stearamide
2. Esters	4–5.3 ppm	Bis(2-ethylhexyl) adipate
		Octadecyl 3-(3,5-di-*tert*-butyl-4-hydroxyphenyl) propionate
3. Alkenes	5–6 ppm	Erucamide
4. Substituted aromatics	7–8.5 ppm	Tris(2,4-di-*tert*-butylphenyl) phosphite
Phenolic antioxidants		Pentaerythritol tetrakis(3-(3,5-di-*tert*-butyl-4-hydroxyphenyl)) propionate
5. Amines, diamines, amides	~2.5 ppm	*N*, *N*-bis(2-hydroxyethyl)-*N*-alkylamine

After Ehret-Henry *et al.* [220]. From J. Ehret-Henry *et al.*, *Food Addit. Contam.*, **9**, 303–314 (1992). Reproduced by permission of Taylor & Francis Ltd (http://www.tandf.co.uk/journals).

rubber vulcanisates were also identified by means of ^{1}H NMR [208]; for completely unknown AO a combination of NMR, MS and IR is required for unambiguous identification.

For forensic purposes 1-NMR spectra of extracts of latex condoms are useful in providing a fingerprint of the additives (positive identification of 15 out of 38 condoms on this basis) [227]. s-NMR is only useful for determining the polymer backbone. Similar forensic analyses for the same purpose have been reported using DCI-MS and FTIR [228]. *Chain-stoppers* (e.g. *t*-butylphenol) in solution process PC have been identified by 1-NMR [229]. 1-NMR is also useful for *troubleshooting* activities. The soluble fraction of plate-out on different tool parts during PVC extrusion, consisting of Irgastab CH-300/BC-29, Irganox 370 and Reoplast 39, was quantitatively determined by ^{1}H NMR [215].

^{13}C 1-NMR spectroscopy is the method of choice for determining the molecular structure of polymers in solution [230]. Polyolefin ^{13}C NMR is mainly quantitative 1D 1-NMR; multiple pulse techniques are used for spectral interpretation. The resolution obtained in ^{13}C NMR spectra of LDPE is an order of magnitude larger than in the corresponding ^{1}H-NMR spectra

(even for 400–500 MHz). Various sophisticated pulse sequences can be used for identifying carbon types in ^{13}C NMR spectra. ^{13}C NMR of LDPE can be used for structural identification (SCB, LCB), detection of chain transfer agents and various unsaturations. LLDPE copolymers are also analysed mainly by ^{13}C NMR [230]; ^{1}H-NMR is used here to detect the different unsaturation species. On the other hand, use of ^{13}C 1-NMR for polymer/additive characterisation is rather limited. Detection of additives in LDPE (e.g. primary and secondary antioxidants and processing aids) is usually carried out via ^{1}H NMR. Detection of additives in HDPE/UHMWPE can also be performed by ^{1}H 1-NMR, provided that the additives are soluble in the NMR solvent. ^{13}C 1-NMR was used for the analysis of ethoxylated amides in HDPE polymer extracts [231]. ^{13}C NMR is applied also for the measurement of alkyl and olefinic end groups and type and amount of comonomer. ^{13}C NMR has equally shown its usefulness in the characterisation of elastomeric components of filled vulcanisates. Werstler [232] has described identification and quantitative analysis of numerous filled, elastomeric vulcanisates by ^{13}C NMR. Before spectral analysis, the samples were extracted with acetone to remove oils and other additives and then solubilised in ODCB, but without removal of the filler.

In principle, ^{13}C 1-NMR is a more suitable technique than ^{1}H NMR for identification and characterisation of extracted *flame retardants* (FRs) as many FRs are partially or totally brominated. However, the solubility of many bromine-containing FRs is often insufficient for ^{13}C NMR experiments in common solvents, such as $CDCl_3$ and tetrachloroethane (TCE), and therefore ^{13}C s-NMR is frequently called in.

A few other successful ^{13}C 1-NMR determinations should be mentioned. Hunt *et al.* [28] used ^{13}C NMR to characterise fractions of extracted analytes of PAG and sorbitan ester samples and identified Irganox 1010. ^{1}H and ^{13}C NMR have been used to identify the main organic components of a breathable diaper backsheet as LLDPE and pentaerythritol tetra-octyl ester (PETO) [233]. The equally present AOs Irganox 1010 and Irgafos 168 were not detected without extraction. Barendswaard *et al.* [234] have reported fully assigned ^{13}C solution spectra of these two antioxidants. Chimassorb 944 in a polyamide matrix can be determined by ^{1}H or ^{13}C 1-NMR using solvents such as formic acid, trifluoroacetic acid or trifluoroethanol [235]. Both ^{1}H and ^{13}C NMR have been used to follow the chemistry of a bis-phenoxidemethylaluminum complex (reaction product of BHT and trimethylaluminum) by exposure in air. Pierre and van Bree [216] also used ^{13}C NMR to

identify epoxidised soybean oil and chlorinated paraffins in PVC extracts. ^{13}C NMR is a good tool to identify alkyl groups of zinc dialkyl dithiophosphates (ZDTP) in *lubricating oils* [236]. Cook and Lehrle [237] have used FTIR and ^{13}C NMR to characterise viscosity index improvers (alkylmethacrylate components) in engine lubricants. Both techniques show the presence of a hydrocarbon side-chain and carbonyl groups. However, neither technique is able to identify and quantitatively analyse the individual monomers, or groups, present in the additives. Methyl silicone, even in small amounts, can readily be determined by ^{13}C NMR. 1D ^{1}H and ^{13}C NMR experiments usually provide sufficient information for the assignment and identification of anionic *surfactants*; phosphate esters of alkylphenylethoxylates can be studied [238]. The assignment of fatty acid salts by means of NMR spectroscopy relies mainly on the resonance frequency of the ^{13}C nucleus of the carbonyl moiety.

The high sensitivity of ^{31}P *NMR spectroscopy* has allowed its application to the study of phosphorous in polymer samples, such as PE [239] and polyester [240–243], containing commercially used levels of phosphorous-based *antioxidants*. In polyolefins, these additive levels vary typically from 300 to 3000 ppm, corresponding to only 20–50 ppm of phosphorous. In particular, ^{31}P NMR spectroscopy has found wide application for the study of the degradation (oxidation, decomposition, hydrolysis) of stabilisers and the identification of their conversion products. Informative spectra can be obtained even at low levels of phosphorous (<1 ppm). In fact, ^{31}P NMR spectroscopy is an ideal method for determining the extent of conversion of phosphite to phosphate because of the high degree of specificity of this technique [216,244–246]. Allen *et al.* [244] successfully used ^{31}P NMR to study the decomposition of Irgafos 168 in PP during γ-irradiation, while Klender *et al.* [245] examined hydrolysis of organophosphorous stabilisers. Also Kenion *et al.* [247] reported a ^{31}P NMR study of the degradation of phosphite antioxidants. Scheirs *et al.* [150] have applied ^{31}P NMR and preparative HPLC-NMR for the characterisation of conversion products formed during degradation of processing antioxidants (Irgafos 168 and Irganox 1076). The ^{31}P NMR resonances of tris(2,4-di-*t*-butylphenyl)phosphite (Irgafos 168), its phosphate and phosphonate derivatives are distinct. 'Pure' Irgafos 168 often contains some phosphate impurities (Scheme 5.3).

^{31}P 1-NMR was also used to follow oxidation of two phosphites to mono- and diphosphates in PP as a function of temperature and number of extrusion passes [248]. The technique has indicated that the antioxidant nonylphenol

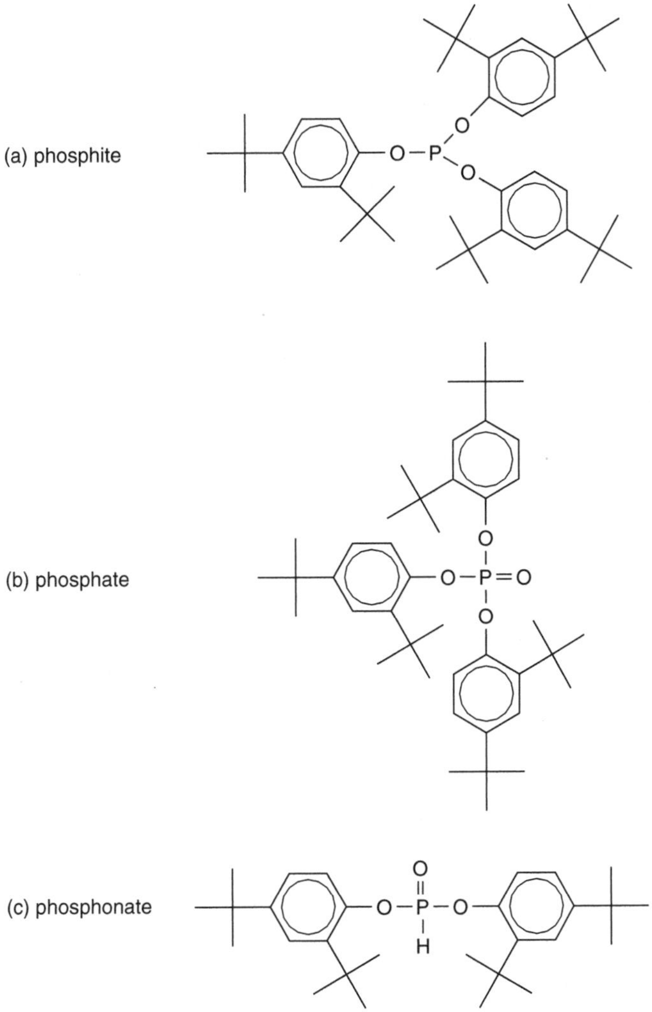

Scheme 5.3 Structures of (a) Irgafos 168, (b) Irgafos 168 phosphate and (c) Irgafos 168 phosphonate (hydrolysis product). After Scheirs *et al.* [150]. Reprinted with permission from J. Scheirs *et al.*, *Advances in Chemistry Series*, **249**, 359–374 (1996). Copyright (1996) American Chemical Society

phosphite had decomposed into various phosphorous-containing degradation products during extrusion of ABS [216]. Phosphonates, *plasticisers* to PVC and polyvinylbutyral, may equally be studied by means of ^{31}P NMR. It is not uncommon to find a dedicated ^{31}P NMR spectrometer being applied for *flame retardant* development. The use of ^{31}P NMR spectroscopy is limited to phosphorous-containing *surfactants*. As the ^{31}P l-NMR chemical shift is insufficient to completely assign the phosphorylated anionic surfactants, ^{31}C NMR spectroscopy is used for their identification.

Fisch [161] has illustrated the use of NMR in support of vinyl stabiliser research. In particular ^{1}H NMR

was applied to di-(2-ethylhexyl)phthalate (DEHP), ^{31}P NMR to the study of phosphite hydrolysis in vinyl stabilisers, ^{13}C NMR to 2-ethylhexylmaleate and butyltin maleate stabilisers, and *^{119}Sn NMR spectroscopy* to monobutyltinsulfide. NMR provides a handle on the composition of *organotin mercaptide stabilisers* and opens the door to correlation of performance with composition. Other examples of NMR applications to vinyl stabilisers include a study of the mechanism of organotin stabiliser action [249] and on the solubility of didecyl phenyl phosphite in rigid vinyl [250]. ^{119}Sn NMR was also used to study the interaction between dibutyltin chloride and dibutyltin dilaurate in

CDCl$_3$ solution; the formation of the monochlorolaurate was observed [251]. Despite the fact that ^{119}Sn is a reasonable sensitive NMR nucleus (Table 5.14) ^{119}Sn l-NMR activity applied to polymer additives is quite limited.

In special circumstances ^{19}F NMR spectroscopy is a useful but laborious analytical tool, e.g. for the investigation of effects of vulcanisation parameters, such as temperature and sulfur concentration, on the formation of decomposition products. In a typical approach Kelm and Gross [252] have used ^{19}F NMR for the study of the decomposition during vulcanisation processing of complex mixtures using fluorine labelling of the molecules of interest. As well known, the most important vulcanisation accelerators of the thiuram type are tetraalkylated thiuram disulfides, such as tetramethylthiuram disulfide (TMTD) and dimethyldiphenylthiuram disulfide. Kelm and Gross [252] synthesised dimethyl-di-(p-fluorophenyl)-thiuram disulfide (FDMPTD), dimethyl-di-(p-fluorophenyl)-thiuram monosulfide (FDMPTM), and zinc-(N-methyl-N-p-fluorophenyl)-dithiocarbamate (FZMPC), as well as a representative of another class of accelerating products, o-fluorophenyl-biguanide (OFPBG), as a derivative of the commercial product o-tolyl-biguanide (OTBG) (Scheme 5.4). Vulcanised rubbers containing these additives were extracted and the soluble parts were analysed by ^{19}F NMR. Formation

of p-fluoro-N-methylaniline (FMA) and other decomposition products was observed. This approach is highly selective and separation techniques are necessary. However, it is based on fluorine-labelled compounds instead of the commercial products. ^{19}F NMR has also been applied to study the effect of plasticisers on LiCF$_3$SO$_3$-containing polymer electrolytes. Use of ^{19}F l-NMR to polymer additives is quite limited.

Use of ^{29}Si NMR can generally be avoided in favour of ^{1}H NMR. For example, Narayan et al. [253] have used proton NMR for the detection of a HALS stabiliser based on silicon technology, after extraction from a PP matrix. ^{17}O l-NMR and s-NMR have been used for the study of PP, EPDM, PIP and NR oxidatively degraded with enriched O$_2$ [204,254].

The application of partial relaxation FTNMR in the determination of unreacted monomers, solvent, water, and additives in polymers, polymer degradation, and functional group and chain structure characterisation has been reviewed [255]. Hummel [95] underestimates the contribution of NMR to polymer/additive analysis.

5.4.1 Multidimensional NMR Spectroscopy

Principles and Characteristics The basic concept of two-dimensional (2D) correlation analysis was born

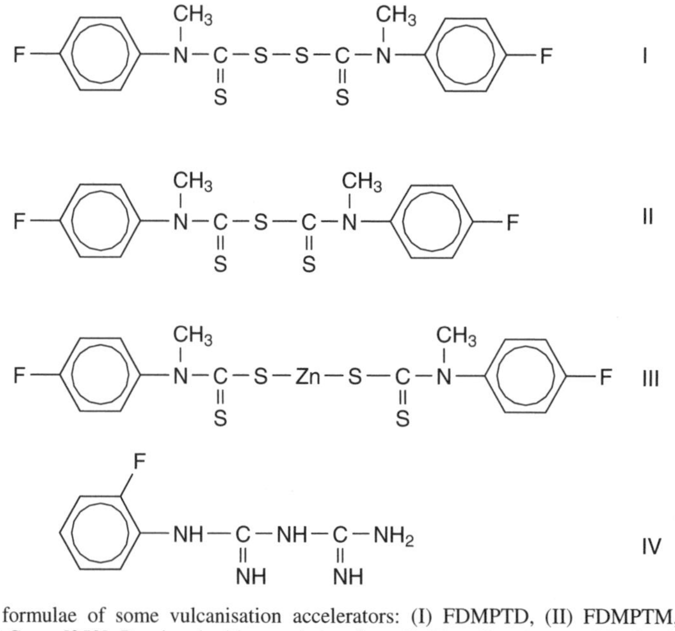

Scheme 5.4 Structural formulae of some vulcanisation accelerators: (I) FDMPTD, (II) FDMPTM, (III) FZMPC and (IV) OFPBG. After Kelm and Gross [252]. Reprinted with permission from *Rubber Chemistry and Technology*. Copyright © (1985), Rubber Division, American Chemical Society, Inc.

in NMR spectroscopy about 30 years ago [256–258]. By dispersing the resonances into a second frequency dimension additional spectral dispersion is achieved. 2D methods were originally developed for use in high-resolution NMR of liquids and have been later adapted to solid-state systems [169]. In the context of 2D spectroscopy, manipulation of spin Hamiltonians, either for the simplification of spectra or for the enhancement of their information content, is of even greater value than for 1D NMR. In the continued development of NMR during the last decades the majority of new methods are in the field of multidimensional NMR. Nowadays, 2D analysis is one of the most important analytical methods in NMR. In conventional magnetic resonance, the spectrum of frequency response $S(\omega)$ is a complex function of a single-frequency variable. In double resonance, a response function $S(\omega_1, \omega_2)$ is measured as a function of two variables. A 2D spectrum represents a signal function $S(\omega_1, \omega_2)$ of two independent frequency variables. This definition does not include stacked plots $S(\tau, \omega)$ of 1D spectra, which are often used in relaxation studies to represent the time evolution of a spectrum after a perturbation.

2D NMR experiments are designed to generate different kinds of frequency information along two axes. Several *strategies* can be employed for this purpose. The general approach adopted in 2D NMR is to apply multiple RF pulses to the sample such that there are two independent variable time intervals in the pulse sequence (Scheme 5.5). In 2D NMR spectroscopy the preparation time establishes the condition of the spin system at the beginning of t_1 and is usually a relaxation delay followed by one or more pulses to start the experiment. The evolution period establishes the second frequency dimension. A series of FID signals are collected with the evolution time t_1 varied systematically to the desired maximum value, t_1 (max). The interval t_1 is usually of the order of microseconds, t_2 of the order of seconds. The evolution period may contain one or more pulses, most commonly a spin-echo sequence. The mixing period is not required in some sequences. Data are acquired during the detection period t_2, for a series of spectra in which t_1 is regularly incremented from 0 to some maximum value, as in a 1D experiment. A 2D data set $S(t_1, t_2)$ is obtained. Double Fourier transformation, with respect to t_2 and then t_1, yields a 2D spectrum $S(\omega_1, \omega_2)$ with two orthogonal frequency (ω) scales. The resulting data are most commonly displayed in a contour format.

A 2D NMR experiment involves a selection of pulses, delays, frequencies, RF phases and amplitudes,

Preparation	Evolution	Mixing	Detection
Δ_1	t_1	Δ_2	t_2

Scheme 5.5 A general 2D NMR pulse sequence

magnetic field gradients, etc. Using the appropriate combination, a spectrum can be produced with selected signals which contain the needed information while removing undesired signals. The key to differences in multidimensional experiments is what nuclei 'feel' during the evolution time (t_1) in terms of pulses, decoupling, pulsed-field gradients, light flashes, etc. The use of field-gradient technology makes it possible to acquire 2D and higher-dimensional spectra very rapidly, greatly reducing experimental times. Field gradients enable improved suppression of unwanted solvent signals, especially from solvent water. The characteristics of several commonly used 2D NMR pulse sequences were listed by Reynolds [177] and others [259].

Part of the power of 2D NMR comes from the ability to provide tremendous spectral dispersion; however, the structural information present from the correlation of frequencies is equally important. The F_2 frequency, Fourier transform of the (t_2) time domain, normally corresponds to a chemical shift axis, but various other combinations of $F_1(\delta^1H, J_{H–H}, J_{C–H}, \delta^{13}C)$ and $F_2(\delta^{13}C, \delta^1H)$ have been described [160] which convey different information. When the Hamiltonian is composed of terms of different physical origin, such as chemical shifts or dipolar or scalar couplings, it is often possible to render spectra more intelligible by separating various interactions in orthogonal frequency domains. The present strength of NMR spectroscopy lies in the application of multidimensional NMR experiments to an analytical problem. In 2D NMR spectra, a *correlation* between peaks can be established that provides detailed *structural or dynamic information*. 1D spectra may be unravelled that are rendered inscrutable because of severe overlap by separating interactions of different physical origin, e.g. chemical shifts and couplings. Information on nearest neighbours and complete coupled spin systems is obtained by means of 2D COSY (correlated spectroscopy) NMR experiments. COSY is a simple sequence, the spectra are rather easy to interpret and require relatively little explanation. Other 2D experiments correlate chemical shifts of nuclei either through bonds (COSY, TOCSY, HMQC) or, in order to obtain 3D structural information, through space (NOESY) [200,201]. NOE is typically employed during the later stages of a structural investigation, when the gross structure of the molecule has been (largely) defined. A knowledge of 2D NMR is fundamental in structural chemistry.

Multidimensional and heteronuclear NMR techniques have revolutionised the use of NMR spectroscopy for the structure determination of organic molecules from small to complex. Multidimensional NMR also allows observation of forbidden multiple-quantum transitions and probing of slow dynamic processes, such as chemical exchange, cross-relaxation, transient Overhauser effects, and spin-diffusion in solids.

Table 5.19 shows the *main benefits* of multidimensional NMR for structure elucidation. The advantages of 2D NMR as compared to its 1D equivalent are: (i) simplification of crowded spectra by spreading resonances into a second dimension; (ii) no need for selective excitation of individual resonances; and (iii) more efficient measurements. 2D NMR techniques facilitate identification of molecules within complex mixtures. Introduction of multidimensional experiments has made the connectivity of carbon and protons much clearer, has much reduced the problem of distinguishing coupling from shift effects by providing extra dimensions for displaying the NMR signal, and has even provided an extra structure-discriminating route. On the other hand, the 2D approach usually requires increased experiment time and computer data storage.

An astonishing number of powerful 2D and higher multidimensional NMR techniques have been developed. The common thread to all of them is to show correlations between certain nuclear properties: chemical shifts, couplings, NOE, and so on [160]. There are quite a few variations of 2D NMR experiments in which properties such as retention time (in HPLC-NMR), distances (imaging) or diffusion coefficients (diffusion ordered spectroscopy) are the variables along one or more axes in the spectra. Multidimensional NMR spectroscopic techniques nowadays play an ever-increasing role [201,202]; data standards for validation are being developed [260].

Multidimensional NMR has extensively been reviewed [206,261–263]; many books deal with the subject [169,197,200–202,264–266]. For experimental aspects of 2D NMR, see Hull [267].

Applications   Useful 2D NMR experiments for identification of *surfactants* are homonuclear proton correlation (COSY, TOCSY) and heteronuclear proton-carbon correlation (HETCOR, HMQC) spectroscopy [200,201]. 2D NMR experiments employing proton detection can be performed in 5 to 20 min for surfactant solutions of more than 50 mM. Van Gorkum and Jensen [238] have described several 2D NMR techniques that are often used for identification and quantification of anionic surfactants. The resonance frequencies of spin-coupled nuclei are correlated and hence give detailed information on the structure of organic molecules.

Ehret-Henry *et al.* [220] have shown that ^{1}H NMR spectra can be used without chromatographic analysis, to shorten the total identification time necessary, and as a fingerprint of all the extractable nonvolatile compounds present in food packaging material (safety control). Figure 5.10 shows a ^{1}H NMR spectrum (in CDCl$_3$ with TMS as internal standard) of a Soxhlet extract of a 35 µm PP film (after solvent evaporation). The assignments of the resonances of Irgafos 168 and its decomposition products were confirmed by a ^{31}P-^{1}H 2D correlation NMR experiment [220].

Since 2D NMR experiments are fast and more informative than regular ^{13}C NMR spectroscopy, they are becoming an alternative to the latter. For quantitative measurements in mixtures, however, 1D NMR spectroscopy with well-resolved resonances must be used. 2D time-domain NMR analysis in the proton spin rotating frame has identified three proton magnetisation fractions in HDPE, corresponding to the crystalline region, chain loops and entangled chain segments, and the amorphous phase. The latter fraction is the most mobile and liquid-like with $T_2 \approx 1$ ms at 120 °C [268].

5.4.1.1   Diffusion-ordered Spectroscopy

Principles and Characteristics   Relevant diffusion processes are interdiffusion driven by concentration

Table 5.19   Benefits of multidimensional NMR

Feature	Multidimensional NMR mode[a]
Resolution of even highly overlapping resonances	J-spectroscopy
Identification of spin-coupled pairs of nuclei	COSY
Identification of all directly bonded and long range heteronuclear pairs	HETCOR, HMQC
Identification of proton pairs interacting magnetically through space	NOESY
Identification of nuclei belonging to separate spin systems	TOCSY
Direct detection of carbon skeletons	2D-INADEQUATE

[a] COSY, homonuclear correlation spectroscopy; HETCOR, heteronuclear chemical shift-correlation spectroscopy; HMQC, heteronuclear multiple quantum coherence; NOESY, nuclear Overhauser and exchange spectroscopy; TOCSY, total correlation spectroscopy; INADEQUATE, homonuclear J-correlated ^{13}C experiment.

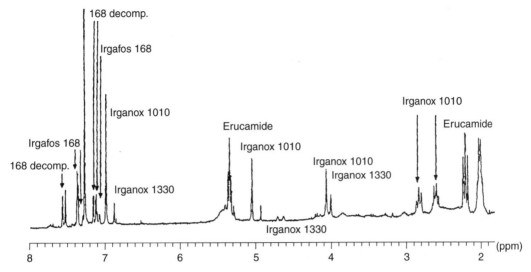

Figure 5.10 ^{1}H NMR spectrum of a Soxhlet extract of a 35 μm PP film. After Ehret-Henry *et al.* [220]. From J. Ehret-Henry *et al.*, *Food Addit. Contam.*, **9**, 303–314 (1992). Reproduced by permission of Taylor & Francis Ltd (http://www.tandf.co.uk/journals)

gradients and self-diffusion driven by Brownian motion. Self-diffusion coefficients differ from interdiffusion coefficients by the impact of the chemical potential. It has been shown that NMR excellently measures the self-diffusion coefficient of solutes in polymer solutions [269], as given by Vrentas and Duda [270]. NMR is about the only method that can measure self-diffusion.

The study of molecular diffusion in solution by NMR methods offers insights into a range of physical molecular properties. Different mobility rates or *diffusion coefficients* may also be the basis for the separation of the spectra of mixtures of small molecules in solution, this procedure being referred to as diffusion-ordered spectroscopy (DOSY) [271] (Figure 5.11). In this 2D experiment, the acquired FID is transformed with respect to t_2 (the acquisition time).

Modern NMR diffusion measurements are all based on the Stejskal–Tanner (S-T) *pulsed field gradient spin-echo experiment* [174] in which a gradient pulse is

used to encode the spatial positions of nuclear spins and a second matched gradient pulse decodes positions. Translation diffusion in the time interval Δ between the gradient pulses results in attenuation of the spin-echo and a Fourier transformation of the half echo permits diffusion rates to be associated with individual lines in the NMR spectrum. The echo attenuation is described by the well-known S-T factor $\exp[-Dq^2(\Delta - \delta/3)]$ [174], where the area of a gradient pulse is $q = \gamma g\delta$, γ is the gyromagnetic ratio, and g and δ are the gradient pulse amplitude and duration, respectively; D is the self-diffusion coefficient of a given component. The time interval Δ between the gradient pulses is typically much greater than the pulse duration δ (typically 100 ms vs. 5–10 μs). Appropriate transformation and display of data for the *diffusion dimension* allows self-diffusion coefficients to be determined for individual resonances.

Pulsed field gradient NMR has become a standard method for measurement of diffusion rates. Stilbs [272] and others have exploited in particular the FT version for the study of mixtures. An added advantage of PFG-NMR is that it can be employed to simplify complex NMR spectra. This simplification is achieved by attenuation of resonances based on the differential diffusion properties of components present in the mixture.

DOSY was developed to provide useful displays of PFG-NMR data sets that incorporate all reliable information about the system under study obtained from PFG-NMR data and prior knowledge. Various pulse sequences for DOSY have been developed [273,274].

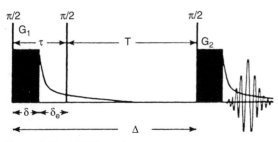

Figure 5.11 DOSY experiment

The DOSY experiment is acquired as a series of 1D NMR spectra in which the entire spectral bandshapes of the individual compounds have been attenuated by varying the gradient strength. A DOSY data set consists of a series of FIDs acquired with various gradient strengths. Several methods have been developed to process the diffusion dimension of DOSY data [275,276]. Once diffusion coefficients have been determined from appropriate data-fitting routines, it is possible to use the data to generate a diffusion dimension with a *2D DOSY spectrum* [277]. One dimension corresponds to the spectroscopic data and the other dimension represents the diffusion domain, i.e. essentially a display of a *'diffusion spectrum'* of each chemical [278]. Sensitivity and resolution of the technique are comparable with those of a conventional 1D spectrum. The correlation between chemical shift and the calculated diffusion coefficient simplifies the analysis of many complex mixtures. For well-resolved resonances, the correct diffusion coefficient is unambiguously determined. The DOSY results demonstrate 'at a glance' qualitative analysis of molecular size distributions in mixtures and, with proper care, quantitative determination of diffusion coefficients [279,280]. With suitable experimental and processing methods, relative accuracies as good as 0.2 % can be obtained for the diffusion coefficients extracted from the spin-echo experiments. The selectivity of high-resolution NMR compensates for the limited resolution in the diffusion dimension to provide a global view of translational dynamics in mixtures.

DOSY is a technique that may prove successful in the determination of additives in mixtures [279]. Using different field gradients it is possible to distinguish components in a mixture on the basis of their diffusion coefficients. Morris and Johnson [271] have developed diffusion-ordered 2D NMR experiments for the analysis of mixtures. PFG-NMR can thus be used to identify those components in a mixture that have similar (or overlapping) chemical shifts but different diffusional properties. Multivariate curve resolution (MCR) analysis of DOSY data allows generation of pure spectra of the individual components for identification. The pure spin-echo diffusion decays that are obtained for the individual components may be used to determine the diffusion coefficient/distribution [281]. Mixtures of molecules of very similar sizes can readily be analysed by DOSY. Diffusion-ordered spectroscopy [273,282], which does not require prior separation, is a viable competitor for techniques such as HPLC-NMR that are based on chemical separation.

Applications   Without authentic samples of the additives, total identification can be difficult using routine NMR correlation methods. However, additives often cover a range of molecular weights and sizes, making them amenable to separation and identification by PFG-NMR. There are two principal regimes in which DOSY techniques find application. On the one hand, DOSY is suitable for low-resolution analysis of mixtures containing species of very different sizes and a broad range of diffusion coefficients (e.g. macromolecules). On the other hand, the high-resolution variant of DOSY is a powerful means for the analysis of complex mixtures of small molecules with relatively similar sizes [277]. Application of DOSY to determine the composition of mixtures is based on differences in the self-diffusion coefficients of the individual compounds. Examples are analysis of complex polymer blends, oligomer analysis, direct additive analysis, determination of molecular weight distributions, study of diffusion processes, etc. [271,279,283].

Jayawickrama *et al.* [284] have recently demonstrated the utility of PFG-NMR as a tool for the identification of those components in mixtures of polymer additives that have similar chemical shifts but different diffusion coefficients. Samples analysed were (Irganox 1330, BHT, Tinuvin P) and (Irganox 1098, Irganox 1330, Tinuvin P) additive solutions. Figure 5.12 shows the DOSY spectrum of (Irganox 1330, BHT, Tinuvin P) with the resonances of each component indicated in the single-pulse ^{1}H NMR spectrum. Diffusion coefficients were extracted from the NMR spectra with DOSY methodology. For well resolved resonances, the correct diffusion coefficients are unambiguously determined. Because the diffusion coefficients of the components of the samples examined are significantly different, most of the resonances in the DOSY spectrum can be identified by simple inspection of the peaks which are aligned with a particular diffusion coefficient, or horizontal row, in the contour plot. The experiments combine the high specificity and information content of NMR spectroscopy with the size selectivity of diffusion coefficients. The added information content of relative molecular size can be of benefit in the analysis of complex mixtures [284].

Linssen and de Vries [285] have examined 1 % di-*t*-butyl-*p*-cresol (DBPC) in low-MW poly(tetrahydrofuran) by means of DOSY (Figure 5.13). DOSY is a powerful tool for the analysis of polydisperse samples and complex mixtures, such as anionic surfactants. It is not to be expected that DOSY will rapidly become a standard tool in polymer/additive analysis.

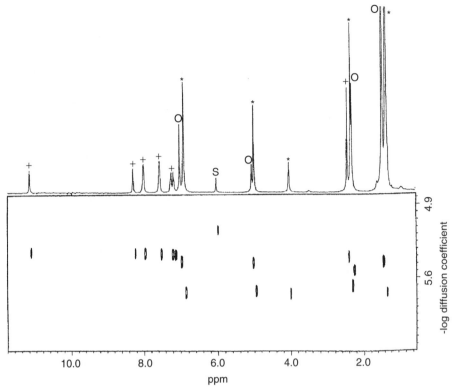

Figure 5.12 ^{1}H DOSY spectrum of (+) Tinuvin P, (*) Irganox 1330 and (o) BHT in TCE-d$_2$ showing resolution of the individual components along the chemical shift and diffusion dimensions. The single pulse ^{1}H NMR spectrum is shown on top of the 2D DOSY plot. After Jayawickrama *et al.* [284]. Reprinted from D.A. Jayawickrama *et al. Magnetic Resonance Chemistry*, **36**, 755–760 (1998). Copyright 1998 © John Wiley & Sons, Ltd. Reproduced with permission

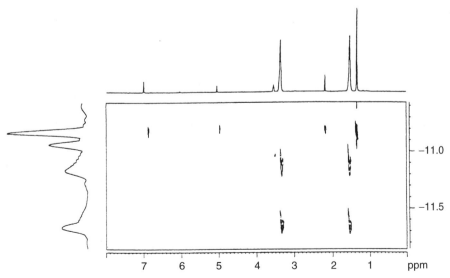

Figure 5.13 DOSY 2D display for 1 % DBPC in PTHF-1500 (horizontal axis: ^{1}H NMR; vertical axis: diffusion coefficients). Reproduced by permission of DSM Research, Geleen

5.5 BIBLIOGRAPHY

5.5.1 General Spectroscopy

M.T.C. De Loos-Vollebregt, *Spectrometrische Analysetechnieken*, Bohn Stafleu Van Loghum, Houten (2004).

G. Gauglitz and T. Vo-Dinh (eds), *Handbook of Spectroscopy*, Wiley-VCH, Weinheim (2003).

S.P. Davis, M.C. Abrams and J.W. Brault, *Fourier Transform Spectrometry*, Academic Press, Orlando, FL (2001).

J. Kauppinen and J. Partanen, *Fourier Transforms in Spectroscopy*, John Wiley & Sons, Ltd, Chichester (2001).

J.C. Lindon, G.E. Tranter and J.L. Holmes (eds), *Encyclopedia of Spectroscopy and Spectrometry*, Academic Press, San Diego, CA (2000).

J.M. Hollas, *High Resolution Spectroscopy*, John Wiley & Sons, Ltd, Chichester (1998).

R. Williams, *Spectroscopy and the Fourier Transform*, VCH Publishers, New York, NY (1996).

P. Suppan, *Chemistry and Light*, The Royal Society of Chemistry, Cambridge (1994).

H. Mark, *Principles and Practice of Spectroscopic Calibration*, John Wiley & Sons, Inc., New York, NY (1991).

5.5.2 Ultraviolet/Visible Spectrophotometry

C. Burgess and T. Frost (eds), *Standards and Best Practice in Absorption Spectrometry*, Blackwell Science, Oxford (1999).

T. Owen, *Fundamentals of Modern UV-Visible Spectroscopy. A Primer*, Hewlett-Packard Corporation, Publ. No. 12-5965-5123E (1996).

M. Thomas, *Ultraviolet and Visible Spectroscopy*, John Wiley & Sons, Ltd, Chichester (1996).

B.J. Clark, T. Frost and M.A. Russell, *UV Spectroscopy: Techniques, Instrumentation and Data Handling*, Chapman & Hall, London (1993).

5.5.3 Infrared Spectroscopy

E. Jiang, *Principles, Experiments and Applications: Based on Research-Grade Nicolet FT-IR Spectrometers*, Thermo Electron, Madison, WI (2003).

D.O. Hummel, *Atlas of Plastics Additives: Analysis of Spectrometric Methods*, Springer-Verlag, Berlin (2002).

H. Günzler and H.-U. Gremlich, *IR Spectroscopy. An Introduction*, Wiley-VCH, Weinheim (2002).

J.M. Chalmers and P.R. Griffiths (eds), *Handbook of Vibrational Spectroscopy*, John Wiley & Sons, Ltd, Chichester (2001), 5 vols.

B. Stuart, *Modern Infrared Spectroscopy*, John Wiley & Sons, Ltd, Chichester (1996).

B.C. Smith, *Fundamentals of Fourier Transform Infrared Spectroscopy*, CRC Press, Boca Raton, FL (1995).

B. Schrader (ed.), *Infrared and Raman Spectroscopy. Methods and Applications*, Wiley-VCH, Weinheim (1994).

J.R. Ferraro and K. Krishnan (eds), *Practical Fourier Transform Infrared Spectroscopy*, Academic Press, San Diego, CA (1990).

A. Fadini and F.M. Schnepel, *Vibrational Spectroscopy: Methods and Applications*, Ellis Horwood, Chichester (1989).

H.A. Willis, J.H. van der Maas and R.G.J. Miller (eds), *Laboratory Methods in Vibrational Spectroscopy*, John Wiley & Sons, Ltd, Chichester (1987).

P.R. Griffiths and J.A. DeHaseth, *Fourier Transform Infrared Spectrometry*, John Wiley & Sons, Ltd, Chichester (1986).

5.5.4 Luminescence Spectroscopy

S.G. Schulman (ed.), *Molecular Luminescence Spectroscopy, Methods and Applications*, John Wiley & Sons, Inc., New York, NY, Parts 1–3 (1985, 1988, 1993).

W.R.G. Baeyens, D. De Keukeleire and K. Korkidis (eds), *Luminescence Techniques in Chemical and Biochemical Analysis*, M. Dekker, New York, NY (1991).

G.G. Guilbault (ed.), *Practical Fluorescence: Theory, Methods and Techniques*, M. Dekker, New York, NY (1991).

5.5.5 Nuclear Magnetic Resonance Spectroscopy

T.N. Mitchell and B. Costisella, *NMR – From Spectra to Structures*, Springer-Verlag, Heidelberg (2004).

Q.T. Pham, R. Petiaud, H. Waton and M.-F. Llauro-Darricades, *Proton and Carbon NMR Spectra of Polymers*, John Wiley & Sons, Ltd, Chichester (2002).

J.D. Roberts, *ABC's of FT-NMR*, University Science Books, Sausalito, CA (2000).

P. Bigler, *NMR Spectroscopy: Processing Strategies*, Wiley-VCH, Weinheim (2000).

T.D.W. Claridge, *High-resolution NMR Techniques in Organic Chemistry*, Pergamon, Amsterdam (1999).

U. Weber and H. Thiele, *NMR Spectroscopy: Modern Spectral Analysis*, Wiley-VCH, Weinheim (1998).

R.S. Macomber, *A Complete Introduction to Modern NMR Spectroscopy*, John Wiley & Sons, Inc., New York, NY (1998).

D. Canet, *Nuclear Magnetic Resonance – Concepts and Methods*, John Wiley & Sons, Ltd, Chichester (1996).

D.M. Grant and R.K. Harris (eds), *Encyclopedia of NMR*, John Wiley & Sons, Inc., New York, NY (1996).

5.6 REFERENCES

1. H.W. Siesler, *Proceedings SPQ-98 (Spectroscopy in Process and Quality Control)*, Advanstar, London (1998).

2. C. Reichardt, *Solvent and Solvent Effects in Organic Chemistry*, Wiley-VCH, Weinheim (2003).

3. Sadtler Research Laboratories, *Infrared Spectra Handbook of Common Organic Solvents*, Philadelphia, PA (1983).

4. J. Workman and A.W. Springsteen (eds), *Applied Spectroscopy. A Compact Reference for Practitioners*, Academic Press, San Diego, CA (1998).

5. W.H. Stevenson III, *Today's Chemist at Work* (10), 47–52 (1998).

6. J.M. Hayes, *Chem. Rev.* **87**, 745 (1987).

7. R. Williams, *Spectroscopy and the Fourier Transform*, VCH Publishers, New York, NY (1996).

8. B.J. Clark, T. Frost and M.A. Russell (eds), *UV Spectroscopy: Techniques, Instrumentation and Data Handling*, Chapman & Hall, London (1993).

9. S. Paul, *Today's Chemist at Work* (10), 21–3 (1998).

10. F. Szabadvery, *History of Analytical Chemistry*, Pergamon Press, New York, NY (1966).

11. J.A. Howell and L.G. Hargis, *Anal. Chem.* **62**, 155R (1990).

12. D.L. Massart, B.G.M. Vandegiste, L.M.C. Buydens, S. de Jong, P.J. Lewi and J. Smeyers-Verbeke, *Handbook of Chemometrics and Qualimetrics*, Elsevier, Amsterdam (1998).

13. C. Burgess and T. Frost (eds), *Standards and Best Practice in Absorption Spectrometry*, Blackwell Science, Oxford (1999).

14. C.T. Cottrell, *Anal. Proc.* **19**, 43–5 (1982).

15. B.P. Chadburn, *Anal. Proc.* **19**, 42–3 (1982).

16. R. Łobiński and Z. Marczenko, *Crit. Rev. Anal. Chem.* **23**, 55–111 (1992).

17. T. Owen, *Fundamentals of Modern UV-Visible Spectroscopy. A Primer*, Hewlett Packard Corporation Publ. No. 12-5965-5123E (1996).

18. *Standard Reference Materials Catalog*, NIST, Gaithersburg, MD, pp. 104–6 (1995).

19. J.P. Hammond, *Spectrosc. Europe* **13** (4), 10–4 (2001).

20. T.D. Harris, *Anal. Chem.* **54**, 741–50A (1982).

21. D.O. Hummel and F. Scholl, *Atlas der Polymer- und Kunststoffanalyse*, C. Hanser Verlag and Verlag Chemie, Munich (1981), Vol. 3.

22. A. Weston and P.R. Brown, *HPLC and CE, Principles and Practice*, Academic Press, San Diego, CA (1997).

23. H.H. Perkampus, *UV Atlas of Organic Compounds*, VCH, Weinheim (1992).

24. J. Scheirs, *Compositional and Failure Analysis of Polymers. A Practical Approach*, John Wiley & Sons, Ltd, Chichester (2000).

25. T.R. Crompton, *Analysis of Polymers. An Introduction*, Pergamon Press, Oxford (1989).

26. J. Urbanski, *Anal. Chem. (Warsaw)* **22**, 749 (1977).

27. T.R. Crompton, in *Polymer Devolatilization* (R.J. Albalak, ed.), M. Dekker, New York, NY (1996), pp. 575–643.

28. T.P. Hunt, C.J. Dowle and G. Greenway, *Analyst* **118** (1), 17–22 (1993).

29. J. Yang, *Proc. Intl. Conf. Pet. Refin. Petrochem. Process.* (X. Hou, ed.), Intl. Acad. Publ., Beijing (1991), Vol. 2, pp. 591–7.

30. J. Haslam, D.C.M. Squirrell, S. Grossman and S.F. Loveday, *Analyst (London)* **78**, 92–106 (1953).

31. D. Munteanu, A. Isfan, C. Isfan and I. Tincul, *Chromatographia* **23**, 7 (1987).

32. S. Tojo and T. Matsumoto, *Chiba-ken Kogyo Shikenjo Kenkyu Hokoku* (1), 15–20 (1987).

33. M.H. Jones and J.T. Woodcock, *Anal. Chem.* **58**, 1145–8 (1986).

34. K. Tôei, S. Motomizu and T. Umano, *Talanta* **29**, 103–106 (1982).

35. Q. Le, *Fenxi Huaxue* **26** (9), 1157 (1998).

36. F.J. Pern, *Angew. Makromol. Chem.* **252**, 195–216 (1997).

37. J.J. Austin, *J. Am. Oil Chem. Soc.* **31**, 424 (1954).

38. W.G. Schwien and H.W. Conroy, *J. Assoc. Off. Anal. Chem.* **48**, 489 (1965).

39. C.S. Sactry, K.E. Rao and U.V. Prasad, *Talanta* **29**, 917 (1982).

40. C.Y. Chu, *Kautsch. Gummi Kunstst.* **41** (1), 33–9 (1988).

41. T.R. Crompton, *Practical Polymer Analysis*, Plenum Press, New York, NY (1993).

42. T.R. Crompton, *Manual of Plastics Analysis*, Plenum Press, New York, NY (1998).

43. J. Haslam and H.A. Willis, *Identification and Analysis of Plastics*, The Iliffe Group, London (1965), p. 307.

44. J. Haslam, H.A. Willis and D.C.M. Squirrell, *Identification and Analysis of Plastics*, John Wiley & Sons, Ltd, Chichester (1983).

45. A.J. Brandolini, J.M. Garcia and R.E. Truitt, *Spectroscopy* **7** (3), 34–9 (1992).

46. P.V.C. Rao, J.V. Prasad and V.J. Koshy, *Ann. Chimica (Rome)* **85**, 171–82 (1995).

47. N.C. Billingham, D.C. Bott and A.S. Manke, *Dev. Polym. Degrad.* **3**, 63–100 (1981).

48. D.C.M. Squirrell, *Analyst* **106**, 1042–56 (1981).

49. S.-J. Kim, *Anal. Sci. Technol.* **13** (3), 282–90 (2000).

50. W. Freitag, *Fresenius' Z. Anal. Chem.* **316** (5), 495–6 (1983).

51. N. Haider and S. Karlsson, *Analyst* **124**, 797–800 (1999).

52. H. Ostromow, *Analyse von Kautschuken und Elastomeren*, Springer-Verlag, Berlin (1981).

53. P. Métois, D. Scholler, J. Bouquant and A. Feigenbaum, *Food Addit. Contam.* **15** (1), 100–11 (1998).

54. N.N. Lytkina, T.M. Egorova and L.N. Mizerovski, *Plast. Massy* (3), 58 (1983).

55. P. Meng, L. Li, C. Yang and G. Song, *Zhongguo Suliao* **13** (9), 73–5 (1999).

56. P.J. Cornish, *J. Appl. Polym. Sci.* **7**, 727 (1963).

57. T. Kawaguchi, K. Ueda and A. Koga, *J. Soc. Rubber Ord. (Japan)*, **28**, 525 (1955).

58. H. Nelissen, *Techn. Rept DSM Research*, Geleen (1998).

59. C.W. Peloso, M.J. O'Connor, S.W. Bigger and J. Scheirs, *Polym. Degrad. Stab.* **62**, 285–90 (1998).

60. C. Stafford, *Anal. Chem.* **34**, 794–6 (1962).
61. J. Soncek and E. Jelinkova, *Analyst (London)* **107**, 623 (1982).
62. L.H. Ruddle and J.R. Wilson, *Analyst (London)*, **94**, 105–9 (1969).
63. A. Magill and A.R. Becker, *J. Pharm. Sci.* **73**, 1663–4 (1984).
64. A. Schaafsma, *Techn. Rept DSM Research*, Geleen (1999).
65. C.L. Hilton, *Anal. Chem.* **32**, 383 (1960).
66. British Standard 2782, Part 4, Method 405B, D (1965).
67. C.L. Hilton, *Anal. Chem.* **32**, 1554 (1960).
68. C.L. Hilton, *Rubber Age (NY)* **84**, 263 (1958).
69. O.N. Kadkin, I.I. Evgen'eva, V.Yu. Mavrin and M.I. Evgen'ev, *Ind. Lab. (Diagn. Mater.)* **65** (10), 637–9 (1999).
70. A.J. Smith, *Anal. Chem.* **36**, 944 (1964).
71. R.H. Campbell and E.J. Young, *Rubber Age (NY)* **100** (3), 71–5 (1968).
72. J. Koch, *Dtsch. Lebensm. Rdsch.* **70**, 209–10 (1974).
73. J. Koch, *Dtsch. Lebensm. Rdsch.* **68**, 401–3 (1972).
74. L. Liu and L. Cai, *Fushun Shiyou Xueyuan Xuebao* **17** (3), 1–10 (1997).
75. J.J. Brandt, *Anal. Chem.* **33**, 1390 (1961).
76. C.A. Parks, *Anal. Chem.* **33**, 140 (1961).
77. G.W. Taylor, *Dyes in History and Archaeology* **7**, 33–9 (1988).
78. A. Navarro Sentanyes, *Rev. Quim. Text.* **106**, 25–32 (1992).
79. A. Shakhnovich and J. Barren, *Proceedings SPE ANTEC 2002*, San Fransisco, CA (2002), pp. 2512–6.
80. C. Haisch and R. Niessner, *Spectrosc. Europe* **14** (5), 10–5 (2002).
81. P. Alessi, A. Cortesi and I. Kikic, *Proceedings GRICU (Gruppo Ricercatori Italiani di Ingegneria Chimica dell'Università)*, Ferrara (1998), pp. 379–82.
82. F.A. Cotton, *Chemical Applications of Group Theory*, John Wiley & Sons, Inc., New York, NY (1971).
83. J.M. Chalmers and G. Dent, *Industrial Analysis with Vibrational Spectroscopy*, The Royal Society of Chemistry, Cambridge (1997).
84. J.W. Cooley and J.W. Tukey, *Math. Comput.* **19**, 297 (1965).
85. A.A. Michelson, *Phil. Mag.* **31**, 338 (1891).
86. P.B. Fellgett, *J. Phys. Radium* **19**, 187 (1958).
87. P. Jacquinot, *Rept Progr. Phys.* **13**, 267 (1960).
88. J. Conne and P. Conne, *J. Opt. Soc. Am.* **56**, 896 (1966).
88a. J.R. Birch and F.J.J. Clarke, *Spectrosc. Europe* **7**(4), 16–22 (1995).
89. F.J.J. Clarke, *Proc. SPIE* **2775**, 6–18 (1996).
90. Available from Division of Electrical Science, NPL, Teddington (UK) (Contact Dr F.J.J. Clarke).
91. G.L. McClure, in *Practical Sampling Techniques for Infrared Analysis* (P.B. Coleman, ed.), CRC Press, Boca Raton, FL (1993), pp. 165–215.
92. W.D. Perkins, in *Practical Sampling Techniques for Infrared Analysis* (P.B. Coleman, ed.), CRC Press, Boca Raton, FL (1993), Ch. 2.
93. S.G. Kazarian, N.M.B. Flichy, D. Coombs and G. Poulter, *Intl Lab.* **32** (2), 28–32 (2002).
94. D.O. Hummel, *Macromol. Symp.* **119**, 65–77 (1997).
95. D.O. Hummel, *Atlas of Plastics Additives. Analysis by Spectrometric Methods*, Springer-Verlag, Berlin (2002).
96. J.A. Sidwell, *The Rapra Collection of Infrared Spectra of Rubbers, Plastics and Thermoplastic Elastomers*, Rapra Technology Ltd, Shawbury (1997).
97. J. Behnke, *The Aldrich Library of FT-IR Spectra*, Aldrich Chemical Co., Milwaukee, WI (1997).
98. D.O. Hummel, *Atlas der Polymer- und Kunststoffanalyse. Bd. 2: Kunststoffe, Fasern, Kautschuk, Harze; Ausgangs- und Hilfsstoffe, Abbauprodukte*, C. Hanser Verlag, Munich (1984).
99. D.O. Hummel (ed.), *Atlas of Polymer and Plastics Analysis: Analytical Methods, Spectroscopy, Characteristic Absorptions, Description of Compound Classes*, John Wiley & Sons, Inc., New York, NY (1988), Vol. 2.
100. D.O. Hummel (ed.), *Atlas of Polymer and Plastics Analysis*, VCH, Weinheim (1991), Vol. 1
101. Sadtler Research Laboratories, *Sadtler Digital FTIR Libraries*, Philadelphia, PA (1989).
102. C.D. Craver, *Infrared Spectra of Plasticizers and Other Additives*, The Coblentz Society, Inc. Kirkwood, MO (1982).
103. D.O. Hummel, *Analysis of Surfactants. Atlas of FTIR Spectra with Interpretations*, Hanser/Gardner, Cincinnati, OH (1996).
104. A.N. Davies, *Spectrosc. Europe* **12** (5), 30–6 (2000).
105. J. Yang, E.J. Hasenoehrl and P.R. Griffiths, *Vib. Spectrosc.* **14**, 1 (1997).
106. P.R. Griffiths and J.A. de Haseth, *Fourier Transform Infrared Spectrometry*, John Wiley & Sons, Inc., New York, NY (1986).
107. F. Rouessac and A. Rouessac, *Chemical Analysis. Modern Instrumental Methods and Techniques*, John Wiley & Sons, Ltd, Chichester (2000).
108. A.J. Ismail, F.R. van de Voort and J. Sedman, in *Instrumental Methods in Food Analysis* (J.R.J. Paré and J.M.R. Bélanger, eds), Elsevier Science, Oxford (1997), pp. 93–139.
109. J. Coates, in *Applied Sectroscopy* (J. Workman and A.W. Springsteen, eds), Academic Press, San Diego, CA (1998), pp. 49–91.
110. T. Hakkarainen, E. Mikkola, J. Laperre, F. Gensous, P. Fardell, Y. Le Tallec, C. Bajocchi, K. Paul, M. Simonson, C. Deleu and E. Metcalfe, *Fire Mater* **24** (2), 101–12 (2000).
111. K.T. Paul, *Cell. Polym.* **14**, 100–17 (1995).
112. B.B. Baker, in *Applied Polymer Analysis and Characterization* (J. Mitchell, ed.), Hanser Publishers, Munich (1987), pp. 329–39.
113. R. Bhardwaj and P. Bakshi, *Proceedings SPE ANTEC '98*, Atlanta, GA (1998), pp. 2219–22.
114. H. Kolnaar, *Techn. Rept DSM Research*, Geleen (1998).
115. J.-Q. Pan, N.C. Liu and W.W.Y. Lau, *Polymer Degrad. Stab.* **62** (1), 165–70 (1998).

116. R.L. Gray and C. Neri, *Proceedings 19th Ann. Intl Conf. on Advances in the Stabilization and Degradation of Polymers*, Luzern (1997), pp. 63–79.

117. Y. Zhang, J.-J. Li and Y.-P. Qian, *Shiyou Huagong* **28**, 48–51 (1999).

117a. M. Blanco, J. Coello, H. Iturriaga, S. Maspoch and E. Bertran, *Appl. Spectrosc.* **49** (6), 747–53 (1995).

118. ASTM D 2124-99. *Standard Test Method for Analysis of Components in Poly(Vinyl Chloride) Compounds Using an Infrared Spectrophotometric Technique*, *Annual Book of ASTM Standards*, ASTM, West Conshohocken, PA (1999), Vol. 08–01.

119. J. Udris, *Analyst (London)* **96**, 130 (1971).

120. R.R. Kozlowski and T.K. Gallagher, *J. Vinyl Add. Technol.* **3**, 249–55 (1997).

121. E. Monroy, N. Wolff, V. Ducruet and A. Feigenbaum, *Analysis* **21**, 221–6 (1993).

122. J. Wang and X. Yang, *Fushun Shiyou Xueyuan Xuebao* **19** (2), 33–9 (1999).

123. T. Kumar, *Analyst* **115**, 1319–22 (1990).

124. P. Fux, *Analyst* **114**, 445–9 (1989).

125. P.-C. Le, *PTS-MS 2022 Report (Aktuelle Analyseverfahren für die Untersuchung von Papier, Kreislauf und Abwasser)*, PTS Analytik-Tage (2000), Paper 3.1–16.

126. S.M. Howdle, J.M. Ramsay and A.I. Cooper, *J. Polym. Sci. B: Polym. Phys.* **32**, 541 (1994).

127. Y.-T. Shieh and T.-H. Tsai, *J. Appl. Polym. Sci.* **69**, 255–61 (1998).

128. M. Reeder, W. Enlow and E. Borkowski, *Factors Influencing the FTIR Analysis of Phosphites in Polyolefins*, Technical Paper F181, GE Specialty Chemicals, General Electric Company, Parkersburg, WV (1989).

129. W.U. Malik, R. Hague and S. PalVerma, *Bull. Chem. Soc. Japan* **36**, 746 (1963).

130. C. Szalkowski and J. Garber, *J. Agr. Food Chem.* **10**, 110 (1962).

131. S. Klingbeil, *PhD Thesis*, University of Göttingen (1995).

132. D. Rendell, *Fluorescence and Phosphorescence Spectroscopy*, John Wiley & Sons, Ltd, Chichester (1987).

133. M.A. Omary and H.H. Patterson, in *Encyclopedia of Spectroscopy and Spectrometry* (J.C. Lindon, ed.), Academic Press, San Diego, CA (2000), pp. 1186–207.

134. J.N. Demas, in *Encyclopedia of Physical Science and Technology* (R.A. Meyers, ed.), Academic Press, San Diego, CA (2001), Vol. 8.

135. L. Zlatkevich (ed.), *Luminescence Techniques in Solid State Polymer Research*, M. Dekker, New York, NY (1989).

136. S.G. Schulman (ed.), *Molecular Luminescence Spectroscopy, Methods and Applications*, John Wiley & Sons, Ltd, Chichester (1985).

137. R. Schuster and H. Schulenberg-Schell, *A New Approach to Lower Limits of Detection and Easy Spectral Analysis*, Agilent Technologies Application Primer No. 5968–9346E (2000).

138. J.N. Miller (ed.), *Standards in Fluorescence Spectrometry*, Chapman & Hall, London (1981).

139. N.S. Allen, *Photochemistry* **28**, 381–452 (1997).

140. M.J. Sonnenschein and C.M. Roland, *Polymer* **31**, 2023 (1990).

141. N. Fukazawa, K. Yoshioka, H. Kukumura and H. Masuhara, *Phys. Chem.* **97**, 6753 (1993).

142. M. Edge, N.S. Allen, R. Wiles, W. McDonald and S.V. Mortlock, *Polymer* **36**, 227 (1995).

143. N.S. Allen, M. Edge, J. Daniels and D. Royall, *Polym. Degrad. Stab.* **62**, 373–83 (1998).

144. M. Edge, R. Wiles, N.S. Allen, W.A. McDonald and S.V. Mortlock, *Polym. Degrad. Stab.* **53**, 141 (1996).

145. N.S. Allen, G. Rivalle, M. Edge, I. Roberts and D.R. Fagerburg, *Polym. Degrad. Stab.* **67**, 325–34 (2000).

146. R. Qi, K. Hu, Q. Zhu, W. Pang and. G. Zhou, *Chem. J. Internet* **2** (3) (2000).

147. J.W. Burdon and N.C. Billingham, *Polym. Mater. Sci. Engng* **62**, 101–5 (1990).

148. D. Phillips and M. Carey, in *Polymer Spectroscopy* (A.H. Fawcett, ed.), John Wiley & Sons, Ltd, Chichester (1996), pp. 369–89.

149. G.F. Kirkbright, R. Narayanaswamy and T.S. West, *Anal. Chim. Acta* **52**, 237 (1970).

150. J. Scheirs, J. Pospíšil, M.J. O'Connor and S.W. Bigger, *Adv. Chem. Ser.* **249**, 359–74 (1996).

151. C.A. Parker and W.J. Barnes, *Analyst* **82**, 606 (1957).

152. E. Cleve and E. Schollmeyer, *Textilveredlung* **30**, 18–21 (1995).

153. A. O'Brien and I. Cooper, *Food Addit. Contam.* **18** (4), 343–55 (2001).

154. H.G. Schild and D.A. Tirrell, *Langmuir* **6**, 1676–9 (1990).

155. R. Verstappen, *Techn. Rept DSM Research*, Geleen (1994).

156. W. Rettig, B. Strehinel, S. Schvadev and H. Seifort (eds), *Applied Fluorescence in Chemistry, Biology and Medicine*, Springer-Verlag, Heidelberg (1999).

157. C.A. Parker and C.G. Hatchard, *Analyst* **87**, 664 (1962).

158. E.M. Purcell, H.C. Torrey and R.V. Pound, *Phys. Rev.* **69**, 37 (1946).

159. F. Bloch, W.W. Hansen and M. Packard, *Phys. Rev.* **70**, 474 (1946).

160. R.S. Macomber, *A Complete Introduction to Modern NMR Spectroscopy*, John Wiley & Sons, Inc., New York, NY (1998).

161. M.H. Fisch, *J. Vinyl Technol.* **12**, 136–41 (1990).

162. R.R. Ernst, *Rev. Sci. Instrum.* **36**, 1689 (1965).

163. A. Street, *LabPlus Intl* **16** (5), 28–31 (2002).

164. W.G. Proctor and F.C. Yu, *Phys. Rev.* **77**, 717 (1950).

165. W.C. Dickinson, *Phys. Rev.* **77**, 736 (1950).

166. J.H. Noggle and R.E. Schirmer, *The Nuclear Overhauser Effect, Chemical Applications*, Academic Press, New York, NY (1971).

167. D. Neuhaus and M.P. Williamson, *The Nuclear Overhauser Effect in Structural and Conformational Analysis*, VCH, Weinheim (1989).

168. S.R. Hartmann and E.L. Hahn, *Phys. Rev.* **128**, 2042 (1962).

169. R.R. Ernst, G. Bodenhausen and A. Wokaun, *Principles of Nuclear Magnetic Resonance in One and Two Dimensions*, Clarendon Press, Oxford (1997).

170. J.B. Grutzner, in *Encyclopedia of Analytical Science* (A. Townshend, ed.), Academic Press, London (1995), pp. 3358–82.

171. A. Kumar and R.C.R. Grace, in *Encyclopedia of Spectroscopy and Spectrometry* (J.C. Lindon, ed.), Academic Press, San Diego, CA (2000), pp. 1643–53.

172. R.R. Ernst, in *Encyclopedia of Nuclear Magnetic Resonance* (D.M. Grant and R.K. Harris, eds), John Wiley & Sons, Ltd, Chichester (1996), pp. 3122–32.

173. R.L. Vold and R.R. Vold, *Progr. NMR Spectrosc.* **12**, 79 (1978).

174. E.O. Stejskal and J.E. Tanner, *J. Chem. Phys.* **42**, 288–92 (1965).

175. L.M. Jackman and F.A. Cotton, *Dynamic NMR Spectroscopy*, Academic Press, New York, NY (1975).

176. A.N. Davies, J. Lambert, R.J. Lancashire and P. Lampen, *Spectrosc. Eur.* **11** (3), 18–20 (1999).

177. W.F. Reynolds, in *Encyclopedia of Spectroscopy and Spectrometry* (J.C. Lindon, ed.), Academic Press, San Diego, CA (2000), pp. 1554–67.

178. A.J. Brandolini and D.D. Hills, *NMR Spectra of Polymers and Polymer Additives*, M. Dekker, New York, NY (2000).

179. C.J. Pouchert and J. Behnke (eds), *Aldrich Library of 75 MHz ^{13}C and 300 MHz ^1H FT-NMR Spectra*, Aldrich Chemical Co., Milwaukee, WI (1992).

180. S. Sasaki, *Handbook of Proton-NMR Spectra and Data*, Academic Press, New York, NY (1996), Vols 1–5.

181. Varian Associates, *High Resolution NMR Spectra Catalogue*, Varian, Palo Alto, CA (1962/3), Vol. 2.

182. Sadtler Research Laboratories, *Sadtler Carbon-13 NMR of Monomers and Polymers*, Philadelphia, PA (1985).

183. R.A. Nyquist, *Interpreting Infrared, Raman and Nuclear Magnetic Resonance Spectra*, Academic Press, Orlando, FL (2001).

184. E.L. Hahn, *Phys. Rev.* **80**, 580 (1950).

185. J.T.W.E. Vogels, *PhD Thesis*, University of Leiden (2002).

186. D.M. Doddrell, D.T. Pegg and M.B. Bendall, *J. Magn. Reson.* **48**, 323–7 (1982).

187. D.G. Gorenstein, *Phosphorus-31 NMR Principles and Applications*, Academic Press, New York, NY (1984).

188. S. Braun, H.-O. Kalinowski and S. Berger, *100 and More Basic NMR Experiments: A Practical Course*, VCH, New York, NY (1995).

189. H. Günther, *NMR Spectroscopy – Basic Principles, Concepts and Applications in Chemistry*, John Wiley & Sons, Ltd, Chichester (1995).

190. A.E. Derome, *Modern NMR Techniques for Chemical Research*, Pergamon Press, Oxford (1995).

191. J.K.M. Sandersad and B.K. Hunter, *Modern NMR Spectroscopy – A Guide for Chemists*, Oxford University Press, Oxford (1994).

192. J.W. Akitt, *NMR and Chemistry – An Introduction to Modern NMR Spectroscopy*, Chapman & Hall, London (1992).

193. C.P. Slichter, *Principles of Magnetic Resonance*, Springer Verlag, Berlin (1990).

194. W. Kemp, *NMR in Chemistry – A Multinuclear Introduction*, Macmillan, London (1986).

195. R.K. Harris, *Nuclear Magnetic Resonance Spectroscopy*, Longman, Harlow (1986).

196. R.J. Abraham, J. Fisher and P. Loftus, *Introduction to NMR Spectroscopy*, John Wiley & Sons, Ltd, Chichester (1988).

197. N. Nakanishi (ed.), *One-dimensional and Two-dimensional NMR Spectra by Modern Pulse Techniques*, Kodansha, Tokyo (1990).

198. G. Bodor, *Structural Investigation of Polymers*, Ellis Horwood, New York, NY (1991).

199. B. Wrachmeyer, *Guide to Multinuclear Magnetic Resonance*, Wiley-VCH, Weinheim (2002).

200. F.J.M. van de Ven, *Multidimensional NMR in Liquids: Basic Principles and Experimental Methods*, VCH Publishers, New York, NY (1995).

201. W.R. Croasmun and R.M.K. Carlson (eds), *Two-dimensional NMR Spectroscopy, Applications for Chemists and Biochemists*, VCH Publishers, New York, NY (1994).

202. H. Friebolin, *Basic One- and Two-dimensional NMR Spectroscopy*, VCH, Weinheim (1991).

203. D.A.W. Wendisch, *Appl. Spectrosc. Rev.* **28** (3), 165–229 (1993).

204. T.M. Alam, M. Celina, D.R. Wheeler, R.A. Assink, R.L. Clough and K.T. Gillen, *Polym. News* **24** (6), 186–91 (1999).

205. K. Saito and T. Ogawa, *Bunseki Kagaku* **49** (1), 3–9 (2000).

206. K. Schmidt-Rohr, *ACS Symp. Ser.* **598**, 184–90 (1995).

207. T.D.W. Claridge, *High-resolution NMR Techniques in Organic Chemistry*, Pergamon, Amsterdam (1999).

208. D.W. Carlson, M.W. Hayes, H.G. Bansaw, A.S. McFadden and A.G. Altenau, *Anal. Chem.* **43**, 1874–6 (1971).

209. R. Sievers, in *Nuclear Magnetic Resonance Shift Reagents* (R. Sievers, ed.), Academic Press, New York, NY (1973).

210. A.D. DeBellis and K.C. Hass, *J. Phys. Chem.* **A 103** (38), 7665–71 (1999).

211. P. Fux, *Analyst* **115** (2), 179–83 (1990).

212. W. Freitag and H. Lind, *Chem. Ind. (London)*, 933 (1980).

213. H.J.M. Bartelink, J. Beulen, E.F.T. Duynstee and E. Konijnenberg, *Chem. Ind.* **5**, 202–4 (1980).

214. P.B. Mansfield, *Chem. Ind. (London)* **28**, 792 (1971).

215. D. Braun and E. Bezdadea, *Angew. Makromol. Chem.* **113**, 77–90 (1983).

216. J. Pierre and J. van Bree, *Kunststoffe* **73**, 319–24 (1983).

217. M. Banks, J. Ebdon and M. Johnson, *Polymer* **35**, 3470–3 (1994).

218. S. Al-Malaika, S. Issenhuth and J. Burdick, *Polym. Degrad. Stab.* **73**, 491–503 (2001).

219. Project AIR-94/025. *Safety and Quality Control of Plastic Materials for Food Contact Materials*, Final Report (1997).

220. J. Ehret-Henry, J. Bouquant, D. Scholler, R. Klink and A.E. Feigenbaum, *Food Addit. Contam.* **9**, 303–14 (1992).

221. A.E. Feigenbaum, J. Bouquant, V.J. Ducruet, J. Ehret-Henry, D.L. Marque, A.M. Riquet, D. Scholler and J.C. Wittmann, *Food Addit. Contam.* **11** (2), 141–54 (1994).

222. Commission of the European Communities, *Draft Synoptic Document 5 on Plastic Materials and Articles Intended to Come into Contact with Foodstuffs*, Brussels (1991).

223. L. Castle, A.J. Mercer and J. Gilbert, *Food Addit. Contam.* **8**, 565–76 (1991).

224. B. van Lierop, L. Castle, A.E. Feigenbaum and A. Boenke, *Spectra for the Identification of Additives in Food Packaging*, Kluwer Academic Publishers, Dordrecht (1998).

225. Bureau Communautaire de Référence, *Spectra for the Identification of Additives in Food Packaging*, Geel (1993).

226. H.A.J. Linssen, *Techn. Rept DSM Research*, Geleen (2000).

227. G.S.H. Lee, K.M. Brinch, K. Kannangara, M. Dawson and M.W. Wilson, *J. Forensic Sci.* **46** (4), 808–21 (2001).

228. R.D. Blackledge and M. Vincenti, *J. Forensic Sci. Soc.* **34** (4), 245–56 (1994).

229. A. Hagenaars, *personal communication* (1999).

230. J.C. Randall, *J. Macromol. Sci., Rev. Macromol. Chem. Phys.* **C29** (2&3), 201–317 (1989).

231. A.H. Sharma, F. Mozayeni and J. Alberts, *Proceedings SPE Polyolefins XI*, Houston, TX (1999), pp. 679–703.

232. D.D. Werstler, *Rubber Chem. Technol.* **53**, 1191 (1980).

233. J. Beulen, H.A.J. Linssen and G.v.d. Velden, *Techn. Rept DSM Research*, Geleen (1997).

234. W. Barendswaard, J. Moonen and M. Neilen, *Anal. Chim. Acta* **283**, 1007–24 (1993).

235. G.v.d. Velden, *personal communication* (2001).

236. Y. Nakata, H. Noda and A. Hyuga, *Sekiyu Gakkaishi* **26** (1), 50–6 (1983).

237. S. Cook and R. Lehrle, *Eur. Polym. J.* **29** (1), 1–8 (1993).

238. L.C.M. van Gorkum and A. Jensen, in *Anionic Surfactants. Analytical Chemistry* (J. Cross, ed.), M. Dekker, New York, NY (1998), pp. 169–208.

239. C.M. Sultany, *Polym. Bull.* **20**, 463–6 (1988).

240. K.R. Carduner and R.O. Carter III, *J. Magn. Reson.* **84**, 361–6 (1989).

241. K.R. Carduner, *Polym. Bull.* **21**, 327–34 (1989).

242. K.R. Carduner, R.O. Carter III, M.F. Cheung, A. Golovoy and H. van Oene, *J. Appl. Polym. Sci.* **40**, 963–75 (1990).

243. M.F. Cheung, K.R. Carduner, A. Golovoy and H. van Oene, *J. Appl. Polym. Sci.* **40**, 977–87 (1990).

244. D.W. Allen, D.A. Leathard and C. Smith, *Chem. Ind.*, 854–5 (1987).

245. G.J. Klender, K.R. Jones and C.W. Calhoun, *Polymer Prepr. (Am. Chem. Soc., Div. Polym. Chem.)* **34** (2), 156–7 (1993).

246. T. Tikuisis and M. Cossar, *Proceedings SPE ANTEC '93*, New Orleans, LA (1993), pp. 270–6.

247. G.B. Kenion, R. Ludicky, J.H. Deitch, M. Leib and D. Doster, *Tappi J.* **77** (6), 199–205 (1994).

248. M.R. Jackupca, D.R. Stevenson and D.L. Stein, *Proceedings Polyolefins X*, Houston, TX (1997), pp. 627–52.

249. R.G. Parker and C.J. Carman, *ACS Symp. Ser.* **169**, 363–73 (1978).

250. T.U. Gevert and S.F. Svenson, *Polymer* **26**, 307–9 (1985).

251. D.W. Allen, J.S. Brooks, R.W. Clarkson, J. Unwin and P.J. Smith, *Polym. Degrad. Stab.* **13** (3), 191–200 (1985).

252. J. Kelm and D. Gross, *Rubber Chem. Technol.* **58**, 37–44 (1985).

253. S. Narayan, R.E. Lee, D. Hallberg and V. Malatesta, *J. Appl. Medical Polymers* **5** (2), 77–9 (2001).

254. T.M. Alam, M. Celina, R.A. Assink, K.T. Gillen and R.L. Clough, *Polym. Prepr. (Am. Chem. Soc., Div. Polym. Chem.)* **38** (1), 784–5 (1997).

255. H. Yu and J. Mi, *Huaxue Shijie* **29** (6), 263–8 (1988).

256. W.P. Aue, E. Bartholdi and R.R. Ernst, *J. Chem. Phys.* **64**, 2229 (1976).

257. J. Jeener, *Ampere International Summer School*, Basko Polje, Yugoslavia (1971).

258. L. Muller, A. Kumar and R.R. Ernst, *J. Chem. Phys.* **63**, 5490 (1975).

259. C. Dybowski, A. Glatfelter and H.N. Cheng, in *Encyclopedia of Spectroscopy and Spectrometry* (J.C. Lindon, ed.), Academic Press, San Diego, CA (2000), pp. 149–58.

260. A.N. Davies, *Spectrosc. Europe* **11** (1), 14–5 (1999).

261. P.L. Rinaldi, in *Encyclopedia of Spectroscopy and Spectrometry* (J.C. Lindon, ed.), Academic Press, San Diego, CA (2000), pp. 2370–81.

262. J.C. Lindon, in *Encyclopedia of Analytical Science* (A. Townshend, ed.), Academic Press, London (1995), pp. 3507–18.

263. G.A. Gray, in *Two-dimensional NMR Spectroscopy* (W.R. Croasmun and R.M.K. Carlson, eds), VCH Publishers, New York, NY (1994), pp. 1–65.

264. K. Schmidt-Rohr and H.W. Spiess, *Multidimensional Solid-state NMR and Polymers*, Academic Press, New York, NY (1994).

265. G.E. Martin and A.S. Zektzer, *Two-dimensional NMR Methods for Establishing Molecular Connectivity*, VCH Publishers, New York, NY (1988).

266. A. Bax, *Two-dimensional Nuclear Magnetic Resonance in Liquids*, Delft University Press, Dordrecht (1984).

267. W.E. Hull, in *Two-dimensional NMR Spectroscopy* (W.R. Croasmun and R.M.K. Carlson, eds), VCH Publishers, New York, NY (1994), pp. 67–456.

268. C. Choi, L. Bailey, A. Rudin and M.M. Pintar, *J. Polym. Sci., Part B: Polym. Phys.* **35** (15), 2551–8 (1997).
269. S. Pickup and F.D. Blum, *Macromolecules* **22**, 3961 (1989).
270. J.S. Vrentas and J.L. Duda, *AIChE J.* **25**, 1 (1979).
271. K.F. Morris and C.S. Johnson, *J. Am. Chem. Soc.* **115**, 4291–9 (1993).
272. P. Stilbs, *Progr. Nucl. Magn. Reson. Spectrosc.* **19**, 1 (1987).
273. P. Stilbs, *Anal. Chem.* **53**, 2135 (1981).
274. D. Wu, A. Chen and C.S. Johnson, *J. Magn. Reson. Ser. A* **115**, 260 (1995).
275. K.F. Morris, P. Stilbs and C.S. Johnson, *Anal. Chem.* **66**, 211 (1994).
276. K. Jankowski. M. Mojski and L. Synoradzki, *Chem. Anal. (Warsaw)* **42**, 567–77 (1997).
277. H. Barjat, G.A. Morris and A.G. Swanson, *J. Magn. Reson.* **131**, 131–8 (1998).
278. E.R. Andrew, *Phil. Trans. R. Soc. London* **A299**, 505 (1981).
279. K.F. Morris and C.S. Johnson, *J. Am. Chem. Soc.* **114**, 3139 (1992).
280. D. Wu, A. Chen and C.S. Johnson, *Bull. Magn. Reson.* **17** (1–4), 21–6 (1995).
281. L.C.M. van Gorkum and T.M. Hancewicz, *J. Magn. Reson.* **130**, 125–30 (1998).
282. G.A. Morris, in *Methods for Structure Elucidation by High Resolution NMR* (G. Batta and K. Köver, eds), Elsevier, Amsterdam (1997).
283. W. Donghui, A. Chien and C.S. Johnson, *J. Magn. Reson. Ser. A* **121**, 88 (1996).
284. D.A. Jayawickrama, C.K. Larive, E.F. McCord and D.C. Roe, *Magn. Reson. Chem.* **36**, 755–60 (1998).
285. H.A.J. Linssen and N.K. de Vries, *Techn. Rept DSM Research*, Geleen (1998).

CHAPTER 6

The Beauty, the Beast and the Oracle

Organic Mass-Spectrometric Methods

Mass spectrometry involves the study of ions in the vapour phase. Mass spectrometers are analytical instruments that convert neutral molecules into gaseous ions and separate those ions according to the ratio of their mass-to-charge (m/z). The location of the mass lines provides a qualitative analysis, and their intensity, mostly measured relative to that of the matrix element or a suitable internal standard, gives a quantitative analysis.

The traditional mass spectrometry laboratory is equipped with top-of-the-line instrumentation, run by a mass spectrometrist for high-performance, nonroutine analyses. The use of mass spectrometry has recently evolved from mass-spectrometrist driven to application driven. In polymer/additive analysis terms this means that the additive specialist/analyst has gained control over the use of instruments such as LC-PB-MS or LC-API-MS. Polymer/additive analysis is served well by both options given in Table 6.1.

Mass spectrometry has a number of features and advantages that can make it a very valuable tool for the identification of organic additives in polymers (Table 6.2). The range of products that can be studied is limited by the ionisation method used and the performance of the mass spectrometer. Mass spectrometry

Additives In Polymers: Industrial Analysis And Applications   J. C. J. Bart
© 2005 John Wiley & Sons, Ltd   ISBN: 0-470-85062-0

Table 6.1 Mass spectrometry in the laboratory

Past: mass-spectrometrist driven
- <50 samples per day: EI, CI, FAB
- Complex instruments run by mass spectrometrists
- Data analysis by mass spectrometrists
- Central MS facilities and many substance classes
- Low throughput, high degree of manual work

Present: applications driven
- Up to 4000 samples per day: ESI, MALDI
- Simple-to-use instruments run by application specialists
- Reasonable automation, application-specific software
- Data analysed by user, certain degree of automation
- Decentralised analyses on specific substance classes

Table 6.2 Main characteristics of mass-spectrometric detection

Advantages
- Powerful
- Selective, sensitive
- Time saving in method development and troubleshooting
- Straightforward fingerprinting
- Reliable
- Small sample amount (fmoles)
- Replaces complex sample pretreatment or separation procedures
- Replaces selective detectors
- Molecular weight determination, absolute MWDs
- Molecular formula (AC-MS)
- Molecular structure (from mass-spectral fragmentation patterns)
- Not an averaging method
- Direct and indirect mixture analysis; purity determination
- Wide choice of sample introduction, mass analysers and ionisation modes in relation to application
- Wide applicability (polarity, thermal stability, volatility, molecular weight)

Disadvantages
- Spectral interpretation
- Expensive equipment
- Operator skills
- Difficult quantitation

provides qualitative and quantitative information concerning the *molecular weight* (molecular weight distribution) and *molecular structure* of organic and inorganic compounds. The technique has relatively high sensitivity, dynamic range and linearity. It can be used as a qualitative analytical tool for *fingerprinting* purposes using widely available mass spectral libraries, as well as for (partial) structural elucidation by identification of molecular fragments. With sufficient precautions and skill, mass spectrometry allows *quantitative analysis* of mixtures of gases or liquids, and in some cases solids in concentrations at the ppb level. This is particularly true for inorganic mass analysis (Section 8.5).

Despite these numerous advantages, mass spectrometry has often been used more as an auxiliary, rather than a primary, identification method for additives in polymers, paints, coatings, etc. Nevertheless, mass spectrometry can be used for direct determination of the composition of unknown admixtures. More difficult is the MS examination of substances of low volatility, as the sample has to be introduced in the gas phase. This requires volatilisation, which often leads to fragmentation.

The analytical chemist is attracted to mass spectrometry mainly by its speed, reliability and use as an almost background-free *detector* for chromatography, which assumes low detection limits. Where the mass spectrometer is not being used as stand-alone equipment, but rather as a detector, the final results are greatly dependent on other – hyphenated – equipment and materials involved, including chromatographic columns; the purity of any gases or solvents used; the quality of standard samples, etc. Clearly, the better the chromatographic separation, the less strain needs to be put on the skills of the mass spectrometer and mass spectrometrist. In organic chemistry MS can readily identify reaction products and by-products, particularly with the help of gas chromatography.

Nowadays, MS is often no longer the analytical bottleneck, but rather what precedes it (sample preparation) and follows it (data handling, searching). Direct mass-spectrometric methods have to compete with the separation techniques such as GC, HPLC and SFC that are commonly used for quantitative analysis of polymer additives. Extract analysis has the general advantage that higher-molecular-weight (less-volatile) additives can be detected more readily than by direct analysis of the polymer compound.

Application of mass spectrometry to analysis of additives in polymers is mainly used as a qualitative tool and relates to three main areas:

(i) direct mass spectral analysis, i.e. with minimal sample preparation;
(ii) indirect methods, i.e. extraction; and
(iii) postcolumn hyphenation (see Chapter 7).

Computerised identification of polymer additives on the basis of MS spectral data has been reported [1].

Inorganic mass spectrometry is described in Section 8.5; related topics are laser mass spectrometry and surface mass spectrometry. In the 37-page review on the analysis of additives by Freitag [2] in 1990, just 13 lines were devoted to mass spectrometry. Later, Bataillard *et al.* [3] still dedicate less than a page to MS, although many regard the technique as one of the most important of its kind available today. This monograph

Table 6.3 Mass spectrometry for polymer/additive analysis

- Small sample amount (μg for direct analysis)
- MW directly derivable from mass spectra
- Molecular formulae from high-resolution mass measurements
- Molecular structures from mass-spectral fragmentation patterns
- Allowance for mixture analysis
- Wide applicability

devotes a much greater share to this technique. Mass spectrometry is not a panacea for all polymer/additive problems although it has developed into a major tool. Table 6.3 illustrates its main assets. Few reviews deal with the use of mass spectrometry for the purpose of polymer/additive analysis [4,5].

6.1 BASIC INSTRUMENTATION

Principles and Characteristics A mass spectrometer consists of various components which are necessary for the formation of ions from molecules, and for their separation and detection (Fig. 6.1). Miniaturisation of MS represents a strategic technology.

The many different kinds of mass spectrometers are generally described according to the types of ionisation sources, mass analysers and detectors (Table 6.4). There are many different ways of introducing samples into the ionisation source. Whereas single-substance samples can be inserted directly by means of a probe, complex mixtures obviously will benefit from a preceding separation step, and this usually involves interfacing chromatography. Interfaces are moving belts, particle beams, etc. (Section 7.3.3.2). The ion source produces a beam of ions, the analyser separates the beam according to the m/z ratio and the detector determines the fraction of the total ion current carried by each of the ions. Except for atomic pressure ionisation (API) sources, mass-spectrometric devices operate under vacuum. Ion production distinguishes gas phase and desorption techniques; molecular and fragment-ions; positive and negative ions; and mono- and multicharged ions. Gas phase techniques are EI, CI (+, PCI; −, NCI)

and FI; desorption techniques are FD, PD, FAB/LSIMS, DCI, LD, ESI, TSP and MALDI. Not all the mass analysers are compatible with all the ionisation methods. Detection in most types of mass spectrometer depends on measurement of the electrons or the secondary ions ejected from a surface by the impact of the ions of interest, usually after electron multiplication.

Consequently, any given mass spectrometric experiment can be performed in many different ways, depending on sample inlet, ionisation type, mass analyser and detector. This choice greatly determines the figures of merit, including mass resolution, speed, abundance, sensitivity, m/z range, and ion transmission. The relative importance of each figure of merit is defined by the analytical problem at hand. It is little wonder, therefore, that with so many different areas of application for mass spectrometry, no single form of mass spectrometry will ever likely supplant all others. Some instrumentation still needs the skills of the experienced mass spectrometrist, e.g. the use of magnetic sector and FTICR instruments.

The mass spectrometer is a mass-flow sensitive device, which means that the signal is proportional to the mass flow dm/dt of the analyte, i.e. the concentration times the flow-rate. It is only now possible to realise the high (theoretically unlimited) mass range and the high-sensitivity multichannel recording capabilities that were anticipated many years ago. Of considerable interest to the problem of polymer/additive deformulation are some of the latest developments in mass spectrometry, namely atmospheric pressure ionisation (API), and the revival of time-of-flight spectrometers (allowing GC-ToFMS, MALDI-ToFMS, etc.).

Many books deal with general mass spectrometry [6–10]. In recent mass spectrometry books, little specific attention has been paid to polymer additives. De Hoffmann and Stroobant [11] and others [12] have traced the history of mass spectrometry.

Applications Mass spectrometry has often been used more as an auxiliary, rather than a primary, identification method for additives in polymers. Table 6.5 shows the suitability of various ionisation modes for oligomer (and polymer) analysis.

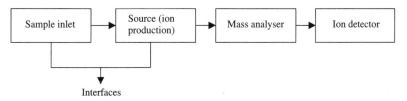

Figure 6.1 Main components of a mass spectrometer

Table 6.4   Basic components of mass spectrometers

Sample inlet	Ionisation sources	Mass analysers	Detectors
GC	Thermal ionisation	Magnetic sector (B)	Faraday cup
LC	Spark source	Double-focusing (EB)	Electron multiplier
SFC	Electron impact (EI)	Reversed geometry (BE)	Magnetic electron multiplier
CE	Photoionisation (PI)	Ion-cyclotron resonance (ICR)	Continuous dynode multiplier
Pyrolysis	Field ionisation (FI)	Quadrupole (Q)	Daly detector
Flow injection	Chemical ionisation (CI)	Quadrupole ion trap (QITMS)	Diode array detector
Solid/liquid probe	Field desorption (FD)	Radio frequency (RF)	Image currents
	Infrared laser desorption (IR-LD)	Time-of-flight (ToF)	Inductive detector
	Desorption chemical ionisation (DCI)	Fourier transform (FTMS)	Microchannel plate
	Fast atom bombardment (FAB)	Triple quadrupole (QQQ)	Photomultiplier
	Plasma desorption (PD)	Four sector (EBEB)	
	Liquid secondary-ion mass spectrometry (LSIMS)	Hybrid (EBQQ)	
	Thermospray (TSP)/plasmaspray (PSP)	Hybrid (EB-ToF, Q-ToF)	
	Electrohydrodynamic ionisation (EHI)	Tandem ToF-ToF	
	Multiphoton ionisation (MPI)		
	Atmospheric pressure chemical ionisation (APCI)		
	Electrospray ionisation (ESI)		
	Ion spray (ISP)		
	Matrix-assisted laser desorption/ionisation (MALDI)		
	Atmospheric pressure photoionisation (APPI)		

Table 6.5   Oligomer analysis by mass spectrometry

Ionisation mode	Oligomer analysis	Polymer MWD	Reference
Direct probe introduction (EI-MS)	+	−	[13]
DCI	+	−	[14]
^{252}Cf PD-MS	+	−	[15]
ESP	+	+	[16]
LSIMS/FAB	+	−	[17]
FD	+	+	[18]
EH	+	+	[19]
LD	+	+	[20]
TSP	+	+	[21]

6.1.1   Inlet Systems

There are several means of introducing samples into the mass spectrometer. The inlet system to be used depends mainly on the nature of the sample (volatility, molecular weight, polarity and thermal stability), and the ionisation method (in particular, the gas pressure in the ion source). The most common inlet systems are:

  (i) cold inlets;
 (ii) hot inlets;
(iii) direct probe inlets;
 (iv) GC inlets;

  (v) inlets for solutions; and
 (vi) total inlet systems.

In cold inlets, gases or compounds that are very volatile at ambient temperature and a pressure of about 10^{-2} Torr are allowed to 'leak' into the mass spectrometer (kept at 10^{-6} Torr). In all-glass heated inlet systems (AGHIS) compounds may be heated to about 300 °C for volatilisation. Compounds that are not sufficiently volatile to be introduced through the cold or hot inlets may be inserted directly into the ion source by means of a probe passing through a vacuum lock. At the low pressures of about 10^{-7}–10^{-5} Torr inside a conventional EI source and with heating, very many compounds are sufficiently volatile to yield good mass spectra. Compounds with relative molecular masses up to about 2000 Da may be measured with a *direct insertion probe*, using common EI/CI methods of ionisation. Direct insertion probes, which are often heated, are also used with milder methods of ionisation, like FAB, allowing much larger molecules to be examined. Relative molecular masses of about 25 000 Da are readily manageable assuming, of course, that the analyser part of the instrument is commensurate with such a mass. The aforementioned inlets are generally used when pure, or nearly pure, compounds are available for analysis.

The mass spectra of mixtures are often too complex to be interpreted unambiguously, thus favouring the separation of the components of mixtures *before* examination by mass spectrometry. Nevertheless, direct polymer/additive mixture analysis has been reported [22,23], which is greatly aided by tandem MS. Coupling of mass spectrometry and a flowing liquid stream involves vaporisation and solvent stripping before introduction of the solute into an ion source for gas-phase ionisation (Section 7.3.3.2). Widespread LC-MS interfaces are thermospray (TSP), continuous-flow fast atom bombardment (CF-FAB), electrospray (ESP), etc. Also, supercritical fluids have been linked to mass spectrometry (SFE-MS, SFC-MS). A mass spectrometer may have more than one inlet (total inlet systems).

6.1.2 Modes of Detection

A mass spectrum is a compilation of ions of measured mass plotted against the measured intensities of the ion signals. The mass scale (given in units of m/z ratio, referenced to the ^{12}C mass of 12.0000 Da exactly) can be measured to a variable degree of accuracy, ranging from integral mass numbers to exact mass measurements to a few mDa. The intensity scale is marked in terms of relative abundance, referred to the most intense ion signal. A conventional mass spectrum is the result of several unimolecular dissociation processes. For unambiguous identification of an analyte, mass spectrometry is one of the most important techniques. For evaluation of mass-spectrometric data, the usual criteria (based on 'expert opinion' rather than founded on statistical evaluation of analytical data) are that four ions should be recorded, the abundances of which should deviate no more than 10 % from a corresponding standard. However, these criteria often prove difficult to meet, in particular for analytes with some of the diagnostic ions being of low mass or low relative abundance. This situation is complicated by the fact that mass spectrometry is available in a variety of modes, GC-MS (EI, CI), GC-MS/MS, LC-MS (ESI, APCI), LC-MS/MS, etc. Harmonisation in this area is very desirable. It is necessary to establish statistically founded *matching criteria* applicable to mass-spectrometric analytical data in order to achieve optimisation of false positive and false negative results in confirmatory analysis.

Comparison with mass-spectral libraries is the easiest way of interpreting mass spectra (and often the only way for non-mass spectroscopists). The NIST/EPA/NIH (NIST 02) electron-ionisation mass spectral library covers 143 000 compounds [24]. Mass-spectrometry search software is available [24a]. However, all mass spectra are not created equal. Consequently, secure comparison of mass spectra for positive compound identification requires attention to the ionisation mode. As the public libraries are far from being complete as regards commercial polymer additives, the creation of proprietary databases is often recommended. In the authors' laboratory a database has been created for some odd 1000 additives [25]. Computer software is also available to assist in the interpretation of series of low-resolution mass spectra, as measured with GC-MS, LC-MS, direct probe insertion or fractionated evaporation of solid mixtures [26]. Application facilitates the analysis of mass spectral fragmentation patterns and isotopic distributions. Mass spectra obtained with EI, CI, FAB, ESI, APCI and other ionisation techniques may be handled.

There are different methods of recording mass spectra:

(i) *Full scanning acquisitions* In this case the mass analyser scans over an appropriate wide m/z range, producing a mass spectrum from which (ideally) molecular ions provide an indication of the molecular weight of the sample. If the sample has fragmented in the ionisation source, then information regarding the molecular structure has been obtained. Full scan mode is used for library matching and structure elucidation. Under appropriate conditions, accurate mass measurements can be carried out.

(ii) *Selected-ion monitoring (SIM) or selected-ion recording (SIR)* For identification of known compounds (targeted analysis) diagnostic fragment ions are all that is needed. The mass analyser will then monitor the specified ions by switching from one to the next. This mode is more sensitive than full scanning (1000-fold increase), because all the available time is spent on the ions of interest rather than monitoring all the ions over a full m/z range.

SIM data acquisitions are usually performed where low levels of components need to be ascertained in complex matrices with a high background ion level. The advantage of this method is that both high sensitivity and high specificity are achieved. Although SIM is sensitive to pg of material, this sensitivity is highly dependent on the matrix and the interferences produced. Frequently, it is only possible to detect ng levels of compound. Both magnetic sector and quadrupole mass spectrometers are used for SIM analyses, as opposed to ToF instruments. Magnetic sector mass spectrometers can even perform SIM at high resolution, whereby the accurate, monoisotopic mass ions are specified and monitored, thus producing highly specific results. High-resolution SIM has not been used frequently in the field of additive analysis.

Obviously, by far the best method of performing SIM is to use a means of sample introduction which generates sample peaks of relatively short peak widths (as in GC or LC) that can be integrated – as opposed to the probe methods of sample introduction which deliver the sample into the ionisation source at a near-constant rate over long periods of time.

6.1.3　Mass Resolution

The main function of the mass analyser is to separate (resolve) the ions formed in the ionisation source on the basis of their mass to charge ratios (m/z). The resolving power of a mass spectrometer is a measure of its ability to separate two ions of any defined mass difference. More precisely, the resolution (R) of a mass analyser is defined as its ability to separate the ion envelopes of two peaks of equal intensity, i.e. the ratio of the mass of a peak (M_1) to the difference in mass between this peak and an adjacent higher mass peak (M_2), i.e.:

$$R = M_1/(M_2 - M_1) \qquad (6.1)$$

This means that a singly charged ion at m/z 500 requires a mass resolution of 500 in order to be separated from another singly charged ion at m/z 501. Resolution is also expressed in ppm ($10^6 \times \Delta M/M$). Low-resolution instruments are defined arbitrarily as instruments that separate unit masses up to m/z 2000. At higher resolution, smaller mass differences may be detected. Resolution above 2000 (10 % valley definition) usually requires access to a magnetic sector instrument, e.g. in the EB, B-IT or EBE ToF hybrid configurations [27,28]. Quadrupole, triple quadrupole, ion trap and conventional ToF mass spectrometers are essentially low-resolution instruments, but recent advances with ToF instruments (notably in the QToF configuration) have led to considerable improvements. High resolving power is needed for exact mass measurement to evaluate the possible chemical compositions of unknown compounds. Resolutions of 2×10^8 have been reported [29]. Fourier-transform instruments are capable of ultra-high-resolution (routine < 2 ppm mass accuracy, externally calibrated). With a modern double-focusing mass spectrometer, it is possible to measure the mass

of an ion to 1 ppm or better and obtain 100 000 or better resolution. Under GC-MS scanning conditions, 5 to 10 ppm mass accuracy is more common, and resolution is set between 2000 and 10 000. For accurate mass analyses using GC-MS, a compromise must be made between the resolution necessary to minimise mass interference and the signal intensity necessary to detect low levels of material.

6.1.4　Isotope Distributions

The *nominal mass* is the mass of an ion with a given empirical formula calculated using the *integer* mass numbers of the most abundant isotope of each element (e.g. $m/z = 249$ could originate from $C_{20}H_9{}^+$, $C_{19}H_7N^+$, $C_{19}H_5O^+$ or $C_{13}H_{19}N_3O_2{}^+$). The *exact mass* is the mass of an ion for a given empirical formula calculated using the *exact* mass of the most abundant isotope of each element (i.e. $C_{20}H_9{}^+$ 249.0704; $C_{19}H_7N^+$ 249.0580; $C_{19}H_5O^+$ 249.0341; $C_{13}H_{19}N_3O_2{}^+$ 249.1479). Atomic mass tables are available [30–33]. Knowledge of the elemental composition of the molecular ion and fragment ions greatly simplifies interpretation. Formula masses for various combinations of C, H, N and O (up to 250 Da) are listed [34].

In general, the mass resolution required for most analyses is such that the singly charged *isotope patterns* of the detected ions are readily discernible. For applications involving molecular weights up to about 1500 Da, this can be provided by magnetic sector, quadrupole, and time-of-flight mass spectrometers.

If one considers Chimassorb 81 (2-hydroxy-4-*N*-octoxy-benzophenone) of molecular formula $C_{21}H_{26}O_3$, a molecular ion ($M^{+\bullet}$) is generated at m/z 326 (nominal mass) in EI conditions with unit resolution set for the analyser in which all the atoms are the lowest mass isotopes (i.e. ^{12}C, ^1H, and ^{16}O). There will also be lower-intensity ions at m/z 327, which correspond to molecules of the same compound in which *one* ^{12}C atom has been replaced by a less-abundant ^{13}C isotope. Ideally, the relative intensities of these two ion peaks relate to the natural abundances of the isotopes (natural abundance of ^{13}C is 1.11 % of that of ^{12}C) multiplied by the number of carbon atoms in the molecule, i.e. the intensity of the m/z 327 ion compared to the m/z 326 ion is equal to 22.2 %. On the basis of the molecular formula and using average atomic masses (Table 6.6) an accurate, but *average molecular weight* of 326.4346 Da results for Chimassorb 81. If unit resolution has been set, this will *not* be the mass of the dominant ion detected by the mass spectrometer. In fact, in this case the dominant ion is $^{12}C_{21}{}^1H_{26}{}^{16}O_3$ with an accurate

Table 6.6 Some frequently encountered atoms with their monoisotopic and average atomic masses

Atom	Isotope	Natural abundance	Monoisotopic mass	Average mass
Hydrogen	^1H	99.985	1.007 825	1.0079
	^2H	0.015	2.01410	
Carbon	^{12}C	98.90	12.000 000	12.0110
	^{13}C	1.10	13.003 355	
Nitrogen	^{14}N	99.63	14.003 074	14.0067
	^{15}N	0.37	15.000 108	
Oxygen	^{16}O	99.762	15.994 915	15.9994
	^{17}O	0.038	16.999 131	
	^{18}O	0.200	17.999 160	
Phosphorous	^{31}P	100	30.973 762	30.9738
Sulfur	^{32}S	95.02	31.972 070	32.0660
	^{33}S	0.75	32.971 456	
	^{34}S	4.21	33.967 866	
	^{36}S	0.02	35.967 080	
Chlorine	^{35}Cl	75.77	34.968 852	35.4527
	^{37}Cl	24.23	36.965 903	
Bromine	^{79}Br	50.69	78.918 336	79.9040
	^{81}Br	49.31	80.916 289	

but *monoisotopic molecular weight* (or exact mass) of 326.1875 Da. For Chimassorb 81, the difference between the average and monoisotopic molecular weight is not great, and indeed, both have the same nominal mass (326 Da). Interpretation of the spectrum would have succeeded regardless of whether the calculations would have been based on monoisotopic or average values; but this is not always the case, especially with high-MW samples (>2000 Da), or those with a pronounced isotope pattern such as those containing halogens or transition metal atoms. As an example, the average and monoisotopic accurate masses for Cyasorb UV 1084 [2,2'-thio-bis-(4-*tert*-octyl-phenolato)-*N*-butyl-amine]-nickel, $C_{32}H_{51}NO_2SNi$, are 572.5194 and 571.2990 Da, respectively, which is a significant difference. A mass spectrum showing a molecular ion at m/z 571 would be consistent with the monoisotopic calculation, but would indicate (mistakenly) that the sample was not Cyasorb UV 1084, on the basis of the average masses. This clearly emphasises the need for calculating correctly the masses of isotopes for correct interpretation of the mass spectra.

Table 6.6 presents a list of some of the most commonly encountered atoms in polymer/additive analysis, together with their monoisotopic and average masses. For the same nominal mass, different exact masses (elemental compositions) do exist. Knowledge of the exact mass of an unknown substance allows its atomic composition to be established. The exact mass of an ion proves the presence of a particular species (compound in a mixture).

Multi-element compounds often show mass spectra in which the molecular ions and fragment-ions are multiplets because some of the elements in the compound have multiple stable isotopes. The distribution of the isotopic signature in mass spectra, calculated with binomial mathematics, proves quite useful for the analysis of flame retardants. Bromine, chlorine and antimony have two stable isotopes in the ratios of 50.69 (^{79}Br): 49.31 (^{81}Br); 75.77 (^{35}Cl): 24.23 (^{37}Cl); and 57.3 (^{121}Sb): 42.7 (^{123}Sb), respectively. The *isotopic distribution* is therefore an important identification tool in low-resolution spectra.

Most mass spectrometers will resolve ions with unit resolution up to at least 2000 Da, and so monoisotopic atomic masses are used in these cases. Above 2000 Da, the resolution should be checked and, if it is insufficient to resolve adjacent isotopes, then average atomic masses can be used in calculations.

6.1.5 Accurate Mass Measurements

Principles and Characteristics Mass spectrometry can provide the accurate mass determination in a direct measurement mode. For a properly calibrated mass spectrometer the mass accuracy should be expected to be good to at least 0.1 Da. Accurate mass measurements can be made at any resolution (resolution matters only when separating masses). For polymer/additive deformulation the nominal molecular weight of an analyte, as determined with an accuracy of 0.1 Da from the mass spectrum, is generally insufficient to characterise the sample, in view of the small mass differences in commercial additives. With the thousands of additives, it is obvious that the same nominal mass often corresponds to quite a number of possible additive types, e.g. NPG dibenzoate, Tinuvin 312, Uvistat 247, Flexricin P-1, isobutylpalmitate and fumaric acid for $m = 312$ Da; see also Table 6.7 for $m = 268$ Da. Accurate mass measurements are most often made in EI mode, since the sensitivity is high, and reference mass peaks are readily available (using various fluorinated reference materials). Accurate mass measurements can also be made in CI

Table 6.7 Calculation of possible atomic compositions[a]

Possible composition	Calculated mass	Deviation
$C_{18}H_{24}N_2$	268.1939	2 ppm
$C_{15}H_{28}N_2S$	268.1973	10 ppm
$C_{13}H_{24}N_4O_2$	268.1899	17 ppm
$C_{16}H_{28}OS$	268.1861	31 ppm

[a] After Lattimer and Harris [36]. Reprinted with permission from *Rubber Chemistry and Technology*. Copyright © (1989), Rubber Division, American Chemical Society, Inc.

and FAB modes, but are considerably more difficult in FI or FD mode. If mass measurement accuracy of 5–10 ppm or better is obtained, elemental compositions of fragment ions can usually be determined or, at the very least, the number of compositions which need to be considered can be reduced drastically. According to ref. [35], with practice, measurements should be achievable to 5 ppm or better. Ion masses measured with an accuracy at the 2–3 ppm level, which requires a mass resolution of 8000–12 000, do provide an unequivocal identification of elemental composition. For example, a molecular ion at m/z 268, measured as 268.1945 with a known accuracy of ± 3 ppm, is readily attributed to $C_{18}H_{24}N_2$ (HPPD) on the basis of Table 6.7.

Resolution does not affect the accuracy of the individual accurate mass measurements when no separation problem exists. When performing accurate mass measurements on a given component in a mixture, it may be necessary to raise the resolution of the mass spectrometer wherever possible. *Atomic composition mass spectrometry* (AC-MS) is a powerful technique for chemical structure identification or confirmation, which requires double-focusing magnetic, Fourier-transform ion-cyclotron resonance (FTICR) or else ToF-MS spectrometers, and use of a suitable reference material. The most common reference materials for accurate mass measurements are perfluorokerosene (PFK), perfluorotetrabutylamine (PFTBA) and decafluorotriphenylphosphine (DFTPP). One of the difficulties of high-mass MS is the lack of suitable calibration standards. Reference inlets to the ion source facilitate exact mass measurement. When appropriately calibrated, ToF mass

spectrometers can provide elemental compositions for both molecular and fragment ions. Mass calibration is most accurate when internal calibrants are employed.

High resolution is used to determine the exact mass of an ion species in a mixture: knowledge of the exact mass of an unknown substance allows its atomic composition to be established. Target analysis exact mass determination proves the presence of a particular ion species (compound) in a mixture. Mass spectrometry is perhaps the only method that can be used to find the empirical formulae of compounds that are not completely pure.

For polymer/additive analyses neither a very high mass range nor an ultrahigh mass resolution is required; isobaric additive ions are not frequent. It is therefore not surprising that tandem sector instruments have not found wide application for polymer/additive identification.

There is rapid growth in the use of accurate mass measurements in the chemical industries. There is equally a clear need for practical guidance in order to obtain robust measurements. At present, LGC coordinates a collaborative study to evaluate the variation in accurate mass measurement across a broad range of instrument types, using an unknown compound of molecular mass of about 450 Da.

Applications   Lattimer [37] has reported high-resolution measurements (measured and calculated mass, atomic composition) of product scans of several PP additives (Irganox 1076/3114, Tinuvin 144/622/770, Goodrite 3150, Chimassorb 944) using a sector mass spectrometer of BEoQ geometry. Figure 6.2 and Table 6.8

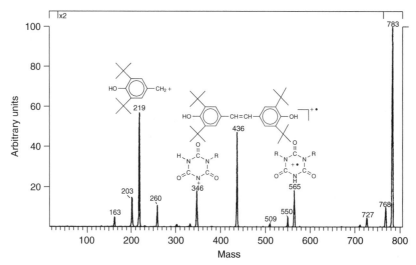

Figure 6.2   Product-ion scan (EI-MS/MS) of M$^{+\bullet}$ 783 (Irganox 3114). After Lattimer [37]. Reprinted from *Journal of Analytical and Applied Pyrolysis*, **26**, R. P. Lattimer, 65–92, Copyright (1993), with permission from Elsevier

Table 6.8 High-resolution mass measurements and atomic compositions of Irganox 3114

Measured mass	Calculated mass	Atomic composition	Ion identity
783.518 8	783.518 6	$C_{48}H_{69}N_3O_6$	(Irganox 3114, $M^{+\bullet}$)
565.351 6	565.351 5	$C_{33}H_{47}N_3O_5$	Fragment ion
436.333 2	436.334 1	$C_{30}H_{44}O_2$	Fragment ion
346.175 6	346.176 6	$C_{18}H_{24}N_3O_4$	Fragment ion
260.164 4	260.165 0	$C_{16}H_{22}NO_2$	Fragment ion
219.174 4	219.174 9	$C_{15}H_{23}O$	Fragment ion

After Lattimer [37]. Reprinted from *Journal of Analytical and Applied Pyrolysis*, **26**, R.P. Lattimer, 65–92, Copyright (1993), with permission from Elsevier.

show the results of the EI-MS/MS product-ion scan of Irganox 3114 ($M^{+\bullet}$).

6.2 ION SOURCES

Principles and Characteristics Ionisation processes are the basis for mass-spectrometric detection. Each of the ionisation techniques occupies its own position in mass spectrometry. The optimum performance of any ionisation method (and therefore the result) will depend critically on the characteristics and reliability of the mass spectrometer. Ionisation may occur in the gas, liquid or condensed phase, and may be either 'hard' or 'soft', i.e. with or without extensive

fragmentation. The main ionisation methods and agents, as classified in Table 6.9, have played a substantial role in polymer/additive analysis. Other ionisation modes, not mentioned in Table 6.9, are used sporadically. For example, electron capture negative ionisation (ECNI) is a very soft and selective ionisation process for electron-accepting molecules. A 250-eV bombardment of argon in the ionisation chamber produces positive ions which decay to produce thermal electrons with a low energy [38]. Luyk *et al.* [39] used the technique to study the pyrolysis products of flame-retardant polymers, such as PS/HBCD. The low-energy electrons attack the pyrolysis products with electron-accepting groups like bromine. The mechanisms by which various ionisation sources operate are still controversial.

Obvious practical requirements are that the ionisation method has to be available; needs to be compatible with the mass spectrometer being used; and able to handle the chemical compound class of investigation. Table 6.10 shows the compatibility of ionisation type, and mass spectrometer.

Electron impact ionisation (EI) stands for extensive fragmentation, but also produces molecular ions. The other ionisation methods shown in Table 6.10 mainly generate quasi-molecular ions for various compound classes. Protonation of organic compounds is one of the most fundamental processes of CI, FAB and ESI mass spectrometry. Apart from electrospray (ESI), which

Table 6.9 Ionisation agents

Ionisation[a] method	Hard ionisation		Soft ionisation				
	Heating	Particle[b] bombardment	Photons	High electric field	Ion/molecule reaction	Spray	Matrix
EI	X	X					
SMB-EI		X					
CI	X				X		
DCI	X				X		
IMR		X			X		
FD/FI	X			X			
EHI				X			
FAB/LSIMS		X					X
TSP	X			X	X	X	
APCI	X			X	X	X	
ESI				X		X	
APPI	X		X		X	X	X
LD			X				
MALDI			X				X
MPI			X				
PD		X					X
HSI	X				X[c]		

[a] Nomenclature (Appendix I).
[b] Electrons, atoms, ions, nuclear fission products.
[c] Rhenium foil.

Table 6.10 Ionisation methods and suitability for different sample classes

Ionisation method	Principal ions detected (+/−)	Mass spectrometer[a]	Sample classes	Approximate mass limit (Da)
Electron impact (EI)	$M^{+\cdot}$ and substantial fragment ions	M, Q, IT, ToF, FTMS	Nonpolar and some polar organics, volatiles	<1000
Chemical ionisation (CI)	MH^+, M (adduct)$^+$ $(M - H)^-$, $M^{-\cdot}$ M (adduct)$^-$ $(M + H)^+$	M, Q, IT, ToF, FTMS	Nonpolar and some polar organics, volatiles	<1 000
Electrospray (ESI)	MH^+, $(M + nH)^{n+}$ $(M - H)^-$ $(M - nH)^{n-}$	M, Q, ToF, IT, FTMS	Polar organics, biopolymers, organometallics, nonvolatiles	<150 000
Atmospheric pressure chemical ionisation (APCI)	MH^+ $(M - H)^-$	M, Q, ToF, FTMS	Polar and some nonpolar organics	<1 500
Atmospheric pressure photoionisation (APPI)	$M^{+\cdot}$, $(M + H)^+$ $MH^{-\cdot}$, $(M - H)^-$	M, Q	Very low-polarity components	<2 000
Fast atom/ion bombardment (FAB/FIB/LSIMS)	MH^+, $(M + H)^+$, $M^{+\cdot}$ $(M - H)^-$, $M^{-\cdot}$ Intact molecular ions	M, Q	Polar organics, proteins, organometallics, nonvolatiles	<20 000[b]
Field desorption/ionisation (FD/FI)	MH^+ (FD), $M^{+\cdot}$ (FI) $(M - H)^-$ (FD), M^- (FI)	M, ToF	Nonpolar and some polar organics, including synthetic polymers	<10 000[b]
Thermospray (TSP)	MH^+, MNH_4^+ $(M - H)^-$	M, Q	Polar and some nonpolar organics	<1 000
Laser desorption ionisation (LDI)	M^+, $(M + Na)^+$ M^-, $(M - H)^-$	IT, FTMS	Organics, organometallics, inorganics, oligomers, polymers[c]	<10 000
Plasma desorption (PD)	MH^+ $(M - H)^-$	ToF, IT, FTMS	Mixture analysis	<50 000
Matrix-assisted laser desorption ionisation (MALDI)	MH^+ $(M - H)^-$	ToF, IT, FTMS	Polar and some nonpolar biopolymers, synthetic polymers	>250 000

[a] M = magnet, Q = quadrupole, ToF = time-of-flight, IT = ion trap, FTMS = Fourier-transform MS.
[b] Depends on the m/z range of the mass spectrometer.
[c] Structural information from fragment ions.

gives rise to multiply charged ions to great extent, only singly charged species are generated in all other methods. Consequently, masses are easily derived from the spectra. *Fragmentations* may be strongly reduced by means of the chemical ionisation (CI) method. Desorption chemical ionisation (DCI) [40], charge-exchange (CE) [41] and ion–molecule reaction (IMR) are variations of the CI method. These techniques have a common feature: the primary ion source is separated from the reaction region, and this allows a better control of both mass and energy distribution of the primary ions. CE-MS instruments have been extensively used for investigating elementary ion-molecule collision processes, but, because of their high cost and complexity, analytical applications have been quite limited.

Typical *gas phase ionisation methods* (vaporisation followed by ionisation) are EI, CI, FI, MPI; desorption/ionisation methods for nonvolatile samples (ions formed in the condensed phase) are FD, PD, SIMS, FAB, DCI, TSP, LD-SIMS, MALD, ESP/ISP.

Ions may be formed in the centre of the source, in the gas phase, or directly on the backing plate (by laser or plasma desorption). For organic compounds that are sufficiently volatile, introduction by vaporisation or by GC through a very small orifice followed by electron impact ionisation is standard procedure, as observed in the case of degradation of polymeric material (e.g. TG-MS, TD-GC-MS, PyMS, PyGC-MS, LPyMS, etc.). If this procedure does not give an unambiguous molecular ion, the next step is chemical ionisation, which usually yields a prominent $[M + H]^+$ peak with little fragmentation. Gas-phase ionisation techniques are difficult for very polar, thermolabile, ionic compounds. *Ion evaporation* is a process in which preformed ions in solution are directly transferred to the gas phase (Fig. 6.3). Although the ion evaporation process is not yet well understood, it is thought to be involved in the ion production in TSP-MS, ESP-MS, EH-MS and certain forms of FD. For liquid-phase ionisation, see Section 7.3.3.2.

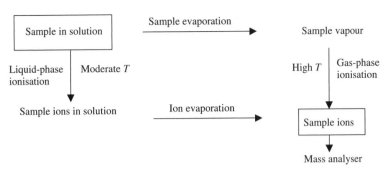

Figure 6.3 Gas-phase vs. liquid-phase ionisation

For non-volatile sample molecules, other ionisation methods must be used, namely *desorption/ionisation* (DI) and nebulisation ionisation methods. In DI, the unifying aspect is the rapid addition of energy into a condensed-phase sample, with subsequent generation and release of ions into the mass analyser. In EI and CI, the processes of volatilisation and ionisation are distinct and separable; in DI, they are intimately associated. In *nebulisation ionisation*, such as ESP or TSP, an aerosol spray is used at some stage to separate sample molecules and/or ions from the solvent liquid that carries them into the source of the mass spectrometer. Less volatile but thermally stable compounds can be thermally vaporised in the direct inlet probe (DIP) situated close to the ionising molecular beam. This DIP is standard equipment on most instruments; an EI spectrum results. Techniques that extend the utility of mass spectrometry to the least volatile and more labile organic molecules include FD, EHD, surface ionisation (SIMS, FAB) and matrix-assisted laser desorption (MALD) as the last resort. Ions may be formed directly (as in SIMS), or neutral species are sputtered from the sample and subsequently ionised in a different process, as in LD-MS. *Plasma sources* (GD, ICP, MIP, PD) are most suitable for inorganic mass spectrometry (Chapter 8).

Table 6.10 reports the main areas of application of the various ionisation methods and the principal ions detected. A breakdown of MS techniques applied to various types of analytes is as follows: thermally stable, low-MW: CI, EI; thermally instable, low-MW: APCI (FIA, LC-MS), ESI; and high-MW: DCI, FD, FAB, LD, ESI (FIA, LC-MS, CZE-MS). Soft ionisation techniques such as FI, FAB and LD are useful for the detection of non-volatile, sometimes oligomeric, polymer additives. Recent developments in ionisation techniques have allowed the analysis of polar, ionic, and high-MW compounds, previously not amenable to mass-spectrometric analysis. Figure 6.4 shows the applicability of various atmospheric pressure ionisation techniques in terms of molar mass and polarity.

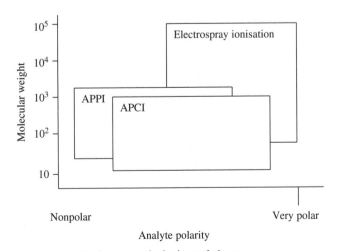

Figure 6.4 Applications of various atmospheric pressure ionisation techniques

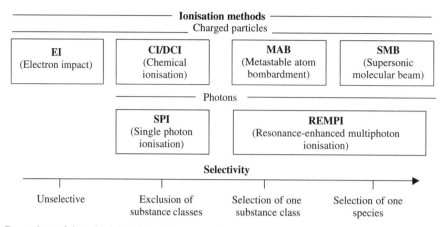

Figure 6.5 Comparison of the selectivity of ionisation methods in mass spectrometry

It is also possible to ionise and fragment selectively by irradiating with *photons* of wavelengths specific to the desorbed species (single photon ionisation, SPI; multiphoton ionisation, MPI; and resonance-enhanced multiphoton ionisation, REMPI). The selectivity of various ionisation methods for on-line mass spectrometry is given in Figure 6.5. REMPI introduces tuneable substance selectivity into mass spectrometry and exhibits two extreme selectivity levels: group-selective REMPI of a group of substances out of one substance class, and isomer-selective REMPI [42]. However, on both levels, remarkably higher selectivity is achieved than for other methods, such as EI, CI or SPI. SPI is also selective in the way that it excludes a substance class because molecules with an ionisation potential higher than the photon energy (>7 eV) are not ionised.

A variety of *combined ion sources* have been described (e.g. EI/FI, EI/CI, EI/CI/FI, FD/EI, FD/FI, TSP/EI, PCI/NCI, APCI/ESI, APCI/ESI/APPI, APCI/ ESI/APEI/APSI/SSI, etc.). Source pressure may range from 10^{-6} Torr (EI, FAB, LSIMS, PD) to 0.1–10 Torr (CI, GD) and atmospheric pressure (APCI, ESI, APPI, ICP).

6.2.1 Electron Impact Ionisation

Principles and Characteristics Electron impact (EI) ionisation is the original ionisation method (1918). Before 1980, mass spectrometry was merely restricted to electron impact (EI), with chemical ionisation (CI) being applied mainly for those samples which resist generation of satisfactory EI data. Nowadays, EI is still a widely used universal and nonselective ionisation method. In EI, the sample is introduced as a vapour

into an ion source, which operates under high vacuum (10^{-5} to 10^{-6} mbar), where molecules are bombarded with relatively high kinetic energy (70 eV) electrons from a heated (tungsten) filament and transformed into positively charged ions. As organic molecules are even-electron species, the process of removing one electron generates a positively charged ion with one unpaired electron ($M^{+\bullet}$), from which the molecular weight of the sample can be inferred directly. Since the ionisation energy of most organic molecules lies in the range of 10 eV, the energy imparted to the molecules during the process not only expels one electron, but excess ionisation energy also causes the charged molecular ion to fragment (hard ionisation). If the electron energy is constant (e.g. 70 eV), the fragmentation pattern is reproducible. Lower ionising energies (10–30 eV), as in *low-voltage electron impact* (LVEI), enhance the relative intensity of molecular ion peaks and reduce the number and relative abundancies of the lower-molecular-weight fragment ions as well as the fragmentation, ionically induced rearrangements, or ion–molecule collisions. There is, however, a marked decrease in sensitivity with decreasing electron energy. For general purposes the *thermal EI ionisation* source temperature is usually set at 180–220 °C.

Supersonic molecular beam (SMB) mass spectrometry (SMB-MS) measures the mass spectrum of vibrationally cold molecules (cold EI). *Supersonic molecular beams* [43] are formed by the co-expansion of an atmospheric pressure helium or hydrogen carrier gas, seeded with heavier sample organic molecules, through a simple pinhole (ca. 100 μm i.d.) into a 10^{-5}-mbar vacuum with flow-rates of 200 mL min^{-1}. In SMB, molecular ionisation is obtained either through improved electron impact ionisation, or through hyperthermal surface ionisation

Table 6.11 Main features of SMB and implications for MS

Main features of SMB	Implications for MS
Vibrational–rotational supercooling	Enhanced EI mass spectral information
Unidirectional motion with hyperthermal kinetic energy (1–20 eV)	Hyperthermal surface ionisation (HSI)
High flow-rate sampling at atmospheric pressure	Fast and ultrafast GC-MS

(HSI) by collision with heated rhenium foil. The main features of SMB are given in Table 6.11. Cold EI with SMB means:

(i) enhanced molecular ion (simultaneous EI + CI);
(ii) improved library search;
(iii) exact isotope information;
(iv) elemental information;
(v) isomer and structural information; and
(vi) tuneable fragmentation.

Enhanced molecular ion implies reduced matrix interference. An SMB-EI mass spectrum usually provides information comparable to field ionisation, but fragmentation can be promoted through increase of the electron energy. For many compounds the sensitivity of HSI can be up to 100 times that of EI. Aromatics are ionised with a much greater efficiency than saturated compounds. Supersonic molecular beams are used in mass spectrometry in conjunction with GC-MS [44], LC-MS [45] and laser-induced multiphoton ionisation followed by time-of-flight analysis [46].

As shown in Table 6.10, EI (and CI) may be used in conjunction with magnetic sector, quadrupole and ion trap mass spectrometers. Introduction of the analyte into an ion source for gas-phase ionisation can be achieved by means of a direct insertion probe (including those of the desorption ionisation type) for solid samples; via a GC interface for reasonably volatile samples in solution; a reservoir (or reference) inlet system for nonpolar, volatile samples and reference materials; or by means of a moving belt or an aerosol beam separator (particle beam) liquid chromatography interface for more polar organic molecules in solution. Under EI conditions it is quite straightforward to admit a reference compound, through the reference inlet, simultaneously with the sample, from either the probe or the GC interface. *Particle-beam* (PB) technology has been used as a method of sample introduction into the MS ion source for EI ionisation of analytes. The PB interface performance has been improved to increase the sensitivity of detection and to minimise

spectral contamination. Since PB-MS counts on gas phase ionisation, the compounds still need to be volatile. Particle-beam MS is only suitable for identification (for apolar components > 50 ng, for polar components > 500 ng). Drawbacks of PB-MS are analyte carry-over or memory effects, signal enhancements, and interferences by high intensity signals of low m/z ions. PB-MS requires high daily maintenance [47]. Mattina [48] has made a plea for a readily switchable PB-API-MS interface on a single mass spectrometer, as later reported by Sakairi and Kato [49]. Electron impact ionisation is traditionally the method of choice for performing accurate mass measurements, although other ionisation methods may be used. With practice, measurements should be achievable to 5 ppm or better. When performing accurate mass measurements on a given component in a mixture, the resolution of the mass spectrometer is raised wherever possible (this is particularly the strength of magnetic sector mass spectrometers). Obviously, the resolution does not affect the accuracy of the individual mass measurements when no separation problem exists. A normal *EI spectrum* can be obtained with a few pg of sample. EI is used in the positive ionisation mode. It is usual to work with at least unit resolution. The following useful information may be obtained from an EI spectrum: molecular ion mass (at the high m/z end of the spectrum), the isotope pattern of the molecular ion region, and fragment-ions. Modern data systems have the means of accessing, searching, and comparing spectra in several commercially available computerised libraries with the acquired spectra in seconds. This is a rapid way of interpreting and confirming EI spectra. It is also possible to extend searching to sub-libraries (e.g. of antioxidants alone, or compounds containing a single chlorine atom). However, the public libraries are rather poor performers with regard to polymer additive EI-MS spectra. Therefore, it is advisable to build up a proprietary database. Library search routines have the advantage of automation; a search method is nowadays included in a general processing routine which is carried out at the conclusion of a GC-MS or LC-MS run.

For the EI ion source, the generated total ion stream is directly proportional to the gas pressure in the impact field, which provides a basic condition for *quantitative analysis*. Compounds can only safely be quantified if influences on the sensitivity of detection, such as ion–molecule reactions and competition in the ionisation process, can be excluded by experimental evidence.

Table 6.12 shows the *main characteristics* of thermal electron impact ionisation. Electron impact can be used for analysing a wide variety of volatile organic

Table 6.12 Main characteristics of thermal electron impact ionisation

Advantages

- Classical spectra
- Library searchable
- Positive compound identification
- Relatively easy to obtain
- Rugged
- Interpretable
- Structural elucidation (unknown compounds)

Disadvantages

- Difficult mixture analysis
- Limited to volatile, low-MW compounds
- Applicability restricted to thermally stable, nonpolar analytes

compounds (up to a molecular weight of ca. 1000 Da), including polymer additives. A disadvantage is that EI does not always give a molecular weight. If identification of a chromatographic peak is the goal, molecular weight information merely defines the boundaries of compound identity. Many compounds have the same molecular weight. If confirmation is the goal, EI provides fragmentation spectra which give additional confidence, provided that standards exist, and can be compared to commercially available reference spectra. EI (and CI) modes have now declined somewhat in favour of other, trendier, ionisation modes, especially those more compatible with direct coupling to LC. PB-MS has been reviewed [48].

Applications The identification of various hydroxybenzophenone *UV absorbers* and dialkyl phthalate *plasticisers* in polymers by MS, IR and UV has been reported [50]. Whereas UVAs could be identified based on their molecular ions detected by EI-MS, only fragment ions were observed for the plasticisers. IR and UV analysis were unsatisfactory for identification purposes. As pointed out by Hayes and Altenau [51] low-voltage mass spectra are very useful in distinguishing between the *antioxidants* PBNA and DBPC and oil signals in acetonitrile extracts of rubber vulcanisates. Generally, small amounts of oil in extract causes considerably interference as the result of the fragmentation pattern of oils (large peaks every 14 mass units, due to the loss of methylene groups). Lehotay *et al.* [52] also analysed polymer extracts employing low-voltage (10 eV) electron ionisation and a direct probe for sample introduction.

Direct desorption MS has been used to identify the nature and amount of *volatiles* produced during processing, i.e. drying, extruding, moulding, etc. of plastic compositions. Programming of the heating in the mass

spectrometer enables process conditions to be simulated. The evolution of perfluoroalkenes from the breakdown of a fluorinated *surfactant*, and alkylphenols and ethoxylated phenols from polyethylene oxide condensates used in the emulsion polymerisation have been detected [53]. EI mass spectra of alkylbenzenesulfonates have been produced through direct introduction of the materials into an EI ion source [54].

The use of the particle-beam interface for introduction of samples into a mass spectrometer (PB-MS), without chromatographic separation, was shown by Bonilla [55] to be a useful method for analysis of *semi-volatile and nonvolatile additives* in PC and PC/PBT blends. The method uses the full power of mass spectrometry to identify multiple additives in a single matrix. The usefulness, speed and simplicity of this approach were illustrated for AOs, UVAs, FRs, slip agents and other additives.

Section 6.4 deals with other EI-MS analyses of samples, i.e. analyses using direct introduction methods (reservoir or reference inlet system and direct insertion probe). Applications of hyphenated electron impact mass-spectrometric techniques for polymer/additive analysis are described elsewhere: GC-MS (Section 7.3.1.2), LC-PB-MS (Section 7.3.3.2), SFC-MS (Section 7.3.2.2) and TLC-MS (Section 7.3.5.4).

6.2.2 Chemical Ionisation

Principles and Characteristics The inherent instability of the molecular ion formed by electron ionisation for certain classes of organic compounds has provided the original impetus behind the development of chemical ionisation (CI) mass spectrometry, first described in 1966 [56]. CI-MS uses the same, or a similar source to that employed by EI ionisation. The analyte is introduced in the ionisation source in its gaseous phase. In this case too, a gas (usually methane, isobutane or ammonia) is introduced into the ionisation chamber, increasing the source pressure considerably. The electron beam emitted from the filament is then exposed to an increased chance of collisions with gas molecules and ions. *Reagent ions* are formed from the gas molecules by collision with the electron beam; it is far more likely that these ions, present in vast excess to the sample molecules, react with the analyte molecules to form product ions, rather than the sample molecules being ionised directly by electron impact. The process of ion–molecule reaction is a less energetic one than in case of EI mode, and molecular fragmentation is much reduced. CI can be operated in both the *positive and negative ionisation modes* (PCI and NCI, respectively), producing either MH^+ or deprotonated molecular anions $(M - H)^-$, with almost no fragment ions, and negatively charged radical molecular ions, $M^{-\bullet}$ (Table 6.10). In both cases, CI is classified as a 'soft' ionisation technique, because the spectra are dominated by quasi-molecular ions, with very little fragmentation.

CI-MS and EI-MS address the same compound class. CI is used mainly when the molecule fragments so completely in EI mode that no $M^{+\bullet}$ ions are observed or when the problem is only knowledge of the molecular weight of the sample component. In fact, EI and CI are usually both carried out on the same sample, as the two ionisation methods produce complementary information of value for the determination of structure and MW of a compound. The detection limits of CI tend to be better than EI, as the latter technique divides the ion current between molecular and fragment ions. A few ng of sample may be detected.

CI and EI are both limited to materials that can be transferred to the ion source of a mass spectrometer without significant degradation prior to ionisation. This is accomplished either directly in the high vacuum of the mass spectrometer, or with heating of the material in the high vacuum. *Sample introduction* into the CI source thus may take place by a direct insertion probe (including those of the desorption chemical ionisation type) for solid samples; a GC interface for reasonably volatile samples in solution; a reference inlet for calibration materials; or a particle-beam interface for more polar organic molecules. This is not unlike the options for EI operation.

In CI, experimental conditions are often chosen such that proton transfer reactions dominate over all other possible reactions, e.g. charge transfer. Methane and isobutane are normally used as reagent gas for nonpolar organics of low proton affinity (e.g. hydrocarbons), which usually protonate to generate MH^+ ions or deprotonate to $[M - H]^-$ ions. Ammonia is useful for more polar organic compounds which will produce either MH^+ or MNH_4^+ ions. For negative CI electron capture operation, ammonia is often the preferred reagent gas, although methane and isobutane can also be used, as ammonia contaminates the source less quickly. Typical CI operating temperatures are 150 to 180 °C (as compared to 180 to 220 °C for EI). In CI conditions, contamination, due to the continual presence of a reagent gas, is much more pronounced than in EI operation.

Table 6.13 lists the *main characteristics* of chemical ionisation. The use of CI overcomes some of the limitations of EI-MS. CI-MS has the advantage of ease of interpretation and of being able to operate at higher input pressures. CI restricts the fragmentation

Table 6.13 Main characteristics of chemical ionisation

Advantages

- Soft ionisation
- Simple spectra (quantitation)
- Molar mass information
- Mixture analysis
- Rugged
- Choice of reagent gas
- Higher selectivity and sensitivity than EI

Disadvantages

- Not library searchable
- No structural elucidation (limited fragmentation)
- Limited to volatile, low-MW compounds
- Applicability restricted to thermally stable, nonpolar analytes

of released volatile compounds. This in turn leads to simple cracking patterns and therefore easy-to-interpret spectra – especially useful in the analysis of mixtures of additives and their degradation products, which often occur in the course of thermal decomposition of polymers. Identification of individual components is clearly favoured by well-developed, non-overlapping, fragmentation patterns.

Harrison [57] distinguishes Brønsted acid and base chemical ionisation, charge exchange (CE) and electron capture chemical ionisation (ECCI).

Among the *separation methods* coupled to EI and CI-MS on-line GC-MS is outstanding. Various groups [58–60] have examined the feasibility of characterising volatiles evolved in TG by CI mass spectrometry.

Numerous articles and reviews [57,61,62] and a book [38] deal with chemical ionisation with permanent gases.

Applications Rudewicz and Munson [63] have used CI-MS for both quantitative and qualitative studies of 0.01–0.1 % additives in 1–2 mg *polypropylene* samples. Using 1.1 % ammonia in methane as a reagent gas $(M + H)^+$ or $(M + NH_4)^+$ ions were obtained with little or no fragmentation of the target compounds such as Ionox 330, Irgafos 168 and Cyasorb UV531, and also very low background from the polymer. Short-term reproducibility of peak areas of 6 % was quoted. Direct insertion probe CI-MS (from 30 to 350 °C) with 1.1 % ammonia in methane as a reagent gas and multiple ion detection (MID) to enhance sensitivity and reproducibility has been used for both qualitative and quantitative studies of Permanax WSP (m/z 217) and the copper deactivator OABH (m/z 295) in very small (150 µg) *polyethylene* samples [64]. Phenolphthalein (PPT, m/z 225) was chosen as the internal standard in view of a similar evaporation temperature as Permanax WSP and OABH and closely related base peaks. Preparation of good calibration standards and the choice of a suitable internal standard are of crucial importance for quantitation of polymer additives. Solutions of pure additives used as standards were found to be unsatisfactory, due to the difference in the evaporation profile between pure additives and those blended in the polymer samples.

In a *polyurethane elastomer* analysed by isobutane CI-MS and CI-MS/MS in TD/Py-CIMS (TD, 20–200 °C; Py, 200–300 °C) mode MDI (diisocyanate), BDO (chain extender), Stabaxol P (stabiliser), AA/ BDO/HDO (cyclic adipate) and residual cyclic ester oligomers were identified [65].

In a study on the identification of organic additives in *rubber vulcanisates* using mass spectrometry, Lattimer *et al.* [22] used direct thermal desorption with three different ionisation methods: EI, CI and FI. Also, rubber extracts were examined directly by four ionisation methods (EI, CI, FD and FAB). The authors did not report a clear advantage for direct analysis as compared to analysis after extraction. Direct analysis was a little faster, but the extraction methods were considered to be more versatile.

Salmona *et al.* [66] used EI and CIMS to identify benzothiazole derivatives leached into injections by rubber plunger seals from disposable syringes. One of the compounds was used as a rubber vulcanisation accelerator, and four others were formed during syringe sterilisation with ethylene oxide. Applications of hyphenated chemical impact mass-spectrometric techniques are described elsewhere: GC-MS (Section 7.3.1.2), for polar and nonpolar volatile organics, SFC-MS (Section 7.3.2.2) and TLC-MS (Section 7.3.5.4).

6.2.2.1 Desorption Chemical Ionisation Mass Spectrometry

Principles and Characteristics The technique called 'desorption CI', also termed 'in-beam', 'direct exposure', 'direct CI', and 'surface ionisation', consists of coating nonvolatile samples on a surface and direct insertion inside an ion source by means of some extended direct insertion probe made of some inert substance (e.g. glass, quartz or Teflon). This replaces the usual practice of vaporising the sample outside the source, which generally results in pyrolysis and gives a much better chance of observing molecular ions in the mass spectra. The technique involves rapid heating in an atmosphere of reagent gas, under the conditions commonly used for chemical ionisation; it has also been used, albeit with less success, with EI ionisation.

To use the DCI probe, 1–2 µL of the sample (in solution) are applied to the probe tip, composed of a small platinum coil, and after the solvent has been allowed to evaporate at room temperature, the probe is inserted into the source. DCI probes have the capability of very fast temperature ramping from 20 to 700 °C over several seconds, in order to volatilise the sample before it thermally decomposes. With slower temperature gradients, samples containing a mixture of components can be fractionally desorbed. The temperature ramp can be reproduced accurately. It is important to use as volatile a solvent as possible, so as to minimise the time required to wait for solvent evaporation, which leaves a thin layer of sample covering the coil. The observed spectrum is likely to be the superposition of various phenomena: evaporation of the sample with rapid ionisation; direct ionisation on the filament surface; direct desorption of ions; and, at higher temperature, pyrolysis followed by ionisation.

The DCI probe is particularly attractive for samples that are susceptible to thermal decomposition, although it can equally well be used as a general means of introducing samples into the ionisation source, i.e. as an alternative to the direct insertion probe. The types of sample which benefit most from DCI probing are higher-molecular-weight, less-volatile compounds, organometallics, and any thermally sensitive compounds [40,67]. DCI is considered to be a soft ionisation technique.

The *main features* of DCI-MS and DCI-MS/MS are given in Table 6.14. DCI has gained rapid popularity because it is relatively simple to adapt to almost any mass spectrometer and gives results similar to FD, but in a more simple manner. It is not a substitute for FD, but it is less expensive and generally produces more fragmentation information than FD. For many compounds, molecular ions will be obtained where conventional solids probes would not do the job. DCI is known for the specificity provided by choosing reagent gas with different proton affinities. The major

Table 6.14 Main features of DCI-MS/MS for polymer/additive analysis

Advantages

- Soft ionisation technique
- Selectivity provided by different reagent gases
- Specificity by collision-induced dissociation
- Applicable to highly polar, thermally labile and nonvolatile molecules
- Ease of operation

Disadvantages

- Short duration of the signal
- No proven record of quantitation

disadvantage of the DCI technique is the short duration of the signal (i.e. 2 s), which in the past was associated with poor reproducibility of mass spectra. Exact mass measurements can be obtained for very fast desorption in a DCI-oaToFMS arrangement.

With the introduction of FAB in 1981, interest in the development of both DCI and FD sharply decreased. Indeed, on highly polar substances FAB provides more valuable results than DCI or FD and a more stable signal. On the other hand, nonpolar substances with high molecular weight are not amenable to FAB, since they are poorly ionised and also they cannot be easily dissolved in the most common FAB matrices. Thus, alternative ionisation methods have to be employed with such compounds. DCI-MS of nonvolatile compounds has been reviewed [40].

Applications Desorption chemical ionisation has proven potential in the analysis of thermally labile, nonvolatile and polar compounds [40,67,68], for the identification of unknown polymers and the study of the thermal degradation mechanisms of polymers. Considering the overall ease of DCI operation, the capability of analysing nonvolatile compounds, and the selectivity provided by choosing different reagent gases, DCI has found surprisingly few practitioners in the analysis of polymer additives.

DCI-MS can be used for the determination of MWD of oligomers in the range 1000–8000 Da after optimisation of a large set of parameters [67]. DCI is adequate for the analysis of high-MW additives (>2000 Da). Figure 6.6 shows the ammonia DCI mass spectrum of Chimassorb 944 (MW 2500–4000); the *m/z* 600 ion is the molecular ion of the monomer, *m/z*

994 corresponds to the dimer losing a triazine group, and a series of ions with a mass interval of a repeating unit of 598 Da are observed.

Chen and Her [23] readily analysed a commercial LDPE/(Irganox 1076, Naugard 524, DOP, oleamide, erucamide) (raincoat) sample by means of *ammonia DCI-MS* (Fig. 6.7) and DCI-MS/MS. The CID spectra of the *m/z* 282/299 and 338/355 ion pairs were indicative of oleamide and erucamide, respectively; *m/z* 663/680 corresponds to oxidised Naugard 524, *m/z* 391 was assigned to the $(M + H)^+$ ion of DOP, and *m/z* 548 to the $(M + NH_4)^+$ ion of Irganox 1076. Other examples of DCI-MS/MS analyses were given: LDPE/(Irganox 1010/1076, Naugard 524, Sandostab P-EPQ, DOP, oleamide, erucamide) (raincoat); HDPE/(Naugard 524, Irganox 1076); and HDPE/(Chimassorb 644, Irganox 1076) (beer crate). *Methane DCI-MS* for polymer/additive analysis of PE extracts has two major drawbacks, namely the observation of pseudomolecular ion(s) and many fragment ions from the additives, and series of ions with a 14-amu interval corresponding to low-MW PE. These matrix ions may obscure ions from additives and make the assignment and selection of precursor ions for MS/MS difficult. On the other hand, ammonia frequently gives very simple CI mass spectra consisting of intense pseudomolecular $(M + H)^+$ and/or $(M + NH_4)^+$ ions with very little fragmentation, because of its high proton affinity. Ammonia DCI-MS is both a 'soft' and selective ionisation technique which ionises the additives preferentially over the matrix. The low-MW polyethylene ions, which were quite intense under the conditions of methane DCI, were almost completely absent in ammonia DCI; thus there was less likelihood of interference.

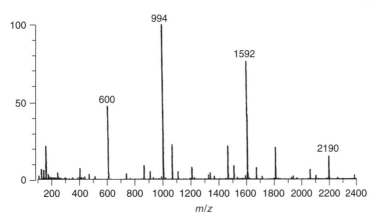

Figure 6.6 Ammonia DCI mass spectrum of Chimassorb 944. After Juo *et al.* [69]. Reprinted from *Analytica Chimica Acta*, **311**, C. Q. Juo *et al.*, 153–64, Copyright (1995), with permission from Elsevier

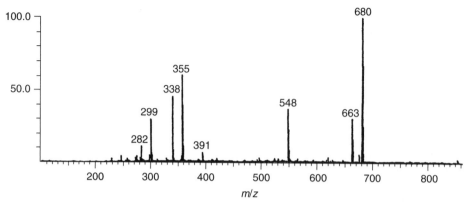

Figure 6.7 Ammonia DCI mass spectrum of an LDPE extract. After Chen and Her [23]. From S.W. Chen and G.R. Her, *Applied Spectroscopy*, **47**, 844–851 (1993). Reproduced by permission of the Society of Applied Spectroscopy

Much greater specificity can be obtained with the use of *tandem mass spectrometry*; the product (daughter)-ion mass spectrum obtained by means of MS/MS provides many structurally characteristic fragment ions and thus much greater confidence in the assignment. Ammonia DCI-MS and DCI-MS/MS were used both to examine toluene extracts of PE containing a wide variety of additives (Irganox 245/259/1010/1024/1076/1098/3114; DLTDP, DSTDP, Naugard 524, Tinuvin P/144/320/326/328/440/622/770, Chimassorb 944, oleamide, erucamide, DOP, DBP). Several toluene extracts of LDPE and HDPE were analysed by ammonia DCI and CID (product-ion spectra), indicating Irganox 1010 (m/z 1194); DSTDP (m/z 700); Naugard 524 (m/z 663, 680); Irganox 1076 (m/z 548); and a compound with the structure as assigned in Scheme 6.1 (m/z 934) in one of the samples [69].

The results thus show that ammonia DCI-MS/MS using a triple quadrupole mass spectrometer is a convenient method for the detection of additives in PE samples. The softness and selectivity provided by ammonia DCI in combination with the specificity provided by CID, demonstrate great potential for identification of additives directly from PE extracts. The utility of DCI in the quantitative analysis of additives has still to be explored. DCI-MS/MS (B/E) with high collision

energy in the analysis of polymer additives is equally a viable method for identifying additives in PE extracts [69]. In comparison with the triple quadrupole instrument, the sector instrument has much higher acceleration energy (3–10 kV vs. 1–250 V) and thus it can produce more structural information. High-energy CID always yielded more fragments than low-energy CID. Desorption chemical ionisation was considered as being the preferred fragmentation technique as not all of the additives in a mixture (with molecular weights <2000 Da) were detected with a harsher ionisation technique. In comparison with FAB, much less fragmentation was observed with DCI using ammonia as a reagent gas. Furthermore, most likely on account of the matrix effect, not all the additives in the extracts were detected by FAB. The softness and the lack of matrix effect make ammonia DCI a better ionisation technique than FAB for the analysis of additives directly from such extracts.

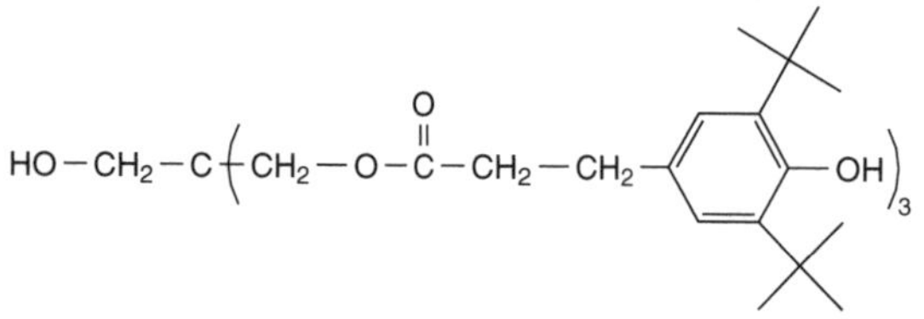

6.2.2.2 Ion–Molecule Reaction Mass Spectrometry

Principles and Characteristics The ion–molecule reaction (IMR) ionisation method belongs to the group of ion beam techniques. The basic structure of these instruments consists of:

(i) a mass-selected primary ion beam source;
(ii) a scattering cell where the ion beam reacts with the sample gas; and
(iii) as mass spectrometer where product ions are mass selected and detected.

In IMR-MS an ion storage section is created, into which ions from a conventional EI ion source are injected and

Scheme 6.1 Structural assignment

where they are trapped. Gas mixtures to be analysed are introduced into this reaction region, and both primary and product ions are detected, after extraction.

Fragmentation can be avoided if ionisation of the neutral constituents to be detected is done by ion–molecule reactions. Ion–molecule reactions occur at low relative kinetic energies between the reactants, so that essentially only the heats of formation of the particles involved govern the reactions, and thus there is usually not enough energy available to cause fragmentation. The choice of *primary ions* is critical for optimising both the selectivity and the sensitivity of IMR-MS measurements. Primary ions should efficiently react with the neutral gas, without producing fragmentation products. A suitable primary ion should have an ionisation energy which is slightly higher than the ones of the neutrals to be detected, so that charge transfer can occur at an appreciable rate. However, it should be low enough in order to avoid fragmentation or dissociative charge transfer due to large amounts of excess energy. For the analysis of complex gas mixtures, several primary ion systems are usually necessary. The gas to be analysed is ionised by reaction with a primary ion beam (typically He^+, Ne^+, Ar^+, Kr^+, Xe^+, CF_3I^+) at low collision energy ($\ll 1$ eV).

Bassi *et al.* [70] have described IMR-MS for on-line gas analysis with a sensitivity of 100 ppb–1 ppm. A mass-selected ion source allows the use of three different primary ion beams (Xe^+, Kr^+ and CF_3I^+), covering the recombination energy range from 10.23 to 14.67 eV. For fast measurements, the change from one primary ion to another can be achieved by a Wien filter. IMR-MS allows quantitative analysis.

Applications In contrast to EI ionisation, ion–molecule reactions in IMR-MS usually avoid fragmentation [71]. This allows on-line multicomponent analysis of complex gas mixtures (exhaust gases, heterogeneous catalysis, indoor environmental monitoring, product development and quality control, process and emissions monitoring) [70]. It should easily be possible to extend the application of the technique to the detection of volatiles in polymer/additive analysis.

6.2.3 Metastable Atom Bombardment

Principles and Characteristics In metastable atom bombardment (MAB), a metastable atom beam, generated by a gun external to the ion volume, is used to bombard the sample. MAB, based on Penning ionisation, offers unique features for gas-phase ionisation. The energy available for ionisation and fragmentation is discrete (excitation energy of the atom, from 8 to 20 eV),

can be fixed by the choice of the rare-gas (He to Xe), and allows the *selective ionisation* of organic compounds. Furthermore, the use of the appropriate rare-gas allows the control of the internal energy deposited in the ions formed, and fragmentation can be made extensively or totally eliminated. MAB sources can control fragmentation. It is possible to generate solely the molecular ion. This possibility offers extremely interesting advantages in terms of sensitivity for MS/MS experiments, because the entire ion current resides in the molecular ion. The ability to generate ions with very low internal energy and reduce fragmentation is very useful for the analysis of labile compounds which show feeble or no molecular ions in EI. The MAB source can be used on magnetic sector instruments as well as quadrupoles and ion traps. MAB can perform mass measurement with the same accuracy as EI, and allows for higher molecular ion abundances. MAB ionisation is particularly useful for mixture analysis.

The MAB ion source offers several advantages over EI for PyMS. By eliminating excessive fragmentation, characteristic of electron ionisation, and by producing highly reproducible mass spectra MAB (Kr) greatly simplifies the analysis of pyrolysis data. Furthermore, MAB ionisation, when combined to MS/MS, provides a useful tool for structural elucidation of pyrolysis products. The ability for selective ionisation can be very useful to reduce the background combination in techniques such as GC-MS, LC-MS or SFC-MS.

Applications TPPy-MAB-MS (Ar, N_2) was recently used for the analysis of bis(1,2,2,6,6-pentamethyl-4-piperidinyl) sebacate in a PUR-based car paint [71a].

6.2.4 Fast Atom Bombardment

Principles and Characteristics In the early mass-spectrometric ionisation techniques, such as EI and CI, the sample needs to be present in the ionisation source *in its gaseous phase*. Volatilisation by applying heat renders more difficult the analysis of thermally labile and involatile compounds, including highly polar samples and those of very high molecular mass. Although chemical derivatisation may be used to improve volatility and thermal stability, many compounds have eluded mass-spectrometric analysis until the emergence of fast atom bombardment (FAB) [72].

FAB-MS is a renewed old technique, first described in 1966 as *molecular beam for solid analysis* (MBSA) [73], and later (1981) further extended with a liquid matrix [72,74]. The FAB experiment is closely related to secondary-ion mass spectrometry (SIMS). In both cases the sample is bombarded with particles of similar kinetic energy. One primary difference is that FAB uses a neutral atom particle beam, while SIMS uses an ion beam. FAB is to be classified as a high-energy beam technique (low keV range) where gas-phase ions are generated by bombarding the sample dissolved or suspended in a liquid matrix, with a beam of energetic particles (usually He, Ar or Xe from an atom gun) (Figure 6.8). Operation of FAB is usually routine and straightforward. In the surface FAB technique [75] the FAB experiment is carried out without any added viscous matrix.

Generally FAB produces protonated, MH^+, or deprotonated, $(M - H)^-$, quasi-molecular ions with a little excess energy which will sometimes produce fragment ions of low intensity. FAB is therefore a 'mild' to 'soft' ionisation technique which produces primarily molecular weight information and some structural information. Positive and negative ionisation mass spectra are produced with equal facility. FAB was originally used with magnetic sector mass spectrometers, but lately mainly with quadrupole mass spectrometers (Table 6.10).

In matrix-assisted particle-desorption techniques (e.g. FAB, SIMS and PDMS), the average energy density pumped into the target surpasses greatly the formation enthalpy not only of the molecular ions of interest, but also of molecular fragments. As a result, a certain fraction of analyte molecules becomes fragmented or otherwise damaged by either direct or indirect action of the primary particles. Desorption of intact, large molecular species is a rather rare event.

Sample preparation for FAB is primarily dissolving the analyte in a liquid matrix. The *matrix* plays an important role in this technique. It is necessary to secure positioning of the sample on the probe tip in a high-vacuum environment until a mass spectrum has been collected. Therefore the matrix (usually a viscous liquid such as glycerol) must be reasonably involatile and have good solvating properties in relation to the sample. In practice, matrices are usually quite polar substances with molecular weights of less than 300 Da. Generally, the matrix constitutes most (usually >75 %) of the resulting mixture that is subjected to the ionisation process that occurs in the vacuum of the mass spectrometer. The choice of matrix is sample related and a multitude of matrices have been reported [77–79]. However, the choice of the ideal matrix for a given FAB experiment can prove to be a difficult task with few guidelines. Nevertheless the choice can be quite critical, as the FAB matrix exerts a profound influence on the ion type(s) produced, ion current intensity, and ultimate fragment ions observed for any given sample. The wrong matrix for a sample can mean no FAB spectrum at all. The liquid matrix permits the surface exposed to the energetic atom beam to be constantly replenished with the sample, and gives a signal lifetime ranging from minutes to over an hour. This increased signal lifetime enables acquisition of accurate mass data, and facilitates a range of tandem mass-spectrometric experiments.

The scope of the matrix is not only to transport and maintain the sample in the high-vacuum region of the source of the mass spectrometer, but also it appears to be necessary for the ion formation process. The matrix is often mixed with small quantities of an electrolyte to improve the results. Matrix and electrolyte tend to complicate FAB and FIB/LSIMS spectra, or at least are viewed by the analyst as essentially complicating the matter.

FAB and FIB/LSIMS produce singly charged ions; a good resolution will not only generate a clear isotope pattern but also an unambiguous, monoisotopic molecular mass. FAB mass spectra are influenced by a variety of experimental parameters, such as the primary beam and matrix used. FAB/LSIMS may give both features of hard ionisation (by means of *particle bombardment*) and of soft ionisation (by means of *matrix*), transferring only about 30 eV to the analyte in the latter case. In fact, the FAB/LSIMS conditions often result in the formation of radical molecular ions $M^{+\bullet}$, and abundant fragment ions which are hard ionisation characteristics, while the soft ionisation characteristic ions including protonated molecules $[M + H]^+$, sodiated molecules $[M + Na]^+$, and dimeric ions such as $[2M + H]^+$, $[M + H + B]^+$ (B: matrix), and $[M + H + L]^+$ (L: ligand) are produced according to

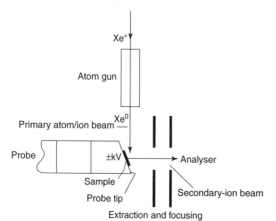

Figure 6.8 Experimental set-up for FAB. After Claeys and Claereboudt [76]. Reprinted from M. Claeys and J. Claereboudt, in *Encyclopedia of Spectroscopy and Spectrometry* (J.C. Lindon, ed.), Academic Press, pp. 505–512, Copyright (2000), with permission from Elsevier

the nature of analytes. The presence of the matrix thus gives rise to a high level of matrix-related ions, high chemical background noise (in particular below m/z 200), poor sensitivity and suppression effects. Glycerol, thioglycerol, triethanolamine, crown ethers, dithiothreitol/dithioerythritol and m-nitrobenzylalcohol, which are the most commonly used matrices, produce a characteristic spectrum which is comprised not only of MH^+ or $(M - H)^-$ ions, but also of polymeric cluster ions. For example, glycerol ions give rise to ions at m/z 93 $[MH^+]$, 185 $[MH + M]^+$, 277 $[MH + M_2]^+$, 369 $[MH + M_3]^+$, etc., up to high masses in positive ionisation mode, and a complementary series starting with m/z 91 $[M - H]^-$, 183 $[(M - H) + M]^-$, 275 $[(M - H) + M_2]^-$, 367 $[(M - H) + M_3]^-$, and so on in negative ionisation mode. If the sample gives rise to ions at any one of these m/z values, then a change of matrix should be considered. Sample-related ions in the region below m/z 200 can be quite difficult to detect with confidence.

Quantitative analysis using FAB is not straightforward, as with all ionisation techniques that use a direct insertion probe. While the goal of the exercise is to determine the bulk concentration of the analyte in the FAB matrix, FAB is instead measuring the concentration of the analyte in the surface of the matrix. The analyte surface concentration is not only a function of bulk analyte concentration, but is also affected by such factors as temperature, pressure, ionic strength, pH, FAB matrix, and sample matrix. With FAB and FIB/LSIMS the sample signal often dies away when the matrix, rather than the sample, is consumed; therefore, one cannot be sure that the ion signal obtained represents the entire sample. External standard FAB quantitation methods are of questionable accuracy, and even simple internal standard methods can be trusted only where the analyte is found in a well-controlled sample matrix or is separated from its sample matrix prior to FAB analysis. Therefore, labelled internal standards and isotope dilution methods have become the norm for FAB quantitation.

The *main characteristics* of FAB-MS are indicated in Table 6.15. FAB ionisation is relatively simple to perform. However, parameter optimisation and data interpretation of the resulting FAB spectra can be complex. Matrix selection for additive analysis is crucial. Solubility of the additives in the matrix is essential for production of viable spectra. FAB/FIB is well suited to organic compounds which exhibit some polarity, and contain either acidic and/or basic functional groups. Compounds with basic groups run well in positive ionisation mode, and those with acidic centres run well in the negative ionisation

Table 6.15 Main features of FAB-MS for polymer/additive analysis

Advantages

- Moderate sensitivity
- Positive- and negative-ion FAB spectral data
- Versatility (matrix choice)
- Direct mixture analysis (somewhat difficult)
- Wide range of compound classes, including relatively high mass
- Applicability for ionic, highly polar or thermally labile compounds
- Ease of use

Disadvantages

- Method optimisation (matrix selection)
- Matrix-related interferences
- High chemical background (noise for m/z <200 Da)
- Analyte–analyte and analyte–matrix interferences
- Difficult quantitation (labelling)
- Applicability to a rather limited range of industrial compound types (less effective for semipolar and nonpolar additives)

mode. Compounds that have both acidic and basic functionalities can be analysed by both ionisation modes. Traditionally, FAB and FIB/LSIMS have been used more in the bio-organic/biomedical area than for any other type of compound. Different samples can have very different responses to the FAB technique, and ion signals can vary significantly for a given concentration of sample. Consequently, analysis of mixtures can present difficulties. A given component of a mixture can suppress other components, while spectra that are consistent with a single-component sample could well have resulted from a mixture. Nevertheless, FAB can potentially provide *direct mixture analysis* without the need for prior fractionation. The intense, long-lived ion current provided by FAB is also well suited to coupling with MS/MS techniques to further facilitate mixture analysis.

FAB has been coupled (both on-line and off-line) with several separation techniques to improve FAB analysis of mixtures (e.g. TLC-FAB-MS, TLC-SIMS). Because FAB and FD provide highly complementary data for many sample types, combination FAB/FD sources have been reported. EI/FD/FAB and EI/FI/FD ion sources have also been mentioned [22], as well as a CI/FAB source [80].

Although introduction of FAB was a milestone in the development of mass-spectrometric ionisation techniques solving many biochemical-related problems, and was and still is a popular technique to use, it appears to be declining since the advent of electrospray.

Reviews of FAB principles, instrumentation and general applications are available, see refs [35,76,79,81,82].

Applications   Early MS work on the analysis of polymer additives has focused on the use of EI, CI, and GC-MS. The major drawback to these methods is that they are limited to thermally stable and relatively volatile compounds and therefore are not suitable for many high-MW polymer additives. This problem has largely been overcome by the development of soft ionisation techniques, such as FAB, FD, LD, etc. and secondary-ion mass spectrometry. These techniques all have shown their potential in the analysis of additives from solvent extract and/or from bulk polymeric material. Although FAB has a reputation of being the most often used soft ionisation method, Johlman *et al.* [83] have shown that LD is superior to FAB in the analysis of polymer additives, mainly because polymer additives fragment extensively under FAB conditions.

Fast atom bombardment (FAB) mass spectrometry has become a widely used technique for the analysis of high-MW and/or thermally labile compounds [84]. Čermák [81] has reviewed the main applications of FAB-MS in qualitative and quantitative analysis of tensides, organic acids and salts, organometallic compounds, inorganic compounds, vulcanised elastomers, synthetic polymers and additives, and aromatic hydrocarbons. Similarly, Cochran [79] has indicated that industrially significant areas as prepolymer and resin characterisation and the analysis of polymer additives (plasticisers, antioxidants, surfactants) are well suited to FAB-MS. In many cases where molecular weight limitations preclude FAB characterisation of the polymer itself, FAB can still be useful for revealing the presence of additives. The typical FAB industrial applications fall into a mass range <1500 Da.

Many additives fragment quite extensively with FAB. The observation of molecular ion and many fragment ions for each additive makes the determination of the number of additives in the extract difficult if not impossible. Moreover, due to the 'matrix effect', not all the additives in polymer extracts are detected under FAB.

FAB-MS has been used for the analysis of *lubricant additives*, thermally labile or involatile organic compounds, such as macromolecules and dyes, and inorganic compounds. *Cationic dyes* and dye intermediates, which are typically acid salts, readily yield preformed ions in the FAB matrix solution. They are also very difficult to address by other MS ionisation methods due to their involatility. Lay and Chang [85] used positive ion FAB to characterise a mixture of amine and ketimine *cross-linking agents* for polymer coatings. Bentz *et al.*

[86] characterised organotin compounds by FAB. Full-scan mode FAB-MS has been utilised for qualitative screening of PVC samples (baby pacifiers) for high levels of *phthalate plasticisers* by direct introduction of the solid PVC material into the mass spectrometer ion source [87]. A small sliver of material was affixed to the target using thioglycerol as a FAB matrix, which was also applied to the surface of the sample in the path of the fast atom beam. The mass spectra showed intense signals for the phthalates, and no other signals except those from the FAB matrix. Several PVC/DEHP samples could thus be screened in an hour, with no sample preparation. Any error or loss of precision associated with sample handling is reduced. Quantitation of DEHP was based on the relative signal of the $[MH]^+$ ions of DEHP, and didecylphthalate (DDP) as an internal reference. The DEHP levels were, within experimental error, equivalent to values determined using the more traditional method, i.e. extraction followed by GC or GC-MS. In principle, potential problems associated with sample homogeneity can also be reduced or eliminated by analysis of solutions containing whole (or even several) pacifiers. Campana and Rose [88] reported that a series of common phthalate plasticisers yield FAB spectra which are similar to their EI spectra. Two amine antioxidants, di-(*t*-octyl)-diphenylamine and phenyl-β-naphthylamine, were identified in rubber extracts using FAB. In both cases an $M^{+\bullet}$ radical cation rather than an $[M + H]^+$ ion was seen in the FAB spectrum.

FAB-MS is highly specific for surfactant species which are difficult to analyse by other methods. The fairly polar *surfactants* represent a class of industrially important compounds for which FAB is highly suited [79]. The fact that cationic surfactants have been proposed as FAB calibration standards attests to their ease of analysis by this technique. Only limited structural information is available from fragment ions in the positive FAB spectra of anionic surfactants. FAB ionisation analysis of anionic surfactants was reported [89], and FAB-MS/MS techniques have been applied to anionic surfactants [90]. Monosubstituted PEG non-ionic surfactant mixtures and isolated oligomeric components have been successfully used as exact mass internal standards in the 700–1200 Da mass range in FAB mass spectrometry [91]. Lattimer [17] has applied FAB to PEGs and PPGs ranging in molecular weight from approximately 400 to 1200 Da.

Industrial samples often consist of a complex mixture of molecular weights and/or functionalities. For their analyses, the fragmentation information in the FAB spectrum is of little use unless the component molecular weights are known. Commercially available FAB/FD combination sources are helpful in this respect. Lattimer *et al.* [92] compared FAB and FD as ionisation techniques for mass-spectrometric analysis of *additive mixtures* (plasticisers, antioxidants, antiozonants, oils and waxes) extracted from rubber compounds. Although both are effective, FD-MS provided only molecular weights, while FAB-MS also gave fragmentation to aid in component assignments. Neither method was useful in quantitative analysis. Some components, particularly minor ones, could be missed in complex mixtures. Although both methods sometimes suppressed certain types of polymer additives when more strongly desorbing species were present, they often complemented each other, with FD-MS and FAB-MS being more effective with nonpolar and polar compounds, respectively.

Sterically hindered phenols and other additives containing thioesters, phosphites, phosphonites and hindered amine moieties were analysed by FAB-MS and LD-FTMS. The laser desorption technique was preferred for analysis of polymer additives because of undesirable fragmentation from FAB experiments [93].

FAB has been used to analyse additives in (un) vulcanised *elastomer systems* [92,94] and FAB matrices have been developed which permit the direct analysis of mixtures of elastomer additives without chromatographic separation. The T-156 triblend vulcanised elastomer additives poly-TMDQ (AO), CTP (retarder), HPPD (antiozonant), and TMTD, OBTS, MBT and N,N-diisopropyl-2-benzothiazylsulfenamide (accelerators) were studied in three matrix solutions (glycerol, oleic acid, and NPOE) [94]. The thiuram class of accelerators were least successful. Mixture analysis of complex rubber vulcanisates without chromatographic separation was demonstrated. The differentiation of matrix ions from sample ions was enhanced by use of high-resolution acquisition.

Recently, Lattimer *et al.* [22,95] advocated the use of mass spectrometry for *direct analysis* of nonvolatile compounding agents in polymer matrices as an alternative to extraction procedures. FAB-MS was thus applied as a means for surface desorption/ionisation of vulcanisates. FAB is often not as effective as other ionisation methods (EI, CI, FI, FD), and FAB-MS is not considered particularly useful for extracted rubber additives analysis compared to other methods that are available [36]. The effectiveness of the FAB technique has been demonstrated for the analysis of a five-component additive mixture [96].

FAB application to *UV stabilisers* in polymers has been discussed in two reports. Layer *et al.* [97] applied FAB together with FD and ESR to study oxidation products of a class of amine UV-stabilisers, 3,3-dialkyldecahydroquinoxalin-2-ones. Riley *et al.* [98] have described a *quantification* procedure to monitor the paint additive Tinuvin 770 in two coating systems (acrylic melamine and a hydroxy ester melamine), and have illustrated the many pitfalls using an external standard, a non-labelled internal standard, and the isotope dilution quantitation method. The external standard method suffered from poor intra-day reproducibility (up to 50 % variation). Small changes in sample matrix affected analyte and internal standard (tricyclohexyl citrate) quite differently, thus also degrading the accuracy of this method. Only by synthesising a deuterium-labelled Tinuvin 770 analogue for use as an internal standard were accurate, matrix-effect-free results obtained. Relatively few studies have been made on the feasibility of quantitative FAB analysis.

The *FAB-MS/MS* approach is a powerful analytical tool for rapid search of additives in extracts: providing not only a fingerprint of the main constituents, but also, through the complementary information obtained by parent- and daughter-ion scan analysis, their structure elucidation. The highly specific FAB-MS/MS technique finds application mainly for fractions of higher polarity. FAB-MS/MS techniques have revealed their use in determining structure and double-bond location in C_2-C_{24} fatty acids [99]. FAB-MS/MS was also used for identification of *n*-octyltin triisooctyl thioglycollate (main component) and tin tetraisooctyl thioglycollate (impurity) [100].

In the deformulation of PE/additive systems by mass spectrometry, much less fragmentation was observed with DCI-MS/MS using ammonia as a reagent gas, than with FAB-MS [69]. FAB did not detect all the additives in the extracts. The softness and the lack of matrix effect make ammonia DCI a better ionisation technique than FAB for the analysis of additives directly from the extracts. Applications of hyphenated FAB-MS techniques are described elsewhere: low-flow LC-MS (Section 7.3.3.2) and CE-MS (Section 7.3.6.1) for polar nonvolatile organics, and TLC-MS (Section 7.3.5.4).

Most recently, little use is being made of FAB-MS for polymer/additive deformulation, possibly as a result of a shift in interest towards electrospray ionisation and difficult quantitation. Practical applications of FAB-MS were reviewed [78,79,81].

6.2.4.1 Liquid Secondary-Ion Mass Spectrometry

Principles and Characteristics In order to achieve higher ion yields than in FAB a caesium (Cs^+) ion gun [101] has been developed; such a fast beam of ions can be used to bombard a sample target with higher energies than a fast beam of atoms (ion energies of up to 35 kV vs. atom beam energies of 8–10 kV). The technique of fast ion bombardment (FIB) is now referred as to liquid secondary-ion mass spectrometry (LSIMS), due to the fact that both the primary and the secondary beams are composed of ions. The term 'liquid' refers to the aforementioned fact that the sample is dissolved in a liquid matrix. The FIB ionisation process appears to be identical in mechanism with the more well-established secondary-ion mass spectrometry (SIMS) ionisation process. However, SIMS is used to examine the surface of a solid sample directly, whereas FIB is generally used to desorb ions from a sample suspended in a liquid matrix like glycerol. The LSIMS spectra generated are usually indistinguishable in nature from FAB spectra, especially if working with low-MW samples (<1000 Da) at reasonable concentrations. As expected, the high-energy caesium ion beam does generally increase the sensitivity of the technique with all samples.

FIB/LSIMS and FAB analyse basically the same type of compounds, with enhanced sensitivity (and sample mass range) for the former technique. From an experimental point of view, the technique requires little expertise to generate useful data quickly.

Mass spectrometers which have the capability of analysing samples by a FAB or FIB/LSIMS-source tend to be fitted with either one alternative or the other; such mass spectrometers have either a magnetic sector or a quadrupole analyser (or hybrid or tandem combinations).

Desorption/ionisation techniques such as LSIMS are quite practical, as they give abundant molecular ion signals and fragmentation for structural information. In the conditions of Jackson *et al.* [96], all the molecular ion and/or protonated molecule ion species were observed in the LSIMS spectrum when only 1 pmol of each additive was placed on the probe tip. However, as mentioned above, in LSIMS/MS experiments the choice of the *matrix* (e.g. NBA, *m*-nitrobenzylalcohol) is very important. Matrix effects can lead to suppression of the generation of molecular ions for some additives. LSIMS is not ideal for the quantitative detection of polymer additives, as matrix effects are very important [96].

LSIMS is a more suitable ionisation technique than FD for analysis of mixtures by means of tandem mass spectrometry, because of the higher ion currents generated from polymer additives using LSIMS. LSIMS/MS experiments may be used in conjunction with FD-MS as a screen to determine class, molecular weight and structure of mixtures of organic polymer additives.

Applications LSIMS spectra are complicated by the fragment ions observed. For example, this turned out to be important in the molecular ion regions of Irganox 3114 and Hostanox O3, because of the series of fragment ions of Irganox 1010 that were observed in mixtures [96].

Applications of hyphenated LSIMS techniques are described elsewhere, e.g. TLC-MS (Section 7.3.5.4).

6.2.4.2 Continuous-Flow Fast Atom Bombardment

Principles and Characteristics Continuous-flow (or dynamic) FAB/FIB [102] and frit FAB/FIB [103] offer a means of introducing samples in solution into a continuous flow of solvent which terminates at the modified FAB/FIB probe tip, and they extend the applicability of FAB. Samples are injected through a conventional HPLC injection valve, or solutions are simply drawn in by the high vacuum in the ionisation source of the mass spectrometer. These very similar techniques are particularly amenable to coupling with HPLC columns, and ionisation of the sample is unchanged with respect to conventional FAB and FIB/LSIMS.

However, the *advantages* of these alternative FAB designs are not just of minor importance, and include improved sensitivity; reduced background signal; more accurate quantification, less analyte discrimination owing to suppression effects; and greater flexibility in operation and coupling to separation methods. Solvents employed are often of the reversed-phase nature (i.e. water, methanol, acetonitrile and combinations thereof) because of the (usually) high polarity of the samples under investigation. However, use of normal-phase solvents such as toluene has also been reported. With standard FAB/FIB, the ionisation source is not heated, which is beneficial to analysis of thermally labile samples. However, in CF-FAB/FIB the temperature required at the probe tip is slightly higher (about 40 °C). A monograph has appeared [104].

6.2.5 Field Ionisation

Principles and Characteristics The pioneering technique of field ionisation (FI) was the first soft ionisation technique, introduced in 1954 [105]. For FI analysis of a reasonably volatile sample, the compound under investigation is volatilised by heat close to the emitter, so that its vapour can condense on to an emitter needle. Hence,

FI is suitable for on-line PyMS or GC-MS; in the latter case, the GC effluent containing the separated components in a flow of heated helium gas is carried directly into the ionisation source. Alternatively, a heated reservoir inlet such as an all-glass heated inlet system (or AGHIS) will allow vaporised samples to exude into the ionisation source. In FI-MS application of a large electric potential (ca. 10 keV) to a metal surface of high curvature results in an intense localised electric field; the gradient produced on the needle tip promotes an electron tunnelling-transfer mechanism which results predominantly in radical molecular ions $M^{+\bullet}$ (soft ionisation) [106]. As no significant internal energy is imparted to the ions, fragmentation is considerably less than that observed by electron impact. The FI source is about 10 times less sensitive than EI. An advantage of FI over EI is that it can be used for samples of much lower volatility and hence molecular weight. Field ionisation does not require reagent gases or large pumping capacities as does CI.

Table 6.16 summarises the *main characteristics* of FI-MS. FI uses high voltages and was once restricted to sensitive double-focusing magnetic sector instruments of relatively high cost. Field ionisation is considered to be the softest ionisation mode. The reproducibility of the non-standard techniques, such as FI-MS and FD-MS, is less well assessed than that of EI-MS. A noticeable drop in FI use occurred after the mid-1980s because of the advent of FAB and other desorption/ionisation methods. FI-MS is only used in a few laboratories worldwide.

FI-MS was reviewed recently [107]; a monograph deals with FI-MS/FD-MS [108].

Applications FI-MS and FD-MS have some attractive features that make them both complementary to other desorption/ionisation methods and unique in their applications. FI has found wide application mainly for the

Table 6.16 Main characteristics of FI-MS

Advantages

- Soft ionisation (fairly high molecular ion abundances)
- Suitable for low-volatility, high-MW analytes (up to ca. 10 000 Da)
- Optimal for nonpolar and slightly polar organics
- Allowance for direct mixture analysis (simple spectra); survey analysis

Disadvantages

- Relatively low sensitivity
- Use of high-voltage source
- Requires high-cost magnetic sector instruments or oaToF-MS
- Limited applicability

analysis of hydrocarbons, which are commonly admitted to the ionisation source via a reservoir inlet system. In these conditions, molecules generate $M^{+\bullet}$ ions with little fragmentation, in contrast to the spectra in EI conditions, which generally generate weak molecular ions accompanied by substantial fragmentation.

Lattimer *et al.* [22,37,65,75] have described direct analysis of rubber or plastic material by field ionisation. Typically, in a diene rubber sample heated in a direct probe (PyFI-MS) from 50 to 750 °C two distinct TIC maxima were observed. At low temperatures (50–400 °C) mainly evaporation of organic additives from the rubber occurs, including fatty acids (MW 256, 284), a *p*-phenylenediamine antiozonant (MW 332), a *t*-octylated diphenylamine antioxidant (MW 393, 505), a *t*-octylphenol/formaldehyde tackifying resin (MW 424, 536) and processing oil ('background' ions) [75]. The second TIC maximum at 400–750 °C represents the evolution of rubber thermal decomposition products (pyrolysates), including isoprene oligomers (MW $68n$, $n = 1$–13). Another study [65] has compared direct mass-spectral identification of residual volatile chemicals, organic additives and degradation products in commercial *elastomer compounds* of unknown composition using 70 eV EI-MS, isobutane CI-MS and FI-MS with thermal desorption and pyrolysis. Residual volatile chemicals and most organic additives were thermally desorbed at lower temperatures (<250 °C), while polymeric components were thermally decomposed (pyrolysed) at higher temperatures (>250 °C). MS/MS and HRMS were carried out to improve the specificity of the analysis. An EPDM bearing was analysed by low-resolution survey FI-MS screening (50–1000 Da region) covering the sample heating range of 70–230 °C; in the same conditions the EI-MS spectrum is considerably more complex, due to the large number of fragment ions from processing oil and other ingredients. Molecular ions using MS/MS and AC-MS pointed to cumyl peroxide curing agent, processing oil, poly-TMDQ, Irganox 1076 and fatty acids/esters. Analysis of a sulfur-cured carbon-black-filled diene rubber V-belt (FI-MS survey scan in sample heating range of 40–300 °C and CI-MS from 70–250 °C) allowed the detection of paraffin wax (C_{20}–C_{45}), unsaturated oil/wax, CHPPD antiozonant, diphenylamine/acetone resin antioxidant, fatty/resin acids, and TBBS accelerator. These examples illustrate the sensitivity, high specificity, and superb mixture analysis capabilities of mass spectrometry.

Direct probe FI-MS of PP samples containing various phosphites was used to study the effect of a sequence of extrusion passes at 260 °C [109].

6.2.6   Field Desorption

Principles and Characteristics Field desorption (FD), invented in 1969 [108,110], is an extension of the field ionisation technique described by Inghram and Gomer [105,106], and was developed for reasons similar to those for the CI and FI modes, namely, to analyse samples that produced no molecular ions by EI. While EI, CI and FI all require ionisation of molecules in the vapour state, this not so for FD. For field desorption analysis of nonvolatile and polar material, the sample needs to be dissolved (or at least suspended) in an appropriate, volatile solvent (such as acetone, dichloromethane, or toluene) to a concentration of ca. $1\,mg\,mL^{-1}$. The sample and solvent are then applied to the emitter wire, either by dipping the wire in the solution or by applying the solution to the wire with a syringe. With the sample applied to the emitter wire, the FD probe can then be inserted into the high-vacuum ion source of the mass spectrometer and exposed to a high-voltage electric field, creating a very high field gradient between emitter and mass spectrometer ion optics. This field gradient causes materials on the emitter to be desorbed and ionised; the ions formed are usually sampled by means of a double-focusing mass spectrometer. Whenever the field alone does not induce sample desorption, a current can be passed through the wire, heating the sample. Typical source temperatures range from $60°$ to $180\,°C$ [108]. Field desorption is a continuous ionisation technique that is not easily pulsed [111].

As there is little excess energy in the soft FD/FI ionisation processes, ions with low internal energy (less than 1 eV) are generated. Molecular and quasi-molecular ions are produced with insufficient internal energy for extensive fragmentation. Usually the major peak represents the $[M + H]^+$ ion. Doubly and triply charged ions may occur. As shown in Table 6.17, cation or anion attachment, thermal emission and proton abstraction are the other *ionisation processes* that can take place in an FD/FI source. Basic, polar organic molecules (e.g. amines) are most likely to be analysed in positive ion mode, where they can form cation attachment ions in the condensed phase with cations such as H^+ (and Na^+). In negative ion mode it is possible for some polar, acidic samples (e.g. carboxylic acids) to ionise by anion attachment, e.g. $(M + Cl)^-$ ions. Thermal emission is observed for (in)organic salts. Most polar organic molecules susceptible to negative ionisation readily produce $(M - H)^-$ ions by proton abstraction.

The two closely related techniques of field desorption [108] and field ionisation [105] are appropriate for

Table 6.17 Ionisation mechanisms in FD/FI techniques

Ionisation mechanism	Technique	Ions formed	Ionisation mode	Type of sample
Field ionisation	FI, FD	$M^{+•}$	+	Nonpolar organics
		$M^{-•}$	−	
Cation attachment	FD/FI	MH^+, MNa^+	+	Basic, polar organics
Anion attachment	FD/FI	$(M + Cl)^-$	−	Acidic, polar organics
Thermal emission	FD/FI	C^+	+	Ionic species[a]
		A^-	−	
Proton abstraction	FD/FI	$(M - H)^-$	−	Acidic, polar organics

[a] Salts, $C^+ A^-$.

the analysis of thermally labile and high-MW samples. FI and FD have emerged as alternative ionisation techniques to complement EI, which is limited to relatively low-MW samples that can be vaporised prior to ionisation, and thus often leads to thermal degradation of involatile or heat-sensitive compounds.

Table 6.18 lists the *main characteristics* of FD-MS. FD is a superior ionisation technique for quantitative analysis, as there are no matrix effects as in LSIMS or MALDI which might suppress the generation of ions from certain additives. However, the technique has some serious drawbacks. The primary difficulty is that FD produces only short-lived, highly variable currents of analyte ions. These analyte ion currents are also very

Table 6.18 Main features of FD-MS

Advantages
- Little sample preparation
- Soft ionisation
- Suitable for nonpolar and some polar organics
- Analysis of thermally labile and high-MW samples
- Rapid survey technique
- Ready insight into the number of components of a mixture (without prior isolation)
- Quantitative analysis

Disadvantages
- Desorption capacity required
- Poor sensitivity
- Moderate reproducibility
- Use of high-voltage source
- No structure elucidation (unless FD-MS/MS)
- Experimentally difficult
- Intrinsically slow technique
- Requires expensive double-focusing mass spectrometer
- Difficult optimisation ('more art than science')
- Restricted use
- Limited applicability

weak (typically two orders of magnitude lower than obtained by LSIMS), and specially designed analysers need to be used to gain useful information. The use of FD in the study of nonpolar polymeric systems (up to 20 kDa) and more polar material (up to 5 kDa) is limited. The approach has not received wider popularity, due to the relatively difficult nature of the experiment compared to other current methods, as well as difficult optimisation, together with the requirement for expensive double-focusing mass spectrometers. In addition, the current that is required limits the mass range. Thermal decomposition can occur for thermally labile systems. The FD-MS experiment is more challenging than LSIMS/MS, because of the low ion currents generated by the former technique and the delicate nature of the emitters. Expert users [107] minimise these difficulties.

FD-MS is a very effective technique for determining molecular weights of thermally labile and nonvolatile compounds, such as polymer additives which do not give good molecular ion spectra during electron impact or chemical ionisation [108]. In order to enhance the structural information of the technique, MS/MS approaches must be used [96]. Hyphenated chromatography–FD/FI-MS techniques appear to be restricted to on-line GC-MS.

Although FD was one of the earliest forms of soft ionisation, poor sensitivity and limited applicability have restricted the impact of the approach in the mainstream of mass spectrometry. More recently, many of the application areas of FD and FI have been appropriated by FAB-MS, which is generally considered to be a technique that requires less expertise; alternatively, laser desorption is frequently being applied. FD-MS is only used in a handful of laboratories worldwide. The technique has recently been reviewed [107], and is subject of various monographs [108,112].

Applications Rather intractable samples, such as organic polymers, are well suited to FD, which avoids the need for volatilisation of the sample. Since molecular ions are normally the *only* prominent ions formed in the FD mode of analysis, FD-MS can be a very powerful tool for the characterisation of polymer chemical mixtures. Application areas in which FD-MS has played a role in the characterisation of polymer chemicals in industry include: chemical identification (molecular weight and structure determination); direct detection of components in mixtures; off-line identification of LC effluents; characterisation of polymer blooms and extracts; and identification of polymer chemical degradation products. For many of these applications, the samples to be analysed are very *complex*

mixtures of chemical compounds. FD-MS has shown successful application to various compound classes, which traditionally can be difficult to analyse by EI (Table 6.10). Mixture analysis is also difficult in FAB, where fragmentation and thermal degradation are extensive. FD generally works more readily for less-polar molecules, while FAB seems best for more polar materials. This feature renders FD and FAB rather complementary in nature for polymer chemical analysis. Thus, in a mixture of diverse polymer chemicals, some components may desorb better by FD, while others will desorb better by FAB.

FD-MS and FAB-MS have been used effectively for identification of organic additives in *rubbers*. Both techniques can be performed quite rapidly, in 5 to 10 min, and provide a complement to IR analysis of compounding additives. Lattimer *et al.* [22,75,92] have described field desorption analysis of rubber extracts. Rubber extracts were also examined directly by EI, CI, FD and FAB [92]. FD/FI-MS was indicated as being most efficient for identifying typical organic additives in rubber vulcanisates. Lattimer [37] has described the identification of Chimassorb 944 by means of FI-MS and FD-MS. Various oligomer series (A: MW $598n + 394$, i.e. 394, 992, 1590, 2188, 2786, 3384; B: MW $598n$, i.e. 598, 1196, 1794, 2392, 2990) were observed in the FD-MS spectrum (Figure 6.9).

FD-MS is also an effective analytical method for direct analysis of many rubber and plastic additives. Lattimer and Welch [113,114] showed that FD-MS gives excellent molecular ion spectra for a variety of polymer additives, including rubber accelerators (dithiocarbamates, guanidines, benzothiazyl, and thiuram derivatives), antioxidants (hindered phenols, aromatic amines), *p*-phenylenediamine-based antiozonants, processing oils and phthalate plasticisers. Alkylphenol ethoxylate surfactants have been characterised by FD-MS [115]. Jackson *et al.* [116] analysed some plastic additives (hindered phenol AOs and benzotriazole UVA) by FD-MS. Reaction products of a *p*-phenylenediamine antiozonant and *cis*-9-tricosene (a model olefin) were assessed by FD-MS [117].

FD-MS by itself provides only limited chemical information. Lattimer *et al.* [92] have also compared the analysis of extracted rubber vulcanisates by means of FD-MS and FAB-MS, using the aforementioned EI/FD/FI/FAB ion source. The systems investigated were neoprene/DOPPD, EPDM/(DOP, PBNA, paraffin wax), neoprene-SBR blend/(DOP, DOPPD, TDBHI). Certain compounds were observed by FD but not by FAB (wax, oil, isocyanurate antioxidant TDBHI). In FAB conditions some polymer additives suppress

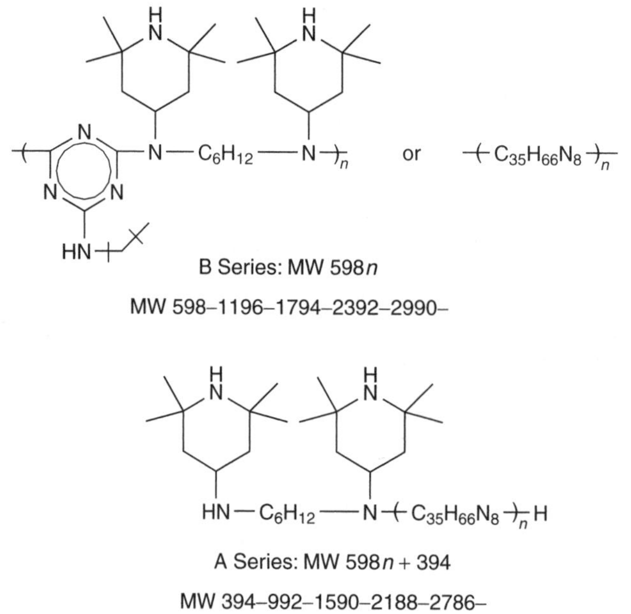

Figure 6.9 Chemical structures of Chimassorb 944 oligomers. After Lattimer [37]. Reprinted from *Journal of Analytical and Applied Pyrolysis*, **26**, R. P. Lattimer, 65–92, Copyright (1993), with permission from Elsevier

the generation from other species. In none of the rubbers were curatives identified, neither by FD nor by FAB. Also neither method was considered to be useful in quantitative analysis, in apparent contrast with the findings for FD–high-energy MS [96]; this mode generates predominantly molecular radical cations $(M^{+\bullet})$, as observed in a mixture of additives. Overall, FD–high-energy MS is a superior ionisation technique for quantitative analysis, as there are no matrix effects which might suppress the generation of ions from certain additives. High-energy FD-MS/MS experiments may be used in conjunction with FD-MS as a screen to determine class, molecular weight and structure of organic polymer additives in *mixtures without prior separation*. Ideally, the resulting spectrum will contain one diagnostic ion for each component.

From the characteristics of the methods, it would appear that FD-MS can profitably be applied to polymer/additive dissolutions (without precipitation of the polymer or separation of the additive components). The FD approach was considered to be too difficult and fraught with inherent complications to be of routine use in the characterisation of anionic surfactants. The technique does, however, have a niche application in the area of nonpolar compound classes such as hydrocarbons and lubricants, compounds which are difficult to study using other mass-spectrometry ionisation techniques.

Off-line coupling of HPLC with FD-MS has been used by several authors [118–121] for the determination of *oligomers*, oligomeric antioxidants (such as poly-TMDQ), ozonation and vulcanisation products. Pausch [122] reported on rubbers, cyclic polyurethane oligomers, as well as on the determination of the molecular weight distribution (up to 5300 Da) and oligomer analysis of polystyrene. Also the components of an aniline–acetone resin were deduced from FD-MS molecular weights [122].

6.2.7 Thermospray Ionisation

Principles and Characteristics Thermospray ionisation (TSP) involves introduction of a relatively high flow (0.2–2 mL min^{-1}) of solvent into the ion source of a mass spectrometer, and is therefore suitable as an interface for HPLC-MS, using standard bore columns. A vaporiser probe (essentially a resistively heated capillary tube of about 100 μm i.d.) acts as a transfer line for taking solvent and solute into the source. The source is heated to prevent condensation of the solvent, and the temperature of the capillary is chosen so as to ensure vaporisation of the solvent. In this way, a vapour jet is generated, which contains small, electrically charged droplets if the solvent is at least partially aqueous and

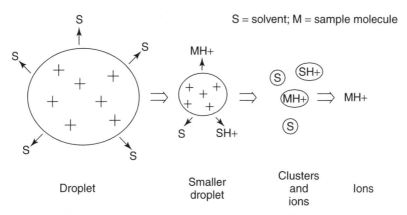

S = solvent; M = sample molecule

Figure 6.10 Production of positive ions by thermospray ionisation. After Ashcroft [35]. From A.E. Ashcroft, *Ionization Methods in Organic Mass Spectrometry*, The Royal Society of Chemistry, Cambridge (1997). Reproduced by permission of The Royal Society of Chemistry

contains an electrolyte (Figure 6.10). The presence of the electrolyte (e.g. ammonium acetate) is essential to enhance the ion yields and sensitivity. The droplets continue to vaporise, becoming smaller in size. Eventually, free (positive and/or negative) ions are expelled from the surface of the droplets. The ions are then analysed by the quadrupole or a magnetic sector analyser of the mass spectrometer. If an electrolyte cannot be used for a particular analysis, an electrical discharge in the source may be employed as an alternative way to effect ionisation. Most thermospray sources are fitted with such a discharge electrode. When this is used, the technique is called filament-on thermospray or *plasmaspray* (PSP) ionisation. In practice, many analyses are a mixture of TSP and PSP ionisation.

Solvents used successfully in TSP operation include water, methanol, acetonitrile, propan-2-ol, dichloromethane and hexane. Any volatile buffer may be employed as an electrolyte, but involatile buffers and inorganic acids such as phosphate salts are to be avoided. This poses some limits on the analysis.

Thermospray ionisation sources are usually outfitted with a quadrupole or magnetic sector mass spectrometer (including hybrids or tandem forms). Thermospray operation allows a reversed-phase solvent system, e.g. a 50:50 (v/v) water-methanol or acetonitrile mix containing 0.1 M ammonium acetate. This ensures compatibility with the 'universal' HPLC procedures available in many industrial research laboratories.

The TSP temperature needs to be high enough to prevent condensation of solvent on the source (typically 200–300 °C). Too high a temperature can lead to the thermal degradation of labile compounds. The *vaporiser temperature*, which is generally set at

150–250 °C, is one of the most critical parameters in TSP operation, and should be optimised for different samples, wherever possible. This is considered to be a considerable drawback in routine operation of unknown polymer/additive extracts. Too low a vaporiser temperature results in the solute and solvent spraying into the ionisation source in their liquid form, without formation of gas-phase ions. Too high a vaporiser temperature causes premature evaporation of the solute and solvent before the outlet of the capillary is reached. This causes an unstable, pulsing ion beam. As ion formation in TSP operation depends very critically on the extent of desolvation and the energy of the nebulised droplets, it is clear that an inappropriate vaporiser temperature will cause loss of sensitivity.

Thermospray (TSP) is another soft ionisation technique which produces predominantly MH$^+$ or (M − H)$^-$ ions, together with some fragmentation. TSP is best suited to the analysis of organic compounds of low molecular mass (<1000 Da) that exhibit some polarity. Polymer additive molecules fall in this wide category.

Thermospray was quite popular before the advent of electrospray, but has now given way to the more robust API techniques, although TSP sources continue to operate. Developed as an LC-MS interface, this technique calls for a continuous flow of sample in solution.

Applications Various surfactant types (ABS, AES, secondary alkane sulfonates, and alkylphenol ethoxysulfates) have been analysed by means of a QQQ using a thermospray source [89]. Other applications of hyphenated thermospray ionisation mass-spectrometric techniques (LC-TSP-MS) are described elsewhere (Section 7.3.3.2).

6.2.8 Atmospheric Pressure Ionisation Techniques

Principles and Characteristics Ion sources for atmospheric pressure ionisation (API) in mass spectrometry were first described in 1958 [123]. The major feature of API-MS is that samples can be routinely introduced into a mass spectrometer at atmospheric pressure, which overcomes the need to cope with traditional ultrahigh vacuum sample introduction systems. This allows interfacing of high-volume liquid introduction techniques. An API ion source (Figure 6.11) generally consists of four parts: a sample introduction device, ion source region (where the ions are generated), ion sampling aperture and ion transfer system (where the pressure difference between the atmospheric pressure source and the high vacuum of the mass analyser is bridged). API comprises various quite different ionisation methods, such as electrospray ionisation (ESI), atmospheric pressure chemical ionisation (APCI) and, more recently, also atmospheric pressure photoionisation (APPI), and others.

Ionisation in an API source can take place in a variety of ways depending on the type of applications, namely by gas-phase ionisation, liquid- and plasma-based ionisation. At present, there are three major application areas of API-MS: air or gas analysis (industrial emissions), on-line LC-MS (largest commercial application), and ICP-MS. A wide variety of sample introduction devices are available for gas analysis by API-MS. For use in ICP-MS, ions are sampled directly from the inductively

coupled plasma. For use in combination with LC-MS and other liquid-introduction techniques, two main types of sample introduction devices are available, i.e. one for APCI and one for ESI. The (most developed) API techniques – APCI and ESI – both:

(i) use thermal pneumatic nebulisation to produce aerosols;
(ii) produce $(M + H)^+$ quasi-molecular ions for MW confirmation;
(iii) use collision-induced dissociation (CID) capability to produce fragmentation; and
(iv) provide MW information on more polar and labile compounds not amenable to PB.

APCI and ESI differ in:

(i) ion generation: corona discharge ionisation (APCI) and ion evaporation (ESI);
(ii) target compounds which can be analysed successfully (molecular weight vs. polarity);
(iii) sensitivity achieved (compound dependent); and
(iv) nature of charged ions (APCI: singly charged; ESI, multiply charged).

API-MS quickly and reliably analyses any size of molecule in almost any matrix, without contamination or degradation. As it is usually not obvious which technique to use, it is therefore often worthwhile analysing the samples of interest with both ionisation techniques and making the final decision empirically.

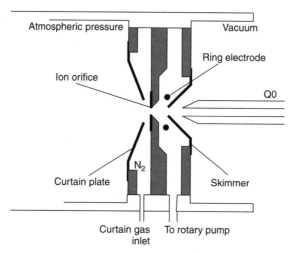

Figure 6.11 Schematic diagram of a typical API source, as for instance used in LC-MS. After Niessen [124]. Reprinted from W.M.A. Niessen, in *Encyclopedia of Spectroscopy and Spectrometry* (J.C. Lindon, ed.), Academic Press, pp. 18–24, Copyright (2000), with permission from Elsevier

Table 6.19 Main characteristics of API-MS

- Sample introduction at atmospheric pressure
- No degradation of thermally labile compounds
- Ultrastable mass calibration
- Reproducible and consistent performance (direct transfer of methods between API systems without re-optimisation)
- Compatibility with buffers and all LC solvents
- Positive- and negative-ion modes
- Precise results, routinely without daily maintenance and cleaning
- Nonclassical MS spectra

Electrospray mass spectrometry

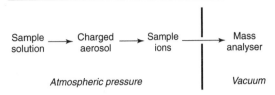

Figure 6.12 Schematics of electrospray mass spectrometry

The *main characteristics* of API-MS are given in Table 6.19. A drawback of API-MS is that the classical MS spectra cannot be used for comparison. API-MS in SIR mode can be used for quantitative assays using standard solutions. API-ToFMS allows high-throughput determination of exact molecular weight and complementary structural information for total sample characterisation. The advantage of direct mass spectrometry using API-MS is that data can be generated in a few minutes and further structural investigation is facilitated by routine tandem mass-spectrometry experimentation. Quadrupole technology supports all modes of tandem MS/MS operation, including product ions, precursor ions, neutral-loss and MRM scans. Benefits include high-energy fragmentation resulting in true MS/MS spectra and superior structural information.

This chapter is mainly limited to *direct API-MS*. For LC-API-MS, see Section 7.3.3.2. APCI, ESI and APPI complement each other in chemical analyses. API-MS was reviewed repeatedly [124,125], and has been the subject of a recent monograph [126].

6.2.8.1 Electrospray Ionisation

Principles and Characteristics Electrospray ionisation (ESI) is an API technique suitable for quadrupole, magnetic sector and ToF instruments. As with all the API techniques, the formation of ions takes place outside the vacuum system of a mass spectrometer. ESI is a desorption technique in which dilute solutions of sample are sprayed into a gas filled chamber. Using a high-velocity nitrogen gas flow, the solution is forced through a syringe held at high potential (usually 2–4 kV), so that the resulting aerosol droplets are charged. The aerosol is then desolvated by a combination of heat, gas flow and vacuum, decreasing the size of the droplets and increasing the density of charge at the surface. When the coulomb repulsion forces overcome the forces of surface tension, the droplets 'explode'. At some stage,

the electric field at the surface of the very small spheres is so high that singly or multiply charged solute ions enter the vapour phase and move to the mass spectrometer (Fig. 6.12). Electrospray thus consists of transferring pre-existing ions from solution to the gas phase for mass analysis, and operates essentially by an 'ion evaporation' process [127]. Actually, electrospray is not an ionisation technique, but rather a charge separation system that results in ions. Doyle *et al.* [128] have demonstrated the process experimentally.

ES ionisation can be pneumatically assisted by a nebulising gas: a variant called *ionspray* (IS) [129]. ESI is conducted at near ambient temperature; too high a temperature will cause the solvent to start evaporating before it reaches the tip of the capillary, causing decomposition of the analyte during ionisation; and too low a temperature will allow excess solvent to accumulate in the sources. Table 6.20 indicates the electrospray ionisation efficiency for various solvents.

The process of ion formation is extremely soft; usually no fragmentation occurs. *Mass spectra* in ESI$^+$ mode are dominated by pseudo-molecular ions (M$^+$, [M + H]$^+$ or [M + Na]$^+$) and cluster ions formed by the addition of one or more solvent molecules. For samples with molecular masses up to ca. 1000 Da (as in case of polymer additives), the ions produced from ESI are similar to those formed by other soft ionisation techniques, namely a protonated molecular ion (MH$^+$) in the positive ionisation mode for basic compounds such

Table 6.20 Electrospray ionisation efficiency

Solution in water:
- High surface tension, coarse droplets, difficult to spray
- Good solvation of ions
- Low sensitivity

Solution in methanol, acrylonitrile:
- Low surface tension, fine droplets, easy to spray
- Less efficient solvation of ions
- High sensitivity

Buffers in HPLC eluent:
- Lower sensitivity at higher buffer concentration

as amines, or a deprotonated molecular ion $(M - H)^-$ in the negative ionisation mode for acidic compounds and compounds with other electronegative groups. However, some adducting may be observed, depending on the sample and its preparation; the affinity for anionisation or cationisation; and the composition of the mobile phase and associated buffers. Generally, under standard electrospray operating conditions, few fragment ions are observed; some form of collision inducement usually achieves fragmentation. Data collection (positive or negative ions) depends on the type of sample to be analysed.

The *main characteristics* of electrospray ionisation are given in Table 6.21. Electrospray ionisation is a suitable technique for producing accurate molecular mass information on a wide range of low-MW samples. ESI-MS is particularly appealing for polar, high-MW samples (more than ca. 1000 Da), where the multiply charged ions formed have m/z values within the range of the spectrometer. However, ESI presents some problems in the identification of unknowns:

(i) mixed mass spectra and/or adducts may result;
(ii) ion chemistry beyond the molecular weight, such as proton-bound dimers (2 M + H), trimers (3 M + H), etc.;

(iii) the protonated molecular ion is not always the base peak; and
(iv) a single (M + H) peak yields no structural information.

ESI mass spectra of *mixtures* are difficult to interpret, because each component produces ions with many different charge states. The most direct and reliable method to solve this problem is to use high-resolution MS and calculate the charge states by measuring the spacing of the isotope peaks. ESI mass spectrometry of (polymeric) mixtures with broad molecular weight distribution benefits from a prior separation that reduces the polydispersity of the analyte.

Fragmentation of the quasi-molecular ions for *structural elucidation* can be achieved in different ways, either by using specific MS/MS techniques, or by varying the electrospray source parameters to induce 'in-source' CID fragmentation. By increasing the voltage on the sampling cone or orifice, the ions are caused to accelerate. They then undergo ion–molecule reactions. The collisions impart energy into the ions, thereby causing fragmentation. *In-source fragmentation* is a straightforward experiment. However, as not only the sample of interest fragments, but also any other components (e.g. residual solvent-related ions) simultaneously present, this method is thus not a specific means of inducing fragmentation. A higher degree of specificity can be achieved in a tandem MS mode by the generation of fragment ions from a specific species, and then by monitoring these fragment ions, the precursor ions, or constant neutral losses. In this way the composition can be established, even when the different species have identical mass.

ESI-MS is the most successful method of coupling a condensed phase separation technique to a mass spectrometer. Because the input to ESI is a liquid, electrospray serves as an *interface* between the mass spectrometer and liquid chromatographic techniques, including SEC and CE (capillary electrophoresis). In LC-MS the flow-rate should lie in the range recommended for the HPLC pump and the mass spectrometer (typically $0.001-1.0$ mL min^{-1}). Recent advances in (nano)electrospray technology include the development of the use of very low solvent flow-rates (30 to 1000 nL min^{-1}) [130,131].

Electrospray mass spectrometry has been reviewed repeatedly [132–135].

Applications Since the first ESI-MS experiments in 1984 [136] the technique has been widely employed to analyse polar molecules ranging from <100 Da up to

Table 6.21 Main characteristics of electrospray ionisation

Advantages

- Ambient temperature operation (does not require heat for desolvation)
- Ideal for thermally labile compounds
- High sensitivity (fmole level)
- Ultrasoft ionisation, generating $[M + H]^+$ or $[M - H]^-$ ions
- Production of multiply charged ions
- Ionisation of large macromolecules
- Effective mass regime for m/z ratios of multiply charged ionic macromolecules
- Molecular weight confirmation
- Appropriate for both volatile and nonvolatile solutes
- Appropriate for ionic/polar analytes
- Hyphenation to chromatographic methods (HPLC, SEC, CE)
- Extended analysis times, allowing multiple MS or MS/MS experiments on as little as 1–2 μL of sample

Disadvantages

- Relatively low LC flow-rates
- Needs to form ions in solution
- Ion suppression in high-salt conditions
- Limited structural information
- Concentration dependent
- Less appropriate for nonpolar analytes

and above 200 000 Da in molecular mass, as a result of the commercial availability of ESI-MS spectrometers. The applicability of electrospray ionisation mass spectrometry (ESI-MS) is generally limited to analytes which are ionic or which can be ionised in solution by acid/base chemistry. The general scope of electrospray applications consists of:

(i) preformed ions;
(ii) polar neutrals that can be ionised (e.g. by ammonia CI); and
(iii) multi-charged ions of (bio)polymers and oligomers.

The corresponding liquid-phase chemistry can be used to promote ion formation by appropriate choice of solvent and pH, salt addition to form $M.Na^+$ or $M.NH_4^+$, and postcolumn addition of reagents. The primary applications of ESI-MS are in the biopolymer field. The phenomenon of routine multiple charging is exclusive to electrospray, which makes it a very valuable technique in the fine chemical and biochemical field, because mass spectrometers can analyse high-molecular-mass samples without any need to extend their mass range, and without any loss of sensitivity. However, with ESI, molecules are not always produced with a distribution of charge states [137]. Nevertheless, this phenomenon somehow complicates the determination of the true mass of the unknown. With conventional low-resolution mass spectrometers, the true mass of the macromolecule is determined by an indirect and iterative computational method.

ESI-MS has found some applications in the mass-spectrometric analysis of *synthetic oligomers and polymers*. This soft ionisation technique affords intact oligomer or polymer ions with a minimal number of fragment ions. The use of ESI-MS for characterisation of synthetic polymers is usually directed towards polar compounds which are more easily ionised, such as PEG oligomers [138]. Widely available ESI-quadrupole mass spectrometers with low resolving power (<1000) are inadequate for the analysis of higher molecular weight polymers. Quadrupole ion traps, state-of-the-art ToF analysers, or magnetic-sector instruments, may give medium mass resolution (≥5 000), and especially an ESI–Fourier-transform mass spectrometer offers high mass accuracy and resolution that may extend the application of the technique to higher polymers. ESI-FTMS has been used to characterise PEG 20 000 [138]. Over 5000 isotope peaks resulting from 47 oligomers in ten charge states were distinguished with resolving powers in excess of 50 000. The ESI interface is also attractive for on-line SEC-MS coupling (Section 7.3.4.2). SEC-ESI-MS has been used for the analysis of octylphenoxy-poly(ethoxy)ethanol, an oligomeric surfactant (Igepal®) [139]. Direct ESI-MS underestimates the sample's polydispersity, compared to SEC [140]. For application of ESI-MS in synthetic polymers without acid or basic functional groups that can be used for ion formation, it is necessary to add (sodium) cations (e.g. to the mobile SEC phase).

Direct ESI^+-MS and ESI-MS/MS in conjunction with ^1H NMR and ^{31}P NMR techniques have been used for the characterisation of commercial aliphatic phosphate/phosphonate oligomers (reaction products of P_2O_5, ethylene oxide and methanol), used as process chemicals [141]. Various phosphorous containing compounds have been examined by direct ESI-MS, such as trixylylphosphate (TXP) and hexachlorophosphazene (HCP), a precursor to many functionalised *flame retardants* [141]. ESI-MS and ESI-MS/MS have also recently been used for the analysis of polymer/additive mixtures. However, there are relatively few reports in the literature of the analysis of *polymer additives* by means of direct electrospray–mass spectrometry. Jackson *et al.* [137] have recently analysed five common organic polymer additives, namely Tinuvin 327, Hostanox O3 and Irganox 1010/1076/3114 by *ESI–low energy MS*. Tinuvin 327 was not suitable for analysis by ESI, presumably more so by APCI. Various factors contribute to the intensity and distribution of the peaks observed in ESI mass spectra, namely the amount of salt (NH_4Cl) added to the analyte to aid ionisation, and the value of the skimmer cone voltage. The molecular ions from the polymer additives were all singly charged species, and were predominantly clusters with an ammonium ion, namely $[M + NH_4]^+$. These ions were also seen by low-energy DCI-MS in a triple quadrupole mass spectrometer [23]. Multiply charged ions were not observed, which is ascribed to the molecular shape and low molecular weight of the polymer additives. The cone voltage at which decomposition is induced depends on the molecular weight of the precursor ion. Fragmentation of low-MW species is induced at low cone voltages, whereas higher-molecular-weight additives require a greater input of internal energy and, therefore, fragment at high values of cone potential. The difference in efficiency of ionisation for certain species is a hindrance to quantitative analysis of mixtures of polymer additives by means of ESI-low energy MS. Mixture analysis has, however, been demonstrated by means of FD–high-energy MS [96] and LC-MB-MS [142].

Jackson *et al.* [137] have also performed CID experiments in an ESI–tandem quadrupole mass spectrometer of Q_1hQ_2 geometry, where *h* is a hexapole collision cell. The skimmer cone voltage and collision energy have a marked effect on the quantitative and qualitative distribution of fragment and molecule ions observed in the *low-energy CID spectra* of polymer additives (Hostanox O3, Emkarate 3020 and Irganox 1010/1076/3114) in the THF solution, generated by ESI [137]. Rearrangements were the predominant pathways for the generation of the fragment ions observed. This is in contrast to the findings in the case of high-energy MS/MS of polymer additives, where direct cleavages of the precursor ion were predominant [96]. It has thus been demonstrated that molecular and structural information can easily be generated for polymer additives by means of ESI–low-energy MS and MS/MS. Complementary data, which aided the confirmation of structures of the additives, were obtained by employing both product ion and precursor ion scanning experiments. A drawback is the fact that the ions observed were highly dependent on the instrumental conditions used.

The low-energy MS/MS spectrum of Irganox 1076 [137] differs considerably from that obtained under high-energy conditions [96]. It is also of interest to note that the low-energy ESI-MS/MS spectrum of Irganox 1010 [137] is comparable with that generated by means of LSIMS-MS under high-energy conditions [143]. This indicates that the internal excitation energy deposited in the LSIMS ionisation process could be similar to that by means of ESI–low-energy CID under the conditions used in both experiments. LSIMS is thought to impart an average of approximately 1–2 eV of internal excitation energy during the ionisation process [144].

Instrumental conditions required for *high-energy MS/MS* on a four-sector instrument are apparently less critical for the different polymer additives, and the results more reproducible [96]. The fragment ions observed in the high-energy product ion spectra were also found to be more characteristic of the class of polymer additive [96]. However, a considerable disadvantage of high-energy in comparison with low-energy experiments is the large difference in instrumentation cost.

Electrospray ionisation has also found applications in inorganic chemistry. In this field, the use of EI-MS has been somewhat limited in its application, because it involves volatilisation of the compound prior to forming ions in the gas phase. Consequently, most studies have involved non-ionic compounds, since ionic species are in general relatively involatile. However, ESI provides a means for transferring ions in solution to the gas phase. Although neutral inorganic and organometallic molecules cannot be directly observed by ESI-MS, they can be detected by prior conversion to closely related ionic species. Typical methods of generating such ions in solution include protonation, quaternisation, metallation, oxidation and chemical reaction.

Applications of hyphenated ESI mass-spectrometric techniques are described elsewhere: LC-ESI-MS (Section 7.3.3.2), SFC-ESI-MS (Section 7.3.2.2) and CE-ESI-MS (Section 7.3.6.1) for polar nonvolatile organics.

6.2.8.2 *Atmospheric Pressure Chemical Ionisation*

Principles and Characteristics Atmospheric pressure chemical ionisation (APCI) has historically been developed before electrospray [125]. APCI is closely related to ESI, in that for both cases ionisation takes place at atmospheric pressure, rather than reduced pressure. A modified electrospray source is used; extraction of ions into the mass spectrometer is similar. In APCI reactant ion spectra are produced by the *corona discharge* method of ionisation, which occurs through an electrical discharge at the tip of a needle inside the source held at high voltage, as shown in Figure 6.13. The source temperature is typically between 120 and 180 °C. Although a high temperature is applied to the probe, most of the heat is used for solvent evaporation and heating the nebuliser gas, with the result that the

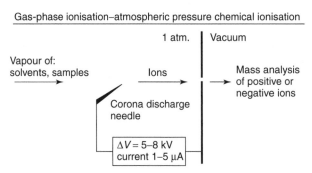

Figure 6.13 Schematic diagram of atmospheric pressure chemical ionisation

thermal effect on the sample is much less than may well be expected. However, with very labile additives, the heated probe may still cause some fragmentation or decomposition. In the atmospheric pressure region around the corona pin, a combination of collisions and charge-transfer reactions generates a chemical ionisation reagent gas plasma. Any target molecules which elute and pass through this region of solvent ions can be ionised by the transfer of a proton to produce $(M + H)^+$ or $(M - H)^-$ quasi-molecular ions that can be transmitted to the mass analyser. The sensitivity of APCI is highly dependent upon the ion–molecule chemistry of both the sample molecules and the so-called reactant ions that are formed with the mobile phase and ambient air. In CI processes take place in a dilute gas phase where particle collisions are less likely and ions have a tendency towards fragmentation. In contrast, when the process takes place at higher pressure, as in APCI, the product ions can either lose internal energy by collisions with inert gas or they can behave as reactant ions themselves. In addition, higher pressures may promote adduct ion formation. Unlike ESI, APCI does not produce multiply charged ions and is therefore not appropriate for analysis of high-MW samples. APCI spectra ideally allow molecular mass determinations of samples up to about 1500 Da.

As for ESI, in order to produce structurally informative fragment ions, in-source cone voltage fragmentation or MS/MS (if a tandem mass spectrometer is available) may be applied. As shown in Table 6.10, APCI is compatible with a variety of mass spectrometers. APCI allows gas-phase and liquid-phase ionisation. APCI tolerates a variety of solvents and buffers, ranging from 100 % aqueous to 100 % organic. The flow-rate of the mobile phase is usually higher in APCI than in ESI (typically $0.2–2\,mL\,min^{-1}$); consequently, the sample takes less time to reach the source, but also the time available for tuning sample-related ions is reduced significantly.

Table 6.22 summarises the *main characteristics* of APCI-MS. Sometimes the heat burden of the APCI interface causes thermal decomposition, which is unwelcome if the requested information is only the molecular weight. On the other hand, studying the thermal fragmentation can provide additional data about the sample. Since the thermal and collisional (CID) fragmentation do not necessarily follow the same pathway, the two methods do complement each other. Therefore, even good use can be made of the thermal decomposition for structure elucidation during APCI experiments [145].

Coupling of APCI-MS/MS to chromatography (typically LC) vastly increases the amount of chemical information obtained from either method alone. The rapid

Table 6.22 Main characteristics of APCI-MS

Advantages

- Produces chemical ionisation mass spectra $(M + H)^+$ at atmospheric pressure
- Soft ionisation
- Gentle vaporisation of the analyte ($<120\,°C$)
- Suitable for labile, neutral to polar compounds
- Most suited for analytes with sufficient proton affinity
- Confirmation tool; molecular weight information
- Fragmentation with CID
- Good sensitivity
- Quantitation feasible
- Easy to use
- High throughput
- Rugged

Disadvantages

- Thermal degradation possible
- High chemical noise at low mass (m/z <150)
- Limited to compounds with molecular weight up to 1500 Da
- Not library searchable

identification of co-evolving compounds by MS/MS, combined with the separating power of the chromatographic method, provides a means of analysing complex mixtures.

APCI-MS/MS has shown the ability to identify and resolve complex products, including isomers, emanating from a TG apparatus [146].

Applications APCI-MS is often more widely applicable than ESI-MS to the analysis of classes of compounds with a low molecular weight, such as basic drugs and their metabolites, antibiotics, steroids, oestrogens, benzodiazepines, pesticides, surfactants, and most other organic compounds amenable to EI. LC-APCI-MS has been used to analyse PET extracts obtained by a dissolution/precipitation procedure [147]. Other applications of hyphenated APCI mass spectrometric techniques are described elsewhere: LC-APCI-MS (Section 7.3.3.2) and packed column SFC-APCI-MS (Section 7.3.2.2) for polar nonvolatile organics.

APCI-MS/MS is not only useful in the analysis of polymers, such as cellulose acetate, but is also of great value in the identification of copolymer substrates, and the various polymer additives such as antioxidants, stabilisers, and plasticisers.

6.2.9 Desorption/Ionisation Methods

Principles and Characteristics Desorption is the term used to describe the transfer of atoms or molecules from a surface-adsorbed state into the gas phase. Mass spectrometrists use the term 'desorption' in a less-rigorous sense to describe a range of techniques which are used to transfer atoms or molecules from the solid state into the gas phase. Mass spectrometry requires ions in vacuum. This process involves formation of ions from volatile compounds (generally EI and CI processes), or desorption of ions into the gas phase, for example, by field desorption (FD) or charged aerosol droplets (thermospray, electrospray, ion spray) for nonvolatile compounds. Compounds of medium volatility and relatively high thermostability, usually in the medium-MW range, can be analysed with thermal desorption techniques, such as direct probe mass spectrometry (DP-MS), see Section 6.4. The advent of *desorption mass-spectrometric methods* has made possible the analysis of compounds which were previously difficult to study with conventional mass spectrometry. In general, desorption/ionisation is carried out by coating the sample on the surface of a suitable metal, and then bombarding the sample with particles of high energy. There are a number of different ways of communicating the energy to the surface molecules, e.g. SIMS, FD, FAB, HSI, PD, LDI and MALDI. In these methods, the desorption event generally results in the removal of several monolayers of material. These desorption techniques have also been coupled with various separation methods for the analysis of thermally labile compounds. The need to volatilise the analyte prior to its ionisation may be bypassed by means of the alternate ionisation techniques of FAB, MALDI, TSI and HSI.

Sample preparation for the common desorption/ionisation (DI) methods varies greatly. Films of solid inorganic or organic samples may be analysed with DI mass spectrometry, but sample preparation as a solution for LSIMS and FAB is far more common. The sample molecules are dissolved in a low-vapour-pressure liquid solvent – usually glycerol or nitrobenzyl alcohol. Other solvents have also been used for more specialised applications. Key requirements for the solvent matrix are sample solubility, low solvent volatility and muted acid – base or redox reactivity. In FAB and LSIMS, the special 'art' of sample preparation in the selection of a solvent matrix, and then manipulation of the mass spectral data afterwards to minimise its contribution, still predominates. Incident particles in FAB and LSIMS are generated in filament ionisation sources or plasma discharge sources.

In the *plasma desorption* (PD) technique, liquid matrices have also been used, but most samples are prepared as a thin-film solid. A solution of the sample in a volatile solvent is sprayed onto a thin support, forming a thin, homogeneous film after solvent evaporation. In PD-MS, which is a radionuclide (^{252}Cf) ionisation technique, ions are formed by a 1-MeV nucleon particle impacting the surface, and are desorbed within 10^{-9} seconds of the impact. The sample ion flux is low, however, and extended signal integration times may be necessary. In matrix-assisted particle-desorption techniques (e.g. FAB, SIMS and PD-MS), the average energy density pumped into the analytes greatly surpasses the formation enthalpy not only of the molecular ions of interest, but also of molecular fragments. As a result, a significant fraction of analyte molecules becomes fragmented or otherwise damaged by either direct or indirect action of the primary particles. The role of a matrix in particle-desorption techniques includes an energy mediation by extending the zone where the energy density is at the level of the molecular-ion formation enthalpy. PD-ToFMS and MALDI-ToFMS couplings are more easily made than couplings of ToF-MS with ions produced in a continuous beam, such as EI, FAB or ESI. PD-MS cannot easily be interfaced on-line with HPLC or configured as a tandem instrument that can provide structural information using collision-induced dissociation (CID). Thermally unstable compounds, which do not give good mass spectra even with FD ionisation, may be subjected to PD-ToFMS. However, the technique has no wide acceptance, due to low mass resolution. The ^{252}Cf-PD method has been largely supplanted by MALDI and ESI-MS, because these methods are more efficient and widely applicable.

In *MALDI*, co-crystallisation of the sample and the energy-absorbing support matrix is sought through careful control of the rate of solvent evaporation from the solution mixture deposited on an inert surface. Lasers commonly used in LD and MALDI instruments generally have pulse widths in the range of 3 to 10 ns, while inexpensive pulsed nitrogen lasers with pulse widths of 250 to 600 ps are also available. In MALDI with UV photons, initial energetic interactions are clearly not of the 'billiard-ball' type, but reflect electronic interactions. With IR photons in MALDI, direct vibrational excitations occur on irradiation. Soft laser desorption/ionisation techniques (MALDI, SELDI, DIOS) are particularly useful for biological macromolecules.

DI methods tend to produce even-electron ions such as protonated molecules $[M + H]^+$, or cationised molecules such as $[M + Na]^+$; these stable ions undergo only a minimum amount of fragmentation. Desorption/ionisation mass-spectral methods (in particular FD, EH and LD) may be used for determining molecular

Table 6.23 Applicability of desorption/ionisation mass-spectrometric techniques

Mass-spectrometric technique	Masses
GC-MS	Volatiles, additives
PyMS, PyGC-MS, DP-MS	Low-MW molecules
DCI, LD, FI, FD, PD	Higher-MW fragments (>1 000 Da)
FAB, LSIMS	Less-volatile chemical species (up to 10^4 Da)
MALDI, SELDI, ESI	High-MW molecules (>10^6 Da)

weight averages in low-MW polymers. Very accurate $\langle M_n \rangle$ and $\langle M_w \rangle$ values can be determined for many types of polymers with average molecular weights of about 5000 Da and less. Table 6.23 compares the applicability of various desorption/ionisation methods with other common mass-spectrometric techniques.

Coupled on-line techniques (GC-MS, LC-MS, MS/MS, etc.) provide for *indirect* mixture analysis, while many of the newer desorption/ionisation methods are well suited for *direct* analysis of mixtures. DI techniques, applied either directly or with prior liquid chromatographic separations, provide molecular weight information up to 5000 Da, but little or no additional structural information. Higher molecular weight (or more labile) additives can be detected more readily in the isolated extract, since desorption/ionisation techniques (e.g. FD and FAB) can be used with the extract but not with the compounded polymer. Major increases in sensitivity will be needed to support imaging experiments with DI in which the spatial distribution of ions in the $x - y$ plane are followed with resolutions of a few tens of microns, and the total ion current obtained is a few hundreds of ions.

Desorption/ionisation MS had been reviewed [148] and is subject of a monograph [149]. PD-MS reviews are also available [150–152].

Applications Bloom identification, an important practical problem in rubber analysis, can best be accomplished by *direct* analysis using desorption/ionisation methods (FD, FI, FAB). Desorption/ionisation techniques such as FD-MS and FAB-MS have also been found to be effective means for analysing polymer extracts [92,113]. FD-MS is particularly useful, since molecular ion abundances are high with respect to fragmentation. DI methods such as FAB [72], LD [83,153] and SIMS [154] have also been applied in the analysis of additives from bulk polymer samples. However, these single-step techniques suffer to varying degrees from matrix interference in the resulting mass spectra.

Several desorption/ionisation methods have been used for the analysis of (mixtures of) dyes: FD-MS [155], SIMS [156], PD-MS [157], FAB-MS [158,159] and LD-MS [160].

Soft desorption/ionisation methods are of course crucial for mass-spectrometric analysis of biological macromolecules.

6.2.10 Photoionisation Techniques

Principles and Characteristics Photoionisation (PI) technology provides a unique combination of capabilities, including: high detection efficiency for many classes of compounds; simplified molecular ion spectra with minimal fragmentation; and large linear dynamic range and minimal ionisation of common solvent or air background. Restricting mass spectrometry to molecular ions plus the lowest energy fragments is analytically valuable.

The basis of photoionisation is to choose a narrow band of energy that lies slightly above the ionisation potentials (IPs) of the molecules of interest, yet below the IPs of common air constituents and solvents. Trace molecules or analytes are then detected without interference from the abundant matrix. Furthermore, fragmentation is minimised because ionisation occurs with very little excess energy. Minimal fragmentation permits screening of all of the important compounds in a multicomponent mixture. Photoionisation also has the advantage of being a near-universal detection system for many classes of compounds; introduces selectivity; and features a large dynamic range and minimal charge-competition effects. Several mass analysers have been used for photoionisation studies, including QMS, B, QIT and ToF-MS. Photoionisation technology is incorporated in a commercially available quadrupole ion trap time-of-flight mass analyser (PI-QIT-ToFMS).

Photons are typically produced from a vacuum discharge lamp (e.g. hydrogen discharge) that generates vacuum – ultraviolet (VUV) photons. If a VUV photon is absorbed by a species with a first ionisation potential (IP) lower than the photon energy, then *single-photon ionisation* (SPI) may occur (see also Figure 6.5) and a molecular ion ($M^{+\bullet}$) is produced. *Multiphoton ionisation* (MPI) is an *n*-photon absorption process leading to ionisation and obeys the condition:

$$n E_{ph} > IP \qquad (6.2)$$

where E_{ph} is the photon energy and IP is the molecular ionisation potential.

Photoionisation detection (PID) enjoys longstanding use with GC [161,162], less so in combination with LC [163]. For GC-PID, the discharge lamp is normally selected such that the energy of the photons is greater than the IP of the analyte but below that of the carrier gas. Early LC-PID methods were substantially similar to GC-PID. *Atmospheric pressure photoionisation* (APPI) detection in LC has recently been introduced and commercialised [164]. At the high collision frequency at atmospheric pressure, species with high proton affinity, and/or low ionisation potentials tend to dominate the positive ion spectra (such as $(M + H)^+$, $M^{+\bullet}$), also depending on solvent choice. Dominant ion types created by APPI are therefore molecular or quasi-molecular. Like APCI, APPI is an ideal complement to electrospray, because of the common interfacing hardware for sampling ions from the atmospheric pressure source into the vacuum. Typical ion currents used in APPI are of the order of 10 nA (cf. 2 μA in APCI). Obviously, the robustness of the APPI source has still to be demonstrated. Post-ionisation reactions may complicate analysis of APPI mass spectra, but it is also possible to exploit these reactions to improve the sensitivity of the method.

APPI and corona discharge–APCI methods were compared [164]. A review of photoionisation and photodissociation methods in MS has appeared [165].

Applications　Tables 6.24 and 6.25 compare various API modes in relation to molar mass and areas of application.

It may be seen that APPI is most suited for very low-polarity components with medium molecular weight (<2000 Da), see also Figure 6.4. APPI enables to distinguish *ortho* and *para* isomers of phenols;

Table 6.24　API modes in relation to molar mass[a]

Analyte	ESI	APCI	APPI
Ionic	xxxx	x	x
Ionisable	xxxx	xx	xx
Polar non-ionic	xxxx	xx	xx
Nonpolar	–	xx	xxx

[a] xxxx, >100 000 Da; xxx, >2 000 Da; xx, >1 000 Da; x, >500 Da.

Table 6.25　Areas of application of various API modes

Analyte	ESI	APCI	APPI
Ionic	xxxx	xx	xx
Ionisable	xxxx	xxx	xxx
Polar non-ionic	xx	xxx	xxx
Nonpolar	x	xx	xxxx

nitrosamine can be detected by dopant APPI, not by direct APPI, APCI or GC-MS.

6.3　MASS ANALYSERS

In mass spectrometers, ions are analysed according to the *m/z* (mass-to-charge) value and not to the mass. While there are many possible combinations of technologies associated with a mass-spectrometry experiment, relatively few forms of mass analysis predominate. They include linear multipoles, such as the quadrupole mass filter, time-of-flight mass spectrometry, ion trapping forms of mass spectrometry, including the quadrupole ion trap and Fourier-transform ion-cyclotron resonance, and sector mass spectrometry. Hybrid instruments intend to combine the strengths of the component analysers.

Mass spectrometers for structure elucidation can be classified according to the method of separating the charged particles. Table 6.26 lists the main mass discriminator types. A useful division is between mass-filtering and mass-dispersing instruments. *Mass filtering*, in which the only ions detected are those with preset *m/z* values, comprises the magnetic sector with an exit slit, and the quadrupole. With scanning mass spectrometers (QMS, magnetic sector) ions in one *m/z* unit are detected, while all others over the full *m/z* range are lost. Improved duty cycles impose a narrow scan range or selected ion monitoring. In *dispersive mass analysers*, ions with different *m/z* values are dispersed along some instrumental parameter ('array detection'). The various array detection instruments employ a different dispersive parameter: space in magnetic sector with diode array detector; frequency in FTICR-MS; *q* parameter in QITMS; and time in ToF-MS. Dispersive instruments can produce full spectra at much greater sample utilisation efficiency than filter instruments, because all ions that enter the mass analyser at any time

Table 6.26　Mass-spectrometric analysers

Mass discriminator[a]	Measured quantity	Kinetic energy
B	Momentum/charge (mv/z)	keV
E	Kinetic energy/charge (mv^2/z)	keV
Q	Mass/charge (m/z)	eV
ToF	Flight time	keV
IT	Frequency	eV
FTICR	Cyclotron frequency	eV

[a] B = magnetic sector; E = electric sector; Q = quadrupole mass filter; ToF = time-of-flight mass spectrometer; IT = ion trap; FTICR = Fourier-transform ion-cyclotron resonance.

Table 6.27 Mass-analyser characteristics (1997 status)

Feature	Sector	QMS	QITMS	ToF-MS	FTICR-MS
Mass resolution	$10^2 - 10^5$	$10^2 - 10^4$	$10^3 - 10^5$	$10^3 - 10^4$	$10^4 - 10^6$
Mass accuracy	1–5 ppm	100 ppm	50–100 ppm	20–100 ppm	1–5 ppm
m/z range	$<10^4$	$<10^4$	7×10^4	$>10^5$	$>10^4$
Dynamic range	10^9	10^7	$10^2 - 10^5$	$10^2 - 10^6$	$10^2 - 10^5$
Abundance sensitivity	$10^6 - 10^9$	$10^4 - 10^6$	10^3	Up to 10^6	$10^2 - 10^5$
Precision	0.01–1 %	0.1–5 %	0.2–5 %	0.1–1 %	0.4–5 %
Efficiency[a]	$<1 \%^e$	<1–95 %	<1–95 %	1–100 %	<1–95 %
Speed[b]	0.1–20 Hz	1–20 Hz	1–30 Hz	$10^1 - 10^4$ Hz	$10^{-3} - 10^1$ Hz
MS^n capability	In tandem	In tandem	Very flexible	MS/MS	Very flexible
Compatibility with ioniser[c]	C	C	P, C	P, C	P, C
Cost	\$\$\$–\$\$\$\$	\$–\$\$	\$–\$\$	\$–\$\$	\$\$\$–\$\$\$\$
Utility requirements[d]	L	B	B	B	L

[a] Transmission × duty cycle.
[b] Time frame of experiment/spectra per second.
[c] P, pulsed; C, continuous.
[d] B, bench-top; L, lab instrument.
[e] Scanning.
After McLuckey [166]. Reprinted from S.A. McLuckey, in *Advances in Mass Spectrometry*, Vol. 14 (K.J. Karjalainen *et al.*, eds), pp. 153–196, Copyright (1998), with permission from Elsevier.

can be detected. Dispersive instruments are also more compatible with pulsed ionisation methods.

Each of the analyser types has a unique set of figures of merit that makes it optimally suited for particular applications (Table 6.27). The main ionisation modes in relation to various mass spectrometers are summarised in Table 6.28.

Mass spectrometers must be regularly tuned or calibrated against a known standard, e.g. perfluorotributy-lamine (PFTBA). The trend is towards miniaturisation ($10 \times 24 \times 14$ in.). A concept for a micro mass spectrometer, with potential applications in process monitoring, has been presented [167]. Mass-spectrometry instrumentation (1997) has been reviewed [166].

Table 6.28 Mass-spectrometer types and compatible ionisation methods

Mass spectrometer[a]	Ionisation interfaces
B	EI, CI, ESI, APCI, APPI, FAB/FIB/LSIMS, FD/FI, TSP
Q	EI, CI, ESI, APCI, APPI, FAB/FIB/LSIMS, (FD/FI), TSP
ToF	EI, CI, ESI, APCI, MALDI, PD, FD/FI, FAB
IT	EI, CI, ESI, ISP, TSP, LDI, MALDI, SIMS, PD, GD
FTICR	EI, CI, ESI, APCI, LDI, MALDI, FAB, FD, PD

[a] B, magnetic sector; Q, quadrupole mass filter; ToF, time-of-flight spectrometer; IT, ion trap; FTICR, Fourier-transform ion-cyclotron resonance.

6.3.1 Sector Analysers

Principles and Characteristics Sector technology is the most mature of all forms of mass analysis. In a magnetic sector analyser (B), a gas stream from the inlet system enters the ionisation chamber (operated at a pressure of about $10^{-6} - 10^{-5}$ Torr) and is bombarded by an electron beam. Ions produced are forced through a first accelerating slit by a weak electrostatic field. A strong electrostatic field then accelerates the ions to their final velocities. The magnetic sector analyser is a direction-focusing device that produces a magnetic field perpendicular to the direction of ion travel. The main requirement is a uniform magnetic field that can be smoothly varied in strength. Ion separation is achieved by magnetic deflection. The path of the ions through the magnet conforms to Equation (6.3).

$$m/z = \mathbf{B}^2 r^2 / 2V \qquad (6.3)$$

where m = the ion mass, z = the ion charge, $\mathbf{B}$ = the magnetic field strength, r = the radius of curvature of the ion path, and V = the accelerating (source) voltage. When ions enter a magnetic field they will follow circular paths of different radii, which depend on their different m/z ratios. Only one circular path at a given magnetic field ($\mathbf{B}$) will allow ions of a particular m/z ratio to reach the detector. Ions with different m/z ratios can be passed in turn to the detector by varying $\mathbf{B}$ appropriately (Figure 6.14). When the ions leave the ionisation source, they have a spread in energy which contributes to their peak widths. One scan of the magnet

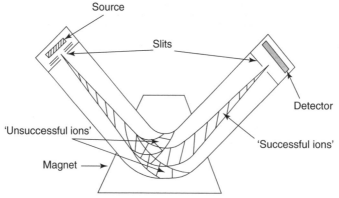

Figure 6.14 Schematic of a magnetic sector analyser

results in the production of one *m/z* spectrum. A *m/z* scan from 12 to 500 may be performed in seconds.

Magnetic sector mass spectrometers accelerate ions to more than 100 times the kinetic energy of ions analysed in quadrupole and ion trap mass spectrometers. The higher accelerating voltage contributes to the fact that ion source contamination is less likely to result in degraded sensitivity. This is particularly important for analysis that requires stable quantitative accuracy.

Magnetic sector mass spectrometers may consist simply of a magnet as an analyser, or (more) frequently, are combined with an electrostatic analyser (ESA, E) before or after the magnetic field. The electrostatic analyser focuses the velocity, and hence the kinetic energy of the ions, but does not mass analyse. The combination of a magnetic and an electrostatic analyser is termed *double focusing*, because of both angular and energy focusing. A double-focusing mass spectrometer is characterised by both high resolution (in excess of 150 000), high sensitivity and high cost. For high resolution accurate mass measurements, double-focusing magnetic sector instruments are invaluable.

The *main characteristics* of sector mass spectrometers are shown in Table 6.29. Magnetic sector mass spectrometers are often considered more difficult to operate than QMS and ToF-MS; the high-voltage source is more demanding to chromatographic interfacing. For figures of merit, see Table 6.27.

Various tandem MS instrument configurations have been developed, e.g. sector instruments, such as CBCE, CBCECB or CECBCE, and hybrid instruments, e.g. BCECQQ (B = magnetic sector analyser, E = electrostatic analyser, C = collision cell, Q = quadrupole mass spectrometer), all with specific performance. Sector mass spectrometers have been reviewed [168].

Table 6.29 Main characteristics of double-focusing sector mass spectrometers

Advantages

- Reproducible 'classical' mass spectra
- High resolution and mass accuracy (elemental composition)
- High sensitivity and dynamic range
- Best quantitative performance of all MS analysers
- Tandem MS experiments
- High-energy CID
- Mature technology

Disadvantages

- Expensive instruments
- Limited scan speeds
- High accelerating voltages (4–8 kV)
- Limited sensitivity at high resolution
- Technically demanding LC-MS coupling
- Complex operation
- Severe vacuum restrictions
- Large size (laboratory instrument)

Applications Sector instruments are applied for niche applications such as high-resolution measurements and fundamental ion chemistry studies. Magnetic sector mass spectrometers remain the instrument of choice in areas of target compound trace analysis, accurate mass measurement and isotope ratio measurement.

Polymer/additive analysis greatly benefits from high-resolution mass data, which often leads to unambiguous identification of (known) additives. However, the investment and operating costs of this instrument do not easily justify its (exclusive) use for the purpose of routine polymer/additive analysis. Analysis of organic polymer additives by means of mass spectrometry is aided by the utilisation of precursor ion and second-generation product ion (MS^3) scanning experiments [169]. A four-sector mass spectrometer was used to determine the fragmentation pathways of these materials under high-energy conditions. Reaction-intermediate scanning and labelling experiments can provide insights into the fragmentation mechanisms of polymer additives. Utilisation of the aforementioned techniques can aid the analysis of additives from polymeric formulations, which are often complex.

6.3.2 Quadrupole Mass Spectrometers

Principles and Characteristics Linear multipoles, such as quadrupoles, hexapoles, octopoles, etc., find wide use in mass spectrometry. The quadrupole mass filter uses four voltage-carrying rods, geometrically arranged as two pairs of parallel rods. Gas-phase ions are introduced into the quad and driven to detection by electrical fields. Ions entering from one end travel with constant velocity in the direction parallel to the poles (z direction), but acquire complex oscillations in the x and y directions by applying to the poles both a direct current (DC) voltage (V_{DC}) and a radiofrequency (RF) voltage (V_{RF}) of opposing polarity, thus giving rise to a quadrupolar field between the rods. There is a 'stable oscillation' that allows a particular ion to pass from one end of the quadrupole to the other without being lost by striking the poles; this oscillation is dependent on the m/z ratio of an ion. Therefore, ions of only a single m/z value will traverse the entire length of the filter at a given set of conditions. In quadrupole instruments there is a trade-off between sensitivity and resolution. The mass resolving property of a quadrupole mass spectrometer (QMS) (ability to separate ions of close masses) is based on path (in)stability of ions in an alternating electrical field. The mass resolution depends on the quad geometry and the voltage settings. Increasing the resolution means that fewer ions will reach the detector. The signal is proportional to the number of ions reaching the detector, and the S/N ratio is related to the time spent to observe each ion and to the resolution. In scan mode, the sensitivity will depend on mass range and scan time. Scanning on a narrow range with low scan speed increases the sensitivity. Increasing the resolution is at the expense of sensitivity, since a smaller number of ions can reach the detector. Quadrupole instruments are limited in their resolution to analyses performed at low (i.e. unity) resolution. A quadrupole mass analyser exhibits a (fairly) constant resolution, expressed as the peak width (in Da) at 10 % of the peak maximum, over the entire mass range used.

A quadrupole mass analyser can be used in different *modes of operation*. In peak hopping mode, only one particular mass is transmitted and the mass analyser is used in the single ion-monitoring (SIM) mode. This mode provides the highest sensitivity for specific ions or fragments. Alternatively, DC-RF voltages scan a specific mass range in order to obtain spectral information. This is called peak scanning. Since a quadrupole electrical field can be adjusted very quickly, owing to the (almost) complete absence of hysteresis, a very fast changeover between masses is guaranteed, so that the largest part of the quadrupole duty cycle can be used for data acquisition.

Table 6.30 lists the *main characteristics* of quadrupole mass spectrometers. QMS is a relatively simple and robust analyser which does not need such a high vacuum as a sector instrument. The maximum admissible pressure at the source of the spectrometer is 10^{-6} mbar in continuous regime and 10^{-5}–10^{-4} mbar during short time intervals. Quadrupole technology assures reproducible and accurate molecular weight measurements day in and day out. For figures of merit, see Table 6.27.

Mass spectrometers with quadrupole analysers are far easier to use than magnetic sector instruments. Quadrupoles are also more easily controlled by computer than are magnetic sectors. This is especially true when peak hopping rather than scanning is desired. Moreover, they are easier to couple than are sector instruments. The high voltage present at the ion source of the latter, combined with the precise positioning required of the source, makes interfacing sector instruments quite difficult. Quadrupole mass spectrometers

Table 6.30 Main characteristics of quadrupole mass spectrometers

Advantages

- High sensitivity
- Sufficiently rapid scanning capability (100 ms per scanning unit)
- Reliability, robustness
- Molecular weight determination
- Good linearity for quantitation
- Simplicity of operation
- Simple vacuum system; high tolerance for pressure
- Ease of chromatographic interfacing
- Relatively accessible cost
- Small size
- Tandem and hybrid instruments

Disadvantages

- Low-resolution, typically 1 Da
- Mass range limited to approximately m/z 4000

present also some advantage over electric sector instruments, by virtue of their cylindrical symmetry, absence of very narrow splits, and relatively high transmission efficiency; however, mass discrimination effects may be larger than in other instruments. Energy and spatial distribution of ions produced in the source and entering the mass analyser are not critical, except for coupling to LC. LC-MS interfaces generally produce ions with a relatively wide energy and spatial distribution.

The quadrupole mass filter is the most abundant mass analyser today and RF-only multipoles are used as transmission devices/collision regions in various instrumental configurations. The mass filter is used extensively as a stand-alone mass analyser and as an analyser in multistage mass spectrometers.

Applications In most polymer/additive analysis applications, a QMS is applied in view of its ease of use, relatively low cost, and coupling with chromatography (Section 7.3). The ability of QMS to cope with large solvent volumes flowing into the ionisation source for extended periods of time and ease of interfacing – both to computers and chromatographs – makes it the choice for multi-user systems, and has facilitated hyphenation with GC, LC and TG. Consequently, QMS are a mainstay of GC-MS, LC-MS and TG-MS.

Quadrupoles are low-resolution MS instruments frequently used for molecular weight determination. QMS provides unit-mass resolution, sufficient dynamic range, good quantitation capabilities, and easy sample introduction without severe vacuum restrictions. The limited mass range (up to 4000 Da) generally does not pose problems in polymer/additive analysis. Some limitations of QMS in polymer research are:

- qualitative analysis of gas mixtures by means of QMS is restricted due to low resolution;
- QMS cannot be used directly for quantitative gas determination and cannot distinguish ions of equal molecular masses (e.g. N_2 and CO);
- QMS requires careful choice of purge gas (in TG) which should not interfere with the expected reaction products; and
- complex gas mixtures as reaction products constitute a problem, especially for gases with low-MW (<40 Da).

6.3.3 Time-of-Flight Mass Spectrometry

Principles and Characteristics Time-of-flight (ToF) mass spectrometry was proposed in 1946, and the first

commercial (linear) ToF-MS was introduced in 1957 based on a Wiley–McLaren focusing design [170]. ToF-MS is essentially mass spectrometry without a magnetic field. ToF mass analysers, which operate in batch mode, provide array detection in which ions of different m/z values are dispersed in time. In fact, ions of different masses are separated by making use of their different velocities after acceleration through a potential (V) of moderately high voltage. The ions are then allowed to drift in a field-free region to the detector. As the ions all have the same translational energy, the lighter ions travel faster than the heavier ones, and therefore separation of the ions depends on the time taken to travel along a fixed distance of the flight tube (i.e. the dispersion element). Since $ze\mathrm{V} = mv^2/2$, the velocity (v) of an ion of mass m is $v = (2ze\mathrm{V}/m)^{1/2}$. The time taken for an ion to reach the detector is $t = (m/z)^{1/2}l/(2e\mathrm{V})^{1/2}$. Since arrival times between successive ions can be less than 10^{-7} s, fast electronics are necessary for adequate resolution. Time-of-flight devices are used with sophisticated ionising methods (FAB, LD and PD).

ToF analysers are able to provide simultaneous detection of all masses of the same polarity. In principle, the mass range is not limited. Time-of-flight mass analysis is more than an alternative method of mass dispersion; it has several special qualities which makes it particularly well suited for applications in a number of important areas of mass spectrometry. These qualities are: fast response time, compatibility with pulsed ionisation events (producing a complete spectrum for each event); ability to produce a snapshot of the contents of the source volume on the millisecond time-scale; ability to produce thousands of spectra per second; and the high fraction of the mass analysis cycle during which sample ions can be generated or collected.

From the very beginning, (linear mode) ToF mass spectrometers have had a reputation as low-resolution instruments, due to the spread in the initial energies of the ions, as well as the spatial distribution of them with unit resolution up to ca. 2000 Da being standard. Although ToF-MS has initially enjoyed great popularity, this was later followed by near-extinction but, more recently, by a dramatic re-emergence. Revival of (redesigned) ToF-MS instrumentation is closely connected with the needs of MALDI as a significant ionisation method; surface analysis by laser probes; PD-MS; elemental analysis by REMPI instruments; ultrafast GC-MS and HPLC-ESI-ToFMS. Important developments in time-of-flight technology, which comprises time-lag focusing [170], reflectron (reToF) systems [171,172] and off-axis or orthogonal geometry with fully accelerated ion beam (oaToF) [173,174], have contributed to the attainment of better mass resolution for *pulsed ionisation techniques*. Relative simplicity and excellent performance in a tandem mass spectrometer result (as applied, for example, in Q-oaToF).

With the resolution being related to the length of the drift tube, a *reflectron* (or ion mirror) is used to extend the flight time while maintaining a small instrument size (Fig. 6.15). The reflectron corrects positional and velocity discrepancies in the acceleration region of the ToF mass spectrometer and removes neutral species which may have resulted from in-flight decay. As a result of the electrostatic reflectron, the spatial spread in the flight direction of ions of given m/z, and thus the detected peak widths, are reduced and the improvement in timing mechanism leads to an increase in resolution. Resolution values in excess of 5000 (FWHM) are now readily available, even allowing accurate mass measurement.

In ToF-MS, the ion source is pulsed to create packets of ions. In the conventional procedure, the system waits for all the ions in a packet to reach the detector before injecting the next packet of ions. Complications arise when ToF-MS is coupled to a *continuous ion source*. Such coupling is therefore often accomplished by the orthogonal extraction approach, in which a segment of the ion stream is accelerated orthogonally by a push-out pulse. However, in this process, up to 95 % of the information contained in the ion steam is lost. Recently, Hadamard transform time-of-flight mass spectrometry (HT-ToFMS) was developed to couple continuous ion

sources to the inherently pulsed nature of time-of-flight measurements [175]. HT-ToFMS features a very fast mass spectral storage rate, enabling its use as a detector for capillary and high-speed separations. Unlike conventional ToF-MS, HT-ToFMS offers a duty cycle of 50 % with a rate of 250 spectra s^{-1} [175]. Several investigators have also employed the reflectron with orthogonal extraction to good advantage.

Tables 6.27 and 6.31 show the *main characteristics* of ToF-MS. ToF-MS shows an optimum combination of resolution and sensitivity. ToF-MS instruments provide up to 40 000 spectra s^{-1}, a mass range exceeding 100 000 (in principle unlimited), a resolution of 5000, and peak widths as short as 200 ms. This is better than quadrupoles and most ion traps can handle. Unlike the quadrupole-type instrument, the detector is detecting every introduced ion (high duty factor). This leads to a 20- to 100-times increase in sensitivity, compared to QMS used in scan mode. The mass range increases quadratically with the time range that is recorded. Only the ion source and detector impose the limits on the mass range. Mass accuracy in ToF-MS is sufficient to gain access to the elemental composition of a molecule. A single point is sufficient for the mass calibration of the instrument. ToF mass spectra are commonly calibrated using two known species, aluminium (27 Da) and coronene (300 Da). ToF is well established in combination with quite different ion sources like in SIMS, MALDI and ESI.

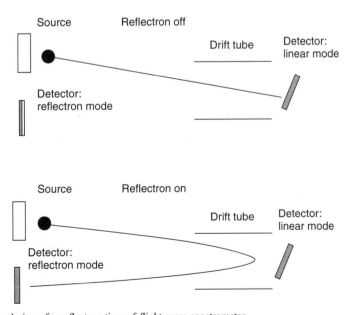

Figure 6.15 Schematic design of a reflectron time-of-flight mass spectrometer

Table 6.31   Main characteristics of ToF-MS

Advantages

- Virtually limitless mass range
- Parallel detection of all the ions over the complete mass range
- Highest spectral generation rate
- Detection of transient signals (narrow chromatographic peaks)
- Compatibility with pulsed ionisation events
- High resolution
- High sensitivity (no ion loss because of scanning)
- Exact mass data, 2–5 ppm
- Fast method development (short run times)
- Little maintenance

Disadvantages

- Low resolution in linear mode
- Strict demands on initial energy and spatial distribution of ions
- High-performance electronics needed
- Footprint (or height: requirements of the flight tube)
- Limited dynamic range of fast digitisers
- Precursor ion selectivity for most MS/MS experiments

State-of-the-art ToF-MS employs reflection lenses and 'delayed extraction' [176] to improve resolution by minimising small differences in ion energies, and in these cases up to 12 000 mass resolution (FWHM, m/z 600) is available. This is sufficient for most modern applications. *Solid probe ToF-MS* (or direct inlet high-resolution mass spectrometry, DI-HRMS) is a breakthrough. DIP-ToFMS is a thermal separation technique. Advantages of DIP-ToFMS are:

(i) reduced sample preparation;
(ii) separation of additives from polymer by probe temperature ramp;
(iii) small sample amounts (generally μg);
(iv) allowance for any sample substrate material (textile, film, wood, metal, concrete);
(v) high speed and sensitivity; and
(vi) exact mass determination (resolution 7500 Da). A disadvantage is that only qualitative information may be obtained.

ToFs can also be used in combination with other mass analysers. Both hybrid sector and quadrupole systems are available. oaToF-MS has been interfaced to a quadrupole mass filter and hexapole gas collision cell, such as to allow recording of mass spectra and product ion spectra with good mass resolution (ca. 10 000), high sensitivity, high mass range (ca. 10 000 Da) and high mass accuracy (<5 ppm) [177,178]. QqToFMS may be fitted with API sources with flow-rates from nL

min^{-1} to mL min^{-1}. Another state-of-the-art instrument is LC-ESI-IT/reToF-MS. The high scan-rate (up to 20 000 scans s^{-1}) of ToF-MS allows for detection of narrow/transient chromatographic signals – typically from on-line GC, GC × GC, LC or CEC. However, ToF-MS can also be used for direct analysis of mixtures. ToF-MS is also actively being pursued in combination with plasma and laser desorption, including MALDI, which require ToF-MS ability to provide a complete spectrum per event and also an unlimited mass range.

The ToF-MS technique was recently reviewed [179–181]. The history of the technique has been traced by ref. [11]; several books have appeared [174,182].

Applications   ToF-MS instruments with specific characteristics are particularly advantageous in the areas of pulsed ionisation events, tandem mass spectrometry (MS/MS), accurate mass determination and chromatographic detection. In view of its extremely fast scanning capability, ToF-MS is useful for real-time monitoring of fast, transient chemical processes and for monitoring of pulsed laser heating of solids. Applications are found in the following techniques: LPy-ToFMS, LDI-ToFMS, L^{2}ToFMS, ToF LMMS, REMPI-ToFMS, MALDI-ToFMS, PD-ToFMS, PyGC-ToFMS, LA-ICP-ToFMS, ToF-SIMS, etc. for *in situ* analysis of (large) organic molecules. ToF-MS relies on prompt ion formation, which essentially occurs during the laser pulse. Organic compounds tend to form ions for a long time after the laser pulse. Reflectron ToF analysers are used in ToF-SIMS, laser desorption, MPI-MS of thermally desorbed species and of sputtered neutrals; an oaToF-MS was employed in pulsed glow-discharge ionisation [183]. Most ToF-MS instruments are being used in conjunction with MALDI. The delayed extraction procedure is particularly useful for MALDI-ToFMS.

The challenge of application of ToF-MS in *chromatographic detection* is the achievement of efficient utilisation of the continuously introduced sample. To be ideal for chromatographic detection, the ionisation efficiency should be high and the sample utilisation efficiency nearly 100 %. Various attempts to improve the duty cycle have been reported [175,184]. oaToF-MS is ideally suited for coupling with fast chromatography and high-efficiency separation techniques (e.g. CZE), because the fast acquisition rates allow a great number of data points to be collected across a narrow peak with high sensitivity and high resolution [185]. The development of oaToF-MS for coupling with continuous ionisation sources has opened up new applications, e.g. ESI-oaToFMS [186] and ICP-oaToF-MS [187], with the object of achieving high sample utilisation efficiencies by using large fractions of the ion beam or by reducing the flight time. API oaToF allows full scan MS data to be generated from sub-mole quantities of material. Collision-induced fragmentation can be produced in the ion source by increasing the sampling cone voltage. The use of oaToF-MS for accurate mass measurement of small molecules is becoming a widespread technique.

ToF-MS plays a significant role in the analysis of high-MW compounds. ESI-ToFMS is suitable for exact mass measurements (2–5 ppm) and for qualitative analysis of oligomers. ToF-MS with different ionisation principles can be used in the area of polymer analysis, e.g. pulsed ion bombardment resulting in secondary ions (SIMS); ionisation by pulsed laser irradiation of the analyte material embedded in a suitable matrix material (MALDI-MS); and GC-ToFMS (with classical EI spectra). ToF-MS with the DIP option (solids insertion probe) equipped with EI and FI sources in the temperature range from ambient to 650 °C has been used for the analysis of UV and EB *curable products* [188]. DIP-ToFMS of a non-UV cured polyesterurethaneacrylate ink readily revealed a reactive diluent, a photoinitiator and polyadipate; in water-based roof paint, acetic acid, styrene, a phthalate plasticiser and propylene glycol as an additive were identified; waterborne polyurethane-based dispersions gave evidence for a diisocyanate, the adipate of 1,6 hexanediol (HD), TEA and BBP as additives; in a pressure-sensitive adhesive, the abietic resin tackifier was readily identified; in textiles composed of aromatic and aliphatic PU fabric coatings, the building blocks of the polymer coating (PEG, PPG and TDI) were

determined together with Santowhite powder and PDMS as additives [188]. In this application, FI is a more successful ionisation mode than EI/CI. Bletsos *et al.* [189] produced ToF-MS spectra of thick films of PDMS and fluorocarbon polymers containing additives in the m/z range up to 4500.

6.3.4 Quadrupole Ion Trap

Principles and Characteristics The quadrupole ion trap mass spectrometer (QITMS) is an ion storage device which consists of three cylindrically symmetrical electrodes of hyperbolic geometry, two end-caps and a ring which form a trapping cavity with a volume of several cm^3 (Figure 6.16). Essentially, the ion storage trap is a spherical configuration of the linear quadrupole mass filter. Ions can be produced in the trap, using an electron impact ioniser, or are generated outside of the trap by various ionisation modes (Table 6.28) and then injected into the trap. Whereas quadrupoles operate at pressures of $<10^{-6}$ Torr, ion traps operate at 10^{-3} Torr of helium. The higher pressure is advantageous for the sensitivity.

Ion traps, attributed to Paul *et al.* [190,191], operate on the basis of first storing ions and then facilitating their detection according to their mass/charge ratio. As a *storage device*, the ion trap acts as an 'electric-field test-tube' for the confinement of gaseous ions. The confining capacity of the ion trap arises from a three-dimensional rotationally symmetrical trapping potential well formed when appropriate potentials are applied to the ion trap electrodes, which allows manipulation of ions under

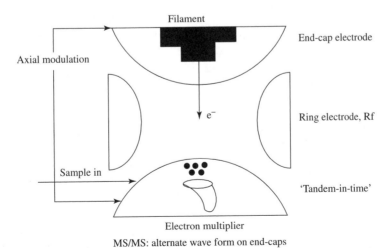

Figure 6.16 Schematic of ion trap analyser (internal ion source)

controlled conditions. The ion trap utilises much more of the total sample (50 %) than quadrupoles (which typically use only 0.1–0.2 % of the ions for analyses). This results in sensitivity down to pg–fg levels. The ion trap functions as a *mass spectrometer* when the trapping field is changed, so that the trajectories of simultaneously trapped ions of consecutive specific mass/charge ratio become sequentially unstable, and ions leave the trapping field in order of m/z ratio. Upon ejection from the ion trap, ions strike a detector and provide an output signal. In the ion trap ionisation, ion storage and ion analysis occur over a defined portion of the mass range. According to Johnson *et al.* [192], with proper attention to the instrumental parameters, QMS and QITMS can yield essentially the same mass spectra for the compounds and conditions studied.

Appropriate use of RF and DC voltages means that some ions can be selectively retained and product ions generated. Some of these ions can then be selected and their product ions generated. In this manner, a fragmentation chain can be established. The ion trap is a typical tandem-in-time mass spectrometer, in which precursor and product ions are created and analysed in the same physical space; ionisation and ion analysis, on the other hand, take place at different times ('MS/MS in time'). The operation can be repeated several times, making it possible to perform MS^n. Ion trap mass spectrometry thus consists of:

(i) generation of ions from analyte molecules, e.g. by means of electrospray ionisation;
(ii) accumulation of the ions in the ion trap;
(iii) isolation of a given ion;
(iv) fragmentation by means of a collision gas; and
(v) accumulation and detection of fragments.

Despite the fact that the ion trap via tandem MS offers more extensive possibilities for identification than QMS, in practice it is not often applied, because routine interpretation of the spectra is often complex and time-consuming.

The *main characteristics* of ion traps are given in Tables 6.27 and 6.32. A side benefit from the trap design is that the ions seldom come into contact with the walls, and thus cleaning is minimised. Ion traps are typically low-resolution systems (1 Da); however, using specific scan methods, higher resolutions can be obtained. The relatively high charge densities found in the ion trap complicate the achievement of high mass accuracy and wide dynamic range; there is a need for optimisation of the collisional activation. The new generation of ion trap mass spectrometers

Table 6.32   Main characteristics of ion trap analysers

Advantages

- Very high sensitivity; trace analysis (fg–pg analyte quantities)
- Selective ion storage for sample matrix clean-up
- Extended m/z range
- Molecular weight determination
- Inherent and efficient MS^n capabilities ($n \leq 12$)
- Targeted compound identification in mixture analysis
- Simple vacuum system (tolerance for high pressures: 10^{-3} Torr)
- Rapid scanning
- Experimental flexibility (interfaceability), convenient use
- Quantitative analysis
- Low cost
- Compact mass analyser

Disadvantages

- Limited dynamic range
- Low–medium resolution
- Relatively low mass accuracy
- Space charge effects and ion–molecule reactions, leading to a poor-quality mass spectrum
- Not all tandem MS experiments are possible (no neutral loss experiments)

combined with an external source (i.e. as APCI, ESI or nanospray [131]) have improved some instrumental properties including:

(i) extended mass range (>70 000 Da);
(ii) resolution (>8000 at 2000 Da); and
(iii) high sensitivity (multisequential MS^n experiments on less than a picomole).

Accurate mass assignment of highly resolved ion-intensity signals remains problematic. QITMS is a relatively simple and compact apparatus (benchtop) and an order of magnitude less expensive than the multistage (triple quad or tandem double-focusing) instruments traditionally used for structural determination.

Instrumental developments concern micro ion traps (sub-mm i.d.) [193], extension of the mass range, mass resolution and capture efficiency for ions generated externally. Fast separations at very low detection levels are possible by means of hybrid QIT/reToF mass spectrometry [194].

QITMS has been reviewed [195–197], and has been the subject of various books [198,199]; a tutorial has also appeared [200].

General Applications   The ion trap can be interfaced with *ionisation methods* that induce ionisation in compounds which are not readily amenable to electron impact due to insufficient vapour pressure and/or modest to high polarity, such as LDI, MALDI, SIMS, GD, ESI, ISP, and TSP. There are several approaches to the introduction of externally generated ions to an ion trap [197]. Energy and spatial distribution of ions produced in the source and entering the mass analyser are not critical. This is important for coupling with LC, because LC-MS interfaces generally produce ions with a relatively wide energy and spatial distribution. The advantages of ion traps with regard to continuing ionisation have also stimulated their combination with laser ionisation.

Commercial ion traps are intended as analysers for *separation methods* (GC, HPLC, SFC, CE). While the original patent for the quadrupole ion trap dates from 1953, the first commercially available ion trap detector (for GC) was produced only in 1982 (by Finnigan). The majority of ion trap instruments in use today are combined with gas chromatography (as GC-MS/MS or GC-CI-MS/MS) but also as LC-MS/MS. The use of an ion trap device for mass analysis and molecular structural analysis, when coupled to liquid inlets, has become a highly important analytical tool. The ultimate signal-to-noise for these tools is highly dependent on the chemical background. Any increase in unwanted ions or droplets can reduce the effective signal-to-noise and the working dynamic range of the mass analyser. The use of an atmospheric pressure ionisation (API) orthogonal aerosol-spraying device can dramatically reduce the unwanted particles and broaden the ion generation process via multiple ionisation modes. Liquid chromatography and capillary electrophoresis combined with electrospray/ion trap mass spectrometry have been reviewed [201].

Ion traps are well suited to tandem MS^n experiments in a single analyser instrument, and allow *structure characterisation* studies (targeted compound identification in mixture analysis). McLuckey *et al.* [202] have defined the conditions that must be met for an MS^n experiment to be feasible in a quadrupole ion trap. Its high MS/MS efficiencies make it an attractive and ultrasensitive detector for organic and inorganic analyses [203–207]. The MS^n capabilities of the ion trap are essential in structural studies on parent or diagnostic fragment ions. Following the introduction of wideband excitation and normalised collision energy in QIT [208], highly reproducible MS/MS spectra suitable for library searching in LC-MS^n are easily obtainable under ESI and APCI conditions [209,210].

Plasticiser contamination from vacuum system O-rings in QITMS systems has been reported [211].

6.3.5 Fourier-Transform Ion-Cyclotron Resonance Mass Spectrometry

Principles and Characteristics Fourier-transform ion-cyclotron resonance (FTICR) was developed in 1971, and although Fourier-transform mass spectrometers (FTMS) now represent fewer than 1 % of all mass spectrometers, they present unique features (Table 6.33). An ICR analyser is a device to determine the mass-to-charge of an ion in a magnetic field by measuring its cyclotron frequency. Several types of FTMS instruments can be found, all composed of four main components: magnet, analyser cell, ultrahigh vacuum system and sophisticated software system. Important primary performance parameters of FTMS improve as the magnetic field strength increases, which explains the trend towards instruments with stronger field strengths; 20 tesla has already been reported [212]. In the analyser cell, the ions are stored, mass analysed and detected. The performance of the FTMS instrument is more sensitive to pressure than other instruments; typically a pressure of 10^{-10} Torr is required. The data system is similar to that used for FTNMR.

The basis for FTMS is *ion-cyclotron motion*. A simple experimental sequence in FTMS is composed of four events: quench, ion formation, excitation and ion detection. Ions are created in or injected into a cubic cell where they are held by an electric trapping potential and a constant magnetic field **B**. Each ion assumes

Table 6.33 Main characteristics of Fourier-transform mass analysers

Advantages

- Ultrahigh mass-resolving power (10^6) and mass accuracy (few ppm)
- High mass range ($>20\,000$ Da)
- High detection efficiency (detection of single ions)
- Real-time monitoring
- Positive and negative ion spectra
- Nondestructive detection (ion stability experiments)
- Adaptability to a wide range of external ion sources
- Inherent tandem MS capabilities
- Relatively compact instrumentation

Disadvantages

- High cost (investment and operational: cooling of superconducting magnets!)
- High vacuum (10^{-10} Torr) hinders coupling with high-pressure ion sources (LC-MS)
- Demanding on computing facilities
- Limited dynamic range (10^2–10^5)
- Limited quantitation capabilities
- Difficult experimentation
- (Permanent) development phase

a cycloidal orbit at its own characteristic frequency, which depends on m/z. The 3D ion trap allows ions to be stored in the high-vacuum ICR analyser cell from seconds to hours, which is many orders of magnitude longer than the residence time of ions in most other types of mass spectrometers. FTICR-MS is the 'big brother' of ITMS. The ion motion in the FTMS cell is complex because of the presence of electrostatic and magnetic trapping fields. After formation by an ionisation event, all trapped ions of a given m/z have the same cyclotron frequency but occupy random positions in the FTMS cell. To detect cyclotron motion, an excitation pulse must be applied to the FTMS cell, so that the ions 'bunch' together spatially into a coherently orbiting ion packet. A radiofrequency pulse brings all of the cycloidal frequencies into resonance simultaneously, to yield a signal as in an interferogram (a time-domain spectrum). This is converted by Fourier transformation to a frequency-domain spectrum, which then yields the conventional m/z spectrum. In this way the various cyclotron frequencies, ν_c (in Hz)

$$\nu_c = z\mathbf{B}/2\pi M \quad \text{(SI units)} \qquad (6.4)$$

are determined as a direct measure of their mass-to-charge ratio m/z.

The ability to selectively excite a particular ion (or group of ions) by irradiating the cell with the appropriate radiofrequencies provides a level of flexibility unparalleled in any other mass spectrometer. The amplitude and duration of the applied RF pulse determine the ultimate radius of the ion trajectories. Thus, by simply turning on the appropriate radiofrequency, ions of a single m/z may be ejected from the cyclotron. In this way, a gas-phase separation of analyte from matrix is achieved. At a fixed radius of the ion trajectories the signal is proportional to the number of orbiting ions. Quantitation therefore requires precise RF control.

Ions of many masses can be detected simultaneously with FTMS. To accomplish *broadband detection*, many frequencies are applied during the excitation event. The ion storage principle allows panoramic registration. The ions do not hit the detector, and they remain available for additional experiments, such as collision-induced dissociation (CID), which can be used to study subsequent daughter generations. Attractive features of FTMS are thus its extensive ion trapping and manipulation capabilities. Targeted ions can be selectively trapped in the ion cell by applying RF pulses to eliminate unwanted ions.

The FTMS instrument operates in a very different fashion from most other types of mass spectrometers. With FTMS, the principal functions of ionisation, mass analysis and ion detection occur in the same space

(the analyser cell), but are spread out in time, whereas with quadrupole and magnetic sector mass spectrometers these events occur simultaneously and continuously, but in different parts of the mass spectrometer. Contrary to magnetic sector or quadrupole mass spectrometers, MS/MS with FTMS is achieved with the same hardware; the experimental sequence is just expanded to include pulse sequences for the mass selection of a precursor ion and for its dissociation. The accurate-mass capability of FTICR-MS generally makes it unnecessary to conduct MS/MS experiments of more than two stages.

In FTICR-MS, ions may be formed either inside the trap (e.g. by electron impact, photoionisation, photodissociation, charge exchange, ion–molecule reaction, CID, etc.) or just outside the ion trap (e.g. laser desorption/ionisation or laser desorption of neutrals followed by electron ionisation, LSIMS with a solids insertion probe, FD, FAB, CF-FAB, PD, API, MALDI, surface-induced desorption, etc.) Desorption/ionisation techniques operate with higher pressures than usually allowed for FTMS operation. The versatility of FTMS follows from the fact that it is an ion trapping instrument. Most *ion sources* work best outside the magnet. Thus, several methods have been developed to guide externally generated ions into an ion trap inside a high-field magnet. The combination of ESI with FTMS is a particularly good match. A solution to the coupling of the inherently pulsed FTICR mass analyser to the inherently continuous electrospray ionisation source in HPLC is to collect and accumulate ions in an octopole ion trap external to the magnet, and then eject the ions into the ICR ion trap for excitation and detection [213]. HPLC–micro ESI-FTICR-MS can detect fmole quantities.

MALDI-FTMS is also capable of spectacular performance. Coupling with MALDI and ESI has driven many developments. *Laser desorption/ionisation* is most popular for analytical FTICR-MS applications. The Nd:YAG laser (266 nm, 4.7 eV) permits both laser desorption (LD) and multiphoton ionisation (MPI), which are effective methods for generating ions. LD is a proven method for emitting large, fragile molecules into the gas phase. MPI offers control of ionisation and fragmentation not possible with EI. Key advantages of using FTMS with MPI are the ability to collect a complete, high-resolution mass spectrum from each laser shot (at variance with scanning instruments), and the ability to trap and study the ions after their formation, e.g. by means of CID. The differences in molar absorptivity of two compounds frequently allow MPI to be used as a method for *selective ionisation* in mixtures in which EI would ionise both compounds. Another important advantage of MPI is that wavelength and power can be used to control the softness of the ionisation.

Fast screening techniques, such as temperature-resolved in-source filament pyrolysis and laser-assisted pyrolysis, benefit from the high cycle time and mass accuracy of FTICR-MS [214]. An additional advantage of FTICR-MS in the study of pyrolysis processes is that MS^n can be readily used for structural identification of desorption and pyrolysis products.

Tables 6.27 and 6.33 illustrate the *main features* of Fourier-transform mass spectrometry. FTMS stands for multipurpose mass spectrometry. As the measurement of frequency is one of the most precise forms of data acquisition, resolution is exceptional, as high as 10^6 (at $m/z = 200$) (dependent upon the sample introduction method). It is possible to zoom from a mass distribution down to a single charge-state and then to its isotope pattern. The sensitivity of a FTMS is much higher than that of conventional mass spectrometers. For low masses (m/z 130) the mass resolution is 1000 times better than for conventional mass spectrometry. A remarkable feature of FTMS is that better mass resolution means better sensitivity. This contrasts with other types of MS, where resolution is increased by selecting a central fraction of the ion beam, thereby sacrificing sensitivity. For singly charged ions of <700 Da, a unique elemental composition can be assigned directly from the measured mass of ca. 1 ppm mass accuracy. With proper mass calibration, the elemental composition of each species can be assigned unambiguously based on sub-ppm mass accuracy. FTMS allows recording of a complete mass spectrum (18–10 000 Da) in 100–500 ms, thus allowing real-time monitoring. FTMS appears to be in a permanent development phase and is not yet mature for routine applications in an industrial environment. Marshall *et al.* [215] have reviewed recent advances in FTICR-MS regarding upper mass limit, mass resolving power, mass accuracy, mass selectivity, sensitivity, and MS^n capability.

FTICR-MS has frequently been reviewed (ref. [215] for sources cited); various dedicated journal special issues are available [216,217], as well as several books [218–221].

Applications Low-field (0.4 T) FTICR-MS with calibration curves determined by collecting 70 eV EI mass spectra has yielded lower limits of detection of the order of tens of picograms (total amount on the solids probe) [222]. Heeren *et al.* [223] performed a rapid microscale analysis with high mass accuracy, by direct temperature resolved desorption and 'in-source' pyrolysis FTICR-MS. The analysis of unknown FR polymer blends, obtained from *household appliances* containing polybrominated biphenyls and diphenyl ethers, tetrabromobisphenol-A and its butylated isomers, and Sb oxide synergists, demonstrated the high resolution and high sensitivity of the system. The polymer samples were screened in broadband mode in the m/z 21–5000 mass range. Series of characteristic isotope patterns for deca-, nona-, octa-, hepta- and pentabromobiphenyl compounds were observed centred around m/z 943 (Br$_{10}$), 879 and 863 (Br$_9$), 801 and 783 (Br$_8$), 721 and 703 (Br$_7$), 641 and 623 (Br$_6$), and 561 and 543 (Br$_5$), respectively. The mass of the individual compounds has been determined with a 7 ppm accuracy [223]. Temperature-resolved FTMS of PS/(Br$_x$DPO, TBBP-A, Sb$_x$O$_y$) under dynamic high-resolution conditions (FWHH 50 000) is able to discriminate between $C_{12}H_2{}^{79}Br_5{}^{81}Br_3O$ (theor. 799.35144 Da, exp. 799.35118 Da, Δppm 0.05) and $C_{12}{}^{79}Br_4{}^{81}Br_4O$ (theor. 799.33344 Da, exp. 799.33323 Da, Δppm −0.26) (Scheme 6.2). The resolution is sufficient to separate the nominally isobaric ions from the Sb$_4$O$_6$ synergist and the *n*-butyl ether derivative of tetrabromobisphenol-A.

Similarly, toluene suspensions of the polystyrene housing of TV sets were examined by means of TPPy-FTMS (300–1200 K) [224]. Diphenylether (DPE) was evidenced by peaks at m/z 141, 142 and 170 and decabromobiphenyl (DBBP) by m/z 943 and 864. Decabromodiphenyl ether (DBDPE) was recognised by thermal degradation products around m/z 800

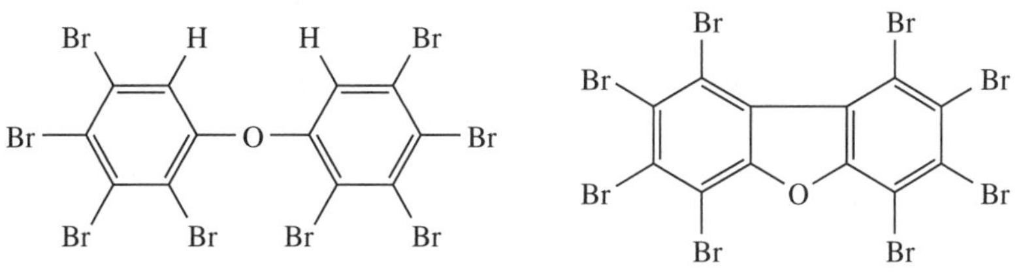

$$C_{12}H_2{}^{79}Br_5{}^{81}Br_3O \qquad C_{12}{}^{79}Br_4{}^{81}Br_4O$$

Scheme 6.2 Structural formulae

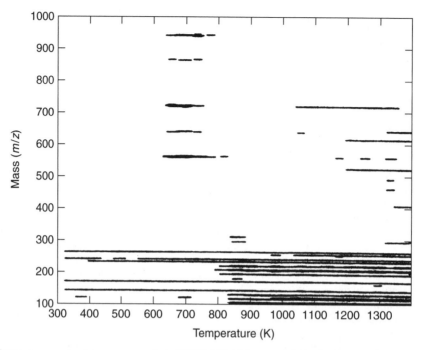

Figure 6.17 Temperature-resolved in-source pyrolysis FTICR-MS of flame-retarded polystyrene (56 spectra with a sampling interval of 1.1 s) from 300 K to 1200 K. After Heeren and Boon [224]. Reprinted from *International Journal of Mass Spectrometry and Ion Processes*, **157/158**, R.M.A. Heeren and J.J. Boon, 391–403, Copyright (1996), with permission from Elsevier

($C_{12}OBr_8^+$), m/z 720 ($C_{12}OBr_7^+$), m/z 640 ($C_{12}OBr_6^+$) and m/z 560 ($C_{12}OBr_5^+$). The temperature-resolved broadband mass spectra (32 000 datapoints each) are shown in Figure 6.17 as a function of their desorption time.

FTICR-MS is capable of powerful mixture analysis, due to its high mass range and ultrahigh mass resolving power. However, in many cases it is still desirable to couple a chromatographic interface to the mass spectrometer for sample purification, preconcentration, and mixture separation. In the example given above, DTMS under HRMS conditions provides the elementary composition. Apart from DTMS, PyGC-MS can be performed to preseparate the mixture of molecules and to obtain the MS spectrum of a purified unknown. Direct comparison with the pure reference compound remains the best approach to obtain final proof.

Kenion *et al.* [225] used LD FTMS for direct analysis of a discoloured adhesive. MALDI-FTICR-MS is a direct way of examining molecular weight distributions. FTMS is not yet widely being used for industrial problem solving, despite long-lasting expectations. Instead, FTICR-MS is the method of

choice for determining ion–molecule reaction pathways, kinetics, equilibria and energetics.

6.3.6 Tandem Mass Spectrometry

Principles and Characteristics Analytical multi-stage mass spectrometry (MS^n) relies on the ability to activate and dissociate ions generated in the ion source in order to identify or obtain structural information about an unknown compound and to analyse mixtures by exploiting two or more mass-separating steps. A basic instrument for the currently most used form, tandem mass spectrometry (MS/MS), consists of a combination of two mass analysers with a *reaction region* between them. While a variety of instrument set-ups can be used in MS/MS, there is a single basic concept involved: the measurement of the m/z of ions before and after a reaction in the mass spectrometer; the reaction involves a change in mass and can be represented as:

$$m_p^+ \longrightarrow m_d^+ + m_n \qquad (6.5)$$

where m_p^+ is the precursor (or parent) ion, m_d^+ is the product (or daughter) ion, and m_n represents one (or more) neutral species. In terms of mass: $m_p = m_d + m_n$. A basic MS/MS experiment usually consists of mass selection of the precursor ion in the first stage of analysis, fragmentation of the precursor ion (by metastable decomposition or by collision-induced dissociation, CID), and mass analysis of the product ions in the second stage of analysis (product-ion scan).

Tandem mass spectrometry is the ultimate problem-solving tool for chemical analysis when enhanced specificity, specificity, selectivity, sensitivity, and/or speed are required, but at a price. This is primarily due to the capacity of MS/MS to obtain spectra of selected precursor ions in complex mixtures. *Advantages* in using MS/MS are:

(i) minimisation (elimination) of sample preparation time (increased speed of analysis, in particular in mixture analysis, as compared to GC-MS, LC-MS, and high-resolution mass analysis (AC-MS));
(ii) direct screening of a complex mixture for a specific target compound;
(iii) (sub)structure identification;
(iv) improvement in specificity of MS/MS vs. MS;
(v) improved sensitivity in terms of signal-to-noise (chemical noise reduction) in comparison with single-stage MS; and
(vi) quantitative studies.

Tandem mass spectrometry provides the highest degree of certainty in analyte identification and may be employed in accordance with recent EU guidelines to obtain data with relevant unambiguity (EC Council Directive SANCO/1805/2000). The sensitivity of MS/MS is typically in the sub-pg range.

MS/MS has especially found analytical applications in combination with soft ionisation techniques, where, without MS/MS, only molecular weight information on the intact molecule is obtained and no fragmentation is observed. MS/MS is then required to achieve structure informative fragmentation. However, there are some *limitations*:

(i) MS/MS is limited to ions up to $m/z \sim 2000$; and
(ii) fragmentation may be insufficient to allow unambiguous structure assignment.

MS/MS experiments are often limited by the amount of sample available, and by the time available to obtain a spectrum.

The principle of MS/MS for direct analysis of a multicomponent system is shown in Figure 6.18, in which the first mass spectrometer (MS I) operates with soft ionisation (FI, FD, CI, LD), and thus produces an ensemble of molecular ions ($M + H^+$, $M - H^+$, or adducts). For identification of molecule ABC only ABC^+ is allowed to enter an interface or fragmentation zone for excitation by collisional activation, laser radiation or surface-induced dissociation. Within the time of one vibration (10^{-13} s), ABC^+ dissociates into fragments characterising the original molecule. They are separated and detected by MS II [226]. Soft ionisation with FI/FD produces low ion yields, which may be insufficient for MS/MS; LVEI (typically at 20 V) can be an alternative. Complete analysis of a multicomponent system is carried out in some 20 min.

In most cases, ion activation in the reaction region or fragmentation zone is applied to increase the internal energy of the ions transmitted from the ion source. The most common means of ion activation in tandem mass spectrometry is *collision-induced dissociation*. CID uses gas-phase collisions between the ion and neutral target gas (such as helium, nitrogen or argon) to cause internal excitation of the ion and subsequent dissociation

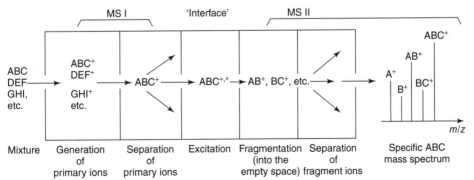

Figure 6.18 Direct MS/MS analysis for the identification of molecule ABC in a mixture. After Schwarz [226]. Reproduced from H. Schwarz, in *Analytiker-Taschenbuch* (R. Borsdorf and H. Günzler, eds), Vol. 8, pp. 199–216, Springer-Verlag, Berlin (1989), by permission of Springer-Verlag, Copyright (1989)

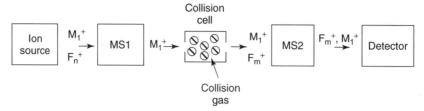

Figure 6.19 Schematic of a CID tandem mass spectrometer where MS1 and MS2 are the mass analysers, M_1^+ is the parent or projectile ion, F_n^+ are fragment ions of M_1^+ generated in the ion source, and F_m^+ are fragment ions formed as a result of CID

Table 6.34 Some MS^2 scan types

Scan mode	MS1	MS2	Application
Product-ion	Selecting m/z	Scanning	To obtain structural information by CID of ions produced in the source
Precursor-ion	Scanning	Selecting m/z	To monitor compounds which in CID give an identical fragment (screening)
Neutral-loss	Scanning $m/z = x$	Scanning $m/z = x - a$	MS1 and MS2 are scanning at a fixed m/z difference to monitor compounds that lose a common neutral species (screening)
Selective reaction monitoring (SRM)	Selecting $m/z = x$	Selecting $m/z = y$	To monitor a specific CID reaction

(Figure 6.19). The alternative application of photo-dissociation offers the possibility of tuning the imparted excess of energy. In this configuration, it is possible to fragment a specific ion isolated from the first mass spectrometer used in the SIM mode. Mixture analysis is possible. Other common modes of MS/MS operation are shown in Table 6.34. These methods are very powerful for identifying unknowns and for providing additional selectivity when searching for known compounds in complex mixtures. The product-ion mode provides an elegant way to obtain more structure information on ions generated by soft ionisation methods. When, instead of compound identification, screening is required, the use of the precursor-ion mode can be very useful

when a common fragment ion is to be expected. An example of the use of the precursor-ion scan mode is monitoring of dialkylphthalate plasticisers by means of the common fragment ion at m/z 149 due to protonated phthalic anhydride. For quantitative analysis, selective reaction monitoring can be applied. In this mode of operation, a precursor ion, selected in the first analyser, is collisionally fragmented and a particular fragmentation reaction is monitored. Because of the enhancement of the selectivity, the determination limit can significantly be improved by means of MS/MS.

In CID, two collision-energy regimes should be considered: low-energy and high-energy collisions, depending on the initial translational energy of the precursor ion upon collision. Low-energy dissociation is often characterised by loss(es) of neutrals from the precursor. This behaviour greatly facilitates interpretation of product-ion mass spectra. High-energy collisions (kV energy) are applicable in magnetic sector instruments as well as in certain applications of postsource decay in ToF instruments, while low-energy collisions are applied in most other systems (triple–quadrupole, ion trap and FTICR). Not surprisingly, the results for the low-energy collisions in a triple quad and those obtained on a sector-quadrupole instrument are usually comparable, but different from the additional fragmentation observed in conditions of high-energy collision.

In single-stage mass spectrometers, CID takes place in the ion source, and is thus sometimes called source CID or *in-source CID*. The advantage of performing CID in single-stage instruments is their simplicity and relatively low cost. The disadvantage is that *all* ions present are fragmented. There is no way to select a specific precursor ion, so there is no sure way to determine which product ions come from which precursor ion.

A most suitable tandem mass spectrometer should:

(i) be able to perform fast scanning high-resolution experiments as required for dynamic sample introduction (e.g. GC, LC);

(ii) not be limited by mass range and analyte concentration dynamic range; and

(iii) permit high-resolution precursor ion selection for MS/MS experiments.

The hybrids QToFMS [185] and BToFMS [28] are getting closest to these requirements. MS/MS experiments can be carried out by means of multiple quadrupoles, multi-sector combinations (B/E or E/B and three- and four-sector) and hybrid (quadrupole/sector) mass spectrometers for the first and third steps, each with specific (dis)advantages, as illustrated elsewhere [227]. In all these cases the two mass spectrometers are physically coupled, with ion separation in different spatial regions of the instrument ('tandem-in-space'). In a different family of mass spectrometers (ion trapping techniques, such as QITMS and FTICR-MS) these stages are performed consecutively within the ion trap itself ('tandem-in-time'). This allows unique MS^n experiments. Moreover, when MS/MS spectrometry is achieved within a single analyser (ion trap), rather than in separate compartments, this represents an appreciable saving in cost. The recent proliferation of ion storage MSs has contributed to the growing interest in multistage experiments.

The data set produced by MS/MS techniques is three-dimensional (precursor m/z, product m/z, intensity). It is particularly desirable to use a dispersive mass analyser for the second stage in a tandem mass spectrometer. The ideal tandem mass analyser – never constructed – would be dispersive in both mass dimensions, thus allowing it to collect the entire MS/MS data set from a single set of source ions [22]. Virtually all the MS/MS systems that have been developed with *dispersive mass analysers* have used a dispersive system for the second mass analyser. The array detection then produces probably the most useful scan – a product spectrum. Collecting a product spectrum for each of the m/z values represented in the precursor spectrum results in the complete MS/MS data set from which all the traditional one-dimensional scans (precursor, product, neutral loss) can be derived. Enke [181] has described MS/MS systems that involve reToF and oaToF analysers as a second stage in tandem mass spectrometry. The extremely powerful hybrid system of QqToF geometry with a reToF instrument allows accurate mass determination (up to 5 ppm), and facilitates interpretation of product-ion mass spectra. Advantages of QToF geometry are simultaneous detection of all ions (high ion collection efficiency, enhanced sensitivity), and orthogonal ToF (high resolution, accurate mass on molecular and fragment ions). The sensitivity and rapid response of

ToF offer a significant advantage over scanning instruments for MS2. Informative spectra can be obtained from fmoles of sample.

Four-sector instruments enable MS/MS with high-resolution selection/analysis for both precursor and product ions. CID is performed with high-energy collisions, i.e. in most cases by single high-energy collisions with helium. The four-sector instruments obviously are complex to operate. Advantages of tandem sector instrument over triple quad tandem mass spectrometers are high mass range and high mass resolution for the separation of isobaric ions from a mixture. Alternatively, a reflectron ToF may be used for MS/MS.

Triple-stage quadrupole instruments, which consist of two mass analysers separated by a quadrupole ion guide (g), hexapole (h) or octopole (o) collision cell (excited by RF only), are among the most widely used instruments for MS/MS (eighth generation in 2003). In the RF-only mode, ions of all masses are transmitted through the quadrupole filter. By adding a gas such as argon to the space between the RF-only rods, ion-neutral collisions may occur. If these are sufficiently energetic, extensive fragmentation of the ions will result; their masses are measured using the final quadrupole mass filter. The instrument can be operated in several ways (normal spectrum, product ions, parent ions, constant loss, SRM) to improve structural analysis. While full scans are being used for qualitative information (molecular weight), selective ion monitoring (SIM) and multiple ion detection (MID) can be used for quantitative information. The main characteristics of QQQ mass spectrometers are given in Table 6.35. Modern LC-QQQ analysis offers separation,

Table 6.35 Main characteristics of modern triple-quad mass spectrometers

Advantages

- High resolution and high sensitivity
- Sensitive target analysis (high specificity)
- Accurate mass
- Quantitative accuracy
- Large dynamic range
- Tolerance for high pressures
- Rapid scanning
- Robust
- Simple to use and maintain
- Bench-top (past: large footprint)

Disadvantages

- Long residence time for a product ion
- Limited to MS^2 experiments
- Expensive

molecular weight determination and specific MS/MS data for components with $\Delta m > 0.2$ Da. Tandem mass spectrometry using the past low-resolution QQQ or IT mass spectrometers has been particularly useful. However, there are several disadvantages associated with these instruments. Lack of high resolving power capabilities, needed for exact mass measurement to evaluate the possible chemical compositions of unknown compounds, has now been overcome for modern triple quads.

Ion trap MS is particularly suited for chemical structure elucidation, as it allows for simultaneous ion storage, ion activation and fragmentation, and product ion analysis. The fragmentation pathway of selected ions and the fragmentation products provide information on the molecular structure. Compared with triple–quadrupole and especially with sector instruments, the ion trap instrument provides more efficient conversion of precursor ion into product ions. However, the CID process via resonance excitation, although quite efficient in terms of conversion yield, generally results in only one (major) product ion in the product-ion mass spectrum. MS/MS with a quadrupole ion trap offers a number of advantages:

(i) mass-selected ions can be accumulated over time;
(ii) all isolated ions can be dissociated, and fragment ions arising from some 90 % of them can be confined; and
(iii) a sequence of several mass-selective operations can be performed, as in MS^n.

The MS^n capability of the ion trap MS system allows isolation of the parent ion of a molecule of interest and its step-wise fragmentation, facilitating interpretation of the product-ion information. In this way, it is possible to investigate the fragmentation pattern of a molecule during ionisation with different methods. Structural isomers can be differentiated if one of their fragment ions is significantly characteristic of the parent molecule. The fragmentation in the ion trap is softer, or more controllable, than that in triple quadrupole instruments. This step-wise fragmentation can be extremely useful in structure elucidation, but is a serious limitation in cases where confirmation of identity should be based on the detection of a number of particular product ions. When GC is interfaced with an ion trap tandem mass spectrometer (GC-MS/MS), individual compounds can be detected at the hundreds of fg level. In an FTICR instrument, which is also an ion trapping device, MS/MS can be performed in a manner similar to MS/MS in ion trap instruments.

Applications The diversity of MS/MS instrumentation offers considerable opportunities for polymer/additive analysis. The best geometry for a particular application depends on a number of factors, including mass resolution in the first and third stages, mass range, sensitivity, available collision energy, the type of information required, data acquisition rate, etc. Polymer science applications of MS/MS comprise:

(i) identification of organic additives in compounded polymers (mixture analysis) without prior chromatographic separation;
(ii) identification of volatile pyrolysates in polymer pyrolysis studies; and
(iii) characterisation of individual oligomers in low-MW polymers.

MS/MS techniques have significant advantages in the areas of impurity identification, background reduction using soft ionisation techniques and extraction of fragmentation information from *complex mixtures*. One important application of MS/MS in polymer/additive analysis is the direct rapid screening of polymeric materials for the presence of additives. Both direct MS/MS and GC-MS/MS have been employed in the quantitative analysis of known compounds in mixtures. The enhanced selectivity and speed of method development are particularly impressive features. Yet, the application of MS/MS to the structural characterisation of polymer additives is still far from widespread [23,37,69,137,228,229]. Low-energy MS/MS experiments were performed by means of sector–quadrupole hybrid [37,228,229] and triple quadrupole [23,137] tandem mass spectrometers. High-energy MS/MS gives fragment-ion peaks that are more characteristic of the class of polymer additives [69,96].

High-MW *oligomeric polymer additives* (e.g. HALS) have been successfully analysed by means of MS, such as Chimassorb 944 with CI-MS/MS [37]; ionisation is most successful with desorption techniques, due to the inherent involatility of these materials. It has been suggested that the most suitable ionisation technique for analysing oligomeric additives is FD [37], but intense distributions have also been observed in the MALDI spectra of the same compounds [230]. Tandem MS must be employed to obtain structural information, as predominantly molecular ions are generated by FD and MALDI ionisation with little fragmentation induced. LSIMS-MS/MS or FD-MS/MS experiments are used in conjunction with FD-MS, as a screen to determine the class, molecular weight and structure of mixtures of organic polymer additives.

Selection of a suitable ionisation method is important in the success of mixture analysis by MS/MS, as clearly shown by Chen and Her [23]. Ideally, only molecular ions should be produced for each of the compounds in the mixture. For this reason, the 'softest' ionisation technique is often the best choice in the analysis of mixtures with MS/MS. In addition to 'softness', selectivity is an important factor in the selection of the ionisation technique. In polymer/additive analysis it is better to choose an ionisation technique which responds preferentially to the analytes over the matrix, because the *polymer extract* often consists of additives as well as a low-MW polymer 'matrix' (oligomers). Few other reports deal with direct tandem MS analysis of extracts of polymer samples [229,231,232]. DCI-MS/MS (B/E linked scan with CID) was used for direct analysis of polymer extracts and solids [69]. In comparison with FAB-MS, much less fragmentation was observed with DCI using NH_3 as a reagent gas. The softness and lack of matrix effect make ammonia DCI a better ionisation technique than FAB for the analysis of additives directly from the extracts. Most likely due to higher collision energy, product ion mass spectra acquired with a double-focusing mass spectrometer provided more structural information than the spectra obtained with a triple quadrupole mass spectrometer.

Tandem MS (DFS equipped with EI/FI/FD source) in conjunction with off-line direct inlet HPLC-UV was used for separation and quantification of *isomeric antioxidants*, $C_{22}H_{30}O_2S$ (MW 358; Scheme 6.3), as antioxidants in THF extracts of surgeons' gloves [232]. Collision activation MS enabled differentiation between the three isomeric structures (Fig. 6.20). Quantification was achieved by chromatographic analysis of the isomeric species, which are not distinguishable by MS. On-line LC-MS facilitates this kind of analysis.

However, for nontargeted analysis, where the class of compound remains unknown, a two-step analysis involving direct analysis by tandem MS followed by HPLC is an advantageous strategy.

Using tandem MS (DFS with EI/FI/FD source), electron impact and collision activation mass spectra of a THF extract of an orthopaedic polymer bandage identified *N*-isopropyl-*N'*-phenyl-*p*-phenylenediamine (IPPD, m/z 226) as a cause for contact dermatitis [232]. FI-MS of the extract of surgeons' gloves indicated thiobis (*t*-butylcresol) (m/z 358; 343, after CID).

Lattimer *et al.* [229] have determined additives in an uncured SBR compound of known composition (Table 6.36) using EI-MS, CI-MS (direct analysis) and FD-MS survey scans (extract analysis, Figure 6.21) on a sector mass spectrometer of BEQQ geometry. All these spectra are rather complex. The FD-MS spectrum shows considerable disturbance by CH_2 splittings. Daughter-ion scans were made on a number of molecular ions ($M^{+\bullet}$ from EI, FI, or FD; MH^+ from CI). EI-MS/MS worked well for those additives that gave intense molecular ions ($M^{+\bullet}$) via this ionisation mode. CI-MS/MS was also satisfactory for most additives. On the other hand, the FI- and FD-MS/MS analyses of these rubber samples were generally not very informative, due to the fact that FI/FD ion currents were too weak and unstable to give daughter-ion spectra with good S/N. It was concluded that the best way to detect and identify typical organic additives in a *compounded rubber* was to use a combination of EI- and CI-MS/MS. Two accelerators, OBTS and DPG, were successfully analysed intact by CI-MS/MS. The MS/MS technique is experimentally rapid compared to other techniques for mixture analysis, such as GC-MS, LC-MS or high-resolution (AC-MS). Similarly, when the selectivity of standard ESI/MS is not adequate, e.g.

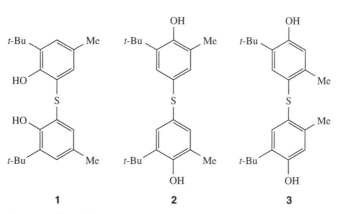

Scheme 6.3 Isomeric *tert*-butyl substituted bisthiophenols

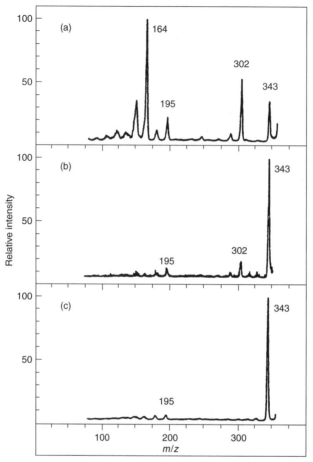

Figure 6.20 Collision activation mass spectra of the EI induced molecular ions m/z 358 of 2,2′-thio-bis-(4-methyl-6-t-butylphenol) (**1**), 4,4′-thio-bis-(6-t-butyl-o-cresol) (**2**) and 4,4′-thio-bis-(6-t-butyl-m-cresol) (**3**) (see Scheme 6.3). After Egsgaard *et al.* [232]. Reprinted from *Trends in Analytical Chemistry*, **11**, H. Egsgaard *et al.*, 164–168, Copyright (1992), with permission from Elsevier

when different species appear at the same m/z value, tandem mass spectrometry has to be employed to obtain complete identification and quantitation [233]. Using product ion, precursor ion or neutral loss scans, the composition can then be established even when the different species have identical mass. ESI-MS/MS allows structural determination of individual molecular species in pmol amounts.

A *five-component mixture* of polymer additives (Irganox 1010/1076/3114, Tinuvin 327, Hostanox O3; 300–1200 Da) in CHCl$_3$ was analysed by means of high-energy mass spectrometry using a four-sector instrument of reverse geometry (B$_1$E$_1$B$_2$E$_2$) and ToF-MS with various ionisation modes (EI, CI, LSIMS, FD, UV-MALDI) [96]. The CID spectra of molecular ions generated by EI, CI, LSIMS and FD were then

Table 6.36 Composition of a rubber compound[a]

Ingredient	phr[c]	Ingredient	phr
Styrene-butadiene rubber[b]	137.5	HPPD	2.0
Zinc oxide	3.0	Refined paraffin wax	2.0
Stearic acid	2.0	Poly-TMDQ	5.0
ISAF-HS carbon-black	72.0	OBTS	2.0
Aromatic processing oil	2.5	DPG	0.1
t-Octylphenol/formaldehyde	2.0	Sulfur	1.75

[a] Total 231.85 phr;
[b] SBR, 23.5 % styrene; oil-extended 27 % oil.
[c] Parts by weight per hundred parts resin.
After Lattimer *et al.* [229]. Reprinted with permission from *Rubber Chemistry and Technology*. Copyright © (1990), Rubber Division, American Chemical Society, Inc.

obtained by means of tandem mass spectrometry. The order of decreasing average internal energy deposited in

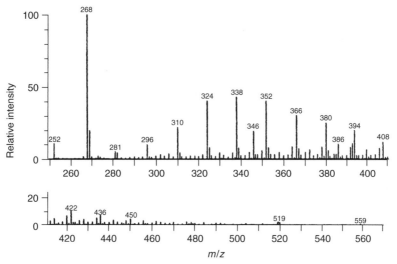

Figure 6.21 Field desorption mass spectrum of the rubber compound (acetone extract analysis) of Table 6.36. After Lattimer *et al.* [229]. Reprinted with permission from *Rubber Chemistry and Technology*. Copyright © (1990), Rubber Division, American Chemical Society, Inc.

the molecular ion is EI > CI, LSIMS > FD, with FD normally being regarded as a very soft ionisation method [234,235]. Higher-mass organic polymer additives (MW > 1000 Da) are not always suited to analysis by means of EI-MS and CI-MS, due to their involatility. Desorption/ionisation techniques such as LSIMS are more practical, as they give more abundant molecular ion signals and fragmentation for structural information. High-energy CID in LSIMS-MS/MS experiments, which generates characteristic ions, is more applicable than low-energy CID to the structural determination of polymer additives. It was observed that high-energy CID LSIMS-MS/MS spectra of Irganox 1010/1076/3114 and Hostanox O3 differ greatly from low-energy CID spectra [23,96]. LSIMS is a more suitable ionisation technique than FD for analysis of mixtures by means of tandem mass spectrometry, because of the higher ion currents generated from polymer additives using LSIMS. On the other hand, LSIMS is not ideal for the quantitative detection of polymer additives, as matrix effects are very important. Field desorption is a good ionisation technique for generating molecular weight information of polymer additives in mixtures. Mainly molecular radical cations (M$^{+\bullet}$) were seen in the FD-MS spectrum of the additive mixture (Fig. 6.22), with little fragmentation. However, the ion currents obtained are typically in the region of two orders of magnitude lower than those obtained by LSIMS ionisation. FD is a superior ionisation technique for quantitative analysis, as there are no matrix effects which might suppress the generation of ions from certain additives.

Moore [231] has used hybrid MS/MS (EBQQ geometry; EI spectra) for the direct analysis of DODPA in *unvulcanised rubber*. Particular advantages of the technique are:

(i) minimal sample preparation;
(ii) ease of interpretation of the high signal-to-noise MS/MS spectra;
(iii) good sensitivity and selectivity for parent ion selection in the double-focusing (EB) primary stage; and
(iv) ability to vary the energy deposited during collisional activation.

Scrivens *et al.* [227] have compared FD-MS/MS experiments on a four-sector instrument of BEBE geometry with those obtained using an oaToF hybrid MS/MS, using complex mixtures of industrial relevance in which conventional MS ionisation techniques such as EI/CI/LSIMS or electrospray have significant limitations. The experiment gives valuable structural details, with the high-energy collisions favouring cleavage rather than rearrangement ions. As expected, the oaToF-MS/MS instrument leads to improved sensitivity over the four-sector machine. Acquisition of both precursor-ion and second-generation product-ion spectra (EI-MS^3 scanning) under high-energy conditions of a four-sector mass spectrometer can aid the analysis of organic additives from polymeric formulations, which are often complex [169]. These techniques were used to determine fragmentation pathways.

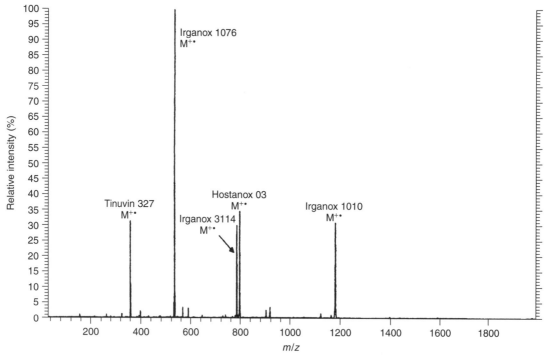

Figure 6.22 FD-MS spectrum of an equimolar mixture of polymer additives. After Jackson *et al.* [96]. Reprinted from A.T. Jackson *et al.*, *Rapid Communications in Mass Spectrometry*, **10**, 1449–1458 (1996). Copyright 1996 © John Wiley & Sons, Ltd. Reproduced with permission

A triple quadrupole mass spectrometer has shown great specificity and potential for identification of additives directly from PE extracts using DCI and low-energy MS/MS [23], although compounds with more rigid structures, such as hindered amines and phosphates, did not produce structurally characteristic fragments allowing a clear structure assignment. This is likely to be due to the low collision energy associated with the triple quadrupole instrument. MS/MS techniques (using QQQ) were used for identification of an *antistatic agent* in a complex mixture of additives (PEG, ethoxylated compounds) in polymer films [100]. Organotin compounds were studied by direct insertion probe QQQ.

Direct polymer compound analysis by soft ionisation, tandem MS/MS and high-resolution (AC-MS) mass spectrometry, has been reviewed [236].

The extent to which fragmentation takes places for each dissociation pathway of the additives depends on the excess of internal energy imparted to the molecular ion. Because the experimental conditions for product-ion mass spectra are strongly instrument dependent, generally not very well standardised and difficult to exchange between instruments from different

manufacturers, there are at present no well-stocked generally applicable *spectral libraries* for MS/MS spectra that can assist in structure elucidation problems. However, this situation is changing (Section 6.3.4).

Numerous applications of MS/MS have been reported, in which the MS/MS instrumentation is either used as a stand-alone instrument for sample introduction by a probe, or in on-line combination with GC or LC. Especially in the latter area, where soft ionisation strategies are frequently applied, MS/MS plays an important role. Identities of various evolved products can be established with more confidence than by using conventional (single-stage) MS analysis. The increasing use of tandem MS (Figure 6.23) reflects the substantial increase in selectivity that can be obtained with this technique. This selectivity is often so high that the work-up procedure can be simplified or omitted. Subjecting products to successive fragmentations (MS^n), as in ion trap MS, gives an extremely powerful tool to obtain the correct information without interferences. Today, the combination of LC with tandem MS is probably the technique in which selectivity comes closest to the meaning of specificity. Tandem mass spectrometry is to be considered as a physical *alternative to chromatography* as a

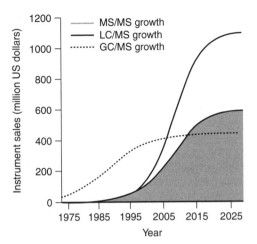

Figure 6.23 Predicted growth of MS/MS. After Willoughby and Sheehan [237]. Reproduced by permission of Global View Publishing

Table 6.37 Mass-spectrometric approaches to polymer/additive analysis

Method	Result
Extract analysis	
Survey scan (EI, CI, FI, FD, FAB)	Low-resolution spectrum of organic additives
Combined chromatography (GC-MS, LC-MS, SEC-MS)	Spectra of chromatographically separated organic additives
Direct analysis	
Survey scan (EI, CI, FI, DI-MS)	Low-resolution spectrum of organic additives and pyrolysates
High-resolution accurate mass analysis	Atomic composition (number and type of atoms) of an organic component
Tandem mass spectrometry (MS/MS)	Two-stage mass analysis (mass spectrum of a selected mass)
Surface analysis (FAB, SIMS)	Spectrum of organic components (bloom) on a rubber surface

separation technique. The high selectivity and resolution of today's MS/MS systems have reduced the need for efficiency of the chromatographic separation for LC/MS analysis. Therefore, shorter columns may be used; an SPE cartridge is sometimes efficient enough to allow direct elution into the MS. In many cases, chromatography is needed only to separate the analytes from the sample matrix, in order to avoid ionisation interferences and contamination of the MS system. SPE provides such a matrix–analyte separation, with the additional benefit of analyte enrichment.

The potential of MS/MS is well documented. Tandem MS has been reviewed [238] and various books deal with the topic [203,237,239].

6.4 DIRECT MASS-SPECTROMETRIC POLYMER COMPOUND ANALYSIS

Principles and Characteristics Mass-spectral analysis methods may be either indirect or direct. Indirect mass-spectral analysis usually requires some pretreatment (normally extraction and separation) of the material, to separate the organic additives from the polymers and inorganic fillers. The mass spectrometer is then used as a detector. Direct mass-spectrometric methods have to compete with separation techniques such as GC, LC and SFC that are more commonly used for quantitative analysis of polymer additives. The principal advantage of direct mass-spectrometric examination of compounded polymers (or their extracts) is speed of analysis. However, quite often more information can be

obtained when a chromatographic step is added to the analytical procedure (Section 7.3). Table 6.37 compares extract and solid analysis of additives by means of mass spectrometry.

The *particle-beam interface* has been used for direct introduction of extracts into the mass spectrometer *without* chromatographic separation [55]. In fact, chromatographic separation is not always essential, especially if structural information is available about the analytes of interest. The main features of this particular approach are:

(i) speed and simplicity: analysis time of 2 min per sample;
(ii) cost effectiveness and environmental friendliness: solvent reduction;
(iii) ease of interpretation: conventional EI mass spectra;
(iv) automation: LC autosampler option; and
(v) applicability: large range of semi-volatile and non-volatile additives.

Direct mass analysis of additives in bulk polymers is in principle an attractive methodology, albeit with many restrictions (Table 6.38). Early MS work has focused on direct thermal desorption of additives from the bulk polymer, followed by EI-MS [22,240], CI-MS [22,63] and FI-MS [22]. However, these traditional approaches are limited to polymer additives that are both stable and volatile at the higher temperatures,

Table 6.38 Main features of direct mass-spectrometric methods

Advantages

- Minimal sample preparation
- No interface
- Various controlled heating modes allowed (isothermal, gradient)
- All modern desorption techniques applicable
- Rapid screening tool
- High sensitivity
- Time-resolved studies

Disadvantages

- Limited to thermally stable and volatile analytes (EI, CI, FI)
- Restrictions on atmospheric conditions
- Low reproducibility (best for in-source)
- Matrix interference (FAB, LSIMS)
- Very small sample amounts (μg level)
- Difficult mixture analysis (ionisation method dependent)
- Difficult detection of minor components
- Difficult quantitation
- Scarcity of reliably mixed polymer formulations
- Costly maintenance of equipment (fouling of ion source)

or can provide meaningful fragment ions at elevated temperatures. With the addition of high-MW, oligomeric organic polymer additives becoming more common, many polymeric materials are no longer suited to analysis by means of GC, LC or SFC. These types of additives can be analysed successfully by means of mass spectrometry [37,230,241], but ionisation is limited to soft desorption/ionisation techniques because of the inherent involatility of these materials.

In principle, the most straightforward way to identify organic additives in a compounded polymer formulation is to heat the material to thermally desorb the volatile compounds, which may then be directed into a mass spectrometer for analysis. Direct mass-spectrometric methods for the study of polymeric materials have no interface; require no sample preparation time; undergo no influence of preparation; may use gradual heating of the sample; and allow application of all modern desorption techniques. In many experimental designs, *neutrals* are desorbed from the extended probes inserted into the ion source. Therefore, rather than being a radically new form of ionisation, the 'in-beam' methods are ways to enhance the volatility of what are normally thought to be nonvolatile compounds. A wide variety of desorption techniques with different conditions of fast and slow heating, inert and activated surfaces, have been described.

In recent years, very effective mass-spectrometric methods have been developed for direct polymer

compound analysis, such as soft ionisation, tandem and high-resolution mass spectrometry. *Direct probe desorption/ionisation methods* comprise:

(i) field desorption (FD);
(ii) secondary-ion mass spectrometry (SIMS);
(iii) ^{252}Cf plasma desorption (PD);
(iv) fast atom bombardment (FAB); and
(v) laser desorption (LD).

However, from the relatively few descriptions of direct mass-spectral analysis of plastics compounds, it follows that this approach still presents some problems. Because direct mixture analysis by means of mass spectrometry without prior fractionation is mostly very complex, *data analysis techniques* are necessary for an effective evaluation of the analytical output. Tas and van der Greef [242] have reviewed the use of direct mass-spectrometric profiling for the analysis of complex systems (e.g. pyrolysates) with emphasis on applications in which pattern recognition techniques have been applied to the evaluation of the resulting profiles. Multivariate data analysis techniques and pattern recognition (PARC) methods are of great importance in the evaluation of mass-spectrometric data of mixtures. Basically, in PARC, profiles are displayed as vectors or points (objects) in a multidimensional space. The dimensions of this space (the number of variable axes) are determined by the number of measurements (variables); in the case of mass spectra, the variables are the mass range scanned or the number of masses monitored. For a detailed observation of similarities and differences between profiles, the detection of groups and trends in the data, display by projection techniques is mandatory. Projection and display techniques such as nonlinear mapping (NLM), hierarchical clustering, principal component analysis (PCA), and discriminant analysis (DA) are frequently used. The main objectives of PCA are:

(i) to reduce the dimensionality of a data set;
(ii) to provide a graphic overview of dominant patterns; and
(iii) to identify significant (and redundant) variables.

Advanced mass spectrometry enables the detection of higher-molecular-weight compounds that can be expected to retain more specific structure information contained in the original complex materials. The application of MS/MS using various scan modes will further extend the capabilities for identification of compounds in complex mixtures. Precursor scan techniques improve insight into the origin of ions in complex pyrolysates and into their routes of formation. Data analysis plays an important role in the detection of trends during processes, in process control, and in quality control of complex base materials and end-products.

The success of direct mass-spectrometric methods generally depends upon *selective volatilisation* of additive components in competition with polymeric fragments. Direct-probe mass spectrometry (DP-MS), also termed direct inlet (DI) or direct insertion probe mass spectrometry (DIP-MS), involves direct insertion of the sample into the mass spectrometer through a vacuum sluice, using a heated sliding rod. For many spectrometers the probe option permits heating samples at temperatures up to 500 °C directly in the ion source, where the additives evaporate according to their boiling point. However, the direct probe inlet has little potential for fractionating samples, unless the components differ widely in volatility. A minimal peak separation is achieved. Some volatile components may be identified directly [243]. Direct-probe mass spectrometry (DP-MS) enables analysis of compounds of medium volatility, and relatively high thermostability, usually in the medium molecular weight range. In such applications, desorption is the dominant process. General experience is that *in vacuo* direct-probe introduction is usually preferable to external heating (continuous flow or molecular leak) devices. Two of the main reasons for this are:

(i) the vacuum allows components to desorb at lower temperatures; and
(ii) there is less chance for components to decompose or condense on walls in transit to the ionisation region [4].

Consequently, higher-molecular-weight (less-volatile) additives can generally be detected more readily with direct-probe introduction.

Characteristics of DP-MS are:

(i) diffusion of volatile additives *in vacuo*;
(ii) no pyrolytic fragmentation; and
(iii) detection of the whole mass stream over temperature (thermogram).

DP-MS suffers from system saturation; sample loads of a few µg are to be used. DP-ToFMS equipped with EI and FI sources is a thermal separation technique for solids which allows exact mass determination (Section 6.3.3). In order to detect and characterise polymer fragments of higher molecular weight, techniques such as DCI, in which the sample is thermally desorbed by the filament on which it is directly deposited, and laser desorption

(LD), in which flash vaporisation of the sample is induced, may be applied. Other techniques which permit detection of less-volatile chemical species are FD (with simultaneous desorption/ionisation of molecules), FAB (with the sample dissolved (dispersed) in a suitable liquid) and SIMS (based on bombardment of a solid surface with high-energy ions). LD-FTICR-MS is superior to FAB-MS for polymer/additive identification because it gives molecular ion fragmentation [83].

In direct insertion techniques, reproducibility is the main obstacle in developing a reliable analytical technique. One of the many variables to take into account is sample shape. A compact sample with minimal surface area is ideal [64]. Direct mass-spectrometric characterisation in the direct insertion probe is not very quantitative, and, even under optimised conditions, mass discrimination in the analysis of polydisperse polymers and specific oligomer discrimination may occur. For nonvolatile additives that do not evaporate up to 350 °C, direct quantitative analysis by thermal desorption is not possible (e.g. Hostanox O3, MW 794). Good quantitation is also prevented by contamination of the ion source by pyrolysis products of the polymeric matrix. For polymer-based calibration standards, the homogeneity of the samples is of great importance. Hyphenated techniques such as LC-ESI-ToFMS and LC-MALDI-ToFMS have been developed for polymer analyses in which the reliable quantitative features of LC are combined with the identification power and structure analysis of MS.

Various methods of analysis exert different thermal stress on a material (Table 6.39). Direct heating in the inlet of a mass spectrometer in order to obtain a mass spectrum of the total pyrolysate is an example of thermochemical analysis. Mass spectrometry has been used quite extensively as a means of obtaining accurate information regarding breakdown products produced upon pyrolysis of polymers. Low residence times allow detection of high masses.

There are two basic approaches to direct mass-spectral analysis of volatile additives, namely constant-temperature heating and temperature-programmed heating (sometimes called thermolysis–mass spectrometry,

Table 6.39 Thermal stress of various analysis methods

Technique	Hot zone	Sample size	Residence time
DP-MS	Ion source	µg	Milliseconds
(TP)PyMS	Expansion chamber	µg	Seconds
GC-MS	Column	100 µg	Seconds
TG-MS	Heated transfer line	2 mg	Seconds
TVA	Heated transfer line	25 mg	Seconds

T-MS). The main direct mass-spectral methods are thermal desorption and pyrolysis mass spectrometry. Several factors favour the efficiency at which volatiles can be removed from a polymeric matrix:

(i) Higher temperatures are more effective: typically 100–300 °C for desorption of volatiles from polymers.
(ii) Heating of the sample close to the decomposition temperature of the polymer.
(iii) Fast removal of desorbed species by heating *in vacuo* or by use of a continuous-flow device.
(iv) Small particle size.

The sensitivity needed for identification of additives is, in general, greater than that needed for identification of volatile pyrolysates.

Thermal desorption (TD) techniques are those where additives are detected at temperatures below the decomposition temperature of the polymer. It is crucial to heat the polymer only to the temperature necessary to vaporise the materials of interest. With instruments that allow the direct probe to be heated independently of the ion source, some degree of fractionation is possible by temperature programming. The advantage of heating a polymer directly inside a mass spectrometry source is that there is little time delay between evolution of pyrolysis products and analysis. Various forms of temperature-programmed heating are TD-MS, TG-MS, vacuum TG-MS [244] or TA-MS in a continuous-flow apparatus [245]. Alternatively, thermal desorption is achieved by heating the sample in a closed vessel held at a constant temperature [240,246]. The atmosphere in which heating takes place can often not easily be varied.

Thermal-programmed solid insertion probe mass spectrometry (TP-SIP-MS) has been proposed [247,248], in which the solid insertion probe consisting of a water-cooled microfurnace enters the mass spectrometer via an airlock. The sample is contained in a small Pyrex tube (i.d. 1 mm, length 20 mm). The TIC trace gives a characteristic evolved gas profile for each compound in a mixture of materials, and the mass spectra associated with each TIC peak give a positive identification of that component as it is vaporised. TP-SIP-MS is appropriate for analysis of small solid particles which are volatile, or produce volatile decomposition products. The technique is a form of evolved gas analysis.

Many of the aforementioned techniques are not appropriate to direct mass-spectrometric analyses of intact high-MW and heat-labile compounds. For such samples, *thermal degradation techniques* (analytical pyrolysis) can be performed to generate more-volatile compounds of lower molecular weight that are amenable

to mass-spectrometric analysis. Pyrolysis techniques are flash-pyrolysis and temperature-programmed pyrolysis mass spectrometry. TPPyMS proves useful in separating organic additives for easier identification. Although direct probes are standard equipment for most mass spectrometers, such devices have found limited use for pyrolysis purposes. In *direct pyrolysis mass spectrometry* (DPyMS) samples are introduced near the ion source of the mass spectrometer via the direct insertion probe for solid materials, and heated at 10 °C min⁻¹ up to 700 °C. Low-MW oligomers and additives contained in polymer samples are evolved at lower temperature with respect to degradation compounds. High-vacuum pyrolysis compounds are quickly removed from the hot reaction zone and ionised. As in DPyMS [249], polymer pyrolysis is achieved very close to the ion source and no transport problems exist, fragments of relatively high mass (up to about 1000–2000 Da) can be detected. This aspect is crucial in defining structures, and is a considerable advantage with respect to other pyrolysis techniques. DPyMS provides finer details on the structure of the primary pyrolysis products, and has become an essential tool in the study of the thermal degradation of polymers [250]. Nevertheless, direct probes have found limited use for pyrolysis purposes for a number of reasons:

(i) possible occurrence of secondary reactions, due to the relatively long residence time of the pyrolysate in the direct-probe cups;
(ii) poor reproducibility compared to conventional pyrolysis;
(iii) relatively low maximum pyrolysis temperature (600 °C); and
(iv) relatively low heating rates.

Direct temperature-resolved pyrolysis-mass spectrometry (DT-MS or TPPyMS) is a fast fingerprint technique for analysis of a broad range of low-MW and polymeric organic materials, as well as inorganic materials. DT-MS is essentially an indirect analysis method, as the molecules are thermally degraded before FTICR-MS analysis. Direct temperature-resolved high-resolution mass spectrometry (DT-HRMS) on sector instruments requires the continuous presence of a calibrant during measurement to assign the mass scale. No internal mass-reference compounds are needed for FTICR-MS. Thermal desorption (TD)-ToFMS for monitoring the thermal decomposition of solid samples placed in the ion source region uses the ability of ToF-MS to give simultaneous multicomponent analysis at high sensitivities [251]. While a thermal separation operates in TPPy-MS, in

TD/Py-REMPI-ToFMS an electronic separation between the components of a mixture is achieved.

Applications Direct analysis of *polymer extracts* by mass spectrometry is a challenging and convenient means for identification of organic additives. Analysis is rapid and generally very specific. Complex mixtures of additives can be examined directly, particularly with a choice of ionisation modes (EI, CI, FD, FAB, etc.). Mass-spectral methods can be used very effectively to complement and extend the results of traditional chromatographic and spectroscopic methods of organic additive analysis. Electron impact (EI-MS) spectra are often difficult to interpret due to the high concentration of processing oils in the extracts and extensive fragmentation of the molecular ions [252]. LVEI-MS has been used for the analysis of PE extracts. Direct MS analysis of polymer extracts has been accomplished using FAB [83,98], LD [83,253], FD [22] and CI [22]. Many examples were given in the previous sections of this chapter.

Analysis of polymer additives by mass spectrometry has, for the most part, been limited to molecular weight determination of the solvent-extracted components [4,254]. Field desorption is a good ionisation technique for generating molecular weight information about polymer additives, lacking the matrix effects that are observed in LSIMS and MALDI spectra [96]. If unknowns are present, structural information is required for identification. This can be resolved by combining soft ionisation techniques with tandem mass spectrometry (MS/MS). The few collision-induced dissociation (CID) experiments reported for organic polymer additives have been generally performed in the low-energy regime (10–200 eV) [23,37,137,228,229]. Precursor-ion scanning techniques have also been utilised for the structural characterisation of polymer additives under low-energy conditions [100,137]. The use of high-energy CID experiments gives fragment ions that are characteristic of the class of polymer additives [69,96]. The difficulties of performing precursor-ion scanning in four-sector instruments have limited the number of applications of these techniques in the high-energy regime. Both precursor-ion [96,255] and MS^3 spectra under high-energy conditions [256] have been reported.

An effective means to facilitate the mass-spectral analysis of rubber acetone extracts is to use desorption/ionisation techniques, such as FD [92,113] and FAB [92]. FAB mass spectra for *rubber extracts* are generally more complex (due to fragment ions) than FD spectra of the same materials. Nevertheless, the FAB spectra are often complementary to FD, since:

(i) the fragment-ions may provide structural information; and
(ii) FAB is generally a more facile ionisation technique for more polar organic additives.

In an acetone extract from a neoprene/SBR hose compound, Lattimer *et al.* [92] distinguished dioctylphthalate (*m/z* 390), di(*t*-octyl)diphenylamine (*m/z* 393), 1,3,5-tris(3,5-di-*t*-butyl-4-hydroxybenzyl)-isocyanurate (*m/z* 783), hydrocarbon oil and a paraffin wax (numerous molecular ions in the *m/z* range of 200–500) by means of FD-MS. Since cross-linked rubbers are insoluble, more complex extraction procedures must be carried out (Chapter 2). The method of Dinsmore and Smith [257], or a modification thereof, is normally used. Mass spectrometry (and other analytical techniques) is then used to characterise the various rubber fractions. The mass-spectral identification of numerous antioxidants (hindered phenols and aromatic amines, e.g. phenyl-β-naphthyl-amine, 6-dodecyl-2,2,4-trimethyl-1,2-dihydroquinoline, butylated bisphenol-A, HPPD, poly-TMDQ, di-(*t*-octyl)diphenylamine) in rubber extracts by means of direct probe EI-MS with programmed heating, has been reported [252]. The main problem reported consisted of the numerous ions arising from hydrocarbon oil in the recipe. In older work, mass spectrometry has been used to qualitatively identify volatile AOs in sheet samples of SBR and rubber-type vulcanisates after extraction of the polymer with acetone [51,246].

The ability of a soft ionisation technique (CI) and collisional activation (MS/MS) to derive chemical information from a complex sample has been illustrated by ref. [258] for the case of dissolution (HCCl$_3$)/precipitation (CH$_3$OH) of HIPS followed by direct probe mass spectrometry of the dried extract. The presence of nonylphenol stabiliser in the plastic was ascertained. Pierre and van Bree [243] used temperature-programmed DIP-MS for identification and quantitative analysis of *antioxidants* in rubbers and plasticisers in polymers, such as α-phenyl-β-naphthylamine, and 1,3-dimethylpentyl-*p*-phenylenediamine. Following up dissolution/precipitation of ABS residual styrene monomers, styrene-acrylonitrile oligomers and antioxidants such as DLTDP were detected. In another ABS compound, nonylphenol (as a decomposition product of trisnonylphenolphosphite) and Cyanox 425 (both antioxidants), the plasticisers DBP and DOP, and 2-phenyl-2-propanol (indicating the use of a cumolhydroperoxide catalyst) were detected. In another study, industrial *coatings* were extracted with dichloromethane for analysis of a common HALS by direct FAB-MS

[98]. The UV light stabiliser bis-(2,2,6,6-tetramethyl-4-piperidinyl)sebacate could be determined quantitatively in the extracts by use of either an isotopically labelled or a nonlabelled internal standard.

Bonilla [55] has used the particle-beam interface for direct qualitative analyses of a number of neat additives (AOs: BHT, Irganox 245/1010/1076, Ronotec 201, Weston TNPP, Ultranox 626, Irgafos 168; UVAs: Tinuvin 234/237/326/328/770/5411; FRs: organic phosphate esters; slip agent: PETS; oligomers: BPA-DA, BPA-BI), additive mixtures, and PC and PC/PBT extracts (injected without preconcentration). Organic molecules with MW > 1000 Da can be analysed by this approach (e.g. PETS, MW >1200 Da; Irganox 1010, MW 1176 Da).

Thermolysis–mass spectrometry is ideal for examining the amount of residual monomer and *processing solvents* present in polymers. In thermolysis, the polymer is heated from room temperature to 200–300 °C, and is then often held isothermally in order to drive off volatile components. Low-temperature pyrolysis (350–400 °C) of PP compounds in direct mass-spectral analysis has shown *volatiles* from PP at every carbon number to masses well above 1000 Da [37].

Direct *solid-state polymer/additive mass analysis* has involved various ionisation modes: EI (Section 6.2.1), CI (Section 6.2.2), DCI (Section 6.2.2.1), FAB (Section 6.2.4), FI (Section 6.2.5), FD (Section 6.2.6) and LD. Survey mass spectra obtained with soft ionisation methods (FI-MS, CI-MS) provide diagnostic overviews of chemical composition. The supplemental tandem (MS/MS) and atomic composition (AC-MS) techniques are used to make specific identifications of various organic ingredients. Direct analysis of polymer systems for more than a few thousand daltons has only just begun. Ionisation methods employed are FD, ESI and MALDI. Solid-probe ToF-MS (or DI-HRMS) is a breakthrough [188].

Yoshikawa *et al.* [240] have reported identification of a variety of commercial additives (BHT, DLTDP, Irganox 1010/1076, Topanol CA, AO 2246, Santowhite, Tinuvin 327, erucamide and oleamide) in PP pellets without prior extraction and further separation using direct heating (from 250 to 350 °C) near the inlet system of a mass spectrometer (EI mode). Up to four additives in commercial samples (0.2 g) were observed by this direct evaporation method (up to 280 °C); pyrolytic products of the base polymer excluded additive analysis below m/z 200. Similarly, antioxidants and *peroxides* in PE and plasticisers and other additives in PVC have been identified by heating the polymers at a temperature high enough to volatilise the additives but below the point at which gaseous polymer degradation products

formed, followed by EI-MS analysis of the gases [259]. DI-FIMS was used to analyse the structures of the oxidation products of the diphosphite Doverphos S-9228 and monophosphite Doverphos 4-HP [109]. TP-SIP-MS was used to identify an Irganox MD 1024 impurity which had caused product failure [248].

DI-MS can be used for temperature-programmed analysis of concentrated Soxhlet extracts or direct solid sampling. With increasing temperature, components with lower volatility are detected. For calibration purposes, test samples of known concentrations may be used. Piringer [260] has shown a linear calibration curve for Irgafos 168 (above 10 ng) in DI-MS analysis. Direct insertion probe (DIP)-MS with a linear temperature programme of 0.5 °C s^{-1} from ambient temperature to 400 °C in the high vacuum of the ion source was used by Bücherl *et al.* [261] for the analysis of additives in *packaging materials*. Direct insertion probe mass spectra of some 100 additives in food packaging have been published [262]. DIP-MS analysis of the *antistatic agent* N, N-bis-(2-hydroxyethyl) alkylamine at 100 °C in the high vacuum of the ion source indicates dominance of saturated C_{14}, C_{16}, C_{18} chains and unsaturated C_{18} chains. More precise characterisation is possible with SFC-MS.

Lay and Miller [87] have reported on screening of PVC samples (in baby pacifiers) for high levels of phthalate *plasticisers*, by direct introduction of the solid material into the mass spectrometer. Quantitation of DEHP was based on the relative signal of the [MH]$^{+}$ ions of DEHP, and didecylphthalate (DDP) as an internal reference. Hayes and Altenau [51] first reported the use of mass spectrometry to directly characterise an SBR sample by observing the antioxidant and *processing oil additives* that were thermally desorbed. Thermal desorption followed by soft ionisation single-stage mass-spectrometric methods allows identification of the various ingredients in elastomers. Lattimer [65] has described the use of TD/EI-MS, TD/(CI, FI-MS), TD/tandem MS and TD/PyMS, using a hybrid mass spectrometer of BEoQ geometry. TD/EI-MS of rubbers is usually hampered by extensive EI fragmentation and intense signals from the processing oil. Direct insertion analysis of vulcanised rubber compounds does not allow for the selective separation of process oils or volatile fragments, and further produces a complex spectrum containing ion fragmentation of the entire compound. Tandem mass spectrometry (MS/MS) is effective for increasing the specificity and sensitivity of detection and identification of additives in direct rubber compound analysis [228,229]. Limitations on the small sample size render difficult detection of the typical low concentrations of volatile additives in conventional rubber compounds. A further problem with the direct insertion method relates to inefficient extraction and isolation of the organic additives, primarily due to carbon-black matrices.

Lattimer *et al.* [22] have reported mass-spectral identification of organic additives in *rubber vulcanisates* of known composition, using direct thermal desorption with three different methods of ionisation (EI, CI, FI). Also three vulcanisates were examined by FAB-MS (without solvent) as a means for surface desorption/ionisation. Acetone, acetonitrile and dichloromethane extracts were examined by EI, CI, FD and FAB-MS. The results, summarised in Table 6.40, show that organic additives can be rapidly identified by various direct mass-spectral approaches. FI/FD is a most efficient vaporisation/ionisation method for identifying typical organic additives in rubber vulcanisates. The analysis can be carried out by using either the untreated rubber (FI mode), or else a solvent extract (FD mode). For unknown samples in which the identity of FD/FI molecular ions is unclear, further analysis using other methods (EI, CI, FAB, MS/MS, GC-MS, LC-MS, high-resolution) can be carried out to gain additional information. For the analyses reported in Table 6.40, no clear advantage was observed for direct analysis as compared to extract analysis. Both approaches gave quite comparable results. For more detailed analysis schemes, however, there is no question that extraction methods are more versatile.

Analysis of solid matter, in particular of rubber vulcanisates, is a classical application of DP-MS [263].

Table 6.40 Optimal methods for detection of organic additives in carbon-black-filled rubber vulcanisates

Component	Direct analysis				Extract analysis			
	EI	CI	FI	FAB	EI	CI	FD	FAB
HPPD	x	x	x	x	x	x	x	x
DOPPD	x	x	x	x	x	x	x	x
DODPA	x	x	x	x	x	x	x	x
Poly-TMDQ	x	x	x	x	x	x	x	x
AO 425	x	x	x	x	x	x	x	x
Stearic acid	x	x	x					
PF resin	x				x			
OBTS		x						
DPG								
TMTD		x						
Processing oil		x					x	
Paraffin wax							x	

Direct insertion of rubber vulcanisate into the ion source of the mass spectrometer is a very fast analysis technique for identification of volatile additives and antioxidants. As already mentioned, a drawback is the interference of numerous intense fragment ions from the hydrocarbon oil in the rubber recipe: numerous peaks due to the oil complicate the spectra of oil masterbatches. Alternatively, the vulcanisate is isothermally heated in a closed vessel and the evolved vapours are introduced into the mass spectrometer ion source via a heated molecular leak. For example, 2-mercaptobenzothiazole sulfenamide accelerator fragments from rubber vulcanisates were detected by closed-vessel heating and introduction into the mass spectrometer via a molecular leak, i.e. by means of DI-EIMS analysis [246]. Lattimer and Harris [4] have examined a triblend rubber vulcanisate/(sulfenamide accelerator, HPPD, poly-TMDQ, palmitic/stearic acid, processing oil) system by means of TD-CI-MS and FI-MS. The accelerator zinc dithiocarbamate (ZDMC) can be unambiguously identified in a vulcanised NR, on the basis of the peak spectrum of the mass trace m/z 304. ZDMC cannot be detected by flash PyGC-MS analysis in the unfragmented state, because of its low thermal stability.

As indicated in Section 6.2.2, DI-CIMS suffers from poor reproducibility. For nonvolatile additives that do not evaporate up to 350 °C, direct quantitative analysis by thermal desorption is not possible. The method depends on polymer formulation standards that are reliably mixed. Wilcken and Geissler [264] described rapid quality control of 1-μg paint samples by means of temperature-programmable DI-EIMS with PCA evaluation.

Direct pyrolysis in the ion source of a mass spectrometer (QMS) was used to analyse PE/(dicumylperoxide, Santonox R) and PVC/DIOP [259]. In-source PyMS is an analytical tool for fast analysis of *flame retardants* in unknown mixtures of polymers [223, 265]. Heeren and Boon [224] used in-source filament pyrolysis FTMS for high-speed, broadband screening of additives in polymeric household appliances.

Also, direct determination of additives by means of laser desorption in solid polymeric materials rather than in polymer extracts has been reported [266]. Takayama *et al.* [267] have described the direct detection of additives on the surface of LLDPE/(Chimassorb 944 LD and Irgafos P-EPQ) after matrix (THAP)-coating. As shown in Scheme 7.13, direct inlet mass spectrometry is also applicable to transfer TLC-MS and TLC-MS/MS analyses without the need for prior analysis. For direct sample introduction a small amount of the selected sample spots on the TLC plate is scratched off and placed on the tile of the direct-inlet sample probe [268].

Relatively few descriptions of direct mass spectral analysis of plastics *compounds* have appeared in the literature [22,37,63,240,243]. Additives in PP were thermally desorbed into a heated reservoir inlet for 80 eV EI-MS analysis [240]. Analysis of additives in PP compounds via direct thermal desorption ammonia CI-MS has been described [269] and direct mass spectrometric oligomer analysis has been reported [21].

Lattimer [37] analysed five PP compounds (Table 6.41) directly in the solid state by programmed TD

Table 6.41 Compositions of polypropylene compounds

Polymer/additive	A	B	C	D	E
Himont 6301[a]	100	100	100	100	100
Irganox 3114	0.10	0.10	0.10	0.10	0.10
Ultranox 626	0.08	0.08	0.08	0.08	0.08
Calcium stearate[b]	0.10	0.10	0.10	0.10	0.10
Tinuvin 770	0.40	0	0	0	0
Tinuvin 144	0	0.40	0	0	0
GoodRite 3150	0	0	0.40	0	0
Tinuvin 622	0	0	0	0.40	0
Chimassorb 944	0	0	0	0	0.40

[a] Polypropylene; contains Irganox 1076 antioxidant. [b] Commercial 'calcium stearate' contains both stearate and palmitate. After Lattimer [37]. Reprinted from *Journal of Analytical and Applied Pyrolysis*, Vol. 26, R. P. Lattimer, pp. 65–92, Copyright (1993), with permission from Elsevier.

(20–300 °C)/PyMS (300–400 °C) using single-stage 70 eV EI, isobutane CI, FI-MS and tandem MS with a hybrid mass spectrometer of BEoQ geometry (resolution ca. 5000; mass accuracy 3 ppm). The general analytical approach to analysis consisted of the use of FI-MS as a screening technique to obtain an overview of the chemical composition of the evolved chemicals, i.e. to determine the number of components and their molecular weights. Supplemental techniques such as MS/MS and AC-MS were then used, in EI and/or CI mode, to elucidate or verify the chemical structures of the various components. FI-MS of the base polymer (Himont 6301) revealed a volatile chemical of MW 530, which an EI-MS/MS product-ion scan identified as Irganox 1076 (added by the manufacturer). Figure 6.2 shows an EI-MS/MS product-ion scan of $M^{+\bullet}$ 783 (Irganox 3114) and Figure 6.24 the CI-MS/MS product-ion scan of MH^+ 481 (Tinuvin 770).

The oligomeric additives Tinuvin 622 and Chimassorb 944 could also be detected by analysis of specific oligomeric components or volatile pyrolysates. In order to determine the nature of the volatile components of Tinuvin 622, the stabiliser was first examined directly by FD-MS. Various oligomer series (MW $283n + 32$; MW $283n + 146$; MW $283n + 201$; MW $283n + 18$; MW $283n + 132$) were observed. PyFI-MS (320–400 °C) revealed other oligomeric series (MW $283n + 183$; MW $283n + 165$; MW $283n$). Probable chemical structures, based on CI-MS/MS and AC-MS,

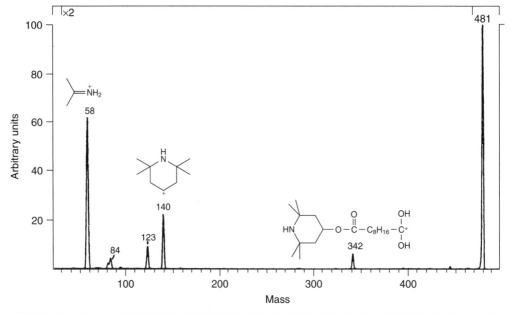

Figure 6.24 Product-ion scan (CI-MS/MS) of MH^+ 481 (Tinuvin 770). After Lattimer [37]. Reprinted from *Journal of Analytical and Applied Pyrolysis*, **26**, R. P. Lattimer, 65–92, Copyright (1993), with permission from Elsevier

were reported. Chimassorb 944 was analysed using FD-MS (Fig. 6.9), CI-MS and CI-MS/MS.

Direct and indirect analysis of polymers and additives using mass spectrometry has been reviewed [4,270]. Kreiner [271] has described accelerator analysis.

6.5   ION MOBILITY SPECTROMETRY

Principles and Characteristics   Ion mobility spectrometry (IMS) is an instrumental technique for the detection and characterisation of organic compounds as vapours at atmospheric pressure. Modern analytical IMS was created at the end of the 1960s from studies on ion–molecule chemistry with mass spectrometers and from ionisation detectors for vapour monitoring. An ion mobility spectrometer (or plasma chromatograph in the original termininology) was first produced in 1970 [272].

Ion mobility spectrometry comprises and is governed by two separate processes:

(i) gas-phase ionisation in air at atmospheric pressure through collisional charge exchanges or ion–molecule reactions; and

(ii) ion characterisation using mobilities of gas-phase ions in a weak electric field.

Ion mobility spectrometry, which is another type of mass analysis, is thus not based on an electromagnetic separation, but rather stands for characterising chemical substances through measurement of gas-phase ion mobilities (effectively size to charge) rather than mass-to-charge ratios (m/z). In practice, a vapour sample is introduced into the reaction region of a drift tube where molecules undergo ionisation. A gas stream counteracts ions moving slowly between the positive and negative ends of the charged chamber. Mobility is determined from the drift velocity attained by ions in the electrostatic field of the drift region at atmospheric pressure. Ion velocities are inversely dependent upon the effective collisional cross-section of an ion, and this makes IMS a kind of molecular size analyser. Phenomena of ion transport in gases have been described [273].

An *ion mobility spectrometer* consists of a sample-introduction device; a drift tube where ionisation and separation of ions takes place; and a detector. Ionisation sources of choice include radioactive sources (e.g. a ^{63}Ni foil), photoionisation methods, corona-spray ionisation, flame ionisation and corona discharge. The most common detection method used to measure the ion current in the IMS drift tube is a circular collector plate that serves as a Faraday cup.

Vapours from nearly all chemical classes or functional groups can be ionised by proton-transfer reactions, ion transfer, charge transfer or dissociative charge transfer. The reagent gas controls the ion chemistry and product-ion formation. As positive ions in IMS are formed through proton-transfer reactions, a reagent gas is used with a proton affinity (PA) slightly below that of the target compound. For example, i-propylamine (PA = 218.6 kcal mole^{-1}) with ethylamine (PA = 217.0 kcal mole^{-1}) as a reagent gas excludes interferences from all compounds with PA <217.0 kcal mole^{-1}. The ions that are formed through APCI processes are generally robust, long-lived, and low-energy species. The ionisation chemistry in the formation of product ions and in reactions with matrices is complex. The special nature and the relative concentration of the product ions depends on the relative gas-phase acidities/basicities and electron/proton affinities of the sample components [274]. Depending on the polarity of the electric field, positive or negative ions can be observed. Product-ion clouds produce current pulses with amplitude proportional to the number of ions, thus yielding chromatogram-like spectra in time. Drift times are typically several milliseconds; single spectra are taken in 10 to 50 ms. Identification of the ions that produce the mobility spectra may be based exclusively on the ion mobilities. More usually, in commercial instruments IMS is coupled to a quadrupole mass spectrometer as a second source of information [275]. There have also been numerous examples of APPI coupled with ion mobility spectrometry (IMS-PID) [276]. Laboratory instruments, field instruments and hand-held instruments are available.

Table 6.42 lists the *main features* of IMS. Detection limits are impressive in IMS as a result of the ion density in air at atmospheric pressure. The response of an ion mobility spectrometer to vapours is concentration dependent, and is quantitative – with minimum detectable levels generally around 10 to 100 pg and linear ranges of about 10^2. The advantages of the atmospheric pressure ionisation in an ion–molecule-reactor are an extremely high ionisation yield and usually no fragmentation. IMS provides molecular data. As IMS does not involve high-vacuum conditions, even highly volatile compounds are easily detected, which in UHV methods (e.g. AES, XPS, etc.) are evaporated before the measurement. IMS is a relatively new technique for the analysis of volatile organic compounds at ultratrace levels with detection limits in the low to sub-picogram range. Ultratrace analytical methods for detection and identification of

Table 6.42   Main characteristics of ion mobility spectrometry

Advantages

- No sample preparation
- Atmospheric pressure chemical ionisation process
- Excellent detection limits (low to sub-pg sensitivity)
- Reasonable selectivity
- Ultratrace molecular identification
- Quantitative behaviour (but limited linear range)
- Simple and reliable instrumentation
- Relatively low cost
- Miniaturisation (palm-size; portability)
- Wide applicability

Disadvantages

- Volatile analysis only
- Limited interpretative and predictive capabilities
- Limited commercial availability
- Poor reproducibility and precision (RSD 5–25 %)
- Limited resolution
- Poor linearity, limited dynamic range
- Matrix effects
- Restricted mixture analysis
- Technical and conceptual foundations not comprehensive
- Extensive method development
- Not a widespread analytical tool

organic compounds are rare. Accurate quantitative determinations from IMS measurements require knowledge about all different organic components of the sample. The current level of understanding and incomplete models for ionisation chemistry in the IMS source fundamentally weaken IMS technology, since interpretive and predictive capabilities are limited.

The IMS response for a compound is strongly dependent on temperature, pressure, analyte concentration/vapour pressure, and proton affinity (or electron/reagent affinity). Pressure mainly affects the drift time, and spectral profiles are governed by concentration and ionisation properties of the analyte. Complex interactions among analytes in a mixture can yield an ambiguous number of peaks (less, equal to, or greater than the number of analytes) with unpredictable relative intensities. IMS is vulnerable to either matrix or sample complexity.

Development of an industrial monitoring application for IMS requires extensive preparatory work, as well as optimal operational conditions for IMS, i.e. the nature of the reagent gas, calibration curves, evaluation of interferents, assessment of reliability. Analysis of mixtures with four or fewer components may be possible, but extension to more complex mixtures should be considered only in special cases, and generally would be unrealistic. Use of preseparators, such as GC columns, is the only known technical approach

to characterisation of large numbers of constituents in complex mixtures.

Ionisation processes in IMS occur in the gas phase through chemical reactions between sample molecules and a reservoir of reactive ions, i.e. the reactant ions. Formation of product ions in IMS bears resemblance to the chemistry in both APCI-MS and ECD technologies. Much yet needs to be learned about the kinetics of proton transfers and the structures of protonated gas-phase ions. Parallels have been drawn between IMS and CI-MS [277]. However, there are essential differences in ion identities between IMS, APCI-MS and CI-MS (see ref. [278]). The limited availability of IMS-MS (or IMMS) instruments during the last 35 years has impeded development of a comprehensive model for APCI. At the present time, the underlying basis of APCI and other ion–molecule events that occur in IMS remains vague. Rival techniques are MS and GC-MS. There are vast differences in the principles of ion separation in MS versus IMS.

Today, there are still only a few data available in the literature for IMS. A compilation of available ion mobilities can be found in ref. [279]. Ion mobility spectrometry has been reviewed [280–282a], and a few monographs are available [278,283].

Applications   Ion mobility spectrometry has found application for military, industrial and forensic purposes. In particular, IMS is used as:

(i) a selective sensor in which the chemistry of ionisation can be tailored for detection of a specific compound or class of compounds;
(ii) a more general analyser for samples of predictable composition; and
(iii) a detector for chromatography.

The ion mobility spectrometer may be viewed as a new generation of *vapour sensor*, or essentially a remote-sensor field monitor (military-grade analytical instrument). A stand-alone IMS can serve as an alarm for toxic gases or vapours. IMS has been used for vapour sensing (threshold sensors), including manufacturing and industrial hygiene [284]. IMS has also been applied for long-term monitoring of toxic vapours in a chemical plant [284]. An (undisclosed) toxic chemical was routinely monitored in a chemical production plant at the 5 ppb-level for over 10 years [284]. Other monitoring applications, including airport security, are described elsewhere [278,285]. Hand-held IMS is capable of detecting TDI, HF, Br_2 and other toxic vapours. A miniature IMS detector with a pulsed corona ionisation source has been described [286].

IMS can be used for chemical analysis of vapours from *electronics packaging* [287]. IMS-QMS has been used to analyse headspace vapours in sealed electronic packages [275,288] and to follow outgassing of polymers [287]. Various types of photoresist solvents, phthalate plasticisers and other polymer additives, such as BHT, were detected. Other applications of IMS in semiconductor technology involve failure analysis; control of the efficiency of cleaning and etching steps; characterisation of process media; and surveillance of the atmosphere of clean rooms.

Hand-held IMS has determined 5 ng to 3.5 μg of dialkylphthalates (DMP, DEP) with an RSD of 6 to 27 % [289]. Only a few attempts were made to apply IMS quantitatively.

IMS is a fast and reasonably accurate way to identify explosives (e.g. TNT, PETN and RDX), as the nitro functional groups in these compounds are electron withdrawing; consequently, the compounds are easily ionised [285]. NQR is another tool for the same purpose. IMS is also applied for the detection of narcotics and warfare gases.

The low numbers of routine applications of IMS have been attributed to the ease of sample overload and to the complexity and uncertainty of ion mobility spectra.

6.6 BIBLIOGRAPHY

6.6.1 Mass Spectrometry (General)

G. Montaudo and R.P. Lattimer (eds), *Mass Spectrometry of Polymers*, CRC Press, Boca Raton, FL (2002).

M.L. Gross and R. Caprioli (eds), *The Encyclopedia of Mass Spectrometry*, Elsevier, Amsterdam (2000–2002), 10 vols.

E. De Hoffmann and V. Stroobant, *Mass Spectrometry: Principles and Applications*, John Wiley & Sons, Ltd, Chichester (2001).

O.D. Sparkman, *Mass Spectrometry Desk Reference*, Global View Publishing, Pittsburgh, PA (2000).

J.C. Lindon, G.E. Tranter and J.L. Holmes (eds), *Encyclopedia of Spectroscopy and Spectrometry*, Academic Press, San Diego, CA (2000).

R. Willoughby and E. Sheehan, *A Global View of MS/MS*, Global View Publishing, Pittsburgh, PA (1999).

J. Barker, *Mass Spectrometry*, John Wiley & Sons, Ltd, Chichester (1998).

K.L. Busch and T.A. Lehman (eds), *Guide to Mass Spectrometry*, VCH Publishers, New York, NY (1997).

R.A.W. Johnston and M.E. Rose, *Mass Spectrometry for Chemists and Biochemists*, Cambridge University Press, Cambridge (1996).

J.R. Chapman (ed.), *Practical Organic Mass Spectrometry*, John Wiley & Sons, Ltd, Chichester (1994).

F.W. McLafferty and F. Tureček, *Interpretation of Mass Spectra*, American Chemical Society, Washington, DC (1993).

H. Budzikiewicz, *Massenspektrometrie. Eine Einführung*, VCH, Weinheim (1992).

F.W. McLafferty and D.B. Stauffer, *Wiley/NBS Registry of Mass Spectral Data*, John Wiley & Sons, Ltd, Chichester (1989).

6.6.2 Mass Spectrometers

R.J. Cotter, *Time-of-Flight Mass Spectrometry. Instrumentation and Applications in Biological Research*, American Chemical Society, Washington, DC (1997).

R.E. March and J.F.J. Todd (eds), *Practical Aspects of Ion Trap Mass Spectrometry*, Vol. I–III, CRC Press, Boca Raton, FL (1995).

P.H. Dawson (ed.), *Quadrupole Mass Spectrometry and Its Applications*, AIP Press, New York, NY (1995).

E.W. Schlag (ed.), *Time-of-Flight Mass Spectrometry and Its Applications*, Elsevier, Amsterdam (1994).

R.J. Cotter (ed.), *Time-of-Flight Mass Spectrometry*, American Chemical Society, Washington, DC (1994).

B. Asamoto (ed.), *FT-ICR/MS: Applications of Fourier Transform Ion Cyclotron Resonance Mass Spectrometry*, VCH Publishers, New York, NY (1991).

A.G. Marshall and F.R. Verdun, *Fourier Transforms in NMR, Optical and Mass Spectrometry: A User's Handbook*, Elsevier Science, Amsterdam (1990).

R.E. March, R.J. Hughes and J.F.J. Todd, *Quadrupole Storage Mass Spectrometry*, Wiley-Interscience, New York, NY (1989).

M.V. Buchanan (ed.), *Fourier Transform Mass Spectrometry: Evolution, Innovation, and Applications*, ACS Symposium Series No. 359, American Chemical Society, Washington, DC (1987).

6.6.3 Ionisation Modes

R.B. Cole (ed.), *Electrospray Ionization Mass Spectrometry: Fundamentals, Instrumentation and Applications*, Wiley-Interscience, New York, NY (1997).

A.E. Ashcroft, *Ionization Methods in Organic Mass Spectrometry*, The Royal Society of Chemistry, Cambridge (1997).

A.G. Harrison (ed.), *Chemical Ionization Mass Spectrometry*, CRC Press, Boca Raton, FL (1992).

R.M. Caprioli (ed.), *Continuous-Flow Fast Atom Bombardment Mass Spectrometry*, John Wiley & Sons, Ltd, Chichester (1990).

L. Prokai, *Field Desorption Mass Spectrometry*, M. Dekker, New York, NY (1990).

P.A. Lyon (ed.), *Desorption Mass Spectrometry*, American Chemical Society, Washington, DC (1985).

6.7   REFERENCES

1. K. Saito, H. Shimizu, Y. Ishida and T. Ooshima, *Bunseki Kagaku* **45** (6), 505–10 (1996).
2. W. Freitag, in *Plastics Additives* (R. Gächter and H. Müller, eds), Hanser Publishers, Munich (1990), Chapter 20.
3. P. Bataillard, L. Evangelista and M. Thomas, in *Plastics Additives Handbook* (H. Zweifel, ed.), C. Hanser Verlag, Munich (2000), pp. 1047–83.
4. R.S. Lattimer and R.E. Harris, *Mass Spectrom. Rev.* **4**, 369–90 (1985).
5. D.O. Hummel, *Atlas of Plastics Additives: Analysis by Spectrometric Methods*, Springer-Verlag, Berlin (2002).
6. J.-P. Hsu, *VG Monographs in Mass Spectrometry* No. 7, VG Organic, Manchester, UK (1993).
7. R.A.W. Johnston and M.E. Rose, *Mass Spectrometry for Chemists and Biochemists*, Cambridge University Press, Cambridge, UK (1996).
8. J.R. Chapman, *Practical Organic Mass Spectrometry*, John Wiley & Sons, Ltd, Chichester (1993).
9. E. Constatin, A. Schnell and M. Thompson, *Mass Spectrometry*, Prentice-Hall, Englewood Cliffs, NJ (1990).
10. F.W. McLafferty and A. Tureček, *Interpretation of Mass Spectra*, University Scientific Books, Mill Valley, CA (1993).
11. E. De Hoffmann and V. Stroobant, *Mass Spectrometry: Principles and Applications*, John Wiley & Sons, Ltd, Chichester (2001).
12. R.G. Cooks, V. Grill and H. Bui, in *Advances in Mass Spectrometry* (E.J. Karjalainen, A.E. Hesso, J.E. Jalonen and U.P. Karjalainen, eds), Elsevier Science, Amsterdam (1998), Vol. 14, pp. 3–39.
13. A.K. Lee and R.D. Sedgwick, *J. Polym. Sci., Polym. Chem. Ed.* **16**, 685 (1978).
14. H.R. Udseth and L. Friedman, *Anal. Chem.* **53**, 29 (1977).
15. B.T. Chait, J. Shpungin and F.H. Field, *Intl J. Mass Spectrom. Ion Proc.* **58**, 121 (1984).
16. M. Dole, C.V. Gupta, L.L. Mack and K. Nakamae, *Polym. Prepr. Am. Chem. Soc., Div. Polym. Chem.* **18** (2), 188 (1977).
17. R.P. Lattimer, *Intl J. Mass Spectrom. Ion Proc.* **551**, 221 (1983/84).
18. R.P. Lattimer, D.J. Harmon and G.E. Hansen, *Anal. Chem.* **52**, 1808 (1980).
19. K.W.S. Chan and K.D. Cook, *Macromolecules* **16**, 1736–40 (1983).
20. R.S. Brown, D.A. Weil and C.L. Wilkins, *Macromolecules* **19**, 1255 (1986).
21. R.P. Lattimer, R.E. Harris and H.-R. Schulten, in *Determination of Molecular Weight* (A.R. Cooper, ed.), John Wiley & Sons, Ltd, Chichester (1989), pp. 391–412.
22. R.P. Lattimer, R.E. Harris, C.K. Rhee and H.-R. Schulten, *Anal. Chem.* **58**, 3188–95 (1986).
23. S.W. Chen and G.R. Her, *Appl. Spectrosc.* **47** (6), 844–51 (1993).
24. S.R. Heller, *Today's Chemist at Work* **8** (2), 45–50 (1999); cf. also NIST O2, National Institute of Standards and Technology, Gaithersburg, MD (2002).
24a. S.E. Stein, *J. Am. Soc. Mass Spectrom.* **5** (4), 316–23 (1994).
25. R. Jennissen, *Techn. Rept DSM Research*, Geleen (1998).
26. K. Varmuza, W. Werther, D. Henneberg and B. Weimann, *Rapid Commun. Mass Spectrom.* **4**, 159 (1990).
27. W. Hüls, H. Münster, R. Pesch and E. Schröder, *Proceedings 44th ASMS Conference on Mass Spectrometry and Allied Topics*, Portland, OR (May, 1996).
28. G. Stilianos, J. Roussis and W. Fedora, *Proceedings 14th International Mass Spectrometry Conference*, Tampere (1997), Paper WePo 109.
29. N.C. Hill, G.M. Alber, L. Schweikhard, T.L. Ricca and A.G. Marshall, *Proceedings 39th ASMS Conference*, Nashville, TN (1991), p. 1501.
30. A.H. Wapstra and G. Audi, *Nucl. Physics* **A432**, 1–54; 55–139 (1985).
31. K. Bos, G. Audi and A.H. Wapstra, *Nucl. Physics* **A432**, 140–84 (1985).
32. A.H. Wapstra, G. Audi and R. Hoekstra, *Nucl. Physics* **A432**, 185–362 (1985).
33. J.R. De Laeter, K.G. Heumann, R.C. Barber, I.L. Barnes, J. Cesario, T.L. Chang and T.B. Coplen, *Pure Appl. Chem.* **63**, 975–90 (1991).
34. R.M. Silverstein and F.X. Webster, *Spectrometric Identification of Organic Compounds*, John Wiley & Sons, Inc., New York, NY (1998).
35. A.E. Ashcroft, *Ionization Methods in Organic Mass Spectrometry*, The Royal Society of Chemistry, Cambridge (1997).
36. R.P. Lattimer and R.E. Harris, *Rubber Chem. Technol.* **62**, 548–67 (1989).
37. R.P. Lattimer, *J. Anal. Appl. Pyrol.* **26**, 65–92 (1993).
38. A.G. Harrison, *Chemical Ionization Mass Spectrometry*, CRC Press, Boca Raton, FL (1992).
39. R. Luyk, J. Pureveen, J.M. Commandeur and J.J. Boon, *Macromol. Chem., Macromol. Symp.* **74**, 235–51 (1993).
40. R.J. Cotter, *Anal. Chem.* **52**, 1589–1606A (1980).
41. A.G. Harrison, in *Ionic Processes in the Gas Phase* (M.A. Almoster Ferreira, ed.), D. Reidel, Dordrecht (1984), p. 23.
42. H.J. Heger, U. Boesl, R. Zimmermann, R. Dorfner and A. Kettrup, *Eur. Mass Spectrom.* **5**, 51–7 (1999).
43. E. Kolodney and A. Amirav, *Chem. Phys.* **82**, 269 (1983).
44. A. Fialkov, A. Gordin and A. Amirav, *Abstracts 7th International Symposium on Hyphenated Techniques in Chromatography and Hyphenated Chromatographic Analyzers (HTC-7)*, Bruges (2002).
45. A. Amirav and O. Granot, *Abstracts 7th International Symposium on Hyphenated Techniques in Chromatography and Hyphenated Chromatographic Analyzers (HTC-7)*, Bruges (2002), Paper 16.2.
46. D.M. Lubman (ed.), *Laser and Mass Spectrometry*, Oxford University Press, New York, NY (1990).
47. J. Hsu, *J. Chromatogr. Libr.* **59**, 345–98 (1996).

48. M.J.I. Mattina, *J. Chromatogr. Libr.* **59**, 325–44 (1996).

49. M. Sakairi and Y. Kato, *J. Chromatogr.* **A794**, 391–406 (1998).

50. S. Tojo and T. Matsumoto, *Chiba-ken Kogyo Shikenjo Kenkyu Hokoku* **1**, 15–20 (1987).

51. M.W. Hayes and A.G. Altenau, *Rubber Age (N.Y.)*, **102** (5), 59–62 (1970).

52. J. Lehotay, J. Daněček, O. Líška, J. Leško and E. Brandšteterová, *J. Appl. Polym. Sci.* **25**, 1943–50 (1980).

53. D.C.M. Squirrell, *Analyst* **106**, 1042–56 (1981).

54. P. Aggozino, L. Ceraulo, M. Ferrugia, E. Caponetti, F. Intravaia and R. Triolo, *J. Colloid Interf. Sci.* **114**, 26–31 (1986).

55. J. Bonilla, *Proceedings SPE ANTEC '95*, Boston, MA (1995), pp. 2475–83.

56. M.S.B. Munson and F.H. Field, *J. Am. Chem. Soc.* **88**, 2621–30 (1966).

57. A.G. Harrison, in *Encyclopedia of Spectroscopy and Spectrometry* (J.C. Lindon, ed.), Academic Press, San Diego, CA (2000), pp. 207–15.

58. S.M. Dyszel, *Thermochim. Acta* **61**, 169 (1983).

59. E. Baumgartner and E. Nachbaur, *Thermochim. Acta* **19**, 3 (1977).

60. T. Tsuneto, I. Murasawa, M. Nagata and Y. Kubota, *J. Anal. Appl. Pyrol.* **33**, 139 (1995).

61. J.H. Bowie, *Mass Spectrom. Rev.* **9**, 349 (1990).

62. M. Vairamani, U.A. Mirza and R. Srinavas, *Mass Spectrom. Rev.* **9**, 235–58 (1990).

63. P. Rudewicz and B. Munson, *Anal. Chem.* **58**, 358–61 (1986).

64. A.I. Vit, M.N. Galbraith, J.H. Hodgkin and D. Yan, *Polym. Degrad. Stab.* **42** (1), 69–73 (1993).

65. R.P. Lattimer, *Rubber Chem. Technol.* **68**, 783–93 (1995).

66. G. Salmona, A. Assaf, A. Gayte-Sorbier and C.B. Airaudo, *Biomed. Mass Spectrom.* **11**, 450–4 (1984).

67. M. Vincenti, E. Pelizzetti, A. Guarini and S. Costanzi, *Anal. Chem.* **64**, 1879 (1992).

68. M.A. Baldwin and F.W. McLafferty, *Org. Mass Spectrom.* **7**, 1353–6 (1973).

69. C.G. Juo, S.W. Chen and G.R. Her, *Anal. Chim. Acta* **311**, 153–64 (1995).

70. D. Bassi, P. Tosi and R. Schögl, *J. Vac. Sci. Technol.* **A16** (1), 114–22 (1998).

71. W. Lindinger, J. Hirben and H. Paretzke, *Intl J. Mass Spectrom. Ion Proc.* **129**, 79–88 (1993).

71a. M. Boutin, *J. Am. Soc. Mass Spectrom.*, in press.

72. M. Barber, R.S. Bordoli, R.D. Sedgwick and A.N. Tyler, *J.C.S. Chem. Commun.* 325–7 (1981).

73. F.M. Devienne and G. Grandclement, *C.R. Acad. Sci.* **262**, 696 (1966).

74. D.J. Surman and J.C. Vickerman, *J.C.S. Chem. Commun.* 324 (1981).

75. R.P. Lattimer, R.E. Harris, C.K. Rhee and H.-R. Schulten, *Rubber Chem. Technol.* **61**, 639 (1988).

76. M. Claeys and J. Claereboudt, in *Encyclopedia of Spectroscopy and Spectrometry* (J.C. Lindon, ed.), Academic Press, San Diego, CA (2000), pp. 505–12.

77. J.L. Gower, *Biomed. Mass Spectrom.* **5**, 191 (1985).

78. M. Takayama, in *New Advances in Analytical Chemistry* (Atta-ur-Rahman, ed.), Harwood Academic Publishers, Amsterdam (2000), pp. 31–76.

79. R.L. Cochran, *Appl. Spectrosc. Rev.* **22** (2–3), 137–87 (1986).

80. J.E. Campana and R.B. Freas, *J.C.S. Chem. Commun.* **21**, 1414 (1984).

81. J. Čermák, *Chem. Listy* **83** (5), 488–505 (1989).

82. M.H. Florêncio, in *New Advances in Analytical Chemistry* (Atta-ur-Rahman, ed.), Harwood Academic Publishers, Amsterdam (2000), pp. 77–96.

83. C.L. Johlman, C.L. Wilkins, J.D. Hogan, T.L. Donovan, D.A. Laude Jr and M.-J. Youssefi, *Anal. Chem.* **62**, 1167–72 (1990).

84. M. Barber, R.S. Bordell, G.J. Elliott, R.D. Sedgwick and A.N. Tyler, *Anal. Chem.* **54**, 645–57A (1982).

85. J.O. Lay Jr. and T.T. Chang, *Abstracts 32nd Annual Conference on Mass Spectrometry and Allied Topics*, San Antonio, TX (1984).

86. B.L. Bentz, P.J. Gale and M.E. Labib, *Abstracts 10th International Mass Spectrometry Conference*, Swansea (1985).

87. J.O. Lay and B.J. Miller, *Anal. Chem.* **59**, 1323A (1987).

88. J.E. Campana and S.L. Rose, *Intl J. Mass Spectrom. Ion Phys.* **46**, 483 (1983).

89. H.T. Kalinoski, in *Anionic Surfactants* (J. Cross, ed.), M. Dekker, New York, NY (1998), pp. 209–50.

90. P.A. Lyon, W.L. Stebbings, F.W. Crow, K.B. Tomen, D.L. Lippstreu and M.L. Gross, *Anal. Chem.* **56**, 8 (1984).

91. M.M. Siegel, R. Tsao, S. Oppenheimer and T.T. Chang, *Anal. Chem.* **62**, 322–7 (1990).

92. R.P. Lattimer, R.E. Harris, D.B. Ross and H.E. Diem, *Rubber Chem. Technol.* **57**, 1013–22 (1984).

93. C.L. Johlman, C.L. Wilkins, J.D. Hogan, T.L. Donovan, D.A. Laude Jr and M.-J. Youssefi, *Anal. Chem.* **62**, 1167–72 (1990).

94. A.J. Deome and P.J. Kane, *Proceedings 32nd Sagamore Army Materials Research Conference (Elastomers Rubber Technology)*, pp. 333–43 (1987).

95. R.P. Lattimer, R.E. Harris, C.K. Rhee and H.-R. Schulten, *Rubber Chem. Technol.* **61**, 639–57 (1989).

96. A.T. Jackson, K.R. Jennings and J.H. Scrivens, *Rapid Commun. Mass Spectrom.* **10**, 1449–58 (1996).

97. R.W. Layer, J.T. Lai, R.P. Lattimer and J.C. Westfahl, *Abstracts 32nd Annual Conference on Mass Spectrometry and Allied Topics*, San Antonio, TX (1984).

98. T.L. Riley, T.J. Prater, J.L. Gerlock, J.E. deVries and D. Schuetzle, *Anal. Chem.* **56**, 2145 (1984).

99. N.J. Jensen, K.B. Tomer and M.L. Gross, *J. Am. Chem. Soc.* **107**, 1863 (1985).

100. D.A. Catlow, M. Johnson, J.J. Monaghan, C. Porter and J.H. Scrivens, *J. Chromatogr.* **328**, 167–77 (1985).

101. W. Aberth, K.M. Straub and A.L. Burlingame, *Anal. Chem.* **54**, 2029 (1982).

102. R.M. Caprioli, T. Fan and J.S. Cottrell, *Anal. Chem.* **58**, 2949 (1986).

103. Y. Ito, T. Takeuchi, D. Ishii and M. Goto, *J. Chromatogr.* **346**, 161 (1985).

104. R.M. Caprioli (ed.), *Continuous-Flow Fast Atom Bombardment Mass Spectrometry*, John Wiley & Sons, Inc., New York, NY (1990).

105. M.G. Inghram and R. Gomer, *J. Chem. Phys.* **22**, 1279 (1954).

106. M.G. Inghram and R. Gomer, *Z. Naturf.* **10a**, 863 (1955).

107. R.P. Lattimer, in *Mass Spectrometry of Polymers* (G. Montaudo and R.P. Lattimer, eds), CRC Press, Boca Raton, FL (2002), pp. 237–68.

108. H.D. Beckey, *Principles of Field Ionization and Field Desorption Mass Spectrometry*, Pergamon Press, London (1977).

109. M.R. Jakupca, D.R. Stevenson and D.L. Stein, *Proceedings Polyolefins X* (SPE, Brookfield, CT, ed.), Houston, TX (1997), pp. 627–52.

110. H.D. Beckey, *Intl J. Mass Spectrom. Ion Phys.* **2**, 500 (1969).

111. H.D. Beckey and D. Schülte, *Z. Instr.* **68**, 302–7 (1960).

112. L. Prokai, *Field Desorption Mass Spectrometry*, M. Dekker, New York, NY (1990).

113. R.P. Lattimer and K.R. Welch, *Rubber Chem. Technol.* **53**, 151–9 (1980).

114. R.P. Lattimer and K.R. Welch, *Rubber Chem. Technol.* **51**, 925–38 (1978).

115. P. Zhu and K. Su, *Org. Mass Spectrom.* **25**, 260 (1990).

116. A.T. Jackson, K.R. Jennings and J.H. Scrivens, *Rapid Commun. Mass Spectrom.* **10**, 1449 (1996).

117. R.P. Lattimer, R.W. Layer and C.K. Rhee, *Rubber Chem. Technol.* **57**, 1023 (1984).

118. R.P. Lattimer, D.J. Harmon and K.R. Welch, *Anal. Chem.* **51**, 1293 (1979).

119. R.P. Lattimer, E.R. Hooser, H.E. Diem and C.K. Rhee, *Rubber Chem. Technol.* **55**, 442 (1982).

120. R.P. Lattimer, E.R. Hooser, H.E. Diem, R.W. Layer and C.K. Rhee, *Rubber Chem. Technol.* **53**, 1170 (1980).

121. J.E.C. Gregg Jr and R.P. Lattimer, *Rubber Chem. Technol.* **57**, 1056 (1984).

122. J.B. Pausch, *Anal. Chem.* **54**, 89–96A (1982).

123. P.F. Knewstubb and T.M. Sugden, *Nature* **181**, 474–5; 1261 (1958).

124. W.M.A. Niessen, in *Encyclopedia of Spectroscopy and Spectrometry* (J.C. Lindon, ed.), Academic Press, San Diego, CA (2000), pp. 18–24.

125. D.I. Carroll, I. Dzidic, E.C. Horning and R.N. Stilwell, *Appl. Spectrosc. Rev.* **17**, 337–406 (1981).

126. Anon., *The API Book*, PE Sciex, Thornhill, Ontario (1989).

127. J.V. Iribarn and B.A. Thomson, *J. Chem. Phys.* **64**, 2287–94 (1976).

128. A. Doyle, D.R. Moffett and B. Vonnegut, *J. Colloid Interf. Sci.* **19**, 136 (1964).

129. T.R. Covey, R.F. Bonner, B.I. Shushan and J.D. Henion, *Rapid Commun. Mass Spectrom.* **2**, 249 (1988).

130. M.S. Wilm and M. Mann, *Intl J. Mass Spectrom. Ion Proc.* **136**, 167 (1994).

131. M.S. Wilm and M. Mann, *Anal. Chem.* **68**, 1–8 (1996).

132. E. Scamporrino and D. Vitalini, in *Modern Techniques for Polymer Characterisation* (R.A. Pethrick and J.V. Dawkins, eds), John Wiley & Sons, Ltd, Chichester (1999), pp. 233–66.

133. K.B. Tomer, M.A. Mosely, L.J. Detering and C.E. Parker, *Mass Spectrom. Rev.* **13**, 431–57 (1994).

134. M. Hamdon and O. Curcuruto, *Intl J. Mass Spectrom. Ion Proc.* **108**, 93–113 (1991).

135. J.B. Fenn, M. Mann, Ch.K. Meng, Sh. Wong and F. Whitehouse, *Mass Spectrom. Rev.* **9**, 37–70 (1990).

136. M. Yamashita and J.B. Fenn, *J. Phys. Chem.* **88**, 4451–9; 4671–5 (1984).

137. A.T. Jackson, A. Buzy, K.R. Jennings and J.H. Scrivens, *Eur. Mass Spectrom.* **2**, 115–27 (1996).

138. P.B. O'Connor and F.W. McLafferty, *J. Am. Chem. Soc.* **117**, 12826 (1995).

139. L. Prokai, *Intl J. Polym. Anal. Charact.* **6**, 379–91 (2001).

140. L. Prokai and W.J. Simonsick Jr, *Rapid Commun. Mass Spectrom.* **7**, 853 (1993).

141. M.J. Taylor, J. Fawcett, S.J. Rumbelow, J.H. Scrivens and C.M. Brookes, unpublished results.

142. J.D. Vargo and K.L. Olson, *Anal. Chem.* **57**, 672 (1985).

143. A.T. Jackson, *Ph.D. Thesis*, University of Warwick (1995).

144. D.H. Williams, A.F. Findeis, S. Naylor and B.W. Gibson, *J. Am. Chem. Soc.* **109**, 1980–6 (1987).

145. P.J. Slegel and T. Karancsi, *Proceedings 14th International Mass Spectrometry Conference*, Tampere, Finland (1997), Paper MoPo 123.

146. B. Shushan, B. Davidson and R.B. Prime, *Anal. Calorim.* **5**, 105 (1984).

147. K.A. Barns, A.P. Damant, J.R. Startin and L. Castle, *J. Chromatogr.* **A712**, 191–9 (1995).

148. K.L. Busch, *J. Mass Spectrom.* **30**, 233–40 (1995).

149. P.A. Lyon (ed.), *Desorption Mass Spectrometry*, ACS Symposium Series, No. 291, American Chemical Society, Washington, DC (1985).

150. R.D. Macfarlane, in *Encyclopedia of Spectroscopy and Spectrometry* (J.C. Lindon, ed.), Academic Press, San Diego, CA (2000), pp. 1848–57.

151. P. Demirev, *Mass Spectrom. Rev.* **14**, 279–308 (1995).

152. R.D. Macfarlane, J.C. Hill and P.W. Geno, *Adv. Mass Spectrom.* **11**, 3–21 (1989).

153. N. Furstenau, F. Hillenkamp and R. Nitsche, *Intl J. Mass Spectrom. Ion Proc.* **31**, 85–91 (1979).

154. M.P. Mawn, R.W. Linton, S.R. Bryan, B. Hagenhoff, U. Jürgens and A. Benninghoven, *J. Vac. Sci. Technol.* **A9**, 307–11 (1991).

155. C.N. McEwen, S.F. Layton and S.K. Taylor, *Anal. Chem.* **49**, 922 (1977).

156. S.D. Richardson, A.D. Thruston, J.M. McGuire and E.J. Weber, *Org. Mass Spectrom.* **28**, 619 (1993).

157. M.U.D. Beug-Deeb, J.A. Bennett, M.E. Inman and E.A. Schweikert, *Anal. Chim. Acta* **218**, 85 (1989).

158. F. Ventura, A. Figueras, J. Caixach, D. Fraisse and J. Rivera, *Fresenius' Z. Anal. Chem.* **335**, 272 (1989).

159. H.S. Freeman, R.B. van Breemen, J.F. Esancy, D.O. Ukponmwan, Z. Hao and W.-N. Hsu, *Text. Chem. Color.* **22**, 23 (1990).

160. M.J. Dale, A.C. Jones, P.R.R. Langridge-Smith, K.F. Costello and P.G. Cummins, *Anal. Chem.* **65**, 793 (1993).

161. J.N. Driscoll, *Am. Lab.*, 71–5 (1976).

162. M.L. Langhorst, *J. Chromatogr. Sci.* **19**, 98–103 (1981).

163. J.N. Driscoll, D.W. Conron, P. Ferioli, I.S. Krull and K.-H. Xie, *J. Chromatogr.* **302**, 43–50 (1984).

164. D.B. Robb, T.R. Covey and A.P. Bruins, *Anal. Chem.* **72**, 3653–9 (2000).

165. J.C. Traeger, in *Encyclopedia of Spectroscopy and Spectrometry* (J.C. Lindon, ed.), Academic Press, San Diego, CA (2000), pp. 1840–7.

166. S.A. McLuckey, in *Advances in Mass Spectrometry* (E.J. Karjalainen, A.E. Hesso, J.E. Jalonen and U.P. Karjalainen, eds), Elsevier Science, Amsterdam (1998), Vol. 14, pp. 153–96.

167. G. Petzold, P. Siebert and J. Müller, in *Micro Total Analysis Systems 2001* (J.M. Ramsey and A. van den Berg, eds), Kluwer Academic Publishers, Dordrecht (2001), pp. 224–6.

168. R. Bateman, in *Encyclopedia of Spectroscopy and Spectrometry* (J.C. Lindon, ed.), Academic Press, San Diego, CA (2000), pp. 2085–92.

169. A.T. Jackson, R.C.K. Jennings, H.T. Yates, J.H. Scrivens and K.R. Jennings, *Eur. Mass Spectrom.* **3**, 113–20 (1997).

170. W.C. Wiley and I.H. McLaren, *Rev. Sci. Instr.* **26**, 1150–7 (1955).

171. B.A. Mamyrin, V.I. Karataev, D.V. Shmikk and V.A. Zagulin, *Sov. Phys. JETP* (English translation) **37**, 45 (1973).

172. B.A. Mamyrin and D.V. Shmikk, *Sov. Phys. JETP* (English translation) **49**, 762 (1979).

173. J.H.J. Dawson and M. Guilhaus, *Rapid Commun. Mass Spectrom.* **3**, 155–9 (1989).

174. R.J. Cotter (ed.), *Time-of-Flight Mass Spectrometry*, American Chemical Society, Washington, DC (1997).

175. R.N. Zare, F.M. Fernández and J.R. Kimmel, *Angew. Chem. Intl Ed. Engl.* **42** (1), 30–5 (2003).

176. K. Tanaka, A. Bowdler and S. Kawabata, *Kratos Analytical Application Note MO122* (1999), Kratos, Japan.

177. H.R. Morris, T. Paxton, A. Dell, J. Langhorne, M. Berg, R.S. Bordoli, J. Hoyes and R.H. Bateman, *Rapid Commun. Mass Spectrom.* **10**, 889–96 (1996).

178. R. Bonner, I. Chernushevich, B. Thomson and T. Covey, *Abstracts 15th Montreux Symposium on Liquid Chromatography/Mass Spectrometry*, Montreux (1998), p. 15.

179. M. Guilhaus, *J. Mass Spectrom.* **30**, 1519 (1995).

180. P. Håkansson, *Braz. J. Phys.* **29** (3), 422–7 (1999).

181. C.G. Enke, in *Advances in Mass Spectrometry* (E.J. Karjalainen, A.E. Hesso, J.E. Jalonen and U.P. Karjalainen, eds), Elsevier Science, Amsterdam (1999), Vol. 14, pp. 197–219.

182. E.W. Schlag (ed.), *Time-of-Flight Mass Spectrometry and Its Applications*, Elsevier, Amsterdam (1994).

183. W. Hang, C. Baker, B.W. Smith, J.D. Winefordner and W.W. Harrison, *J. Anal. At. Spectrom.* **12**, 143 (1997).

184. R. Grix, R. Kutscher, G. Li, U. Grüner and H. Wollnik, *Rapid Commun. Mass Spectrom.* **2**, 83–5 (1988).

185. R.H. Bateman, R.S. Bordoli, A.J. Gilbert, J.B. Hoyes and H.R. Morris, *Proceedings 44th ASMS Conference on Mass Spectrometry and Allied Topics*, Portland, OR (1996), p. 796.

186. O.A. Mirgorodskaya, A.A. Shevchenko, I.V. Chernushevich, A.F. Dodonov and A.I. Miroshnikov, *Anal. Chem.* **66**, 99 (1994).

187. D.P. Myers, G. Li, P.P. Mahoney and G.M. Hieftje, *J. Am. Soc. Mass Spectrom.* **6**, 400 (1995).

188. F. De Boever, *Micromass 2001, New Products Seminar and Users Event*, Amsterdam (2001).

189. I.V. Bletsos, D.M. Hercules, J.H. Magill, D. VanLeyen, F. Niehuis and A. Benninghoven, *Anal. Chem.* **60**, 938–44 (1988).

190. W. Paul and H. Steinwedel, *US Pat.* 2,939,952 (June 7, 1960).

191. W. Paul, *Angew. Chem. Intl Ed. Engl.* **29**, 739 (1990).

192. J.V. Johnson, R.A. Yost, P.E. Kelley and D.C. Bradford, *Anal. Chem.* **62**, 2162–72 (1990).

193. W.B. Whitten, J. Moxom, P.T.A. Reilly and J.M. Ramsey, in *Micro Total Analysis Systems 2001* (J.M. Ramsey and A. van den Berg, eds), Kluwer Academic Publishers, Dordrecht (2001), pp. 210–12.

194. J.-T. Wu, M.G. Qian, M.X. Li, K. Zheng, P. Huang and D.M. Lubman, *J. Chromatogr.* **A794**, 377–89 (1998).

195. R.E. March, in *Encyclopedia of Spectroscopy and Spectrometry* (J.C. Lindon, ed.), Academic Press, San Diego, CA (2000), pp. 1000–9.

196. P.S.H. Wong and R.G. Cooks, *Curr. Separ.* **16**, 85–92 (1997).

197. R.E. March, in *Advances in Mass Spectrometry* (E.J. Karjalainen, A.E. Hesso, J.E. Jalonen and U.P. Karjalainen, eds), Elsevier Science, Amsterdam (1998), Vol. 14, pp. 241–78.

198. R.E. March, R.J. Hughes and J.F.J. Todd (eds), *Quadrupole Storage Mass Spectrometry*, Wiley-Interscience, New York, NY (1989).

199. R.E. March and J.F.J. Todd (eds), *Practical Aspects of Ion Trap Mass Spectrometry*, CRC Press, Boca Raton, FL (1995), 3 vols.

200. R.E. March, *J. Mass Spectrom.* **32**, 351 (1997).

201. M.E. Bier and J.C. Schwartz, in *Electrospray Ionization Mass Spectrometry* (R.B. Cole, ed.), John Wiley & Sons, Ltd, Chichester (1997), Chapter 7.

202. S.A. McLuckey, G.L. Glish and G.J. van Berkel, *Intl J. Mass Spectrom. Ion Proc.* **106**, 213 (1991).

203. K.L. Busch, G.L. Glish and S.A. McLuckey, *Mass Spectrometry/Mass Spectrometry. Techniques and Applications of Tandem Mass Spectrometry*, VCH, Weinheim (1988).

204. R.G. Cooks, G.L. Glish, S.A. McLuckey and R.E. Kaiser, *Chem. Engng News* **69** (12), 26–30, 33–41 (1991).

205. B.A. Eckenrode, S.A. McLuckey and G.L. Glish, *Intl J. Mass Spectrom. Ion Proc.* **106**, 137 (1991).

206. D.E. Goeringer, W.B. Whitten, J.M. Ramsey, S.A. McLuckey and G.L. Glish, *Anal. Chem.* **64**, 1434 (1992).

207. J.F.J. Todd and A.D. Penman, *Intl J. Mass Spectrom. Ion Proc.* **106**, 1 (1991).

208. L.L. Lopez, P.R. Tiller, M.W. Senko and J.C. Schwartz, *Rapid Commun. Mass Spectrom.* **13**, 663 (1999).

209. M. Sanders, *Proceedings 47th ASMS Conference on Mass Spectrometry and Allied Topics*, Dallas (1999), pp. 1813–14.

210. C. Baumann, M.A. Cintora, M. Eichler, E. Lifante, M. Cooke, A.M. Prryborowska and J.M. Halket, *Rapid Commun. Mass Spectrom.* **14**, 349 (2000).

211. K.M. Verge and G.R. Agnes, *J. Am. Soc. Mass Spectrom.* **13** (8), 901–905 (2002).

212. C.L. Hendrickson, J.J. Drader, D.A. Laude Jr and A.G. Marshall, *Rapid Commun. Mass Spectrom.* **10**, 1829–32 (1996).

213. M.W. Senko, C.L. Hendrickson, M.R. Emmett, S.D.-H. Shi and A.G. Marshall, *J. Am. Soc. Mass Spectrom.* **8**, 970–6 (1997).

214. R.M.A. Heeren, G.J. van Rooij, P.B. O'Connor, M.C. Duursma and J.J. Boon, *Proceedings 14th International Mass Spectrometry Conference*, Tampere (1997), Paper ThPo 016.

215. A.G. Marshall, C.L. Hendrickson and M.E. Emmett, in *Advances in Mass Spectrometry* (E.J. Karjalainen, A.E. Hesso, J.E. Jalonen and U.P. Karjalainen, eds), Elsevier Science, Amsterdam (1998), Vol. 14, pp. 221–39.

216. C.L. Wilkins (ed.), *Fourier Transform Mass Spectrometry*, Special Issue in *Trends Anal. Chem.* **13**, 223–51 (1994).

217. A.G. Marshall (ed.), *Fourier Transform Ion Cyclotron Resonance Mass Spectrometry*, Special Issue in *Int. J. Mass Spectrom. Ion Proc.* **137/138**, 410 (1996).

218. M.V. Buchanan (ed.), *Fourier Transform Mass Spectrometry: Evolution, Innovation, and Applications*, American Chemical Society, Washington, DC (1987).

219. A.G. Marshall and F.R. Verdun, *Fourier Transforms in NMR, Optical, and Mass Spectrometry: a User's Handbook*, Elsevier, Amsterdam (1990).

220. B. Asamoto (ed.), *FT-ICR/MS: Analytical Applications of Fourier Transform Ion Cyclotron Resonance Mass Spectrometry*, VCH Publishers, New York, NY (1991).

221. R. Williams, *Spectroscopy and the Fourier Transform*, VCH Publishers, New York, NY (1996).

222. M.C. Nyman, J.J. Ferra, J. Perez, H.I. Kenttamaa and E.R. Blatchley III, *Proceedings 14th International Mass Spectrometry Conference*, Tampere (1997), Paper WePo 107.

223. R.M.A. Heeren, C.G. de Koster and J.J. Boon, *Anal. Chem.* **67** (21), 3965–70 (1995).

224. R.M.A. Heeren and J.J. Boon, *Intl J. Mass Spectrom. Ion Proc.* **157/158**, 391–403 (1996).

225. G.B. Kenion, R. Ludicky, J.H. Deitch, M. Leib and D. Doster, *Tappi J.* **77** (6), 199–205 (1994).

226. H. Schwarz, in *Analytiker-Taschenbuch* (R. Borsdorf and H. Günzler, eds), Springer-Verlag, Berlin (1989), Vol. 8, pp. 199–216.

227. J.H. Scrivens, T. Jackson, H.T. Yates, M. Taylor, R. Bateman and M. Green, *Proceedings 14th International Mass Spectrometry Conference*, Tampere (1997), Paper ThPo 119.

228. R.P. Lattimer, *Rubber Chem. Technol.* **61**, 658 (1988).

229. R.P. Lattimer, H. Münster and H. Budzikiewicz, *Rubber Chem. Technol.* **63**, 298–307 (1990).

230. K. Ng, B. Piatek and H. Tong, *Proceedings 42nd ASMS Conference on Mass Spectrometry and Allied Topics*, Chicago, IL (1994).

231. C. Moore, *Spectrosc. Intl J.* **7**, 1–10 (1989).

232. H. Egsgaard, E. Larsen, W.B. Pedersen and L. Carlsen, *Trends Anal. Chem.* **11** (4), 164–8 (1992).

233. X. Han and R.W. Gross, *J. Am. Soc. Mass Spectrom.* **6**, 1202–10 (1995).

234. A.T. Jackson, K.R. Jennings and J.H. Scrivens, *Rapid Commun. Mass Spectrom.* **10**, 1459–62 (1996).

235. W. Brand and K. Levsen, *Intl J. Mass Spectrom. Ion Phys.* **35**, 1 (1980).

236. P.J. Fordham. J.W. Gramshaw, H.M. Crews and L. Castle, *Food Addit. Contam.* **12** (5), 651–9 (1995).

237. R. Willoughby and E. Sheehan, *A Global View of MS/MS*, Global View Publishing, Pittsburgh, PA (1999).

238. W.M.A. Niessen, in *Encyclopedia of Spectroscopy and Spectrometry* (J.C. Lindon, ed.), Academic Press, San Diego, CA (2000), pp. 1404–10.

239. F.W. McLafferty, *Tandem Mass Spectrometry*, John Wiley & Sons, Inc., New York, NY (1983), p. 6.

240. T. Yoshikawa, K. Ushimi, K. Kimura and M. Tamura, *J. Appl. Polym. Sci.* **15**, 2065–72 (1971).

241. K. Ng, B. Piatek and C. Shimanskas, *Proceedings 41st ASMS Conf. on Mass Spectrometry and Allied Topics*, San Francisco, CA (1993).

242. A.C. Tas and J. van der Greef, *Trends Anal. Chem.* **12**, 60–6 (1993).

243. J. Pierre and J. van Bree, *Kunststoffe* **73**, 319–24 (1983).

244. G.J. Mol, R.J. Gritter and G.E. Adams, in *Applications in Polymer Spectroscopy* (E.G. Brame, ed.), Academic Press, New York, NY (1978), pp. 257–77.

245. J.G. Moncur, A.B. Campa and P.C. Pinoli, *J. High Resolut. Chromatogr., Chromatogr. Commun.* **5**, 322–3 (1982).

246. A.S. Hilton and A.G. Altenau, *Rubber Chem. Technol.* **46**, 1035–43 (1973).

247. P.A. Barnes and G.M.B. Parkes, *J. Thermal. Anal.* **39**, 607–18 (1993).

248. P.A. Barnes, G.M. Parkes and P. Sheridan, *J. Therm. Anal.* **42**, 841–54 (1994).

249. H.-R. Schulten and R.P. Lattimer, *Mass Spectrom. Rev.* **3**, 231 (1984).

250. G. Montaudo, *Trends Polym. Sci.* **4** (3), 81–6 (1996).

251. D. Price, D. Dollimore, N.S. Fatemi and R. Whitehead, *Thermochim. Acta* **42**, 323–32 (1980).

252. D.W. Carlson, M.W. Hayes, H.C. Ransaw, R.S. McFadden and A.G. Altenau, *Anal. Chem.* **43**, 1874–6 (1971).

253. B. Asamoto, J.R. Young and R.J. Citerin, *Anal. Chem.* **62**, 61 (1990).

254. J.H. Scrivens, *Adv. Mass Spectrom.* **13**, 447 (1995).

255. J.H. Scrivens, K. Rollins, R.C.K. Jennings, R.S. Bordoli and R.H. Bateman, *Rapid Commun. Mass Spectrom.* **6**, 272 (1992).

256. K.D. Ballard, S.J. Gaskell, R.C.K. Jennings, J.H. Scrivens and R.G. Vickers, *Rapid Commun. Mass Spectrom.* **6**, 553 (1992).

257. H.L. Dinsmore and D.C. Smith, *Anal. Chem.* **20**, 11–24 (1948).

258. J.C. Tou, D. Zakett and V.J. Caldecourt, in *Tandem Mass Spectrometry* (F.W. McLafferty, ed.), John Wiley & Sons, Inc., New York, NY (1983), pp. 435–50.

259. F. Janssen, *Electrotechniek* **57** (2), 96–100 (1979).

260. O.-G. Piringer, *Verpackungen für Lebensmittel*, VCH, Weinheim (1993).

261. T. Bücherl, A. Gruner and N. Palibroda, *Packag. Technol. Sci.* **7**, 139–54 (1994).

262. B. van Lierop, L. Castle, A. Feigenbaum and A. Boenke, *Spectra for the Identification of Additives in Food Packaging*, Kluwer Academic Publishers, Dordrecht (1998).

263. J.B. Pausch, R.P. Lattimer and H.L.C. Meuzelaar, *Rubber Chem. Technol.* **56**, 1031–44 (1983).

264. H. Wilcken and M. Geissler, *Shimadzu Application Training Seminar*, Duisburg (1996).

265. R. Luyk, H.A.J. Govers, G.B. Eykel and J.J. Boon, *J. Anal. Appl. Pyrol.* **20**, 303 (1991).

266. W.H. McClennen, J.M. Richards, H.L.C. Meuzelaar, J.B. Pausch and R.P. Lattimer, *Polym. Mater. Sci. Engng* **53**, 203–207 (1985).

267. S. Takayama, H. Takeguchi and A. Kawabata, *Proceedings Workshop on Mass Spectrometry of Polymers*, Catania (1999), Paper P10.

268. J.J. Monaghan, W.E. Morden, T. Johnson, I.D. Wilson and P. Martin, *Rapid Commun. Mass Spectrom.* **6**, 608–15 (1992).

269. H.L. Friedman, *J. Macromol. Chem.* **A1**, 57 (1967).

270. J.H. Scrivens and A.T. Jackson, *NATO ASI Ser., Ser.* **C521**, 201–34 (1999).

271. J.G. Kreiner, *Rubber Chem. Technol.* **44**, 381–96 (1971).

272. F.W. Karasek, *Res. Dev.* **21**, 34 (1970).

273. E.A. Mason and E.W. McDaniel, *Transport Properties of Ions in Gases*, John Wiley & Sons, Inc., New York, NY (1973).

274. F.W. Karasek and G.E. Spangler, *J. Chromatogr. Libr.* **20**, 377 (1981).

275. K. Budde, *Proceedings Electrochem. Soc. (Anal. Tech. Semicond. Mater. Process Charact.)* **90–11**, 215–26 (1990).

276. C.S. Leasure, M.E. Fleischer, G.K. Anderson and G.A. Eiceman, *Anal. Chem.* **58**, 2142–7 (1986).

277. F.W. Karasek, *Res. Dev.* **21**, 25 (1970).

278. G.A. Eiceman and Z. Karpas, *Ion Mobility Spectrometry*, CRC Press, Boca Raton, FL (1994).

279. C. Shumate, R.H. St Louis and H.H. Hill Jr, *J. Chromatogr.* **373**, 141 (1986).

280. J.E. Roehl, *Appl. Spectrosc. Rev.* **26**, 1 (1991).

281. R.H. St Louis and H.H. Hill, *CRC Crit. Rev. Anal. Chem.* **21**, 321 (1990).

282. H.H. Hill, W.F. Siems, R.H. St. Louis and D.G. McMinn, *Anal. Chem.* **62**, 1201A (1990).

282a A.B. Kanu and H.H. Hill, *LabPlus Intl* **18** (2), 20–26 (2004).

283. T.W. Carr (ed.), *Plasma Chromatography*, Plenum Press, New York, NY (1984).

284. R.J. Dam, in *Plasma Chromatography* (T.W. Carr, ed.), Plenum Press, New York, NY (1984), pp. 177.

285. D. Filmore, *Today's Chemist at Work*, 31–36 (March 2002).

286. J. Xu, W.B. Whitten and J.M. Ramsey, in *Micro Total Analysis Systems 2001* (J.M. Ramsey and A. van den Berg, eds), Kluwer Academic Publishers, Dordrecht (2001), pp. 337–8.

287. T.W. Carr, *Natl Bur. Standards (US), Spec. Publ.* **519**, 697 (1979).

288. T.W. Carr, *Proceedings, International Reliability Physics Symposium*, San Francisco, CA (April 1979).

289. E.J. Poziomek and G.A. Eiceman, *Environ. Sci. Technol.* **26**, 1313–18 (1992).

CHAPTER 7

The Lame, the Blind and the Broker

Multihyphenation and Multidimensionality in Polymer/Additive Analysis

Additives In Polymers: Industrial Analysis And Applications J. C. J. Bart
© 2005 John Wiley & Sons, Ltd ISBN: 0-470-85062-0

426 MULTIHYPHENATION AND MULTIDIMENSIONALITY IN POLYMER/ADDITIVE ANALYSIS

For the characterisation of additives and polymer fractions (e.g. oligomers), a variety of techniques are employed to obtain information about quality and quantity. Starting from the sample preparation procedures (SFE, SPE, derivatisation, etc.) the structure-based separations and detection may be off-line (e.g. MALDI-ToFMS), semi on-line (e.g. evaporative FTIR) or on-line (e.g. UV, ESI-MS, LS). The combination of a single analytical column and a selective detector often does not suffice for the recognition and quantification of all analytes of interest in complex samples. *Multidimensionality* has come into focus because of co-elution of peaks; distrust of the identification of analytes on the basis of retention values; the need to incorporate sample preparation in the total analytical procedure; and the more stringent demands made concerning the identification and/or confirmation of analytes of interest (often at trace levels). Multidimensional techniques for sample preparation, separation, detection, identification and quantification of analytes are frequently subdivided into *coupled-column techniques*, which comprise on-line sorbent-based sample preparation, heartcut operations and comprehensive separation techniques, and *hyphenations* where the separation module is coupled with a spectroscopic detector which provides structural information. This terminology is not always strictly adhered to, and, moreover, does not cover all possible combinations of techniques. In addition, second-order dimensionality has been achieved by combining hyphenation and coupled-column approaches in one set-up. Hirschfeld [1] has coined the term 'hyphenated techniques' to describe systems that consist of directly interfaced instruments, usually with extensively developed automation. The main step of all hyphenated techniques

is frequently the separation step, but not necessarily so. Hyphenation of separation techniques can occur at the front end or outlet side (Scheme 7.1). While postcolumn hyphenation is a must, precolumn hyphenation should be kept simple (and be validated). Multidimensional techniques are generally more application-specific than single analytical methods (*cf.* Table 7.1). The degree of sophistication of a method very much depends on the complexity of the sample matrix and on how many selectivity-generating steps have to be introduced. In hyphenated techniques the selectivity can stand tougher samples.

With the many possible modes of hyphenation in the sample preparation-separation-identification (multidash) sequence, the analyst is potentially on dangerous ground. Setting up an on-line technique is not simply the coupling of two well-established subtechniques, but will generally require adaptation and subsequent optimisation, as well as profound knowledge of the underlying principles. It is not the new hyphenation which matters, but its practical analytical advances. By simultaneously obtaining various sets of data (e.g. separation data and spectral data) on the same sample, hyphenated analytical techniques have become powerful tools for analysis, providing much more useful information than that obtained by operating the techniques independently. Table 7.2 lists the main characteristics of hyphenated techniques. Hyphenations serve a wide variety of purposes, such as:

(i) identification of unknowns;
(ii) compositional drift determination studies;
(iii) deformulation of competitive products; and
(iv) quantitative analysis.

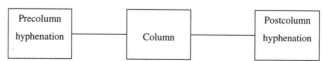

Scheme 7.1 Precolumn and postcolumn hyphenation

Table 7.1 Some selected principles of hyphenated methods

Concern of coupling	Principle(s)	Example(s)
Sample preparation	Extraction	SFE-SFC-MS
Trace enrichment, cleanup	Solid-phase extraction	SPE-GC
Separation	Chromatography	GC-MS, GC-FTIR, HPLC-AAS, SFC-NMR
Volatilisation	Pyrolysis	PyGC-MS, PyGC-FTIR, ETV-ICP-MS
Sample transportation	Flow analysis	FIA-AAS
Detection	MSD, ESD, FTIR	GC-MS, MS-MS, HPLC-FTIR, HPLC-ICP-MS
Simultaneous analysis	Thermal analysis	TG-DSC
Microanalysis	Microscopy, laser ablation	Laser OES, LA-ICP, FIM-MS
Analyte desorption	Laser ablation, thermal desorption	LA-AAS, LA-ICP-MS, TD-GC-MS

Table 7.2 Main characteristics of hyphenated techniques

Advantages

- On-line sample preparation
- Reduced sample handling steps
- Speed of analysis (time savings)
- Highly automated procedures (up to total automation from sample preparation to analysis)
- Improved sensitivity (lower detection limits)
- Improved selectivity (easier data handling)
- Improved reproducibility
- Reduced risk for cross-contamination, analyte loss and decomposition
- Multicomponent analysis in a single shot
- Improved reliability of analytical data
- Information multiplied, results simplified

Disadvantages

- Interface construction
- Demanding method development (simultaneous optimisation)
- Mutual interferences
- Suboptimal operating conditions
- Operational complexity

Although the advantages of coupling techniques are considerable, a hyphenated technique is only as good as its *interface*. In particular, in order to eliminate mutual interferences, suboptimal performance is often an inevitable compromise. An important goal in hyphenation is robustness. Separation science and spectroscopy or mass spectrometry are very specialised research areas. When applying a hyphenated technique, an analyst expects the other half just to work. Hyphenated techniques are data intensive and require the use of computers to control acquisition, storage, display and manipulation of data. Typically, a chromatographic system coupled to a fast-scanning spectrometric detector (e.g. diode array device or mass spectrometer) records full spectra at regular intervals over the entire chromatogram. The obtained 3D data provide a rich source of information that can be analysed by appropriate chemometric techniques. Thus, the analytical potential of hyphenated systems is greatly enhanced in combination with mathematical tools for extracting information from a 3D data analysis. The new generation of analytical techniques, such as GC-HRMS, LC-DAD, CE-MS, LC-MS, LC-NMR can be used with much higher efficiency when combined with pattern recognition methods [2]. This holds, for example, for impurity profiling.

The 'ideal' interface is rare. Table 7.3 lists the qualities required for an ideal interface from a chromatographic point of view [3]. Nowadays, hyphenation goes a long way towards *total analysis systems* (e.g. HPLC-UV-NMR-MS), especially in the pharmaceutical industry. Such 'magic-wand' systems are by no means a panacea for all analytical problems; they are more likely to be confined to niche applications. Multihyphenation and multidetector monitoring set their own

Table 7.3 Qualities of the ideal chromatographic interface

- Maintenance of chromatographic integrity (concentration and distribution); trueness
- Absence of thermal degradation of analytes within the interface
- Compatibility with different fluids under pressure programming conditions and with modifiers
- Highly versatile chromatographic separations
- High analyte transport efficiency
- Highly sensitive technique
- Ease of automation
- High sample throughput
- Reliable, low-maintenance interface

Table 7.4 Problems in multidimensional chemical analysis

- Different sensitivities of the tandem instruments
- Solvent choice
- Overall instrumental reliability (subtechniques, interfaces, 'leaks')
- Data handling
- Human resources (various operators for subtechniques)
- Cost

limits, mainly in terms of practical problems (Table 7.4). Examples are differences in sensitivity (UV, MS vs. FTIR, NMR) and solvent characteristics.

If total analysis systems are not exactly the best option for polymer/additive analysis, then neither are chip-based hyphenations. The reasons here are completely different: minute sample size (homogeneity problems for more traditional samples), and (theoretical) limits to chromatography and concentration-based sensitivity, etc.

This chapter deals mainly with (multi)hyphenated techniques comprising wet sample preparation steps (e.g. SFE, SPE) and/or separation techniques (GC, SFC, HPLC, SEC, TLC, CE). Other hyphenated techniques involve thermal-spectroscopic and gas or heat extraction methods (TG, TD, HS, Py, LD, etc.). Also, spectroscopic couplings (e.g. LIBS-LIF) are of interest. Hyphenation of UV spectroscopy and mass spectrometry forms the family of laser mass-spectrometric (LAMS) methods, such as REMPI-ToFMS and MALDI-ToFMS. In REMPI-ToFMS the connecting element between UV spectroscopy and mass spectrometry is laser-induced REMPI ionisation. An intermediate state of the molecule of interest is selectively excited by absorption of a laser photon (the wavelength of a tuneable laser is set in resonance with the transition). The excited molecules are subsequently ionised by absorption of an additional laser photon. Therefore the ionisation selectivity is introduced by the resonance absorption of the first photon, i.e. by UV spectroscopy. However, conventional UV spectra of polyatomic molecules exhibit relatively broad and continuous spectral features, allowing only a medium selectivity. Supersonic jet cooling of the sample molecules (to $5-50\,K$) reduces the line width of their vibronic peaks in the UV absorption spectra, increasing the spectroscopic selectivity. Under these conditions, well-structured spectral features (often as sharp as those found in IR spectroscopic fingerprints) can be observed in the molecular UV spectra. The very high sensitivity and selectivity of REMPI-ToFMS allows detection of volatiles in headspace. Multidimensionality has been reviewed [4].

7.1   PRECOLUMN HYPHENATION

Principles and Characteristics   Because of the limited selectivity of extraction, a chromatographic analysis is almost always needed. Recently, a fair amount of progress has been made regarding the front end of the total analysis procedure, namely the integration of sample preparation (this being the analytical bottleneck) and separation. The idea behind such systems is to perform sample extraction, cleanup and concentration as an integral part of the analysis in a closed system. Scheme 7.2 shows the main procedures related to sample preparation for chromatographic analysis.

Recovery procedures have traditionally involved some form of solvent, gas or heat extraction from the bulk sample matrix. Some of these lend themselves to precolumn hyphenation (e.g. SFE, TD, Py, HS), as opposed to others (e.g. Soxhlet, ultrasonics). Extraction of additives should not be considered as an isolated step, because it may strongly influence the subsequent chromatographic separation. The success of an analysis may very often depend more on the extraction procedure than on the chromatographic separation. In hyphenation there should be compatibility between the sample preparation and subsequent chromatographic analysis.

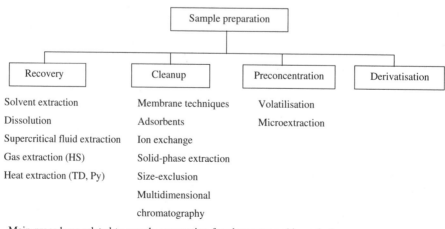

Scheme 7.2   Main procedures related to sample preparation for chromatographic analysis

In rare instances, recovery of an analyte from the sample matrix is so efficient that all unwanted matrix components are left behind. Usually, however, *cleanup* is necessary to eliminate unwanted co-extracted matrix components from the sample extract. Coupled column techniques (e.g. SPE, SEC and other multidimensional chromatographic methods) are useful for this purpose. The goals of cleanup are to achieve removal of matrix interferences; concentration or dilution of the analyte; and preparation of the analyte in a form most suited to the needs of the final chromatographic techniques. The ultimate success of the analysis is often critically dependent on that of the cleanup step. In the usual analytical methods, additives are extracted using large quantities of solvents first, followed by concentrating the resulting solution to make the analysis possible. Suitable *concentration methods* are SFE, SPME, CIS and CT. Use of SFE eliminates the use of large quantities of solvents. Chromatography may also be coupled to *derivatisation* (e.g. in a PTV injector). The main reason for derivatising a sample, prior to separation by GC, is to render highly polar materials sufficiently volatile, and to allow their elution at high temperatures without thermal decomposition or molecular change.

Coupling of sample extraction/preparation steps with chromatographic analysis has the potential of achieving very rapid, sensitive and cost-effective analyses. In general, on-line extraction/separation methods are most successful when the extraction fluid is of the same phase as the chromatographic carrier fluid. In principle there are two different modes of coupling a sample preparation method to a separation technique, as discussed in Section 3.3.3.1. One is the on-line mode (Table 7.5), connecting the two techniques directly with tubing and valves. An ideal on-line extraction/GC method would combine the ability of liquid solvent extraction to extract efficiently a broad range of analytes, with the ability of gas-phase extraction methods (e.g. headspace analysis) to rapidly and efficiently transfer the extracted analytes to the gas chromatograph. In the off-line mode, the extract is usually transferred from the collector to the analytical column through robotic interaction. Both modes have distinct, but different, advantages.

Table 7.5 On-line precolumn chromatographic techniques

Sample preparation	GC	SFC	HPLC	TLC
SFE	+	+	+	+
SPE	+	+	+	+
SPME	+		+	
Pyrolysis	+			
Derivatising	+		+	

The solvents used in extraction may affect subsequent chromatography. Precolumn hyphenation is considered to be much more difficult and critical than the more developed postcolumn hyphenation. Sample preparation and injection are considerable bottlenecks in terms of ruggedness.

The most common precolumn chromatographic techniques discussed here are SFE (Section 3.4.2), SPE (Section 3.5.1) and SPME (Section 3.5.2). However, sampling methods such as thermal desorption, pyrolysis and headspace (Section 4.2.2) may also be classified in this category.

Supercritical fluid extraction (SFE) has found application in combination with various chromatographic techniques. There are several modes by which SFE can be applied to the preparation of samples for chromatography [5]. The analyte may be separated selectively from the interfering matrix components (Scheme 7.3a), but is more usually co-extracted (Scheme 7.3b), and may subsequently be analysed by an appropriate instrumental technique (on-line or off-line) (Scheme 7.3c). SFE used in isolation is generally not selective enough to isolate specific solutes from the extraction matrix without further cleanup or resolution from co-extracted species prior to qualitative and quantitative analysis. In more elaborate methods, SFE is coupled with sorbent technology. Scheme 7.3d shows SFE extraction of both analyte and interfering components, followed by fractionation of the analyte from interfering solutes. In Scheme 7.3e the interfering co-extracted components from the SFE step are permanently isolated on the sorbent cartridge.

One of the attractive features of SFE with CO_2 as the extracting fluid is the ability to directly couple the extraction method with subsequent analytical methods (both chromatographic and spectroscopic). Various modes of on-line analyses have been reported, and include continuous monitoring of the total SFE effluent by MS [6,7], SFE-GC [8–11], SFE-HPLC [12,13], SFE-SFC [14,15] and SFE-TLC [16]. However, interfacing of SFE with other techniques is not without problems. The required purity of the CO_2 for extraction depends entirely on the analytical technique used. In the *off-line mode* SFE takes place as a separate and isolated process to chromatography: extracted solutes are trapped or collected, often in a suitable solvent for later injection on to chromatographic instrumentation. Off-line SFE is inherently simpler to perform, since only the extraction parameters need to be understood, and several analyses can be performed on a single extract. Off-line SFE still dominates over on-line determinations of additives—an

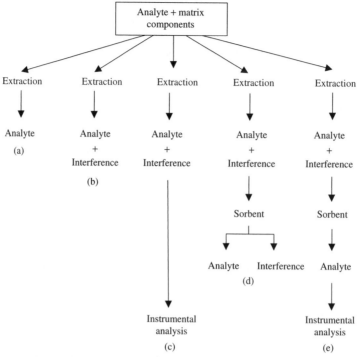

Scheme 7.3 Generalised supercritical fluid methods for extraction and analysis with and without separation of interfering components. After King [5]. Reproduced from the *Journal of Chromatographic Science*, by permission of Preston Publications, a Division of Preston Industries, Inc.

important reason being the need for representative sample sizes. Off-line SFE also offers the greatest capability for method development.

In the on-line process, SFE and chromatography are coupled to form an integrated process: following SFE the extracted species are passed directly to the chromatograph, usually via a trap or sample loop and a valve switching device. *On-line SFE* collection requires minimal sample handling between the extraction and chromatographic steps. The major advantage of on-line approaches is the potential to transfer every extracted analyte molecule to the chromatographic system. Depending on the samples, the interface may have to be optimised for the analytes and for the matrices. Some keywords are: restrictor temperature and plugging, trap temperature, loss of solutes, trap dimensions, band broadening, peak focusing and memory effects. On-line techniques are preferred when maximum sensitivity is needed from small samples. Analyte sensitivity enhancement obtained by on-line interfacing can be significant (i.e. 1000-fold or greater). Sample loadability of the extraction vessel is relatively small (1–3 g) because of the analyte capacity of the analytical chromatographic column and detector. In comparison,

the sample size in off-line SFE can be much larger (>5–10 g) and the extraction can be achieved with a faster flow-rate (0.5–7 mL min^{-1} compressed) as compared to on-line SFE. The latter technique generally requires that the chromatographic system be used for sample collection during the SFE step, thus reducing the number of chromatographic runs that can be performed during a set period of time. As the extraction solvent is easily removed, while the analytes of interest are trapped for subsequent separation, trace analysis is possible with on-line GC or SFC. SFE has a distinct advantage in multidimensional systems, as the initial separation method, except in analysis of dilute aqueous samples (which usually require removal of water by solid-phase extraction as the first sample preparation step). In direct coupling, the solvent peak is eliminated and the analysis of compounds that elute with the solvent becomes possible. SFE selectivity is determined mainly by solute solubility in the supercritical extraction fluid; therefore competition of SFE-chromatography with multidimensional chromatographic techniques is difficult. SFE has also been hyphenated to multidimensional chromatographic techniques such as LC-GC and SFC-GC. The inflexibility of SFE coupled

Table 7.6 Main features of coupled SFE-chromatographic techniques

Advantages

- Combined sample preparation and analysis
- Reduction in number of sample handling steps
- Increased sample throughput
- High recoveries (generally better than Soxhlet or TD techniques)
- High extraction rates
- Automated operation
- Processing of thermolabile compounds
- (Limited) manipulation of selectivity (variation of SF solvent power)
- Sample concentration (by decompression prior to chromatographic analysis)

Disadvantages

- Mainly qualitative data
- Detectabilities at ppm–ppb levels
- Complicated interfaces
- Blocking of capillaries (by cryogenic cooling or precipitation)
- Method development effort

Table 7.7 Advantages of on-line solid-phase extraction–chromatography couplings

- Rapid sample preparation mode
- Preconcentration (manifold sample enrichment)
- Sample cleanup
- Minimal sample preparation efforts
- No sample losses during work-up
- Orthogonality (e.g. SPE-GC, SPE-CE)

to chromatographic separation techniques has limited their continued usage. In view of the elimination of organic solvents, on-line SFE techniques are particularly attractive to industries whose current sample preparation techniques require solvents that are undergoing increased regulation (i.e. chlorinated solvents). Table 7.6 summarises the main characteristics of coupled SFE-chromatographic techniques.

In *sorption methods*, compounds are partitioned from the sample into the bulk of a polymer retaining phase. Sorptive techniques comprise solid-phase extraction (SPE), solid-phase microextraction (SPME) and stir bar sorptive extraction (SBSE) (Section 3.5.3). SPE represents the most commonly used sample preparation technique nowadays for cleanup and sample preconcentration. *SPE-chromatographic couplings* can be off-line (cartridge), at-line (robotic interface), on-line (transfer line) or in-line (complete integration). For nonroutine samples, the normal mode of operation for SPE is off-line, i.e. the analyte(s) are preconcentrated and then analysed separately. Stand-alone SPE, including such steps as eluate collection, evaporation, reconstitution and sample transfer for injection, can be a time-consuming sample preparation technique. An automated on-line SPE system would be used as an alternative rapid sample preparation mode for large numbers of routine samples and process monitoring. In on-line SPE, the purified analytes are desorbed from the SPE column directly into the analytical system, thus significantly reducing cycle time. For specific purposes, *on-line SPE* systems have been

coupled directly to a range of analytical instrumentation, such as HPLC, SFC, GC (if PTV technique can be used), CE, ICP, spectrophotometry, etc. Advantages of SPE-chromatographic couplings are indicated in Table 7.7.

Benefits of on-line SPE include elimination of eluate collection, evaporation, and processing in a completely closed system with protection from contact with hazardous solvents, light and oxygen. On-line SPE offers ruggedness, excellent precision and high sample throughput, and is ideally suited to perform a fast matrix–analyte separation prior to e.g. LC-MS or direct MS analysis.

Whereas SPE is a sample cleanup method, *SPME* is essentially a solvent-free sampling method. Stir bars in hyphenated SBSE-TDS-CGC configuration for product control analysis are a powerful tool for the extraction and analysis of organic compounds in aqueous matrices.

Also, on-line coupling of pressurised hot-water extraction (PHWE) with chromatography (GC, NPLC and RPLC) has been reported [17]. In even more advanced approaches, two extraction methods have been combined in order to achieve high selectivity and sensitivity, e.g. PHWE-MMLLE-GC [18], where microporous membrane liquid–liquid extraction (MMLLE) serves as a trapping device concentrating and cleaning the extract. Other such examples are MAP™-SFE and MAE-MAHS-SPME, i.e. on-line microwave-assisted extraction followed by solid-phase microextraction in headspace assisted by microwave irradiation [19]. Also, ASE®-ASPEC has been proposed in which automated sample preparation with extraction cartridges (ASPEC) [20] purifies, fractionates and concentrates ASE® extracts by SPE [21]. Fully automated ASE®-ASPEC-HPLC-MS has been reported [21].

Sample handling in chromatography has been reviewed [22], also specifically for SFE [23,24]; several textbooks have appeared [25,26].

General applications The majority of reported analytical chromatographic applications of SFE to date have been concerned with the coupling of SFC (Section 7.1.1.4), using very small extraction cells, with capillary and packed column SFC instruments to effect sequential SFE-SFC separation schemes. Applications of on-line SFE-chromatography have been reviewed [23].

7.1.1 Chromatographic Sampling Methods

Principles and Characteristics Gas chromatographic sampling varies with sample type: e.g. SFE, TD, TG, Py and HS for solids; and SFE, LLE, SPE, HS, LC, SDE, for liquids. Sample introduction methods for GC are commonly classified as primary methods (HS, PT, LLE, SFE, SPE) and secondary methods (column coupled separations: GC, LC, SFC). The difference between these two groups is that in the primary techniques the number of compounds injected onto the column may be very large (depending on the sample), whereas secondary techniques provide the ability to select the number of compounds transferred to a second-stage column. *On-line sample preparation-GC techniques* may also be subdivided into solvent-free techniques and those requiring introduction of some 10–100 μL of an organic solvent into the GC. The latter category comprises LVI-GC (including SPE-GC), NPLC-GC and RPLC-GC. Although NPLC-GC is markedly successful, limitations are the heartcut nature of the set-up and the restricted use of NPLC separations. RPLC-GC is an analytical technique of minor interest, due to problems created by most typical RPLC eluents. Within the LVI-based subgroup, SPE-GC is the favoured option.

As to the solvent-free subgroup, SFE can be coupled to a wide range of chromatographic techniques, including GC, SFC, HPLC, SEC and TLC. On-line coupling of SFE to GC and SFC was considered to be relatively straightforward, since SCFs will decompress into gases which are inherently compatible with GC and SFC. As a means of sample introduction to GC, on-line SFE is an alternative to conventional syringe injection, to HS, PT, TD and Py. In comparison to these introduction systems, SFE has the potential to encompass a wide range of volatile to nonvolatile analytes. SFE introduces extracted analytes to the GC in the gas phase, without the problems related to introducing large volumes of liquid solvents. Yet, on-line SFE-GC has not become the promised viable approach, due to various operational and design problems and the unexpectedly complicated nature of supercritical fluid–analyte–matrix interactions. On the other hand, SPME can be coupled rather easily with GC, and sorption is rapid. SPME can be used to concentrate volatile or nonvolatile components from liquid, solid or gaseous sampling. The SPME device not only combines extraction and concentration, but also directly transfers the adsorbed compounds into a GC injector. Coupling to GC, GC-MS, split/splitless and on-column injection or desorption of the analytes in an SPME-HPLC interface have been described. Headspace solid-phase microextraction (HS-SPME) shortens the time of extraction and facilitates analysis of solid samples, provided that the analytes are volatile. As mentioned in Section 4.2, GC is also well equipped with a variety of injection modes.

Sampling methods in SFC are far more restricted than in the case of GC (Section 4.3 and Table 7.5). Not surprisingly, supercritical fluid extraction is an obvious choice.

An ideal extraction method for subsequent *LC analysis* requires compatibility of the extraction solvent with the mobile phase used in the chromatographic separation of additives. In case of incompatibility, the additive-containing extract has to be concentrated almost or even to dryness for subsequent dilution or redissolving in an appropriate solvent, usually the major component of the mobile phase. Extract concentration is often also required because of the low additive content in the polymer samples. On-line extraction and chromatographic analysis have also been carried out using a multiple-solvent gradient and a special guard column [27]. The principle of the method is to dissolve the sample in a good solvent and precipitate or suspend the polymer sample in the HPLC instrument with a nonsolvent. This precipitate or suspension will absorb on to a guard column. By adding a solvent to the adsorbed suspension with a gradient system, the monomers, additives and polymers will redissolve and elute from the column [28]. The general situation of sample pretreatment/introduction in LC is superior to GC if aqueous samples are considered. Use of disposable or reusable SPE cartridges, combined on-line with a (reversed-phase) LC system, enables efficient analyte enrichment from 10–100 mL water samples, without analyte breakthrough. On-line coupling of SFE to LC has received less attention than SFE-GC.

The number of reports on *on-line TLC analysis* of extracts is quite limited. Stahl [16,29] described a device for supercritical extraction with deposition of the fluid extract on to a moving TLC plate. On-line SFE-TLC provides rapid and simple insight into the extraction performance. Its strength is that the extract is deposited on a plate, which means that detection is a static process. Limitations of SFE-TLC are that quantification is difficult, and that the stability of components on the support material or in the presence of oxygen may be a problem. For additives in beverages (such as benzoic and sorbic acid preservatives) an SPE-TLC method has been described [30].

Overviews of sample preparation for chromatographic separations are available [31–33]. See also Table 4.15. Trapping methods for GC were critically reviewed [34].

Applications The potential of a variety of direct solid sampling methods for in-polymer additive analysis by GC has been reviewed and critically evaluated, in particular, static and dynamic headspace, solid-phase microextraction and thermal desorption [33]. It has been reported that many more products were identified after SPME-GC-MS than after DHS-GC-MS [35]. Off-line use of an amino SPE cartridge for sample cleanup and enrichment, followed by TLC, has allowed detection of 11 synthetic colours in beverage products at sub-ppm level [36]. SFE-TLC was also used for the analysis of a vitamin oil mixture [16].

Sandra and David [37] have reported on validation studies for SHS-HSGC, Py-HSGC, on-line LVI-GC, on-line SPME-GC, on- and off-line SPE-GC, on-line and off-line derivatisation-GC, SBSE-TD-GC, and PTV-LC(SEC)-GC.

7.1.1.1 On-line SFE–GC Coupling

Principles and Characteristics Gas chromatography is the most widely applied analytical tool for volatile organics, and therefore on-line coupled SFE-GC is likely to be also of interest for such compounds. Indeed, most of the initial coupling of SFE to chromatographic methods was obtained with GC. Since there is a high compatibility between solubility in CO_2 and volatility in GC, this trend is likely to continue to be developed into highly automated robotic systems. The vast majority of GC separations utilise capillary (e.g. 250–320 μm i.d.) columns, and most SFE-GC work relates to capillary GC.

Conventional introduction systems in GC allow gaseous or liquid samples to be introduced in a narrow band. In case of SFE, the solvent is a liquefied gas when introduced into the interface of an on-line SFE-GC. After depressurisation the volume will expand about 10^3 times. A large amount of gaseous solvent must be separated and removed from the solutes in the interface. Cryogenic focusing should be applied to prevent band spreading. The choice of fluids for SFE-GC is limited by the need for volatility (as the depressurised fluid) in addition to the general characteristics of SFE fluids. Highly automated analytical systems for SFE-GC were developed in the late 1980s. Multivessel on-line SFE-GC was described [38]. Figure 7.1 shows a schematic diagram of a typical on-line SFE-GC.

Coupled (or on-line) SFE-GC requires four essential steps:

(i) efficient extraction of the analytes from the sample matrix;
(ii) transfer of the analytes from SFE to GC;

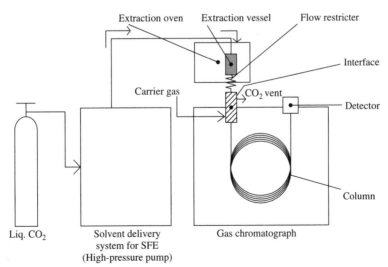

Figure 7.1 Schematic diagram of a typical on-line SFE-GC. After Maeda and Hobo [39]. Reprinted from T. Maeda and T. Hobo, in *Hyphenated Techniques in Supercritical Fluid Chromatography and Extraction* (K. Jinno, ed.), pp. 255–274, Copyright (1992), with permission from Elsevier

(iii) depressurisation of the SCF to the gaseous state and removal from the chromatographic system; and

(iv) refocusing of the extracted analytes in a narrow band for good chromatographic peak shapes and efficient separations.

A variety of approaches has been used to couple SFE extraction with capillary GC [40,41] since the first reports of on-line SFE-GC in 1986 [8,42]. Three *general approaches* to SFE-GC are outstanding:

(i) use of a sample loop and valve to introduce a heartcut of the flowing SFE extract to the GC [43];

(ii) use of a cryogenic or sorbent trap external to the GC to collect the SFE analytes from the depressurised fluid [11]; and

(iii) direct injection (split/splitless, PTV or cold trap-injection), in which the analytes are directly transferred to the GC from the SFE cell by the extraction fluid, without the need for intermediate valves or traps.

Recently, much interest has been shown in the use of a PTV injector as an interface between the SFE pressure restrictor and the GC column. Various SFE-GC interfaces have been discussed [39,41,44], with emphasis on the use of restrictor heating and cryogenic focusing. Choice and optimisation of SFE-GC methods were discussed by Hawthorne [40]. Direct coupling of SFE with GC in an on-line approach is conceptually straightforward, assuming quantitative deposition of the extracted analyte into the chromatographic inlet. This technique is most successful, particularly for obtaining quantitative results. For quantitative SFE-GC, a calibration method is required in which the calibration standard is introduced into the GC system in the same way as the real sample. Split SFE-GC techniques can accommodate larger samples than on-column techniques (<100 mg), which is of particular advantage for heterogeneous samples [45]. On-column SFE yields maximum sensitivity, since all of the extracted analytes are transferred into the GC column. Split SFE-GC extends the range of samples that can be analysed by (on-column) SFE-GC techniques, because nonvolatile matrix components are prevented from being transferred into the GC column; split SFE-GC is more suitable for larger sample sizes and high analyte concentrations. External analyte traps are advantageous when the sample requires the use of modified supercritical fluids.

Selectivity in on-line SFE-GC is regulated by varying extraction temperature and pressure, and by the use of alternate SCFs (CO_2, N_2O, Xe), modifiers, and different adsorbents (for removal of interfering analytes). Calibration of both on-column and split SFE-GC is accomplished by either adding a standard solution to an inert matrix in the extraction cell, or by injecting the standard. Both on-column and split SFE-GC methods yield good quantitative reproducibilities (2–10 % RSD for replicate SFE-GC analyses). Hawthorne *et al.* [45] have described quantitative split SFE-GC-MS analysis.

SFE-GC is an attractive approach to coupling the extraction, concentration and chromatographic steps for the analysis of samples containing analytes that can be analysed using capillary GC. Often it is difficult to identify all the components which are extracted from samples by FID alone. This is a particular problem when the sample history and/or the identity of the compounds of interest are not known. When SFE-GC is combined to powerful spectroscopic detectors, unique data can be obtained, allowing their use as routine tools in the analytical laboratory. For positive identification of components of interest, multihyphenated techniques such as SFE-GC-AED, SFE-GC-MS, SFE-GC-FTIR-MS are employed [46].

On-line SFE coupled to GC or SFC, according to the thermal stability of the analytes, are both very competitive with classical methods of analysis in terms of sensitivity and analysis time. Since all of the extracted analytes are transferred to the GC system, much higher method sensitivities can be obtained. Several modes of operation are possible utilising on-line SFE-GC, including: quantitative extraction of all analytes from a sample matrix; quantitative extraction and concentration of trace analytes; selective extractions at various solvating powers to obtain specific fractions; and periodic sampling (multiple-step extractions) of the effluent at various pressures for qualitative characterisation of the sample matrix.

Off-line SFE is inherently simpler for the novice to perform, since only the SFE (and analyte collection) step needs to be understood. In off-line SFE further cleanup or a pretreatment step can be employed to eliminate interferences. With off-line SFE, sensitivities are limited by the fact that only about 1 μL of the collection solvent is generally injected into the GC. The daily sample throughput can be higher using off-line SFE, since SFE-GC requires that the GC be used for a sample collection device (rather than performing chromatographic separations) during the SFE extraction, whereas several off-line extracts can be loaded into an autosampler for unattended GC analysis.

Table 7.8 Main characteristics of on-line SFE-GC coupling

Advantages

- Sample extraction, concentration, separation and detection in 1 h
- No sample handling between extraction and separation (no contamination)
- No modification of the gas chromatograph
- High sensitivity (quantitative collection of extracted analytes)
- Reduced possibilities for degradation and loss of analyte (extraction at relatively low temperatures)
- No interference of modifiers with target analyte determination
- Reliable quantitation (using TDM and HS-GC)
- Class-selective extractions (multiple SFE-GC analyses at different extraction pressures)
- Various operating modes
- Fractionation of complex samples
- Automation

Disadvantages

- Analyte solubility in an SCF (limited for higher-MW and polar compounds)
- Sample range (fundamentally limited by GC, see Figure 7.2)
- Complete extraction needed for quantitative analysis
- Small sample size (<100 mg for on-column); heterogeneity problems
- Complicated optimisation
- Small extraction cell volumes
- Co-extraction of undesirable compounds
- Need for SFC- or high-purity CO_2 grade

Table 7.8 summarises the *main characteristics* of on-line SFE-GC-D (D = FID, MS, NPD, ECD). SFE-GC-FID/MS requires SFC-grade CO_2, while it is essential to use the high-purity CO_2 grade for SFE-GC-ECD. SFE-GC appears to have advantages when relatively volatile analytes (e.g. hexane) need to be determined, particularly when such species are difficult to collect using off-line methods. The selectivity obtainable with the wide range of solvent powers available with SFE provides some potential for fractionation of complex samples and isolation of specific analytes from a matrix. For uncharged relatively apolar compounds, which can be dissolved in $scCO_2$, SFE shows several advantages over liquid extraction techniques. However, for more polar analytes, the extraction efficiency depends strongly on the extraction conditions (p, T, modifier).

Liu *et al.* [9] have overcome some of the most common drawbacks of on-line SFE-GC devices [10,11], namely:

(i) relatively long analysis time;

(ii) need for complete extraction for quantitative analysis (occasionally impossible due to extremely strong adsorptive interactions); and

(iii) small extraction cell volumes (15–300 μL) to avoid column overloading.

Large cell dimensions require either longer extraction times than is practical, or high extraction flow-rates. Small sample sizes create problems with regard to the homogeneity of the material samples. These authors have developed a high-speed, thermal desorption modulated (TDM) SFE-GC-FID with a large extraction cell (1 cm × 10 mm i.d.) which permits simultaneous sample extraction and analysis. In this device, samples of extract are introduced as sharp injection pulses to the high-speed GC at 10-s intervals, where they are separated while extraction proceeds. With this system, the extract is sampled continuously, thereby allowing almost real-time monitoring of the extraction of selected components. The system generates much pertinent data in a very short time, and is useful in designing and optimising extraction processes. Quantitative results could be extrapolated before extraction was complete, from curves obtained by signal averaging of segmented chromatograms using Bartle's 'hot-ball' model [47]. The total amounts of analytes present in the original sample matrix were derived from the equation:

$$m_o = m_1 + m_2^2/(m_2 - m_3) \qquad (7.1)$$

where m_o is the amount of analyte in the original sample matrix, and m_i is the amounts of analyte extracted within time t_i. Compared to the thermal desorption method, on-line SFE-GC can recover higher-molecular-weight molecules without using high temperatures. Figure 7.2 shows the molecular ranges of substances which can be handled by chromatography and SFE.

Modifiers can be used very effectively in on-line SFE-GC to determine the concentration levels of the respective analytes. This presents an advantage in terms of the use of modifiers in SFE, since they appear as solvent peaks in GC separations and do not interfere with the target analyte determination. Although on-line SFE-GC is a simple technique, its applicability to real-life samples is limited compared to off-line SFE-GC. As a result, on-line SFE-GC requires suitable sample selection and appropriate setting of extraction conditions. If the goal is to determine the profile or matrix composition of a sample, it is required to use the fluid at the maximum solubility. For trace analysis it is best to choose a condition that separates the analytes from the matrix without interference. However, present SFE-GC techniques are not useful for samples

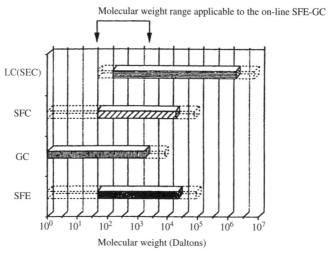

Figure 7.2 Molecular weight range applicable to chromatography and SFE. After Maeda and Hobo [39]. Reprinted from T. Maeda and T. Hobo, in *Hyphenated Techniques in Supercritical Fluid Chromatography and Extraction* (K. Jinno, ed.), pp. 255–274, Copyright (1992), with permission from Elsevier

that contain high concentrations of extractable, but not 'GC-able' matrix components. To increase the range of analytes which can be investigated by SFE-GC, on-line derivatisation of 'non-GC-able' components has been envisaged [48].

Instrumentation and applications of on-line SFE-GC have been reviewed repeatedly [9,10,23,40,49–51].

Applications   Off-line SFE-GC-MS has been used for the analysis of additives in polyurethanes [52]. Marín *et al.* [53] have used off-line SFE-GC to determine the detection limits for DOP and DBP in flexible PVC. The method was compared with Soxhlet liquid extraction.

On-line SFE-GC finds use especially in petroleum-related applications [54], but has also been applied to polymer additives [47,55]. PBT polymers were extracted at 200 bar and 55 °C for the determination of carbonic acid diphenyl esters and other *volatiles*, using on-line SFE-GC-MS [47]. Extraction of entrained volatiles is a quality test for some polymers. SFE-GC-FTIR-MS has been employed to reveal the cause of odour of a smelly hose (a plasticiser) [56]. SFE-GC can also profitably be used for the determination of *residual solvents* in polymers such as benzene, toluene and *o*-xylene [57]. *Oligomers* of PE (up to 1000 Da) were determined by GC after supercritical fluid extraction [58].

The depth distribution of *light stabilisers* in coatings has been studied in 1–3 mg microtomed slices, by means of SFE-GC with ToF-SIMS and nitrogen thermionic detection, as well as by direct ToF-SIMS analysis; results were in good agreement [59]. As the SFE effluent

is totally compatible for on-line introduction into a GC, SFE-GC is an effective way to deal with light-sensitive or very volatile target analytes [60]. SFE-GC-AED has been used profitably for the determination of *organotin compounds* [61] and alkyl lead compounds without interferences from co-extracted components. Applications of SFE-GC have been reviewed [24,51].

7.1.1.2   On-line SPE–GC Coupling

Principles and Characteristics   Although early published methods using SPE for sample preparation avoided use of GC because of the reported lack of cleanliness of the extraction device, SPE-GC is now a mature technique. Off-line SPE-GC is well documented [62,63] but less attractive, mainly in terms of analyte detectability (only an aliquot of the extract is injected into the chromatograph), precision, miniaturisation and automation, and solvent consumption. The interface of SPE with GC consists of a transfer capillary introduced into a retention gap via an on-column injector. Automated SPE may be interfaced to GC-MS using a PTV injector for large-volume injection [64]. LVI actually is the basic and critical step in any SPE-to-GC transfer of analytes. Suitable solvents for LVI-GC include pentane, hexane, methyl- and ethylacetate, and diethyl or methyl-*t*-butyl ether. Large-volume PTV permits injection of some 100 μL of sample extract, a 100-fold increase compared to conventional GC injection. Consequently, detection limits can be improved by a factor of 100, without the need for preconcentration. (For a 1-mL sample, the detection limit is typically of the order of 0.1 ppb.) Alternatively, the sample volume to be processed can be reduced while retaining a high sensitivity. Reduction in sample volume not only reduces the overall sample processing time, but also enables automation to be applied to a wider range of SPE methods. Automation of SPE-GC allows higher sample throughputs, with minimal eluate evaporation and maximum analyte recovery. Such a system is ideally suited to trace analysis. *On-line SPE-LVI-PTV-GC-MS* without packing material in the liner, with fast early solvent vaporisation and concentration of the target compounds at the inlet of the GC column, handles 20-mL samples and avoids degradation resulting from catalytic effects of the packing material and desorption by packing particles [65]. The main characteristics of on-line SPE-LVI-PTV-GC are given in Table 7.9.

The inherent advantages of on-line SPE-GC for the analysis of aqueous samples are well known [66] and comprise the use of an integrated and closed system in which the SPE extract is transferred from the sample preparation module to the autosampler via an autosampler vial.

A recent extension of the scope of SPE-GC and SPE-GC-MS concerns the use of AED detection with its multielement detection capability and unusually high selectivity. Hankemeier [67] has described *on-line SPE-GC-AED* with an on-column interface to transfer 100 μL of desorbing solvent to the GC. The fully on-line set-up is characterised by detection limits of $5-20\,ng\,L^{-1}$, because of quantitative transfer of the analytes from the SPE to the GC module. On-line coupling of SPE with GC is more delicate than SPE-LC, because of the inherent incompatibility between the aqueous part of the SPE step and the dry part of the GC system.

Applications On-line SPE-GC and SPE-GC-MS couplings find wide application for sample cleanup of biological, environmental and industrial analysis of *aqueous samples* [67]. SPE-GC-AED/MS is ideally suited for the (nontarget) screening of hetero-atom-containing compounds in aqueous samples, and allows confirmation plus identification in one run [68]. Specific applications of hyphenated SPE-GC systems for polymer/additive analysis were not identified.

7.1.1.3 On-line SPME–GC Coupling

Principles and Characteristics As mentioned already (Section 3.5.2) solid-phase microextraction involves the use of a micro-fibre which is exposed to the analyte(s) for a prespecified time. GC-MS is an ideal detector after SPME extraction/injection for both qualitative and quantitative analysis. For SPME-GC analysis, the fibre is forced into the chromatography capillary injector, where the entire extraction is desorbed. A high linear flow-rate of the carrier gas along the fibre is essential to ensure complete desorption of the analytes. Because no solvent is injected, and the analytes are rapidly desorbed on to the column, minimum detection limits are improved and resolution is maintained. On-line coupling of conventional fibre-based SPME coupled with GC is now becoming routine. Automated SPME takes the sample directly from bottle to gas chromatograph. Split/splitless, on-column and PTV injection are compatible with SPME. SPME can also be used very effectively for sample introduction to fast GC systems, provided that a dedicated injector is used for this purpose [69,70].

The *main characteristics* of on-line SPME-GC coupling are given in Table 7.10. Although the SPME sampling regime itself (room-temperature operations) is not harsh, other parts of the analytical system (such as the

Table 7.9 Main features of on-line SPE-LVI-PTV-GC

Advantages

- Total transfer of extracted species (closed system), cleanup, concentration
- No analyte losses by evaporation
- Decreased risk of sample contamination
- No degradation of target compounds
- High sensitivity
- Reduced sample-processing time
- Automation
- On-site use

Disadvantages

- Requires great differences in polarity
- Analysis of aqueous samples

Table 7.10 Main characteristics of on-line SPME-GC coupling

Advantages

- Solvent-free sample introduction
- Excellent chromatography
- Simplicity of experimental set-up; relatively problem-free
- Relative ease of method development and optimisation

Disadvantages

- Smaller application range than SPE-GC
- Lower analyte detectability and precision than SPE-GC
- Unsuitable for thermolabile compounds

GC injection port) may well be. This was clearly demonstrated by a comparison between cryotrapping/direct injection, cryotrapping/SPME, and solid-phase (Tenax-GC) extraction for sampling of odorous sulfur compounds [71]. Thermally labile compounds are likely to break down in the GC injection port/column/transfer line. Instead of SPME-GC, the recently developed SPME-HPLC [72] might be more applicable to analysis of such thermally unstable compounds.

SPME-GC and SPE-GC are complementary rather than competitive techniques. SPE requires at least 100 µL, of which only a small fraction is actually used for subsequent GC or HPLC analysis. This problem is addressed by SPME. Although SPME-GC is an attractive technique, its precision (RSDs, 3–20 %) is often not as good as in SPE-GC (RSDs, 1–10 %).

Membrane extraction with a sorbent interface (MESI) is a relatively new sampling/sample preparation technology, which has not yet been commercialised [73]. Górecki and Pawliszyn [70] coupled micro-GC to solid-phase microextraction (SPME) and membrane extraction with a sorbent interface (MESI). In this arrangement, the membrane, permeable to small organic molecules, acts as a selective barrier, allowing some molecules to pass from the outside to the carrier gas stream, while blocking others. After crossing the membrane the molecules are concentrated on the sorbent trap, which is periodically flash-heated to desorb the analytes and inject them in the GC column. SPME and membrane extraction with a sorbent interface (MESI) [73] are both techniques capable of concentrating the analytes from a variety of matrices and introducing them – in the form of a narrow band – to a GC column, without additional focusing steps. The most important aspect of MESI is its ability to concentrate volatile analytes selectively on-line before fast separation. This can result in dramatically improved sensitivity. Even with a preconcentration time of only 1 min, the sensitivity of MESI was higher by more than two orders of magnitude compared with direct injection of a gas sample. The portability of SPME makes it an ideal tool for rapid field screening of a multitude of samples. SPME is a batch method, well suited for screening, whereas MESI enables semicontinuous analysis – ideal for monitoring of both gaseous and liquid streams. Potential applications of this technique include process analysis. An impact on polymer/additive analysis is less likely.

Applications   Coupling of solid-phase *micro*-extraction and a *micro*-GC (separation times of 15 sec) is suitable for rapid field screening and potentially useful for process analysis. Odours at ppt level can be analysed by

hyphenated SPME-GC-MS [74]. A total of 18 low-MW products formed during thermo-oxidation of PA6.6 at 100 °C, in addition to similar products originating from the lubricants, were extracted by SPME and identified by GC-MS [75]. Exempt coating solvents have been sampled by means of HS-SPME and analysed by GC-FID/MS after spiking with acetone-d_6, methylacetate-d_3 and/or *m*-chlorobenzotrifluoride [76]. Potter and Pawliszyn [77] reported the combined approach of SPME-GC-QITMS for direct analysis of extracts of *volatile organics* in water. SPME-GC is well suited to water toxicity studies, and has been used for the analysis of 4-nonylphenol, a metabolite of many alkylpolyethoxylate surfactants and detergents, in water [78]. SPME-GC was also used to determine DEHP in an immersed plastic solution [79]. *Metal speciation* by means SPME-GC-ICP-MS has been reported [80].

7.1.1.4   On-line SFE–SFC Techniques

Principles and Characteristics   An important reason for pursuing off-line SFE-SFC is the need for representative sample sizes. On the other hand, small sample sizes have an advantage in avoiding the heavy constructions needed for large-sample high-pressure applications. Also, small samples allow a high linear velocity in the extractor, reducing the extraction time, and diminish build-up of polymeric material in restrictors and pumping. On-line SFE-SFC-UV coupling was first reported in 1985 [81], and can be performed in both static and dynamic mode. Direct transfer of solutes from SFE to SFC using scCO$_2$ is a major attraction of using SFE directly coupled to SFC. On-line SFE has most often been coupled with both packed and capillary SFC (SFE-pSFC, SFE-cSFC) and with capillary GC (SFE-CGC). Coupling of SFE to SFC is achieved either in parallel or in series. The parallel connection, which requires independent SFE and SFC systems, is more suitable for process monitoring, as opposed to the analytical nature of the connection in series.

SFE-SFC operation consists of extraction, trapping, and transfer to the column for chromatography. Concentration of the solutes of interest before sample introduction to SFC may be achieved by:

(i)   repeated evaporation of scCO$_2$;
(ii) trapping of the solutes; or
(iii) use of on-line cryogenic trapping.

The latter system allows quantitation [82]. Cryotrapping requires very low temperatures, because volatile analytes are partially or completely lost, especially at long extraction times. An alternative to cryotrapping is a technique using a sorbent. Once extraction and trapping are complete, the analytes are removed from the trap with the supercritical fluid. Koski and Lee [83] have discussed solute focusing in this second extraction step. Transfer of the analytes from the trap to the column may be aided by elevating the trap temperature to quickly desorb analytes from the trap surface. The main points for improvement in on-line SFE-SFC are trapping and transfer. Polar compounds are usually recovered in higher yields from open tubular column traps than from sorbent-filled traps, due to better deactivation of the former. Generally, the success of on-line SFE-SFC depends greatly on the trapping technique used.

In SFE-SFC, the same fluid that is used for extraction is likewise used for chromatography. Even though the fluid is decompressed and exists as a gas in the interface, where extracted analyte is focused, raising the pressure and the temperature of the chromatographic mobile phase converts it into an SCF once again for SFC. SFE-SFC couplings are simple and easy to build; commercial equipment is available. Cotton *et al.* [84] and others [44] have described on-line SFE-SFC instrumentational arrangements, from relatively simple to complicated. However, these arrangements have a constraint in the quantitative analysis of components in the sample. Extraction performed in stop-flow mode does not allow thorough extraction. In order to extract as much as possible, the fluid flow should be continuous and the extract should be trapped by means of an adsorbent. Direct coupling of SFE to SFC usually involves a valve switching device. A thermal modulation interface has been developed for SFE-cSFC [85]. Various detectors have been reported in the on-line SFE-SFC combination, such as MSD, ELSD, FID, UVD, FTIR, NPD, SCD [81,86]. Very pure SFE-SFC grade solvents (99.999 %) enable fluids such as CO_2 to be linked directly to several detection systems, such as FID, ECD, UV, FTIR, MS and AED. On-line SFE-SFC-FID is capable of separating and detecting numerous polymer additives using a single method. On-line SFE-SFC-FTIR and -MS are discussed in Sections 7.3.2.1 and 7.3.2.2, respectively.

SFE can be combined with several forms of SFC, i.e. with conventional packed columns (1–4.6 mm i.d.: packed-column SFC or pSFC), with capillary columns (10–250 μm i.d.; capillary SFC or cSFC), and recently with packed capillary columns (200–530 μm i.d., 3–10 μm particles; packed capillary SFC or pc-SFC).

Since cSFC has good chromatographic properties for most of the additives used, *SFE-cSFC coupling* has become a widely used 2D technique, albeit mostly

Table 7.11 Main characteristics of on-line SFE-cSFC

Advantages

- Sensitivity
- Reproducibility
- Small sample size
- High linear velocity (reduced extraction time)
- Growth potential

Disadvantages

- Long equilibrium time required (for static mode)
- Considerable method development
- Heterogeneity problems
- Difficult quantitation (sample size, CO_2 flow-rate control)
- Not suitable for aqueous samples (operational problems)
- Interference of co-extracted compounds

in off-line mode in view of heterogeneity problems. According to Greibrokk [23], coupling SFE with either SFC or LC for determination of additives in polymers has considerable growth potential, combined with improved sampling methods, smaller samples and automated equipment.

The main characteristics of on-line SFE-cSFC are given in Table 7.11. As the whole extract is transferred to the analytical column, low detectable quantities are possible. Nevertheless, quantitation of the SFE-cSFC system parameters is difficult to achieve. Effects from changes in CO_2 flow-rates on quantitative measurements were discussed [87]. An accurate and practical method for the measurement of CO_2 gas flow-rates at the system's restrictor outlets was developed using an ELCD detector [87]. For on-line SFE-cSFC under static conditions, the sample to be extracted is held in a cell under pressure to equilibrate for a certain period of time before an aliquot of the extract is introduced into the cSFC column. The major disadvantage is the long equilibrium time necessary. Static SFE-cSFC with aliquot sampling can be used for basic qualitative investigations. Dynamic SFE-cSFC operates principally by continuously exposing the analytes to a fresh stream of SCF and trapping, thus accumulating the extracted components. Various trapping methods have been described [84]. For successful analysis by coupled SFE-cSFC various parameters must be optimised. The efficiency of intermediate trapping must be considered. The nature of the analyte is crucial, while the presence of co-extracted, interfering compounds demands either selectivity during extraction, or trapping on an absorbent from which selective desorption into the SFC column is possible. The sample size must avoid overloading of the SFC column. The conditions for efficient SFC analysis must be optimised, preferably off-line. Not all samples are suitable for SFE-cSFC. If an analyte can be analysed by GC, then SFE-GC should be the preferred technique. The presence of water in the matrix presents considerable problems by freezing in the SFE-cSFC flow path. While direct extraction of aqueous media has been described in SFE-pSFC [88,89], it proved necessary to separate water from the SCF containing phenol analytes by means of a phase separator. The smaller capacities of capillary columns have rendered SFE-cSFC on aqueous samples difficult if not impossible. The presence in the matrix of interfering compounds that are soluble in the SCF may also pose problems.

On-line SFE-pSFC coupling was first reported by Sugiyama *et al.* [81] using packed HPLC columns, and was later reviewed by Anderson [90]. The major advantages of the technique are summarised in Table 7.12. High selectivity towards the solutes of interest can be achieved in SFE-pSFC as a result of the extraction selectivity provided by pressure and temperature parameters and the possible use of modifiers in scCO$_2$, with additional selectivity provided by pSFC through the wide range of commercially available column packings. A main challenge lies in the analysis of more polar compounds. The sample capacity of packed columns makes the technique valuable for fraction collection of solutes for additional investigation. The robust back-pressure regulators used in pSFC – as compared to capillary restrictors in cSFC – help to make SFE-pSFC more stable and reproducible than SFE-cSFC [90]. Packed column SFC provides lower resolution than cSFC, but allows a greater sample capacity and a rapid analysis time.

Most of the SFE-SFC devices developed are designed to obtain qualitative results. However, various quantitative analyses of polymer additives have been reported [82,87,91–93]. The ability to remove the SCF is particularly important when SFE is coupled on-line

to chromatographic techniques. This can be particularly true for trace analysis, when it is useful to get the analytes in as concentrated a form as possible in order that they can be measured with sufficient sensitivity by the chosen analytical technique.

On-line SFE–packed capillary SFC is an interesting development in comparison with SFE-cSFC, because of the higher loadability and shorter analysis times. In comparison with SFE-pSFC, the advantages are a lower pressure drop, higher efficiency (theoretical number of plates) and lower flow-rates, resulting in easier interfacing with FID or MS instruments.

On-line SFE-SFC modes present several distinct *advantages* that are beyond reach of either technique when used separately (Table 7.13). An obvious advantage of SFE is that it is an ideal way to introduce a sample into an SFC system. Because the injection-solvent is the same as the mobile phase, in this respect the criteria for a successful coupling of different techniques are fulfilled [94], i.e. the output characteristics from the first instrument and the input characteristics of the second instrument are compatible. Supercritical fluid techniques can separate high-MW compounds; are significantly faster than classical Soxhlet extractions; and require less heat and solvent. SFE-SFC techniques are versatile,

Table 7.13 Advantages and limitations of on-line SFE-SFC techniques

Advantages

- Compatibility of extraction solvent and chromatographic mobile phase
- On-line automation of sample preparation and chromatographic analysis steps
- Cleanup, fractionation and concentration prior to high-resolution chromatography
- High reproducibility
- Increased productivity
- Preparation with minimal sample contamination
- High sensitivity
- Small sample requirements (0.5 mg)
- Quantitative analysis
- Trace analysis capability
- Versatility
- Wide application range (recommended for reactive and thermally labile analytes)
- Commercial equipment

Disadvantages

- Small SFC-user community
- Historical instrumental problems (plugging with larger samples, carry-over effects)
- Severe method development (large number of experimental variables)
- Restrictions on sample nature (excluding aqueous samples and polar compounds)
- Poorly defined application area for SFC

Table 7.12 Main characteristics of on-line SFE-pSFC

Advantages

- One-step analysis
- High selectivity
- Rapid extraction and direct analysis of unstable and oxidation sensitive solutes
- Quantitative analysis
- Suitability for trace analysis (quantitative transfer of solutes between extractor and chromatograph)
- Low contamination potential (for trace-level solutes)
- High loading capacity
- Use of modifiers
- Wide range of detectors
- Reproducibility
- Automation

Disadvantages

- Relatively low resolution

and their solvating power can be manipulated through changes in pressure and by addition of modifiers. When the sample contains thermally labile or reactive compounds, SFE-SFC is recommended. The procedure is excellent for thermally unstable polymer additives in commercial plastics or for fatty acids and triglycerides in foods, etc., which cannot readily be analysed by GC without derivatisation. SFE-SFC complements SFE-GC when target analytes are thermally labile or beyond the volatility range of GC. An additional advantage of on-line SFE-SFC over on-line SFE-HPLC and SFE-GC is that it is unlikely that sample constituents which are insoluble in the mobile phase will be introduced into the column. On-line SFE-SFC allows for small sample sizes to be used, as compared to off-line studies. Nevertheless, off-line is the simpler system from the experimental point of view. In SFE-SFC, SFE eliminates the injection problems that plague quantitative SFC analysis.

In the mid-to-late 1990s, SFC became an established technique, although only holding a niche position in the analytical laboratory. The lack of robust instruments and the inflexibility of such systems has led to the gradual decline of SFE-SFC. Only a small group of industrial SFE-SFC practitioners is still active. Also the application area for SFC is not as clearly defined as for GC or HPLC. Nevertheless, polymer additives represent a group of compounds which has met most success in SFE-SFC. The major *drawbacks* of SFE-SFC are the need for an optimisation procedure for analyte recovery by SFE (Section 3.4.2), and the fair chance of incompatibility with the requirements of the chromatographic column. The mutual interference of SFE and SFC denotes non-ideal hyphenation.

Many reviews have dealt with SFE-SFC coupling [23,24,95–101], some specifically with SFE-cSFC [84], and others with SFE-pSFC [90]. See also Section 4.3.

Applications SFE-SFC solves problems in such diverse areas as polymers/monomers, oils/lubricants, foods, pharmaceuticals, natural products, specialty chemicals, coatings, surfactants and others. Off-line SFE–SFC survives alongside on-line determinations of additives, because of the need for representative sample sizes. Off-line SFE-SFC was used for extraction of AOs from PP [102]. In cases where the analyst wishes to perform further analysis on the extracted species, it is useful to be able to isolate the extract from the solvent. The ability to remove the solvent easily is particularly important when SFE is coupled on-line to chromatographic techniques, but is equally important for trace analysis when it is useful to concentrate

the analyte as much as possible. SFE-SFC coupling has initially extensively been used for the determination of *low-MW components and additives* in polymers [15,38,82,103–106]. On-line SFE-SFC has found applications for extraction and separation of antioxidants [15], plasticisers and low-MW oligomers from thermoplastics [107]. Levy *et al.* [105] reported on-line SFE-SFC analysis of PE/(BHT, Irgafos 168, Irganox 1010/1035/1076/1330, erucamide) films. Also, on-line SFE-SFC analyses of LDPE/(BHT/BHEB, Isonox 129, Irganox 1010/1076) and of PE/(Irgafos 168, Cyasorb UV3346, Cyanox 1790) have been described [38]. As most of the additives in *polyolefins* are of medium polarity and soluble in scCO$_2$, SFC is suitable for their analysis. The Suprex Corporation [108] has reported a variety of SFE-SFC-FID and SFE-SFC-UV analyses of polymer additives: LLDPE/(Tinuvin 326, Irganox 1010), HDPE/(Irganox 1010/1076, Irgafos 168, Ultranox 626, Ethanox 398, Weston 399, Cyanox 1790) and HDPE/(Irgafos 168, Irganox 1010/1076/1330). On-line SFE-SFC with CO$_2$ has been reported for PE and PP/(Irganox 1010, BHT, erucamide, Tinuvin 770, Irgafos 168, Isonox 129, DLTDP) [92], PE and PS/(BHT/BHEB, Isonox 129, Irganox 1010/1076, Irgafos 168, Cyasorb UV3346, Cyanox 1790, stearylstearamide, HBCD and erucamide) [82], PU/(Amgard TCEP/TMCP/V6 and Termolin 101) [109], as well as for the determination of other low-MW organics in the following systems: nylon/(cyclic trimer and ethylbis-stearamide) [15] and PET/cyclic trimer [103]. On-line SFE-cSFC (CO$_2$ only) was used for PE/erucamide acid amide [110] and PE/Irganox [110]. On-line SFE-SFC was also applied for the analysis of PE/(Irganox 1076, TNPP, Weston 618), PS/EBS, polymer film/(BHT, Irganox 1076, stearamide, erucamide), styrofoam/-HBCD, ethafoam/(stearyl stearamide, Irganox 1010) and LDPE pellet/(BHT/BHEB, Isonox 129, Irganox 1010/1076) [104]. Hirata *et al.* [111] have described the use of on-line SFE-pcSFC for the extraction of PE film.

On-line SFE-SFC-FID of an HDPE/(Irgafos 168, Irganox 1010) extract shows co-extracted C$_{10}$ to C$_{30}$ polyethylene oligomers interfering with antioxidants in the case of FID detection (Figure 7.3); when the UV detector is used, the oligomeric species are not detected, as they contain few chromophoric functional groups (Figure 7.4) [112]. This exemplifies the importance of selecting the appropriate detection configuration for analysis. Cotton *et al.* [15] described SFE-cSFC conditions which extracted additives selectively and separated them from *co-extracted oligomers*.

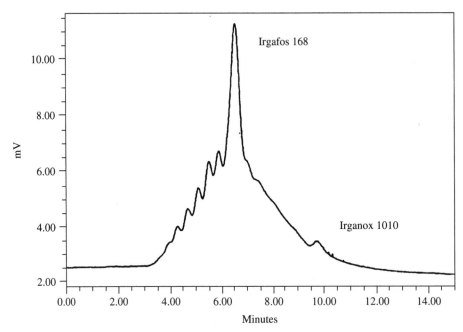

Figure 7.3 SFE-SFC-FID chromatogram of an HDPE/(Irgafos 168, Irganox 1010) extract. After Tikuisis and Dang [112]. Reprinted with permission from T. Tikuisis and V. Dang, in *Plastics Additives, An A−Z Reference* (G. Pritchard, ed.), Chapman & Hall, London, pp. 80−94. (1998) Copyright CRC Press, Boca Raton, Florida

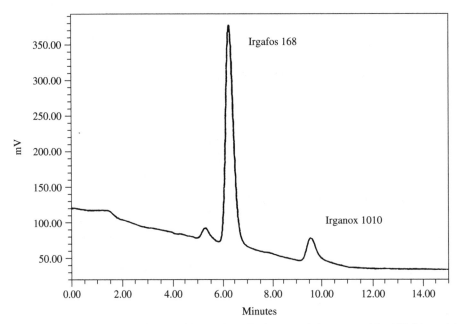

Figure 7.4 SFE-SFC-UV chromatogram of an HDPE/(Irgafos 168, Irganox 1010) extract. After Tikuisis and Dang [112]. Reprinted with permission from T. Tikuisis and V. Dang, in *Plastics Additives, An A−Z Reference* (G. Pritchard, ed.), Chapman & Hall, London, pp. 80−94 (1998) Copyright CRC Press, Boca Raton, Florida

The applicability of on-line SFE-cSFC for the analysis of polymer additives has also been demonstrated for several PP samples. In extracting PP film utilising SFE-SFC, restrictor plugging was mentioned as a problem with larger samples [20]. Bücherl et al. [113] compared Soxhlet extraction–SFC with Soxhlet extraction–HPLC and on-line SFE-SFC of biaxially stretched PP/(Irganox 1010/1076, Irgafos 168) film for food contact applications. It was noticed that the SFE extraction efficiency is lower than that of the more drastic Soxhlet extraction, which also extracts very high-MW substances not relevant in food migration processes. SFE-SFC is suitable as a screening test. Bücherl et al. [113] have also analysed PP/(Irganox 1010/1076, Irgafos 168, antistatic) by means of SFE-SFC-FID, also revealing various oligomers.

On-line SFE-pSFC-FID, using formic or acetic acid modified CO_2 as an extraction solvent, was used to analyse a dialkyltin mercaptide stabiliser in *rigid PVC* (Geon 87444) [114]. Hunt et al. [115] reported off-line SFE-pSFC-UV analysis of PVC/(DIOP, chlorinated PE wax, Topanol CA), using methanol as a modifier. Individual additives are unevenly extracted at lower pressures and temperatures, where extraction is incomplete. Topanol CA, the most polar of the three PVC additives studied, could not be fully extracted in the time-scale required (15–20 min), even at the highest CO_2 temperature and pressure obtainable. However, methanol-modified CO_2 enhances extraction of Topanol CA. PVC film additives (DEHP, fatty acids, saturated and aromatic hydrocarbons) were also separated by off-line SFE-preparative SFC, and analysed by PDA and IR [116].

Flame retardants in *polyurethane foams* were determined by SFE-SFC [117]. Off-line SFE-SFC-FID was used for the analysis of additives in polyurethanes [52], and on-line SFE-SFC for extraction of additives from isocyanate formulations [107].

On-line SFE-SFC-ELSD analysis of the *textile and fibre finish* components, butyl stearate/palmitate/myristate, was reported [118]. Similarly, Kirschner et al. [119] used on-line SFE-pSFC for fibre finish analysis; on-line SFE-SFC was also instrumental in the analysis of the total composite finish on a commercial textile thread [120].

Various on-line SFE-SFC applications for the analysis of polymer additives and *oligomers* have been reported, such as: additives in PE [121]; polymer additives and oligomers in PP; nylon pellets and PEEK granules [15]; and oligomers in PET films [103], PS [122] and PMMA [123], sometimes in quantitative fashion [15,103]. SFE-SFC has identified the cyclic trimer tris-(ethylene terephthalate) in commercially available PET film [103]. Also, SFE-SFC extraction and

separation of acrylic acid dimers, trimers, etc., from adhesive formulations of low-profile additives from cured thermoset composites has been reported [107].

Linear calibration curves for *quantitation* by means of SFE-SFC may be constructed by extracting standards spiked on to quartz wool in the extraction vessel. SFE-pSFC calibration curves for BHT, Isonox 129, Irganox 1076 and Irganox 1010 added to LDPE at various spike levels were reported [82]. Ashraf-Khorassani et al. [104] used on-line SFE-CT-pSFC with a cryofocusing trap (CT) for quantification of additives in a variety of polymeric matrices: PS/*N*, *N*-ethyl-bis-stearamide (EBS), PE/(TNPP, Irganox 1076, Weston 618), LDPE/(BHT/BHEB, Irganox 1076), PE/(stearyl stearamide, Irganox 1010), styrofoam/(HBCD, phthalocyanine blue), LDPE/(Irgafos 168, Cyasorb UV3348, Cyanox 1790) and LDPE/(BHT/BHEB, Isonox 129, Irganox 1076/1010). Addition of a cryofocusing trap permits focusing of low-level analytes into discrete narrow bands, allowing quantitative and reproducible results with better detection limits. After extraction and collection in the trap, all of the extracted components are directly injected on to the chromatographic column. Table 7.14 gives a selection of the quantitative results, which are quite satisfactory. The same holds for the quantitative determination of 1.45 % dimethyltin mercaptide in rigid PVC (RSD 2.9 %, $n = 6$), by means of on-line SFE-pSFC-FID [114].

Other quantitative SFE–cryogenic trapping SFC-FID analyses have been reported for additives in LDPE [82].

Table 7.14 Experimental SFE-SFC concentrations of different additives from various polymers

Additive	SFE-SFC calculated concentration (ppm)	Expected concentration (ppm)
BHT, BHEB (in LDPE)	510	540
Isonox 129 (in LDPE)	220	210
Irganox 1076 (in LDPE)	250	265
Irganox 1010 (in LDPE)	250	240
Irgafos 168 (in LDPE)	200	205
Cyasorb UV3346 (in LDPE)	1000	1050
Cyanox 1790 (in LDPE)	110	120
Stearyl stearamide (in PE)	95	100
Irganox 1010 (in Ethafoam)	310	300
HBCD (in styrofoam)	120	140
BHT (in a polymer film)	240	250
Irganox 1076 (in polymer film)	200	185
Stearamide (in polymer film)	520	510
Erucamide (in polymer film)	95	100

After Ashraf-Khorassani et al. [104]. Reproduced from the *Journal of Chromatographic Science*, by permission of Preston Publications, a Division of Preston Industries, Inc.

On-line SFE-SFC has also been used for the quantification of erucamide and antioxidants in PE [110]. Cotton *et al.* [15] have reported quantitative extraction of additives from PP at five different extraction pressures, at a constant flow-rate and temperature. Below 50 atm, extraction was negligible; between 50 and 200 atm, Tinuvin 326 and 770 were extracted, along with small quantities of oligomers. Higher pressures lead to the extraction of all the additives present, with the integrated peak areas conforming well to the actual concentrations.

Linear calibration curves were obtained for three additives by direct SFC and SFE-SFC from an inert matrix, and from PP samples [93]. Quantitative determination was linear over a wide range of analyte content. The duration of extraction and analysis did not exceed 75 min. In order to analyse a mixture of additives, which is fundamental for industrial applications, experimental conditions corresponded to a compromise. PE and PP samples from several manufacturers were extracted by on-line SFE-pSFC-FID by Ryan *et al.* [92]. Extraction efficiencies exceeding 92 % were obtained for 50 mg samples for ten different additives ranging from BHT (218 Da) to Irganox 1010 (1178 Da). Additives were typically extracted under relatively mild conditions. Oligomers are extracted at much higher pressures. It is thus possible to selectively extract either oligomers or additives. Accurate quantitation was reported. Baner *et al.* [87] have used monitoring of CO_2 gas flow-rates as a basis for quantitative analysis of Irganox 1010 and 1076 in PP and of triacetin (TA; plasticiser) in the biodegradable polymer Biopol, by means of on-line SFE-cSFC-FID. Quantitation of total extractables in polymers by SFE appears to be a viable screening method for the total amount of potential migrants in polymers. The observed high uncertainty for the method (variation coefficients of about 25 %) was ascribed to the small sample sizes (0.17 g PP) used, and to the uncertainty in the FID calibration factors. Quantitative SFE-cSFC coupling requires further development. The method works well only for substances with similar CO_2 solubilities. Bias against higher-MW additives should also be eliminated.

On-line SFE-SFC method development for validated quantitative analysis of PP/(Irganox 1010/1076, Tinuvin 327) has been reported [93]. SFE conditions required optimisation of extraction time and pressure, matrix type (particle or film) and matrix parameters (particle size, film thickness, sample weight). About 30 % of extracts were lost during collection. Very poor recoveries (20–25 %) were reported from ground samples (particle size 100 μm < d < 160 μm) and additive-dependent recoveries of 45–70 % for 30-μm-thick films. Bücherl

et al. [124] have used on-line SFE-cSFC-FID/MS for simultaneous quantitative determination (FID) and identification of unknown extractable substances (EI-MS) in packaging materials. Residue monomers, oligomers, additives, lacquers, adhesive gums and sealing agents are important sources of such substances. SFE-cSFC has also been used for extraction and quantitation of additives in ground PE and PP samples [92]. Quantitative analysis of polymer additives, including Irganox 1010/1076, Irgafos 168, Tinuvin 326/770/P, stearamide and triacetin, by on-line SFE-SFC has been reported by various authors [15,87,92,104,109]. High extraction recoveries (92–96 %) were mentioned. Other quantitative analyses of antioxidants in PE have appeared [106].

The minute sample sizes allowed in SFE-SFC analysis (typically 0.5 mg; cf. the approximate weight of 30 mg for a single pellet), which is several orders of magnitude smaller than the sample weights used in GC, HPLC or IR analysis (5–10 g), allows us to perform additive dispersion studies on a pellet-to-pellet basis [106].

Anderson [90] foresees a bright future for SFE-pSFC, with applications in the polymer industry. However, it does not appear that routine SFE-SFC quantitation is an easy matter. SFE, SFC and SFE-SFC applications have been described in various reviews and books (Sections 3.4.2, 4.3 and Bibliography).

7.1.1.5 SFE–HPLC Couplings

Principles and Characteristics There are many compounds which are not amenable to GC analysis, and in these cases on-line SFE-HPLC coupling can be a useful tool. On-line SFE-HPLC coupling was first reported in 1983 [13]; a commercial instrument became available in 1988 [125,126]. An SFE-HPLC interface is less straightforward than in the case of either GC or SFC. The main problem lies in coupling a sample preparation technique where gas is produced in the interface to a chromatographic technique with a liquid mobile phase. In SFE-HPLC, a gas must be introduced into an LC column, or the extract must be deposited on an interface while the decompressed CO_2 gas is vented. The recirculating mode of SFE with an in-line sampling valve provides an easy means of coupling SFE with HPLC (Figure 7.5).

SFE-HPLC consists of various steps. Extraction and collection of the analytes are closely related to each other, but each is controlled by separate variables. For this reason, the joint optimisation of the steps is difficult to perform, and they must be studied independently

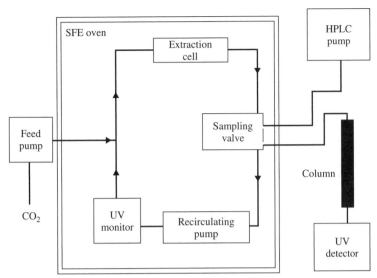

Figure 7.5 Schematic representation of a coupled SFE-HPLC system employing a recirculating extraction manifold interfaced to HPLC via a sample injection valve. After Lynch [54]. Reprinted from T. Lynch, in *Chromatography in the Petroleum Industry* (E.R. Adlard, ed.), *J. Chromatography Library*, **56**, 269–303, Copyright (1995), with permission from Elsevier

(often using factorial design) [12]. Experimental variables comprise static and dynamic SFE (with or without a modifier), an off-line and on-line coupling mode, column characteristics and a mobile phase. Only systems that use solid sorbent traps allow for the use of dynamic as well as static SFE. Dynamic extraction linked with HPLC is particularly useful where maximum sensitivity is required. Hirata and Okamoto [127] introduced a compromise between on- and off-line HPLC analysis, using a microbore column packed with silica. Instead of connecting the trap on-line to the HPLC, it was removed from the extractor and placed in the HPLC system, just before the column, and used as an injection loop.

Use of FID and SCD are compatible with SFE-HPLC, since they are flame-based and unaffected by gases in the mobile phase. Unfortunately, SCD can only be used with micro-HPLC (column i.d. $<320\,\mu m$), which requires miniaturised equipment not commonly found in most analytical laboratories. When following SFE with HPLC analysis using a spectroscopic detector, a medium-purity grade is usually sufficient.

In combinations of SFE and HPLC with more polar compounds, the need for modified CO_2 is obvious. On-line SFE-HPLC is particularly helpful for those samples that are:

(i) light- and/or air-sensitive;
(ii) at trace level; and
(iii) present in limited quantities.

Trace analysis is particularly attractive for SFE-HPLC since quantitative transfer of all analytes extracted to the chromatographic system becomes possible. At present, on-line SFE-HPLC appears to be feasible for qualitative analysis only; quantitation is difficult due to possible pump and detector precision problems. Sample size restrictions also appear to be another significant barrier to using on-line SFE-HPLC for quantitative analysis of 'real' samples. On-line SFE-HPLC has therefore not proven to be a very popular hyphenated sample preparatory/separation technique. Although on-line SFE-HPLC has not been quantitatively feasible, SFE is quite useful for quantitative determination of those analytes that must be analysed by off-line HPLC, and should not be ruled out when considering sample preparatory techniques. In most cases, all of the disadvantages mentioned with the on-line technique (Table 7.15) are eliminated. On- and off-line SFE-HPLC were reviewed [24,128].

The configuration can be expanded by adding other sample preparation instruments to facilitate automating other preparative steps that may intervene between SFE and the analytical instrument, e.g. solvent exchange, internal standard addition, serial dilutions for calibration curve generation, SPE for further cleanup of the extract output by SFE, derivatisation of components within the SFE extract, and many other (currently) manual–human intervention techniques.

The recirculating mode of SFE has also been coupled successfully to commercially HPLC-MS systems, with

Table 7.15 Main features of on-line SFE-HPLC

Advantages

- Reliable
- Fast
- Environmentally friendly
- Trace analysis

Disadvantages

- Method development (factorial design)
- Not quantitatively feasible
- Sample size restrictions

various interfaces (MB, TSP, PB) [54]. The PB/TSP combination on the same instrument allows excellent flexibility, as the particle beam gives EI-type spectra and therefore fragmentation data, and the thermospray ionisation mode provides molecular weight data. This permits quite rapid collection of much structural information on unknowns.

Applications   Off-line SFE-HPLC appears to be applicable and quantitative for a variety of samples in many 'real'-world matrices. The main challenge lies in the use of this technique for the more polar compounds. Quantitative off-line SFE-SFC-UV analysis of HDPE/Ethanox 330 was described after extensive method development (varying modifiers, modifier concentration, temperature) [129]. Soxhlet extraction and SFE-RPLC-UV of PE samples were compared [127]. A sample size (inhomogeneity) problem was pointed out when a SFE reproducibility study was performed on five 3-mg samples of PE. This points to limits

in the usefulness of miniaturisation of analytical tools for polymer/additive analysis. Figure 7.6 demonstrates the inhomogeneity for 0.5–0.7-mg PE samples determined by SFE-HPLC according to ref. [127]. Hirata and Okamoto [127] examined additives in PE and PP by means of off-line SFE-μHPLC; after decompression the analytes were collected in a microtrap filled with silica.

Reports of on-line SFE-HPLC are rare, perhaps because the majority of analytes that have been extracted using SFE can be separated using either GC or SFC. On-line SFE-HPLC is often used to monitor extraction efficiencies. SFE-HPLC optimised for temperature (120 °C), pressure (384 bar), SCF flow and modifier (methanol) has been used for the quantification of Irganox 1010 and Irgafos 168 extracted from PP. In this case Thilén and Shishoo [12] varied three SFE parameters for optimisation of the extraction efficiency, and five parameters for the collection efficiency, see Figures 7.7 and 7.8. Despite these efforts, low recoveries were observed (Table 7.16). This was attributed to problems associated with the compounding process, and not to uncertainties in the extraction and analytical method.

Daimon *et al.* [130] used on-line micro SFE-HPLC for the analysis of PP/(BHT, Tinuvin 326, Irganox 1010). A disadvantage is that the technique requires the extraction of very small sample sizes, in order that extracts will not overload the HPLC system.

SFE-LC-MS has given unique information for the examination of polymer additives and their breakdown products [131]. Lynch *et al.* [46] studied PE/(Santonox R, Cyasorb UV531, Irganox 3114/1010/1076) by means of SFE-LC-TSP-MS and SFE-LC-PB-MS. This system has been shown to give less degradation in polymer

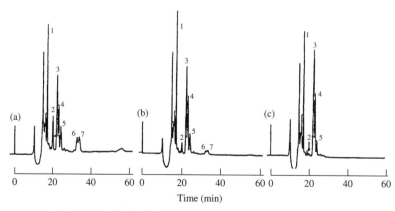

Figure 7.6 Heterogeneous distribution of additives in (a) 0.728 mg, (b) 0.580 mg, and (c) 0.494 mg PE samples determined by SFE-HPLC, according to Hirata and Okamoto [127]. Reprinted from Y. Hirata and Y. Okamoto, *Journal of Microcolumn Separations*, **1**, 46–50, Copyright © (1989, John Wiley & Sons, Inc.). This material is used by permission of John Wiley & Sons, Inc.

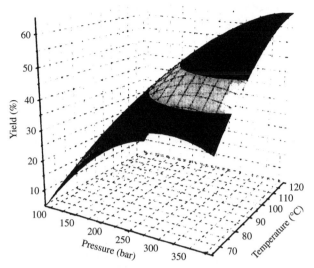

Figure 7.7 SFE recoveries of Irganox 1010 from PP as a function of pressure and time (modifier: 2 % MeOH; CO$_2$ flow, 4 mL min^{-1}). After Thilén and Shishoo [12]. Reprinted from M. Thilén and R. Shishoo, *Journal of Applied Polymer Science*, **76**, 938–946, Copyright © (2000, John Wiley & Sons, Inc.). This material is used by permission of John Wiley & Sons, Inc.

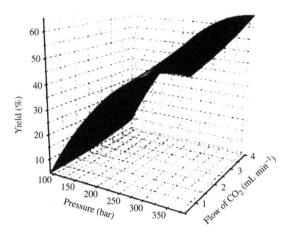

Figure 7.8 The effect of flow of scCO$_2$ on the recovery of Irganox 1010 from PP by means of SFE (temperature 120 °C; modifier, 2 % MeOH). After Thilén and Shishoo [12]. Reprinted from M. Thilén and R. Shishoo, *Journal of Applied Polymer Science*, **76**, 938–946, Copyright © (2000, John Wiley & Sons, Inc.). This material is used by permission of John Wiley & Sons, Inc.

Table 7.16 Recoveries from SFE optimisations[a]

Analyte	Optimisation[b]	Recovery (ppm)	Recovery (%)	SD (%)
Irganox 1010	1	1570	62.6	2.8
Irgafos 168	1	1850	74.1	2.0
Irganox 1010	2	1450	57.2	5.0
Irgafos 168	2	1720	68.7	2.9

[a] Results are from four replicate tests.
[b] Optimisation of extraction efficiency (1) and collection efficiency (2). After Thilén and Shishoo [12]. Reprinted from M. Thilén and R. Shishoo, *Journal of Applied Polymer Science*, **76**, 938–946, Copyright © (2000, John Wiley & Sons, Inc.). This material is used by permission of John Wiley & Sons, Inc.

additives than was seen when an off-line SFE with collection in solvent was employed.

7.1.1.6 On-line SPE–HPLC Coupling

Principles and Characteristics Off-line SPE utilises disposable cartridges or discs containing 50 to 2000 mg of sorbent, whereas on-line SPE-HPLC utilises so-called precolumns which have obligatory small dimensions. The SPE column frequently contains a low-efficiency sorbent which performs a preseparation of the sample, after which the analyte-containing fraction is directed on to a second high-efficiency column for separation and quantitation of the analytes of interest. For SPE followed by HPLC, the outlet of the precolumn is switched directly into the mobile phase flow path of the HPLC system, via a high-pressure valve. This situation is often referred to as column-switching liquid chromatography. This process eliminates off-line sample transfers, including eluate collection, evaporation, reconstitution and injection, thereby significantly reducing the cycle time. On-line SPE-HPLC coupling has become routine since 1995, and is now a robust technique. This is explained by good compatibility of the HPLC aqueous mobile phase with SPE of aqueous (biological or environmental) samples. SPE has usually been coupled on-line with HPLC systems using UV/VIS detectors (SPE-LC-UV). Despite the high sensitivity of SPE-LC-UV, the selectivity of the technique is limited, which is a considerable disadvantage when analysing complex samples. Fluorescence and electrochemical detectors have also been used, as well as diode array detectors (SPE-LC-DAD) and mass spectrometers (SPE-LC-MS). On-line SPE is ideally suited to perform a fast matrix–analyte separation prior to LC-MS or direct MS analysis.

Organic compounds can be determined rapidly by means of SPE-LC-(tandem)-MS, using two short columns [132]. However, combination of a strong SPE sorbent with a short analytical LC column, a popular combination for fast generic SPE-LC-MS, may result in poor chromatography if the LC eluent is too weak for efficient (rapid) desorption of the extract from the SPE sorbent. This problem may be overcome by *thermally assisted desorption* for on-line SPE (TASPE) [133]. Being a chromatographic process, SPE is a function of temperature. Few examples of thermally assisted desorption for on-line SPE can be found in the literature [134]. Temperature control enhances both speed and performance of SPE and on-line desorption to the analytical column. The principles and advantages of TASPE-LC have been described [133]. Increased desorption temperature leads to shorter SPE time and better SPE-LC efficiency. Hyphenated SPE-LC-MS or SPE-MS is rapidly gaining popularity, because it provides total automation, enhanced sensitivity and a major reduction of analyst time. For on-line SPE-LC-MS, SPE must be at least as fast as (fast) LC-MS.

The advantages of on-line SPE are shown in Table 7.17. Using on-line SPE-LC systems, a higher degree of automation, with a lower amount of sample and low detection limits, is usually obtained.

Klink [135] recently discussed sample preparation procedures for LC-MS. SPE can be so well integrated into the concept of LC-MS, that in many automated applications no clear distinction exists between SPE and LC [135]. In on-line LC-MS mode, the possibilities for changing the eluent are rather limited, because of the tolerance of the eluent for the interface. Moreover, the conventional gradient mode may lead to strong fluctuations in the response of the MS detector. Here the off-line mode, using SPE for concentration followed by selective elution, enables very far-reaching preseparation, due to the differences in the polarity of the eluents applied and their mixtures. Although the overall benefits of SPE for LC-MS applications are positive, extracts

Table 7.17 Main features of on-line SPE-HPLC(-MS) coupling

Advantages

- Reduction in sample-transfer steps
- Elimination of analyte collection, evaporation, reconstitution and injection
- Superior performance, reduced costs
- Closed system (light and air protection)
- Multiple choice of media and phases (including cartridges, columns and discs)
- Quantitative sample transfer
- Minor use of sample, solvent and time (compared to LC)
- Increased sample throughput
- Good method development resources available
- Trace enrichment
- Advanced automation

Disadvantages

- Limited cleanup and selectivity
- Method development may be complex

may contain fine particles which cause pressure build-up in HPLC columns.

Development and validation of protocols for solid-phase extraction coupled to LC and LC-MS have been described [136]. Hennion [137] has reviewed on-line SPE-HPLC coupling followed by various detection modes.

Applications With the current use of soft ionisation techniques in LC-MS, i.e. ESI and APCI, the application of MS/MS is almost obligatory for confirmatory purposes. However, an alternative mass-spectrometric strategy may be based on the use of oaToF-MS, which enables accurate mass determination at 5 ppm. This allows calculation of the elemental composition of an unknown analyte. In combination with retention time data, UV spectra and the isotope pattern in the mass spectrum, this should permit straightforward identification of unknown analytes. Hogenboom *et al.* [132] used such an approach for identification and confirmation of analytes by means of on-line SPE-LC-ESI-oaToFMS. Off-line SPE-LC-APCI-MS has been used to determine fluorescence whitening agents (FWAs) in surface waters of a Catalan industrialised area [138]. Similarly, Alonso *et al.* [139] used off-line SPE-LC-DAD-ISP-MS for the analysis of industrial textile waters. SPE functions here mainly as a *preconcentration device.*

On-line SPE-LC-ISP-MS/MS (with run times of 1–3 min) is widely used in pharmaceutical, clinical, and combinatorial chemistry, as well as in environmental monitoring, whenever rapid target MS analysis in a complex matrix is required. Thermally assisted desorption (TAD) increases the recovery of analytes significantly [134]. Baltussen *et al.* [140] also used on-line SPE-HPLC-MS after sorptive preconcentration.

7.1.1.7 On-line SPME–HPLC Coupling

Principles and Characteristics Solid-phase microextraction (SPME) has been exploited most effectively when coupled to GC, although more recently it has also been used for HPLC. Chen and Pawliszyn [141] first introduced the technology of coupling standard SPME fibre sampling to narrow-bore HPLC in 1995. Adsorption of analytes on to the SPME fibre is the same for both GC and HPLC; the difference is the means of desorption. Unlike GC, in case of HPLC no hot injector is available for thermal desorption of the analytes from the fibre. For SPME-HPLC, the analytes are desorbed by the eluent solvents of the HPLC system. This requires a separate interface. Both a loop and an 'in tube' interface have been described. Conventional HPLC injection ports are unsuitable for SPME, but can easily be modified by replacing the injection loop with a specially designed desorption chamber. The interface allows mobile phase to contact the SPME fibre; remove the adsorbed analytes, and deliver them to the column for separation (dynamic desorption). This is necessary because SPME is an equilibrium process. When analytes are more strongly adsorbed to the fibre, the fibre can be soaked in mobile phase before the material is injected on to the column (static desorption). For nonvolatile and thermally unstable organic analytes, SPME-HPLC provides a simple alternative to LLE-HPLC or SPE-HPLC methods, and eliminates the need for derivatisation before SPME-GC can be used.

The main characteristics of on-line SPME-HPLC(-MS) are shown in Table 7.18. Most of the SPME fibres are compatible with HPLC solvents. SPME combined with HPLC provides a means by which simple, rapid concentration of analytes can be achieved together with a means of introduction of the concentrated analytes to the HPLC system. This eliminates the need for larger injection volumes, and avoids derivatisation if the analytes were to be detected by GC. An advantage of the SPME method over LLE methods is the absence of a solvent peak in chromatograms obtained after extraction by SPME. SPME is not suitable for organic solutions. As SPME is a microextraction technique, coupling to μ HPLC may be envisaged.

Applications The scientific literature on this relatively new approach is still quite limited. SPME-HPLC-MS is suitable for quantitation of polar and semipolar organic compounds from aqueous solutions.

Table 7.18 Main features of on-line SPME-HPLC(-MS) coupling

Advantages

- Time-saving direct sampling from solutions/slurries (SPME) and solids (HS-SPME)
- Solvent-free extraction
- Analyte concentration
- Reproducible results from mg sample amounts
- High selectivity and identification capacity (MS^n option)
- Applicable to thermally unstable and nonvolatile compounds
- Choice and tuning of MS ionisation conditions define selectivity and sensitivity of the hyphenated technique
- Automation

Disadvantages

- Requires a special interface
- Characterisation of matrix influences requires determination of equilibrium times
- Desorption process requires appropriate solvents with sufficient desorption capacity
- Not suitable for organic solutions
- Expensive consumables

It is important that any method for surfactant analysis maintains the same oligomer distribution in the extracted samples. LLE and SPE are generally combined with chromatographic methods for separation and resolution of non-ionic surfactants into their ethoxamers. An alternative is the use of SPME-HPLC, recently reported by Chen and Pawliszyn [141]. Alkylphenol ethoxylate surfactants such as Triton X-100 and various Rexol grades in water were determined by means of SPME-NPLC-UV (at 220 nm) [142]. Detection limits for individual alkylphenol ethoxamers were at low ppb level.

7.2 COUPLED SAMPLE PREPARATION – SPECTROSCOPY/SPECTROMETRY

Principles and Characteristics Extraction or dissolution methods are usually followed by a separation technique prior to subsequent analysis or detection. While coupling of a sample preparation and a chromatographic separation technique is well established (Section 7.1), hyphenation to spectroscopic analysis is more novel and limited. By elimination of the chromatographic column from the sequence precolumn–column–postcolumn, essentially a chemical sensor remains which ensures short total analysis times (1–2 min). Examples are headspace analysis via a sampling valve or direct injection of vapours into a mass spectrometer (TD-MS; see also Section 6.4). In such cases high reliance is likely to be placed on chemometrics.

Sometimes, the combined selectivities afforded by both SFE and spectroscopic methods may render intermediate separations moot. In particular, use of CO_2 facilitates direct coupling to either a mass spectrometer or an IR spectrophotometer. On-line monitoring of SFE analytes may also be performed by means of ultraviolet (UV) or flame ionisation detection (FID). UV detection requires the presence of a chromophore in the extracted analytes. This illustrates one of the other advantages of using scCO$_2$, as it is transparent down to about 190 nm. By using absorbance detection (in contrast to FID), spectra giving functional group information can be obtained [143]. As fibre-optic monitor systems are available for wavelengths in the UV and NIR range, this technique may become increasingly important in coming years. However, other spectroscopic methods are also used in conjunction with SFE extraction, such as NMR [144] or fluorescence detection [145].

In SFE-FTIR there is no column, therefore every extracted analyte will elute simultaneously. *Off-line SFE-FTIR* is readily achieved by trapping the extracted analytes in an IR-suitable solvent (i.e. perchloroethylene) and observing the IR absorbance of the analyte-containing extract through a liquid cell. On-line analyses, however, are inherently more intricate, due to the difficulties of trapping analytes and quantitatively transferring them into the FTIR. Ikushima *et al.* [146] have reported development of an *on-line SFE-FTIR* system. In direct SFE-FTIR the analyte may be transferred via a stainless-steel transfer line directly to the FTIR flow-cell. On-line SFE-FTIR without intermediate trapping eliminates solvent use and results in a fairly rapid and reproducible analysis technique. The adaptability of FTIR detection for a variety of SFE protocols (e.g. static, dynamic, static–dynamic, direct analysis, dynamic–cryotrap) has been described [147]. Common on-line SFE-IR methods use expensive IR transparent windows with poor optical properties. On the other hand, an inexpensive fibre-optic FTIR accessory for on-line SFE-IR shows excellent optical qualities (transmission window 5000–800 cm^{-1}), and can be used for both qualitative and quantitative work [148].

The ability of SFE-FTIR to perform a variety of extraction methods is a definite advantage, especially for the study of complex mixtures containing analytes of varying solubility. For analytes which are readily solubilised in CO_2, direct dynamic and direct static–dynamic SFE-FTIR methods are quite successful. Elimination of the trapping process reduces both analysis time and potential analyte loss arising from incomplete trapping and/or recovery. Analytes which are not as readily solubilised by scCO$_2$, however, pose more of a problem, and dynamic extraction with cryotrapping may be necessary so that the analyte(s) may be exhaustively extracted first and then transferred into the FTIR as a whole. Disregarding matrix effects, results from cryotrapping SFE-FTIR can vary, depending upon trap temperature and CO_2 density, as well as analyte solubility.

Table 7.19 summarises the *main features* of on-line SFE-FTIR. The technique is particularly useful for those applications where chromatographic separation of the extract is not required. Exclusion of the column provides substantial time savings. In many cases, direct transfer into the FTIR flow-cell is feasible without the necessity of intermediate trapping. Direct SFE to FTIR interfacing provides an alternative to liquid extraction–IR methods that require hazardous solvents. Solvent extraction–IR and SFE-FTIR methods often show differences. This is not too surprising, as organic solvents frequently tend to extract components from a matrix more vigorously than scCO$_2$, and thus remove more of the oligomer and organic components present in the sample. Quantification using internal standards is usually done directly from the Gram–Schmidt reconstruction. Co-extractives may pose problems, since

Table 7.19 Main features of on-line flow-cell SFE-FTIR

Advantages

- No loss of extracted volatile analytes
- No use of flow restrictor
- Time savings (exclusion of column)
- Nondestructive technique
- Allowance for dual detection (post-flow-cell multihyphenation)
- No environmentally hazardous solvents
- Compatibility with solvent extraction data

Disadvantages

- Limited to analytes soluble in a supercritical fluid (usually scCO$_2$)
- Limitations on the use of mobile phases and modifiers
- Complex analysis for multicomponent extraction (no analyte separation)
- Many variables for method optimisation
- Use of very small (eventually nonrepresentative) sample sizes
- Need for high-pressure, small-volume flow-cell
- Limited experience (few research groups)
- Need for intermediate trapping in case of slow kinetics of extraction
- Background interference (CO_2 absorption: 3504–3822 cm^{-1}, 2137–2551 cm^{-1}, <900 cm^{-1})
- Need for dedicated FTIR (for on-line system)
- No exhaustive extraction
- Limited quantitation

the quantities of such components will vary greatly with the quantity of sample used. Error values stated (RSD 5 to 20 %) are for the inclusive method (sample method and analysis combined). Direct observation of the extraction process with FTIR was reported [149]. SFE-FTIR coupling has been reviewed [147].

As is well known, $scCO_2$ swells and plasticises glassy polymers for 'solvent-free' incorporation of additives. It is possible to change the degree of swelling of such a polymer, and consequently its free volume, merely by changing the density of $scCO_2$. This phenomenon offers the possibilities of adjusting or 'tuning' the amount of solute incorporated into the polymer phase, and also of influencing the process of diffusion of the solute within CO_2-swollen polymer. *In situ* FTIR spectroscopy under supercritical conditions allows the measurement of the partitioning of a solute or co-solvent between the SCF phase and the polymer phase, and can reveal specific intermolecular interactions [150]. Applications of $scCO_2$ impregnation of glassy polymers include dyeing processes [150].

On-line SFE-NMR coupling was also reported [151,152]. SFE provides some degree of separation by means of solubility and affinity to the matrix. This offers the possibility of transferring analytes directly from the extraction into the NMR probe. Drawbacks in the acquisition of SFE-NMR and SFC-NMR spectra are the elongated spin-lattice relaxation times T_1 of protons and the pressure dependence of ^1H NMR chemical shifts [153].

Smith and Udseth [154] first described *SFE-MS* in 1983. Direct fluid injection (DFI) mass spectrometry (DFI–MS, DFI-MS/MS) utilises supercritical fluids for solvation and transfer of materials to a mass-spectrometer chemical ionisation (CI) source. Extraction with $scCO_2$ is compatible with a variety of CI reagents, which allow a sensitive and selective means for ionising the solute classes of interest. If the interfering effects of the sample matrix cannot be overcome by selective ionisation, techniques based on tandem mass spectrometry can be used [7]. In these cases, a cheaper and more attractive alternative is often to perform some form of chromatography between extraction and detection. In SFE-MS, on-line fractionation using pressure can be used to control SCF solubility to a limited extent. The main features of on-line SFE-MS are summarised in Table 7.20. It appears that the direct introduction into a mass spectrometer of analytes dissolved in supercritical fluids without on-line chromatography has not actively been pursued.

The high selectivity and resolution of today's MS systems have reduced the need for efficiency of the separation for LC-MS analysis. In many cases, only

Table 7.20 Main characteristics of on-line SFE-MS

Advantages

- Great reduction in sample handling and preparation
- High-speed detection
- Rapid screening analytical capability (MS/MS)

Disadvantage

- Overloading of the mass spectrometer

a separation between analytes and sample matrix is sufficient for reliable quantitation. Therefore, with an efficient SPE cleanup, the LC separation can be omitted. SPE provides such a matrix–analyte separation, with the additional benefit of analyte enrichment, and avoids ionisation interferences and contamination of the MS system. In *on-line SPE-MS*, the purified analytes are desorbed from the SPE cartridge and transferred totally and directly into the MS. The integrated cleanup, trace enrichment and LC separation on a single SPE cartridge (packed with small sorbent particles) coupled to mass spectrometry for fast and selective analysis has great potential for fully automated, fast target analysis using various matrices. The major benefits of on-line SPE-MS include total automation, high precision, and enhanced sensitivity compared to liquid–liquid extraction and off-line SPE. Hyphenated (SPE)-LC-MS^n is a well-accepted technology, in particular in the pharmaceutical laboratory for high-throughput determinations. For fast SPE-LC-MS^n, it is important that the SPE process keeps pace with the (LC)MS^n analysis. Direct interfacing of SPE to high-sensitivity detectors (MS/MS), with SPE replacing the HPLC column, is gaining acceptance [155].

SPME with infrared detection (SPME-IR) is a useful alternative to SPME-GC. As opposed to the GC application, which uses a cylinder of solid or liquid coated on a fibre, the IR application utilises a small square of polymer film, which is more compatible with IR spectroscopic sample handling. Once the organics are extracted into the polymer film (using a similar procedure to that with the syringe device), IR absorbance spectroscopy is used for quantitation. *SPME-IR* is also sensitive, selective, fast and environmentally friendly (reusable films); the SPME-IR procedure does not require the use of organic extraction solvents. SPME-IR instrumentation is inexpensive and readily portable, and optical measurements are inherently simple to perform [156]. SPME coupled with laser desorption mass spectrometry (LDMS) is a fast tool for the detection of a variety of analytes at ppb concentrations [157]. *Direct SPME-MS^n*, omitting the GC separation step, may be used for targeting a specific analyte amongst a limited number of compounds. Combined with a rapid microwave distillation step, direct SPME-MS^n significantly reduces analysis time.

Apart from the aforementioned sample preparation techniques (SFE, SPE and SPME), other sample collection modes are coupled directly to spectroscopy (e.g. fast pyrolysis and fast thermolysis-FTIR) and spectrometry (e.g. LD-ITMS).

Applications Supercritical fluid extraction of ABS composites has been used to separate the halogenated *flame retardants* TBBA, TBPA and HBCD; IR spectroscopy was used to identify the organic flame retardants [158]. On-line flow-cell SFE-FTIR (without trapping) of *fibre/finishing oils* was reported with extraction being conducted as static, dynamic, static-dynamic or dynamic-cryotrap [149]. The detection limit for *n*-tetracosane (C_{24}) from Celite was 74 ng. The same procedure without chromatographic separations has been used for (multicomponent) finish analysis of different polymer fibres: polyurethane (PDMS oil; 98.3 % extractable), polyamide (glycerol triesters, alcohol ethoxylates, alcohol polyethylene oxide–ethylene oxide surfactants, phosphites, PEG derivatives, fatty acid soaps; 91.9 % extractable) and Kevlar/aramid (substituted phenol, C_8–C_{18} triglycerides, sorbitol derivatives; 89 % extractable) (RSDs of 5–9 %) [119,147]. Quantitative recovery of finish oils was reported. Certain finish components show low solubility in scCO$_2$. Also, online quantitation of another complex finishing system, composed of fatty acids, phosphorous-containing antioxidants, fatty acid esters, ethoxylate/propoxylate block copolymers, Sorbitol® and Sorbitan® derivatives, on a commercial polyester fibre (Dacron®), was achieved with FTIR [159]. A calibration plot was generated on finish standards spiked on to pre-extracted finish-free fibres and the per cent finish-on-yarn was calculated for the polyester fibre sample. However, the drawbacks of the approach are the use of very small (and possibly not representative) sample sizes; limitations to the use of modifiers; and the need for a dedicated FTIR. At variance to the 'softer' SFE method, Soxhlet extraction also removes low-MW material from the polymer itself. Liescheski *et al.* [160] used SFE-FTIR to characterise polyurethane fibre finishes. The method can also be used for *extraction dynamics studies*, as the solubility of analytes in CO$_2$ can be monitored directly by FTIR for varying extraction conditions. Also, SFE-DFI-MS can be used to monitor supercritical fluid extractions.

Partitioning of *azo-dyes* between PMMA and scCO$_2$ has been measured by means of *in situ* FTIR and UV/VIS spectroscopy. The use of CO$_2$ to incorporate dyes into polymers has potential in the textile industry and in the production of advanced polymeric materials for nonlinear optical (NLO) applications [150].

As in the case of SFE-FTIR, for SFE-NMR the direct monitoring of the supercritical fluid extraction process is also an application. Fractions can be collected and examined with off-line chromatography. However, NMR allows the direct detection of samples which are extracted, and provides sufficient spectral dispersion for independent detection of several different compounds, even if the extraction profiles overlap. Generally, the concentrations and total amount of sample in SFE are higher than in SFC. As a result, monitoring of the extraction process is also possible using NMR instruments at lower fields, thus providing an easy and simple way to better understand an extraction process. Figure 7.9 shows the expanded continuous-flow ^{1}H NMR spectrum of diallylphthalate.

Solid-phase microextraction (SPME) is compatible with high-speed separations. SPME-IR has been applied to VOCs in soil [156]. Górecki and Pawliszyn [69] reported separation of 28 volatile organic compounds listed in EPA method 624, with ITMS detection, in less than 150 s.

7.3 POSTCOLUMN HYPHENATION

Principles and Characteristics Chromatography and spectroscopy are totally orthogonal techniques, which provide complementary information about the identity of the components and their concentration in a sample. Consequently, their on-line combination (spectroscopic investigation of the separated sample components) is a most powerful approach. In a chromatographic separation, there always exists the possibility that two or more components are eluted as one peak. Likewise, different compounds may present highly similar spectra. The probability of correctly identifying an unknown component increases greatly when it exhibits both the correct elution time and a good spectral match. The high degree of compound selectivity possible through the combination of chromatography with detectors providing structural information, has been the driving force behind the development of effective hyphenated techniques (Table 7.21).

Combined chromatographic–spectroscopic techniques allow complex multicomponent data to be obtained in a single experiment. Essentially, the analysis of monomers, additives, oligomers and polymer can be performed in one step on-line. Postcolumn hyphenation, which comprises spectroscopic detectors, sniffing, fraction collection or heartcutting, is well

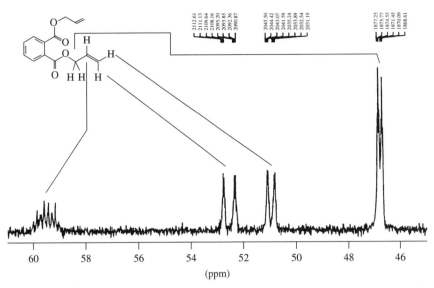

Figure 7.9 Continuous-flow ^{1}H NMR spectrum (400 MHz) of diallylphthalate in supercritical carbon dioxide (flow-rate 2 mL min^{-1}, pressure approximately 136 bar, and temperature 321 K). After Richter *et al.* [153]. Reprinted with permission from B.E. Richter *et al.*, *Analytical Chemistry*, **68**, 1033–1039. Copyright (1996), American Chemical Society

Table 7.21 Main hyphenated chromatographic–spectroscopic techniques

UV/VIS, FTIR	Raman	NMR	MS
GC	HPLC	GC	GC
SFC	CE	SFC	SFC
HPLC	TLC	HPLC	HPLC
TLC		SEC	SEC
SEC		CE	TLC
			CE/CEC

Table 7.22 Main postcolumn hyphenated techniques used in polymer/additive analysis

Separation technique	Hyphenated postcolumn detection						
	UV	FTIR	R	MS	NMR	AED	ICP-AES/MS
GC	+	+		+		+	+
SFC	+	+		+	+		+
HPLC	+	+		+	+		+
TLC		+	+	+			
SEC		+		+			

developed and no major problems remain. The main postcolumn chromatographic techniques with impact on polymer/additive analysis are given in Table 7.22, but other useful chromatography–detector combinations are equally operative. Unfortunately, some excellent spectroscopic detectors, like Fourier-transform infrared via light-pipe interface and atomic emission detection, are no longer commercially available.

Table 7.22 shows that most current emphasis is on FTIR and MS couplings. Chromatographic interfaces with IR spectrometers have served greatly to popularise the technique of vibrational spectroscopy. *Infrared chromatographic interfaces*, including GC, HPLC, SFC, TLC and SEC, have served to separate complex mixtures, identify compounds, and make quantitative measurements – with the detection limits in some cases reaching pg levels. Figure 7.10 shows some typical optical arrangements. HPLC-FTIR, SFC-FTIR and TLC-FTIR are not as sensitive as GC-FTIR, but are more appropriate for analyses involving nonvolatile mixture components. In GC-FTIR and SFC-FTIR analyte-deposition-based methods are more sensitive and versatile than flow-cell-based techniques. Indeed, mutually similar interfaces have been developed for the coupling of GC [161], SFC [162], and LC [163,164] with FTIR, in which the column effluent is deposited directly on a moving IR-transparent substrate, and transmission spectra are recorded under an FTIR microscope. Various reviews and books deal with coupling of chromatography (GC, HPLC, SEC, SFC) to FTIR [165–167], including applications [168].

When interfacing any detection method to a separation process, the integrity of the separation must be maintained. The resolution between two components, which is gained during a highly efficient separation, can

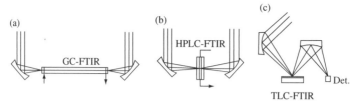

Figure 7.10 Typical optical arrangements employed for detection of (a) vapour phase; (b) liquid phase; and (c) solid chromatographic phases. After White [167]. Reprinted from R. White (ed.), *Chromatography/Fourier Transform Infrared Spectroscopy and Its Applications* Marcel Dekker Inc., New York, NY (1990), by courtesy of Marcel Dekker Inc.

be degraded significantly by a poorly designed interface. Yet, some interface designs are close to ideal with few mutual interferences, and a high sensitivity that enables a low limit of quantification in the assay. Ideal detectors used in chromatography should be universal, sensitive, quantitative and able to deliver detailed structural information. No single detector fulfils all of these criteria (Section 4.1) and this provides the driving force to the use of *multihyphenated systems* (e.g. cSFC-FTIR-MS). For example, in the analysis of polymeric samples, FTIR is extremely useful in providing identification of additives and impurities, and also in identifying the type of oligomers present. It cannot, however, be easily used to provide information on the mass distribution of the oligomers. On the other hand, while MS is ideal for providing molecular mass data, it often fails to differentiate between isomers, as opposed to FTIR. In most instances, IR spectroscopy alone cannot provide unequivocal mixture−component identification. For this reason, chromatography−FTIR results are often combined with retention indices or mass-spectral analysis to improve structure assignments (e.g. GC-FTIR-MS). Differences in FTIR and MS sensitivity limit joint analysis applications to components present in moderate to high concentrations. It follows that futuristic multihyphenated chromatospectrometric systems, such as SFC-UV/FTIR/MS/MS/FID [169] or current LC-UV-NMR-MS [170] may easily lead to more frustration than results. With tandem detection, a nondestructive technique is coupled first, e.g. it is possible to couple both IR and MS to the same gas chromatograph, producing GC-FTIR-MS (rather than GC-MS-FTIR). Multidetector technology can be applied to provide a means of validating analyser performance.

Hyphenation of chromatographic separation techniques (SFC, HPLC, SEC) with *NMR spectroscopy* as a universal detector is one of the most powerful and time-saving new methods for separation and structural elucidation of unknown compounds and molecular compositions of mixtures [171]. Most of the routinely used NMR flow-cells have detection volumes between 40

and 180 µL for conventional separations with analytical columns, and recent designs employ detection volumes in the order of 200 nL for capillary separations. The low flow-rates used in capillary chromatography permit the use of deuterated solvents.

Great success in identifying the constituents in a mixture has been obtained by combining high-resolution separation techniques with *mass spectrometry* as a detection device. Mass spectrometry is the detector of choice for the chromatographer. Because MS data are far more specific than UV (or even a diode array) less time is spent on developing chromatography. MS can distinguish components that co-elute – even if the UV spectra are identical – and structural information can distinguish components – even if they have the same molecular weight! Molecular weight information complements LC-UV and accelerates additive analysis. Introduction of various chromatography/mass-spectrometer interfaces has greatly facilitated the analysis of ionic or thermally labile compounds. The power of combining MS detection with any separation technique is that it provides a second dimension of separation. From a chromatographic point of view, the 'ideal' interface to mass spectrometry should possess several qualities:

(i) chromatographic integrity should be maintained;
(ii) various ionisation methods should be allowed;
(iii) high analyte transport efficiency and ready ionisation, for high sensitivity;
(iv) no thermal degradation of the analyte in the interface;
(v) compatibility with fluids under pressure programming, for chromatographic versatility; and
(vi) high reliability.

Problems arise in interfacing column chromatographic techniques to a mass spectrometer: from the difference in material flow requirements between the two instruments and the desire to generate information about the sample without interference from the mobile phase in which it is diluted. The most favourable case occurs for gas chromatography. Mass spectrometers can accept column flow-rates of $1-2\,mL\,min^{-1}$ into the ion source. This is compatible with the optimum flow-rates of some types of open tubular columns in GC, but much less than the typical flow-rates of packed columns. The gas burden of the source is a bigger problem for LC and SFC. A liquid chromatograph operating at a flow-rate of $1.0\,mL\,min^{-1}$ produces nearly $200-1000\,mL\,min^{-1}$ of vapour at STP, of which the mass spectrometer can accept only $1-2\,mL\,min^{-1}$.

The ideal mass-spectrometric interface should allow for a range of ionisation methods (Tables 7.23 and 7.24). The ionisation of organic molecules for use with chromatographic outlets include EI, CI, APCI for samples that can be vaporised prior to ionisation; alternative ionisation techniques using TSP, ESP or FAB are needed for labile, high-MW or ionic samples.

Table 7.23 Combinations of separation techniques and mass discrimination[a]

Separation	Ionisation[b,c]	Mass discrimination
GC	EI[d]	Sector (B, E)
SFC	CI[d]	Time-of-flight (ToF)
HPLC	FI[d]	Quadrupole (Q)
SEC	TI	Ion trap (IT)
IC	Desorption[e]	FTICR
TLC	Nebulisation[f,g]	
CE	Photon ionisation	

[a] Most couplings are technically feasible, see ref. [172].
[b] Ionisation methods are reviewed by refs [173,174].
[c] For ion-activation processes in MS/MS, see ref. [175] and table 2 of ref. [176].
[d] Ionisation of molecules in the vapour state.
[e] Comprises FAB, L-SIMS, SIMS, PD, MALD, desorption CI (DCI) and FD.
[f] Includes fluid introduction techniques such as ICPI, TSP, ESP/ISP, and HNI/APCI.
[g] With HPLC-MS interfaces, removal of eluent and ionisation are combined in one step.

High analyte transport efficiency and ready ionisation should occur, thus producing a highly sensitive technique. The ultimate sensitivity and speed of analysis available with MS detection can depend greatly on the preceding separation process. The ability to perform reliably and consistently liquid phase separations, which are capable of generating chromatographic peaks 1 sec or less in width, has rapidly matured in recent years. Chromatography–MS couplings have been reviewed [177]. The specifications of hyphenated mass spectrometry are readily available (*www.hypenms.nl*).

This chapter also deals in particular with chromatographic detection by *atomic plasma spectrometry* and *plasma mass spectrometry* (AED, MIP, ICP). With the application of such detectors, metal-specific signals can be obtained – thus the information content of a separation increases significantly. The major objectives of interfaced chromatography–atomic plasma source emission spectrometry (C-APES) are:

(i) to monitor eluates for their elemental composition with high elemental sensitivity;
(ii) to determine the selected element with high selectivity over co-eluted elements;
(iii) to tolerate non-ideal chromatographic resolution from complex matrices;
(iv) to detect molecular functionality through derivatisation with element-tagged reagents;
(v) to detect a number of elements simultaneously for empirical and molecular formula determination; and
(vi) to enable trace-elemental analysis with species selectivity.

Four types of atomic spectrometry have been interfaced for chromatographic detection, namely AAS, FES, AFS and APES. Ebdon *et al.* [178] have discussed coupling of HPLC with AAS. HPLC-FAAS is relatively insensitive. Application of HPLC-GFAAS or

Table 7.24 Sample types in relation to chromatographic interfaces, ionisation methods and mass spectrometers

Sample classes	Chromatographic interfaces	Ionisation methods	Mass spectrometers[a]
Polar organic compounds	GC, cSFC	CI	B, Q
	LC, CE	ESI	B, Q, ToF
	pSFC, LC	APCI	B, Q
	Low-flow LC, CE	FAB/FIB/LSIMS	B, Q
	None	FD/FI, DCI	B, (Q)
	pSFC, LC	TSP	B, Q
Nonpolar organic compounds	GC, PB-LC	EI	B, Q, IT, ToF
	GC, cSFC	CI	B, Q, ToF
	None	FD/FI, DSC	B, (Q)
Oligomers	None	FD/FI	B, (Q)
Polymers	None	MALDI	ToF, IT

[a] B = magnet; IT = ion trap; Q = quadrupole; ToF = time-of-flight.

HPLC-QFAAS is also problematical. Most development of atomic plasma emission in HPLC detection has been with the ICP and to some extent the DCP, in contrast with the dominance of the microwave-induced plasmas as element-selective GC detectors. An integrated GC-MIP system has been introduced commercially. Significant polymer/additive analysis applications are not abundant for GC and SFC hyphenations. Wider adoption of plasma spectral chromatographic detection for trace analysis and elemental speciation will depend on the introduction of standardised commercial instrumentation to permit interlaboratory comparison of data and the development of standard methods of analysis which can be widely used.

Gas and liquid chromatography directly coupled with atomic spectrometry have been reviewed [178,179], as well as the determination of trace elements by chromatographic methods employing atomic plasma emission spectrometric detection [180]. Sutton *et al.* [181] have reviewed the use and applications of ICP-MS as a chromatographic and capillary electrophoretic detector, whereas Niessen [182] has briefly reviewed the applications of mass spectrometry to hyphenated techniques.

7.3.1 (Multi)hyphenated GC Techniques

Among chromatographic (GC, SFC, HPLC, TLC) couplings to FTIR, *GC-FTIR* is the more sensitive; for this purpose various interface designs have been developed such as light-pipe and trapping systems. GC-FTIR is essentially a powerful tool for the analysis of vapours, but has found limited use only in polymer/additive analysis, not unlike the fate of other GC-FTIR devices with various front-end accessories (TD, Py, HS). GC-FTIR is in competition with GC-MS. The limits of quantitation are approximately 10, 100 and 20 ng mL^{-1} for GC-MS (SIM), light-pipe IR and DD-IR, respectively [183]. GC–Raman and μGC-UV are not used for polymer/additive analysis. Similarly, GC-NMR coupling with direct sample trapping, first reported in 1965 [184], and on-line GC-NMR coupling [185] have also made no inroads.

Gas chromatography is a most favourable case for interfacing to a *mass spectrometer*, as the mobile phases commonly used do not generally influence the spectra observed, and the sample, being in the vapour phase, is compatible with the widest range of mass-spectral ionisation techniques. The primary incompatibility in the case of GC-MS is the difference in operating pressure for the two hyphenated instruments. The column outlet in GC is typically at atmospheric pressure, while source pressures in the mass spectrometer range from 2 to 10^{-5} Torr for CI and EI, respectively. Column flow-rates of 1–2 mL min^{-1} (adjusted to STP) into the ion source are compatible with modern MS vacuum systems. This is also the optimum flow-rate range for open tubular capillary columns of conventional dimensions. Coupling such columns to an MS, therefore, presents the fewest problems. Low-resolution mass spectrometry is routinely used in GC-MS applications. High-resolution measurements provide accurate elemental composition data for all ions in the spectra, or for a single ion such as the molecular ion. This chapter illustrates that only GC-MS contributes significantly to polymer/additive analysis by means of hyphenated gas-chromatographic methods. GC-MS is a widely used analytical tool; various polymer additive protocols have been developed, embracing HTGC-MS, GC-HRMS and fast GC-MS with a wide variety of front-end devices (SHS, DHS, TD, DSI, LD, Py, SPE, SPME, PTV, etc.). Ionisation modes employed are mainly EI, CI (for gases) and ICPI (for liquid and solid samples). GC-FTIR-MS is proposed as a method of choice for volatile mixture analysis; serial flow GC-FTIR-MS instruments have reached commercial maturity.

GC-AAS has found late acceptance because of the relatively low sensitivity of the flame; graphite furnaces have also been proposed as detectors. The quartz tube atomiser (QTA) [186], in particular the version heated with a hydrogen–oxygen flame (QF), is particularly effective [187] and is used nowadays almost exclusively for GC-AAS. The major problem associated with coupling of GC with AAS is the limited volume of measurement solution that can be injected on to the column (about 100 μL). Virtually no GC-AAS applications have been reported. As for *GC-plasma source techniques* for element-selective detection, GC-ICP-MS and GC-MIP-AES dominate for organometallic analysis and are complementary to PDA, FTIR and MS analysis for structural elucidation of unknowns. Only a few industrial laboratories are active in this field for the purpose of polymer/additive analysis. GC-AES is generally the most helpful for the identification of additives on the basis of elemental detection, but applications are limited mainly to tin compounds as PVC stabilisers.

Finally, we can note that GC or GC-MS in general are capable of analysing only 20 % of the chemicals in use worldwide.

7.3.1.1 Gas Chromatography–Spectroscopy

Principles and Characteristics Gas chromatography/infrared spectrometry systems couple the separation capability of GC with the selectivity for functional groups and identification power of FTIR in order to analyse multicomponent organic mixtures in small elution volumes with real-time spectral display. GC-FTIR yields 3D information: retention time, wavelength and absorption. GC samples for IR examination were originally obtained by trapping the solute vapour at low temperatures and then examining the fractions collected off-line [188]. Material in a GC-CT-IR trap may be transferred to an infrared gas cell for examination in the vapour phase; to a suitable micro cell (as a liquid); or may be condensed on a cold surface as a solid for examination by conventional spectroscopic techniques. The first publications on the GC-CT-IR coupling appeared in the late 1970s.

FTIR has significant advantages over 'classical' IR, mainly related to the capability for recording a spectrum in a very short time interval. This allows FTIR to be hyphenated with GC (and other separation techniques). An ideal interface between chromatography and FTIR couples both techniques without sacrificing the performance of either. Specific interface designs depend on many factors, including the physical state of the matter to be analysed. The three predominant interface techniques for *GC-FTIR* are flow-through cell (light-pipe), matrix isolation (MI) and cold trapping (or direct deposition, DD), see Figure 7.11. In GC-FTIR the mobile phases commonly used are transparent in the mid-IR region, favouring direct coupling via a flow-through cell. Mobile-phase elimination techniques are used primarily to increase sensitivity. The simplest

(and most commonly used) interface for GC-FTIR is a long-pathlength *light-pipe* connection to the GC by a length of heated capillary tubing and coated inside with a reflecting material (gold). The light-pipe acts as an on-line gas cell and provides on-the-fly FTIR vapour-phase spectra. The light-pipe interface is characterised by poor sensitivity (high ng range). Typical sample quantities required to yield identifiable spectra using a light-pipe with an i.d. of 1 mm are between 5 and 25 ng. Low-volume (90 μL) light-pipe and scan rates up to 20 spectra ensure no loss of resolution for analytical capillary columns.

Matrix isolation (of sample molecules in a matrix of solid argon at about 12 K) yields IR spectra with narrow lines and detailed fine structures [189]. However, the need for special matrix isolation libraries for computer searching has limited general acceptance of this approach. The detection limits of a matrix isolation instrument are in the hundreds of pg range. A device based on the matrix isolation principle was introduced commercially in 1984 [190], but is ultimately no longer available. The third procedure is based on '*trapping*' of the GC eluting materials on a moving, subambient cooled surface (usually a ZnSe plate), which transports the condensed sample into the focus of an FTIR microscope, and provides near real-time spectra and chromatogram reconstruction [191,192]. The CGC-μFTIR interface is similar to that reported for SFC-FTIR [193]. Disadvantages are crystallisation of water and organic solvent on the ZnSe plate. The cost of the trapping system is considerably lower than that of matrix-isolation GC-FTIR. Detection limits in the sub-ng to mid-pg range are now claimed for deposition methods of HRGC-FTIR [191]. GC-cryotrapping-IR has potential for trace-level analysis when combined with LVI and SPE [194]. Nearly all chromatography–FTIR measurements requiring rapid data acquisition have employed MCT detectors.

The success of *spectral identification* depends on the appropriate reference spectra for comparison. IR measurement of eluates that are at slightly subambient temperature is advantageous considering that the large databases of condensed-state spectra may be searched. Spectra measured by matrix-isolation GC-FTIR have characteristically narrow bandwidths compared with the spectra of samples in the condensed phase near ambient temperature or in the gas phase. In addition, the relative intensities of bands in the spectra of matrix-isolated samples often change compared with either gas- or condensed-phase spectra [195]. GC-FTIR spectra obtained by direct deposition match well with the corresponding reference spectra in standard phase

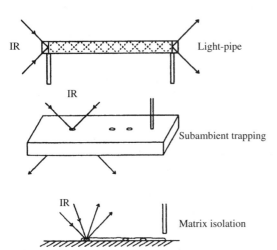

Figure 7.11 Gas chromatography–IR interfaces. After White [166]. Reprinted from R.L. White, in *Encyclopedia of Spectroscopy and Spectrometry*, Academic Press, J.C. Lindon (ed.), pp. 288–293, Copyright (2000), with permission from Elsevier

(KBr) databases. *Quantitative analysis* by GC-FTIR is complicated by many uncertainties associated with both the chromatography and spectroscopy [196]. Bulk property detectors (e.g. TCD, FID, etc.) can be used for quantitative analysis when mixture components are known, but provide little structural information for unknown mixture components. Both integrated absorbance and Gram–Schmidt vector methods have been used for the quantitative analysis of mixture components in GC-FTIR.

The main characteristics of GC-FTIR are collected in Table 7.25. GC-FTIR is a powerful tool for the analysis of vapours, but otherwise has lost favour in comparison to GC-MS (in analogy to PyGC-FTIR vs. PyGC-MS). GC-FTIR can provide information about complex mixtures that are unobtainable with GC-MS (and vice versa). GC-FTIR allows isomer identification, as opposed to GC-MS. However, the latter technique is strong in homologue identification and molecular weight determination. The sensitivity of GC-FTIR coupling is lower than GC-MS coupling.

In chromatography–FTIR applications, in most instances, IR spectroscopy alone cannot provide unequivocal mixture-component identification. For this reason, chromatography–FTIR results are often combined with retention indices or mass-spectral analysis to improve structure assignments. In GC-FTIR instrumentation the capillary column terminates directly at the light-pipe entrance, and the flow is returned to the GC oven to allow in-line detection by FID or MS. Recently, a multihyphenated system consisting of a GC, combined with a cryostatic interfaced FTIR spectrometer and FID detector, and a mass spectrometer, has been described [197]. Obviously, *GC-FTIR-MS* is a versatile complex mixture analysis technique that can provide unequivocal and unambiguous compound identification [198,199]. Actually, on-line GC-IR, with

Table 7.25 Main characteristics of GC-FTIR coupling

Advantages

- Real-time analysis
- Functional group selectivity
- Identification power (depending upon appropriate databases)
- Nondestructive analysis
- Good spectroscopic resolution
- Numerous sample introduction methods
- Multihyphenation (FID, MS)

Disadvantages

- Fairly low sensitivity (ng–pg range)
- Sample volatility (in CGC-μFTIR device)
- Interdetector competition (MS)

---

adsorption of the solute in a cold pack tube, was first described in a GC-IR-MS arrangement [188]. GC-FTIR-MS interfacing has been achieved by splitting of the column effluent between the IR and MS spectrometers in either a parallel or serial configuration. Commercial instruments are now capable of performing routine mixture analysis at the 10–100-ng component levels. Differences in FTIR and MS sensitivity have limited joint analysis applications to components present in moderate to high concentrations. The initial mismatch in sensitivity of the two detectors is now minimised.

There is a need for increased chromatography-FTIR sensitivity to extend IR analysis to trace mixture components. GC-FTIR-MS was prospected as the method of choice for volatile complex mixture analysis [167]. HPLC-FTIR, SFC-FTIR and TLC-FTIR are not as sensitive as GC-FTIR, but are more appropriate for analyses involving nonvolatile mixture components. Although GC-FTIR is one of the most developed and practised techniques which combine chromatography (GC, SFC, HPLC, SEC, TLC) and FTIR, it does not find wide use for polymer/additive analysis, in contrast to HPLC-FTIR.

In addition to standard liquid injection there are many GC accessories which can provide different methods of sample introduction to the column, such as HS, SPE, SFE, TD, TG, Py, etc. Examples of such GC-FTIR devices are TD-GC-FTIR (with a cryostat interface) and PyGC-FTIR.

Standard practices for GC-IR analysis have been described (ASTM E 1642-94). Griffiths [200] has discussed GC-FTIR designs. Sample preparation methods for hyphenated infrared techniques, in particular GC-FTIR, have been reported [201]. The technique has been reviewed repeatedly [167,183,201–204]; a monograph [205] has appeared.

Miniaturised *GC-UV* equipment consisting of an injector, an 8-cm-long GC separation column and parallel light-pipe, has been coupled to a powerful UV spectrophotometer capable of scanning from 168 to 330 nm, with access to an expanding database of UV vapour phase spectra [206]. The highly sensitive GC-UV system is particularly suited for the detailed analysis of a wide range of gaseous samples; liquid samples can be analysed by direct injection, while a thermal desorption module allows the analysis of solid samples.

Applications   Off-line GC-FTIR using a trap device has been used for the determination of a wide range of polymer additives (with highest boiling constituents below 250 °C) in amounts down to 0.01 % with an accuracy of 5 to 10 % [207]. According to Haslam *et al.* [208], the technique could not be applied effectively to substances with a b.p. much in excess of 120 °C.

On-line GC-FTIR is a powerful tool for analysis of *vapours*, and is being used for the study of out-gassing phenomena (in TD-GC-FTIR configurations). Headspace GC-FTIR is a technique that has received little attention, but has great utility in the analytical field for *odours* and volatile gases which are evolved from sample matrices [209]. GC-FTIR has also been used in combination with HPLC-FTIR to study the ageing process of acrylic structural adhesives [210]. Relatively high boiling range samples can be analysed without compromising spectroscopic performance; mixed phthalates with boiling points between 300 °C and 500 °C, eluting from the GC, pass through a GC-IR light-pipe without condensing. Wilkins *et al.* [211] reported analysis of a commercial lacquer thinner by GC-FTIR-MS, detecting 30 components. Rau and McGrattan [212] have investigated the thermally induced decomposition of EVA copolymers by a combination of TG-FTIR and GC-FTIR measurements of the trapped gases.

As to the main limitation of MS vs. FTIR detection, namely the inability to distinguish closely related isomers, this rarely plays a role in additive analysis. Notable examples of isomeric additives are the bifunctional stabilisers $C_{22}H_{30}O_2S$ as 4,4'-thio-bis-(6-t-butyl-m-cresol), 2,2'-thio-bis-(4-methyl-6-t-butylphenol) and 4,4'-thio-bis-(2-methyl-6-t-butylphenol) (Section 6.3.6), the bisphenolic antioxidants $C_{23}H_{32}O_2$ (Plastanox 2246 and Ethanox 720) and the phenolic antioxidants $C_{15}H_{24}O$ (nonylphenol and di-t-butyl-p-cresol).

A GC-IR-MS system with library search capability has been used to effectively identify the pyrolysis products of polybutadiene and the antioxidant additive 2,6-di-t-butyl-4-methylphenol [199]. Paper for food packaging was analysed by P&T (at 100 °C) combined with μGC-UV. No specific applications of μGC-UV to polymer/additive analysis have as yet been reported.

7.3.1.2 Gas Chromatography–Mass Spectrometry

Principles and Characteristics The most commonly used and widely known technique coupled to gas chromatography is mass spectroscopy. In early couplings, with packed GC columns being the major problem encountered, elimination of the carrier gas at relatively high flow-rates (25 mL min^{-1}), resulted in a significant loss of sensitivity. With the advent of open tubular devices, the column flow could be passed directly into the mass spectrometer. GC-MS is typically

used in verification or identification of unknown components. GC-MS analysis is particularly difficult in the off-line mode. On-line GC-MS, initiated in 1957 [213], can only be performed on volatile analytes or those that can be derivatised. The main reason for success in GC-MS is that the effluent is compatible with classical gas-phase ionisation methods. It is estimated that there are over 30 000 bench-top GC-MS systems worldwide. Figure 7.12 shows indicative purchase prices (2001) of GC-MS instruments

As may be seen from Table 7.26, a great variety of *commercial GC-MS systems* are currently available [215]. GC-MS systems differ widely in characteristics: injection systems (headspace, fluid injector, solids probe, pyrolyser), ionisation techniques (EI, SMB-EI, PCI, NCI, ECD-NCI, FAB, FI, CID, HSI, ICPI), mass spectrometer (QMS, QQQ, QITMS, ToF-MS, B), mass range ($450 < m/z < 4000$), and resolution (up to 5000 for B and ToF-MS). The ease of coupling GC systems with relatively low-cost *mass spectrometers* makes GC the method of choice for numerous analyses. In low-resolution GC-MS the two important types of instrument are the ultratrace full scan instrument (e.g. ITMS) and most other quadrupole mass spectrometers using full scan for large amounts of analyte and selected ion monitoring (SIM) for detecting very small amounts ($<1-10$ ng). Both techniques have their supporters [216]. Detection limits of GC-MS instruments vary considerably with mass-spectral acquisition mode. The better sensitivity due to large-volume injection to the GC allows MS in scan mode to be used for detection. With SIM, a limited number of ions are monitored during a selected time interval. The presence of the analyte is determined by 'diagnostic' ions at the correct retention time and with the correct abundance ratio (between certain limits). Quadrupole mass spectrometers can be used with GC open tubular columns and microbore LC columns. With the ion trap mass spectrometer the whole mass spectrum is stored for each point of the chromatogram (e.g. one spectrum per second). Subsequently, full-scan identification of the analyte by library search may be performed with the data system while recording a new acquisition. By using tandem MS the matrix is eliminated and S/N is greatly improved as compared to SIM or the full-scan mode. For QITMS a full-scan identification of components at (at least) the 10 pg level is claimed. The practical aspects of GC-QITMS systems have been discussed [217].

ToF-MS does not scan, the ions are pulsed into the analyser and the electronics accumulate 40 000 spectra per second, which are averaged and saved up to 500 times per second. GC-oaToFMS instruments are

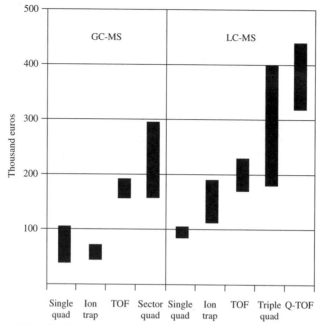

Figure 7.12   Purchase costs (2001) of GC-MS and LC-MS instruments. After Niessen [214]. Reproduced by permission of Bèta Publishers

currently available in two types: designed either to be capable of exact mass measurement or to give fast spectral acquisition rates (for fast GC). GC-oaToFMS or HRGC-MS can measure masses to an accuracy of <5 ppm, which is equivalent to <2 mDa at mass 400.

Coupling of high-resolution double-focusing mass spectrometers (e.g. of EBE geometry) to gas chromatographs is now well established, and benchtop single-ion monitoring systems, which rival the sensitivities of more complex instrumentation, have turned GC-MS into a routine tool. Implementation of fast, efficient collision-activated dissociation (CAD) methods allows acquisition of diagnostic fragmentation patterns for multicomponent mixtures. MS/MS is applied for structure elucidation.

HRGC as well as HPLC and SFC all require transportation of the separated analytes from a high-pressure zone (the column) to a low-pressure zone (the ion source), without loss of sample or chromatographic integrity. High-resolution capillary columns require only a modest carrier gas flow ($0.1–3.0\,\mathrm{mL\,min^{-1}}$) that can flow directly into the ionisation source without upsetting the vacuum. The normal 30 m, 0.25 mm i.d. fused silica capillary column emerges from the GC oven directly into the GC *interface*, and terminates within the ionisation source. Microbore columns (<0.1 mm) are also being

Table 7.26   Overview of some commercial GC-MS couplings

Mass spectrometer	m/z range	Ionisation techniques	Optionals
QMS	800	EI, PCI, NCI	Spectral libraries
QMS	450	EI	Spectral libraries
QMS	1 024	EI, PCI, NCI	
DFS	2 500	EI, PCI, NCI, FAB	MS/MS, resolution 5 000
ToF-MS	1 500	EI, PCI, NCI, FI	Resolution 5 000
QMS	1 200	EI, PCI, NCI	LVI, 10^5 dynamic range
QMS	700/900	EI, PCI, NCI	DIP, Py, spectral libraries
QMS	1 000/1 500	EI, PCI, NCI	
QITMS	1 000	EI, PCI, NCI	MS/MS
QMS/QQQ	2 500/4 000	EI, PCI, NCI	
QITMS	650	EI, CI	MS/MS
ToF-MS	1 500	SMB-EI, HSI	>100 scans s^{-1}

used. The main task of the GC-MS interface, which is kept as short as possible, is to heat the column evenly to a desired temperature in order to prevent condensation of (high boiling) eluates before they reach the ionisation source (no decomposition of thermolabile and/or reactive compounds by active surfaces should occur). The GC-MS interface temperature is important, and should be set to be at least at the highest GC oven temperature used, and possibly 10 or 20 °C higher. If too low, then the emerging chromatographic peaks will be too broad and give tailing; if too high then a noisy baseline results, with risk of column bleeding. The gas load entering the ion source must be within the pumping capacity of the mass spectrometer. Typical conventional GC-MS analysis requires 10–40 min. Using low pressure (LPGC) with 10-m columns, the optimum velocity is a factor of three to five higher than normally used. Analysis of samples with a high content of matrix residues will lead to rapid contamination of the ion source and a corresponding loss of sensitivity after a few measurements.

In *high temperature (HT)-GC-MS* and PyGC-MS experiments, special attention should be given to the stability of the column. GC columns can lose some of the stationary phase ('bleeding') when heated up to the maximum operating temperature; the thicker the stationary phase, the more column bleed may be expected. When coupled on-line to a mass spectrometer, the stationary phase may foul the ion source, which leads to rapid decay in sensitivity and detection of usually siloxane-related mass peaks at m/z 207, 281, 355, etc. HTGC-MS coupling was discussed by ref. [218].

There is no universal interface that can facilitate the common forms of *ionisation*, such as EI or CI, FAB or LSIMS used in conjunction with chromatography. In GC-MS tandem systems, ion sources in general use are EI, CI and ICPI (for liquid and solid samples). The ionisation source should be optimised and calibrated for EI, CI$^+$, or CI$^-$ analyses using reference and standard compounds introduced from either the reference inlet or by use of a direct insertion probe. Classical *electron ionisation* yields little useful information about structures which are either very stable or very labile. For the first, the few fragment ions present in the upper mass range do not contain sufficient information about the structure – and the usually very abundant molecular ion cannot compensate for this. For the second, only masses in the lower mass range are present, and these are not very informative. The molecular ion is either of low intensity or completely absent. *GC-CIMS* allows selection of the reagent gas (methane, isobutane or ammonia), depending on the application. Compounds with electronegative substituents and unsaturation can

be expected to have a large electron capture cross-section and thus work particularly well in the negative ion mode. Amirav *et al.* [219–221] used a continuous supersonic molecular beam (SMB) to separate carrier gas and analyte molecules from a GC capillary; this is made possible by spatial separation of small and large molecules within the supersonic beam. Nonselective ionisation techniques and quadrupole mass spectrometry were used. With SMB, EI becomes a soft ionisation method with a dominant M$^+$ peak. Vibrational cooling attainable in SMB enables improved mass-spectral characterisation of isomers (Scheme 7.4). Improved isomer characterisation is of importance to fast GC-MS, as it provides a mass-spectral channel of information that is compatible with fast GC even without isomer chromatographic separation.

Hyperthermal surface ionisation (HSI) is an ultrasensitive tuneable selective ion source [222,223] which is based on the very effective ionisation of various hyperthermal molecules upon their scattering from a surface with a high work function, such as rhenium oxide. Molecule-surface electron transfer constitutes the major and most important HSI mechanism for GC-MS.

GC-MS analysis is schematically indicated in Figure 7.13. In GC-MS *data collection*, account should be taken of the fact that GC peaks eluting from a capillary column are only a few seconds wide, which imposes fast scan times. As most compounds analysed by GC-MS are low-MW (<800 Da), a relatively short

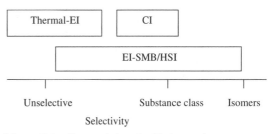

Scheme 7.4 Characteristics of cold electron impact

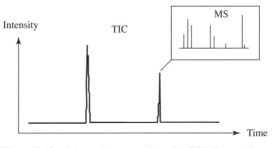

Figure 7.13 Schematic presentation of a GC-MS analysis

scan range is sufficient; scan speeds of one second or even faster are sustainable on modern mass spectrometers. For good chromatographic practice, the peaks should be well separated, symmetrical in shape and present no tailing. The spectra obtained should be of a suitable quality for library searching. However, not all mass spectra are created equal. Consequently, secure comparison of mass spectra for positive compound identification requires attention to the ionisation mode. Traditional interpretation of GC-MS data is based often only on mass-spectral analytical parameters and public (NIST/EPA/NIH, Wiley Registry of Mass Spectral Data or Pfleger/Maurer/Weber), user-created or private, mass-spectral databases using AMDIS. The low effectivity of commercial library search for state-of-the art polymer additives is well known, and the end-users are the best at creating their own files (see ref. [224]) for verification purposes. In this respect, GC-MS protocols are rather labour intensive. In order to increase the reliability and correctness in identification, obviously, the GC retention times (retention indices, RI) should be considered. However, RI databases for standard nonpolar liquid phases include data for only some 35 000 organic compounds, i.e. far less than the 275 000 EI mass spectra in the aforementioned libraries. Isolation of pure spectra in GC-MS analysis by means of mathematical methods has been reported [225]. The mass spectra of the analytes studied must contain significant structural information. If this is not so, one should try other ionisation techniques. GC-MS analysis of complex mixtures can yield chromatographic data containing several hundred resolvable peaks. Powerful software is required to produce an unambiguous identity for every peak, particularly for the case of very similar spectra.

A good way to test the operation of the GC and the GC-MS interface is to run a standard sample. For this purpose the so-called Grob test mixture [226] is suitable.

Table 7.27   Main characteristics of generic GC-MS coupling

Advantages

- High sensitivity (low pg levels in PCI scan mode; 1 fg in NCI SIM mode)
- Analysis of (semi)volatile complex mixtures
- Separation and positive identification of unknown components (fingerprinting capability)
- Isotopically labelled internal standards
- Interference-free analyte quantification (SIM)
- Well-established technique
- Wide variety of front-end accessories
- Highly sophisticated trace analysis tool
- Powerful data systems
- Universal and specific
- Wide applicability

Disadvantages

- No unambiguous structural elucidation
- Unacceptable quantification in scan mode
- Modest speed (20–40 min, unless fast GC)
- Not suited for polar or thermolabile compounds
- Relatively low resolving power (QMS)

Table 7.27 shows the *main characteristics* of conventional GC-MS. In general, GC-MS cannot unambiguously elucidate the structure of a completely unknown substance, although this might be possible employing a high-resolution mass spectrometer. The addition of retention time, accurate mass determination, and CID of selected ions further increases the power of GC-MS for the identification of unknown compounds (Scheme 7.5). Due to the added selectivity of MS/MS, a sharp, well-resolved peak is available for quantitative analysis. Quantitation based on MS/MS daughter ions yields accurate results, even in the presence of interferences. An IR spectrum simultaneously produced in a GC-FTIR-MS system helps significantly in structure elucidation. Complex hyphenated systems such as TD-CT/Py/GC-MS/FTIR have been devised for the same purpose.

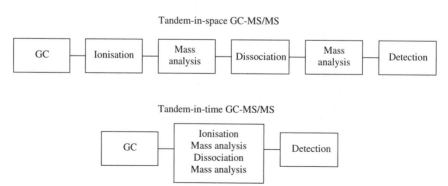

Scheme 7.5   Comparison of tandem-in-time (QQQ) and tandem-in-space (QIT) MS/MS

Because the sensitivity of the GC-MS is exceptionally good, reaching down to ppb level in a 1-mg (1-μM) sample, the technique is irreplaceable in the analysis of complex mixtures that have some volatility. However, the analysis of polar or higher-MW compounds with limited volatility encounters difficulties using GC-MS. Fewer than 10 % of all organic substances are suitable for direct GC-MS analysis [227]. The volatility problem can be addressed by *derivatisation* or PyGC techniques. Derivatisation procedures are employed in GC-MS to render the sample components volatile and suitable for use with a simple direct-inlet MS interface; by use of derivatisation techniques another 10–15 % of all compounds will become GC-MS compatible. For useful derivatives in GC-MS, see also ref. [228]. It should not be expected that samples with low volatility due to their high molecular weight can be rendered volatile by derivatisation. Also, larger molecules have a lower capability to form ions in the MS (compound dependent), and most MS instruments have limitations regarding the maximum m/z value that can be analysed. GC-MS has largely taken over from sulfur-, phosphorous-, nitrogen- and halogen-selective detectors, which suffer from matrix effects and have limitations to their usefulness.

The use of stable, narrow-bore columns and fast oven temperature programmes result in analysis times of 5–15 min, while maintaining the final resolution obtained with conventional GC approaches. *Fast GC-MS* (5–200 s) requires the matching of several technologies:

(i) the choice of a sample introduction method capable of producing a narrow injection bandwidth;
(ii) the choice of an appropriate column size; and
(iii) a fast detector.

Mass spectrometers for fast GC-MS using narrow-bore columns are typically QMS [229], ToF-MS [230], ITMS [231] and magnetic sector types [232]. Fast GC-MS relies on the use of high-performance, small internal diameter, short GC column lengths without loss in separation efficiency, high heating and cooling rates, as well as high head pressures (4–9 bar), higher split ratio than conventional capillary columns and fast scan rates. Pressure programs for the carrier gas should be used to guarantee an optimum flow through the column. As peaks are considerably narrower (typically 0.3 s wide for 0.25 mm i.d. columns) due to the shorter retention times, fast data acquisition and improved data processing are necessary. This can be achieved by automated peak finding and sample comparison algorithms, automated mass-spectral

deconvolution, enhanced quantitation algorithms and custom report design software. Rather than using multivariate data analysis (MDA) techniques [233] the use of SMB-MS [221] appears to be most practical for fast GC-MS.

Fast GC-MS can be carried out with short lengths of column with diameters ranging from i.d. 0.05–0.15 mm to 0.53 mm. Most GC-MS instruments are designed to cope with flow-rates below 1–2 mL min^{-1}, which means column diameters need to be relatively small, so as not to overload the mass-spectrometer vacuum system. Typical fast GC-MS methods include 'open split' [234], 'subambient' [235] and 'time-compressed' methods. *Time-compressed chromatography* (TCC) is fast GC-MS based on microbore fast GC [236] and ToF fast mass spectrometry [230]. Ultrafast GC-ToFMS has been developed, and is commercially available [237]. TCC fast GC-MS suffers from significant difficulties; does not necessitate ToF mass analysis; is of limited optimal use; and can be replaced by other approaches [238]. Flow programming [239] is a known method for reducing the GC analysis time. The high flow-rate encountered with SMB-MS enables flow programming with flow-rate ratios of up to 150 mL min^{-1}, in marked contrast with the limited flow programming flow-rates ratio in conventional GC-MS. The features of supersonic molecular beams were used to establish an alternative method for fast GC-MS. In *GC-SMB-MS* the flow from the GC column is supplemented with a make-up flow, and the gas expands supersonically through a nozzle at atmospheric pressure into a region of low pressure. The supersonic expansion results in reduction of the internal energy of the molecules due to vibrational cooling, reducing the probability of fragmentation. Supersonic GC-MS shows enhanced molecular ions, combined with tuneable fragmentation and library compatibility, extended isomer and structural information, exact isotope abundance and elemental information. EI with SMB is claimed to be the ideal ion source for GC-MS [240] because of:

(i) ultrasensitive and selective hyperthermal surface ionisation (HSI);
(ii) fast GC-MS; and
(iii) broad applicability, including thermally labile and relatively nonvolatile large molecules.

As to the optimum *column diameter* for fast GC-MS, convincing evidence [238] shows that:

(i) a megabore column (0.53 mm i.d.) is the optimal choice for ultrafast GC;

(ii) a narrow-bore column (0.25 mm i.d.) is a good choice where subminute analysis time is not required and extra GC resolution is needed; and

(iii) the microbore column (<0.1 mm i.d.) is of limited use.

Fast GC-MS analysis with microbore columns does not lend itself to trace analysis. It is ideally suited, however, to multicomponent mixtures where ultimate sensitivity is not so important. Contrary to fast GC, with fast GC-MS analysis the mass spectrometer and ionisation methods play a major role in the overall separation capability. Enhanced mass spectrometric separation is more useful than the modest gain achieved with the microbore column. Use of a short, high flow-rate megabore column thus provides a superior alternative to practical fast GC-MS, especially when coupled with mass spectrometry in supersonic molecular beams. Megabore columns offer additional benefits such as allowing on-column injection, which, when coupled with a high flow-rate, provides an effective handling capability for thermally labile compounds. In cases where better chromatography is desirable, a short standard narrow-bore column (0.25 mm i.d.) with a column flow-rate of 5–10 mL min^{-1} can be employed. In short, the gain in resolution with a microbore column is small, SMB-MS compensates for the loss of GC resolution and fast GC-MS is faster with megabore columns. The ability to use conventional quadrupole or ion trap analysers with their limited scan speed is one of the most significant advantages of fast megabore GC-SMB-MS. This is made possible through their capability for handling very high gas flow-rates through a short megabore column. With SMB, splitless injection is performed without any cryofocusing. Fast GC-SMB-HSI-ToFMS was also described [241], and the advantage of reduced analysis time over the HPLC-fluorescence technique for trace analyte detection was pointed out.

At present, supersonic GC-MS represents the ultimate performance in GC-MS, as shown in Table 7.28. One of the major advantages of fast GC-MS is its ability to handle a larger group of thermally labile compounds that usually dissociate in the GC injector, column, or MS ion source [238]. The high carrier-gas flow-rate correspondingly reduces the molecular residence time and dissociation in the injector. Combined with thinner column films and reduced column length, this also minimises dissociation in the GC column. Fast GC-SMB-MS seems to offer the ultimate handling capability of thermally labile compounds [221]. As GC-SMB-MS allows a considerable lowering of the elution temperature, by about 200 °C, the size of

Table 7.28 Main features of fast GC-SMB-MS

Advantages

- Reduced sample preparation time
- Ultrafast total analysis time (from sample to result)
- Unique information (cold EI, HSI)
- Selective ionisation, isomer information
- High concentration detection sensitivity (splitless, HSI)
- Large-volume injection
- Flexibility and ease of use (wide choice in operational parameters)
- Short columns (3–4 m, 0.53 or 0.25 mm i.d.)
- High flow-rates (4–200 mL min^{-1})
- Increased sample throughput
- Suitable for difficult sample preparation problems (DSI)
- Compatibility with various mass analysers (QMS, QITMS, ToF-MS)
- Ultimate handling capability of thermally labile compounds
- Laser desorption injection
- Automation

Disadvantages

- Postprocessing required for quantitation
- Restricted polarity range
- Restricted to volatiles (MW <1 000 Da)

compounds amenable for GC-MS analysis is almost doubled. Many compounds which are too thermolabile for conventional GC-MS can be analysed by ultrafast GC-SMB-ToFMS [240].

An alternative fast GC-MS mode, *low-pressure GC-MS*, is based on the high optimal velocity for the carrier gas obtained when a separation is performed under reduced pressure. The pressure required for LPGC-MS is 120 kPa, generating a flow of 1 mL He min^{-1}. The main advantages of low-pressure GC-MS are shown in Table 7.29.

GC-IMR-MS is based on gentle ionisation of gas molecules by ion–molecule reactions (IMR) [242]. Such reactions between reaction ions and sample gases produce a significantly smaller excess of energy than does electron impact ionisation. Thus, IMR provides a

Table 7.29 Main advantages of LPGC-MS

- Faster analysis (a factor of three to five for temperature-programmed, up to a factor of 10 for isothermal runs)
- Higher sensitivity (increase by factor of five)
- Compatibility with existing injection techniques and ion trap technology
- Low elution temperatures (less column bleed)
- High loadability (film thickness 0.1–1.0 μm)
- Forgiving (10 m, 0.53 mm i.d. capillaries)
- Quantitative analysis
- Increased sample throughput

'soft' ionisation that greatly reduces, and in many cases completely eliminates, fragmentation of molecules in the sample gas. Several different reaction ions (Kr and Xe) are used for IMR (Section 6.2.2.2). GC-IMR-MS covers molecular components up to m/z 500 and allows precise analysis of complex gas mixtures into the sub-ppm range.

GC-MS has been reviewed [203,204] with particular attention being paid to additive analysis [243]; GC-MS interfaces were reported by Oehme [227]. For GC-REMPI-ToFMS, see Section 7.5. Various pertinent monographs are available [227,228,244,245].

Applications GC-MS is widely used in thermoplastics problem-solving. The presence of unknown components is a recurring problem, both during production and in end products. Such components often cause unwanted side-effects such as discoloration, smell, loss of specific properties and side reactions. In these cases it is important to identify and quantify components quickly and effectively. Complex problems such as unwanted product discoloration can usually be addressed by means of a multidisciplinary approach.

GC-MS is limited to those additives that are both thermally stable and reasonably volatile (i.e. boiling points below about 300 °C). However, in the GC sense many organic additives are unstable or not very volatile, such as rubber curatives, higher-molecular-weight stabilisers, oligomeric materials. The use of short GC columns, supersonic molecular beam techniques, low stationary phase loadings, and high-temperature phases extend the applicability of GC-MS to some higher-MW materials. HT-GC-MS (up to 380 °C) has been used to separate additives up to MW 1200 (Irganox 1010) [246]. As a result, GC and GC-MS are widely used in the qualitative and quantitative analysis of polymer and rubber additives. General applications of GC-MS are listed in Table 7.30.

Various *GC-MS protocols* for additive analysis have been developed. Kawamura *et al.* [247] have presented

Table 7.30 General applications of GC-MS

- Identification/quantification of residual monomers/solvents/volatile additives
- Analysis of processing off-gases, exhaust gases, combustion
- Monitoring of working environment conditions
- Identification/quantification of off-odour/taint components
- Identification/quantification of degradation products/contaminants
- Trace analysis
- Survey of competitor products
- Pyrolysis of polymeric materials

the total ion chromatogram by GC-MS of 53 additives in PE. Up to 89 components were separated in other simultaneous determination methods [248].

GC-MS is well suited for the determination of *monomers*, such as α-methylstyrene in styrene-based polymers [249], or other low-MW compounds produced from the processing of rubbers and plastics. Variants of GC-MS such as HS-GC-MS, CT-GC-MS, TD-GC-MS are ideal for identifying *residual volatiles* in polymers, including monomers, monomer impurities, oxidative degradation products, solvents, odorants or irritants, etc. Residual VCM from 0.02 to 0.1 ppm in PVC has been determined by GC-MS monitoring [250,251]. Maleic acid in unsaturated polyester resins has been determined by aminolysis with a primary aliphatic amine followed by GC-MS analysis [252]. Long-term properties of GFR polyester composites were studied by GC-MS and HS-GC-MS using multivariate data analysis of low-MW products [253]. By means of partial least-squares (PLS) modelling it was possible to estimate the degradation time directly from the quantity of 13 identified degradation products. GC-MS has also been used for the determination of the by-products 1,3-dichloro-2-propanol and 3-chloro-1,2-propanediol of the polyamidoamine-epichlorohydrin wet-strength resins for paper [254].

GC-MS and GC-AED techniques were used for the direct analysis of used tyre vacuum pyrolysis oil [255]. Antioxidants and antiwear additives (0.25–5 wt% DODPA, α-NPA, TCPs, TPP, IPPs) in *lubricating synthetic oils*, essentially esters of branched-chain alcohols such as pentaerythritol, neopentylglycol and trimethylolpropane, were determined by means of GC-SIM-MS using diphenylamine (DPA) as an internal standard [256]; similarly, TCPs, TPP, IPPs, IIPs and I2P were quantitatively analysed by GC-FPD using triethylphosphate (TEP) as an internal standard. RSD values of 3–6 % were reported for GC-SIM-MS, and 7–9 % for GC-FPD.

Polymer *extracts* are frequently examined using GC-MS. Pierre and van Bree [257] have identified nonylphenol from the antioxidant TNPP, a hindered bisphenol antioxidant, the plasticiser DOP, and two peroxide *catalyst residues* (cumol and 2-phenyl-2-propanol) from an ABS terpolymer extract. Tetramethylsuccinodinitrile (TMSDN) has been determined quantitatively using specific-ion GC-MS in extracts of polymers prepared using azobisisobutyronitrile; TMSDN is highly volatile. Peroxides (e.g. benzoyl or lauroylperoxide) produce acids as residues which may be detected by MS by methylation of the evaporated extract prior to GC-MS examination [258]. GC-MS techniques are not applicable to all compounds, especially thermolabile compounds such as N-nitrosamides or the non-volatile NOC.

GC-MS was used to correlate the *flame retardancy* of various PC, PPE, PBT and PS grades containing DBDPE (Saytex 102E), $Mg(OH)_2$, red P and TPP during combustion [259]. GC-NCI-MS has been exploited to determine a variety of brominated flame retardants (PBDEs, TBBP-A, HBCD, PBBs) [260]. Decabromodiphenylether (DBDPE) (MW 959) was analysed by means of GC-QMS using tris(perfluoroheptyl)-s-triazine (TPFHST) as a calibration compound (MW 1185) [261]. Formation of PBDDs and PBDFs during pyrolysis of several polymer blends containing BFRs was also analysed with GC-MS [262], and the products of the pyrolysis in air of CPVC/ABS and PVC/ABS blends with and without the *smoke suppressant* Fe(OOH) were studied with the same technique [263].

Maehara and Yamada [248] have developed a simultaneous GC-MS analysis method for 89 polymer additives, including 34 PVC *plasticisers*. GC-MS has provided a means for analysis of mixtures of C_1 to C_{10} alkylbenzylphthalates [264]. A mixture of 10 phthalates has been separated and detected by MS-FID. Exposure to phthalate esters via indoor air inhalation was monitored by sampling air in a charcoal tube, followed by toluene extraction and GC-MS analysis (with internal standardisation) and GC-FPD analysis (with external calibration) [265]. Kumar [266] has published a mass-spectral guide for quick identification of dialkylphthalates in complex mixtures. Generally, mass spectra of branched phthalates are grossly similar to those of straight-chain phthalates. In these conditions, it is recommended that reference standards of both samples be chromatographed so that characterisation of branched phthalates can be based on both retention time and mass spectra. Recently, the analysis of adipic acid ester plasticisers by GC-MS was reported [267]. Frequently encountered artefacts in GC-MS typically arise from BHT and phthalates [268]. The γ-radiolysis products of dibutylphthalate (DBP) were analysed by GC, GC-MS and PyGC-MS [269]. A quantitative GC-MS assay with accurate mass selected ion monitoring was developed for leachates and extracts of the neurotoxic plasticiser n-butylbenzenesulphonamide (NBBS) commonly used in polyamides; the $[^{13}C_6]$ NBBS isotopomer was used as an isotope-labelled internal standard. The mass difference between NBBS and its stable isotopomer is 6 Da. Quantitation was based on the ratio of the chromatographic peak areas for ions m/z 170.0276 and 176.0477, as well as m/z 141.0010 and 147.0211.

A GC-SIM-MS method was developed for the determination of *UV stabilisers* in PET bottles using BHT, Chimassorb 81, Cyasorb UV24/5411, Tinuvin P/326/327 as standard compounds and benzophenone as an internal standard; linear calibration curves in the range of 0.4–100 pg were obtained [270]. Breakdown products of the *antioxidant* tetrakis-[methylene-3-(3,5-di-butyl-4-hydroxy-phenyl)-propionate]methane (Irganox 1010) were studied by GC-MS by various authors [271,272]. Irganox 1010 (MW 1177) is difficult to be analysed by GC because of low volatility. However, Me-3-(3,5-di-t-butyl-4-hydroxyphenyl)propionate (MP) is formed as a transesterification product during methanol extraction at high temperature. The content of Irganox 1010 in a polyester resin can thus be determined by GC-MS via the low-MW methyl ester derivative after thermal extraction with methanol [272]. Similarly, Irganox 1076 in HIPS has been determined by GC-MS (LOD 0.1 ppm) after ultrasonic extraction and transesterification with methanol [273]. During extraction from a polymer, additives may inadvertently be subjected to (further) oxidation. GC-MS analysis has revealed oxidation of Irgafos 168 to a phosphate in HDPE also under processing conditions or multipass extrusion. The extent of conversion of the phosphite can be quantified by using GC as well as ^{31}P NMR spectroscopy [274]. MS, fluorimetry and SFC are other useful techniques for studying the degradation of Irgafos 168. MS is very specific for detecting the presence of the phosphate degradation product because of an abundant ion peak at m/z 316. By using FTICR-MS and ToF-SIMS Asamoto *et al.* [275] and Mawn *et al.* [276] observed an m/z 662 ion and attributed this ion to the Irgafos 168 phosphate.

Dilettato *et al.* [271] have identified coextracted low-mass *oligomers* by GC-MS. In fact, GC-MS is an excellent technique for oligomer analysis, since each linear oligomeric hydrocarbon fraction can also be assigned a carbon number. At variance to Ziegler–Natta catalysts, metallocene-based catalysts produce polyolefins without much formation of wax-like oligomers. GC-MS can be used to study the chemical composition of waxes. Oligomers in polyolefins can be analysed by GC up to a MW of about 400 Da. Many other low-MW oligomeric additives can only be analysed using GC-MS by means of pyrolysis, i.e. when fragmented.

Degradation products of LDPE/(BHT, Chimassorb 944) after long-term exposure to compost, water and air (chemical hydrolysis at pH 5 and pH 7) at room temperature were examined by GC-MS [277]; the structural changes in the LDPE film were monitored by DSC and SEC. Among the 79 low-MW degradation products identified by GC-MS the main components were linear and branched alkanes, alkenes, alcohols, ketones and esters. The degradation and transformation products of Chimassorb 944 were identified, as well as the interaction products of Chimassorb 944 with BHT (processing antioxidant). The traditional strategy for determining additive residues is labour intensive, typically involving extraction and generic matrix cleanup prior to GC with selective detection. Nowadays, the selectivity of MS methodology has removed the need for SPE cleanup stages. Exact mass GC-MS using oa-ToF can be exploited to calculate elemental composition, and therefore to predict structure and identity of unknowns or confirm target compounds. Exact mass measurement also enhances selectivity and sensitivity, as characteristic mass spectrograms can be extracted with very narrow mass windows.

Mixtures of (isomeric) long-chain 4-alkylphenols – suspected endocrine disruptors – are directly used in the production of alkylphenolpolyethoxylates, giving rise to some 200 components in the case of the non-ionic *surfactant* Triton-X. LC-ESI-MS has been used for the determination of the total free alkylphenol content in extracts and wastewater, whereas GC-MS in SIM mode is the method of choice for determining each free alkylphenol in mixtures composed of five to 30 compounds [278]. GC-CIMS with ammonia reagent was used to identify and quantitate nonylphenolpolyethoxy carboxylate (NPEC) metabolites of nonylphenolpolyethoxylate surfactants in effluents and river water samples [279]. Ammonia as a reagent gas gave intense ammonia–molecular ion adducts for each NPEC, without secondary fragmentation. Because calibration curves for the NPECs were not linear, an external calibration curve was prepared for each group of samples prior to analysis.

A good example of the use of high-sensitivity GC-MS is the identification of trace compounds causing *odour and taint* produced in foodstuffs packaged in plastic materials (cf. flavour analysis in food science). Printing and coating solvents in plastic bag headspace have been identified by GC-MS as causes of odour or taint [258]. Off-odour release from HDPE resin pellets containing a Vitamin E formulation (CF-120) consisting of Vitamin E, glycerol, PEG-400 and GMC was evaluated by a sensor panel and GC-MS analysis [280]; GMC was the major off-odour contributor. Aroma and solvent barrier properties of multilayer PA6 flexible packaging were analysed by passing a gas carrier stream through a test cell where a permeant chamber is separated from the detector gas stream by a sheet of barrier film. The permeant was then monitored in the gas stream by GC or GC-MS [281]. The organoleptic

performance of extrusion coating LDPE for aseptic liquid packaging applications has been determined by means of HS-GC-MS, RPLC and a taste panel [282]. As the off-taste of film samples is similar to the off-taste of synthetic water samples containing 1.5 ppb aldehydes and ketones and 100–200 ppb acids, it is possible to control the organoleptic quality of LDPE by means of GC analysis.

The photoproducts of UV degradation of enhanced degradable LDPE (with photosensitisers and biodegradable filler) were identified by SPME-GC-MS [283]. Over 100 hydrocarbons, ketones, carboxylic acids, ketoacids, lactones, alcohols, esters and other compounds were detected. Degradation products from the additives, which were used to increase the degradation, were found. With the current high-performing analytical technique, the risk of over-analysis is great. Virus et al. [284] identified up to 118 impurities in a US extracted polymer by means of GC-MS/MS.

The overwhelming majority of substances identified by GC-MS in solvent extracts of plastics intended for food contact are oligomers, additives, and adventitious contaminants, with little evidence for polymerisation aids [285]. Balafas et al. [286] have analysed 136 Australian *food packaging materials* for the presence of six phthalate esters (DEP, DOP, DBP, DMP, BBP, DEHP) and one adipate ester (DEHA) using Soxhlet extraction (recoveries from 95–131 %) and GC-MS. The highest concentrations of plasticisers were detected in printed PE materials; hence the ultimate source of plasticisers may have been the printing inks. Fatty acid methyl esters (FAME) were determined by fast GC-MS [287]. Hydrogenation of unsaturated fatty acids can take place in the GC-MS interface if hydrogen is used as the carrier gas [288]. For monitoring the migration of acetyltributylcitrate (ATBC) from vinylidene chloride-*co*-vinyl chloride films into food during microwave cooking Castle et al. [289] used isotope dilution GC-MS with [^{2}H$_4$]ATBC as a deuterated internal standard.

Contaminants in *recycled plastic packaging waste* (HDPE, PP) were identified by MAE followed by GC-MS analysis [290]. Fragrance and flavour constituents from first usage were detected. Recycled material also contained aliphatic hydrocarbons, branched alkanes and alkenes, which are also found in virgin resins at similar concentration levels. Moreover, aromatic hydrocarbons, probably derived from additives, were found. Postconsumer PET was also analysed by Soxhlet extraction and GC-MS; most of the extracted compounds (30) were thermally degraded products of additives and polymers, whereas only a few derived from the original contents (soft drinks) [291]. Quantitation was carried out using GC-MS (SIM) and deuterated internal standards; most compounds score below the FDA food contact threshold of 215 ppb. SPME was suggested as an alternative, more rapid, mode of extraction, especially for the analysis of volatile contaminants that are more likely to migrate into soft drinks. A similar study of contaminants in recycled PET was carried out by means of HS-GC-MS and extraction followed by GC-MS [292].

Also, polymer additives in *medical plastics* have been analysed by means of GC-MS [293]. 4,4′-Methylenebis(2-chloroaniline) (MBOAC), used in the polymer industry to cure urethane elastomers and epoxy resins, has been analysed in body fluids by means of GC-MS [294]. GC-MS has also been used to examine extracted PP/Triclosan (sanitiser) [243].

Most *dyes*, including sulfonated azo dyes, are nonvolatile or thermally unstable, and therefore are not amenable to GC or gas-phase ionisation processes. Therefore, GC-MS techniques cannot be used. GC-MS and TGA were applied for the identification of acrylated polyurethanes in *coatings* on optical fibres [295]. Although GC-MS is not suited for the analysis of polymers, the technique can be used for the study of the products of pyrolysis in air, e.g. related to smoke behaviour of CPVC/ABS and PVC/ABS blends [263].

Organotin compounds (*biocides* and polymer stabilisers) in water were analysed by means of HS-SPME-GC and GC-MS (SIM) [296].

Increasing reliance on mass spectrometry as the universal detector for GC has not solved all the problems of additive identification. Isomer identification is impossible (except for REMPI technology), but is hardly an issue in additive analysis.

(a) Sampling Techniques for GC-MS

Principles and Characteristics With regard to the front end of GC-MS instrumentation, polymer/additive analysis may take advantage of direct injection (split/splitless, cool on-column, PTV, etc.) and many accessories such as SHS, DHS, P&T, TD, CT, DSI, solid probe, LD, Py, SPE, SPME, and SFE (Figure 7.14), some being more specific for solid sampling. With regard to GC, conventional methods rely on syringe-based liquid dispensing inside the GC injector. Table 7.31 summarises the *injection techniques* used routinely and the area in which they are most applicable. The availability of large-volume injector with PTV means that sensitivity may be extended by at least two orders of magnitude.

Of course, to be able to use the direct injection method of sample introduction, the analyte or the polymer system must be soluble in a solvent. Other methods of sample introduction need to be considered in order to eliminate the involatile material from the chromatographic separation. These have become extremely effective in the analysis of matrices such as polymers.

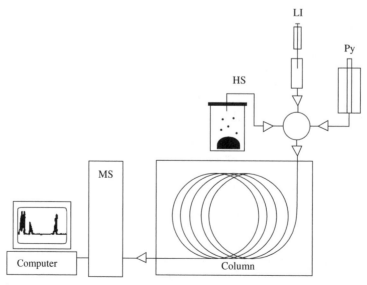

Figure 7.14 Schematic set-up of a GC-MS system with various injectors (HS, headspace; LI, liquid injector; Py, pyrolyser)

Table 7.31 Overview of direct injection techniques used in GC-MS analysis

Introduction technique	Column type	Selected mode	Sample type	Comments
Split/splitless	Capillary	Split Pulsed splitless Splitless Pulsed splitless	High concentration High concentration Low concentration Low concentration	New technique (>2 μL) Useful with large injections (>2 μL) High sensitivity
Cool on-column	Capillary	Ramped temperature Track oven	Low concentration or thermally labile	Minimal sample discrimination and decomposition
Purged packed	Packed Large-bore capillary	N/a	N/a	Reasonable if good peak resolution is not required
Programmed temperature vaporisation	Capillary	Split Pulsed split Splitless Pulsed splitless Solvent vent	High concentration High concentration Low concentration Low concentration Low concentration	

After Golby [243]. Reproduced by permission of Rapra Technology Ltd.

Most introduction systems are related to extraction of volatiles through the application of heat. Many systems involve cryogenic trapping of volatiles prior to GC-MS analysis. Meyer [297] has described a *vacuum distillation* system integrated with the GC via a cold-trapping tube. Ligon and George [298] have used a continuous-flow thermal desorption system in which volatiles from the heated polymer are trapped on to a liquid-nitrogen-cooled section at the head of the GC column. Similar devices have been described by others [299,300]. A *direct sample introduction* (DSI) device has been described which is based on sample introduction and controlled vaporisation in a disposable mini-vial inside the GC injector [220]. As the nonvolatile matrix residue is retained, the need for sample cleanup is eliminated but sample preconcentration is not provided. Legislation has been tightened up to the extent that in many cases very low concentrations can hardly be measured with the (very sensitive) GC-MS without preconcentration techniques and/or large-volume liquid injection (LVI).

In *static headspace sampling* [301,302] the polymer is heated in a septum-capped vial for a time sufficient for the solid and vapour phases to reach equilibrium (typically 2 hours). The headspace is then sampled (either manually or automatically) for GC analysis, often followed by FID or NPD detection. Headspace sampling is a very effective method for maintaining a clean chromatographic system. Changing equilibrium temperature and time, and the volumes present in the headspace vial can influence the 'sensitivity' of the static headspace system. SHS-GC-MS is capable of analysing volatile compounds in full scan with ppb level

detection. Large-volume headspace injection is a useful tool for lowering detection limits to ppt level in full scan. Combination of full scan and SIM can lower detection limits even more.

Dynamic headspace GC-MS involves heating a small amount of the solid polymer sample contained in a fused silica tube in a stream of inert gas. The volatile components evolved on heating the sample are swept away from the sample bulk and condensed, or focused on a cryogenic trap before being introduced onto the chromatographic column via rapid heating of the trap. The technique can be used qualitatively or quantitatively; DHS-GC-MS is considered to be well suited towards routine quantitative analysis.

Purge-and-trap GC-MS is a slightly different approach to the aforementioned techniques. It has wide applicability and has the advantage of being able to handle larger or more representative samples. The technique is not usually associated with thermally heating the sample, although most modern instruments do provide this option. It involves passing an inert gas, usually the carrier gas directly through the sample, if it is a liquid, or the sample atmosphere if it is a solid matrix. The purge volatiles are trapped in much the same way as in the dynamic headspace technique, before being introduced on to the column. The advantages of this technique are that, because larger sample sizes can be used and the atmosphere or liquids can be purged for long periods of time, very low detection limits can be obtained (ppt). This is especially useful for odour analysis in polymeric foams and films. A disadvantage of the existing *preconcentration techniques*, such as purge-and-trap and large-volume liquid injection, is that not only the analyte but also the matrix is enhanced, which can lead to many problems in combination to GC-MS. Moreover, with existing LVI injection techniques the most volatile compounds cannot easily be analysed, as they are often lost during the extraction step. Purge-and-trap was up to now often the only alternative. A disadvantage of purge-and-trap is cross-contamination of samples.

SFE-GC-MS is particularly useful for (semi)volatile analysis of thermo-labile compounds, which degrade at the higher temperatures used for HS-GC-MS. Vreuls *et al.* [303] have reported in-vial liquid–liquid extraction with subsequent large-volume on-column injection into GC-MS for the determination of organics in water samples. Automated in-vial *LLE-GC-MS* requires no sample preparation steps such as filtration or solvent evaporation. On-line *SPE-GC-MS* has been reported [304]. Smart *et al.* [305] used thermal extraction-gas chromatography–ion trap mass spectrometry (*TE-GC-MS*) for direct analysis of TLC spots. Scraped-off material was gradually heated, and the analytes were thermally extracted. This thermal desorption method is milder than laser desorption, and allows analysis without extensive decomposition.

The use of focused or slightly defocused laser light for sample desorption and volatilisation seems to be the ideal injection method for ultrafast GC-MS. With laser desorption injection, chromatography is the limiting time step. Sample preparation is eliminated through the ability to reproducibly desorb and inject a very small sample amount that does not require further cleanup. Laser desorption injection can uniquely provide an additional dimension of spatial information for 2D surface chemical mapping. Laser desorption injection is especially suitable for the organic analysis of surfaces. While most of the laser desorption schemes are based on laser desorption of samples that are placed inside the mass-spectrometer vacuum chamber, in *LD-GC-MS* the laser desorption unit is mounted on the ultrafast GC-MS injector unit [221]. Laser desorption fast GC-MS requires the high flow-rate of the megabore SMB approach. Amirav *et al.* [223,306,307] have reported development of an atmospheric-pressure excimer-laser desorption fast-GC-MS inlet to uniquely desorb and inject a variety of untreated samples and matrices into the fast GC in a helium-purge atmospheric environment.

Applications   HS-GC-MS was used to identify odour in a manufacturing plant as an acetal [308] and to analyse a colour body problem [308a]. HS-GC-MS and GC-MS were both used for *failure analysis*; blister space

on a circuit board was sampled with a syringe [309]. DHS-GC-MS was used to analyse the *curing system* of a vulcanised butyl rubber (IIR): CBS, TMTD, TETD, MBT (sulfur accelerators), BHT (AO), sulfur, DEP, and DOP [243]. The same technique was also used in a study of *migration* of ε-caprolactam from a PA6 cooking utensil into food [243]. A student experiment for the determination of *VOCs* from food-grade LDPE by DHS-Tenax-GC-MS has been reported [310]. Off-line PT GC-MS was used for the determination of 22 volatiles in the ppt–ppb range in the raw materials of plastic food packaging [311].

On-line SFE-GC-MS was used for the analysis of organic extractables from human hair [312]. Van Lieshout *et al.* [313] described GC-MS analysis of an SFE extract of an (ABS) impact-modified PC/PBT blend identifying Ionol CP, Dressinate, cyclic PBT trimer, Irganox 1076 and Irganox PS 800. TD-GC-MS was used in the development of flame retardants, and for the analysis of fire debris [314]. The application of laser desorption fast GC-MS analysis was employed in the analysis of DOP on a stainless-steel surface [221].

(b) Hyphenated IMS Technology

Principles and Characteristics   In chromatographic techniques, ion mobility spectrometry serves as a sophisticated detector, enhancing the reliability of identification of the chromatographs and adding another dimension to the analysis. Hyphenated IMS technology comprises coupling to GC, LC, SFC and CE, but the compelling need for these methods is not obvious. The role for IMS laboratory instrumentation is unclear when so many capable detectors are already commercially available. In field analysis, however, IMS may come to occupy a significant position as a sophisticated detector for portable gas chromatographs.

An ion mobility spectrometer offers to prospective users an attractive detector for a GC, from the perspective of detection limits and specificity. A mobility spectrometer, even with low resolution, allows interrogation of compound identities and imparts better specificity than the electron-capture detector. When gaseous analytes are delivered individually to IMS, the mobility spectrum contains information for identification, provided that operating conditions are kept constant for the unknown and reference spectra. The connection of a GC column to an ion mobility spectrometer is convenient, due to the compatibility of pressures. GC-IMS coupling simplifies the ionisation chemistry in IMS. The main technical problems arise from interferences (matrix effect) and from the reproducibility of the quantitative response of IMS. Few data regarding precision are available for GC-IMS, but these range from 15 to 25%. No GC manufacturer has adopted IMS as an option. The large selection of well-recognised detectors for GC almost excludes substantial changes in the situation.

GC-IMS-MS instruments are ideally suited for laboratory studies, as a complex mixture can be separated; ionisation in relatively clean systems can take place; and the identity of the ions can be studied and verified by mass spectrometry [315]. However, the cost of such systems is quite prohibitive, and their complexity confines their utilisation to the laboratory. In GC-IMS-MS, the gas chromatograph is used to preseparate the components of the sample, with the IMS used as its detector. The ions that constitute the mobility spectrum are then further characterised by MS.

IMS has also been proposed as a qualitative detector for LC [316,317].

GC-IMS has also been reviewed [318].

Applications GC-IMS is applied only in a few laboratories. Most applications have been directed toward environmental analyses. GC-IMS is used in niche areas, such as high-speed air-quality monitoring (on board space stations) and detecting chemical warfare agents. Snyder *et al.* [319] have described a hyphenated field-portable hand-held GC-IMS device, which was applied to the separation of phosphate (TMP, TEP)/phosphonate (DMMP, DEMP, DIMP, DEEP) mixtures. A mixture of four phosphonate analytes can be successfully resolved with a small GC-IMS device in under 8 s.

Failures to address key technical concerns have caused interest to wane in GC-IMS and IMS.

7.3.1.3   GC–Plasma Source Detection

Principles and Characteristics Most common GC detectors (TCD, FID, IRD and PID) are nonselective; others are limited to specific elements such as NPD (N, P) or FPD (S, P). Multielement-selective detectors in GC include optical detectors that measure absorption, emission or fluorescence radiation and mass detectors (Section 4.2). In practice, atomic emission methods prevail over absorption- and fluorescence-based techniques. Atomic emission detection coupled to GC selectively detects most of the elements which may be present in the organic and inorganic compounds which can be eluted from a gas chromatograph [255,320,321]. The combination of a gas chromatograph with an atomic spectrometer is dominated by inductive coupling and microwave induction plasma emission generation (Scheme 7.6). General requirements for a sample analysable by GC-ESD (element-selective detection) include a small volume (1–25 μL) of a solution of nonpolar thermally stable species in a volatile nonpolar solvent (or gas phase).

Plasmas compare favourably with both the chemical combustion flame and the electrothermal atomiser with respect to the efficiency of the excitation of elements. The higher temperatures obtained in the plasma result in increased sensitivity, and a large number of elements can be efficiently determined. Common *plasma sources* are essentially He MIP, Ar MIP and Ar ICP. Helium has a much higher ionisation potential than argon (24.5 eV vs. 15.8 eV), and thus is a more efficient ionisation source for many nonmetals, thereby resulting in improved sensitivity. Both ICPs and He MIPs are utilised as emission detectors for GC. Plasma-source mass spectrometry offers selective detection with excellent sensitivity. When coupled to chromatographic techniques such as GC, SFC or HPLC, it provides a method for elemental speciation. Plasma-source detection in GC is dominated by GC-MIP-AES

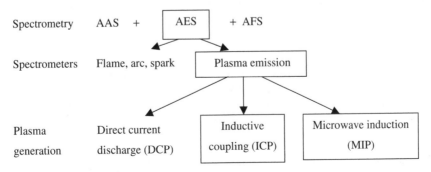

Scheme 7.6 GC-plasma source techniques

and GC-ICP-MS. No commercial MIP-MS systems are available.

In MIP, an argon or helium plasma is initiated and sustained in a microwave cavity that focuses power from a microwave source, typically at 2.45 GHz, into a sample contained in a discharge cell made of quartz [322]. The sample, mixed with pure helium make-up gas, enters the plasma, which causes the molecules to become atomised; the atoms are raised to electronically excited states or are ionised, and as they return to lower energy levels, photons are emitted at wavelengths characteristic for each particular element. The light emitted is analysed by a diode array spectrometer (171 to 800 nm). *Microwave-induced plasma* is not nearly as energetic as inductively induced plasma, and so reagent gases are employed to help energy transfer between the plasma atoms and molecules, and the sample atoms. For this reason, a number of reagent gases (such as $N_2/10\%$ CH_4; H_2, O_2) are usually made available for the plasma torch. The main advantages of using MIPs as alternative plasma sources in chromatographic detection are principally the reduced gas flows and power consumption needed to sustain the plasma [323]. Microwave-induced plasmas suffer from certain drawbacks, one being the easy manner in which the plasma can be extinguished in the presence of excess organic material. The limited thermal energy of MIP sources results in difficulties with sample desolvation and atomisation. Plasma quenching often occurs in tandem systems, as a result of the elution of the solvent peak from the gas chromatograph. In GC-MIP-AES, solvent venting is necessary to prevent plasma instability.

The use of a microwave plasma to excite emission from a GC effluent was first described in 1965 [324]; it has been extensively evaluated as an atomic emission detector for GC [325]. GC-MIP-AES has been commercially available since 1989. Various GC-MIP-AES interfaces have been described [326,327]. The general term *GC-AED* has become widely used for the microwave–plasma-band system described by Sullivan and Quimby [328]. GC-AED has become increasingly popular because of its selectivity for, in principle, all elements. In spite of some interelemental interferences and molecular structure effects, GC-MIP provides a universal and selective multi-element system for organic and inorganic analysis. It can differentiate between halogens and detect oxygenated and deuterated compounds. It is a relatively inexpensive detector that can be used for a preliminary rapid scan of a complex mixture to indicate elemental compounds. Oxygen is probably the most difficult element for which to achieve high selectivity, due to background levels of oxygen from entrained air, water and gas impurities. Various oxygen-selective

GC detectors are available: O-FID, FTIR and MIP-AES. The latter has many of the characteristics of an ideal oxygen-selective detector for GC. MIP-AES successfully replaces O-FID for oxygen-containing compounds, ECD for halogen-containing compounds, and FPD for sulfur-containing analytes. Although the sensitivity provided by AED is lower than that achieved with ECD, this method has to be considered as an advancement when separation problems occur [329]. The main disadvantage of ECD in comparison with AED is the selectivity.

Typical detection limits for GC-AED are $0.1\,pg\,s^{-1}$ for organometallics, $0.2\,pg\,s^{-1}$ for carbon and $1\,pg\,s^{-1}$ for sulfur. Determination of the elemental composition of unknowns provides highly desirable information additional to that obtained from GC-MS. For several elements (e.g. N), detectability using AED is worse compared to NPD or ECD. However, this drawback can be overcome by means of LVI-GC-AED [67]. Additionally, the detector response per mass unit of an element is fairly compound-independent, although this statement does not go unchallenged [330]. While the technique cannot be used uncritically to find empirical formulae, it does offer robust and very useful element ratio data on unknown compounds. In favourable cases, where hetero-atoms are present, and practical assumptions can be made, the empirical formula can be shown to be one of several possibilities. For qualitative analysis, area ratios are calibrated directly from those of two elements of an internal standard. Although GC-AED reduces the need for high resolution from the column, as any interfering substances that do not contain the specific element of interest are not displayed in single-element monitoring, co-elutions adversely affect the determination of low-concentration compounds, due to problematic background correction.

The most important advantages of MIP-AES as an analytical technique for GC detection of metals and metalloids are indicated in Table 7.32. MIP-AES is one of the most powerful analytical tools for selective detection in GC, and is potentially quantitative [331]. Elemental figures of merit for GC-MIP detection have been reported [332]. Microwave-induced plasmas have found much greater use in GC than in HPLC interfacing. Reviews on empirical and molecular formula determination by GC-MIP have been published [332,333].

Several studies have focused on the use of low-pressure and atmospheric-pressure MIP-MS as a detector for GC [334]. Atmospheric-pressure *GC-MIP-MS* systems have some limitations with regard to analysing many low-mass elements (P, S, Cl, etc.). Low-pressure MIP-MS is better equipped for this purpose. Both non-metals and metals have been analysed by GC-MIP-MS.

Table 7.32 Main characteristics of GC-MIP-AES

Advantages

- High sensitivity (sub-pg range)
- High elemental selectivity
- Versatility (easy tuning to a particular wavelength)
- Simultaneous multi-element analysis
- Wide dynamic range (four to five decades)
- Potentially quantitative (robust element ratios)
- Broad application area

Disadvantages

- Limited thermal energy
- Low tolerance to the introduction of even a relatively small amount of sample
- Empirical formula determination dependent on elemental response factors
- Rather complex and costly
- Commercially available

Phosphorous- and sulfur-containing compounds separated by GC have been determined with both He and N_2 MIP-MS systems with low ng to pg detection limits [335]. Halogenated compounds and organotins can be determined by GC-MIP-MS, at low to sub-pg detection limits.

Although the data generated by a MIP emission detector for GC are similar in many ways to those produced by a mass-selective detector, the information provided is often complementary [68]. An AES detector provides:

(i) equal detector response for all components, as well as excellent linearity, which allows accurate relative quantitation of mixtures;
(ii) the possibility of complete characterisation of the elemental composition of all eluted materials; and
(iii) element selectivity.

MSD provides molecular weight, fragmentation information and mass selectivity. Also, simultaneous GC-MS/MIP-AES has been described, using both a low-pressure and an atmospheric-pressure splitter [336]. The combination of MS and AED data sets provides the potential for application to a wide range of analytical problems, such as screening for the presence of hetero-atom-containing analytes (AED), identification and confirmation (MS) and quantification (MS, AED). On-line LVI-GC-AED/MS (dual detection) has been described with small (i.e. less than 0.5 s) differences in retention time of a compound with AED and MS detection [67]. The dual-hyphenation set-up largely eliminates data-interpretation problems caused by small differences in retention time, or retention indices and is,

therefore, a distinct improvement over a purely conventional GC-AED plus GC-MS approach.

The *inductively coupled argon plasma* (ICP) discharge is the most widely used general analytical emission spectrochemical source. The basic difference between the ICP and MIP torch is the method of energy transfer, the physical arrangement of both torches being fundamentally similar. In addition, the ICP torch may not provide good spectra for the common elements, but is very sensitive to elements of higher atomic weight and, in particular, the metallic elements. The ICP torch is often employed with a capillary column system, but can also be used with packed columns, provided that argon is used as the carrier gas. Although energy transfer between the plasma and the elements being examined is more efficient with ICP sources, the power requirement is, nevertheless, considerably greater than that needed by the microwave torch. Most modern ICP atomic emission spectrometers utilise a diode array sensor.

ICP is not quite a suitable emission source for GC detectors, due to very low sensitivity (ng range), because of the large dilution of the GC effluent with plasma gases [337]. On the other hand, direct-current plasma (DCP), which can also be sustained in argon at atmospheric pressure, provides detection limits in the low pg range [338]. However, MIP is the best excitation source to be coupled with GC. Although the inductively coupled argon plasma has been little used for GC, e.g. as GC-ICP-AES [339], it is finding favour as a mass-spectral ion source in *GC-ICP-MS* mode. The use of ICP-MS as a chromatographic detector was first described in the late 1980s. The versatility of ICP-MS is reflected by the variety of front-end analytical arrangements and sample introduction systems used, suitable to cope with samples in gaseous, liquid or solid form. The use of ICP-MS as a chromatographic detector is well established [337,340,341]. ICP-MS has been hyphenated to all forms of chromatographic techniques for speciation analyses, including GC, LC, SFC and CE. In coupling GC with atomic spectrometry, the chromatographic column ideally runs directly into the source. Such a set-up prevents sample loss, peak broadening and tailing. Alternatively, a heated transfer line is used [342]. The analyte is also more efficiently ionised in the plasma, as it is already in the vapour form, thus requiring no desolvation and vaporisation upon aspiration into the ICP. Various GC-ICP-MS interface designs have been described [343–348]; see Fig. 7.15. No commercial GC-ICP-MS transfer lines are available. As a chromatographic detector, ICP-QMS is not necessarily the best choice; GC-ICP-ToFMS has been explored [349].

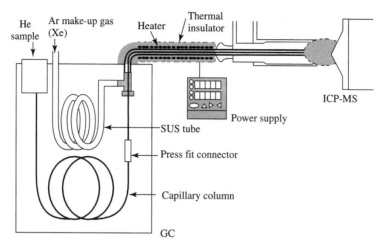

Figure 7.15 Schematic of the GC-ICP-MS interface. Reproduced by permission of Agilent Technologies

GC-ICP-MS is mainly used in relation to gaining speciation information. In the case of organometallic speciation with GC-ICP-MS, the most abundant isotope of the different elements, leading to the most sensitive signals (e.g. ^{120}Sn, ^{202}Hg and ^{208}Pb), can be chosen. ^{126}Xe originating from Xe, doped to the carrier gas, can act as an internal standard for instrumental factors during GC analysis. GC-ICP-MS provides sensitive element-selective information as well as atomic ratios and empirical formulae [350]. Whereas MIP-MS and MIP-AES are equivalent in terms of sensitivity, ICP-MS gains two decades over ICP-AES. In comparison with GC-MIP-AES, detection limits with GC-ICP-MS are at least a factor of 10 superior; typical detection limits for organo-Sn, -Hg and -Pb are in the order of 10–100 fg (absolute as metal). Gaseous sample introduction techniques generally provide better detection limits than liquid introductions (as in HPLC-ICP-MS), with levels ranging from pg to sub-pg.

Table 7.33 reports the main characteristics of GC-ICP-MS. Since both GC and ICP-MS can operate independently and can be coupled within a few minutes by means of a transfer line, hyphenation of these instruments is even more attractive than GC-MIP-AES. GC-ICP-MS is gaining popularity, probably due to the fact that speciation information is now often required when analysing samples. Advantages of GC-ICP-MS over HPLC-ICP-MS are its superior resolution, resulting in sharper peak shapes and thus lower detection limits. GC-ICP-MS produces a dry plasma: when the separated species reach the ICP they are not accompanied by solvent or liquid eluents. This reduces spectral interferences. Variations on the GC-ICP-MS

Table 7.33 Main characteristics of GC-ICP-MS coupling

Advantages

- Multi-element and elemental speciation analyses
- Approximately 100 % sample transport efficiency from GC to plasma
- Excellent analytical recoveries
- High ionisation efficiencies of vapour-phase analytes
- Superior detection limits
- Ultratrace detection of organically bound metals
- Wide linear dynamic range (six decades)
- Less isobaric interferences than LC
- Ability to perform isotope dilution analysis

Disadvantages

- Analyte condensation in interface
- Not suitable for nonvolatile compounds
- Limited number of applications

theme are HTGC-ICP-MS, HG-CT-GC-ICP-MS, GC-HR-ICP-MS, GC-DFS-ICP-MS with low- and high-temperature (410 °C) transfer lines.

Microwave plasma detection has been reviewed [351], also in relation to GC [352,353]. Coupling of chromatography (GC, SFC, HPLC) and capillary electrophoresis (CE) with ICP-MS and MIP-MS detectors has also been reviewed [181,334,335]. Various specific GC-ICP-MS reviews have appeared [334,337,345,346,354,355].

Applications Since the introduction of commercial GC-MIP-AES systems, organometallic analysis can be performed routinely, e.g. for speciation of organotin compounds used as PVC stabilisers [356]. The hyphenated GC-AED technique can be used to improve the efficiency of additive analysis. While on the one hand

AED serves as a universal detector (on the basis of the universal lines C-193, C-496 and C-486), on the other hand, it is also a specific detector. AED is capable of generating valuable information on the element content of the eluted components, which can be of significant help in the identification of the additive type (i.e.: P in peroxide decomposers, S in thiosynergists, N in HALS, Cl in UV-stabilisers, etc.). Various authors have applied GC-AED for *additive analysis* [328,357–360]. GC-AED was used to calculate empirical formula of a variety of compounds, including diethylphthalate ($C_{12}H_{14}O_4$) [320] and fatty ester methyl esters [359]. AED is effective in the tasks of specific element monitoring and component pattern recognition [361]. Although the results are promising, since element-specific information can be obtained from additives (which is of great help in identifying additives in unknown samples), the signal-to-noise ratio of AED for the identification of additives on the basis of element (S, Cl, P or N) detection is much lower as compared with the standard means of detection (FID), and should be improved. Nevertheless, AED has been used successfully to monitor a variety of additives in polymer extracts. The element-specific chromatograms provide fast information about the types of additives. Because of the high resolution of CGC and the selectivity and sensitivity of AED, complex mixtures of polymer additives can be characterised. According to Verdurmen *et al.* [360], GC-AED is not yet a routine tool for additive analysis in an industrial research laboratory, but may be used for specific problems in combination with mass spectrometry.

GC-AED employing microwave plasma emission has been used for determining *tin compounds* applied in the plastics and paints industry, and in agriculture [356]. Most tin compounds are highly polar; consequently, they must be derivatised before separation on a GC column. David and Sandra [362] have illustrated the applicability of capillary GC-AED in the analysis of a supercritical fluid extract of a rubber. Organotin compounds have also been analysed by means of GC-MIP-MS [363]. GC-MIP-MS with a microplasma (MP) ion source has been used for the detection of negatively charged halogen atoms, as an alternative to GC-ECD [364]. GC-AED in combination with GC-MS was applied in the characterisation of used-tyre vacuum pyrolysis oil, containing organic additives [255].

ICP-MS is being used more frequently in combination with a front-end separation technique such as GC, as a specific and highly selective detector for a variety of *speciation applications*. GC-ICP-MS is a powerful technique for speciation of organometals, as it combines the high resolving power of capillary GC with the high sensitivity and selectivity of ICP-MS [365–367]. The technique performs particularly well in the speciation of organic compounds of trace elements, notably Sn, Hg, Pb, As, Ge, Se, Co and Ni [368]. Owing to its high sensitivity, detection limits are significantly better than those of GC-AAS and GC-MIP-AES. For nonpolar organometallic compounds, GC-ICP-MS is an excellent detector, which has been applied to both packed column low-resolution GC (LRGC) and capillary high-resolution GC (HRGC). Achievable LODs for the elements investigated were reported as being in the low- to mid-pg L^{-1} range. The improvements compared with HPLC-ICP-MS are a result of the fact that nearly 100 % (as against 2–4 %) of the analyte reaches the plasma, and that no plasma power is used for desolvation. GC-ICP-MS was used for the speciation of organotin compounds [345] and other organometallic species [365], including tetraethyllead; detection limits of 0.1–1.0 pg are quoted [347]. GC-ICP-MS was also used for analysing standards of nonpolar organometallic compounds of P, As, Sb and Sn (in concentrations of 0.1 to 100 ng L^{-1}). HG-CT-GC-ICP-MS was applied to the determination of butyltin compounds (MBT, DBT, TBT) [369].

On the whole, the applications of plasma-source emission detection to GC in the field of polymer/additive analysis are limited. The same holds for GC–atomic absorption spectrometry [370].

7.3.2 (Multi)hyphenated SFC Techniques

When the separation stage of a hyphenated system is SFC, special interfacing problems exist between the SFC and the spectrometer, due to the unique properties of supercritical fluids. Phase changes, varying sample introduction rates, mobile-phase compatibility, mobile-phase elimination, integrity of the SFC separation, and detection are all problems with SFC hyphenated analytical methods. Capillary columns are directly compatible with MS and FTIR detectors. With greater supercritical mobile-phase flow, packed columns are more compatible with liquid detectors traditionally used for HPLC, such as UV or PDA. As in the case of precolumn hyphenation, it is therefore useful to clearly distinguish hyphenation involving cSFC and pSFC.

Both flow-cell and solvent elimination SFC-FTIR are useful, in particular for thermolabile components. This hyphenated technique requires a compromise between chromatographic and spectroscopic requirements. Its use is limited. Also, the application of SFE-SFC-FTIR to polymer/additive analysis is not widespread. SFC-NMR spectroscopy is a fairly specialised hyphenated technique, which is not widely available, but offers interesting prospects for application. SFC-MS is a specialised technique with a complicated coupling (column and ionisation type, interface, mass spectrometer), which is not in routine general use and has found only limited applicability for polymer/additive analysis. For this purpose, on-line SFE-SFC-MS appears to be more popular than SFE-SFC-FTIR. SFC is still in its infancy as far as speciation analysis of metal-containing additives is concerned. Introduction of carbon-containing gases into plasmas adversely affects the excitation properties of the plasma. Easy hyphenation of pSFC to MIP-AES has been reported. ICP-AES has been coupled to both pSFC and cSFC. SFC-ICP-MS requires a rather complicated instrumental design. In terms of detection methods, SFC is a separation technique with higher potential than LC in hyphenated techniques. Hyphenated systems involving supercritical fluids have been reviewed [371].

7.3.2.1 SFC–FTIR Hyphenation

Principles and Characteristics Fourier-transform infrared detection in SFC is attractive because it can offer structural information about the analytes [372]. The coupling was introduced in 1983 [373]. Various approaches have been advanced:

(i) Matrix isolation (MI) methodologies.
(ii) Coupling with a high-pressure 'closed' liquid-phase flow-cell through which the column effluent is

passed for transmission or ATR FTIR measurement (on-line mode).

(iii) Continuous mobile phase elimination with deposition of the eluites on an appropriate substrate prior to transmission or reflectance IR measurement (off-line or nearly on-line modes, *cf.* Figs 7.16 and 7.17).

Consequently, the major *experimental options* for the analyst are packed or capillary SFC, mobile phase with/without modifier and off-line or on-line mode, namely direct deposition (DD-SFC-FTIR) vs. flow-cell. Both small-bore packed columns and narrow-bore open-tubular columns have been used for SFC-FTIR analysis using a pressure-stable, thermostated, flow-cell or solvent elimination interfaces.

The FTIR detector is constrained by two major problems: mid-IR absorption by most chromatographically compatible mobile phases, and relatively low FTIR sensitivity compared to several other detectors. Since SFC analytes are nonvolatile, mobile phases are easily eliminated. This is advantageous, as absorbance of the mobile phase in the flow-cell set-up and changes in the mobile phase spectrum with applied pressure render background correction difficult, particularly if modifiers are used. Actually, practically all SFC-FTIR investigations have dealt with 100 % CO_2. $scCO_2$ is a suitable *mobile phase* for SFC-FTIR, because of its IR transparency. Nevertheless, some important spectral regions are obscured because of strong CO_2 absorption. Supercritical xenon has been used as an alternative mobile phase for cSFC-FTIR, both from a chromatographic point of view and by virtue of its spectral transparency [375]. Xenon affords several unique advantages, including:

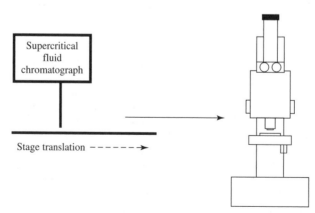

Figure 7.16 Schematic representation of off-line SFC-FTIR. After deposition of the eluites on to a moving ZnSe substrate the window is moved to the focus of a stand-alone FTIR microscope, where the spectrum of each spot is measured with the plate stationary. After Griffiths *et al.* [374]. Reprinted from P.R. Griffiths *et al.*, in *Hyphenated Techniques in Supercritical Fluid Chromatography and Extraction* (K. Jinno, ed.), pp. 83–101, Copyright (1992), with permission from Elsevier

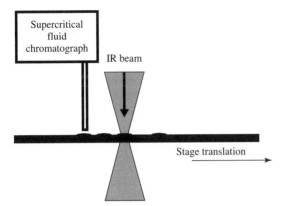

Figure 7.17 Schematic representation of on-line SFC-FTIR. Each eluite passes through the IR beam a few seconds after deposition. After Griffiths *et al.* [374]. Reprinted from P.R. Griffiths *et al.*, in *Hyphenated Techniques in Supercritical Fluid Chromatography and Extraction* (K. Jinno, ed.), pp. 83–101, Copyright (1992), with permission from Elsevier

(i) no need for spectral background subtraction;

(ii) detectability of OH-stretching vibrations ($3500\,cm^{-1}$ region) and aromatic CH-deformations ($<800\,cm^{-1}$); and

(iii) closer matching to library spectra for spectra obtained in Xe than in CO_2.

scN_2O (nitrous oxide) is unsuitable for flow-cell SFC-FTIR.

In SFC-MI-FTIR, a cryogenic deposition surface is maintained at 150 K, instead of the 10 K needed for GC-MI-FTIR. Detection limits of 100 ng are reported for this interface [189]. *Flow-cell FTIR detection* in SFC is similar to GC-FTIR, although the smaller internal diameters of SFC columns have imposed different cell designs. The infrared transparency of CO_2 permits the use of flow-cells with pathlengths 10–100 times greater than those employed for HPLC-FTIR. For on-line detection the flow-cell must be able to withstand high pressure (up to 4000 psi), have IR transparent windows (typically 2 mm thick) and exhibit a volume considerably smaller than the peak volume. The cell volumes in on-line SFC-FTIR are inevitably large compared with those employed in SFC with UV detection, because of the constraints imposed by spectroscopic considerations. The IR flow-cell dimensions must be designed according to the column type in use. The flow-cell design in cSFC-FTIR is an inevitable compromise between the chromatographic and spectroscopic requirements, and thus does not qualify as an ideal hyphenation mode. The requirements for an optimised flow-cell for cSFC have been given [376]. Raynor *et al.* [377] have described a

Table 7.34 Main characteristics of hyphenated flow-cell SFC-FTIR

Advantages

- Nondestructive
- Real-time monitoring
- Quantitative potential
- Multihyphenation

Disadvantages

- High-pressure cell
- Limited choice of mobile phases
- Narrow IR wavelength windows
- Compromise between chromatographic and spectroscopic requirements
- Difficult spectral searching
- Relatively low sensitivity

flow-cell interface for use with a 50-μm i.d. open tubular column. Nanogram detection levels are possible in flow-cell SFC-FTIR. Also a light-pipe SFC-FTIR flow-cell has been developed [378]. The main features of this coupling are given in Table 7.34.

As the entire effluent stream is monitored, all sample components are detected. High-pressure IR flow-cells allow real-time monitoring. Flow-cell SFC-FTIR has the benefit of being a nondestructive technique, permitting placement of a second detector (such as FID or MS) post-flow-cell for added data collection. The main disadvantage of on-line SFC-FTIR is blackout of certain regions of the spectrum because of mobile-phase absorption (unless scXe is used). The fraction of the spectrum that is lost depends on the complexity, polarity and symmetry of the molecules comprising the mobile phase. Various important regions of the spectrum may be lost in flow-cell SFC-FTIR, such as 3850–$3475\,cm^{-1}$ (O—H stretch), 2575–$2040\,cm^{-1}$ (C≡N stretch) and 800–$600\,cm^{-1}$ (aromatic C—H bending). This situation deteriorates for pSFC, where it is common practice to add organic modifiers to CO_2. If methanol is added to CO_2 at a concentration of 5 %, most of the useful IR spectrum is lost. Spectral searching of flow-cell SFC-FTIR spectra will only yield unambiguous results if the reference spectrum is measured under the same SCF conditions.

The problems discussed above may be circumvented by eliminating the mobile phase before measuring the spectra of the eluites, as first demonstrated by Shafer *et al.* [379] for pSFC-FTIR. Each eluite was deposited on a moving glass plate, on which a layer of powdered KCl or KBr had been laid down from methanol slurry for diffuse reflectance spectroscopy (SFC-DRIFTS). *Solvent elimination SFC-FTIR* after deposition of the eluites on to a moving ZnSe substrate is quite straightforward; the window is moved to the focus of a stand-alone FTIR microscope, where the spectrum of each spot is measured with the plate stationary (Figure 7.16). Measurements made when the eluites pass through the focused IR beam immediately after deposition (DD-SFC-FTIR) are referred to as on-line measurements (Figure 7.17). The spectra recorded can be referenced to conventional condensed-phase libraries for identification. A commercial interface, the LC-Transform™, delivers the separation process eluents via a spray nozzle; the carrying solvents are evaporated during the spray process, and the resulting chromatogram is deposited as a continuous track on an IR-transparent (e.g. Ge, ZnSe) flat sample collection disc for scanning of the track in a stand-alone microspectrometer [380]. The LC-Transform™ is particularly effective when coupled to SFC. The CO_2 flashes as it enters the terminal low-pressure region, and the solutes spray out of the capillary tip and deposit directly on the sample collection disc. It is important to deposit the analytes in an area as small as possible (preferably <100 μm) to build up thickness and enhance absorbance, by analogy to DD-GC-FTIR and HPLC-FTIR. Consequently, it is preferable that solutes are deposited at room temperature as solids rather than liquids, as the latter tend to spread over a larger area of window surface. Evaporative SFC-FTIR interfaces are equally applicable to separations effected on capillary or 1-mm i.d. packed columns using modified or unmodified mobile phases. No solvent interferences occur. The minimum quantity of any analyte that can be identified in real-time varies from several hundred pg for strongly absorbing compounds to some tens of ng for very weak absorbers [379–381]. The technique is particularly useful for organic compounds that are too nonvolatile to be separable by GC. The main features of evaporative SFC-FTIR are summarised in Table 7.35.

As most polymers contain relatively few additives, the high resolving power of capillary SFC columns is

Table 7.35 Main characteristics of solvent-elimination SFC-FTIR

Advantages

- High sensitivity
- Access to condensed-phase library reference spectra
- Use of a variety of mixed mobile phases
- Flexibility
- No need for dedicated FTIR microspectrometer
- Nanogram identification limits

Disadvantages

- Loss of volatile compounds during mobile-phase evaporation
- Co-deposition of mobile-phase impurities

not required, and separations in less than 10 min may be achieved by pSFC (in comparison with 40–60 min for cSFC). Up to 1990, most DD-SFC-FTIR measurements involved cSFC separations; deposition from packed columns with reduced diameter is a more difficult, yet feasible, matter. SFE-SFC-FTIR [86] and SFE-SFC-UV [14] have also been reported.

Infrared coupling to SFC has further developed into the use of two or more complementary detectors, such as cSFC-FTIR-MS [382] and cSFC-UV-FTIR-FID. Nevertheless, SFC-FTIR appears not to have developed the same importance as SFC-MS (*cf.* Section 7.3.2.2).

Development and application of various chromatographic interfaces to infrared spectrometers have been reviewed [167], more particularly also for SFC-FTIR [167,372,383,384] and on-line cSFC-FTIR [385]. Taylor and Calvey [386] have reviewed high-pressure flow-cell cSFC-FTIR and pSFC-FTIR. Griffiths *et al.* [374] reviewed SFC-FTIR involving elimination of the mobile phase.

Applications Both solvent-elimination and flow-cell SFC-FTIR procedures have proved useful in the analysis of a variety of mixtures of high-MW, reactive and thermolabile compounds [383,386]. SFC-FTIR microspectrometry is a technique with considerable potential for polymer/additive characterisation. Accurate identification of unknown additives was achieved by means of cSFC-FTIR, using a microscope accessory and the solvent elimination interface [381]. Spots (ca. 200 μm in diameter) associated with peaks of erucamide and Irganox 1010 were collected at the solvent elimination interface on a KBr disc and analysed in the FTR microscope [383]. Raynor *et al.* [381] showed how 21 common *polymer additives* with a variety of chemical types with molecular weight from 225 to 1178 Da could be separated by SFC on a nonpolar capillary column in a single chromatogram; good-quality FTIR spectra could be obtained on sample quantities of the order of 100 ng deposited on KBr discs with solvent elimination. The vapour pressure of many SFC eluites is sufficiently low that the deposits last for several days on the substrate, without being lost through vaporisation. As methanol-modified CO_2 cannot be used directly with either FID or with a flow-cell FTIR apparatus (due to background noise) the pure modifier was spiked directly into the extraction vessel. The method has been used successfully for the analysis of unknown additives in PP samples, and has attracted significant industrial interest. SFC-FTIR is especially advantageous in the analysis of unknown mixtures of additives in which two or more components have similar retention times, such as Tinuvin 144 and a degradation product of Irgafos 168 [381].

SFE has been linked to cSFC-FTIR to analyse *antioxidants*. cSFC-FTIR/FID with solvent elimination on a KBr disc was reported for qualitative analysis of a PP/(Tinuvin P/440, Irganox 1010/3114, erucamide) extract [381]. Good-quality FTIR spectra could be obtained from nonvolatile components deposited at levels of the order of 100 ng. Smith *et al.* [380] separated, detected, and identified Irganox 1076 by SFC-IR, using a solvent-elimination interface originally designed for LC-FTIR. The optimum sheath conditions for depositing this antioxidant with MeOH-modified scCO$_2$ were determined. Off-line SFE-SFC-FTIR (flow-cell) was used for the analysis of PE/(Irganox 1010, Irgafos 168, Irgafos 168 phosphate, BHT) [121]. Both scCO$_2$ and Soxhlet extract large amounts of low-MW polymer.

Flow-cell SFC-FTIR has been successfully employed in the study of UV curing *inks and coatings*, which are mixtures of reactive diluents, thermally labile photoinitiators and reactive oligomers [387]. Solvent-elimination cSFC-FTIR of dimethylsiloxane *oligomers* [193] and solvent-elimination pSFC-FTIR of styrene and methyl phenyl siloxane oligomers [388] were reported. Phenyl-ethoxy-acrylate oligomers were analysed successfully, employing in-line cSFC-flow FTIR-FID using a CO$_2$ mobile phase [377].

Yang *et al.* [389] rapidly distinguished compounds extracted from *paper*, using on-line SFE-SFC-FTIR in conjunction with principal component analysis. The quantitative determination of the surfactant mixture Triton X-100 and other complex oligoether *surfactants* by means of cSFC-FTIR flow-cells has been reported [390,391]. Practical applications of SFC-FTIR include the determination of nonvolatile compounds from microwave-susceptible packaging that may migrate into heated food. Another application is the analysis of *fibre finishes* on fibre/textile matrices.

On-line SFE-pSFC-FTIR was used to identify extractable components (additives and monomers) from a variety of nylons [392]. SFE-SFC-FID with 100 % CO$_2$ and methanol-modified scCO$_2$ were used to quantitate the amount of residual caprolactam in a PA6/PA6.6 copolymer. Similarly, the more permeable PS showed various additives (Irganox 1076, phosphite AO, stearic acid – *ex* Zn-stearate – and mineral oil as a melt flow controller) and low-MW linear and cyclic oligomers in relatively mild SCF extraction conditions [392]. Also, antioxidants in PE have been analysed by means of coupling of SFE-SFC with IR detection [121]. Yang [393] has described SFE-SFC-FTIR for the analysis of polar compounds deposited on polymeric matrices, whereas Ikushima *et al.* [394] monitored the extraction of higher fatty acid esters. Despite the expectations, SFE-SFC-FTIR hyphenation in on-line additive analysis of polymers has not found widespread industrial use. While applications of SFC-FTIR and SFC-MS to the analysis of additives in polymeric matrices are not abundant, these techniques find wide application in the analysis of food and natural product components [395].

7.3.2.2 SFC–MS Hyphenation

Principles and Characteristics SFC-MS is a sensitive coupled technique that can be selective or universal; it was first mentioned in 1978 [396]. Further developments are given in Table 7.36. It is used in an on-line mode with 'open cell' gas-phase interfaces, where the mobile phase is decompressed to low pressures. SFC presents a number of features which allow for easier coupling with MS than other chromatographies. In practice, however, SFC-MS coupling did not turn out to be as easy as expected, a fact which can be ascribed to the problems met in the adiabatic expansion of the mobile phase and the effects of pressure gradients in the ion

Table 7.36 SFC-MS interfaces, ionisation techniques and mass analysers

Interface	Ionisation type	Mass analyser
Molecular-beam	Chemical (CI)	Quadrupole (Q, QQQ)
Direct-fluid-injection (DFI)[a]	Electron impact (EI)	Double-focusing sector (DFS)
Thermospray (TSP)[a]	Charge exchange (CE)	Time-of-flight (ToF)
Particle-beam (PB)[a]	Electron-capture (ECNI)	Fourie-transform (FT)
Moving belt (MB)[a]	Atmospheric pressure (API)	Ion trap detector (ITD)
Plasma desorption (PD)[b]	Secondary ion (SI)	
Supercritical fluid-injection (SFI)[c]	Fast atom bombardment (FAB)	

[a] On-line.
[b] Off-line.
[c] Direct introduction to MS without prior chromatography.

source on the MS performance [397]. On the other hand, since SFC is typically performed without nonvolatile buffers, problems encountered with such buffers in LC are not faced by SFC. The other important factor in SFC-MS is column type. The gas burden for capillary columns is much less than for packed columns, which affects the choice of interface used to couple SFC to MS. While a typical mass-spectrometer vacuum system can readily handle the total gas volume resulting from a capillary column, the high flow-rates for conventional packed columns require an enrichment device (such as MB or PB) to remove the excess gas from the expanding supercritical fluid. The choice of column type (i.e. flow-rate) often dictates which interface, which ionisation methods, and even which types of mass spectrometer may or may not be used (Table 7.36).

The chromatographic and mass spectrometric choices facing the analyst in coupling SFC and MS successfully, namely: injection method; column; type of flow restrictor and mass spectrometer; ionisation method and type of vacuum system, have been described [398]. In SFC-MS coupling, the restrictor plays a major role, as the expansion behaviour to a large extent determines the overall performance of the SFC-MS system and defines the range of applications.

Many SFC-MS interfaces have been constructed [399–401], not unlike LC-MS interfaces. The *interface design* must prevent adsorption or thermal decomposition before the compound reaches the ion source. In the combination of any high-pressure process with MS, efficient ion production and transmission from the source to the MS analyser is necessary. This is particularly challenging in the case of SFC-MS, as most SFC with an unmodified mobile phase is performed with pressure/density-programmed elution, i.e. the pressure within the SFC system changes during the course of the separation. Two types of SFC-MS interfaces are commonly distinguished: direct-coupling and mobile-phase-elimination. A variety of *direct-coupling interfaces* have been described, including SFC-TSP-MS [402], SFC-DFI-MS [403] and SFC-API-MS [404]. SFC-TSP-MS can accept the entire SFC effluent from $scCO_2$, even for conventional 4.6 mm i.d. packed columns, but has low sensitivity and the thermospray mass spectra are often difficult to interpret. Little work has been performed with this interface, as reviewed by Combs *et al.* [3]. In the direct-fluid-injection (DFI) interface, the entire effluent from the column is introduced into the ion source ionisation volume through a flow restrictor. DFI interfaces are mechanically simple, low-cost devices which allow freedom in the choice of the detection mode. A drawback is that the performance of the mass

spectrometer in the EI mode is adversely affected by the introduction of relatively large quantities (1–5 mL/min) of the mobile phase into the ion source. DFI interfaces provide relatively simple cSFC-MS operation, as long as detection limits of <10–100 ppm are not required. Detection limits of high flow-rate (100–200 µL/mm) SFC-DFI-MS interfaces are in the range of a few tens of pg in the SIM mode, but are complex and not commercially available.

Various *'transport' type interfaces*, such as SFC-MB-MS and SFC-PB-MS, have been developed. The particle-beam interface eliminates most of the mobile phase using a two-stage momentum separator; with the moving-belt interface, the column effluent is deposited on a belt, which is heated to evaporate the mobile phase. These interfaces allow the chromatograph and the mass spectrometer to operate independently. By depositing the analyte on a belt, the flow-rate and composition of the mobile phase can be altered without regard to a deterioration in the system's performance within practical limits. Both EI and CI spectra can be obtained. Moving-belt SFE-SFC-MS^n has been described.

The most common interfaces for MS coupling to cSFC and pSFC are given in Table 7.37. There is no universal, ideal SFC-MS interface. Mobile-phase-eliminating and direct-coupling interfaces are compared in Table 7.38.

Whereas GC-EI-MS and LC-EI-MS are both well-established analytical tools, SFC-EI-MS has been more elusive. Some of the cSFC-EI-MS spectra are actually not true EI spectra but rather CE-EI spectra [405,406]. With the DFI interface, the effluent is introduced directly into an EI or CI ion source – the latter being quite common. As the SFC pressure varies over the course of the separation, so does the flow of mobile phase into the ion source. Using SFC-DFI-MS, the aprotic CO_2 acts as a typical reagent molecule for charge exchange (CE) with solute molecules. Charge exchange spectra are obtained which are usually comparable to true EI spectra (*cf.* the CO_2 charge exchange spectrum of Irgafos 168 [407]) and a library search can be performed. The sensitivity of cSFC-EI-MS systems is in the pg range for scan data, which is comparable to GC-EI-MS. pSFC-EI-MS was recently described [408,409]; artefact-free

Table 7.37 Most common SFC-MS interfaces

Interface	cSFC	pSFC
Direct-fluid-injection	+	+
Moving belt		+
Thermospray		+
Particle-beam		+
API source	+	+

Table 7.38 Comparison of mobile-phase-elimination and direct-coupling SFC-MS interfaces

Mobile-phase-eliminating interfaces

Advantages

- Good separating capacity
- MS operates in absence of mobile phase-derived ions
- Ease of identification (EI, CI)
- High flow-rates (MB interface)

Disadvantages

- Low sensitivity
- No identification of nonvolatile and polar molecules
- Expensive, cumbersome, dated (MB interface)
- Limited applicability

Direct-coupling interfaces

Advantages

- Good separating capacity
- Better detection limits
- Molecular mass information of thermally unstable components (CI)
- Searchable spectra (EI-like charge exchange spectra)

Disadvantage

- Mass spectra influenced by mobile-phase properties

EI-MS spectra can be obtained, comparable to those generated under GC-EI-MS conditions.

Chemical ionisation is the most frequently used *ionisation technique* in SFC-MS. It provides better sensitivity compared with charge exchange. All standard CI reagent gases can be used in SFC-MS. In the absence of a modifier, CO_2 can act as reagent gas giving real molecular ion spectra (eq. 7.2):

$$CO_2^{+\bullet} + M \longrightarrow CO_2 + M^{+\bullet} \qquad (7.2)$$

Alternatively, with a polar modifier such as MeOH, which acts as a proton-donor reagent gas, and under APCI conditions, $[M + H]^+$ ions are observed. CI provides information about the molecular mass of the sample. In contrast to EI, the fragmentation patterns in CI are not available from spectrum libraries. Hence, identification of unknown components is problematic (*cf.* the methanol CI spectrum of Irgafos 168 [407]). Calibration with a polysiloxane compound has been used for SFC analyses using ammonia CI as reagent gas in the CI ionisation source. If very mild ionisation conditions are applied, such as APCI, almost no fragmentation occurs, and structure elucidation is difficult unless tandem MS is used. On-line SFC-API-MS has recently emerged as a highly promising technique [404,410]. Most SFC-API-MS work has concerned capillary columns in view of the fact that API sources can accommodate high flow-rates.

A limitation in the use of API sources results from the frequent application of mobile-phase composition programming in pSFC. Pinkston *et al.* [411] have compared electrospray and electron impact for open-tubular and packed-column SFC-MS. Direct on-line coupling of SFC to FAB/MS (as well as SFC-ELSD) is also very promising to detect components which give no response in a UV detector [412].

Early work on *cSFC-MS coupling* for the separation of nonpolar, high-MW components like synthetic oligomers was based on a direct introduction method; this straightforward approach culminated in the late 1980s and then declined (Figure 1 of ref. [413]). The low flow-rates delivered by capillary columns in cSFC-DFI-MS ($1-5\,\mathrm{mL\,min^{-1}}$, as liquid) are easily accommodated by the vacuum equipment of a standard mass spectrometer. cSFC-DFI-MS^n using a double-focusing magnetic sector (DFS) mass spectrometer and EI ionisation has been described; the advantages of the system are high resolution and an expanded mass scale, whereas the complexity of interfacing with the chromatographic inlet system due to the high potential ($6-10\,\mathrm{kV}$) of the ion source is a disadvantage. The major application areas of this hyphenated system are: the analysis of low concentrations of low molecular mass; of thermally unstable compounds in complex mixtures; and of high-MW compounds that are not amenable to GC (e.g. waxes, surfactants, polymer additives, etc.). For the latter class of compounds, spectra obtained in HPLC-MS mostly lack the structural information provided by electron-impact spectra. However, cSFC-MS with direct fluid introduction (DFI) interfaces suffers from some inherent limitations:

(i) variable ionisation conditions during analysis;
(ii) carrier fluid flow variation during density/pressure programming;
(iii) the need for high pressures (high flow-rates) for high-MW compounds;
(iv) losses in sensitivity in EI mode; and
(v) charge exchange (CE) with $CO_2^{+\bullet}$ ions [413].

cSFC-MS coupling has often been found to be problematic due to the strong cooling effect of the expanding SCF carrier. CI has produced detection limits that are superior to those of EI: the latter requires high dilution of the decompressed fluid to the requisite low pressures for EI. The low capacity of cSFC columns has been claimed to be responsible for the lack of sensitivity of cSFC-MS systems. Capillary columns in SFC-MS ($50\,\mu\mathrm{m}$ i.d.) are limited to $10\,\mathrm{ng}$ per component (easy column overloading). A commercial DFI interface [414] has been used most for cSFC-MS. cSFC-API-MS, reported in 1993 [415], has only sparingly been used, probably on account of the unavailability of commercial interfaces. Yet, introducing samples to the ion source in an SCF mobile phase allows the ionisation process to be better controlled than if solvents or buffers had been used. Sjöberg and Markides [416] have reported cSFC-API-MS with interchangeable ESI and APCI mass spectrometry. The resolution and sensitivity of this system are comparable to those of cSFC-FID. All SFC-MS reviews [3,398,417,418] point to the promise of API methods. Design and optimisation of cSFC-APCI-ToFMS were described by Lazar *et al.* [419], who stressed the importance of a great many experimental factors. Sjöberg [420] has compared APCI vs. ESI interface performance in cSFC-API-MS. ESI is a 'milder' ionisation technique compared to APCI. pSFC is more useful for routine use, robust, and easier to couple to MS than cSFC.

Flow limitations restrict application of the DFI interface for *pSFC-MS coupling*. pSFC-DFI-MS with electron-capture negative ionisation (ECNI) has been reported [421]. The flow-rate of eluent associated with pSFC (either analytical scale – 4.6 mm i.d. – or microbore scale: 1–2 mm, i.d.) renders this technique more compatible with other LC-MS interfaces, notably TSP and PB. There are few reports on workable pSFC-TSP-MS couplings that have solved real analytical problems. Two interfaces have been used for pSFC-EI-MS: the moving-belt (MB) [422] and particle-beam (PB) interfaces [408]. pSFC-MB-MS suffers from: mechanical complexity of the interface; decomposition of thermally labile analytes; problems with quantitative transfer of nonvolatile analytes; and poor sensitivity (low ng range). The PB interface is mechanically simpler but requires complex optimisation and poor mass transfer to the ion source results in a limited sensitivity. Table 7.39 lists the main characteristics of pSFC-PB-MS. Jedrzejewski

Table 7.39 Main characteristics of pSFC-PB-MS

Advantages

- Unique EI spectra (library searchable)
- Very fast and efficient separations
- Allowance for thermally labile and polar compounds
- Allowance for high-MW analytes of limited volatility
- High flow-rate capability
- Mechanically simple interface

Disadvantages

- Need for a sophisticated optimisation scheme
- Need for removal of extensive gas loads
- Limited sensitivity (ng range)

and Taylor [408] have indicated the need for extensive optimisation (modifier concentration, chromatographic conditions, instrumental settings). The goals of maximum sensitivity and efficient separation are in conflict. The particle beam is an acceptable choice in cases where sensitivity, volatility and analyte polarity are not an issue.

Atmospheric pressure ionisation methods, such as ESI and APCI, have been most commonly used to interface pSFC and MS [410,415,423–426]. pSFC-API-MS devices are designed for accepting high flow-rates. Use of a pressure-regulating fluid interface is required: the pSFC flow-rate must be within a range dictated by the interface capillary and the acceptable fluid flow into the ion source. Essentially, the coupling requires that the chromatographic conditions are adjusted. This is then not the most ideal mode of hyphenation. Consequently, no existing pSFC-MS interface appears to have all the qualities of an ideal interface at this time. pSFC-API-MS is expected to gain broader acceptance.

Jedrzejewski and Taylor [408] have evaluated microbore pSFC-PB-MS. Microbore separations can take full advantage of the simple DFI design, and are still being pursued. Packed-column SFC-MS was recently reviewed [13]. The two techniques, cSFC and pSFC, are complementary: compounds in complex mixtures may be more easily identified with cSFC-MS, while pSFC-MS may be more suitable for target component analysis.

Mass spectrometers involved in SFC-MS coupling are reportedly QMS, QQQ, FTMS, QITMS, ToF-MS and sector MS. Interfacing SFC with Fourier-transform mass spectrometers, which require relatively low analyser pressures ($<10^{-7}$ Torr) has the potential for high-resolution mass spectra [427–429]. By coupling the SFC to an ion trap, MS/MS and MS^3 data can be generated during a single analysis. Complementary MS/MS and pseudo MS^3 data can be generated on a QQQ mass spectrometer, using source-induced CID followed by ion selection and MS/MS analysis [410].

Table 7.40 summarises the *general characteristics* of on-line SFC-MS. The method is potentially most useful for thermally labile and involatile compounds that are unsuitable for GC-MS. Because the MS instrument is the main source of information, the reproducibility of the retention and the separation selectivity are much less important than for other SFC applications. As a result, mass spectroscopists do not feel restrained by the limits on reproducibility, which slowed the uptake of SFC by chromatographers. Method development should not be underestimated. Practical problems are associated with interfacing and the effect of the expanding

Table 7.40   Main characteristics of on-line SFC-MS

Advantages

- Commercial interfaces (limited)
- Evaporation of SCF facilitated in comparison to most LC solvents
- Retention time reproducibility less crucial than for SFC
- Variety of ionisation modes
- Suitable for thermally labile and involatile compounds
- Quantitative analysis

Disadvantages

- Relatively low sensitivity
- Capillary column overloading
- Practical problems with interface
- Flow-rate restrictions
- High maintenance requirements
- No mass information for $m/z < 50$
- Method development
- Non-ideal hyphenation.

SCF carrier freezing and blocking the flow restrictor. Some interfaces lack sensitivity [430]. SFC-MS lacks information at m/z values below 50, due to the high abundance of the CO_2 peak. SFC-MS is not as promising as was originally thought.

For sample identification, especially for complex mixtures, SFC-MS coupling gives better results than DIP. In addition to the mass-spectral data, when using this method the retention times can also be used for identification. The great disadvantage of DIP is the incomplete peak separation, because separation can be achieved only by the use of different evaporation temperatures. An important difference to GC-MS is

interfacing is that in SFC-MS the flow restrictor must be heated in order to prevent solute precipitation. Careful choice of the temperatures of the column, transfer line and probe tip, and of the position of the pressure restrictor in the SFC-MS interface allow successful analysis of components difficult to analyse using GC-MS and LC-MS. SFC-MS has several potential advantages over HPLC-MS, namely, shorter analysis times and favourable chromatographic resolution and sensitivity. SFC-ESI-MS allows a much wider range of flows than those of conventional HPLC-ESI-MS, since CO_2 can act as a nebulising gas. Nevertheless, it would appear that SFC-MS has not yet reached the levels of exploitation that were anticipated. This is ascribed to the fact that SFC-MS is considered to be difficult to use and requires excessive maintenance.

Bücherl *et al.* [431] have described direct SFE-cSFC-HRMS/FID coupling (Figure 7.18) with reduced effective flow of CO_2 into the mass spectrometer and significant loss of sample molecules (except for some low-MW compounds).

Satisfactory performance of the *SFE-SFC-HRMS* instrumentation (resolution 1200) was only possible after optimisation (temperatures, restrictor and quartz tube positions, flow characteristics and sample transfer conditions). Mass spectra obtained for Irganox 1010/1076/1330 and Irgafos 168/P-EPQ by SFC-HRMS were identical with those obtained by use of DIP [431]. However, the sensitivity of the SFE-SFC-MS interface is low (at best 4 % of that obtained with sample introduction via DIP). An enormous amount of sample is lost in all parts of the coupling system (SFE, SFC and

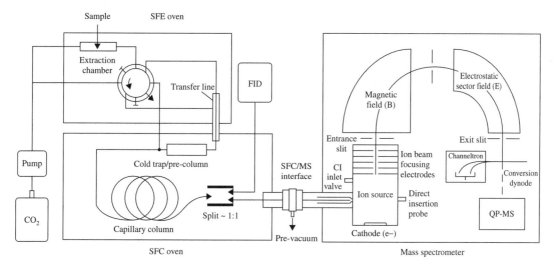

Figure 7.18   Schematic diagram of an SFE-cSFC-HRMS/FID system. After Bücherl *et al.* [431]. From T. Bücherl *et al.*, *Journal of High Resolution Chromatography*, **17**, 765–769 (1994). © Wiley-VCH, 1994. Reproduced by permission of Wiley-VCH

interface). SFC-ELSD-MS split coupling has also been reported [432].

SFC-MS coupling has repeatedly been reviewed [3, 397,398,413,417,418,433–437] and cSFC-MS detection [438] and SFC-DFS [439] were considered specifically. SFC-MS interface technology was recently reviewed by Niessen [440].

Applications SFC coupled to MS is one of the most successful applications of SFC, which range from natural products and polymer additives to forensic examination of controlled drugs [398,437,441,442]. SFC-MS spectra show an increased abundance of ions in the low mass range (100–200 Da). Apparently, strong fragmentation occurs under SFC conditions. The SFC-MS spectra are found to be independent of the experimental conditions and show a good similarity to direct insertion probe (DIP) spectra [403]. EI has generally been indicated as a practical method of ionisation for the polymer additives investigated; CI with isobutane or methane showed almost no spectra characteristics and, with ammonia, the CI sensitivity was very low. SFC-MS is applied especially to the analysis of synthetic polymers and non-ionic surfactants in industrial laboratories. The suitability of SFC-MS for the analysis of thermally labile materials has been emphasised [421,443].

The application range of cSFC-DFI-MS (Table 7.41) appears to be restricted either to the analysis of low-MW substances or to problems related to high-MW samples where low detection limits are not needed [124,444,445]. The analysis of *surfactants* [446] by SFC-MS is frequently performed to demonstrate the feasibility of newly developed interface technology for practical applications. A rugged cSFC-MS method has been developed for the analysis of ethoxylated alcohols (AEs), which are non-ionic surfactants incorporated into a wide variety of industrial and consumer products [447]. cSFC-DFI-DFS was used for the analysis of low-MW, thermally unstable peroxides, and the higher-MW surfactants Triton X-100 and

Dobanol Ethoxylate [443]. At least 16 Triton units with mass 910 were observed. A study of the reactions of amines and amine derivatives with scCO$_2$ using cSFC-MS was also reported [448]. Both cSFC-APCI-MS and cSFC-ESI-MS of PEG 600 and PPG 425 were described [416]. Direct insertion probe (DIP) methodology was used for the structure analysis of the *antistatic agent N,N-bis(2-hydroxyethyl)alkylamine*. When analysed by SFC-MS coupling, the same sample could be separated into six components. The alkyl chains consist of saturated C$_{12}$, C$_{14}$, C$_{16}$ and C$_{18}$ chains and of C$_{18}$ chains with one double bond where 18:1 and 16:0 chains dominate.

pSFC-PB-MS was used to analyse a synthetic mixture of Irganox 1076/1093/1098/MD-1024 [409]. These compounds are not amenable to GC-MS techniques because of their molecular weights and limited volatility. Mackay and Smith [52] have used the analysis of *antioxidants* (BHT) and plasticisers in biocompatible polyurethanes to further illustrate the feasibility of on-line SFE-SFC-MS. Consistent and representative mass spectra were obtained in an analysis that took less than an hour to complete. Some of the additives could not be analysed by GC-MS. Off-line SFE-GC-MS and SFE-SFC-FID were also used to characterise the extracted material. cSFC-EI/CI-MS has been used for the analysis of standard compounds [449] and additives in real products such as microwave packaging [450]. On-line SFE-SFC-MS (sector field) with simultaneous quantitative detection by FID, is a rapid and easy analysis method for polymer homologues and additives in packaging materials [124]. The main aim of this coupling is not an improvement in the detection limit by using highly sophisticated techniques, but to gain quick and reliable information for the identification of unknowns. The electron impact determination limit for Irganox and Tinuvin compounds in a polymer is approximately 5–100 ppm for a 2-mg sample. The application illustrates how structural work and routine polymer analysis can be done with this time-saving method.

Table 7.41 Applications of direct cSFC-DFI-MS

Analytes	Ionisation method[a]	Mass analyser	Detection limits	Reference
Ethoxylated alcohols, poly(dimethylsiloxane)s, alkylethoxysulfates	PCI (1 % NH$_3$ in CH$_4$)	Hybrid MS (QMS/QITMS)	– – –	[444]
Non-ionic surfactants	PCI + CID	QQQ	MS-MS, product-ion scan	[445]
Polymer additives in packaging materials	EI	DFS	10–100 ng (full scan)	[124]

[a] PCI, positive chemical ionisation; EI, electron impact; CID, collisionally induced dissociation.

Table 7.42 Composition of Irgafos P-PEQ

Component	Share (%)	Component	Share (%)
BHT	18.8	Oxidised 4,4'-mono-P-EPQ	0.8
Irgafos 168	26.2	3,3'-P-EPQ	1.3
4,3'-mono-P-EPQ	1.3	4,3'-P-EPQ	16.4
4,4'-mono-P-EPQ	8.8	4,4'-P-EPQ	27.4

After Bücherl *et al.* [124]. Reprinted from T. Bücherl *et al.*, *Packaging Technological Science*, **7**, 139–154 (1994). Copyright 1994 © John Wiley & Sons, Ltd. Reproduced with permission.

Irgafos P-EPQ, used as a processing stabiliser, is a mixture of eight compounds (Figure 7.19 and Table 7.42), as determined by cSFC-MS and DIP-MS. Figure 7.19 shows the SFC chromatogram of Irgafos P-EPQ; the missing component, BHT, was too volatile to be trapped in the cryotrap of SFE-SFC-FID/MS equipment [124]. The quantitative composition of Irgafos P-EPQ was determined by means of SFC-FID. Comparable results have been achieved by HPLC analysis [451].

cSFC-DIP-MS was also used to demonstrate the applicability of an optimised system in terms of sensitivity, signal-to-noise ratio and quality of mass-spectral data in the *quantitative determination* of Irgafos 168, Irgafos 168 phosphate and Irganox 1076 (with Tinuvin 770 as an internal standard) in industrial samples of polymeric materials [403]; CE-EI

spectra were obtained. Carrott *et al.* [452] have studied the performance of 25 pure additives (Irganox 245/1010/1035/1075/1330/1425/3114/PS 802, Irgafos 168, BHT, HMBP, Tinuvin 327/328/384/440/622/770/ 1130, Topanol CA, Synprolam, Chimassorb 944, Cyasorb UV1164/UV531, oleamide and erucamide) and of PE/(Irgafos 168, Irgafos 168 phosphate, Irganox 1010) by means of pSFC-APCI-MS. Standard mixtures of additives could be separated in less than 15 min with a high degree of resolution. Five additives were not eluted: Chimassorb 944, Tinuvin 770, Irganox 1425, Tinuvin 622 and Synprolam, not even at 300 bar and at 20 % modifier concentration. Except for the hindered phenol type AOs, typically the Irganox-based additives (which showed extensive fragmentation), the additives generated good spectra in the positive-ion mode. Negative ionisation is ideal for Irganox species, as these compounds then provide intense molecular ion peaks and informative fragmentation [452]. The pSFC-MS technique is linear over a wide concentration range ($0.05-25\,\mu g\,mL^{-1}$) and pg LODs with positive-ion APCI (single ion monitoring) were determined for Tinuvin 327 (68 pg) and Irganox 1010 (390 pg). In a real PE sample, unidentified products were observed (possibly degradation products of Irganox 1010). pSFC-APCI-MS (double-focusing sector) has been used for the analysis of PEGs and PSs [423]. It appears that pSFC-APCI-MS is a powerful method for identification of polymer additives, provided that a library of mass

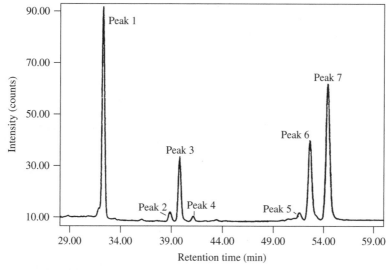

Figure 7.19 SFC chromatogram of Irgafos P-EPQ: Peaks (1) Irgafos 168; (2) 4,3'-mono-P-EPQ; (3) 4,4'-mono-P-EPQ; (4) oxidised 4,4'-mono-P-EPQ; (5) 3,3'-P-EPQ; (6) 4,3'-P-EPQ; (7) 4,4'-P-EPQ. After Bücherl *et al.* [124]. Reprinted from T. Bücherl *et al.*, *Packaging Technological Science*, **7**, 139–154 (1994). Copyright 1994 © John Wiley & Sons, Ltd. Reproduced with permission

spectra of such additives using this technique is available. *On-line SFE-cSFC-MS* studies on polymer additives have been reported [124,431]. SFE-SFC-EIMS is an interesting niche approach.

SFC-MS appears to be a rather specialised technique which as yet finds limited routine application.

7.3.2.3 SFC-NMR Spectroscopy

Principles and Characteristics SFC-NMR is a newly developed technique [453–455]. Although RPLC-NMR is now established as a routine method in analytical laboratories, some problems arise from HPLC solvents which are not favourable for NMR spectroscopy. Instead, scCO$_2$ is a solvent which provides nearly ideal properties for ^{1}H NMR spectroscopy: solvent signal suppression is not necessary, and unrestricted observation of the whole spectral range is possible: similar to ^{1}H NMR spectroscopy of deuterated solvents. Therefore, 2D experiments in stopped-flow mode are more easily implemented than with eluents that produce strong background signals. Even SFC with a modifier is feasible, because the added amounts are normally in a concentration range which allows the use of deuterated solvents; otherwise, only for some regions of the spectrum can no information be obtained [456]. Consequently, NMR is no longer restricted by the conditions dictated by the chromatographic separation (ideal hyphenation).

SFC-NMR is available from 200 to 800 MHz, and is suitable for all common NMR-detected nuclei. SFC/SFE-NMR requires dedicated probe-heads for high pressure (up to 350 bar) and elevated temperature (up to 100 °C). SFC-NMR is carried out with conventional packed columns, using modifier, pressure and temperature gradients. The resolution of ^{1}H NMR spectra obtained in SFE-NMR and SFC-NMR coupling under continuous-flow conditions approaches that of conventionally recorded NMR spectra. However, due to the supercritical measuring conditions, the ^{1}H spin-lattice relaxation times T_1 are doubled.

The *main characteristics* of on-line SFC-NMR are given in Table 7.43. SFC-NMR allows rapid and selective separation of larger sample volumes than LC-NMR. SFC-NMR provides nearly the same possibilities as LC-NMR, but at the expense of greater technical effort. The quality of HPLC-NMR and SFC-NMR spectra is equivalent. SFC-NMR is affected by drifting signals arising from the gradient conditions (pressure, density). A disadvantage of SFC-NMR spectroscopy includes the pressure dependence of the chemical shifts; the increase in T_1 of the system (compared to that in the liquid state) increases the experiment time. Albert [152]

Table 7.43 Main characteristics of SFC-NMR spectroscopy

Advantages

- NMR-transparent SCFs
- Spectroscopy not dictated by chromatographic conditions
- Broad applicational area

Disadvantages

- Need for pressure-proof NMR detection cell
- Pressure-dependent chemical shifts
- Expensive equipment

has described direct on-line coupling between SFE, SFC and high-field ^{1}H NMR.

Applications Albert *et al.* [455] have shown continuous-flow SFC-NMR spectra of five *plasticisers* (DEP, DNPP, DPP, BBP, DNBP). On-flow and stopped-flow pSFC-NMR of synthetic mixtures of phthalates were reported [457]. The feasibility of SFC-NMR coupling has been demonstrated with real-life applications [458]. Figure 7.20 shows a reconstruction of an extraction profile from a PVC tube [152]. The profiles of the integral aromatic proton signals between 7.2 and 8.2 ppm and the ester protons at 4.42 ppm display the relative concentration of the extracted phthalate as a function of the proceeding extraction. The structure of the extracted phthalate could be assigned to DEHP (Figure 7.21).

Figure 7.22 shows the ^{1}H NMR chromatogram (contour plot) of the separation of a 10 % phthalate mixture in CH$_2$Cl$_2$. The spectrum is almost free from interferences; the NMR resolution is excellent, and it is possible to identify all plasticisers even at concentrations as low as 2 %, which corresponds to 60 µg per component. In contrast, in on-line HPLC-^{1}H NMR separation the regions between 3.9–3.3 and 1.9–1.7 ppm are completely obscured by solvent signals.

On-line SFC-^{1}H NMR has also been applied for the investigation of *monomeric acrylates* [456]; SFC–NMR and HPLC-NMR were compared for both these acrylates [456] and phthalates [459]. Direct SFC–NMR coupling offers the advantage that the recorded continuous-flow ^{1}H NMR spectrum is not obscured by solvent signals.

SFC-NMR is still a fairly specialised system available in few laboratories. However, wider application of the technique is expected [152].

7.3.2.4 SFC–Plasma Source Detection

Principles and Characteristics SFC has been coupled to MIP-AES, MIP-MS, ICP-AES and ICP-MS

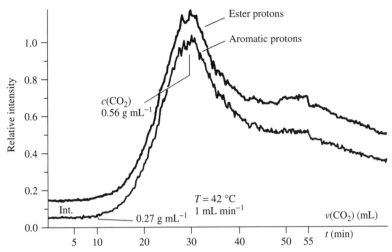

Figure 7.20 Extraction profile of a phthalate from PVC. After Albert [152]. Reprinted from *Journal of Chromatography*, **A785**, K. Albert, 65–83, Copyright (1997), with permission from Elsevier

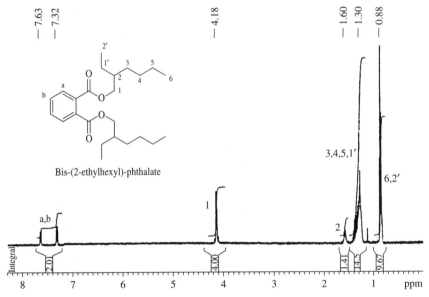

Figure 7.21 Stopped-flow ^{1}H NMR spectrum (400 MHz) of bis(2-ethylhexyl)phthalate from PVC with supercritical CO_2. After Albert [152]. Reprinted from *Journal of Chromatography*, **A785**, K. Albert, 65–83, Copyright (1997), with permission from Elsevier

detectors, thereby producing high selectivity. Regardless of the *plasma* used, the source must fulfil certain basic criteria in order to function as an acceptable elemental detector for SFC. The plasma must be able to promote the transitions of interest (metals as well as nonmetals) and be unaffected (in terms of operating characteristics) by the introduction of CO_2 and modifying agents into the plasma. These operating characteristics are excitation temperature, electron number density and stability of the plasma. The criterion of being unaffected by the introduction of CO_2 is the most difficult to fulfil. Packed-column flow-rates are in the order of $60–120$ mL min^{-1} of CO_2, while capillary column flows are a factor of 10 less. Introduction of carbon-containing gases into Ar plasmas (ICP and MIP) affects the excitation properties of the plasma adversely [460,461]. In SFC-ICP

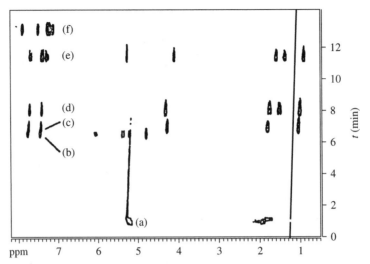

Figure 7.22 ¹H NMR chromatogram (contour plot, 400 MHz) of an SFC separation of five plasticisers: (a) CH₂Cl₂, (b) diallyl, (c) di-*n*-propyl, (d) di-*n*-butyl, (e) benzyl-*n*-butyl, and (f) diphenylphthalate. After Albert [152]. Reprinted from *Journal of Chromatography*, **A785**, K. Albert, 65–83, Copyright (1997), with permission from Elsevier

detection, a degradation in plasma energy is noted when CO₂ is introduced into the plasma at capillary column flow-rates. Additionally, the limited amount of energy of the Ar discharge is inadequate to excite nonmetal species. To overcome the shortcomings of the Ar plasma as an elemental detector for SFC, helium-based plasmas have been employed [462–466]. In terms of the effects of mobile-phase-induced perturbations on the characteristics of helium plasmas, SFC presents a situation intermediate to those of GC-MIP and LC-MIP.

As SFC provides gaseous sample introduction to the plasma and thus near-100 % analyte transport efficiency, coupling SFC with plasma mass spectrometry offers the potential of a highly sensitive, element-selective chromatographic detector for many elements. Helium high-efficiency *microwave-induced plasma* has been proposed as an element-selective detector for both pSFC and cSFC [467,468]; easy hyphenation of pSFC to AED has been reported [213].

While most preliminary SFC-plasma coupled techniques employed microwave-induced plasmas (MIPs), the use of ICP-MS is now increasing [469]. An advantage of microcolumn *SFC-ICP hyphenation* is the significantly reduced flow-rates of microcolumns compared with those of conventional columns. Both pSFC-ICP-AES [470,471] and cSFC-ICP-AES [472] were described. In the case of elemental detector selectivity (e.g. AES) complete chromatographic resolution is not required. The detector possesses linearity over several orders of concentrative magnitude. Minimum detectable quantities for nonmetals range from sub to low ng mL⁻¹.

SFC-ICP-MS requires rather expensive and complicated instrumental design [473,474]. Interfacing the SFC restrictor with the ICP torch follows different approaches for pSFC and cSFC [469]. Polar modifiers, however, do not have a serious deleterious effect on the ICP plasma, which enables the polarity of the mobile phase to be changed with no significant loss of sensitivity or resolution. This enables analysis of compounds which are too polar for adequate separation with pure CO₂ as the mobile phase. SFC is still in its infancy as far as speciation analysis of metal-containing additives is concerned.

Applications   SFC can separate organometallic compounds which are nonvolatile and thermally labile. Since many organometallic compounds are thermally unstable, there remains an advantage of the SFC approach over GC-based techniques. Helium MIP-MS has been used as an element-selective detector for SFC of halogenated compounds [475]. SFC-ICP-MS shows high potential for the determination of ultratrace levels of organometallic and nonmetal compounds such as tetraalkyltin compounds [474]. cSFC-ICP-MS [476,477] couplings sensitively and selectively detect heavier elements (organotin and -lead compounds). Various organotin compounds (e.g. tetraphenyltin, TPT) were analysed by SFC-ICP-MS with detection limits in the low to sub-pg range [478]. The impact of these techniques on polymer/additive analysis is as yet very small.

7.3.3 (Multi)hyphenated HPLC Techniques

In principle, liquid chromatography is better suited for polymer/additive analysis than gas chromatography as it can separate much higher-molecular-weight materials (up to MW 1500–2000 Da) and can successfully handle thermally unstable additives (such as accelerators). The most common detection mode is, of course, HPLC-UV [479]. UV detection offers excellent sensitivity, but depends on the presence of a chromophore in the sample and has little or no capability of identifying unknown compounds. The *LC-UV coupling* is hardly considered as a tandem instrument. LC-UV tandem systems are basically multiwavelength detectors. The multiwavelength dispersion detector can only monitor a separation at one wavelength, but has the higher resolution. The diode array detector monitors UV absorption simultaneously at all wavelengths, but the number of diodes limits its resolution. The tandem LC-UV system is used to monitor a separation at the optimum wavelength; test the purity of a peak from its spectrum; or compare the spectrum of a peak to that of a reference. An advanced detection technique for HPLC is the photodiode array with light-pipe flow cell (with optical path of 5 cm), allowing a new level of UV sensitivity and detection of components which cannot be detected by conventional detectors. HPLC-UV-ELSD coupling (separation–identification–quantification) is a valid deformulation scheme for additive-containing polyolefins. Tandem systems employing *fluorescence spectroscopy* help in identifying eluted compounds; selectively monitor a specific type of sample; or enhance the sensitivity of the system to a particular compound. In modern instruments the change in excitation and emission wavelengths can be programmed to suit the elution pattern of the eluents. LC-UV and LC-FL tandem systems were reviewed [204].

Mobile-phase opacity is primarily responsible for the lack of general-purpose *HPLC-FTIR interfaces* possessing sensitivity comparable to GC-FTIR. Various HPLC-FTIR interfaces have been developed for all HPLC categories except ion exchange. Eluent components of HPLC are commonly collected with an evaporative interface and are analysed by IR microscopy with low ng range sensitivity; Raman microscopy is equally applicable. *On-line HPLC-NMR* offers considerable advantages over the off-line mode, but is still a rather specialised technique which is not widely accessible. Few polymer/additive problems have been resolved this way; NMR sensitivity is gradually improving.

On-line LC-MS undoubtedly is a more important and versatile identification technique than LC-FTIR. However, there is no single 'universal' LC-MS interface available: every interface has its specific limitations with regard to flow-rate and composition of the LC eluent, polarity and molecular mass of the analytes, and/or ionisation technique(s) that can be used. For the non-mass spectroscopist, LC-MS developments have been a rather confusing matter. The developments of 30 years of LC-MS can be summarised as follows:

(i) Reduction of interface space (MBI, DLI, TSP, CF-FAB, MAGIC, PB, TMD, ESP/ISP, HNI, APCI, SSI, CF MALDI) to API techniques (electrospray, turbo-ionspray, nanospray, APCI, APPI, HNI).

(ii) Replacement of mass spectrometers from scanning type (sector, QMS) to pulsed-source mass spectrometers (IT, ToF, Q-ToF, FTMS).

(iii) Flow-rate compatibilisation (covering the 30 nL-2 mL min^{-1} range).

(iv) From 'Spray and Pray' to robust devices.

LC-MS is now a nature technology and operation of an LC-MS system is no longer the realm of an MS specialist. The proper choice of the LC-MS mode to be used in a specific situation depends on analyte class, sample type and problem (detection, confirmation, identification). On-line LC-MS is used more for specialised applications than for general polymer or rubber compound analysis. This derives from the fact that LC-MS method development (column, solvent system, solvent programme, ionisation mode) is rather time consuming. LC-MS (in particular with API interface) enables analysis of a wide range of polar and nonvolatile compounds which cannot be analysed by GC (*cf.* Scheme 7.7).

The liquid chromatograph can be employed to separate different species, and *atomic spectrometric techniques* can identify and confirm the presence of specific elements eluted at a particular retention time. The eluent from a liquid chromatography column is ideally suited for direct injection into an atomic spectrometer (flame, graphite furnace or ICP/MIP) by means of a simple nebuliser. LC has been coupled (in)directly with different forms of atomic spectrometer (FAAS, GFAAS, LEAFS). Flame atomic spectrometry detection for HPLC does not offer good sensitivity for the most part. However, GFAAS systems have been interfaced to HPLC because of their good detection limits, and have been used for speciation [480]. Applications of LC-LEAFS and LC-GFAAS in polymer/additive analysis are mainly restricted to the determination of some organotin compounds with sensitivities down to 60 pg [204].

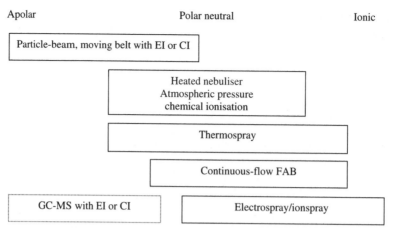

Scheme 7.7 Comparative analyte ranges for the major LC-MS interfaces

7.3.3.1 Liquid Chromatography–Fourier-transform Infrared Couplings

Principles and Characteristics Deformulation of resin compositions poses a difficult analytical challenge, due to the mixture of compounds that are present within the solid polymer matrix. As interpretation of IR spectra of mixtures is often very difficult, on-line HPLC-FTIR coupling is of great importance. For the practical association of the liquid chromatograph with the IR spectrometer, the problems are twofold. Firstly, either the solvent needs to be removed before measurement, or an infrared transparent solvent must be employed. Unfortunately, of the few IR-transparent solvents, many are not compatible with the mobile-phase requirements of the liquid chromatograph. Secondly, the IR spectrometer is generally less sensitive than UV or MS and, consequently, the LC-IR system is put at a disadvantage. Thus larger samples may be needed, and this may be unacceptable to the chromatographic system. Efficient coupling of LC and FTIR therefore requires a sophisticated interface. Various LC-FTIR interfaces have been proposed [163,481–491]. Instrumental variables in LC-FTIR hyphenation are many: column-type, mobile-phase and flow-rate (LC), flow-cell, solvent-elimination and spray-type (interface), and substrate and detection mode (FTIR). A general-purpose LC-FTIR interface is not available; the various operational approaches serve different purposes. Mobile-phase interferences pose problems for both the (on-line) flow-cell and (semi on-line) mobile-phase elimination approaches.

The *flow-cell interface* in HPLC-FTIR, first reported in 1975 [492], is the most straightforward. Flow-cells consist of two IR-transparent windows (KBr for non-aqueous, and AgCl or ZnSe for aqueous solvents) separated by a spacer. The thickness of the spacer determines the pathlength and volume of the flow-cell. The pathlength of the cell is generally less than 0.2 mm for organic solvents, and less than 0.03 mm for water–organic solvent mixtures. Sensitivity is thus quite low, and only major components are detected specifically. Other disadvantages are that little information about the solute can be obtained in spectral regions where the mobile phase shows appreciable absorption. Common HPLC mobile phases exhibit characteristic infrared spectra; therefore, mobile-phase absorptivity varies with wavelength. The optimum HPLC-FTIR pathlength at a given wavelength is inversely proportional to the absorptivity of the mobile phase at that wavelength. Mixture–component infrared spectra are obtained from flow-cell measurements by subtracting mobile-phase contributions from acquired spectra.

RPLC is barely compatible with the flow-cell approach. Solvent programming capabilities (gradient elution) in HPLC are not easily implemented with flow-cell interfaces, because reference spectra for a wide range of solvent mixtures must be measured in order to effectively subtract mobile-phase contributions from eluent spectra. Solvent programming represents a challenge for mobile-phase elimination interfaces as well. Solvent removal conditions at the start of the separation may not be appropriate for solvent mixtures eluting later in the solvent programme. In order to minimise problems associated with eluent absorption, the choice of solvents in flow-cell LC-FTIR is generally limited to chlorinated alkanes or deuterated solvents. Due to the small optical pathlength, the absolute detection limits in on-line LC-FTIR are in the (high) μg range, which frequently implies that analyte concentrations of $1-10\,g\,L^{-1}$ have to be injected to obtain identifiable spectra. Despite its poor detection limits, flow-cell HPLC-FTIR can be useful for the specific and quantitative detection of major constituents of mixtures.

Fixed pathlength transmission flow-cells for aqueous solution analysis are easily clogged. *Attenuated total reflectance* (ATR) provides an alternative method for aqueous solution analysis that avoids this problem. Sabo *et al.* [493] have reported the first application of an ATR flow-cell for both NPLC and RPLC-FTIR. In micro-ATR-IR spectroscopy coupled to HPLC, the trapped effluent of the HPLC separation is added dropwise to the ATR crystal, where the chromatographic solvent is evaporated and the sample is enriched relative to the solution [494]. Detection limits are not optimal. The ATR flow-cell is clearly inferior to other interfaces.

Microbore HPLC-FTIR detection limits are about 10 times lower than analytical-scale HPLC-FTIR detection limits. The lowest reported LC-FTIR detection limits are approximately $100-1000$ times higher than the best GC-FTIR detection limits. The main characteristics of flow-cell HPLC-FTIR are summarised in Table 7.44. Because of mobile-phase interferences, flow-cell HPLC-FTIR is considered as a powerful tool only for the specific detection of major components but is otherwise a method of limited potential, and SFE-SFC-FTIR has been proposed as an alternative [391].

The obvious alternative for the in-line flow-through cell in HPLC-FTIR is *mobile-phase elimination* ('transport' interfacing), first reported in 1977 [495], and now the usual way of carrying out LC-FTIR, in particular for the identification of (minor) constituents of complex mixtures. Various spray-type LC-FTIR interfaces have been developed, namely, thermospray (TSP) [496], particle-beam (PB) [497,498], electrospray (ESP) [499] and pneumatic nebulisers [486], as compared by Somsen *et al.* [500]. The main advantage of the *TSP-based*

Table 7.44 Main characteristics of flow-cell HPLC-FTIR

Advantages

- Simple and low cost
- Specific and quantitative detection of major components
- Routine application

Disadvantages

- Small optical pathlength
- Narrow spectral window (solvent interferences)
- Limited choice of mobile phase
- Incompatible with RPLC
- Low sensitivity (high µg range)
- Restricted applicability (special-purpose method)

systems is that relatively high flow-rates (0.5–1 mL/min) of both organic and aqueous eluents can be handled, and conventional-size LC can be used. The high temperature of the TSP may induce analyte losses by evaporation or thermal degradation. Moderate FTIR sensitivity is achieved. Jansen [483] has reported a direct deposition LC-FTIR device consisting of thermospray (50–180 °C), moving belt and diffuse reflectance optics accessories, which was applied to the analysis of additives in polymers with detection limits in the 100-ng range. Wood [498] studied *LC-PB-FTIR* using dioctyldiphenylamine as a model compound. Although PB-FTIR analysis of compounds at the ng level has been indicated, the reported sample quantities are usually in the (high) µg range. The modest analyte detectability is related to the low efficiency (5–10 %) of analyte transfer in the PB interface. The feasibility of *electrospray* (ESP) nebulisation as a means of coupling micro-LC and FTIR was also indicated [499], but further studies on µLC-ESP-FTIR have been limited.

Gagel and Biemann [486] reported a *nebuliser-based* LC-FTIR method which involves continuous deposition of the effluent from a narrow-bore NPLC column on a rotating IR-reflective disc. The solvent-elimination FTIR interface has lately been perfected [484] by having the chromatographic effluent pass into a nozzle, which sprays the mobile phase as a tightly focused jet on to the surface of a continuously rotated sample-collection disc. A commercial equivalent of this interface (LC Transform™) is compatible with the major LC modes (RP, NP, SEC) and other separatory methods (SFC, FFF, TREF). Reported benefits of this fraction collector are: low-to-mid ng range sensitivity; control of mobile-phase flow; automatic sample collection; ready adaptation to RP and NP chromatography modes; and compactness [501,502]. Thus, the LC chromatogram is actually stored, enabling efficient FTIR measurements, since both the spectral resolution and the signal-to-noise ratio strongly depend on the available analysis time. A limitation of this HPLC-FTIR method is that only non-volatile solutes can be analysed. Volatile components tend to simply evaporate off the collection-disc surface before the system can yield useful spectra. Fortunately, these same samples are good candidates for GC-FTIR methods. Later, Somsen *et al.* [503] modified a spray-jet interface, originally developed for on-line LC-TLC for coupling of narrow-bore RPLC and FTIR [163]; the effluent was deposited on to a ZnSe substrate for analysis by FTIR transmission microscopy, and an identification limit of ca. 10 ng was quoted. Transmission measurements using a ZnSe substrate appear to be preferable to measurement in R-A mode on an aluminium mirror.

The solvent-elimination interface offers various advantages over the flow-cell approach:

(i) recording of eluent interference-free full mid-IR spectra of the deposited compounds independently from the LC conditions;
(ii) free selection and optimisation of mobile-phase composition and elution method, virtually independently of the spectrometer requirements;
(iii) full exploitation of the sensitivity of the FTIR spectrometer;
(iv) coverage of a relatively large part of the chromatographic peak within the IR beam; and
(v) off-line recording and 'postrun' signal averaging.

The main *advantages* of mobile-phase elimination techniques (Table 7.45) derive from high enrichment factors attained. However, 100 % yield can be obtained only for nonvolatile eluents. The solvent-elimination approach features increased sensitivity and enhanced spectral quality: two important conditions for reliable identification of (low-level) sample constituents. For mobile-phase elimination LC-FTIR high-purity solvents are required. Solvent-elimination LC-FTIR is more complicated than on-line FTIR detection. The interface should adequately evaporate the eluent, maintaining chromatographic resolution during the deposition process. In this respect, the LC flow-rate, the composition of the eluent and the nature of analytes and substrate are important factors. Since enrichment is achieved by selective evaporation, the solute must be significantly less volatile than the mobile phase. The method is not suitable for use with an aqueous mobile phase, due to the high surface tension and latent heat of evaporation for water.

Table 7.45 Main characteristics of solvent-elimination HPLC-FTIR

Advantages

- High enrichment
- Better sensitivity (low-to-mid ng range) than flow-cell HPLC-FTIR
- Relatively high flow-rates
- No solvent interferences
- Spectral quality for unambiguous identification of low-level components
- Compatibility with many separatory modes
- Spectral storage
- Automation

Disadvantages

- Analysis of nonvolatile solutes only
- Need for high-purity solvents
- Less suitable for RPLC

In solvent-elimination LC-FTIR, basically three types of substrates and corresponding IR modes can be discerned, namely, powder substrates for diffuse reflectance (DRIFT) detection, metallic mirrors for reflection–absorption (R-A) spectrometry, and IR-transparent windows for transmission measurements [500]. The most favourable solvent-elimination LC-FTIR results have been obtained with IR-transparent deposition substrates that allow straightforward transmission measurements. Analyte morphology and/or transformation should always be taken into consideration during the interpretation of spectra obtained by solvent-elimination LC-FTIR. Dependent on the type of substrate and/or size of the deposited spots, often special optics such as a (diffuse) reflectance unit, a beam condenser or an FTIR microscope are used to scan the deposited substances (typical diameter of the FTIR beam, $20 \mu m$).

As the vast majority of LC separations are carried out by means of gradient-elution RPLC, solvent-elimination RPLC-FTIR interfaces suitable for the elimination of aqueous eluent contents are of considerable use. RPLC-FTIR systems based on TSP, PB and ultrasonic nebulisation can handle relatively high flows of aqueous eluents $(0.3-1 \, mL \, min^{-1})$ and allow the use of conventional-size LC. However, due to diffuse spray characteristics and poor efficiency of analyte transfer to the substrate, their applicability is limited, with moderate (100 ng) to unfavourable $(1-10 \mu g)$ identification limits (mass injected). Better results (0.5–5 ng injected) are obtained with pneumatic and electrospray nebulisers, especially in combination with ZnSe substrates. Pneumatic LC-FTIR interfaces combine rapid solvent elimination with a relatively narrow spray. This allows deposition of analytes in narrow spots, so that FTIR transmission microscopy achieves mass sensitivities in the low- or even sub-ng range. The flow-rates that can be handled directly by these systems are $2-50 \, \mu L \, min^{-1}$, which means that micro- or narrow-bore LC (i.d. 0.2–1 mm) has to be applied.

For microbore HPLC, with a flow of less than $100 \, \mu L \, min^{-1}$, off-line LC-FTIR has been developed using *matrix isolation techniques*. The solutes are deposited on a moving IR salt window [504] or on a rotating plated disc [486], and are measured afterwards with the aid of a FTIR microscope or a reflectance accessory. FTIR detection was first applied to the analysis of microbore HPLC eluent by Teramae and Tanaka [505]. In microbore HPLC-FTIR the amount of mobile phase required for separation is much less than for conventional scale HPLC. This simplifies both flow-cell and mobile-phase elimination interfaces. Flow-cell

Table 7.46 LC-FTIR interface detection limits

Interface	NPLC	RPLC	SFC	SEC
Transmission flow-cell	50 ng[a]	20 μg	50 ng	50 ng[a]
ATR flow-cell	1 mg	1 mg	–	–
DRIFTS	10 ng[a]	1 μg	50 ng	1 μg[a]
Mobile phase elimination	100 ng[a]	100 ng[a]	–	1 μg[a]

[a] Microbore HPLC-FTIR.
After White [167]. Reprinted from R. White (ed.), *Chromatography/Fourier Transform Infrared Spectroscopy and Its Applications*, Marcel Dekker Inc., New York, NY (1990), by courtesy of Marcel Dekker Inc.

microbore HPLC-FTIR may employ deuterated solvents such as $CDCl_3$, D_2O or CD_3CN. Detection limits for NP and RP microbore HPLC-FTIR solvent-elimination interfaces are dictated by the concentration of impurities in the mobile phase; the useful dynamic range is limited by the sample capacity of microbore columns. Microbore NPLC-FTIR mobile-phase elimination interfaces work well when solvent is easily evaporated. This is not the case for RP solvents that typically contain water.

Table 7.46 shows the LC-FTIR interface detection limits. Detection limits approaching those for GC-FTIR light-pipe interfaces have been reported for flow-cell HPLC-FTIR when IR-transparent mobile phases are employed. For both the moving-belt and thermospray LC-MS couplings the detection limits are in the ng range. Selective evaporation consisting of fraction collection followed by *DRIFT* identification achieves a detection limit of 100 ng.

On-line HPLC-FTIR now provides a fast and discriminative technique for the identification and confirmation of organic molecules. The IR detector can be used as a qualitative tool, whereas the UV-VIS detector provides quantitative data. Combining these techniques to mass spectrometry, which yields information on the molecular weight of compounds and fragments, creates a highly effective analytical data set. Unlike the solutes used in conventional flow detectors, those analysed by the static HPLC-FTIR disc remain available and accessible for further analysis, e.g. by means of MS. Using a 1- or 5-μL syringe, analysts can add a small drop of appropriate solvent on to a deposit and withdraw a sample into the syringe for further analysis, which includes MS studies of the collected solutes.

LC-IR using *surface-enhanced IR absorption spectroscopy* (SEIRAS) was recently designed in order to develop a highly sensitive and rapid analysis method for polymer additives [506]. The method, which consists of spraying the LC eluents on to a metal film of Ag on a BaF_2 substrate, allows an enhancement factor of about 90.

LC-FTIR has been reviewed [204,507]. Various mobile-phase elimination designs were discussed by White [167]. Resolution of complex LC-FTIR spectroscopy data was described [508]. A general overview of flow-cell based IR detection and of early solvent-elimination interfaces for LC-FTIR has recently appeared; solvent-elimination RPLC-FTIR interfaces have also been described [500].

Applications Jansen [483] and others [509] showed the feasibility of solvent-elimination LC-FTIR in the analysis of polymer additives using the TSP-based coupling method. Somsen *et al.* [503] identified additives in a test mixture (Irganox 1076, Ionol 220, Tinuvin 327/328, Cyasorb UV531), in PVC/(Irganox 1076, *N, N*-bis(hydroxyethyl)alkylamine, oleamide) and in PP/GMS by solvent-elimination based coupling of RPLC and FTIR with deposition on a moving ZnSe window (Figure 7.23).

Dwyer [484] identified *antioxidants and UV absorbers* such as Irgafos 168, Irganox 1010/1076/3114, and Cyasorb UV531 in a (hexane:CH_2Cl_2 1:1) extract of PP using a commercially available NPLC-solvent elimination FTIR interface. The LC Transform™ is extremely useful in quickly providing the analyst with detailed knowledge of polymer formulations. Solvent elimination HPLC-UV-FTIR (LC Transform™), as tested for some 20 polymer additives in the 220–1178 Da molecular weight range, typically achieves low- to mid-ng detection limits [194]. Only BHT and Ultranox 626 were not successfully analysed. Optimisation of the procedure was described; for non-UV absorbing additives, ELSD was used. Figure 7.24 compares UV (at 280 nm) and ELSD detection with the Gram–Schmidt reconstruction (GSR) for the RPLC separation of nine polymer additives. Direct comparison between ELSD and GSR clearly shows a loss in resolution when using the LC Transform™ interface.

The minimum quantity of additives to be deposited in order to accurately and spectrometrically identify a compound is 0.5–1.0 μg (depending on the specific absorbance region). The data indicate that the detection limits for polymer additives, which are virtually ideal analytes because of their high molecular weight and powdered form, fall in the low- to mid-ng range. The limit of identification of the LC-FTIR method (about 10 ng injected) allows determination of additives at levels commonly met in polymers. This was confirmed by Cosemans *et al.* [510], who tested the interface with two solutions of a total of 21 polymer additives (with molecular weights from 220 to 1178 Da), most of which were positively analysed by gradient (CH_3CN/H_2O) RPLC-FTIR (LC Transform™). The test solutions contained

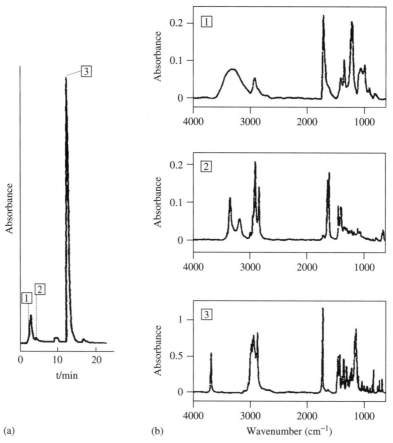

Figure 7.23 (a) LC-UV chromatogram (275 nm) of a PVC sample extract; (b) FTIR spectra of peaks. Legend: (1) monoesterified *N*,*N*-bis(hydroxyethyl)alkylamine, (2) oleamide and (3) Irganox 1076. After Somsen *et al.* [503]. From G.W. Somsen *et al.*, *Analyst*, **121**, 1069–1074 (1996). Reproduced by permission of The Royal Society of Chemistry

the following additives (MW in parenthesis) in concentrations from 5 to 15 mg 100 mL^{-1}: (1) Tinuvin P (225), Irganox 245 (586), DBPC (220), Plastanox 2246 (341), Cyanamide 162 (369), Cyasorb UV531 (326), Tinuvin 326 (316), Goodrite 3114 (784), Ionox 330 (774), Irgafos 168 phosphate (662), Irganox 1063 (514), Irganox 1076 (530), Irgafos 168 (646); (2) Santonox (359), Santowhite Powder (382), Topanol CA (544), Tinuvin 120 (438), Tinuvin 327 (358), Irganox 1010 (1178), Irganox 1063 (514), and Mixxim BB/100 (658). The deposit diameter was found to depend upon the nature of the additive, its concentration and the eluent composition. DBPC (220 Da) and Tinuvin P (225 Da), as well as DTBP (206 Da), were too volatile for optimal deposition. Irganox 1076 leads to a discontinuously spread spot which precludes a good FTIR spectrum. For method development purposes, commercial PP, PA6, PC and ABS extracts with known additive packages were

analysed by LC-FTIR. High-concentration components, which are not detected by PDA (220 nm), can contaminate low-concentration additive deposits. Additive deposits of polyether ester and polycarbonate samples are easily contaminated with oligomer deposits; on the other hand, *polyolefin oligomers* do not contaminate the additive deposits.

RPLC-FTIR of an Irganox 1010, Irganox PS800 containing polyolefin extract was reported with a thermospray/moving belt/DRIFT interface [483]; detection limits of 100 ng were reported for this experimental device. LC-TSP-FTIR has also been used for the identification of other antioxidants, as shown in Figure 7.25 [500].

Temperature-programmed packed capillary liquid chromatography (TP-CLC), coupled off-line to solvent elimination FTIR (LC Transform™) has recently been used for gradient separations of Irganox 1010/1076/3114 dissolved in DMF with LOD of about 40 ng [511]. Low

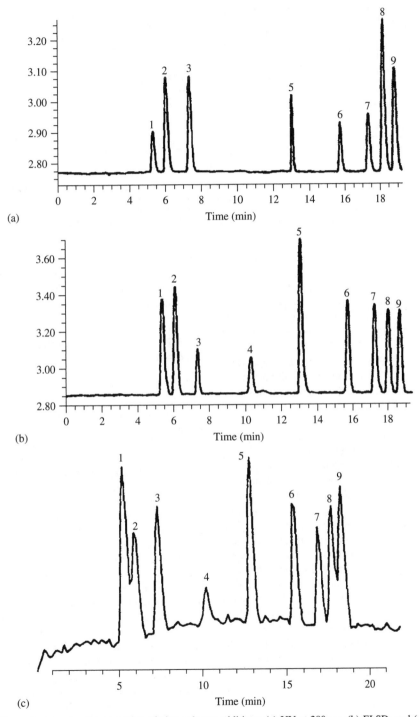

Figure 7.24 Chromatograms for the separation of nine polymer additives: (a) UV at 280 nm; (b) ELSD; and (c) GSR. Legend: 1, Irganox 1098; 2, Santowhite; 3, Lowinox; 4, Kemamide; 5, Irganox 259; 6, Irganox 1010; 7, Ethanox 330; 8, Tinuvin 328; 9, Tinuvin 327. After Jordan and Taylor [194]. Reproduced from the *Journal of Chromatographic Science*, by permission of Preston Publications, a Division of Preston Industries, Inc.

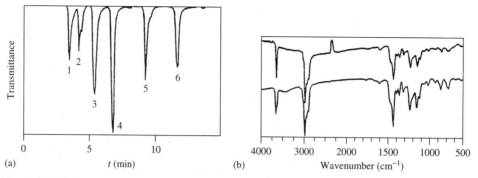

Figure 7.25 (a) LC-TSP-FTIR functional group $(3100–2800\,cm^{-1})$ chromatogram of phenolic antioxidants. Legend: (1) Irganox 3114 and 1035; (2) Irganox 1010; (3) Irganox 1330; (4) Irganox 565; (5) Irganox 1076; (6) Irgafos 168. (b) LC-TSP-FTIR spectrum (top) and standard FTIR spectrum (bottom) of Irganox 1330. After Somsen *et al.* [500]. Reprinted from *Journal of Chromatography*, **A856**, G.W. Somsen *et al.*, 213–242. Copyright (1999), with permission from Elsevier

flow-rates of $1–10\,\mu L\,min^{-1}$ could be handled by adaptation of the pneumatic nebuliser. The LC-UV-FTIR configuration is particularly well suited for the separation of the components (including several positional isomers) of Irgafos P-EPQ, see Figures 7.26 and 7.27. The compound 3,3'-P-EPQ was not found, most probably due to its low abundance. The *TP-CLC-FTIR* combination is quite interesting in being related to temperature-programmed GC-FTIR, with the basic difference that nonvolatile substances, including oligomers, are separated. With true capillaries (i.d. <1 mm), only isocratic mobile phases can be used.

Successful combination of a chromatographic procedure for separating and isolating additive components with an on-line method for obtaining the IR spectrum enables detailed compositional and structural information to be obtained in a relatively short time frame, as shown in the case of additives in PP [501], and of a *plasticiser* (DEHP) and an aromatic phenyl phosphate *flame retardant* in a PVC fabric [502]. RPLC-TSP-FTIR with diffuse reflectance detection has been used for *dye analysis* [512]. The HPLC-separated components were deposited as a series of concentrated spots on a moving tape. HPLC-TSP-FTIR has analysed polystyrene samples [513,514]. The LC Transform™ has also been employed for the identification of a stain in carpet yarn [515] and a contaminant in a multiwire cable [516]. HPLC-FTIR can be used to maintain consistency of raw materials or to characterise a performance difference.

Sudo *et al.* [506] have reported the detection of 10 ng of triphenylphosphate by means of LC-SEIRAS; the method is used for the analysis of additives in commercial polypropylene.

Compared with other LC detection modes (UV, MS), the use of IR detection in LC is still rather limited.

Worldwide, some 500 LC-FTIR users are registered. A review on the determination of rubber and plastic chemicals by liquid chromatography–spectroscopy has appeared [494].

7.3.3.2 Liquid Chromatography–Mass Spectrometry

Principles and Characteristics As pointed out elsewhere, LC is incurably blind, whereas the lameness of MS may only be cured with great effort [517]. Their handicaps are largely overcome if they are put to use together. A mass spectrometer is an ideal detector for HPLC analysis, since it offers universality, selectivity and sensitivity (only an order of magnitude less sensitive than LIF detection). Off-line LC-MS coupling, first described in 1973 [518], exploits many advantages of coupling of a mass spectrometer to an HPLC. However, on-line LC-MS provides volatility enhancement not possible with conventional MS inlets, and allows nonvolatile and thermally sensitive compounds to be handled. LC-MS is one of the most important analytical techniques. LC-MS hyphenation combines an instrument that operates in the condensed phase with another that operates under vacuum. Potential advantages of on-line LC-MS coupling are summarised in Table 7.47. For a given LC-MS combination, multi-signal analysis allows us to determine which mode produces the best response for a particular analyte. Actually, MS is more than just a detection method. It is in fact, another separation technique, orthogonal to HPLC; it relies on a different physical property of the analyte to effect separation. As mass spectrometry is *not* a universal detector, it is often not sufficient just to rely on LC-MS. Interpretation of the spectral data in LC-MS, and even in

Abundance	Chemical name	Chemical structure
36–46 %	'4,4'-P-EPQ'	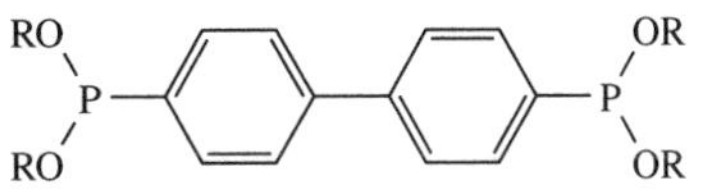
17–23 %	'4,3'-P-EPQ'	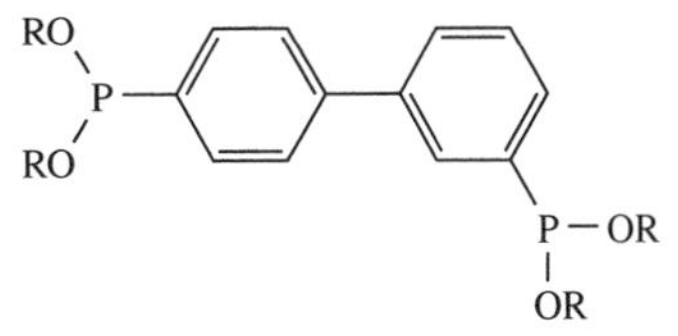
2–5 %	'3,3'-P-EPQ'	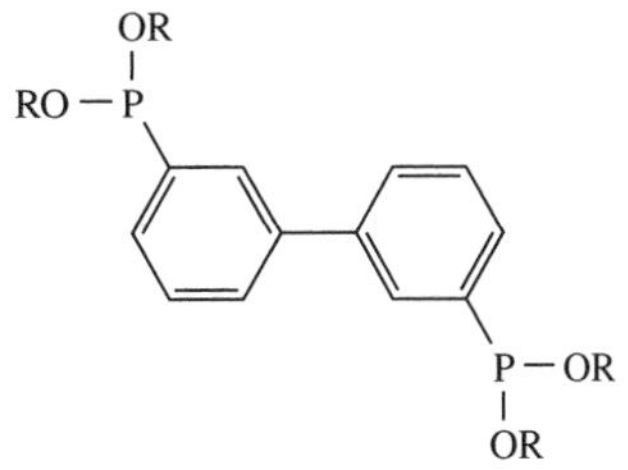
2–5 %	'Oxidised P-EPQ'	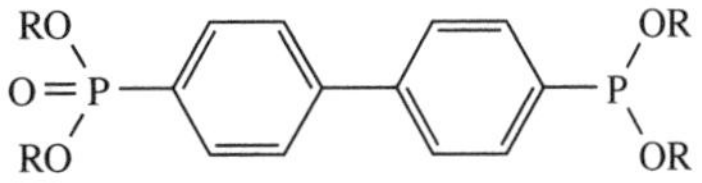
11–19 %	'Mono-P-EPQ'	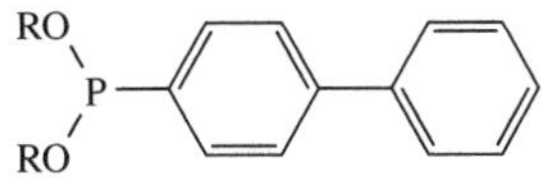
9–18 %	'Irgafos 168'	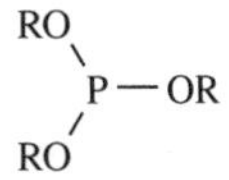
		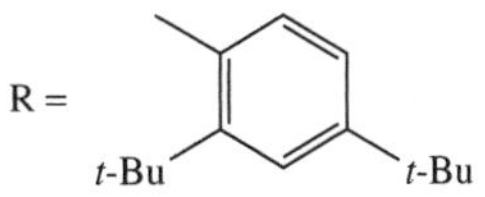

Figure 7.26 Structure, name and abundance of the main components of Irgafos P-EPQ, according to the manufacturer (Ciba-Geigy Ltd). After Bruheim *et al.* [511]. From I. Bruheim *et al.*, *Journal of High Resolution Chromatography*, **23**, 525–530 (2000). © Wiley-VCH, 2000. Reproduced by permission of Wiley-VCH

LC-MS[n], generally requires some background information on the nature of the solutes. Compared with GC-MS with EI/CI, LC-MS does not offer the same identification possibilities, because of the different ionisation mechanisms. Nevertheless, LC-MS has become an invaluable tool to selectively quantify solutes, and to confirm structures or to elucidate structural characteristics. A drawback of LC-MS is that measurable organic compounds are very limited compared with compounds separable by LC alone. LC-MS places considerable constraints on conditions that may be used for the LC side. Use of LC-MS is generally more rewarding for large series than for isolated problem-solving cases. Changes in LC methods required for modern LC-MS systems generally involve changes in sample preparation and solution chemistry to:

(i) ensure adequate analyte concentration;
(ii) maximise ionisation through careful selection of solvents and buffers; and

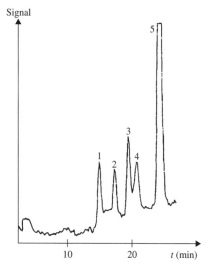

Signal

10   20 t (min)

Figure 7.27 Constructed Gram–Schmidt chromatogram of a temperature-programmed packed-capillary LC separation of 4.8 μg Irgafos P-EPQ dissolved in DMF; temperature programme: 50 °C for 8 min, 4 °C min^{-1} up to 140 °C. Legend: 1, mono-P-EPQ; 2, Irgafos 168; 3, 4,3′-P-EPQ; 4, oxidised 4,4′-P-EPQ; 5, 4,4′-P-EPQ. After Bruheim *et al.* [511]. From I. Bruheim *et al.*, *Journal of High Resolution Chromatography*, **23**, 525–530 (2000). © Wiley-VCH, 2000. Reproduced by permission of Wiley-VCH.

(iii) minimise the presence of compounds that compete for ionisation or suppress the signal through gas-phase reactions.

LC-MS is essentially a *two-stage process*, composed of liquid introduction and analyte ionisation involving transfer of the analyte from the condensed phase to the gas phase. This process involves aerosol formation. This is accomplished by disruption of the bulk liquid structure with an energy source sufficient to overcome the surface energy of the solvent, such as heat (vaporisation), pneumatic nebulisation (shearing off droplets from a liquid stream), or an electric potential (droplet creation in a field). Consequently, the common components of a hyphenated LC-MS system are: sample introduction–separation–analyte transport–ionisation–mass filtering–detection of ions–acquisition and control–mass spectrum. Most of these steps allow multiple choices. Consequently, a broad variety of chromatography–mass-spectrometric devices have been developed. To interface HPLC with MS the following features are essential:

(i) high enrichment of solute to solvent;
(ii) high collection and transference efficiency;
(iii) minimum band spreading; and
(iv) analysis of samples with low volatility.

Table 7.47 General characteristics of on-line LC-MS

Advantages

- Alternative method for introducing samples into MS source
- Universal detection for LC (total ion current) from <800 Da (EI) to 10^5 Da (ESI)
- Analysis of thermolabile compounds, not amenable to GC-MS
- Analysis of nonvolatile analytes
- No analyte derivatisation needed
- Low detection limits (<10^{-12} g)
- Variety of ionisation processes
- Multisignal analysis (+ and − ionisation; SIM or scanning; fragmenter voltage)
- Unambiguous analyte identification (LC-MS, LC-MS^n)
- Assessment of peak purity
- Quantitative with authentic standard (good linearity over two to three decades)
- Reduced need for sample cleanup strategies (high throughput)
- Wider applicability than GC-MS
- Rugged, user-friendly
- Reliable, versatile

Disadvantages

- Not always the most economical solution
- Poor understanding of fragmentation of even-electron ions
- Variety of ionisation modes required to cover full polarity range
- Occasional serious optimisation problems
- Possible matrix effects (polar compounds)
- Lack of structural information (exception: EI spectra)
- No MS^n reference
- Low sensitivity of many detection systems

The *major difficulties* met in combining LC and MS are:

(i) apparent flow-rate incompatibility (introduction of 1 mL min^{-1} of a liquid effluent – corresponding to over 1000 mL min^{-1} gas flow – from a conventional LC column into the high vacuum of the mass spectrometer);
(ii) ionisation of nonvolatile and/or thermally labile analytes;
(iii) solvent composition incompatibility (frequent use of nonvolatile mobile phase additives in LC separation development).

While the first two difficulties have been overcome there is no general solution available for the problem of incompatibility of mobile-phase composition. LC-MS systems are more complicated than GC-MS, as the eluted substances are mostly involatile, co-eluted with solvent, and frequently not efficiently ionised by EI or CI processes. Solutions to the problem are various, including surface ionisation (SIMS, FAB, FD, HSI,

PD, LDI, MALDI, etc.) and laser desorption. Recent developments in LC-MS, mainly atmospheric ionisation and off-axis spraying, have made the technique as reliable, versatile and easy to use as capillary GC-MS.

Different options are available for LC-MS instruments. The vacuum system of a mass spectrometer typically will accept liquid flows in the range of $10-20\,\mu L\,min^{-1}$. For higher flow-rates it is necessary to modify the vacuum system (TSP interface), to remove the solvent before entry into the ion source (MB interface) or to split the effluent of the column (DLI interface). In the latter case only a small fraction ($10-20\,\mu L\,min^{-1}$) of the total effluent is introduced into the ion source, where the mobile phase provides for chemical ionisation of the sample. The currently available commercial LC-MS systems (Table 7.48) differ widely in characteristics: mass spectrometer (QMS, QQQ, QITMS, ToF-MS, B, B-QITMS, QToF-MS), mass range (m/z 25 000), resolution (up to 5000), mass accuracy (at best <5 ppm), scan speed (up to $13\,000\,Da\,s^{-1}$), interface (usually ESP/ISP and APCI, nanospray, PB, CF-FAB). There is no single LC-MS interface and ionisation mode that is readily suitable for all compounds

of interest. LC-MS is the single most widely publicised analytical technique with major recent developments (nL min^{-1} flow-rates), ionisation modes (ESI, APCI, SSI) and mass spectrometers (QMS, QITMS, ToF-MS, QQQ, QToF-MS). Capital investments (2001) for the main commercial instruments are shown in Figure 7.12.

Three general *strategies* for LC-MS interfacing were developed:

(i) solvent removal/vaporisation/ionisation (MB, PB);
(ii) direct ionisation (CF-FAB); and
(iii) mobile-phase nebulisation/soft desolvation/ionisation (DLI, TSP, ESP, APCI, SSI).

LC-MS inlet probes support all conventional HPLC column diameters from <1 mm to 4.6 mm. In LC-MS the mobile phase must be eliminated, either before entering or from inside the mass spectrometer, so that the production of ions is not adversely affected. The problem of removing the solvent is usually overcome by direct-liquid-introduction (DLI), mechanical transport devices, or particle beam (PB) interfaces. The main disadvantages of *transport devices* are that column

Table 7.48 Overview of some commercial LC-MS couplings

Mass spectrometer	m/z range	Interface(s)	Optionals
QMS	50–3 000	(nano) ESI, APCI, APPI	
QITMS	50–2 200, 4 000	(nano) ESI, APCI, APPI	
ToF-MS	10–20 000	(nano) ESI, APCI	
QMS	3 000	ISP, APCI	Nanospray
QQQ	1 800, 3 000	ISP, APCI, APPI	MS/MS, nanospray
QToF-MS	12 000	ISP, APCI, MALDI	Mass accuracy <5 ppm
ToF-MS	25 000	ISP, APCI	Mass accuracy <5 ppm
QITMS	3 000, 6 000	ESI, APCI, SIMS	MS^n ($n < 4$, $n < 12$); nanospray
reToF-MS	20 000	ESI, APCI	Mass accuracy <5 ppm; resolution 2 000, nanospray
B	1 500	APCI, ESI, CF-FAB	DFS, resolution 5 000
QITMS	2 000	ESI, APCI, SSI	MS^n
ToF-MS	>20 000	ESI, APCI	Mass accuracy <5 ppm (m/z 150–400); nanoflow
QQQ	1 600, 4 000	ESI, APCI	MS/MS, nanoflow
QToF-MS	>20 000	ESI, APCI, MALDI	Mass accuracy <5 ppm (m/z 150–900); resolution 5000; nanoflow
QMS	1 500, 2 000	ESI, APCI	
QMS	1 500	ESI, APCI	Solids probe
QITMS	2 000, 4 000	ESI, APCI	MS/MS, MS^n ($n < 11$); nanospray
QQQ	4 000	ESI, APCI, FAB	MS/MS, solids probe; nanospray
QQQ	1 500	ESI, APCI	High resolution
QMS	2 000, 4 000	ESI, (APCI)	
QMS	1 000	Thermabeam^{TM} EI	Searchable spectra

Table 7.49 LC-MS interfaces, ionisation modes and mass spectrometers

Interface	Ionisation modes	Mass spectrometers
Moving-belt (MB)	EI, CI, FAB	QMS, B
Direct-liquid-introduction (DLI)	CI	QMS, QITMS
Thermospray (TSP)	EDI-CI, CI, EI	QMS, B, QITMS, QQQ
CF-FAB	FAB, LSIMS	QMS
Particle-beam (PB)	EI, CI, FAB, (L)SIMS	QMS, QITMS, B
Electrospray (ESP)/ionspray (ISP)	ESI	QMS, B, FTMS, QITMS, ToF-MS, QQQ, QToF-MS
Heated nebuliser	APCI, APPI	QMS, QITMS, ToF-MS, B-QITMS, QToF-MS, QQQ
Pneumatic nebuliser (PN)	MALDI	ToF-MS
Sonic spray (SSI)	ESI-like	QITMS
Supercritical molecular beam (SMB)	EI, HSI	QMS, ToF-MS

flow-rates are restricted to a maximum of about 1 mL min^{-1}, and eluates containing polar modifiers produce a drastic reduction in sensitivity. The latter means that many applications based on RP separations are not suited to this technique. It is generally agreed that microbore columns should be used in interfacing techniques that do not use a transport device, i.e. DLI techniques.

LC-MS interfaces generally produce ions with a relatively wide energy and spatial distribution. Table 7.49 lists the main LC-MS interface types. The most important types of contemporary LC-MS interfaces are direct inlet systems: PB, TSP, API, ICPI and MIP (the latter two for plasma source detection, *cf.* Section 7.3.3.5). Three main types of LC-MS coupling systems are usually distinguished:

  (i) removal of the solvent leaving the solute;
 (ii) reduced-flow systems; and
(iii) differential selection of solute ions.

Earlier LC-MS systems used interfaces that either did not separate the mobile-phase molecules from the analyte molecules (DLI, TSP) or did so before ionisation (PB). The analyte molecules were *then* ionised in the mass spectrometer under vacuum, often by traditional EI ionisation. These approaches are successful only for a very limited number of compounds. On the other hand, in atmospheric pressure ionisation, the analyte

molecules are ionised *first*, at atmospheric pressure. The analyte ions are then mechanically and electrostatically separated from neutral molecules. Spray formation, which is necessary to disrupt the HPLC flow, requires some energy in the form of heat and/or gas flow.

The mobile phase in LC-MS may play several roles: active carrier (to be removed prior to MS), transfer medium (for nonvolatile and/or thermally labile analytes from the liquid to the gas state), or essential constituent (analyte ionisation). As LC is often selected for the separation of involatile and thermally labile samples, ionisation methods different from those predominantly used in GC-MS are required. Only a few of the ionisation methods originally developed in MS, notably EI and CI, have found application in LC-MS, whereas other methods have been modified (e.g. FAB, PI) or remained incompatible (e.g. FD). Other ionisation methods (TSP, ESI, APCI, SSI) have even emerged in close relationship to LC-MS interfacing. With these methods, ion formation is achieved within the LC-MS interface, i.e. during the liquid- to gas-phase transition process. LC-MS ionisation processes involve either gas-phase ionisation (EI), gas-phase chemical reactions (CI, APCI) or ion evaporation (TSP, ESP, SSI). Van Baar [519] has reviewed ionisation methods (TSP, APCI, ESI and CF-FAB) in LC-MS.

In practice, the most appropriate ionisation method for a given type of sample must be considered, and the LC requirements should then be checked on any experimental incompatibility (flow-rate range, buffers and matrices, etc.) influencing ionisation method, chromatographic resolution and sensitivity. Sample enrichment prior to analysis and exchange of the widely used LC buffer systems against a suitable MS solvent is advantageous. In order to keep LC-MS coupling as simple as possible, LC methods are preferred which do not require buffers, electrolytes or matrix modifications. In this respect, TSP and FAB/FIB/LSIMS ionisation sources are often not the preferred choice. The current most popular ionisation modes in LC-MS are ESI and APCI; for polymer/additive analysis, mainly EI and APCI. ESI and APCI are largely incompatible with nonpolar compounds. LC-PB-MS is (nearly) obsolete, whereas LC-APPI-MS still needs to show its potential. Alternatives are LC-SMB-MS and GC-SMB-MS (but both are noncommercial). Table 7.50 will help to select the most adequate LC-MS configuration for a particular application. Table 7.51 outlines the main features of multisignal analysis in LC-MS.

Two LC-MS systems were developed, based on nearly full *removal of the solvent*: moving-belt [520] and particle-beam [521]. Mechanical transport devices

Table 7.50 Pathfinder of LC-MS techniques

Application	Ionisation mode	Mass analyser
• Verification, determination and quantification of known molecules	API, in-source CID	QMS
• Structure confirmation	API or PB-EI	QMS, QQQ, QITMS
• Library identification of nontargeted nonpolar molecules	PB-EI	QMS
• Elemental composition determination	API, high-resolution analyser	ToF-MS, B
• Structural analysis of high-mass molecules	API, high-resolution analyser, MS/MS	ToF-MS, QToF-MS/MS, QQQ
• High-mass molecules, MWD	ESP, high mass range analyser	ToF-MS
• High throughput quantitation of known molecules in complex matrices	API, MS/MS	QQQ

Table 7.51 Multisignal analysis in LC-MS

Mode	Information
+ ve or − ve ionisation	Determination of best analysis mode; confirmation of molecular weight: $[M + H]^+$ and $[M + H]^-$
Selected ion monitoring	Sensitivity for quantitation of low-level target compounds
Scanning	Complete spectra for high-level compounds and unknowns
Fragmenter voltage	Molecular ions (mass confirmation), fragmentation (structure determination)

reduce the solvent and concentrate the solute in a region removed from the ion source of the MS. Arpino [522] has reviewed the *moving-belt interface* (MBI), the most successful in these types of interfaces. The MBI device consists of an endless continuously moving Kapton®

ribbon, which transports the column effluent from the LC column outlet towards the MS ion source. During transport, the mobile phase is removed by gentle heating and evaporation of the solvent under reduced pressure (analyte-enrichment approach). Inside the source the solute is flash-vaporised from the belt into the EI/CI source. Alternatively, FAB ionisation may be used. However, in these cases, material sticking to the belt; thermal degradation during the evaporation phase; and mechanical complexity, are unfavourable. The main features of LC-MB-MS are given in Table 7.52.

LC-MB-MS has routinely been used for polymer/additive analysis [523]. Due to problems associated with this interface (e.g. sensitivity and analyte desorption), it is not considered to be a general analytical tool; its use was mainly relegated to specific cases. Few studies have been published on the MB interface, now extinct.

Particle-beam technology has become the generic designation for relatively simple LC-MS (and SFC-MS) sampling approaches for the introduction of solvent-depleted solute particles into the ion source through a momentum separator. In the particle-beam interface, gas-phase analyte enrichment is pursued (Figure 7.28). In the original LC-PB-MS design (MAGIC: monodisperse aerosol generating

Table 7.52 Main features of LC-MB-MS

Advantages

- Free choice of EI, CI and FAB ion sources.
- Compatible with normal column flow-rates and solvents.
- Suitable for semivolatile, relatively nonpolar and/or thermally labile analytes
- Most universal

Disadvantages

- Fairly complex and technically demanding operation
- Low sensitivity
- Surface adsorption/decomposition
- Unsuitable for trace analysis (high chemical background)

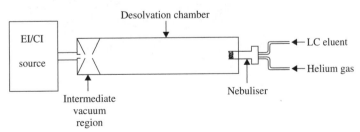

Figure 7.28 Typical particle-beam interface. After Ashcroft [524]. From A.E. Ashcroft, *Ionization Methods of Organic Mass Spectrometry*, The Royal Society of Chemistry, Cambridge (1997). Reproduced by permission of The Royal Society of Chemistry

interface for chromatography) [525,526], the eluent from the HPLC is sprayed through a heated nebuliser into an evaporation chamber. The vapour passes through a two-stage momentum separator, where the lighter solvent molecules are removed by differential pumping, leaving a stream of uncharged gas-phase solute particles to pass into the EI/CI ion source. Based on advanced particle–beam technology, the Thermabeam™ mass detector (TMD) vaporises and removes the chromatographic eluent, then ionises the analyte particles by EI. Thermal nebulisation and use of sheath gas creates a fine aerosol. In the momentum separator, heat and pressure reduction expand the volume of the aerosol and remove the chromatographic eluent, resulting in dry analyte particles with the preheated spray. TMD is a TSP-PB hybrid. The aerosols from a thermal source have a smaller particle distribution than the monodisperse aerosols. The mechanistic understanding of the fundamental processes underlying the PB interface performance is rather poor, but comprises pneumatic nebulisation with aerosol formation, followed by desolvation with solute enrichment, separation, solute vaporisation, concentration of the analytes into a beam, and EI/CI ionisation. Although the chance for thermal decomposition is minimised, it is not completely eliminated. The PB interface is similar to the MB interface, in that it is able to decouple chromatography from MS analysis. LC-PB-MS generates EI spectra with a reproducible fragmentation pattern of a molecule, which can be searched against commercial databases, industry standards or user-created compound libraries. This type of ionisation is well suited for qualitative analysis where extensive structure-informative fragmentation is highly desirable. Analyte detectability is rather poor in PB-MS, and it is difficult sometimes to determine the molecular weight of an unknown compound from the mass spectra, due to the fragmentation. Figure 7.29 shows an LC-PB-MS retention time–mass plot of a polymer additive extract.

LC-PB-MS is especially suited to NPLC systems. RPLC-PB-MS is limited to low-MW (<500 Da) additives. For higher masses, LC-API-MS (combined with tandem MS and the development of a specific mass library) is necessary. Coupling of LC via the particle-beam interface to QMS, QITMS and magnetic-sector instruments has been reported. In spite of the compatibility of PB-MS with conventional-size LC, microbore column (i.d. 1–2 mm) LC-PB-MS has also been developed. A well-optimised PB interface can provide a detection limit in the ng range for a full scan mode, and may be improved to pg for SIM analyses.

Table 7.53 shows the *main characteristics* of LC-PB-MS. Of all LC-MS interface methods, LC-PB-MS comes closest to GC-MS (Scheme 7.7). The particle beam is an acceptable choice in cases where sensitivity, volatility and analyte polarity are not an issue. Usually, the function of UV is added to LC-PB-MS; this allows the investigation of peak homogeneity. Drawbacks of LC-PB-MS are the low sensitivity and the nonlinearity

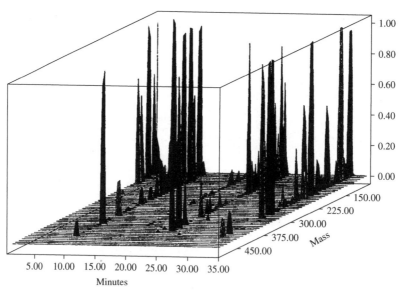

Figure 7.29 LC-PB-MS retention time–mass plot of a polymer additive extract. Courtesy of F. Ploeg, Waters/Micromass, Etten-Leur. Reproduced by permission of Waters Corporation

Table 7.53 Main characteristics of LC-PB-MS

Advantages

- Mechanically simple
- Reliable
- Interface compatibility: 100 % aqueous to 100 % organic
- Little spectral interference from the mobile phase
- Positive identification of unknown compounds (library searchable EI/CI spectra)
- Suitable for specific applications (e.g. polymer additives)
- Broad application

Disadvantages

- Sensitivity to eluent changes
- Less suitable for RPLC mode
- Low efficiency, limited sensitivity (ca. 5 ng in full scan mode)
- Limited mass range (<800 Da for TMD)
- Restricted to thermally stable molecules with reasonable vapour pressure
- Not suitable for highly polar analytes
- Complex optimisation

of the response. The low sensitivity, which is due to the low analyte transfer efficiency in the interface (as little as 10–12 % of the analyte in the liquid stream passes into the MS source), may be overcome by trace-enrichment procedures coupled on-line to LC. Nevertheless, highly polar and thermally labile analytes cause problems. LC-PB-MS is usually limited to compounds with a molecular mass below 1000 Da. The highest molecular mass observed by LC-PB-MS is about 2000 Da for styrene oligomers [527]. Advances in nebulisation techniques (i.e. use of an oscillating capillary nebuliser, OCN) have recently overcome the problem of the nonlinear response experienced with PB-MS. The system is more tolerant to NP (organic) solvents than to RP (aqueous)-based ones. This may be a drawback for coupling of many 'universal' RPLC methods. The use of high water content mobile phases lowers interface sensitivity, and thus a higher concentration of organic solvent should be preferred. A reduced mobile phase flow-rate, possibly in the µL min^{-1} range, as in microflow LC-PB-MS, enhances sensitivity and reduces the chance for thermal decomposition [528]. The difficulties in interfacing an ionisation method that comprises the use of a reasonably fragile filament in a high vacuum with a chromatographic technique eluting about 1 mL min^{-1} should not be underestimated.

Recently, a revival of electron ionisation LC-MS has been proposed by development of *LC-SMB-MS* [529]. In this system, the LC output (50–250 µL min^{-1}) is vaporised at atmospheric pressure and expanded from a supersonic nozzle into the vacuum system as neutral molecules. The vaporised solvent serves as the supersonic molecular beam carrier gas. The vibrationally cold molecules in the SMB are ionised by 70-eV electrons and mass analysed. Thermally labile molecules that are not amenable to GC can be analysed successfully. The characteristics of LC-SMB-MS are:

(i) enhanced molecular ion and structural, isomer and isotope mass-spectral information; and
(ii) suitable for the standard range of APCI compounds and nonpolar compounds.

Isotope abundance and elemental information is exhibited in cold EI mass spectra, due to the enhanced molecular ion and the ionisation of isolated molecules without CI-type reactions. Supersonic LC-MS provides a linear response, unlike LC-PB-MS. LC-SMB-MS is expected to compete with APPI, APCI and PB LC-MS modes (Scheme 7.8 and Figure 6.4).

Jones *et al.* [530] addressed the difficulties of validating LC-PB-MS and LC-TSP-MS methods with inter-laboratory studies; RSD values exceeded 4 %. Stable-isotope-labelled analogues that co-elute with the target analytes provide the best means for coping with enhancement or degradation in signal due to co-eluting components. For tuning, the spectrometer in LC-PB-MS PFK is often used.

In *reduced-flow LC-MS systems*, the solvent flow into the spectrometer is reduced to a level where the pumping system can cope. Essentially, three such systems have been developed: direct-liquid-introduction (DLI), flowing FAB [531] and electrospray [532]. An alternative approach to belt transport interfacing is to deliver the column eluate directly into the MS source and use CI techniques. Methods based on this principle are called *direct-liquid-injection systems*, which are comprised of capillary flow restrictors, diaphragms,

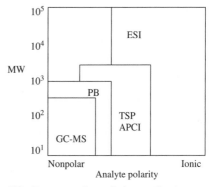

Scheme 7.8 Representation of the application ranges of LC-MS interfaces

nebulisers and microbore HPLC. DLI rarely achieves sufficient desolvation for EI, and only if it is combined with microbore LC. CI techniques have to be used in view of the amount of liquid entering the MS. This is a disadvantage, because CI spectra generally provide limited structural information. An advantage of LC-CIMS systems is that they are ideally suited to the use of polar solvents, as these promote protonation of the solute and production of the quasi-molecular ion (MH$^+$). CI DLI interfaces are suited to applications requiring RP separations. Another advantage of these systems is that the technique is amenable to the analysis of nonvolatile and thermally labile compounds. A major problem with DLI systems is that most mass spectrometers can only accept liquid flow-rates of less than $50 \mu L \, min^{-1}$, which is considerably less than those associated with conventional 4 mm i.d. HPLC columns (i.e. $1–2 \, mL \, min^{-1}$). This leads to a sensitivity problem. The best method of performing DLI LC-MS is thus to use microbore columns operating at flow-rates of $5–50 \, \mu L \, min^{-1}$. Such systems allow 100 % transfer of injected material to the MS, and although only sub-μg amounts of solute can be handled by these columns, the detection limits are 100 times greater than achieved by most other HPLC-MS interfaces. Nevertheless, DLI interfaces are far from ideal, and few studies deal with this interface, which never properly caught on. Both MB and DLI interfaces are no longer in use.

In a continuous-flow or *dynamic fast atom bombardment* (CF-FAB) interface column flow-rates are restricted to about $5–15 \, \mu L \, min^{-1}$ by the rate of solvent removal from the target. Most column types can be used in conjunction with continuous CF-FAB/FIB/LSIMS. Microbore (i.d. 1 mm) and capillary (i.d. 0.3 mm) columns operate optimally at $<50 \, \mu L \, min^{-1}$ and $<5 \, \mu L \, min^{-1}$, respectively, and are suitable for direct coupling. For larger i.d. HPLC columns (e.g. 2.1 and 4.6 mm), the flow needs to be split as it cannot pass directly through to the CF-FAB/FIB probe. In LC-FAB-MS a small liquid stream, mixed with an appropriate FAB matrix solvent, e.g. glycerol, flows through a narrow-bore fused-silica capillary towards either a stainless-steel frit (frit-FAB) or a gold-plated FAB target, where solvent evaporation takes place. If it is not possible to have the matrix in the mobile phase, then it can be introduced by postcolumn addition. Ions are generated by bombardment of the liquid film by fast atoms or ions, i.e. the common ionisation process in FAB or LSIMS. Both isocratic and gradient elution are allowed; RP and NP solvents are suitable (e.g. water, acetonitrile, methanol, toluene, and mixtures thereof, and electrolytes in low concentration). The major

advantages of the LC-CF-FAB-MS are high sensitivity, high mass capability ($>6000 \, Da$) and applicability to thermally unstable and/or involatile polar compounds, especially biopolymers. Use of LSIMS, FAB and LDI in LC-MS is confined to relatively high-molecular-weight solutes. LC-FAB-MS is particularly suitable to analytes that do not yield other mass spectra (EI, CI, etc.). CF-FAB is characterised by a widely variable response. For a particular application, the observed sensitivity depends on the wettability of the probe tip, the temperature of the probe, and the composition and flow-rate of the mobile phase. Matrix interferences are possible. It is often advisable to install a UV detector either in-line (i.e. in between column outlet and CF-FAB/FIB probe), or simultaneously (i.e. using a splitting device to direct one part of the eluent to the continuous-flow probe and another part to the alternative detector). cHPLC-CF-FAB-MS^n coupling has also been described [533].

The *electrospray ionisation* (ESI) process is based on ion evaporation phenomena [534]. For electrospray (ESP), the mobile phase is forced through a metal capillary needle at a high electrical potential. Electrospray nebulisation results in an aerosol of small highly charged droplets of one polarity. This is at variance with thermospray and heated nebuliser interfaces, which generate both positively and negatively charged droplets. The excess charge resides at the surface. Upon solvent evaporation, the electric field at the droplet surface increases and the multiply charged droplets will subdivide into a fine spray of smaller droplets. Nitrogen gas is used to help desolvate this aerosol. The use of gas-assisted nebulisation has made *LC-ESP-MS* into one of the most commonly used interfaces. When LC and MS are coupled via electrospray ionisation, the disparate flow ranges of the two techniques (usually $0.2–2.0 \, mL \, min^{-1}$ for LC and $0.2–2.0 \, \mu L \, min^{-1}$ for ESI-MS) must be reconciled. A wide variety of high and low flow-rate interface devices are available for LC-ESP-MS [535]. If a 4.6-mm HPLC column is used with a mobile phase flowing at $1 \, mL \, min^{-1}$, one may opt to send only a portion of the eluent through to the ESI source, and the remainder to an alternative detector. The electrospray ionisation efficiency is solvent dependent. Solutions in water show high surface tension, coarse droplets, difficult spraying, good solvation of ions and low sensitivity, while solutions in methanol and acetonitrile are characterised by low surface tension, fine droplets, easy spraying, less-efficient solvation of ions and high sensitivity. Contaminants in LC-ESI-MS may be in the gas phase – notably plasticisers (e.g. DBP m/z 279, 296, 301; DOP m/z 391, 408, 413) from plastic hoses or floor covering, solvents from floor waxes (e.g. carbitol m/z 135, 152, 157), or caprolactam (from the steel-braided nylon connection between gas cylinder and pressure reducing station). Typical liquid-phase contaminants are detergents, surfactants, buffers and additives.

The *ionspray* (ISP, or pneumatically assisted electrospray) LC-MS interface offers all the benefits of electrospray ionisation with the additional advantages of accommodating a wide liquid flow range (up to $1\,mL\,min^{-1}$) and improved ion current stability [536]. In most LC-MS applications, one aims at introducing the highest possible flow-rate to the interface. While early ESI interfaces show best performance at $5-10\,\mu L\,min^{-1}$, ionspray interfaces are optimised for flow-rates between 50 and $200\,\mu L\,min^{-1}$. A gradient capillary HPLC system ($320\,\mu m$ i.d., $3-5\,\mu L\,min^{-1}$) is ideally suited for direct coupling to an electrospray mass spectrometer [537]. In sample-limited cases, nano-ISP interfaces are applied which can efficiently be operated at sub-$\mu L\,min^{-1}$ flow-rates [538,539]. These flow-rates are directly compatible with micro- and capillary HPLC systems, and with other separation techniques (CE, CEC).

ESI and APCI are soft ionisation techniques which usually result in quasi-molecular ions such as $[M + H]^+$ with little or no fragmentation; molecular weight information can easily be obtained. However, experimental conditions can also be chosen in such a way that a sufficiently characteristic pattern is obtained, allowing verification [540]. ESI is amenable to thermally labile and nonvolatile molecules. Both ESI and APCI are much more sensitive than PB and very well suited for quantitative analysis, but less so for unknown samples. The choice among the two is usually determined by the application. Recently, nanoscale LC-ESI-MS has been developed [541]. The nano-electrospray ion source offers the highest sensitivity available for LC-MS (atto- to femtomole range) and can also be used as an off-line ion source.

In LC-ESP-MS configurations, a variety of mass spectrometers has been used: QMS, QITMS, magnetic sector, oaToF-MS, FTMS, QToF-MS/MS, QQQ. LC-ESP-MS is most suitable for quadrupoles. In conjunction with QMS ESI can also measure high-mass/high-charge ratio molecules. LC-ESI-oaToFMS affords high performance in terms of sensitivity, resolution (4000–5000 FWHM from 400 to 4000 *m/z*), a wide mass range (up to 10 000 *m/z*), fast acquisition rates (several mass spectra), and is suitable for fast chromatographic analyses [542]. LC-ESI-oaToFMS shows mass measurement errors <3 ppm for MS data and <5 ppm for MS/MS data. As ESP, like the TSP interface, is a dedicated ion source, special attention must be paid to tuning and calibration of the instrument.

Table 7.54 Main characteristics of LC-ESI-MS

Advantages

- Allows low flow-rates (preferably $5-10\,\mu L\,min^{-1}$)
- Fast LC-MS for short HPLC columns (<10 cm)
- Compatibility with capillary HPLC, RPLC and NPLC
- Soft ionisation (molecular mass information)
- Amenable to thermally labile, nonvolatile, very polar and ionic compounds
- High resolution, wide mass range
- Most sensitive LC-MS technique
- Concentration sensitive detection (microbore LC for best sensitivity)
- Wide linear range
- Quantitative analysis
- Robust, ease of operation
- Worldwide spread

Disadvantages

- No structural information (verification of known compounds only)
- Multiple charging may complicate data interpretation
- Special attention required to tuning and calibration
- Interferences (gas-phase and liquid-phase contaminants)

Table 7.54 summarises the *main characteristics* of LC-ESI-MS. Analytes may be macromolecular, including polymers. Smaller, volatile molecules can also be analysed easily by electrospray. For compounds with molecular weights below 1 kDa, ionspray is more popular than electrospray. Analytes must be capable of forming ions in solution. Molecules without ionisable groups will not run well in electrospray. ESP can form multiple-charge ions, which increases the upper mass limit (>150 kDa). This capability is significant for macromolecules. Electrospray is a concentration technique with a linear range of approximately $10^{-9}-10^{-4}$ M; its limits of detection are routinely in the low pg or even fg range. Brereton [543] described the use of chemometrics in the analysis of LC-ESI-MS/DAD data.

A group of techniques employing *differential selection of solute ions* relies on nebulisation and ionisation of the eluent, with some discrimination of ion selection in favour of the solute. Main representatives are APCI [544] and thermospray [545]. In a *thermospray interface* a supersonic jet of vapour and small droplets is generated out of a heated vaporiser tube. Controlled, partial vaporisation of the HPLC solvent occurs before it enters the ion source. Ionisation of nonvolatile analytes takes place by means of solvent-mediated CI reactions and ion evaporation processes. Most thermospray sources are fitted with a discharge electrode. When this is used, the technique is called *plasmaspray* (PSP) or discharge-assisted thermospray. In practice, many analyses are a mixture of TSP and PSP ionisation, especially those employing on-line RPLC-MS with gradient elutions where the first part of the run conforms to TSP, and the latter part to PSP. Thermospray operation requires careful optimisation of a variety of mostly interrelated, experimental parameters. Typical detection limits for LC-TSP-MS analysis are in the range of 1–10 ng for full-scan analysis.

Thermospray (TSP) is a soft ionisation technique which produces predominantly MH^+ or $(M - H)^-$ quasimolecular ions, together with some fragmentation. TSP is best suited to the analysis of organic compounds of molecular mass (<1200 Da) that exhibit some polarity. Besides FD and FAB ionisation methods, TSP-MS and, more recently, ESI-MS have become the most commonly used determination methods for polar compounds. Polymer additive molecules fall in this wide category. Thermospray was quite popular before the advent of electrospray, but has now given way to the more robust API techniques, although some TSP sources continue to operate. TSP is tolerant of both polar and nonpolar solvents, which means that, when coupled to HPLC, both NP and RP separations can be carried out with isocratic and gradient elutions. Both TSP and PSP ionisation work best with flow-rates of ca. 1 mL min^{-1}, and allow for coupling to standard-bore chromatography. Many HPLC separations in polymer/additive analysis rely on gradient elution, and, as the optimum TSP operating conditions change substantially with respect to the mobile-phase composition, the change in solvent can be a cause of instability. An increase in the organic content of the mobile phase generally leads to a decrease in sensitivity [545a], and many such analyses may require PSP operation.

Table 7.55 Main characteristics of LC-TSP-MS

Advantages

- Compatible with high flow-rates (0.2–2 mL min^{-1}); handles total eluate volumes
- Can be used with NPLC and RPLC
- Retrofitting to most MS
- Soft ionisation (molecular mass information)
- Suitable to polar and involatile molecules as well as ions
- Rugged, reliable, simple to operate

Disadvantages

- Rather elaborate optimisation of experimental parameters; low user-friendliness
- Not suited to all analytes (insufficient ion yields)
- Limited molecular mass range (preferably <1500 Da)
- Decomposition of thermolabile compounds
- Limited structural information
- Lack of robustness
- No longer commercially available

Table 7.55 lists the *main characteristics* of LC-TSP-MS. Thermospray suffers from several limitations that are problematic for routine and consistent use. The upper molecular weight limit for TSI is low, and a lack of sensitivity is also inherent, making trace analysis difficult. As TSP produces predominantly molecular mass information, frequently a multiple analyser mass spectrometer, such as a QQQ or hybrid instrument in LC-TSP-MS^n fashion is necessary in order to obtain more structural information for unambiguous compound identification. Another disadvantage is the inability to perform EI because of a high source pressure. Particle-beam interfaces solve this problem. Thermospray is a disappearing interface type for LC-MS, and it is now being replaced by atmospheric-pressure interfacing techniques. Thermospray nebulisation continues to be used in nebulisation for ICP-MS.

LC-APCI-MS is a derivative of discharge-assisted thermospray, where the eluent is ionised at atmospheric pressure. In an *atmospheric pressure chemical ionisation* (APCI) interface, the column effluent is nebulised, e.g. by pneumatic or thermospray nebulisation, into a heated tube, which vaporises nearly all of the solvent. The solvent vapour acts as a reagent gas and enters the APCI source, where ions are generated with the help of electrons from a corona discharge source. The analytes are ionised by common gas-phase ion–molecule reactions, such as proton transfer. This is the second-most common LC-MS interface in use today (despite its recent introduction) and most manufacturers offer a combined ESI/APCI source. LC-APCI-MS interfaces are easy to operate, robust and do not require extensive optimisation of experimental parameters. They can be used with a wide variety of solvent compositions, including pure aqueous solvents, and with liquid flow-rates up to 2 mL min^{-1}.

APCI is a soft ionisation mode that does not induce much fragmentation, although informative fragmentation can be induced by increasing the cone voltage setting. Consequently, LC-APCI-MS spectra are not searchable in commercial databases. APCI mass spectra are not as predictable as EI spectra. For this reason, only user-generated spectral libraries can be searched, usually in both the APCI$^+$ and APCI$^-$ modes. In HPLC-APCI-MS quantification is performed by HPLC, and mass spectrometry is used to verify the findings. In combination with HPLC retention time, reliable verification of the compounds can be performed. APCI is best used with flow-rates >1 mL min^{-1}, and is suitable for RPLC coupling. Various mass spectrometers have been employed in conjunction to LC-APCI-MS, namely QMS, QITMS, ToF-MS, QQQ, QToF-MS and

Table 7.56 Main characteristics of LC-APCI-MS

Advantages

- Ease of use
- Very rugged; robust
- No extensive optimisation of experimental parameters
- Wide variety of solvent compositions (incl. aqueous)
- High flow-rates (up to $2\,mL\,min^{-1}$); fast LC-MS for short columns
- Heater temperature not critical, only on gradients
- Ultrahigh sensitivity
- Relatively high mass range (2 000 Da)
- Suitable for fragile compounds (noncovalent complexes and thermally labile species)
- Verification of known compounds (ultrasoft ionisation)
- Wide range of analyte polarity (even with NPLC)
- General purpose (wide applicability)

Disadvantages

- No searchable commercial databases
- Requires volatile buffers

B-QITMS. LC-APCI-ToFMS is still in an early stage of development. LC-API-MS is more general purpose than LC-PB-MS; these techniques are complementary.

Table 7.56 summarises the *main characteristics* of LC-APCI-MS. APCI is appropriate for the analysis of small molecules up to approximately 2000 Da. Since APCI is dependent on the proton affinities of analytes, it can be used to selectively analyse compounds of interest, even in complex matrices. APCI also tends to produce less cluster ions and can accommodate higher HPLC flow-rates. Heated nebuliser–APCI is ideal for moderately labile, neutral to polar compounds. The system provides low detection limits in the 1–10 ng range in full scan acquisition and in the 0.1–0.5 ng range in MID, which are superior by a factor of at least 50 to all other LC-MS interfaces available. Excellent linearity over two orders of magnitude is achieved. APCI-MS in the negative mode is about a factor less sensitive than in the positive mode. The negative mode shows predominantly $[M - H]^-$ and is therefore very well suited for obtaining reliable molecular weights. This fact is used for the search on 'new' additives or reaction products. Controllable, reproducible orifice-induced fragmentation (OIF) or CID/MS provides qualitative information that helps characterise chemical structure. Mixed mode analysis allows molecular-weight and structural information to be obtained in a single analysis. Sensitivity and fragmentation obtained by LC-APCI-MS depends on probe temperature and extraction voltage.

HPLC-APCI-MS data are found to be most useful when the chemical identity of additives in a polymer is already known (verification); otherwise tandem MS

is required. HPLC-UV-APCI-MS with in-line UV detection (PDA) between the HPLC column and MS (Figure 7.30) provides the following information for structural assignments: retention time comparisons (HPLC), high-resolution UV spectra (1.2 nm optical resolution), peak homogeneity, quantitation (UV); molecular weight and fragmentation pattern (MS). In this experimental set-up, there is an offset of several seconds in retention times of the eluting peaks on the UV output compared with the MS output. HPLC-FTIR similarly provides additional identification, which is based on molecular structure. Because identifying a substance by MS requires knowledge about its possible molecular structure, the two techniques form a highly complementary system of identification.

Commercially available *atmospheric pressure ionisation* (API) systems usually combine the two fundamentally different processes, namely gas-phase ionisation (APCI) and liquid phase ionisation (ESI), and allow 100 ms switching between ESI^+, ESI^-, $APCI^+$ and $APCI^-$ The term ESCI™ or electrospray chemical ionisation has been coined for this purpose. Table 7.57 lists the main characteristics of LC-API-MS. API techniques give mainly molecular weight information. The fragmentation obtained by CID is useful, but depends on the applied voltages and on the instrument geometry. Building mass-spectra libraries is possible at the users' level, but it is difficult to build large libraries similar to electron impact ionisation in the case of GC-MS or LC-PB-MS.

API techniques differ according to probe design, ionisation process and application field. API technology is compatible with a broad range of liquid separation techniques, including RPLC, NPLC, SFC, IC and CZE (see also Table 7.58). LC-API-MS allows analysis of a wide range of polar thermolabile and nonvolatile compounds which cannot be analysed by GC. In API techniques, mobile-phase composition is a crucial factor for ionisation: acidic pH for positive ions (4–5 is preferred; acetic acid, formic acid); basic pH for negative ions (8–9 is preferred; NH_4OH, DEA, TEA); and eventually postcolumn addition of reagent to adjust the eluent pH for better ESI detectability. The most common polar solvents used with API techniques are (mixtures of) water, methanol, acetonitrile and propan-2-ol, with 1:1 (v/v) water/acetonitrile or water/methanol being used frequently. At present, LC-API-MS is the most powerful interface, followed at a distance by LC-PB-MS.

Soft ionisation modes, such as API, which leave the (pseudo)molecular ion intact without much fragmentation, offer more sensitivity, and are ideal for quantitative work at low levels (e.g. breakdown products). With the use of soft ionisation techniques in LC-MS, tandem MS

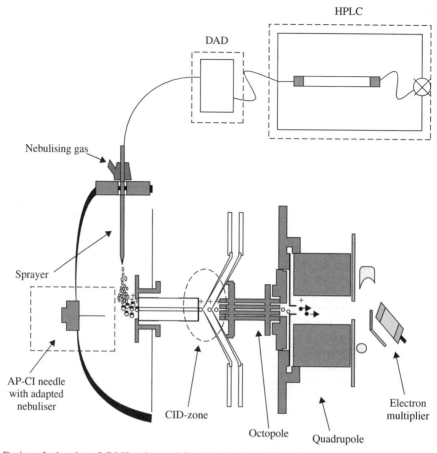

Figure 7.30 Design of a benchtop LC-MS orthogonal flow introduction device

is almost obligatory for confirmatory purposes. Otherwise, a mass-spectrometric strategy may be based on oaToF-MS with accurate mass determination at 5 ppm (at a sensitivity an order of magnitude higher than that of QMS or magnetic sector). Accurate monoisotopic mass determination of the molecular ion, as carried out by means of LC-oaToFMS, results in highly practical compound assignments (also for trace components). This technology, which needs no compromise between resolution and sensitivity, has apparently not yet been applied to polymer/additive analysis and would probably require the APCI mode. LC-API-ToFMS features are given in Table 7.59. Enke [546] has emphasised enhanced chromatographic resolution through deconvolution of total-ion chromatograms obtained with modern ToF-MS analysers. Where the increased analytical resolution afforded by the high data density is not needed, a shorter, low-resolution chromatographic run can be used, and optimisation can be accelerated, since the

compounds need not be chromatographically separated to yield all the desired results.

Capillary HPLC (usually SEC or LCCC), coupled semi off-line with MALDI-ToFMS for oligomer and end-group characterisation, may be operated using collection modules precoated with matrix material.

Sonic spray ionisation (SSI) is a room-temperature LC-MS interfacing technique first developed in 1994, in which a liquid flow is sprayed from the tip of a capillary under atmospheric pressure, with a gas flow coaxial to the capillary [547]. When the gas velocity is of the order of the speed of sound, charged droplets are formed because of the shear stress of the sonic gas flow. SSI is a compromise between ESI and APCI, and does not require a heated nebuliser (as in APCI), or application of high voltage (as in ESI) to the sprayer. As SSI forms charged droplets without heating, it is ideally suited for analysis of thermally labile compounds. Furthermore, ionisation without additional high voltages increases the

Table 7.57 Main features of LC-API-MS

Advantages

- Widest range of analytes (diverse polarities and molar masses)
- Chromatographer-friendly
- Wide range of flow-rates (50 nL min^{-1} to >1 mL min^{-1}; preferably <10 μL min^{-1})
- Easy to use
- High sample throughput
- Small sample amount required
- Robust, high reliability
- Ionisation at atmospheric pressure
- Soft ionisation modes: APCI, ESI, APPI (+ ve and − ve)
- Quantitation over wide linear dynamic range (over three orders of magnitude)
- Attomole detection limits
- High-resolution MS/MS
- Decreasing method development time
- 3 000 Da mass range
- Mature and versatile
- Wide applicability range

Disadvantages

- Relatively expensive
- Cone voltage-dependent fragmentations
- No commercial databases

Table 7.58 Compatibility of LC and API-MS technologies

Chromatographic mode	ESI	APCI
RPLC	+++	++
NPLC	+	+++
Ion pair	++	++
Ion exchange	+	+

Table 7.59 Main features of LC-API-ToFMS

- Parallel mass analyser (high sensitivity)
- High resolution (>6 000, FWHM)
- Wide mass range
- High mass accuracy (<5 ppm; selectivity and elemental composition based on monoisotopic masses)
- High dynamic range (exceeding four decades)
- Differentiates between isobaric compounds
- Fast chromatography and MS scan rate (5 000 Da s^{-1})
- APCI + ESI (flexibility)
- Total integration: LC-UV-MS
- Improved peak definition (quantitation)
- Automation, ease of use
- In-source CAD (full structural characterisation)

flexibility in the choice of mobile phases. Ionisation is ESI-like. Ultrasonically assisted electrospray nebulisation interfaces are now commercially available in combination with API sources [548]. SSI-MS has been coupled to HPLC [549], as well as CE [550], and is readily combined over a wide range of flow-rates and mobile phases.

A new development concerns *atmospheric pressure photoionisation* (APPI). APPI is a recently introduced API technique for LC-MS [551]. Exchangeable APCI/ESI/APPI ion sources are now commercially available. The APPI source allows both gas and liquid introduction, and the analyte is nebulised into a heated probe. From an instrumental point of view, APPI and APCI ion sources are similar: in the PhotoSpray™ source, the corona discharge needle of APCI is substituted by a VUV lamp source (usually Kr, IP = 10.6 eV). This energy level is high enough to ionise many classes of organic compounds, but low enough to minimise ionisation of air and common HPLC solvents. Relatively low ionisation energy means that the APPI source causes minimal fragmentation and typically generates molecular ions or protonated molecules and little adducts. As ionisation is achieved with photons instead of electrons as for APCI, APPI is not limited by gas phase acid-base chemistry. Nevertheless, initial LC-APPI-MS experiments of polyaromatics showed a poor response. However, for analytes that do not readily ionise, the use of ionisable dopants such as toluene (IP = 8.83 eV) and acetone (IP = 9.70 eV), which are co-vaporised in the auxiliary gas, provides a means to overcome the low sensitivity previously associated with LC-PID [551]. Dopant photoions react to completion with solvent and analyte molecules present in the ion source. In APPI two processes may take place:

(i) charge transfer (via the dopant); or
(ii) solvent transfer.

In the latter case the dopant is ionised, interacts with the solvent and, subsequently, solvent clusters interact with the analyte. Molecular and protonated molecular ions are observed, indicating that ionisation can occur via proton (toluene) and electron transfer (acetone).

Table 7.60 shows the *main characteristics* of LC-APPI-MS. Using APPI, at an LC flow-rate of 200 μL

Table 7.60 Main characteristics of LC-APPI-MS

Advantages

- Soft ionisation (+ ve and − ve ions)
- Complementary to ESI and APCI
- More sensitive than APCI
- Compatible with NPLC and RPLC
- Suitable for low-polar compounds
- No derivatisation necessary

Disadvantage

- Robustness to be demonstrated (new technology)

Table 7.61 Comparison of API performance in LC-MS

Matrix effect	ESI > APCI ≥ APPI
Sensitivity to polar compounds	ESI > APCI ≥ APPI
Sensitivity to low-polar compounds	APPI >> APCI > ESI
Chemical background	ESI > APCI > APPI
Susceptibility to mobile-phase composition	APPI > ESI > APCI

min^{-1}, it is possible to obtain analyte signal intensities eight times as high as those obtainable with a commercially available corona discharge-APCI source [551]. First indications are that the range of compounds that can be efficiently ionised by APPI closely follows that of APCI – a finding that points toward the conclusion that the ion–molecule reactions responsible for analyte ionisation in each source ultimately follow similar pathways. As shown in Figure 6.4, APPI offers more sensitive detection for some compound classes compared to APCI – notably highly polar solutes. Target application areas are pharmaceuticals, clinical analysis, food and environmental analysis and polymer/additive analysis. The three API techniques are compared in Table 7.61.

Another *multi-API interface* has been described [552], but not commercialised, which realises two key techniques, namely the combination of EI mode (as an atmospheric-pressure spray with electron impact ionisation, APEI) with APCI, APSI, ESI and SSI modes in one LC-MS system. LC-MS using the APEI mode is very useful for the analysis of hydrocarbons, aromatic compounds and other compounds with relatively high volatility and thermal stability. The APSI (atmospheric-pressure spray ionisation) interface is almost the same as thermospray, but the mass spectra are very different. An all-purpose single LC-multi-API-MS system is of great interest because of the ease of replacement of five interfaces, which can deal with a wide variety of organic compounds from hydrocarbons with low polarity to proteins with high polarity. It is very difficult

to predict exactly what type of organic compounds can be measured by each mode. LC-multi-API-MS thus provides:

(i)   the APEI mode for low-polarity organic compounds (e.g. aromatics);
(ii) the APCI mode for relatively polar compounds with nitrogen and/or oxygen atoms which are neutral in solution;
(iii) the APCI mode for organic compounds with high cation affinity; and
(iv) the ESI or SSI modes for polar analytes which exist as, or are associated with, ions in solution.

Conversion of LC to LC-MS methods has been discussed [553], and Taylor [554] has dealt with *column technology* for LC-MS analysis. Vissers [555] has discussed on- and off-line coupling of microcolumn LC to MS. The choice of an HPLC column depends on the separation to be performed, the HPLC solvent delivery systems available, and (indirectly) on the ionisation technique. Solvent flow-rates for different HPLC couplings are given in Table 7.62. If the probe and source are not suited to the delivered flow-rates, the eluent needs to be split after the HPLC column. Conversely, a column of less than 2.1 mm i.d. will probably not have a sufficiently high flow-rate for APCI operation. In RP chromatography, users sometimes must modify HPLC conditions to allow compatibility with MS. Fast-gradient elution LC-QToF is a useful tool in the structural characterisation of low concentration components in complex mixtures [556]. Interfacing LC with MS is becoming more practical with the miniaturisation of HPLC equipment. As LC-MS hyphenation employs a very mass-selective detector, the chromatographic separation requirements do not always demand high resolution; consequently, short columns (2 to 10 cm) packed with small particles (2 and 3 μm) may suffice. Narrow-bore LC-MS generates narrower

Table 7.62 Solvent flow-rates in LC-MS methods

Interface	Ionisation source	Optimum flow-rates[a]	Notes
PB	EI	0.5–2 mL min⁻¹	Higher flow-rates acceptable only for a low aqueous content in the mobile phase
ESP	ESI	30 nL–1 mL min⁻¹	Wide range of solvents acceptable (most universal)
APCI probe	APCI	50 μL–2 mL min⁻¹	Wide range of solvents acceptable
CF-FAB	FAB/FIB/LSIMS	1–10 μL min⁻¹	Matrix needed
TSP probe	TSP	0.5–2 mL min⁻¹	Wide range of solvents acceptable Needs presence of an electrolyte

[a] min–max.

Table 7.63 Summary of interface types and chemistry in LC-MS couplings

Type	Ionisation	Species formed	Comments
MB	EI, CI, FAB	$M^{\bullet+}$, $M + H^+$, $M - H^-$	Insensitive, not robust
PB	EI, CI	$M^{\bullet+}$, $M + H^+$, $M - H^-$	Insensitive, good for NP
DLI	EI, CI	$M^{\bullet+}$, $M + H^+$, $M - H^-$	Not robust, superseded by ESP
CF-FAB	FAB	$M + X^+$, $M - H^-$	Low flow essential, requires matrix
ESP	ESI	$M + X^+$, $M - H^-$	Very sensitive, excellent for biopolymers
$-$ ve ion APCI	ECNI	$M^{\bullet-}$	Very specific, little used, superseded by APCI
TSP	Thermal CI	$M + X^+$, $M - H^-$	Well used, superseded by APCI and ESP
Corona discharge	APCI	$M + X^+$, $M - H^-$	Well used, very popular
SSI	ESI	$M + X^+$, $M - H^-$	New, ideal for thermally unstable compounds
API	APPI	$M^{\bullet+}$, $M + H^+$, $M - H^-$	New

peaks, which means better sensitivity (taller peaks) for the same injection mass, or smaller injections for the same peak size when compared with conventional 4.6-mm i.d. columns. Combination of LC with MS is facilitated by using a LC system with microbore columns (i.d. 1 mm) instead of conventional columns (i.d. 4–5 mm). Flow-rates in LC systems with microbore columns ($10–50\,\mu L\,min^{-1}$) enable the total effluent to be introduced into the ion source. This increases the sensitivity of the method, compared to systems which require splitting of the effluent.

Table 7.63 lists LC-MS interface types and the species formed in the various ionisation processes.

Mass analysers used in LC-MS hyphenation are indicated in Table 7.64. On-line coupling of LC with magnetic-sector instruments (with in-source fragmentation with MS^n capability) is problematic because the high flow-rates of liquid with high-voltage sources lead to breakdown and noisy spectra (sparking in the source). Therefore, combination of a magnetic-sector instrument with most high-pressure LC-MS interfaces is considered as being far from ideal. ToF-MS is particularly useful in applications where either high-mass analytes are to be analysed or where high scan speeds are required (fast chromatography). ToF-MS in LC-MS can only be used as a universal detector, because no ion selection can be performed. While the oaToF system provides a 20–100-fold improvement in sensitivity compared to a scanning QQQ system, and the ability to mass analyse very narrow peaks at high flow-rates, the QToF hybrid offers unsurpassed performance in terms of specificity (high mass accuracy) and sensitivity (using low-energy CID). This greatly facilitates the identification of unknowns. One key advantage of FTMS systems over other high-performance mass spectrometers is the ability to obtain accurate mass measurements on both parent

Table 7.64 Main characteristics of currently available mass analysers for LC-MS

Mass analyser	Advantages of LC-MS	Disadvantages of LC-MS
QMS	Simple operation and maintenance; low cost; constant resolution (Δm); linear m/z scale; low ion accelerating voltages	Low resolving power (up to 4000); low m/z accuracy ($\sim$0.1 units); slow acquisition rate (1 spectrum s^{-1}); maximum m/z range $\sim$3000
QITMS	Simple operation and maintenance; low-cost MS/MS; constant resolution (Δm); linear m/z scale; low ion accelerating voltages; very high sensitivity	Low resolving power (up to 4000); low m/z accuracy ($\sim$0.1 units), slow acquisition rate (10 spectra s^{-1}); maximum m/z range $\sim$6000
ToF-MS	Simple operation; high acquisition rate ($>$100 spectra s^{-1}); high m/z accuracy ($\sim$0.01 units); m/z range $>$ 10 000; high resolving power ($>$5000)	High vacuum and high accelerating voltages; QToF-MS for true MS/MS; resolution (Δm) decreases with increasing mass
B	MS/MS performance; high resolving power (up to 50 000); high m/z range ($>$10 000), very high m/z accuracy (0.001 units)	Very high cost; complex operation and maintenance; low acquisition rate (1–10 spectra s^{-1})

and MS/MS product-ions. The features of state-of-the-art LC-FTMS (based on an ion trap-FTMS hybrid) are unsurpassed, with resolving-power values of $10^5–10^6$ and mass accuracies below 1 ppm. The drawbacks of LC-FTMS are difficult coupling, problems with fast signals (narrow LC peaks) and demanding computing facilities. Readers without access to LC-FTMS [557] need to manage with suboptimal general analytical solutions. FTMS and magnetic-sector instruments are much more expensive, and fit into specialised LC-MS niches.

Although use of *LC-MS/MS* in the polymer/additive analysis field may seem attractive because of high sensitivity; high mass resolution and good mass measurement accuracy; good specificity; advanced structural identification power and characterisation of unknown (reaction) products; as well as ultimate detection limits and high throughput, the cost of LC-MS/MS will usually constitute a deterrent for most such analytical laboratories unless open-shop facilities are available. The LC-MS^n domain was once dominated by triple-quad mass spectrometers, especially for high-sensitivity targeted quantitative analysis, but other (hybrid) systems are now commercially available, such as QToF-MS, and B-QITMS, etc. LC-QToFMS^n is the current state-of-the-art LC-MS device (± 3 ppm in MS, ± 5 ppm in MS/MS). LC-MS^n allows full-scan MS/MS sensitivity for structural elucidation, and is essential for selectivity in the analysis of complex matrices. Verification of unknown additives can be based on accurate mass rather than identification of a fragmentation pattern. HPLC-API-MS/MS is extensively used for rapid quantitative method development and analyte identification (mainly in the pharmaceutical industry). It is to be considered that such LC-MS (MS) spectra differ much more than EI spectra. This is caused by experimental parameters such as different ionisation methods, gases, solvents, collision offset voltages, etc. In many cases it is relatively straightforward for a trained mass spectroscopist to interpret an MS/MS spectrum from any mass spectrometer if he/she knows the structure! Thanks to recent developments in ion-trap technology [558], highly reproducible MS/MS spectra (ESI, APCI), suitable for library searching, can now be obtained. A small library of such product-ion mass spectra (1000 compounds) includes dyes but no polymer additives. Mass spectra obtained from multiple stages of MS (MS$^+$) are also included.

Much LC-MS work is carried out in a qualitative or semi-quantitative mode. Development of *quantitative LC-MS* procedures for polymer/additive analysis is gaining attention. When accurate quantitation is necessary, it is important to understand in depth the experimental factors which influence the quantitative response of the entire LC-MS system. These factors, which include solvent composition, solvent flow-rate, and the presence of co-eluting species, exert a major influence on analyte mass transport and ionisation efficiency. Analyte responses in MS procedures can be significantly affected by the nature of the organic modifier used in the RPLC

eluent. This may create problems in quantitative analysis. The suitability of the ELS detector for use as a concentration detector in LC-MS has been recognised. The principle of operation of ELSD involves evaporation of the volatile eluent to leave particles or nonvolatile solute, which scatter light. This means that it has the same eluent requirements as mass spectrometry. Parallel HPLC-UV-ELSD-MS is in use. LC-ELSD-MS is applied in high-throughput screening.

The sensitivity of LC-MS can be improved in an *SPE-LC-MS configuration* with on-line trace-enrichment. Actually, it can even be advantageous to replace the SPE-LC part (typically involving a 25-cm-long LC column and a run time of 30–50 min) by a single, short column (SSC). The SSC column (as short as 1 cm and packed under high pressure) then serves for both trace enrichment *and* separation, with a short run time. On-line and off-line SPE-LC-MS/MS allow high-throughput quantitative LC-MS/MS analyses.

A good starting point for LC-MS *method development* is a 50×2.1 mm Xterra MS C$_{18}$ column (3.5 µm); injection volume 5 µL; mobile phase: H$_2$O/ACN 90/10 to 10/90 v/v; 250 µL min^{-1} gradient; 50 °C. Generic protocols and smart automated (or zero) method development is coming within reach [559].

Many excellent reviews on the development, instrumentation and applications of LC-MS can be found in the literature [560–563]. Niessen [440] has recently reviewed interface technology and application of mass analysers in LC-MS. Column selection and operating conditions for LC-MS have been reviewed [564]. A guide to LC-MS has recently appeared [565]. Voress [535] has described electrospray instrumentation, Niessen [562] reviewed API, and others [566,567] have reviewed LC-PB-MS. For thermospray ionisation in MS, see refs [568,569]. Nielen and Buytenhuys [570] have discussed the potentials of LC-ESI-ToFMS and LC-MALDI-ToFMS. Miniaturisation (reduction of column i.d.) in LC-MS was recently critically evaluated [571]. LC-MS/MS was also reviewed [572]. Various books on LC-MS have appeared [164,433,434,573–575], some dealing specifically with selected ionisation modes, such as CF-FAB-MS [576] or API-MS [577].

Applications Major application areas of LC-MS technology are pharmaceuticals (up to (LC-PDA-ESP)$_4$-MS configurations), environmental protection (metal speciation), food industry, biotechnology, petrochemicals and consumer products. Many of the additives used in plastics production are insufficiently volatile to be analysed by GC-MS, and are more readily analysed by LC-MS. Similarly, some oligomers are not readily analysed by GC-MS, but are amenable to analysis by LC-MS. LC-MS has been in use for the characterisation and identification of additives in plastic materials since 1985 [578,579]. Yet, LC-MS has struggled to enter routine laboratories within the polymer industry, although LC-MS/MS has recently been used for the analysis of plastic additives [271,580–584].

Capillary HPLC-MS has been reported as a confirmatory tool for the analysis of synthetic dyes [585], but has not been considered as a general means for structural information (degradant identification, structural elucidation or unequivocal confirmation); positive identification of minor components (trace component MW, degradation products and by-products, structural information, thermolabile components); or identification of degradation components (MW even at 0.01 % level, simultaneous mass and retention time data, more specific and much higher resolution than PDA). Successful application of LC-MS for additive verification purposes relies heavily and depends greatly on the quality of a MS library. Meanwhile, MB, DLI, CF-FAB, and TSP interfaces belong to history [440].

Off-line coupling of LC and MS for additive analysis has received a fair amount of attention in the past. Combination of LC with FD-MS has been particularly attractive, since the two techniques are inherently compatible in terms of molecular-weight range and types of materials that can be analysed [494,586,587]. The off-line procedure requires that individual components be trapped in vials as they elute from the LC. The rubber antioxidant poly(2,2,4-trimethyl-1,2-dihydroquinoline) has been studied by *off-line LC-FDMS*: over 40 components in eight oligomeric series were identified [587,588]. Similarly, a series of rubber tackifying resins (alkylphenol/formaldehyde condensation products) was examined [589]. LC and FD-MS have also been used to characterise degradation (oxidation and ozonation) products of two common rubber *p*-phenylene-diamine antiozonants [590,591]. Off-line RPLC-FDMS with accurate mass measurement has already been used by Lattimer *et al.* [587] in 1979, for the analysis of oligomers of PS (up to 18 monomer units) and poly-TMDQ (up to eight units). Electron impact ionisation is quite limited in its ability to analyse higher-mass oligomers, because of extensive thermal decomposition and fragmentation in the ion source.

RPLC-UV-MB-MS (dual detection in series) with methane CI mass spectra (*m/z* 200 to 1200) was first used for identification of AOs and UVAs [523,578]; ng quantities of a nine-component additive mixture (BHT, Santowhite Powder, Topanol CA, Cyasorb UV531/5411, Irganox 1010/1076, Ionox 330, DSTDP) were detected as standards and extracts from PP; RSD of 1.6–3.7 % for the standard addition method with absorbance detection and about 4–10 % for MS detection were reported. DSTDP, which is non-absorbing at 280 nm, could not be detected with the UV detector; however, it was easily identified by LC-MS. Irganox 1010 was not detected by CI-MS, but rather on the basis of chromatographic retention. Similarly, acetonitrile extracts of automotive component mouldings (PP/(BHT, Irganox 1076, DOP, palmitic and stearic acid, octadecanol)) were examined. The procedure enables additives to be identified and determined in PP in amounts down to 2 mg (0.01 % in polymer) with an accuracy of ±5 to 10 %. LC-MB-MS with chemical ionisation was also used for the determination of thermally labile *N*-nitroso compounds [592]. Spray deposition on to a moving belt surface in LC-MS has been used for the determination of phenolic and other additives in polymer systems. The technique is unsuitable for trace analysis.

Direct liquid injection (DLI) has been used even less. Hirter *et al.* [579] have reported the early analysis of a synthetic antioxidant mixture (Irganox 1010/1076/1098) by means of μRPLC-DLI-QMS with CI. In early studies, the HPLC effluent was vaporised by laser radiation [593]; both EI and solvent-mediated CI spectra were obtained in the on-line mode from analytically difficult molecules. However, the instrumentation was complex; the sensitivity was not as good as that obtained by GC-MS; and thermal decomposition was observed with other compounds. This direct introduction approach with enrichment was used for the analysis of phthalates.

LC-TSP-MS without tandem mass capabilities has only met with limited success for additive analysis in most laboratories. Thermospray ionisation was especially applied between 1987 and 1992 in combination with LC-MS for a wide variety of compound classes, e.g. dyes (Fig. 7.31). Thermospray, particle-beam and electrospray LC-MS were used for the analysis of 14 commercial azo and diazo dyes [594]. No significant problems were met in the LC-TSP-MS analysis of neutral and basic azo dyes [594,595], at variance with that of thermolabile sulfonated azo dyes [596,597]. LC-TSP-MS^n has been used to elucidate the structure of Basic Red 14 [598]. The applications of LC-TSP-MS and LC-TSP-MS^n in dye analysis have been reviewed [599].

For analysis of surfactants, i.e. detection, identification and quantification, LC-TSP-MS and MS/MS are also qualified methods for substance-specific information [600–602]. A mixture of non-ionic surfactants, comprising nonylphenol ethoxylates [C_9H_{19}-(C_6H_4)-O-(CH_2-CH_2-O)$_m$-H], anionic surfactants and PEG, was

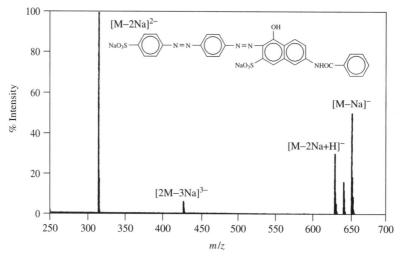

Figure 7.31 Negative-ion thermospray mass spectrum of the disulfonated azo dye Direct Red 81 (mobile phase contains 10 mmol L^{-1} ammonium acetate). After Niessen [568]. Reprinted from W. Niessen, in *Encyclopedia of Spectroscopy and Spectrometry* (J.C. Lindon, ed.), Academic Press, pp. 2353–2360, Copyright (2000), with permission from Elsevier

analysed by LC-TSP-MS [603]. LC-TSP-MS/MS is necessary for identification in cases with compounds of identical m/z ratio present in the sample. LC-TSP-MS spectra (with postcolumn addition of ammonium acetate solution) are dominated by molecular ions of the type $[M + NH_4]^+$ and $[M + H]^+$. This method can quantify concentrations of linear primary alcohol ethoxylates at levels of 25 to 100 ppb for total alcohol ethoxylates, and less than 3 ppb for individual homologues [601]. LC-TSP-MS was also used for the determination of butyltins in marine samples [604]. The applications of thermospray ionisation have been reviewed [605].

HPLC-CF-FAB-MS has been used in the analysis of cationic surfactants and commercial fabric softeners [606,607]. On-line NPLC-CF-FAB-MS (magnetic sector), with accurate mass and tandem mass spectrometry, has been used for the determination of ditallowdimethylammonium chloride (DTDMAC), a cationic softener used as the active ingredient in fabric softeners [606]. DTDMAC is a mixture of various combinations of tallow ($C_{14,16,18}$) alkyl homologues. The range of applications of LC-CF-FAB-MS is expanded by the ability to analyse the higher-molecular-weight cationic surfactants that exhibit poor solubility in polar solvents and, consequently, are not amenable to RPLC analysis. CF-FAB-MS is especially useful in the analysis of highly polar and labile compounds, where the electrospray interface may be a useful alternative.

Although the *electrospray interface* is a powerful tool, especially in the analysis of highly polar

and ionic compounds, ESI is less suitable for LC-MS analysis of polymer additives. LC-ESI-MS has been applied to organotin compounds [608,609], (sulfonated) azo dyes [610,611], and surfactants (including alkylphenol ethoxylates) [612–614], mainly in relation to environmental problems. μLC-ESI-QITMS was used for separation and detection of dibutyltin (DBT) and other organotin compounds leached from potable water PVC pipes [615]. LC-ESI-MS has been employed for the identification of DBP and DEHP in river sediments [616]. LC-MS techniques are widely used for analysis of dyes: a class of chemicals of widely differing structure and properties. LC-ESI-MS greatly enhances the ability to characterise increasingly complex, polar and nonvolatile dyestuffs. The sensitivity and specificity (by CID in the electrospray transport or by MS/MS) achieved with LC-ESI-MS enables monitoring for trace levels of dyes in mixtures – a necessity for production process control and environmental monitoring.

Electrospray has been successful for numerous azo dyes that are not ionic salts. Several anthraquinone dyes have been analysed by LC-ESI-MS [552]. Electrospray achieves the best sensitivity for compounds that are precharged in solution (e.g. ionic species or compounds that can be (de)protonated by pH adjustment). Consequently, LC-ESI-MS has focused on ionic dyes such as sulfonated azo dyes which have eluded analysis by particle-beam or thermospray LC-MS [594,617,618]. Techniques like LC-PB-MS and GC-MS, based on gasphase ionisation, are not suitable for nonvolatile components such as sulfonated azo dyes. LC-TSP-MS on

Table 7.65 Application of LC-MS techniques in characterisation of dye classes[a]

Dye class	LC-PB-MS	LC-TSP-MS	LC-ESI-MS
Sulfonated azo	−	±	+
Cationic	−	±	+
Azo (disperse)	±	+	+
Azo (solvent)	±	+	+
Anthraquinone	±	±	±

[a] In order of increasing success for MS analysis in terms of sensitivity and specificity (−, ±, +).
After Yinon *et al.* [599]. Reprinted from *Journal of Chromatography Library*, **59**, J. Yinon *et al.*, 187–218, Copyright (1996), with permission from Elsevier.

a single quadrupole system usually results in single-ion spectra, lacking structural information for compound confirmation. LC-ESI-MS probably offers the best combination of sensitivity and specificity (CID in the electrospray transport region). However, electrospray sensitivity is often reduced for nonpolar dyes that do not have sites of (de)protonation to form cations or anions for positive or negative ion detection. Applications of electrospray/ionspray LC-MS in dye analysis are several [618a] and have been reviewed [599]; see also Table 7.65. Microbore HPLC-ESI-MS was used for quantitative analysis of cationic surfactants (DTDMAC, DEEDMAC) using external standards. LS-ESI-ToFMS (resolution 5000, m/z range 70–10 000) is well suited to determine the accurate mass of an extracted surfactant. Mengerink [619] compared the performance of ESI and APCI interfaces in the analysis of some 32 additives (AOs, UVAs, plasticisers, whiteners). At variance with the APCI interface which created enough ions for an acceptable response, the electrospray interface showed low ionisation efficiency in the absence of an electrolyte in the mobile phase. LC-ESI-QITMS has also been employed for analysis of a mixture of neutral PEG molecules (mean MW of 200, 400, 600 and 1000) and charged aminopolyethylene glycols (mean MW 350 and 750) [620]. Pereira and Games [621] used LC-ESI-QITMS and LC-APCI-QITMS for the determination of sulfonamides. APCI gave better sensitivity than ESI, and detection limits were in the low ppb range. LC-ESI-MS in its various forms (SEC-ESI-ToFMS, GPEC®-ESI-ToFMS) is well suited for polymer analysis. GPEC®-MS was used for oligomer analysis (up to $n = 20$) of dipropoxylated bisphenol A/adipic acid with parallel UV and ESI-ToFMS detection [570].

A variety of 17 polyolefin additives (Tinuvin 328/P, Irganox 1010/1076/1330/3114, Irgafos 168, Isonox 129, Cyasorb UV531, MD 1024, BHT, BEHB, Lowinox 44B25, Succonox 16/18, Naugard 445, erucamide) were

separated and identified by means of RPLC-PDA-TMD-MS (dual detection), by matching against a user-built and an external library [622]. Peak purity was controlled. LC-UV-TMD-MS was also used for the analysis of erucamide from PE packaging film with UV/VIS for peak identification, verification of peak homogeneity and quantification [623]. Commercial dyes were subjected to LC-TMD-QQQ [624].

LC-PB-MS is a typical example of a niche application for polymer/additive analysis. Yinon *et al.* [599] have reviewed the applications of LC-PB-MS in dye analysis. Measurements of dyes with LC-TSP-MS were found to be more sensitive by two or three orders of magnitude than with LC-PB-MS. LC-PB-MS has been applied to the study of the transformation products of the antioxidant Irganox 1330 in food-contact polymers subjected to electron-beam irradiation [625,626]. Recently, Allen *et al.* [625,626] and others [623,627] have applied LC-PB-MS for additive analysis. Cosemans-Craeghs [628] has compared LC-PB-MS, LC-APCI-MS, LC-ESI-MS and LC-TMD-MS for polymer additive analysis of seven test solutions containing 36 different additives. The mean detection limit of LC-PB-MS is about 1 ng, but is compound class dependent. LC-PB-MS has been applied to azo dyes [624]. The ionisation technique used in particle-beam interfaces (EI) generally does not fulfil the criterion of producing predominantly the molecular ions that are required for oligomer and polymer applications. Nevertheless, in experiments on polymer extracts, it was possible to identify both additives and oligomers. LC-PB-MS has been used for the analysis of styrene oligomers (up to 18-mer, MW 1930 Da) [527]. Particle-beam approaches are more effective than bulk vaporisation using solids insertion probe sample introduction for EIMS analysis (limit: 11-mer, MW 1202 Da).

Various LC-PB-MS and LC-APCI-MS comparisons have been reported on polymer additive extracts [540, 563,629,630]. The complementary character of the EI and APCI modes was confirmed. Yu *et al.* [630] compared LC-PB-MS and RPLC-UV-APCI-MS for detection and identification of unknown additives (in the 252 to 696 Da range) in an acetonitrile extract from PP (containing Irganox 1076, Naugard XL-1 and a degradation product, NC-4, 3-(3,5-di-*t*-butyl-4-hydroxyphenyl) propanoic acid, 7,9-di-*t*-butyl-1-oxaspiro [4,5] deca-6,9-diene-2,8-dione and octadecanol-1). Comparison was based on EI data (identification of chemical structure), APCI⁻ (MW information; CID spectrum with limited fragmentation) and PDA (210 nm). The components were identified by EI and confirmed by APCI⁻ (with better sensitivity and linearity); MS and PDA showed

Table 7.66 Polymer additives evaluated by means of LC-PB-MS and LC-APCI-MS

Additive[a]	Chemical formula	Molecular weight
DGEBA	$C_{21}H_{24}O_4$	340.2
Ethanox 720	$C_{23}H_{32}O_2$	340.3
Plastanox 2246	$C_{23}H_{32}O_2$	340.3
Tinuvin 312	$C_{18}H_{20}N_2O_3$	312.2
Uvistat 247	$C_{20}H_{24}O_3$	312.2
Nonylphenol	$C_{15}H_{24}O$	220.2
DBPC	$C_{15}H_{24}O$	220.2
Topanol L	$C_{22}H_{30}O_2S$	358.2
Santonox R	$C_{22}H_{30}O_2S$	358.2
Triphenyl phosphate	$C_{18}H_{15}O_4P$	326.1
Cyasorb UV531	$C_{21}H_{26}O_3$	326.2
Cyasorb UV3638	$C_{22}H_{12}N_2O_4$	368.1
Tinuvin 315	$C_{22}H_{28}N_2O_3$	368.2
Plastanox 425	$C_{25}H_{36}O_2$	368.3
Ultranox 626 diphosphate	$C_{33}H_{50}O_8P_2$	636.3
Irganox 1098	$C_{40}H_{64}N_2O_4$	636.5

[a] Structural isomer pairs: Ethanox 720 and Plastanox 2246; Topanol L and Santonox R.

comparable chromatograms. The hydrocarbon component was only detected by EI. APCI⁻ exhibited better sensitivity and linearity than EI. In using LC-APCI-MS it is useful to know where one is looking for. Cosemans-Craeghs [540] has compared the identification power of LC-PB-MS (Thermabeam™) and LC-APCI-MS for 31 additives (some of equal molecular mass, see Table 7.66), distributed in six test samples, and of various polymer extracts. LC-PB-MS allows ready identification of additives with the same molecular weight and structural isomers. APCI-MS, operated at medium orifice voltage (50 V), gives (de)protonated molecules in the positive/negative ion mode, radical molecular ions as well as fragment-ions, and also permits differentiation of additives of identical molecular weight (even structural isomers). As LC-APCI-MS covers a wider mass range (up to 1400–1500 Da) with positive and negative spectra, the technique is preferred over LC-PB-MS. However, for identification of unknown components by means of LC-APCI-MS, either a spectral library or a set of fragmentation rules must be built up. LC-PB-MS is not very suitable for extremely volatile samples, which tend to be pumped away along with the solvent, or for thermally labile compounds, because of the source temperature (200 to 250 °C). Nevertheless, LC-PB-MS in combination with PDA, most recently in LC-DAD-MS^n configuration, is used extensively for polymer/additive analysis both by industrial additive producers, the polymer industry and end users [623,625–627]. Out of the 150 LC-PB-MS units worldwide, some 40 % also deal with additive problems.

Analysis of potential endocrine disrupters, such as alkylphenols (degradation products of non-ionic surfactants), diphenolalkanes (cumylphenol, bisphenol-A, bisphenol-F) and benzophenones (from the paper industry) with RPLC-SSI-QITMS in the negative ion mode has been described [631].

Analysis of plastic monomers, (PET) oligomers and additives in food contact materials, and measurement of migration of foods and food simulants using *LC-APCI-MS* (and ICP-MS and SEC-FTIR), has been reported by Jickells [632]. RPLC-APCI⁺-MS has been used for analysis of the migration potential of the oligomeric fraction in virgin and recycled PET plastics used for food and beverage packaging [633]. Quantitative analysis (sub ppm) is possible by means of LC-API-MS^n (e.g. 10 ppb octylphosphonic acid in food). LC-APCI-MS has also been used for the characterisation of natural antioxidants (from rosemary, sage, and barley) [634]. LC-APCI-MS may cause in-source oxidation of several compounds, such as Santonox R, Ultranox 626, DGEBA and Irgafos 168. APCI spectra of Irganox-type compounds reveal loss of *t*-butyl groups by a retro-Friedel–Crafts catalysed reaction [540], as observed when positive and negative mode spectra are examined, as in the case of Irganox 1010 and Irganox 1076.

The degradation products of Irganox 1010, (MA₄, *cf.* Scheme 7.9) from extrusion of PE in mild processing conditions were characterised using LC-APCI-MS [635]. Various decomposition reactions of Irganox 1010 may be envisaged:

(i) Hydrolysis of ester groups:

$$MA_4 \longrightarrow MA_3OH \longrightarrow MA_2(OH)_2 \longrightarrow$$
$$MA(OH)_3$$

(ii) Dealkylation (loss of *t*-butyl groups)
(iii) Loss of di-*t*-butylphenol.

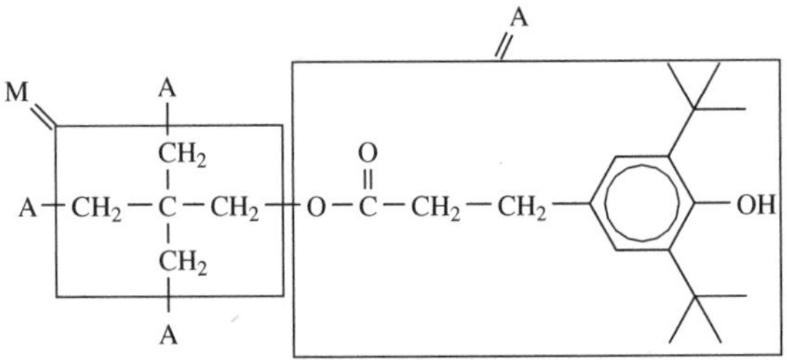

Scheme 7.9   Structure of Irganox 1010, M(A)₄

Table 7.67   Hydrolysis of Irganox 1010[a]

$n = t\text{-Bu}$ loss	$[MA_4-H]^-$	$[MA_3OH-H]^-$	$[MA_2(OH)_2-H]^-$	$[MA(OH)_3-H]^-$
0	**1175**	**915**	**655**	393
1	**1119**	**859**	**599**	337
2	**1063**	**803**	543	281
3	**1007**	**747**	487	
4	**951**	**691**	431	
5	**895**	635		
6	**839**	579		
7	783			
8	727			

[a] Theoretical and experimentally observed (bold) m/z values.

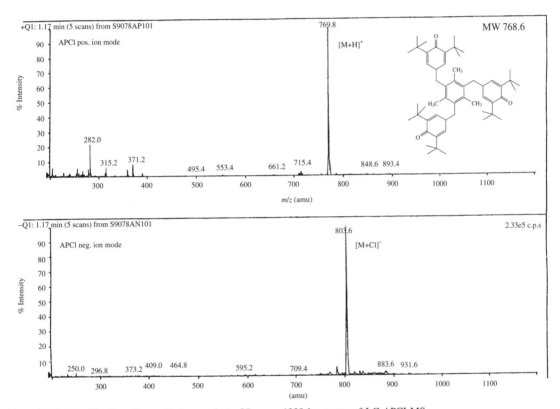

Figure 7.32   Identification of an oxidation product of Irganox 1330 by means of LC-APCI-MS

LC-APCI-MS confirms these degradation pathways. Table 7.67 shows that one or two di-t-butyl-4-hydroxy-phenylpropionate side-chains are lost by hydrolysis of the ester groups.

Figure 7.32 shows the identification of an oxidation product of Irganox 1330 by means of APCI-MS. LC-APCI-MS/MS (high-resolution sector field-ion trap hybrid) has also been used for the analysis (elemental composition and structure) of Irganox PS 802 [636].

The retention behaviour of oligomers in RPLC/-NPLC-APCI-MS (35 to 1500 Da) has been described [637]. In isocratic chromatography only a limited number n of oligomers can be separated, as the retention usually increases excessively at high n, so that gradient elution is necessary for successful separation of samples with broader molar mass distribution. RPLC-APCI-UV/QITMS was used for the analysis of accelerators (CBS, MBT, MBTS, BT) extracted

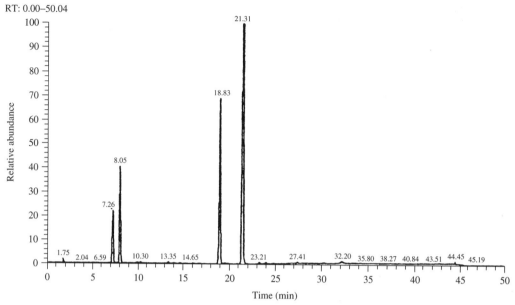

Figure 7.33   RPLC-UV chromatogram at 275 nm of a solution of BT ($t = 7.26$ min), MBT ($t = 8.05$ min), CBS ($t = 18.98$ min) and MBTS ($t = 21.31$ min) in dioxane/acetonitrile. After Mosch and Giese [638]. Reproduced by permission of the authors

Table 7.68   Results of quantification by LC-APCI$^+$-MS

Parameter	NC-4	Naugard-XL	Octadecanol-1	Irganox 1076
m/z monitored (Da)	415.2	473.2	294.3	530.5
Detection limit (ppb)	30.4	14.4	27.2	12.8
Linear range (ppb)	30.4–7600	14.4–3600	27.2–6800	12.8–3200
% in original sample (w/w)	0.07	0.04	0.4	0.06
Correlation coefficient (R^2)	0.997	0.998	0.999	0.979

(dioxane/acetonitrile) from rubbers [638]. In the experimental conditions chosen, LC-MS detected only two accelerators, as opposed to RPLC-UV (Fig. 7.33). LC-API-QITMS has also been used for the determination of alkylphenols down to ng mL^{-1} levels [639]. Some typical quantitative LC-APCI$^+$-MS results [630] are shown in Table 7.68. LC-APCI$^+$-MS was also used for the quantification of tributyltin [608]. DFI-APCI-MS analysis of DEP in SIR mode ($m/z = 223$) was reported [640].

There is no single LC-MS interface that is ideally suited for all compounds of interest to analytical chemists. It is evident that LC-APCI-MS and LC-PB-MS are currently the LC-MS methods most frequently used for polymer/additive analysis. The two techniques are compared in Table 7.69. When PB *and* API interfacing techniques are used, much more structural information can be obtained, and unambiguous identification

of unknowns – i.e. analysis of non-target next to target analytes – becomes possible. This was recognised previously [641]. Although the complementarity of LC-PB-MS (EI data; qualitative analysis) and LC-API-MS (APCI data; quantitative analysis) has been established satisfactorily for the case of polymer/additive analysis [630], few industrial users beyond Waters/Micromass are likely to benefit from this instrumental synergism (multi-API-MS users [552] would also benefit). As LC-APCI-MS yields information from both the positive and negative ionisation mode, covers a greater mass range, and is very well suited for quantitative analysis, it has an edge over LC-PB-MS. However, the technique appears to be unduly selective (i.e. highly dependent upon the operating conditions). In scan mode it is desirable to have adequate sensitivity for *all* additives, as for LC-PB-MS. By combining EI and APCI, positive chemical

Table 7.69 Comparison of LC-PB-MS and LC-APCI-MS in polymer/additive analysis

Characteristic	PB	APCI
Spectral features	Molecular mass data; specific fragmentation patterns	Relatively simple spectra: (de)protonated species, radical molecular ions, fragment ions; pos./neg. mode
Databases	Commercially available (EI)	Only user-generated (instrument dependent)
Mass range	500–800 Da[a]	1 400–1 500 Da[a]
Sensitivity	PB < APCI	APCI > PB
General fragmentation rules	Applicable	Applicable
Specific fragmentation rules	Unknown	Unknown
Analytical result	Compound identification	Compound verification; quantitation

[a] Analyte dependent.

identification with sensitive quantitation of an unknown compound can be obtained.

Whereas the components of (known) test mixtures can be attributed on the basis of APCI$^{+/-}$ spectra, it is quite doubtful that this is equally feasible for unknown (real-life) extracts. Data acquisition conditions of LC-APCI-MS need to be optimised for existing universal LC separation protocols. User-specific databases of reference spectra need to be generated, and knowledge about the fragmentation rules of APCI-MS needs to be developed for the identification of unknown additives in polymers. Method development requires validation by comparison with established analytical tools. Extension to a quantitative method appears feasible. Despite the current wide spread of LC-API-MS equipment, relatively few industrial users, such as ICI, Sumitomo, Ford, GE, Solvay and DSM, appear to be somehow committed to this technique for (routine) polymer/additive analysis.

LC-tandem MS was recently used for polymer/additive characterisation. In cases of soft ionisation processes (e.g. ESI, APCI, etc.), MS/MS is often necessary to confirm the ionic species. QITMS has the potential to improve the detection limits for organotin analysis compared to QMS. HPLC-UV and LC-API-MS/MS have been employed for the characterisation of the products of photodegradation of benzotriazole-based UV absorbers (Tinuvin P/328/900) under mild conditions [642]. Among the photoproducts identified by tandem MS were: benzotriazole, phenolic fragments and solvent–UVA adducts. ICI uses LC-API-MS^n for routine analyses of high-MW polymer additives [581–583]. Recently, Kashiwagi et al. [563] have reviewed LC-MS developments in the application of polymer/additive analysis, and compared four LC-MS interfaces.

7.3.3.3 LC–NMR Spectroscopy

Principles and Characteristics In most cases, off-line LC-NMR techniques are laborious and time-consuming, including working up the samples for redissolving in deuterated solvents, and subsequent recording of NMR spectra. On-line separation has many advantages over conventional off-line separation. Although LC-NMR was introduced in the early 1980s, implementation of *on-line LC-NMR* has been hampered by the low sensitivity of NMR at the nL scale; the limited dynamic range; the need for deuterated solvents; lack of suitable techniques for suppression of the solvent signal; and cost. Meanwhile, many of the major technical obstacles have been overcome. Hyphenated experiments now enable simultaneous efficient separation and NMR detection with structural identification in a closed system [643–645].

LC-NMR hyphenation consists of a liquid chromatograph (autosampler, pump, column and oven) and a classical HPLC detector. The flow of the detector is brought via an interface to the flow-cell NMR probe. Using commercial NMR flow-cells with volumes between 40 and 180 μL, in connection with microbore columns or packed capillaries, complete spectra have been provided from 1 nmol of sample. These micro-cells allow expensive deuterated solvents to be used, and thus eliminate solvent interference without excessive cost. The HPLC eluent can be split in order to allow simultaneous MS detection.

LC-NMR can be operated continuously ('on-flow') or discontinuously ('stopped-flow'). The optimum flow-rate in *continuous-flow NMR* is a compromise between best resolution and sensitivity. The sensitivity in NMR measurements has been increased significantly by:

(i) an increase in magnetic field strengths (600 to 800 MHz);
(ii) use of filtering techniques;
(iii) use of NMR micro-cells (60 μL volume); and
(iv) increase in flow-rate.

These developments have made it much easier to obtain well-resolved spectra in the on-flow mode [646]. Most current LC-NMR methods use low-flow-rates/long analysis times to increase signal-to-noise ratios, but do so at the expense of chromatographic performance. The shortcomings of continuous-flow NMR are poor sensitivity due to limited measuring time with each individual analyte, the flow-rate dependency of the NMR line width, and the relatively large required analyte concentrations (high mM level). The on-flow detection limit in capillary HPLC-NMR (column diameter <0.5 mm; 500 MHz) is in the low μmolar range. When >30 μg amounts of an analyte are present, on-line NMR acquisition becomes practical. Further, substantial band broadening and loss of chromatographic resolution is observed in NMR flow-through cells. In *stopped-flow mode*, the residence time of a component in the magnetic field increases, and lower detection levels (down to ng level) are attainable [647]. In stopped-flow mode, decisions based on other analytical methods (mostly UV/DAD) are required. However, many compounds have no UV absorbance, structure specificity is low, and UV cannot be used quantitatively (a large UV signal can be a small amount, and vice versa). Other available methods are RI (not possible with gradient elution; low sensitivity, low structure specificity), radioactivity detection (needs radioactive labelling; high specificity), ECD or ELSD (low structure specificity) or FTIR (low sensitivity, small observable window).

Eluents suitable for LC-NMR must be clean from an NMR standpoint; minimal UV absorbance is of less-critical importance. Eluents are generally selected to have as few NMR signals as possible, since it will usually be desirable to suppress these signals for optimal results. Therefore, CH_3CN is best suited for RP-chromatography. In HPLC-^{1}H NMR coupling, the use of deuterated compounds (except D_2O) is generally restricted in view of cost. Use of the proton-free eluent CCl_4 confines the separation to the adsorption mode. This problem can be overcome in SFC-NMR. RPLC-NMR coupling is limited by suppression of solvent signals. For RP chromatography, D_2O can be used in place of H_2O, and also provides a deuterium lock. Likewise, for NPLC or SEC $CDCl_3$ can be used instead of $CHCl_3$. A deuterium level of about 1 % is sufficient for most purposes. For LC-NMR using nonstandard solvents, high-quality solvent suppression is an important factor. Suppression of multiple eluent signals can be performed, e.g. by

(i) presaturation;
(ii) non-excitation;
(iii) pulsed field gradient (PFG)-based methods;
(iv) filtering methods; or
(v) post-acquisition data processing.

The requirements of on-flow NMR allow only the use of a few such *solvent suppression methods*. Ideally, the solvent suppression method should eliminate solvent resonances without disturbing solute signals; apply to multiple resonances; allow for changing frequencies of the solvent signals in gradient LC systems; and yield a uniform signal strength to allow integration [648]. Efficient solvent suppression in the continuous-flow and in the stopped-flow mode can be performed by application of a NOESY-type presaturation scheme. In addition, the observable regions in the spectrum can be further 'cleaned' by ^{13}C-decoupling to remove the satellites in the ^{1}H spectrum of the eluent. The problem with solvent suppression is sensitivity (400–500 ng). Although highly efficient solvent-suppression methods are now available, and deuterated solvents are no longer strictly mandatory for HPLC-NMR, it is still convenient to prepare mobile phases from acetonitrile/D_2O, methanol-d_4/D_2O, or acetonitrile-d_3; superheated water is also a suitable eluent for RPLC [649].

Table 7.70 lists some of the *main features* of LC-NMR hyphenation. At 11.7 T (500 MHz for ^{1}H) a 4 mm LC-NMR flow-probehead readily provides a detection limit of ca. 5 μg for on-flow (1 mL min^{-1}) and 150 ng for stopped-flow (in 3 h) for a typical 350-Da substance. Miniaturisation and hyphenation of NMR to various capillary-based microanalytical systems (LC, CZE) was described [650].

On-line coupling of separation techniques to NMR has recently been reviewed [459,651–653], and solvent suppression methods in NMR spectroscopy in particular [654].

Table 7.70 Main on-line LC-NMR features

Advantages

- Sample requirements (few μg)
- Detection limits in μg–ng range
- Suitable for all common NMR nuclei (^{1}H, ^{19}F, ^{31}P, ^{13}C, etc.)
- Multiple solvent suppression
- No need for individual component isolation
- LC-peak autodetection (UV, MS)
- On-flow NMR detection of isocratic and gradient LC runs
- Stopped-flow for longer 1D and 2D measurements
- RP-, NP-, and special phase chromatography
- Efficient separation with unequivocal structural peak assignment
- Rapidly developing technology

Disadvantages

- Interfering solvents
- Impurities of HPLC-grade solvents
- Not yet a widely accessible routine method (emerging technique)
- Expensive equipment

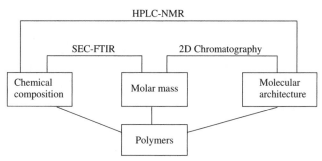

Scheme 7.10 Polymer analysis by hyphenated techniques

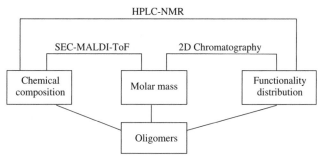

Scheme 7.11 Oligomer analysis by hyphenated techniques

Applications ⋅ Off-line RPLC-^{1}H NMR was used for the study of trimethylolpropane-triacrylate (TMPTA), a cross-linking agent for polymers and adhesives [494]. Not surprisingly, on-line LC-NMR (and LC-NMR-MS) techniques find applications mostly for studies of pharmaceuticals, drug metabolism and food chemistry. Most of the applications reported were performed on ^{1}H and ^{19}F, but ^{31}P and ^{13}C are also useful nuclei for HPLC-NMR. On-line LC-NMR coupling provides also excellent structural information on complex polymer systems (Schemes 7.10 and 7.11). Running a single on-line experiment allows information on the number of components, their chemical composition and substitution pattern to be obtained. LC-NMR applied to polymer analysis yields degree of branching, configuration, end-groups, and sometimes also the absolute molecular mass.

In cases where 2D NMR experiments are insufficient for a complete analysis of anionic surfactant mixtures, LC-NMR may provide better information. Characterisation of fatty alcohol ethoxylate (FAE) based *oligomeric surfactants* by on-line 2D (GCOSY, TOCSY and Homo 2DJ) stopped-flow HPLC-^{1}H NMR has been described [655,656]. The analysis of a typical mixture comprising three components (PEG and PEOs with different end-groups) is shown in Figure 7.34. In this representation, the ^{1}H NMR frequency domain is in the

horizontal dimension, and the chromatographic separation time is in the vertical dimension. The results are summarised in Table 7.71.

From the NMR spectra extracted from the contour plot at the corresponding chromatographic peak maxima it appears that the sample components are polyethylene oxides with different end-groups. Analysis of the aliphatic and aromatic parts of the spectrum gives evidence about the chemical structure of the end-groups, including their linearity or branching. LC-NMR and LC-MS were also used for characterisation of the chemical heterogeneity in a polymeric alkyl terminated PEO surfactant [657]. On-line RPLC-^{1}H NMR has been applied for the analysis of oligostyrenes [658].

LC-^{1}H NMR has been used as a purity determination method for phthalic acid esters, as illustrated for DBP [659].

Thermal degradation of Irganox 1076 in air was studied by means of HPLC-UV/VIS and by preparative HPLC-NMR. At 180 °C cinnamate and dimeric oxidation products are formed, and at 250 °C de-alkylation products are observed [660]. On-line LC-NMR hardly covers a real need in polymer/additive analysis, as the off-line option is mostly perfectly adequate for that purpose.

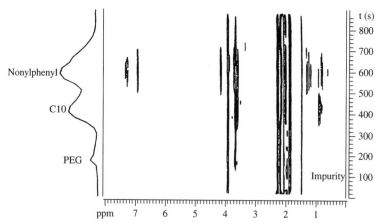

Figure 7.34 Contour plot of chemical shift vs. retention time, and reconstructed chromatogram of an FAE-based surfactant (Table 7.71). After Schlotterbeck *et al.* [655]. Reproduced from G. Schlotterbeck *et al.*, *Polym. Bull.*, **38**, 673–679 (1997), by permission of Springer-Verlag, Copyright (1997)

Table 7.71 α-Hydroxy-ω-alkoxy(aryloxy) FAE structural information obtained by on-line LC-NMR[a]

Components	Alkoxy(aryloxy) end-group
PEG	–
C_{10}-PEO	$-(CH_2)_5-CH(CH_3)-CH(CH_3)_2$
Nonylphenyl-PEO	$-C_6H_4-C_9H_{19}$ isomeric mixture

[a] FAE, fatty alcohol ethoxylate; PEG, poly(ethylene glycol); PEO, α-hydroxy-ω-alkoxy(aryloxy) polyethylene oxide.
After Schlotterbeck *et al.* [655]. Reproduced from G. Schlotterbeck *et al.*, *Polym. Bull.*, **38**, 673–679 (1997), by permission of Springer-Verlag, Copyright (1997).

7.3.3.4 Hypernated LC Techniques

Principles and Characteristics Traditional analytical approaches include off-line characterisation of isolated components, and the use of several chromatographic separations, each optimised for a specific spectroscopic detector. Neither LC-NMR nor LC-MS alone can always provide complete structure determinations. For example, MS may fail in assigning an unequivocal structure for positional isomers of substituents on an aromatic ring, whereas NMR is silent for structural moieties lacking NMR resonances. Often both techniques are needed.

Multiple hyphenation ('hypernation') provides comprehensive spectroscopic information from a single separation. The first doubly hyphenated HPLC-NMR-MS appeared in 1995 [661], and its value is now accepted; meanwhile fully integrated *on-line LC-NMR-MS* and MS^n systems (QMS, QITMS) are commercially available. On-line LC-NMR-MS coupling is by no means trivial. For example, the sensitivity of NMR is limited, while MS is incompatible with non-volatile buffers. The choice of a compatible solvent is restricted. Deuterated solvents, attractive for LC-NMR, seriously complicate interpretation of the MS spectra obtained. Moreover, the HPLC and MS instruments cannot be positioned too close to the NMR magnet. Finally, accurate determination of delay times between the NMR and MS signals is needed to avoid incorrect signal correlations. Although expensive, HPLC-NMR-MS may be the best way to unequivocally characterise complex mixtures.

NMR/MS has also been hyphenated to LC-SPE and LC-DAD modules. Sample enrichment and exchange of the HPLC mobile phase with an NMR suitable solvent is advantageous. LC-SPE-NMR/MS gains up to a factor of four in LC-NMR S/N for a single injection. No deuterated solvents are needed for separation and trapping. Optimisation of the separation procedure is less critical than for HPLC-UV.

Wilson *et al.* [662–665] have described various prototype systems for *total organic analysis devices*. It has proved technically feasible to obtain UV, IR, NMR and MS spectra (together with atomic composition based on accurate mass determination) following RPLC separation. The fully integrated approach offers the benefit that one chromatographic run is required, thus ensuring that all of the spectrometers observe the same separation. Such multiple hyphenations might favour the analysis of complex mixtures for both confirmation of identity and structure determination (should this represent a cost-effective approach). Table 7.72 illustrates the main features of on-flow multiple LC hyphenation.

Scheme 7.12 shows some tested configurations of integrated total spectroscopy laboratories. Couplings may be parallel or in series; other examples are LC-NMR-MS and GC-FTIR-MS. Wilson [664] has

Table 7.72 Main characteristics of prototype on-flow LC hypernation

Advantages

- One chromatographic run
- Fully integrated system
- Rapid and efficient chromatographic and full spectroscopic data collection on major components in mixtures
- High potential analytical power
- Management appeal

Disadvantages

- Overcomplex multiple splitting of eluates
- Limitations on eluent choice (minimisation of solvent interferences)
- Different spectroscopic time-scales
- Need for optimisation (solvent compatibility, flow-rates, peak broadening, etc.)
- Complex operation
- Downgrading of sensitivity
- Robustness, reliability, vulnerability
- Capital intensive

described the use of a *HPLC-UV-NMR/MS* system with parallel operation of NMR and MS. Placing the mass analyser after the NMR spectrometer (in-line) results in pressurisation of the NMR flow probe. Relatively few chromatographic compromises are needed to implement successful HPLC-NMR and HPLC-NMR-MS experiments. RPLC with conventional columns and flow-rates of $1 \, mL \, min^{-1}$ are used routinely. The UV detector, which can be either a variable wavelength instrument or a diode array spectrometer is a very convenient means of monitoring the separation. On-line LC-UV-NMR/QITMS with split flow (NMR:MS = 95:5) is an advanced spectroscopy laboratorium under full automation, with LC-peak selection, positive and negative ionisation, automatic MS fragmentation, 2D-NMR, and multiple NMR solvent suppression. Various mass spectrometers (QMS, QQQ, ITMS, ToF-MS) with different ionisation modes (EI, API) have been coupled. MS data evaluation on one LC peak comprises detection

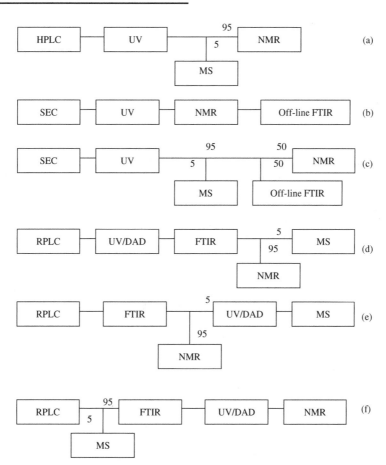

Scheme 7.12 Layout of several tested prototype HPLC hypernations. Splitting ratios are indicated

Table 7.73   Main features of HPLC-UV-NMR/MS

Advantages

- Accommodates RPLC with few chromatographic compromises
- Optimisation of the separation procedure not critical
- High structure specificity
- High sensitivity
- Fragmentation information
- Complete structure determinations
- Complex mixture analysis

Disadvantages

- Complex and capital-intensive instrumentation
- Constraints on compatible solvent composition
- Solvent-dependent ionisation
- Deuterium exchange

of molecular ions, fragmentations (MS/MS), isolation and further fragmentations (MS/MS/MS, etc.), and a check for peak consistency (overlap) [170]. Prototype versions including ToF-MS provide the ability to determine molecular formulae via accurate mass determination [663]. Table 7.73 shows the main benefits and limitations of LC-UV-NMR/MS.

FTIR in multiply hyphenated systems may be either off-line (with on-line collection of peaks) [666,667] or directly on-line [668,669]. Off-line techniques may be essential for minor components in a mixture, where long analysis times are required for FT-based techniques (NMR, IR), or where careful optimisation of the response is needed. In an early study a prototype configuration comprised SEC, a triple quadrupole mass spectrometer, off-line evaporative FTIR with splitting after UV detection; see Scheme 7.12c [667]. Off-line IR spectroscopy (LC Transform™) provides good-quality spectra with no interferences from the mobile phase and the potential for very high sensitivity. Advanced approaches consist of an HPLC system incorporating a UV diode array, FTIR (using an ATR flow-cell to obtain on-flow IR spectra), NMR and ToF-MS.

Where the use of multiple spectroscopic analysis on a single HPLC separation is an advantage, the benefit of using the simplest possible mobile phase for separations is manifest. While selecting compatible *solvent systems* for NMR and MS is sometimes complex, addition of IR (even off-line) places even more constraints on solvent composition. For SEC-NMR-IR CDCl$_3$ is a suitable eluent [666]; for RPLC-FTIR-UV-NMR-MS D$_2$O-CD$_3$CN is recommended. Superheated D$_2$O has been proposed as the mobile phase [670].

A major difficulty is also the difference in the sensitivities of the detectors, with NMR and IR being of relatively low sensitivity as compared to UV/DAD and

MS. Consequently, the layout of Scheme 7.12e has been proposed as an improvement over Scheme 7.12d [671]. Some 50–100 µg of each analyte on the column is needed to obtain identifiable spectra (structure dependent). Sensitivity can be improved to the low µg or even ng range in miniaturised flow NMR systems.

Hyphenated techniques, in which chromatographic separations are allied with powerful spectroscopic techniques, are gradually eliminating the need for isolating unknowns in a pure form before identification. The possibilities of multiple hyphenation are by no means exhausted, and other types of spectrometers (e.g. fluorescence) could also be included. Although the concatenation of chromatography to an array of spectroscopic detectors does not present overwhelming difficulties, it is still not clear whether the benefits of hypernated systems outweigh the costs of assembling such an array of instruments.

Wilson [664,672] has reviewed multiple hyphenation of liquid chromatography.

Applications   Ideally, multiply hyphenated systems should be assembled rapidly in response to real need. Access to these means is restricted to a few laboratories only. Multiple LC hyphenations have been used to analyse test mixtures of polymer additives; see Table 7.74. The relative ease with which SEC-UV using CDCl$_3$ as a solvent can be coupled to on-line ^{1}H NMR and an 'in series' off-line FTIR (Scheme 7.12b), has been shown for a mixture of polymer additives (BHT, Irganox 1076, DIOP) [666]. Figure 7.35 shows representative spectra for on-flow NMR and MS and off-line FTIR of 2,6-di-*t*-butyl-4-methoxyphenol.

Table 7.74   LC hypernations and applications to additive test mixtures

Multiple hyphenation	Lay-out	Testing	Reference(s)
HPLC-UV-NMR-MS	Scheme 7.12a	–	[664,672]
SEC-NMR-IR[a]	Scheme 7.12b	[b]	[666]
SEC-UV-MS-NMR-IR[a]	Scheme 7.12c	[b]	[664,667]
FIA-IR-UV(DAD)-NMR-ToFMS	–	–	[663]
RPLC-IR-UV(DAD)-NMR-ToFMS	Scheme 7.12d,e	–	[668,671]
RPLC-IR-UV(DAD)-NMR-MS	Scheme 7.12f	[c]	[665,673]

[a] Off-line.
[b] BHT, Irganox 1076, DIOP.
[c] Irganox 245, BHA, BHT, bisphenol A, Topanol CA.

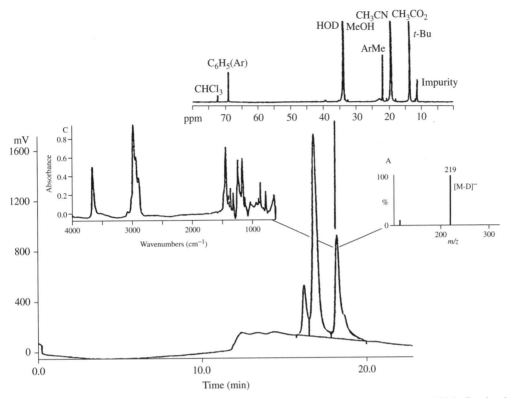

Figure 7.35 HPLC(SEC)-UV-MS-NMR-IR analysis applied to polymer additives. After Wilson [664]. Reprinted with permission from I.D. Wilson, *Analytical Chemistry*, **72**, 534–542A. Copyright (2000) American Chemical Society

A five-compound model test mixture of polymer additives (Irganox 245, BHA, BHT, bisphenol A, Topanol CA) in CD$_3$CN in amounts ranging from ca. 230 to 900 µg on the column were tested with the hypernated configuration of Scheme 7.12f, using an ATR-FTIR flow-cell and on-flow 500 MHz ^{1}H NMR [673]. In a case of considerable technical overkill, the system was also used to identify a suspected polymer additive as being BHT.

Future work needs to be directed toward applications to real samples rather than model test mixtures. The proof of the pudding is in the eating!

7.3.3.5 *Liquid Chromatography–Plasma Source Detection*

Principles and Characteristics Plasma source techniques are more widely used in connection with liquid chromatography than atomic absorption spectrometry (see Section 7.3.3). ICP is a natural complement to liquid chromatography, and HPLC-ICP procedures

have been quite widely adopted. Liquid flow-rates used in most LC techniques are of the order of 0.1 to 1 mL min^{-1}, which are comparable to conventional liquid flow-rates for direct aspiration of solutions into the ICP. In addition, the LC mobile phase elutes from the column at atmospheric pressure, which is ideally suitable for the ICP-MS sample introduction. The direct injection nebuliser (DIN) is particularly useful for interfacing an LC system with an ICP system [674,675]. Of course, the dead volumes introduced by nebuliser and spray chamber must not degrade the resolution gained by the LC separation.

ICP is intolerant to the solvents commonly used in LC development. Consequently, most *LC-ICP-AES* systems have been employed with ion-exchange columns, as this separation process largely involves aqueous mobile phases that are amenable to the ICP-AES instrument. Use of acetonitrile or THF in the mobile phase has usually evoked a change in interface design to accommodate the different solvents. The advantages of LC-ICP-AES include multi-element detection and the ability to obtain real-time chromatograms. LC-ICP-AES provides sub-ppm detection levels, and slightly better values for μLC-ICP-AES [676]. However, this is not the sensitivity necessary for trace-element speciation of real samples. The greater sensitivity of ICP-MS over ICP-AES can provide this ability. Suyani *et al.* [677] compared ICP-AES and ICP-MS as detection methods for organotin speciation, using cation-exchange and ion-pair HPLC. ICP-MS detection limits obtained for organotin compounds separated by ion-pair and ion-exchange HPLC ranged from 400 to 1000 pg. Absolute detection limits obtained by HPLC-ICP-MS are approximately three orders of magnitude lower (better) than those obtained with ICP-AES.

Plasma source mass spectrometry provides a means to analyse samples for various elements at sub-ng to pg levels by combining the ability of a plasma to efficiently ionise elements with the sensitivity of a mass spectrometer. The physical coupling of LC to ICP-MS is straightforward. Various liquid chromatographic modes (reversed phase, reversed phase ion-pair, micellar, ion-exchange, size-exclusion and chiral LC) have been interfaced to ICP-MS [679]. Microbore (1.0 mm i.d.) and minibore (2.0 mm i.d.) LC columns are considered to be most amenable for ICP-MS detection. Figure 7.36 shows a schematic of an LC-ICP-MS system. Resolved LC analyte fractions are introduced via a nebuliser (pneumatic, thermospray, ultrasonic) or direct injection into a high-temperature argon plasma, where they are decomposed, atomised and ionised. The resulting ions are transported through a sampling interface into a mass spectrometer. Coordination of the chromatographic retention time for each peak with elemental abundance in the MS enables calculation of species concentration. Spectral display routines allow overlay of UV and ICP-MS spectra for method development and elucidation.

Table 7.75 summarises the main features of *LC-ICP-MS*. Various mass analysers (QMS, ITMS, ToF-MS,

Table 7.75   Main features of LC-ICP-MS

Advantages

- Simplicity of interface
- Applicable to wide variety of LC modes
- Wide variety of stationary and mobile phases
- Compatibility of flow-rates
- Multi-element and multi-isotope capability
- Low levels of detection for many elements
- Rugged, versatile, routine
- Trace-elemental analysis with species selectivity
- No need for high preconcentration factors

Disadvantages

- Most expensive element-specific detector
- Isobaric and polyatomic interferences (10 000 mass resolution separates interferences)
- Need for certified reference materials (validation)

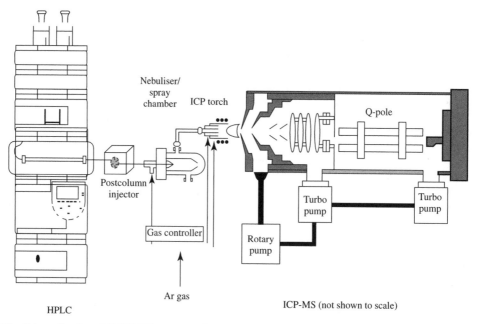

Figure 7.36   Schematic of an LC-ICP-MS system. After DeNicola and Caruso [678]. Reproduced by permission of Reed Elsevier

sector-field) have been used in connection with LC-ICP-MS; QMS is standard. In LC-ICP-MS coupling the loading of organic modifier must be regulated so that the plasma is not extinguished; oxygen may be added to the sample aerosol. LC-ICP-MS presents serious problems with carbon deposition on the sampler, and skimmer wear. μLC-ICP-MS and μCZE-ICP-MS have been described [675]. LC-ICP-MS and IC-ICP-MS studies have been reported by various authors [341]. The ICP-MS ion source is more commonly employed in LC-MS than in GC-MS.

Few authors have coupled HPLC to MIP-AES [680–682], despite the need for a nonmetal element-specific HPLC detector. Advantages of *HPLC-He MIP-MS* include improved detection for the halogens. The primary disadvantage is the inability of the plasma to tolerate aerosol introduction.

LC combined with ICP-MS and MIP-MS detectors has recently been reviewed [181,335,337,679,683–685].

Applications RPLC-ICP-AES was used for speciation and quantification of polar, low-MW *silanols* [686]. Cr(III, VI) can be determined by IC-ICP-OES at the ppt level. However, many HPLC applications for organometallic compounds demand the use of gradient elution, and a high flow-rate combined with gradient elution results in unstable plasma conditions.

HPLC-ICP-MS is a very useful tool in analysing metal-bearing compounds. In particular, HPLC-ICP-MS has been used repeatedly for separation and speciation of *organotin compounds* [632,677,687–690]. RPLC-ICP-MS has been employed extensively for the analysis of environmental and general analytical samples, notably for the speciation of butyltin compounds which are used extensively in PVC production [691]. The toxic effect of ions such as $Sn(C_4H_9)_3^+$ on marine species is a major environmental concern. Other chromatographic modes (cation exchange and ion-pair) have also been used widely for tin speciation with ICP-MS detection [677,692]. A comparison of GC-ICP-MS and HPLC-ICP-MS for the analysis of organotin compounds is available [693]. Organotin separations by HPLC offer the advantage that derivatisation is not required. Also, RPLC-ICP-MS analysis (Cr(III) and Cr(VI)) of azo dyes (Acid Blue 158/193) has been reported [694]. *Halogenated compounds* can be detected with HPLC-MIP-MS.

7.3.4 Hyphenated SEC Techniques

Both SEC-FTIR and SEC-MS are mainly used to provide information on the chemical composition of polymers. Although analysis of additives by means of *evaporative SEC-FTIR* has been described, it is not in use as a routine analytical tool for this purpose. *SEC-MS* in its various forms, such as μSEC-MALDI-ToFMS, SEC-ESI-FTMS or GPEC®-ESI-ToFMS, is useful for oligomer profiling. On-line *SEC-^{1}H NMR* (500 MHz) was applied to the analysis of PMMA polymers and chloral oligomers [695]. Multivariate curve resolution (MCR) can be applied to the analysis of SEC-NMR experiments where the separation is insufficient to resolve individual components in mixtures in both the elution and spectroscopic domain [696]. Applications of *SEC-ICP-MS* [697,698] and *SEC-ICP-AES* [686] for polymer/additive analysis are very limited. The latter technique was used for the determination of nonpolar, high-MW PDMS polymers.

7.3.4.1 Evaporative SEC–FTIR Coupling

Principles and Characteristics As there are considerable limitations in the use of SEC for low-MW compounds (limited resolution, insensitivity of detectors), it is not surprising that hyphenated SEC techniques have been called in to improve the separation and identification power of the technique. Use of FTIR as a detector for SEC provides the ability to simultaneously characterise the molecular weight and molecular weight distribution, and to identify and quantify IR-active functional groups. This enables characterisation of the concentration of IR-active functional groups as a function of molecular weight. Direct interfacing of SEC and FTIR has involved in-line flow-cells and solvent evaporation interfaces which can remove the solvent prior to FTIR spectral analysis. Low-volume flow-cells offer continuous monitoring of the eluates, with little loss of chromatographic resolution. However, application is limited, due to the strong infrared absorption inherent in SEC eluents throughout much of the mid-IR spectral range. Moreover, the concentrations present in SEC are usually inadequate for IR analysis. Until recently, the evaporative interface offered the only practical method for coupling FTIR to SEC; however, a commercial HTSEC-FTIR interface is now available, with temperature control on both flow-cell and line.

Both a pneumatic heated nozzle system [487] and an ultrasonic nozzle/vacuum system [699] have been described for removing the troublesome solvent in order to simplify IR analysis. The former system (LC Transform™) has been commercialised [700], and allows full use of the mid-IR spectral range by providing analyte films free from solvent interference. The evaporative interface removes the SEC mobile phase at the exit of the column, and deposits the eluting polymer as a continuous film or as a series of discrete films on a substrate. A SEC-FTIR sampling configuration of transmission-reflection on a rear-surface-aluminised germanium disc provides good signal-to-noise IR spectra which resemble conventional transmission spectra [701]. Transparent Ge discs allow FTIR identification in transmission mode. This increases the sensitivity significantly, so that low-level components, which go unnoticed in reflectance mode, are detected and identified.

Cheung *et al.* [702] have evaluated various solvent evaporative high-temperature SEC-FTIR interfaces. This detection approach was initially employed only for qualitative analysis, but is recently also being used quantitatively. For that purpose the polymer film quality generated by the interface is of critical importance (thickness effects). Table 7.76 lists the *main features* of evaporative SEC-FTIR for polymer analysis.

The technique offers advantages over alternative techniques (e.g. preparative SEC fractionation followed by IR). SEC-FTIR is much more suitable to quantitation of additives than SEC-MALDI-ToFMS. Combining SEC-FTIR with SEC-MALDI offers a more powerful tool than when either is used alone.

Applications Applications of SEC-FTIR include quantitative analysis of copolymers [701]; product deformulation of hot melt adhesives; characterisation of polymer compositional heterogeneity; analysis of complex mixtures of urethane oligomers; and eventually also the identification and quantitative analysis of polymer additives

Table 7.76 Assets of evaporative SEC-FTIR for polymer analysis

Advantages

- Simultaneous characterisation of MW, MWD and comonomer concentrations, as $f(\text{MW})$
- Speed, simplicity
- Adaptable to copolymer systems
- No matrix or solvent interference
- Good sensitivity (low ng)
- Potentially quantitative (calibration with ^{1}H NMR)
- Use of existing equipment

Disadvantages

- Operator-intensive
- Nontrivial uniform analyte deposition on collection disc
- Only mean chemical composition per polymer fraction
- Database required
- Limited multihyphenation capabilities
- Off-line

Table 7.77 Analysis of synthetic polymers using SEC-FTIR and SEC-MALDI-ToFMS

FTIR	MALDI-ToFMS
• Determination of component variation in copolymers	• Endgroup determination
• Measurement of components in mixtures	• Repeat units of homopolymers and copolymer (up to 20 kDa)
• Reverse engineering of competitor products	• Mass measurements to 10^6 Da
• Identification of additives and impurities	• Accurate determination of $M_{\text{w}}/M_{\text{n}}$ for low-MW (<20 kDa)

and impurities. Table 7.77 compares the application areas of SEC-FTIR and SEC-MALDI-ToFMS.

SEC-FTIR yields the average polymer structure as a function of molecular mass, but no information on the distribution of the chemical composition within a certain size fraction. SEC-FTIR is mainly used to provide information on MW, MWD, CCD, and functional groups for different applications and different materials, including polyolefins and polyolefin copolymers [703–705]. Quantitative methods have been developed [704]. Torabi *et al.* [705] have described a procedure for quantitative evaporative FTIR detection for the evaluation of polymer composition across the SEC chromatogram, involving a post-SEC treatment, internal calibration and PLS prediction applied to the second derivative of the absorbance spectrum.

SEC-FTIR is widely used for *adhesive analysis* [704,706]. Evaporative SEC-FTIR has been used for the study of adhesives composed of a high-MW acrylic fraction, a medium-MW PVAc fraction, SBR in the low-MW fraction and a trace amount of dilaurylthiodipropionate (DLTDP) in the very low-MW region [706]. The suitability of a moving ZnSe window for SEC-FTIR was demonstrated by analysing oligomers in a PS standard mixture [503].

Jickells [632] has considered the use of off- and on-line SEC-FTIR for quantitative analysis of *polymer additives* such as DOP. Meyer-Dulheuer [707] has described the separation of various additives (Irganox 1010, Irgastab BZ 530/CH 55, Tinuvin 326) from PVC, by means of SEC-FTIR with solvent evaporation. Absolute concentrations of 10 ng Irganox 1010 could easily be detected. Dissolutions of PVC/DEHP and PC/PETS (release agent) were analysed by SEC-FTIR using a thermospray/moving belt/DRIFT interface [483]. The detection limits of the method are in the 100 ng range, depending on the IR sensitivity and volatility of the solutes. This is not extremely sensitive.

SEC-FTIR with principal factor analysis has been used for the evaluation of 120 FTIR spectra of 120 SEC fractions of thermally and radiolytically aged multicomponent systems consisting of Estane 5703, a nitroplasticiser (bis-2,2-dinitropropylacetal/bis-2,2-dinitropropylformal 1:1) and Irganox 1010 [708].

7.3.4.2 On-line SEC–MS Techniques

Principles and Characteristics Problems connected with sample preparation, ionisation and detector efficiency can lead to errors in the quantitation of mass averages and MWD in the case of ESI-MS and MALDI-MS. Coupling of SEC with MS makes it possible to overcome these difficulties. SEC-MS has developed since the early 1990s. Two methods are currently outstanding: on-line SEC-ESI-MS (QMS or FTMS) and semi on-line SEC-MALDI-ToFMS [709].

It has been demonstrated that ESI-MS is compatible with the SEC conditions applied to the routine analysis of synthetic oligomers and polymers, and that the coupling offers specific benefits in terms of accurate molecular weight calibration [710]. The widely accepted desorption model for electrospray ionisation, which is an ultrasoft ionisation technique, relies on the existence of preformed ions in solution [711]. As most synthetic polymers contain no acidic or basic functional groups that can be utilised for ion formation through acid–base equilibria, cationisation is a preferred technique for producing gaseous ions from synthetic oligomers and polymers by desorption ionisation with a minimal number of fragment ions (e.g. in the case of *SEC-ESI-MS*). For the study of synthetic polymers by means of SEC-ESI-MS, cationisation is necessary to produce $M[Na^+]$, $M[2Na^+]$ and $M[3Na^+]$ pseudomolecular ions. SEC-ESI-MS can give accurate analysis of oligomers and low-MW polymers, but the molecular weight range is rather limited. Polymer species up to 3000 Da can be successfully analysed by SEC-ESI-MS. For higher-molecular-weight polymers, MALDI-ToFMS after off-line SEC with fraction collection is now the method of choice.

The *benefits* of SEC-ESI-MS are:

(i) separation of species according to molecular size;
(ii) rapid resolution of the MWD (narrow 'cuts');
(iii) use of a small fraction (ca. 1%) of the effluent;
(iv) allowance for alternative parallel detection (DRI, spectrophotometric); and
(v) calibration of SEC without the use of chemically unrelated external calibrants.

SEC-ESI-FTMS combines the size separation based technique of SEC with one of the most powerful mass spectrometric techniques of FTMS offering high mass accuracy (ppm), ultrahigh resolving power ($>10^6$) and the capability to perform tandem mass spectrometry. The technique enables generation of oligomer elution profiles, which can be used for accurate calibration of standard SEC data. Coupling of SEC to ESI-MS is further described in ref. [710].

Hyphenation of SEC and MALDI-ToFMS is very attractive for the characterisation of polymer samples. Since the preparation of MALDI-MS samples requires that analyte and matrix be co-crystallised in an intimate mixture, an off-line (or semi on-line) approach is mandatory. *SEC-MALDI-ToFMS* can be used to obtain an absolute calibration curve for SEC, or SEC is used as one of the sample preparation steps for MALDI [712]. In the former case, small fractions with low polydispersity (PD) are collected after SEC separation. Molecular-weight estimates provided by MALDI-MS agree with the values obtained by conventional techniques only in the case of samples with narrow MWD [713,714]. SEC-MALDI-MS coupling circumvents many of the problems encountered in SEC. It is then no longer necessary to use generic SEC standards for the universal calibration curve, and the SEC curve can be calibrated against absolute molecular weights specifically for a given polymeric material. In the SEC-MALDI-MS interface (e.g. LC Transform™) the SEC components are deposited on to a stainless steel foil which has been precoated with matrix material. The chromatogram is 'painted' on to the matrix surface in a reproducible manner, as a continuous track. On completion of the collection phase, the matrix foil is transferred to a nondedicated MALDI-MS. The procedure has considerable advantages over the manual dried drop technique, and is the only sampling approach capable of the systematic automation of SEC-MALDI experiments and of securing an even analyte distribution. The interface acts as a fraction collector, evaporator, mixer and spotter in one. In semi on-line SEC-MALDI-ToFMS, uniform preparation of the matrix is no longer an issue, and sample spectra are more representative of the entire sample. The polydispersity PD (M_w/M_n) problem has also disappeared: for every fraction collected by SEC, PD is approximately equal to one. Table 7.78 summarises the main characteristics of SEC-MALDI-ToFMS.

There is no need for perfect separation in the eluting peak; MALDI can improve the resolution of chromatography. Semi on-line SEC-MALDI-ToFMS makes allowance for the separation of polymer, oligomers and additives. Ion suppression has been noticed for

Table 7.78　Main features of SEC-MALDI-ToFMS

Advantages

- Semi on-line
- Uniform matrix deposits
- Simple and reproducible sample preparation
- No restrictions on polydispersity
- No need for perfect chromatographic separation
- Automated sample analysis

Disadvantages

- Cost of interface
- Some loss in sensitivity
- No quantitation

Table 7.79　Current limitations of SEC-MALDI-ToFMS

- Selection of the matrix
- Manual data acquisition
- Oligomers limited to 20 kDa
- No coupling of laser and some sample types (e.g. PO)
- Detector discrimination (favours low mass)
- Suppression of some ions

bromine (e.g. in FR materials). It should be stressed that although additives are observed by SEC-MALDI-ToFMS, the technique is most appropriate for the polymer part. The current limitations of the technique are outlined in Table 7.79. Combination of SEC-FTIR and SEC-MALDI-ToFMS is more powerful than either used alone.

MALDI does not replace chromatography (e.g. SEC) for molecular weight distributions. For PD >1.4 it is necessary to separate in order to reduce the polydispersity. Chromatography is needed to enhance MALDI.

Applications　SEC-ESI-FTMS has been applied to the characterisation of butylmethacrylate (BMA) and butylacrylate (BA) *oligomers* [715]. Detection of sodiated mixed polyester oligomers in excess of 3 kDa by means of SEC-ESI-FTMS has also been reported [716]. ESI-FTMS provides absolute molecular weight determination for polymers that have been separated by SEC. Simonsick and Prokai [717] have used SEC-ESI-MS as a tool for polymer characterisation. Cross-linkers, additives and stabilisers and coalescing solvents contained in complex waterborne acrylic–methacrylic *automotive coating formulations* were analysed in a single experiment and readily differentiated from the acrylic macromonomers (MW <3 kDa). SEC-ESI-MS has also found application in the separation of a commercial polydisperse mixture of the *surfactant* octylphenoxypoly(ethoxy)ethanol,

$C_8H_{17}C_6H_4O(CH_2CH_2O)_nCH_2CH_2OH$ [710]. Selected-ion chromatograms of sodiated species allowed for monitoring of the elution profiles of individual oligomers, and were used for calibration of SEC. GPEC®-APCI-MS analysis of PEG monomethylether 750 was reported [564]. Also copolymers have been characterised by GPEC®-ESI-MS [718]. SEC-MALDI-MS is an exciting way of determining the copolymer distribution. The technique details the oligomers present.

SEC-UV-MS-NMR-IR (LC Transform™) has been used for the analysis of a model test mixture of additives composed of BHT, Irganox 1076 and DIOP; see Section 7.3.3.4 [667].

7.3.5　Hyphenated TLC Techniques

Linking TLC with a tandem instrument differs from combining GC or LC with an appropriate spectrometer. Hyphenation of planar chromatographic techniques represents a niche application compared to HPLC-based methods. Due to the nature of the development process in TLC, the combination is often considered as an *off-line in situ procedure* rather than a truly hyphenated system. True in-line TLC tandem systems are not actually possible, as the TLC separation must be developed before the spots can be monitored. It follows that all TLC tandem instruments operate as either fraction collectors or off-line monitoring devices. Various elaborate plate extraction procedures have been developed. In all cases, TLC serves as a cleanup method.

It is not necessary to extract the analyte from the adsorbent prior to subjecting it to structural analysis. Dedicated commercial probes for *in situ* coupling of TLC and MS, FTIR and Raman are available. Given the nondestructive nature of many of the hyphenated techniques, it is possible to perform multiple experiments on the same TLC spot. In a sense this is analogous to multiple hyphenated chromatographic techniques. There is no need to repeat the chromatographic separation a second time to allow an additional spectroscopic analysis to be performed. In addition, the plate can be physically transferred to the spectrometer. This enables the analyst to choose from a variety of postchromatographic characterisation techniques, and also avoids commitment to permanent expensive and complex instrumentation for occasional problems. Nonetheless, the combination of TLC with vibrational spectroscopy in order to identify unknown chromatographic spots suffers from some severe limitations.

Many methods can be used for spot location and sample identification, including visual analysis, UV/VIS,

F, reflectance IR, NIRS, Raman, NMR, optical and electron microscopy techniques, AES, XPS, radio-imaging methods and mass spectrometry in many varieties. Scanning of 2D chromatograms using laser-based indirect fluorometric detection and photoacoustic detection, and FAB and LSIMS, has been demonstrated. Sample positioning and manipulation is central to each of these methods. The ability of the TLC plate to effectively act as a storage device for the separation, enabling chromatography and spectrometry to be performed at locations distant from each other in both time and space, provides a degree of flexibility not present in directly coupled systems.

The most common off-line monitoring system is that of scanning densitometry. Scanning the TLC plate for UV absorption after development and drying has already been discussed (Section 4.4.1.2). TLC migration data (R_F values), even if combined with *in situ* UV/VIS spectra, are seldom sufficiently selective to enable unambiguous identification of analytes. For this purpose, more selective detection techniques are necessary, notably MS, NMR [719], FTIR, Raman or fluorescence (line-narrowing) spectroscopy. Amongst these, TLC-MS and TLC-FTIR are the most popular methods, followed by TLC-Raman. Of these techniques, MS is generally considered the most useful. The amount of sample that can be loaded on to a TLC plate is often insufficient for analysis by IR or NMR. TLC using plates coated with 250 μm absorbent is an excellent technique for separating quantities of up to 100 mg of additive mixtures into their individual components. Preparative TLC (2-mm-thick plates) handles larger quantities (up to 1 g), as needed for full characterisation by NMR. TLC tandem systems are tedious to use and sometimes difficult to operate. Serious consideration should be given to the alternative use of LC-MS and LC-FTIR. TLC tandem systems are seen as the last resort–to be used only under very special circumstances [204].

Recently, various excellent reviews have been published on hyphenated and multidimensional TLC techniques [720–722].

7.3.5.1 TLC–Fluorescence Spectroscopy

Principles and Characteristics Many of the planar chromatography methods rely on fluorescence detection to achieve the required identification limits; exploitation of sensitive and selective derivatisation reactions is of considerable importance. Most TLC scanning densitometers can be operated in the fluorescence mode and are able to record *in situ* excitation spectra of TLC

spots [723]. Spots may naturally fluoresce or have moieties chemically bonded to them, which induce fluorescence. Quenching of a fluorescent background is another, even more common and popular approach. In that case the matrix of the plate includes a fluorescing (inorganic) compound which, in turn, has its fluorescence suppressed by the analyte.

Fluorescence-based detection is nondestructive, in that the photons absorbed and re-emitted do not consume the sample; long integration times can provide an extraordinary high level of sensitivity. Interpretable spectra can be recorded from sub-ng amounts. Although not suitable for direct characterisation, common fluorescence scanning densitometers can be used to considerably improve the confidence of retention-based identification of fluorescent compounds. Electronic scanning densitometers using vidicon tubes or charge-coupled device (CCD) cameras have been developed mainly for TLC image analysis [724–730]. Such detectors can also be used to very rapidly acquire fluorescence emission spectra of TLC spots [731–733]. Fluorescence spectrometry is used generally only as a quantitative detection tool in relation to TLC. The use of *conventional fluorescence spectroscopy* and fluorescence line narrowing spectroscopy (FLNS) as an aid to compound identification was described [721]. FLNS requires cryogenic temperatures (<30 K) and is a slow scanning technique unsuitable for direct coupling to column liquid chromatography.

TLC-XRF is the designation of the *in situ* TLC analytical method coupled with XRF. The subject has not been widely investigated since its introduction in 1985 by Toda *et al.* [734]. Only a few attempts have been made to combine TLC and XRF, mainly because XRF could not easily give 2D images. Elemental imaging mapping under atmospheric conditions is now possible using X-ray tubes with a spot size of about 100 μm in diameter. Combination of TLC with XRF may be accomplished without any interface [735]. The technique enables direct visualisation of elements developed on a TLC plate. As XRF is nondestructive, it is possible to use it with another method to analyse, re-extract from the plate, or colour with reagent after measurement. Individual elements on TLC plates may be detected even under conditions of poor separation or reproducibility. In addition to inorganic compounds, organic compounds including electronegative elements can be detected simultaneously that cannot always be measured satisfactorily by competitive techniques. Phenolic halogen-containing compounds can distinctly be detected by individual elemental imaging. Detection limits are 1 nmol μL^{-1} for Mn, Co, and Fe,

$10\,nmol\,\mu L^{-1}$ for Cu, Cl and I, and $100\,nmol\,\mu L^{-1}$ for Br [735].

Rather high concentrations are needed to afford *in situ* TLC-XRF elemental images. TLC-XRF is mainly used for element qualification rather than quantification. The major difficulty in using XRF for detection in TLC is the strong emission background of the (cellulose and silica gel) support, which seriously interferes with particular spectral regions. In practice, TLC plates with moderately low background, i.e. 0.5-mm-thick cellulose plates and 0.25-mm-thick silica gel plates, are used [735].

Applications   A TLC-fluorimetric method has been described for the determination of 2,4- and 2,6-diaminotoluene in methanol extracts of flexible polyurethane foams [736]. The precision of this method at the 20 ppm level is ±30 %.

7.3.5.2   TLC–Infrared Techniques

Principles and Characteristics   Both mid-IR (2.5–50 μm) and near-IR (0.8–2.5 μm) may be used in combination to TLC, but both with lower sensitivity than UV/VIS measurements. The infrared region of the spectrum was largely ignored when the only spectrometers available were the dispersive types. Fourier-transform instruments have changed all that. Combination of TLC and FTIR is commonly approached in two modes:

(i)   transfer TLC-FTIR, involving transfer of analytes from the TLC plate to an IR-transparent substrate, prior to FTIR measurement; and

(ii)   *in situ* TLC-FTIR, i.e. the separated compounds are analysed directly on the TLC plate [737].

In both cases, either conventional FTIR transmission or diffuse reflection detection may be used. Because TLC and the postspectroscopic evaluation are not linked directly, few compromises have to be made with regard to the choice of the solvent system employed for separation. Chromatographic selectivity and efficiency are not influenced by the needs of the detector. The TLC plate allows the separation to be made in a different site from the laboratory where the separated analytes are evaluated. The fact that the sample is static on the plate, rather than moving with the flow of a mobile phase, also puts less demand on the spectrometer. The popularity of TLC-IR derives in part from its low cost.

Until the introduction of *in situ* HPTLC-FTIR [738], the combination of TLC and UV spectroscopy was the only *in situ* coupling method available in TLC. A great *advantage* of TLC-FTIR over TLC-UV coupling is of course detection (including quantification) of non-UV absorbing substances on TLC plates, which makes the method applicable to almost all kinds of substance (nearly all chemical compounds show IR absorption). The quality of TLC-FTIR spectra is sufficient for identification of unknown substances and discrimination between closely related substances [739]. TLC-FTIR provides lower detection limits, does not damage eluates, and does not alter chromatographic resolution. *Disadvantages* arise because of the background interference of the stationary phase. It is not possible to record all wavelengths equally well. In comparison with the KBr technique, there are fewer absorption bands, and only the region between $3550\,cm^{-1}$ and $1370\,cm^{-1}$ can be evaluated on commercially available plates [740]. Moreover, using IR spectrometry, *in situ* measurements of spots on TLC plates are associated with spectral distortions caused by interaction of the analyte and the strongly absorbing stationary phase.

When considering libraries of spectra for identification purposes, the effect of sample preparation on spectral characteristics is also important. Two FTIR sampling methods have been adopted for IR analysis of TLC eluates in the presence of a stationary phase, namely DRIFTS [741] and PAS [742], of comparable sensitivity. It is to be noted that *in situ* TLC-PA-FTIR and TLC-DRIFT spectra bear little resemblance to KBr disc or DR spectra [743,744]. This hinders spectral interpretation by fingerprinting. For unambiguous identification, the use of a reference library consisting of TLC-FTIR spectra of adsorbed species is necessary.

Targeted *quantitative determinations* can be made at a wavelength that is specific for the particular compound. Obviously, if a substance does not have IR absorbance bands in the limited region from 3100 to $1600\,cm^{-1}$, it is not possible to employ direct quantification. In diffuse reflection, quantification of weakly absorbing samples on a non-absorbing matrix of infinite layer thickness is only possible in Kubelka–Munk units. These prerequisites are not met under the conditions used for *in situ* measurements on ordinary TLC plates, because the matrix also sorbs in the IR region and the layer thickness is limited (ca. 200 μm). Different ways of quantitative determinations by direct TLC-FTIR coupling have been introduced. As a universal method, evaluation of peak areas in Gram–Schmidt chromatograms, containing changes over the whole spectral range, has proved to be appropriate and practical [745]. On the other hand, the use of the peak area of window chromatograms is selective but of lower precision. The different possibilities of quantitative determinations by direct TLC-FTIR coupling were compared [746]. Relative standard deviations between 1.3 and 6.1 % may be achieved. When comparing methods, not only the precision but also the time requirement and method selectivity should be considered.

Physically removing material from the spot (including a portion of the stationary phase) to a proper medium is a simple way to couple TLC with a spectrometric technique. The main reason for using sample *transfer TLC-FTIR* is to avoid the strong IR-absorbance of the TLC adsorbent by extracting the analytes via a solvent to an IR-transparent pellet or powder. This allows measurement of the full spectra $(4000-500\,cm^{-1})$ at a reasonable sensitivity using conventional FTIR transmission or diffuse reflection detection and direct comparison of absorbance spectra. Dissolving the analyte in a suitable solvent does not require special equipment. Standard, nondedicated spectrometers can be used. However, transfer methods are usually laborious and have potential for loss, decomposition and/or contamination of the analyte. It is often also difficult to completely maintain chromatographic resolution during a transfer process. Numerous transfer TLC-FTIR methods have been reported [721,744,747–750]. A sample-transfer accessory was designed to transfer eluates from the TLC plate to a non-absorbing substrate prior to IR analysis [751]. With this TLC-FTIR interface, eluates are removed from a conventional TLC plate, deposited on KCl powder, and analysed by DRIFTS. *Diffuse reflectance* is the most efficient method for FTIR analysis of solutes deposited on highly scattering stationary phases. Chalmers *et al.* [752] have used off-line TLC-DRIFT, which was based upon an elaborate solute extraction procedure with transfer of material from the TLC plate to a KCl pellet. Following a transfer procedure involving scraping off, extraction, centrifugation, filtration and concentration, good-quality DRIFT spectra of small amounts (100 ng) of analyte separated by TLC were obtained [744]. Other transfer methods for TLC-FTIR have been described, in which the analyte is eluted from the plate without scraping off the adsorbent, which simplifies sample preparation work. In one such approach, a powder layer of KBr is used to coat a TLC plate. The zones are eluted off the plate with an appropriate solvent, thereby moving them into the powder layer. The pattern of the separated zones is thus directly transferred on to the KBr layer [753]. Diffuse reflectance is then employed to measure the spots. Other approaches include ATR [754] or isolation of the spots

and use of IR on them [749]. The eluate transfer method is the method of choice for TLC-FTIR techniques.

In the alternative *in situ TLC-FTIR* approach, the separated compounds are analysed by FTIR directly on the TLC plate. Percival and Griffiths [737] first reported *in situ* FTIR detection of spots on a TLC plate. In *in situ* methods, the spots on the TLC plate are usually first localised by UV light or by staining (using a second plate) prior to IR measurement. By using GC-IR software it is possible to obtain chromatograms from a TLC separation. The major problem encountered in *in situ* TLC-FTIR is the strong absorption background of the sorbent, which can obscure spectral features and limit analyte detectability to about 100 ng per chromatogram zone, i.e. about ten times higher than that of UV densitometry [746]. Diffuse reflectance infrared detection (DRIFT) is the most commonly used FTIR mode for *in situ* TLC-FTIR following early work by Fuller and Griffiths [741]. By preparation of the TLC plate on an IR-transparent material, such as silver chloride, FTIR spectra from TLC spots can be obtained directly using transmission spectroscopy [737].

Kovar *et al.* [738–740] have greatly contributed to the development and optimisation of on-line HPTLC-FTIR coupling. The principle of *on-line HPTLC-FTIR* coupling [743] depends on scanning the plate, fixed on to a computer-controlled x, y stage, with an IR beam in a DRIFT unit. The first such IR densitometer for TLC was developed by Glauninger *et al.* [755]. Application of commercial spectral libraries for (automatic) analyte identification is limited as material deposited from solution, as a thin layer, produces a DR spectrum whose relative band intensities can be markedly different from the DR spectrum of solid material prepared by grinding [744]. Also, use of DRIFT as an *in situ* detection method in TLC may cause serious interferences in particular spectral regions. DRIFT spectra of TLC spots contain parts where sensitivity is high and appropriate analyte information can be obtained, and other parts where the signal-to-noise ratio is poor and only minimal information can be extracted. Various conventional TLC phases, such as silica, alumina, cellulose and reversed-phase materials, can be used in combination with scanning DRIFT [756]. Silica gel absorbs strongly in the regions from 3700 to $3100\,cm^{-1}$ and from 1600 to $800\,cm^{-1}$, obscuring possible analyte absorptions at these frequencies. In those cases, sample transfer to a KCl pellet is appropriate.

To improve the performance of TLC-FTIR coupling, optimised sorbent layers are required which allow substantial reduction in identification limits. The full potential of the technique is yet to be realised. Lack of sensitivity and spectral distortions associated with *in situ* DRIFTS and PA-FTIR detection methods severely restrict the utility of these techniques. Even though the sample transfer technique requires more effort to obtain infrared spectra, this effort is well worth the lower detection limits and increased spectral information gained. Resolution degradation is not a problem for *in situ* detection methods, and 100 % yields are easily obtained when sample transfer is not required. On-line HPTLC-FTIR coupling has been reviewed [745], as well as TLC-DRIFT analysis [744].

The interferometer in an FTIR spectrometer modulates light of the source with a wavelength-dependent frequency. Therefore, in combination with a photoacoustic cell, an FTIR spectrometer can yield photoacoustic IR spectra, which can be applied for identification purposes. The use of a PAS cell in analysing spots from TLC plates has been reported [742,757–759]. Qualitative and quantitative information on TLC samples can be obtained by dispersive and laser PAS at UV/VIS wavelengths [760,761]. Most densitometric measurements of TLC plates are performed by reflectance scanning densitometry in the UV/VIS spectral region. The advantage of scanning thin-layer plates by *photoacoustic spectroscopy* is that it can be used effectively with very black or highly absorbing samples [742]. However, the results of densitometric reflectance measurements strongly depend upon the depth distribution of the analysed material. In principle, PAS is considered to be a more reliable technique for quantification of TLC plates than reflectance densitometry, which can only measure concentrations at the surface of a TLC plate. Non-uniform distribution of concentration of the analyte within the sorbent is the main reason for low reproducibility in TLC measurements. On the whole, PA-FTIR and DRIFTS are both not particularly well suited for *in situ* characterisation of TLC separated spots. The quality of photoacoustic spectra is generally not as good as that obtained by diffuse reflectance spectrometry. In both cases, a significant proportion of the 'fingerprint' region is obscured due to the presence of the substrate. Another disadvantage is that PA-FTIR spectra of *in situ* TLC spots bear little resemblance to KBr disc spectra [743]. Consequently, meaningful comparisons can only be made with a library of difference spectra of adsorbed substances on silica gel.

Also, *NIRS detection* in reflectance mode can be used for TLC [762]. The main reason for studying the applicability of NIRS as an *in situ* detection tool in TLC is that adsorbents such as silica gel have no strong absorption in the NIR region; background interferences are thus expected to be small. The fact that the major bands in NIRS are present in all organic species obviates the need for visualising agents in TLC-NIR. Scanning of the path of elution at a wavelength common to all the expected species (e.g. a C–H band in the 2200 nm region) will locate the eluted materials. NIRS is suitable for analysis of compounds lacking UV absorption, so that detection is possible without prior derivatisation. Since in analogy to TLC-FTIR spectra NIR spectra in reflectance mode are affected by analyte-adsorbent interactions, again creation of a spectral library of adsorbed species is required. This must be considered as a major deterrent. Plate-thickness variation and adsorption of water vapour on to the plate are major complications for combination of NIRS to TLC [763]. Mustillo and Ciurczak [764] have reviewed the development and role of near-infrared detection in TLC.

TLC-FTIR is now in a rapid growth phase with commercial instrumentation. Somsen *et al.* [721] discussed applications of TLC in combination with FTIR, NIRS and PA-FTIR. TLC-FTIR has been reviewed [167].

Applications Identification of polymer additives by TLC-IR is labour intensive and comprises extraction, concentration of extracts, component separation by TLC on silica, drying, removal of spots, preparation of KBr pellets and IR analysis. The method was illustrated with natural rubber formulations, where *N*-cyclohexyl-2-benzothiazyl sulfenamide, IPPD and 6PPD antioxidants, and a naphthenic plasticiser were readily quantified [765]. An overview of polymer/additive type compounds analysed by *transfer TLC-FTIR* is given in Table 7.80.

Crompton [770] has described extraction of scraped TLC spots (Cyasorb UV531 and Ionol CP), followed by off-line IR analysis of KBr discs. Chalmers *et al.* [752] have used off-line TLC-DRIFT for extracted PP/(Irganox 1330/1010, Topanol OC, erucamide). Volatility of Topanol OC on the particle surfaces of the KCl substrate highlights a limitation of the off-line TLC-DRIFTS technique.

Applicability of *in situ TLC-FTIR* is shown in Table 7.81. Nonylphenol (NP) and alkylphenol ethoxylates $(NP(EO)_n, n = 3,9,14)$ were detected by means of TLC-FTIR [771].

The applicability of alternative *photothermal densitometric techniques*, such as PAS, for characterisation of TLC plates with particular emphasis on the in-depth distribution of compounds in the sorbent, has been investigated [776]. No specific applications for polymer/additive systems appear to have been reported so

Table 7.80 Compounds analysed by transfer TLC-FTIR[a]

Compound types	Transfer method[b]	FTIR mode	Detection limit (μg)	Reference(s)
Dyes	Scrape + extraction	DRIFT	0.01[c]	[744]
	Scrape + Wick–Stick	Transmission	n.d.[d]	[749]
	In situ extraction + Wick–Stick	Transmission	1[c]	[766]
	Chromalect	DRIFT	1[c]	[767]
Phenols	Scrape + extraction	Transmission	2[c]	[747,768]
Phthalates	Scrape + extraction	DRIFT	0.05[c]	[748]
Polymer additives	Scrape + Wick–Stick	DRIFT	1[c]	[750]
	In situ extraction	DRIFT	10	[752]

[a] Abbreviations: n.d. = not determinable.
[b] Transfer methods: Chromalect = using TLC-FTIR accessory of Analect; Scrape + extraction = spot scrape-off followed by solvent extraction and evaporation on KBr; Scrape + Wick–Stick = spot scrape-off followed by Wick–Stick technique [769]; *in situ* extraction + Wick–Stick = *in situ* spot extraction followed by Wick–Stick technique; *in situ* extraction = *in situ* spot extraction followed by evaporation on KCl.
[c] Estimate value.
[d] Analysed amount of substance not stated.
After Somsen *et al.* [721]. Reprinted from *Journal of Chromatography*, **703**, G.W. Somsen *et al.*, 613–665. Copyright (1995), with permission from Elsevier.

Table 7.81 Compounds analysed by *in situ* TLC-FTIR

Compounds	Sorbent[a]	FTIR mode	Detection limit (μg)	Reference
Dyes	SiO_2, Al_2O_3	Transmission	0.2[b]	[737]
	Al_2O_3	Transmission	0.01	[772]
	SiO_2	DRIFT	1[b]	[741]
	ZrO_2	DRIFT	0.3	[773]
	ZrO_2	DRIFT	0.01	[774]
Phthalates	SiO_2, RP	DRIFT	1	[748]
Surfactants	SiO_2	DRIFT	50[b]	[775]

[a] RP, reversed-phase.
[b] Estimated on the basis of reported data.
After Somsen *et al.* [721]. Reprinted from *Journal of Chromatography*, **703**, G.W. Somsen *et al.*, 613–665. Copyright (1995), with permission from Elsevier.

far. There are also relatively few reported applications of *FT-NIRS* as an *in situ* detection method for TLC (see ref. [777]).

7.3.5.3 TLC–Raman Spectroscopy Techniques

Principles and Characteristics The prospects of Raman analysis for structural information depend upon many factors, including sample scattering strength, concentration, stability, fluorescence and background scattering/fluorescence from the TLC substrate. *Conventional dispersive Raman spectroscopy* has been considered as a tool for *in situ* analysis of TLC spots, since most adsorbents give weak Raman spectra and minimal interference with the spectra of the adsorbed species. Usually both silica and cellulose plates yield good-quality conventional Raman spectra, as opposed to polyamide plates. Detection limits for TLC fractions

by the conventional Raman technique are of the order of about $10\,\mu g$ per mm^2 (sample dependent) [778]. Although conventional, visibly excited Raman spectroscopy is a more sensitive detector for stable, nonfluorescent samples than FT Raman, it cannot be used with either fluorescor-doped or iodine-stained TLC plates, and in many cases sample degradation or fluorescence renders it less reliable or even useless. Also, the micro-Raman system illuminates only a small area, and therefore no signal is obtained from most of the volume of the TLC fraction, reducing sensitivity and further promoting sample damage. Despite the absence of sorbent interferences, the practical use of conventional Raman spectroscopy in TLC is thus very restricted, most probably because Raman scatter is inherently weak [778,779].

Introduction of NIR-exciting *FT-Raman spectrometers* [780] has prompted reassessment of RS as a more promising *in situ* detection method for TLC fractions than TLC-FTIR [779]. Advantages of near-IR excitation with a Nd:YAG laser (1064 nm) are avoidance of fluorescence and photo-induced sample damage, excellent spectral reproducibility, ease of operation, and low cost of the instrument relative to conventional dispersive spectrometers. While FT Raman spectroscopy can be used to analyse fractions *in situ* on TLC plates, the spectral quality depends critically upon the Raman scattering strength of the compound and its concentration in the eluted spot. A disadvantage of FT-RS instruments is that they are intrinsically less sensitive than modern multichannel dispersive Raman spectrometers, since the Raman intensity is inversely proportional to the wavelength to the fourth power. Similarly to the IR case, compared with reference spectra of pure compounds,

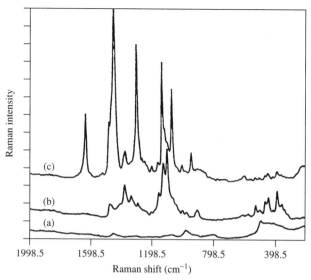

Figure 7.37 Comparison of spectra of different TLC plates: (a) silica gel; (b) cellulose; and (c) polyamide 11. After Everall *et al.* [779]. From N.J. Everall *et al.*, *Applied Spectroscopy*, **46**, 597–601 (1992). Reproduced by permission of the Society for Applied Spectroscopy

small frequency shifts, broadening and different relative intensities of peaks are observed in *in situ* Raman spectra, owing to strong interaction of the adsorbates with silica gel. This renders fingerprint identification more difficult. Nevertheless, *in situ* Raman is suitable for identification of totally unknown substances, by reference to a Raman atlas, provided that the sample concentration on the plate is high enough [781]. A specific Raman atlas for polymer additives is lacking. The detection limit depends critically upon the size of the eluted spot. A major disadvantage of the technique is also that it samples only a small (ca. 1 mm^2) area of the eluted spot, compared with the DRIFTS method, which samples the contents of the entire spot. Reliable identification by FT–Raman requires some 100–200 μg of components in a mixture. For FT–Raman the silica-plate exhibits the weakest background interference (Fig. 7.37); cellulose and polyamide 11 are unsuitable for *in situ* detection of adsorbed species. FT–Raman spectroscopy is a more generally applicable system for *in situ* analysis of fractions on TLC plates than conventional RS, especially if one takes into account recent technical progress.

The sensitivity limitations of TLC-FT-Raman spectroscopy may be overcome by applying the SERS effect [782]. Unlike infrared, a major gain in Raman signal can be achieved by utilising surface activation and/or resonance effects. *Surface-enhanced Raman* (SER) spectra can be observed for compounds adsorbed on (rough) metallic surfaces, usually silver or gold colloids [783,784], while *resonance Raman* (RR) spectra

are obtained when the frequency of the exciting radiation coincides with an electronic absorption band of the analyte [785]. Under conditions of dispersive SER or RR, Raman cross-sections can be 10^2–10^6-fold larger than in conventional Raman spectroscopy, i.e. from a level of several μg mm^{-2} for a TLC spot, to sub ng–pg level [786–790]. To use surface-enhanced Raman spectroscopy (SERS) in TLC, the spots are sprayed with a silver hydrosol prior to Raman excitation [788]. The technique is not routinely applicable as an analytical tool, since not all molecules give good SERS spectra; moreover, SERS spectra are not identical to normal, unenhanced Raman spectra, making identification more difficult. With *surface-enhanced resonance Raman scattering* (SERRS) the enhancement factor is even as much as 10^8 to 10^9, which permits *in situ* vibrational spectrometry down to pg amounts [787,790,791]. This requires matching the excitation laser with the absorption properties of the compound, combined with application of a silver sol to the substance adsorbed on the layer.

The use of near-IR-laser excited FT-SERS eliminates the disturbing fluorescence of impurities found with visible excitation, and provides SERS enhancement factors that are about 20 times larger than those found for excitation at 514.5 nm [792]. For a strong Raman scatterer (fluorene), a typical detection limit of 500 ng is found for a 3-mm diameter spot. For weak scatterers, the detection limits may be in the high-μg region, which means that some compromise between chromatographic resolution and sample capacity is needed. With the high-sensitivity enhancement provided by TLC-FT-SERS, no compromise of chromatographic performance by using HPTLC plates is necessary.

Since SERS and SERRS are substance specific, they are ideal for characterisation and identification of chromatographically separated compounds. SE(R)R is not, unfortunately, as generally applicable as MS or FTIR, because the method requires silver sol adsorption, which is strongly analyte-dependent. SE(R)R should, moreover, be considered as a qualitative rather than a quantitative technique, because the absolute activity of the silver sol is batch dependent and the signal intensity within a TLC spot is inhomogeneously distributed. TLC-FTIR and TLC-RS are considered to be more generally applicable methods, but much less sensitive than TLC-FT-SERS; FT-Raman offers μm resolution levels, as compared to about 10 μm for FTIR. TLC-Raman has been reviewed [721].

Applications Sollinger and Sawatzki [793] have reported the use of TLC-Raman for routine applications, e.g. TLC of hydroxybenzenes (including hydroquinone and pyrogallol) on conventional, silica gel and specific Raman–TLC plates (coated with spherical silica gel). Databases were used for identification of substances. Typical detection limits were in the low μg region per application, Micro-Raman spectrometry has been employed in analysing TLC fractions from polymer additives within a detection limit of 1–5 μg mm^{-2} for a TLC spot [778]. Highly Raman active compounds, like *optical brighteners* [794] can be detected without surface-enhanced scattering on specially modified silica gel plates. The identification limit is about 25 ng for these substances and about 100 ng for dyes. Analyses of *polymer additives* (Topanol OC, Irganox 1010, erucamide) by means of TLC-Raman, TLC-FT–Raman and TLC-DRIFTS using silica, cellulose and polyamide plates were compared with 514 nm excitation (Figure 7.38) [779]. Reasonable-quality spectra were obtained from sample loadings equivalent to about 3 μg mm^{-2} in the most favourable case. The concentration of the eluted spot is the limiting factor in this approach since the Raman experiment samples only about 1 mm^2 of the total sample area. Background Raman features from the TLC adsorbent ultimately obscure the spectrum of very dilute loadings of additives.

TLC–Raman laser microscopy ($\lambda = 514$ nm) in conjunction with other techniques (IR microscopy, XRF and HPLC-DAD-ESI-MS) has been used in the analysis of a yellow *impurity* in styrene attributed to reaction of the polymerisation inhibitor *t*-butylcatechol (TBC) and ammonia (from a washing step) [795]. Although TLC-FT-Raman did not allow full structural characterisation, several structural elements were identified. Exact mass measurement indicated a $C_{20}H_{25}O_3N$ compound which was further structurally characterised by ^1H and ^{13}C NMR.

SERS can be used to characterise ng and pg amounts of solutes on colloidal silver-treated HPTLC plates using

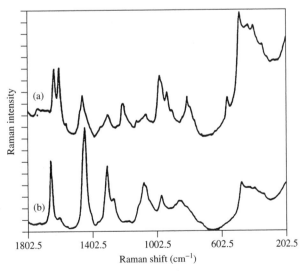

Figure 7.38 TLC-FTRS of erucamide (a) and Irganox 1010 (b), 200 μg each. After Everall *et al.* [779]. From N.J. Everall *et al.*, *Applied Spectroscopy*, **46**, 597–601 (1992). Reproduced by permission of the Society for Applied Spectroscopy

an Ar-ion or He-Ne laser [779,789]. A laser Raman microprobe in combination with HPTLC-SERS has allowed measurement of spots down to 1 μm in size, containing sub-ng quantities of material. Koglin [789] has demonstrated that TLC-SERS can be used to study the adsorption behaviour of cationic surfactants on silica gel. Clearly, SERS is a very interesting method for obtaining structural information about very small amounts of compounds separated by TLC.

Somsen *et al.* [796] have reported the use of SERR spectroscopy for the *in situ* selective determination and semi-quantitative analysis of structurally similar *dyes* separated by TLC. The limits of identification of the TLC-SERRS method (ca. 5 ng applied) were sufficient for acquisition of spectra of impurities present in the certified dye standards. SERRS may also be used for *in situ* identification of highly fluorescent molecules on HPTLC plates.

7.3.5.4 TLC–Mass-spectrometric Methods

Principles and Characteristics With sample spots in a developed thin-layer chromatogram, there are no constraints on the operation of the mass spectrometer. Depending on the analytical information required, either low- or high-resolution mass-spectral data can be recorded, and both positive- and negative-ion mass spectra can be obtained from the same sample spot.

Various *methodological approaches* are possible to obtain mass spectra from substances separated by TLC (Scheme 7.13):

(i) Manual and cost-effective removal of the stationary phase from the zone of interest followed by solvent extraction (in much of the early TLC-MS work) or direct introduction into the ion source of the spectrometer (used for additive screening and verification). Stable, volatile analytes can be determined by TD-MS after volatilisation from a heatable probe tip, leaving behind the chromatographic matrix [797,798]. Less-volatile, or unstable, analytes can be ionised by FAB/LSIMS after mixing the stationary phase with a suitable liquid matrix [799,800], by ^{252}Cf PD [801] or by LD followed by laser ionisation mass spectrometry, directly from TLC plates [802]. Although such materials can also be ionised by field desorption (FD), the elution procedure is still required to transfer the TLC fraction to the FD emitter. The methods are destructive of both sample and chromatogram, are characterised by high sample loading, μg-level detection limits and high backgrounds [803–805]. Unless there are only a few components on the plate which are of interest, excising chromatographic zones and transfer of the separated material from the TLC plate to the mass spectrometer is a laborious, time-consuming procedure which is also subject to contamination.

(ii) Use of a micro-solvent device that enables electrospray (ES) ionisation to be performed on compounds separated by TLC [806].

(iii) Instrumental methods employing probes or plate scanners (*cf.* ref. [721]). In one such plate scanner, the analytes are locally thermally evaporated

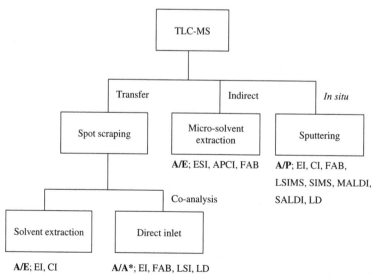

Scheme 7.13 TLC-MS modes. **A**, additive; **A***, adsorbent; **E**, eluent; **P**, plate

(by a small H_2-O_2 flame, incandescent lamp or a pulsed CO_2 laser) into a stream of gas, which carries them into the source of the mass spectrometer for chemical ionisation [807,808]. The process should be carried out without decomposition or condensation in the transfer lines. Other drawbacks are insensitivity and high temperature [808,809]. Alternatively, the chromatogram is placed intact within the source housing, and a spatially resolved 1D or 2D organic map of the surface is obtained by plate scanning [810]. Since time is not a factor in the TLC-MS detection system, any spot in the 2D chromatogram can be investigated in any order. Particle-induced desorption techniques (FAB, LSIMS, MALDI) or LD allow direct analysis of samples from within the chromatographic matrix [811]. Brown *et al.* [812] have developed several scanning TLC-MS systems based on a liquid SIMS approach.

The ability to perform mass spectrometry on analytes directly on the (HP)TLC sorbent removes the need to recover them prior to analysis. This greatly reduces the amount of work required to confirm identity. As TLC methods are developed for larger and more nonvolatile sample molecules, the use of sputtering methods will undoubtedly become increasingly important.

Fast-scanning quadrupole and magnetic-sector instruments provide scan speeds (for integral mass resolution) of $0.1\,s\,decade^{-1}$. A mass spectrum measured from mass 10 to mass 1000 would require 0.2 s for scanning the analyser. Other mass-analysis devices, such as ion traps or FTICR instruments, also have a time function in scanning. ToF mass spectrometers do not scan, but do require a pulsed ionisation method and time for passage of separate ion packets through a flight tube. These latter instruments have not been widely used for TLC-MS.

As evident from Scheme 7.13, most modern ionisation techniques have been used for TLC-MS, and no single ionisation method is used exclusively with TLC-MS. Various ionisation methods may be applied that avoid the need to evaporate the sample into an EI or CI source; these are based in particular on sputtering (FAB, SIMS) or laser desorption. Several sputtering methods of ionisation do not require the use of a liquid matrix, e.g. TLC-SIMS [797]. Recent developments include the use of matrix-assisted laser desorption ionisation (MALDI) and surface-assisted laser desorption ionisation (SALDI). It is obvious that TLC-MS is complemented with TLC-MS^n [800] and TLC-HRMS techniques. Table 7.82 lists the general characteristics of TLC-MS.

The *advantages* of TLC-MS are derived mostly from the well-known characteristics of TLC (Section 4.4.1.2),

Table 7.82 General characteristics of TLC-MS

Advantages

- High information power (structural information, molecular mass)
- No physical interface
- Wide variety of ionisation modes
- Modest cost
- No time factor
- Imaging

Disadvantages

- Analytes must be desorbed/ionised
- Relatively high detection limits
- Low sensitivity for polar compounds
- Large background interferences
- Few commercial interfaces
- TLC-MS development lags behind TLC-FTIR and LC-MS

extended through the high information power of MS for each spot, In direct TLC-MS analyses overlapping zones can clearly be identified. No matter what the mechanism for transfer of the separation achieved on the plate to the mass spectrometer, the resulting mass spectra can be obtained with high quality and relatively good sensitivity from $<1\,\mu g$ of material, using either on-line or off-line TLC-MS [813,814]. At present, unequivocal results can be obtained with amounts of substance of about 100 ng per chromatographic zone [721,815,816]; this is expected to improve. For true detection and identification, 2D spatial imaging of the layer and sample transfer to the vapour phase without degrading the chromatographic resolution or mass spectrometer sensitivity, are required.

Wide use of TLC-MS is hampered by the lack of commercially available interfaces. This also restricts automation and high throughput. Commercial direct insertion probes for scanning TLC-MS are available [811]. Compared with on-line LC-MS operation, TLC-MS hyphenation is much less highly developed.

The use of ionisation techniques such as EI and CI for TLC stationary phases has generally been limited to relatively nonpolar and thermally stable molecules. Polar involatile compounds, separated on silica gel, generally strongly adsorb on to the matrix, and decompose when heat is applied for volatilisation [817]. Use of less-adsorbent phases, such as polyamide, is particularly useful for *TLC-EIMS* work, because the analytes are not as strongly adsorbed to this phase and do not require high probe temperatures [818,819]. For compounds that are not suitable candidates for TLC-EIMS, FAB can be employed. Chemical ionisation, although suitable for TLC-MS, appears to have been little used.

As indicated in Scheme 7.13 for *in situ* identification by means of TLC-MS, a variety of ionisation techniques are used. Most important are FAB and LSIMS, in which organic molecules are sputtered from surfaces by the impact of an energetic particle beam (velocity about $10^5 \, \mathrm{m \, s^{-1}}$), and laser desorption (LD) [794,820–823]. As the sample ions formed by these *'soft' ionisation* methods are usually the same even-electron ions, such as $(M + H)^+$ formed in CIMS, the spectra are relatively simple, with few or no fragment ions against a considerable background; the structural information that can be derived is generally limited and analyte detectability is compromised (detection limits 10–100 ng) [721,811,816]. Increased molecular ion fragmentation may be obtained by use of tandem MS [824,825].

Both *FAB-MS and LSIMS* require application of an appropriate liquid matrix to the stationary phase. Chang *et al.* [826] have devised a simple technique for transferring TLC spots to a FAB probe. Continuous-flow FAB/FIB has been coupled to TLC, by using a probe which holds a complete TLC plate, pretreated with glycerol, inside the ionisation source. The TLC plate can then be moved so that all the components are passed through the line of the fast atom or ion beam [827]. The TLC analyte will thus be presented to the mass spectrometer together with the stationary phase, any organic phase-modifier present, binding agents, fluorescent indicators and eventually (high-boiling) residual solvents from the mobile phase. Mass spectra obtained from a single spot taken from a silica gel TLC plate will consist of analyte ions together with ions from the excipients. In FAB-MS or LSIMS analysis, the problem is exacerbated by the formation of matrix and salt adduct ions.

Table 7.83 lists the main characteristics of TLC-FAB-MS/LSIMS. A key difference between EI/CI and FAB/LSIMS/LD is the fact that sampling in FAB and LSIMS is from a specified location that corresponds to the impact footprint of the primary particle beam. The natural compatibility of FAB, LSIMS and LD with the direct mass-spectrometric analysis of TLC plates is readily apparent. Most mass-spectrometric measurements are destructive in nature, but FAB and LSIMS are surface-sensitive techniques in which the material actually consumed in the analysis is sputtered only from the top few microns of the sample spot. The underlying bulk is not affected, and can be used for further probing. The major limitation of TLC-FAB depends on the capability of the compounds to produce a good spectrum.

Table 7.83 Main features of HPTLC-FAB-MS/LSIMS

Advantages

- Sampling of volatile and nonvolatile analytes
- Suitable for polar and/or thermally unstable substances
- Small samples
- Short analysis time
- No contamination
- Surface-sensitive
- Spatial resolution (spot size 0.01 to 1 mm^2)
- Nondestructive

Disadvantages

- High background
- Limited fragmentation and structural information (molecular ion)
- High detection limits (10–100 ng)

It appears that use of FAB has declined somewhat in favour of LSIMS [828]. Off-line TLC-LSIMS^n data contain increased structural information [824]. Mostly, LSIMS procedures involve removal of the stationary phase from the plate for subsequent mass-spectroscopic analysis, but scanning motorised probes are also available. The most sophisticated developments in TLC-MS scanners allow 2D *imaging* of TLC plates using LSIMS [810,811,816]. Optimal use of such devices has required considerable ingenuity to overcome certain of the technical problems associated with LSIMS. A potential problem in TLC-LSIMS is that background and matrix peaks can be large and may obscure those of the analyte (especially when the latter is present in low concentration). Using MS/MS techniques can circumvent such problems. In contrast, ESI-MS data do not suffer from background problems to the same extent.

On-line *TLC-ToFSIMS coupling* is well suited for direct detection and identification of separated substances on TLC plates [829–832]. A vapour-deposited noble-metal layer, preferably 0.2-μm-thick Ag or Au, is necessary for successful SIMS analysis in order to ensure a conducting surface and a high secondary-ion yield. The TLC plate may be placed directly in the SIMS ion source. The organic molecules are easily identified by their quasi-molecular ions like $[M + H]^+$, $[M + Ag]^+$, etc. and/or by characteristic fragments. ToF-SIMS is applicable even if only extremely small amounts of sample material are available, with LODs of 30 ng per spot. TLC-SIMS, which is suitable for quantitative analysis of thermally unstable compounds of low volatility [833], provides only surface sensitivity [834]. Successful imaging of samples directly from TLC plates with high spatial and mass resolution and good sensitivity was demonstrated using a ToF mass spectrometer [835]. Focusing liquid metal ion guns allow a spot size of $10 \, \mu m$ diameter. TLC-SIMS coupling is an active research area [836].

Laser desorption has been used for direct desorption of sample molecules from TLC plates since the early years of TLC-MS. In *direct laser desorption*, the photon energy must be absorbed by the components of the chromatogram or by the sample itself. For this reason most early work used infrared lasers.

Laser microprobe MS (LMMS) can be used for direct analysis of normal-phase HPTLC plates [802,837]. Kubis *et al.* [802] used polyamide TLC plates; polyamide does not interfere with compound identification by the mass spectrum, owing to its low-mass fragment-ions ($m/z < 150$). LMMS is essentially a surface analysis technique, in which the sample is ablated using a Nd-YAG laser. The UV irradiation desorbs and ionises a microvolume of the sample; the positive and negative ions can be analysed by using a ToF mass spectrometer. The main characteristics of TLC-LMMS are indicated in Table 7.84 [838].

Automated organic and elemental ion mapping of TLC plates by LMMS techniques, without focus correction, has been reported [802,839]. One of the early TLC-MS scanners used laser desorption combined with CI detection [807,808]. The use of laser desorption mass spectrometry (LDMS), in connection with TLC separations, allows sampling of a very small area of a spot (ca. $5 \, \mu m$). In this way spot homogeneity can be determined (e.g. in the case of overlapping components), and also leaves the bulk of the material unaffected for further study. An important advantage

Table 7.84 Main characteristics of TLC-LMMS

Advantages

- Direct analysis (no analyte removal from plate, no liquid matrix)
- Extreme sensitivity (10 ppm for elements; 0.5 ng detection of sampled mass)
- Structural and relative molecular mass information
- High spatial resolution (laser beam diameter $1 \, \mu m$)
- Examination of chemical heterogeneities within a single TLC spot
- High measurement speed (10 min)
- Analysis of organic and inorganic molecules; elemental information

Disadvantages

- High cost (not a viable proposition for TLC spot analysis alone)
- Sample charging
- Non-informative spectra from nonpolar compounds (contain low-mass structurally specific ions but no molecular species)
- High background

of LD is that no liquid matrix of the type required for FAB/LSIMS (e.g. glycerol) or extraction solvent is needed in order to obtain spectra. This enables spots and bands on the TLC plate to be examined and imaged without loss of spatial resolution. However, UV-based LD is characterised by poor reproducibility and significant fragmentation [840]. This problem can partly be circumvented by the use of IR laser desorption, followed by multiphoton ionisation (MUPI).

To compensate for the perceived problems of LD, the use of MALDI/SALDI has been considered. The LOD of *TLC-MALDI-MS* is about 1 ng (for localised signals) [841,842]. Unlike LD, MALDI analysis of TLC plates requires special means of adding the matrix to sample molecules already separated within the chromatographic matrix. It is important that any matrix materials employed do not interfere with the mass spectra obtained. The key to successful TLC-MALDI-ToFMS is hence the preparation of the TLC plate prior to mass spectrometry. Gusev *et al.* [843] have demonstrated the use of MALDI for the direct analysis of TLC plates; the method of plate preparation involves pressing a previously prepared layer of matrix crystals into the TLC plate. Bristow and Creaser [842] have reported TLC-MALDI-MS of dyestuffs using plates prepared by spraying the matrix solution on to the plate with a TLC reagent sprayer. Alternative methods to the matrix pressing technique have been developed and were compared [841,842]. Direct quantification using MALDI has been reported [844]. TLC combined with surface-assisted laser desorption ionisation (TLC-SALDI-MS) has also recently been described [845].

Zhu and Yeung [846] have described quantitative TLC by *laser pyrolysis scanning* (LPS). Pyrolysis is used here as a universal mechanism for transferring species to a detector. In this procedure, a TLC plate after separation of the analytes is irradiated with a CO_2 CW infrared laser to produce a high-temperature spot. The analyte is thus pyrolysed and swept into an FID or ECD detector by a carrier gas. The length of the TLC plate, enclosed in a pyrolysis cell, is scanned. TLC-LPS should also allow one to identify the separated compounds from the gas-phase fragments. For example, by connecting to a gas chromatograph, it is ready for pyrolysis–GC fingerprint identification. The LPS technique is limited to silica gel and alumina plates. Reversed-phase plates, cellulose plates and plates with an organic binder are not suitable for FID, because of the high organic contents. Also, low-boiling-point solvents must be used for easy solvent evaporation in LPS-FID to minimise the background. Use of TLC-LPS, which depends on TLC plates with low background for FID or ECD, is limited.

The main limitation of *in situ* TLC-MS (sputtering) techniques, namely significant background contributions and limited fragment-ion information, can be overcome by tandem MS. In analogy to TLC-MS two basic approaches to TLC-MS^n may be distinguished, namely 'manual' and 'instrumental'. The former case involves manipulation of the sample, is readily implemented, and requires no additional equipment or interface. Instrumental *TLC-MS/MS* uses a dedicated probe, by which the track of interest is moved through the ion source of the mass spectrometer. Mass-spectrometry data can then be acquired from each component in turn. Several types of instrument capable of performing MS/MS are available, with the two most common types being the triple quadrupole and the magnetic sector/quadrupole hybrid. In any screening analysis, the ability to use high mass resolution MS to identify and confirm ion empirical formulae is becoming increasingly important. MS/MS experiments with multisector or FTICR mass spectrometers help in characterising the sample ions sputtered from a chromatogram.

The mass spectral benefits of TLC-MS^n have been described [800]. The quality of the spectra is generally very good. Compared with TLC-MS, the MS/MS technique offers the advantage that matrix interferences from ions derived from either co-chromatographing contaminants or the general 'background' are absent, and the fragmentation provides diagnostic ions, enabling unambiguous identification and structure determination to be performed. The common methods of ionisation (EI, CI and FAB, LSIMS) used in conjunction with the MS/MS ability of the tandem mass spectrometer ensure that the analyst can obtain molecular weight and structural information. TLC coupled with FAB-MS/MS and EI-MS/MS is therefore a facile and rapid technique for identification of a range of polar and nonpolar substances separated by TLC, without the need to recover the analytes prior to mass spectrometry. Apart from the mass-spectral advantages indicated before, TLC, when coupled with a tandem mass spectrometer, also offers some major chromatographic advantages over HRGC-MS and HPLC-MS (*cf.* Section 4.4.1.2). Due to the limited availability of MS/MS instruments, relatively few groups have made use of TLC-MS/MS.

The direct coupling of TLC with MS was reviewed, including discussion of LD FTMS, TLC-SIMS, and methods for interfacing TLC and MS [810,847]. Busch [810,811,815,816,848,849] and others [721,850,851] have repeatedly reviewed TLC-MS.

Applications   TLC-MS can be applied to qualitative and quantitative analysis of a very wide range of analytes (at least as wide as HPLC-based methods) and to samples that might well be considered quite unsuitable for other separation techniques. TLC-MS has been applied to dyes [827]. In connection to TLC screening of additives direct inlet (DI) MS is a suitable tool for identification which requires only very small amounts of matter (μg). Squirrell [258] has used HRMS for the examination of a UV-absorbing additive extracted from a PET composition. The material, with a TLC R_f value as Cyasorb UV531, exhibited UV/VIS characteristics differing from this compound. IR, UV, NMR and low-resolution MS gave evidence for a benzophenone derivative for which a $C_{28}H_{20}O_7$ structure (relative molecular mass 468.1209 Da) was postulated, at variance to synthetic considerations ($C_{29}H_{24}O_6$, 468.1573 Da). HRMS gave evidence for a relative molecular mass of 468.1544 Da, showing $C_{29}H_{24}O_6$ to be the correct formula, with confirmation being obtained from the masses of the fragment ions (Fig. 7.39).

A mixture of polymer additives (Topanol O, Irganox 1076), separated on silica gel, has been detected by means of TLC-EIMS/MS [800]. According to Ramaley *et al.* [807,808] more-polar compounds are more difficult to detect, with acids and alcohols giving low sensitivity and amino acids being undetectable. At higher molecular mass (typically above 300 Da), and decreased volatility, detection becomes problematic. Various authors [852] have used TLC-FAB-MS. High-performance TLC plates were used for separation of mixtures of amine antioxidants (in gas oils) and surfactants in mixed (cationic, non-ionic and anionic) systems, followed by extraction of the sample spots with the thio-glycerol liquid matrix typically used in FAB; detection levels below 20 ng μL^{-1} [853]. Reversed-phase TLC on aluminium-backed plates coupled with FAB-MS and LSIMS has been used for identification and separation of 27 *food dyes* (consisting of 12 permitted for use in food and 15 dyes unlawful in Japan) using C_{18}-modified silica gel as adsorbent and two mobile phases. The method was successfully applied for the identification of unlawful dyes in imported foods [799,854–857].

There have been some examples of the use of LDMS applied to the analysis of compounds separated via TLC, although not specifically dealing with polymer additives [852]. Dewey and Finney [838] have described direct TLC-spectroscopy and TLC-LMMS as applied to the analysis of *lubricating oil additives* (phenolic and amine antioxidants, detergents, dispersants, viscosity index improvers, corrosion inhibitors and metal deactivators). Also a series of general organics and ionic surfactants were analysed by means of direct normal-phase HPTLC-LMMS [837]. Novak and Hercules [858] have

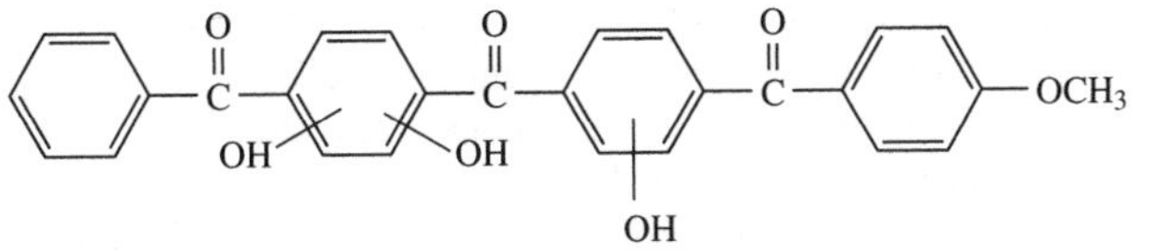

$C_{28}H_{20}O_7$ Relative molecular mass = 468.1209

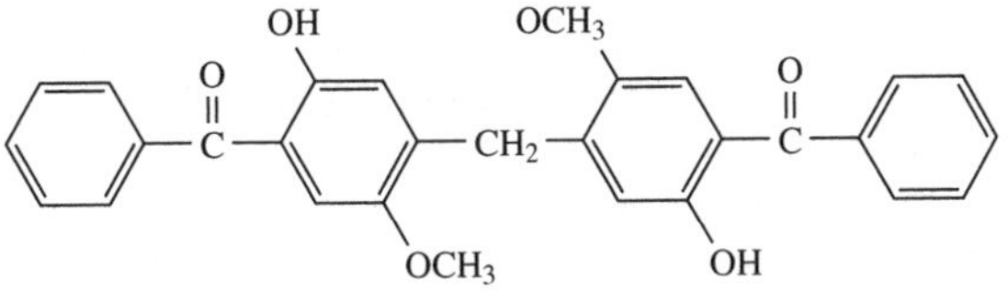

$C_{29}H_{24}O_6$ Relative molecular mass = 468.1573

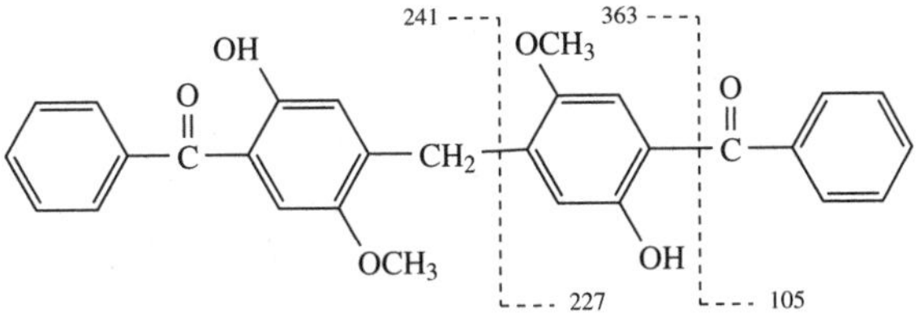

Formula	Theoretical mass	Measured mass
$C_{29}H_{24}O_6$	468.1573	468.1544
$C_{22}H_{19}O_5$	363.1232	363.1244
$C_{15}H_{13}O_3$	241.0865	241.0945
$C_{14}H_{11}O_3$	227.0708	227.0703

Figure 7.39 Structural assignments of a benzophenone derivative based on low- and high-resolution mass spectrometry. After Squirrell [258]. From D.C.M. Squirrell, *Analyst*, **106**, 1042–1056 (1981). Reproduced by permission of The Royal Society of Chemistry

used a LAMMA®-1000 laser microprobe (Nd:YAG laser, 265 nm) in combination with a ToF mass analyser, to examine dyes separated by means of TLC and to produce 2D images of triphenylmethane dyes on a polymer surface. TLC-MS^n applications are limited [820], and usually refer to normal-phase chromatographic separations on silica gel. Most TLC-MS^n applications use FAB or LSIMS as the initial ionising technique. However, EI ionisation (70 eV) has been applied to a mixture of polymer additives partially resolved by TLC on silica gel using an ethyl acetate-hexane solvent system [800]. MS and MS^n data were acquired using a tandem mass spectrometer of BEqQ geometry. The appropriate zone of silica was removed from the plate and packed into the probe tip, and the analytes were then slowly thermally desorbed. Good MS^n product-ion spectra were obtained for the $[M - CH_3]^+$ ions at m/z 205 for Topanol O and m/z 515 for Irganox 1076, respectively. Representative TLC-EIMS and TLC-EIMS^n spectra for Irganox

1076 were reported [800]. Wilson and Morden [820] have described practical HPTLC-MS/MS applications for compounds containing plasticisers.

7.3.6 Hyphenated CE Techniques

Hyphenation in capillary electrophoresis is still in its infancy. Critical aspects of CE hyphenation include the minute volumes of sample injected (typically a few nL) and small flow-rates (in the order of nL min^{-1}). Interfaces are not commercially available. *CZE-UV* can be used for the analysis of higher polyamide oligomers in HFIP solution [859]. A solvent elimination design with nebuliser has been described for *CE-FTIR* and CEC-FTIR coupling; absolute detection limits are hundreds of pg [860]. An advantage of CE-FTIR is that analytes may be detected and identified without derivatisation. *CE(C)-NMR* [861–863] is advancing rapidly.

Mass spectrometry is an attractive detector for CE, that can provide high sensitivity, excellent selectivity for unambiguous confirmatory analysis, and structural information for the identification of new compounds [864]. Coupling is readily achieved, owing to the compatibility of capillary flow-rates with MS. A number of *CE-MS interfaces*, including coaxial sheath-flow, liquid junction and sheathless arrangements, have been developed. ESI-MS is the most popular ionisation mode, but FAB-MS has also been reported (e.g. as CEC-ESI-MS [865] and CEC-FAB-MS), as well as CE-EIMS [866]. CEC offers particular advantages in coupling to MS, since narrow chromatographic peaks produce a higher mass flux. The combination of extremely efficient chromatography with the unrivalled specificity of MS results in a very powerful analytical technique. The method allows handling of extremely small sample quantities. Obviously, CZE-MS/MS is an expensive method of analysis [867].

Capillary electrophoresis is a technique that offers potential benefits for elemental speciation analysis. Various papers describe on-line *CE-ICP-MS* and *CE-ICP-AED couplings* [181,868–871], which require due attention to the flow-rate (typically of the order of $\mu L\,min^{-1}$) and low sample volume of a typical CE separation. The aim is quantitative elemental speciation in less than 1 min, with detection limits in the low $ng\,mL^{-1}$ to sub-ng mL^{-1} range for various sample types. Speciation studies are of primary interest for CE-ICP-MS, as only ions, complexes and molecules containing the same element need to be separated. On-line CE-ICP-MS (CZE mode) for selenium speciation has been described [871]. The possibility of detecting more than 20 elements simultaneously is the most important advantage of *PIXE*, which has recently been coupled to CE [872]. In comparison to CE-ICP-MS, higher detection limits and higher uncertainties of the measured signals were observed.

7.3.6.1　Hyphenated CE–MS Techniques

Principles and Characteristics　The main reasons for hyphenating MS to CE are the almost universal nature of the detector, its sensitivity and the structural information obtainable, including assessment of peak purity and identity. As CE is a liquid-phase separation technique, coupling to the mass spectrometer can be achieved by means of (modified) LC-MS interfaces. Because of the low flow-rates applied in CE, i.e. typically below $100\,nL\,min^{-1}$, a special coupling device is required to couple CE and the LC-MS interface. Three such devices have been developed, namely a

coaxial coupling, a liquid junction coupling and a direct coupling. Also, a variety of ionisation methods have been attempted since the first reported CE-MS coupling [873]: electrospray, ion spray and atmospheric pressure chemical ionisation, as well as continuous-flow fast atom bombardment – usually as modifications of low flow-rate LC-MS interfaces. Two additional mass-spectrometric ionisation techniques have been coupled off-line with CE, namely ^{252}Cf plasma desorption [874] and MALDI [875].

ESI is the method of choice interfacing CE and MS [876]. It both ionises and desolvates the analytes of interest, and provides good yields of the molecular ion for high sensitivity with polar, fragile and thermally labile analytes. Conventional ESI normally accepts $\mu L\,min^{-1}$ flow-rates. Therefore, interfacing CE with ESI requires a make-up liquid junction to accommodate the $nL\,min^{-1}$ flow-rates of CE. The development in nanospray ESI presages an ideal coupling between CE and ESI-MS. *CE-ESI* has been coupled to QMS [873,877,878], ITMS [879,880], magnetic sector [881], FTICR-MS [882] and ToF-MS [883–885]. Several technical details have to be carefully optimised to make such interfaces routine. *CE-ESI-MS* is still considered to be an immature technique, mainly due to current technical and chemical limitations. CE-ESI interfacing must accommodate high voltage (HV) at the separating capillary and connection to another HV delivering instrument. Volatile buffers (as in HPLC-ESI coupling) are not electrospray compatible. When coupled to ESI-MS, information on the solute's structure and molecular weight can be obtained rapidly. On-line CE-ion spray MS has also been reported [886].

Although electrospray ionisation has advantages for studies of ionic and very polar molecules, APCI is the preferred ionisation method for many other compounds – in that it provides better sensitivity and has advantages for quantitative studies. *CE-APCI-MS* studies were conducted on both QQQ and QITMS instruments [887].

The combination of CE with continuous-flow fast atom bombardment (*CF-FAB-MS*) requires the use of an interface, because of the incompatibility of the CF-FAB process and CE for liquid flow [888]. The CF-FAB source requires a solvent, usually water/glycerol (95–5 v/v), which is maintained at a steady flow-rate of $2–15\,mL\,min^{-1}$. Flow-rate in CE does not exceed $1\,nL\,min^{-1}$.

Variations may also be introduced at the CE end of the coupling. CZE is the most widely practised mode of CE. Nanospray CZE-ESI-MS [889],

CZE-APCI-MS [887] and CZE-CF-FAB-MS^n [533] couplings have been described. CEC-MS coupling offers:

(i) unique selectivity for polar and ionised solutes;
(ii) low flow-rate;
(iii) low peak volume of analytes; and
(iv) short analysis time.

The combination of CE and MS enables high-efficiency separations of a wide range of solutes. As with any CE technique the running buffer determines the quality of the separation. In practice, therefore, most CE-MS separations have been performed in an acidic buffer. Unfortunately, these conditions severely constrain the range of samples that can be successfully analysed by CE-MS. CE-MS requires a small volume of a concentrated sample $(10^{-3}\,\text{mole}\,\text{L}^{-1})$. Sensitivity is reported as <1 pmol for SIM and 10 pmol for full mass spectra. Very little quantitative work has been performed with CE-MS techniques.

CE-MS, CE-MS^n and CEC-MS have raised a great deal of attention recently, because of their universality, specificity and sensitivity. CE-MS and CEC-MS are useful for adding an extra selectivity dimension to detect co-eluted molecules of different masses and obtaining information about molecular mass and structure. CE-MS has frequently been reviewed [433,434,440,873,876,890–896].

Applications CE is very tolerant of complex sample matrices, and therefore its on-line combination with MS provides for highly selective detection of compounds in low-volume but otherwise complex mixtures. Sulfonated azo and other *textile dyes* at the 3 pmol level (SIM) have been analysed by CE-MS using a liquid junction coupling to an ionspray interface [880] and a coaxial electrospray interface [897]. CEC-ESI-MS has been used for neutral azo- and anthraquinone-based textile dyes and impurity testing [865]. In CZE-MS, the first separation is on the basis of the analyte's charge-to-size ratio (CZE), and then on the basis of its mass-to-charge ratio (MS). Most of the CZE-ESI-MS work uses the positive ion mode. CEC-APCI-MS has been used for analysis of low-molecular-mass *organic and inorganic anions* [898].

The range of applications of CE-MS is still rather limited [899]. Few 'real' unknown samples have been analysed by CE-MS. In particular, CE-MS activities for synthetic polymer additive analysis purposes are not abundant. On the other hand, ITP and ITP-CE separations of food additives on a chip have been reported [900].

7.4 MULTIDIMENSIONAL CHROMATOGRAPHY

Principles and Characteristics In some cases, analysis using an appropriate combination of a single separation and detection method is not satisfactory, and it becomes necessary to utilise a combination of separation methods and/or multidetector monitoring. This approach is termed multidimensional, or coupled chromatography and is meant to describe a specific sequential combination of chromatographic procedures.

Typical one-dimensional separation techniques are GC, LC, SEC, CE, etc. Separations in single-column systems employing high-resolution capillary columns are primarily based on column efficiency, whereas, in 2D systems, the source of separation power is the difference in selectivity between two or more stationary phases. The term 'two-dimensional chromatography' has its origin in flat-bed chromatography. Systems such as SEC-VIS, SEC-MALLS are considered as being 1^+ dimensional separation methods. In combination with size separation NMR and IR provide a higher dimension ($1\frac{1}{2}$–2D systems).

Multidimensional techniques are preferred when a very large number of theoretical plates are required for separation; when a large number of samples of a similar kind are to be analysed in the shortest possible time; and for analysis of complex mixtures which contain only a few adjacent components of interest. For such separation problems, unidimensional techniques are inherently inefficient, as only a small fraction of the column is actually in use at any given time. *Multimodal separations* employ two or more chromatographic methods in series, such as on-line LC-GC or SFC-GC coupling. Multidimensional or multimodal separations can be performed on-line or off-line, particularly in liquid chromatography. In general, these systems are designed to separate selected components in a primary system, which are then transferred in some manner (preferably on-line) to a secondary, fundamentally different separating system, where further resolution is attainable. Constituents of no interest go to waste. Both methods aim at: solving selectivity problems; improving the resolution of parts of a complex mixture; affecting trace enrichment of selected analytes; or increasing sample throughput by employing techniques such as heartcutting, backflushing, foreflushing, cold trapping or recycle chromatography. Heartcutting methods are coupled systems designed to isolate one or only a few analytes of interest from a complex matrix. A system of this type has a peak capacity typical of a single-dimension system. *Comprehensive multidimensional systems* are based upon the eventual transfer of *all*, or a representative portion of all, analytes composing the initially injected sample to all further separation modes. Comprehensive *multi*-dimensional separation methods are needed for a truly comprehensive characterisation of complex polymers. The most relevant comprehensive techniques are GCxGC, LCxGC and SFCxGC. For nomenclature and conventions, see ref. [900a].

The objective of combined analytical separations is to obtain nonredundant information from independent systems. The success of all multidimensional methods in chromatography is dependent on the creation of complementary separation mechanisms, applied in a sequential manner, to enhance the separation capacity of the system. For techniques to be complementary to each other, the acquired data should be orthogonal. A multidimensional system is commonly defined as a system in which:

(i) the components of a mixture are subjected to two or more different separation steps (mechanisms); and
(ii) components separated in any single step remain always separated until completion of the separative operation [901].

Multidimensional methods thus involve a combination of single mechanisms and systems. In any multidimensional (usually 2D) approach, it is desirable that each dimension be as 'pure' as possible in terms of selectivity of the separation mechanism. In comprehensive 2D separations, the precision (or chromatographic resolution) becomes a limiting factor and is ultimately determined by the quality of the separation in both dimensions.

In theory, multidimensional separation methods can be developed by combining almost any of the different chromatographic mechanisms or phases, electrophoretic techniques, or field-flow fractionation subtechniques. The potential variety of multidimensional combinations is exceedingly large. In practice, however, there are far more single-dimensional separations than multidimensional ones. A simple tandem arrangement does not satisfy the second criterion, and is not considered to be multidimensional. Examples would include the use of a short column packed with a reversed-phase material used for the first separation, followed by a separation on a longer column packed with the same phase and using the same or a similar mobile phase. Simply joining two different columns in series rarely offers much advantage in HPLC, except for SEC where columns of different pore sizes linked in series are used to optimise the separation of a sample with wide molecular weight distribution. Another useful application of simply joining two columns in series is the simultaneous separation of anions and cations using tandem anion- and cation-exchange columns in series. A 2D planar TLC system clearly satisfies the first criterion for multidimensional separation, provided that component displacement along the two axes is governed by different mechanisms.

There are two general types of multidimensional chromatography separation schemes: those in which the effluent from one column flows directly on to a second column at some time during the experiment, and those in which some type of trap exists between the two columns to decouple them (off-line mode). The purpose of a trap is often to allow collection of a fixed eluate volume to reconcentrate the analyte zone prior to the second separation step, or to allow a changeover from one solvent system to another. The use of *off-line multidimensional techniques* (conventional sample cleanup) with incompatible mobile phases, is common in the literature, and replacing these procedures with automated on-line multidimensional separations will require continuous development efforts.

Multidimensional chromatography requires particular attention to *sample preparation*, interface configuration and detection. Whenever two such chromatographic operations are preceded by a solvent extraction step or on-line sorbent-based sample preparation, these should be considered as part of the multidimensional systems, adding its own dimension to yield a dimensionality of three. Various extraction techniques have been combined with multidimensional chromatography, such as SFE, SPE, PHWE and MMLLE (see also Section 7.1.1). SPE-GC, SPE-CE are orthogonal techniques (i.e. based on different chromatographic methods). In principle, analysis of a sample by multicolumn chromatography is relatively simple. Two columns of different selectivity can be connected. If each column is connected to a separate detector, two sets of retention data are obtained simultaneously. Retention data on the two columns are correlated for the same sample components. A multidimensional system is easily automated once the appropriate chromatographic conditions are established.

Effective multidimensional separation requires not only a multiplicity of separation stages, but also that the integrity of separation achieved in one stage must be carried through to the others (fidelity). Separation modes need to be (made) compatible. The compatibilities of each column with each mobile phase must be considered carefully. An example is a system in which the cleanup column is a size-exclusion chromatography column compatible with organic-containing mobile phases and water, and the analytical column is a reversed-phase HPLC column (SEC-RPLC). In some cases, no change in mobile phase is required (e.g. GC-GC, SFC-SFC).

Table 7.85 Multidimensional chromatography

First dimension	Second dimension					
	GC	SFC	HPLC	SEC	TLC	CE/CEC
GC	+		+		+	
SFC	+	+	+		+	
HPLC	+	+	+	+	+	+
SEC	+	+	+		+	
TLC	+				+	

In SFE-GC, LC-CEC, LC-SFC and LC-GC, increasing difficulty (in this order) is experienced in the change of mobile phase.

The prevailing aspect of the first dimension separation is miniaturisation, in order to keep the injection volume in the second dimension sufficiently low. The most important aspect of the second dimension is speed, in order to maintain the separation achieved in the first dimension as much as possible.

Table 7.85 gives an overview of the main multidimensional chromatographies. Multidimensional success stories are GC-GC, LC-LC and LC-CE. Various multidimensional approaches may be distinguished:

(i) 2D planar separations (using the two right-angle dimensions of a continuous surface); and
(ii) coupled column separation (in which the effluent from one column is shunted to subsequent columns, which are subject to new separative conditions).

The greatest strength of 2D planar methods is that they distribute components widely over a 2D space of high peak capacity.

Table 7.86 compares single-column and multidimensional chromatographic methods. Multidimensional approaches have attracted attention for their capability to solve problems related to sample preparation, separation and identification of analytes in complex samples. Sample cleanup, concentration and fractionation are performed as an integral part of the analysis in a closed system. In multidimensional methods, none of the sample material is wasted, and sensitivity is increased accordingly. Multidimensional separation methods can lead to exceptionally high peak resolution, particularly when directly orthogonal techniques are combined [902]. The strength of coupled-column technology is that columns can be coupled in flexible arrangements and used to focus on specific components. With two parameters characteristic of the component, one can greatly multiply the certainty of identifying the component relative to the 1D system in which only one parameter is available. Disadvantages of the noncoupled column approach are

Table 7.86 Main characteristics of single-column and multidimensional chromatographic systems

Single-column systems	Multidimensional systems
Advantages	
• Low cost	• Minimal sample preparation
• Wide availability	• Sample cleanup
• Relative simplicity	• No contamination
• Ease of operation	• Reduced total analysis time (greater efficiency)
	• High peak resolution
	• Several sets of independent retention data
	• Flexibility
	• Greatly increased selectivity
	• Improved reliability and repeatability
	• Analysis of very complex mixtures
Disadvantages	
• Limited resolving power	• Capital investment
• Limited information	• Complex instrumentation (complicated interfaces)
	• Operator expertise necessary
	• Mobile phase (in)compatibilities
	• Retention time window drift
	• Restrictions on detectors (related to mobile phase flow-rate)
	• Difficult quantitative analysis
	• Time-consuming method development
	• Few commercial sources

typically the time involved (solvent evaporation is often the slowest step), and potential problems with contamination and quantitative recovery of the analyte from the first separation. Refocusing on-line is often faster and more convenient than solvent evaporation off-line. For a multidimensional system, quantitation is highly dependent on the technique used to transfer components from one separating system to another. Very rigorous control of a variety of factors, including flow-rates, temperatures and switching times, is necessary. Applying on-line multidimensional chromatography to routine analyses requires a better understanding of both instrumentation and the separation processes than do conventional off-line techniques.

Giddings [901] and others [903,904] have reviewed the use of multiple dimensions in analytical separations.

General applications While capillary separation methods produce peak capacities, n, numbered in hundreds, many real-world mixtures (e.g. in the petroleum industry) require values of $n \approx 10^4$. This can only be achieved by coupling of chromatographic columns, either with the same or different mobile phases. Multidimensional separation procedures are especially important for complex samples of numerous components where a high degree of resolution is required (typically in the petroleum industry).

The beauty of 2D gel electrophoresis as a separation technique is the orthogonality of the two separation dimensions: separation by charge (isoelectric point) in the first dimension, and separation by size in the second dimension. Two-dimensional gel electropheresis is the core separation technique for proteomics, along with HPLC (for preparative isolation).

When the complexity or 'dimensionality' of a sample increases (with zero dimensionality corresponding to a pure compound), hyphenated and multidimensional techniques soon become indispensable. This also certainly holds true for polymers with variations in molecular weight, chemical composition, functional groups or end-groups, branching, etc. Separation may require five dimensions. Perhaps the most common use of multidimensional separations is the pretreatment of a complex matrix in an off-line mode. In this, the first column serves as a filter to remove sample matrix components which might foul the analytical column. In polymer/additive analysis, extracted oligomers often contribute greatly to the complexity of the problem. Reported applications of multidimensional chromatographic techniques in polymer/additive analysis have mainly involved LC-GC, LC-LC, SEC-GC, SEC-LC and SEC-TLC. Multidimensional techniques were used in the comprehensive analysis of migrants from food packaging materials [904a]. For recent applications, see ref. [904b].

7.4.1 Multidimensional Gas Chromatography

Principles and Characteristics Multidimensional gas chromatography (MDGC) is widely used, due to the mobile-phase compatibility between the primary and secondary separating systems, which allows relatively simple coupling with less-complicated interfaces. In its simplest form, 2DGC can be carried out in the off-line mode. The most elementary procedure involves manual collection of effluent from a column, followed by re-injection into another column of a different selectivity (e.g. from an apolar to a polar column). Selecting proper GC-column combinations is critical. In on-line mode, the interface in MDGC must provide for the quantitative transfer of the effluent from one column

to the next, without altering the composition of the transferred sample or degrading the resolving power of the second column. Cold trapping (cryogenic refocusing) is a term used in MDGC to describe the refocusing of the analytes at the head of the second column after transfer from the first. It reconcentrates the transferred fraction into a narrow band prior to re-injection on the second column to maintain resolution. In MDGC the second column is generally a conventional dimension column. Sensitivity improvement is normally not achieved, and only a small increase in overall separation capacity may be realised. The specialised technology is favoured by a wide availability of sensitive and selective detectors. Selective detectors such as FTIR and $MS(MS^n)$ have been added to multidimensional GC [905].

Objectives for MDGC comprise short analysis times, high resolution, increase in peak capacity, improved determination of trace components, and avoidance of high column temperatures for the elution and venting of low-volatility components which are not of interest for the analysis. It is necessary that the components of interest be sufficiently volatile to be transported in the gas phase. MDGC lacks variation of the selectivity depending on the mobile phase, and is somewhat limited by the selectivity differences which are obtained using common stationary phases [906]. The primary drawback of MDGC is that it can only be applied to a limited range of zones of a primary column separation, i.e. only to a few parts of a sample. A multidimensional GC system may be used in various *operational modes*, such as solvent-flush, back-flush and heartcut. 2DGC based on heartcutting or sample cleanup techniques invariably uses only a part of the sample, but gains its power by rejecting the portion of the sample that is not of interest to the analyst. A narrow cut from the first column results in less potential interference for separation on the second column. Obviously, heartcut methods are rarely used when it is necessary to completely analyse every single component in a sample, as is usually necessary in polymer/additive analysis. Separations with transfer of eluate cuts from a first to a coupled second column should only be called multidimensional if additional new and significant sets of qualitative (e.g. relative retention) as well as the related better quantitative data are achieved. Basically, this can only be the case if either the polarity of the stationary phase or the temperatures for the separation systems in the coupled columns are different. In complex mixtures, the (GC-)MS spectrum is not sufficient for identification.

GC-TLC is not particularly difficult, but has been little used since its inception in the mid-1960s [907–912]. In most instances TLC was used to either confirm the identity of a GC peak or as a test of peak homogeneity. These problems are now generally more conveniently solved by GC-MS.

As for *sample preparation*, SPE-GC has become more popular than NPLC-GC. Aqueous samples are not compatible with NPLC-GC, while RPLC-GC has never become a success. SPE-GC-(tandem)MS and SPE-GC-AED systems have demonstrated excellent performance. SPME is an equilibrium technique while SPE affords exhaustive extraction of the analytes. Laser desorption injection in LD-GC-MS can uniquely provide an additional dimension of spatial information for 2D surface chemical mapping [221].

Comprehensive two-dimensional gas chromatography (GCxGC) is emerging as a new, powerful separation technique. In GCxGC two 'orthogonal' 1D GC separations are applied to an *entire* sample. The sample is first separated on a normal-bore nonpolar capillary GC column in the programmed temperature mode. All of the effluent of this first column is then focused in very many, very narrow fractions (band width <10 ms) at regular, short intervals, and subsequently re-injected on to a second (semi)polar capillary column, which is short and narrow to allow for very rapid separations. The resulting chromatogram has two time axes (retention on each of the two columns) and a signal intensity corresponding to the peak height. Characteristic GCxGC chromatograms related to solute structure can be obtained by using selective GC detectors (including elemental-selective detectors). The detector should permit a high sampling rate (>100 Hz), e.g. FID, μECD, or ToF-MS. In comprehensive gas chromatography, contour plots ($r_t - r_t$) substitute for chromatograms.

GCxGC can compete with the most complex GC-MS techniques. For quantitative analysis, GCxGC is far superior. Comprehensive 2DGC is fast and provides a complete characterisation of the entire sample in a single analysis. Peak overlap is much less likely than in conventional GC. The two independent separations provide an inherent validation of peak assignments. The method is applicable to analytes with boiling points up to 400 °C. The performance of GCxGC is considerably extended beyond that of MDGC. GCxGC achieves Giddings' vision of continuous multidimensional gas chromatography [902,913].

GCxGC is characterised by a stunning peak capacity (retention plane rather than retention time, i.e. the product of the peak capacity of each individual column) and high resolution; GCxGC tunes selectivity by using two columns with different polarities, while sensitivity is increased significantly by 10 to 50 times. For complex mixtures, comprehensive 2DGC can obtain very large

peak capacities (thousands of peaks) and high information throughput. GCxGC is proven for qualitative screening. Using retention-time standardisation, accurate and precise quantitative analysis can be achieved if time and cost are no problem. Competition to GCxGC arises from unidimensional multidetection GC with adequate data processing, where complete chromatographic resolution is often no longer required.

Important issues are orthogonality of dimensions, compatibility of methods, ease of interfacing, analysis time in each dimension, robustness and reproducibility. Important attributes of GCxGC are the timed modulator, detector and software package. Really comprehensive GCxGC generates up to 1 gigabyte of data per analysis. By choosing a specific detector, e.g. an NPD or ECD, or evaluating only a specific target (mass or 2D-location), the amount of data generated is kept manageable but the analysis is much less comprehensive. Data evaluation and quantification are still bottlenecks. 2DGC is mainly a research tool. Chemometrics is being applied [914,915]. Coupling of ToF-MS with comprehensive GC (GCxGC) results in a unique 3D separation of analytes. State-of-the-art *GCxGC-ToFMS* can collect up to 500 spectra per second. The technique is especially suitable for analysis of *very complex* mixtures and extracts. Further development of GCxGC requires accepted modulation procedures [916], quantitative analysis, validation studies and significant applications.

MDGC and GCxGC have recently been reviewed [917–921a].

Applications Multidimensional GC has been used for many years in many different operational modes, such as solvent-flush, back-flush and heartcut, using fully automated multiple transfers. Heartcut 2D methods, especially GC-GC and LC-GC, are only applicable to the determination of volatile, thermally stable solutes. MDGC is typically more suitable for the analysis of highly complex hydrocarbon mixtures than for a limited number of additives (volatiles, oligomers) in polymers. Typical areas of application of MDGC are hydrocarbon, environmental and food analysis with involvement of the separation of isomers and enantiomers. Application of GC-LC for the determination of oligomers in soluble polymers may be envisaged, providing that analyte concentrations are high enough.

GCxGC is most often applied for target analysis. Main application areas for GCxGC are to be found where unresolved peaks are the norm, i.e. for atmospheric organics (e.g. urban air), food and flavour (organoleptics), (petro)chemical and forensic analyses, and for trace-level studies. The peak capacities of a single-column chromatographic system do not match analytical needs in petroleum, natural product and biological chemistry, at variance with the problems in polymer/additive analysis where unresolved peaks are not the norm. Application of GCxGC to food samples offers a new tool to enforce stricter legislation. European legislation (Directive 02/657/EC) requires lower quantitation limits in food (e.g. 10 ppb of pesticide). Often this cannot be obtained in GC-MS and HR-MS, due to low signals and interfering matrices, and therefore will be a driving force for GCxGC. GCxGC is being used in operating refineries, in oil pricing, in phenolic additives analysis, and in solvent purity checks. GCxGC is also used in the characterisation of synthetic macromolecules [922] and can be helpful in identifying catalyst killers. A major application is the TD-GCxGC of smelly polymers. Py-GCxGC is being explored for polymers and additives. Applications of GCxGC have been reviewed [923].

It seems only a matter of time until one-dimensional GC analyses are not accepted any more in demanding cases, as they are not sufficiently specific.

7.4.2 Multidimensional Supercritical Fluid Chromatography

Principles and Characteristics Many samples are intrinsically so complex that they cannot be resolved by a single chromatographic method. In other instances, the complexity of the analytical problem is increased, as in case of extraction of additives from polymers, often leading to co-extraction of oligomers. Although cSFC has a high resolving power, the efficiency of a single capillary column is often inadequate for the separation of closely related compounds and complex mixtures, e.g. oligomers.

Multidimensional SFC offers advantages similar to other coupled chromatographic systems for the analysis of complex samples. Because of the characteristics of the supercritical fluids, SFC can play an important role in combining two chromatographic techniques without severe problems. The application of CO_2, as a common mobile phase for both normal-phase and reversed-phase chromatography, is a simple alternative to many LC-LC methods. Often, the use of SFC in a multidimensional system (SFC-SFC, SFC-GC and LC-SFC) avoids many of the problems associated with the interfacing of two dissimilar mobile phases. Substituting open-tubular SFC for the capillary dimension of a coupled column system allows a wider range of samples to be analysed, with only a moderate sacrifice in chromatographic efficiency.

SFC-SFC is more suitable than LC-LC for quantitation purposes, in view of the lack of a suitable mass-sensitive, universal detector in LC. Group quantitation can be achieved by FID. The ideal SFC-SFC system would consist of a short (10–30 cm) packed-capillary primary column, interfaced to a long (5–10 m) open-tubular column, but such a combination is difficult to realise, due to the different flow-rates required for each column type. Coupled SFC-SFC is often configured with a solute concentration device prior to valve switching on to the SFC. The main approaches to this concentration stage are the use of absorbent material or cryofocusing. Davies *et al.* [924] first introduced two-dimensional cSFC (cSFC-cSFC), and its use has been reported [925,926].

Alternatively, the LC dimension of LC-GC may be replaced by packed-column SFC, in order to improve the compatibility between two mobile phases and to allow the FID to be used for both separations. Because of the relatively nonpolar nature of $scCO_2$, *SFC-GC* is particularly recommended as a substitute for many normal-phase LC-GC analyses. The techniques developed for solvent evaporation at the LC-GC interface are often not required in SFC-GC, because the solutes are deposited at the front of the GC column when CO_2 decompresses into a gas at the end of the SFC column.

SFC-TLC is largely unexplored. Stahl [927] developed a device for supercritical fluid extraction with deposition of the fluid extracts on a moving TLC plate. Wunsche *et al.* [928] have described an automated apparatus for direct pSFC-TLC coupling. Compared to collecting the effluent from the SFC in decompression vessels, the direct deposition of the effluent on the TLC plate leads to significant losses of analytes. Multidimensional SFC has been reviewed [929].

Applications On-line pSFC-GC has been applied to the analysis of fossil fuels, such as group-type separations of high-olefin gasoline (saturates, olefins and aromatics) [930]. No significant applications concerning polymer/additive analysis can be mentioned.

7.4.3 Multidimensional Liquid Chromatography

Principles and Characteristics The use of a liquid chromatographic separation as a means of preparing samples for subsequent analysis by another chromatographic separation is well established. The goal of such 'cleanup' separations is to reduce the complexity of the original sample matrix by separating a fraction of that matrix from the analyte. Often, the separation scheme used in such cases is truly multidimensional, since the mode of the preliminary separation step is different from that of the analytical separation which follows. Techniques involving these various sample preparation schemes prior to analysis by chromatographic separation may be off-line multicolumn separations, where sample preparation and analytical columns are not physically coupled, and there is some manual manipulation of the sample between columns. Typical examples of such multidimensional separation schemes include sample preparation with a normal-phase separation on a glass column packed with silica gel and eluted with organic solvent mixtures, followed by evaporation of the solvent from the collected fraction(s) containing the analyte(s), redissolution of the fraction(s) in an appropriate solvent and analysis by RPLC or GC.

The very different selectivities of the various modes of liquid chromatography, such as absorption, partition, ion-exchange, size-exclusion and affinity chromatography, offer a much richer number of possibilities for multidimensional separations of truly different selectivity than does GC. As shown in Table 7.85, *multidimensional liquid chromatography* (MDLC) comprises a number of techniques, basically LC-GC, LC-LC, LC-SFC, LC-SEC, and LC-CE (in particular for samples of biological origin), and LC-TLC. For example, microcolumn LC has been interfaced with TLC-FTIR [931], conventional HPLC [932], microcolumn LC [933], GC [934], SFC [935] and CZE [936] as the second dimension. MDLC is well suited to solving selectivity problems, and offers a very high separating power. The challenges in applying the technique are in efficiently coupling the columns for transfer of analytes, and in maintaining mobile phase/column compatibility. Primary uses of MDLC are minimisation of separation time for the analysis of complex mixtures, cleanup and trace enrichment. The literature abounds with descriptions of multidimensional LC methods. Applications are numerous, and comprise sampling of a section of the chromatogram for further resolution (heartcut); on-column concentration (the use of a weaker eluent to reconcentrate analytes prior to separation); or the use of columns having stationary phases with similar characteristics but differing phase ratios. Not all column types are compatible. Typical examples of compatible systems are SEC-NPLC, RPLC-IEC and IEC-RPLC.

The use of LC as a sample pretreatment of GC has received considerable attention. The separation mechanisms in LC and GC are complementary, providing a powerful combined tool for the separation of complex mixtures. In *LC-GC* the primary LC column typically achieves separation by class, and the GC column fractionates members within the class [937,938]. LC-GC combines the selectivity of liquid chromatography with the efficiency and sensitivity of gas chromatography, yielding relatively high peak capacities. The choice between on-line and off-line LC-GC is not trivial. Off- and at-line LC-GC offer greater flexibility of instrumentation, and might be economically more viable. Disadvantages lie in exposure of the sample to air and light, and losses caused by adsorption of the analytes on glassware. Fully on-line automated LC-GC is the technique of choice when large series of samples have to be analysed. Matching the two technologies is not straightforward, since the two separation techniques operate in phases which are in two different physical states. LC-GC coupling requires that a liquid mobile phase is converted into a GC compatible vapour phase.

LC-GC transfer techniques are essentially the following:

(i) direct introduction (by means of auto-injectors or autosamplers);
(ii) eluent feed (introduction of the LC effluent directly by the LC pump); and
(iii) gas transfer [906].

The first reported on-line LC-GC coupling was by Majors [939], using direct introduction. The major limitation of this approach is the liquid volume ($1\,\mu L$) that can be introduced directly into the capillary GC column, considering the obvious incompatibility with conventional LC columns that operate at flow-rates of $1-3\,mL\,min^{-1}$. Consequently, only a small fraction of a selected peak can be sampled for introduction into the high-resolution GC column, rendering quantitative analysis difficult. One approach to quantitative eluent transfer is to miniaturise the LC system. LC-GC has gained popularity, especially with LC microcolumns ($0.1-1.0\,mm$) to reduce eluent volume. The reduction of the internal diameter of the LC column utilised for multidimensional chromatographic applications to microcolumn dimensions introduces various advantages and minimises the problems encountered when large volumes of liquid are introduced into capillary GC columns. The solutes are diluted much less, and much larger sections of the LC chromatogram can be introduced into the capillary GC, allowing quantitative transfer of the components of interest, and resulting in greater reproducibility and better quantitation. Small transfer fractions ($40\,\mu L$) may be injected into the GC system. In μLC-CGC, the primary microcolumn is used

Table 7.87 Main features of directly coupled LC-GC

Advantages

- Minimal sample preparation
- Powerful tool for preseparating complex samples (LC)
- Separation of by-products from interested fraction (narrow windows)
- Complete transfer (quantitation)
- Improved sensitivity and selectivity (enhanced separation efficiency)
- Reproducibility
- Use of detection modes incompatible with LC
- More straightforward interfacing to MS than for LC
- Ease of automation
- Speed of analysis
- Reduced solvent consumption

Disadvantages

- Dependence on component volatility
- Restricted GC column/eluent choice
- Interface durability
- Need for thermally stable solutes
- Demanding method optimisation
- Difficult technique
- Cost

as a highly efficient cleanup step, or a chemical class fractionation; GC carries out the final separation more efficiently. The sensitivity of the analysis is improved dramatically, typically 400–1000 times.

Table 7.87 shows the *main features* of on-line micro LC-GC (see also Table 7.86). The technique allows the high sample capacity and wide flexibility of LC to be coupled with the high separation efficiency and the many selective detection techniques available in GC. Detection by MS somewhat improves the reliability of the analysis, but FID is certainly preferable for routine analysis whenever applicable. Some restrictions concern the type of GC columns and eluent choice, especially using LC columns of conventional dimensions. Most LC-GC methods are normal-phase methods. This is partly because organic solvents used as eluents in NPLC are compatible with GC, making coupling simpler. RPLC-GC coupling is demanding; water is not a suitable solvent for GC, because it hydrolyses the siloxane bonds in GC columns. On-line RPLC-GC has not yet become routine. LC-GC technology is only applicable to compounds that can be analysed by GC, i.e. volatile, thermally stable solutes. LC-GC is appropriate for complex samples which are difficult or even impossible to analyse by a single chromatographic technique. Present LC-GC methods almost exclusively apply on-column, loop-type or vaporiser interfaces (PTV).

The main reason for pursuing on-line LC-GC is that HPLC provides far better resolution than conventional techniques of *sample preparation* (e.g. involving

cartridges) and automation, which thus greatly reduces manual sample preparation work. With solid samples, LC-GC cannot be used without extraction of the analytes from the solid matrix into a suitable solvent. Coupling of the extraction system directly to LC-GC allows the whole analysis to be made in a closed system even for solid samples, and the reliability of the analysis is improved. Off-line SFE and microcolumn LC-GC have been performed [940]. There have been applications reported in the area of microcolumn LC-GC using NP, RP and size-exclusion modes. Also, on-line coupling of SFE with μLC-GC has been described [934]. The advantages of on-line SFE-LC-GC are reduced sample preparation, increased resolving power, high sensitivity, quantitation, precision, minimisation of analyte loss and sample contamination, and automation potential. Less than 0.1 mL of organic solvent is required for the entire analytical procedure. The main characteristics of *SFE-LC-GC-MS* are reported in Table 7.88 [941].

Also, subcritical (hot/liquid) water can be used as a mobile phase for packed-column RPLC with solute detection by means of FID [942]. In the multidimensional on-line PHWE-LC-GC-FID/MS scheme, the solid sample is extracted with hot pressurised water (without the need for sample pretreatment), and the analytes are trapped in a solid-phase trap [943]. The trap is eluted with a nitrogen flow, and the analytes are carried on to a LC column for cleanup, and separated on a GC column using the on-column interface. The closed PHWE-LC-GC system is suitable for many kinds of sample matrices and analytes. The main benefit of the system is that the concentration step is highly efficient, so that the sensitivity is about 800 times better than that obtained with traditional methods [944]. Because small sample amounts are required (10 mg), special attention has to be paid to the homogeneity of the sample. The system is

Table 7.88 Main characteristics of SFE-LC-GC-MS

Advantages

- Quick derivatisation (SFE)
- Selective extraction (SFE)
- Effective cleaning-up (LC)
- Removal of derivatisation reagent (LC)
- Removal of unreacted acids (SFE/LC)
- Low detection limit (LC-GC)
- Low contamination risk (on-line)
- Powerful identification (MS)
- Environmentally friendly

Disadvantages

- Complicated construction
- No routine analysis
- Analysis time 1–2 h

best suited for samples in which the amounts of analytes are very small. Less-sensitive detectors may be used. Young *et al.* [945] have reported SWE-WRPLC separations of hydrophobic compounds, in which subcritical water extraction (SWE) is combined with water-only reversed-phase liquid chromatography (WRPLC).

In *comprehensive LCxGC*, on-line coupling of LC with GC is established, in which fractions of the complete LC analysis are transferred to the GC system. The key to successful LCxGC (or LC-GC) is the large volume sampling step that forms the bridge between LC and GC. A conventional LC column, generally operated at a flow of 1 mL min^{-1}, will provide peaks 1 to 2 min wide. As a consequence, fractions of approximately 250 µL will have to be transferred into the GC system to preserve the resolution obtained in the first-dimension LC separation, which is required in LCxGC. When microcolumns are used in LC, the entire peak volume can be transferred to the GC, and reversed-phase systems can be used without great problems.

In on-line *multidimensional HPLC* (MDHPLC) two relatively high-efficiency columns are coupled in an instrument, via the use of valves, traps and other means. In LC-LC the precolumn is used for sample cleanup and prefractionation, before introduction of the fraction of interest to the analytical column. Much of the instrumentation for MDHPLC is the same as that in conventional 'one-dimensional' experiments. However, the additional complexity of MDHPLC experiments leads to greater difficulties than those found in conventional HPLC:

(i) development of two LC separations;
(ii) compatibility of both columns with all the mobile phases (alternatively: solvent changeover); and
(iii) valve switching (timing).

Trapping systems in MDHPLC usually involve the quantitative transfer of an analyte from one column to another, with an opportunity for additional analyte manipulation between columns. Some manipulations may include trapping the analyte zone in a sample loop of known volume, reconcentration of the analyte zone, or a change in the nature of the solvent system. The column flow-rate for the first column must be generally far less or similar to that of the second column. To bring about a significant improvement in resolution, the retention mechanisms for the two columns should be complementary. The mobile phase for the first column must be a weak solvent for the sample transferred to the second column, as well as being miscible with the mobile phase for the second column. On-line LC-LC can involve interfacing normal-phase and reversed-phase LC

Table 7.89 Main features of on-line MDHPLC

Advantages

- Increased analytical throughput
- Reduced operator effort
- Increased selectivity
- Separation of highly polar and nonvolatile compounds
- Sample cleanup
- Analysis of very complex mixtures
- Additional information (in comparison to conventional LC)

Disadvantages

- Instrumental and operational design
- Operator expertise
- Mobile-phase (in)compatibilities
- Relatively low sensitivity
- Cost

eluents – an approach which requires an intermediate solvent exchange step. Effective trapping techniques are important for trace analysis, since quantitatively collecting the analyte eluting from a cleanup column and refocusing it into a narrow band, which gives the narrowest (i.e. tallest) peaks in the analytical separation, maximises the sensitivity of the analysis. The detector chosen for MDHPLC should be one which shows minimal response to changes in mobile-phase flow-rate.

Table 7.89 lists the *main characteristics* of MDHPLC (see also Table 7.86). In MDHPLC the mobile-phase polarity can be adjusted in order to obtain adequate resolution, and a wide range of selectivity differences can be employed when using the various available separation modes [906]. Some LC modes have incompatible mobile phases, e.g. normal-phase and ion-exchange separations. Potential problems arise with liquid-phase immiscibility; precipitation of buffer salts; and incompatibilities between the mobile phase from one column and the stationary phase of another (e.g. swelling of some polymeric stationary-phase supports by changes in solvents or deactivation of silica by small amounts of water).

Different HPLC techniques that can be coupled are GPEC® (for the separation according to chemical composition and functional type), SEC (separation on molecular size), and LCCC (for separation according to functional type). The SEC-GPEC®, LCCC-SEC, and NP-GPEC®/RP-GPEC® couplings have been successfully applied. For SEC-GPEC®, a molar mass-chemical composition distribution (MMCCD) is obtained, and molar mass-functional type distributions (MMFTD) can be obtained when SEC is coupled with LCCC or GPEC®. Another example of MDLC is the coupling of ion-exclusion and ion-exchange separation modes. Also, higher-dimensional systems such as LC-LC-GC-FID are in use [946].

Various authors have described on-line *LC-SFC coupling* [947,948]. Coupling of LC to SFC with conventional-size LC columns, where only a small fraction of the peak of interest is transferred to the SFC, allows only for qualitative results, and does not address the need for improved sensitivity in cSFC. Cortes *et al.* [948] have described relatively large-volume sample introductions (>10 μL) into cSFC, using microcolumn LC in the first dimension. LVI-LC-cSFC provides enhanced sensitivity compared with conventional cSFC injection techniques. LC-cSFC is expected to be of utility in the characterisation of complex samples, and in the determination of components which are thermally labile; do not contain significant chromophores; or do not have sufficient volatility to be analysed by GC.

Direct coupling between gravity-flow column liquid chromatography (CC) and TLC had already been reported in 1969 [949]. Boshoff *et al.* [950] first mentioned direct HPLC-TLC coupling. HPLC-HPTLC couplings have been described by various authors [951–953]; a simple apparatus for the cleanup of extracts isolated by SPE has been described [954]. Separation by adsorption chromatography takes place preferentially, as a result of hydrogen bonding or dipole–dipole interactions. Further separation within a polarity group can then be achieved either two-dimensionally or off-line, by partition chromatography on TLC plates. In *LC-TLC coupling*, the effluent from a narrow-bore LC column is deposited on a TLC plate after a normal- or reversed-phase separation, without serious loss of chromatographic information [952,955,956]. The interface is a fused capillary which connects the column outlet to a commercial spray jet assembly for TLC [957,958]. The column eluent is nebulised by mixing with (heated) nitrogen gas and sprayed as an aerosol on to the layer. At flow-rates typical for narrow-bore columns (5–100 μL min^{-1}), the whole column eluent can be applied to the layer. Because of the reduced flow-rate, microcolumn (i.d. 0.53 mm) HPLC has great promise for coupling to TLC, e.g. as μHPLC-TLC-FLD [953]. For higher column flow-rates, a splitter is required. The restricted flow capacity of the spray jet interface can be overcome by adding an on-line SPE step to concentrate column fractions.

In HPLC-TLC coupling, the crucial aspect is the maintenance of the chromatographic integrity during the deposition process. The chromatogram is preserved after LC separation, and is available for further separation and/or investigation. LC-TLC coupling increases the separation efficiency, and allows detection modes which are incompatible with LC (e.g. spectroscopic techniques

such as fluorescence line-narrowing, phosphorescence and surface-sensitised spectroscopy, which generally require immobilised analytes) [953]. The usefulness of TLC plates as a LC buffer memory for fluorescence emission and excitation experiments has been clearly demonstrated. The attraction of LC-TLC is its ability to separate multiple samples simultaneously, and thus to allow fractions of similar composition to be identified and grouped together for further analysis. LC-TLC has also received some attention as a solvent-elimination interface for spectroscopic identification using DRIFT, fluorescence or SERRS. In μRPLC-TLC-DRIFT (*in situ*) the limited spectral window is a drawback [959]. LC-TLC presents a high resolving power. The technique has reached a reasonable level of maturity. The principles of HPLC-TLC have been described [722].

HPLC and CE may also be coupled to give an on-line multidimensional set-up [960]; CZE and HPLC are highly orthogonal systems. On-line HPLC-GC has been reviewed repeatedly [49,906,961–963], as well as MDHPLC [964]; various textbooks are available [965,966].

Applications If an extract needs further cleanup, it is possible to couple it with multidimensional chromatographic techniques such as LC-LC or LC-GC. The first chromatographic step can then be used for the on-line *cleanup* and concentration of the extract, and the second one for the final separation. Large-volume, on-column injection (LVI-COC) is particularly useful for coupled LC-GC in which 100–350 μL fractions of eluent from the NPLC cleanup separation step are transferred on-line to the GC column. For example, on-line removal of high-MW interfering material, such as polymers from a polymer/additive dissolution, can be achieved easily by using SEC before the fraction containing additives is transferred to the GC.

On-line NPLC-GC-FID and/or FTIR analysis has been used in discriminating between *paraffin waxes* and paraffin oils present in, or migrating between, food packaging and food simulants; FID was used for quantitation [967]. In a typical application, on-line coupled LC-GC-FID has also been used for the analysis of food contamination by *mineral oil* from printed cardboard [968]. The technique has revealed that many foods are contaminated with mineral oil products. Grob *et al.* [969] have determined mineral oil in canned food by on-line LC-LC-GC-FID. DEHP was determined in salad oil by means of conventional LC-GC [970]. HPLC-GC-MS/MS (ion trap) can serve highly useful purposes in areas of applications such as impurity identification in complex mixtures. The advantages are easier and more reliable analysis (disturbing compounds are removed in the LC step); increased sensitivity due to LC-LC or LC-GC (large fractions can be transferred to GC); and a closed system (no contamination, minimisation of sources of errors). On-line PHWE-LC-GC with FID or MS detection was used to determine *brominated flame retardants* in sediments, as an alternative to Soxhlet and GC-MS [971]. After extraction with pressurised hot water, the analytes were adsorbed in a solid-phase trap. In this arrangement, LC is used for very selective cleanup and concentration of the extract; final analysis is carried out by GC.

Coupled LC-LC can separate high-boiling petroleum residues into groups of saturates, olefins, aromatics and polar compounds. However, the lack of a suitable mass-sensitive, universal detector in LC makes quantitation difficult; SFC-SFC is more suitable for this purpose. Applications of multidimensional HPLC in *food analysis* are dominated by off-line techniques. MDHPLC has been exploited in trace component analysis (e.g. vitamin assays), in which an adequate separation for quantitation cannot be achieved on a single column [972]. LC-LC-GC-FID was used for the selective isolation of some key components among the irradiation-induced olefinic degradation products in food, e.g. dienes and trienes [946].

Partial resolution of the various isomers of pentaery-thritol tetrastearate (PETS) was achieved using LVI-LC-cSFC [948]. LC-CE has mainly been applied to samples of biological origin. Micro RPLC has been combined with alumina TLC to separate an eight-compound mixture [959]; the complete TLC plate could be imaged by FTIR, and pure spectra of all individual compounds could be obtained.

Using MS detection relaxes the constraints on LC resolution, because additional separation occurs in the mass domain. In principle, LC-MS may yield a complete 2D distribution of a polymer according to chemical composition and molar mass. If MS detection is employed, the efficient cleaning in the LC step makes it possible to use total ion monitoring and even to identify unknown compounds from the sample. As extracts often contain interfering compounds, mass spectrometry in selective ion mode is a practical detector. Fully automated multidimensional LC-MS-MS-MS systems are available.

Two-dimensional liquid-chromatographic separations are also of great potential interest in *polymer analysis*. After separating macromolecules, according to only one type of heterogeneity, by one experiment, there is no chance to get a correlation between different types of heterogeneity, unless two LC experiments are hyphenated on-line. In LC-SEC polymers are separated in the first dimension, according to chemical composition or functionality type distribution – using gradient-elution or isocratic LC; in the second dimension the resulting fractions are separated according to size by SEC. LC-SEC is an excellent separation method for the simultaneous determination of the functionality-type distribution and MWD of polymers [973]. A disadvantage is the required separation time. LC-SEC may be equipped with a broad choice of detectors (RI, UV, ELSD, MALLS, MS, FTIR).

SEC in combination with multidimensional liquid chromatography (LC-LC) may be used to carry out *polymer/additive analysis*. In this approach, the sample is dissolved before injection into the SEC system for prefractionation of the polymer fractions. High-MW components are separated from the additives. The additive fraction is collected, concentrated by evaporation, and injected to a multidimensional RPLC system consisting of two columns of different selectivity. The first column is used for sample prefractionation and cleanup, after which the additive fraction is transferred to the analytical column for the final separation. The total method (SEC, LC-LC) has been used for the analysis of the main phenolic compounds in complex pyrolysis oils with minimal sample preparation [974]. The identification is reliable because three analytical steps (SEC, RPLC and RPLC) with different selectivities are employed. The complexity of pyrolysis oils makes their analysis a demanding task, and careful sample preparation is typically required.

7.4.3.1 Multidimensional Size-exclusion Chromatography

Principles and Characteristics An obvious extension of size-exclusion chromatography (SEC) is combination with another chromatographic technique. In multidimensional size-exclusion chromatography (SEC-GC, SEC-HPLC, SEC-CZE) SEC achieves a gross separation between high-MW components (to be disposed of) and low-MW components (to be separated further by means of a secondary technique). Important advantages of the fully on-line coupling of sample preparation and GC analysis are the considerable time saving, the improved detection limits, and the reduced risk of losing unstable analytes. Cortes *et al.* [975] have described on-line μ*SEC-GC-MS* for rapid identification of polymer additives, including antioxidants, plasticisers, lubricants, flame retardants waxes, and UV stabilisers, in solutions

Table 7.90 Main features of μSEC-GC analysis of polymer dissolutions

Advantages

- Reduced sample handling and preparation time
- Potential for full automation
- Minimisation of extraction losses
- Quantitative analysis
- Polymer *and* additive analysis
- High analyte detectability

Disadvantages

- Upper temperature limit of GC column
- Polymer dissolution required
- Adequate separation of polymer and additives needed
- Not suitable for high-MW additives

of the polymeric materials. The method is expected to be applicable to a wide range of polymeric materials, as long as they can be dissolved and an adequate separation of the polymer and additive fractions can be obtained in the preliminary step (microcolumn SEC); this separation is necessary to prevent the introduction of nonvolatile (polymeric) components into the capillary GC-MS system.

Table 7.90 shows the *main characteristics* of the polymer/additive analysis procedure, consisting of dissolution followed by μSEC-GC. An advantage of this approach is the minimisation of the extraction losses inherent in conventional sample preparation techniques. The determination of higher-molecular-weight additives is restricted, due to the upper interface temperature attainable in the GC-MS system. The usefulness of the tool for quantitative analysis has been ascertained [948,976].

Most LC-GC methods apply vaporiser, loop-type or on-column interfaces, a trapping loop or PTV. The maximum volume of 500–800 μL, that can be handled by an on-column interface, makes it less suitable for SEC-GC when using conventional SEC columns. The loop interface permits transfer of volumes up to 10 mL, and only the temperature during transfer has to be optimised. The main disadvantage of the loop-type interface is that it can be used only for high-boiling analytes [977]. Compared with conventional LC-GC systems, the volume of the fraction of interest eluted from the SEC column is fairly large, typically several mL. Since the whole fraction rather than an aliquot is introduced into the GC system, analyte detectability is high. This allows small sample sizes (few mg). While SEC is often performed using columns with a large inner diameter (>7 mm), for on-line SEC-GC it is advantageous to use lower i.d. columns (4.6 mm). In this way, lower fraction volumes must be transferred.

Requirements for ensuring SEC compatibility with GC were discussed [978]. A potential problem in LVI is the repeatability of retention times.

For the analysis of nonvolatile compounds, on-line coupled microcolumn *SEC-PyGC* has been described [979]. Alternatively, on-line μSEC coupled to a conventional-size LC system can be used for separation and quantitative determination of compounds, in which volatility may not allow analysis via capillary GC [976]. An automated SEC-gradient HPLC flow system for polymer analysis has been developed [980]. The high sample loading capacity available in SEC makes it an attractive technique for intermediate sample cleanup [981] prior to a more sensitive RPLC technique. Hence, this intermediate step is especially interesting for experimental purposes whenever polymer matrix interference cannot be separated from the peak of interest. Coupling of SEC to RPLC is expected to benefit from the miniaturised approach in the first dimension (no broadening). Development of the first separation step in *SEC-HPLC* is usually quite short, unless problems are encountered with sample/column compatibility.

Also, *SEC-cSFC* has been reported [947]. Due to the nature of the interface and the use of conventional-sized SEC columns, only small portions of the SEC fractions can be analysed by cSFC (i.d. <100 μm); packed-capillary SFC would be an obvious improvement to avoid overloading. Fujimoto *et al.* [931] have reported the combined power of *μSEC-TLC-DRIFT* (*in situ*). Because of the reduced flow-rate, microcolumn HPLC has great promise for this type of coupling.

Applications   Multidimensional SEC techniques can profitably be applied to soluble polymer/additive systems, e.g. PPO, PS, PC – thus excluding polyolefins. A fully automated on-line sample cleanup system based on SEC-HRGC for the analysis of additives in polymers has been described, as illustrated for PS/(200–400 ppm Tinuvin 120/327/770, Irgafos 168, Cyasorb UV531) [982]. In this process, the high-MW fractions are separated from the low molecular masses. SEC is often used as a sample cleanup for on-line analysis of additives in food extracts; these analyses are usually carried out as on-line LVI-SEC-GC-FPD.

Cortes *et al.* [975] have used on-line μSEC-CGC for rapid determination of a great variety of *additives* in an emulsion ABS-PVC blend, HIPS and a styrene–acrylate–ethylene rubber polymer. These systems are difficult to analyse, because of the high levels of insolubles such as fillers, pigments, or rubber modifiers. The additives were separated from the polymer fraction in a polymer/additive dissolution using μSEC, and were transferred on-line to a capillary GC-MS. In this process, the way in which components of interest are transferred from the LC system to the GC column is crucial. Liquid solvent should not reach the capillary column stationary phase. Ideally, the action of the microcolumn SEC should exclude the polymer matrix and cause total permeation of the additives. On-line μSEC-GC/LC analysis of dissolved polymer samples (e.g. PC homopolymer or ABS terpolymer) yields quantitative additive analysis, as opposed to the dissolution/precipitation technique where additives may exhibit solubility dependence [948,976]. μSEC-GC has been used widely, as in the quantitative analysis of a dissolution of HIPS/(Ionol, palmitic/stearic acid, Irganox 1076, Irgafos 168) [983]. Perkins *et al.* [304] have reported the development of an on-line analysis method for 100 ppm polymer additives in a 500 μL SEC fraction in DCM by means of at-column GC. High-MW compounds (1200 Da) were determined at low concentrations (total elution time: 27 min; repeatability 3–7 %).

SEC-GC-FID, according to Figure 7.40, has been used to carry out the simultaneous determination of the polymer average molecular masses and molar mass distribution and the concentration of additives [984]. The effluent was split and adsorbed on PTV packing material before GC analysis. The choice of PTV

packing material is crucial. Test samples consisted of THF dissolutions of PVC/(DMP, DEP, DPP, DIBP, DNBP, BBP, DEHP) with *n*-docosane as an internal standard, PVC/(DIPA, DIBA, DEHA, DBS, BEHA) and PVC/(BHT, bisphenol-A, Sumilizer MDP-S/BBM-S, Santonox R). Repeatability analysis indicated RSD values of 0.5–2.0 % for the phthalates. The linearity range of the calibration curves for each of the phthalates differed, because of their different adsorptive force to the packing material of PTV. On-line μSEC-PyGC has been used to characterise a styrene–acrylonitrile copolymer in terms of the average polymer composition as a function of molecular size [979].

The literature reports various (multidimensional) chromatographic approaches involving SEC and LC operating on dissolved polymer/additive mixtures. Floyd [985] has used microbore (1 mm i.d.) SEC-RPLC for the quantitative analysis of Tinuvin P in a cellulose acetate solution in THF, after separation of the polymeric and additive fractions; total analysis time about 30 min. Relative accuracy and precision of 3 % and 1.5 % were quoted. SEC-RPLC was also used to determine the styrene level in polystyrene crystals [986]. Additives in copolymers have been separated in a SEC/C$_{18}$ system [987]. Chlorohydrin mixtures may be analysed by RPLC, but not in the presence of polymer. Thus, SEC

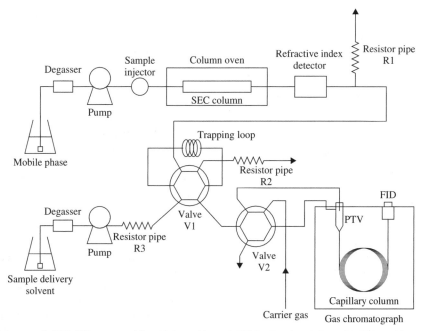

Figure 7.40 Scheme of SEC-GC system. After Kobayashi *et al.* [984]. Reprinted from N. Kobayashi *et al.*, *Journal of Microcolumn Separations*, **12**, 501–507, Copyright © (2000, John Wiley & Sons, Inc.). This material is used by permission of John Wiley & Sons, Inc.

was used to separate material for subsequent examination by RPLC. In particular, open-column SEC was performed in order to separate the additive and its transformation products from heat-treated PVC stabilised with [1-^{14}C]-TETO [988]. However, the low-MW fraction still contained substantial amounts of PVC oligomers with molecular weights up to about 2000 Da [989]. Consequently, HP-SEC was necessary in order to obtain a sample containing essentially only TETO and its derived chlorohydrins. The indicated fraction was trapped and analysed by RPLC. A combination of chromatographic techniques was used here, but HP-SEC performed an essential function in obtaining a clean sample for RPLC; it is doubtful whether this could have been achieved differently. Nerín *et al.* [990,991] used SEC-NPLC for the separation of a dissolved polymer fraction (diverted to waste) from the analyte-containing fraction. µSEC-LC has recently been used for the analysis of photocrosslinkers in Cyclotene 4000 series advanced electronic resins [992]. BHT (MW 220) and BHA (MW 180) were separated from vegetable oil (MW >600) by means of SEC, and separated from each other by RPLC [993].

Cotter *et al.* [994] have used on-line SEC-LC for the *simultaneous analysis* of polymers and low-MW additives. Staal *et al.* [27] have described the dissolution of polycarbonate and polysulfone in THF and precipitation on to a C$_{18}$ guard column. Separation of polymer and additives was achieved using gradient elution from water–THF (50:50 vol. %) as nonsolvent to 100 % THF. The additives elute first, followed by the oligomers and polymer. Additives grafted on to polymers may be characterised by SEC-HPLC cross-fractionation using ELS detection [995]. SEC-GPEC® coupling allows a mass-chemical composition distribution (MMCCD) to be obtained, whereas a molar mass-functional type distribution (MMFTD) of polymers may be derived from SEC-LCCC.

Fujimoto *et al.* [931] have taken advantage of the increase in chromatographic resolution provided by combining µSEC and TLC. The SEC chromatogram of a mixture of four polymer additives (Irganox 1010, AO 2246, Cyasorb UV9, oleamide), which could not be separated completely by µSEC alone because of the similarity of molecular sizes (AO 2246 and oleamide) (*cf.* Fig. 7.41), was deposited continuously on to a moving silica plate, which was subsequently developed in a direction perpendicular to the direction of deposition, thereby resolving all of the analytes (Figure 7.42). *In situ* DRIFT spectra of the individual spots permitted the identification of each additive at a limit of detection of less than 8 µg, which is considerably better than for photoacoustic (50 µg per

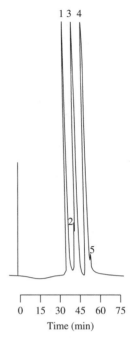

Figure 7.41 Microcolumn size-exclusion separation of a four-component mixture using THF as the mobile phase and UV detection at 230 nm. Legend: (1) Irganox 1010; (2) oleamide; (3) AO 2246; (4) Cyasorb UV9; (5) sample solvent (CHCl$_3$). After Fujimoto *et al.* [931]. Reprinted from *Journal of Chromatography*, **438**, C. Fujimoto *et al.*, 329–337, Copyright (1988), with permission from Elsevier

component) and sample transfer DR measurements (20–50 µg per component). The procedure may be extended to polymer/additive dissolutions.

7.4.4 Multidimensional Thin-layer Chromatography

Principles and Characteristics Multidimensionality in planar chromatography takes a broader context than is common for column separations. For TLC, various modes of multidimensional separations are used:

(i) unidirectional continuous development (with change of solvent);
(ii) unidirectional multiple development (repeated elution in the same direction);
(iii) sequential 2D development (with redevelopment with a different separation mechanism in the orthogonal direction) under conditions of capillary-controlled or forced flow of the mobile phase,

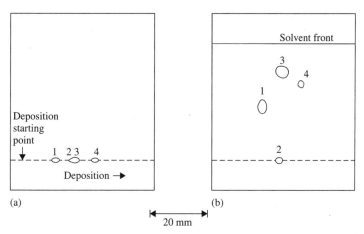

Figure 7.42 Appearance of chromatographic spots. (a) After HPLC separation and deposition on the TLC plate; (b) after development of the chromatogram with benzene. After Fujimoto *et al.* [931]. Reprinted from *Journal of Chromatography*, **438**, C. Fujimoto *et al.*, 329–337, Copyright (1988), with permission from Elsevier

mobile-phase gradients and system automation (mainly for qualitative analysis);

(iv) coupling of a second chromatographic technique to TLC (i.e. GC-TLC, HPLC-TLC); and

(v) use of TLC as a sample preparation method (i.e. TLC-GC, TLC-HPLC after spot transfer) [996,997].

Two-dimensional separations in planar chromatography are rather trivial to perform. All unidimensional multiple development techniques employ successive repeated development of the layer in the same direction, with removal of the mobile phase between developments. The main variants are multiple chromatography and incremental multiple development. The basis for automated multiple development (AMD) is the automation of unidimensional, incremental, multiple development with a reverse solvent strength gradient [998]. 2D TLC finds limited use, and is mainly a qualitative technique.

Research in coupled chromatographic systems with TLC as one separation dimension is limited. Although interfacing column chromatographic methods such as GC, LC or SFC as one dimension and planar chromatography as the second is not very common, it offers some unique possibilities for enhancing both separation capacity and detection. Interfacing GC or HPLC to TLC is technically not particularly difficult. In most cases TLC was used to confirm the identity of a GC peak. This problem is now solved by GC-MS. The driving force for interfacing TLC to GC, HPLC or SFC is low, because of the perception that column systems provide superior separation capacity, and in view of the difficulty of quantifying the TLC chromatograms obtained. However, the

thin-layer plate acts as an efficient storage detector, especially where on-line detection lacks adequate sensitivity or selectivity. Coupling of column and TLC separation allows the sample components to be further investigated free of time constraints. This is advantageous for sample components that require derivatisation for convenient or selective detection; for sequential scanning using different detection principles or to preserve and transport the separation to different locations for detection; and for applications employing solid-phase spectroscopic identification techniques.

In a *total spot injection* technique to analyse TLC spots by GC, the spot is scraped off and introduced in a special injection port of the GC [999]. Alternatively, the scraped-off TLC zone can be analysed by *TD-GC-MS* [305]. Compared to other thermal methods and laser desorption [804,809,858,1000] thermal extraction (up to 315 °C) is milder, and can desorb thermally stable volatile and semivolatile compounds from silica gel without extensive decomposition. At 315 °C, most thermally stable compounds will not undergo a pyrolysis reaction in an inert gas atmosphere. This method provides an effective means for quantitative and qualitative analysis of compounds separated on TLC plates. The detection limit can be extended to the sub-ng/spot level. Analysis of separated TLC spots cut from the plate has also been carried out by *PyGC* [1001,1002]. Zhu and Yeung [1000] used direct coupling of TLC to GC by low-power *laser desorption* (TLC-LDGC), and they observed considerable fragmentation of the analytes into decomposition products. Background peaks from the sorbent, coupled with pyrolysis products generated from the analytes, limit the usefulness of this approach.

In both cases, GC fingerprint libraries must be built before quantitative analysis can be routinely carried out. In analysis of QTLC by *laser pyrolysis scanning* (LPS), the TLC plates are placed in a chamber after development, and were irradiated with an IR laser to produce a high temperature at the location of the spot. The analyte is swept by a carrier gas to a GC, and detected with FID or ECD. The technique combines the separation power of TLC and the detection modes of GC [846].

Overpressure layer chromatography (OPLC) can be employed as a cleanup method for HPLC, or used for prefractionation. OPLC-HPLC coupling has been described by ref. [1003].

The greatest strength of 2D planar methods is that they distribute components widely over a 2D space of high peak capacity. Multidimensional TLC development has the advantages of requiring simple equipment; is compatible with scanning densitometry for solute identification and quantitation; and enables exploitation of the spot reconcentration mechanism.

Postcolumn hyphenated and multidimensional TLC techniques have been reviewed [721,996,1004].

Applications   AMD has been used for the study of the PEG oligomer distribution [1005]. No specific application concerning polymer/additive analysis appears to have been reported.

7.5   MULTIDIMENSIONAL SPECTROSCOPY

Principles and Characteristics   A new level of understanding is achieved when several analytical techniques are combined in a hyphenated approach. On the other hand, in a simple experiment in which spectroscopic analysis is performed as a function of time, concentration, or other additive properties, the output will be *multidimensional*, and a number of independent variables will determine its dimensions. If

such an experiment is conducted by varying spatial coordinates, frequency, or other domains, the situation changes. Multidimensional experiments greatly advance our understanding of structure–property relationships in polymeric materials.

The separation of interactions by 2D spectroscopy can be compared with 2D chromatography. In a one-dimensional thin layer or paper chromatogram, the separation of the constituents by elution with a given solvent is often incomplete. Elution with a second solvent in a perpendicular direction may then achieve full separation. In NMR spectroscopy, the choice of two solvents is replaced by the choice of two suitable (effective) Hamiltonians for the evolution and detection periods which allow unique characterisation of each line.

The general experimental approach used in *2D correlation spectroscopy* is based on the detection of dynamic variations of spectroscopic signals induced by an external perturbation (Figure 7.43). Various molecular-level excitations may be induced by electrical, thermal, magnetic, chemical, acoustic, or mechanical stimulations. The effect of perturbation-induced changes in the local molecular environment may be manifested by time-dependent fluctuations of various spectra representing the system. Such transient fluctuations of spectra are referred to as *dynamic spectra* of the system. Apart from time, other physical variables in a generalised 2D correlation analysis may be temperature, pressure, age, composition, or even concentration.

Two-dimensional spectroscopy is a rather novel concept, and a powerful tool in analysing spectra. The advantages of generalised 2D correlation spectroscopy are:

(i) enhancement of spectral resolution by spreading spectral peaks over the second dimension (simplification of complex spectra);

(ii) identification of various inter- and intramolecular interactions through selective correlation of peaks;

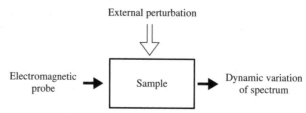

Figure 7.43   General conceptual scheme to obtain a 2D correlation spectrum by inducing selective time-dependent spectral variations with an external perturbation (mechanical, electrical, chemical, magnetic, optical, thermal, etc.). After Noda [1006]. From I. Noda, *Applied Spectroscopy*, **44**, 550–561 (1990). Reproduced by permission of the Society for Applied Spectroscopy

(iii) probing the specific order of the spectral intensity changes; and

(iv) unambiguous band assignment based upon the correlations between various bands [1006].

Synchronous 2D correlation spectra represent coupled or related changes of spectral intensities, while asynchronous correlation spectra represent independent or separate variations [1007]. The 2D cross-correlation analysis enhances similarities and differences of the variations of individual spectral intensities, providing spectral information not readily accessible from 1D spectra.

Although the idea of generating 2D correlation spectra was introduced several decades ago in the field of NMR [1008], extension to other areas of spectroscopy has been slow. This is essentially on account of the time-scale. Characteristic times associated with typical molecular vibrations probed by IR are of the order of picoseconds, which is many orders of magnitude shorter than the relaxation times in NMR. Consequently, the standard approach used successfully in 2D NMR, i.e. multiple-pulse excitations of a system, followed by detection and subsequent double Fourier transformation of a series of free-induction decay signals [1009], is not readily applicable to conventional IR experiments. A very different experimental approach is therefore required. The approach for generation of *2D IR spectra* defined by two independent wavenumbers is based on the detection of various relaxation processes, which are much slower than vibrational relaxations but are closely associated with molecular-scale phenomena. These slower relaxation processes can be studied with a conventional IR spectrometer with the use of a simple time-resolved technique. 2D IR spectra are especially suited for elucidating various chemical interactions among functional groups. The type of information contained in a dynamic spectrum is determined by the selection of the perturbation (e.g. migration, drawing, aggregation, etc.).

The formal approach of 2D correlation analysis to time-dependent spectral intensity fluctuations has been extended to UV, Raman [1010], and near-IR spectroscopy [1011–1014]; 2D fluorescence is upcoming.

Main advances in ESR spectroscopy have recently come about by adding new dimensions to basic 1D ESR [1015]. Dimensions such as time and radiofrequency radiation have either created new spectroscopies or enriched one-dimensional forms. Examples are pulsed ESR and ENDOR.

Generalised 2D correlation spectroscopy is powerful in exploring complicated NIR spectra [1016]. Both 2D IR [1017] and 2D NIR correlation spectroscopy [1018] have been reviewed; for *n*D NMR, see refs [1019,1020].

Figure 7.44 shows the 2D UV chromatogram (RPLC-UV/VIS (DAD)) for a five-compound test mixture of polymer additives [662]. Any spectral data collected during hyphenated chromatography–spectroscopy measurements can be readily transformed into 2D correlation spectra.

Multidetector chromatographic–spectroscopic methods are frequently also described in terms of multidimensionality. In GC detection, both FTIR and AED are used in parallel with MS, i.e. after a split. GC-FTIR-MS systems are commercially available, at variance to GC-AED-MS. The latter can be applied in the analysis

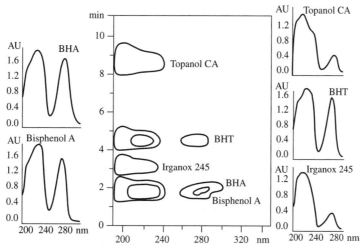

Figure 7.44 2D UV chromatogram (254 nm) for a five-compound test mixture. After Louden *et al.* [662]. Reproduced from D. Louden *et al.*, *Anal. Bioanal. Chem.*, **373**, 508–515 (2002), by permission of Springer-Verlag, Copyright (2002)

of organometallic compounds. Another example of a *multidetection strategy* is the combination of GC, UV spectroscopy and MS (3D technique). There are various designs for a GC-MS or GC-UV interface: by effusive gas inlet, continuous supersonic beams and pulsed supersonic beams.

The probability of co-eluting components in the conventional GC-MS separation of complex mixtures is quite high. GC-ToFMS is well placed for complex mixture analysis. Deconvolution software makes it possible to distinguish almost co-eluting components on the basis of mass spectra. One way to combine UV spectroscopy and ion formation (and thus mass spectrometry) is by resonance-enhanced multiphoton ionisation (REMPI) with wavelength-tuneable lasers. This *3D technique* (GC retention time, excitation laser UV wavelength, mass) allows unperturbed GC resolution, high UV selectivity (isomer selection) and fast, well-resolved mass-selective monitoring [1021]. As a highly selective 3D trace analysis technique, GC-UV-MS greatly reduces time-consuming sampling 'cleanup' procedures.

Monitoring of GC eluents by UV/ToF-MS detection may be carried out using either selective REMPI or unselective LEI [1021]. Laser-enhanced electron ionisation (LEI) combined with ToF-MS (GC-LEI-ToFMS) produces ag levels of sensitivity for organotin [1022]. A major advantage of laser-induced electron ionisation is that the ionisation method is switched very rapidly and easily from unselective (LEI) to highly selective (REMPI) ionisation. The use of a selective ionisation technique, such as MPI, increases target analyte specificity. Both GC-MPI-FTMS [1023] and GC-REMPI-ToFMS [1024] have been described. Very good molecular cooling is indispensable for optimum selectivity of gas-phase UV spectroscopy and REMPI. GC-jet-REMPI-MS is well suited for analysis of target compounds in complex matrices. *GC-REMPI-ToFMS* is a concept for coupling GC-MS with high-resolution UV spectroscopy as an additional analytical dimension to reach ultrahigh selectivity [1025]. A capillary GC is interfaced with a dynamic gas sample cell, where laser−analyte interaction takes place. In GC-REMPI-ToFMS, the chromatographic eluent is expanded repetitively (20 Hz) as short supersonic jet pulses (∼150 μs in duration) into the vacuum of the mass spectrometer. Under jet-cooled conditions, UV absorption spectroscopy (i.e. REMPI spectroscopy) becomes highly compound selective. Using GC-REMPI-MS, the chemical species discrimination can be pushed far above the usual limits of chromatographic separation, simply by virtue of the ionisation process [1024]. In order to be effective, the laser wavelength should be able to ionise the analytes of interest. It is surmised that a tuneable UV laser would allow the selective determination of a given class and possibly a subclass of compounds. The multidimensional nature of the approach comes from the ability to perform on-line monitoring following analytical processes and products from any of these events:

(i) photoelectrons by means of a total electron capture detector (TECD);
(ii) photo-ions via ToF-MS; and
(iii) laser-induced fluorescence (LIF) by the optical emission probe.

Detection limits in the 100 fg range can be obtained with a tuneable UV laser working at a wavelength of maximum absorption for the compounds of interest. Continuous supersonic beams require high gas loads and combination with a pulsed ionisation technique (e.g. REMPI) is unfavourable in terms of sensitivity. Pulsed valves are a better approach for a GC-UV-MS interface [1021].

Applications The potential use of 2D correlation spectroscopy is very wide [1007]. Most multidimensional techniques arise from the correlation of frequency domains in the presence of external perturbations, as in NMR. For applications of multidimensional NMR spectroscopy and NMR diffusion measurements, see Sections 5.4.1 and 5.4.1.1.

Dynamic IR spectroscopy coupled with 2D correlation analysis can provide insights into submolecular interactions in blends and compounds [1017]. 2D IR spectroscopy allows identification of specific interactions between components in polymer mixtures. While blends and copolymers have been studied [1026], no reports on compounds have yet appeared. Applications of 2D IR spectroscopy to polymeric materials have been reviewed [1017,1026].

For applications of multidetector methods, see Section 7.3.3.4 in particular.

7.6 BIBLIOGRAPHY

7.6.1 General

E.R. Malinowski, *Factor Analysis in Chemistry*, John Wiley & Sons, Ltd, Chichester (2002).
Z. Deyl (ed.), *Directory of Hyphenated Techniques*, Elsevier, Amsterdam (1994).

7.6.2 Multihyphenation and Multidimensionality

U.A.Th. Brinkman (ed.), *Hyphenation. Hype and Fascination*, Elsevier, Amsterdam (1999).

R.P.W. Scott, *Tandem Techniques*, John Wiley & Sons, Ltd, Chichester (1997).

E.J. Karjalainen and U.P. Karjalainen, *Data Analysis for Hyphenated Techniques*, Elsevier Science, Amsterdam (1996).

U.A.Th. Brinkman and H. Poppe (eds), *Multidimensionality. Hyphenation and Coupled-Column Techniques, J. Chromatography* **A703** (1995).

T. Provder, M.W. Urban and H.G. Barth (eds), *Hyphenated Techniques in Polymer Characterization*, ACS Symposium Series No. 581, American Chemical Society, Washington, DC (1994).

7.6.3 Precolumn Hyphenation

M. Caude and D. Thiébaut, *Practical Supercritical Fluid Chromatography and Extraction*, Harwood Academic Publishers, Amsterdam (1999).

R.M. Smith and S.B. Hawthorne (eds), *Supercritical Fluids in Chromatography and Extraction*, Elsevier Science Publishers, Amsterdam (1997).

F.A. Settle (ed.), *Supercritical Fluid Chromatography and Extraction*, Prentice Hall, Upper Saddle River, NJ (1997).

M.S. Verrall (ed.), *Supercritical Fluid Extraction and Chromatography*, John Wiley & Sons, Ltd, Chichester (1996).

S.A. Westwood (ed.), *Supercritical Fluid Extraction and Its Use in Chromatographic Sample Preparation*, Blackie A&P, London (1993).

K. Jinno (ed.), *Hyphenated Techniques in Supercritical Fluid Chromatography and Extraction*, Elsevier Science Publishers, Amsterdam (1992).

B. Wenclawiak, *Analysis with Supercritical Fluids: Extraction and Chromatography*, Springer-Verlag, Berlin (1992).

7.6.4 Postcolumn Hyphenation

S. Down and J. Halket (eds), *LC/MS Applications Database Online*, HD Science Ltd, Nottingham (2003).

R. Willoughby, E. Sheehan and S. Mitrovich, *A Global View of LC/MS*, Global View Publishing, Pittsburgh, PA (2002).

R.E. Ardrey, *LC-MS. An Introduction*, John Wiley & Sons, Ltd, Chichester (2002).

K. Alberti, *On-line LC-NMR and Related Techniques*, John Wiley & Sons, Ltd, Chichester (2002).

W.M.A. Niessen (ed.), *Current Practice of Gas Chromatography–Mass Spectrometry*, M. Dekker, New York, NY (2001).

W.M.A. Niessen, *Liquid Chromatography–Mass Spectrometry*, M. Dekker, New York, NY (1999).

M. Oehme, *Practical Introduction to GC-MS Analysis with Quadrupoles*, Hüthig Verlag, Heidelberg (1998).

M. McMaster and C. McMaster, *GC/MS: Practical User's Guide*, Wiley-Interscience, New York, NY (1998).

W.M.A. Niessen and R.D. Voyksner, *Current Practice in Liquid Chromatography–Mass Spectrometry*, Elsevier, Amsterdam (1998).

R. Houriet (ed.), *HPLC Liquid Chromatography: Major Mass Spectrometers for GC/MS and LC/MS*, EDP Sciences, Les Ulis (1998); in: *Analusis* **26** (2) (1998).

R. Smits (ed.), *Proceedings 5th International Symposium on Hyphenated Techniques in Chromatography and Hyphenated Chromatographic Analysers, J. Chromatogr.* **A819** (1–2) (1998).

H.-J. Hübschmann, *Handbuch der GC/MS. Grundlagen und Anwendung*, Wiley-VCH, Weinheim (1996).

F.G. Kitson, B.S. Larsen and C.N. McEwen, *Gas Chromatography and Mass Spectrometry. A Practical Guide*, Academic Press, San Diego, CA (1996).

W.M.A. Niessen and J. van der Greef, *Liquid Chromatography–Mass Spectrometry*, M. Dekker, New York, NY (1992).

P.C. Uden (ed.), *Element-Specific Chromatographic Detection by Atomic Emission Spectroscopy*, American Chemical Society, Washington, DC (1992).

C.W. Huang, *Interfacing Supercritical Fluid Chromatography with Double Focusing Mass Spectrometry*, UMI, Ann Arbor, MI (1990).

R. White (ed.), *Chromatography/Fourier Transform Infrared Spectroscopy and Its Applications*, M. Dekker, New York, NY (1990).

A.L. Yergey, C.G. Edmonds, I.A.S. Lewis and M.L. Vestal, *Liquid Chromatography–Mass Spectrometry: Techniques and Applications*, Plenum Press, New York, NY (1990).

W. Herres, *HRGC-FT-IR: Capillary Gas Chromatography–Fourier Transform Infrared Spectroscopy. Theory and Applications*, Hüthig Publishers, Heidelberg (1987).

7.6.5 Multidimensional Chromatography

L. Mondello, A.C. Lewis and K.D. Bartle (eds), *Multidimensional Chromatography*, John Wiley & Sons, Ltd, Chichester (2001).

K. Grob, *On-line Coupled LC-GC*, Hüthig Publishers, Heidelberg (1991).

H.J. Cortes (ed.), *Multidimensional Chromatography*, M. Dekker, New York, NY (1990).

7.6.6 Multidimensional Spectroscopy

I. Noda and Y. Ozaki, *Two-Dimensional Correlation Spectroscopy*, Wiley-Interscience, New York, NY (2004).

H. Friebolin, *Basic One- and Two-Dimensional NMR Spectroscopy*, VCH, Weinheim (1991).

R.R. Ernst, G. Bodenhausen and A. Wokaun, *Principles of Nuclear Magnetic Resonance in One and Two Dimensions*, Clarendon Press, Oxford (1997).

F.J.M. van de Ven, *Multidimensional NMR in Liquids: Basic Principles and Experimental Methods*, VCH, New York, NY (1995).

M.W. Urban and T. Provder, *Multidimensional Spectroscopy of Polymers. Vibrational, NMR, and Fluorescence Techniques* ACS Symposium Series No. 598, American Chemical Society, Washington, DC (1995).

W.R. Croasmun and R.M.K. Carlson (eds), *Two-Dimensional NMR Spectroscopy, Applications for Chemists and Biochemists*, VCH, New York, NY (1994).

K. Schmidt–Rohr and H.W. Spiess, *Multidimensional Solid-State NMR and Polymers*, Academic Press, New York, NY (1994).

K. Nakanishi (ed.), *One-Dimensional and Two-Dimensional Spectra by Modern Pulse Techniques*, Kodansha, Tokyo (1990).

7.7 REFERENCES

1. T. Hirschfeld, *Anal. Chem.* **52**, 297A (1980).
2. E.J. Karjalainen and U.P. Karjalainen, *Data Analysis for Hyphenated Techniques*, Elsevier Science, Amsterdam (1996).
3. M.T. Combs, M. Ashraf-Khorassani and L.T. Taylor, *J. Chromatogr.* **A785**, 85–100 (1997).
4. U.A.Th. Brinkman and H. Poppe (eds), *Multidimensionality. Hyphenation and Coupled-Column Techniques*, *J. Chromatogr.* **A703** (1995).
5. J.W. King, *J. Chromatogr. Sci.* **27**, 355–64 (1989).
6. R.D. Smith and H.R. Udseth, *Fuel* **62**, 466–9 (1983).
7. H.T. Kalinoski, H.R. Udseth, B.W. Wright and R.D. Smith, *Anal. Chem.* **58**, 2124–9 (1986).
8. S.B. Hawthorne and D. Miller, *J. Chromatogr. Sci.* **24**, 258–64 (1986).
9. Z.Y. Liu, P.B. Farnsworth and M.L. Lee, *J. Microcol. Sep.* **4**, 199–208 (1992).
10. B.W. Wright, S.R. Frye, D.G. McMinn and R.D. Smith, *Anal. Chem.* **59**, 640–4 (1987).
11. S.B. Hawthorne, M.S. Kreiger and D.J. Miller, *Anal. Chem.* **60**, 472 (1988).
12. M. Thilén and R. Shishoo, *J. Appl. Polym. Sci.* **76**, 938–46 (2000).
13. K.K. Unger and P. Roumeliotis, *J. Chromatogr.* **282**, 519–26 (1983).
14. K. Sugiyama, M. Saito, T. Hondo and A. Senda, *J. Chromatogr.* **332**, 107–16 (1985).
15. N.J. Cotton, K.D. Bartle, A.A. Clifford, S. Ashraf, R. Moulder and J. Dowle, *J. High Resolut. Chromatogr.* **14**, 164–8 (1991).
16. E. Stahl, *J. Chromatogr.* **142**, 15 (1977).
17. T. Hyötyläinen and M.-L. Riekkola, *LC.GC Europe* **15** (5), 298–306 (2002).
18. K. Kuosmanen, T. Hyötyläinen, K. Hartonen and M.-L. Riekkola, *Abstracts 7th International Symposium on Hyphenated Techniques in Chromatography and Hyphenated Chromatographic Analyzers (HTC-7)*, Bruges (2002), Paper P86.
19. Y.I. Chen, Y.-S. Su and J.-F. Jen, *Abstracts 7th International Symposium on Hyphenated Techniques in Chromatography and Hyphenated Chromatographic Analyzers (HTC-7)*, Bruges (2002), Paper P25
20. *Application Note, The ASPEC System: Automated Sample Preparation with Extraction Columns,* Gilson Medical Electronics, Middleton, WI (1989).
21. M. Papagiannopoulos and A. Mellinthin, *Abstracts 7th International Symposium on Hyphenated Techniques in Chromatography and Hyphenated Chromatographic Analyzers (HTC-7)*, Bruges (2002). Paper P120.
22. K. Robards, P.R. Haddad and P.E. Jackson, *Principles and Practice of Modern Chromatographic Methods*, Academic Press, London (1994).
23. T. Greibrokk, *J. Chromatogr.* **A703**, 523–35 (1995).
24. R.W. Vannoort, J.-P. Chervet, H. Lingeman, G.J. De Jong and U.A.Th. Brinkman, *J. Chromatogr.* **505**, 45–77 (1990).
25. S.A. Westwood (ed.), *Supercritical Fluid Extraction and its Use in Chromatographic Sample Preparation*, Blackie A&P, London (1993).
26. J. Pawliszyn (ed.), *Sampling and Sample Preparation for Field and Laboratory*, Elsevier, Amsterdam (2003).
27. W.J. Staal, P. Cools, A.M. van Herk and A.L. German, *Chromatographia* **37** (3–4), 218–20 (1993).
28. W.J. Staal, P. Cools, A.M. van Herk and A.L. German, *Separation Technol.* 857 (1994).
29. E. Stahl and W. Schiltz, *Z. Anal. Chem.* **280**, 99 (1976).
30. M.C. Smith and J. Sherma, *J. Planar Chromatogr.* **8**, 103–6 (1995).
31. S.K. Poole, T.A. Dean, J.W. Oudsema and C.F. Poole, *Anal. Chim. Acta* **236**, 3 (1990).
32. Special issue on *Sample Preparation for Chromatographic Analysis, Anal. Chim. Acta* **236** (1) (1990).
33. J.C.J. Bart, *Polym. Test* **20**, 729–40 (2001).
34. C.J. Koester and R.E. Clement, *Crit. Rev. Anal. Chem.* **24**, 263–316 (1993).
35. M. Hakkarainen, A.-C. Albertsson and S.J. Karlsson, *Environ. Polym. Degrad.* **5** (2), 67–73 (1997).
36. S.M. Dugar, J.N. Leibowitz and R.H. Dyer, *J. AOAC Intl* **77** (5), 1335–7 (1994).
37. P. Sandra and F. David, *Abstracts 7th International Symposium on Hyphenated Techniques in Chromatography and Hyphenated Chromatographic Analyzers (HTC-7)*, Bruges (2002), Paper P21.1.
38. J.M. Levy and M. Ashraf-Khorassani, in *Hyphenated Techniques in Supercritical Fluid Chromatography and Extraction* (K. Jinno, ed.), Elsevier, Amsterdam (1992), pp. 197–223.
39. T. Maeda and T. Hobo, in *Hyphenated Techniques in Supercritical Fluid Chromatography and Extraction* (K. Jinno, ed.), Elsevier, Amsterdam (1992), pp. 255–74.

40. S.B. Hawthorne, in *Supercritical Fluid Extraction and Its Use in Chromatographic Sample Preparation* (S.A. Westwood, ed.), Blackie, London (1993), pp. 65–86.

41. M.L. Lee and K.E. Markides (eds), *Analytical Supercritical Fluid Chromatography and Extraction*, Chromatography Conferences Inc., Provo, UT (1990).

42. S.B. Hawthorne and D.J. Miller, *J. Chromatogr. Sci.* **24**, 527 (1986).

43. J.M. Levy and J.P. Guzowski, *Fresenius' Z. Anal. Chem.* **330**, 207 (1988).

44. M.-L. Riekkola, P. Manninen and K. Hartonen, in *Hyphenated Techniques in Supercritical Fluid Chromatography and Extraction* (K. Jinno, ed.), *J. Chromatogr. Libr. Ser.* **53**, 275–304 (1992).

45. S.B. Hawthorne, D.J. Miller and J.J. Langenfeld, *J. Chromatogr. Sci.* **28**, 2–8 (1990).

46. T.P. Lynch, R.E.A. Escott, P.G. McDowell, I. Roberts and M.J. Carrott, *Proceedings 4th International Symposium on Supercritical Fluid Chromatography and Extraction*, Cincinnati, OH (1992), p. 179–80.

47. K.D. Bartle, A.A. Clifford, S.B. Hawthorne, J.J. Langenfeld, D.J. Miller and R. Robinson, *J. Supercrit. Fluids* **3**, 143–9 (1990).

48. R. Hillmann and K. Bachmann, *J. High Resolut. Chromatogr.* **17**, 350 (1994).

49. E.C. Goosens, D. de Jong, G.J. de Jong and U.A.Th. Brinkman, *Chromatographia* **47**, 313–45 (1998).

50. M.D. Burford, K.D. Bartle and S.B. Hawthorne, *Adv. Chromatogr.* **37**, 163–204 (1997).

51. S.B. Hawthorne, *Anal. Chem.* **62**, 633A (1990).

52. G.A. Mackay and R.M. Smith, *J. Chromatogr. Sci.* **32**, 455–60 (1994).

53. M.L. Marín, A. Jiménez, J. López and J. Vilaplana, *J. Chromatogr.* **A750**, 183–90 (1996).

54. T.P. Lynch, in *Chromatography in the Petroleum Industry* (E.R. Adlard, ed.), *J. Chromatogr. Libr.* **56**, 269–303 (1995).

55. S. Schmidt, L. Blomberg and T. Wännman, *Chromatographia* **28**, 400–4 (1989).

56. T.P. Lynch, unpublished results (2000).

57. *Suprex Corporation Application Chromatogram SFE-65* (1989).

58. M.M. Robson, K.D. Bartle, A.A. Clifford and T.P. Lynch, *Proceedings 7th International Symposium on Supercritical Fluid Chromatography and Extraction*, Indianapolis, IN (1996), Paper D-18.

59. F. Andrawes, T. Valcarcel, G. Haacke and J. Brinen, *Anal. Chem.* **70** (18), 3762–5 (1998).

60. J.M. Levy, A.C. Rosselli, E. Storozynsky, R. Ravey, L.A. Dolata and M. Ashraf-Khorassani, *LC.GC* **10** (5), 386–91 (1992).

61. V. Lopez-Avila, Y. Liu and W.F. Beckert, *J. Chromatogr.* **785**, 279–88 (1997).

62. H. Kobayashi, K. Ohyama, N. Tomiyama, Y. Jimbo, O. Matano and S. Goto, *J. Chromatogr.* **643**, 197 (1993).

63. J. Beltran, F.J. Lopez and F. Hernandez, *Anal. Chim. Acta* **283**, 297 (1993).

64. S. Öllers, M. van Lieshout, H.-G. Janssen and C.A. Cramers, *LC-GC Intl* **7**, 435–9 (1997).

65. R. Sasano, T. Hamada, M. Kurano and M. Faruno, *J. Chromatogr.* **A896**, 41–9 (2000).

66. J.J. Vreuls, G.J. de Jong, R.T. Ghysen and U.A.Th. Brinkman, *J. AOAC* **77**, 306 (1994).

67. Th. Hankemeier, *Ph.D. Thesis*, Free University of Amsterdam (2000).

68. Th. Hankemeier, J. Rozenbrand, M. Abhadur, J.J. Vreuls and U.A.Th. Brinkman, *Chromatographia* **48**, 273 (1998).

69. T. Górecki and J. Pawliszyn, *Anal. Chem.* **67** (18), 3265–74 (1995).

70. T. Górecki and J. Pawliszyn, *LC-GC Intl* **12** (2), 123–7 (1999).

71. S. Ferary and J. Auger, *J. Chromatogr.* **A750**, 63 (1996).

72. R. Eisert and J. Pawliszyn, *Anal. Chem.* **69**, 3140 (1997).

73. M.J. Yang and J. Pawliszyn, *LC.GC Intl* **9** (5), 283–96 (1996).

74. *Supelco Application Note 147*, Supelco Inc., Bellefonte, PA (1998).

75. M. Groning and M. Hakkarainen, *J. Chromatogr.* **A932**, 1–11 (2001).

76. R.R. Bodrian, *J. Coat. Technol.* **72**, 69–74 (2000).

77. D.W. Potter and J. Pawliszyn, *J. Chromatogr.* **625**, 247 (1992).

78. K.K. Chee, M.K. Wong and H.K. Lee, *J. Microcol. Sep.* **8** (2), 131–6 (1996).

79. Z.-L. Liu, C.-H. Li and C.-Y. Wu, *Fenxi Shiyanshi* **21** (1), 71–3 (2002).

80. T. de Smaele, L. Moens, R. Dams and P. Sandra, in *Applications of Solid Phase Microextraction* (J. Pawliszyn, ed.), The Royal Society of Chemistry, Cambridge (1999), pp. 296–310.

81. K. Sugiyama, M. Saito, T. Hondo and M. Senda, *J. Chromatogr.* **332**, 107–16 (1985).

82. M. Ashraf-Khorassani and J.M. Levy, *J. High Resolut. Chromatogr.* **13**, 742–7 (1990).

83. I.J. Koski and M.L. Lee, *J. Microcol. Sep.* **3**, 481 (1991).

84. N.J. Cotton, M.W. Raynor and K.D. Bartle, in *Supercritical Fluid Extraction and Its Use in Chromatographic Sample Preparation* (S.A. Westwood, ed.), Blackie, London (1993), pp. 87–111.

85. S. Mitra and N.K. Wilson, *J. Chromatogr. Sci.* **28**, 182 (1990).

86. R.J. Skelton, C.C. Johnson and L.T. Taylor, *Chromatographia* **21**, 4 (1986).

87. L. Baner, T. Bücherl, J. Ewender and R. Franz, *J. Supercrit. Fluids* **5**, 213–9 (1992).

88. D. Thiébaut, J.P. Chervet, R.W. Vannoort, G.J. De Jong, U.A.Th. Brinkman and R.W. Frei, *J. Chromatogr.* **477**, 151 (1989).

89. J. Hedrick and L.T. Taylor, *Anal. Chem.* **61**, 1986 (1989).

90. I.G.M. Anderson, in *Supercritical Fluid Extraction and Its Use in Chromatographic Sample Preparation* (S.A. Westwood, ed.), Blackie, London (1993), pp. 112–44.

91. M. Ashraf-Khorassani, M.L. Kumar, D.J. Koebler and G.P. Williams, *J. Chromatogr. Sci.* **28**, 599–604 (1990).

92. T.W. Ryan, S.C. Jocklovich, J.C. Watkins and E.J. Levy, *J. Chromatogr.* **505** (1), 273–82 (1990).

93. F. Martial, J. Huguet and C. Brunel, *Polym. Intl* **48** (4), 299–306 (1999).

94. K.D. Bartle, M.P. Burke, A.A. Clifford, I.L. Davies, J.P. Kithinji, M.W. Raynor, G.F. Shilstone and A. Williams, *Eur. Chromatogr. News* **2**, 12 (1988).

95. L. Taylor, in *Supercritical Fluid Chromatography and Extraction* (F.A. Settle, ed.), Prentice Hall, Upper Saddle River, NJ (1997), pp. 183–97.

96. A.A. Clifford, in *Supercritical Fluid Extraction and Chromatography* (M.S. Verrall, ed.), John Wiley & Sons, Ltd, Chichester (1996), pp. 241–57.

97. P.J. Sandra, A. Kot, A. Medvedovici, F. David and L. Toribio, in *Supercritical Fluids in Separation Sciences* (O. Kaiser, ed.), Incom, Düsseldorf (1997), pp. 279–89.

98. L.T. Taylor, *J. Chromatogr. Sci.* **35**, 374–82 (1997).

99. H.M. NcNair, *Am. Lab.* **29**, 10–13 (1997).

100. S.H.Y. Wong, in *Supercritical Fluid Chromatography* (S.H.Y. Wong and I. Sunshine, eds), CRC Press, Boca Raton, FL (1997), pp. 51–69.

101. M. Saito, T. Hondo and M. Senda, in *Multidimensional Chromatography. Techniques and Applications* (H.J. Cortes, ed.), M. Dekker, New York, NY (1990), pp. 331–57.

102. N.J. Cotton, K.D. Bartle, A. Clifford and C.J. Dowle, *J. Appl. Polym. Sci.* **48**, 1607–19 (1993).

103. K.D. Bartle, T. Boddington, A.A. Clifford, N.J. Cotton and C.J. Dowle, *Anal. Chem.* **63**, 2371–7 (1991).

104. M. Ashraf-Khorassani, D.S. Boyer and J.M. Levy, *J. Chromatogr. Sci.* **29**, 517–21 (1991).

105. J.M. Levy, A.C. Rosselli, E. Storozynsky, R.M. Ravey, L.A. Dolata and M. Ashraf-Khorassani, *Am. Lab.* **10**, 386–91 (1992).

106. T. Tikuisis and M. Cossar, *Proceedings SPE ANTEC '93*, New Orleans, LA (1993), p. 270.

107. *Ashland Chemical Analytical Services & Technology Bulletin 2531*, Ashland Specialty Chemical Company, Dublin, OH (1998).

108. *Suprex Corporation Application Chromatograms SFE-24, SFE-73, SFE-74, SFE-75* (1989); *SFE-94* (1991).

109. G.A. Mackay and R.M. Smith, *HRC – J. High Resolut. Chromatogr.* **18**, 607–609 (1995).

110. H. Engelhardt, J. Zapp and P. Kolla, *Chromatographia* **32**, 527 (1991).

111. Y. Hirata, F. Nakata and M. Horihata, *J. High Resolut. Chromatogr., Chromatogr. Commun.* **11**, 81 (1988).

112. T. Tikuisis and V. Dang, in *Plastics Additives. An A–Z Reference* (G. Pritchard, ed.), Chapman & Hall, London (1998), pp. 80–94.

113. T. Bücherl, A.L. Baner and O.-G. Piringer, *Dtsch. Lebensm.-Rdsch.* **89** (3), 69–71 (1993).

114. J.W. Oudsema and C.F. Poole, *J. High Resolut. Chromatogr.* **16** (3), 198–202 (1993).

115. T.P. Hunt, C.J. Dowle and G. Greenway, *Analyst* **116**, 1299–1304 (1991).

116. T. Imahashi, Y. Yamauchi and M. Saito, *Bunseki Kagaku* **39** (2), 79–85 (1990).

117. G.A. MacKay and R.M. Smith, *Anal. Proc.* **29**, 463 (1992).

118. M.J. Drews, R.P. Paradkar, J. Helvey and K. Ivey, *Abstracts 8th Intl. Symposium on Supercritical Fluid Chromatography and Extraction*, St. Louis, MO (1998), L-21.

119. C.H. Kirschner, S.L. Jordan, L.T. Taylor and P. Seemuth, *Anal. Chem.* **66**, 882–7 (1994).

120. F. Höfler and G. Alt, *Proceedings 11th International Symposium on Capillary Chromatography*, Monterey, CA (1990).

121. R.C. Wieboldt, K.D. Kempfert and D.C. Dalrymple, *Appl. Spectrosc.* **44**, 1028–34 (1990).

122. D. Leyendecker, F.P. Schmitz and E. Klesper, *Chromatographia* **23**, 38 (1987).

123. K. Ute, N. Miyatake, T. Asada and K. Hatada, *Polym. Bull.* **28**, 561 (1992).

124. T. Bücherl, A. Gruner and N. Palibroda, *Packag. Technol. Sci.* **7**, 139–54 (1994).

125. W.G. Engelhart and A.G. Gargus, *Am. Lab.* **20**, 30 (1988).

126. J.B. Nair and J.W. Huber, *LC.GC* **6**, 1071 (1988).

127. Y. Hirata and Y. Okamoto, *J. Microcol. Sep.* **1**, 46–50 (1989).

128. A.L. Howard and L.T. Taylor, in *Supercritical Fluid Extraction and Its Use in Chromatographic Sample Preparation* (S.A. Westwood, ed.), Blackie, London (1993), pp. 146–68.

129. A.M. Pinto and L.T. Taylor, *J. Chromatogr.* **A811**, 163–70 (1998).

130. H. Daimon, Y. Katsura, T. Kondo and Y. Hirata, *Proceedings 12th International Symposium on Capillary Chromatography* (K. Jinno, P. Sandra and T. Hanai, eds), Kobe (1990), p. 456.

131. T.P. Lynch, in *Applications of Supercritical Fluids in Industrial Analysis* (J.R. Dean, ed.), Blackie, Glasgow (1993), Chapter 8.

132. A.C. Hogenboom, W.M.A. Niessen, D. Little and U.A.Th. Brinkman, *Abstracts 15th Montreux Symposium on Liquid Chromatography/Mass Spectrometry*, Montreux, Switzerland (1998), Paper 46.

133. A. Duinkerken, A. Schellen, O. Halmingh, M. van Gils and B. Ooms, *LC.GC Europe* **13** (3), 182–6 (2000).

134. T. Renner, D. Baumgarten and K.K. Unger, *Chromatographia* **45**, 199–205 (1997).

135. F. Klink, *LC.GC Europe* **13** (6), 396–409 (2000).

136. J.A.B. Ooms, G.J.M. van Gils, A.R. Duinkerken and O. Halmingh, *Intl Lab.* **30** (6), 18–23 (Nov. 2000).

137. M.-C. Hennion, *J. Chromatogr.* **A856**, 3–54 (1999).

138. M. Castillo, M.R. Boleda, F. Ventura and D. Barceló, *Abstracts 15th Montreux Symposium on Liquid Chromatography/Mass Spectrometry*, Montreux, Switzerland (1998).

139. M.C. Alonso, M. Castillo and D. Barceló, *Abstracts 15th Montreux Symposium on Liquid Chromatography/Mass Spectrometry*, Montreux, Switzerland (1998).

140. E. Baltussen, H. Snijders, H.-G. Janssen, P. Sandra and C.A. Cramers, *J. Chromatogr.* **A802**, 285–95 (1998).

141. J. Chen and J. Pawliszyn, *Anal. Chem.* **67**, 2530–3 (1995).

142. A.A. Boyd-Boland and J. Pawliszyn, *Anal. Chem.* **68**, 1521–9 (1996).

143. S.A. Liebman, E.J. Levy, S. Lurcott, S. O'Neill, J. Guthrie, T. Ryan and S. Yocklovich, *J. Chromatogr. Sci.* **27**, 118 (1989).

144. U. Braumann, H. Händel, K. Albert, R. Ecker and M. Spraul, *Anal. Chem.* **67**, 930–5 (1994).

145. D.C. Dunham, M.S. Desmarais, T. Breid and J.W. Hills, *Anal. Lett.* **28**, 2561–74 (1995).

146. Y. Ikushima, N. Saito, K. Hatakeda, S. Ito, M. Arai and K. Arai, *I & EC Res.* **31**, 574–8 (1992).

147. L.T. Taylor and S.L. Jordan, *J. Chromatogr.* **A703**, 537–48 (1995).

148. D.C. Tilotta, D.L. Heglund and S.B. Hawthorne, *Am. Lab.* **28**, 36R–36T (1996).

149. C.H. Kirschner and L.T. Taylor, *Anal. Chem.* **65**, 78–83 (1993).

150. S.G. Kazarian, M.F. Vincent, B.L. West and C.A. Eckert, *J. Supercrit. Fluids* **13**, 107–12 (1998).

151. U. Braumann, H. Händel, K. Albert, R. Ecker and M. Spraul, *Anal. Chem.* **67**, 930–5 (1995).

152. K. Albert, *J. Chromatogr.* **A785**, 65–83 (1997).

153. B.E. Richter, B.A. Jones, J.L. Ezzell, N.L. Porter, N. Avdalovic and C. Pohl, *Anal. Chem.* **68**, 1033–9 (1996).

154. R.D. Smith and H.R. Udseth, *Anal. Chem.* **55**, 2266–72 (1983).

155. G.D. Bowers, C.P. Clegg, S.C. Hughes, A.J. Harker and S. Lambert, *LC.GC Mag. Sep. Sci.* **15**, 48 (1997).

156. D.C. Stahl and D.C. Tilotta, in *Applications of Solid Phase Microextraction* (J. Pawliszyn, ed.), The Royal Society of Chemistry, Cambridge (1999), pp. 625–37.

157. M.E. Cisper, W.L. Earl, N.S. Nogar and P.H. Hemberger, *Anal. Chem.* **66**, 1897–1901 (1994).

158. E. Marioth, G. Bunte and Th. Härdle, *Polym. Recycl.* **2** (4), 303–8 (1996).

159. S.L. Jordan, L.T. Taylor and P.D. Seemuth, *Proceedings 5th International Symposium on Supercritical Fluid Chromatography and Extraction*, Baltimore, MD (1994), Paper E-9.

160. P.B. Liescheski, P.D. Seemuth and T.O. Trask, *Text. Res. J.* **66**, 436–41 (1996).

161. P. Jackson, G. Dent, D. Carter, D.J. Schofield, J.M. Chalmers, T. Visser and M. Vredenbregt, *J. High Resol. Chromatogr.* **16**, 515 (1993).

162. K.L. Norton and P.R. Griffiths, *J. Chromatogr.* **A703**, 383 (1995).

163. G.W. Somsen, R.J. van de Nesse, C. Gooyer, U.A.Th. Brinkman, N.H. Velthorst, T. Visser, P.R. Kootstra and A.P.J.M. de Jong, *J. Chromatogr.* **552**, 635 (1991).

164. R. Willoughby, E. Sheehan and S. Mitrovich, *A Global View on LC.MS*, Global View Publishing, Pittsburgh, PA (1998).

165. J.M. Chalmers and G. Dent, *Industrial Analysis with Vibrational Spectroscopy*, The Royal Society of Chemistry, Cambridge (1997).

166. R.L. White, in *Encyclopedia of Spectroscopy and Spectrometry* (J.C. Lindon, ed.), Academic Press, San Diego, CA (2000), pp. 288–93.

167. R. White (ed.), *Chromatography/Fourier Transform Infrared Spectroscopy and Its Applications*, M. Dekker, New York, NY (1990).

168. G. Jalsovszky, in *Encyclopedia of Spectroscopy and Spectrometry* (J.C. Lindon, ed.), Academic Press, San Diego, CA (2000), pp. 282–7.

169. K. Anton, in *Encyclopedia of Analytical Science* (A. Townshend, ed.), Academic Press, London (1995), pp. 4856–62.

170. M. Spraul, U. Braumann, M. Godejohann and M. Hoffmann, *Pre-ENC Bruker NMR Symposium*, Orlando, FL (2000).

171. K. Albert, M. Dachtler, T. Glaser, H. Händel, T. Lacker, G. Schlotterbeck, S. Strohschein, L.-H. Tseng and U. Braumann, *J. High Resolut. Chromatogr.*, **22**(3), 135–43 (1999).

172. A.L. Burlingame, R.K. Boyd and S.J. Gaskell, *Anal. Chem.* **66**, 634R (1994).

173. D.F. Hunt, *Intl J. Mass Spectrom. Ion Phys.*, **45**, 111 (1982).

174. F.H. Field, *Adv. Mass Spectrom.* **10**, 271 (1986).

175. S.R. Horning, M.E. Bier, R.G. Cooks, G. Brusini, P. Traldi, A. Guiotto and R. Rodighiero, *Biomed. Environ. Mass Spectrom.* **18**, 927 (1989).

176. R.A. Yost and D.D. Fetterolf, *Mass Spectrom. Revs* **2**, 1 (1983).

177. W.M.A. Niessen, in *Encyclopedia of Spectroscopy and Spectrometry* (J.C. Lindon, ed.), Academic Press, San Diego, CA (2000), pp. 293–300.

178. L. Ebdon, S.J. Hill and R.W. Ward, *Analyst* **112**, 1–16 (1987).

179. L. Ebdon, S.J. Hill and R.W. Ward, *Analyst* **111**, 1113–38 (1986).

180. P.C. Uden, in *Determination of Trace Elements* (Z.B. Alfassi, ed.), VCH, Weinheim (1994), pp. 425–60.

181. K. Sutton, R.M.C. Sutton and J.A. Caruso, *J. Chromatogr.* **A789**, 85–126 (1997).

182. W.M.A. Niessen, in *Encyclopedia of Spectroscopy and Spectrometry* (J.C. Lindon, ed.), Academic Press, San Diego, CA (2000), pp. 843–9.

183. N. Ragunathan, K.A. Krock, N. Klawun, T.A. Sasaki and C.L. Wilkins, *J. Chromatogr.* **A856**, 349–97 (1999).

184. E.G. Brame, *Anal. Chem.* **37**, 1183–4 (1965).

185. J. Buddrus and H. Herzog, *Org. Magn. Reson.* **15**, 211–3 (1981).

186. J.C. van Loon and B. Radziuk, *Can. J. Spectrosc.* **21** (2), 46–50 (1976).

187. L. Ebdon, R.W. Ward and D.A. Leathard, *Analyst* **107**, 129–43 (1982).

188. R.P.W. Scott, I.A. Fowliss, D. Welti and T. Wilkens, *Gas Chromatography 1966* (A.B. Littlewood, ed.), The Institute of Petroleum, London (1966), p. 318.

189. J.H. Rayner, M.A. Moseley, E.D. Pellizzari and G.R. Vely, *J. High Resolut. Chromatogr., Chromatogr. Commun* **11**, 209 (1988).

190. J.F. Schneider, G.T. Reedy and D.G. Ettinger, *J. Chromatogr. Sci.* **23**, 49–53 (1985).

191. S. Bourne, A.M. Haefner, K.L. Norton and P.R. Griffiths, *Anal. Chem.* **62**, 2447 (1990).

192. R. Fuoco, K.H. Shafer and P.R. Griffiths, *Anal. Chem.* **58**, 3249–54 (1986).

193. S.L. Pentoney, K.H. Shafer and P.R. Griffiths, *J. Chromatogr. Sci.* **24**, 230–5 (1986).

194. S.L. Jordan and L.T. Taylor, *J. Chromatogr. Sci.* **35**, 7–13 (1997).

195. S. Bourne, G.T. Reedy, G. Coffey and D. Mattson, *Am. Lab.* (Fairfield, CT) **16** (6), 90–101 (1994).

196. D.F. Gurka, I. Farnham, B.B. Potter, S. Pyle, R. Titus and W. Duncan, *Anal. Chem.* **61**, 1584–9 (1989).

197. N. Ragunathan, A.T. Sasaki, K.A. Krock and C.L. Wilkins, *Anal. Chem.* **66** (21), 3751 (1995).

198. M.J.D. Low and S.K. Freeman, *J. Agric. Food Chem.* **16**, 525 (1968).

199. W.P. Duncan, *Am. Lab.* **20** (8), 40–6 (1988).

200. P.R. Griffiths, *Chemical IR Fourier Transform Spectroscopy*, Wiley-Interscience, New York, NY (1975).

201. G.L. McClure, in *Practical Sampling Techniques for Infrared Analysis* (P.B. Coleman, ed.), CRC Press, Boca Raton, FL (1993), Chapter 7.

202. P.R. Griffiths, S.L. Pentoney, A. Giorgetti and K.H. Shafer, *Anal. Chem.* **58**, 1349A (1986).

203. S.A. Liebman and T.P. Wampler, in *Pyrolysis and GC in Polymer Analysis* (S.A. Liebman and E.J. Levy, eds), M. Dekker, New York, NY (1985), pp. 53–148.

204. R.P.W. Scott, *Tandem Techniques*, John Wiley & Sons, Ltd, Chichester (1997).

205. W. Herres, *HRGC-FTIR: Capillary Gas Chromatography–Fourier Transform Infrared Spectroscopy, Theory and Applications*, Hüthig Publishers, Heidelberg (1987).

206. Inscan Inc., *Lab. Products Intl* **12** (1), 5 (1998).

207. D.M.W. Anderson, *Analyst (London)* **84**, 50 (1959).

208. J. Haslam, A.R. Jeffs and H.A. Willis, *Analyst (London)* **86**, 44 (1961).

209. R.J. McGorrin, T.R. Pofahl and W.R. Croasmun, *Anal. Chem.* **59**, 1109A–12A (1987).

210. G.A. Luoma and R.D. Rowland, *J. Chromatogr. Sci.* **24**, 210 (1986).

211. C.L. Wilkins, G.N. Giss, R.L. White, G.M. Brissey and E.C. Onyiniuka, *Anal. Chem.* **54**, 2260 (1982).

212. A. Rau and B.J. McGrattan, *Abstracts Pyrolysis '98*, Neuherberg/Munich (1998), p. 79.

213. J.C. Holmes and F.A. Morrell, *Appl. Spectrosc.* **11**, 86 (1957).

214. W.M.A. Niessen, *Chem. 2 Wkbld.* **17**, 29–34 (8 Sept 2001).

215. W.M.A. Niessen, *Chem. 2 Wkbld.* **8**, Lab. 6–9 (24 Apr 1999).

216. A. Vermoesen, J. Vercammen, C. Sanders, D. Courtheyn and H.F. De Brabander, *J. Chromatogr.* **564**, 385 (1991).

217. N.A. Yates, M.M. Booth, J.L. Stephenson and R.A. Yost, in *Practical Aspects of Ion Trap Mass Spectrometry* (R.E. March and J.F.J. Todd, eds.), CRC Press, Boca Raton, FL (1995), Vol. 3, p 121.

218. W. Blum and R. Aichholz, in *Encyclopedia of Analytical Science* (A. Townshend, ed.), Academic Press, London (1995), pp. 1833–44.

219. S. Dagan and A. Amirav, *Intl J. Mass Spectrom. Ion Proc.* **133**, 187 (1994).

220. A. Amirav and S. Dagan, *Eur. Mass Spectrom.* **3**, 105–11 (1997).

221. A. Amirav, S. Dagan, T. Shahar, N. Tzanani and S.B. Wainhaus, *Adv. Mass Spectrom.* **14**, 529–62 (1998).

222. A. Danon and A. Amirav, *J. Phys. Chem.* **93**, 5549 (1989).

223. A. Amirav, *Org. Mass Spectrom.* **26**, 1–17 (1991).

224. D.O. Hummel, *Macromol. Symp.* **119**, 65–77 (1997).

225. M. Statheropoulos, E. Smaragdis, N. Tzamtzis and C. Georgakopoulos, *Anal. Chim. Acta* **331**, 53–61 (1996).

226. K. Grob, G. Grob and K. Grob, *J. Chromatogr.* **156**, 1 (1978).

227. M. Oehme, *Practical Introduction to GC-MS Analysis and Quadrupoles*, Hüthig Publishers, Heidelberg (1998).

228. F.G. Kitson, B.S. Larsen and C.N. McEwen, *Gas Chromatography and Mass Spectrometry, A Practical Guide*, Academic Press, San Diego, CA (1996).

229. C.C. Grimm, S.W. Lloyd and L. Munchausen, *Intl Lab.* 10A–10F (July 1996).

230. M.I. Yavor, B. Hartmann and H. Wollnik, *Intl J. Mass Spectrom. Ion Process.* **130**, 223–6 (1994).

231. P.G. Van Ysacker, H.-G.M. Janssen, H.M.J. Snyders, P.A. Leclercq and C.A. Cramers, *J. Microcol. Sep.* **5**, 413–19 (1993).

232. P.A. Leclercq, H.M.J. Snyders, C.A. Cramers, K.H. Maurer and U. Rapp, *J. High Resolut. Chromatogr.* **12**, 652 (1989).

233. W. Windig, W.H. McClennen and H.L.C. Meuzelaar, *Chemometr. Intell. Lab. Syst.* **1**, 151–65 (1987).

234. D. Henneberg, U. Henrichs and G. Schomburg, *J. Chromatogr.* **112**, 343 (1975).

235. S.A. Rossi, J.V. Johnson and R.A. Yost, *Biol. Mass Spectrom.* **21**, 420 (1992).

236. C.P.M. Schutjes, E.A. Vermeer, J.A. Rijks and C.A. Cramers, *J. Chromatogr.* **253**, 1 (1982).

237. S.C. Davis, A.A. Makarov and J.D. Hughes, *Rapid Commun. Mass Spectrom.* **13**, 237–41 (1999).

238. A. Amirav, N. Tzanani, S.B. Wainhaus and S. Dagan, *Eur. Mass Spectrom.* **4**, 7–13 (1998).

239. S. Peters and R. Sacks, *J. Chromatogr.* **30**, 187 (1992).

240. A. Amirav, A. Gordin and A. Fialkov, *Abstracts 7th International Symposium on Hyphenated Techniques*

in *Chromatography and Hyphenated Chromatographic Analyzers (HTC-7)*, Bruges (2002), Paper P4.

241. S.C. Davis, A.A. Makarov and J.D. Hughes, *Rapid Commun. Mass Spectrom.* **13**, 247–50 (1999).

242. D. Bassi, P. Tosi and R. Schögl, *J. Vac. Sci. Technol.* **A16** (1), 114–22 (1998).

243. A. Golby, in *Development in Polymer Analysis and Characterisation*, Rapra Technology Ltd., Shawbury (1999), Paper 7.

244. W.M.A. Niessen (ed.), *Current Practice of Gas Chromatography–Mass Spectrometry*, M. Dekker, New York, NY (2001).

245. M. McMaster, *GC/MS. A Practical User's Guide*, John Wiley & Sons, Inc., New York, NY (1998).

246. F. David, L. Vanderroost and P. Sandra, *Proceedings 13th International Symposium on Capillary Chromatography*, Riva del Garda, Italy (1991).

247. Y. Kawamura, K. Watanabe, K. Sayama, Y. Takeda and T. Yamada, *Shokuhin Eiseigaku Zasshi* **38** (5), 307–18 (1997).

248. T. Maehara and T. Yamada, *Shokuhin Eiseigaku Zasshi* **40** (3), 189–97 (1999).

249. S. Tan, T. Takashi and T. Okada, *J. AOAC* **74**, 815–8 (1991).

250. T. Baba, *Shokuhin Eiseigaku Zasshi* **18** (6), 500–3 (1977).

251. S.G. Gilbert, J.R. Giacin, J.R. Morano and J.D. Rosen, *Packag. Dev. Syst.* **5**, 20 (1975).

252. F. Herrmann, Z. Dušek and P. Matoušek, *J. Appl. Polym. Sci.: Appl. Polym. Symp.* **48**, 503–10 (1991).

253. M. Hakkarainen, G. Gallet and S. Karlsson, *Polym. Degrad. Stab.* **64**, 91–9 (1999).

254. L. Boden, M. Lundgren, K.-E. Stensio and M. Gorczynski, *J. Chromatogr.* **A788** (1/2), 195–203 (1997).

255. S. Mirmiran, H. Pakdel and C. Roy, *J. Anal. App. Pyrol.* **22** (3), 205–15 (1992).

256. M. Bernabei, R. Seclì and G. Bocchinfuso, *J. Microcol. Sep.* **12** (11), 585–92 (2000).

257. J. Pierre and J. van Bree, *Kunststoffe* **73**, 319–24 (1983).

258. D.C.M. Squirrell, *Analyst* **106**, 1042–56 (1981).

259. K. Takeda and T. Nemoto, *Recent Advances in Flame Retardant Polymeric Materials* (Business Communications Co., ed.) **10**, 79–87 (1999).

260. S. Brandsma, J. van Hesselingen, P. Leonards and J. de Boer, *Varian Chromatogr. News* **1**, 18 (2002).

261. K. Usukara, T. Seko and N. Onda, *Abstracts IUPAC International Congress on Analytical Sciences 2001*, Tokyo (2001), Paper 4024.

262. R. Luyk, *Ph.D. Thesis*, University of Amsterdam (1993).

263. L.F. Lu, D. Price, G.J. Milnes, P. Carty and S. White, *Polym. Degrad. Stab.* **64** (3), 601–603 (1999).

264. E. Dziwinski, J. Hetper and R. Kolodenny, *J. Chromatogr.* **288** (1), 221–6 (1984).

265. T. Otake, J. Yoshinaga and Y. Yanagisawa, *Environ. Sci. Technol.* **35** (15), 3099–3102 (2001).

266. R. Kumar, *Am. Lab.* **31** (22), 32–5 (1999).

267. Y. Kotsuka and S. Suzuki, *Kawasaki-shi Kogai Kenkyusho Nenpo* **22**, 19–22 (1996).

268. H.-J. Hübschmann, *Handbuch der GC/MS, Grundlagen und Anwendung*, Wiley-VCH, Weinheim (1996).

269. Z. Zhong and Y. Fu, *Fushe Yanjiu Yu Fushe Gongyi Xuebao* **19** (4), 247–53 (2001).

270. M. Monteiro, C. Nerín, C. Rubio and F.G.R. Reyes, *J. High Resolut. Chromatogr.* **21** (5), 317–20 (1998).

271. D. Dilettato, P.J. Arpino, K. Nguyen and A. Bruchet, *J. High Resolut. Chromatogr.* **14** (5), 335–42 (1991).

272. K. Kurihara and F. Tanoue, *Bunseki Kagaku* **49** (3), 205–208 (2000).

273. S.-K. Yu, *Shiyou Huagong* **30** (10), 792–5 (2001).

274. J. Scheirs, J. Pospíšil, M.J. O'Connor and S.W. Bigger, *Adv. Chem. Ser.* **249**, 359–74 (1996).

275. B. Asamoto, J.R. Young and R. Citerin, *Anal. Chem.* **62**, 61 (1990).

276. M.P. Mawn, R.W. Linton, S.R. Bryan, B. Hagenhoff, U. Jurgens and A. Benninghoven, *J. Vac. Sci. Technol.* **A9**, 1307 (1991).

277. N. Haider and S. Karlsson, *Polym. Degrad. Stab.* **74**, 103–112 (2001).

278. R. Espejo, K. Valter, M. Simona, Y. Janin and P. Arrizabalaga, *Abstracts 7th International Symposium on Hyphenated Techniques in Chromatography and Hyphenated Chromatographic Analyzers (HTC-7)*, Bruges (2002), Paper P43.

279. J.A. Field and R.L. Reed, *Environ. Sci. Technol.* **30** (12), 3544–50 (1996).

280. Y.C. Ho and K.L. Yam, *Polym. Plast. Technol. Eng.* **38** (1), 19–41 (1999).

281. T. Imai, B.R. Harte and J.R. Giacin, *J. Food Sci.* **55**, 158–61 (1990).

282. M.N.G. Storm van Leeuwen and G. Wullms, *Proceedings 6th European Polymers, Laminations and Coatings Symposium*, Copenhagen (1999), pp. 199–207.

283. F. Khabbaz, A.-C. Albertsson and S. Karlsson, *Polym. Degrad. Stab.* **61**, 329–42 (1998).

284. E.D. Virus, I.A. Revelsky, A.I. Revelsky and I.N. Glazkov, *Abstracts 7th International Symposium on Hyphenated Techniques in Chromatography and Hyphenated Chromatographic Analyzers (HTC-7)*, Bruges (2002), Paper P171.

285. P.J. Fordham, J.W. Gramshaw and L. Castle, *Food Addit. Contam.* **18** (5), 461–71 (2001).

286. D. Balafas, K.J. Shaw and F.B. Whitfield, *Food Chem.* **65**, 279–87 (1999).

287. P. Gerhards, *Shimadzu Application Note GC-MS* 40 (1998).

288. G.C. Jamieson, *J. High Resolut. Chromatogr.* **5**, 632–3 (1982).

289. L. Castle, J. Gilbert, S.M. Jickells and J.W. Gramshaw, *J. Chromatogr.* **437**, 281–6 (1988).

290. W. Camacho and S. Karlsson, *Polym. Degrad. Stab.* **71** (1), 123–34 (2001).

291. L.M. Konkol, R.F. Cross, I. Harding and E. Kosior, *Abstracts 7th International Symposium on Hyphenated*

Techniques in Chromatography and Hyphenated Chromatographic Analyzers (HTC-7), Bruges (2002), Paper P80.

292. B. Raffael and C. Simoneau, *Abstracts 7th International Symposium on Hyphenated Techniques in Chromatography and Hyphenated Chromatographic Analyzers (HTC-7)*, Bruges (2002), Paper 15.3.

293. N.P.H. Ching, G.N. Jham, C. Subbarayan, D.V. Bowen, A.L.C. Smit, C.E. Grossi, R.G. Hicks, F.H. Field and T.F. Nealon, *J. Chromatogr. Biomed. Appl.* **222**, 171–7 (1981).

294. K. Jedrzejczak and V.C. Gaind, *Analyst* **117**, 1417 (1992).

295. D. Fan, Y. Lin, J. Hsiao, J. Shieh, S. Huang and S. Lin, *Proceedings SPE ANTEC '92*, Detroit (1992), pp. 1130–8.

296. M.-R. Lee and C.-C. Chou, *Abstracts 7th International Symposium on Hyphenated Techniques in Chromatography and Hyphenated Chromatographic Analyzers (HTC-7)*, Bruges (2002), Paper P95.

297. J.A. Meyer, *J. Chromatogr.* **99**, 709–20 (1974).

298. W.V. Ligon and M.C. George, *J. Polym. Sci.: Polym. Chem. Ed.* **16**, 2703–9 (1978).

299. J.G. Moncur, T.E. Sharp and E.R. Byrd, *J. High Resolut. Chromatogr., Chromatogr. Commun.* **4**, 603–11 (1981).

300. R.J. Lloyd, *J. Chromatogr.* **284**, 357–71 (1984).

301. H. Hachenberg and A.P. Schmidt, *Gas Chromatographic Headspace Analysis*, Heyden, London (1977).

302. R.P. Lattimer and J.B. Pausch, *Am. Lab.* **12** (8), 80–8 (1980).

303. R.J.J. Vreuls, E. Romijn and U.A.Th. Brinkman, *J. Microcol. Sep.* **10** (7), 581–8 (1998).

304. R. Perkins, D. Nicholas and R. Sasano, *www.ATAS-int.com* (2000).

305. X. Chen and R.B. Smart, *J. Chromatogr. Sci.* **30**, 192–6 (1992).

306. S. Dagan and A. Amirav, *J. Am. Soc. Mass Spectrom.* **6**, 120–31 (1995).

307. A. Amirav and S. Dagan, *Intl. Lab.* **17A**–17L (March 1996).

308. D.C. Shaker and H.C. Kim, *Proceedings SPE ANTEC '98*, Atlanta, GA (1998), pp. 2114–8.

308a. G.B. Kenion, R. Ludicky, J.H. Deitch, M. Leib and D. Doster, *Tappi J.* **77**(6), 199–205 (1994).

309. M. Ezrin and G. Lavigne, *Proceedings SPE ANTEC '91*, Montreal (1991), pp. 2230–3.

310. S.C. Hodgson, R.J. Casey, J.D. Orbell and S.W. Bigger, *J. Chem. Edu.* **77** (12), 1631–3 (2000).

311. U.M. Pagnoni, *Chrompack News* **22** (2), 12–13 (1995).

312. B.A. Benner, *Proceedings 8th International Symposium on Supercritical Fluid Chromatography and Extraction*, St. Louis, MO (1998), L-39.

313. M.H.P.M. van Lieshout, H.-G. Janssen, C.A. Cramers, M.J.J. Hetem and H.J.P. Schalk, *J. High Resolut. Chromatogr.* **19**, 193–9 (1996).

314. R. Rella, A. Sturaro, G. Parvoli, D. Ferrara and L. Doretti, *LC.GC Europe* **15** (9), 603–9 (2002).

315. M.J. Cohen and F.W. Karasek, *J. Chromatogr. Sci.* **8**, 330 (1970).

316. F.W. Karasek and D.W. Denney, *Anal. Lett.* **11**, 993 (1973).

317. S.H. Kim, F.W. Karasek and S. Rokushika, *Anal. Chem.* **50**, 152 (1978).

318. G.A. Eiceman and Z. Karpas, *Ion Mobility Spectrometry*, CRC Press, Boca Raton, FL (1994).

319. A.P. Snyder, C.S. Harden, A.H. Brittain, M.-G. Kim, N.S. Arnold and H.L.C. Meuzelaar, *Anal. Chem.* **65**, 299 (1993).

320. J. Sullivan and B. Quimby, *Anal. Chem.* **62**, 1034–43 (1990).

321. B. Quimby, V. Giarrocco and J. Sullivan, *J. High Resolut. Chromatogr.* **15**, 705–9 (1992).

322. P.C. Uden, in *Selective Detectors* (R.E. Sievers, ed.), John Wiley & Sons, Inc., New York, NY (1995), pp. 143–69.

323. L.K. Olson and J.A. Caruso, *Spectrochim. Acta* **49B**, 7 (1994).

324. A.J. McCormack, S.C. Tong and W.D. Cook, *Anal. Chem.* **37**, 1470 (1965).

325. C. Bradley and J.W. Carnahan, *Anal. Chem.* **60**, 858 (1988).

326. C.I.M. Beenakker, *Spectrochim. Acta* **32B**, 173 (1977).

327. B.D. Quimby and J.J. Sullivan, *Anal. Chem.* **62** (10), 1027–34 (1990).

328. J.J. Sullivan and B.D. Quimby, *J. High Resolut. Chromatogr.* **12** (5), 282–6 (1989).

329. S. Slaets, F.C. Adams and F. Laturnus, *LC.GC Intl* 580–6 (Sept 1998).

330. C. Webster and M. Cook, *J. High Resolut. Chromatogr.* **18**, 319 (1995).

331. W. Yu, Y. Huang and Q. Ou, *ACS Symp. Ser.* **479**, 44–61 (1992).

332. P.C. Uden, Y. Yoo, T. Wang and Z.B. Cheng, *J. Chromatogr.* **468**, 319 (1989).

333. A.H. Mohamad and J.A. Caruso, in *Detectors for Gas Chromatography* (J.C. Giddings, E. Grushka and P.R. Brown, eds), M. Dekker, New York, NY (1987), pp. 191–227.

334. F.A. Byrdy and J.A. Caruso, in *Selective Detectors* (R.E. Sievers, ed.), John Wiley & Sons, Inc., New York, NY (1995), pp. 171–207.

335. L.K. Olson, D.T. Heitkemper and J.A. Caruso, *ACS Symp. Ser.* **479**, 288–308 (1992).

336. D.B. Hooker and J. DeZwaan, *ACS Symp. Ser.* **479**, 132–51 (1992).

337. S.J. Hill, M.J. Bloxham and P.J. Worsfold, *J. Anal. At. Spectrom.* **8**, 499 (1993).

338. I.S. Krull, K.W. Panaro, J. Noonan and D. Erickson, *Appl. Organomet. Chem.* **3**, 295–308 (1989).

339. P.O. Duebelbeis, S. Kapila, D.E. Yates and S.E. Manahan, *J. Chromatogr.* **351**, 465–73 (1986).

340. N.P. Vela, L.K. Olson and J.A. Caruso, *Anal. Chem.* **65**, 585A (1993).

341. H. Klinkenberg, *On the use of Inductively Coupled Plasma Mass Spectrometry in Industrial Research, Ph.D. Thesis*, University of Amsterdam (1995).

342. J. Poehlman, B.W. Pack and G.M. Hieftje, *Intl Lab.* 26–9 (July 1999).

343. A.W. Kim, M.E. Foulkes, L. Ebdon, S.J. Hill, R.L. Patience, A.G. Barwise and S.J. Rowland, *J. Anal. At. Spectrom.* **7**, 1147 (1992).

344. E.H. Evans and J.A. Caruso, *J. Anal. At. Spectrom.* **8**, 427 (1993).

345. J.C. van Loon, L.R. Alcock, W.H. Pinchin and J.B. French, *Spectrosc. Lett.* **19**, 1125 (1986).

346. G.R. Peters and D. Beauchemin, *J. Anal. At. Spectrom.* **7**, 965 (1992).

347. G. Pritzl, F. Stuer-Lauridsen, L. Carlsen, A.K. Jensen and T.K. Thorsen, *Intl J. Environ. Anal. Chem.* **62**, 147–59 (1996).

348. M. Yamanaka and O.F.X. Donard, *Agilent Technologies Application Note* (Apr. 2000).

349. A.M. Leach, M. Heisterkamp, F.C. Adams and G.M. Hieftje, *J. Anal. At. Spectrom.* **15**, 151–5 (2000).

350. N.S. Chong and R.S. Houk, *Appl. Spectrosc.* **41**, 66 (1987).

351. A. de Wit and J. Beens, *J. Chromatogr. Libr.* **56**, 159–200 (1995).

352. T.H. Risby and Y. Talmi, *CRC Crit. Rev. Anal. Chem.* **14** (3), 231 (1983).

353. D. Deruaz and J.-M. Mermet, *Analusis* **14** (3), 107–18 (1986).

354. W.G. Pretorius, L. Ebdon and S.J. Rowland, *J. Chromatogr.* **646**, 369 (1993).

355. B. Lalère, J. Szpunar, H. Budzinski, P. Garrigues and O.F.X. Donald, *Analyst* **12**, 2665 (1995).

356. R. Łobiński, W.M.R. Dirkx, M. Ceulemans and F.C. Adams, *Anal. Chem.* **64**, 159 (1992).

357. F. David and P. Sandra, *LC.GC Intl* **5** (12), 22–6 (1994).

358. Y. Zeng and P. Uden, *J. High Resolut. Chromatogr.* **17**, 217–22 (1994).

359. P.L. Wylie, J.J. Sullivan and B.D. Quimby, *J. High Resolut. Chromatogr.* **13**, 499–506 (1990).

360. E. Verdurmen, R. Verstappen, J. Swagten, H. Nelissen, G. Heemels and J.C.J. Bart, *Abstracts 4th Intl Symposium on Hyphenated Techniques in Chromatography and Hyphenated Chromatographic Analyzers (HTC-4)*, Bruges (1996), Poster E38.

361. F.C.-Y. Wang, *Anal. Chem.* **71**, 2037–45 (1999).

362. F. David and P. Sandra, *LC.GC* **11** (4), 282–7 (1993).

363. H. Suyani, J.T. Creed, J.A. Caruso and R.D. Satzger, *J. Anal. At. Spectrom.* **4**, 777–82 (1989).

364. C. Brede, S. Pedersen-Bjergaard, E. Lundanes and T. Greibrokk, *J. Anal. At. Spectrom.* **15**, 55–60 (2000).

365. A. Prange and E. Jantzen, *J. Anal. At. Spectrom.* **10**, 105 (1995).

366. T. De Smaele, L. Moens, R. Dams and P. Sandra, *LC.GC Intl* **9**, 138 (1996).

367. F. Adams, M. Ceulemans and S. Slaets, *LC.GC Europe* **14** (9), 548–65 (2001).

368. T. De Smaele, L. Moens and R. Dams, in *Plasma Source Mass Spectrometry* (G. Holland and S.D. Tanner, eds), The Royal Society of Chemistry, Cambridge (1997), pp. 109–23.

369. E. Segovia García, J.I. García Alonso and A. Sanz-Medel, in *Plasma Source Mass Spectrometry* (G. Holland and S.D. Tanner, eds), The Royal Society of Chemistry, Cambridge (1997), pp. 182–91.

370. B. Radziuk, Y. Thomassen, L.R.P. Butler, J.C. van Loon and Y.K. Chau, *Anal. Chim. Acta* **108**, 31 (1979).

371. P. Sandra, A. Kot, A. Medvedovici and F. David, in *Practical Supercritical Fluid Chromatography and Extraction* (M. Caude and D. Thiébaut, eds), Harwood Academic Publishers, Amsterdam (1999), pp. 283–320.

372. K. Jinno, *Chromatographia* **23**, 55 (1987).

373. K.H. Shafer and P.R. Griffiths, *Anal. Chem.* **55**, 1939 (1983).

374. P.R. Griffiths, K.L. Norton and A.S. Bonanno, in *Hyphenated Techniques in Supercritical Fluid Chromatography and Extraction* (K. Jinno, ed.), Elsevier, Amsterdam (1992), pp. 83–101.

375. S.B. French and M. Novotny, *Anal. Chem.* **58**, 164–6 (1986).

376. R.C. Wieboldt, G.E. Adams and D.W. Later, *Anal. Chem.* **60**, 2422 (1988).

377. M.W. Raynor, A.A. Clifford, K.D. Bartle, C. Rayner, A. Williams and B.W. Cook, *J. Microcol. Sep.* **1**, 101 (1989).

378. J.W. Jordan and L.T. Taylor, *J. Chromatogr. Sci.* **24**, 82 (1986).

379. K.H. Shafer, S.L. Pentoney and P.R. Griffiths, *Anal. Chem.* **58**, 58 (1986).

380. S.H. Smith, S.L. Jordan, L.T. Taylor, J. Dwyer and J. Willis, *J. Chromatogr.* **764**, 295–300 (1997).

381. M.W. Raynor, K.D. Bartle, I.L. Davies, A. Williams, A.A. Clifford, J.M. Chalmers and B.W. Cook, *Anal. Chem.* **60**, 427–33 (1988).

382. M. Carrott, M. Kaplan, S. Bajic and G. Davidson, *Abstracts 5th International Symposium on Supercritical Fluid Chromatography and Extraction*, Baltimore, MD (1994), C-11.

383. K.D. Bartle, A.A. Clifford and M.W. Raynor, in *Hyphenated Techniques in Supercritical Fluid Chromatography and Extraction*, Elsevier, Amsterdam (1992), pp. 103–27.

384. G. Davidson and T.J. Jenkins, *Spectrosc. Europe* **4**, 32 (1992).

385. Ph. Morin, in *Practical Supercritical Fluid Chromatography and Extraction* (M. Caude and D. Thiébaut, eds), Harwood Academic Publishers, Amsterdam (1999), pp. 179–200.

386. L.T. Taylor and E.M. Calvey, in *Hyphenated Techniques in Supercritical Fluid Chromatography and Extraction* (K. Jinno, ed.), Elsevier, Amsterdam (1992), pp. 65–82.

387. R. Fuoco, S.L. Pentoney and P.R. Griffiths, *Anal. Chem.* **61**, 2212 (1989).

388. C. Fujimoto, Y. Hirata and K. Jinno, *J. Chromatogr.* **332**, 47 (1985).

389. J. Yang, E.J. Hasenoehrl and P.R. Griffiths, *Vib. Spectrosc.* **14**, 1–8 (1997).

390. T.J. Jenkins, M. Kaplan and G. Davidson, *Proceedings 5th International Symposium on Supercritical Fluid Chromatography and Extraction*, Baltimore, MD (1994), Paper C-10.

391. T.J. Jenkins, M. Kaplan, M.R. Simmonds, G. Davidson, M.A. Healy and M. Poliakoff, *Analyst* **116** (12), 1305–11 (1991).

392. S.L. Jordan, L.T. Taylor, P.D. Seemuth and R.J. Miller, *Text. Chem. Color.* **29** (2), 25–32 (1997).

393. J. Yang, *Diss. Abstr. Intl* **B53** (12, pt. 1), 6271 (1993).

394. Y. Ikushima, N. Saito, K. Hatakeda and S. Ito, in *Supercritical Fluid Processing for Food Biomaterials* (S.S.H. Rizvi, ed.), Blackie, New York, NY (1994), pp. 244–54.

395. E.M. Calvey, *Semin. Food Anal.* **1** (2), 133 (1996).

396. L.G. Randell and A.L. Wahrhaftig, *Anal. Chem.* **50**, 1703 (1978).

397. P.J. Arpino, F. Sadoun and H. Virelizier, *Chromatographia* **36**, 283–8 (1993).

398. J.D. Pinkston and T.L. Chester, *Anal. Chem.* **67**, 650–6A (1995).

399. D.M. Sheeley and V.N. Reinhold, *J. Chromatogr.* **474**, 83 (1989).

400. H.T. Kalinoski and L.O. Hargiss, *J. Chromatogr.* **474**, 69 (1989).

401. E.C. Huang, B.J. Jackson, K.E. Markides and M.L. Lee, *Anal. Chem.* **60**, 2715 (1988).

402. J. Via and L.T. Taylor, *Anal. Chem.* **66**, 1385 (1994).

403. R. van Leuken, M.A.A. Mertens, H.-G.M. Janssen, P. Sandra, G. Kwakkenbos and R. Deelder, *J. High Resolut. Chromatogr.* **17**, 573–6 (1994).

404. D. Thomas, P.G. Sim and F.M. Benoit, *Rapid Commun. Mass Spectrom.* **8**, 105 (1994).

405. T.L. Chester, J.D. Pinkston, D.P. Innis and D.J. Bowling, *J. Microcol. Sep.* **1**, 182 (1989).

406. E.R. Baumeister, C.D. West, C.F. Ijames and C.L. Wilkins, *Anal. Chem.* **63**, 251 (1991).

407. M.A.A. Mertens, P.A. Leclercq and H.-G.M. Janssen, *LC.GC Intl* **8** (6), 328–32 (1995).

408. P.T. Jedrzejewski and L.T. Taylor, *J. Chromatogr.* **A677**, 365–76 (1994).

409. P.O. Edlund and J.D. Henion, *J. Chromatogr. Sci.* **27**, 274–82 (1989).

410. D.G. Morgan, K.L. Harbol and N.P. Kitrinos, *J. Chromatogr.* **A800**, 39–49 (1998).

411. J.D. Pinkston, T.R. Baker and C.A. Smith, *Proceedings 7th International Symposium on Supercritical Fluid. Chromatography and Extraction*, Indianapolis, IN (1996), Paper L-28.

412. M. Takeuchi and T. Saito, *J. High Resolut. Chromatogr.* **14**, 347–51 (1991).

413. P.J. Arpino and P. Haas, *J. Chromatogr.* **A703**, 479–88 (1995).

414. R.D. Smith, J.C. Fjeldsted and M.L. Lee, *J. Chromatogr.* **247**, 231 (1982).

415. L.N. Tyrefors, R.X. Moulder and K.E. Markides, *Anal. Chem.* **65**, 2835–40 (1993).

416. P.J.R. Sjöberg and K.E. Markides, *J. Chromatogr.* **A785**, 101–10 (1997).

417. U. Petersson and K.E. Markides, *J. Chromatogr.* **734**, 311–18 (1996)

418. P.J. Arpino, *Adv. Mass Spectrom.* **13**, 151–75 (1995).

419. I.M. Lazar, M.L. Lee and E.D. Lee, *Anal. Chem.* **68**, 1924–32 (1996).

420. P.J.R. Sjöberg, *Abstracts 15th Montreux Symposium on Liquid Chromatography/Mass Spectrometry*, Montreux Switzerland (1998), p. 70.

421. C. Brede and E. Lundanes, *J. Chromatogr.* **712**, 95–101 (1995)

422. E.D. Ramsey, D.E. Games, J.R. Perkins and J.R. Startin, *J. Chromatogr.* **464**, 353 (1989).

423. K. Matsumoto, S. Nagata, H. Hattori and S. Tsuge, *J. Chromatogr.* **605**, 87–94 (1992).

424. J.D. Pinkston, *Abstracts 8th International Symposium on Supercritical Fluid Chromatography and Extraction*, St. Louis, MO (1998), E-20.

425. F. Sadoun, H. Virelizier and P.J. Arpino, *J. Chromatogr.* **647**, 351–9 (1993).

426. T.R. Baker and J.D. Pinkston, *J. Am. Soc. Mass Spectrom.* **9**, 498–509 (1998).

427. E.D. Lee, J.D. Henion, R.B. Cody and J.A. Kinsinger, *Anal. Chem.* **59**, 1309 (1987).

428. D.A. Laude, S.L. Pentoney, P.R. Griffiths and C.L. Wilkins, *J. Chromatogr.* **59**, 2283 (1987).

429. H.H. Hill and D.A. Atkinson, in *Hyphenated Techniques in Supercritical Fluid Chromatography and Extraction* (K. Jinno, ed.), Elsevier, Amsterdam (1992), Chapter 2.

430. R.D. Smith, H.T. Kalinoski and H.R. Udseth, *Mass Spectrom. Revs.* **6**, 445 (1987).

431. T. Bücherl, A. Eschler, A. Gruner, N. Palibroda and E. Wolff, *J. High Resolut. Chromatogr.* **17**, 765–9 (1994).

432. J.-P. Mercier, P. Chaimbault, A. Salvador and M. Dreux, *Abstracts HPLC '98*, St. Louis, MO (1998), P-0542-T.

433. W.M.A. Niessen, *Liquid Chromatography–Mass Spectrometry*, M. Dekker, New York, NY (1999).

434. W.M.A. Niessen and J. van der Greef, in *Liquid Chromatography–Mass Spectrometry. Principles and Applications*, Marcel Dekker, New York, NY (1999), Chapter 11.

435. J.D. Pinkston, in *Analysis with Supercritical Fluids: Extraction and Chromatography* (B. Wenclawiak, ed.), Springer-Verlag, Berlin (1992), Chapter 9.

436. J.D. Pinkston, in *Practical Supercritical Fluid Chromatography and Extraction* (M. Caude and D. Thiébaut, eds), Harwood Academic Publishers, Amsterdam (1999), pp. 161–78.

437. S.V. Olesik, *J. High Resolut. Chromatogr.* **14**, 5 (1991).

438. J.D. Pinkston and D.J. Bowling, in *Hyphenated Techniques in Supercritical Fluid Chromatography and Extraction* (K. Jinno, ed.), Elsevier, Amsterdam (1992), pp. 25–46.

439. C.W. Huang, *Interfacing Supercritical Fluid Chromatography with Double Focusing Mass Spectrometry*, UMI, Ann Arbor, MI (1990).

440. W.M.A. Niessen, *J. Chromatogr. Libr.* **59**, 3–70 (1996).

441. J. Cousin and P.J. Arpino, *Analusis* **14**, 215 (1986).

442. A.J. Berry, D.E. Games and J.R. Perkins, *J. Chromatogr.* **36**, 147 (1986).

443. M.A.A. Mertens, H.-G.M. Janssen, C.A. Cramers, W. J. L. Genuit, G.J. van Velzen, H. Dirkzwager and H. van Binsbergen, *J. High Resolut. Chromatogr.* **19**, 17–22 (1996).

444. J.D. Pinkston, T.E. Delaney, K.L. Morand and R.G. Cooks, *Anal. Chem.* **64**, 1571 (1992).

445. H.T. Kalinoski and L.O. Hargiss, *J. Am. Soc. Mass Spectrom.* **3**, 150 (1992).

446. H.T. Kalinoski and R.D. Smith, *Anal. Chem.* **60**, 529 (1988).

447. C.A. Smith and J.D. Pinkston, *Proceedings 8th International Symposium on Supercritical Fluid Chromatography and Extraction*, St. Louis, MO (1998), Paper E-18.

448. J.D. Pinkston, R. Hentschel and C.A. Smith, *Proceedings 5th International Symposium on Supercritical Fluid Chromatography and Extraction*, Baltimore, MD (1994), Paper E-6.

449. P.J. Arpino, D. Dilettato, K. Nguyen and A. Bruchet, *J. High Resolut. Chromatogr.* **13**, 5 (1990).

450. E.M. Calvey, T.H. Begley and J.A.G. Roach, *J. Chromatogr. Sci.* **33**, 61 (1995).

451. G. Ligner, *Dtsch. Lebensm.-Rdsch.* **6**, 178–81 (1993).

452. M.J. Carrott, D.C. Jones and G. Davidson, *Analyst* **123** (9), 1827–33 (1998).

453. K. Albert and U. Braumann, *Spec. Publ. R. Soc. Chem.* **163**, 86–93 (1995).

454. L.A. Allen, T.E. Glass and H.C. Dorn, *Anal. Chem.* **60**, 390 (1988).

455. K. Albert, U. Braumann, L.-H. Tseng, G. Nicholson, E. Bayer, M. Spraul, M. Hofmann, C. Dowle and M. Chippendale, *Anal. Chem.* **66**, 3042–6 (1994).

456. K. Albert, U. Braumann, R. Streck, M. Spraul and R. Ecker, *Fresenius' Z. Anal. Chem.* **352**, 521–8 (1995).

457. U. Braumann and M. Spraul, *Proceedings 8th International Symposium on Supercritical Fluid Chromatography and Extraction*, St. Louis, MO (1998), Paper L-19.

458. U. Braumann, H. Händel, S. Strohschein, M. Spraul, G. Krack, R. Ecker and K. Albert, *J. Chromatogr.* **A761**, 336–40 (1997).

459. K. Albert, M. Dachtler, T. Glaser, H. Händel, T. Lacker, G. Schlotterbeck, S. Strohschein, L.-H. Tseng and U. Braumann, *J. High Resolut. Chromatogr.* **22** (3), 135–43 (1999).

460. C.B. Motley, M. Ashraf-Khorassani and G.L. Long, *Appl. Spectrosc.* **43**, 737 (1989).

461. M.W. Blades and B.L. Caughlin, *Spectrochim. Acta* **40B**, 579 (1985).

462. R.J. Skelton, P.B. Farnsworth, K.E. Markides and M. Lee, *Anal. Chem.* **61**, 1815 (1989).

463. D. Luffer, L. Galante, L. David, M. Novotny and G. Hieftje, *Anal. Chem.* **60**, 1365 (1988).

464. J.W. Carnahan, G.K. Webster and L. Zhang, *Abstracts 16th Annual Meeting, Federation of Analytical Chemistry and Spectroscopist Societies*, Chicago, IL (1989), No. 517.

465. Q. Jin, G. Hieftje, F. Wang and D.M. Chambers, *Abstracts 1990 Winter Conference on Plasma Spectrochemistry*, St. Petersburg, FL (1990), No. S-8.

466. M. Wu and J.W. Carnahan, *Appl. Spectrosc.* **44**, 673 (1990).

467. D.R. Luffer and M. Novotny, in *Hyphenated Techniques in Supercritical Fluid Chromatography and Extraction*, Elsevier, Amsterdam (1992), pp. 171–96.

468. G.K. Webster and. J.W. Carnahan, in *Element-Specific Chromatographic Detection by Atomic Emission Spectroscopy* (P.C. Uden, ed.), American Chemical Society, Washington, DC (1992), pp. 25–43.

469. J.M. Carey and J.A. Caruso, *Trends Anal. Chem.* **11**, 287 (1992).

470. K. Jinno, in *Hyphenated Techniques in Supercritical Fluid Chromatography and Extraction* (K. Jinno, ed.), Elsevier, Amsterdam (1992), pp. 151–70.

471. K. Jinno, H. Yoshida, H. Mae and C. Fujimoto, in *Element-Specific Chromatographic Detection by Atomic Emission Spectroscopy* (P.C. Uden, ed.), American Chemical Society, Washington, DC (1992), Chapter 13.

472. J.W. Olesik and S.V. Olesik, *Anal. Chem.* **59**, 796 (1987).

473. E. Blake, M.W. Raynor and D. Cornell, *J. High Resolut. Chromatogr.* **18**, 33 (1995).

474. W.-L. Shen, N.P. Vela, B.S. Sheppard and J.A. Caruso, *Anal. Chem.* **63**, 1495–6 (1991).

475. L.K. Olson and J.A. Caruso, *J. Anal. At. Spectrom.* **7**, 993 (1992).

476. N.P. Vela and J.A. Caruso, *J. Chromatogr.* **641**, 337 (1993).

477. J.M. Carey, N.P. Vela and J.A. Caruso, *J. Anal. At. Spectrom.* **7**, 1173 (1992).

478. L.K. Olson and J.A. Caruso, *Proceedings 5th International Symposium on Supercritical Fluid Chromatography and Extraction*, Baltimore, MD (1994), Paper 14.

479. B.J. Clark, T. Frost and M.A. Russell (eds), *UV Spectroscopy: Techniques, Instrumentation and Data Handling*, Chapman & Hall, London (1993).

480. G. Mortensen, B. Pedersen and G. Pritzl, *Appl. Organomet. Chem.* **9**, 65 (1995).

481. K.S. Kalasinsky, J.A.S. Smith and V.F. Kalasinsky, *Anal. Chem.* **57**, 1969–74 (1985).

482. C. Fujimoto and K. Jinno, *Anal. Chem.* **64**, 476–81A (1992).

483. J.A. Jansen, *Fresenius' Z. Anal. Chem.* **337**, 398–402 (1990).

484. J. Dwyer, *Am. Lab.* 22–6 (May 1994).

485. K. Jinno and C. Fujimoto, *J. High Resolut. Chromatogr., Chromatogr. Commun.* **4**, 532 (1981).

486. J.J. Gagel and K. Biemann, *Anal. Chem.* **58**, 2184–9 (1986).

487. J.J. Gagel and K. Biemann, *Anal. Chem.* **59**, 1266–72 (1987).

488. D. Keuhl and P.R. Griffiths, *J. Chromatogr. Sci.* **17**, 471 (1979).

489. D. Keuhl and P.R. Griffiths, *Anal. Chem.* **52**, 1394 (1980).

490. J.W. Hellgeth and L.T. Taylor, *Anal. Chem.* **59**, 295 (1987).

491. A.H. Dekmezian and T. Morioka, *Anal. Chem.* **61**, 458–61 (1989).

492. K.L. Kizer, A.W. Mantz and L.C. Bonar, *Am. Lab.* **7** (5), 85 (1975).

493. M. Sabo, J. Gross, J. Wang and I.E. Rosenberg, *Anal. Chem.* **57**, 1822 (1985).

494. J.B. Pausch, *Anal. Chem.* **54**, 89–96A (1982).

495. P.R. Griffiths, *Appl. Spectrosc.* **31**, 497 (1977).

496. P.R. Griffiths and C.M. Conroy, *Adv. Chromatogr.* **25**, 105 (1986).

497. R.M. Robertson, J.A. de Haseth, J.D. Kirk and R.F. Browner, *Appl. Spectrosc.* **42**, 1365 (1988).

498. D.J. Wood, *Spectrosc. Intl* **2** (3), 36 (1990).

499. M.W. Raynor, K.D. Bartle and B.W. Cook, *J. High Resolut. Chromatogr.* **15**, 361 (1992).

500. G.W. Somsen, C. Gooyer and U.A.Th. Brinkman, *J. Chromatogr.* **A856**, 213–42 (1999).

501. LabConnections, *Application Note 10: Analysis of Polymer Additives*, LabConnections, Marlborough, MA (1995).

502. J. Sidwell, *Polymer Testing '96* (Rapra Technology Ltd., ed.), Shawbury (1996), Paper 6.

503. G.W. Somsen, E.J.E. Rozendom, C. Gooyer, N.H. Velthorst and U.A.Th. Brinkman, *Analyst* **121**, 1069–74 (1996).

504. K. Jinno, C. Fujimoto and Y. Hirata, *Appl. Spectrosc.* **36**, 67 (1982).

505. N. Teramae and S. Tanaka, *Spectrosc. Lett.* **13**, 117 (1980).

506. E. Sudo, Y. Esaki and M. Sugiura, *Bunseki Kagaku* **50** (10), 703–7 (2001).

507. D. Warren Vidrine, in *Fourier Transform Infrared Spectroscopy* (J.R. Ferraro and L.J. Basile), Academic Press, San Diego, CA (1979), Vol. 2, p. 129.

508. F.C. Sanchez, T. Hancewicz, B.G.M. Vandegiste and D.L. Massart, *Anal. Chem.* **69**, 1477–84 (1997).

509. A.M. Robertson, D. Littlejohn, M. Brown and C.J. Dowle, *J. Chromatogr.* **588**, 15 (1991).

510. J. Cosemans, S. Bremmers and J.C.J. Bart, *Tech. Rep. DSM Research*, Geleen (1997).

511. I. Bruheim, P. Molander, E. Lundanes, T. Greibrokk and E. Ommundsen, *J. High Resolut. Chromatogr.* **23** (9), 525–30 (2000).

512. M.A. Mottaleb and D. Littlejohn, *Anal. Sci.* **17**, 429–34 (2001).

513. M.A. Mottaleb, *Anal. Sci.* **15**, 57 (1999).

514. M.A. Mottaleb, B.G. Cooksey and D. Littlejohn, *Fresenius' Z. Anal. Chem.* **358**, 536–8 (1997).

515. LabConnections, *Application Note AN-14*, LabConnections Inc., Marlborough, MA (1995).

516. LabConnections, *Application Note AN-15*, LabConnections Inc., Marlborough, MA (1995).

517. P.J. Schoenmakers and C.G. de Koster, in *Recent Applications in LC-MS, LC.GC Europe Special Edn* pp. 38–43 (Nov. 2002).

518. H.R. Schulten and H.D. Beckey, *J. Chromatogr.* **84**, 315–20 (1973).

519. B.L.M. van Baar, *J. Chromatogr. Libr.* **59**, 71–133 (1996).

520. W.H. McFadden, H.L. Schwartz and S. Evens, *J. Chromatogr.* **122**, 389 (1976).

521. P.C. Winter, D.B. Perkins, W.K. Williams and F.F. Browner, *Anal. Chem.* **60**, 489 (1988).

522. P.J. Arpino, *Mass Spectrom. Rev.* **8**, 35 (1989).

523. J.D. Vargo and K.L. Olson, *J. Chromatogr.* **353**, 215–24 (1986).

524. A.E. Ashcroft, *Ionization Methods of Organic Mass Spectrometry*, The Royal Society of Chemistry, Cambridge (1997).

525. P.C. Winkler, D.D. Perkins, W.K. Williams and R.F. Browner, *Anal. Chem.* **60**, 489 (1988).

526. R.C. Willoughby and R.F. Browner, *Anal. Chem.* **56**, 2625 (1984).

527. G.G. Jones, R.E. Pauls and R.C. Willoughby, *Anal. Chem.* **63**, 460–3 (1991).

528. T. Takeuchi, D. Ishii, A. Saito and T. Ohki, *J. High Resolut. Chromatogr., Chromatogr. Commun.* **5**, 91–2 (1982).

529. A. Amirav and O. Granot, *Abstracts 7th International Symposium on Hyphenated Techniques in Chromatography and Hyphenated Chromatographic Analyzers (HTC-7)*, Bruges (2002), Paper 16.2.

530. T.L. Jones, L.D. Betowski and V. Lopez-Avila, *Trends Anal. Chem.* **13**, 333 (1994).

531. R.M. Caprioli, T. Fan and J.S. Cottrell, *Anal. Chem.* **58**, 2949 (1986).

532. M. Yamashita and J.B. Fenn, *J. Chem. Phys.* **88**, 4451 (1984).

533. M.A. Moseley, L.J. Deterding, K.B. Tomer and J.W. Jorgenson, *J. Chromatogr.* **480**, 197 (1989).

534. J.V. Iribarne, P.J. Dziedzic and B.A. Thomson, *Intl J. Mass Spectrom. Ion Phys.* **50**, 331 (1983).

535. L. Voress, *Anal. Chem.* **66**, 481–6A (1994).

536. E.D. Lee, W. Mück, J.D. Henion and T.R. Covey, *Biomed. Environ. Mass Spectrom.* **18**, 844 (1989).

537. S.A. Cohen, T. Dourdeville, D. Della Rovere and J. Holyoke, *Abstracts 22nd International Symposium on High Performance Liquid Phase Separations and Related Techniques*, St. Louis, MO (1998), Paper P-0418-T.

538. M.S. Wilm and M. Mann, *Intl J. Mass Spectrom. Ion Proc.* **136**, 167 (1994).

539. M.S. Wilm and M. Mann, *Anal. Chem.* **68**, 1–8 (1996).

540. J. Cosemans-Craeghs, *Techn. Report DSM Research*, Geleen, Netherlands (1999).

541. J.N. Alexander, G.A. Schultz and J.B. Poli, *Rapid Commun. Mass Spectrom.* **12**, 1187 (1998).

542. J.F. Banks, E. Gulcicek and C. Whitehouse, *Proceedings 14th International Mass Spectrometry Conference, Tampere, Finland (1997)*, Paper LaPo 001.

543. R. Brereton, *Proceedings Solutions for Scientists Symposium* (Advanstar Communications, ed.), London (1999).

544. A.P. Bruins, T.R. Covey and J.D. Henion, *Anal. Chem.* **59**, 2642 (1987).

545. C.R. Blakley and M.L. Vestal, *Anal. Chem.* **55**, 750–4 (1983).

545a. D.J. Liberato and A.L. Yergey, *Anal. Chem.*, **58**, 6 (1986).

546. C.G. Enke, in *Advances in Mass Spectrometry* (E.J. Karjalainen, A.E. Hesso, J.E. Jalonen and U.P. Karjalainen, eds), Elsevier Science, Amsterdam (1999), Vol. 14, pp. 197–219.

547. A. Hirabayashi, M. Sakairi and H. Koizumi, *Anal. Chem.* **66**, 4557 (1994); **67**, 2878 (1995).

548. J.F. Banks, J.P. Quinn and C.M. Whitehouse, *Anal. Chem.* **66**, 3688 (1994).

549. D.A. Volmer, *LC.GC Europe* **13** (11), 838–47 (2000).

550. Y. Hirabayashi, A. Hirabayashi and H. Koizumi, *Rapid Commun. Mass Spectrom.* **13**, 712–5 (1999).

551. D.B. Robb, T.R. Covey and A.P. Bruins, *Anal. Chem.* **72**, 3653–9 (2000).

552. M. Sakairi and Y. Kato, *J. Chromatogr.* **A794**, 391–406 (1998).

553. J.L. Merdink and J.W. Dolan, *LC.GC Europe* **13** (11), 806–11 (Nov. 2000).

554. T. Taylor, *Proceedings Solutions for Scientists Symposium* (Advanstar Communications, ed.), London (1999).

555. J.P.C. Vissers, *J. Chromatogr.* **A856**, 117–43 (1999).

556. M.J. Lee, S. Monté, J. Sanderson and N.J. Haskins, *Rapid Commun. Mass Spectrom.* **13**, 216–21 (1999).

557. B.E. Winger, D.C. Tutko, K.D. Henry and J.E. Campana, *Proceedings 14th International Mass Spectrometry Conference*, Tampere, Finland (1997), Paper FrOr02.

558. L.L. Lopez, P.R. Tiller, M.W. Senko and J.C. Schwartz, *Rapid Commun. Mass Spectrom.* **13**, 663 (1999).

559. D. Vrielink, A. Schellen, E. Koster and B. Ooms, *Intl Lab.* **32** (7A), 14–5 (2002).

560. W.M.A. Niessen and A.P. Tinke, *Anal. Chem.* **A703**, 37–57 (1995).

561. J. Slobodník, B.L.M. van Laar and U.A.Th. Brinkman, *J. Chromatogr.* **A703**, 81–121 (1995).

562. W.M.A. Niessen, *J. Chromatogr.* **A794**, 407–35 (1998).

563. T. Kashiwagi, O. Hiroaki, M. Katoh and H. Takigawa, *Sumitomo Kagaku (Osaka)* (2), 71–85 (1993).

564. P. Sandra, G. Vanhoenacker, F. Lynen, L. Li and M. Schelfant, in *Guide to LC-MS* (R. Smits, ed.), *LC.GC Europe Special Issue* (Dec. 2001), pp. 8–21.

565. R. Smits (ed.), *Guide to LC-MS, LC.GC Europe Special Issue* (Dec. 2001).

566. A. Cappiello, *Mass Spectrom. Rev.* **15**, 283–96 (1996).

567. C.S. Creaser and J.W. Stygall, *Analyst* **118**, 1467 (1993).

568. W.M.A. Niessen, in *Encyclopedia of Spectroscopy and Spectrometry* (J.C. Lindon, ed.), Academic Press, San Diego, CA (2000), pp. 2353–60.

569. P.J. Arpino, *Mass Spectrom. Rev.* **9**, 631–69 (1990).

570. M.W.F. Nielen and F.A. Buytenhuys, *LC.GC Europe* **14** (2), 82–8 (2001).

571. P. Myers and K.D. Bartle, in *Recent Applications in LC-MS, LC.GC Europe Special Edn*, pp. 44–7 (Nov. 2002).

572. J. Wieling, *Chromatographia* **55** (Suppl.), S107–13 (2002).

573. A.L. Yergey, G.C. Edmonds, I.A.S. Lewis and M.L. Vestal, *Liquid Chromatography/Mass Spectrometry Techniques and Applications*, Plenum Press, New York, NY (1990), pp. 1–306.

574. W.M.A. Niessen and R.D. Voyksner, *Current Practice in Liquid Chromatography–Mass Spectrometry*, Elsevier, Amsterdam (1998).

575. M.A. Brown (ed.), *Liquid Chromatography/Mass Spectrometry Applications in Agricultural, Pharmaceutical and Environmental Chemistry*, ACS Symposium Series No. 420, American Chemical Society, Washington, DC (1990).

576. R.M. Caprioli (eds), *Continuous-flow Fast Atom Bombardment Mass Spectrometry*, John Wiley & Sons, Inc., New York, NY (1990).

577. Anon., *The API Book*, PE Sciex, Thornhill, ON (1989).

578. J.D. Vargo and K.L. Olson, *Anal. Chem.* **57**, 672–5 (1985).

579. P. Hirter, H.J. Walther and P. Dätwyler, *J. Chromatogr.* **323** (1), 89–98 (1985).

580. H. Egsgaard, E. Larsen, W.B. Pedersen and L. Carlsen, *Trends Anal. Chem.* **11** (4), 164–8 (1992).

581. A.T. Jackson, C.K. Jennings, H.T. Yates, J.H. Scrivens and K.R. Jennings, *Eur. Mass Spectrom.*, **3**, 113–20 (1997).

582. A.T. Jackson, K.R. Jennings and J.H. Scrivens, *Rapid Commun. Mass Spectrom.* **10**, 1449–58 (1996); 1459–62 (1996).

583. A.T. Jackson, A. Buzy, K.R. Jennings and J.H. Scrivens, *Eur. Mass Spectrom.* **2**, 115–27 (1996).

584. C.G. Juo, S.W. Chen and G.R. Her, *Anal. Chim. Acta* **311**, 153–64 (1995).

585. W.C. Brumley, C.M. Brownrigg and A.H. Crange, *J. Chromatogr.* **A680**, 635 (1994).

586. H.-R. Schulten, *J. Chromatogr.* **251**, 105–28 (1982).

587. R.P. Lattimer, D.J. Harmon and K.R. Welch, *Anal. Chem.* **51**, 1293–6 (1979).

588. R.P. Lattimer, E.R. Hooser and P.M. Zakriski, *Rubber Chem. Technol.* **53**, 346–56 (1980).

589. R.P. Lattimer, E.R. Hooser, H.E. Diem and C.K. Rhee, *Rubber Chem. Technol.* **55**, 442–55 (1982).

590. R.P. Lattimer, E.R. Hooser, H.E. Diem, R.W. Layer and C.K. Rhee, *Rubber Chem. Technol.* **53**, 1170–90 (1980).

591. R.P. Lattimer, E.R. Hooser, R.W. Layer and C.K. Rhee, *Rubber Chem. Technol.* **56**, 431–9 (1983).

592. I.G. Beattie, D.E. Games, J.R. Startin and J. Gilbert, *Biomed. Mass Spectrom.* **12**, 616 (1985).

593. C.R. Blakley, M.J. McAdams and M.L. Vestal, *Adv. Mass Spectrom.* **8**, 1616 (1980).

594. R. Straub, R.D. Voyksner and J.T. Keever, *J. Chromatogr.* **627**, 173 (1992).

595. J. Yinon and J. Saar, *J. Chromatogr.* **586**, 73 (1991).

596. M.A. McLean and R.B. Freas, *Anal. Chem.* **61**, 2054 (1989).

597. A. Groeppelin, M.W. Linder, K. Schellenberg and H. Moser, *Rapid Commun. Mass Spectrom.* **5**, 203 (1991).

598. L.D. Betowski and J.M. Ballard, *Anal. Chem.* **56**, 2604 (1984).

599. J. Yinon, L.D. Betowski and R.D. Voyksner, *J. Chromatogr. Libr.* **59**, 187–218 (1996).

600. L. Prokai, B.-H. Hsu, H. Farag and N. Bodor, *Anal. Chem.* **61**, 1723 (1989).

601. K.A. Evans, S.T. Dubey, L. Kravetz, I. Dzidic, J. Gomulka, R. Müller and J.R. Stork, *Anal. Chem.* **66** (5), 699 (1994).

602. H.F. Schröder, *J. Chromatogr. Libr.* **59**, 263–324 (1996).

603. J.D. Reynolds, D. Stubbs and M. Barriger, *Proceedings 39th ASMS Conference on Mass Spectrometry and Allied Topics*, Nashville, TN (1991), p. 1354.

604. W.R. Cullen, G.K. Eigendorf, B.U. Nwata and A. Takatsu, *Appl. Organomet. Chem.* **4**, 581 (1990).

605. P.J. Arpino, *Mass Spectrom. Rev.* **11**, 3–40 (1992).

606. D.L. Lawrence, *J. Am. Soc. Mass Spectrom.* **3**, 575–81 (1992).

607. K.A. Evans, S.T. Dubey, L. Kravetz, I. Dzidic, J. Gomulka, R. Müller and J.R. Stork, *Anal. Chem.* **66** (5), 699 (1994).

608. K.W.M. Siu, G.J. Gardner and S.S. Berman, *Anal. Chem.* **61**, 2320 (1989).

609. T.L. Jones and L.D. Betowski, *Rapid Commun. Mass Spectrom.* **7**, 1003 (1993).

610. P.O. Edlund, E.D. Lee, J.D. Henion and W.L. Budde, *Biomed. Environ. Mass Spectrom.* **18**, 233 (1989).

611. H.-Y. Lin and R.D. Voyksner, *Anal. Chem.* **65**, 451 (1993).

612. D.D. Popenoe, S.J. Morris, P.S. Horn and K.T. Norwood, *Anal. Chem.* **66**, 1620 (1994).

613. K.B. Sharrard, P.J. Marriott, M.J. McCormick, R. Colton and G. Smith, *Anal. Chem.* **66**, 3394 (1994).

614. P.L. Ferguson, C.R. Iden and B.J. Brownawell, *Anal. Chem.* **72** (8), 4322–30 (2000).

615. T.L. Jones-Lepp, K.E. Varner and B.A. Hilton, *Appl. Organomet. Chem.* **15** (12), 933–8 (2001).

616. M. Möder and P. Popp, in *Applications of Solid Phase Microextraction* (J. Pawliszyn, ed.), The Royal Society of Chemistry, Cambridge (1999), pp. 311–26.

617. A.P. Bruins, L.O.G. Weidoff, J.D. Henion and W.L. Budde, *Anal. Chem.* **59**, 2647 (1987).

618. M. Holcapek, P. Jandera and J. Prikyl, *Dyes & Pigm.* **43** (2), 127–37 (1999).

618a. M. Trojanowicz, J. Orska-Gawryś, I. Surowiec, B. Szostek, K. Urbaniak-Walczak, J. Kehl and M. Wróbel, *Stud. Conserv.* **49**, 115–30 (2004).

619. Y. Mengerink, *Tech. Rep. DSM Research*, Geleen (1998).

620. J. Palmgrén, E. Toropainen, S. Auriola and A. Urtti, *Abstracts 7th International Symposium on Hyphenated Techniques in Chromatography and Hyphenated Chromatographic Analyzers (HTC-7)*, Bruges (2002), Paper P119.

621. L.M. Pereira and D.E. Games, *Proceedings 14th International Mass Spectrometry Conference*, Tampere, Finland (1997), Paper MoPo120.

622. Waters Corp., *Polyolefin Additives Determination with UV and Mass Detection*, (Integrity System Publications), Waters Corp., Milford, MA (n.d.)

623. B.J. Altepeter, *LaborPraxis* **19** (5), 90–2 (1995).

624. J. Yinon, T.L. Jones and L.D. Betowski, *J. Chromatogr.* **482**, 75 (1989).

625. D.W. Allen, M.R. Clench, A. Crowson and D.A. Leathard, *J. Chromatogr.* **629** (2), 283–90 (1993).

626. D.W. Allen, M.R. Clench, A. Crowson, D.A. Leathard and R. Saklatrala, *J. Chromatogr.* **A679**, 285–97 (1994).

627. Waters Corp., *Thermabeam Mass Detector* (Integrity System Publications), Waters Corp., Milford, MA (1997).

628. J. Cosemans-Craeghs, *Tech. Rep. DSM Research*, Geleen, Netherlands (1997).

629. K. Yu, E. Block, V. Mykytin and M. Balogh, *Proceedings International GPC Symposium '98*, Phoenix, AZ (1998), pp. 211–32.

630. K. Yu, E. Block and M. Balogh, *LC.GC* **18** (2), 162–178 (2000); *LC.GC Europe* **12** (9), 577–87 (1999).

631. T. Benijts, R. Dams, W. Lambert and A.P. de Leenheer, *Abstracts 7th International Symposium on Hyphenated Techniques in Chromatography and Hyphenated Chromatographic Analyzers (HTC-7)*, Bruges (2002), Paper 18.2

632. S.M. Jickells, *Abstracts 4th International Symposium on Hyphenated Techniques in Chromatography and Hyphenated Chromatographic Analyzers (HTC-4)*, Bruges (1996), B12.

633. K.A. Barns, A.P. Damant, J.R. Startin and L. Castle, *J. Chromatogr.* **A712**, 191–9 (1995).

634. M.N. Maillard, G. Giampaoli and M.E. Cuvelier, *Talanta* **43** (3), 339–47 (1996).

635. H. Henderickx and C. de Koster, *Tech. Rep. DSM Research*, Geleen, Netherlands (1999).

636. N. Palibroda, O.-G. Piringer and J. Brandsch, *Proceedings 14th International Mass Spectrometry Conference*, Tampere, Finland (1997), Paper ThPo104.

637. P. Jandera, M. Holčapek and L. Kolářová, *Intl J. Polym. Anal. Charact.* **6**, 261–94 (2001).

638. A. Mosch and U. Giese, *Proceedings DIK Workshop Moderne Prüfmethoden – von Rohstoffen über Mischungen bis zum Produkt*, Hannover (2000).

639. R.D. Voyksner and J.T. Keever, *Abstracts 15th Montreux Symposium on Liquid Chromatography/Mass Spectrometry*, Montreux, Switzerland (1998).

640. M.E. Alvarez-Piñeiro, M.J. López de Alda-Villaizán, P. Paseiro-Losada and M.A. Lage-Yusty, *J. High Resolut. Chromatogr.* **20** (6), 321–4 (1997).

641. M.J.I. Mattina, *J. Chromatogr. Libr.* **59**, 325–44 (1996).

642. M.A. Dearth, T.J. Korniski and J.L. Gerlock, *Polym. Degrad. Stab.* **48**, 111–20 (1995).

643. K. Albert, *J. Chromatogr.* **A703**, 123 (1995).

644. K. Albert, G. Schlotterbeck, U. Braumann, H. Händel, M. Spraul and G. Krack, *Angew. Chem. Int. Ed. Engl.* **34**, 1014 (1995).

645. L.H. Tseng, K. Albert, M. Spraul and U. Braumann, *Proceedings 23rd Intl Symposium on High Performance Liquid Phase Separations and Related Techniques HPLC '99*, Granada, Spain (1999).

646. H. Pasch and W. Hiller, *Macromolecules* **29**, 6556–9 (1996).

647. N. Watanabe and E. Niki, *Proc. Jpn Acad.* **54** (Ser. B), 194–9 (1978).

648. D.A. Laude and C.L. Wilkins, *Anal. Chem.* **59**, 546–51 (1987).

649. R.M. Smith, O. Chienthavorn, I.D. Wilson, B. Wright and S.D. Taylor, *Anal. Chem.* **71**, 4493–7 (1999).

650. T.L. Peck and J.V. Sweedler, in *Micro Total Analysis Systems 2001* (J.M. Ramsey and A. van den Berg, eds), Kluwer Academic Publishers, Dordrecht (2001), pp. 417–9.

651. S.A. Korhammer and A. Bernreuther, *Fresenius' Z. Anal. Chem.* **354**, 131–5 (1996).

652. A. Nordon, C.A. McGill and D. Littlejohn, *Analyst* **126**, 260–72 (2001).

653. J.C. Lindon, J.K. Nicholson and I.D. Wilson, *Adv. Chromatogr.* **36**, 315–82 (1996).

654. M. Liu and X. Mao, in *Encyclopedia of Spectroscopy and Spectrometry* (J.C. Lindon, ed.), Academic Press, San Diego, CA (2000), pp. 2145–52.

655. G. Schlotterbeck, H. Pasch and K. Albert, *Polym. Bull.* **38**, 673–9 (1997).

656. W. C. Lenhart, T. L. Matochik, F. M. Michaels and M. Nair, *Abstracts HPLC'98*, St. Louis, MO (1998), P-0530-T.

657. T.C. Schunk, W.C. Lenhart, W.F. Smith and F.A. Fox, *Abstracts HPLC'98*, St. Louis, MO (1998), P-0541-M.

658. H. Pasch, W. Hiller and R. Haner, *Polymer* **39**, 1515–23 (1998).

659. T. Ihara, T. Saito, Y. Shimizu, R. Iwasawa, M. Kinoshita, S. Kinugasa and A. Nomura, *Abstracts 7th Intl. Symposium on Hyphenated Techniques in Chromatography and Hyphenated Chromatographic Analyzers (HTC-7)*, Bruges (2002), P68.

660. J. Scheirs, J. Pospíšil, M.J. O'Connor and S.W. Bigger, *Adv. Chem. Ser.* **249**, 359–74 (1996).

661. F.S. Pullen, A.G. Swanson, M.J. Newman and D.S. Richards, *Rapid Commun. Mass Spectrom.* **9**, 1003–6 (1995).

662. D. Louden, A. Handley, E. Lenz, I. Sinclair, S. Taylor and I.D. Wilson, *Anal. Bioanal. Chem.* **373**, 508–15 (2002).

663. D. Louden, A. Handley, S. Taylor, E. Lenz, S. Miller, I.D. Wilson and A. Sage, *Analyst* **125**, 927–32 (2000).

664. I.D. Wilson, *Anal. Chem.* **72**, 534–42A (2000).

665. D. Louden, A. Handley, R. Lafont S. Taylor, I. Sinclair, E. Lenz, T. Orton and I.D. Wilson, *Anal. Chem.* **74**, 288–94 (2002).

666. M. Ludlow, D. Louden, A. Handley, S. Taylor and I.D. Wilson, *Anal. Commun.* **36**, 85–7 (1999).

667. M. Ludlow, D. Louden, A. Handley, S. Taylor, B. Wright and I.D. Wilson, *J. Chromatogr.* **A857**, 89–96 (1999).

668. D. Louden, A. Handley, S. Taylor, E. Lenz, S. Miller, I.D. Wilson and A. Sage, *Anal. Chem.* **72**, 3922–6 (2000).

669. D. Louden, A. Handley, S. Taylor, E. Lenz, S. Miller, I.D. Wilson, A. Sage and R. Lafont, *J. Chromatogr.* **A910**, 237–46 (2001).

670. E. Lenz, S. Taylor, I. Sinclair, C. Collins, I.D. Wilson, D. Louden and A. Handley, *Abstracts 7th International Symposium on Hyphenated Techniques in Chromatography and Hyphenated Chromatographic Analyzers (HTC-7)*, Bruges (2002), P98.

671. D. Louden, A. Handley, S. Taylor, E. Lenz, S. Miller, I.D. Wilson, A. Sage and R. Lafont, *J. Chromatogr.* **A910**, 237–46 (2001).

672. I.D. Wilson, *J. Chromatogr.* **A892**, 315–27 (2000).

673. D. Louden, A. Handley, E. Lenz, I. Sinclair, S. Taylor and I.D. Wilson, *Anal. Bioanal. Chem.* **373**, 508–15 (2002).

674. K.E. LaFreniere, V.A. Fassel and D.E. Eckels, *Anal. Chem.* **59**, 879 (1987).

675. A. Tangen, R. Trones, T. Greibrokk and W. Lund, *J. Anal. At. Spectrom.* **12**, 667–70 (1997).

676. K.S. Jinno, S. Nakanishi and T. Nagoshi, *Anal. Chem.* **56** (11), 1977–9 (1984).

677. H. Suyani, J.I. Creed, T. Davidson and J.A. Caruso, *J. Chromatogr. Sci.* **27**, 139–43 (1989).

678. K. DeNicola and J.A. Caruso, *LabPlus Intl* **17** (1), 14–17 (2003).

679. K.L. Sutton and J.A. Caruso, *J. Chromatogr.* **A856**, 243–58 (1999).

680. H.A.H. Billet, J.P.J. van Dalen, P.J. Schoenmakers and L. de Galan, *Anal. Chem.* **55**, 847 (1983).

681. K.G. Michlewicz and J.W. Carnahan, *Anal. Lett.* **20**, 1193 (1987).

682. L. Zhang, J.W. Carnahan, R.E. Winans and P.H. Neill, *Anal. Chem.* **61**, 895 (1989).

683. H. Goenaga Infante and F.C. Adams, in *Guide to LC-MS* (R. Smits, ed.), *LC.GC Europe Special Issue* (Dec. 2001), pp. 31–41.

684. S.L. Bonchid-Cleland, H. Dong and J.A. Caruso, *Am. Lab.* **27**, 34N (1995).

685. N.P. Vela and J.A. Caruso, *J. Anal. At. Spectrom.* **8**, 787 (1993).

686. S.B. Dorn and E.M.S. Frame, *Analyst* **119** (8), 1687–94 (1994).

687. C. Rivas, L. Ebdon and S.J. Hill, *J. Anal. At. Spectrom.* **11**, 1147 (1996).

688. B. Fairman and R. Wahlen, *Spectrosc. Europe* **13** (5), 16–22 (2001).

689. C.F. Harrington, G.K. Eigendorf and W.R. Cullen, *Appl. Organomet. Chem.* **10**, 339 (1996).

690. H. Suyani, D. Heitkemper, J.T. Creed and J.A. Caruso, *Appl. Spectrosc.* **43**, 962–7 (1989).

691. X. Dauchy, R. Cottier, A. Batel, R. Jeannot, M. Borsier, A. Astruc and M. Astruc, *J. Chromatogr. Sci.* **31**, 416 (1993).

692. J.W. McLaren, K.W.M. Siu, J.W. Lam, S.N. Willie, P.S. Maxwell, A. Palepu, M. Koether and S.S. Berman, *Fresenius' Z. Anal. Chem.* **337**, 721 (1990).

693. R. Wahlen, *LC.GC Europe* **15** (10), 670–7 (2002).

694. G.K. Zoorob and J.A. Caruso, *J. Chromatogr.* **A773**, 157–62 (1997).

695. K. Hatada and K. Ute, *Polym. Mat. Sci. Engng* **62**, 332 (1990).

696. L.C.M. van Gorkum and T.M. Hancewicz, *J. Magn. Reson.* **130**, 125–30 (1998).

697. S.C.K. Shum and R.S. Houk, *Anal. Chem.* **65**, 2972 (1993).

698. K. Takatera and T. Watanabe, *Anal. Chem.* **65**, 3644 (1993).

699. A.H. Dekmezian, T. Morioka and C.E. Camp, *J. Polym. Sci.* **28B**, 1903 (1990).

700. J.N. Willis and L. Wheeler, *Appl. Spectrosc.* **47**, 1128 (1993).

701. T.C. Schunk, S.T. Balke and P. Cheung, *J. Chromatogr.* **A661**, 227–38 (1994).

702. P. Cheung, S.T. Balke, T.C. Schunk and T.H. Mourey, *J. Appl. Polym. Sci.: Appl. Polym. Symp.* **52**, 105–24 (1993).

703. L.M. Wheeler and J.N. Willis, *Appl. Spectrosc.* **47**, 1128–30 (1993).

704. J.L. Dwyer, *Proceedings SPE ANTEC '99*, New York, NY (1999), pp. 2440–6.

705. K. Torabi, S.T. Balke and T.C. Schunk, *Proceedings International GPC Symposium '98*, Phoenix, AZ (1998), pp. 167–88.

706. R.J. Papez, *Proceedings International GPC Symposium '98*, Phoenix, AZ (1998), pp. 207–10.

707. T. Meyer-Dulheuer, *Proceedings 4. Würzburger Tage der instrumentellen Analytik in der Polymertechnik*, SKZ, Würzburg (1999), Paper M.

708. J.R. Schoonover and S. Zhang, *Polym. Prepr. Am. Chem. Soc., Div. Polym. Chem.* **42** (1), 381–2 (2001).

709. H.M. Burger, H.M. Muller, D. Seebach, K.O. Bornsen, M.Schar and H.M.Widmer, *Macromolecules* **26** (18), 4783–90 (1993).

710. L. Prokai and W.J. Simonsick, *Rapid Commun. Mass Spectrom.* **7**, 853–6 (1993).

711. J.V. Iribarn and B.A. Thomson, *J. Chem. Phys.* **64**, 2287–94 (1976).

712. X. Lou, J.L.J. van Dongen and E.W. Meyer, *Abstracts 6th International Symposium on Hyphenated Techniques in Chromatography and Hyphenated Chromatographic Analyzers (HTC-6)*, Bruges (2000), P47.

713. P.M. Lloyd, K.G. Suddaby, J.E. Varney, E.Scrivener, P.J. Derrick and D.M. Haddleton, *Eur. Mass Spectrom.* **1**, 293–300 (1995).

714. G. Montaudo, D. Gorozzo, M.S. Montaudo, C. Puglisi and F. Samperi, *Macromolecules* **28**, 7983–9 (1995).

715. R.A. Hutchinson, M.C. Grady and W.J. Simonsick, *Preprints International Symposium on Free Radical Polymerization: Kinetics and Mechanism*, Il Ciocco, Italy (2001).

716. W.J. Simonsick, D.J. Aaserud and M.C. Grady, *Proc. Am. Chem. Soc. Div. Polym. Mat.: Sci. Engng* **78**, 52 (1998).

717. W.J. Simonsick and L. Prokai, in *Chromatographic Characterization of Polymers, Hyphenated and Multidimensional Techniques* (T. Provder, H.G. Barth and M.W. Urban, eds), American Chemical Society, Washington, DC (1995), Chapter 4.

718. F.A. Buytenhuys, H.J.F.M. van de Ven, M. Nielen, P.J.C.H. Cools, B. Klumperman and A.L. German, *Abstracts International GPC Symposium '98*, Phoenix, AZ (1998), p. 31.

719. I.D. Wilson, M. Spraul and E. Humpfer, *J. Planar Chromatogr.* **10**, 217–19 (1997).

720. C.F. Poole, *J. Chromatogr.* **A856**, 399–427 (1999).

721. G.W. Somsen, W. Morden and I.D. Wilson, *J. Chromatogr.* **703**, 613–65 (1995).

722. J.-P. Salo and H. Salomies, *Farmaseuttinen Aikakauskirja DOSIS* **12**, 152–64 (1996).

723. R.E. Allen and L.J. Deutsch, in *TLC: Quantitative Environmental and Clinical Applications* (J.C. Touchstone and D. Rogers, eds.), John Wiley & Sons, Inc., New York, NY (1980), p. 348.

724. M.L. Gianelli, D.H. Burns, J.B. Callis, G.D. Christian and N.H. Andersen, *Anal. Chem.* **55**, 858 (1983).

725. D.H. Burns, J.B. Callis and G.D. Christian, *Trends Anal. Chem.* **5**, 50 (1988).

726. S. Ebel and W. Wuthe, in *Proceedings 3rd International Symposium on Instrumental HPTLC* (R.E. Kaiser, ed.), Bad Dürkheim (1985), p. 381.

727. V.A. Pollak and J. Schulze-Clewing, *J. Chromatogr.* **437**, 97 (1988).

728. V.A. Pollak and J. Schulze-Clewing, *J. Planar Chromatogr.* **3**, 104 (1990).

729. V.A. Pollak, A. Doelemeyer, W. Winkler and J. Schulze-Clewing, *J. Chromatogr.* **59**, 241 (1992).

730. S.M. Brown and K.L. Busch, *J. Planar Chromatogr.* **5**, 338 (1992).

731. M.L. Gianelli, J.B. Callis, N.H. Andersen and G.D. Christian, *Anal. Chem.* **53**, 1357 (1981).

732. J.A. Cosgrove and R.B. Bilhorn, *J. Planar Chromatogr.* **2**, 362 (1989).

733. S.E. Ebel and W. Windmann, *J. Planar Chromatogr.* **4**, 171 (1991).

734. S. Toda, M. Horibe, A. Okubo, S. Yamazaki and N. Hamada, *Proceedings XXIVth Colloquium Spectroscopicum Internationale*, Garmisch-Partenkirchen, Germany (1985), Vol. 4, pp. 780–1.

735. M. Ohnishi-Kameyama and T. Nagata, *Anal. Chem.* **70** (9), 1916–20 (1998).

736. J.L. Guthrie and R.W. McKinney, *Anal. Chem.* **49**, 1676 (1977).

737. C.J. Percival and P.R. Griffiths, *Anal. Chem.* **47**, 154 (1975).

738. G. Glauninger, K.-A. Kovar and V. Hoffmann, in *Software Entwicklung in der Chemie 3, Proceedings 3rd Workshop on Computational Chemistry* (G. Gauglitz, ed.), Springer-Verlag, Berlin (1989), pp. 171–80.

739. K.-A. Kovar, H.K. Enblin, O.R. Frei, S. Rienas and S.C. Wolff, *J. Planar Chromatogr.* **4**, 246–50 (1991).

740. S. Bayerbach, G. Gauglitz, K.-A. Kovar and W. Pisternick, in *Dünnschicht-chromatographie in Memoriam Professor Dr. Jork* (R.E. Kaiser, W. Günther, H. Gunz and G.W. Wulff, eds), InCom Sonderband, Düsseldorf (1995), pp. 1–13.

741. M.P. Fuller and P.R. Griffiths, *Anal. Chem.* **50**, 1906 (1978).

742. L.B. Lloyd, R.C. Yeates and E.M. Eyring, *Anal. Chem.* **54**, 549–52 (1982).

743. R.L. White, *Anal. Chem.* **57**, 1819–22 (1985).

744. M.P. Fuller and P.R. Griffiths, *Appl. Spectrosc.* **34**, 533–9 (1980).

745. S.A. Stahlmann, *J. Planar Chromatogr.* **12**, 5–12 (1999).

746. O.R. Frey, K.-A. Kovar and V. Hoffmann, *J. Planar Chromatogr.–Mod. TLC* **6**, 93–9 (1993).

747. A. Otto, U. Bode and H.M. Heise, *Fresenius' Z. Anal. Chem.* **331**, 376 (1988).

748. S.G. Bush and A.J. Breaux, *Mikrochim. Acta* **1**, 17 (1988).

749. K.O. Alt and G. Szekely, *J. Chromatogr.* **202** (1), 151 (1980).

750. J.M. Chalmers and M.W. Mackenzie, *Appl. Spectrosc.* **39**, 634 (1985).

751. K.H. Shafer, J.A. Herman and H. Bui, *Am. Lab.* **20** (2), 142 (1988).

752. J.M. Chalmers, M.W. Mackenzie, J.L. Sharp and R.N. Ibett, *Anal. Chem.* **59**, 415–18 (1987).

753. H. Yamamoto, K. Wada, T. Tajima and K. Ichimura, *Appl. Spectrosc.* **45**(2), 253 (1991).

754. B.N. Tarasevich and M.V. Polyakova, *Vestn. Mosk. Univ. Ser. 2: Khim.* **22** (20), 188 (1981).

755. G. Glauninger, K.-A. Kovar and V. Hoffmann, *Fresenius' J. Anal. Chem.* **338**, 710 (1990).

756. P.R. Brown and B.T. Beauchemin, *J. Liq. Chromatogr.* **11**, 1001 (1988).

757. S.L. Castleden, C.M. Elliott and G.F. Kirkbright, *Anal. Chem.* **51**, 2152 (1979).

758. V.A. Fishman and A.J. Bard, *Anal. Chem.* **53**, 102 (1981).

759. A. Hagman and S. Jacobsson, *J. Chromatogr.* **395**, 271 (1987).

760. K. Imaeda, K. Ohsawa, K. Uchiyama and S. Nakamura, *Anal. Sci.* **3**, 11 (1987).

761. K. Ohsawa, K. Uchiyama, Y. Yoshimura, A. Takasuka, K. Mibe, K. Tamura, Y. Ohtani and K. Imaeda, *Anal. Sci.* **6**, 589 (1990).

762. E.W. Ciurczak, *Proceedings 8th International NIRA Symposium*, Tarrytown, NY (1985).

763. A. Fong and G.M. Hieftje, *Appl. Spectrosc.* **48**, 394–9 (1994).

764. D.M. Mustillo and E.W. Ciurczak, *Appl. Spectrosc. Rev.* **27**, 125–41 (1992).

765. R. de C.L. Dutra, *Polim.: Cienc. Tecnol.* **6** (2), 26–31 (1996).

766. T. Iwaoka, S. Tsutsumi, K. Tada and F. Suzuki, *Sankyo Kenkyusho Nenpo* **40**, 39–46 (1988).

767. K.M. Shafer, P.R. Griffiths and W. Shu-Qin, *Anal. Chem.* **58**, 2708 (1986).

768. U. Bode and H.M. Heise, *Mikrochim. Acta* **1**, 143 (1988).

769. H.R. Garner and H. Packer, *Appl. Spectrosc.* **22**, 122 (1967).

770. T.R. Crompton, *Analysis of Polymers, An Introduction*, Pergamon Press, Oxford (1989).

771. H. Hellmann, *Fresenius' Z. Anal. Chem.* **321** (2), 159–62 (1985); **322** (1), 42–6 (1985).

772. M.M. Gomez-Taylor and P.R. Griffiths, *Appl. Spectrosc.* **31**, 528 (1977).

773. M.D. Danielson, J.E. Katon, S.P. Bouffard and Z. Zhu, *Anal. Chem.* **64**, 2183 (1992).

774. S.P. Bouffard, J.E. Katon, A.J. Sommer and N.D. Danielson, *Anal. Chem.* **66**, 1937 (1994).

775. N. Buschmann and A. Kruse, *Commun. J. Com. Esp. Deterg.* **24**, 457 (1993).

776. I. Vovk, M. Franko, J. Gibkes, M. Prošek and D. Bicanic, *J. Planar Chromatogr.* **10** (4), 258–62 (1997).

777. H. Yamamoto, O. Yoshikawa, M. Nakatani, F. Tsuji and T. Maeda, *Appl. Spectrosc.* **45**, 1166 (1991).

778. J. Von Czarnecki and H.W. Hiemesch, *Actual. Chim.* **4**, 55–56 (1980).

779. N.J. Everall, J.M. Chalmers and I.D. Newton, *Appl. Spectrosc.* **46**, 597–601 (1992).

780. B. Chase, *Anal. Chem.* **14**, 881A (1987).

781. C. Petty, *Spectrochim. Acta* **49**, 645–55 (1993).

782. A. Rau, *J. Raman Spectrosc.* **24**, 251–4 (1993).

783. R.L. Garrell, *Anal. Chem.* **61**, 401–11A (1989).

784. J.J. Laserna, *Anal. Chim. Acta* **283**, 607–22 (1993).

785. M.D. Morris and D.J. Wallen, *Anal. Chem.* **51**, 182–92A (1979).

786. D.W. Armstrong, L.A. Spino, M.R. Andrias and E.W. Findsen, *J. Chromatogr.* **369**, 227 (1986).

787. E. Koglin, *J. Planar Chromatogr.* **2**, 194–7 (1989).

788. E. Koglin, *J. Mol. Struct.* **173**, 369–76 (1988).

789. E. Koglin, *J. Planar Chromatogr.* **6**, 88–92 (1993).

790. E. Koglin, *J. Planar Chromatogr.* **3**, 117–20 (1990).

791. G.W. Somsen, S.K. Coulter, C. Gooyer, N.H. Velthorst and U.A.Th. Brinkman, *Anal. Chim. Acta* **349**, 189 (1997).

792. S. Keller, T. Loechte, B. Dippel and B. Schrader, *Fresenius' Z. Anal. Chem.* **346**, 863 (1993).

793. S. Sollinger and J. Sawatzki, *GIT Fachz. Lab.* **1**, 14–18 (1999).

794. I.D. Wilson and W. Morden, *J. Planar Chromatogr.* **4**, 226 (1991).

795. C. de Koster, D. Wienke, J. Cosemans and H. Linssen, *Tech. Rep. DSM Research*, Geleen, Netherlands (1999).

796. G.W. Somsen, P.G.J.H. ter Riet, C. Gooyer, N.H. Velthorst and U.A.Th. Brinkman, *J. Planar Chromatogr.* **10**, 10–17 (1997).

797. S.E. Unger, A. Vincze, R.G. Cooks, R. Chrisman and L. D. Rothman, *Anal. Chem.* **53**, 976–81 (1981).

798. D. Dekker, *J. Chromatogr.* **168**, 508–11 (1979).

799. H. Oka, Y. Ikai, F. Kondo, N. Kawamura, J. Hayakawa, K. Masuda, K. Harada and M. Suzuki, *Rapid Commun. Mass Spectrom.* **6**, 89 (1992).

800. J.J. Monaghan, W.E. Morden, T. Johnson, I.D. Wilson and P. Martin, *Rapid Commun. Mass Spectrom.* **6**, 608–15 (1992).

801. R.D. Macfarlane and D.F. Togerson, *Science* **191**, 920 (1976).

802. A.J. Kubis, K.V. Somayajula, A.G. Sharkey and D.M. Hercules, *Anal. Chem.* **61**, 2516–23 (1989).

803. K. Heyns and H.F. Grützmacher, *Angew. Chem. Intl Ed. Engl.* **1**, 400 (1962).

804. G.J. Down and S.A. Gwyn, *J. Chromatogr.* **103**, 208–10 (1975).

805. R. Kraft, A. Otto, A. Makower and G. Etzoki, *Anal. Biochem.* **113**, 193–6 (1981).

806. R.M. Anderson and K.L. Busch, *J. Planar Chromatogr.* **11**, 336 (1998).

807. L. Ramaley, M.E. Nearing, M.-A. Vaughan, R.G. Ackman and W.D. Jamieson, *Anal. Chem.* **55**, 2285 (1983).

808. L. Ramaley, M.-A. Vaughan and W.D. Jamieson, *Anal. Chem.* **57**, 353 (1985).

809. R. Kaiser, *Chem. Br.* **5**, 54–61 (1969).

810. K.L. Busch, *J. Planar Chromatogr.* **5**, 72 (1992).

811. K.L. Busch, in *Handbook of Thin-Layer Chromatography* (J. Sherma and B. Fried, eds), M. Dekker, New York, NY (1996), Chapter 9.

812. S.M. Brown, H. Schurz and K.L. Busch, *J. Planar Chromatogr.* **3**, 222 (1990).

813. C.F. Poole and S.K. Poole, *Anal. Chem.* **66**, 27A (1994).

814. S. Nyiredy, in *Chromatography* (E. Heftmann, ed.), Elsevier, Amsterdam (1992), pp. A109–50.

815. K.L. Busch, *Trends Anal. Chem.* **6**, 95–100 (1987).

816. K.L. Busch, *Trends Anal. Chem.* **11**, 314–24 (1992).

817. J. Henion, G.A. Maylin and B.A. Thomson, *J. Chromatogr.* **271**, 107 (1983).

818. R. Kraft, A. Otto, H.-J. Zopfl and G. Etzold, *Biomed. Environ. Mass Spectrom.* **14**, 1 (1987).

819. R. Kraft, D. Buttner, P. Frank and G. Etzold, *Biomed. Environ. Mass Spectrom.* **14**, 5 (1987).

820. I.D. Wilson and W. Morden, *LC.GC Intl* **12** (2), 72–80 (1999).

821. S.M. Brown and K.L. Busch, *J. Planar Chromatogr.* **4**, 189 (1991).

822. P. Martin, W. Morden, P. Wall and I.D. Wilson, *J. Planar Chromatogr.* **5**, 255 (1992).

823. R. Lafont, C.J. Porter, E. Williams, H. Read, E.D. Morgan and I.D. Wilson, *J. Planar Chromatogr.* **6**, 421 (1993).

824. I.D. Wilson, R. Lafont, R.G. Kingston and C.J. Porter, *J. Planar Chromatogr.* **3**, 359–61 (1990).

825. W. Morden and I.D. Wilson, *J. Planar Chromatogr.* **9**, 84–91 (1996).

826. T.T. Chang, J.O. Lay and R.J. Francel, *Anal. Chem.* **56** (1), 109–11 (1984).

827. G.W. Somsen, W. Morden and I.D. Wilson, *J. Chromatogr.* **A703**, 613–65 (1995).

828. K.L. Busch, J.O. Mullis and R.E. Carlson, *J. Liq. Chromatogr.* **16**, 1695 (1993).

829. Y. Kushi and S. Handa, *J. Biochem.* **98**, 265 (1985).

830. L. Merschel, W. Sichtermann, N. Buschmann and A. Benninghoven, *Proceedings SIMS X* (A. Benninghoven, B. Hagenhoff and H.W. Werner, eds), John Wiley & Sons, Ltd, Chichester (1996), pp. 787–90.

831. N. Buschmann, L. Merschel and S. Wodarczak, *Tenside, Surfactants, Deterg.* **33**, 16–20 (1996).

832. A. Oriňák, R. Oriňaková, A. Benninghoven, V. Andruch, M. Justinová and K. Jacenková, *Abstracts 7th International Symposium on Hyphenated Techniques in Chromatography and Hyphenated Chromatographic Analyzers (HTC-7)*, Bruges (2000), Paper P118.

833. K. Banno, M. Matsuoka and R. Takahashi, *Chromatographia* **32**, 179–81 (1991).

834. G.C. DiDonato and K.L. Busch, *Anal. Chem.* **58**, 3231–2 (1986).

835. K.L. Busch, J.A. Mullis and J.A. Chakel, *J. Planar Chromatogr.* **5**, 9–15 (1992).

836. S. Takahashi (to Hitachi), *UK Pat. Appl.* 2,187,327 (Aug. 27, 1986); *C.A.* **106**: 224094d.

837. R.W. Finney and H. Read, *J. Chromatogr.* **471**, 389–96 (1989).

838. C.R. Dewey and R.W. Finney, *Anal. Proc. (London)* **27** (5), 125–7 (1990).

839. Z.A. Wilk and D.M. Hercules, *Anal. Chem.* **59**, 1819–25 (1987).

840. A.I. Gusev, A. Proctor, Y.I. Rabinovich and D.M. Hercules, *Anal. Chem.* **67**, 1805 (1995).

841. S. Mowthorpe, M.R. Clench, A. Cricelius, D.S. Richards, V. Parr and L.W. Tetler, *Rapid Commun. Mass Spectrom.* **13**, 264–70 (1999).

842. A.W.T. Bristow and C.S. Creaser, *Rapid Commun. Mass Spectrom.* **9**, 1465 (1995).

843. A.I. Gusev, O.J. Vasseur, A. Proctor, A.G. Sharkey and D.M. Hercules, *Anal. Chem.* **67**, 4565–70 (1995).

844. A.J. Nicola, A.I. Gusev and D.M. Hercules, *Appl. Spectrosc.* **50**, 1479–82 (1996).

845. Y.-C. Chen, J. Shiea and J. Summer, *J. Chromatogr.* **A826**, 77 (1998).

846. J. Zhu and E.S. Yeung, *J. Chromatogr.* **463** (1), 139–45 (1989).

847. K.L. Busch, *J. Mass Spectrom.* **30**, 233–40 (1995).

848. K.L. Busch, J.O. Mullis and R.E. Carlson, *J. Liq. Chromatogr.* **16**, 1713 (1993).

849. K.L. Busch, *Chromatogr. Sci.* **55**, 183–209 (1991).

850. I.D. Wilson and W. Morden, *J. Planar Chromatogr.* **9**, 84 (1996).

851. I.D. Wilson, *J. Chromatogr.* **A856**, 429–42 (1999).

852. K.L. Busch, S.M. Brown, S.J. Doherty, J.C. Dunphy and M.V. Buchanan, *J. Liq. Chromatogr.* **13**, 2844–69 (1990).

853. K.H. Bare and H. Read, *Analyst* **112**, 433–6 (1987).

854. H. Oka, Y. Ikai, J. Hayakawa, K. Masuda, K. Harada, M. Suzuki, V. Martz and J.D. MacNiel, *J. Agric. Food Chem.* **41**, 410 (1993).

855. K. Masuda, K. Harada, H. Oka, N. Kawamura and M. Yamada, *Org. Mass Spectrom.* **24**, 74 (1989).

856. K. Harada, K. Masuda, M. Suzuki and H. Oka, *Biol. Mass Spectrom.* **20**, 522 (1991).

857. H. Oka, Y. Ikai, T. Ohno, N. Kawamura, J. Hayakawa, K. Harada and M. Suzuki, *J. Chromatogr.* **674**, 301 (1994).

858. F.P. Novak and D.M. Hercules, *Anal. Lett.* **18**, 503–18 (1985).

859. Y. Mengerink, Sj.v.d. Wal, H.A. Claessens and C.A. Cramers, *J. Chromatogr.* **A871** (1/2), 259–68 (2000).

860. J.A. De Haseth, R. Todebush and L.-T. He, *Abstracts 7th International Symposium on Hyphenated Techniques in Chromatography and Hyphenated Chromatographic Analyzers (HTC-7)*, Bruges (2002), Paper 4.2.

861. K. Pusecker, J. Schewitz, P. Gfrörer, L.-H. Tseng, K. Albert and E. Bayer, *Abstracts HPLC'98*, St. Louis, MO (1998), p. 161.

862. K. Albert, *Angew. Chem. Intl Ed. Engl.* **34** (6), 641–2 (1995).

863. D.L. Olson, M.E. Lacey and J.V. Sweedler, *Anal. Chem.* **70** (7), 257–64A (1998).

864. W.C. Brumley, *J. Chromatogr.* **603**, 267–72 (1992).

865. G.A. Lord, D.B. Gordon, L.W. Tetler and C.M. Carr, *J. Chromatogr.* **700A**, 27–33 (1995).

866. D. Birkett, *Chem. Br.* 22–7 (July 2002).

867. J.-Y. Zha, P. Thibault, T. Tazawa and M.A. Quilliam, *J. Chromatogr.* **A781**, 555–64 (1997).

868. J.W. Olesik, J.A. Kinzer and S.V. Olesik, *Anal. Chem.* **67**, 1 (1995).

869. Y. Liu, V. Lopez-Avila, J.J. Zhu, D.R. Wiederin and W.F. Beckert, *Anal. Chem.* **67**, 2020–5 (1995).

870. Q. Lu, S.M. Bird and R.M. Barnes, *Anal. Chem.* **67**, 2949 (1995).

871. B. Michalke, *LC.GC Europe*, 36–41 (Jan. 2000).

872. C. Vogt, J. Vogt and H. Wittrisch, *J. Chromatogr.* **A727**, 301 (1996).

873. J.A. Olivares, N.T. Nguyen, C.R. Yonker and R.D. Smith, *Anal. Chem.* **59**, 1230–2 (1987).

874. B. Sundqvist and R.D. MacFarlane, *Mass Spectrom. Rev.* **4**, 421–60 (1985).

875. M. Karas, U. Bahr and U. Giessman, *Mass Spectrom. Rev.* **10**, 335–58 (1991).

876. G.A. Ross, *LC.GC Europe* **14** (1), 45–9 (2001).

877. E.D. Lee, W. Mück, J.D. Henion and T.R. Covey, *J. Chromatogr.* **458**, 313–21 (1988).

878. P. Thibault, S. Pleasance and M.V. Laycock, *J. Chromatogr.* **542**, 483–501 (1991).

879. T.E. Wheat, K.A. Lilley and W.S. Schnute, *Proceedings 45th ASMS Conference on Mass Spectrometry and Allied Topics*, ASMS, Palm Springs, CA (1997).

880. E.D. Lee, W. Mück, J.D. Henion and T.R. Covey, *Biomed. Environ. Mass Spectrom.* **18**, 253–7 (1989).

881. J.R. Perkins and K.B. Tomer, *Anal. Chem.* **66**, 2835 (1994).

882. S.A. Hofstadler, J.H. Wahl, R. Bakhtair, G.A. Anderson, J.E. Bruce and R. Smith, *J. Am. Soc. Mass Spectrom.* **5**, 894 (1994).

883. L. Fang, R. Zhang, E.R. Williams and R.N. Zare, *Anal. Chem.* **66**, 3696 (1994).

884. J.F. Banks and T. Dresch, *Anal. Chem.* **68**, 1480 (1996).

885. J.F. Banks, E. Gulcicek and C. Whitehouse, *Proceedings 14th International Mass Spectrometry Conference*, Tampere, Finland (1997), Paper LaPo001.

886. E.D. Lee, W. Mück, J.D. Henion and T.R. Covey, *Biomed. Environ. Mass Spectrom.* **18**, 844 (1989).

887. Y. Takada, M. Sakaira and H. Koizumi, *Anal. Chem.* **67**, 1474 (1995).

888. N.J. Reinhold, W.M.A. Niessen, U.R. Tjaden, L.G. Gramberg, E.R. Verhey and J. van der Greef, *Rapid Commun. Mass Spectrom.* **3**, 348 (1989).

889. R. D. Smith, J. A. Olivares, N. T. Nguyen and H. R. Udseth, *Anal. Chem.* **60**, 436 (1988).

890. G. Choudhary, A. Apffel, H. Yin and W. Hancock, *J. Chromatogr.* **A887** (1–2), 85–101 (2000).

891. J. Ding and P. Vouros, *Anal. Chem.* **71** (1), 378–85A (1999).

892. J.F. Banks, *Electrophoresis* **18** (12/13), 2255–66 (1997).

893. W.C. Brumley and W. Winnik, *J. Chromatogr. Libr.* **59**, 481–527 (1996).

894. J. Cai and J.D. Henion, *J. Chromatogr.* **A703**, 667 (1995).

895. R.D. Smith, J.H. Wahl, D.R. Goodlett and S.A. Hofstadler, *Anal. Chem.* **65**, 574A (1993).

896. K.B. Tomer, L.J. Deterding and C.E. Parker, *Adv. Chromatogr.* **35**, 53–99 (1993).

897. L.W. Tetler, P.A. Cooper and C.M. Carr, *Rapid Commun. Mass Spectrom.* **8**, 179 (1994).

898. W. Ahrer and W. Buchberger, *Abstracts 15th Montreux Symposium on Liquid Chromatography/Mass Spectrometry*, Montreux (1998), p. 32.

899. V. Spikmans, N.W. Smith, M.G. Tucker, R. Horsten and M. Mazereeuw, *LC.GC Europe*, 486–94 (July 2000).

900. R. Bodor, M. Zuborova, E. Olvecka, V. Majadova, M. Masar, D. Kaniansky and B. Stanislauski, *J. Sep. Sci.* **24** (9), 802–9 (2001).

900a. P. Schoenmakers, P. Marriott and J. Beens, *LC.GC Europe* **16** (6), 335–9 (2003).

901. J. Giddings, in *Multidimensional Chromatography* (H.J. Cortes, ed.), M. Dekker, New York, NY (1990), pp. 1–27.

902. J.C. Giddings, *Anal. Chem.* **56**, 1258A (1984).

903. I.L. Davies, K.E. Markides, M.L. Lee, M.W. Raynor and K.D. Bartle, *J. High Resolut. Chromatogr., Chromatogr. Commun.* **12**, 193 (1989).

904. H.J. Cortes, *Chromatogr. Sci.* **45**, 211 (1989).

904a. K. Grob, unpublished results (2004).

904b. Special Issue on *Recent Applications in Multidimensional Chromatography, LC.GC Europe* (November 2003).

905. S.G. Claude and R. Tabacchi, in *Proceedings 8th International Symposium on Capillary Chromatography* (P. Sandra, ed.), Hüthig Publishers, Heidelberg (1987), p. 564.

906. H.J. Cortes, in *Multidimensional Chromatography* (H.J. Cortes, ed.), M. Dekker, New York, NY (1990), pp. 251–99.

907. I.C. Nigam, M. Sahasrabudhe and L. Levi, *Can. J. Chem.* **41**, 1535–9 (1963).

908. J. Janák, *J. Gas Chromatogr.* **1** (10), 20–3 (1963).

909. J. Janák, *J. Chromatogr.* **15**, 15–28 (1964).

910. J. Janák, I. Klimes and K. Hana, *J. Chromatogr.* **18**, 270–7 (1965).

911. R. Kaiser, *Z. Anal. Chem.* **205**, 284–98 (1964).

912. H.C. Curtius and M. Muller, *J. Chromatogr.* **32**, 222 (1968).

913. J.C. Giddings, *J. Chromatogr.* **703**, 3–15 (1995).

914. R.E. Synovec, *Abstracts 7th International Symposium on Hyphenated Techniques in Chromatography and Hyphenated Chromatographic Analyzers (HTC-7)*, Bruges (2002), Paper 12.2.

915. V. van Mispelaar, H.-G. Janssen, A. Tas and P. Schoenmakers, *Abstracts 7th International Symposium on Hyphenated Techniques in Chromatography and Hyphenated Chromatographic Analyzers (HTC-7)*, Bruges (2002), Paper 12.1.

916. J.B. Philips and E.B. Ledford, *Field Anal. Chem.* **1**, 23 (1996).

917. W. Bertsch, *J. High Resolut. Chromatogr.* **22**, 647–65 (1999); **23**, 167–81 (2000).

918. W. Bertsch, in *Multidimensional Chromatography* (H.J. Cortes, ed.), M. Dekker, New York, NY (1990), pp. 74–144.

919. C.F. Poole and S.K. Poole, *Chromatography Today*, Elsevier, Amsterdam (1991).

920. G. Schomburg, *J. Chromatogr.* **A703**, 309–25 (1995).

921. J.B. Phillips and J. Xu, *J. Chromatogr.* **A703**, 327–34 (1995).

921a. J.V. Hinshaw, *LC-GC Europe* **17** (2), 86–95 (2004).

922. B. Winniford, T. McCabe, J. Luong, K. Sun, B. Gerhart and H.J. Cortes, *Abstracts 7th International Symposium on Hyphenated Techniques in Chromatography and Hyphenated Chromatographic Analyzers (HTC-7)*, Bruges (2002), Paper 8.1.

923. M. Pursch, K. Sun, B. Winniford, H.J. Cortes, A. Weber, T. McCabe and J. Luong, *Anal. Bioanal. Chem.* **373** (6), 356–67 (2002).

924. I.L. Davies, B. Xu, K.E. Markides, K.D. Bartle and M.L. Lee, *J. Microcol. Sep.* **1**, 71–84 (1989).

925. S. Hishimoto and Y. Hirata, *Abstracts 8th International Symposium on Supercritical Fluid Chromatography and Extraction*, St. Louis, MO (1998), L-20.

926. L.Q. Xie, K.E. Markides, M.L. Lee, N.K. Hollenberg, G.H. Williams and S.W. Graves, *Chromatographia* **35**, 363–71 (1993).

927. E. Stahl, *J. Chromatogr.* **142**, 15 (1977).

928. L. Wunsche, U. Keller and I. Flament, *J. Chromatogr.* **552**, 539 (1991).

929. I.L. Davies, K.E. Markides, M.L. Lee and K.D. Bartle, in *Multidimensional Chromatography* (H.J. Cortes, ed.), Marcel Dekker, New York, NY (1990), pp. 301–30.

930. J.M. Levy, J.P. Guzowski and W.E. Huhak, *J. High Resolut. Chromatogr., Chromatogr. Commun.* **10**, 337–41 (1987).

931. C. Fujimoto, T. Morita and K. Jinno, *J. Chromatogr.* **438**, 329–37 (1988).

932. T. Takeuchi, M. Asai, H. Haraguchi and D. Ishii, *J. Chromatogr.* **499**, 549–56 (1990).

933. L.A. Holland and J.W. Jorgenson, *Anal. Chem.* **67**, 3275–83 (1995).

934. H.J. Cortes, L. Green, C. Shayne and M. Robert, *Anal. Chem.* **63**, 2719 (1991).

935. R. Moulder, K.D. Bartle and A.A. Clifford, *Analyst* **116**, 1293 (1991).

936. A.V. Lemmo and J.W. Jorgenson, *Anal. Chem.* **65**, 1576–81 (1993).

937. J.A. Apffel and H. McNair, *J. Chromatogr.* **279**, 139 (1983).

938. H.J. Cortes, C.D. Pfeiffer and B.E. Richter, *J. High Resolut. Chromatogr., Chromatogr. Commun.* **8**, 469 (1985).

939. R.E. Majors, *J. Chromatogr. Sci.* **18**, 571 (1980).

940. R.M. Campbell, D.M. Meunier and H.J. Cortes, *J. Microcol. Sep.* **1**, 302–308 (1989).

941. M. Shimmo, T. Hyötyläinen, K. Hartonen and M.-L. Riekkola, *J. Microcol. Sep.* **13**, 202–10 (2001).

942. D.J. Miller and S.B. Hawthorne, *Anal. Chem.* **69**, 623–7 (1997).

943. T. Hyötyläinen, K. Kuosmanen, M. Shimmo, T. Andersson, K. Hartonen and M.-L. Riekkola, *Abstracts 7th International Symposium on Hyphenated Techniques in Chromatography and Hyphenated Chromatographic Analyzers (HTC-7)*, Bruges (2002), Paper 5.1.

944. T. Hyötyläinen, T. Andersson, K. Hartonen, K. Kuosmanen and M.-L. Riekkola, *Anal. Chem.* **72**, 3070–6 (2000).

945. T.E. Young, S.C. Ecker, R.E. Synovec, N.T. Hawley, J.P. Lomber and C.M. Wai, *Talanta* **45**, 1189–99 (1998).

946. M. Biedermann, K. Grob, D. Fröhlich and W. Meier, *Z. Lebensm. Unters. Forsch.* **195**, 409–16 (1992).

947. I.S. Lurie, *LC.GC* **6**, 1066–7 (1988).

948. H.J. Cortes, R.M. Campbell, R.P. Himes and C.D. Pfeiffer, *J. Microcol. Sep.* **4**, 239–44 (1992).

949. J.H. van Dijk, *Fresenius' Z. Anal. Chem.* **247**, 262 (1969).

950. P.R. Boshoff, B.J. Hopkins and V. Pretorius, *J. Chromatogr.* **126**, 35 (1976).

951. W.R.G. Baeyens and B. Lin Lang, *J. Planar Chromatogr.* **1**, 198–213 (1988).

952. J.W. Hofstraat, M. Engelsma, R.J. van de Nesse, C. Gooijer, N.H. Velthorst and U.A.Th. Brinkman, *Anal. Chim. Acta* **186**, 247–59 (1986).

953. J.W. Hofstraat, M. Engelsma, R.J. van de Nesse, U.A.Th. Brinkman, C. Gooijer and N.H. Velthorst, *Anal. Chim. Acta* **193**, 193–207 (1987).

954. J. Bladek, *J. Planar Chromatogr.* **6**, 495 (1993).

955. D.E. Jänchen and H.J. Issaq, *J. Liq. Chromatogr.* **11**, 1941 (1988).

956. E. Müller and H. Jork, *J. Planar Chromatogr.* **6**, 21 (1993).

957. O.R. Queckenberg and A.W. Frahm, *J. Planar Chromatogr.* **6**, 55 (1993).

958. H.-J. Stan and F. Schwarzer, *J. Chromatogr.* **A819**, 35 (1998).

959. C. Fujimoto, T. Morita, K. Jinno and K.H. Shafer, *J. High Resolut. Chromatogr.* **11**, 810–14 (1988).

960. S.F.Y. Li, *Capillary Electrophoresis, Principles, Practice and Applications*, Elsevier, Amsterdam (1992).

961. T. Hyötyläinen and M.-L. Riekkola, *LC.GC Europe* **13** (9), 658–64 (2000).

962. K. Grob, *J. Chromatogr.* **A703**, 265–76 (1995).

963. K. Grob, *Chimia* **45**, 109 (1991).

964. H.J. Cortes and L.D. Rothman, in *Multidimensional Chromatography* (H.J. Cortes, ed.), M. Dekker, New York, NY (1990), pp. 219–50.

965. K. Grob, *On-line Coupled LC-GC*, Hüthig Publishers, Heidelberg (1991).

966. H.J. Cortes (ed.), *Multidimensional Chromatography*, M. Dekker, New York, NY (1990).

967. J. Simal-Gándara, M. Sarria-Vidal and R. Rijk, *J. AOAC Intl* **83** (2), 311–19 (2000).

968. C. Droz and K. Grob, *Z. Lebensm. Unters. Forsch.* **A205** (3), 239–41 (1997).

969. K. Grob, M. Huber, U. Boderius and M. Bronz, *Food Addit. Contam.* **14**, 83–8 (1997).

970. B. Pacciarelli, E. Muller, R. Schneider, K. Grob, W. Steiner and D. Frohlich, *J. High Resolut. Chromatogr., Chromatogr. Commun.* **11**, 135 (1988).

971. K. Kuosmanen, T. Hyötyläinen, K. Hartonen and M.-L. Riekkola, *J. Chromatogr.* **A943**, 113–22 (2002).

972. R.E. Majors, *Liq. Chromatogr. HPLC Mag.* **2**, 358 (1984).

973. A van de Horst and P.J. Schoenmakers, *Abstracts 7th International Symposium on Hyphenated Techniques in Chromatography and Hyphenated Chromatographic Analyzers (HTC-7)*, Bruges (2002), Paper P163.

974. T. Andersson, T. Hyötyläinen and M.-L. Riekkola, *J. Chromatogr.* **A896**, 343–9 (2000).

975. H.J. Cortes, B.M. Bell, C.D. Pfeiffer and J.D. Graham. *J. Microcol. Sep.* **1** (6), 278–88 (1989).

976. H.J. Cortes, G.E. Bormett and J.D. Graham. *J. Microcol. Sep.* **4** (1), 51–7 (1992).

977. K. Grob Jr and J.M. Stoll, *J. High Resolut. Chromatogr., Chromatogr. Commun.* **9**, 518 (1986).

978. B. Jongenotter and H.-G. Janssen, *LC.GC Europe* **15** (6), 338–57 (2002).

979. H.J. Cortes, G.L. Jewett, C.D. Pfeiffer, S. Martin and C. Smith, *Anal. Chem.* **61** (9), 961–5 (1989).

980. T.C. Schunk and F.A. Fox, *Abstracts 22nd International Symposium on High Performance Liquid Phase Separations and Related Techniques*, St. Louis, MO (1998), P-0642-T.

981. M.J. Shepherd and J. Gilbert, *J. Chromatogr.* **218**, 703 (1981).

982. J. Blomberg, P.J. Schoenmakers and N.v.d. Hoed, *J. High Resolut. Chromatogr.* **17** (6), 411–14 (1994).

983. H.J. Cortes, unpublished results (1999).

984. N. Kobayashi, H. Arimoto and Y. Nishikawa, *J. Microcol. Sep.* **12** (9), 501–7 (2000).

985. T.R. Floyd, *Chromatographia* **25** (9), 791–6 (1988).

986. E. El Mansouri, N. Yagoubi and D. Ferrier, *J. Chromatogr.* **A771**, 111–18 (1997).

987. E.L. Johnson, R. Gloor and R.E. Majors, *J. Chromatogr.* **149**, 571 (1978).

988. M.J. Shepherd and J.J. Gilbert, *J. Chromatogr.* **178**, 435–41 (1979).

989. J. Gilbert, M.J. Shepherd and M.A. Wallwork, *J. Chromatogr.* **193**, 235 (1980).

990. C. Nerín, J. Salafranca, J. Cacho and C. Rubio, *J. Chromatogr.* **A690**, 230 (1995).

991. C. Nerín, J. Salafranca and J. Cacho, *Food Addit. Contam.* **13**, 243 (1996).

992. D.W. Patrick, D.A. Strand and H.J. Cortes, *J. Sep. Sci.* **25** (8), 519–26 (2002).

993. R.E. Majors and E.L. Johnson, *J. Chromatogr.* **167**, 17 (1978).

994. R.L. Cotter, R.J. Limpert and C. Deluski, *Am. Lab.* **19** (2), 54–62 (1987).

995. M. Augenstein and M. Stickler, *Makromol. Chem.* **191** (2), 415–28 (1990).

996. C.F. Poole and S.K. Poole, in *Multidimensional Chromatography* (H.J. Cortes, ed.), M. Dekker, New York, NY (1990), pp. 29–73.

997. M. Zakaria, M.-F. Gonnord and G. Guiochon, *J. Chromatogr.* **271**, 127 (1983).

998. K. Burger, *Fresenius' Z. Anal. Chem.* **318**, 228 (1984).

999. W.J. De Klein, *Fresenius' Z. Anal. Chem.* **249**, 81–3 (1970).

1000. J. Zhu and E.S. Yeung, *Anal. Chem.* **61**, 1906–10 (1989).

1001. S.J. Lyle and M.S. Tehrani, *J. Chromatogr.* **236**, 25–30 (1982).

1002. A.J. Al-Sayegh and S.J. Lyle, *J. Anal. At. Pyrol.* **14**, 323–30 (1986).

1003. E. Minscovics, M. Garami and E. Tyihak, *J. Planar Chromatogr.* **4**, 299–303 (1991).

1004. C.F. Poole and S.K. Poole, *J. Chromatogr.* **A703**, 573–612 (1995).

1005. M.T. Belay and C.F. Poole, *J. Planar Chromatogr.* **4**, 424 (1991).

1006. I. Noda, *Appl. Spectrosc.* **44** (4), 550–61 (1990).

1007. I. Noda, *Appl. Spectrosc.* **47** (9), 1329–36 (1993).

1008. W.P. Aue, E. Bartholdi and R.R. Ernst, *J. Chem. Phys.* **64**, 2229 (1976).

1009. A. Bax, *Two Dimensional Nuclear Magnetic Resonance in Liquids*, Reidel, Boston, MA (1982).

1010. K. Ebihara, H. Takahashi and I. Noda, *Appl. Spectrosc.* **47**, 1343 (1993).

1011. I. Noda, G.M. Story, A.E. Dowrey, R.C. Reeder and C. Marcott, *Macromol. Symp.* **119**, 1 (1997).

1012. Y. Ozaki, Y. Liu and I. Noda, *Appl. Spectrosc.* **51**, 526 (1997).

1013. Y. Liu, Y. Ozaki and I. Noda, *J. Phys. Chem.* **100**, 7326 (1996).

1014. M. Shimoyama, T. Ninomiya, K. Sano, Y. Ozaki, H. Higashiyama, M. Watari and M. Tomo, *J. Near Infrared Spectrosc.* **6**, 317–24 (1998).

1015. R. Bramley, in *Encyclopedia of Inorganic Chemistry* (R.B. King, ed.), John Wiley & Sons, Ltd, Chichester (1993), Vol. 3, pp. 1062–81.

1016. Y. Ozaki and I. Noda, *J. Near Infrared Spectrosc.* **4**, 85 (1996).

1017. C. Marcott, A.E. Dowrey and I. Noda, *Anal. Chem.* **66**, 1065–78A (1994).

1018. Y. Ozaki and Y. Wang, *J. Near Infrared Spectrosc.* **6**, 19–31 (1988).

1019. R.R. Ernst, G. Bodenhausen and A. Wokaun, *Principles of Nuclear Magnetic Resonance in One and Two Dimensions*, Clarendon Press, Oxford (1997).

1020. J.N.S. Evans, *Biomolecular NMR Spectroscopy*, Oxford University Press, Oxford (1995).

1021. R. Zimmermann, Ch. Lermer, K.W. Schramm, A. Kettrup and U. Boesl, *Eur. Mass Spectrom.* **1**, 341–51 (1995).

1022. S.M. Colby, M. Stewart and J.P. Reilly, *Anal. Chem.* **62**, 2400 (1990).

1023. M.P. Chiarelli and M.L. Gross, in *Lasers and Mass Spectrometry* (D.M. Lubman, ed.), Oxford University Press, New York, NY (1990), pp. 272–90.

1024. R.L.M. Dobson, A.P. D'Silva, S.J. Weeks and V.A. Fassel, *Anal. Chem.* **58**, 2129–37 (1986).

1025. R. Zimmermann, U. Boesl, H.-J. Heger, E.R. Rohwer, E.K. Ortner, E.W. Schlag and A. Kettrup, *J. High Resolut. Chromatogr.* **20**, 461–70 (1997).

1026. I. Noda, A.E. Dowrey and C. Marcott, in *Modern Polymer Spectroscopy* (G. Zerbi, ed.), Wiley-VCH, Weinheim (1999), pp. 1–32.

CHAPTER 8

Are you in your element?

Inorganic and Element Analytical Methods

Additives In Polymers: Industrial Analysis And Applications   J. C. J. Bart
© 2005 John Wiley & Sons, Ltd   ISBN: 0-470-85062-0

Although the polymer industry is often considered a bulk industry, the development and production of the different types, grades and compounds requires advanced technology and an appreciable amount of R&D. In support of the industrial development phase, production control and quality management, the polymer industry needs to determine more than 60 elements in polymeric materials, in concentrations ranging from per cent down to ppt levels.

Elemental analysis plays an important role throughout the life-cycle of a polymeric material and in product development. In the R&D phase, elemental analysis is used in the study of catalytic systems, needed for the production of most polymers, in the determination of catalytic residues in finished products, for the analysis of additives (type and amount), the most important being inorganic fillers, plasticisers, flame retardants, UV-stabilisers and thermal stabilisers, and for the determination of pigments and dyes. In the production phase, elemental analysis is used for on-line or off-line inspection of product specification, usually within the framework of a QA system, that is ISO 9000, EN 29 000, EN 45 001, etc. Particular attention is required for materials to be used for food-contacting and medical purposes. In service conditions, elemental analysis may be called in to deal with contamination problems. In the recycling stage, especially for the polymer-to-polymer route, elemental analysis is used for quality assurance, to monitor the element content that should remain within given production norms or – as is the case with some heavy metals – to check compliance with legal regulations. Conservative estimates of the total number of elemental analyses (determination of an element in a polymer base material) carried out in the research and production labs in the European polymer industry amount to 1–2 million year^{-1}, at an approximate cost of fifty million euros [1]. In all these cases, sampling needs to be considered carefully, in particular, as it is well known that many polymer compounds are intrinsically heterogeneous.

Table 8.1 shows some selected inorganic components in polymeric matrices. The broad variety of elements contained in polymers may be classified into three product-oriented categories:

(i) elements being an intrinsic part of the monomers used in polymer manufacture (e.g. Cl in PVC);
(ii) elements occurring in the additive package; and
(iii) adventitious impurities.

Elements of type (i) are present in the percentage level, those of type (ii) are typically in the percentage to ppm range, whereas elements of type (iii) may be found on ppb level. Figure 8.1 shows the generally expected concentration levels of various classes of additives in polymers. Percentage range concentrations are often indicative of additives such as fillers, flame retardants, pigments, etc. Minor metallic constituents are usually either catalyst residues (Al, Ti, Fe, Cr, Si, V, etc.) or adventitious metallic impurities left in the polymer after manufacture (processing equipment traces, corrosion products). Another source of traces of metals in polymers are neutralising chemicals added in the final stages of manufacture.

As for legally permitted elemental concentrations in some polymeric products, both CONEG (Coalition of Northeastern Governors)/USA and the European Packaging and Packaging Waste Directive 94/62 EC regulations are concerned with the problems of packaging waste and the currently permitted heavy-metal content in packaging. Regulated impurities in food packaging material (permitted upper limits for extractable percentage of impurities in colorants) in the EU are Pb 0.0100, As 0.0100, Hg 0.0050, Cd 0.0100, Zn – , Se 0.0100, Ba 0.0100, Cr 0.1000, Sb 0.0500 and aromatic amines 0.0500 %. However, in both regulations it is no longer just the bioavailable metal, but the total metal content in coloured/printed packages which matters. As from 30 June 2001, the sum of four elements (Pb, Cd, Cr(VI) and Hg) must not exceed 100 mg kg^{-1}. It is obvious

Table 8.1 Selected inorganic components in polymer/additive matrices

Function	Chemical composition[a]	Elemental composition
Hydroperoxide decomposer	Dialkyldithiocarbamates, dialkyldithiophosphates	P, Ni, S, Zn
Free-radical scavenger	n-Butylamine nickel-2,2′-thio-bis-(4-t-octylphenolate)	P, Ni, S
Heat stabiliser	Organotin mercaptides/sulfides/carboxylates; antimony mercaptides; metal carboxylates; lead stearate/phosphite/phthalate/sulfate	S, Sb, Sn, Ba, Ca, Cd, Mg, Sr, Zn P, Pb, S
Co-stabiliser	Organic phosphites	P
Plasticiser	Trialkyl phosphates	P
Acid acceptor	Metallic soaps (stearates)	Ca, Zn
Flame resistance	Al_2O_3, antimony oxides, boron compounds, halogen compounds, phosphate esters, metal hydrates, magnesium compounds, tin compounds, molybdenum compounds, silicones	Al, B, Br, Cl, Mo, P, Sb, Si, Sn, Zn
Filler/reinforcement[b,c]	Carbonates, glass fibres, $Al(OH)_3$, kaolin, talc, silica, wollastonite, glass spheres, mica	Al, Ca, Fe, Mg, K, Na, S, Si, Zr
Active filler[c]:		
Abrasion	SiC	Si
Friction	MoS_2	Mo, S
Slip	Mullite, calcium silicate, $ZrSiO_4$	Al, Ca, Si, Zr
Electrical properties	Al, Zn, Cu, Ni, Al_2O_3, $BaTiO_3$	Al, Ba, Cu, Ni, Ti, Zn
Magnetic properties	Barium ferrite, iron oxide	Ba, Fe
Thermal properties	Beryllium oxide	Be
Reinforcement	Metallic fibres, MgO, ZrO	Fe, Mg, Zr
Weathering (UV) stability	ZnO	Zn
Adhesion promotion	Chromium complexes, silanes, titanates, zirconium aluminates	Al, Cr, Si, Ti, Zr
Radiation protection	$BaSO_4$, PbO, ZrO	Ba, Pb, S, Zr
Pigment	TiO_2, ZnS (white); Fe_2O_3, Cd (S, Se), Pb (Cr, Mo, S) O_4 (red); ultramarine, Co (Al, Cr)$_2O_4$ (blue); Cr_2O_3, (Co, Ni, Zn)$_2$ (Ti, Al) O_4 (green); Fe_3O_4 (brown); (Ti, Ba, Sb) O_2 (yellow)	Al, Ba, Cd, Co, Cr, Fe, Mo, Ni, Pb, S, Sb, Se, Ti, Zn
Biocide	10, 10′-oxy-bis-phenoxarsin, tributyltin oxide, metallic soaps	As, Cu, Hg, S, Sb, Sn, Zn
Nucleating agent	Mineral fillers, pigments, metals, metal oxides, phosphates, BN, NaF, Na-benzoate, Ca-stearate/montanate	Al, B, Ca, Cd, Co, Cr, F, K, Mg, Na, P, S, Sb, Si, Ti

[a] Selective examples.
[b] Inert fillers/extenders increase polymer bulk and reduce costs.
[c] Active fillers produce specific improvements in properties.

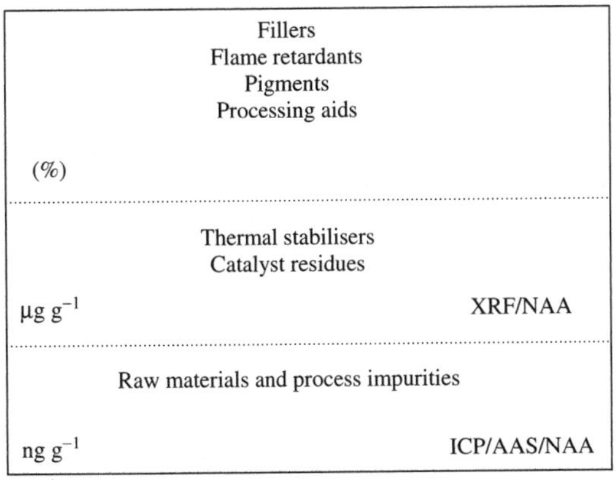

Figure 8.1 Typical scale of elemental concentration levels in relation to additive functionality

that considerable analytical effort is needed to show compliance with these regulations.

8.1 ELEMENT ANALYTICAL PROTOCOLS

Although elemental analysis (i.e. the determination of elements ranging from H to U) in a polymeric material is very common practice in the polymer industry, among polymer processors, research and application laboratories, and end-users, the analytical methods and protocols used are widely different and are not harmonised within the EC, let alone worldwide. Among the currently established 344 ISO methods for plastics, there are no 'ISO-approved' methods available for the elemental analysis of polymer material [2]. Only recently, the European Committee for Standardisation (CEN), Technical Committee CEN/TC 249, issued a draft on the determination of cadmium in plastics (Draft prEN 1122). Hence, the effective application of an element analysis method is largely depending on in-house analytical skills. This situation does not favour the mutual interpretation of analytical results between industrial, governmental and other parties. Until recently, mostly 'home-made' reference materials, i.e. polymer materials that are thoroughly analysed using different techniques or over longer times or that have been prepared on a weight basis, were used to validate these analytical methods. However, such in-house certification is an expensive, time-consuming and experience-intensive matter that is only done for products with major sales. To improve the accuracy and precision of the currently used analytical measurement protocols, certified reference materials (CRM) are ideal tools. However, CRMs of this type are rather few. Actually, the effective introduction of QA systems requires *internationally accepted certified reference materials*. The elements can be added to the polymer as an organic or inorganic compound, with the difference that an inorganic compound is insoluble, rather immobile but discrete in particle size and thus may give rise to a higher inhomogeneity. Organic, soluble compounds may be mixed more homogeneously with the matrix, but are more mobile and thus prone to migration, leading to a lower stability. An optimum balance needs to be found.

The Polymer Elemental Reference Material (PERM) project [1] has stimulated the exchange of knowledge and expertise on elemental analysis in polymer material among major industrial polymer laboratories, governmental, private and university laboratories. Apart from producing CRMs, development and intercomparison of analytical methods and protocols for the elemental analysis of polymer materials is much needed. In one of the few round-robins, with 23 participating laboratories, nine different element analytical techniques were used: XRF, ICP-AES. FAAS, GF-SS-ZAAS, GFAAS. ICP-MS, IPAA, INAA and PAA; 85–88 % of the values were within ±10 % deviation from the certified value [3].

8.1.1 Element Analytical Pretreatment Protocols

Elemental analysis can be categorised in various ways:

  (i) bulk vs. surface;
 (ii) macro vs. micro (spatially resolved);
(iii) single-element vs. multi-element;

 (iv) destructive vs. non-destructive;
  (v) chemical vs. instrumental;
 (vi) major element vs. (ultra)trace element; and
(vii) speciation vs. no chemical bonding information.

The choice of method for a particular application should take into account this categorisation, as well as the precision and accuracy required. This chapter mainly deals with bulk analysis and speciation.

Destructive methods require removal of the polymer matrix before the elements can be transferred to a liquid phase. *Matrix removal methods* are relatively indifferent with respect to the form of the material. Polymer raw material is produced as granulate, pellets, powder, platelets, flakes, foils, bars, rods or fibres. Polymer end material can take up any form. Matrix removal can be done in several ways by: burning off the organic matrix in a crucible; dissolving the polymer in a suitable organic solvent (Section 3.7); or by dissolving the polymer in a mixture of appropriate acids, eventually with the addition of heat (e.g. on a hot plate, or using microwave destruction) and pressure. The main problems associated with matrix removal methods are well known. There always exists the uncertainty that the elements to be determined are lost because of their intrinsic volatility (e.g. F, Cl, Br, Hg) and the possibility of contamination, stemming from the chemicals used, the instruments or from the laboratory environment (e.g. Na, Ca, Fe, Cr). Thus, reliable and accurate results are not easily obtained.

The most frequently used methods for elemental analysis in plastics (certainly in the past) deal with *digestions* of some kind. Also, some derivatisation methods (e.g. hydride generation for element analysis, or the equivalent TMAH treatment for molecular analysis) may be used to generate volatile species which are more easily separated from each other by chromatography. Derivatisation reactions are often far from being well controlled.

For destructive measuring methods, a CRM would serve as a reference to check the recovery of a particular matrix removal procedure. This is especially important for 'open' destructions at atmospheric pressure. Alternatively, isotope dilution methods may be used; once isotopic equilibrium is established, loss of analyte does not affect the analysis result. Isotope dilution techniques are only available in a few specialised laboratories. Another type of problem is encountered in pressurised methods oxidising the matrix in a closed vessel or bomb. Due to the large amounts of gas (CO_2, NO_x, SO_2) evolving from samples with a high organic matrix content, an excessive pressure build-up occurs that prohibits the use of large sample aliquots. Typical amounts are between 250–500 mg of sample, which is very low in case of mixed polymer waste. A CRM may serve here as a reference for measurement variabilities that are to be expected in relation to a particular sample material and its associated *inhomogeneities*.

After matrix removal, samples can be measured using various techniques, such as AAS, AES, ICP, etc. Traditional chemical analysis methods, involving separation and gravimetric, titrimetric or polarographic determination of the elements, are being replaced by a wide selection of instrumental methods.

Nondestructive radiation techniques can be used, whereby the sample is probed as it is being produced or delivered. However, the sample material is not always the appropriate shape or size, and therefore has to be cut, melted, pressed or milled. These handling procedures introduce similar problems to those mentioned before, including that of sample homogeneity. This problem arises from the fact that, in practice, only small portions of the material can be irradiated. Typical nondestructive analytical techniques are XRF, NAA and PIXE; micro-destructive methods are arc and spark source techniques, glow discharge and various laser ablation/desorption-based methods. On the other hand, *direct solid sampling* techniques are also not without problems. Most suffer from matrix effects. There are several methods in use to correct for or overcome matrix effects:

   (i) dilution;
  (ii) matrix matching;
 (iii) use of internal standards;
  (iv) standard addition;
   (v) chemical separation; and
  (vi) isotope dilution.

Analyte dilution sacrifices sensitivity. Matrix matching can only be applied for simple matrices, but is clearly not applicable for complex matrices of varying composition. Accurate correction for matrix effect is possible only if the IS is chosen with a mass number as close as possible to that of the analyte element(s). Standard addition of a known amount of the element(s) of interest is a safe method for samples of unknown composition and thus unknown matrix effect. Chemical separations avoid spectral interference and allow preconcentration of the analyte elements. Sampling and sample preparation have recently been reviewed [4].

8.1.2 Elemental Analysis Methods

Principles and Characteristics Inorganic components in organic polymers serve a wide variety of purposes, as shown in Table 8.1. It is readily apparent from this table that multi-element analytical techniques are the best approach for the inorganic deformulation of polymer/additive matrices. For this purpose, a broad selection of techniques are available nowadays, both in solution and in the solid state, as indicated in Table 8.2, including emission spectrometric methods, usually after some form of sample treatment or separation step (e.g. ICP-AES, GC-AED, PyGC-AED), but also directly in the solid state (GD-OES and SSS), absorption spectrometry (e.g. AAS), fluorescence spectroscopy (XRF) and mass spectrometry (e.g. ICP-MS, LA-ICP-MS). Techniques are available for bulk analysis (e.g. AAS, ISE, XRF, ICP-AES, ICP-MS, etc.), but also for surface analysis (e.g. XPS, LA-ICP-AES, LA-ICP-MS) and elemental distribution analysis (e.g. EPMA, iXPS, iSIMS). With the exception of the surface analytical techniques, most elemental analysis methods are assumed to yield quantitative results after some form of calibration. NAA and PIXE are the exceptional absolute trace methods, which can be called in for validation of other analytical procedures.

Table 8.2 categorises the various elemental analysis methods as to their application (atomic range, aggregation state, sampling area). The table also indicates whether the method is supposed to be applied to an additive separated from the polymer or not, either in solution (after extraction), or in the solid state. Various element-sensitive detectors are worth considering in combination with chromatography (e.g. GC-AED and PyGC-AED). The main analytical techniques used for elemental bulk analysis in industrial research are XRF, FAAS, GFAAS, ICP-AES, ICP-MS and NAA. On the whole, good agreement on results is frequently wanting.

A variety of techniques have been used for the determination of *nonmetallic elements*, ranging from nondestructive techniques such as XRF, to classical techniques based on digestion of the sample followed by an appropriate analytical finish. Typical traditional quantitative determinations of light elements in polymers are Dumas, Kjeldahl (N); Wurzschmitt, Carius (S); Wurzschmitt, Volhard, Schöniger (Cl). Other methods may be found in ref. [5]. The analysis of elemental carbon, hydrogen, and nitrogen (CHN) is of limited use for additive analysis in polymers, as it constitutes an insufficient basis for the identification of polymer additives. Yet, a few situations can be envisaged in which such an analysis is indicated. While the classical chemical approaches for determining *nitrogen* quantitatively are highly reliable, sample processing by chemical dissolution or combustion in pure oxygen is required prior to analysis. Consequently, chemically intractable materials are thus

Table 8.2 Inorganic analysis of additives in polymeric material

Technique	Aggregation state	Applicability		Legend
		Atomic range[a]	Sampling area	
CHN analyser	Solution/solid	C, H, N	Bulk	3
FAAS	Solution	Wide (68)	Bulk	1
GFAAS	Solid	Wide (50)	Bulk	3
ICP-AES	Solution	Wide (73)	Bulk	1
SSS-AES	Solid	Wide	Bulk	3
GD-OES	Solid	Wide (88)	Bulk	3
XRF	Solution/solid	Wide ($Z \geq 4$)	Bulk	3
GD-MS	Solid	All	Bulk	3
SSMS	Solid	Wide (Li-U)	Bulk	3
ICP-MS	Solution	Wide (76)	Bulk	1
MIP-MS	Solution	Wide	Bulk	1
NAA	Solution/solid	Wide (70)	Bulk	3
PIXE	Solid	Wide ($Z \geq 6$)	Bulk	3
ISE	Solution	Limited (10)	Bulk	2
AES	Solid	Wide ($Z \geq 3$)	Bulk	3
XPS	Solid	Wide ($Z \geq 3$)	Surface	3
SIMS	Solid	Wide (All)	Surface	3
SNMS	Solid	Wide (All)	Surface	3
RBS	Solid	Wide ($Z \geq 10$)	Surface	3
LA-ICP-AES	Solid	Wide (73)	Surface, micro-analysis	3
LA-ICP-MS	Solid	Wide (76)	Surface, micro-analysis	3
EPMA	Solid	Wide ($Z \geq 4$)	Surface, imaging, micro-analysis	3
CE	Solution	Restricted	Bulk	1
GC-AED	Solution	Limited	Bulk	1
PyGC-AED	Solid	Limited	Bulk	3
LC-ICP-MS	Solution	Wide (76)	Bulk	3

1 Analysis after separation of polymer and additive (in solution).
2 Analysis without separation of polymer and additive (in solution).
3 Intact (in-polymer) analysis.
[a] In parentheses: number of elements or elemental range.

analysed with difficulty. With the exception of 14 MeV NAA, nondestructive (instrumental) methods are insufficiently accurate ($\pm 2\%$). In conjunction with some background knowledge of the sample at hand, information about the total nitrogen number may be useful, especially for the class of nitrogen containing hindered amine light stabilisers, although obviously not in nylons. Nitrogen analysis may thus also be useful in troubleshooting for verification of the extraction efficiency. While nitrogen analysis does not suffer from some of the matrix effects of other in-polymer analysis techniques (e.g. XRF), it does require a moderate amount of time per analysis. CHN is rarely used in a production environment, and is predominantly a tool used in support and research laboratories.

Obtaining accurate results for *sulfur determination* at ppm level is illustrative of a problematic element. Trace sulfur analysis is an analytical challenge. Sulfur cannot be determined by conventional AAS, since its resonance lines are located in the vacuum ultraviolet. Determination by ICP-MS is plagued by interferences from oxygen ($^{16}O^{16}O^+$, $^{16}O^{16}O^1H^+$, $^{16}O^{18}O^+$) and argon (^{36}Ar), and requires high resolution. The limit of detection of NAA for this element is only about $30\,\mu g\,g^{-1}$. Moreover, sample dissolution is not straightforward for sulfur as the target element, as many sulfur compounds are volatile and a common reagent to dissolve plastics, sulfuric acid, cannot be used. As sulfur is a very common element, the risks of contamination are high. Direct (total) sulfur analysers are based on combustion of the samples in an oxygen-rich atmosphere at high temperature, passing the combustion gas through a filter and moisture trap, followed by subsequent detection of SO_2 by different techniques, including UV-fluorescence or IR detection. This allows sulfur determinations down to $0.35\,\mu g\,g^{-1}$ (RSD 20%) [6]. However, total sulfur, as determined by such combustion techniques, provides no useful information as to the type or source of the sulfur. Electrothermal vaporisation as a means of sample introduction for ICP-MS produces a dry plasma and, therefore, all interferences from polyatomic ions containing oxygen are significantly reduced. ETV-ICP-MS is a (costly) alternative for sulfur determination (as SO_2), with considerable advantages over dedicated sulfur-specific analysers based on combustion with oxygen at high temperatures. The traditional GC sulfur detector is flame photometry (FPD). More recently, chemiluminescence detection of sulfur (SCD) as a finish to high-resolution capillary chromatography has evolved as the detector of choice for extremely low concentrations. A general lack of reference materials with a sulfur concentration below the $5 \, \mu g \, g^{-1}$ level is noticed.

Trends in element analysis are multi-element (survey) analysis, lower concentration levels, micro/local element analysis and speciation (coupling with chromatography). An overview of the determination of elements in polymeric materials is available [7]. Reviews on sample preparation for trace analysis are given in refs [8–10]. Quality assurance of analytical data in routine elemental analysis has been discussed [11]. Organic analysis is obviously much more requested in relation to polymer/additive matrices than elemental analysis.

Applications Mika *et al.* [12] have described two simple indirect methods for determination of HALS in polyolefins, based on quantitative microanalysis of nitrogen in the sample by hydrogenolytic transformation of nitrogen into ammonia. In the first method, a Raney-nickel catalyst was in contact with the sample, and ammonia formed in the reduction was determined photometrically. In the second procedure, the nickel-based catalyst was used as packing for the reduction tube, and the resulting ammonia was titrated. None of the common stabilisers, including phosphites and thioesters, interferes with the analysis. However, no other nitrogen-containing additive should be present, as, otherwise, erroneous results are generated.

8.2 SAMPLE DESTRUCTION FOR CLASSICAL ELEMENTAL ANALYSIS

Destructive solid sample preparation methods, such as digestion and mineralisation, are well known as they have been around for some time; they are relatively cheap and well documented [13–15]. Decomposition of a substance or a mixture of substances does not refer so much to the dissolution, but rather to the conversion of slightly soluble substances into acid- or water-soluble (ionogenic) compounds (chemical dissolution).

This involves a chemical change. It is required that any decomposition procedure should alter the original environment of the sample into a digest, i.e. a solution in which the analyte is distributed homogeneously. More specific conditions set to a decomposition technique are [4]:

(i) guarantee of the integrity of the procedure with respect to the analyte;
(ii) removal of residual matrix components which interfere in the detection;
(iii) possibility of adjustment of the oxidation state of the analyte; and
(iv) safety.

Decomposition and dissolution processes are generally subject to fairly large sources of error, namely:

(i) incomplete dissolution of the analytes;
(ii) losses by volatilisation;
(iii) solvent contamination; and
(iv) contamination from the vessel walls.

The analyst's requirements to digestion procedures are given in Table 8.3.

Figure 8.2 shows the main steps in sample preparation for elemental analysis. Sample preparation involves numerous steps from sample collection to sample presentation, as a homogeneous solution for instrumental analysis. Sample preparation can involve combinations of the following: drying of the sample, leaching, extraction, digestion of the matrix, postdigestion chemistry, analytical separation, solvent removal, and exchange. Sample preparation is a time-consuming, error-prone and difficult step in an analysis. More recently, various preparation techniques have revolutionised sample preparation. Nevertheless, sample preparation methods, such as extractions or digestions, are still a limiting factor in analytical science. Without fast and reliable digestion techniques, modern analytical methods such as AAS, ICP-AES, DCP-AES and ICP-MS would not give correct results in the determination of elements in complex samples, because solubilisation of analytes and

Table 8.3 Requirements to matrix digestion procedures

- High speed
- High quality (total digestion, no contamination, no removal of unstable compounds)
- Safety
- Validated methods (CRMs, proof of correct digestion, documentation)
- Total solution for different materials
- Digestion overnight

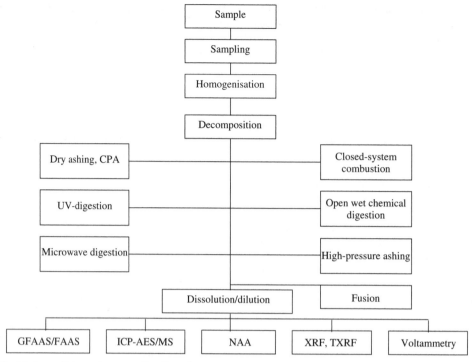

Figure 8.2 Modern solid sample preparation flow-chart

removal of interfering molecules are essential for correct analyses. Moreover, many ultratrace detection techniques, such as ICP-MS, rely on homogeneous samples for analysis. To transform solid samples into homogeneous solutions, dissolution methods using concentrated mineral acids at elevated temperatures and pressures are the most common sample preparation methods in use.

Decomposition methods are usually classified as melt decompositions, wet decompositions (with liquid decomposing agents) and dry decompositions by combustion. Sample decomposition methods are varied, and involve open and closed systems (at low and high pressure), UV and thermal activation, low or high temperature, and use of conventional convective or microwave heating. Table 8.4 lists the main sample decomposition methods for trace-element determination.

Modern decomposition techniques (ashing, wet decomposition) are nowadays usually performed in closed vessels made of inert and very pure materials (such as borosilicate glass, PTFE, PFA, quartz, glassy carbon) at high temperature and pressure. A recent development in wet decomposition at elevated T, p is based upon microwave heating, again mainly in closed systems. Mineral acid decompositions typically use HNO_3 as the main oxidising acid, often in combination

Table 8.4 Decomposition methods for elemental analysis

Combustion:

1. In open systems
   - Dry ashing
   - Low-temperature ashing (LTA, CPA)

2. In closed systems
   - Flask combustion
   - Combustion bombs

3. In dynamic systems

Wet digestion:

1. In open systems (thermal and UV digestion)
2. In closed systems (pressure digestion)
   - with conventional heating (HPA)
   - with microwave heating (MOD)

Fusion

with non-oxidising acids such as HF and HCl. The widespread use of the halogen-based acids in digestion procedures increases the complexity of sample processing, especially for detection of analytes at ultratrace levels. Trace contaminants in digestion techniques are often due to the reagents used, the ambient air and the composition of the vessel walls that come into contact with

Table 8.5 Instrumental devices for decomposition methods

Methodology	Device(s)
Melt decomposition	Parr bomb
Wet chemical thermal decomposition	Tölg pressure bomb, wet incineration automat, Kjeldahl digestion, high-pressure incinerator (Knapp)
Microwave decomposition	Microwave oven
Combustion	Calorimetric bomb (Berthelot), oxygen flask (Schöniger), Wickbold apparatus, cold plasma incinerator, micro-Dumas combustion (CHN analyser), sulfur-specific analysers (S, C, N, O), pyrochemiluminescence
UV decomposition	UV decomposition device

the digestion solution. Table 8.5 shows the main commercially available devices for decomposition methods.

Although the need for complete decomposition is often stressed (see also Table 8.3), not all detection techniques demand the same degree of mineralisation. Table 8.6 classifies analytical techniques according to the amount of mineralisation that they need [4]. Ideally, a purely instrumental approach is the only way to prevent losses and contamination due to decomposition. Choosing a decomposition mode simply to be able to meet the requirements of the detection technique is an incomplete approach. The choice of decomposition should primarily be directed by both the matrix and element of interest.

A monograph on mineral acid decomposition in inorganic analysis is available [16], as well as an acid decomposition site (*http://www.sampleprep.duq.edu/sampleprep*).

8.2.1 Combustion Analysis

The basis of all combustion techniques is the removal of carbon as CO_2 and hydrogen as H_2O. When determining

Table 8.6 Mineralisation requirements of detection techniques

Degree of mineralisation	Detection techniques
No	NAA, PIXE, SS-ZGFAAS
Partial	GFAAS, ICP-MS, ICP-AES, FAAS
Complete	ISE, DPASV, DPCSV, HG-AAS, IC, RNAA, IDMS

the carbon and hydrogen contents of a sample, the CO_2 and H_2O produced are measured, but, when determining other elements, these are allowed to be lost to the atmosphere, whilst the elements to be determined are converted into a convenient form for measurement.

8.2.1.1 Combustion in Open Systems

Principles and Characteristics Dry ashing is defined as any sample pretreatment process involving heating of a sample (usually at 400–800 °C) in the open atmosphere or in a controlled flow of gas, but not in a closed container, with or without the addition of other materials, in order to completely remove the organic constituents ('burning off'), and to form a solid inorganic ash. This ash can subsequently be dissolved in acid, thus enabling the determination of elemental species by AAS, ICP, colorimetry, voltammetry, ion chromatography or any other suitable elemental analytical technique.

Dry ashing procedures for elemental and ash content analysis fall into the following categories:

- in the atmosphere, with or without the use of an ashing aid;
- in an oxygen-enriched atmosphere;
- in a microwave furnace;
- low-temperature ashing with an oxygen plasma;
- combined dry ashing–wet ashing procedures.

Dry ashing uses the combustion of the sample with oxygen at high temperatures or excited oxygen/oxygen radicals at low temperatures. The source of oxygen here is from the air of from flasks, and not from mineral acids. Optimum ashing conditions for various polymeric matrices (PA, PS, PVC and PTFE) have been reported [17], as well as recommended ashing conditions per element [18]. The main characteristics of dry ashing for elemental analysis are given in Table 8.7.

Ashing is routinely used in the sample preparation of transition metals, but losses of volatile analytes such as halogens, S, P, Si, As, Se, Cd or Zn can occur.

In the *Wickbold method*, solid samples are vaporised in an oxygen stream and fed into an oxyhydrogen flame, which burns in a cooled quartz tube. The combustion products are condensed here, or are captured in an absorption solution as gaseous materials. Although combustion in a Wickbold apparatus is a quick and effective method for destroying organic material of all types, incomplete destruction may occur [19]. In special digestion vessels, known as *cold-plasma ashers* (CPA),

Table 8.7   Main features of dry ashing

Advantages

- Simplicity and cheapness
- Low blanks (reduced contamination)
- Absence of hazardous chemicals (cf. wet ashing)
- Sample size (up to 1000 g)

Disadvantages

- Low recoveries (no protection against volatilisation losses)
- Low reliability
- Long ashing times
- Two-stage process (ash product needs dissolution)
- Difficult dissolution of ashed materials
- Cross-contamination between successive samples (vessel wall effects)

excited oxygen (a plasma) is generated under reduced pressure, using a high-frequency field, whereby mainly oxygen atoms with a short lifetime appear. Oxidation is achieved at relatively low temperature, and easily volatile matter (such as compounds of As, Sb, Se and Te) can be quantitatively detected in the process. The procedure is slow. Anderson [18] has described the general principles of dry ashing, and has compared it with alternative sample dissolution techniques.

Dry ashing competes with other techniques for sample dissolution of organic (and biological) species for elemental analysis, such as wet ashing in open or closed vessels, oxygen flask techniques, combustion tube techniques, and microwave digestion techniques in open or sealed vessels. Dry ashing for elemental analysis has now largely been displaced by wet ashing because of its disadvantages, and should not be used unless there is good reason for so doing.

Dry ashing procedures are also used to quantify the involatile inorganic component of a sample (the 'ash content'), see ASTM D 5630-94. When the ashing is complete, the ash is simply weighed. Carbon-black (CB) quantification is generally carried out using nitrogen-blanketed tube furnaces in which the polymer matrix is burned away, leaving behind CB and some ash. Nitrogen

is used as an inert blanket, so that the polymer is ashed without oxidising the carbon-black. The carbon-black content determination of polymers is described in ASTM D 1506-94b. Alternatives for gravimetric CB determination by muffle furnace ashing are microwave furnace ashing or TG analysis.

Applications   Quantitative dry ashing (typically at 800 °C to 1200 °C for at least 8 h), followed by acid dissolution and subsequent measurement of metals in an aqueous solution, is often a difficult task, as such treatment frequently results in loss of analyte (e.g. in the cases of Cd, Zn and P because of their volatility). Nagourney and Madan [20] have compared the ashing/acid dissolution and direct organic solubilisation procedures for stabiliser analysis for the determination of phosphorous in tri-(2,4-di-*t*-butylphenyl)phosphite. Dry ashing is of limited value for polymer analysis. Crompton [21] has reported the analysis of Li, Na, V and Cu in polyolefins. Similarly, for the determination of Al and V catalyst residues in polyalkenes and polyalkene copolymers, the sample was ignited and the ash dissolved in acids; V^{5+} was determined photo-absorptiometrically and Al^{3+} by complexometric titration [22].

The simplest identification procedure for inorganic fillers is incineration, with subsequent detection of the inorganic elements. During ashing, several pigments, fillers and inorganic additives may undergo compositional changes (Table 8.8).

Inorganic fillers can be quantified gravimetrically by furnace ashing or TG analysis. The standard procedure for ashing polymers is described in ASTM D 5630-94, which makes provision for the use of microwave furnaces and quartz crucibles. As pointed out elsewhere [23], the determination of filler content by furnace ashing or TG can be misleading in certain circumstances (e.g. as a result of a reaction or decomposition). In the case of the removal of the polymer matrix by ashing to isolate the filler component, care must be taken to ensure that the moderate ashing temperature does not induce

Table 8.8   Transformations of inorganic additives after ashing (4 h at 650–950 °C)

Additive	Ashing temperature (°C)	End-product	Additive	Ashing temperature (°C)	End-product
MnO_2	950	Mn_3O_4	PbO_2	650, 950	PbO
CdS	650	$CdSO_4$ (+CdO)	$PbSO_4$	950	PbO
CdS	950	CdO	PbS	650	$PbSO_4$ (+PbO)
CdSe	650, 950	CdO	PbS	950	PbO

After Affolter [7]. Reproduced by permission of Huthig GmbH.

modification of the fillers. Fillers such as glass fibre, talc and chalk are unaffected at moderate ashing temperatures up to 500 °C. Once the filler is isolated, it can then be analysed by FTIR spectroscopy in a KBr pellet. The fibre content of GFR-composites is generally determined by gravimetry, after furnace ashing and acid washing. For the determination of glass-fibre length the polymeric matrix can be ashed at low temperature by means of activated oxygen in a low-temperature asher (LTA). The ashing process of some polymers, such as ABS and LCP, is quite lengthy (up to weeks). The fibre-length distribution can be measured by sieve analysis of the liberated glass fibres. An alternative non-invasive technique for measurement of monofilament lengths is CLSM.

Heinrichs *et al.* [24] have determined 37 elements in plastic waste after ashing in an oxygen plasma, followed by acid digestion in an autoclave. Cold-plasma ashing is used for destroying filters made of cellulose or polymers, as applied to collect dust samples. Even PTFE filters can be ashed by this method. Low-temperature ashing, in which the oxygen in the atmosphere over the sample to be ashed is activated by exposure to an electric discharge and/or UV light, has been used for the determination of trace metals in PP, PS, PVC and PET [25].

With few exceptions, dry ashing is required for the analysis of rubber, followed by dissolution of the residue in an acid mixture that is suitable for the analyte and the matrix.

8.2.1.2 Combustion in Closed Systems

Principles and Characteristics Combustion analysis is used primarily to determine C, H, N, O, S, P, and halogens in a variety of organic and inorganic materials (gas, liquid or solid) at trace to per cent level, e.g. for the determination of organic-bound halogens in epoxy moulding resins, halogenated hydrocarbons, brominated resins, phosphorous in flame-retardant materials, etc. Sample quantities are dependent upon the concentration level of the analyte. A precise assay can usually be obtained with a few mg of material. Combustions are performed under controlled conditions, usually in the presence of catalysts. Oxidative combustions are most common. The element of interest is converted into a reaction product, which is then determined by techniques such as GC, IC, ion-selective electrode, titrimetry, or colorimetric measurement. Various combustion techniques are commonly used.

Except for dry incineration in a platinum crucible, oxygen-induced decomposition requires special apparatus in order to avoid the loss of volatile substances. Combustion of organic materials may be achieved with the aid of oxygen under increased pressure in a calorimetric or oxygen bomb (Berthelot). Combustion techniques are limited by the inability to always obtain complete combustion, especially when dealing with polymeric and highly fluorinated materials and the high blanks for chlorine and to a lesser extent sulfur when using either Schöniger or Parr® Oxygen Bomb combustions. A limitation of organic elemental analysis is the inability to obtain homogeneous sampling of many industrial materials when only 1–2 mg is weighed out. Various Parr® bombs for preparing samples for chemical analysis are available:

(i) oxygen combustion bombs for converting oxygen-combustible samples into water-soluble forms rapidly, with complete recovery of all trace constituents;

(ii) acid digestion bombs for dissolving samples in strong acids or alkalis in chemically resistant vessels at elevated temperatures and pressure (Section 8.2.2.1); and

(iii) microwave digestion bombs for treating samples with strong acids or alkalis in metal-free vessels using microwave energy for rapid digestion (Section 8.2.2.2).

Francis [26] has recently described some modern methods for element determinations, including the oxygen flask technique and the determinations of C, H, O, N (Kjeldahl), halogens, S, P and F (ion-selective electrode).

(a) Organic Elemental (C, H, N, O) Analysis

Simultaneous analysis of C, H, and N in liquid or solid organic materials may be carried out after combustion in an oxygen atmosphere at temperatures up to 1000 °C. The gaseous combustion products (CO_2, H_2O, N_2) are flushed with a carrier gas (helium) through a reductant, and quantification is obtained by GC. For analysis of O, the combustion is performed in helium over platinised carbon. Carbon monoxide formed is converted to CO_2 by passage over CuO, and measured in the same manner as for analysis of C. A 2–3 mg sample is required.

In *Micro Dumas combustion* (CHN analysis) the sample is vaporised and carried by a stream of CO_2 over nickel oxide at 1000 °C to oxidise the sample to CO_2, H_2O and N_2. Nickel reduces nitrogen oxides in the heated combustion tube. Carbon monoxide, formed by reduction of CO_2 by nickel, is oxidised by passage through hopcalite at 110 °C. Traces of unoxidised methane are completely oxidised by passage through specially prepared copper oxide at 700 °C. Any carbonaceous sample residue, which might retain nitrogen, is completely oxidised by passing oxygen over the heated residue. The mixture of CO_2 and H_2 is collected over 30 % KOH, which absorbs CO_2, and the residual gas (N_2) is measured by displacing and weighing an equal volume of mercury. The Micro Dumas combustion procedure is capable of determining organic nitrogen in amounts down to 1 %, in a wide range of polymers. This method gives results which do not differ from the true nitrogen content of the sample by more than 0.2 %. Replicate analyses agree within 0.1 %.

(b) Schöniger Oxygen Flask Combustion

In this method, which has proved to be important, a weighed organic sample placed in a platinum basket is ignited in pure oxygen in a thick-walled 1-litre conical flask equipped with a solid glass stopper. The gases formed are absorbed by a solution in the flask, resulting in inorganic ions (F^-, Cl^-, Br^-, I^-, SO_4^{2-}, etc.), which are then quantitatively determined by an appropriate spectrophotometric or titrimetric method. Sample weight is limited to less than 200 mg (generally less than 20 mg). Oxygen flask combustion, according to Schöniger [27], offers advantages when readily volatilised elements such as halogens, Se, S, P, B, F, Hg, As and Sb are to be determined in an organic matrix. Woodget [28] has recently summarised the recommended methods for elemental analysis by oxygen flask combustion.

(c) Parr® Oxygen Bomb

Pressure or melt decompositions with an oxidising agent, such as sodium peroxide, may be carried out in a closed, pressure-resistant container such as a Parr® bomb. Oxygen combustion bombs are used for determining sulfur and other elements in solid or liquid combustible materials, using a procedure which converts samples rapidly to soluble forms with complete recovery of all constituents. This technique is similar to the Schöniger oxygen flask combustion, except that up to 1 g of sample (pellet or film) can be ignited at 40 atm of pure oxygen in a stainless-steel vessel. The technique has better sensitivity (detection limits of 20 ppm Cl and S), and is more suitable for volatile samples. Vessel wall adsorption may occur.

(d) Pyro-hydrolysis

Combustion pyro-hydrolysis is based on the chemical sequence:

$$RF + H_2O + O_2 \xrightarrow{900-1100\,°C} HF + \text{oxides}$$

$$HF + \text{oxides} \xrightarrow{\text{buffer}} NaF + H_2O$$

In the Antek® Fluoride Analyzer, a pyrolysis furnace is combined with an ion-specific electrode cell (ISE). Table 8.9 compares this specific analyser to a conventional combustion bomb.

Applications In combustion techniques for total organic carbon and nitrogen, a few mL of sample are added to an ampoule containing 25 % H_3PO_4 and $KHSO_4$. The ampoule is flame sealed and placed in an autoclave for one hour. The persulfate oxidises any organic carbon to CO_2. IR measures the CO_2 after opening the ampoule inside the instrument. The method has a detection limit of 1 ppm carbon. Inorganic bicarbonate/carbonate can be measured by omitting the persulfate. Crompton [30] has described the qualitative and quantitative analysis of various elements (F, Cl, Br, I, N, S, P) in polymers, using oxygen flask combustion, followed by spectrophotometric or titrimetric procedures. The same author [31] has described several applications of oxygen flask combustion followed by titration for the determinations of 2 to 80 % of total Cl, Br and I in polymers, of Cl (from 5 to 500 ppm) in polymers containing Cl, S and/or P, and/or F and of traces of S (>50 ppm) in polyalkenes. Determination of flame retardants often relies on the quantitative determination of Cl and Br by

Table 8.9 Advantages and disadvantages of two combustion techniques

Specific ion analyser	Combustion bomb
• Automated method (electronic calibration and data handling)	• Wet chemical method (manual calibration, data collection, calculations)
• Rapid analyses (15 min per sample)	• Slow analyses
• Equipment and start-up costs relatively high	• Equipment and start-up costs relatively low
• High maintenance	• Low maintenance
• Specific	• Versatile (can be used for other ions by changing electrodes)

After Amos *et al.* [29]. Reproduced by permission of Carl Hanser Verlag GmbH & Co.

oxygen flask combustion methods [32]. Schöniger oxygen flask combustion–turbidimetry was used for determining total chlorine in amounts down to 5 ppm in chlorobutyl rubber and other chlorine-containing polymers [33]. The technique has found application for the determination of chlorine in PVC [34] and phosphorous in polyolefins [35]. Fluoride was determined by photofluorimetric titration [32]. The oxygen flask method (ISO 6528-1 (1992) and BS 7164-23.1 (1993)) also determines the total sulfur content in rubbers.

Micro amounts of sulfur in polymer are usually determined by oxygen flask combustion, sodium peroxide fusion in a metal bomb followed by titration [30], pyroluminescence [36] or ICP-AES. An oxygen flask combustion photometric titration procedure capable of determining total sulfur in polymers in amounts down to 50 ppm was reported. The repeatability of the sulfur determination in polyolefins in the oxygen flask is ±40 % at 50 ppm level, improving to ±2 % at the 1 % level [21]. Crompton [31] has also combined Schöniger flask combustion with a colorimetric procedure for the determination of phosphorous in polymers in various concentration ranges (0.01 to 2 %, 2 to 13 %).

In a round-robin of two chlorine-containing elastomers (with 1 and 10 % Cl) and two flame-retarded thermoplasts (with 10 % Br) halogens were determined according to oxidative digestion (Schöniger, Parr, Wickbold) followed by argentometric titration or ion chromatographic analysis of the halogenide (13 participating laboratories) [37]. The intra- and interlaboratory confidence levels amounted to 10 % for halogen concentrations of 10 %.

Nitrogen in polymers may be determined by micro-Dumas combustion [31] or pyrochemiluminescence [38]. The Micro Dumas method is available in a commercially produced CHN analyser. N-analysis is typically industrially applied to Atmer 163 in PP, slip agents in film or Chimassorb 944 in PE. Quality control of additive dispersion in LDPE greenhouse film formulations composed of N-containing Chimassorb 944 (HALS), S-containing Cyasorb UV1084 (Ni quencher) and N, S-free Chimassorb 81 and Irganox 1010 (antioxidants) was carried out by means of nitrogen and sulfur analysis using a CHSN elemental analyser based on combustion [39].

A PET oligomer isolation method has utilised chloroform extraction in a Parr bomb lined with a Teflon-TFE fluoro-carbon resin [40]. The analytics of *fluoropolymer processing aids* (combustion analysis, XRF, FTIR, ^{19}F NMR, OM) have recently been described [29]. Combustion analysis (Parr® Oxygen Bomb Calorimeter) can be used for quantitative analysis

of such PPAs in polyolefinic matrices. The matrix is completely combusted, while the processing aid is converted into HF; the F⁻ concentration is then measured with a specific ion electrode (Antek® Fluoride Analyzer). The method is very sensitive and capable of accurately measuring processing-aid concentrations down to 50 ppm in PE in about one hour [41]. Parr® bomb analysis was also used to verify fluoropolymer processing additives in masterbatches [42].

8.2.2 Wet Matrix Digestion

Mineral acid dissolution is an important sample preparation process for instrumental analysis, as it liberates element ions into a solution that can be directly introduced into an analytical instrument. For quantitative analysis, most instruments require a solution.

Depending upon the nature, origin, quantity of sample, elements of interest and concentration range, different techniques of wet digestion have to be adopted. Sample preparation methods range from open vessels (e.g. a platinum container), to pressurised closed glass (Carius) tubes, steel-jacketed Teflon bombs, open and closed vessel microwave-assisted acid digestion systems, at low pressure (<10 atm) and at moderate pressure (10−80 atm). Dissolution assemblies like a Kjeldahl flask, Parr® bomb, crucibles, quartz and platinum vessels, PTFE containers and specially designed apparatus are utilised for wet matrix digestion [43]. Whenever destruction of organic matter with oxidising acid leads to violent reaction, dry ashing may offer a more suitable method.

8.2.2.1 Conventional Decompositions

Principles and Characteristics Wet ashing is the hydrolysis and oxidation of the sample using one or more mineral acids. It comprises the decomposition of the matrix by oxygen, released at elevated temperatures by the acid(s). The primary chemicals reported to have been used in wet ashing (eventually as mixtures) are HNO_3, H_2SO_4, $HClO_4$, H_3PO_3, H_3PO_4, HF and H_2O_2. Besides these, many so-called ashing aids have been used, such as $Mg(NO_3)_2$, KNO_3, $NaNO_3$, HBr, HCl, HCOOH and H_3PO_4. Ideally, a reagent should dissolve the sample completely. The main properties and fields of applications of acids used for digestion have been reported [44]. Perchloric acid is the strongest oxidising agent for organic matter, but carries a high danger of explosion. Nitric acid is much safer, but is not a sufficiently strong oxidising agent to completely mineralise organic materials in open systems. Wet incineration automates have been described which are suitable for decompositions of organic matrices with HF, H_2SO_4, HNO_3 and $HClO_4$, as well as for the determination of nitrogen by the Kjeldahl method with H_2SO_4 decomposition [19]. *Kjeldahl nitrogen analysis* is a classical measurement in analytical chemistry, which has been used extensively over the past 120 years and has recently been excellently reviewed [45].

Successful wet chemical digestions depends on the following influences:

- appropriate consideration of chemistry using small volumes of high-purity acids or acid mixtures;
- resistant reaction vessels made of temperature-stable and (where appropriate) pressure-resistant pure materials; and
- high temperature for rapid, more complete reactions. Chemical rates roughly double for each $10\,°C$ of temperature increase.

In some elemental determination methods, such as polarography/voltammetry, it is necessary to destroy the organic materials completely. Oxidative decomposition, with the addition of hydrogen peroxide and UV radiation, may then be used. In contrast to the wet-chemical thermal decomposition processes in open or pressurised vessels, *UV digestion* requires only small amounts of acid in addition to H_2O_2, without a significant increase in temperature. This keeps the contamination resulting from heavy-metal levels in the reagents at a low level. Usually, high-pressure mercury lamps with a high intensity, as well as a great radiant flux, are used. The sample solutions are located in quartz vessels. Since the sample solution heats up only slightly in an open vessel (largely avoided by cooling), there is no loss of volatile substances. Several quartz glass reaction vessels can be arranged around the UV emitter in the centre. Highly reactive chemical radicals are produced by UV radiation, and ozone is generated. In the secondary reactions, organic substances that bind to heavy metals (i.e. they form complexes and therefore evade voltammetric or photometric analysis) are degraded [19]. UV digestion of a sample is directly proportional to the UV intensity and irradiation time. It is inversely proportional to the organic substance concentration. The digestion is also directly proportional to the temperature of the sample.

Wet decomposition in *open vessels* allows large sample sizes and achieves low detection limits, but may give rise to systematic errors due to:

- contamination caused by reagents and container material;
- losses of elements caused by adsorption on the surface of the vessel or by reaction with the vessel material; and
- losses of elements by volatilisation.

During wet digestion carried out with the application of solubilising and/or oxidising digestion reagents in *closed systems*, pressure builds up. The forerunners of these methods were the Carius digesters. As compared with open-vessel systems, the favoured pressure decomposition (actually a misnomer) has several advantages for trace-element analysis (Table 8.10).

To fully exploit the advantage of nitric acid, pressure decomposition systems have to be used that permit application of HNO_3 at temperatures above the boiling point. Temperatures above $300\,°C$ are required for complete digestion of organic materials with HNO_3 alone in closed systems.

Two significant sample preparation innovations that were considered leading-edge technologies in the 1980s are the *high-pressure asher* (HPA) system and the alternative microwave sample preparation systems, designed by Knapp and Grillo [46,47]. Modern HPA technology has been described in ref. [48]. HPA devices, which can operate up to $320\,°C$ at maximum pressures of 130 bar (1920 psi), make it possible to perform acid decompositions at $300\,°C$ and generate practically carbon-free solutions. Other similar devices, such as the Tölg pressure decomposition containers, have a lower temperature limit of $170\,°C$. The decomposition times in the Tölg bomb and in the high-pressure incinerator lie between one and three hours. Table 8.11 shows the main features of high-pressure ashing. A complete methods collection for HPA is available; HPA digestion parameters for PVC, PP and paint pigments were given [49]. Optimal decomposition parameters for PP and PVC in HPA are also readily available [17].

Table 8.10 Advantages of pressure decomposition for trace-element analysis

- No evaporative losses of elements
- Shorter reaction times (by a factor of three to 10), and improved decomposition because of the high temperature (above the boiling point of the reagent)
- Use of HNO_3 (available in very high purity) as a reagent for most digestions
- Lower blank values because of reduced reagent quantities
- No contamination from external sources
- Automation (increased throughput)

Table 8.11 Main characteristics of high-pressure ashing

- Minimum acid requirements (2.5 mL for 0.15 g organics)
- Single acid digestion
- Minimum organic residue
- No need for time-consuming predigestion steps
- Complete sample mineralisation
- No loss of elements
- Reduced risk of contamination
- Automated sample preparation
- Maximum operator safety

Woittiez and Sloof [4] have compared the (dis)-advantages of a dozen dry and wet ashing procedures. For trace-element analysis, low-temperature dry ashing, microwave-assisted wet digestion, quartz-lined high-pressure wet ashing and low-pressure dry ashing with reflux seem to be the best choices. From the economical point of view, microwave-assisted wet digestion and Schöniger combustion score best.

Applications Sulfur and phosphorous ions generated from the thioester and/or phosphite type AOs originally present in a plastic sample can be quantified after wet digestion using specific ion electrodes or ion chromatography (IC). Nitrogenous-based AOs (e.g. aryl amines) present in a plastic sample can be measured by traditional Kjeldahl analysis or pyrolysis/chemiluminescence, using an automated nitrogen analyser [50]. Crompton [31] has reported Kjeldahl digestion–boric acid titration methods for the determination of *organic nitrogen* in polyolefins in amounts between 0.002 and 75 %, and from 1 to 90 %. The Kjeldahl digestion procedures described quantitatively decompose amines, amino compounds, amino acids, amides, nitriles and their simple derivatives, and also many refractory nitrogen compounds. Quantitative decomposition of nitrogen-containing polymers, e.g.

styrene–acrylonitrile copolymers and polyacrylonitrile, may be achieved. The same author has also reported a Kjeldahl digestion–indophenol blue spectrophotometric procedure (at 625 nm) for the determination of 0.002–75 % organic nitrogen in polymers (reproducibility ±5 %). Forms of nitrogen for which accurate results are not usually obtained include those with −N−N (e.g. diazo) and N−O (e.g. nitro) linkages, and some resistant heterocyclic structures. One of the problems with the classical Kjeldahl digestion procedure for determination of organic nitrogen in polymers is connected with the form in which the sample exists [21]. For fine powder, or a very thin film, acid digestion may be adequate to enable quantitative extraction of the relevant substance from the polymer. Reliable nitrogen contents can be obtained for fine polymer powders using the Kjeldahl digestion procedure with a mixture of concentrated H_2SO_4, K_2SO_4 and an oxidation catalyst. However, low nitrogen results are found for polymers in larger granular form; for the analysis of such samples, classical microcombustion techniques are recommended. Another reason why the Kjeldahl sulfuric acid digestion procedure is of difficult application for the digestion of polymer powders, and certainly for polymer granules in the determination of metals, is that the Kjeldahl digestion catalyst, usually a mercury, copper or selenium salt, interferes in all likelihood with methods for the determination of traces of metals in digests. Chimassorb 944 has been analysed by the determination of the nitrogen content in the polymer by combustion or Kjeldahl methods [51]. The Kjeldahl method presumes the absence of other N-containing compounds. Other methods to determine nitrogen in polymers are micro-Dumas combustion [31] and pyrochemiluminescence [38].

Table 8.12 shows a selection of elemental analyses of polymers after oxidative wet digestion. Crompton [31] has described the determination of 0.1 ppm of copper in polyalkenes by means of wet digestions of

Table 8.12 Analysis of polymers after oxidative wet digestion

Analyte	Polymer	Digestion mixture	Technique
Al, Ca, Sb	PVC	H_2SO_4–H_2O_2/Kjeldahl	FAAS
Al, Pb	PVC	H_2SO_4–H_2O_2/Kjeldahl	FAAS
As, Ba, Hg	Various	HNO_3, 8 h 240 °C/pressure	Various
Cd	Various	HNO_3–H_2SO_4/pressure	FAAS
Cd, Pb, Sn, Zn	PVC	H_2SO_4–H_2O_2	FAAS
Cd, Sb, Sn	PVC	H_2SO_4–H_2O_2/Kjeldahl	FAAS
Mg, Pb	PVC	H_2SO_4–H_2O_2/Kjeldahl	FAAS
Sb	PVC	HCl–HNO_3–$HClO_4$–H_2SO_4	HG-AAS
Sn	PVC	H_2SO_4–H_2O_2	GFAAS
Sn	Various	HNO_3–HCl/high pressure	GFAAS

Adapted from Welz and Sperling [52]. From B. Welz and M. Sperling, *Atomic Absorption Spectrometry*, Wiley-VCH, Weinheim (1999). Reproduced by permission of Wiley-VCH.

the matrix with aqueous alcoholic magnesium nitrate first and with HNO_3/H_2SO_4 next, followed by complexation and photometric determination of coloured copper diethyl dithiocarbamate at 432 nm. *Phosphorous* in polymers is also frequently analysed either by oxygen flask methods; by digestion with concentrated sulfuric acid–perchloric acids followed by spectrophotometry; or by mineralisation in nitric–perchloric acid with subsequent titration or photometric determination [31]. For example, the micro-determination of phosphorous in polymers has been carried out by $H_2SO_4/HClO_4$ digestion and colorimetric measurement at 430 nm of a yellow phosphovanadomolybdate complex formed by reaction of the digest with ammonium vanadate/molybdate. This method is capable of determining total phosphorous in polymers in amounts down to 0.5 % and up to 20 %, with an accuracy of ±5 %.

ASTM D 3171-90 describes a standard test method for *fibre content* determination of resin–matrix composites by matrix digestion in a hot medium, such as HNO_3 or H_2SO_4/H_2O_2 [53].

Determination of *flame retardants* often relies on the quantitative determination of P (wet digestion and precipitation as ammonium phosphomolybdate), oxygen flask combustion methods for Cl and Br, or XRF. Korenaga [54] has determined traces of arsenic in acrylic fibres containing Sb_2O_3 as a fire retardant, by AAS analysis of the concentrated $HNO_3/HClO_4/H_2SO_4$ digest of the polymer at 193.7 nm. The procedure was calibrated against standards of arsenic-free acrylic fibre and pure As_2O_3.

Besecker *et al.* [55] have recently described a simple and optimised HPA digestion technique for PE/PP blends using nitric acid; 23 elements were readily determined by ICP-AES with detection limits in the low ppb range and 5 % RSD. Recoveries were in the 90–100 % range for all elements. The complete sample preparation process takes about three hours. Harsh acid mixtures are not required; generally HNO_3 is the acid of choice in ICP-AES analysis. As an alternative to the conventional digestion of recycled polymeric matrices for consumer electronics in the determination of additives containing Ti, Zn, Br, Cd, Sn, Sb and Pb, dissolution of the thermoplastic waste followed by TXRF is considerably less time-consuming [56].

Method development for high-pressure ashing and closed microwave digestion was reported for wet oxidation and extraction of Pb, Cd, Cr and Hg from various *food packaging materials* [57]. Use of HPA resulted in the highest median recoveries of the spiked elements (Pb and Cd, 92 %; Cr, 97 %; Hg, 83 %). The use of In as an internal standard improved the accuracy

of results from both ICP-MS and ICP-AES. Wet digestions are typically also used for inorganic species determinations in foods. Oxidative UV photolysis was used in the trace determination of metals in food or plant materials [58]. An interlaboratory comparison with remarkable laboratory reliability (RSD = 21.1 % for 52 ± 11 ppm) was reported for the determination of Cd in *wrapping paper* [59]. Digestion methods used by the 13 participants were pressure digestion in a PTFE bomb with sub-boiled HNO_3, HPA, microwave and open digestion; detection methods applied were ETAAS, ASV and potentiometry.

Sulfated ash is the inorganic residue remaining when organic substance is sulfatised in the presence of sulfuric acid. Alkali metals, alkaline earths and some other metals are converted into sulfates. Nadkarni *et al.* [60] have recently discussed the limitations of reliability and reproducibility of the *sulfated ash test method* (ASTM D 874-96), which has long been used as a quantitative measure of the ash-forming metallic constituents in unused lubricating oils and additive concentrates. This method provides predictable and chemically understood salts when additives based on the older additive technology (primarily Ba, Ca, Zn and P) are ashed. New additive technology has introduced other elements, notably Mg, B, etc., which result in the formation of nonstoichiometric oxides, (pyro)phosphates, etc. in addition to metal sulfates. Because of the complexity of the salts formed, and the poor interlaboratory reproducibility, it is not possible to predict the sulfated ash values from the elemental compositions for additive packages. With the newer additive technology, ASTM D 874 has lost its usefulness as a tool to either control blending operations or as a quick test to check additive treat levels for meeting product specification. In view of the high degree of accuracy and precision of the tests available for measuring metals and nonmetals, it is recommended that the actual metal and nonmetal concentrations be used in the specifications, rather than the unpredictable and error-prone sulfated ash values [60].

8.2.2.2 Microwave-assisted Decompositions

Principles and Characteristics Instead of thermal initiation, microwave decomposition may be of use for sample preparation involving combustion or acid digestion. The advantages over thermal initiation lie in the shorter time needed (minutes instead of hours). Microwave oven digestion (MOD) systems are not analytical instruments. Functionally, they are chemical reaction systems for acid decomposition of samples, producing solutions suitable for introduction into common analytical instrumentation.

Abu-Samra *et al.* [61] published the first paper on microwave-assisted decomposition in 1975, and the first compendium on microwave sample preparation was edited in 1988 [14]. Introduction of high-temperature/pressure devices in the early 1990s has given a further impulse to the development of microwave-assisted decomposition [8]. Microwave energy as a heat source for digestion/ashing procedures has become an important link between the time-consuming sample decomposition techniques and the fast determination methods, up to the extent that completely automated on-line digestion/determination techniques are now available (with AAS, ICP-AES or ICP-MS).

Modern commercial instrumentation for microwave-assisted decomposition operates at a frequency of 2.45 GHz. The corresponding energy is insufficient to rupture molecular bonds directly; the observed processes are rotational excitation of dipoles and molecular motion associated with the migration of ions. Therefore, the designation 'microwave-*assisted* decomposition' is highly appropriate. In microwave heating, energy is directly transferred to absorbing molecules of both sample and reagents (vessel walls are microwave-transparent). Because mineral acids in aqueous solution are all ionised, polar and charged, microwave absorption by these reagents is very high. The control of acid decomposition methods benefits from the consistent, predictable and reproducible energy absorption. Decomposition times are reduced as a result of the rapid heating process of the sample/acid solution. Modern microwave oven digestion systems are capable of processing samples at a rate of 20–30 per hour, and can be operated unattended, thus maximising the follow-up analytical operation potential.

Pressure conditions generated during microwave acid digestion are a result of microwave heating, which raises acid temperature and vapour pressure, and accumulation of gaseous decomposition products (CO_2, NO_x, SiF_4) of the reaction inside the vessel. *Pressure digestion* allows the sample to be essentially isolated from the laboratory environment, thus minimising both loss of volatile analytes and contamination from external particles. Moreover, digestion takes place at higher temperatures compared to digestion at atmospheric pressure, because of the boiling-point effect. The high temperature also has a strong influence on the oxidising power of the digesting acids. Not surprisingly, it is necessary to distinguish between low-pressure (<10 bar) and high-pressure systems (<80 bar). The latter systems are particularly suitable for trace analysis in thermally resistant matrices. Various microwave-assisted matrix digestion techniques may be distinguished [62]:

- low-pressure systems with home appliance microwave ovens;
- commercial low-pressure microwave systems;
- high-pressure microwave systems;
- nonpressurised microwave systems; and
- dry ashing in a microwave oven.

For vessel types, see Table 3.23. Most commercially available systems for microwave-assisted sample preparations are based on closed-vessel devices that offer either a pressurised or nonpressurised mode. The rate of a digestion reaction is controlled primarily by temperature, and only indirectly by pressure. Closed-vessel microwave dissolution systems are limited only by the temperature and pressure safety tolerances of the reaction vessel and the microwave absorption characteristics of the solution. Matusiewicz [63] has described low- and high-pressure microwave acid digestion vessels. The results of microwave-assisted digestion are evaluated on the basis of the following criteria: completeness of matrix destruction; reproducibility with respect to the analyte determination; freedom from trace losses and contamination; ease of handling; and expenditure of time. The microwave-assisted techniques allow samples to decompose rapidly (ca. 10 min); total experimental time including preparation, cooling, transfer and cleaning, amounts to 45–60 min for microwave-assisted high-pressure ashing, i.e. three or four times faster than conventional high-pressure ashing. Nonpressurised microwave ashing takes up an intermediate position. Cold-plasma ashing is very slow (10–20 h) [62]. On-line digestions are an important option for nonpressurised techniques.

Microwave dissolution is superior to hot-plate and block-digestion wet acid-digestion methods. Microwave-enhanced reaction methods in sample preparation for elemental analysis are robust and relatively low cost. Calibration as a means of transferring microwave procedures is only as good as the ability to reproduce the microwave field precisely. The ability to reproduce reaction conditions of a calibration-controlled method is typically ±5–10 °C or more. Kingston *et al.* [64,65] have outlined the necessary descriptors for documenting a microwave method. Williams and Haswell [66] have highlighted the advantages of microwave-enhanced flow systems over batch processes, e.g. in the field of sample dissolution for elemental analysis. The main characteristics of the various microwave-assisted digestion techniques are given below.

(a) Low-Pressure Systems with Home Appliance Microwave Ovens

In view of insufficient safety precautions for chemical operations, these systems are generally not recommended for microwave-assisted digestions of polymeric material.

(b) Commercial Low-Pressure Microwave Systems

High-temperature/low-pressure inorganic digestions are an area of application that has benefited from recent advances in vessel and sensor design. The inert properties of Teflon and its resistance to acid attack make it the material of choice for microwave pressure-vessel construction. Improved commercial systems offer additional safety precautions and improved facilities for pressure and/or temperature control. Also, the distribution of microwave radiation inside the oven cavity is fairly homogeneous. Low-pressure systems allow decomposition temperatures of about $180\,°C$. However, for many matrices, such temperatures are not sufficient to guarantee the complete ashing of thermoresistant sample components.

(c) High-Pressure Microwave Systems

Closed vessels capable of performing digestions at elevated pressures were introduced in the mid-1970s. The development of closed containers – allowing microwaves to pass through and heat the digestion mixture directly – means that much higher temperatures can be achieved, resulting in faster dissolution than conventional atmospheric pressure techniques. Vessel or sample container technology has driven the advancement of microwave-based equipment. High-pressure/high-temperature equipment is designed to support digestion up to a pressure of 150 bar, and typically in large volumes of solvent (100 mL). Temperatures substantially higher than in low-pressure microwave systems can be reached (up to $300\,°C$), resulting in more thorough digestion – even of difficult matrices. For the complete decomposition of a matrix, the working temperature should be held for at least 10 minutes. Digestions in closed quartz vessels are about 100 times faster than by hot plate. The advantage of pressurised microwave digestion (PMD) with precise T, p control is that nearly every sample matrix can be decomposed completely in a single-stage process. Modern microwave digestion systems allow simultaneous decomposition of six to 12 vessels. As the power is shared between the vessels, fewer vessels

Table 8.13 Characteristics of high-pressure microwave systems

Advantages

- Short digestion and cooling times
- High sample throughput (automation)
- No volatilisation of elements
- Ease of use
- Enhanced oxidation potential (superheated HNO_3 digestion)
- High-quality digestion
- Low consumption of high-purity reagents (low blank values)
- T power vs. time
- Accurate analysis (matrix free)

Disadvantages

- Potential safety problems
- Limited amount of sample (typically 0.5–1 g)
- Limited temperature range
- Difficulty in obtaining complete dissolution
- Not very suitable for organic compounds
- No possibility of adding reagents (single step procedure)
- Memory effects from external sources (porous PTFE walls)
- Difficult cleaning

have advantages. Knapp *et al.* [67] have described several designs for pressure-controlled microwave-assisted wet digestion systems. The *main characteristics* of commercial closed, pressurised microwave digestion vessels are given in Table 8.13.

Matusiewicz [63,68] has reported the development of a high-pressure, high-temperature, focused-microwave-heated acid (HNO_3) digestion system. This microwave technique requires only about 3 % of the time necessary for the thermal high-pressure (HPA) technique. The technique of microwave heating samples in sealed containers to speed up acid digestion has been in widespread use for the past few years [69,70].

(d) NonPressurised Microwave Systems

Advantages of the use of atmospheric pressure devices over pressurised vessels are [44]:

- safety;
- possibility of sequential additions of reagents during digestion;
- precise control of the energy;
- wider range of vessel materials;
- allowance for complete dryness of the digest;
- handling of large sample size; and
- automation.

Ambient pressure systems are the best option for safety reasons, because no overpressure can occur. Open vessel systems allow much larger sample sizes to achieve lower detection limits. Moreover, nonpressurised microwave digestion is suitable for on-line decompositions in continuous-flow systems [71]. Nonpressurised microwave systems are limited by a low maximum temperature, which cannot exceed the ambient-pressure boiling point of the acid (or the acid mixture). The oxidising power of nitric acid is insufficient at low temperatures (ca. 120 °C). Other disadvantages are the danger of contamination by the laboratory environment, and that of trace losses, especially if mercury is the analyte. Losses of organometallic compounds of arsenic, antimony and tin may also occur if such digestion is performed without precautions.

Mermet [44] has described focused-microwave-assisted atmospheric pressure acid digestion technology, including the application to the Kjeldahl nitrogen determination in organic materials. *Microwave-assisted Kjeldahl technology* leads to shorter digestion time and high recovery, and is advantageous in comparison with the time-consuming conventional procedure. The rapid digestion is possible by using the efficient energy coupling associated with microwave energy and the high-temperature boiling acid (H_2SO_4, bp 340 °C), which allows elevated temperatures (>300 °C) to be used in the procedure. The atmospheric pressure microwave-assisted digestion method has been developed for the combined analysis of total phosphorous and Kjeldahl nitrogen in complex matrices [72].

(e) On-line Microwave-enhanced Processes

Technology for microwave-digestion flow systems is available for sample digestion, dissolution and detection, as well as organic synthesis. Burguera *et al.* [73] have described the coupling of microwave sample dissolution to FAAS by using flow-injection analysis (FIA) methodology. The impetus behind such developments is to further reduce the labour-intensive process of batch microwave sample dissolution by using a flow system, thus increasing sample throughput and facilitating automation. On-line microwave sample dissolution has also been coupled with ICP-AES/MS [74].

(f) Dry Ashing in a Microwave Oven

The use of microwave digestion−wet oxidation overcomes sample preparation problems for many polymer-based materials. However, this will result in a reduction in sensitivity compared with an ashing procedure, because of dilution. Use of an aqueous phase is not necessary while heating with microwaves. Dry ashing in a microwave oven (up to 1200 °C in 2 min) is a valid alternative to conventional dry ashing in a muffle furnace. The advantages compared to conventional dry ashing techniques are striking: time saving, low energy consumption and user friendliness. Memory effects are often lower in microwave-assisted techniques than in conventional ashing techniques, which may be due to the shorter contact time in the microwave vessels.

Postdigestion evaporation procedures using both conventional and microwave heating methods have been compared [75]. The use of microwave energy to perform evaporations (instead of the conventional methods) is advantageous. Volatile inorganic species are recovered completely using microwave-assisted evaporation, which is in contrast to the losses experienced using hot-plate evaporation. Microwave-assisted evaporation is a new tool for trace inorganic analysis, where losses due to volatilisation of many inorganic species are no longer a concern. Reviews on commercially available microwave-assisted digestion systems are given in refs [62,69,70,76−79].

Applications Sample preparation capabilities for AAS, ICP, ICP-MS are greatly improved by advanced microwave technology with high-pressure vessels (up to 300 °C and 1500 psi), which provide:

(i) fast, complete, and reproducible digestions;
(ii) high throughput and rapid sample turnaround;
(iii) improved recoveries;
(iv) retention of volatile analytes; and
(v) elimination of cross-contaminants.

Numerous microwave applications have been published on decomposition, fusion, dry and wet mineralisation, ashing and extraction. Knapp *et al.* [67] have reported decomposition efficiencies of over 96 % for PE, PVC, PS and PB, using PMD. Boron in polyolefins was determined after high-pressure microwave digestion followed by ICP-MS [80].

Microwave digestion with H_2SO_4/HNO_3 was used for the determination of monoester tin *heat stabilisers* for PVC [81]. A method for high-pressure (75 bar) closed-vessel microwave digestion in HNO_3 has been developed for several polymers (HDPE, PS, PE/PP, PE precursor), which eliminates the need for harsh acid mixtures (including H_2SO_4 and HF) [82]. Using ICP-AES analysis, 1−18 elements were determined in the low ppb range with precision better than 5 % RSD. The accuracy of the method was determined by performing spike recoveries; recoveries amounted to 90−100 % for all elements.

Plastic scrap, such as household waste for use as an auxiliary fuel in rotary kilns in the cement industry, requires regular analysis for several metals that may cause environmental problems or affect the cement product quality. These metals originate from colouring agents, fillers and stabilisers. The difficulties of the determination arise mainly from the sample preparation step. Mixed plastic waste is a difficult sample. Due to different origins, this material shows great inhomogeneity; a strongly changing composition; and an unpredictable reaction behaviour, depending on the kind and quantity of additives. Some additives form very reactive intermediate products during wet ashing procedures, leading to violent reactions. According to Zischka *et al.* [83] accurate determination of heavy metals in plastic scrap is strongly influenced by the selection of the sample digestion method. These authors have developed a universally applicable closed-vessel microwave-assisted digestion method for safe and reliable routine determination of Cd, Cr and Pb in plastic scrap [83]. An interlaboratory test on the determination of Cd, Cr and Pb in PVC and PUR using different microwave digestions has been reported [83a].

Extended applications of microwave furnaces include open-vessel digestion, hydrolysis, ashing of organic and inorganic materials, determination of loss on ignition (LOI) and residue on ignition (ROI), drying and high-temperature reactions (synthesis). Table 8.14 compares some typical ashing times (conventional vs. microwave).

Table 8.14 Typical ashing times

Material	Conventional (min)	Microwave (min)	Time savings (%)
Polyester (filled)	480	15	97
Polyethylene (unfilled)	30	5	83
Polyethylene (% carbon-black)	30	7	77
Polypropylene	30	5	83

Ref.: ASTM D 5630-94 Ash Content in Thermoplastics; ASTM D 1506-97 Carbon-Black/Ash Content.

As microwave sample preparation has evolved, standard microwave procedures have been developed and approved by numerous standard methods organisations (ASTM, AOAC International, EPA, etc.), see ref. [64]. Examples are standard test methods for carbon black/ash content (ASTM Method D 1506-97), lead analysis in direct paint samples (ASTM Method E 1645-94), etc. Table 8.15 shows some microwave ashing references (detection: weight). A French AFNOR method utilises the atmospheric pressure single-mode microwave method as an alternative sample preparation procedure for Kjeldahl nitrogen determination [84]. The performance of a microwave-assisted decomposition for rapid determination of glass fibre content in plastics for QC has been described [85].

Residual moisture content of samples can play a major role in reproducible sampling of the matrix, as well as in the interaction of the microwave energy with the sample.

8.2.3 Fusion Methods

Principles and Characteristics Fused-salt media or fluxes will decompose most substances (usually inorganic) at the high temperature required for their use ($300-1000\,^\circ$C) and the high concentration of reagent brought into contact with the sample (up to tenfold excess). The production of a clear melt signals completion of the decomposition. After cooling, the melt is dissolved. Various reagents may be used for the fusion of polymer samples prior to the determination of metals [17]. Fusion with sodium carbonate is useful for polymers which release acidic vapors upon ignition. The acid is trapped in the solid alkaline reagent. Sodium bisulfate forms a very high-temperature melt, in which the oxides of many metallic elements dissolve and are put in a chemical form suitable for subsequent analysis. Sodium peroxide is a reagent for the fusion of polymer samples preparatory to analysis of metals such as zinc, and nonmetals such as chlorine and bromine.

The addition of fluxes increases the risk of raising the blank value, owing to the amount of flux required for a

Table 8.15 Microwave ashing references

Matrix	Application	Ashing vessel	Conditions	Ref.
HDPE	TiO_2	Porcelain crucible	$3-4$ min at $800\,^\circ$C	[86]
HDPE/TiO_2	Residual ash	Porcelain crucibles or quartz fibre crucibles	15 min at $800\,^\circ$C	[87]
HDPE/C	Residual ash	Porcelain crucibles or quartz fibre crucibles	12 min at $600\,^\circ$C	[87]
LDPE	TiO_2	Porcelain crucible	$1.5-5$ min at $800\,^\circ$C	[86]
LDPE	SiO_2	Porcelain crucible	$3-4$ min at $800\,^\circ$C	[86]
PP	TiO_2	Porcelain crucible	$3-4$ min at $800\,^\circ$C	[86]

successful fusion. In addition, the final aqueous solution obtained from fusion will have a high salt concentration, which may cause difficulties in subsequent analysis steps. The high temperatures required for a fusion increase the danger of volatilisation losses. These disadvantages make fusion a less than ideal technique for extreme trace-element determination. Nowadays, such analyses are likely to be carried out by modern nondestructive physical techniques.

Applications Basic methods for the determination of halogens in polymers are fusion with sodium carbonate (followed by determination of the sodium halide), oxygen flask combustion and XRF. Crompton [21] has reported fusion with sodium bicarbonate for the determination of traces of chlorine in PE (down to 5 ppm), fusion with sodium bisulfate for the analysis of titanium, iron and aluminium in low-pressure polyolefins (at 1 ppm level), and fusion with sodium peroxide for the complexometric determination using EDTA of traces of bromine in PS (down to 100 ppm). Determination of halogens in plastics by ICP-MS can be achieved using a carbonate fusion procedure, but this will result in poor recoveries for a number of elements [88]. A sodium peroxide fusion-titration procedure is capable of determining total sulfur in polymers in amounts down to 500 ppm with an accuracy of $\pm 5\%$ [89].

Micro tin in PP and lubricating oil was analysed spectrophotometrically after isolation of the analyte from the samples with $KHSO_4$ as a fusing agent, $(NH_4)_2S_2O_8$ as an oxidant and Na_2SO_4 as an auxiliary ashing agent [90].

8.3 ANALYTICAL ATOMIC SPECTROMETRY

Principles and Characteristics Sources for analytical atomic spectrometry include arc, spark, flame, graphite furnace, low-pressure discharges (GD), plasma discharges (ICP, DCP, MIP), and laser plume. These sources are either continuum sources, which emit radiation over a relatively broad spectral range, or line sources emitting a few discrete lines only. Examples of continuum sources are deuterium lamps with a useful continuous spectrum from 160 to 375 nm, and a tungsten filament lamp in the wavelength region between 320 and 2500 nm. Low-pressure arc lamps, hollow-cathode discharge tubes and electrodeless discharge lamps are common line sources which emit narrow spectral lines. Sources are used directly as emission sources; as atomisers for atomic absorption or atomic fluorescence; or

for ion production in connection with optical and mass-spectrometric detection. The most widely used wavelength selection device is the échelle monochromator, which uses high diffraction orders and large angles of diffraction with resolution of around 0.015 nm, compared to 0.2 nm for a typical AAS monochromator.

In this chapter, only *atomic spectrochemical methods* are discussed. Atomic spectra are line spectra, and are specific to the absorbing or emitting atoms (elements), i.e. the spectra contain information on the atomic structure. Each spectral line can be regarded as the difference between two atomic states:

$$E_2 - E_1 = \Delta E = h\nu \qquad (8.1)$$

where E_1 and E_2 are the energies of the two levels. Three forms of radiative transition are possible between the energy levels E_2 and E_1:

(i) spontaneous emission for the transition from a higher excited state to a lower state;
(ii) induced emission for the transition from a higher excited state to a lower state stimulated by external radiation of the corresponding frequency; and
(iii) absorption of radiation (the reversed process to induced emission) with a corresponding transition of a lower state to a higher excited state.

Atomic spectra are much simpler than the corresponding molecular spectra, because there are no vibrational and rotational states. Moreover, spectral transitions in absorption or emission are not possible between all the numerous energy levels of an atom, but only according to selection rules. As a result, emission spectra are rather simple, with up to a few hundred lines. For example, absorption and emission spectra for sodium consist of some 40 peaks; for elements with several outer electrons, absorption spectra may be much more complex and consist of hundreds of peaks.

Absorbed energy may be rapidly lost to the surroundings by collisions, allowing the system to relax to the ground state. Alternatively, the energy is re-emitted a few milliseconds later, in a process called fluorescence (AFS). In emission spectrochemical methods, the radiation emitted by the analyte following nonradiational excitation is measured. The main spectrochemical techniques are those involving thermal excitation processes occurring in a flame, or a plasma or electrical excitation using a direct current arc or a high-voltage spark (Table 8.16). Not all possible atomic spectrometric measurement modes are used to the same extent. For instance, AFS and GFAES are rarely used in practice.

Table 8.16   Classification of atomic spectral methods

Atomisation method	Typical atomisation temperature (K)	Basis for method[a]
Flame	1 700–3 150	A, E, F
Electrothermal	1 200–3 000	A, F
Inductively coupled argon plasma	6 000–8 000	E, F
Direct-current argon plasma	6 000–10 000	E
Electric arc	4 000–5 000	E
Electric spark	40 000?	E

[a] A, absorption; E, emission; F, fluorescence spectroscopy.

Both emission and absorption spectra are affected in a complex way by variations in *atomisation temperature*. The means of excitation contributes to the complexity of the spectra. Thermal excitation by flames (1500–3000 K) only results in a limited number of lines and simple spectra. Higher temperatures increase the total atom population of the flame, and thus the sensitivity. With certain elements, however, the increase in atom population is more than offset by the loss of atoms as a result of ionisation. Temperature also determines the relative number of excited and unexcited atoms in a source. The number of unexcited atoms in a typical flame exceeds the number of excited ones by a factor of 10^3 to 10^{10} or more. At higher temperatures (up to 10 000 K), in plasmas and electrical discharges, more complex spectra result, owing to the excitation to more and higher levels, and contributions of ionised species. On the other hand, atomic absorption and atomic fluorescence spectrometry, which require excitation by absorption of UV/VIS radiation, mainly involve resonance transitions, and result in very simple spectra.

Lasers produce spatially narrow and very intense beams of radiation, and lately have become very important sources for use in the UV/VIS and IR regions of the spectrum. Dye lasers (with a fluorescent organic dye as the active substance) can be tuned over a wavelength range of, for instance, 20–50 nm. Typical solid-state lasers are the ruby laser (0.05 % Cr/Al_2O_3; 694.3 nm) and the Nd: YAG laser (Nd^{3+} in an yttrium aluminium garnet host; 1.06 µm).

The ideal *spectrochemical detector* has a high sensitivity, a high signal-to-noise ratio and constant response over a considerable range of wavelengths, fast response time and minimal output signal in the absence of illumination. The most widely used detector in the UV region is the photomultiplier tube (PMT). A diode-array multichannel spectrometer uses a photodiode-array detector, consisting of an arrangement of 1000 or more silicon photodiodes. With such diode-array detectors placed along the focal plane of a monochromator, all wavelengths of interest can be monitored simultaneously. GFAAS is now equipped with solid-state detectors.

As shown in Table 8.17, there is considerable overlap of capabilities between element analytical methods. A general understanding of the basic principles of the various techniques is necessary for an informed choice of the best technique. The atomic spectrometry techniques used most are ETA-AAS, ICP-AES and ICP-MS.

Dymott [91] has recently compared FAAS, GFAAS, ICP-AES and ICP-MS. As a rule, ICP-MS produces the best *detection limits* (typically 1–10 ppt), followed by GFAAS (usually in the sub-ppb range), then ICP-AES (in the order of 1–10 ppb) and finally FAAS (in the sub-ppm range). Flame techniques do not offer the sensitivity required for the analysis of many solids. For FAAS, *short-term precision* is in the range of 0.1–1.0 %. Long-term precision depends on the spectrometer optics; double-beam types are capable of long-term precision of 1–2 %, whereas single-beam types are typically in the 10 % range. Primarily because of difficulties in injecting small volumes, GFAAS short-term precision is generally in the range 0.5–5 %. Long-term precision is totally dependent on the tube type and service condition. ICP-AES short-term precision is reasonably good, around 0.1–2 %, and, even over periods of hours, should be no worse than 1–5 %. Simultaneous spectrometers generally have good short- and long-term precision compared with sequential systems. Short-term for ICP-MS is in the range 0.5–2 %, with long-term precision around the 5 % level. FAAS and GFAAS have limited *dynamic ranges*, of the order of only 10^2–10^3, therefore solutions must be held in a narrow range of concentrations. ICP-AES has considerably wider dynamic range, up to 10^5, making it a more suitable technique for highly concentrated samples, or samples with wide concentrations of analyte elements. ICP-MS typically operates at much lower concentration levels, so that ranges up to 10^8 can be achieved.

Practically all classical methods of atomic spectroscopy are strongly influenced by interferences and matrix effects. Actually, very few analytical techniques are completely free of interferences. However, with atomic spectroscopy techniques, most of the common interferences have been studied and documented. Interferences are classified conveniently into four categories: chemical, physical, background (scattering, absorption) and spectral. There are virtually no *spectral interferences* in FAAS; some form of background correction is required. Matrix effects are more serious. Also GFAAS shows virtually no spectral interferences, but

Table 8.17 Summary of characteristics of elemental analysis techniques

Feature	FAAS	GFAAS	ICP-AES	ICP-MS
Detection limits	Very good for some elements	Excellent for some elements	Very good for all elements	Excellent for most elements
Sample throughput	10–15 s per element	3–4 min per element	6–60 elements/min	All elements in 2–5 min
Dynamic range	10^3	10^2	10^5	10^5–10^8
Analytical range (% to)	10 ppb–1 000 ppm	10 ppt–1 ppm	10 ppt–1 000 ppm	1 ppt–1 000 ppm
Precision				
Short-term	0.1–1.0 %	0.5–5 %	0.1–2 %	0.5–2 %
Long-term	Two-beam 1–2 % One-beam <10 %	1–10 % (tube lifetime)	1–5 %	<5 %
Interferences				
Spectral	Very few	Very few	Many	Few
Chemical (matrix)	Many	Very many	Very few	Some
Physical (matrix)	Some	Very few	Very few	Some
Mass effects	None	None	None	Many
Isotopic	None	None	None	Many
Dissolved solids in solution	0.5–5 %	>20 % (slurries)	1–20 %	0.1–0.4 %
Elements applicable to	68+	50+	73	76
Sample volumes required	Large (mL)	Very small (μL)	Medium (mL)	Small (μL or mL)
Semiquantitative analysis	No	No	Yes	Yes
Isotopic analysis	No	No	No	Yes
Ease of use	Very easy	Moderately difficult	Easy	Moderately difficult
Method development	Easy	Difficult	Moderately easy	Difficult
Unattended operation	No	Yes	Yes	Yes
Capital costs	Low	Medium to high	High	Very high
Running costs	Low	Medium	High	High
Cost per elemental analysis	Low	High	Medium	Medium

After Dymott [91]. Reproduced by permission of Reed Elsevier.

background effects do exist on a major scale (compensation by means of Zeeman background correction). *Matrix effects* can be very serious, and generally occur in the vapour phase. Spectral interferences are fairly common in ICP-AES (use of alternate analytical lines is recommended); background effects require correction. Chemical matrix effects are fairly minimal. Spectral interference problems in ICP-MS are fairly insignificant. Much interference can now be overcome by using collision cell technology or high-resolution magnetic sector ICP-MS. Background effects are negligible. Chemical matrix effects can be a major problem in ICP-MS, resulting in severe limitations in the use of the system. Internal standardisation techniques are called for. There is a continuing search for methods that are either interference-free, or that incorporate interference-removal steps prior to the analytical measurement. For figures of merit of analytical atomic spectroscopy methods, see also ref. [92].

Atomic spectroscopy has been reviewed [92]; a recent update is available [93]. An overview of sample introduction in atomic spectrometry is available [94]. Several recent books deal with analytical atomic spectrometry [95–100].

Table 8.18 Applications of sources of atomic spectrometry

Source	Spectrometry (sample[a]), concentration range[b]
Flame	AES (l), t; AAS (l), t; AFS (l), ut
Arc (dc)	AES (l, s), t; MS (s), ut
Spark	AES (s), t; MS (s), ut
GD	AES (s, l, g), t; AAS (s, l, g), t; AFS (s, l, g), ut; MS (s, g), ut
ICP	AES (l), t; AFS (l), t; MS (l), ut
MIP	AES (g), t; MS (g), ut
Furnace	AES (l, s), t; AAS (l, s), ut
Laser	AES (s), m; AAS (s), t; AFS (s), ut; MS (s), ut

[a] s, solid; l, liquid; g, gas.
[b] m, minor; t, trace; ut, ultratrace.

Applications The general application areas of various forms of atomic spectrometry are given in Table 8.18.

As to expected performance, Figure 8.3 shows some (conservative) values for lateral and in-depth resolutions for various physical methods of analysis. Recent developments already allow better performance, e.g. a lateral resolution of about 0.1 μm with liquid metal ion sources in SIMS and small beams of 150 μm in diameter in μXPS.

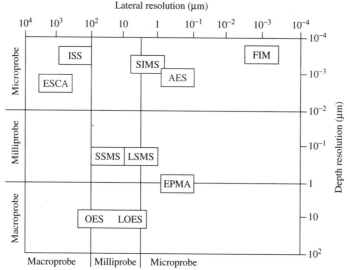

Figure 8.3 Comparison of lateral and depth resolutions of various physical methods. For acronyms, see Appendix I. After Ramendik *et al.* [101]. Reprinted from G. Ramendik *et al.*, in *Inorganic Mass Spectrometry* (F. Adams *et al.*, eds), John Wiley & Sons, Inc., New York, NY, pp. 17–84, Copyright © (1988, John Wiley & Sons, Inc.) This material is used by permission of John Wiley & Sons, Inc.

8.3.1 Atomic Absorption Spectrometry

Principles and Characteristics Atomic absorption spectrometry (AAS) was first described in 1955 [102,103]. In AAS, the analyte is presented as an atomic vapour, and radiation of the right wavelength (selected based on the element to be analysed) is passed through it to excite atoms from the ground state to an excited electronic level. This process is accompanied by attenuation of the incident radiation at a particular frequency, and can occur only when the energy difference between the two levels matches exactly the energy of the photons. The energy transitions correspond to radiation in the UV/VIS regions of the electromagnetic spectrum. As only atoms in the ground state are involved in this process, ionisation must be kept to a minimum.

In principle, all elements can be determined by AAS, since the atoms of any element can be excited and are therefore capable of absorption. The limitations lie practically only in the field of instrumentation. Measurements below 200 nm in the vacuum UV range are difficult, owing to the incipient absorption of atmospheric oxygen. With modified instruments and a shielded flame or a graphite furnace, it is possible to determine such elements as iodine at 183.0 nm, sulfur at 180.7 nm, and phosphorous at 177.5 nm, 178.3 nm and 178.8 nm.

Table 8.19 shows the *main characteristics* of AAS. Originally, a hollow-cathode was used as *excitation*

Table 8.19 General features of atomic absorption spectrometry

Advantages

- High specificity and selectivity
- High detection ability
- Relatively simple operation
- Suitable for ultratrace contents through to major constituents
- Sequential and near-simultaneous multi-element AAS
- Quantitation (relative technique)
- Speciation analysis
- Mature technology
- Automation

Disadvantages

- Time-consuming sample preparation step
- Method development
- Chemical and physical matrix effects
- Competitive edge

source, and the atom reservoir was a combustion flame, but alternative atomisation methods have been developed, such as the electrodeless discharge lamp (EDL), graphite furnace, hydride generation and cold vapour techniques. Hollow-cathode lamps (HCLs) from commercial sources are available for about 40 elements. Moreover, various multi-element sources (e.g. Na/K, Ca/Mg, Cu/Zn, Al/Ca/Mg, etc.) are on the market. Ideally, of course, the atomiser for AAS should yield complete atomisation of the element of interest, irrespective of the sample matrix. Both flame and nonflame atomisers are less than ideal atomisers for AAS. In both cases, the sample is generally introduced as an aqueous solution. Samples can also be introduced into a graphite furnace as a slurry or even as a solid. Thermal energy is then used to evaporate the solvent and to dissociate the sample without extensive ionisation.

Conventional AAS measures one element in all samples, before switching to the next element. Multi-element AAS, for the most part, is limited to two to six elements. Fast sequential AAS with four multi-element hollow-cathode lamps can measure up to 15 elements in near-simultaneous manner. With fast sequential lamp selection, all elements in each sample are measured before progressing to subsequent samples, providing complete results for each sample in minutes. Replacing the traditional hollow-cathode lamps used in most atomic absorption instruments with semiconductor diode lasers would allow the user to perform background-corrected multi-element AAS with high selectivity, a large dynamic range, and internal standardisation. However, so far, diode lasers are less ideal spectroscopic sources – being not continuously tuneable. AAS is available with single- and double-beam optics. The stability of measurements depends largely on the beam configuration.

In AAS, the sensitivity is directly proportional to the number of atoms in the ground state. Optimisation of the operating conditions is an essential part of the development of analytical methods using AAS. The number of absorbing atoms in AAS does not stand in simple relation to the concentration of the analyte in the sample. Various phase changes take place on the way to the formation of gaseous atoms. In a flame, these comprise nebulisation, desolvation, volatilisation of aerosol particles and, finally, dissociation of the molecules. Quantitative modelling of these processes carries a high level of uncertainty. Given the less than quantitative nebulisation/atomisation process and interferences, it is not surprising that changes in the intensity of the detector signal cannot be directly related to analyte concentration in the sample. Like most other spectrometric techniques, AAS is a relative technique, in which the relationship between quantity or mass of the analyte and the measured value is determined by *calibration*, i.e. comparison with a standard. Calibration techniques in use for conventional AAS are external standard and standard addition methods. Fast sequential AAS allows internal standardisation (IS) and fast semiquantitative analysis. IS corrects for the extinction of the analytes, according to the extinction of the internal standard, but not for chemical interferences. RSD values

of < 2 % are quoted. A condition is that the elements are measured in rapid succession. In general, depending on the analyte and the analytical line used in the AAS technique, calibration curves in AAS are only linear for absorbance values up to about 0.5–0.8. Absolute calibration is a desirable aim.

Any difference in the behaviour of the analyte atoms in the sample and in the standard implies an *interference*. AAS using a line source for excitation suffers little spectral interference. Background interference in AAS is more important. This nonspecific absorption is caused by:

(i) light absorption by free molecules from matrix components which have not dissociated (molecular absorption); and/or
(ii) light scattering by solid or liquid particles of matrix substances which have not completely evaporated.

To overcome this problem, three different correction techniques have been proposed; the Zeeman technique offers the most advantages.

In *flame AAS* (FAAS), either an air/acetylene or a nitrous oxide/acetylene flame is used to evaporate the solution and dissociate the sample into its component atoms. FAAS has found wide application, especially in the past, because of its simplicity, the low cost of the equipment and its versatility in excitation of various elements of different nature. In FAAS, little spectral interference occurs and, moreover, these interferences are well identified and usually easily controlled. FAAS has similar detection limits to ICP-AES, namely $0.1–100\,\mu g\,L^{-1}$. Under usual conditions, the relative error of flame absorption analysis is of the order of 1 % to 2 %. FAAS is a single-element technique that is ideal when one or a few elements must be determined in a large number of samples. The disadvantages are that refractory elements such as B, Mo, Ta, V, W and Zr cannot be determined with good sensitivity, because the flame temperatures are not high enough to atomise a large fraction of such analytes. When several elements are to be determined in each sample, fast sequential FAAS is quicker than conventional ICP-AES (Table 8.30). Only liquid samples can be analysed, and relatively large sample volumes (a few mL) are required; flame cells are not very convenient for the direct atomisation of solid samples. Semiquantitative analysis, which is very common in ICP-AES, is impossible in conventional FAAS, due to the measurement order (element per element), but is possible in fast sequential FAAS, due to complete analysis per sample (sample per sample).

Standard deviations for fast sequential FAAS are about 2%. FAAS has had relatively little improvement in detection limits and performance over the last 15 years.

In the application of *electrothermal AAS* (ETAAS) at least a dozen atomisation systems have been developed [104]. The most advantageous designs are probably those that use a graphite support (platform, boat, microboat and cup) for both sample weighing and atomisation. The *graphite furnace* consists of an atomisation system whereby relatively small volumes of sample (a few μL or μg) are introduced into a resistively heated graphite tube (typically 3–5 cm long, 3–8 mm i.d.) through which the optical beam of the spectrometer is passed [105]. The latest graphite furnace design is the transversely heated graphite furnace with an integrated L'vov platform. Liquid samples, slurries and solid samples can be analysed. In a drying step, the solvent is evaporated at a temperature just above its boiling point; in the ashing step, the temperature is raised to remove organic matter and as many volatiles of the matrix as possible to separate the analyte from all interfering components. The atomisation step at high temperatures (2000 to 3000 °C), which lasts for a few seconds, is followed by a cleaning step to remove leftover residues. The element of interest is thus atomised, and the absorption of the source radiation by the atomic vapour is measured. GFAAS equipped with a CCD camera enables continuous recording of images of the events taking place inside the tube.

Table 8.20 shows the *main features* of solid sampling in atomic absorption spectrometry. The most important advantages of electrothermal vaporisation (ETV) compared to conventional solution nebulisation are the low sample consumption ($< 100 \mu L$), higher transport efficiency and lower level of MO^+ ions formed when aqueous solutions are analysed. A severe drawback of the method is the relatively short transient signal of 1–2 s produced by ETV [106]. Electrothermal atomisers offer the advantage of unusually high sensitivity for small volumes of sample. Typically, sample volumes between 0.5 and 10 μL are sufficient: under these circumstances, absolute detection limits typically lie in the range of 10^{-10} to 10^{-13} g of analyte. However, quantitative transfer of accurately weighed microsamples into electrothermal atomiser tubes is not trivial [107]. With ETAAS it is possible to make standardless elemental analysis (through the Lambert–Beer law) with the highest sensitivity for aqueous solutions [108]. An instrumental analytical method can be defined as standardless when a sufficiently reliable equation exists from which the concentration or quantity of the analyte can be calculated from a single measurement.

Table 8.20 Main characteristics of solid sampling in atomic absorption spectrometry

Advantages

- High sensitivity
- No sample pretreatment (dissolution)
- Low sample consumption
- Standardless elemental analysis

Disadvantages

- Labour intensive (no automatic weighing, problematic solid sample insertion)
- Different weighing for each replicate
- Sample inhomogeneity
- Difficult calibration (as opposed to liquid samples)
- Autosampler accuracy
- Less-effective use of chemical modifiers (as compared to liquid samples)
- High matrix concentration (pronounced spectral interferences, high background absorbances)
- Few possibilities for dilution
- Inferior precision (as compared to liquid samples)
- Low analytical range
- Furnace contamination
- Relatively slow
- Extensive method development
- Difficult determination of refractory metals
- Formation of refractory carbides

The major advantage of GFAAS is that its detection limits (0.01–10 μg/L) are 10–100 times better than for FAAS or ICP-AES. The technique is thus appropriate for the determination of low concentrations in limited sample volumes. Routinely, a precision of 3–10% can be obtained for GFAAS, as compared to ca. 0.5% for FAAS, ca. 1.5% for ICP-AES, and 2–3% for ICP-MS. Accuracy depends on the quality of standards, the presence of interferences, extents of contamination, etc. Analysis of solid materials can be challengingly difficult. As no dissolution is used, and none of the sample matrix is removed before the introduction of the sample into the furnace, this can result in severe vapour-phase chemical interferences. Matrix effects in GFAAS are now largely under control, using a combination of platform technology, background correction, matrix modification, etc., but the technique still requires considerable training and method development. GFAAS shares with FAAS the disadvantages that refractory element performance is relatively limited, and that it is essentially a single-element technique. Another disadvantage is the low analytical range, usually less than two orders of magnitude. Furnace methods are slow – typically requiring several minutes per element. Recently, GFAAS has been developed into a simultaneous, multi-element technique by optically combining and transmitting a portion of the output of several hollow cathode lamps through the graphite tube. Measurements are made for selected groups of elements. GFAAS is now equipped with solid-state detectors. Modern high-performance AA spectrometers feature automated atomiser exchange, enabling switching between flame and furnace modes. Alternatives to the slow GFAAS are modern simultaneous ICP-AES and ICP-MS techniques, also in terms of sensitivity. Electrothermal atomisation is ordinarily applied only when flame or plasma atomisation provides inadequate detection limits.

Solid sampling Zeeman atomic absorption spectrometry (SS-ZAAS), AAS with direct Zeeman background correction (ZBGC), which is a variant of the classical GFAAS technique, was introduced commercially in 1979 [109]. This method of background correction is based on the splitting of atomic spectral lines into several components, under the influence of a magnetic field applied to the spectral source (Zeeman effect). ZBGC compensates for nonspecific absorption and structured background, and allows for virtually interference-free GFAAS [109a]. The optical configuration associated with the polarised Zeeman background correction method combines the benefits of true double-beam operation with background correction, by measuring two polarised components of a single beam passing through the sample. The polarised Zeeman method is a true double-beam method, since it makes use of a sample and reference beam, but, in this case, sample and reference beams are simply differently polarised components of the same spectral light source. Zeeman atomic absorption is not free from problems [17]. The limitation of SS-ZAAS is that only about 20 elements can be measured. Recently the SR (high-speed-self-reversal) background correction technique has been announced, which competes with the well-known Zeeman systems.

Cold vapour (CV) and *hydride generation* (HG) techniques were developed for the determination of Hg, As, Se, Bi, etc. at the ng L^{-1} level. Further developments are a multi-element system, employing a continuum source and a high-resolution monochromator [110]. The flame technique, the graphite furnace, hydride generation, and cold vapour techniques are nowadays of equal significance. These techniques are applied in trace, ultratrace and nanotrace analysis. Each of these techniques has its own atomiser, its own specific mechanisms of atomisation and interference, and preferred field of application. Organic liquids are frequently analysed by FAAS and GFAAS, aqueous solutions by HG-AAS and CV-AAS, and solid samples by GFAAS.

Atomic absorption spectrometry is also being used in the hyphenated mode. Parris *et al.* [111] described

GC-GFAAS interfacing. HPLC-GFAAS has also been reported.

In industrial research laboratories, AAS (in particular FAAS) is no longer being used to the same extent as in the past, despite the aforementioned important improvements in AAS technology. More rapid, multi-element methods have gradually taken over, such as ICP-AES, ICP-MS and NAA. However, the determination of one element is faster with AAS than with an ICP technique. Also, ICP-AES does not supersede GFAAS in terms of sensitivity and selectivity.

Multi-element AAS has been reviewed [112], as well as ETAAS [104] and instrumental aspects of GFAAS [113]. Various monographs on analytical atomic absorption spectrometry are available [52,96,114,115], and on GFAAS [116] and ETAAS [117] more in particular.

Applications A limited number of papers refer to the use of AAS in relation to polymer/additive deformulation. Elemental analysis of polymers and rubbers by AAS may be carried out after dissolution in an organic solvent (Table 8.21), after oxidative wet digestion (Table 8.12), after dry ashing (Table 8.22) or directly in the solid state (Table 8.23).

Nagourney and Madan [20] have considered both AAS and ICP-AES as reliable measurement techniques for the determination of metal components in mixed-metal/phosphite stabiliser systems in PVC. For reasons given elsewhere (Section 8.3.2.4), in this case ICP-AES was considered the technique of choice for most metal stabiliser determinations, while AAS remains a useful method to corroborate the ICP-AES results. For the determination of tin in *rigid PVC* by means of HG-AAS, the main effort has been to develop a sample digestion procedure [118]. Tin and Ti from a PVC potable

Table 8.21 Analysis of polymers after dissolution in an organic solvent

Analyte(s)	Polymer	Solvent	Technique
Ca, Cu, Fe, K, Na, Si	PI	Cyclohexanone	GFAAS
		N-methylpyrrolidone	
Cd	PVC	DMSO-HNO$_3$	GFAAS
Cd, Pb	PVC	DMF	FAAS
Cd, Pb	PVC	Acetone and others	GFAAS
Sb	Various	DMF–HCl	HG-AAS
Sn	PVC	DMF	FAAS/GFAAS
Sn	PVC	DMF–Br$_2$	HG-AAS
Sn	PVC	THF	GFAAS
Zn	Rubber	Ethylacetonate	FAAS

Adapted from Welz and Sperling [52]. From B. Welz and M. Sperling, *Atomic Absorption Spectrometry*, Wiley-VCH, Weinheim (1999). Reproduced by permission of Wiley-VCH

Table 8.22 Analysis of rubbers by FAAS after dry ashing

Analyte	Ashing temperature	Treatment of the ash
Cr	600 °C	Solution in HNO₃
Cu	550 °C	Extraction with HCl, treatment with H₂SO₄—HF
Fe	550 °C	Heated with HCl
Mn	550 °C	Heated with HCl, treatment with H₂SO₄—HF
Pb	650 °C	Treatment with formic acid, solution in aqua regia
Zn	550 °C	Solution in H₂SO₄—HNO₃

Adapted from Welz and Sperling [52]. From B. Welz and M. Sperling, *Atomic Absorption Spectrometry*, Wiley-VCH, Weinheim (1999). Reproduced by permission of Wiley-VCH.

water pipe were analysed by GFAAS; Ca and Mg were analysed in an air–acetylene flame [119]. ETAAS was used for the determination of lead in PVC [120]. Braun and Richter [121] have used AAS, CE and complexometric EDTA titrations for the determination of heavy metals (Pb, Sn, Ba, Cd, Zn, Ca) in PVC stabilisers, such as Irgastab 17A and 17MOK-N after transfer to an aqueous phase. Complexometric titrations were not found to be highly accurate in the concentration range of interest; AAS and CE require standard solutions for quantitative determination, although they are more precise and faster than complexometry. Although the classic titration gives acceptable results for the metal content in the samples without large apparatus, CE and AAS have several advantages compared with chemical methods. FAAS has also been used to measure trace concentrations of Mg, Pb [122] and Ca, Al, Sb [123] in PVC samples containing a high concentration of

alkaline earths. Rabadán *et al.* [118,124] determined Sn in organotin stabilisers and in PVC, by HG-AAS.

Total lead and arsenic contents and other heavy elements in *food packaging*, toys and consumer goods were measured by direct determination of the colorant, using XRF, or by indirect determination using AAS [125]. When critical values were found, extraction conditions according to the revised Norm EN 73-3-1988 were used. Traces of Ni, Cu, Zn, Fe, Mn, Pb, Cd and Cr (0.03–0.15 ppm) in polymers were determined after ashing the polymer and examination of a nitric acid extract of the ash by AAS at specific wavelengths [31]. The method was calibrated against standard solutions of the heavy metals in nitric acid. Mercury in various food packaging materials was determined by closed microwave (high-pressure) ashing followed by CV-AAS; use of SnCl₂ as a reducing agent resulted in considerably better accuracy than use of sodium borohydride reagent [57].

Haslam *et al.* [32] reported the determination of Al in polyolefins by AAS. Typical AAS tests on *rubber compounds* involve several steps. The sample is combusted, and the resulting ash is dissolved in distilled de-ionised water. The solution is then used for AAS [126]. AAS or EDS can also be used for element analysis of filler particles. In order to determine the uniformity of tin compounds in polychloroprene after milling and pressing, Hornsby *et al.* [127] have ashed various pieces from one composition. After fusion of the residue with sodium peroxide and dissolution in HCl, the Sn content was determined by means of AAS. Typical industrial AAS measurements concern the determination of Ca in Ca stearate, Zn in Zn stearate, Ca- and Zn stearate in PE, Ca and Ti in PE film or Al and V in rubbers.

Table 8.23 Applications of direct solid sampling in ETAAS

Sample	Element	Atomiser[a]	Comments[b]	Reference
Fibre and plastic paper	Cu, Fe, Mn, Si	Graphite tube	–	[129]
Paper	Cd, Hg, Cu, Cr	Boat	–	[130]
	10 elements	CRA	–	[131]
Plastic	Pb	CRA	RSD = 15 %	[132]
Textile fibres	Cu, Cd, Mn, Rb	Cup-in-tube	RSD = 3.1–12 %	[133]
Plastic film, PVC	Cr, Pb, Cu	Cup-in-tube	RSD = 4–6 %	[134]
Polyester fibre	Au	Graphite tube	DL = 0.02 μg g⁻¹	[135]
Polyesters	Sb, Mn	Graphite cup	RSD = 6–14 %	[107]
Polyethylene	Cr, Cu, Ni, Pb	Boat	–	[136]
	Cd	Graphite tube, L'vov platform	RSD = 1.1–1.4 %	[137]
Polypropylene	Ti, Al, Mg	Graphite furnace	DL = 12 ng g⁻¹	[138]
Polymers	Cu, Cr, Fe	Graphite tube	–	[139]
PVC air filter	Be	CRA	–	[140]
PVC	Pb	Graphite tube	RSD = 1.4–10 %	[120]
ABS, PA6, PVC	Cd	Graphite boat	–	[141]

[a] CRA, carbon rod atomiser
[b] RSD, relative standard deviation; DL, detection limit.

Table 8.24 Comparison of methods; cadmium content in $\mu g\,g^{-1}$ (ppm)

Material	XRF	Colorimetry[a]	FAAS[b]	SS-ZAAS[c]
ABS (green)	$\gg$75	10 100	10 000	>10 000
ABS (blue)	$\gg$75	2550	2320	2050 ± 250
ABS (beige)	$\gg$75	670	680	810 ± 160
PA6 + GF (blue)	$\ll$75	<5	4	<5
PA6 + GF (green)	>75	310	300	300 ± 25
PA6 + GF (brown)	$\gg$75	690	700	800 ± 150
PVC, soft (blue)	$\gg$75	670	600	530 ± 50
PVC, soft (brown)	$\gg$75	690	600	580 ± 30

[a] After decomposition and dithizone method [143].
[b] After decomposition [143].
[c] Calibration with aqueous standards.
After Rühl [141]. Reproduced from W.J. Rühl, *Fresenius' Z. Anal. Chem.*, **322**, 710–712 (1985), by permission of Springer-Verlag, Copyright (1985).

Table 8.22 shows some rubber analyses by FAAS after dry ashing. The concentration of Rh in polymers was measured by FAAS [128]. The accuracy of 10–20 % was in agreement with a dissolution procedure; the precision obtained for direct solid analysis was between 10 and 20 %. Due to the relatively high analyte content of lead in paint, the determination is mostly performed by FAAS. Typical digestion procedures include dry ashing, wet and microwave digestion.

Table 8.23 collects together some typical ETAAS analyses of polymer formulations; see also ref. [141a]. GFAAS has also been applied for the determination of additive elements in lubricating oils [52]. Solid-sampling GFAAS and NAA are preferred analytical tools for the analysis of mg samples, also in relation to RM production.

Within the framework of the PERM project [1], a microhomogeneity study of the doped material in two concentration ranges was carried out by SS-ZAAS on three pilot elements (Pb, Cd and Hg), to investigate the dissolution of the dopants in the base material. SS-ZAAS is very suitable for the determination of *microhomogeneity*, as the minimum sample mass that can be tested is that of about one pellet. Cadmium was selected, as this element can be compared to the results of the CRM 'Cd in PE'; Hg and Pb were selected because of their ease of determination by SS-ZAAS. For each of the three elements, 20 analyses were carried out on material from five different batches. From these data, homogeneity and a minimum sample mass were determined. Earlier, Janssen *et al.* [136] had reported the determination of Ni, Cu, Pb and Cr in PE by means of direct ZAAS. These materials were subsequently used as synthetic reference materials for QC (stabiliser and pigment contents). It was recently pointed out that LA-ICP-MS using a UV laser is a more suitable tool for studying heterogeneity of polymers than is SS-ZAAS [142].

Table 8.24 compares SS-ZGFAAS results for product control of Cd in automotive polymers with conventional methods [141]. SS-ZAAS was also used for QC purposes to determine Cu, Mn, Fe, Cd and Pb in adhesive tapes [144].

As to the analysis of trace elements in *paper, cardboard* and raw materials for the production of paper, high concentration elements such as Cu, Fe or Ti can easily be determined by FAAS; Cd and Pb are frequently analysed by GFAAS. Cadmium in pulp and paper was determined by AAS after pressurised digestion with nitric acid [145]. An interlaboratory comparison of Cd in wrapping paper was reported, mainly based on pressure digestion in PTFE bombs with sub-boiled nitric acid, followed by ETAAS [59]. For wrapping paper used for foodstuffs, next to the total content of toxic heavy metals, the soluble or leachable fraction is of particular interest.

Synthetic fibres can contain residues of catalysts or stabilisers, flame retardants, or antibacterials (e.g. organotin compounds) [146]. The sensitivity of FAAS is mostly adequate for the determination of trace elements in *textile fibres*. Völlkopf *et al.* [133] used direct solids sampling to determine Cd, Cu, Mn and Rb in textile fibres such as polyester and polyamide, by GFAAS. Traces of As in acrylic fibres containing Sb_2O_3-based fire retardants were analysed by FAAS at 193.7 nm after acid digestion [54]. The procedure was calibrated against standards of free acrylic fibre and pure As_2O_3.

Atomic absorption spectrometry in general leads to some 1000 papers yearly.

8.3.2 Atomic Emission Spectrometry

Principles and Characteristics Under appropriate conditions, all elements can be made to emit characteristic spectra. Unfortunately, no single excitation source excites all elements in an optimal way. The *ideal emission source* should be both an ideal atomiser and an ideal excitation source, and should also allow control of the excitation energy by the user, in order to optimise the source conditions for the determination of elements with both a low and a high excitation energy. In practice, the detection power of optical spectroscopy methods in which volatilisation, atomisation and excitation are made by the same source is a compromise between the requirements of atomisation (high energy) and excitation (low energy). Using lasers, these steps may be optimised separately (laser ablation/laser spectroscopy). For minimisation of background emission and other interferences, atomisation and excitation should occur in an inert chemical environment. The source should further provide reproducible atomisation and excitation conditions, in order to obtain precise and accurate single-element and multi-element results. Plasma atomic emission sources – inductively coupled plasma (ICP), direct current plasma (DCP) and microwave-induced plasma (MIP) – have led to a revival of great importance, especially for trace-element determination, and closely approach the ideal emission source, quite contrary to techniques using classical sources (flame, arc, and spark), which have a rather limited field of application. ICP and DCP equipment has seen the most widespread analytical applications; microwave plasmas have not caught on to the same extent.

There are two kinds of discharge commonly used – the pulsed radiofrequency (r.f.) spark and the triggered, low-voltage arc discharge. As most instruments use the former, it follows that most research has focused on the r.f. spark source, in which a pulsed 1–MHz r.f. voltage of up to 100 kV is fed to the electrodes in vacuum. *Arc ablation* has long been used for production of an aerosol at the surface of electrically conducting samples. Electrical arc sources for atomic emission spectrometry (AES) use currents between 5 and 30 A and burning voltages of 20–60 V between the electrodes, for an arc gap of a few millimetres. Arc emission requires a somewhat lengthy sample preparation, but the analysis time is quite short. With the direct current (d.c.) arc, 70 to 80 elements can be excited; because of its limited precision and accuracy the major use of the arc is in qualitative and semiquantitative analysis. The *spark* allows higher precision and accuracy, and is applied mainly in quantitative determinations, especially for rapid multi-element analysis of solid samples, as required for process control, e.g. in the steel industry. When sparks are used, the ablation of electrically conducting solids is less dependent on matrix composition. This applies both to high-voltage sparks (up to 10 kV) and to medium-voltage sparks, as used

for emission spectrometric analysis of metals. Arcs and sparks, however, are non-ideal sources, because of complex spectra with significant background, severe matrix effects, lack of reproducibility, etc. Not surprisingly, therefore, other excitation sources, such as the laser microprobe and reduced pressure discharge (e.g. glow discharge), have been developed. Other new excitation principles in atomic emission spectrometry (AES) are high-frequency (ICP) or microwave plasmas (MIP). The detection limits of ICP-AES and MIP-AES lie between those of FAAS and GFAAS.

Plasma sources were developed for emission spectrometric analysis in the late-1960s. Commercial inductively coupled and d.c. plasma spectrometers were introduced in the mid-1970s. By comparison with AAS, *atomic plasma emission spectroscopy* (APES) can achieve simultaneous multi-element measurement, while maintaining a wide dynamic measurement range and high sensitivities and selectivities over background elements. As a result of the wide variety of radiation sources, optical atomic emission spectrometry is very suitable for multi-element trace determinations. With several techniques, absolute detection limits are below the ng level.

The main detectors used in AES today are photomultiplier tubes (PMTs), photodiode arrays (PDAs), charge-coupled devices (CCDs), and vidicons, image dissectors, and charge-injection detectors (CIDs). An innovative CCD detector for AES has been described [147]. New developments are the array detector AES. With modern multichannel échelle spectral analysers it is possible to analyse any luminous event (flash, spark, laser-induced plasma, discharge) instantly. Considering the complexity of emission spectra, the importance of spectral resolution cannot be overemphasised. Table 8.25 shows some typical spectral emission lines of some common elements. Atomic plasma emission sources can act as chromatographic detectors, e.g. GC-AED (see Chapter 4).

Table 8.25 Principal spectral emission lines, minimum detectable limits, and linear dynamic range of some commonly monitored elements in GC-AED

Element	Emission wavelength (nm)	Minimum detectable level (pg/s)[a]	Linear dynamic range
Carbon	193	1	2×10^4
Nitrogen	174	50	2×10^4
Sulfur	181	2	1×10^4
Oxygen	777	120	5×10^3
Chlorine	479	40	1×10^4

[a] (S/N) = 2.

After Sullivan [148]. Reproduced by permission of Hewlett Packard.

Applications Atomic emission spectrometry has been used for polymer/additive analysis in various forms, such as flame emission spectrometry (Section 8.3.2.1), spark source spectrometry (Section 8.3.2.2), GD-AES (Section 8.3.2.3), ICP-AES (Section 8.3.2.4), MIP-AES (Section 8.3.2.6) and LIBS. Only ICP-AES applications are significant. In hyphenated form, the use of element-specific detectors in GC-AED (Section 4.2) and PyGC-AED deserves mentioning.

8.3.2.1 Flame Emission Photometry

Principles and Characteristics Flame emission instruments are similar to flame absorption instruments, except that the flame is the excitation source. Many modern instruments are adaptable for either emission or absorption measurements. Graphite furnaces are in use as excitation sources for AES, giving rise to a technique called electrothermal atomisation atomic emission spectrometry (ETA AES) or graphite furnace atomic emission spectrometry (GFAES). In flame emission spectrometry, the same kind of interferences are encountered as in atomic absorption methods. As flame emission spectra are simple, interferences between overlapping lines occur only occasionally.

Flame emission photometry is used mainly for the determination of alkali metals and some easily excited elements (Na, K, Li, Ca, etc.). This is related to the fact that the number of excited atoms in the flame decreases exponentially with increasing excitation energy. Moreover, at variance to AAS, where the sensitivity is directly proportional to the number of atoms in the *ground* state, the sensitivity of AES increases with an increasing number of atoms in the *excited* state.

Applications The use of flame photometry (AES) in the analysis of plastics materials is very limited. A sodium neutralising aid (0.02 %) in HDPE and PP has been determined by flame photometry of a nitric acid solution of dope ashed polymer residues and ash blending with carbon containing 0.1 % Pd [31]. Calibration was achieved against known blends of sodium carbonate in $MgSO_4$. In the particular case, flame photometry after ashing at 500 °C in the presence of an ashing aid gives quantitative recovery, as shown by comparison with NAA (nondestructive method). Lithium (0.2 ppm) in polyalkenes and copolymers has been determined by ashing of the polymer and analysis of the ash in dilute nitric acid solution by flame photometry at 670.9 nm [31].

8.3.2.2 Arc and Spark Emission Spectrometry

Principles and Characteristics Arc and spark discharges have widely been used as excitation sources for qualitative and quantitative emission spectrometry since the 1920s; commercial instruments became available during the 1940s.

Spark-source spectrometry is an analytical technique for the analysis of electrically conducting materials, where the sample is one of the electrodes. If the two electrodes are in contact with the surface of an insulating material, a high-frequency spark is generated on the surface. The conductivity of the plasma channel becomes highest near the surface of the nonconducting material and solid material is vaporised. The radiation contains emission lines of the insulating material. In spark discharge [149], sample atomisation takes place in localised overheated spots hit by the discharge, where the sample surface is locally melted. Unlike glow discharge, the composition of the flux of atoms released into the discharge is *different* from the composition of the sample. Therefore, analysis of different matrices with a single calibration is not possible with spark sources. If the sample is structurally inhomogeneous, atomisation by spark becomes even more complex. Spark and arc discharges exhibit a high sensitivity, large applicability and only a few interferences, but they do not yield a stable ion population. Glow discharge is such a stable, low-energy ion source [150]. There are fundamental differences in the operation of the two types of discharge, spark source and Grimm-type glow discharge – both important spectral sources for analysis of solid conductive matrices by optical emission spectrometry, as well as in their excitation and noise characteristics [151]. However, from the point of view of the applications to bulk elemental analysis, it is sample atomisation which makes the essential difference between both sources.

Spark sources are especially important for *metal analysis*. To date, medium-voltage sparks (0.5–1 kV) often at high frequencies (1 kHz and more), are used under an argon atmosphere. Spark analyses can be performed in less than 30 s. For accurate analyses, extensive sets of calibration samples must be used, and mathematical procedures may be helpful so as to perform corrections for matrix interferences. In arc and spark emission spectrometry, the spectral lines used are situated in the UV (180–380 nm), VIS (380–550 nm) and VUV (<180 nm) regions. Atomic emission spectrometry with spark excitation is a standard method for production and product control in the metal industry.

Golloch *et al.* [152,153] first described the use of the *sliding spark source* (SSS) for the analytical detection of metals in polymers. Intense optical emission is observed when positionally stable high-current surface sparks supplied by a pulse-generator with definite discharge parameters (max. 800 A per pulse; electrode material Cu, Ag) are sliding over compact nonconductive materials such as plastics. Substrate thermal vaporisation, ionisation and excitation processes in the surface discharge plasma induce emission corresponding to the neutral and ionic states. The emission spectra (185–510 nm, spectral resolution <1.5 nm), obtained with fibre optics linked to a diode array detector, gave good results for the determination of metals in PVC. In a further development, a sliding spark source, which generates discharges using up to 2 kJ of stored energy, and a CCD spectrometer, have been used for ablation and excitation of elements with sufficient intensities even when the polymer substrate does not contain chlorine [154]. The method allows direct *in situ* analysis of nonconductive material without prior sample preparation. In sliding spark technology, the detection time is about one second, and the system can recognise plastics irrespective of their pigmentation, working with a wide range of thermoplastics, chlorinated and brominated polymers, heavy-metal additives, fillers and glass-fibre reinforcements [154a].

Applications Spark-source atomic emission and mass spectrometry have been used for routine analysis of solids, particularly for quality assurance and comparative work. As with GD-MS, spark sources are restricted to samples that are, to some extent, electrically conducting, or that can be made conducting by grinding to a powder and mixing with a high-purity conducting binder.

The only really noteworthy application stems from the recent development of sliding spark spectrometry with emission detection (SSS-AES). Sliding spark spectrometry, which is similar to laser-PES, has been used in *polymer waste sorting* for identification of PVC and environmentally undesirable additives [155]. SSS-AES provides a rapid simultaneous multi-element identification system for PVC, Cl-containing waste plastics and flame-retardant thermoplastics within about one second, due to the occurrence of emission lines of reactive fillers, inorganic pigments, and the increased volatility in the presence of chlorine [152,153]. Examples are PVC (ca. 55 wt %), chlorinated PE (15–20 wt %) and a FR-ABS sample containing 10 wt % of flame retardant with unknown chlorine concentration [153,154] Detection of several Cl emission lines without line interferences allows the easy identification of PVC. A limit of detection of 0.5 wt % was obtained for quantitative evaluation of Cl. SSS-AES was also used to detect metals in non-chlorine-containing polymers [154]. Figure 8.4 shows part of the sliding spark emission spectra of an ABS sample containing Cd and Zn.

The technique has also been used for the qualitative analysis of polymer samples containing pigments or fillers. PE and PP samples containing TiO_2 (rutile), ZnS, $BaSO_4$, $CaCO_3$, $PbSO_4$, talc ($H_2Mg_3Si_4O_{12}$) and wollastonite ($CaSiO_3$) were successfully analysed with SSS-AES [154]. Sliding spark spectrometry allows rapid survey analysis. Detection of metal containing additives,

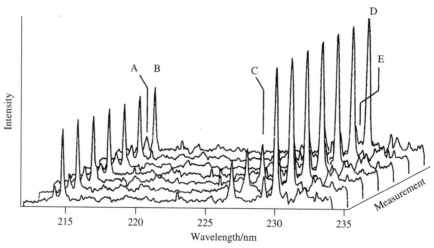

Figure 8.4 Reproducibility of sliding spark emission spectra taken at different sites on the surface of an ABS sample containing Cd and Zn. A: Zn I 213.86 nm; B: Cd II 214.44 nm; C: Cd II 226.50 nm; D: C III 229.68 nm; E: Cd I 228.80 nm. After Golloch and Siegmund [154]. Reproduced from A. Golloch and D. Siegmund, *Fresenius' Z. Anal. Chem.*, **358**, 804–811 (1997), by permission of Springer-Verlag, Copyright (1997)

halogens and flame retardants in polymers by means of SSS is an attractive alternative to other direct solid sampling methods. Limits of detection are about 0.1 wt % for elements if chlorine-free polymers can be obtained. Application of a CCD spectrometer enables fast detection of various emission lines, including background subtraction and use of an internal standard. Quantification is possible if the matrix is known.

In order to validate sliding spark spectrometry results, plastic material was collected and the element concentration was determined via AAS after digestion. The samples were used as calibration standards. Additional standards were obtained by manufacturing known amounts of additives in the polymer matrix. Calibrations were made for Cd, Cr, Pb, Zn, Sb, Si and Ti in chlorine-free polymers; Al, Ba, Ca, Cd, Pb, Sn, Ti, Zn in PVC; chlorine (as PVC); and bromine in polyurethane (PUR). A calibration curve for Br as a flame retardant in PUR is shown in Figure 8.5.

8.3.2.3 Glow-discharge Spectrometry

Principles and Characteristics Glow discharge (GD) is a general term for fairly low-power electrical discharges in reduced-pressure systems (e.g. fluorescent tubes and neon lights). A glow discharge stands for a thin plasma of low power density, where ionised and energetically excited atoms make up a very small fraction of the total atom number density. It is formed when a cell, consisting of a grounded anode and a cathode, is filled with a gas (typically argon) at low pressure (0.1–10 mbar). A potential difference (of the order of 1 kV) is then applied between the electrodes, which creates gas breakdown (i.e. splitting of the gas into positive ions and electrons). The Ar^+ ions are accelerated towards the cathode (the sample), causing emission of electrons upon bombardment. The ions collisionally sputter neutral atoms from the sample surface into the adjacent GD plasma. As these sputtered atoms diffuse through the GD plasma, the collisions with electrons, ions and metastable atoms excite and ionise the sample atoms. Excitation collisions (and the subsequent decay, with emission of light characteristic of the species) are responsible for the 'glow' discharge, which is essentially a self-sustaining plasma. The electrodes of the cell can be mounted in different geometries (coaxial cathode, planar diode, hollow cathode lamp, hollow cathode plume, Grimm or Marcus source) [156].

The main parameters for monitoring and controlling the source are current, potential and pressure. Glow discharge can be operated in three different electrical operation modes, with either a direct current (d.c.), radiofrequency (r.f.) powered or pulsed-glow discharge system. Most GD devices are operated in the continuous mode, at a power of a few watts. The glow discharge can also be 'boosted', by combining it either with a laser, a graphite furnace, a microwave discharge, magnetic fields or external gas jets, in order to improve the analytical results. In *direct-current (d.c.) glow discharge*, which is the most common type, the electrodes must be electrically conducting. In order to initiate the discharge, the voltage across the electrodes must exceed a minimum value (500–1100 V). The d.c. analysis is limited to conductive or semiconductive samples, to very thin nonconductive layers ($<2 \mu m$), and to flat samples larger than 20 mm diameter. R.f. GD-OES has eliminated these limitations. The major advantage of glow discharge devices energised by a *radiofrequency* (r.f.) alternating electric field (MHz; delivered power 10–100 W) is their ability to sputter nonconducting materials. The major fields of applications of r.f. glow discharges to date have been in bulk analysis, but there is an increasing interest in depth profile analysis of nonconducting surface layers such as paints. *Pulsed GD systems*, which operate in short repetitive pulses ($10–10^3$ Hz), can be run at high peak power (typically 20 kW), even if just for microseconds. Pulsed operation of GD allows more experimental control parameters; operates with separate atomisation and excitation/ionisation steps; achieves enhanced photon and ion yield, and allows time-resolved

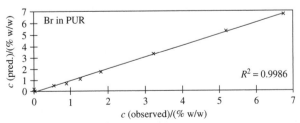

Figure 8.5 Sliding spark emission calibration curve of Br in PUR. After Golloch and Siegmund [154]. Reproduced from A. Golloch and D. Siegmund, *Fresenius' Z. Anal. Chem.*, **358**, 804–811 (1997), by permission of Springer-Verlag, Copyright (1997)

spectroscopy. Both pulsed d.c. and r.f. discharges have been described. Pulsed discharges offer arguably the greatest reward for MS experiments, particularly ToF-MS. For the differences between r.f.- and d.c.-glow discharge, see ref. [156a].

Glow discharge is essentially a simple and efficient way to generate atoms. Long known for its ability to convert solid samples into gas-phase atoms, GD techniques provide ground-state atoms for atomic absorption or atomic fluorescence, excited-state atoms for atomic emission, and ionised atoms for MS [158]. Commercial instrumentation has been developed for all these methods, except for GD-AFS and pulsed mode GD.

The use of GD for optical emission is based on sputtering. The material to be analysed serves as the cathode of the glow discharge. In GD-OES, sputtering and atomic emission are combined. GD-OES uses the low-pressure, hollow-anode glow discharge (Grimm source), combined with one or more optical spectrometers (polychromator or monochromator). Usable emission lines are available for all elements (including hydrogen). Payling and Larkins [159] have published 961 000 optical emission lines for 88 elements. Combined d.c./r.f. GD-OES instruments have been developed. The r.f. GD-OES sources have demonstrated desirable figures of merit (Table 8.26), typically yielding one or two orders of magnitude better detection limits for conducting sample matrices than the commercial spark source. GD-OES can analyse practically any solid material which is vacuum-compatible. In most work on depth profile analysis by GC-OES, the GD sources used are based on the design of Grimm [160], which requires flat samples.

Table 8.27 shows the *main characteristics* of r.f. GD-OES, which is a valuable technique for the direct analysis of solids. Related techniques are GD-AAS, GD-AFS, GD-MS, arc-OES, spark-OES, FANES, ICP-OES and LA-OES. Glow discharge is considered as a growth area, although the true analytical potential of glow-discharge spectrometry (GDS) remains largely

Table 8.27 Main features of r.f. GD-OES

Advantages

- Relatively inexpensive; low operating costs
- Easy to operate, versatile
- Little specimen preparation; large samples (20 cm)
- Direct, rapid spectrochemical analysis (relatively simple spectra)
- No ultrahigh vacuum
- High sample throughputs, routine analysis
- Exceptional figures of merit
- Detectability: 10^{14} at. cm^{-2}
- All elements (survey analysis)
- Direct bulk trace analysis of both conducting and nonconducting sample materials (more problematic)
- Matrix-independent emission yields
- Surface and quantitative depth profile (QDP) analysis of multimatrix materials (nm to $100 \, \mu m$)
- Ability to sputter nonconducting materials (not for d.c.-GD)
- Relatively high sputtering rate
- Temporal resolution (pulsed GD)
- Imaging capabilities
- Mature technique

Disadvantages

- Weak excitation/ionisation source
- Solids-only technique
- Problems with complex matrices
- No chemical bonding information
- Regular-shaped large area samples (Grimm sources)
- Vacuum compatibility
- Needs new developments
- No general agreement on best quantitative schemes

unexplored. Table 8.28 compares GDS to some competing techniques. An intrinsic phenomenon of glow discharge, that distinguishes it from most other spectroscopic sources, is a natural separation of the initial sampling step (sputter atomisation) from the subsequent analytical processes (e.g. excitation and ionisation). Consequently, matrix and interelement effects are small or nondetectable. Quantitative analysis can thus be achieved against a few or even one external standard that does not necessarily have the same (or a very similar) matrix composition. The approximation of matrix-independent emission yields in GD-OES is the basis for quantitation.

GDS instruments are viable alternatives to the traditional arc and spark-source spectroscopies for bulk metals analysis. Advantages of GDS over surface analysis methods such as AES, XPS and SIMS are that an ultrahigh vacuum is not needed and the sputtering rate is relatively high. In surface analysis, GD-OES, AES, XPS and SIMS will remain complementary techniques. GD-OES analysis is faster than AES (typically 10 s vs. 15 min). GD-OES is also 100 times more sensitive than

Table 8.26 Analytical figures of merit of GD-OES

Figures of merit	Value
Lower limit of detection (ppm)	1–100
Minimum detectable number of atoms (atoms cm^{-2})	$10^{13}–10^{15}$
Minimum information depth	1 nm
Relative instrumental depth resolution	<10 %
Penetration rate	1–100 nm s^{-1}
Short-term precision (major and minor elements)	<1 % RSD

After Bengtson [157]. From A. Bengtson, in *Surface Characterization. A User's Sourcebook* (D. Brune *et al.*, eds), Wiley-VCH, Weinheim (1997), Chapter 14. Reproduced by permission of Wiley-VCH.

Table 8.28 Competing techniques

Technique	Principal advantage	Principal disadvantage
• **Direct solid bulk analysis**		
Arc/spark OES	Simple	Nonlinear
XRF	Well developed	Light elements
• **Depth profiling**		
SIMS	High sensitivity	Quantification
AES/XPS	True surface	Slow
• **Combined bulk analysis/depth profiling**		
GD-MS	Sensitivity	Slow
SNMS	Quantification	Slow

After Bengtson *et al.* [161]. Reprinted from A. Bengtson *et al.*, in *Glow Discharge Optical Emission Spectrometry* (R. Payling *et al.*, eds), pp. 3–11. Copyright © 1997 John Wiley & Sons, Limited. Reproduced with permission.

AES and can detect hydrogen, which AES cannot do. On the other hand, AES can provide some chemical bonding information.

Several recent reviews deal with GD-OES [157,162, 163] and pulsed glow discharge [164,165]. Several books on GDS have appeared [158,166,167].

Applications Although GD-OES is most popular in metals research, it can also be applied to prepainted steel, coatings and polymers. Examples of analysis of coatings are:

(i) on metal sheets: surface treatment (anticorrosion);
(ii) on glass: thin layers (>1 nm);
(iii) on organic materials; and
(iv) in paints: identification of pigments.

GD-OES is helpful in identifying surface quality problems in both the automotive and aerospace industries.

Baudoin *et al.* [168,169] first presented qualitative *depth profiles* of lacquer and polymer coatings by means of r.f. GD-OES. Quantitative depth profiles were successively obtained by Payling *et al.* [170] on prepainted metal coated steel. Samples comprised a (rutile) pigmented silicone-modified polyester topcoat over a polymer primer, on top of an aluminium–zinc–silicon alloy coated steel substrate. With GD-OES in r.f. mode, it was possible to determine the depth profile through the polymer topcoat, polymer primer coat, metal alloy coating, and alloy layer binding to the steel substrate with a total depth of 50 μm, all in about 60 min on the one sample. GD-OES depth profiles of unexposed and weathered silicone-modified polyesters were also reported [171]. Radiofrequency GD-OES has further been used to determine mixed TiO_2 and Fe_2O_3 *pigment loadings* in polyester/melamine–formaldehyde polymers [172]. The elemental compositions of the pigmented polymer coatings were derived (RSD ~5 %). During sputtering, some loss of polymer may occur; very soft polymers may be problematic. Qualitative depth profiles of pigmented polyester coatings on galvanised steel, and a quantitative depth profile of titanium in one of the coatings, have been presented [173]. Use of GDS has also been reported for tyre and surface analysis [174]. GD-OES does not rank amongst the most wanted tools for polymer/additive analysis.

8.3.2.4 Inductively Coupled Plasma–Atomic Emission Spectrometry

Principles and Characteristics A plasma, which is an electrically neutral, highly ionised gas composed of ions, electrons and neutral particles, has significantly higher gas temperatures and less-reactive chemical environments than flames. The plasma gas is a stream of inert gas, usually argon, at atmospheric pressure, which is energised with high-frequency electromagnetic fields (r.f. or microwave energy), or with direct current. The plasma is maintained inside an arrangement of three open concentric quartz tubes (the so-called plasma torch). The plasma torches used today in both ICP-AES and ICP-MS, are almost exclusively based on the original Fassel design [175,176].

In ICP-AES and ICP-MS, sample mineralisation is the Achilles' heel. *Sample introduction systems* for ICP-AES are numerous: gas-phase introduction, pneumatic nebulisation (PN), direct-injection nebulisation (DIN), thermal spray, ultrasonic nebulisation (USN), electrothermal vaporisation (ETV) (furnace, cup, filament), hydride generation, electroerosion, laser ablation and direct sample insertion. Atomisation is an essential process in many fields where a dispersion of liquid particles in a gas is required. Pneumatic nebulisation is most commonly used in conjunction with a spray chamber that serves as a droplet separator, allowing droplets with average diameters of typically <10 μm to pass and enter the ICP. Spray chambers, which reduce solvent load and deal with coarse aerosols, should be as small as possible (micro-nebulisation [177]). Direct injection in the plasma torch is feasible [178]. Ultrasonic atomisers are designed to specifically operate from a vibrational energy source [179].

The sample aerosol is introduced into the centre of the plasma, where it is exposed, for a few milliseconds, to the harsh conditions of the discharge, where discharge temperatures of 5000–9000 K are observed. This leads to complete desolvation, vaporisation and dissociation of all sample material introduced, as well as to ionisation and UV/VIS emission phenomena. In combination with a high-quality spectrometer, plasma sources approach the characteristics of the ideal atomic emission source. The emission line intensities are proportional to the concentration of each element in the sample.

The use of an ICP as a spectroscopic excitation source [180,181] has been beneficial to the field of inorganic elemental analysis [175,176]. Boumans [182] and Montaser and Golightly [183] have described the fundamental properties of ICPs. A significant number of papers has been published concerning the fundamental properties and characteristics of the ICP, making this atmospheric discharge probably the most extensively researched excitation and ionisation source. All these studies show that the structure and excitation and ionisation mechanisms occurring in the ICP are exceedingly complex. Atoms formed in an ICP can be ionised and excited by various mechanisms, including thermal excitation/ionisation, Penning and charge transfer ionisation/excitation. In this area, there is a considerable gap in knowledge, in particular regarding the influence of operating parameters such as gas flow-rates and r.f. power on excitation.

Instrumental *ICP-AES options* are characterised by simultaneous measurement sequences (speed, robustness) or sequential arrangements (resolution, flexibility, wavelength range) with radial, axial or dual plasma viewing geometries (Figure 8.6). Advantages of the traditional radial plasma are robustness (few matrix effects, less fouling), low detection limits (ppm) and cost. Axial viewing increases the pathlength; axial plasmas are suitable for trace analysis by enhancing the sensitivity by a factor of five to 10, but at a cost of increased

interferences. Typical detection limits in axial configuration are $0.01-1.0\,\mu g\,L^{-1}$, as opposed to $1-10\,\mu g\,L^{-1}$ in radial configuration. Axial detection limits are comparable to GFAAS.

ICP-AES has first developed from the use of a photographic plate to single-channel detectors such as a (circular) photomultiplier tube (PMT), with considerable loss of information. By combining an échelle spectrometer with full-frame 2D CCD and CID detectors and association of linear CCD arrays in commercial ICP-AES systems, fast acquisition of the entire UV/VIS spectra is now possible. The use of CCD/CIDs requires extensive data processing (chemometrics), but marks essentially a renaissance of photographic plate ICP-AES. Cutting-edge *technology* for CCD simultaneous ICP-AES consists of échelle optics and special (custom-designed) chip detectors (as digital photographic plates), providing complete flexibility in (multiple) wavelength selection (167–785 nm) to avoid spectral interferences; a built-in wavelength library; simultaneous background measurement; increased linear dynamic range (ppb to %); high resolution (FWHM 0.007 nm); low detection limits (sub-ppb); and high speed (10 seconds' integration time for all elements). The fastest ICP-AES measures 73 elements in just 35 s [147]. Scanning CCD revolutionises ICP by combining the flexibility of traditional sequential ICP systems with the performance and reliability of simultaneous systems, and allows collection of any element in any order, one at a time. The benefits of the availability of entire spectra are:

(i) improved flexibility in analytical line selection (avoiding spectral interferences);
(ii) selection of lines with the same ICP operating parameter optimisation;

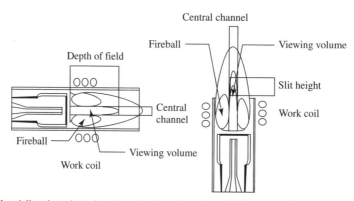

Figure 8.6 Horizontal axially viewed torch and vertically oriented, radially viewed plasma configuration. Reproduced by permission of Varian, Inc.

(iii) use of several lines with various sensitivities to extend the dynamic range;

(iv) use of several atomic and ionic lines to verify possible matrix effects;

(v) qualitative and semiquantitative analysis;

(vi) confirmation of results at alternative wavelengths (including far-UV for halogens); and

(vii) fast diagnostics [184].

Wavelength database libraries of >32 000 analytical lines can be used for fast screening of the échellogram. Such databases allow the analyst to choose the best line(s) for minimum interferences, maximum sensitivity and best dynamic range. Further extension of the wavelength range (from 120 to 785 nm) is desirable for alkali metals, Cl, Br, Ga, Ge, In, B, Bi, Pb and Sn, and would allow measurement of several emission lines in a multivariate approach to spectral interpretation [185].

For ICP system control, daily testing of four elements (Ba, Mg, Zn and Ar) has been proposed [186]. The concentration of each element is determined from measured intensities via calibration with standards. For standardisation, use is made of reference solutions. Calibration of ICP-AES has been critically evaluated [187].

ICP-AES and ICP-MS analyses are hampered in almost all cases by the occurrence of sample *matrix effects*. The origins of these effects are manifold, and have been traced partly to physical and chemical aerosol modifications inside sample introduction components (nebulisation effects). Matrix effects in ICP-AES may also be attributed to effects in the plasma, resulting from easily ionised elements and spectral background interferences (most important source of systematic errors). Atomic lines are usually more sensitive to matrix effects than are ionic lines. There exist several options to overcome matrix interferences in multi-element analysis by means of ICP-AES/MS, namely:

(i) matrix matching (most effective, but time-consuming);

(ii) separation of analyte from matrix;

(iii) internal standardisation (requires several internal standards; compromises multi-element analysis);

(iv) standard additions (spiking; requires addition of several elements at once; can result in additional interferences); and

(v) better understanding of plasma features (best option).

Cyclonic spray chambers greatly reduce matrix effects.

Table 8.29 shows the *main characteristics* of ICP-AES as a fast multi-element technique. Analytical figures of merit for ICP emission spectrometers are

Table 8.29 Main characteristics of ICP-AES

Advantages

- Liquid sampling, small sample size (few mL); solids (after dissolution, laser ablation)
- Simultaneous or sequential multi-elemental determination capability (up to 100 ppm)
- Flexibility (all wavelengths)
- Atomic and ionic emission lines
- Low detection limits for most metallic elements and some nonmetals (sub-ppb–100 ppb; tens of pg to ng)
- Wide linear dynamic range (four to five decades)
- Almost complete elemental coverage (>70 metallic/nonmetallic elements); more limited for simultaneous ICP-AES
- Accuracy (10 % or better with simple standards; 0.5 % with appropriate techniques)
- Precision (0.2–0.5 %)
- Analytical speed, high sample throughput, robust
- Bulk analysis (unless LA)
- High tolerance to matrix effects
- Allowance for transient signals (e.g. ETV)
- Semiquantitative scanning; quantitation (standards)
- Linear calibration (over approximately five decades)
- Compact (in case of CCD-ICP-AES)
- Relative ease of operation
- Mature technology

Disadvantages

- Complex spectra (spectroscopic interferences)
- Less suitable for solid sampling
- Destructive
- Matrix effects (correctable)
- Memory effects (cross-contamination: nebuliser)
- Torch devitrification
- Expensive equipment and use
- High argon consumption

Table 8.30 Analytical speeds

Technique	Number of elements	Analysis time per sample (min)
Conventional FAAS	8	4.7
Conventional ICP-AES	8	2.6
Fast sequential FAAS	8	2.0
Simultaneous ICP-AES	8	1.1

Reproduced by permission of Varian, Inc.

available [188]. Table 8.30 compares typical analytical speeds for various related techniques.

Conventional ICP-AES has similar detection limits to FAAS (although inferior to those of ICP-MS) and is much faster when many elements are determined in the same sample. The detection limits of modern, fast ICP-AES are equal to those of conventional, slow GFAAS. Table 8.31 compares the detection power of various atomic emission spectrometries. The greater

Table 8.31 Power of detection of optical atomic emission spectrometric methods

Method	Detection limits	
	Relative	Absolute
Flame AES	$1\,ng\,mL^{-1}$ (alkali)	–
DC arc AES	$0.1-2\,\mu g\,g^{-1}$	$1-20\,ng$
Spark AES	$1-5\,\mu g\,g^{-1}$	$10-100\,ng$
GD-AES	$1-5\,\mu g\,g^{-1}$	$1-5\,ng$
Laser AES	$1-10\,\mu g\,g^{-1}$	$1-10\,pg$
ICP-AES	$0.01-5\,ng\,mL^{-1}$	$0.1-10\,ng$ (ETV)
DCP-AES	$1-20\,ng\,mL^{-1}$	–
MIP-AES	$2-40\,ng\,mL^{-1}$	$0.05-0.5\,ng$

linear dynamic range of ICP-AES compared with AAS is also extremely important if samples with widely varying concentrations are analysed. ICP-AES can analyse some nonmetallic elements (P, S) which are difficult for AAS. However, some volatile elements (B, I, Os, Hg) are problematic in ICP. Refractory elements can be determined with good sensitivity by ICP-AES, because of the high temperatures reached in the plasma. ICP-AES is more expensive than FAAS (e.g. because of the complex monochromator needed).

Today, ICP-AES is an indispensable inorganic analytical tool. However, because of the high plasma temperature, ICP-AES suffers from some severe spectral interferences caused by line-rich spectra of concomitant matrix elements such as Fe, Al, Ca, Ni, V, Mo and the rare-earth elements. This is at variance with AAS. The spectral interference can of course be minimised by using a (costly) high-resolution spectrometer. On the other hand, the high temperature of the ICP has the advantage of reducing chemical interferences, which can be a problem in AAS.

ICP-based analysis techniques and spectrometer technology were reviewed recently [147,189–191]. An up-to-date monograph on ICP-AES is available [192]. A useful Internet site for the ICP community is *http://listserv.syr.edu/archives/plasmachem-l.html*.

Applications Over the last 20 years, ICP-AES has become a widely used elemental analysis tool in many laboratories, which is also used to identify/quantify emulsifiers, contaminants, catalyst residues and other inorganic additives. Although ICP-AES is an accepted method for elemental analysis of *lubricating oils* (ASTM D 4951), often, unreliable results with errors of up to 20 % were observed. It was found that viscosity modifier (VM) polymers interfere with aerosol formation, a critical step in the ICP analysis, thus affecting the sample delivery to the plasma torch [193]. Modifications

of the ASTM D 4951 ICP-AES procedure were developed. DIN 38 406 describes the determination of 24 elements by means of ICP-AES [194].

Nagourney and Madan [20] have discussed methods for the atomic spectroscopic determination of Ba, Cd, Zn and P in *vinyl stabilisers*, in particular ICP-AES. In-house products were used as preferred reference materials. Both AAS and ICP-AES are reliable measurement techniques for this problem. For metal levels usually exceeding 0.5 %, AAS determinations require multiple dilutions to achieve a final concentration within the narrow linear working range. Barium determinations by AAS require chemical addition to minimise ionisation interferences. Phosphorous determinations by AAS are impractical, owing to the insensitivity in a flame and the time-consuming nature of electrothermal AAS determinations. The wider linearity, increased phosphorous sensitivity, and sequential multi-element capability of ICP-AES makes it the technique of choice for most metal stabiliser determinations; AAS can be used to verify the ICP-AES results. This conclusion applies to determinations in both aqueous and organic media. Comparison of ashing/acid dissolution and direct organic solubilisation procedures for these stabiliser analyses denotes a 12 % loss of phosphorous in tri-(2,4-di-*t*-butylphenyl) phosphite in the former case, possibly due to formation of a volatile acid species. Organic dissolution and ICP-AES measurement showed excellent agreement with the expected value, making it the preferred method.

The metal element contents in several thermal stabilisers for PVC were determined by ICP-AES with a detection limit of $0.004-0.07\,\mu g\,mL^{-1}$, RSD of 1.1–3.8 %, and recovery of 99.6–102.0 % [195]. In a study of weathering of PVC by Carlsson *et al.* [196], all the siding samples used were first ashed in air at up to 500 °C (constant weight). The ash was then analysed by ICP-AES after dissolution in aqua regia. Ash residues were also characterised by SEM. Because volatile tin compounds could have been lost during ashing, some PVC samples were digested in concentrated H_2SO_4, and then dissolved in HCl (2 M).

Typical industrial ICP-AES measurements are Al, Ca, Co, Cr, Fe, K, Na, Ni, P (S), Sn and Si in polycarbonate; Al, Ca, Cd, Co, Cu, Fe, K, Mg, Mn, Mo, Na, Ni, Pb, Y and Zn in polyether foams; and Ag, Al, B, Ba, Be, Ca, Cd, Co, Cr, Cu, Fe, K, Mg, Mn, Mo, Ni, P, Pb, Si, Sn, Te, Ti, Tl, V, Zn in industrial wastewater management [197].

A simple, rapid and accurate ICP-AES method was developed for the analysis of Ca, Mg, Fe, Al, Ti in polypropylene [198]. With optimised operating parameters and analytical wavelength, RSD was below

5%, with a sample recovery of 91–107%. ICP-AES was also used for the determination of As in PP [199]. Trace amounts of Ti and Al (*catalyst residues*) in PP were determined by AAS and ICP-AES in the presence of Na [200]. Solubilisation of polymeric matrices is often complicated. With ICP-AES, the detection limit and standard deviations are lower, and the analysis time is considerably shorter since only 1.0 g of sample needs to be mineralised, as compared to 10 g for AAS (RSD 8.5%). ICP-AES analysis of PP containing Al, Ba, Ca, Fe, Li, Mg, Na, P, Ti and Zn has been compared for pneumatic and ultrasonic nebulisers [179]. Precision for the determinations made by USN-ICP-AES is generally better than for those by PN-ICP-AES. Other routine measurements frequently carried out industrially by means of ICP are Cr, Cu, Fe, Mg, Mn and Ni in UHMWPE, and Cu and I in nylons after dissolution in HCOOH/HCl. ICP-AES has also been used for the determination of Ge in PET [201].

Sodium, K, Ca, Mg, Zn and Fe and inorganic chlorides are the most widely occurring contaminants in solvent dyes for engineering thermoplastics such as PC, and need to be contained to within 250 ppm (specification; ICP verification for QC purposes) [202].

ICP-AES analysis has been reported for the determination of sulfur in various copolymers and rubbers (e.g. as curing agents) after Schöniger mineralisation [203], and substitutes for the classic iodometric method, which involves converting S into SO_2, is very time-consuming and is complicated by interferences from halogenated polymers and nitrilic rubbers. ICP-AES has also been used to determine ultratrace amounts of cerium (0.032% ± 0.001) in paint after ashing; the calibration curve extended over a broad analytical dynamic range for Ce (0.2–700 ppm) [204]. Elemental analyses by means of ICP-AES have been reported to characterise a membrane, which was prepared from polyethyleneimine by cross-linking with dibromoalkane, followed by alkylation [205].

ETV-ICP-AES can be used to determine sulfur catalyst residues in bisphenol-A; a drawback is serious contamination of the graphite [206].

Challenges of ICP-AES/MS for polymer/additive analysis are the lack of CRMs. A monograph describes applications of ICP-AES [182].

8.3.2.5 Direct-current Plasma Discharge

Principles and Characteristics A direct-current plasma (DCP) jet is a flowing, gas-stabilised electrical discharge that is maintained by a core consisting of a continuous direct-current arc which is formed between a tungsten cathode and two graphite anodes [207]. The sample is nebulised with Ar gas. The major advantages of the three-electrode DCP are its good stability; its ability to handle organic as well as aqueous solutions; and its acceptance of solutions with a relatively high solids content. The DCP has been successfully applied to many analyses, achieving low detection limits, good precision, and accuracy. It is not as widely accepted or used as ICP, and the analytical performance characteristics are generally somewhat inferior to ICP. Excitation temperatures can reach 6000 K; sample volatilisation is not complete, however, because residence times in the plasma are relatively short. In addition, this can lead to problems for samples that contain elements with high excitation energy. Another disadvantage of the DCP is the small region where optimal line-to-background intensity ratios occur.

DCP-AES can be used for high-viscosity matrices, slurries, etc. Organic solvents and acids can be handled without problems. Sample preparation is simpler than for ICP. Operating costs are much lower than for ICP-AES. Table 8.32 compares DCP-AES to ICP-AES and FAAS; Table 8.33 shows typical detection limits. DCP and its applications were reviewed [208].

Applications The principal plasma emission sources that have been applied for detection in GC have been

Table 8.32 Comparison of FAAS, DCP-AES and ICP-AES

Feature	FAAS	DCP-AES	ICP-AES
Range of sample types		xx	
Ease of use		x	
Detection limits		x	xx
Dynamic range		x	xx
Freedom from interferences		x	xx
Element coverage		x	xx
Sample throughput		xx	xx
Consumable costs	x	x	
Instrument cost	xx	x	

Table 8.33 Typical detection limits ($\mu g\,L^{-1}$) of FAAS, DCP-AES and ICP-AES

Element	FAAS	DCP-AES	ICP-AES	Element	FAAS	DCP-AES	ICP-AES
Al	30	15	5	P	4 000	100	40
B	600	10	2	S	–	–	40
Cu	3	4	2	Si	200	10	5
K	2	10	60	Zr	1 500	5	0.8
Mn	2	2	2				

MIP and DCP. ICP has been little used for GC, but ICP and DCP have been used effectively as HPLC detectors.

8.3.2.6   Microwave-induced Plasma

Principles and Characteristics The major drawbacks of ICP with argon as the support gas lie in numerous isobaric polyatomic ion interferences and in the lack of sufficient energy to ionise halogens and nonmetals to the necessary extent. With these weaknesses of ICP in mind, the possibility of generating microwave-induced plasmas with alternative gases to argon is of interest.

Microwave-induced plasmas are electrodeless discharges operated in a noble gas at rather low power (usually <200 W). The use of helium, nitrogen, and argon MIPs has been explored. A microwave-induced plasma (MIP) is initiated by a spark of a Tesla coil, not unlike ICP. The electrons generated oscillate in the microwave field and gain sufficient kinetic energy to ionise the support gas by collision. Microwave plasmas are operated at 1–5 GHz. Usually, a microwave frequency of 2450 MHz is used, because of the commercial availability of microwave generators of this frequency. Gas kinetic temperatures are of the order of 2000–3000 K. As a result, atomisation frequencies are not much better than those of common flames. Microwave plasmas are highly ionising. Ground-state atoms, ions, and molecules and excited state species, as well as electrons, are present in high abundance in microwave plasmas.

In a typical MIP-MS instrument, the ICP portion is replaced with one of a variety of microwave discharge sources, usually a fairly standardised (modified) Beenakker cavity connected to a microwave generator. The analytical MIP at intermediate power (<500 W) is a small and quiet plasma source compared with the ICP. The mass spectrometer needs no major modifications for it to be interfaced with the MIP. With MIP used as a spectroscopic radiation source, typically consisting of a capillary (1 mm i.d.), a power of 30–50 W and a gas flow below 1 L min^{-1}, multi-element determinations are possible. By applying electrodeposition on graphite electrodes, ultratrace element determinations are within reach, e.g. pg amounts of Hg.

By far the greatest advantage and driving force for the MIP is the ability to use *helium* as the plasma gas. The advantages of helium are its monoisotopic nature; high ionisation potential; and increased electrical resistivity, heat capacity, and thermal conductivity compared with those of argon. Using He as a working gas, S, P and halogens can be detected. A microwave-induced helium plasma (MIP) operating at very low power (1–10 W) is particularly valuable for excitation of nonmetals such as halogens and chalcogens [209]. The most common use for He-MIP has been as an ionisation source for the element-specific detection of volatile analytes at low- to sub-ppb levels, typically in a GC-MIP-MS configuration, but also in combination with liquid chromatography. Commercially available GC-MIP-AES is very useful for speciation. A microplasma is best, because He is very costly. SFC-MIP has also been used for optical emission studies [210,211].

Applications A method for multi-element determination of major elements in commercial and in-house prepared polymer/additive formulations by MIP-AES after microwave digestion with nitric acid has been reported [212]. The precision obtained varied between 2 and 4.5 %, depending on the element determined.

Helium MIP-MS is also a very sensitive detection system for *organotin compounds*, such as tetraethyltin (TET), tetrabutyltin (TBT), triethyltin bromide (TET-Br), tripropyltin chloride (TPT-Cl), tributyltin chloride (TBT-Cl), and others, separated by CGC [213]. Detection limits at sub-pg levels were achieved, and linear dynamic ranges of at least three orders of magnitude were obtained.

8.3.3   Atomic Fluorescence Spectrometry

Principles and Characteristics Atomic fluorescence spectrometry (AFS) is based on excitation of atoms by radiation of a suitable wavelength (absorption), and detection and measurement of the resultant de-excitation (fluorescence). The only process of analytical importance is resonance fluorescence, in which the excitation and fluorescence lines have the same wavelength. Non-resonance transitions are not particularly analytically useful, and involve absorption and fluorescence photons of different energies (wavelength).

In AFS, the analyte is introduced into an atomiser (flame, plasma, glow discharge, furnace) and excited by monochromatic radiation emitted by a primary source. The latter can be a continuous source (xenon lamp) or a line source (HCL, EDL, or tuned laser). Subsequently, the fluorescence radiation is measured. In the past, AFS has been used for elemental analysis. It has better sensitivity than many atomic absorption techniques, and offers a substantially longer linear range. However, despite these advantages, it has not gained the widespread usage of atomic absorption or emission techniques. The problem in AFS has been to obtain a sufficiently stable, reproducible and high-intensity light source that completely covers the useful electromagnetic spectrum (190–800 nm).

AFS instruments are mainly used to detect the *vapour-forming elements*, such as those that form hydrides (As, Bi, Ge, Pb, Se, Sb, Sn and Te). AFS is less prone to spectral interferences than either AES or AAS. Detection limits in AFS are low, especially for elements with high excitation energies, such as Cd, Zn, As, Pb, Se and Tl. In recent years, the use of AFS has been boosted by the production of specialist equipment that is capable of determining individual analytes at very low concentrations (at the ng L^{-1} level). The analytes have tended to be introduced in a gaseous form. AFS methods and instrumentation have been reviewed [214–216], see also ref. [17].

Applications   No reference to polymer/additive problem solving by AFS has been recorded.

8.3.4   Direct Spectrometric Analysis of Solid Samples

Principles and Characteristics   Solid samples may be analysed by means of digestion and slurry techniques. However, the usefulness of these methods is limited. Digestion is a relatively slow and time-consuming process; samples may sometimes be difficult to digest; one has to ensure that all elements of interest are stable in the solution; the blank of chemicals used might bring problems with the detection limit, and grinding solids into a powder fine enough for the slurry technique can be prone to contamination. Analytical chemistry faces great challenges regarding direct analysis of solid matter. Direct solid sampling (SS) obviates the need to dissolve samples before measurement, and avoids wet oxidation processes by means of microwave digestion or other traditional dry or wet ashing procedures. Direct analysis of solid materials using atom-spectrometric methods is on its way to becoming routine in industrial analysis. Table 8.34 gives the main requirements for an *ideal direct solid sampling* method.

Advantages and disadvantages of performing the direct analysis of solid samples without prior dissolution or digestion are given in Table 8.35. In direct analysis, particular attention should be paid to influences from the matrix, sample inhomogeneities, calibration and measurement uncertainty. As the sample is not diluted, the matrix concentration may be high, which can lead to chemical interferences and high background absorbance.

Table 8.34   Requirements of an ideal direct solid sampling method for elemental analysis

- Simultaneous multi-element analysis
- Applicable to a wide range of sample compositions
- Wide dynamic range of concentrations
- Suitable for small and large sample sizes
- Simple changing of samples
- Simple operation
- Reproducible sampling
- Accurate measurement
- From microanalysis (local) to distribution analysis (1D and 2D) and macroanalysis (bulk)
- Suitable for standardless quantification
- Acceptable costs of instrumentation and maintenance

Adapted from Moenke-Blankenburg [217]. From L. Moenke-Blankenburg, in *Lasers in Analytical Atomic Spectroscopy* (J. Sneddon *et al.*, eds), VCH Publishers, New York, NY (1997), pp. 125–195. Reproduced by permission of Wiley-VCH.

Table 8.35   General characteristics of analytical solid sampling procedures

Advantages

- No sample pretreatment (avoids wet decomposition)
- Higher detection power (substantial dilution in solubilisation procedures)
- Reduced total analysis time (high sample throughput)
- Low susceptibility to contamination or analyte loss due to volatilisation, sorption or incomplete solubilisation
- Minimum number of analytical process steps
- Avoidance of corrosive or toxic reagents (hazards)
- Allowance for analysis of very small quantities
- Provision for microdistribution and homogeneity check
- More economical

Disadvantages

- Relatively poor repeatability precision (sample inhomogeneity; µg range)
- Difficult introduction of a solid sample into the atomiser
- Sample weighing for each replicate measurement
- Repeated sampling procedure for each element (GFAAS)
- Difficult dilution
- Markedly lower precision than for solution analysis
- Chemical interferences and high background absorbance
- Sample residues (contamination)
- Semiquantitative (complex analyte addition technique)
- Calibration problems (few suitable calibration samples)

The influence of matrix concomitants often cannot be recognised or quantified. Progress in background correction techniques (e.g. direct Zeeman–AAS [218]), furnace techniques, microweighing, and electronic signal processing have gradually made possible the elimination or suppression of interferences that had previously hampered the application of direct solid sampling. Sample inhomogeneity in small samples may affect precision. Standard deviations ranging from 5 to 30 % are generally observed, depending on the weight of the sample [219]. Adding matrix modifiers to solids is generally more difficult and less effective than in solution.

Different analytical procedures have been developed for direct atomic spectrometry of solids applicable to inorganic and organic materials in the form of powders, granulate, fibres, foils or sheets. For sample introduction without prior dissolution, a sample can also be suspended in a suitable solvent. Slurry techniques have not been used in relation to polymer/additive analysis. The required amount of sample taken for analysis typically ranges from 0.1 to 10 mg for analyte concentrations in the ppm and ppb range. In direct solid sampling method development, the mass of sample to be used is determined by the sensitivity of the available analytical lines. Physical methods are direct and relative instrumental methods, subjected to matrix-dependent physical and nonspectral interferences. Standard reference samples may be used to compensate for systematic errors. The minimum difficulties cause INAA, SNMS, XRF (for thin samples), TXRF and PIXE.

Table 8.36 lists the main classical and newer approaches to solid sampling for elemental analysis. Little work on the introduction of solids into flames has been reported, because of problems of sample delivery and the relatively low source temperature. In arc and spark emission and in laser ablation as a sampling technique, the ablated sample material cannot be determined exactly. The limitations of arc or spark ablation are the poor precision and the instability of the plasma when excessive sample quantities are introduced. Nonconducting solids of materials rendered conducting by pelletising with a conductive material (copper or graphite) can be sampled by arc or spark erosion techniques. GFAAS is the best adapted of the spectrometric methods for use with direct solid sampling. Instruments for solid sampling AAS are commercially available. Direct electrothermal atomisation of solids requires a device for background correction, usually based on the Zeeman effect. SS-GF-ZAAS analysis is restricted to solids and metals volatilised at temperatures reached in the electrothermal atomiser; severe matrix effects due to nonspecific background absorption make standardisation difficult. Few commercial AAS instruments are equipped for solid sampling. To calibrate a method, use is generally made of CRMs with a matrix similar to the sample. Solid sample analysis using GFAAS requires no complete matching of the sample and calibrant material.

Application of XRF as a solid sampling method for multi-element determination is restricted to materials with relatively high contents; the direct determination of Pb with NAA is difficult because of physical problems. Laser evaporation and radiofrequency sputtering of electrically nonconducting materials, particularly in the bulk form, are important for atomic spectrometric analysis. Local surface areas are selected, using a microscope, to sample just the surface layer or to probe deeply into the bulk substrate. Laser ablation suffers from rather poor precision; this can be overcome by using an internal standard, for example, one of the major matrix components. Several variations of atomic spectrometry using laser vaporisation have been reported. Solids can also be analysed directly using various techniques coupled to ICP-MS: electrothermal vaporisation, laser ablation (as well as arc/spark ablation), direct sample insertion, fluidised bed introduction, etc. These techniques are even more interesting for samples which are difficult to dissolve. All these approaches complicate the optimisation process.

Hybrid systems which have been used to determine metals in solids include ETV-ICP, spark–MIP, spark–flame, arc–flame, laser–flame, laser–ETA and laser–DCP. ETV-ICP-MS is a solid sampling process which is automated, multi-element with high detection power, and amongst the best currently achieved.

Direct solid sample analysis is still mostly a *subsidiary method*, confined to specific analytical tasks, rather than truly complementary to traditional analysis via solutions. Solid sampling is not standard in routine

Table 8.36 Main element analytical solid sampling techniques and typical sample mass

SS-AES (0.1–1 mg)	NAA (1 mg–10 g)	SSMS (0.1–25 mg)
GD-AES (n.g.)	PIXE (0.1–10 mg)	GD-MS (1–100 µg)
SSS-AES* (n.g.)	µPIXE (>0.1 ng)	ETV-ICP-MS* (1 mg)
GFAAS (µg)	RBS (0.1–10 mg)	LA-ICP-MS* (10 ng–100 µg)
SS-GF-ZAAS* (µg)		SIMS (n.g.)
ETV-ICP-AES* (n.g.)		SNMS (n.g.)
LA-ICP-AES* (10 ng–100 µg)		
L-PES* (1 ng–1 µg)		
XRF (1–100 mg)		
µ-SR-XRF (>0.1 ng)		
TEM-EDS (>1 fg)		

* New technique.
n.g. = not given.

and monitoring laboratories. No single method can be used for the direct analysis of every sample.

Modern trace analysis is interested in detailed information about the distribution of elements in microareas and their chemical binding forms (speciation). The limited sample mass implies methods with absolute detection limits as high as possible. Use of the sputtering process as a sampling technique localises the analytical zone at the outer layers of a solid, and allows analysis to progress into the interior.

Solid and slurry sampling using the graphite furnace have been reviewed [220,221]. A monograph dealing with solid sample analysis is available [222].

Applications Specific applications of the direct spectrometric analysis methods of solid samples of Table 8.36 are given under the specific headings. One investigation that is practically only possible by direct solids analysis is checking the *homogeneity of polymers* [136,137]; this is of significance for reference materials and for quality control. A method for the assessment of microhomogeneity should meet various requirements [223]:

(i) no sample pretreatment;
(ii) precise analysis of metal representative samples (<1 mg); and
(iii) speed (large number of analyses).

Obviously, direct SS-GF-ZAAS fulfils these criteria. A carefully performed homogeneity study using SS-GF-ZAAS was reported for the VDA CRMs (Cd in polyethylene) [137,223,224]. Solid sampling also allows analysis of high-purity materials (blank-free, low limits of analysis).

Rühl [141] has reported on very large scale SS-GFAAS and XRF analysis of Cd in polymers for *product control* purposes in the automobile industry (approximately 20 000 parts per year) as a reaction on a Swedish law for environmental protection (upper limit of 75 ppm). Another sample of direct SS-GFAAS in industry is the control of all raw materials, processing steps, and products for adhesive tapes for the content of Cu, Mn, Fe, Cd and Pb, which act as a rubber poison by catalytic effects [144].

8.4 X-RAY SPECTROMETRY

X-ray spectrometric techniques can be classified by their excitative mode as follows:

- X-ray tube excitation;
- particle excitation (protons or heavy charged particles, PIXE; electrons, SEM and electron microprobe);
- synchrotron radiation (SR); high brilliance; and
- radioisotope excitation (^{55}Fe, ^{109}Cd or ^{241}Am; in portable spectrometers).

The latter X-ray source requires more calibration, for obvious reasons. On encountering matter, X-rays (discovered by Röntgen in 1895) may pass through unaffected or undergo reflection (TXRF), refraction, diffraction (XRD), polarisation, elastic and inelastic scattering (SAXS), photoelectric absorption (XRF, XAS, XPS) and electron–positron pair production (above 1.02 MeV, i.e. for very hard X-rays). In other excitation modes, X-rays may be generated and emitted (EPMA, PIXE, EXEFS). Electron beams or charged particles are the usual excitation sources that can generate X-rays. X-ray radiation may cause radiation damage.

X-ray spectrometry (XRS) encompasses a broad field of techniques, ranging from elemental analysis (XRF, XPS, PIXE, EPMA), without or with imaging capabilities (iXPS, iPIXE), to structural analysis of the ordered state (XAS, XES, XRD) or disordered state (XAS, XES, SAXS). Analytical methods that employ X-rays can be categorised in techniques that (only) use X-rays to excite sample atoms (e.g. XPS) and those that employ the information contained in the energy and number of X-rays emitted by a material under investigation (e.g. XRF, XRD, EPMA, PIXE). These methods yield structural, elemental or chemical information, derived from a fairly large sample volume (bulk analysis), or originating from a well-defined (and microscopically) small location (microanalysis, surface analysis). Both XRD and WDXRF are governed by Bragg's law: $2d \sin \theta = n\lambda$, where λ stands for the wavelength, n the order of reflection, d the interplanar spacing and θ the diffraction angle, but for quite different reasons. In the case of long-range order, the specific scattering angles carry information on the periodical ordering dimensions, and the intensities contain information on the location of the electrons within that order. No assumption regarding order is made to enable absorption experiments (XAS) and only short-range information is derivable. Also small-angle X-ray scattering (SAXS) can be used for the study of disordered materials, such as amorphous fractions, glasses, etc. Modern crystallographic instrumentation is highly advanced.

Many of the aforementioned techniques comprise experimentally important derivatives based on alternative X-ray sources (SR-XRF), detection modes

(EDXRF vs. WDXRF, sequential vs. simultaneous XRF), geometry (TXRF, SEXAFS, glancing angle), theoretical interpretations (EXAFS, XANES), microanalysis (μXRF, EPMA), hyphenation or on-line analysis. There is sufficient scope for polymer/additive analysis by means of XRS tools.

X-ray spectrometry is regularly being reviewed [225, 226]. An overview of X-ray techniques is available [227]. A handbook of X-ray spectrometry has appeared [228]. XRS techniques generate some 5000 literature references yearly.

8.4.1   X-ray Fluorescence Spectrometry

Principles and Characteristics   X-ray fluorescence (XRF) is based on the interaction (absorption/emission) of X-rays with matter. When matter is irradiated with photons of sufficiently high energy, an electron from an inner shell (usually K or L) of one of the atoms present may be ejected, leading to an electronically excited ion. An electron from a higher energy level may rapidly fill the vacancy created, and the difference in energy between the two levels is generally released in the form of an X-ray photon. These photons, which have a narrow energy bandwidth, are specific for the particular electron transition, and therefore characteristic of the ionised element. The electron transitions from high to low energy states are limited by a set of selection rules based on the four quantum numbers of each electron. Conventional X-ray spectroscopy is largely confined to the region from 0.01 nm (U $K\alpha$) to 2 nm (F $K\alpha$). As the electron displacement involves the inner shells, the energy of the characteristic X-ray lines is largely independent of the valency of the element (i.e. no measurable chemical shift).

In a competitive process, the excess energy can be dissipated by emission of a second or Auger electron from an outer shell of the atom, leaving it in a doubly ionised excited state. The relative importance of AES and XRF depends upon the atomic number (Z) of the element involved. High Z values favour fluorescence, whereas low Z favours AES. This fact, taken together with X-ray absorbance in air, makes XRF into a method which is not very sensitive for elements with atomic numbers below $Z \approx 10$. Measurements of solid samples are normally made under vacuum, as the absorption of air renders analysis of elements lighter than Ti impossible.

The *sample format* used for XRF measurement is typically compressed fine particle size powder (2–10 mm thick, 20–50 diameter), moulded film or

Table 8.37   General features of X-ray fluorescence spectrometry

Sample requirements	$\leqslant 5.0$ cm in diameter; 10–100 mg
Sample form	Thin film, solid, multilayers, liquid
Destructiveness	Sample preparation mostly unavoidable
Measurement environment	Air
Sampling depth	Normally ca. 10 μm; few tens of Å in TXRF
Depth profiling	Only by using variable-incidence X-rays
Range of elements	All but low-Z elements: H, He and Li; multi-element
Spectral interferences	Presence of neighbouring elements; matrix effects; no memory effects
Standardisation	Calibration standards; standardless quantitation
Dynamic range	Sub-ppm to 100 %
Minimum detection limit	Some hundred ng
Precision	For major elements: 0.05 % relative
Accuracy	±1 % for composition, ±3 % for thickness
Lateral resolution	Down to 10 μm using a micro beam (SR-XRF)
Chemical bond information	Only via soft X-ray spectra
Main use	Element identification and quantification; determination of thickness
Instrument cost	Relatively inexpensive (low power) to very expensive (high power spectrometers)
Analysis cost	Low

plaque, fused pellets or pills (diluted with a flux to overcome inter-element effects). For quantitative analyses, a flat and homogeneous sample surface is needed. Powder samples have to be pulverised to fineness, where absorption in single grains can be neglected. The *general features* of XRF are given in Table 8.37. X-ray fluorescence allows direct analysis of the solid, with high sample throughput. The sensitivity of XRF depends on the atomic number of the element considered, energy and intensity of the incident radiation, instrument geometry, reflectivity of the analysing crystal (WDXRF) and detector efficiency. Sensitivity limitations are felt particularly for low-Z elements. Soft X-rays are absorbed by air. XRF is reliable, inexpensive, fast and efficient, but nonspecific. It is an analytical workhorse. Inhomogeneity of the sample, including particle-size effects, is one of the main sources of systematic error in XRF.

XRF spectrometry typically uses a polychromatic beam of short-wavelength X-radiation to excite lines with longer wavelength characteristics from the sample to be analysed. *XRF equipment* consists of an X-ray source, sample holder and detection system, which can be either wavelength-dispersive (WD) or energy-dispersive (ED). XRF instrumentation now ranges from portable XRF spectrometers to benchtop energy dispersive systems and high-performance wavelength dispersive spectrometers. Most XRF work (99 %) is carried out with ordinary Rh tube excitation. Most conventional WDXRF instruments use a high power (2 to 4 kW) X-ray Bremsstrahlung source, whereas ED spectrometers use either a high or low power (0.5 to 1.0 kW) primary source. Also X-ray tubes with a rotating anode (up to 18 kW) are in use. Due to their lower beam intensities, application of radionuclides is restricted to energy-dispersive systems. For radionuclides used as primary sources in EDXRF, see ref. [229]. The brightest excitation source is synchrotron radiation. Better detection limits can be achieved with high-power systems. Breitländer Silikatglas Eichproben XRF-SUS are well suited for direct drift correction.

The basic function of the spectrometer is to separate the polychromatic beam of radiation coming from the specimen in order that the intensities of each individual characteristic line can be measured. In principle, the wide variety of instruments (WDXRF and EDXRF types) differ only in the type of source used for excitation, the number of elements which they are able to measure at one time and the speed of data collection. Detectors commonly employed in X-ray spectrometers are usually either a gas-flow proportional counter for heavier elements/soft X-rays (useful range: $E < 6$ keV; 1.5–50 Å), a scintillation counter for lighter elements/hard X-rays ($E > 6$ keV; 0.2–2 Å) or a solid-state detector (0.5–8 Å).

In recent years, XRF has seen the development of some powerful new variations. Improved light element (Be, B, C, N, O) analysis by WDXRF is based on ultrathin (75 μm) high-transmission end-window X-ray tubes; close coupling between anode and sample; 4-kW X-ray generators at high (150 mA) currents; very coarse collimators; multilayer analyser crystals; very thin (<1 μm) detector windows and multiple curved crystal optics. Modern configurations comprise the use of more than one detector/monochromator system. Depending on the monochromator crystal(s) chosen, different parts of the periodic system can be covered (e.g. from F to Fe, or from Al to U). A trend is also observed towards the use of parallel beam instead of focusing geometry (for better intensity in grazing angle incidence). Portable XRF spectrometers (8 kg) are now also based on X-ray tube excitation (as used for alloy analysis by UN inspectors in Iraq).

XRF instruments are either single-channel or *sequential* and multichannel or quasi-simultaneous. The sequential spectrometer scans the radiation emitted by the sample by changing the angle sequentially. For different wavelength regions, different analyser crystals must be used to fulfil Bragg's equation. As for each wavelength separate counting has to be carried out with crystal and detector set at the appropriate angles (θ and 2θ), the measurement time is relatively high (depending on θ scan range). With a sequential spectrometer, a wide range of different elements may be determined. High-performance sequential XRF now allows analysis of up to 84 elements of the periodic table, with a choice of some ten crystals allowing optimisation for each region of the X-ray spectrum.

Simultaneous spectrometers consist of various combinations of analyser crystals and detectors, arranged around the sample at fixed angle settings. Use of a multi-channel X-ray spectrometer with simultaneous determination of up to 24 elements can considerably increase the analysis speed (a few seconds to a few minutes).

Wavelength-dispersive spectrometers employ diffraction by a single crystal to separate characteristic wavelengths emitted by the sample. Three types of WD instruments are in use, namely sequential, simultaneous and sequential–simultaneous WDXRF. X-rays emerging from the sample and angularly dispersed by the diffracting crystal are measured with a suitable detector, usually a gas-flow proportional counter for long wavelengths and a NaI(Tl) scintillation detector for short wavelengths.

Table 8.38 shows the *main features* of WDXRF. With WDXRF, the detectability of light elements (down to Be) is optimised by using X-ray tubes and detectors with ultrathin windows and widely spaced (focusing) analyser crystals. As the intensities per analytical line in WDXRF are two orders of magnitude higher quantitative determination is much faster than with EDXRF. A totally unknown sample may be measured for 73 elements in about 14 min. The big advantage of WD spectrometers is their excellent wavelength-to-energy resolution, especially in the low-energy region and the high count-rate range (10^6 cps) in operation. Consequently, WDXRF has few problems with spectral interference. With the current high resolution, it is possible to distinguish sulfide/sulfate and phosphite/phosphate. WDXRF is generally accepted as having the best precision of all the commonly used instrumental techniques.

WDXRF data processing with increasing accuracy of analysis (from left to right) is summarised in Figure 8.7. XRF is used to identify/quantify fillers, stabilisers,

Table 8.38 Main features of wavelength-dispersive X-ray fluorescence spectrometry (WDXRF)

Advantages

- Simple sample preparation
- Solid or liquid samples
- Ease of use
- Nondestructive (unless oddly shaped)
- Speed, ruggedness
- Wide element coverage (Be−U); multicomponent analysis
- Widest dynamic range (0.1 ppm to 100 %) of any spectrometric technique
- Sequential, simultaneous and sequential–simultaneous capabilities
- High resolution
- High precision (short- and long-term repeatability)
- Accuracy (normally 0.1 % relative)
- Various quantitation modes (incl. standardless analysis)
- Automation
- Mature technology

Disadvantages

- Severe matrix effects
- Need for rather large specimens (surface area of several cm^2)
- Inflexible in terms of sample handling
- Time-consuming sequential mode

pigments, flame retardants, and contaminants, resolves plant and customer problems, is used to evaluate competitive products and provides customer service.

Energy-dispersive spectrometry (EDS) is a technique of X-ray spectroscopy that is based on the simultaneous collection and energy dispersion of characteristic X-rays. Typical ED detectors are thermoelectrically cooled semiconductors (usually operated at 77 K), PIN diodes, Si drift chambers, CdZnTe (CZT), and Peltier-cooled detectors with 240 eV resolution. Counting times are typically in the range of 100–10 000 s (about 10–30 times that of WDXRF), but this is the total measurement time for multi-element determination. Generally, the energy resolution is much lower with ED detectors than with diffraction-based WD systems; the energy resolution of the Si(Li) detector is less than 145 eV at 5.9 keV. In EDXRF, peak overlap is thus much more common than in WDXRF, e.g. between the $K\alpha$ and $K\beta$ lines of elements with atomic numbers $Z + 1$ and Z. Also, as there are no pure background regions between the peaks, background correction is more difficult, and rather sophisticated mathematical treatment of the spectra is required. EDS spectrometers are most frequently attached to electron column instruments (SEM, TEM).

Table 8.39 shows the main features of EDXRF. EDXRF is not able to detect the fine structure of the K, L, M, etc. lines. EDXRF is used for applications which require measurement of a limited number of elements, and where the resolution and ultralow detection limits of wavelength-dispersive systems are not necessary. For example, EDXRF has been used as a rapid screening technique for the determination of Br and Sb in plastic recyclate at a LOD of 5 ppm [230]; the method was validated by means of NAA [231]. Conventional EDXRF systems and benchtop units have a limited detection capability for low-Z-elements and cannot directly measure fluorine in processing aids.

In EDXRF, trends are for miniaturisation, development and optimisation of high-resolution room-temperature detectors and extension of the application range towards the determination of light elements.

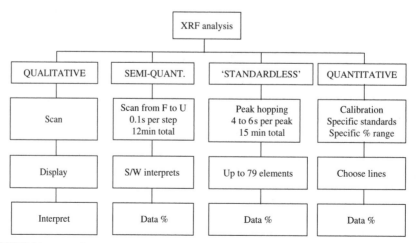

Figure 8.7 WDXRF data processing

Table 8.39 Main features of energy-dispersive X-ray fluorescence spectrometry (EDXRF)

Advantages

- Analysis 'as is', or minimal sample preparation
- Rapid survey analysis of solids, powders, composites, films, coatings, pastes, slurries and liquids
- Nondestructive (analysis of odd-shaped specimens)
- Wide element coverage (Na–U)
- Detection limits: 100–200 ppm for $Z > 11$
- Accuracy (nominally 4–5 % relative; for concentrations >5 at. %)
- Sampling depth ($> 0.02 \mu$m)
- Various quantitation modes (incl. standardless analysis)
- Imaging/mapping (in SEM, EPMA and STEM)
- Lateral resolution (0.5–1 μm bulk; 1 nm for thin samples in STEM)
- Suitable for production sites (benchtop instruments)
- Mature technology

Disadvantages

- Severe matrix effects
- Relatively poor resolution
- Relatively poor sensitivity
- No chemical bonding information
- Sample size limitations (SEM, EPMA, XRF; 3 mm diameter thin foils in TEM)

Table 8.40 Comparisons between WDXRF and EDXRF

Feature	WDXRF	EDXRF
Detection mode	Wavelength	Energy
Detectors	Scintillation, proportional	Si(Li), Si PIN diode
Element coverage	Be–U	Na–U
Fine structure	Yes	No
kV/mA combinations	Up to 70 kV/120 mA (max. 4.2 kW)	Up to 50 kV, 2 mA (max. 50 W)
Sensitivity	High	Low
Spectral resolution	15 eV	175 eV
Spatial resolution	Better	–
Miscellaneous	Stable	Compact

Table 8.40 compares the main characteristics of WDXRF and EDXRF. Multidispersive XRF combines the benefits of the WDXRF technique for routine elemental analysis with the complete flexibility offered by EDXRF for nonroutine analysis. Clearly, modern XRF instrumentation is rather varied, ranging from simple benchtop EDXRF equipped with a low-power X-ray tube and high-resolution proportional counter for some key elements, to 4 kW simultaneous multichannel spectrometers with 28 fixed element channels for

analysis in just a few seconds and full-scan-based standardless analysis for elements O–U within 2 min.

Both WDXRF and EDXRF lend themselves admirably to quantitative analysis, since there is a relationship between the wavelength or energy of a characteristic X-ray photon and the atomic number of the element from which the characteristic emission line occurs. The fluorescence intensity of a given element is proportional to the weight fraction. Emitted fluorescence radiation is partly absorbed by the matrix, depending on the total mass absorption coefficient:

$$\mu = \sum_i \mu_i w_i \tag{8.2}$$

where μ_i and w_i are the mass absorption coefficient and mass fraction of element i, respectively. Apart from the absorption effect, an enhancement effect may also operate. XRF intensities are very rarely directly proportional to concentration. Consequently, in the conversion of net line intensity to analyte concentration, it is necessary to correct for any absorption and/or enhancement effects. *Matrix effects* in XRF, which are a result of attenuation of the incident and fluorescent X-rays in the sample, are essentially well understood. Whenever the attenuation differs for a sample and a standard, systematic errors result. For thin films, XRF intensity–composition–thickness equations derived from first principles can be used for the precision determination of composition and thickness. This holds also for each individual layer of multiple-layer films. In thin-film techniques the amount of sample in the irradiated spot has to be known. For infinite-thickness samples, quantitative evaluation of element concentrations in XRF is a rather more complicated matter. One has to take into account several factors related to the physics of the process (e.g. cross-sections, penetration depths), spectrometer characteristics (e.g. geometry, detector properties) and sample (mass attenuation, inter-element effects). The details of quantitative analysis are beyond the scope of this section, but can be found elsewhere [232–235]. Only the essentials are given here.

Quantitative XRF analysis has developed from specific to universal methods. At the time of poor computational facilities, methods were limited to the determination of few elements in well-defined concentration ranges by statistical treatment of experimental data from reference material (linear or second order curves), or by compensation methods (dilution, internal standards, etc.). Later, semi-empirical influence coefficient methods were introduced. Universality came about by the development of fundamental parameter approaches for the correction of total matrix effects (absorption and enhancement). Fundamental parameters and semi-empirical methods are frequently combined [232]. Mathematical equations on fluorescence, absorption and enhancement, and the possible mathematical approaches to quantification of multi-element specimens, with empirical, semi-empirical and fundamental influence coefficients (FIC), are given by Lachance and Claisse [232]. The term 'influence coefficient' is used to quantify the total matrix effect of element j on analyte i.

As XRF is not an absolute but a comparative method, sensitivity factors are needed, which differ for each spectrometer geometry. For quantification, matrix-matched standards or matrix-correction calculations are necessary. Quantitative XRF makes ample use of calibration standards (now available with the calibrating power of some 200 international reference materials). Table 8.41 shows the *quantitative procedures* commonly employed in XRF analysis. Quantitation is more difficult for the determination of a single element in an unknown than in a known matrix, and is most complex for all elements in an unknown matrix. In the latter case, full qualitative analysis is required before any attempt is made to quantitate the matrix elements.

Conventional XRF analysis uses calibration by regression, which is quite feasible for known matrices. Both single and multi-element *standards* are in use, prepared for example by vacuum evaporation of elements or compounds on a thin Mylar film. Comparing the X-ray intensities of the sample with those of a standard, allows quantitative analysis. Depending on the degree of similarity between sample and standard, a small or large correction for matrix effects is required. Calibration standards and samples must be carefully prepared; standards must be checked frequently because of polymer degradation from continued exposure to X-rays. For trace-element determination, a standard very close in composition to the sample is required. This may be a certified reference material or a sample analysed by a primary technique (e.g. NAA). Standard reference material for rubber samples is not commercially available. Use can also be made of an internal standard,

where a known amount of an element with characteristic radiation excited and absorbed to (almost) the same extent as for the analyte element is added to sample and standard. For this purpose, a $Z \pm 1$ element, as compared to the analyte element, is commonly used. In this way, a correction is made for the different matrix effects of sample and standard. Other frequently applied methods for correction of matrix effects rely on dilution and on standard addition, where sample preparation is essentially as for the internal standard method. Dilution may be achieved by fusion with a flux (borax or pyrosulfate). It is vital that all standards and unknowns be presented to the spectrometer in a reproducible and identical manner.

For complicated samples where matrix or inter-element effects are present, a linear calibration curve may not be valid, and one should consider using an empirical model for concentration correction. This usually requires a large set of standards of similar composition to the unknown, which generally makes analysis rather impractical. Inter-element effects can be calculated from a basic knowledge of physical parameters in combination with the appropriate use of samples of known composition, pure elemental standards or composite standards.

An approach to reducing XRF matrix effects, without requiring a large number of standards, is to present the sample to the spectrometer as a thin film (negligible absorption/enhancement effects). The main problem with this approach is that the limits of detection are significantly degraded, due to the small amount of sample analysed. This latter limitation can largely be eliminated by using TXRF (Section 8.4.1.1).

Alternatively, *fundamental parameter methods* (FPM) may be used to simulate analytical calibrations for homogeneous materials. From a theoretical point of view, there is a wide choice of equivalent fundamental algorithms for converting intensities to concentrations in quantitative XRF analysis. The fundamental parameters approach was originally proposed by Criss and Birks [239]. A number of assumptions underlie the application of theoretical methods, namely that the specimens be thick, flat and homogeneous, and that, for calibration purposes, the concentrations of *all* the elements in the reference material be known (having been determined by alternative methods). The classical formalism proposed by Criss and Birks [239] is equivalent to the fundamental influence coefficient formalisms (see ref. [232]). In contrast to empirical influence coefficient methods, in which the experimental intensities from reference materials are used to compute the values of the coefficients, the fundamental influence coefficient approach calculates

Table 8.41 Quantitative procedures in XRF analysis

Single-element methods	Internal standardisation
	Standard addition
	Use of scattered source radiation
Multiple-element methods	Type standardisation
	Use of influence coefficients [236,237]
	Fundamental parameter techniques [238]

the values from theory, i.e. without prior knowledge of the elemental concentrations in the specimens being analysed, and confirms their validity by using them to linearise a scatter of experimental intensity data. Analysis based on fundamental parameter methods incorporates an iterative process, which involves making initial estimates of specimen composition which are gradually refined during the process. In the application to polymers, a slight disadvantage is that the method generally requires knowledge of light element concentrations (particularly C:H ratios) which cannot normally be measured using XRF. Universal standardless analyses for infinite thickness samples are applied when specific matching standards are not available; when complicated calibration procedures are required; or when samples can only be obtained in small quantities or as irregular shapes.

There are two possible ways of XRF analysis used in fundamental parameter methods, namely analysis with and without standards. The intensity of the measured characteristic radiation I_{im} is related to the calculated intensity of radiation I_{ical}

$$I_{im} = k_i \times k_o \times I_{ical} \qquad (8.3)$$

where k_i is a coefficient depending on the spectrometer parameters and k_o is a constant depending on the way of normalisation of the X-ray tube spectrum. The values for the product of coefficients $k_i \times k_o$ for different wavelengths can be obtained for different elements with standard samples of known composition [239]. In another approach, XRF analysis with fundamental parameters is carried out by determining the k_i coefficient as a function of the wavelength, i.e. taking into account the influence of the spectrometer parameters on the X-ray beam emitted from the sample.

The use of fundamental parameters is attractive for various reasons. They impose fewer restrictions on the number of standards required for analysis. This simplifies the standardisation protocol for maintaining a XRF system, and permits greater flexibility in dealing with different types of materials. Intensity/concentration algorithms of the fundamental type, i.e. without recourse to the use of standards, have gradually developed [238–240] and are now widely available [241]. Functionality and quality of XRF software have reached a very high level, with a large variety of evaluation procedures and correction models for quantitative analysis, and calculation of fundamental parameter coefficients for effective matrix corrections. Nevertheless, there is still a need for accuracy improvement of fundamental parameters, such as the attenuation functions.

XRF nowadays provides accurate concentration data at major and low trace levels for nearly all the elements in a wide variety of materials. Hardware and software advances enable on-line application of the fundamental approach in either classical or influence coefficient algorithms for the correction of absorption and enhancement effects. Vendors' software packages, such as QuantAS® (ARL), SSQ (Siemens), X40, IQ$^+$ and SuperQ (Philips), are precalibrated analytical programs, allowing semiquantitative to quantitative analysis for elements in any type of (unknown) material measured on a specific X-ray spectrometer without standards or specific calibrations. The basis is the fundamental parameter method for calculation of correction coefficients for matrix elements (inter-element influences) from fundamental physical values such as absorption and secondary fluorescence. UniQuant® (ODS) calibrates instrumental sensitivity factors (k values) for 79 elements with a set of standards of the pure element. In this approach to inter-element effects, it is not necessary to determine a calibration curve for each element in a matrix. Calibration of k values with pure standards may still lead to systematic errors for unknown polymer samples. UniQuant® provides semiquantitative XRF analysis [242].

It is possible to determine components in complex EPs where matrix effects can be severe. For example, zinc (as zinc borate), chlorine (as dechlorane flame retardant), antimony (as oxide) and fibre-glass have been determined in nylon using just one standard. Many users have refined the universal precalibrated programmes for standardless XRF and made them more efficient for matrix correction by using variable correction coefficients. OilQuant® offers possibilities for analysing polymers [243]. Software packages usually provide:

(i) a quantitative analysis in the absence of standards (e.g. applicable to mixed plastic waste polymers);
(ii) a quantitative analysis with the highest accuracy if standards are available (see also Figure 8.7); and
(iii) analysis of small and/or odd shaped samples.

International or in-house standards in combination with fundamental parameters software, lead to the same accuracy as conventional analysis using regression analysis of standards. Provided that accurate standards are available, the main factors that determine the accuracy of XRF are the matrix absorption correction and (in the case of EDXRF) the spectrum evaluation programme, i.e. correction for spectral overlap and background.

Quantitation in XRF was reviewed [225], as well as standardless analytical methods using an extended FPM for the determination of sample compositions [244,245].

Various significant *advancements in XRF technology* are total-reflection X-ray fluorescence (TXRF, see Section 8.4.1.1), micro-XRF and synchrotron radiation XRF (SR-XRF). Among the others, synchrotron radiation offers continuous radiation up to very high energies (>120 keV) with high intensity and brightness. Brightness (in photons s^{-1} mm^{-2} $mrad^{-2}$ per 0.1 % band pass) for conventional and SR X-ray sources are 10^8 and 10^{21}, respectively. Third-generation SR excitation sources allow use of the *K* series of emission lines in analysis of the heavy elements [246]. Typically, 0.01 ppm W can be determined in a 500-s measurement. Sub-μm beam spots have been demonstrated using SR. It is possible to conduct *SR-XRF* with spatial resolutions of 0.1−1 μm and with LODs in the 10−100-ppb (sub-pg level) range. The main limitation of SR-XRF is the same as for PIXE, but to a greater extent, namely availability. The potential of SR-XRF undoubtedly lies in its microbeam capability.

XRF has also been hyphenated to various chromatographic techniques, cf. TLC-XRF (Section 7.3.5.1). For process XRF, the stream interface is a simple by-pass flow; the window material that allows the X-rays to enter the product stream−a thin film of polycarbonate−confines pressure and temperature.

The literature on XRF is abundant. Recent general reviews are refs [235,237]; for sample preparation see ref. [247]. EDXRF was specifically dealt with in ref. [248] and an excellent X-ray detector overview is available [225]. Several recent XRF monographs have appeared [233,249,249a], also covering TXRF [250] and quantitative XRF [232,251].

Applications X-ray fluorescence is widely used for direct examination of polymeric materials (analysis of additives, catalyst residues, etc.) from research to recycling, through production and quality control, to troubleshooting. Many problems meet the concentration range in which conventional XRF is strong, namely from ppm upwards. Table 8.42 is merely indicative of the presence of certain additive classes corresponding to elemental analysis; element combinations are obviously more specific for a given additive. It should be considered that some 60 atomic elements may be found in polymeric formulations. The XRF technique does not provide any structural information about the analytes detected; the technique also has limited utility when

Table 8.42 Elements and corresponding additives present in commercial polymers

Element(s)	Additive(s)
Al	Talc, catalyst residue, flame retardant
As	Biocide
B	Flame retardant, smoke suppressant
Br, Cl	Flame retardant
Ca	Stearate processing aid, filler, heat stabiliser
Cr	Pigment, catalyst residue
Cu	Pigment, biocide
F	Processing aids
Hg	Biocide
Mg	Talc, flame retardant
Mo	Smoke suppressant
P	Phosphite stabiliser, plasticiser, flame retardant
S	Vulcanising agent, dye, biocide
Sb	Flame retardant, biocide
Se	Pigment
Si	Silica, filler, catalyst residue
Sn	Flame retardant, pigment, stabiliser, biocide
Ti	Pigment, catalyst residue
Zn	Stearate processing acid, heat stabiliser, pigment, smoke suppressant, biocide

more than one analyte is present in a sample containing the same element.

Due to the different working principles of WDXRF and EDXRF, the applications differ strongly (Table 8.43). Simultaneous WDXRF with ten channels (elements) and increased sensitivity for the low atomic number elements (e.g. a few ppm of phosphorous in a low atomic number matrix) has been used for QC of polymer granules [252]. To detect elements at trace levels (ppm−ppt), generally the special XRF modes, mainly EDXRF techniques, are applied like TXRF, SR-XRF or μXRF. Detection limits with SR-XRF are now at the attogram level.

Raw or used oils are usually analysed for additive-element content, including Ba, Zn, Mn, Ca, P and Cl, plus naturally occurring elements including S, N, Ni and Na [253]. Typical elements determined in paints by XRF are Zn, P, Sn, Pb, Al, Ti, Cd, Cr, Co, Ba, Ca and Si; in rubbers they are mostly Zn, Mg, Si, S, Ca and Fe. Because of sample preparation complications

Table 8.43 A selection of applications of X-ray spectrometry

Field	Method(s)
Plastic materials	EDXRF, WDXRF
Ga in PUR foam	WDXRF
Textile fibres	TXRF
Thin-film characterisation	XRF, EDXRF
Multilayers	EDXRF
Art conservation (paintings, etc.)	EDXRF, TXRF

in the elemental analysis of *rubbers*, nondestructive analysis techniques such as XRF, PIXE, SEM-EDS and LIESA® have been introduced into the rubber industry, but the literature on this subject is scarce. Kump *et al.* [254] have reported quantitative multi-element analysis of pulverised rubber samples by XRF utilising ^{109}Cd and ^{55}Fe excitation radioactive sources, single-element thick standards and k_o-NAA validation. An accuracy of 6 % or better can be achieved. Techniques like AAS and ICP-AES with complicated and, in many cases, rather uncontrollable sample preparation processes are less convenient, although in many cases they are more sensitive.

As already indicated in Scheme 2.12, XRF is profitably used for general screening of polymer formulations on inorganic components, before and after extraction. In the case of several PVC blends, such screening has indicated the presence of Cl, Ca, Ti, Cr, Fe, Zn, Mo, Cd, Sn, Sb and Pb [255]. It is well known that X-ray radiation may cause radiation damage, such as coloration of PVC samples during XRF analysis.

Houben [256] has compared the determination of *flame-retardant* elements Br, P, S, K, Cl and F in polycarbonate using commercial (X40 and UniQuant®) software. For the X40 method, a calibration line for each element in PC or PC/ABS blends was mapped for the conversion of intensities to concentrations. With the universal UniQuant® method, sensitivity factors (ks) were calibrated with pure standards. The X40 method turned out to be more reliable than UniQuant® for the determination of FRs in PC and PC/ABS blends, even in the case of calibration of k values with PC standards. Standard errors of 5 % were achieved for Br, P, S and K, and 20 % for Cl and F; the latter element could not be determined by means of UniQuant® (Table 8.44). GFR PC cannot be quantified with these two methods, because of the heterogeneous nature of the composites. Other difficult matrices for XRF analysis are PBT, PS and PP compounds containing both BFRs and Sb_2O_3 (10–30 wt %) due to self-absorption of Sb and interelement effects.

Although XRF is not a surface tool, it can be effective in differentiating the composition of bulk and surface

Table 8.44 Detection limits (ppm) for FR elements in PC and PC/ABS blends

Method	Br	P	S	K	F
X40	12.3	2.0	9.2	5.1	100
UniQuant®[a]	18.6	5.3	8.6	3.1	n.d.

[a] Reference: element in PC standards.
n.d. Not determinable.
After Houben [256].

of plastic mouldings. One way is to remove the top layer by rubbing it with an abrasive paper or fine (30 μm) diamond on a polyester backing. Alternatively, by careful selection of a characteristic line with a relatively small critical depth, measurements can be made which only relate to the very surface of a sample. Validation of these results is difficult matter. Surface effects on flame-retardant nylon mouldings were reported; see Table 8.45 [252]. XRF was also used to determine the content of 2-(alkylstannylthio)acetates in the surface of 1.1-mm-thick PVC after roll milling and hot pressing [257].

Mineral *fillers* such as dolomite, chalk, talc and kaolin originating from different geographical locations have specific elemental compositions. Their spectra can therefore be used to identify the source of supply [258]. WDXRF is also in use for fingerprinting of the composition of *E-glass fibres* in polyesters and polyamides (as glass pearls) for source classification and traceability purposes [259]. Although the composition of E-glass is very specific, it still encompasses a given range for the various elements (Table 8.46). Obviously, differences in the organic sizing of glass fibres cannot be traced through this method.

EDXRF has been used for comparison of the elemental composition of some 100 *carpet fibres* (both automotive and residential) for forensic source classification [260]. SEM-EDS is too insensitive for this

Table 8.46 Composition of E-glass (wt %)

SiO_2	52–56	MgO	0–5
CaO	16–25	$Na_2O + K_2O$	0–2
Al_2O_3	12–16	TiO_2	0–1.5
B_2O_3	5–10	Fe_2O_3	0–0.8

Table 8.45 Surface effects on flame-retardant nylon mouldings

Feature	Analyte line					
	Si $K\alpha$	Cl $K\alpha$	Zn $L\alpha$	Zn $K\alpha$	Sb $L\alpha$	Sb $K\alpha$
Wavelength (Å)	7.13	4.73	12.25	1.44	3.44	0.47
Critical depth (μm)	30	70	8	580	60	13 000

After Warren [252]. From P.L. Warren, *Analytical Proceedings*, **27**, 186–187 (1990). Reproduced by permission of The Royal Society of Chemistry.

purpose [261]. Similarly, XRF has also been used for forensic purposes to distinguish yellow PE bags of different origin on the basis of Pb and Ti concentrations [262]. XRF is quite suitable for the characterisation of *pigments* and siccatives, and quality control (e.g. Sn content in various products). For nondestructive determination of pigments in easel paintings and medieval manuscripts, both μXRF and TXRF (Section 8.4.1.1) are frequently used.

As XRF is least suitable for use with elements of low atomic number, primary *antioxidants* which contain either oxygen or nitrogen as the active element cannot be analysed by XRF. Secondary AOs, however, which contain either phosphorous or sulfur as the active element, can be readily determined by this method. As the XRF technique is a nonspecific method, the same calibration curve for a given element (e.g. phosphorous) can be employed for a variety of phosphite antioxidants. For example, a calibration curve for the phosphorous content of unknown samples was based on PE/Ultranox 626 with variable additive levels [50]. Obviously, knowledge of the type of AO formulated into the sample is required for conversion of the phosphorous content to the AO concentration. In this respect, XRF analysis is most suited for QC type applications, although it can also be employed as an effective screening tool to determine the presence of secondary antioxidants in unknown samples. In Ultranox® 2795P (blend of Ultranox® 210/626 and Ca stearate), the P and Ca contents were measured by means of XRF, while Ultranox 210/626 were quantified by SEC [263]. In principle, for process control purposes Irganox B blends (Irganox 1010/1076 with Irgafos 168 in various ratios) can quantitatively be determined by means of XRF measurement of one element (P) only. This requires a high accuracy for this low-Z element, and presumes that the blend composition is rigorously stoichiometric. In practice, the approach is not completely satisfactory. Alternatives are direct in-polymer UV, mid-IR or NIR analysis. XRF has also been used for the quantitative analysis of metal-based heat stabilisers in PVC [141].

The *antiblock agent* Sipernat 44 ($Na_2O \cdot Al_2O_3 \cdot 2SiO_2 \cdot 4H_2O$) in LDPE can be determined by Al and Si measurement after ashing. Quantitative XRF analysis sets limits on the Si/Al ratio. Determination of P, S, Ca, Zn and Cu in additives in *lubricating oils* was carried out using on-line EDXRF and fundamental parameter software [264]; comparison was made with wet chemical analysis (P, S) and AAS (Ca, Zn, Cu).

Uncommon elements may be used as 'tracer' compounds. '*Tracers*' are being used increasingly by polymer manufacturers as a rapid means of screening complaint polymer samples, in order to ascertain whether the polymer has been produced by a particular manufacturer. Tracers are often based on innocuous metal stearates, such as strontium, lithium and bismuth. Since these elements are seldom encountered in other common plastics additives, they can be used as unique identifiers for certain grades of polyethylene. XRF analysis is ideal for rapidly detecting these tracer compounds. The 1σ precision value for each of these elements is approximately 1 ppm for a 100 s measurement.

Table 8.47 shows the available options for the analysis of *polymer processing aids*, namely combustion and instrumental methods. The best method is dependent on PPA type, the level to be measured, and the available equipment (see also Section 8.2.1.2). Fluoropolymer processing aid concentrations can be determined by WDXRF configured to measure either fluorine or a tracer, and by EDXRF to analyse a tracer [29]. Calibration curves are required. At present, EDXRF or benchtop XRF units cannot directly measure fluorine. For resin or masterbatch producers who prefer to make on-line XRF measurements of processing aid concentrations (to letdown levels of 50–100 ppm), processing aids that contain a tracer (usually $BaSO_4$) are available. The analysis time is less than two minutes.

In relation to the European Standard EN 71/3 concerning *toy safety* [266]. Gimeno Adelantado *et al.* [267] have proposed a simultaneous multi-elemental preconcentration approach to increase the sensitivity of the technique for the elements of interest (Sb(III), Ba, Cd, Cr(III), Hg, Pb and As(III)). As the organic matrix of the

Table 8.47 Analysis of polymer processing aids

Method	Concentrates	End-use levels[d]	Testing time
Oxygen bomb (Parr®)[a]	+	+	75 min
Pyro-hydrolysis[a,b]	+	+	15 min
Oxygen flask (Schöniger)	+	+	10 min
WDXRF[a]	+	+	25 min
Pulse ¹⁹F NMR[a]	+		<10 min
FTIR/NIR[a]	+		n.g.
DSC[c]	+		n.g.
¹H NMR[c]	+	+	1 h

[a] Fluoro polymer processing aid
[b] Combustion (pyro-hydrolysis) with fluoride specific electrode detection
[c] PEO
[d] Letdown levels
After Dewitte [265]. Reproduced by permission of the author.

sample is totally destroyed with molten $NaOH/AgNO_3$ and converted into an aqueous solution of the species to be determined, aqueous solution reference species can be used for calibration. The results obtained were comparable with an AAS method.

Metals can enter the polymer system at the polymer production stage (e.g. catalyst residues), in a subsequent compounding stage (e.g. additives), or during production of the final consumer article (e.g. adhesives, ink). Out of the ten heavy metals according to Ferguson [268], namely As, Bi, Cd, Hg, In, Pb, Se, Sb, Te and Tl, only Cd and Pb are sometimes present in certain plastics packaging articles. Typical sources of metals are pigments (Cd, Pb, Zn), stabilisers (Cd) and lubricants (Zn). Reliable literature references on *metal concentrations* in plastics packaging materials and in mixed plastics waste (MPW), containing more than 65 wt % PE, PP and 10 wt % PS, are relatively few [269]. Mark [269] has examined the concentration ranges of metals in used plastics articles, such as bottles, films, technical and other nonpackaging articles – providing data relevant to both recycling and recovery. A maximum content of 1000 ppm can be estimated for MPW, if all metals are taken into account. Lead, Zn and Cu account for 50 % of the total content. The total amount of neurotoxic metals which are specific to plastics does not exceed 300 to 400 mg kg^{-1}. Metals not found in MPW at levels higher than a few ppm are Hg, As, In, Te, Sb, Tl and Bi. Many of the aforementioned elements are found in or on other packaging materials, such as paper, and are not specific to polymers. The levels of metals found in MPW were in compliance with legislative limits.

Reliable sampling of MPW is critical for the reliability and validity of the analytical results. Sampling while shredding, after sorting and of baled material have been discussed [269]. As for sample preparation for EDXRF, both extrusion blending and a grinding and densification procedure are used to produce a homogeneous composite sample of solid material before pressing the typical plaques for EDXRF. Sample preparation for chemical analysis is mainly carried out by oxidation in an acid medium at an elevated temperature. A pressurised bomb, microwave oven decomposition or plasma-induced ash formation are all used for acid digestion. The different methods of oxidation of plastics have not been validated in a very rigorous way, and different treatment times and temperature levels have been used.

The high sensitivity of XRF for elements of higher atomic weight (e.g. Br, Cl) is of advantage in the identification of plastics in *postconsumer waste*, and is being used in practice only for separating 'PVC' bottles from other plastics. Using a combination of X-ray and transmission detectors greatly extends the scope of a system, enabling it to deal with the most common types of plastics entering the waste stream from retail packaging sources (PE, PP, PET, PS, PVC). In industrial-scale systems for separating PVC bottles from other plastics, XRF and IR detection are used in conjunction, namely XRF for PVC and IR to separate PET, HDPE, PP and PS. Although NIR systems may also be used, a basic drawback with NIRS is that it is not well adapted to detection of plastics containing a large amount of fillers and/or certain types of FRs, or to compounds that are grey or black. Other pertinent spectrometry technologies include sliding spark and laser-induced emission (LIESA).

In the framework of the European PERM project (development of CRMs), it was necessary to monitor homogeneity both on the micro-level (between individual pellets, mg-amounts) and the macro-level (between bottles, g-amount). As XRF typically uses g-amounts, this technique is quite suitable for a macrohomogeneity study by means of multi-element analysis on tablets of hot-pressed pellets (amount about 10 g each). In this way, all elements of interest could be monitored for their homogeneity within one batch and between different batches. For the development of CRMs, an element stability study is also needed, in view of the fact that elements, depending on their physico-chemical properties, may migrate throughout the material, even evaporate or change chemical form. To test element stability, aged pellets are to be re-analysed by XRF for elemental loss.

XRF is widely used in *production control*, e.g. for the determination of catalyst residues in polyolefins [270]. Table 8.48 shows the typical results of a standardless analysis of PP. Industrial XRF analyses of additives in polymers (QC) are usually carried out by either calibration curve or monitor programming. Use of calibration curves allows measuring the full additive concentration range, and is not restricted to a given additive. Consequently, there is no need to implement a new calibration curve upon adjustments of additive or additive contents. For a calibration curve, at least

Table 8.48 Standardless XRF analysis of PP (ppm levels)

Element	Certificate	UniQuant®
Al	160	180
Si	480	390
Cl	55	87
Ca	140	150
Ti	49	56

three calibration standards are needed. A disadvantage is a slightly lower accuracy in comparison to monitor programming. In fact, any systematic deviation from the curve adds to the inaccuracy. Monitor analytical programmes are designed for a particular polymer grade. The monitor or reference standard is almost identical to the sample, and thus overcomes matrix and particle-size effects and renders calibration lines unnecessary. Monitor samples are based on product specification. Monitor analysis is well suited for accurate analysis, and is incidentally also the cheapest method. For a given product portfolio, three monitor samples are necessary. The accuracy of the monitor method is only dependent on the standard deviation of the measurement and the accuracy of the chemical analysis. Although the monitor programme has the higher accuracy, it is optimal only in the target area. However, the method has to be upgraded upon every change in additive, and is thus not suitable for a larger range of concentrations. The whole range of technical product grades can be analysed by selecting a suitable array of monitors. Typical industrial XRF analyses are the determination of Al, Zn, Ca, Cl, Ti and Fe in HDPE; of Al, Mg, Zn, Ca, Cl, Ti, Fe, S, P and Si in PP; of Zn and S in XPE; of Cd, Pb and Sb in ABS; and of Cu, K and I in nylons. Some typical lower limits of detection (3σ) of elements in PET using Peltier-cooled detection (PCD) are 8.9 (P), 1.5 (Ti), 1.1 (Mn), 0.6 (Co) and 5.9 ppm (Sb) [271].

XRF can also be used for measuring the thickness of very thin metallic coatings. The applications of XRF have been reviewed [253]. For further applications of XRF in in-polymer additive analysis, see also ref. [23].

8.4.1.1 Total-reflection X-ray Fluorescence

Principles and Characteristics   Another significant XRF development is total-reflection X-ray fluorescence (TXRF). One of the major problems that inhibits obtaining good detection limits in small samples is the high background due to scatter from the sample substrate support material. This can be overcome by using a very small angle of incidence, less than the critical angle, so that total reflection of the exciting radiation occurs [272]. The specimen is deposited upon a polished plate that is optically flat. The most common carrier materials are quartz, silicon, germanium, glassy carbon and Perspex acrylic plastic. Due to the low glancing angle, the primary radiation barely penetrates into the plate (ca. 3 nm) and there is very little scattering. The emitted X-rays are detected by a Si(Li) detector

placed immediately above the sample, and are further processed as in EDXRF. In TXRF, which is in fact a variant of EDXRF with substantially lower background, absolute detection limits of a few pg can be achieved for many elements.

TXRF is most applicable to liquid samples, but success has also been achieved with direct analysis of some solids, e.g. very thin sections of organic tissue and polymer film. Alternatively, small amounts of solid material can be analysed by TXRF after acid digestion.

Table 8.49 shows the most important *characteristics* of TXRF. Advantages are reduced spectral background and high detection efficiency. The technique allows measurement of pg amounts or concentrations in the range of a few tenths of a ppb, without preconcentration. As the specimen is only a few µg thick, generally one is not faced with the rather complicated matrix effects as encountered with thick samples. Consequently, quantification is quite straightforward. Usually a single element not present in the sample is added as an internal standard, and a calibration curve is established which is valid for all matrices. TXRF systems are now moving from laboratory to industry.

The hypersensitivity of TXRF to the solid surface roughness allows for surface morphology diagnosis and the determination of element distribution along the heterogeneous surface. Glancing-angle X-ray techniques, which take advantage of the restricted evanescent wave penetration into the sample, provide surface sensitivity comparable to the photoelectron-based surface investigation tools such as AES and XPS. Unlike these latter techniques, a vacuum environment is not necessary for glancing-angle X-ray techniques, which

Table 8.49   Main characteristics of total-reflection X-ray fluorescence

Advantages

- Ease of specimen preparation
- No vacuum requirements
- Ability to handle mg quantities of material
- Primary analytical method
- Surface-sensitive method
- Multi-element capability (ppt level)
- Detectability (10^{10} at. cm^{-2})
- High sensitivity (pg amounts)
- Quantitative
- Absence of matrix effects

Disadvantages

- Only liquid and thin film sampling
- Surface roughness
- Need for flat sample

greatly enhance the applicability of the technique to surface problems.

Taniguchi and Ninomiya [273] and Ninomiya *et al.* [274] have reviewed TXRF as an inherently surface-sensitive, nondestructive and cost-saving method in the analysis of trace elements and other microcomponents in polymers and other materials. An overview of sources, samples and detectors for TXRF is available [275].

Applications   Some typical applications of TXRF are quantitative microanalyses of samples in the 10–100 µg range, forensic applications, microinclusions, environmental samples like rain-water, sea-water, etc.

Metal derivatives (Ti, Zn, Cd, Sn, Sb, Pb) and bromine from additives in recycled thermoplasts from consumer *electronic waste* were determined by dissolving the samples in an organic solvent, followed by TXRF analysis [56]. The procedure proved considerably less time-consuming than conventional digestion of the polymer matrix. Results were validated independently by INAA.

TXRF has also been used for the characterisation of single, colourless *textile fibres* (polyesters, modified cellulose and wool), yielding a fingerprint trace-element pattern, suitable for forensic purposes [276,277]. Sample preparation involved dissolution/predigestion in HNO_3 and matrix removal (O_2 cold plasma).

TXRF is frequently used for *contamination control* and ultrasensitive chemical analysis, in particular in relation to materials used in semiconductor manufacturing [278,279], and metallic impurities on resin surfaces, as in PFA sheets [279,280]. TXRF has been used by Simmross *et al.* [281] for the quantitative determination of cadmium in the four IRMM polyethylene reference materials (VDA-001 to 004). Microsamples (20–100 µg) from each reference material were transferred by hot pressing at 130 °C as 3 µm thin films straight on to quartz glass discs commonly used for TXRF analysis. The results obtained were quite satisfactory (Table 8.50). Other reports of the *forensic application* to plastic materials by TXRF have appeared [282], including a study of PE films by elemental analysis [283].

TXRF offers the possibility of a simple, multielemental analysis of practically all elements composing inorganic *pigments*, with the advantage relative to some other analytical techniques that just a few µg of sample material are sufficient to obtain a fingerprint of the pigment. The technique was applied to identify the origin of tiny, hard, particle-like spots of unknown material, formed on the surface of the original varnish of

Table 8.50   Polyethylene CRMs from the Institute for Reference Materials and Measurements (IRMM)

IRMM label	Manufacturers' label	Cd content (mg kg^{-1})	
VDA-001	Sicolen Yellow 09/16493	40.9 ± 1.2^a	39.6^b
VDA-002	Sicolen Orange 28/16494	75.9 ± 2.1	75.2
VDA-003	Sicolen Red 39/16495	197.9 ± 4.8	192.3
VDA-004	Sicolen Bordeaux 49/16496	407 ± 12	392.5

^{a} Certified value.
^{b} TXRF ($\pm 6\%$).

cars [284]. The key elemental composition and impurity pattern of medieval manuscript pigments can provide information on the origin of the artist's material; for this purpose, TXRF with microsampling by means of (destructive but invisible) cotton-wool rubbing is often used in conjunction with portable Raman spectroscopy (MRS) with a 780 nm diode laser (no microsampling necessary) [285–287]. Pigment identification by means of TXRF alone is of course limited by the fact that only elements with $Z > 13$ can be detected; moreover, some pigments show the same key elements. In combination, TXRF reveals quantitative element analytical data and Raman spectroscopy molecular information. Identification is then based on comparison with a pigment library (400 entries).

TXRF has recently also been used to discriminate between green PE garbage bags, based on Ti, Pb, Cr and Cu analysis (forensic evidence) [288]. XRF provided a more effective method of discrimination when compared to IR spectrometry or DSC.

8.4.2 Particle-induced X-ray Emission Spectrometry

Principles and Characteristics   Particle-induced X-ray emission spectrometry (PIXE) is a high-energy ion beam analysis technique, which is often considered as a complement to XRF. PIXE analysis is typically carried out with a proton beam (proton-induced X-ray emission) and requires nuclear physics facilities such as a Van der Graaff accelerator, or otherwise a small electrostatic particle accelerator. As the highest sensitivity is obtained at rather low proton energies (2–4 MeV), recently, small and relatively inexpensive tandem accelerators have been developed for PIXE applications, which are commercially available. Compact cyclotrons are also often used.

PIXE is not a true nuclear analytical method, because it is based on the interaction of fast ions with the electron clouds of the atoms. In fact, PIXE is based on *X-ray emission*. The specimen to be analysed is irradiated by accelerated, charged particles (usually protons or occasionally heavier ions) and electrons are ejected from the innermost shells in atoms of the sample. An X-ray quantum is emitted when an electron from an outer shell fills a vacancy. The X-rays are analysed with a suitable spectrometer. The method is intrinsically multivariate, because the X-rays form a spectrum. The atomic numbers of the elements involved are determined by X-ray spectrometry, and no information about the isotopic composition is obtained.

In 1970, Johansson *et al.* [289] first showed that a combination of excitation with 2–MeV protons and X-ray detection with a Si (Li) detector, permitting energy-dispersive detection (5–25 keV) with high efficiency, constituted a very powerful method for multi-elemental analysis of trace elements. A PIXE spectrum is usually quite complicated. The highest sensitivity is obtained for $20 < Z < 40$ and $Z > 75$ in the analysis of trace elements in a matrix of light elements and with thin samples of low mass. PIXE handles essentially thick specimens (flat and smooth), as well as thin specimens (foils). Since PIXE analysis *in vacuo* demands conducting specimens, graphite is added as a binding material. With thin specimens, such as a plastic foil, and a beam area of a few mm^2, the amount probed by the proton beam can be 10^{-4} g or even lower. Analysis depth is quoted as 50 μm (depending on particle energy and on sample composition), with depth and lateral resolutions of 10 and 1 μm down to 0.1 μm, respectively. With a spatial resolution of a few μm, the absolute detection limits are of the order of 10^{-15}–10^{-16} g.

PIXE is a primary analytical technique, like NAA, and permits absolute determinations of concentrations. The basis for *quantitative PIXE* is, as in all X-ray methods, that there exists a relationship between the net peak area of an X-ray line in the spectrum and the amount of element in the sample. One of two methods can be applied to calibration:

(i) use of thin-film standards, i.e. a relative method; or
(ii) use of fundamental physical parameters in combination with an experimentally determined efficiency curve.

A correction for matrix effects is usually required. Difficulties in PIXE quantitation may be relieved by complementary information from RBS [290], or FTIR and elastic backscattering (EBS) analysis [291]. FTIR can give a rough estimation of the elemental composition, while EBS or RBS can deliver information on the major-element composition.

Table 8.51   Main features of PIXE

Advantages

- Ease of specimen handling
- Relatively nondestructive (local radiation damage with overdose)
- Primary analytical technique
- Multi-element technique ($Z \geq 6$)
- High sensitivity ($f(Z)$)
- Low background
- Trace-element analysis (0.1–1 ppm level)
- Quantitative
- Good accuracy (5 %) and precision (1 %)
- Well-understood physical process
- Macro- and microanalysis; microtomographic analysis
- Fast (min)
- Automatable
- 1D and 2D imaging
- Mature technique

Disadvantages

- Unsuitable for the lightest elements and for rare earths
- Need for particle accelerator
- Vacuum atmosphere
- Limited beam current (radiation damage)
- Relatively complex spectrum processing (spectral interferences)
- No chemical state information
- Expensive
- Need for associated specialist expertise

Table 8.51 shows the *main features* of PIXE. The most important advantage of PIXE is the possibility of detecting more than 20 elements simultaneously. PIXE is most suitable for the determination of trace elements in a light matrix, where the limit of detectable concentration is of the order of 0.1–1 ppm. The absolute detection limit is a few pg. PIXE is characterised by straightforward matrix effects, and allows chemical shift measurements. Spectral interferences cannot be avoided. The interaction between light ions of MeV energy and matter is relatively nondestructive. PIXE may damage a specimen in the case of too large an accumulated dose. The risk is, of course, more serious for very sensitive materials (e.g. paper, parchments).

With charged-particle microprobes, the samples must be stained and thinned to improve both contrast and signal-to-noise ratio; coated with a thin conducting layer to reduce charging effects and improve spatial resolution; and be in vacuum to maintain the charged-particle beams. Finally, information on the chemical state of the detected elements is difficult to obtain using techniques based on charged particles.

PIXE has been compared to other focusing X-ray beam techniques [292]. Although the measurement part of PIXE and EDXRF is exactly the same, the spectra have a different appearance, due to different excitation processes. In XRF, creation of a vacancy in an inner shell is most efficient for an energy of the exciting photon just above the binding energy of the electron. The intensity of the characteristic lines therefore decreases with decreasing atomic number Z, also reflecting the decreasing fluorescence yield. In PIXE, ionisation cross-sections are highest for light elements, and decrease with increasing Z. Therefore, the peaks of the lighter elements are much larger than for the heavy elements. Also, the physical processes causing background in EDXRF and PIXE spectra are different. Background in XRF originates from scattering of the incident X-rays from the source. In PIXE, the main source of background is secondary-electron Bremsstrahlung (SEB) from the deceleration in the sample of ejected electrons. Table 8.52 emphasises some other differences between the two techniques. If an analytical problem can be solved using XRF, there is little sense in using PIXE.

PIXE was originally employed as a bulk analytical technique (macro-PIXE). Since proton beams can be generated with much higher brilliance (i.e. particle density in the beam) than that of photon beams produced by X-ray tubes, macro-PIXE is ideally suited for analysing very small amounts of material. Focusing of the proton beam to sub-μm or μm dimensions with currents of the order of 100 pA leads to microbeam PIXE and the proton microprobe, variously known as the *nuclear microprobe* (NMP) or nuclear microscope. Micro-PIXE is a combination of the microbeam technique with PIXE analysis, and is analogous to the electron microprobe. The differences are that the electron gun is replaced by a proton accelerator, and that the MeV protons are rather more difficult to focus than keV electrons; a proton microprobe is therefore considerably more complicated and expensive than its electron counterpart. *μPIXE* is a trace-element analysis technique. It is the ratio of characteristic lines to continuum intensity, rather than the actual characteristic X-ray yield per unit beam charge, that is very different from the EPMA situation. The great advantage of μPIXE compared with the electron microprobe is that the detection limits are better by two or three orders of magnitude. The reason is the lower continuous Bremsstrahlung background induced by ions in the sample material. PIXE is the most sensitive of the analytical methods that can be combined with the microbeam technique. The spatial resolution of the proton and electron microprobes is comparable (1 μm). Microbeam PIXE overcomes some of the drawbacks of other techniques, such as low sensitivity (EPMA) or low spatial resolution (XRF).

The proton beam can be scanned across the surface of the specimen in the same fashion as in an electron microprobe, and thus it is possible to obtain concentrations as a function of position. With μPIXE in the scanning mode, the spatial distribution of trace elements can be studied by 1D and 2D imaging (iPIXE), i.e. line-scan and pixel mapping. Cutting-edge NMPs allow imaging at high resolution and sensitivities below 100 ppb [292a]. The combination of high spatial resolution (1–2 μm) and deep beam penetration into a sample (many tens of μm) means that NMP can produce high-resolution images of features deep in a sample. Unlike a focused electron beam, the nuclear microprobe does not lose its focus as it penetrates the specimen. NMP makes it possible to produce quantitative trace-element images in real-time. Microanalytical techniques such as PIXE, micro-PIXE, solid sampling AAS, electron- and laser-microprobe techniques, analyse (or consume) sample masses from 1–2 mg down to 1 pg. *Nuclear microscopy* (iPIXE), which can determine the concentrations of major and trace elements, represents a powerful

Table 8.52 Comparison between PIXE and XRF

Feature	PIXE	Conventional XRF
Specimen quantity	Only mg required	Gram quantities may be required
Sample thickness	Only 10–15 mg cm^{-2}	Thicker samples needed (often 2 mm or more)
Lowest detectable concentration (fractional)	10^{-6}–10^{-7}	$\sim 10^{-8}$
Portability	Yes (source instead of accelerator)	Yes (γ source instead of X-ray tube)
Complexity of technology	Relatively complex	Relatively simple
Nonconducting specimens	Coating with conducting film required	No coating needed
Volatile specimens	Necessitates bringing proton beam into air	No problems
Spatial scanning	Possible (iPIXE)	Possible (iXRF)
Microbeam	Possible (μPIXE)	Possible (μXRF)

array of high-energy ion microbeam techniques. Development of more efficient detectors is desirable. As for hyphenation, a PIXE detector was used in connection with capillary electrophoresis for metal-specific and sensitive detection [293].

PIXE has been reviewed repeatedly [294–296]. Several monographs have appeared [297,298].

Applications   The main application fields of PIXE are earth science, air pollution studies (aerosol analysis), mineralogical studies, forensic science, arts and archaeology. In the external-beam PIXE technique, the proton beam is taken out to ambient air. This mode finds application in the analysis of art objects (paintings, books, etc.).

PIXE, in conjunction with SR-XRF and XAS, has been used to study friction wear and dissolution of orthopaedic implant systems by analysing the distribution and chemical state of the metallic elements (Fe, Cr, Ni and Ti) in the tissues around a failed implant [300]. Russell *et al.* [301] have reported comparisons of trace-element content (Ca, Fe, Cu, Zn) in different polymer foil types (PP, polyimide, polyester, PC, PS). Polycarbonate Kimfol was identified as the purest substrate then available (Ca 3.6 ± 1.1, Fe 0.3 ± 0.1, Cu 0.08 ± 0.05, Zn 0.14 ± 0.05 ng cm^{-2}). Brissaud *et al.* [299] have determined the concentration profile of colour pigments in model paintings using PIXE at six different proton energies. Various 30–50-μm-thick layers of acrylic paints were examined: white (Ca, Ti), blue (Si, Cu), yellow (S, Zn, Cd, Ba) and orange (S, Ca, Ti, Fe, Zn, Ba), as well as their superpositions. Figure 8.8 shows the Zn, Cu and Cd concentrations for a multilayer made of orange, blue and yellow pigments. The order of layer deduced from the profiles corresponds to the order of deposition. The spatial resolution of each colour layer is poor. Analytical polymer/additive problems have so far not been too difficult or perhaps impossible to solve with means other than PIXE.

PIXE has also been used for pigment analysis in furniture and interior painting [301a]. The growing use of μPIXE and associated beam techniques in art and archaeometry is noticeable [302,303]. Even delicate materials such as paper and parchment are unaffected by microbeams.

The applications of high-energy ion microbeams have been collected in a monograph [304].

8.4.3   X-ray Absorption Spectrometry

Principles and Characteristics   In X-ray absorption spectroscopy (XAS) the absorption coefficient (μ)

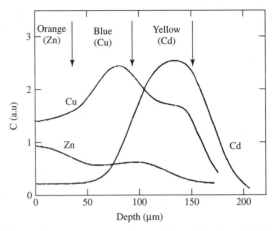

Figure 8.8   Mean concentration profiles of Zn, Cu and Cd in multilayer acrylic paint, consisting of orange deposited on blue, both on yellow. The arrows define the three paint layer limits as they were evaluated through weighing. After Brissaud [299]. Reprinted from *Nuclear Instruments and Methods in Physics Research*, **B117**, I. Brissaud *et al.*, 179–185, Copyright (1996), with permission from Elsevier

of a material is measured as a function of the incident photon energy. When the energy of the incident X-ray reaches the binding energy of an electronic level of one of the constituting elements of the sample, absorption increases abruptly (absorption threshold), and a core level photoelectron is ejected. Beyond the *absorption edge*, there are oscillations in the absorption coefficient μ, referred to as an absorption fine structure. The oscillations are due to the scattering of the photoelectron wave function by the surrounding neighbours of the excited atom. Outgoing and backscattered waves interfere constructively or destructively, thus giving rise to oscillations of the absorption coefficient.

The whole region around the absorption edge is referred to as X-ray absorption fine structure (XAFS), and is divided into two regions, the X-ray absorption near-edge structure (*XANES*) (also called near-edge X-ray absorption fine structure, NEXAFS), extending about 40 eV beyond the edge, and the extended X-ray absorption fine structure (*EXAFS*) extending typically 1000 eV beyond the edge. In addition, there can be pre-edge features in the spectrum, which arise from excitations of the core electron to higher states within the atom, and can fingerprint the valency state of the atom. All three spectral features provide structural information about the target atom (the absorber), the atom which is emitting the photoelectron. The strong oscillations just beyond the absorption edge (XANES), which arise from multiple scattering of photoelectrons by atoms in a local cluster around the absorbing atom, provide information on the geometrical arrangement of these near-neighbour atoms. The weaker structure observed from 40 eV beyond the edge (EXAFS), which represents interference between the outgoing photoelectron wave and waves following backscattering from neighbouring atoms, hides information on interatomic distances, coordination number and atomic types. EXAFS is therefore an element-selective technique and a probe of local order around the excited atom, which allows us to study disordered or amorphous matter. One can always unconditionally choose the element whose environment is being probed, by measuring the absorption spectra at a selected edge.

A XAS experiment involves the irradiation of a sample with a tuneable source of monochromatic X-rays, usually from a synchrotron facility (high brilliance). Third-generation synchrotrons have sufficient intensity to observe XAFS spectra up to 100 keV. Nevertheless, laboratory-scale XAFS spectroscopy is of importance, despite the vast availability of synchrotron beam time [305].

Table 8.53 shows the *main features* of XAS. The advantages of EXAFS over diffraction methods are that the technique does not depend on long-range order, hence it can *always* be used to study local environments in amorphous (and crystalline) solids and liquids; it is atom specific and can be sensitive to low concentrations of the target atom (about 100 ppm). XAS provides information on interatomic distances, coordination numbers, atom types and structural disorder; and oxidation state by inference. Accuracy is 1–2 % for interatomic distances, and 10–25 % for coordination numbers.

X-ray absorption spectroscopy is a reliable and routine method to investigate the local atomic environments of selected elements in materials, and provides useful

Table 8.53 Main features of X-ray absorption spectroscopy

Advantages

- Nondestructive
- Wide applicability (all aggregation states)
- Element specific (for Li to U)
- Short-range order (local environment around absorber)
- High sensitivity
- High reliability
- Chemical shifts
- Depth profiling (glancing incidence angles)
- Theoretically well understood
- Relatively low cost/experiment

Disadvantages

- Need for synchrotron radiation facility
- Expensive equipment

'fingerprints' for individual compounds. In particular, XAS is applicable to the highly dispersed state, where the information power of XRD breaks down.

XANES spectroscopy is also the basis of chemically sensitive *X-ray imaging*, as well as qualitative and quantitative microspectroscopy [306]. μXANES is attractive for chemical analysis, with its spatial resolution down to 10 μm. Variations on the theme are surface EXAFS (SEXAFS), grazing incidence XAS and *in situ* time-resolved XAS investigations. Grazing angle XAFS can be used for the study of ultrathin multilayer systems.

EXAFS has recently been reviewed [307]; XAS and NEXAFS monographs are also available [308,309].

Applications EXAFS spectroscopy is obviously well suited to speciate and quantify the state of heavy metals, e.g. in soils [310]. Similarly, it allows differentiation of ZnO and Zn stearate on the basis of the X-ray absorption structure of zinc.

Zinc dialkyldithiophosphates (ZDDPs), which act as *antiwear additives* in lubricating oils and were postulated to exist in various molecular forms (monomer, dimer or neutral form, and basic form), were studied by multi-edge (Zn K-, P K- and S K-) XAS for structural assessment [311]. Grazing incidence absorption spectroscopy measurements have provided evidence for breakdown of the ZDDP molecule following its adsorption on to a steel substrate surface [312]. XANES and CEMS were used to study the interaction of perfluoropolyalkyl ether (PFPAE) additives with Fe-based alloys [313].

Friction wear and dissolution of *orthopaedic implant systems* were examined by PIXE, SR-XRF and XAS [300]. From XAFS analysis, it appeared that the chemical state of Fe had changed as a consequence of dissolution in the organic tissue.

Elemental distribution and chemical state of ppm metal impurities can be measured using synchrotron-based X-ray fluorescence (μXRF) and X-ray absorption spectroscopy (μXAS), both with a 1–2 μm² spatial resolution [314].

XANES, which can be used to determine molecular structure and orientation of chemisorbed molecules on well-characterised single-crystal surfaces and is able to discriminate between the same atoms in different bonding situations, has been used to examine the supramolecular organisation adopted by the dye Reactive Red 3:1 physisorbed and chemisorbed on to cotton and cellophane substrate materials [315]. A distinct difference in the nature of the *dye/cotton interaction* was observed for different preparative methods. The mode by which dyes interact with fabrics is of great commercial importance in the textile industry. Some requirements are colour fastness, resistance to chemical attack (by detergent components), etc.

Grazing angle XAS techniques (XANES) can be applied to ultrathin film systems [316]. Selected NEXAFS, XPS and FTIR spectroscopy results were obtained for plasma-polymerised films with different monomers (styrene, acetylene, ethylene and butadiene) [317].

XAS has widely been used for catalyst studies [308], and also in relation to polymers and catalyst residues.

8.4.4 X-ray Diffraction

Principles and Characteristics In X-ray powder diffraction (XRD) a collimated beam of X-rays, with wavelength $\lambda \sim 0.5-2\,\text{Å}$, is incident on a sample placed in a capillary tube of low-absorption glass, and is diffracted by the crystalline phases in the specimen according to Bragg's law ($\lambda = 2d \sin \theta$, where d is the spacing between atomic planes in the crystalline phase). The intensity of the diffracted X-rays is measured as a function of the diffraction angle 2θ and the specimen's orientation. Capillary sample spinning improves the diffraction statistics and reduces preferred orientation. The diffraction pattern is used to identify the specimen's crystalline phases and to measure its structural properties. Diffraction utilises the interference of the radiation scattered by atoms in an ordered structure, and is therefore limited to studies of materials with long-range order. X-ray diffraction analysis is an extremely simple and reliable method for qualitative identification of crystalline inorganic and organic substances.

Crystallography is an advanced discipline [318]. Modern crystallography has been developed since the discovery of X-ray diffraction in 1912 from the original basis laid down by classical crystallographers. One of the beauties of this modern discipline, while it can be somewhat mathematical, is the universal use of standardised notations and conventions, as developed through the International Union of Crystallography (IUCr).

The diffractometer has gradually evolved in terms of maximum power of sealed X-ray tubes, rotating anodes, new X-ray optics, better detector efficiency, position-sensitive detection and, lately, real-time multiple-strip (RTMS) fast X-ray detection, which replaces a single detector by an integrated array of parallel detectors to provide an up to 100-fold increase in efficiency compared with traditional detectors without compromise on resolution. Time-resolved powder diffraction is possible using 2D detectors. XRD is increasingly been recognised as a powerful tool for process control.

Modern *powder diffraction technology* in $\theta-\theta$ geometry makes use of ceramic X-ray tubes, high-resolution Peltier-cooled Si(Li) solid-state detectors with low-angle performance, high sensitivity and S/N ratio, and an automatic sample changer. Peltier-cooled detectors offer a superior energy resolution of 300 eV (cf. 600 (3000) eV for a scintillation detector with/without monochromator, respectively) and remove fluorescence radiation and the need for traditional $K\beta$ filtration. Peaks with d-spacing as large as 200 Å can be observed using $\text{Cu}K\alpha$ radiation. Vertical and horizontal goniometers with $\theta-2\theta$ and $\theta-\theta$ operation are available, eventually sharing a common X-ray source. Diffraction patterns of samples can be obtained from $3-75°$ 2θ. Full diffraction patterns with the intensity and resolution for Rietveld analysis can be collected in two minutes.

Synchrotron radiation X-ray powder diffraction (SR-XRD) is a fine experimental tool (not just for routine measurements), allowing μXRD microbeam crystallography (small spot size: $50-100\,\mu\text{m}^2$) in view of the high brilliance of the source, and 2D mineralogical mapping. *Single-crystal XRD* has advanced greatly, with powerful CCD diffractometers and software for solving the phase problem [319,320]. Small-molecule CCD single-crystal X-ray crystallography with minimal human intervention is now so rapid that results are obtained in many cases before NMR or MS analyses are completed. Large databases are available (CSD, ICSD).

Knowledge of the crystal structure is a prerequisite for the rational understanding of many properties of crystalline materials, such as morphology, photostability and dissolution rate. For many chemical compounds, it is often difficult or even impossible to grow single crystals large enough for structure elucidation by single-crystal diffraction techniques. High-quality powder diffraction data are, in general, far easier to obtain. However, structure solution from powder diffraction data is more difficult.

Table 8.54 indicates the *main characteristics* of X-ray powder diffraction. Powder samples of 1 mg or greater can be analysed. The principal advantage is that substances (e.g. $BaSO_4$) can be identified directly, and not indirectly via their elements (e.g. Ba and S). This is particularly advantageous if both silica and silicates (kaolin, talc) are present. Detection limits are matrix dependent, but ca. 3 % in a two-phase mixture, with synchrotron radiation ca. 0.1 %.

XRD allows standard materials characterisation. A set of diffraction angles and relative intensities is characteristic of each individual chemical compound. By

Table 8.54 Main characteristics of X-ray powder diffraction

Advantages

- Polycrystalline samples (powders and solids)
- Simple sample preparation
- Nondestructive (for most materials)
- Phase identification (ICDD file)
- Structural analysis (atomic level)
- Various geometries ($\theta-\theta$, $\theta-2\theta$)
- Solid-state technology detection
- High speed; high-throughput analysis
- High-energy resolution (300 eV Peltier cooled; 0.04° 2θ FWHM)
- Low-angle performance ($\theta = 0.24°$ for goniometer radius of 500 mm)
- Qualitative and quantitative (multiphase) analysis
- Microfocus (ca. 10 μm lateral resolution)
- Maintenance-free
- Automation (sample changer)
- Advanced discipline

Disadvantages

- Only applicable for crystalline materials
- Ionising radiation

comparison of data measured on samples of unknown composition, with the values found in card files or on data carriers (CD-ROM) (e.g. ASTM Powder Diffraction File, ICDD data bank), identification of the components can be carried out. Such data banks of single-phase X-ray powder diffraction patterns contain line sets of more than 80 000 different crystallographic phases. For calibration, mixtures of known chemical and crystallographic composition are examined. Computer-based search and match routines permit positive identification of unknowns. *Crystallographic analysis* traditionally comprises two main tasks: indexing of a powder pattern and unit-cell refinement. Rietveld analysis is used for the simulation of a powder diagram from structural data; for standardless quantitative phase analysis; and for refinement of crystal structure data from powder measurements, and hence the determination of molecular configuration and conformation. Quantitative determinations of the content of a single or several crystalline components are usually achieved by comparative measurements (with an 'internal' or 'external' standard). It is necessary to randomise the orientation of the sample and to select suitable, texture-free reflections, since preferred orientation can lead to large errors in quantitative X-ray analysis.

The main use of XRD consists of the identification and quantitative analysis of crystalline phases, accurate determination of lattice parameters and atomic arrangements, determination of the degree of crystallinity and crystalline size, orientation and strain. Specialised

uses are defect imaging and characterisation of atomic arrangements in amorphous materials and multilayers, concentration profiles with depth and film thickness measurements. XRD is a bulk technique with a probing depth of a few μm; it is not a typical technique for depth profiling.

Selected area diffraction (SAD) combined with microscopy is an important supplementary tool to X-ray diffraction in crystal structure analysis. SAD has the additional advantage of giving the correlation between morphology and crystal structure whenever single crystals are too small for single-crystal X-ray analysis.

The analytical power of XRF and XRD has lately been combined in an integrated XRF/XRD system, in which XRD powder measurements are examined for phase identification and Rietveld analysis on the basis of element concentrations. Process analysis, a former stronghold of XRF, can now be performed by high-speed XRD, which is supported by XRF element-analytical data.

Overviews [321] and monographs [322,323] dealing with X-ray powder diffractometry are available.

Applications The general applications of XRD comprise routine phase identification, quantitative analysis, compositional studies of crystalline solid compounds, texture and residual stress analysis, high- and low-temperature studies, low-angle analysis, films, etc. Single-crystal X-ray diffraction has been used for detailed structural analysis of many pure polymer additives (antioxidants, flame retardants, plasticisers, fillers, pigments and dyes, etc.) and for conformational analysis. A variety of analytical techniques are used to identify and classify different crystal polymorphs, notably XRD, microscopy, DSC, FTIR and NIRS. A comprehensive review of the analytical techniques employed for the analysis of polymorphs has been compiled [324]. The Rietveld method has been used to model a mineral-filled PPS compound [325].

Inorganic fillers in plastics compositions are usually in a very finely divided form and, as such, are ideal for powder XRD study. A sample size of a few mg gives a good pattern in 1 or 2 h. Crystalline *mineral fillers* can usually be observed directly in the complete polymeric formulation, in concentrations exceeding about 1%. Combined XRD/XRF studies are favoured [326]. A mineral filler is easy to identify in a compound in the absence of other fillers.

For the identification of mineral fillers in EPs, XRD was compared with various other analytical techniques (Table 8.55). Both IR and XRF allowed identification

Table 8.55 Potential of various techniques for the analysis of mineral fillers in polyamides

Filler	Wet chemical[a]	XRF	IR[b]	RS	XRD
Talc	2	1,2	1	1	1
Clay	2	1,2	1	0	0
Mica	2	1,2	1	1	1
Wollastonite	2	1,2	1	1	1
Chalk	2	1,2	1	1	1
E-glass	2	1,2	0	0	0

[a] Hydrolysis with HCl and fuming with HF, in combination with IR.
[b] μFTIR and ATR-FTIR.
0 = identification and quantification not possible
1 = identification possible
2 = quantification possible
After Klok *et al.* [258].

of the mineral fillers by spectral comparison, namely on the basis of specific absorption bands (IR) and intensity ratios of the inorganic elements detected (XRF). Although XRD is the preferred technique for identifying crystalline mineral fillers, it cannot be used for amorphous fillers, such as glass, or for crystalline fillers heated to very high temperature. XRF can quantify mineral fillers by means of a determination of the ash residue. In the determination of mineral fillers, TGA (total ash content), XRF (ash composition) and XRD (phase analysis) complement each other. Wet chemical methods, hydrolysis with HCl, and fuming off of the insoluble material using HF, yield quantitative results. Comparison with the ash residue content, in combination with an IR analysis of the residues obtained, yields information on the presence of other polymers, impact modifiers and flame retardants. The XRD signals of clay interfere with those of polyamides.

XRD analysis on sulfated ash residue of new additive packages in *lubricating oils* has indicated the presence of nonstoichiometric mixed salts of the elements Mg, Ca, Zn, P and B [60].

With little experience in XRD, one can recognise most of the common fillers in original *rubbers*, such as calcite, clay, talc, red iron oxide, zinc oxide, titanium dioxide, quartz and gypsum [326]. As in almost all other forms of analysis, the most satisfactory results are achieved by examining the filler in the purest form in which it can be obtained. Thus, it may be desirable to separate the filler from the matrix. Ashing is such a concentration method, and has been used for subsequent identification of inorganic fillers. Diffraction of rubber ash can give additional information, particularly with compounds which undergo changes; contain amorphous fillers such as silica; or have low minor constituents, whose pattern could be lost in the background of the original rubber compound. However, study of

the ash alone is not sufficient, since the ashing process could change many rubber components into some other forms. Similarly, pyrolysis can be used to quantitatively determine volatiles, mineral fillers (as residues) and carbon-black in rubbers and elastomers [327,328]. Mixtures of fillers can be analysed, but obviously the difficulty of interpretation increases with the number of phases. Determination of minor inorganic components is challenging. XRF/XRD has been used for the measurement of total ash weight for mixtures of clay, chalk and TiO_2 from *paper* samples [329].

XRD and TEM are the primary and complementary tools in the characterisation of polymer–clay *nanocomposites* [330]. Other frequently used techniques are DSC and NMR. There are several terms used to describe the dispersion of clays in polymers. A delaminated (or exfoliated) system is considered to be one where the clay layers are well dispersed in the system, with the individual silicate layers being spaced far apart; hence, no reflections are observed in XRD. Intercalated systems exist when polymer chains have entered between the clay layers, but have not pushed them apart to give a delaminated structure. In an intercalated system, the well-ordered multiclay layered structures will give an XRD pattern with a *d*-spacing larger than that of the original clay. In an immiscible system, the clay layers never separate when mixed with the polymer. These large stacks give an XRD pattern which shows no change in *d*-spacings when compared to the original clay. Basically, this is not a nanocomposite, but rather a filler in a polymer. These three categorisations have been primarily based on XRD results. XRD, however, does have some shortcomings, in that it cannot differentiate between delaminated and disordered immiscible systems. In addition, there are systems that are not cleanly intercalated or delaminated, as suggested by XRD. As disorder increases in the sample, XRD becomes a less-definitive technique than TEM. Order dependence and the inability to differentiate between types of dispersions are a major drawback of XRD. For XRD, the ease of analysis and the ability to measure accurately the *d*-spacing between clay layers are key strengths, most appropriate for intercalated nanocomposites.

XRD was used to investigate the spacings of silicate layers of montmorillonite (from 1.9 to 4 nm) in PP/montmorillonite (MMT) nanocomposites prepared by *in situ* graft-intercalation in the presence of acrylamide [331]. Similarly, XRD and TEM were used to study the dispersibility of PP/MMT nanocomposites prepared by melt intercalation using organo-montmorillonite and conventional twin screw extrusion [332]. Various delaminated and intercalated polymer (PA6, PA12, PS,

PP-*g*-MA) silicate nanocomposites and intercalated thermoset silicate nanocomposites for flame-retardant applications were characterised by XRD and TEM [333]. XRD, TEM and FTIR were also used in the study of 1D CdS nanoparticle-poly(vinyl acetate) nanorod composites prepared by hydrothermal polymerisation and simultaneous sulfidation [334]. The CdS nanoparticles were well dispersed in the polymer nanorods. The intercalation of polyaniline (PANI)-DDBSA (dodecylbenzenesulfonate) into the galleries of organo-montmorillonite (MMT) was confirmed by XRD, and significantly large *d*-spacing expansions (13.3–29.6 Å) were observed for the nanocomposites [335].

DeBellis *et al.* [336] have studied the conformational preference of 3′ substituents in 2-(2′-hydroxyphenyl)benzotriazole (BZT) *UV absorbers*, in relation to the photopermanence in coatings. A rotomer of BZTs, in which the intramolecular hydrogen-bond has been disrupted (Figure 8.9), has been implicated in its photodegradation. The position of this equilibrium depends upon the polarity and H-bonding ability of the medium in which the UVA operates. Ghiggino *et al.* [337] have used principal-component analysis to resolve the absorption spectrum of Tinuvin P and 5-sulfonated Tinuvin P into separate components due to the planar and nonplanar forms. Each form makes significant contributions to the long- and short-wavelength absorptions (i.e. the resolved spectra overlap considerably). However, there is no fundamental reason to expect a similar situation for in-coating behaviour. The superior photostability of a 3′-alpha-cumyl substituted BZT has been rationalised in terms of the pronounced conformational preference of this group, as supported by molecular dynamics (MD) simulations, solution NMR and solid-state X-ray crystallographic data (Figure 8.10).

XRD can be used for studies of *polymorphic transformations* in bloom formation. In combination with DTA, XRD has been used to study polymorphic transformations of Ca stearate and Ca stearate monohydrate [338]. Full pattern refinement has been used for quantitative analysis of mixtures of crystalline

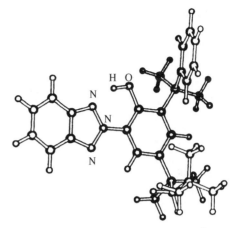

Figure 8.10 Solid-state structure of Tinuvin® 928 [2-(2′-hydroxy-5′-*tert*-octylphenyl)benzotriazole]. After DeBillis *et al.* [336]. Reprinted with permission from A.D. DeBellis *et al.*, *ACS Symposium Series*, **805**, 453–467. Copyright (2002), American Chemical Society

components in melamine-based *flame retardants* [339]. Accuracies of 1 % or better can readily be obtained.

XRD can be applied for the identification of organic and inorganic *pigments* [340]. Diffraction data for a selection of organic pigments used in paints have been listed [340,341]. Another useful source of reference on XRD data of synthetic dyes and pigments is ref. [342]. Some problems related to XRD of organic pigments (OP) are:

(i) low scattering power;
(ii) weak XRD spectra for intensely coloured, low-concentration OPs; and
(iii) low crystallinity [341].

Pigments in stratigraphic section may be analysed by TXRF (elemental composition), μRaman spectroscopy (functional groups) or XRD (structural analysis). At variance to conventional XRD, SR-XRD allows acquisition of patterns on thin stratigraphic sections

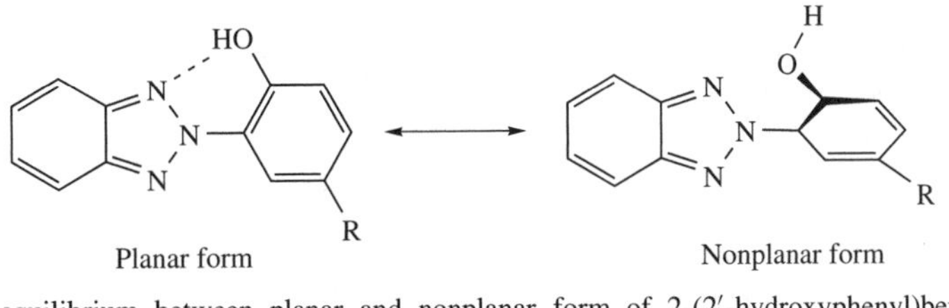

Planar form   **Nonplanar form**

Figure 8.9 Rotomeric equilibrium between planar and nonplanar form of 2-(2′-hydroxyphenyl)benzotriazole UVAs. After DeBillis *et al.* [336]. Reprinted with permission from A.D. DeBillis *et al.*, *ACS Symposium Series*, **805**, 453–467. Copyright (2002), American Chemical Society

(about $100\,\mu m$), from substrate to weathered outermost layer [343]. This allows the determination of the spatial distribution of pigments forming different layers.

Small-angle X-ray scattering (SAXS) data have made it possible to deduce the localisation of organic additives (pigments) in the bulk of isotactic polypropylene (iPP) [344]. This work has confirmed that the additives are located in the amorphous phase, in spite of their crucial influence on the formation of the crystalline phase of iPP. SAXS has also been used to study the 3D structure of different carbon-black aggregates, and silica-filled SBR rubber compounds [345].

The industrial applications of X-ray diffraction have been described [346].

8.5 INORGANIC MASS SPECTROMETRY

Principles and Characteristics Inorganic mass spectrometry is the most universal and most sensitive instrumental method for simultaneous determination of practically all elements. Although the development of the spark source in 1934 has essentially marked the beginning of inorganic mass spectrometry, it was not until 1958 that the first commercial spark-source mass spectrometer (SSMS) became available. Despite this early start, the use of mass spectrometry for inorganic analysis has been very limited compared to organic applications. The latest evolution in measurement technology has resulted in the gradual uprooting of inorganic optical spectrometric techniques by inorganic mass spectrometry. Both ICP-AES and GFAAS are declining in use in comparison to ICP-MS. The sensitivity of ICP-MS exceeds that of GFAAS.

Mass spectrometry is the only *universal multi-element method* which allows the determination of all elements and their isotopes in both solids and liquids. Detection limits for virtually all elements are low. Mass spectrometry can be more easily applied than other spectroscopic techniques as an absolute method, because the analyte atoms produce the analytical signal themselves, and their amount is not deduced from emitted or absorbed radiation; the spectra are simple compared to the line-rich spectra often found in optical emission spectrometry. The resolving power of conventional mass spectrometers is sufficient to separate all isotope signals, although expensive instruments and skill are required to eliminate interferences from molecules and polyatomic cluster ions.

For mass-spectrometric analysis, a number of ionisation methods are available, differing in the type

of evaporation and atomisation of the sample material (thermal evaporation, laser ablation, electron beam, ion bombardment, etc.). Inorganic mass spectrometry for (ultra)trace-element determination consists of various plasma mass-spectrometric methods (such as SSMS, GD-MS, LMMS, (LA-)ICP-MS) and a variety of other methods (TIMS, AMS, RIMS). Currently, there is no universal ion source that is capable of providing chemical data similar to plasma, EI and CI characteristics simultaneously and independently of one another. Switched ion sources (hot and cool plasma) are under development [346a]. Table 8.56 lists the characteristics of the *ideal PS-MS method*, still beyond reach. The early available technique, SSMS, has a somewhat limited application, being suitable mainly for conducting solid samples, such as metals, semiconductors and minerals, In addition to SSMS, isotope dilution thermal ionisation MS (ID-TIMS) [347] has received some attention since the 1950s. Despite the existence of these techniques, progress in mass spectrometry for inorganic analysis has been slow, mainly because:

(i) SSMS has always been somewhat cumbersome and lengthy, with limited throughput;
(ii) inorganic samples are usually more difficult to volatilise and to ionise than organic samples, and require more powerful ion sources; and
(iii) some inorganic compounds are difficult to dissociate.

During the production of ions from a solid, a number of complex, interconnected, and obscure processes take place. When multiple ionisation occurs, the spectra obtained are complex.

Mass-spectrometric techniques may be divided into two types: those in which the ions are formed directly, as in ICP-MS and SIMS, and those in which particles are

Table 8.56 List of desiderata of plasma-source mass spectrometry

• Universal ion source	• No sample preparation
• Low detection limits	• Macro, micro and transient samples
• Excellent precision ($<0.1\,\%$ RSD)	• Amenable to speciation
• Broad dynamic range ($>10^7$)	• Spatial resolution in solid samples
• No spectral or matrix interferences	• Rapid ($10\,s$ per sample)
• Complete, simultaneous elemental coverage	• Inexpensive (initial and continuing)
• Isotope analysis capability	• Simple, automated, compact
• Absolute (standardless) analysis	

Table 8.57 Comparison of various mass-spectrometry methods for elemental analysis of solids

Feature	SSMS	GD-MS	SNMS	SIMS
Metals and semiconductors	Yes	Yes	Yes	Yes
Insulators	After mixing with conductor	After mixing with conductor	Directly	Charge compensation
Bulk analysis	Yes	Yes	Yes	(Yes)
Surface layer/depth profiling	Difficult	Yes	Yes	Yes
Lateral resolution	ca. 100 μm	–	1–5 μm	1–5 μm
Elemental coverage	Large	Large	Large	Large
RSF range[a]	Small	Small	Small	Wide
Dynamic range	Limited by photoplates (graded exposures)	Very wide	Wide	Wide
Sensitivity	ng g^{-1}	ng g^{-1}	μg g^{-1}	ng g^{-1} (not all elements)
Speed of analysis	Slow	Fast	Fast	Moderate
Main MS interferences	Multiply charged ions	'Argides', dimers	'Argides', dimers	Polyatomic ions
Ease of analysis	Skilled interpretation required	Good	Good	Skilled operation required

[a] RSF, relative sensitivity factors.

sputtered from the sample and subsequently ionised in a different process, as in SNMS, LA-ICP-MS and GD-MS. Table 8.57 compares various mass-spectrometric methods for the elemental analysis of solids.

Nowadays, in *multi-element bulk trace analysis* MS competes mainly with optical and XRF spectrometry. Although these methods are more widely used, MS has strengthened its application via ICP-MS with its robust ion source; convenient ways of sample introduction; and low and high resolution. Thermal ionisation MS (TIMS) has become more versatile and, with isotope dilution standardisation in solid targets, remains as a reference method. Glow-discharge MS can be used for the analysis of both conductive and nonconductive materials. The role of spark-source MS (SSMS) is now quite limited. The detection power of various inorganic mass-spectrometric techniques is roughly as follows: SNMS (ppm), r.f.-GD-MS (10 ppb), SIMS, LIMS and SSMS (ppb), d.c.-GD-MS (100 ppt), LA-ICP-MS (10 ppt), ICP-MS (aqueous solution, ppq).

The main advantages of plasma-source mass spectrometry (PS-MS) over other analytical techniques, such as PS-AES and ETAAS, are the possibilities of quantitative isotope determination and isotope dilution analysis; the rapid spectral scanning capability of the mass spectrometer; and semiquantitative determinations to within a factor of two or three. Several *labelling methods* are used for the quantification of analytes present in complex mixtures. In these methods, the sample is spiked with a known quantity of a labelled form of the analyte. Both stable and radioactive isotopes are used as labels. Isotope ratios may be measured by IDMS or NMR techniques.

Flow-injection (FI) on-line analyte preconcentration and matrix removal techniques greatly enhance the performance of atomic spectrometry [348]. By using USN with membrane desolvation (MDS) as the interface, FI sorbent extraction can be directly coupled with ICP-MS for the analysis of organic solutions [349].

As to the distribution of trace elements, e.g. contaminants residing at the *surface* or segregated at interfaces or in inclusions, electron probe X-ray micro analysis and AES are the standard techniques, but both are limited by their poor detection power. Mass-spectrometric techniques are more sensitive by several orders of magnitude, and SIMS in particular is playing an increasingly important role in solving technical problems. Laser mass-spectrometric methods are based on non-resonant laser ionisation, or laser ablation and resonant laser ionisation. Whereas LMMS involves ionisation of evaporated and atomised sample material in a laser microplasma under high-vacuum conditions, LA-ICP-MS uses the evaporation of sample material by a (de)focused laser beam in an inert gas atmosphere (e.g. Ar), under normal pressure and post-ionisation in a gas discharge of an inductively coupled plasma. In RIMS one or more dye lasers are tuned precisely to the wavelength required for the excited states and ionisation of evaporated atoms, in order to achieve a highly selective ionisation of analyte.

Noticeable trends in inorganic mass spectrometry are speed, simultaneity, and fewer problems from isobars and tuned plasma conditions. Inorganic MS techniques used for inorganic trace analysis have been reviewed [350]. Various monographs deal with inorganic mass spectrometry [351,352] and plasma-source mass spectrometry in particular [353,354].

Applications   Table 8.58 shows the main fields of application of inorganic mass spectrometry. Mass-spectrometric techniques find wide application in inorganic analysis, and are being used for the determination of elemental concentrations and of isotopic abundances; for speciation and surface characterisation; for imaging and depth profiling. Solid-state mass spectrometry is usable as a quantitative method only after calibration by standard samples.

The artificial separation between organic and inorganic mass-spectrometric methods is now narrowing, as shown by *speciation* studies (Section 8.8). Plasma-source MS (PS-MS), mainly as ICP-MS and MIP-MS, has been particularly effective when applied to speciation analysis. Direct speciation is also possible with electrospray MS (ESI-MS).

Gijbels [355] has reviewed the elemental analysis of high-purity solids by mass spectrometry. For applications of plasma-source mass spectrometry, see ref. [353].

8.5.1   Spark-Source Mass Spectrometry

Principles and Characteristics   The original idea of spark-source mass spectrometry (SSMS) is due to Dempster [356], long before the first commercial instruments. In spark-source MS, atomisation and ionisation

Table 8.58 Fields of application of inorganic mass spectrometry

Topic	Methods	Reference
Trace and ultratrace analysis	SSMS, GD-MS, LA-ICP-MS, (IDMS)	Chapter 8
Surface analysis[a]	SIMS, SNMS, GD-MS, LA-ICP-MS	–
Isotope analysis	TIMS, ICP-MS, SIMS, (IDMS)	Section 8.5.4
Speciation[b]	ICP-MS, MIP-MS, ESI-MS, IDMS	Section 8.8

[a] Lateral element distribution, surface contamination, depth profiling.
[b] Hyphenated to chromatographic separation techniques.

are achieved by applying *in vacuo*, a high-frequency (ca. 1 MHz) potential difference of 20–100 kV between the electrodes, one or both of which must be prepared from the sample material. The gaseous positively charged ions formed are accelerated into the mass spectrometer by a d.c. potential. The wide initial kinetic energy of ions formed in the spark (some keV) requires application of a double-focusing mass spectrometer. Time-of-flight and quadrupole analysers have also been used in special applications, but with poor resolution.

Table 8.59 shows the *main characteristics* of SSMS. SSMS is unsurpassed when it comes to obtaining a general view of impurity element concentrations in a sample – covering the whole periodic table (including elements such as C, N, O, Cl and P). Spark-source mass spectrometry is superior to NAA, in that the detection limits are constant for all elements. In comparison with nine other techniques, SSMS detects the greatest number of elements to the lowest detection limit, and for solid samples it gives the lowest detection limit [357]. Because the process of ion production in SSMS shows approximately equal sensitivity for all elements, it is possible to use the technique for semiquantitative analysis, without regard to the actual changes in sensitivity that occur. For quantitative analysis, however, the ionisation processes need to be understood and controllable. Standardless spark-source

Table 8.59 Main characteristics of spark-source mass spectrometry

Advantages

- Simple sample preparation only (without chemical pretreatment)
- Very efficient method for multi-element analysis of conducting solid samples
- Simultaneous detection of nearly all elements with high specificity
- Relatively uniform sensitivity for all sample types (Li to U)
- Applicable to major, minor and trace elements (detection limits 0.1 ppm)
- High absolute sensitivity

Disadvantages

- Sampling requirement: ca. 25 mg
- Less suitable for insulators; no direct analysis of solutions
- Poor lateral and in-depth resolution
- Poor accuracy (15–30 %)
- Long analysis time
- Electrical charging effects
- Complex and expensive instrumentation
- Outdated instrumentation
- Declining technique
- Niche applications only

mass-spectrometry analysis has been reported [358]. The procedure is severely limited in a practical sense, in that five or six internal standards with widely differing properties are necessary to achieve quantitative multi-element analysis. SSMS is characterised by relatively few interferences, but its ion yield is erratic and has a wide energy spread.

SSMS can be classified among the milliprobe techniques (Figure 8.3), i.e. it is a unique link between microprobe techniques and macroanalytical methods that are characterised by poor lateral and in-depth resolutions (as in OES), or that have no lateral resolution whatsoever (as in NAA). Also, the achievable precision and accuracy are poor, because of the irreproducible behaviour of the r.f. spark. Whereas analysis of metals, semiconductors and minerals is relatively simple and the procedures have become standardised, the analysis of nonconducting materials is more complex and generally requires addition of a conducting powder (e.g. graphite) to the sample [359]. Detection limits are affected by the dilution, and trace contamination from the added components is possible. These problems can be overcome by the use of lasers [360]. Coupled with isotope dilution, a precision of 5 % can be attained for SSMS.

SSMS has always been somewhat impractical and lengthy, because of the vacuum pumping required, and the need for integration of spectra over time – to overcome the influence of instabilities of the spark source. Investment and running costs are high, and throughput is usually only one sample per day. As commercial equipment is no longer available, the spark source used for inorganic analysis is rated as a declining, almost extinct technique which has been superseded by the more efficient GD-MS and ICP-MS techniques. Reviews on spark-source MS are available [101,355,361,362].

Applications Real applications of spark-source MS started on an empirical basis before fundamental insights were available. SSMS is now considered obsolete in many areas, but various unique applications for a variety of biological substances and metals are reported. Usually, each application requires specific sample preparation, sparking procedure and ion detection. SSMS is now used only in a few laboratories worldwide. Spark-source mass spectrometry is still attractive for certain applications (e.g. in the microelectronics industry). This is especially so when a *multi-element survey analysis* is required, for which the accuracy of the technique is sufficient (generally 15–30 % with calibration or within an order of magnitude without). SSMS is considered to be a sensitive survey method for the analysis of surfaces and thin films [363]. Spark-emission spectrometry enables rapid and simultaneous determination of many elements in metals, including C, S, B, P, and even elements such as N and O. SSMS has been used for the analysis of trace levels of certain elements causing discoloration in TiO_2 [364]. The applications of SSMS have been reviewed [101]. Spark-source mass spectrometry has sporadically been used for the analysis of polymers and plastics [365,366], but no real significant contributions to polymer/additive analysis can be reported.

8.5.2 Glow-Discharge Mass Spectrometry

Principles and Characteristics The glow discharge is a simple, two-electrode device filled with a noble gas to a pressure of about 0.1–10 mm Hg. A few hundred volts applied across the electrodes cause breakdown of the gas and formation of ions, electrons and other species that make the GD useful in analytical chemistry. Glow discharge has been known for over 60 years as an *ion source* for mass spectrometry [367]. In recent years, the glow discharge has been developed as a more stable, low-energy alternative ion source. The first commercial GD-MS (utilising a double-focusing mass spectrometer) was introduced in 1983 [368]. Quadrupole-based GD-MS is now most commonly used [369], but ToF-MS has also been considered. From an instrumental point of view, the ICP source in ICP-MS can easily be replaced with a GD source [370].

Table 8.60 shows the *main features* of GD-MS. Whereas d.c.-GD-MS is commercial, r.f.-GD-MS lacks commercial instruments, which limits spreading. Glow discharge is much more reliable than spark-source mass spectrometry. GD-MS is particularly valuable for studies of alloys and semiconductors [371]. Detection limits at the ppb level have been reported for GD-MS [372], as compared to typical values of 10 ppm for GD-AES. The quantitative performance of GD-MS is uncertain. It appears that ± 5 % quantitative results are possible, assuming suitable standards are available for direct comparison of ion currents [373]. Sources of error that may contribute to quantitative uncertainty include: sample inhomogeneity, spectral interferences, matrix differences and changes in discharge conditions.

Glow-discharge methods need new developments or will go into decline. The greatest needs for GD methods are:

(i)   increase in detection limits (10^2–$10^3 \times$);
(ii) convenient methods for nonconductors;

Table 8.60   Main features of GD-MS

Advantages

- Direct solid sampling (bulk metal, thin film, compacted powder, solution residues)
- Response to both metallic and nonmetallic elements (survey elemental scans)
- Minimal matrix effects
- $ng\,g^{-1}$ detection limits attainable
- Sensitivities generally similar for most elements
- Good precision
- Isotopic information
- Mass spectra much simpler than optical emission spectra

Disadvantages

- Slow
- Low source pressure (0.1–10 Torr)
- No chemical information
- Electronic charging effects
- Relatively expensive
- No commercially available instrumentation for r.f.-GD-MS

(iii) reduction of spectral interferences; and

(iv) new application areas (e.g. polymers, organics).

Tables 8.57 and 8.61 compare the performance of GD-MS to other inorganic mass-spectrometry techniques, including LA-ICP-MS which acts as a strong competitor. GD and laser ablation techniques offer the possibility to obtain information about the distribution of analyte within the sample, but on quite a different scale. Pulsed GD requires ToF-MS [374].

Several reviews deal with GD-MS [156,355,369,373, 375]. A recent monograph is available [166].

Table 8.61   Comparison between GD-MS and LA-HR-ICP-MS

Feature	Glow discharge	Laser ablation
Spatial resolution:		
Lateral	2 mm	<5 μm
Depth	10–100 nm	~1 μm
Quantification	External standard	Internal standardisation
Semiquantitative	+	−
Matrix dependence	Weak	Strong
Analysis speed	Slow	Fast
Accuracy	Good	Matrix dependent
Materials:		
(Semi)conductors	+	+
Nonconductors	+[a]	+
Depth profiling	+	−

[a] After compacting with a graphite or metal matrix.

Applications   GD-MS is until now mainly a multi-element tool for direct bulk analysis of *conducting samples* (metals and alloys) and the analysis of thin layers of solid samples. Glow-discharge spectrometry finds numerous fields of application for direct analysis, and needs further development, for example regarding the investigation of nonmetals and nonmetallic coatings by means of high-frequency lamps, and the procedures for calibration in case of depth profiling as well as microdistribution analysis. Up to now, reference materials for these purposes have been rare. GD-MS has been applied to the analysis of insulators after compacting a graphite or metal matrix. A great reduction in r.f.-GD-MS applications has been noticed recently. No applications in the polymer/additive area have been recorded.

8.5.3   Inductively Coupled Plasma–Mass Spectrometry

Principles and Characteristics   The need for simple spectra, adequate spectral resolution, and low detection limits, as the basic requirements for a successful inorganic analytical technique, led to the consideration of using a combination of a mass spectrometer as a detector and a high-pressure plasma as an ion source [376–378]. In fact, since an ICP exhibits temperatures of 5000–9000 K, its use as an ion source for mass spectrometry was almost obvious, featuring both relative absence of matrix interferences and high first-degree ionisation yields. It was already in an early stage of development of ICP as an excitation source for AES, that there were indications that its characteristics made it suitable to serve as an ion source for mass spectrometry. Namely, the majority of the prominent lines used in ICP-AES spectra originate from the singly charged ion ground state. Apparently, the ICP efficiently ionises most elements introduced to single charged positive ions (M^+), so that the plasma can be used as an ion source for the mass spectrometer. Technically, ICP-MS was born at the moment when continuous ion extraction from the bulk plasma without any intervening boundary layer proved to be possible and commercial ICP-MS instruments were first launched in 1983 [379]. A schematic diagram of such an instrument is given in Figure 8.11.

In Figure 8.12, the basic set-up of an ICP-MS instrument is presented as a block diagram, consisting of a sample introduction system, the inductively coupled argon plasma (ICP) and the mass-specific detector. By far the most commonly applied sample introduction technique is a pneumatic nebuliser, in which a stream of argon (typically $1\,L\,min^{-1}$), expanding with high

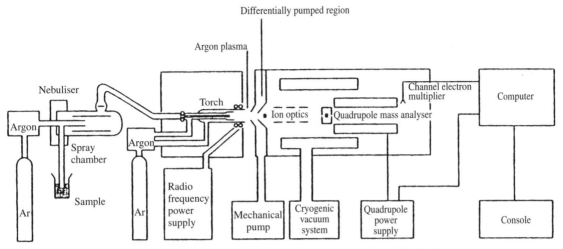

Figure 8.11 Schematic diagram of an ICP-MS instrument. Reproduced by permission of Perkin-Elmer

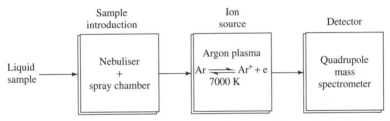

Figure 8.12 Block diagram of an ICP-MS. After Klinkenberg [380]. Reproduced by permission of the author

velocity from an orifice, converts a stream of liquid into an aerosol of droplets with a wide spread in diameter. Nebulisers are almost exclusively used in conjunction with a spray chamber that serves as a droplet separator, allowing droplets with diameters of typically $<10\,\mu$m to pass. The sample introduction system is the Achilles' heel of all ICP-MS instruments. In fact, only about 2 % of the sample solution actually reaches the plasma in a conventional system. The large variety of nebulisers used for the introduction of liquid samples reflects the diversity of analytical problems encountered. The choice of nebuliser depends on matrix constitution (high dissolved solid content, slurry, organic), amount of sample available (nebulisers with a limited uptake rate such as micro-concentric and direct injection) and sensitivity (ultrasonic, thermospray, hydraulic high-pressure and direct-injection nebulisers). The ultrasonic nebuliser (USN) coupled to a desolvation system is highly efficient.

The front-end arrangements used nowadays in conjunction with ICP-MS are tailored to the analytical problem at hand. Sample introduction techniques which

can be easily adapted to ICP-MS are flow injection (FI), liquid chromatography (LC), laser ablation (LA) and electrothermal vaporisation (ETV). The sample material introduced accordingly into the spray chamber is exposed to the hostile conditions of the ICP, and is successively desolvated, vaporised and dissociated, whereafter the atoms are ionised and/or excited, thus making the ICP a very effective ion source for *inorganic mass spectrometry*. The dwell time in the ICP equals a few milliseconds. Sample dissociation is extremely efficient and few, if any, molecular fragments of the original sample pass into the mass spectrometer. Most elements are ionised to a large extent (>95 %), especially those with first degree ionisation potentials below 8 eV (Table 8.65), thus making the ICP a very effective ion source for inorganic MS. The argon plasma operates at atmospheric pressure, and is sampled by means of a differentially pumped interface consisting of two orifices, called the sampler and the skimmer.

It is obvious from the history of ICP-MS that the *interface* is of crucial importance. Within the interface, conditions are converted from the high

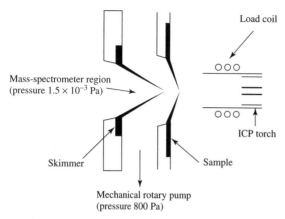

Figure 8.13 Schematic overview of the interface between ICP and MS. Reproduced by permission of Perkin-Elmer

temperature and atmospheric pressure of the ICP to the room-temperature and high-vacuum conditions necessary for mass-spectrometry detection (Figure 8.13). Ion sampling should take place without changing the composition of the bulk plasma. Inside the mass spectrometer, the ion beam is focused and transmitted by a series of electrostatic lenses, termed the ion optics. Hereafter, the ion beam enters a mass analyser. The function of the mass analyser in ICP-MS is to separate and allow the passing of a very specific narrow mass region (<1 amu) while rejecting all others. The vast majority of ICP-MS instruments is equipped with quadrupole mass analysers. The mass-resolving properties of quadrupoles are based upon path stability or instability of ions within the quadrupole device. For our purposes, the exclusive existence of singly charged species (isotopes) is presumed, since these are the species normally detected by ICP-MS.

Table 8.62 shows the *main characteristics* of ICP-MS, which is widely used in routine analytical applications. The ICP ion source has several unique advantages: the samples are introduced at atmospheric pressure; the degree of ionisation is relatively uniform for all elements; and singly charged ions are the principal ion product. Theoretically, 54 elements can be ionised in an ICP with an efficiency of 90 % or more. Even some elements that do not show ionic emission lines should be ionised with reasonable efficiency (namely, As, 52 % and P, 33 %) [381]. This is one of the advantages of ICP-MS over ICP-AES. Other features of ICP-MS that make it more attractive than ICP-AES are much lower detection limits; ability to provide isotopic ratio information and to offer isotope dilution capabilities for quantitative analysis; and clean and simple spectra. The

Table 8.62 Main features of ICP-MS

Advantages

- Diversity in sample introduction systems (nebulisation, coupling, solid sampling)
- Small liquid sample size (few mL)
- Sequential and simultaneous multi-element ultratrace detection
- Isotope ratio and dilution determinations
- Low detection limits for most elements (sub-ppt for solutions; ppb for solids)
- Wide dynamic range (up to 10 decades)
- Almost complete elemental coverage (isotopic fingerprinting)
- Spectral simplicity
- Semiquantitative scanning capability; quantitation
- Accuracy (0.2 % isotopic; 5 % or better quantitative and 20 % or better semiquantitative)
- Fast (high sample throughput; 10 s per isotope)
- Versatile; robust, excellent short- and long-term stability
- User friendliness
- Matrix tolerance
- Mature technology

Disadvantages

- Troublesome sample preparation (except for LA solid sampling)
- Lack in harmonisation of ICP operating conditions
- Not suitable for micro and transient sampling
- Poor sample introduction efficiency
- Destructive
- Polyatomic ion interferences (for unit mass resolution ICP-QMS)
- Low concentration of dissolved solids (clogging)
- Carbon deposition in torch and cones
- Mass discrimination
- 1–5 % RSD
- Detection limits Z dependent
- Speciation with complications
- No direct spatial resolution
- Expensive equipment
- High argon consumption

linear dynamic range for many elements is such that major, minor and trace levels may all be determined at once. Detection limits are superior to (almost) any other trace-element analytical technique, including GFAAS, and are even better in high-resolution instruments. Compared to ICP-AES, ICP-MS has a limited capability in the determination of very high analyte concentrations (more than hundreds of milligrams per litre). ICP-MS as a detector for organometallic compounds is superior to detection methods such as AAS and MIP-AES.

Whilst qualitative survey scans are useful in the preliminary analytical investigations of a problem, the extent to which such signals can be quantified is of greater importance. Unless the sample matrix is simple, *internal standardisation* (or frequent calibration) will be required for quantitative analysis using external calibration with a series of standard solutions. For the best results, an element with properties similar to that of the analyte is needed (in order to behave similarly in ICP-MS), so several internal standards may be required for a multi-elemental analysis spanning the entire mass range. Single- and multi-element solutions are available as inorganic reference standards for ICP-MS. The ultimate internal standard is an enriched isotope of the element to be analysed [382].

Although the detection limit of an ICP-MS is about 1 ppt, the device is rather inefficient in the transport of the ions from the plasma to the analyser (interface efficiency of about 1 %). The influence of the ICP-MS sampling cone is still to be worked out. Introduction of organic solvents into an ICP-MS decreases the sensitivity, due to excessive solvent loading of the plasma.

ICP-MS presents various shortcomings as compared to the requirements of an ideal PS-MS technique (Tables 8.62 and 8.56). Simultaneous detectors, as in ToF-MS or array-detector atomic mass spectra (ADAMS), offer several advantages in terms of sensitivity, precision, LOD (50 ppq), resolving power and sample throughput. PS-ToFMS and ICP-ADAMS are still in their infancy.

In recent years several new instruments have been developed based on different mass-spectrometer principles. Two different categories of *ICP-MS instruments* are currently commercially available: low-resolution instruments (using either QMS, ITMS or ToF-MS) and focusing high-resolution instruments (DFS, FTMS). Selected specifications for these two categories are shown in Table 8.63. Both the quadrupole-based and the double-focusing instruments allow a sequential multi-element measurement, whereas ICP-ToFMS allows

simultaneous multi-element detection. Bench-top ICP-oaToFMS provides fairly high resolution (4000), high speed and mass detection, without compromising mass range. ICP-ToFMS is almost ideally suited for measurement of fast transient signals, e.g. in ETV and laser ablation applications with analysis of the entire mass range in less than 30 μs. For the different ICP-ToFMS instrument geometries, see ref. [382a]. The ion trap is also an attractive alternative mass analyser for coupling with an ICP [383]. The 3D quadrupole ion-trap mass spectrometer (3DQMS) produces a 3D quadrupolar potential field within the cavity, trapping the ions in a stable oscillating trajectory within the cell. Ion storage in the trap under conditions of a collision cell leads to the advantageous and complete destruction of polyatomic interfering molecules [383]. When polyatomic ions collide with collision or reaction gases such as He, H_2 or NH_3 they are dissociated into their component atoms or ions. However, individual m/z values are measured sequentially, resulting in a loss of sampling speed and duty cycle. Moreover, the restricted dynamic range of the ITMS (10^4–10^6 ions) can limit its dynamic range. Detection levels are in the ppt range for most elements. Also, FTMS has been coupled to ICP for elemental analysis [384]. Although this combination can eliminate most isobaric interferences and offers simultaneous multi-elemental measurements, its cost, ultrahigh vacuum requirements, restricted ion-storage capability and lengthy measurement time have limited its routine application. Every instrument has its place and particular use. A quadrupole will always have better single-element sensitivity than the multi-element, simultaneous ToF-MS.

Recent advances in ICP-MS technology also comprise other forms of collision-cell technology (CCT) and multicollector systems for high-precision isotope ratio determinations. Yet, ICP-MS is considered as having reached a level of stagnation in development [385].

Several authors [386,387] have discussed the spectroscopic and nonspectroscopic (matrix) interferences in ICP-MS. ICP-MS is more susceptible to nonspectroscopic matrix interferences than ICP-AES [388–390]. *Matrix interferences* are perceptible by suppression and (sometimes) enhancement of the analyte signal. This enhanced susceptibility has to be related to the use of the mass spectrometer as a detection system. In fact, since both techniques use the same (or comparable) sample introduction systems (nebulisers, spray chambers, etc.) and argon plasmas (torches, generators, etc.), it is reasonable to assume that, as far as these parts are concerned, interferences are comparable. The most severe limitation of ICP-MS consists of polyatomic

Table 8.63 Typical specifications of ICP-MS instruments

Feature	Type of instrument	
	Low resolution	High resolution
Resolution	1 amu[a]	300–10 000 amu[b]
Mass range	1–300 amu	1–500 amu
Orders of linear dynamic range	5–9	9
Short-term precision (%)	1–3	2
Detection limit (ng L^{-1})[c]	<1–10	<0.01–1 ($R =$ 300–500)

[a] Peak width at 10 % from the peak maximum.
[b] 10 % valley between two adjacent peaks.
[c] Based on three times the standard deviation of the blank.

interferences on the elemental signals, originating from argon and/or the sample matrix. *Spectral interferences* appear in ICP-MS when partial or complete overlap occurs with the analyte isotope peak. In that case, the net analyte signal cannot be isolated from the interferent. Isobaric interferences usually do not pose a serious problem to ICP-MS, because in most cases a non-interfered isotope is available or an inter-element correction can be carried out. The formation of polyatomic species still forms the most intractable ICP-MS interference and also the least fundamentally investigated [391]. Polyatomic ions having the general form $(M_x)_n^+$, wherein M_x represents the constituent isotope(s) and n the number of constituent isotopes per polyatomic species, can be composed of all isotopes present in or added to the argon plasma. Some typical examples of polyatomic interferences are shown in Table 8.64.

The almost continuous confrontation with polyatomic spectral interferences poses a serious problem in ICP-QMS analysis, because detection limits are degraded and the measured intensity is incorrectly attributed to the analyte ion. The problem of spectral interferences

mainly exists in the region below m/z 120. Spectral interferences can be reduced by various means:

(i) high mass resolution (ICP-HRMS);
(ii) cavity trapping of polyatomic ions (3DQMS);
(iii) collision/reaction cell technology (CCT/CRC); and
(iv) cold plasma.

With He *all* argon-based polyatomic interferences are eliminated, enabling elements such as K, Ca, Cr, Fe, Cu, As and Se to be analysed at ppt levels in one run, without the matrix-induced limitations of cool plasmas or the need for hydride generation. Table 8.64 shows the mass resolution (R) necessary to separate typical interferences. In ICP-HRMS *high mass resolution* typically goes up to 12 000. Figure 8.14 shows the layout of ICP-HRMS instruments equipped with a double-focusing magnetic sector mass analyser (BE geometry). Use of ICP-HRMS enables the separation of isotopes from interfering peaks and hence the correct attribution of mass peaks to both isotopes and interfering species. Second-generation ICP-HRMS equipped with a dynamic reaction system using low reactive gases and a DFS mass spectrometer has a resolution some 1000 times higher than that of conventional ICP-MS and allows 100 % rejection of interferences such as $^{40}Ar^{16}O^+$ and $^{56}Fe^+$, or $^{40}Ar^{35}Cl^+$ and $^{75}As^+$ [392]. ICP-HRMS is commonly used for high-precision isotope ratio determinations. Although the increase in resolution is paid for by a certain loss of sensitivity, ICP-HRMS is the most sensitive ICP-MS; using a high-efficiency nebuliser (USN) detection limits down to 1 pg L^{-1} for uninterfered elements have been achieved [393].

Collision/reaction cell technology (CCT/CRC) is a cheaper alternative for reducing the impact of interferences than is high-resolution magnetic sector ICP-MS.

Table 8.64 Mass resolution (R) necessary to separate typical interferences

Nuclide	Molecular ion	Resolution
$^{31}P^+$	$^{14}N^{16}O^1H^+$	968
$^{31}P^+$	$^{15}N^{16}O^+$	1 458
$^{32}S^+$	$^{16}O_2^+$	1 801
$^{56}Fe^+$	$^{40}Ar^{16}O^+$	2 502
$^{63}Cu^+$	$^{40}Ar^{23}Na^+$	2 790
$^{64}Zn^+$	$^{32}S_2^+$	4 261
$^{75}As^+$	$^{40}Ar^{35}Cl^+$	7 775
$^{80}Se^+$	$^{40}Ar_2^+$	9 688

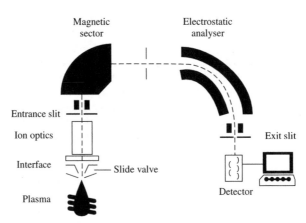

Figure 8.14 Layout of an ICP-HRMS instrument. After Jakubowski and Stuewer [386]. Reproduced by permission of the authors

ICP-CRC-MS has made the technique applicable to a much wider range of sample types than was previously possible. Common ICP-MS spectral interferences reduced by the closely related dynamic reaction cell (DRC) technology are (polyatomic species/interfered analyte): $^{12}C^{15}N$ and $^{12}C^{14}N^{1}H/^{27}Al$, $^{38}Ar^{1}H/^{39}K$, $^{40}Ar/^{40}Ca$, $^{35}Cl^{16}O/^{51}V$, $^{35}Cl^{16}O^{1}H/^{52}Cr$, $^{36}Ar^{16}O/^{52}Cr$, $^{40}Ar^{12}C/^{52}Cr$, $^{38}Ar^{16}O^{1}H/^{55}Mn$, $^{40}Ar^{16}O/^{56}Fe$, $^{40}Ar^{16}O^{1}H/^{57}Fe$, $^{40}Ar^{35}Cl/^{75}As$ and $^{40}Ar^{40}Ar/^{80}Se$. DRC is a much lower-energy device than the collision cell. Also hexapole collision cells remove spectroscopic interferences by 'argides' (argon ions and all molecular ions with argon atoms). In an alternative reaction cell development, octopoles are used. Octopoles have higher ion transmission than hexapoles or quadrupoles, but are much smaller. Consequently, the operating pressure is higher and very high reaction efficiencies are achieved.

'Cold plasma' with reduced temperature is another way to cope with the most annoying problems from interferences, even in the case of low-resolution instruments [394]. The effect consists of weaker ionisation conditions coming close to chemical ionisation [395]. In particular, argides are reduced by orders of magnitude in comparison to conventional ICP operation. However, at lower plasma temperatures, evaporation of analyte material is considerably reduced. Reducing the plasma temperature also has a dramatic effect on the ionisation (and therefore sensitivity) of many elements. Table 8.65 shows the ion population as a function of plasma temperature and ionisation potential. As a result, the cold plasma technique is only advantageous for a rather small number of elements and applications.

Solid samples may be introduced into the ICP-MS by means of electrothermal volatilisation (ETV), arc, spark and laser ablation as front-end arrangements. Coupling of a graphite furnace to ICP-MS results in a method that combines the suitability of the graphite furnace for handling solid samples with the detection power of ICP-MS. Among the advantages of ETV as a sample introduction technique for ICP-MS are: high transport efficiency compared to conventional solution nebulisation,

and lower partial pressure of oxygen in the plasma, which consequently reduces the MO^{+}/M^{+} ratio. The sample only has to be vaporised – not atomised – and therefore separation of the species on the basis of their volatility is possible. The possibility of on-line separation with *ETV-ICP-MS* of fractions in a sample with different volatility has been demonstrated [106]. When compared to (i) solid sampling GFAAS and (ii) XRF, respectively, (i) the multi-element capability, and (ii) the possibility of obtaining accurate results without using a matrix-matched solid standard and the very low sample consumption (1 mg for ETV-ICP-MS vs. 10 g for XRF) seem to be the most important advantages of ETV-ICP-MS. Solid-sampling ETV-ICP-MS/AES allows simultaneous oligo-element homogeneity determinations in very small samples (mg). ETV-ICP-MS is still in its infancy, and efforts should be focused on the maximum number of elements that can be determined, taking into account the short duration of the transient signals involved.

LA-ICP-MS allows microscopic profiling of solid samples. Using a UV or IR laser, elemental measurements can be made with a degree of lateral resolution (typically $20-50\,\mu m$) limited only by the laser spot size [396]. LA-ICP-MS makes it possible to obtain some information relating to element concentration with respect to sampling depth. The depth probed is $1-10\,\mu m$ per laser pulse. Depth profiling and imaging/mapping are possible. Direct multi-elemental ultratrace analysis of both conductors and nonconducting materials can be performed with both UV and IR LA-ICP-MS. Both LA and speciation analysis generate transient signals, for which ToF-MS is the best detection mode. ICP-ToFMS has been used for the detection of transient signals by laser ablation [397]. Drawbacks are lower sensitivity and increased spectral overlap. Solids or thin films are, however, more usually digested into solution prior to analysis.

ICP-MS has been hyphenated to all forms of chromatographic techniques for speciation analysis, including GC, LC, SFC and CE, where it is used as a *chromatographic detector*. For this purpose, ICP-QMS is not necessarily the best choice. Experimental GC-ICP-MS interfaces have been described [398]. GC-ICP-ToFMS has been explored [399]. Detection limits for several organotin and organolead compounds were in the low fg region, with a dynamic range exceeding six orders of magnitude. Liquid chromatography is the easiest of the chromatographic separation techniques to couple with ICP-MS, principally because the flow-rates used with LC are of the order of 0.1 to $1.0\,mL\,min^{-1}$. In addition, the LC mobile phase elutes

Table 8.65 Ion population (%) as a function of plasma temperature and ionisation potential

Element	IP (eV)	Plasma temperature			
		5 000 K	6 000 K	7 000 K	8 000 K
Na	5.14	90.0	98.9	99.8	99.9
Al	5.98	56.2	94.5	99.1	99.8
Sb	8.64	0.3	9.0	57.6	90.9
Se	9.75	0.0	1.1	17.8	66.6
Hg	10.43	0.0	0.3	6.5	42.6

from the column at atmospheric pressure, which is ideally suitable for the ICP-MS sample introduction (see also Section 7.3.3.5). General aspects of ICP-MS and different fields of application have been reviewed by refs [400,401] and refs [402–404], respectively. The analytical capabilities of modern ICP-MS were recently critically evaluated [405]. Several monographs are available [192,353,406–408].

Applications ICP-MS has become the technique of choice for the determination of elements in a wide range of liquid samples at concentrations in the ng L^{-1} to $\mu g\,L^{-1}$ range. Typical applications of ICP-MS are: multi-element analysis of liquids (even with high solid contents); element speciation by hyphenation to chromatographic techniques; continuous on-line gas analysis; multi-element trace analysis of polymers; and trace analysis in high-purity materials. ICP-MS is routinely used for quality control purposes.

Analysis of plastic monomers, (PET) oligomers and additives in food contact materials and measurement of migration of foods and food simulants using ICP-MS, APCI-MS and SEC-FTIR has been reported by Jickells [409]. Fordham *et al.* [410] have reported (semi)quantitative trace-element analysis for five polymer samples (LDPE, HDPE, PP, PET and PS), using solution nebulisation ICP-MS after microwave sample digestion using nitric/sulfuric acid. Results were validated using in-house reference materials prepared by doping the polymers with a cocktail of metal stearates (Al, Cr, Co, Mg, Pb and Zn) and Sb_2O_3 via the melt, and by comparing to NAA. Polymer reference materials were analysed by LA-ICP-MS, for homogeneity.

Colourless PE cling-film and coloured PE garbage bags from different manufacturing origins were characterised by ICP-MS after microwave digestion (HNO_3/H_2O_2) [411]. Black bags were discriminated by differences in concentrations of four elements (Mg, Al, Fe and Ba); orange bags were distinguished using Mg, Cr, Mn, Fe and Pb concentrations and their lead isotope

ratios. For polymer films, which contain a sufficiently high concentration of lead additives, lead isotope ratios can also be used as an alternative method for sample discrimination. The ability to distinguish and match such plastic samples is of forensic value. Solution nebulisation (after microwave digestion) and LA-ICP-HRMS for elemental comparison of whole packaging tapes (elements Co, Cr, Cu, Mn, Ni, V and Zn) and the adhesive layer have been reported [412]. Also, in this case ICP-HRMS serves forensic purposes, as it can discriminate between tape production batches. LA-ICP-HRMS facilitates both the sample preparation and the sampling of small, clean tape areas. Various relevant organometal species such as alkyl- and/or phenyltin and various lead compounds exist as ionic non- or insufficiently volatile species. This requires derivatisation by hydride generation or ethylation with $NaBEt_4$.

Application to *solid polymer/additive formulations* is restricted, for obvious reasons. SS-ETV-ICP-MS (cup-in-tube) has been used for the simultaneous determination of four elements (Co, Mn, P and Ti) with very different furnace characteristics in mg-size PET samples [413]. The results were compared to ICP-AES (after sample dissolution) and XRF. Table 8.66 shows the very good agreement between the various analytical approaches. The advantage of directly introducing the solid sample in an ETV device is also clearly shown by the fact that the detection limit is even better than that reported for ICP-HRMS. The technique also enables speciation of Sb in PET, and the determination of various sulfur species in aramide fibres. ETV offers some advantages over the well-established specific sulfur analysers: very low sample consumption; the possibility of using an aqueous standard for calibration; and the flexibility to carry out the determination of other analytes. The method cannot be considered as very economic.

The direct determination of sulfur in bisphenol A at trace levels by SS-ETV-ICP-MS has been described using ^{34}S as the isotope selected [206]. Table 8.67

Table 8.66 Determination of Co, Mn, P and Ti in PET[a]

Method	n	Co	Mn	P	Ti
ETV-ICP-MS[b]	7	99.3 ± 3.9 (4.2)	53.5 ± 3.2 (6.4)	46.6 ± 3.9 (9.1)	90.6 ± 5.6 (6.7)
ETV-ICP-MS[c]	7	99.0 ± 5.9 (6.4)	54.1 ± 1.5 (3.0)	46.2 ± 3.6 (8.3)	93.9 ± 6.3 (7.2)
XRF	4	99.8 ± 0.8 (0.5)	53.8 ± 0.8 (0.9)	45.8 ± 3.0 (4.1)	92.0 ± 0.4 (0.3)
ICP-AES	4	100.8 ± 3.8 (2.4)	55.2 ± 2.2 (2.5)	42.8 ± 2.5 (3.7)	94.6 ± 3.8 (2.5)

[a] Mean values (ppm) for *n* determinations; uncertainties as 95 % confidence intervals and RSD % (in brackets).
[b] Standard addition with aqueous solution.
[c] External calibration using solid standards.
After Resano *et al.* [413]. From M. Resano *et al.*, *Journal of Analytical Atomic Spectrometry*, **15**, 389–395 (2000). Reproduced by permission of The Royal Society of Chemistry.

Table 8.67 Determination of sulfur in bisphenol A samples[a]

Method	Calibration	Results (ppm)	RSD (%)
SS-ETV-ICP-MS	Aqueous standards	1.96 ± 0.13	8.8
	Standard solutions	1.93 ± 0.17	8.2
Sulfur-specific analyser[b]	Liquid organic standards	2.13 ± 0.48	14.0
Sulfur-specific analyser[c]	Liquid organic standard	2.20 ± 0.42	15.3
SS-ETV-ICP-MS	Aqueous standards	0.293 ± 0.017	8.2
	Standard solutions	0.286 ± 0.031	9.9
Sulfur-specific analyser[b]	Liquid organic standards	0.355 ± 0.095	25.6

[a] Two samples.
[b] Detection mode: microcoulometric titration.
[c] Detection mode: UV fluorescence.
After Resano *et al.* [206]. From M. Resano *et al.*, *Journal of Analytical Atomic Spectrometry*, **16**, 793–800 (2001). Reproduced by permission of The Royal Society of Chemistry.

Table 8.68 Limits of detection attainable for sulfur analysis using different methods

Technique	LOD (ng g^{-1})	Reference(s)
SS-ETV-ICP-MS[a]	4	[206]
Solution-ICP-MS[b]	16 000	[206]
Solution-ETV-ICP-MS	50	[414]
Solution-ETV-ICP-MS[c]	4	[415]
Solution-ICP-HRMS	10	[416]
Desolvation membrane-ICP-HRMS	0.01	[416]
Specific sulfur analyser[d]	190	[413]

[a] Calculated considering the introduction of 3 mg of sample.
[b] Calculated considering the dissolution of 0.25 g in 100 mL.
[c] 5 % N_2 added as oxygen scavenger.
[d] Detection mode: microcoulometric titration.

Table 8.69 Hierarchy of analytical methods with respect to accuracy

Analytical data	Analytical method
True value	No method known
Definitive value	Definitive method, e.g. IDMS
Reference method value	Reference method
Assigned value	Routine method

After Heumann [420]. Reprinted from K.G. Heumann, in *Inorganic Mass Spectrometry* (F. Adams *et al.*, eds), John Wiley & Sons, Inc. New York, NY, pp. 301–376, Copyright © (1988, John Wiley & Sons, Inc.). This material is used by permission of John Wiley & Sons, Inc.

shows that the accuracy of SS-ETV-ICP-MS and sulfur-specific analysers is equal, but precision seems to be better for the former technique. Table 8.68 compares the detection limits reported in the literature for sulfur analysis using different methods.

LA-ICP-MS allows quick simultaneous oligo-element *homogeneity determinations* in mg samples of polymeric material. Coupling of ICP-MS to chromatographic techniques provides *element speciation* capabilities, especially as a detector for LC. Kingston *et al.* [417] have described a 'speciated' technique for the determination of Cr(III) and Cr (VI) by HPLC-ICP-MS.

A monograph on ICP-MS applications has appeared [418].

8.5.4 Isotope Dilution Mass Spectrometry

Principles and Characteristics There are basically two ways of obtaining accurate results in micro and trace analyses, namely either use of standard reference materials (if available) for calibrating analytical methods, or else use of highly accurate methods. Table 8.69

shows a hierarchy of analytical methods with respect to accuracy. '*Scientific knowledge is a body of statements of varying degrees of certainty – some most unsure, some nearly sure, but none absolutely certain*' [419]. There is no known analytical method that is able to product 'true values' in all cases. Definitive values can be obtained by definitive methods, such as isotope dilution mass spectrometry (IDMS).

Isotope dilution (ID) is a technique for the quantitative determination of element concentrations in a sample, on the basis of isotope ratios [382]. An important prerequisite for isotope dilution is the availability of two stable isotopes, although in some cases the use of long-lived radionuclides allows the application range to be further extended [420].

Isotope dilution mass spectrometry (IDMS) can be applied with most of the ionisation methods used in mass spectrometry to determine isotope ratios with greater or lesser accuracy. For calibration by means of isotope dilution, an exactly known amount of a spike solution, enriched in an isotope of the element(s) to be determined, is added to an exactly known amount of sample. After isotopic equilibration, the isotope ratio for the mixture is determined mass spectrometrically. The attraction of IDMS is its potential simplicity; it relies only on the measurement of ratios. The accurate measurement of large isotope ratios challenges the dynamic measurement range of modern mass spectrometers. From:

(i) the measured isotope ratio of the blend of sample and spike;
(ii) the masses of sample and spike (determined most accurately by weighing);
(iii) the natural isotopic composition of the analyte element (or in the case of natural variations in the isotopic abundances, the experimentally determined isotope ratio in the sample); and
(iv) the isotopic composition of the spike

the analyte concentration can be calculated with very good accuracy. The equations used for this calculation can be found elsewhere [420].

Precise and accurate isotope measurements are a major requirement for IDMS. Table 8.70 summarises the *main characteristics* of IDMS. The high precision and accuracy can be attributed to the fact that no quantitative isolation of the elements to be determined is necessary. A stable isotope used as a spike is the best internal standard to apply. A major advantage of isotope dilution is that once isotopic equilibrium is established, loss of analyte does not affect the analysis result, so that highly reliable results can be obtained. The detection limit of IDMS essentially depends on the element to be determined, on the matrix and on the blank. Therefore, the optimum ionisation method for each element to be determined must be selected. Two major disadvantages of IDMS are its destructive nature and the fact that if the matrix is not a homogeneous solution the sample must be chemically treated in order to carry out the spiking process. The cost of spike isotopes is usually no longer a limiting factor.

Table 8.70 Main characteristics of IDMS

Advantages

- Precise and accurate analysis
- Nonquantitative isolation of the analyte
- Ideal internal standardisation
- Mono-, oligo- and multi-element analyses
- High sensitivity with low detection limits

Disadvantages

- Destructive
- Suitable for elements with more than one isotope
- Requires relatively stable isotopic pattern in nature
- Need for chemical sample preparation
- Time-consuming
- Relatively expensive

The isotope dilution method can be used for the measurement of molecules or elemental species (about 60 elements have stable isotopes). This approach allows ultratrace analysis because, contrary to radioactive labelling where the measurement relies on detecting atoms that decay during the period of measurement, *all* of the labelled atoms are measured.

Mono- and oligoelement analyses can be carried out by IDMS with thermal ionisation; multi-element determinations can be carried out with mass-spectrometric methods, notably ICP-MS. When using *thermal ionisation mass spectrometry* (TIMS), where the ionisation agent is heat, the quantitative determination will almost invariably be carried out by the isotope dilution technique. In thermal ionisation isotope dilution MS (ID-TIMS or TI-IDMS), a sample is placed on a metal filament, which is heated in a vacuum in the MS source to give ions, which are then separated in the mass spectrometer and detected [347]. Filament loading procedures are element-specific, and separation and purification of the element are necessary for highly accurate determinations. For ID-TIMS the analyte elements have to be separated from the matrix. Thermal processes produce both positive and negative ions. For quantitation, an amount of the element to be determined, enriched in one of its isotopes, is added to the sample. In ID-TIMS, the detection capability can be limited either by the variation of the analyte content of the procedure blanks, or by the precision of the isotope ratio measurements for the spike solution. TIMS is a suitable method for the determination of precise and accurate isotopic ratios, with excellent precision of better than 0.1 %, but is rather impractical for routine application, as extensive sample preparation is required. The technique is available only in a limited number of specialised laboratories. Until recently, TIMS was considered the ideal method for isotopic analysis. However, this role is gradually being taken over by ICP-HRMS, in particular in multicollector mode.

In *isotope dilution inductively coupled plasma-mass spectrometry* (ID-ICP-MS) the spike, the unspiked and a spiked sample are measured by ICP-MS in order to determine the isotope ratio. Using this technique, more precise and accurate results can be obtained than by using a calibration graph or by standard addition. This is due to elimination of various systematic errors. Isotopes behave identically in most chemical and physical processes. Signal suppression and enhancement due to the matrix in ICP-MS affects both isotopes equally. The same holds for most long-term instrumental fluctuations and drift. Accuracy and precision obtained with ID-ICP-QMS are better than with other ICP-QMS calibration strategies [421]. In general, the agreement with certified values is excellent. The precision of the determination of isotopic compositions is of the order of 1 % when the respective isotopic concentrations do not differ by more than one order of magnitude. ID-ICP-MS is more generally available than ID-TIMS. Application of the isotope dilution methodology with ICP-MS requires less-complicated sample pretreatment, and thus offers higher sample throughput. For (ultra)trace determination, the latter feature results in much less risk of contamination than with ID-TIMS. The precision of the results is comparable with that of ID-TIMS, which is a more lengthy method, requiring isolation of the element of interest. ICP-QMS cannot measure such extreme isotope ratios at TIMS. Detection limits of another multi-element analysis technique, ID-SSMS, are in the range 0.1–1 ppm [422].

Heumann *et al.* [420,423] have recently given overviews of plasma MS for isotope ratio measurements. A review of the ID-SSMS method can also be found in the literature [422].

Applications The application of the isotope dilution technique is especially useful in carrying out precise and accurate micro and trace analyses. The most accurate results in mass spectrometry are obtained if the isotope dilution technique is applied (RSDs better than 1 % in trace analysis). Therefore, application of IDMS is especially recommended for calibration of other analytical data, and for certification of standard reference materials. The technique also finds application in the field of isotope geology, and is used in the nuclear industry for quantitative isotope analysis.

ID-TIMS has been used as an oligo-element method for the determination of photographically relevant trace elements (Cr 40–100 ng g^{-1}, Cd < 3 ng g^{-1}, Pb 10 ng g^{-1} to 5.5 µg g^{-1}) in AgCl emulsions after spiking using the isotope ratios ^{52}Cr/^{53}Cr, ^{114}Cd/^{116}Cd and ^{206}Pb/^{208}Pb [424]. In this analysis, Ag$^+$ was selectively removed by precipitation as AgCl. Results for Pb and Cr were in excellent agreement with ID-ICP-QMS. Comparison between ID-TIMS and ID-ICP-MS also made clear that sample inhomogeneity limits the between-sample precision.

ID-TIMS has frequently been used for *certification* of Standard Reference Materials by NIST (*ex* NBS) and IRMM (*ex* BCR), as in case of the Polymer Elemental Reference Material (PERM) project for the production of PE standards with certified concentrations of eight elements, namely Cd, Cr, Hg, Pb, As, S, Cl and Br [1]. Also certification of the VDA-001 to -004 PE/Cd CRMs has taken advantage of both ID-TIMS [224,425] and

ID-ICP-MS [425,426]. Certified values for the 4 PE/Cd materials (VDA-001 to -004) with nominal Cd concentrations of 0.35 to 3.6 mmol kg^{-1} were established by IDMS, using an enriched ^{111}Cd spike [427]. The results did not show significant differences between bottles at the 300-mg level. For the sample with the highest Cd content (VDA-004; 407 ppm) the measurement reproducibility was worse, which stands in relation to heterogeneity problems, as confirmed by GF-ZAAS results.

SI-traceable reference values for Cd, Cr and Pb amount contents in HDPE samples from the PERM project [1] were determined by IDMS used as a primary method of measurement [425], see Table 8.71. Cadmium and Pb were measured with ICP-MS, and Cr with PTI-MS after decomposition of the polymer matrix using a high-pressure asher (HPA). There is good agreement between the IDMS-derived data and the official certified values. The results of the classical IDMS approach [428] and of 'on-line' IDMS methodology [429] for the determination of Cr, Cd, Hg and Pb in PERM materials [1] were in agreement with each other and with the overall mean of the interlaboratory exercise [426]. 'On-line' IDMS is based on a simple instrumental set-up, whereby enriched material is automatically and continuously added on-line to both calibrant and sample solutions, shortly before their introduction into the ICP-MS. In this approach (resembling external calibration), a curved calibration graph is used as a reference. The classical IDMS approach involves no calibration graph.

The use of ICP-MS for the measurement of isotope ratios is becoming increasingly more important. ID-ICP-MS has been used for the determination of trace elements in other CRMs, such as Pb in BCR 60, BCR 61, BCR 62 and BCR 320 [430]; Cu in copper ore RMs [431]; 11 trace elements (Cr, Ni, Zn, Sr, Mo, Cd, Sn, Sb, Ti, Pb and U) in a marine sediment reference material [421]; and Pb in CRM 983 (from NIST) [430,432].

Concern over the migration of plasticisers from packaging materials has led to various studies, amongst which monitoring levels of various phthalates from coatings of regenerated cellulose film [433] and of acetyltributyl citrate (ATBC) from vinylidene chloride/vinyl

Table 8.71 SI-traceable amount contents (ppm) in PERM materials

Element	BCR CRM-680	BCR CRM-681
Cd	141.9 ± 5.6	22.04 ± 0.72
Cr	113.4 ± 2.7	17.50 ± 0.32
Pb	106.3 ± 2.0	13.69 ± 0.33

After Vogl *et al.* [425]. Reproduced from J. Vogl *et al.*, *Accred. Qual. Assur.*, **5**, 314–324 (2000), by permission of Springer-Verlag, Copyright (2000).

chloride copolymers [434] in food, using *isotope dilution GC-MS* to fulfil the requirements of high specificity, high quantitative precision and accuracy.

8.6   RADIOANALYTICAL AND NUCLEAR ANALYTICAL METHODS

Principles and Characteristics   Various nuclear analytical techniques, such as neutron activation analysis (NAA), isotope dilution analysis (IDA), ion beam analysis methods (PIXE, RBS), γ-ray spectrometry (Mössbauer analysis) and radiotracer methods, are all employed for polymer/additive analysis, even though on a restricted scale. With *radiochemical methods*, the radiation emitted from radioactive atomic nuclei is used as analytical information. If the radioactive nuclei are generated in the analytical sample itself (using a neutron source), we are dealing with an activation analysis (Section 8.6.1). Radiochemical analysis or tracer analysis uses radioactive nuclides as highly sensitive measurable indicators. Radionuclides may be produced as a result of a nuclear reaction between an activating species (neutrons, γ-photons, charged particles, etc.) and isotope(s) of the element of interest. The radiation emitted by these radionuclides is measured with a suitable detector after a variable decay time depending on the half-life of the nuclide. In isotope dilution analysis, a small, exactly defined amount of the radioactive material is added to the analytical sample.

Although following similar nuclear reaction schemes, *nuclear analytical methods* (NAMs) comprise bulk analysing capability (neutron and photon activation analysis, NAA and PAA, respectively), as well as detection power in near-surface regions of solids (ion-beam analysis, IBA). NAMs aiming at the determination of elements are based on the interaction of nuclear particles with atomic nuclei. They are nuclide specific in most cases. As the electronic shell of the atom does not participate in the principal physical process, the chemical bonding status of the element is of no relevance. The general scheme of a nuclear interaction is:

$$a + A \longrightarrow B^* + b \quad \text{(prompt reaction)} \quad (8.4)$$

$$B^* \xrightarrow{(\lambda)} C + c \quad \text{(delayed reaction)} \quad (8.5)$$

or $A(a,b)B^*$ in the physics notation, where A, B, C denote atomic nuclei and a, b, c particles or γ-quanta; λ is a decay constant. The basis of the nuclear analytical method for chemical analysis is the measurement of the amount of either light or heavy product produced in a known flux of projectiles for a known length of time. Nuclear reactions such as (n, α) or (n, p) are possible

only with high-energy neutrons, whereas (n, γ) reactions predominate with low-energy neutrons. Since many elements are activated simultaneously, radioactivity measurements in either prompt or delayed analysis yield both the identity of the radioactive nuclides and their amounts. Almost only γ-ray emitters are used in activation analysis. Prominent features of NAMs are sensitivity, selectivity, multi-element determination and linear calibration over a wide dynamic range.

Nuclear analytical methods have been exploited for elemental analysis of polymers, because of their quantitative character, generally high sensitivity (ng g^{-1} range) and depth resolution in the nm range. Moreover, the polymer matrix does not act as a chemical or physical interference. Also, the risk of sample inhomogeneity is minimised, since samples with masses as high as 1 g can be analysed. A disadvantage is the radiation damage induced by the primary beam. However, this damage can be either monitored or suppressed by sample cooling, and both quantification and depth profiling are still possible without sputtering techniques. As NAMs require expensive apparatus (nuclear reactor; accelerator in radioactive control areas) their availability is restricted to a small number of suitably equipped institutions.

Prominent NAMs are NAA, PAA, RBS and PIXE. Actually, PIXE is not a true nuclear analytical method, because it is based on the interaction of fast ions with the electron clouds of the atoms. A suitable combination of NAMs allows all elements of the periodic table to be studied. PAA is especially suitable for the determination of light elements, whereas NAA, RBS and PIXE detect medium and heavy elements very well.

The characteristics of radiochemical methods are well known [435]. An overview of the determination of elements by nuclear analytical methods has appeared [436]. Some selected reviews of nuclear methods of analysis are available: charged particle activation analysis [437,438], instrumental neutron activation analysis [439–441] and ion-beam analysis [442].

Applications   Radiotracer measurements, which combine high sensitivity and specificity with poor spatial resolution, have been used for migration testing. For example, studies have been made on HDPE, PP and HIPS to determine effects of manufacturing conditions on migration of AOs from plastic products into a test fat [443]. Labelled antioxidant was determined radioanalytically after 10 days at 40 °C. Acosta and Sastre [444] have used radioactive tracer methods for the determination of styrene ethyl acrylate in a styrene ethyl acrylate copolymer.

Neutron activation analysis (NAA) is a supreme technique for elemental analysis (Section 8.6.1). Other nuclear analytical techniques, such as PIXE (Section 8.4.2) and RBS, also find application in investigations of diffusion processes [445].

Fractional separation of tin compounds used as stabilisers in PVC was based on substoichiometric isotope dilution analysis [446]. The tin compounds were isolated by extraction and complexed with salicylideneamino-2-thiophenol, followed by controlled addition of γ-irradiated tributyl tin oxide and measurement of the γ-activity. PVC containing a nominal 0.63 % $(Bu)_3Sn$ analysed 0.614 ± 0.016 % in nine determinations.

Activation analysis, the application of radiotracers and other radiochemical methods in innovative trace analysis are indispensable, first of all in the preparation of standard reference samples.

8.6.1 Activation Analysis

Principles and Characteristics In *neutron activation analysis* (NAA) the sample is irradiated by neutrons. The principal reaction in NAA is:

$$n + A \longrightarrow B^* + \gamma \qquad (8.6)$$

also written as $A(n,\gamma)B^*$. There are three types of neutron sources used traditionally for activation analysis: reactors, accelerators and isotope sources. The most common is the nuclear reactor, in which neutrons are produced by fission of uranium. In reactor NAA (n, γ) reactions predominate because of the large cross-sections with *low-energy neutrons* (thermal neutrons). A (n, γ) reaction consists of neutron capture, leading to an excited state of the nucleus formed (an isotope of the original element, but with higher mass number) followed by prompt emission (within ca. 10^{-14} s) of a high-energy photon (ca. 1 MeV) to yield the radionuclide or a stable nuclide in its ground state. Typically, γ-rays result either from transitions among activated nuclear energy states of the compound nucleus or from radioactivated decay products.

In terms of atomic spectrometry, NAA is a method combining excitation by nuclear reaction with delayed de-excitation of the radioactive atoms produced by emission of ionising radiation (β, γ, X-ray). Measurement of delayed particles or radiations from the decay of a radioactive product of a neutron-induced nuclear reaction is known as simple or delayed-gamma NAA, and may be purely instrumental (INAA). The γ-ray energies are characteristic of specific indicator radionuclides, and their intensities are proportional to the amounts of the various target nuclides in the sample. NAA can thus

provide both qualitative and quantitative information regarding the elemental composition of a wide variety of samples. NAA is one of the simplest techniques for trace-element analysis. Samples are inserted in PE or PP cans for short irradiation, or in a quartz vial contained in an aluminium can for long irradiation. Each sample is transported to and from the reactor by means of a pneumatically operated tube. Analysis is typically with a multichannel analyser, so that a number of different elements can be determined simultaneously.

The number of target atoms in the sample may be calculated directly from the basic equation of activation analysis:

$$A = n\phi\sigma(1 - e^{-\lambda t})(e^{-\lambda d}) \qquad (8.7)$$

where A = absolute activity of the sample (disintegrations s^{-1}), n = number of atoms of the target nuclide, ϕ = neutron flux density (n cm^{-2} s^{-1}), σ = reaction cross-section (cm^2), λ = decay constant for the product radionuclide (s^{-1}), t = time of irradiation (s), and d = time of decay before counting (s). It is possible to use the above equation to obtain elemental concentrations directly from the measured peak area ('absolute' activation analysis). However, such an absolute method is seldom applied, because of uncertainties in the physical nuclear constants (such as σ, λ) and difficulties in the determination of some experimental parameters (such as the neutron flux). It is usually found to be more convenient to use a *comparative method* whereby a reference with a known content of the considered element is also irradiated. The practice of multi-element INAA is simplified by the use of standard reference materials (SRMs) and synthetic multi-element standards (a few reports only), which are irradiated simultaneously with the samples.

The k_0-NAA method has been developed to overcome the labour-intensive and time-consuming work of preparing multi-element standards when routine multi-element or panoramic analyses are required [447]. It is intended to be an absolute technique in which uncertain nuclear data are replaced by a composite nuclear constant, the k_0-factor, which has been determined experimentally for each radionuclide with high accuracy. This k_0 is given by:

$$k_0 = A_r{}^*\theta \ \sigma_0\gamma/A_r\theta^*\sigma^*\gamma^* \qquad (8.8)$$

where A_r is the relative atomic mass, θ is the abundance of the isotope yielding the radioactive isotope, σ_0 is the thermal neutron cross-section and γ is the relative intensity of the gamma radiation in the decay scheme of the radionuclide; *denotes the comparator. This method uses 'composite nuclear constants' (generalised

k_0 values) for the elements concerned and several flux and flux-ratio monitors are irradiated together with the samples. For determination of the k_0-factor, gold was used as a comparator, which was co-irradiated as dilute Au–Al wire. This Au–Al wire should be co-irradiated with each sample. The k_0 factor contains only well-defined, invariable nuclear constants and no terms relating to the experimental conditions. The actual determination of k_0 factors is not trivial. The k_0 factors for the relevant gamma lines of about 120 analytically interesting radionuclides have been determined and reported [448]. Simonits et al. [449] have detailed the theory and principles of the k_0-NAA method. A user-oriented outline of the k_0 method has been published [448]. It is estimated that the overall contribution of the application of k_0-NAA to the uncertainty of the analytical result is approximately 4 %.

Table 8.72 lists the *main characteristics* of NAA. NAA allows solid samples, thus avoiding sample dissolution, with the attendant problems of: difficult or incomplete dissolution of some matrices; possible losses of some elements by volatilisation; or precipitation or adsorption losses. NAA also does not need any laborious procedures such as ashing or digestion. Neutron activation analysis measures two-thirds of the periodic table, including transition metals, halogens, lanthanides and platinum-group metals. NAA provides multi-elemental capability, with sensitivity varying irregularly from one element to the next. The sensitivity of various elements for NAA varies greatly. Favourable elements such as Mn or Eu can be determined in sub-ng quantities, while the method is less sensitive by about six orders of magnitude for Fe or Pb. A problem with NAA is that the signal is measured against a background of Bremsstrahlung. INAA can determine element levels differing by seven to eight orders of magnitude with about the same and very high precision, and inhomogeneities as low as 1 % can be detected. Detection limits in the ng g^{-1} region for 42 elements for a 1.0 g sample of PE by NAA were reported [450]. Activation analysis is most suited to the multi-elemental analysis of minor, trace and ultratrace elements in small samples.

The sources of uncertainty in NAA analysis are well understood, and can be derived in advance, modelled and assessed experimentally. There are two main kinds of interferences in the calculation of trace-element concentrations by INAA. The first one is formation of the same radionuclide from two different elements. Another kind of interference is from two radionuclides having very close γ lines. Whenever interferences occur, the radionuclide of interest can be carried through a post-irradiation radiochemical separation without the danger of contamination.

Table 8.72 Main features of neutron activation analysis

Advantages

- Instrumental (quasi nondestructive) bulk analysis technique
- Minimal sample preparation (limited contamination problem)
- Suitable for intractable (even large) samples
- Simultaneous multi-elemental capability (F to U, except for Pb, Bi, P and Tl)
- Ultratrace method
- Quantitative and standardless (absolute method; no calibration curves)
- Wide applicability (gas, liquid, solid, slurries, dispersions; organic, inorganic)
- Sample amount (from mg to kg)
- Rapid
- High sensitivity for many elements (ppb level)
- Very high specificity
- Reasonably high precision (1–5 %)
- Linear measuring range (from sub-pg g^{-1} up to %-level)
- Nearly matrix independent
- Few well-defined interferences, low blank values
- Reference technique ('primary ratio method')
- Low cost
- Automation (no exposure to radiation for operator)
- Mature technique

Disadvantages

- Analysis cycle time (irradiation, γ-spectrometry, data evaluation: 8 h to 14 days)
- No speciation analysis
- No lateral resolution
- Need for neutron source (often not on-site)
- Sample is rendered radioactive
- Nuclear expertise required
- Contamination during sample preparation stage (crushing)
- Limited use for routine laboratory

Because neutron and γ-radiation are highly penetrating, the method is virtually matrix-independent. NAA is a bulk technique. NAA can also be a selective technique, because appropriate choice of experimental parameters (such as irradiation and decay times), and use of neutrons of varying energies, can discriminate against unwanted elements.

Activation analysis is based on a principle different from that of other analytical techniques, and is subject to other types of systematic error. Although other analytical techniques can compete with NAA in terms of sensitivity, selectivity, and multi-element capability, its potential for blank-free, matrix-independent multi-element determination makes it an excellent reference technique. NAA has been used for validation of XRF and TXRF.

Two practical problems in the analysis of plastics by NAA, namely contamination during sample preparation (especially sample crushing in relation to ultratrace analysis) and destruction of the sample capsule due to pressure build-up by the gases formed during neutron irradiation, have been addressed [451]. Other limitations of the technique include time restrictions dependent on the half-lives of the radionuclides measured. Rapid analysis based on short-lived isotopes is possible only in exceptional cases. For many elements, the turnaround time is long (up to weeks), which hampers routine analysis. Although the method may be labour intensive, this can be reduced by automation. Being a nuclear process, NAA does not provide information on the chemical form of the element (speciation) and also has no *in situ* microprobe capability and only limited autoradiography.

NAA is a mature field. The principles are well understood. No dramatic improvements are expected in either instrumentation or methodology that would revolutionise the method. However, access to a nuclear reactor is becoming a serious concern, in view of the ageing and shutdown of many research reactors.

The most probable reactions occurring in *photon activation analysis* (PAA) are:

$$\gamma + A \longrightarrow B^* + n \tag{8.9}$$

$$B^* \longrightarrow C + \gamma' \tag{8.10}$$

where the photons emitted in the delayed decay are detected and used as analytical signal. Typical PAA reactions for light elements (with $t_{1/2}$, min) are: $^{12}C(\gamma,n)^{11}C$ (20.3), $^{14}N(\gamma,n)^{13}N$ (9.96), $^{16}O(\gamma,n)^{15}O$ (2.03) and $^{19}F(\gamma,n)^{18}F$ (109.7). The characteristics of PAA closely match those of NAA. As photons and neutrons are highly penetrating, both PAA and NAA are bulk analysis methods. PAA is particularly well suited for determination of light elements (C, N, O) and the halogens, and has a good detection capability for most elements beyond Na. PAA determines a wide variety of elements ($Z \geq 6$, including Ti, Ni, Nb, Tl, Pb and Bi which are not accessible to NAA). In the certification of reference materials, instrumental PAA is a valuable complement to any sample dissolving method.

INAA was recently reviewed [452,453]. Various books deal with radioanalytical chemistry and nuclear methods of analysis [435,454–458]. A nuclear spectrometry handbook is available [459].

Applications   NAA is a powerful technique for ultratrace analysis of metallic impurities in plastics. PE is one of the most important canning materials for NAA, mainly on account of its high purity and low price. The cross-section of PE for thermal neutrons amounts to

$0.33 \times 10^{-24}\,cm^{-2}$, so that the material assumes only a little activity after irradiation in a nuclear reactor. INAA has been used for the determination of some 20 trace elements in LDPE (prepared noncatalytically and preferred for encapsulation in NAA) [460]. A large scatter of element contents was found in both granules and foils from the same batch, indicating a heterogeneous distribution for the majority of the elements present as impurities in PE. NAA is used in relation to local aggregations of *catalyst residues* and additives, and is a useful tool for homogeneity testing. NAA is widely used by various polyolefin producers in the determination of catalyst residues (Ti, Al, Mg, Sb, Cl, Br, F, Na, Mn, Zn). Traces of chromium in polyolefins originating from Phillips-type polymerisation catalysts have been determined by NAA [461].

Using k_0-NAA, the concentrations of seven trace elements (Na, Mg, Al, Cl, Ti, Mn and Zn) in HDPE household plastics were determined with short-time irradiation [462]. Similarly, NAA has been used for the determination of 50 ppm silica catalyst support in HDPE powder. Silicon is activated by the reaction $^{28}Si(n, p)^{28}Al$ [463]. The silicon concentration is measured by counting the 1.78 MeV γ-ray emission from the decay of ^{28}Al. A Conostan 5000 ppm Si standard was used for calibration. NAA has also been used for the analysis of sodium in polyolefins; in this case the γ activity, due to the radioactive isotope produced in the nuclear reaction $^{23}Na(n, \gamma)^{24}Na$, is measured. The procedure is sensitive and specific, and does not involve a preliminary sample ashing step. The sodium contents obtained by NAA are usually significantly higher (on average +12%) than by ashing at 650 °C followed by flame photometry [464]. Appreciable losses of sodium can occur during high-temperature ashing of polyolefins containing up to 500 ppm of this element. NAA is very useful for determining oxygen in polymers at all concentration levels.

Lee *et al.* [451] have described procedures to overcome several practical problems in the analysis of plastics by NAA; more than 30 elements were analysed in eight PE and six PP materials in the range of sub-ppb to per cent. Table 8.73 shows the close correspondence of the results for quantitative determination of Mg in PP by means of NAA and LA-ICP-MS [465]. Also, a comparison of NAA and EDXRF for the determination of Ba, Cr, Fe and Ti in polycarbonate was reported [465a].

Several in-house PVC compositions containing ten elements (Al, Ba, Ca, Cd, Mg, Na, Pb, Sb, Sn and Ti) were examined by NAA; the high chlorine content masks the determination of Al, Ca, Mg and Ti [466]. Also, the high content of Sb masked the other elements.

Table 8.73 Quantitative determination of Mg in PP by NAA and LA-ICP-MS

Sample	Mg content (ppm)	
	NAA	LA-ICP-MS
A	13	11
B	6.1	6.3
C	24	23
D	13	17
E	31	29

After Marshall *et al.* [465]. From J. Marshall *et al.*, *Journal of Analytical Atomic Spectrometry*, **6**, 145–150 (1991). Reproduced by permission of The Royal Society of Chemistry.

Thus PVC seems to cause severe matrix effects. In PA4.6 Cu and I were determined simultaneously [466a].

Neutron activation analysis is particularly useful for the analysis of trace halogens, and is unique for fluorine. On the other hand, phosphorous is particularly troublesome, because its radioactive isotope, ^{32}P, does not emit gamma rays. Yet, NAA has been proposed as a routine technique for measuring the phosphorous content in polymers [467]. Phosphorous may be determined by INAA with reactor activation and beta counting, as the (n, γ) reaction on phosphorous produces ^{32}P (half-life 14.3 days), a pure beta-emitter. Corrections must be applied for sample thickness, spectral interferences (caused mainly by Sb, Zn and Br) and nuclear interferences (S and Cl). A fairly large amount of polymer is required (200 mg) to constitute a representative sample. A detection limit of $2 \mu g g^{-1}$ was attained. However, the method is less suited for quality control, as short-lived interfering nuclides can be eliminated only by waiting about ten days before beta counting. Constantinescu *et al.* [468] measured phosphorous and sulfur in various polymeric matrices, through the use of cyclotron-accelerated particles. NAA has also been used for the determination of migration of phosphorous-based *additives* from food packaging material into food-simulating solvents [469]. The main drawback of this approach is activation of other elements to radionuclides that interfere with Cerenkov counting for ^{32}P.

In the field of RM *certification*, NAA represents a major analytical technique. It possesses unique quality assurance and self-verification aspects. Not surprisingly, therefore, NAA has been used to certify NIST standard reference materials [470]. By analogy, NAA has also been instrumental in analysing the EC polymer reference materials within the framework of the PERM project [1]. NAA was also used to validate a TXRF procedure for the determination of additives containing Ti, Zn, Br, Cd, Sn, Sb and Pb [56].

Dutch law has indicated NAA as the technique of choice for the analysis of polymer waste (recycling). INAA has been recommended as the analysis technique for the determination of cadmium in industrial products.

8.7 ELECTROANALYTICAL TECHNIQUES

Principles and Characteristics A substantial percentage of chemical analyses are based on electrochemistry, although this is less evident for polymer/additive analysis. In its application to analytical chemistry, electrochemistry involves the measurement of some electrical property in relation to the concentration of a particular chemical species. The electrical properties that are most commonly measured are: potential or voltage, current, resistance or conductance charge or capacity, or combinations of these. Often, a material conversion is involved and therefore so are separation processes, which take place when electrons participate on the surface of electrodes, such as in polarography. Electrochemical analysis also comprises currentless methods, such as potentiometry, including the use of ion-selective electrodes.

Analytical methods based upon oxidation/reduction reactions include oxidation/reduction titrimetry, potentiometry, coulometry, electrogravimetry and voltammetry. Faradaic oxidation/reduction equilibria are conveniently studied by measuring the potentials of electrochemical cells in which the two half-reactions making up the equilibrium are participants. *Electrochemical cells*, which are galvanic or electrolytic, reversible or irreversible, consist of two conductors called electrodes, each of which is immersed in an electrolyte solution. In most of the cells, the two electrodes are different and must be separated (by a salt bridge) to avoid direct reaction between the reactants.

The main electroanalytical techniques are: electrogravimetry, potentiometry (including potentiometric titrations), conductometry, voltammetry/polarography, coulometry and electrochemical detection. Some electroanalytical techniques have become very widely accepted; others, such as polarography/voltammetry, less so. Table 8.74 compares the main electroanalytical methods.

For polymer/additive analysis, electrogravimetry, potentiometry, conductometry and voltammetry have never played a major role. Because of many complications, which can arise by the use of *conductometry* for complicated matrices (such as most polymeric compounds), the technique is not extensively applied in this field. Conductometric measurements are mostly

Table 8.74 Differences between electroanalytical methods

Parameter	Potentiometry	Voltammetric methods	Coulometric methods
Current development	Vanishingly small	Yes	Yes
Polarisation	Absent	Complete concentration polarisation	Minimisation/compensation
Chemical effect	Undetectable	Minimal	Quantitative, conversion into another state
Cell nature	Galvanic	Three electrode system	Electrolytic
Final measurement	Cell potential	Current	Product mass or quantity of charge

useful in analytical applications if, in the course of an analytical procedure, there are relatively large changes in conductivity. Therefore, conductivity is principally used as an indication method in titrimetry. Polarography did find relatively wide application in the past. The application of coulometry is essentially restricted to the important Karl Fischer moisture determination method. By using electrogravimetry, materials are quantitatively determined via complete cathodic or anodic deposition (as metals or metal oxides) on electrodes, and subsequent weighing.

Advantages of electroanalytical methods are:

(i) the relatively inexpensive equipment;
(ii) a small laboratory space requirement;
(iii) the ability to determine (ultra)trace analyte levels using polarographic, voltammetric and amperometric techniques;
(iv) the extended automation at minimal cost;
(v) the speed of analysis; and
(vi) the way that they are unaffected by turbidity, colour, or suspensions.

Nevertheless, these methods no longer enjoy great popularity. In general, electrochemical methods are only recommended where a simpler method is not available.

The main contribution of electroanalytics to polymer/additive analysis lies in electrochemical detection

(Section 4.1). Electroanalytical chemistry is described in a monograph [471].

Applications Electrochemical techniques, while lacking the wide elemental range and long linear response of some atomic and mass spectrometry technologies, offer a valuable alternative in a number of specific applications, and have particular advantages for direct speciation and anion determination. In ion chromatography, amperometric, potentiometric and conductometric detection is widely used [472], see also Section 4.4.2.5.

Crompton [21] has reviewed the use of electrochemical methods in the determination of phenolic and amine antioxidants, organic peroxides, organotin heat stabilisers, metallic stearates and some inorganic anions (such as bromide, iodide and thiocyanate) in the 1950s/1960s (Table 8.75). The electrochemical detector is generally operated in tandem with a universal, nonselective detector, so that a more general sample analysis can be obtained than is possible with the electrochemical detector alone.

Ostromow [328] has described the use of conductometry for the analysis of extracts from elastomers and rubbers, such as the determination of various vulcanisation accelerations: dithiocarbamates, thiurams (tetramethylthiuramdisulfide, tetramethylthiurammonosulfide), 2-mercaptobenzothiazole, diphenylguanidine

Table 8.75 Electrochemical methods in polymer/additive analysis

Additive(s)	Matrix	Method
Vulcanisation accelerators	Rubbers	Polarography
Phenolic, amine-type AOs	Polymers	Polarography
Phenols, cresols, amines	Polymers	Voltammetry
Antioxidants	Polymers	Differential cathode-ray polarography
Cresols, quinols	Polyester acrylates	Anodic voltammetry
Cresols, phenols	PVC	Anodic voltammetry
Organic peroxides	Polystyrene	Cathode-ray polarography
Sn	PVC	Polarography
Metallic (Cd, Zn, Ba) stearates	Polymers	Alternating current polarography
Cd, Pb, Zn salt	PVC	Polarography

After Crompton [21]. Reprinted from T.R. Crompton, *Analysis of Polymers. An Introduction*, Pergamon Press, Copyright (1989), with permission from Elsevier.

and *o*-tolylbiguanide. More recent applications are given in Sections 8.7.1–8.7.3. The application of electrochemical methods to polymer/additive analysis has not been very extensive, as opposed to food chemistry [473]. Electrochemical methods in polymer analysis have recently been reviewed [474].

8.7.1 Potentiometric Methods

Principles and Characteristics Analytical methods that are based upon potential measurements are termed potentiometric methods, or potentiometry. Potentiometric measurements are based on the determination of a voltage difference between two electrodes plunged into a sample solution (mostly aqueous or selected non-aqueous solvents) under null current conditions. The external reference electrode (ERE) is a half-cell having a known electrode potential that remains constant and is independent of the composition of the analyte solution. By convention, the reference electrode is always treated as the anode in potentiometric measurements. The indicator electrode, which is immersed in a solution of the analyte, has a potential that varies in a known way with variations in the concentration of an analyte. The standard hydrogen electrode (SHE) is considered internationally as the primary reference electrode, but is somewhat troublesome to maintain and use. Common EREs are the saturated KCl calomel electrode (SCE) and the Ag/AgCl/KCl electrode. The indicator electrode is composed of an internal reference electrode (IRE) bathed in a reference solution that is physically separated from the sample by a membrane. As the design of such selective electrodes depends on the nature of the internal reference electrode, the internal solution and the membrane, they exhibit variable specificity. The measurement of potentials between indicator and reference electrodes is made under conditions of minimal flow of current, so that ohmic drop and electrode reactions are of no real significance, and the voltage measured is essentially identical to the cell potential capability. Most indicator electrodes used in potentiometry are highly selective in their responses. The most familiar potentiometric devices are glass electrodes for the measurement of pH. The potentiometric response tends to be slow, so potentiometry is used infrequently in analysis [475].

Electrodes which only respond to certain free (not bound) measured ions are called *ion-selective electrodes* (ISE). The term is now usually applied to all potentiometric measuring electrodes that are capable of providing data concerning the concentration or activity of a particular chemical ion or species. Although no indicator electrode is absolutely specific in its response, a few are now available that are remarkably selective. Indicator electrodes are mainly of three types: metallic, membrane and ion-selective field effect transistors (ISFET). Ion-selective electrodes are available for F^-, Cl^-, Br^-, I^-, CN^-, SCN^-, Cd^{2+}, Cu^{2+}, Pb^{2+}, Ag^+/S^{2-}, each with specific interferences. Gas-sensing probes for CO_2, NO_2, H_2S, SO_2, HF, HCN and NH_3 are available. These electrodes are used in direct ionometry, or as indicator electrodes for many measurements involving titrimetry and complexometry.

Most measurements include the determination of ions in aqueous solution, but electrodes that employ selective membranes also allow the determination of molecules. The sensitivity is high for certain ions. When specificity causes a problem, more precise complexometric or titrimetric measurements must replace direct potentiometry. According to the Nernst equation, the measured potential difference is a measure of the activity (rather than concentration) of certain ions. Since the concentration is related to the activity through an appropriate activity coefficient, calibration of the electrode with known solution(s) should be carried out under conditions of reasonable agreement of ionic strengths. For quantitation, the standard addition method is used.

One of the most fruitful uses of potentiometry in analytical chemistry is its application to titrimetry. Prior to this application, most titrations were carried out using colour-change indicators to signal the titration endpoint. A *potentiometric titration* (or indirect potentiometry) involves measurement of the potential of a suitable indicator electrode as a function of titrant volume. The information provided by a potentiometric titration is not the same as that obtained from a direct potentiometric measurement. As pointed out by Dick [473], there are advantages to potentiometric titration over direct potentiometry, despite the fact that the two techniques very often use the same type of electrodes. Potentiometric titrations provide data that are more reliable than data from titrations that use chemical indicators, but potentiometric titrations are more time-consuming.

Recently, specific ion analysers have been developed, consisting of a pyrolysis furnace with an ion-specific electrode cell and a sample handling system. These systems are alternatives to conventional combustion bombs (Section 8.2.1.2). The advantages of such specific ion analysers include automation, electronic calibration and data handling, and rapid analyses; the disadvantages are relatively high equipment and start-up costs, high maintenance costs, and the fact that the analysis is restricted to the ion for which it is designed.

Electrochemical methods such as potentiometry allow analyses up to $\mu g\,L^{-1}$ quantities, or, with methods such as voltammetry, they extend into the micro-trace range. Table 8.74 compares potentiometry to other electroanalytical techniques. Potentiometry and ion-selective electrodes are described in various books [476–480].

Applications Potentiometry finds widespread use for direct and selective measurement of analyte concentrations, mainly in routine analyses, and for endpoint determinations of titrations. Direct potentiometric measurements provide a rapid and convenient method for determining the activity of a variety of cations and anions. The most frequently determined ion in water is the hydrogen ion (pH measurement). Ion chromatography combined with potentiometric detection techniques using ISEs allows the selective quantification of selected analytes, even in complex matrices. The sensitivity of the electrodes allows sub-ppm concentrations to be measured.

Potentiometry is used in the control of foodstuffs (analysis of NO_3^-, F^-, Br^-, Ca^{2+}, etc. in drinks) and in the analysis of chloride in pulp and paper. Applications in polymer/additive analysis are very limited. In a study of the behaviour of several aliphatic and aromatic bromo-compounds on heating to 500 °C to evaluate their flame retardancy potential for PET, the Br released during heating was collected in a $1\,\mu M$ $NaBr–5M\,NaNO_2$ absorption solution, and determined potentiometically using a Br–selective electrode [481]. Motonaka *et al.* [482] have determined traces of Mn(II) with an ion-selective electrode consisting of a membrane of manganese dibenzyldithiocarbamate in silicone rubber. Ion-selective electrodes in non-aqueous media were used for effective analytical process control in the preparation of organotin heat stabilisers for PVC, to determine organic and inorganic chlorides (as well as mercaptan sulfur), which are the main control parameters of the technology [483]. Quality control of starting materials, intermediates and final products was elaborated. A method for selective determination of monoester–tin compounds by potentiometric titration using a Cu-selective electrode was developed, and a potentiometric determination of Sn(IV) was adapted to control total tin content in heat stabilisers for PVC [81]. Potentiometric titration using ISE has also been used in the determination of organic sulfur [484], and for the analytical control (Cl) of synthetic fibre production [485].

An example of multidisciplinary analysis is the study of the performance of antistatic and slip additives in polyolefins, as investigated by means of XPS,

PA-FTIR, IR, TLC, NMR, potentiometry and chemiluminescence [486]. Analysis of fluoropolymer-based processing aids with a specific ion (F^-) analyser can be carried out within 15 min per sample [29].

Potentiometric titration procedures with sodium methoxide have been reported for non-sulfur-containing organotin compounds in solvent extracts of polymers, and for phenolic antioxidants with sodium isopropoxide in pyridine medium [21]. Organotin compounds in solvent extracts of PVC can be determined by potentiometric and manual titration procedures [487,488].

In process chromatography, potentiometry may be used to monitor selected ions or pH, as these values change over the course of the gradient.

8.7.2 Voltammetric Methods

Principles and Characteristics Voltammetric methods are electrochemical methods which comprise several current-measuring techniques involving reduction or oxidation at a metal–solution interface. Voltammetry consists of applying a variable potential difference between a reference electrode (e.g. Ag/AgCl) and a working electrode at which an electrochemical reaction is induced ($Ox + ne^- \longrightarrow Red$). Actually, the experimental set-up is a three-electrode system in which the current is measured between the working electrode and the auxiliary, while the applied potential is measured between the working electrode and the reference electrode. The most widely used working electrodes are vitreous carbon, platinum, gold, silver and mercury. Heyrovski's original voltammetric method employing the hanging mercury dropping electrode (HMDE) is called polarography.

The key factor in voltammetry (and polarography) is that the applied potential is varied over the course of the measurement. The voltammogram, which is a current-applied potential curve, $I = f(E)$, corresponds to a voltage scan over a range that induces oxidation or reduction of the analytes. This plot allows identification and measurement of the concentration of each species. Several metals can be determined. The limiting currents in the redox processes can be used for quantitative analysis; this is the basis of voltammetric analysis [489]. The methods are based on the direct proportionality between the current and the concentration of the electroactive species, and exploit the ease and precision of measuring electric currents. Voltammetry is suitable for concentrations at or above ppm level. The sensitivity is often much higher than can be obtained with classical titrations. The sensitivity of voltammetric methods can be greatly enhanced by the inclusion of a preconcentration step in which the analyte is accumulated at the electrode. The analytical signal is generated by a subsequent stripping step.

Stripping voltammetry (SV) is an extension of polarography for trace-metal analysis, and can be used for the simultaneous quantitation of several metal ions. The importance of *cyclic voltammetry* (CV), in which the potential is continuously varied in a cyclic manner, lies more in its ability to provide a quick overview of the electrochemical properties of a compound than in its applications to quantitative analysis. Voltammograms can be obtained by measuring the current output for a fixed concentration of analyte, using a series of discrete potentials covering the desired range. Use of the instrumentation under flowing conditions is referred to as *hydrodynamic voltammetry*. Amperometry may be regarded as a form of hydrodynamic voltammetry, in which the potential has been selected and the species measured are changing. Table 8.74 summarises the main differences of voltammetric methods as compared to other electroanalytical methods.

Table 8.76 shows the *main characteristics* of voltammetry. Trace-element analysis by electrochemical methods is attractive due to the low limits of detection that can be achieved at relatively low cost. The advantage of using standard addition as a means of calibration and quantification is that matrix effects in the sample are taken into consideration. Analytical responses in voltammetry sometimes lack the predictability of techniques such as optical spectrometry, mostly because interactions at electrode/solution interfaces can be extremely complex. The role of the electrolyte and additional solutions in voltammetry are crucial. Many determinations are pH dependent, and the electrolyte can increase both the conductivity and selectivity of the solution. Voltammetry offers some advantages over atomic absorption. It allows the determination of an element under different oxidation states (e.g. Fe^{2+}/Fe^{3+}).

Table 8.76 Main characteristics of voltammetry

Advantages

- Oligo and trace-element analysis tool
- Refined, clean and simple technique
- Little or no sample preparation
- Rapid
- High sensitivity (ppm level)
- No separation needed
- Low-cost, portable equipment

Disadvantages

- pH dependency
- Complex interactions at electrode/solution interfaces

Voltammetry has been adapted to HPLC (when the mobile phase is conducting), and CE as a *detection technique* for electroactive compounds. In this usage, the voltammetric cell has been miniaturised (to about 1 μL) in order not to dilute the analytes after separation. This method of amperometric detection in the pulsed mode is very sensitive. However, this device makes it possible to detect few analytically important molecules besides phenols, aromatic amines and thiols.

Special electrochemical *sensors* that operate on the principle of the voltammetric cell have been developed. The area of chemically modified solid electrodes (CMSEs) is a rapidly growing field, giving rise to the development of new electroanalytical methods with increased selectivity and sensitivity for the determination of a wide variety of analytes [490]. CMSEs are typically used to preconcentrate the electroactive target analyte(s) from the solution. The use of polymer coatings showing electrocatalytic activity to modify electrode surfaces constitutes an interesting approach to fabricate sensing surfaces useful for analytical purposes [491].

For the theory of voltammetry/polarography, the reader is referred to refs [473,476].

Applications Barendrecht [492] has described the voltammetric determination of phenolic and aminotype antioxidants. Also, the applicability of a PVC-graphite electrode as an amperometric detector in a flow-injection system using a wall-jet flow cell for the determination of propylgallate (PG), octylgallate (OG) and *tert*-butylhydroxyanisole (BHA), has been described [493]. The voltammetric cell used contained the PVC-graphite working electrode, an Ag/AgCl/3M KCl reference electrode, and a glassy carbon counter electrode. Cyclic voltammetric analysis of octylgallate (OG) in food by means of a polypyrrole electrode modified with nickel phthalocyanine in a FIA system has been reported [494]. Anodic stripping voltammetry (ASV) has been used to determine Ionol (2,6-di-*t*-butyl-*p*-cresol) and quinol in polyester acrylates, as well as Ionol and 4,4-isopropylidenediphenol in PVC [495]. Other similar analyses are reported in Table 8.75. Inorganic pigments from paintings and polychromed sculptures immobilised in polymer (Palaroid B72) film modified electrodes have been identified by CV and anodic and cathodic stripping differential pulse voltammetry (ASDPV/CSDPV) techniques [496]. The results were found to be in agreement with PLM, SEM/EDX, XRD and FTIR analysis.

At one time, voltammetry (particularly classical polarography) was an important tool for the determination of inorganic ions and certain organic species in aqueous solutions. However, these analytical applications have been supplanted by various atomic spectroscopy methods, and voltammetry has ceased to be important in analysis, except for certain special applications. Voltammetric methods are available for rapid determination (in less than 10 min) of up to four trace metals (e.g. Cu, Cd, Pb, Zn) in marine waters in a single voltammetric sweep [497].

8.7.2.1 Polarographic Methods

Principles and Characteristics Contrary to potentiometric methods that operate under null conditions, other electrochemical methods impose an external energy source on the sample to induce chemical reactions that would not otherwise occur spontaneously. It is thus possible to analyse ions and organic compounds that can either be reduced or oxidised electrochemically. Polarography, which is a division of voltammetry, involves partial electrolysis of the analyte at the working electrode.

Polarography refers to the use of mercury electrodes, which have a constantly renewed surface in application. This includes the dropping mercury electrode (DME) and the static mercury drop electrode (SMDE). In a polarographic experiment, the indicator electrode is a mercury droplet growing at the end of a narrow glass capillary. There are usually two other electrodes in the cell containing the sample: an auxiliary electrode (often a platinum wire) and a reference electrode (e.g. a saturated calomel electrode). Mercury has several specific advantages that make it attractive as an electrode material for quantitative analysis. Among the voltammetric methods, polarography yields the most reproducible results, because of its frequently renewed electrode/solution interface. However, the use of readily oxidisable mercury limits the method mostly to reductions. In all polarographic methods, the current is proportional to the concentration C^* of the electroactive species, rather than to the total amount present in the sample. This is a consequence of the local nature of polarographic electrolysis. Because polarography records an interfacial reaction, it is often subject to interference by strongly adsorbed species. Adsorbate interference affects all voltammetric methods.

Classical polarography is not optimised for analytical sensitivity. This inefficiency has been remedied in *pulse polarography*. Instead of applying a steadily increasing voltage on the mercury droplet, in pulse polarography the voltage is pulsed. The advantage of this approach is increased sensitivity and better distinction between analytes differing only by a few tens of mV. Pulse polarography has two major modes of operation: normal pulse polarography (NPP) and differential pulse polarography (DPP). Various polarographic techniques are thus used: classical DC (with DME and SMDE modes) and pulse and differential pulse polarography. The detection limit for DC/DME is about 1 ppm, as compared to 10 ppb for stripping voltammetry. DPP and square-wave polarography have long supplanted the classical linear-scan method, because of their greater sensitivity, convenience and selectivity. In the emulsion formation polarographic technique, a metal is released from an organometallic sample by means of a mineralisation treatment, transfer into an aqueous phase and subsequent emulsification [498]. Surfactants can be used to give sufficiently high concentrations without suppressing the polarographic peaks.

Polarisation titrations are often referred to as amperometric or biamperometric titrations. It is necessary that one of the substances involved in the titration reaction be oxidisable or reducible at the working electrode surface. In general, the polarisation titration method is applicable to oxidation–reduction, precipitation and complexation titrations. Relatively few applications involving acid/base titration are found. Amperometric titrations can be applied in the determination of analyte solutions as low as 10^{-5} M to 10^{-6} M in concentration.

The ascent of spectroscopic methods has reduced the relative importance of polarography as an analytical method, although the high sensitivity of square-wave and pulse polarography, and of stripping voltammetry, often still gives them an edge for several environmentally important metals, such as cadmium and lead. Polarography is sometimes an alternative to atomic absorption methods but otherwise is outclassed.

The basic techniques of polarography were recently summarised by Crompton [499].

Applications As the basic process of electron transfer at an electrode is a fundamental electrochemical principle, polarography can widely be applied. Polarography can be used to determine electroreductible substances such as monomers, organic peroxides, accelerators and antioxidants in solvent extracts of polymers. Residual amounts of *monomers* remain in manufactured batches of (co)polymers. For food-packaging applications, it is necessary to ensure that the content of such monomers is below regulated level. Polarography has been used for a variety of monomers (styrene, α-methylstyrene, acrylic acid, acrylamide, acrylonitrile, methylmethacrylate) in

Table 8.77 Polarographic methods for the determination of monomers

Monomer	Polymer	Method	Reference(s)
α-Methylstyrene	Polystyrene	Solvent solution polarography	[500,501]
Styrene	Polystyrene	Solvent solution polarography	[500–502]
Methylmethacrylate	Polymethacrylates	Polarography	[503,504]
Acrylonitrile	Styrene–acrylonitrile	Polarography	[505]
Styrene, acrylonitrile	Styrene–acrylonitrile	Solution in DMF, polarography	[502,506–508]
Styrene	Polyester laminates, styrene–butadiene acrylics	Polarography	[509]
Acrylamide, acrylic acid	Polyacrylamide	DPP	[510]

various polymers. Table 8.77 gives a non-exhaustive list of monomers determined by various polarographic methods. Other applications are listed in Table 8.75. Most of this work is dated (1950–1980).

Styrene and acrylonitrile monomers in SAN were determined in a single polarographic analysis [502], with lower limits of detection of 20 and 2 ppm, respectively. The styrene monomer contents determined by UV spectroscopy at 292 nm in CCl_4 solution were found to be in good agreement with the polarographic results [499]. Acrylonitrile could not be verified in the same way. Residual acrylonitrile (ACN) monomer in SAN production samples has been determined polarographically directly in DMF solution [506]. This procedure without using combustive or separative techniques avoids indirect measurement based on extraction or reprecipitation procedures. Styrene monomer does not interfere with the polarographic determination of ACN in DMF. The procedure for the determination of acrylamide monomer in polyacrylamide involves extraction, removal of interfering cationic and anionic species, and polarographic reduction [510]. Acrylic acid is polarographically distinguishable from acrylamide in a neutral medium, whereas ethylacrylate interferes. The detection limit of acrylamide monomer by polarography is less than 1 ppm.

Classical polarography was found to be more applicable to rubber accelerators than to phenolic and amine type antioxidants. *Antioxidants* have been studied with various types of electrodes (DME, graphite, platinum) [492,511–514]. A procedure has been described involving the conversion of an antioxidant into a polarographically reducible form [515]. Polarography has also been applied to the analysis of organic *vulcanisation accelerators* (e.g. Vulkacit AZ/F/P/Thiuram) [516]. Mocker *et al.* [517–519] have used polarography for raw material control and for the determination of accelerators in batches of unvulcanised rubbers and in elastomers. Examples include the polarographic determination of 2-mercaptobenzothiazole and

2-mercaptobenzimidazole, and of *N*-cyclohexyl-2-benzothiazylsulfenamide in elastomers. Polarography was considered to be more applicable to rubber accelerators than to antioxidants [519]. Nevertheless, the polarography of some amines and phenols has been studied [520–522]. *Organic peroxides* (benzoyl peroxide, *p-tert*-butylperbenzoate and lauryl peroxide) in PS have also been determined by polarography [31,523]. Cadmium, Pb and Zn salts of *fatty acids* were determined by polarography [524], and Cd, Zn and Ba stearates or laurates in PVC were determined by alternating current polarography [525].

Pulse polarography is widely applicable to the analysis of inorganic substances (*metallic cations, inorganic anions*), including metallic components in polymers. In the majority of cases it is necessary to subject the samples to wet-ashing, or microwave digestion under pressure, to destroy all organic matter. A polarographic method has been developed for determining traces of Ti in HDPE. Sensitivity is improved by the use of catalytic oxidation of Ti^{3+} by reaction with $KClO_3$. The method detects ca. 0.0002 5 % Ti [526]. Tin can be determined in PVC by polarography and AAS [527]. Emulsion formation polarography was used for the determination of cobalt (the most reactive drier metal used in varnishes [498]) and for the differential pulse polarographic determination of Pb^{2+} as lead naphthenate in paint driers and varnishes [528]. Also, the polarographic determination of free sulfur in an acetone extract has been reported [529]. Proske [530] has described the polarographic determination of free sulfur in vulcanisates. Polarographic sulfate determination in a vulcanisate has also been reported [531]. Polarisation titrations are applicable to phenols and phosphates. Mercaptans in PMMA were determined by amperometric titration with a standard silver solution [532].

It will not have escaped the attention of the careful reader that most references concerning the applications of polarography to polymer/additive analysis are rather dated.

8.7.3 Coulometric Methods

Principles and Characteristics The three related electroanalytical methods: electrogravimetry, potentiostatic coulometry, and amperostatic coulometry or coulometric titrimetry, differ from potentiometry in several regards (Table 8.74). The word 'coulometry' implies measurement of charge (in coulombs); in practice, almost all modern coulometric measurements measure the current, and integrate this over time to calculate the corresponding charge. Coulometric methods at constant current or constant potential are based on the use of a reagent that can react with or convert the analyte quantitatively to a different oxidation state. Coulometric and gravimetric methods share the common advantage that the proportionality constant between the quantity measured and the analyte mass is derived from theory and accurately known physical constants, thus eliminating the need for calibration standards. The process of coulometry yields the number of analyte equivalents involved in the sample under test. This will lead to a quantitative determination of the analyte in the sample. The current efficiency must be 100 % for the method to be used.

Coulometry and *amperometry* can be distinguished by the extent to which the analyte undergoes a Faradaic reaction at the working electrode, namely complete and partial, respectively. Coulometry is essentially high-efficiency amperometry with working electrodes of large surface area. Successful coulometric or amperometric detection can result only if the applied potential is chosen correctly.

Perhaps the most useful application of coulometry is in *coulometric titrations*, in which the titrant is not added to the solution but is generated internally from a previously added precursor reagent. For example, Ce(IV) can be generated from Ce(III)-containing solutions in concentrated H_2SO_4, and used to titrate Fe(II) or Ti(III). The amount of titrant that must be generated to reach the equivalence point is then measured coulometrically. Coulometric titrations are similar to other titrimetric methods, in that analyses are based on measuring the combining capacity of the analyte with a standard reagent. A typical example is the generation of I_2 between two platinum electrodes immersed in an oxygen-free solution of KI. The generating reaction at the anode is:

$$2I^- - 2e \longleftrightarrow I_2 \qquad (8.11)$$

The generation of iodine coulometrically at the anode has an extensive application in the *Karl Fischer* (KF) technique of water determination. The current state-of-the-art of the coulometric KF titration is given by ref. [533].

Coulometric methods are as accurate and precise as conventional gravimetric and volumetric procedures and, in addition, are readily automated. In contrast to gravimetric methods, coulometric procedures are usually rapid, and do not require that the product of the electrochemical reaction be a weighable solid. The methods are moderately sensitive, and offer a reasonably selective means for separating and determining a number of ions.

The techniques of voltammetry/polarography, atomic absorption, ICP, etc., have in most cases supplanted the coulometric approach for the determination of inorganic analytes. Coulometry and the use of coulometry in food analysis have recently been reviewed [473,476].

Applications The coulometric Karl Fischer titration is a widely used *moisture determination* method (from ppm to 100 %). In the presence of water, iodine reacts with sulfur dioxide through a redox process, as follows:

$$I_2 + SO_2 + 2H_2O \longrightarrow H_2SO_4 + 2\,HI \qquad (8.12)$$

In the presence of methanol as a solvent, the reaction is:

$$CH_3OSO_2H + I_2 + H_2O \rightleftharpoons CH_3OSO_3H + 2HI \qquad (8.13)$$

Before the Karl Fischer coulometric determination of water can be made, solid components that are not soluble must either be ground into powders or extracted with anhydrous solvents. Crompton [31] has reported the determination of 0.2 % water in cellulose after extraction with methanol. Its use in the plastics industry requires the addition of an accessory pyrometry oven. The Karl Fischer determinations involve heating a plastic resin sample (usually less than 5 g) in the pyrolysis oven, and transfer of the resulting vapors to the reagent vessel, followed by electrical coulometry of the water in the vapour stream. Karl Fischer pyrometry–coulometry presents several disadvantages. The fragility of the apparatus requires situation in an analytical laboratory setting, often at a considerable distance from the moulding floor. Transporting dried engineered resins (which are highly hygroscopic) without contamination is problematic. Accuracy at low moisture levels requires careful control of the background moisture in the KF system. In addition, because coulometry does not start until the moisture rate exceeds some preset level (typically $0.5\,\mu g\,s^{-1}$) a portion of the water vapour signal is lost and Karl Fischer results typically have a negative bias at low moisture levels [534]. The consumables

of Karl Fischer titration are relatively expensive. Raisanen [535] has described a new headspace-based device,
which overcomes the limitations of Karl Fischer and
LOD techniques.

The water (moisture) content can rapidly and accurately be determined in polymers such as PBT, PA6,
PA4.6 and PC via coulometric titration, with detection
limits of some 20 ppm. Water produced during heating of
PET was determined by Karl Fischer titration [536]. The
method can be used for determining very small quantities of water ($10 \mu g - 15 mg$). Certified water standards
are available. Karl Fischer titrations are not universal. The method is not applicable in the presence of
H_2S, mercaptans, sulfides or appreciable amounts of
hydroperoxides, and to any compound or mixture which
partially reacts under the conditions of the test, to produce water [31]. Compounds that consume or release
iodine under the analysis conditions interfere with the
determination.

Conductometric titrations offer several advantages
compared with potentiometric titration methods, such as
better precision and better differentiation of the basic
components in polymers, but they are more laborious.
ASTM D 4928-96 is an established KF method for the
determination of water in crude oils.

Microcoulometric titration is used as the detection
mode in some commercial sulfur-specific analysers.
Sulfur in PP and waxes (range from 0.6 to 6 ppm S)
were determined by means of an oxidative coulometric
procedure [537]. The coulometric electrochemical array
detector was used for determining a variety of synthetic
phenolic antioxidants (PG, THBP, TBHQ, NDGA,
BHA, OG, Ionox 100, BHT, DG) in food and oils [538].

8.8 SOLID-STATE SPECIATION ANALYSIS

Principles and Characteristics   The fastest growing
area in elemental analysis is in the use of hyphenated
techniques for speciation measurement. Elemental speciation analysis, defined as the qualitative identification
and quantitative determination of the individual chemical forms that comprise the total concentration of an
element in a sample, has become an important field
of research in analytical chemistry. Speciation or the
process yielding evidence of the molecular form of an
analyte, has relevance in the fields of food, the environment, and occupational health analysis, and involves
analytical chemists as well as legislators. The environmental and toxicological effects of a metal often depend
on its forms. The determination of the total metal content

does not provide this information. Environmental legislation now stipulates maximum allowable levels of toxic
species rather than total elemental concentration. Speciation analysis bridges the gap between inorganic and
organic analysis, and constitutes one of the most challenging analytical fields. Trace-element measurements
need to be correlated with selective detection. The sensitivity of molecular-specific detection techniques is far
from reaching that of element-specific detection. Both
metals and nonmetals may be speciated; therefore compounds containing elements such as the halogens may be
separated. Speciation studies are mainly concerned with
Hg, Sn, Pb, As, Se, Sb, Cr and Fe. *Chemical speciation* refers to various chemical states within one phase,
whereas physical speciation of a trace element refers
to the distribution over various coexisting phases. Spatial speciation deals mostly with surface layers in solids
(depth profiling, lateral profiling).

Speciation involves a number of discrete analytical
steps comprising the extraction (isolation) of the analytes from a solid sample, preconcentration (to gain
sensitivity), and eventually derivatisation (e.g. for ionic
compounds), separation and detection. Various problems
can occur in any of these steps. The entire analytical procedure should be carefully controlled in such a way that
decay of unstable species does not occur. For speciation analysis, there is the risk that the chemical species
can convert so that a false distribution is determined. In
general, the accuracy of the determinations and the traceability of the overall analytical process are insufficiently
ensured [539].

As in all other analytical activities, *sample preparation* is crucial. Methods are required to extract molecular species under conditions mild enough to maintain the molecular integrity of the species (e.g. MAE,
SPME, ASE®). Extraction recoveries and decomposition cannot be controlled through recovery studies with
spiked samples [540]. Avoidance of chemical pretreatment is possible if direct methods such as solid sampling
can be adopted (e.g. LA-ICP-MS, GD-MS). There are
few *diagnostic tools* able to trace the conversion or
degradation of species during sampling, storage and the
measurement step, since conventional speciation methods can only measure the species concentrations in the
final solutions at the time of measurement. Kingston
et al. [541] have proposed speciated isotope dilution
mass spectrometry (SIDMS) to remedy this situation,
by spiking the sample at each step with enriched stable
isotopes of the same species. Speciated isotope dilution
enables the quantitative correction of species transformation during analysis. SIDMS has the potential to be
used as a diagnostic tool to validate other methods and

to certify speciated standards. Lack of quality control of speciation often affects the compatibility of data within and between laboratories. One of the limitations in using certified organometallic reference materials for speciation analysis is their short shelf-life. Many organometallic compounds are prone to oxidation and/or decomposition in storage, and therefore often require recertification. This can be overcome by developing appropriate standards and methods for combining HPLC-ICP-MS speciation measurement with isotope dilution mass spectrometry (IDMS). The SM&T programme of the European Commission has launched collaborative projects to improve the state-of-the-art speciation analysis in Europe [542]. Although CRMs have been prepared for the validation of speciation analysis methods, there is a general lack of suitable reference materials.

Key issues in speciation analysis are:

(i) a selectivity sufficient to determine the chemical forms of the element of interest; and
(ii) a high sensitivity for ultratrace measurement.

Speciation essentially means a high degree of selectivity (appropriately handled by multidimensional chromatography). Atomic spectroscopic techniques including AAS, ETAAS, PS-AES and PS-MS offer the high sensitivity required for ultratrace measurements, but generally lack the selectivity necessary to identify different species of the same element. Coupling of a sensitive element-specific detector to a chromatographic (or electrophoretic) separation technique (Tables 8.78 and 8.79) has great promise in speciation analysis. The main disadvantage of GC is the need for derivatisation of analytes prior to analysis, as the native species are usually ionic and lack the necessary volatility and thermal stability. Liquid chromatography is considerably more difficult to hyphenate on-line with the most interesting specific detectors than is gas chromatography.

Since the concentrations of the various compounds or oxidation states in which trace elements can occur are always lower than the total content of the analyte, speciation analysis is normally an ultratrace determination in the ng L^{-1} range for solutions and the ng g^{-1} range for

Table 8.78 Most frequently coupled speciation techniques

Separation technique	Element-selective detection techniques
GC	AAS (FAAS, GFAAS, QFAAS)
HPLC	AES (ICP-AES, MIP-AES, DCP-AES)
SFC	FPD, ECD, FID
CE	AFS
FI	MS (EI, PI, API), ICP-MS, IDMS
	LEI

Table 8.79 Main techniques for speciation analysis

Element-specific detection	Molecular-specific detection
GC-AES	GC-MS
GC-FPD	HPLC-FLD
GC-AAS	HPLC-MS[a]
GC-ICP-AES, GC-MIP-AES	
GC-ICP-MS, GC-MIP-MS	
HPLC-AAS	
HPLC-ICP-AES	
HPLC-ICP-MS	
SFC-ICP-MS	
CZE-ICP-MS	

[a] ESI or APCI-MS.

solids. The requirements placed on the detector are thus correspondingly high. The detector sensitivity for speciation varies per element. Detectors used for speciation analysis are either element specific (e.g. AAS) or nonspecific (e.g. FID, FPD, ECD). Classical detectors after LC or GC separation have been applied for the determination of some chemical forms of elements, e.g. FPD or FID detection for TBT. The early *interfaces* for speciation, GC-FAAS and HPLC-AAS, do have a number of limitations. GC-AAS is limited to samples which are thermally stable and have favourable gas solution partition coefficients. Various atomic spectroscopic techniques have been applied as detectors for HPLC (e.g. FAAS, GFAAS, FAF). Flame techniques give poor detection, due to the short residence time in the flame and the low-temperature environment. HPLC-FAAS generally does not provide the sensitivity required for many real samples. GFAAS exhibits useful detection limits, but is not capable of monitoring HPLC effluent continuously, because time is required to run through the several furnace cycles. Similarly, ETAAS cannot be applied in a continuous (on-line) mode; moreover, the necessary manipulations caused by the off-line character of the method may increase considerably the risks of errors. GC-AES is limited to volatile and thermally stable compounds. HS-GC-AED is an adequate speciation tool.

Plasmas have an important place in speciation analysis. With the introduction of GC-MIP, GC-ICP-MS and HPLC-ICP-MS, it has become possible to monitor both the metallic and nonmetallic components of a single sample simultaneously at the low detection levels required. Advantages of CGC-ICP-MS comprise high transport efficiency, few isobaric interferences, stable plasma operation, good chromatographic resolution and pg range detection limits. GC-ICP-ToFMS generates high expectations. ICP-AES and ICP-MS afford continuous on-line real-time monitoring for metal species separated by HPLC. Detection limits are affected greatly by the chromatographic conditions (anion-exchange chromatography, reversed-phase chromatography, etc.) and are species dependent. For elemental speciation by LC-ICP-MS, see ref. [543].

With the exception of GC-MIP-AES there are no commercial instruments available for speciation analysis of organometallic species. Recently, a prototype automated *speciation analyser* (ASA) for practical applications was described [544,545], which consists of a P&T system (or focused microwave-assisted extraction), multicapillary GC (MC-GC), MIP and plasma emission detection (PED). MCGC-MIP-PED provides short analysis times (<1 min) and a 5–10 pg absolute detection limit.

Mass spectrometry can be specific in certain cases, and would even allow on-line QA in the isotope dilution mode. MS of molecular ions is seldom used in speciation analysis. API-MS allows compound-specific information to be obtained. APCI-MS offers the unique possibility of having an element- and compound-specific detector. A drawback is the limited sensitivity of APCI-MS in the element-specific detection mode. This can be overcome by use of on-line sample enrichment, e.g. SPE-HPLC-MS. The capabilities of ESI-MS for metal speciation have been critically assessed [546]. Use of ESI-MS in metal speciation is growing. Houk [547] has emphasised that neither ICP-MS (elemental information) nor ESI-MS (molecular information) alone are adequate for identification of unknown elemental species at trace levels in complex mixtures. Consequently, a plea was made for simultaneous use of these two types of ion source on the same liquid chromatographic effluent.

For chemical speciation, X-ray absorption spectroscopy is another supreme tool taking advantage of its electronic and structural information power. Also, REMPI-MS is outstanding in its selectivity for molecular species. Radioanalytical methods have also been used for speciation analysis [548]. Microscopical speciation analysis requires SSIMS or LMMS [549].

Table 8.80 shows the present status of speciation methodology. For trace-metal speciation, atomic absorption detectors feature a relatively high absolute detection limit (10 pg level), as compared to the 0.1 to 1 pg sensitivity level for molecular ion MS techniques as well as for MIP-AES. The detection limit of LEI-ToFMS is in the attogram range. Speciation has been reviewed [550]. Various monographs deal with speciation analysis [542,551,552].

Applications   Speciation analysis is particularly important in plant and animal biochemistry and nutrition (food/food supplements), clinical biochemistry, industrial chemistry and environmental chemistry. In the

Table 8.80   Status of speciation methodology

+ Well-understood measurement part of the analytical process
+ Many interlaboratory exercises
+ Availability of matrix CRMs
+ Performing research instrumentation (GC-ICP-ToFMS)
− General lack of commercial instruments (except GC-MIP-AES)
+ Tailor-made instrumentation (ASA prototype)
− Rather complex methodologies (specialist rather than routine)
− Methods not legally enforceable (extraction, complex matrices, compounds' integrity)
− Insufficient accuracy and traceability of overall analytical process

latter field the analyte of interest is usually known, at variance with bio-inorganics. In health studies, speciation is of growing importance. Suitable certified organometallic speciation standards (>99 % pure) are under development, as well as validated methods for Sn, As, Hg, Se and Pb [553].

Organotin compounds have numerous applications (as biocides, catalysts, stabilisers, etc.). Speciation information at ng levels is usually required. In studies on the technology of obtaining *heat stabilisers* for PVC, it was found that for evaluation of the stabiliser quality it is not enough to determine the total tin content, but it is necessary to know the proportion of mono- and diesterchlorotin compounds in the mixture [554]. Methods of tin determination were elaborated which permit speciation of monoester tins, diester tins and inorganic tin compounds in intermediates formed in the process of obtaining stabilisers. As monoesterchlorotin compounds (MMET and MBET) form chelates with EDTA, as opposed to diesterchlorotins (DMET and DBET), they may be selectively determined by potentiometric titration with the use of Cu-ISE after formation of a complex with EDTA in an alcoholic–aqueous medium. Use of microwave digestion and an automatic system for potentiometric determination of the content of tin, monoesterchlorotins, chlorine, mercaptan sulfur and light organic compounds assures analytical control of organotin heat-stabiliser technology for PVC. The analysis is also of importance from an economical point of view, since tin is the most expensive material in this technology.

Other techniques used for *organotin speciation* comprise GC-FAAS, GC-GFAAS, GC-ICP-MS, HPLC-FAAS, HPLC-GFAAS, HPLC-DCP (after continuous on-line hydride generation), HPLC-ICP-AES, HPLC-ICP-MS, etc. [555]. Whereas ICP-AES does not provide an adequate response for ng levels of tin, ICP-MS can detect sub-ng to pg levels. GC-ICP-ToFMS has been used for speciation analysis of organotin and organolead compounds (tetramethyltin, tetraethyltin and tetraethyllead) [556]. With fast GC separation (<3 min, peak width 1 s) the (fast) transient signals produced were analysed by ToF-MS without many of the problems encountered with scanning mass spectrometers. Femtogram limits of detection for each compound were quoted with a dynamic range exceeding six orders of magnitude.

For RPLC-ICP-MS speciation of butyltin compounds, the absolute detection limits of 0.15 ng (as tin) for tributyltin and 0.24 ng for di- and monobutyltin were reported [557]. Of the systems incorporating LC-ICP-MS, a number of chromatographic mechanisms have been employed. Ion-exchange, ion-pair and micellar LC systems tend to involve the use of nonvolatile additives in the mobile phase, which is incompatible with API-MS instrumentation. LC-ICP-MS and ion-pair chromatography-based LC-ICP-MS were used for speciation and detection of organotin compounds (TMT-Cl, TBT-Cl, TPhT-Ac) [558]. Detection limits obtained by ICP-MS are three orders of magnitude lower than those obtained with ICP-AES. LC-APCI-MS and LC-ICP-MS were compared for speciation of organotin compounds (DBT, TBT, DPhT, TPhT) [559]. While the use of ICP-MS has the advantage of superior sensitivity coupled with selectivity, LC-API-MS provides molecular information on the intact organotin species. Thus, each technique can be used as a means of validation of the other, giving the analyst greater confidence in the speciation of organotin compounds. Epler *et al.* [560] have described the use of LEI as a detector for speciation of organolead compounds using LC, and have reported detection limits comparable with those which can be achieved using ICP-MS (sub-ng mL^{-1}). Vela and Caruso [561] have compared ICP-MS and FID detection of alkyltin compounds separated using cSFC. Limits of detection for SFC-ICP-MS (about 0.2–0.8 pg) were more than an order of magnitude better than for SFC-FID. SFC and SFE are still in their infancy as far as applications to speciation analysis are concerned.

Kolasa *et al.* [562] have reported changes in the degree of oxidation of chromium (from Cr^{6+} to Cr^{3+}) in the course of probe mineralisation of PE for AAS analysis. HPLC-ICP-MS has been used as a selenium-specific detector [563]. Other *selenium speciation* work by ICP-MS has been reported [564]. Numerous other examples of speciation analysis have been described for the most appropriate techniques (Chapter 7).

Speciation analysis of organometal compounds by means of GC-MIP-AES and GC-ICP-MS has been reviewed [565], as has as metal speciation by HPLC

[566]. Applications of ICP-MS as a chromatographic detector in speciation studies have been reported [567].

8.9 BIBLIOGRAPHY

8.9.1 Sampling and Sample Preparation

J.C. Lindon, G.E. Tranter and J.L. Holmes (eds), *Encyclopedia of Spectroscopy and Spectrometry*, Academic Press, San Diego, CA (2000).

P. Gy, *Sampling for Analytical Purposes*, John Wiley & Sons, Ltd, Chichester (1998).

H.M. Kingston and S.J. Haswell (eds), *Microwave-Enhanced Chemistry*, American Chemical Society, Washington, DC (1997).

M. Stoeppler (ed.), *Sampling and Sample Preparation*, Springer-Verlag, Berlin (1997).

P. Gy, *Sampling of Heterogeneous and Dynamic Materials Systems. Theories of Heterogeneity, Sampling and Homogenisation*, Elsevier, Amsterdam (1992).

Z. Sulcek and P. Povondra, *Methods of Decomposition in Inorganic Analysis*, CRC Press, Boca Raton, FL (1989).

H.M. Kingston and L.B. Jassie (eds), *Introduction to Microwave Sample Preparation: Theory and Practice*, American Chemical Society, Washington, DC (1988).

R. Bock, *A Handbook of Decomposition Methods in Analytical Chemistry*, International Textbook Company, London (1979).

8.9.2 Atomic Spectrometry

M. Cullen (ed.), *Atomic Spectroscopy in Elemental Analysis*, Blackwell Publishing, Oxford (2003).

J. Nölte, *ICP Emission Spectrometry. A Practical Guide*, Wiley-VCH, Weinheim (2003).

R.K. Marcus (ed.), *Glow Discharge Plasmas in Analytical Spectroscopy*, John Wiley & Sons, Inc., New York, NY (2002).

J.A.C. Broekaert, *Analytical Atomic Spectrometry with Flames and Plasmas*, Wiley-VCH, Weinheim (2000).

D. Beauchemin, D.C. Grégoire, D. Günther, V. Karanassios, J.-M. Mermet and T.J. Woods (eds), *Discrete Sample Introduction Techniques for Inductively Coupled Plasma Mass Spectrometry*, Elsevier, Amsterdam (2000).

B. Welz and M. Sperling, *Atomic Absorption Spectrometry*, Wiley-VCH, Weinheim (1999).

K.W. Jackson (ed.), *Electrothermal Atomization for Analytical Atomic Spectroscopy*, John Wiley & Sons, Ltd, Chichester (1999).

L. Ebdon, E.H. Evans, A. Fisher and S. Hill (eds), *An Introduction to Analytical Atomic Spectrometry*, John Wiley & Sons, Ltd, Chichester (1998).

D.J. Butcher and J. Sneddon, *A Practical Guide to Graphite Furnace Atomic Absorption Spectrometry*, John Wiley & Sons, Inc., New York, NY (1998).

G. Schlemmer and B. Radziuk, *A Laboratory Guide to Graphite Furnace Analytical Atomic Spectroscopy*, Springer-Verlag, Berlin (1998).

R. Payling, D.G. Jones and A. Bengtson (eds), *Glow Discharge Optical Emission Spectrometry*, John Wiley & Sons, Ltd, Chichester (1997).

U. Kurfürst (ed.), *Solid Sample Analysis: Direct and Slurry Sampling using GF-AAS and ETV-ICP*, Springer-Verlag, Berlin (1997).

J.R. Dean, *Atomic Absorption and Plasma Spectroscopy*, John Wiley & Sons, Ltd, Chichester (1997).

J.W. Robinson, *Atomic Spectroscopy*, M. Dekker, New York, NY (1996).

J. Sneddon, T.L. Thiem and J.-I. Lee (eds), *Lasers in Analytical Atomic Spectroscopy*, VCH Publishers, New York, NY (1996).

M. Parkany, *The Use of Recovery Factors in Trace Analysis*, The Royal Society of Chemistry, Cambridge (1996).

C. Vandecasteele and C.B. Block, *Modern Methods for Trace Element Determination*, John Wiley & Sons, Ltd, Chichester (1993).

L.H.J. Lajunen, *Spectrochemical Analysis by Atomic Absorption and Emission*, The Royal Society of Chemistry, Cambridge (1992).

S.J. Haswell (ed.), *Atomic Absorption Spectrometry – Theory, Design and Applications*, Elsevier Science, Amsterdam (1991).

J. Sneddon (ed.), *Sample Introduction in Atomic Spectroscopy*, Elsevier Science, Amsterdam (1990).

P.W.J.M. Boumans, *Inductively Coupled Plasma Emission Spectroscopy*, Vols I and II, John Wiley & Sons, Inc., New York, NY (1987).

8.9.3 X-ray Spectrometry

K. Tsuji, R. van Grieken and J. Injuk (eds), *X-Ray Spectrometry. Recent Technological Advances*, John Wiley & Sons, Ltd, Chichester (2004).

K. Tsuji, *Advances in X-Ray Spectrometry*, John Wiley & Sons, Ltd, Chichester (2003).

D.J. Dyson, *X-Ray and Electron Diffraction Studies in Materials Science*, Maney Publishing, Leeds (2003).

R.E. van Grieken and A.A. Markowicz, *Handbook of X-ray Spectrometry*, M. Dekker, New York, NY (2002).

K.H.A. Janssens, F.C.V. Adams and A. Rindby (eds), *Microscopic X-Ray Fluorescence Analysis*, John Wiley & Sons, Ltd, Chichester (2000).

E. Lifshin (ed.), *X-ray Characterization of Materials*, Wiley-VCH, Weinheim (1999).

R. Jenkins, *X-Ray Fluorescence Spectrometry*, Wiley-Interscience, New York, NY (1999).

F.H. Chung, *Industrial Applications of X-Ray Diffraction*, M. Dekker, New York, NY, (1999).

V.E. Buhrke, R. Jenkins and D.K. Smith (eds), *A Practical Guide for the Preparation of Specimens for X-Ray Fluorescence and X-Ray Diffraction Analysis*, Wiley-VCH, New York, NY (1998).

G.R. Lachance and F. Claisse, *Quantitative X-Ray Fluorescence Analysis, Theory and Application*, John Wiley & Sons, Ltd, Chichester (1998).

M.M. Woolfson (ed.), *An Introduction to X-Ray Crystallography*, Cambridge University Press, Cambridge (1997).

A. Michette and S. Pfauntsch (eds), *X-Rays: The First Hundred Years*, John Wiley & Sons, Inc., New York, NY (1996).

H. Saisho and Y. Gohshi, *Applications of Synchrotron Radiation to Materials Analysis*, Elsevier Science, Amsterdam (1996).

R. Jenkins and R.L. Snyder, *Introduction to X-Ray Powder Diffractometry*, John Wiley & Sons, Ltd, Chichester (1996).

R. Klockenkämper, *Total-Reflection X-Ray Fluorescence Analysis*, John Wiley & Sons, Inc., New York, NY (1996).

S.A.E. Johansson, J.L. Campbell and K.G. Malmqvist (eds), *Particle-Induced X-Ray Emission Spectrometry (PIXE)*, John Wiley & Sons, Inc., New York, NY (1995).

T. Yamanaka, *Material Analysis Using Powder X-Ray Diffraction*, Kodansha, Tokyo (1993).

J. Stöhr, *NEXAFS Spectroscopy*, Springer-Verlag, New York, NY (1992).

B. Dziunikowski, *Energy Dispersive X-Ray Fluorescence Analysis*, Elsevier, Amsterdam (1989).

D.C. Koningsberger and R. Prins (eds), *X-Ray Absorption, Principles, Applications. Techniques of EXAFS, SEXAFS and XANES*, John Wiley & Sons, Inc., New York, NY (1988).

K.L. Williams, *Introduction to X-ray Spectrometry*, Allen & Unwin, London (1987).

E.P. Bertin, *Introduction to X-ray Spectrometry Analysis*, Plenum, New York, NY (1985).

8.9.4 Inorganic Mass Spectrometry

R. Thomas, *Practical Guide to ICP-MS*, CRC Press, Boca Raton, FL (2004).

S. Nelms (ed.), *Inductively Coupled Plasma Mass Spectrometry Handbook*, Blackwell Publishing, Oxford (2004).

J.R. de Laeter, *Applications of Inorganic Mass Spectrometry*, Wiley-Interscience, New York, NY (2001).

H.E. Taylor, *Inductively Coupled Plasma Emission and Mass Spectrometry, Practices and Techniques*, Academic Press, San Diego, CA (2001).

G. Holland and S.D. Tanner (eds), *Plasma Source Mass Spectrometry, The New Millennium*, The Royal Society of Chemistry, Cambridge (2001).

S.J. Hill (ed.), *Inductively Coupled Plasma Spectrometry and Its Applications*, Sheffield Academic Press, Sheffield (1999).

A. Montaser (ed.), *Inductively Coupled Plasma Mass Spectrometry: From A to Z*, Wiley-VCH, New York, NY (1998).

E.H. Evans, J.J. Giglio, T.M. Castillano and J.A. Caruso, *Inductively Coupled and Microwave Induced Plasma Sources for Mass Spectrometry*, The Royal Society of Chemistry, Cambridge (1995).

G. Holland and A.N. Eaton, *Applications of Plasma Source Mass Spectrometry*, Part I and II, The Royal Society of Chemistry, Cambridge (1991/1993).

K.E. Jarvis, A.L. Gray and R.S. Houk, *Handbook of Inductively Coupled Plasma Mass Spectrometry*, Blackie, London (1992).

A. Montaser and D.W. Golightly, *Inductively Coupled Plasmas in Analytical Atomic Spectrometry*, VCH Publishers, New York, NY (1992).

K.E. Jarvis, A.L. Gray, J.G. Williams and I. Jarvis, *Plasma Source Mass Spectrometry*, The Royal Society of Chemistry, Cambridge (1990).

A.R. Date and A.L. Gray, *Applications of Inductively Coupled Plasma Mass Spectrometry*, Blackie, Glasgow (1989).

F. Adams, R. Gijbels and R. van Grieken, *Inorganic Mass Spectrometry*, John Wiley & Sons, Inc., New York, NY (1988).

8.9.5 Nuclear Analytical Methods

Z. Alfassi (ed.), *Non-Destructive Elemental Analysis*, Blackwell Publishing, Oxford (2001).

J. Kantele and J. Aysto, *Handbook of Nuclear Spectrometry*, Academic Press, London (1995).

Z.B. Alfassi and C. Chung (eds), *Prompt Gamma Neutron Activation Analysis*, CRC Press, Boca Raton, FL (1995).

D.J. Winefordner and I.M. Kolthoff (eds), *Activation Spectrometry in Chemical Analysis*, Wiley-Interscience, New York, NY (1991).

W.D. Ehmann and D.E. Vance (eds), *Radiochemistry and Nuclear Methods of Analysis*, Wiley-Interscience, New York, NY (1991).

S.J. Parry, *Activation Spectrometry in Chemical Analysis*, John Wiley & Sons, Inc., New York, NY (1991).

C. Segebade, H.-P. Weise and G.J. Lutz, *Photon Activation Analysis*, Walter de Gruyter, Berlin (1988).

8.9.6 Trace-element Analysis

J.R. Dean, *Methods for Environmental Trace Analysis*, John Wiley & Sons, Ltd, Chichester (2003).

M. Sargent and G. MacKay, *Guidelines for Achieving Quality in Trace Analysis*, The Royal Society of Chemistry, Cambridge (1995).

Z.B. Alfassi (ed.), *Determination of Trace Elements*, VCH, Weinheim (1994).

8.9.7 Electroanalysis

F. Scholz (ed.), *Electroanalytical Methods, Guide to Experiments and Applications*, Springer-Verlag, Berlin (2002).

G. Henze, *Polarographie und Voltammetrie*, Springer-Verlag, Berlin (2001).

M.R. Smyth and J.G. Vos, *Analytical Voltammetry*, Elsevier, Amsterdam (1992).

8.9.8 Speciation Analysis

R. Cornelis, J. Caruso, H. Crews and K. Heumann (eds), *Handbook of Elemental Speciation*, John Wiley & Sons, Ltd, Chichester (2003/4), 2 vols.

L. Ebdon, L. Pitts, R. Cornelis, H. Crews, O.F.X. Donard and Ph. Quevauviller (eds), *Trace Element Speciation for Environment, Food, Health*, The Royal Society of Chemistry, Cambridge (2001).

J.A. Caruso, K.L. Sutton and K.L. Ackley (eds), *Elemental Speciation – New Approaches for Trace Element Analysis*, Elsevier, Amsterdam (2000).

Ph. Quevauviller, *Methods Performance Studies for Speciation Analysis*, The Royal Society of Chemistry, Cambridge (1998).

I.S. Krull (ed.), *Trace Metal Analysis and Speciation*, Elsevier, Amsterdam (1991).

J.R. Kramer and H.E. Allen (eds), *Metal Speciation: Theory, Analysis and Application*, Lewis, Chelsea (1988).

G.R. Batley (ed.), *Trace Element Speciation: Analytical Methods and Problems*, CRC Press, Boca Raton, FL (1987).

8.10 REFERENCES

1. A. Lamberty, W. van Borm and Ph. Quevauviller, *Fresenius' Z. Anal. Chem.* **370**, 811–18 (2001).
2. *International Standards Organization Catalogue*, ISO, Geneva (1994).
3. A. Lamberty, P. De Bièvre and A. Götz, *Fresenius' Z. Anal. Chem.* **345**, 310–13 (1993).
4. J.W. Woittiez and J.E. Sloof, in *Determination of Trace Elements* (Z.B. Alfassi, ed.), VCH, Weinheim (1994), pp. 93–107.
5. A. Krause, A. Lange and M. Ezrin, *Plastics Analysis Guide. Chemical and Instrumental Methods*, Hanser Publishers, Munich (1983).
6. D. Lawrenz, *Phys. Stat. Sol.* **A167**, 373 (1998).
7. S. Affolter, *Kautsch. Gummi Kunstst.* **44**, 739–43 (1991).
8. L. Dunemann, J. Begerow and A. Bucholski, in *Ullmann's Encyclopedia of Industrial Chemistry*, VCH, Berlin (1994), Vol. B5, p. 65.
9. F.E. Barton II and W.R. Windham, *J. Assoc. Off. Anal. Chem.* **71**, 1162–7 (1988).
10. H.M. Kingston, P.J. Walter, S. Chalk, E. Lorentzen and D. Link, in *Microwave-Enhanced Chemistry* (H.M. Kingston and S.J. Haswell, eds), American Chemical Society, Washington, DC (1997), pp. 223–349.
11. A. Schramel and B. Michalke, in *Proceedings GSF Symposium on Analytical Chemical Quality Assurance, 1993* (A. Kettrup and E. Flammenkamp, eds), Ecomed, Landsberg (1995), pp. 11–21.
12. V. Mika, L. Preisler and J. Sodomka, *Polym. Degrad. Stab.* **28**, 215–25 (1990).
13. A.A. Samra, J.S. Morris and S.R. Koirtyohann, *Anal. Chem.* **47**, 1475 (1975).
14. H.M. Kingston and L.B. Jassie (eds), *Introduction to Microwave Sample Preparation*, American Chemical Society, Washington, DC (1988).
15. H.-M. Kuss, *Fresenius' Z. Anal. Chem.* **343**, 788 (1992).

16. Z. Sulcek and P. Povondra, *Methods of Decomposition in Inorganic Analysis*, CRC Press, Boca Raton, FL (1989).

17. C. Vandecasteele and C.B. Block, *Modern Methods for Trace Element Determination*, J. Wiley & Sons, Ltd, Chichester (1993).

18. R. Anderson, in *Encyclopedia of Analytical Science* (A. Townshend, ed.), Academic Press, London (1995), pp. 4489–95.

19. G. Schwedt, *The Essential Guide to Analytical Chemistry*, John Wiley & Sons, Ltd, Chichester (1997).

20. S.J. Nagourney and R.K. Madan, *J. Test. Eval.* **19** (1), 77–82 (1991).

21. T.R. Crompton, *Analysis of Polymers. An Introduction*, Pergamon Press, Oxford (1989).

22. A.J. Smith, *Anal. Chem.* **36**, 944 (1964).

23. J. Scheirs, *Compositional and Failure Analysis of Polymers. A Practical Approach*, John Wiley & Sons, Ltd, Chichester (2000).

24. H. Heinrichs, H.-J. Brumsack and W. Schultz, in *6. Colloquium Atomspektrometrische Spurenanalytik* (B. Welz, ed.), Bodenseewerk Perkin-Elmer, Überlingen (1991), pp. 769–74.

25. H. Narasaki and K. Umezawa, *Kobunshi Kagaku* **29**, 438–42 (1972).

26. H.J. Francis, in *Applied Polymer Analysis and Characterization* (J. Mitchell, ed.), Hanser Publishers, Munich (1987).

27. W. Schöniger, *Mikrochim. Acta*, 123 (1955); 869 (1956).

28. B. Woodget, in *Encyclopedia of Analytical Science* (A. Townshend, ed.), Academic Press, London (1995), pp. 4495–4502.

29. S.E. Amos, G.M. Giacoletto, J.H. Horns, C. Lavallée and S.S. Woods, in *Plastics Additives Handbook* (H. Zweifel, ed.), Hanser Publishers, Munich (2000), pp. 553–84.

30. T.R. Crompton, *Practical Polymer Analysis*, Plenum Press, New York, NY (1993).

31. T.R. Crompton, *Manual of Plastics Analysis*, Plenum Press, New York, NY (1998).

32. J. Haslam, H.A. Willis and D.C.M. Squirrell, *Identification and Analysis of Plastics*, John Wiley & Sons, Ltd, Chichester (1983).

33. J.Z. Falcon, J.L. Love, L.J. Gaeta and A.G.G. Altenau, *Anal. Chem.* **47**, 171 (1975).

34. K. Tanaka and T. Morikawa, *Kagaku to Kogyo (Osaka)*, **48**, 387 (1974); *C.A.* **82**, 98 702 (1974).

35. H. Narasaki, K. Miyagi and A. Unno, *Bunseki Kagaku* **22** (5), 541 (1973).

36. D. Coz and K. Baranwal, *Rubber World*, 30–35 (Jan. 1999).

37. S. Affolter, A. Ritter and M. Schmid, *Ringversuche und polymeren Werkstoffen*, EMPA, St. Gallen, Switzerland (2000).

38. H.A. Hernandez, *Intl Lab.* 84 (Sept 1981).

39. S. Halim Hamid, *Polym. Mater. Sci. Eng.* **82**, 25–6 (2000).

40. W.R. Hudgins, K. Theurer and T. Mariani, *J. Appl. Polym. Sci.: Appl. Polym. Symp.* **34**, 145–55 (1978).

41. *Dynamar Polymer Processing Additive Technical Information, Parr Bomb Analytical Method for Determining Total Organic Fluorine Concentration in Polyethylene*, Dyneon, Oakdale, MN (n.d.).

42. S.S. Woods and A.V. Pocius, *Proceedings SPE ANTEC 2000*, Orlando, FL (2000), pp. 191–208.

43. A.D. Sawant, in *Encyclopedia of Analytical Science* (A. Townshend, ed.), Academic Press, London (1995), pp. 4503–11.

44. J.-M. Mermet, in *Microwave-Enhanced Chemistry. Fundamentals, Sample Preparation and Applications* (H.M. Kingston and S.J. Haswell, eds), American Chemical Society, Washington, DC (1997).

45. H.A. McKenzie, *Trends Anal. Chem.* **13**, 138 (1994).

46. G. Knapp and A. Grillo, *Am. Lab.* **4**, 76 (1986).

47. G. Knapp, *Microchim. Acta* **2**, 445 (1991).

48. R.T. White, P. Kettisch and P. Kainrath, *At. Spectrosc.* **19** (6), 187–92 (1998).

49. Anton Paar GmbH, *HPA-S List of Applications*, Graz (n.d.).

50. T. Tikuisis and V. Dang, in *Plastics Additives. An A–Z Reference* (G. Pritchard, ed.), Chapman & Hall, London (1998), pp. 80–94.

51. Ciba-Geigy, *Chimassorb 944 Determination in Polypropylene, High Density Polyethylene (HDPE) and Low Density Polyethylene (LDPE) by the Total Nitrogen Content, Analytical Method Code No. KC65/1*, Basel (1980).

52. B. Welz and M. Sperling, *Atomic Absorption Spectrometry*, Wiley-VCH, Weinheim (1999).

53. ASTM D 3171-90, *Standard Test Method for Fiber Content of Resin–Matrix Composites by Matrix Digestio, Annual Book of ASTM Standards*, ASTM, West Conshohocken, PA (1998), Vol. 15.03.

54. T. Korenaga, *Analyst* **106**, 40 (1981).

55. K.D. Besecker, C.B. Rhoades, B.T. Jones and K.W. Barnes, *At. Spectrosc.* **19** (6), 193–7 (1998).

56. H. Fink, U. Panne, M. Theisen, R. Niessner, T. Probst and X. Lin, *Fresenius' Z. Anal. Chem.* **368** (2–3), 235–9 (2000).

57. L. Perring, M.-I. Alonso, D. Andrey, B. Bourqui and P. Zbinden, *Fresenius' Z. Anal. Chem.* **370** (1), 76–81 (2001).

58. P.L. Buldini, S. Cavalli and A. Mevoli, *J. Chromatogr.* **A739**, 167 (1996).

59. P. Ostapczuk, G. Knezevic, L. Matter and G. Steinle, *Fresenius' Z. Anal. Chem.* **345**, 308–9 (1993).

60. R.A. Nadkarni, R.R. Ledesma and G.H. Via, *Soc. Automot. Eng. (Spec. Publ.)*, **SP-1121**, 129–34 (1995).

61. A. Abu-Samra, J.S. Morris and S.R. Koirtyohann, *Anal. Chem.* **47**, 1475 (1975).

62. J. Begerow and L. Dunemann, in *Sampling and Sample Preparation* (M. Stoeppler, ed.), Springer-Verlag, Berlin (1997), pp. 155–66.

63. H. Matusiewicz, in *Microwave-Enhanced Chemistry* (H.M. Kingston and S.J. Haswell, eds), American Chemical Society, Washington, DC (1997), pp. 353–69.

64. P.J. Walker, S. Chalk and H.M. Kingston, in *Microwave-Enhanced Chemistry* (H.M. Kingston and S.J. Haswell, eds), American Chemical Society, Washington, DC (1997), pp. 55–221.

65. F.A. Settle, P.J. Walker, H.M. Kingston, M.A. Pleva, T. Snider and W. Boute, *J. Chem. Inf. Comput. Sci.* **32**, 349–53 (1992).

66. K.E. Williams and S.J. Haswell, in *Microwave-Enhanced Chemistry* (H.M. Kingston and S.J. Haswell, eds), American Chemical Society, Washington, DC (1997), pp. 401–22.

67. G. Knapp, F. Panholzer, A. Schalk and P. Kettisch, in *Microwave-Enhanced Chemistry* (H.M. Kingston and S.J. Haswell, eds), American Chemical Society, Washington, DC (1997), pp. 423–51.

68. H. Matusiewicz, *Anal. Chem.* **66**, 751 (1994).

69. L.B. Jassie and H.M. Kingston, in *Introduction to Microwave Sample Preparation: Theory and Practice* (H.M. Kingston and L.B. Jassie, eds), American Chemical Society, Washington, DC (1988), pp. 1–6.

70. R. Sturgeon and H. Matusiewicz, *Progr. Anal. Spectrosc.* **12**, 21–39 (1989).

71. V. Karanassios, F.H. Li, B. Liu and D.S. Salin, *J. Anal. At. Spectrom.* **6**, 457 (1991).

72. L. Collins, S.J. Chalk and H.M. Kingston, *Anal. Chem.* **68**, 2610–14 (1996).

73. M. Burguera, J.L. Burguera and O.M. Alarcón, *Anal. Chim. Acta* **179**, 351 (1986).

74. R.E. Sturgeon, S.N. Willie, B.A. Methven, J.W.H. Lam and H. Matusiewicz, *J. Anal. At. Spectrom.* **10**, 981 (1995).

75. D.D. Link and H.M. Kingston, *Anal. Chem.* **72**, 2908–13 (2000).

76. H.M. Kingston and S.J. Haswell (eds), *Microwave-Enhanced Chemistry. Fundamentals, Sample Preparation and Applications*, American Chemical Society, Washington, DC (1997).

77. M. Bettinelli, in *Encyclopedia of Analytical Science* (A. Townshend, ed.), Academic Press, London (1995), pp. 4511–18.

78. L. Dunemann, *Nachr. Chem. Tech. Lab.* **39** (10), M1 (1991).

79. H.M. Kingston and L.B. Jassie, in *Introduction to Microwave Sample Preparation: Theory and Practice* (H.M. Kingston and L.B. Jassie, eds), American Chemical Society, Washington, DC (1988), pp. 93–154.

80. P. Knape, personal communication (2000).

81. K. Jankowski, A. Jerzak and L. Synoradzki, *Chem. Anal. (Warsaw)* **43** (3), 427–36 (1998).

82. K.D. Besecker, C.B. Rhoades, B.T. Jones and K.W. Barnes, *At. Spectrosc.* **19** (2), 55–9 (1998).

83. M. Zischka, P. Kettisch and P. Kainrath, *At. Spectrosc.* **19** (6), 223–7 (1998).

83a. A. Ritter, E. Michel, M. Schmid and S. Affolter, *Polym. Test* **23**, 467–74 (2004).

84. Association Française de Normalisation, *Kjeldahl Nitrogen Determination with Microwave Sample Preparation*, Rept NF V03-100 (1992).

85. U. Sengutta, *Plastverarbeiter* **53** (7), 62–3 (2002).

86. C.C. Budke and D.G. McFadden, *Plast. Compd* **13** (2), 65–71 (March/Apr. 1990).

87. S.E. Carr and C.R. Moser, *Proceedings Intl Coal Testing Conference*, Lexington, KY (1990), pp. 52–6.

88. J. Marshall and J. Franks, *Proceedings XVIth FACSS Mtg*, Chicago (1989), Paper No. 843.

89. A.S. Colson, *Analyst (London)* **113**, 791 (1963).

90. Y. Wang, L. Liu and X. Zhang, *Yejin Fenxi* **21** (2), 11–13 (2001).

91. T. Dymott, *LabPlus Int.* 22–24 (Sept./Oct. 2000).

92. J.A.C. Broekaert, in *Ullmann's Encyclopedia of Industrial Chemistry*, VCH, Berlin (1994), Vol. B5, pp. 559–652.

93. J.R. Bacon, J.S. Crain, L. van Vaeck and J.G. Williams, *J. Anal. At. Spectrom.* **14**, 1633–59 (1999).

94. *Spectrochim. Acta* **50B** (4–7), (June 1995).

95. J.A.C. Broekaert, *Analytical Atomic Spectrometry with Flames and Plasmas*, Wiley-VCH, Weinheim (2000).

96. J.R. Dean, *Atomic Absorption and Plasma Spectroscopy*, John Wiley & Sons, Ltd, Chichester (1997).

97. J.W. Robinson, *Atomic Spectroscopy*, M. Dekker, New York, NY (1996).

98. L.H.J. Lajunen, *Spectrochemical Analysis by Atomic Absorption and Emission*, The Royal Society of Chemistry, Cambridge (1992).

99. J. Sneddon (ed.), *Sample Introduction in Atomic Spectroscopy*, Elsevier Science, Amsterdam (1990).

100. S.J. Haswell (ed.), *Atomic Absorption: Theory, Design and Applications*, Elsevier Science, Amsterdam (1991).

101. G. Ramendik, J. Verlinden and R. Gijbels, in *Inorganic Mass Spectrometry* (F. Adams, R. Gijbels and R. van Grieken, eds), John Wiley & Sons, Inc., New York, NY (1988), pp. 17–84.

102. A. Walsh, *Spectrochim. Acta* **7**, 108 (1955).

103. C.T.J. Alkemade and J.M.W. Milatz, *J. Opt. Soc. Am.* **45**, 583 (1955).

104. C. Bendicho and M.T.C. de Loos-Vollebregt, *J. Anal. At. Spectrom.* **6**, 353–74 (1991).

105. B.V. L'vov, *Spectrochim. Acta* **17**, 761–70 (1961).

106. P. Richner, D. Evans, C. Wahrenberger and V. Dietrich, *Fresenius' Z. Anal. Chem.* **350**, 235–41 (1994).

107. W.J. Price, T.C. Dymott and P.J. Whiteside, *Spectrochim. Acta* **35B**, 3–10 (1980).

108. G. Torsi, P. Reschiglian, C. Locatelli. F.N. Rossi and D. Melucci, *Spectrosc. Europe* **10** (2), 16–21 (1998).

109. U. Kurfürst and K.-H. Grobecker, *Laborpraxis* **5**, 28–31 (1981).

109a. M.T.C. De Loos-Vollebregt and L. de Galan, *Progr. Anal. At. Spectrosc.*, **8**, 47–81 (1985).

110. J.M. Harnly, *Anal. Chem.* **51**, 2007 (1979).

111. G.E. Parris, W.R. Blair and F.E. Brinckman, *Anal. Chem.* **49**, 378–86 (1977).

112. K.S. Farah and J. Sneddon, *J. Appl. Spectrosc. Rev.* **30** (4), 351 (1995).

113. B.E. Erikson, *Anal. Chem.*, **72**, 543–546A (2000).
114. L. Ebdon, E.H. Evans, A. Fischer and S. Hill (eds), *An Introduction to Analytical Atomic Spectrometry*, John Wiley & Sons, Ltd, Chichester (1998).
115. J.E. Cantle (ed.), *Techniques and Instrumentation in Analytical Chemistry*. Vol. 5. *Atomic Absorption Spectrometry*, Elsevier, Amsterdam (1982).
116. D.J. Butcher and J. Sneddon, *A Practical Guide to Graphite Furnace Atomic Absorption Spectrometry*, John Wiley & Sons, Inc., New York, NY (1998).
117. K.W. Jackson, *Electrothermal Atomization for Analytical Atomic Spectrometry*, John Wiley & Sons, Ltd, Chichester (1999).
118. J. Rabadán, J. Galbán and J. Aznárez, *At. Spectrosc.* **14**, 95–8 (1993).
119. G.R. Dietz, J.D. Banzer and E.M. Miller, *J. Vinyl Technol.* **1** (3), 161–3 (1979).
120. M.A. Belarra, I. Lavilla, J.M. Anzano and J.R. Castillo, *J. Anal. At. Spectrom.* **7**, 1075–8 (1992).
121. D. Braun and E. Richter, *Kautsch. Gummi Kunstst.* **50**, 696 (1997).
122. M.A. Belarra, J.M. Anzano, F. Gallarta and J.R. Castillo, *J. Anal. At. Spectrom.* **2**, 77–9 (1987).
123. M.A. Belarra, J.M. Anzano, F. Gallarta and J.R. Castillo, *J. Anal. At. Spectrom.* **1**, 141 (1986).
124. J.M. Rabadán, J. Galbán, J.C. Vidal and J. Aznárez, *J. Anal. At. Spectrom.* **5**, 45 (1990).
125. *Ciba Technical Bulletin 3H, Colorants for Use in Food Packaging, Toys and Consumer Goods*, Ciba, Basel (Nov. 1998).
126. C.Y. Chu, *Kautsch. Gummi Kunstst.* **41** (1), 33–9 (1988).
127. P.R. Hornsby, P.A. Mitchell and P.A. Cusack, *Polym. Degrad. Stab.* **32** (3), 299–312 (1991).
128. R. Kannipayoor and J.C. van Loon, *Spectrosc. Lett.* **20**, 871 (1987).
129. J.D. Kerber, A. Koch and G.E. Peterson, *At. Absorpt. Newsl.* **12**, 104 (1973).
130. G. Knezevic and U. Kurfürst, *Fresenius' Z. Anal. Chem.* **322**, 717–8 (1985).
131. P.J. Simon, B.C. Glessen and T.R. Copeland, *Anal. Chem.* **49**, 2285 (1977).
132. P. Girgis-Takla and I. Chroneos, *Analyst* **103**, 122 (1978).
133. U. Völlkopf, R. Lehmann and D. Weber, *J. Anal. At. Spectrom.* **2**, 455–8 (1987).
134. G.R. Carnrick, B.K. Lumas and W.B. Barnett, *J. Anal. At. Spectrom.* **1**, 443–7 (1986).
135. J.D. Kerber, *At. Absorpt. Newsl.* **10**, 104 (1971).
136. A. Janssen, B. Brückner, K.-H. Grobecker and U. Kurfürst, *Fresenius' Z. Anal. Chem.* **322**, 713–16 (1985).
137. J. Pauwels, C. Hofmann and K.-H. Grobecker, *Fresenius' Z. Anal. Chem.* **345**, 475–7 (1993).
138. I.G. Kalaydjieva, *Fresenius' Z. Anal. Chem.* **371**, 394–5 (2001).
139. E.L. Henn, *Anal. Chim. Acta* **73**, 273–81 (1974).
140. I. Havezov and B. Tamnev, *Fresenius' Z. Anal. Chem.* **291**, 127 (1978).
141. W.J. Rühl, *Fresenius' Z. Anal. Chem.* **322**, 710–12 (1985).
141a. F.J. Langmyhr and G. Wibetoe, *Progr. Anal. At. Spectrosc.* **8**, 193–256 (1985).
142. A.M. Dobney, A.J.G. Mank, K.-H. Grobecker, P. Conneely and C.G. de Koster, *Anal. Chim. Acta* **423**, 9–19 (2000).
143. B.E. Saltzan, *Anal. Chem.* **25**, 493 (1953).
144. R. Erb, *Fresenius' Z. Anal. Chem.* **322**, 719–20 (1985).
145. G. Knezevic and O. Töppel, *Das Papier* **45** (6), 285–7 (1991).
146. C. Tonini, *Tinctoria* **77**, 358 (1980).
147. M.B. Knowles, T.T. Nham and S.J. Carter, *Intl Lab.* **8**, 25–32S (1998).
148. J.J. Sullivan, *Application Note 228–108*, Hewlett Packard (1989), pp. 1–7.
149. K. Laqua, *Emissionsspektroskopie*, in *Ullmanns Encyklopädie der Technischen Chemie*, Verlag Chemie, Weinheim (1980), Bd. 5.
150. W.W. Harrison and B.L. Bentz, *Progr. Anal. Spectrosc.* **11**, 53–110 (1988).
151. Z. Weiss, in *Glow Discharge Optical Emission Spectrometry* (R. Payling, D.G. Jones and A. Bengtson, eds), John Wiley & Sons, Ltd, Chichester (1997), pp. 729–34.
152. T. Seidel, A. Golloch, H. Beerwald and G. Böhm, *Fresenius' Z. Anal. Chem.* **347**, 92–102 (1993).
153. A. Golloch and T. Seidel, *Fresenius' Z. Anal. Chem.* **349**, 32–5 (1994).
154. A. Golloch and D. Siegmund, *Fresenius' Z. Anal. Chem.* **358**, 804–11 (1997).
154a. Anon., *Chem. Anl. Verfahren (4)*, 48–50 (1998).
155. A. Golloch and T. Seidel, *Gleitfunkenspektrometrie zur Kunststoffsortierung. Schnelle Identifizierung von PVC und umweltrelevanter Additieven*, IFAT, Munich (1993).
156. A. Bogaerts and R. Gijbels, *Spectrochim. Acta* **53B**, 1–42 (1998).
156a. K.A. Marshall, T.J. Casper, K.R. Brushwyler and J.C. Mitchell, *J. Anal. At. Spectrom.* **18**, 637–45 (2003).
157. A. Bengtson, in *Surface Characterization. A User's Sourcebook* (D. Brune, R. Hellborg, H.J. Whitlow and O. Hunderi, eds), Wiley-VCH, Weinheim (1997), Chapter 14.
158. B. Chapman, *Glow Discharge Processes*, John Wiley & Sons, Inc., New York, NY (1980).
159. R. Payling and P. Larkins, *Optical Emission Lines of the Elements*, John Wiley & Sons, Ltd, Chichester (2000).
160. W. Grimm, *Spectrochim. Acta* **23B**, 443 (1968).
161. A. Bengtson, D.G. Jones and R. Payling, in *Glow Discharge Optical Emission Spectrometry* (R. Payling, D.G. Jones and A. Bengtson, eds), John Wiley & Sons, Ltd, Chichester (1997), pp. 3–11.
162. A. Bogaerts, M. van Straaten and R. Gijbels, *J. Appl. Phys.* **77**, 5 (1994).
163. R. Payling, in *Glow Discharge Optical Emission Spectrometry* (R. Payling, D.G. Jones and A. Bengtson,

eds), John Wiley & Sons, Ltd, Chichester (1997), pp. 20–47.

164. W.W. Harrison, C. Yang and E. Oxley, *Anal. Chem.* **73**, 480–7A (2001).

165. N. Jakubowski, D. Steuwer and V. Vieth, *Anal. Chem.* **59**, 1825 (1987).

166. R.K. Marcus (ed.), *Glow Discharge Plasmas in Analytical Spectroscopy*, John Wiley & Sons, Inc., New York, NY (2002).

167. R. Payling, D.G. Jones and A. Bengtson (eds), *Glow Discharge Optical Emission Spectrometry*, John Wiley & Sons, Ltd, Chichester (1997).

168. J.L. Baudoin, M. Chevrier, P. Hunault and R. Passetemps, *Spec. Publ., R. Soc. Chem.* **128**, 177 (1993).

169. J.L. Baudoin, M. Chevrier, B. Herman, R. Passetemps and P. Hunault, *Spectra Anal.* **22**, 47–50 (1993).

170. R. Payling, D.G. Jones and S.A. Gower, *Surf. Interface Anal.* **20**, 959 (1993).

171. D.G. Jones and R. Payling, in *Glow Discharge Optical Emission Spectrometry* (R. Payling. D.G. Jones and A. Bengtson, eds), John Wiley & Sons, Ltd, Chichester (1997), pp. 661–7.

172. D.G. Jones, R. Payling, S.A. Gowner and E.M. Boge, *J. Anal. At. Spectrom.* **9**, 369–73 (1994).

173. F. Canpont, *PhD. Thesis*, Université Claude Bernard, Lyon (1993).

174. P. Bourrain, D. Guicherd and L. Peeters, *Wire J. Intl* **18**, 44–8 (1985).

175. V.A. Fassel, *Pure Appl. Chem.* **49**, 1533 (1977).

176. R.H. Scott, V.A. Fassel, R.N. Kniseley and D.E. Nixon, *Anal. Chem.* **46**, 75 (1974).

177. S. Maestre, J. Mora, J.-L. Todolí and A. Canals, *J. Anal. At. Spectrom.* **14**, 61–7 (1999).

178. M. Haldimann, M. Baduraux, A. Eastgate, P. Froidevaux, S. O'Donovan, D. Von Gunten and O. Zoller, *J. Anal. At. Spectrom.* **16**, 1364–9 (2001).

179. R.I. Botto, *J. Anal. At. Spectrom.* **8**, 51–7 (1993).

180. S. Greenfield, I.L. Jones and C.T. Berry, *Analyst* **89**, 713 (1964).

181. R.H. Wendt and V.A. Fassel, *Anal. Chem.* **37**, 920 (1965).

182. P.W.J. Boumans, *Inductively Coupled Plasma Emission Spectroscopy*, John Wiley & Sons, Inc., New York, NY (1987), 2 vols.

183. A. Montaser and D.W. Golightly, *Inductively Coupled Plasmas in Analytical Atomic Spectroscopy*, VCH, New York, NY (1992).

184. J.-M. Mermet, *Proceedings IUPAC Intl Congress on Analytical Sciences 2001 (ICAS 2001)*, Tokyo (2001), Paper 3D02.

185. J.-M. Mermet, *Spectrochim. Acta* **B56**, 1657–72 (2001).

186. J.-M. Mermet, *Spectrochim. Acta* **B48**, 743–55 (1993).

187. J.-M. Mermet, *Spectrochim. Acta* **B49**, 1313–24 (1994).

188. J.-M. Mermet and E. Poussel, *Appl. Spectrosc.* **49**, 12–18A (1995).

189. A. Montaser, J.A. McLean, H. Liu and J.-M. Mermet, in *Inductively Coupled Plasma Mass Spectrometry* (A. Montaser, ed.), Wiley-VCH, New York, NY (1998), pp. 1–31.

190. J.-M. Mermet, in *Inductively Coupled Plasma Mass Spectrometry and Its Applications* (S.J. Hill, ed.), Sheffield Academic Press, Sheffield (1999), pp. 35–70.

191. J.A.C. Broekaert, in *Determination of Trace Elements* (Z.B. Alfassi, ed.), VCH, Weinheim (1994), Chapter 6.

192. H.E. Taylor, *Inductively Coupled Plasma Emission and Mass Spectrometry. Practices and Techniques*, Academic Press, San Diego, CA (2001).

193. J.G. Bansal and F.C. McElroy, *Soc. Automot. Eng. (Special Publ.)*, **SP-996**, pp. 61–8 (1993).

194. DIN 38406, *Bestimmung der 24 Elemente Ag, Al, B, Ba, Ca, Cd, Co, Cr, Cu, Fe, K, Mg, Mn, Mo, Na, Ni, P, Pb, Sb, Sr, Ti, V, Zn and Zr durch Atomemissionsspektrometrie mit induktiv gekoppeltem Plasma (ICP-AES) (E22)*, Beuthe Verlag, Berlin (1987).

195. W. Fan, S. Wan and Q. Su, *Suliao Gongye* **28** (1), 27–8 (2000).

196. D.J. Carlsson, M. Krzymien, D.J. Worsfold and M. Day, *J. Vinyl Addit. Technol.* **3**, 100–106 (1997).

197. J. Claessens (Bayer, Antwerp), personal communication (2001).

198. Z.-B. Yang, *Shiyou Huagong* **30** (8), 635–7 (2001).

199. Z. Lei, J. Tao and M. Zhang, *Guangpu Shiyanshi* **15** (4), 84–6 (1998).

200. D. Oktavec and J. Lehotay, *At. Spectrosc.* **14**, 103–5 (1993).

201. T. Nakahara and T. Wasa, *Bull. Univ. Osaka Prefect., Ser.* **A41**, 13–21 (1992).

202. A. Shakhnovich and J. Barren, *Proceedings SPE ANTEC 2002*, San Francisco, CA (2002), pp. 2512–6.

203. G. DiPasquale and B. Casetta, *At. Spectrosc.* **5** (5), 209–10 (1984).

204. K.L. Wong, *Anal. Chem.* **53**, 2148–9 (1981).

205. J. Nagaya, A. Tanioka and K. Miyasaka, *J. Appl. Polym. Sci.* **48** (8), 1441–7 (1993).

206. M. Resano, M. Verstraete, F. Vanhaecke, L. Moens and J. Claessens, *J. Anal. At. Spectrom.* **16**, 793–800 (2001).

207. R.J. Decker, *Spectrochim. Acta* **35B**, 19–31 (1980).

208. J. Sneddon, *Spectroscopy* **4** (3), 26 (1989).

209. A.M. Bilgic, E. Voges, U. Engel and J.A.C. Broekaert, *J. Anal. At. Spectrom.* **15**, 579 (2000).

210. C. Motley and G.L. Long, *J. Anal. At. Spectrom.* **5**, 477 (1990).

211. L.K. Olson and J.A. Caruso, *J. Anal. At. Spectrom.* **7**, 993 (1992).

212. K. Jankowski, A. Jerzak, A. Sernicka-Poluchowicz and L. Synoradzki, *Anal. Chim. Acta* **440** (2), 215–21 (2001).

213. H. Suyani, J.T. Creed and J.A. Caruso, *J. Anal. At. Spectrom.* **4**, 777 (1989).

214. S.J. Hill and A.S. Fisher, in *Encyclopedia of Spectroscopy and Spectrometry* (J.C. Lindon, ed.), Academic Press, San Diego, CA (2000), pp. 50–5.

215. S. Greenfield, *Trends Anal. Chem.* **14**, 435–42 (1995).

216. D.J. Butcher, *Spectroscopy* **8** (2), 14 (1993).

217. L. Moenke-Blankenburg, in *Lasers in Analytical Atomic Spectroscopy* (J. Sneddon, T.L. Thiem and Y.-I. Lee, eds), VCH Publishers, New York, NY (1997), pp. 125–95.

218. T. Hadeishi and R. McLaughlin, *Fresenius' Z. Anal. Chem.* **322**, 657–9 (1985).

219. U. Kurfürst, *Analytiker Taschenbuch*, Springer-Verlag, Berlin (1991), p. 190.

220. U. Kurfürst, in *Solid Sample Analysis* (U. Kurfürst, ed.), Springer-Verlag, Berlin (1998), pp. 129–90.

221. M. Stoeppler and U. Kurfürst, in *Solid Sample Analysis* (U. Kurfürst, ed.), Springer-Verlag, Berlin (1997), pp. 247–318.

222. U. Kurfürst (ed.), *Solid Sample Analysis*, Springer-Verlag, Berlin (1998).

223. J. Pauwels, C. Hofmann and C. Vandecasteele, *Fresenius' Z. Anal. Chem.* **348**, 418–21 (1994).

224. J. Pauwels, A. Lamberty, P. De Bièvre, K.-H. Grolecker and C. Bauspiess, *Fresenius' Z. Anal. Chem.* **349**, 409–11 (1994).

225. I. Szalóki, S.B. Török, C.-U. Ro, J. Injuk and R.E. van Grieken, *Anal. Chem.* **72** (12), 211–33 (2000).

226. S.B. Török, J. Lábár, M. Schmeling and R.E. van Grieken, *Anal. Chem.* **70** (12), 495–517R (1998).

227. K. Janssens and F. Adams, in *Encyclopedia of Analytical Science* (A. Townshend, ed.), Academic Press, London (1995), pp. 5574–93.

228. R.E. van Grieken and A.A. Markowicz (eds), *Handbook of X-Ray Spectrometry: Methods and Techniques*, M. Dekker, New York, NY (1993).

229. U. Kramar, in *Encyclopedia of Spectroscopy and Spectrometry* (J.C. Lindon, ed.), Academic Press, San Diego, CA (2000), pp. 2467–77.

230. M. Riess, *Proceedings 3rd Würzburger Tage der instrumentellen Analytik in der Polymertechnik*, SKZ, Würzburg (1997).

231. B. Danzer, *Dissertation*, Erlangen–Nürnberg University (1996).

232. G.R. Lachance and F. Claisse, *Quantitative X-Ray Fluorescence Analysis. Theory and Application*, John Wiley & Sons, Ltd, Chichester (1998).

233. R. Jenkins, *X-Ray Fluorescence Spectrometry*, Wiley-Interscience, New York, NY (1999).

234. A.A. Markowicz and R.E. van Grieken, in *Handbook of X-Ray Spectrometry* (R.E. van Grieken and A.A. Markowicz, eds), M. Dekker, New York, NY (1993), p. 339.

235. U. Myint, J. Tölgyessy and K. Kristiansen, in *Surface Characterization. A User's Sourcebook* (D. Brune, R. Hellborg, H.J. Whitlow and O. Hunderi, eds), Wiley-VCH, Weinheim (1997), Chapter 7.

236. M.J. Beattie and R.M. Brissey, *Anal. Chem.* **26**, 980 (1954).

237. R. Jenkins, in *X-Ray Characterization of Materials* (E. Lifshin, ed.), Wiley-VCH, Weinheim (1999), pp. 171–209.

238. J. Sherman, *Spectrochim. Acta* **7**, 283 (1955).

239. J.W. Criss and L.S. Birks, *Anal. Chem.* **40**, 1080–6 (1968).

240. W.K. de Jongh, *X-Ray Spectrom.* **2**, 151 (1973).

241. Omega Data Systems BV, *Uniquant*® 5 (2001).

242. ThermoARL, *Introduction to UniQuant*® 5, Ecublens (2001).

243. A. Bühler, *Petro Industry News, 34–35* (Nov./Dec. 2001); *Hydrocarb. Engng (7/6)*, 55–8 (June 2002).

244. I. Szalóki, A. Somogyi, M. Braun and A. Tóth, *X-Ray Spectrom.* **28**, 399–405 (1999).

245. R.A. Barrea and R.T. Mainardi, *X-Ray Spectrom.* **27**, 111–16 (1998).

246. I. Nakai and Y. Terada, *Proceedings ICAS 2001*, Tokyo (2001), Paper 4E12.

247. A.B. Blank and L.P. Eksperiandova, *X-Ray Spectrom.* **27**, 147–60 (1998).

248. R.P. Pettersson and E. Selin-Lindgren, in *Surface Characterization. A User's Sourcebook* (D. Brune, R. Hellborg, H.J. Whitlow and O. Hunderi, eds), Wiley-VCH, Weinheim (1997), Chapter 8.

249. V.E. Buhrke, R. Jenkins and D.K. Smith (eds), *A Practical Guide for the Preparation of Specimens for X-Ray Fluorescence and X-Ray Diffraction Analysis*, Wiley-VCH, New York, NY (1998).

249a. P. Brouwer, *Theory of XRF*, PANanalytical Almelo, The Netherlands (2003).

250. R. Klockenkämper, *Total-Reflection X-Ray Fluorescence Analysis*, John Wiley & Sons, Inc., New York, NY (1996).

251. R. Tertian and F. Claise, *Principles of Quantitative X-Ray Fluorescence Analysis*, Heyden & Sons, London (1982).

252. P.L. Warren, *Anal. Proc.* **27**, 186–7 (1990).

253. C. Streli, P. Wobrauschek and P. Kregsamer, in *Encyclopedia of Spectroscopy and Spectrometry* (J.C. Lindon, ed.), Academic Press, San Diego, CA (2000), pp. 2478–87.

254. P. Kump, M. Nečemer, B. Smodiš and R. Jačimovič, *Appl. Spectrosc.* **50** (11), 1373–7 (1996).

255. B. Ribar and Z. Skrbic, *Polimeri (Zagreb)* **7**(12), 357–9 (1986).

256. E. Houben, *Tech. Rep. DSM Research*, Geleen, Netherlands (1997).

257. C. Jurriaan, M. van den Heuvel, G. Hoentjen and A.K. de Kreek, *Makromol. Chem., Rapid Commun.* **5** (4), 235–8 (1984).

258. G. Klok, J. Rousseau, H. Goertz and H. Nelissen, *Tech. Rep. DSM Research*, Geleen, Netherlands (1998).

259. L. Steffanie and H. Vanwersch, *Tech. Rep. DSM Research*, Geleen, Netherlands (1999).

260. R.D. Koons, *J. Forensic Sci.* **41** (2), 199–205 (1996).

261. S.G. Ryland, *J. Forensic Sci.* **31** (4), 1314–29 (1986).

262. Y. Nir-El, *J. Forensic Sci.* **39**, 758 (1994).

263. G. Harm, J. Wittenauer and P. Roscoe, *Proceedings Polymer Additives '95*, Chicago, IL (1995).

264. D.J. Leland, D.E. Leyden and A.R. Harding, *Adv. X-Ray Anal.* **32**, 39–44 (1989).

265. G. Dewitte, *Proceedings European Plastic Additives Technical Experts Conference (EPATEC '98)*, Vaalsbroek (1998), Paper 2.

266. Association Française de Normalisation, *European Standard EN71/3, Safety of Toys, Part III, Toxicity of Toys*, Paris (1982).

267. J.V. Gimeno Adelantado, V. Peris Martinez, F. Bosch Reig, M.T. Doménech Carbó and F. Bosch Mossi, *Anal. Chim. Acta* **276**, 39–45 (1993).

268. J.E. Ferguson, *The Heavy Metals: Chemistry, Environmental Impact and Health Effect*, Pergamon Press, Oxford (1990).

269. F.E. Mark, *Metals in Source Separated Plastics Packaging Waste*, APME, Brussels (1996).

270. L.G. Meisinger, *Chem. Plants Proc. (3)*, 46–7 (2000).

271. L. Käselitz and B. Gade, unpublished results (2002).

272. W. Michaelis, J. Knoth, A. Prange and H. Schwenke, *Adv. X-Ray Anal.* **28**, 75 (1984).

273. K. Taniguchi and T. Ninomiya, *Tetsu to Hagane* **76** (8), 1228–36 (1990).

274. T. Ninomiya, S. Nomura and K. Taniguchi, *Hyomen Kagaku* **11** (3), 189–94 (1990).

275. R.S. Hockett, *Adv. X-Ray Anal.* **39**, 767–9 (1997).

276. A. Prange, U. Reus, H. Böddeker, R. Fischer and F.-P. Adolf, *Anal. Sci.* **11**, 483 (1995).

277. A. Prange, U. Reus, H. Böddeker, R. Fischer and F.-P. Adolf, *Adv. X-Ray Chem. Anal. Jpn* **26s**, 1 (1995).

278. H. Ryssel, L. Frey, N. Streckfuss, R. Schork, F. Kroninger and T. Falter, *Appl. Surf. Sci.* **63** (1–4), 79–87 (1993).

279. K. Ohtani, K. Ihara and T. Ohmi, *Proceedings Electrochemical Society (Contamination, Control and Defect Reduction in Semiconductor Manufacturing)* **92** (2), 361–74 (1992).

280. K. Ohtani, T. Ohmi and K. Ihara, *J. Electrochem. Soc.* **140** (8), 2244–9 (1993).

281. U. Simmross, R. Fischer, F. Düwel and U. Müller, *Fresenius' Z. Anal. Chem.* **358** (4), 541–15 (1997).

282. T. Ninomiya, S. Nomura and K. Taniguchi, *X-sen Bunseki no Shinpo* **19**, 227–35 (1988).

283. J. Buscaglia, *PhD. Thesis*, City University, New York; *Diss. Abstr. Intl* **B60** (4), 1575 (1999).

284. M. Nečemer, P. Kump and B. Stropnik, *Acta Chim. Slov.* **43** (1), 61–6 (1996).

285. B. Wehling, P. Vandenabeele, L. Moens, R. Klockenkämper, A. von Bohlen, G. van Hooydonk and M. de Reu, *Mikrochim. Acta* **130**, 253–60 (1999).

286. P. Vandenabeele, B. Wehling, L. Moens, B. Dekeyzer, B. Cardon, A. von Bohlen and R. Klockenkämper, *Analyst* **124**, 169–72 (1999).

287. R. Klockenkämper, A. von Bohlen, L. Moens and W. Devos, *Spectrochim. Acta* **48B** (2), 239–46 (1993).

288. F. Duwell, R. Fischer, T. Schonberger, U. Simmross and D. Weis, *Proceedings Meeting of the International Association of Forensic Scientists*, Tokyo (1996).

289. T.B. Johansson, K.R. Axelsson and S.A.E. Johansson, *Nucl. Instr. Meth.* **84**, 141 (1970).

290. V. Potocek and N. Toulhoat, *Nucl. Instr. Meth.* **B109/110**, 197–202 (1996).

291. W.M. Kwiatek, J. Leki, E.M. Dutkiewicz and C. Paluszkiewicz, *Nucl. Instrum. Meth.* **B114**, 345–9 (1996).

292. S.A.E. Johansson and J.L. Campbell, in *Particle-Induced X-Ray Emission Spectrometry* (S.A.E. Johansson, J.L. Campbell and K.G. Malmqvist, eds), John Wiley & Sons, Ltd, Chichester (1995), pp. 419–34.

292a. C. Ryan, *Materials World*, **10** (6), 23–4 (2002).

293. H. Wittrisch, S. Conradi, E. Rohde, J. Vogt and C. Vogt, *J. Chromatogr.* **A781**, 407–16 (1997).

294. W. Maenhout, *Nucl. Instr. Meth.* **B49**, 518 (1990).

295. W. Maenhout, *Scan. Microsc.* **4**, 43 (1990).

296. N.F. Mangelson and M.W. Hill, *Scan. Microsc.* **4**, 63 (1990).

297. S.A.E. Johansson, J.L. Campbell and K.G. Malmqvist (eds), *Particle-Induced X-Ray Emission Spectrometry (PIXE)*, John Wiley & Sons, Inc., New York, NY (1995).

298. S.A.E. Johansson and J.L. Campbell, *PIXE: A Novel Technique for Elemental Analysis*, John Wiley & Sons, Ltd, Chichester (1998).

299. I. Brissaud, G. Lagarde and P. Midy, *Nucl. Instr. Meth. Phys. Res.* **B117**, 179–85 (1996).

300. A.M. Ektessabi, S. Shikine, M. Hamdi, N. Kitamura, M. Rokkum and C. Johansson, *Int. J. PIXE* **10** (1/2), 37–45 (2000).

301. S.B. Russell, C.W. Schulte, S. Faiq and J.L. Campbell, *Anal. Chem.* **53**, 571 (1981).

301a. M. Brenner, J.-O Lill, M. Ström and P. Tunander, *Stud. Conserv.* **49**, 99–106 (2004).

302. K.G. Malmqvist, in *Particle-Induced X-Ray Emission Spectrometry (PIXE)* (S.A.E. Johansson, J.L. Campbell and K.G. Malmqvist, eds), John Wiley & Sons, Inc., New York, NY (1995), pp. 367–417.

303. C. Oger, J. Allart and G. Weber, *Proceedings Art 2002*, Antwerp (2002).

304. F. Watt and G.W. Grime (eds), *Principles and Applications of High-Energy Ion Microbeams*, Hilger, Bristol (1987).

305. K. Sakurai and X. Guo, *Spectrochim. Acta* **B54**, 99–107 (1999).

306. S.G. Urquhart, A.P. Hitchcock, A.P. Smith, H.W. Ade, W. Lidy, E.G. Rightor and G.E. Mitchell, *J. Electron Spectrosc. Relat. Phenom.* **100**, 119–35 (1999).

307. P. Le Fèvre, H. Magnan and D. Chandesris, *Rev. Metall./Cah. Inf. Tech.* **96** (9), 1041–55 (1999).

308. D.C. Koningsberger and R. Prins (eds), *X-Ray Absorption, Principles, Applications. Techniques of EXAFS, SEXAFS and XANES*, John Wiley & Sons, Inc., New York, NY (1988).

309. J. Stöhr, *NEXAFS Spectroscopy*, Springer-Verlag, New York, NY (1992).

310. A. Manceau, J.C. Harge, G. Sarret, J.L. Hazemann, M.C. Boissett, M. Mench, Ph. Cambier and R. Prost, *Colloq., Inst. Nucl. Rech. Agron.* **85**, 99–120 (1997).

311. E.S. Ferrari, K.J. Roberts and D. Adams, *Wear* **236**, 246–58 (1999).

312. E.S. Ferrari, K.J. Roberts, M. Sansone and D. Adams, *Wear* **236**, 259–75 (1999).

313. J.N. Cutler, J.H. Sanders and G. John, *Tribol. Lett.* **4** (2), 149–54 (1998).

314. S.A. McHugo, A.C. Thompson, C. Flink, E.R. Weber, G. Lamble, B. Gunion, A. MacDowell, R. Celestre, H.A. Padmore and Z. Hussain, *J. Cryst. Growth* **210**, 395–400 (2000).

315. G.P. Hastie, J. Johnstone and K.R. Roberts, *J. Mater. Sci. Lett.* **17**, 1223–5 (1998).

316. D.T. Jiang and E.D. Crozier, *Can. J. Phys.* **76**, 621–43 (1998).

317. I. Retzko, J.F. Friedrich, A. Lippitz and W.E.S. Unger, *J. Electron Spectrosc. Relat. Phenom.* **121** (1–3), 111–29 (2001).

318. J.-J. Rousseau, *Basic Crystallography*, John Wiley & Sons, Ltd, Chichester (1998).

319. S. Byrams, *Today's Chemist at Work* **8** (12), 16–20 (1999).

320. D. Watkin, *Today's Chemist at Work* **9** (10), 22–4 (2000).

321. S. García-Granda and M.A. Salvadó, in *Encyclopedia of Analytical Science* (A. Townshend, ed.), Academic Press, London (1995), pp. 5560–74.

322. T. Yamanaka, *Material Analysis Using Powder X-Ray Diffraction*, Kodansha, Tokyo (1993).

323. R. Jenkins and R.L. Snyder, *Introduction to X-Ray Powder Diffractometry*, John Wiley & Sons, Ltd, Chichester (1996).

324. T.L. Threlfall, *Analyst* **120**, 2435 (1995).

325. R.W. Morton, D.E. Simon, J.F. Geibel, J.J. Gislason and R.L. Heald, *Adv. X-Ray Anal.* **44**, 103–9 (2001).

326. F. Mersch and R. Zimmer, *Kautsch. Gummi Kunstst.* **39**, 427–32 (1986).

327. F. Endter, *Kautsch. Gummi* **8**, WT 302–(1955).

328. H. Ostromow, *Analyse von Kautschuken und Elastomeren*, Springer-Verlag, Berlin (1981).

329. S. Sturm, *Adv. Instrum.* **40**, 1487–97 (1985).

330. A.B. Morgan, J.W. Gilman and C.L. Jackson, *Proceedings ACS Div. Polym. Mater.: Sci. Eng.* **82**, 270–1 (2000).

331. X. Liu, J. Fan, Q. Li, X. Zhu and Z. Qi, *Gaofenzi Xuebao* **5**, 563–7 (2000).

332. Q. Zhang, Q. Fu, L. Jiang and Y. Lei, *Polym. Intl* **49** (12), 1561–4 (2000).

333. J.W. Gilman, T. Kashiwagi, M. Nyden, J.E.T. Brown, C.L. Jackson, S. Lomakin, E.P. Giannelis and E. Manias, in *Chemistry and Technology of Polymer Additives* (S. Al-Malaika, A. Golovoy and C.A. Wilkie, eds), Blackwell Science, Oxford (1999), pp. 249–65.

334. J.-H. Zeng, J. Yang, Y. Zhu, Y.-F. Liu, Y.-T. Qian and H.-G. Zheng, *Chem. Commun.* (15), 1332–3 (2001).

335. W. Jia, E. Segal, D. Kornemandel, Y. Lambot, M. Narkis and A. Siegmann, *Synth. Met.* **128** (1), 115–20 (2002).

336. A.D. DeBellis, R. Iyengar, N.A. Kaprinidis, R.K. Rodebaugh and J. Suhadolnik, *ACS Symp. Ser.* **805**, 453–67 (2002).

337. K.P. Ghiggino, A.D. Scully and I.H. Leaver, *J. Phys. Chem.* **90**, 5089 (1986).

338. R.D. Vold, J.D. Grandine and M.J. Vold, *J. Colloid Sci.* **3**, 339–61 (1948).

339. A. Braam, *Tech. Rep. DSM Research*, Geleen, Netherlands (2001).

340. D.O. Hummel, *Atlas of Plastics Additives: Analysis by Spectrometric Methods*, Springer-Verlag, Berlin (2002).

341. C.J. Curry, D.F. Rendle and A. Rogers, *J. Forensic Sci. Soc.* **22**, 173–8 (1982).

342. A. Whitaker, in *The Analytical Chemistry of Synthetic Dyes* (K. Venkataraman, ed.), John Wiley & Sons, Inc., New York, NY (1977).

343. N. Salvadó, T. Pradell, E. Pantos, M.Z. Papis, J. Molera and M. Vendrell, *Proceedings Art 2002*, Antwerp (2002).

344. A. Wlochowicz and J. Garbarczyk, *Polym. Intl* **34** (3), 253–6 (1994).

345. J. Fröhlich, St. Kreitmeier and D. Göritz, *Kautsch. Gummi Kunstst.* **51**, 370–5 (1998).

346. F.H. Chung, *Industrial Applications of X-Ray Diffraction*, M. Dekker, New York, NY (1999).

346a. V. Majidi, C. Lewis, C. Hassell and W. Hang, *Proceedings Colloquium Spectroscopicum Internationale XXXIII*, Granada, Spain (2003), p. 82.

347. J.D. Fassett and P.J. Paulsen, *Anal. Chem.* **61**, 643A (1989).

348. Z. Fang, *Flow Injection Separation and Preconcentration*, VCH, Weinheim (1993).

349. R. Ma, C.W. McLeod and K.C. Thompson, in *Plasma Source Mass Spectrometry. Developments and Applications* (G. Holland and S.D. Tanner, eds), The Royal Society of Chemistry, Cambridge (1997), pp. 192–201.

350. J.S. Becker and H.-J. Dietze, *Spectrochim. Acta* **53B**, 1475 (1998).

351. J.R. de Laeter, *Applications of Inorganic Mass Spectrometry*, Wiley-Interscience, New York, NY (2001).

352. F. Adams, R. Gijbels and R. van Grieken, *Inorganic Mass Spectrometry*, John Wiley & Sons, Inc., New York, NY (1988).

353. G. Holland and S.D. Tanner (eds), *Plasma Source Mass Spectrometry. Developments and Applications*, The Royal Society of Chemistry, Cambridge (1997).

354. K.E. Jarvis, A.L. Gray, J.G. Williams and I. Jarvis, *Plasma Source Mass Spectrometry*, The Royal Society of Chemistry, Cambridge (1990).

355. R. Gijbels, *Talanta* **37**, 363 (1990).

356. A.J. Dempster, *Proc. Am. Philos. Soc.* **75**, 755 (1935).

357. A.M. Ure, *Trends Anal. Chem.* **1**, 314 (1982).

358. G.I. Ramendik, *Int. J. Mass Spectrom. Ion Proc.* **172** (3), 221 (1998).

359. A.M. Andreani, J.C. Brun, J.P. Mer, A. Fillot and R. Stefani, *Méthodes Phys. Anal.* **7**, 258 (1971).

360. K.P. Jochum, L. Matus and H.M. Seufert, *Fresenius' Z. Anal. Chem.* **331**, 136–9 (1988).

361. K.P. Jochum, *PTB – Ber. ThEx – Phys. – Tech. – Bundesanst.*, **PTB-ThEx-3**, 36 (1997).

362. J.R. Bacon and A.M. Ure, *Analyst* **109**, 1229–34 (1984).

363. R.E. Honig, *Thin Solid Films* **31**, 89 (1976).

364. R.L. Beveridge, *Appl. Spectrosc.* **27**, 271 (1973).

365. G.I. Ramendik, V.I. Dershiev and Yu.V. Vasyuta, *Zh. Anal. Khim.* **34**, 1016 (1979).

366. M. Aoki and K. Kurosaki, *Kobunshi Kagaku* **29**, 442 (1972).

367. F.W. Aston, *Mass Spectra and Isotopes*, Longmans, New York, NY (1942).

368. J.E. Cantle, E.F. Hall, C.J. Shaw and P.J. Turner, *Intl J. Mass Spectrom. Ion Phys.* **46**, 11 (1983).

369. A. Raith, in *Glow Discharge Optical Emission Spectrometry* (R. Payling, D.G. Jones and A. Bengtson, eds), John Wiley & Sons, Ltd, Chichester (1997), pp. 735–55.

370. U. Greb, L. Rottmann and M. Hamester, *Finnigan MAT Appl. Note No. 8* (1995).

371. W.W. Harrison, C.M. Barshick, J.A. Klingler, P.H. Ratliff and Y. Mei, *Anal. Chem.* **62**, 943–9A (1990).

372. J.E. Cantle, E.F. Hall, C.J. Shaw and P.J. Turner, *Intl J. Mass Phys.* **46**, 11–13 (1983).

373. W.W. Harrison, in *Inorganic Mass Spectrometry* (F. Adams, R. Gijbels and R. van Grieken, eds), John Wiley & Sons, Inc., New York, NY (1988), pp. 85–123.

374. Y. Su, P. Yang, Z. Zhou, X. Wang, F. Li, B. Huang, J. Ren, M. Chen, H. Ma and G. Zhang, *Spectrochim. Acta* **53B**, 1413–20 (1998).

375. A. Bogaerts, in *Encyclopedia of Spectroscopy and Spectrometry* (J.C. Lindon, ed.), Academic Press, San Diego, CA (2000), pp. 669–76.

376. A.L. Gray, *J. Anal. At. Spectrom.* **1**, 403 (1986).

377. A.R. Date and A.L. Gray, *Analyst* **106**, 1255 (1981).

378. R.S. Houk, V.A. Fassel, G.D. Flesch, H.J. Svec, A.L. Grag and C.E. Taylor, *Anal. Chem.* **52**, 2283 (1980).

379. D. Douglas, *Can. Res.* **16**, 55–60 (1983).

380. H. Klinkenberg, *On the Use of Inductively Coupled Plasma Mass Spectrometry in Industrial Research PhD. Thesis*, University of Amsterdam (1995).

381. R.S. Houk, *Anal. Chem.* **58**, 97A (1986).

382. H.P. Longerich, *At. Spectrosc.* **10**, 112 (1989).

382a. S.J. Ray and G.M. Hieftje, *J. Anal. At. Spectrom.* **16**, 1206–16 (2001).

383. D.W. Koppenaal, C.J. Barinaga and M.R. Smith, *J. Anal. At. Spectrom.* **9**, 1053 (1994).

384. K.E. Milgram, F.M. White, K.L. Goodmer, C.H. Watson, D.W. Koppenaal, C.J. Barinaga, B.H. Smith, J.D. Winefordner, A.G. Marshall, R.S. Houk and J.R. Eyler, *Anal. Chem.* **69**, 3714 (1997).

385. R. Łobiński, *Spectrochim. Acta* **53B**, 177 (1998).

386. N. Jakubowski and D. Stuewer, in *Plasma Source Mass Spectrometry* (G. Holland and S.D. Tanner, eds.), The Royal Society of Chemistry, Cambridge (1997), pp. 298–312.

387. E.H. Evans and J.J. Giglio, *J. Anal. At. Spectrom.* **8**, 1 (1993).

388. J.A. Olivares and R.S. Houk, *Anal. Chem.* **58**, 20 (1986).

389. J.J. Thompson and R.S. Houk, *Appl. Spectrosc.* **41**, 801 (1987).

390. E.H. Evans and J.A. Caruso, *Spectrochim. Acta* **47B**, 1001 (1992).

391. A.A. van Heuzen and N.M.M. Nibbering, *Spectrochim. Acta* **46B**, 1013 (1993).

392. N. Bradshaw, E.F.H. Hall and N.E. Sanderson, *J. Anal. At. Spectrom.* **4**, 801–3 (1989).

393. S.I. Yamasaki, A. Tsumura and Y. Takaku, *Mikrochem. J.* **49**, 305 (1994).

394. S.J. Jiang, R.S. Houk and M.A. Stevens, *Anal. Chem.* **60**, 1217 (1988).

395. S.D. Tanner, *J. Anal. At. Spectrom.* **10**, 905 (1995).

396. B. Masters and B.L. Sharp, *Proceedings European Winter Conference on Plasma Spectrochemistry*, Cambridge (1995), p. 99.

397. P.P. Mahoney, G. Li and G.M. Hieftje, *J. Anal. At. Spectrom.* **11**, 401 (1996).

398. M. Yamanaka and O.F.X. Donard, *Agilent Technologies Application Note* (Apr. 2000).

399. A.M. Leach, M. Heisterkamp, F.C. Adams and G.M. Hieftje, *J. Anal. At. Spectrom.* **15**, 151–5 (2000).

400. G. Horlick, *Spectroscopy* **7**, 22 (1992).

401. R.S. Houk, *Acc. Chem. Res.* **27**, 333 (1994).

402. I. Jarvis and K.E. Jarvis, *Chem. Geol.* **95**, 1 (1992).

403. I.B. Brenner and H.E. Taylor, *Crit. Rev. Anal. Chem.* **23**, 355 (1992).

404. R.M. Barnes, *Anal. Chim. Acta* **283**, 115 (1993).

405. D.A. Solyom, O.A. Grøn, J.H. Barnes and G.M. Hieftje, *Spectrochim. Acta* **B56**, 1717–29 (2002).

406. A. Montaser (ed.), *Inductively Coupled Plasma Mass Spectrometry: From A to Z*, VCH Publishers, New York, NY (1997).

407. S.J. Hill (ed.), *Inductively Coupled Plasma Spectrometry and Its Applications*, Sheffield Academic Press, Sheffield (1999).

408. E.H. Evans, J.J. Giglio, T.M. Castillano and J.A. Caruso, *Inductively Coupled and Microwave Induced Plasma Sources for Mass Spectrometry*, The Royal Society of Chemistry, Cambridge (1995).

409. S.M. Jickells, *Proceedings 4th International Symposium on Hyphenated Techniques in Chromatography and Hyphenated Chromotographic Analyzers (HTC-4)*, Bruges (1996), Paper E05.

410. P.J. Fordham, J.W. Gramshaw, L. Castle, H.M. Crews, D. Thompson, S.J. Parry and E. McCurdy, *J. Anal. At. Spectrom.* **10**, 303–9 (1995).

411. K.E. Nissen, J.T. Keegan and J.P. Byrne, *Can. J. Anal. Sci. Spectrosc.* **43** (4), 122–8 (1998).

412. A.M. Dobney, W. Wiarda, P. de Joode and G.J.Q. van de Peyl, unpublished results (2001).

413. M. Resano, M. Verstraete, F. Vanhaecke, L. Moens and A. van Alphen, *J. Anal. At. Spectrom.* **15**, 389–95 (2000).

414. H. Naka and D.C. Grégoire, *J. Anal. At. Spectrom.* **11**, 359 (1996).

415. L.L. Yu, W.R. Kelly, J.D. Fassett and R.D. Vocke, *J. Anal. At. Spectrom.* **16**, 140 (2000).

416. T. Prohaska, C. Latkoczy and G. Stingeder, *J. Anal. At. Spectrom.* **14**, 1501 (1999).

417. H.M. Kingston, D. Huo, Y. Lu and S. Chalk, *Spectrochim. Acta* **B53** (2), 299 (1998).

418. A.R. Date and A.L. Gray, *Applications of Inductively Coupled Plasma Mass Spectrometry*, Blackie, Glasgow (1989), p. 169.

419. R.P. Feynman, *What Do You Care What Other People Think?*, HarperCollins, London (1993), p. 245.

420. K.G. Heumann, in *Inorganic Mass Spectrometry* (F. Adams, R. Gijbels and R. van Grieken, eds.), John Wiley & Sons, Inc., New York, NY (1988), pp. 301–76.

421. J.W. McLaren. D. Beauchemin and S.S. Berman, *Anal. Chem.* **59**, 610 (1987).

422. I. Cornides, in *Reviews on Analytical Chemistry* (L. Niinisto, ed.), Akademiai Kiado, Budapest (1982), p. 105.

423. K.G. Heumann, S.M. Gallus, G. Radlinger and J. Vogl, *J. Anal. At. Spectrom.* **13** (9), 1001 (1998).

424. F. Vanhaecke, J. Diemer, K.G. Heumann, L. Moens and R. Dams, *Fresenius' Z. Anal. Chem.* **362**, 553–7 (1998).

425. J. Vogl, D. Liesegang, M. Ostermann, J. Diemer, M. Berglund, C.R. Quétal, F.D.P. Taylor and K.G. Heumann, *Accred. Qual. Assur.* **5**, 314–24 (2000).

426. A. Dobney, H. Klinkenberg, F. Souren and W. van Borm, *Anal. Chim. Acta* **420** (1), 89–94 (2000).

427. A. Götz, A. Lamberty and P. De Bièvre, *Intl J. Mass Spectrom. Ion Proc.* **123**, 1–6 (1993).

428. P. De Bièvre and G.H. Debus, *J. Nucl. Instrum. Methods* **32**, 224 (1965).

429. H. Klinkenberg, W. van Borm and F. Souren, *Spectrochim. Acta* **B51**, 139 (1996).

430. M.J. Campbell, C. Vandecasteele and R. Dams, in *Applications of Plasma Source Mass Spectrometry* (G. Holland and A.N. Eaton, eds.), The Royal Society of Chemistry, Cambridge (1991), p. 130.

431. E.S. Beary, K.A. Brletic, P.J. Paulsen and J.R. Moody, *Analyst (Cambridge)* **112**, 441–4 (1987).

432. M.J. Campbell, *PhD. Thesis*, University of Ghent (1991).

433. L. Castle, A.J. Mercer, J.R. Startin and J. Gilbert, *Food Addit. Contam.* **5**, 9 (1988).

434. L. Castle, J. Gilbert, S.M. Jickells and J.W. Gramshaw, *J. Chromatogr.* **437**, 281–6 (1988).

435. S.J. Parry, *Activation Spectrometry in Chemical Analysis*, John Wiley & Sons, Inc., New York, NY (1991).

436. H.-P. Weise, W. Görner and M. Hedrich, *Fresenius' Z. Anal. Chem.* **369**, 8–14 (2001).

437. J. Hoste and K. Strijckmans, *J. Trace Microprobe Tech.* **5**, 53 (1987).

438. K. Strijckmans and C. Vandecasteele, *Anal. Chim. Acta* **195**, 141 (1987).

439. W.D. Ehmann and D.E. Vance, *CRC Critical Rev. Anal. Chem.* **20**, 405–43 (1989).

440. R. Dybczynski, *Chem. Anal. (Warsaw)* **30**, 749 (1985).

441. L. Kosta, *Fresenius' Z. Anal. Chem.* **324**, 649 (1986).

442. B.L. Doyle, *Microbeam Anal.* **21**, 15 (1986).

443. W. Freytag and K. Figge, *Kunststoffe* **74** (8), 441–4 (1984).

444. J.L. Acosta and R. Sastre, *Rev. Plast. Mod.* **29** (224), 212–5 (1975).

445. M.P. de Jong, M.J.A. de Voigt, L.J. van Yzendoorn and H.H. Brongersma, *Bull. Mater. Sci.* **22** (3), 687–90 (1999).

446. H. Imura and N. Suzuki, *Anal. Chem.* **55** (7), 1107 (1983).

447. A. Simonits, F. De Corte and J. Hoste, *J. Radioanal. Chem.* **24**, 31 (1975).

448. F. De Corte, A. Simonits, A. De Wispelaene and J. Hoste, *J. Radioanal. Nucl. Chem.* **113**, 145–61 (1987).

449. A. Simonits, L. Moens, F. De Corte, A. De Wispelaene, A. Elek and J. Hoste, *J. Radioanal. Chem.* **60**, 461 (1980).

450. G. Kennedy, *ACS Symp. Ser.* **440**, 128 (1989).

451. K.-Y. Lee, S.-K. Shim, Y.-Y. Yoon, Y.-S. Chung and G.-H. Lee, *J. Radioanal. Nucl. Chem.* **241** (1), 129–34 (1999).

452. Z.B. Alfassi, in *Determination of Trace Elements* (Z.B. Alfassi, ed.), VCH Publishers, New York, NY (1994), pp. 253–307.

453. R. Dams, in *Ullmann's Encyclopedia of Industrial Chemistry*, VCH, Berlin (1994), Vol. B5, pp. 689–704.

454. Z.B. Alfassi and C. Chung, *Prompt Gamma Neutron Activation Analysis*, CRC Press, Boca Raton, FL (1995).

455. W.D. Ehmann, D.E. Vance, J.D. Winefordner and I.M. Kolthoff (eds.), *Radiochemistry and Nuclear Methods of Analysis*, Wiley-Interscience, New York, NY (1991).

456. J.D. Winefordner and I.M. Kolthoff (eds.), *Activation Spectrometry in Chemical Analysis*, Wiley-Interscience, New York, NY (1991).

457. Z.B. Alfassi (ed.), *Activation Analysis*, CRC Press, Boca Raton, FL (1989).

458. J. Tölgyessy and M. Kyrš, *Radioanalytical Chemistry*, Ellis Horwood, Chichester (1989).

459. J. Kantele and J. Aysto, *Handbook of Nuclear Spectrometry*, Academic Press, London (1995).

460. J. Kučera and L. Soukal, *J. Radioanal. Chem.* **80** (1–2), 121–7 (1983).

461. S. Haukka, *Analyst* **116**, 1055 (1991).

462. W.H. El-Abbady, *Al-Azhar Bull. Sci.* **10** (1), 181–6 (1999).

463. D.R. Battiste, J.P. Butler, J.B. Cross and M.P. McDaniel, *Anal. Chem.* **53**, 2232 (1981).

464. T.R. Crompton, *Eur. Polym. J.* **4**, 473–96 (1968).

465. J. Marshall, J. Franks, I. Abell and C. Tye, *J. Anal. At. Spectrom.* **6**, 145–50 (1991).

465a. B. Holynska, C.G. De Koster, J. Ostachowicz, L. Samek and D. Wegrzynek, *X-Ray Spectrom.* **29**, 291 (2000).

466. M. Hemmerlin, J.-M. Mermet, M. Bertucci and P. Zydowicz, *Spectrochim. Acta* **B52**, 421–30 (1997).

466a. D.A.W. Bossus and C.J.L.X. Hünen, *Tech. Rep. DSM Research*, Geleen, Netherlands (1988).

467. J. St.-Pierre and G. Kennedy, *J. Radioanal. Nucl. Chem.* **234** (1–2), 51–4 (1998).

468. B. Constantinescu, E. Ivanov, G. Pascovici and D. Plostinaru, *J. Radioanal. Nucl. Chem. Lett.* **155**, 25 (1991).

469. T.D. Lickly, T. Quinn, F.A. Blanchard and P.G. Murphy, *Appl. Radiat. Isot.* **39** (6), 465–70 (1988).

470. R.R. Greenberg, *J. Radioanal. Nucl. Chem.* **113**, 233 (1987).

471. B.H. Vassos and G.W. Ewing, *Electroanalytical Chemistry*, John Wiley & Sons, Inc., New York, NY (1983).

472. W.W. Buchberger and P.R. Haddad, *J. Chromatogr.* **A789**, 67–83 (1997).

473. J.G. Dick, in *Instrumental Methods in Food Analysis* (J.R.J. Paré and J.M.R. Bélanger, eds.), Elsevier, Amsterdam (1997), pp. 267–365.

474. B. Philipp and H. Anders, in *Applied Polymer Analysis and Characterization* (J. Mitchell, ed.), Hanser Publishers, Munich (1987), pp. 127–42.

475. R.D. Rocklin, *J. Chromatogr.* **546**, 175 (1991).

476. F. Rouessac and A. Rouessac, *Chemical Analysis. Modern Instrumental Methods and Techniques*, John Wiley & Sons, Ltd, Chichester (2000).

477. A. Evans, *Potentiometry and Ion-Selective Electrodes*, John Wiley & Sons, Inc., New York, NY (1987).

478. V.M. Maxwell, *Ion-Selective Electrodes*, Oxford University Press, Oxford (1988).

479. E.P. Serjeant, *Potentiometry and Potentiometric Titrations*, John Wiley & Sons, Inc., New York, NY (1984).

480. J. Kortya and K. Stulik, *Ion-Selective Electrodes*, Cambridge University Press, New York, NY (1984).

481. M. Day and D.M. Wiles, *J. Fire Sci.* **1** (4), 255–70 (1983).

482. J.S. Motonaka, S. Ikeda, N. Tanaka and H. Nishioka, *Chem. Eng. News* 56 (Jan. 1985).

483. K. Jankowski, M. Mojski and L. Synoradzki, *Chem. Anal. (Warsaw)* **42**, 567–7 (1997).

484. J. Kalous, D. Brázdová and K. Vytřas, *Anal. Chim. Acta* **283**, 645 (1993).

485. L.N. Bykova, N.A. Kazaryan, N.S. Chernova, I.P. Shakhova and N.M. Kvashi, in *Proceedings ISE Conference, Budapest, 5–9 Sept. 1977* (E. Pungor, ed.), Akademiai Kiado, Budapest (1978).

486. A.H. Sharma, F. Mozayeni and J. Alberts, *Proceedings Polyolefins XI*, SPE, Houston, TX (1999), pp. 679–703.

487. J. Udris, *Analyst* **96**, 130 (1971).

488. A. Groagova and M. Pribyl, *Fresenius' Z. Anal. Chem.* **234**, 423 (1968).

489. R. de Levie, *Principles of Quantitative Chemical Analysis*, McGraw-Hill, New York, NY (1997).

490. A.R. Guadalupe and H. Abruna, *Anal. Chem.* **57**, 142 (1985).

491. A.R. Hillman, in *Reactions and Applications of Polymer Modified Electrodes, Electrochemical Science and Technology of Polymers* (R.G. Linford, ed.), Elsevier, London (1987), p. 241.

492. E. Barendrecht, *Anal. Chim. Acta* **24**, 498 (1961).

493. M. Luque, A. Ríos and M. Valcárcel, *Anal. Chim. Acta* **395** (1–2), 217–23 (1999).

494. A. Guadarrama, C. De la Fuente, J.A. Acuna, M.D. Vazquez, M.L. Tascon and P. Sanchez Batanero, *Quim. Anal. (Barcelona)* **18** (2), 209–16 (1999).

495. V.V. Budyina, V.G. Marinin, Yu.V. Vodzinskii, A.I. Kalinina and I.A. Korschunov, *Zavod. Lab.* **36**, 1051 (1970).

496. A. Doménech-Carbó, M.T. Doménech-Carbó, M. Moya-Moreno, J.V. Gimeno-Adelantado and F. Bosch-Reig, *Anal. Chim. Acta* **407**, 275–89 (2000).

497. J. Bruce, *Intl Lab.* **32** (6A), 18–19 (2002).

498. J. García-Antón and J.L. Guiñón, *Analyst* **110**, 1365 (1985).

499. T.R. Crompton, in *Polymer Devolatilization* (R.J. Albalak, ed.), M. Dekker, New York, NY (1996), pp. 575–643.

500. V. Novak, *J. Chem. Prumsyl.* **22**, 298 (1972).

501. R.M. Podzeeva, U.I. Lukhovitskii and V.L. Forpov, *Zavod. Lab.* **37**, 168 (1971).

502. T.R. Crompton and D. Buckley, *Analyst (London)* **90**, 76–82 (1965).

503. A.I. Katelin, V.N. Komleva and L.N. Mal'kova, *Proizvod. Khim. Prom.-St.* **11**, 62 (1977).

504. T.R. Crompton and L.W. Meyers, *Plasti Polym.* 205 (June 1968).

505. G. Schwoetzer, *Fresenius' Z. Anal. Chem.* **260**, 10 (1972).

506. G.C. Claver and M.E. Murphy, *Anal. Chem.* **31**, 1682 (1959).

507. F.R. Mayo, F.M. Lewis and C. Walling, *J. Am. Chem. Soc.* **70**, 1529 (1948).

508. P. Shapras and G.C. Clover, *Anal. Chem.* **36**, 2282 (1964).

509. V. Novak and J. Seidl, *J. Chem. Prumsyl.* **28**, 186 (1978).

510. S.R. Betso and J.D. McLean, *Anal. Chem.* **48**, 766 (1976).

511. V.F. Gaylor, P.J. Elving and A.L. Conrad, *Anal. Chem.* **25**, 1078 (1953).

512. V.F. Gaylor, A.L. Conrad and J.H. Landerl, *Anal. Chem.* **29**, 224, 228 (1957).

513. J.F. Hedenberg and H. Freiser, *Anal. Chem.* **25**, 1355 (1953).

514. J.W. Hamilton and A.L. Tappel, *J. Am. Oil Chem. Soc.* **40**, 52 (1963).

515. C. Budke, D.K. Bannerjee and G.D. Miller, *Anal. Chem.* **36**, 523 (1964).

516. G. Proske, *Kautschuk* **16**, 13–17 (1940).

517. F. Mocker, *Kautsch. Gummi* **11** (10), WT281–92 (1958); **12** (16), WT155–9 (1959).

518. F. Mocker, *Kautsch. Gummi* **13** (4), WT91–108 (1960); **13** (7), WT187–94 (1960).

519. F. Mocker and J. Old, *Kautsch. Gummi* **12** (7), WT190–1 (1959); **15** (5), WT143–6 (1962).

520. R.N. Adams, *Rev. Polarogr.* **11**, 71 (1963).

521. A. Zweig, E. Lancaster, M.T. Neglia and W.H. Jura, *J. Am. Chem. Soc.* **86**, 413 (1964).

522. F.J. Vermillion and T.A. Pearl, *J. Electrochem. Soc.* **111**, 1392 (1964).

523. E.J. Kuta and F.W. Quackenbusch, *Anal. Chem.* **32**, 1069 (1960).

524. E. Schroeder and S. Malz, *Dtsch. Farben Z.* **5**, 417 (1958).

525. L.M. Mal'kova, A.I. Kalanin and E.M. Derepletchikova, *Zh. Anal. Khim.* **27**, 1924 (1972).

526. P. Kondziela and K. Czaja, *Zesz. Nauk.–Wyzska Szk. Pedagog. im. Powstancow Slask Opulu (Ser.) Chem.* **6**, 115–20 (1984); *C.A.* **101**, 182074 (1984).

527. H. Fassy and P. Lalet, *Chim. Anal. (Paris)* **52**, 1281–4 (1970).

528. K. Peltonen, *Analyst* **111**, 819 (1986).

529. F. Mocker and J. Old, *Kautsch. Gummi* **14** (10), WT301 (1961).

530. G. Proske, *Kautsch. Gummi* **1** (12), 339–43 (1948).

531. F. Mocker, J. Old and G. Walther, *Kautsch. Gummi* **15** (7), WT197 (1962).

532. J. Haslam, D.C.M. Squirrell, S. Grossman and S.F. Loveday, *Analyst (London)* **78**, 92–106 (1953).

533. G. Robertson, *Petro Industry News*, pp. 14–15 (June/July 2001).

534. Cosa Instrument Corp., *Using the CA-06/VA-06* (Oct. 1994), p. 4.

535. W.R. Raisanen, *Proceedings SPE ANTEC '97*, Toronto (1997), pp. 3603–5.

536. D.C.M. Squirrell, *Analyst* **106**, 1042 (1981).

537. I.J. Oita, *Anal. Chem.* **55** (14), 2434–6 (1983).

538. D.R. McGabe and I.N. Acworth, *Am. Lab.* **30** (13), 18B–D (1998).

539. F. Adams, *Accredit. Qual. Assur.* **3**, 308 (1998).

540. Ph. Quevauviller, *J. Chromatogr.* **A750**, 25 (1996).

541. H.M. Kingston, D. Huo, H. Boylan and Y. Han, *Proceedings IUPAC International Congress on Analytical Sciences 2001 (ICAS 2001)*, Tokyo (2001), Paper 2D01.

542. Ph. Quevauviller, *Methods Performance Studies for Speciation Analysis*, The Royal Society of Chemistry, Cambridge (1998).

543. K. DeNicola and J.A. Caruso, *LabPlus Intl* **17** (1), 14–17 (2003).

544. B. Rosenkranz, Ph. Quevauviller and J. Bettmer, *Intl Lab.* **30** (3), 20–6 (2000).

545. F. Adams, *Proceedings 6th International Symposium on Hyphenated Techniques in Chromatography and Hyphenated Chromatographic Analyzers (HTC-6)*, Bruges (2000).

546. J.W. Olesik, J.A. Kinzer, E.J. Grunwald, K.K. Thaxton and S.V. Olesik, *Spectrochim. Acta* **B53** (2), 239 (1998).

547. R.S. Houk, *Spectrochim. Acta* **B53**, 267–71 (1998).

548. H.A. Das, in *Determination of Trace Elements* (Z.B. Alfassi, ed.), VCH, Weinheim (1994), pp. 461–542.

549. L. Van Vaeck, A. Adriaens and F. Adams, *Spectrochim. Acta* **53B** (2), 367–78 (1998).

550. G.M. Hieftje, *Spectrochim. Acta* **53B**, 165 (1998).

551. L. Ebdon, L. Pitts, R. Cornelis, H. Crews, O.F.X. Donard and Ph. Quevauviller (eds.), *Trace Element Speciation for Environment, Food, Health*, The Royal Society of Chemistry, Cambridge (2001).

552. I.S. Krull, *Trace Metal Analysis and Speciation*, Elsevier, Amsterdam (1991).

553. B. Fairman and R. Wahlen, *Spectrosc. Europe* **13** (5), 16–22 (2001).

554. K. Jankowski, A. Jerzak and L. Synoradzki, *Chem. Anal. (Warsaw)* **43** (3), 427–36 (1998).

555. A. Morabito and Ph. Quevauviller, *Spectrosc. Europe* **14** (4), 18–23 (2002).

556. A.M. Leach, M. Heisterkamp, F.C. Adams and G.M. Hieftje, *J. Anal. At. Spectrom.* **15**, 151 (2000).

557. X. Dauchy, R. Cottier, A. Batel, R. Jeannot, M. Borsier, A. Astruc and M. Astruc, *J. Chromatogr. Sci.* **31**, 416–21 (1993).

558. H. Suyani, J. Creed, T. Davidson and J. Caruso, *J. Chromatogr. Sci.* **27**, 139–43 (1989).

559. S. White, T. Catterick, B. Fairman and K. Webb, *J. Chromatogr.* **A794**, 211–18 (1998).

560. K.S. Epler, T.C. O'Haver and G.C. Turk, *J. Anal. At. Spectrom.* **9**, 79–82 (1994).

561. N.P. Vela and J.A. Caruso, *J. Chromatogr.* **641**, 337–45 (1993).

562. D. Kolasa, K. Samsonowska and K. Zorawska, *Chemik* **51** (4), 100–1 (1998).

563. W. Goessler, D. Kuehnelt, C. Schlagenhaufen, K. Kalcher, M. Alegaz and K.J. Irgolic, *J. Chromatogr.* **A789**, 233–45 (1997).

564. S.M. Bird, H. Ge, P.C. Uden, J.F. Tyson, E. Block and E. Denoyer, *J. Chromatogr.* **A789**, 349–59 (1997).

565. F. Adams, *Proceedings Solutions for Scientists Symposium*, Advanstar Communications, London (1999).

566. H. De Beer, *Metal Speciation by High Performance Liquid Chromatography*, PhD. Thesis, Rand Afrikaans University, South Africa (1993).

567. K. Sutton, R.M.C. Sutton and J.A. Caruso, *J. Chromatogr.* **A789**, 85–126 (1997).

CHAPTER 9

Academic science always works

Direct Methods of Deformulation of Polymer/Additive Dissolutions

In the preceding chapters, polymer/additive deformulation moved mainly along the lines of extraction and dissolution–precipitation, i.e. *after* separation of the high-MW fraction of the sample from the additives, followed by single or hyphenated (multidimensional) chromatographic–spectroscopic and – mass-spectrometric techniques. This rather elaborate analytical approach was considered to be an *indirect* method. Cracking the code of a dissolved polymer/additive sample, *without* separation of the polymeric and additive parts prior to the onset of molecular characterisation procedures has hardly been touched on. In this chapter we explore the feasibility of this more direct *in situ* approach (quick and dirty). Indeed, although separation of polymer and additives has the advantage of removing possible mutual interferences in the examination of the components, this separation is not always necessary. An example is the case of polyacrylamide as an additive incorporated into polyvinyl alcohol [1]. Direct multicomponent polymer/additive analysis may be carried out in dissolution or in the solid state (Table 9.1). None of these methods is straightforward.

Solvents play an important role in polymer/additive analysis, namely for extraction of additives and dissolution of polymeric material, as a chromatographic liquid and as a window in spectroscopy. A solvent should generally have the following properties:

(i) be clear and colourless;
(ii) be volatile without leaving a residue;
(iii) have good long-term resistance to chemicals;
(iv) have neutral reaction;
(v) have a slight or pleasant smell;
(vi) be anhydrous;
(vii) have constant physical properties according to the manufacturers' specification;
(viii) low toxicity;
(ix) be biologically degradable; and
(x) be as inexpensive as possible [2].

Depending on the application, one or more of these properties might prevail. For example, a high-volatility

Table 9.1 Main hurdles in direct methods of multicomponent polymer/additive analysis

Procedure	Hurdle
Dissolution/chromatography	Solubility; chromatographic resolution
Dissolution/spectroscopy	Solubility; selectivity, sensitivity
Dissolution/mass spectrometry	Solubility; quantitation
Spectroscopy	Chemometrics
Mass spectrometry	Mass spectroscopist
Thermal analysis	Identification
Pyrolysis	Quantitation
Laser desorption	Quantitation

Additives In Polymers: Industrial Analysis And Applications J. C. J. Bart
© 2005 John Wiley & Sons, Ltd ISBN: 0-470-85062-0

Table 9.2 Main features of dissolution methods for polymer/additive deformulation

Advantages

- No grinding of polymer required
- Quantitative additive recovery (if soluble)
- Relatively fast
- Wide applicability

Disadvantages

- Solubility dependency
- Not universal
- Restricted selection of solvents (suitable for host polymer and additives *and* deformulation technique)
- Relatively low sensitivity
- Labour and solvent intensive
- Additive volatility (transformation)
- Co-presence of oligomeric fraction
- Difficult automation
- Unsuited for QC

solvent can evaporate rapidly so as to leave the dissolved substance (e.g. in a paint film, for deposition, etc.). Dissolution of polymers was discussed in Section 3.1.3, and can be accelerated by microwave heating [3]. The solubility of a polymer depends not only on its chemical nature but also on molecular weight, branching, degree of cross-linking, tacticity, etc. A more extended summary of the solubility of plastics is given in ref. [4].

Table 9.2 reports the *main characteristics* of dissolution methods for the purpose of additive analysis. Dissolution of polymeric materials can overcome the extraction resistance of some additive classes (e.g. high-MW additives); the host polymer matrix could also have a relatively high degree of crystallinity, which limits the extraction solvent in effectively permeating the polymer and solvating the analytes. The dissolution mode is fast in comparison to several extraction procedures. Solvent choice for additives is easier than for polymer–additive combinations, and is even more complicated in relation to the follow-up analytical characterisation technique; it is frequently difficult to find a single solvent. Dissolution may be a problem, in particular for engineering plastics (e.g. LCPs, blends) or cross-linked polymers. Consequently, the dissolution procedure is not a universal method for polymer/additive analysis: inorganics such as fillers and pigments usually do not dissolve. The main advantage of the dissolution procedure over the previously described dissolution/precipitation mode (Section 3.7) is the possibility of quantitation (of the solubilised additives). In the latter case, this may be compromised due to coprecipitation of the additives. Evaluation of the influence of the polymer on the retention of additives

requires knowledge of the chemistry to comprehend the diversity of chemical and physical interactions. It is important to understand how these may affect solubility. This discussion is beyond the scope of this book. A disadvantage of the dissolution method is the co-presence of the oligomeric fraction. Dissolution methods also impair sensitivity, because of the dilution effect on account of the polymer solubility. Moreover, if the resultant solution is not transparent, spectrophotometric techniques may be difficult to apply. As a consequence, the range of analytical tools that can be used for analysis of additives and polymer in solution is more limited: chromatography (mainly SEC, see Section 9.1.1), spectroscopy (mainly NMR, see Section 9.2.1), spectrometry (mainly MALDI-ToFMS, see Section 9.3.1) for organics, and ICP-AES and polarography for inorganics (see also Sections 8.3.2.4 and 8.7.2.1). Also, in-source pyrolysis methods have been used for the examination of polymer dissolutions.

Claver and Murphy [5] and others [6] have described polarographic determinations of residual styrene and acrylonitrile contents in styrene–acrylonitrile copolymers in DMF solution of the polymer, i.e. without using combustive or separative techniques. The method was employed for plant support. Emulsion formation polarography has been used for the determination of cobalt in varnishes [7], and of lead in paint as driers and varnishes [8]. Previous polarographic methods for the determination of metals in varnishes or lacquers generally required the prior destruction of the organic matter by ashing or acid digestion, which is time-consuming and tedious. Dissolution of PVC/(mixed-metal/phosphite stabiliser) systems in an appropriate organic solvent (2-butoxyethanol), followed by ICP-AES measurement for the determination of Ba, Cd, Zn and P, has been reported by ref. [9].

Selective dissolution of the polymer may be used industrially to separate polymer from additives for recycling purposes. However, separation of PPE from its additives (CB, talc, mica) in integrated circuit board scrap by means of trichloroethylene would not seem to meet industrial requirements (toxicity, cost) [10].

9.1 CHROMATOGRAPHIC METHODS

Principles and Characteristics Of the chromatographic methods discussed in Chapter 4 essentially only SEC and HPLC are used to some extent for the analysis of dissolutions containing both macromolecular and additive components. SPE is a useful device for working

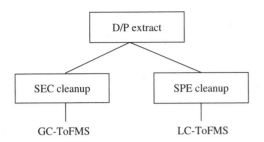

Scheme 9.1 Cleanup procedures for dissolved polymer/additive formulations

up of polymer/additive dissolutions: the apolar polymer is retained on the C_{18} sorbent, while analytes may be eluted (see Scheme 9.1). In the fractionation of dissolutions, it is advantageous to make use of the differences in polarity and affinity of the components with the sorbent. For simultaneous screening of many analytes, advantage may be taken of the high full scan sensitivity of ToF-MS; exact mass, retention time and isotope ratio information is gathered rapidly.

In principle, TLC is also suited for the analysis of oligomers, polymers and polymer/additive solutions. *Thin-layer chromatography* permits analysis of some complex types of polydispersity, such as those of composition and molar mass for copolymers, and those of molar mass and end-groups for oligomers [11]. In analogy with the use of TLC (Section 4.4.1.2) and hyphenated TLC techniques (Section 7.3.5) for the analysis of polymer/additive extracts, one might expect that it is equally possible to separate a dissolved polymer/additive sample by means of TLC, without prior precipitation of the polymer from the solution. By direct application of the polymer solution to the TLC plate and appropriate choice of the eluent, the polymer is not eluted – as opposed to the additives. In practice, however, attempts to chromatograph polymer solutions directly by TLC are often unsuccessful, since a smear of the large excess of polymer over the full length of the chromatogram generally masks or distorts the additive spots.

Applications Most plasticisers can be removed almost quantitatively from PVC by Soxhlet extraction or direct refluxing with diethyl ether for several hours. With the dissolution (THF)/reprecipitation (*n*-hexane) method, PVC compounds can be isolated. However, reliable results are also obtained by using a short cut in which a PVC compound (0.3 g) is dissolved in THF (10 g), insoluble material is allowed to settle and then the THF solution *including* PVC is injected on to the GC column [12]. Although direct gas chromatographic analysis

of THF dissolutions of plasticised PVC wire cable products was reported as a simple and accurate means for quality control [13], the obvious drawbacks of this procedure are corrosion (due to loss of HCl) and fouling of the column packing; this can be obviated by regular replacement of the first few centimetres of the packing, or by combination with SPE. By calibration against standard compositions, the method is capable of giving quantitative results. Use of SEC-UV or SEC-RI is more indicated.

Using a HFIP/H_3PO_4 gradient, it is possible to directly analyse a PA6/(S-EED, Irganox 1098, Chimassorb 944, Cyasorb UV1164/3364) sample quantitatively after dissolution of the polymer in HFIP [14]. No sample pretreatments, such as precipitation or extraction at elevated temperatures, are needed. In the absence of UV-absorbance of the polyamide at higher wavelengths, the stabilisers can selectively be determined at an optimum wavelength (S-EED, Chimassorb 944 and Cyasorb UV3364 at 247 nm; Irganox 1098 at 278 nm; Cyasorb UV1164 at 347 nm). The procedure is extendable to other polyamides and polyesters. Gradient separation of oligomers, additives and polymers in a PC blend by multiple solvent extractions on a HPLC column has been reported [15].

Lawrence and Ducharme [16] have described a fast, simplified method for the detection of fluorescent whiteners in polymers, in which the polymer dissolution was applied directly to the thin layer. Similarly, using fluorescent indicator-containing layers, the presence of small amounts of UV light-absorbing antioxidants in polymers could be determined.

9.1.1 Size-Exclusion Chromatography

Principles and Characteristics With the relatively low resolution power of SEC for low-MW compounds, the qualifying features of direct SEC on polymer/additive dissolutions are to be found at the opposite end of the molecular weight scale for additives, namely oligomeric amd polymeric as well as grafted additives, which are all difficult or impossible to access by means of conventional extraction methods. The low resolving power also favours quantitative *groupwise additive analysis*. Application of SEC for additive analysis is essentially based on removal of the polymer matrix, instead of extracting the additives. The polymer elutes from the SEC column first, and is diverted to waste; the fraction containing the analytes is then directed to an analytical chromatographic column. SEC may thus be used as a simple and inexpensive fractionation tool of additives. Dissolution followed by flash SEC (30 min) is faster than

Table 9.3 Application of SEC to polymer/additive dissolutions

Advantages
- No need for additive extraction
- Chromatographic separation of polymer and additive(s)
- Sample cleanup
- Simultaneous polymer characterisation (MW, MWD) and additive analysis
- Additive screening
- Simple chromatographic fingerprint
- Analysis of oligomeric and polymeric additives
- Groupwise additive analysis
- Quantitative (group) analysis
- Speed

Disadvantages
- Solvent restrictions
- Polymer solubility limitations
- Limited resolving power
- Low additive concentration in polymer solution
- High detection limit at small sample loadings

any extraction method. The *main characteristics* of the method are given in Table 9.3. The simultaneous determination of polymer analytical data (MW, MWD) and additive-related data is shared with liquid NMR (structural information), see Section 9.2.1.

As shown in Section 7.4.3.1, multidimensional size-exclusion chromatography (SEC-GC, μSEC-GC-MS, SEC-PyGC, SEC-HPLC, and SEC-TLC) is used for polymer/additive analysis. In these approaches SEC achieves a gross separation between high-MW components (to be disposed of) and low-MW components, which are then further separated. The main features of μSEC-GC analysis of polymer dissolutions are given in Table 7.90. On-line μSEC-GC/LC analysis of a dissolved polymer sample yields quantitative analysis, as opposed to the dissolution/precipitation technique where additives may exhibit solubility dependence [17,18]. Cortes *et al.* [19] have described on-line coupled micro-column SEC-GC-MS for direct identification of additives in polymers. For the analysis of nonvolatile compounds, on-line coupled micro-column SEC-PyGC has been reported [20]. Another way of separating additives from the polymer after dissolution was explored by Nerín *et al.* [21,22], who linked a HPSEC column with a normal-phase HPLC column. Polymer/additive analysis may also be carried out by means of off-line SEC combined with MDLC [23], see Section 7.4.3. Furthermore, advantage has been taken of the increase in chromatographic resolution by the combination of SEC and TLC [24].

Applications   Direct analysis of polymer/additive dissolutions and of *lubricants* containing antioxidants

[25] may profitably be carried out by means of SEC, as also illustrated elsewhere (Section 4.4.2.3). Polyurethane samples can be readily dissolved using hot DMF, and the solution can then be analysed directly for antioxidant content by HT-SEC, employing an in-line guard column to filter out the polymer. In this way, BHT, Irganox 245 and Tinuvin 328 in *polyurethane* were directly determined in (HT)SEC separation after dissolution rather than extraction [26]. In order to overcome problems with extraction procedures (incomplete extraction, degradation) Mady *et al.* [27] have used direct determination of additives in both ambient and high-temperature soluble polymers, using SEC and HT-SEC, respectively. For analysis of additives in ambient-temperature soluble polymers (PS, PVC, polyols, urethanes, oils, waxes and silicones) the entire sample was dissolved in organic solvents such as THF, and the solution was injected directly into the chromatograph. Antioxidants such as BHT, Irganox 565/1010/1076 and Ethyl 330, and UVAs such as Tinuvin 144/328/P, were analysed in some of the above polymers.

For samples that meet the solubility requirements of the SEC approach, analyses were also reported for additives in polymers such as PVC and PS [28,29]. Direct SEC analysis of *PVC additives* such as plasticisers and thermal stabilisers in dissolution mode has been described [28,30,31]. In the analysis of a dissolved PS sample using a SEC column of narrow pore size, the group of additives was separated on a normal-phase column after elution of the polymer peak [21]. Column-loading capacity of HPSEC for the analysis of additives, their degradation products and any other low-MW compounds present in plastics has been evaluated for PS/HMBT, PVC/TNPP and PVC/TETO (glyceryl tri[1-^{14}C] epoxyoleate) [31]. It was shown that HPSEC can be used to separate low-MW compounds from relatively large amounts of polymers without serious loss of resolution of the additives; the technique has also been used for the group analysis of chlorohydrin transformation products of the TETO model compound [32].

For analysis of additives in polymers such as polyesters, nylons and polyolefins (e.g. PP/(Tinuvin 328, BHT) and LLDPE/(BHT, Irganox 1076, Cyasorb UV531, Goodrite 3114)), which are soluble at high temperature, HT-SEC/RI, UV, FTIR was used. These systems were analysed by using a non-UV absorbing mobile phase to accommodate the UV detector. UV detection was essential to attain the detectability required for the low concentrations of additives present in commercial polymers, using decalin as a solvent. Table 9.4 shows the additive detectability by using three different detectors.

Table 9.4 Additive detectability

Additive	Detector		
	RI	FTIR ($1740\,cm^{-1}$)	UV (280 nm)
Irganox 1076	25 µg	20 µg	1.0 µg
Irganox 1010	150	20	–
Tinuvin P	15	>100	0.2
Tinuvin 328	20	>100	0.2
Tinuvin 900	10	>100	0.2
BHT	35	–	0.3
Ethyl 330	40	>100	0.3
Goodrite 3114	–	–	0.3

After Mady *et al.* [27]. Reproduced by permission of the authors.

SEC is a useful technique for determining hydrolytic and UV degradation in outdoor applications of polyesters and nylons by monitoring simultaneously the MWD of the polymer and the effects on additives. Lederer and Amtmann [33] used hot TCB dissolution of extruded PVC window profiles in combination with HT-SEC for the direct and simultaneous determination of the common organic components except cross-linked impact modifiers. Using double RI/IR detection, a 'one-shot' chromatographic analysis of the PVC matrix and of all the soluble organic and metal organic PVC additives could be accomplished. It was possible to analyse the whole organic fraction without prior precipitation of PVC. Pasch *et al.* [34] have reported SEC-RI/UV of polymers containing UV stabiliser units, namely DHPVC-*g*-UVA and statistical copolymers of styrene, methylmethacrylate and UV stabilisers (benzophenones, phenyl and naphthyl benzotriazoles). The distribution of the UV stabiliser units along different graft copolymer fractions can be visualised using the UV detector at 313 nm. SEC profiles indicate that, in all cases, part of the UV stabiliser units is located in the polymer bulk, whereas a second part is located in the oligomer and monomer region – obviously as unreacted UV stabiliser molecules. SEC-ELSD is also used for the analysis of polymers and polymer additives [35].

Cortes *et al.* [18] have quantitatively determined polymer additives in a *polycarbonate* homopolymer and an ABS terpolymer. In that case, a multidimensional system consisting of a microcolumn SEC was coupled on-line to either capillary GC or a conventional LC system. The results show losses of certain additives when using the conventional precipitation approach. An at-column GC procedure has been developed for rapid determination (27 min) of high-MW additives (ca. 1200 Da) at low concentrations (100 ppm) in 500-µL SEC fractions in DCM for on-line quality control (RSD of 2–7 %) [36]. Also, SEC-NPLC has been used for the analysis of additives in dissolution of polymeric

materials such as *polystyrene* [21,22]. SEC-TLC was employed to separate a mixture of four polymer additives which could not be separated completely by µSEC alone [24]. A SEC chromatogram was deposited continuously on to a moving silica plate, which was subsequently developed in a direction perpendicular to the direction of deposition, thereby resolving all of the analytes. *In situ* DRIFT spectra of the individual spots permitted identification of each additive at a limit of detection of about 1 µg. The procedure may be extended to polymer/additive dissolutions.

Staal *et al.* [15] have described the dissolution of polycarbonate and *polysulfone* in THF and precipitation on to a C_{18} guard column. Separation of polymer and additives was achieved using gradient elution from water–THF (50:50 vol. %) as nonsolvent to 100 % THF. The additives elute first, followed by the oligomers and polymer.

A single SEC-RI-UV run has provided data on both the MWD of a *rubber* and the concentrations of stabilisers [37]. Kuo *et al.* [38] have used various liquid chromatographic techniques (SEC, NPLC, RPLC, IC) for the characterisation of ingredients in *coatings and inks*. In new coating systems based on low-MW polymers, oligomers and reactive additives which produce high-MW cross-linked products, knowledge of the MW and MWD of polymer components, chemical composition and identity of oligomers and additives is essential for the optimisation of product performance.

SEC has sometimes been used with off-line IR spectroscopy for the detection of polymer additives, such as dioctylphthalate, as well as on-line [39]. Dissolutions of PVC/DEHP and of PC/pentaerythritoltetrastearate (release agent) were analysed by SEC-FTIR using the thermospray/moving belt/DRIFT interface [40]. The detection limits of the method were in the 100 ng range, depending on the IR sensitivity and volatility of the solutes. This is not extremely sensitive.

Occasionally there is the need for simultaneous determination of MW, MWD of polymers and identification/quantification of additives [38]. This was the case for polymer *and* additive analysis of SBR/(softeners, flavour agents, stabilisers) (chewing gum) [41]. The many constituents of the SBR portion of the sample were not resolved, since adjacent components were similar in size. It should be stressed, however, that the need for simultaneous determination of the molecular weight of polymers and the identification/quantification of additives is exceptional rather than the rule. The determination of molecular weight distributions by SEC has indicated that *oligomer fractions* analysed by dissolution and (Soxhlet) extraction methods may differ essentially [42].

9.2 SPECTROSCOPIC TECHNIQUES

In Chapter 5 it was argued that the prospects of multicomponent additive analysis of polymer extracts by means of UV and FTIR are not bright. Therefore, it should not be expected that this is improving for polymer/additive dissolutions. On the contrary, for such systems, essentially only NMR spectroscopy has led to significant results, although the number of pertinent reports is much restricted even here.

UV spectrophotometry can be used for the analysis of polymer/additive mixtures in solution, provided that the following restrictions are taken into account:

(i) Only polymer additives, which are absorbers of UV radiation, can be determined. The majority of stabilisers have an aromatic structure, so that they are UV-active and show a distinct UV spectrum.

(ii) The polymeric material should be UV transparent (e.g. polyethylene). Other polymers, such as styrenics, absorb strongly in the UV region, making it difficult or impossible to observe any signals caused by such additives. UV spectrophotometry cannot be used for the detection of additives for polymers absorbing above about 250 nm (e.g. ABS).

(iii) Many solvents (aromatics, esters, ketones, DMF, $CHCl_3$, etc.) do not allow UV detection at low wavelengths (<250 nm); also, in ethers, the UV cut-off may be dramatically influenced by contaminations, which are formed by oxidation, such as peroxides in THF, etc., or by stabilisers. Table 3.50 gives an overview of common solvents for polymer/additive systems. Semicrystalline polymers are not easily soluble.

(iv) UV spectrophotometry is not highly specific, and cannot easily be used to identify unknown additives or to analyse multicomponent systems.

(v) The UV method does not perform well in the presence of other interferences (e.g. pigments).

The scope of UV analysis of dissolved polymer/additive matrices is thus quite restricted and mainly limited to special cases in which the additive package is known, e.g. the determination of Irganox 1098 in GFR-PA4.6 after dissolution in H_2SO_4/HNO_3. Fibreoptic dissolution analysis by means of a UV diode array spectrometer is well known. In comparison to IR spectroscopy, UV spectrophotometry is better equipped to provide quantitative data.

Infrared spectroscopy, which is recognised as an analytical technique with high selectivity and fingerprinting ability for molecular compounds, can be used for the analysis of polymer/additive mixtures in solution, provided that the following restrictions are taken into account:

(i) IR absorption at realistic additive concentrations (e.g. antioxidants from 250 ppm to 2 %) may be obscured by absorption from the polymer itself. Spectral subtraction of an appropriate reference polymer can remove matrix interferences; however, this may be impaired, as additive-free material is not always available.

(ii) Many polymer solvents (Table 3.50), as well as most HPLC mobile-phase systems, are opaque in the mid-IR region. Where the solvent is a strong absorber in the IR spectral region, the spectra of solutes are easily masked. (For the same reason, on-line LC-FTIR coupling using a flow-cell is generally hampered by eluent interferences, and solvent-elimination LC-FTIR is to be preferred).

(iii) The FTIR spectrum may contain absorbances from any other additive in the polymer which causes interference.

(iv) Infrared spectroscopy is mainly used as a qualitative tool, as opposed to UV spectrophotometry.

Also, the usefulness of IR analysis of dissolved polymer/additive matrices is thus quite restricted. When the specific additive used is known; when only qualitative results are required; and when the polymer matrix is appropriate, FTIR can provide a much easier, faster alternative to NMR spectroscopy. A rare example of additive analysis in dissolution by means of infrared has been reported for PE/oleamide [43].

9.2.1 Nuclear Magnetic Resonance Spectroscopy

Principles and Characteristics   The qualifying features for the application of solution NMR to extracts of polymeric materials have already been outlined in Section 5.4. For NMR spectroscopy, which is a powerful analytical tool for identification and quantification, extraction of additives from the polymer is not required. Recent NMR developments suggest various possibilities for *direct* additive analysis:

(i) Direct measurement (^{1}H NMR) using a high-resolution A/D converter, i.e. detection of minor signals (corresponding to concentrations of about 100 ppm; $\geq$40 MHz NMR).

(ii) Use of field gradients and suppression of the main signal (restricted to singlets, mainly PE).

(iii) Selection of one spin system (selective Total Correlation Spectroscopy); sensitivity to be assessed.

(iv) A combination of gradient and CH-correlation, e.g. Heteronuclear Multiple Quantum Coherence Total Correlation Spectroscopy (HMQC-TOCSY), with high selectivity and application to small amounts (20 μg) in 'bulk' (20–100 mg).

With the low-frequency ^{1}H l-NMR spectrometers of the past, analysis of low-concentration additives such as antioxidants was quite problematic, in view of lack of sensitivity. Moreover, in low-frequency ^{1}H NMR most polymer solvents may easily interfere with the solute resonances. In higher-field, high-resolution NMR (500 MHz), solute interferences may be overcome by means of selective signal suppression. However, this is no longer a necessary strategy for the present 900-MHz NMR generation, where both sensitivity and resolution are adequate and quantification is facilitated.

In this chapter we illustrate a direct method of characterisation of polymer/additive dissolutions by means of (500 MHz) NMR, which takes advantage of selective signal suppression allowing elimination of unwanted signals, such as the ca. $10^5 \times$ more intensive PE signal. The most effective approach to solvent suppression is the destruction of the net solvent magnetisation by pulsed field gradients (PFGs). Patt and Sykes [44] and others have proposed pulse sequences for efficient suppression. Solvent suppression methods have recently been compared [45,46,46a]. Obviously, suppression hinders quantification with reference to the polymer peak. The recently developed area of LC-NMR often favours the use of protonated solvents for reasons of economy, making robust solvent suppression essential.

Requirements for l-NMR analysis of a polymer/additive solute are essentially the same as for additive extracts, namely:

(i) usually ^{1}H NMR;
(ii) solubility in an NMR solvent;
(iii) no interaction of NMR solvent and additives; and
(iv) separation of additive signals from those of NMR solvent and polymer.

Quite obviously, a disadvantage of the classical sample preparation technique, consisting of dissolving a sample in a solvent, which may eventually lead to volatilisation and degradation of the additives, is not totally eliminated (see Section 3.7). Actually, the solvent choice is more restrictive (Table 9.5). In fact, NMR for polymer/additive dissolutions is feasible only in cases of a common solvent for polymer and additives, compatible

Table 9.5 Selected NMR solvents for polymers[a,b]

Polymer	Solvent(s)	Dissolution temperature ($^\circ$C)
Polyethylene	TCB, TCE-d$_2$, 2:1 v/v mixture of TCB: 1,2-dichlorobenzene-d$_4$ or benzene-d$_6$, hexachlorobutadiene, 4:1 v/v mixture of *cis*-decaline: *p*-dioxane-d$_8$	100–160
Polypropylene	TCB, TCE-d$_2$	110–150
Ethylene/propylene copolymers	TCE-d$_2$	90–110
Ethylene/vinylacetate copolymer	TCE-d$_2$	>100
Polystyrene	TCB, TCE-d$_2$, THF-d$_8$, chloroform-d$_1$	25–130
Polyvinylchloride	THF-d$_8$, DMF-d$_7$, TCB/TCE-d$_2$	25–140
Polyoxymethylene	*p*-Chlorophenol	~170
Polyethylene terephthalate	TFA-d$_1$, TCE-d$_2$, HFIP	>120
Polyester resins	Chloroform-d$_1$, acetone-d$_6$, DMSO-d$_6$	Ambient
Polyamides	Formic acid, TFA-d$_1$, trifluoroethanol-d$_3$, H$_2$SO$_4$, DMF-d$_7$ (for semi-aromatics)	60–70
Polymethylmethacrylate	Chloroform-d$_1$, benzene-d$_6$, TCE-d$_2$	>80
Polyvinylacetate	TCE-d$_2$	25–60
Polyvinyl alcohol	D$_2$O, DMSO-d$_6$	25–60
Polyurethane	DMF-d$_7$, DMSO-d$_6$, TCE-d$_2$	60
Polyacrylonitrile	DMSO-d$_6$	~100
Poly(diphenylene ether) sulfones	DMSO-d$_6$, chloroform-d$_1$	~100
Styrene–butadiene copolymer	Chloroform-d$_1$	Ambient

[a] See Appendix I for acronyms.
[b] For properties of deuterated solvents, see Claridge [47].

with the NMR requirements. This is the case for poly-olefins, PET, PBT, PVC, poly(ester-*block*-ether) copoly-mers, EPDM, SBR, etc., but is more difficult or impossi-ble for polyamides, fluoroelastomers, Teflon and others. Taking appropriate precautions, e.g. using *cis*-decalin-d_{18} instead of proteo-*cis*-decalin, may largely eliminate interfering solvent resonances.

For direct l-NMR analysis of polymer/additive anal-ysis, a few recommendations may be made. It is good practice to carry out the sample preparation in a nitro-gen atmosphere, followed by sealing. Experimental conditions must be such that no measurable amount of polymer degradation or loss of stabiliser occurs during analysis. It has been reported that several of the solvents commonly employed in l-NMR analysis of polyolefins are unsatisfactory because of observed degradation of the stabilisers during NMR measurements at high tem-perature [48]. Such degradation is found in polyolefin solutions of both TCB and hexachlorobutadiene, at vari-ance with a 4:1 v/v mixture of *cis*-decaline and *p*-dioxane-d_8. By its very nature, dissolution results in analyte dilution, even if contained (typical conditions: 5 mg sample in 0.6 mL solvent). For extruded poly-mer samples, a number of pellets should preferably be sampled (scraped), so as to minimise sample-to-sample variations in additive concentration.

The *main characteristics* of the application of NMR to polymer/additive dissolutions are given in Table 9.6. NMR solvent–additive interactions should be avoided. This may be problematic for some systems, such as polyamides, where the polymer solvent (e.g. formic

Table 9.6 Application of NMR to polymer/additive dissolutions

Advantages

- Primary analytical tool
- No need for additive extraction
- Additive screening (soluble additives only)
- Simple spectroscopic fingerprint
- Simultaneous analysis of polymer and additives
- Analysis of low- and high-molecular weight additives
- Quantitative analysis with internal standard (for ^{1}H or ^{31}P NMR) or polymer (^{1}H NMR)
- ^{1}H and diffusion resolution
- Various nuclei (^{1}H, ^{19}F, ^{31}P)

Disadvantages

- Solvent restrictions (matrix solubility; additive interaction; NMR requirements)
- Long dissolution times
- Diluted solutions
- Moderate sensitivity
- Need for signal suppression techniques (not for >500 MHz NMR generation)
- Measurement times (field strength dependent)

acid) does not dissolve the additives. The disadvantages of the method are long dissolution times at high T and fairly long measurement times (16 h for PE solution on a 300-MHz instrument but only 1 h at 500 MHz). No information is gathered about certain additives (e.g. insoluble metal soaps or very low-concentration components such as optical whiteners). The technique is well suited for fairly low additive concentrations (ca. 100–500 ppm), but difficult for traces (<5 ppm).

It appears that purification of commercially avail-able solvents is sometimes required for the complete elimination of impurity resonances. Occasionally, these impurities may be turned into advantage, as in the case of $C_2D_2Cl_4$ where the (known) C_2DHCl_4 content may be used as an internal standard for *quantitation*. Thus, removal of every impurity peak is not always essential for identification and quantitative analysis of stabilis-ers in PE. Determination of the concentration of addi-tives in a polymer sample can also be accomplished by incorporation of an internal NMR standard to the dissolution prepared for analysis. The internal standard (preferably aromatic) should be stable at the temperature of the NMR experiment, and could be any high-boiling compound which does not generate conflicting NMR resonances, and for which the proton spin-lattice relax-ation times are known. 1,3,5-Trichlorobenzene meets the requirements for an internal NMR standard [48]. The concentration should be comparable to that of the ana-lytes to be determined.

Applications   Literature on the application of l-NMR to polymer/additive dissolution is scarce. Schilling and Kuck [48] first described a direct method for char-acterisation and quantification of additives in solu-tions of polymers by means of high-field (11.7 T), high-resolution ^{1}H NMR with selective signal sup-pression. While stabiliser degradation was observed in 1,2,4-trichlorobenzene and hexachlorobutadiene solu-tions, additives and polymer were very stable (over a period of 3 h) in a mixture of 80 % *cis*-decalin and 20 % *p*-dioxane-d_8. This mixture is also a good solvent for polyolefins, and provides a stable deuterium lock signal for the NMR spectrometer. In this solvent mixture, small changes in stabiliser concentration were observed after 16 h at 115 °C or after several weeks at 20 °C. Exam-ination of HDPE/(0.18 wt % Irganox 1010, 0.20 wt % Irganox MD 1024) in *cis*-decalin: *p*-dioxane-d_8 (4:1 v/v) at 115 °C was carried out with effective suppression of the polymer resonance and partial suppression of the large solvent peaks of *cis*-decalin. Careful sam-ple preparation and exclusion of oxygen are required to prevent stabiliser loss during NMR measurement.

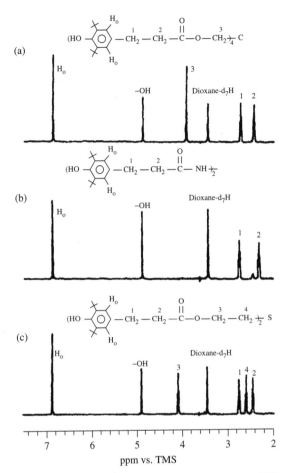

Figure 9.1 500 MHz ^{1}H NMR spectra of (a) Irganox 1010, (b) Irganox MD 1024, and (c) Irganox 1035 in a 80:20 (v/v) mixture of *cis*-decalin: *p*-dioxane-d_8 at 115 °C. After Schilling *et al.* [48]. Reprinted from *Polymer Degradation and Stability*, **31**, F.C. Schilling *et al.*, 141–152, Copyright (1991), with permission from Elsevier

Identification of *phenolic stabilisers* by high-resolution NMR is quite simple (although not always unique), particularly at the high magnetic fields available in modern spectrometers. The proton spectra of three common stabilisers, Irganox 1010/MD 1024/1035, are shown in Figure 9.1. Clearly, in HDPE samples studied at field strength of 11.7 T resonances unique to each additive can readily be observed [48]. Quantitative analysis was carried out using a standard solution of the additives and unstabilised polymer. The NMR spectrum of the standard was recorded with a basic set of experimental parameters. This procedure removed the need for obtaining NMR data under conditions of full proton relaxation. Nonetheless, it is desirable to obtain NMR

data under conditions of maximum experimental sensitivity. This requires knowledge of the proton spin-lattice relaxation times.

Schilling and Kuck [48] have also reported 500-MHz ^{1}H-NMR spectra of PE cable material containing unknown amounts of unidentified stabilisers (Figure 9.2). Resonances of Irganox 1010 and MD 1024 could readily be traced. Since the spectrum was recorded under identical conditions as in the standard sample, the level of the stabilisers could be determined (ca. 0.2 wt. %). The proposed NMR method (high field, high resolution with selective signal suppression) can be employed not only to verify and quantify additives in fresh, extruded and aged polyolefin products, but also in other polymer systems provided that a suitable solvent is available.

From the list of most popular NMR nuclei (^{1}H, ^{13}C, ^{19}F and ^{31}P), it is obvious that NMR of dissolved polymer/additive matrices also holds promise for direct analysis of phosphorous-containing additives. This was

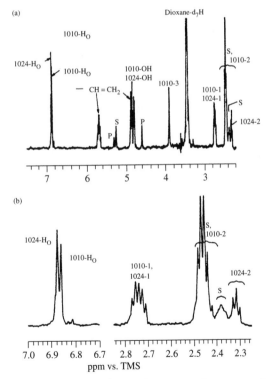

Figure 9.2 500 MHz ^{1}H NMR spectra of a commercial stabilised PE product; solvent and temperature as in Figure 9.1. See Figure 9.1 for structural designations (S and P refer to residual solvent and polymer resonances, respectively). After Schilling *et al.* [48]. Reprinted from *Polymer Degradation and Stability*, **31**, F.C. Schilling *et al.*, 141–152, Copyright (1991), with permission from Elsevier

shown for the characterisation of transformation products of several phosphorous-containing polymer additives in PE and a styrene-butadiene copolymer, dissolved in appropriate solvents (see Table 9.5) [49]. An advantage of ^{31}P NMR is the absence of interference of the matrix. On the other hand, in comparison with ^{1}H NMR, ^{31}P NMR performs less well with regard to structure elucidation (many protons vs. few phosphorous nuclei). Moreover, as relaxation times are long, the quantification is difficult.

Spectroscopic characterisation of additives and their degradation products is very useful in understanding the chemistry that occurs during normal polymer processing operations. Generation of free-radical species in polymers by the combination of heat and oxygen is of greatest concern during processing operations up to 300 °C, such as extrusion, film-blowing, or moulding, which convert base resins into useful products. Whereas only incidental oxygen is present in extruders and most moulding machines, blown-film lines expose molten polymer to the atmosphere. Protection of the polymer against the results of direct oxygen attack by trapping radicals is provided by primary antioxidants (PAOs), such as hindered phenolics. The use of secondary antioxidants (SAOs), such as phosphites and phosphonites, is well established. Some phosphorous-containing SAOs, varying considerably in hydrolytic and thermal stability, are shown in Figure 9.3. Additives A (Naugard 524) and C (Irgafos P-EPQ) are hydrolytically stable, whereas Irgafos P-EPQ is not very resistant to thermal degradation. While TNPP (additive B) is not very stable hydrolytically, it does exhibit thermal stability comparable to tris(2,4-di-*t*-butylphenyl)phosphite (additive A).

Brandolini *et al.* [49] have described spectroscopic characterisation by means of ^{31}P NMR (at 161.9 MHz),

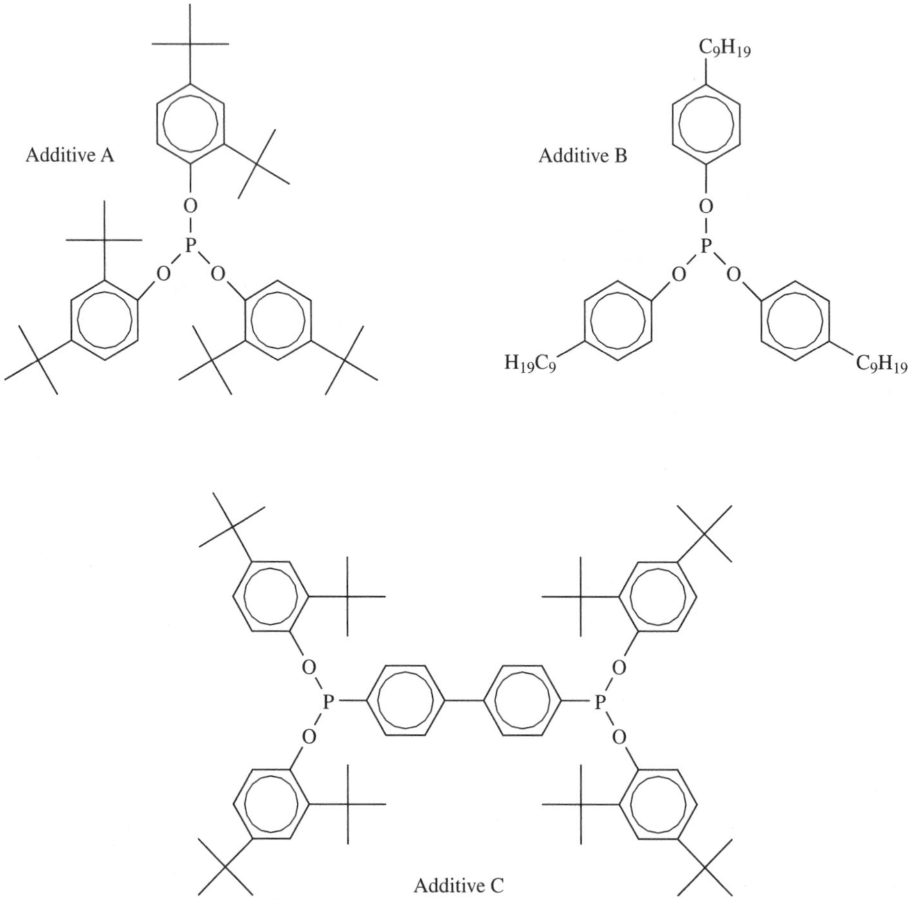

Figure 9.3 Chemical structures for the phosphorous-containing secondary antioxidants Naugard 524 (A); tris(nonylphenyl)phosphite (B); and Irgafos P-EPQ (C). After Brandolini *et al.* [49]. Reprinted from *Spectroscopy*, **7**(3), A.J. Brandolini *et al.*, 34–39, Copyright (1992), with permission from IOS Press

FTIR and UV of the degradation products of these *phosphorous-containing polymer additives*. The ^{31}P NMR chemical shifts are very sensitive to changes in chemical structure; the NMR experiment is responsive enough to allow analysis at commercially relevant concentrations (300–3000 ppm). Interpretable spectra of additive solutions could be obtained in a relatively short time (700–800 transients, or about 2 h). For the observation of low phosphorous levels (20–50 ppm) in polymers, overnight or weekend runs are frequently necessary (5 000–20 000 transients). When additive A is heated at 140 °C in air for 4 h the ^{31}P NMR spectrum changes considerably. The degradation products are readily assigned to phosphate di- and triesters. Although hydrolysis is usually a major concern for phosphite antioxidants, this is not a problem for additive A. After compounding of additive A in a styrene–butadiene block copolymer, or after extrusion of PE, ^{31}P NMR shows little degradation. On the other hand, ^{31}P NMR spectra of TNPP at 140 °C in air for 4 h denote significant formation of phosphoric acid and three phosphate ester components. When extruded at the unrealistically low temperature of 140 °C for a commercial process, the phosphonite additive C has degraded considerably. Irgafos P-EPQ is more thermally labile than either A or B, and hydrolytically stable. Although ^{31}P NMR provides a highly specific analysis of the extent of additive degradation, the experiments are quite time-consuming (at least in the given conditions), and require sufficient sample. To differing degrees, IR and UV spectrophotometries, mass spectrometry and thermal analysis are also useful for studying the oxidative reactions of these compounds. If only general information is needed, or if the sample's formulation is well characterised, FTIR is a rapid and simple technique for studying degradation. UV spectrophotometry is not particularly useful for the additives of Figure 9.3.

De Vries and Beulen [50] have carried out direct quantitative measurements of HDPE/(1000 ppm Irganox 1076, Zn stearate), HDPE/(1000 ppm Irganox 1076), and VLDPE/(1500 ppm Irganox B 220, 200 ppm Irganox 1076), dissolved in the NMR solvent $C_2D_2Cl_4$ at 120 °C by means of 300-MHz ^{1}H NMR using signal suppression. Observed proton resonances (PE 1.28 ppm; Irgafos 168 7.05, 7.23, 7.37 ppm; Irganox 1076 2.55, 2.85, 4.0 and 6.96 ppm; Irganox 1010 4.00, 6.92 ppm) allowed the direct determination of aromatic additives in PE (without prior extraction). Quantification at a level of approximately 100 ppm or higher was possible, provided that the degree of deuteration of $C_2D_2Cl_4$ (internal reference) is known. For aliphatic additives (in the δ 2–4 ppm shift range) different experimental conditions

(pulse angle) are needed, and quantification is more difficult. A drawback of the method is the relatively long measurement time (from 14 h on a 300-MHz NMR to only two hours at 500 MHz).

As many bromine-containing *flame retardants* do not dissolve in common NMR solvents (typically $CDCl_3$ and tetrachloroethane), ^{1}H l-NMR can not generally be applied and ^{13}C s-NMR may then be called in. However, in favourable circumstances, e.g. for FR 1025 (poly-pentabromobenzylacrylate, Ameribrom) in PBT (Tribit 1500 GN 30), direct ^{1}H l-NMR in $C_2D_2Cl_4$ of the fraction insoluble in HFIP can be used, in view of the unique resonance position of the benzylacrylate fragment in FR-1025.

Results presented show that high-field, high-resolution NMR with or without selective signal suppression can be used to identify type and quantity of several additives in dissolved polymeric samples (but no very low concentration or inorganic additives). From one 400-MHz ^{1}H l-NMR spectrum of LLDPE, m-PE or a LLDPE/m-PE blend, information can be obtained about type and amount of unsaturation and type and amount of stabilisers (250–1000 ppm level); at 600 MHz information can be obtained even in a short time. Although the usefulness of NMR analysis of dissolved polymer/additive matrices is subject to restrictions, application of the technique is worth considering. Yet, little activity is noticed. High-resolution l-NMR (600–800 MHz) equipped with a high-temperature accessory (125 °C), if available, in combination with DOSY (physical separation) and/or suppression of polymer and solvent signals may profitably be used for additive screening, as an alternative to LC-NMR (chemical separation). The problem with solvent suppression is sensitivity (400–500 ng for a 600-MHz instrument) which still compares poorly to mass spectrometry. In cases where the specific additive used is known, the polymer matrix is appropriate, and when only qualitative results are required, FTIR is usually a much easier and faster alternative to NMR spectroscopy. For unique identification of unknown additives, MS is to be preferred.

For diffusion NMR, which is another approach to the separation of polymer and additive signals by application of field gradients, see Section 5.4.1.1.

9.3 MASS-SPECTROMETRIC METHODS

Mass analysis of polymer/additive extracts (i.e. without the polymeric component, see Chapter 6) is obviously an easier task for additive identification than mass analysis of polymer/additive dissolutions. Yet, it is of interest to explore the chances of the latter approach and to identify the most promising routes. Section 6.2 suggests the different approaches for extracted and dissolved polymer/additive samples in relation to mass spectrometry. Ionisation methods such as DCI, FAB, FD, LDI (including MALDI) and ICPI are suitable entries for dissolved samples.

Desorption chemical ionisation (DCI) mass spectrometry has been used for detecting additives extracted from polymers [51,52] by a solvent as volatile as possible. To use the DCI probe, $1-2\,\mu L$ of the sample, in solution, are applied to the probe tip, composed of a small platinum coil, and after the solvent has been allowed to evaporate at room temperature, the probe is inserted into the source. The sample is then subject to fast temperature ramping. DCI does not seem to be the most suitable mass-spectrometric method for analysis of dissolved polymer/additive matrices, because:

(i) with a few exceptions, the solvents used for dissolution of polymer/additive matrices (Table 3.50) are not very volatile;
(ii) in this case DCI is not a valid alternative to the direct insertion probe.

On the basis of the results by Chen and Her [51] it may be expected, however, that the low-proton-affinity PE matrix will be barely detectable, and thus not interfering, in ammonia DCI-MS of a dissolved PE/additive sample.

Whereas the use of conventional *fast atom bombardment* (FAB) in the analysis of polymer/additive *extracts* has been reported (see Section 6.2.4), the need for a glycerol (or other polar) matrix might render FAB-MS analysis of a dissolved polymer/additive system rather unattractive (high chemical background, high level of matrix-, solvent- and polymer-related ions, complicated spectra). Yet, in selected cases the method has proved quite successful. Lay and Miller [53] have developed an alternative method to the use of sample extraction, cleanup, followed by GC in the quantitative analysis of PVC/DEHP with plasticiser levels as typically found in consumer products (ca. 30%). The method relied on addition of the internal standard didecylphthalate (DDP) to a THF solution of the PVC sample with FAB-MS quantitation based on the relative signal levels of the [MH]$^+$ ions of DEHP and DDP obtained from full-scan spectra, and on the use of a calibration curve (intensity ratio m/z 391/447 vs. mg DEHP/mg DDP). No FAB-matrix was added. No ions associated with the bulk of the PVC polymer were observed. It was

reasoned that selective ionisation of the analyte and reference standard, rather than bulk PVC, occurs because of the high molecular weight of the PVC molecule. The method, based on the analysis of solutions, reduces or eliminates potential problems with sample homogeneity. Direct FAB-MS (with thioglycerol matrix) was used for qualitative screening purposes.

As *field desorption* (FD) refers to an experimental procedure in which a solution of the sample is deposited on the emitter wire situated at the tip of the FD insertion probe, it is suited for handling lubricants as well as polymer/additive dissolutions (without precipitation of the polymer or separation of the additive components). Field desorption is especially appropriate for analysis of thermally labile and high-MW samples. Considering that FD has a reputation of being difficult to operate and time consuming, and in view of recent competition with laser desorption methods, this is probably the reason that FD applications of polymer/additive dissolutions are not frequently being considered by experimentalists.

DIP-ToFMS is theoretically another option for the separation of additives from polymer in dissolutions using a probe temperature ramp. However, the technique also allows direct handling of solid substrate material, which is even more convenient. The technique has profitably been used for the analysis of non-UV cured ink, revealing diluent, photo-initiator and polymer [54].

It is of particular interest that *MALDI-ToFMS*, which is not the most obvious choice for the analysis of *additives* in solution, finds practical application for rapid screening of polymer/additive dissolutions.

9.3.1 MALDI-MS Analysis of Polymer/Additive Dissolutions

Principles and Characteristics Although it might appear that MALDI-ToFMS should perform particularly well only for the polymer part of polymer/additive systems, the technique also yields useful information about *additives* contained in UV-insensitive polymers, such as polyolefins. The latter materials are hardly an insignificant part of the total polymer market!

Lasers have advanced the analytical use of mass spectrometers to characterise additives in polymers, and routine application of MALDI is no longer limited to high molecular masses only. MALDI can now clearly produce isotopically resolved mass spectra of small molecules (<800 Da) in an L-ToF instrument, which can be used successfully for the characterisation of molecules of different chemical classes. High mass resolution with an improvement of mass accuracy to

Table 9.7 Features of MALDI-ToFMS analysis of polymer/additive dissolutions

Advantages

- Very fast sample preparation and measurement (20 samples on a sample slide in 1 h)
- Possibility of measuring mass spectra of complex sample mixtures directly (without extensive separation or sample cleanup)
- Easy identification of additives by mass number (high resolution; ±0.1 Da)
- No interference (high resolution) with the polymer matrix (in particular for UV-insensitive polymers)
- Broad mass range
- Absence of any fragmentation (on linear instruments)
- Suited for mixture analysis
- Identification of various additives by one procedure
- High sensitivity (0.05 wt % and less)

Disadvantages

- Quick identification only of known additives
- Matrix effects
- Difficult quantitation
- Restricted applicability to UV-sensitive polymers

0.1 Da of the monoisotopic masses, high sensitivity (down to $20 \, \text{fmol} \, \mu\text{L}^{-1}$), and the absence of any fragmentation, are important advantages for a L-ToF system. The *main characteristics* of MALDI-ToFMS, as applied directly to polymer/additive dissolutions, are summarised in Table 9.7.

Applications MALDI-ToFMS is at its best as a rapid screening technique for quick identification of known additives. However, this screening is rendered slightly more complicated by the fact that MALDI-ToFMS spectra of pure additives and of additives in the presence of excess macromolecules are not always identical (matrix effect) [55]. For unknown additives, the relation MALDI-ToFMS spectrum−chemical structure is not easily established, and the use of FD or MALDI-MS/MS is then needed. As MALDI-MS shows a sensitivity difference for the various additives, it cannot easily quantify them unless the analytes are very similar. For differentiation of additives with the same mass number (e.g. Tinuvin 315 and Cyasorb UV3638 with $m/z = 368$) high resolution is required, as provided by delayed extraction MALDI-ToFMS.

Meyer-Dulheuer [55] has analysed the *pure* additives (phenolic antioxidants, benzotriazole UV stabilisers and HALS compounds) of Table 9.8 in THF solutions by means of MALDI-ToFMS. As it turns out, polar molecules in the mass range of below 800 Da, which have a high absorption coefficient at the laser wavelength used, can often be measured without any matrix [55,56]. In this case, there is no matrix-assisted laser desorption and ionisation (MALDI) process any more. It is a simple laser desorption/ionisation (LDI) process. The advantage of this method is a matrix-free mass spectrum with the same mass resolution as in the MALDI case,

Table 9.8 Characterisation of additives in polymers by means of MALDI-ToFMS

Polymer	Additive	Functionality[a]	Solvent	Matrix modification[b]	Minimum detection limit[c]
PP	Hostanox O 3	AO	Toluene	−	0.05 wt %
PP	Hostavin N 20	L-HALS	Toluene	−	0.01
PP	Irganox 3114	AO	Toluene	−	0.10
PP	Tinuvin 765	L-HALS	Toluene	Dithranol	0.20
PP	Hostavin N 20/N 30	L/H-HALS	Toluene	Dithranol	0.01
HDPE	Hostanox O 3	AO	Toluene	−	0.05
HDPE	Hostavin N 20	L-HALS	Toluene	−	0.01
HDPE	Hostavin N 30	H-HALS	Toluene	Dithranol	0.20
HDPE	Tinuvin 765	L-HALS	Toluene	Dithranol	0.20
HDPE	Tinuvin 320/ Hostavin N 30	UV/H-HALS	Toluene	Dithranol	n.d.
PA6	Hostavin N 20	L-HALS	Trifluoroethanol	−	0.05
PA6	Irganox B 1171	AO	Trifluoroethanol	−	0.01
PA6	Tinuvin 320	UV	Trifluoroethanol	Dithranol	0.30
PA6	Tinuvin 350	UV	Trifluoroethanol	Dithranol	0.10
PMMA	Irganox 3052 FF	AO	THF	Dithranol	ND
PMMA	Tinuvin 213	UV	THF	Dithranol	ND
PMMA	Tinuvin 571	UV	THF	Dithranol	ND

[a] AO, phenolic antioxidant; UV, benzotriazole UV stabiliser; L-HALS, low-MW HALS; H-HALS, high-MW HALS.
[b] Dithranol or 1,8,9-trihydroxyanthracene.
[c] n.d., not determined; ND, not detectable.
After Meyer-Dulheuer [55]. Reproduced by permission of H. Pasch.

but, in addition, fragmentation peaks may occur as a result of higher laser intensities. The need for the use of a laser-light absorbing matrix depends on the chemical structure of the analyte of interest. Whereas a direct laser desorption process can determine low-molecular-weight laser-light absorbing additives, for higher-molecular-weight or photolytically labile compounds the use of a matrix is necessary.

Phenols and benzotriazoles absorb in the range of the laser light used (337 nm), are sufficiently stable for direct laser desorption, and can be analysed without a light-absorbing matrix. Identification takes place on the basis of the molecular ions $[M + H]^+$ (see Figure 9.4 for Tinuvin 350), $[M + Na]^+$ or $[M + K]^+$ (Na and K derive from natural sources); some fragmentation was observed for Irganox 3052 FF. The Irganox B 1171 blend shows $[M + Na]^+$ and $[M + K]^+$ adducts of its components (Figure 9.5). Tinuvin 320 shows some fragmentation peaks. The MALDI-ToFMS spectra of Tinuvin 213 show three peak series of $[M + Na]^+$

ions with mass increments of 44 Da, suggesting three series of polyethylene oxide (PEO) oligomers with benzotriazole end-groups, see Figures 9.6 and 9.7, and Table 9.9. Component analysis of additives such as Tinuvin 213 and blends (Irganox B 1171, etc.) is possible with MALDI-ToFMS, but is often difficult by other means.

Even HALS compounds which absorb weakly at 337 nm can be analysed directly *without matrix assistance*, with the exception of the high-MW Hostavin N 30 (ca. 1500 Da), which fragments by direct laser desorption; ionisation of intact molecules occurs only in the presence of a (dithranol) matrix. Direct laser desorption leads only to noncharacteristic, low-MW fragments. Hostavin N 20 leads to $[M + H]^+$, $[M + Na]^+$, $[M + K]^+$ and some fragmentation peaks. MALDI-ToFMS of Tinuvin 765, which consists of a mono- and bifunctional sterically hindered amine, only shows the adduct peaks of the bifunctional amine; apparently, the monofunctional amine is not ionisable.

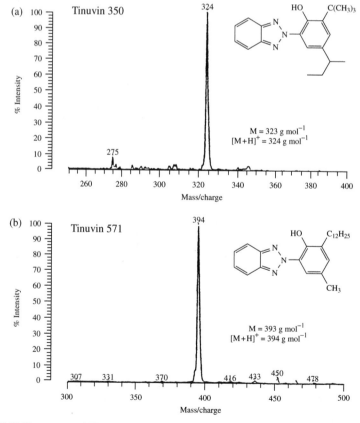

Figure 9.4 MALDI-ToFMS spectrum of Tinuvin 350. After Meyer-Dulheuer [55]. Reproduced by permission of H. Pasch

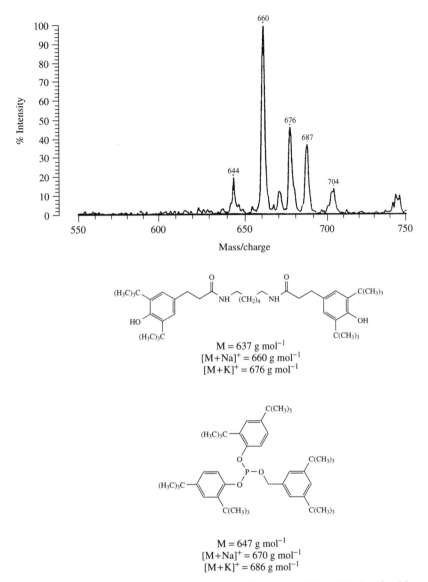

Figure 9.5 MALDI-ToFMS spectrum of Irganox B 1171 (Irganox 1098/Irgafos 168 1:1 blend). After Meyer-Dulheuer [55]. Reproduced by permission of H. Pasch

Jackson *et al.* [57] have compared various ionisation techniques (EI, CI, LSIMS, FD, UV MALDI) of a five-component mixture of polymer additives consisting of Irganox 1010/1076/3114, Tinuvin 327 and Hostanox O3 (see also Section 6.3.6). Using a silver trifluoroacetate solution, predominantly $(M + Ag)^+$ molecular ions were generated by means of UV MALDI ionisation with little fragmentation, often together with some clusters (Figure 9.8), which makes assignment of unknowns more complex than for analysis by

means of FD. Analysis by means of UV MALDI-MS is rapid, and the experiments are simple, but the instrumental capability for CID experiments is far from being widely available. UV MALDI-ToFMS results were not very reproducible, with the distribution of relative intensities of the cationated molecular ions varying with each acquisition [57]. Also, matrix effects were prominent in MALDI experiments and this – added to the lack of reproducibility (due to sample preparation) and the complication introduced by the presence of various adducts of

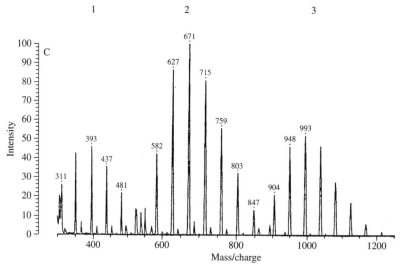

Figure 9.6 MALDI-ToFMS spectrum of Tinuvin 213. After Meyer-Dulheuer [55]. Reproduced by permission of H. Pasch

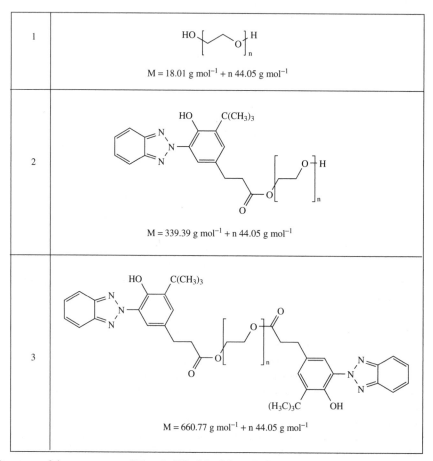

Figure 9.7 Structures of the components of Tinuvin 213. After Meyer-Dulheuer [55]. Reproduced by permission of H. Pasch

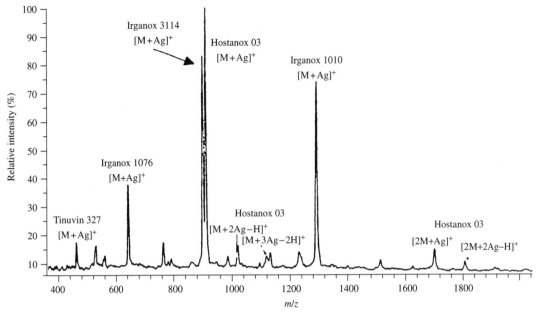

Figure 9.8 UV MALDI-MS spectrum of a mixture of polymer additives. After Jackson *et al.* [57]. Reprinted from A.T. Jackson *et al.*, *Rapid Communications in Mass Spectrometry*, **10**, 1449–1458 (1996). Copyright 1996 © John Wiley & Sons, Ltd. Reproduced with permission

Table 9.9 Attribution of mass peaks of the series 1, 2 and 3 of Tinuvin 213 (Figure 9.6)

n	1 $[M + Na]^+$	2 $[M + Na]^+$	3 $[M + Na]^+$
4	–	539	–
5	–	582	904
6	–	627	948
7	349	671	993
8	393	715	1 036
9	437	759	1 080
10	481	803	1 124
11	526	847	1 168

After Meyer-Dulheuer [55]. Reproduced by permission of H. Pasch.

some additives with silver – indicates that at the present time MALDI is unsuitable for quantitative analyses of additive mixtures.

MALDI-FTMS has been used for characterisation of Jeffamine D-2000 and nylon-6 oligomers (MW 25 000) [58]. The mass resolving power ($m/\Delta m = 200\,000$) and mass accuracy (a few ppm) of FTMS were sufficient to mass-resolve the individual components of complex mixtures of homologous series of oligomers differing in chemical composition and end-group composition. MALDI-FTMS analysis of Jeffamine D-2000, R_1-$(C_3H_6O)_x$-R_2, shows four homologous series of oligomers, each with distinct end-group composition.

Meyer-Dulheuer [55] and Pasch *et al.* [59] have described the use of MALDI-ToFMS for the direct determination of polar additives in polymeric matrices of different polarity. For this purpose, various model *polymer/additive dissolutions* were eventually matrix-modified with dithranol and deposited on a sample holder without further treatment; there was no need to add alkali-metal ions for adduct formation. The systems examined were PP/(Hostanox O 3, Hostavin N 20, Hostanox SE 10, Irganox 3114, Tinuvin 765), PP/(Hostavin N 20, Hostavin N 30), HDPE/(Hostanox O 3, Hostavin N 20, Hostavin N 30, Tinuvin 765), PMMA/(Irganox 3052 FF, Tinuvin 213, Tinuvin 571), HDPE/(Tinuvin 320, Hostavin N 30), and PA6/(Hostavin N 20, Irganox B 1171, Tinuvin 320, Tinuvin 350), all charged with 0.01–1 wt % of additives. All additives investigated showed high absorption in the UV/VIS range, at variance with the polymer. Laser irradiation leads to energy absorption by the analyte and the matrix, causing desorption of additives and matrix molecules. As the additives are ionised preferentially due to their high polarity, it is possible to determine them directly in the presence of the polymers.

Table 9.8 shows the results of MALDI-ToFMS experiments of the selected additives in PP, HDPE, PA6 and PMMA. The observed detection limits are in the technical range of interest. Identification of the additives

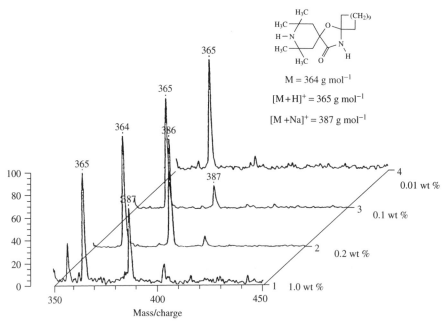

Figure 9.9 MALDI-ToFMS spectra of HDPE/Hostavin N 20. After Meyer-Dulheuer [55]. Reproduced by permission of H. Pasch

in PP and HDPE is possible, either by direct or (dithranol) matrix-assisted laser desorption. In general, phenols and benzotriazoles with high absorption coefficient at the laser light wavelength (337 nm) can be determined by direct laser desorption in polymer/additive dissolutions. PP/Tinuvin 765 required matrix-assistance (in contrast to pure Tinuvin 765) probably because of the highly apolar nature of PP. In the MALDI-ToFMS spectra of the (matrix-modified) PP/additive and HDPE/additive dissolutions, no mass peaks of the polymeric matrix were observed (see Figure 9.9 for HDPE/Hostavin N 20).

In spite of the high polarity of PA6, identification of additives was also feasible in formulations of PA6/additive dissolutions, although with decreased sensitivity. Hostavin N 20, Irganox B 1171, Tinuvin 320 and Tinuvin 350 can be determined in PA6 in technical concentrations, although the sensitivity is less than for nonpolar polymers, such as polyolefins. This was tentatively explained as follows. In a nonpolar polymer matrix, the electronically excited polar additive molecule can easily be desorbed. In the polar polyamide matrix, desorption of the additives is hindered by strong polar interactions (e.g. hydrogen bridges) between the excited analytes and the polymer matrix. This hinders selective desorption of the additives by laser irradiation. However, in a polymer/additive matrix-modified solution, evaporated to dryness, the interactions between the polar

additive and PA6 are already greatly inferior to those in a compounded polymer. Selective desorption of $[M + H]^+$ is not possible for PA6/Hostavin N 20, apparently in view of hydrogen bonding between the amide groups of the polymer and the additive. PA6/Tinuvin 320 leads to more fragmentation than in the case of pure Tinuvin 320. For PA6/Tinuvin 350, one observes the $[M - CH_3]^+$ fragment instead of the $[M + H]^+$ molecular ion peak for pure Tinuvin 350.

For PMMA/additive dissolutions, it was not possible to identify any additive characteristic mass peaks, either by direct laser desorption or with matrix-assistance (dithranol, DHBA or sinapinic acid, 4-hydroxy-3,5-dimethoxy-cinnamic acid). This has again been ascribed to very strong interaction between PMMA and additives, which suppresses desorption of additive molecules. Also, partial depolymerisation of photolytically labile PMMA by laser irradiation may play a role, which leads to saturation of the detector by PMMA fragment-ions and disappearance of additive mass peaks below noise level. Meyer-Dulheuer [55] has also reported MALDI-TOFMS analysis of a coating/2-ethylhexyldiphenylphosphate sample. *Quantitative determination* of the additives by means of MALDI-ToFMS proved impossible. Possibly the development of reproducible (automated) sample handling procedures or thin films might overcome this problem.

It is concluded that MALDI-ToFMS is a suitable method for direct analysis of low-MW additives in complex polymeric materials (in dissolution), in particular as a rapid screening technique (within 0.5 h). However, in order to turn this method into a general tool for identification and quantitation, considerably more work needs to be done. Identification of additives in polymeric matrices by means of MALDI-ToFMS would greatly benefit from reference libraries of additives contained in such matrices. This is not unlike the situation observed for ToF-SIMS.

9.4 REFERENCES

1. F.-C. Wang, *J. Chromatogr.* **A753** (1), 101–8 (1996).
2. D. Stoye and W. Freitag (eds), *Paints, Coatings and Solvents*, Wiley-VCH, Weinheim (1998).
3. G.N. LeBlanc, *Am. Lab.* **32** (18), 32–7 (2000).
4. T. R. Crompton, *Practical Polymer Analysis*, Plenum Press, New York, NY (1993).
5. G.C. Claver and M.E. Murphy, *Anal. Chem.* **31**, 1682 (1959).
6. T.R. Crompton and D. Buckley, *Analyst* **90**, 76 (1965).
7. J. García-Antón and J.L. Guiñón, *Analyst* **110**, 1365 (1985).
8. K. Peltonen, *Analyst* **111**, 819 (1986).
9. S.J. Nagourney and R.K. Madan, *J. Test. Eval.* **19** (1), 77–82 (1991).
10. S.M. Reynolds and D.F. Ober, *Proceedings SPE ANTEC 2002*, San Francisco, CA (2002), pp. 3652–3.
11. E.S. Gankina and B.G. Belinskii, in *Handbook of Thin-Layer Chromatography* (J. Sherma and B. Fried, eds), M. Dekker, New York, NY (1991), pp. 807–62.
12. A.S. Wilson, *Plasticisers. Principles and Practice*, The Institute of Materials, London (1995).
13. A.A. Krychkov and A.S. Rysev, *Plast. Massy* **1**, 71–3 (1990).
14. Y. Mengerink, *Tech. Rep. DSM Research*, Geleen, Netherlands (2001).
15. W.J. Staal, P. Cools, A.M. van Herk and A.L. German, *Chromatographia* **37** (3–4), 218–20 (1993).
16. A.H. Lawrence and D. Ducharme, *J. Chromatogr.* **194**, 434–6 (1980).
17. H.J. Cortes, G.E. Bormett and J.D. Graham, *J. Microcol. Sep.* **4** (1), 51–7 (1992).
18. H. J. Cortes, R. M. Campbell, R. P. Himes and C. D. Pfeiffer, *J. Microcol. Sep.* **4**, 239–44 (1992).
19. H.J. Cortes, B. Bell, C.D. Pfeiffer and J.D. Graham, *J. Microcol. Sep.* **6**, 278–88 (1989).
20. H.J. Cortes, G.L. Jewett, C.D. Pfeiffer, S. Martin and C. Smith, *Anal. Chem.* **61** (9), 961–5 (1989).
21. C. Nerín, J. Salafranca, J. Cacho and C. Rubio, *J. Chromatogr.* **A690**, 230 (1995).
22. C. Nerín, J. Salafranca and J. Cacho, *Food Addit. Contam.* **13**, 243 (1996).
23. T. Andersson, T. Hyötyläinen and M.-L. Riekkola, *J. Chromatogr.* **A896**, 343–9 (2000).
24. C. Fujimoto, T. Morita and K. Jinno, *J. Chromatogr.* **438**, 329 (1988).
25. V.S. Greene and V.J. Gatto, *Abstracts International GPC Symposium '98*, Phoenix, AZ (1998), p. 41.
26. N.H. Mady, R. Liu and J. Viczkus, *Proceedings SPI Ann. Tech. Mark. Conference, 30th (Polyurethanes)*, Toronto (1986), p. 332.
27. N.H. Mady, R. Liu and J. Karliner, *Proceedings SPE ANTEC '84*, New Orleans (1984), pp. 323–5.
28. M.J. Shepherd and J. Gilbert, *J. Chromatogr.* **178**, 435 (1979).
29. N. Yagoubi, C. Mur, A. Baillet and D. Baylocq-Ferrier, *Analusis* **19** (8), 252–6 (1991).
30. K. Sreenivasan, *J. Liq. Chromatogr.* **9**, 2425 (1986).
31. M. J. Shepherd and J. Gilbert, *J. Chromatogr.* **218**, 703–13 (1981).
32. M.J. Shepherd and J. Gilbert, *Eur. Polym. J.* **17**, 285 (1981).
33. K. Lederer and I. Amtmann, *J. Liq. Chromatogr.* **13** (9), 1865–75 (1990).
34. H. Pasch, A. Al-Mobasher, S. Attari, K.F. Shuhaibar and F.A. Rasoul, *J. Appl. Polym. Sci.: Appl. Polym. Symp.* **45**, 209–26 (1990).
35. E. Meehan and S. O'Donohue, *Abstracts International GPC Symposium '98*, Phoenix, AZ (1998), p. 24.
36. R. Perkins, O. Nicolas and R. Sasano, *www.ATAS-int.com* (2000).
37. D.W. Brown and D.A. Loucks, *Proceedings ACS Rubber Division Spring Meeting. (133rd)*, Dallas, TX (1988), Paper 68.
38. C. Kuo, M.D. Griesen and D. Robinson, *Abstracts 214th ACS National Meeting*, Las Vegas, NV (Sept 1997), American Chemical Society, Washington, DC (1997), pp. 7–11.
39. S. Jickells, *Proceedings 4th International Symposium on Hyphenated Techniques in Chromatography and Hyphenated Chromatographic Analysers (HTC-4)*, Bruges (1996), E05.
40. J.A.J. Jansen, *Fresenius' Z. Anal. Chem.* **337** (4), 398–402 (1990).
41. J.L. Ekmanis, *Polym. Prepr. (Am. Chem. Soc., Div. Polym. Chem.)* **27** (2). 270–1 (1986).
42. H. El Mansouri, N. Yagoubi, D. Scholler, A. Feigenbaum and D. Ferrier, *J. Appl. Polym. Sci.* **71** (3), 371–5 (1999).
43. J. Haslam, H.A. Willis and D.C.M. Squirrell, *Identification and Analysis of Plastics*, John Wiley & Sons, Ltd, Chichester (1983).
44. S.L. Patt and B.D. Sykes, *J. Chem. Phys.* **56** (6), 3182–4 (1972).
45. W.S. Price, *Ann. Repts. NMR Spectroscopy* **38**, 289–354 (1999).
46. W. E. Hull, in *Two-Dimensional NMR Spectroscopy* (W. R. Croasmun and R.M.K. Carlson, eds.), VCH Publishers, New York, NY (1994), pp. 67–456.

46a. M. Liu and X. Mao, in *Encyclopedia of Spectroscopy and Spectrometry* (J.C. Lindon, ed.), Academic Press, San Diego, CA (2000), pp. 2145–52.

47. T.D.W. Claridge, *High-Resolution NMR Techniques in Organic Chemistry*, Pergamon, Amsterdam (1999).

48. F.C. Schilling and V.J. Kuck, *Polym. Degrad. Stab.* **31**, 141–52 (1991).

49. A. J. Brandolini, J. M. Garcia and R. E. Truitt, *Spectroscopy* **7** (3), 34–9 (1992).

50. N.K. de Vries and J. Beulen, *Tech. Rep. DSM Research*, Geleen, Netherlands (1999).

51. S.W. Chen and G.R. Her, *Appl. Spectrosc.* **47** (6), 844–51 (1993).

52. C.G. Juo, S.W. Chen and G.R. Her, *Anal. Chim. Acta* **311** (2), 153–64 (1995).

53. J.O. Lay and B.J. Miller, *Anal. Chem.* **59**, 1323A (1987).

54. F. de Boever, at *Micromass 2001 New Products Seminar & Users Event*, Amsterdam (2001).

55. T. Meyer-Dulheuer, *Diplomarbeit*, Technical University of Darmstadt (1997).

56. K.R. Lykke, D.H. Parker, P. Wurz, J.E. Hunt, M.J. Pellin, D.M. Gruen and J.C. Hemminger, *Anal. Chem.* **64**, 2797 (1992).

57. A.T. Jackson, K.R. Jennings and J.H. Scrivens, *Rapid Commun. Mass Spectrom.* **10**, 1449–58 (1996).

58. C.G. de Koster, *Tech. Rep. DSM Research*, Geleen, Netherlands (1996).

59. H. Pasch, T. Meyer-Dulheuer and M. Resch, *Kautsch. Gummi Kunstst.* **51**, 782 (1998).

CHAPTER 10

Much learning does not teach understanding (Heraclitus)

A Vision for the Future

If one wishes to predict the future of additive analysis in polymers, it is relevant to consider the prospects of further evolution of polymeric and additive materials; the influence of legislation and environment; instrumental developments and currently unsolved problems. It then becomes clear that additive analysis stands a fair chance remaining in use for some time, certainly in a strongly competitive environment, which will require improved product design specifications, quality assurance and research for new applications. As ideal production environments are rare, customer complaints will also require continuous attention. Government regulations are another reason for continuous analytical efforts.

The reader should be aware of the many pitfalls accompanying them at each step of the total analytical process. Careful consideration of the characteristics of the various instrumental methods described may help to avoid such pitfalls. It should be noticed that polymer/additive analysis is hardly ever the rate-determining step in the decision process, even though speed is highly desirable. Apart from process control, polymer/additive analysis will not easily justify the introduction of high-speed robotised analytical techniques.

Not surprisingly, most of the work reported in the preceding chapters was carried out in *industrial research* laboratories. However, much of the underlying fundamental theory has an academic origin. One of the largest current threats is the gradual disappearance of analytical departments in many (technical) universities, which will lead to a rapid slow-down in the understanding of the principles of new techniques which are applicable to industrial problem solving.

Additives In Polymers: Industrial Analysis And Applications   J. C. J. Bart
© 2005 John Wiley & Sons, Ltd   ISBN: 0-470-85062-0

10.1 TRENDS IN POLYMER TECHNOLOGY

Today's polymers (thermoplasts, thermosets, rubbers) are in various stages of their life-cycle, from inception (high-temperature polymers, such as PEI, PAI, PES, TPE copolymers), growth (HDPE, LLDPE, PP, PET, EPs), maturity (LDPE, PS, ABS) to ageing (SBR, PVC). The spectacular growth in polymer use that occurred during the past half-century primarily involved commodity polymers such as PE, PP, PS, PVC, PET, polybutadiene, and polyamides, which have found uses in packaging, transport, consumer products, electrical appliances, wires and cables, building and construction, leisure, etc. Besides the more obvious engineering applications for thermoplastics and elastomers, polymers are used extensively in medical and pharmaceutical technology, food technology, cosmetics and homecare products, environmental applications and other areas. The 'Plastic Age' started in 1979 when the volume of plastic produced exceeded that of steel. World production amounts to over 140 Mt (1999), with about an equal share for Europe, Asia and USA, and 5 % ROW; growth expectations of 5.3 % y^{-1} will lead to an increase in world consumption to 250 Mt y^{-1} by 2010.

The global production capacities for thermoplastics (2000) are indicated in Tables 10.1 and 10.2.

The four largest classes of synthetic polymers (PE, PP, PVC and PET) make up about 80 % of the world market. About 60 % of the production of polymers supplies structural materials to the market (packaging 41 %, building components 20 %, electric insulation 9 %, automobile parts 7 %, agriculture 2 %, miscellaneous

Table 10.1 Major plastics consumption (kilotonnes)

Polymer	1990[a]	1998[b]	2000[b]	2001[a]	AAGR[c]
PE	30 440	45 582	49 693	51 580	4.9
LDPE	13 755	–	–	16 866	1.9
HDPE	11 599	–	–	21 996	6.0
LLDPE	5 086	–	–	12 718	8.7
PP	12 500	23 456	26 655	27 929	7.6
PET	11 695	20 705	24 025	25 670	7.4
PA	4 620	5 525	5 890	5 925	2.3
PVC	18 460	23 656	25 840	27 004	3.5
PS	7 440	11 039	11 750	12 129	4.5
ABS/SAN	n.g.	4 513	5 019	n.g.	5.4[d]
PC	628	1 135	1 380	1 785	9.9
PMMA	n.g.	848	882	n.g.	2.0[d]
POM	285	525	585	638	7.6

[a] After Scheidl [1].
[b] After Scheidl [2].
[c] Average annual growth (%) in the 1990–2001 period.
[d] Average annual growth (%) in the 1998–2000 period.
n.g. not given.

Table 10.2 Global production capacities for thermoplastics (2000)

Thermoplastic	Global production (Mt y^{-1})	Growth rate (%)	Plant capacity (kt y^{-1})[e]
Commodities[a]	140	2–5	400
Engineering plastics[b]	7	6–8	50–100
High-performance polymers[c]	0.25–0.3	8–20	<50
Specialty polymers[d]	Embryonic	–	–
Global thermoplasts	148	5.3	–

[a] PE, PP, PET, PVC, PS, ABS, PMMA, ASA.
[b] PC, PPO, PBT, PA, POM.
[c] PEI, PES, PSU, LCP, PEEK.
[d] Functional polymers PAN, PPV, nanocomposites.
[e] Typical size of a world-scale production plant.

21 %), while 40 % serve as functional materials (as additives, processing aids, adhesives, coatings, viscosity regulators, lubricants, etc.).

It is of interest to consider the diverse consumer segments in more detail. ABS, PS and PP account for some 50 % of the *electrical appliances'* market. As for automotives, plastics consumption in European cars amounts to some 10 wt % (1998 level) with the following average breakdown: PP 46 %, PUR 11 %, PA 10 %, ABS 10 %, PE 8 %, PVC 4 %, PET/PBT 3 %, others 8 %; the average US car has about 257 pounds of plastic (2001), with an expected growth to 314 pounds by 2009. With current engines, a decrease in the weight of gasoline-burning cars by 300 kg is required to conform to the CO_2 Kyoto Protocol. This implies extended use of light materials [3]. A concept car with all-plastic bodywork is currently under development by Chrysler. Long-fibre-reinforced thermoplastics (LFRT) are one of the highest growth plastic material areas. Long-glass-fibre-reinforced PP (LGFPP) is the most prevalent product, used in a wide range of automotive applications, primarily to replace metal. Typical business items in the *automotive* industry are shown in Table 10.3. PVC and elastomers have emerged as prime automotive material substitution targets; olefinic and styrenic TPEs also promise of rapid growth. A life-cycle assessment of PVC is available [4].

Packaging trends involve lower cost, more functionality, volume growth, shorter product life-cycles and increased value. Rigid plastic packaging now dominates markets which were previously the exclusive realms of materials such as glass and metal, despite the fact that some technical aspects of plastic are still inferior to those of such traditional materials, e.g. the failure of plastics to withstand high temperatures without deforming, or barrier properties of a standard comparable to that

Table 10.3 Business items in the automotive industry

- Cost reduction
- Voluntary heavy-metal controls vs. imposed regulatory prohibition
- Restricted substances (phase-out of Cr^{6+} in metal coatings)
- Non-asbestos mineral fibres (friction materials)
- Competitive green image issues (e.g. for FRs)
- Recyclate materials (contaminant content)
- Moulded-in-colour parts
- Vehicle interior air quality (organics emission from components)
- Restricted substance management and recycling system (RSMR)
- Material selection (mono-materials, soft touch, nonsensitising, environmentally friendly)
- Weight reduction (improved fuel economy)

Table 10.4 Business drivers in the polymer industry

- Mature industry (e.g. polyolefins); lower overall growth
- Replacement market
- Increased globalisation (shift from developed to developing regions)
- Growing economies
- Redefinition of core business; identification of key areas (consolidation)
- Industry restructuring (portfolio swapping, marketing agreements)
- Economy-of-scale (companies, plants)
- Reduction of standard-grade products (niche products to independent compounders)
- Focus on production methods
- Cost pressure; focus on price
- Emphasis on environmental and recycling issues (HSE friendliness)
- Consumer demand for safe products
- Higher product performance expectations (market requirements)
- Application competence
- Cost of new product development
- Technological advancements
- Rapid changes in distribution (e-commerce; Internet polymer auction sites)
- Internal efficiencies
- Decreasing technical service support

of glass. The highest growth in flexible packaging will be in the convenience and ready-to-prepare foods, electronics and games, and medical sectors. Plastics in packaging have dramatically improved food safety and reduced spoilage. Because of the packaging industry's need for product differentiation, improved functionality, and material minimisation, it offers the opportunity for new and technologically improved superior materials. Current developments in packaging materials are lighter weight packages with increased barrier properties to moisture, oxygen, other gases and aromas (longer shelf-life for packed products), lower-gauge films, biodegradability, edibility, microwaveability, and easy and economical recycling [5].

The use of polymers in *medical technology* continues to grow rapidly. In the *cable industry*, the main polymers are PVC, polyolefins and halogenated elastomers; PVC seems set to remain. In *building and construction*, polymeric materials are used extensively for piping and conduit, cladding and profiles, and insulation. Unlike the uncertainties over some of the additive halogenated FR chemicals, the use of reactive brominated species in unsaturated polyesters and polyurethanes is growing.

Table 10.4 shows the main trends affecting the overall polymer business. However, differentiation according to polymer type is necessary (Table 10.5). The plastics industry still has a highly fragmented structure, and consolidation will continue in order to meet the demands of global competition, until there are no further incremental efficiencies to be gained. At the present time, the plastics industry is plagued by low profit margins and surplus capacity.

With the increasing costs of new product development, industry concentrates on specific materials and chemistries, and on meeting specific regulatory requirements. Technological developments (Table 10.6)

Table 10.5 Market drivers for different polymer types

Polymer type	Driving force(s)
Commodities	Bulk, consolidation, optimisation of existing processes; cost-driven, merging businesses
Engineering plastics	Technology driven, process/production cost driven
High-performance polymers	Technology driven
High-end specialties	R&D driven

are affected by the increasing costs of R&D and the difficulty in establishing patent rights. As a result, companies are becoming increasingly involved in licensing, technological joint ventures and R&D partnering. *Polyolefins* have recently undergone a relatively rapid change of technology, driven by new polymerisation catalysts that provide enhanced polymers and opportunities to broaden product capability and lower the operating costs of existing processes. Today's very large global polyolefins industry is based on the use of only five monomers: ethylene, propylene, butene-1, hexene-1 and octene-1. The introduction of metallocene and other single-site catalyst (SSC) technologies has made possible a new usage of cheap co-monomers (such as styrene, norbornene and carbon monoxide), higher catalyst activities (i.e. lower

Table 10.6 Technological challenges and threats

Opportunities:

- Replacement of traditional materials by plastics
- Materials with less environmental impact
- Barrier and hot-fill packaging technologies
- Shorter business lifetimes
- Postconsumer plastics recycling
- Innovations (e.g. ease of processing)
- Miniaturisation and user-friendliness (E&E)
- Use of SFs in polymer processing
- Nanotechnology
- Replacement of obsolete technology

Threats:

- Degrading of industrial scientific organisations
- Trimming of corporate, long-term research
- No time for innovation (<10 % of researcher's time[a])
- R&D partnering
- Recruitment
- Interdisciplinary polymer science education

[a] O'Rourke and Carroll [8].

catalyst residues), new process/comonomer combinations, and a tremendous broadening of the product range capabilities of all polyolefin processes (high pressure, solution, slurry and gas phase) [6]. The very specific nature of SSCs provides close control over polymerisation, leading to a very narrow MWD and compositional distribution, and virtually eliminates the soluble low-MW by-product fraction of polymers. Metallocene technology for polyolefins affects additive packages for antioxidants, antiblock/slip agents, nucleating agents, processing aids and lubricants, and leads to new problems in polymer/additive analysis. Late-transition-metal catalysts (LTMCs) are another revolutionary development.

Significant progress has been made in adapting the properties of *engineering plastics* by modifying them with additives, fillers, reinforcements, impact modifiers, processing aids, etc. Industrial efforts have been directed to styrenics (ABS/SAN, SMA), engineering plastics for general use (PA, PET, PBT, PTT, PC, POM, MPPO, etc.), high-performance EPs (PPS, PSU, PES, PEI, PAR, fluorurated and silicon-containing polymers) and advanced super-polymers (aramids, PI, PAI, PEEK, PEK, PPSU, LCP). EP business will show more and more bulk character; additive knowhow will further spread over this business column. A few blend systems are outstanding, such as ABS/PC, ABS/PA, ABS/PVC, PC/PET, PC/PBT, PPO/PS, PPO/PA and PSU/ABS. *Polymer blends* now constitute a significant fraction of the commercially available polymer products in the marketplace. Some are based on miscibility between certain components (e.g. PPO/PS), but most have multiple phases (e.g. PC/ABS). As the demands on the performance of reinforced plastics increase, a need for high-strength fibres is becoming more evident. As for composites, which offer high stiffness and strength, low weight and excellent chemical and corrosion resistance, about 70 % are thermoset – the remaining being thermoplastic. It is expected that the trend will reverse in the near future, in view of the recyclable properties of the latter. Higher tensile strength, tougher, more abrasion-resistant composites are now available.

In order to sustain solid growth in the new market environment, three factors are considered to be key [7]:

(i) application of more efficient processes to monomer and polymer synthesis;
(ii) modified functional polymers for novel market-driven applications; and
(iii) reuse of polymeric materials (intelligent polymer recycling).

The EU Directive on Packaging and Packaging Waste sets a recovery target for used packaging of 25 % by 2001 and 50–65 % by 2005. It is likely that recycling of postconsumer plastics will require substantial polymer/additive analyses.

'Hot' traditional market segments are currently the automotive sector (nanocomposites), beer bottles, and packaging (smart, functional, strong). The plastics sector is shifting emphasis away from commodity products to applications such as ICT, energy supply and biomaterials. High-technology industries (aerospace, microelectronics, biomedical, membranes, etc.) are the driving force for the development of new polymeric systems combining specific functional properties with thermal stability. High-performance polymers in advanced technologies comprise high-modulus/high-strength temperature-stable polymers; high-temperature polymer blend materials for micro- and nanoelectronics; polymers with special architectures (block-copolymers, star polymers, hyperbranched polymers); high-temperature membranes; composites for aviation and space; and chemical modifications of high-performance polymers. Whereas purity is a prerequisite for microelectronics, optoelectronics and biomedical applications, numerous other high-tech users require modified polymers that can withstand a broad range of processing and application conditions. In-service reliability is a key differentiating factor for new materials or new components. The market requires robust products. The selection of additives is critical, and is often a proprietary form of knowledge. Especially in small niche markets the use of additives can provide a cheaper and effective solution to achieving the necessary chemical and physical property improvements. It is general expectation that R&D activities will focus on new processes, new catalysts, and reduction of product portfolio complexity. Product innovations, such as metallocene polyolefins, will be more exceptional in the future. Possible exceptions might be found in the development of environmentally degradable polymers (EDPs). Nanocomposite technology is expected to lead to new commercial applications, in the automotive industry (structural applications), medical/biomedical, E&E (flame retardancy), packaging (barrier properties), coatings, and more. For successful development of nanocomposites, it is necessary to be able to properly characterise the materials, understand the level of compatibility required between the clay/tube and polymer, and produce consistent nanocomposite repeatability.

Massive *material substitutions* have taken place, e.g. starting with the replacement of flax and other cellulosics for synthetic noncellulosic textile fibres since the 1940s, followed by polymeric fibres (aliphatic polyamides) with special properties (stiffness and heat resistance) in the 1950s and 1960s. Material substitution is not only a matter of science and technology, but can also have political origin. The addition of desirable properties such as flame retardancy, resistance to corrosion and mechanical resistance, high- and low-temperature resistance, biocompatibility, differential permeability (to gases or liquids) and specific electrical properties to polymeric materials has opened new frontiers. Polymer-blend technology and mineral- and glass-fibre reinforcement technology have brought polymer materials into automotive and aerospace applications (e.g. high-impact polyamide and PP blends). The last decade or so has witnessed the commercial introduction of a whole new group of polyolefins, called plastomers. Distinctions between plastic and rubber materials have become blurred with the wide range of polymers and blends using polyolefin technology. Other examples of newly introduced materials are Shell's aliphatic polyketone Carilon (later withdrawn) and Corterra polyester (polytrimethylene terephthalate, PTT) and Showa–Denko's poly-*N*-vinylacetamide.

Obviously, there exists severe *interplastics competition*, e.g. PP vs. ABS, clarified PP vs. PS, PA, PVC, HDPE and PS (Table 10.7). A wide range of cross-linked and thermoplastic elastomer applications, from footwear to automotive parts and toothbrushes, are adopting new metallocene-catalysed polyolefin elastomers (POEs). These low-density copolymers of ethylene and octene were first accepted as impact modifiers for TPOs, but now displace EPDM, (foamed) EVA, flexible PVC, and olefinic thermoplastic vulcanisates (TPVs). Interpolymer competition may also result from

Table 10.7 Intermaterial competition

Traditional materials	New materials
Wood, glass, metal, etc.	Polymers, LGFPP
Cost-effective EPs (ABS, PS, PA)	Polyolefins
Flexible PVC	m-PE plastomers, TPE
Elastomers	m-PO
PS (packaging applications)	m-PP
PVC (bottles)	PET
ASA	Hivalloy PP resins (UV resistance)
PC (aircraft interiors)	PEI, PES, PEEK (inherent flame retardancy)

improvements in additive technology. Developments in the past two decades have demonstrated continuous improvement in processability and performance of filled polyolefin compounds, pushing out quite a few competitive exotic polymers. An example is the fire-retardant formulation of PP (with high light stability, absence of blooming and good processability) for high-performance applications (appearance parts, fibres) replacing FR-ABS [9]. Such reformulations redefine the areas of competition for PP versus other polymers. Future materials will be characterised by controlled complexity in order to achieve their functionality.

The ability to downgauge, decrease part weight, improve barrier properties and reach new levels of product performance are propelling polyolefins into new markets previously dominated by other plastics. The high growth rate in PP production capacity is mainly being driven by the ability of PP to replace other resins on a cost/performance basis. For example, functionalisation of PP by incorporation of acrylic functionality has extended its weatherability performance. Interpolymer competition will have a significant impact on the amount and type of additives used.

10.2 TRENDS IN ADDITIVE TECHNOLOGY

Additives are avenues for breaking into new markets and enhancing the value of existing polymers. By definition, chemical additives depend upon plastics for their existence. Factors that affect polymer consumption are absolutely vital considerations for the additive industry. Typically, the new technology of metallocene-based polyolefins has a significant impact on various classes of additives. Recently, both plastics and additives have moved to a new stage in their life-cycles, resulting in many changes. Although rapid growth is over, polymer additives still present an 'above-average' growth potential, despite intensified global competition. In both industries there is greater emphasis on cost, and in some cases demand on increased service. Investments are shifted from new products to process development. Fewer new products are being introduced, and those that are introduced generally have more of an evolutionary than a revolutionary character. The use of additives can provide a cheap and effective solution to new material development, in particular in niche markets. Even if the number of additives is only increasing slowly, the number of *possible combinations* of existing products is enormous, so that sophisticated solutions to problems are still always possible. There are now far more categories of additives than there were 25 years ago (Appendix II). The chemistry of additives is often extremely complex, and their choice can be bewildering. Murphy [10] has recently reviewed new developments in additives.

After a decade of sluggish R&D in the 1980s, the *additive business* in the 1990s was characterised by an upswing. Health and environmental concerns, and the resulting regulations and restrictions, combined with the processors' demands for cost-effective solutions and more demanding applications were important driving forces for innovation by additives suppliers. All additives are required to perform more efficiently at ever lower loadings and to perform and remain stable over ever wider processing and application conditions. With better understanding of the mechanism by which the specific functions are achieved, products have been steadily refined, so that today the actual amount of additive used in plastics has been reduced, in some cases, to fractions of a percentage. In recent years, developments in additives have been aimed at safer use, minimising dose level, prolonging stabilising action, preventing migration, and minimising undesirable colours of the polymer.

Other possible incentives for additive development are problems or deficiencies with existing additives and the need to find a use for a by-product or expand the use of a product. For example, packaging materials may be extended with temperature indicators, O_2-absorbers or food additives. Decreased volatility, increased heat stability, and increasing environmental considerations during processing and in use, are other goals of new product developments. These combined efforts result in many new product offerings. Only large producers are able to afford the R&D associated with testing the new additives.

The US$16 billion plastics additives industry (2000) is expected to increase from a volume of 17 billion lbs (1998) to 21 billion lbs by 2004 [11]. The plastics additives market is particularly dynamic in China (from

Table 10.8   Drivers for plastic additive growth

Positive:

- Resin growth (volume)
- Emerging economies
- Resin diversification (catalyst technology)
- New applications (e.g. demand for fluorescent, pearlescent and other brilliant pigments; introduction of photoconductive elements in polymer-based electro-optical devices)
- Improved tailor making (e.g. separation between a functional and compatible part)
- Multifunctional additives
- Adding functionality to (packaging) materials
- Cost reduction (in specialties industry)

Negative:

- Formulation optimisation (economy of use)
- Increased effectiveness
- Cost reduction (in polymer industry)
- New concepts (e.g. unconventional additive-free problem solving)
- Less antagonism (by separation of additives)

Neutral:

- Resin technology
- Interpolymer competition
- Customer-oriented specialty product forms
- Environmental/regulatory changes

Table 10.9   Apparent supplier strategies for chemical additives for plastics

• Structural changes ('urge to merge')	• Additive packages (materials handling)
• New pricing strategies	• Blending (on specification)
• Cost improvements	• New supply sourcing
• Globalisation (global plastic customers)	• Internet trading sites
• End market focus	
	• Environmental/regulatory compliance
• Core competences	• Shareholder value
• Product equivalents	• Consultancy (to users)

US$2.4 billion in 2003 to US$3.5 billion in 2007) [11a]. Factors affecting plastic additive growth are given in Table 10.8, and supplier strategies in Table 10.9. Demand for additives will become more standardised. Consolidation, commoditisation, cost demands and service requirements push polymer and additive producers towards a global presence. Lower growth (related to polymer growth) and lower volume due to inherently higher-performing polymer structures are compensated by the need for additives for tailored polymer properties and consultancy to the polymer industry regarding effective additive packages. Four additive classes (antioxidants, flame retardants, light stabilisers and impact modifiers) constitute over 80 % of the materials used in higher growth plastics. Environmental and regulatory pressure may be summarised by a few keywords: product forms (pellets, etc.), ecolabels ('Blue Angel', 'White Swan'), recycling and biodegradability. Additive manufacturers are faced with the imperative to develop products that not only enhance the overall processability and functionality of the plastic and rubber end products, but satisfy environmental and health standards as well. Governmental regulations are both a threat and an opportunity.

Intermaterial competition affects the additive business. For example, if metallocene polyolefins displace PVC markets, then the demand for heat stabilisers and plasticisers might decrease, while the need for products such as fluoropolymer processing aids, antiblocks and slip additives might increase. At present, PVC as the main user of additives is under threat as never before. However, 'green' PVC, containing more environmentally friendly FRs and plasticisers as well as non-heavy-metal stabilisers, is now environmentally acceptable.

The additive business is characterised by rapid change, competitive pressures, and continuous developments in *technology* (Table 10.10). Recently, fragrance has been added to masterbatch technology.

Table 10.10 Technical trends in polymer additives

Opportunities:

- Better dosage of traditional grades (pellets vs. powder)
- One-pack systems
- Better specs
- Reduction in concentration variability
- New product developments
- Additives with new functions (e.g. UVAs as FRs, or FRs as thermal stabilisers)
- Additives operating at an earlier stage during polymer degradation
- Multifunctional formulations
- Increased interaction between additive and polymer
- Functionalisation of polymer backbone (anchoring)
- Polymeric additives, polymer-bound (grafted) additives
- Environmentally friendly additives
- Plastic waste management
- Nanocomposites (for polar thermoplastics)
- Reactive additives
- Intelligent additives (go where action is needed)
- Surface HALS
- Avoidance of antagonism between additives
- Patent expirations
- Lower analytical costs

Threats:

- Pressure on brominated flame retardants (Europe)
- Miniaturisation
- New polymer developments
- e-beam processing

Table 10.11 Technical trends for some additive classes

Fillers:	*Plasticisers:*
• Surface modifications	• Replacement of plasticised PVC
• Nanoclays	• Processing aids
• Surface properties (scratch, . . .) of polymer parts	• Food-contact and healthcare applications
Pigments and colorants:	*Heat/UV stabilisers:*
• New pigments	• Synergy
• Pigment/polymer interactions	• New environment (radiation, . . .)
• Novel effects	• Food-contact and healthcare applications

Fragrance encapsulated in a polyurethane microcapsule can be applied to a fabric by standard textile finishing processes. Target markets for scent-enduring formulations include automotive interiors and packaging for health and beauty aids and cosmetics. Hyperbranched polymers, as multifunctional nanocarriers, may be used for performing boosting applications in paper-coating formulations and textile dyeability [12,13]. In new materials science concepts, the role of additives is high (e.g. in thermoplastic elastomers). The technological trends and outlooks differ for various additive classes (Table 10.11), see ref. [14].

The trend continues toward use of synthetic or vegetable-oil based rather than animal-fat derived additives, as a result of concern about BSE. Some companies also pursue this interest for kosher-certified applications. There is a growing trend in additives technology to develop formulations specifically designed for a particular performance, often to solve narrowly focused problems. Additives for plastics are being used more aggressively. Not only is there now a greater selection among existing alternatives, there are also more ways to extend the capabilities of additives beyond their traditional users. Additives are increasingly expected to expand their functionality, e.g. they should not only provide a protective barrier for the packaged product, but must also be involved in the total design of both the package and its contents.

As argued above, it is not reasonable to expect that new polymer developments will easily stay away from the use of additives. In this respect e-beam processing of plastics constitutes an alternative to chemical additives, but is very limited in scope [15]. Table 10.12 gives the current status for several additive groups.

10.2.1 Advances in Additives

It is beyond the scope of this book to report extensively on recent advances in additives. The examples given merely serve to trigger the analyst to keep abreast

Table 10.12   Additive group status

- *Impact modifiers:*

  Products: Acrylic, EPDM/EPR, MBS/ABS/MABS
  Driving forces: Growth in PVC consumption (construction applications)
  Threats: Negative sentiments PVC

- *Organic peroxides:*

  Products: Wide variety
  Driving forces: Growth of resin production (PE, PVC, thermosets)
  Threats: New technology (metallocenes) displacing (reducing) needs

- *Plasticisers:*

  Products: Mainly phthalates and trimellitates
  Driving forces: Growth in PVC consumption
  Threats: Interpolymer competition (PVC replacement)

- *Antioxidants:*

  Products: Phenolic primary AOs, organophosphite secondary AOs
  Driving forces: Global production of polyolefins
  Threats: None

- *Light stabilisers:*

  Products: Hindered amine technology, benzotriazoles, benzophenones
  Driving forces: Consumer demand; most resin systems
  Threats: Reducing volume, surface coating, higher-performing resins

- *Heat stabilisers:*

  Products: Organotin, mixed metal, lead-based
  Driving forces: Growth in PVC consumption (construction applications)
  Threats: PVC image

- *Flame retardants:*

  Products: Brominated and phosphorous-based products, specialty inorganics
  Driving forces: Demands for safer end-use
  Threats: Government regulations; shift to alternative chemistries or resins

of new developments in order to be well equipped in providing solutions for increasingly sophisticated analytical problems. Developments in *stabiliser technology* have been significant in recent decades. New developments aim at improvements such as better maintenance of melt flow-rates, lower initial colour and better colour maintenance, higher processing temperature performance, enhanced long-term thermal and light stability, inhibition of gas fade discoloration, enhanced additive compatibility, resistance to interactions with other additives, reduced taste and odour, and the suppression of visual imperfections (gels), more product forms and better price/performance ratio [16]. For UV stabilisers noticeable developments are:

(i)   UV stabilisation of pre-coloured exterior body parts (instead of painting) [17];
(ii) increase in HALS, decrease in nickel stabilisers;
(iii) new synergists for HALS; and
(iv) more product forms (blends).

Most of the development in the additives field in the last 20 years came from the evolution of the so-called

'secondary structure' of additives, i.e. development of stabilisers with higher-MW, better compatibility, lower volatility, etc., i.e. all the properties directly connected with the physical aspects of polymer stabilisation. Some new trends in polymer stabilisation are given in Table 10.13. With increasingly demanding market requirements, selection of the stabilisation package becomes more critical.

Table 10.13   New trends in polymer stabilisation

Chemistry	Advances
Hydrolytically stable phosphates	Process stabilisation, colour retention
Lactone chemistry	Improved melt processing, colour improvers
Hydroxylamines	Reducing phenolic discoloration; alternative to traditional phenolic AO based stabilisation
NOR HALS	Reduced co-additive interactions; acidic polymer applications; flame retardancy

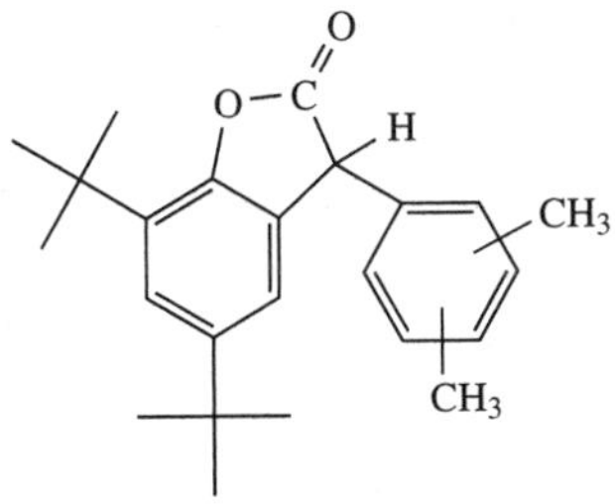

Figure 10.1 Chemical structure of Lactone HP-136 (Ciba Specialty Chemicals)

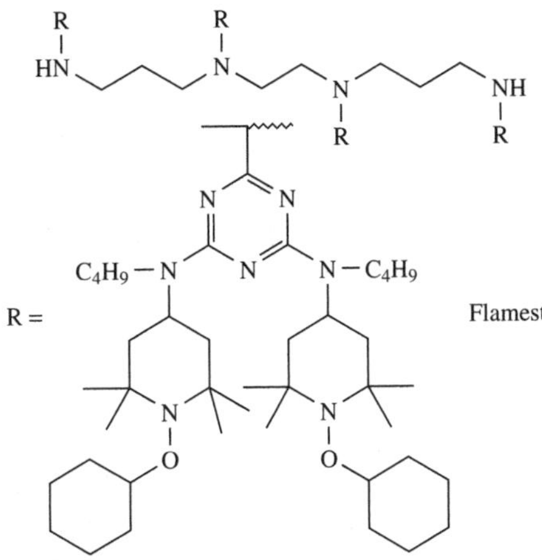

Figure 10.2 Chemical structure of a UV-stable nonhalogenated *N*-alkoxy-hindered amine flame-retardant additive (Flamestab NOR 116) for polyolefins (Ciba Specialty Chemicals)

New developments are hydroxylamines and lactones (for processing stability), which operate at an earlier stage during stabilisation. Lactone (benzofuranone) chemistry has been identified as commercially viable, and marks a revolutionary advance in comparison to hindered phenols and phosphites [18]. New lactone chemistry (Figure 10.1) provides enhanced additive compatibility, reduced taste and odour (organoleptics), resistance to irradiation-induced oxidation, and inhibition of gas fade discoloration. The commercial introduction of fundamentally new types of stabilisers for commodity and engineering polymers is not expected in the near future.

The evolution of HALS technology to meet the requirements of emerging polyolefin markets and applications is a story of continuous improvement and structure/property optimisation [19–21]. Current trends in the industry are toward low-volatility, extraction-resistant, hindered phenolic AOs, such as oligomeric polysiloxane based, high-MW antioxidants

(Figure 3.22) [22]. Another typical example of a new, evolutionary high-MW hindered amine stabiliser (HMW HAS), with new levels of light stability and reduced interaction with pigments, is Chimassorb 2020 (Fig. 3.25e).

Advanced extraction-resistant long-term stabilisers were described by Schmutz *et al.* [23] and some hydrolysis-resistant high-performance phosphite processing stabilisers (Cyanox 2704, Doverphos S-9228) were introduced. Other new developments concern alkoxyamine HALS (NOR-HALS), such as the low-MW (liquid) Tinuvin 123-S (Ciba Specialty Chemicals) and Amfine's polymeric LA-900, both low-basicity amines designed to resist performance degradation in acidic environments, e.g. in agricultural films or applications containing halogenated FRs. Similarly, Cyasorb UV6435 (Cytec Industries) offers enhanced resistance to agricultural chemicals. Recent innovations in light stabilisation have broadened the palette of colours available in various applications (e.g. in plastic shutters). A great variety of other stabilisers has recently been introduced, such as the hydroxyphenyl benzotriazole (BZT) UVA Cyasorb UV5326 for indirect food contact, and novel light stabilisers of the di-*t*-butyl-hydroxybenzoate class (dual role as thermal and light stabiliser), both Cytec Industries. Ethanox 330 antioxidant (Albemarle) is the only high-MW phenolic antioxidant sanctioned by FDA for indirect food contact use in all polymers. Recently, other new classes of antioxidants (aminoglycoside and β-lactam antibiotics) for POs and PVC were described [24,25]. GLCC has introduced an expanded range of BZT UV stabilisers [26], as well as nonvolatile but mobile silicon-based HALS (Uvasils).

Other activities concern additives for *recyclates* [27]. Recyclates behave differently from virgin material. Damage, such as chemical changes, carbonyl groups, hydroperoxides, double bonds and discoloration, is introduced into the recyclate during processing and the lifetime of the plastic article, and the fluctuating composition of recyclates including the presence of contaminants (other polymers, metal traces, contact media, (in)organic impurities) is responsible for a modified degradation and processing behaviour. Compositional analysis, including additive analysis, will be necessary.

Continuous *technological improvement* can further be illustrated by the case of rigid PVC applications with a succession of technologies based on tin mercaptides, mixed Ba/Cd carboxylates, lead phosphite, Ba/Zn carboxylates, Ca/Zn carboxylates and, finally, tin maleates. Lead-free PVC stabilisers have assumed higher significance through increasing ecological awareness [28]. Metal-free thermal stabilisers for PVC have been developed, such as pyrimidinediones [29], organic thioles [30] and polyols [31,32] and Mg,Al hydrotalcite [33]; see also ref. [34]. New trends in polymer stabilisation were reviewed recently [35,36].

Flame retardancy is another active field of product innovation. Some 150 to 200 flame retardants (FRs) cover most of the requirements of the market (1997 value: US$2.2 billion or 14% of the plastics additives market). The markets for *flame retardants* are growing relative to average industrial markets (Table 1.8), with mineral flame retardants growing twice as fast as other FRs. Because the range of applications is diverse, so are the types of FRs; end-product specifiers are the major driving force behind innovation. The requirements of a fibre manufacturer are particularly demanding [9], now being met by Reoflam® FG-372 (FMC). Other industry drivers here are the gradual phasing out of brominated materials, coupled with increased demand for halogen-free systems. In addition, more stringent fire regulations require systems that offer a slower heat-release rate in fires, along with low smoke generation, low FR additive toxicity, smoke toxicity and corrosion. In the fire safety area, the national regulatory activities in the EC are uncoordinated, causing increasing disharmony [37]. A Dutch ban on TBBP-A, bis(2,3-dibromopropylether) (FR-720), is imminent. Some polybromobiphenyls (penta, octa) are no longer being produced (end 2004) or are replaced (e.g. Firemaster 550). Pentabromodiphenyl oxide also is proposed for a ban in Europe in 2004. The controversy over the use of BFRs has driven business-machine OEMs serving global markets to favour non-controversial FR systems. In some countries, brominated materials are still widely used in TV manufacturing. New brominated FRs continue to be introduced, e.g. tris(tribromophenyl)cyanurate (FR-245) for styrene copolymers and tris(tribromoneopentyl)phosphate (Dead Sea Bromine) aimed at use with PP, and the brominated polystyrene Saytex HP-7010 (Albemarle) for PBT, PET and polyamides [38].

The threatened bans on halogen FRs have not materialised in North America. Still, the nonhalogens are making inroads into selected areas [39]. In Europe, nonhalogen FRs are thriving, and the halogen-based additives are under increasing restrictions because of environmental perceptions, corrosiveness and toxicity of the smoke produced. The technological challenge is to achieve acceptable fire retardancy with nonhalogen materials without negatively affecting the mechanical and physical properties of the compound. Search for nonhalogen FRs has been successful in several polymer systems, e.g. Clariant's APP-based Exolit AP

and Bayer's DMPP-based Levagard VP SP 51009 flame retardants. The serendipitously discovered novel non-heavy-metal, nonhalogenated FR synergists are based on *N*-alkoxy amine (NOR) chemistry [40,41] (Figure 10.2).

Other recently reported systems are the nonhalogen phosphorous flame retardant FP-500 (Amfine) for PC/ABS and PPO blends, the reactive phosphite Struktol PD3710 (Schill & Seilacher) for epoxies [42], and nonhalogen, nonphosphorous FRs for the cable industry. Mechanistic studies on polymer combustion and fire retardance are necessary to develop nontoxic, environmentally friendly, new FRs (e.g. intumescent systems) with high effectiveness. On the other hand, high-throughput approaches have been proposed [43]. Halogenated materials, usually in combination with antimony oxide, function by quenching radical reactions in the gaseous phase. Hydrated fillers, like $Al(OH)_3$ and $Mg(OH)_2$ simply release water on heating. Intumescent FR systems (usually phosphate-based), which are particularly suited to olefinic polymers, function by developing an expanded char layer on the surface of the material exposed to heat. Halogen-free FR additives based on phosphorous function by developing a protective char. Environmentally safe fire-proofing agents have recently focused on (in)organic Si compounds, nanocomposites, borates, hydroxides and melamine derivatives. Development of FRs is tending towards synergistic systems, in which several means of retardance are used, e.g. combinations of phosphorous with halogens. Flame-retardant one-pack systems are now on the market (e.g. Fyrebloc, GLCC). Also in this field, interadditive competition (e.g. between phosphorous, nitrogen, silicon and metal oxide/hydroxide FR systems) is both an opportunity and a threat.

Use of higher-performance polymers may suppress the demand for FRs in business machines. Flame retardancy has recently been reviewed [39], as well as the technology of halogen-free FR additives [44]. The future technology of polymer flame retardancy has been described [45].

Technological advances are also being reported for other additives. For *plasticisers*, migration is a main point of concern, with the result that certain phthalate types have been gradually phased out and replaced by high-MW polymeric types, such as ethylene copolymers. Mesamoll (Bayer) is a phthalate-free plasticiser. PVC is being reformulated from (DOP, Pb, Sb_2O_3) to (no DOP, no Pb, ATH/MDH, low Sb_2O_3) [46]. Although the literature occasionally reports the evaluation of novel materials as external plasticisers for PVC, few if any of these are likely to be commercialised. The costs of the obligatory tests for HSE effects now constitute a large additional development cost.

Replacement of heavy-metal-based *pigments* is really possible via organic pigments for the coloration of polyolefins [47]. Some recent advancements are based on high-dispersibility quinacridone, diketopyrrolopyrrole (DPP), anthraquinone, isoindolinone, benzimidazolone, etc., and liquid colour concentrates. Recently introduced new pigments are Ciba Irgacolor Yellow 2 GTF and Ciba Cromophtal Yellow 3 RLR. Typical inorganic powder pigments are, for example, bismuth vanadate, Cr/Sb/Ti oxide, Mn/Sb/Ti oxide, Cr/Fe/Zn oxide, Co/Al oxide, etc.

In *filler* design, the control of particle size and particle-size distribution (from micro to meso and nano) is increasingly exploited for the preparation of polymer composites. Incentives are the increase of interface area, and hence, the volume fraction of the polymer–filler interfacial zone, and better homogeneity of the particle distribution. Other advancements consist of *in situ* generation of filler particles by hydrolysis of inorganic/organic hybrid precursors [48,49], surface modification, encapsulation, multifunctional fillers and improved compounding technology. Current materials of interest are engineered mineral fibres, graphite nanotubes/nanoplatelets/nanofibres and conductive fillers.

Other new product developments have concerned *peroxide initiators* for unsaturated polyesters [50]. Chlorinated fluorocarbons have been virtually removed from mould release agents and rigid polyurethane insulation foams. Fluoropolymer-based *processing aids* (such as vinylidene/hexafluoropropylene copolymer with/without polyethylene oxide) have been introduced [51]. Some recent developments also include a permanent (nonmigratory) *antistat* (Irgastat P22) and very efficient antifog additives (Atmer 651, Atmer 691). Polymeric antistatic agents and their utilisation were recently reviewed [52]. Millad 3988 (Milliken Chemical) is a new nucleating and *clarifying agent* for PP. Recently, new very effective *biocides* (e.g. Irgosan PA) were also introduced. New *dispersion aids* for pigment concentrates (Ceridust-waxes, Clariant) were commercialised, see also ref. [53].

Many new products tend to be reactive rather than additive, are polymeric, and are more precisely designed to meet the requirements of the polymer manufacturer or formulator. Low-MW additives face two major problems:

(i) evaporation during high-temperature moulding and extrusion; or
(ii) migration to the surface and extraction.

Especially in recent years, there has been a tendency to develop stabilisers with higher-MW (>2000 Da) to prevent loss under severe conditions of application. *Polymeric additives* for polymers, including impact modifiers, flexibilisers, antistatic agents, and processing aids, have been reviewed [54].

In another approach, additive functionality is attached to the polymer. This may be achieved by:

(i) specific polymer design;
(ii) random co-polymers;
(iii) functionalisation;
(iv) grafting; or
(v) reactive processing.

Whatever the approach, the result is a difficult-to-analyse system. Such options suit polymer producers better than additive suppliers. Aromatic polymers (PPO) have been mentioned as char-forming FRs. *Polymeric UV absorbers*, blended in proper proportions with commercial plastics, have potential use as stabilisers for fibres, films and coatings. Several monomeric stabilisers containing a vinyl group were homopolymerised and used as stabilisers for PE, PVC, acrylates, polystyrene, cellulose acetate and several vinyl polymers [55].

The use of copolymers is essentially a new concept free from low-MW additives. However, a *random copolymer*, which includes additive functions in the chain, usually results in a relatively costly solution; yet industrial examples have been reported (Borealis, Union Carbide). Locking a flame-retardant function into the polymer backbone prevents migration. Organophosphorous functionalities have been incorporated in polyamide backbones to modify thermal behaviour [56]. The materials have potential for use as fire-retardant materials and as high-MW fire-retardant additives for commercially available polymers. The current drive for incorporation of FR functionality *within* a given polymer, either by blending or copolymerisation, reduces the risk of evolution of toxic species within the smoke of burning materials [57]. Also, a UVA moiety has been introduced in the polymer backbone as one of the co-monomers (e.g. 2,4-dihydroxybenzophenone-formaldehyde resin, DHBF).

Great Lakes has reported that *functionalisation* with graftable moieties results in a product which can be chemically bound to a polysiloxane backbone, e.g. Silanox MD. Functionalisation of polysiloxanes with HALS (polymer-bound HALS, P-HALS) and phenolic antioxidants has been described [22]. Functionalised polysiloxanes (Figure 3.23) exhibit high stabilisation activity in critical applications such as PP fibres and PE cables [58].

Grafting may be achieved by using chemical reactions (e.g. decomposition of azo groups, chain transfer initiation, redox methods, etc.), photo-induction, γ radiation-induction or via the so-called xanthate method [59]. Grafting reactions alter the physical and mechanical properties of the polymer used as a substrate. Grafting differs from normal chemical modification (e.g. functionalisation of polymers) in the possibility of tailoring material properties to a specific end use. Photostabilisation of PE and PP can be achieved as a result of the grafting of 2-hydroxy-4-(3-methacryloxy-2-hydroxy-propoxy) benzophenone using γ radiation [60]. The 2-(2-hydroxyvinylphenyl)2-benzotriazole has been used as a grafting monomer for atactic PP, E-P copolymers, ethylene/vinyl acetate, PMA, and PMMA. Clariant has introduced a reactive HALS (PR-31), which can be grafted in sunlight on the surface of polyolefins after migration from the bulk (surface-bonded additive functionality). This 'chemical anchoring' alleviates blooming and plate-out in processing, and gives a better performance than oligomeric HALS at equal loadings [61]. Reactive impact modifiers can be chemically grafted on to the matrix polymer. Grafting of plasticising side groups on to PVC is a more novel route to plasticisation, which avoids the limitations related to plasticiser migration [62]. Polyurethane-bound stabilisers have been described [63]. There has been comparatively little commercialisation of the grafting process. Reasons are partly economic and partly technical, namely, the concurrent formation of homopolymers in most cases and the lack of reproducibility in largely heterogeneous reactions [64]. In addition, there is the difficulty of controlling the grafted side chains in molecular weight distribution. The graft products are usually characterised by a variety of methods (gravimetry, IR, thermal analysis, pyrolysis, SEM, swelling measurements, determination of MW and MWD, dielectric relaxation, etc.) [64].

The advantages in adopting a *reactive processing* approach to polymer stabilisation include high retention of antioxidants in the polymer matrix (no volatilisation) leading to efficient stabilisation; reduced risk of migration of AOs into the human environment; cost effectiveness and attractiveness of the option of producing highly modified concentrates for use as conventional additives in the same or other polymers. Site-specific AO grafts can be targeted for premium performance of speciality niche products. The trend toward reactive products is also noticed for impact modifiers, antistats and flame retardants. As to the latter, emphasis is both on char-forming polymers, intumescent additives and monomeric reactive FRs, which are bound chemically

Table 10.14 Multifunctional additives for polymers

Multiple functionalities		
Flame retardant	Smoke suppressor	
Flame retardant	Plasticiser	
Flame retardant	Lubricant	
Flame retardant	Thermal stabiliser	
Flame retardant	Thermal stabiliser	Light stabiliser
Plasticiser	Thermal stabiliser	
UV absorber	Thermal stabiliser	
UV absorber	Antioxidant	
UV stabiliser	Antioxidant	Processing aid
UV stabiliser	Flame retardant synergist	
UV stabiliser	Flame retardant	Smoke suppressor
Antioxidant	Thermal stabiliser	
Antioxidant	Thermal stabiliser	Thickening agent
Antioxidant	Plasticiser	
Melt stabiliser	Light stabiliser	
Nucleating agent	Optical whitener	
Slip agent	Antiblocking agent	
Pigment	Filler	
Filler	Nucleating agent	
Metal deactivator	Antioxidant	Flame retardant
Antiplasticiser	Processing aid	
Compatibiliser	Carrier resin	

with the matrix polymer to prevent leaching, volatilisation and blooming. Development of reactive systems may well be restricted to niche markets.

Some of the latest developments centre around the synergy which can be obtained by using particular additives in combination, and focus on the multifunctional properties of certain compounds [65]. There are no definite dividing lines among the polymer additives of various types. Plasticisers may serve another function as lubricant, processing aid, impact modifier, thermal stabiliser, or flame retardant. For example, triphenylphosphate acts as a plasticiser and flame retardant in HIPS/PC blends [66]. *Multifunctional additives* are recently being introduced, (Table 10.14), e.g. Nylostab S-EED (melt, light and long-term heat stabiliser, fibre dyeability) in polyamide fibre manufacture, and Lowinox MD 24 (metal deactivator and antioxidant). Liu *et al.* [67] have reported a series of bifunctional stabilisers (UV-HALS, UV-AO and HALS-AO) containing *ortho*-hydroxy benzophenone, hindered amine or hindered phenol groups, which are effective in protecting PP against photo-oxidation and thermal oxidation. Other examples are hindered amine thermal stabilisers (HATS), such as Hostavin N-30, which serves both as an antioxidant and a UV stabiliser. Also, other multifunctional additives provide UV and thermal stability to polyolefins. Various other polyfunctional stabilisers for polymers have been mentioned [68,69]. In the stabiliser field, we are still waiting for a product that can provide good UV light stability, thermal stability and process stability simultaneously. Hyperbranched polymers that can act as a nanocarrier for many functional entities offer the prospect of tailor-made and multifunctional performance plastics additives [70]. Multifunctional additives in plastic recycling have been reviewed [71]. Multifunctional additives for rubber compounds have been discussed [72].

Other noteworthy developments are carrier materials, such as Stamypor (DSM) and Accurel (AKZO), for production of concentrates with liquid or low-melting additives and reactants (see Section 1.2.1). The biggest growth area for *additive carriers* is coming from liquid peroxides and silanes, due to related health and safety issues for shopfloor staff. The NOR HALS stabiliser Tinuvin 123-S (a non-interacting, low-MW liquid) for TPO, PP and some blends is delivered in a solid carrier (Accurel).

Emphasis is also being given to on 'green' or eco polymer additives, such as sorbitol, vitamin E, natural fibres, etc. [32,73,74].

Information about the market introduction of new additives is easily accessible by means of various annual reports in *Plastics Engineering* (e.g. refs [75,76]), *Modern Plastics International* [77], *Plastics Additives and Compounding* (e.g. ref. [78]) or otherwise (*Additives for Polymers*, etc.), as well as regular conferences such as AddCon, AddPlast and SPE meetings. For business opportunities for 2002–2006, see ref. [79].

10.3 ENVIRONMENTAL, LEGISLATIVE AND REGULATORY CONSTRAINTS

In this time of rapidly changing social attitudes toward the stewardship of the *environment* and natural resources, health and safety concerns, there is an increased demand for 'safer' plastics and 'more friendly' additives. Increasing hygienic and eco-toxicological concerns comprise health-risk claims, consumer perceptions, 'green' competition, recyclability and workplace safety. A key challenge for the 21st century is decreasing the environmental impact of manufacturing processes. Environmental (and economical) constraints determine a shift away from the use of (large amounts of) solvents, especially halogen-containing ones, see Table 10.15. The Montreal Protocol aims at phasing out chlorofluorocarbons, CH_3CCl_3 and CCl_4 (blowing agents and solvents). Governmental regulations have already coerced many to switch to

Table 10.15 The EPA's voluntary toxics release inventory

- Heavy-metal (Cd, Cr, Pb, Hg, Ni) compounds
- Cyanides
- Benzene, toluene, xylene
- CCl_4, $CHCl_3$, CH_2Cl_2
- Methyl ethyl ketone
- Methyl isobutyl ketone
- Tetrachloroethylene
- Trichloroethylene
- Trichloroethane

solvents of lower toxicity. Examples are supercritical fluid polymerisation media, solvent-less powder coatings, environmentally friendly spray-on-mould release agents, SFE, SFC, etc.

Other environmental issues are directed in particular to heavy metals, halogens, phthalates, volatile organic components (VOCs), and waste disposal. These concerns have been translated into *legislative and regulatory constraints* and high waste-disposal costs. The new European Chemical Policy (REACH) proposes sweeping changes for the regulation and testing of chemicals.

All additives are subject to some form of regulatory control through general health and safety-at-work legislation. Particulate and gaseous emissions (VOCs) are regulated this way. Particular attention has also been given to flame and smoke toxicity. Other wide-ranging legislation covers the use, handling and 'fitness for purpose' of all chemicals, in particular for sensitive uses such as food and water contact, medical uses or toys. EU Directives exist for materials in contact with foodstuffs. The laws and regulations that govern packaging procedures have grown increasingly complex, up to overregulation. The Landfill Directive has implications for the disposal of plastics products, recovery and recycling [80]. Table 10.16 summarises the response of polymer/additive formulators to legislative and commercial pressures.

Table 10.16 Response of polymer/additive formulators to legislative and commercial pressures

Domain	Action(s)
Health	Nontoxic or hazardous ingredients; supply forms
Safety	Reduction or elimination of solvents
Environment	Polymer identification; additive risk assessments
Performance	Multifunctionality; improved physical performance and processability
Production	Additive 'packages', masterbatches, blends (formulations)
Economics	Lower concentrations
Commercial	New product development efforts
Legislation	Additional method development and optimisation
Quality	QA schemes (precision, accuracy)

10.3.1 Trends in Manufacturing, Processing and Formulation

As stressed in Sections 10.1 and 10.2, business concerns in the polymer and additive industries express the need for regulatory compliance, with emphasis on health, safety and environmental (HSE) issues, including postconsumer recycling. Changes in technology are driven partly by the desire to produce plastics which are ever more closely specified for particular purposes. Materials development must take place within an increasingly (and understandably) strict environment which regulates the potential hazards of chemicals in the workplace (production stage); of materials in contact with foodstuffs and users (consumer stage); and (the ultimate criterion) the long-term impact of additives on the environment when the product is recycled or otherwise comes to final disposal (postconsumer stage). Life-cycle and risk assessments may play an important role in the realistic evaluation of product use. Stricter regulatory regimes stimulate additive innovation.

In recent years there has been increasing concern that some of the additives which are apparently most effective in their function might be hazardous to the health of people working with them, or may be pollutants to the environment. The analyst should be aware of toxicity issues connected to polymer additives. Some doubt is being expressed over the long-term viability of some additive classes, in particular halogen-containing flame retardants, plasticisers, as well as heavy metals in lubricants (metallic stearates), pigments and heat and light stabilisers. Although there are hardly any bans on materials, or hard scientific reasons, industry has taken steps forward on the environmental front. An area that has experienced a revolutionary change in the field of plastics has been the replacement of *heavy-metal pigments*, such as cadmium and chromium by organic chemicals. Lead-based pigments are due to follow them. Brilliant plastics materials with high colour saturation are now based on organic pigments, such as Cromophtal DPP Red BP, replacing cadmium reds; 13805 Yellow, 14342 Orange and 15151 Red masterbatches for PE already replace lead chromate based colours. Quinacridone pigments are other organic alternatives for some heavy-metal colours.

A similar situation pertains in the PVC stabiliser sector, where environmental restrictions have acted as an incentive to PVC *reformulation* (replacement of the commonly used liquid and solid Ba/Cd and solid Pb stabiliser systems). Producers of heat stabilisers for PVC nowadays strive to create products that are heavy-metal-free and provide strong weatherability and stain resistance. Cadmium has gone already, and lead, although not under specific legislative pressure, is being replaced in wires and cables and some pipe and profile applications, as calcium/zinc variants have become more versatile and stable. Tin is an alternative to cadmium and lead in rigid PVC applications [81]. For rubbers, development of zinc-free sulfur vulcanisation is an important issue.

Regulations and standards have historically driven the *flame-retardant chemicals* industry. The normal fire-, smoke- and toxicity related standards have been joined by environmental standards caused by the alleged environmental impact of halogens and antimony. Suitable replacements have not been found for these materials in all cases. In the 1990s the major emphasis on FRs has been in developing products to reduce smoke density, toxicity and corrosivity, with the final aim of arriving ultimately at a situation where all the ingredients in a polymer are nontoxic during smoke generation. The FR sector has still to contend with threats from outside bodies over some classes of brominated materials. Restrictions on the use and marketing of certain substances may be foreseen, notably with respect to PBDEs. When tested to the very stringent German Dioxin Ordinance the bromine-containing flame retardant Reoflam® PB-370 (FMC) does not contain or generate any detectable levels of brominated dioxins or dibenzofurans. The claim that it is difficult or impossible to recycle polymers containing BFRs has also been disproven by Reoflam® PB-370, which has been compounded, moulded, reground, and then put through the cycle again (up to 16 times) with no significant changes in melt flow-rate, flame-retardant performance, impact strength or electrical properties [9]. End-of-life recyclability of HIPS and PBT using ethylene 1,2-bis(tetrabromophthalimide) (EBTBP) as the flame retardant has been discussed [82]. Using kneading technology with countercurrent extraction, FRs can be separated from used plastics while maintaining the original physical properties [83].

The past decade has also seen a continued interest in new halogen-free FR additives for polymers, both because of legislative restrictions and on technical grounds. The technology of new ecologically friendly halogen-free flame-retardant additives for polymeric systems is varied, and comprises polymer char formers and low-melting glass systems, etc. [44,84]. It is likely that the most objected-to HFRs will decline in favour of other less challengeable halogen compounds such as higher-MW ones (more difficult to analyse!). Various new halogen-free FRs (arylphosphates and ammonium polyphosphates) have recently been introduced. Although the nonhalogen issue is still very important, halogen compounds are still plentiful. An annotated bibliography on the *halogen controversy* in flame retardancy [85] is available [86]. If halogens constitute an environmental problem, fire certainly is. For the current regulatory status of FRs, see ref. [86a].

There have been suspicions about possible adverse health effects from DOP/DEHP and other *phthalates* for three decades. The issue still persists. Many toy manufacturers have been persuaded to stop using soft PVC in toys. The EU has recently banned DEHP and DBP (as from September 2004) on political rather than scientific grounds.

Workplace safety has been taken care of by the reworking of some classes of additives into more environmentally acceptable forms. Some trends are the increased use of additive concentrates or masterbatches and the replacement of powder versions by uniform pellets or pastilles which release less dust and flow more easily. Moreover, the current move to multicomponent formulations of stabilisers and processing aids in a low- or nondusting product also takes away the risk of operator error, aids quality control, ISO protocols and good housekeeping. An additional benefit is more homogeneous incorporation of the additives in the polymeric matrix.

Suppliers in the plastics additive field have actively pursued actions for new formulations that offer equivalent performance without regulatory threat. Yet, in the future, even greater impact of regulatory agencies on the chemical industry may be expected (e.g. new EU automotive recycling regulations as from 2005). Government regulations as to postconsumer recyclates (PCR), although likely to be sparse in the US, will lead to the need for further technical developments in the field of polymer/additive analysis. The demand for plastics to function under stricter regulatory regimes and in wider or more 'punishing' markets stimulates additive R&D activity. Additional method development and optimisation will lead to more environmentally safe and friendly additives, not unlike the biological antioxidant vitamin E [87], and presumably heavy-metal-free, halogen-free, phthalate-free, with replacement of Sb_2O_3 by Sb_2O_5, etc. Analysts of additive packages will undoubtedly notice these trends.

10.4 ANALYTICAL CONSEQUENCES

Polymer/additive analysis means permanent development: new polymers, new additives and new chemistry, new regulations and specifications, new analytical technologies and understanding. The analytical specialist should be aware of these developments (see also Table 1.15). Progress in polymer/additive analysis is a combination of few instrumental breakthroughs and many evolutions in mature techniques. The rapid development of automated instrumentation over the past 15 years has heralded a renaissance in analytical chemistry, and offers more reliable and rapid forms of analyte detection.

Some of the challenges facing the industrial laboratory are: limited resources, cost containment, productivity, timeliness of test results, chemical safety, spent chemicals disposal, technician capability, analytical capability, disappearing skills, and reliability of test results. The present R&D climate in the chemical industry is one of downsizing at corporate level (lean and mean), erosion of boundaries between basic and applied science, and polymer science and analytical chemistry as 'Cinderella' subjects. Difficult chemical analyses are often run by insufficiently skilled workers (a managerial issue).

Polymer/additive analysis is a typical industrial analytical problem, and indeed not one of the easiest or least important ones. Requirements set to industrial analytical expertise vary from new analytical approaches for product innovation, to service-oriented problem solving (combination of analytical expertise and specific product knowledge), and cost-efficient analysis of a few grades (plant service) (Scheme 10.1). Reported prospects set the instrumental trends in the polymer industry (Table 10.17). For traditional quality laboratories this translates into:

(i)   new laboratory functions;
(ii) at-line, on-line, in-line measurements;
(iii) changes in technology used; and
(iv) computer networking and process integration.

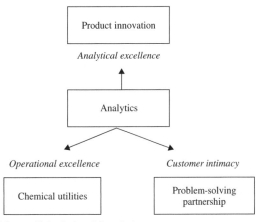

Scheme 10.1 Industrial analytics

Table 10.17 Instrumental trends in polymer industry

- R&D laboratories as 'knowledge' facilitators
- Traditional quality laboratories closer to production
- Need for universal analysis methods and standardisation
- Minimisation of sample preparation
- Shift from traditional testing procedures to push-a-bottom analysers
- Global need for service and support

Far more interesting analyses are being carried out on a daily basis than are ever being published (out of confidentiality or lack of time). Therefore, this book only reflects the accessible part of global efforts.

Industry demands tools that are essentially 'Faster Better Cheaper' as well as:

- more selective, sensitive and rapid (high productivity);
- reliable, verifiable, consistent and producing timely test results;
- robust, reproducible and traceable;
- user-friendly (short method development times; black-boxes);
- allowing fast payback;
- having preset methodology and being standardised;
- being total solutions for different materials; and
- on-line.

General trends in analytical chemistry are given in Table 10.18. The basic needs in polymer/additive analysis were already given in Table 1.10.

Continuous developments in polymer technology (Section 10.1), additive technology (Section 10.2) and legislation (Section 10.3) determine the continuous

Table 10.18 Trends in analytical chemistry

- More complex samples (more interest in 'minor' components)
- More sophisticated questions (low levels, speciation, etc.)
- *In situ* analysis (no separation, 'quick and dirty')
- Simplification of sample preparation
- Shift towards dedicated analytical equipment
- More spectroscopy and sensors
- (Multi)hyphenation
- Space- and time-resolved analyses
- Single molecule/atom detection
- Microanalytics (miniaturisation)
- System integration
- More information per analysis
- High information content (data management)
- Greater role of software
- Reduced total analysis time (operator and instrumental; HTE)
- Improved reproducibility and repeatability
- Validation services

evolution of additive-containing polymeric materials which are offered for analytical scrutiny. Current problems in material science are often complex and require multimethod solutions. Regulations are requiring more, not less, product testing. For complex analytical problem solving, expertise is key; industry needs both all-rounders with product and process knowledge, as well as (a few) specialists with the time to follow external developments and knowhow transfer. It is often difficult to quantify the benefits of a well-staffed analytical laboratory. It is clear that sophisticated technologies are not sufficient by themselves to solve complex problems. At present, little expertise development takes place in industry; instead, key experts monitor new outside developments. Subcritical expertises are shared with third parties or outsourced. For quality assurance, an internal expert is needed.

Today's analytical methods are generally sufficiently selective and sensitive, but industrial analytical chemists are faced with changes in analytical demands:

- sustainable development, quality control, on-line/at-line process analysis;
- fragmentation of applications, specialised applications; and
- high throughput, user-friendliness, on-line and *in situ* monitoring and screening, performance vs. cost of ownership.

Miniaturisation of scientific instruments, following on from size reduction of electronic devices, has recently been hyped up in analytical chemistry (Tables 10.19 and 10.20). Typical examples of miniaturisation in sample preparation techniques are micro liquid–liquid extraction (in-vial extraction), ambient static headspace and disc cartridge SPE, solid-phase microextraction (SPME) and stir bar sorptive extraction (SBSE). A main driving force for miniaturisation is the possibility to use MS detection. Also, standard laboratory instrumentation such as GC, HPLC [88] and MS is being miniaturised. Miniaturisation of the LC system is compulsory, because the pressure to decrease solvent usage continues. Quite obviously, compact detectors, such as ECD, LIF, UV (and preferably also MS), are welcome.

Miniaturisation is advantageous and desirable in many cases where compatibility is an issue (e.g. in

Table 10.19 Miniaturisation in separation sciences

- Column diameter	- Instrument integration (chips)
- Injection system	- Particle size
- Detection system	- Sample size

Table 10.20 Advantages and constraints of miniaturisation

Advantages

- Operation close to theoretical limits
- Heat dissipation (easier temperature control)
- Reduced analysis volumes; minimal peak volumes
- Mass sensitivity
- Higher efficiencies
- Compatibility with detection system
- Energy and solvent management; less waste
- Lower analysis costs
- Portability
- Speed

Constraints

- Operational difficulties (trained analyst)
- Sample capacity
- Robustness (high maintenance)
- Sensitivity (when not sample limited)

multidimensional chromatography). Reduction of peak volumes simplifies transfer to secondary separation systems, but the sensitivity can suffer. The major drawback of these Lilliputian techniques is poor concentration limits of detection. The challenge of using these techniques for solving problems related to concentration limits of detection will require that analysts pay closer attention to all portions of the separation process, namely better sample preparation, preconcentration and sample input techniques; better and more sensitive detection techniques; and techniques to improve the separation selectivity and molecular sensitivity of the molecules with which they are dealing. Miniaturisation in the μTAS direction [89] is not expected to play a role in polymer/additive analysis, which is not a sample-limited problem and where portability for out-of-lab monitoring is not an issue. An important benefit of μTAS would be the ability to completely standardise an analytical protocol on a chip.

Miniaturisation is tied up to *high-speed analysis*. Some potential high sample throughput methods for use in polymer/additive analysis are SPE, HPTLC, SPE-TLC and μSEC-LS. Although high-throughput screening (HTS) is fashionable, it is again not expected to have much impact on polymer/additive problem solving activities where (rightly) no huge amounts of samples are available. Data processing is as much a bottleneck for HTS as sample extraction is for conventional analysis. In both cases automation is needed. Factors that are likely to be more effective in terms of speed in polymer/additive analysis include:

(i) autosamplers (to permit operation when the facility is closed);
(ii) efficient sample handling techniques;

(iii) fast separation configurations (e.g. GC or LC);
(iv) optimisation programmes (to reduce method development time);
(v) separation simulation software (to translate methods between different columns and supports);
(vi) adequate databases; and
(vii) expert systems.

Solvent management is important, not only as a direct cost, but because the use of organic solvents poses a threat to both health and environment. Moreover, concentrating sample extracts and changing solvents in a method sequence are generally time consuming and sources of method variability. This dictates the use of small samples requiring only a few mL of solvent for extraction, and favours the use of SPE for isolation, concentration and solvent exchange.

Breakthroughs in *computerisation* now permit new opportunities for automation, greater reliability, and lowered limits of detection with improved precision and accuracy. Computer-assisted identification of additives included in polymers by combining MS, ^{1}H-NMR, ^{31}P-NMR and LC-MS data has been described [90]. Polymer/additive analysis also faces a growing mass of increasingly complex data. For example, in images, each pixel is characterised by a complete spectrum. Spectra are often measured in two and higher dimensions (e.g. in high-resolution NMR). For the application of computers to R&D of polymer additives, see ref. [91].

In some cases, analytical black boxes are the requirement (e.g. for high-throughput); in other cases hyphenation of analytical black boxes, which generate large amounts of data, inevitably means a move away from the technique specialist to the *data management* specialist. Reliance on automation means that validation is a problem. Polymer/additive analysis is steadily becoming more mathematised. Multivariate approaches are being adopted across a range of fields to reveal sample patterns, hidden phenomena, and variable relationships in seemingly complex data. Chemometrics is used for optimisation of the mobile phase in HPLC, in NIRS [92], in PyGC-MS [93], etc. It is a reasonable expectation that the role of chemometrics will increase. Advanced chromatographic and spectroscopic data handling, network remote control, graphical evaluation, user-definable queries with databases, chromatogram/spectra overlays, custom report generation and laboratory information management systems (LIMS) are high on the list of priorities in the industrial laboratory. Data management faces two interests: for instrument firms, differentiation from the competition is an important driving force; for the chemical industry, software represents a major cost.

Optimisation of software (generic format) would be auspicious, but even no generally accepted norms are available. Specialised MS, FTIR and ToF-SIMS computer-based industrial *additive libraries* are the necessary support for analytical developments. Web-based data mining tools are also starting to become available.

A caveat is that neophytes are so highly focused on instrumental methods and chemometrics that they are likely to neglect basic analytical skills in 'wet laboratory' assays.

10.4.1 General Analytical Tool Development

In the preceding chapters did we succeed in cracking the code of additive packages both qualitatively and quantitatively? Of course, the answer depends greatly on the complexity of the problem, but also on the general approach, which consists of two stages:

(i) isolation of the additive molecules (atoms) from the polymeric matrix; and
(ii) identification and quantification.

Isolation may occur by liquid–solid interaction (extraction, dissolution) or heat (thermal, pyrolytic, laser). Extraction methods easily handle qualitative screening for low- to medium-MW compounds; fail for high-MW components or polymer-bound functionalities; and are less reliable quantitatively (analyte dependent). When applicable, dissolution methods suffer from sensitivity, because of the dilution effect on account of the polymer. In-polymer analysis performs well for qualitative screening, but is as yet not strongly performing for quantitative analysis, except for some specific questions.

In previous chapters the reader has been provided with a guided sightseeing tour through a garden of categorical analytical tools with some being seasoned, withered or even downtrodden, others flourishing in exuberance or under oppression, and a few in bud. Table 10.21 marks the breakthrough of the most important analytical tools for the problem areas that are the subject of this monograph; see also ref. [94] for the history of analytical instrumentation in the 20th century. It may readily be seen that the generally available tool-box is growing steadily; in the past two decades the pace of analytical diversification and innovation has been higher than ever before, not least thanks to cheap computing power. This is even in spite of the fact that software development lags behind hardware development. On the other hand, some analytical tools for polymer/additive

Table 10.21 Milestones of analytical development

Period	Analytical methods that developed during that period
<1900	Chromatography, emission spectroscopy, PAS, DTA, OM
1900s	IR[a], MS, UV microscopy
1910s	TG, XRD
1920s	US, MW, far-IR, RS, SS, AES, near-field microscopy
1930s	NPLC, SSMS, TEM
1940s	MW heating, PC, GC, LC, UV[a], IR (double-beam)[a], NMR, ESR, RS[a], MS[a], PyIR, PyMS, SIMS, ToF, XRF[a], TEM[a]
1950s	GC[a], RPLC, PyGC, FTIR, Michelson's interferometer, ATR, NMR[a], μIR, ESR[a], Mössbauer, QMS, QITMS, ToF-MS[a], SSMS[a], FID, GC-MS, TG-DTA[a], AAS, XPS, EPMA[a]
1960s	TVA, SFC, HS-GC, SEC, HPLC, L, LR, FT-NMR, LR-NMR, CL, CIMS, FD, QMS[a], GC-MS[a], CuPy[a], Py-FIMS, PyGC-MS, TG-MS, LMMS, DSIMS[a], DSC, XPS[a], AES[a], SEM/EDX, ICP, GDS
1970s	MW[a], IC[a], FTIR[a], NIR, ¹³C NMR[a], NMRI, PyFT-IR[a], CuPyMS[a], GC-FTIR, LD, FTICR, PDMS, LMMS[a], SSIMS[a], PDA, SEM[a], STEM[a], ESEM, PIXE
1980s	MAE[a], SPE, SFE, cSFC[a], FT-Raman, NMRI[a], TSP, FAB, PB, ESI-MS, QITMS[a], PD-MS[a], PyGC-FTIR, GC-QITMS[a], SFC-MS, ICP-MS[a], LA-ICP-MS, rfGD-OES[a], MALDI-MS, LDMS, TG-FTIR[a], ELSD, STEM[a], STM[a], AFM[a]
1990s	SPME[a], SPE-GC, SPE-HPLC, CL[a], ICP-ToFMS[a], LC-API-MS[a], HRTG[a], MDSC[a], MTGA[a], TMDSC[a], μTA[a], (multi)hyphenated techniques[a]

[a] Analytical instruments' funding dates.

deformulation have disappeared in favour of others (various LC-MS substitutions, evaporative LC-FTIR instead of in-flow HPLC-FTIR, etc.), while some techniques are changing gradually (e.g. from LC via HPLC to nanoflow HPLC), or are subject to diversification (cSFC, pSFC). Yet, most newly developed methods do coexist and compete with 'older' methods.

Some analytical methods are highly mature (NAA, XRD, XRF, XAS), the theory is well assessed, and just instrumental and incremental improvements (more intense sources, better detectors) may be expected. However, in many other areas the sharply increasing power of analytical instrumentation (with regard to both hardware and software) and its transformation into tools for in-process control (such as NIRS, LR-NMR, etc.) are most appropriately considered as breakthroughs.

For the industrial chemist, faced with efficiency requirements, full awareness of the rapidly changing analytical environment is essential; budget constraints (and not only these) determine the need for a well-balanced choice of method and equipment.

Table 10.22 Basis for selection of analytical technique

- Sample preparation requirements
- Range of applicability
- Sensitivity and detection limit
- Analytical precision and reproducibility
- Adequate calibration concepts
- Analytical concentration range[a]
- Multifunctionality

- Extensive and comprehensive automation
- Specificity

- Process controllability
- Control on analytical interferences
- Difficulties with contamination
- Operator skills' requirements

- Total analytical time

- Instrument cost
- Equipment configured to meet customer needs

- Maximum performances, rugged equipment
- Standardisation

- Quality assurance and certification
- Complete documentation of experiments

- Result recording conforming to ISO and GLP
- Network compatibility
- User-friendly software

- Integrated databases for continuous overview
- Workplace safety

[a] Quality control (ppm), trace analysis (ppb), ultratrace analysis (ppt–ppq).

Table 10.23 Future instrumental developments

- Evolutionary instrument market (rather than revolutionary new forms of analysis)
- Shift in emphasis (from petrochemicals to biochemicals and genomics)
- Multidimensional technology (comprising integrated, automated data processing)
- Visualisation
- Single-molecule and -atom analysis
- Information management
- Miniaturisation (up to LOC); microarrays
- Prevailing fashion
- General instrument industry slowdown

Table 10.22 shows some of the parameters that should be taken into consideration when selecting an analytical technique for a specific application. Ideally, the analyst requires a general system which does not need to be optimised for each single analyte and is matrix independent. Nonstandard samples need to be analysed almost routinely. It is probably for this reason that methods such as SFE, MAE, LC-MS and SFE-SFC (to name a few) do not find the expected general application, in spite of some very attractive features. Instrumentation developed under ISO 9000 control, which uses state-of-the-art computer technology, operating platform and user-friendly application software programs, is highly favoured. Some actual trends in polymer/additive deformulation make it possible to predict the (near) future. In this respect we can notice some relevant developments.

10.4.2 Future Trends in Polymer/Additive Analysis

Progress in polymer/additive analysis has closely mirrored the changes in technology in both the polymer and additive industries during the past decades. Whilst the pharmaceutical and biochemical industries

exploded in new technologies, polymer/additive analysis progressed with its own evolution. Advances in this field are concomitant with general instrument development (Table 10.23). The driving forces of change in polymer/additive analysis are:

(i) R&D with the highly knowledge-based use of a multiplicity of advanced techniques;
(ii) problem-solving capabilities with routine but increasingly sophisticated analysis; and
(iii) process control (QC).

Obviously, the scope of additive analysis for R&D purposes (product innovation and understanding of additive performance), quality control, troubleshooting and competitor product analysis differ (Scheme 10.2). Product knowledge (see Sections 10.1 and 10.2) is particularly desirable for the latter two activities.

The additive analysis reported has been largely confined to conventional polymers (polyolefins, polycondensates, PS, PVC, etc.) Very little work, if any, has been reported on advanced engineering plastics. Similarly, also relatively little research activity has focused on additives in acrylics or blends.

The chances to predict successfully the future on the basis of the past are fewer than ever, because of the current great pace of change (the doubling of knowledge in 15 years) and the appearance of new inventions. Describing the past is more reliable than forecasting the future. For example, the prospects expressed a decade ago [95] of the use of SFE on-line with the various chromatographic techniques and their interfacing to spectroscopic techniques, with SFE-GC-UV-IR-MS and SFE-SFC-UV-IR-MS being used commonly where a sample is automatically placed in the extractor with no further contact until the final report, has certainly not yet come true. Yet, environmental (and economical) constraints determine a shift away from the use of (large

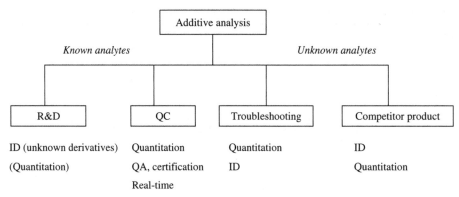

Scheme 10.2   Scope of polymer/additive analysis

amounts of) solvents, especially halogen-containing ones. Consequently, for the future one may expect:

(i) more SFE, ASE® and MAE rather than Soxhlet extractions;
(ii) greater emphasis on microflow chromatographic techniques; and
(iii) a greater share for in-polymer analysis.

The latter development, both qualitatively and quantitatively, requires more substantial advancements to be made.

For new analytical techniques to prosper, they must have demonstrated applications to real-world samples, with outstanding figures of merit relative to competing approaches. Table 10.24 opposes the prospects of conventional separation procedures and advanced *in situ* analyses by the currently most qualifying techniques. Lab-on-a-chip (LOC) devices are unlikely to be robust enough to cope with the moderately complex (i.e. 'dirty') matrices that are real-life samples. Industrial chemists need to avoid a lot of work for every analyte and every matrix. Obstacles to solid analysis are relatively poor sensitivities, narrow linear dynamic ranges and unavailability of solid standards. The trend

Table 10.24   Comparison of general features for bulk polymer/additive analysis

Feature	Separation procedures	*In situ* analysis
Matrix interference	No (except eventually D/P)	Yes (FAB, SIMS, PyGC-MS; not for L^2MS)
Selectivity	Universality rare (except for MAE-HPLC-ELSD/UV)	Yes (L^2MS, MS^n)
Co-additive interferences	No	Yes
Quantification	Feasible	Very difficult (e.g. L^2MS)
Sample preparation	Yes	No
Identification	General reference libraries (commercial, proprietary)	Specific reference libraries (usually lacking, e.g. TLC-MS)
	Retention factors	Deductive reasoning (e.g. LMMS); PCA
Analytical complexity	Limited	High
Instrumental cost	Relatively low	Medium–high (e.g. Raman, NMR, PyFTIR)
Survey analysis	TLC	Limited (e.g. LMMS, L^2MS)
Experimental effort		
Time	High	Relatively low
Interpretative effort	Low (generalist)	High (specialist)
Analyte integrity	Eventually compromised	Verifiable
Selective detection	Very limited	Occasionally

towards simple sample preparation procedures favours techniques such as HS, NIRS, Raman, and sensor technology with respect to AAS and ICP. Where understanding is lacking, libraries are efficient substitutes for thought. In this respect, polymer/additive analysis is poorly equipped for many techniques. The major threat is the development of high-performing additive-free polymeric material. Many of the driving forces in the polymer additive industry will affect polymer/additive analytical activities.

The increasing complexity of polymer/additive analysis may be illustrated by the evolution of hindered amine stabilisers (HAS) in the last three decades, namely from low-MW Tinuvin 770 to the oligomeric products Tinuvin 622, Chimassorb 944 and their blends (e.g. Tinuvin 783); high-MW Chimassorb 119; synergistic mixtures of low- and high-MW HAS; and, most recently, to the HMW HALS Chimassorb 2020 (end-capped Chimassorb 944) showing premium performance (excellent UV and long-term heat-stabilising effect, low volatility, excellent toxicity profile, no pigment interactions, high extraction resistance), see Figure 3.25. In-polymer additive analyses such as PyGC-MS are required for the identification of such products. Determination of high-MW (oligomeric and polymeric) additives, grafted additive functions and reactive additives is still in its infancy. Demands for higher sensitivity (detection at levels <100 ppm) and resolution increase with the advent of more complex materials and tighter specifications, and new demands are placed on analytical techniques for higher performance.

10.4.2.1 Sample Preparation

Minimisation of sample preparation is the main bottleneck in polymer/additive analysis. The importance of sample preparation increases with miniaturisation of the separation techniques. However, there is no point in improving instrumentation when the true sources of errors in measurement are sampling, sample inhomogeneity or sample instability.

It is apparent from Chapter 3 that *new sample preparation technologies* generally are faster, more efficient and cost effective; more easily automated and safer; use smaller amounts of sample and less organic solvent; provide better recovery; and meet or exceed precision and accuracy compared to traditional sample preparation techniques. Conventional methods of the analysis of additives in polymers are mostly based on the separation of the polymer matrix and additives by means of extraction. Many extraction principles are

being applied to polymer/additive analysis. The ageing, time-consuming and economic Soxhlet method is still widely used against all the odds, but is now slowly being replaced by the considerably more expensive SFE, MAE and ASE® techniques, which reduce solvent use and sample preparation time. The equally low-cost ultrasonication method finds few followers. Although SFE presents various advantages and is useful for extraction of thermally labile substances, a significant drawback is the need for optimisation of *many* operational parameters, notably solvent modifier and pressure. This extent of method development contrasts with the requirements for routine application, and makes a real breakthrough in analytical SFE uncertain. Microwave-assisted extraction is less complicated than SFE, but also requires optimisation to some extent; the need for a dielectric medium limits application, and safety (high p, T) is an important issue. Relatively few applications in the field of polymer/additive microwave extraction have been reported. Accelerated solvent extraction needs even less parameter optimisation (mainly selection of the extraction solvent), but more experience has to be gained with this promising technique. Thus, although in the last decade many advances have been reported in the extraction field, there is still no clearly developed view on the actual power of many new techniques (SFE, MAE, ASE®) for additive extraction *without* complicating (and undesirable) optimisation procedures which are *analyte dependent*. It takes time for new extraction methods to become commonplace in the analytical laboratory. Reasons are the need for equivalence testing, mastering of new skills, method development and payback of capital investment. In many cases, extraction methods are more versatile than direct mass-spectrometric approaches [96]. It is generally advisable to verify extraction results spectroscopically (before and after extraction). A real *breakthrough* would be fast and quantitative extraction of *all* analytes in a matrix independent fashion. An alternative sample preparation route, involving dissolution, presents other troubles (solvent choice; insolubility of inorganics; dilution) and is therefore not universal. The recent development of advanced extraction-resistant additives (see ref. [23]) determines the need for alternative, in-polymer, additive analysis techniques. In view of the heterogeneity of polymeric materials, there is limited scope for microsampling.

Analyte stability is another source of concern; the chemical identity should be preserved all through the analytical process. This is not always guaranteed. Analytes may be oxidised or hydrolysed during extraction. Compound integrity can also suffer in the chromatographic column, through heat-induced decay. Extraction is an extreme form of chromatography. Recently, the boundaries between extraction and chromatography are blurring, as shown by:

- the relation between GC, SFC and HPLC;
- the use of superheated/subcritical water for extraction and chromatography; and
- the role of enhanced fluidity solvents and pressurised fluid extractions.

Techniques such as SPE simplify the labour of sample preparation, increasing its reliability, and eliminating the cleanup step by using more selective extraction procedures.

Integration of sample preparation and chromatography by on-line coupling aims at reduction of analysis time. It is apparent from Section 7.1 that these hyphenated techniques are not yet contributing heavily to the overall efficiency of polymer/additive analysis in industry. On-line SFE-SFC requires considerable method development, and MAE-HPLC is off-line. Enhancement of sensitivity for trace analysis requires appropriate sample preparation and preconcentration schemes, as well as improved detection systems.

10.4.2.2 Separation Methods

Deformulation of multicomponent additive packages may be achieved by chemical means (extraction, chromatography), physically (e.g. tandem MS fragmentation; DOSY-NMR, self-diffusion; LR-NMR, relaxation time), mathematically (e.g. PCA), or in combination. While polymer analysis cannot do *without* polymer separation, it is true that polymer/additive analysis still greatly benefits from the separation of the additives from the polymeric matrix. As polymer/additive deformulation of real-life samples still relies heavily on extraction and dissolution, the procedure is logically followed by a mix of well worked-out chromatographic techniques, all endowed with a variety of specific detectors. The chromatographic methods are conceptually rather simple, and have become a ubiquitous part of quantitative chemical analysis. General trends in chromatography may be summarised as follows: microchromatography, miniaturisation, microfluidics–chromatography, hyphenation, unified chromatography, trace analysis and speciation.

The current status of chromatography is shown in Table 10.25. Since reducing separation time is a major issue, there is a pronounced trend toward shorter columns filled with small particles. The current trends for lower flow (micro- and nano-LC) columns, and great strides to achieve (ultra-) fast chromatographic

Table 10.25 Current status in chromatography

Gas chromatography:

- Evolutionary
- New injection types
- Large-volume analysis
- Narrow-bore GC(-MS)
- High-temperature GC
- Selective stationary phases
- Fast analysis
- Comprehensive GC
- Multicapillary columns
- Micro systems technology

Liquid chromatography:

- Standardisation
- Temperature-programming
- Miniaturisation
- Microcolumn LC
- New phases (incl. monoliths)
- Multidimensionality
- HTLC

Supercritical fluid chromatography:

- Rapid analysis
- Extended GC
- Niche applications

separation, are aspects which will not fail to be greatly appreciated by industrial research chemists, and will soon certainly have active followers for polymer/additive deformulation. Rapid chromatographic analysis may be achieved by using highly porous monolith columns. Multicapillary columns are as yet too unreliable. New techniques for reducing separation time, such as the use of flow gradients or a combination of flow and solvent gradients, have not been widely adopted in polymer/additive analysis. More often, in order to increase sample throughput, part of the chromatographic resolution is sacrificed, while still fully maintaining the sensitivity by using highly selective detection techniques such as tandem MS and NMR. Downsizing and integration of chemical processes is in principle expected to lead to huge gains in performance, speed, size, throughput and automation. However, as polymer/additive analysis is usually not sample limited, the scale of the separation methodology does not need to be reduced. Important time-savings in chromatography may be achieved by using fewer, more generalised methods, e.g. one GC method and one HPLC method, thus reducing regular calibration and validation efforts.

Gas chromatography in its various forms (GC, HS-GC, HSGC, HTGC, GC-MS, PyGC-MS) is widely used for additive analysis of low- to medium-MW species. As shown in Section 4.1, although some detection methods

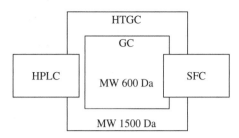

Figure 10.3 Competitive chromatographic techniques

(e.g. FID to GC, PDA to HPLC) are work-horses with excellent records for verification and quantification, new appearances are being made (often in hyphenated form with FTIR and MS). The resolving power and sensitivity of modern *gas chromatography* is absolutely amazing, and some liquid chromatographies are rapidly developing along similar lines. For routine application in an industrial research laboratory, the most outstanding amongst these is HTGC. Many potential HPLC and SFC applications (medium-MW) fall in the working range of HTGC (up to 420 °C), see Figure 10.3. Perfecting the throughput, such as a rapid (120 °C min^{-1}) in-vial extraction and large-volume cold on-column injection (LVI-GC), dramatically shortens the total analysis time, thus increasing lab productivity and lowering the cost of ownership. Use of narrow-bore GC for additive analysis is still largely unexplored. Evaluation requires choice of the optimal column geometry and temperature gradient to achieve the desired detection limits and instrumental response times. Redesigned high-speed GC may be expected. GC-AED lacks a real breakthrough. Speciation analysis of organometallic polymer additives is limited to GC-MIP-AES and GC-ICP-MS.

Liquid chromatography in its various guises (NPLC, RPLC, µHPLC, HTLC, SEC) is the single most widely used technique for polymer/additive analysis. The ever-increasing demand on chemical laboratories to increase sample throughput has provided the impetus for HPLC column manufacturers to introduce new stationary phases and a range of column geometries to meet the requirements of speed, high sensitivity and reduced sample availability. Use of LC-PDA-ELSD greatly overcomes the need for GC. It remains to be demonstrated, however, that HPLC-ELSD with analyte-specific optimisation is experimentally more advantageous than carrying out both a HTGC and a HPLC run. Temperature-programmed HPLC was developed specifically as a robust tool for the analysis of difficult oligomeric polymer additives. LVI-µHTLC can be used for the determination of very low additive concentrations (0.6 ppm Irganox 1076 in LDPE) [97].

As µHPLC is likely to be much less robust, conventional HPLC equipment will stay.

Advances in *size-exclusion chromatography*, coupled with refractive index, absorption, viscosity, and light-scattering detectors, and MALDI-ToFMS, have made it possible to accurately determine molecular weight distribution (oligomer profiling), even at the relatively low values of polymeric additives (up to about 5000 Da). Advances in column design, e.g. high-resolution PS/DVB columns ($>10^5$ plates m^{-1}) mean that SEC can provide a valuable alternative to conventional HPLC techniques for the separation of small molecules.

Despite the fact that *thin-layer chromatography* is simple and cheap, it is slowly declining for the purpose of polymer/additive analysis, yet is valuable for special applications: separation of components which are not easily released by GC or LC columns, fast screening and preparative chromatography. Modern instrumental developments originate here from different application areas (pharmaceutical and biomedical analysis). Fuller appreciation of the advantages of TLC in combination with advanced spectroscopic characterisation, by both chromatographers and spectroscopists, may lead to a revival of these techniques in both routine analysis and for problem solving. However, this would require development of the potential of TLC by instrument manufacturers, resulting in the commercial production of dedicated instrumentation. TLC tandem systems are often seen as the last resort, to be used only under very special circumstances [98].

The sensitivity of *electrophoretic methods* extends to the attomole range, which is particularly useful for scarce samples (seldom a problem in additive analysis) and degradation products.

Alternatives to off- and on-line chromatography are desirable in order to avoid the time-consuming chromatographic step, and are imperative for polymer-bound additive functionalities.

10.4.2.3 Spectroscopy

In polymer/additive analysis, spectroscopic methods are used for studying both molecular and atomic composition, usually as a detector for chromatographic techniques. Application of spectroscopic techniques to molecular additive analysis depends on the nature of the sample and its complexity (Table 10.26). Application of the intrinsically simple monocomponent analyses by means of UV/VIS and FTIR is rather exceptional for real-life samples. Most industrial samples are complex. It is in the area of multicomponent analysis that most

Table 10.26 Application of spectroscopic methods in polymer/additive analysis

Sample	Monocomponent analytes	Multicomponent analytes
Solution	Straightforward	Hyphenation (detector)
Melt/solid	Straightforward	Mathematical model

progress has been made. Nevertheless, this remains a difficult matter.

In polymer/additive deformulation (of extracts, solutions and in-polymer), spectroscopic methods (nowadays mainly UV, IR and to a lesser extent NMR followed at a large distance by Raman) play an important role, and even more so in process analysis, where the time-consuming chromatographic techniques are less favoured. Some methods, as NMR and Raman spectrometry, were once relatively insensitive, but seem poised to become better performing. Quantitative polymer/additive analysis may benefit from more extensive use of 600–800 MHz 1-NMR equipped with a high-temperature accessory (soluble additives only).

Powerful signal-processing analytical techniques, such as multiwavelength and derivative spectrophotometry not only offer many advantages regarding elimination of interferences and multicomponent determinations, but have also facilitated the resolution of complex mixtures of organic constituents with highly overlapped spectra, contributing to more effective use of separation techniques. Sophisticated computer-implemented mathematical methods, such as Fourier transformation, have led to the improvement of many spectroscopic methods, and have spawned entirely new methods, such as FT-IR, FT-NMR and FTICR-MS. Likewise, the introduction of matrix methods has led to completely new ways of analysing data, as in near-infrared spectroscopy.

Raman spectroscopy has enjoyed a dramatic improvement during the last few years: the interference by fluorescence of impurities is virtually eliminated. Up-to-date near-infrared Raman spectrometers now meet most demands for a modern analytical instrument concerning applicability, analytical information and convenience. In spite of its potential abilities, Raman spectroscopy has until recently not been extensively used for real-life polymer/additive-related problem solving, but does hold promise. Resonance Raman spectroscopy exhibits very high selectivity. Further improvements in spectrophotometric measurement detection limits are also closely related to advances in laser technology. Apart from Raman spectroscopy, areas in which the laser is proving indispensable include molecular and fluorescence spectroscopy. The major use of lasers in analytical atomic

spectrometry is in four areas: laser sampling, laser-enhanced ionisation spectrometry, laser-excited atomic fluorescence spectrometry (LEAFS) and laser-induced breakdown (emission) spectroscopy [99].

As for miniaturisation, examples are UV/VIS spectrometers using a linearly variable filter rather than a diffraction grating as the wavelength separation device (matchbox size), and double-focusing mass sensors (postage stamp size).

10.4.2.4 Mass Spectrometry

Various analytical methods have made quantum leaps in the last decade, not least on account of superior computing facilities which have revolutionised both data acquisition and data evaluation. *Major developments* have centred around mass spectrometry (as an ensemble of techniques), which now has become a staple tool in polymer/additive analysis, as illustrated in Chapters 6 and 7 and Section 8.5. The impact of mass spectrometry on polymer/additive analysis in 1990 was quite insignificant [100], but meanwhile this situation has changed completely. Initially, mass spectrometrists have driven the application of MS to polymer/additive analysis. With the recent, user-friendly mass spectrometers, additive specialists may do the job and run LC-PB-MS or LC-API-MS. The constant drive in industry to increase speed will undoubtedly continuously stimulate industrial analytical scientists to improve their mass-spectrometric methods.

Mass spectrometry combines exquisite sensitivity with a precision that often depends more on the uncertainties of sampling and sample preparation than on those of the method itself. Mass spectrometry is a supreme *identification and recognition* method in polymer/additive analysis through highly accurate masses and fragmentation patterns; quantitation is its weakness. Direct mass spectrometry of complex polymeric matrices is feasible, yet not often pursued. Solid probe ToF-MS (DI-HRMS) is a breakthrough. Where used routinely, mass spectrometrists are usually still in charge. At the same time, however, costs need to be watched.

There are as yet no signs that the development of mass spectrometry is slowing down. This has consequences for the analyst. Whereas in the past the practitioner of polymer/additive analysis primarily benefited from a good knowledge of chromatography, it is now essential that such analytical scientists are conversant with current MS theory and practice. Advances in MS are due to new conceptual design, new ionisation methods (e.g. switched plasma sources) and software developments. In particular, the diversity of MS/MS instrumentation offers considerable opportunities for polymer/additive analysis (Section 6.3.6). In this respect, the FAB-MS/MS approach is a powerful analytical tool for a rapid search of additives in extracts. The highly specific technique finds application mainly for fractions of higher polarity.

Some new developments are additive analysis by DIP-ToFMS and MALDI-ToFMS, and the direct determination of additives by PyGC-MS. MALDI-ToFMS may be used as a quick screening method of known additives in UV-insensitive polymers. For rapid screening purposes, various reference libraries – under development – might be of considerable help, such as MALDI-ToFMS, PyGC-MS and ToF-SIMS spectra of additives in polymeric matrices (as opposed to spectra of the pure compounds). In the application of mass-spectrometry techniques to polymer/additive analysis, some crucial information is still lacking. For example, the sensitivity of MALDI-ToFMS is known, at variance with that of PyGC-MS; quantification with MALDI-ToFMS seems to be impossible without considerable improvements in sample handling. Considerably more work needs to be done to turn MALDI-ToFMS of polymer/additive dissolutions into a valid analytical tool.

The use of high-MW, oligomeric, organic polymer additives (typically HALS) is becoming more common in many polymeric materials, and these compounds still elude analysis by means of GC, SFC or LC. These additives have been successfully analysed by means of mass spectrometry [101–103], but ionisation is limited to desorption techniques (FD, LD), because of the inherent involatility of the high-MW materials. It was suggested previously that the most suitable ionisation technique for analysing oligomeric additives is FD [101], but intense distributions have also been observed in MALDI spectra of the same compounds [103]. Tandem mass spectrometry must be employed to obtain structural information, as FD and MALDI generate often predominantly molecular ions with little induced fragmentation. LSIMS-MS/MS or FD-MS/MS experiments are therefore often used in conjunction with FD-MS, as a screen to determine the class, molecular weight and structure of mixtures of organic polymer additives. MALDI-ToFMS has potential as a *rapid screening* device in polymer/additive analysis, especially in cases of 'inert' matrices (such as polyolefins). Selective polymer/additive analysis may be carried out by means of L^2ToFMS, by careful choice of the ionising laser wavelength. Resonance-enhanced laser mass spectrometry is a new 2D method for selective detection of trace quantities of organic

compounds in mixtures. Application of LDIOS (laser desorption/ionisation on silicon) for oligomeric additives might prove useful for this analytically rather intractable group of compounds.

Although often used as a qualitative (identification) tool, MS may act as a quantitative inorganic mass detector. Quantification of organic analytes often takes place in combination with chromatography or in tandem MS mode. It should be realised that mass spectrometry is certainly not a panacea for all polymer/additive problems, although it is developing into a major tool for this purpose.

10.4.2.5 Hyphenated Techniques

The mass spectrometer is mainly used as a mass detector in chromatography (GC, SFC, HPLC, SEC, TLC). With the great variety of interfaces, ionisation modes and mass spectrometers, chromatography–mass spectrometry is highly diversified. High-resolution separations combined with accurate mass measurements and element-sensitive detection (MIP, ICP) have been reported.

The use of GC-MS in polymer/additive analysis is now well established. Various *GC-based polymer/additive protocols* have been developed, embracing HTGC-MS, GC-HRMS and fast GC-MS with a wide variety of front-end devices (SHS, DHS, TD, DSI, LD, Py, SPE, SPME, PTV, etc.). Ionisation modes employed are mainly EI, CI (for gases) and ICPI (for liquid and solid samples). Useful instrumental developments are noticed for TD-GC-MS. GC-SMB-MS is a fast *analytical* tool as opposed to fast chromatography only [104]. GC-ToFMS is now about to take off. GC-REMPI-MS represents a 3D analytical technique based on compound-selective parameters of retention time, resonance ionisation wavelength and molecular mass [105].

SFC-based methods still need to show their potential, in spite of past great promise. pSFC-APCI-MS is a powerful method for identification of polymer additives, provided that a library of mass spectra of polymer additives using this technique is available. SFC-MS appears less performing than originally announced; nevertheless, SFE-SFC-EIMS is an interesting niche approach to additive analysis. On the other hand, we notice the lack of real breakthrough in SFE-SFC-FTIR.

It is considerably more difficult to achieve on-line *LC hyphenation* than with GC. Sample size restrictions are a significant barrier to using some techniques, such as on-line SFE-HPLC, for quantitative analysis of 'real' samples. Application of LC-MS to polymer/additive analysis has been much more troubled than that of GC-MS. Two key directions have emerged from 25 years of LC-MS developments, namely LC-PB-MS (EI mode) and LC-APCI-MS, but both with limitations. The former is a niche application for polymer/additive analysis, the latter is apparently LC-MS design dependent. A general approach for quantitative polymer/additive analysis, based on the use of LC-PB-MS in conjunction with LC-APCI-MS by Waters Corp. [106], is unlikely to find many practitioners, despite the fact that this combination undoubtedly offers considerable potential for confirmation, positive identification and quantification (retention times, library matching, APCI$^-$, APCI$^+$, UV). On the other hand, a multi-API interface (comprising easily replaceable APCI, ESI, SSI, APEI and APSI) has been described [107], which realises the two desired key techniques, namely the combination of EI mode (as an atmospheric pressure spray with electron impact ionisation, APEI) with APCI, APSI, ESI and SSI modes in one LC-MS system. All-purpose LC-multi-API-MS provides:

(i) the APEI mode for low-polarity organic compounds (e.g. aromatics);
(ii) the APCI mode for relatively polar compounds with nitrogen and/or oxygen atoms which are neutral in solution;
(iii) the APCI mode for organic compounds with high cation affinity; and
(iv) the ESI or SSI modes for polar analytes which exist as ions in solution or are associated with ions in solution [108].

This development is particularly promising, as it is usually very difficult to exactly describe what type of organic compounds can be measured by each mode. LC-SMB-MS is expected to substitute APPI, APCI and PB LC-MS modes [109]. Fast LC-oaToFMS/ELSD allows speed, high resolution, and automated exact mass measurement. μHTLC-ESI-MS is suggested as being a useful analytical tool, e.g. for the structural evaluation of polymeric HAS (in analogy to μHTLC-ELSD studies). A drawback of LC-MS is that measurable organic compounds are very limited compared with separable compounds by LC alone. A much desired breakthrough is a sensitive mass-proportional LC detector. New hyphenated techniques requiring evaluation for polymer/additive analysis are LC-MS/MS, in particular LC-QToFMS/MS (for accurate mass determination and molecular recognition), HPLC-NMR and HPLC-NMR-MS. Other multiply coupled systems (e.g.

HPLC-PDA-MS) are already being used. Although HPLC-NMR-MS provides a very powerful approach for compositional and structural analysis, it by no means represents the limit of what is possible in terms of hyphenation. On-line extraction and the attachment of multiple detectors (e.g. IR, F) make the technique even more powerful. Other 'analytical laboratories' such as TG-DTA-DSC-FTIR, TD-CT/Py/GC-MS/FTIR and HPLC-UV/NMR/IR/MS have been put to work, but do not represent practical solutions for routine polymer/additive analysis.

Evaporative LC-FTIR is rapidly gaining industrial acceptance as a useful tool in low-MW additive analysis. HPLC has also been coupled with various element-selective detectors. There is significant demand for *speciation* information for many elements, and the separation ability of chromatography coupled to ICP-MS offers the analyst a versatile tool for such studies. It is apparent that ICP-MS is increasingly being employed for chromatographic detection. Several modes of GC, SFC, LC and CE have been hyphenated with ICP-MS for improved detection limits compared to other traditional methods of detection such as UV-VIS spectroscopy. Inorganic speciation deserves more attention.

SFE-GC and SFC-GC may replace normal-phase LC-GC. SFC-SFC may be adopted for a wide range of applications, particularly for solutes that are not amenable to GC and as a replacement for some coupled LC-LC separations. However, no applications to additives in polymers can be mentioned.

In many cases, the current approach to hyphenation of two (or more) techniques, typically a combination of a separation method and an identification technique (spectroscopic or spectrometric), is still not totally satisfactory. This is especially the case when the optimum operating conditions of both techniques are compromised in their combination. In that respect, any proposed improvement is welcome. Multihyphenated techniques, although fancy, usually become quite complicated, so as to require dedicated analysts. In relation to Scheme 10.2, it should be realised that hyphenated techniques are costly and complex to run; they are most useful for unknown analytes.

10.4.2.6 In-polymer Analysis

Despite well-deserved attention and considerable efforts, direct polymer/additive analysis (without separation) has not been turned into routinely workable concepts. Table 10.27 shows the main approaches.

Some general characteristics of direct solid-state polymer/additive analysis methods have already been

Table 10.27 *In situ* polymer/additive analysis techniques

- Direct probe MS
- Spectroscopy (UV/VIS, IR, R, NIRS, NMR, Mössbauer, etc.)
- Thermal analysis (TG, DSC, DTA)
- Desorption techniques (SHS, DHS, TD, LD)
- Pyrolysis

given in Table 10.24. Cost-saving (direct) analysis of additive packages on intact polymeric matrices requires either (hyphenated) mass-spectrometric or spectroscopic techniques, where *in situ* spectroscopic methods are of limited use only, except for well-defined systems (e.g. in-process streams). Direct in-polymer/additive analysis heavily relies on thermal methods (DTMS and TD-GC-MS), pyrolysis (PyGC-MS) and laser desorption (LDI-MS), all preferably with mass-spectrometric detection.

Although the feasibility of *direct probe MS* for the analysis of additives in complex polymeric matrices has been demonstrated (Section 6.4), application is limited, difficult and requires above-average mass-spectroscopic expertise. Direct desorption in the MS probe is usually limited to screening of volatile components. *Direct multicomponent spectroscopic analysis* has other hurdles to overcome (UV/VIS: lack of spectral discrimination; IR/R: functional-group recognition only, with no discriminative power for additives with similar functionalities; NIRS: unsuitable for R&D problems; NMR: sensitivity).

Thermo-analytical methods (DSC, TG) may be used as a quick tool in quality assurance, incoming goods and damage analysis. μTA is not expected to have much impact in polymer/additive analysis. At present, TG-based methods are more suitable for the higher-concentration flame-retardant and filler analysis market segment. Conventional thermal analysis (TG, DSC, DTA) does not allow identification unless hyphenated. TG-MS may develop into a useful system for rapid screening of intact molecules (now up to about 500 Da). Provided that there is proper commercial development, this technique can yield results superior to those of direct probe MS, as a (high-resolution) thermal separation step is involved. However, in terms of molecular mass, the performance will not be superior to LC-PB-MS, and the possibility of thermal degradation is inherent to the technique. Whereas the future probably will put more emphasis on TG-MS, TG-FTIR is expected to decline. Round-robins contribute to the validation of analytical methods, e.g. comparison between TGA methods and solvent extraction [110,111]. For the analysis of complex solids, controlled *thermal desorption* techniques in combination with GC-MS are useful tools in problem solving, and are efficient ways to collect and

concentrate trace amounts of volatile components. The use of chemosensors is very limited.

Analytical pyrolysis is useful for industrial QC of organic polymer systems. Pyrolysis in its most extended form, PyGC-MS, allows qualitative analysis by comparison with existing additive libraries. This method seems on the verge of rapidly gaining the industrial analyst's trust for low-concentration additive analysis. Yet, the prospects of quantification by means of PyGC-MS are still to be evaluated in full and do not appear to be a trivial matter. Pyrolysis is particularly appealing for polymer-bound additive functionalities, graft polymers, impact modifiers and other rather intractable samples. However, in this case, matrix effects (which compromise quantitation) are unavoidable. Conceptually of course, pyrolysis methods lack analytical elegance: analytes are first broken down experimentally, leaving the analyst with the difficult task of molecular reconstruction. Combined multistep thermal desorption–pyrolysis of solid samples can be carried out by means of PTV injection to HTGC-FID [112]. The system can be used to obtain well-separated fractions of low-MW components (monomers, solvents), additives and pyrolysis products.

Laser desorption methods (such as LD-ITMS) are indicated as cost-saving real-time techniques for the near future. In a single laser shot, the LDI technique coupled with Fourier-transform mass spectrometry (FTMS) can provide detailed chemical information on the polymeric molecular structure, and is a tool for direct determination of additives and contaminants in polymers. This offers new analytical capabilities to solve problems in research, development, engineering, production, technical support, competitor product analysis, and defect analysis. Laser desorption techniques are limited to surface analysis and do not allow quantitation, but exhibit superior analyte selectivity.

All available methods (TG-MS, PyGC-MS and LDI-MS) suffer from difficult *quantitation*, although for different reasons. In TG-MS, selective volatilisation may not reflect the composition in the solid; the quantitation problem of PyGC-MS requires assessment of the importance of matrix effects. Laser ablation methods cannot easily be calibrated. Quantitation is simplified in case of dual detection (MS for identification, FID for quantitation). A general drawback of many direct methods, which allow only small sampling volumes, is granule-to-granule variations.

Apart from the problems addressed regarding highly efficient, routinely and generally applicable bulk analysis of relatively low-MW organic additives in manufactured polymers and spatially resolved analysis, other analytical problems of considerable complexity (HMW, grafting, incorporation in the polymer backbone, reactive systems, etc.) will require more attention in the near future, as well as the use of direct in-polymer analysis. Systems such as those described by Zhu and Weil [113], namely EPDM/(PPO, melamine, kaolin), where PPO acts as a char-forming polymeric additive, require a high dose of analytical ingenuity for characterisation where the well-established extraction/dissolution procedures fail. Direct analysis, still in an early stage of development for relatively low-MW additives, is the remaining option. The future should see a greater contribution from in-polymer analysis. Development of a general, reliable, qualitative and quantitative, in-polymer additive analysis technique would be a real breakthrough, but is most probably utopian.

10.4.2.7 Spatially Resolved Polymer/Additive Analysis

The continually increasing sensitivity of analytical instruments makes it possible to probe smaller samples. For smaller volumes, surface properties become more important. Surface analysis is a rather new and rapidly developing field. Analytical difficulties increase with the degree of heterogeneity, from homogeneous to surface treated, coated, layered, continuously varying composition to totally heterogeneous.

The scope of *UV microscopy* in polymer/additive analysis is confined to the study of UV stabilisers (ageing, migration). The greatest potential inherent in high-resolution *infrared microscopy* is the ability to obtain spectra from sub-μm regions of samples that can aid in identification and characterisation of materials (problem solving). In the field of on-line microanalysis there appear to be equally good prospects for dispersive *Raman micro-probes* using NIR excitation beyond 1000 nm and linear array detectors with good sensitivity for the investigation at the microscopic level or for remote analysis by means of optical fibres of samples which fluoresce under visible illumination. With these 2D sensors the powerful techniques already developed in the visible region, like confocal line imaging, could be extended to the NIR region. Fibre-guide technology makes the analyst independent of the location of the samples. Dispersive Raman may be considered for the study of all types of heterogeneous materials.

Surface characterisation methods, both elemental (e.g. XPS, AES) and molecular (e.g. ToF-SIMS), are gaining in importance, in view of the active development of surface-modification technology to render fillers of all types more acceptable to the matrix and improve interfacial bonding, e.g. surface-modified rubber particles as a reinforcing, elastomeric filler [114]. ToF-SIMS can also be employed profitably for problems related to molecular diffusion, migration and blooming (specific ToF-SIMS in-polymer additives libraries are now available). On the other hand, laser desorption techniques, such as LD-ITMS, which offer prospects for spatially resolved additive analysis, are more suitable for impurity detection in industrial troubleshooting.

The marriage of spectroscopy with imaging has recently led to a plethora of new techniques, such as iXPS, iSIMS, NMRI, ESRI, ICL, etc. *NMR imaging of a mobile material within rigid solids is relatively simple.* NMRI has been applied to polymers [115,116]. *ESRI* has been developed into a method for the study of diffusion processes, and for spatial and spectral profiling of radicals formed during polymer degradation. Fast imaging techniques may be expected to be in future greater demand.

10.4.2.8 In-process Analysis

Many traditional laboratory/off-line methods are now moving in the direction of in-process applications. On-line GC had already been introduced in the 1950s. Using chip- and microsystem technology, μGC is now being introduced, which achieves analysis times of 30 seconds and is therefore suitable for quality control. SPME-μGC is potentially useful for process analysis.

The shift from lab to line goes hand in hand with spectroscopisation. Spectroscopy has witnessed a revolution, with the development of miniature spectrometers using optical fibres coupled with advanced software based on chemometrics (enabling on-line measurements). Fibre-optics technology favours the speed of the analytical process, where the place of the sample generation and the analytical laboratory do not coincide. Key market drivers for *process spectroscopy* are technology improvement, a trend toward low-maintenance systems and a shift from process GC to MS and IR. Various spectroscopic probes have actively been used in in-process analysis of additives in polymeric materials (mid-IR > UV/VIS > NIRS > Raman). Spectroscopic sensors and multivariate computational techniques allow extraction of information from complex systems, for the purpose of process monitoring, feedback and quality assurance. Amongst in-process analysis techniques, the application of mid-IR with PMD evaluation is of great interest [117]. NIRS has shown good potential for selected additive analysis in process control. Where extensive model building is required (as in the case of NIRS) spectroscopic techniques are less suitable for R&D purposes. There are some signs of diminishing efforts in the field of in-process polymer/additive analysis, which probably stands in relation to the difficulty of multicomponent analysis.

In the field of in-process analysis, analytical NMR applications also constitute a growth area – and also in relation to additives. This stems from the fact that the method makes it possible to use chemical analytical data in polymer quality control. Robust tools for hostile chemical plant environments are now available. The field of process analytical chemistry has been pushed to the forefront of the partnership between industry and academia.

10.4.2.9 Quantitative Analysis

As may be obvious from previous chapters, quantitative analysis requires more substantial advancements to be made than qualitative analysis (library-based fingerprinting, screening, identification, recognition). For many polymer/additive problems, the classical methods are usually sensitive enough, and sophisticated instrumental methods are available, allowing analytical chemists to probe samples for components at much lower concentration levels.

Quantitative analysis of multicomponent additive packages in polymers is difficult subject matter, as evidenced by results of round-robins [110,118,119]. Sample inhomogeneity is often greater than the error in analysis. In procedures entailing extraction/chromatography, the main uncertainty lies in the extraction stage. Chromatographic methods have become a ubiquitous part of quantitative chemical analysis. Dissolution procedures (without precipitation) lead to the most reliable quantitative results, provided that total dissolution can be achieved; follow-up SEC-GC is molecular mass-limited by the requirements of GC. Of the various solid-state procedures (Table 10.27), only TG, SHS, and eventually Py, lead to easily obtainable accurate quantitation.

More attention should be devoted to the quantitative determination of analytes in real-life samples. The accuracy of the determinations and the traceability of the overall analytical process are insufficiently ensured [120,121]. As no primary methods are available for the purpose, this necessarily implies the use of certified reference materials.

10.4.2.10 Quality Assurance

At the present time, when customers and public opinion are demanding the accountability of laboratories

carrying out analyses related to socially sensitive areas, such as recycling, product safety, environmental pollution, etc., the importance of harmonising protocols for quality assurance (QA) schemes is obvious. In this respect the chemical industry has arrived at a full and detailed internal evaluation. Internal QA schemes can then form the basis for third-party assessment. Industry is rapidly adopting the ISO 9000 series of standards for quality assurance and quality management (QM). In this framework, laboratories must be able to validate the precision of analytical measurements by an unbroken and fully documented chain from these measurements upstream to the source. It is equally important that the quality of the databases used is impeccable.

Quality assurance considerations lead to the need for appropriate reference materials, and their consistent and effective use to monitor the precision and accuracy of laboratory analyses. In this context, *certified reference materials* (CRMs), now still largely lacking in the polymer/additive area, play an important role. In previous years, some attempts have been undertaken to prepare some inorganic CRMs (VDA and PERM projects), but this is highly insufficient when we consider that some 60 elements are used in polymer/additive formulations. The lack of CRMs for organic compounds in polymeric matrices is an even more serious handicap. Nagourney and Madan [122] have demonstrated that intermediate or finished in-house materials can be utilised successfully as QA reference materials. Good QC of polymer/additive formulations as yet has not been achieved.

The globalisation of trade and the growing importance of analytical tools determine an increasing demand for international *standardisation* of the latter. The development of transnational industrial sectors urgently requires international standards for analytical methods. Such internationally accepted analytical protocols as yet are almost completely lacking in the field of polymer/additive analysis. *Validation* and/or collaborative studies that demonstrate satisfactory method performance are needed to move many of the research methods to a regulatory level, from which we all benefit. This monograph illustrates the difficulty of reaching international consensus on analytical tools, which are very diverse and complementary, all with a proper application area.

10.4.3 Analytical Challenges

The determination of the most appropriate method for a given analytical problem is the first demanding task. Routine polymer/additive analysis, where the nature and approximate concentrations of the components are known, is difficult enough. Even more skill and experience are required for more difficult cases:

(i) determination of 'reactive' type additives with formation of transformation and/or interaction products;
(ii) colour body analysis;
(iii) exposed samples;
(iv) recycled samples; and
(v) migration problems.

Identification and resolution of *colour body problems* is especially demanding, because the colour bodies are often unstable and present at very low concentrations (subppm). Typical analytical tools used here are extraction followed by HPLC-DAD or LC-MS. Handling exposed samples always presents a challenge in analysis, as chemical reactions may have taken place during polymer degradation. To determine the original concentration of the additive prior to ageing requires more deductive analysis. Understanding analytical results from *recycled samples* poses an even greater challenge, because there is usually no information about the history of the plastic parts in the recyclate. Polymer cross-linking may well lead to incomplete extraction, and direct techniques are often even less suited. For illustrative examples, see ref. [123].

Polymer samples that are either insoluble, infusible, highly filled, highly foamed or in a suspension, all present an analytical challenge. Some examples of polymer samples that are difficult to analyse are brittle polymers, highly cross-linked samples (elastomers, thermosets), highly carbon-black or mineral-filled polymers, glass-reinforced composites, high-impact polyblends, foams, multilayer films, fibres, surface-modified polymers, viscous samples (e.g. prepregs), liquid paints, large mouldings, etc. [124].

The difficulty of analysing *high-MW additives* is best exemplified for Chimassorb 944 and related oligomeric HALS compounds, where a battery of analytical tools has been used over the years (PyGC, HPLC-FID/UV, SEC-UV, RPLC-UV, D/P-HPLC, etc.), ending up with μHTLC-ELSD [125], see Section 4.4.2.2. The theoretically unlimited possibilities of tailoring the molecular weight of silicon-bound stabilisers [22], the functionality and structural characteristics, represent a considerable challenge for improved end-use performance, as well as in analytical terms. For case studies of *extraction resistance* (high-MW additives, polymer-bound functionalities, additive adsorption, insoluble additives), see Section 3.6.3. Oil-extended SBR represents a demanding extraction matrix, due to the large amount of total extractables (up to 40 %), as well as the wide variety of compounds present in the sample (extender oils, organic

acids, AOs, cross-linking agents, binders, accelerators, carbon-black, fillers). In this field, an additional analytical challenge is the determination of thermolabile low-MW additives such as nitrosamines in cross-linked rubbers. A recent difficult problem has been the analysis of nanoscopic fillers.

One of the greatest challenges in the laboratory is to prove that accurate results have been generated for unknown samples.

10.4.4 Polymer/additive Analysis at the Extremes

The range of techniques which can now be used routinely in an industrial laboratory has probably never been wider. Polymer/additive analysis takes advantage of all the instrumental possibilities, up to their limits (Table 10.28). For example, LVI-μHTLC can be used for the determination of very low additive concentrations (0.6 ppm Irganox 1076 in LDPE) [97]. However, polymer/additive analysis is not a sample-limited field where advancements such as nano-LC (capillary internal diameter $<100\,\mu$m)/nanoflow electrospray interface/QToFMS, allowing flow-rates of $150\,$nL min^{-1} and MS/MS spectra of as little as 1 fmole per component, excite the analyst. On the other hand, high-performance polymer/additive analysis does benefit from on-line simultaneous analysis, fast screening, high sensitivity and high resolution. Typical high-throughput sample techniques were given in Section 10.4. The limits of the usefulness of miniaturisation of the analytical tools for polymer/additive analysis have been pointed out repeatedly (e.g. for SFE-HPLC, CE-ESI-MS). Although integration of the overall analytical process – sample preparation, analyte separation, and detection/quantification – in one total chemical analysis system (TAS) is well under way, the results are not always encouraging (e.g. SFE-SFC-MS), despite the prospective advantages. For most analysts, on-line LC-NMR and LC-NMR-MS are currently still beyond the limits of this field. More immediate use can be expected of LC-multi-API-MS (APEI, APCI, APPI). In view of the generally limited sample streams for polymer/additive analysis, ultrafast GC (separation in seconds) is equally beyond practical needs.

Table 10.29 shows the variability of *operating conditions* in various polymer/additive analysis techniques.

With improving detector performance, the smaller can be the sample size and, consequently, the more rapid the sample pretreatment. However, as shown repeatedly in quantitative analysis, small sample sizes (several mg) face homogeneity problems and set a

Table 10.28 Limits of instrumental equipment for polymer/additive analysis

Feature	Upper limit	Lower limit
Temperature	HT-GC, HT-PTV, HTLC, HT-SEC, ICP-AES, ICP-MS, MIP-AES, MIP-MS, LD, LIP-AES, HTA	SFE, SFC, SWE, LTA
Pressure	SFE, PHWE, PFE, HPSE, PLE, PMD, HPLC, OPLC, HPPLC, PD-SEC, (H)PDSC, HP-OIT, HPSEM	SIMS, XPS, AES, TEM, SEM
Speed	SFE, SPE, MAE, ASE®, HSGC, HSLC, HSTLC, SEC-LS, SPE-TLC, EDXRF, EPMA, EN	Soxhlet, D/P
Wavelength	US	UV/VIS, mid-IR, NIR
Voltage	CE	LVEI, LV (SEM)
Performance	HPLC, HPSEC, HPTLC, HPCE	–
Sensitivity	F, LVI-μHTLC	–
Selectivity	ESD, LD, REMPI, MS, ISE, AAS, RRS, SERS	FID
Specificity	LC-MS[a]	UV, FTIR[b]
Resolution	HR-CGC, GC-ToFMS, HRMS, HR-ICP-MS, MS/MS, DI-HRMS, HRLEELS, HRTGA	LR-NMR, LRRS
Spatial resolution	LA-ICP-MS, LMMS, LIBS, LDI, LTA, SAX	NMRI
Molecular mass	ESI, MALDI	PB, TG-MS, TD
Fragmentation	EIMS	CI, FI, FAB, LD, API
Dimensionality	MHE, MDGC, GC-MAB-MS, MDHPLC, TLC, MS^n, L^{2}MS, LPyMS/MS, MPI	1D
		2D (XPS, AES, SIMS)
Orthogonality	Chromatography–spectroscopy, NPLC-RPLC	–
Column technology	MC-GC	Monocapillary
Equipment size	SR, NAA, sector MS	μGC, μHPLC, CE, μECD, μTCD
Instrumental complexity	On-line LC-UV-NMR-QITMS, LC-multi-API-MS, GC-AES/MS, multihyphenation, multidimensionality	Off-line
Cost	FTICR-MS, 900 MHz NMR	Soxhlet, SPME, SPE
Technique	Cutting-edge	Obsolete

[a] Substance-specific.
[b] Substance class-specific.

Table 10.29 Extreme operating parameters of polymer/additive analysis techniques

Feature	High	Low
Volume	Gas phase, LVI, LV HSI, LVS, PLC	Liquid/solid phase, CIS, CT, SPME, nano-ESI
Solvent use	Soxhlet, MAE, ASE®, HPLC	SPME, SHS, DHS, DIMS, Py, TD, LD
Temperature variation	TPD, TPPy, HTGC, HTLC	CT, CIS
Time	Simultaneous, dual detection	Sequential
Speed	Real-time	Soxhlet
Wavelength dependency	Selectivity, sensitivity	–
Destructiveness	Extraction, dissolution, chromatography	NDE, NDT, LD
Spatial resolution	Scanning, SEM, STEM	OM, NMRI
Remoteness	On-line	Off-line
Angular dependency	TRXRF	XRF
Modulation	MTDSC, MTDTA, MTGA	DSC, DTA

limit to the usefulness of miniaturised analytical tools for polymer/additive analysis.

Table 10.30 describes the general characteristics of additive analysis problems. For polymer/additive analysis, the analyst now has at its disposal powerful tools with a proven record of being efficient in industrial problem solving. Historically, the analysis of polymer/additive packages has experienced a long series of tool development, as may be seen by comparing classic textbooks covering the literature up to 1970 [126], 1985 [127,128] with the current text. At the present pace of development, this monograph may well be outdated in less than a decade. The future of in-polymer additive analysis will be strongly affected by developments in many instrumental disciplines. Factors that inhibit the breakthrough of new technologies include the availability of commercial instrumentation; the ability to significantly improve an analysis by objective measures and by solving real-world problems; minimal learning requirements; and the substantial reduction of the labour, time and cost involved in the analysis [129]. Looking back over the past 20 years, various proposed polymer/additive technologies have not met

Table 10.30 Complexity of the polymer/additive analytical problem

Feature	High	Low
Sample size	SPE, SPME, µSEC, TG, Py, µTA, µFTIR, µRS, µXPS, STM, AFM, LA-AES, MRS	No sample-limitation, extraction technology, preparative scale
Sample nature	Heterogeneous	Homogeneous
Mass	Several 1000 Da	Low 100 Da
Surface	XPS, AES, SIMS, SE(R)RS	Bulk methods
Nature of analyte(s)	Oligomers, high-MW, thermally instabile, involatile, polymer-bound, multicomponent, surface modifications, trace level, speciation	Monomers, mono-component, high concentrations, low-MW
Selectivity	Matrix-dependent	No interferences

such requirements. The persistence of Soxhlet extraction (as from 1879) is illustrative; the application of TG-MS is hampered by the learning requirements. Some other techniques such as SFC and CZE face an uncertain future in this area. It is equally of interest to note the tools that did not survive the test of time. 'Wet' analytical chemistry is less fashionable nowadays in polymer/additive analysis. The same holds for classical gravimetric, colorimetric, and volumetric analysis and various electroanalytical techniques (now mainly existing only as detectors). Other examples of declining or *obsolete techniques* are TVA, paper chromatography, and various mass-spectrometric methods (e.g. moving belt, thermospray, FI, FAB), including SSMS (no longer commercially viable). GD-MS is struggling for survival. Other techniques have evolved, such as LC to HPLC and µHPLC, TLC to HPTLC, IR to FTIR, in-flow LC-FTIR to evaporative LC-FTIR, etc. Currently available methods that are under-utilised are image analysis for quantification of results; chemiluminescence for early detection of oxidative processes; and acoustic emission for subvisible mechanical damage measurement.

In preceding chapters we have indicated which tools are nowadays being used routinely or currently are under development. General trends are higher sensitivity, more information, and faster and further automation. Automatic analyses are nice (sample in, report out), but interactive analysis tools are better. It is not realistic to expect the need for more analyses. Some future needs are more reliable quantitation, reference materials and simplification of data management. A particular problem in additive analysis concerns accuracy and traceability. In many cases, extractable rather than total concentration is determined. There are still many quantitative analytical methods waiting to be developed. It is here that the field will advance. Table 10.31 lists some proposed *(r)evolutionary developments* in polymer/additive analysis.

For the use-inspired analytical chemist, it is crucial to stay on top of the latest advances in science

Table 10.31 Future directions in polymer/additive analysis[a]

- Accelerated dissolution of polymers (3)
- Application of PHWE (2)
- Development of organic CRMs (3)
- Ultrahigh pressure HPLC (3)
- Application of GC-REMPI-ToFMS (2)
- Combined thermal desorption/pyrolysis using PTV-HTGC-FID/ToFMS (1)
- Use of vacuum TG-MS and HR-TG-MS-PCA (2)
- Reliable quantitation (3)
- Application of LC-SMB-MS (2)
- Use of LC-multi-API-MS (APCI, APEI, APPI) (2)
- UV Raman microspectroscopy (2)
- Raman spectroscopy (1)
- PA-FTIR for surface profiling (2)
- Quantification by chemiluminescence methods (3)
- NMRI (2)
- Direct quantitative MS probe for complex analyte mixture (5)
- Simplification of data management (4)
- HPLC-UV/NMR/IR/MS (6)

[a] In increasing order of difficulty: trivial (1), pragmatic (2), ambitious (3), unrealistic (4), impossible (5); overkill (6).

to catch the technical edge, yet to stay away from the hype, which is not needed in polymer/additive analysis (CE, CEC, µTAS, ultra microanalysis). Novel techniques are currently under development that aim at enhancing analyte extraction efficiencies, improve cleanup processes and permit the analysis of additives in polymers *in situ*. These include laser-excited fluorescence, thermal-desorption mass spectrometry (TD-MS) and remote laser-induced fluorescence (RLIF). Trends in optical microanalysis (UV/VIS, near-IR, mid-IR) are near-field microscopy (sub-µm to nm), near-field microspectroscopy (IR, R; from 1 µm to 200 nm), surface-enhanced infrared analysis (SEIRA) and multilayer depth profiling by new ATR and PAS approaches. Laser-acoustics can be used for testing thin coatings (10 nm) on good sound-guiding substrates; the method is nondestructive, easy-to-use, quick (30 s) and requires no sample preparation. Polymer/additive analysis may benefit from techniques such as HRTG-MS-PCA, UV Raman microspectroscopy and NMRI. Current *cutting-edge techniques* are TD-CIS-GC-MS, GC-REMPI-ToFMS, LC-multi-API-MS, LC-QToFMS, L^2ToFMS, MALDI-FTMS, MPI-FTMS, MPI-ToFMS, PS-MS, speciation analysis, iSIMS, NSOM, μTA and EN. Generally, the more complex a technique, the more application specific it is. GC-REMPI-ToFMS offers selectivity, sensitivity and mass accuracy. L^2ToFMS can differentiate isomeric UV stabilisers. L^2MS using REMPI allows for highly sensitive and selective detection of target molecules directly in the solid state, without interference of the polymeric matrix. Post-ionisation at various wavelengths is an extremely powerful technique. These techniques are not simply toys for analytical champions, but point to an interesting future for polymer/additive analysis.

10.4.5 Advanced Polymer/Additive Deformulation Schemes

After careful reading up to this page, the reader might easily feel more uncomfortable than before consulting this book. Being highly interdisciplinary, polymer/additive analysis is not plain sailing. One could wonder whether there is an *ideal method* for polymer/additive analysis. The obvious answer is that there is none. Polymer/additive analysis is a problem far away from the mathematically and physically well-described relationship between a crystal and diffraction, where XRD is the structural analytical tool *par excellence*, providing a virtually exact *total solution* in the greatly over-determined condition of a diffraction pattern with respect to the structural parameters to be established. Polymer/additive analysis is not in such a favourable condition. No litmus paper solution can be expected here. It is usually necessary to use several analytical tools to confirm with certainty the composition of an additive system. Any physico-chemical technique (GC, HPLC, UV, IR, NMR, MS, etc.) reveals here only part of a complex reality. Nevertheless, in specific cases, a simple UV measurement of a thin film containing a UV absorber can turn out to be the ideal method. Clearly, a UV spectrum is usually not sufficiently selective for unambiguous attribution of an unknown component in admixtures, whereas an IR spectrum yields (other) fragmentary information; generally, a retention time is equally too thin a basis for appropriate assignment, and MS data contain much additional useful information. In combination, however, unambiguous assignment may be arrived at. It is the task of the analytical chemist to

Table 10.32 Requirements for the ideal polymer/additive analysis technique

• All-in-one-solution	• Appropriate fit to application
• Universal	• Rapid analysis times
• Ideal sample	• High sensitivity for all additive classes
• Suitable for solid and liquid probes	• Analyte specific
• No sample preparation/separation	• Quantitative (standardless)
• Solubility not required	• Good reproducibility; traceability
• Suitable for low- and high-MW analytes	• Low cost of ownership
• Mass accuracy (elemental composition)	• No environmental impact
• No matrix interferences, no thermal stress	• Commercial instrumentation
• Limited method development effort	• Access to general databases

choose the *optimal solution* within given boundaries of available instrumental possibilities.

Table 10.32 is a shortlist of the characteristics of the ideal polymer/additive analysis technique. It is hoped that the ideal method of the future will be a reliable, cost-effective, qualitative and quantitative, in-polymer additive analysis technique. It may be useful to briefly compare the two general approaches to additive analysis, namely conventional and in-polymer methods. The classical methods range from inexpensive to expensive in terms of equipment; they are well established and subject to continuous evolution; and their strengths and deficiencies are well documented. We stressed the hyphenated methods for qualitative analysis and the dissolution methods for quantitative analysis. Lattimer and Harris [130] concluded in 1989 that there was no clear advantage for direct analysis (of rubbers) over extract analysis. Despite many instrumental advances in the last decade, this conclusion still largely holds true today. Direct analysis is experimentally somewhat faster and easier, but tends to require greater interpretative difficulties. Direct analysis avoids such common extraction difficulties as:

(i) variable extraction efficiencies of different solvents; and
(ii) the evaporative loss of more-volatile components (e.g. accelerator fragments and residual monomers/solvents).

Direct analysis also provides the opportunity to identify both the organic additives (by thermal desorption) and polymer matrix components (by pyrolysis) in TPPy experiments. The two approaches are not competitive but complementary.

If we consider only a few of the general requirements for the ideal polymer/additive analysis techniques (e.g. no matrix interferences, quantitative), then it is obvious that the choice is much restricted. Elements of the ideal method might include LD and MS, with reference to CRMs. Laser desorption and REMPI-MS are moving closest to direct selective sampling; tandem mass spectrometry is supreme in identification. Direct-probe MS may yield accurate masses and concentrations of the components contained in the polymeric material. Selective sample preparation, efficient separation, selective detection, mass spectrometry and chemometric deconvolution techniques are complementary rather than competitive techniques. For elemental analysis, LA-ICP-ToFMS scores high.

Few laboratories in the world (probably none) will have access, directly or indirectly, to the many polymer/additive tools mentioned in previous chapters. Actually, there would even appear to be little need for such a broad diversification in a single dedicated laboratory, especially in industry, where return-on-investment is an important factor in any decision-making process aiming at the introduction of yet another emerging analytical technique. Many tools are not mutually exclusive (e.g. Soxhlet, SFE, MAE, ASE®, or GC, SFC, HPLC, TLC). Yet, some are definitely to be preferred to others, in the particular circumstances of a given laboratory, as in the case of a laboratory which has to deal with very general questions (e.g. R&D) or just with more specialised or standardised ones (plant support laboratory, in-process analysis). Whereas the high end of the market (R&D) might need a four-sector MS, LC-MS, and s-NMR, the low end (process control) might prefer LR-NMR, NIRS and ASE®. In all cases, equally sophisticated (new) analytical techniques are needed for efficient analysis.

The best method or the most suitable combination of methods can be discussed only in regard to the actual analytical problem. The ideal method for polymer analysis in an industrial environment is often essentially that practical one which identifies and quantitates the desired components at the lowest acceptable total cost for the customer, compatible with the desired accuracy and time constraints. Three examples may illustrate the necessary *pragmatic trade-off*. Despite being old methods, classical polymer/additive analysis techniques, based on initial additive separation from the polymer matrix through solvent extraction methods followed by preconcentration, still enjoy great popularity. This

time-consuming procedure may introduce problems, due to variable extraction efficiencies. Although this is analytically unsatisfactory, it is nevertheless often tolerable in industrial problem solving. On the other hand, after extraction, LC-QToFMS/MS is in principle a very desirable method, providing a highly accurate molecular mass in combination with a fragmentation pattern (for recognition), and is particularly appealing in the case of high throughput for unknowns, but otherwise not cost-effective for low sample throughput. Finally, if one were to come to the conclusion that one's problem might preferably be solved with PIXE or FTICR-MS, one would be well advised to step back and reconsider the choice. It is quite likely that more easily accessible techniques would suit the case. However you solve the problem, use your whole tool-box. Many well-defined problems can actually be solved using rather simple means. UV spectrophotometry performs quite well for the quantitative determination of a UV absorber in polyolefins. However, the problem becomes rapidly more complicated when more than one UV absorbing component is present, and therefore the method does not provide a generally applicable solution to the problem.

This textbook essentially provides a survey of state-of-the art methods for polymer/additive analysis and their inherent *limitations*, in order to make the analyst aware of the realistic expectations of their equipment ('fitness for use').

It is generally difficult to identify developments with high potential where interferences do not preclude general application. To ensure the *relevance* of a method, its application to real sample analysis must be demonstrated. The accuracy of an analytical method should be confirmed by an independent method, or by the analysis of certified reference materials. Detailed comparative studies of the method developed with other well-established methods for polymer/additive analysis are not frequent in the analytical literature. Nevertheless, some examples may be found in Section 3.6. Improvements in analytical techniques are reasonably sought in sample preparation and in hyphenated chromatographic techniques. However, greatest efficiency is often gained from the use of databases rather than accelerated extraction or hyphenation.

Where necessary, *analytical artefacts* have been stressed, as in the case of chromatography, with the examples of stabilisers and other polymer additives [131]. Pitfalls have also been described, as, for example, interferences in GC-MS between matrix components present at high concentrations, and analytes in low concentration. These interferences may lead to false-positive and -negative results, and false quantification.

This may often be avoided either by using slightly modified methods (e.g. different columns, reagents) or by the use of a completely different physico-chemical technique, based on other principles. This reduces the chance of missing out crucial information.

Various techniques have been introduced which still lack specific applications in polymer/additive analysis, but which may reasonably be expected to lead to significant contributions in the future. Examples are LC-QToFMS, LC-multi-API-MS, GC-ToFMS, Raman spectroscopy (to a minor extent), etc. Expectations for DIP-ToFMS [132], PTV-GC-ToFMS [133] and ASE® are high. The advantages of SFC [134,135], on-line multidimensional chromatographic techniques [136,137] and laser-based methods for polymer/additive analysis appear to be more distant. Table 10.33 lists some *innovative polymer/additive analysis protocols*. As in all endeavours, the introduction of new technology needs a champion.

Two quantitative dissolution schemes have been proposed, consisting of SEC with GC [137] or LC [143] follow-up. The former protocol is in regular industrial use (but mostly as a local application only), the latter requires testing for polymer/additive analysis. The operational MAE + HPLC-ELSD/UV scheme lacks universality – being designed for polyolefins and quantifies extractables only [141]. Miniaturised LC in combination with temperature programming (LVI-TP-pcLC-UV) is a powerful technique for the determination of very low levels of additives in polyolefin extracts [97]. Gradient HPLC-ELSD/PDA is not universal [142]. The PHWE-LC-GC method offers a challenging opportunity for polymer/additive analysis for polar analytes. Taylor's

unified sample preparation mode for extraction of various matrix/analyte combinations (polar and nonpolar), based on a hybrid SFE/ESE® system [149], needs exposure to real-life industrial samples for further verification. PTV-HTGC-ToFMS is currently being tested for polymer/additive analysis [133]. Other possibilities (to be tested) are based on thermal methods: TG screening (for definition of pyrolysis temperature) followed by rapid PyGC-MS screening (for identification), and vacuum TG-MS (for quantification of significant masses) with multivariate analysis; no oligomeric interferences are to be expected here. The above list of concepts on how to analyse polymer additives is extensive, but nevertheless incomplete. It represents a rich source of ideas, from which the best should be chosen, combined and further elaborated.

Rather than being a scientific field on its own, polymer/additive analysis is an expertise supposed to serve as a tool for problem solving. It is necessary to make a selection. Not all of the many analytical protocols presented can become standard methods. It will be essential for the future of polymer/additive analysis that progress is made towards accepted techniques and broad use. The success of a technique requires extreme simplicity.

The keen observer may have noticed that modern methods of additive analysis in polymers overlap with those in allied areas (Figure 10.4): rubber [150,151], paints [152,153] and coatings [152], adhesives [153], inks [153], food [154], impregnated paper and faced paperboard [155], etc. Clearly, tool-boxes will differ to some extent, as do the analytes.

Table 10.33 Advanced polymer/additive analysis protocols[a]

- Dissolution + μSEC-HTGC-MS [136–138]
- GC-SMB-MS [139]
- GC-JET UV-REMPI/LEI-reToFMS [140]
- PTV-HTGC-ToFMS [133]
- MAE + HPLC-ELSD/UV [141]
- MAE/PLE + LVI-TP-pcLC-UV [97]
- LVI-μHTLC-ELSD/UV [97,125]
- Gradient HPLC-ELSD/PDA [142]
- Dissolution + SEC + LC-LC [143]
- PHWE-LC-GC [144,145]
- LC-SMB-EI-MS [146]
- LC-APCI-MS, LC-QToFMS, LC-multi-API-MS [107]
- HPLC(SEC)-UV-MS-NMR-IR [147]
- Fast PyGC-MS (for QC)
- TD/Py-REMPI-ToFMS
- TG, PyGC-ToFMS, vac TG-MS + MVA
- MAE + SPME-MS^n (for rapid odour analysis)

[a] After Bart [148]. Reprinted from *Polymer Degradation and Stability*, **82**, J.C.J. Bart, 197–205, Copyright (2003), with permission from Elsevier.

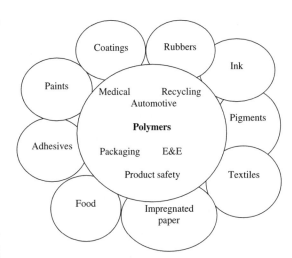

Figure 10.4 Extended additive analysis

10.5 EPILOGUE

Large-volume standardised products are entering the market, together with sophisticated niche products, while the technical service needed for specialty grades is under pressure. At the same time, the technological possibilities for the improvement of polymeric materials (intended as a suitable combination of polymer and additives with desired properties) are still enormous, including process improvements for better and quicker incorporation of fillers and pigments and consistency of processing. With today's industrial laboratories being under mounting pressure to maintain high productivity to meet performance and to reduce costs, more incremental developments, rather than radical developments, may be foreseen. In these times of globalisation, stronger competition, degradation of scientific organisations, mobility, flexibility, redundancies, outsourcing, restructuring, concentrations, consolidation, divestures, mergers and joint-ventures in the polymer industry, laboratory analysis cannot fail to pass through a period of rapid evolution. Some of the challenges posed to the industrial laboratory are summarised in Table 10.34.

On the whole, nowadays, the analytical chemist is being asked far more sophisticated questions than they ever were in the past. These concern more complex polymer/additive systems, lower additive concentrations, polymeric additives, grafted additive functionalities, distribution, migration, imaging and more reliable quantitation (spatially resolved or not). The trend for more detail is not likely to stop easily, in view of the ever-increasing quality awareness of the modern customer, and this determines the need for sophisticated, state-of-the-art tools. Experience learns that profit flows to organisations that insist on the highest standards in every aspect of business, including analytical chemistry. Analytical chemistry is critically important in product and process development processes and beyond.

Table 10.34 Challenges facing the industrial laboratory

• Increasingly competitive environment	• Chemical safety
• Limited resources, staff turnover	• Spent chemicals' disposal
• Changes in job responsibility	• Changes in technology and regulations
• Knowledge management	• Technician capability
• Cost containment	• Analytical capability
• Productivity gains	• Reliability of test results (quality)
• Timeliness of test results	• Interdisciplinarity

At universities at the turn of last century, analytical chemistry was the very heart of the chemistry curriculum: in fact, chemistry was *synonymous* with analytical chemistry. In the 1920s organic chemistry took a fair share; the 1930s saw the upsurge of physical chemistry, the 1940s a renewed emphasis on inorganic chemistry, the 1950s polymer chemistry with strong development of materials science in later years; in the 1960s organometallic chemistry came in focus, in the 1970s quantum chemistry was at the forefront, the 1980s saw a great interest in surface chemistry, and in the 1990s supramolecular chemistry was developed. In most of these areas analytical chemistry played a vital role. Despite these contributions, nowadays *academic analytical chemistry* is struggling for survival in many industrialised countries. Actually, Robinson [156], some 35 years ago feared that academic analytical chemistry was close to extinction. Findeis and Wilson [157] also considered analytical chemistry as a fading discipline. The image of analytical chemistry is certainly a troubled one for some [156–159]. It is a branch of chemistry that suffers from high fragmentation, and many chemists pretend to be able to perform their own analysis. With the high complexity of today's chemical systems, this may only be justified in part. It is actually surprising that the discipline of analytical chemistry has not been more fashionable in the past decade or so, which saw the renaissance of analytical opportunities. The visible disappearance of analytical chemistry from the undergraduate curriculum in chemistry will not fail to translate itself into future problems downstream and affect the profession. This textbook should give unbelievers food for thought. Both the possibilities for application and the interrelationships between methods of analytical chemistry are usually far from obvious for the neophyte.

The strength of analytical chemistry is its applicability to problem solving, preferably at the right level (speed, efficiency, required accuracy, selectivity, etc.). Not surprisingly, therefore, industry requires and has provided strong analytical support. This monograph shows that *applied industrial analytical chemistry* is still very much alive. Analysis of polymeric materials is typically an area in which the analytical chemist is faced directly with problem-solving aspects. The search for an optional analysis method renders the analytical activities very challenging, and requires the mind of a detective. However, if polymer/additive analysis is to survive as an industrial expertise, it should not be a drain on the finances, but a source of cost savings for the company concerned.

Progress in the polymer industry has always come from two sources: academic institutions *and* industrial laboratories. This will certainly remain the case for the future, but even more so, as, at present, the industry focus is almost entirely on applied research and development. As a consequence, the polymer industry must rely more than ever on the fundamental research done at academic institutions.

Gazing in the polymer/additive crystal ball appears to indicate that the following future developments affecting polymer/additive analysis will take place:

(i) new polymeric structures (mainly in terms of better control on polymeric structures, less so in terms of entirely new polymers);
(ii) slowing down of the pace of development of new additives;
(iii) niche markets for polymer producers (grafted additives);
(iv) better-performing instrumental methods;
(v) greater complexity of analytical problems;
(vi) greater impact of quality assurance schemes; and
(vii) diminishing academic analytical support.

Despite the fact that polymer/additive analysis is a growth area (Figure 1.2), and taking into account the aforementioned considerations, it could well be that the near future will see a more modest increase beyond the current 400 research papers per year on the topic of polymer/additive deformulation. This book has provided a roadmap to polymer/additive analysis. It is up to the users to make ingenious use of the (necessarily limited) equipment at their disposal for their specific problem-solving cases.

Folks, that's all, the challenge remains at the bottom line:

The more you know, the more you forget.
The more you forget, the less you know.
The less you know, the less you forget.
The less you forget, the more you know.

10.6 BIBLIOGRAPHY

A.J. Peacock, *Handbook of Polyethylene*, M. Dekker, New York, NY (2000).

Special issue *Today's Chemist at Work* (March 15, 1999); special issue *Anal. Chem.* (March 1, 1999).

J.W. Gooch, *Analysis and Deformulation of Polymeric Materials*, Plenum Press, New York, NY (1997).

J.R.J. Paré and J.M.R. Bélanger (eds), *Instrumental Methods in Food Analysis*, Elsevier, Amsterdam (1997).

L.J. Calbo, *Handbook of Coatings Additives*, M. Dekker, New York, NY (1987).

10.7 REFERENCES

1. K. Scheidl, *Proceedings Specialty Plastic Films (SP '97)*, Zürich (1997).
2. K. Scheidl, *Proceedings ETP '98 (Engineering Thermo Plastics)*, Zürich (1998), pp. 1–5.
3. M. Costes, *Proceedings Plastics Trends '99*, Brussels (1999).
4. Department of Environment, Transport and the Regions (UK), *Life Cycle Assessment of Polyvinyl Chloride and Alternatives*, DETR, London (2001).
5. S.J. Risch, *ACS Symp. Ser.* **753**, 1–7 (2000).
6. N. Kashiwa and S. Kojoh, *Preprints IUPAC Polymer Conference on the Mission and Challenges of Polymer Science and Technology (IUPAC-PC 2002)*, Kyoto (2002), pp. 691–2.
7. D. Freitag, *Preprints IUPAC 37th International Symposium on Macromolecules*, Gold Coast (1998), p. 996.
8. R. O'Rourke and C. Carroll, *ECN Process Rev.* pp. 21–3 (July 2002).
9. D.L. Buszard, *Proceedings Polypropylene '98*, Zürich (1998).
10. J.S. Murphy, *The Additives for Plastics Handbook*, Elsevier Advanced Technology, Oxford (1996, 2001).
11. Townsend Tarnell Inc., *Chemical Additives for Plastics – 1999*, TTI, Mount Olive, NJ (1999).
11a. BRG Townsend Inc., *Plastics Additives in China, 2003*, Mount Olive, NJ (2003).
12. C. Bruens, R. Nieland and D. Stanssens, *Polym. Polym. Comp.* **7** (8), 581–7 (1999).
13. DSM Venturing & Business Development, *Profile Sheet 7, Hybrane®*, DSM, Heerlen (2002).
14. H. Zweifel (ed.), *Plastics Additives Handbook*, Hanser Publishers, Munich (2000).
15. M. Stern, *Proceedings SPE ANTEC 2000*, Orlando, FL (2000), pp. 1772–6.
16. W. Voigt and R. Todesco, *Polym. Degrad. Stab.* **77**, 397–402 (2002).
17. U. Kanja, H. Ohleier and P. Wetzel, *Kunststoffe* **88** (3), 362–6 (1998).
18. R.E. King, *Proceedings SPE ANTEC '99*, New York, NY (1999), pp. 2266–73.
19. D. Horsey, M. Bonora and P. Holbein, *Proceedings Polyolefins XI*, Houston, TX (1999), pp. 541–53.
20. S.B. Samuels, J.M. Eng and A.H. Wagner, *Proceedings Polyolefins XI*, Houston, TX (1999), pp. 521–39.
21. J. Malík, L. Avár and M. Zäh, *Proceedings International Conference on Additives in Polyolefins*, Houston, TX (1999), pp. 17–34.
22. R.L. Gray, R.E. Lee and C. Neri, *Proceedings Polyolefins X*, Houston, TX (1997), pp. 599–613.
23. Th. Schmutz, E. Kramer, H. Zweifel and G. Dörner, *J. Elastomers Plast.* **30** (1), 55–67 (1998).
24. R.F. Grossman and M.B. Detweiler, *Proceedings Polyolefins 2001*, Houston, TX (2001), pp. 483–7.
25. R.F. Grossman and M.B. Detweiler, *J. Vinyl Addit. Technol.* **7** (1), 24–5 (2001).

26. R.E. Lee, V. Malatesta, R.M. Riva, M. Angaroni and C. Neri, *Proceedings 9th International Conference on Additives 2000*, Clearwater Beach, FL (2000).

27. R. Pfaendner, *Kunststoffe* **89** (7), 76–9 (1999).

28. M. Schiller, G. Zuschnig, A. Egger and C. Videler, *Proceedings PVC 99*, Brighton (1999), pp. 370–6.

29. P.J. Clucas and T. Hjertberg, *Preprints IUPAC 37th International Symposium on Macromolecules*, Gold Coast (1998), p. 373.

30. W.H. Starnes, *Abstracts Second International Conference on Polymer Modification, Degradation and Stabilisation (MoDeSt2)*, Budapest (2002).

31. J. van Haveren and D.S. van Es, *Kunststof Rubber* **1**, 33–4 (2003).

32. J. van Haveren, D.S. van Es, J. Steenwijk and C. Boeriu, *Proceedings 11th International Conference on Additives 2002*, Clearwater Beach, FL (2002).

33. Süd-Chemie, *Application Note Sorbacid®, Heavy Metal Free PVC Stabilisation*, Süd-Chemie AG, Moosburg, Germany (2001).

34. J. Murphy, *Plastics Addit. Compd* **1** (5), 12–17 (1999).

35. P. Solera and G. Capocci, *Polym. Polym. Compos.* **7**, 521–36 (1999).

36. J.R. Pauquet, *Plast. Rubber Compos. Proc. Appl.* **27** (1), 19–24 (1998).

37. M. Papez, *Proceedings Flame Retardants '94* (The British Plastics Federation, ed.), London (1994), pp. 9–20.

38. D. De Schrijver, *Proceedings 3rd European Additives and Colors Conference* (J.M. Liégois, ed.), SPE Benelux, Antwerp (2003), pp. 46–51.

39. V. Wigotsky, *Plast. Eng.* **57** (2), 22–33 (2001).

40. R. Srinivasan, A. Gupta and D. Horsey, *Proceedings International Conference on Additives in Polyolefins*, Houston, TX (1998), pp. 69–83.

41. M.V. Troutman, *Proceedings 11th International Conference on Additives 2002*, Clearwater Beach, FL (2002).

42. S. Sprenger and R. Utz, *Proceedings AddCon World '98*, London (1998), Paper 18.

43. R.M. Nyden and J.W. Gilman, *Recent Adv. Flame Retard. Polym. Mater.* **11**, 19–24 (2000).

44. J. Davis, *Proceedings AddCon '95*, Basel (1995), Paper 3.

45. K. Takeda, *Kogaku Kogyosha* **50** (8), 577–85 (1999).

46. R. Schmidt, *Proceedings AddCon World '98*, London (1998), Paper 17.

47. M.P. Ansias, *Pitture Vernici, Eur. Coat.* **75** (12–13), 39–41 (1999).

48. H. van Damme, *Preprints Euro-Fillers'01*, Łódź, Poland (2001), Paper L-01.

49. I. Soma, *Purasuchikkusu* **49** (10), 20–7 (1998).

50. A. Maltha, *Proceedings Organic Peroxides Symposium* (AKZO Nobel), Apeldoorn (1998), Paper 5.

51. S.E. Amos, *Proceedings Polyolefins X*, Houston, TX (1997), pp. 133–43.

52. H. Sumi, *Purasuchikkusu* **49** (10), 34–40 (1998).

53. N. Sale, *Proceedings 3rd European Additives and Colors Conference* (J.M. Liégois, ed.), SPE Benelux, Antwerp (2003), pp. 117–23.

54. M. Glotin and A. Bouilloux, *Caoutch. Plast.* **76** (782), 29–30 (1999).

55. J.S. Parmar and R.P. Singh, in *Handbook of Engineering Polymeric Materials* (N.P. Cheremisinoff, ed.), M. Dekker, New York, NY (1997), pp. 399–409.

56. J.W. Wheeler, Y. Zhang and J.C. Tebby, *Spec. Publ. R. Soc. Chem.* **224**, 252–65 (1998).

57. J.W. Wheeler, Y. Zhang and J.C. Tebby, *Proceedings 6th European Meeting on Fire Retardancy of Polymeric Materials*, Lille (1997), pp. 90–1.

58. R.L. Gray and C. Neri, *Proceedings 19th Annual International Conference on Advances in the Stabilization and Degradation of Polymers* (V.A. Patsis, ed.), Luzern, Switzerland, (1997), pp. 63–79.

59. H. Hatakeyama and B. Ranby *Cellul. Chem. Technol.* **9**, 583–96 (1975).

60. F. Ranogajec, M. Mlinac and I. Dvornik, *Radiat. Phys. Chem.* **18** (11), 511 (1981).

61. R.D. Leaversuch, *Mod. Plast. Intl* **10**, 65 (1995).

62. A.S. Wilson, *Plasticisers – Selection, Application and Implications, Rapra Review Report No. 88*, Rapra Technology Ltd, Shawbury (1996).

63. D. Munteanu and C. Bolcu, *Proceedings AddCon World '98*, London (1998), Paper 23.

64. E.M. Abdel-Bary and E.M. El-Nesr, in *Handbook of Engineering Polymeric Materials* (N.P. Cheremisinoff, ed.), M. Dekker, New York, NY (1997), pp. 501–15.

65. L. Johansson, *Proceedings AddCon World '98*, London (1998), Paper 3.

66. F.C.-Y. Wang, *J. Chromatogr.* **A883**, 199–210 (2000).

67. N. Liu, J.-Q. Pan and W.W.Y. Lau, *Polym. Degrad. Stab.* **62**, 307–14 (1998).

68. K. S. Minsker, R. Z. Biglova, E. Ya. Davydov and G. E. Zaikov, *Plast. Massy* **4**, 25–7 (1998).

69. W.D. Habicher, B. Pawelke, I. Bauer, K. Yamaguchi, C. Kosa, S. Chmela and J. Pospíšil, *J. Vinyl Addit. Technol.* **7** (1), 4–18 (2001).

70. M. Holmes, *Plastics Addit. Compd* **1** (6), 18–19 (1999).

71. H. Herbst and R. Pfaendner, *Proceedings ARC '98 (Annual Recycling Conference)*, SPE, Brookfield, CT (1998), pp. 205–14.

72. B. Adhikari, T.K. Khanra and S. Maiti, *J. Sci. Ind. Res.* **51**, 429–43 (1992).

73. K. Iwahashi, *Fain Kemikaru* **30** (1), 38–47 (2001).

74. J. van Haveren, D.S. van Es and M.J.A. van den Oever, *Proceedings 3rd European Additives and Colors Conference* (J.M. Liégois, ed.), SPE Benelux, Antwerp (2003), pp. 73–6.

75. D. Johnson, *Plast. Eng.* pp. 38–50 (Nov. 2000).

76. D. Johnson, *Plast. Eng.* **55** (11), 30–47 (1999).

77. Special Report, Chemicals and Additives '98, in *Mod. Plast. Intl* **28** (9), 76–140 (1998).

78. Anon., *Plast. Addit. Compd* **3** (10), 12–29 (Oct. 2001).

79. BRG Townsend Inc., *Plastic Additives 2002*, BRG Townsend, Mt. Olive, NJ (2002).

80. P.W. Dufton, *Proceedings AddCon World '98*, London (1998), Paper 12.

81. M. Cuilleret, *Proceedings AddCon World '98*, London (1998), Paper 11.

82. R.B. Dawson and S.D. Landry, *Proceedings International SAMPE Symposium and Exhibition (Society for the Advancement of Material and Process Engineering)*, Long Beach, CA (2002), Vol. 42, pp. 989–1000.

83. Matsushita Electric, *www.matsushita.co.jp*

84. G.E. Zaikov and S.M. Lomakin, *Intl J. Polym. Mater.* **41** (1–2), 153–69 (1998).

85. IUPAC White Book on Chlorine (R.-P. Martin and G.J. Martens, eds), *Pure Appl. Chem.* **68** (9) (1996).

86. E.D. Weil, *Proceedings 10th Annual BCC Conference on Flame Retardancy*, Stamford, CT (1999).

86a. R.B. Dawson, S.D. Landry and V. Steukers, *Proceedings 15th Annual BCC Conference on Flame Retardancy* Stamford, CT (2004).

87. Y.C. Ho, S.S. Young and K.L. Yam, *J. Vinyl Addit. Technol.* **4** (2), 139–50 (1998).

88. A. Weston and P.R. Brown, *HPLC and CE, Principles and Practice*, Academic Press, San Diego, CA (1997).

89. A. Manz, N. Graber and H.M. Widmer, *Sens. Actuators* **B1**, 244–8 (1990).

90. K. Saito and T. Ogawa, *Bunseki Kagaku* **49** (1), 3–9 (2000).

91. N. Inui, M. Sasaki, M. Yamazaki and K. Yanagi, *Nippon Gomu Kyokaishi* **66** (6), 356–67 (1993).

92. P. Geladi and E. Dåbakk, *J. Near Infrared Spectrosc.* **3**, 119–32 (1995).

93. A.C. Tas and J. van der Greef, *Trends Anal. Chem.* **12**, 60–6 (1993).

94. M.S. Lesney, in Special Issue *Today's Chemist at Work* (March 1999), pp. 41–9.

95. S.A. Westwood (ed.), *Supercritical Fluid Extraction and Its Use in Chromatographic Sample Preparation*, Blackie A&P, London (1993).

96. R.P. Lattimer, R.E. Harris, C.K. Rhee and H.-R. Schulten, *Anal. Chem.* **58**, 3188–95 (1986).

97. P. Molander, K. Haugland, D.R. Hegna, E. Ommundsen, E. Lundanes and T. Greibrokk, *J. Chromatogr.* **A864**, 103–9 (1999).

98. R.P.W. Scott, *Tandem Techniques*, John Wiley & Sons, Ltd, Chichester (1997).

99. J. Sneddon and Y.-I. Lee, *Spectrosc. Lett.* **30** (7), 1417–27 (1997).

100. W. Freitag, in *Plastics Additives* (R. Gächter and H. Müller, eds), Hanser Publishers, Munich (1990), Chapter 20.

101. R.P. Lattimer, *J. Anal. Appl. Pyrol.* **26**, 65–92 (1993).

102. K. Ng, B. Piatek and C. Shimanskas, *Proceedings 41st ASMS Conference on Mass Spectrometry and Allied Topics*, San Francisco, CA (1993).

103. K. Ng, B. Piatek and H. Tong, *Proceedings 42nd ASMS Conference on Mass Spectrometry and Allied Topics*, Chicago, IL (1994).

104. S. Amirav, S. Dagan, T. Shahar, N. Tzanani and S.W. Wainhaus, *Adv. Mass Spectrom.* **14**, 529–62 (1998).

105. C. Weikhardt, U. Boesl and E.W. Schlag, *Anal. Chem.* **66**, 1062 (1994).

106. K. Yu, E. Block and M. Balogh, *LC.GC Europe*, pp. 577–87 (Sept 1999).

107. M. Sakairi and Y. Kato, *J. Chromatogr.* **A794**, 391–406 (1998).

108. J. Yinon, L.D. Betowski and R.D. Voyksner, *J. Chromatogr.* **59**, 187–218 (1996).

109. A. Amirav and O. Granot, *Abstracts 7th Intl. Symposium on Hyphenated Techniques in Chromatography and Hyphenated Analyzers (HTC-7)*, Bruges (2002), Paper 16.2.

110. S. Affolter, A. Ritter and M. Schmid, *Ringversuche an polymeren Werkstoffen*, EMPA, St. Gallen (2000).

111. S. Affolter, M. Schmid and B. Wampfler, *Kautsch. Gummi Kunstst.* **52**, 519–28 (1999).

112. M.P.M. van Lieshout, H.-G. Janssen, C.A. Cramers and G.A. van den Bosch, *J. Chromatogr.* **A764**, 73–84 (1997).

113. W. Zhu and E.D. Weil, *J. Appl. Polym. Sci.* **67** (8), 1405–14 (1998).

114. B.D. Bauman, in *Plastics Additives, An A–Z Reference* (G. Pritchard, ed.), Chapman & Hall, London (1998), pp. 584–9.

115. S. Blackband and P. Mansfield, *J. Phys. C; Solid State Phys.* **19**, L49–52 (1986).

116. L.A. Weisenberger and J.L. Koenig, *Macromolecules* **23**, 2445–53 (1990).

117. K.W. van Every and M.J. Elder, *Proceedings SPE ANTEC '96*, Indianapolis, IN (1996), pp. 2400–5.

118. S. Affolter and M. Schmid, *Intl J. Polym. Anal. Charact.* **6**, 35–57 (2000).

119. J.C.J. Bart, M. Schmid and S. Affolter, *Anal. Sci.* **17**, i729–32 (2001).

120. H. Marchandise, *Fresenius' Z. Anal. Chem.* **345**, 82 (1993).

121. F. Adams, *Accred. Qual. Assur.* **3**, 308 (1998).

122. S.J. Nagourney and R.K. Madan, *J. Test. Eval.* **19** (1), 77–82 (1991).

123. P. Bataillard, L. Evangelista and M. Thomas, in *Plastics Additives Handbook* (H. Zweifel, ed.), Hanser Publishers, Munich (2000), pp. 1047–83.

124. J. Scheirs, *Compositional and Failure Analysis of Polymers. A Practical Approach*, John Wiley & Sons, Ltd., Chichester (2000).

125. R. Trones, *Development of Packed Capillary High Temperature Liquid Chromatography Utilizing Light Scattering and Inductively Coupled Plasma Mass Spectrometry Detection*, PhD. Thesis, University of Oslo (1999).

126. J. Haslam, H.A. Willis and D.C.M. Squirrell, *Identification and Analysis of Plastics*, Heyden, London (1980).

127. T.R. Crompton, *Practical Polymer Analysis*, Plenum Press, New York, NY (1993).

128. T.R. Crompton, *Manual of Plastics Analysis*, Plenum Press, New York, NY (1998).

129. R. Stevenson, K. Mistry and I. Krull, *Am. Lab.* **30** (16), 16A (1998).

130. R.P. Lattimer and R.E. Harris, *Rubber Chem. Technol.* **62**, 548–67 (1989).

131. B.S. Middleditch and A. Zlatkis, *J. Chromatogr. Sci.* **25** (12), 547–51 (1987).

132. F. De Boever, *Micromass 2001, New Products Seminar and Users Event*, Amsterdam (2001).

133. A. Geeve, unpublished results (2002).

134. C. Berger, *Chromatogr. Sci. Ser.* **75**, 301–48 (1998).

135. K. Anton, M. Bach, C. Berger, F. Walch, G. Jaccard and Y. Carlier, *J. Chromatogr. Sci.* **32** (10), 430–8 (1994).

136. H.J. Cortes, G.E. Bormett and J.D. Graham, *J. Microcol. Sep.* **4** (1), 51–7 (1992).

137. H.J. Cortes, B.M. Bell, C.D. Pfeiffer and J.D. Graham, *J. Microcol. Sep.* **1**, 278–88 (1989).

138. H.J. Cortes, G.L. Jewitt, C.D. Pfeiffer, S. Martin and C. Smith, *Anal. Chem.* **61** (9), 961–5 (1989).

139. S. Dagan and A. Amirav, *Intl J. Mass Spectrom. Ion Proc.* **133**, 187 (1994).

140. R. Zimmermann, Ch. Lermer, K.W. Schramm, A. Kettrup and U. Boesl, *Eur. Mass Spectrom.*, 341–51 (1995).

141. B. Marcato and M. Vianello, *J. Chromatogr.* **A869**, 285–300 (2000).

142. P.G. Alden and M. Woodman, *Proceedings Intl GPC Symposium '98*, Phoenix, AZ (1998), pp. 437–57.

143. T. Andersson, T. Hyötyläinen and M.-L. Riekkola, *J. Chromatogr.* **A896**, 343–9 (2000).

144. T. Hyötyläinen, K. Kuosmanen, M. Shimmo, T. Andersson, K. Hartonen and M.-L. Riekkola, *Abstracts 7th International Symposium on Hyphenated Techniques in Chromatography and Hyphenated Chromatographic Analyzers (HTC-7)*, Bruges (2002), Paper 4.3.

145. K. Kuosmanen, T. Hyötyläinen, K. Hartonen and M.-L. Riekkola, *J. Chromatogr.* **A943**, 113–22 (2002).

146. A. Amirav and O. Granot, *Abstracts 7th International Symposium on Hyphenated Techniques in Chromatography and Hyphenated Chromatographic Analyzers (HTC-7)*, Bruges (2002), Paper 16.2.

147. I.D. Wilson, *Anal. Chem.* **72**, 534–42A (2000).

148. J.C.J. Bart, *Abstracts Second International Conference on Polymer Modification, Degradation and Stabilisation (MoDeSt2)*, Budapest (2002); *Polym. Degrad. Stab.* **82**, 197–205 (2003).

149. M. Ashraf-Khorassani, S. Hinman and L.T. Taylor, *J. High Resolut. Chromogr.* **22** (5), 271–5 (1999).

150. W. Hofmann, *Rubber Technology Handbook*, Hanser Publishers, Munich (1996).

151. C.Y. Chu, *Kautsch. Gummi Kunstst.* **41** (1), 33–9 (1988).

152. D. Stoye and W. Freitag (eds), *Paints, Coatings and Solvents*, Wiley-VCH, Weinheim (1998).

153. J.W. Gooch, *Analysis and Deformulation of Polymeric Materials – Paints, Plastics, Adhesives and Inks*, Plenum Press, New York, NY (1997).

154. J.R.J. Paré and J.M.R. Bélanger (eds), *Instrumental Methods in Food Analysis*, Elsevier, Amsterdam (1997).

155. M.M. van Deursen, *PhD. Thesis*, Technical University of Eindhoven, Eindhoven (2002).

156. J.W. Robinson, *Anal. Chem.* **40** (11), 33–6A (1968).

157. A.F. Findeis and M.K. Wilson, *Anal. Chem.* **42**, 27–31A (1970).

158. H. Cherdron, *J. Polym. Sci.: Polym. Symp.* **75**, 119–24 (1993).

159. R. Smits, *LC.GC Europe* **13** (7), 460–2 (2000).

List of Symbols

ACRONYMS OF TECHNIQUES

AAM	Analyte addition method
AA(S)	Atomic absorption (spectrometry)
AC-MS	Atomic composition mass spectrometry
ADAMS	Array-detector atomic mass spectrometry
AED	Atomic emission detection
AES	(1) Atomic emission spectrometry; (2) Auger electron spectroscopy
AFID	Phosphorous/nitrogen-selective alkali/flame ionisation detector
AFM	Atomic force microscopy
AFS	Atomic fluorescence spectrometry
AGHIS	All-glass heated inlet system
AMD	(1) Automated multiple development (chamber); (2) Atmospheric mass detector; (3) Advanced method development
AMS	Accelerator mass spectrometry
AOTF	Acousto-optical tuneable filter
AOTS	Acousto-optical tuneable spectrometer/scanning
AP	Atom probe
APCI	Atmospheric pressure chemical ionisation
APEI	Atmospheric pressure spray with electron impact ionisation
APES	Atomic plasma emission spectrometry
API	Atmospheric pressure ionisation
AP MALDI	Atmospheric pressure MALDI
APPI	Atmospheric pressure photo-ionisation
APSI	Atmospheric pressure spray ionisation
AS	Atomic spectroscopy
ASA	Automated speciation analyser
AS(DP)V	Anodic stripping (differential pulse) voltammetry
ASE®	Accelerated solvent extraction
ASEC	Aqueous size-exclusion chromatography
ASGDI	Atmospheric sampling glow discharge ionisation
ASPEC	Automated sample preparation with extraction cartridges
ASV	Anodic stripping voltammetry
ATR	Attenuated total reflectance
B	Magnetic sector analyser
BET	Brunauer, Emmett and Teller (surface area)
BPC	Bonded-phase chromatography
C	Collision cell
CA	Collision activation
CAD	Collision-activated dissociation
C/AED	Chromatography–atomic emission detection
C-APES	Chromatography–atomic plasma emission spectrometry
CapSEC	Capillary size-exclusion chromatography
CC	(1) Cryogenic collection (trap); (2) Column chromatography
CCD	Charge-coupled device
CC-GC	Capillary column gas chromatography
CCIGC	Capillary column inverse gas chromatography
CCT	Collision cell technology
CDS	Chromatographic data system
CDT	Corona discharge treatment
CE	(1) Capillary electrophoresis; (2) Charge exchange (ionisation)

Additives In Polymers: Industrial Analysis And Applications   J. C. J. Bart
© 2005 John Wiley & Sons, Ltd   ISBN: 0-470-85062-0

CEC	Capillary electrochromatography	CV-AAS	Cold (mercury) vapour atomic absorption spectrometry
CED	Coulometric electrochemical detector		
CEMS	Conversion electron Mössbauer spectroscopy	CV-AFS	Cold vapour atomic fluorescence spectroscopy
CF(D)	Conventional fluorescence (detection) (see F, FLD)	CZE	Capillary zone electrophoresis
		CZT	CdZnTe (detector)
CF-FAB MS	Continuous-flow fast atom bombardment mass spectrometry	DAD	Diode-array detector
CF-MALDI	Continuous-flow MALDI	DALLS	Dual-angle laser light scattering
Cf PD	^{252}Cf plasma desorption	DCI	(1) Direct chemical ionisation; (2) Desorption chemical ionisation
CGC	Capillary gas chromatography		
CGE	Capillary gel electrophoresis	DCP	Direct-current (argon) plasma
CHA	Concentric hemispherical analyser	DCP-AES	Direct-current plasma atomic emission spectrometry
cHPLC	Capillary high-performance liquid chromatography		
		DD	(1) Dipole−dipole interactions; (2) Decoupling/double resonance (high power ^{1}H decoupling); (3) Direct deposition
CI	Chemical ionisation		
CID	(1) Collision-induced dissociation; (2) Charged-injection device		
		DE	Delayed extraction
CIEF	Capillary isoelectric focusing	DEI	Desorption electron ionisation
CI-MS, CIMS	Chemical ionisation mass spectrometry	DEPT	Distortionless enhancement of polarisation transfer
CIS	Cooled injection system		
CITP	Capillary isotachophoresis	DFI	Direct-flow injection/direct-fluid introduction
CL	Chemiluminescence		
CLC	Capillary liquid chromatography	DF-ICP-MS	Double-focusing ICP-MS
CLND	Chemiluminescent nitrogen detector	DFS	Double-focusing sector
CLSM	Confocal laser scanning microscopy (see LSCM)	DHS	Dynamic headspace
		DI	(1) Desorption/ionisation; (2) Direct inlet
CMA	Cylindrical mirror analyser		
CMSE	Chemically modified solid electrode	DID	Discharge ionisation detector
CNG	Constant neutral gain	DI-MS	Direct inlet mass spectrometry
CNL	Constant neutral loss	DIN	Direct-injection nebuliser
COC	Cool on-column	DIP	Direct inlet (insertion) probe
COC-SVE	Cool on-column with solvent vapour exit	DLI	(1) Direct laser ionisation; (2) Direct liquid introduction; (3) Direct liquid interface
COSY	Correlation spectroscopy (NMR)		
		DMA	Dynamic mechanical analysis
CP	(1) Cross-polarisation; (2) Conducting polymer (chemosensor)	DMAE	Dynamic microwave-assisted extraction
		DME	Dropping mercury electrode
CPA	Cold-plasma asher	DMI	Difficult matrix introduction (GC)
CPAA	Charged particle activation analysis	DOSY	Diffusion ordered spectroscopy
CRC	Collision/reaction cell	DP	(1) Differential pressure (viscosity detector); (2) Density profiling
CRYSTAF	Crystallisation analysis fractionation		
CSDPV	Cathodic stripping differential pulse voltammetry		
		D/P	Dissolution/precipitation
cSFC	Capillary column SFC	DPASV	Differential pulse anodic stripping voltammetry
CSV	Cathodic stripping voltammetry		
CT	(1) Cold trap, cryotrapping; (2) Computed X-ray tomography	DPCSV	Differential pulse cathodic stripping voltammetry
CuPy	Curie point pyrolysis	DP-MS	(*In vacuo*) direct probe mass spectrometry
CV	Cyclic voltammetry		

DPP	Differential pulse polarography	EDXRF	Energy-dispersive X-ray fluorescence
DP(S)V	Differential pulse (stripping) voltammetry	EGA	Evolved gas analysis
DPyMS	Direct pyrolysis mass spectrometry	EGC	Evolved gas collection
		EGD	Evolved gas detection
DR	Diffuse reflectance	EGP	Evolved gas profile
DRC	(1) Dynamic rate control;	EHD, EHI	Electrohydrodynamic ionisation
	(2) Dynamic reaction cell	EI	Electron ionisation/impact
DRI	Differential refractive index (detector) (see RI(D))	EI-MS, EIMS	Electron impact mass spectrometry
		ELCD	Electrolytic conductivity (Hall) detector (see HECD)
DRIFTS	Diffuse reflectance infrared Fourier-transform spectroscopy	ELP-DAD	Extended light path diode-array detector
DS	Derivative spectroscopy		
DSC	Differential scanning calorimetry	ELS(D)	Evaporative light scattering (detector)
DSI	(1) Direct sample introduction;	EM	Electron microscopy
	(2) Dirty sample introduction	em	emission
DT	Differential trapping	EMD	Evaporative mass detection
DTA	Differential thermal analysis	EMP	Electron microprobe
DTD	Direct thermal desorption	emr	Electromagnetic radiation
DTGS	Deuterated triglycine sulfate (detector)	EN	Electronic nose
		ENDOR	Electron nuclear double resonance
DTMA	Differential thermomechanical analysis	EPC	Electronic pressure control
		EPMA	Electron-probe microanalysis
DT-MS	Direct temperature-resolved mass spectrometry	EPR	Electron paramagnetic resonance (see ESR)
DV	Differential viscosimeter	ERE	External reference electrode
		ESA	Electrostatic analyser
E, ESA	Electric sector analyser, electrostatic analyser	ESCA	Electron spectroscopy for chemical analysis (see XPS)
EA	Elemental analysis	ESCI™	Electrospray chemical ionisation
EBS	Elastic backscattering	ESD	Element-selective detector
EC	Electrochemical (analyser)	ESE®	Enhanced solvent extraction
ECCI	Electron capture chemical ionisation	E-SEM, ESEM	Environmental scanning electron microscopy
ECD	(1) Electron-capture detector;	ES(I), ESP	Electrospray (ionisation)
	(2) Electrochemical detector	ESIMS	Electrospray ionisation mass spectrometry
ECL	Electrochemical luminescence		
ECNI	Electron capture negative ionisation	ESR	Electron spin resonance (same as EPR)
ED	(1) Electrochemical detection;	ESRI	Electron spin resonance imaging
	(2) Energy dispersive	ETAAS, ET-AAS	Electrothermal (atomisation) atomic absorption spectrometry
EDAX®, EDX	Energy-dispersive X-ray spectrometry	ETA AES	Electrothermal atomisation atomic emission spectrometry
EDI-CI	Electric discharge-induced chemical ionisation	ETA LEAFS	Electrothermal atomisation laser-excited atomic fluorescence spectrometry
EDL	Electrodeless discharge lamp		
EDS	(1) Energy-dispersive spectrometry;		
	(2) Electron diffraction spectroscopy	ETV	Electrothermal vaporisation
ED-SEC	Electro-drive size-exclusion chromatography	EVA	Evolved volatile analysis
EDXRA	Energy-dispersive X-ray analysis (SEM)	ex	Excitation

EXAFS	Extended X-ray absorption fine structure	FT-IMS	Fourier-transform ion mobility spectrometry
EXEFS	Extended X-ray emission fine structure	FT-IR, FTIR	(Fourier-transform) infrared spectroscopy
		FTIR-μS	FTIR-microspectroscopy (see μFTIR)
F, FL	Fluorescence (detector)		
FAAS	Flame atomic absorption spectrometry	FT LMMS	Fourier-transform laser-microprobe mass spectrometry
FAB	Fast atom bombardment	FTMS	Fourier-transform mass spectrometry
FAB-SSIMS	Fast atom bombardment static secondary ion mass spectrometry	FTNMR	Fourier-transform NMR
FAES	Flame atomic emission spectrometry	FT-RS, FTRS	Fourier-transform Raman spectroscopy
FAF	Flame atomic fluorescence spectrometry	GA	Gas analysis
		GC	Gas chromatography
FANES	Furnace atomic nonthermal emission spectrometry	GCxGC	Comprehensive gas chromatography
FD	Field desorption	GC-MS	Gas chromatography–mass spectrometry
FD-MS, FDMS	Field desorption mass spectrometry	GCO	Gas chromatography–olfactometry
FES	Flame emission spectrometry	GC-SNIFF	Sniffing port gas chromatography (see GCO)
FEXTRA	Feststoff Extraction		
FFF	Field-flow fractionation	GD-(MS)	Glow-discharge (mass spectrometry)
FF-TLC	Forced-flow TLC	GD-OES	Glow-discharge-optical emission spectrometry
FI	(1) Field ionisation; (2) Flow injection	GDS	Glow-discharge spectrometry
FIA	Flow-injection analysis	GE	Gradient elution
FIB	Fast ion bombardment	GEC	Gradient-elution chromatography
FID	(1) Flame ionisation detector; (2) Free induction decay (NMR)	GE-HPLC	Gradient-elution high-performance liquid chromatography
FIM	Field ion microscopy	GFAAS	Graphite furnace atomic absorption spectrometry
FI-MS, FIMS	Field ionisation mass spectrometry		
FIR	Far infrared	GFAES	Graphite furnace atomic emission spectrometry
FL	Fluorescence, fluorometry		
FLD	Fluorescence detector	GLC	Gas–liquid chromatography
FLNS	Fluorescence line-narrowing spectroscopy	GPC	Gel permeation chromatography (see SEC)
FMASE	Focused microwave-assisted Soxhlet extraction	GPEC®	Gradient polymer elution chromatography
FMC	Flow microcalorimetry	GSC	Gas–solid chromatography
FMW	Focused microwave-enhanced extraction	GSSMS	Gliding spark source mass spectrometry
FNAA	Fast neutron activation analysis		
FPA	Focal plane array (detector)	h	Hexapole
FPD	Flame photometric detector	HC(L)	Hollow cathode (lamp)
FSCD	Fluorine-induced sulfur chemiluminescence detector	HDID	Helium discharge ionisation detector
		HECD	Hall electrolytic conductivity detector (see ELCD)
FSCE	Free solution capillary electrophoresis		
		He(HE)MIP	Helium (high-efficiency) microwave-induced plasma
FSD	Fourier self-deconvolution		
FTD	Flame thermoionic detector	HEN	High-efficiency nebuliser
FTICR	Fourier-transform ion-cyclotron resonance	HETCOR	Heteronuclear chemical shift-correlation spectroscopy

HG-AAS	Hydride generation AAS	HT-SEC, HTSEC	High-temperature size-exclusion chromatography
HG-ICP-AES	Hydride generation ICP-AES		
HID	Helium ionisation detector	HT-ToFMS	Hadamard transform time-of-flight mass spectrometry
HMDE	Hanging mercury dropping electrode		
HMQC-TOCSY	Heteronuclear multiple quantum coherence-total correlation spectroscopy	IAA	Instrumental activation analysis
		IAM	Indoor air monitoring
		IBA	Ion-beam analysis
HNI	Heated nebuliser ionisation	IC	(1) Ion chromatography;
HPA	High-pressure asher		(2) Interaction chromatography
HPCE	High-performance capillary electrophoresis	ICL	Imaging chemiluminescence
		ICP(I)	Inductively coupled plasma (ionisation process)
HPDSC	High-pressure DSC		
HPLC	High-performance liquid chromatography	ICP-AES	Inductively coupled plasma–atomic emission spectrometry
HP-OIT	High-pressure oxidative induction time	ICP-DRC-MS	Dynamic reaction cell ICP-MS
		ICP-MS	Inductively coupled plasma–mass spectrometry
HPPLC	(1) High-pressure planar liquid chromatography;		
	(2) High-performance precipitation liquid chromatography	ICP-OES	Inductively coupled plasma–optical emission spectrometry
		ICR	Ion-cyclotron resonance
HPSE	High-pressure solvent extraction	ID	Identification
HPSEC	High-performance size-exclusion chromatography (see SEC)	ID(A)	Isotope dilution (analysis)
		ID-GC-MS	Isotope dilution gas chromatography–mass spectrometry
HPSEM	High-pressure SEM		
HPTLC	High-performance thin-layer chromatography	ID-ICP-MS	Isotope dilution–inductively coupled plasma–mass spectrometry
HR-CGC, HRGC	High-resolution (capillary) gas chromatography	IDMS	Isotope dilution mass spectrometry
		ID-TIMS	Isotope dilution–inductively coupled plasma–mass spectrometry (see TI-IDMS)
HR-ICP-MS	High-resolution inductively coupled plasma mass spectrometry		
HRLEELS	High-resolution low-energy electron loss spectroscopy		
		IEC	Ion-exchange chromatography
HRMS	High-resolution mass spectrometry	IGC	Inverse gas chromatography
HRTG	High-resolution thermogravimetry	IMMS	Ion mobility mass spectrometry
HS-GC	Headspace gas chromatography	IMR-MS	Ion-molecule reaction mass spectrometry
HSGC	High-speed gas chromatography		
HSI	Hyperthermal surface ionisation	IMS	(1) Ion mobility spectrometry;
HSLC	High-speed liquid chromatography		(2) Infrared microspectroscopy (see µFTIR)
HSSE	Headspace sorptive extraction		
HS-SPME	Headspace solid-phase microextraction	INAA	Instrumental neutron activation analysis
HSTLC	High-speed TLC	INADEQUATE	Homonuclear J-correlated ^{13}C experiment (NMR)
HT	High temperature		
HTA	High-temperature asher	IPC	Ion-pair chromatography
HTE	High-throughput experimentation	iPIXE	Imaging particle-induced X-ray emission
HT-GC, HTGC	High-temperature gas chromatography		
		IR	Infrared
HT-GPC	High-temperature gel permeation chromatography (see HT-SEC)	IRD	Infrared detector
		IRE	(1) Internal reflection element;
HTLC	High-temperature liquid chromatography		(2) Internal reference electrode
		IRED	Infrared emitting diode
HTS	High-throughput screening	IRMS	Isotope ratio mass spectrometry

IRS	Internal reflectance spectroscopy (see ATR)	LEI	Laser-enhanced ionisation
ISE	Ion-selective electrode	LI	(1) Laser ionisation; (2) Liquid injection
ISFET	Ion-selective field effect transistor	LIBS	Laser-induced breakdown spectroscopy
iSIMS	Imaging secondary ion mass spectrometry	LIESA®	Laser-induced emission spectral analysis (see LIBS)
IS(P)	Ionspray		
IT(D)	Ion trap (detector)	LIF(S)	Laser-induced fluorescence (spectroscopy)
ITMS	Ion trap mass spectrometry		
ITP	Isotachophoresis	LLC	Liquid–liquid chromatography (or partition chromatography)
iXPS	Imaging X-ray photoelectron spectroscopy		
		LLE	Liquid–liquid extraction
		LLPS	Liquid–liquid phase separation
KF	Karl Fischer (coulometry)	LMMS	Laser microprobe mass spectrometry
		LMS	Laser mass spectrometry (see LAMS)
L	Laser		
LA	Laser ablation	L²MS	Two-step laser mass spectrometry
LA-AES	Laser ablation–atomic emission spectrometry (see LIBS)	l-NMR	Liquid nuclear magnetic resonance
		LOC	Lab-on-a-chip
LAC(CC)	Liquid adsorption chromatography (at critical conditions)	LPGC	Low-pressure gas chromatography
		LPS	Laser pyrolysis scanning
LAES	Laser ablation–emission spectrometry (see LIBS)	LR	Laser Raman
		LRGC	Low-resolution gas chromatography
LA(L)LS	Low-angle (laser) light scattering	LR-NMR	Low-resolution NMR
LAMMA®	Laser microprobe mass analysis (see LMMS)	LRRS	Low-resolution Raman spectroscopy
		LS	Light scattering
LAMS	Laser(-assisted) mass spectrometry; (resonance-enhanced) laser mass spectrometry (see REMPI)	LSC	Liquid–solid chromatography
		LSCM	Laser scanning confocal microscopy (see CLSM)
LAN	Local area network	LSD	Light-scattering detector
LA-OES	Laser ablation–optical emission spectrometry (see LIBS)	LSE	Liquid–solid extraction
		LSIMS, LSI-MS	Liquid secondary ion mass spectrometry
LC	Liquid chromatography		
LCCC	Liquid chromatography under critical conditions	LTA	Low-temperature asher
		lToF-MS	Linear ToF-MS
LCEC	Liquid chromatography with electrochemical detection	L²ToFMS	Laser-desorption laser-photoionisation ToF-MS
LC-ICP-MS	Liquid chromatography–inductively coupled plasma–mass spectrometry	LVEI	Low-voltage electron ionisation
		LV HSI	Large–volume headspace injection
		LVI	Large–volume injection
LCT	LC-Transform™	LVS	Large–volume sampling
LD	Laser desorption		
LDI	Laser desorption/ionisation	MAB	Metastable atom bombardment
LD-IMS	Laser desorption–ion mobility spectrometry	MAE	Microwave-assisted extraction
		MAGIC™	Monodisperse aerosol generating interface for chromatography
LDIOS	Laser desorption/ionisation on silicon		
		MALD(I)	Matrix-assisted laser desorption/ionisation
LDMS	Laser desorption mass spectrometry		
LE	Laser excitation	MA(L)LS	Multiple-angle (laser) light scattering
LEAFS	Laser-excited atomic fluorescence spectrometry		
		MAP	Microwave-assisted process
LED	Light emitting diode	MAP™	Microwave-assisted Process™

MAP-HS	Microwave-assisted gas-phase extraction	MSD	Mass selective detector
MAP-SFE	MAP™ under supercritical fluid conditions	MSEGA	Mass spectrometric evolved gas analysis
MAS	Magic-angle spinning	M-SPE	Miniaturised solid-phase extraction
MASE	Microwave-assisted solvent extraction	MTDSC	Modulated temperature DSC
		MTDTA	Modulated temperature DTA
MB(I)	Moving belt (interface)	MTGA™	Modulated thermogravimetric analysis
MBSA	Molecular beam for solid analysis	MUPI	Multiphoton ionisation (see MPI(s))
MC	Mass chromatograph	MW	Microwave
MCC	Multicapillary column	MWSPE	Microwave-assisted solid-phase extraction
MC-GC	Multicapillary gas chromatography	MXRF	Micro X-ray fluorescence (see μXRF)
MC-ICPMS	Multicollector ICP-MS		
MCP	Microchannel plate	μFTIR	Micro Fourier-transform infrared (see FTIR-μS)
MCT	Mercury−cadmium−telluride (detector)	μLC	Micro liquid chromatography
MDGC	Multidimensional gas chromatography	μRS	Micro Raman spectroscopy (see MRS)
MDHPLC	Multidimensional high-performance liquid chromatography	μSEC	Micro size-exclusion chromatography
MDLC	Multidimensional liquid chromatography	μTA	Micro thermal analysis
		μTAS	Miniaturised total analysis system
MDS	Membrane desolvator	μXAS	Micro X-ray absorption spectroscopy
MDSC™	Modulated differential scanning calorimetry	μXPS	Micro X-ray photoelectron spectroscopy
MECC, MEKC	Micellar electrokinetic capillary chromatography	μXRF	Micro X-ray fluorescence
MESI	Membrane extraction with sorbent interface	NAA	Neutron activation analysis
		NACE	Non-aqueous capillary electrophoresis
MFI	Melt-flow index		
MHE	Multiple headspace extraction	NAM	Nuclear analytical method
MI	Matrix isolation	NAMS	Neutron activation mass spectrometry
MID	Multiple ion detection		
MIKES	Mass-analysed ion kinetic energy spectrometry	NARP	Nonaqueous reversed-phase
		N-BPC	Normal bonded-phase chromatography
MIM	Multiple ion monitoring		
MIP	Microwave-induced plasma	NCD	Nitrogen chemiluminescence detector
MMLLE	Microporous membrane liquid−liquid extraction	NCI	Negative chemical ionisation
MOD	Microwave-oven digestion	NDE	Nondestructive evaluation
MP	(1) Mobile phase;	NDT	Nondestructive testing
	(2) Microplasma;	NELC	Nonexclusion liquid chromatography
	(3) Modulus profiling	NEXAFS	Near-edge X-ray absorption fine structure (see XANES)
MPI(S)	Multiphoton ionisation (spectroscopy)	NIR(A)	Near-infrared reflectance (analysis)
MR	Magnetic resonance		
MRM	Multiple reaction monitoring	NIRS	Near-infrared spectroscopy
MRS	Micro Raman spectroscopy	NLM	Nonlinear mapping
MS	Mass spectrometry	NLO	Nonlinear optics
MS^n	Multiple-stage mass spectrometry; tandem mass spectrometry	NMP	Nuclear microprobe
		NMR	Nuclear magnetic resonance

NMRI	Nuclear magnetic resonance imaging	PDA	Photodiode array (detection)
NOE	Nuclear Overhauser effect/enhancement	PDECD	Pulse discharge electron-capture detector
NP	Normal phase	PD-MS, PDMS	Plasma-desorption mass spectrometry
NPC	Normal-phase chromatography		
NPD	Nitrogen phosphorous detector or thermoionic detector	PDPI	Photodissociation–photoionisation
NP-GPEC®	Normal-phase gradient polymer elution chromatography	PDPID	Pulse discharge photoionisation detector
NP-HPLC, NPLC	Normal-phase liquid chromatography	PD-SEC	Pressure-drive size-exclusion chromatography
NPP	Normal pulse polarography	PEC	Pressurised flow capillary electrochromatography
NQR	Nuclear quadrupole resonance	PED	Plasma emission detection
NSOM	Near-field scanning optical microscopy (see SNOM)	PES	Photoelectron spectroscopy
		PFE	Pressurised fluid extraction
		PFG-NMR	Pulsed-field gradient nuclear magnetic resonance
o	Octopole		
oaToF	Orthogonal acceleration time-of-flight	PFPD	Pulsed flame photometric detector
		PFT	Pulse Fourier-transform (spectrometers)
OCN	Oscillating capillary nebuliser		
OES	Optical emission spectrometry	PFT-NMR	Pulse Fourier-transform NMR
O-FID	Oxygen-specific flame ionisation detector	PGSE	Pulsed gradient spin-echo (NMR)
		PHWE	Pressurised hot water extraction
OIF	Orifice-induced fragmentation	PI	(1) Photoionisation;
OL	Oxyluminescence		(2) (Laser) post-ionisation;
OM	Optical microscopy		(3) Plasma-ionisation
OPLC	Overpressured layer chromatography	PID	(1) Photon-induced dissociation;
O-SCD	Ozone-induced sulfur chemiluminescence detector		(2) Photoionisation detection
		PIGE	(1) Particle-induced γ-ray emission;
OSM	One-step microwave-assisted extraction		(2) Proton-induced γ-ray spectrometry
OTC	Open-tubular columns		
OTLC	Open-tubular liquid chromatography	PIL	Plasma-induced luminescence
ot-SFC	Open-tubular SFC	PI-MS, PIMS	Photoionisation mass spectrometry
OTT	Open-tubular trapping	PI-ToF	Photoionisation time-of-flight
		PIXE	(1) Particle-induced X-ray emission;
PA	Photoacoustics		(2) Proton-induced X-ray emission
PAA	Photon activation analysis	PLC	Preparative layer chromatography
PA-FTIR	Photoacoustic Fourier-transform infrared	PLE	Pressurised liquid extraction
		PLM	Polarised light microscopy (see PM)
PARC	Pattern recognition		
PA(S)	Photoacoustic (spectroscopy)	PLOT	Porous-layer open-tubular column
PA-UV	Photoacoustic UV spectrophotometry	PM	Polarisation microscopy (see PLM)
		PMD	Pressurised microwave digestion
PB	Particle beam	PME	Polymer modified electrode
PC	Paper chromatography	PMT	Photomultiplier tube
PCA	Principal component analysis	PN	Pneumatic nebulisation
PCD	Peltier-cooled detection	PPIP	Packed purged injection port
PCI	Positive chemical ionisation	PRLC	Precipitation–redissolution liquid chromatography
pcSFC	Packed-capillary supercritical fluid chromatography		
PD	^{252}Cf plasma desorption	PS-AES	Plasma source–atomic emission spectrometry

PSD	(1) Position-sensitive detector; (2) Photon-stimulated desorption; (3) Post-source decay	RIMS	Resonance ionisation mass spectrometry
PSE	Pressurised solvent extraction	RLIF	Remote laser-induced fluorescence
pSFC	Packed column SFC	RLVI	Rapid large-volume injection
PS-MS	Plasma-source mass spectrometry	RNAA	Radiochemical neutron activation analysis
PSP	Plasma spray	RP	Reversed phase
PT, P&T	Purge-and-trap	RPC	(1) Reversed-phase chromatography; (2) Rotation planar chromatography
PTI-MS	Positive thermal ionisation–mass spectrometry	RP-HPLC, RPLC	Reversed-phase (high-performance) liquid chromatography
PTV	Programmed temperature vaporising (inlet)	RPTLC	Reversed-phase TLC
Py	Pyrolysis	RR(S)	Resonance Raman (scattering)
PyFTIR	Pyrolysis-Fourier transform infrared	RS	Raman scattering/spectroscopy
PyGC	Pyrolysis gas chromatography	S/SL	Split/splitless (capillary)
PyGC-MS	Pyrolysis gas chromatography–mass spectrometry	SAD	Selected area diffraction
		SALDI	Surface-assisted laser desorption/ionisation
PyMS	Pyrolysis mass spectrometry	SALS	Small-angle light scattering
		SAX	Selected area XPS
Q	Quadrupole mass filter	SAXS	Small-angle X-ray scattering
q	Quadrupole ion guide	SBSE	Stir bar sorptive extraction
QDP	Quantitative depth profiling	SCD	(1) (Flame) sulfur chemiluminescence detector; (2) Segmented charge-coupled device
QF	Quartz tube flame atomiser		
QFAAS	Quartz furnace atomic absorption spectrometry		
Q-ICP-MS	Quadrupole ICP-MS (also ICP-QMS)	SCE	Saturated calomel (reference) electrode
QITMS	Quadrupole ion trap mass spectrometer	SCOT	Support-coated open-tubular column
QMS	Quadrupole mass spectrometer	SD	Steam distillation
QQQ, QqQ	Triple quadrupole analyser	SDE	Simultaneous distillation extraction
QTA	Quartz tube atomiser	SDME	Static mercury drop electrode
QTLC	Quantitative thin-layer chromatography	SEB	Secondary electron Bremsstrahlung
		SEC	Size-exclusion chromatography (see GPC)
QToF	Quadrupole time-of-flight (mass spectrometry)		
		SEEC	Size-exclusion electrochromatography
R	(Normal) Raman	SEIRAS	Surface-enhanced infrared absorption spectroscopy
RALLS	Right-angle laser light scattering		
RBS	Rutherford backscattering spectroscopy	SELDI	Surface-enhanced laser desorption ionisation
RCD	Redox chemiluminescence detector	SEM	Scanning electron microscopy
		SE(R)RS	Surface-enhanced (resonance) Raman spectroscopy
REMPI	Resonance enhanced multiphoton ionisation		
		SEXAFS	Surface extended X-ray absorption fine structure
reToF	Reflectron time-of-flight		
r.f.-GD-AES	Radiofrequency powered glow discharge atomic emission spectrometry	SF	Shake-flask
		SFC	Supercritical fluid chromatography
		SFE	Supercritical fluid extraction
RI(D)	Refractive index (detector)	SFI	Supercritical fluid injection

SFT	Supercritical fluid techniques	STM	Scanning tunnelling microscopy
SHE	Standard hydrogen electrode	SV	Stripping voltammetry
SHS	Static headspace	SVE	Solvent vapour exit
SI	Soft ionisation	SWC	Superheated water chromatography
SIDMS	Speciated isotope dilution mass spectrometry	SWE	Subcritical and/or superheated water extraction
SIM(-MS)	Selected-ion monitoring (single ion monitoring) mass spectrometry	TAD	Thermally assisted desorption
SIMS	Secondary ion mass spectrometry	TAHM	Thermally assisted hydrolysis and methylation (*cfr.* THM)
SI-MS	Soft ionisation mass spectrometry		
SIR	Selected-ion recording (equivalent to SIM)	TALLS	Triple-angle laser light scattering
		TAS	Total analysis system
SLM	Supported liquid membrane (extraction)	TASPE	Thermally assisted on-line solid-phase extraction
SMB	Supersonic molecular beam	TCC	Time compressed chromatography
SMDE	Static mercury drop electrode	TCD	Thermal conductivity detector
SM-SPE	Semimicro solid-phase microextraction	TCT-GC-MS	Thermal desorption cold trap injection GC-MS
SNIFF	Sniffing port	TD	(1) Thermal desorption;
s-NMR	Solid-state nuclear magnetic resonance		(2) Thermodilatation
		TDM	Thermal desorption modulator
SNMS	Sputtered and/or secondary neutral mass spectrometry	TD-MS	Thermal desorption mass spectrometry
SNOM	Scanning near-field optical microscopy (see NSOM)	TEA	(1) Thermal evolution analysis; (2) Thermoelectric analysis;
SP	Stationary phase		(3) Thermal energy analyser
SPE	(1) Solid-phase extraction;	TECD	Total electron-capture detector
	(2) Single-pulse excitation (NMR)	TE-GC-MS	Thermal extraction GC-MS
SPE-PT	Solid-phase extraction pipette tips	TEM	Transmission electron microscopy
SPI	Single photon ionisation	TG(A)	Thermogravimetry, thermogravimetric analysis
SPME	Solid-phase microextraction		
SPM-GC	Simultaneous pyrolysis methylation–gas chromatography	TGS	Triglycine sulfate (detector)
		THM	Thermally assisted hydrolysis and methylation
SR	(1) Specular reflectance;		
	(2) Synchrotron radiation	THM-GC-MS	Thermally assisted hydrolysis and methylation GC-MS
SRM	Selected reaction monitoring		
SR-XRF	Synchrotron radiation X-ray fluorescence	TI	Thermal ionisation
		TID	Thermoionic detector
SS	(1) Solid sampling;	TI-IDMS	Thermal ionisation–isotope dilution mass spectrometry (see ID-TIMS)
	(2) Spark source		
SS-AES	Sliding spark-source atomic emission spectrometry	TIMS	Thermal ionisation mass spectrometry
SSC	Single, short column		
SSI	Sonic spray ionisation	TLC	Thin-layer chromatography
SSMS	(1) Spark-source mass spectrometry;	TLE	Thin-layer electrochromatography
	(2) Solid-state mass spectrometry	TLF	Time-lag focusing
SSS	Sliding (gliding) spark source and/or spectrometry	TMD	ThermaBeam™ mass detector
		TMDSC	Temperature modulated DSC (see MTDSC)
SS-ZAAS	Solid sampling Zeeman atomic absorption spectrometry		
		T-MS	Thermolysis-mass spectrometry
STEM	Scanning transmission electron microscopy	TOCSY	Total correlation spectroscopy
		ToF	Time-of-flight

ToFMS, ToF-MS	Time-of-flight mass spectrometry	XAFS	X-ray absorption fine structure
ToF-SIMS	Time-of-flight secondary ion mass spectrometry	XANES	X-ray absorption near-edge structure (see NEXAFS)
TP	Thermal programming	XAS	X-ray absorption spectroscopy
TPLC	Temperature-programmed liquid chromatography	XES	X-ray emission spectrometry
TPPy	Temperature-programmed pyrolysis	XPS	X-ray photoelectron spectroscopy (see ESCA)
TREF	Temperature rising elution fractionation	XRD	X-ray diffraction
		XRF	X-ray fluorescence
TRS	Time-resolved spectroscopy	XRS	X-ray spectrometry
TRXRF	Total-reflection X-ray fluorescence (see TXRF)	ZAAS	Zeeman atomic absorption spectrometry
TS	Thermospray (see TSP)	ZBGC	Zeeman background correction
TSD	(1) Thermoionic specific detector; (2) Thermally stimulated discharge	ZE	Zone electrophoresis
TSD-GC-MS	Thermally stimulated desorption GC-MS	ZETAAS	Zeeman electrothermal atomic absorption spectrometry
TSM	Two-step microwave-assisted solvent extraction	ZGFAAS	Zeeman graphite furnace atomic absorption spectrometry
TSP	Thermospray (see TS)	3DQMS	3D Quadrupole ion trap mass spectrometer
TSQ	Triple-stage quadrupole mass spectrometer		
TTP	Temperature–time profile		
TVA	Thermal volatilisation analysis		
TXRF	Total-reflection X-ray fluorescence (see TRXRF)		

CHEMICAL NOMENCLATURE

Polymers and Products

UHV	Ultra-high vacuum		
UPLC	Ultra performance liquid chromatography	ABS	Acrylonitrile-butadiene-styrene terpolymer
US	Ultrasound	aPP	Atactic polypropylene
USN	Ultrasonic nebulisation	ASA	Acrylonitrile-styrene-acrylic ester copolymer
UTLC	Ultra-thin layer chromatography		
UV	Ultraviolet	BAN	Butadiene-acrylonitrile copolymers
UVD	UV/visible detector	BR	Butadiene rubbers, polybutadienes
VI	Volatiles interface	CCD	Chemical composition distribution
VIS	Visible	ClPE, CPE	Chlorinated polyethylene
VPD-AAS	Vapour deposition atomic absorption spectrometry	CPO	Chlorinated polyolefin
		CPVC	Chlorinated poly(vinyl chloride)
VUV	Vacuum ultraviolet	CR	Polychloroprene (chloroprene rubber)
VWD	Variable wavelength detector		
WCOT	Wall-coated open–tubular column	DHBF	2,4-Dihydrobenzophenone – formaldehyde resin
WD	Wavelength dispersive		
WDS	Wavelength dispersive spectrometry	EDP	Environmentally degradable polymer and plastic
WDXRF	Wavelength dispersive X-ray fluorescence	EO-PO	Oxyethylene-oxypropylene copolymers
WRPLC	Water-only reversed-phase liquid chromatography	EP	(1) Engineering plastic; (2) Epoxide resin

EPDM	Ethylene-propylene-diene rubber, ethylene-propylene terpolymer, poly(ethylene-*co*-propylene-*co*-3,5-ethylidene norbornene)	MPW	Mixed plastic waste
		MUF	Melamine urea formaldehyde resin
EPM	Ethylene-propylene copolymer	NBR	Acrylonitrile-butadiene rubber, nitrile rubber
EPR	Ethylene-propylene rubber		
EVA	Ethylene-vinyl acetate copolymer, poly(ethylene-*co*-vinyl acetate)	NR	Natural rubber; polyisoprene
EVAL	Ethylene-vinyl alcohol copolymer, poly(ethylene-*co*-vinyl alcohol)	PA	Polyamide
		PA6/6.6	Polyamide 6/6.6
EVOH	Poly(ethylene-*co*-vinyl alcohol)	PAG	Poly(alkylene glycol)
		PAI	Polyamidimide
		PAN	Polyacrylonitrile
FRP	Fibre reinforced polymer	PAn, PANI	Polyaniline
FR-PET	Flame retarded PET	PAO	Poly(alpha-olefin)
FTD	Functionality type distribution	PAR	Polyarylate
		PB, P1B	Polybutene-1
GFR	Glass-fibre reinforced	PBMA	Poly(butyl methacrylate)
		PBT	Poly(butylene terephthalate)
		PC	Polycarbonate
HDPE	High-density polyethylene	PCR	Post-consumer recyclate
HECO	Heterophasic ethylene-propylene copolymer	PDMS	Polydimethylsiloxane
		PE	Polyethylene
HIPS	High-impact polystyrene	PECT	Poly(ethylene-*co*-1,4-cyclohexane-dimethylene terephthalate)
HSPE	High-strength polyethylene		
		PEEK	Poly(etheretherketone)
IIR	Isobutylene-isopropene rubber; poly(isobutene-*co*-isoprene)	PEG	Poly(ethylene glycol)
		PEI	Polyethylene imine, polyetherimide
iPP	Isotactic polypropylene	PEK	Polyether ketone
IPS	Impact polystyrene	PEMA	Poly(ethyl methacrylate)
IR	Isoprene rubber; poly(*cis*-1,4-isoprene)	PEO	Poly(ethylene oxide); α-Alkoxy-ω-hydroxy polyethylene oxide
		PES	Polyether sulfone
LCP	Liquid crystalline polymer	PET, PETP	Poly(ethylene terephthalate)
LDPE	Low-density polyethylene	PFA	Perfluoroalkoxy-modified tetrafluoroethylene; Perfluoroalkoxy vinyl ether
LFRT	Long fibre reinforced thermoplastic		
LGFPP	Long glass-fibre reinforced polypropylene		
		PFPAE	Perfluoropolyalkyl ether
LLDPE	Linear low-density polyethylene	P-*g*-A, P*g*A	Additive-grafted polymer
		PI	(1) Polyimide; (2) Polyisoprene
MABS	Methyl methacrylate-acrylonitrile-butadiene-styrene		
MBS	Methyl methacrylate-butadiene-styrene terpolymer	PIB	Polyisobutylene
		PIP	Polyisoprene
MDPE	Medium-density polyethylene	PMA	Poly(methacrylate)
MLDPE	Medium low-density polyethylene	PMMA	Poly(methyl methacrylate)
MM(CC)D	Molecular mass (chemical composition) distribution	PO	Polyolefins
		POE	(1) Polyolefin elastomer; (2) Polyoxyethylene
MMFTD	Molecular mass functional type distribution		
		Poly-TMDQ	Poly(2,2,4-trimethyl-1,2-dihydroquinoline)
m-PE	Metallocene polyethylene		
m-PO	Metallocene polyolefin	POM	Poly(oxymethylene)
m-PP	Metallocene polypropylene	PP	Polypropylene
MPPO	Modified polyphenylene oxide	PP-*co*-PE	Ethylene/propylene copolymer

PP-*g*-MA	Polypropylene-graft-maleic anhydride	Z/N	Ziegler–Natta
PPA	Polyphthalamides		
PPE	Poly(phenylene ether)		
PPG	Poly(propylene glycol)		
PPO	Poly(phenylene oxide); poly(2,6-dimethylphenylene oxide)		

Additives/Chemicals

PPS	Polyphenylene sulfide
PPSU	Polyphenylsulfone
PPV	Polyphenylene vinylene
PS	Polystyrene
PSDVB	Poly(styrene-divinylbenzene)
PSF	Polystyrene foam
PSU	Polysulfone
PTFE	Poly(tetrafluoroethylene)
PTHF	Poly(tetrahydrofuran)
PTT	Polytrimethylene terephthalate
PUF	Polyurethane foam
PU(R)	Poly(urethane)
PVA	Poly(vinyl alcohol) (see PVAL, PVOH)
PVAc	Poly(vinyl acetate)
PVAL	Poly(vinyl alcohol) (see PVA, PVOH)
PVC	Poly(vinyl chloride)
PVDC	Poly(vinylidene chloride)
PVDF	Poly(vinylidene fluoride) (see PVF2)
PVF	Poly(vinyl fluoride)
PVF2	Poly(vinylidene fluoride) (see PVDF)
PVOH	Poly(vinyl alcohol) (see PVA, PVAL)

ACN	Acrylonitrile (see AN)
ADC	Azodicarbonamide
AE	Ethoxylated alcohol
AIBN	2,2′-Azobisisobutyronitrile
AM	Acrylamide
AN	Acrylonitrile (see ACN)
AO	(1) Antioxidant; (2) Active oxygen
AP	Alkylphenol
APEO	Alkylphenol polyethoxylate
APP	Ammonium polyphosphate, $(NH_4PO_3)_n$
ATBC	Acetyltributyl citrate
ATH	Alumina trihydrate
BA	(1) Blowing agent; (2) Butylacrylate
BADGE	Bisphenol-A diglycidyl ether
BBP	Butylbenzylphthalate
BCP	Butylcyclohexyl phthalate
BDO	1,4-Butanediol
BDP	Bis(diphenyl)phosphate
BEHA	(1) *N*,*N*-Bis-(2-hydroxyethyl) alkyl (C_8-C_{18}) amine; (2) Bis(2-ethylhexyl)azelate
BEO	Brominated epoxy oligomer
BFR	Brominated flame retardant
BHA	Butylated hydroxyanisole; *t*-butyl-4-methoxy-phenol
BHEB	Butylhydroxyethyl benzene
BHT	(1) Butylhydroxytoluene; (2) *β*-Hydroxytoluene
BIO-S	Biostabiliser
BMA	Butyl methacrylate
BOP	Benzyloctylphthalate
BOPD	Bis(2-ethylhexyl)peroxydi-carbonate
BPA	Bisphenol-A
BPA-BI	Bisphenol-A bisimide
BPA-DA	Bisphenol-A dianhydride
BPP	Bispyrene propane
BQ	Benzoquinones
Br_xBB	Bromobiphenyl
Br_{10}DPO	Decabromodiphenyl ether
BT	Benzothiazole
BZT	2-Hydroxybenzotriazoles

SAN	Styrene-acrylonitrile copolymer
SBR	Styrene-butadiene rubber; poly(butadiene-*co*-styrene)
SBS	Styrene hydrogenated butadiene-styrene terpolymer
SMA	Styrene-maleic anhydride copolymer
SMC	Sheet moulding compound
sPS	Syndiotactic PS
SR	Synthetic rubber
ST-DVB	Cross-linked styrene-divinylbenzene

TPE	Thermoplastic elastomer
TPO	Thermoplastic olefin
TPU	Thermoplastic polyurethane
TPV	Thermoplastic vulcanisate

UF	Urea formaldehyde resin
UHMWPE	Ultrahigh-molecular–weight polyethylene

VLDPE	Very low-density polyethylene

CaSt	Calcium stearate	DEHP	Di(2-ethylhexyl)phthalate
CB	(1) Chain-breaker;	DEMP	Diethyl methylphosphonate
	(2) Carbon-black	DEP	Diethylphthalate
CBA	Chemical blowing agent	DEVP	Diethyl vinyl phosphonate
4-CBA	4-Carboxybenzaldehyde	DFTPP	Decafluorotriphenylphosphine
CB-D	Chain-breaking donor	DG	Dodecyl gallate
CBS, CZ	N-Cyclohexyl-2-benzothiazole	DGEBA	Diglycidyl ether of bisphenol A
	sulfenamide	DHA	Di-n-hexyl adipate
CFC	Chlorofluorocarbon	DHBA	2,5-Dihydroxybenzoic acid
Ch	Chromophore		(gentisic acid)
CHPPD	N-cyclohexyl-N'-phenyl-p-	DHDPM	4,4'-Dihydroxydiphenylmethane
	phenylene-diamine	DHP	Dihexylphthalate
CRM	Certified reference material	DIBA	Diisobutyladipate
CRS	Chemical reference substances	DIBP	Diisobutylphthalate
CTA	Crude terephthalic acid	DIDP	Diisodecylphthalate
CTAB	Cetyltrimethylammonium	DIHP	Diisoheptylphthalate
	bromide	DIMP	Diisopropyl methylphosphonate
		DINP	Diisononylphthalate
DAP	Diallylphthalate	DIOA	Diisooctyladipate
DBBA	3,6-Di-t-butyl-4-hydroxybenzyl	DIOP	Diisooctylphthalate
	acrylate	DIPA	Diisopropyladipate
DBBP	Decabromobiphenyl (see Br$_x$BB)	DIUP	Diisoundecylphthalate
DBDPE	Decabromodiphenylether	DLO	Diffusion-limited oxidation
DBDPO	Decabromodiphenyloxide (see	DLTDP	Dilaurylthiodipropionate
	Br$_{10}$DPO)	DLTP	Dilaurylthiopropionate
DBET	Dibutylester chlorotin	DMA	(1) Dimethyladipate;
DBP	Dibutylphthalate		(2) 1,3-Dimethyladamantane;
DBPA	Dibutylphthalate absorption		(3) Dimethylacrylamide;
	(test)		(4) Dimethylacetamide
DBPC	Di-tert-butyl-p-cresol	DMET	Dimethylester chlorotin
DBS	(1) Di-n-butylsebacate;	DMF	N,N-Dimethylformamide
	(2) Dibromostyrene;	DMIP	Dimethylisophthalate
	(3) 1,2,3,4-Di-p-	DMMP	Dimethyl methylphosphonate
	methylbenzylidene sorbitol;	DMOP	Dimethyl o-phthalate
	(4) Sodium dodecyl benzene	DMP	Dimethylphthalate
	sulfonate;	DMS	(1) Dimethyl sebacate;
	(5) Dibenzylsulfide		(2) Dimethylsilicone
DBT	Dibutyltin	DMSO	Dimethylsulfoxide
DBTDL	Di-n-butyltin dilaurate	DMT	Dimethylterephthalate
DCHP	Dicyclohexylphthalate	DNA	Dinonyladipate
DCM	Dichloromethane	DNBP	Di-n-butylphthalate
DCPD	Dicyclopentadiene	DNDP	Di-n-decylphthalate
DDBSA	Dodecylbenzenesulfonic acid	DNHP	(1) Di-n-hexylphthalate;
DDP	Didecylphthalate		(2) Di-n-heptylphthalate
DEA	Diethylamine	DNNP	Di-n-nonylphthalate
DeBP	Decylbenzylphthalate	DNOP	Di-n-octylphthalate
Deca-BDE	Decabromodiphenylether (see	DNP	Dinonylphthalate
	DBDPO)	DNPD	N,N'-Di-β-naphthyl-p-
DEEDMAC	Diethylester dimethylammonium		phenylenediamine
	chloride	DNPT	Dinitroso pentamethylenetetramine
DEEP	Diethyl ethylphosphonate	DOA	Dioctyladipate
DEHA	Di(2-ethylhexyl)adipate	DODPA	Di(t-octyl)diphenylamine

DOP	Dioctylphthalate
DOPPD	Dioctyl-*p*-phenylene diamine
DOS	Dioctylsebacate
DPA	Diphenylamine
DPE	Diphenylether
DPG	1,3-Diphenylguanidin
DPhT	Diphenyltin
DPP	(1) Diphenylphthalate;
	(2) Dipropylphthalate;
	(3) Diketopyrrolopyrrole
DPPD	*N*,*N*'-Diphenyl-*p*-phenylenediamine
DPPH	Diphenylpicrylhydrazyl
DSTDP	Distearyl 3,3'-thiodipropionate
DTBP	(1) 2,4-Di-*t*-butylphenol;
	(2) Di-*t*-butylperoxide
DTDMAC	Ditallowdimethylammonium chloride
DTMHP	Di(trimethylhexyl)phthalate
DUP	Diundecylphthalate
DVB	Divinylbenzene
E	Ethylene
EA	(1) Ethyl acrylate;
	(2) Extrusion aid
EBA	*N*,*N*'-Ethylene-bis-stearamide
EBS	Ethyl-bis-stearamide
EBTBP	Ethylene 1,2-bis(tetrabromophthalimide)
EDTA	Ethylenediamine-tetraacetic acid
EDTMQ	6-Ethoxy-1,2-dihydro-trimethylquinoline
EHA	2-Ethylhexyl acrylate
EHDPP	2-Ethylhexyl diphenylphosphate
ENB	Ethylidene-norbornene (C$_9$)
EO	Ethylene oxide, oxirane
EPO	Epoxidised vegetable oil
ES	External standard
ESBO, ESO	Epoxidised soybean oil
ETA	Ethanol−toluene azeotrope
FAE	Fatty alcohol ethoxylates
FAME	Fatty acid methyl esters
FB	Fluidised bed
FDMPTD	Dimethyl-di-(*p*-fluorophenyl)-thiuram disulfide
FDMPTM	Dimethyl-di-(*p*-fluorophenyl)-thiuram monosulfide
FMA	*p*-Fluoro-*N*-methylaniline
FR	Flame retardant
FWA	Fluorescence whitening agent

FZMPC	Zinc-(*N*-methyl-*N*-*p*-fluorophenyl)-dithiocarbamate
GDS	Glycerol distearate
GF	(1) Glass fibre;
	(2) Glass-filled
GFR	Glass fibre reinforced
GMS	Glycerol monostearate
HALS	Hindered amine light stabiliser
HAS	Hindered amine stabiliser
HATS	Hindered amine thermal stabiliser
HBCD	Hexabromocyclododecane
HB 307	Mixture of synthetic triglycerides
HCP	Hexachlorophosphazene
HD	Hydroperoxide-decomposing antioxidant
HDO	1,6-Hexanediol
HFIP	1,1,1,3,3,3-Hexafluoroisopropanol; hexafluoropropan-2-ol
HFR	Halogenated flame retardant
HMD(A)	Hexamethylenediamine
HMDI	1,6-Hexamethylene di-isocyanate
HMTA	Hexamethylenetetramine
HMW	High-molecular-weight
HPMP	Hydroxy-pentamethylpiperidine
HPPD	*N*-(1,3-Dimethylbutyl)-*N*'-phenyl-*p*-phenylenediamine
HQS	8-Hydroxyquinoline-5-sulfonic acid
HRP	Horseradish peroxidase
HS	Heat stabilised
IIP	Di-(2-propyl)phenylphenyl-phosphate
IM	Impact modifier
I2P	Bis(2-propyl)phenyldiphenyl-phosphate
IPA	Isopropylalcohol
IPP	(2-Propyl)phenyldiphenyl-phosphate
IPPD	*N*-Isopropyl-*N*'-phenyl-*p*-phenylene diamine
IS	Internal standard
L	Ligand
LHSC	Lower halogenated subsidiary colours
LMW	Low-molecular-weight
LS	Light stabiliser
LTMC	Late transition metal catalyst

M	Modifier	ODCB	*o*-Dichlorobenzene
M$^{+\bullet}$	Molecular ion	ODPA	Octylated diphenylamine
MA(H)	Maleic anhydride	ODS	Octadecyl-modified silica gel
MB	Masterbatch	OFPBG	*o*-Fluorophenyl-biguanide
MBET	Monobutylester chlorotin	OG	Octylgallate
MBI	2-Mercaptobenzimidazole	OP	(1) Octylphenol; (2) Organic
MBOAC	4,4′-Methylenebis(2-chloroaniline)		pigment
MBT	(1) 2-Mercaptobenzothiazole;	OTBG	*o*-Tolyl-biguanide
	(2) Monobutyltin	OTC	Organic tin compound
MBTS	Bismercaptobenzothiazole and/or		
	2,2′-dibenzothiazyl disulfide	P	Polar
MD	Metal deactivator	p	Proton
MEA	Monoethanolamine	PA	(1) Polymer additive;
MEHP	Mono(2-ethylhexyl)phthalate		(2) Proton affinity
MEK	Methylethylketone	PABA	*p*-Aminobenzoic acid
MH$^+$	Protonated molecular ion	PAH	Polynuclear aromatic hydrocarbon
(M-H)$^-$	Deprotonated molecular ion	PAO	Primary antioxidant
MHB	Methyl-*p*-hydroxybenzoate	PBA	Physical blowing agent
MIBK	Methylisobutylketone	PBB	Polybrominated biphenyls
MMA	Methylmethacrylate	PBDE, PBDPE	Polybrominated diphenylethers
MMET	Monomethylester chlorotin	PBNA	*N*-Phenyl-*β*-naphthyl amine
MMT	Montmorillonite	PCB, PCBP	Polychlorinated biphenyls
MP	(1) Melamine pyrophosphate;	PeBDPO	Pentabromodiphenyl oxide (see
	(2) Me 3-(3,5-di-*tert*-butyl-4-		penta-BDE)
	hydroxyphenyl) propionate;	Penta-BDE	Pentabromodiphenyl ether (see
	(3) Mobile phase (modifier)		PeBDPO)
MPEG	Methoxy polyethylene glycol	PER	Pentaerythritol
MTBE	Methyl-*t*-butylether	PERM	Polymeric elemental reference
			material
n	Neutron	PETN	Pentaerythritol tetranitrate
NBA	*m*-Nitrobenzylalcohol	PETO	Pentaerythritoltetraoctyl ester
NBBS	*n*-Butylbenzenesulfonamide	PETS	Pentaerythritoltetrastearate
NDB	No-dust blend	PFK	Perfluorokerosene
NDGA	Nordihydroguaiaretic acid	PFTBA	Perfluorotributylamine
NOC	*N*-Nitroso compounds	PG	*n*-Propylgallate
	(nitrosamines)	P-HALS	Polymer-bound hindered amine
NOR-HALS	Alkoxyamine HALS		light stabiliser
NO$_x$	Nitrogen oxides	PHPAA	*p*-Hydroxyphenylacetic acid
NP	(1) *p*-Nonylphenol; (2) Nonpolar	PO	Propylene oxide
*α*NPA	*α*-Naphthylphenylamine	PPA	(1) Polymer processing additive;
NPOE	2-Nitrophenyloctylether		(2) Poly(1,2-propylene adipate)
		PPT	Phenolphthalein
OABH	*N*,*N*′-Dibenzaloxalyldihydrazide	5-PT	5-Phenyltetrazole
	(Eastman)	PTC	Phase transfer catalyst
OB, OBA	Optical brightener, optical	PVI	Pre-vulcanisation inhibitor
	brightening agent		
OBPA	Oxy-bis-phenoxarsine	Q	Quencher
OBS	Organic based stabiliser	QM	Quinone methide
OBSH	4,4′-Oxy-bis(benzene sulfonyl		
	hydrazide)	RDX	Research Department Explosive
OBTS	*N*-Oxydiethylene-2-benzothiazyl-		(Cyclotrimethylenetrinitranine)
	sulfenamide	RM	Reference material

SAO	Secondary antioxidant
SC	Supercritical
scCO$_2$	Supercritical CO$_2$
SCF	Supercritical fluid (see SF)
SDN6S	Sodium 2-(N-dodecylamino)-naphthalene-6-sulfonate
S-EED	1,3-Benzene dicarboxamide; N,N'-bis-2,2,6,6-tetramethylpiperidynyl
SF	Supercritical fluid (see SCF)
SI	Secondary ion
SRM®	Standard Reference Material, registered trademark (NIST)
SSC	Single site catalyst
TA	(1) Terephthalic acid; (2) Triacetin
TAC	Triallylcyanurate
TAIC	Triallylisocyanurate
TBAH	Tetrabutylammonium hydroxide
TBBA, TBBP-A	Tetrabromobisphenol-A
TBBP-S	Tetrabromobisphenol-S-bis-(2,3-dibromopropyl ether)
TBC	t-Butyl catechol
TBHQ	t-Butylhydroquinone
TBP	(1) 2,4,6-Tribromophenol; (2) Tributyl phosphate
TBPA	t-Butylperoxyacetate
TBPB	t-Butylperoxybenzoate
TBPE	1,2-Bis(tribromophenoxy)ethane
TBPO	t-Butylperoxy-2-ethyl-hexanoate
TBPP	t-Butylperoxypivalate
TBT	Tributyltin
TBT-Cl	Tributyltin chloride
TCB	1,2,4-Trichlorobenzene
TCE	(1) 1,1,2,2-Tetrachloroethane; (2) Trichloroethylene
TCP	Tricresylphosphate
TCPP	Trichlorophenoxy phenol
TDBHI	1,3,5-Tris(3,5-di-t-butyl-4-hydroxybenzyl)isocyanurate
TDI	2,4-Toluene diisocyanate
TEA	Triethylamine
TEGDMA	Triethyleneglycol dimethacrylate
TEHTM	Tris(2-ethylhexyl)trimellitate
Tenax	Absorbent charcoal
TEP	Triethylphosphate
TET	(1) Triethyltin; (2) Tetraethyltin
TETD	Tetraethylthiuramdisulfide
TETHM	Tri-2-ethylhexyl trimellitate
TETO	Glyceryl tri [1-^{14}C]epoxyoleate

TFA	Trifluoroacetic acid
TFE	Tetrafluoroethylene
TGS	Triglycine sulfate
THBP	2,4,5-Trihydroxybutyrophenone
thd	Tris-(2,2,6,6-tetramethyl-3,5-heptane dionato)
THF	Tetrahydrofuran
TMAH	Tetramethylammonium hydroxide
TMDQ	2,2,4-Trimethyl-1,2-dihydroquinoline
TMP	(1) 2,2,6,6-Tetramethylpiperidine; (2) Trimethylphosphate
TMPTA	Trimethylolpropanetriacrylate
TMS	Tetramethylsilane (internal standard)
TMSDN	Tetramethylsuccinodinitrile
TMSN	Tetramethylsuccinonitrile
TMT	Tetramethyltin
TMT-Cl	Trimethyltin chloride
TMTD	Tetramethylthiuram disulfide
TNPP	Tris (nonylphenyl) phosphite
TNT	Trinitrotoluene
TOP	Trioctyl phosphate
TOTM	Trioctyl trimellitate
TPA	3,3-Di(thio)-propionic acid
TPFHST	Tris(perfluoroheptyl)s-triazine
TPhT	Triphenyltin
TPhT-Ac	Triphenyltin acetate
TPP	(1) Triphenyl phosphate; (2) Triphenylphosphine
TPT	Tripropyltin
TPT-Cl	Tripropyltin chloride
TSH	p-Toluene sulfonyl hydrazide
TSS	p-Toluene sulfonyl semicarbazide
TTP	Tritolyl phosphate
TXP	Trixylylphosphate
UDP	Undecylphthalate
UVA	UV absorber
VM	Viscosity modifier
VNP	Very nonpolar
VOCs	Volatile organic compound(s)
VP	Very polar
YAG	Yttrium aluminium garnet
ZBEC	Zinc benzyldiethyldithiocarbamate
ZDDP, ZDTP	Zinc dialkyl dithiophosphates
ZDEC	Zinc-N-diethyldithiocarbamate
ZDMC	Zinc-N-dimethyldithiocarbamate
ZnSt	Zinc stearate

6PPD	*N*-1,3-Dimethylbutyl-*N'*-phenyl-*p*-phenylenediamine

PHYSICAL AND MATHEMATICAL SYMBOLS

A	Ampere, unit of electric current
A	(1) Mass number of a nucleus; (2) Absorbance; (3) Area
Å	Ångstrom, unit of wavelength, $1\,\text{Å} = 10^{-8}\,\text{cm}$
a, ag	Atto (10^{-18}), attogram
AC	Alternating current
a.m.u, amu	Atomic mass unit
AU	Absorbance unit
B	Magnetic field strength
B_0	Static magnetic field (flux density)
B$_0$	External (applied) magnetic field amplitude (NMR)
B$_1$	Oscillating magnetic field vector of the observing channel (NMR)
b.p.	Boiling point
C	(1) Degrees Centigrade; (2) Coulomb
C, *c*	(1) Concentration or molar concentration; (2) Thermal capacity
c	Velocity of light
CED	Cohesive energy density
Ci	Curie
CLS	Classical least-squares
COF	Coefficient of friction
CP	Curie-point
cP	Centipoise
cps	Counts per second
CV	Coefficient of variation
CW	Continuous wave (laser)
D	(1) Debye; (2) Diffusion
D	(1) Diffusion coefficient; (2) Distribution ratio
d	(1) Diameter, thickness; (2) Density; (3) Diffusion path length; (4) Interplanar spacing of crystal
d_p	Particle diameter

DA	Discriminant analysis
Da	Dalton or atomic mass unit
d.c., DC	Direct current
DL	Detection limit
E	Electrical field strength
E	(1) Energy (in eV); (2) Potential
e	Unit charge of an electron
e$^-$	Electron
EM	Electromagnetism
em	Emission wavelength used in fluorescence detection
EMSA	Electron microscope surface area
ES(TD)	External standardisation (calibration)
eV	Electron volt ($23.06\,\text{kcal}\,\text{mol}^{-1}$)
ex	Excitation wavelength used in fluorescence detection
f	(1) Function (general); (2) Fractional free-volume; (3) Flow-rate; (4) Volume fraction
f, fg, fmol	Femto (10^{-15}); femtogram, femtomole
FIC	Fundamental influence coefficient
FP(M)	Fundamental parameter (method) (see XRF)
fs	Femtosecond ($10^{-15}\,\text{s}$)
FT	Fourier transform
ft	foot
FWHH, FWHM	Full-width at half-height/maximum
G	(1) Gauss unit of magnetic field strength; (2) Giga (10^9)
G	(1) Free enthalpy (Gibbs free energy); (2) Geometric term
g	(1) Gram; (2) Gradient pulse amplitude
GSR	Gram−Schmidt reconstruction
Gy	Gray
H	Hildebrand parameter
H	(1) Enthalpy; (2) Surface loss parameter
h, hr	Hour
h	Planck's constant
HETP	Height equivalent to a theoretical plate
HR	High resolution
HT	High temperature
HV	(1) High voltage; (2) High vacuum

Hz	Hertz, unit of frequency (cycles per second)	m	(1) Nuclear spin quantum number; (2) Mass of atom or ion
$h\nu$	Photon energy in eV	$\langle M_n \rangle$	Number average molecular weight
		$\langle M_w \rangle$	Weight average molecular weight
I	(1) Magnetic spin of a nucleus, angular momentum quantum number (integer or half-integer); (2) Current; (3) Intensity	MAU	Milliabsorbance unit
		MCR	Multivariate curve resolution
		MD	(1) Mahalanobis distance; (2) Molecular dynamics
		MDA	Multivariate data analysis
I_o	Intensity of incident light	MFI, MI	Melt flow index
i.d.	Internal diameter	mg, mmol, mL	Milligram, millimole, millilitre (10^{-3})
ILS	Inverse least squares		
in.	inch	MH	Mark–Houwink (constants)
IP	Ionisation potential	mil	0.001 inch
IS(TD)	Internal standardisation (calibration)	mLOD	Mass limit of detection
		m.p.	Melting point
		MPa	Mega Pascal
J	Joule, a unit of energy	MW	Molecular weight
J	Spin coupling constant (NMR)	MWD	Molecular weight distribution
		m/z	Mass-to-charge ratio
K	Kelvin	N	(1) Newton; (2) Normal
K	(1) Partition coefficient or equilibrium constant; (2) Force constant of a bond; (3) Reduced ion mobility	N	(1) Number of neutrons in a nucleus; (2) Noise; (3) Column plate number
k	(1) Kilo (10^3); (2) Boltzmann constant	n	Refractive index
k	Wave vector	n	(1) Number of components; (2) Number of measurements
k	(1) Molar absorption coefficient; (2) Retention factor; (3) Thermal conductivity; (4) Instrumental sensitivity factor	ND	Not detectable
		ng, nm, nmol	Nanogram, nanometre, nanomole (10^{-9})
k_m	Mass transfer coefficient	ns	Nanosecond (10^{-9} s)
k_o	Composite nuclear constant		
k_r	Reaction rate constant	p	Pico (10^{-12})
		p	(1) Pressure; (2) Vapour pressure
L	Litre	Pa	Pascal
L	Length (column length)	PC	Personal computer
l	Pathlength	PCA	Principle component analysis
lb	pound (0.453 kg)	PCR	Principle component regression
LOD	(1) Limit of detection; (2) Loss on drying	PD	Polydispersity
		PDA	Principal discriminant analysis
LOQ	Limit of quantitation	PFG	Pulsed field gradient
LSR	Least-squares regression	pg, pmol	Picogram, picomole (10^{-12})
		phr	Parts by weight per hundred parts of resin
M	(1) Molarity (mol L^{-1}); (2) Mega (10^6)	PLS(R)	Partial least-squares (regression)
M	(1) Atomic or molecular weight; (2) Adsorption constant at two interfaces	PMD	Principle components/Mahalanobis distance discriminant analysis
		ppb	Parts per billion
m	(1) Milli; (2) metre	ppm	Parts per million
		ppq	Parts per quadrillion

ppt	Parts per trillion	T_1, T_{1r}	Nuclear spin-lattice (longitudinal) relaxation time; in the rotating frame
ps	Picosecond (10^{-12} s)		
psi	Pounds per square inch	T_2	Nuclear spin-spin (transverse) relaxation time
q	(1) Wave vector;	t	ton
	(2) Internuclear distance	t	(1) Time(s);
q	Normal coordinate		(2) Layer thickness
		t_o	Dead time
R	Isotope ratio	$t_{1/2}$	Half-life time
R, R_s	(1) Universal gas constant;	t_p	Pulse width
	(2) Recovery factor (extraction);	t_R	Retention time
	(3) Resolution	TRT	Temperature-rise time
R_f	Retention value (planar chromatography)	TV	Television
R^2	Square of the multiple correlation coefficient	u	Carrier gas velocity
		UHMW	Ultra high molecular weight
r	(1) Reaction rate;	UHV	Ultra high vacuum
	(2) Internuclear distance;	UV-A	UV wavelength range 315–380 nm
	(3) Radius	UV-B	UV wavelength range 280–315 nm
RF, r.f.	Radiofrequency	UV-C	UV wavelength range 200–280 nm
rms	Root mean square		
RRT	Relative retention time	V	Volt
RSD, r.s.d.	Relative standard deviation	V	Volume; molar volume
RSF	Relative sensitivity factor	v	(1) Velocity;
r.t.	Room temperature		(2) Recoil velocity
r_t	Retention time (see t_R)	v/v	Volume/volume (solution concentration)
S	(1) Sensitivity factor;	W	Watt, measure of RF power
	(2) Solubility	w/w	Weight/weight (solution concentration)
S	(1) Spin quantum number;		
	(2) Selectivity;	x	Molar fraction (in general)
	(3) Solubility coefficient;	x, y, z	Cartesian coordinates
	(4) Response function		
s	Second	$Y\%$	Extraction yield
s	Scattering coefficient		
SD	Standard deviation	Z	(1) Atomic number;
SI	Système International d'Unités		(2) Number of ions
SML(T)	Group specific migration limit	z	Zepto (10^{-21})
S/N, SNR	Signal-to-noise ratio	z	(1) Axis of $\mathbf{B}_o$, the external (applied) magnetic field;
SPC	Statistical process control		
SR	Self-reversal (background correction)		(2) Ion charge;
S-T	Stejskal–Tanner (NMR)		(3) Depth
S/W	Software		

T	(1) Tesla, unit of magnetic field strength (10^4 Gauss);
	(2) Tera (10^{12})
T	(1) Absolute temperature (K);
	(2) Transmittance
T_g	Glass transition temperature
T_m	Melting temperature

PHYSICAL AND MATHEMATICAL GREEK SYMBOLS

α	Particle
β	(1) Bohr magnetron;
	(2) Proportionality constant;
	(3) Particle

γ | (1) Gyromagnetic ratio of a nucleus; (2) Gamma ray

Δ | (1) Shift or difference (e.g. ΔE, energy difference); (2) Symbol for heat

ΔE_{vap} | Evaporation heat

ΔH_p | Latent heat of vaporisation

ΔR | Resolution

ΔS | Entropy of mixing

δ | (1) Chemical shift (ppm relative to a reference); (2) Solubility parameter; (3) Duration of the gradient pulse

$\delta_d, \delta_p, \delta_h$ | Solubility parameter based on nonpolar or pure dispersive, polar and hydrogen bonding interactions

ε | (1) Molar absorption coefficient; (2) Dielectric constant

ε^* | Complex dielectric constant

ε' | Real part of complex dielectric constant

ε'' | Imaginary part of complex dielectric constant (dielectric loss)

ε^o | (1) Solvent strength parameter; (2) Permittivity of free space

η | Viscosity of medium

θ | (1) Angle between internuclear vector and $\mathbf{B}_o$; (2) Incident angle; (3) Bragg angle; (4) Isotope abundance

λ | (1) Wavelength, unit Å; (2) Decay constant

μ | (1) Magnetic moment of a nucleus; (2) Molecular dipole moment; (3) Micro (10^{-6}); (4) Reduced mass of a system; (5) Absorption coefficient

μm | Micron

ν | (1) Frequency (Hz); (2) Velocity

ρ | Density, unit g cm^{-3}

σ | (1) Standard deviation; (2) Nuclear shielding constant; (3) Thermal neutron cross-section

τ | Time constant (detector)

ϕ | (1) Neutron flux; (2) Volume fraction of solute (ϕ_1) and polymer (ϕ_2) in a mixture

χ | Flory–Huggins polymer–solvent interaction parameter

ω | (1) Angular velocity (rad s^{-1}); (2) Light modulation frequency

GENERAL ABBREVIATIONS

AATCC | American Association of Textile Chemists and Colorists (Research Triangle Park, NC)

ACS | American Chemical Society (Washington, DC)

AFNOR | Association Française de Normalisation; French Association for Standardisation (Paris, France)

AMDIS | Automated Mass Spectral Deconvolution and Identification System (NIST)

AOAC | Association of Official Analytical Chemists International (Arlington, VA)

AP | Asian Pacific

ASTM | American Society for Testing and Materials (West Conshohocken, PA)

BCC | Business Communications Co. Inc. (Norwalk, CT)

BCR | Bureau Communautaire de Référence; European Commission DG XII Community Bureau of Reference (Geel, Belgium); now IRMM

BS | British Standards

BSE | Bovine spongioform encephalopathy

CAS | Chemical Abstracts Service (USA)

CCDC | Cambridge Crystallographic Data Centre (Cambridge, England)

CDC | Corporate Development Consultants

CEC | Commission of the European Communities (Brussels, Belgium)

CEN | Comité Européen de Normalisation; European Committee for Standardisation (Brussels, Belgium)

CFR | Code of Federal Regulations (USA)

CI, C.I. | Colour Index

CITAC | Co-operation on International Traceability in Analytical Chemistry

CONEG | Coalition of Northeastern Governors (USA)

CSD | Cambridge Structural Database (CCDC, Cambridge, UK)

DAM | Draft Amendment (ISO)

dB | Database

DETR | Department of Environment, Transport and the Region (London, UK)

DIN	(1) Deutsches Institut für Normung, German Institute on Standardisation (Berlin, Germany); (2) Deutsche Industrie Normen (German Industrial Standards)	MSDS	Material Safety Data Sheet
		NBS	National Bureau of Standards (now NIST)
DIS	Draft International Standard (ISO)	NIH	National Institutes of Health (Bethesda, MD)
DKI	Deutsches Kunststoff Institut (Darmstadt, Germany)	NIST	National Institute of Standards and Technology (formerly NBS) (Gaithersburg, MD)
EC	European Community	NPL	National Physics Laboratory (Teddington, UK)
EEC	European Economic Community		
EEE, E&E	Electrical and Electronic Equipment	OEM	Original Equipment Manufacturer
EN	European Norm		
ENV	European Prestandard	PQ	Performance Qualification
EPA	Environmental Protection Agency (USA)	PTAI	Phillip Townsend Associates Inc. (Houston, TX)
EU	European Union		
EUCAP	European Collection of Automotive Paints	QA	Quality Assurance
		QC	Quality Control
EURACHEM	Association of European Chemical Laboratories (Lisbon, Portugal)	R&D	Research and Development
		REACH	Registration, Evaluation and Authorization of Chemicals (EU)
FAAM	Food Additives Analytical Manual	RoHS	Restrictions on Hazardous Substances
FDA	Food and Drug Administration (USA)	ROW	Rest of World
		RSMR	Restricted Substance Management and Recycling System
GDP	Gross domestic product		
GLCC	Great Lakes Chemical Corporation (West Lafayette, IN)	SM&T	Standards, Measurements and Testing Programme, EU (formerly BCR)
GLP	Good Laboratory Practice	SOP	Standard Operating Procedure
		SPE	Society of Plastic Engineers (Brookfield, CT)
HSE	(1) Health, Safety and Environment; (2) Health & Safety Executive (Sudbury, UK)		
		TM	Trademark
		TTI	Townsend Tarnell Inc. (Mount Olive, NJ); now Townsend's Polymer Services & Information (T-PSI)
ICDD	International Center for Diffraction Data (Swarthmore, PA)		
ICSD	Inorganic Crystal Structure Database (FIZ, Karlsruhe, Germany)	UN	United Nations
IRMM	Institute for Reference Materials and Measurements (Geel, Belgium)	US	United States
		USEPA	United States Environmental Protection Agency
ISO	International Organization for Standardization (Geneva, Switzerland)		
IUCr	International Union of Crystallography	VDA	Verband der Automobilindustrie, German Federation of Car Industry (Frankfurt, Germany)
IUPAC	International Union of Pure and Applied Chemistry		
LGC	Laboratory of the Government Chemist (formerly NPL) (Teddington, UK)	WELAC	Western European Accreditation Cooperation
LIMS	Laboratory Information and Management Systems	WTC	World Trade Center (New York, NY)

Functionality of Common Additives Used in Commercial Thermoplastics, Rubbers and Thermosetting Resins

Accelerators

Accelerate chemical, photochemical, biochemical reactions or processes, e.g. cross-linking or degradation of polymers. Also called promoters, co-catalysts. Refer usually to the cure process in thermosetting resins.

Acid Scavengers

Neutralise acidity (e.g. deriving from acidic catalyst residues) effectively. Commonly salts of weak organic acids (metal stearates) or inorganic bases (hydrotalcite). Also called antiacids.

Activators

Chemical compounds used with a catalyst to permit polymerisation at room temperature (also called accelerators or promoters).

Adhesion Promoters

Improve adhesion of dissimilar materials such as polymers to inorganic substrates. Also called primers. Primers generally contain a multifunctional chemically reactive species capable of acting as a chemical bridge. In theory, any polar functional group in a compound may contribute to improved bonding to mineral surfaces. However, only a few organofunctional silanes have the balance of characteristics required in successful adhesion promotion: polymer reactivity, covalent bonding to inorganic surfaces, hydrolysability, and other important properties such as water solubility, surface activity, and wetting control. Silane adhesion promoters are now used in several systems, including urethane, epoxy, PVC plastisol, acrylic, and latex coatings. High levels of adhesion in various co-extrusion and metal bonding applications are also achieved with adhesive resins consisting of polyolefins with incorporated functional groups (hot melt adhesives, tackifiers).

Antiblocking Agents

Blocking refers to the adhesion force between two flat surfaces. When plastic film is produced, there is a tendency for contacting layers to stick together, or 'block', making the separation of film difficult. Reduced blocking is achieved by antiblocking agents (mostly inorganic minerals such as talc, diatomaceous earth and silica, or primary fatty acid mixtures) causing surface roughness. Antiblocks are often used in combination with slip additives.

Antifogging Agents

'Fogging' is formation of small water droplets (visible condensation) on the surface of a polymer film. Undesirable effects may result from fog formation, such as reduction of clarity and dripping. Incorporation of antifogging agents eliminates the reduction of transparency by migration to the surface and increases the polymer surface critical wetting tension. This results in

Additives In Polymers: Industrial Analysis And Applications   J. C. J. Bart
© 2005 John Wiley & Sons, Ltd   ISBN: 0-470-85062-0

a film on the surface rather than droplets. Glycerol esters act as antifogging agents. Application areas are agricultural and food-packaging film. For food-wrapping films the selected additives must conform with national regulations.

Antimicrobials

Natural or synthetic, mostly low-MW molecules that reduce or suppress microbe populations in plastics. Specialised products (both organic and inorganic) which tend to be used most in plasticised materials such as PVC and polyurethanes, as well as in rubber articles. The main cause of microbial growth is the additives themselves: plasticisers, starch fillers, lubricants, thickening agents and oils.

See Preservatives.

Antioxidants

Inhibit autoxidation of organic materials by interfering with free radical reactions that lead to incorporation of oxygen into macromolecules in a chain mechanism consisting of two interacting cyclical processes (Scheme II.1).

The primary cycle (A) can be interrupted by electron acceptors (oxidising agents) and electron donors (reducing agents). *Chain-breaking (CB) antioxidants* or *primary AOs* interrupt the primary oxidation cycle and act as radical scavengers, reacting rapidly with propagating radicals, ROO• and R•. Chain-breaking donor (CB-D) antioxidants generally form the basis of heat-stabilising systems for polyolefins. Compounds classified as antifatigue or antiflexcrack inhibitors are listed among CB antioxidants. Typical CB antioxidants are hindered phenolics and secondary aromatic amines. Phenolics generally resist staining and discoloration; however, they can form quinone structures with conjugated double bonds. Important commercial primary antioxidants are IPPD and DPPD (antidegradants for rubbers, e.g. in tyres), Irganox 1010, Irganox 1076, Goodrite 3114, Cyanox 2246, Topanol O, BHT and (more recently) Vitamin E (for polyolefins). *Preventive antioxidants* or *secondary AOs* interrupt the second oxidative cycle (B) by preventing or inhibiting generation of free radicals. The most important preventive mechanism is hydroperoxide decomposition (PD) leading to inert and inactive oxidised end-products and thereby eliminating formation of alkoxy and hydroxy radicals. Secondary antioxidants include hydroxylamines and phosphite compounds (e.g. Irgafos 168), which

Mechanisms of antioxidant action.

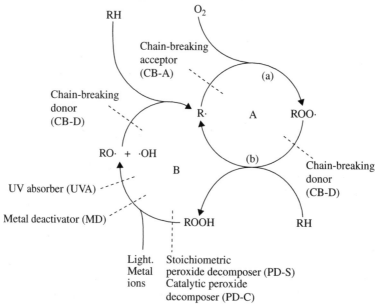

Scheme II.1 Mechanism of antioxidant action. After Grassie and Scott [1]. From N. Grassie and G. Scott, *Polymer Degradation and Stabilisation*, Cambridge University Press, Cambridge (1988). Reproduced by permission of Cambridge University Press

are nondiscolouring and FDA-approved for many indirect food applications. They are used in synergistic combination with primary AOs. Secondary AOs are most effective at elevated temperatures such as those encountered in processing and protect both the polymer and the primary AOs from degradation. TNPP, Weston 618, Ultranox 624 and Sandostab PEPQ are other important examples. Organophosphite secondary AOs also trap peroxyl and alkoxyl radicals thereby eliminating further oxidative chain reactions. Chemically, AOs should preferably be hydrolytically stable, colour stable and stable to high temperatures. Superior AOs for long-term use in polymers have MW > 500 Da. Chemical classes of compounds in use as AOs are naturally occurring phenolic compounds, phospholipids, hydroxy or amino carboxylic acids, sterically hindered phenols, secondary aromatic amines, organic sulfides, metal dialkyl dithiophosphates or carbamates, trivalent phosphorous compounds, quinoline derivatives, and hindered piperidines.

Three factors determine the effectiveness of antioxidants in polymers, namely: (i) intrinsic molar activity; (ii) substantivity in the polymer; and (iii) solubility in the polymer. Multifunctional AOs combine multiple functions in one molecule. Sterically hindered amine stabilisers (HAS), such as Chimassorb 944, Tinuvin 622 and Tinuvin 783 are prime examples.

Polymer-bound antioxidants must be molecularly dispersed (i.e. infinitely soluble) and cannot be physically lost from the substrate. High-MW phenolic AOs are preferred for applications requiring FDA approval, minimal discoloration, and long service life at high temperatures. Antioxidants are used for protection of polymers, plastics, elastomers, foods, fuels and lubricants.

Antiozonants

Protect elastomeric materials against deterioration by ozone that is either naturally generated by electrical discharge and solar radiation or produced in urban areas as a consequence of pollutant emissions. Ozone attacks double bonds in polymeric compounds and causes cracking along stress lines of a material. An effective antiozonant either reacts rapidly with ozone or forms a protective barrier on the surface of the product. The choice of antiozonants depends on the processing characteristics, nondiscolouring and staining requirements of the end-product, and conditions of use and cost. There is little known about the mechanism of antiozonant action.

Antiradiation Agents

Retard degradation processes through γ radiation. Phenyl-β-naphthyl amine (PBN) is a commercial agent widely used in the rubber industry; polyaniline (PAn) is also effective.

Antiskinning Agents

Control skin formation as a result of autoxidation of alkyd or oil paint exposed to air. Representative classes of compounds are oximes and antioxidants.

Antislip Agents

Minimise the tendency of plastics films to slip past each other when it would be undesirable.

See Slip agents.

Antistatic Agents

Reduce or eliminate the tendency of plastics to retain electrostatic charges. The exceptionally low electrical conductivity of most plastics leads to charge build-up on the surface, with consequent dust attraction affecting appearance and performance of the end-product, static attraction between separate articles, handling problems during transport, storage and packing, and electrical discharge causing fire or explosion.

Antistatic agents are ionic substances, such as inorganic salts or organic materials, which attract water molecules. Outstanding materials are derived from the fatty amide and amine chemistry, such as ethoxylated alkyl amines, glycerolmonostearate, fatty alkanolamines, and sodium alkylsulfonates ($C_{12}-C_{16}$ alkyl groups are preferred because of low loss by evaporation).

'External' antistatics, applied to finished products as wash coatings, act as a functional surfactant at the polymer/air interface, and have an immediate but short term effect over a broad range of polymers without any potential compatibility problems. They perform by taking up humidity from the air, thus increasing the conductivity and therefore reducing build-up of static changes. They are generally ineffective in extremely dry environments. Classical or *'internal'* *antistatic agents* are incorporated into the polymer and produce long-term protection against static charge accumulation. The efficiency of antistatics may be

evaluated directly, by measuring the rate at which an electrical charge is dissipated, or indirectly, by the electrical conductivity of the material. Antistatics play a role in the practical problem of a static-free polymer formulation for supermarket plastic bags that you can actually open.

Biocides

Protect a plastic material against attack by bacteria, fungi, algae, moulds, etc., especially when additives such as plasticisers are present. Biocides, which come in several types – fungicides, bactericides, etc. – are by definition toxicologically active and therefore attract the attention of regulators. The ideal antimicrobial agent is toxic to the organisms it will be used against; compatible; stable; soluble or dispersible in the product composition; nonstaining; noncorrosive; unaffected by physical conditions of the product; penetrating; economic; registered; having minimal environmental impact. Antimicrobials may be both inorganic and organic. The most widely used and traditional antimicrobial is oxybisphenoxarsine (OBPA); other biocides worth mentioning are trichlorophenoxyphenol (TCPP) or Triclosan® (Ciba Specialty Chemicals) and Zeneca's Vanquish® products. Metallic soaps of copper, zinc, mercury and tin are also effective biocides for a variety of applications. Most plastics do not require protection against microorganisms. In fact there is probably as much interest in promoting biodegradable behaviour as in inhibiting it. Application areas are PVC and polyurethane foams.

See Antimicrobials.

Blowing Agents

Generate a gas chemically or by simple evaporation during polymer processing, which effects expansion or foaming the polymer creating a cellular structure during processing. Blowing (or foaming) agents can be compressed gases or volatile liquids or compounds that decompose at processing temperatures to form a gas. *Chemical blowing agents* (CBAs) liberate gas as a result of chemical reactions, including thermal decomposition. Decomposition of CBAs can be either endothermic or exothermic. The basic types of industrial CBAs are organic (azo compounds, e.g. ADC; hydrazine derivatives, e.g. TSH, OBSH; sulfonyl semicarbazides, e.g. TSS; tetrazoles, e.g. 5-PT; *N*-nitroso compounds, e.g. DNPT) and in a few cases inorganic (bicarbonates/carbonates). CBAs find application for PVC (50 %),

PE/PP (25 %), rubbers (including EVA) (20 %) and other thermoplastics (5 %). *Physical blowing agents* (PBAs) liberate gases as a result of physical processes (evaporation, desorption) at elevated temperatures or reduced pressures. PBAs are mostly volatile liquids, freons, aliphatic hydrocarbons or solid blowing agents.

Bonding Agents

Typically used for the coating of polyester and polyamide substrates with plasticised PVC. Chemically: one-component bonding agents (e.g. aromatic polyisocyanurate) and two-component bonding agents (e.g. aliphatic polyisocyanate); liquids.

See Pot-life extenders.

Carbon-Blacks

Impart a variety of desirable characteristics to plastics, including colour, protection from UV radiation, and conductivity. Properties and uses of each carbon-black type can vary, e.g. furnace black comes in high abrasion, fast extrusion, high modulus, general purpose, semireinforcing and other grades. Black colloidal carbon fillers are made by the partial combustion and/or thermal cracking of natural gas, oil, or another hydrocarbon. Carbon-blacks are widely used as a filler and pigment in PVC, phenolic resins, and polyolefins, increase resistance to UV light and electrical conductivity and sometimes act as a cross-linking agent.

Catalysts

Chemical compounds (usually an organic peroxide) which initiate polymerisation of a resin (also called hardeners).

Clarifying Agents

Essentially nucleating agents which cause crystallisation to take place in several locations at once, starting a large number of small spherulites which give rise to less light scattering and greater light transmission. Give finished parts (typically of injection moulded polypropylene) improved aesthetics through enhanced see-through clarity and surface gloss. In addition, faster process cycles and physical properties are commonly obtained.

Performance criteria of clarifying agents are clarity, organoleptics, processability and nucleation. A typical third generation clarifying agent is the sorbitol acetal Millad® 3988 (3,4-dimethyl dibenzylidene sorbitol).

Colorants (Dyes, Pigments)

Colorants for plastics are commonly chemically classified as inorganic white and coloured pigments, organic pigments, polymer-soluble organic colorants (dyes), carbon pigments (carbon-black, graphite) and inorganic effect pigments (e.g. metal flakes, mica-based pearl effects). Act either by absorbing parts of the spectrum and reflecting other parts (solid pigments, dyes), or by transmitting only certain wavelengths (transparent colours). Pigments have by far the greater significance in the plastics industry, and dyes in the textile industry. Pigments are virtually insoluble in plastics, whereas are soluble. The term pigment is linked with a specific particle size range ($\sim$0.01 to $\sim$1 μm). Coloration with pigments requires a dispersion, coloration with dyes involves a dissolving process. The colour change resulting from coloration of plastics is based on the wavelength-dependence and/or scattering (reflectance) of light. Dyes can only absorb light and not scatter it. These colorants are, therefore, transparent. The optical effect of pigments is based on reflectance and coloured and opaque resins result. With all coloured pigments that selectively absorb and reflect, the shade is influenced by particle size. No reflectance occurs when the particle sizes are very small.

See Dyes, Pigments.

Compatibilisers

Usually polymeric substances of appropriate chemical structure and morphology which promote the miscibility of incompatible materials. Block copolymers are especially useful surfactants at the polymer/polymer interface because the two blocks can be made up from molecules of the individual polymers to be mixed. Typical compatibilisers in polymer blends are: LDPE-*g*-PS in PE/PS; CPE in PE/PVC; acrylic-*g*-PE, -PP, -EPDM in polyolefin/PA; and maleic-*g*-PE, -PP, -EPDM, -SEBS in polyolefin/polyesters.

Corrosion Inhibitors

Used to kill rest active catalyst residues, e.g. metal stearate and hydrotalcite.

Coupling Agents

Improve adhesion between two dissimilar phases in a composite material, e.g. a resin matrix/reinforcement interface. Coupling agents are used in trace quantities to modify a surface in such a way that bonding occurs with another kind of surface, e.g. a mineral and a polymer. Create a better dispersion of the mineral and favour agglomerate breakdown. A most common use of silane coupling agents is in adhering inorganic materials such as glass to organic materials such as polymers.

See Adhesion promoters.

Cross-linking Agents

Provide lateral bonds within the skeletal polymer backbone to improve compressive strength, abrasion resistance, and thermal stability. The process of cross-linking existing linear macromolecules is analogous to vulcanisation of natural rubber. Major polymers to be cured with cross-linking peroxides are natural rubber (NR) and the synthetic analogues (SBR, BR, IR). Others include polyethylene (PE), copolymers of ethylene, propylene and a diene (EPDM), nitrile polymers (NBR) as well as silicone and fluor polymers. The resulting material is insoluble and infusible, making chemical analysis of the extent of reaction extremely difficult. Determination of the extent of cross-linking frequently relies on the measurement of changes in physical, electrical and mechanical properties.

See Compatibilisers, Vulcanisation accelerators.

Curing Agents

Consist of a range of chemicals which promote cross-linking; can initiate cure by catalysing ('catalysts', hardeners, initiators), speed up and control cure (activators, promoters) or perform the opposite function (inhibitors) producing thermosetting compounds and specialised thermoplastics (e.g. peroxides in polyesters, or amines in epoxy formulations). The right choice of a cure system is dependent on process, process temperature, application and type of resin.

See Activators, Catalysts, Initiators.

Defoaming Agents

Remove trapped gases from liquid mixes during compounding.

Degradation Additives

Degrade plastics (solid waste) by one of four processes: biodegradation, photodegradation, chemical degradation, and hydrodegradation.

See Photosensitisers, Pro-oxidants.

Diluents

Reduce resin viscosity and facilitate processing. Refer frequently to epoxy resins.

Dispersing Agents

Aid in the uniform dispersion of additives. Make powdered solids (e.g. particulate fillers with high energy and hydrophilic surface) more compatible with polymers by coating their surfaces with an adsorbed layer of surfactant in the form of a dispersant. Surface coating reduces the surface energy of fillers, reduces polymer/filler interaction and assists dispersion. Filler coatings increase compound cost. Fatty acids, metal soaps, waxes and fatty alcohols are used as dispersants commonly in concentrations from 2 to 5 wt %.

See Coupling agents, Fillers.

Driers

Are used to accelerate autoxidation and hardening of oxidisable coatings. Metal soaps, used as paint driers, can be made from a variety of carboxylic acids, including the commercially important naphthenic and 2-ethyl hexanoic acids, tall oil, fatty acids, neodecanoic and isononanoic acid. Cobalt is unquestionably the most active drier metal available. Metallic driers such as cobalt naphthenate or octoate and zinc salts can interact with UVAs, HALS, or AOs.

Dyes

Impart a bright and intense colour to a substrate by a process which at least temporarily destroys any crystal structure of the colouring substances. Dyes are transparent and easy to disperse and process. Dyes are incompatible with polyolefins, having a tendency to bleed and plate out. Due to the solubility and lack of migration resistance, dyes are rarely used in moulding resins.

See Colorants.

Elastifying Agents

Promote improved elasticity and cold temperature flexibility. Used in cables, self-adhesive products, road-marking materials, etc.

Emulsifiers

Anionic emulsifiers, such as alkali salts of fatty acids, can be applied in anionic latices. Are particularly important as wetting and foaming agents in latex technology. Poly(glycol) ethers act as nonionogenic emulsifiers.

Exotherm Modifiers

Reduce the maximum temperature reached during an exothermic cross-linking reaction.

Extenders

Are used in combination with primary plasticisers to lower cost.

See Fillers, Plasticisers, Process oils.

Fibres

Improve mechanical properties or reinforce polymers. A fibre is a unit of matter of relatively short length, characterised by a high ratio of length to thickness or diameter. Fibrous additives may be polymeric, carbonaceous, metallic, glassy or ceramic in nature. Glass fibres provide short (staple, chopped, milled) or continuous fibre reinforcement, and are used widely in both thermosets and thermoplastics for increased strength, dimensional stability, thermal stability, corrosion resistance, and dielectric properties. Used in moulding compounds, spray-up processes, die moulding, lay-up, and other lamination processes. Glass fibres are often surface modified, e.g. with coupling agents, to improve bonding with polymer matrix or to impart special properties such as electrical conductivity (by metal coating). The fibre length of glass is typically reduced to 50–500 μm during processing.

Fillers

Particulate additives (usually low-cost of fibrous, lamellar or spherical type) are designed to alter the physical, mechanical, electrical or other properties of a polymer or to extend a resin. Fillers may be classified as nonreinforcing, semi-reinforcing and reinforcing or as carbon blacks and light coloured fillers. Commodity minerals, such as ground limestone, are relatively inexpensive and used mostly as extenders. Specialty minerals (e.g. talc, calcium carbonate, diatomaceous earth) are usually reinforcing fillers. These are inherently of small particle size and often chemically surface modified. Organic fillers may be of natural origin (e.g. wood flour) or synthetic (including fluoropolymer spheres and milled polymer waste). They might increase the flammability and decrease the moisture resistance of plastics. Spherical filler particles (micro-spheres), usually hollow, are used to reduce weight in a product without much adverse effect on mechanical properties.

See Reinforcements.

Flame Retardants

Substances applied to or incorporated in a combustible material (e.g. organic polymers, nylon, vinyl and rubber, etc.) to reduce flammability. Act by retarding ignition, control/douse burning, reduce smoke evolution. Slow down or interrupt the self-sustained combustion cycle when the heat-flux is limited. Flame retardants (FRs) improve the combustion behaviour and alter the combustion process (cool, shield, dilute, react) so that decomposition products will differ from non-flame retarded articles. FRs are usually divided into three classes:

- *Halogenated additives.* Most widely used, work in the gas phase above the burning polymer surface by chemically interrupting the flame propagation mechanism. Halogenated FRs, especially brominated, have been targeted for substitution being perceived as an environmental hazard by persistence and bio-accumulation; when heated they can form toxic by-products (CO, smoke).
- *Intumescent additives.* React with the polymer substrate to produce a char layer which forms an effective barrier between heat source and oxygen and

the fuel derived from pyrolysis of the polymer. The efficiency with which a polymer can be flame retarded is related to its inherent tendency to char.
- *Spumific additives.* Decompose at the combustion temperature of the polymer to produce inert gases, such as CO_2 and H_2O, which dilute the combustion gases and hinder burning.

Flame retardants can reduce processability and interfere with other additives. Flame retardants are either reactive or additive. Reactive flame retardants (used mainly in thermosets) are chemically bound to the polymer; additive FRs (used primarily in thermoplastics) are physically mixed with the resin during or after polymerisation. Flame retardants are generally designed to provide a particular level of resistance to ignition or flame spread. The ideal FR additive is colourless, easily blended, compatible with the substrate (no blooming and plate-out), has no deleterious effects on (mechanical) properties of the substrate, allows all finished article colours, remains effective throughout the service life of the products, is thermally and light stable, resistant to ageing and hydrolysis, odourless, does not cause corrosion, is highly effective in small quantities, generates a very low amount of smoke, has minimal toxicological effects, is economic and recyclable. This goal is still to be reached. Nearly all the present-day commercial FRs and smoke suppressants can be divided into several chemical groups:

- Antimony and other inorganic compounds (tin, molybdenum, aluminium, magnesium, iron, boron, with ATH accounting for about 40 % in volume of FR shipments in Europe).
- Halogenated (usually bromine or chlorine) compounds (inhibit free radical flame reactions).
- Phosphorous compounds (toxicity problems).
- Melamine derivatives.

Major disadvantages of inorganic FRs are their relatively low decomposition temperature and requirement of a large fraction of material to give sufficient flame retardant performance in most polymers.

Flame retardants cause the plastic material to be safer in use in its final form. They are only used when the marketplace requires it in a wide range of applications spanning the construction, moulded parts, sealants, coatings and textile industries. High levels (up to 25 %) of FR are required in polypropylene – one of the most difficult plastics to render flame retardant – to meet standards for certain applications, such as in the electronics industry. Between 150 and 200 FRs have been designed to cover most of the market requirements.

Foaming Agents

Generate or release gases into selected areas of the fluid polymers (used almost exclusively to control foaming processes).

See Blowing agents.

Free-radical Scavengers

Function by inhibiting degradation of a resin which has already formed free radicals reducing them to stable, unreactive products, and stopping any further decomposition. Hindered phenols are able to inhibit oxidation by scavenging most of the oxygen-centred free radicals, such as $^{\bullet}OH$, $RO^{\bullet}$, and $ROO^{\bullet}$ before they are able to attack the polymeric substrate to form an alkyl radical. Commercially, Hindered Amine Light Stabilisers (HALS) are the most important light stabilisers which act (partly) as free radical scavengers.

See Antioxidants, Light stabilisers.

Friction Agents

Lubricate the surface of a plastics component in contact with another (especially metal) part.

See Slip agents.

Hardeners

See Catalysts.

Heat Stabilisers/Processing Stabilisers

Prevent polymers from degrading thermally through radical formation and oxidation during processing (often up to $50-60\,^{\circ}C$ above the melting point of the polymer), but also in application. Heat stabilisers act by stopping oxidation, or by attacking the decomposition products of oxidation. Phosphite/phosphonites are most effective stabilisers during processing, protecting both polymer and primary antioxidants. Other melt processing stabilisers are hindered phenols, hydroxylamines and lactones.

Heat stabilisers for PVC act by HCl scavenging and include organotins, mixed metal salt blends, and lead compounds. The latter account for nearly 64 % of volume (in 1994), followed by barium/cadmium and organotin compounds. Cadmium-based heat stabilisers are rapidly being replaced due to environmental concerns. Barium/zinc and calcium/zinc compounds show a high growth rate. It is expected that methyltin stabilisers will soon dominate the growing PVC pipe market.

Homogenising Agents

Assist the widely dissimilar ingredients used in a rubber compound to coalesce and mix into a homogeneous uniform processable mass. Homogenisers are low-MW polymeric resin blends. The homogenising resin blend contains portions that are compatible with aliphatic, naphthenic and aromatic parts of the elastomers in a blend and higher-MW homologues of the plasticisers. They have a wetting effect. Fatty acid derivatives and phenolic resins are used.

See Dispersing agents.

Impact Modifiers

Improve both impact strength and rigidity of thermoplastics by using up the energy of crack propagation. Elastomers are prototypical toughening additives. Examples of high-polymeric impact modifier/thermoplastic matrix systems are EVA, CPE and MBS in PVC, EP(D)M and SBS in PA, and acrylic rubbers in polyesters.

Inhibitors

Retard or prevent chemical reactions (such as cure).

See Scorch retardants.

Initiators

Usually organic peroxides acting as initiators for radical polymerisation (acrylics, styrenics, PVC, LDPE).

Light Stabilisers

Reduce or eliminate reactions caused mainly by UV or visible light which would otherwise impair the stability of polymers in outdoor use. Stabilisers in general ensure safe processing and long-term application of polymers inherently sensitive to oxidation processes. UV radiation in the range of 280–400 nm causes most of the degradation in polymers due to breakdown of chemical bonds in a polymer in a process called photodegradation. This process ultimately causes cracking, chalking, colour changes and loss of physical properties, such as impact strength, tensile strength, elongation, etc.

A typical degradative sequence is:

$$R \longrightarrow R^* \longrightarrow R^{\bullet} \qquad (II.1)$$

where active stabilisers are UV screeners/absorbers for R, energy quenchers for the UV photoexcited state R^* and radical scavengers for the free radicals $R^{\bullet}$. Another degradative sequence is:

$$R^{\bullet} \xrightarrow{O_2} ROO^{\bullet} \xrightarrow{RH} ROOH + R^{\bullet} \qquad (II.2)$$

with radical scavengers for $R^{\bullet}$ and $ROO^{\bullet}$, and hydroperoxide decomposers for ROOH as active stabilisers. These various stabilisers prevent degradation according to different mechanisms.

Much effort has been devoted to optimise the practical utility of stabilisers and their physical properties to ensure better permanency in the host polymer and ecological acceptability in agreement with legislative requirements applied strictly in some specific areas of use (in contact with food or other sensitive organic materials). Compatibility with the resin is more important in light stabilisers than in antioxidants, since they are used at higher concentrations. Compatibility of the generally polar light stabilisers is more difficult to achieve with nonpolar resins such as polypropylene. Major current products include benzophenones, benzotriazoles, hindered amines, nickel-containing compounds and others. A trend influencing the market includes the introduction of polymer bound light stabilisers. The choice of stabilisers for the light stabilisation of fibres and multifilaments is usually restricted to high-MW HAS

Hindered amine light stabilisers (HALS, e.g. Tinuvin 770, Tinuvin 622, Chimassorb 944, etc.) function by way of their highly efficient ability to act as free radical scavengers and peroxide decomposers, hence their ability to remove rapidly the species formed on light exposure that lead to oxidative degradation. The active form of the tetramethylene piperidine moiety – the nitroxyl radical – reacts with all radicals intervening in the degradation of polyolefins. Polymers stabilised with hindered amines are resistant to photodegradation even after the starting hindered amines have been

completely consumed. This is due to the fact that hindered amine oxidation products such as hydroxylamines and hydroxylamine ethers are also inhibitors of photo-oxidation. Both are capable of trapping peroxy radicals.

More than 200 users annually consume over US$ 55 million of light stabilisers. IR absorbers (e.g. amine, antimony salt mixture) protect against IR radiation; applicable in PC, PMMA, PVC, etc.

See Free-radical scavengers, Peroxide decomposers, Quenchers, Ultraviolet absorbers, Ultraviolet screening agents.

Lubricants

Improve flow characteristics of polymers, facilitate processing of plastic resins and act as functional surfactants at polymer/air or polymer/metal interfaces. Lubricants may be internal (incorporated into the host resin during production or compounding) or external (placed between the sliding surfaces before or during operation). Internal lubricants reduce melt viscosity and friction or increase slipperiness between polymer particles before melting. External lubricants are coated on the resin pellets or sprayed on the moulds or dies. Act at the external interface of the polymer and reduce friction between polymer melt and metal surfaces of the process equipment preventing a polymer from sticking to the mould or the machinery by migration to the surface during processing or use, with formation of a separating layer.

Saturated hydrocarbons (waxes), fatty acids, metal soaps, fatty acid amides and esters (primarily C_{16}–C_{18}) act as internal lubricants, fluoro elastomers as external lubricants. Many other polymer additives, e.g. antistatic agents, antifogs, antioxidants, UV stabilisers, etc., act as lubricants in the barrel of the extruder once they are in the liquid form.

See Antiblocking agents, Release agents, Slip agents.

Mastication Agents

Lower the viscosity of natural rubber as a result of breaking of the molecular chains to enable problem-free compounding with saving of cost and time. Without the presence of oxygen, mastication would not be possible. The consequence of mastication is a reduction of average molecular weight.

See Peptisers, Processing aids.

Metal Deactivators

Retard efficiently oxidation of polymers catalysed by metal impurities. Function by chelation. Effective metal deactivators are complexing agents which have the ability to co-ordinate the vacant orbitals of transition metal ions to their maximum co-ordination number and thus inhibit co-ordination of hydroperoxides to metal ions. Main use of stabilisation against metal-catalysed oxidation is in wire and cable applications where hydrocarbon materials are in contact with metallic compounds, e.g. copper.

Nucleating Agents

Provide nuclei for crystal growth and control spherulite formation in crystallisable polymers. Addition of nucleating agents produces more uniform spherulites and a structure with higher degree of crystallinity, can improve transparency, speeds up moulding cycle time. Nucleating agents are either finely divided inorganic materials (CB, silica, kaolin, talc, molybdenum disulfide) or (much preferred) organic materials (alkali metal salts of aromatic carboxylic acids, sorbitol derivatives). Need to have melting temperatures chosen with an appreciation of the processing conditions in which they will be used, and are accordingly divided into melt-sensitive and melt-insensitive nucleating agents.

See Clarifying agents.

Odour Modifiers

Mask an undesirable odour or add a desirable one. Their value depends on the persistence of their action.

Optical Brighteners

Improve the initial colour of plastics (correct discolorations), enhance whiteness (produce brilliant white end-use articles) and increase the brilliancy of coloured articles. Optical brighteners are special fluorescent organic substances which absorb UV radiation (360 to 380 nm) re-emitting it as visible blue light and compensating the unwanted yellow cast of polymer substrates. Requirements are lightfastness, thermal and migration stability (compatibility), low volatility, homogeneous distribution in the finished article. Chemical classes: benzoxazoles and coumarin derivatives (100 to 500 ppm). Usually, fluorescent whiteners are introduced by dry-blending with the plastics material (powder or pellets).

Parting Agents

See Release agents.

Peptisers

Generally radical acceptors or oxidation catalysts, which effectively remove free radicals formed during milling and mixing procedures. Inter-macromolecular action leads to reduction of the entanglements between polymer molecules. Chemically: activated zinc soaps.

See Mastication agents.

Peroxide Decomposers

Act to decompose hydroperoxides into stable molecules such as alcohols and ethers, before they can react with light to form free radicals. Main chemical classes are trivalent phosphorous compounds and thio-synergists (esters of thiodipropionic acid). Sulfur-based organic antioxidants decompose hydroperoxides by non-radical reactions. Typical peroxide decomposers are Irgafos 168, Ultranox 626, Irganox PS 800 and others.

See Light stabilisers, (Secondary) antioxidants.

Photosensitisers

Increase the photodegradation rate by absorbing light and initiating chemical changes.

Pigments

Give self-colouring, opaque and transparent optical effects. Pigments are generally insoluble in the plastic in which crystal or particulate structure is retained to some degree; colour results from the dispersion of fine particles ($\sim 0.01-1\,\mu$m) throughout the resin. Some pigments have the additional value of conferring magnetic or corrosion protection properties. Leaf-like effect pigments (metallic or pearlescent) act through a combination of reflection and scattering.

See Colorants.

Plasticisers

Soften and render products more flexible by reduction of the brittleness of the end-product. Also used to modify viscosity and improve the flow characteristics and processability of a polymer. Are designed to space out the polymer molecules, facilitating their movements and leading to enhanced flexibility (lower modulus) and ductility. Plasticisers may play a dual role as stabilisers or cross-linkers. Performance criteria are compatibility, plasticising efficiency, processability and permanence.

Plasticisers are classified according to their solvating power and migration characteristics. Selection of plasticisers requirements are: (i) 'solvation' of the polymer at an economic rate at the compounding stage; and (ii) no exudation of the system through volatilisation (at high temperatures) and blooming or bleeding (at lower temperatures) during the service life of the plasticised polymer. The migration rate of the plasticiser depends to a large extent on the relative differences in solubility parameters between polymer, plasticiser and environment. Most plasticisers are monomeric non-volatile organic liquids or low-melting-point solids, such as dioctyl phthalate or stearic acid; polymeric plasticisers (e.g. phthalic and adipic polyester) are usually very viscous liquids. Mostly used in PVC (80 % market share), mainly as phthalate esters of C_8, C_9 and C_{10} alcohols.

Pot-life Extenders

Promote pot-life, e.g. for PVC plastisols and isocyanate systems or catalysed UF and MUF core resins, avoiding loss of the resin and creation of environmental and disposal problems by extended interruptions in production.

Preservatives

Protect against mould, mildew, fungi, and bacterial growth. Should be migratory to replace consumed material. Biocides are most commonly used in PVC due to its plasticiser content and use in outside applications.
See Antimicrobials, Biocides.

Processing Aids

Function by coating the metal surface of a die, effectively changing the interfacial properties between metal and polymer melt. Counter processing problems by lowering the melt viscosity or improving mixing and melt homogeneity. Reduce gel formation during extrusion. Processing aids are liquids during the moulding process, become a part of the material and reduce moulding cycle time. Polymer processing additives (PPAs) are classified according to their use into antistatics, antiblocking and assembling agents, dispersants, homogenisers, lubricants, mastication aids (or peptisers), reinforcing agents and tackifiers. The major chemical classes are dominated by oleochemical derivatives and hydrocarbons (oils, waxes and resins). Fluoropolymer-based additives may act as PPA to improve the processing of polymers such as LDPE, LLDPE, HDPE, PP and others. High-polymeric processing aids (acrylates) or flow improvers are used in PVC.
See Lubricants.

Process Oils

Facilitate pre-vulcanisation processing, increase softness, extensibility and flexibility of the vulcanised end-product. The rubber processing industry consumes large quantities of materials which have a plasticising function; complex mixtures (paraffinic, naphthenic, aromatic) of mineral hydrocarbon additives, used with the large tonnage natural and synthetic hydrocarbon rubbers, are termed process oils. Because of the complexity of these products, precise chemical definition is usually not attempted. If the inclusion of an oil results in cost reduction it is functioning as an extender. The term plasticiser is commonly reserved for synthetic liquids used with the polar synthetic rubber.

Promoters

See Activators.

Pro-oxidants

Increase the oxidation rate of polymers, e.g. metal ions which increase the hydroperoxide decomposition rate. Photodegradation and thermal degradation are enhanced by transition metal ion containing pro-oxidants, such as iron dithiocarbamate (as opposed to nickel dithiocarbamate, which acts as a photo-antioxidant).

Quenchers

Act physically by stabilising polymers degrading by photolysis. Photons absorbed by chromophores (Ch) present in the plastics material raise the energy level of the chromophore. In the excited state the chromophore must lose excess energy to stabilise again. Excited-state quenchers (Q) accept energy from such an excited-state chromophore which then reverts to the stable ground state. The quencher dissipates the excess energy as heat, fluorescence, or phosphorescence, methods that do not result in resin degradation. Quenchers intercept the energy before any molecular bonds of the polymer chain are broken.

For energy transfer to occur from an excited chromophore (donor) to the quencher (acceptor), the latter has to have lower lying energy states than the donor. The transfer can be either a long-range energy transfer in which the distance between chromophore and quencher (d_{Q-Ch}) may be up to 10 nm, collisional, or an exchange energy transfer. Nickel quenchers are used in agricultural films.

See Light stabilisers.

Reinforcements

Intended to improve physical properties, reinforcements enhance the dimensional stability of materials, increase impact resistance, and improve tensile strength. The distinction between fillers and reinforcements is sometimes vague. Classification according to use.

See Fillers.

Reinforcing Fibres

See Fibres.

Release Agents

De-moulding of a plastics object may be facilitated by means of an internal additive, coating or sheet. Mould release agents can be either internal (mixed into the compound and migrating to the surface) or external (applied to the mould surface). Internal release agents are used in moulding compounds, including injection moulded plastics. Some internal release agents also function as lubricants and antistatic/antiblocking agents. External mould release agents (e.g. Carnauba wax) are for use with thermosetting resins, mainly in open moulds and low throughput operations. Also known as parting agents.

See Lubricants.

Scorch Inhibitors/Retardants

The standard approach to reducing scorch (e.g. discoloration resulting from oxidation) involves addition of one or more antioxidants. Scorch retardants prevent premature decomposition of peroxides and cross-linking of polymers (pre-vulcanisation).

See Vulcanisation retardants.

Shrinkage Modifiers

Control shrinkage after moulding. Any filler will decrease shrinkage; most commonly used are silica, clay, calcium carbonate, alumina, talc, powdered metals and lithium aluminium silicate.

Slip Agents

Slip for plastic films is defined as low surface friction which permits a layer of film to slide easily over another layer of film or over machine surfaces in film-fabricating and packaging equipment. Slip agents provide surface lubrication to films and sheets, reduce tack immediately after processing and facilitate processing on high-speed processing equipment. They act as a functional surfactant at the polymer/air interface, have limited compatibility with the polymer and exude to the surface, providing a coating that reduces the coefficient of friction. The amount on the surface is related to the ease of slip. They can perform secondary functions, e.g. act as antistatic agents. They are available in liquid or solid form. The liquids generally tend to be silicone-based (e.g. PDMS) which provides excellent lubricity to the surface. Typical solid slip additives are waxes (e.g. Carnauba wax, PE waxes) or PTFE (Teflon). Slip additives for film extrusion are primary fatty amides with variable total carbon chain length, such as oleamide ($C_{18}^=$, from animal sources), erucamide ($C_{22}^=$, from rape seed-oil), stearamide (C_{18}) and behenamide (C_{22}). All of these amides are readily available for commercial purposes but none are commercially available in pure form, i.e. with only one fatty component. The amides are originally uniformly dispersed in the resin but after extrusion they 'bloom' to the surface at varying rates. Oleamide provides faster response than erucamide probably because it is a smaller molecule and thus can migrate to the surface faster to provide the slip. Slip agents in films are always used in combination with antiblocking agents. Stearamide and behenamide show lower effectiveness as slip agents than oleamide and erucamide because of their saturation; their antiblocking effectiveness is much better.

Smoke Suppressants

Change the nature of the polymer combustion process to reduce smoke formation.

See Flame retardants.

Stabilisers

Prevent degradation of polymers by chemical reactions. Stabilisers are mono-, bi- or polyfunctional. Polyolefin stabilisers represent a major proportion of the stabiliser market.

See Antioxidants, Light stabilisers.

Surfactants

'Surface active agents' (surfactants) are active (adsorb) at surfaces and reduce surface tensions. Surfactants work because they are amphiphilic: they have opposing solubility tendencies in one molecule, such as a hydrocarbon chain and a polar end. Because of this disparity in solubility, they tend to form concentration gradients at dissimilar phase interfaces. Surfactant additives are classified according to the interface at which they are active.

Surfactants used as lubricants are added to polymer resins to improve the flow characteristics of the plastic during processing; they also stabilise the cells of polyurethane foams during the foaming process. Surfactants are either nonionic (e.g. fatty amides and alcohols), cationic, anionic (dominating class; e.g. alkylbenzene sulfonates), zwitterionic, hetero-element or polymeric (e.g. EO-PO block copolymers). Fluorinated anionic surfactants or 'super surfactants' enable a variety of surfaces normally regarded as difficult to wet. These include PE and PP: any product required to wet the surface of these polymers will benefit from inclusion of fluorosurfactants. Surfactants are frequently multicomponent formulations, based on petro- or oleochemicals.

See Antistatic agents, Foaming agents, Lubricants, Slip agents, Wetting agents.

Tackifiers

In the rubber industry tackiness means the ability of uncured rubber compounds to stick together under moderate pressure. Tackifiers improve adhesion and joint strength at contact areas, should not affect processing characteristics of the unvulcanised compound and not provide the vulcanisates with surface tack.

See Processing aids.

Thickening Agents

Increase viscosity, e.g. magnesium oxide in unsaturated polyester resins increases the viscosity so much that a liquid resin becomes a tackfree solid, suitable for making sheet moulding compound (SMC). A relatively high viscosity is also desirable for latex compounds used to coat textiles.

Thixotropic Agents

Modify the dependence of viscosity on shear rate, producing low viscosity at high rates, and vice versa.

Ultraviolet Absorbers

Are used especially in outdoor applications to preserve polymers against harmful radiation which would otherwise cause degradation (e.g. colour change in opaque resins). Like UV screeners, UV absorbers are designed to interrupt the first step of the degradation process by absorption of light. However, unlike pigments used as screeners, UV absorbers are (nearly) colourless additives, which absorb light very intensely in the UV-A and UV-B regions of the spectrum (280 to 380 nm). The (physical) protection mechanism of UV absorbers is based essentially on absorption of the harmful radiation and its dissipation in a manner that does not lead to photosensitisation, i.e. dissipation of heat. These compounds must be very photostable and should not absorb in the visible region of the spectrum to avoid discoloration of the polymeric material or coating. For long-term protection, a UV absorber must be resistant to oxidative processes that could degrade it to nonabsorbing species. Low-MW hydroxybenzophenones (e.g. Chimassorb 81; UV absorption in the 260–350 nm range, stable to high heat temperatures, excellent compatibility with many polymers) and hydroxyphenylbenzotriazoles (e.g. Tinuvin 327; main absorption capacity between 290 nm and 400 nm) are the most extensively used UV absorbers. The stabilising effectiveness of some UV absorbers goes beyond the pure light filter effect. There is evidence that 2-hydroxyphenylbenzotriazoles and 2-hydroxybenzophenones are multifunctional additives capable of absorbing light and of quenching photo-excited chromophores in a variety of polymers.

A fundamental disadvantage of UVAs is the fact that they need a certain absorption depth (sample thickness) for good protection of a plastic. Therefore, no protection of surfaces (crazing) and only limited protection are observed in thin samples, e.g. fibres or films. UV absorbers are very effective in UV absorbing substrates, such as polystyrenes, polyesters and polyolefins. There is a demand for more effective UV stabilisers than the currently available products with the following inherent problems: limitation of load and loss by bleed-out due to low resin compatibilities, loss by sublimation during processing or extraction, migration over time in end-use products. The greatest need includes thin film, fibres and coatings applications where there exists a high surface area to volume ratio. Since phenolic and HALS AOs, unlike UVAs, do not depend on the thickness of the sample for effectiveness, they are especially useful for such substrates. Polymer-bound UV stabilisers, such as NORBLOC 7966, provide permanent placement of the benzotriazole/benzophenone in the polymer. Ideal organic UV absorbers in sunscreens should strongly absorb UV-A (315–380 nm) and UV-B (280–315 nm) light, be nontoxic, nonstaining, odourless, waterproof, sweatproof and cheap. Commercial products in UV absorbing lotions are *p*-aminobenzoic acid (PABA), PABA ester, oxybenzone, octocrylene, octyl salicylate, octyl methoxy cinnamate.

See Light stabilisers, Ultraviolet screening agents.

Ultraviolet Screening Agents

Act by screening from UV light thereby effectively inhibiting the photo-excitation of light absorbing species present in the polymer; perhaps the oldest form of UV protection for polymers and coatings. UV screening can, of course, be effected by some pigments added to a formulation to provide opacity or translucence. The most effective is carbon-black, but its application is obviously limited. Titanium dioxide and zinc oxide also act as UV screeners by absorbing or reflecting UV light. However, use of opaque pigments alone has a number of significant drawbacks, such as unwanted colour (not desirable for end use), loss of transparency (because of the particulate nature of pigments), poor surface protection and photodegradative effects (some screeners that absorb UV light become sites for degradation).

See Light stabilisers, Ultraviolet absorbers.

Viscosity Modifiers

Improve thermal and mechanical shock, increase elongation, and obtain higher impact strength and flexibility in a material.

Vulcanisation Accelerators/Retardants

Increase the rate of the cross-linking action with sulfur considerably; allow for lower sulfur content to achieve optimum vulcanisate properties. Organic accelerators (e.g. thiuram, dithiocarbamate, etc.) are of major importance. In some cases it is necessary to retard the onset of vulcanisation to assure sufficient processing safety. The antioxidant 2-mercaptobenzimidazole (MBI) acts as a retarder for most accelerators.

See Cross-linking agents, Scorch inhibitors.

Wear Additives

Improve the ability of a base polymer to resist wear in dynamic applications. Common wear additives are PTFE and silicone oil.

Wetting Agents

Wet out solid substrates. Act as a functional additive at the polymer/air interface, e.g. filler particle surfaces, and help their uniform dispersion in a polymer matrix without agglomeration.

See Surfactants.

Table II.I gives references for a variety of plastic, rubber and coatings additives and polymerisation aids.

REFERENCES

1. N. Grassie and G. Scott, *Polymer Degradation and Stabilisation*, Cambridge University Press, Cambridge (1988).
2. M. Ash and I. Ash, *Handbook of Plastic and Rubber Additives*, Gower Publishing Ltd, Aldershot (1995).
3. W. Hofmann, *Rubber Technology Handbook*, Hanser Publishers, Munich (1996).
4. J. Murphy, *The Additives for Plastics Handbook*, Elsevier Science Ltd, Oxford (1996, 2001)
5. M. Domininghaus, *Die Kunststoffe und ihre Eigenschaften*, VDI Verlag, Düsseldorf (1992).
6. T.J. Henman (ed.), *World Index of Polyolefine Stabilizers*, Kogan Page, London (1982).

Table II.1 Classes of commercial plastic, rubber and coatings additives and polymerisation aids (including references to databases by trade name, chemical, function, application and manufacturer)

Additive class	References
Abrasives	[2]
Accelerator activators	[3]
Accelerators	[2,4–8]
Acid scavengers – Acid neutralisers – Antiacids	[2,9–11]
Activators	[2,12]
Adhesion promoters – Bonding agents	[2,5,13–17]
Antiblocking agents	[2,4,5,9–11,13,18,19]
Anticoagulants	[2]
Anticorrosives	[15,19]
Antidegradants	[2,6,12,20]
Antiflexcracking agents	[2,12]
Antifoaming agents	[2,12,19]
Antifogging agents	[2,4,9,11,17,21]
Antifouling agents – Antibacterials – Antiallergics – Antimicrobials – Biostabilisers – Biocides – Fungistats – Germicides	[2,3,5,9,11,13,14,19,22–29]
Antioxidants	[2,4,5–7,9–11,13,18,20,22,23,30–39]
Antiozonants	[2,3,6,7,12,30]
Antiscorch agents – Scorch retardants	[2,9]
Antisettling agents	[8,15]
Antiskinning agents	[8,40]
Antistatics	[2,4,5,9,10,11,13,18,19,22,23,41,42]
Binders – Bonding agents	[8]
Biodegradation promoters	[9]
Blocking agents	[2,4,11]
Blowing agents – Blow promoters – Foaming agents	[2–5,9–14,18,22,23,41,43–46]
Carbon-blacks	[2,3,9,12,47]
Catalysts	[2,4]
Chain extenders	[2]
Chelators	[2,13]
Clarifiers	[2,4]
Coagulants	[2,12]
Colorants – Dyes – Pigments	[1–5,7–11,13,14,22,23,37,41,48–56]
Compatibilisers	[2,4,9,23,57]
Conductive fillers	[4,9,58]
Corrosion inhibitors	[8,15]
Coupling agents	[2,4,5,11,14,18,23,37,41,59]
Creaming agents	[2]
Cross-linking agents	[2–5,7,11,12,22,60]
Curing agents	[2,4,8,9,18,23,60]
Deactivators – Metal deactivators	[2,6,9,10,13,22,31,32,61]
Defoamers	[2,8,15]
Deodorants – Odorants – Fragrances	[2,3,9,14]
Detackifiers – Antitack agents	[2]
Diluents – Viscosity depressants	[2,9,62]
Dispersants	[2,4,8,9,12,17,19,63,64]
Driers – Dehydrating agents – Desiccants – Siccatives	[2,8,40]
Emulsifiers – Co-emulsifiers	[2,5,12,65]
Fatigue agents	[3]
Fibres	[2,4,9,13,19,37,66,67]
Fillers – Extenders	[2–4,8–14,22,37,41,62,67–75]
Flame retardants – Fire retardants – Smoulder retardants – Smoke-suppressants	[2,4,5,7,9–11,13,14,18,22,23,37,41,48,72,76–84a]
Flexibilisers	[2]
Flow improvers	[85]

(continued)

Friction agents	[4]
Gellants	[2]
Hardeners	[2,3,5,8]
Heat sensitisers	[2]
Homogenisers	[2,12]
Impact modifiers	[2,4,5,9,13,18,22,23,37,48,72,86]
Inhibitors – Scorch inhibitors	[2,5,6,9]
Initiators	[2,5,17,60]
Intermediates	[2]
Light stabilisers – IR absorbers – UV absorbers	[2,4,5–8,10,11,13,18,20,23,32,41,87–90]
Lubricants	[2,4,5,9–13,18–20,22,23,37,41,91–93]
Mastication aids	[3,12]
Modifiers	[2,4,5,9,13,23,48,55,84,94]
Nucleating agents	[2,4,5,9–11,18,22,41,55]
Optical brighteners – Whiteners	[2,9,11,22,37]
Peptisers	[2,3,12]
Plasticisers – Softeners	[2–5,7,9,12–14,18–20,22,23,37,41,48,84,92,95–97a]
Pot-life extenders	[98]
Preservatives	[2,4,12,18,41,64,99]
Processing aids	[2–4,9,12–14,22,23,32,34,37,48]
Process oils – Extender oils – Mineral oils	[3,12,14]
Reclaiming agents	[2]
Reducing agents	[2]
Reinforcements	[2,4,9–11,13,22,37]
Release agents – Mould release agents	[2–5,9,17,19]
Retarders	[2]
Shortstops – Terminators	[2]
Shrinkage modifiers	[4,9]
Slip agents	[2,4,5,10,11,15,19,100]
Softeners	[3]
Solubilisers	[2,63]
Solvents	[2,8,101]
Stabilisers	[2,4–6,8–12,18,20,22,23,32,37,41,72,84,88,92,95,102–104]
Surfactants	[2,4,9,14,105–118]
Suspending agents	[2]
Synergists	[2]
Tackifiers	[2,12,14]
Thermal stabilisers	[34]
Thickeners – Thixotropes – Viscosity modifiers	[4,8,9,12,13,15,17,95,119]
Toughening agents	[9,120]
Vulcanising accelerators/retarders	[2,3,12]
Waxes	[2,17,121]
Wear additives	[41]
Wetting agents	[2,8,9,12]

7 T.R. Crompton, *Practical Polymer Analysis*, Plenum Press, New York, NY (1993).

8 W. Freitag and D. Stoye (eds), *Paints, Coatings and Solvents*, Wiley-VCH, Weinheim (1998).

9 G. Pritchard, *Plastics Additives. An A–Z Reference*, Chapman & Hall, London (1998).

10 C. Maier and T. Calafut, *Polypropylene. The Definitive User's Guide and Databook*, Plastics Design Library, Norwich, NY, 1998.

11 H. Zweifel (ed.), *Plastics Additives Handbook*, Hanser Publishers, Munich (2000).

12 H.-W. Engels, H.-J. Weidenhaupt, M. Abele, M. Pieroth and W. Hofmann, in *Ulmann's Encyclopedia of Industrial Chemistry*, Weinheim (1993), Vol. A23, pp. 365–420.

13 L. Mascia, *The Role of Additives in Plastics*, Edward Arnold Publishers, London (1974).

14 C.H. Hepburn, *Rubber Compounding Ingredients – Need, Theory and Innovation, Part II. Processing, Bonding, Fire Retardants, Rapra Review Report Nr. 97*, Rapra Technology Ltd, Shawbury (1997).

15 L.J. Calbo, *Handbook of Coatings Additives*, M. Dekker, New York, NY (1987).

16 I. Skeist, *Handbook of Adhesives*, Van Nostrand-Reinhold, Princeton, NJ (1990).

17 E.W. Flick, *Plastics Additives*, Noyes Publications, Park Ridge, NJ (2001), 3 vols.

18 Phillip Townsend Associates Inc., *Chemical Additives for Plastics*, Mount Olive, NJ (1997).

19. A.L. McKenna, *Fatty Amides*, Witco Chemical Corporation, Memphis, TN (1982).

20. U. Schönhausen, *Positive List I. Additives for Plastics, Elastomers and Synthetic Fibres for Food Contact Applications*, Ciba-Geigy Ltd, Basel (1993).

21. ICI Surfactants – Polymer Additives Group, *Antifog Agents for Agricultural and Food Packaging Films. Brochure 90-5E* (1998).

22. R. Gächter and H. Müller (eds), *Plastics Additives Handbook*, Hanser Publishers, Munich (1993).

23. P. Dufton, *Functional Additives for the Plastics Industries*, Rapra Technology Ltd, Shawbury (1998).

24. A.L. Eilender and R.A. Oppermann, in *Handbook of Coatings Additives* (L.J. Calbo, ed.), M. Dekker, New York, NY (1987), pp. 177–224.

25. D. Nichols, *Plast. Addit. Compd.* **4** (12), 14–7 (2002).

26. J. Simmons, *Plast. Addit. Compd.* **2** (10), 20–3 (2000).

27. J. Markarian, *Plast. Addit. Compd.* **4** (12), 18–21 (2002).

28. M. Ash and I. Ash, *The Index of Antimicrobials*, Gower Publishing Ltd, Aldershot (1996).

29. E. Lück and M. Jager, *Antimicrobial Food Additives*, Springer-Verlag, Berlin (1997).

30. M. Ash and I. Ash, *The Index of Antioxidants and Antiozonants*, Gower Publishing Ltd, Aldershot (1997).

31. S. Al-Malaika, in *Handbook of Engineering Polymeric Materials* (N.P. Cheremisinoff, ed.), M. Dekker, New York, NY (1997), pp. 105–8.

32. P.P. Klemchuk, in *Oxidation Inhibition in Organic Materials* (J. Pospíšil and P.P. Klemchuk, eds), CRC Press, Boca Raton, FL (1989), Vol. I, pp. 11–32.

33. J. Pospíšil, in *Oxidation Inhibition in Organic Materials* (J. Pospíšil and P.P. Klemchuk, eds), CRC Press, Boca Raton, FL (1989), Vol. I, pp. 33–59.

34. F. Gugumus, in *Oxidation Inhibition in Organic Materials* (J. Pospíšil and P.P. Klemchuk, eds), CRC Press, Boca Raton, FL (1989), Vol. I, pp. 61–172.

35. C. Neri, *Chim. Ind. (Milan)* **79** (9), 1223–32 (1997).

36. R.A. Larson, *Naturally Occurring Antioxidants*, Lewis Publishers, Boca Raton, FL (1997).

37. J.T. Lutz (ed.), *Thermoplastic Polymer Additives: Theory and Practice*, M. Dekker, New York, NY (1989).

38. K. Schwarzenbach, B. Gilg, D. Müller, G. Knobloch, J.-R. Pauquet, P. Rota-Graziosi, A. Schmitter, J. Zingg and E. Kramer, in *Plastics Additives Handbook* (H. Zweifel, ed.), Hanser Publishers, Munich (2000), pp. 1–139.

39. G. Scott, *Antioxidants in Science, Technology, Medicine and Nutrition*, Albion Publishing, Chichester (1997).

40. R. Hurley, in *Handbook of Coatings Additives* (L.J. Calbo, ed.), M. Dekker, New York, NY (1987), pp. 485–509.

41. B. Stocker, *World Plastics Rubber Technol.* 33–40 (1998).

42. H. Sumi, *Purasuchikkusu* **49** (10), 34–40 (1998).

43. J.M. Methven, *Foams and Blowing Agents, Rapra Review Report No. 25*, Rapra Technology Ltd, Shawbury (1990).

44. F.A. Shutov, in *Handbook of Polymeric Foams and Foam Technology* (D. Klempner and K.C. Frisch, eds), Hanser Publishers, Munich (1991), pp. 375–408.

45. *Proceedings Blowing Agent Systems: Formulations and Processing*, Rapra Technology Ltd, Shawbury (1998).

46. F.A. Shutov, *Progr. Rubber Plast. Technol.* **5** (4), 289 (1989).

47. J.-B. Donnet, R.C. Bansal and, M.-J. Wang, *Carbon Black*, M. Dekker, New York, NY (1993).

48. J. Edenbaum (ed.), *Plastics Additives and Modifiers Handbook*, Chapman & Hall, London (1996).

49. G. Buxbaum (ed.), *Industrial Inorganic Pigments*, Wiley-VCH, Weinheim (2001).

50. K. Hunger (ed.), *Industrial Dyes*, Wiley-VCH, Weinheim (2002).

51. H.M. Smith (ed.), *High Performance Pigments*, Wiley-VCH, Weinheim (2002).

52. W. Herbst and K. Hunger, *Industrial Organic Pigments*, VCH, Weinheim (1997).

53. I.R. Wheeler, *Metallic Pigments in Polymers*, Rapra Technology Ltd, Shawbury (1999).

54. *Pigment Blacks for Plastics, Degussa Technical Bulletin No. 40*, Frankfurt am Main (1998).

55. Associazione Italiana di Scienza e Tecnologia delle Macromolecole (AIM), *Additivi per Materiali Polimerici (Atti del XXIV Convegno Scuola Mario Farina, Gargnano)*, Pacini Editore, Pisa (2002).

56. R.A. Charvat (ed.), *Coloring of Plastics*, SPE, Brookfield, CT, (2003).

57. S. Datta and D.J. Lohse, *Polymeric Compatibilizers: Uses and Benefits in Polymer Blends*, C. Hanser Verlag, Munich (1996).

58. B. Davis, in *Industrial Applications of Surfactants* (D.R. Karsa, ed.), The Royal Society of Chemistry, London (1987), pp. 307–17.

59. R.N. Rothon, *The Coupling Agent Index*, Intertech, Portland, MN (2000).

60. AKZO-NOBEL, *Organic Peroxides Symposium*, Apeldoorn (1998).

61. M.G. Chan, in *Oxidation Inhibition in Organic Materials* (J. Pospíšil and P.P. Klemchuk, eds), CRC Press, Boca Raton, FL (1989), Vol. I, pp. 225–46.

62. M. Ash and I. Ash, *Handbook of Fillers, Extenders and Diluents*, Ashgate Publishing, Endicott, NY (1998).

63. M. Ash and I. Ash, *Dispersants, Solvents and Solubilizers*, Edward Arnold Publishers, London (1988).

64. R.J. Kane (ed.), *Paper Coating Additives*, TAPPI Press, Atlanta, GA (1995).

65. A.D. James, P.H. Ogden and J.M. Wates, in *Industrial Applications of Surfactants* (D.R. Karsa, ed.), The Royal Society of Chemistry, London (1987), pp. 250–68.

66. A.R. Bunsell (ed.), *Fiber Reinforcement for Composite Materials*, Elsevier Science, Amsterdam (1988).

67. B.N. Rothon, *Adv. Polym. Sci.* **139**, 67–107 (1999).

68. H.S. Katz and J.V. Milewski (eds), *Handbook of Fillers for Plastics*, Chapman & Hall, London (1987).

69. J.S. Falcone Jr, in *Kirk-Othmer Encyclopedia of Chemical Science and Technology* (M. Howe-Grant, ed.), John Wiley & Sons, Inc., New York, NY (1992).

70. J. Jancar (ed.), *Mineral Fillers in Thermoplastics. I. Raw Materials and Processing*, Springer-Verlag, Berlin (1999).

71. G. Wypych (ed.), *Handbook of Fillers*, ChemTec, Toronto (1999).

72. S. Al-Malaika, A. Golovoy and C.A. Wilkie (eds), *Chemistry and Technology of Polymer Additives*, Blackwell Science, Oxford (1999).

73. T.J. Pinnavaia and G.W. Beall, *Polymer–Clay Nanocomposites*, John Wiley & Sons, Inc., New York, NY (2001).

74. R.N. Rothon, *Particulate Fillers for Polymers, Rapra Review Report No. 141*, Rapra Technology Ltd, Shawbury (2002).

75. G. Pritchard, *Novel and Traditional Fillers for Plastics: Technology and Market Developments, Rapra Technology Industry Analysis Report*, Rapra Technology Ltd, Shawbury (1999).

76. M. Ash and I. Ash, *The Index of Flame Retardants*, Gower Publishing Ltd, Aldershot (1997).

77. C.J. Hilado, *Flammability Handbook for Plastics*, Technomic Publishing Company, Lancaster, PA (1998).

78. J.H. Troitzsch, *Flame Retardants*, C. Hanser Verlag, Munich (1996).

79. J. Green, *Flammability and Flame Retardants in Plastics, Rapra Review Report, Vol. 4 (6)*, Rapra Technology Ltd, Shawbury (1991).

80. P.W. Dufton, *Fire-additives and Materials*, Rapra Technology Ltd, Shawbury (1995).

81. J.H. Troitzsch, *Plastics Flammability Handbook*, C. Hanser Publishers, Munich (2004).

82. A.F. Grand and C.A. Wilkie (eds), *Fire Retardancy of Polymeric Materials*, M. Dekker, New York, NY (2000).

83. A.R. Horrocks and D. Price (eds), *Fire Retardant Materials*, Woodhead Publishers, Cambridge (2001).

84. J.R. LeBlanc, *Digest of Polymer Additives*, Specialized Technology Resources Inc., Enfield, CT (2000).

84a. M. Xanthos, *Functional Fillers for Plastics*, Wiley-VCH, Weinheim (2005).

85. K.D. Böhme, in *Plastics Additives* (R. Gächter and H. Müller, eds), C. Hanser Publishers, Munich (1993), pp. 481–524.

86. J.T. Lutz and D.L. Dunkelberger, *Impact Modifiers for PVC. The History and Practice*, John Wiley & Sons, Inc., New York, NY (1992).

87. J.S. Parmar and R.P. Singh, in *Handbook of Engineering Polymeric Materials* (N.P. Cheremisinoff, ed.), M. Dekker, New York, NY (1997), pp. 399–409.

88. P.J. Schirmann and M. Dexter, in *Handbook of Coatings Additives* (L.J. Calbo, ed.), M. Dekker, New York, NY (1987), pp. 225–69.

89. A. Valet, *Light Stabilizers for Paints*, C.R. Vincentz, Hannover (1996).

90. F. Gugumus, in *Plastics Additives Handbook* (H. Zweifel, ed.), Hanser Publishers, Munich (2000), pp. 141–425.

91. E. Richter, *Plast. Addit. Compd* **2** (11), 14–21 (2000).

92. U. Schönhausen, *Positive List II, Additives Used in the PVC Industry for Food Contact Applications*, Ciba-Geigy Ltd, Basel (1990).

93. I. Ash and M. Ash, *Handbook of Lubricants, Synapse Information Resources*, Endicott, NY (2001).

94. J.T. Lutz and R.F. Grossman (eds), *Polymer Modifiers and Additives*, M. Dekker, New York, NY (2001).

95. M. Ash and I. Ash, *Plasticizers, Stabilizers and Thickeners: What Every Chemical Technologist Wants to Know*, Edward Arnold Publishers, London (1989), Vol. 3.

96. A.S. Wilson, *Plasticisers – Selection. Applications and Implications, Rapra Review Report No. 88*, Rapra Technology Ltd, Shawbury (1996).

97. A.S. Wilson, *Plasticisers: Principles and Practice*, The Institute of Materials, London (1995).

97a. G. Wypych, *Handbook of Plasticizers*, ChemTech Publishers, Toronto (2004).

98. J. Ridnell, *Proceedings Asian Intl Laminates Synposium*, TAPPI Press, Atlanta, GA (1997), pp. 99–102.

99. D. Stoye and W. Freitag (eds), *Paints, Coatings and Solvents*, Wiley-VCH, Weinheim (1998).

100. A. Maltby and R.E. Marquis, *J. Plast. Film Sheeting* **14** (2), 111–20 (1998).

101. J. Brandrup and E.H. Immergut (eds), *Polymer Handbook*, John Wiley & Sons, Inc., New York, NY (1999).

102. J. Sedlář, in *Oxidation Inhibition in Organic Materials* (J. Pospíšil and P.P. Klemchuk, eds), CRC Press, Boca Raton, FL (1989), Vol. II, pp. 1–28.

103. F. Gugumus, in *Oxidation Inhibition in Organic Materials* (J. Pospíšil and P.P. Klemchuk, eds), CRC Press, Boca Raton, FL (1989), Vol. II, pp. 29–162.

104. C. Kröhnke and F. Werner, *Stabilizers for Polyolefins, Rapra Review Report No. 132*, Rapra Technology Ltd, Shawbury (2001).

105. M.R. Porter, *Handbook of Surfactants*, Chapman & Hall, London (1994).

106. M. Ash and I. Ash, *Handbook of Industrial Surfactants*, Gower Publishing Ltd, Aldershot (1997).

107. J. Cross (ed.), *Anionic Surfactants: Analytical Chemistry*, M. Dekker, New York, NY (1998).

108. E. Jungermann (ed.), *Cationic Surfactants*, M. Dekker, New York, NY (1970).

109. D.C. Cullum (ed.), *Introduction to Surfactant Analysis*, Blackie, Glasgow (1994).

110. T.M. Schmitt, *Analysis of Surfactants*, M. Dekker, New York, NY (1992).

111. J. Cross (ed.), *Nonionic Surfactants – Chemical Analysis*, M. Dekker, New York, NY (1987).

112. G. Hollis (ed.), *Surfactants Europe*, The Royal Society of Chemistry, Cambridge (1995).

113. H.W. Stache (ed.), *Anionic Surfactants – Organic Chemistry*, M. Dekker, New York, NY (1996).

114. B. Jönsson, B. Lindman, K. Holmberg and B. Kronberg, *Surfactants and Polymers in Aqueous Solution*, John Wiley & Sons, Ltd, Chichester (1998).

115. D.R. Karsa (ed.), *Industrial Applications of Surfactants*, The Royal Society of Chemistry, London (1987).

116. D.O. Hummel, A. Baum, H. Zimmermann and A. Solti, *Analysis of Surfactants*, Hanser Gardner Publishers, Cincinnati, OH (1996).

117. K.R. Lange (ed.), *Surfactants*, Hanser Gardner Publishers, Cincinnati, OH (1999).

118. D.O. Hummel, *Handbook of Surfactant Analysis*, John Wiley & Sons, Ltd, Chichester (1999).

119. G. Davison and D.R. Skuse (eds), *Advances in Additives for Water-based Coatings*, The Royal Society of Chemistry, Cambridge (1999).

120. A.A. Collyer, *Rubber Toughened Engineering Plastics*, Chapman & Hall, London (1994).

121. W.H. Pushaw III, in *Handbook of Coatings Additives* (L.J. Calbo, ed.), M. Dekker, New York, NY (1987), pp. 271–80.

Specimen Polymer Additives Product Sheets

Additives In Polymers: Industrial Analysis And Applications   J. C. J. Bart
© 2005 John Wiley & Sons, Ltd   ISBN: 0-470-85062-0

PAO-B01 (PO-AS code A004).

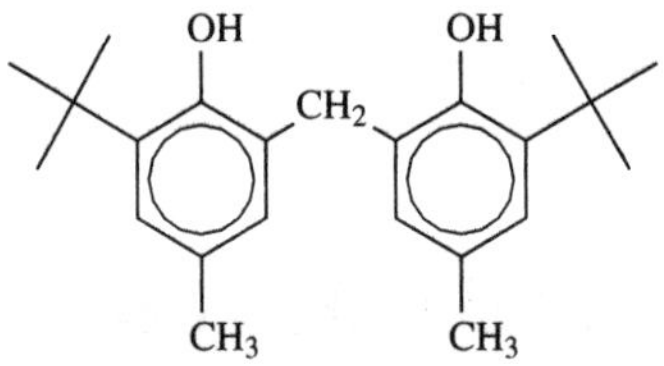

Chemical name(s):	2,2′-Methylenebis (6-*t*-butyl-4-methylphenol)
	2,2′-Bis (6-*t*-butyl-*p*-cresyl)methane
	2,2′-Methylenebis (6-*t*-butyl-*p*-cresol)
	Phenol, 2,2′-methylene-bis-[6-(1,1-dimethylethyl)-4-methyl-]
	2,2′-Methylenebis (4-methyl-6-*t*-butylphenol)

Molecular formula:	$C_{23}H_{32}O_2$
Molecular weight:	340
CAS:	119-47-1
EINECS:	204-327-1

Physical properties:	Melting point: 120–133 °C

Solubilities (20 °C, %/w): Methanol: 30, Ethanol: 45, Acetone: 45, Ethyl acetate: 50, Chloroform: 45, Heptane: 5. Insoluble in water

Trade name	Supplier	Remarks
AKROCHEM AO 235	AKROCHEM	
ANTIOXIDANT BKF	BAYER	
AO 2246	MARUBENI CORPORATION	
ASM ZKF	BAYER	
CAO-14	PMC SPECIALTIES	
CAO 5	PMC SPECIALTIES	
CYANOX 2246	CYTEC	FORMER AMERICAN CYANAMID
HOSTANOX 04	CLARIANT	FORMER HOECHST
IRGANOX 2246	CIBA SPECIALTY CHEMICALS	
KEMINOX 246	CHEMIPRO KASEI KAISHA	
LOWINOX 22M46	GREAT LAKES	FORMER LOWI
MBP 5	GREAT LAKES	FORMER SFOS
OXICHEK 114	FERRO CORPORATION	
PLASTANOX 2246	CYTEC	FORMER AMERICAN CYANAMID
SUMILIZER MDP-S	SUMITOMO CHEMICALS	
RALOX 2246	RASCHIG	
RALOX 46	RASCHIG	
SANTOWHITE PC	FLEXSYS	FORMER MONSANTO
SEENOX 224M	CHEMIPRO KASEI KAISHA	
ULTRANOX 246	GENERAL ELECTRIC	FORMER BORG WARNER
VANOX 2246	R.T. VANDERBILT	
VANOX MBPC	R.T. VANDERBILT	
VULKANOX BKF	UNIROYAL	
YOSHINOX 2246	YOSHITOMI	

General remarks:
LD50: 4880 mg/kg

Figure III.1 Product sheet of the phenolic antioxidant 2,2′-methylene-bis-(6-*tert*-butyl-4-methylphenol). After Gijsman [1]. Reproduced by permission of P. Gijsman

AA-08 (PO-AS code B008).

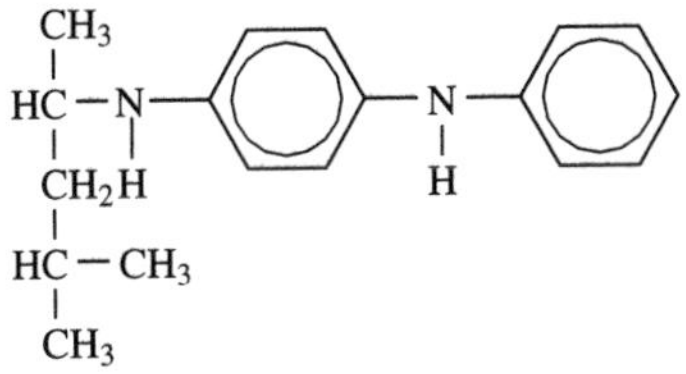

Chemical name(s):	N-1,3-dimethyl-butyl-N'-phenyl-paraphenylene-diamine
Molecular formula:	$C_{18}H_{24}N_2$
Molecular weight:	268
CAS:	793-24-8
EINECS:	
Physical properties:	Melting point: 46 °C

Solubilities: Soluble in methylene chloride, acetone, ethyl acetate, ethanol. Insoluble in water.

Trade name	Supplier	Remarks
AKROCHEM PD-2	AKROCHEM	
ANTOZITE 67P	R.T. VANDERBILT	
PERMANAX 6PPD	FLEXSYS	FORMER AKZO NOBEL
SANTOFLEX 13	FLEXSYS	FORMER MONSANTO
VALKANOX 4020	BAYER	

Figure III.2 Product sheet of the aromatic amine N-1,3-dimethylbutyl-N'-phenyl-paraphenylene-diamine. After Gijsman [1]. Reproduced by permission of P. Gijsman

PM01 (PO-AS code C001).

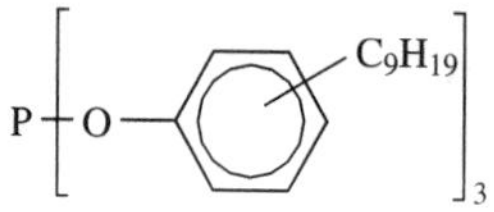

Chemical name(s):	Trisnonylphenyl phosphite

Molecular formula:	$C_{45}H_{69}O_3P$
Molecular weight:	688
CAS:	26523-78-4
EINECS:	247-759-6

Physical properties:	Liquid

Solubilities: Soluble in most organic solvents. Insoluble in water

Trade name	Supplier	Remarks
ALKANOX TNPP	GREAT LAKES	FORMER ENICHEM
ADK STAB 1178	ASAHI DENKA KOGYO	
ADK STAB 329	ASAHI DENKA KOGYO	
AREPHOS P39	ARECOM	
CHEL 39	REAGENS	
DOVERPHOS 4	DOVER	
DOVERPHOS HP-4HR	DOVER	CONTAINS 0.75% TIPA*
GARBEFIX OS101	GREAT LAKES	FORMER SFOS
IRGAFOS TNPP	CIBA SPECIALTY CHEMICALS	
IRGASTAB CH55	CIBA SPECIALTY CHEMICALS	
JP-351	JOHOKU CHEMICAL	
LANKROMARK LE109	AKCROS CHEMICALS	FORMER LANKRO CHEMICALS
LOWINOX TNPP	GREAT LAKES	FORMER LOWI
MARK 1178	ASAHI DENKA KOGYO	
MIANTO P-800	MIWON COMMERCIAL	
NAUGARD 512	UNIROYAL CHEMICAL	
NAUGARD P	UNIROYAL CHEMICAL	
NAUGARD PHR	UNIROYAL CHEMICAL	
NAUGARD TNPP	UNIROYAL CHEMICAL	
PHOSCLERE 315	FLEXSYS	FORMER AKZO
POLYGARD	NAUGATUCK	
SANTOWHITE TNPP	FLEXSYS	FORMER MONSANTO
SUMILIZER TNPP	SUMITOMO CHEMICALS	
TNPP	MARUBENI CORPORATION	
TNPP	DEER POLYMER	
WESTON 399	GENERAL ELECTRIC	CONTAINS 0.75% TIPA*, FORMER BORG WARNER
WESTON TNPP	GENERAL ELECTRIC	FORMER BORG WARNER
WYTOX 312	UNIROYAL CHEMICAL	

*TIPA = TRIISOPROPYLAMINE

Figure III.3   Product sheet of trisnonylphenylphosphite. After Gijsman [1]. Reproduced by permission of P. Gijsman

S-M01 (PO-AS code D002).

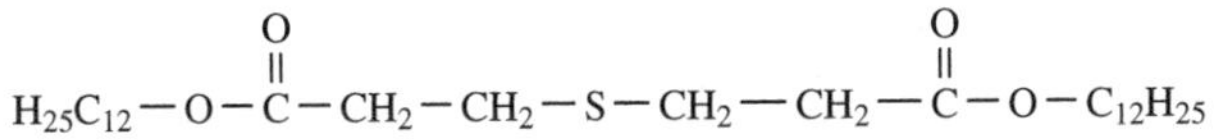

$$H_{25}C_{12}-O-\overset{\overset{\displaystyle O}{\|}}{C}-CH_2-CH_2-S-CH_2-CH_2-\overset{\overset{\displaystyle O}{\|}}{C}-O-C_{12}H_{25}$$

Chemical name(s):	Dilauryl thiodipropionate
	Didodecyl 3,3′-thiodipropionate
	Thiobis(dodecyl propionate)
	Thiodipropionic acid dilauryl ester
	Bis(dodecyloxycarbonylethyl) sulfide

Molecular formula:	$C_{30}H_{58}O_4S$
Molecular weight:	514
CAS:	123-28-4
EINECS:	204-614-1

| Physical properties: | Melting point: | 39–41 °C |
| | Boiling point: | 240 °C (1 mm Hg) |

Solubilities: Soluble in toluene, acetone, ether, chloroform. Slightly soluble in alcohols, ethyl acetate. Insoluble in water

Trade name	Supplier	Remarks
AFLUX 32	RHEIN CHEMIE	
ARENOX DL	ARECOM	
ARGUS DLTDP	ASAHI DENKA KOGYO	
CARSTAB DLTDP	MORTON	
CHIMOX 12	CHIMOSA	
CYANOX LTDP	CYTEC	FORMER AMERICAN CYANAMID
DLTDP	MARUBENI CORPORATION	
EVANSTAB 12	EVANS	
HOSTANOX SE1	CLARIANT	FORMER HOECHST
HOSTANOX SE3	CLARIANT	FORMER HOECHST
IRGANOX PS800	CIBA SPECIALTY CHEMICALS	
LANKROMARK DLTDP	AKCROS CHEMICALS	FORMER LANKRO CHEMICALS
LOWINOX DLTDP	GREAT LAKES	FORMER LOWI
MIANTOL L	MIWON COMMERCIAL	
NAUGARD DLTDP	UNIROYAL CHEMICAL	
NONOX DLTP	ZENICA	FORMER ICI
PLASTANOX LTDP	CYTEC	FORMER AMERICAN CYANAMID
SEENOX DL	SUN	
SUMILIZER TLP	SUMITOMO CHEMICALS	

Figure III.4 Product sheet of the thiosynergist dilaurylthiodipropionate. After Gijsman [1]. Reproduced by permission of P. Gijsman

BP-01 (PO-AS code F001).

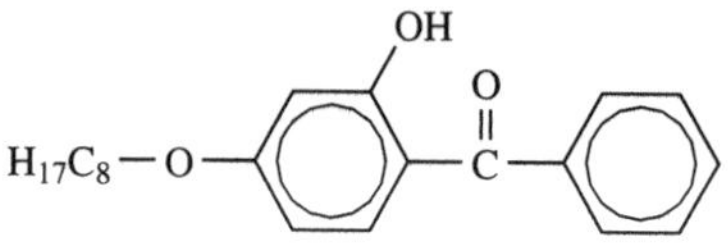

Chemical name(s):	2-Hydroxy-4-*n*-octoxybenzophenone
	2-Hydroxy-4-(octyloxy) benzophenone
	[2-Hydroxy-4-(octyloxy)phenyl]phenylemethanone

Molecular formula:	$C_{21}H_{26}O_3$
Molecular weight:	326
CAS:	1843-05-6
EINECS:	217-421-2

| Physical properties: | Melting point: | 45–48 °C |

Solubilities (30 °C, %/w): Methanol: 3, Ethyl acetate: 67, Methyl ethyl ketone: 65, Toluene: 77

Trade name	Supplier	Remarks
ADUVEX 248	WARD BLENKINSOP	
ANTIUV P	GREAT LAKES	FORMER SFOS
ARELITE BP81	ARECOM	
CARSTAB 700	CINCINNATI MILACRON	
CHIMASSORB 81	CIBA SPECIALTY CHEMICALS	
CYAGARD UV531	CYTEC	FORMER AMERICAN CYANAMID
CYASORB UV531	CYTEC	FORMER AMERICAN CYANAMID
FERRO AM300	FERRO CORPORATION	
HOSTAVIN ARO8	CLARIANT	FORMER HOECHST
KEMISORB 12	CHEMIPRO KASEI KAISHA	
LANKROMARK LE285	AKCROS CHEMICALS	FORMER LANKRO
LOWILITE 22	GREAT LAKES	FORMER LOWI
MARK 1413	ASAHI DENKA KOGYO	
MISORB BP-12	MIWON COMMERCIAL	
PALMAROLE LS.BO.087	PALMAROLE	
SANDUVOR 3035	CLARIANT	FORMER SANDOZ
SEESORB 102	SUN	
SUMISORB 130	SUMITOMO CHEMICALS	
ULTRA V300	FLEXSYS	FORMER MONSANTO
UVASORB 3C	3V	FORMER SIGMA
UVCHEK AM300	FERRO CORPORATION	
UVINUL 3008	BASF	
UVINUL 408	BASF	
UVISOL BP12	UVISOL	
VIOSORB 130	POLYCHEMICALS	

Figure III.5 Product sheet of the UV absorber 2-hydroxy-4-*n*-octoxybenzophenone. After Gijsman [1]. Reproduced by permission of P. Gijsman

MS-02 (PO-AS code l002).

$$H_2N - C_4H_9$$

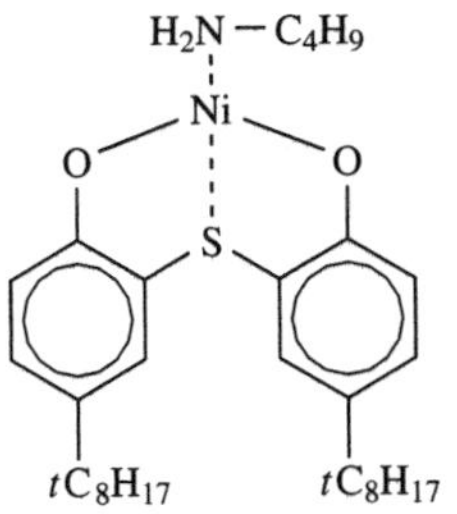

| Chemical name(s): | 2,2′-Thiobis (4-t-octylphenolato)-n-butylamine nickel(II) |

Molecular formula:	$C_{32}H_{51}O_2NiS$
Molecular weight:	572
CAS:	14516-71-3
EINECS:	

| Physical properties: | Melting point: | 275–281 °C |

Solubilities (20 °C, g/100 mL): Acetone: <1, Ethyl acetate: 1.7, Methanol: <1, Hexane: 28, Chloroform: >100

Trade name	Supplier	Remarks
CHIMASSORB N705	CIBA SPECIALTY CHEMICALS	
CYASORB UV1084	CYTEC	
KEMISORB 6NB	CHEMIPRO KASEI KAISHA	
NI 1	CHIMOSA	
QUENCHER 84	GREAT LAKES	
SEESORB 612NH	GREAT LAKES	
UVASORB NI	3V	

Figure III.6 Product sheet of the nickel quencher 2,2′-thio-bis-(4-$tert$-octylphenolato)-n-butylamine nickel(II). After Gijsman [1]. Reproduced by permission of P. Gijsman

LM-HALS-01 (PO-AS code K002).

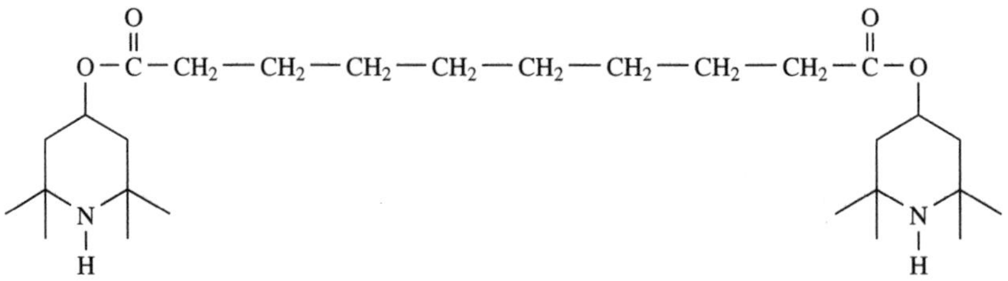

Chemical name(s):	Bis (2,2,6,6-tetramethyl-4-piperidinyl) sebacate
	Decanedioic acid, bis (2,2,6,6-tetramethyl-4-piperidinyl) ester
	Bis (2,2,6,6-tetramethyl-4-piperidyl) sebacate
	Bis (2,2,6,6-tetramethyl-4-piperidinyl) decanedioate

Molecular formula:	$C_{28}H_{52}N_2O_4$
Molecular weight:	481
CAS:	52829-07-9
EINECS:	258-207-9

Physical properties:	Melting point:	80–85 °C

Solubilities (20 °C, g/100 mL): Methanol: 38, Trichloromethane: 46, Acetone: 19, Ethanol: 50, Hexane: 5, Methylene chloride: 56.

Trade name	Supplier	Remarks
ADK STAB LA-77	PALMAROLE	
ARELITE HA70	ARECOM	
HALS 77	TRAMACO	NOWADAYS PALMAROLE
HALS LS770	MARUBENI CORPORATION	
JF-90	JOHOKU CHEMICAL	
LOWILITE 77	GREAT LAKES	FORMER LOWI
MARK LA77	ASAHI DENKA KOGYO	
MIHALS	MIWON COMMERICAL	
SANOL LS770	SANKYO	
SUMISORB 577	SUMITOMO CHEMICALS	
TINUVIN 770	CIBA SPECIALTY CHEMICALS	
TOMISORB 77	YOSHITOMI	
UVASEB 770	GREAT LAKES	
UVASORB HA77	3V	
UVISOL 077	UVISOL	

Figure III.7 Product sheet of the low-MW hindered amine light stabiliser di-(2,2,6,6-tetramethyl-4-piperidinyl) sebacate. After Gijsman [1]. Reproduced by permission of P. Gijsman

HM-HALS-01 (PO-AS code K003).

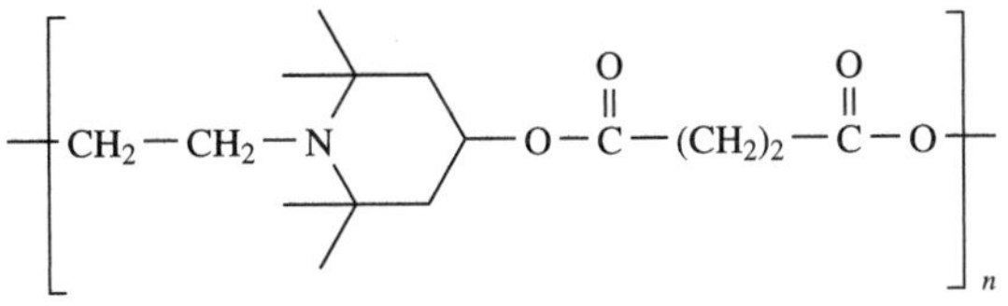

Chemical name(s):	Dimethyl succinate polymer with 4-hydroxy-2,2,6,6-tetramethyl-1-piperidine ethanol	

Molecular formula:	$(C_{31}H_{54}N_2O_9)_n$
Molecular weight:	>3100-4000
CAS:	65447-77-0
EINECS:	

Physical properties:	Melting point:	55–70 °C

Solubilities (20 °C, %/w): Hexane: <0.01, Acetone: 2, Ethyl acetate: 5, Chloroform: 40, Methylene chloride: 40, Methanol: 0.1, Water: <0.01.

Trade name	Supplier	Remarks
TINUVIN 622	CIBA SPECIALTY CHEMICALS	

Figure III.8 Product sheet of the high-MW hindered amine light stabiliser Tinuvin 622. After Gijsman [1]. Reproduced by permission of P. Gijsman

REFERENCES

1. P. Gijsman, *Stabiliser Trade Name–Chemical Structure Index*, DSM Research, Geleen (1998).

Index

Additives In Polymers: Industrial Analysis And Applications   J. C. J. Bart
© 2005 John Wiley & Sons, Ltd   ISBN: 0-470-85062-0

With kind thanks to Indexing Specialists (UK) Ltd of Hove for compilation of this index